COMPREHENSIVE POLYMER SCIENCE

IN 7 VOLUMES

COMPREHENSIVE POLYMER SCIENCE

The Synthesis, Characterization, Reactions & Applications of Polymers

CHAIRMAN OF THE EDITORIAL BOARD

SIR GEOFFREY ALLEN, FRS

Unilever Research and Engineering, London, UK

DEPUTY CHAIRMAN OF THE EDITORIAL BOARD

JOHN C. BEVINGTON

University of Lancaster, UK

Volume 3

Chain Polymerization I

VOLUME EDITORS

GEOFFREY C. EASTMOND

University of Liverpool, UK

ANTHONY LEDWITH

Pilkington plc Group Research, UK

SAVERIO RUSSO

Università di Sassari, Italy

PIERRE SIGWALT

Université Pierre et Marie Curie, Paris, France

PERGAMON PRESS

OXFORD · NEW YORK · BEIJING · FRANKFURT
SÃO PAULO · SYDNEY · TOKYO · TORONTO

U.K.	Pergamon Press plc, Headington Hill Hall, Oxford OX3 0BW, England
U.S.A.	Pergamon Press, Inc., Maxwell House, Fairview Park, Elmsford, New York 10523, U.S.A.
PEOPLE'S REPUBLIC OF CHINA	Pergamon Press, Room 4037, Qianmen Hotel, Beijing, People's Republic of China
FEDERAL REPUBLIC OF GERMANY	Pergamon Press GmbH, Hammerweg 6, D-6242 Kronberg, Federal Republic of Germany
BRAZIL	Pergamon Editora Ltda, Rua Eça de Queiros, 346, CEP 04011, Paraiso, São Paulo, Brazil
AUSTRALIA	Pergamon Press (Australia) Pty Ltd., P.O. Box 544, Potts Point, NSW 2011, Australia
JAPAN	Pergamon Press, 5th Floor, Matsuoka Central Building, 1-7-1 Nishishinjuku, Shinjuku-ku, Tokyo 160, Japan
CANADA	Pergamon Press Canada Ltd., Suite No. 271, 253 College Street, Toronto, Ontario, Canada M5T 1R5

Copyright © 1989 Pergamon Press plc

First edition 1989

Library of Congress Cataloging-in-Publication Data

Comprehensive polymer science.

Includes index.
Contents: v. 1. Polymer characterization/volume editors, Colin Booth & Colin Price—v. 2. Polymer properties/volume editors, Colin Booth & Colin Price—[etc.]—v. 7. Specialty polymers & polymer processing/volume editor, Sundar L. Aggarwal.
1. Polymers and polymerization
I. Allen, G. (Geoffrey), 1928– .   II. Bevington, J. C.
QD381.C66   1988   547.7   88-25548

British Library Cataloguing in Publication Data

Comprehensive Polymer Science.

Vol. 3: Chain polymerization
1. Polymers
I. Allen, Geoffrey   II. Bevington John C.
III. Eastmond, Geoffrey C.
547.7

ISBN 0-08-036207-9 (vol. 3)
ISBN 0-08-032515-7 (set)

Printed in Great Britain by A. Wheaton & Co. Ltd., Exeter

Contents

Preface

It is only 60 years since Staudinger's model of the molecular nature of a polymer was becoming universally accepted and the physical states of rubbers, plastics and fibres understood. Unfortunately, for some time many academic chemists continued not to appreciate the full significance of polymerization reactions and physicists tended to regard polymeric materials as inevitably being of indeterminate composition and unamenable to study by conventional physical methods.

Nevertheless, in the 1930s the foundations were laid for the understanding of the main polymerization mechanisms. An industry based on synthetic rubbers, plastics and fibres was soon established. In World War II it played a major strategic role and afterwards grew to be one of the main elements of the heavy chemicals industry. It became recognized that synthetics may be superior to natural materials in their properties and that they may be used for completely new purposes.

Alongside the production of well-defined materials there grew the ability to characterize the structure of polymer molecules and to understand the relationships between methods of preparation and subsequent treatment, structure and properties, both chemical and physical. As a result, a vast literature of polymer science and technology has been generated and four Nobel prizes awarded specifically for contributions to polymer science. Add to this the fact that many biological molecules, including polypeptides, enzymes, antibodies, carbohydrates and so on, are polymers of varying degrees of complexity, then the universality of polymers in the physical and biological sciences and technologies forms a dominant modern theme.

Comprehensive Polymer Science is a series of volumes designed to set down the structure of this vast subject in such a way that researchers and teachers of polymer science and workers in associated fields can find an authoritative and comprehensive account of the topic of immediate interest. That topic is set out in a framework of related subjects. The text is focused on synthetic polymers with little reference to biological macromolecules *per se* but the science underpins both physical and biological systems.

To ensure that the wide coverage is maintained at an authoritative level, more than 250 authors from 20 countries have been enlisted. Their contributions have been organized into a series of major themes:

Volume 1	Polymer Characterization
Volume 2	Polymer Properties
Volumes 3–5	Polymerization Mechanisms
Volume 6	Polymer Reactions
Volume 7	Specialty Polymers & Polymer Processing

Because of the wide coverage the editors were presented with a particularly difficult decision with regard to symbols and nomenclature. The latter does not follow strictly the recommendations of IUPAC nor are symbols consistent throughout the whole work. However, usage in a particular chapter is consistent with the practice in the current literature. Thus a reader will be able to frame new publications in the context of the information presented in this series of volumes.

We should like to acknowledge the way in which the staff at the publisher, particularly Dr Colin Drayton (who initially proposed the project), Dr Helen McPherson and their editorial team, have supported the editors and authors in their endeavour to produce a text that is both complete and up-to-date and that will appeal to industrial and academic researchers alike. *Comprehensive Polymer Science* is a milestone in the literature of the subject in terms of coverage, clarity and a sustained high level of presentation.

GEOFFREY ALLEN
London

JOHN C. BEVINGTON
Lancaster

Contributors to Volume 3

Dr T. Aida
Department of Synthetic Chemistry, Faculty of Engineering, University of Tokyo, Hongo, Bunkyo-ku, Tokyo 113, Japan

Professor W. J. Bailey
Department of Chemistry, University of Maryland, College Park, MD 20742, USA

Professor C. H. Bamford
Institute of Medical and Dental Bioengineering, University of Liverpool, PO Box 147, Liverpool L69 3BX, UK

Dr C. A. Barson
Department of Chemistry, University of Birmingham, PO Box 363, Birmingham B15 2TT, UK

Dr J. K. Beasley
1 Deer Run Drive, Spring Valley, Wilmington, DE 19807, USA

Professor J. C. Bevington
Department of Chemistry, University of Lancaster, Bailrigg, Lancaster LA1 4YA, UK

Dr N. C. Billingham
School of Chemistry and Molecular Sciences, University of Sussex, Falmer, Brighton BN1 9QJ, Sussex, UK

Dr S. Boileau
Laboratoire de Chimie Macromoléculaire, Collège de France, Place Marcelin Berthelot 11, F-75231 Paris Cedex 05, France

Professor B. Boutevin
Directeur de Recherche au CNRS, Laboratoire de Chimie Appliquée, Université des Sciences et Techniques de Languedoc, Ecole Nationale Supérieure de Chimie, 8 rue Ecole Normale, F-34075 Montpellier Cedex, France

Professor Dr D. Braun
Deutsches Kunststoff-Institut, Schlossgartenstrasse 6R, D-6100 Darmstadt, Federal Republic of Germany

Dr S. Bywater
Chemistry Division, National Research Council, Montreal Road Laboratories, Ottawa, Ontario K1A 0RG, Canada

Professor J. M. G. Cowie
Department of Chemistry, University of Stirling, Stirling FK9 4LA, UK

Dr W. K. Czerwinski
Deutsches Kunststoff-Institut, Schlossgartenstrasse 6R, D-6100 Darmstadt, Federal Republic of Germany

Dr M. P. Dreyfuss
Michigan Molecular Institute, 1910 W St Andrews Road, Midland, MI 48640-2696, USA

Dr P. Dreyfuss
Michigan Molecular Institute, 1910 W St Andrews Road, Midland, MI 48640-2696, USA

Professor M. Fontanille
Université de Bordeaux 1, Laboratoire de Chimie des Polymères Organiques, Institut du Pin, 351 cours de la Libération, F-33405 Talence Cedex, France

Professor E. J. Goethals
Laboratory of Organic Chemistry, State University of Ghent, 281 Krijgslaan (S-4), B-9000 Ghent, Belgium

Professor A. E. Hamielec
Department of Chemical Engineering, McMaster Institute for Polymer Production Technology, McMaster University, Hamilton, Ontario L8S 4L7, Canada

Dr C. J. Hamilton
Speciality Materials Group, Department of Applied Chemistry, University of Aston, Aston Triangle, Birmingham B4 7ET, UK

Professor T. Higashimura
Department of Polymer Chemistry, Faculty of Engineering, Kyoto University, Yoshida Honmachi, Sakyo-ku, Kyoto 606, Japan

Professor Y. Imanishi
Department of Polymer Chemistry, Faculty of Engineering, Kyoto University, Yoshida Honmachi, Sakyo-ku, Kyoto 606, Japan

Professor S. Inoue
Department of Synthetic Chemistry, Faculty of Engineering, University of Tokyo, Hongo, Bunkyo-ku, Tokyo 113, Japan

Mr R. Jérôme
Université de Liège, Laboratoire de Chimie Macromoléculaire et de Catalyse Organique, Bâtiment B6, 4000 Sart Tilman (Liège 1), Belgium

Dr Y. Kawakami
Department of Synthetic Chemistry, Faculty of Engineering, Nagoya University, Furo-cho, Chikusa-ku, Nagoya 464, Japan

Professor J. P. Kennedy
Institute of Polymer Science, University of Akron, Akron, OH 44325, USA

Professor Dr H. R. Kricheldorf
University of Hamburg, Institute for Technical and Macromolecular Chemistry, Bundesstrasse 45, D-2000 Hamburg 13, Federal Republic of Germany

Dr P. Kubisa
Polish Academy of Sciences, Centre of Molecular and Macromolecular Studies, 90362 Lodz, Boczna 5, Poland

Dr J. F. MacGregor
Department of Chemical Engineering, McMaster Institute for Polymer Production Technology, McMaster University, Hamilton, Ontario L8S 4L7, Canada

Dr M. K. Martin
Specialty Chemicals Division, 3M Company, 3M Center, St Paul, MN 55144-1000, USA

Professor K. Matyjaszewski
Department of Chemistry, Carnegie-Mellon University, 4400 Fifth Avenue, Pittsburgh, PA 15231, USA

Dr G. Moad
Division of Chemical Polymers, CSIRO, GPO Box 4331, Melbourne 3001, Victoria, Australia

Dr H.-U. Moritz
Institut für Technische Chemie, Technische Universität Berlin, Sekr TC3, Strasse des Juni 135, 1000 Berlin 12, Federal Republic of Germany

Dr A. H. E. Müller
Institut für Physikalische Chemie, Johannes-Gutenberg-Universität Mainz, Welderweg 15, D-6500 Mainz, Federal Republic of Germany

Professor Dr O. Nuyken
Universität Bayreuth, Makromolekulare Chemie I, Postfach 101251, Universitätsstrasse 30, Gebäude NW 2, D-8580 Bayreuth, Federal Republic of Germany

Professor K. F. O'Driscoll
Department of Chemical Engineering, University of Waterloo, Waterloo, Ontario N2L 3G1, Canada

Dr S. D. Pask
Bayer AG, 4047 Dormagen, Federal Republic of Germany

Professor S. Penczek
Polish Academy of Sciences, Centre of Molecular and Macromolecular Studies, 90362 Lodz, Boczna 5, Poland

Dr A. Penlidis
Department of Chemical Engineering, McMaster Institute for Polymer Production Technology, McMaster University, Hamilton, Ontario L8S 4L7, Canada

Professor Y. Pietrasanta
Laboratoire de Chimie Appliquée, Université des Sciences et Techniques de Languedoc, Ecole Nationale Supérieure de Chimie, 8 rue Ecole Normale, F-34075 Montpellier Cedex, France

Dr R. E. Putnam
100 Alden Avenue, Marietta, 0H 45750, USA

Professor Dr K.-H. Reichert
Institut für Technische Chemie, Technische Universität Berlin, Sekr TC3, Strasse des Juni 135, 1000 Berlin 12, Federal Republic of Germany

Dr E. Rizzardo
Division of Chemical Polymers, CSIRO, GPO Box 4331, Melbourne 3001, Victoria, Australia

Dr J. M. Rooney
Sun Chemical Company, 631 Central Avenue, Carlstadt, NJ 07072, USA

Professor A. Rudin
Institute for Polymer Research, University of Waterloo, Waterloo, Ontario N2L 3G1, Canada

Professor G. Sauvet
Université Paris-Nord, Centre Scientifique et Polytechnique, Laboratoire de Recherches sur les Macromolécules, Unité Associée au CNRS, Avenue Jean-Baptiste Clément, F-93430 Villetaneuse, France

Dr M. Sawamoto
Department of Polymer Chemistry, Faculty of Engineering, Kyoto University, Yoshida Honmachi, Sakyo-ku, Kyoto 606, Japan

Dr J. Šebenda
Institute of Macromolecular Chemistry, Czechoslovak Academy of Sciences, Heyrovského nam 2, 16206 Prague 6, Czechoslovakia

Professor P. Sigwalt
Laboratoire de Chimie Macromoléculaire, Université Pierre et Marie Curie, Tour 44, Couloir 44-54, 1er Etage, 4 Place Jussieu, F-75252 Paris Cedex 05, France

Dr S. Slomkowski
Polish Academy of Sciences, Centre of Molecular and Macromolecular Studies, 90362 Lodz, Boczna 5, Poland

Dr D. H. Solomon
Division of Chemical Polymers, CSIRO, GPO Box 4331, Melbourne 3001, Victoria, Australia

Dr M. Stickler
Röhm GmbH Chemische Fabrik, Kirschenallee, Postfach 4242, D-6100 Darmstadt 1, Federal Republic of Germany

Dr Y. Y. Tan
Polymer Chemistry Laboratory, State University of Groningen, Nijenborgh 16, 9747 AG Groningen, The Netherlands

Professor Ph. Teyssié
Université de Liège, Laboratoire de Chimie Macromoléculaire et de Catalyse Organique, Bâtiment B6, 4000 Sart Tilman (Liège 1), Belgium

Dr B. J. Tighe
Speciality Materials Group, Department of Applied Chemistry, University of Aston, Aston Triangle, Birmingham B4 7ET, UK

Professor D. A. Tirrell
Polymer Science and Engineering Department, University of Massachusetts, Amherst, MA 01003, USA

Professor T. Tsuruta
Faculty of Engineering, Department of Industrial Chemistry, Science University of Tokyo, Kagurazaka 1-3, Shinjuku-ku, Tokyo 162, Japan

Contents of All Volumes

Volume 4   Chain Polymerization II

Volume 5   Step Polymerization

1

General Concepts

JOHN M. G. COWIE

University of Stirling, UK

1.1 GENERAL CONCEPTS OF CHAIN REACTIONS

Formation of polymer from monomer, or monomers, is traditionally divided into two broad reaction classes: step (or condensation) polymerizations and chain growth (or addition) polymerizations. The distinction between these processes can be made on mechanistic grounds as they exhibit quite dissimilar reaction kinetics which lead to different distributions of species as a function of the extent of reaction, and, depending on the detailed mechanism, the molecular weight distributions of the polymers formed.

In contrast to small molecule systems, it is a distinctive feature of polymeric molecules that they can have essentially the same chemical composition but widely differing molecular weights. This is a consequence of the fact that they are composed of different numbers of the same repeating unit and, in general, within a given polymer sample there will exist a wide range of molecular weights. The term degree of polymerization is used to define the number of monomer units in a given chain and when used to characterize a polymer sample will of necessity be an average value. The number of chains of different degrees of polymerization in a given polymer sample (or alternatively its molecular weight distribution, MWD) reflects the chemistry of the polymerization process, and a knowledge of the MWD is important not only because of the relationships between the molecular weight and properties (Volume 1) but also because of the insight such knowledge provides towards an understanding of reaction mechanisms (Volume 3, Chapter 4 and other chapters relating to specific mechanisms).

A classical step growth reaction involves a series of monomer + monomer, monomer + oligomer, monomer or oligomer + macromolecule, and macromolecule + macromolecule reactions, whereas addition reactions proceed by monomer + macromolecule reactions alone. The general features and the detailed aspects of step growth polymerizations are described in Volume 5 of this series.

In this chapter some common but essential features of chain growth polymerizations are described to provide an understanding of the subject in broad outline as an introduction to the more specialized chapters.

Addition polymerizations are known to take place *via* a chain reaction which is a self-sustaining molecular reaction maintained by the fact that the product of one stage initiates a further reaction step, which in turn regenerates more product capable of causing further reactions. The polymerization thus involves a large number of identical and successive events, initiated by some triggering mechanism, which eventually lead to the production of long chain-like molecules.

For a successful polymerization, the chain reaction must be started by first providing a suitable active centre capable of reacting with the monomers present and in doing so regenerate the active centre during each addition step thereby allowing the reaction to continue. Unlike a nuclear fission reaction, addition polymerization is not a branching chain reaction. It can be regulated by the controlled production of initiating active centres which only regenerate themselves singly at each step and are susceptible to deactivation in a variety of ways. Thus an addition polymerization has three clearly defined reaction stages: (i) initiation of the reaction; (ii) propagation and chain growth by means of a chain reaction; and (iii) termination of the growing chain.

1.1.1 Initiation

1.1.1.1 Free radical initiation

The active centre which initiates and propagates the chain can be either a free radical, an ion or a radical ion. The free radical is perhaps the most commonly used type of initiator and is produced by decomposition of a suitable molecule, either thermally or by irradiation (see also Volume 3, Chapter 8). In general the active species is produced by homolytic fission of a σ bond to produce two free radicals (equation 1). These free radicals must be capable of transferring their activity to the monomer but must also be produced at a reasonable rate at the temperature selected for the polymerization reaction. Useful groups of compounds able to act as initiators are: (i) peroxides (organic or inorganic); (ii) azo compounds; (iii) hydroperoxides and peresters; and (iv) organometallics. Benzoyl peroxide is typical of the group (i) initiators and reacts as shown in equation (2), where two identical radicals are produced by rupture of the O—O bond, which has a low dissociation energy. The rate of decomposition will of course be temperature dependent and it has been suggested as a useful guide that an initiator decomposition rate constant in the range $k = 10^{-5}$ to 10^{-6} s^{-1} will produce radicals at suitable rates for polymerization. This immediately defines a useful temperature range for each initiator. For benzoyl peroxide the rate equation for thermal decomposition is $k = 10^{14} \exp(-29.9 \text{ kJ}/RT)$ s^{-1} and this gives an optimum temperature range of 330–350 K. Examples of other suitable peroxides are shown in Table 1.

$$\text{I} \longrightarrow 2\text{R}\cdot \tag{1}$$

$$\underset{\underset{O}{\|}}{\text{PhC}}\text{—O—O—}\underset{\underset{O}{\|}}{\text{CPh}} \longrightarrow 2\underset{\underset{O}{\|}}{\text{PhC}}\text{—O}\cdot \tag{2}$$

The archetypal azo initiator is 2,2'-azobisisobutyronitrile (AIBN), which decomposes thermally to produce two identical free radicals with the liberation of nitrogen (equation 3). While the C—N bond has a relatively high dissociation energy, the driving force in this facile decomposition appears to be the liberation of the extremely stable N_2 molecule. As with the peroxides, the thermal decomposition of azo compounds is temperature dependent and the useful temperature ranges for some of these are given in Table 1.

$$\underset{\underset{CN}{|}}{\text{Me}_2\text{C}}\text{—N}\!\!=\!\!\text{N—}\underset{\underset{CN}{|}}{\text{CMe}_2} \longrightarrow 2\underset{\underset{CN}{|}}{\text{Me}_2\text{C}}\cdot \; + \; N_2 \tag{3}$$

Table 1 Some Radical Initiator Decomposition Rate Equations and the Corresponding Optimum Temperature Range for Use

Initiator	Rate equations (s^{-1})	Useful temperature range (K)
RC(O)OO(O)CR		
R = Et	$k = 10^{14}\exp(-146 \text{ kJ}/RT)$	382–402
R = Bu^t	$k = 6.3 \times 10^{15}\exp(-156.8 \text{ kJ}/RT)$	377–395
RN=NR		
R = Me$_2$C(CN)	$k = 3 \times 10^{15}\exp(-128.7 \text{ kJ}/RT)$	313–333
R = PhCHMe	$k = 1.3 \times 10^{15}\exp(-152.6 \text{ kJ}/RT)$	378–398
R = Me$_2$CH	$k = 5 \times 10^{13}\exp(-170.5 \text{ kJ}/RT)$	453–473

Combination of these two different initiating groups in one molecule produces a bifunctional initiator, *e.g.* (**1**). Structures such as these can be used to prepare block copolymers by stimulating the decomposition of one type of group in the presence of a monomer while leaving the other group intact for subsequent reaction with a second monomer. Thus thermal decomposition of the azo group at ~ 340 K allows reaction with monomer A to produce a macroinitiator terminated with the more stable peroxide group. This can then be decomposed catalytically at room temperature in the presence of a different monomer B to produce a block copolymer. Alternatively, the peroxide can be used first, followed by the azo group. The exact structure and block sequence of the block copolymer formed will depend on the termination mechanism, as will be seen later, after each stage in the reaction.

$$Bu^tOO-\overset{\overset{\displaystyle O}{\|}}{C}-R-N{=}N-R-\overset{\overset{\displaystyle O}{\|}}{C}-OOBu^t$$

(**1**)   di-*t*-butyl-4,4′-azobis-4-cyanoperoxyvalerate

The low temperature decomposition of peroxides can be catalyzed by a tertiary amine as shown in equation (4) or when involved in a redox reaction (see also Volume 3, Chapter 9), *e.g.* equation (5). In both cases only one radical is produced. Other low temperature initiations can be effected by some organometallic compounds. Thus silver alkyls tend to be unstable and will decompose at or below ambient temperature to produce single radicals (equation 6). Also useful in this respect are the alkyl boron compounds (R_3B), which are thought to react in the presence of oxygen according to Scheme 1. Whereas the group (i) and (ii) initiators tend to produce two identical radical fragments, the hydroperoxides and peresters yield two different radical fragments which may have different reactivities, *e.g.* as in equations (7) and (8). It should be noted that the hydroperoxides will also take part in redox reactions such as shown in equation (6). Certain vinyl monomers, such as styrene and methyl methacrylate, may be susceptible to thermal initiation (equation 9; see also Volume 3, Chapter 10), but this is not a particularly successful method for most monomers. A more useful alternative is to irradiate the monomer with high energy electromagnetic radiation such as X-rays or γ-rays where there is hydrogen transfer between two monomers to produce two radicals (equation 10; see Volume 4, Part 3).

$$Ph\overset{\overset{\displaystyle O}{\|}}{C}-O-O-\overset{\overset{\displaystyle O}{\|}}{C}Ph \;+\; R_3N \longrightarrow R_3N^+\bar{O}-\overset{\overset{\displaystyle O}{\|}}{C}Ph \;+\; PhCO\cdot \tag{4}$$

$$Fe^{2+} \;+\; H_2O_2 \longrightarrow Fe^{3+} \;+\; OH^- \;+\; OH\cdot \tag{5}$$

$$AgEt \longrightarrow Ag \;+\; Et\cdot \tag{6}$$

$$R_3B \;+\; O_2 \longrightarrow R_2BOOR$$

$$R_2BOOR \;+\; 2R_3B \longrightarrow R_2BOBR_2 \;+\; R_2BOR \;+\; 2R\cdot$$

Scheme 1

$$Ph-\overset{\overset{\displaystyle Me}{|}}{\underset{\underset{\displaystyle Me}{|}}{C}}-O-OH \longrightarrow Ph-\overset{\overset{\displaystyle Me}{|}}{\underset{\underset{\displaystyle Me}{|}}{C}}-O\cdot \;+\; OH\cdot \tag{7}$$

cumyl peroxide

$$Ph\overset{\overset{\displaystyle O}{\|}}{C}-O-OBu^t \longrightarrow Ph\overset{\overset{\displaystyle O}{\|}}{C}O\cdot \;+\; Bu^tO\cdot \tag{8}$$

6-butyl perbenzoate

$$CH_2{=}\underset{\underset{\displaystyle X}{|}}{CH} \longrightarrow \cdot CH{=}\underset{\underset{\displaystyle X}{|}}{CH} \;+\; H\cdot \tag{9}$$

$$2CH_2{=}\underset{\underset{\displaystyle X}{|}}{CH} \xrightarrow{\;\gamma\text{-rays}\;} CH_2{=}\underset{\underset{\displaystyle X}{|}}{C}\cdot \;+\; Me\underset{\underset{\displaystyle X}{|}}{CH}\cdot \tag{10}$$

High energy radiation can lead to the formation of ionic species and, in the presence of suitable monomers, may initiate ionic polymerization. A more effective method is to use a photosensitive initiator which absorbs UV radiation and decomposes to produce radicals (see Volume 4, Chapter 20), *e.g.* equation (11). This method has the advantage that specific temperatures need not be used, radical production starts immediately the reaction is exposed to the radiation and stops when it is switched off. Photoinitiation can, in specialized systems, create initiating ions.

$$PhC\text{—}CPh \xrightarrow{\ hv\ } PhC\cdot \ + \ PhCH\cdot \tag{11}$$

with benzoin

Active species of various types may also be created electrochemically (see Volume 4, Chapter 24).

1.1.1.2 Ionic initiation

Chain reactions can also be initiated by anions or cations but the mechanisms differ for each and both differ from the radical processes. The ionic initiators are much more specific towards the type of monomer but can initiate polymerization in some for which radical initiation is ineffective.

Monomers which are suitable for anionic initiation are those with an electron-withdrawing substituent attached to the double bond. These include phenyl, nitrile and carboxylic groups which help to stabilize the anion. The reaction is a nucleophilic attack of the monomer by the carbanions which either results in transfer of an electron to the monomer or actual addition of the carbanion moiety. Thus the reaction of sodium metal with naphthalene produces a radical anion (green) which reacts with styrene to produce a styryl radical ion (red) (equation 12). This may dimerize to produce a dianion and chain growth takes place from both ends of this active species.

$$Na^+ \left[\bigcirc\!\bigcirc \right]^{\cdot -} + \ CH_2\!\!=\!\!CH \longrightarrow \ \cdot CH_2\text{—}CH^- Na^+ \ + \ \bigcirc\!\bigcirc \tag{12}$$

with Ph (under first CH_2) and Ph (under second).

Organometallic compounds are also effective and butyllithium will add onto a monomer to produce the active species which is again an ion pair (equation 13). Other initiators which can be used are Grignard reagents and alkali metal suspensions (see also Volume 3, Part 3).

$$Bu^- Li^+ \ + \ CH_2\!\!=\!\!CH \longrightarrow \ BuCH_2CH^- \ Li^+ \tag{13}$$

with Ph.

Monomers with electron-donating substituents are most susceptible to initiation by carbenium or oxonium ions as they are prone to electrophilic attack on the double bond. Classical protonic acids (HCl, H_2SO_4, $HClO_4$) can act as initiators but more useful are the Lewis acid–base combinations where a Lewis acid, such as BF_3, reacts first with a coinitiator (H_2O) to produce the active species $[BF_3OH]^-H^+$ (equation 14), which can react with a monomer to produce a carbenium ion. Other possible initiator–coinitiator combinations are $AlCl_3/RCl$, $AlCl_3/H_2O$, $TiCl_4/MeOH$ or $SbCl_5$ alone, *etc.* (see also Volume 3, Part 4).

$$H^+[BF_3OH]^- \ + \ CH_2\!\!=\!\!\underset{Me}{\overset{Me}{C}} \longrightarrow \ Me\text{—}\underset{Me}{\overset{Me}{C^+}} \ [BF_3OH] \tag{14}$$

In both anion- and cation-initiated polymerizations there is a counter (gegen) ion present which can play a significant role in the subsequent propagation reactions.

In addition to the three classical initiating systems for chain reaction polymerizations, other initiating systems also exist. For example, group transfer polymerization, which involves a controlled Michael addition to certain monomers susceptible to anionic polymerization is a recent development where initiation is effected by silylketylacetals (see Volume 4, Chapter 10).

Transition-metal-based initiators are also important (see Volume 4, Part 1) and form the basis of coordination polymerizations, including the classic Ziegler–Natta systems. More recently metathesis reactions, using transition metal catalysts, have been developed.

Although not a chain polymerization in the classic sense, active species in the form of radicals or ions may be generated in plasmas (see Volume 4, Chapter 21).

1.1.2 Propagation

1.1.2.1 Radical (see also Volume 3, Chapter 6)

Once initiation has taken place, growth of the polymer chain is effected by the repetitive addition of a monomer to the active centre with regeneration of that centre. Although the lifetime of an individual radical centre is short, the propagation steps occur many times in rapid succession to build a polymer chain comprising 10^2 to 10^6 monomer units before termination takes place. For vinyl monomers all of these reaction steps can be generalized by equation (15) and each can be characterized by the same rate constant. Normally the addition is head-to-tail, as shown, for both steric and electronic reasons. While head-to-head, or tail-to-tail, additions can and do take place, radical (I) is thermodynamically more stable than radical (II) (equation 16), and any unfavourable interactions between adjacent R groups on formation of the polymer are also minimized.

$$\text{\textbackslash\textbackslash}CH_2\!-\!\underset{\underset{R}{|}}{CH}\cdot \; + \; CH_2\!\!=\!\!\underset{\underset{R}{|}}{CH} \longrightarrow \text{\textbackslash\textbackslash}CH_2\!-\!\underset{\underset{R}{|}}{CH}\!-\!CH_2\!-\!\underset{\underset{R}{|}}{CH}\cdot \tag{15}$$

(I)

$$\text{\textbackslash\textbackslash}CH_2\!-\!\underset{\underset{R}{|}}{CH}\cdot \; + \; CH_2\!\!=\!\!\underset{\underset{R}{|}}{CH} \longrightarrow \text{\textbackslash\textbackslash}CH_2\!-\!\underset{\underset{R}{|}}{CH}\!-\!\underset{\underset{R}{|}}{CH}\!-\!CH_2\cdot \tag{16}$$

(II)

Other possible structures can be formed when dienes are polymerized due to the presence of two double bond sites in the monomer unit. The possible addition mechanisms can be illustrated using the unsymmetrical diene 2-methyl-1,3-butadiene as the model (Scheme 2). This can be polymerized to give four distinct structures, if stereoregularity is ignored for the moment. Dienes can undergo 1,4-addition reactions with the retention of one unsaturated group per monomer unit in the main chain and this also leads to *cis–trans* isomerism. These different structural arrangements also change the properties of the polymer; thus, natural rubber, for which this is the basic building block, is made up of the *cis*-1,4-addition polymer exclusively, whereas gutta percha has the *trans*-1,4-addition form. However, each double bond is also capable of acting as a vinyl group so that 1,2-addition and 3,4-addition are also possible, both of which can be arranged in specific tactic forms as will be seen later. For symmetrical dienes, *e.g.* 1,3-butadiene, these two modes of addition become indistinguishable. As many diene polymerizations result in a mixture of structures in any one polymer chain, control can be difficult and the synthesis of structurally pure samples requires the use of specific catalysts.

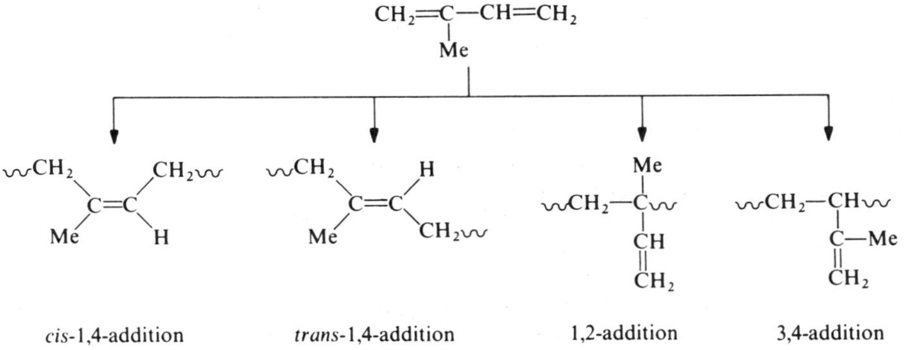

$$CH_2\!\!=\!\!\underset{\underset{Me}{|}}{C}\!-\!CH\!\!=\!\!CH_2$$

| *cis*-1,4-addition | *trans*-1,4-addition | 1,2-addition | 3,4-addition |

Scheme 2

1.1.2.2 *Ionic*

The mechanism of anionic polymerization involves the insertion of a monomer between the counterion and the carbanion on the terminal unit of the growing chain, followed by nucleophilic attack of the anion on the monomer double bond (equation 17; see Volume 3, Part 3). For cationic propagation the mechanism is similar, only now there is an electrophilic attack by the terminal carbenium ion on the monomer (see Volume 4, Part 4).

$$
\text{\textasciitilde CH}_2\text{—CH}^-\quad G^+ \longrightarrow \text{\textasciitilde CH}_2\text{—CH—CH}_2\text{—CH}^-\quad G^+ \tag{17}
$$

$$
\underset{R}{|}\qquad\qquad \underset{R}{|}\qquad \underset{R}{|}
$$

$$
\text{CH}_2\text{=CH}
$$

$$
\underset{R}{|}
$$

These reactions occur in rapid succession but the rate will be influenced by the positioning of the counterion relative to the carbanion or carbenium ion. This will depend on the solvent used, which will determine the state of solvation of the ion pair.

Cyclic monomers such as lactones, lactams, cyclic amines and cyclic ethers, which polymerize by a ring-opening mechanism, can be encouraged to do so by ionic initiation. The tendency to form linear high polymer may depend on the size of the ring, thus a five-membered lactone does not form polymer but a six-membered lactone will polymerize. However, this is not a general feature and the behaviour can depend on the chemical nature of the ring, *e.g.* oxolane, a five-membered ring, is readily polymerized cationically.

Two general mechanisms for propagation have been proposed and are shown in equations (18) and (19). In the first of these, there is cleavage of the ring by the initiator to form an ionic or zwitterionic end group, followed by attack on another ring which is opened to regenerate the ionic end. The second mechanism suggests the initiator forms a complex which is the active species.

$$\longrightarrow \longrightarrow \longrightarrow \text{polymer} \tag{18}$$

$$\longrightarrow \longrightarrow \longrightarrow \text{polymer} \tag{19}$$

Some cyclic compounds such as cyclohexane and 1,4-dioxane, but especially those which are stabilized by their aromaticity (benzene, 1,3,5-triazine, borazines) have not yet been polymerized, but the common inorganic polymers, polysiloxanes and polyphosphazenes can be prepared from cyclic precursors (see Volume 4, Part 5).

Many polymerizations are conducted in homogeneous systems, in a solution of monomer or another solvent, and it is from studies of such systems that an understanding of reaction mechanisms has been established. Other polymerizations, however, are heterogeneous either by virtue of the insolubility of the initiators (as in Ziegler catalysis) or because of the insolubility of the polymer in its monomer or some other polymerizing medium. In such cases the reaction kinetics are complicated and may differ according to the type of system under investigation. Thus if a polymer precipitates during a reaction the propagating species (usually radicals) may be trapped, still active, in the precipitated polymer in the absence of monomer. In some cases the precipitated polymer may absorb monomer and the polymerization can then continue inside and outside the polymer phase (see Volume 4, Chapter 16). Alternatively, polymerizations may be performed deliberately in an emulsion or some other dispersed system (see Volume 4, Part 2).

1.1.3 Termination

1.1.3.1 Bimolecular termination

In free-radical-initiated polymerizations, chain termination occurs by reaction with other free radicals in the system, or with impurities, if present. The growing chain can be terminated by interaction with initiator fragments but a more likely process is the bimolecular reaction between two active chains. This can occur in two different ways.

(a) Combination. This involves the formation of a σ bond when two free radical sites collide. A single chain is created from this mechanism with a head-to-head placement at the site of combination (equation 20).

(b) Disproportionation. Here there is hydrogen abstraction from one growing chain end by the radical end of another to give two chains, one with an unsaturated terminal unit and the other a saturated chain end (equation 21).

In both cases there is mutual termination of two active kinetic chains and no further chain reaction involving these radicals can occur.

$$\text{\textasciitilde}CH_2-\dot{C}H + H\dot{C}-CH_2\text{\textasciitilde} \longrightarrow \text{\textasciitilde}CH-CH\text{\textasciitilde} \tag{20}$$
$$\qquad\quad | \qquad\quad | \qquad\qquad\qquad\qquad | \quad |$$
$$\qquad\quad R \qquad\quad R \qquad\qquad\qquad\qquad R \quad R$$

$$\text{\textasciitilde}CH_2-\overset{\text{Me}}{\underset{R}{\dot{C}}} + \overset{\text{Me}}{\underset{R}{\dot{C}}}-CH_2\text{\textasciitilde} \longrightarrow \text{\textasciitilde}CH_2-\overset{\text{Me}}{\underset{R}{C}}H + \overset{\text{Me}}{\underset{R}{C}}=CH\text{\textasciitilde} \tag{21}$$

1.1.3.2 Chain transfer

Other mechanisms have been proposed in which one kinetic chain is terminated by interaction with another molecular species which is not a free radical. The other species may be solvent molecules, monomers, another polymer chain backbone or a molecule susceptible to an abstraction reaction. In general the radical removes an atom or molecular fragment from the interacting molecule to leave another radical species (equation 22). If this is sufficiently reactive to initiate further chain growth then the rate of the reaction is unaffected but the chain length of the product is reduced. If the new radical produced is less reactive towards the monomer or totally unreactive then the reaction is retarded or perhaps even inhibited completely.

$$\text{\textasciitilde}CH_2-\underset{R}{\dot{C}}H + AB \longrightarrow \text{\textasciitilde}CH_2-\underset{R}{C}HA + B\cdot \tag{22}$$

In the presence of very active chain transfer agents, or high concentrations of transfer agents, chain transfer may become a dominant process and lead to the formation of large numbers of short polymer chains (telomers) with end groups derived from the chain transfer agent. This process is described as telomerization (see Volume 3, Chapter 14). If these end groups are potentially reactive, then the terminal functionality can be used in chain extension reactions and block copolymer formation.

1.1.3.3 Ionic systems

In ionic polymerizations the termination reactions are not as clearly defined as in the radical processes. Anionic reactions in particular have no bimolecular termination mechanism and in the absence of impurities the active end remains active indefinitely. In practical terms they can be terminated when water, carbon dioxide, alcohols or other agents are added to the system, and reactive terminal groups may be incorporated into the polymer chains if a suitable choice of the reagent is made. Chain transfer reactions are not a common feature in anionic polymerizations, although transfer to solvent has been identified in certain systems.

For cationic polymerizations the situation is somewhat different and several termination and chain transfer reactions have been identified.

(a) Ionic rearrangement. This is a spontaneous anion–cation recombination or anion splitting, which results in the loss of the active carbenium ion end group of the growing polymer chain.

(b) Transfer. This can occur with or without termination of the kinetic chain. Proton or hydride ion transfer to a monomer results in termination of one chain but with reinitiation of a second when the receiving monomer is transformed into a new carbenium ion. Other transfer reactions with substances such as water, tertiary amines, ethers or esters tend to produce new inactive cationic species at the terminal unit which are incapable of further growth, thereby terminating the kinetic chain completely.

1.1.4 Equilibrium

For a polymerization reaction to occur the free energy difference, ΔG, between the monomer and the polymer must be negative and, although independent of the nature of the chain carrier, it will depend on the availability of a suitable kinetic route. The size and magnitude of ΔG will be determined by the enthalpy change ΔH and the entropy change ΔS through the well known relation

$$\Delta G \;=\; \Delta H \;-\; T\Delta S \tag{23}$$

In a polymerization reaction the gain in vibrational and rotational entropy tends to be counterbalanced by a loss of external rotational entropy, but there is a net loss of translational entropy when the monomer molecules are covalently bonded together to form a chain. This loss of entropy due to connectivity is the major contribution to ΔS which, with values typically of the order of $100\,\mathrm{J\,mol^{-1}}$, is negative and unfavourable towards polymerization. However, for vinyl monomers, where there is conversion of a π to a σ bond, the polymerization is a highly exothermic process with ΔH values ranging from $-34\,\mathrm{kJ\,mol^{-1}}$ to $-160\,\mathrm{kJ\,mol^{-1}}$. Under normal temperature conditions ΔH will easily be the dominating term and ensure that ΔG is negative, thereby favouring the polymerization. Only at high temperatures will the $T\Delta S$ term become significant and eventually dominate, thereby rendering ΔG positive and unfavourable. This implies that at some temperature $\Delta G = 0$ and $\Delta H = T\Delta S$, *i.e.* equilibrium conditions prevail.

The influence of temperature on polymerization reactions can also be assessed from kinetic considerations. As an addition polymerization can be described as a basic three-step process, initiation, propagation and termination, a kinetic scheme can be developed which accounts for all three stages and arrives at rate expressions for each step with corresponding rate constants k_d, k_p and k_t respectively. As will be seen later, for a free radical polymerization this kinetic analysis shows that the rate of polymerization is proportional to $k_p(k_d/k_t)^{1/2}$ and the degree of polymerization is proportional to $k_p/(k_d k_t)^{1/2}$, so the temperature dependence of polymerization can be assessed using these relations. If it is assumed that each rate constant follows an Arrhenius dependence

$$k_i \;=\; A\exp(-E_i/RT) \tag{24}$$

where E_i is the activation energy for each step, it follows that

$$k_p(k_d/k_t)^{1/2} \;=\; A_p(A_d/A_t)^{1/2}\exp-[E_p \;+\; E_d/2 \;-\; E_t/2] \tag{25}$$

and the overall activation energy for a polymerization reaction, E_a, is given by

$$E_a \;=\; E_p \;+\; E_d/2 \;-\; E_t/2 \tag{26}$$

Approximate values of the individual activation energies can now be assigned to estimate E_a. For a thermally decomposed radical initiator, E_d lies typically in the range $120-170\,\mathrm{kJ\,mol^{-1}}$, for most vinyl monomers E_p values are between 20 and $45\,\mathrm{kJ\,mol^{-1}}$, and for the termination step $E_t \sim 8-20\,\mathrm{kJ\,mol^{-1}}$. This means that E_a is approximately $85\,\mathrm{kJ\,mol^{-1}}$ and that the rate of polymerization should at least double for each 10 K rise in temperature. A similar analysis for the degree of polymerization shows that E_a is now negative and implies that the polymer chain length will decrease as the temperature of polymerization increases. The logical conclusion is that, as the temperature rises, a temperature will be reached above which no polymer will be formed.

It has been observed by Dainton and Ivin[1] that the polymerization rate does increase as the temperature is raised but that it eventually reaches a maximum after which it decreases and eventually becomes zero. The temperature at which this occurs is a characteristic of the monomer but also depends on conditions of monomer concentration and pressure.

These observations can be explained by recognizing that the propagation step is reversible and that depropagation can occur (equation 27).

$$\sim\!\!M_n^{\displaystyle \cdot} \quad + \quad M \underset{k_{dp}}{\overset{k_p}{\rightleftharpoons}} \sim\!\!M_{n+1}^{\displaystyle \cdot} \tag{27}$$

The rate expression is then

$$-d[M]/dt \quad = \quad R_p \quad = \quad (k_p[M] \quad - \quad k_{dp})[M\cdot] \tag{28}$$

and at equilibrium, where the overall polymerization rate R_p is zero, the equilibrium constant K is

$$K \quad = \quad (k_p/k_{dp}) \quad = \quad 1/[M_e] \tag{29}$$

where $[M_e]$ is the equilibrium monomer concentration. The temperature at which this equilibrium occurs is the ceiling temperature, T_c, and can be described schematically as the intersection of the propagation and depropagation rate curves (see Figure 1). Hence above T_c it is impossible for any polymeric material to form.

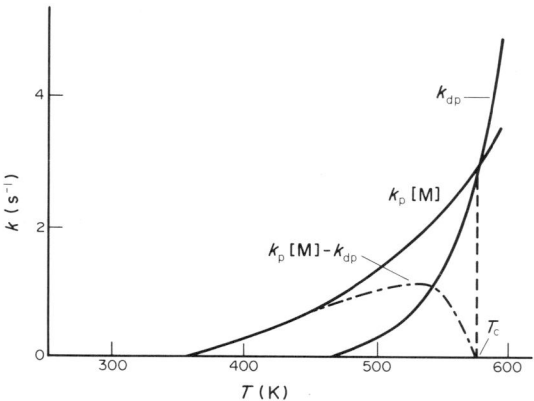

Figure 1 The temperature dependence of $k_p[M]$ and k_{dp} for styrene (after Dainton and Ivin[1])

The thermodynamic significance of T_c can be related to this kinetic analysis by again using the Arrhenius expressions for the rate constants. At equilibrium

$$A_p\exp(-E_p/RT_c)[M_e] \quad = \quad A_{dp}\exp(-E_{dp}/RT_c) \tag{30}$$

and it follows that

$$T_c \quad = \quad (E_p \quad - \quad E_{dp})/\{R\ln(A_p/A_{dp}) \quad + \quad R\ln[M_e]\} \tag{31}$$

The difference in activation energies for the forward and reverse reaction is simply the enthalpy change for the polymerization ΔH_p, so equation (31) becomes

$$T_c \quad = \quad \Delta H_p/\{R\ln(A_p/A_{dp}) \quad + \quad R\ln[M_e]\} \tag{32}$$

The equilibrium constant can be related to the standard free energy change $\Delta G°$, and it follows from equation (33)

$$\Delta G° \quad = \quad -RT\ln K \quad = \quad RT\ln[M_e] \tag{33}$$

that $R\ln(A_p/A_{dp}) = \Delta S°$ the standard entropy change. Equation (33) can be recast as

$$T_c \quad = \quad \Delta H_p/(\Delta S_p° \quad + \quad R\ln[M_e]) \tag{34}$$

This shows that the ceiling temperature is a function of the free monomer concentration and that for a given monomer concentration there will be a specific ceiling temperature at which the particular $[M_e]$ will be in equilibrium with polymer chains. Thus for any polymerization temperature an

Table 2 Equilibrium Monomer Concentrations for 298 K
Estimated from Equation (34)

Monomer	$[M_e]$ (mol dm^{-3})
Vinyl acetate	10^{-9}
Styrene	10^{-6}
Methyl methacrylate	10^{-3}
α-Methylstyrene	2.6

equilibrium monomer concentration can be estimated, as shown in Table 2, for a selected temperature of 298 K.

As ΔH_p is negative, a rise in temperature will cause $[M_e]$ to increase, thus at 405 K methyl methacrylate has a value of $[M_e] \approx 0.5$ mol dm^{-3}, whereas α-methylstyrene will not polymerize at all. Ceiling temperatures then refer to a given monomer concentration and it is more convenient to refer it to a standard state. This can either be referred to pure liquid monomer or a concentration of 1 mol dm^{-3}; typical examples for pure liquid monomers are given in Table 3. While the ceiling temperature alters with monomer concentration, it is also sensitive to pressure. As ΔH and ΔS are both negative, an increase in T_c is obtained if $-\Delta S$ can be decreased. This can be achieved either by raising the monomer concentration (in a solution polymerization) or by decreasing the volume change normally observed during polymerization. Experimental data show that there is a linear dependence of $\log T_c$ on pressure and the Clapeyron–Clausius equation

$$\frac{\mathrm{d}T_c}{\mathrm{d}P} = \frac{T_c \Delta V}{\Delta H} \tag{35}$$

is applicable. Typical values for the rate of increase of T_c with pressure are 0.17 K MPa^{-1} for α-methylstyrene and 0.2 K MPa^{-1} for tetrahydrofuran.

Table 3 Ceiling Temperatures Based on Pure Liquid Monomer as the
Standard State

Monomer	T_c *(pure liquid monomer)* (K)
Tetrafluoroethylene	853
Styrene	583
Methyl methacrylate	493
Thioacetone	368
Tetrahydrofuran	353
α-Methylstyrene	334
Acetaldehyde	242

It should be noted that whereas the preceeding discussion has been cast in terms of free radical polymerizations, the thermodynamic argument is independent of the nature of the active species. Consequently, the analysis is equally valid for ionic polymerizations. A further point to note is that for the concept to apply an active species capable of propagation and depropagation must be present. Thus inactive polymer can be stable above the ceiling temperature for that monomer but the polymer will degrade rapidly by a depolymerization reaction if main chain scission is stimulated above T_c (see Volume 6, Chapter 15).

1.1.5 Heats of Polymerization

It was noted earlier that addition polymerization is an exothermic process and the change in enthalpy is typically in the range 34–160 kJ mol^{-1}. The particular values differ for each monomer and are influenced by several factors, *viz*: (a) the energy difference between monomer and polymer resulting from resonance stabilization of the double bond by the substituent or by conjugation; (b) steric strains in the polymer imposed on the new single bonds by substituent interactions; and (c) polar or secondary bonding effects.

The most important factors are (a) and (b) and while the general observation is that the higher the resonance stabilization in the monomer the less exothermic the reaction, it is believed that steric factors have the greatest effect on ΔH_p. Thus the unusually high steric strains generated on forming poly(α-methylstyrene) are caused by interactions between the phenyl rings and α-methyl units and result in low values for ΔH_p and T_c. These also seem to be responsible for the facile unzipping degradation reaction. Polymerization of tetrafluoroethylene, on the other hand, is a highly exothermic reaction, and produces a polymer with little steric strain. Some values for ΔH_p are shown in Table 4, which have been measured calorimetrically and can be compared favourably with data calculated from equation (34).

Table 4 Heats of Polymerization for Selected Monomers

Monomer	$-\Delta H_p{}^a$ (kJ mol^{-1})	$-\Delta H_p{}^b$ (kJ mol^{-1})
α-Methylstyrene	35.2	34.1
Isobutene	52.7	—
Formaldehyde	54.3	58.5
Methyl methacrylate	58.1	56.0
Ethyl methacrylate	58.9	60.2
Styrene	68.5	—
Vinyl acetate	89.0	—
Vinyl chloride	95.7	—
Tetrafluoroethylene	155.5	—

[a] Calorimetric.
[b] Equation (34).

The large heat of polymerization can have serious practical consequences, especially when polymerizations are rapid, and can even lead to thermal explosions. To avoid these defects the rate of the process must be controlled or other practical expediencies adopted. Heat removal is particularly problematic in bulk polymerizations taken to high conversions as the reaction mixtures become very viscous and efficient stirring becomes difficult.

The generation of dangerous hot spots in the reaction may be avoided by keeping path lengths for heat loss low, by performing polymerizations in solution, or providing a heat sink by carrying out polymerizations in emulsions and dispersions (Volume 4, Part 2) where a large volume of inert liquid phase is present.

1.1.6 Polymer Microstructure

When an asymmetric vinyl monomer (CH_2=CHR) is polymerized, or a diene undergoes 1,2- or 3,4-addition, each substituted backbone chain atom is a chiral centre and there are two possible stereoregular placements for the monomer when it adds onto the growing chain. These centres do not necessarily exist in the monomer but are formed when the monomer is joined to the chain and the stereoregularity of the polymer is determined by the way the incoming monomer orients itself with respect to the growing chain end. If there is no control over the monomer orientation as it is attached to the chain, then each of the two possible placements occur at random and chains formed have irregular sequences of steric placements. The polymer formed in this way is called an *atactic* chain. Most radical-initiated polymerizations lead to the formation of atactic polymers. This can be attributed to the fact that in the growing chain the terminal unit is a radical in a planar sp^2 configuration, which can rotate freely. As the incoming monomer adds to the chain it forms a new radical and in doing so locks the penultimate unit into its steric configuration relative to the rest of the chain, but with freedom of bond rotation the fixing of the penultimate unit orientation is a random process.

Under certain conditions closer control of the orientation of the incoming monomer is possible. In ionic polymerizations the position of the counterion can assist and if the terminal unit acts as an ion pair, a weak four centred coordination complex may form between the incoming monomer and the ion pair. This holds the monomer always in the same orientation relative to the chain end so that it adds in the same way each time forming a stereoregular chain. Ziegler–Natta catalysts can also be used to control monomer orientation but the mechanism differs from the ionic polymerizations. Using some of these systems it is possible to synthesize polymers with only one type of stereoregular placement. If all the chiral atoms along the chain have the same steric configuration, then the polymer is said to be *isotactic*. In the extended planar zigzag representation, shown in Figure 2(a), all the R groups are located on the same side of the plane bisecting the chain backbone longitudinally. If on the other hand each successive substituted atomic centre has a regularly alternating configuration, then the chain is *syndiotactic*. In this case the planar zigzag form shows that the R group alternates from one side of the plane to the other in a regular fashion as shown in Figure 2(b).

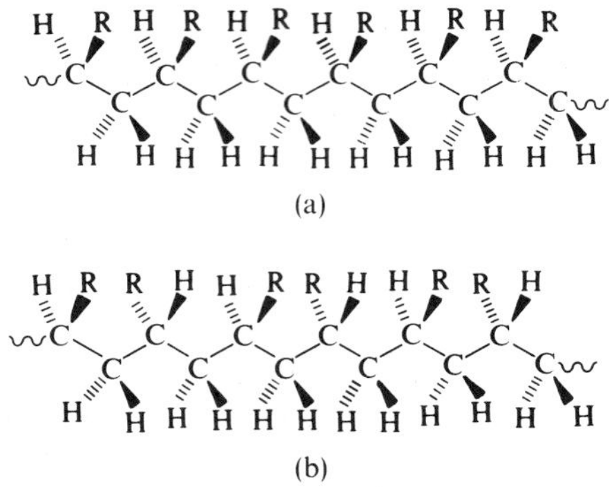

(a)

(b)

Figure 2 (a) Zigzag structure showing isotactic placings; (b) zigzag structure showing syndiotactic placings

As described earlier, dienes can be polymerized by 1,2- or 3,4-addition in which the diene behaves like an asymmetric vinyl monomer and these can be prepared in isotactic, syndiotactic and atactic forms.

None of the above systems yield optically active polymers, but in some cases of ring-opening polymerization, where a substituted heterocyclic monomer is involved, heterotactic or isotactic structures which rotate polarized light can be formed (Scheme 3). If a pure enantiomer of propylene oxide is used, then an isotactic polymer is formed which is optically active. If racemic mixtures of the monomer are used, the reaction is more complex and may polymerize randomly to produce atactic material. However, there are two other possibilities: the reaction may show preference for one enantiomer over another, producing a chain enriched in one form, or, alternatively, the growing chain may prefer to incorporate only the same type of enantiomer thereby forming an all $d(+)$ or all $l(-)$ chain. The former is a *stereoelective* reaction whereas the latter is a *stereoselective* polymerization.

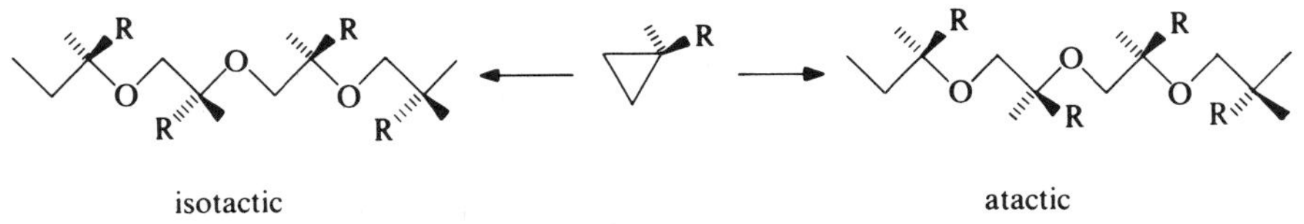

isotactic   atactic

Scheme 3

1.1.7 Network Formation[2,3]

In free-radical-initiated addition polymerization reactions using bifunctional monomers it is normally expected that the product will be a linear polymer. This is not always found to be the case as side reactions can lead to branching or even network formation. If, during the propagation, there is chain transfer from a growing chain A to an inactive chain B which involves hydrogen abstraction from chain B, then a new radical site is created on B capable of initiating further chain growth or reacting with a second radical site. This will form either a branched molecule (equation a; Scheme 4). or a crosslink (equation b; Scheme 4). When the transfer reactions produce single branched molecules only, this is insufficient to lead to an infinite network structure. If two growing branches combine to form a crosslink, or reaction (b), Scheme 4 occurs, there is an increased probability of network formation but this is more likely if an alternative crosslinking reaction is also available to the system or if dienes are involved in the reaction. In the latter case radical attack at the site of the double bond results in attachment of one chain to the other with the concurrent generation of a radical on an adjacent carbon atom from which a new chain can grow as represented in equation (36)

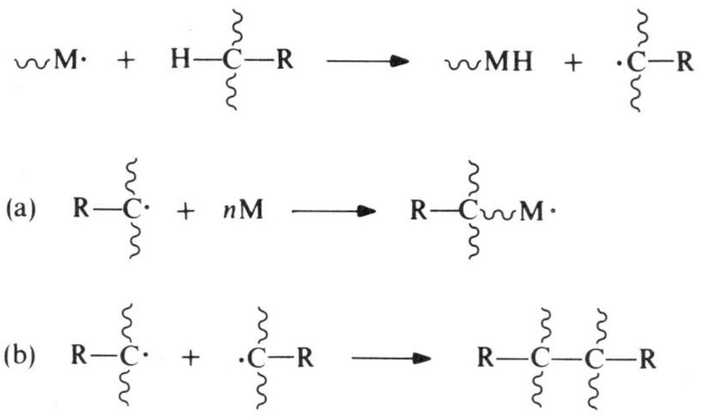

$$\text{~~M}\cdot \; + \; \text{H}-\overset{\xi}{\underset{\xi}{\text{C}}}-\text{R} \;\longrightarrow\; \text{~~MH} \; + \; \cdot\overset{\xi}{\underset{\xi}{\text{C}}}-\text{R}$$

$$(a) \qquad \text{R}-\overset{\xi}{\underset{\xi}{\text{C}}}\cdot \; + \; n\text{M} \;\longrightarrow\; \text{R}-\overset{\xi}{\underset{\xi}{\text{C}}}\text{~~M}\cdot$$

$$(b) \qquad \text{R}-\overset{\xi}{\underset{\xi}{\text{C}}}\cdot \; + \; \cdot\overset{\xi}{\underset{\xi}{\text{C}}}-\text{R} \;\longrightarrow\; \text{R}-\overset{\xi}{\underset{\xi}{\text{C}}}-\overset{\xi}{\underset{\xi}{\text{C}}}-\text{R}$$

Scheme 4

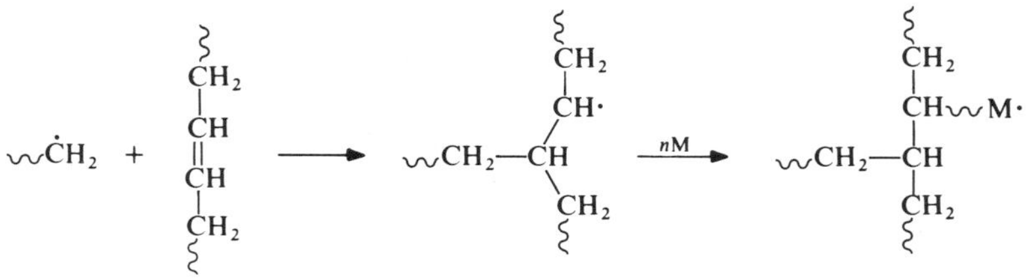

$$\text{~~}\dot{\text{C}}\text{H}_2 \; + \; \begin{array}{c}\xi\\ \text{CH}_2\\ |\\ \text{CH}\\ ||\\ \text{CH}\\ |\\ \text{CH}_2\\ \xi\end{array} \;\longrightarrow\; \text{~~CH}_2-\begin{array}{c}\xi\\ \text{CH}_2\\ |\\ \text{CH}\cdot\\ |\\ \text{CH}\\ |\\ \text{CH}_2\\ \xi\end{array} \;\xrightarrow{\,n\text{M}\,}\; \text{~~CH}_2-\begin{array}{c}\xi\\ \text{CH}_2\\ |\\ \text{CH~~M}\cdot\\ |\\ \text{CH}\\ |\\ \text{CH}_2\\ \xi\end{array} \qquad (36)$$

(or reaction a, Scheme 4). This in effect creates a tetrafunctional unit in the chain and the continued reaction can lead to the eventual formation of an infinite network.

Polydienes and diene-containing copolymers are susceptible to crosslinking by oxygen, which eventually leads to their embrittlement and degradation. In the classical vulcanization of rubber, a more controlled crosslinking is achieved by adding sulfur which reacts with the unsaturated groups and forms sulfur bridges between the molecules (see Volume 6, Chapter 4). Network formation can also be assisted by adding a reagent such as dicumyl peroxide to the polydienes but the efficiency of this reaction has been found to depend on the type of the elastomer. More generally, crosslinking can be effected by γ-irradiation of a linear polymer sample or by adding divinyl compounds to the polymerization mixture, *e.g.* divinylbenzene or 1,2-ethanediol dimethacrylate which act as tetra-functional units.

Crosslinking processes are important in practice and are used to provide mechanical stability to naturally occurring rubbers and other elastomers. The reaction is also used to cure polymers in the formation of surface coatings, crosslinked beads and polymer support reagents.

The crosslinking reaction is then a progressive build up of larger and larger molecules, starting from the chains initially present (primary chains) which are joined together forming bigger crosslinked species. These in turn are linked to each other as the numbers of primary chains are depleted and, if conditions are right, an infinite network is formed. This can occur quite abruptly during the course of the reaction at the gel point, where the reaction mixture consists of large insoluble three dimensional network structures and a sol fraction comprising smaller crosslinked units which are still soluble and have yet to be incorporated into the larger structures. Thus there is a changing population of molecular sizes and structures throughout the reaction, which continues beyond the gel point as the remaining sol units are linked into the gel phase.

Clearly, it is of interest to be able to follow the gelation and predict the occurrence of the gel point. This can be achieved by using simple arguments, developed by Flory, which apply equally well to network formation by crosslinking of preformed chains or by using multifunctional monomers in the polymerization reaction.

Consider a polymer sample containing initially x-mers, all of the same length, undergoing an ideal crosslinking reaction in which each time a crosslink forms the total number of molecules in the system is reduced by one. It is also assumed that there is no intramolecular crosslinking and that the intermolecular reaction is a random process. The crosslink density ρ can be defined as the probability of any monomeric unit being the site of a crosslink and is expressed as

$$\rho \; = \; 2\Phi N_0 \qquad (37)$$

where Φ is the number of crosslinks formed and N_0 is the total number of monomer units in the system. For primary molecules, with a number average degree of polymerization P_n, the crosslinking reaction begins with a primary chain A being linked to a second primary chain B, but if an infinite network is to be formed then a certain critical number of crosslinks ε must be formed on chain B. This can be derived from the total number of units available on chain B for crosslinking and the probability of a unit reacting, *i.e.*

$$\varepsilon = \rho(P_n - 1) \tag{38}$$

This shows that the formation of an infinite network will only be possible if there are sufficient crosslinks formed to ensure that $\varepsilon = 1$. A pair of crosslinked chains can be regarded as being equivalent to a tetrafunctional group in a non-linear step growth reaction, so at least one crosslink bond per primary chain is needed for an infinite network to form. This means that critical conditions are given by $\varepsilon = 1$ and

$$\rho_c = \frac{1}{(P_n - 1)} \approx \frac{1}{P_n} \tag{39}$$

thereby suggesting that the larger the primary molecules the fewer the number of crosslinks which are required for gelation.

A polymer sample with a completely monodisperse molecular weight distribution is rather unusual and equation (40) should be derived for a polydisperse sample. As the probability of finding a crosslink point on a long primary chain is greater than on a short one, it is more realistic to use the weight fraction of x-mer, W_x, rather than the mole fraction when calculating ε and so

$$\varepsilon = \rho W_x(x - 1) \tag{40}$$

If the expectations for all the species in the distribution are considered then an average ε can be written as

$$\bar{\varepsilon} = \sum_i \varepsilon_i = \sum_i \rho(W_x)_i(x_i - 1) \tag{41}$$

and

$$\bar{\varepsilon} = \rho \sum_i [(W_x)_i x_i - (W_x)_i] \tag{42}$$

Remembering that $\sum_i (W_x)_i x_i = \bar{P}_w$, the weight average degree of polymerization, and that $\sum_i (W_x)_i = 1$ it follows that

$$\bar{\varepsilon} = \rho \bar{P}_w \tag{43}$$

and equation (43) remains valid for any molecular weight distribution. Thus for a random crosslinking process $\varepsilon = 1$ when the sample is monodisperse but if there is a 'most probable' distribution where $(\bar{P}_w/\bar{P}_n) = 2$ then the critical value is $\varepsilon = 0.5$.

A modern theoretical treatment of crosslinking is presented later in these volumes (Volume 5, Chapter 8; Volume 6, Chapter 8).

1.1.8 Thermoplastics and Thermosets

As addition polymerization can produce linear, branched and crosslinked materials, it is necessary to understand how the properties change with structure. The commonly used characteristic parameters which act as guides to polymer behaviour are the glass transition temperature, T_g, and the melting point T_m, where $T_m > T_g$.

If, for simplicity, the behaviour of a totally amorphous linear polymer is to be considered, it will possess a T_g but no T_m. Below the T_g the polymer behaves like a rigid, glassy material which, at the molecular level, is indicative of the fact that the chains are essentially immobile and incapable of large scale cooperative Brownian motion. When the sample is heated, it expands but remains a rigid plastic until a specific temperature is reached, *e.g.* T_g, above which there is a rapid change in sample behaviour. The chains are now able to move relatively freely, the sample softens and behaves initially as a tough leather-like material. With a further increase in temperature the chains become more mobile until they eventually flow like a viscous liquid. On cooling below the T_g, chain mobility is again lost and the polymer regains its hard glassy qualities. This process of thermal cycling—hardening–softening–hardening—is infinitely reversible (assuming that no degradation takes place) and a material which behaves in this way is called a *thermoplastic*. Thus if one wishes to impart a certain shape to a thermoplastic material, it can be heated above its T_g where it softens sufficiently to be moulded into shape and on cooling below T_g the shape is 'locked in' and retained. This shape can be changed simply by reheating, altering and cooling again.

It should be noted, however, that the magnitude of the T_g will depend on the relative flexibility of the polymer chain and can vary from as low as ~ 148 K for poly(dimethylsiloxane) to as high as 445 K for poly(α-methylstyrene) or even higher for some condensation polymers. The value will determine the likely end use of the polymer, *i.e.* a low T_g is necessary for elastomeric behaviour and a high T_g is suitable for rigid engineering plastics; these criteria refer of course to the temperature range in which the materials will be used.

Polymers which tend to crystallize show significant property changes when compared with totally amorphous material. The crystalline regions tend to act like crosslinks between the amorphous regions and while the effect on properties is less noticeable below T_g, crystallinity can lead to an increase in T_g compared with the totally amorphous counterpart. It will also increase the modulus of the sample in the temperature region between T_g and T_m. The effect of this can be seen in polyethylene where the T_g is subambient but the polymer can act as a tough leathery material up to just below the melting point because of the high level of crystallinity. Heating above T_m leads to the melting of the crystallites and a highly viscous liquid is obtained. These materials are also thermoplastic as the changes are reversible and they are capable of undergoing the hardening–softening recycling process.

In semi-crystalline polymers the crystalline regions act as thermally labile crosslinks which can break and reform repeatedly. When the crosslink is a covalent bond, which can be broken at high temperature but is unlikely to reform, there is a significant change in physical properties. Chemical crosslinking of a polymer results in an increase in T_g caused by the restrictions in chain motion by the crosslinks which tie the chains together. As the crosslink density increases, so too does the T_g and this may rise to temperatures which exceed the onset of degradation in the sample.

When a thermoplastic material is heated above its T_g, the chains are able to move quite freely, which means that the material softens and can flow like a viscous liquid, but in a crosslinked material this is no longer possible. Now the chains are restricted by crosslinking and have only limited spacial movement relative to their neighbours to which they are bound. It is impossible for chains to slip past one another and so liquid flow cannot take place in these systems. Thus in an infinite network, chain movement is insufficient above T_g, or for high crosslink densities T_g may be too high to reach, for the material to soften sufficiently to allow moulding like the thermoplastics. Crosslinked systems are called *thermoset* polymers, *i.e.* once the crosslinking reaction has taken place the material is essentially thermally intractable (although it can be shaped by machining). This illustrates the importance of being able to predict the gel point in a crosslinking reaction, for once this has been reached, the polymer cannot be dissolved in solvents or softened sufficiently by heat, thereby restricting the ability to fabricate the material after the formation of an infinite network.

1.2 GENERAL REFERENCES

1. F. S. Dainton and K. J. Ivin, *Q. Rev., Chem. Soc.*, 1958, **12**, 61.
2. P. J. Flory, 'Principles of Polymer Chemistry', Cornell University Press, New York, 1953.
3. A. Charlesby, 'Atomic Radiation and Polymers', Pergamon Press, Oxford, 1960.

2
Copolymerization

ARCHIE E. HAMIELEC, JOHN F. MACGREGOR and
ALEX PENLIDIS
McMaster University, Hamilton, Ontario, Canada

2.1 INTRODUCTION

Copolymerization permits the synthesis of an almost unlimited range of polymers and is often used, therefore, to obtain a better balance of properties for the commercial application of polymeric materials. Copolymers may be synthesized by chain growth and step growth condensation polymerization processes. This section deals exclusively with chain growth copolymerization, in which one or more types of monomer add to an active center located on a growing (or live) polymer chain. The term 'copolymerization' includes the simultaneous polymerization of two or more types of monomer; however, the terms terpolymerization and multicomponent polymerization are often used to indicate polymerization of three types of monomer and three or more types of monomer, respectively. Chain growth copolymerizations may be done using various active centers, including free radical, ionic and Ziegler–Natta processes. There are four basic copolymer structures: random, alternating, block and graft. Random copolymers have relatively random distributions of the two monomer units along the polymer chain. Alternating copolymers have the two monomer units, M_A and M_B, occurring in an alternating fashion ($M_A M_B M_A M_B M_A M_B$). A block copolymer is a linear copolymer with one or more long uninterrupted sequences of each type of monomer unit ($M_A M_A M_A M_A M_A M_B M_B M_B M_B M_B$). A graft copolymer is a branched copolymer with the main chain and branches having different compositions. This section deals exclusively with random and alternating copolymers.

Although the reactivity of a monomer in copolymerization cannot be predicted from its behavior in homopolymerization, the terminal model (also called 'simple copolymerization kinetics') can predict, with acceptable accuracy, the detailed microstructure of copolymer chains and the consumption rates of the monomers in many instances. In some cases it may be necessary to use more sophisticated models to predict details of the polymer chain microstructure accurately. Copolymerizations may be performed using heterogeneous as well as homogeneous processes (such as solution, bulk, suspension, emulsion, slurry and gas phase processes) and using batch, semibatch and continuous-flow reactors. It should be noted, however, that the monomer concentrations local to the active centers control the copolymerization, not the average concentrations in the reactor. It has recently been observed that, with heterogeneous Ziegler–Natta catalysts, it is possible to obtain, simultaneously, families of copolymer molecules with quite different chain microstructures. This is probably due to the multisite nature of the catalysts, with the propagation constants varying from site to site. A similar explanation can be used to explain the large polydispersities ($\bar{M}_w / \bar{M}_n$) which are obtained for polyalkenes using heterogeneous Ziegler–Natta catalysts. If the multisite nature of these catalysts are properly accounted for, the terminal model may be valid for these systems.

For excellent introductions to chain growth copolymerization, one can refer to the recent texts by Odian[1] and Rudin,[2] and to the review by Wittmer.[3]

2.2 SIMPLE COPOLYMERIZATION KINETICS — TERMINAL MODEL

2.2.1 Introduction

The basic hypothesis of simple copolymerization kinetics is that the reactivity of an active center only depends upon the monomer unit in the copolymer chain on which the active center is located. Therefore, for a binary copolymerization, the growth of copolymer chains, microstructure development and monomer consumption are uniquely described by four propagation reactions. One should state, in addition, that the copolymer chains must have reasonably large molecular weights. This is to ensure that the statistics for monomer addition to active centers are valid, and that the amount of monomer consumed in reactions other than propagation is negligible. This is often called the long chain approximation. When dealing with copolymer oligomers these hypotheses may not be valid. The model which is based on simple copolymerization kinetics is called the terminal model. This model uses the four propagation steps outlined in equations (1) to (4)

$$P_{m,n,A} + M_A \xrightarrow{k_{AA}} P_{m+1,n,A} \tag{1}$$

$$P_{m,n,A} + M_B \xrightarrow{k_{AB}} P_{m,n+1,B} \tag{2}$$

$$P_{m,n,B} + M_A \xrightarrow{k_{BA}} P_{m+1,n,A} \tag{3}$$

$$P_{m,n,B} + M_B \xrightarrow{k_{BB}} P_{m,n+1,B} \tag{4}$$

where $P_{m,n,A}$ is a live copolymer chain with m units of monomer A (M_A) and n units of monomer B (M_B) bound in the chain, with the active center located on monomer A. The active center may be at the end of the chain or anywhere along the polymer backbone. Values of propagation constants and reactivity ratios may be different if the active centers are nonterminal rather than terminal. The first subscript on the propagation constants denotes the location of the active center, and the second subscript denotes the monomer reactant. For a binary copolymerization there are two types of active center and the total concentrations of these are given by

$$[P_A] = \sum_{m=1}^{\infty} \sum_{n=1}^{\infty} [P_{m,n,A}] \tag{5}$$

$$[P_B] = \sum_{m=1}^{\infty} \sum_{n=1}^{\infty} [P_{m,n,B}] \tag{6}$$

$$[P] = [P_A] + [P_B] \tag{7}$$

It is implied in equations (1)–(4) that the rate constants are independent of chain length. Reactivity ratios for this simple binary copolymerization are given by

$$r_A = k_{AA}/k_{AB} \tag{8}$$

$$r_B = k_{BB}/k_{BA} \tag{9}$$

Reactivity ratios define the relative reactivities of radical centers towards the comonomers. For linear binary copolymer chains, the number of times M_A follows M_B in the chain equals the number of times M_B follows $M_A \pm 1$. Therefore, one can equate R_{AB} to R_{BA} for linear polymer chains (other than oligomers) with a small error, where R_{AB} and R_{BA} are the instantaneous rates of formation of units in the copolymer chains where M_B follows M_A, and M_A follows M_B, respectively. For branched copolymer chains the errors are larger, but still acceptable. Application of this equality gives

$$k_{AB}[P_A][M_B] = k_{BA}[P_B][M_A] \tag{10}$$

$$\Phi_A = k_{BA}[M_A]/(k_{BA}[M_A] + k_{AB}[M_B]) = k_{BA}f_A/(k_{BA}f_A + k_{AB}f_B) \tag{11}$$

where Φ_A is the fraction of active centers of type A and f_A and f_B are the mole fractions of unreacted monomers A and B.

2.2.2 Instantaneous Copolymer Composition Equation

The specific polymerization rates of monomers A and B are given by

$$R_{PA} = k_{AA}[P_A][M_A] + k_{BA}[P_B][M_A] \tag{12}$$

$$R_{PB} = k_{AB}[P_A][M_B] + k_{BB}[P_B][M_B] \tag{13}$$

Dividing equation (12) by equation (13), using equation (10), and introducing reactivity ratios defined by equations (8) and (9), one can readily derive the following form of the instantaneous copolymer composition equation

$$F_A/F_B = R_{PA}/R_{PB} = \{(r_A[M_A] + [M_B])[M_A]\}/\{([M_A] + r_B[M_B])[M_B]\} \tag{14}$$

where F_A and F_B are overall mole fractions of monomer A and monomer B in the copolymer produced instantaneously (*i.e.* in a very small time interval).

It should be pointed out that, in general, all of the copolymer molecules produced instantaneously may not have the same composition, thus requiring the use of overall values for F_A and F_B. This subject will be dealt with more fully in Section 2.2.3. Equation (14) was originally derived by Alfrey and Goldfinger,[4] May and Lewis,[5] and Wall.[6] An equivalent expression to equation (14) which is often used is as follows.

$$F_A = \frac{(r_A - 1)f_A^2 + f_A}{(r_A + r_B - 2)f_A^2 + 2(1 - r_B)f_A + r_B} \tag{15}$$

In principle, equations (14) and (15) can be used to predict the overall composition of copolymer produced instantaneously, given the reactivity ratios and the concentrations of unreacted monomer, or they can be used to estimate the reactivity ratios, given F_A at various f_A levels. This subject is dealt with more fully in Section 2.2.6. A graph of equation (15) for various binary reactivity ratio pairs, showing the magnitude and direction of compositional drift for a batch reactor, is shown in Figure 1. An azeotropic point is also shown (a point where composition of copolymer and unreacted monomer is the same, and no compositional drift occurs).

2.2.3 Instantaneous Bivariate Distribution of Composition and Chain Length for Linear Copolymer Chains

When applying equations (14) and (15), one usually assumes that all of the copolymer molecules produced instantaneously have the same composition. Since the molecular weight of a copolymer is finite, the compositions, as well as the chain lengths of the individual polymer molecules, cannot all be identical. Therefore, for copolymer chains produced instantaneously, there is a bivariate distribution of composition and chain length. This distribution was first derived by Stockmayer,[7] who showed that, although the variance in composition was small for high molecular weight copolymers,

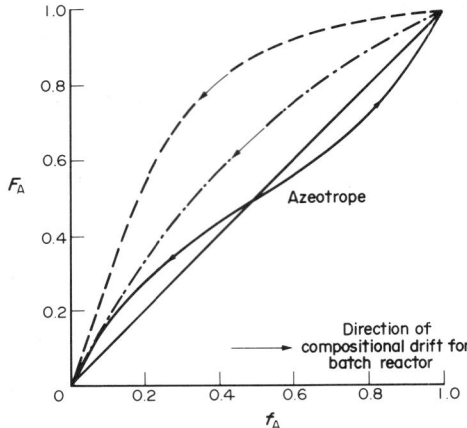

Figure 1 Graph of the instantaneous copolymer composition equation (equation 15) for various binary reactivity ratio pairs, showing magnitude and direction of compositional drift for a batch reactor and an azeotropic point; — — —, $r_A = 10$, $r_B = 0.5$; — · — · —, $r_A = 2$, $r_B = 0.5$; and ———, $r_A = 0.5$, $r_B = 0.5$

it could be significant for low molecular weight copolymers. The instantaneous bivariate distribution for linear copolymer chains is as follows

$$W(l,y) = [(l/\lambda^2)(1 - \rho + \rho l/2\lambda)\exp(-l/\lambda)]\{[1/(2\pi\sigma^2)^{1/2}]\exp(-y^2/2\sigma^2)\} \tag{16}$$

where $W(l,y)\,dl\,dy$ is the weight fraction of copolymer having chain length and composition deviation in the ranges l to $l + dl$ and y to $y + dy$. λ is the number average chain length of the growing chains, often called the kinetic chain length. (When chain transfer reactions are significant, the number average degree of polymerization of the propagating chains and the kinetic chain length are not equal.) ρ is the fraction of growing chains which are terminated by combination. The quantities y, σ^2 and K are given by

$$y = y_A - F_A \tag{17}$$

$$\sigma^2 = F_A(1 - F_A)K/l \tag{18}$$

$$K = (1 - 4F_A(1 - F_A)(1 - r_A r_B))^{1/2} \tag{19}$$

where y_A is the mole fraction of monomer A in the copolymer and, as defined earlier, F_A is the overall mole fraction of monomer in copolymer chains produced instantaneously.

It is convenient to introduce the two dimensionless parameters, τ and β, where

$$\tau = \frac{k_{td}[P]}{k_p[M]} + \frac{k_{fm}}{k_p} + \frac{k_{fT}[T]}{k_p[M]} \tag{20}$$

$$\beta = \frac{k_{tc}[P]}{k_p[M]} \tag{21}$$

λ and ρ are related to τ and β by

$$\lambda = 1/(\tau + \beta) \tag{22}$$

$$\rho = \beta/(\tau + \beta) \tag{23}$$

For binary copolymerization, the pseudokinetic rate constants are related to elementary kinetic rate constants by

$$k_{tc} = k_{tc_{AA}}\Phi_A^2 + 2k_{tc_{AB}}\Phi_A\Phi_B + k_{tc_{BB}}\Phi_B^2 \tag{24}$$

$$k_{td} = k_{td_{AA}}\Phi_A^2 + 2k_{td_{AB}}\Phi_A\Phi_B + k_{td_{BB}}\Phi_B^2 \tag{25}$$

$$k_{fm} = k_{fm_{AA}}\Phi_A f_A + k_{fm_{AB}}\Phi_A f_B + k_{fm_{BA}}\Phi_B f_A + k_{fm_{BB}}\Phi_B f_B \tag{26}$$

$$k_p = k_{AA}\Phi_A f_A + k_{AB}\Phi_A f_B + k_{BA}\Phi_B f_A + k_{BB}\Phi_B f_B \tag{27}$$

$$k_{fT} = k_{fT_A}\Phi_A + k_{fT_B}\Phi_B \tag{28}$$

where k_{fm} and k_{fT} are pseudokinetic rate constants for chain transfer to monomer and to chain transfer agent T, respectively.

Equation (16) shows that, for copolymer chains having a chain length l, the composition follows a Gaussian distribution with mean F_A and variance σ^2, defined by equations (18) and (19). For ideally random copolymers, $r_A r_B = 1$ (see Section 2.2.5.2) and, hence, $K = 1$. The variance is at a maximum at the mid point copolymer composition $F_A = 0.5$, and varies inversely with chain length. For high molecular weight copolymers, it may be possible to set the variance to zero with negligible consequences in polymer microstructure calculations. Equation (16) then reduces to the simpler form

$$W(l) = (\tau + \beta)(\tau + \beta/2(\tau + \beta)l)l \exp(-(\tau + \beta)l) \tag{30}$$

where $W(l)dl$ is the weight fraction of copolymer of composition F_A and chain length l produced instantaneously. Stockmayer[7] has already pointed out that, for batch and semibatch reactors, equations (14) and (16) may be integrated to find the bivariate distribution of chain length and composition for the copolymer molecules accumulated in the reactor over a finite time period.

The appropriate starting points for the calculation of copolymer chain composition and molecular weight distribution for linear copolymers are equations (15) and (16).[8]

2.2.4 Integrated Copolymer Composition Equation

Monomer conversion X is related to the overall mole fraction F_A and to the mole fraction of unreacted monomer f_A in a batch reactor by the following integral equation

$$\ln(1 - X) = \int_{f_{A0}}^{f_A} \frac{df_A}{(F_A - f_A)} \tag{31}$$

where monomer conversion X is given by

$$X = (N_{A0} + N_{B0} - N_A - N_B)/(N_{A0} + N_{B0}) \tag{32}$$

where N_A and N_B are the number of moles of monomers A and B in the reactor at any given time.

Meyer and Lowry[9] substituted for F_A using equation (15), and then integrated equation (31) analytically for constant r_A and r_B and obtained

$$X = 1 - (f_A/f_{A0})^\alpha((1 - f_A)/(1 - f_{A0}))^\beta((f_{A0} - \delta)/(f_A - \delta))^\gamma \tag{33}$$

where

$$\alpha = r_B/(1 - r_B) \tag{34}$$

$$\beta = r_A/(1 - r_A) \tag{35}$$

$$\gamma = (1 - r_A r_B)/((1 - r_A)(1 - r_B)) \tag{36}$$

$$\delta = (1 - r_B)/(2 - r_A - r_B) \tag{37}$$

Reactivity ratios have a small temperature dependence, therefore equation (33) is often used for nonisothermal batch copolymerization. Equation (33) is also often used with experimental X *vs.* f_A data to estimate reactivity ratios. This subject is considered in Section 2.2.6.

2.2.5 Copolymer Microstructure — Reactivity Ratios and Sequence Length Distributions

2.2.5.1 *Monomer sequence length distributions*

We will now derive expressions for monomer sequence length distributions in binary copolymerization. The following probabilities may be defined: P_{AA} = probability that a growing chain with an M_A end adds another M_A monomer, and P_{AB} = probability that a growing chain with an M_A end adds another M_B monomer, with similar definitions for P_{BB} and P_{BA}. The above events are the only ones possible, therefore the probabilities P_{AA} and P_{AB} must give unity when added together. The same is true for P_{BB} and P_{BA}. The probability that exactly n units of M_A in series may be found in a growing chain is

$$N(M_A, n) = P_{AA}^{n-1}(1 - P_{AA}) = P_{AB}P_{AA}^{n-1} \tag{38}$$

and, since

$$\sum_{n=1}^{\infty} N(M_A, n) = 1 \tag{39}$$

where $N(M_A,n)$ is the fraction of M_A sequences of length n. The probabilities may be related to reactivity ratios as follows

$$P_{AA} = k_{AA}[P_A][M_A]/(k_{AA}[P_A][M_A] + k_{AB}[P_A][M_B])$$

$$= \left(\frac{r_A[M_A]}{[M_B]}\right)\Bigg/\left(\frac{r_A[M_A]}{[M_B]} + 1\right) \tag{40}$$

In a similar manner

$$P_{AB} = 1\Bigg/\left(\frac{r_A[M_A]}{[M_B]} + 1\right) \tag{41}$$

$$P_{BB} = \left(\frac{r_B[M_B]}{[M_A]}\right)\Bigg/\left(\frac{r_B[M_B]}{[M_A]} + 1\right) \tag{42}$$

$$P_{BA} = 1\Bigg/\left(\frac{r_B[M_B]}{[M_A]} + 1\right) \tag{43}$$

Using these relationships, $N(M_A,n)$ can be expressed in terms of f_A as

$$N(M_A,n) = \alpha^{n-1}/(1+\alpha)^n \tag{44}$$

$$\text{where} \quad \alpha = r_A f_A/(1 - f_A) \tag{45}$$

A similar relationship may be derived for $N(M_B,n)$

$$N(M_B,n) = \beta^{n-1}/(1+\beta)^n \tag{46}$$

$$\text{where} \quad \beta = r_B f_B/(1 - f_B) \tag{47}$$

Average sequence lengths, $\bar{N}(M_A)$ and $\bar{N}(M_B)$, are given by

$$\bar{N}(M_A) = \sum_{n=1}^{\infty} nN(M_A,n) = 1 + \alpha \tag{48}$$

$$\bar{N}(M_B) = 1 + \beta \tag{49}$$

These expressions, which apply for infinitely long chains, were originally derived by Alfrey and Goldfinger,[4] Wall[6] and, more recently, by other workers.[10-17] Galvan and Tirrell[17] have derived an expression for the average sequence length which is a function of chain length. The calculations show that, for finite chains, the infinite chain assumption introduces the maximum error when there are long sequences of one of the monomers. They point out that this can be important for copolymers such as linear low density polyethylenes which contain low levels of comonomers, 1-butene, 1-hexene or 1-octene. Again, it should be pointed out that these conclusions may have to be modified to account for the multisite character of heterogeneous Ziegler–Natta catalysts.

In batch copolymerization, the sequence length distribution for monomer A in the final co-polymer product is given by

$$\bar{N}(M_A,n) = \int_0^{X_A} \frac{N(M_A,n)}{\bar{N}(M_A)} dX \Bigg/ \int_0^{X_A} \frac{dX_A}{\bar{N}(M_A)} \tag{50}$$

A similar expression may be written for monomer B.

2.2.5.2 *Random copolymers*

When the reactivity ratio product, $r_A r_B$, equals unity, then the probabilities P_{AA} and P_{BA} are equal or, in other words, the probability that an M_A unit follows an M_A unit in the copolymer equals the probability that it follows an M_B unit. This condition must be satisfied for an ideal random copolymer to form. Random copolymers are formed when the active center adds to either monomer with equal facility (*i.e.* $k_{AA} = k_{AB}$, $k_{BB} = k_{BA}$ and $r_A = r_B = 1$), but this is a special case. Free radical copolymerization of ethylene and vinyl acetate provides a random copolymer with the composition of the copolymer equal to the composition of unreacted monomers.[2] The more common situation, where $r_A r_B \sim 1$ for random free radical copolymerization, includes styrene–butadiene ($r_A r_B = 1.1$) and vinyl choride–vinyl acetate ($r_A r_B = 0.9$).

2.2.5.3 *Alternating copolymers*

When crosspropagation rate constants (k_{AB} and k_{BA}) are much larger than their respective homopropagation rate constants (k_{AA} and k_{BB}), and the reactivity ratios approach zero, one obtains an alternating copolymer with $F_A = F_B = 0.5$. Such copolymerizations occur in free radical systems when the two monomer types have opposite polarities.[2] The reactivity of a polar monomer can be considerably enhanced in copolymerization with a monomer of opposite polarity. Maleic anhydride does not homopolymerize under normal free radical polymerization conditions; however, styrene–maleic anhydride copolymers synthesized at moderate temperatures are almost purely alternating ($r_A = r_B = 0$). The complex participation model has often been used to explain the mechanism of synthesis of alternating copolymers (see Section 2.3 and Volume 4, Chapter 23).

2.2.6 Determination of Binary Reactivity Ratios

The determination of reactivity ratios with small confidence intervals requires sensitive analytical techniques, careful planning of experiments and the use of statistically valid methods of estimation. Unfortunately, most of the binary reactivity ratio data published to date were not calculated using statistically valid methods, and confidence intervals were not usually given. The reactivity ratios are highly correlated, and there are often major disagreements in estimated values from different laboratories. In this section, some of the analytical techniques which show promise, the design of experiments to obtain the most precise estimates, and statistically valid methods for estimating reactivity ratios are discussed. Later, in Section 2.3.4, some statistical methods for discriminating between the terminal and penultimate models are also discussed.

2.2.6.1 *Experimental techniques*

The starting point for the experimental measurement of reactivity ratios is, usually, either the instantaneous equations (equations 14 and 15) or the integrated equation (equation 33), with which composition data may be used at high conversions. The instantaneous equations are used at low monomer conversions. Most copolymerizations are done in batch reactors, either at low conversions or up to high conversions, using a range of initial monomer compositions (f_{A0}). Methods for selecting appropriate initial monomer compositions to yield reactivity ratios with the smallest confidence intervals are discussed in Section 2.2.6.3. There are two methods for the determination of the copolymer composition: (i) measurement of the concentrations of unreacted monomers, with the use of mass balances to calculate the mole fractions of the monomers in the accumulated copolymer; and (ii) direct measurement of the composition of the accumulated copolymer. In either case it is necessary to separate the monomers from the copolymer, as both unreacted monomers and polymer contain the same functional groups. Monomer concentrations are usually measured by gas chromatography or high performance liquid chromatography. These analytical methods usually provide adequate accuracy and relatively short analysis times.[18-22] Although direct measurements of copolymer composition (using IR, UV, NMR spectroscopy and other techniques) are usually less accurate, they may be necessary when side reactions consume significant amounts of monomer. These measurement techniques can, of course, provide information on monomer sequence lengths in the copolymer, and this kind of information may be necessary in order to discriminate between copolymerization models.[24-27]

Further discussion of experimental techniques and sampling analysis ('sequential' *vs.* 'quenching') may be found in recent work by Hautus *et al.*[22,23] Another, but less common, reactor type used to measure reactivity ratios is the well-mixed continuous-stirred tank reactor operated under steady-state conditions (CSTR). With this type of reactor, various monomer compositions are used in the feed stream. The instantaneous composition equations (equations 14 and 15) can now be used at all conversion levels. To ensure molecular mixing, and to maintain reaction mixture viscosities at reasonable levels, inert solvents and moderate monomer conversions are normally employed.

2.2.6.2 *Statistically valid methods for estimation of reactivity ratios*

Traditional methods for estimating reactivity ratios[5,28-30] are based on, first, transforming the instantaneous copolymer composition equation into a form that is linear in the parameters r_A and

r_B, for example

$$(f_A/f_B)(F_B/F_A - 1) = -r_B + r_A(f_A/f_B)^2(F_B/F_A) \tag{51}$$

and then estimating the reactivity ratios by graphical plotting or by linear least-squares. However, these approaches, aside from requiring that the instantaneous copolymer composition equation be valid, are statistically unsound, because the 'independent variable' $(f_A/f_B)^2(F_B/F_A)$ contains errors, and the dependent variable $(f_A/f_B)(F_B/F_A - 1)$ does not have constant variance. Both of the latter assumptions are necessary for linear least-squares to be a statistically valid estimation method. As a result, it has been found[31,32] that they often lead to very poor estimates with misleading confidence intervals. The common use, in the past, of these statistically invalid estimation procedures is one of the reasons for the wide variation in reactivity ratios reported by different researchers.

It should be mentioned here that the Kelen and Tudos[30] method, although statistically invalid, can be used to obtain at least 'good initial' r_A and r_B estimates, provided the experiments have been suitably designed. More details are given by McFarlane *et al.*[33] and Laurier *et al.*[34]

Currently, reactivity ratios are usually estimated using procedures based on the statistically valid error-in-variables model (EVM), or on its modifications.[18,19,35-40] These methods allow one to take into account properly all the sources of experimental error. No groups of variables are considered to be 'independent' and free of error, or 'dependent' with 'constant' error. All measured variables (*e.g.* $[M_j]^m, j = A, B$ comonomers) are considered on an equal basis, and represented as coming from some true (but unknown) value $[M_j]^t$ which is contaminated with measurement error, that is

$$[M_j]_i^m = [M_j]_i^t + \varepsilon_i \tag{52}$$

To illustrate the EVM method, consider the integrated copolymer composition equation (33). Suppose N experiments were performed, and, for each experiment ($i = 1, 2, \ldots, N$), the values of the initial comonomer concentrations, $[M_{A0}]_i$ and $[M_{B0}]_i$, are measured with error variances σ_{A0}^2 and σ_{B0}^2, and covariance σ_{AB0}^2, and the final comonomer concentrations $[M_A]_i$, $[M_B]_i$ are measured independently of the initial concentrations and with variances σ_A^2, σ_B^2, and covariance σ_{AB}^2. The values of the variances will depend on the measurement methods used, and their values can be estimated from replicated experiments.

The copolymer composition equation (33) can be rewritten in terms of the true comonomer concentrations as

$$f\{[M_{A0}]^t, [M_{B0}]^t, [M_A]^t, [M_B]^t, r_A, r_B\} = 1 - X - \left(\frac{f_A}{f_{A0}}\right)^\alpha \left(\frac{1-f_A}{1-f_{A0}}\right)^\beta \left(\frac{f_{A0}-\delta}{f_A-\delta}\right)^\gamma = 0 \tag{53}$$

where

$$X = 1 - \frac{[M_A]^t + [M_B]^t}{[M_{A0}]^t + [M_{B0}]^t} \tag{54}$$

$$f_{j0} = \frac{[M_{j0}]^t}{\sum\limits_{j=1}^{2} [M_{j0}]^t}, \quad j = A, B \tag{55}$$

Since the only measurements one has of these comonomer concentrations are contaminated with error (equation 52), the functional relationship (equation 53) will not hold exactly if the true concentrations $[M_j]_i^t$ are replaced by their measurements $[M_j]_i^m$, that is

$$f\{[M_{A0}]_i^m, [M_{B0}]_i^m, [M_A]_i^m, [M_B]_i^m, r_A, r_B\} = \varepsilon_i \tag{56}$$

where ε_i represents the propagated measurement error in $f_i\{\quad\}$ of equation (53).

Approximate EVM estimates of the reactivity ratios can be obtained by minimizing the weighted sum of squares expression

$$\sum_{n=1}^{N} \frac{\{f\{[M_{A0}]_i^m, [M_{B0}]_i^m, [M_A]_i^m, [M_B]_i^m, r_A, r_B\}\}^2}{\text{Var}(\varepsilon_i)} \tag{57}$$

where $\text{Var}(\varepsilon_i)$ is the estimate of the variance of the propagated error ε_i based on a first order Taylor series expansion, that is:

$$\text{Var}(\varepsilon_i) \doteq P_{A0_i}^2 \sigma_{A0}^2 + P_{B0_i}^2 \sigma_{B0}^2 + 2P_{A0_i} P_{B0_i} \sigma_{AB0}^2 + P_{A_i}^2 \sigma_A^2 + P_{B_i}^2 \sigma_B^2 + 2P_{A_i} P_{B_i} \sigma_{AB}^2 \tag{58}$$

where P_{A0_i}, P_{B0_i} and P_{B_i} are the partial derivatives of the functional relationships $f_i\{\ \}$ in equation (53) with respect to the variables $[M_{A0}]$, $[M_{B0}]$, $[M_A]$ and $[M_B]$, respectively, evaluated at the measured values of the comonomer compositions for the *i*th experiment, that is

$$P_{j_i} = \left[\frac{\partial f_i\{\ \}}{\partial [M_j]}\right]_m \tag{59}$$

Analytical expressions for these derivatives are given in Patino-Leal *et al.*[39] for a slightly different form of the copolymer equation from that used here. This approximate EVM procedure is sometimes referred to as 'the error propagation method'. From the analysis of many sets of copolymer composition data, it has been found[39] that this method gives estimates of the reactivity ratios that are very close (usually to within 0.1%) to the estimates given by the exact EVM procedures.[19,22,23,40] It is widely used, therefore, as it is both simpler to use and sufficiently accurate. The estimates obtained by these EVM approaches are superior to those obtained by arbitrary graphical or arbitrary least-squares procedures.

Although EVM procedures require more information than arbitrary least-squares procedures, namely on the measurement error variances and covariances, the point estimates of the reactivity ratios obtained by these procedures have been shown to be very insensitive to changes in the covariance values. The great improvement provided by the EVM estimation methods apparently comes from simply accounting correctly for the major sources of error.

In addition to obtaining point estimates of the reactivity ratios, it is important to obtain some idea of the amount of uncertainty associated with these estimates. An approximate 95% joint confidence region for the true values of r_A and r_B is given by the region contained in the ellipsoid

$$h_{AA}(r_A - \hat{r}_A)^2 + 2h_{AB}(r_A - \hat{r}_A)(r_B - \hat{r}_B) + h_{BB}(r_B - \hat{r}_B)^2 \leq \chi^2_{0.95,2} \tag{60}$$

where h_{AA}, h_{BB} and h_{AB} are the elements of the Hessian matrix evaluated at the final point estimates $(\hat{r}_A, \hat{r}_B)$, and $\chi^2_{0.95,2}$ is obtained from tables of the chi-squared distribution with two degrees of freedom. An estimate of the Hessian matrix is usually available from the nonlinear estimation routine used to minimize equation (57). Since the Hessian is proportional to the error variances used in the EVM procedure, the joint 95% confidence region will be sensitive to these values, even though the point estimates are not. A recent comprehensive evaluation of the Fineman–Ross method and EVM for the estimation of binary reactivity ratios has been made by Chee and Ng.[41] The EVM method is by far the superior method, with the Fineman–Ross method giving errors often greater than 200%.

Extensions of these EVM methods to the estimation of reactivity ratios in terpolymerizations have been discussed by Duever *et al.*[42,43]

2.2.6.3 *Design of experiments for precise estimation of reactivity ratios*

When performing experiments to obtain data for the estimation of reactivity ratios, one would like to obtain data containing the maximum amount of information possible. This involves ideas on the statistical design of experiments, where a design consists of the choice of the experimental conditions at which one is to perform the experiments and collect the data. In order to estimate the reactivity ratios in the copolymer composition (equations 33 and 53), this involves the choice of a set of conditions $(f_{j0}, f_j)_i$, $i = 1, 2 \ldots, N$, at which to perform and analyze N experiments.

One measure of the precision of the estimates obtained from any experimental design is the area of the joint confidence region, given in equation (60) for the EVM procedure. In order to minimize the area of this confidence region, it can be shown that one should choose the experimental conditions in such a way as to maximize the determinant of the Hessian matrix H, that is

$$\max \quad |H| \tag{61}$$

$$(f_A, f_{A0})_i = 1, 2, \ldots, N$$

An estimate of the Hessian matrix is given by

$$H \doteq X^T V^{-1} X \tag{62}$$

where X is the $(N \times 2)$ Jacobian matrix of partial derivatives of the functional copolymer equation (53) with respect to the reactivity ratios, i.e. $(\partial f\{\ \}/\partial r_j)$, evaluated at the best estimates of the reactivity ratios, and at the chosen experimental settings $(f_{A0}, f_A)_i$, $i = 1, 2, \ldots, N$, and V is the diagonal covariance matrix of the propagation errors or residuals defined in equation (58). It should be noted that, because of the nonlinear nature of the equations, the Hessian matrix is dependent upon the reactivity ratios, *i.e.* it appears that one needs to know these ratios prior to designing an experiment to estimate them. In practice, one uses the best estimates available in order to carry out the design.

These optimal design procedures are usually used only as a method of determining the experimental regions which contain most information on the reactivity ratios for various combinations of (r_A, r_B). The use of these procedures in designing experiments for the closely related problem of parameter estimation in vapor–liquid equilibrium experiments, based on EVM methods, has been discussed by Sutton and MacGregor.[37] Tidwell and Mortimer[31,32] have discussed their use in the design of copolymer reactivity ratios experiments based on the instantaneous copolymer composition equations (equations 14 and 15) and on least-squares estimation. The design procedure involves the choice of both the initial monomer composition (f_{A0}) and the final composition (f_A) or the final conversion (X) for each experiment. It can be shown that the design will lead to two optimal experimental regions for estimating the two reactivity ratios. For example, with styrene–acrylonitrile copolymerization $(r_A = 0.36, r_B = 0.08)$ these methods show that one of the experimental conditions must be very low styrene composition $(f_{A0} \simeq 0.03)$ in order to obtain good estimates of r_B. If several runs are not performed at these conditions, then very wide confidence regions result. In general, one will also obtain much higher precision by using the integrated copolymer equation (equation 33), and running the experiment to higher conversion.[22] This is true because, for low conversion runs, the experimental errors in (f_{A0}, f_A) are large compared to the difference $(f_{A0} - f_A)$ used to estimate the copolymer composition. Similarly, very high conversion experiments (particularly in a CSTR) provide little information, because the copolymer composition depends mostly on the monomer feed composition (f_{A0}) and is insensitive to the reactivity ratios.

2.2.7   Multicomponent Copolymerization

Multicomponent copolymerization is the term used to designate the chain growth copolymerization of three or more types of monomer. When n types of monomer are simultaneously copolymerized there are n types of active center. This assumes that the propagation reaction rate constants do not vary spatially as with heterogeneous Ziegler–Natta catalysts. The specific rate of consumption of monomer of type j and the total specific rate of consumption of monomer are given by

$$R_{p_j} = \left(\sum_{i=1}^{n} k_{ij} [P_i] \right) [M_j] \tag{63}$$

$$R_p = \sum_{j=1}^{n} R_{p_j} \tag{64}$$

The mole fraction of monomer j in the copolymer produced instantaneously is given by

$$F_j = R_{p_j}/R_p = \left(\sum_{i=1}^{n} k_{ij}\Phi_i \right) f_j \bigg/ \left(\sum_{i=1}^{n} \sum_{j=1}^{n} k_{ij}\Phi_i f_j \right) \tag{65}$$

where Φ_i is the fraction of active centers of type i and f_j is the mole fraction of unreacted monomer j.

Application of the stationary-state hypothesis for active centers of type j gives

$$\left(\sum_{i=1}^{n} k_{ji} f_i \right) \Phi_j = \left(\sum_{i=1}^{n} k_{ij}\Phi_i \right) f_j, \quad j = 1, \ldots, m; \ i \neq j \tag{66}$$

The composition of the copolymer formed instantaneously from n monomer types can be expressed in terms of the unreacted monomer mole fractions and $n(n-1)$ binary reactivity ratios. This has already been demonstrated for binary copolymerization; for terpolymerization the instantaneous copolymer composition equations take the form[2]

$$F_A/F_C = \frac{f_A(f_A r_{BC} r_{CB} + f_B r_{CA} r_{BC} + f_C r_{CB} r_{BA})(f_A r_{AB} r_{AC} + f_B r_{AC} + f_C r_{AB})}{f_C(f_A r_{AB} r_{BC} + f_B r_{AC} r_{BA} + f_C r_{AB} r_{BA})(f_C r_{CA} r_{CB} + f_A r_{CB} + f_B r_{CA})} \tag{67}$$

$$F_B/F_C = \frac{f_B(f_A r_{CB} r_{AC} + f_B r_{AC} r_{CA} + f_C r_{AB} r_{CA})(f_B r_{BA} r_{BC} + f_A r_{BC} + f_C r_{BA})}{f_C(f_A r_{AB} r_{BC} + f_B r_{AC} r_{BA} + f_C r_{AB} r_{BA})(f_C r_{CA} r_{CB} + f_A r_{CB} + f_B r_{CA})} \tag{68}$$

where F_A, F_B and F_C are the overall mole fractions of monomers A, B and C bound in the terpolymer produced instantaneously.

2.2.8 Copolymer Composition Control

2.2.8.1 Compositional drift in batch copolymerization

In the absence of an azeotrope, and when one monomer is more reactive than the other in a binary batch copolymerization (*e.g.* $r_A > 1$ and $r_B < 1$), the instantaneous copolymer composition will decrease in monomer A with increase in conversion. The extent of composition drift, which leads to a copolymer heterogeneous in composition, depends on the ratio of reactivity ratios r_A/r_B (increasing with any increase in r_A/r_B), the initial monomer composition (f_{A0}) and the monomer conversion (X). A copolymer which is heterogeneous in composition usually has inferior properties, therefore industrial processes have been developed to reduce composition heterogeneity. These processes are usually semibatch, but are sometimes continuous. There are, however, certain semibatch emulsion polymerization processes where heterogeneous copolymers are produced through the latex particles to achieve certain property improvements.[44]

2.2.8.2 Monomer feed policies in semibatch copolymerization

There are two basic monomer feed policies which may be used in semibatch copolymerization to minimize compositional drift. Effective commercial processes are usually based on one or a combination of these feed policies. The two feed policies are discussed briefly herein; additional detail can be found elsewhere.[8,45] The definitions of policies 1 and 2 for a binary copolymerization are given below.

Policy 1: All of the slower monomer, and sufficient of the faster monomer (to give the desired copolymer composition F_A), are added to the reactor at time zero. Thereafter, the faster monomer is fed to the reactor with a time-varying feedrate to maintain N_A/N_B (the ratio of the number of moles of monomers A and B in the reactor) and F_A constant with time.

Policy 2: A charge of monomers A and B at the desired monomer concentration levels (to give the desired F_A) is added to the reactor at time zero. Thereafter, monomers A and B are fed to the reactor with time-varying feedrates to maintain $[M_A]$, $[M_B]$ and F_A constant with time.

The equations which need to be solved in order to determine the required feedrates to produce a homogeneous copolymer in a semibatch process are as follows

$$dN_A/dt = -\hat{\Phi}_A[P]N_A + F_{A,in} \tag{69}$$

$$dN_B/dt = -\hat{\Phi}_B[P]N_B + F_{B,in} \tag{70}$$

$$\frac{dV}{dt} = M_{mA}F_{A,in}/\rho_{mA} + M_{mB}F_{B,in}/\rho_{mB} + (M_{mA}\hat{\Phi}_A[P]N_A$$
$$+ M_{mB}\hat{\Phi}_B[P]N_B)/\rho_p - \hat{\Phi}_A[P]N_A/\rho_{mA} - \hat{\Phi}_B[P]N_B/\rho_{mB} \tag{71}$$

where

$$\hat{\Phi}_A = (k_{AA}\Phi_A + k_{BA}\Phi_B) \tag{72}$$

$$\hat{\Phi}_B = (k_{AB}\Phi_A + k_{BB}\Phi_B) \tag{73}$$

and are constants for constant F_A. The initial conditions are

$$t = 0, \ N_A = N_{A0}, \ N_B = N_{B0}, \ V = V_0 \tag{74}$$

and the constraints are

Policy 1 $$F_{B,in} = 0, \ d(N_A/N_B)/dt = 0 \tag{75}$$

Policy 2 $$d[M_A]/dt = d[M_B]/dt = 0 \tag{76}$$

Given the time variation of the total polymer radical concentration ($[P]$), one can readily solve for $F_{A,in}$ and $F_{B,in}$ using the above equations. The time variation of $[P]$ depends on the initiator feed policy, if chemical initiation is employed.

With policy 1, the monomer concentrations are continuously falling, as with a batch copolymerization, and as a consequence, lower levels of long chain branching and crosslinking are obtained than with policy 2, if reactions which produce long branches are significant. A limiting form of policy 2 ($[M_A]$ and $[M_B] \sim 0$, or monomer-starved feed) can be convenient, as the composition of the copolymer is the same as the composition of monomers in the feed stream. A knowledge of reactivity ratios is thus not required. In addition, the low monomer levels preclude a runaway polymerization. A possible major disadvantage is excessive levels of long chain branching and crosslinking.

2.3 ALTERNATIVE COPOLYMERIZATION KINETIC MODELS

2.3.1 Deviations from the Terminal Model

Deviations from the simple copolymerization model have been noted for various comonomer pairs and for various polymerization systems, including free radical,[46] anionic and heterogeneous Ziegler–Natta. Bywater[47] has considered the applicability of the terminal model in anionic polymerization and made the following observations. Two difficulties may arise. 'In hydrocarbon solvents with lithium and sodium based initiators, $[P_i]$ is not the total concentration of growing polymer chains ending in monomer unit-i but, due to self-association phenomena, only that part in an active form. The reactivity ratios determined are, however, unaffected by the association phenomena. As each reactivity ratio refers to a common active center, the effective concentration of active species is reduced equally for both monomer types. In polar solvents, such as THF, association apparently does not occur, but there will be two types of active centers, one an anion and the other an ion-pair.' If these phenomena are properly accounted for, the reactivity ratios will not depend on concentration and possibly the terminal model will be valid.

Recent measurement of reactivity ratios for two heterogeneous Ziegler–Natta catalysts[48,49] have shown them to have a significant concentration dependence. To account for this apparent deviation from the terminal model, the existence of more than one type of active site on the catalyst surface has been suggested.[48,49] Terminal models which account for multiple active site types have been proposed.[50,51] Usami et al.[52] discuss the effect of multiple active site types on molecular weight and short chain branching distribution in the synthesis of linear low density polyethylene by heterogeneous Ziegler–Natta catalysts.

In order to explain some deviations from the terminal model in free radical polymerization, Merz et al.[53] introduced the concept of the penultimate effect. In other words, the reactivity of a radical center may depend on the type of adjacent monomer unit bound in the copolymer chain. It has been shown by Hill et al.[55] that model discrimination (e.g. the terminal model vs. the penultimate model) requires detailed measurements of the copolymer microstructure, and that data on changes in overall composition are insufficient for this purpose. They compared the terminal model with the penultimate and complex-participation models for the copolymerization of styrene and acrylonitrile, and found that the penultimate model gave significantly better predictions of copolymer microstructures. Bartlett and Nozaki[54] observed that solutions of some comonomers exhibited a yellow coloration, which was attributed to the formation of donor–acceptor complexes between the two monomers. It was suggested that this complex participates in the propagation reactions and causes deviations from the terminal model. These copolymers generally showed a high degree of alternation of the comonomers bound in the chains. On the other hand, other authors have shown that highly alternating copolymers are often formed, even though the donor–acceptor complexes were, apparently, not present during the polymerization.[56] Some details of the penultimate and complex-participation models will be given later in this section. Deviations from the terminal model can also occur when depropagation reactions are significant, which can happen when polymerizations are performed near ceiling temperatures.

2.3.2 Penultimate Model

The penultimate model as employed by Hill et al.[55] for the free radical copolymerization of styrene (S) and acrylonitrile (AN) is described herein. The propagation reactions are

$$\text{SS} \cdot + \text{S} \xrightarrow{\ k_{SSS}\ } \text{SS} \cdot \tag{77}$$

$$\text{SS} \cdot + \text{AN} \xrightarrow{\ k_{SSAN}\ } \text{SAN} \cdot \tag{78}$$

$$\text{ANS·} + \text{S} \xrightarrow{k_{\text{ANSS}}} \text{SS·} \tag{79}$$

$$\text{ANS·} + \text{AN} \xrightarrow{k_{\text{ANSAN}}} \text{SAN·} \tag{80}$$

$$\text{SAN·} + \text{AN} \xrightarrow{k_{\text{SANAN}}} \text{ANAN·} \tag{81}$$

$$\text{SAN·} + \text{S} \xrightarrow{k_{\text{SANS}}} \text{ANS·} \tag{82}$$

$$\text{ANAN·} + \text{AN} \xrightarrow{k_{\text{ANANAN}}} \text{ANAN·} \tag{83}$$

$$\text{ANAN·} + \text{S} \xrightarrow{k_{\text{ANANS}}} \text{ANS·} \tag{84}$$

with reactivity ratios

$$r_{\text{SS}} = k_{\text{SSS}}/k_{\text{SSAN}}, \; r_{\text{ANAN}} = k_{\text{ANANAN}}/k_{\text{ANANS}} \tag{85}$$

$$r_{\text{ANS}} = k_{\text{ANSS}}/k_{\text{ANSAN}}, \; r_{\text{SAN}} = k_{\text{SANAN}}/k_{\text{SANS}} \tag{86}$$

The instantaneous copolymer composition equation for the penultimate model is given by

$$F_{\text{S}}/F_{\text{AN}} = \frac{1 + r_{\text{ANS}}(f_{\text{S}}/f_{\text{AN}})(r_{\text{SS}}f_{\text{S}} + f_{\text{AN}})/(r_{\text{ANS}}f_{\text{S}} + f_{\text{AN}})}{1 + r_{\text{SAN}}(f_{\text{AN}}/f_{\text{S}})(r_{\text{ANAN}}f_{\text{AN}} + f_{\text{S}})/(r_{\text{SAN}}f_{\text{AN}} + f_{\text{S}})} \tag{87}$$

2.3.3 Complex-participation Model

Seiner and Litt[57] were the first to report a mathematical analysis of the complex-participation model. The Seiner–Litt model, in its most general form, allows the participation of both monomers and donor–acceptor complex in the propagation reactions, and is thus not limited to pure alternating copolymerization. Hill *et al.*[56] have reviewed the use of this model, covering investigations which have ranged from the theoretical treatments of various methods of analysis of copolymerization data, to studies of systems which may proceed *via* complex participation.

In its simplest form, the generalized complex-participation model can be described by eight propagation reactions and an equilibrium reaction forming the complex from the monomers. The complex-participation model as employed by Hill *et al.*[55] for the copolymerization of styrene and acrylonitrile is given below. The propagation reactions include the same four which apply in the terminal model, plus four additional ones, which are

$$\text{S·} + \overline{\text{ANS}} \xrightarrow{k_{\text{S}\overline{\text{ANS}}}} \text{S·} \tag{88}$$

$$\text{S·} + \overline{\text{SAN}} \xrightarrow{k_{\text{S}\overline{\text{SAN}}}} \text{AN·} \tag{89}$$

$$\text{AN·} + \overline{\text{ANS}} \xrightarrow{k_{\text{AN}\overline{\text{ANS}}}} \text{S·} \tag{90}$$

$$\text{AN·} + \overline{\text{SAN}} \xrightarrow{k_{\text{AN}\overline{\text{SAN}}}} \text{AN·} \tag{91}$$

The equilibrium reaction for complex formation is

$$\text{S} + \text{AN} \underset{}{\overset{K}{\rightleftharpoons}} \overline{\text{SAN}} \tag{92}$$

and the appropriate reactivity ratios for this model are

$$r_{\text{S}} = k_{\text{SS}}/k_{\text{SAN}}, \; r_{\text{AN}} = k_{\text{ANAN}}/k_{\text{ANS}} \tag{93}$$

$$P_{\text{S}} = k_{\text{S}\overline{\text{SAN}}}/k_{\text{S}\overline{\text{ANS}}}, \; P_{\text{AN}} = k_{\text{AN}\overline{\text{ANS}}}/k_{\text{AN}\overline{\text{SAN}}} \tag{94}$$

$$S_{\text{S}} = k_{\text{S}\overline{\text{ANS}}}/k_{\text{SAN}}, \; S_{\text{AN}} = k_{\text{S}\overline{\text{ANS}}}/k_{\text{ANS}} \tag{95}$$

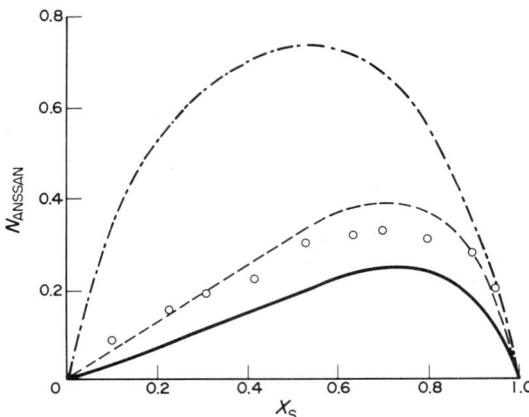

Figure 2 A comparison of the predictions of the number fraction of styrene sequences of length two (N_{ANSSAN}) at various styrene mole fractions in styrene/acrylonitrile copolymers (X_S) by the terminal, penultimate and complex-participation models with experimental ^{13}C NMR measurements after Hill *et al.*[55] —, terminal model; –––, penultimate model; –·–·–, complex participation model; and ○, experimental data

2.3.4 Model Discrimination

The penultimate and complex-participation models require the estimation of six additional parameters over those required in the terminal model. Given the extreme influence that both the choice of experimental design and the statistical estimation methods have on the reactivity ratio estimates in the terminal model, it is to be expected that the use of efficient designs and EVM-type estimation methods will be even more crucial for obtaining efficient estimates of the greater number of parameters in these models.

Recent work by Duever *et al.*[43] on the use of Monte Carlo estimation methods for estimating the parameters in the penultimate model appears to be promising. Hill *et al.*[55] have pointed out that composition data alone cannot be used to discriminate between these models, since most of the models can adequately fit these data. Information on the copolymer structure or sequence length distribution is needed. This can be obtained from ^{13}C NMR in the form of triad and tetrad fractions. Hill *et al.*[55] have shown that, for the copolymerization of styrene and acrylonitrile, these structural data indicate the superiority of the penultimate model. Figure 2 shows a comparison of the predictions of the number fraction of styrene sequences of length two in the copolymerization of styrene and acrylonitrile by the terminal, penultimate and complex-participation models, with experimental measurements based on ^{13}C NMR. For further details, and comparisons for different sequence lengths for both styrene and acrylonitrile, refer to Hill *et al.*[55] A reasonable approach to modeling copolymerizations would appear to be to use the terminal model for composition and rate calculations, but to resort to the penultimate or complex-participation model if copolymer microstructure is important and the terminal model is inadequate for this purpose.

2.4 REFERENCES

1. G. Odian, 'Principles of Polymerization', McGraw-Hill, New York, 1970.
2. A. Rudin, 'The Elements of Polymer Science and Engineering', Academic Press, New York, 1982.
3. P. Wittmer, *Makromol. Chem., Suppl.*, 1979, **3**, 129.
4. T. Alfrey and G. Goldfinger, *J. Chem. Phys.*, 1944, **12**, 205.
5. F. R. Mayo and F. M. Lewis, *J. Am. Chem. Soc.*, 1944, **66**, 1594.
6. F. T. Wall, *J. Am. Chem. Soc.*, 1944, **66**, 2050.
7. W. H. Stockmayer, *J. Chem. Phys.*, 1945, **13**, 199.
8. A. E. Hamielec and J. F. MacGregor, in 'Polymer Reaction Engineering', ed. K. H. Reichert and W. Geiseler, Hanser, New York, 1983, p. 21.
9. V. E. Meyer and G. G. Lowry, *J. Polym. Sci., Part A*, 1965, **3**, 2843.
10. F. P. Price, *J. Chem. Phys.*, 1962, **36**, 209.
11. F. P. Price, in 'Markov Chains and Monte Carlo Calculations in Polymer Science', ed. G. G. Lowry, Dekker, New York, 1970.
12. C. Tosi and G. Catinella, *Makromol. Chem.*, 1970, **137**, 211.
13. W. H. Ray, *J. Macromol. Sci., Rev. Macromol. Chem.*, 1972, **8**, 1.
14. G. S. Georgiev, *J. Macromol. Sci., Chem.*, 1976, **A10**, 1081.

15. G. S. Georgiev, *Makromol. Chem.*, 1979, **180**, 1277.
16. G. S. Georgiev, *Polym. Bull. (Berlin)*, 1983, **10**, 527.
17. R. Galvan and M. Tirrell, *J. Polym. Sci., Polym. Chem. Ed.*, 1986, **24**, 803.
18. L. H. Garcia-Rubio, M. G. Lord, J. F. MacGregor and A. E. Hamielec, *Polymer*, 1985, **26**, 2001.
19. R. Van der Meer, H. N. Linssen and A. L. German, *J. Polym. Sci., Polym. Chem. Ed.*, 1978, **16**, 2915.
20. F. L. M. Hautus, A. L. German and H. N. Linssen, *J. Polym. Sci., Polym. Lett. Ed.*, 1985, **23**, 311.
21. M. Johnson, T. S. Karmo and R. R. Smith, *Eur. Polym. J.*, 1978, **14**, 409.
22. F. L. M. Hautus, H. N. Linssen and A. L. German, *J. Polym. Sci., Polym. Chem. Ed.*, 1984, **22**, 3487.
23. F. L. M. Hautus, H. N. Linssen and A. L. German, *J. Polym. Sci., Polym. Chem. Ed.*, 1984, **22**, 3661.
24. A. Rudin, K. F. O'Driscoll and M. S. Rumack, *Polymer*, 1981, **22**, 740.
25. K. F. O'Driscoll, L. T. Kale, L. H. Garcia-Rubio and P. M. Reilly, *J. Polym. Sci., Polym. Chem. Ed.*, 1984, **22**, 2777.
26. L. H. Garcia-Rubio and N. Ro, *Can. J. Chem.*, 1985, **63**, 253.
27. L. H. Garcia-Rubio, N. Ro and R. D. Patel, *Macromolecules*, 1984, **17**, 1998.
28. M. Fineman and S. D. Ross, *J. Polym. Sci.*, 1950, **5**, 259.
29. D. Braun, W. Bendlein and G. Mott, *Eur. Polym. J.*, 1973, **9**, 1007.
30. T. Kelen and F. Tudos, *J. Macromol. Sci., Chem.*, 1975, **A9**, 1.
31. P. W. Tidwell and G. A. Mortimer, *J. Polym. Sci., Part A.*, 1965, **3**, 369.
32. P. W. Tidwell and G. A. Mortimer, *J. Macromol. Sci., Rev. Macromol. Chem.*, 1970, **C4**, 281.
33. R. C. McFarlane, P. M. Reilly and K. F. O'Driscoll, *J. Polym. Sci., Polym. Chem. Ed.*, 1980, **18**, 251.
34. G. C. Laurier, K. F. O'Driscoll and P. M. Reilly, *J. Polym. Sci., Polym. Symp.*, 1985, **72**, 17.
35. M. J. Box, *Technometrics*, 1970, **12**, 219.
36. H. I. Britt and R. H. Luecke, *Technometrics*, 1973, **15**, 233.
37. T. L. Sutton and J. F. MacGregor, *Can. J. Chem. Eng.*, 1977, **55**, 602.
38. B. Yamada, M. Itahashi and T. Otsu, *J. Polym. Sci., Polym. Chem. Ed.*, 1978, **16**, 1719.
39. H. Patino-Leal, P. M. Reilly and K. F. O'Driscoll, *J. Polym. Sci., Polym. Lett. Ed.*, 1980, **18**, 219.
40. H. Patino-Leal and P. M. Reilly, *Technometrics*, 1981, **23**, 221.
41. K. K. Chee and S. C. Ng, *Macromolecules*, 1986, **19**, 2779.
42. T. A. Duever, K. F. O'Driscoll and P. M. Reilly, *J. Polym. Sci., Polym. Chem. Ed.*, 1983, **21**, 2003.
43. T. A. Duever, K. F. O'Driscoll and P. M. Reilly, submitted to *J. Polym. Sci., Polym. Chem. Ed.*, 1987.
44. D. R. Bassett and A. E. Hamielec, *ACS Symp. Ser.*, 1981, **165**.
45. A. E. Hamielec, J. F. MacGregor and A. Penlidis, 'Multicomponent Free-radical Polymerizations in Batch, Semi-Batch and Continuous Reactors', Symposium Proceedings, IUPAC, Genoa, May 1987. 'IUPAC International Symposium on Free Radical Polymerization Kinetics and Mechanisms'.
46. C. Walling and E. R. Briggs, *J. Am. Chem. Soc.*, 1945, **67**, 1774.
47. S. Bywater, in 'Comprehensive Chemical Kinetics', ed. C. H. Bamford and C. F. H. Tipper, Elsevier, New York, 1976, vol. 15, p. 53.
48. M. Kakugo, Y. Naito, K. Mizunuma and T. Miyatake, *Macromolecules*, 1982, **15**, 1150.
49. Y. Doi, R. Ohnishi and K. Soga, *Makromol. Chem., Rapid Commun.*, 1983, **4**, 169.
50. C. Cozewith and G. Ver Strate, *Macromolecules*, 1971, **4**, 482.
51. K. W. McLaughlin and C. A. J. Hoeve, *Polym. Prepr., Am. Chem. Soc., Div. Polym. Chem.*, 1986, **27**, 246.
52. T. Usami, Y. Gotoh and S. Takayama, *Polym. Prepr., Am Chem. Soc., Div. Polym. Chem.*, 1986, **27**, 110.
53. E. Merz, T. Alfrey and G. Goldfinger, *J. Polym. Sci.*, 1946, **1**, 75.
54. P. D. Bartlett and K. Nozaki, *J. Am. Chem. Soc.*, 1946, **68**, 1495.
55. D. J. T. Hill, J. H. O'Donnell and P. W. O'Sullivan, *Macromolecules*, 1982, **15**, 960.
56. D. J. T. Hill, J. H. O'Donnell and P. W. O'Sullivan, *Prog. Polym. Sci.*, 1982, **8**, 215.
57. J. A. Seiner and M. Litt, *Macromolecules*, 1971, **4**, 308.

3

Block and Graft Copolymers

JOHN M. G. COWIE
University of Stirling, UK

3.1 INTRODUCTION

The incorporation of two different monomers, A and B, into a polymer chain in a statistical fashion leads to copolymers with properties which depend on the composition of the product and are normally intermediate between those of the parent homopolymers. The lengths of the sequences are determined by the relative reactivities of the two monomers but, on average, are usually short (Volume 3, Chapters 2, 15). In the limit, single monomers may alternate regularly in the chain and these are known as alternating copolymers (Volume 4, Chapter 22). On the other hand, the monomers can be combined in a more regular fashion, either by linking extended linear sequences of one to linear sequences of the other by end-to-end addition to give block copolymers, or by attaching chains of B at points on the backbone chain of A, forming a branched structure known as a graft copolymer. Possible structures, within this latter group, that can be formed are illustrated in Figure 1, showing different arrangements of the blocks, the size of which may vary from a few monomers up to several thousand. In the graft copolymers the branch chains are normally distributed at random along the backbone of the primary chain, but more regularly branched structures can be formed which are called comb-branch polymers, where the branching can be, in the limit, at every second atom in the main chain backbone. While these structures have been illustrated with a two-monomer system, both block and graft copolymers can be constructed from statistical copolymer sequences.

The properties of block and graft copolymers are determined by the length of the blocks and their chemical composition. The latter factor will determine whether or not the two different types of block are miscible with one another and even though covalent bonding of one dissimilar block to another enhances miscibility, these structures usually tend to segregate and form heterogeneous systems. This can produce a variety of morphologies depending on the relative block lengths, the molecular weight of the complete molecule, the continuity of one phase compared with the second and the method of preparing the sample. A schematic representation of some possible morphologies is illustrated in Figure 2, showing the different distributions of the two phases. The biphasic nature means that the material will exhibit some of the characteristics of each separate homopolymer, and this can be advantageous. One example may serve to illustrate the point. A triblock (ABA) copolymer can be synthesized with the A blocks composed of material with a high glass transition temperature T_g and a B block which is an elastomeric polymer with a low T_g. At ambient temperatures the A blocks are in the glassy state and anchor the ends of the elastomeric block, thereby acting like crosslinks, which will be stable up to the T_g of the glassy polymer. The material behaves like a crosslinked elastomer but has the advantage over traditionally vulcanized rubber in that the 'glassy crosslinks' are thermally labile and the material can be readily processed simply by

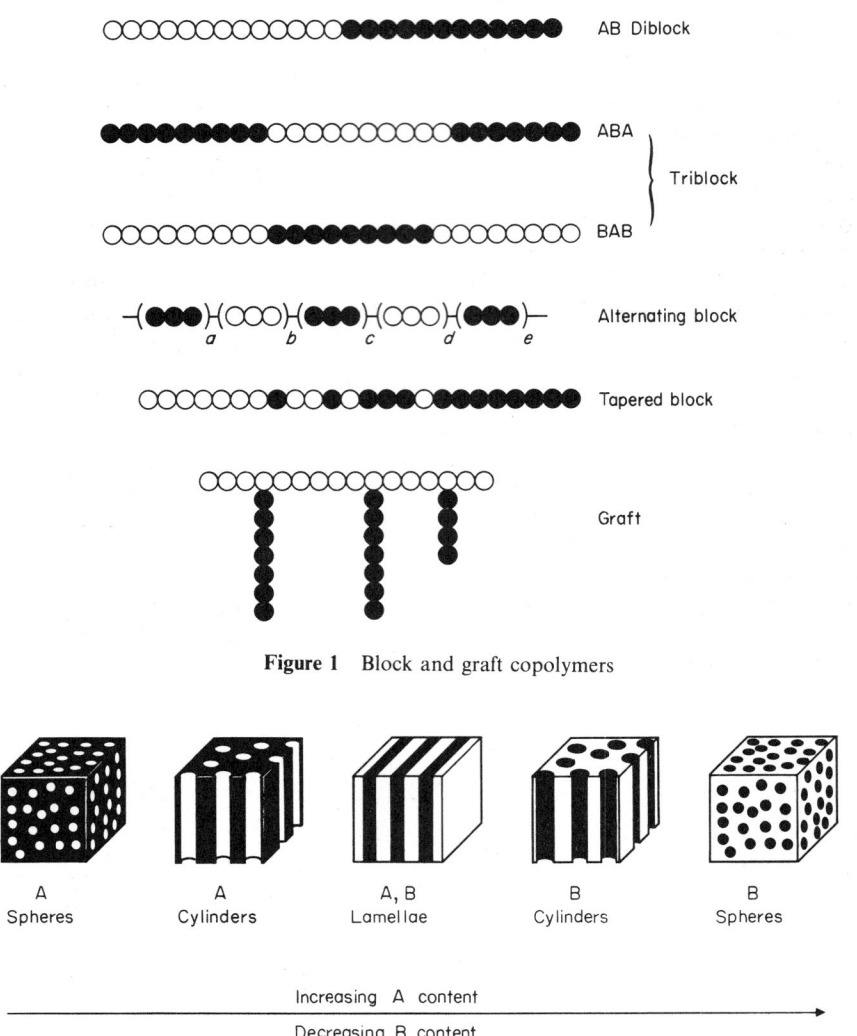

Figure 1 Block and graft copolymers

Figure 2 Structure of an AB block copolymer with, from left to right, an increasing chain length of the A block. (1) spheres of A in a continuous matrix of B; (2) cylinders of A in a continuous matrix of B; (3) lamellae; (4) and (5) reversal of the continuous phase to component A

heating above the higher T_g. The elastomeric properties are regained on cooling as the A blocks vitrify and phase separate again. These materials are called *thermoplastic elastomers*. The best elastomeric behaviour is obtained when the B block forms a continuous phase in the matrix and this can be controlled by altering the relative block lengths of A to B. It can also be controlled by casting the materials from different solvents if the liquid chosen is a better solvent for one block rather than the other. Thus the modulus for poly[styrene-*block*-(ethylene-*co*-butene)-*block*-styrene] triblock copolymer can be varied by casting films from different solvents as shown in Figure 3.

When tetrahydrofuran is used, which is a better solvent for the polystyrene block than the centre block, the polystyrene coils are more effectively dispersed and form the continuous phase in the matrix. This produces a high modulus in the plateau region between the two glass transitions. When heptane is used, the reverse is true: the polystyrene chains are now in a contracted form, whereas those of the centre block are expanded and form the dispersed phase. Neither of these give equilibrium morphologies as they are controlled by the interactions between the individual blocks and the solvent rather than between the blocks themselves. A pressed film (solvent free) should be closer to this equilibrium state and the modulus behaviour for this sample is found to lie between the two extremes exhibited by the solvent cast samples.

In the subsequent sections the principles of the main procedures which are most commonly used in the synthesis of block and graft copolymers are outlined. Detailed descriptions of these and other processes may be found in the subsequent specialized chapters.

3.2 BLOCK COPOLYMER SYNTHESIS

Block copolymers can be synthesized by sequential addition reactions using: (i) ionic initiators where an active site is kept 'alive' on the end of the initial block, which is then capable of initiating chain growth of a second monomer on the end of the first chain; (ii) coupling of different blocks with functional terminal units, either directly or through a reaction involving a small intermediate molecule; and (iii) bifunctional radical initiators where a second potentially active site is incorporated at one end of the first chain grown, which can initiate at a later stage a new chain from the macroradical produced.

The anionic method is particularly successful in preparing well-defined block copolymers by making use of the observation that there is no easily discernible termination step and, if kept free from impurities, the 'living' carbanionic end groups can be used to initiate the polymerization of a second monomer.

The main limitation to the method is that the anion of one monomer must be able to initiate the polymerization of a second monomer and this may not always be the case. Thus polystyryl lithium can initiate the polymerization of methyl methacrylate to give an (A—B) diblock but, because of its relatively low nucleophilicity, the methyl methacrylate anion cannot initiate styrene propagation. Best results are achieved when two monomers of high electrophilicity are used, *e.g.* styrene (St) with butadiene (Bd) or isoprene and (A—B—A) triblocks can be formed as shown in equations (1a) and (1b). The triblock can also be prepared by coupling the two carbanions using an organic dihalide (equation 2) and other coupling agents such as phosgene or dichlorodimethylsilane are equally effective. This method can also be used to prepare radial blocks with multifunctional compounds as illustrated with silicon tetrachloride in equation (3).

$$(PSt)_x^- \; Li^+ \quad + \quad yBd \quad \longrightarrow \quad (PSt)_x(PBd)_y^- \; Li^+ \tag{1a}$$

$$(PSt)_x(PBd)_y^- \; Li^+ \quad + zSt \quad \longrightarrow \quad (PSt)_x(PBd)_y(PSt)_z^- \; Li^+ \tag{1b}$$

$$(PSt)_w(PBd)_x^- \; Li^+ \quad + \quad ClCH_2Cl \quad + \quad Li^+ \; {}^-(PBd)_y(PSt)_z \quad \longrightarrow \quad (PSt)_w(PBd)_{x+y}(PSt)_z \quad + \quad 2LiCl \tag{2}$$

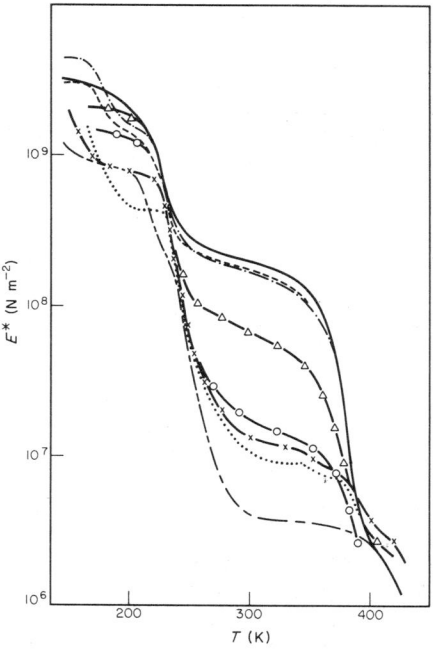

Figure 3 Temperature dependence of the complex modulus E^* for poly[styrene-*block*-(ethylene-*co*-butene)-*block*-styrene] films cast from various solvents; (———) tetrahydrofuran, (—·—·—) chloroform, (---) toluene, ○ cyclohexane, △ bromobutane, × pressed, (·····) cyclohexane, and (————) *n*-heptane (reproduced from *Macromolecules*, 1979, **12**, 52 with permission from the American Chemical Society)

$$4 \; \underset{\text{PSt} \quad \text{PBd}}{\text{\textasciitilde\textasciitilde\textasciitilde}}^- \text{Li}^+ \; + \; \text{Cl}\!-\!\underset{\text{Cl}}{\overset{\text{Cl}}{\text{Si}}}\!-\!\text{Cl} \; \longrightarrow \; \text{\textasciitilde\textasciitilde Si\textasciitilde\textasciitilde} \tag{3}$$

An interesting consequence of the marked differences in reactivity ratios found in some of the anionic systems is that in a mixture of monomers pure blocks of one can be obtained without incorporation of the second monomer. In styrene/butadiene mixtures the latter reacts most rapidly and can be almost completely polymerized before the styrene begins to react. As the butadiene becomes depleted, styrene is incorporated progressively until it is the only monomer left and the remaining chain grown is purely polystyrene. This produces a 'tapered' diblock copolymer.

Triblock copolymers can be constructed if a bifunctional initiator is generated as when sodium naphthalene is used with styrene or α-methylstyrene. Radical anions are formed, which combine to give a dianion and growth can then take place from both ends. Addition of a second monomer then yields a triblock structure.

3.2.1 Transformation Reactions

Potentially there are greater numbers of monomers which are suitable for cationic polymerizations than for anionic but the cationic method is less successful in block copolymer synthesis because, in many systems, the existence of a living carbocationic species is doubtful. Consequently, the involvement of carbocations in block copolymer synthesis tends to be limited to mixed reactions, *e.g.* the coupling of poly(tetrahydrofuran) cations with polystyryl anions to give an (A—B) diblock (equation 4). A more versatile approach is to use a transformation reaction in which one type of active terminal species is converted into a second type. Two general reactions have been identified: (i) a terminal unit anion–cation transformation by a two-electron oxidation process (equation 5); and (ii) carbanion to free radical conversion, which is a one-electron oxidation step (equation 6).

$$\text{\textasciitilde\textasciitilde CH}_2\!-\!\overset{+}{\text{O}}\underset{\text{PF}_6^-}{\bigcirc} \; + \; \underset{\text{Na}^+}{^-\text{CHCH}_2\text{\textasciitilde\textasciitilde}} \; \longrightarrow$$

$$\text{\textasciitilde\textasciitilde}\!\left(\text{poly THF}\right)\!\text{\textasciitilde\textasciitilde CH}_2\text{O}\!-\!\left(\text{CH}_2\right)_{\!4}\!\text{\textasciitilde\textasciitilde}\!\left(\text{PSt}\right)\!\text{\textasciitilde\textasciitilde} \; + \; \text{NaPF}_6 \tag{4}$$

$$\text{\textasciitilde\textasciitilde C}^- \; \longrightarrow \; \text{\textasciitilde\textasciitilde C}^+ \; + \; 2e^- \tag{5}$$

$$\text{\textasciitilde\textasciitilde C}^- \; \longrightarrow \; \text{\textasciitilde\textasciitilde C}\cdot \; + \; e^- \tag{6}$$

In the anion–cation transformation reaction, the anionically generated 'living' polymer chain is end capped with a halide, producing a chain which can be isolated for subsequent reaction. This can be used to initiate a cationic polymerization of a suitable monomer by activating the end with a silver or lithium salt according to the general scheme shown in equations (7a)–(7c). Halides may not

$$\text{\textasciitilde\textasciitilde M}_1^-\text{Li}^+ \; + \; \text{BrRBr} \; \longrightarrow \; \text{\textasciitilde\textasciitilde M}_1\text{RBr} \; + \; \text{LiBr} \tag{7a}$$

$$\text{\textasciitilde\textasciitilde M}_1\text{RBr} \; + \; \text{Ag}^+\text{Y}^- \; \longrightarrow \; \text{\textasciitilde\textasciitilde M}_1\text{R}^+\text{Y}^- \; + \; \text{AgBr} \tag{7b}$$

$$\text{\textasciitilde\textasciitilde M}_1\text{R}^+\text{Y}^- \; + \; n\text{M}_2 \; \longrightarrow \; \text{\textasciitilde\textasciitilde M}_1\text{\textasciitilde\textasciitilde M}_2^+\text{Y}^- \tag{7c}$$

always be the best terminating agents and Grignard reagents have been used for this purpose with much greater success.

The reverse cation–anion transformation is also feasible and involves the end capping of the carbenium ion with a species capable of further reaction with an alkyllithium: one example of this is demonstrated in equations (8a)–(8c).

$$\sim\!\!\sim\!\!M_1^+\,Y^- \quad + \quad RNH_2 \quad \longrightarrow \quad \sim\!\!\sim\!\!M_1NRH \quad + \quad HY \tag{8a}$$

$$\sim\!\!\sim\!\!M_1NRH \quad + \quad R'Li \quad \longrightarrow \quad \sim\!\!\sim\!\!M_1NR^-Li^+ \quad + \quad R'H \tag{8b}$$

$$\sim\!\!\sim\!\!M_1NR^-Li^+ \quad + \quad nM_2 \quad \longrightarrow \quad \sim\!\!\sim\!\!M_1N\!\!\sim\!\!\sim\!\!M_2^-Li^+ \tag{8c}$$

Anion–radical transformations can be effected in a number of ways but one must always begin with the carbanion-terminated chain.

(a) This chain can be end capped with a halide perester (Scheme 1), which provides a chain with a potential radical-forming site at one end. Thermal decomposition of this group in a second-stage reaction, in the presence of another monomer, generates an alkoxy macroradical from which to grow the second block but also produces a second radical fragment likely to produce some homopolymer as a contaminant.

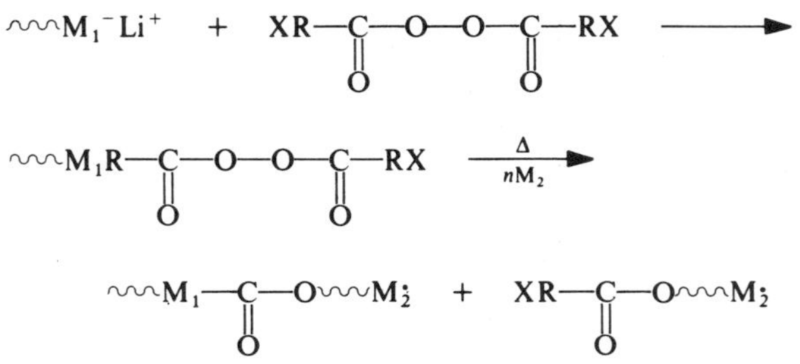

Scheme 1

(b) An alternative route involves end capping to produce a terminal hydroxyl followed by reaction with trichloroacetyl isocyanate (equation 9). This new, reactive end group can be used to initiate the growth of a second block *via* the photoreduction method proposed by Bamford (Scheme 2), where magnesium or rhenium carbonyls are excited by UV or visible radiation and extract a chlorine atom from the terminal unit, thereby creating a radical site. As only one radical is formed, this is a much 'cleaner' reaction compared with (a); however, block lengths are more difficult to control in both these radical reactions and the exact structure of the product formed can depend on the mechanism of the termination reaction as will be seen later.

$$\tag{9}$$

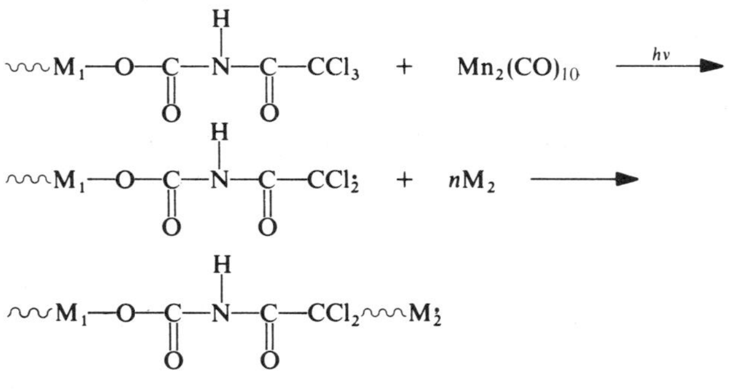

Scheme 2

3.2.2 Coupling Reactions

It is clear from the foregoing that polymer chains can be synthesized with functional groups in the α or the ω position, or both. If two different types of block are functionalized, they can be linked together to form copolymers.

Anionic polymerizations can be terminated by addition of another molecule which will introduce an ω-functional group in the chain. Excess carbon dioxide or cyclic anhydrides lead to terminal carboxylic groups, while addition of excess phosgene produces an acid chloride function. Similarly, isocyanates generate ω-amide functions and lactones yield ω-hydroxyl groups.

The 'Inifer' process developed by Kennedy can be used to functionalize vinyl monomers *via* a cationic route by initiating a polymerization with an alkyl halide/boron trichloride mixture $\{R^+ BCl_4^-\}$. The termination by transfer to an alkyl halide leaves a halide-terminated polymer. This can be transformed to a hydroxyl terminal unit *via* the sequence (i) dehydrohalogenation, (ii) hydroboration, (iii) oxidation and hydrolysis (Scheme 3). These ω-functional blocks may be coupled to form diblock copolymers using standard reaction techniques, *e.g.* diisocyanates will couple ω-hydroxy and/or ω-amine blocks together. Direct reactions can also occur and ω-acid chlorides combine readily with ω-hydroxy units.

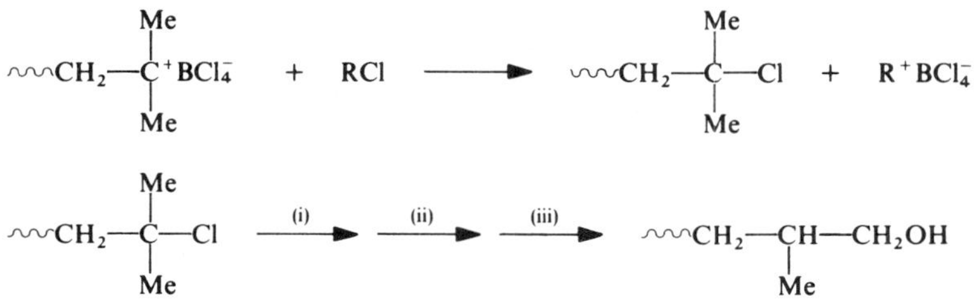

$$\text{CH}_2\!-\!\underset{\underset{\text{Me}}{|}}{\overset{\overset{\text{Me}}{|}}{\text{C}}}{}^{+}\text{BCl}_4^- \quad + \quad \text{RCl} \quad \longrightarrow \quad \text{CH}_2\!-\!\underset{\underset{\text{Me}}{|}}{\overset{\overset{\text{Me}}{|}}{\text{C}}}\!-\!\text{Cl} \quad + \quad R^+\text{BCl}_4^-$$

$$\text{CH}_2\!-\!\underset{\underset{\text{Me}}{|}}{\overset{\overset{\text{Me}}{|}}{\text{C}}}\!-\!\text{Cl} \quad \xrightarrow{\text{(i)}}\;\xrightarrow{\text{(ii)}}\;\xrightarrow{\text{(iii)}}\quad \text{CH}_2\!-\!\underset{\underset{\text{Me}}{|}}{\text{CH}}\!-\!\text{CH}_2\text{OH}$$

Scheme 3

Macroazonitriles can be employed and structures based on (**1**) in equation (10), with either diol, acid, or acid chloride terminal functions, are preferred. Functionalized chains can be linked to the azo compounds, then the azo group can be decomposed thermally to produce radical sites for further chain growth. This type of reaction will be dealt with more fully in the next section.

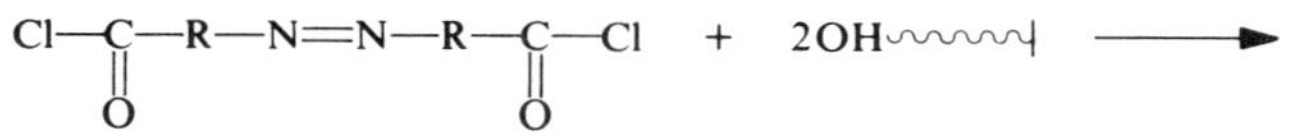

$$\underset{\underset{\text{O}}{\|}}{\text{Cl}\!-\!\text{C}}\!-\!\text{R}\!-\!\text{N}\!=\!\text{N}\!-\!\text{R}\!-\!\underset{\underset{\text{O}}{\|}}{\text{C}}\!-\!\text{Cl} \quad + \quad 2\text{OH}\text{\textasciitilde\textasciitilde\textasciitilde} \quad \longrightarrow$$

(**1**)

$$\text{\textasciitilde\textasciitilde}\text{O}\!-\!\underset{\underset{\text{O}}{\|}}{\text{C}}\!-\!\text{R}\!-\!\text{N}\!=\!\text{N}\!-\!\text{R}\!-\!\underset{\underset{\text{O}}{\|}}{\text{C}}\!-\!\text{O}\text{\textasciitilde\textasciitilde}$$

(10)

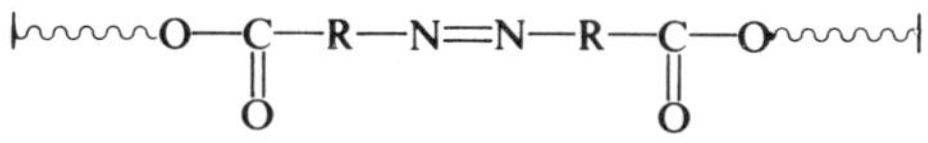

3.2.3 Bifunctional Radical Initiators

Block copolymers produced wholly by radical reactions are much more difficult to control and mixed products may result, depending on whether termination is by combination or disproportionation. A typical set of reaction pathways is illustrated in Scheme 4, using a sequential initiator such as di-*t*-4,4′-azobis(4-cyanoperoxyvalerate) or a similar derivative. Stage one in the reaction is the thermal decomposition of the azo group at temperatures of between 310 K and 330 K in the presence of the first monomer. The product isolated at the end of this reaction is a mono- or bifunctional macroinitiator and this can be activated subsequently either in bulk, solution or emulsion at low temperatures by using a reducing agent such as tetraethylenepentamine to catalyze the perester decomposition. When carried out in the presence of a second monomer, this will produce block copolymers and probably some homopolymer. This reaction procedure can be reversed and the azo group reacted in the second stage. The main disadvantage of this method is that a mixture of block structures may be obtained and it is essential to know the termination mechanism if any degree of control is to be exercised.

3.3 GRAFT COPOLYMER SYNTHESIS

In the synthesis of graft copolymers, the sites at which the second and subsequent blocks are attached to the first are no longer terminal units but are at positions along the backbone of the first

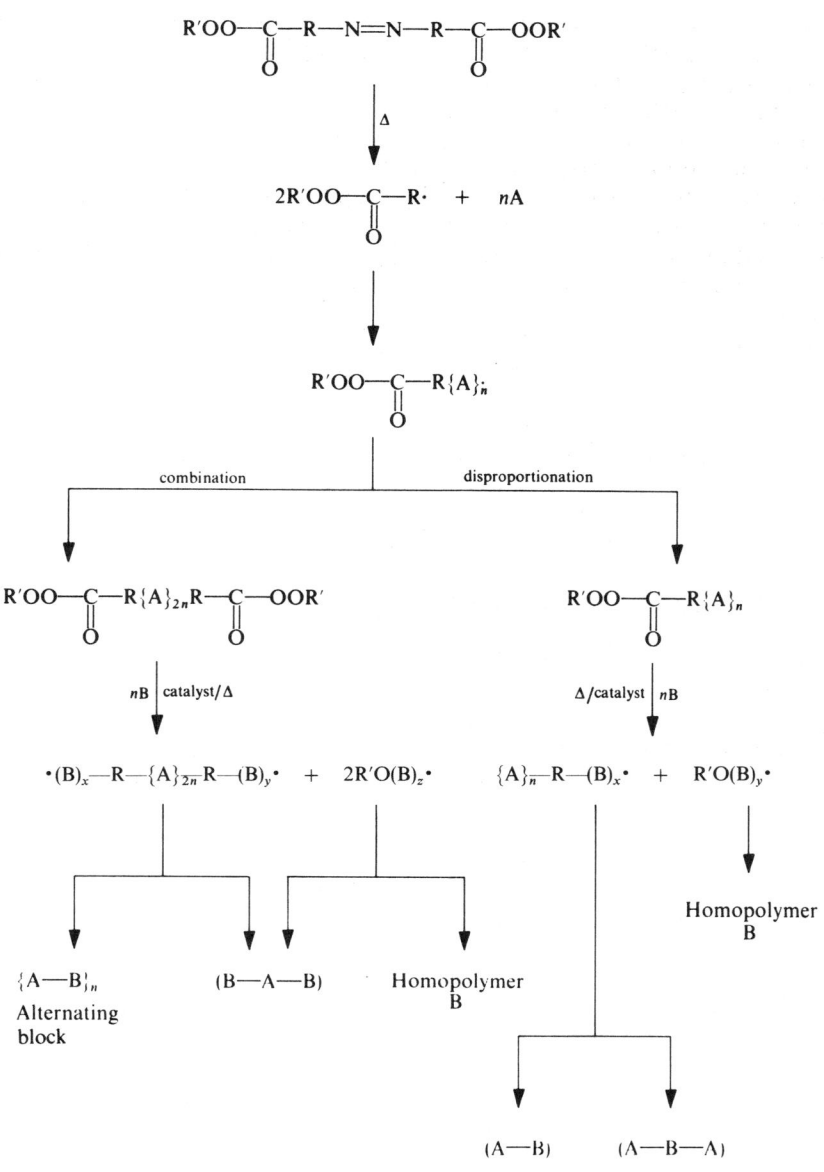

Scheme 4

chain. There are three main techniques for preparing graft copolymers: (a) grafting 'from'; (b) grafting 'onto'; and (c) *via* macromonomers.

The grafting 'from' procedure requires active sites to be created on the polymer chain capable of initiating the growth of other chain branches comprising a second monomer. Free radical sites can be formed by direct or mutual radiation with γ-rays of a polymer in the presence of the second monomer. This is a simple method but can also lead to homopolymer formation.

Preirradiation of a polymer in the presence of oxygen leads to formation of peroxy groups on the polymer which are relatively stable and this allows the polymer to be isolated and stored for further reaction. Polymers prepared in this way can be heated in the presence of a second monomer: the peroxy groups decompose to produce radical sites and the grafting process can take place. This method has been used to prepare poly(styrene-*graft*-acrylonitrile). The free radical approach can be used in other ways. Graft copolymers are formed when a chain transfer to preformed polymer can be effected with a second monomer present in the reaction mixture, but this depends on the radical source, *e.g.* methyl acrylate can be grafted to natural rubber when benzoyl peroxide is the initiator but is much less likely to do so with azobisisobutyronitrile. The effectiveness of this grafting 'from' technique is a function of the reactivity and polarity of the radical site and the monomer. Alternatively, radical-forming sites can be introduced into the chain backbone by *in situ* modification of some monomer units or by copolymerization. Thus, a polymer with trihalide groups pendant to the chain can be activated in the presence of a second monomer, using a reaction similar to that shown in Scheme 2, thereby forming a graft rather than a block copolymer. This type of reaction may also lead to a crosslinked structure if termination of the radical by combination predominates (Scheme 5).

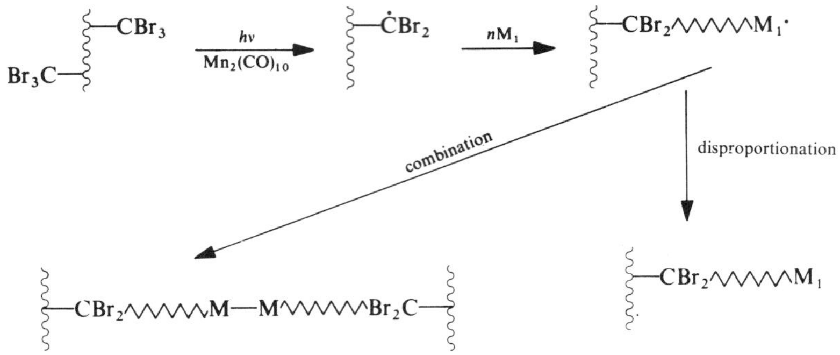

Scheme 5

By altering the number of active sites on the backbone the number and distribution of grafted chains can be controlled. The length of each graft will depend on both the rate of initiation and the monomer concentration, but the mechanism by which termination of the growing radical takes place will determine the ratio of branches to crosslinks in the system. If this is exclusively by combination, then the occurrence of crosslinking will be high. However, the network formation can be modified by addition of a chain transfer agent which will produce a mixture of branches and crosslinks but also some homopolymer. In such systems the amount of chain transfer agent added will determine the ratio of branches to crosslinks.

A similar mixture of structures, but little of the contaminating homopolymer, will be obtained when a second monomer is used whose radicals terminate partly by disproportionation and partly by combination.

Finally, photodegradation of pendant ketones results in the formation of radical sites capable of initiating a graft, but like many of the other radical techniques there is also a tendency for homopolymerization to occur.

Anionic sites suitable for grafting 'from' reactions can be introduced by metallation, involving the complexation of a hydrocarbon polymer by organolithium compounds. The reaction is assisted by complexing the lithium first with tetramethylenediamine, which acts as a solvating base. Aromatic chlorine is readily exchanged for lithium, which then acts as an initiator for the anionic polymerization of suitable monomers.

Grafting 'onto' methods involve having sites on the main chain which can be attacked by a growing second chain, thereby linking the two by covalent bonding. In anionic polymerizations the linking of two chains by phosgene creates a polymeric ketone which can react with a second chain as shown in Scheme 6. Other electrophilic functional groups are effective in this reaction, *e.g.* ester, nitrile, anhydride, *etc.*, and can be used as grafting sites by growing carbanions such as the polystyryl

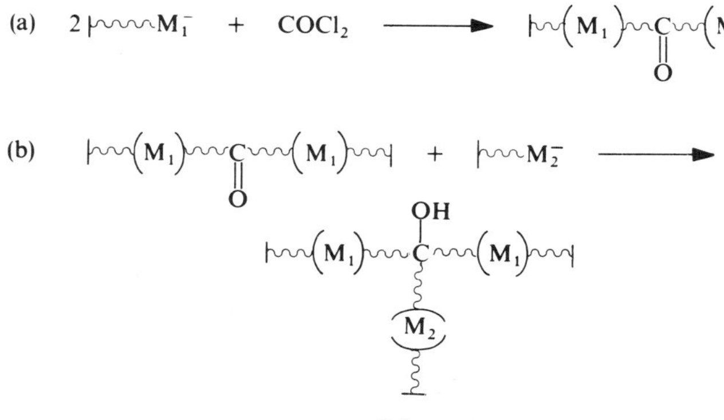

Scheme 6

ion. Many of these grafting techniques produce random branching along the primary chain, also a distribution of branch lengths. A more controlled grafting procedure can be used to produce regular comb-branch structures and this is achieved either by polymerization of macromonomers or by using a polymer analogous reaction on a suitable backbone.

In the first case macromonomers can be prepared by functionalizing a short chain with a vinyl unit. A typical reaction is shown in equation (11) and other methods have been reported. Polymerization of these monomers produces a well-defined graft structure with branches located at regular intervals along the chain. If the starting macromonomers have a uniform length, then the branches will also be regular, but mixed lengths can also be prepared. Copolymerization with another monomer will alter the regularity of the branching points but will maintain the uniformity of branch length.

$$\text{CH—OH} + \text{CH}_2{=}\text{CH} \longrightarrow \text{CH}_2{=}\text{CH—C—O—CH} \tag{11}$$
$$\underset{R}{|} \qquad \underset{COCl}{|} \qquad\qquad \underset{O}{\overset{\|}{}} \quad \underset{R}{|}$$

Poly(acid chloride)s have been used for polymer analogous reactions and ω-functionalized units can be condensed at these sites to produce structures similar to those obtained using macromonomers.

3.4 INTERPENETRATING POLYMER NETWORKS

Block and graft copolymer formation can be regarded as methods of mixing two kinds of polymer molecules where the components are bound together by covalent bonding rather than relying simply on physical mixing. Intermediate between these two extremes are the interpenetrating networks (IPN), which constitute a class of two-phase polymer systems where the components are held in the

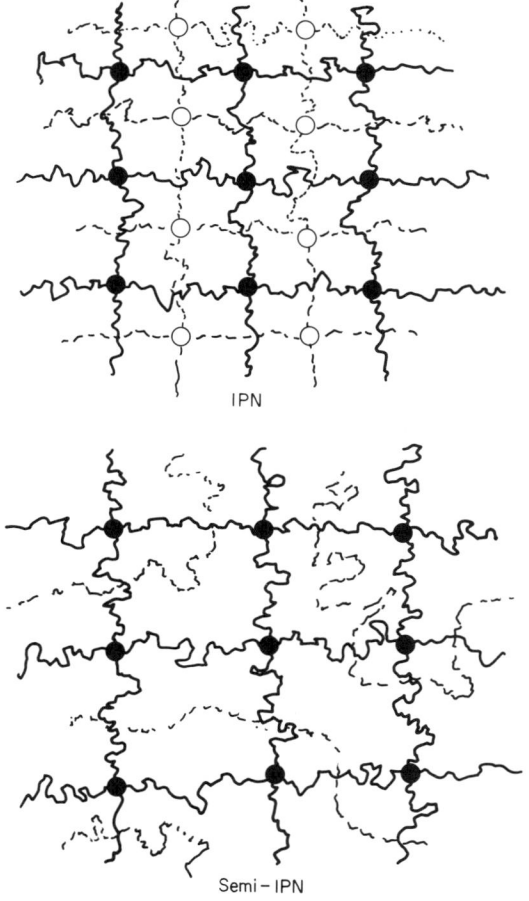

Figure 4 Interpenetrating and semi-interpenetrating polymer networks

mixture because they are physically trapped in the form of intertwined networks. Two examples of the different types are shown schematically in Figure 4.

The class can be subdivided on the basis of the method of formation and the term IPN is usually applied to structures made sequentially where the starting point is a preformed network of polymer A. This is then exposed to a mixture of the second monomer B and a crosslinking agent, both of which are absorbed by the first network causing it to swell. An *in situ* polymerization then results in the formation of a network of polymer B, which interpenetrates the polymer A network. The overall reaction thus produces a primary network through which a secondary network is woven and the resulting matrix appears to form a quasi-single-phase material. This can sometimes behave like a one-phase intimately mixed system, but in the majority of systems the characteristics of a two-phase mixture are exhibited. The extent of miscibility of the two components will be an important parameter in determining the morphology of the IPN. If the two polymers are miscible, the networks should be continuous, but if they are immiscible the level of interpenetration will be much lower as phase separation will occur during the polymerization process.

Continuity of phase is maintained in a network of polymer A, when swollen with monomer B, but as the polymerization of B takes place, a phase separation process will occur, which takes place early in the reaction for incompatible pairs and progressively later in the polymerization for more miscible polymer combinations. This will create phase-separated domains in the structure and the tendency is then for polymer B to have less phase continuity than polymer A. Reversing the order of polymerization should also reverse the phase continuity.

However, the thermodynamic interactions between the components will not be the only factor determining the final morphology and the ability of the chains to undergo phase separation will also depend on the physical restrictions, such as chain entanglements, and viscosity effects experienced during the polymerization reaction. Just as the distribution of phases can be controlled to some extent in block copolymers by the type of solvent used, so too the extent to which phase separation occurs in IPN systems will be a balance between the thermodynamic interactions and the kinetic effects.

Other methods can be used to prepare these interpenetrating systems. If the two crosslinking reactions are carried out simultaneously in the same reaction mixture, the product is called a simultaneous interpenetrating network (SIN). These can be prepared if one network is formed by a chain growth mechanism and the other by a step growth mechanism. A third group, called interpenetrating elastomeric networks (IEN), can be formed by taking two polymer latex emulsions which are then crosslinked in a single reaction.

Semi-IPNs can be prepared by omitting the crosslinking agent for monomer B, so that after polymerization of B the network A now contains linear chains of B which are trapped in the structure by entanglements.

4

Molecular Weight Distributions

NORMAN C. BILLINGHAM

University of Sussex, Brighton, UK

4.1 INTRODUCTION

Scientific study of the chemistry of small molecules is dominated by the law of constant composition and the axiom that all molecules of any pure compound have the same molecular weight; measurement of molecular weight (usually by mass spectrometry) is one of the standard starting points in structure determination in organic chemistry. However, the concept of a single, definitive molecular weight presents great problems when the substances concerned are high polymers. We have seen in earlier chapters that chain polymerization reactions involve the repeated addition of monomer units to a limited number of active centres in a chain propagation reaction. In typical batch polymerizations new active centres are generated continuously, the monomer concentration changes during the reaction as the monomer is consumed and active centres may be removed by chain termination reactions or transferred from one molecular chain to a new one by a chain transfer reaction. It is only under very special conditions that all polymer chains grow to the same length; with very few exceptions the statistical nature of polymerization reactions produces chains which have varying numbers of repeat units, so that the product contains a distribution of molecular chain lengths and is said to be polydisperse. It is not possible to characterize such a polymer by a single molecular weight and the chain lengths can only be specified completely by a molecular weight distribution.

Since the molecular weight distribution of a polymer reflects the mechanistic and kinetic history of its synthesis, the measurement of the distribution is an important tool for the study of polymerizaton mechanisms and also for the study of degradation reactions, involving scission of the polymer chains. Additionally, polymers which are identical to standard methods of chemical analysis may differ in molecular weight distribution, with consequent differences in physical and mechanical properties. For example, a small proportion of very long chains in a polymer may lead to a large increase in melt viscosity, as long chains readily become entangled; such increases in viscosity lead to unwanted changes in processing behaviour. Conversely, small proportions of very short chains may behave as plasticizers, leading to a reduction of the glass transition temperature. Yield, fracture and orientation are also sensitive to the molecular weight distribution. Effects of this kind are often quite subtle and characterization of molecular weight distribution can be important in both the study and the control of physical properties.

In this section we consider the basic definitions of molecular weight distribution in chain polymerization, emphasizing particularly the relationship of molecular weight distribution to the mechanisms of the polymerization reaction. Other more fundamental aspects of distribution functions, relevant to the characterization of polymers, are discussed in Volume 1, Chapter 3. In the last 20 years the rapid development of chromatographic methods for studying molecular size distributions in polymers has made molecular weight distributions much more accessible. Although chromatographic methods are very good for qualitative comparison of polymer samples, conversion of a chromatographic elution curve into a true molecular weight distribution is still a very complex process and chromatographic methods can easily be very fast and efficient ways of getting wrong answers. For many purposes it is either necessary or convenient to be content with one or more molecular weight averages. The definition of such averages and their relation to the parent distribution are also discussed below.

4.2 MOLECULAR WEIGHT DISTRIBUTIONS AND AVERAGES

Because of the statistical nature of polymerization reactions, complete characterization of molecular size in a polymer requires the measurement of its molecular weight distribution. Distribution data are most conveniently presented in the form of a histogram or, more commonly, a continuous curve. The ordinate of the distribution curve can be any quantity which is a direct or indirect measure of chain length, such as molecular weight, radius of gyration, limiting viscosity number or chromatographic elution volume. The abscissa can be any quantity which measures polymer concentration independently of molecular weight, such as number or weight of polymer molecules or the refractive index of their solution. Clearly the distribution can be presented in many forms. In practice we choose to define the distribution in such a way that it can be compared with calculations from a proposed reaction mechanism. First it is convenient to define some of the required symbols.

(i) A polymer chain with i repeat units is said to have a degree of polymerization of i and its molecular weight is M_i; obviously M_i is the product of i and the molecular weight of the repeat unit.

(ii) The number of molecules of size M_i in a sample is denoted by N_i and their number fraction ($= N_i/\Sigma N_i$) is denoted by n_i.

(iii) Similarly, the total weight of molecules of length i is denoted by W_i and their weight fraction ($= W_i/\Sigma W_i$) is denoted by w_i.

(iv) Unless otherwise stated, the symbol Σ denotes summation over all real molecular species, *i.e.* over the range $i = 1$ to $i = \infty$.

Note that n_i is the mole fraction of molecules with degree of polymerization i and that the number and weight of polymer molecules with i units are related by $W_i = N_i M_i/N_A$, where N_A is Avogadro's constant.

One of the simplest ways to present a distribution is to plot N_i or n_i against M_i, giving the number or frequency distribution. If n_i is chosen, then the existence of the frequency distributon is expressed by equation (1).

$$n_i = n(M_i) \tag{1}$$

The mean μ of such a distribution will be given by equation (2)[1,2] and the variance σ^2 is given by equation (3).

$$\mu = \int_0^\infty M_i n(M_i)\,di \tag{2}$$

$$\sigma^2 = \int_0^\infty (i-\mu)^2 n(M_i)\,di = \Sigma(i-\mu)^2 n(M_i) \tag{3}$$

Since the frequency function $n(M_i)$ defines the fraction of molecules of size M_i, it is a normalized function and governed by equation (4).

$$\Sigma n(M_i) = 1 \tag{4}$$

Although the value of i must be integral, it is typically much greater than one and the discrete function $n(M_i)$ can usually be replaced by the continuous function $n(M_i)dM_i$; the normalization condition then becomes equation (5), where the lower limit of integration can be set to zero because $n(0)$ is very small or zero. The use of continuous functions is often convenient when dealing with calculation of distributions from reaction mechanisms.

$$\int_0^\infty n(M_i)dM_i = 1 \tag{5}$$

The normalized frequency distribution is convenient for many purposes and often the easiest to predict. It places equal statistical weight on all polymer molecules, irrespective of their chain lengths, so that a molecule with $i=2$ and a molecule with $i=10^5$ each count equally as one molecule in the distribution. In many circumstances this is not particularly realistic since the property of interest may depend on the weight of the molecules rather than just how many there are. An alternative way of representing the distribution is to assign statistical weightings to the molecules according to their chain length and to plot W_i or w_i against M_i or i. This yields the weight distribution of the polymer with the same rules as before defining normalization. Obviously the frequency and weight distributions are interrelated and the relationship is given by equations (6) and (7).

$$w(M_i) = M_i n(M_i)/\Sigma M_i n(M_i) \tag{6}$$

$$n(M_i) = \frac{w(M_i)}{M_i} \left/ \Sigma \frac{w(M_i)}{M_i} \right. \tag{7}$$

Any peak in the number distribution function $n(M_i)$ must satisfy equation (8). The corresponding condition for the weight distribution can be expressed *via* equation (4) in the form of equation (9), so that at a peak in the number distribution dw_i/dM_i is equal to $n(M_i)$ and must be positive. For a weight distribution with a single peak this can only be true for values of M_i below the peak value so that the peak of a frequency distribution must lie below the corresponding peak in the weight distribution.

$$dn_i/dM_i = \dot{n}(M_i) = 0 \tag{8}$$

$$dw_i/dM_i = M_i \dot{n}(M_i) + n(M_i) = 0 \tag{9}$$

Distribution curves presented in the ways so far described are often termed differential distributions. An alternative way of presenting the data is to plot either $\overset{i}{\underset{1}{\Sigma}} n_i$ or $\overset{i}{\underset{1}{\Sigma}} w_i$ against M_i. This procedure gives integral or cumulative curves where any point on the curve gives the total number or weight fraction of molecules having molecular weights up to and including M_i. Figure 1 shows typical number, weight and integral curves for the molecular weight distribution of a polymer.

Because of the problems of measuring molecular weight distributions accurately, it is often necessary to measure one or more average values. In principle the average molecular weight may be defined in many ways, each giving a different value unless the polymer is monodisperse. For most purposes the choice is dictated by experimental convenience. There are two very commonly used averages.

(i) The number average molecular weight, $\bar{M}_n$, defined by equation (10). This average is measured by methods which determine the number of polymer molecules present in a given sample weight, *e.g.* colligative properties or end-group analysis (see vol. 1). It corresponds to the mean value of the number distribution and can be related to the weight distribution data by equation (11).

$$\bar{M}_n = \frac{\Sigma N_i M_i}{\Sigma N_i} = \Sigma n_i M_i \tag{10}$$

$$\bar{M}_n = \frac{\Sigma W_i}{\Sigma W_i/M_i} \tag{11}$$

(ii) The weight average molecular weight, $\bar{M}_w$, defined by equation (12). This average is measured by methods, typically light scattering, in which the measurement is sensitive to the weight of the

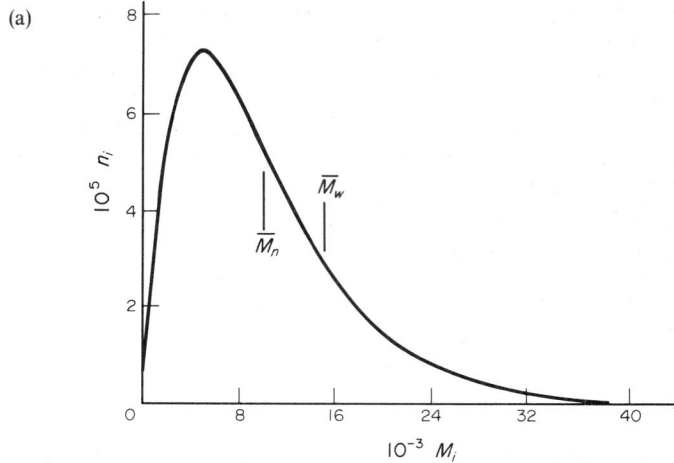

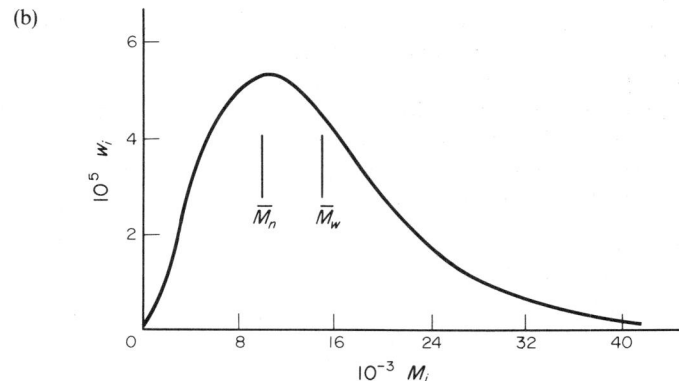

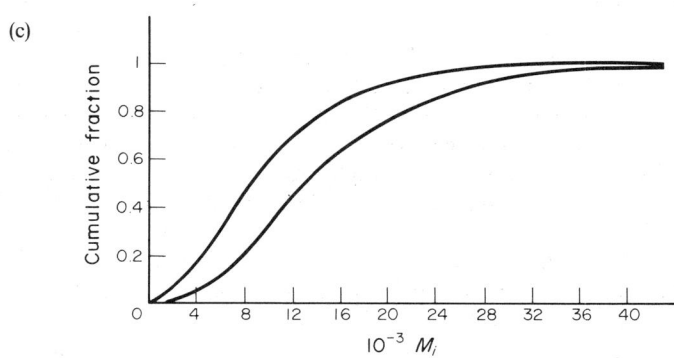

Figure 1 Typical molecular weight distribution curves for a polymer: (a) number or frequency distribution; (b) corresponding weight distribution; and (c) cumulative or integral curves corresponding to (a) (upper curve) and (b) (lower curve). Number and weight averages are indicated

polymer molecule. It corresponds to the mean of the weight distribution and is related to the number data by equation (13).

$$\bar{M}_w = \frac{\Sigma W_i M_i}{\Sigma W_i} = \Sigma w_i M_i \tag{12}$$

$$\bar{M}_w = \frac{\Sigma N_i M_i^2}{\Sigma N_i M_i} \tag{13}$$

It is possible to define a whole series of averages in terms of the parameters of the number and weight distributions, by equation (14), where $n = 0$ yields $\bar{M}_n$ and $n = 1$ yields $\bar{M}_w$. For $n = 2$ and $n = 3$ the averages are the z and $z + 1$ average respectively; the z average is important in the analysis of sedimentation of polydisperse polymers in the ultracentrifuge. Higher averages can be defined (see

Volume 1, Chapter 3) but are close to impossible to measure with precision since they depend on increasing powers of M and thus an ever more precise definition of the high molecular weight end of the distribution. Fractional values of n are possible but fairly pointless unless there is a sensible experimental method for measurement of the resulting average. An example of a more complex average is the viscosity average, defined by equation (15) where α is the exponent in the relationship of limiting viscosity number $[\eta]$ to molecular weight (equation 16) and is dependent upon solvent and upon temperature (see Volume 1, Chapter 3).

$$\bar{M} = \frac{\Sigma N_i M_i^{n+1}}{\Sigma N_i M_i^n} \tag{14}$$

$$\bar{M}_v = \left[\frac{\Sigma N_i M_i^{1+\alpha}}{\Sigma N_i M_i} \right]^{1/\alpha} \tag{15}$$

$$[\eta] = KM^\alpha \tag{16}$$

If the definitions of number and weight averages are substituted into the definition of the variance of the distribution (equation 3), it is easily shown that equation (17) holds, from which equation (18) follows directly and, since σ can only be positive or zero, it follows that the ratio $\bar{M}_w/\bar{M}_n$ must be greater than unity, except for a monodisperse polymer, for which it is equal to unity; this polydispersity ratio is often quoted as a convenient measure of the breadth of the molecular weight distribution. It ranges from values close to 1.0 for the best synthetic monodisperse polymers up to values of above 100 for some polymers prepared on coordination catalysts or for branched polymers. Since precisely the same arguments can be applied to all averages, it follows that increases in the exponent n in equation (14) cause the value of the corresponding average to increase.

$$\sigma^2 = (\bar{M}_w \bar{M}_n) - \bar{M}_n^2 \tag{17}$$

$$\frac{\bar{M}_w}{\bar{M}_n} = 1 + \frac{\sigma^2}{\bar{M}_n^2} \tag{18}$$

Although the ratio $\bar{M}_w/\bar{M}_n$ is frequently quoted as a measure of the range of molecular sizes present in a sample, it does not uniquely define the molecular weight distribution. At the same time a ratio $\bar{M}_w/\bar{M}_n$ close to unity does not imply very high monodispersity; a mixture of 20 mol% of polymer with $i = 50$ into a polymer with $i = 100$ would give $\bar{M}_w/\bar{M}_n = 1.05$.

The important feature of number and weight distributions and averages is that they place different importance on different parts of the molecular weight distribution. Small molecules contribute to the number distribution on equal terms with large molecules. In contrast they contribute very much less to the total weight of the molecules; a very small weight fraction of small molecules can correspond to a high number fraction. The point is illustrated in Table 1 which shows the effect on $\bar{M}_w$ and $\bar{M}_n$ if small amounts of a second polymer of higher or lower molecular weight are added to a sample of molecular weight 10^5. Addition of high molecular weight polymer changes $\bar{M}_w$ rapidly, leaving $\bar{M}_n$ unaffected; addition of low molecular weight material has the opposite effect.

For any distribution function it is possible to define a series of quantities which are the moments of the distribution. In the case of a polymer, M_i cannot have negative values and the moments of the number distribution Q_n are defined by equation (19), where n is zero or a positive integer. The zero moment of the number distribution is equal to the total concentration of polymer molecules in the

Table 1 Effect of Addition of a Second Monodisperse Polymer to a Monodisperse Polymer of Molecular Weight 10^5

MW added	w_i	n_i	$\bar{M}_n$	$\bar{M}_w$
10^7	0.01	0.0001	100 999	199 000
10^7	0.1	0.0011	110 987	1 090 000
10^6	0.01	0.001	100 908	109 000
10^6	0.1	0.011	109 890	190 000
10^3	0.01	0.503	50 251	99 010
10^3	0.1	0.917	9174	90 100
10^2	0.01	0.910	9099	99 001
10^2	0.1	0.991	991	90 010

system and the first moment is equal to the concentration of monomer units so that $\bar{M}_n = Q_1/Q_0$. Similarly the other averages for which n is an integer in equation (14) are related to the ratios of successive moments of the distribution, so that $\bar{M}_w = Q_2/Q_1$, $\bar{M}_z = Q_3/Q_2$, *etc.* Equation (19) is important in that it allows the computation of average molecular weights from a known distribution. In principle the relationship between a distribution and its moments is uniquely defined, so that a set of average molecular weights can be used to calculate the molecular weight distribution.[3] Although this method has been discussed and applied, it is rarely possible to determine a sufficient number of averages with sufficient precision for it to be reliable; if the polymer is well enough characterized for this approach to be good, then there are enough data to allow proper calibration of a chromatographic system.

$$Q_n = \Sigma n_i M_i^n \tag{19}$$

4.3 MOLECULAR WEIGHT DISTRIBUTIONS AND REACTION MECHANISM

4.3.1 Introduction

We have seen that the most useful distribution functions for description of the molecular weight distribution of a polymer are the number distribution $n(M_i)$ and the weight distribution $w(M_i)$. For a homopolymer, calculation of the distribution function expected for any particular reaction mechanism depends upon evaluating either $n(M_i)$ or $w(M_i)$ as a function of M_i. In other words we need to calculate the number (or mole) fraction of molecules of length i or their weight fraction, the two being interrelated. The underlying principle that makes this possible is that the number of polymer molecules under consideration is very large and will be governed by statistical laws. Thus if the probability of forming a chain containing exactly i units can be calculated for a given reaction mechanism, then it will equal the frequency with which such chains appear in the molecular weight distribution.

Detailed analyses of the molecular weight distributions expected for different types of mechanism are presented in, for example, Volume 3, Chapter 18. At this point we introduce the basic concepts only.

We have seen earlier (Volume 3, Chapter 1) that the general mechanism of formation of a linear homopolymer by a chain reaction involves four distinct types of reaction. Initiation reactions generate new active centres to which the monomer molecules add sequentially in a propagation step. The propagation may be interrupted by chain transfer or by chain termination. Chain transfer stops the growth of a given molecular chain whilst passing the kinetic chain to a new molecule by reinitiating and forming a new active centre. In contrast, chain termination involves the interruption of both the molecular chain and the kinetic chain. The exact form of the distribution of chain lengths depends upon the relative importance of these reactions and their relative rates; some of the important cases are dealt with below.

4.3.2 Perfect Polymerization: The Poisson Distribution

The simplest polymerization to describe is also the most difficult to achieve in reality. It is that in which both chain termination and chain transfer are eliminated. Once an active centre is generated in an initiation event, propagation continues without transfer until the monomer is exhausted. Because there is no mechanism for deactivation of an active centre, polymerizations of this type are said to be 'living'. Living polymerization requires an active centre whose reactivity is highly specific; it must be reactive enough to propagate but completely inert to chain transfer or termination in the reaction system. Anionic polymerizations of a few monomers are the best known examples of this type[4] although living polymerizations involving cations, coordination complexes and group-transfer mechanisms are also known (see Volume 3, Chapters 25 and 39; Volume 4, Chapter 10).

The simplest example of a living polymerization occurs when all of the initiation takes place simultaneously, as is the case when a monomer is instantaneously mixed with a population of rapidly reacting initiator. In this case all of the active centres are generated at once and each has an identical probability of propagation. If the conversion is high enough that a significant number of monomer molecules is polymerized by each active centre, and if the propagation is irreversible, then the molecular weight distribution is equivalent to the distribution of a large number of objects into a smaller number of identical categories in which they become fixed. The result is a Poisson distribution for which the number distribution of chain lengths is given by equation (20), where γ is the average number of monomer units added to each active centre. The corresponding weight distribution is given by equation (21). For all practical purposes the number and weight distributions are identical. The corresponding number and weight average degrees of polymerization are given by equations (22) and (23), so that the polydispersity ratio is given by equation (24) and $\bar{i}_w$ is virtually the same as $\bar{i}_n$ for large values of γ. Thus polymers following the Poisson distribution are very close to being monodisperse, certainly as close as can be achieved by simple chain growth statistics and much closer than can be achieved by fractionation of a polydisperse sample. Figure 2 shows the typical form of the Poisson distribution for a polymer with an average degree of polymerization of 100 units.

$$n_i = e^{-\gamma}\gamma^{i-1}/(i-1)! \tag{20}$$

$$w_i = ie^{-\gamma}\gamma^{i-1}/(i-1)!(\gamma+1) \tag{21}$$

$$\bar{i}_n = 1+\gamma \tag{22}$$

$$\bar{i}_w = 1+\gamma+\gamma/(1+\gamma) \approx 1+\bar{i}_n \tag{23}$$

$$\bar{i}_w/\bar{i}_n = 1+1/\gamma \tag{24}$$

It is interesting to comment on the effects of conversion during a batch polymerization reaction in a living system. Once polymerization is initiated the total number of active centres remains constant throughout. As monomer is consumed in a batch reactor, the parameter γ increases, but the distribution function is unchanged. Most importantly, the same parameter γ applies to all polymer molecules in the system. Thus the isolated polymer always has a Poisson distribution but the average molecular weight increases with conversion, even though consumption of monomer results in a decrease in the rate of polymerization. Essentially the same is true of a reactor continuously fed with monomer, where the average chain length will increase with the amount of monomer supplied but the distribution remains the same.

4.3.3 Random Polymerization: The Most Probable Distribution

Living polymerization represents the closest approach to perfection of control over molecular weight distribution which can be achieved by chain growth of a polymer molecule from a single active centre in one direction. It is extremely difficult to achieve. A much more common situation is typified by a polymerization in which initiation occurs continuously and there is a chain termination or transfer reaction. In such reactions new chains are being generated and removed continuously.

Let us assume that active centres are generated at a constant rate in a reactor in which the monomer concentration is maintained constant by continuous injection of monomer and that

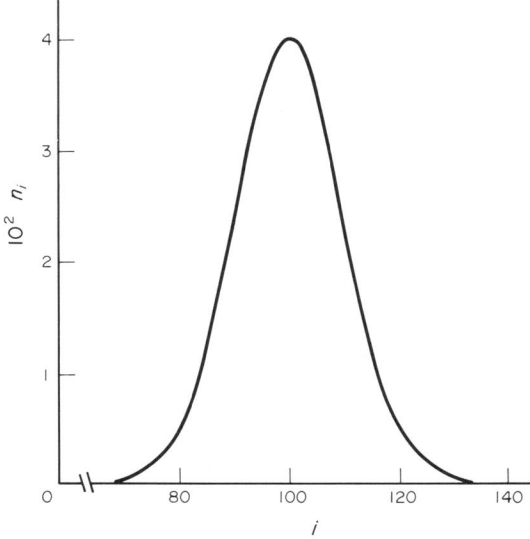

Figure 2 Poisson frequency distribution of chain lengths for a polymer with $\bar{i}_n = 100$. The weight distribution is essentially indistinguishable from this

propagating chains are stopped in a first-order process which may be either a chain transfer step, or in the case of a radical centre, disproportionation. In such a case an active centre can only do one of two things. It may propagate by the addition of a monomer or it may deactivate. If the probability of propagation is ϕ, then the probability of deactivation is $1-\phi$ and these two probabilities do not change with time. For a chain to grow to contain i units it must propagate exactly $i-1$ times and then deactivate. The distribution of chain lengths is analogous to the tossing of a coin which is biased to one side; the number distribution is given by equation (25) and the weight distribution by equation (26). This distribution is illustrated in Figures 3 and 4; it is an important case because it also arises when polymer is formed by step polymerization (see Volume 5), by random exchange reactions between polymer chains and by random scission of a higher molecular weight linear polymer. The function is a binomial distribution. Its application to polymers has been described by Schulz,[5] and by Flory[6] and it often known as the Schulz–Flory or the 'most probable' distribution.

$$n_i = \phi^{i-1}(1-\phi) \tag{25}$$

$$w_i = i\phi^{i-1}(1-\phi) \tag{26}$$

The most probable distribution is very much wider than the Poisson case although it reduces to the Poisson distribution as i tends to ∞ and ϕ tends to 1.[1] The average molecular weights are given by equations (27), (28) and (29), so that the ratios for high molecular weight polymers ($\phi \approx 1$) are $\bar{i}_z : \bar{i}_w : \bar{i}_n = 3:2:1$. Note in particular the very marked difference between the number and weight

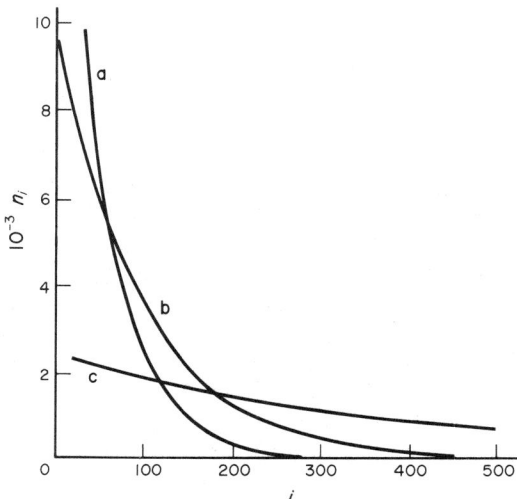

Figure 3 The most probable frequency distribution of chain lengths for polymers: (a) $\bar{i}_n = 50$; (b) $\bar{i}_n = 100$; and (c) $\bar{i}_n = 400$ (adapted from Peebles[2])

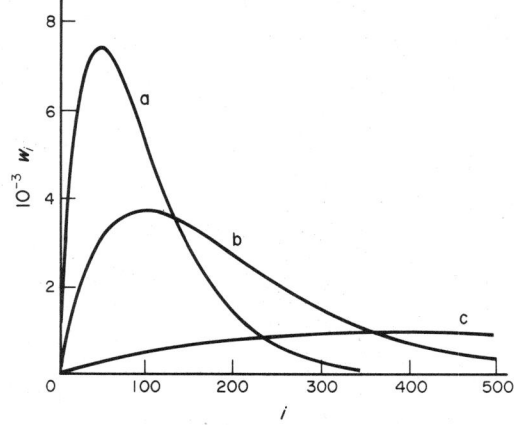

Figure 4 The most probable weight distribution of chain lengths for polymers corresponding to the frequency distributions in Figure 3: (a) $\bar{i}_n = 50$; (b) $\bar{i}_n = 100$; (c) $i_n = 400$ (adapted from Peebles[2])

distributions at the low molecular weight end of the curves; the polymer contains a large number fraction of small molecules but they represent a small weight fraction. The maximum in the weight distribution function lies at $\bar{i}_n$ and the inflexion point at $\bar{i}_w$.

$$\bar{i}_n = 1/(1-\phi) \tag{27}$$

$$\bar{i}_w = (1+\phi)/(1-\phi) \tag{28}$$

$$\bar{i}_z = (1+4\phi+\phi^2)/(1-\phi)(1+\phi) \tag{29}$$

In order for high molecular weight polymer to be formed, equation (27) requires that ϕ be close to unity. In such a case we can make the approximation of equation (30), which allows equations (25) and (26) to be rewritten as equations (31) and (32). For this reason, and because of the exponential appearance of the frequency distribution, the most probable distribution is also often known as the exponential distribution function.

$$\bar{i}_n = 1/(1-\phi) \approx -1/\ln \phi \tag{30}$$

$$n_i \bar{i}_n = \exp(-1/\bar{i}_n) \tag{31}$$

$$w_i \bar{i}_n = (i/\bar{i}_n)\exp(-1/\bar{i}_n) \tag{32}$$

For specific cases which fit the conditions required for the most probable distribution, ϕ is calculable from the kinetics of the reaction and this calculation has been discussed in detail by Bamford *et al.*[7] For the case of addition polymerization, ϕ is determined by the relative rates of propagation and deactivation of chains. For an active centre which deactivates by chain transfer, or termination with a species S, it is given by equation (33), where k_p and k_{tr} are the rate constants for the propagation and transfer respectively. For a free radical reaction with both transfer and termination by disproportionation, equation (29) becomes equation (34), where k_t is the rate constant for termination and I the rate of initiation. Substitution from either equation (33) or (34) into equation (27) yields kinetic expressions for $\bar{i}_n$ which are identical with those derived by kinetic arguments.

$$\phi = k_p[M]/(k_p[M]+k_{tr}[S]) \tag{33}$$

$$\phi = k_p[M]/(k_p[M]+k_{tr}[S]+(Ik_t)^{1/2}) \tag{34}$$

4.3.4 Deviations from Ideal Distributions

So far we have looked at two cases where the molecular weight distribution can be predicted rather simply. Polymerizations which are instantaneously initiated and free of termination and transfer lead to a Poisson distribution, whereas those which have continuous initiation and a random termination or transfer step lead to a most probable distribution, at least under conditions where the initiation and propagation rates remain constant. Both of these situations are ideals which can be approached in practice with careful control.

In the general case the mechanisms may be more complicated and the rates of the individual reaction steps can change with conversion, both because of changes in concentration of species and because of changes in rate constants. In principle the distribution reflects the history of the polymerization and can be related to the mechanism where this is well enough understood. Goodrich[8] has discussed in general the problem of predicting molecular weight distributions from kinetic schemes. In practice there is no single, universal treatment but a number of generalizations are possible. In the following sections we give a qualitative outline of some of the specific cases which can have practical relevance.

4.3.4.1 Instantaneous initiation with termination or transfer

Szwarc[4] has emphasized the practical conditions required for synthesis of a polymer with a Poisson distribution of molecular weights. All chains must be initiated simultaneously, termination and transfer must be excluded, the reaction conditions must be spatially homogeneous, the rate of propagation must be independent of chain length and all propagation steps must be identical and irreversible. These conditions are extremely difficult to achieve. In particular, increasing the molecular weight of the polymer requires either a higher monomer concentration or a lower active centre concentration. Since the monomer concentration is limited by the need to keep the solution fluid and homogeneous, higher molecular weights imply reduced active centre concentrations and it becomes increasingly harder to eliminate termination by impurities.

Szwarc[4] and Peebles[2] both give exhaustive treatments of the effects of impurity termination in living polymerization. The precise effect depends on the exact reaction. If impurities added with the monomer very rapidly kill a proportion of the active centres, then the distribution remains narrow but the average molecular size is higher than expected. Termination which is slower relative to propagation causes the deactivation of some chains at the same time as others continue to propagate and the molecular weight distribution becomes broadened in a complex way which depends on whether termination occurs at a constant rate or at a rate which varies with conversion.[2] An understanding of these effects is fundamental in attempts to synthesize monodisperse polymers.

The effect of chain transfer on a living polymerization with rapid initiation has been analyzed by Coleman *et al.*[9] and by Orofino and Wenger.[10] This situation is common in cationic polymerizations where initiation is often very rapid, termination slow or nonexistent and the reaction dominated by chain transfer to monomer (Volume 3, Part 4). The effect is to broaden the distribution because although all active centres are generated simultaneously, molecular chains are generated randomly. If the transfer reaction is sufficiently rapid relative to the propagation, then the polymer produced in a batch reactor is expected to have a most probable distribution.

4.3.4.2 Living polymerization with slow initiation

The effect of slow initiation of a living polymerization has been thoroughly analyzed by Gold;[11] this situation may apply in anionic, group transfer or coordination polymerizations. It can be modelled qualitatively by imagining a batch reactor containing a fixed amount of monomer into which are injected very small aliquots of initiator which reacts instantaneously. The instantaneous distribution for chains generated by each aliquot will be a Poisson one but each group of chains will have a different value of γ as it is initiated in a lower monomer concentration than the previous group. The molecular weight distribution can thus be visualized as a convolution of an infinite number of Poisson distributions with different values of γ, with the individual distributions weighted in a way determined by the kinetics of initiator addition. In principle we could use this approach to generate any required distribution to order.

This situation can be realized in a reactor which contains a fixed initial concentration of monomer whose living polymerization is initiated by slow injection of initiator. Provided that the rate of initiation is rapid, the molecular weight distribution depends upon the profile of the initiation, which can be controlled to give narrow or extremely wide distributions. Procedures of this type have been carried out by using electrochemical methods to generate anionic initiators.[12] The same situation can be realized by using mechanical injection of the initiator solution by a pump and also by the use of controlled monomer addition to a continuous tubular reactor; control of molecular weight distribution in this way has been discussed by Meira and Johnson.[13] This approach is limited only by the few monomers which can be polymerized with the required strict control.

In the more familiar case of a batch reactor where all of the initiator is added at the beginning but reacts relatively slowly, the extent of broadening depends upon the ratio k_p/k_i but does not approach the most probable distribution; the distribution is characterized by a long tail to low molecular weight and a rather sharp cut off at the high molecular weight end of the frequency distribution (effectively corresponding to the molecular weight of the chains initiated at the beginning of the reaction). This distribution is illustrated in Figure 5.

4.3.4.3 Free radical polymerization with combination

In the particular case of free radical polymerization it is possible for chains to be deactivated by combination reactions, which lead to chains becoming coupled together. This would be expected to have the effect of making short chains less probable. In the specific case where termination is by combination, there is no transfer reaction and the monomer concentration remains constant the distribution is given by equations (35) and (36).[7] The number distribution is compared with the most probable case in Figure 6. The absence of short chains is obvious and expected. The distribution is narrower and $\bar{i}_w/\bar{i}_n = 1.5$. In reality most radical polymerizations have a mixture of termination mechanisms and also are subject to chain transfer; the resulting distribution is thus the appropriately

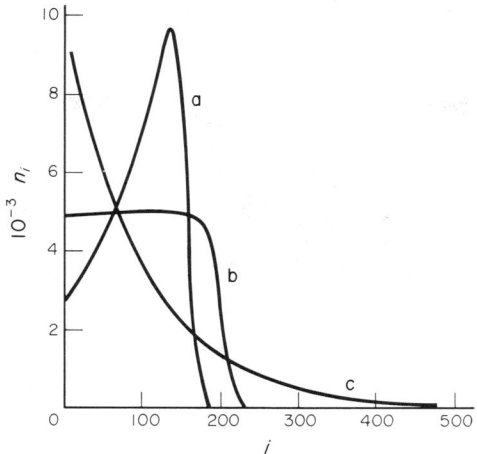

Figure 5 Frequency distribution for polymers with $\bar{i}_n = 100$: (a) living polymerization, no transfer, slow initiation with $k_p/k_i = 10^2$; (b) living polymerization, no transfer, slow initiation with $k_p/k_i = 10^4$; and (c) most probable distribution (adapted from Gold[11])

weighted sum of equations (25) and (35).

$$n_i = (-\ln \phi)^2 i \tag{35}$$

$$w_i = ((-\ln \phi)^3 i^2 \phi^i)/2 \tag{36}$$

4.3.4.4 *Free radical polymerization with monomer consumption*

In introducing the most probable distribution, we considered a polymerization in which the rate of initiation and the monomer concentration are kept constant so that ϕ remains constant throughout the polymerization. In fact this situation is rare, certainly in laboratory-scale reactions, and a much more likely case is that of a batch reaction, where the rate of initiation may be constant but the monomer concentration decreases with conversion as the monomer is used up. If the propagation is irreversible and termination is by disproportionation or transfer, then the most probable distribution represents the instantaneous distribution; chains which are formed in the early stages of the reaction will have the most probable distribution corresponding to the value ϕ at that point. Once a chain is terminated it remains as part of the polymer but plays no further part in the

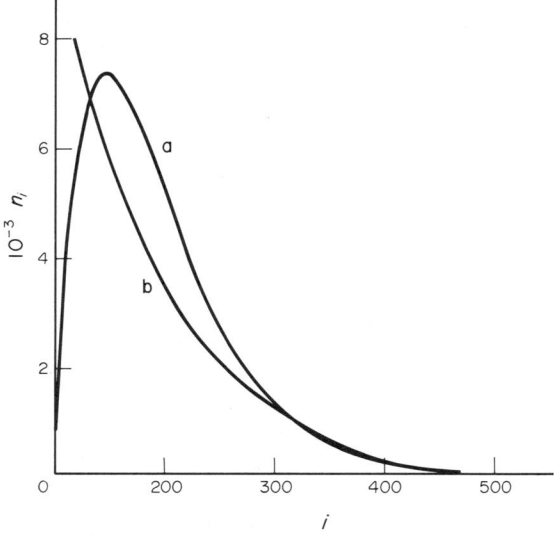

Figure 6 Effect of termination by combination on frequency distribution of a polymer with $\bar{i}_n = 100$; (a) termination by combination, and (b) most probable distribution

reaction. Thus as the concentration of monomer decreases, and the value of ϕ falls, the molecular weight distribution remains most probable but the average molecular weight of the polymer falls. For this situation the most probable distribution represents the instantaneous distribution of the polymer being formed at any point; the overall distribution is effectively the convolution of an infinite series of most probable distributions. The same general considerations will apply to a reaction in which the initiator is consumed and the rate of initiation decreases with conversion, although in this case the average molecular weight will tend to rise with conversion.

Note the contrast between this situation and the Poisson distribution of a living polymer, discussed in Section 4.2. In the living polymerization case the value of γ increases with conversion but the same value applies to all polymer molecules at any one time; the average molecular weight increases with conversion and the molecular weight distribution does not change. In the more general case the value of ϕ changes with conversion but polymer molecules are formed with the distribution appropriate to the conversion at which they are generated. In the most typical case, where the initiation rate is constant but the monomer is used up, the average chain length falls with conversion.

Mathematically the distribution functions for cases where the parameter ϕ is not constant are often very complex and depend on the specific details of the reaction. Peebles[2] and Bamford *et al.*[7] have discussed these cases in great detail. As an example, Figure 7 shows the predicted weight distributions for a batch polymerization with constant initiation rate and varying conversion when termination is by transfer or disproportionation.[7] The effect is not very significant below 25% conversion but above this the shift towards lower molecular weight is very noticeable. The ratio $\bar{i}_w/\bar{i}_n$ is 2.0 at the beginning of the reaction, changes little over the first 25% conversion and rises to 2.5 at about 80% conversion. Even in this general case the distribution is not particularly wide. The very wide distributions found in some polymerizations with coordination catalysts (where $\bar{M}_w/\bar{M}_n$ can be 50 or above) arise mainly from a spread of active centres of different reactivity so that k_p is not a constant. Wide distributions can also be formed under conditions which lead to values of k_p or k_{tr} which vary with conversion, *e.g.* if the viscosity of the polymerization mixture changes significantly with conversion. They also arise if the transfer reactions can lead to branching, as is discussed below.

4.4 EMPIRICAL DISTRIBUTION FUNCTIONS

4.4.1 Introduction

We have seen that there are two extreme cases where the molecular weight distribution of a polymer can be predicted rather simply but that the majority of real cases lead to more complicated distributions. Examination of the figures so far presented will show that the distribution functions for random reactions are all rather similar despite being generated from different equations. This similarity has led to the proposition that the distribution can be approximated by empirical

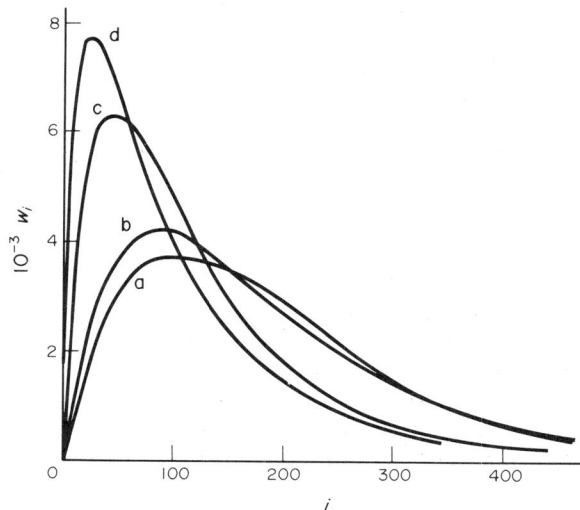

Figure 7   Effect of conversion on the weight distribution from a polymerization with $\bar{i}_n = 100$ at low conversion. Conversions are: (a) 0.1%; (b) 0.25%; (c) 0.75%; and (d) 0.95% (adapted from Peebles[2])

functions which are mathematically tractable and can be easily integrated to give averge molecular weights. Empirical functions are important in the analysis of data from polymer fractionation experiments; they are also useful in comparison of distributions and in some curve-fitting procedures for the analysis of chromatographic fractionation data. In the following sections we briefly introduce some of the more important functions.

4.4.2 The Schulz Exponential Distribution

This is a two-parameter function which is convenient as a model for many polymers. It takes the form described by equations (37) and (38),[14,15] where ψ and n are parameters. The averages are given by equations (39) and (40), so that the polydispersity is given by equation (41).

The Schulz distribution is a limiting form of the binomial distribution and includes a number of other distributions as special cases. For $n=1$ it reduces to the most probable distribution and for $n=2$ it becomes the instantaneous distribution expected for radical polymerization with termination by combination. As n increases, the distribution narrows and for $n=\infty$ it becomes the Poisson distribution.

$$n_i = e^{-i\psi} i^{n-1} \psi^i / \Gamma(n) \tag{37}$$

$$w_i = e^{-i\psi} \psi^{n+1} i^n / \Gamma(n+1) \tag{38}$$

$$\bar{i}_n = n/\psi \tag{39}$$

$$\bar{i}_w = (n+1)/\psi \tag{40}$$

$$\bar{i}_w/\bar{i}_n = 1 + 1/n \tag{41}$$

4.4.3 The Log-normal Distribution

The normal or Gaussian distribution function describes the symmetrical distribution of individuals about the mean. Although it can be used as an approximation to the Poisson distribution or for other narrow distributions, as for example in fractionated polymers, it cannot be used to describe wide distributions because its symmetry about the mean would require negative chain lengths. One way of getting round this problem is to assume that the logarithm of the chain length has a Gaussian distribution. The resulting function is given by equations (42) and (43), where i_0 is the mean and β is the variance of the distribution of $\log i$. The averages are then given by equations (44) and (45). This distribution function is the log-normal distribution, sometimes referred to as the Lansing–Kraemer distribution.[16] It has been a popular model distribution function, mainly because of the ease with which the constants $\bar{i}_0$ and β can be determined from fractionation data, using graph paper with probability scaling.[2] However, its use has been criticized.[17]

$$n_i = \frac{1}{i\beta\sqrt{2\pi}} \exp\left(\frac{-(\log i/i_0)^2}{2\beta^2}\right) \tag{42}$$

$$w_i = \frac{1}{\bar{i}_n \beta\sqrt{2\pi}} \exp\left(\frac{-(\log i/i_0)^2}{2\beta^2}\right) \tag{43}$$

$$\bar{i}_n = \bar{i}_0 \exp(-\beta^2/2) \tag{44}$$

$$i_w = \bar{i}_0 \exp(\beta^2/2) \tag{45}$$

4.4.4 The Generalized Exponential Distribution

The generalized exponential distribution is a three-parameter function, given by equations (46) and (47), where m, k and y are constants. In the case where $m=1$ and $y=0.5$ this is the conventional χ^2 distribution. It becomes the most probable distribution for $k=m=1$, the Schulz exponential distribution for $m=1$ and the log-normal distribution for $m=0$. With the availability of computers, fitting of the three-parameter equation to experimental data has become more feasible and more popular. Some of the problems in doing so have been discussed by Jakes and Saudek.[18]

$$n_i = my^{k/m} i^{k-1} [\exp(-yi^m)]\Gamma(k/m) \tag{46}$$

$$w_i = my^{(k+1)/m} i^k [\exp(-yi^m)]/\Gamma[(k+1)/m] \tag{47}$$

4.5 MORE COMPLEX CASES: BRANCHED POLYMERS AND COPOLYMERS

4.5.1 Introduction

We have so far considered only the simplest type of chain reaction polymerization, that in which there is a single monomer and the polymer produced is linear. Even within this limitation, the distribution of molecular sizes is a complex problem in the general case, although it can be simplified. In practice the situation is often even more complex for two reasons. Firstly, adventitious chain transfer reactions or deliberately added polyfunctional monomers may lead to chain branching or cross-linking. Secondly, the polymerization may involve more than one monomer, leading to copolymers. Both of these possibilities lead to new problems which are discussed in more detail in the appropriate chapters of this work. At this stage we briefly illustrate the nature of the problems.

4.5.2 Chain Branching

Chain branching can arise in a chain polymerization for a number of reasons. In the polymerization of a conventional difunctional monomer the commonest cause of branching is chain transfer reactions with the polymer which result in generation of a new initiating centre on an existing polymer chain. If the active centre is a radical and termination is by combination, then cross-linking may also occur. These cases typically lead to relatively low levels of branching. Conversely, branching may arise from the use of monomer molecules which contain more than one double bond capable of polymerization. Such reactions typically lead to high levels of branching and to cross-linking.

The detailed calculation of the molecular weight distribution for a branching system depends upon the particular mechanism involved. Branching by chain transfer alone cannot form a cross-linked network unless there is some combination mechanism. Bamford and Tompa discussed the case of branching by chain transfer to polymer with first-order termination.[3] This is also related to the case of graft copolymerization onto a preformed polymer chain. In a batch polymerization a polymer molecule initially formed at low conversion will gain more branches than one formed later, partly because it is in the system for longer and partly because the chain length of polymers formed at the beginning is higher. The distribution function is complex but the effect is that the molecular weight distribution becomes considerably broadened and shifted to higher molecular weight.

Polymerization of multifunctional monomers allows cross-linking to occur and has been discussed by Flory[19] and by Bamford *et al.*[7] In the early stages the polymerization yields linear molecules which contain pendant functionalities whose subsequent reaction causes cross-linking. This situation can be treated as the cross-linking of an initial population of preformed polymer chains and will eventually lead to coupling of chains to form an infinite network. At this point the system consists of a mixture of individual polymer chains with an infinite network and gelation occurs. The probability of formation of an infinite network, q, can be expressed by equation (48), where x is the probability that any unit of the primary chain has entered into cross-linking. If q is less than unity then there is no gelation, whereas a value greater than unity gives a gel. The critical condition for gelation is thus $q = 1$ and the critical value of x for gelation, x_c, is given by equation (49). This is a very important conclusion because it implies that gelation of long chains occurs very easily. If we denote the number of cross-links per primary molecule by γ, then γ is equal to $x\bar{i}_n$ and its value at the gel point is given by equation (50). Thus for a homogeneous primary polymer gelation occurs when $\gamma = 1$. Broadening of the distribution favours gelation and the critical value of γ is 0.5 for a most probable distribution of the primary molecules. This means that a simple polymerization will produce a gel if there is one cross-link for every two polymer chains. Stockmayer[20] has evaluated the gel point and the changes in molecular weight distribution for the coupling of polymers of different initial distribution. He points out that the distribution becomes extremely heterogeneous; coupling of molecules decreases $\bar{M}_n$ but increases $\bar{M}_w$ and the distribution rapidly broadens. The theory of gelation is dealt with in more detail in Volume 5, Chapters 2, 4 and 8.

$$q = x(\bar{i}_w - 1) \tag{48}$$

$$x_c = 1/(\bar{i}_w - 1) \approx 1/\bar{i}_w \tag{49}$$

$$\gamma_c = \bar{i}_n/\bar{i}_w \tag{50}$$

4.5.3 Copolymers

It is common in polymer chemistry to modify the properties of a homopolymer by the introduction of a second monomer to give a copolymer. In principle the second monomer may be introduced randomly into the polymer chain or may be present with some degree of ordering. The limiting cases of ordered copolymers are block, graft and alternating copolymers.

The simplest copolymerization is that in which two monomers are mixed and polymerized simultaneously with termination by transfer or disproportionation. This case is considered by Bamford *et al.*[7] and by Stockmayer.[21] The polymer formed in any small increment of reaction is expected to be heterogeneous in respect of both composition and chain length. The distribution of composition irrespective of chain length is a normal distribution, whilst the distribution of chain length irrespective of composition is expected to have the most probable form. In any real polymerization the problem is that the two monomers are not included in the copolymer in the ratio of their concentrations in the monomer mixture. Differences in reactivity cause one monomer to be consumed faster than the other so that in a batch reaction there is a slow drift in the monomer composition with conversion. At the same time the consumption of monomers leads to the expected changes in molecular weight distribution. The result is that the general case for a random copolymer is a complex distribution of both composition and molecular size, which is both difficult to predict and difficult to characterize.

If a copolymerization is carried out under conditions which lead to a strictly alternating copolymer, then it will behave as if it were the reaction of a single monomer and the normal equations apply. Similarly a random grafting to a preformed polymer chain behaves like a branching homopolymerization as discussed above, although some methods of synthesis can allow close control over both the length and distribution of the grafted chains. The specific case of a block copolymer formed by coupling of two preformed homopolymers has recently been analyzed by Yan and Yuan.[22] If the number of distributions of the two prepolymers are n(i) and n(j) respectively, then the block copolymer is characterized by the two-dimensional distribution function of equation (51), where n(ij) is the number fraction of chains containing i units of type 1 and j units of type 2. The corresponding averages are related to those of the homopolymers by equations (52) and (53), where x is the weight fraction of polymer molecules of type 1 in the polymer.

$$n(ij) = n(i)n(j) \tag{51}$$

$$\bar{M}_n = \bar{M}_n^1 + \bar{M}_n^2 \tag{52}$$

$$\bar{M}_w = \frac{x\bar{M}_w^1 + (1-x)\bar{M}_w^2}{1 - 2x(1-x)} \tag{53}$$

4.6 REFERENCES

1. See, for example, H. Mulholland and C. R. Jones, 'Fundamentals of Statistics', Butterworths, London, 1968.
2. L. H. Peebles, Jr., 'Molecular Weight Distributions in Polymers', Interscience, New York, 1971.
3. C. H. Bamford and H. Tompa, *Trans. Faraday Soc.*, 1954, **50**, 1097.
4. M. Szwarc, 'Carbanions, Living Polymers and Electron-Transfer Processes', Interscience, New York, 1968.
5. G. V. Schulz, *Z. Phys. Chem., Abt. B.*, 1935, **30**, 379.
6. P. J. Flory, *J. Am. Chem. Soc.*, 1936, **58**, 1877.
7. C. H. Bamford, W. G. Barb, A. D. Jenkins and P. F. Onyon, 'The Kinetics of Vinyl Polymerization by Radical Mechanisms', Butterworths, London, 1958, chap. 7.
8. F. C. Goodrich, in 'Polymer Fractionation', ed. M. J. R. Cantow, Academic Press, New York, 1967, chap. F.
9. B. D. Coleman, F. Gornick and G. Weiss, *J. Chem. Phys.*, 1963, **39**, 3233.
10. T. A. Orofino and F. Wenger, *J. Chem. Phys.*, 1961, **35**, 532.
11. L. Gold, *J. Chem. Phys.*, 1958, **28**, 91.
12. B. L. Funt and D. R. Richardson, *J. Polym Sci., Part A-1*, 1970, **8**, 1055; 1971, **9**, 2171.
13. G. R. Meira and A. F. Johnson, *Polym. Eng. Sci.*, 1981, **21**, 415.
14. G. V. Schulz, *Z. Phys. Chem., Abt. B*, 1939, **43**, 25.
15. B. H. Zimm, *J. Chem. Phys.*, 1948, **16**, 1099.
16. W. D. Lansing and E. O. Kraemer, *J. Am. Chem. Soc.*, 1935, **15**, 1369.
17. A. M. Kotliar, *J. Polym. Sci., Part A*, 1964, **2**, 4303, 4327.
18. J. Jakes and V. Saudek, *Makromol. Chem.*, 1986, **187**, 2223.
19. P. J. Flory, 'Principles of Polymer Chemistry', Cornell University Press, Ithaca, NY, 1953, chap. 9.
20. W. H. Stockmayer, *J. Chem. Phys.*, 1943, **11**, 45; 1944, **12**, 125.
21. W. H. Stockmayer, *J. Chem. Phys.*, 1945, **13**, 199.
22. D. Yan and C. Yuan, *Makromol. Chem., Rapid Commun.*, 1987, **8**, 83.

5

General Experimental Methods in Polymerization

MANFRED STICKLER
Röhm GmbH, Darmstadt, FRG

5.1 PURIFICATION OF REAGENTS

Commercially available monomers may contain a number of impurities or additives that must be carefully removed prior to polymerization. These impurities are often by-products of the monomer synthesis, *e.g.* divinylbenzene and ethylbenzene, in the case of styrene.[1] Since many vinyl monomers react readily with oxygen to form peroxy compounds, these autoxidation products and their thermal or photochemical decomposition products must also be accounted for (*e.g.* for methyl methacrylate, the alternating copolymer of the monomer with oxygen and formaldehyde and methyl pyruvate derived from it).[2-5] Furthermore, monomers may contain oligomers and polymeric material. To prevent unintentional polymerization during storage, stabilizers such as phenols[6] and amines[7] are used as additives. Reaction products of these stabilizers or inhibitors with monomer, oxygen or free radicals derived therefrom can also be present in a monomer. Finally, dissolved oxygen and moisture may be present.

The method of purification of a given monomer naturally depends on its chemical nature, the impurities that can be expected according to its preparation and, last but not least, on the type of polymerization reaction that one intends to perform (*i.e.* whether free radicals or ionic species are to be used as active centres of chain growth). Generally, the usual methods of organic chemistry can be applied.[8-11] Owing to the character of a chain reaction, polymerizations are extremely sensitive towards impurities and thus all these operations must be performed very rigorously. This applies especially to so-called living polymerizations.[12]

The purification process may be controlled using highly sensitive analytical procedures, such as gas chromatography or mass spectrometry. However, since absolutely pure chemical compounds do not exist, the purity of a monomer is always defined functionally. A monomer may be said to be pure if different methods of preparation and repeated purification lead to identical experimental results, *e.g.* identical analytical data (UV, IR and mass spectra and gas chromatograms) or, more stringently, to identical behaviour in kinetic investigations.

The removal of stabilizer from liquid monomers is usually accomplished by extraction with either dilute alkali or acid, depending on the type of inhibitor (phenol or amine). After subsequent washing with distilled water, the monomer is dried with an appropriate drying agent[10] and then fractionated by distillation, usually in an inert atmosphere of nitrogen or argon and under reduced pressure. It may be advantageous to add a small amount of copper stearate during distillation, since this compound destroys peroxides very efficiently.[13] The distillate is collected in a reservoir, which may be sealed under an inert gas atmosphere for storage. If storage of freshly distilled monomer for a short time is necessary, this should be carried out in the dark at temperatures as low as possible.

However, especially for kinetic experiments, monomers should be used immediately after their purification. For longer storage times, the purified monomer should once more be stabilized.[8]

The removal of inhibitor may be performed more conveniently by adsorption chromatography. With this technique, the monomer is run directly into the still through a column filled with activated alumina. This is a highly efficient method for the simultaneous removal of peroxides.[14,15] However, in the case of ester monomers such as acrylates or methacrylates, one should be aware that activated alumina may hydrolyze the ester group. The corresponding free acids and alcohols produced from such a hydrolysis can be eliminated by the addition of calcium hydride to the still. Refluxing the monomer over calcium hydride also destroys traces of water.[12,15]

The final purification of a monomer is most often performed by prepolymerization.[16] For this purpose, the reservoir with the monomer is attached to a high vacuum system. Residual dissolved oxygen is then first removed by several cycles of freezing and thawing.

Several methods of initiation have been proposed for inducing the subsequent prepolymerization. A photochemical initiation without added sensitizer has been proposed as the most convenient technique.[17,18] However, since the unsensitized photopolymerization of many monomers is very slow, a prolonged exposure to UV radiation with short wavelength is needed. Under these conditions, volatile by-products may be formed which might unintentionally pollute the monomer again.[15,17-19]

The addition of thermal initiators may also cause trouble. Thus, according to the results of Dulog et al.,[14] after the prepolymerization of methyl methacrylate (MMA) with azobis(isobutyronitrile) (AIBN), it is impossible to obtain the residual monomer completely free from traces of initiator by either distillation or adsorption chromatography.

Fortunately, many vinyl monomers do polymerize at elevated temperatures even in the absence of initiator. In most cases, these 'thermal' polymerizations are due to trace impurities and are not truly spontaneous thermal polymerizations as is the case with styrene or MMA. However, at least with free-radical polymerizations, such 'thermal' prepolymerizations should yield a monomer of probably the highest purity, since it is believed that all impurities which might disturb a free-radical polymerization are eliminated by this pretreatment.[15,16]

In some cases, a prepolymerization can be achieved even at room temperature if, after the purification of the monomer by distillation, it is stirred for some days with a small amount of activated alumina[15] or with a molecular sieve.[15,20]

Besides the more conventional methods, such as fractionated crystallization, solid monomers can be purified to a high degree by zone melting (provided they are stable at their melting point). A sublimation in high vaccum may also be adequate.

Rectification at lower temperatures is useful for the purification of monomers which are gaseous at ambient temperatures. Traces of humidity and of oxygen are removed from these monomers with the help of molecular sieves, e.g. the so-called BTS catalyst produced by the BASF AG[21] or, more conveniently and much more rigorously, with Oxisorb of Messer Griesheim GmbH.[22]

Solvents used in polymerization experiments are purified following the well-known methods of organic chemistry;[23] the same holds for catalysts or any other additive used in a given recipe.

5.2 PREPARATION OF THE REACTION SET-UP

Besides rigorous purification of the reagents, the careful selection and cleaning of the reaction vessel is of great importance for polymerization systems. Numerous types of laboratory polymerization reactors have been described in the literature; some simple apparatuses for the polymerization of liquid, gaseous and solid monomers have been presented in the review of Kern and Schulz.[8]

Polymerizations with extremely low concentrations of active chain-ends, e.g. spontaneous thermal polymerizations or cases involving living chain carriers, and particularly kinetic investigations of these systems, require special care in cleaning the apparatus used. It is then desirable to carry out the final steps of the purification of the reagents on a high-vacuum line where greased stop-cocks should be avoided as far as possible and where breakseals or, at least, vacuum-tight Teflon valves should be used instead.[12,15]

At the beginning of such an experiment, any oxygen is removed from the reactor by several cycles of evacuation and flushing with oxygen-free nitrogen. Moisture adhering very obstinately to the glass surface can be removed by the 'flaming in vacuum' technique. With ionic polymerizations, even such rigorous cleaning can be insufficient. Thus, washing of the whole reaction vessel with a solution of the organometallic initiator to be investigated is often used.[12]

Furthermore, one should always be aware that the surface of a reaction vessel may contribute to the chain-initiation or chain-termination processes in the bulk of such a reaction system. Under these circumstances, the rate of polymerization may depend on the material and/or the surface-to-volume ratio of the vessel. To overcome these influences, the surface of the polymerization reactor may by modified, *e.g.* by silylation, in the case of glass vessels, or by conditioning.[15]

In industrial practice, instead of the somewhat more complicated and laborious breakseal method, reactors equipped with self-sealed rubber gaskets together with hypodermic syringes are widely used for polymerization reactions sensitive towards oxygen and humidity. This technique offers more flexibility and expediency.[10]

The design of a polymerization vessel may have a significant influence, especially with kinetic investigations. Polymerizations are highly exothermic and, even at relatively low degrees of monomer conversion, the viscosity of the reaction mixture may become very high. In rapidly polymerizing systems, the dissipation of the heat produced is a severe problem. Here, a favourable suface-to-volume ratio of the reaction vessel and the additional possibility of effective stirring is advantageous. Otherwise it is impossible to carry out the reaction under isothermal conditions. This problem has recently been discussed in detail by Thiele *et al.*[24]

Another point which should be considered when selecting a reactor is the half-life of the monomer conversion of a given polymerization. The time taken to mix and heat up the reagents and, if necessary, the time needed for quenching the reaction and for withdrawing samples, should always be negligible when compared with this half-life of monomer consumption. With free-radical polymerizations, which under normal conditions are relatively slow reactions, a stirred-tank reactor generally fulfills this prerequisite quite well. Such a simple laboratory reactor has recently been designed for precision measurements of low-conversion polymerization kinetics.[25] More sophisticated devices for the study of anionic polymerizations with half-lives of the order of about 10 min are shown in the excellent monograph of Szwarc.[12] For even shorter half-lives, automatically controlled reactors are needed. Warzelhan *et al.*[26] presented such a stirred-tank reactor, equipped with magnetic valves, which is claimed to be suitable for the kinetic study of polymerizations with half-life periods as small as 2 s (*cf.* Figure 6 in Volume 3, Chapter 38).

Studies of even faster reactions, with half-lives in the region of 0.05 to 2 s, require adoption of flow techniques. The capillary flow-method was introduced in 1923 by Hartridge and Roughton;[27] modern versions have been developed by the schools of Szwarc[12,28] and Schulz[29,30] to investigate anionic polymerization kinetics; *cf.* also Volume 3, Chapter 38 for further information.

The stopped-flow method, originally invented by Chance and co-workers to study enzymatic reactions,[31,32] may also be adopted for fast polymerization reactions; details and a discussion of the advantages and shortcomings of these two flow techniques can be found elsewhere.[12]

With fast polymerization reactions, the stirred-flow reactor proposed by Denbigh and co-workers[33-35] may also be used.[12] When working with small reactor volumes (about 10 ml), the residence time is reduced considerably and a steady state of the flow system is approached within a few minutes.

5.3 ISOLATION AND PURIFICATION OF POLYMERS

In some cases (*e.g.* with kinetic experiments), it is desirable to stop a polymerization prior to complete conversion of the monomer. This may be in order to study the properties of the polymer at different degrees of conversion or to avoid undesired side reactions which occur predominantly at high contents of polymer (*e.g.* chain branching).

In the simplest case, rapid cooling of the reaction mixture will lead to a quenching of the polymerization. Another possibility is to pour the sample into a nonsolvent, thus precipitating the polymer and at the same time highly diluting the residual monomer and the catalyst. It is more effective to stop the reaction by the addition of some inhibitor or a compound that destroys the catalyst. The selection of such an additive naturally depends on the type of polymerization. Free-radical polymerizations can be stopped with the help of stable free radicals, such as 2,2'-diphenyl-1-picrylhydrazyl (DPPH) or 2,2,6,6-tetramethylpiperidine-*N*-oxyl (TEMPO). With anionic or cationic polymerizations, the addition of water, acids or bases is recommended.

Isolation is especially easy if a polymer is insoluble in its own monomer (*e.g.* acrylonitrile). Then filtration or centrifugation can be used to separate the polymer from the heterogeneous reaction mixture. If the polymer remains soluble, one may strip residual monomer and other volatile constituents from the reaction mixture with the help of a vacuum distillation, or the polymer can be precipitated. With the former procedure, nonvolatile compounds (*e.g.* initiator residues) will, of course, remain together with the polymer, and monomer and solvent inclusions can scarcely be avoided.

Thus, precipitation is generally the best way to isolate polymer. The precipitating liquid should be miscible with the monomer and the solvent(s) used as the reaction medium, and it should be a good solvent for the initiator and all other additives used with the given polymerization recipe. Furthermore, the solvents used for the precipitation should be volatile enough to be easily removed when drying the polymer in high vacuum at elevated temperatures. A compilation of solvents and nonsolvents for polymers can be found in the monograph of Gnamm and Fuchs.[36]

The precipitation of a polymer is affected by the concentration of the polymer solution, the rate and method of mixing this solution with the nonsolvent, the temperature and, of course, by the chemical nature and thermodynamic quality of solvent and nonsolvent. If the polymer solution is too concentrated, the polymer precipitates as a gummy material which includes residual monomer and solvent. If, on the other hand, the polymer solution is too dilute, the precipitate may be so finely disperse that it will not settle down and this makes handling very awkward. With such colloidal systems, the addition of a small amount of dilute acetic acid or a dilute solution of sodium chloride may help to overcome these difficulties and to bring the polymer into an agglomerated and manageable form. Sometimes cooling is sufficient to break down the colloidal system; in any case, vigorous stirring is essential. Optimal conditions will be found by trial.

To work at an acceptable viscosity, solutions of very high molecular weight polymers must be strongly diluted for an efficient precipitation fractionation. To avoid handling huge solvent volumes, it is convenient to use the spraying technique for the precipitation in these cases.[37]

Polymer precipitation should be quantitative. Normally, low molecular weight material is the most soluble and consequently might be lost in the precipitation procedure. It is reported that PMMA with a degree of polymerization less than about 50 is still soluble in methanol.[38] Under special circumstances, *e.g.* with end-group analysis, such a loss of low molecular weight polymer (even if it can hardly be detected by weighing) will lead to substantial errors and stronger precipitants should be looked for (*e.g.* petroleum ether for the precipitation of low molecular weight PMMA).

Purification of polymers can be achieved by repeated cycles of precipitation; with crosslinked materials, swelling and extraction is possible. In both cases it is advantageous to change the polarity of solvents and nonsolvents in the subsequent purification steps.

A careful method of general applicability is that of freeze drying, where benzene, dioxane or water are most often used as solvents. With water-soluble polymers, dialysis may also be used to remove low molecular weight impurities and, at the same time, increase the polymer concentration.[39]

Polymers obtained by one or other of the foregoing isolation and purification methods are normally wet, and the removal of the last traces of solvent and precipitant is often extremely difficult, because of solvent inclusions and/or particular interactions between polymer and solvent molecules. Drying is conventionally performed by heating the material in a high vacuum oven at temperatures slightly above the glass transition temperature of the polymer. The time required for obtaining weight constancy can be minimized by reducing the particle size of the solid. Freeze drying may be very advantageous, as it results in a polymer with a porous structure, enabling the easy release of included solvent residues.

5.4 REFERENCES

1. R. H. Boundy and R. F. Boyer, 'Styrene, Its Polymers, Copolymers, and Derivatives', Reinhold, New York, 1952.
2. C. E. Barnes, R. M. Elofson and G. D. Jones, *J. Am. Chem. Soc.*, 1950, **72**, 210.
3. G. V. Schulz and G. Henrici, *Makromol. Chem.*, 1956, **18/19**, 437.
4. F. R. Mayo and A. A. Miller, *J. Am. Chem. Soc.*, 1958, **80**, 2493.
5. C. H. Bamford and P. R. Morris, *Makromol. Chem.*, 1965, **87**, 73.
6. C. Moureu and C. Dufraisse, *C. R. Hebd. Seances Acad. Sci.*, 1922, **174**, 259.
7. I. L. Wolk (Phillips Petroleum Co.) *US Pat.* 2 407 861 (1946) (*Chem. Abstr.*, 1947, **41**, 617).
8. W. Kern and R. C. Schulz, in 'Houben-Weyl, Methoden der Organischen Chemie', 4th edn., ed. E. Müller, Thieme, Stuttgart, 1961, vol. XIV/1, p. 25.
9. G. M. Burnett, in 'Techniques of Organic Chemistry', 2nd edn., ed. A. Weissberger, Interscience, New York, 1963, vol. VIII/2, p. 1139.
10. D. Braun, H. Cherdron and W. Kern, 'Techniques of Polymer Synthesis and Characterization', Wiley-Interscience, New York, 1972.
11. S. R. Sandler and E. Karo, 'Polymer Syntheses', Academic Press, New York, 1977.
12. M. Szwarc, 'Carbanions, Living Polymers and Electron Transfer Processes', Interscience, New York, 1968.

13. G. M. Burnett, 'Mechanism of Polymer Reactions', Interscience, New York, 1954.
14. L. Dulog, W. Vogt and W. Kern, *Makromol. Chem.*, 1966, **97**, 75.
15. M. Stickler and G. Meyerhoff, *Makromol. Chem.*, 1978, **179**, 2729.
16. P. D. Bartlett and K. Nozaki, *J. Am. Chem. Soc.*, 1946, **68**, 2377.
17. C. H. Bamford and M. J. S. Dewar, *Proc. R. Soc. London, Ser. A*, 1949, **197**, 356.
18. C. H. Bamford and M. J. S. Dewar, *Discuss. Faraday Soc.*, 1947, **2**, 310.
19. M. H. Mackay and H. W. Melville, *Trans. Faraday Soc.*, 1949, **45**, 323; *Trans. Faraday Soc.*, 1950, **46**, 63.
20. T. H. Bates, J. V. F. Best and T. F. Williams, *Nature (London)*, 1960, **188**, 469.
21. M. Schütze, *Angew. Chem.*, 1958, **70**, 697.
22. M. Schubert and R. Erdmann, *Gas Aktuell*, 1971, **1**, 20.
23. J. Riddick and W. B. Bunger, in 'Techniques of Chemistry', 3rd edn., A. Weissberger, Wiley-Interscience, New York, 1970, vol. II.
24. R. Thiele, W. Hänisch and G. Weickert, *Plaste Kautsch.*, 1983, **30**, 181.
25. M. Stickler, *Makromol. Chem., Macromol. Symp.*, 1987, **10/11**, 17.
26. V. Warzelhan, G. Löhr, H. Höcker and G. V. Schulz, *Makromol. Chem.*, 1978, **179**, 2211.
27. M. Hartridge and F. J. W. Roughton, *Proc. R. Soc. London, Ser. A*, 1923, **104**, 376.
28. C. Geacintov, J. Smid and M. Szwarc, *J. Am. Chem. Soc.*, 1962, **84**, 2508.
29. R. V. Figini, H. Hostalka, K. Hurm, G. Löhr and G. V. Schulz, *Z. Phys. Chem. (Wiesbaden)*, 1965, **45**, 269.
30. G. Löhr, B. J. Schmidt and G. V. Schulz, *Z. Phys. Chem. (Wiesbaden)*, 1972, **78**, 177.
31. B. Chance, *J. Franklin Inst.*, 1940, **229**, 455, 613, 737.
32. B. Chance, Q. H. Gibson, R. H. Eisenhardt and K. K. Lonberg-Holm, 'Rapid Mixing and Sampling Techniques in Biochemistry', Academic Press, New York, 1964.
33. B. Stead, F. M. Page and K. G. Denbigh, *Discuss. Faraday Soc.*, 1947, **2**, 263.
34. K. G. Denbigh and F. M. Page, *Discuss. Faraday Soc.*, 1954 **17**, 145.
35. E. F. Caldin, 'Fast Reactions in Solution', Blackwell, Oxford, 1964.
36. H. Gnamm and O. Fuchs, 'Lösungsmittel und Weichmachungsmittel', 8th edn., Wissenschaftliche Verlagsgesellschaft mbH, Stuttgart, 1980, vol. II.
37. R. Schulz and A. Sabel, *Makromol. Chem.*, 1954, **14**, 115.
38. G. Henrici-Olivé and S. Olivé, *Adv. Polym. Sci.*, 1961, **2**, 496.
39. J. John, *Nature (London)*, 1959, **183**, 1055.

6

Overall Mechanisms

JOHN C. BEVINGTON
University of Lancaster, UK

6.1 INTRODUCTION

This chapter presents an overview of radical polymerization in homogeneous systems. It shows how the component reactions fit together and how the overall kinetics can be related to mechanistic and other features of the separate steps. Some specific examples are quoted but subsequent chapters should be consulted for detailed information.

An idealized kinetic treatment is presented first and is followed by discussion of aspects of each of the elementary reactions. Special attention is given to systems involving thermal and photochemical initiators; direct thermal or photochemical polymerization without added initiator is not considered. The final section of the chapter refers to the model compound approach to the elucidation of processes occurring during radical polymerization.

Reactions involving free radicals as chain carriers are the most generally applicable and usually the most easily performed processes for making high polymers from monomers containing carbon–carbon unsaturation. Detailed study of polymerizations in homogeneous systems, where all the reactants and also the polymer radicals and molecules are soluble, is essential for a proper understanding of polymerizations of other types including those using emulsions or suspensions of monomers, which are of great importance for the large scale production of polymers.

Some emphasis is placed on reactions which give rise to abnormal groupings, including end groups. In high polymers, these groupings correspond to a small fraction by weight of the whole material but they cause interruptions in structural regularity. They can be produced in any of the reactions which together make up the whole process of polymerization. In many cases, abnormal groups are important as possible sites for instability of a polymer; their role has been discussed in general[1] and for poly(vinyl chloride) in particular.[2]

6.2 KINETIC TREATMENTS

6.2.1 Rates of Polymerization

If various assumptions are made, the overall rate, R_p, of a homogeneous radical polymerization can be related to the concentrations of monomer and initiator, and the velocity constants of the elementary reactions. Rigorous treatments have been developed[3] but the essentials of the argument can be presented quite simply.

If reactive radicals are generated in pairs by thermolysis of an initiator, I, the rate of initiation, R_i, of polymerization is given by equation (1) in which k_d is the velocity constant for the dissociation of I and f (the efficiency of initiation) is, for many combinations of monomer and initiator, effectively constant over quite wide ranges of concentrations (see Volume 3, Chapter 8). If radicals are formed in pairs by a photochemical process in which a molecule, I, is activated directly or indirectly, R_i is given by equation (2); $\mathscr{I}$ is the intensity of the incident photochemically active light and k depends on the absorption coefficient of the absorbing species. In kinetic work, there should be only weak absorption otherwise radical production is not uniform through the system.

$$R_i = 2f k_d [\text{I}] \tag{1}$$

$$R_i = 2f k \mathscr{I} [\text{I}] \tag{2}$$

Initiating radicals can be generated singly in redox reactions of the type shown in equation (3) (see Volume 3, Chaptder 9); in such cases, all the primary radicals initiate the growth of polymer chains unless the concentration of monomer is very low.

$$\text{H}_2\text{O}_2 + \text{Fe}^{2+} \longrightarrow \cdot\text{OH} + \text{OH}^- + \text{Fe}^{3+} \tag{3}$$

The general growth (or propagation) reaction represented in equation (4) is assigned a velocity constant k_p, which is taken as being independent of the size of the reacting radical and independent of the medium, *i.e.* the concentration of the monomer and the nature of any diluent. $P_n\cdot$ represents an active propagating polymer chain with n monomer units. The reaction is quite strongly exothermic with ΔH of the order of $-80\,\text{kJ}\,\text{mol}^{-1}$ for most monomers; it involves an increase in order of the system and so ΔS is negative (see Volume 3, Chapter 1). The reverse of reaction (4) is referred to as depropagation with a velocity constant k_{-p}; it becomes more and more significant as the temperature is raised because the activation energy E_{-p} is greater than E_p. For some monomers, including α-methylstyrene, depropagation becomes very pronounced at those temperatures (say 40–70 °C) where radical polymerizations are commonly performed (see Volume 3, Chapter 1). At one time, it was thought that such monomers could not be polymerized by a radical mechanism although ionic polymerizations were known to be possible at comparatively low temperatures. It is now known that α-methylstyrene engages readily in radical polymerization if the conditions are properly selected.

$$\text{P}_n\cdot + \text{M} \longrightarrow \text{P}_{n+1}\cdot \tag{4}$$

Growth occurs predominantly by head-to-tail addition so that, for a monomer $\text{CH}_2{=}\text{CHX}$, the growing radical is represented as $\text{P}_n\text{CH}_2\text{CHX}\cdot$ and the resulting macromolecule has a very regular structure (see Section 6.4.1).

The rate of consumption of monomer is expressed by equation (5) in which $[\text{P}\cdot]$ is the total concentration of radicals of all sizes capable of growth and $[\text{M}]$ is the concentration of monomer. In most systems, $[P\cdot]$ is likely to be in the region of 10^{-8} mol l^{-1} and so it cannot be measured reliably by ESR.

$$R_p = -\frac{d[M]}{dt} = k_p [P\cdot] [M] \tag{5}$$

Termination of the growing chains occurs by the interaction of any pair of radicals in the system to give inactive products. The rate of removal of radicals, in polymerization kinetics, often is given by equation (6), in which k_t is supposed to be independent of the sizes of the interacting radicals and not to vary during the polymerization (see Volume 3, Chapters 11 and 12). The termination may occur by combination (also referred to as coupling) of radicals, giving one macromolecule containing a head-to-head linkage —$CH_2CHXCHXCH_2$—; the rate of formation of macromolecules is $\frac{1}{2}k_t[P\cdot]^2$. Another type of interaction, referred to as disproportionation, is possible; a hydrogen atom is transferred from one radical to the other as in equation (7). Two polymer molecules are formed, one containing an unsaturated end group; in this case, the rate of formation of polymer molecules is $k_t[P\cdot]^2$ (see Volume 3, Chapter 8).

$$R_t = k_t[P\cdot]^2; k_t = k_{tc} + k_{td} \tag{6}$$

$$P_nCH_2CHX\cdot + P_mCH_2CHX\cdot \longrightarrow P_nCH_2CH_2X + P_mCH=CHX \tag{7}$$

Some authors[3,4] have expressed R_t as $2k_t[P\cdot]^2$, the actual value of k_t being half that of k_t in equation (6). Whenever quoted values of k_t and the velocity constants for other steps are considered, it is essential to note which of the two expressions for R_t has been adopted.

In homogeneous radical polymerizations, the average interval between the generation of an initiating radical and termination of the derived polymer radical (*i.e.* the average life time of a growing radical) is very much less than the period over which R_p is measured. In most cases, only a small fraction of the initiator decomposes during this period so that R_i does not change appreciably; further, little of the monomer is consumed and the contraction accompanying polymerization is insufficient to affect concentrations significantly. It is supposed that a steady state is established early in the polymerization, the values of R_i and R_t becoming almost equal and remaining so. From equations (2) and (6), the steady concentration of growing radicals, $[P\cdot]_s$, is given by equation (8) so that the steady rate of polymerization, $R_{p,s}$, can be expressed by equations (9)–(11); the stationary rate is usually written simply as R_p and a stationary state is applied. These equations are regarded as the 'ideal' rate expressions for homogeneous radical polymerizations; they form the basis for determination of $k_p/k_t^{\frac{1}{2}}$.

$$[P\cdot]_s = (R_i/k_t)^{\frac{1}{2}} \tag{8}$$

$$R_{p,s} = k_p(R_i/k_t)^{\frac{1}{2}} [M] \tag{9}$$

$$R_{p,s} = (2fk_dk_p^2/k_t)^{\frac{1}{2}} [I]^{\frac{1}{2}} [M] \tag{10}$$

$$R_{p,s} = (2fk \, \mathscr{I} \, k_p^2/k_t)^{\frac{1}{2}} [I]^{\frac{1}{2}} [M] \tag{11}$$

6.2.2 Effects of Temperature

The variation with temperature of $R_{p,s}$ for fixed $[M]$ and $[I]$ depends on the activation energies associated with the dissociation of initiator, E_d, the growth reaction, E_p, and the termination process, E_t; it is supposed that f does not vary appreciably with temperature. The composite activation energy for the overall polymerization, E_{pol}, is deduced from equation (10) and is given by equation (12). The value assigned to E_t allows for any shift with temperature of the balance between combination and disproportionation; both reactions have low values of E but that for the latter is slightly the larger. Typical values of E_d, E_p and E_t are 140, 30 and 10 kJ mol^{-1} respectively, so that E_{pol} is expected to be in the region of 95 kJ mol^{-1}. The relationship between R_p and T is disturbed when the temperature is raised sufficiently for depropagation to become significant. Initiators of the usual types dissociate at rates ordinarily regarded as suitable over only a limited range of temperatures, say 30 °C.

$$E_{pol} = \frac{1}{2}E_d + E_p - \frac{1}{2}E_t \tag{12}$$

Redox initiators have quite low activation energies, typically about 40 kJ mol^{-1}, leading to values of E_{pol} of about 45 kJ mol^{-1}. Therefore redox systems are effective at comparatively low temperatures and over quite wide ranges. The values of E_{pol} are even smaller for systems involving photochemical initiation without any contribution from thermal processes; since E_d can be removed from equation (12), E_{pol} is typically about 25 kJ mol^{-1}.

6.2.3 Molecular Weights of Polymers (see also Volume 3, Chapter 18)

The kinetic chain length, $\bar{v}$, in a polymerization is defined as the average number of growth reactions which follow an initiation process; normally, the number average value is considered. In the scheme involving initiation, growth and termination, $\bar{v}$ is the number of monomeric units in the average polymer radical at termination; it is related to the average degree of polymerization, $\overline{DP}$, of the polymer by equation (13). The value of q is two for termination entirely by combination since two radicals unite to give a single molecule, but it is unity if disproportionation is the sole mode of termination. If initiation produces an end group R— which can be determined in the polymer by an analytical procedure, $\bar{v}$ can be found as (number of monomeric units in a sample of polymer)/ (number of R— end groups in that sample), whatever the relative importances of the two types of termination.

$$\overline{DP}_n = q\,\bar{v}_n \tag{13}$$

In the steady state, $\bar{v}$ is related to $R_{p,s}$ and R_i (and therefore to R_t) by equation (14). Use of equations (1) and (10) leads to equation (15); substitution of a value of $[P\bullet]_s$ from equation (8) leads to equations (16)–(18), in the last of which $k_{t,c}$ and $k_{t,d}$ are the velocity constants for termination by combination and disproportionation respectively.

$$\bar{v} = R_{p,s}/R_i = R_{p,s}/R_t \tag{14}$$

$$\bar{v} = k_p[P\bullet][M]/k_t[P\bullet]^2 \tag{15}$$

$$\overline{DP}_n = 2(k_p^2/2fk_dk_t)^{\frac{1}{2}}[I]^{-\frac{1}{2}}[M] \quad \text{Exclusively combination} \tag{16}$$

$$\overline{DP}_n = (k_p^2/2fk_dk_t)^{\frac{1}{2}}[I]^{-\frac{1}{2}}[M] \quad \text{Exclusively disproportionation} \tag{17}$$

$$\overline{DP}_n = \frac{k_p[M]}{(2fk_d[I])^{\frac{1}{2}}} \frac{(k_{tc}+k_{td})^{\frac{1}{2}}}{[(k_{tc}/2)+k_{td}]} \tag{18}$$

Bimolecular termination involves any pair of growing radicals and so the reaction must lead to a distribution of molecular weights, MWD, in the polymer. Provided that $\overline{DP}$ is high and $[M]$ and $[I]$ remain essentially unchanged during polymerization, $\overline{DP}_w/\overline{DP}_n$ is 1.5 for combination and 2.0 for disproportionation (see also Volume 3, Chapters 4 and 18).[5]

6.2.4 Non-stationary Phases (see also Volume 3, Chapter 7)

Equations (9)–(11) can be used to determine of $k_p/k_t^{\frac{1}{2}}$. The additional information needed for evaluation of k_p and k_t separately can be obtained from experiments in which R_i is suddenly changed. This requirement is responsible for the special importance of photoinitiation in kinetic studies of radical polymerizations; the generation of radicals can be started or stopped virtually instantaneously by opening or closing the shutter of the lamp. Abrupt changes in R_i are not possible if radicals are formed by thermal processes since there must be continuous variation in the rate of production while the temperature is changing.

Consider a system capable of polymerization and suppose that production of initiating radicals is started abruptly at a rate R_i. The value of $[P\bullet]$ builds up from zero towards the stationary value $[P\bullet]_s$; during this period, R_p rises from zero towards $R_{p,s}$. Correspondingly, a sudden cessation in the production of radicals is followed by decreases in $[P\bullet]$ and R_p towards zero values. The periods during which $[P\bullet]$ and R_p are changing are referred to as non-stationary phases. Their durations cannot be defined precisely but ordinarily R_p rises close to $R_{p,s}$ or alternatively falls almost to zero within about one minute. Methods have been devised and used for following the polymerization during these phases;[6] in the main, they depend upon heating effects associated with the propagation process.

Detailed treatments of the non-stationary phases show how k_p/k_t can be evaluated.[7] Using values of $k_p/k_t^{\frac{1}{2}}$, it is then possible to evaluate k_p and k_t separately. Some of the methods for experimental investigation of non-stationary phases are applicable at any stage in the polymerization, even when the system contains a fairly large amount of dissolved polymer so that its viscosity is high.

Important variants on the methods for studying the non-stationary phases depend upon the use of intermittent illumination in the rotating sector technique[8] or the newer technique of spatially intermittent illumination;[9] in both cases, R_p is measured as a function of the intervals between the light periods (see also Volume 3, Chapter 7). When properly applied, they lead to values of k_p/k_t. A very useful critical survey[10] has been provided of the techniques which can lead to values of k_p and

k_t. It includes discussion of a promising procedure[11] in which a polymerizing system is illuminated with periodic laser light flashes; consideration of $\overline{DP}$ and MWD of the polymer produced in such a system allows direct determination of k_p.

Although there is still uncertainty about values of k_p and k_t even for much studied monomers such as styrene and methyl methacrylate, it is clear that they depend upon the nature of the monomer and also upon temperature and pressure. At 60 °C and normal pressure, typical values (in $l\,mol^{-1}\,s^{-1}$) for k_p lie between 100 and 4000 and for k_t between 3×10^6 and 7×10^7.

6.2.5 Limitations of the Idealized Scheme

The scheme considered up to this point has been used for quite a long time and has been moderately successful. For several monomers, k_p and k_t have been evaluated and to some extent the results have been correlated according to the principles of physical organic chemistry. Substantial differences between the results from different laboratories and from alternative procedures are disturbing since most of them cannot be explained in terms of experimental deficiencies. Plainly, critical re-examination of the scheme is needed to discover whether or not some important factors have not been allowed for.

The idealized treatment makes no allowance for the fact that some substances affect quite markedly R_p and $\overline{DP}$ of the resulting polymer although they apparently cannot be involved in initiation, propagation or termination. It is necessary to postulate additional reactions such as chain transfer (see Section 6.5) to account for these effects and then to realize that these reactions may occur even in systems where their effects are less apparent.

There are many examples of deviations from 'ideal behaviour' as represented by equation (10); frequently, but not always, they can be satisfactorily explained by variations in the values of k_d, k_p and k_t without invoking additional chemical processes. These variations are generally associated with alterations in the nature of the reaction medium. Subtle but important effects may arise from preferential solvation of growing macroradicals, leading to non-uniform distribution of reactants, even in systems which are seemingly homogeneous (see Section 6.4.4).

6.3 INITIATION

6.3.1 Rates and Efficiencies of Initiation

In principle, R_i can be determined in several ways but none of them is entirely satisfactory. The subject is discussed here for the production of radicals by thermal dissociation of substances, such as peroxides and azo compounds.

The value of k_d for an initiator can be measured by monitoring either its disappearance or the appearance of a stable product. The measurements should be made for very dilute solutions in selected solvents to reduce possible effects of induced decomposition caused by attack of radicals on molecules of the initiator (see Section 6.5.5). The value of k_d generally depends upon the nature of the solvent so that it is not certain that a value found using an inert solvent is applicable to a system containing monomer. These solvent effects may arise from differences between the changes in solvation on passing from reactant to transition state in the various media, with consequent effects on the activation parameters $\Delta H^{\ddagger}$ and $\Delta S^{\ddagger}$ (see Volume 3, Chapter 8).[12]

It is necessary to consider also the efficiency of initiation, *i.e.* f in equation (1); usually the radicals which actually initiate polymerization enter the polymer and form end groups. A cage effect operates for dissociations in the liquid phase whether or not monomer is present; the separation of the radicals is impeded by the surrounding molecules so that their geminate interaction is favoured. The cage effect is significant in connection with the value of f only if the 'waste' products of the geminate reaction are distinguishable from the parent molecule, as in Scheme 1; in other cases the initiator is regenerated. It is unlikely that it will ever be possible to make reliable calculation of f from first principles, but it can be measured by comparing the amount of initiator decomposed during a polymerization with the amount included in the polymer in the form of end groups.[13] For azonitriles, values of f in the region of 0.7 have been confirmed by direct determination of the 'waste' products formed during polymerization.

$$Me_2C(CN)N{=\!=}NC(CN)Me_2 \longrightarrow 2\,Me_2C(CN)\cdot\ +\ N_2$$

$$2Me_2C(CN)\cdot \longrightarrow Me_2C(CN)C(CN)Me_2\ \ or\ \ Me_2CHCN\ +\ CH_2{=\!=}C(CN)Me$$

Scheme 1

Direct determination of fk_d may be possible by studying the decomposition of the initiator in an inert solvent containing excess of a radical scavenger (see Section 6.7.2); the rate of consumption of scavenger is measured, possibly by spectrophotometry.[14] The importance of geminate interaction of radicals when the scavenger is present in the solution is assumed to be the same as when the primary radicals are captured by monomer to initiate polymerization. In many cases, there is unfortunately doubt about the exact stoichiometry of the reaction between the scavenger and primary radicals (see Section 6.7.2).

Another method using a radical scavenger depends upon measurement of inhibition periods during polymerization (see Section 6.7.1).[15] The total number of 'available' radicals generated during inhibition is directly related to the initial amount of scavenger in the system. It is necessary again to make assumptions about the stoichiometry of the radical/scavenger reaction. It may also be difficult to define the exact duration of inhibition because, when [scavenger] has fallen to a very low value, there is competition between it and the monomer for the capture of primary radicals.

An alternative approach to the evaluation of R_i depends upon finding $\bar{v}$ for a polymerization and then applying equation (14). For this purpose, measurements of $\overline{DP}_n$ can be used with equation (13) but information is needed on the balance between combination and disproportionation and also on the frequency of transfer reactions (see Section 6.5.1). Another procedure for finding $\bar{v}$ is based upon analysis of the polymer for end groups derived from the initiator;[13] it requires no assumptions about the termination process, except in the rather rare cases where primary radical termination (see Section 6.6.3) is significant, and it avoids problems associated with all transfer reactions except transfer to initiator (see Section 6.5.5) but allowance can be made for that process.[16] The procedure requires some knowledge of the actual initiation reaction and the compositions of end groups derived from the initiator.

6.3.2 Dead-end Polymerization

The discussion so far has involved the assumption that [I] does not change appreciably during the polymerization. If this is not so, R_i declines continuously through the reaction and indeed may fall effectively to zero, so that R_p also becomes very small. Such systems are referred to as dead-end polymerizations.[17] It has been shown that, knowing any two of the quantities f, k_d and $k_p/k_t^{\frac{1}{2}}$, the third can be determined by kinetic analysis of the reaction. The continuous decreases in R_i and R_p lead to gradual changes in $\bar{v}$, $\overline{DP}$ and MWD. Essentially instantaneous measurements are possible for R_p but measurements of $\overline{DP}$ and MWD must refer to the whole polymer formed up to a certain point in the reaction, *i.e.* they are integral values; the more useful differential values, corresponding to the polymer being formed at a particular stage in the reaction, can only be deduced from the variations in integral values.

6.4 GROWTH REACTIONS

6.4.1 General Considerations

The growth reaction represented in equation (4) normally proceeds by head-to-tail addition largely because the radical —$CH_2CHX\cdot$ is stabilized relative to the radical —$CHXCH_2\cdot$, to an extent depending on the nature of the substituent X. An occasional head-to-head reaction (equation 19) is possible, especially during polymerization at comparatively high temperature; it is likely to be followed immediately by a tail-to-tail addition (equation 20) to produce again the more stable macroradical and to give a three-unit section represented as —$CH_2CHXCHXCH_2CH_2CHX$—. Head-to-head placements, although rare, may be important for certain polymers, notably poly(vinyl chloride), because of the likelihood that they could act as nuclei for reactions having deleterious effects on the material. For this reason and others, there has been interest in radical polymerization of vinyl chloride at quite low temperatures in order to produce polymers possessing very regular structures without blemishes.

$$PCH_2CHX\cdot \;+\; CHX{=}CH_2 \longrightarrow PCH_2CHXCHXCH_2\cdot \tag{19}$$

$$PCH_2CHXCHXCH_2\cdot \;+\; CH_2{=}CHX \longrightarrow PCH_2CHXCHXCH_2CH_2CHX\cdot \tag{20}$$

Head-to-head placements in poly(vinyl acetate) have received much attention partly because they are rather more likely to occur in that polymer than in many others (the substituent —OCOMe being rather less effective than say —Ph in stabilizing the growing radical) and partly because techniques are available for their detection. The classical method of working[18] requires conversion of poly(vinyl acetate) to poly(vinyl alcohol) which is then treated with periodic acid. The reagent cleaves the central bond in the diglycol unit —CH(OH)—CH(OH)— so that the concentration of head-to-head groups in the original polymer can be deduced from any reduction in $\overline{DP}$. The periodic oxidation leads to the formation of aldehydic end groups.[19] The quantity of acetaldehyde is appreciably greater than expected for a statistical distribution of head-to-head linkages along the polymer chains, indicating a tendency for them to occur close to chain ends; this feature might affect the reliability of results deduced from changes in $\overline{DP}$ of the polymer as a whole. An alternative and newer procedure requires detailed examination of the NMR spectra of samples of poly(vinyl alcohol); it is possible to find the number of adjacent methylene groups, corresponding to the tail-to-tail reactions which follow head-to-head additions.[20]

Information on the frequency of head-to-head growth is available for only a few monomers. The process is comparatively important for unsaturated fluorocarbons, such as vinylidene fluoride,[21] but seems to be particularly rare for monomers such as $CH_2\text{=}CXY$ in which the groups X and Y are fairly large; in the case of methyl methacrylate for example, there must be steric interference to the close approach of the head of a macroradical to the head of a monomer molecule. If a particular monomer can be polymerized by both radical and ionic mechanisms, then head-to-head irregularities are even less likely to be formed in the materials produced by ionic polymerizations.

Structural rearrangements are rather less likely in radical polymerizations than in cationic processes and it is usually possible to predict with confidence the nature of the repeating units in the macromolecular chains. Suggestions that certain monomers can give rise to abnormal units, such as —$CH_2CH\text{=}C\text{=}N$— in polyacrylonitrile, lack conclusive evidence. The overall kinetics of polymerization are affected noticeably only if the occasional unusual addition gives a radical having a reactivity that is very different from that of the normal growing radical; this situation seems to occur with vinyl benzoate and vinylferrocene (see Section 6.6.1).

Certain special monomers, including cyclic ketene acetals[22] and derivatives of vinyl cyclopropane,[23] undergo ring-opening polymerizations (equations 21 and 22). In some cases, ring-opening and 'normal' additions are of comparable frequency; each such polymerizing system therefore contains two distinct types of macroradicals which must have different reactivities towards the monomer so that kinetic complexities can be expected (see Volume 3, Chapter 22).

$$P\bullet \ + \ CH_2\text{=}C \underset{O}{\overset{O}{\Big\langle}}_{Ph} \qquad \longrightarrow \qquad PCH_2COOCH_2CHPh\bullet \tag{21}$$

$$P\bullet \ + \ CH_2\text{=}CH \!\!\!\bigtriangleup_X \qquad \longrightarrow \qquad PCH_2CH\text{=}CHCH_2CHX\bullet \tag{22}$$

For monomers such as acrylic anhydride which contain two unconjugated C=C bonds, intramolecular growth may be prominent (equation 23), its importance at a particular temperature depending upon [monomer] (see Volume 4, Chapter 23).[24] The uncyclized monomeric units contain pendant unsaturated groups which may subsequently engage in growth processes, perhaps leading to crosslinking. Even when the conversion is so low that reaction of pendant groups can be ignored, the system contains more than one type of growing radical, *viz.* the reactant and the products shown in equation (23) (see also Volume 4, Chapter 23).

$$PCH_2\overset{\bullet}{C}HCOOCOCH\text{=}CH_2 \longrightarrow PCH_2\overline{CHCOOCO\underset{\bullet}{C}HCH_2} \quad or \quad PCH_2\overline{CHCOOCOCHCH_2}\bullet \tag{23}$$

Radical polymerization of a conjugated diene gives polymer containing several types of monomeric unit; structures (1), (2) and (3) are found in polybutadiene. These polymerizations are not stereospecific (see also Volume 3, Chapter 1). The monomers $CH_2\text{=}CHX$ and $CH_2\text{=}CXY$, each of which gives rise to essentially only one type of monomeric unit, in principle at least can be converted

$$\begin{array}{ccc}
\text{—CH}_2\text{—CH—} & \text{—CH}_2\diagdown_{\text{CH=CH}}\diagup^{\text{CH}_2\text{—}} & \text{—CH}_2\diagdown_{\text{CH=CH}}\diagdown_{\text{CH}_2\text{—}} \\
\quad|\quad & & \\
\text{CH} & & \\
\;\|\; & & \\
\text{CH}_2 & & \\
\\
\textbf{(1)} & \textbf{(2)} & \textbf{(3)}
\end{array}$$

into isotactic and syndiotactic polymers but radical polymerizations do not produce strictly stereoregular polymers; this statement may need modification if the polymerizing system contains certain additives (see Section 6.4.4). For methyl methacrylate and other monomers having bulky substituents, there is limited steric control of the growth process leading to a tendency towards syndiotacticity which generally increases as the temperature of polymerization is reduced.

Steric effects can be significant in growth reactions and they are the main cause of the comparatively low heats of polymerization for monomers such as methyl methacrylate and more especially α-methylstyrene. The failure of stilbene, esters of cinnamic acid and many other α,β-disubstituted monomers to engage in homopolymerization under ordinary conditions is also mainly attributed to steric effects. Rather surprisingly, dialkyl fumarates give polymers of quite high molecular weight although the isomeric dialkyl maleates do not engage in radical polymerization unless the system contains a *cis–trans* isomerization catalyst as well as a source of initiating radicals.[25] Both k_p and k_t for the fumarate esters are small (values of 0.015 and 16 $41\,\text{mol}^{-1}\,\text{s}^{-1}$ respectively have been quoted for the diethyl ester) but the ratio is such that high polymer can be produced at a reasonable rate (see also Volume 3, Chapter 17).

The assumption that k_p for the growth reaction shown in equation (4) is independent of the number of monomeric units already incorporated in the radical is thought to be sound except for very small growing radicals (see Volume 3, Chapter 1). In the early stages of the growth of a chain, the reactive site is quite close to an end group; this group may well be very distinct structurally from the monomeric units and so it may exert quite a different electronic effect upon the reactive site and so modify its reactivity. Any effect of this type is expected to become negligibly small as soon as several monomeric units have been added and indeed there is no convincing evidence for an effect of radical size on k_p for larger macroradicals.

Electron spin resonance (ESR) spectroscopy has been used successfully in studies of several types for polymer systems, including the generation of radicals in solid polymers by means of high energy and UV radiations. Unfortunately the steady concentrations of growing radicals during actual polymerizations are usually too low for their detection by most spectrometers and clearly therefore detailed structural information about the radicals cannot be obtained. The concentrations can be enhanced by the use of special systems such as highly viscous media and those in which polymer is precipitated as it is formed so that the rate of termination is greatly reduced. There have also been accounts[26,27] of the direct observation by ESR of the propagating radicals in emulsion polymerizations. Flow methods and spin-trapping techniques can be applied effectively in conjunction with ESR for obtaining information about growing radicals.

Developments in ESR instrumentation, in particular cavity design, have led to notable improvement in sensitivity to say $10^{-7}\,\text{mol}\,l^{-1}$. Examples of applications now include direct examination of the growing radicals in polymerizations of triphenylmethyl methacrylate[28] and dialkyl itaconates[29] in what might be termed 'normal' systems. There has been a report[30] of a study of the radicals present during the polymerization of MMA initiated by benzoyl peroxide; it was concluded that there are two stable conformations of the growing radical, interconverting by rotation about the C_α—C_β bond. Further improvement in sensitivity by a factor of 10 is regarded as feasible and it would open up many possibilities for application of ESR to radical polymerizations in fluid systems; thus, determination of the stationary concentration of radicals coupled with the use of equation (5) would allow direct determination of k_p.

6.4.2 Copolymerization (see also Volume 3, Chapters 2, 15 and 16)

The simplest effective treatment of the radical copolymerization of monomers M_A and M_B supposes that the system contains only two types of growing radicals, referred to as $P_A\cdot$ and $P_B\cdot$. A radical $P_A\cdot$ has an M_A unit at its reactive end and it is assumed that all such radicals have identical reactivities so that no distinction is drawn between, for example, growing radicals —M_A—M_A—$M_A\cdot$ and —M_B—M_B—$M_A\cdot$; similar considerations apply to radicals $P_B\cdot$. This assumption means that a binary radical copolymerization involves four growth reactions (Scheme 2) and so-called monomer reactivity ratios, r_A and r_B, are defined as shown. The scheme is described as based on the neglect of any effects of penultimate groups upon the reactivity of a macroradical and clearly corresponds to an approximation.

A relationship can readily be established between the ratio of the numbers of M_A and M_B units in a binary copolymer and $[M_A]/[M_B]$ in the feed mixture of monomers (equation 24). Except in special cases, one of the monomers is consumed preferentially so that there are continuous changes in the composition of the feed and that of the copolymer. In academic studies of copolymerizations, it is usual to allow only very low conversions so that the effects of the changes in compositions are

$$P_A\cdot \;+\; M_A \longrightarrow P_A\cdot \qquad \text{Velocity constant} \;=\; k_{AA}$$

$$P_A\cdot \;+\; M_B \longrightarrow P_B\cdot \qquad\qquad\qquad\qquad k_{AB}$$

$$P_B\cdot \;+\; M_B \longrightarrow P_B\cdot \qquad\qquad\qquad\qquad k_{BB}$$

$$P_B\cdot \;+\; M_A \longrightarrow P_A\cdot \qquad\qquad\qquad\qquad k_{BA}$$

$$r_A \;=\; k_{AA}/k_{AB} \qquad r_B \;=\; k_{BB}/k_{BA}$$

Scheme 2

minimized; in large scale preparations, when high conversions must be achieved, monomers are added to the system in a controlled manner so that the compositions are maintained at the required values. Equation (24) is referred to as an instantaneous equation; the effects of the drifts in the compositions can be allowed for in integrated equations. It is possible also to elaborate the treatment to allow for effects of penultimate groups on reactivities of macroradicals.

$$(M_A/M_B)_{copolymer} \;=\; \frac{[M_A]}{[M_B]} \cdot \frac{r_A[M_A] \;+\; [M_B]}{[M_A] \;+\; r_B[M_B]} \tag{24}$$

If there are no effects of penultimate groups, the copolymerization of n monomers involves n^2 growth reactions. Each of these reactions occurs in a binary copolymerization and so, in principle at least, it is possible to predict the composition of a polycomponent copolymer from knowledge of the composition of the feed and of the characteristics of the appropriate binary systems.

Equation (24) provides the basis for determination of monomer reactivity ratios; these quantities can be deduced also from detailed information on the distribution of the types of monomeric units in copolymers. Extensive compilations of these ratios are available[31] but some of the published results are, for various reasons, somewhat unreliable.

If k_{AA} is known for the homopolymerization of M_A, then values of k_p for the cross-propagations of $P_A\cdot$ with other monomers can be found from the appropriate monomer reactivity ratios. Even if absolute values of the velocity constants for the cross-propagations are not found, relative values can be very useful. The value of k_p for a homopolymerization depends upon the reactivities of both radical and monomer and they cannot be separated; from studies of copolymerization, it is possible to compare the reactivities of monomers towards selected polymer radicals and to unravel the factors governing the reactivities of monomers. The reactivities of macroradicals towards a particular monomer can also be compared but, for this purpose, k_p for each of the homo-polymerizations is required; reliable information is sparse.

By and large, studies of copolymerizations have confirmed the view that monomers of high reactivity give rise to polymer radicals of low reactivity. Quantitative information on copolymerizations is needed for the development of the 'Q and e' and 'patterns' treatments for the prediction of behaviour in radical polymerizations (see also Volume 3, Chapter 17).[32]

Many ingenious schemes have been proposed and tested for the preparation of block and graft copolymers by methods involving radical processes. The difficulties usually encountered with the methods are associated with contamination of the products with homopolymers, with distributions of kinetic chain lengths in radical polymerizations and with uncertainties about transfer and termination processes. One of the more successful methods depends upon the use of so-called macromers; these materials can be represented by a formula $CH_2{=}CHR$ where R represents a chain consisting of repeating units. A macromer can be copolymerized, usually radically, with a more usual type of monomer to give what is effectively a polymer with grafts (see also Volume 6, Chapters 9 and 12).

6.4.3 Diffusion Control

The rates at which macroradicals interact in solution are governed by the rates of diffusion processes (see Section 6.6.2). Propagation reactions are much less susceptible to diffusion control for two reasons; first, the actual chemical reaction is considerably slower than that in termination and, second, diffusion of the small monomer molecule is much faster than that of a macroradical. Diffusion control of propagation may operate at very high conversions when the system is so viscous that diffusion, even of the monomer, is seriously impeded.

The growth reaction is accelerated by application of high pressure since it is a process of association and the volume of activation, $\Delta V_p^{\ddagger}$, is negative. The effect is not cancelled by decrease in rate of diffusion of monomer caused by the marked increase in viscosity accompanying increase in pressure, at least in the early stages of reaction;[33] it is predicted, however, that diffusion control of propagation will become evident at rather lower conversions than for polymerizations performed at normal pressures (see also Volume 3, Chapter 21).

When a polymerization is performed in a medium which is not a solvent for the polymer being produced, polymer radicals are included in the precipitate. The extent to which further growth of trapped radicals is impeded depends upon the physical form of the precipitate, in particular the extent to which it is swollen by the monomer; generally the effect on growth reactions is much less than that on termination processes involving pairs of radicals (see Section 6.6.2 and Volume 4, Chapter 16).

6.4.4 Effects of the Medium

Radical polymerizations in general are less sensitive than ionic processes to the chemical character of the medium and many of them can be performed in aqueous systems. Some solvents, *e.g.* carbon tetrachloride, may act as reactive transfer agents (see Section 6.5.1) and others, *e.g.* aromatic nitro compounds, as retarders (see Section 6.7.3); many others, apparently not involved in chemical processes, have smaller but significant effects on R_p and $\overline{DP}$. Probably the most common of these influences arise from effects of solvation on the activation parameters ΔH^+ and ΔS^+ for propagation, as already mentioned in connection with the dissociation of initiators in solution.

Solvation must be considered in another connection also. Physical studies of solutions of polymers in mixtures of solvents show that commonly there is preferential solvation of macromolecules by one of the components of the mixture; the effect must occur also for macroradicals, including their reactive sites. Consequently, for growing radicals in a mixture of diluent and monomer, the effective concentration of monomer in the immediate vicinity of the radical sites may differ significantly from the concentration averaged over the whole system. This non-uniformity would inevitably cause deviation from the ideal rate equation, in particular the kinetic order with respect to monomer.[34] A similar effect might arise in a copolymerization so that the effective ratio of the concentrations of the monomers in the vicinity of a growing chain could differ from that expected on the basis of uniform distribution of the components of the system. It has been shown[35] for several pairs of monomers that quite marked dependence of apparent monomer reactivity ratios on the nature of the diluent can be fully and readily explained by effects of preferential solvation.

It is conceivable that an explanation of the type being discussed can account for differences between the monomer reactivity ratios found for systems rich in M_A and those for systems in which M_B is abundant. Differences of this type are usually attributed to penultimate group effects (see Section 6.4.2); it should not be supposed that these effects can be neglected but they may not be the sole cause for the apparent failure of equation (24) to apply over the whole range of values of $[M_A]/[M_B]$.

Another explanation of kinetic abnormalities in radical polymerizations has been advanced, based on the use of thermodynamic activities in place of concentrations in rate equations.[36] It is likely that this treatment can be related to that based upon preferential solvation.

Certain types of template polymerization might be regarded as extreme cases of non-uniform distribution of reactants, leading to very special effects (see also Volume 3, Chapter 19).

Explanations of other types also have been proposed to account for effects of the medium in radical polymerizations. Complexing of solvent with radicals has been considered as a process which could be in competition with complexing of monomer with radicals.[37,38] Another treatment is based on the possible involvement in polymerization of vibrationally excited radicals (so-called hot radicals) having reactivities different from those of normal radicals.[39]

Much larger effects of solvents are found for ionizable monomers. The ionized and unionized species are regarded as quite distinct in reactivity so that any change affecting the extent of ionization must modify the kinetic characteristics of the overall polymerization. It has been found that the polymerizations of acrylic and methacrylic acids are affected by the pH of the medium (see Volume 3, Chapter 20).[40]

The early discovery[41] of the marked effects of lithium chloride upon the polymerization of acrylonitrile in dimethylformamide has been followed by many accounts of profound influences of Lewis acids upon certain radical polymerizations and copolymerizations.[42] The effects include modifications of overall rates, tacticities of products and monomer reactivity ratios in copolymerizations. In some cases, the Lewis acid can cause alkenes, such as isobutene, to participate readily in alternating radical copolymerizations with acrylic and methacrylic monomers (see Volume 4, Chapter 22). These additives must exert their effects by complexing with monomers and/or radicals. Many schemes have been proposed and ideas on the so-called non-classical growth reactions have been reviewed.[43]

6.5 TRANSFER REACTIONS (see also Volume 3, Chapter 13)

6.5.1 General Features

An important deficiency of the scheme considered in Section 6.2 is its neglect of a process known as chain transfer. Certain additives, *e.g. n*-butanethiol, at quite low concentrations cause marked reduction in $\overline{DP}$ of polymer produced under particular conditions. An atom or group of atoms is first transferred from the additive to the polymer radical as in Scheme 3; the radical derived from the so-called transfer agent then reacts with monomer to re-initiate polymerization. On this basis, chain transfer, unless very frequent, has no effect on $\bar{v}$ and R_p but interrupts the growth of the molecular chain. The simple relationship (equation 13) between $\bar{v}$ and $\overline{DP}$ is invalidated and q may become appreciably less than one.

$$P_n\cdot + BuSH \longrightarrow P_nH + BuS\cdot$$
$$BuS\cdot + M \longrightarrow P_1\cdot$$

Scheme 3

The process of transfer introduces end groups of new types. There may also be modification in the MWD; for a system in which termination occurs by combination, $\overline{DP}_w/\overline{DP}_n$ shifts from 1.5 towards 2.0 as transfer increases in frequency.

A transfer constant is defined as k_f/k_p, where k_f refers to the first step in Scheme 3. Essentially there is competition between the abstraction process and the growth reaction; k_f/k_p can be found from examination of the relationship between [transfer agent] and $\overline{DP}$ for the polymer. The magnitude of k_f/k_p depends on the natures of both transfer agent and monomer, and on the temperature, since the activation energies E_f and E_p are generally different; transfer constants do not vary greatly with pressure.[44] For *n*-butanethiol with styrene at 60 °C, k_f/k_p is approximately 20; it and other reactive transfer agents are used in practice for limiting $\overline{DP}$ of the polymer produced in certain cases. Much smaller transfer constants are found for substances commonly used as diluents in radical polymerizations; thus for benzene with styrene at 60 °C, k_f/k_p is in the region of 0.02×10^{-4}.

The so-called 'patterns treatment' uses the value of k_f for the reaction of a polymer radical with toluene as a measure of the inherent reactivity of the radical even in propagation processes (see Volume 3, Chapter 17).[32]

6.5.2 Degradative Transfer

Transfer agents for which k_f/k_p is comparatively large generally cause a reduction in R_p; the process is then referred to as degradative transfer. Hydrogen abstraction occurs readily if the X—H bond in a transfer agent is comparatively weak, meaning that the product radical X· is quite stable and therefore unreactive. Some of the X· radicals engage in processes other than re-initiation (see Scheme 4) and there is then reduction in $\bar{v}$ and in R_p as well as in $\overline{DP}$. If this effect is very pronounced, with few of the X· radicals being consumed in re-initiation, then the additive is better described as a retarder. In an extreme case where the additive is so effective that it reacts with primary radicals and the consumption of monomer is suppressed, it is described as an inhibitor. Appreciable consumption of monomer begins only when the additive has been consumed (see Section 6.7.1).

$$X\cdot + P_n\cdot \longrightarrow XP_n \quad \text{*or* products of disproportionation}$$
$$2X\cdot \longrightarrow X_2 \quad \text{*or* products of disproportionation}$$

Scheme 4

6.5.3 Transfer to Monomer

Transfer to monomer is pronounced for only a few monomers, *e.g.* allyl acetate and isopropenyl acetate; with these monomers, the process is degradative because of the stability and low reactivity of the product radical and a severe limitation is placed on R_p for the polymerization and on DP for the resulting polymer. The failure of propylene to form high polymers by radical reactions, even at the high pressures used successfully with ethylene, is due to the ready occurrence of degradative transfer to monomer giving the allyl radical $CH_2{=}CHCH_2\cdot$.

For styrene and other common monomers at say 60 °C, the transfer constant to monomer is so small that on average the reaction occurs once for about 10^4 to 10^5 growth steps. The process, however, sets an upper limit to the value of $\overline{DP}$ which can be achieved even if the conditions are such that $\bar{v}$ is very great. This case is unusual in homogeneous polymerizations except for those performed at high pressures.

The first stage in transfer is usually represented as in Scheme 3, the polymer radical abstracting an atom from a particular site in the molecule. Other reactions may be possible; in the case of the thiol, the hydrogen atoms of the alkyl group may be involved to a very limited extent. When transfer occurs with an unsaturated molecule such as a monomer. an alternative process, equation (25), may take place with an atom being transferred from the radical to the monomer. Very pronounced effects of some organometallic compounds (Q) on $\overline{DP}$ for the polymerization of methyl methacrylate have been interpreted in terms of catalyzed transfer to monomer, represented formally in Scheme 5.[45]

$$P_nCH_2CHX\cdot \ + \ Q \ \longrightarrow \ P_nCH{=}CHX \ + \ \dot{Q}H$$

$$\dot{Q}H \ + \ CH_2{=}CHX \ \longrightarrow \ Q \ + \ MeCHX\cdot$$

Scheme 5

$$P_nCH_2CHX\cdot \ + \ CH_2{=}CHX \ \longrightarrow \ P_nCH{=}CHX \ + \ MeCHX\cdot \tag{25}$$

6.5.4 Transfer to Polymer

Attack by a growing radical on a polymer chain leads to structural branching (see Scheme 6) with consequent effects on physical properties of the polymer. Transfer of this type is most likely in the later stages of polymerization when the concentration of polymer is high. Its effects on R_p and $\overline{DP}_n$ are small but there is broadening of the MWD since the largest molecules are those most likely to be involved in the transfer and therefore to be enlarged. Direct characterization of transfer to polymer is difficult; it is usual to draw analogies with the reactions of small molecules selected to represent the monomeric units in the polymer, *i.e.* to adopt the model compound approach (see Section 6.8).

$$P_nCH_2CHXP_m \ + \ P_jCH_2CHX\cdot \ \longrightarrow \ P_nCH_2\dot{C}XP_m \ + \ P_jCH_2CH_2X$$

$$P_nCH_2\dot{C}X\,P_m \ + \ CH_2{=}CHX \ \longrightarrow \ \underset{\underset{\overset{|}{\overset{CH_2}{\underset{\overset{|}{CHX}}{}}}}{}}{P_nCH_2CXP_m}$$

Scheme 6

Scheme 6 represents an intermolecular process; intramolecular transfer is most likely when the cyclic transition state corresponds to about six atoms. This reaction occurs readily during the high pressure radical polymerization of ethylene; it leads to short branches even in the earliest stages of polymerization.[46]

6.5.5 Transfer to Initiator

Growing radicals may attack molecules of the initiator. If the process is pronounced, it is better described as induced decomposition of the initiator. It can be represented formally by Scheme 7. The initiator becomes incorporated in the form of end groups but, from the point of view of initiation, it is wasted so that the efficiency (f in equation 1) is depressed.

$$P_n\cdot \ + \ X_2 \ \longrightarrow \ P_nX \ + \ X\cdot$$

$$X\cdot \ + \ M \ \longrightarrow \ P_1\cdot$$

Scheme 7

Initiators which are very effective in transfer are referred to as inifers.[47] They include some organosulfur compounds such as tetramethylthiuram disulfide $Me_2NCSSSCSNMe_2$. Substances of this type can be used for the production of polymers of quite low molecular weight with special end groups. These inifers are usually responsible for degradative transfer and the relationships between R_p and [I] do not conform to equation (10).

While considering transfer to initiator, mention might be made of primary radical transfer. A radical derived from the initiator might engage in a reaction such as that in equation (26) so that a fragment from the initiator is not actually incorporated as an end group; there is clear evidence that this process can occur in some cases.[48]

$$Me_3CO \cdot \ + \ CH_2{=}CRMe \longrightarrow Me_3COH \ + \ CH_2{=}CRCH_2 \cdot \tag{26}$$

6.6 TERMINATION

6.6.1 First-order Termination

The treatment in Section 6.2.1 was based upon removal of radicals in pairs; it can readily be adapted to allow for deactivation of polymer radicals singly; this type of termination occurs when the system contains a retarder, X, functioning in either of the ways referred to in Section 6.5.2 or Section 6.7.1. Equations (6) and (9) are modified so that R_t is given by $k'_t[P\cdot][X]$ and $R_{p,s}$ by $k_p R_i[M]/k'_t[X]$ leading to equation (27) for polymerization initiated by thermal dissociation of I. The kinetic order with respect to I is 1.0 instead of 0.5; for systems in which both types of termination occur, an effective order between 0.5 and 1.0 is expected. When the monomer itself is a powerful degradative transfer agent, $[X]^{-1}$ in equation (27) must be replaced by $[M]^{-1}$ so that $R_{p,s}$ becomes independent of $[M]$.

$$R_p \ = \ (2fk_dk_p/k'_t)\,[I]\,[M]\,[X]^{-1} \tag{27}$$

It appears that a special type of unimolecular termination occurs with vinyl benzoate.[49] There is occasional abnormal addition to give a highly stabilized radical incapable of further growth; the reaction is therefore effectively a termination involving a single polymer radical. A similar effect occurs with vinylferrocene.[45]

6.6.2 Diffusion Control (see also Volume 3, Chapter 12)

The interaction of macroradicals in solution is a diffusion-controlled process so that bimolecular termination is governed by physical rather than purely chemical factors. The first indications to this effect came with the discovery that, at some point in many homogeneous radical polymerizations, there is pronounced auto-acceleration accompanied by the production of polymer of comparatively high $\overline{DP}$.[51] It was postulated that, when sufficient polymer has accumulated in the system for the viscosity to reach a certain level, k_t begins to fall because of restricted diffusion although the initiation and growth processes are hardly affected. To maintain the balance between R_i and R_t, $[P\cdot]_s$ must rise and R_p increases correspondingly; according to equation (14), there is an increase in $\bar{v}$ leading to increase in $\overline{DP}$ and a consequent broadening of MWD for the whole polymer. The term 'gel effect' is widely used.

It is now recognized that bimolecular termination is diffuson controlled, even in the first stages of polymerization, and that k_t is related inversely with the viscosity of the medium.[52] Translational diffusion of macroradicals is not the only process to be considered; even if the centres of gravity of two such radicals are close, rotational movement is needed to bring the reactive points together. These movements are influenced by the viscosity of the medium and by the physical characteristics of the polymer. A form of diffusion control therefore operates at all stages although onset of the gel effect may signify a change from control by rotational diffusion to control by translational diffusion.[53,54] It now appears that, at any stage in a polymerization, k_t decreases with radical size, tending towards a limiting value;[55] most probably the effect can be correlated with diffusion control.

Bimolecular termination is an associative process so that $\Delta V^{\ddagger}$ must be negative; the reaction would therefore be expected to be more rapid at high pressures, with perhaps a shift towards combination in any competition between that process and disproportionation. In fact, k_t falls with rising pressure because the very marked increase in viscosity causes slower diffusion of the macroradicals.[33]

Rates of copolymerizations can be explained in terms of diffusion control of termination in a way which is much more satisfactory than older explanations.[56]

A special type of diffusion control occurs in systems where the polymer comes out of solution as it is formed. The precipitate contains 'buried' polymer radicals which are almost immobile so that reactions between macroradicals are prevented. Much of the work on this effect has been done with acrylonitrile (see Volume 4, Chapter 16).

6.6.3 Primary Radical Termination

It has been supposed that in some systems an appreciable proportion of the primary radicals reacts with macroradicals instead of with monomer. The process is most likely to occur if [M] is low and primary radicals are generated quickly. Detailed kinetic treatments have been given[57,58] but the essential feature is the prediction of an order for the overall polymerization of less than 0.5 with respect to I, as actually found in some cases. Direct evidence for primary radical termination comes from detailed examination of end groups derived from the initiator.[59]

Primary radical termination must be less affected by diffusion than interactions of pairs of large radicals. When the viscosity of the system is high, movements of the small primary radicals are much less impeded than those of macroradicals; it is expected therefore that primary radical termination is more pronounced at high conversions and in polymerizations at high pressures.

6.7 RETARDATION AND INHIBITION

6.7.1 General Considerations

Degradative transfer (see Section 6.5.2) causes reduction in R_p because of inefficiency in the re-initiation step but a process of another type also can cause retardation. If a growing radical adds to a reagent to give a product with very low reactivity towards monomer, effectively the kinetic chain is terminated prematurely and R_p is reduced. Substances of various classes function in this way; they include aromatic nitro compounds and quinones but each case requires separate consideration. It is quite common for several processes to operate in parallel for a particular retarder although they are indistinguishable kinetically. Thus for *p*-benzoquinone, some evidence indicates clearly that radical attack occurs at an oxygen atom (equation 28) but there is also evidence that the reaction can occur at a carbon atom (equation 29) and in other ways also; it is quite likely that the competition between the alternative reactions is governed to some extent by the nature of the radical being considered.

$$P\cdot \ + \ O{=}\!\!\left\langle\!\!\bigcirc\!\!\right\rangle\!\!{=}O \ \longrightarrow \ PO{-}\!\!\left\langle\!\!\bigcirc\!\!\right\rangle\!\!{-}O\cdot \tag{28}$$

$$P\cdot \ + \ \underset{O}{\overset{O}{\bigcirc}} \ \xrightarrow{\text{by steps}} \ \underset{\overset{.}{O}}{\overset{OH}{\underset{}{\bigcirc}}}\!P \tag{29}$$

Inhibitors can be regarded as especially effective retarders, capturing primary radicals or very small polymer radicals so efficiently that no production of polymer can be detected until the concentration of the additive has fallen almost to zero. Inhibitors are of great practical importance for stabilization of monomers during storage. They may be added to the monomer in the still during purification by distillation, to prevent polymerization during heating; clearly the inhibitors used in such cases should be involatile otherwise the distillate becomes contaminated. Inhibitors can be used to stop polymerizations which for some reason have become so fast as to be uncontrollable. As mentioned in Section 6.3.1 and referred to in more detail in Section 6.7.2, inhibitors can be used to good effect in studies of the initiation process in radical polymerizations.

There have been extensive studies of kinetic aspects of inhibition and retardation in radical polymerization and the subject has been comprehensively reviewed;[60] it is an area to which the ideas connected with so-called hot radicals have been applied (see Section 6.4.4).

An important feature of retarded polymerizations is the suppression, partial or complete, of termination by interaction of pairs of growing radicals and its replacement by reactions in which single polymer radicals are deactivated. Equations such as (6) and (8) to (11) therefore need modification; for the overall polymerization, the kinetic order with respect to initiator is increased from 0.5 towards unity. Consumption of retarder during polymerization can cause difficulties in kinetic approaches because the balance between terminations involving one and two growing radicals is likely to shift as the concentration of retarder gradually decreases; in fact, the stationary state treatment cannot be properly applied to systems in which the rate of polymerization changes significantly with time.

When an additive is so effective in deactivating initiating radicals or very small growing radicals that it can be classified as an inhibitor and if its concentration is not excessively high, there comes a time when its concentration has fallen so much that some conversion of monomer to polymer can be detected. Ideally the rate of polymerization then builds up to the value expected for a reaction proceeding in the absence of the additive; it may be necessary to make allowance for the consumption of initiator during the inhibition period. The duration of the transitional period depends upon the reactivity of the inhibitor, the change from zero to full rate of polymerization being more abrupt for the more reactive inhibitors.

Inhibitors have been used to measure rates of initiation (see Section 6.3.1); it is supposed that the rate of production of initiating radicals is unaffected by the presence of the inhibitor and that there is a simple relationship between the number of radicals generated and the number of molecules of inhibitor originally present and consumed during the inhibition period. For this purpose, the inhibition period is commonly defined by considering a conversion *vs.* time plot and extrapolating its linear portion to give an intercept on the time axis. Problems obviously can arise in connection with the transitional period; it has been shown by kinetic analysis[61] that, for many purposes, it is more satisfactory to consider the time required for the polymerization to attain 0.648 of its maximum rate and to use that time in the calculation of the rate of initiation.

6.7.2 Stabilized Radicals

Highly stabilized radicals form an interesting group of inhibitors; the best known is diphenyl-picrylhydrazyl (DPPH) (**4**), which is widely used as a standard in ESR spectroscopy. The mechanism of the action of DPPH has been much studied and it is certainly not direct attachment to the radical being scavenged. There is an added complication in that the products formed from DPPH are aromatic nitro compounds which function as retarders (Figure 1); this secondary reactivity of DPPH must operate to some extent even during the period of inhibition so that one 'molecule' of DPPH becomes effectively equivalent to slightly more than one primary radical (see Section 6.3.1). Similar effects arise with some other stabilized radicals; the uncertainties seem unavoidable because those structural features which are responsible for the stabilization of the radical also provide additional sites for interaction with reactive radicals. A commercially available stabilized radical, which seems to be rather more satisfactory than DPPH as a scavenger, is 2,2,6,6-tetramethylpiperidine-*N*-oxyl (**5**), commonly referred to as TEMPO; related substances also have been used.

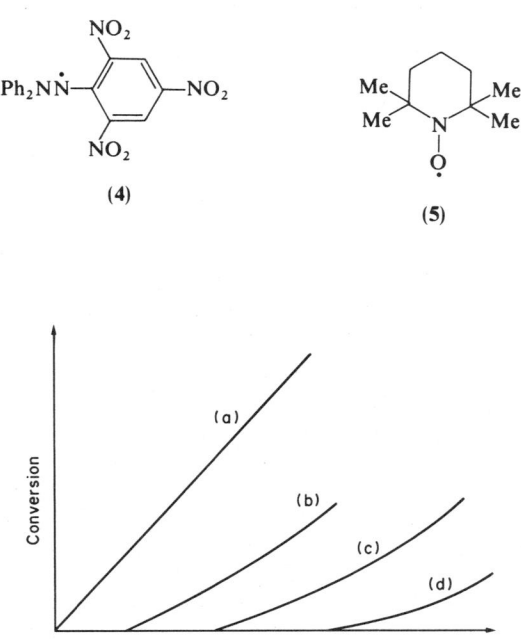

Figure 1 Sketch showing the effects of an inhibitor which becomes converted into a retarder: line (a) refers to a polymerization without inhibitor; lines (b), (c) and (d) refer to polymerizations with increasing [inhibitor], producing lengthening inhibition period and more pronounced retardation

Although TEMPO is very reactive towards carbon-centred radicals, it does not react with those in which the unpaired electron is on an oxygen atom; this difference forms the basis of a powerful method for study of the initiating processes for the important group of initiators giving rise to oxygen-centred radicals such as hydroxy, benzoyloxy and *t*-butoxy.[62] Suppose that the primary radicals are generated in a system containing monomer and TEMPO; the oxygen-centred radicals react with monomer to give carbon-centred radicals just as they would if TEMPO were absent and polymerization were proceeding normally. These new radicals react however with the scavenger to give stable products which can be identified and determined, so providing information on the relative importances of alternative reactions of the oxygen-centred radicals with the monomer. For the *t*-butoxy radical with methyl methacrylate, the principal process is the expected attachment of the radical to the tail of the molecule of monomer; abstraction of hydrogen to give a substituted allylic radical occurs to an appreciable extent and there is a minor contribution from hydrogen abstraction involving the ester group of the monomer (Scheme 8). In other cases, the occurrence of a significant amount of head initiation has been detected.

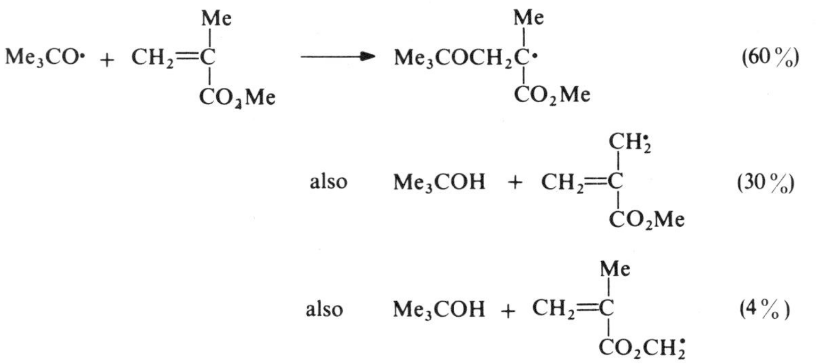

$$Me_3CO\cdot \ + \ CH_2{=}\underset{\underset{CO_2Me}{|}}{\overset{\overset{Me}{|}}{C}} \longrightarrow Me_3COCH_2\underset{\underset{CO_2Me}{|}}{\overset{\overset{Me}{|}}{C}}\cdot \qquad (60\%)$$

$$also \quad Me_3COH \ + \ CH_2{=}\underset{\underset{CO_2Me}{|}}{\overset{\overset{CH_2^\cdot}{|}}{C}} \qquad (30\%)$$

$$also \quad Me_3COH \ + \ CH_2{=}\underset{\underset{CO_2CH_2^\cdot}{|}}{\overset{\overset{Me}{|}}{C}} \qquad (4\%)$$

Scheme 8

Initiation through hydrogen abstraction (*i.e.* primary radical transfer) would give rise to an end group which is not an initiator fragment (see Section 6.5.5), with serious implications for procedures for measuring kinetic chain lengths by end group analysis (see Section 6.3.1), deducing rates and efficiencies of initiation and assessing the relative importances of combination and disproportionation. It must be noted, however, that the trapping experiments do not by themselves establish that a radical such as $CH_2{=}C(CO_2Me)CH_2\cdot$ actually reacts with monomer to initiate growth of a polymer chain.

6.7.3 Organic Retarders and Inhibitors

Marked retardation can arise in copolymerizations if M_A is reactive and M_B is rather unreactive so that the corresponding polymer radicals $P_A\cdot$ and $P_B\cdot$ are respectively unreactive and reactive (see Section 6.4.2). The growth reaction between $P_B\cdot$ and M_A occurs readily but the other cross-propagation ($P_A\cdot$ and M_B) involves two relatively unreactive species and is slow. An effect is very pronounced if styrene (M_A), even in very small amounts, is present in vinyl acetate during polymerization.

Polynuclear aromatic hydrocarbons reduce R_p for the radical polymerizations of many monomers; inhibition is observed for comparatively unreactive monomers such as vinyl acetate, giving radicals of high reactivity. Many of these hydrocarbons possess high affinity for methyl radicals, giving products which are stabilized to an appreciable extent. It is believed that aromatic hydrocarbons are responsible for the difficulty in obtaining samples of monomeric *N*-vinylcarbazole which show reproducible rates of polymerization. In cases of this type, a final purification by 'prepolymerization' is normally very effective; the procedure involves polymerization of part of the monomer so that any residual inhibitor or retarder is consumed.

As mentioned in Section 6.7.1, quinones are rather effective in reducing the rates of radical polymerizations but it is improbable that a unique mechanism is generally applicable. Relationships between the effectiveness of a quinone and its redox potential have been proposed.[63]

Aromatic nitro compounds have great influences upon the rates of radical polymerizations. Extensive studies have been made on the effects of substituents upon the reactivities of these compounds and the results have been analyzed in terms of the Hammett and Taft equations.[64] As stated in Section 6.7.2, the nitro groups in DPPH are responsible for at least some of the complexities found when that substance is used as a radical scavenger.

Compounds of several other classes have been used and studied as inhibitors and retarders; they include nitroso compounds of various types, phenols and aromatic aldonitrones, but there are many uncertainties about the courses of the reactions by which they exert their effects. Certain phenols are extensively used as stabilizers for monomers with which they do not react directly; there has been disagreement on the role of molecular oxygen in these systems and it has been claimed that the additive is effective only if oxygen also is present.

For some systems, quite simple and plausible schemes can be constructed but usually they are found not to be completely satisfactory. Thus diphenylpicrylhydrazine is expected to give up hydrogen to a radical very readily according to equation (30) to yield the stabilized radical DPPH, which would then rapidly deactivate a second reactive radical.[65] The substituted hydrazine might therefore be expected to be equivalent to two reactive radicals but this is not exactly so, partly because of the presence of nitro groups in the substituted hydrazine and in the derived DPPH. Rather similarly, an aromatic aldonitrone is expected to react with a radical according to equation (31), but then to react with a second radical.[66] These aldonitrones undergo a side reaction with monomer, by dipolar cycloaddition; clearly it must destroy the simple relationship between the number of additive molecules consumed and the number of radicals removed from the system.

$$P_n\cdot \; + \; Ph_2NNHC_6H_2(NO_2)_3 \; \longrightarrow \; P_nH \; + \; DPPH \qquad (30)$$

$$P_n\cdot \; + \; \underset{\displaystyle ArCH=\overset{\textstyle O}{\overset{\uparrow}{N}}-Ph}{} \; \longrightarrow \; \underset{\displaystyle \underset{P_n}{|}}{ArCH-\overset{\textstyle \overset{\cdot}{O}}{\overset{|}{N}}-Ph} \qquad (31)$$

6.7.4 Inorganic Compounds

Certain ionic compounds can function as retarders or inhibitors. Iron(III) chloride, for example, acts as an inhibitor for styrene and as a retarder for acrylonitrile or methacrylonitrile. The velocity constant for the process represented by equation (32) depends upon the nature of the polymer radical and upon that of X; with methyl methacrylate, for example, the salicylate is a weak retarder whereas the bromide is an inhibitor.[67]

$$P_n\cdot \; + \; FeX_3 \; \longrightarrow \; P_nX \; + \; FeX_2 \qquad (32)$$

An intramolecular electron transfer process is believed to operate in the unimolecular termination during the radical polymerization of vinylferrocene.[50] The proposed mechanism involves electron transfer between the growing end of the polymer radical and the iron atom in the terminal monomeric unit.

Oxygen exerts a retarding or even inhibiting effect on many radical polymerizations. In a sense, it acts as a comonomer (see equation 33) to give a relatively unreactive radical. In some cases, copolymers of fairly low molecular weight are included among the products; these copolymers are of course peroxides which later may initiate polymerization. It is to avoid the effects of oxygen that radical polymerizations are normally performed under strictly anaerobic conditions.

$$P_n\cdot \; + \; O_2 \; \longrightarrow \; P_nOO\cdot \qquad (33)$$

6.8 MODEL REACTIONS

Some of the reactions involved in the overall process of radical polymerization are difficult to study experimentally; in several such cases, use has been made of 'model reactions' in which a polymer radical or a polymer molecule is represented by an appropriately chosen small radical or molecule. The reactions of the small species can usually be characterized, perhaps in terms of the yields of products from competing reactions; it is then supposed that the macroradicals or macromolecules behave similarly. The model approach can be useful qualitatively or at best semi-quantitatively, but it is doubtful whether the deductions made from it should ever be regarded as completely justified.

The procedure can be illustrated by considering its application to the problem of the relative importances of combination and disproportionation during termination (see Section 6.2.1); thus the growing radical of methyl methacrylate (6) is likened to the radical (7) which is produced by the dissociation of the azo ester (8). The yields of the products from (7) can be determined; they give a direct indication of the relative importances of the two types of interradical reactions.[68] The relationship between (6) and (7) is such that a polymer chain in the former is represented by a hydrogen atom in the latter; this representation might be regarded as a gross approximation.

$$
\begin{array}{cccc}
 & \text{Me} & \text{Me} & \\
 & | & | & \\
\text{P}_n\text{CH}_2\text{C}\cdot & \text{Me}\!-\!\text{C}\cdot & \text{Me}_2\text{C}\!-\!\text{N}\!=\!\text{N}\!-\!\text{CMe}_2 \\
 & | & | & | \qquad\quad | \\
 & \text{CO}_2\text{Me} & \text{CO}_2\text{Me} & \text{CO}_2\text{Me}\quad\text{CO}_2\text{Me} \\
 & & & \\
\textbf{(6)} & \textbf{(7)} & \textbf{(8)} &
\end{array}
$$

The model approach has been used in connection with branching in vinyl polymerization as a consequence of transfer to polymer (see Section 6.5.4). In this case, the monomeric unit, in say polystyrene, is represented by either ethylbenzene or isopropylbenzene. The transfer constants for these hydrocarbons in the polymerization of styrene are known and they are taken as indicating the reactivity in tranfser of the monomeric unit in the polymer. An improved procedure uses larger models representing two or more adjacent monomeric units.

Fairly successful attempts have been made to compare the behaviour of small radicals in initiation with that of polymer radicals in the growth reactions of copolymerization. For example, the reactions with various monomers of the radical $\text{Me}_2\text{C(CN)}\cdot$, derived from azobis(isobutyronitrile), have been compared with the reactions of the poly(methacrylonitrile) radical with the same monomers in copolymerizations.[69] Similarly $\text{MeCHPh}\cdot$ has been used as a model for the polystyrene radical.[70, 71, 72]

6.9 REFERENCES

1. D. H. Solomon, *J. Macromol. Sci., Chem.*, 1982, **A17**, 337.
2. T. Hjertberg and E. M. Sörvik, *J. Macromol. Sci., Chem.*, 1982, **A17**, 983.
3. G. C. Eastmond, in 'Comprehensive Chemical Kinetics', ed. C. H. Bamford and C. F. H. Tipper, Elsevier, Amsterdam, 1976, vol. 14a, chap. 1.
4. G. M. Burnett, 'Mechanism of Polymer Reactions', Interscience, New York, 1954, p. 98.
5. Ref. 4, p. 76.
6. W. I. Bengough and H. W. Melville, *Proc. R. Soc. London, Ser. A*, 1955, **225**, 330; 1955, **230**, 429.
7. P. E. M. Allen and C. R. Patrick, 'Kinetics and Mechanisms of Polymerization Reactions', Ellis Horwood, Chichester, 1974, p. 136.
8. Ref. 7, p. 155.
9. K. F. O'Driscoll and H. K. Mahabadi, *J. Polym. Sci., Polym. Chem. Ed.*, 1976, **14**, 869.
10. M. Stickler, *Makromol. Chem., Macromol. Symp.*, 1987, **10/11**, 17.
11. O. F. Olaj, I. Bitai and G. Gleixner, *Makromol. Chem.*, 1985, **186**, 2569.
12. C. Walling, 'Free Radicals in Solution', Wiley, New York, 1957, p. 37.
13. J. C. Bevington, *Trans. Faraday Soc.*, 1955, **51**, 1392.
14. C. E. H. Bawn and S. F. Mellish, *Trans. Faraday Soc.*, 1951, **47**, 1216.
15. C. H. Bamford, W. G. Barb, A. D. Jenkins and P. F. Onyon, 'The Kinetics of Vinyl Polymerization by Radical Mechanisms', Butterworths, London, 1958, p. 50.
16. J. C. Bevington and T. D. Lewis, *Polymer*, 1960, **1**, 1.
17. A. V. Tobolsky, *J. Am. Chem. Soc.*, 1958, **80**, 5927.
18. P. J. Flory and F. S. Leutner, *J. Polym. Sci.*, 1948, **3**, 880; 1950, **5**, 267.
19. R. L. Adelman and R. C. Ferguson, *J. Polym. Sci., Polym. Chem. Ed.*, 1975, **13**, 393.
20. D. W. Ovenall, *Macromolecules*, 1984, **17**, 1458.
21. R. E. Cais and N. J. A. Sloane, *Polymer*, 1983, **24**, 179.
22. W. J. Bailey, S.-R. Wu and Z. Ni, *J. Macromol. Sci., Chem.*, 1982, **A18**, 973.
23. I. Cho and K. D. Ahn, *J. Polym. Sci., Polym. Chem. Ed.*, 1979, **17**, 3169.
24. G. B. Butler and A. Matsumoto, *J. Polym. Sci., Polym. Lett. Ed.*, 1981, **19**, 167.
25. T. Otsu, *Makromol. Chem., Macromol. Symp.*, 1987, **10/11**, 235.
26. M. J. Ballard, R. G. Gilbert, D. H. Napper, P. J. Pomery and J. H. O'Donnell, *Macromolecules*, 1984, **17**, 504.
27. W. Lau, D. G. Westmoreland and R. W. Novak, *Macromolecules*, 1987, **20**, 457.
28. Y. Kuwae and S. Nozakura, *Polym. J.*, 1981, **13**, 919.
29. T. Sato, S. Inui, H. Tonaka, T. Ota, H. Kamachi and K. Tonaka, *J. Polym. Sci., Polym. Chem. Ed.*, 1987, **25**, 637.
30. H. Kamachi, Y. Kuwae, M. Kohno and S. Nazakura, *Polym. J.*, 1982, **14**, 749.
31. R. Z. Greenley, *J. Macromol. Sci., Chem.*, 1980, **A14**, 445.
32. A. D. Jenkins, *Makromol. Chem., Macromol. Symp.*, 1987, **10/11**, 1.
33. K. E. Weale, in 'Reactivity, Mechanism and Structure in Polymer Chemisty', eds. A. D. Jenkins and A. Ledwith, Wiley, London, 1974, p. 162.
34. J. Pavlinec, J. Jergušová and Š. Florián, *Eur. Polym. J.*, 1982, **18**, 279.
35. H. J. Harwood, *Makromol. Chem., Macromol. Symp.*, 1987, **10/11**, 331.

36. E. A. Lissi and M. Moya, *Eur. Polym. J.*, 1980, **16**, 185.
37. C. H. Bamford, in 'Molecular Behaviour and the Development of Polymeric Materials', ed. A. Ledwith and A. M. North, Chapman and Hall, London, 1975, p. 55.
38. G. Henrici-Olivé and S. Olivé, *Makromol. Chem.*, 1966, **96**, 221.
39. F. Tüdös, T. Kelen and T. Földes Berezhnikh, *J. Polym. Sci., Polym. Symp.*, 1975, **50**, 109.
40. G. Blauer, *Trans. Faraday Soc.*, 1960, **56**, 606.
41. C. H. Bamford, A. D. Jenkins and R. Johnston, *Proc. R. Soc., London, Ser. A*, 1957, **241**, 364.
42. J. C. Bevington and J. R. Ebdon, in 'Developments in Polymerisation — 2', ed. R. N. Haward, Applied Science, London, 1979, p. 26.
43. V. A. Kabanov, *Makromol. Chem., Macromol. Symp.*, 1987, **10/11**, 193.
44. Ref. 33, p. 167.
45. N. S. Enikolopyan, B. R. Smirnov, G. V. Ponomarev and I. M. Belgovskii, *J. Polym. Sci., Polym. Chem. Ed.*, 1981, **19**, 879.
46. D. J. Cutler, P. J. Hendra, M. E. A. Cudby and H. A. Willis, *Polymer*, 1977, **18**, 1005.
47. T. Otsu, M. Yoshida and A. Kuriyma, *Polym. Bull. (Berlin)*, 1982, **7**, 45.
48. R. D. Grant, E. Rizzardo and D. H. Solomon, *Makromol. Chem.*, 1984, **185**, 1809.
49. M. Litt and V. Stannett, *Makromol. Chem.*, 1960, **37**, 19.
50. M. H. George and G. F. Hayes, *J. Polym. Sci., Polym. Chem. Ed.*, 1975, **13**, 1049.
51. R. G. W. Norrish and R. R. Smith, *Nature (London)*, 1942, **150**, 336.
52. A. M. North, 'The Kinetics of Free Radical Polymerization', Pergamon Press, Oxford, 1966, p. 73.
53. B. W. Brooks, *Proc. R. Soc. London, Ser. A*, 1977, **357**, 183.
54. D. T. Turner, *Macromolecules*, 1977, **10**, 221.
55. K. Ito and K. F. O'Driscoll, *J. Polym. Sci., Polym. Chem. Ed.*, 1979, **17**, 3913.
56. A. M. North, in 'Reactivity, Mechanism and Structure in Polymer Chemistry', ed. A. D. Jenkins and A. Ledwith, Wiley, London, 1974, p. 152.
57. H. K. Mahabadi and K. F. O'Driscoll, *Makromol. Chem.*, 1977, **178**, 2629.
58. P. C. Deb and S. Ray, *Eur. Polym. J.*, 1978, **14**, 607.
59. G. Moad, D. H. Solomon, S. R. Johns and R. I. Willing, *Macromolecules*, 1982, **15**, 1188.
60. G. C. Eastmond, 'Comprehensive Chemical Kinetics', ed. C. H. Bamford and C. F. Tipper, Elsevier, Amsterdam, 1976, vol. 14A, pp. 127–152.
61. C. H. Bamford, A. D. Jenkins and R. Johnston, *Proc. R. Soc. London, Ser A*, 1957, **239**, 214.
62. D. H. Solomon and G. Moad, *Makromol. Chem., Macromol. Symp.*, 1987, **10/11**, 109.
63. P. D. Bartlett, G. S. Hammond and H. Kwart, *Discuss. Faraday Soc.*, 1947, **2**, 342.
64. F. Tüdös, I. Kende and M. Azor, *J. Polym. Sci., Part A*, 1963, **1**, 1369.
65. J. C. Bevington and N. A. Ghanem, *J. Chem. Soc.*, 1958, 2254.
66. S. Szakács, T. Földes-Berezsnich, F. Tüdös and L. Jókay, *Eur. Polym. J.*, 1979, **15**, 295.
67. E. R. Entwistle, *Trans. Faraday Soc.*, 1960, **56**, 284.
68. A. M. North and D. Postlethwaite, in 'Structure and Mechanism in Vinyl Polymerization', ed. T. Tsuruta and K. F. O'Driscoll, Dekker, New York, 1969, p. 109.
69. J. C. Bevington, T. N. Huckerby and N. W. E. Hutton, *J. Polym. Sci., Polym. Chem. Ed.*, 1982, **20**, 2655.
70. D. A. Cywar and D. A. Tirrell, *Macromolecules*, 1986, **19**, 2908.
71. K. Ito, *Polymer*, 1985, **26**, 1253.
72. J. C. Bevington, D. A. Cywar, T. N. Huckerby, E. Senogles and D. A. Tirrell, *Eur. Polym. J.*, 1988, **24**, 699.

7

Experimental Techniques

MANFRED STICKLER
Röhm GmbH, Darmstadt, FRG

7.1 INTRODUCTION

As with any kinetic investigation, the ultimate aim of free-radical polymerization kinetics is to establish, for a given polymerization system, the reaction mechanism and to measure the reaction rate constants associated with its individual elementary reaction steps. The free-radical polymerization of vinyl monomers is one of the most extensively and thoroughly investigated organic reactions. Early kinetic studies were made almost 50 years ago. They revealed the chain nature of the reaction mechanism with its three essential elementary steps: chain initiation, propagation and termination.

The basic equation of an ideal free-radical polymerization relates the rate of polymerization R_p with the concentrations of the reactants, *i.e.* monomer concentration [M] and initiator concentration [I] (equations 1 and 2). The overall rate constant of polymerization K combines the elementary rate constants of the chain initiation $f k_d$ (f = cage efficiency; k_d = decomposition rate constant of the initiator), the chain propagation k_p and the chain termination k_t.

$$R_p = -\frac{d[M]}{dt} = K[M][I]^{1/2} \tag{1}$$

$$K = \left(\frac{2fk_d k_p^2}{k_t}\right)^{1/2} \tag{2}$$

A second experimentally observable quantity is the number-average degree of polymerization $\bar{P}_n$. It is related to the rate of polymerization *via* the well-known Mayo equation (equation 3). Here, $C_M = k_{fm}/k_p$ and $C_S = k_{fs}/k_p$ are (relative) chain-transfer constants for the monomer and for any other compound S present in the reaction system (*i.e.* solvent, initiator, chain-transfer agent, polymer); $\lambda = k_{td}/(k_{td} + k_{tc})$ is the fraction of disproportionation in the termination reaction.

$$\frac{1}{\bar{P}_n} = \frac{1 + \lambda}{2} \frac{k_t}{k_p^2} \frac{R_p}{[M]^2} + C_M + C_S \frac{[S]}{[M]} \tag{3}$$

It is important to note that equations (1)–(3) combine the instantaneous variables [M], [I], R_p and $\bar{P}_n$. Thus, they are only valid for vanishingly small increments of monomer conversion. However, with a few exceptions, most kinetic experiments yield cumulative quantities, *e.g.* the monomer conversion x and the cumulative number-average degree of polymerization $\bar{P}_n$. Thus, it is desirable to use, as far as possible, integrated relationships instead of equations (1) and (3) when analyzing kinetic data. Such integrated relationships have been derived for several simple free-radical polymerization systems.[1]

Discussions about general experimental methods of reaction kinetics have been published in the literature.[2-4] There are also several reviews or monographs available which deal specifically with experimental methods of free-radical polymerization kinetics,[5-11] including the treatise of steady-state and nonsteady-state kinetics,[5-8,11] kinetics at advanced degrees of monomer conversion[9] and heterogeneous polymerization systems.[10] Experimental techniques for living polymerizations are collected in the book by Szwarc[12] (*cf.* Volume 3, Chapter 38 for further details). For a discussion of the purification of reagents, preparation of the reaction set-up and isolation and purification of polymers see also Volume 3, Chapter 5.

7.2 MEASURING RATES OF POLYMERIZATION

Rates of polymerization which, according to equation (1), will allow us to estimate K, can be obtained by differentiating time–conversion data $x(t)$. The direct measurement of time seldom limits the accuracy of a rate determination, but precautions must be observed in timing a polymerization reaction. The time required for sampling or for initiating and quenching the reaction is likely to introduce a much larger uncertainty than is inherent in the performance of a good clock. This is important for rapid polymerizations, such as those with half-lives of monomer conversion of less than about one hour. Kinetic experiments at low degrees of conversion often belong to this category of reactions. It is common practice to prepare the reaction mixture at a low temperature for thermally initiated polymerizations, and then increase its temperature rapidly to the desired value. The time required for this process depends on a number of factors: the heat capacity and thermal conductivity of the reaction mixture, the size, shape and material of the reaction vessel, the temperature gradient, *etc.* It is necessary, therefore, to know the time lag of a given experimental set-up.

Estimates of monomer conversion may be determined in two different ways: by directly measuring the concentration of monomer (by chemical analysis or by collecting and weighing the polymer produced) or, alternatively, by indirectly measuring monomer conversion with the help of physical properties of the reaction system. The discontinuous analytical procedures are often laborious and time consuming. To avoid this expenditure of time and effort, indirect methods of conversion measurement without sampling are frequently used. It should be stressed, however, that the accuracy of even the most precise of these indirect methods cannot be better than the accuracy of the standard analytical procedure that was used to calibrate the method.

7.2.1 Dilatometry

The polymerization of a liquid monomer is generally accompanied by a decrease in volume. Therefore dilatometry is a convenient indirect way of recording the progress of monomer conversion. As the equipment is relatively simple and inexpensive, the method has been widely used. Numerous types of dilatometers have been reported in the literature.[6,9,11,13] They include instruments used to study heterogeneous systems (*e.g.* emulsion polymerizations[14] or reactions where the polymer is insoluble in its own monomer[15-17]) as well as those used to investigate polymerizations under high pressure.[18] Several types of dilatometers with automatic recording of the capillary heights have also been described.[11,18,19]

The application of conventional dilatometers to polymerization reactions is limited by two sources of error. If the monomer itself is used as a recording liquid for the volume decrease, the increasing viscosity of the reaction mixture soon causes a distortion of the meniscus in the capillary, which makes accurate reading impossible. In addition, the dissipation of the heat of reaction by either convection or conduction is hindered by the viscous solution. Dilatometers with mercury as a recording liquid are often used to overcome these difficulties.[6,11,20] There should be no need to emphasize that a dilatometer resembles a highly sensitive thermometer and thus requires a thermostat with a very high degree of temperature constancy (variations less than ± 0.01 K).

As it is an indirect method for measuring conversion, dilatometry needs calibration. This does not only mean determination of the parameters of the measuring device (*i.e.* capillary radius and volume of the dilatometer) but also and especially the volume contraction coefficient for a given monomer/polymer system. There have been several attempts at predicting volume contraction coefficients, but without any convincing success to date.[21-24] Systematic experimental studies are rather rare, even for the most familiar monomers, and have not supplied conclusive results.[20,25] With copolymerizations, the situation is even more uncertain.[26-31] Generally, a linear relationship between volume contraction and monomer conversion is assumed. It was shown that this assumption might not hold up to complete conversion for the bulk polymerization of methyl methacrylate.[32]

7.2.2 Density Measurement

With the development of a sensitive digital density-measuring device (DMA),[33,34] the density as an intrinsic physical property of monomer/polymer mixtures may also be determined with high precision. This technique may also be applied to polymerizing systems.[11,35-37] Due to the low volume of the measuring tube (≈ 0.5 ml), this method is especially valuable where only small amounts of monomer are available. Furthermore, the small volume allows a rapid heating-up of the reaction mixture and thus the time-lag is minimal. Calibration is done with solutions of the polymer in its monomer. Automatic data acquisition is easily performed.[11]

Several types of DMA cells are available (*e.g.* from Anton Paar, Graz, Austria). The external cell DMA 11 can be used for photopolymerizations and, with the DMA 512HT, experiments at high temperatures and pressures are possible (maximum values 150 °C and 4×10^7 Pa). These might be valuable options for the investigation of polymerizations under more industrial conditions.

7.2.3 Measurement of Refractive Index or Dielectric Constant

The progress of monomer conversion can also be followed by measuring changes in the refractive index of the polymerizing reaction mixture. This method is quite similar to the dilatometric technique in many aspects. With both methods, a high precision can be obtained in principle; both are sensitive to small temperature variations. While volume contractions for a complete conversion are usually about 20%, the change in refractive index is considerably less ($\approx 10\%$). Thus, in experiments where high resolution is needed, interferometers should be used. However, in general, the measurement of refractive index has not achieved the popularity of dilatometry.

Instruments for continuously monitoring the refractive index of a polymerizing system are worth mentioning, however. The availability of fibre-optic devices and opto-electronic systems has also promoted interesting innovations in the last few years for measurement of the refractive index.[11]

Closely related to the refractive index is the dielectric constant, and the relative changes of these two physical properties during a polymerization are of about the same magnitude. Although modern electronic equipment has considerably increased the accuracy of dielectric constant measurement, this method has only been used in a few cases (especially with nonstationary polymerization kinetics[38]).

7.2.4 Conversions from Light-scattering and Viscosity Measurements

The measurement of viscosity and of the intensity of scattered light are both methods that are extremely sensitive to the presence of very small amounts of polymer. In addition, with both techniques, the measured quantity is dependent not only on polymer concentration but also on the molecular weight of the product. Thus evaluation of kinetic data from such experiments can be rather complicated.[6]

In earlier studies, nonsteady-state polymerizations were investigated[39-44] in order to obtain the rate constants k_p and k_t. More recently, static and dynamic light-scattering measurements have been applied simultaneously to polymerizing systems. Since the polymer concentration was determined concurrently *via* Raman spectroscopy in these experiments, very detailed information was obtained not only about the reaction kinetics but also about the molecular parameters (average molecular weight $\overline{M}_w$, molecular weight distribution, radius of gyration) and the dynamics (hydrodynamic radius, diffusion coefficient) of the polymer produced.[45] Although more complicated than usual standard techniques in free-radical polymerization kinetics, this type of light-scattering analysis shows a unique potential which will surely be exploited in future work.

7.2.5 Spectroscopic Methods

Raman spectroscopy, mentioned above, is one of many spectroscopic methods which can be used for kinetic investigations. The absorption of radiation is a specific property of the absorbent molecule; both the magnitude of the absorption coefficient and the wavelength at which absorption maxima occur are characteristic of the absorbent molecule and only to a lesser extent of its physical state.

Among the spectroscopic methods available, those operating in the UV or IR regions of the spectrum are the most convenient and have a long history. Sophisticated mathematical methods of data analysis have been reported, even for more complex reaction systems,[46] and Fourier transform infrared spectroscopy (FTIR), a powerful improvement on the classical IR technique, is becoming more and more familiar. Raman spectroscopy can be used alone or as a complementary analytical tool to the IR method.

Under appropriate conditions, other spectroscopic techniques, *e.g.* fluorescence spectroscopy or NMR analysis, may give the desired kinetic information. We shall not go into further details here.

7.2.6 Chromatographic Methods

Chromatographic analysis can be used very efficiently to study polymerization kinetics. This is especially the case with complex reaction mixtures, where the possibility of a separation prior to detection and quantification is needed.

With volatile and thermally stable monomers, gas chromatography (GC) is a very convenient technique. Modern modifications such as capillary GC have enormous separation capabilities and head-space GC allows the easy determination of low-boiling monomers in a matrix of compounds with a high boiling-point. With high-boiling or thermally labile compounds, liquid chromatography (LC) should be used instead of GC.[47]

By far the most important application of chromatographic methods is the determination of residual monomer. However, the accuracy of this method is generally too low to determine the small changes in concentration of the monomer during the study of initial polymerization kinetics. Thus, higher conversions of the reactants are generally needed and only integrated rate equations are appropriate.

On the other hand, the measurement of polymer produced may be performed with very high sensitivity and good precision, even at extremely low conversions, with LC methods. Under appropriate conditions (especially with careful column selection), both monomer conversion and molecular weight distribution of the polymer can be obtained in a single run *via* size-exclusion chromatography (SEC).

7.2.7 Thermoanalytical Methods

Polymerizations are accompanied by the release of a considerable reaction enthalpy. This heat evolution can be used for monitoring the progress of the reaction. There exist two basically different types of experiment: the calorimetric experiments which measure the (integral) production of heat, and those experiments where the output signal of the instrument is proportional to the time derivative of the heat production (dH/dt).

The most familiar method of the latter type is differential scanning calorimetry (DSC). As the rate of heat production is proportional to the rate of polymerization, DSC is one of the few techniques in experimental polymerization kinetics where reaction rates and not concentration/time data are directly obtained. Conversion histories are then readily obtained by integration of these rate data. This procedure is inherently more accurate than the reverse one of evaluating rates from the slope of a conversion curve.

DSC measurements are typically run dynamically (*cf.* ref. 48); isothermic operation is also possible.[49, 50] In the case of copolymerization, the heat of reaction may vary with conversion, thus preventing a simple reduction of DSC data. Methods to overcome this difficulty have been proposed.[51] Calorimetric experiments may be conducted isothermally or adiabatically. The latter type is of particular importance for nonstationary polymerization kinetics (*cf.* Sections 7.5.1 and 7.5.2).

The calorimetric analysis of free-radical polymerization kinetics has, in most cases, been looked upon from the point of view of reaction engineering. In this way, with the help of an energy balance equation, the reaction enthalpy and the polymerization rate can be evaluated with appreciable accuracy from the measurement of only a few temperatures and the mass flow of the monomer. Studies of this type will also give kinetic parameters at advanced degrees of monomer conversion and are thus of the uttermost industrial importance.[52-54]

7.2.8 Other Physicochemical Methods

Several other methods for the estimation of monomer conversion with the help of physico-chemical properties of the reaction system have been reported in the literature. Some of them seem to be (at least nowadays) only of academic interest (*e.g.* the measurement of diamagnetic suscept-ibility[55] or Brillouin spectroscopy[56]); with others, the situation and the potential have not as yet been fully investigated. Thus, as an example, the measurement of the velocity of sound was proposed for the study of polymerization kinetics,[57] but other authors observed that ultrasonic waves may change kinetic parameters, *cf.* ref. 58. Future work should clarify these discrepancies.

7.2.9 Gravimetry

Having discussed in the preceding sections various methods of an indirect measurement of monomer conversion, we shall now finally come to the most direct way of obtaining conversion data for polymerizations: the gravimetric determination of the polymer produced.

Generally, sealed tubes are used for this purpose and the polymer is isolated by precipitation. This technique has been criticized as being too laborious, extremely clumsy[6, 7] and susceptible to several errors, *e.g.* inclusion of monomer and/or solvent with the precipitated polymer, loss of low molecular weight material, uncontrolled polymerization conditions for the different sealed tubes within one series, *etc.* On the other hand, gravimetry needs no calibration, which saves time and in addition avoids any systematic error.

Considering this latter very important advantage of gravimetry, together with the drawbacks of the sealed-tube technique, led the author to the development of a highly improved gravimetric method in his laboratory. The sealed tubes were (for low-conversion polymerizations) replaced by a small tank reactor which, besides other details, is equipped with a thermostatted jacket and a stirrer. This stirred tank reactor (STR) allows an absolutely isothermal conduction of the polymerization which is controlled with a thermometer. Before starting the polymerization by the sudden addition of initiator, the reactor and the monomer are purged with nitrogen. Thus any dissolved oxygen is removed and the start of the reaction is very well defined.

After appropriate time intervals, samples are withdrawn from the STR with the help of a Teflon microvalve. The polymerization is quenched by the addition of an efficient inhibitor and the conversions are then determined by precipitating and weighing the polymer. A more detailed description of the instrument and its manipulation is given elsewhere.[11]

It was shown that very accurate time/conversion data can be obtained with the STR. In addition, at each conversion level, molecular weights or other polymer properties of interest (*e.g.* tacticities, end-group content) can be measured. Finally, one should mention that the STR can also be operated in a semibatch or continuous mode.

7.3 MEASURING RATES OF INITIATION

Measuring monomer conversions of a steady-state free-radical polymerization yields only the overall polymerization constant K. To get information about the individual rate constants of the elementary reactions, additional experiments are needed. Measuring rates of initiation is the first step to resolve K into fk_d and k_p^2/k_t. If the decomposition of the initiator is also investigated, the separation of f and k_d is possible.

7.3.1 Decomposition Rate of Initiators

Numerous methods have been reported in the literature for the direct measurement of initiator decomposition.[11] Iodometry, IR analysis or HPLC coupled with UV detection may be used with peroxides. The volumetric measurement of evolved nitrogen is a technique applicable to azo-initiators. Further methods include UV spectroscopy, polarography or DSC.

It has been claimed in the literature that an interesting feature of the latter technique is the possibility of obtaining the complete Arrhenius law of decomposition from a single dynamic experiment. A criticism of the DSC methods has been given recently, however.[11]

It was shown that with polymerizations where $k_d t \ll 1$ is fulfilled ($t = $ polymerization time), the decomposition rate constant k_d may be determined solely from time/conversion data.[1]

7.3.2 Rates of Initiation

Owing to the cage effect, the rate of chain initiation is generally less than twice the rate of production of primary radicals. This is taken into consideration by introducing an efficiency factor f (with $f \leq 1$). A direct way of estimating f is to analyze quantitatively the side products generated in the cage reaction. Modern separation techniques (HPLC) and methods for the elucidation of the structures of these by-products (*e.g.* NMR) have greatly facilitated this task (*cf.* ref. 59).

However, other methods, such as the trapping of primary radicals with the help of inhibitors, have most often been used to evaluate rates of initiation. The theory of inhibition kinetics has been given in detail by Bagdasar'yan.[8]

Expecting simple stoichiometry, the use of stable free radicals as inhibitors has been thought to be superior to that of molecule inhibitors such as nitro compounds or quinones. 2,2-diphenyl-1-picrylhydrazyl (DPPH) has certainly been the most popular stable free radical for the study of inhibition kinetics. Other compounds suggested include triphenylverdazyl, Banfield's radical or 2,2,6,6-tetramethylpiperidine-*N*-oxyl (TEMPO). In contrast to DPPH, which is suspected of undergoing side reactions with some monomer systems (for a more detailed discussion see ref. 11), TEMPO and its derivatives seem to be really ideal inhibitors. Fascinating details about initiation pathways have been obtained with these scavengers.[60]

The assay of the polymer for the initiator fragments present as polymer end-groups is a further way of evaluating rates of chain initiation in free-radical polymerizations. To avoid primary radical termination, the concentration of the initiator should be as low as possible in such experiments. Generally, polymers with rather high molecular weights are produced under these circumstances. Thus, the initiator fragments constitute only a very small fraction of the total weight of the polymer and highly sensitive analytical techniques are needed for the estimation of the initiator-fragment concentration in the polymer.

While sophisticated chemical methods, such as Palit's dye-partition technique,[61,62] have been used for end-group assay, radiochemical tracer analysis has more general applicability.[63-65] This method requires no knowledge of the mode of the termination reaction (disproportionation or combination); however, an essential condition is that initiation should occur solely by addition to the double bond of the monomer and not by hydrogen-abstraction reactions. Primary radical termination or chain transfer to the initiator, or both, give rise to complications but might be accounted for quantitatively.

Besides radiochemical labelling, ^{1}H NMR has been used for determining initiator residues incorporated into polymers during their preparation.[66] Both these methods only give information about the number of such labels in a polymer molecule; they are not sensitive to the environment of the initiator-derived functionality. Using ^{13}C-enriched initiators together with ^{13}C NMR has recently been shown to be an excellent tool for investigating the fate of the initiator fragments, thereby getting very detailed information about not only the initiator efficiency f but also the nature of the possible side reactions of the primary radicals.[65,67]

7.4 KINETIC PARAMETERS FROM MOLECULAR WEIGHTS AND MOLECULAR WEIGHT DISTRIBUTIONS

Free-radical polymerization is a statistical process and the molecular weight distribution (MWD) and the average molecular weights of the polymer produced are functions of the probability of chain propagation p. This propagation probability is, on the other hand, closely related to the mechanism and kinetic parameters of the polymerization. Thus, *vice versa*, the determination of the MWD should allow quantitative conclusions to be drawn about p and the kinetic parameters. This possibility has been recognized since the early days of polymerization kinetics, and the theoretical MWDs derived for a large variety of polymerization mechanisms have been compiled.[68]

The technique of current choice for measuring the MWD of polymers is size-exclusion chromatography (SEC). SEC has contributed greatly to our knowledge about the effect of polymerization reaction variables on product characteristics. However, with the increasing use of this method, it has become apparent that (besides several other problems not mentioned here) it has a poor sensitivity in the low and high molecular weight regions. This may lead to considerable uncertainties and thus restricts the value of MWD data for kinetic analysis.[11, 25, 69]

The combination of number-average degrees of polymerization with rates of polymerization yields information about chain-transfer constants (*cf.* equation 3).[70] The chain-transfer constant of the monomer, C_M, is determined with the highest accuracy under conditions where the bimolecular termination reaction between two macroradicals is only of negligible importance and where chain transfer to the monomer is by far the overwhelming process of production of dead polymer. To a very good approximation, equation (3) is then reduced to $1/\bar{P}_n = 2/\bar{P}_w = C_M$, and the more accurately accessible $\bar{P}_w$ can be used instead of $\bar{P}_n$.[71]

The second kinetic parameter derived from a Mayo plot is the product $0.5(1 + \lambda)(k_t/k_p^2)$. Thus, once k_p^2/k_t is known, the fraction of disproportionation termination λ can be calculated. However, the factor $0.5(1 + \lambda)$ has frequently been neglected in the literature when analyzing kinetic and molecular weight data with the help of the Mayo equation. Values of k_p^2/k_t estimated in this way can be wrong by up to a factor of 2.

7.5 HOW TO OBTAIN THE INDIVIDUAL RATE CONSTANTS OF CHAIN PROPAGATION AND CHAIN TERMINATION

The measurement of the stationary polymerization rate and of the average degree of polymerization discussed above will only yield the ratio k_p^2/k_t but not the individual values of the reaction rate constants for chain propagation k_p and chain termination k_t. The estimation of these latter quantities is generally based on kinetic experiments conducted under nonstationary polymerization conditions. Such experiments give us the ratio k_p/k_t which can be combined with k_p^2/k_t deduced from stationary experiments to calculate finally k_p and k_t. Several types of nonstationary or pseudo-stationary polymerization experiments have been developed; a survey of methods is given in the monographs of Burnett[6] and Bagdasar'yan.[8]

7.5.1 Method of Initial Nonstationary Kinetics

With this technique, the very beginning of a polymerization with the rise of the polymer radical concentration from zero to its stationary value is observed. During this pre-effect, the monomer conversion increases according to equation (4),[8] where τ is the average lifetime of the polymer radicals, which is typically of the order of magnitude of 1 s. For times $t \gg \tau$, equation (4) reduces to the simple form of equation (5), which may be used to plot x *vs.* t thus to calculate τ.

$$x = (k_p/k_t) \ln[\cosh(t/\tau)] \tag{4}$$

$$x = (k_p/k_t) [(t/\tau) - \ln 2] \tag{5}$$

7.5.2 Post-polymerization Method

If, with a stationary photopolymerization system, illumination is suddenly stopped, the concentration of the polymer radicals will decrease owing to the termination reaction. The additional conversion during this post-polymerization period (after-effect) is given by equation (6).[8] Equation

(6) considers only the bimolecular termination mechanism; suitable corrections for additional first-order termination by impurities or for a possible thermal initiation have been calculated. However, it is better to eliminate these uncontrolled factors by using a continuous weak, but controlled, background illumination.[8]

$$x = (k_p/k_t) \ln[1 + (t/\tau)] \qquad (6)$$

A combination of the pre- and after-effect is the method of brief illumination, where a short impulse of intensive light is superimposed on a background illumination.[8]

The study of these nonstationary periods of polymerization, the durations of which are often only a few seconds, requires sophisticated automatic apparatus for measuring very small monomer conversions. To have a really instantaneous beginning or cessation of radical production, only photochemical or radiochemical initiation of the polymerization is useful. Melville and co-workers have developed a series of such techniques. Most of these methods involve adiabatic reaction conditions, where the temperature of the system is measured directly or *via* some physical property of the system which depends on temperature, *i.e.* volume,[72] refractive index[73] or dielectric constant;[74] *cf.* refs. 8 and 9.

7.5.3   Rotating-sector Method

An alternative means of studying nonstationary polymerization is intermittent illumination. This is by far the most widely used technique for the determination of the lifetime τ of polymer radicals. The principle of this method is to use a series of nonstationary phases of polymerization by periodic interruption of a photochemical initiation. Thus many pre- and after-effects are combined. The interruption of the initiating light-beam is achieved mechanically with the help of a rotating disc with sector-shaped apertures.

This principle was developed by Briers, Chapman and Walters[75] and applied to polymerization systems for the first time by Burnett and Melville[76] and Bartlett and Swain.[77] The theory has been reviewed in detail by Burnett,[6,7] including corrections for a steady dark reaction, mixed first- and second-order chain terminations, nonuniform light absorption and the penumbra effect (trapezoidal light impulses). More recently, an extension to copolymerization has been reported.[78,79] Details of rotating-sector apparatuses are also given in the reviews of Burnett;[6,7] modern electronical controls of the light-chopper unit have been described by Nagy *et al.*[80]

When compared with the pre-effect, the sector technique is said to be somewhat less accurate, since several independently measured quantities have to be combined to evaluate τ. Under carefully controlled conditions, an error of about 20% has recently been assigned to the lifetime as obtained by intermittent illumination;[80] this would lead to errors of about 40 and 20% for k_t and k_p, respectively.

7.5.4   Flow Methods

A flow method for the determination of k_p/k_t was proposed by Funt and Collins.[81] The principle of this technique is comparable to that of brief illumination (*cf.* Section 7.5.2): the reaction mixture flowing through a tube passes a small irradiated zone where chain initiation takes place. In the subsequent dark zone of the tube, post-polymerization occurs. This after-effect is quenched by running the polymerizing system into a solution of an inhibitor. The residence time in the dark zone is varied by changing the length of the corresponding part of the tube. Measuring the polymer yield as a function of this residence time allows the evaluation of τ.

A very similar experimental technique was described earlier by Goldfinger and Heffelfinger,[82] while Penchev[83] used a capacity flow method to get k_p/k_t.

7.5.5   Spatially Intermittent Polymerization

A linear analogue to the rotating-sector method was recently designed by O'Driscoll and Mahabadi.[84] A solution of monomer and photoinitiator flows through a dark tubular reactor which is covered with an aperture plate with regularly spaced slots through which UV radiation shines. The alternating dark and light regions produce a spatially intermittent polymerization system (SIP).

With the rotating-sector method, only small amounts of polymer can normally be obtained for further investigations (*e.g.* determination of MWD) and the produced material shows the cumulative properties corresponding to the conversion at the end of the experiment, typically about 5%. Owing to its continuous mode of operation, the SIP reactor, on the other hand, may produce large quantities of polymeric material at a fixed conversion level, which might be as low as 0.1%. In this way, it is even possible to study the conversion dependence of τ.

In addition, with the SIP reactor, both k_p^2/k_t and k_p/k_t can be measured under identical conditions and with comparable precision. Thus k_p and k_t data obtained with this method are claimed to be highly reliable.

7.5.6 Viscosity Method

The extreme sensitivity of viscosity measurements also allows their application to the study of the nonstationary periods of a polymerization. The viscosity method developed by Bamford and Dewar[39] differs from all other methods discussed so far in that the intensity of the illumination is here needed only in relative units and that it gives the values for all the elementary reaction rate constants directly. Furthermore, its experimental technique is comparatively simple.

However, the theoretical background of the viscosity method is rather complicated.[6, 39] The original theory was developed for photochemical polymerizations with diradicals as chain carriers and termination exclusively by disproportionation. A modified version with monoradical initiation was given later on.[85]

7.5.7 ESR Spectroscopic Method

ESR spectroscopy was introduced to free-radical polymerization kinetics by Fischer.[86] ESR is known to be possibly the best method for the determination of k_p and k_t, since, in principle, the stationary concentration of free-radicals can be measured directly. Unfortunately, this stationary concentration of growing radicals is (under normal conditions of low-conversion polymerizations) too small to be detected by currently available commercial ESR spectrometers. Thus, only specially modified instruments will yield the desired information.[87, 88] Besides the estimation of k_p *via* measurement of the steady-state concentration of polymer radicals and the stationary rate of polymerization, it should also be possible to obtain k_t by some nonstationary experiment. Up to now, however, scan rates of ESR spectrometers are too slow and termination rate constants could only be obtained under special conditions with low-conversion systems.[88]

With higher monomer conversions, the situation is much better; owing to the gel effect, the termination rate constant decreases and the concentration of the macroradicals rises greatly. Under such favourable conditions, propagation and termination rate constants have been obtained just recently[89, 90] and we can surely expect further exciting results in the near future.

7.5.8 Laser Flash Initiation Methods

During the past decade, laser light sources appeared as very convenient tools in photochemistry and photophysics. The key advantages of a laser-initiated chain reaction are: (i) the reaction system is prepared in a well-defined initial state on a time scale which is extremely short when compared with the subsequent kinetics; (ii) the full temporal evolution of the chain is monitored in real time; and (iii) the radical concentrations can be varied over several orders of magnitude by control of the intensity of the laser pulse.[91]

The potential of flash photolysis for the estimation of k_p/k_t had already been indicated in 1968,[92] but it is only recently that this technique has gained its proper attention in free-radical polymerization kinetics. Olaj and co-workers[93, 94] developed an experimental method where the polymerizing system is subjected to periodic laser light flashes; thus a pseudo-stationary state is established similar to that of the rotating-sector method or spatially intermittent polymerization. The mathematical treatment of the kinetics is very much facilitated, however. Moreover, the analysis of the MWD of the polymer produced allows the direct determination of k_p without any reference to k_p/k_t.[95]

By using a powerful exciplex laser, Buback and co-workers studied the time-resolved polymerization, where radical growth induced by a laser pulse of 10 ns width was measured *via* FTIR spectroscopy on a time scale of micro- and milli-seconds.[96,97] Values of k_p/k_t and $k_t[P\cdot]_0$ are obtained from such experiments ($[P\cdot]_0$ is the concentration of macroradicals just after the laser pulse). If, in addition, the quantum yield for the production of radicals is known (or measured by an independent experiment), the individual constants k_p and k_t can be calculated.

7.6 REFERENCES

1. M. Stickler, *Makromol. Chem.*, 1979, **180**, 2615.
2. R. Livingston, in 'Techniques of Organic Chemistry', 2nd edn., ed. A. Weissberger, Interscience, New York, 1961, vol. VIII/1, chap. III, p. 55.
3. E. F. MacNichol, Jr., in 'Techniques of Organic Chemistry', 2nd edn., ed. A. Weissberger, Interscience, New York, 1961, vol. VIII/1, chap. IV, p. 89.
4. R. Huisgen, in 'Houben-Weyl, Methoden der Organischen Chemie', 4th edn., ed. E. Müller, Thieme, Stuttgart, 1955, vol. III/1, p. 99.
5. L. Küchler, 'Polymerisationskinetik', Springer, Berlin, 1951.
6. G. M. Burnett, 'Mechanism of Polymer Reactions', Interscience, New York, 1954.
7. G. M. Burnett, in 'Techniques of Organic Chemistry', 2nd edn., ed. A. Weissberger, Interscience, New York, 1963, vol. VIII/2, p. 1139.
8. Kh. S. Bagdasar'yan, 'Theory of Free-Radical Polymerization', Israel Program for Scientific Translations, Jerusalem, 1968.
9. G. P. Gladyshev and K. M. Gibov, 'Polymerization at Advanced Degrees of Conversion', Israel Program for Scientific Translations, Jerusalem, 1970.
10. H. Oelmann and H. Fuerst, *Plaste Kautsch.*, 1976, **23**, 638.
11. M. Stickler, *Makromol. Chem., Macromol. Symp.*, 1987, **10/11**, 17.
12. M. Szwarc, 'Carbanions, Living Polymers and Electron Transfer Processes', Interscience, New York, 1968, p. 18.
13. P. H. Plesch, *Int. Lab.*, 1986, October, 18.
14. C. J. Lyons and E. Elbing, *J. Macromol. Sci., Chem.*, 1982, **A17**, 113.
15. K. Marquardt and P. Mehnert, *Angew. Makromol. Chem.*, 1973, **28**, 177.
16. A. Garton and M. H. George, *Polymer*, 1975, **16**, 934.
17. M. H. George and A. Garton, *J. Macromol. Sci., Chem.*, 1977, **A11**, 1389.
18. T. Sasuga and M. Takehisa, *J. Macromol. Sci., Chem.*, 1978, **A12**, 1307.
19. V. D. McGinniss and R. M. Holsworth, *J. Appl. Polym. Sci.*, 1975, **19**, 2243.
20. G. V. Schulz and G. Harborth, *Angew. Chem.*, 1947, **A59**, 90.
21. F. S. Nichols and R. G. Flowers, *Ind. Eng. Chem.*, 1950, **42**, 292.
22. S. Loshaek and T. G. Fox, *J. Am. Chem. Soc.*, 1953, **75**, 3544.
23. A. Adicoff and R. Yee, *J. Appl. Polym. Sci.*, 1976, **20**, 2473.
24. R. Zana, *J. Polym. Sci., Polym. Phys. Ed.*, 1980, **18**, 121.
25. G. C. Eastmond, *Makromol. Chem., Macromol. Symp.*, 1987, **10/11**, 71.
26. J. C. Bevington, H. W. Melville and R. P. Taylor, *J. Polym. Sci.*, 1954, **14**, 463.
27. P. Wittmer, *Angew. Makromol. Chem.*, 1974, **39**, 35.
28. D. Braun and G. Disselhoff, *Polymer*, 1977, **18**, 963.
29. D. Braun, G. Disselhoff and F. Quella, *Makromol. Chem.*, 1978, **179**, 1239.
30. G. Disselhoff, *Polymer*, 1978, **19**, 111.
31. Y.-D. Ma, T. Fukuda and H. Inagaki, *Polym. J.*, 1983, **15**, 673.
32. D. Panke and W. Wunderlich, *Makromol. Chem.*, 1973, **167**, 351.
33. O. Kratky, H. Leopold and H. Stabinger, *Z. Angew. Phys.*, 1969, **27**, 273.
34. H. Leopold, *Elektronik*, 1970, **9**, 297.
35. W. Jaeger, Ch. Wandrey, G. Reinisch and K.-J. Linow, *Faserforsch. Textiltech.*, 1978, **29**, 647.
36. B. Trathnigg, *Makromol. Chem.*, 1980, **181**, 1979.
37. B. Trathnigg, *Angew. Makromol. Chem.*, 1980, **88**, 127.
38. C. M. Burrell, T. G. Majury and H. W. Melville, *Proc. R. Soc. London, Ser. A*, 1951, **205**, 309.
39. C. H. Bamford and M. J. S. Dewar, *Proc. R. Soc. London, Ser. A*, 1948, **192**, 309, 329.
40. Yu. S. Cherkinskii, *Polym. Sci. USSR (Engl. Transl.)*, 1977, **19**, 524.
41. S. G. Kulichikhin and A. Ya. Malkin, *Polym. Sci. USSR (Engl. Transl.)*, 1980, **22**, 2093.
42. A. Ya. Malkin, S. G. Kulichikhin, D. N. Emel'yanov, I. E. Smetania and N. V. Ryabokon', *Polymer*, 1984, **25**, 778.
43. I. M. Bel'govskii, M. A. Markevich and N. S. Yenikolopyan, *Polym. Sci. USSR (Engl. Transl.)*, 1964, **6**, 958.
44. G. N. Kornienko, A. Chervenka, I. M. Bel'govskii and N. S. Yenikolopyan, *Polym. Sci. USSR (Engl. Transl.)*, 1969, **11**, 3072.
45. B. Chu and D. Lee, *Macromolecules*, 1984, **17**, 926.
46. H. Mauser, 'Formale Kinetik', Bertelsmann Universitätsverlag, Düsseldorf, 1974.
47. F. Eisenbeiß, E. Dumont and H. Henke, *Angew. Makromol. Chem.*, 1978, **71**, 67.
48. G. O. R. Alberda van Ekenstein and Y. Y. Tan, *Eur. Polym. J.*, 1982, **17**, 839, 1061.
49. K. Horie, I. Mita and H. Kambe, *J. Polym. Sci., Part A-1*, 1968, **6**, 2663; *J. Polym. Sci., Part A-1*, 1969, **7**, 2561.
50. J. R. Ebdon and B. J. Hunt, *Anal. Chem.*, 1973, **45**, 804.
51. D. H. Sebastian and J. A. Biesenberger, *J. Macromol. Sci., Chem.*, 1981, **A15**, 553.
52. H. M. Andersen, *J. Polym. Sci., Part A-1*, 1966, **4**, 783.
53. B. Hentschel, *Chem.-Ing.-Tech.*, 1979, **51**, 823.
54. H. Weber and U. Guggisberg, in 'Thermal Analysis, Proceedings of the International Conference on Thermal Analysis', ed. H. G. Wiedemann, Birkaeuser Verlag, Basel, 1980, p. 461.
55. J. Farquharson, *Trans. Faraday Soc.*, 1936, **32**, 219; *Trans. Faraday Soc.*, 1937, **33**, 824.

56. D. A. Jackson and J. R. Stevens, *Molecular Physics*, 1975, **30**, 911.
57. F. Dinger, P. Hauptmann and R. Säuberlich, *Plaste Kautsch.*, 1982, **29**, 681; *Plaste Kautsch.*, 1983, **30**, 546.
58. K. F. O'Driscoll and A. U. Sridharan, *J. Polym. Sci., Polym. Chem. Ed.*, 1973, **11**, 1111.
59. S. Bizilj, D. P. Kelly, A. K. Serelis, D. H. Solomon and K. E. White, *Aust. J. Chem.*, 1985, **38**, 1657.
60. D. H. Solomon and G. Moad, *Makromol. Chem., Macromol. Symp.*, 1987, **10/11**, 109.
61. S. R. Palit, *Makromol. Chem.*, 1959, **36**, 89; *Makromol. Chem.*, 1960, **38**, 96.
62. D. Pramanick and B. Chakraborty, *Colloid Polym. Sci.*, 1981, **259**, 995.
63. J. C. Bevington, *Fortschr. Hochpolym.-Forsch.*, 1960, **2**, 1.
64. G. Ayrey, *Chem. Rev.*, 1963, **63**, 645.
65. J. C. Bevington, *Makromol. Chem., Macromol. Symp.*, 1987, **10/11**, 89.
66. K. Hatada, T. Kitayama and H. Yuki, *Makromol. Chem., Rapid Commun.*, 1980, **1**, 51.
67. G. Moad, D. G. Solomon, S. R. Johns and R. I. Willings, *Macromolecules*, 1982, **15**, 1188; *Macromolecules*, 1984, **17**, 1094.
68. L. H. Peebles, Jr., 'Molecular Weight Distributions in Polymers', in 'Polymer Reviews', Interscience, New York, 1971, vol. 18.
69. A. Rudin, *Makromol. Chem., Macromol. Symp.*, 1987, **10/11**, 273.
70. G. Henrici-Olivé and S. Olivé, *Fortschr. Hochpolym.-Forsch.*, 1960, **2**, 496.
71. M. Stickler and G. Meyerhoff, *Makromol. Chem.*, 1978, **179**, 2729.
72. S. W. Benson and A. M. North, *J. Am. Chem. Soc.*, 1958, **80**, 5625.
73. N. Grassie and H. W. Melville, *Proc. R. Soc. London, Ser. A*, 1951, **207**, 285.
74. C. M. Burrell, T. G. Majury and H. W. Melville, *Proc. R. Soc. London, Ser. A*, 1951, **205**, 309.
75. F. Briers, D. L. Chapman and E. Walters, *J. Chem. Soc.*, 1926, 562.
76. G. M. Burnett and H. W. Melville, *Nature (London)*, 1945, **156**, 661.
77. P. D. Bartlett and C. G. Swain, *J. Am. Chem. Soc.*, 1945, **67**, 2273.
78. J. P. Marano, Jr., L. H. Shendalman and C. A. Walker, *J. Polym. Sci., Part A-1*, 1970, **8**, 3461.
79. T. Fukuda, Y.-D. Ma and H. Inagaki, *Macromolecules*, 1985, **18**, 17.
80. A. Nagy, D. Szalay, T. Földes-Bereznich and F. Tüdös, *Eur. Polym. J.*, 1983, **19**, 1047.
81. B. L. Funt and E. Collins, *J. Polym. Sci.*, 1958, **28**, 97.
82. G. Goldfinger and C. Heffelfinger, *J. Polym. Sci.*, 1954, **13**, 123.
83. P. Penchev, *Makromol. Chem.*, 1976, **177**, 413.
84. K. F. O'Driscoll and H. K. Mahabadi, *J. Polym. Sci., Polym. Chem. Ed.*, 1976, **14**, 869.
85. G. Dixon-Lewis, *Proc. R. Soc. London, Ser. A*, 1949, **198**, 510.
86. H. Fischer, *Adv. Polym. Sci.*, 1968, **5**, 463.
87. S. E. Bresler, E. N. Kozbekov, V. N. Fornichev and V. N. Shadrin, *Makromol. Chem.*, 1972, **157**, 167; *Makromol. Chem.*, 1974, **175**, 2875.
88. M. Kamachi, M. Kohno, Y. Kuwae and S. Nozakura, *Polym. J.*, 1982, **14**, 749; *Polym. J.*, 1985, **17**, 541.
89. M. J. Ballard, R. G. Gilbert, D. H. Napper, P. J. Pomery, P. W. O'Sullivan and J. H. O'Donnell, *Macromolecules*, 1986, **19**, 1303.
90. M. J. Ballard, D. H. Napper, R. G. Gilbert and D. F. Sangster, *J. Polym. Sci., Polym. Chem. Ed.*, 1986, **24**, 1027.
91. D. J. Nesbitt and S. R. Leone, *J. Chem. Phys.*, 1981, **75**, 4949.
92. Yu. L. Spirin and T. S. Yatsimirskaya, *Teor. Eksp. Khim.*, 1968, **4**, 546.
93. O. F. Olaj, I. Bitai, H. F. Kauffmann and G. Gleixner, *Oesterr. Chem.-Ztg.*, 1983, **84**, 264.
94. O. F. Olaj, I. Bitai and G. Gleixner, *Makromol. Chem.*, 1985, **186**, 2569.
95. O. F. Olaj, I. Bitai and F. Hinkelmann, *Makromol. Chem.*, 1987, **188**, 1689.
96. M. Buback, H. Hippler, J. Schweer and H.-P. Vögele, *Makromol. Chem., Rapid Commun.*, 1986, **7**, 261.
97. H. Brackemann, M. Buback and H.-P. Vögele, *Makromol. Chem.*, 1986, **187**, 1977.

8

Azo and Peroxy Initiators

GRAEME MOAD and DAVID H. SOLOMON
CSIRO, Melbourne, Australia

8.1 INTRODUCTION

The most common form of initiator for radical polymerization is a compound which acts as a thermal source of radicals. Typically this is an azo or peroxy initiator. A particular initiator will have a convenient decomposition rate over a relatively narrow temperature range. Accordingly an initiator is often categorized according to its half-life at a given temperature and many such tabulations can be found (see, for example, ref. 1). The initiators may also be used at lower temperatures when they form one component of a photochemical or redox initiation system.

Whatever mode of radical generation is employed, recent studies show that radical sources must also be classed according to the types of radicals formed and their suitability for use with the particular monomers, solvent and other agents present in the reaction medium.[2]

We shall consider only the more recent developments in this area, covering earlier work only to the extent necessary to provide an appropriate background, and refer the reader to other works for additional information.[2-9] The redox chemistry of the initiators and aspects of their behaviour in aqueous or heterogeneous media (*e.g.* during emulsion polymerization) are covered in other chapters of this volume.

Within these bounds we attempt to provide a critical survey of the chemistry of azo and peroxy initiators which is relevant to the initiation of polymerization. We describe the various types of azo and peroxy initiators, paying particular attention to the nature of radicals produced and the factors which influence the initiator efficiency. We consider the selectivity shown by oxygen-centred and carbon-centred radicals in their reactions with monomers and other components of the polymerization medium. Finally, we detail some of the methods which have been applied in characterizing the initiation process.

8.2 AZO INITIATORS

Two classes of azo compounds will be distinguished: the dialkyldiazenes (**1**) and the hyponitrites (**2**).

$$-\overset{|}{\underset{|}{C}}-N{=}N-\overset{|}{\underset{|}{C}}-\qquad\qquad-\overset{|}{\underset{|}{C}}-ON{=}NO-\overset{|}{\underset{|}{C}}-$$

$$\textbf{(1)}\qquad\qquad\qquad\textbf{(2)}$$

8.2.1 Dialkyldiazenes

These azo compounds are alkyl derivatives of diazene (diimide; HN=NH).[10] The kinetics and mechanism of the thermal and photochemical decomposition of these materials has been comprehensively reviewed by Engel.[11] While all aliphatic azo compounds are photochemical sources of radicals, those with half-lives appropriate for use as thermal initiators are generally tertiary and have resonance-stabilizing groups (*e.g.* nitrile, aryl, carboxyalkyl) in the α positions. The azo compounds in most common usage as polymerization initiators are the azonitriles: 2,2'-azobis(2-methylpropanenitrile) [better known as azobis(isobutyronitrile) or AIBN (**3**)] (see Section 8.2.2), 1,1'-azobis(1-cyclohexanenitrile) (**4**), 4,4'-azobis(4-cyanovaleric acid) (**5**), *etc.*

$$\underset{\textbf{(3)}}{\overset{\displaystyle Me\qquad\qquad Me}{Me-\overset{\overset{\displaystyle |}{|}}{\underset{\underset{\displaystyle CN}{|}}{C}}-N{=}N-\overset{\overset{\displaystyle |}{|}}{\underset{\underset{\displaystyle CN}{|}}{C}}-Me}}\qquad\qquad \underset{\textbf{(4)}}{\underset{\underset{\displaystyle CN\quad CN}{}}{\bigcirc\!\!-N{=}N-\!\!\bigcirc}}\qquad\qquad \underset{\textbf{(5)}}{\overset{\displaystyle Me\qquad\qquad Me}{HO_2CCH_2CH_2-\overset{\overset{\displaystyle |}{|}}{\underset{\underset{\displaystyle CN}{|}}{C}}-N{=}N-\overset{\overset{\displaystyle |}{|}}{\underset{\underset{\displaystyle CN}{|}}{C}}-CH_2CH_2CO_2H}}$$

While the majority of dialkyldiazenes in use as initiators are symmetrical, unsymmetrical azo compounds, for example triphenylmethylazobenzene (**6**), also find application in special circumstances (see Volume 3, Chapter 10).[12]

$$Ph_3C-N{=}N-\bigcirc$$

$$\textbf{(6)}$$

Many factors are important in determining the dependence of thermolysis rates (k_d) on structure. The stability of the generated radicals appears to be a factor.[13] Values of k_d for simple dialkyldiazenes R–N=N–R increase in the series where R is aryl, primary, secondary, tertiary, allyl and are dramatically accelerated by α substituents capable of delocalizing the free spin of the incipient radical.[13] However, steric factors are also important[14] and for alicyclic azo compounds a good correlation exists between k_d and ground state strain.[15] Additional factors become important for bicyclic and other conformationally constrained azo compounds.[16] However, such compounds are seldom used as polymerization initiators and further discussion on this subject is beyond the scope of this chapter. For a discussion of the various influences and a comprehensive tabulation of thermolysis rates, readers are referred to Engel's review.[11]

One important fact to note regarding the photochemistry of dialkyldiazenes is that the main light-induced reaction is *trans–cis* isomerization. This process has a moderate quantum yield ($\phi \sim 0.5$).[11] In general, aliphatic *cis* azo compounds have at best a transient existence under typical reaction conditions. Thus, the rate of photoisomerization approximates the rate of initiator disappearance. Where the *cis* isomer is thermally stable, quantum yields for initiator decomposition are low ($\phi < 0.1$).[11]

The efficiency for generation of 'useful' radicals from azo compound initiators is low, typically 50–70%.[17–19] The remaining radicals are lost as cage recombination products.

The amount of cage reaction is a function of the viscosity of the reaction medium. Quoted initiator efficiencies almost without exception apply to zero or low conversion, yet during polymerization the medium viscosity increases dramatically with conversion and the initiator efficiency will drop accordingly. For a styrene (S) polymerization in 50% (v/v) benzene initiated by AIBN, the 'instantaneous' initiator efficiency varies from 76% at low conversion to 20% at 90–95% conversion.[20] The common assumption that the rate of initiation ($k_d f$) is invariant with conversion cannot be supported.

For azo compounds where the α positions are not fully substituted, rearrangement to the corresponding hydrazone may also reduce the initiator efficiency (Scheme 1).[6] Such azo compounds are also susceptible to an induced decomposition mechanism involving initial abstraction of an α hydrogen.[6]

Scheme 1

The rate of decomposition of azo compounds may be modified by transition metal salts (*e.g.* Cu^{2+}).[7,21] However, solvent effects, while observable, are small[11,21,22] and there is little evidence for radical-induced decomposition except as noted above (see also Section 8.2.2).[11]

8.2.2 Azobis(isobutyronitrile)

While some details of the kinetics of radical production from AIBN remain to be unravelled,[20,23] its decomposition mechanism and behaviour as a polymerization initiator are largely understood (see Scheme 2).

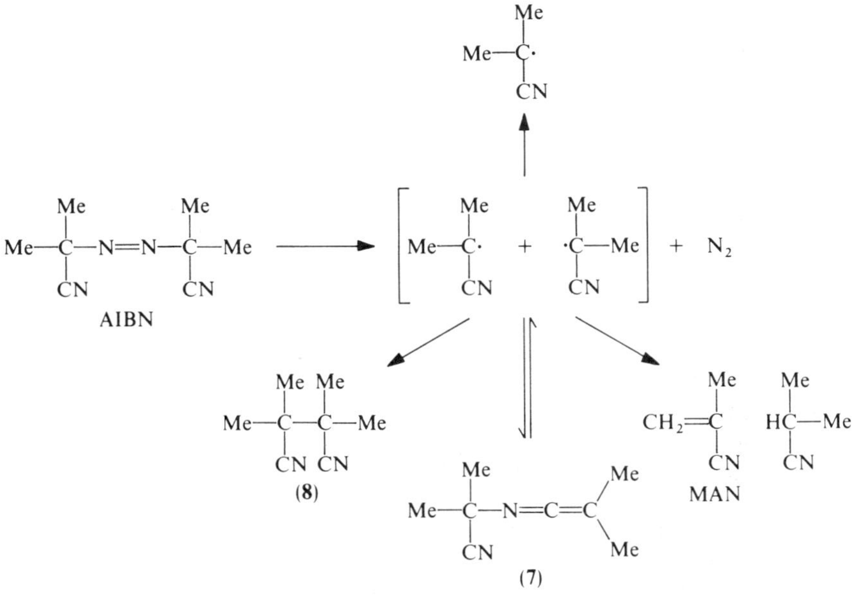

Scheme 2

While the initial breakdown of certain azo compounds is proposed to occur in two stages[11] and, in particular, the diazenyl radical $Me_2(CN)CN=N\cdot$ has been suggested to have transient existence at $-196\,°C$,[24] for AIBN decomposition in solution, nitrogen loss is sufficiently rapid that the diazenyl radicals cannot be trapped or initiate polymerization. All observed reactions are attributable to the cyanoisopropyl radical.

Cage recombination, which reduces the efficiency of initiation, also produces a variety of byproducts (Scheme 2). The major product of cage recombination is the ketenimine (7).[25] Suggestions[26,27] that (7) might be formed by a non-radical mechanism have been discounted.[20] The ketenimine (7) is itself thermally unstable and reverts to cyanoisopropyl radicals at a rate similar to that for AIBN thermolysis.[28] Thus, the formation of this material (7) complicates any analysis of the kinetics of initiation.[23] A second, and perhaps more important, concern is the potential reactivity of ketenimine (7) under polymerization conditions, for example as a transfer agent. This possibility has not yet been investigated.

Tetramethylsuccinodinitrile (8) appears inert under polymerization conditions but concern has been raised regarding its toxicity and it must be removed from polymers used in food contact applications.[10,29]

Methacrylonitrile (MAN) readily copolymerizes.[30-32] The copolymerized MAN may impair the thermal stability of polymers. For the case of polystyrene (PS) initiated with AIBN such groups have been proposed as weak links which rupture during the initial stages of thermal degradation.[33]

Dialkyldiazenes are often preferred over other (peroxide) initiators because of their lower susceptibility to induced decomposition. The importance of transfer to initiator during polymerizations initiated by AIBN has been the subject of some controversy.[34-38] However, recent studies[32] of S polymerization initiated by ^{13}C-labelled AIBN demonstrate conclusively that transfer to initiator has little importance in that system. Other explanations for those irregularities in polymerization kinetics previously attributed to transfer to initiator will have to be considered.

Nonetheless, there are clearly complications associated with the use of AIBN. Some of these may be avoided through the use of alternate azo initiators. Recently, azobis(methyl isobutyrate) (see Section 11.3.3) has seen use in laboratory studies of polymerization.[39]

8.2.3 Hyponitrites

The hyponitrites (2), alkyl esters of hyponitrous acid ($HON=NOH$), are low temperature sources of alkoxy radicals[40-43] or acyloxy radicals[44] (Scheme 3). Tertiary hyponitrites are not susceptible to induced decomposition; however, the same is not true of primary and secondary hyponitrites.[45,46] While di-t-butyl (9) and dicumyl hyponitrites (10) have proved convenient sources of t-butoxy and cumyloxy radicals respectively in the laboratory, the commercial utilization of hyponitrites is limited by their availability.

$$Me_3CON=NOCMe_3 \rightarrow Me_3CO\cdot \quad {}^{N_2} \quad \cdot OCMe_3$$

(9)

Scheme 3

$$Me_2PhCON=NOCPhMe_2$$

(10)

The hyponitrites are somewhat more efficient than the dialkyldiazenes (see above) with respect to radical generation.[40] However, a proportion of radicals is lost through cage reaction with formation of the corresponding peroxide in the case of tertiary hyponitrites[44,47,48] or ketone plus alcohol (disproportionation) in the case of hyponitrites with α hydrogens.[49] This proportion shows direct dependence on the viscosity of the reaction medium.[48] Approximately 5% of radicals are lost through cage recombination when dicumyl hyponitrite is decomposed in bulk MMA or S at $60\,°C$.[47] The product of cage recombination, dicumyl peroxide, is also a source of cumyloxy radicals although it is stable at the temperatures at which hyponitrites are usually employed.

Dialkyl hyponitrites show only weak absorption at $\lambda > 290$ nm[46] and their photochemistry is a largely neglected area. The triplet-sensitized decomposition of these materials has, however, been investigated by Mendenhall et al.[46]

8.3 PEROXIDES

Peroxides are the most widely used of all initiators and their general chemistry[1,6,51] and behaviour as initiators of radical polymerization[1,52,53] have been reviewed. Readers are referred in particular to Swern's trilogy[54] for a most comprehensive coverage of the literature through 1970.

Peroxides may be employed as thermal or photochemical initiators, or be used as a component in a redox system (see Section 8.2). In the former two instances all reactions should ideally ensue from unimolecular homolysis of the relatively weak O–O bond. However, various forms of induced and nonradical decomposition complicate the kinetics of radical generation and reduce the initiator efficiency.[51,55]

Six classes of peroxides will be distinguished. These are: (a) diacyl peroxides (**11**); (b) peroxy-dicarbonates (**12**); (c) peresters (**13**); (d) hydroperoxides (**14**); (e) dialkyl peroxides (**15**); and (f) inorganic peroxides (hydrogen peroxide, persulfate (**16**), peroxydiphosphate).

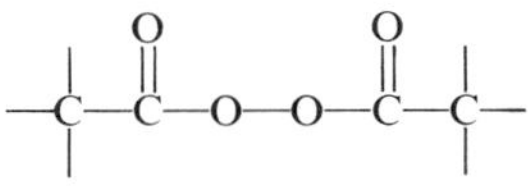

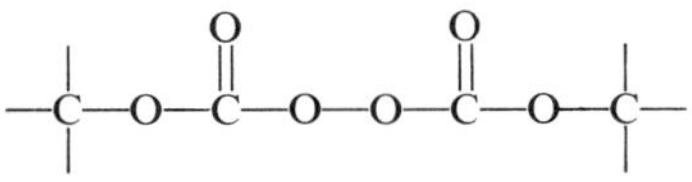

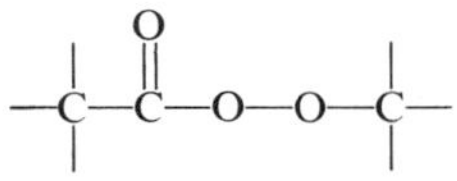

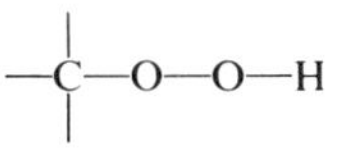

8.3.1 Diacyl Peroxides

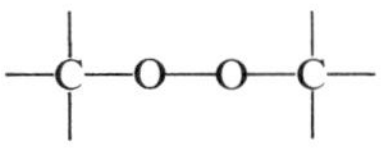

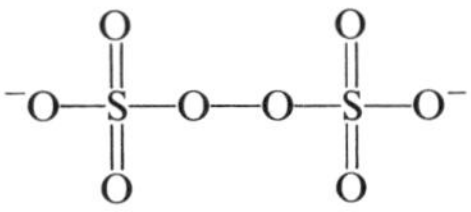

Diacyl peroxides (**11**) decompose to acyloxy radicals which may lose carbon dioxide to afford aryl or alkyl radicals as illustrated below for dibenzoyl peroxide (BPO) (Scheme 4).

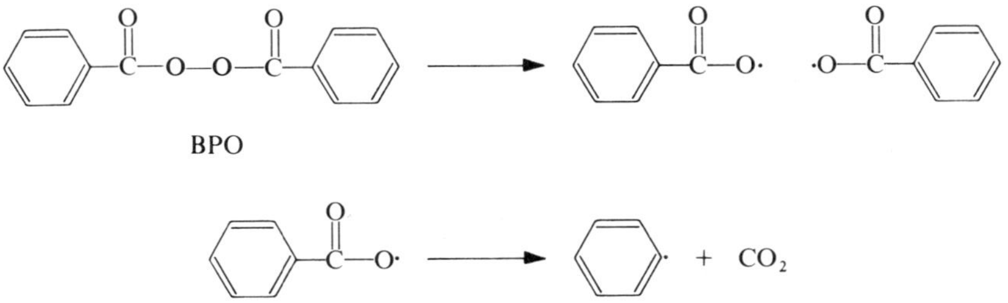

Scheme 4

Of the vast range of diacyl peroxides reported only two see widespread usage as initiators of radical polymerization. These are BPO and didodecanoyl (dilauroyl) peroxide (LPO). The lower diacyl peroxides (*e.g.* diacetyl peroxide, **17**) are very susceptible to induced decomposition and cannot be conveniently handled in a pure state.

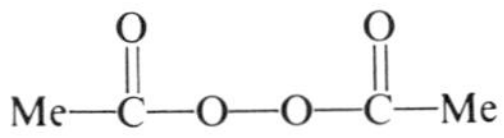

Accordingly, the reactions of diacyl peroxides will be discussed in terms of the chemistry of BPO and LPO. However it should be noted that the rates of decomposition of the diacyl peroxides and the rates of β scission of the acyloxy radicals formed are dependent on structure.[6,55]

The rate of β scission of thermally generated benzoyloxy radicals is slow compared to the rate of escape from the solvent cage. Thus, in the presence of a reactive substrate (*e.g.* a monomer) the production of benzoyloxy radicals from BPO (at low conversions of monomer) can be almost 100% efficient.[56] The only significant cage reaction is the reformation of BPO[57] and little phenyl benzoate and/or biphenyl are formed.[18,58]

In contrast, aliphatic acyloxy radicals generally give complete loss of carbon dioxide before cage escape and have at best transient existence.[6,59] These initiators should be considered as sources of alkyl rather than acyloxy radicals. However, ester end groups may still arise in the polymers initiated by LPO[60] through transfer to initiator to which diacyl peroxides are particularly susceptible (Scheme 5).

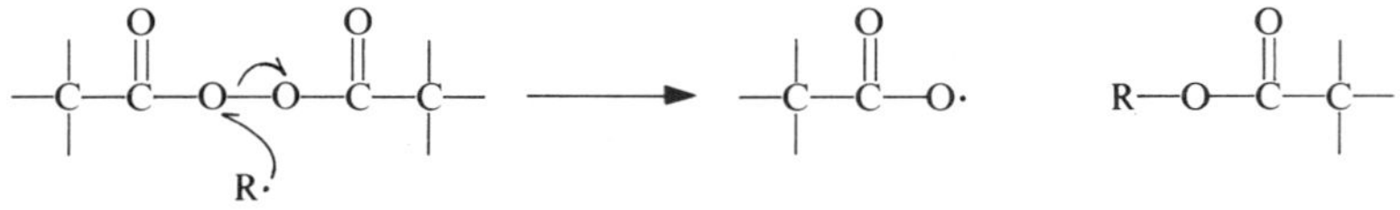

Scheme 5

A further consequence of the rapidity of β scission is that the production of radicals from aliphatic diacyl peroxides is <100% efficient. Significant yields of esters and alkanes are formed by cage recombination.

Photochemically generated benzoyloxy radicals apparently undergo more rapid loss of carbon dioxide than those produced thermally.[61] It has been suggested that during the photodecomposition of BPO, β scission occurs in concert with O–O bond rupture.[62] As a result the proportion of phenyl radicals produced is much higher and initiator efficiency is substantially <100% due to formation of phenyl benzoate and biphenyl in the solvent cage.

Non-radical decomposition to an ion pair either by direct heterolysis or by interaction with a nucleophile is also possible. However, initial reaction in this way does not preclude the eventual formation of radicals from the products by what may be written as an overall redox initiation. A well-known example is the decomposition of BPO in the presence of amines (see Volume 3, Chapter 9). Certain monomers, for example N-vinylcarbazole and N-vinylimidazole, also induce the decomposition of BPO.[8,63–65]

The rates of decomposition of the diacyl peroxides show marked solvent dependence.[55] This is largely attributable to the equally marked susceptibility to induced decomposition. It is therefore not surprising that transfer to initiator (Scheme 5) is a major complication in polymerizations initiated by BPO[56,66] and other diacyl peroxides. It has been shown[32,56] that during S polymerization initiated by 0.1 M BPO at 60 °C, as many as 75% of chains are terminated by transfer to initiator or primary radical termination (see Section 8.6.2) between 30% and 75% conversion. This accounts for the polymer having a much narrower molecular weight distribution than that prepared with AIBN initiation under similar conditions.[32]

8.3.2 Dialkyl Peroxydicarbonates

Dialkyl peroxydicarbonates (12) have been reported as low temperature sources of alkoxy radicals.[67] However, it is established that for primary and secondary esters the rate of loss of carbon dioxide is slow[55,68,69] compared to the rate of addition to most monomers or reaction with other substrates. They should, therefore, be considered a source of alkoxycarbonyloxy radicals (Scheme 6). The uses of peroxydicarbonates as initiators of polymerization have been reviewed.[70,71]

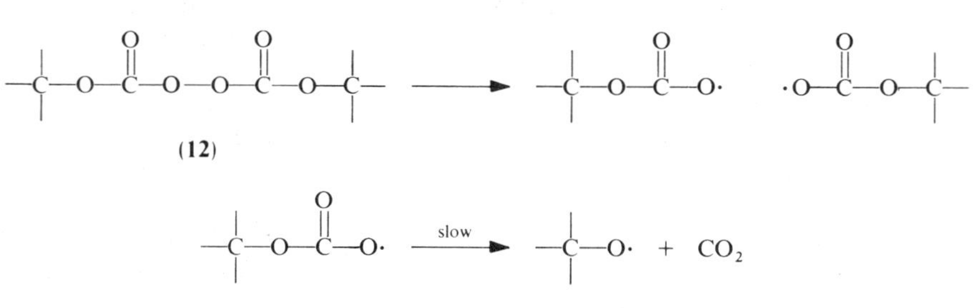

(12)

Scheme 6

The most common peroxydicarbonates are the diisopropyl (18) and dicyclohexyl esters (19). These initiators have the disadvantage of being extremely susceptible to induced decomposition and as a consequence their rates of decomposition show pronounced dependence on the nature of the reaction medium and their concentration.[1] Induced decomposition may involve a mechanism analogous to that described for diacyl peroxides (Scheme 5).[55] A more important pathway for primary and secondary peroxydicarbonates involves initial abstraction of an α hydrogen (Scheme 7).[72]

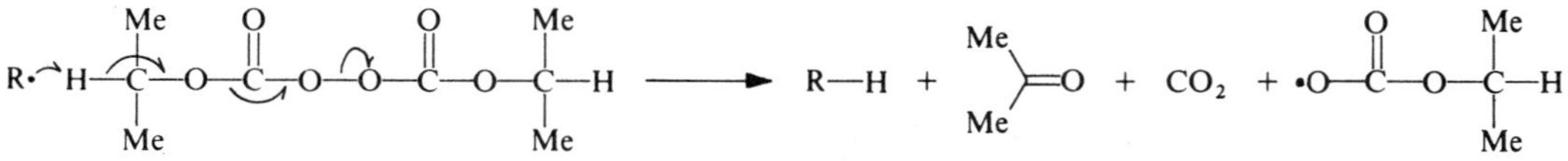

(18)

Scheme 7

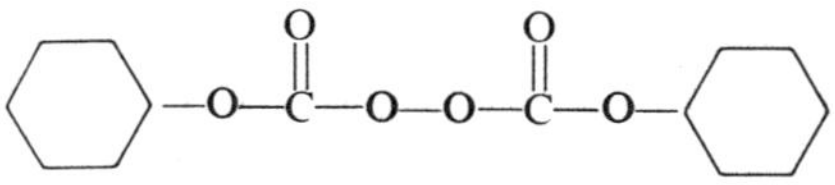

(19)

8.3.3 Peresters

The syntheses and chemistry of peresters (13) have been reviewed by Singer.[73] They are sources of alkoxy and acyloxy radicals (Scheme 8). Those used as initiators are usually *t*-butyl esters, *e.g.* (20) and (21).

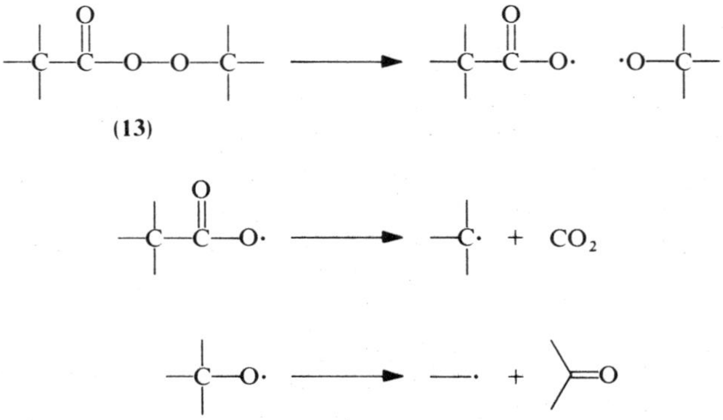

(13)

Scheme 8

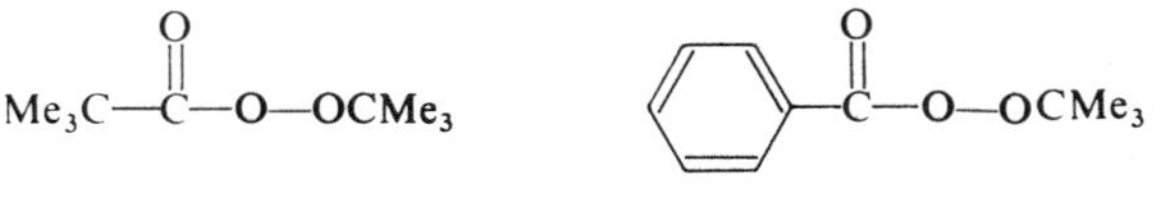

(20)   (21)

The rates of decomposition and β scission are very much dependent on the perester structure. However, with few exceptions (*e.g.* cyclopropanoyloxy radical) the aliphatic acyloxy radicals do not have sufficient lifetime to initiate polymerization or be trapped by efficient scavengers. For certain aliphatic peresters, as with the corresponding diacyl peroxides, there is evidence for concerted two-bond cleavage producing an alkoxy and an alkyl radical directly.[6,44]

Peresters of tertiary alcohols are generally less susceptible to induced decomposition than the corresponding diacyl peroxides (see above). The most common peresters are the cumyl and *t*-butyl esters.

Esters of peroxalic acid are sources of alkoxy radicals (Scheme 9).[74] Decomposition is proposed to take place by concerted three-bond cleavage. Di-*t*-butyl peroxalate (22) is widely used as a clean, low temperature source of *t*-butoxy radicals in the laboratory.

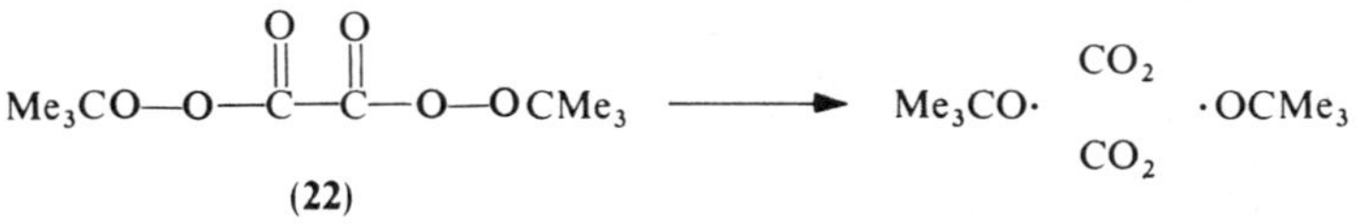

(22)

Scheme 9

Peresters are seldom used as photoinitiators since photodecomposition requires light of 250–300 nm, a region where many monomers also absorb. This situation may be improved by incorporating a suitable chromophore into the molecule (*e.g.* perester (23) has $\lambda_{max} = 366$ nm and ϕ near unity[75]), or through the use of sensitizers.

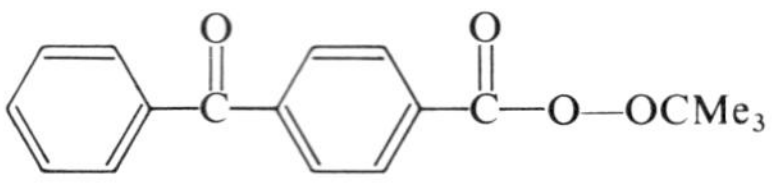

(23)

8.3.4 Alkyl Hydroperoxides

Alkyl hydroperoxides (14) are high temperature sources of alkoxy and hydroxy radicals.[76] They are more commonly used as one component of a redox system (see Volume 3, Chapter 9). The most common initiators of this class are cumyl (24) and *t*-butyl hydroperoxides (25) and a range of ketone peroxyacetals (*e.g.* 26).

The hydroperoxy hydrogen may be readily abstracted and these initiators are efficient transfer agents (Scheme 10).[76] The initiator efficiency is often low and polymers prepared with a hydroperoxide initiator may possess a proportion of potentially labile alkylperoxy end groups.

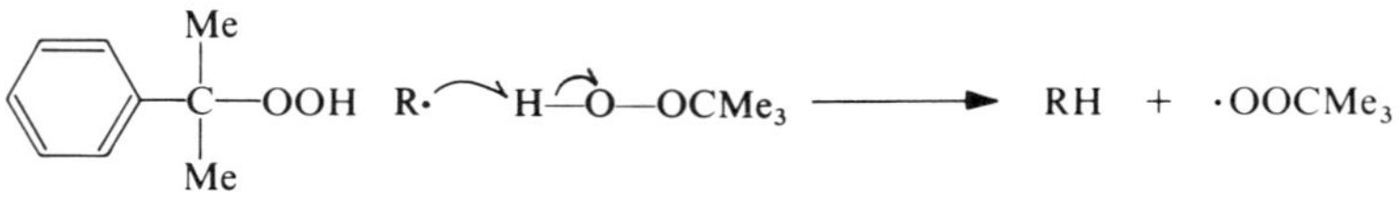

(24)   (25)

Scheme 10

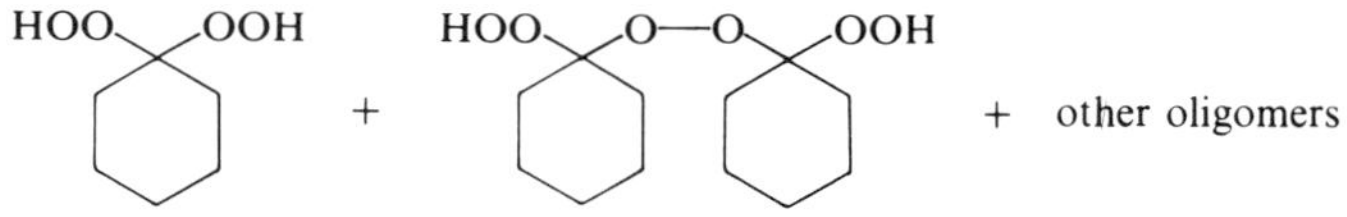

(26)

8.3.5 Dialkyl Peroxides

Dialkyl peroxides (15) are employed as high temperature sources of alkoxy radicals. Those most often encountered are dicumyl (27) and di-*t*-butyl peroxides (28), sources of cumyloxy and *t*-butoxy radicals respectively. The chemistry of tertiary dialkyl peroxides is generally not complicated by induced or ionic decomposition mechanisms. However, primary and secondary dialkyl peroxides may degrade by a variety of induced and non-radical decomposition mechanisms and are seldom used.

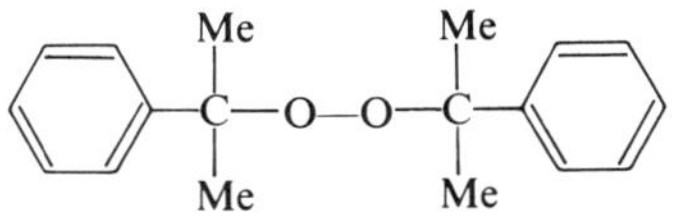

(27)   (28)

8.3.6 Inorganic Peroxides

Inorganic peroxides (hydrogen peroxide, persulfate (16), see Scheme 11, peroxydiphosphate) have limited solubility in organic media but are useful initiators in aqueous media or in heterogeneous systems (*e.g.* in emulsion). These materials are also often used as a component in redox initiation systems (see Chapter 9).

(16)

Scheme 11

8.4 MULTIFUNCTIONAL INITIATORS

These initiators contain two or more radical-generating functions within the one molecule. They are used to achieve higher molecular weight polymers and higher degrees of conversion, and in the production of block and graft copolymers.[77-79]

There are two main types of multifunctional initiator. The first, represented by the azo peroxide (29),[80] include low molecular weight compounds of well-defined structure, typically containing three thermolabile centres.[81] The second type includes polymeric azo or peroxy compounds in which the radical-generating functions may be side chains (*e.g.* 30) or part of the backbone (*e.g.* 31).[82-84] The applications of polymeric azo compounds have recently been reviewed by Nuyken and Weidner.[85,86]

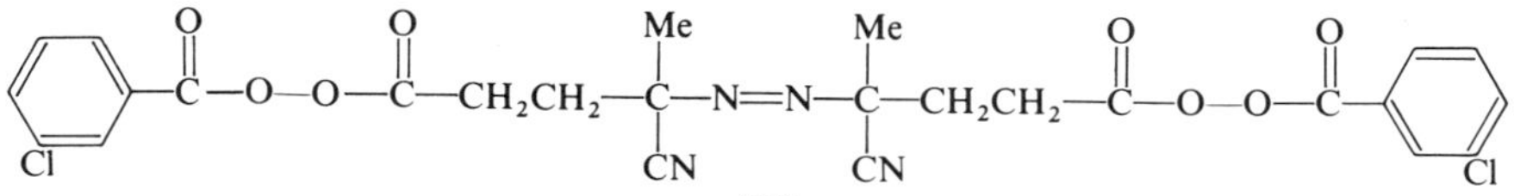

(29)

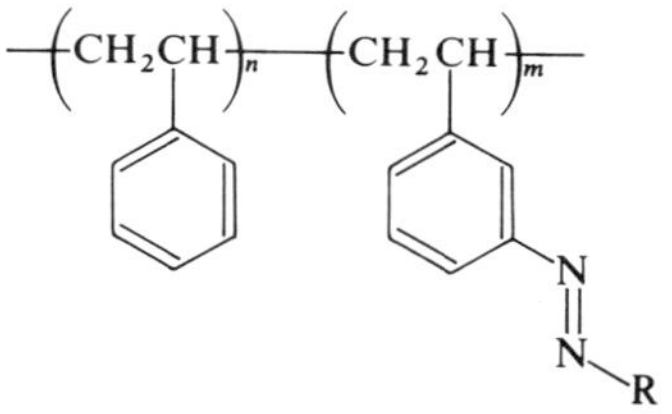

(30) 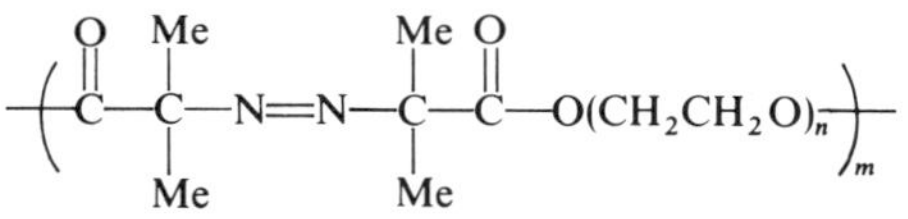   (31)

One further type of multifunctional initiator deserves mention. These are α-hydroperoxydiazenes (*e.g.* 32)[87] and derived peresters.[88] These initiators are good low temperature sources of alkyl and hydroxy or acyloxy radicals (Scheme 12) and intermediate species have at best transient existence. The α-hydroperoxydiazenes, like other hydroperoxides (see Section 8.3.4), are susceptible to induced decomposition.[88]

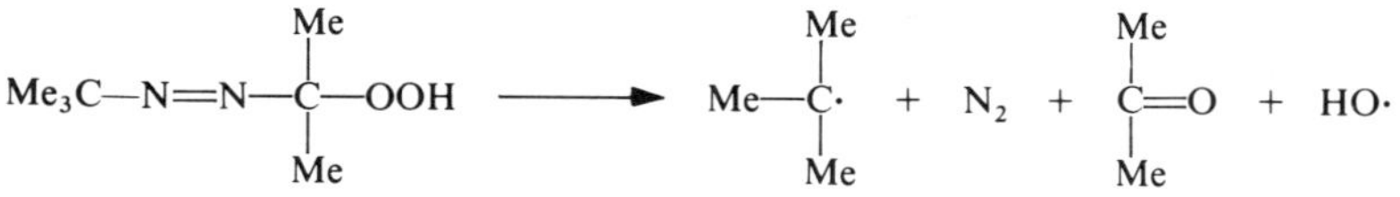

(32)

Scheme 12

8.5 OXYGEN-CENTRED RADICALS

This section details the chemistry of oxygen-centred radicals generated during the initiation process. These species are the first-formed species in the homolysis of peroxides and hyponitrites. An excellent compilation of absolute and relative rate data for reactions of oxygen-centred radicals[89] covering the literature through 1982 is available. Accordingly, in this section we concentrate on the specificity of the reactions of oxygen-centred radicals with monomers and other components of the polymerization medium.

8.5.1 Alkoxy Radicals

Alkoxy radicals are frequently encountered initiating species, both in polymerizations and as the subject of laboratory studies. Most studies have concentrated on *t*-butoxy and to a lesser extent cumyloxy radicals and relatively little attention has been paid to other alkoxy radicals.

8.5.1.1 *t-Butoxy radicals*

The reactions of *t*-butoxy radicals are amongst the most studied of all radical processes. Pioneering studies by Walling,[90] Ingold[91] and others[92] have shown that *t*-butoxy radicals have a marked propensity for hydrogen atom abstraction and established the reactivities of common solvents and other substrates.

Kunitaki and Murakami[93] and Sato and Otsu[94] employed ESR and spin trapping (see Section 8.8.1.1) to show that for the case of monomers possessing an α methyl substituent, abstraction from that group was competitive with addition to the double bond. Subsequent investigations by Encina *et al.*[95] and Rizzardo *et al.*[96–102] have shown that the interaction of *t*-butoxy radicals with monomers bearing sp^3 hydrogens inevitably produces a mixture of initiating species (*e.g.* Schemes 13 and 14). Indeed, with simple alkenes[101] and vinyl ethers[103] abstraction predominates over addition.

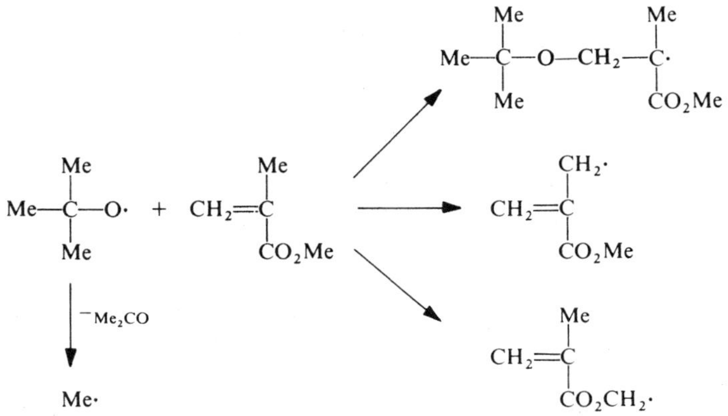

Scheme 13

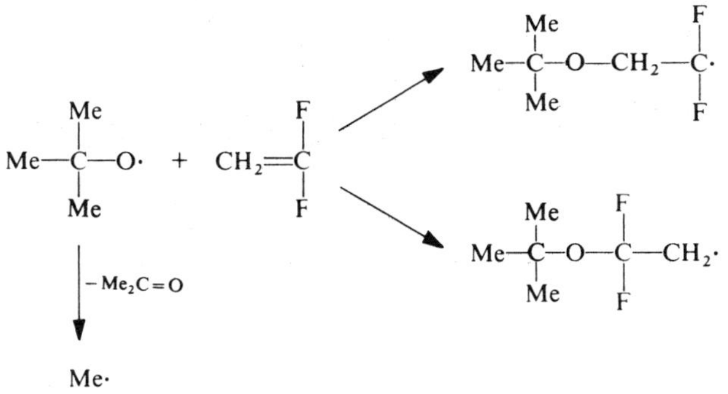

Scheme 14

This behaviour must be taken into account in studies of polymerization mechanisms which rely on end group determination.[104] Clearly, those chains initiated by abstraction products will not contain an initiator residue and some data in the literature will need to be reinterpreted in this light.[105]

Abstraction from monomer also gives rise to unsaturated initiating species and consequently to potentially reactive double bonds at the chain ends. These may copolymerize, leading to the formation of grafts or crosslinks, particularly if the polymerization is taken to high conversion.

The relative importance of hydrogen abstraction and double bond addition is a function of the particular reaction conditions employed. Higher temperatures have been shown to favour abstraction over addition for the reaction of *t*-butoxy radicals with α-methylstyrene.[95] However, the opposite trend was observed with isobutylene.[106,107] The ratio of hydrogen abstraction *vs.* double bond addition has also been shown to be solvent dependent,[108,109] the former reaction being of greater importance in more polar solvents. However, this propensity of *t*-butoxy radicals for hydrogen abstraction also has other, perhaps more important, ramifications for polymerizations carried out in solution.[110] Two solvents which see widespread usage in industry, toluene and 2-butanone, are excellent hydrogen· donors and polymers prepared in these solvents can be anticipated to have predominantly solvent-derived end groups.[2,111,112] Even benzene, a solvent sometimes used in laboratory studies because of its low reactivity toward radicals, affords a measurable yield of products from hydrogen abstraction.[111-113] At high conversions the reactivity of the preformed polymer must also be considered.[114-116]

As well as indicating a need for care in solvent selection, this behaviour offers a potential means of controlling the nature of the end groups in polymerizations initiated by *t*-butoxy radicals. The solvent-derived radicals (usually alkyl) are likely to be more selective in their reactions with monomers (carbon-centred radicals react principally by addition, see Section 8.6). Thus by appropriate choice of solvent, complications stemming from the formation of the unsaturated end groups (*e.g.* in MMA polymerizations, see above) may be eliminated.

Studies of the relative reactivity of *t*-butoxy radicals, in abstraction from substituted toluenes[117,118] or addition to substituted styrenes,[119] point to their being slightly electrophilic. However, Sato and Otsu[94] found that the order of reactivity of *t*-butoxy radicals towards a series of monomers was different from that of more electrophilic species (benzoyloxy radicals). He concluded that, for reactions of *t*-butoxy radicals, product radical stability is important in determining reactivity. Cuthbertson *et al.*[102] examined the reactions of *t*-butoxy radicals towards fluoroalkenes and found a pattern of reactivities more characteristic of a nucleophilic species. The factors which determine reactivity of radicals towards fluoroalkenes have recently been discussed by Arnaud *et al.*[120]

When considered in relation to other oxygen-centred radicals (*e.g.* hydroxy, benzoyloxy), *t*-butoxy and other alkoxy radicals (see below) show a high degree of regiospecificity in adding to carbon–carbon double bonds (see Table 1). Significant amounts of head addition are seen only with the haloalkenes (*e.g.* Scheme 14),[102,106] simple alkenes[101] and vinyl acetate.[98]

The case of vinylidene fluoride (and other fluoroalkenes) deserves special mention.[102,106] *t*-Butoxy radicals afford predominantly head addition with this monomer (Scheme 14). The influence of steric factors is expected to be small. Polar factors and considerations of product radical stability suggest that tail addition should be preferred. An explanation for the apparently anomalous behaviour may lie with the relative strengths of the bonds being formed in head *vs.* tail addition (CF_2-O *vs.* CH_2-O).[119] α-Fluorine substitution is known to stabilize a C–O bond. The C–O bond dissociation energies in CF_3-O-CF_3 and CF_3-OH are greater by 22 and 18 kcal mol^{-1} (1 cal = 4.184 J) respectively than those in CH_3-O-CH_3 and CH_3-OH.[121]

t-Butoxy radicals also undergo unimolecular fragmentation producing acetone and methyl radicals (see Section 8.6.1). Under typical reaction conditions (60–120 °C) β scission is competitive with most bimolecular reactions involving *t*-butoxy radicals.[111] The precise amount of β scission observed is dependent on the reactivity of the particular monomer(s) and the reaction conditions and is of greater significance in more polar solvents and at higher temperatures.

8.5.1.2 *Cumyloxy radicals*

The behaviour of cumyloxy radicals generally mirrors that of *t*-butoxy radicals, the principal difference being that the rate of β scission (affording acetophenone and methyl radicals) is significantly greater.[122,123] Thus the fraction of methyl radical initiation of S or MMA polymerization at 60 °C is greater by a factor of six.[47]

8.5.1.3 *Primary and secondary alkoxy radicals*

Relatively few studies have been concerned with the reactions of primary and secondary alkoxy radicals (isopropoxy, methoxy, *etc.*) with monomers. When considered in relation to *t*-butoxy or cumyloxy radicals, the most notable feature of the chemistry of primary and secondary alkoxy radicals is a reduced tendency to abstract hydrogen or to undergo β scission.[106,124-126]

Table 1 Rates and Selectivities for Reaction of Carbon-centred and Oxygen-centred Radicals

| Radical | Reactivity Relative to Styrene | | | | | | | | k | Ref. |
	MS^a	S^b	VAC^c	MMA^d	MA^c	MAN^c	AN^c	Toluene	$(1\,mol^{-1}\,s^{-1})$	
Bu^{t}O•^{e}	1.3 (83/0/17/0)	1.0 (100/0/0)	0.06 (80/15/5)^i	0.28 (64/0/32/4)	0.06 (84/2/15)	0.12 (73/0/27)^i	0.05 (100/00/)^i	0.19^i	1×10^6	98–100, 108, 109
Bu^{t}O•^{f}	0.23 (>99)	1.0	0.009	0.09 (61/0/39/—)	0.05	0.08 (68/0/32)	—	0.02	—	94
Bu^{t}O•^{g}	1.8 (90/—/10/—)	1.0 (100/—/<1)	0.3 (80/—/20)	0.3 (60/—/40/—)	0.7 (58/—/42)	—	—	—	—	95
PhMe$_2$CO•^{i}	—	1.0 (100/0/0)	—	0.1 (71/0/26/3)	—	—	—	—	—	97
HO•^{j}	1.2 (83/3/5/9)	1.0 (84/8/8)	—	0.63 (87/6/5/2)	0.34 (80/17/3)	0.96	0.27	—	1.19×10^{10}	149
HO•^{k}	—	—	—	1.0	—	—	—	—	—	147
PhCO$_2$•^{e}	—	1.0 (80/6/14)	0.26 (76/24/0)	0.11 (93/7/<1/0)	0.02 (83/17/0)	—	0.02 (99/1/0)	—	7.4×10^3	99, 130, 131
PhCO$_2$•^{l}	—	1.0	0.36	0.12	0.05	—	<0.05	—	1.0×10^9	129
SO$_4^-$•^{k}	—	—	—	1.0	—	0.38	0.08	—	—	147
Me•^{m}	0.86	1.0	0.04	1.8	1.3	2.7	2.2	0.0004	—	170, 171
n-C$_6$H$_{11}$•^{n}	0.60	1.0	—	—	3.47	7.53	—	—	1.5×10^5	175
c-C$_6$H$_{11}$•^{o}	0.93	1.0	—	5.0 (100/0/0/0)	6.7 (99.8/0.2/0)	13.3	24.0	—	—	218
Bu^t•^{p}	—	1.0 (>95)	0.03 (>95)	>7.6	>7.6	—	7.6 (>95/)	0.00005	1.32×10^5	176, 177
Me$_2$CCN•^{i}	—	1.0 (100/0/0)	40	0.56	0.03^h	—	—	0.0	—	181, 188
Ph•^{e}	—	1.0 (100/0/1)	0.3 (>95)	1.2 (100/0/<1/0)	0.8 (97/3/0)	—	(100/0)	0.01	—	100, 111
Ph•^{q}	—	1.0	—	1.6	—	—	—	0.01	1.1×10^8	199
Ph•^{r}	3.57	1.0	0.38	3.17	2.93	3.58	3.13	—	—	198
Ph•^{l}	1.24	1.0	0.23	1.78	0.78	2.46	—	—	—	197
Ph•^{s}	—	1.0	≥0.08	1.7	—	—	0.8	—	—	196
Ph•^{t}	—	1.0	—	1.24	0.92	1.35	—	—	—	195

^{a} Tail addition/head addition/abstraction/aromatic substitution. ^{b} Tail addition/head addition/abstraction. ^{c} Tail addition/head addition/aromatic substitution. ^{d} Tail addition/head addition/abstraction. ^{e} Tail addition/head addition/abstraction/abstraction α methyl/abstraction ester methyl. ^{f} 60 °C, acetone. ^{g} 25 °C, p-xylene. ^{g} 20 °C, benzene. ^{h} Ethyl acrylate. ^{i} 60 °C, bulk. ^{j} 60 °C, cyclohexane. ^{k} 25 °C, water. ^{l} 60 °C, benzene. ^{m} 5-Hexenyl, 69 °C, acetonitrile/acetic acid. ^{n} 25 °C, acetonitrile/acetic acid. ^{o} 20 °C, methylene chloride. ^{p} 27 °C, 2-propanol. ^{q} 25 °C, freon 113. ^{r} 60 °C, CCl$_4$. ^{s} 60 °C, benzene/DMF. ^{t} 25 °C water/acetone.

8.5.2 Acyloxy Radicals

Only aromatic acyloxy (from diacyl peroxides and peresters) and alkoxycarbonyloxy (from peroxydicarbonates) radicals will be considered. Aliphatic acyloxy radicals have at best transient existence and do not have a sufficient lifetime to enable direct reaction with monomers.

8.5.2.1 *Benzoyloxy radicals*

The reactions of benzoyloxy radicals with simple alkenes was investigated by Kochi[127,128] and the relative reactivities of monomers have been established by Bevington *et al.*[129] More recently, the selectivity of the reactions of benzoyloxy radicals with common monomers has been extensively studied by Moad *et al.*[56,99,100,112,130−132] Benzoyloxy radicals show remarkably poor regio-specificity when adding to carbon–carbon double bonds and invariably give both head and tail addition.[99,100,130,133] They also display a marked propensity for aromatic substitution (Scheme 15).[99] On the other hand, in relation to alkoxy radicals they show little tendency to abstract hydrogen[100] (compare data in Table 1).

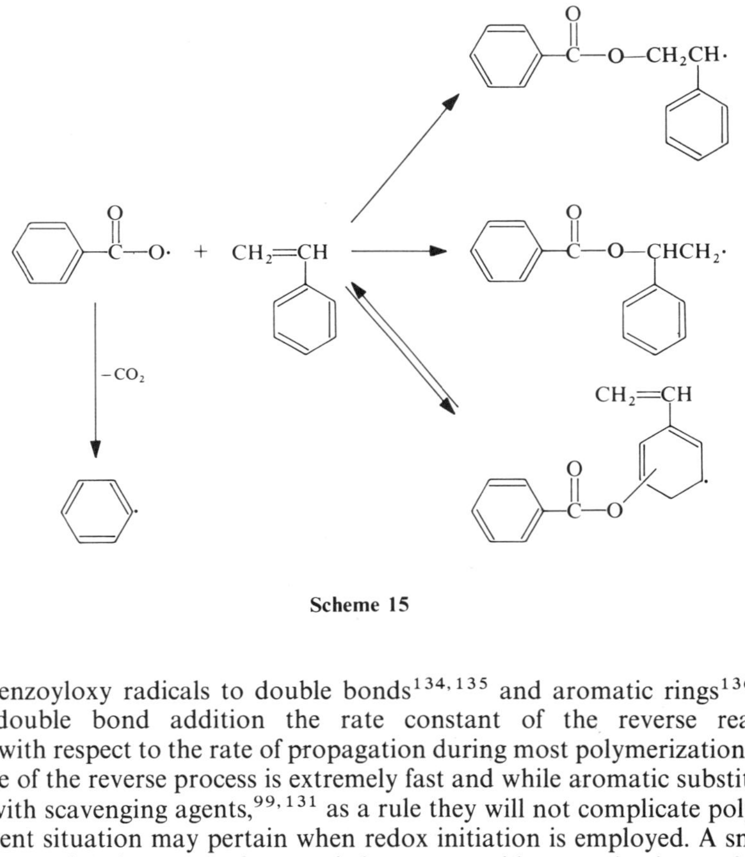

Scheme 15

Additions of benzoyloxy radicals to double bonds[134,135] and aromatic rings[136] are reversible reactions. For double bond addition the rate constant of the reverse reaction is slow ($k = 10^2$–$10^3 \, s^{-1}$) with respect to the rate of propagation during most polymerizations. For aromatic substrates the rate of the reverse process is extremely fast and while aromatic substitution products may be trapped with scavenging agents,[99,131] as a rule they will not complicate polymerizations.[56] However, a different situation may pertain when redox initiation is employed. A small proportion (with relation to normal end groups) of aromatic benzoate residues can be detected in PS prepared with benzoyl peroxide, though it is likely that these arise through attack on PS.[56,137]

The rate of β scission of benzoyloxy radicals is such that in most polymerizations initiated by these radicals both phenyl and benzoyloxy end groups will be formed. While a number of studies have attempted to derive the rate of this process a definitive value is not yet available.[100] The rate of β scission is dependent on whether the radical is generated photochemically or thermally[62] but otherwise appears to be comparatively insensitive to reaction conditions and, in particular, solvent changes. This is evidenced by the finding that similar relative reactivities are obtained from direct competition experiments[132] as from studies on individual monomers when β scission is used as a clock reaction.[129,130] It follows that a reliable value for the rate constant for β scission would enable the absolute rates of initiation by benzoyloxy radical to be estimated and the importance of primary radical termination to be assessed.[138,139]

8.5.2.2 *Alkoxycarbonyloxy radicals*

The chemistry of these radicals parallels that of the aroyloxy radicals (*e.g.* benzoyloxy, see above). The isopropoxycarbonyloxy radical undergoes a facile reaction with aromatic substrates (*e.g.* toluene) by reversible aromatic substitution.[140,141] Thus reaction with S affords ring substitution as well as the expected double bond addition.[142] The isopropoxycarbonyloxy also shows little tendency to abstract hydrogen.[127,143] For example with MMA, hydrogen abstraction, while observed, is a minor pathway ($\leq 1\%$).[143]

In the above systems, no products attributable to the reactions of isopropoxy radicals are observed,[127,143] indicating that the rate of β scission is slow relative to addition to monomers or other substrates.[69]

8.5.3 Hydroxy Radicals

The transient radicals produced in reactions of hydroxy radicals with vinyl monomers in aqueous solution have been determined directly by EPR[144–146] or UV spectroscopy.[147,148] These studies have demonstrated that hydroxy radical reactions with monomers and other species proceed at or near the diffusion-controlled limit[147] and show a lack of specificity.[145,146,148]

More recently, Grant *et al.*[150] have applied a radical-trapping technique (see Section 8.8.1.4) to carry out a quantitative analysis of the reactions of hydroxy radicals with a range of vinyl and α-methylvinyl monomers in organic media. Their results are summarized in Table 1. Even though resonance and steric factors should combine to favour 'normal' tail addition, substantial yields of head addition, aromatic substitution and abstraction are obtained. However, it is notable that the relative amount of abstraction *vs.* addition is significantly less than is obtained with *t*-butoxy radicals under similar conditions (see Table 1).

Trends in the relative reactivity of monomers are explicable in terms of the electrophilicity of the hydroxy radical.[150]

8.5.4 Sulfate Radical Anion

The sulfate radical anion may interact with a carbon–carbon double bond in two ways (Scheme 16).[150] It may add to the double bond (pathway a) or it may accept an electron from the monomer to generate a radical cation, which may in principle occur either by direct electron transfer (pathway b) or by an addition elimination sequence (pathway a then c). The radical cation may, in principle, propagate by cationic and/or radical mechanisms. However, in aqueous media the radical cation is rapidly hydrated to give a hydroxyl adduct. Radical cation formation is facilitated by acidic conditions.[150]

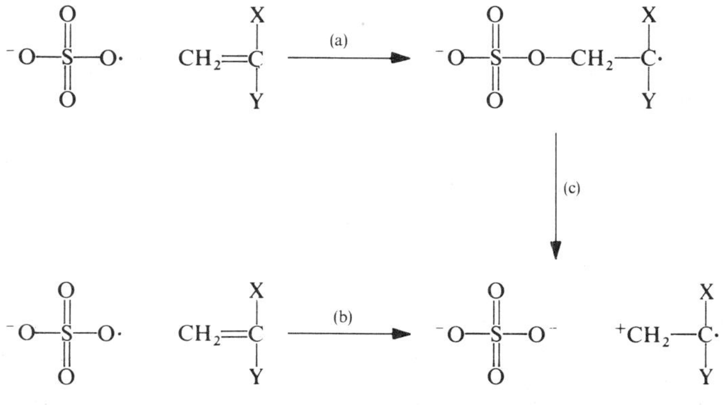

Scheme 16

There is some controversy surrounding the exact mechanism of initiation of S polymerization by the sulfate radical anion. Experiments conducted by Ledwith and Russell[151] and more recently by Citterio *et al.*[152,153] have shown no evidence of an intermediate sulfate adduct and these workers have proposed that reactions of the sulfate radical anion with electron rich alkenes and S derivatives proceed by pathway (b) over a wide range of pH and reaction conditions. However, other workers[154] have rationalized similar data by allowing the initial formation of a sulfate adduct. Attempts to detect an intermediate in the reaction of sulfate radical anion with S by UV[155] or with cyclohexene by ESR and conductivity measurements[156] clearly point to the addition being the major pathway. Moreover, PS formed with persulfate initiation can, depending on reaction conditions, possess a high proportion of sulfate end groups.[157-160] Therefore, while it is not possible at this stage to exclude the possibility that single electron transfer might be involved (pathway b) in some circumstances, the bulk of available evidence is consistent with the view that the initiation of S polymerization involves the initial formation of a sulfate adduct (pathway a) and that where a radical cation is formed, it is produced by pathway (c).

It is generally accepted that reaction with electron deficient monomers involves initial addition of the sulfate radical anion to the monomer. Reactions of the sulfate radical anion with acrylic acid derivatives have been shown to give rise to the sulfate adduct (pathway a) under neutral or basic conditions, but under acidic conditions afford the radical cation, probably by an addition elimination process (pathway c). The sulfate radical anion shows little tendency to abstract hydrogen from methyl-substituted acrylic acids.[150]

It has also been suggested that the sulfate radical anion may be converted to the hydroxyl radical in aqueous solution. The major evidence for this is the formation of a proportion of hydroxy end groups in some polymerizations. However, the hydrolysis of the sulfate radical anion is slow $(k \approx 10^7 \, M^{-1} \, s^{-1})$ compared with its rate of reaction with most monomers $(k = 10^8 - 10^9 \, M^{-1} \, s^{-1})$[147] and should only compete in circumstances where the monomer concentration is very low. The formation of hydroxy end groups in polymerizations initiated by sulfate radical anions can be accounted for by the hydration of an intermediate cation radical or by the hydrolysis of an initially formed sulfate adduct either during the polymerization or subsequently. The hydroxy radical and sulfate radical anion show quite different selectivity in their reactions with unsaturated substrates. Most notably, the sulfate radical anion has a somewhat lower propensity for abstracting hydrogen than does the hydroxyl radical.

8.6 CARBON-CENTRED RADICALS

Carbon-centred radicals may be produced directly from azo compounds or indirectly from peroxides as scission products of the initially formed acyloxy or alkoxy radicals. Alkyl radicals, when considered in relation to oxygen-centred radicals, show a high degree of regiospecificity in their reactions with vinyl and α-methylvinyl monomers.

Recent reviews by Giese,[161] Tedder and Walton,[162-165] Ruchardt[166] and Beckwith[167] have discussed the various factors which determine the rate and selectivity of addition and substitution processes involving carbon-centred radicals while Lorand[168] and Asmus and Bonifacic[169] have provided a compilation of rate data covering the literature through 1982.

8.6.1 Simple Alkyl Radicals

The reactions of methyl radicals with monomers, solvents, inhibitors and other materials have been extensively studied principally by Szwarc et al.[170,171] (solution) and Pryor et al.[172] (gas phase) from the point of view of establishing the relative reactivities of these substrates.

The rates of hydrogen abstraction and aromatic substitution by methyl radicals are small with respect to those for addition to the double bond of all common monomers with the exception of simple alkenes.[168,169] Nevertheless, since these reactions do occur they must be borne in mind for polymerizations carried to high conversion.

The interaction of representative primary, secondary and tertiary alkyl radicals (n-hexyl, cyclohexyl, t-butyl) with monomers (acrylates and S) and other alkenic substrates has been extensively studied by Giese et al.[161] These species show a high degree of regiospecificity in their reactions with acrylates (see Table 1). Significant amounts of head addition are however observed with β-alkylacrylates[173] and the proportion can be correlated with the steric size of the β substituent.

Other studies aimed at establishing absolute rate constants for the reactions of alkyl radicals with monomers have been reported by Minisci et al. (n-heptyl,[174] isopropyl,[174] 5-hexenyl[175]) and Fischer[176-177] (t-butyl). Some of their data are reported in Table 1.

Geise and Meister[178] established the relative nucleophilicity of primary, secondary and tertiary alkyl radicals in addition reactions by determining their reactivity towards a series of substituted styrenes (radical, ρ^+): $Me_3C\cdot$, 1.1; $c\text{-}C_6H_{11}\cdot$, 0.68; $n\text{-}C_6H_{13}\cdot$, 0.45.

8.6.2 Cyanoisopropyl Radicals

The resonance-stabilized cyanoisopropyl radicals can in principle react either at carbon or at nitrogen (Scheme 17). These species may undergo self-reaction (Scheme 2), or react with other radical species (see Volume 3, Chapter 11) or organometallic reagents[179] to produce a ketenimine derivative. However, as yet no examples of a ketenimine being derived in an addition reaction with either monomers[27,32] or spin traps[180] have been reported.

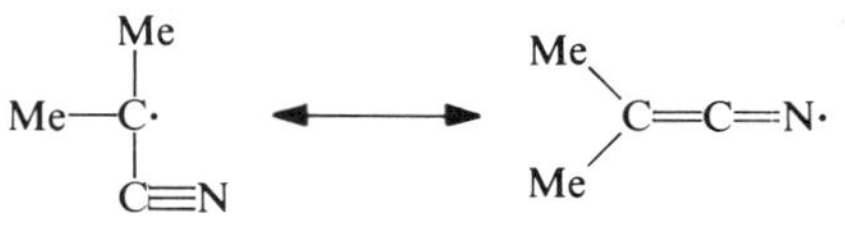

Scheme 17

Cyanoisopropyl radicals are comparatively selective in their reaction with monomers (*e.g.* S, MMA) and generally afford exclusively tail addition. However, Bevington *et al.*[181,182] have recently shown by end group determination that cyanoisopropyl radicals give *ca.* 10% head addition with vinyl acetate (VAC) at 60 °C and that the proportion of head addition increases with increasing temperature. It should be noted that similar end groups are formed by primary radical termination as are formed by head addition and this complicates an accurate assessment of the importance of these processes.

Cyanoisopropyl radicals also have little tendency to abstract hydrogen from monomer, solvent or polymer.[32] Note however that cyanoisopropyl radicals, like other carbon-centred radicals, react with oxygen at diffusion-controlled rates.[183] In incompletely degassed media, abstraction products, peroxide linkages and other defect structures may arise through the intermediacy of a peroxy radical (Scheme 18).[30]

Scheme 18

Recent studies by Bevington *et al.*[181,182,184-189] have established the relative reactivity of a variety of monomers towards cyanoisopropyl radicals and clearly demonstrate the electrophilic character of this species. However, there are as yet no absolute rate data for the reactions of cyanoisopropyl radicals with substrates. A number of reports (see for example refs. 190–193) suggest that primary radical termination is important during polymerizations initiated by AIBN. If this is the case then the rate constants for addition to monomers[191] must be substantially lower than those for other alkyl radicals (see Table 1). However, for the case of S polymerization, it has recently been demonstrated[32] that the process has little importance except where very high rates of radical generation are employed.[192] It is worth noting also that model studies show that primary radical termination between cyanoisopropyl radicals and polystyryl radicals involves a significant amount of disproportionation (see Section 11.4). This result is significant for those measuring the extent of primary radical termination by end group determination.[190]

8.6.3 Aryl Radicals

Relative reactivities of monomers and other substrates towards phenyl radicals have been obtained by a variety of methods and representative data are reported in Table 1.[100,111,193-196] Direct comparison of the numbers between studies is complicated since the reaction conditions differ. Scaiano and Stewart[197] have determined the absolute rate constants for the attack of phenyl radicals on a variety of substrates including MMA ($k(25\,°C) = 1.8 \times 10^8$), S ($k(25\,°C) = 1.1 \times 10^8$) and BPO ($k(25\,°C) = 2.1 \times 10^7$).

Aryl radicals may attack aromatic rings (*e.g.* S[99]) or abstract hydrogen (*e.g.* MMA[100]) in competition with addition to a monomer double bond. However for the cases studied the selectivity shown for double bond addition is high and other products account for $\leq 1\%$ of the total (see Table 1).[99,100] The degree of specificity for tail *vs.* head addition is also very high. Significant head addition is seen only where tail addition is retarded by steric factors (*e.g.* methyl crotonate,[100] β-substituted methyl vinyl ketones[200]).

8.7 REACTIVITY OF MONOMERS TOWARDS RADICALS

A knowledge of the reactivity of radicals towards monomers, and the factors which determine such reactivity, is vital to the understanding of polymerization and copolymerization. This section is concerned specifically with the reactivities of monomers towards the initiator-derived radicals; reactivities of monomers and propagating radicals are considered in Chapter 17 of this volume.

Absolute rate constants are important in defining initiator efficiency and calculating the importance of such processes as primary radical termination by computer simulation.[191]

However, for many purposes relative rate constants will suffice and they are generally more reliable than absolute rate constants. The data are used for estimating the nature and distribution of end groups by kinetic simulation.[2] The relative rates of the various initiation and termination processes can also have a dramatic influence on the compositional heterogeneity and molecular weight distribution of low molecular weight copolymers.[201]

A series of rules whereby the course and relative rates of radical addition reactions can be predicted has been devised by Tedder[165] and Giese.[161] These rules indicate that while regiospecificity is mainly determined by steric factors, the rates of addition of radicals to 1-substituted and 1,1-disubstituted alkenes should be determined largely by polar influences. Tedder[165] has indicated that stability of the incipient radical may be of consideration where the substituent has π orbitals capable of delocalizing the free spin (*e.g.* Ph, $-CH=CH_2$; note that many monomers fit into this category). However, it is clear from the data in Table 1 and that presented by Giese[161] that, for the more polar radicals, polar factors are indeed the major influence.

However, while rules may be employed qualitatively to predict overall trends in reactivity, precise values must still be determined by experiment. For the most common initiating species many data are available,[90,168,169] although significant gaps still exist. Furthermore, with radicals which are only slightly electrophilic (or nucleophilic),[202] for example *t*-butoxy radical, it is clear that other than polar factors become important in determining reactivity (see Section 8.5.1.1).

8.8 TECHNIQUES

There are many difficulties associated with examining radical–monomer reactions by direct examination of a polymer sample. Principal among these is the low concentration of the initiator residues. Typically they comprise 0.01–1% of a given sample. Recent advances in instrument sensitivity and new techniques, particularly in the field of NMR spectroscopy, allow initiator residues to be observed in favourable circumstances for polymers of modest molecular weight. However, in general they are essentially undetectable by conventional procedures.

To overcome these problems two basic approaches have been employed. The first involves isolation of the reaction of interest by employing a reagent to trap the first-formed adduct. This involves conducting the polymerization in the presence of an appropriate inhibitor. The second involves labelling a radical such that residues derived from it and incorporated into a polymer can be more readily identified (see also Section 11.5).[17,185,203]

8.8.1 Radical Trapping

Trapping techniques rely on the radical of interest being more reactive towards monomer (or less reactive towards the radical trap) than the monomer-derived radicals involved in subsequent propagation steps. The majority of polymerization inhibitors can be employed with varying degrees of effectiveness. The other essential is that the trapped products be stable under the polymerization conditions. A variety of reagents have been employed.[204] The most widely applied reagents include spin traps, transition metal salts, metal hydrides and nitroxides. The advantages and limitations of these will be considered in the following sections.

8.8.1.1 Spin traps

The term spin trapping is reserved for describing radical-trapping reactions in which the free spin is retained in the trapped product (*i.e.* the product is a radical).[205,206] The reagents most commonly employed are nitrones and nitroso compounds and the technique then involves EPR detection of the *relatively* stable nitroxides formed by the trapping of more transient radicals (Scheme 19). The application of this method to the study of radical polymerization was first described by Chalfont et al.[207] who employed 2-methyl-2-nitrosopropane (33) as a trap for the study of S polymerization initiated by *t*-butoxy radicals. Since that time Kunitake and Murakami,[93] Sato, Otsu *et al.*,[94] Kamachi *et al.*[95] and Bevington *et al.*[180] have applied this trap (33) in the study of a wide range of initiating systems.

$$R\cdot \ + \ O{=}N{-}CMe_3 \ \longrightarrow \ \underset{(33)}{R{-}\overset{\textstyle O\cdot}{\underset{|}{N}}{-}CMe_3}$$

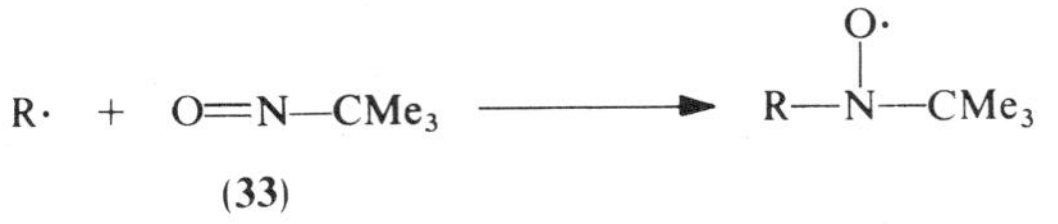

Scheme 19

Initiation has also been studied using phenyl *t*-butyl nitrone (34) as a spin trap.[93,180,209] However, since the radical centre is one carbon removed from the trapped radical (Scheme 20), the EPR spectrum is less sensitive to the nature of that radical and there is consequently greater difficulty in resolving signals and in interpreting the results.[93]

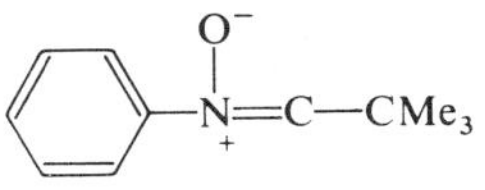

(34)

Scheme 20

The spin-trapping technique has the advantage that it is experimentally simple to carry out (no product isolation is required) and that it is compatible with a wide range of initiating systems. However, there are major limitations in the use of this technique, particularly when quantitative results are required. Not all radicals are trapped at equal rates or with equal efficiency. Moreover, many nitroxides are not stable under the reaction conditions. In particular, they react with radicals at or near diffusion-controlled rates and they can also undergo β scission either to regenerate the trapped radical or to form a new radical. Nitroxide stability is strongly dependent on the nature of the trapped species.

Finally, side reactions involving the trap and the monomer may give rise to products which complicate the interpretation of the EPR spectra. For example, nitroso compounds react with α-methylvinyl monomers by an ene reaction,[143] *t*-butyl radicals are produced by thermal or photochemical decomposition of 2-methyl-2-nitrosopropane and are trapped as di-*t*-butyl nitroxide; other processes are described in recent reviews.[205] Many of these complications can be avoided or allowed for by carrying out appropriate control experiments.

The application of a more thermally and photochemically stable nitroso compound, 2,4,6-tri-*t*-butylnitrosobenzene (35) in the study of polymerization, has been described by Savedoff and Ranby[210] and more recently by Lane and Tabner.[133]

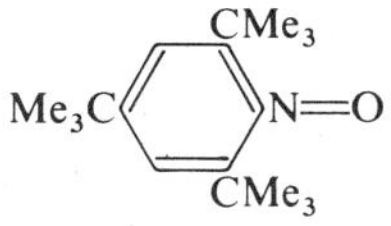

(35)

8.8.1.2 *Transition metal salts*

Many transition metal salts are inhibitors of polymerization. Those used as radical traps are usually in higher oxidation states and react with radicals by ligand transfer or electron transfer to afford products which can be determined by conventional analytical techniques (*e.g.* Scheme 21).[204,211] Transition metal salts include copper(II) (*e.g.* $Cu(OAc)_2$, $CuCl_2$, $Cu(SCN)_2$)[212,174,175] and iron(III) salts (*e.g.* $FeCl_3$).[213] A limitation of these trapping agents is that they show selectivity for nucleophilic radicals (*e.g.* those derived from addition to S). The reduced form of the metal salt may play a role in inducing the decomposition of a peroxide initiator.

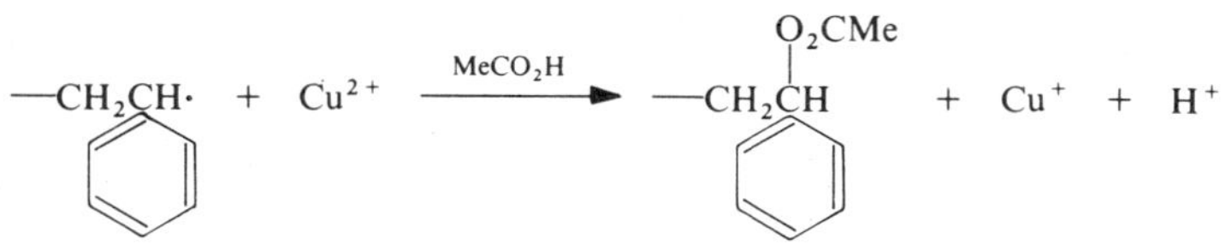

Scheme 21

The reduction of radicals by transition metal salts (*e.g.* Ti^{3+}) may also be employed as a trapping reaction (*e.g.* Scheme 22).[200] These reagents are selective for electrophilic radicals (*e.g.* those derived by tail addition to acrylic monomers or alkyl vinyl ketones).

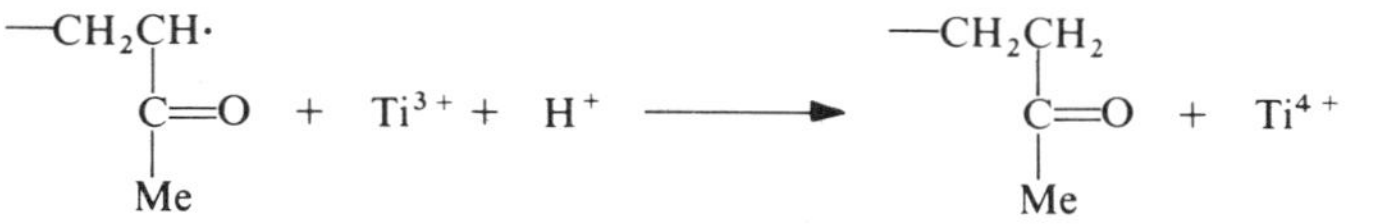

Scheme 22

The dependence of the rate of oxidation/reduction of radicals on radical structure means that the various products from the reaction of an initiating radical with monomers will not all be trapped with equal efficiency and as a consequence complex mixtures can arise.

A further limitation is the low solubility of many transition metal salts in organic media.

8.8.1.3 *Metal hydrides*

A method for studying radical reactions in which alkyl radicals and a mercury(II) hydride radical trap are produced on reduction of an alkyl mercury(II) salt with sodium borohydride was developed by Hill and Whitesides[214] and first applied to the study of initiation[178,215-219] by Giese *et al.* (Scheme 23).[204]

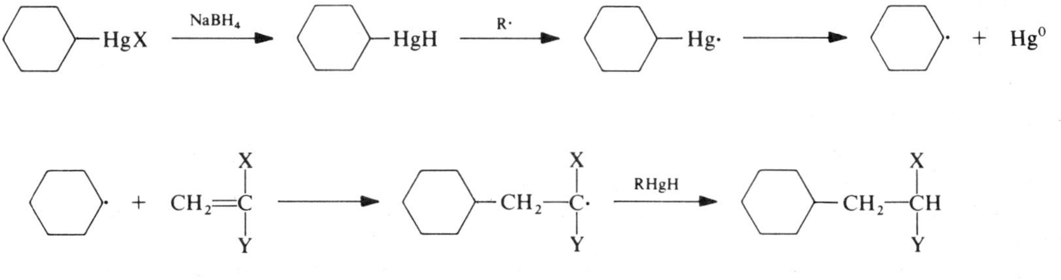

Scheme 23

Electrophilic radicals are scavenged at a substantially greater rate than nucleophilic radicals. Thus conditions can be found such that an excellent yield of adducts of simple alkyl radicals (*e.g.* *n*-hexyl, cyclohexyl, *t*-butyl, alkoxyalkyl) to, for example, acrylic monomers are obtained. With less electron deficient monomers, for example S, oligomerization may result.[215] Another consequence of the metal hydride traps being selective for electrophilic radicals is that the products from the reaction of a radical with a given monomer may not all be trapped with equal efficiency. Other possible complications in the utilization of this method have been discussed by Russell *et al.*[220]

A related radical-trapping technique involves the use of Group IV hydrides as trapping reagents.[204] The reduction of alkyl halides by stannyl[221] or germyl[222] radicals affords alkyl radicals which may react with a substrate (*e.g.* monomer) in competition with trapping by the hydride. This technique has seen widespread use in the study of intramolecular radical reactions.[204] One limitation with the use of the Group IV hydrides as radical traps in the study of polymerization is that the traps may themselves add monomer, albeit reversibly.[223]

8.8.1.4　*Nitroxides*

A well-known feature of the chemistry of nitroxides (*e.g.* **36–38**) is that they rapidly combine with carbon-centred radicals to afford alkoxyamines.[224,225] This reaction, which occurs at or near diffusion-controlled rates,[226–228] has been employed to detect radical intermediates in organic reactions[229–231] and also in the identification of primary radicals produced from photo-initiators.[232] Rizzardo and Solomon[96] applied this knowledge to develop a versatile technique for examining the initiation step of polymerization (see Scheme 24). The method relies on the initiator-derived radicals either not reacting or reacting only slowly with the nitroxide while the propagating radicals are efficiently scavenged to yield stable alkoxyamines. The technique is thus most suited to the study of the reactions of heteroatom-centred (*t*-butoxy,[96–102,108,109] cumyloxy,[47] isopropoxy, benzoyloxy,[99,131,132] isopropoxycarbonyloxy,[143] hydroxy,[149] butanethiyl[233]) and more reactive carbon-centred radicals (methyl,[99,111] undecyl,[234] phenyl[99,100,111]). Major advantages of this method over other trapping techniques are that typical conditions for solution/bulk polymerization can be employed and that a very wide range of initiating systems can be examined. The application of the technique is greatly facilitated by the use of a nitroxide possessing a UV chromophore (*e.g.* **37**, **38**).[235]

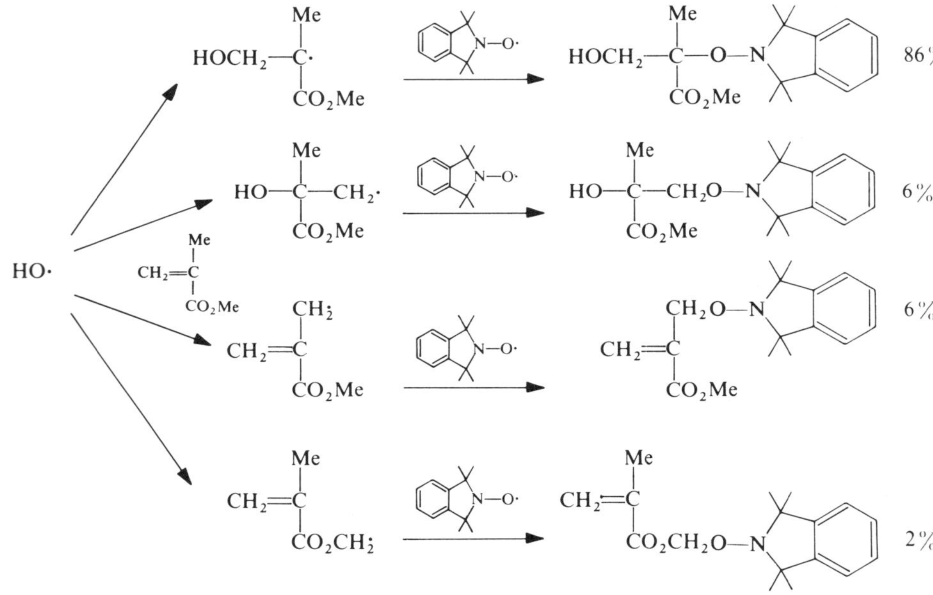

Scheme 24

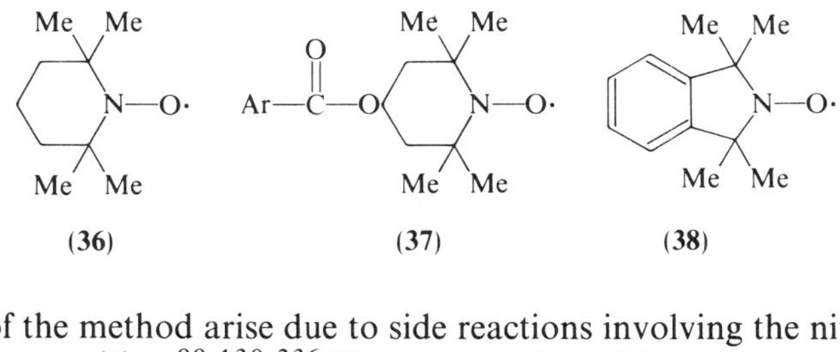

(**36**)　　　　　　　　(**37**)　　　　　　　　(**38**)

Some limitations of the method arise due to side reactions involving the nitroxide. These include induced initiator decomposition.[99,130,236] However, such problems can usually be avoided by the correct choice of nitroxide and reaction conditions or allowed for by conducting appropriate control experiments. It should also be noted that the trapping of radicals by nitroxides is a reversible reaction[237,238] and this may limit the temperature range in which the technique can be employed.[111] Finally, care must also be taken in examining photoinitiation since nitroxides may abstract hydrogen,[239,240] react by photoelectron transfer[241] or undergo β scission[242-244] under the influence of UV light.

8.8.2 Nuclear Magnetic Resonance

The sensitivity of modern NMR allows initiator residues to be determined directly in polymer samples when the signals are discrete from those due to the backbone carbons.[31,245,246] In some cases special pulse sequences may be applied to suppress NMR signals due to backbone carbons or hydrogens, thus allowing obscured end group resonances to be observed.[32,247]

Selective labelling may either enhance the visibility of the initiator residues or suppress signals due to the backbone. A technique based on ^{13}C labelling of the initiator and ^{13}C NMR has been developed as a means of examining the mechanism of the initiation process[20,32,56,185] and for determining the relative reactivity of monomers towards initiating radicals.[132,185] Hatada et al.[248,249] have shown that for polymers prepared from perdeuterated monomer the initiator residues may be conveniently determined by ^1H NMR.

The advantage of the NMR methods over alternative labelling techniques (*e.g.* radiolabelling) is that NMR signals are extremely sensitive to the environment of the initiator residue. For example, by ^{13}C NMR it is possible to distinguish those end groups formed by head addition, tail addition to monomer, transfer to initiator and primary radical termination.[32,56] Recent studies[185] have demonstrated the sensitivity of end group signals to the nature of the attached and more remote monomer units. For example, the labelled carbons of the end group in PS prepared with AIBN-α-^{13}C are sensitive to the configuration of carbons up to six monomer units removed.[56] Residual initiator and the initiator-derived products will usually also give rise to discrete signals in the NMR spectrum (see also Section 11.5.4).

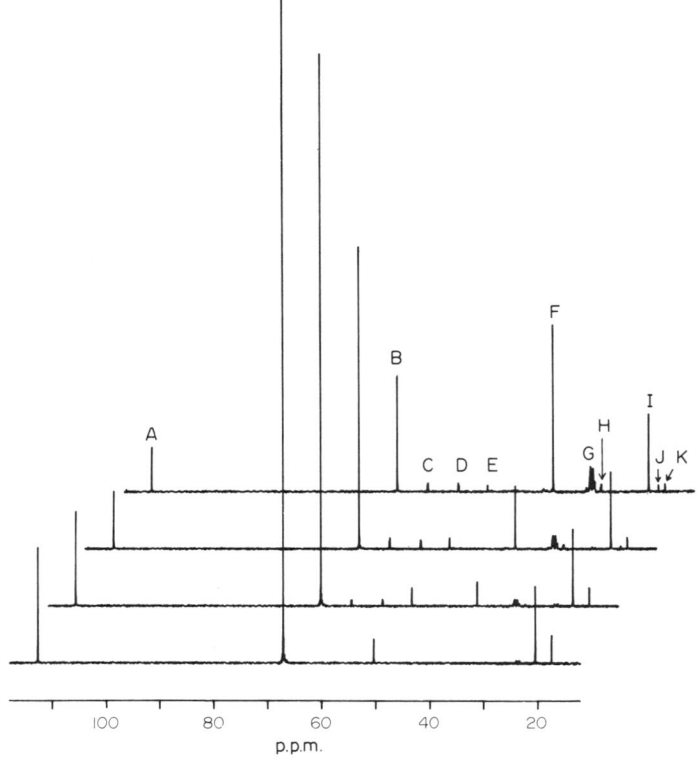

Figure 1 62.9 MHz ^{13}C NMR spectra recorded during copolymerization of styrene (1.6 M) and methyl methacrylate (0.4 M) in toluene/benzene-d^6 initiated by azobis(isobutyronitrile-α-^{13}C) (0.05 M) at 70 °C. From bottom to top the spectra were obtained after 120, 360, 870 and 1840 min. Signals are assigned as follows (refer to Scheme 2):(A) CH_2=CHPh; (B) $Me_2C(CN)$-N=$NC(CN)Me_2$; (C) Me_2C=C=$NC(CN)Me_2$; (D) Me_2C=C=$NC(CN)Me_2$; (E) CH_2=$CMe(CO_2CH_3)$; (F) $Me_2C(CN)$-$C(CN)Me_2$; (G) $Me_2C(CN)CH_2CHPh$–; (H) $Me_2C(CN)CH_2CMe(CO_2Me)$–; (I) $PhCH_3$; (J) $Me_2CH(CN)$; and (K) CH_2=C-$(CH_3)(CO_2Me)$

The power of the technique is illustrated by the example shown in Figure 1.[250] From one experiment and without need to isolate any materials it is possible to determine (a) the total fate of the initiator as a function of conversion (initiator efficiency, nature and amount of by-products); (b) the chain ends (reactivity of primary radicals towards monomers, head *vs.* tail addition, *etc.*); (c) the rate of polymerization; and (d) the number average molecular weight [(number of end groups)/ (monomer used)].

8.9 REFERENCES

1. C. S. Sheppard and V. R. Kamath, *Polym. Eng. Sci.*, 1979, **19**, 597.
2. D. H. Solomon, P. Cacioli and G. Moad, *Pure Appl. Chem.*, 1985, **57**, 985.
3. C. Walling, 'Free Radicals in Solution', Wiley, New York, 1957, p. 55.
4. J. C. Bevington, 'Radical Polymerization', Academic Press, London, 1961, p. 29.
5. K. F. O'Driscoll and P. Ghosh, in 'Stucture and Mechanism in Vinyl Polymerization', ed. T. Tsuruta and K. F. O'Driscoll, Dekker, New York, 1969, p. 60.
6. T. Koenig, in 'Free Radicals', ed. J. K. Kochi, Wiley–Interscience, New York, 1973, vol. 1, p. 113.
7. G. C. Eastmond, in 'Comprehensive Chemical Kinetics' ed. C. H. Bamford and C. F. H. Tipper, Elsevier, Amsterdam, 1976, vol. 14a, p. 22.
8. J. C. Bevington and J. R. Ebdon, in 'Developments in Polymerization', Applied Science, London, 1979, vol. 2, p. 1.
9. M. K. Mishra, *J. Macromol. Sci., Rev. Macromol. Chem.*, 1981, **20**, 149.
10. C. S. Sheppard in 'Encyclopaedia of Polymer Science and Engineering', 2nd edn., Wiley, New York, 1985, vol. 2, p. 143.
11. P. S. Engel, *Chem. Rev.*, 1980, **2**, 99.
12. T. Otsu and T. Tazaki, *Polym. Bull. (Berlin)*, 1986, **16**, 277.
13. J. W. Timberlake, in 'Substituent Effects in Radical Chemistry', ed. H. G. Viehe, Z. Janousek and R. Merenyi, Reidel, Dordecht, 1986, p. 271.
14. C. Rüchardt, *Top. Curr. Chem.*, 1980, **88**, 1.
15. W. Duisman and C. Rüchardt, *Tetrahedron Lett.*, 1974, 4517.
16. R. A. Firestone, *J. Org. Chem.*, 1980, **45**, 3604.
17. G. Ayrey, *Chem. Rev.*, 1963, **63**, 645.
18. J. K. Fink, *J. Polym. Sci., Polym. Chem. Ed.*, 1983, **21**, 1445.
19. T. Fukuda, Y. Ma and H. Inagaki, *Macromolecules*, 1985, **18**, 17.
20. G. Moad, E. Rizzardo, D. H. Solomon, S. R. Johns and R. I. Willing, *Makromol. Chem., Rapid Commun.*, 1984, **5**, 793.
21. J. Barton and E. Borsig, 'Complexes in Free-Radical Polymerization,' Elsevier, Amsterdam, 1988, p. 118.
22. H. Tanaka, K. Fukuoka and T. Ota, *Makromol. Chem., Rapid Commun.*, 1985, **6**, 563.
23. W. Barbe and C. Rüchardt, *Makromol. Chem.*, 1983, **184**, 1235.
24. P. B. Ayscough, B. R. Brooks and H. E. Evans, *J. Phys. Chem.*, 1964, **68**, 3889.
25. A. B. Jaffe, K. J. Skinner and J. M. McBride, *J. Am. Chem. Soc.*, 1972, **94**, 8510.
26. J.-C. Roy, J. R. Nash, R. R. Williams, Jr. and W. H. Hamil, *J. Am. Chem. Soc.*, 1956, **78**, 519.
27. J. C. Bevington, T. N. Huckerby and N. W. E. Hutton, *Eur. Polym. J.*, 1982, **18**, 963.
28. G. S. Hammond, O. D. Trapp, R. T. Keys and D. L. Neff, *J. Am. Chem. Soc.*, 1959, **81**, 4878.
29. H. Ishiwata, T. Inoue and K. Yoshihira, *J. Chromatogr.*, 1986, **370**, 275.
30. J. C. Bevington and H. G. Troth, *Trans. Faraday Soc.*, 1962, **58**, 186.
31. W. H. Starnes, Jr., I. M. Plitz, F. C. Schilling, G. M. Villacorta, G. S. Park and A. H. Saremi, *Macromolecules*, 1984, **17**, 2507.
32. G. Moad, D. H. Solomon, S. R. Johns and R. I. Willing, *Macromolecules*, 1984, **15**, 1094.
33. C. N. Cascaval, S. Straus, D. W. Brown and R. E. Florin, *J. Polym. Sci., Polym. Symp.*, 1976, **57**, 81.
34. J. A. May, Jr. and W. B. Smith, *J. Phys. Chem.*, 1968, **72**, 2993.
35. W. A. Pryor and T. R. Fiske, *Macromolecules*, 1969, **2**, 62.
36. G. Ayrey and A. C. Haynes, *Makromol. Chem.*, 1974, **175**, 1463.
37. J. G. Braks and R. Y. M. Huang, *J. Appl. Polym. Sci.*, 1978, **22**, 3111.
38. R. D. Athey, *J. Polym. Sci., Polym. Chem. Ed.*, 1977, **15**, 1517.
39. M. Stickler, *Makromol. Chem.*, 1986, **187**, 1765.
40. L. Dulog and P. Klein, *Chem. Ber.*, 1971, **104**, 895, 902.
41. H. Kiefer and T. G. Traylor, *Tetrahedron Lett.*, 1966, 6163.
42. E. M. Y. Quinga and G. D. Mendenhall, *J. Org. Chem.*, 1985, **50**, 2836.
43. J. D. Druliner, P. J. Krusic, G. F. Lehr and C. A. Tolman, *J. Org. Chem.*, 1985, **50**, 5838.
44. T. W. Koenig, in 'Organic Free Radicals', ed. W. A. Pryor, American Chemical Society, Washington, 1978, p. 315.
45. G. D. Mendenhall and L. W. Cary, *J. Org. Chem.*, 1975, **40**, 1646.
46. G. D. Mendenhall, L. C. Stewart and J. C. Scaiano, *J. Am. Chem. Soc.*, 1982, **104**, 5109.
47. E. Rizzardo, A. K. Serelis and D. H. Solomon, *Aust. J. Chem.*, 1982, **35**, 2013.
48. H. Kiefer and T. G. Traylor, *J. Am. Chem. Soc.*, 1967, **89**, 6667.
49. G. D. Mendenhall and E. M. Y. Quinga, *Int. J. Chem. Kinet.*, 1985, **17**, 1187.
50. O. L. Magelli and C. S. Sheppard, in 'Organic Peroxides', ed. D. Swern, Wiley–Interscience, New York, 1970, vol. 1, p. 1.
51. R. Hiatt, in 'Organic Free Radicals', ed. W. A. Pryor, American Chemical Society, Washington, 1978, p. 315.
52. K. F. O'Driscoll, in 'Organic Peroxides', ed. D. Swern, Wiley–Interscience, New York, 1970, vol. 1, p. 609.
53. H. Seidl and G. Luft, *J. Macromol. Sci., Chem.*, 1981, **15**, 1.
54. D. Swern (ed.), 'Organic Peroxides', Wiley–Interscience, New York, 1970, 1971, vols. 1–3.
55. R. Hiatt, in 'Organic Peroxides', ed. D. Swern, Wiley-Interscience, New York, 1970, vol. 2, p. 799.
56. G. Moad, D. H. Solomon, S. R. Johns and R. I. Willing, *Macromolecules*, 1982, **15**, 1188.
57. J. C. Martin and J. H. Hargis, *J. Am. Chem. Soc.*, 1969, **91**, 5399.

58. C. G. Swain, L. J. Schaad and A. J. Kresge, *J. Am. Chem. Soc.*, 1958, **80**, 5313.
59. D. E. Falvey and G. B. Schuster, *J. Am. Chem. Soc.*, 1986, **108**, 7419.
60. Y. N. Anisimov, S. S. Ivanchev and A. I. Yurzhenko, *Polym. Sci. USSR (Engl. Transl.)*, 1967, **9**, 773.
61. D. Lefort and J. Y. Nedelec, *Tetrahedron*, 1980, **36**, 3199.
62. L. Grossi, J. Lusztyk and K. U. Ingold, *J. Org. Chem.*, 1985, **50**, 5882.
63. J. C. Bevington, C. J. Dyball and J. Leech, *Makromol. Chem.*, 1979, **180**, 657.
64. T. Sato, M. Abe and T. Otsu, *Makromol. Chem.*, 1977, **178**, 1259; J. C. Bevington, C. J. Dyball and J. Leech, *Makromol. Chem.*, 1977, **178**, 2741.
65. B. B. Dambatta and J. R. Ebdon, *Eur. Polym. J.*, 1986, **22**, 783.
66. N. G. Podosenova, E. G. Zotikov, O. P. Bovkunenko and V. I. Mel'nichenko, *J. Appl. Chem. USSR (Engl. Transl.)*, 1979, **53**, 1513.
67. H. C. McBay and O. Tucker, *J. Org. Chem.*, 1954, **19**, 869.
68. S. G. Cohen and D. B. Sparrow, *J. Am. Chem. Soc.*, 1950, **72**, 611.
69. D. E. Van Sickle, *J. Org. Chem.*, 1969, **34**, 3446.
70. M. Yamada, K. Kitagawa and T. Komai, *Plast. Ind. News*, 1971, **17**, 131.
71. W. A. Strong, *Ind. Eng. Chem.*, 1964, **56**, 33.
72. E. F. J. Duynstee, M. L. Esser and R. Schellekens, *Eur. Polym. J.*, 1980, **17**, 1127.
73. L. A. Singer, in 'Organic Peroxides', ed. D. Swern, Wiley–Interscience, New York, 1970, vol. 1, p. 265.
74. P. D. Bartlett, E. P. Benzing and R. E. Pinock, *J. Am. Chem. Soc.*, 1960, **82**, 1762.
75. S. N. Gupta, I. Gupta and D. C. Neckers, *J. Polym. Sci., Polym. Chem. Ed.*, 1981, **19**, 103.
76. R. Hiatt, in 'Organic Peroxides', ed. D. Swern, Wiley–Interscience, New York, 1970, vol. 2, p. 1.
77. Y. N. Barantsevich and S. S. Ivanchev, *Polym. Sci. USSR (Engl. Transl.)*, 1983, **25**, 2341.
78. K. F. O'Driscoll and J. C. Bevington, *Eur. Polym. J.*, 1985, **21**, 1039.
79. C. Simionescu, E. Comanita, M. Pastravanu and S. Dumitriu, *Prog. Polym. Sci.*, 1986, **12**, 1.
80. C. Simionescu, K. G. Sik, S. Dumitriu, E. Comanita and M. Pastravanu, *Polym. Bull. (Berlin)*, 1986, **15**, 503.
81. B. Z. Gunesin and I. Piirma, *J. Appl. Polym. Sci.*, 1981, **26**, 3103.
82. B. Hazer and B. M. Baysal, *Polymer*, 1986, **27**, 961.
83. Y. Yagci, U. Tunca and N. Biçak, *J. Polym. Sci., Polym. Lett. Ed.*, 1986, **24**, 49.
84. X.-Y. Qiu, W. Ruland and W. Heitz, *Angew. Makromol. Chem.*, 1984, **125**, 69.
85. O. Nuyken and R. Weidner, *Adv. Polym. Sci.*, 1986, **73**, 145.
86. O. Nukyen, in 'Encyclopaedia of Polymer Science and Engineering', 2nd edn., Wiley, New York, 1985, vol. 2, p. 158.
87. R. D. Grant, E. Rizzardo and D. H. Solomon, *J. Chem. Soc., Chem. Commun.*, 1984, 867.
88. A. S. Nazran and J. Warkentin, *J. Am. Chem. Soc.*, 1982, **104**, 6405.
89. J. A. Howard and J. C. Scaiano, 'Landoldt–Börnstein, New Series, Radical Reaction Rates in Solution', ed. H. Fischer, Springer–Verlag, Berlin, 1984, vol. 13, part d.
90. C. Walling, *Pure Appl. Chem.*, 1967, **15**, 69.
91. K. U. Ingold, *Pure Appl. Chem.*, 1967, **15**, 49.
92. J. K. Kochi, in 'Free Radicals', ed. J. K. Kochi, Wiley–Interscience, New York, 1973, vol. 2, p. 665.
93. T. Kunitake and S. Murakami, *J. Polym. Sci., Polym. Chem. Ed.*, 1974, **12**, 67.
94. T. Sato and T. Otsu, *Makromol. Chem.*, 1977, **178**, 1941.
95. M. V. Encina, M. Rivera and E. A. Lissi, *J. Polym. Sci., Polym. Chem. Ed.*, 1978, **16**, 1709.
96. E. Rizzardo and D. H. Solomon, *Polym. Bull. (Berlin)*, 1979, **1**, 529.
97. P. G. Griffiths, E. Rizzardo and D. H. Solomon, *J. Macromol. Sci., Chem.*, 1982, **17**, 45.
98. P. G. Griffiths, E. Rizzardo and D. H. Solomon, *Tetrahedon Lett.*, 1982, **23**, 1309.
99. G. Moad, E. Rizzardo and D. H. Solomon, *Macromolecules*, 1982, **15**, 909.
100. G. Moad, E. Rizzardo and D. H. Solomon, *Aust. J. Chem.*, 1983, **36**, 1573.
101. M. J. Cuthbertson, E. Rizzardo and D. H. Solomon, *Aust. J. Chem.*, 1983, **36**, 1957.
102. M. J. Cuthbertson, E. Rizzardo and D. H. Solomon, *Aust. J. Chem.*, 1985, **38**, 315.
103. H.-G. Korth and R. Sustmann, *Tetrahedron Lett.*, 1985, **26**, 2551.
104. E. Rizzardo and D. H. Solomon, *J. Macromol. Sci., Chem.*, 1979, **13**, 1005.
105. S. M. Aliwi and C. H. Bamford, *J. Chem. Soc., Faraday Trans. 1*, 1977, **73**, 776.
106. I. H. Elson, S. W. Mao and J. K. Kochi, *J. Am. Chem. Soc.*, 1975, **97**, 335.
107. C. Walling and W. Thaler, *J. Am. Chem. Soc.*, 1961, **83**, 3877.
108. R. D. Grant, P. G. Griffiths, G. Moad, E. Rizzardo and D. H. Solomon, *Aust. J. Chem.*, 1983, **36**, 397.
109. R. D. Grant, E. Rizzardo and D. H. Solomon, *Makromol. Chem.*, 1984, **185**, 1809.
110. J. K. Allen and J. C. Bevington, *Proc. R. Soc. London, Ser. A.*, 1961, **262**, 271.
111. D. S. Bednarek, G. Moad, E. Rizzardo and D. H. Solomon, *Macromolecules*, 1987, submitted.
112. D. S. Bednarek, G. Moad, E. Rizzardo and D. H. Solomon, 'Polymer 85 Preprints', RACI, Polymer Division, Melbourne, 1985, p. 49.
113. J. L. Brokenshire and K. U. Ingold, *Int. J. Chem. Kinet.*, 1971, **3**, 343.
114. E. A. Lissi and A. Leon, *J. Polym. Sci., Polym. Chem. Ed.*, 1979, **17**, 3023.
115. N. Ohto, E. Niki and Y. Kamiya, *J. Chem. Soc., Perkin Trans. 2*, 1977, 1416.
116. E. Niki and Y. Kamiya, *J. Chem. Soc., Perkin Trans. 2*, 1975, 1221.
117. H. Sakurai and A. Hosomi, *J. Am. Chem. Soc.*, 1967, **89**, 458.
118. C. Walling and J. A. McGuinness, *J. Am. Chem. Soc.*, 1969, **91**, 2053.
119. M. Jones, G. Moad, E. Rizzardo and D. H. Solomon, *J. Org. Chem.*, Submitted.
120. R. Arnaud, R. Subra, C. V. Barone, F. Lelj, S. Olivella, A. Solé and N. Russo, *J. Chem. Soc., Perkin Trans. 2*, 1986, 1517.
121. B. E. Smart, in 'Molecular Structure and Energetics', ed. J. F. Liebman and A. Greenberg, VCH: Deerfield Beach, Florida, 1986, vol. 3, p. 141.
122. C. Walling and A. Padwa, *J. Am. Chem. Soc.*, 1963, **85**, 1593.
123. A. Baignee, J. A. Howard, J. C. Scaiano and L. C. Stewart, *J. Am. Chem. Soc.*, 1983, **105**, 6120.
124. C. Walling and R. T. Clark, *J. Am. Chem. Soc.*, 1974, **96**, 4530.
125. E. Abuin, C. Mujica and E. Lissi, *Rev. Latinoam. Quim.*, 1980, **11**, 78.

126. P. Brun and B. Waegell, in 'Reactive Intermediates', Plenum Press, New York, 1983, vol. 3, p. 367.
127. D. J. Edge and J. K. Kochi, *J. Am. Chem. Soc.*, 1973, **95**, 2635.
128. J. K. Kochi, *J. Am. Chem. Soc.*, 1962, **84**, 1572.
129. J. C. Bevington, D. O. Harris and M. Johnson, *Eur. Polym. J.*, 1965, **1**, 235.
130. G. Moad, E. Rizzardo and D. H. Solomon, *Makromol. Chem., Rapid Commun.*, 1982, **3**, 533.
131. G. Moad, E. Rizzardo and D. H. Solomon, *J. Macromol. Sci., Chem.*, 1982, **17**, 51.
132. G. Moad, E. Rizzardo and D. H. Solomon, *Polym. Bull. (Berlin)*, 1984, **12**, 471.
133. J. Lane and B. J. Tabner, *J. Chem. Soc., Perkin Trans. 2*, 1984, 1823.
134. L. R. C. Barclay, D. Griller and K. U. Ingold, *J. Am. Chem. Soc.*, 1982, **104**, 4399.
135. A. L. J. Beckwith and C. B. Thomas, *J. Chem. Soc., Perkin Trans. 2*, 1973, 861.
136. J. Saltiel and H. C. Curtis, *J. Am. Chem. Soc.*, 1971, **93**, 2056.
137. L. H. Garcia Rubio, N. Ro and R. D. Patel, *Macromolecules*, 1984, **17**, 1998.
138. K. C. Berger, P. C. Deb and G. Meyerhoff, *Macromolecules*, 1977, **10**, 1075.
139. P. C. Deb and S. K. Kapoor, *Eur. Polym. J.*, 1979, **15**, 961.
140. M. E. Kurz, E. M. Steele and R. L. Vecchio, *J. Org. Chem.*, 1974, **39**, 3331.
141. T. Nakata, K. Tokumaru and O. Simamura, *Bull. Chem. Soc. Jpn*, 1970, **43**, 3590.
142. M. J. Cuthbertson, unpublished results.
143. M. J. Cuthbertson, G. Moad, E. Rizzardo and D. H. Solomon, *Polym. Bull. (Berlin)*, 1982, **6**, 647.
144. H. Fischer and G. Giacometti, *J. Polym. Sci., Polym. Symp.*, 1967, **16**, 2763.
145. H. Fischer, *Z. Naturforsch., Teil A*, 1964, **19**, 866.
146. H.-K. Roth and P. Wünsche, *Acta Polym.*, 1981, **32**, 491.
147. P. Maruthamuthu, *Makromol. Chem., Rapid Commun.*, 1980, **1**, 23.
148. N. A. McAskill and D. A. Sangster, *Aust. J. Chem.*, 1985, **38**, 2137.
149. R. D. Grant, E. Rizzardo and D. H. Solomon, *J. Chem. Soc., Perkin Trans. 2*, 1985, 379.
150. R. O. C. Norman, P. M. Storey and P. R. West, *J. Chem. Soc. B*, 1970, 1088.
151. A. Ledwith and P. J. Russell, *J. Polym. Sci., Polym. Lett. Ed.*, 1975, **13**, 109.
152. A. Citterio, C. Arnoldi, C. Giordano and G. Castaldi, *J. Chem. Soc. Perkin Trans. 1*, 1983, 891.
153. C. Arnoldi, A. Citterio and F. Minisci, *J. Chem. Soc., Perkin Trans. 2*, 1983, 531.
154. W. E. Fristad and J. R. Peterson, *Tetrahedron*, 1984, **40**, 1469.
155. N. A. McAskill and D. F. Sangster, *Aust. J. Chem.*, 1979, **32**, 2611.
156. O. P. Chawla and R. W. Fessenden, *J. Phys. Chem.*, 1975, **79**, 2693.
157. N. Misra and B. M. Mandal, *J. Polym. Sci., Polym. Lett. Ed.*, 1985, **23**, 63.
158. A. K. Banthia, B. M. Mandal and S. R. Palit, *J. Polym. Sci., Polym. Chem. Ed.*, 1977, **15**, 945.
159. N. Misra and B. M. Mandal, *Macromolecules*, 1984, **17**, 495.
160. N. N. Ghosh and B. M. Mandal, *Macromolecules*, 1986, **19**, 19.
161. B. Giese, *Angew. Chem., Int. Ed. Engl*, 1983, **22**, 753.
162. J. M. Tedder and J. C. Walton, *Acc. Chem. Res.*, 1976, **9**, 183.
163. J. M. Tedder and J. C. Walton, *Tetrahedron*, 1980, **37**, 701.
164. J. M. Tedder, *Tetrahedron*, 1982, **38**, 313.
165. J. M. Tedder, *Angew. Chem., Int. Ed. Engl.*, 1982, **21**, 401.
166. C. Rüchardt, *Top. Curr. Chem.*, 1980, **88**, 1.
167. A. L. J. Beckwith, *Tetrahedron*, 1981, **37**, 3073.
168. J. P. Lorand, in 'Landoldt–Börnstein, New Series, Radical Reaction Rates in Solution', ed. H. Fischer, Springer–Verlag, Berlin, 1984, vol. 13, part a, p. 135.
169. K.-D. Asmus and M. Bonifacic, in 'Landoldt–Börnstein, New Series, Radical Reaction Rates in Solution', ed. H. Fischer, Springer–Verlag, Berlin, 1984, vol. 13, part b.
170. M. Szwarc, *J. Polym. Sci.*, 1955, **16**, 367.
171. L. Herk, A. Stefani and M. Szwarc, *J. Am. Chem. Soc.*, 1961, **83**, 3008.
172. W. A. Pryor, D. L. Fuller and J. P. Stanley, *J. Am. Chem. Soc.*, 1972, **94**, 1632.
173. B. Giese and S. Lachhein, *Angew. Chem., Int. Ed. Engl.*, 1981, **20**, 967.
174. T. Caronna, A. Citterio, M. Ghirardini and F. Minisci, *Tetrahedron*, 1977, **33**, 793.
175. A. Citterio, A. Arnoldi and F. Minisci, *J. Org. Chem.*, 1979, **44**, 2674.
176. H. Fischer, in 'Substituent Effects in Radical Chemistry', ed. H. G. Viehe, Z. Janousek and R. Merenyi, Reidel, Dordecht, 1986, p. 123.
177. K. Münger and H. Fischer, *Int. J. Chem. Kinet.*, 1985, **17**, 809.
178. B. Giese and J. Meister, *Angew. Chem., Int. Ed. Engl.*, 1977, **16**, 178.
179. Z. M. Dzhabiyeva, P. Ye Matkovskii, Ye. L. Pechatnikov and N. A. Byrikhina, *Polym. Sci. USSR (Engl. Transl.)*, 1985, **27**, 2416.
180. J. C. Bevington, P. F. Fridd and B. J. Tabner, *J. Chem. Soc., Perkin Trans. 2*, 1982, 1389.
181. J. C. Bevington, T. N. Huckerby and S. C. Varma, *Eur. Polym. J.*, 1986, **22**, 427.
182. J. C. Bevington, E. N. J. Heseltine, T. N. Huckerby and S. C. Varma, *J. Polym. Sci., Polym. Chem. Ed.*, in press.
183. B. Maillard, K. U. Ingold and J. C. Scaiano, *J. Am. Chem. Soc.*, 1983, **105**, 5095.
184. J. C. Bevington, J. R. Ebdon, T. N. Huckerby and N. W. E. Hutton, *Polym. Commun.*, 1982, **23**, 163.
185. J. C. Bevington, J. R. Ebdon and T. N. Huckerby, *Eur. Polym. J.*, 1985, **21**, 685.
186. J. C. Bevington, T. N. Huckerby and B. J. Hunt, *Br. Polym. J.*, 1985, **17**, 43.
187. J. C. Bevington and T. N. Huckerby, *Macromolecules*, 1985, **18**, 176.
188. J. C. Bevington, T. N. Huckerby and N. W. E. Hutton, *Eur. Polym. J.*, 1984, **20**, 525.
189. C. A. Barson, J. C. Bevington and T. N. Huckerby, *Polym. Bull. (Berlin)*, 1986, **16**, 209.
190. W. A. Pryor and T. H. Fiske, *Macromolecules*, 1969, **2**, 62.
191. W. A. Pryor and J. H. Coco, *Macromolecules*, 1970, **3**, 500.
192. H. K. Mahabadi and K. F. O'Driscoll, *Makromol. Chem.*, 1977, **178**, 2629.
193. P. C. Deb and I. D. Gaba, *Makromol. Chem.*, 1978, **179**, 1559.
194. W. Konter, B. Bömer, K.-H. Köhler and W. Heitz, *Makromol. Chem.*, 1981, **182**, 2619.

195. S. C. Dickerman, I. S. Megna and M. M. Skoultchi, *J. Am. Chem. Soc.*, 1959, **81**, 2270.
196. J. C. Bevington and T. Ito, *Trans. Faraday Soc.*, 1968, **64**, 1329.
197. W. A. Pryor and T. R. Fiske, *Trans. Faraday Soc.*, 1969, **65**, 1865.
198. Y. A. Levin, A. G. Abul'khanov, V. P. Nefedov, M. S. Skorobogatova and B. E. Ivanov, *Dokl. Chem. (Engl. Transl.)*, 1972, **235**, 728.
199. J. C. Scaiano and L. C. Stewart, *J. Am. Chem. Soc.*, 1983, **105**, 3609.
200. A. Citterio, F. Minisci and E. Vismara, *J. Org. Chem.*, 1982, **47**, 81.
201. M. N. Galbraith, G. Moad, D. H. Solomon and T. H. Spurling, *Macromolecules*, 1987, **20**, 675.
202. R. J. Elliot and W. G. Richards, *J. Chem. Soc., Perkin Trans. 2*, 1982, 943.
203. S. R. Palit and B. M. Mandal, *J. Macromol. Sci., Rev. Macromol. Chem.*, 1968, **2**, 225.
204. B. Giese, 'Radicals in Organic Synthesis: Formation of Carbon–Carbon Bonds', Pergamon Press, Oxford, 1986.
205. M. J. Perkins, *Adv. Phys. Org. Chem.*, 1981, **17**, 1.
206. E. G. Janzen, *Acc. Chem. Res.*, 1971, **4**, 31.
207. G. R. Chalfont, M. J. Perkins and A. Horsfield, *J. Am. Chem. Soc.*, 1968, **90**, 7141.
208. M. Kamachi, Y. Kuwae and S. Nozakura, *Polym. Bull. (Berlin)*, 1981, **6**, 143.
209. J. C. Bevington, B. J. Tabner and P. F. Fridd, *Rev. Roum. Chim.*, 1980, **25**, 947.
210. L. G. Savedoff and R. Ranby, *Polym. Prepr., Am. Chem. Soc., Div. Polym. Chem.*, 1978, **19**, 629.
211. F. Minisci, *Acc. Chem. Res.*, 1975, **8**, 165.
212. J. K. Kochi, *J. Am. Chem. Soc.*, 1962, **84**, 774.
213. Y. K. Chong, E. Rizzardo and D. H. Solomon, *J. Am. Chem. Soc.*, 1983, **105**, 7761.
214. C. L. Hill and G. M. Whitesides, *J. Am. Chem. Soc.*, 1974, **96**, 870.
215. B. Giese and J. Meister, *Chem. Ber.*, 1977., **110**, 2588.
216. B. Giese, G. Kretzschmar and J. Meixner, *Chem. Ber.*, 1980, **113**, 2787.
217. B. Giese and J. Meixner, *Angew. Chem., Int. Ed. Engl.*, 1980, **19**, 206.
218. B. Giese and J. Meixner, *Chem. Ber.*, 1981, **114**, 2138.
219. B. Giese and R. Engelbrecht, *Polym. Bull. (Berlin)*, 1984, **12**, 55.
220. G. A. Russel. W. Jiang, S. S. Hu and R. K. Khanna, *J. Org. Chem.*, 1986, **51**, 5499.
221. B. Giese, J. A. Gonzalez-Gomez and T. Witzel, *Angew. Chem., Int. Ed. Engl.*, 1984, **23**, 69.
222. P. Pike, S. Hershberger and J. Hershberger, *Tetrahedron Lett.*, 1985, **26**, 6289.
223. R. Sommer and H. G. Kuivilla *J. Org. Chem.*, 1968, **33**, 802.
224. E. G. Rozantsev and V. D. Sholle, *Synthesis*, 1971, **4**, 190.
225. H. G. Aurich and W. Weiss, *Top. Curr. Chem.*, 1976, **59**, 65.
226. A. L. Aleksandrov, E. M. Pliss and V. F. Shuvalov, *Bull. Akad. Sci. USSR, Div. Chem. Sci. (Engl. Transl)*, 1979, **28**, 2262.
227. P. Schmid and K. U. Ingold, *J. Am. Chem. Soc.*, 1978, **100**, 2493.
228. A. L. J. Beckwith, V. M. Bowry, M. O'Leary, G. Moad, E. Rizzardo and D. H. Solomon, *J. Chem. Soc., Chem. Commun.*, 1986, 1003.
229. W. K. Robbins and R. H. Eastman, *J. Am. Chem. Soc.*, 1970, **92**, 6077.
230. A. T. Bottini and L. J. Cabral, *Tetrahedron Lett.*, 1977, **18**, 615.
231. L. M. Lawrence and G. M. Whitesides, *J. Am. Chem. Soc.*, 1980, **102**, 2493.
232. H. J. Hageman and T. Overeem, *Makromol. Chem., Rapid Commun.*, 1981, **2**, 719.
233. G. Moad, unpublished data.
234. G. Moad, E. Rizzardo and D. H. Solomon, *Tetrahedron Lett.*, 1981, **22**, 1165.
235. P. G. Griffiths, G. Moad, E. Rizzardo and D. H. Solomon, *Aust. J. Chem.*, 1983, **36**, 397.
236. S. V. Rykov and V. D. Sholl, *Bull. Akad. Sci. USSR, Div. Chem. Sci. (Engl. Transl.)*, 1971, **20**, 2238.
237. D. W. Grattan, D. J. Carlsson, J. A. Howard and D. M. Wiles, *Can. J. Chem.*, 1979, **57**, 2834.
238. E. Rizzardo, *Chem. Aust.*, 1987, **54**, 32.
239. J. F. W. Keana, R. J. Dinerstein and F. Baitis, *J. Org. Chem.*, 1971, **36**, 209.
240. L. J. Johnston, M. Tencer and J. C. Scaiano, *J. Org. Chem.*, 1986, **51**, 2806.
241. D. R. Anderson, J. S. Keute, H. L. Chapel and T. H. Koch, *J. Am. Chem. Soc.*, 1979, **101**, 1904.
242. D. R. Anderson and T. H. Koch, *Tetrahedron Lett.*, 1977, **18**, 3015.
243. J. F. W. Keana and F. Baitis, *Tetrahedron Lett.*, 1968, 365.
244. J. M. Coxan and E. Patsalides, *Aust. J. Chem.*, 1982, **35**, 509.
245. K. Hatada, V. Terawaki, T. Kitayama, M. Kamachi and M. Tamaki, *Polym. Bull. (Berlin)*, 1981, **4**, 451.
246. T. Kashiwagi, A. Inaba, J. E. Brown, K. Hatada, T. Kitayama and E. Masuda, *Macromolecules*, 1986, **19**, 2160.
247. S. R. Johns, E. Rizzardo, D. H. Solomon and R. I. Willing, *Makromol. Chem.*, 1983, **4**, 29.
248. K. Hatada, T. Kitayama and H. Yuki, *Makromol. Chem., Rapid Commun.*, 1980, **1**, 51.
249. K. Hatada, T. Kitayama and E. Masuda, *Polym. J. (Tokyo)*, 1986, **18**, 395.
250. G. Moad, E. Rizzardo and D. H. Solomon, 'Polymer 85 Preprints', RACI, Polymer Division, Melbourne, 1985, p. 46.

9
Redox Initiators

CLEMENT H. BAMFORD
University of Liverpool, UK

9.1 INTRODUCTION

Peroxy compounds, both organic and inorganic, became well-estabilished initiators of free-radical polymerization during the period 1920–1940, but a new chapter in their history opened in 1940 with the discovery that addition of small quantities of a reducing agent greatly increased the rate of initiation.[1-4] An interesting review by Bacon[4] describes the early history of redox initiation. Initiators composed of mixtures of oxidizing and reducing agents, now termed redox initiators, have become very important, particularly in industry where they are used extensively in low-temperature emulsion polymerization, for example in synthetic rubber production. The activation energies for radical generation by redox processes are often low, of the order $40 \, \text{kJ mol}^{-1}$ (compared to $130 \, \text{kJ mol}^{-1}$, approximately, for simple thermal dissociation into initiating radicals) so that they are particularly suitable for low-temperature polymerizations.

The term reduction activation has also been applied to the mechanism of initiation by these systems.[1]

The essential feature in redox initiation is a single-electron transfer, generating free radicals which are sufficiently active to initiate. Commonly, the electron transfer takes place between two separate molecular species, but intramolecular transfers yielding radicals are known and will be discussed later.

Numerous redox initiators are known and many 'recipes' are current in emulsion polymerization technology. An early comprehensive account was published by Bovey *et al*;[5] later articles have been written by Duck[6] and by Vanderhoff[7] and a useful summary has been provided by Cooper.[8] The nature of the redox process would lead us to expect, correctly, that derivatives of transition metals (both ionic and covalent) would be frequently occurring components. In such circumstances the electron transfer step produces one radical, together with a change in the oxidation state of the metal. In systems free from transition metals both oxidizing and reducing components yield radical products so that radicals are generated in pairs. The possibility of geminate-type recombination in solvent cages[9] therefore arises with the latter but not the former type. Frequently the exact mechanistic details of redox initiation are not clear.

It is possible to classify redox systems in various ways; in the following we shall divide them according to their solubilities in water and organic liquids. Both thermal- and photo-initiation are considered.

9.2 AQUEOUS REDOX INITIATORS INCORPORATING TRANSITION METAL DERIVATIVES

The classic and undoubtedly most-studied redox system is a mixture of hydrogen peroxide and a ferrous salt. As long ago as 1894 Fenton[10] discovered the powerful oxidizing properties of such mixtures. Subsequent studies by Haber and Willstätter[11] and by Haber and Weiss[12] led to the proposal of a mechanism based on $\cdot OH$ and $HO_2\cdot$ radicals which, in essence, has been confirmed by later work.[13-15] The currently accepted component reactions are shown in Scheme 1. In the presence of a vinyl monomer the $\cdot OH$ radicals formed in reaction (a) initiate polymerization and all these radicals may be scavenged if the concentration of monomer is sufficiently high. The molar yield of ferric ion, per mol of hydrogen peroxide decomposed, then falls to one, from a maximum of two in the absence of monomer. Under these conditions the rate of initiation is determined by reaction (a) in Scheme 1; this has been found[14] to have a rate coefficient of $4.45 \times 10^8 \exp(-4733/T)$ $mol^{-1} dm^3 s^{-1}$.

$$Fe^{2+} + H_2O_2 \rightarrow Fe^{3+} + OH^- + \cdot OH \text{ (or } FeOH^{2+} + \cdot OH) \tag{a}$$

$$Fe^{2+} + \cdot OH \rightarrow Fe^{3+} + OH^- \tag{b}$$

$$\cdot OH + H_2O_2 \rightarrow H_2O + HO_2\cdot \tag{c}$$

$$Fe^{2+} + HO_2\cdot \rightarrow Fe^{3+} + HO_2^- \tag{d}$$

$$Fe^{3+} + HO_2\cdot \rightarrow Fe^{2+} + H^+ + O_2 \tag{e}$$

Scheme 1

Peroxy compounds in general and reducing ions other than ferrous (*e.g.* Ti^{3+}, Ag^+, Cu^+) may also partake in similar reactions, which are essentially one-electron transfers with concomitant rupture of the $-O-O-$ bond.[13,16,17] Thus with an organic diperoxide the primary process is represented by equation (1). When no radical scavenger is present the radical product reacts with further ferrous salt, as in equation (2), but the radicals may be intercepted by a monomer, with initiation of polymerization.

$$R-O\overset{\frown}{-}O-R \rightarrow R-O\cdot + {}^-O-R \tag{1}$$
$$\underset{Fe}{}$$

$$Fe^{2+} + \cdot OR \rightarrow Fe^{3+} + {}^-OR \tag{2}$$

Similarly with persulfates the analogous reactions in equations (3) and (4) occur.[17]

$$Fe^{2+} + {}^-O_3S-O-O-SO_3^- \rightarrow Fe^{3+} + {}^-O_3S-O\cdot + SO_4^{2-} \tag{3}$$

$$Fe^{2+} + {}^-O_3S-O\cdot \rightarrow Fe^{3+} + SO_4^{2-} \tag{4}$$

'Acid salts' of persulfates (peroxymonosulfates) ($^-O_3S-O-OH$) in the presence of reducing metal ions such as Ag^+, Mn^{2+}, Fe^{2+} or Co^{2+} are, as would be anticipated, active redox initiators.[18,19]

The interaction of hydroperoxides and ferrous salts is essentially similar to the redox processes above and is represented by equation (5).

$$Fe^{2+} + RO-OH \rightarrow Fe^{3+} + {}^-OH + \cdot OR \tag{5}$$

The radical species generated is alkyloxy rather than hydroxyl, a fact which has important implications for the synthesis of block or graft copolymers by processes based on equation (5). Hydroperoxides are frequently used as components of redox systems in emulsion polymerization and their reactions have been examined in some detail; for details see refs. 5 and 17. Walling[20] has tabulated rate coefficients for reactions of the type (5) and Orr and Williams[21] have discussed efficiencies of initiation.

The use of diacyl peroxides as components of redox initiations was pioneered by Kern and his colleagues,[3] who postulated reactions similar in principle to equations (1) and (2) leading to acyloxy radicals. Cu^+ is particularly effective[22] in decomposing diacyl peroxides and peresters and the processes occurring are exemplified in equation (6). This reaction, a convenient source of acyloxy radicals, is useful in the acyloxylation of alkenes.

$$(PhCO-O)_2 + Cu^+ \rightarrow PhCO-O-\overset{\cdot}{O}-OCPh + Cu^{2+} \tag{6}$$
$$\downarrow$$
$$PhCO-O\cdot + {}^-OOCPh$$

Sulfur analogues of peroxides undergo related redox reactions. Kolthoff *et al.*[23] suggested that in the disulfide–ferrous citrate complex system radicals are generated by reaction (7). In the presence of strongly reducing ions such as Ti^{3+}, Cr^{2+} or V^{2+}, hydroxylamine is reduced to $\cdot NH_2$ radicals and Davis, Evans and Hagginson[24] showed that such combinations will initiate polymerization. According to Evans, Baxendale and Cowling[25] hypobromous acid and Fe^{2+} interact to generate bromine atoms which can initiate reaction (8).

$$Fe^{2+}(\text{complex}) + RS\text{–}SR \rightarrow Fe^{3+}(\text{complex}) + RS^- + \cdot SR \tag{7}$$

$$Fe^{2+} + HOBr \rightarrow Fe^{3+} + HO^- + Br\cdot \tag{8}$$

Although in the examples presented above the transition metal ion is the reducing component of the redox system there are many instances in which ions in their higher oxidation states (often complexed) function as the oxidizing components. One-electron transfers in such systems have been recognized since the early work of Haber and Willstätter[11] and Bäckström,[26] who interpreted the autoxidation of sulfites catalyzed by cupric salts in terms of the radical ion $\cdot SO_3^-$ (as in reaction 9).

$$Cu^{2+} + SO_3^{2-} \rightarrow Cu^+ + \cdot SO_3^- \tag{9}$$

Redox reactions between Co^{3+} and a variety of organic substrates were studied by Bawn and White.[27] Oxidation by ceric ion provides a well-known technique for generating radical sites on polymer molecules for the synthesis of block and graft copolymers. The method is applicable to many substrates with labile protons such as alcohols,[28] amides and urethanes (see equations 10, 11 and 12).[29]

$$\text{wwCH}_2\text{—CHww} + Ce^{4+} \longrightarrow \text{wwCH}_2\text{—CHww} + Ce^{3+} + H^+ \tag{10}$$
$$\overset{|}{\text{OH}} \qquad\qquad\qquad\qquad \overset{|}{\text{O}\cdot}$$

$$\text{wwNHCOww} + Ce^{4+} \longrightarrow \text{ww}\dot{\text{N}}\text{COww} + Ce^{3+} + H^+ \tag{11}$$

$$\text{wwNHCOOww} + Ce^{4+} \longrightarrow \text{ww}\dot{\text{N}}\text{COOww} + Ce^{3+} + H^+ \tag{12}$$

A detailed monograph published by Hebeish and Guthrie[30] describes the extensive work on grafting to cellulose which has been carried out. In these systems radicals may also be generated by interaction of Ce^{4+} with the vinyl monomer and there is generally some termination by Ce^{4+} (as in reaction 13). Hebeish and Guthrie[30] also discuss redox initiation involving Mn^{4+} and Mn^{3+}.

$$R_r\cdot + Ce^{4+} \rightarrow P_r + Ce^{3+} + H^+ \tag{13}$$

Photoinitiated redox reactions in aqueous solution are well known. Dainton and James[31] studied the photodecomposition of water in the hydration shells of some cations M^{n+} and anions A^{n-} and postulated the following processes (equations 14–16), concluding that the hydrogen atoms formed are capable of initiating polymerization. Many ion pairs have been shown to photoinitiate; some details are given in ref. 17.

$$M^{n+}(H_2O) + hv \rightarrow M^{(n+1)+} + {}^-OH + H\cdot \tag{14}$$

$$A^{n-}(H_2O) + hv \rightarrow A^{(n-1)-} + {}^-OH + H\cdot \tag{15}$$

$$A^{n-}H_3O^+ + hv \rightarrow A^{(n-1)-} + H_2O + H\cdot \tag{16}$$

9.3 AQUEOUS REDOX INITIATORS WITHOUT TRANSITION METALS

Photoinitiation of polymerization by dyes is a redox process first reported by Oster[32] in 1954, and subsequently developed by Oster and his colleagues.[33, 34] Free-radical polymerization may be photoinitiated by many dyes in the presence of mild reducing agents and low concentrations of oxygen. Oster *et al.* proposed[35] that the dye is first reduced to the leuco form which is later oxidized with free-radical formation. A second suggestion, originating from Delzenne *et al.*[36] is that interaction between the photoexcited dye and oxygen produces hydrogen peroxide, which then undergoes a redox reaction with the reducing agent. A review of the subject has been written by Oster and Yang.[37]

The high rates of polymerization obtainable with the aid of dye photoinitiation have been utilized in rapid imaging processes for displays and holography. One system employs an aqueous solution of barium acrylate or acrylamide with a photoredox initiator of a phenothiazine dye and benzene sulfinate ions. Little *et al.*[38] have made a detailed study of the closely related system methylene blue (D^+Cl^-), benzene sulfinateand RSO_2^- and have interpreted their results in terms of the mechanism of Scheme 2.

$$D^+ + h\nu \rightarrow D^+(^1S) \tag{a}$$

$$D^+(^1S) \rightarrow D^+(^1T) \tag{b}$$

$$D^+(^1S) + RSO_2^- \rightarrow D^+ + RSO_2^- \tag{c}$$

$$D^+(^1T) \rightarrow D^+ \tag{d}$$

$$D^+(^1T) + RSO_2^- \rightarrow D\cdot + RSO_2\cdot \tag{e}$$

$$D^+(^1T) + RSO_2^- \rightarrow D^+ + RSO_2^- \tag{f}$$

$$D\cdot + RSO_2\cdot \rightarrow D^+ + RSO_2^- \tag{g}$$

$$2D\cdot + H_2O \rightarrow D^+ + DH + OH^- \tag{h}$$

$$RSO_2\cdot + M \rightarrow R_1\cdot \tag{i}$$

Scheme 2

Photon absorption (a) by the dye molecule takes the latter into the first excited state ^1S which can then undergo a radiationless conversion (b) to the triplet ^1T. The ^1S and ^1T states may be deactivated by sulfinate (equations (c), (f), respectively), but while the singlet does not oxidize the sulfinate the triplet is able to do so, giving rise to the redox reaction (e) and generating sulfinate radicals. These radicals enter into several destruction reactions (g), (h), and also initiate polymerization (equation i). The presence of sulfinate groups in the final polymer has been demonstrated. In agreement with the proposed redox process (e) is the observation that introduction of electron-withdrawing groups into the *para* position of the sulfinate ion reduces the rate coefficient of the reaction. In practice, the compounds (1) and (2) may replace sulfinates with advantage, *e.g.* a longer active life.[39]

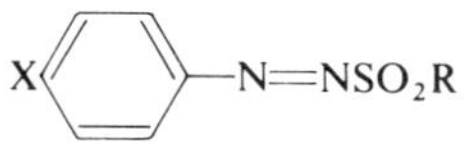

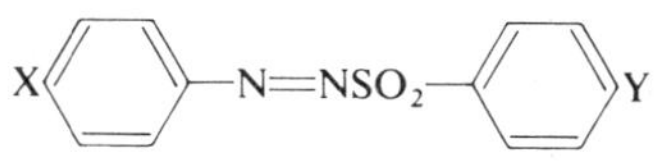

Many combinations of oxidizing and reducing agents free from transition metals can function as redox initiators. Single electron transfer in these systems leads to two free radicals, both of which may, in principle, initiate polymerization, although the behaviour in any given system depends on the radical and monomer reactivities. Indeed, it cannot be assumed that reaction between oxidizing and reducing agents will necessarily lead to radical generation.[20]

Redox initiators of the type we are considering are used in industrial emulsion polymerization.[5-8] A list of common components is given in Table 1. Often the detailed chemistry involved is somewhat obscure. Some typical interactions are presented in equations (17a, b, c): persulfates with metabisulfites (a) and thiosulfates (b) and cumene hydroperoxide with thiosulfates (c).

$$S_2O_8^{2-} + S_2O_5^{2-} \rightarrow SO_4^-\cdot + SO_4^{2-} + S_2O_5^-\cdot \tag{17a}$$

$$S_2O_8^{2-} + S_2O_3^{2-} \rightarrow SO_4^-\cdot + SO_4^{2-} + S_2O_3^-\cdot \tag{17b}$$

$$PhCMe_2OOH + S_2O_3^{2-} \rightarrow PhCMe_2O\cdot + S_2O_3^-\cdot + {}^-OH \tag{17c}$$

Table 1 Components for Redox Initiators

Oxidizing agents	Potassium persulfate, potassium ferricyanide, potassium permanganate, ceric sulfate, *t*-butyl hydroperoxide, cumene hydroperoxide, pinane hydroperoxide, di-isopropylbenzene hydroperoxide
Reducing agents	Sodium hyposulfite, sodium metabisulfite, sodium sulfide, sodium thiosulfate, hydrazine hydrate

Rodriguez and his co-workers have examined the polymerization of acrylamide initiated by the pairs: persulfate + thiosulfate;[40] persulfate + metabisulfite;[41] and chlorate + sulfite;[41].

9.4 NON-AQUEOUS REDOX INITIATORS

9.4.1 Peroxide Tertiary Amine Systems

The interaction of organic peroxides with tertiary amines is probably the most familiar example of this class. Bartlett and Nozaki[42] discussed the kinetics of the reactions with primary, secondary and tertiary amines, but much of the detailed information comes from the work of Horner and his collaborators.[43-47] An interesting survey has been written by Walling.[20] The system dibenzoyl peroxide + dimethylaniline has been studied extensively; according to Horner and co-workers the main reactions are those in Scheme 3.

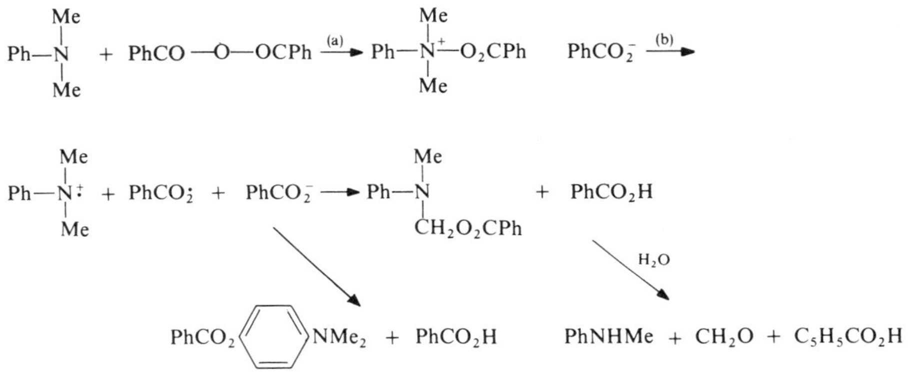

Scheme 3

The quaternary salt first formed (equation a) decomposes as shown in equation (b), generating benzoyloxy radicals. In the presence of a monomer these radicals initiate polymerization; Horner and Schwenk[43] noted that polystyrene prepared with this initiator contained negligible nitrogen, but significant quantities of oxygen which could be recovered as benzoic acid by hydrolysis. When no monomer is present the major products are benzoic acid, monomethylaniline and formaldehyde.

Other reducing agents are also able to enhance the rate of initiation by dibenzoyl peroxide including sulfinic acids, α-ketols, formic acid, thiols and hydrazines.[17] Addition of soluble metal derivatives (such as naphthenates) increases further the activities of these systems.[4,17]

Other diacyl peroxides and hydroperoxides may react similarly.[43] Hirai and his colleagues have used the redox interaction between tri-*n*-butylborane and dibenzoyl peroxide (molar ratio 2:1) to initiate radical polymerization at $-50\,°C$[48,49] and $-20\,°C$.[50]

Free-radical polymerization may also be initiated by alkylboranes in the presence of oxygen.[17,51,52]

9.4.2 Initiators Based on Transition Metal Complexes

Transition metal complexes, typically carbonyls, hexaaryl isocyanides, triphenylphosphines and triphenyl phosphites, are members of an extensive class of initiating systems, thermal and photo-chemical, which are effective over a wide range of temperatures and are particularly useful for some synthetic purposes. The complexes alone are not initiators of free-radical polymerization (although, of course, some of them induce other types of polymerization) but require the presence of a coinitiator. Coinitiators may be of two types: (1) an organic halide and (2) an unsaturated species such as an alkene with electron-attracting groups, acetylene or a substituted acetylene. The reactions involved are quite different in the two cases and we shall discuss them separately.

9.4.2.1 *Type 1 initiation*

The use of metal carbonyls in the presence of high concentrations of carbon tetrachloride for the synthesis of telomers of ethylene was reported by Freydlina and Belyavskii[53] in 1959. Bamford and Finch[54] showed that the reaction is a free-radical one and that initiation involves both carbonyl and halide. Subsequently Bamford and his colleagues have made extensive studies of these initiators. Ref. 55 describes early work and later investigations are reported below.

The kinetic behaviour of all the systems investigated follows a common pattern, of which important features are outlined below.

(a) *Square root behaviour.* At constant halide concentration the rate of polymerization is proportional to the square root of the concentration of the metal derivative, as expected for an uncomplicated free-radical polymerization. Deviations from this behaviour appear at sufficiently high metal complex concentrations, the exact conditions being determined by the nature of the metal and the halide. As a rule these deviations are not caused by retardation, since $k_p k_t^{-\frac{1}{2}}$ retains its normal value for the monomer concerned under the prevailing conditions, but arises from inhibition. That is, an intermediate in the initiation mechanism interferes with radical generation. The phenomenon has been termed 'self-inhibition'; it is discussed further in ref. 55. Systems showing self-inhibition include $Mo(CO)_6/CCl_4$, $Mo(CO)_6/CHCl_3$, $Cr(CO)_6/CCl_4$, $Co_4(CO)_{12}/CCl_4$; often with halides of higher activity self-inhibition is lower, thus for $Mo(CO)_6/CCl_3CO_2Et$, $Mo(CO)_6/CBr_4$ it is negligible.

(b) *Halide dependence.* This is an important feature. As [halide] is increased (from zero) the rate of polymerization increases from zero and eventually achieves a plateau value; in this 'plateau region' the rate of polymerization has an order zero in [halide]. The exact shape of the 'halide curve' depends on the natures of the metal, halide, solvent and other species which may be present[55] (see c). Thus, a very active halide gives rise to a curve which ascends sharply from the origin; it will appear that the form of the curve at low [halide] (the 'sharpness') is of practical importance.

(c) *Addition of free ligand.* Often the rate of initiation is depressed by addition of the free ligand (*i.e.* carbon monoxide in the case of metal carbonyls), to an extent depending on the nature of the metal. For example, $Mn_2(CO)_{10}$ is relatively insensitive while $Ni(CO)_4$ is highly sensitive to the presence of carbon monoxide. This effect is a further example of true inhibition. In the presence of the added ligand the sharpness of the halide curve is reduced.

These and other observations on the behaviour of the initiators have enabled a general picture of their chemistry to be constructed. Radical generation is essentially an electron transfer process from transition metal to halide, the former assuming a higher oxidation state and the latter generally cleaving into an anion and radical, see equation (18) where M represents the metal atom. Electron transfer to a halide Cl–R may follow two routes, shown in equations (19a, b), the relative importance of reactions (a), (b) depending *inter alia* on the electron affinity of R. When $R = CCl_3$, route (a) predominates,[56] the radicals being exclusively $\cdot CCl_3$; CBr_4 behaves in a like manner.[57] With trichloroacetic acid and derivatives route (a) predominates with Mo, Mn, Pt complexes, but with Ni compounds (b) is comparably important.[58] It would thus appear, for example, that the anion derived from ethyl trichloroacetate (3) can associate more readily with Ni than with Mo, Mn or Pt,

$$R^- = Cl_2C \overset{\cdots}{=\!=} C - OC_2H_5$$
$$\underset{O}{\big|}$$

(3)

so that route (b) is favoured. The interaction of nickel carbonyl with α-bromoketones in DMF solution provides another example of the occurrence of route (b).[55,59] N-Haloamides,[60] imides[60] and urethanes[29,61,62] are very active coinitiators in these processes and yield the corresponding nitrogen-centred radicals. Many of the results described above have been obtained by use of ^{14}C-labelled halides;[56,57] another technique, involving the use of polymeric coinitiators, has also been utilized and leads to similar conclusions.[55]

$$M^0 + Cl\text{–}CCl_3 \rightarrow M^I Cl^- + \cdot CCl_3 \tag{18}$$

$$e^- + Cl\text{–}R \begin{cases} \longrightarrow Cl^- + R\cdot & (19a) \\ \longrightarrow Cl\cdot + R^- & (19b) \end{cases}$$

The kinetic features, especially the form of the halide dependence, has been interpreted to imply the existence of a rate-determining step which does not involve the halide.[54,55] Two obvious possibilities, S_N1 ligand scission and S_N2 ligand replacement by monomer (m) or other species, are both encountered, and are typified in equations (20) and (21). Ref. 55 lists the natures of the ligand exchange encountered in different systems, so far as these are known.

$$S_N1 \quad Ni\{P(OPh)_3\}_4 \rightleftharpoons Ni\{P(OPh)_3\}_3 + P(OPh)_3 \tag{20}$$

$$S_N2 \quad Mo(CO)_6 + m \rightleftharpoons m \cdots Mo(CO)_5 + CO \tag{21}$$

The reactions accompanying the S_N1 ligand scission in tetrakis(triphenyl phosphite)nickel(0) in equation (20) in the presence of carbon tetrachloride in monomer solution (m = methyl methacrylate or styrene) are presented in Scheme 4.[63,64] The forward reaction in equation (a) is potentially rate determining and independent of the nature and concentration of the monomer, and the equilibrium in (b) is rapidly established and lies well over to the right. The net results of reactions (a) and (b) is, therefore, a ligand exchange with the monomer; the product $Ni\{P(OPh)_3\}_3(MMA)$ is more reactive towards CCl_4 than is $Ni\{P(OPh)_3\}_4$ and enters into the redox process (c). This gives rise to a Ni^I derivative which may participate in further redox reactions, e.g. (d).

$$Ni\{P(OPh)_3\}_4 \underset{k_2}{\overset{k_1}{\rightleftharpoons}} Ni\{P(OPh)_3\}_3 + P(OPh)_3 \tag{a}$$

$$Ni\{P(OPh)_3\}_3 + MMA \overset{K}{\rightleftharpoons} Ni\{P(OPh)_3\}_3(MMA) \tag{b}$$

$$Ni\{P(OPh)_3\}_3(MMA) + CCl_4 \overset{k_3}{\rightarrow} \cdot CCl_3 + Ni^I \tag{c}$$

$$Ni^I + CCl_4 \overset{k_4}{\longrightarrow} \cdot CCl_3 + Ni^{II} \tag{d}$$

Scheme 4

Direct kinetic observations[63,64] made on the changes in absorption spectra accompanying ligand exchange have permitted evaluation of k_1, k_2/K and K_0, the equilibrium constant for the overall process shown in equation (22) for m = methyl methacrylate or styrene. At $25\,°C$ $k_1 = 2.5 \times 10^{-4}\,s^{-1}$ with an activation enthalpy of $111\,kJ\,mol^{-1}$. This rate coefficient controls the rate of initiation for [halide] in the plateau region, and it is worth noting for comparison that the rate of thermal decomposition of azobis(isobutyronitrile) has approximately the same value in toluene at $86\,°C$.

$$Ni\{P(OPh)_3\}_4 + m \overset{K_0}{\rightleftharpoons} Ni\{P(OPh)_3\}_4 m + P(OPh)_3 \tag{22}$$

Both redox processes (c) and (d) generate initiating radicals and have been studied,[64,65] with different halides and methyl methacrylate and styrene as monomers, by observations on polymerization kinetics. In most of this work the appropriate differential equations were integrated numerically without assumption of stationary state conditions for any species containing a nickel atom. Stationary state conditions were applied to radical intermediates. As is apparent from Table 2, the rate of oxidation of Ni^0 to Ni^I measured by k_3 (reaction (c), Scheme 4) varies widely from one halide to another when methyl methacrylate is the monomer. The second redox stage (d) appears from the values of k_4 to be relatively slow except in the cases of CCl_4 and CBr_4; with CCl_4 the second oxidation may be more rapid than the first and has little influence on the kinetics.[63] The high reactivity of bromo derivatives compared to the corresponding chlorine compounds is noteworthy. Bamford and Sakamoto[65] conclude that both electron affinity of the halide and solvation of the transition complex play important roles in determining rate coefficients and activation parameters.

In styrene solution the second stage of oxidation by CCl_4 is much slower than in methyl methacrylate;[64] when CBr_4 is the halide this stage is rapid and the two redox processes cannot easily be separated. The kinetics of the oxidation of Ni^I by CCl_4 in styrene are complicated; partial dimerization of the species possibly occurs and the dependence of the rate on $[CCl_4]$ is not simple first order, so that no values for k_4 are quoted in Table 2.

The high reactivity of bromo derivatives extends to the $\cdot CBr_3$ radical, which can interact with reducing nickel species in a fashion reminiscent of a similar process[66] with $Mo(CO)_6$.

The thermodynamic behaviour of these systems has been discussed.[64] The Ni–monomer bond formed in reaction (b) of Scheme 4 is stronger for methyl methacrylate than for styrene; in the latter instance the dissociation energy is apparently less than 20% of that of the Ni–P bond broken in reaction (a).

Table 2 Absolute Rate Coefficients

Halide	T (°C)	k_3 (mol^{-1} dm^3 s^{-1})	k_4 (mol^{-1} dm^3 s^{-1})
Ni{P(OPh)$_3$}$_4$. Monomer: methyl methacrylate[63,65]			
CHBr$_3$	38	5.5×10^{-3}	1.5×10^{-4}
CBr$_4$	38	30	1
CH$_2$Br$_2$	25	3×10^{-6}	$< 10^{-6}$
	38	15×10^{-6}	$< 10^{-6}$
CCl$_3$CO$_2$Et	38	1.5	0
CHCl$_2$CO$_2$Et	38	4.0×10^{-3}	0
CHCl$_3$	25	4.5×10^{-5}	0
	38	1.43×10^{-4}	0
CH$_2$Cl$_2$	25	1×10^{-6}	0
	38	5.0×10^{-6}	0
CCl$_4$	6	1.1×10^{-1}	$> k_3$
	16	1.4×10^{-1}	$> k_3$
	25	1.4×10^{-1} (mean)	$> k_3$
Ni{P(OPh)$_3$}$_4$. Monomer: styrene[64]			
CCl$_4$	6.5	3×10^{-1}	
	15.5	6.3×10^{-1}	
	25	1.19	

Table 2 (*Continued*)

Halide		CCl$_4$		CBr$_4$		CCl$_3$CO$_2$Et
	T (°C)	$10^3 k_3$ (mol^{-1} dm^3 s^{-1})	T (°C)	$10^3 k_3$ (mol^{-1} dm^3 s^{-1})	T (°C)	$10^3 k_3$ (mol^{-1} dm^3 s^{-1})
Mo(CO)$_6$. Monomer: methyl methacrylate[70]						
	41.7	2.10	23.0	4.4	36.0	1.60
	49.0	4.47	42.0	40	42.0	2.70
	72.8	43.6	52.0	200	49.0	6.80
	80.0	79.4[a]	72.8	1260	69.2	22.0
			80.0	2510[a]	80.0	145[a]

	T (°C)	$10^2 k_2'$ (s^{-1})	k_2 (mol^{-1} dm^3 s^{-1})
	41.7	0.20	
	49.0	0.35	
	60.0	0.75	
	72.8	1.60	
	80.0	2.40[a]	456

[a] Extrapolated.

Systems in which the rate-determining reaction in the plateau region is an S_N2 ligand exchange are typified by molybdenum carbonyl (*cf.* equation 21),[67,68] which has received most detailed study. The major component reactions are presented in Scheme 5 for CCl$_4$ as halide.

$$\text{Mo(CO)}_6 + \text{m} \underset{k_2}{\overset{k_1}{\rightleftharpoons}} \text{m} \cdots \text{Mo(CO)}_5 + \text{CO} \qquad \text{(a)}$$
$$(4)$$

$$\text{m} \cdots \text{Mo(CO)}_5 \xrightarrow{k_2'} \text{inactive products} \qquad \text{(b)}$$

$$\text{m} \cdots \text{Mo(CO)}_5 + \text{CCl}_4 \xrightarrow{k_4} (5) + \text{CO} \qquad \text{(c)}$$

$$(5) \xrightarrow{k_4} \cdot\text{CCl}_3 + \text{Mo}^\text{I} \text{ derivative} \qquad \text{(d)}$$

Scheme 5

Reactions (c) and (d) constitute the redox process. Under stationary conditions Scheme 5 leads to relations (23), (24) for the rate of initiation $\mathscr{I}$ (assuming initiation by all $\cdot CCl_3$ radicals generated) and the rate of polymerization R_p, respectively. Equation (24) is consistent with the kinetic features (a)–(c). Note that plateau conditions correspond to equation (25) and that the plateau rate is given by equation (26).

$$\mathscr{I} = k_1 [Mo(CO)_6][m] \frac{[CCl_4]}{k_2'/k_3 + (k_2/k_3)[CO] + [CCl_4]} \tag{23}$$

$$R_p = k_p k_t^{-1/2} k_1^{1/2} [Mo(CO)_6]^{1/2}[m]^{3/2} \left\{ \frac{[CCl_4]}{k_2'/k_3 + (k_2/k_3)[CO] + [CCl_4]} \right\}^{1/2} \tag{24}$$

$$[CCl_4] \gg k_2'/k_3 + (k_2/k_3)[CO] \tag{25}$$

$$R_{p,\,plateau} = k_p k_t^{-1/2} k_1^{1/2} [Mo(CO)_6]^{1/2}[m]^{3/2} \tag{26}$$

The second term in the denominator of equation (24) accounts for the inhibition by added ligand. The sharpness S of the halide curve may be defined as the reciprocal of the halide concentration corresponding to a rate of polymerization $\varphi R_{p,\,plateau}$ which is a designated fraction of the plateau value, φ being constant and < 1. From equations (24) and (26) we can derive equation (27) from which the sharpness should increase with the activity of the halide (k_3) and should decrease in the presence of added ligand, as found experimentally [(b), (c), respectively]. The fact that the rate-determining step in the plateau region involves monomer has two consequences: (i) the rate of polymerization has an order 3/2 in monomer concentration (in the absence of active solvents, see below) and (ii) there is some degree of monomer selectivity, the rate of initiation depending on the character of the monomer. For example, methyl methacrylate appears more active than styrene in this respect. Some (coordinating) solvents are able to replace monomer in the ligand exchange; in these circumstances their concentrations appear in the rate equation. Examples of such solvents are ethyl acetate,[67] dioxane[68] and acetic anhydride.[68]

$$S = \left(\frac{1}{\varphi^2} - 1 \right) \frac{k_3}{k_2' + k_2[CO]} \tag{27}$$

Self-inhibition, which is encountered with some $Mo(CO)_6$/halide systems at higher concentrations, has been attributed to interaction between intermediates (4) and (5) leading to inactive products.[69]

The redox reaction (c), Scheme 5, has been studied by Bamford and Sakamoto.[70] Complex (4) was prepared photochemically in monomer solution and halide added; the progress of the reaction was then followed spectrophotometrically. Absolute values of k_2, k_2' and k_3 were thus obtained and are set out in Table 2. The radical yield of the reaction at low $[Mo(CO)_6]$ was found to be unity in the plateau region in accordance with Scheme 5.

The expected Mo^I derivative formed in reaction (d) undergoes a further secondary oxidation yielding Mo^V, which may be identified from its ESR spectrum.[71] The relation between ESR signal strength and reaction time is sigmoid; for short times the rate of development of free spins is much less than the rate of radical generation measured by polymerization kinetics, suggesting that the initial product (Mo^I) is undetected by ESR. With CCl_4 the secondary oxidation to Mo^V is relatively slow and yields two further radicals; thus some steps in the oxidation $Mo^I \rightarrow Mo^V$ occur without radical generation and may, for example, involve a two-electron transfer. With CBr_4 at $80\,°C$ secondary oxidation is rapid and the two steps cannot readily be separated.[57]

The relatively high reactivity of carbon tetrabromide in these process is evident from Table 2.

Activities of metal derivatives and halides

A list of the initiating activities of many transition metal derivatives may be found in ref. 55, which also presents the temperature dependence for selected cases.

Generally the derivatives of Group VIII metals are most active, for example, the nickel compounds $Ni\{P(OPh)_3\}_4$, $Ni(CO)_2\{PPh_3\}_2$ and $Ni(CO)_4$ initiate readily at room temperatures.[72,73] The activities of Group VIIA derivatives are rather low at temperatures below $80\,°C$; those of Group VIA are higher, especially for molybdenum, the most active metal in the group, $Mo(CO)_6$ having an activity comparable to azobis(isobutyronitrile). Thus it is possible to choose a metal derivative which is suitable for initiation in any part of the temperature range normally used in polymerization.

As already pointed out, halide activity is increased by a change from chlorine to bromine, as exemplified in Table 2. Fluorides and iodides are inactive. Multiple substitution in the halide also increases the activity (Table 2), as does location of the halogen atom at an allylic or benzylic position. The presence of an electron-attracting group such as carboxyl in the molecule often increases the activity. These factors closely parallel those which determine the rate of capture of thermal electrons by a halide as might be anticipated from equation (19).[74] N-Haloamides, imides and urethanes are highly active and yield the corresponding nitrogen-centred radicals.

(i) Photoinitiation

Most of the systems described can photoinitiate free-radical polymerization. Koerner von Gustorf and colleagues[75-77] described early observations, mostly on iron compounds, and suggested mechanisms closely related to those already described[55] for thermal reactions. Strohmeier and co-workers[78] reported photoinitiation of polymerization of ethyl acrylate and vinyl chloride by transition metal carbonyls and related derivatives in the presence of carbon tetrachloride but advanced no mechanisms.

Manganese and rhenium carbonyls ($Mn_2(CO)_{10}$, $Re_2(CO)_{10}$) are probably the most useful photoinitiators of this type. The systems $Mn_2(CO)_{10}/CCl_4$ and $Re_2(CO)_{10}/CCl_4$ were first investigated as photoinitiators by Bamford et al.[79,80] Long wave limits of absorption are situated at approximately 460 and 380 nm, respectively, and photoinitiation occurs up to these wave lengths. Both systems exhibit halide curves similar to those found in the thermal reactions; the halide curves for $Re_2(CO)_{10}$ are remarkably sharp, those for $Mn_2(CO)_{10}$ become increasingly sharp as the light intensity is reduced. The quantum yield for initiation in the plateau region is effectively unity for the in-source reaction (see below).

Both carbonyls can give rise to unexpectedly long photo-after-effects when used as photoinitiators with common monomers, which may extend to periods of hours. With $Mn_2(CO)_{10}$ the long-lived after-effect requires the presence of a coordinating additive such as acetylacetone,[81] but in the case of $Re_2(CO)_{10}$ no special additive is required[80] and it is supposed that the vinyl monomer behaves in this way. These prolonged after-effects, which must not be confused with normal photo-after-effects arising from the finite rate of bimolecular termination, are useful in practice (see below).

The view was early expressed[80] that photodissociation of the carbonyl yields two dissimilar fragments, only one of which reacts rapidly with the coinitiator while the second regenerates the original carbonyl. This suggestion, which has persisted in many subsequent studies,[82-85] offers the simplest explanation of the overall stoichiometry, which is expressed by the equation (28).

$$Mn_2(CO)_{10} + CCl_4 \xrightarrow{h\nu} Mn(CO)_5Cl + \tfrac{1}{2}Mn_2(CO)_{10} + \cdot CCl_3 \qquad (28)$$

It is not supported by flash photolysis studies of $Mn_2(CO)_{10}$ in non-polar liquids, although a fairly recent report[86] suggests that two primary processes occur giving $2Mn(CO)_5$ and $Mn_2(CO)_9$ + CO respectively. In view of the important effect played by the solvent, the writer believes[61,62] that Scheme 6 represents the most satisfactory mechanism of photoinitiation by these carbonyls in coordinating media. Here s represents a coordinating molecule, which for $Re_2(CO)_{10}$ may be monomer.

$$Mn_2(CO)_{10} + h\nu \underset{a}{\rightarrow} Mn_2(CO)_{10}^* \overset{s}{\underset{b}{\rightarrow}} (CO)_5Mn\text{-}s\text{-}Mn(CO)_5$$

$$(CO)_5Mn\text{-}s\text{-}Mn\text{-}(CO)_5 \underset{c}{\rightarrow} (CO)_5Mn\text{-}s + Mn(CO)_5 \underset{d}{\rightarrow} Mn_2(CO)_{10} + s$$

$$(6)$$

$$(CO)_5Mn\text{-}s + CCl_4 \underset{e}{\rightarrow} Mn(CO)_5Cl + s + \cdot CCl_3 \text{ (slow)}$$

$$(CO)_5Mn + CCl_4 \underset{f}{\rightarrow} Mn(CO)_5Cl + \cdot CCl_3 \text{ (fast)}$$

Scheme 6

By virtue of the intervention of the solvent or coordinating additive two dissimilar species are formed on dissociation, having different reactivities. The coordinated species (6) reacts slowly (e) with the coinitiator and is responsible for the prolonged after-effect, while the other fragment $(CO)_5Mn$ reacts rapidly (f) and is responsible for most of the in-source initiation. Competition for the fragments $(CO)_5Mn\text{-}s$ and $(CO)_5Mn$ between the second-order recombination (d) and the redox reactions (e), (f) in Scheme 6, accounts for the observation referred to earlier that the sharpness of the halide curves increases with decreasing light intensity. Scheme 6 is also consistent with the experimental value close to unity for the quantum yield of radical generation of the in-source reaction.[79, 80]

(ii) Applications

These initiators have the practical advantage of consisting of components which are mostly stable at room temperatures. For example, tetrakis{triphenyl phosphite} nickel(0), which, in the presence of carbon tetrachloride, initiates below ambient temperatures and does not decompose rapidly below its melting point (145 °C) in the solid state.

Many of the derivatives (*e.g.* $Mn_2(CO)_{10}$) may be readily purified by sublimation in vacuum.

Type 1 initiation by the systems under discussion provides a synthetic route to polymers having specified terminal groups, by virtue of the fact that with a coinitiator R–X (X = Cl, Br) the initiating radical R• becomes a terminal group in the polymer chain (see equation 19a). The final polymer molecule therefore carries one or two such residues, depending on whether the chain terminates by disproportionation or combination, respectively. This procedure is of wide application since the coinitiator may be almost any type of molecule containing a halogen substitutent or into which the latter may be introduced.[87] For example, bioactive species have been incorporated as terminal groups[29, 88] and Eastmond *et al.*[89] have grafted polymer to glass surfaces in this way.

When the coinitiator is a polymer containing active halogen the same reactions lead to block[90] or graft[91-96] formation, as typified in Scheme 7. This presents the simplest cases, in which termination of the chains of m^2 is by disproportionation. With combination, the product in (a) would be a three-block species, while in (b) crosslinks instead of simple grafts would be formed, so that network formation and gelation would ensue. Measurement of gel times in such systems and application of Flory's gelation theory[97] have been used to estimate the ratio of combination to disproportionation in the terminaton of m^2 chains. The occurrence of chain transfer to solvent or monomer, initiator decay or formation of unattached primary radicals delay the gel time and introduce errors into the simple treatment. Equations for the gel time when these phenomena are significant have been developed.[92, 93]

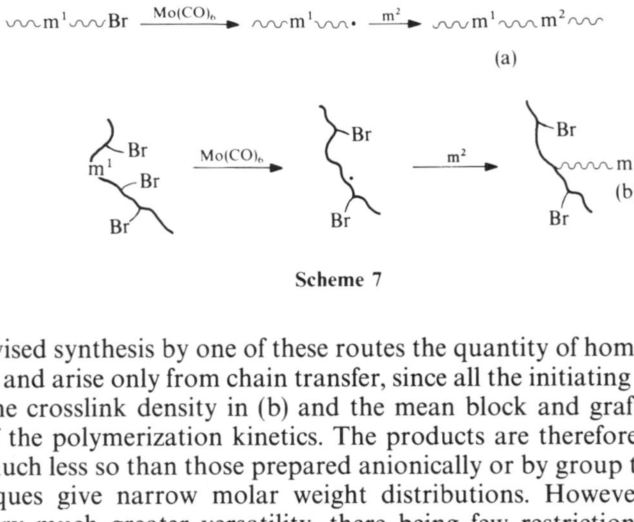

Scheme 7

In a properly devised synthesis by one of these routes the quantity of homopolymer of m^2 formed should be minimal, and arise only from chain transfer, since all the initiating radicals are attached to polymer chains. The crosslink density in (b) and the mean block and graft lengths are calculable from knowledge of the polymerization kinetics. The products are therefore relatively well characterized, although much less so than those prepared anionically or by group transfer polymerization, since these techniques give narrow molar weight distributions. However, the syntheses under discussion have very much greater versatility, there being few restrictions on the nature of the coinitiator and the monomer m^2. Note that no metal becomes chemically bound to the polymeric products in these reactions.

The wide range of applications of the techniques may be illustrated by their use in the synthesis of block copolymers in which each block is an alternating copolymer,[98] the synthesis of block copolymers of polypeptides and vinyl polymers[99] and, in the biomaterials field, to the grafting of polymer chains carrying antiplatelet agents to poly(ether urethanes).[29, 88, 100] Variants of the reactions in Scheme 7 have been studied in connection with the formulation of new (negative) photoresists and presensitized lithographic plates.[101]

Eastmond and his colleagues[102] have investigated the details of the component reactions, notably the termination process, and have made interesting morphological studies[103] with the products.

The reactions in Scheme 7 may be elaborated in some obvious ways, *e.g.* by incorporating halogen atoms in both terminal groups of the polymeric coinitiator molecule to obtain products with a larger number of blocks per chain. A recent report by Bamford *et al.*[104] points out the value of (readily available) halogen-containing isocyanates, *e.g.* chloromethyl, 2-chloroethyl, trichloromethyl, trichloroacetyl isocyanates in extending these syntheses. The highly reactive isocyanate functions couple readily to many different types of groups encountered in polymers as main chain, side chain or terminal units, *e.g.* –OH, –COOH, –NH₂, –OOCNH–, thereby introducing halogen atoms which can act as coinitiators.

The syntheses outlined may be carried out thermally or photochemically; often photochemical initiation with the aid of $Mn_2(CO)_{10}$ or $Re_2(CO)_{10}$ is expeditious and allows the reactions to be preformed under mild conditions. 'Photo-after-effect grafting'[61] is convenient to apply to preformed materials at ambient temperatures. After irradiation of a solution of $Mn_2(CO)_{10}$ or $Re_2(CO)_{10}$ in monomer containing the coordinating additive (if necessary) to form the long-lived intermediate (**6**), Scheme 6, the liquid is poured (in vacuum) on to the substrate to be grafted (the coinitiator); grafting then proceeds thermally. This technique avoids the difficulty of irradiating an irregular solid uniformly as required for in-source grafting, and it is suitable for delicate materials. For example, it has been used for grafting to a poly(ester-urethane) vascular prosthesis.[61,104]

Arene chromium tricarbonyls ($arCr(CO)_3$) have been studied as Type 1 photoinitiators.[105–107] The order of quantum yields ($\lambda = 365$ nm) with methyl methacrylate as monomer, fluorobenzene > chlorobenzene > benzene > toluene > anisole is probably the same as the order of ar–Cr bond energies.[106]

9.4.2.2 *Type 2 initiation*

Recognition of the possibility of this mode of initiation started from the observation[83] that pure liquid tetrafluoroethylene at $-93\,^\circ C$ containing a low concentration of manganese or rhenium carbonyls polymerizes rapidly on irradiation. (A thermal reaction between tetrakis(triphenyl phosphite)nickel(0) and maleic anhydride had been reported earlier[108] to generate radicals; this process, although probably Type 2, is inefficient). It soon became clear[83–85] that C_2F_4 is an effective coinitiator at ambient temperatures with common vinyl monomers. The dependence of rate of polymerization on $[C_2F_4]$ is of the form of the familiar halide curve; however, initiation does not involve halide abstraction and the polymeric products have end groups carrying metal atoms. The structure of the terminal portions of the chains of poly(methyl methacrylate) prepared with $Mn_2(CO)_{10}/C_2F_4$ photoinitiation are shown in (**7**). Such end groups and those from $Re_2(CO)_{10}$ are readily detected by their IR absorption bands near 2100 cm^{-1}.[85,109–111] The initiating radicals are therefore represented by (**8**); radical formation is thus the result of *addition* of a metal-containing fragment to the coinitiator.

$$(CO)_5Mn-CF_2CF_2CH_2\overset{\overset{\displaystyle Me}{|}}{\underset{\underset{\displaystyle CO_2Me}{|}}{C}}\!\!\sim \qquad (CO)_5Re-\overset{\overset{\displaystyle R^1}{|}}{\underset{\underset{\displaystyle R^2}{|}}{C}}-\overset{\overset{\displaystyle R^3}{|}}{\underset{\underset{\displaystyle R^4}{|}}{C}}\!\!\cdot \qquad (CO)_5Re-\overset{\overset{\displaystyle R^1}{|}}{\underset{\underset{\displaystyle R^2}{|}}{C}}-\overset{\overset{\displaystyle R^3}{|}}{\underset{\underset{\displaystyle R^4}{|}}{C}}\!\!\cdot$$

$$(7) \qquad\qquad\qquad (8) \qquad\qquad\qquad (9)$$

Coinitiators other than fluoroalkenes have been identified. Types have already been listed (Section 9.4.2) and activities are presented in Table 3.[109,111] The data illustrate the importance of structural factors. To show high activity a Type 2 coinitiator must be able to form strong bonds with the transition metal. In the alkenes in Table 3 the bond strength is enhanced by the presence of electronegative substituents such as fluorine, carbonyl and nitrile. However, steric factors are also very important.[111] Thus, there is much steric compression in (**9**) with $R^1 = R^2 = CO_2Me$ or CO_2Et, consequently $C(CO_2Et)_2=C(CO_2Et)_2$ is a feeble coinitiator, with quantum yield of initiation $\varphi = 0.06$. If $R^1 = H$ the compression is relieved, and $CH(CO_2Me)=C(CO_2Me)_2$ is active with $\varphi = 0.74$ (Table 3). Finally, $Re_2(CO)_{10}$ is more active than $Mn_2(CO)_{10}$. This is also likely to be a bond energy effect, since Re–C is generally much stronger than Mn–C.[112]

The mechanism of Type 2 photoinitiation by $Mn_2(CO)_{10}$ and $Re_2(CO)_{10}$ has been discussed in refs. 84, 85, 113.

Type 2 is not confined to photochemical processes. Manganese pentacarbonyl chloride is an interesting thermal initiator which is active with coinitiators of both types.[114] In methyl methacrylate (MMA) solution at $60\,^\circ C$, when the compound is approximately ten-fold as active as azobis-(isobutyronitrile), it has been proposed that the initiating species is $Mn^0(MMA)(CO)_3$ derived by

Table 3 Activities of Type 2 Coinitiators: Quantum Yields of Initiation[109,111]

Additive (concentration 0.2 mol dm^{-3})	$Re_2(CO)_{10}$ ($\lambda = 365$ nm)	$Mn_2(CO)_{10}$ ($\lambda = 435.8$ nm)
$CH_2=CHCO_2Me$	0	0
cis-$CH(CO_2Et)=CH(CO_2Et)$	0.18	0
trans-$CH(CO_2Et)=CH(CO_2Et)$	0.77	0
$CH_2=C(CO_2Et)_2$	0.14	0
$CH(CO_2Me)=C(CO_2Me)_2$	0.74	0
$C(CO_2Et)_2=C(CO_2Et)_2$	0.06	0
$CH_2=CHCN$	0.07	0
trans-$CHCN=CHCN$	0.76	0
$CF_2=CF_2$[(a)]	1.0	0.9
$CF_2=CFCl$[(a)]	0.87	0.72
$CF_2=CCl_2$[(a)]	0.65	0.02
$CF_3CF=CF_2$	1.0	0.08
$CF_2=CFCF=CF_2$	1.0	0
(fluorobenzene)	0	0
(fluoro-$CF=CF_2$ benzene)	1.0	0
$CH\equiv CH$[b]	0.90	0
$C(CO_2H)\equiv C(CO_2H)$[a]	0.98	0.42
$C(CO_2Me)\equiv C(CO_2Me)$[a]	0.98	0.61

[a] concentration 0.1 mol dm^{-3}. [b] concentration 0.4 mol dm^{-3}.

disproportionation of the thermally formed dimer (**10**), which then initiates in the two modes as shown in equations (30) and (31).

$$(CO)_3Mn \underset{Cl}{\overset{MMA\ Cl}{\diamondsuit}} Mn(CO)_3 \longrightarrow Mn^0(MMA)(CO)_3 + MnCl_2 + MMA + 3CO \qquad (29)$$

(**10**)

$$Mn(MMA)(CO)_3 + CCl_4 \xrightarrow{-2CO} Mn(CO)_5Cl + MMA + {}^\bullet CCl_3 \qquad (30)$$

$$Mn(MMA)(CO)_3 + C_2F_4 \xrightarrow{2CO} (CO)_5MnCF_2\dot{C}F_2 + MMA \qquad (31)$$

The activation energy for radical generation in the plateau region (*i.e.* that of reaction 29) is approximately 109 kJ mol^{-1} and the process is likely to be monomer selective.

(i) Applications

Polymers such as (**7**) are normally prepared at or below ambient temperatures. At higher temperatures, for example 100 °C, they become unstable and decompose with generation of macroradicals (equation 32). Consequently if the decomposition is carried out in the presence of a monomer a block copolymer is formed, consisting, in general, of chains of polymers of the two monomers linked by a unit of the coinitiator.[115] In view of these properties, macromolecules typified by (**7**) have been termed 'macroinitiators'.[115] Radical generation (equation 32) is retarded by carbon monoxide and it is possible that the primary step is the scission of CO.[116]

$$(\mathbf{7}) \rightarrow Mn(CO)_5 + {}^\bullet CF_2CF_2CH_2\overset{\overset{\displaystyle Me}{|}}{\underset{\underset{\displaystyle CO_2Me}{|}}{C}}\!\!\sim\!\!\sim \qquad (32)$$

The concept of macroinitiators has been extended to include multifunctional species carrying side chains with $Mn(CO)_5$ terminations, suitable for the synthesis of graft copolymers.[117]

As with Type 1, there are few restrictions on the natures of the monomers which can be used in these reactions, so that Type 2 forms a versatile technique for synthesis of block and graft copolymers by free-radical methods. However, so far the practical applications are less well developed than those of Type 1.

9.4.2.3   *Initiation by metal chelates*

Both cationic[118] and anionic[119] polymerizations initiated by chelates have been reported. The first demonstration of the thermal generation of radicals from chelates and the initiation of polymerization was due to Arnett and Mendelsohn,[120] who showed that manganese(III) acetylacetonate ($Mn^{III}(acac)_3$) has a much higher activity than corresponding derivatives of the other metals examined. Kastning *et al.*[121] extended these findings by noting that the rate of polymerization of styrene at 110 °C is proportional to $[Co^{III}(acac)_3]^{\frac{1}{2}}$, as expected for a free-radical mechanism. These workers found that '... the ligand combines with the polymer', but doubted whether there was a change in the oxidation state of the metal.

Bamford and Lind,[122] working with $Mn^{III}(acac)_3$ and methyl methacrylate and styrene as monomers, confirmed that the square root relation holds at 80 °C and showed from measurements of $k_p k_t^{-\frac{1}{2}}$ that no retardation occurs. Reduction to the Mn^{II} state was indicated by the magnetic susceptibilities of the manganese products. Further, the rate of initiation was found to depend on the nature of the solvent and it was concluded[55,122] that radical generation is preceded by coordination of monomer or solvent to the metal as shown in Scheme 8, reactions (a), (b) (s represents a coordinating solvent). The concomitant rupture of an Mn–O bond (k_1 or k_2) is considered rate determining so that the rate of radical formation (assumed to equal the rate of initiation $\mathscr{I}$) is given by

$$\mathscr{I} = [Mn(acac)_3](k_1[m] + k_2[s]) \tag{33}$$

The order in [m] of the overall polymerization reaction therefore lies between 1.5 for an inert solvent (*e.g.* benzene) and 1.2, approximately, for a coordinating solvent (*e.g.* ethyl acetate). The primary radical is derived from the ligand by a redox process (c) or (d) with Mn^{III} and so becomes attached to the polymer as a terminal group (*cf.* Kastning *et al.*).[121] For Cu^{II} ethylacetoacetate and styrene, Nishikawa and Otsu[123] reported a rate of polymerization proportional to $[styrene]^n [chelate]^{\frac{1}{2}}$, n lying between 1.3 and 2.4 depending on the solvent.

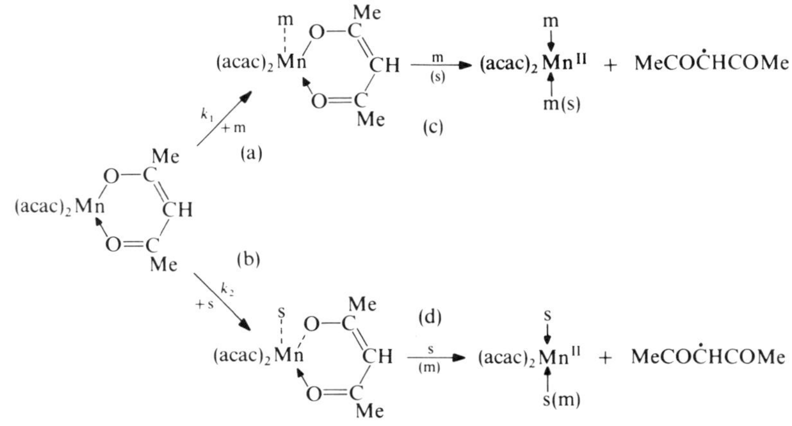

Scheme 8

The influence of substituents in the acetylacetonate ligand has been examined. Manganese(III) 1,1,1-trifluoroacetylacetonate ($Mn^{III}(facac)_3$) shows remarkable monomer selectivity, apparently initiating polymerization only of monomers with electron-attracting groups.[122] Thus it initiates free-radical polymerization of methyl methacrylate and acrylonitrile rapidly, but is inactive towards styrene and vinyl acetate. Scheme 9 shows the mechanism advanced to account for this behaviour; heterolytic dissociation of Mn–O (a) is followed by coordination of monomer to the anion (b) and subsequent electron transfer to the adjacent Mn^+ (c). Detachment of the ligand and initiation then proceed as in (d).

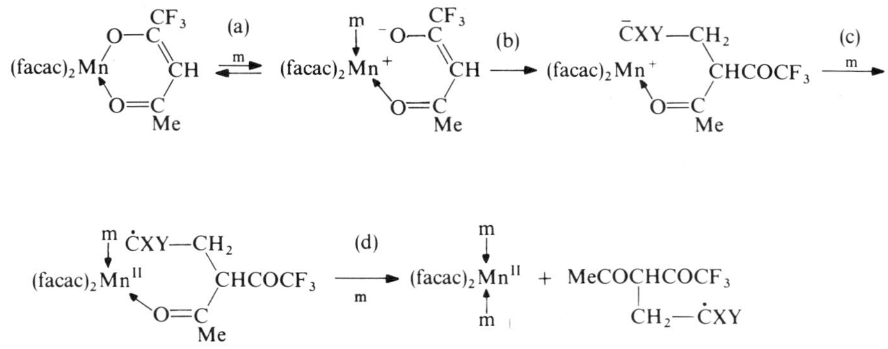

Scheme 9

Many different types of additive have been reported to 'activate' metal chelates,[121] including dimethyl sulfoxide (DMSO),[124] pyridine, piperidine and 2,6-dimethylpiperidine,[125] and carbonyl compounds.[126] Kaeriyama[126] concluded that the effectiveness of ketones depends on their nucleophilicity. Bamford and Ferrar[127] examined the kinetics of initiation in the presence of electron donors such as DMSO and various amines at 25 °C. These latter greatly increase the rate of initiation, with DMSO or 1,2-diaminopropane up to 1000-fold, so that the systems become effective as room temperature initiators. At the same time monomer selectivity appears, of the type described above but less extreme. Probably both non-selective and selective components contribute; in the former the electron donor may behave as in (b), Scheme 8, while in the latter it gives rise to dipolar species as in Scheme 9. The primary radicals are derived from the ligand so that substituents in the latter are incorporated as terminal groups in the polymer chains.

Thermal initiation by the following chelates of Mn^{III} has also been reported:[122] tris(1,3-diphenyl-1,3-propanediono), tris(1,3-di-p-chlorophenyl-1,3-propanediono) and tris(3-phenyl-2,4-pentanediono).

Photoinitiation by chelates has received some attention. Quantum yields of photoinitiation by $Mn^{III}(acac)_3$ and $Mn^{III}(facac)_3$ are small, and no monomer selectivity is encountered with the later.[127] Aliwi and Bamford[128] have studied chelates of vanadium(V). Chloroxobis(2,4-pentanediono)-vanadium(V) (10) initiates in the near UV ($\lambda = 365$ nm) with low quantum yield ($\sim 2 \times 10^{-2}$) according to equation (34).[129]

$$V^V O(acac)_2 Cl + h\nu \rightarrow V^{IV} O(acac)_2 + {}^\bullet Cl \qquad (34)$$

$$\underline{}$$

(10)

The activity is greatly increased in the presence of electron donors: thus with DMSO (10% v/v) the quantum yield at $\lambda = 365$ nm is 0.59 and both chlorine and DMSO residues are found as terminal groups.[130] The related chelate methoxo-oxobis(8-quinolyloxo)vanadium(V) (11) also photoinitiates at $\lambda = 365$ nm, generating ${}^\bullet CH_2OH$ and $MeO^\bullet$ radicals and forming the V^{IV} chelate (12)[131] (equation 36). Photoactive red polymers incorporating V^V chelates have been described.[132,133]

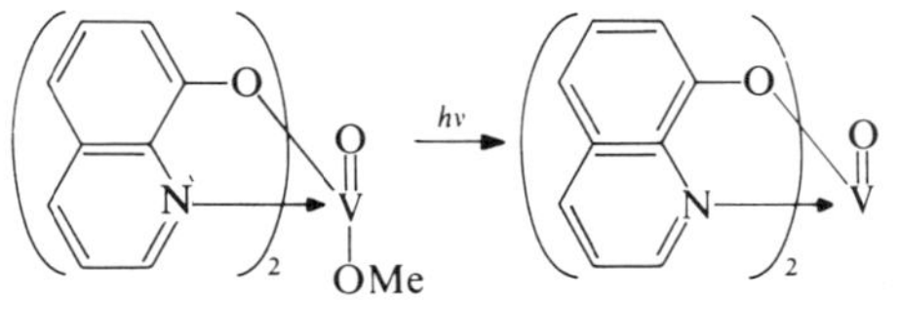

(11)   **(12)**

$$+ {}^\bullet CH_2OH + MeO^\bullet \qquad (35)$$

9.5 REFERENCES

1. R. G. R. Bacon, *Trans. Faraday Soc.*, 1946, **42**, 140.
2. L. B. Morgan, *Trans. Faraday Soc.*, 1946, **42**, 169.
3. W. Kern, *Makromol. Chem.*, 1948, **1**, 199, 209, 229, 249; 1948, **2**, 48.
4. R. G. R. Bacon, *Q. Rev., Chem. Soc.*, 1955, **9**, 287.
5. F. A. Bovey, I. M. Kolthoff, A. I. Medalia and E. J. Mecham, 'Emulsion Polymerization', Interscience, New York, 1955.
6. E. W. Duck, in 'Encyclopedia of Polymer Science and Technology', Wiley, New York, 1966, Vol. 5, p. 801.
7. J. W. Vanderhoff, in 'Vinyl Polymerization', part II, ed. G. E. Ham, Marcel Dekker, New York, 1969, chap. 1.

8. W. Cooper, in 'Reactivity Mechanism and Structure in Polymer Chemistry', eds. A. D. Jenkins and A. Ledwith, Wiley, London, 1974, chap. 8
9. R. M. Noyes, in 'Progress in Reaction Kinetics', ed. G. Porter, Pergamon, Oxford, 1961, chap. 5; 'Encyclopedia of Polymer Science and Technology', Interscience, New York, 1965, vol. 2.
10. H. J. W. Fenton, *J. Chem. Soc.*, 1894, **65**, 899.
11. F. Haber and R. Willstätter, *Ber. Dtsch. Chem. Ges.*, 1931, **64**, 2844.
12. F. Haber and J. Weiss, *Proc. R. Soc. London, Ser. A.*, 1934, **147**, 332.
13. N. Uri, *Chem. Rev.*, 1952, **50**, 375.
14. W. G. Barb, J. H. Baxendale, P. George and K. R. Hargrave, *Trans. Faraday Soc.*, 1951, **47**, 462.
15. J. H. Baxendale, M. G. Evans and G. S. Park, *Trans. Faraday Soc.*, 1946, **42**, 155.
16. J. H. Merz and W. A. Waters, *Discuss. Faraday Soc.*, 1947, **2**, 179.
17. G. C. Eastmond in 'Comprehensive Chemical Kinetics', eds. C. H. Bamford and C. F. H. Tipper, Elsevier, Amsterdam, 1976, vol. 14A, chap. 1.
18. R. K. Samal, R. R. Das, M. C. Nayak, G. V. Suryanarayana, G. Panda and D. P. Das, *J. Polym. Sci., Polym. Chem. Ed.*, 1981, **19**, 2751.
19. R. K. Samal, M. C. Nayak and D. P. Das, *Eur. Polym. J.*, 1982, **18**, 313.
20. C. Walling, 'Free Radicals in Solution', Wiley, New York, 1957.
21. R. J. Orr and H. L. Williams, *J. Am. Chem. Soc.*, 1955, **77**, 3715; *Can. J. Chem.*, 1952, **30**, 985.
22. J. K. Kochi and P. J. Krusic, *Spec. Publ.-Chem. Soc.*, 1970, **24**, 147.
23. I. M. Kolthoff, W. Stricks and N. Tanaka, *J. Am. Chem. Soc.*, 1955, **77**, 5215.
24. P. Davis, M. G. Evans and W. C. E. Higginson, *J. Chem. Soc.*, 1951, 2563.
25. M. G. Evans, J. H. Baxendale and D. J. Cowling, 'The Labile Molecule', *Discuss. Faraday Soc.*, 1947, **2**, 206.
26. H. L. J. Bäckström, *Z. Phys. Chem. (Leipzig)*, 1934, **B25**, 122.
27. C. E. H. Bawn and A. G. White, *J. Chem. Soc.*, 1951, 331, 339, 343.
28. G. Mino, S. Kaizerman and E. Rasmussen, *J. Polym. Sci.*, 1958, **31**, 242; 1958, **38**, 393, 523; *J. Am. Chem. Soc.*, 1959, **81**, 1494.
29. C. H. Bamford, I. P. Middleton, Y. Sataka and K. G. Al-Lamee, in 'Advances in Polymer Synthesis', eds. B. M. Culbertson and J. E. McGrath, Plenum Press, New York, 1986, p. 291.
30. A. Hebeish and J. T. Guthrie, 'The Chemistry and Technology of Cellulosic Copolymers,' Springer, Berlin, 1981.
31. F. S. Dainton and D. G. L. James, *Trans. Faraday Soc.*, 1958, **54**, 649.
32. G. Oster, *Nature (London)*, 1954, **173**, 300.
33. G. Oster and Y. Mizutani, *J. Polym. Sci.*, 1956, **22**, 177.
34. G. K. Oster, G. Oster and J. R. Nussbaum, *Polym. Prepr., Am. Chem. Soc., Div. Polym. Chem.*, 1961, **1**, 290.
35. G. K. Oster, G. Oster and G. Prati, *J. Am. Chem. Soc.*, 1957, **79**, 595.
36. G. Delzenne, S. Toppet and G. Smets, *J. Polym. Sci.*, 1960, **48**, 347.
37. G. Oster and N.-L. Yang, *Chem. Rev.*, 1968, **68**, 125.
38. M. J. Little, J. D. Margerum, A. M. Lackner and C. T. Petrusis, *J. Phys. Chem.*, 1971, **75**, 3066.
39. G. A. Delzenne, H. K. Peeters and U. L. Laridon, IUPAC Symposium, Leuven, June, 1972.
40. J. P. Riggs and F. Rodriguez, *J. Polym. Sci., Part A-1*, 1967, **5**, 3167.
41. F. Rodriguez and R. D. Givey, *J. Polym. Sci.*, 1961, **55**, 713.
42. P. D. Bartlett and K. Nozaki, *J. Am. Chem. Soc.*, 1947, **69**, 2299.
43. L. Horner and E. Schwenk, *Angew. Chem.*, 1949, **61**, 411; *Justus Liebigs Ann. Chem.*, 1949, **566**, 69.
44. L. Horner and C. Betzel, *Justus Liebigs Ann. Chem.*, 1953, **579**, 175.
45. L. Horner, *J. Polym. Sci.*, 1955, **18**, 438.
46. L. Horner and K. Sherf, *Justus Liebigs Ann. Chem.*, 1951, **573**, 35.
47. L. Horner and W. Kirmse, *Justus Liebigs Ann. Chem.*, 1955, **597**, 48.
48. H. Hirai, *J. Macromol. Sci., Chem.*, 1975, **A9**, 883.
49. H. Hirai and M. Komiyama, *J. Polym. Sci., Polym. Chem. Ed.*, 1975, **13**, 2419.
50. M. Komiyama and H. Hirai, *J. Polym. Sci., Polym. Chem. Ed.*, 1976, **14**, 307.
51. F. S. Arimoto, *J. Polym. Sci.*, 1966, **A4**, 275.
52. C. H. Bamford, in 'Molecular Behaviour and the Development of Polymeric Materials', eds. A. Ledwith and A. M. North, Chapman and Hall, London, 1975, chap. 2.
53. R. K. Freydlina and A. B. Belyavskii, *Dokl. Akad. Nauk SSSR*, 1959, **127**, 1027; *Izv. Akad. Nauk SSSR, Ser. Khim.*, 1961, **1**, 177.
54. C. H. Bamford and C. A. Finch, *Proc. R. Soc. London, Ser. A*, 1962, **268**, 553.
55. C. H. Bamford, in 'Reactivity Mechanism and Structure in Polymer Chemistry', eds. A. D. Jenkins and A. Ledwith, Wiley, London, 1974, chap. 3.
56. C. H. Bamford, G. C. Eastmond and V. J. Robinson, *Trans. Faraday Soc.*, 1964, **60**, 751.
57. C. H. Bamford, G. C. Eastmond and F. J. T. Fildes, *Proc. R. Soc. London, Ser. A*, 1972, **326**, 453.
58. C. H. Bamford, G. C. Eastmond and D. Whittle, *Polymer*, 1969, **10**, 771.
59. E. Yoshisato and S. Tsutumi, *J. Am. Chem. Soc.*, 1968, **90**, 4488.
60. C. H. Bamford, F. J. Duncan, R. J. W. Reynolds and J. D. Seddon, *J. Polym. Sci., Part C*, 1968, **23**, 419.
61. C. H. Bamford and I. P. Middleton, *Eur. Polym. J.*, 1983, **19**, 1027.
62. C. H. Bamford, I. P. Middleton, IUPAC Macro '83 (Bucharest, Romania), Plenary and Invited Lectures, Part 2, p. 168.
63. C. H. Bamford and E. O. Hughes, *Proc. R. Soc. London, Ser. A*, 1972, **326**, 469.
64. C. H. Bamford and E. O. Hughes, *Proc. R. Soc. London, Ser. A*, 1972, **326**, 489.
65. C. H. Bamford and I. Sakamoto, *J. Chem. Soc., Faraday Trans. 1.*, 1974, **70**, 330.
66. C. H. Bamford, G. C. Eastmond and F. J. T. Fildes, *Chem. Commun.*, 1970, 146; *Proc. R. Soc. London, Ser. A*, 1972, **326**, 431.
67. C. H. Bamford, R. Denyer and G. C. Eastmond, *Trans. Faraday Soc.*, 1965, **61**, 1459.
68. C. H. Bamford, R. Denyer and G. C. Eastmond, *Trans. Faraday Soc.*, 1966, **62**, 688.
69. C. H. Bamford, G. C. Eastmond and W. R. Maltman, *Trans. Faraday Soc.*, 1966, **62**, 2531.
70. C. H. Bamford and I. Sakamoto, *J. Chem. Soc., Faraday Trans. 1.*, 1974, **70**, 344.

71. C. H. Bamford, G. C. Eastmond and F. J. T. Fildes, *Proc. R. Soc. London, Ser. A*, 1972, **326**, 431.
72. C. H. Bamford and K. Hargreaves, *Trans. Faraday Soc.*, 1967, **63**, 392.
73. C. H. Bamford, G. C. Eastmond and P. Murphy, *Trans. Faraday Soc.*, 1970, **66**, 2598.
74. K. R. Wilson and D. R. Herschbach, *Nature (London)*, 1965, **208**, 182.
75. E. Koerner von Gustorf and F. W. Grevels, *Fortschr. Chem. Forsch.*, 1969, **13**, 366.
76. E. Koerner von Gustorf, M. C. Henry and C. DiPietro, *Z. Naturforsch., Teil B*, 1966, **21**, 42.
77. E. Koerner von Gustorf, M.-J. Jun and G. O. Schenck, *Z. Naturforsch., Teil B*, 1963, **18**, 503.
78. W. Strohmeier and P. Hartmann, *Z. Naturforsch., Teil B*, 1964, **19**, 882; W. Strohmeier and H. Grubel, *Z. Naturforsch., Teil B*, 1967, **22**, 98, 553.
79. C. H. Bamford, P. A. Crowe and R. P. Wayne, *Proc. R. Soc. London, Ser. A*, 1965, **284**, 455.
80. C. H. Bamford, P. A. Crowe, J. Hobbs and R. P. Wayne, *Proc. R. Soc. London, Ser. A*, 1966, **292**, 153.
81. C. H. Bamford and J. Paprotny, *Polymer*, 1972, **13**, 208.
82. C. H. Bamford, *Pure Appl. Chem.*, 1973, **34**, 173.
83. C. H. Bamford and S. U. Mullik, *Polymer*, 1973, **14**, 38.
84. C. H. Bamford and S. U. Mullik, *J. Chem. Soc., Faraday Trans. 1*, 1973, **69**, 1127.
85. C. H. Bamford and S. U. Mullik, *J. Chem. Soc. Faraday Trans. 1*, 1975, **71**, 625.
86. S. P. Church, H. Hermann, F.-W. Grevels and K. Schaffner, *J. Chem. Soc., Chem. Commun.*, 1984, 785.
87. See A. K. Alimoglu, C. H. Bamford, A. Ledwith and S. U. Mullik, *Macromolecules*, 1977, **10**, 1081.
88. C. H. Bamford, I. P. Middleton, K. G. Al-Lamee, J. Paprotny and Y. Satake, *Polym. J.* 1987, **19**, 475.
89. G. C. Eastmond, C. Nguyen-Huu and W. H. Piret, *Polymer*, 1980, **21**, 598.
90. C. H. Bamford, *Eur. Polym. J., suppl.*, 1969, 1.
91. C. H. Bamford, R. W. Dyson and G. C. Eastmond, *J. Polym. Sci., Part C*, 1967, **16**, 2425.
92. C. H. Bamford, R. W. Dyson and G. C. Eastmond, *Polymer*, 1969, **10**, 885.
93. C. H. Bamford, R. W. Dyson, G. C. Eastmond and D. Whittle, *Polymer*, 1969, **10**, 759.
94. C. H. Bamford, G. C. Eastmond and D. Whittle, *Polymer*, 1969, **10**, 771.
95. J. Ashworth, C. H. Bamford and E. G. Smith, *Polymer*, 1972, **13**, 57; *Pure Appl. Chem.*, 1972, **30**, 25.
96. C. H. Bamford and G. C. Eastmond, in 'Recent Advances in Polymer Blends, Grafts and Blocks', ed. L. H. Sperling, Plenum Press, New York, 1974, p. 162.
97. P. J. Flory, 'Principles of Polymer Chemistry', Cornell University Press, Ithaca, New York, 1953.
98. C. H. Bamford and Xio-zu Han, *Polymer*, 1981, **22**, 1299.
99. Y. Imanishi, M. Tanaka and C. H. Bamford, *Int. J. Biol. Macromol.*, 1985, **7**, 89.
100. C. H. Bamford, in 'New Trends in the Photochemistry of Polymers', eds. N. S. Allen and J. S. Rabek, Elsevier, 1985, chap. 8, p. 129.
101. H. M. Wagner and M. D. Purbrick, *J. Photogr. Sci.*, 1981, **29**, 230.
102. G. C. Eastmond, *Pure Appl. Chem.*, 1981, **53**, 657.
103. G. C. Eastmond and D. G. Phillips, *Polymer*, 1979, **20**, 1501.
104. C. H. Bamford, I. P. Middleton, J. Paprotny and K. G. Al-Lamee, *Br. Polym. J.*, 1987, **19**, 269.
105. C. H. Bamford, K. G. Al-Lamee and C. J. Konstantinov, *J. Chem. Soc., Faraday Trans. 1*, 1977, **73**, 1406.
106. C. H. Bamford and K. G. Al-Lamee, *J. Chem. Soc., Faraday Trans. 1*, 1984, **80**, 2175.
107. C. H. Bamford and K. G. Al-Lamee, *J. Chem. Soc., Faraday Trans. 1*, 1984, **80**, 2187.
108. C. H. Bamford and E. O. Hughes, *J. Chem. Soc., Faraday Trans. 1*, 1972, **68**, 1474.
109. C. H. Bamford and S. U. Mullik, *Polymer*, 1976, **17**, 225.
110. C. H. Bamford and S. U. Mullik, *J. Chem. Soc., Faraday Trans. 1*, 1976, **72**, 368.
111. C. H. Bamford and S. U. Mullik, *J. Chem. Soc., Faraday Trans. 1*, 1977, **73**, 1260.
112. D. Lalage, S. Brown, J. A. Connor and H. A. Skinner, *J. Organomet. Chem.*, 1974, **81**, 403.
113. S. M. Aliwi, C. H. Bamford and S. U. Mullik, *J. Polym. Sci., Polym. Symp.*, 1975, **50**, 33.
114. C. H. Bamford and S. U. Mullik, *J. Chem. Soc., Faraday Trans. 1*, 1976, **72**, 2218.
115. C. H. Bamford and S. U. Mullik, *Polymer*, 1976, **17**, 94.
116. A. K. Alimoglu, C. H. Bamford, A. Ledwith and S. U. Mullik, *Vysokomol Soedin. Ser. A*, 1979, **21**, 2403.
117. C. H. Bamford and S. U. Mullik, *Polymer*, 1978, **19**, 948.
118. Y. Nishikawa and T. Otsu, *Makromol. Chem.*, 1969, **128**, 276.
119. Y. Nishikawa, T. Otsu and S. Watanuma, *Makromol. Chem.*, 1968, **115**, 278.
120. E. M. Arnett and M. A. Mendelsohn, *J. Am Chem. Soc.*, 1962, **84**, 3821, 3824.
121. E. G. Kastning, H. Naarmann, H. Reis and C. Berding, *Angew. Chem., Int. Ed. Engl.*, 1965, **4**, 322.
122. C. H. Bamford and D. J. Lind, *Proc. R. Soc. London, Ser. A*, 1968, **302**, 145.
123. Y. Nishikawa and T. Otsu, *Kogyo Kagaku Zasshi*, 1969, **72**, 1836.
124. D. J. Lind, Thesis, University of Liverpool, 1967.
125. K. Uehara, Y. Kataoka, M. Tanaka and N. Murata, *Nippon Kagaku Kaishi*, 1969, **72**, 754.
126. K. Kaeriyama, *Kenkyu Hokoku-Sen'i Kobunshi Zairyo Kenkyusho*, 1970, **92**, 1.
127. C. H. Bamford and A. N. Ferrar, *Proc. R. Soc. London, Ser. A*, 1971, **321**, 425.
128. C. H. Bamford and A. N. Ferrar, *J. Chem. Soc., Faraday Trans. 1*, 1972, **68**, 1243.
129. S. M. Aliwi and C. H. Bamford, *J. Chem. Soc., Faraday Trans. 1*, 1974, **70**, 2092.
130. S. M. Aliwi and C. H. Bamford, *J. Chem. Soc., Faraday Trans. 1*, 1975, **71**, 52.
131. S. M. Aliwi and C. H. Bamford, *J. Chem. Soc., Faraday Trans. 1*, 1975, **71**, 1733.
132. S. M. Aliwi and C. H. Bamford, *Polymer*, 1977, **18**, 375.
133. S. M. Aliwi and C. H. Bamford, *Polymer*, 1977, **18**, 381.

10

Other Initiating Systems

GRAEME MOAD, EZIO RIZZARDO and DAVID H. SOLOMON
CSIRO, Melbourne, Australia

10.1 THERMAL INITIATION

10.1.1 Introduction

A purely thermal polymerization is one in which monomer is converted to polymer by thermal energy alone. Self-initiated or spontaneous homo- and co-polymerization has been reported for many monomers and monomer pairs. Homopolymerization generally requires substantial thermal energy whereas copolymerization between certain electron-acceptor and electron-donor monomers can occur at ambient temperature. The occurrence of true thermal polymerization can be difficult to establish since trace impurities in the monomers or reaction vessel often prove to be the actual initiators. In most cases of self-initiated polymerization, the identity of the initiating radicals and the mechanisms by which they are formed remain obscure.

10.1.2 Homopolymerization

A purely thermal homopolymerization to high molecular weight polymer has only been demonstrated unequivocally for styrene (S) and some of its derivatives, and for methyl methacrylate (MMA). For certain monomers, for example, 2-vinylpyridine, 2-vinylthiophene, 2-vinylfuran[1] and dimethylaminoethyl methacrylate,[2] the evidence for self-initiation is perhaps less rigorous.

Carefully purified styrene polymerizes by a free radical mechanism at a rate of 0.1% per hour at 60 °C and 2% per hour at 100 °C.[3] At 180 °C, 80% conversion of monomer to polymer is achieved in approximately 40 minutes.[4] Polymer production is accompanied by the formation of several styrene dimers and trimers amounting to around 2% by weight. The dimer fraction consists largely of *cis*- and *trans*-1,2-diphenylcyclobutanes (6) while the stereoisomeric tetrahydronaphthalenes (4) are the main constituents of the trimer fraction.[5]

The radicals responsible for the initiation of polymerization have been identified as 1-phenylethyl (3) and 1-phenyl-1,2,3,4-tetrahydronaphthalenyl (2) by examining the end groups of styrene oligomers generated in polymerizations retarded by FeCl$_3$/DMF.[6] The most plausible mechanism for the formation of radicals (2) and (3) was first proposed by Mayo in 1961 (Scheme 1). It involves the formation of a Diels–Alder dimer (1) of styrene [(4 + 2) cycloaddition] followed by transfer of a hydrogen atom from dimer to a styrene molecule (molecule-assisted homolysis).[7] To account for certain features of the polymerization kinetics, it has been postulated that radical production proceeds exclusively through the isomer of (1) in which the phenyl group is axial. Both isomers of (1) can give rise to trimers (4),[5] presumably by an ene reaction between (1) and styrene. Trimers (4) can also form by cage combination of radicals (2) and (3).

Diphenylcyclobutanes (6) are believed to arise *via* 1,4-diradicals (5). Indeed it is not impossible that diradicals (5) are a second minor source of initiation.[1] Diradicals have been implicated in the thermal initiation of 2,3,4,5,6-pentafluorostyrene where transfer of a fluorine atom from Diels–Alder dimer to monomer would be highly unlikely (because of high C–F bond strength) and for monomers which cannot form a Diels–Alder adduct (*e.g.* MMA).[1,8]

The self-initiated thermal polymerization of methyl methacrylate occurs at a much slower rate than that of styrene at the same temperature. At 90 °C, styrene polymerizes about 70 times faster.[1] Several dimers and trimers have been isolated from the reaction mixture. These include *cis*- and *trans*-dimethyl 1,2-dimethyl-cyclobutane-1,2-dicarboxylate which were shown by kinetic investigations to arise *via* 1,4-diradical intermediates.[9] On the basis of this, and in the absence of plausible alternatives, it is tempting to assume that the diradicals are also responsible for the initiation of polymerization.

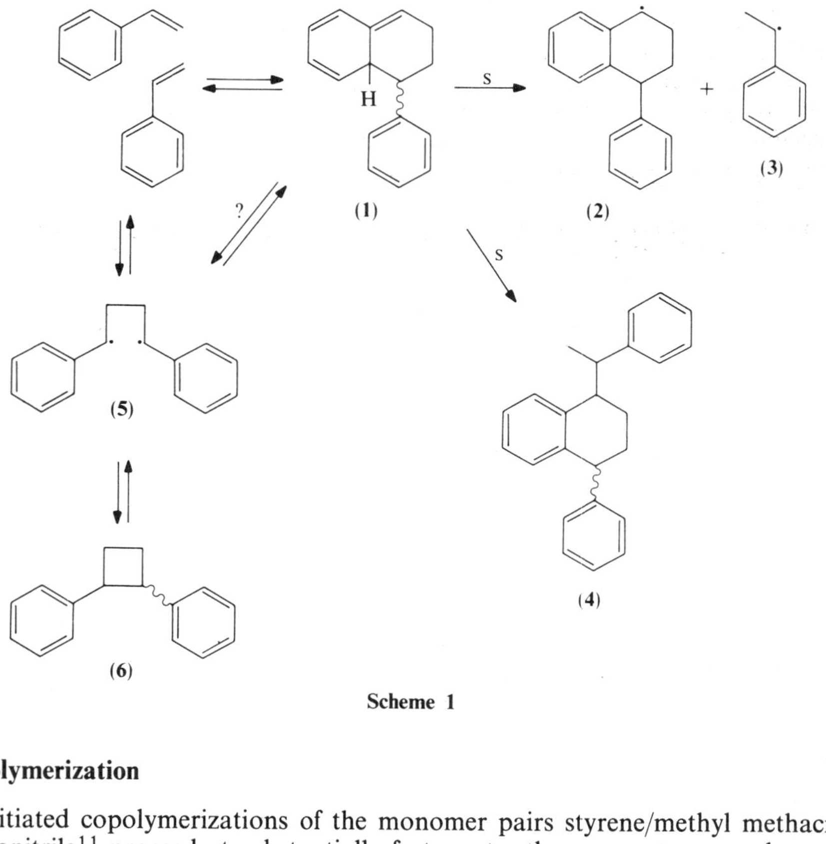

Scheme 1

10.1.3 Copolymerization

The self-initiated copolymerizations of the monomer pairs styrene/methyl methacrylate[10] and styrene/acrylonitrile[11] proceed at substantially faster rates than pure styrene polymerization. For styrene/acrylonitrile the mechanism of initiation is believed to be analogous to that of pure styrene (Scheme 1) but with acrylonitrile acting as the dienophile in the formation of the Diels–Alder adduct.[11]

In the cases where one monomer is a strong electron donor and the other a strong electron acceptor the spontaneous polymerization may exhibit free radical character, ionic character, or both, and the products can be either homopolymers of one or both monomers, or copolymers. The polymerizations which result in alternating copolymers are believed to proceed by a radical mechanism. Examples are provided by styrene with maleic anhydride,[12] vinylidene cyanide,[13] or dimethyl 1,1-dicyanoethane-2,2-dicarboxylate;[14] *p*-methoxystyrene with trimethyl ethylenetricarboxylate[14] or dimethyl cyanofumarate;[15] 1,2-dimethoxyethylene with maleic anhydride;[16] and vinyl sulfides with a range of electron-accepting monomers.[17] In the last cases free radical species have been detected by means of a spin trapping method.

A number of mechanisms have been proposed to explain the initiation processes. These include electron transfer followed by proton transfer to give two monoradicals,[17] hydrogen abstraction by a charge-transfer complex from solvent,[12] and formation of a diradical from a charge-transfer complex.[18]

Hall[14] has developed a unifying concept based on tetramethylenes (reasonance hybrids of 1,4-diradical and zwitterionic limiting structures) to rationalize all donor–acceptor polymerizations. The tetramethylenes may have predominant zwitterionic character or predominant diradical character depending on the nature of the substituents.[14,19] The former are viewed as initiators of ionic homopolymerization and the latter as initiators of alternating free radical copolymerization.

However, in spite of considerable effort, unequivocal proof for any of the proposed mechanisms is still lacking.

10.2 INIFERTERS

10.2.1 Introduction

The term iniferter (*ini*tiator — trans*fer* agent — chain *ter*minator) was coined by Otsu *et al.*[20] to describe a class of initiators with the following attributes: (1) One (or both) of the radicals formed on initiator decomposition is long lived and is unable (or slow) to initiate polymerization. Thus, primary radical termination is the major pathway for the cessation of chain growth. The initiator may additionally be susceptible to radical-induced decomposition (transfer to initiator). Ideally, these will be the only termination mechanisms. (2) The bond to the end group formed by the aforementioned mechanisms is thermally or photochemically labile and undergoes reversible homolysis to regenerate the propagating radical. (3) Primary radical termination occurs exclusively by combination. Transfer to initiator occurs exclusively by group transfer.

The ideal polymerization mechanism is summarized in Scheme 2.

$$\text{\textit{Initiation}: } R\!-\!R' \rightarrow R\cdot + R'\cdot$$

$$\text{\textit{Propagation}: } R\cdot + M \rightarrow R\!-\!M\cdot \rightarrow \rightarrow R\!\!-\!\!(M)_n\!M\cdot$$

$$\text{\textit{Termination}: } R\!\!-\!\!(M)_n\!M\cdot + R'\cdot \rightleftarrows R\!\!-\!\!(M)_n\!M\!-\!R'$$

$$R\!\!-\!\!(M)_n\!M\cdot + R\!-\!R' \rightarrow R\!\!-\!\!(M)_n\!M\!-\!R' + R\cdot$$

Scheme 2

The above requirements render the iniferters unsuitable for obtaining high molecular weight polymers since, with the initiators described to date, rates of initiation and polymerization are slow. To date, the applications of these materials have been in pseudo-living radical polymerization and in the synthesis of block and graft copolymers (see Volume 6).

10.2.2 Photochemical Iniferters

Otsu *et al.* have reported that disulfides, including diphenyl disulfide PhS–SPh (7)[21] and dithiuram disulfide (8),[22] act as photochemical iniferters. With these initiators transfer to initiator (by group transfer) may be an important mechanism for the termination of polymer chains. However, the end groups formed are indistinguishable from those formed by primary radical termination (Scheme 2).

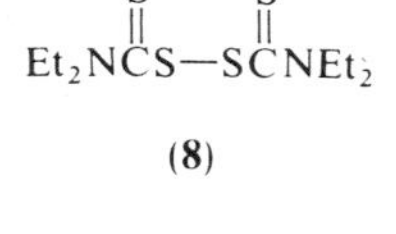

$$\underset{\text{Et}_2\text{NCS}-\text{SCNEt}_2}{\overset{\displaystyle \overset{\text{S}}{\|} \qquad \overset{\text{S}}{\|}}{}}$$

(8)

For the dithiocarbamates, studies on model compounds have shown that the end groups formed by addition to monomer (primary dithiocarbamates) are much less susceptible to photodissociation than benzyl or tertiary derivatives.[23] Therefore, the only 'living' ends are those formed by primary radical termination or transfer to initiator.

The disulfide iniferters (*e.g.* **7, 8**) suffer from the disadvantage that both of the radicals produced on decomposition of the initiator (or subsequent decomposition of the macroinitiator) may initiate polymerization. This reduces the steady state concentration of the sulfur centred radicals, and increases the likelihood that termination by the reaction of two propagating radicals will compete with primary radical termination. Consequently, there is a slow loss of 'living' ends.

Benzyl and tertiary thiocarbamates may be used directly as photoiniferters.[23] Photodissociation of the C–S bond affords a reactive alkyl radical (to initiate polymerization) and a less reactive sulfur centred radical (to undergo primary radical termination). While initiation by the sulfur centred radical cannot be completely avoided, the use of these compounds allows substantially better control of the polymerization process and greater scope for controlling polymer architecture. Otsu *et al.* have demonstrated the use of the mono-, di- and multi-functional initiators, (**9**), (**10**) and (**11**) in the production of AB and ABA block copolymers and star polymers respectively.[24] Copolymers of (**12**) were used successfully to initiate graft polymerization.[25]

(9)   (10)

(11)   (12)

10.2.3 Thermal Iniferters

Many stable radicals undergo combination with more reactive radicals at or near diffusion controlled rates to form relatively weak bonds and yet react with monomers (initiate polymerization) only slowly. Initiators which generate such species have application as thermal iniferters.

10.2.3.1 Hexasubstituted ethanes

The use of 1,2-disubstituted-1,1,2,2-tetraphenylethanes (*e.g.* **13**) was first reported by Braun *et al.*[26,27] Related systems have been examined by Otsu *et al.*[28] and Crivello *et al.*[29]

The following example is illustrative. Thermal decomposition of the pinacol derivative (**13**) at 70 °C in MMA affords diphenylalkyl radicals which initiate polymerization and take part in primary radical termination to form oligomers of the general structure (**14**). The bond formed in the latter reaction is thermally labile and the oligomers (**14**) are effective initiators of further polymerization. Loss of 'living' ends must occur as the diphenylalkyl radical produced from the macroinitiator also adds monomer. Otsu *et al.*[30] have indicated that a further problem with these iniferters is loss of 'living' ends through primary radical termination by disproportionation.

(13)   (14)

The rates of decomposition of hexasubstituted ethanes[31] and the derived macroinitiators vary according to the degree of steric crowding about the C–C bond undergoing homolysis. It is probably as a consequence of this that these iniferters appear most suited for the polymerization of 1,1-disubstituted monomers (*e.g.* MMA, α-methylstyrene).[32]

Iniferters based on silylated pinacols have been investigated by Crivello *et al.* Of particular note is the cyclic compound (**15**) which undergoes thermolysis to afford a diradical thus offering a route to ABA block copolymers.[33]

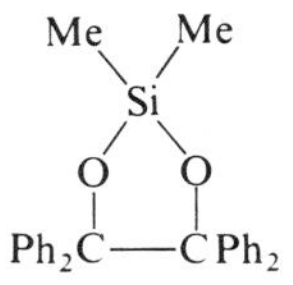

(15)

10.2.3.2 *Triphenylmethylazobenzene*

Otsu and Tazaki[34] have examined the use of triphenylmethylazobenzene, $Ph_3CN=NPh$ (**16**) as an iniferter. The phenyl radical initiates polymerization and the triphenylmethyl radical acts as a radical trap. The chemistry of the macroinitiators so formed is analogous to that described above and the use of (**16**) suffers from the same disadvantages as the hexasubstituted ethanes since triphenylmethyl radical may initiate polymerization, albeit slowly,[35] or undergo disproportionation with the propagating radical.

10.2.3.3 *Alkoxyamines*

Rizzardo *et al.*[36,37] pioneered the use of alkoxyamines (*e.g.* **17**) as iniferters. These compounds are readily prepared by decomposing an alkyl radical source [*e.g.* azobis(isobutyronitrile) (AIBN)] in the presence of a nitroxide.

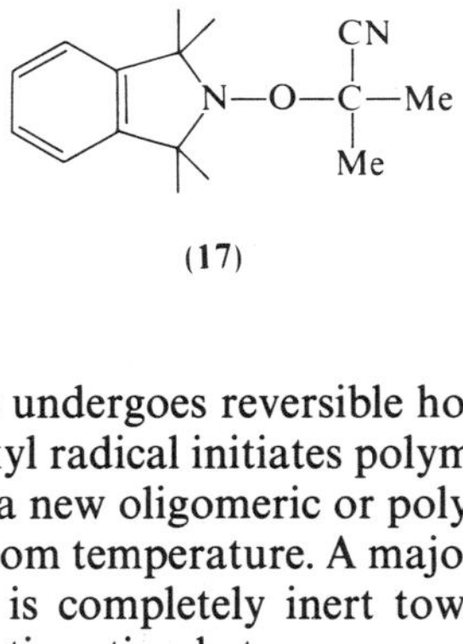

(17)

The C–O bond of the alkoxyamine undergoes reversible homolysis on heating to afford an alkyl radical and a stable nitroxide. The alkyl radical initiates polymerization and the nitroxide combines with the propagating radical to form a new oligomeric or polymeric alkoxyamine initiating species. The latter are stable and isolable at room temperature. A major advantage of the alkoxyamines over other iniferters is that the nitroxide is completely inert towards most monomers under normal polymerization conditions. Disproportionation between propagating species and the nitroxide, and consequent loss of 'living' ends, is not observed except with tertiary radicals (*e.g.* that from MMA) and then is only a minor pathway.[38]

The rate of homolysis of the alkoxyamines (R–ON<) is sensitive to steric compression around the C–O bond. Thus the rates of homolysis vary with the structure of both the nitroxide fragment (*e.g.* decreasing in the series **18** > **19** > **20** > **21**) and the alkyl radical fragment (3° > 2° > 1°).

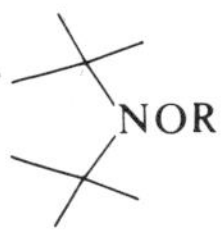

(18) 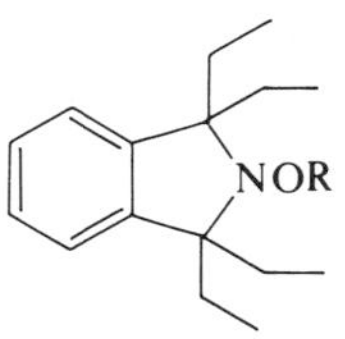   (19) 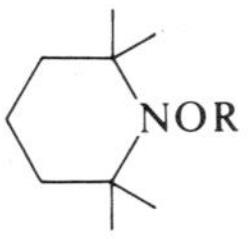   (20) 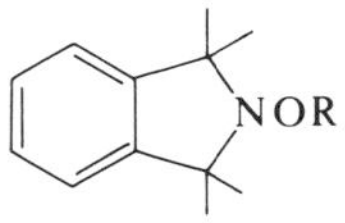   (21)

10.3 REFERENCES

1. W. A. Pryor and L. D. Lasswell, *Adv. Free-Radical Chem.*, 1975, **5**, 27.
2. M. D. Shalati and R. M. Scott, *Macromolecules*, 1975, **8**, 127.
3. R. H. Boundy and R. F. Boyer, 'Styrene, its Polymers, Copolymers and Derivatives', Reinhold, New York, 1952.
4. A. W. Hui and A. E. Hamielec, *J. Appl. Polym. Sci.*, 1972, **16**, 749.
5. K. Kirchner and K. Riederle, *Angew. Makromol. Chem.*, 1983, **111**, 1 and references therein.
6. Y. K. Chong, E. Rizzardo and D. H. Solomon, *J. Am. Chem. Soc.*, 1983, **105**, 7761.
7. W. A. Pryor, *ACS Symp. Ser.*, 1978, **69**, 33.
8. W. A. Pryor, M. Iino and G. R. Newkome, *J. Am. Chem. Soc.*, 1977, **99**, 6003.
9. J. Lingnau, M. Stickler and G. Meyerhoff, *Eur. Polym. J.*, 1980, **16**, 785.
10. C. Walling, *J. Am. Chem. Soc.*, 1949, **71**, 1930.
11. K. Kirchner and H. Schlapkohl, *Makromol. Chem.*, 1976, **177**, 2031.
12. M. Matsuda and K. Abe, *J. Polym. Sci., Part A-1*, 1968, **6**, 1441.
13. J. K. Stille and D. C. Chung, *Macromolecules*, 1975, **8**, 83.
14. H. K. Hall, Jr., *Angew, Chem., Int. Ed. Engl.*, 1983, **22**, 440.
15. H. K. Hall, Jr., A. B. Padias, A. Pandya and H. Tanaka, *Macromolecules*, 1987, **20**, 247.
16. T. Kokubo, S. Iwatsuki and Y. Yamashita, *Makromol. Chem.*, 1969, **123**, 256.
17. T. Sato, M. Abe and T. Otsu, *J. Macromol. Sci., Chem.*, 1981, **A15**, 367.
18. N. G. Gaylord and A. Takahashi, *Adv. Chem. Ser.*, 1969, **91**, 94.
19. K. Jug, *J. Am. Chem. Soc.*, 1987, **109**, 3534.
20. T. Otsu and M. Yoshida, *Makromol. Chem., Rapid Commun.*, 1982, **3**, 127.
21. T. Otsu, M. Yoshida and A. Kuriyama, *Polym. Bull. (Berlin)*, 1982, **7**, 45.
22. T. Otsu and A. Kuriyama, *Polym. Bull. (Berlin)*, 1984, **11**, 135.
23. T. Otsu and A. Kuriyama, *J. Macromol. Sci., Chem.* 1984, **A21**, 961.
24. T. Otsu and A. Kuriyama, *Polym. Bull. (Berlin)*, 1984, **11**, 135; A. Kuriyama and T. Otsu, *Polym. J.*, 1984, **16**, 511; T. Otsu and A. Kuriyama, *Polym. J.*, 1985, **17**, 97.
25. T. Otsu, K. Yamashita and K. Tsuda, *Macromolecules*, 1986, **19**, 287.
26. A. Bledzki and D. Braun, *Makromol. Chem.*, 1981, **182**, 1047.
27. A. Bledzki, D. Braun and K. Titzschkau, *Makromol. Chem.*, 1983, **184**, 745.
28. T. Otsu, A. Matsumoto and T. Tazaki, *Polym. Bull. (Berlin)*, 1987, **17**, 323.
29. J. V. Crivello, J. L. Lee and D. A. Conlon, *J. Polym. Sci., Polym. Chem. Ed.*, 1986, **24**, 1251.
30. T. Otsu and T. Tazaki, *Polym. Bull. (Berlin)*, 1987, **17**, 127.
31. C. Rüchardt, *Top. Curr. Chem.*, 1980, **88**, 1.
32. A. Bledzki and D. Braun, *Polym. Bull. (Berlin)*, 1986, **16**, 19.
33. J. V. Crivello, J. L. Lee and D. A. Conlon, *Polym. Bull. (Berlin)*, 1986, **16**, 95.
34. T. Otsu and T. Tazaki, *Polym. Bull. (Berlin)*, 1986, **16**, 277.
35. G. Moad, E. Rizzardo and D. H. Solomon, *Macromolecules*, 1982, **15**, 909.
36. E. Rizzardo, *Chem. Aust.*, 1987, **54**, 32.
37. D. H. Solomon, E. Rizzardo and P. Cacioli, *Eur. Pat. Appl.* EP135280 (*Chem. Abstr.*, 1985, **102**, 221 335q).
38. D. S. Bednarek, G. Moad, E. Rizzardo and D. H. Solomon, *Macromolecules*, 1988, in press.

11

Chemistry of Bimolecular Termination

GRAEME MOAD and DAVID H. SOLOMON
CSIRO, Melbourne, Australia

11.1 INTRODUCTION

In the context of this work the term bimolecular termination is taken to refer to termination which involves two propagating radicals. This chapter is therefore concerned with the chemistry of radical–radical reactions.

By way of introduction we summarize factors known to influence radical–radical reactions involving low molecular weight species. We then consider termination during homopolymerization of the more common monomers and provide a critical survey of the literature on both model systems and polymerizations. The section on copolymerization is necessarily brief due to a lack of experimental data. Finally, we consider the techniques used to characterize termination, pointing out their advantages and limitations.

11.2 RADICAL–RADICAL REACTIONS

Radical–radical reactions may take several pathways (Scheme 1): (a) combination which usually, but not invariably, takes place by a simple head-to-head coupling of radicals. Exceptions[1] include the formation of quinone methide derivatives from the combination of substituted benzyl radicals (see Section 11.3.2) and the formation of ketenimines from α-cyanoalkyl radicals (see Section 11.3.5); (b) disproportionation which involves the transfer of a β-hydrogen from one radical to another;

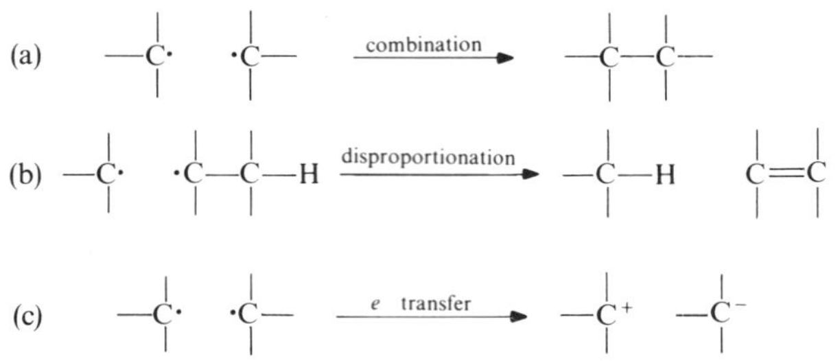

(a)   —Ċ· ·Ċ— $\xrightarrow{\text{combination}}$ —Ċ—Ċ—

(b)   —Ċ· ·Ċ—Ċ—H $\xrightarrow{\text{disproportionation}}$ —Ċ—H   C=C

(c)   —Ċ· ·Ċ— $\xrightarrow{e \;\; \text{transfer}}$ —Ċ⁺   —Ċ⁻

Scheme 1

(c) electron transfer to afford an ion pair. This pathway is rare for reactions involving only carbon-centred radicals and will not be considered further in this chapter.

The last comprehensive review of small molecule radical–radical reactions appeared in 1973.[2] The pathways observed depend mainly on the structure of the radicals involved and show only small variation with such factors as the nature of the reaction medium, temperature, pressure, *etc.*

For carbon-centred radicals both combination and disproportionation are usually observed. Early theories rationalized the outcome of radical–radical reactions in terms of a single four-centre transition state or intermediate which could lead to either disproportionation or combination.[3,4] These theories were discounted by the observation of an admittedly small dependence of k_{td}/k_{tc} on temperature, pressure and solvent and it is now generally recognized that combination and disproportionation must be considered as two separate reactions.[2] This view is also supported by theoretical studies.[5-8]

For a given series of radicals, there is an increase in the amount of disproportionation with an increase in the number of β-hydrogen atoms. However, there is no direct correlation between k_{td}/k_{tc} and the number of β-hydrogens and it is evident that other than statistical factors are involved.[2,9]

The discrepancies have been rationalized in terms of the greater sensitivity of combination to steric factors. For simple alkyl radicals a correlation has been found between $\log (k_{td}/k_{tc})$ (after statistical correction) and Taft steric parameters.[10]

The importance of steric factors on k_{td}/k_{tc} can be gauged by considering the radicals (1) and (2). The self-reaction of cumyl radicals (1) affords predominantly combination (*cf.* Table 1) while the radicals (2), in which a β hydrogen is replaced by a *t*-butyl group, give predominantly disproportionation.[11]

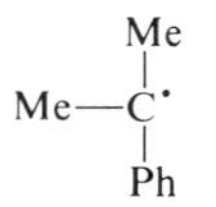

$$\text{Me}\!-\!\overset{\displaystyle \text{Me}}{\underset{\displaystyle \text{Ph}}{\overset{|}{\underset{|}{\text{C}^{\bullet}}}}} \qquad\qquad \text{Me}\!-\!\overset{\displaystyle \text{CH}_2\text{Bu}^t}{\underset{\displaystyle \text{Ph}}{\overset{|}{\underset{|}{\text{C}^{\bullet}}}}}$$

(1) 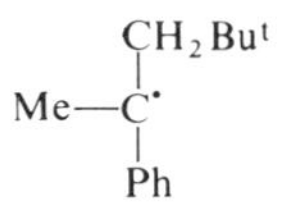   (2)

In extreme cases suitably bulky substituents at the radical centre can render a radical persistent. A well-known example is the di-*t*-butyl methyl radical (3).[12] This radical (3) cannot decay by disproportionation since there are no β hydrogens. The trineopentylmethyl radical (4) is another example of a persistent radical. In this case both disproportionation and combination are substantially retarded by steric factors (see below).[13]

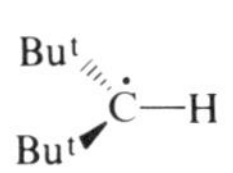

$$\overset{\displaystyle \text{Bu}^t}{\underset{\displaystyle \text{Bu}^t}{}}\!\!\overset{\bullet}{\text{C}}\!-\!\text{H} \qquad\qquad \overset{\displaystyle \text{Bu}^t\text{CH}_2}{\underset{\displaystyle \text{Bu}^t\text{CH}_2}{}}\!\!\overset{\bullet}{\text{C}}\!-\!\text{CH}_2\text{Bu}^t$$

(3) 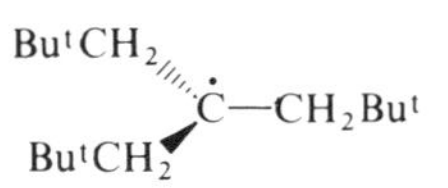   (4)

Selectivity for transfer of axial *vs.* equatorial hydrogens is observed when conformationally biased cyclohexyl radicals (*e.g.* 5) disproportionate. This behaviour has been explained in terms of disproportionation being subject to stereoelectronic control.[14] The proposal is that the C–H bond which can achieve best overlap with the orbital bearing the unpaired spin is transferred preferentially. The process is thus mechanistically analogous to β scission where a similar requirement applies.[15]

Table 1 Relative Importance of Combination *vs.* Disproportionation in Model Systems

System	·Structure	Temperature (°C)	k_{td}/k_{tc}	Refs.
S	(7)	20	0.073[a]	44
S	(8)	20	0.141[a]	44
S	(7)	80	0.081	44
S	(8)	80	0.146	44
S	(9)	80	0.156	44
S	(9) or (10)	80	0.157	47
S	(7)	118	0.097	43
S	(8)	118	0.107	43
S	(9) or (10)	160	0.082	47
α-Methylstyrene	(1)	20–60	0.05[a]	20
α-Methylstyrene	(1)	55	0.1[a]	50
MMA	(12)	70–90	0.78	26
MMA	(12)	90	0.62	19
MMA	(12)	115	0.61	19
MMA	(12)	140	0.60	19
MMA	(12)	165	0.59	19
Ethyl methacrylate	(13)	80	0.72	26
Butyl methacrylate	(14)	80	1.17	26
MAN	(15)	80	0.1[a,b]	74–76
MAN	(17)	80	0.1	47

[a] Unsymmetrical coupling product is also formed but is not included in ratio shown. [b] Simple β-substituted derivatives give similar values of k_{td}/k_{tc} [see A. Ueda and S. Nagai, *Kobunshi Ronbunshu*, 1986, **46**, 331 (*Chem. Abstr.*, 1986, **105**, 115451b)]

Stereoelectronic effects have also been invoked in explaining the resistance of the trineopentyl-methyl (**4**)[16] and triisopropylmethyl (**6**)[17] radicals towards decay by disproportionation. In the case of (**6**) the β hydrogens are constrained to lie in the nodal plane of the *p* orbital due to buttressing between the methyls of adjacent isopropyls.[9]

In view of these results, it is also reasonable to question whether stereoelectronic factors (and steric hindrance to adoption of the required conformation) might also influence the reactivity of other, less conformationally constrained, radicals (*e.g.* polymer chains) towards disproportionation.

The presence of α substituents which delocalize the free spin (*e.g.* Ph–, CN–) favours combination over disproportionation[2,18,19] (see also Table 1). A completely satisfactory explanation for this effect has yet to be proposed. However, it seems reasonable[20] that if overlap between the semi-occupied orbital and the breaking C–H bond favours disproportionation, then substituents which delocalize the free spin will serve to diminish the interaction and thus disfavour disproportionation.

Values of k_{td}/k_{tc} for simple alkyl radicals are sensitive to reaction conditions.[20-21] However, the effects are generally small. For example, values of k_{td}/k_{tc} for *t*-butyl radicals show small variation with temperature (disproportionation decreases with increasing temperature)[22] and medium viscosity, (disproportionation increases with increasing viscosity).[21,22] Note, however, that influences of pressure and viscosity on k_{td}/k_{tc} can be substantial for pairwise generated radicals which undergo self-reaction within the solvent cage.[22]

11.3 HOMOPOLYMERIZATION

11.3.1 Combination *vs.* Disproportionation

Since the mode of termination plays an important part in determining the molecular weight distribution and may lead to the incorporation of defect structures into polymer chains, a knowledge of k_{td}/k_{tc} is vital to the understanding of structure–property relationships. Unsaturated linkages at the ends of polymer chains, as may be formed by disproportionation, have long been thought to contribute to polymer instability and it has recently been demonstrated that both head-to-head linkages and unsaturated ends are weak links during the thermal degradation of poly(methyl methacrylate) (PMMA).[24,25] Polymer chains with unsaturated ends may also be reactive during polymerization. Copolymerization of such macromonomers is one of the possible mechanisms for the formation of long chain branches.[26–28] These macromonomers may also function as transfer agents.[28]

A knowledge of k_{td}/k_{tc} is important in designing polymer syntheses. For example, in the preparation of block copolymers by free radical processes[29–31] where ABA or AB blocks may be produced depending on whether termination involves combination or disproportionation respectively. Finally, the relative importance of combination and disproportionation is also important in the interpretation of polymerization kinetics and, in particular, in the derivation of absolute rate parameters (see Chapter 16 in this volume).

11.3.2 Styrene and Derivatives

Numerous studies over the last four decades have sought to elucidate the mechanism of termination during styrene (S) polymerization (Scheme 2) and most have concluded that termination involves predominantly combination.[32–42] However, distinction between a k_{td}/k_{tc} of 0.0 and one of (say) ≤ 0.2 is difficult given the precision achievable even with the most modern techniques and instrumentation. Nonetheless, many workers have taken 'predominantly combination' to mean 'exclusively combination'. This extrapolation is unjustified.

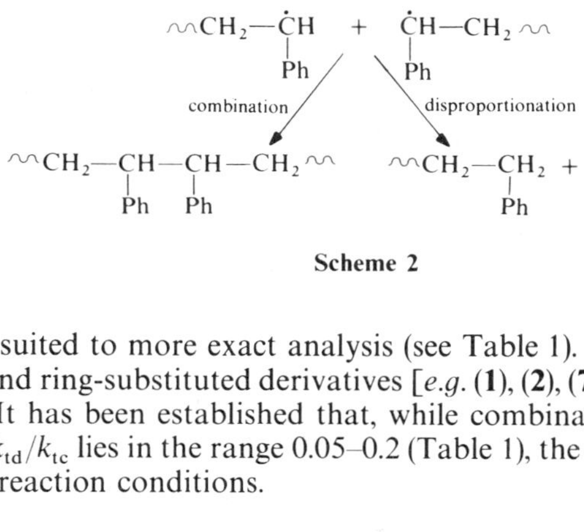

Scheme 2

Model systems are suited to more exact analysis (see Table 1). The self-reactions of phenylethyl radicals and of α-, β- and ring-substituted derivatives [e.g. (1), (2), (7)–(10)] have been investigated by several groups.[43–47] It has been established that, while combination is dominant, disproportionation does occur and k_{td}/k_{tc} lies in the range 0.05–0.2 (Table 1), the precise value being dependent on radical structure and reaction conditions.

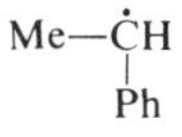

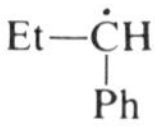

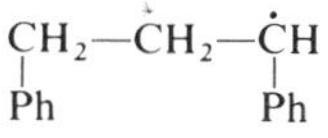

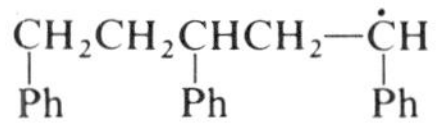

It is noteworthy that the extent of disproportionation shows only a marginal increase with increasing chain length for oligostyryl radicals [in the series (7)–(10)].[44,47] This result suggests that for polystyryl radicals k_{td}/k_{tc} should be similar or only slightly greater than that observed in the self-reaction of the 1,3,5-triphenylpentyl radicals (10), see Table 1.

When the radicals (7) or (8) are generated photochemically at 20 °C there is also evidence,[44] by way of the aromatized product [e.g. (11)], of the reversible formation of a quinonoid species by α-*ortho* coupling (Scheme 3). It is suggested[44] that the nonobservance of α-*para* coupling under these conditions or of either coupling reaction when phenylethyl radicals are generated thermally, may be a reflection of the facility of the reverse reaction at higher temperatures (80 vs. 20 °C) rather than an indication that such reactions do not occur. An analogous reversible unsymmetrical coupling process takes place between benzyl radicals[48] and between cumyl radicals (1).[20,49,50] In these cases there is evidence for both α-*ortho* and α-*para* coupling.

The temperature dependence of k_{td}/k_{tc} for self-reaction of 1,3-diphenylpropyl (9) or 1,3,5-triphenylpentyl radicals (10) is small but measurable with k_{td}/k_{tc} varying from 0.16 at 80 °C to 0.08 at 160 °C.[47]

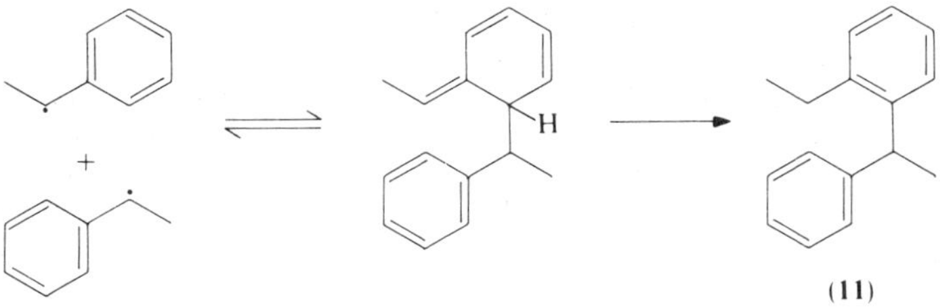

(11)

Scheme 3

Olaj *et al.*[51] were the first in recent times to suggest that termination of S polymerization involves a significant fraction of disproportionation. They analyzed the molecular weight distribution of PS samples prepared at temperatures in the range 20–90 °C and estimated k_{td}/k_{tc} to be *ca.* 0.2. More recently they[52] have analyzed the molecular weight distribution of PS samples prepared with photoinitiation at 60 and 85 °C and estimated the values for k_{td}/k_{tc} under these conditions to be *ca.* 0.5 and 0.67 respectively, thus indicating that k_{td}/k_{tc} has a substantial temperature dependence. These findings are supported by the work of Berger and Meyerhoff[53,54] and Stickler.[55] The former[53,54] conducted a detailed study of S polymerization initiated with radiolabelled AIBN and concluded that k_{td}/k_{tc} ranged from 0.168 at 30 °C to 0.468 at 62 °C and 0.663 at 80 °C. Their[53,54] kinetic analysis has however been criticized since they did not allow for chain length dependence of the termination rate constants (see Chapter 11 in this volume).

Furthermore, Dawkins and Yeadon[36] have recently discussed the problems associated with estimating k_{td}/k_{tc} on the basis of polydispersity measurements and determined that k_{td}/k_{tc} should be 'substantially smaller' than suggested by Berger and Meyerhoff.[53,54] The data from the latter studies, at least at higher temperatures (≥ 60 °C), are also inconsistent with model studies, which predict that k_{td}/k_{tc} should have only a small temperature dependence and, indeed, should decrease with increasing temperature (see above and Table 1), and with other values of k_{td}/k_{tc} based on end group determination (see above).

Moad *et al.* have employed ^{13}C NMR to define and quantify the end groups in samples of PS prepared at 60 °C and with either ^{13}C-labelled BPO[56] or AIBN[57] as initiator. They were able to show that under the conditions employed there are 1.7 ± 0.2 initiator-derived end groups per molecule, a result which indicates some disproportionation. However, further work is required (better signal-to-noise spectra) for a precise value of k_{td}/k_{tc}.

The influence of ring substituents (*p*-Cl, *p*-OMe) on k_{td}/k_{tc} has been investigated by Ayrey *et al.*[35] who determined that disproportionation was facilitated by a *p*-OMe substituent. However, this result also is contrary to the findings of model studies[43] (on substituted phenylethyl radicals) which show that electron donating substituents (*p*-OMe) slightly disfavour combination.

In summary, studies on model systems show that, while combination is the dominant termination mechanism during S polymerization, disproportionation should occur to a small yet significant extent. However, further studies are required to firmly establish the precise value of k_{td}/k_{tc} for S polymerization particularly at temperatures ≥ 80 °C, the nature and extent of temperature dependence, and reconcile the studies on model systems and polymerizations.

11.3.3 Alkyl Methacrylates

The termination reaction in MMA polymerization (Scheme 4) has been investigated using a wide range of techniques (see Table 2). While it is generally accepted that both combination and disproportionation are important, there is considerable discrepancy in the values of k_{td}/k_{tc} obtained in the various studies. In some cases the differences may be attributed to variations in the way molecular weight data are interpreted[58–60] or to the failure to allow for other modes of termination under the polymerization conditions. In other cases reasons for the discrepancies are less clear. Model studies are more clear cut. Trecker and Foote[19] examined the self-reaction of 1-methoxy-carbonyl-1-methylethyl radicals (**12**) and, more recently, Bizilj *et al.*[26] have re-examined this system and reported studies on mono- and oligo-meric models for methyl, ethyl and *n*-butyl methacrylate termination [*e.g.* (**12**–**14**)]. Their findings on k_{td}/k_{tc} are reported in Table 1. For the three systems, combination and disproportionation are of similar importance. Particular note should be made of the fact that the ratio k_{td}/k_{tc} shows only a marginal temperature dependence (see Table 1).[19,26]

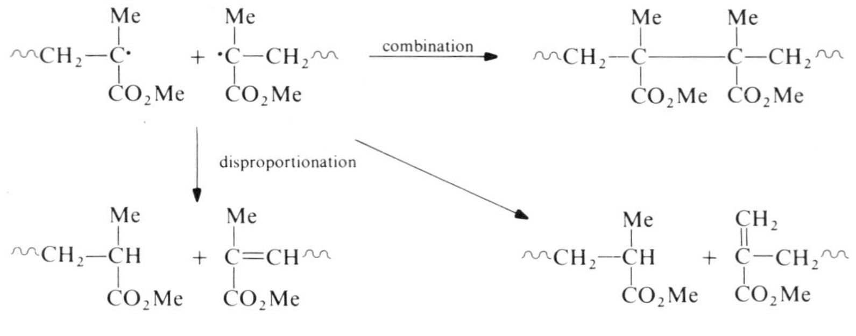

Scheme 4

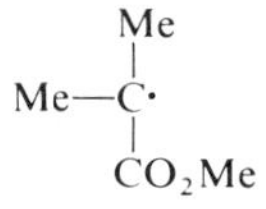

(12) 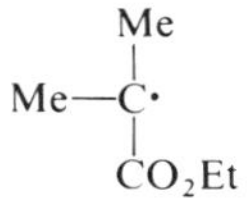   (13) 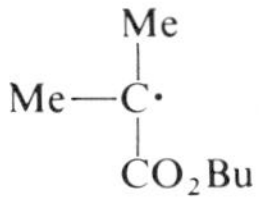   (14)

Table 2   Values of k_{td}/k_{tc} as a Function of Temperature for Methyl Methacrylate Polymerization

Temperature (°C)	37	60	63	64	Ref. 70	62	71	55	25
−25						0.14			
0	1.50					0.50			
15						0.76			
25	2.12			2.0					
30						1.18			
40			0.45						
45					∞				
60	5.67	1.35	0.75	2.7	2.62		2.57	0.44	1.28
80			1.32	4.0					

A number of studies report that the value of k_{td}/k_{tc} in MMA polymerization is dependent on the polymerization conditions. Boudevska *et al.*[61] have proposed that the preferred termination mechanism during MMA polymerization is solvent dependent and that disproportionation is favoured in benzene. Four independent studies[37,62–64] have examined the temperature dependence of the termination mechanism for MMA polymerization. While these studies do not agree on precise values for k_{td}/k_{tc}, all suggest a substantial temperature dependence for k_{td}/k_{tc} and, moreover, indicate that the extent of disproportionation increased with increasing temperature (see Table 2). These results are at variance with model studies (see above) which show that k_{td}/k_{tc} decreases, albeit marginally, with increasing temperature.[19]

An early report[65] indicated that the self-reaction of 1-methoxycarbonyl-1-methylethyl radicals (12), like cyanoisopropyl radicals (15) (see Section 11.1.3.5), afforded an additional compound which was thermally unstable but unidentified. This suggests[66] the possibility that unsymmetrical C–O coupling might occur in competition with normal C–C coupling. In this context, it is noteworthy that a precedent for a reversible unsymmetrical C–O coupling mode has been established for the case where normal, C–C coupling, is sterically very hindered.[67] However, the more recent studies[19,26,29] on reactions of radicals (12) and related species provide no supporting evidence for the existence of such a pathway.

Bizilj *et al.*[26] have demonstrated that there is significant selectivity in disproportionation of oligomeric methacrylyl radicals. Abstraction of a methyl hydrogen (to generate a terminal methylene

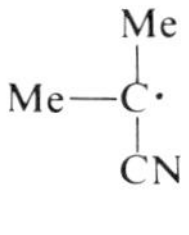

(15)

group) is preferred *ca.* 80-fold over abstraction of a methylene hydrogen (to afford an internal double bond — refer Scheme 4). The most obvious explanation for this behaviour is that the methyl hydrogens are more sterically accessible than the methylene hydrogens.

Given the preference for transfer of methyl *vs.* methylene hydrogens when oligomeric or polymeric (see below) species disproportionate, the model studies [involving radicals (12–14)] should tend to overestimate k_{td}/k_{tc}. They have six *vs.* three methyl hydrogens which might be abstracted. Yet most studies on PMMA have yielded values of k_{td}/k_{tc} higher than for the simple models (compare Tables 1 and 2). The explanation almost certainly lies with steric factors (see Section 11.2) which favour disproportionation for polymeric or oligomeric species. An examination of dimeric or larger models is required to clarify this situation. Bizilj *et al.*[26] have indicated that $k_{td}/k_{tc} \leq 1.80$ for the dimeric model.

In view of the inconsistencies in measured values of k_{td}/k_{tc} from the various studies on MMA polymerization (Table 2), it would clearly be desirable to determine the termination-derived ends in PMMA directly. Hatada *et al.*[25,68] have shown that unsaturated ends in PMMA may be identified and quantified by ^{1}H NMR. Unfortunately, the technique does not allow either saturated ends or head-to-head linkages to be determined simultaneously. Nonetheless, this study did confirm the preference for transfer of a methyl *vs.* a methylene hydrogen in disproportionation indicated by the model studies.[26]

Values of k_{td}/k_{tc} for polymerizations of ethyl and *n*-butyl and other methacrylates have been determined.[69–71] Their behaviour is similar to MMA to the extent that both combination and disproportionation are important.

11.3.4 Alkyl Acrylates

The termination mechanism during methyl acrylate polymerization (20–60 °C, benzene) has been variously determined to be predominantly disproportionation[33,70] or predominantly combination.[69,71,72] Ayrey *et al.*[71] have shown that transfer reactions are important during the polymerization and this may have led to erroneous conclusions being drawn in some of the earlier studies. They concluded that termination was almost exclusively by combination.

Bamford *et al.*[69] (25 °C) and Fehérvári *et al.*[73] (50 °C, DMF) have confirmed this finding and determined that bimolecular termination of propagating radicals from higher acrylate esters also involves predominantly combination.

11.3.5 Methacrylonitrile

Cyanoisopropyl radicals (15) (see Scheme 5), the simplest model for the propagating radical of methacrylonitrile (MAN), have been extensively studied (see Sections 8.2.1 and 8.62 in this volume).

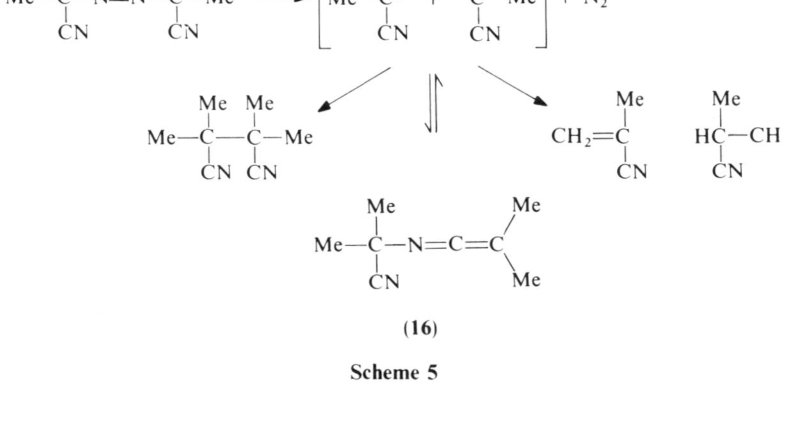

Scheme 5

Whereas methacrylyl radicals [(12–14), see above] give substantial disproportionation, with (15) combination is dominant. Disproportionation accounts for only 10% of radicals.[74,75] A similar finding emanated from a recent study by Serelis and Solomon[76] on the reactions of oligo(MAN) radicals. They found k_{td}/k_{tc} to have little if any dependence on the oligomer chain length ($n \leq 4$). As with (12–14), disproportionation involves preferential abstraction of a methyl hydrogen and the chains will, therefore, possess a potentially reactive terminal methylene.[26]

Cyanoisopropyl radicals (15) undergo unsymmetrical C–N coupling in competition with C–C coupling and, indeed, the former reaction which leads to a ketenimine (16) is the major pathway (Scheme 5).[77] The preferential formation of the ketenimine (16) probably reflects the importance of polar and steric influences.[5] However, the ketenimine (16) is thermally unstable and a source of cyanoisopropyl radicals (15). This reaction is not unique to monomeric radicals. A ketenimine (18) is also a major product from the reaction of the 'dimeric' MAN radical (17) with cyanoisopropyl radicals (15) (see Scheme 6).[78]

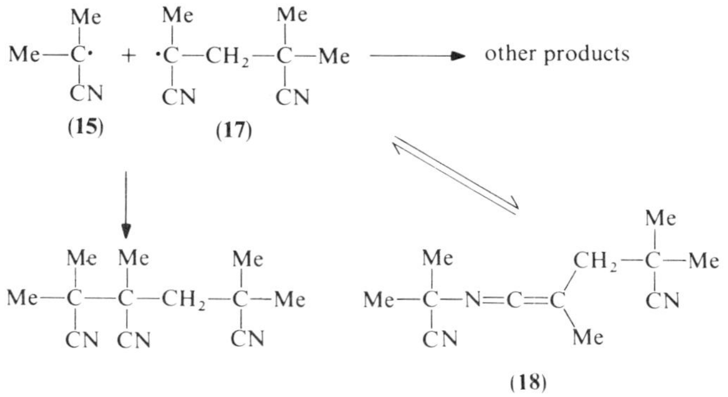

Scheme 6

Termination of MAN polymerization has been seldom studied. Bamford et al.[69] have examined MAN polymerization (DMSO, 25 °C) by the gelation technique and have estimated that termination occurs predominantly by disproportionation ($k_{td}/k_{tc} = 1.86$) in apparent contradiction of the model studies[74–76] (see above). However, note should be made of the somewhat different reaction conditions.

11.3.6 Acrylonitrile

Acrylonitrile polymerization has been examined by the gelation technique (25 °C, DMSO) and it has been suggested on the basis of these experiments that termination occurs predominantly by combination.[69]

This agrees with most[79] earlier studies[80–83] on polymers prepared in various solvents (H₂O, DMF) and at a range of temperatures (10–90 °C).

11.3.7 Vinyl Acetate

A number of reports[33,70,84] have appeared suggesting that termination during vinyl acetate (VAC) polymerization occurs exclusively or substantially by disproportionation. However, the majority of these investigations did not allow for the occurrence of other termination mechanisms, in particular transfer to monomer and polymer, which are extremely important during VAC polymerization.[85,86]

These problems were addressed by Bamford et al.[69] who used the gelation technique to show that bimolecular termination occurs predominantly by combination (25 °C).

11.3.8 Vinyl Chloride

Studies on vinyl chloride (VC) polymerization, like those involving VAC, are complicated by the fact that typically only a small proportion of termination events may involve radical–radical reactions and studies[87,88] on the termination mechanism which do not make due allowance for this should be viewed circumspectly.

Park and Smith[89] recognized this and attempted to allow for such factors in their examination of the termination mechanism during VC polymerization at 30 and 40 °C in chlorobenzene. They determined the initiator-derived ends in PVC prepared with radiolabelled AIBN and concluded that $k_{td}/k_{tc} = 3.0$. However, recent studies suggest the data may need to be re-evaluated.[90,91]

Atkinson *et al.*[91] have applied the gelation technique and propose that termination involves predominantly combination.

11.3.9   Conclusions

Precise numbers for k_{td}/k_{tc} are not yet available for most polymerizations. The discrepancies between studies of model systems and polymerizations are in many cases substantial and are not readily explicable in terms of current theories. The discrepancies are most evident in data for higher reaction temperatures (see, in particular, the discussion on S (Section 11.1.3) and MMA (Section 11.1.4) above).

It is tempting to attribute problems in reconciling model studies with actual polymerizations to difficulties in interpreting the results (polymerizations are often complicated by other termination pathways, in particular transfer reactions, which, while not fully characterized, must be allowed for — see Section 11.5) or to some undefined inadequacy in the model (see Section 11.5.5). It is evident that much more effort needs to be devoted to this area.

Despite these problems it is possible to make the following tentative generalizations regarding the mechanism of bimolecular termination; (a) termination of polymerizations involving vinyl monomers occurs predominantly by combination; (b) termination of polymerizations of α-methylvinyl monomers involves a measurable amount of disproportionation; (c) during disproportionation of radicals bearing an α-methyl substituent (for example, those derived from MMA) there is a high degree of selectivity for transfer of a hydrogen from the α-methyl group rather than the methylene group; (d) within a series of (vinyl or) α-methylvinyl monomers k_{td}/k_{tc} appears to decrease according to the ability of the substituent to stabilize a radical centre in the series $Ph < CN < CO_2R < Me$.

11.4   COPOLYMERIZATION

Bimolecular termination in copolymerization is a very complex process. For a multicomponent copolymerization involving n monomers there are $n(n+1)/2$ possible pairs of radicals which may interact.[58] In each case reaction may involve either combination or disproportionation or both. In the case of disproportionation between unlike radicals there are two directions for hydrogen atom transfer. Thus, there are a minimum of $n(3n+1)/2$ reactions to be considered. This number may be multiplied further if there is more than one pathway for combination and/or disproportionation or if penultimate unit effects are important.

For these reasons the present discussion is limited to binary copolymerization. The possible products from bimolecular termination of S–MMA copolymerization are shown in Scheme 7. Even for a binary system there are significant impediments associated with determining k_{td}/k_{tc} for the various reactions involved by examining copolymerizations or copolymers directly and relatively few systems have been studied in detail. Overall values for k_{td}/k_{tc} can, of course, be estimated by end group determination using the same techniques as employed for characterizing homopolymerization. However, interpretation of these data requires a knowledge of the relative importance of cross- and homo-termination.

Analysis of copolymerization kinetics usually requires (implicitly or explicitly) an assumption to be made regarding the relative rate constants for cross-termination and the two homotermination processes. The various treatments used to describe the kinetics of copolymerization are discussed in Chapter 16 in this volume.

If the extents of cross-termination *vs.* homotermination are known and it is presumed that penultimate group and medium effects are unimportant then characterizing cross-termination is simplified considerably since k_{td}/k_{tc} for the two homotermination processes are then the same as during the corresponding homopolymerizations.[37,70]

Serelis *et al.*[92] have studied a number of model systems with the aim of estimating k_{td}/k_{tc} for pairs of unlike radicals generated from the appropriate unsymmetrical azo compounds (see Section 11.5.5).

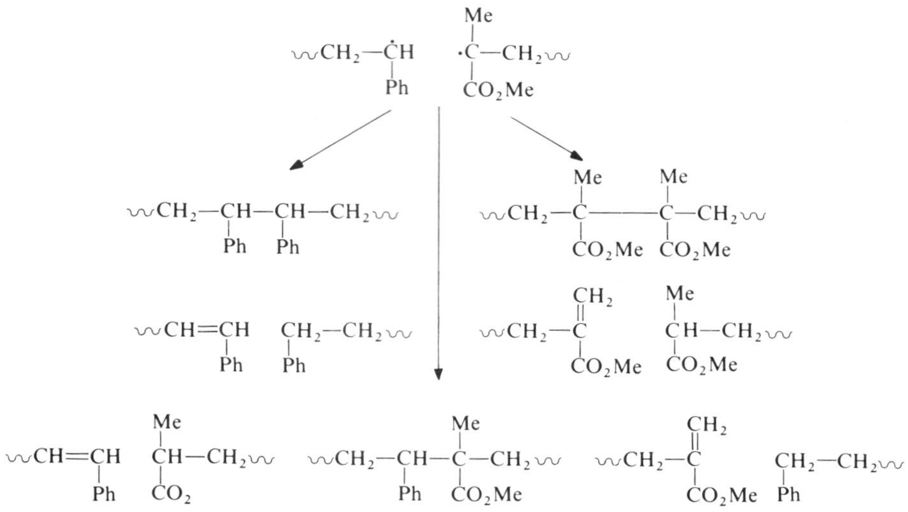

Me
wwCH₂—ĊH   ·Ċ—CH₂ww
        | |
        Ph   CO₂Me

wwCH₂—CH—CH—CH₂ww   wwCH₂—C———C—CH₂ww
        | | | |
        Ph   Ph   CO₂Me CO₂Me

                                          Me   Me

wwCH=CH   CH₂—CH₂ww   wwCH₂—C   CH—CH₂ww
       | || |
       Ph   Ph   CO₂Me   CO₂Me
                                   CH₂   Me

Me   Me   CH₂

wwCH=CH   CH—CH₂ww   wwCH₂—CH—C—CH₂ww   wwCH₂—C   CH₂—CH₂ww
       | | | | ||
       Ph   CO₂   Ph   CO₂Me   CO₂Me   Ph

Scheme 7

The reaction between radicals (9) and (12) may be considered a model for cross-termination in S–MMA copolymerization.[47] The value for k_{td}/k_{tc} (90 °C) was determined as 0.56 and it was also established that, in disproportionation, transfer of a hydrogen from radical (9) to radical (12) was *ca.* 5.1 times more prevalent than transfer in the reverse direction [*i.e.* from (12) to (9)]. Values of k_{td}/k_{tc} (90 °C) for the self-reaction of these radicals are 0.13 and 0.75 for (9)[47] (see Section 11.3.2) and (12)[26] (see Section 11.3.3) respectively.

For the cross-reaction radicals (9) and (15), a model for cross-termination in S–MAN polymerization, k_{td}/k_{tc} (90 °C) was 0.61 and, in disproportionation, hydrogen transfer from (9) to (15) was *ca.* 2.2 times more frequent than transfer from (15) to (9).[47] Both self-reactions involve predominantly combination. Values of k_{td}/k_{tc} (80 °C) are 0.18 and 0.05 for radicals (9) (see Section 11.3.2) and (15) (see Section 11.3.5) respectively. A point to note is that the values k_{td}/k_{tc} for homotermination give no guide as to the value for k_{td}/k_{tc} in cross-termination.

The value of k_{td}/k_{tc} (80 °C) in the cross-reaction for the two methacrylyl models (12) and (14) (1.22) is similar to that seen in the self-reaction of (14) (1.12).[92] The value of k_{td}/k_{tc} (80 °C) for the self-reaction of (12) is 0.79. There is a small preference (*ca.* 1.4 fold) for the transfer of hydrogen from the butyl ester (14) to the methyl ester (12). Data on k_{td}/k_{tc} in primary radical termination are also relevant. Heitz *et al.*[93] report that primary radical termination between cyanoisopropyl radicals (15) and oligoethylene radicals involves substantial disproportionation. Bimolecular termination of primary alkyl radicals involves largely combination.[2] Heitz *et al.* have also indicated that oligo-styrene radicals combine with cyanoisopropyl radicals (15) (98 °C, toluene).[93] This result is incon-sistent with the findings of the above-mentioned model study on S–MAN copolymerization. However, Moad *et al.*[57] have recently re-examined this system and found it to be much more complicated than originally indicated and, in particular, have shown that some disproportionation does occur. Finally Barton *et al.*[95] have indicated that primary radical termination between poly(butyl methacrylate) radicals and cyanoisopropyl radicals involves largely disproportionation.

Another important finding to come from the three model studies of Serelis *et al.*[92,96] is that a preference for cross-termination over homotermination exists. The rate difference between cross-termination and homotermination is zero or small in the case of reactions of radicals (14) and (12).[92] For the S–MMA and S–MAN models they estimate that the rate constant for cross-termination is at least two to three fold greater than that for the faster of the two homotermination processes.[47,96] The preferences are, however, substantially less than are calculated by analyzing copolymerization data according to the classical 'chemical control' model (see Chapter 16 in this volume).[96,97]

Ito also has conducted model studies aimed at determining the relative rates of cross-termination over homotermination in S–MMA[98] and S–MAN[99] copolymerization. He came to the conclusion that cross-termination was not favoured over homotermination ($\phi \sim 1$) (see also Chapter 16 in this volume). However, errors introduced through the assumptions inherent in the kinetic analysis and by presuming that termination by disproportionation does not occur[99] are likely to be large.

11.5 TECHNIQUES

There are three basic methods for examining the termination reaction. These involve: (a) evaluating the kinetics of polymerization; (b) analyzing polymer samples — the molecular weight distribution and the type and number of end groups per chain are dependent on the termination mechanism; (c) examining model systems.

Methods (a) and (b) suffer from the limitation that they require some foreknowledge of the details of the polymerization mechanism. In particular, transfer reactions must be absent or characterized sufficiently well that their influence can be gauged.

Method (c) requires care in choosing the model and selecting reaction conditions. A very simple model cannot be expected to behave exactly the same as a polymeric species. Nonetheless, because model systems are more suited to exact analysis they can offer a valuable insight into the types of reaction to expect in termination of polymerization.

It is usually desirable to use a combination of techniques to obtain a complete and accurate picture.

11.5.1 Polymerization Kinetics

The mode of termination can be calculated by comparing the kinetic chain length (the ratio of the rate of propagation to the rate of initiation or termination) with the measured molecular weight.[39,40,51,55] This method lacks favour because of the large errors inherent in rate measurements.

11.5.2 Gelation

A method developed by Bamford *et al.*[69] for determining k_{td}/k_{tc} is known as 'the gelation technique' and involves measuring the time required for gelation when a polymeric halo compound [*e.g.* poly(vinyl trichloroacetate)]/$Mn_2(CO)_{10}$/hv is used to initiate polymerization. Under these conditions termination by combination will give rise to a crosslink while disproportionation will lead to graft formation. The gelation time is then a measure of the degree of crosslinking effected and hence k_{td}/k_{tc}. In addition to the usual assumptions, concerning absence of, or allowing for, competing termination pathways, it is necessary to know the initiator efficiency and rate of initiation. In the original work[69] the results were calibrated by reference to S polymerization for which k_{td}/k_{tc} was taken to be 0. In view of more recent studies, which show that disproportionation is significant (see Section 11.3.2), the data may require minor adjustment.

11.5.3 Molecular Weight Distribution

Another method of obtaining k_{td}/k_{tc} relies on a precise evaluation of the molecular weight distribution.[36,41] The mode of termination has a substantial influence on the molecular weight distribution with the polydispersity $(\bar{M}_w/\bar{M}_n) = 2.0$ if termination occurs exclusively by disproportionation and 1.5 if termination is by combination. This method is only applicable to very low conversion polymers since $\bar{M}_w/\bar{M}_n$ is conversion dependent. The molecular weight distribution may be measured by a variety of techniques including GPC[36,41] and ultracentrifugation.[38]

11.5.4 End Group Analysis

One of the classical methods for the evaluation of k_{td}/k_{tc} involves the quantitative determination of the polymer end groups and number average molecular weight. The errors inherent in this technique are often large. Extreme precision is required both in end group determination, as they comprise only a very small fraction of a typical polymer, and in measuring the molecular weight.

Usually the initiator-derived ends are determined since these may be readily labelled for detection by chemical,[100-103] radiochemical[58,104] or spectroscopic (NMR,[105] IR,[106] UV[107]) methods. The NMR method (see Section 8.8.2) offers substantial advantages in determining initiator-derived end groups since initiator residues incorporated into the polymer by chain initiation can be distinguished from unchanged initiator, end groups formed by transfer to initiator or copolymerization of initiator by-products.[56,57] Ideally, termination by combination will give rise to chains with two, and disproportionation to chains with one, initiator residue per molecule. The value of k_{td}/k_{tc} can therefore be calculated directly by applying the relationship: $k_{td}/k_{tc} = (2 - x)/2(x - 1)$; where x is the number of initiator fragments per molecule.

With the advent of NMR instrumentation with the requisite high sensitivity, direct detection of initiation- and termination-derived end groups is feasible in circumstances where there is no interference from signals due to the polymer chain.[25,68]

In evaluating the results it is necessary to allow for termination by other processes (*i.e.* transfer, inhibition, primary radical termination) and other side reactions (*e.g.* chain branching). In circumstances where transfer to monomer, solvent, *etc.* are significant k_{td}/k_{tc} will be overestimated.

If initiator residues are determined it is also necessary to ensure that end groups formed by all possible initiation pathways are detectable. It is also necessary to allow for or measure the contribution of other mechanisms whereby initiator fragments might be introduced into the polymer. These include: transfer to initiator; primary radical termination; and copolymerization of initiator by-products. If the latter processes are not allowed for, k_{td}/k_{tc} will be underestimated.

11.5.5 Model Studies

Determination of k_{td}/k_{tc} by direct analysis of a polymerization or the resultant polymer often requires data on other aspects of the polymerization mechanism that is not readily available. For this reason, it is often more appropriate to investigate the termination mechanism, in the first instance, by examining the self-reaction of low molecular weight models. Evaluation of k_{td}/k_{tc} is simplified and competing side reactions are more readily detected and allowed for. The adequacy of the models may, in some cases, be questioned. These studies, nonetheless, offer a valuable starting point since when conducted carefully, they provide unambiguous results and clearly demonstrate types of reactions that are likely to occur in a polymerization. In circumstances where there is a dramatic difference between polymerization and model there is clearly a need to reconcile the difference (*e.g.* in MMA and S polymerization at higher temperatures — see above). If a substantial chain length dependence for k_{td}/k_{tc} could be demonstrated unambiguously it would be an important finding.

Alkyl radicals can be conveniently generated from azo compounds or from aliphatic diacyl peroxides (see Section 8.1 in this volume). Simple models are readily available (for example, AIBN[74-76] for MAN polymerization and azobis(methyl isobutyrate)[26] for MMA polymerization). More complex models which enable the influences of chain length to be examined or the study of copolymerization often require some synthetic expertise.

A complication with this method is (co)polymerization of the unsaturated disproportionation product. While this may be seen as an advantage in some cases[74] in that it allows the examination of the reactions of oligomeric radicals, it severely complicates the product analysis.

Such problems may be circumvented by conducting the experiment in the presence of an inhibitor,[26,92] the concentration of which is chosen such that all radicals which escape the solvent cage are trapped, and all reactions of the initiator-derived radicals with other species are thus eliminated. In these circumstances, the cage recombination products may be examined as a model for the termination. In media of low viscosity there is no evidence that values for k_{td}/k_{tc} are different for cage and encounter reactions.[22]

ACKNOWLEDGEMENT

We are grateful to Dr A. K. Serelis for allowing us to communicate his unpublished data on models for termination in copolymerization and for providing a survey of the literature on termination.

11.6 REFERENCES

1. W. P. Neumann and R. Stapel, in 'Substituent Effects in Radical Chemistry', ed. H. G. Viehe, Z. Janousek and R. Merenyi, Reidel, Dordrecht, Netherlands, 1986, p. 271.
2. M. J. Gibian and R. C. Corley, *Chem. Rev.*, 1973, **73**, 441.
3. J. A. Kerr and A. F. Trotman-Dickenson, *Prog. React. Kinet.*, 1961, **1**, 105.

4. J. N. Bradley and B. S. Rabinovitch, *J. Chem. Phys.*, 1962, **36**, 3498.
5. T. Minato, S. Yamabe, H. Fujimoto and K. Fukui, *Bull. Chem. Soc. Jpn.*, 1978, **51**, 1.
6. M. Imoto, S. Sakai and T. Ouchi, *J. Chem. Soc. Jpn.*, 1985, 97.
7. S. W. Benson, *Acc. Chem. Res.*, 1986, **19**, 335.
8. J. J. Dannenberg and B. Baer, *J. Am. Chem. Soc.*, 1987, **109**, 292.
9. C. Rüchardt, *Top. Curr. Chem.*, 1980, **88**, 1.
10. H.-D. Beckhaus and C. Rüchardt, *Chem. Ber.*, 1977, **110**, 878.
11. G. Fraenkel and M. J. Geckel, *J. Chem. Soc., Chem. Commun.*, 1980, 55.
12. D. Griller and K. U. Ingold, *Acc. Chem. Res.*, 1976, **9**, 13.
13. K. Schreiner and A. Berndt, *Tetrahedron Lett.*, 1973, 3411.
14. A. L. J. Beckwith and C. J. Easton, *J. Chem. Soc., Perkin Trans. 2*, 1983, 661.
15. A. L. J. Beckwith, *Tetrahedron*, 1981, **18**, 3073.
16. K. Schlüter and A. Berndt, *Tetrahedron Lett.*, 1979, 929.
17. D. Griller, S. Içli, C. Thankachan and T. T. Tidwell, *J. Chem. Soc., Chem. Commun.*, 1974, 913.
18. K. U. Ingold, in 'Free Radicals', ed. J. K. Kochi, Wiley, New York, 1973, vol. 1, p. 1.
19. D. J. Trecker and R. S. Foote, *J. Org. Chem.*, 1968, **33**, 3527.
20. S. F. Nelsen and P. D. Bartlett, *J. Am. Chem. Soc.*, 1966, **88**, 137.
21. H.-H. Schuh and H. Fischer, *Helv. Chim. Acta*, 1978, **61**, 2463.
22. D. D. Tanner and P. M. Rahimi, *J. Am. Chem. Soc.*, 1982, **104**, 225.
23. R. C. Neuman, Jr. and M. E. Frink, *J. Org. Chem.*, 1983, **48**, 2430.
24. P. Cacioli, G. Moad, E. Rizzardo, A. K. Serelis and D. H. Solomon, *Polym. Bull.*, 1984, **11**, 325.
25. T. Kashiwagi, A. Inaba, J. E. Brown, K. Hatada, T. Kitayama and E. Masuda, *Macromolecules*, 1986, **19**, 2160.
26. S. Bizilj, D. P. Kelly, A. K. Serelis, D. H. Solomon and K. E. White, *Aust. J. Chem.*, 1985, **38**, 1657.
27. C. H. Bamford and E. F. T. White, *Trans. Faraday Soc.*, 1958, **54**, 268.
28. P. Cacioli, D. G. Hawthorne, R. L. Laslett, E. Rizzardo and D. H. Solomon, *J. Macromol. Sci., Chem.*, 1986, **23**, 839.
29. H. Tanaka, T. Kagawa, T. Sato and T. Ota, *Macromolecules*, 1986, **19**, 934.
30. G. C. Eastmond and J. Grigor, *Makromol. Chem., Rapid Commun.*, 1986, **7**, 375.
31. C. H. Bamford, G. C. Eastmond, J. Woo and D. H. Richards, *Polymer*, 1982, **23**, 643.
32. K. F. O'Driscoll and J. C. Bevington, *Eur. Polym. J.*, 1985, **21**, 1039.
33. C. H. Bamford and A. D. Jenkins, *Nature (London)*, 1955, **176**, 78.
34. G. M. Burnett and A. M. North, *Makromol. Chem.*, 1964, **73**, 77.
35. G. Ayrey, F. G. Levitt and R. J. Mazza, *Polymer*, 1965, **6**, 157.
36. J. V. Dawkins and G. Yeadon, *Polymer*, 1979, **20**, 981.
37. J. C. Bevington, H. W. Melville and R. P. Taylor, *J. Polym. Sci.*, 1954, **12**, 449; 1954, **14**, 463.
38. J. Hakozaki and N. Yamada, *J. Chem. Soc. Jpn.*, 1967, **70**, 1560 (*Chem. Abstr.*, 1968, **68**, 96 263r).
39. F. R. Mayo, R. A. Gregg and M. S. Matheson, *J. Am. Chem. Soc.*, 1951, **73**, 1691.
40. D. H. Johnson and A. V. Tobolsky, *J. Am. Chem. Soc.*, 1952, **74**, 938.
41. J. G. Braks and R. Y. M. Huang, *J. Appl. Polym. Sci.*, 1978, **22**, 3111.
42. G. Henrici-Olivé and S. Olivé, *J. Polym. Sci.*, 1960, **48**, 329.
43. M. J. Gibian and R. C. Corley, *J. Am. Chem. Soc.*, 1972, **94**, 4178.
44. G. Gleixner, O. F. Olaj and J. W. Breitenbach, *Makromol. Chem.*, 1979, **180**, 2581.
45. J. R. Shelton and C. K. Liang, *J. Org. Chem.*, 1973, **38**, 2301.
46. C. G. Overberger and A. B. Finestone, *J. Am. Chem. Soc.*, 1956, **78**, 1638.
47. A. K. Serelis, unpublished data.
48. H. Langhals and H. Fischer, *Chem. Ber.*, 1978, **111**, 543.
49. K. J. Skinner, H. S. Hochster and J. M. McBride, *J. Am. Chem. Soc.*, 1974, **96**, 4301.
50. R. C. Neuman, Jr. and M. J. Amrich, Jr., *J. Org. Chem.*, 1980, **45**, 4629.
51. O. F. Olaj, J. W. Breitenbach and B. Wolf, *Monatsh. Chem.*, 1964, **95**, 1646.
52. O. F. Olaj, H. F. Kauffmann, J. W. Breitenbach and H. Bieringer, *J. Polym. Sci., Polym. Lett. Ed.*, 1977, **15**, 229.
53. K. C. Berger and G. Meyerhoff, *Makromol. Chem.*, 1975, **176**, 1983.
54. K. C. Berger, *Makromol. Chem.*, 1975, **176**, 3575.
55. M. Stickler, *Makromol. Chem.*, 1979, **180**, 2615.
56. G. Moad, D. H. Solomon, S. R. Johns and R. I. Willing, *Macromolecules*, 1982, **15**, 1188.
57. G. Moad, D. H. Solomon, S. R. Johns and R. I. Willing, *Macromolecules*, 1984, **17**, 1094.
58. J. C. Bevington, 'Radical Polymerization', Academic Press, London, 1961.
59. P. W. Allen, G. Ayrey, F. M. Merret and C. G. Moore, *J. Polym. Sci.*, 1956, **22**, 549.
60. G. Ayrey and C. G. Moore, *J. Polym. Sci.*, 1959, **36**, 41.
61. H. Boudevska, C. Brutchkov, S. Platchkova and J.-P. Pascault, *Makromol. Chem.*, 1981, **182**, 3257.
62. J. G. Braks, G. Mayer and R. Y. M. Huang, *J. Appl. Polym. Sci.*, 1980, **25**, 449.
63. G. V. Schulz, G. Henrici-Olivé and S. Olivé, *Makromol. Chem.*, 1959, **31**, 88.
64. C. H. Bamford, G. C. Eastmond and D. Whittle, *Polymer*, 1969, **10**, 771.
65. J. S. Mackie and S. Bywater, *Can. J. Chem.*, 1957, **35**, 570.
66. J. C. Bevington and H. G. Troth, *Trans. Faraday Soc.*, 1962, **58**, 186.
67. W. P. Neumann and R. Stapel, *Chem. Ber.*, 1986, **119**, 3422.
68. K. Hatada, T. Kitaiyama and E. Masuda, *Polym. J.*, 1986, **18**, 395.
69. C. H. Bamford, R. W. Dyson and G. C. Eastmond, *Polymer*, 1969, **10**, 885.
70. A. K. Chaudhuri and S. R. Palit, *J. Polym. Sci., Part A-1*, 1968, **6**, 2187.
71. G. Ayrey and A. C. Haynes, *Eur. Polym. J.*, 1973, **9**, 1029.
72. G. Ayrey, M. J. Humphrey and R. C. Poller, *Polymer*, 1977, **18**, 840.
73. A. Fehérvári, E. B. Gyevi, T. Földes-Bereznich and F. Tüdos, *J. Macromol. Sci., Chem.*, 1982, **18**, 431.
74. A. F. Bickel and W. A. Waters, *Recl. Trav. Chim. Pays-Bas*, 1950, **69**, 1490.
75. W. Barbe and C. Ruchardt, *Makromol. Chem.*, 1983, **184**, 1235.
76. A. K. Serelis and D. H. Solomon, *Polym. Bull.*, 1982, **7**, 39.

77. A. B. Jaffe, K. J. Skinner and J. M. McBride, *J. Am. Chem. Soc.*, 1972, **94**, 8510.
78. G. Moad, D. H. Solomon and R. I. Willing, in preparation.
79. Y. Tsuda, *Kobunshi Kagaku*, 1960, **17**, 364.
80. C. H. Bamford, A. D. Jenkins and R. Johnston, *Trans. Faraday Soc.*, 1959, **55**, 179.
81. W. Bracke, J. A. Empen and C. S. Marvel, *Macromolecules*, 1968, **1**, 465.
82. B. E. Bailey and A. D. Jenkins, *Trans. Faraday Soc.*, 1960, **56**, 903.
83. J. C. Bevington and D. E. Eaves, *Trans. Faraday Soc.*, 1959, **55**, 1777.
84. B. L. Funt and W. Pasika, *Can. J. Chem.*, 1960, **38**, 1865.
85. D. C. Bugada and A. Rudin, *Polymer*, 1984, **25**, 1759.
86. F. F. Vercauteren and W. A. B. Donners, *Polymer*, 1986, **27**, 993.
87. F. Danusso, G. Pajar and D. Sianesi, *Chim. Ind. (Milan)*, 1959, **41**, 1170 (*Chem. Abstr.*, 1961, **55**, 20 499a).
88. G. Talamini and G. Vidotto, *Chim. Ind. (Milan)*, 1964, **46**, 16 (*Chem. Abstr.* 1964, **60**, 10 804a).
89. G. S. Park and D. G. Smith, *Makromol. Chem.*, 1970, **131**, 1.
90. W. H. Starnes Jr., I. M. Plitz, F. C. Schilling, G. M. Villacorta, G. S. Park and A. H. Saremi, *Macromolecules*, 1984, **17**, 2507.
91. W. H. Atkinson, C. H. Bamford and G. C. Eastmond, *Trans. Faraday Soc.*, 1970, **66**, 1446.
92. D. P. Kelly, A. K. Serelis, D. H. Solomon and P. Thompson, *Aust. J. Chem.*, 1987, **40**, 1631.
93. W. Güth and W. Heitz, *Makromol. Chem.*, 1976, **177**, 1835.
94. W. Konter, B. Bömer, K.-H. Köhler and W. Heitz, *Makromol. Chem.*, 1981, **182**, 2619.
95. J. Barton, I. Capek, V. Juranicova and S. Reidel, *Makromol. Chem. Rapid. Commun.*, 1986, **7**, 521.
96. G. Moad, A. K. Serelis, D. H. Solomon and T. H. Spurling, *Polym. Commun.*, 1984, **25**, 240.
97. C. Walling, *J. Am. Chem. Soc.*, 1949, **71**, 1930.
98. K. Ito, *Polymer*, 1985, **26**, 1253.
99. K. Ito, *J. Polym. Sci., Polym. Chem. Ed.*, 1978, **16**, 2725.
100. S. R. Palit and B. M. Mandal, *J. Macromol. Sci., Rev. Macromol. Chem.*, 1968, **2**, 225.
101. N. N. Ghosh and B. M. Mandal, *Macromolecules*, 1986, **19**, 19.
102. E. Rizzardo and D. H. Solomon, *J. Macromol. Sci., Chem.*, 1979, **13**, 997.
103. E. Rizzardo and D. H. Solomon, *J. Macromol. Sci., Chem.*, 1979, **13**, 1005.
104. G. Ayrey, *Chem. Rev.*, 1960, **63**, 645.
105. J. C. Bevington, J. R. Ebdon and T. N. Huckerby, *Eur. Polym. J.*, 1985, **21**, 685.
106. Yu. N. Anisimov, S. S. Ivanchev and A. I. Yurzhenko, *Polym. Sci. USSR (Engl. Trans.)*, 1967, **9**, 773.
107. L. H. Garcia Rubio, N. Ro and R. D. Patel, *Macromolecules*, 1984, **17**, 1998.

12

Kinetics of Bimolecular Termination

KENNETH F. O'DRISCOLL
University of Waterloo, Ontario, Canada

12.1 INTRODUCTION

Before two polymeric free radicals react to terminate their chain growth, they must first undergo a series of diffusional steps. These diffusional steps can be separated into two different types: the *translational* diffusion, during which the two chains enter into collision with each other and the *segmental* diffusion, during which the two chain ends approach each other and ultimately enter into a small volume element in which reaction can take place. This process, shown schematically in Figure 1, was first described by Benson and North[1] and has come to be recognized as a necessary basis for describing the complex behavior of the termination process in free radical polymerization.

When the concentration of polymer chains, propagating or terminated, is low, the segmental diffusion process will control the rate of termination. As the polymer concentration increases, translational diffusion becomes rate controlling. Thus, as a polymerization advances to moderate conversion, the increase in rate often observed (variously called the Trommsdorf[2], Norrish–Smith[3] or gel effect) can be ascribed to the decrease in the rate constant for termination, caused by the change in the diffusion process which controls the rate. Finally, at very high conversion, propagating chains are quite immobile and their termination occurs by a process termed 'reactive diffusion', a unique form of translational diffusion where the addition of monomer units serves to bring two (otherwise immobile) chain ends into reactive proximity. Reactive diffusion is discussed in more detail below.

Regardless of which of these processes is rate controlling, the overall process occurring when two radicals (P) terminate each other can be mechanistically described by

$$P + P \xrightarrow{k_T} \{PP\} \xrightarrow{k_S} (PP) \xrightarrow{k_C} P\text{–}P \tag{1}$$

Figure 1 Schematic representation of two polymeric free radicals diffusing together, undergoing segmental motion and then terminating by combination

where the rate constants subscripted T, S and C refer, respectively, to the net rate constants for the translational and segmental diffusion processes, and for the chemical reaction step. The encounter collision of the two coils is $\{P\,P\}$, the encounter collision of the two radical chain ends is (PP) and the dead polymer is P–P. (The latter is treated as though the chains had combined, but it could equally have been treated as though a disproportionation reaction had occurred.) The overall rate constant for the termination reaction, k_t, can be defined by either the appearance of the product (P–P) or disappearance of the two propagating radicals. Following conventional kinetic treatment

$$-(1/2)\mathrm{d}[P]/\mathrm{d}t = \mathrm{d}[P–P]/\mathrm{d}t = k_t[P]^2 \tag{2}$$

whence

$$-\mathrm{d}[P]/\mathrm{d}t = 2k_t[P]^2 \tag{3}$$

The overall rate constant for termination has been defined according to equation (3) by many authors (mostly North American). However, the reader is advised that many authors have defined the rate constant by

$$-\mathrm{d}[P]/\mathrm{d}t = k_t[P]^2 \tag{4}$$

Clearly the rate constant defined by equation (4) is just twice the numerical value of that defined by equation (3) and one must make allowance for this in using literature values for the overall termination rate constant and numbers derived from it. In this chapter k_t is defined by equation (4).

Although k_t is referred to as a rate 'constant', it must be appreciated that it can and does change during the course of a polymerization, even if it is isothermal. As a result of the existence of the three sequential diffusion processes outlined above, the whole description of the termination kinetics can become quite complicated for a polymerization taking place over a wide range of polymer concentrations. To simplify this description somewhat, the following discussion is broken down into three separate categories: very low conversion (*ca.* 0 to 10% polymer in solution), the gel effect region (*ca.* 20 to 80%) and very high conversion (*ca.* >90%). As will be seen, this division is quite arbitrary and the exact boundaries between these regions depends on the molecular weight of the polymer and on those reaction parameters which affect polymer mobility — solvent, chain stiffness, temperature, *etc.*

12.2 k_t AT LOW CONVERSIONS

The ratio of rate constants k_p^2/k_t appears frequently in any analytical description of free radical polymerization kinetics and can often be determined from experimental information on steady state polymerizations where, for example, rate or molecular weight is measured. Typically, k_t values of $10^{7\pm1}\,\mathrm{l\,mol^{-1}\,s^{-1}}$ are observed. However, to determine the absolute value of k_t it is necessary to carry out an analysis of a non-steady state polymerization. In such an experiment the average lifetime of a propagating chain, τ, can be measured and related to the ratio k_p/k_t. This numerical value can, in turn, be used with the separately measured rate of initiation, R_i, or k_p^2/k_t to obtain separately the numerical values of k_p and k_t.

A number of techniques have been used to conduct experiments under non-steady state conditions. Perhaps the best known is the rotating sector technique,[4] which depends on intermittent photochemical initiation of polymerization in which the illumination period is short compared to the time necessary to obtain steady state conditions. In the rotating sector experiment, the lifetime is given by

$$\tau = k_p[M]/k_tR_p \tag{5}$$

An analogue of the rotating sector experiment, called spatially intermittent polymerization,[5] provides a sample adequate for both conversion and molecular weight measurements. This technique is conducted in a continuous flow tubular reactor, in which the monomer and photo-initiator pass alternately through dark and light regions. The large sample which can be collected obviates the need for a separate experimental determination of R_i or k_p^2/k_t. The equations describing this experiment are formally identical to those of the rotating sector, with the period of rotation being replaced by the ratio of linear flow velocity to length.

Other techniques which have been used include non-steady state measurements of viscosity,[6] or of volume changes by dilatometry[7] and steady state measurements of emulsion polymerization rate.[8] The latter determines k_p when the number of particles is known and the number of radicals per particle is assumed to be 1/2. Recent work which measures this quantity directly by electron spin resonance (ESR)[9] is discussed below.

The non-steady state experiments have, collectively, given data which exhibit enormous scatter, even though they have been restricted to conversion levels below 1%. Figure 2 shows one such collection[10] of rate constants for styrene. It is probable that the scatter is caused, at least in part, by the existence of a chain length dependence of k_t. The lifetime determination is done under one set of conditions and the k_p^2/k_t measurement under another, where the chain length is different. This experimental problem is circumvented in the spatially intermittent polymerization technique, by which it was unequivocally shown that a chain length dependence of k_t does exist as illustrated in Figure 3.[5]

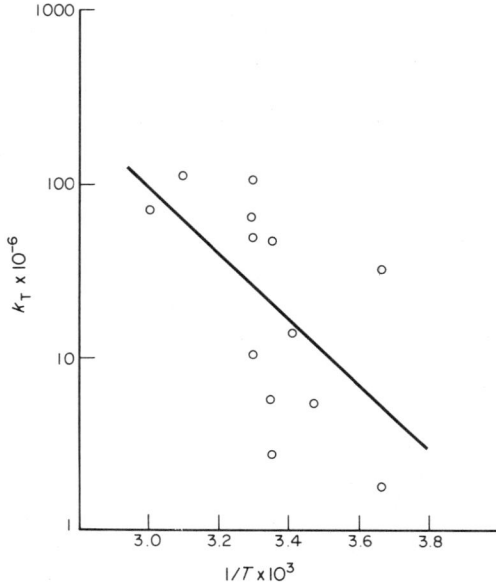

Figure 2 Arrhenius plot of k_t data as collected in ref. 10

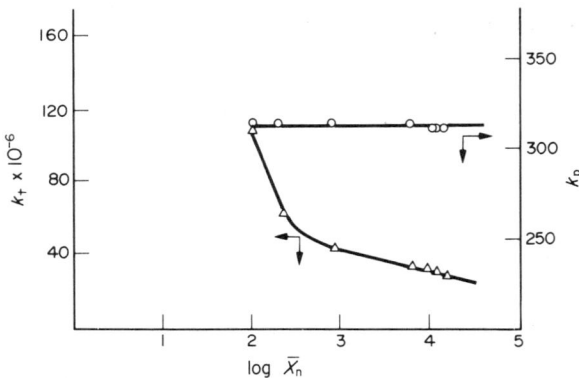

Figure 3 Effect of chain length on k_p and k_t for methyl methacrylate

A theoretical description of the chain length dependence of k_t can be found in the work of Mahabadi and O'Driscoll.[11] Using the model embodied in equation (1) and considering chain flexibility, chain size, excluded volume effects and polymer–solvent interactions, they showed that the *average* termination rate constant, defined by

$$k_t = \sum_{n=1}^{\infty} \sum_{m=1}^{\infty} (k_{t(n,m)}[P_n][P_m]/[P])$$

(6)

could be described as a product of two functions

$$k_t = F_1 F_2 \tag{7}$$

where F_1 is a function of the segmental friction coefficient, ξ, and temperature, and F_2 is a function of the linear chain expansion coefficient, α, and the lengths of the two reacting chains, N_i and N_j

$$F_1 \sim T/\xi \tag{8}$$

$$F_2 = (\alpha_i \alpha_j)^{-1.3}[1 - 0.37(\alpha_i \alpha_j)^{-0.37} \times \{1 - 4(2/3\pi)^{1/2}(N_0)^{1/2}(\alpha_i \alpha_j)^{-1/2}(N_i N_j)^{-1/4}\}] \tag{9}$$

Assuming $N_i = N_j = X_n$ (the average chain length), these equations predict a chain length dependency

$$k_t \sim X_n^a \tag{10}$$

where a is predicted to be -0.5 for short chains (length *ca.* 100 or less), falling off to -0.1 for large chains (length > 1000). They also predict an effect of segment size and solvent viscosity (as they affect ξ) and solvent quality (as expressed by α) which have been experimentally verified.

The chain length dependency is not noticed in most polymerizations at low conversions and high degrees of polymerization because the -0.1 power is so small. However, when dealing with low molecular weights (as for example in the case of measuring chain transfer constants or the rate constant for primary radical termination, or when doing precise work at ordinary molecular weights) the chain length effect can be large enough to distort experimental results and conclusions unless it is explicitly considered. Yasukawa and Murakami[12] showed by computer simulation that a -0.1 power dependence of k_t on chain length would produce an apparent order of polymerization, with respect to initiator, of 0.45 instead of the expected 0.5. This was verified by Stickler,[13] who also showed that the termination rate constant for a monomeric radical was 50 times greater than the average for the polymeric radicals.

As another example of the possible importance of chain length dependence, Figure 4[14] shows the primary radical termination as calculated for styrene, with and without consideration for chain length dependence of k_t. The curvature in the plot where chain length dependence was neglected was originally attributed to the possibility of disproportionation of styrene in the termination reaction. (It had previously been accepted from a great deal of evidence that styrene terminates only by combination.) The plot without curvature shows that, when chain length dependence is included, disproportionation need not be invoked, and the termination of styrene is indeed by combination.

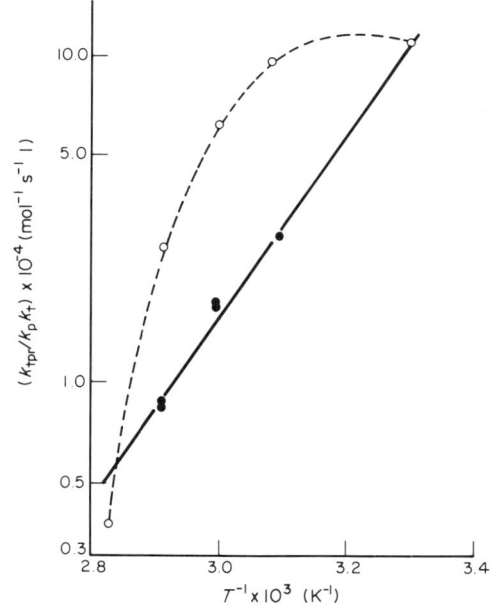

Figure 4 Calculated variation of the characteristic constant for primary radical termination with temperature in experiments done at variable ($\bigcirc$) and constant ($\bullet$) chain lengths

A particular effect of solvent quality on k_t is caused by the influence of polymer concentration on polymer size at low polymer concentrations. Rosen[15] showed experimentally that small amounts of polymer in solution would actually increase the value of k_t. He reasoned that the polymer in solution would cause the polymeric radical to find itself in a thermodynamically poorer solvent compared to the same radical in pure solvent. Coils would be smaller because of this. Therefore, two colliding chains would have a smaller volume to explore by segmental diffusion before reacting, and would take less time to react. Mahabadi and O'Driscoll[16] expanded the theoretical treatment of α to show quantitatively that the rate enhancement observed at *ca.* 1 to 10% polymer in solution was consistent with this reasoning. Horie and Mita[17] went on to show the limitation of this approach as the solvent quality becomes sufficiently poor to make it difficult for the chains to explore their combined coils. Experimental observation of this phenomenon in bulk polymerization has been made, both with respect to rate and degree of polymerization.[18] It should be noted that the effect is small, seldom amounting to more than a 5 or 10% change in the expected rate.

The most significant conclusions to be drawn from this theoretical treatment are: (i) any factor which decreases the segmental friction coefficient will increase k_t; (ii) any factor which decreases the reacting coil size will increase k_t; and (iii) any careful kinetic study which changes the chain length or solvent character must take into consideration the effects embodied in the previous two points.

While equation (7) gives an exact relation between the important physical parameters and the experimentally observable average termination rate constant, it is an awkward expression of the chain length dependence. Numerous workers have suggested and used an equation of the form

$$k_{t(n,m)} = k_{t0}(nm)^{-a} \tag{11}$$

where $k_{t(n,m)}$ is the rate constant for termination between two radicals of lengths m and n and k_{t0} is a constant. Mahabadi[19] has shown that, if the exponent a is treated as a constant, the relation between the observed k_t and X_n is

$$k_t = k_{t,0}\, g(a)\, X_n^{-2a} \tag{12}$$

where $g(a)$ involves the gamma function of a. Experimental verification of the latter equation has often been found.

12.3 k_t AT INTERMEDIATE CONVERSIONS: THE GEL EFFECT

The gel effect is commonly accepted as beginning in that portion of a bulk or solution polymerization where the reaction rate increases and the molecular weight of the polymer formed also increases. Any complete model of the gel effect must describe not only the increased rate (and molecular weight) during the gel effect, but also a smooth transition between the low conversion rate and the gel effect. To that end it is useful to consider the question: when does the gel effect begin? Some authors have used simple conversion–time plots, and taken the upward curvature from linearity to indicate the onset of the gel effect; others have used the deviation from linearity of a plot of $\ln(R_p)/[M]$. Given the decrease in rate caused by the presence of polymer, it has also been suggested that the minimum in rate which can be observed defines the onset of the gel effect. As will be seen in the following discussions, most models employ an empirical or theoretical means of specifying the onset of the gel effect in order to trigger a 'switch' from low conversion kinetics to gel effect kinetics.

Burnett and Duncan[20] were the first to postulate a set of equations to describe the gel effect. They suggested the existence of a critical concentration of polymer, which is dependent on the polymer molecular weight, and above which the propagating chains are immobile and incapable of terminating each other. This approach gave a simple set of equations which were in accord with experimental observations. The equations demanded the existence of a 'switch', *i.e.* a concentration at which chains switched from being mobile and capable of terminating to being immobile and incapable of terminating. Furthermore, no allowance was made for the fact that small growing chains might terminate before they became immobile.

Fifteen years later, Cardenas and O'Driscoll[21] postulated that two populations of propagating chains existed in a solution at a given concentration: those large enough to be entangled and those small enough to be freely mobile. The transition between the two chain sizes was established with a concept,[22] useful in viscosity measurements, that a critical constant, K_c, exists, which is equal to the product of polymer concentration and chain length, and which determines the minimum chain size for entanglement. Assigning a rate constant $k_{t,e}$ to reaction between the entangled species, the

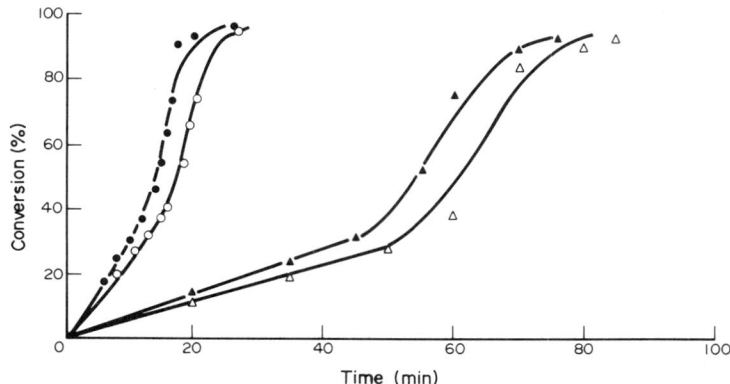

Figure 5 Bulk polymerization of methyl methacrylate at 70 °C ($\triangle$, $\blacktriangle$) and 90 °C ($\bigcirc$, $\bullet$); solid lines are the model[21] predictions

normal k_t to reaction between the unentangled and a geometric mean of the two for reaction between the unentangled and the entangled chains, it was shown that a two parameter model (K_c and $k_t/k_{t,e}$) accurately predicted rate and molecular weight behavior over a wide range of conversions, initiator concentrations and temperature for methyl methacrylate (MMA) in bulk. This is illustrated in Figure 5 for MMA at two temperatures and two initiator concentrations.

In addition to the successful description of the rate and molecular weight to be expected in polymerization, the model also led to a simple understanding of the reason why some monomers show a stronger gel effect than others. Put simply, it depends on the ratio k_p^2/k_t. A monomer having a large value of such a ratio and polymerizing at a given rate will produce, in a given time, the same amount of polymer but with a higher molecular weight than will a different monomer, polymerizing at the same rate, with a lower value for that ratio. This explains why, for example, styrene has such a small gel effect compared to MMA.

In spite of its ability to describe bulk polymerization, the Cardenas and O'Driscoll model failed to adequately describe polymerization in solution and, most importantly, was not intended to describe very high conversion. The success of the model coupled with its limitations led to several other models which took different physical approaches.

One such approach to a useful description of the gel effect was taken by Hamielec and Marten[23] who introduced the concept of free volume into a semi-empirical model. They postulated a relation between the weight average molecular wight, M_w and the free volume of the reacting system, V_f

$$k_t \sim M_w^{-1.75} \exp(-A/V_f) \tag{13}$$

The exponent of molecular weight was empirically chosen, A is a constant and the concentration dependence of k_t is contained in V_f. This approach and the approach of O'Driscoll and Cardenas fitted the same data equally well.

In a series of papers, Soh and Sundberg[24] combined the entanglement and free volume concepts so that, using a switch at some critical chain length, the value obtained was

$$k_{t(n,m)} = k_{t,0}(X_n/X_{n,c})^{-2.4} \exp(-A/V_f) \tag{14}$$

This approach gave a good fit to bulk polymerization data for six different monomers.

At the same time, the work of de Gennes with scaling and reptation concepts[25] was showing great promise for describing physical phenomena of polymers in solution, particularly diffusion. Tulig and Tirrell[26] developed a model for k_t using the concepts of reptation and a critical concentration, C^*, above which chains interpenetrate, and a second critical concentration, C^{**}, above which reptative behavior dominates. For chains of length N, C^{**} is defined as $K_c/N^{0.5}$, and above that concentration

$$k_t = k_{t,\min} + k_t^{**}(N^{**}/N)^2(C^{**}/C)^{1.75} \tag{15}$$

where the ** superscripts refer to values at the transition concentration and $k_{t,\min}$ is a minimum value for the rate constant. This works quite well at concentrations ranging up to *ca.* 70%, but in that range another switch had to be introduced (C^{***}) and another formulation for the rate constant was needed (see below).

In several publications, Ito[27] has also used scaling and reptation concepts and has proposed equations similar to those of Tulig and Tirrell. De Gennes has discussed bimolecular diffusion-controlled reactions in general and the termination reaction of polymerization in particular in a pair of papers.[28] He concluded that k_t should vary inversely with the square of chain length when reptation controls the process.

In all the gel effect modeling described above, the authors sought to explain the deviations from the behavior expected for classical free radical polymerization kinetics by changes in k_t alone. In fact, it has long been recognized that initiator efficiency and even the propagation rate constant, k_p, are reduced by increasing viscosity. Based on a series of papers on the kinetics of emulsion polymerization, it has been shown[9] that it is possible to use ESR measurements to obtain chain radical concentrations in emulsion particles. This enables the determination of rate constants with few assumptions. In this manner, it has been demonstrated[9] that k_t can be adequately described by the Soh and Sundberg model.[24]

This experimental development is particularly interesting since the technique involves seeded emulsion polymerizations, and it is quite possible to experimentally generate a given average chain length radical in the presence of a different chain length polymeric matrix. A similar approach in bulk or solution polymerization might be possible using the pulsed laser technique pioneered by Buback and co-workers[29] in high pressure ethylene polymerization. These could be powerful tools for testing models, especially when it is recognized that most models have used an average molecular weight of the polymerizing system, which drifts with conversion, to describe the gel effect on the rate constants of a separate distribution of radical chain lengths.

12.4 k_t AT VERY HIGH CONVERSIONS

The description of a polymerization approaching very high conversion, *ca.* 90% or more, is of great industrial importance. However, judging by the published literature, it has (until recently) received very little detailed attention in a kinetic sense. Any description of the polymerization process at very high conversion will necessarily include, not only diffusion in viscous media, but also diffusive behavior in a glass, the validity of steady state assumptions and the possibility of depropagation becoming kinetically important. Such considerations imply that one cannot consider the termination rate constant in isolation, but rather that it is necessary to consider it together with the propagation rate constant, the initiator rate constant and its efficiency, and with the relation between the polymerizing systems' glass transition temperature, T_g, the conversion level and the temperature of polymerization as well as the Gibbs free energy of polymerization (*i.e.* reversibility).

The most coherent approach to this problem has been made by Stickler and his colleagues[30] who have utilized a combination of the free volume[23] approach for k_p with the reactive diffusion concept of Schulz for k_t.[31]

A theoretical consideration of the course of a bulk polymerization carried out at a temperature below the polymer's T_g suggests that a polymerizing mixture would become glassy before reaching 100% conversion. The glass temperature of the reacting mixture may be calculated from free volume considerations to be determined by a weighted sum, computed from the volume fractions of each species in the reaction mixture and their glass transition temperatures. In this glassy state, it has been proposed that polymerization will cease, because monomer is unable to diffuse to a live chain end.

Wunderlich and Stickler have presented[32] experimental evidence to show that MMA does not follow this simple idea. Figure 6 shows that the 'limiting conversion' observed at 80 °C, well below the T_g of PMMA, was a moderately strong function of initiator concentration. Further study revealed that polymerization continued, albeit at a very low rate, without cessation after the monomer–polymer reaction mixture had become glassy. From a practical, as opposed to a theoretical, point of view it is reasonable to ignore this very slow rate of polymerization (which is not understood) in order to model the industrially important final stage of polymerization. The following treatment does so, and explicitly includes k_p as well as k_t, but does not include depolymerization or the influence of high conversion on the initiation process.

Under conditions where the polymer concentration is very high, it might be expected that the reduced mobility of chains described in the gel effect becomes so great that they may be considered truly immobile. If that is the case, ordinary bimolecular reactions as illustrated by equation (1) can no longer occur. Instead, we must recognize that the active ends of two propagating chains can move through space by adding monomer, and so come into the same small volume element where they can terminate each other. This is termination by reactive diffusion and the rate constant for the process

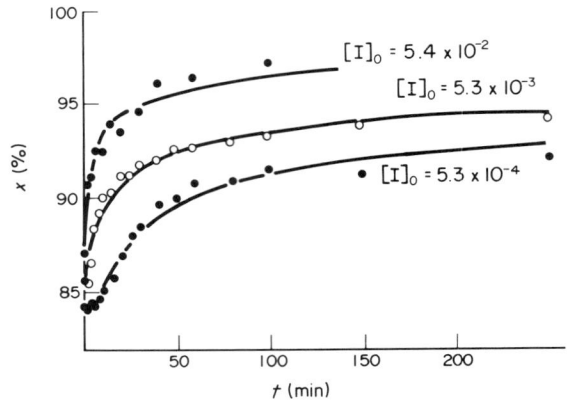

Figure 6 Influence of initiator concentration on limiting conversion of methyl methacrylate at 80 °C; solid lines are the model predictions[32]

can be written as proportional to the diffusion coefficient, D_R

$$k_{t,R} = C D_R \tag{16}$$

where the constant C contains known geometric factors and

$$D_R = (n_s l_0^2/6) k_p[M] \tag{17}$$

where l_0 and n_s are the length and number of monomers in a statistical segment of polymer.

Because the termination process is, under these conditions, defined in terms of the propagation rate constant, a precise description of k_p at high conversion is needed. At low conversions, the propagation reaction is quite fast and chemically controlled with a rate constant k_p of the order of magnitude of 10^2 to 10^3. As the polymerization becomes more and more viscous, it is not unreasonable to suppose that the addition of monomer might be limited by its diffusivity to the immobile chain end. The rate constant is then given, in terms of the free volume theory, by

$$k_p \doteq k_{p,0} \exp[-V^*(1/V_f - 1/V_{f,0})] \tag{18}$$

where V^* is an empirical constant and $V_{f,0}$ and V_f are the free volumes of monomer and reacting mixture, respectively. The latter can be calculated from the work of Kelley and Bueche.[33]

The empirical constant V^* may be regarded as a critical fraction of the free volume required for monomer to diffuse to the chain end. Fitting of data to the above equations gave $V^* = 0.35$, independent of temperature and initiator type or concentration.

This model was combined with that of Hamielec and Marten[23] to produce a comprehensive description of the kinetics of bulk polymerization over a wide conversion range, extending to very high conversion. The major additions to the equations presented above were the introduction of two 'switches', $V_{f,cr1}$ and $V_{f,cr2}$ — critical free volumes of the reaction mixture at which the gel effect and the very high conversion equations are used. $V_{f,cr1}$ replaces $V_{f,0}$ in equation (18) above and V^* was assigned the value 1.0. With four 'fixed' parameters and two adjustable ones this then gave an excellent fit to both rate and molecular weight data.

Other approaches, such as using reptation theory, have been less successful in describing very high conversion. So it must be regarded as an important area in which the lack of general knowledge and insufficient data combine to make it necessary that more work be done.

12.5 TERMINATION IN COPOLYMERIZATION

In copolymerization, the termination reaction is often discussed in terms of the three different reactions possible when two different chain ends combine

$$\left.\begin{array}{l} P_A + P_A \longrightarrow \\ P_B + P_B \longrightarrow \\ P_A + P_B \longrightarrow \end{array}\right\} \text{dead polymer} \tag{19}$$

where P_A and P_B represent active chains having terminal units of monomers A and B respectively. Rate constants for the three steps are usually designated k_{tAA}, k_{tBB} and k_{tAB} respectively.

The problem with this approach to describing the termination reaction in copolymerization is that the diffusion-controlled processes which operate in homopolymerization also operate in copolymerization. Designating rate constants in terms of the end unit of the chain is, therefore, the equivalent of assuming that that unit determines the speed of the rate-controlling diffusion process. This is physically incorrect, but, so far, no one has given a quantitative treatment which is correct as an alternative in all cases. What follows then is a brief summary of those treatments which have been given. The subject has been reviewed by Chiang and Rudin.[34]

The well-known 'ϕ factor', defined by

$$\phi = k_{tAB}/(k_{tAA} k_{tBB})^{0.5} \tag{20}$$

was originally introduced to describe rates of copolymerization.[35] It can be determined from rates of copolymerization as a function of composition. Often the ϕ values are much greater than unity. Since ϕ is the ratio of the cross termination to the geometric mean of the homotermination reaction rate constants, values of ϕ that are very different from unity imply that chemical effects (*e.g.* electrostatic interactions between groups of differing polarity) cause a marked increase in the cross termination reaction. Since the reaction is diffusion controlled, chemical changes cannot be expected to cause an increase in the rate. For this reason, the ϕ factor must be regarded as an empirical parameter which may be useful for describing the rate of copolymerization, but which does not have any physical significance at the molecular level.

An extension of the ϕ factor has been developed by Russo *et al.*,[36] who have attributed the segmental diffusion which is rate controlling to the two end units of the reacting chain, *i.e.* the last four carbon atoms. This leads to what is essentially a penultimate unit effect in the termination step, and necessitates the determination of a larger number of parameters than is usually justified by the data. It has, however, been successful in describing some systems,[36] but not others.[37]

Treating the termination process as the average of reactions between chains which have many different diffusive speeds was suggested first by Atherton and North,[38] who postulated the simple expression

$$k_t = F_1 k_{tAA} + F_2 k_{tBB} \tag{21}$$

proposing that the average k_t was merely the average of the homotermination k_t values weighted by the copolymer compositions F_1 and F_2. This was extended by Ito and O'Driscoll,[37] who averaged the segmental friction coefficients.

All of the above has been done on systems of two monomers at low conversion. No one has yet treated binary copolymerizations taken to high conversion where the gel effect is important, or multicomponent polymerizations with three or more monomers, Also, all these treatments have assumed that the simple Mayo–Lewis model of copolymerization propagation is applicable. Recent work by Fukuda *et al.*[39] has demonstrated, by experimental determination of k_t in copolymerization (using the rotating sector technique), that the Mayo–Lewis model may be inapplicable for styrene–methyl methacrylate. The apparent large maximum in k_t as composition is varied may be an artifact

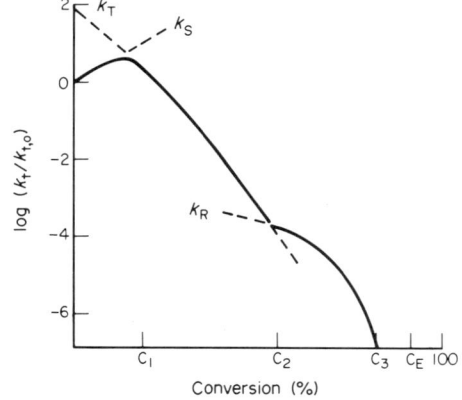

Figure 7 Schematic variation of the termination rate 'constant' with conversion in a bulk polymerization. The symbols are defined in the text

of the model being assumed; they suggested a penultimate unit effect in copolymerization for this system might make the experimental k_t data easily understandable.

12.6 CONCLUSIONS

From the work described above, it would appear that calling the parameter k_t a rate 'constant' is a gross abuse of the meaning of the word. Not only does k_t begin to increase as soon as polymer appears in the reaction mixture, but at about 10% conversion it changes from being controlled by segmental diffusion, k_S, to being controlled by translational diffusion, k_T. As the polymer concentration increases further, the combined effect of more polymer and higher molecular weight polymer further depresses k_T, until it is so slow that reactive diffusion, k_R, becomes dominant. This is shown schematically in Figure 7 where the numbers on the ordinate are very approximate. The conversion level C_1 is the onset of the gel effect, C_2 the onset of reactive diffusion and C_3 the limiting conversion, at which the system becomes glassy and beyond which little polymerization is observed. Should C_3 be higher than the equilibrium monomer concentration, C_E, the schematic representation would need revision to indicate that the apparent value of k_t would go to zero at C_E

12.7 REFERENCES

1. S. W. Benson and A. M. North, *J. Am. Chem. Soc.*, 1962, **84**, 935.
2. E. Trommsdorf, H. Kohle and P. Lagally, *Makromol. Chem.*, 1948, **1**, 169.
3. R. G. W. Norrish and R. R. Smith, *Nature (London)*, 1942, **150**, 336.
4. G. M. Burnett and H. W. Melville, *Nature (London)*, 1945, **156**, 661.
5. K. O'Driscoll and H. K. Mahabadi, *J. Polym. Sci., Polym. Chem. Ed.*, 1976, **14**, 869.
6. M. J. S. Dewar and C. H. Bamford, *Nature (London)*, 1946, **158**, 380.
7. S. W. Benson and A. M. North, *J. Am. Chem. Soc.*, 1958, **80**, 5625.
8. M. Morton, P. Saltiello and H. Landfield, *J. Polym. Sci.*, 1952, **8**, 111.
9. M. J. Ballard, R. G. Gilbert, D. H. Napper, P. J. Pomery, P. W. O'Sullivan and J. H. O'Donnell, *Macromolecules*, 1986, **19**, 1303.
10. R. Korus and K. O'Driscoll, in 'Polymer Handbook', ed. J. Brandrup and E. Immergut, Wiley, New York, 1975, p. 11.
11. H. K. Mahabadi and K. O'Driscoll, *J. Polym. Sci., Polym. Chem. Ed.*, 1977, **15**, 283.
12. T. Yasukawa and K. Murakami, *Polymer*, 1980, **21**, 1423.
13. M. Stickler, *Makromol. Chem.*, 1986, **187**, 1765.
14. H. K. Mahabadi and K. O'Driscoll, *Makromol. Chem.*, 1977, **178**, 2629.
15. W. A. Ludwico and S. L. Rosen, *J. Polym. Sci., Polym. Chem. Ed.*, 1976, **14**, 2121.
16. H. K. Mahabadi and K. O'Driscoll, *Macromolecules*, 1977, **10**, 55.
17. K. Horie and I. Mita, *Macromolecules*, 1977, **11**, 1175.
18. J. Dionisio, H. K. Mahabadi and K. O'Driscoll, *J. Polym. Sci., Polym. Chem. Ed.*, 1979, **17**, 1891.
19. H. K. Mahabadi, *Macromolecules*, 1985, **18**, 1319.
20. G. M. Burnett and G. L. Duncan, *Makromol. Chem.*, 1962, **51**, 154.
21. J. Cardenas and K. O'Driscoll, *J. Polym. Sci., Polym. Chem. Ed.*, 1976, **14**, 883.
22. R. Porter and J. Johnson, *Chem. Rev.*, 1966, **66**, 1.
23. F. L. Marten and A. E. Hamielec, *J. Appl. Polym. Sci.*, 1982, **27**, 489.
24. S. K. Soh and D. C. Sundberg, *J. Polym. Sci., Polym. Chem. Ed.*, 1982, **20**, 1299, 1315, 1331, 1345.
25. P. G. de Gennes, 'Scaling Concepts in Polymer Physics', Cornell University Press, Ithaca, NY, 1979.
26. T. Tulig and M. Tirrell, *Macromolecules*, 1981, **14**, 1501.
27. K. Ito, *Polym. J.*, 1980, **12**, 499.
28. P. G. de Gennes, *J. Chem. Phys.*, 1982, **76**, 3316, 3322.
29. M. Buback, H. Hippler, J. Schweer and H. Vogele, *Makromol. Chem., Rapid Commun.*, 1986, **7**, 261.
30. M. Stickler, D. Panke and A. E. Hamielec, *J. Polym. Sci., Polym. Chem. Ed.*, 1984, **22**, 2243.
31. G. V. Schulz, *Z. Phys. Chem. (Frankfurt Main)*, 1956, **8**, 290.
32. W. Wunderlich and M. Stickler, *Polym. Prepr., Am. Chem. Soc., Div. Polym. Chem.*, 1984, **25**, (2), 7.
33. F. N. Kelley and F. Bueche, *J. Polym. Sci.*, 1961, **50**, 549.
34. S. S. M. Chiang and A. Rudin, *J. Macromol. Sci., Chem.*, 1975, **A9**, 237.
35. H. W. Melville, R. Noble and W. F. Watson, *J. Polym. Sci.*, 1947, **2**, 229.
36. G. Bonta, B. M. Gallo and S. Russo, *J. Chem. Soc., Faraday Trans. 1*, 1975, **71**, 1727.
37. K. Ito and K. F. O'Driscoll, *J. Polym. Sci., Polym. Chem. Ed.*, 1979, **17**, 3913.
38. J. N. Atherton and A. M. North, *Trans. Faraday Soc.*, 1962, **58**, 2049.
39. T. Fukuda, Y. Ma and H. Inagaki, *Macromolecules*, 1985, **18**, 17.

13
Chain Transfer

C. ANTHONY BARSON

The University of Birmingham, UK

13.1 INTRODUCTION

The principal features of chain-transfer reactions in radical polymerization have been described in Volume 3, Chapters 1 and 3. Although the principles of such reactions were proposed in 1937,[1] the first chemical evidence for a specific transfer process followed a few years later.[2] Chain transfer involves the reaction of a polymer radical $P_n\cdot$ with some other species XA, known as the transfer agent, which is present in the polymerizing system. A labile atom or group of atoms X is transferred from XA to $P_n\cdot$, with the generation of a new radical $A\cdot$, as shown in equation (1). $A\cdot$ may then reinitiate polymerization by reaction with monomer M, as shown in equation (2). XA can be monomer, initiator, solvent, dissolved additive or terminated polymer; X is commonly hydrogen or a halogen. Although the kinetic chain is continued, by the preservation of a radical centre, the molecular chain is terminated.

$$P_n\cdot + XA \rightarrow P_nX + A\cdot \tag{1}$$

$$A\cdot + M \rightarrow AM\cdot \text{ (or } P_1\cdot) \tag{2}$$

If transfer proceeds according to equations (1) and (2), the mean degree of polymerization $(\overline{DP})$ is reduced, except where XA is terminated polymer. Since the kinetic chain is continued, the overall rate of polymerization remains largely unchanged, unless there is appreciable reduction in the reactivities at the radical sites. The process thus described is regarded as conventional transfer.

When the reactivity of $A\cdot$ with monomer molecules is appreciably less than that of $P_n\cdot$, the concentration of $A\cdot$ increases and the chance of its taking part in termination reactions also increases. This leads to a decrease in the concentration of reactive species which can be involved in propagation, with a consequent decrease in the rate of polymerization. In addition, the kinetic-chain length is decreased. This type of reaction is known as degradative chain transfer.

If k_f is the rate constant for the step shown in equation (1) and k_p is the rate constant for the normal propagation step in the polymerization, then the ratio k_f/k_p is the transfer constant C for the polymer radical with species XA.

13.2 EFFECT OF TRANSFER ON THE DEGREE OF POLYMERIZATION

The basic steps for a polymerization undergoing transfer are given in Scheme 1. M, I and S represent monomer, initiator and solvent (or dissolved additive), respectively, and $P_n\cdot$ represents a polymer radical containing n units of monomer. The radicals $Y\cdot$ and $I\cdot$, derived from the initiator by dissociation or transfer, respectively, are not necessarily identical (see Section 13.4.1).

$$\text{Initiation}\begin{cases} I \xrightarrow{k_d} 2Y\cdot \\[2ex] Y\cdot + M \xrightarrow{k_i} P_1\cdot \end{cases}$$

$$\text{Propagation } P_n\cdot + M \xrightarrow{k_p} P_{n+1}\cdot$$

$$\text{Transfer to monomer } P_n\cdot + M \xrightarrow{k_{rM}} P_n + M\cdot$$

$$\text{Transfer to initiator } P_n\cdot + I \xrightarrow{k_{rI}} P_n + I\cdot$$

$$\text{Transfer to solvent (or dissolved additive) } P_n\cdot + S \xrightarrow{k_{rS}} P_n + S\cdot$$

$$\text{Transfer to polymer } P_n\cdot + P_m \xrightarrow{k_{rP}} P_n + P_m\cdot$$

$$\text{Reinitiation by transfer radicals}\begin{cases} M\cdot + M \xrightarrow{k_{pM}} P_2\cdot \\[2ex] I\cdot + M \xrightarrow{k_{pI}} P_1\cdot \\[2ex] S\cdot + M \xrightarrow{k_{pS}} P_1\cdot \\[2ex] P_m\cdot + M \xrightarrow{k_{pP}} P_{m+1}\cdot \end{cases}$$

$$\text{Termination}\begin{cases} P_n\cdot + P_m\cdot \xrightarrow{k_{tc}} P_{n+m} \\[2ex] P_n\cdot + P_m\cdot \xrightarrow{k_{td}} P_n + P_m \end{cases}$$

Scheme 1

If all of the polymer molecules are linear, $\overline{DP}$ is defined by equation (3). This may also be expressed in terms of rates, as in equation (4). Provided that propagation accounts for most of the consumption of monomer, the rate of incorporation of monomer units is given by the rate of propagation. The rate of formation of polymer chain-ends depends on initiation, termination and transfer reactions. Each initiation step in Scheme 1 produces two chain-ends. Each transfer step, with the exception of transfer to polymer, produces two additional chain-ends, provided that all the radicals formed in transfer steps are consumed by reinitiation or by termination with polymer radicals. Termination by combination does not increase the number of chain-ends further, but each termination step, by disproportionation, produces two more chain-ends.

$$\overline{DP} = \frac{\text{number of monomer units polymerized}}{\text{(number of polymer chain-ends)}/2} \tag{3}$$

$$\overline{DP} = \frac{\text{rate of incorporation of monomer units}}{\text{(rate of formation of polymer chain-ends)}/2} \tag{4}$$

If termination is exclusively by combination, the value of $\overline{DP}$ is given by equation (5), where f is the efficiency of initiation; $P\bullet$ refers to a polymer radical incorporating any number of monomer units $\left([P\bullet] = \sum\limits_{n=1}^{\infty} [P_n\bullet]\right)$. If termination is exclusively by disproportionation, an additional term appears in the denominator (equation 6). Equation (6) may be rearranged into the form of equation (7), where the last term is omitted if termination is exclusively by combination. The individual transfer constants C_M, C_I and C_S are defined in Scheme 2. Since, under stationary state conditions for $[P\bullet]$, $2fk_d[I] = 2k_t[P\bullet]^2$ and the rate of polymerization (R_p) is given by $R_p = k_p[P\bullet][M]$, equation (7) may be rewritten as equation (8). In this equation, $\alpha = 0$ or 1, dependent on whether termination is exclusively by combination or disproportionation, respectively, and k_t is the appropriate rate constant for termination.

$$\overline{DP} = \frac{k_p[P\bullet][M]}{fk_d[I] + k_{fM}[P\bullet][M] + k_{fI}[P\bullet][I] + k_{fS}[P\bullet][S]} \tag{5}$$

$$\overline{DP} = \frac{k_p[P\bullet][M]}{fk_d[I] + k_{fM}[P\bullet][M] + k_{fI}[P\bullet][I] + k_{fS}[P\bullet][S] + k_{td}[P\bullet]^2} \tag{6}$$

$$\frac{1}{\overline{DP}} = \frac{fk_d[I]}{k_p[P\bullet][M]} + C_M + C_I\frac{[I]}{[M]} + C_S\frac{[S]}{[M]} + \frac{k_{td}[P\bullet]}{k_p[M]} \tag{7}$$

$$\frac{1}{\overline{DP}} = \frac{(1+\alpha)k_t R_p}{k_p^2[M]^2} + C_M + C_I\frac{[I]}{[M]} + C_S\frac{[S]}{[M]} \tag{8}$$

$$C_M = k_{fM}/k_p; \quad C_I = k_{fI}/k_p; \quad C_S = k_{fS}/k_p$$

Scheme 2

Equation (8) forms the basis for the determination of the values of the individual transfer constants. If the value of $\overline{DP}$ in the absence of transfer is defined by $\overline{DP}_0$, equation (9) results. This equation is commonly known as the Mayo equation, although transfer to monomer and to initiator were omitted when it was first derived.[3] For the Mayo equation to be valid, the conditions need to be such that concentrations do not change appreciably. The concentrations are instantaneous values, as is that of $\overline{DP}$. The equation becomes invalid if degradative chain transfer takes place (see Section 13.4.6). A modified form of the equation[4] can be employed to take account of changes in the rate of polymerization. The Mayo equation can also be modified by the inclusion of an extra term to take account of primary radical termination;[5] use of this modified equation gives an insight into the importance of such termination in the reactions under investigation.

$$\frac{1}{\overline{DP}} = \frac{1}{\overline{DP}_0} + C_M + C_I\frac{[I]}{[M]} + C_S\frac{[S]}{[M]} \tag{9}$$

Upon equating the rates of initiation and termination under stationary state conditions for $[P\bullet]$, $[I]$ in equation (8) may be expressed in terms of R_p, leading to equation (10).

$$\frac{1}{\overline{DP}} = \frac{(1+\alpha)k_t R_p}{k_p^2[M]^2} + C_M + C_I\frac{k_t R_p^2}{fk_d k_p^2[M]^3} + C_S\frac{[S]}{[M]} \tag{10}$$

13.3 DETERMINATION OF TRANSFER CONSTANTS

13.3.1 General

By plotting $1/\overline{DP}$ against R_p (see equation 10), it is possible to deduce the types of transfer which are occurring. For example, curvature in the plot indicates the presence of transfer to initiator (Figure 1). To simplify the evaluation of individual transfer constants, it is helpful to avoid transfer to solvent (or dissolved additive). If it is necessary to carry out the polymerization in the presence of a solvent, the latter should be chosen so that it is 'inert'. Confirmation of the absence of significant

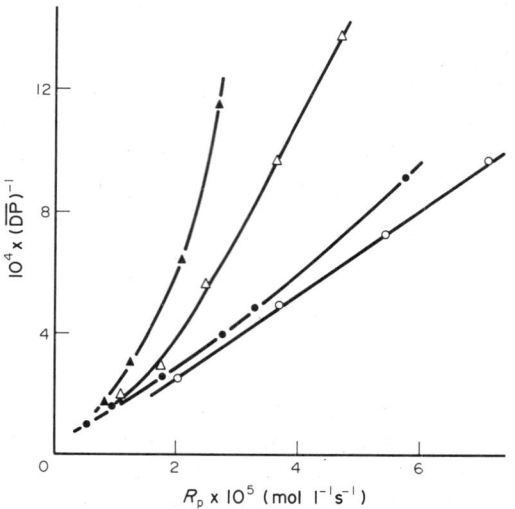

Figure 1 Dependence of $1/\overline{DP}$ on R_p for polymerizations of styrene at 60 °C, initiated by azobis(isobutyronitrile) ($\bigcirc$), dibenzoyl peroxide ($\bullet$), cumene hydroperoxide ($\triangle$) and *t*-butyl hydroperoxide ($\blacktriangle$) (from data of B. Baysal and A. V. Tobolsky, *J. Polym. Sci.*, 1952, **8**, 529)

transfer to solvent is obtained by showing that the value of the intercept at $R_p = 0$, in the plot of $1/DP$ against R_p, is independent of the concentration of the solvent.

The values of individual transfer constants, except in the case of transfer to polymer, can be determined by application of equation (10). The details of such procedures, as they are applied to each type of transfer, are described later under the relevant headings.

Another general method for determining transfer constants takes into account the way in which a fragment of transfer agent is incorporated as an end-group into the polymer by the reinitiation step in equation (2). A sufficiently sensitive method is required to measure the incorporation of these end-groups. A method successfully employed to follow transfer in this way uses radiotracers, in which group A of the transfer agent is radioactively labelled.[6,7] A particular advantage of using labelled transfer agents is that the method is generally unaffected by simultaneous transfer reactions involving unlabelled substances. If A· is consumed by processes other than the reinitiation in equation (2), the rate of displacement of group A in equation (1) exceeds its rate of incorporation into polymer in equation (2); under such circumstances, the transfer constant will be underestimated. Errors may arise if the mechanism of transfer differs from that in equations (1) and (2), since the labelled group A might not then become incorporated into the polymer. Errors may also arise if, in addition to transfer, the transfer agent may be incorporated through copolymerization. Careful control of radiochemical purity is essential to avoid mistakes in interpreting which chemical species is incorporated into the polymer. For example, as a result of radiotracer experiments, benzene was formerly thought to copolymerize with vinyl acetate (VAC),[8] but subsequent work[9] showed that the amount of incorporation of labelled benzene was explained by transfer alone. The error in the earlier work probably arose when a polymerizable radiochemical impurity, the precursor acetylene, was mistaken for labelled benzene when measuring the activity in the polymer. Since a hydrogen atom is commonly the species transferred to the propagating polymer radical, care must also be exercised when tritium is incorporated into the transfer agent[6] to avoid kinetic measurements being subject to appreciable isotope effects.

13.3.2 Determination of C_M

In the absence of transfer to solvent (or dissolved additive), equation (10) simplifies to equation (11), which may also be written in the form of equation (12). Since curvature in a plot of $1/\overline{DP}$ against R_p indicates transfer to initiator, the extrapolation of the linear portion, where such transfer is insignificant, to the $1/\overline{DP}$ axis yields the value of C_M (Figure 1).

$$\frac{1}{\overline{DP}} = \frac{(1+\alpha)k_t R_p}{k_p^2 [M]^2} + C_M + C_I \frac{k_t R_p^2}{f k_d k_p^2 [M]^3} \tag{11}$$

$$\frac{1}{\overline{DP}} = \frac{(1+\alpha)k_t R_p}{k_p^2 [M]^2} + C_M + C_I \frac{[I]}{[M]} \tag{12}$$

13.3.3 Determination of C_I

To evaluate C_I, equation (11) can be rearranged into the form of equation (13). When the left-hand side of equation (13) is plotted against R_p, k_t/k_p^2 is determined from the intercept at $R_p = 0$. If the value of fk_d is known from other data, the slope of the graph yields C_I. Alternatively, if the value of C_M is negligible, equation (12) may be simplified and rearranged to give equation (14). A plot of the left-hand side of equation (14) against [I]/[M] yields a straight line with a slope of C_I.

$$\left(\frac{1}{\overline{DP}} - C_M\right)\frac{1}{R_p} = \frac{(1+\alpha)k_t}{k_p^2[M]^2} + C_I\frac{k_t R_p}{fk_d k_p^2[M]^3} \tag{13}$$

$$\frac{1}{\overline{DP}} - \frac{(1+\alpha)k_t R_p}{k_p^2[M]^2} = C_I\frac{[I]}{[M]} \tag{14}$$

If a radiolabelled initiator is used, the rate at which initiator fragments are incorporated into the polymer (R_{inc}) may be determined.[10] The position of the labelling in the initiator molecule must be such that the value of R_{inc} obtained only takes account of initiator incorporated into the polymer by normal initiation or transfer processes. R_{inc} can then be expressed by equation (15), which leads to equation (16), where R_i is the rate of initiation. If the concentration of labelled initiator is kept constant and R_p is varied by the addition of a second unlabelled initiator, a plot of R_{inc} (determined from radiotracer analysis of the polymer) against R_p is linear. R_i and C_I may be found from the intercept and slope, respectively. The above treatment assumes that the transfer process destroys the ability of initiator fragments incorporated into the polymer to dissociate into radicals capable of promoting further initiation.

$$R_{inc} = R_i + 2k_{fI}[P\bullet][I] \tag{15}$$

$$R_{inc} = R_i + 2C_I[I]R_p/[M] \tag{16}$$

13.3.4 Determination of C_S

Values of C_S were first derived by application of the Mayo equation.[3] If transfer to monomer and initiator are both negligible, equation (9) requires that a plot of $1/\overline{DP}$ against [S]/[M] is linear with a slope of C_S and with an intercept having the value of $1/\overline{DP}_0$. It is essential that $\overline{DP}_0$ in equation (9) remains constant throughout the determination, if accurate values of C_S are to be obtained. By reference to equation (8), it can be seen that $\overline{DP}_0$ is only constant if $R_p/[M]^2$ is constant. The latter can be achieved by making appropriate adjustments to [I] throughout the experimental measurements. To eliminate the term caused by transfer to initiator in equation (9), an initiator which shows negligible transfer may be used, e.g. azobis(isobutyronitrile) (AIBN). The slope of the plot of $1/\overline{DP}$ against R_p then gives the value of C_S. If $R_p/[M]^2$ is not kept constant, $[(1/\overline{DP}) - (1+\alpha)k_t R_p/k_p^2[M]^2]$ can be plotted against [S]/[M], yielding a straight line of slope C_S (see equation 8).

Another method, suitable for determining C_S when the solvent has a high reactivity, makes use of the ratio of the rates of consumption of solvent and monomer. Provided that the molecular chains are long and that transfer to monomer is negligible, it is permissible to divide the rate equation for transfer to solvent by that for propagation, yielding equation (17). If the rates of consumption of solvent and monomer can be monitored, either directly or from their incorporation into the polymer, a plot of the ratio of these rates against [S]/[M] has a slope of C_S. This method assumes more importance in cases where it is difficult to obtain accurate values for $\overline{DP}$ by conventional methods, such as when the value of $\overline{DP}$ is small. In such cases, however, it is necessary to take account of the consumption of monomer by the reinitiation step in Scheme 1. Equation (18) takes this into account; under stationary state conditions for [S•], it simplifies to equation (19), which may be plotted to yield C_S.

$$\frac{d[S]/dt}{d[M]/dt} = \frac{k_{fS}[P\bullet][S]}{k_p[P\bullet][M]} = C_S\frac{[S]}{[M]} \tag{17}$$

$$\frac{d[M]/dt}{d[S]/dt} = -\frac{k_p[P\bullet][M] + k_{pS}[S\bullet][M]}{k_{fS}[P\bullet][S]} \tag{18}$$

$$\frac{d[M]/dt}{d[S]/dt} = \frac{k_p[P\bullet][M] + k_{fS}[P\bullet][S]}{k_{fS}[P\bullet][S]} = \frac{[M]}{C_S[S]} + 1 \tag{19}$$

To minimize one of the problems caused by very reactive transfer agents, a method of 'moderated' copolymerization has been proposed.[11] The aim is to copolymerize the particular monomer (M_B) in low concentration with a 'moderating' monomer (M_A) which gives negligible transfer to the solvent and which is present in very much higher concentrations. Since M_A is now the principal monomer in the mixture, polymer radicals terminating in M_B are rare. Thus copolymerization leads to values of $\overline{DP}$ which can be measured quite easily. The transfer constant for M_B with solvent can be determined even when very high values of this constant would, in cases of homopolymerization, lead to very small values of $\overline{DP}$. It therefore circumvents the situation where the polymer radical is normally so short that its radical end is influenced by the close proximity of the group at the other end.

In determining C_S by the above methods, it is important that the solvent should act only as a simple diluent and should not influence R_p; thus the solvent must not take part in copolymerization. It is also important that the rate constants for the individual steps in the polymerization are not influenced by the ratio $[S]/[M]$; thus the value of f must remain constant within the chosen range of $[S]/[M]$. Since retardations accompanying degradative chain transfer cause increased complexity, these must also be absent (see Section 13.4.6). Furthermore, the Mayo equation is invalid in cases where the solvent causes precipitation of the polymer.

13.3.5   Determination of C_P

Since transfer to polymer does not change the number of polymer molecules in the system, studies of $\overline{DP}$ are not helpful in studying such transfer to polymer. Whilst $\overline{DP}$ is unchanged, the transfer mechanism results in the formation of branches on the polymer chain and leads to a broadening of the molecular weight distribution (MWD). The number of branches per monomer molecule polymerized (ρ) is related to the transfer constant C_P and the extent of reaction (p) by equation (20).[12] A method has been suggested for determining the values of MWD resulting from transfer to polymer;[13] when termination is solely by disproportionation, the distributions of linear and branched polymers are found to be interrelated. Formulae have also been derived[14] for determining the rate of formation of branching centres and its relationship to MWD.

$$\rho = -C_P[1 + (1/p)\ln(1-p)] \qquad (20)$$

To measure transfer to polymer *via* the extent of branching, monomer can be polymerized in the presence of preformed polymer. Provided that it is possible to distinguish chemically between the preformed polymer and the monomer, the fraction of branches in the polymer can be determined.[15] Radiotracer labelling of the preformed polymer or of the monomer provides a very sensitive method of analysis for calculating the number of branches and their average length.[16]

A further method of measuring transfer to polymer utilizes model compounds. It is assumed that a monomer unit in a polymer chain has a similar reactivity to a model compound; in the model, hydrogen atoms or methyl groups are substituted for the polymer chains at each end of the monomer unit or units.[17]

13.4   IMPORTANCE OF DIFFERENT TYPES OF TRANSFER

13.4.1   Transfer to Initiator

An extensive list of values of C_I is published elsewhere.[18] If, as a result of transfer to initiator, a molecule is destroyed as a potential initiator when incorporated into the polymer, R_i will decrease more rapidly than in the absence of transfer. The increased rate of consumption of initiator amounts to its induced decomposition by reaction with polymer radicals $P\cdot$. Early work indicated that there was no appreciable transfer to AIBN in the polymerization of styrene (S).[19,20] In contrast with dibenzoyl peroxide (BPO) and other peroxides in the polymerization of S,[21] AIBN shows linear behaviour in a Mayo-type plot (see Figure 1). For this reason it has usually been preferred as an initiator. More recent work, however, has led to values of 0.012 to 0.16 for C_I in the polymerization of S at 60 °C[22,23] and of 0.02 with methyl methacrylate (MMA) at the same temperature.[24]

BPO and a large number of its derivatives have been recognized for many years as transfer agents.[25] The values of C_I for several of these peroxides in polymerization with S at 70 °C are listed in Table 1; the *para* substituents are listed in order of increasing electronegativity; this generally appears to increase the rate of transfer. With such peroxides, transfer destroys the potential of the fragments for further initiation by cleaving the O—O bond, as shown in equation (21). The radical produced by this step is identical to the primary radical from the decomposition of the initiator. It can reinitiate in an identical manner, either directly or *via* the phenyl or substituted phenyl radical formed by the loss of carbon dioxide. In normal initiation, both aroyloxy and aryl groups are incorporated into the polymer, in proportions determined by the relative probability that aroyloxy groups can react with monomer or, alternatively, dissociate to aryl groups. Increasing transfer with such initiators therefore increases the ratio of aroyloxy fragments to aryl fragments in the polymer.[26] Primary radical transfer (see Section 13.4.5) leads to a similar increase.

$$P_n{\cdot} + RCO_2OCOR \rightarrow P_nOCOR + RCO_2{\cdot} \tag{21}$$

Table 1 Transfer Constants C_I for Substituted Dibenzoyl Peroxides with Styrene at 70 °C[25]

Substituent	Para position	C_I Meta position	Ortho position
t-Butyl	*ca.* 0		
Methoxy	0.074		
Methyl	0.003		0.175
(Hydrogen)	(0.075)	(0.075)	(0.075)
Acetate	0.187		
Fluorine	0.219	0.246	0.40
Chlorine	0.216	0.346	1.91
Bromine	0.193	0.465	2.17
Iodine	0.293	0.262	
Cyanide	0.804		
Nitro	7.4	6.2	

From the studies of substituted dibenzoyl peroxides,[25] it is evident that electron-withdrawing substituents increase the probability of transfer. Since the rate of primary dissociation of the peroxide is reduced by such substituents, higher concentrations need to be used to achieve similar rates of polymerization, a factor which increases the probability of transfer even further. A particular peroxide may also show marked differences in reactivity towards different propagating radicals. For instance, bis(2,3,4,5-tetrachlorobenzoyl) peroxide behaves normally in MMA and VAC, whereas there is marked transfer in S.[25] Hydroperoxides also readily take part in transfer,[21] but, in this case, the mechanism involves the scission of the O—H bond rather than the O—O bond.[27]

Many sulfur-containing initiators undergo transfer very readily and frequently cause degradative chain transfer. Such substances are often called 'inifers' (initiator–transfer agents), a term which was first applied in cationic polymerization.[28] Because of the practical difficulty of distinguishing between transfer and termination with common initiators, the term 'iniferter' (initiator–transfer agent–terminator) has been proposed.[29] In the case of diphenyl disulfide, (PhS)$_2$, it was suggested that transfer proceeds according to equation (22).[30] Thus it produces a different radical from that formed by the primary dissociation of the initiator, believed to be PhS•. The disulfide radical is very unreactive, leading to degradative chain transfer. Another group of sulfur-containing initiators which have been studied quite extensively are the tetraalkylthiuram disulfides.[30] This group also forms a disulfide radical on transfer, as in equation (23).

$$P_n{\cdot} + PhSSPh \rightarrow P_nPh + PhSS{\cdot} \tag{22}$$

$$P_n{\cdot} + (R_2NCSS)_2 \rightarrow P_nCSNR_2 + R_2NCS.SS{\cdot} \tag{23}$$

Sulfur-containing initiators and other iniferters provide the possibility for systems of living radical polymerization in homogeneous solution. In Scheme 3, a polymer chain, formed as the result of transfer, dissociates into a polymer (1) with a radical chain-end and a small radical (2); the latter may be sufficiently stable not to initiate a new polymer chain. A radical polymerization of this type can then proceed *via* a living mechanism, according to Scheme 3.[31] S and MMA have been photo-polymerized in homogeneous solution by a living radical mechanism using several sulfur-containing iniferters. Diphenyl disulfide, dibenzoyl disulfide, tetraethylthiuram disulfide, benzyl diethyldithiocarbamate and 2-phenylethyl diethyldithiocarbamate all promote such polymerizations.[29,32] The polymerizations are accompanied by a marked retardation, and the number-average molecular weights increase as a function of time. Scheme 3 illustrates how monomer is incorporated into the polymer in blocks. Use can be made of this phenomenon in preparing multicomponent block copolymers.[33]

$$-CH_2CHRX \rightleftharpoons -CH_2CHR \cdot + X \cdot \xrightarrow{nCH_2=CHR} -(CH_2CHR)_nCH_2CHR \cdot + X \cdot$$

$$\text{(1)} \qquad \text{(2)}$$

$$\rightleftharpoons -(CH_2CHR)_nCH_2CHRX \xrightarrow{mCH_2=CHR} -(CH_2CHR)_{n+m}CH_2CHRX$$

Scheme 3

13.4.2 Transfer to Monomer

The values of C_M for a selection of the more common monomers are given in Table 2. A more comprehensive list is published elsewhere.[18] In comparison with other types of transfer these constants are relatively small, with the result that such transfer is not usually a major factor in limiting the molecular weights of polymers. The small values may result from the fact that the abstraction of a hydrogen atom often requires the breaking of strong vinyl C–H bonds.

Table 2 Monomer Transfer Constants $C_M{}^a$

Monomer	*Temperature* (°C)	$10^4 C_M$
Acrylamide	60	0.6
Acrylonitrile	60	0.26–0.3
Allyl acetate	80	176–700
Allyl chloride	80	1600
Methacrylonitrile	60	5.81
Methyl acrylate	60	0.036–0.325
Methyl methacrylate	60	0.07–0.18
Styrene	60	0.07–1.37
Vinyl acetate	60	1.75–2.8
Vinyl chloride	60	10.8–12.8

a L. J. Young, in 'Polymer Handbook', ed. J. Brandrup and E. H. Immergut, Interscience, New York, 1975, p. II-57.

As Table 2 shows, allylic monomers are, relatively, more susceptible to this type of transfer; it is usually accompanied by retardation. It has been shown[34] that, in the polymerization of allyl acetate, the molecular weights of the polymers are small and independent of R_p, which is first order in [I]. When deuterated monomer, $CH_2=CHCD_2O_2CMe$, was used, R_p and $\overline{DP}$ both increased.[35] It was concluded that the transfer proceeds according to equation (24). The retardation observed is a consequence of resonance stabilization of the transfer radicals between the structures $CH_2=CH\dot{C}HX$ and $\cdot CH_2CH=CHX$, resulting in inefficiency in reinitiation.

$$P_n \cdot + CH_2=CHCH_2O_2CMe \rightarrow P_n + CH_2=CH\dot{C}HO_2CMe \qquad (24)$$

It has been shown that, when allylic monomers are polymerized in the presence of modifying agents, such as complexing agents, Lewis or protonic acids, both R_p and $\overline{DP}$ are increased.[36-38] The kinetics have been studied in relation to the nature of the functional group X in the structure $CH_2=CHCH_2X$.[39] Such functional groups determine the relative strengths of the C–H bonds at the α position and thus control the rate of transfer. The increase in R_p in the presence of complexing agents is connected with the high probability for addition of the transfer radical to the double bond of the monomer; the latter is determined by the nature of the complexing agent. The lowest rate of abstraction of hydrogen atoms at the α position is observed for monomers with the highest basicity, where interaction of the functional group with the complexing agent is strongest.

In the case of monomers with α methyl groups, abstraction of a hydrogen atom leads to the possibility of resonance stabilization of the product radical between $CH_2=CX\dot{C}H_2$ and $\cdot CH_2CX=CH_2$. Transfer involving the formation of such radicals might not be readily followed by reinitiation, thus leading to retardation. Although MMA does not show such retardation, isopropenyl acetate has been shown to undergo degradative chain transfer.[40,41] With MMA, the high monomer reactivity may mask the effect of any resonance stabilization, allowing efficient reinitiation after transfer.

Because of the relatively large value of C_M in the case of VAC, transfer to monomer readily occurs with the formation of the species $CH_2=CHO_2C\dot{C}H_2$. This can reinitiate polymerization with its radical end and also take part in propagation with another propagating radical.[42] The latter process results in branching, although the branches may be readily removed by hydrolysis (see also Section 13.4.4).

A recent development in transfer to monomer has been to promote such transfer by the use of a third agent. If, instead of the reinitiation step in Scheme 1, the transfer radical $S\cdot$ undergoes further transfer, as in equation (25), the transfer agent is regenerated and the overall process is one of catalyzed transfer to monomer. Such a process has considerable potential over other methods used to control the MWD of polymers; it does not introduce foreign end-groups into the polymer and can be regulated by controlling the transfer agent. The first reported example of a high efficiency catalyst for transfer to monomer involved the use of a cobalt complex of hematoporphyrin tetramethyl ether in the polymerization of MMA.[43] When applied to the polymerization of S, the same complex did not promote such strong transfer.[44] Cobaloxime, synthesized from cobalt(II) acetate and dimethylglyoxime, has also been used as a catalytic transfer agent; in this case, the transfer constants were shown to be dependent on chain length.[45]

$$S\cdot + M \rightarrow S + P_1 \cdot \tag{25}$$

13.4.3 Transfer to Solvent or Dissolved Additive

Transfer occurs with a wide range of compounds. The transfer constants of a selection of these are given in Table 3. A comprehensive list of values is published elsewhere.[18] The magnitude of the constants varies over a very wide range and much effort has been expended in attempting to correlate magnitude with molecular structure. In general, the order of reactivity depends on the stabilization of the radical formed in transfer, in much the same way as in copolymerization.

Table 3 Values of C_S for Some Transfer Agents with Polystyryl Radicals at 60 °C[a]

Transfer agent	$10^4 C_S$
Benzene	0.018–0.04
Cyclohexane	0.024–0.063
Toluene	0.105–2.05
Ethylbenzene	0.67–2.7
t-Butylbenzene	0.04–0.06
1-Chlorobutane	0.04
1-Bromobutane	0.06
1-Iodobutane	1.85
Butyl alcohol	0.06–1.6
Acetone	4.1
Chloroform	0.5–3.4
Diphenylamine	0.9
Triethylamine	7.1–7.5
Triphenylmethane	3.5
Butyl disulfide	24
2-Methyl-2-propanethiol	31 000–46 000
1-Butanethiol	210 000–250 000
Carbon tetrachloride	87–148
Carbon tetrabromide	17 800–4 200 000
Bromotrichloromethane	650 000
p-Benzoquinone	2 270 000

[a] L. J. Young, in 'Polymer Handbook', ed. J. Brandrup and E. H. Immergut, Interscience, New York, 1975, p. II-57.

A detailed treatment of the factors which control reactivity in transfer is to be found elsewhere[46,47] (see also Volume 3, Chapter 17).

Aliphatic hydrocarbons, because of their strong C–H bonds, are not very susceptible to transfer. But a systematic study of the effects of substituents in aromatic hydrocarbons, on their ability to transfer in the polymerization of S, showed a wide variation in C_S.[48,49] Benzene, because of the strength of the C–H bond, is not very susceptible to transfer. Benzylic hydrogen atoms are found to be increasingly susceptible to transfer as their degree of substitution is increased. This results from the decreasing strength of the C–H bonds in such circumstances, together with an increased possibility of resonance stabilization of the product radical from the transfer step. The ability to transfer is, however, lost if substitution eliminates the benzylic hydrogen atom, as in the case of *t*-butylbenzene.[48] Increased resonance stabilization of the transfer radical also permits ready transfer in the case of diphenylmethane and triphenylmethane.[48]

A number of transfer agents containing halogens and oxygen were used in some early studies on the molecular basis for reactivity in transfer.[49] Increasing transfer results from an increasing polarizability of the atom transferred, which is in turn determined by the degree and type of substitution. Semi-quantitative assessments of reactivity were attempted by applying the Q and e scheme which had been applied earlier to copolymerization. This met with some success, since it was able to predict the order of reactivity of a series of monomers with carbon tetrabromide as transfer agent[50] (see also Volume 3, Chapter 17).

Halocarbons readily undergo transfer reactions, since a halogen atom can be abstracted from such compounds. Carbon tetrabromide has a larger value of C_S than carbon tetrachloride, since the latter has a stronger carbon–halogen bond. The readiness to transfer results from the possibility of resonance stabilization of the product trihalocarbon radicals, involving the non-bonding electrons on the halogen atoms. The transfer constant in the polymerization of S with carbon tetrachloride is found to be dependent on the concentrations of monomer and of solvent. The latter dependence can be explained by the hot-radical theory[51] (see also Volume 3, Chapter 6).

Because of the very great readiness of some halocarbons to transfer, the $\overline{DP}$ of the product can be limited to very small numbers (see also Volume 3, Chapter 14). Since the propagating radicals never contain many monomer units, the average transfer constants are very much influenced by the rate constants of the monomeric, dimeric and trimeric radicals. It has been shown that the transfer constants can vary in magnitude appreciably as the growing radical increases in size from monomeric to trimeric or tetrameric.[52−55] In the telomerization of MMA with carbon tetrabromide at 30 °C, it increases by a factor of about two[56] and with bromotrichloromethane by about four times.[57] When S is telomerized with bromotrichloromethane at 30 °C, the transfer constant increases about sixty-fold as the growing radical increases from monomeric to trimeric.[58]

As already described in Section 13.4.1, many sulfur-containing organic compounds are strong transfer agents. Thiols normally undergo transfer by hydrogen abstraction.[59] Disulfides are stronger transfer agents than monosulfides, due to the susceptibility of attack on the S–S bond. Chlorophosphines have been found to act as transfer agents in the polymerization of S[60] and of MMA.[61,62] Trichlorophosphine, dichloro(phenyl)phosphine and chlorodiphenylphosphine show varying susceptibility to transfer in the polymerization of S in the order $PClPh_2 > PCl_2Ph > PCl_3$; this was attributed to the difference in stabilities arising from the phenyl groups in the radical transition states and also to the polar nature of the chlorophosphines. In the case of MMA similar trends were observed, but R_p was increased, possibly due to complex formation between the transfer agent and the monomer. When trichlorophosphine oxide and phenylphosphonic dichloride are compared, an increase in R_p is again observed with MMA; the phenyl-substituted reagent is more susceptible to transfer.[63]

The method of 'moderated' copolymerization[11] has been applied to some systems where large amounts of transfer cause experimental difficulties when using conventional methods for determining transfer constants (see Section 13.3.4). Since the method depends on suppressing the overall effect of transfer by using a large molar ratio of a second monomer, it is capable of revealing penultimate group effects in transfer reactions.[64] Such effects are quite marked with carbon tetrabromide at 60 °C. Styryl radicals with S, methyl acrylate (MA) or MMA penultimate groups give transfer constants of 337, 302 and 60, respectively.[65] The large reduction in transfer constant when MMA is the penultimate unit appears to result from steric hindrance between the α methyl group and the incoming bromine atom. This view is supported when the transfer agent is 1-butanethiol. In this case,[66] there is very little difference between the transfer constants when MA and MMA are the penultimate groups to a terminal styryl radical. It is therefore concluded that the rate of transfer with carbon tetrabromide is not determined solely by the terminal group, but can be reduced considerably by an α methyl group on the penultimate unit.

13.4.4 Transfer to Polymer

Since transfer to polymer only becomes significant at higher concentrations of polymer, it is unlikely to present difficulties at the lower concentrations used for studying other types of transfer. When polymerizations are taken to higher conversion, however, its effect cannot be ignored. It can produce significant effects on the physical properties of the polymers formed, because of the structural changes brought about by an increase in branching. Methods dependent on the shapes of polymer molecules, such as measurements of physical properties, can be used to study the branching, provided that the branches are not very short.

Because of the difficulties involved in measuring C_p, there are fewer reliable data in the literature than for other transfer constants. Values vary over a wide range, even for common polymeric radicals with common polymers. A comprehensive list of values is to be found elsewhere.[18]

In the polymerization of ethylene, both long and short branches are formed (see Volume 3, Chapter 21). The long branches arise from intermolecular transfer and are only formed in significant quantities at higher conversions. The short branches occur, however, at very low conversions.[67] This leads to the conclusion that they are formed as the result of intramolecular transfer, in which there is a 'back-biting' involving a transient six-membered cyclic structure at the radical chain-end.[68,69] Once the transfer radical has added to monomer, there is a high probability of further 'back-biting' transfer, because of the close proximity of the new radical to the main chain. This results in the short side-branches themselves becoming branched, leading to the formation of clusters of very short side-chains.[70] Confirmation of the size and location of side branches in polyethylene formed under high pressure has been found by using ^{13}C NMR and nitric acid degradation.[71] It was concluded that the ethyl and butyl side groups are excluded from crystalline zones. Models have been used to predict that other intermediate length branches can also occur.[72]

In the polymerization of VAC, in addition to the branching caused by transfer to monomer (see Section 13.4.2), transfer to polymer causes further branching. If hydrogen abstraction occurs from the methyl group of the pendant acetoxy group, long branches are formed which may be hydrolyzed.[42] When abstraction occurs at the hydrogen in the α position, long branches are also formed, but these cannot be hydrolyzed. On the basis of transfer to model compounds, it is predicted that transfer to polymer occurs mainly by the latter abstraction process, forming non-hydrolyzable branches.[73] Short branches may also be formed by an intramolecular 'back-biting' transfer process involving transients with six-membered rings,[42] analogous to those occurring in the polymerization of ethylene.

13.4.5 Primary Radical Transfer

In much the same way as transfer agent and monomer compete for reaction with polymer radicals, they can also compete for reaction with primary radicals from the dissociation of initiator. When transfer occurs with the latter radicals, it is known as primary radical transfer. Thus a transfer step, as in equation (26), competes with that involving addition of monomer to Y• in Scheme 1. What happens subsequently to A• determines the effect such transfer has on the overall polymerization, as it does in transfer involving polymer radicals.

$$Y\bullet + XA \rightarrow YX + A\bullet \qquad (26)$$

If A• initiates polymerization efficiently, R_p is unaffected. If A• is less efficient in this respect, R_p will be reduced. If A• is unreactive, it may lead to degradative transfer. This reduces the overall rate of initiation as well as the rate of initiation determined by measuring the rate of incorporation of initiator fragments. Any determination of a transfer constant which depends on measuring the incorporation of transfer agent into the polymer will lead to high values, unless account is taken of the incorporation of transfer agent by primary radical transfer. Only by extrapolation to zero concentration of transfer agent can the effect of primary radical transfer be eliminated;[47] the latter decreases in importance as the concentration of transfer agent is reduced.

A detailed mathematical analysis taking account of non-ideality resulting from primary radical transfer has been proposed.[74] This allows all relevant transfer constants to be obtained from kinetic data alone, without recourse to molecular weight measurements or to tracer techniques.

13.4.6 Degradative Chain Transfer

Degradative chain transfer results from the low reactivity of a radical formed in a transfer step. If the efficiency of the reinitiation (or initiation in the case of primary radical transfer) is low, the radicals may be consumed by other processes, such as termination steps. This leads to a reduction in R_p and therefore to retardation. If this effect is pronounced, it is preferable to consider the transfer agent as a retarder, although the distinction between the two terms is rather arbitrary.

The reduction in R_p with degradative chain transfer causes an increase in the order with respect to initiator from 0.5 to values closer to 1.0, depending on the efficiency of reinitiation. The resultant order can be related to this efficiency (F), to R_p and to the corresponding rate in the absence of transfer agent ($R_{p,0}$) by equation (27).[75] A reduction in the monomer reactivity with the transfer radical is usually accompanied by an increase in the reactivity of the corresponding polymer radical. Under such circumstances, F is reduced and the chances of degradative chain transfer are increased. Degradative chain transfer is therefore most likely to occur when the tendency to transfer is greatest. Thus triphenylmethane gives an order with respect to initiator of almost 1.0 in the polymerization of VAC, with considerable retardation.[76] In the comparable polymerizations of S and MMA, R_p is almost unchanged and the order with respect to initiator is close to 0.5.[6]

$$\text{Order with respect to initiator} = 0.5[1 + (1-F)(1-R_p/R_{p,0})] \tag{27}$$

Similar behaviour is observed when tertiary amines are used as transfer agents with the same three monomers; VAC is the only monomer of the three where significant retardation is observed.[77] The amount of retardation for the monomers is in the order VAC > MMA > S, which is consistent with the generally accepted order of monomer reactivity. The radicals formed by the abstraction of a hydrogen atom in transfer are, therefore, apparently stabilized and interact with the least reactive monomer comparatively slowly. The situation regarding reactivity may therefore be compared with that in copolymerization, where the least reactive monomers form the most reactive polymer radicals and *vice versa* (see also Volume 3, Chapter 17).

Degradative chain transfer involving solvent or added transfer agent causes the order with respect to [M] to increase above unity. As [S]/[M] increases, termination can occur increasingly by steps other than those in Scheme 1. This reduces [$P_n \cdot$], but, since $R_p = k_p[P_n \cdot][M]$, the order with respect to [M] increases. Orders as high as 2.5, in the case of the polymerization of VAC initiated by BPO in the presence of pyridine, have been observed.[78]

13.5 REFERENCES

1. P. J. Flory, *J. Am. Chem. Soc.*, 1937, **59**, 241.
2. J. W. Breitenbach and A. Maschin, *Z. Phys. Chem., Abt. A*, 1940, **187**, 175.
3. F. R. Mayo, *J. Am. Chem. Soc.*, 1943, **65**, 2324.
4. E. A. Lissi, *J. Macromol. Sci., Chem.*, 1981, **A15**, 447.
5. P. C. Deb and G. Meyerhoff, *J. Polym. Sci., Polym. Phys. Ed.*, 1974, **12**, 2163.
6. J. C. Bevington and H. G. Troth, *Trans. Faraday Soc.*, 1962, **58**, 2005.
7. J. C. Bevington and H. G. Troth, *Trans. Faraday Soc.*, 1963, **59**, 1348.
8. W. H. Stockmayer and L. H. Peebles, *J. Am. Chem. Soc.*, 1953, **75**, 2278.
9. J. W. Breitenbach, G. Billek, E. Faltlhansl and E. Weber, *Monatsh. Chem.*, 1961, **92**, 1100.
10. J. C. Bevington and T. D. Lewis, *Polymer*, 1960, **1**, 1.
11. C. H. Bamford, *J. Chem. Soc., Faraday Trans. 1.*, 1976, **72**, 2805.
12. P. J. Flory, *J. Am. Chem. Soc.*, 1947, **69**, 2893.
13. B. L. Gutin, *Vysokomol. Soedin., Ser. A*, 1978, **20**, 620.
14. N. G. Taganov, *Vysokomol. Soedin., Ser. A*, 1981, **23**, 2772.
15. R. B. Carlin and N. E. Shakespeare, *J. Am. Chem. Soc.*, 1946, **68**, 876.
16. J. C. Bevington, G. M. Guzman and H. W. Melville, *Proc. R. Soc. London, Ser A*, 1954, **221**, 437, 453.
17. D. Lim and O. Wichterle, *J. Polym. Sci.*, 1958, **29**, 579.
18. L. J. Young, in 'Polymer Handbook', ed. J. Brandrup and E. H. Immergut, Interscience, New York, 1975, p. II-57; G. C. Eastmond, in 'Comprehensive Chemical Kinetics', ed. C. H. Bamford and C. F. H. Tipper, Elsevier, Amsterdam, 1976, vol. 14A, p. 153.
19. D. H. Johnson and A. V. Tobolsky, *J. Am. Chem. Soc.*, 1952, **74**, 938.
20. N. G. Saha, U. S. Nandi and S. R. Palit, *J. Chem. Soc.*, 1958, 12.
21. B. Baysal and A. V. Tobolsky, *J. Polym. Sci.*, 1952, **8**, 529.
22. W. A. Pryor and T. R. Fiske, *Macromolecules*, 1969, **2**, 62.
23. J. A. May, Jr. and W. B. Smith, *J. Phys. Chem.*, 1968, **72**, 2993.
24. G. Ayrey and A. C. Haynes, *Makromol. Chem.*, 1974, **175**, 1463.
25. W. Cooper, *J. Chem. Soc.*, 1952, 2408.
26. J. C. Bevington, J. Toole and L. Trossarelli, *Makromol. Chem.*, 1959, **32**, 57.
27. C. Walling and Y.-W. Chang, *J. Am. Chem. Soc.*, 1954, **76**, 4878.
28. J. P. Kennedy, *J. Macromol. Sci., Chem.*, 1979, **A13**, 695.
29. T. Otsu and M. Yoshida, *Makromol. Chem., Rapid Commun.*, 1982, **3**, 127.
30. T. Ferington and A. V. Tobolsky, *J. Am. Chem. Soc.*, 1958, **80**, 3215.
31. T. Otsu, M. Yoshida and T. Tazaki, *Makromol. Chem., Rapid Commun.*, 1982, **3**, 133.

32. T. Otsu, M. Yoshida and A. Kuriyama, *Polym. Bull. (Berlin)*, 1982, **7**, 45.
33. T. Otsu and M. Yoshida, *Polym. Bull. (Berlin)*, 1982, **7**, 197.
34. P. D. Bartlett and R. Altschul, *J. Am. Chem. Soc.*, 1945, **67**, 816.
35. P. D. Bartlett and F. A. Tate, *J. Am. Chem. Soc.*, 1953, **75**, 91.
36. M. N. Masterova, L. I. Andreyeva, V. P. Zubov, L. S. Polak and V. A. Kabanov, *Vysokomol. Soedin., Ser. A*, 1976, **18**, 1957.
37. V. P. Zubov, Ye. S. Garina, V. F. Kornil'eva, M. N. Masterova, V. A. Kabanov and L. S. Polak, *Vysokomol. Soedin., Ser. A*, 1973, **15**, 100.
38. V. F. Kornil'eva, M. N. Masterova, Ye. S. Garina, V. P. Zubov, V. A. Kabanov, L. S. Polak and V. A. Kargin, *Vysokomol. Soedin., Ser. A*, 1971, **13**, 1830.
39. V. P. Zubov, M. V. Kumar, M. N. Masterova and V. A. Kabanov, *J. Macromol. Sci., Chem.*, 1979, **A13**, 111.
40. R. Hart and G. Smets, *J. Polym. Sci.*, 1950, **5**, 55.
41. N. G. Gaylord and F. R. Eirich, *J. Polym. Sci.*, 1950, **5**, 743.
42. H. W. Melville and P. R. Sewell, *Makromol. Chem.*, 1959, **32**, 139.
43. N. S. Enikolopyan, B. R. Smirnov, G. V. Ponomarev and I. M. Belgovskii, *J. Polym. Sci., Polym. Chem. Ed.*, 1981, **19**, 879.
44. B. R. Smirnov, V. D. Plotnikov, B. V. Ozerkovskii, V. P. Roshchupkin and N. S. Enikolopyan, *Vysokomol. Soedin., Ser. A*, 1981, **23**, 2588.
45. A. F. Burczyk, K. F. O'Driscoll and G. L. Rempel, *J. Polym. Sci., Polym. Chem. Ed.*, 1984, **22**, 3255.
46. C. H. Bamford, W. G. Barb, A. D. Jenkins and P. F. Onyon, 'The Kinetics of Vinyl Polymerization by Radical Mechanisms', Butterworths, London, 1958.
47. J. C. Bevington, 'Radical Polymerization', Academic Press, New York, 1961.
48. R. A. Gregg and F. R. Mayo, *Discuss. Faraday Soc.*, 1947, **2**, 328.
49. R. A. Gregg and F. R. Mayo, *J. Am. Chem. Soc.*, 1953, **75**, 3530.
50. N. Fuhrman and R. B. Mesrobian, *J. Am. Chem. Soc.*, 1954, **76**, 3281.
51. T. Földes-Bereznich, M. Szesztay, E. Boros Gyevi and F. Tüdös, *J. Macromol. Sci., Chem.*, 1981, **A16**, 977.
52. F. R. Mayo, *J. Am. Chem. Soc.*, 1948, **70**, 3689.
53. F. M. Lewis and F. R. Mayo, *J. Am. Chem. Soc.*, 1954, **76**, 457.
54. H. W. Melville, J. C. Robb and R. C. Tutton, *Discuss. Faraday Soc.*, 1953, **14**, 150.
55. W. I. Bengough and R. A. M. Thomson, *Trans. Faraday Soc.*, 1960, **56**, 407.
56. C. A. Barson and R. Ensor, *Eur. Polym. J.*, 1977, **13**, 113.
57. C. A. Barson and R. Ensor, *Eur. Polym. J.*, 1977, **13**, 53.
58. C. A. Barson, R. A. Batten and J. C. Robb, *Eur. Polym. J.*, 1974, **10**, 97.
59. L. A. Wall and D. W. Brown, *J. Polym. Sci.*, 1954, **14**, 513.
60. H. Uemura, T. Taninaka and Y. Minoura, *J. Polym. Sci., Polym. Lett. Ed.*, 1977, **15**, 493.
61. H. Uemura, T. Taninaka and Y. Minoura, *J. Polym. Sci., Polym. Chem. Ed.*, 1978, **16**, 41.
62. T Ogawa and J. Gallego, *Eur. Polym. J.*, 1978, **14**, 825.
63. T. Taninaka, H. Uemura, K. Iwamoto and Y. Minoura, *J. Polym. Sci., Polym. Chem. Ed.*, 1978, **16**, 1549.
64. C. H. Bamford and S. N. Basahel, *J. Chem. Soc., Faraday Trans. 1*, 1978, **74**, 1020.
65. C. H. Bamford, S. N. Basahel and P. J. Malley, *Pure Appl. Chem.*, 1980, **52**, 1837.
66. C. H. Bamford and S. N. Basahel, *J. Chem. Soc., Faraday Trans. 1*, 1980, **76**, 112.
67. A. G. Morrell, *Discuss. Faraday Soc.*, 1956, **22**, 152.
68. A. H. Willbourn, *J. Polym. Sci.*, 1959, **34**, 569.
69. J. C. Woodbrey and P. E. Ehrlich, *J. Am. Chem. Soc.*, 1963, **85**, 1580.
70. K. Casey, C. T. Elston and M. K. Phibbs, *J. Polym. Sci., Polym Lett. Ed.*, 1964, **2**, 1053.
71. D. J. Cutler, P. J. Hendra, M. E. A. Cudby and H. A. Willis, *Polymer*, 1977, **18**, 1005.
72. W. L. Mattice and F. C. Stehling, *Macromolecules*, 1981, **14**, 1479.
73. J. T. Clarke, R. O. Howard and W. H. Stockmayer, *Makromol. Chem.*, 1961, **44–46**, 427.
74. P. C. Deb, *Eur. Polym. J.*, 1980, **16**, 759.
75. P. W. Allen, F. M. Merrett and J. Scanlan, *Trans. Faraday Soc.*, 1955, **51**, 95.
76. J. C. Bevington and H. G. Troth, *Trans. Faraday Soc.*, 1963, **59**, 127.
77. C. H. Bamford and E. F. T. White, *Trans. Faraday Soc.*, 1956, **52**, 716.
78. P. Ghosh and G. Mukhopadhyay, *Eur. Polym. J.*, 1979, **15**, 141.

14

Telomerization

BERNARD BOUTEVIN and YVES PIETRASANTA
Ecole Nationale Supérieure de Chimie, Montpellier, France

14.1 INTRODUCTION

Telomerization reactions, in contrast to polymerization reactions, are generally regarded as reactions which lead to oligomers having very low molecular weights and even monoaddition compounds. For the first time, in 1946, Handford[1] defined telomerization as the reaction between a compound XY called the telogen and one or several molecules of a polymerizable species M called the taxogen, under polymerization conditions. Telomerization leads to the formation of telomers $X(M)_nY$, $1 < n < 10$, according to the reaction in equation (1).

$$XY + nM \xrightarrow{\text{catalyst}} X(M)_nY \tag{1}$$

The products can be classified as intermediate between organic monomeric and macromolecular compounds. The scope of this topic was first outlined by Starks in 1974.[2]

We note that the catalysts which have been used especially are the traditional generators of free radicals: organic initiators, redox systems and UV, γ (^{60}Co) or electron beams. More recently other initiators have been used for syntheses, *e.g.* initiation transfer agents called 'inifers'.[3] Yet the radical and redox initiators are the most commonly used and are considered in more detail below.

14.2 MECHANISMS AND KINETICS OF TELOMERIZATION REACTIONS

The mechanisms of telomerization are described by the following equations (2)–(8).

14.2.1   Radical Telomerization

Concerning the radical telomerization using an initiator I, such as a peroxide, a peracid or a diazoic compound, we get

Initiation
$$I \xrightarrow{k_d} 2R_1 \cdot \tag{2}$$

$$R_1 \cdot + XY \rightarrow RY + X \cdot \tag{3}$$

$$X \cdot + M \rightarrow X-M \cdot \tag{4}$$

Propagation
$$XM_n \cdot + M \xrightarrow{k_p} XM_{n+1}^{\bullet} \tag{5}$$

Chain Transfer
$$XM_n \cdot + XY \xrightarrow{k_f} X(M)_nY + X \cdot \tag{6}$$

Termination
$$X \cdot + X \cdot \xrightarrow{k_t} X_2 \tag{7}$$

$$XM_n \cdot + XM_p \cdot \rightarrow \text{polymer} \tag{8}$$

Concerning radical telomerization, we have used the formula established by Mayo[4] for radical polymerization. Thus, the instantaneous degree of polymerization $(\overline{DP_n})_i$ is related to the concentration of monomer (M), telomer (T) and initiator (I) and to the appropriate rate constants (equation 9) when we consider only transfer to the telogen, which is the most important step in telomerization.

$$\frac{1}{(\overline{DP_n})_i} = \frac{(2fk_dk_t)^{\frac{1}{2}}}{k_p} \frac{[I]^{\frac{1}{2}}}{[M]} + C_T \frac{[T]}{[M]}$$

$$\text{where } C_T = \frac{k_f}{k_p} \tag{9}$$

To calculate the real or cumulative $\overline{DP_n}$ we have to know the variations of monomer, initiator and telogen concentrations with time. Variation of monomer is given by Tobolsky's equation (equation 10)[5], and that of telogen (equation 12) was obtained by O'Brien.[6]

$$\log \frac{[M]_0}{[M]} = \frac{k_p}{(k_t)^{-\frac{1}{2}}} \left(\frac{2f}{k_d}\right)^{\frac{1}{2}} [I]_0^{\frac{1}{2}} \left(1 - e^{\frac{-k_dt}{2}}\right) \tag{10}$$

$$[I] = [I]_0 e^{-k_dt} \tag{11}$$

$$\log \frac{[T]_0}{[T]} = C_T \log \frac{[M]_0}{[M]} \tag{12}$$

Thus we get[7]

$$(\overline{DP_n})_{cum}^i = \frac{i}{\displaystyle\sum_{j=0}^{j=i-1} \frac{1}{(\overline{DP_n})_i}} \tag{13}$$

from

$$(\overline{DP_n})_{cum}^i = \frac{[M]_0 - [M]}{[T]_0 - [T]} \tag{14}$$

It is also possible to get the values of the cumulation weight average degree of polymerization $(\overline{DP_w})_{cum}$. In equation (15) we consider only the variation of $(\overline{DP_w})_{cum}$ with monomer concentration. It is not the real value of $(\overline{DP_w})$ since we consider that in each interval i of the rate of conversion in monomer, the reaction is isomolecular (see Table 1).

$$(\overline{DP_w})_{cum} = \sum_{j=0}^{j=i-1} \frac{(DP_n)_i}{i} \tag{15}$$

For instance, let us consider the telomerization of vinyl acetate with chloroform initiated by benzoyl peroxide at 60 °C. The instantaneous $\overline{DP_n}$ is given by equation (16) derived from equation (9).

$$\frac{10^3}{(\overline{DP_n})_i} = 5.8 \frac{[I]^{-\frac{1}{2}}}{[M]} + 15 \frac{[CHCl_3]}{[M]} \tag{16}$$

Table 1 Variations of Degrees of Polymerization with the Degree of Conversion in Monomer

Yields %	0	20	40	60	80	100
time (min)	0	33	84	135	249	514
$(\overline{DP}_n)_i$	65	52	40	27	13	1
$(\overline{DP}_n)_{cum}$	65	58	50	41	29	5
$(\overline{DP}_w)_{cum}$	65	58	56	52	47	40
Evolution of the polydispersity[7]	1.00	1.01	1.11	1.26	1.63	8.00

Table 2 Transfer Constants (C_{Me}) and Termination Constants (C_{CCl_4}) of Several Catalysts and Telogen in Redox Telomerization

	C_{Me}	C_{CCl_4}
$CF_2=CFCl/CCl_4/Fe^{2+}$	75	0.0130
$CH_2=CHCO_2Et/CCl_4/Fe^{2+}$	8	0.0001
$CH_2=CHCO_2Et/CCl_4/Cu^+$	800	0.0001

We note that the second term, which is due to transfer to chloroform, is 75 times greater than the term due to bimolecular termination (Table 2).

14.2.2 Redox Telomerization

For the telomerization initiated by a redox catalyst, the mechanism is generally described as in equations (17)–(21).

Initiation
$$RCCl_3 + Me^{n+} \xrightarrow{k_i} [Me(Cl)]^{(n+1)+} + RCCl_2 \cdot \tag{17}$$

$$RCCl_2 \cdot + M \rightarrow RCCl_2 M \cdot \tag{18}$$

Propagation
$$RCCl_2 M_n \cdot + M \xrightarrow{k_p} RCCl_2 M_{n+1}^{\cdot} \tag{19}$$

Chain Transfer
$$RCCl_2 M_n \cdot + RCCl_3 \xrightarrow{k_f} RCCl_2 M_n Cl + RCCl_2 \cdot \tag{20}$$

Termination
$$RCCl_2 M_n \cdot + [Me(Cl)]^{(n+1)+} \xrightarrow{k_t} RCCl_2 M_n Cl + Me^{n+} \tag{21}$$

The first redox telomerizations had been conducted by Asscher and Vofsi[8] using salts of transition metals, such as iron or copper, as catalysts. Yet Freidlina[9–10] used metal carbonyl.[9] The surveys performed from 1965 to 1970 led to several hypotheses about the mechanism of redox reaction. An interesting model had been suggested by Svezdin et al.[11] who indicated that all the redox telomerizations were carried out in a coordination sphere around the metal as we can see in Scheme 1.

More recently, new metallic complexes have been used as catalysts to improve the yield and the selectivity of the reaction in both chain length and molecular structure. Among them, we can notice the metal carbonyls of Fe,[12] Mn,[13–15] Cr;[16] and metallic complexes such as $RuCl_2(PPh_3)_3$,[17] $Pt(PPh_3)_4$, $RhCl(PPh_3)_3$,[18] $Ni[P(OPh)_3]_4$.[19]

Using the same method as employed for free radical initiation, the kinetics of the redox telomerization have been studied (see equations 22–24).[20–21]

$$\frac{1}{(\overline{DP}_n)_i} = \phi C_{Me} C + C_{RCCl_3} R \tag{22}$$

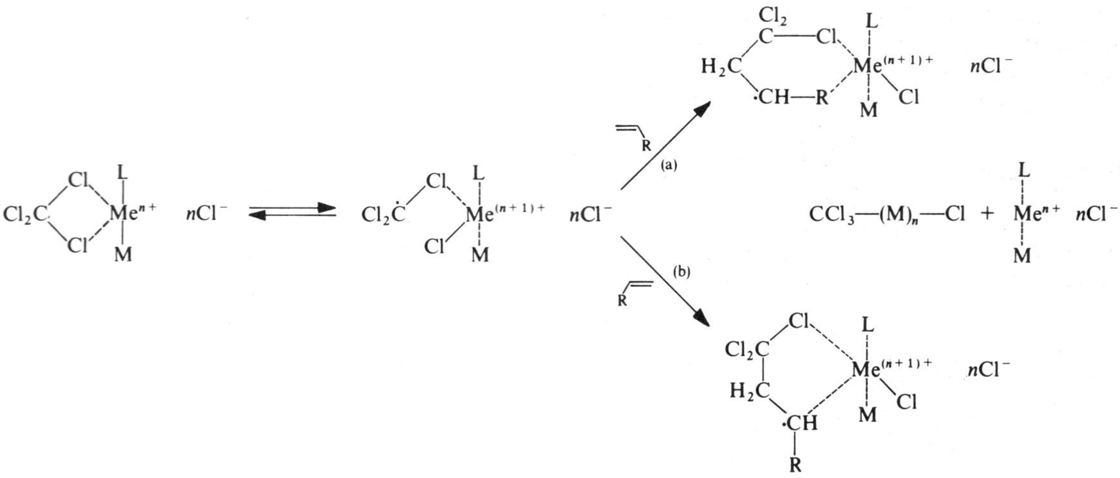

Scheme 1   Svezdin mechanism of redox telomerization: (a) $R = CN$, CO_2R', $n = 1$; (b) $R = H$, Cl, alkyl, $n = 1\text{--}3$

where

$$C = \frac{[Me^{n+}]}{[M]}, \; R = \frac{[RCCl_3]}{[M]}, \; C_{Me} = \frac{k_t}{k_p}, \; \phi = \frac{[Me^{(n+1)+}]}{[Me^{(n+1)+}]_0}, \; C_{RCCl_3} = \frac{k_{f,RCCl_3}}{k_p}$$

$$[RCCl_3] = [RCCl_3]_0 \exp[1 - (1-\phi)fk_i[Me^{(n+1)+}]_0 t] \tag{23}$$

$$\log \frac{[M]_0}{[M]} = \frac{R_0}{\phi C_{Me} C_0} \left\{ 1 - \exp\left[(1-\phi)fk_i[Me^{(n+1)+}]_0 t \right] \right\} \tag{24}$$

In this case, termination of the $[Me(Cl)]^{(n+1)+}$ complex is much greater than the chain transfer on $RCCl_3$. For instance, for the telomerization of ethyl acrylate with CCl_4 using $FeCl_3$/benzoin as catalyst, if we plot $1/(\overline{DP}_n)_0$ *vs.* C_0 and then *vs.* R_0 we obtain in the first case a straight line whose slope gives a high value of C_{Me}, whereas in the second case the slope is C_{CCl_4} which is very small compared with C_{Me}. Several values of transfer and termination constants are detailed in Table 2. It is most insteresting to note the difference between the C_{Fe} and C_{Cu} values. Using copper compounds as the catalyst leads to products with low $\overline{DP}_n$, very often with one to one adducts.

All these relations can be used when values of $\overline{DP}_n$ are greater than 5, because there is no variation of the rate constants k_t, k_p, k_f with degrees of polymerization greater than this value.[23] On the contrary, when $\overline{DP}_n$ is smaller than 5 the $\overline{DP}_n$ values must be calculated by the molecular distribution (*i.e.* the molar fraction F_n *vs.* n) and this distribution is called 'Telomer'. In radical telomerization the equation was proposed by David and Gosselain[24] and in redox catalysis we have equation (25).

$$F_n = \frac{\phi C_{Me}^n \cdot C_0}{\displaystyle\prod_{i=0}^{i=n} (1 + \phi C_{Me}^i \cdot C_0)} \tag{25}$$

14.3   MONOMERS AND TELOGENS USED IN TELOMERIZATION

14.3.1   Monomers

All the monomers involved in radical polymerization can be used in telomerization. Concerning the chain lengths, for a given telogen, smaller lengths are generally obtained by redox catalysis rather than by radical initiation.

Thus, first with allylic derivatives, monoaddition products are essentially obtained, whatever the catalyst used. Second, with vinyl monomers (vinyl chloride, styrene, vinyl acetate, *etc.*) and dienes, the monoaddition compound is obtained with copper or metal complex catalysts, otherwise a distribution of adducts with one to ten monomers is observed. Moreover with a free radical initiator it is possible to produce a $\overline{DP}_n$ up to 100. Thirdly, the acrylic monomers result in a large range of degrees of polymerization; with free radical initiation we usually obtain high degrees of polymerization. Only telogens with high transfer constants such as thiols (0.2 to 2)[25] and bromo-compounds (CCl_3Br; 0.05 to 0.1) enable us to obtain low degrees of polymerization. In contrast, with redox catalysis the degrees of polymerization are lower and a wider range than of those for other monomers is obtained, depending on the telogen used (CCl_4, CCl_3CF_3 and CCl_3CO_2R).[27]

14.3.2 Telogens

Many telogens may be used in telomerization. They are chosen in a different way depending on either the process used or how the required telogen cleaves. The procedure,[8] the cleavage of the telogen and the products obtained are described in Table 3.

Table 3 Different Methods of Cleavage in the Telogens

Bond cleaved	Method	Ref.
C–I	$C_6F_{13}I + CF_2{=}CFCl \xrightarrow[\text{initiation}]{\text{radical}} C_6F_{13}{\leftarrow}CF_2CFCl{\rightarrow}_n I$	28
C–Br	$CCl_3Br + CF_2{=}CFCl \xrightarrow[\text{initiation}]{\text{radical}} CCl_3{\leftarrow}CF_2CFCl{\rightarrow}_n Br$	29
C–Cl	$CCl_3H + CF_2{=}CFCl \xrightarrow{\text{FeCl}_3} HCCl_2{\leftarrow}CF_2CFCl{\rightarrow}_n Cl$ $CCl_4 + CF_2{=}CFCl \xrightarrow{\text{RuCl}_2(\text{Ph}_3)_3} CCl_3CF_2CFCl_2$ $CCl_3F + CF_2{=}CFCl \xrightarrow{\text{CuCl}} CFCl_2CF_2CFCl_2$	30
C–F	$CCl_3F + CF_2{=}CFCl \xrightarrow{\text{AlCl}_3} Cl_3CCF_2CF_2Cl$	31
C–H	$CCl_3H + CH_2{=}CHOCOMe \xrightarrow[\text{initiation}]{\text{radical}} CCl_3{\leftarrow}CH_2CH{\rightarrow}_n H$ $\qquad\qquad\qquad\qquad\qquad\qquad\qquad\quad\; |$ $\qquad\qquad\qquad\qquad\qquad\qquad\qquad\; OCOMe$	32
S–H	$HO_2CCH_2SH + CH_2{=}CHCO_2C_{12}H_{25} \xrightarrow[\text{initiation}]{\text{radical}} HO_2CCH_2S{\leftarrow}CH_2CH{\rightarrow}_n H$ $\qquad\qquad\qquad\qquad\qquad\qquad\qquad\qquad\qquad\qquad\qquad\quad |$ $\qquad\qquad\qquad\qquad\qquad\qquad\qquad\qquad\qquad\qquad CO_2C_{12}H_{25}$	33
P–H	$\begin{array}{c}\text{EtO}\\ \diagdown\\ O{\leftarrow}PH\\ \diagup\\ \text{EtO}\end{array} + CF_2{=}CFCl \xrightarrow[\text{initiation}]{\text{radical}} \begin{array}{c}\text{EtO}\\ \diagdown\\ O{\leftarrow}P{\leftarrow}CF_2CFCl{\rightarrow}_n{-}H\\ \diagup\\ \text{EtO}\end{array}$	34
Si–H	$MeSiCl_2H + CH_2{=}CHC_6F_5 \xrightarrow{\text{H}_2\text{PtCl}_6} MeSiCl_2CH_2CH_2C_6F_5$	35
N–H	$C_7F_{15}CH_2O_2CCH{=}CH_2 + H\overline{NCH_2{-}CH_2} \xrightarrow[\text{initiation}]{\text{radical}} C_7F_{15}CH_2O_2CC_2H_4\overline{NCH_2{-}CH_2}$	36

Thus we can see that telomerization reactions permit the formation of a large variety of mono- or poly-molecular products which contain silicon, phosphorus, nitrogen, sulfur and very often halide atoms.

14.3.3 Ring-opening Telomerization

Of late years, the notion of telomerization has been given a very special extended meaning in ring-opening polymerizations.

(i) For THF[63]

$$(CH_2{=}CHCO)_2O + HSbF_6 \text{ (or } CF_3SO_3H) + THF \rightarrow CH_2{=}CHCO{\leftarrow}OCH_2CH_2CH_2CH_2{\rightarrow}OCOCH{=}CH_2 \quad (26)$$

(ii) For 'phosphonitrilic chlorides' with PBr_5[64] we have $PBr_4{\leftarrow}P{=}NCl_2{\rightarrow}_n Br$.

(iii) For cyclosiloxanes by cationic initiation[65]

$$(Me_2SiO)_3 + CF_3SO_3H \rightarrow CF_3SO_3 (SiO)_n H \xrightarrow[\text{(ii) Me}_3\text{SiOCOMe}]{\text{(i) MeCO}_2\text{H}} Me_3SiO (SiO)_n COMe \tag{27}$$

with Me groups on the Si atoms.

(iv) For epoxides[66]

$$CF_2\!\!-\!\!CFCF_2CF_2X + C_3F_7OCFCF_2O^- Cs^+ \longrightarrow C_3F_7OCF(CF_3)CF_2O (CF(CF_2CF_2X)CF_2O)_n CF(CF_2CF_2X) COF$$

$$X = Br, Cl \tag{28}$$

14.4 APPLICATIONS OF THE REACTIONS OF TELOMERIZATION

Several different applications may be distinguished according to whether or not the telomers possess reactive functions.

14.4.1 Non-functional Telomers

These are oligomers of a similar nature to the corresponding polymers and yet their chain extremities are very well identified. Traditionally, the telomers have been prepared as intermediates for the synthesis of amino acids,[1] *i.e.* telomers of ethylene with carbon tetrachloride by chemical transformation of the two extremities of the chain. Subsequently, telomers have been synthesized in order to be used as non-extractable additives of the corresponding polymers. Nevertheless, the main industrial application concerns the fluorinated products. Thus, the synthesis of perfluoroalkyl iodides is performed by the reaction of C_2F_5I with C_2F_4 to lead to iodides with chains containing about eight to twenty carbon atoms. These compounds are precursors for the synthesis of surfactants[37] and the company Atochem has mainly prepared these precursors in this manner.

The other interesting application concerns the fabrication of chlorofluorinated oils (Kelf or Voltalef). Initially, Barnhart[38] prepared them by adding sulfuryl chloride to chlorotrifluoroethylene (equation 29).

$$SO_2Cl_2 + CF_2\!\!=\!\!CFCl \xrightarrow[\text{initiation}]{\text{radical}} Cl (CF_2\!\!-\!\!CFCl)_n Cl \tag{29}$$

More recently, the Oxy company has prepared these kinds of oils from telomers $CCl_3 (CF_2\!\!-\!\!CFCl)_n Cl$[30,39] obtained by the reaction of ClF_3 on the extremities of the chain[40] to increase the thermal stability.

14.4.2 Monofunctional Telomers

Monofunctional telomers may be obtained by chemical transformation of the end of the chain, but it is also possible to prepare them using a functional telogen as illustrated in the example of equation (30).

$$CH_2\!\!=\!\!CHCl + Cl_3CCO_2H \xrightarrow{\text{catalyst}} HO_2CCCl_2 (CH_2\!\!-\!\!CHCl)_n Cl \tag{30}$$

Using $FeCl_3 \cdot 6H_2O$ as catalyst, molecular weights lower than 1000 are obtained,[41] but with free radical initiators they may even reach 20000.[42]

Using this approach, Ito *et al.*[43] synthesized macromonomers by reacting glycidyl methacrylate and telomers of stearyl methacrylate with thioglycolic acid (equation 31). This method of telomerization competes with the preparation of macromonomers *via* ionic processes.[44] These macromonomers find applications in the synthesis of graft copolymers used as emulsifying agents for blends of PE/PVC[45] and PE/PS.[46] Recently the Arco company has proposed the use of polystyrene macromers as thermoplastic additives in photocuring or electron beam curing formulations.

$$CH_2\!\!=\!\!CMeCO_2CH_2CH\!\!-\!\!CH_2 + HO_2CCH_2S (CH_2CMeCO_2R)_n H \xrightarrow{NR_3}$$

$$CH_2\!\!=\!\!C(Me)CO_2CH_2CHOHCH_2O_2CCH_2S (CH_2CMeCO_2R)_n H \tag{31}$$

As we have mentioned before, monofunctional fluorinated surfactants can be prepared by transformation of $C_nF_{2n+1}I$ compounds; the latter telomers are used as telogen in a second telomerization,[47-48] the so called bistelomerization (equation 32).

$$R_FI + CH_2=CH\cdot(CH_2)_{n-2}CO_2R \xrightarrow[\text{ii, Zn/EtOH}]{\text{i, radical telomerization}} R_F\cdot(CH_2)_n CO_2R \tag{32}$$

The Akzo company has synthesized 'long chain ramified' monocarboxylic acids, by telomerization of acetic anhydride with α-alkenes catalyzed by manganese(III) acetate,[49] commercialized under the brand name Kortacids (equation 33)

$$\tag{33}$$

The metallic salts of these acids such as calcium salts present interesting properties of solubility: for example they are soluble in mineral oil and lead to solutions which have a low viscosity with Newtonian flow and which enable the friction coefficients to be decreased drastically.

14.4.3 Difunctional or Telechelic Telomers

These telomers are very interesting because they may be used as monomers in polycondensation reactions and they are prepared by several different methods briefly described here:

(i) by chemical transformation of the two end groups of the telomers;[50]

$$CCl_3\cdot(CF_2CFCl)_n Cl \xrightarrow[\text{ii, oleum}]{\text{i, AlCl}_3} HO_2C\cdot(CF_2CFCl)_{n-1}CF_2CO_2H \tag{34}$$

(ii) from non-conjugated dienes as taxogens and functional telogens,[51a]

$$Cl_2(O)PCCl_3 + CH_2=CH\cdot(CH_2)_2CH=CH_2 \rightarrow (Cl_2OPCCl_2CH_2CHClCH_2)_2 \tag{35}$$

(iii) from conjugated dienes with allylic alcohol and a special catalyst;[51b]

$$CH_2=CHCH=CH_2 + CH_2=CHCH_2OH \xrightarrow[\text{Alkaline earth}]{\text{Rh(HO}_3)_3} \text{telechelic diol (1,4-}trans\text{-polybutadiene)} \tag{36}$$

(iv) by monoaddition of a functional telogen on the two extremities of a ditelogen compound;[52]

$$CCl_3CH_2CCl_3 + CH_2=CH\cdot(CH_2)_8CO_2Me \rightarrow CH_2\cdot(CCl_2CH_2CHCl\cdot(CH_2)_8CO_2Me)_2 \tag{37}$$

(v) from difunctional disulfides as telogen;[53]

$$OC=NC_6H_4S\text{-}SC_6H_4N=CO + CH_2=CHPh \xrightarrow[\text{initiation}]{\text{radical}} OC=NC_6H_4S\cdot(CH_2CH)\text{-}SC_6H_4N=CO \tag{38}$$

(vi) from telogens with weak bonds (equation (39) where R = −CN, −OPh, −OSiMe$_3$),[54]

$$\tag{39}$$

(vii) with a 'ditelogen involvement' by a cationic mechanism,[3]

$$\tag{40}$$

(viii) and finally, the functional group may be produced by the initiator and termination either by duplication[55] or by transfer to the initiation (the product in reaction (41) is called 'PBHT' by the Arco company).[56,57]

$$H_2O_2 + CH_2{=}CHCH{=}CH_2 \rightarrow HO{-}(CH_2CH{=}CHCH_2{-})_x{-}(CH_2CH{-})_y{-}OH \qquad (41)$$

$$HO{-}(CH_2CH_2O{-})_y\overset{S}{\overset{\|}{C}}{-}S{-}S{-}\overset{S}{\overset{\|}{C}}{-}(OCH_2CH_2{-})_x OH + CH_2{=}CH \rightarrow$$

$$HO{-}(CH_2CH_2O{-})_y\overset{S}{\overset{\|}{C}}{-}S{-}(CH_2CH{-})_n{-}S{-}\overset{S}{\overset{\|}{C}}{-}(OCH_2CH_2{-})_x{-}OH \qquad (42)$$

$$\underset{Et}{\overset{Et}{>}}N{-}\underset{S}{\overset{\|}{C}}{-}S{-}S{-}\underset{S}{\overset{\|}{C}}{-}N\underset{Et}{\overset{Et}{<}} \;+\; CH_2{=}\underset{Ph}{\overset{}{CH}} \longrightarrow \left[\underset{Et}{\overset{Et}{>}}N{-}\underset{S}{\overset{\|}{C}}{-}S{-}\left(CH_2{-}\underset{Ph}{\overset{}{CH}}\right)_n\right]_2 \qquad (43)$$

14.4.4 Multifunctional Telomers

Multifunctional compounds are prepared from a functional monomer by multiadditions. One mainly uses acrylic monomers as starting materials to obtain different chosen functional groups (acid, alcohol, epoxide, amine, amide, *etc.*) inside chains. When one uses radical initiation the telogen is either a thiol or a bromocompound such as CCl_3Br. When redox catalysis is involved, CCl_4 or $RCCl_3$ species are used as telogens.

Numerous chemical transformations of polyacrylic telomers have been carried out in order to obtain interesting industrial applications such as:

(a) the preparation of additives for antifouling paints[58] where the telomers of acrylic acid with various thiols are esterified by tributyltin oxide;
(b) the grafting of drugs onto these same telomers[59] in order to increase their efficiency.

In these two cases, the covalent bond between the substrate and the active compound gives to the molecule the desired delaying effect.

Concerning the synthesis of photocrosslinkable compounds,[60] acrylic derivatives have been grafted onto multifunctional substrates in order to introduce more than two crosslinkable double bonds on these oligomers (equation 44).

$$C_{12}H_{25}SH \;+\; CH_2{=}CHCO_2CH_2CH{-}CH_2 \xrightarrow{\text{AIBN}} C_{12}H_{25}S{-}(CH_2CH{-})_n{-}H \xrightarrow[\text{NR}_3]{\text{acrylic acid}}$$

$$C_{12}H_{25}S{-}(CH_2CH)_{\overline{n-x}} \qquad (CH_2CH)_x{-}H$$

(44)

Varnishes have thus been obtained which are cured either by UV radiation or by electron beams, in about 10^{-1} s.[61] The residual epoxides, acids or alcohol groups are responsible for good adhesive properties on substrates such as aluminum.

Finally another application concerns the synthesis of surfactants.[62] These compounds are especially desirable when they are fluorinated because they present good efficiency, even when they are used in very small amounts (equation (45) where $10 < n < 100$, $C_{R_FI} \simeq 0.05$ and $R = -C_nF_{2n+1}$).

$$R_F{-}I + CH_2{=}CHCONH_2 \xrightarrow[\text{initiation}]{\text{radical}} R_F{-}(CH_2CH{-})_n I \qquad (45)$$

Other telogens and hydrophilic monomers have also been used: $C_nF_{2n+1}{-}C_2H_4{-}SH$ with acrylamide ($C_T = 0.6$); $C_nF_{2n+1}{-}C_2H_4{-}I$ with *N*-vinylpyrrolidone ($C_T = 0.05$); $C_nF_{2n+1}{-}C_2H_2{-}O_2C{-}CCl_3$ with acrylamide by redox catalysis $C_T \simeq 1.75$.

14.5 CONCLUSION

Telomerization by radical initiation or redox catalysis represents an interesting means of synthesis in both organic and macromolecular chemistry. It results in some simple or polyfunctional compounds which contain heteroatoms such as F, Cl, P, N, S, Si. The knowledge of polymerization kinetics from rate constants and transfer coefficients enables the preparation of macromolecules with a predictable degree of polymerization and a polydispersity as low as possible. The potential for industrial applications of this kind of compound is thus very wide and is still increasing.

14.6 REFERENCES

1. W. E. Handford, *US Pat.* 2 396 786 (1946) (*Chem. Abstr.*, 1946, **40**, 3628).
2. C. M. Starks, in 'Free Radical Telomerization', Academic Press, New York, 1974.
3. J. P. Kennedy, R. Santos and M. Walters, *Polym. Bull.(Berlin)*, 1984, **11**, 261.
4. F. R. Mayo, *J. Am. Chem. Soc.*, 1943, **65**, 2324.
5. A. V. Tobolsky, *J. Am. Chem. Soc.*, 1958, **80**, 5927.
6. J. L. O'Brien and F. Gornick, *J. Am. Chem. Soc.*, 1955, **77**, 4757.
7. B. Boutevin, Y. Pietrasanta and G. Bauduin, *Makromol. Chem.*, 1985, **186**, 283.
8. M. Asscher and D. Vofsi, *J. Chem. Soc.*, 1961, 2261.
9. R. Kh. Freidlina and C. E. Chukovskaya, *Synthesis*, 1974, 477.
10. R. Kh. Freidlina, E. Ts. Chukovskaya and B. A. Enolin, *Dokl. Akad. Nauk SSSR*, 1964, **159**(6), 1346.
11. V. L. Svezdin and G. A. Domrachev, *Dokl. Akad. Nauk SSSR*, 1971, **198**(1), 108.
12. T. T. Vasil'eva, L. F. Germanova, V. I. Dostovolova, B. V. Nelyvbin and R. Kh. Freidlina, *Izv. Akad. Nauk SSSR*, 1983, **12**, 2759.
13. M. A. Moskalenko, A. B. Terent'ev and R. Kh. Freidlina, *Izv. Akad. Nauk SSSR*, 1982, **6**, 1260.
14. M. A. Moskalenko and A. B. Terent'ev, *Izv. Akad. Nauk SSSR*, 1983, **10** 2322.
15. M. A. Moskalenko, A. B. Terent'ev and R. Kh. Freidlina, *Izv. Akad. Nauk SSSR*, 1981, **3**, 637.
16. N. A. Grior'ev, Tumanskaya and R. Kh. Freidlina, *Izv. Akad. Nauk SSSR*, 1981, **7**, 1515.
17. H. Matsumoto, *J. Synth. Org. Chem.*, 1978, **12**, 1059.
18. A. Behr, *Aspects Homogeneous Catal.*, 1984, **5**, 3.
19. Y. Inone, S. Ohno and H. Hashimoto, *Chem. Lett.*, 1978, 367.
20. B. Boutevin and Y. Pietrsanta, *Makromol. Chem.*, 1985, **186**, 817.
21. B. Boutevin and Y. Pietrasanta, *Makromol. Chem.*, 1985, **186**, 831.
22. B. Boutevin, C. Maubert, Y. Pietrasanta and P. Sierra, *J. Polym. Sci.*, 1981, **19**, 511.
23. B. Boutevin, M. Maliszewicz and Y. Pietrasanta, *Makromol. Chem.*, 1985, **186**, 1467.
24. C. David and P. A. Gosselain *Tetrahedron* 1962, **70**, 639.
25. B. Boutevin, H. Snoussi and M. Taha, *Eur. Polym. J.*, 1985, **21**(5), 445.
26. G. Bauduin, B. Boutevin, J. P. Mistral and L. Sarraf, *Makromol. Chem.*, 1985, **186**, 1445.
27. B. Boutevin, M. Maliszewicz and Y. Pietrasanta, *Makromol. Chem.*, 1982, **183**, 2333.
28. R. N. Haszeldine and B. R. Steele, *J. Chem. Soc.*, 1953, 1592.
29. R. L. Enrenfeld, *US Pat.*, 1 778 375 (1957), (*Chem. Abstr.*, 1957, **51**, 11 759).
30. B. Boutevin and Y. Pietrasanta, *Eur. Polym. J.*, 1976, **12**, 219.
31. A. Posta, and O. Paleta, *Collect Czech. Chem. Commun.*, 1966, **31**, 2389.
32. G. Bauduin, D. Bondon, J. Martel, Y. Pietrasanta and B. Pucci, *Makromol. Chem.* 1981, **182**, 2589.
33. C. Bonardi, B. Boutevin, Y. Pietrasanta and M. Taha, *Makromol. Chem..* 1985, **186**, 261.
34. J. A. Bittles and R. M. Joyce, *US Pat.* 2 559 754 (1951) (*Chem. Abstr.*, 1952, **46**, 1026).
35. B. Boutevin, Y. Pietrasanta and B. Youssef, *J. Fluorine Chem.*, 1986, **31**(1), 57.
36. G. S. Tesoro, *US Pat.* 3 719 698 (1973) (*Chem. Abstr.*, 1973, **79**, 20 250).
37. L. Lichtenberger, *Chim. Ind. (Paris)*, 1971, **104**, 815.
38. W. S. Barnhart *US Pat.* 2 837 580 (1958) (*Chem. Abstr.*, 1959, **53**, 618g).
39. S. Saran Mohan (Occidental Chemical Corp) *Eur. Pat. Appl.*, 93 580 (1983) (*Chem. Abstr.*, 1984, **100**, 68 933).
40. S. Saran Mohan (Occidental Chemical Corp) *Eur. Pat. Appl.*, 93 579 (1983) (*Chem. Abstr.*, 1984, **100**, 68 969).
41. B. Boutevin, Y. Pietrasanta and M. Taha, *Makromol. Chem.*, 1982, **183**, 2985.
42. B. Boutevin, Y. Pietrasanta and M. Taha, *Makromol. Chem.*, 1982, **183**, 2977.
43. K. Ito, N. Usami and Y. Yamashita, *Macromolecules*, 1980, **13**, 216.
44. R. Milkovich, *Polym. Prepr., Am. Chem. Soc., Div. Polym. Chem.*, 1980, **21**, 40.
45. B. Boutevin, Y. Pietrasanta, M. Taha and T. Sarraf, *Polym. Bull.(Berlin)*, 1985, **14**, 14.
46. B. Boutevin, Y. Pietrasanta and T. Sarraf, *Angew. Makromol. Chem.*, 1987, **148**, 195.
47. H. Jaeger (Ciba Geigy A.G.) *Ger. Pat.* 2 142 056 (1972) (*Chem. Abstr.*, 1971, **77**, 125 942).
48. N. O. Brace, *J. Org. Chem.*, 1962, **27**, 4491.
49. H. Van Breberode, *J. Am. Chem. Soc.* 1984, **61**(2) 247.
50. B. Boutevin and Y. Pietrasanta, *Eur. Polym. J.*, 1975, **12**, 231.
51. (a) M. Corallo and Y. Pietrasanta, *Phosphorus Sulfur*, 1977, **3**, 359; (b) W. Heitz, K. Klauss and W. Mehnert, *Makromol. Chem.*, 1976, **177**(5), 1625.
52. A. Battais, B. Boutevin, Y. Pietrasanta and T. Sarraf, *Makromol. Chem.*, 1982, **183**, 2359.
53. T. Dkano, M. Katayama and I. Shinohara, *J. Appl. Polym. Sci.*, 1978, **22**, 369.
54. A. Bledzki, *Kunststoffe* 73, 1983, **3**, 156.
55. J. Brossas, *Inf. Chim.* 1974, **128**, 185.

56. De Soto (J. M. Zimmerman, J. J. Krajewski and G. K. Noren), *Eur. Pat. Appl.* 0161 502 (1985) (*Chem. Abstr.*, 1986, **104**, 131 609).
57. M. Imoto, T. Otsu and J. Yonezawa, *Makromol. Chem.*, 1960, **36**, 93.
58. A. Battais, B. Boutevin and Y. Pietrasanta, (DRET) *Fr. Pat.* 81 13080 (1981) (*Chem. Abstr.*, 1983, **98**, 217 326).
59. G. Bauduin, J. M. Bessière, D. Bondon, J. Martel and Y. Pietrasanta, *Makromol. Chem.*, 1981, **182**, 3491.
60. G. Bauduin, B. Boutevin, W. J. Deiss and Y. Pietrasanta, (Pechiney Ugine Kuhlmann), *Fr. Pat.* 83 06 0666 (1983) (*Chem. Abstr.*, 1985, **102**, 97 058).
61. B. Boutevin, W. J. Deiss and Y. Pietrasanta, 'Radcure', Bâle, 1985, pp. 1–13.
62. B. Boutevin, Y. Pietrasanta, A. Lantz and M. Taha (Atochem), *Fr. Pat.* 84 198 34 (1984) (*Chem. Abstr.*, 1987, **106**, 69 170).
63. H. J. Kreb and W. Heitz, *Makromol. Chem., Rapid Commun.*, 1981, **2**, 427.
64. F. Yamada, T. Yasui and I. Shinohara, *J. Macromol. Sci., Chem.*, 1981, **A15**, 585.
65. M. Scibiorek and J. Choinowski, *Eur. Polym. J.*, 1981, **17**, 413.
66. T. I. Ito, J. Kaufman, R. H. Kratzer, J. G. Nakahara and K. J. L. Paciorek, *J. Fluorine Chem.*, 1979, **14**, 93.

15

Copolymer Composition

DAVID A. TIRRELL

University of Massachusetts, Amherst, MA, USA

15.1 INTRODUCTION

The successful synthesis of new materials *via* free radical copolymerization requires a thorough understanding of the factors that control the structures of copolymer chains. Of the primary structural variables used to describe polymer chains, two (copolymer composition and comonomer sequence distribution) are unique to copolymers. The present chapter examines the compositions and sequences of copolymers prepared *via* free radical processes.

The products of free radical copolymerizations are, with few exceptions, determined by the kinetics, rather than the thermodynamics, of the chain growth process. The problem of predicting copolymer composition and sequence then reduces to the writing of a set of differential equations that describe the rates at which each of the two monomers enters the copolymer chain by attack of the growing macroradical. This requires a kinetic model of the copolymerization process, and several such models have been described in the copolymerization literature. The following sections examine the fundamental bases of the most important of these models, and assess the degree to which such models can account for experimentally observed copolymerization behavior.

15.2 TERMINAL MODEL

15.2.1 Theory

The standard kinetic treatment of free radical copolymerization was introduced in 1944, in papers contributed independently by Mayo and Lewis,[1] by Alfrey and Goldfinger,[2] and by Wall.[3] Following earlier suggestions by Dostal,[4] by Norrish and Brookman,[5] and by Jenckel,[6] Mayo and Lewis described their experimental work on the radical copolymerization of styrene and methyl methacrylate in terms of a model in which the rate constant for addition of each monomer was assumed to be dependent on the identity of the terminal unit on the growing chain. Four elementary propagation steps were then considered (equations 1–4).

$$\text{wwM}_A\cdot \ + \ M_A \ \xrightarrow{\ k_{AA}\ } \ \text{wwM}_A M_A\cdot \tag{1}$$

$$\text{wwM}_A\cdot \ + \ M_B \ \xrightarrow{\ k_{AB}\ } \ \text{wwM}_A M_B\cdot \tag{2}$$

$$\text{wwM}_B\cdot \ + \ M_A \ \xrightarrow{\ k_{BA}\ } \ \text{wwM}_B M_A\cdot \tag{3}$$

$$\text{wwM}_B\cdot \ + \ M_B \ \xrightarrow{\ k_{BB}\ } \ \text{wwM}_B M_B\cdot \tag{4}$$

By writing differential equations that describe the rates of disappearance of monomers M_A and M_B, and by assuming steady state concentrations of the radical centers $M_A\cdot$ and $M_B\cdot$, one arrives at a simple expression that relates the ratio of monomers in the copolymer $(d[M_A]/d[M_B])$ to the concentrations of monomers in the feed mixture $([M_A]$ and $[M_B])$ (equation 5).

$$\frac{d[M_A]}{d[M_B]} \ = \ \frac{[M_A]}{[M_B]} \frac{r_A[M_A] \ + \ [M_B]}{[M_A] \ + \ r_B[M_B]} \tag{5}$$

The parameters r_A and r_B are *reactivity ratios* defined as

$$r_A \ = \ k_{AA}/k_{AB} \quad \text{and} \quad r_B \ = \ k_{BB}/k_{BA} \tag{6}$$

A detailed derivation of equation (5) is given in Chapter 2 of this volume.

Analysis of the terminal model is readily extended to the predictions of copolymer sequence distribution. Sequence distributions are most generally and conveniently specified in terms of the number fractions of uninterrupted sequences of a given monomer (M_A or M_B) that are of a particular length. The number fraction is of course identical to the probability that a given uninterrupted sequence, selected at random, is of that length. Consider a sequence of M_A units of length x. Such a sequence arises in the terminal model when a growing macroradical terminating in $M_A\cdot$ adds $(x-1)$ M_A units followed by an M_B. The probability that such a sequence forms is obtained as the product of the probabilities of each of the independent steps that lead to the sequence. Thus the number fraction of sequences of M_A of length x (N_x^A) is given as

$$N_x^A \ = \ P_{AA}^{x-1} P_{AB} \tag{7}$$

where P_{AA} is the probability that $\text{wwwM}_A\cdot$ adds M_A

$$P_{AA} \ = \ \frac{k_{AA}[M_A\cdot][M_A]}{k_{AA}[M_A\cdot][M_A] \ + \ k_{AB}[M_A\cdot][M_B]} \ = \ \frac{r_A}{r_A \ + \ [M_B]/[M_A]} \tag{8}$$

and $P_{AB}(=1-P_{AA})$ is the probability that $\text{wwwM}_A\cdot$ adds M_B. The lengths of M_B sequences are determined in similar fashion, such that

$$N_x^B \ = \ P_{BB}^{x-1} P_{BA} \ = \ \frac{r_B}{r_B \ + \ [M_A]/[M_B]} \tag{9}$$

Thus equations (8) and (9) allow calculation of the comonomer sequence distribution from a knowledge of the terminal model reactivity ratios and the monomer feed composition. Use of equations (5), (8) and (9) is of course restricted to conditions under which the monomer feed composition is fixed. In practical terms, this requires the use of low monomer conversions in order to avoid serious errors arising from compositional drift as M_A and M_B enter the copolymer chain at different rates. Use of the integrated form of equation (5) has been recommended in order to relax the low conversion restriction (*cf.* Chapter 2 of this volume).

15.2.2 Experiment

The terminal model has proven remarkably successful in correlating a large body of copolymerization data. Greenley[7] has provided the most recent tabulation of free radical reactivity ratios, in which some 900 ratios were recalculated, *via* the equations of Kelen and Tudos,[8,9] from the original experimental data. Table 1 lists some representative reactivity ratios for several of the most important vinyl monomers.

Table 1 Monomer Reactivity Ratios in Radical Copolymerization[a]

M_A	r_A	M_B	r_B	*Ref.*
Acrylonitrile	0.030–0.100	Butadiene	0.10–0.45	10–14
	7.00	Ethylene	0.00	15
	6.00	Maleic anhydride	0.00	16
	0.14	Methyl methacrylate	1.32	17
	0.00–0.17	Styrene	0.29–0.55	18–27
	4.05–5.51	Vinyl acetate	0.040–0.060	28, 29
	2.55–4.00	Vinyl chloride	0.020–0.070	30–34
Butadiene	0.50–0.75	Methyl methacrylate	0.027–0.32	35–37
	1.35–1.83	Styrene	0.37–0.84	14, 37–44
	8.80	Vinyl chloride	0.04	45
Ethylene	0.40	Maleic anhydride	0.00	46
	0.050	Styrene	14.9	47
	0.13–0.88	Vinyl acetate	0.72–3.74	48–53
	0.020–0.34	Vinyl chloride	0.96–4.38	54–62
Maleic anhydride	0.010–0.020	Methyl methacrylate	3.10–6.36	63–66
	0.000–0.020	Styrene	0.000–0.097	67–76
	0.40–0.67	Vinyl chloride	0.040–0.100	77
Methyl methacrylate	0.22–0.64	Styrene	0.28–0.62	78–97
	22.2–28.6	Vinyl acetate	0.030–0.070	98–100
	8.99	Vinyl chloride	0.070	101
Styrene	18.8–60.0	Vinyl acetate	0.010–0.16	102–104
	12.4–25.0	Vinyl chloride	0.005–0.160	32, 105–108
Vinyl acetate	0.24–0.98	Vinyl chloride	1.03–2.30	101, 102, 109–113

[a] Values shown represent a range selected from a listing of reactivity ratios provided by R. Z. Greenley as a personal communication to the author.

The data in Table 1 provide several important insights regarding the current status of our understanding of radical copolymerization. First, the observed reactivity patterns may be divided into a rather small number of classes, which may be distinguished from one another on the basis of the magnitudes of r_A and r_B. The most useful classification is as given below.

(i) $r_A \approx r_B \approx 1$ ($k_{AA} = k_{AB}$; $k_{BA} = k_{BB}$). Neither radical center shows substantial preference for either M_A or M_B, so that the relative rates of monomer consumption are determined only by the relative monomer concentrations in the feed mixture. Equation (5) simplifies to

$$\frac{d[M_A]}{d[M_B]} = \frac{[M_A]}{[M_B]} \tag{10}$$

and the copolymer and monomer feed compositions are thus identical. Of the copolymerization systems listed in Table 1, butadiene–styrene ($r_A = 1.35$–1.83; $r_B = 0.37$–0.84) and vinyl acetate–vinyl chloride ($r_A = 0.24$–0.98; $r_B = 1.03$–2.30) approach this pattern most closely. Each shows small but significant deviations, however, so that compositional drift with conversion becomes an important practical concern.

(ii) $r_A \approx r_B \approx 0$ ($k_{AA} \approx k_{BB} \approx 0$). Each of the radical centers shows a strong preference for cross-propagation. In the extreme case, the copolymer is perfectly alternating and of 1:1 composition, regardless of the composition of the monomer feed mixture. Equation (5) simplifies to

$$\frac{d[M_A]}{d[M_B]} = \frac{[M_A] [M_B]}{[M_B] [M_A]} = 1 \tag{11}$$

The copolymerization of maleic anhydride ($r_A = 0.00$–0.02) with styrene ($r_B = 0.00$–0.097) behaves in this manner.

(iii) $r_A > 1$; $r_B < 1$ ($k_{AA} > k_{AB}$; $k_{BA} > k_{BB}$). Each of the radical centers prefers to add M_A, so that the copolymer is always enriched in M_A relative to the feed. This situation arises frequently in radical copolymerization, and many examples may be found in Table 1. A special case is that in which $r_A r_B = 1$ ($k_{AA}/k_{AB} = k_{BA}/k_{BB}$); *i.e.* in which both active centers show the *same* preference for addition of one of the monomers. Equation (5) then becomes

$$\frac{d[M_A]}{d[M_B]} = r_A \frac{[M_A]}{[M_B]}$$ (12)

This behavior is often termed *ideal copolymerization*.

(iv) $r_A < 1$; $r_B < 1$ ($k_{AB} > k_{AA}$, $k_{BA} > k_{BB}$). Each of the radical centers prefers cross-propagation, but the preference is not absolute. This results in a tendency toward alternation, which grows stronger as r_A and r_B approach zero. The copolymerization of acrylonitrile ($r_A = 0.00$–0.17) with styrene ($r_B = 0.29$–0.55) provides a good example. A characteristic of such copolymerizations is the existence of the so-called *azeotropic composition*, at which the copolymer and feed compositions are equal. This situation arises when

$$\frac{d[M_A]}{d[M_B]} = \frac{[M_A]}{[M_B]}$$ (13)

which requires that

$$\frac{r_A[M_A] + [M_B]}{[M_A] + r_B[M_B]} = 1$$ (14)

so that

$$\frac{[M_A]}{[M_B]} = \frac{1 - r_B}{1 - r_A}$$ (15)

at the azeotropic point. The azeotropic composition is of some practical significance in that at this point compositional drift with conversion may be neglected. This allows batch copolymerizations to be run to high conversions without the introduction of substantial compositional heterogeneity into the product (*cf.* Chapter 2 of this volume).

The patterns of copolymerization behavior discussed above are summarized most succinctly in the form of composition curves in which copolymer composition (*e.g.* as mole fraction M_A in the copolymer) is plotted as a function of the monomer feed composition (as mole fraction M_A in the feed). Figure 1 shows schematic composition curves for each of the four classes of copolymerizations described above. Such composition curves, calculated *via* equation (5), reproduce in satisfactory fashion the experimental compositions obtained in the vast majority of radical copolymerizations. Thus the terminal model is extraordinarily useful as a context in which to describe the compositions of copolymers prepared from vinyl monomers of widely varying structure and reactivity. On the other hand, the model is not as successful in the prediction of comonomer sequence distribution[114] and overall copolymerization rate.[115]

A second point that becomes apparent on examination of Table 1 is that the experimentally reported reactivity ratios, even for these most common vinyl monomers, span some range. It is not at

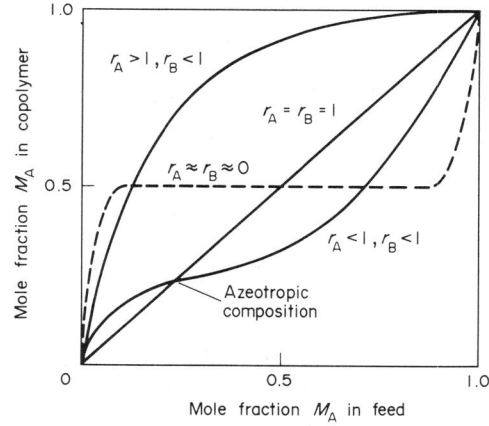

Figure 1   Copolymer composition (mole fraction M_A in copolymer) as a function of monomer feed composition (mole fraction M_A in feed) for various reactivity ratio combinations

all uncommon for the highest and lowest reported values to differ by a factor of two for a given copolymerization, and much larger variation may be found. The sources of experimental and statistical uncertainty in determining reactivity ratios are discussed in Chapter 2 of this volume.

15.3 PENULTIMATE MODEL

The fundamental assumption of the terminal model, *i.e.* that the reactivity of the growing radical is determined only by the identity of the last-added monomer unit, is equivalent to the assumption that the relative rates of monomer addition are insensitive to substitution at positions more remote than that β to the radical center. Remote substituent effects are well known in organic chemistry, and it is plausible that copolymerization reactivity ratios should be affected by units that precede the terminal residue on the propagating macroradical. Merz, Alfrey and Goldfinger[116] suggested in 1946 that a proper description of the propagation step should take into account four distinct active centers, which are defined by the identities of their terminal and *penultimate* units

$$\sim\!\!M_A M_A\cdot \ + \ M_A \ \xrightarrow{\ k_{AAA}\ } \ \sim\!\!M_A M_A M_A\cdot \tag{16}$$

$$\sim\!\!M_A M_A\cdot \ + \ M_B \ \xrightarrow{\ k_{AAB}\ } \ \sim\!\!M_A M_A M_B\cdot \tag{17}$$

$$\sim\!\!M_B M_A\cdot \ + \ M_A \ \xrightarrow{\ k_{BAA}\ } \ \sim\!\!M_B M_A M_A\cdot \tag{18}$$

$$\sim\!\!M_B M_A\cdot \ + \ M_B \ \xrightarrow{\ k_{BAB}\ } \ \sim\!\!M_B M_A M_B\cdot \tag{19}$$

$$\sim\!\!M_A M_B\cdot \ + \ M_A \ \xrightarrow{\ k_{ABA}\ } \ \sim\!\!M_A M_B M_A\cdot \tag{20}$$

$$\sim\!\!M_A M_B\cdot \ + \ M_B \ \xrightarrow{\ k_{ABB}\ } \ \sim\!\!M_A M_B M_B\cdot \tag{21}$$

$$\sim\!\!M_B M_B\cdot \ + \ M_A \ \xrightarrow{\ k_{BBA}\ } \ \sim\!\!M_B M_B M_A\cdot \tag{22}$$

$$\sim\!\!M_B M_B\cdot \ + \ M_B \ \xrightarrow{\ k_{BBB}\ } \ \sim\!\!M_B M_B M_B\cdot \tag{23}$$

The copolymer composition is determined by the relative rates of monomer consumption

$$\frac{d[M_A]}{d[M_B]} = \frac{k_{AAA}[M_A M_A\cdot][M_A] \ + \ k_{BAA}[M_B M_A\cdot][M_A] \ + \ k_{ABA}[M_A M_B\cdot][M_A] \ + \ k_{BBA}[M_B M_B\cdot][M_A]}{k_{AAB}[M_A M_A\cdot][M_B] \ + \ k_{BAB}[M_B M_A\cdot][M_B] \ + \ k_{ABB}[M_A M_B\cdot][M_B] \ + \ k_{BBB}[M_B M_B\cdot][M_B]}$$

Assumption of steady state concentrations of each of the four radical centers leads to equation (25) for the copolymer composition

$$\frac{d[M_A]}{d[M_B]} = \frac{1 \ + \ \dfrac{r_{BA}X(r_{AA}X \ + \ 1)}{r_{BA}X \ + \ 1}}{1 \ + \ \dfrac{r_{AB}(r_{BB} \ + \ X)}{X(r_{AB} \ + \ X)}} \tag{25}$$

where $X = [M_A]/[M_B]$ and the reactivity ratios are defined as

$$r_{AA} \ = \ \frac{k_{AAA}}{k_{AAB}} \qquad r_{BA} \ = \ \frac{k_{BAA}}{k_{BAB}} \qquad r_{AB} \ = \ \frac{k_{ABB}}{k_{ABA}} \qquad r_{BB} \ = \ \frac{k_{BBB}}{k_{BBA}} \tag{26}$$

Prediction of monomer sequence lengths by the penultimate model is conceptually identical to that described previously for the terminal model. The probability (P_{BAA}) that an $\sim\!\!M_B M_A\cdot$ chain end adds M_A is

$$P_{BAA} \ = \ \frac{k_{BAA}[M_B M_A\cdot][M_A]}{k_{BAA}[M_B M_A\cdot][M_A] \ + \ k_{BAB}[M_B M_A\cdot][M_B]} \ = \ \frac{[M_A]}{[M_A] \ + \ [M_B]/r_{BA}} \tag{27}$$

and

$$P_{AAA} = \frac{[M_A]}{[M_A] + [M_B]/r_{AA}} \tag{28}$$

A sequence consisting of an isolated M_A unit arises only when an $\sim\sim M_B M_A \cdot$ chain end adds M_B, so the number fraction of M_A sequences that are of length one is

$$N_1^A = 1 - P_{BAA} \tag{29}$$

For longer sequences, enumeration of the required propagation steps leads to

$$N_x^A = P_{BAA} P_{AAA}^{x-2}(1 - P_{AAA}) \tag{30}$$

Equations (25), (29) and (30) for the penultimate model are thus equivalent to equations (5) and (8) developed previously for the terminal model; in each case, knowledge of the copolymerization reactivity ratios allows calculation of copolymer compositions and sequences as functions of the ratio of monomer concentrations in the feed.

15.4 COMPLEX PARTICIPATION MODEL

Radical copolymerizations of electron rich alkenes with electron poor alkenes are anomalous in several respects. Such monomer pairs often afford alternating copolymers over the entire range of feed compositions,[117] and one often observes in such systems a marked sensitivity of the overall copolymerization rate to temperature, solvent and monomer concentration.[118,119] Butler and co-workers have also noted anomalies in the stereochemistry[120-122] and regiochemistry[123] of certain copolymerizations of electron rich and electron poor alkenes.

A mechanistic scheme that accounts for this behavior invokes the participation of 1:1 alkenic electron donor–acceptor (EDA) complexes in the propagation step. Specifically, it is proposed that the 1:1 complex $(\overline{M_A M_B})$ competes with free monomers for the growing chain end. Modification of the terminal model in this way requires consideration of eight propagation steps

$$\sim M_A\cdot + M_A \xrightarrow{k_{AA}} \sim M_A M_A\cdot \tag{31}$$

$$\sim M_A\cdot + M_B \xrightarrow{k_{AB}} \sim M_A M_B\cdot \tag{32}$$

$$\sim M_B\cdot + M_A \xrightarrow{k_{BA}} \sim M_B M_A\cdot \tag{33}$$

$$\sim M_B\cdot + M_B \xrightarrow{k_{BB}} \sim M_B M_B\cdot \tag{34}$$

$$\sim M_A\cdot + \overline{M_A M_B} \xrightarrow{k_{A\overline{AB}}} \sim M_A M_A M_B\cdot \tag{35}$$

$$\sim M_A\cdot + \overline{M_B M_A} \xrightarrow{k_{A\overline{BA}}} \sim M_A M_B M_A\cdot \tag{36}$$

$$\sim M_B\cdot + \overline{M_A M_B} \xrightarrow{k_{B\overline{AB}}} \sim M_B M_A M_B\cdot \tag{37}$$

$$\sim M_B\cdot + \overline{M_B M_A} \xrightarrow{k_{B\overline{BA}}} \sim M_B M_B M_A\cdot \tag{38}$$

if radical additions to each 'side' of the complex are regarded as distinct, and a complexation equilibrium

$$M_A + M_B \underset{}{\overset{K}{\rightleftharpoons}} \overline{M_A M_B} \tag{39}$$

An analysis of this mode, which predicts copolymer composition and sequence as functions of the feed composition, has been provided by Hill and co-workers.[124] The mole ratio of M_A to M_B in the copolymer is given as

$$\frac{d[M_A]}{d[M_B]} = \frac{(1 - P_{BB})(P_{AB} + P_{A\overline{AB}}) + (1 - P_{AB})(P_{BA} + P_{B\overline{BA}})}{(1 - P_{BA})(P_{AB} + P_{A\overline{AB}}) + (1 - P_{AA})(P_{BA} + P_{B\overline{BA}})} \tag{40}$$

where the transition probabilities are defined as

$$\left.\begin{array}{ll}
P_{BB} = r_B[M_B]/\Sigma M_B & P_{BA} = [M_A]/\Sigma M_B \\[2mm]
P_{B\overline{AB}} = s_B[\overline{M_A M_B}]\Sigma M_B & P_{B\overline{BA}} = s_B q_B[\overline{M_A M_B}]/\Sigma M_B \\[2mm]
P_{AB} = [M_A]/\Sigma M_A & P_{AA} = r_A[M_A]/\Sigma M_A \\[2mm]
P_{A\overline{AB}} = s_A q_A[\overline{M_A M_B}]/\Sigma M_A & P_{A\overline{BA}} = s_A[\overline{M_A M_B}]/\Sigma M_A
\end{array}\right\} \tag{41}$$

with

$$\Sigma M_A = [M_A] + r_A[M_A] + s_A[\overline{M_A M_B}][1 + q_A] \tag{42}$$

and

$$\Sigma M_B = r_B[M_B] + [M_A] + s_B[\overline{M_A M_B}][1 + q_B] \tag{43}$$

The reactivity ratios in this formulation are defined as

$$r_A = k_{AA}/k_{AB} \qquad r_B = k_{BB}/k_{BA} \tag{44}$$

$$q_A = k_{A\overline{AB}}/k_{A\overline{BA}} \qquad q_B = k_{B\overline{BA}}/k_{B\overline{AB}} \tag{45}$$

$$s_A = k_{A\overline{BA}}/k_{AB} \qquad s_B = k_{B\overline{AB}}/k_{BA} \tag{46}$$

so that equation (40) specifies the copolymer composition in terms of monomer concentrations and seven parameters (six reactivity ratios and the complexation equilibrium constant, K).

Sequence information can be calculated in the usual manner, *i.e.* as the number fraction of sequences of either monomer of length x. For M_A, the number fraction of sequences of length x is

$$N_x^A = \frac{p_x^A}{\displaystyle\sum_{m=1}^{\infty} p_m^A} \tag{47}$$

where

$$p_1^A = P_B \frac{[(1 - P_{AA})(P_{AB} + P_{A\overline{BA}})(P_{BA} + P_{B\overline{BA}}) + P_{B\overline{AB}}(P_{A\overline{AB}} + P_{AB})]}{P_{A\overline{AB}} + P_{AB}} \tag{48}$$

and

$$p_x^A = P_B P_{AA}^{x-2} \frac{[(1 - P_{AA})(P_{BA} + P_{B\overline{BA}})(P_{AB} P_{AA} + P_{A\overline{BA}} P_{AA} + P_{A\overline{AB}})]}{(P_{A\overline{AB}} + P_{AB})} \tag{49}$$

The quantity P_B is the probability of selecting an M_B unit that entered the chain either as free M_B or *via* the reactions shown in equations (35) and (37). However, P_B need not be evaluated, since this quantity may be eliminated from the expression (equation 47) for N_x^A. Equations (40) and (47) thus allow calculation of copolymer composition and sequence as functions of the monomer feed composition.

15.5 OTHER COPOLYMERIZATION MODELS

The terminal, penultimate and complex participation models have been discussed widely in the copolymerization literature. Two additional models — the complex dissociation model and the depropagation model — have not been considered as extensively, but each is physically plausible and each has been analyzed in sufficient detail that compositions and sequences may be calculated. These models are outlined briefly here; the reader is directed to the original papers for a thorough description.

15.5.1   Complex Dissociation Model

Tsuchida and Tomono[125] suggested in 1971 that EDA complexes may take part in radical copolymerizations not by adding to the chain end in a concerted fashion, but rather by delivering only one of the two complexed monomers. Thus the terminal model must be modified by consideration of four new propagation steps

$$\sim\sim M_A\cdot \; + \; \overline{M_A M_B} \; \longrightarrow \; \sim\sim M_A M_A\cdot \; + \; M_B \tag{50}$$

$$\sim\sim M_A\cdot \; + \; \overline{M_A M_B} \; \longrightarrow \; \sim\sim M_A M_B\cdot \; + \; M_A \tag{51}$$

$$\sim\sim M_B\cdot \; + \; \overline{M_A M_B} \; \longrightarrow \; \sim\sim M_B M_A\cdot \; + \; M_B \tag{52}$$

$$\sim\sim M_B\cdot \; + \; \overline{M_A M_B} \; \longrightarrow \; \sim\sim M_B M_B\cdot \; + \; M_A \tag{53}$$

Hill and co-workers[126] have provided an analysis of this kinetic scheme, and have demonstrated the calculation of copolymer composition and sequence according to this model.

15.5.2   Copolymerization with Depropagation

Most radical copolymerizations are strongly exothermic and effectively irreversible. Near the ceiling temperature, however, the influence of depropagation must be considered. Lowry[127] in 1960 developed a general theory that predicts copolymer composition for systems in which the addition of one of the two monomers is reversible. O'Driscoll and co-workers[128] subsequently derived composition equations equivalent to but more general than those of Lowry, and provided expressions for sequence distributions as well.

15.6   EVALUATION OF COPOLYMERIZATION MODELS

Three experimental approaches have been used to evaluate the theoretical treatments of copolymerization discussed above. The majority of such studies have compared measured compositions and sequences with those predicted by each of the kinetic models. But, in fact, the determination of sequence distributions is still a considerable technical challenge, so that many investigators have been limited to composition measurements alone. More recently, measurements of absolute rate constants and trapping experiments on simple model radicals have been brought to bear on questions of copolymerization mechanism.

15.6.1   Composition and Sequence

It was recognized early in the study of radical copolymerization that sequence distribution should be more sensitive than composition to the details of the chain growth process. In their original paper on the penultimate model,[116] Merz, Alfrey and Goldfinger pointed out that it is not in measurements of composition, but rather 'in the length of the [comonomer] sequences that the effect of the monomer in the chain preceding the free radical chain end would become noticeable . . .'. Berger and Kuntz[129] in 1964 analyzed this problem quantitatively for several hypothetical and several real copolymerizations. They showed, for example, that the compositions predicted by the terminal and penultimate models would be indistinguishable over a 1000-fold variation in $[M_A]/[M_B]$, for a system in which the terminal model reactivity ratios are $r_A = 0.1$ and $r_B = 0.9$ and the penultimate model parameters are $r_{AA} = 0.94$, $r_{BA} = 0.01$, $r_{BB} = 0.9$ and $r_{AB} = 5$. Thus quite large penultimate effects ($r_{AA}/r_{BA} = 94$) can be masked. On the other hand, much more modest effects are readily apparent in the predicted sequence distributions. Analysis of the copolymerization of styrene (M_A) and maleic anhydride (M_B) according to the terminal ($r_A = 0.0227$, $r_B = 0$) or penultimate ($r_{AA} = 0.017$, $r_{BA} = 0.063$, $r_{AB} = r_{BB} = 0$) models leads to rather different predictions. In the terminal model analysis, the number fraction of styrene residues isolated between maleic anhydride units is 0.23; in the penultimate model analysis, 0.09. Thus a penultimate effect of a factor of four leads to sequence predictions markedly different from those of the terminal model.

More recently, Hill and co-workers have analyzed the bulk copolymerization of styrene and acrylonitrile in terms of the composition and sequence predictions of the terminal, penultimate and complex participation models.[114] They find that all three models reproduce the experimental composition data rather well, although the penultimate and complex models offer statistically

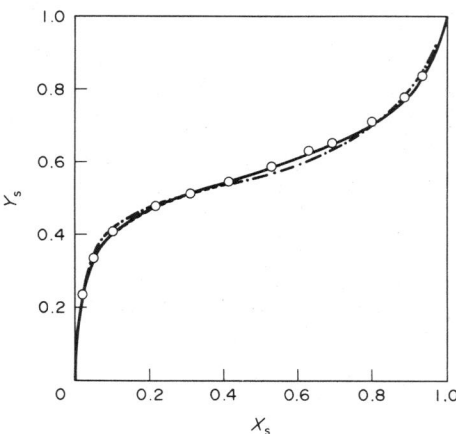

Figure 2 Copolymer composition curve for the copolymerization of acrylonitrile and styrene in bulk at 60 °C. Y_S = mole fraction styrene in copolymer; X_S = mole fraction styrene in comonomer feed. ○, experimental data; ——, penultimate model and complex model with no restriction; − − − −, complex model with equilibrium constant for complexation fixed at 0.52; — · — · —, terminal model (reprinted with permission of The American Chemical Society from ref. 114)

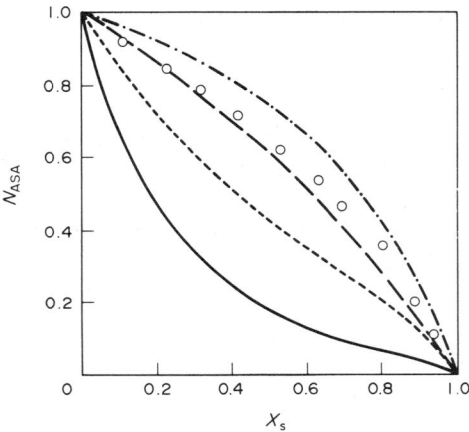

Figure 3 Number fraction of styrene sequences of length one (N_{ASA}) in copolymers of acrylonitrile and styrene, as a function of the mole fraction of styrene in the monomer feed mixture (X_S). ○, experimental; ——, complex model with no restriction; − − − −, complex model with equilibrium constant for complexation fixed at 0.52; — —, penultimate model; — · — · —, terminal model (reprinted with permission of The American Chemical Society from ref. 114)

significant improvements over the terminal model. Figure 2 shows the experimental compositions as well as the best-fit predictions of each of the three models. The predictions are remarkably similar, and data of very high precision are required for model discrimination. The sequence predictions of the three models are quite different, however, and allow a clear distinction between the penultimate and complex kinetic schemes (Figure 3). Hill and co-workers conclude that the bulk copolymerization of styrene and acrylonitrile is best described by a penultimate model with $r_{SS} = 0.23$, $r_{AS} = 0.63$, $r_{SA} = 0.09$ and $r_{AA} = 0.04$.

Determination of copolymer sequence has also provided insight into the effects of solvent on radical copolymerization. Harwood and co-workers have noted that copolymerizations that involve ionic, highly polar or hydrogen-bonding monomers are subject to large solvent effects (*i.e.* the composition curves for such copolymerizations vary dramatically with solvent).[130] Furthermore, sequence distributions determined in such systems are inconsistent with the predictions of *any* of the conventional kinetic schemes, if one uses reactivity ratios determined from the relation between copolymer composition and monomer feed composition. On the other hand, comparison of the sequence distributions of copolymers of identical composition (but prepared in different solvents from feeds of different $[M_A]/[M_B]$) shows them to be identical. Harwood concludes that the conditional probabilities governing monomer addition (and therefore the reactivity ratios) must be independent of solvent, and that the role of the solvent is to influence the relative concentrations of monomers available to the growing chain end. Harwood has provided convincing evidence for this behavior in the copolymerizations of styrene with acrylic acid, methacrylic acid and acrylamide, and in the copolymerization of vinylidene chloride with methacrylonitrile.

15.6.2 Measurements of Absolute Rate Constants

Each of the copolymerization models predicts not only composition and sequence, but also the overall propagation rate constant, k_p, as a function of monomer feed composition. Fukuda and co-workers[115] used a rotating sector technique to determine k_p over a range of feed compositions for the copolymerization of styrene and methyl methacrylate. They found large and systematic deviations from the predictions of the terminal model, but were able to reproduce their experimental observations by assuming a small penultimate effect. This is an intriguing observation in view of the 'classic' nature of styrene/methyl methacrylate copolymerization, and suggests that absolute rate measurements may prove particularly powerful in probing copolymerization mechanisms.

15.6.3 Model Reactions

The mechanistic assumptions of the terminal, penultimate and complex participation models may be evaluated *via* trapping experiments with simple alkyl radicals. For example, the penultimate model, as applied to the copolymerization of monosubstituted alkenes, implies that the selectivity of the attacking radical should be sensitive to the nature of the substituent that lies γ to the radical center

$$\sim\sim\sim \underset{\underset{\gamma}{|}}{CH_2}CH\underset{\underset{\alpha}{|}}{CH_2}CH\cdot \qquad (1)$$

Tirrell and co-workers have determined the relative rates of addition of acrylonitrile and styrene (k_A/k_S) to a series of γ-substituted propyl radicals,[131] and report that a γ-cyano group depresses the relative affinity of the radical center for acrylonitrile by a factor of 3.5. This 'penultimate effect' is remarkably consistent with those inferred by Hill and co-workers *via* composition and sequence analyses,[130] and lends support to the penultimate model as a physically realistic description of the copolymerization of styrene and acrylonitrile. Analogous measurements of k_A/k_S for the 1-phenyl-ethyl[132] and 1-cyanoethyl[133] radicals are consistent with this view.

Trapping experiments have also been applied to the evaluation of the complex participation model. The hypothesis that alkenic EDA complexes add in concerted fashion to alkyl radicals — the fundamental assertion of the model — is subject to direct experimental test, as shown in equations (54)–(57)

$$R\cdot \begin{cases} \xrightarrow{M_A} RM_A\cdot \xrightarrow{T} RM_AT \qquad (54) \\ \qquad\qquad\qquad\quad (2) \\ \xrightarrow{M_B} RM_B\cdot \xrightarrow{T} RM_BT \qquad (55) \\ \qquad\qquad\qquad\quad (3) \\ \xrightarrow{\overline{M_AM_B}} RM_AM_B\cdot \xrightarrow{T} RM_AM_BT \qquad (56) \\ \qquad\qquad\qquad\qquad\quad (4) \\ \xrightarrow{\overline{M_BM_A}} RM_BM_A\cdot \xrightarrow{T} RM_BM_AT \qquad (57) \\ \qquad\qquad\qquad\qquad\quad (5) \end{cases}$$

T = radical trap

The radical of interest, $R\cdot$, is generated in the presence of M_A, M_B and a radical trap, T. If $R\cdot$ undergoes concerted complex addition, trapping of the simple alkene adducts $RM_A\cdot$ and $RM_B\cdot$ will not be observed. Determination of the yields of products (2) and (3) then allows an estimate of the maximum extent to which the complex participates in the consumption of M_A and M_B. Jones and Tirrell have reported trapping experiments on the 1-butyl radical in its reactions with *N*-phenyl-maleimide and two donor alkenes (2-chloroethyl vinyl ether and styrene).[134, 135] In each case, simple addition of *N*-phenylmaleimide was the dominant reaction; no evidence for concerted addition of the EDA complex was obtained.

15.7 REFERENCES

1. F. R. Mayo and F. M. Lewis, *J. Am. Chem. Soc.*, 1944, **66**, 1594.
2. T. Alfrey, Jr. and G. Goldfinger, *J. Chem. Phys.*, 1944, **12**, 205.
3. F. T. Wall, *J. Am. Chem. Soc.*, 1944, **66**, 2050.
4. H. Dostal, *Monatsch. Chem.*, 1936, **69**, 424.
5. R. G. W. Norrish and E. F. Brookman, *Proc. R. Soc. London, Ser. A*, 1939, **171**, 147.
6. E. Jenckel, *Z. Phys. Chem., Abt. A*, 1942, **190**, 24.
7. R. Z. Greenley, *J. Macromol. Sci., Chem.*, 1980, **14**, 445.
8. T. Kelen and F. Tudos, *J. Macromol. Sci., Chem.*, 1975, **9**, 1.
9. F. Tudos, T. Kelen, T. Foldes-Bereznich and B. Turcsanyi, *J. Macromol. Sci., Chem.*, 1976, **10**, 1513.
10. W. H. Embree, J. M. Mitchell and H. L. Williams, *Can. J. Chem.*, 1951, **29**, 253.
11. H. Asai and T. Imoto, *J. Polym. Sci., Part B*, 1964, **2**, 553.
12. F. T. Wall, R. W. Powers, G. D. Sands and G. S. Stent, *J. Am. Chem. Soc.*, 1948, **70**, 1031.
13. I. Sakurada, T. Okada, S. Hatakeyama and F. Kimura, *J. Polym. Sci., Part C*, 1963, **1**, 1233.
14. H. R. Hennery-Logan and R. V. V. Nichols, quoted by F. R. Mayo and C. Walling, *Chem. Rev.*, 1950, **46**, 191.
15. V. F. Gromov, P. M. Khomikovskii and A. D. Abkin, *Vysokomol. Soedin.*, 1961, **3**, 1015.
16. F. R. Mayo, F. M. Lewis and C. Walling, *J. Am. Chem. Soc.*, 1948, **70**, 1529.
17. C. Simionescu, N. Asandei and A. Liga, *Makromol. Chem.*, 1967, **110**, 278.
18. W. M. Ritchey and L. E. Ball, *J. Polym. Sci., Part B*, 1966, **4**, 557.
19. F. M. Lewis, C. Walling, W. Cummings, E. R. Briggs and W. J. Wenisch, *J. Am. Chem. Soc.*, 1948, **70**, 1527.
20. A. Chapiro and A.-M. Jendrychowska-Bonamour, *J. Polym. Sci., Part C*, 1963, **1**, 1211.
21. K. W. Doak, *J. Am. Chem. Soc.*, 1950, **72**, 4681.
22. S. Tazuke and S. Okamura, *J. Polym. Sci., Part A-1*, 1968, **6**, 2907.
23. K. R. Hennery-Logan and R. V. V. Nichols, quoted by F. R. Mayo and C. Walling, *Chem. Rev.*, 1950, **46**, 191.
24. K. W. Doak, M. A. Deahl and I. H. Christmas, '137th ACS Meeting Abstract Papers', Cleveland, Ohio, April 1960, vol. 1, no. 1, p. 151.
25. N. L. Zutty and R. D. Burkhart, *Adv. Chem. Ser.*, 1962, **34**, 52.
26. J. Asakura, M. Yoshihara, Y. Matsubara and T. Maeshima, *J. Macromol. Sci., Chem.*, 1981, **15**, 1473.
27. Yu. D. Semchikov, L. A. Smirnova, T. E. Knyazeva, S. A. Bulgakova, G. A. Voskoboinik and V. I. Sherstyanykh, *Vysokomol. Soedin., Ser A*, 1984, **26**, 704.
28. F. R. Mayo, C. Walling, F. M. Lewis and W. F. Hulse, *J. Am. Chem. Soc.*, 1948, **70**, 1523.
29. T. Alfrey, J. Bohrer, H. Haas and C. Lewis, *J. Polym. Sci.*, 1950, **5**, 719.
30. N. Ashikari and A. Nishimura, *J. Polym. Sci.*, 1958, **31**, 250.
31. E. C. Chapin, G. E. Ham and R. G. Fordyce, *J. Am. Chem. Soc.*, 1948, **70**, 538.
32. B. R. Thompson and R. H. Raines, *J. Polym. Sci.*, 1959, **41**, 265.
33. E. W. Rugeley, T. A. Field and G. H. Fremon, *Ind. Eng. Chem.*, 1948, **40**, 1724.
34. Kh. U. Usmanov, A. A. Yul'chibaev and Kh. Yuldasheva, *Uzb. Khim. Zh.*, 1967, **11**, 27.
35. C. Walling and J. A. Davison, *J. Am. Chem. Soc.*, 1951, **73**, 5736.
36. M. F. Margaritova and V. A. Raiskaya, *Tr. Mosk. Khim.-Tekhnol. Inst. im. DI Mendeleeva*, 1953, **4**, 37.
37. F. M. Lewis, C. Walling, W. Cummings, E. R. Briggs and W. J. Wenisch, *J. Am. Chem. Soc.*, 1948, **70**, 1527.
38. R. D. Gilbert and H. L. Williams, *J. Am. Chem. Soc.*, 1952, **74**, 4114.
39. I. Crescentini, G. B. Getchele and A. Zanella, *J. Appl. Polym. Sci.*, 1965, **9**, 1323.
40. E. J. Meehan, *J. Polym. Sci.*, 1946, **1**, 318.
41. R. J. Orr and H. L. Williams, *Can. J. Chem.*, 1952, **30**, 108.
42. R. J. Orr, *Polymer*, 1961, **2**, 79.
43. T. Imoto, Y. Ogo, S. Goto and T. Mitani, *Kogyo Kagaku Zasshi*, 1966, **69**, 1371.
44. J. M. Mitchell and H. L. Williams, *Can. J. Res.*, 1949, **27**, 35.
45. G. V. Tkachenko, P. H. Khomikovskii, A. D. Abkin and S. S. Medvedev, *Zh. Fiz. Khim.*, 1957, **31**, 242.
46. R. A. Terteryan and V. S. Khrapov, *Vysokomol. Soedin., Ser A*, 1983, **25**, 1850.
47. R. A. Terteryan and S. D. Livshits, *Sb. Nauchn. Tr.-Vses. Nauchno-Issled. Inst. Pererab. Nefti*, 1982, **41**, 118.
48. R. D. Burkhart and N. L. Zutty, *J. Polym. Sci., Part A*, 1963, **1**, 1137.
49. B. Erusalimsky, N. Tumarkin, F. Duntoff, S. Lyubetzky and A. Goldenberg, *Makromol. Chem.*, 1967, **104**, 288.
50. M. Ratzch, W. Schneider and D. Musche, *J. Polym. Sci., Part A-1*, 1971, **9**, 785.
51. A. L. Gol'denberg and S. G. Lubetskii, *Dokl. Akad. Nauk SSSR*, 1968, **179**, 900.
52. A. L. German and D. Heikens, *J. Polym. Sci., Part A-1*, 1971, **9**, 2225.
53. R. Van der Meer and A. L. German, *J. Polym. Sci., Polym. Chem. Ed.*, 1979, **17**, 571.
54. R. D. Burkhart and N. L. Zutty, *J. Polym. Sci., Part A*, 1963, **1**, 1137.
55. F. I. Duntov and B. L. Erusalimskii, *Vyskomol. Soedin.*, 1965, **7**, 1075.
56. F. I. Duntov, A. L. Gol'denberg, M. A. Litvinova and B. L. Erusalimskii, *Vysokomol. Soedin., Ser. A*, 1967, **9**, 1920.
57. B. L. Erusalimskii, N. Tumarkin, F. Duntoff, S. Lyubetzky and A. Gol'denberg, *Makromol. Chem.*, 1967, **104**, 288.
58. A. Misono and Y. Uchida, *Bull. Chem. Soc. Jpn.*, 1966, **39**, 2458.
59. T. Otsu, Y. Kinoshita and A. Nakamachi, *Makromol. Chem.*, 1968, **115**, 275.
60. T. Otsu, T.-C. Lai, Y. Khinoshita, A. Nakamachi and M. Imoto, *Kogyo Kagaku Zasshi*, 1968, **71**, 904.
61. A. Misono, Y. Uchida and K. Yamada, *Bull. Chem. Soc. Jpn.*, 1967, **40**, 2366.
62. C. E. Wilkes, J. C. Westfahl and R. H. Backderf, *Polym. Prepr., Am. Chem. Soc., Div. Polym. Chem.*, 1967, **8**, 386.
63. D. C. Blackley and H. W. Melville, *Makromol. Chem.*, 1956, **18**, 16.
64. A. M. North and D. Postlethwaite, *Polymer*, 1964, **5**, 237.
65. M. C. DeWilde and G. Smets, *J. Polym. Sci.*, 1950, **5**, 253.
66. E. Tsuchida, T. Shimomura, K. Fujimora, Y. Ohtani and I. Shinohara, *Kogyo Kagaku Zasshi*, 1966, **70**, 1230.
67. T. Alfrey and E. Lavin, *J. Am. Chem. Soc.*, 1945, **67**, 2044.
68. T. L. Ang and H. J. Harwood, *Polym. Prepr., Am. Chem. Soc., Div. Polym. Chem.*, 1964, **5**, 306.
69. M. B. Huglin, *Polymer*, 1962, **3**, 335.

70. K. Uehara, T. Nishi, T. Tsuyuri, F. Tamura and N. Murata, *Kogyo Kagaku Zasshi*, 1967, **70**, 750.
71. C. H. Bamford and W. G. Barb, *Discuss. Faraday Soc.*, 1953, **14**, 208.
72. C. B. Chapman and L. Valentine, *J. Polym. Sci.*, 1959, **34**, 319.
73. E. Tsuchida, Y. Ohtani, H. Nakadai and I. Shinohara, *Kogyo Kagaku Zasshi*, 1967, **70**, 573.
74. T. V. Sheremeteva and G. N. Larina, *Dokl. Akad. Nauk SSSR*, 1965, **162**, 1323.
75. K. Noma, M. Niwa and K. Iwasaki, *Kobunshi Kagaku Zasshi*, 1963, **20**, 646.
76. C. C. Price and J. G. Walsh, *J. Polym. Sci.*, 1951, **6**, 239.
77. E. Tsuchida and T. Kawagoe, *Enka Biniiru To Porima*, 1968, **8**, 21.
78. K. Uehara, T. Nishi, T. Tsuyuri, F. Tamura and N. Murata, *Kogyo Kagaku Zasshi*, 1967, **70**, 750.
79. T. Ito and T. Otsu, *J. Macromol. Sci., Chem.*, 1969, **3**, 197.
80. F. T. Wall, R. E. Florin and C. J. Delbecq, *J. Am. Chem. Soc.*, 1950, **72**, 4769.
81. V. P. Zubov, L. I. Valuev, V. A. Kabanov and V. A. Kargin, *J. Polym. Sci., Part A-1*, 1971, **9**, 833.
82. E. L. Madruga, J. San Roman and M. A. Del Puerto, *J. Macromol. Sci., Chem.*, 1979, **13**, 1105.
83. C. Simionescu, B. C. Simionescu and S. Ioan, *J. Macromol. Sci., Chem.*, 1985, **22**, 765.
84. C. C. Price and J. G. Walsh, *J. Polym. Sci.*, 1951, **6**, 239.
85. E. J. Arlman, H. W. Melville and L. Valentine, *Recl. Trav. Chim. Pays-Bas*, 1949, **68**, 945.
86. F. T. Wall, quoted by F. R. Mayo and C. Walling, *Chem. Rev.*, 1950, **46**, 191.
87. A. P. Sheinker and A. D. Abkin, *Tr. Tashk. Konf. Mirnomu Ispol'z. At. Energ.*, 1961, **1**, 395.
88. T. Otsu, T. Ito and M. Imoto, *Kogyo Kagaku Zasshi*, 1966, **69**, 986.
89. L. S. Luskin and R. J. Myers, in 'Encyclopedia of Polymer Science and Technology,' Interscience, New York, 1964, vol. 1, p. 246.
90. F. M. Lewis, C. Walling, W. Cummings, E. R. Briggs and F. R. Mayo, *J. Am. Chem. Soc.*, 1948, **70**, 1519.
91. A. Guyot and J. Guillot, *J. Chem. Phys.*, 1964, **61**, 1434.
92. J. B. Kinsinger, J. S. Bartlett and N. H. Rauscher, *J. Appl. Polym. Sci.*, 1962, **6**, 529.
93. R. H. Wiley and B. Davis, *J. Polym. Sci.*, 1962, **62**, S-132.
94. F. M. Lewis, F. R. Mayo and W. F. Hulse, *J. Am. Chem. Soc.*, 1945, **67**, 1701.
95. H. Fujihara, K. Yamazaki, Y. Matsubara, M. Yashihara and T. Maeshima, *J. Macromol. Sci., Chem.*, 1979, **13**, 1081.
96. K. F. O'Driscoll, L. T. Kale, L. H. Rubio and P. M. Reilly, *J. Polym. Sci., Polym. Chem. Ed.*, 1984, **22**, 2777.
97. S. A. Chen and L. C. Tsai, *Makromol. Chem.*, 1986, **187**, 653.
98. J. C. Bevington and M. Johnson, *Eur. Polym. J.*, 1968, **4**, 669.
99. J. N. Atherton and A. M. North, *Trans. Faraday Soc.*, 1962, **58**, 2049.
100. Min Szu-Kwei and Chen Ho Chu, *Hua Hsueh Hsueh Pao*, 1957, **23**, 262.
101. P. Argon, T. Alfrey, J. Bohrer, H. Haas and H. Wechsler, *J. Polym. Sci.*, 1948, **3**, 157.
102. F. R. Mayo, C. Walling, F. M. Lewis and W. F. Hulse, *J. Am. Chem. Soc.*, 1948, **70**, 1523.
103. T. Nakata, T. Otsu and M. Imoto, *J. Polym. Sci., Part A*, 1965, **3**, 3383.
104. Yu. D. Semchikov, L. A. Smirnova, T. E. Knyazeva, S. A. Bulgakova, G. A. Voskoboinik and V. I. Sherstyanykh, *Vysokomol. Soedin., Ser. A*, 1984, **26**, 704.
105. E. C. Chapin, G. E. Ham and R. G. Fordyce, *J. Am. Chem. Soc.*, 1948, **70**, 538.
106. K. W. Doak, *J. Am. Chem. Soc.*, 1948, **70**, 1525.
107. J. Ulbricht, J. Giesemann and M. Gebauer, *Angew. Makromol. Chem.*, 1968, **3**, 69.
108. G. V. Tkachenko, V. S. Etlis, L. V. Stupen and L. P. Kofman, *Zh. Fiz. Khim.*, 1959, **33**, 25.
109. N. Grassie, I. C. McNeill and I. F. McLaren, *J. Polym. Sci., Part B*, 1965, **3**, 897.
110. T. Imoto, Y. Ogo and H. Nakamoto, *Bull. Chem. Soc. Jpn.*, 1968, **41**, 543.
111. C. S. Marvel, G. D. Jones, T. W. Mastin and G. L. Schertz, *J. Am. Chem. Soc.*, 1942, **64**, 2356.
112. K. Hayashi and T. Otsu, *Makromol. Chem.*, 1969, **127**, 54.
113. T. Kimura and K. Yoshida, *Kagaku to Kogyo (Osaka)*, 1953, **27**, 288.
114. D. J. T. Hill, J. H. O'Donnell and P. W. O'Sullivan, *Macromolecules*, 1982, **15**, 960.
115. T. Fukuda, Y. D. Ma and H. Inagaki, *Macromolecules*, 1985, **18**, 17.
116. E. Merz, T. Alfrey and G. Goldfinger, *J. Polym. Sci.*, 1946, **1**, 75.
117. J. M. G. Cowie, (ed.), 'Alternating Copolymers', Plenum Press, New York, 1985.
118. D. J. T. Hill, J. H. O'Donnell and P. W. O'Sullivan, *Prog. Polym. Sci.*, 1982, **8**, 215.
119. M. Yoshimura, H. Mikawa and Y. Shirota, *Macromolecules*, 1978, **11**, 1085.
120. K. G. Olson and G. B. Butler, *Macromolecules*, 1983, **16**, 707.
121. K. G. Olson and G. B. Butler, *Macromolecules*, 1984, **17**, 2480.
122. K. G. Olson and G. B. Butler, *Macromolecules*, 1984, **17**, 2486.
123. G. B. Butler, K. G. Olson and C. L. Tu, *Macromolecules*, 1984, **17**, 1884.
124. R. E. Cais, R. G. Farmer, D. J. T. Hill and J. H. O'Donnell, *Macromolecules*, 1979, **12**, 835.
125. E. Tsuchida and T. Tomono, *Makromol. Chem.*, 1971, **141**, 265.
126. D. J. T. Hill, J. H. O'Donnell and P. W. O'Sullivan, *Macromolecules*, 1983, **16**, 1295.
127. G. G. Lowry, *J. Polym. Sci.*, 1960, **17**, 463.
128. J. A. Howell, M. Izu and K. F. O'Driscoll, *J. Polym. Sci., Part A-1*, 1970, **8**, 699.
129. M. Berger and I. Kuntz, *J. Polym. Sci., Part A*, 1964, **2**, 1687.
130. H. J. Harwood, *Makromol. Chem., Macromol. Symp.*, 1987, **10/11**, 331.
131. S. A. Jones, G. S. Prementine and D. A. Tirrell, *Macromolecules*, 1986, **19**, 2908.
132. D. A. Cywar and D. A. Tirrell, *Macromolecules*, 1986, **19**, 2908.
133. G. S. Prementine and D. A. Tirrell, *Macromolecules*, 1987, **20**, 3034.
134. S. A. Jones and D. A. Tirrell, *Macromolecules*, 1986, **19**, 2080.
135. S. A. Jones and D. A. Tirrell, *J. Polym. Sci., Polym. Chem. Ed.*, 1987, **25**, 3177.

16

Rates of Copolymerization

DIETRICH BRAUN and WOJCIECH K. CZERWINSKI
Deutsches Kunststoff-Institut, Darmstadt, FRG

16.1 INTRODUCTION

Binary copolymerizations have been investigated for more than 40 years. Relations between monomer feed and copolymer composition were developed in the early 1940s by Alfrey, Goldfinger, Mayo, Lewis and Wall. In comparison with the many thousands of binary systems described in the literature since then using the so-called copolymerization equation and by reactivity ratios, there are rather few investigations of the kinetics and the rates of binary free-radical copolymerizations. The copolymerization of ternary monomer systems has received even less attention until now; the number of publications describing ternary polymerizations is not much higher than a hundred. Most authors primarily discussed problems concerning the composition of terpolymers and there is little information about the rates of ternary copolymerizations.[1]

16.2 EXPERIMENTAL METHODS

Rates of copolymerization can be measured by the usual methods, *e.g.* gravimetry of the formed polymer or consumption of monomers by gas chromatography. Rather simple and very useful are dilatometric measurements (see Volume 3, Chapter 5). The correlation between shrinkage and conversion in a polymerization reaction is given by[2]

$$U = \frac{100(\Delta V)}{KV}$$

where U is the conversion, V the initial volume, ΔV the decrease in volume and K the so-called conversion factor. K can be evaluated from this equation by direct gravimetric determination of the conversion for a polymerization of known shrinkage. A second method is based on the densities of monomer (ρ_M) and polymer (ρ_P)

$$K = (\rho_P - \rho_M)/\rho_P$$

The two equations are valid for homocomponent as well as multicomponent polymerizations, though in the latter case K is dependent on the composition of the monomer mixture. For binary copolymerizations of monomers M_A, M_B, Wittmer[3] developed the equation

$$K = K_{AA} B_{AA} + K_{BB} B_{BB} + K_{AB} B_{AB}$$

where B_{AA}, B_{BB} and B_{AB} are the frequencies of bonds $M_A M_A$, $M_B M_B$ and $M_A M_B$, and proved its validity for the system ethyl acrylate/styrene. Braun *et al.*[4] described the dilatometric behaviour of the three binary copolymerizations of the monomers benzyl methacrylate (BMA), styrene (S) and methyl methacrylate (MMA) by this relation.

In a ternary system of monomers M_A, M_B and M_C there may result bonds of the type $-M_A M_A-$, $-M_B M_B-$ and $-M_C M_C-$ as well as $-M_A M_B-$, $-M_A M_C-$ and $-M_B M_C-$ or the reverse combinations of these.

According to Wittmer's theory, the ii steps (i = A, B, C) are associated with the conversion factors of the homopolymerizations K_{ii} and the ij steps (i, j = A, B, C; i $\neq$ j) with the alternating conversion factors K_{ij} which can be evaluated from the three binary copolymerizations. The overall conversion factor in a terpolymerization will thus be combined from six increments[5]

$$K = K_{AA} B_{AA} + K_{BB} B_{BB} + K_{CC} B_{CC} + K_{AB} B_{AB} + K_{AC} B_{AC} + K_{BC} B_{BC}$$

The B_{ii} and B_{ij} values are the frequencies of the single bonding types in the terpolymer. These bonding frequencies can be evaluated from the nine possible binary reactions in a terpolymerization of the monomers M_A, M_B and M_C.[6] Thus by knowing the conversion factors K_{ii} for the homopolymerizations, K_{ij} for the binary copolymerizations and the B values, the overall conversion factor for a binary copolymerization or a terpolymerization can be calculated. The experimental and theoretical conversion factors for the ternary system BMA/S/MMA have been compared by Disselhoff.[5]

In binary systems where one of the comonomers does not homopolymerize, complete conversion is not attainable. In this case the relation between conversion and dilatometric constant K must be modified and an appropriate expression has been derived and experimentally verified for the system diethyl maleate (DEM)/styrene.[7]

16.3 BINARY COPOLYMERIZATION

16.3.1 Classical Rate Model

Several kinetic models for binary copolymerizations are described in the literature. The classical rate model developed by Melville *et al.*[8] is based on the assumption that the reactivity of the growing radical is controlled by the reactivity of the terminal chain component only. Under these conditions the following scheme of propagation steps is valid

$$-M_A{}^\bullet + M_A \xrightarrow{k_{pAA}} -M_A{}^\bullet \tag{1}$$

$$-M_A{}^\bullet + M_B \xrightarrow{k_{pAB}} -M_B{}^\bullet \tag{2}$$

$$-M_B{}^\bullet + M_A \xrightarrow{k_{pBA}} -M_A{}^\bullet \tag{3}$$

$$-M_B{}^\bullet + M_B \xrightarrow{k_{pBB}} -M_B{}^\bullet \tag{4}$$

Under steady state conditions and assuming that equation (5) is valid

$$k_{pAB}[-M_A{}^\bullet][M_B] = k_{pBA}[-M_B{}^\bullet][M_A] \tag{5}$$

Melville, Noble and Watson[8] derived the following rate equation for R_{cop}, the overall rate of copolymerization

$$R_{cop} = \frac{r_A[M_A]^2 + 2[M_A][M_B] + r_B[M_B]^2)R_i^{1/2}}{(r_A^2 \delta_A^2[M_A]^2 + 2\phi\delta_A\delta_B r_A r_B[M_A][M_B] + r_B^2 \delta_B^2[M_B]^2)^{1/2}} \tag{6}$$

where
$$r_i = \frac{k_{pii}}{k_{pij}}; \qquad \delta_i = \frac{k_{tii}^{1/2}}{k_{pii}};$$

cross termination constant
$$\phi = \frac{k_{tAB}}{(k_{tAA} k_{tBB})^{1/2}};$$

and rate of initiation
$$R_i = 2k_d f [I]$$

Some rational simplifications of this model have been proposed by Abkin.[9] The termination rate is assumed to be constant for various monomer feed compositions and therefore only the initiation rates for the homopolymerizations are required. Under these conditions, time-consuming initiation rate measurements in the monomer mixtures can be omitted.

16.3.2 Model of Diffusion-controlled Termination

Many experimental investigations have shown that these rather simple assumptions can only be applied in special cases. Therefore other authors have discussed more complex models. As the so-called cross termination rate constant in equation (6) was found not to be constant in many experimental investigations, several authors took the termination steps into greater consideration. North and Postlethwaite[10] have postulated a reaction scheme with diffusion control

$$-M_A\cdot + \cdot M_B - \underset{k_{-1}}{\overset{k_1}{\rightleftharpoons}} -M_A\cdot\cdot M_B- \qquad (7)$$

$$-M_A\cdot\cdot M_B - \underset{k_{-2}}{\overset{k_2}{\rightleftharpoons}} -M_A{:}M_B- \qquad (8)$$

$$-M_A{:}M_B - \overset{k_c}{\longrightarrow} P \qquad (9)$$

Here the growing radicals diffuse towards one another (reaction 7), with sequential motion permitting a direct approach of reactive centres (reaction 8), culminating in a termination step (reaction 9) to give the product P.

The overall rate constant of termination (k_t) depends on the viscosity η_i of the reaction medium

$$k_t = \frac{k_t^0}{\eta_L} \qquad (k_t^0 = \text{proportionality factor}) \qquad (10)$$

Applying this assumption, the rate equation can be modified as follows

$$R_{hom} = \frac{k_p}{(k_t^0)^{0.5}} \eta_L^{0.5} (2k_d f[I])^{0.5} [M] \qquad \text{for homopolymerizations} \qquad (11)$$

$$R_{cop} = \frac{(r_{ij}[M_i]^2 + 2[M_i][M_j] + r_{ji}[M_j]^2) R_i^{0.5}}{\left(\dfrac{k_t^0}{\eta_L}\right)^{0.5} \left(\dfrac{[M_i]}{k_{pij}} + \dfrac{[M_j]}{k_{pji}}\right)} \qquad \text{for copolymerizations (Atherton and North[11])} \qquad (12)$$

By substituting $d_i = (k_t)^{0.5}/k_{pij}$ and $d_j = (k_t)^{0.5}/k_{pji}$ it follows that

$$R_{cop} = \frac{(r_{ij}[M_i]^2 + 2[M_i][M_j] + r_{ji}[M_j]^2) R_i^{0.5} \eta_L^{0.5}}{d_i[M_i] + d_j[M_j]} \qquad (13)$$

The parameters d_i and d_j are calculated as fitting factors for given values of R_{cop}, η_L, r_{ij} and r_{ji}.

16.3.3 Penultimate Model

The diffusion rate of the growing chain end depends not only on the viscosity of the medium but also on the conformational characteristics of the radical chain end. Based on this effect, a reaction model has been developed assuming, as a first approximation, that the length of the rearranging segment is given by the last four carbon atoms (penultimate effect). With this simplification Merz,

Alfrey and Goldfinger[12] derived a modified copolymer composition equation. A rate equation corresponding to this model was developed by Russo and Munari[13,14] and has been successfully applied to describe the copolymerization rate of several binary systems. If this assumption is made, then 10 different reactions must be considered[13]

$$-M_i-M_j^\bullet + \cdot M_i^\bullet - M_j - \xrightarrow{\; k_{tijij} \;} P$$

$$-M_i-M_j^\bullet + \cdot M_j - M_j - \xrightarrow{\; k_{tijjj} \;} P \tag{14}$$

etc.

Using the simplification[13] (for example)

$$k_{tijij} = 2(k_{tijji} \times k_{tjiij})^{0.5}$$

$$k_{tijjj} = 2(k_{tijji} \times k_{tjjjj})^{0.5} \tag{15}$$

etc.

expression (16) for the rate of copolymerizations may be obtained

$$R_{cop} = \frac{[M_A + M_B]R_i^{0.5}}{x+1} \times \frac{r_{AB}x^3 + 3r_{AB}x^2 + 2x + r_{AB}r_{BA}x + r_{BA}}{r_{AB}^2 x^2 \delta_A + r_{AB}x\delta_{BA} + \dfrac{r_{BA}x(r_{AB}r_{BA}x+1)}{r_{BA}+x}\delta_{AB} + \dfrac{r_{BA}^2(r_{AB}x+1)}{r_{BA}+x}\delta_B} \tag{16}$$

for $x = [M_A]/[M_B]$, r_{AB}, r_{BA}, δ_A and δ_B are defined in equation (6) and

$$\delta_{AB} = (k_{tABBA})^{0.5}/k_{pAA}, \; \delta_{BA} = (k_{tBAAB})^{0.5}/k_{pAA}.$$

Recent reexamination of the classical binary model by Inagaki *et al.*[15-19] shows however that the rates of free-radical copolymerizations are not yet comprehensively understood in terms of the elementary processes of propagation and termination. Measurements of absolute rate constants k_p and k_t reveal that particularly the value of the constant k_p is entirely different from what the model predicts. The authors therefore indicate the necessity of reinterpretation of the propagation step in copolymerization reactions.

16.3.4 Supplementary Concepts

Each of the growing radicals may react in a different energy state before giving up its reaction heat. Assuming only the lowest (cold) and highest (hot) extreme states the 'theory of hot radicals' was developed.[20] The theory leads to the conclusion that the rate constants of chain propagation reactions, and hence the reactivity ratios, may depend on the composition of the monomer mixture or on the dilution.

In some cases the classical binary model may not be valid, *e.g.* if the monomers are capable of association, owing to secondary valency forces. In this regard the donor–acceptor interaction between molecules has a very remarkable effect. Models taking into account the interactions between radical and monomer or solvent[21] or monomer–monomer[22] (complex models) were developed. For the second case a rather complicated rate equation[23,24] results. A simpler method by Georgiev and Zubov[25] was successfully applied by many authors[26-29] to describe the reaction rate of some binary systems exhibiting significant monomer–monomer complex formation.

16.3.5 Some Experimental Results

For the verification of rate equations for binary copolymerizations some experimental results are presented below.

As Figure 1 shows, the polymerization reactivities of styrene and *N*-vinylpyrrolidone (NVP) are very different. The copolymerization can be described satisfactorily with good accuracy by means of the classical rate equation (6). This is due to the large difference in the reactivity of the two monomers involved, which under practical conditions means that only a few active chains with NVP units as radical end groups are present; therefore one kind of termination reaction (with S at the growing chain ends) predominates, which is in agreement with the prerequisite of this model.

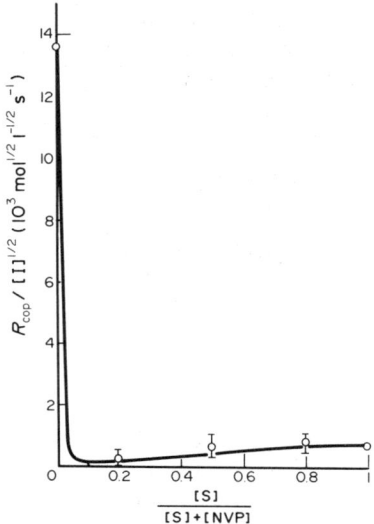

Figure 1 Copolymerization rate of the system S/NVP as a function of the monomer feed (60 °C, AIBN), showing ○, measured points and —, curve calculated with equation (6) and $\phi=0.033$, $r_{AB}=20.51$ and $r_{BA}=0.030$

For the system S/MMA the reactivities of the two monomers are also quite different. But, as shown in Figure 2, the copolymerization rate cannot be exactly described by means of the classical model (fine dashed line). This fact is well known from the literature and was explained by Russo and Munari[13] as an effect of the penultimate monomer unit. Another possibility for describing this system is to use variable cross-termination parameters ϕ in the classical model (solid line in Figure 2). The change of ϕ with the monomer feed is also indicated in Figure 2 (long dashed line).

As can be seen in Figure 3, the reaction rate of the system NVP/MMA cannot be described by means of the classical copolymerization model with either $\phi=1$ or $\phi=0.0002$ as the best value (dashed lines). Here, interactions between NVP and MMA can be observed using ¹H and ¹³C NMR spectroscopy.[30] The observed shifts can be explained[31,32] by an interaction between the carbonyl carbon atom of MMA and the nitrogen atom of NVP.

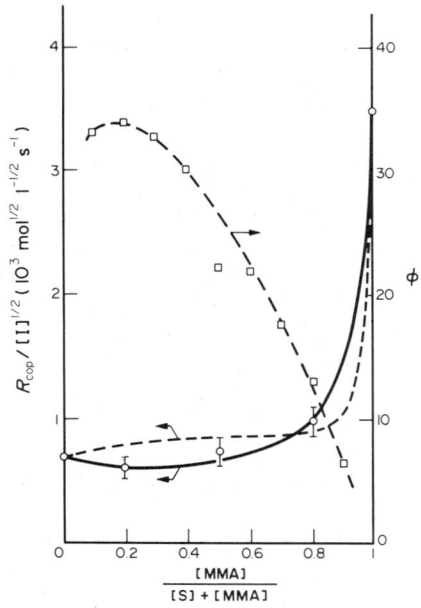

Figure 2 Copolymerization rate of the system S/MMA and the value of ϕ (equation 6) as a function of the monomer feed (60 °C, AIBN): ○, measured values; ---, calculated with equation (6) and $\phi=15.73$; ——, calculated with equation (6) and variable ϕ values ($r_{AC}=0.522$, $r_{CA}=0.482$), and -□-□-□, ϕ values calculated from the reaction rate

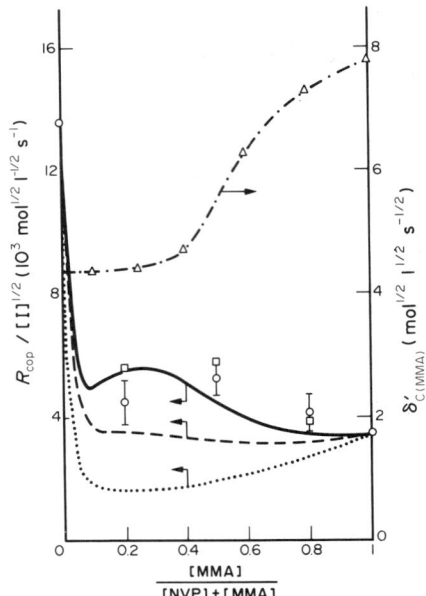

Figure 3 Copolymerization rate of the system NVP/MMA and δ'_C as a function of the monomer feed (60 °C, AIBN): $\bigcirc$, measured values; $\square$, literature data[33] (same reaction conditions); ...., calculated with equation (6) and $\phi = 1$; — —, calculated with equation (6) and $\phi = 0.0002$ as best value (nonlinear estimation); ——, calculated with equation (6), $\phi = 3$ and variable δ'_C values ($r_{BC} = 0.103$, $r_{CB} = 4.163$); and $\triangle$ —·—·, δ'_C, calculated with $2k_d f = 0.913 \ 10^{-5} \mathrm{s}^{-1}$

Starting from earlier investigations[33] on the MMA homopolymerization in *N*-methylpyrrolidone (NMP) as a nonpolymerizable model of NVP, a series of polymerizations was carried out with increasing dilution, where the monomer concentrations were varied over a wide range. From the values of the overall rate constant k_{cop} the ratio $\delta = k_{tii}^{0.5}/k_{pii}$ was calculated, supposing the initiation rate constant $2k_d f$ to have the same value as in bulk homopolymerization.

This condition is not exactly fulfilled.[34] The experimental values therefore are indicated as δ'_r. The change of $\delta'_{C(MMA)}$ with increasing dilution is represented in Figure 3 (dotted-dashed line) for [NMP] = [NVP]. Since this dependence is very pronounced the variation was introduced as a function of the ratio [NVP]:[MMA] into the classical rate equation (6). The resulting rate curve fits the experimental points quite well, confirming the importance of the monomer–monomer interaction in the copolymerization kinetics of this system.

Reactivity ratios for the system DEM/MMA were determined as $r_{DEM} = 0$ and $r_{MMA} = 370$.[35] From these values it follows that DEM participates only to a very small extent in the propagation step; in practice it is not homopolymerizable. Starting from this fact, the copolymerization of the system DEM/MMA may be regarded, to a very good approximation, as homopolymerization of MMA in DEM as solvent. Under these conditions the copolymerization rate can be interpreted by equation (11). Taking into account the viscosity of the reaction medium which changes between $\eta_L = 0.37$ cP for MMA and $\eta_L = 1.33$ cP for DEM with this simple model, a rather good agreement between the measured and the calculated values results.[35]

The copolymerization rates for the systems DEM/S and DEM/MMA show quite similar features.[36] However, from the reactivity ratios of the system DEM/S ($r_{DEM} = 0.001$ and $r_S = 6.592$) it results that DEM is more frequently incorporated into polymer chains during the copolymerization with styrene than in the system DEM/MMA. Therefore, the copolymerization rate can be better interpreted by means of a binary copolymerization model using the change of the viscosity of the reaction medium from $\eta_L = 0.47$ cP for S to $\eta_L = 1.34$ cP for DEM, which is nearly the same interval as for DEM/MMA. For this reason the reaction model of Atherton and North[11] (equation 12) may be used to describe the measured rates. The fitting factors d_A and d_C were estimated with equation (13) as 135.7 and 186.6 $(\mathrm{mol} \ s \ cP \ l^{-1})^{0.5}$ respectively.[36]

16.4 TERNARY COPOLYMERIZATION

As mentioned above, up to now in the literature there have been no systematic investigations of the rates of ternary copolymerizations. From the experiences with binary copolymerization systems, for an exact evaluation of the rate equation for ternary systems a knowledge of the initiation rates seems to be necessary; in this regard a linear interpolation of the initiation rate for binary and ternary monomer systems is not sufficient. Since the required values are missing, the applicability of the classical ternary rate equation of Blackley, Melville and Valentine[1] is essentially unknown.

16.4.1 Classical Rate Model for Three Homopolymerizable Monomers

For three homopolymerizable monomers the propagation steps of a ternary copolymerization are much more complicated than in the case of binary copolymerizations as the propagation is composed of nine reactions

$$-M_A\bullet + M_A \xrightarrow{k_{pAA}} -M_A\bullet \tag{17}$$

$$-M_A\bullet + M_B \xrightarrow{k_{pAB}} -M_B\bullet \tag{18}$$

$$-M_A\bullet + M_C \xrightarrow{k_{pAC}} -M_C\bullet \tag{19}$$

$$-M_B\bullet + M_A \xrightarrow{k_{pBA}} -M_A\bullet \tag{20}$$

$$-M_B\bullet + M_B \xrightarrow{k_{pBB}} -M_B\bullet \tag{21}$$

$$-M_B\bullet + M_C \xrightarrow{k_{pBC}} -M_C\bullet \tag{22}$$

$$-M_C\bullet + M_A \xrightarrow{k_{pCA}} -M_A\bullet \tag{23}$$

$$-M_C\bullet + M_B \xrightarrow{k_{pCB}} -M_B\bullet \tag{24}$$

$$-M_C\bullet + M_C \xrightarrow{k_{pCC}} -M_C\bullet \tag{25}$$

On this basis a ternary rate equation was derived by Melville *et al.*[1] assuming the validity of the steady state condition

$$R_{cop} = \frac{\left\{\displaystyle\sum_{A,B,C}\left([M_A]r_{AB} + [M_B] + [M_C]\frac{r_{AB}}{r_{AC}}\right)\Delta_A\right\}R_i^{1.2}}{\left\{\displaystyle\sum_{A,B,C}D_{AA} + 2\sum_{A,B,C}D_{AB}\right\}^{1.2}} \tag{26}$$

where

$$\sum_{A,B,C}D_{AA} = \delta_A^2 r_{AB}^2\Delta_A^2 + \delta_B^2 r_{BC}^2\Delta_B^2 + \delta_C^2 r_{CA}^2\Delta_C^2; \quad \sum_{A,B,C}D_{AB} = \phi_{AB}\delta_A\delta_B r_{AB}r_{BC}\Delta_A\Delta_B + \phi_{BC}\delta_B\delta_C r_{BC}r_{CA}\Delta_B\Delta_C + \phi_{CA}\delta_C\delta_A r_{CA}r_{AB}\Delta_C\Delta_A$$

$$\Delta_A = \begin{vmatrix} \dfrac{r_{BC}}{r_{BA}}[M_A] & [M_A] \\[2ex] -\left\{\dfrac{r_{BC}}{r_{BA}}[M_A] + [M_C]\right\} & \dfrac{r_{CA}}{r_{CB}}[M_B] \end{vmatrix}$$

$$\Delta_B = -\begin{vmatrix} -\left\{[M_B] + \dfrac{r_{AB}}{r_{AC}}[M_C]\right\} & [M_A] \\[2ex] [M_B] & \dfrac{r_{CA}}{r_{CB}}[M_B] \end{vmatrix} \qquad \Delta_C = \begin{vmatrix} -\left\{[M_B] + \dfrac{r_{AB}}{r_{AC}}[M_C]\right\} & \dfrac{r_{BC}}{r_{BA}}[M_A] \\[2ex] [M_B] & -\left\{[M_C] + \dfrac{r_{BC}}{r_{BA}}[M_A]\right\} \end{vmatrix}$$

Melville *et al.* have verified this equation for the monomer system S/p-methoxystyrene/MMA. With the aid of interpolated initiation rate values good agreement between the calculated and the measured rate was obtained for three different monomer feeds.

The assumptions of Abkin[9] concerning initiation and termination steps are particularly interesting in the case of ternary copolymerizations. They allow considerable reduction in the number of experiments required to test the applicability of the classical model. In this connection Braun *et al.*[37] have applied Abkin's simplifications to the classical ternary reaction model as a first approximation.

If λ_{nm} is the ratio of the initiation rate constants (k_i) for monomers n and m

$$\lambda_{nm} = \left(\frac{k_{in}}{k_{im}}\right)^{0.5} \quad \text{and} \quad R_{p,n} = \frac{R_{p,\text{homo},n}}{[I]^{0.5}}$$

where $R_{p,\text{homo},n}$ is the absolute homopolymerization rate of the monomer M_n at the initiator concentration $[I]$, the following rate equation results

$$R_{cop} = -\frac{d[M]}{dt} = \frac{R_{p,C}T^{1/2}}{[M]_0^{3/2}} \left\{ \frac{\dfrac{a}{r_{CB}}[M_A][M_B] + \dfrac{[M_A]}{r_{CA}}Z}{ab[M_A][M_B] - XZ}\left(W + \frac{e[M_B]Y}{Z}\right) + \frac{[M_B]Y}{r_{CB}Z} + U \right\} \quad (27)$$

$$T = \lambda_{AC}^2[M_A] + \lambda_{BC}^2[M_B] + [M_C]; \qquad U = \frac{[M_A]}{r_{CA}} + \frac{[M_B]}{r_{CB}} + [M_C]; \qquad W = c[M_A] - b[M_B] + d[M_C];$$

$$X = \frac{\lambda_{AC}[M_A]}{r_{CA}R_{p,C}} + \frac{[M_B]}{r_{AB}R_{p,C}} + \frac{[M_C]}{r_{AC}R_{p,C}}; \qquad Y = a[M_A] + f[M_B] + g[M_C]; \qquad Z = \frac{[M_A]}{r_{BA}R_{p,C}} + \frac{\lambda_{BC}[M_B]}{r_{CB}R_{p,B}} + \frac{[M_C]}{r_{BC}R_{p,C}}$$

$$a = \frac{1}{r_{BA}R_{p,C}} - \frac{\lambda_{BC}}{r_{CA}R_{p,B}}; \qquad b = \frac{1}{r_{AB}R_{p,C}} - \frac{\lambda_{AB}}{r_{CB}R_{p,A}}; \qquad c = \frac{\lambda_{AC}}{r_{CA}R_{p,A}} - \frac{1}{R_{p,A}};$$

$$d = \frac{\lambda_{AC}}{R_{p,A}} - \frac{1}{r_{AC}R_{p,C}}; \qquad e = \frac{\lambda_{AC}}{r_{CA}R_{p,A}} - \frac{1}{r_{AB}R_{p,C}}; \qquad f = \frac{1}{R_{p,C}} - \frac{\lambda_{BC}}{r_{CB}R_{p,B}}; \qquad g = \frac{1}{r_{BC}R_{p,C}} - \frac{\lambda_{BC}}{R_{p,B}}$$

For the calculations using equation (27) the following constants are required: monomer reactivity ratios r_{ij}, the three individual homopolymerization rates $R_{p,n}$ and only three initiation rate constants k_{in}. The results correspond directly to the values divided by $[I]^{0.5}$.

16.4.2 Model of Diffusion-controlled Termination

The ternary copolymerization kinetics can be simplified if one of the comonomers is not able to homopolymerize. Assuming diffusion control of chain termination[10,11] (equation 10), the following ternary rate equation can be developed for the case when M_A cannot homopolymerize

$$R_{cop} = \{2k_d f[I]\eta_L\}^{0.5} \times \frac{(S_A' + S_B' + S_C')}{(V_A + V_B + V_C)} \quad (28)$$

$$S_A' = \{[M_B] + R[M_C]\}\Delta_A' \qquad S_B' = \left\{\frac{[M_A]}{r_{BA}} + [M_B] + \frac{[M_C]}{r_{BC}}\right\}\Delta_B' r_{BC} \qquad S_C' = \left\{\frac{[M_A]}{r_{CA}} + \frac{[M_B]}{r_{CB}} + [M_C]\right\}\Delta_C' r_{CA}$$

$$R = \frac{k_{pAC}}{k_{pAB}}$$

$$V_A = d_A \Delta_A' R \qquad V_B = \delta_B^0 \Delta_B' r_{BC} \qquad V_C = \delta_C^0 \Delta_C' r_{CA}$$

$$d_A = \frac{(k_t^\circ)^{0.5}}{k_{pAC}} \qquad \delta_B^0 = \frac{(k_t^0)^{0.5}}{k_{pBB}} \qquad \delta_C^0 = \frac{(k_t^0)^{0.5}}{k_{pCC}}$$

$$\Delta_A' = \begin{vmatrix} \dfrac{r_{BC}}{r_{BA}}[M_A] & [M_A] \\ -[M_C] & [M_A] + \dfrac{r_{CA}}{r_{CB}}[M_B] \end{vmatrix} \qquad \Delta_B' = \begin{vmatrix} \dfrac{r_{CA}}{r_{CB}}[M_B] & [M_A] \\ -[M_B] & [M_B] + R[M_C] \end{vmatrix}$$

$$\Delta'_C = \begin{vmatrix} R[M_C] & [M_B] \\ \\ -[M_C] & [M_C]+\dfrac{r_{BC}}{r_{BA}}[M_A] \end{vmatrix}$$

Considering the penultimate effect for monomers M_B and M_C according to Russo *et al.*,[13,14] equations (14) and (15) with the steady state condition yield equation (29)

$$R_i = R_t = \{(k_{tBBBB})^{0.5}[-M_B-M_B{}^\bullet] + (k_{tBCCB})^{0.5}[-M_B-M_C{}^\bullet] + (k_{tCCCC})^{0.5}[-M_C-M_C{}^\bullet] + (k_{tCBBC})^{0.5}[-M_C-M_B{}^\bullet]\}^2 \quad (29)$$

In a system with a nonhomopolymerizable monomer M_A the following three assumptions may be made: (a) $[-M_A-M_A{}^\bullet]=0$; (b) the radicals $-M_B-M_A{}^\bullet$, $-M_C-M_A{}^\bullet$, $-M_A-M_B{}^\bullet$ and $-M_A-M_C{}^\bullet$ do not show any penultimate effect; and (c) the viscosity of the medium is influence on the termination of all growing radicals. Under these conditions equation (29) becomes

$$R_i = R_t = \frac{1}{\eta_L}[(k^0_{tBBBB})^{0.5}[-M_B-M_B{}^\bullet] + (k^0_{tCBBC})^{0.5}[-M_C-M_B{}^\bullet] + (k^0_{tCCCC})^{0.5}[-M_C-M_C{}^\bullet] + (k^0_{tBCCB})^{0.5}[-M_B-M_C{}^\bullet]$$
$$+(k^0_t)^{0.5}\{[M_A{}^\bullet]+[-M_A-M_B{}^\bullet]+[-M_A-M_C{}^\bullet]\}]^2 \quad (30)$$

With

$$k_{AB}[-M_A{}^\bullet]: k_{BC}[-M_B{}^\bullet]: k_{CA}[-M_C{}^\bullet]=\Delta_A: \Delta_B: \Delta_C \quad (31)$$

$$\frac{(-M_i-M_j{}^\bullet)}{(-M_j-M_j{}^\bullet)} \simeq \frac{k_{ij}(M_i{}^\bullet)}{k_{jj}(M_j{}^\bullet)} \quad (32)$$

and with equations (28) and (30) the final rate equation (33) results

$$R_{cop} = \{2k_d f[I]\eta_L\}^{0.5} \times \frac{(S'_A + S'_B + S'_C)}{\left(J+\dfrac{L}{M}+\dfrac{N}{P}\right)} \quad (33)$$

S'_i are defined in equation (28)

$$J = d_A R\Delta'_A \qquad M = \Delta'_A + r_{BC}\Delta'_B + \frac{r_{CA}}{r_{CB}}\Delta'_C \qquad P = r\Delta'_A + \Delta'_B + r_{CA}\Delta'_C$$

$$L = (\delta^0_B(r_{BC}\Delta'_B+\Delta'_A) + \delta^0_{CB}\frac{r_{CA}}{r_{CB}}\Delta'_C)\,r_{BC}\Delta'_B \qquad N = (\delta^0_C(r_{CA}\Delta'_C+R\Delta'_A)+\delta^0_{BC}\Delta'_B)r_{CA}\Delta'_C$$

where d_A, δ^0_B and δ^0_C are defined in equation (28),

$$\delta^0_{BC} = \frac{k^0_{tBCCB}}{k_{pCC}} \quad \text{and} \quad \delta^0_{CB} = \frac{k^0_{tCBBC}}{k_{pBB}}$$

16.4.3 Some Experimental Results

The results of rate measurements of the polymerization of the system S/NVP/MMA are shown in Figure 4. As can be seen, the homopolymerization rates of these three monomers are very different. The binary systems S/NVP and S/MMA as well as nearly all investigated ternary systems polymerize with a very similar rate to that of the styrene homopolymerization, *i.e.* in this case the slowest polymerizing component determines the overall rate of the ternary system.

The experimentally measured values can be interpreted by means of the classical rate equation (26). All necessary kinetic constants have been determined by homopolymerization and binary copolymerization experiments. ϕ_{AC} and ϕ_C were taken as functions of the ratios [S]:[MMA] and [NVP]:[MMA] respectively. The measured and calculated rate values are summarized in Table 1; the comparison shows very good agreement between the calculated and the experimentally determined rates. Therefore in this case the classical rate equation can be applied.

An interpretation with the aid of the simplified ternary model (equation 27) yields only approximate rates (Table 1). It is, however, possible in this case to estimate quickly the copolymerization rate for different monomer feeds without any knowledge of the initiation rates of the monomer mixtures.

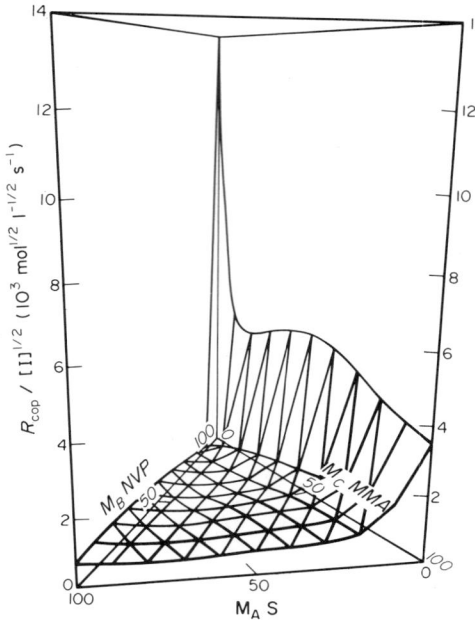

Figure 4 Experimentally determined ternary copolymerization rate of the system S/NVP/MMA as a function of the monomer feed (60 °C, AIBN, in bulk)

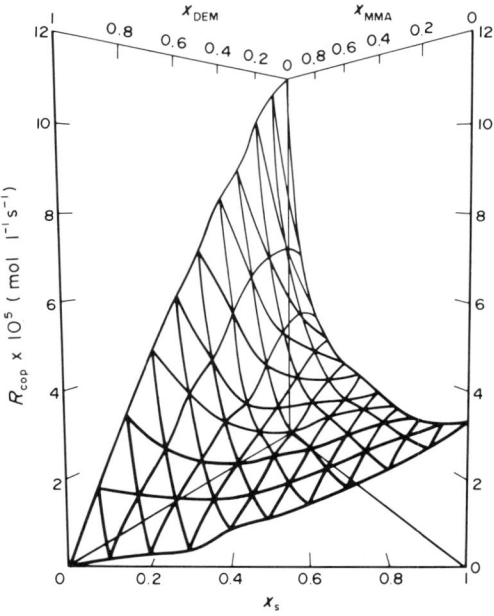

Figure 5 Experimentally determined ternary copolymerization rate of the system DEM/MMA/S as a function of the monomer feed (60 °C, AIBN, in bulk)

In a similar way, the experimental rates for the system DEM/MMA/S are shown in Figure 5.[38] Also, in this case, the component with the slowest homopolymerization rate (DEM) determines the copolymerization rate of the ternary system over a wide range of monomer feeds.

For the interpretation of the experimental results both equations (28) and (33) were used; measured and calculated values are compared in Table 2.

As can be seen, equation (28) can describe the experimentally determined rates satisfactorily. Nevertheless, considering a penultimate effect in the case of the binary system S/MMA in the applied

Table 1 Measured and Calculated Polymerization Rates of the Ternary System
S/NVP/MMA

System I *S:NVP:MMA* (mol %)	$R_{cop}/[AIBN]^{0.5} \times 10^4$ (mol$^{0.5}$ l$^{-0.5}$ s$^{0.5}$) $60^\circ C$ in bulk, AIBN measured	calculated (equation 26)	calculated (equation 27)
10 : 10 : 80	16.672	13.778	25.084
40 : 10 : 50	9.727	6.941	15.676
80 : 10 : 10	6.982	6.773	8.057
20 : 20 : 60	9.311	7.870	18.259
40 : 20 : 40	7.745	6.626	13.108
60 : 20 : 20	6.310	5.940	9.184
10 : 40 : 50	11.803	9.316	16.868
50 : 40 : 10	5.105	4.849	6.052
10 : 60 : 30	9.183	6.666	10.794
30 : 60 : 10	4.315	3.579	4.903
10 : 80 : 10	5.236	3.180	4.348

Table 2 Measured and Calculated Polymerization Rates of the Ternary System
DEM/MMA/S

System II *DEM: MMA:S* (mol %)	$R_{cop} \times 10^5$ (mol l^{-1} s^{-1}) $60^\circ C$ in bulk, AIBN measured[a]	calculated (equation 28)	calculated (equation 33)
10.3 : 9.8 : 79.9	2.69	3.66	2.70
45.6 : 9.7 : 44.7	1.45	2.00	1.49
29.4 : 10.1 : 10.5	0.65	0.49	0.43
23.0 : 23.0 : 54.0	2.42	3.88	2.24
33.6 : 33.1 : 33.3	2.37	4.00	2.14
55.8 : 22.5 : 21.7	1.58	1.83	1.18
9.6 : 45.4 : 45.0	3.36	7.38	3.31
22.9 : 53.8 : 23.3	3.84	·5.93	2.81
44.8 : 45.0 : 10.2	2.98	2.94	1.74
9.3 : 81.3 : 9.4	5.83	7.78	4.36

[a] $[AIBN] = 2 \times 10^{-3}$ [mol l^{-1}]

rate model (equation 33) permits polymerization rates of ternary systems to be described even more precisely.

16.5 CONCLUSIONS

In this chapter some systematic investigations on the rate of several binary and ternary copolymerization systems are presented and compared with theoretically developed rate equations, based on various kinetic models. The systems discussed and the rather good agreement between calculations and experimental results permit the conclusion that the rates of binary and ternary copolymerizations may be interpreted on the basis of Melville's model, but that it is necessary to take into consideration some deviations which result from some simplifications in the model. The systems presented here also show that Melville's model must be adapted to the individual nature of the monomers, especially in the case where big differences exist in the reactivities of the monomers involved or when there are specific interactions between them.[39]

16.6 REFERENCES

1. D. C. Blackley, H. W. Melville and L. Valentine, *Proc. R. Soc., London, Ser. A*, 1954, **227**, 10.
2. G. V. Schulz and G. Harborth, *Angew. Chem., A*, 1947, **59**, 90.
3. P. Wittmer, *Angew. Makromol. Chem.*, 1974, **39**, 35.

4. D. Braun and G. Disselhoff, *Polymer*, 1977, **18**, 963.
5. G. Disselhoff, *Polymer*, 1978, **19**, 111.
6. G. Ham, (ed.), 'Copolymerization', Interscience, New York, p. 6, 38.
7. D. Braun and G. Cei, *Makromol. Chem.*, 1987, **188**, 189.
8. H. W. Melville, B. Noble and W. F. Watson, *J. Polym. Sci.*, 1947, **2**, 229.
9. A. D. Abkin, *Dokl. Akad. Nauk SSSR*, 1950, **75**, 403.
10. A. North and D. Postlethwaite, in 'Structure and Mechanism in Vinyl Polymerization', ed. T. Tsuruta and K. O'Driscoll, Dekker, New York, 1959, p. 99.
11. J. Atherton and A. North, *Trans. Faraday Soc.*, 1962, **58**, 2049.
12. E. Merz, T. Alfrey and G. Goldfinger, *J. Polym. Sci.*, 1946, **1**, 75.
13. S. Russo and S. Munari, *J. Macromol. Sci., Chem.*, 1968, **2**, 1321.
14. G. Bonta, B. M. Gallo and S. Russo, *J. Chem. Soc., Faraday Trans. 1*, 1975, **71**, 1727; *Polymer*, 1975, **16**, 429.
15. T. Fukuda, Y.-D. Ma and H. Inagaki, *Polym. Bull.*, 1983, **10**, 288.
16. T. Fukuda, Y.-D. Ma and H. Inagaki, *Makromol. Chem., Suppl.*, 1985, **12**, 125.
17. T. Fukuda, Y.-D. Ma and H. Inagaki, *Polym. J.*, 1982, **14**, 705.
18. T. Fukuda, Y.-D. Ma and H. Inagaki, *Macromolecules*, 1985, **18**, 17.
19. Y.-D. Ma, T. Fukuda and H. Inagaki, *Macromolecules*, 1985, **18**, 26.
20. F. Tüdös, T. Kelen and T. Földes-Berezsnikh, *J. Polym. Sci., Polym. Symp.* 1975, **50**, 109.
21. G. Henrici-Olive and S. Olive, *Makromol. Chem.*, 1962, **58**, 188; 1963, **68**, 219; *Z. Phys. Chem.*, 1966, **47**, 286; 1966, **48**, 35, 51.
22. D. J. T. Hill, J. H. O'Donnell and P. W. O'Sullivan, *Prog. Polym. Sci.*, 1982, **8**, 215.
23. D. Braun and W. Czerwinski, *Makromol. Chem.*, 1983, **184**, 1071.
24. P. C. Deb and G. Meyerhoff, Polymer, 1985, **26**, 629.
25. G. S. Georgiev and V. P. Zubov, *Eur. Polym. J.*, 1978, **14**, 93.
26. B. Tizianel, C. Caze and C. Loucheux, *J. Macromol. Sci., Chem.*, 1985, **22**, 1477.
27. K. Fujimori, *Polym. Bull.*, 1985, **13**, 459.
28. K. Fujimori, P. P. Organ, M. J. Costigan and I. E. Craven, *J. Macromol. Sci., Chem.*, 1986, **23**, 647.
29. K. Fujimori and A. S. Brown, *Polym. Bull.*, 1986, **15**, 223.
30. D. Braun and W. K. Czerwinski, *Makromol. Chem.*, 1987, **188**, 2389.
31. T. Ishida, S. Kondo and K. Tsuda, *Makromol. Chem.*, 1977, **178**, 3221.
32. K. Tsuda, S. Kondo, K. Yamashita and K. Ito, *Makromol. Chem.*, 1984, **185**, 81.
33. D. Braun, G. Disselhoff and F. Quella, *Makromol. Chem.*, 1981, **182**, 2951.
34. M. Kamachi, D. J. Liaw and S. Nozakura, *Polym. J.*, 1981, **13**, 41.
35. D. Braun and G. Cei, *Makromol. Chem.*, 1986, **187**, 1699.
36. D. Braun and G. Cei, *Makromol. Chem.*, 1986, **187**, 1713.
37. F. Quella, W. K. Czerwinski and D. Braun, *Makromol. Chem.*, 1987, **188**, 2403.
38. D. Braun and G. Cei, *Makromol. Chem.*, 1987, **188**, 171.
39. D. Braun and W. K. Czerwinski, *Makromol. Chem., Makromol. Symp.*, 1987, **10/11**, 415.

17

Reactivities in Free-radical Polymerization

CLEMENT H. BAMFORD
University of Liverpool, UK

17.1 INTRODUCTION

Free-radical addition to the alkenic double bonds of vinyl monomers, and radical abstraction processes from monomers or other components of the reaction mixture, form the basis of free-radical polymerization. In homopolymerization, regiospecificity of addition determines the nature of the linkages between units of the polymer chains, *i.e.* whether the units are linked head–tail, tail–tail, *etc.*, while rates of addition and abstraction (transfer) reactions have influences on polymer molar masses. Abstraction processes may also be responsible for the existence of imperfections such as branch points or double bonds in the polymer chains.

In this chapter we discuss the structural features influencing addition and abstraction reactions of propagating radicals. Clearly, reactions of both primary and propagating radicals are important in polymerization; the former have been considered elsewhere[1] and will only be referred to here for purposes of illustration or emphasis.

17.2 PROPAGATING RADICAL–MONOMER REACTIONS

The alternative of a head or tail addition to a vinyl monomer $CH_2=CXY$ gives rise to four types of propagation reaction, as shown in equations (1)–(4):

$$\sim\!\sim\!\sim CH_2\dot{C}XY + CH_2=CXY \rightarrow \sim\!\sim\!\sim CH_2CXYCH_2\dot{C}XY \qquad (1)$$

(1)

$$\text{\small $\sim\!\sim$}CH_2\dot{C}XY + CXY{=}CH_2 \rightarrow \text{\small $\sim\!\sim$}CH_2CXYCXY\dot{C}H_2 \qquad (2)$$

$$\text{\small $\sim\!\sim$}CXY\dot{C}H_2 + CXY{=}CH_2 \rightarrow \text{\small $\sim\!\sim$}CXYCH_2CXY\dot{C}H_2 \qquad (3)$$

$$\text{\small $\sim\!\sim$}CXY\dot{C}H_2 + CH_2{=}CXY \rightarrow \text{\small $\sim\!\sim$}CXYCH_2CH_2\dot{C}XY \qquad (4)$$

Both reactions (1) and (3) lead to radical products of the same type as the initial radicals, and, if propagation occurred exclusively by either or both of these, polymers with a regular head-to-tail (1,3) structure would be formed. On the other hand, reactions (2) and (4) in alternation would give rise to head-to-head and tail-to-tail units.

In 1935, Staudinger and Steinhoffer[2] demonstrated the predominance of head-to-tail structures in polystyrene by destructive distillation of the polymer. The products were found to have phenyl substituents on alternate carbon atoms in the molecules, with no such substituents on the adjacent carbons. Subsequently, overwhelming evidence has accumulated indicating essentially similar structures for many vinyl polymers;[3-6] in particular, high resolution nuclear magnetic resonance (NMR) analysis can be used to indicate the basic structure and also allows imperfections to be identified. The evidence from polymer structures, therefore, strongly favours reactions (1) or (3).

Electron spin resonance (ESR) studies of the propagating radicals have also been made.[7] Radical concentrations in polymerizing systems are generally rather low ($\sim 10^{-7}$ mol dm^{-3}) and could not be readily detected with the early spectrometers; consequently, observations were carried out on radicals trapped in solid polymers,[7] or in viscous media, *i.e.* under conditions precluding rapid radical termination. The pioneering work of Fischer and his colleagues[8-10] demonstrated that, under appropriate conditions in flowing systems, radical concentrations high enough for detailed ESR analysis could be obtained. Spectrometer design has now developed to an extent which permits well-resolved spectra to be obtained from propagating radicals under stationary conditions during polymerization. A supersensitive ESR spectrometer was described in 1972 by Bresler *et al.*[11,12] By means of a specially designed balance cavity, the absorption of microwave power was transformed into the rotation of the plane of polarization of microwave oscillations, so that the spectrometer had analogies with an optical polarimeter. It was claimed that the apparatus was sensitive to DPPH in benzene at a concentration of 2×10^{-10} mol dm^{-3}. Kamachi *et al.*[13] have also reported highly resolved ESR spectra, using a spectrometer with a specially designed TM$_{110}$ cavity, stated to have sensitivity an order of magnitude higher than the conventional TE$_{011}$ mode cavity. Spin-trapping has also been used to elucidate the nature of the adduct between primary radicals and monomers.[1] All these investigations leave little room for doubt that, in general, the propagating radical has basically the structure (1) and is formed by head-to-tail addition. The evidence both from polymer and radical structures, therefore, strongly favours reaction (1) as the major propagation process. Primary radicals usually show a preference for tail addition, as explained elsewhere,[1] but other types of reaction, notably hydrogen abstraction, are also commonly encountered, especially with oxygen-centred radicals. In normal propagation it is unusual to find radicals other than those which are carbon based.

In spite of the general predominance of reaction (1), 'mistakes' occur leading to imperfections (*e.g.* head–head addition) in the polymers, with a probability depending on the nature of the monomer. This topic will be referred to later.

17.2.1 Factors Influencing Rates and Regioselectivity of Addition

Early ideas of the orientation of radical addition to alkenes (*i.e.* the regiospecificity) were taken from organic chemistry, in which the electronic theory was very successful in interpreting mechanisms of reactions involving ionic intermediates. Mayo and Walling and their colleagues[14,15] emphasized the importance of resonance stabilization and of polar contributions to the transition states of radical additions in determining rates of addition. These views followed naturally from the interpretation of the 'abnormal' addition of hydrogen bromide to unsymmetrical alkenes—the so-called anti-Markovnikov addition—advanced by Kharasch and his colleagues[16] and by Hey and Waters[17] who, in 1937, independently recognized the free-radical character of the addition. It was proposed that the chain-carrying radical adds to the alkene so as to give the most stable radical intermediate.

The reactivities of monomers towards a series of reference polymer radicals may be compared with the aid of copolymerization data. If the rate coefficients for addition of radical $\text{\small $\sim\!\sim$}A\cdot$ to monomers A and B are k_{AA} and k_{AB}, respectively, the reactivity ratio r_A is defined as k_{AA}/k_{AB}, so that $1/r_A = k_{AB}/k_{AA}$ is a measure of the rate coefficient for addition of $\text{\small $\sim\!\sim$}A\cdot$ to monomer B. Studying a series of different B monomers allows the reactivities of the monomers towards $\leadsto$A• to be compared. Furthermore, by using different types of reference radical $\leadsto$A•, the influence of the properties of the latter may be examined. Mayo and Walling[14] adopted this procedure and produced an extensive list which was later enlarged somewhat by Walling.[15] These compilations show that different reference radicals give essentially similar sequences of monomer reactivities, and reveal the generally high reactivities of monomers such as styrene, butadiene and, to a smaller extent, acrylonitrile, which form radicals generally considered to be resonance stabilized. Relative radical reactivities may also be compared with the help of copolymerization data; since $k_{AB} = k_{AA}/r_A$, knowledge of the absolute value of k_{AA}, together with r_A, allows evaluation of k_{AB} and so provides the desired information. As expected, the stabilized radicals show lower reactivity. There is, therefore, an inverse relation between the reactivities of monomers and their derived radicals, the most reactive monomers giving the least reactive radicals and *vice versa*. Mayo and Walling[14] found that monomer reactivities vary with the substituent X in $CH_2=CHX$ in the sequence $Ph > CH_2=CH- > MeCO- > -CN > -CO_2R > Cl > -CH_2Z > MeCO_2- > -OR$. This sequence was interpreted as corresponding to decreasing 'resonance stabilization' of the derived radicals of type $\leadsto CH_2\dot{C}HX$. Although a given conjugating substituent stabilizes both monomer and radical by electron delocalization, the effect on the radical is much greater.[15]

Mayo and Walling[14] also demonstrated the existence of an 'alternating tendency' in copolymerization, that is, radical $\leadsto$A• tends to react preferentially with monomer B and *vice versa*, when the monomers are of comparable reactivity. As was noted by early workers,[18,19] this is most easily explained in terms of polar effects; in fact, it is possible to arrange monomers in a polarity series determined by the electron-withdrawing or electron-donating properties of their substituent groups which accords well with the alternation data.[15]

These theories using resonance stabilization and polar factors are in good agreement with the experimental findings that the chain-carrying radical in polymerization has structure (1), since this is the most stable species which can be formed, and head–tail addition (reaction 1) is favoured by polar effects. Formation of an unstabilized $\leadsto \dot{C}H_2$ radical as in reactions (2) and (3) is less favourable energetically.

Since these early proposals were formulated, a considerable amount of quantitative data has accumulated on small radical addition reactions to alkenes and abstraction reactions, and it is of interest to consider the impact of this on our ideas of free-radical polymerization. Data are often discussed with relation to four features: (i) stabilization of radical reactants and products by electron delocalization, (ii) steric effects, (iii) polar effects, and (iv) the strength of the new bond being formed, addition occurring in a manner which gives the strongest bond. Combinations of these are usually required to rationalize the experimental observations[20] and the analysis can rarely be carried out quantitatively. The strength of the new bond (iv) is a function of (i), (ii) and (iii), but specific influences may also operate; for example, the presence of halogen substituents in the alkene leads to weaker bonds and a radical would, therefore, tend to add at the site free from halogen, in the absence of other factors. In polymerizations, addition to the substituted carbon atom of an alkenic monomer $CH_2=CXY$ is uncommon and this feature is, therefore, less in evidence; however, its occasional occurrence may be of practical significance in introducing weak links into the chains.

Free-radical additions are normally strongly exothermic (compare Table 4) and their transition states are, therefore, likely to be closer to the reactants than to the products. The rather small value of the secondary isotope effects observed when the alkene is substituted with deuterium at the reaction centre has been interpreted as supporting this picture of the reaction.[21,22] In these circumstances, steric interactions (ii) and polar effects (iii) would be expected to have considerable influence on the rates of reaction, while stabilization of the radical product would be less important. Note, however, that detailed calculations of Strausz *et al.*[23] do not support the simple interpretation of the magnitude of the secondary isotope effect given above.

In discussing data on small radical addition, regiospecificity and rates of addition will be considered together; both are intimately bound up with the same structural factors and cannot profitably be separated.

17.2.1.1 Stabilization of radicals

Tedder, Walton and their colleagues[20,24,25] have amassed much data on the activation parameters and the regioselectivity for addition of methyl and substituted methyl radicals to unsymmetrical alkenes (see Table 1). Tedder[26] has argued that the preferential addition of radicals to the unsubstituted carbon (C-2) is not primarily a result of resonance stabilization of the radical

Table 1 Relative Rates $(2k_2/k_e)^a$ and Orientation Ratios [(2):(1)] for Addition of Methyl, Trifluoromethyl and Trichloromethyl Radicals to Monosubstituted Ethylenes $(C(2)H_2=C(1)HX)^{24}$

X	$\cdot CH_3$		$\cdot CF_3$		$\cdot CCl_3$	
	$2k_2/k_e$	Orientation ratio [(2):(1)]	$2k_2/k_e$	Orientation ratio [(2):(1)]	$2k_2/k_e$	Orientation ratio [(2):(1)]
H	1.0	1:1	1.0	1:1	1.0	1:1
Me	0.7	1:0.15	2.3	1:0.1	4.0	1:0.07
F	0.9	1:0.20	0.48	1:0.09	0.7	1:0.08
Cl	4.2	1:>0.01	1.3	1:>0.01	2.5	1:>0.01
Br	—	—	1.2	1:>0.01	—	—
CF$_3$	0.9	1:0.33	0.40	1:>0.02	0.9	1:>0.01
CN^b	34.0	—	0.72	1:>0.01	4.5	1:>0.01
MeO	—	—	2.1	1:>0.01	—	—
CH$_2$=CH–	8.1	1:0.01	20.0	1:>0.01	—	—

^{a} k_2 = rate of addition to C(2) and k_e = the rate of addition to ethylene. ^{b} Data for ethyl radicals.

adduct, but arises from deactivation of the substituted carbon (C-1). Typical data upon which this argument is based are presented in Table 1.[24] In all the reactions tabulated, addition takes place preferentially at the (unsubstituted) tail of the alkene, but frequently the rates are lower than those of addition to ethylene. If resonance stabilization were a dominating factor, substituents should enhance the rate and not reduce it. A more satisfactory explanation of the results is that, in these instances, the regioselectivity is governed by steric effects (steric compression) which retard head addition.

Similar conclusions may be adduced from activation parameters. As an illustration we consider the addition of trichloromethyl radicals to the $CH_2=$ ends of the following alkenes: $CH_2=CH_2$, $CH_2=CHMe$, $CH_2=CHF$, $CH_2=CFMe$, $CH_2=CHCl$ and $CH_2=CF_2$. The activation energies for addition are quite similar—close to 26.8 kJ mol^{-1}, except for addition to the highly polar $CH_2=CF_2$ for which $E_A = 32.2 \text{ kJ mol}^{-1}$. Since stabilization of the radical adducts would reduce the activation energy, it has been argued[26] from these data that such stabilization, which would be expected to vary from one system to another, cannot be an important parameter. While this may be true in the present systems, the general argument is weakened by the fact that the extents of resonance stabilization anticipated are not large. The converse type of argument has also been used[26] against the validity of the Mayo–Walling picture, *viz.* that, in the additions of $\cdot CCl_3$ to $CH_2=CH_2$ and the head ends of $CH_2=CHF$ and $CH_2=CF_2$, the unpaired spin in the product is located in each case on $-\dot{C}H_2$, yet the observed activation energies are 26.4, 35.2 and 47.7 kJ mol^{-1}. These data evidently indicate that factors other than resonance stabilization contribute to the activation energy, in these cases, mainly polar influences (see below).

Inspection of Table 1 shows that additions of $\cdot CH_3$ and $\cdot CF_3$ proceed much more rapidly to butadiene than to ethylene, and this may be attributed to electron delocalization in the adduct radical. Butadiene differs from the other alkenes in the table in that the substituent $-CH=CH_2$ has π orbitals which can overlap the half-filled orbitals in the radical and so confer stabilization. This is the only alkene in Table 1 in which resonance stabilization of the adduct radical has a marked effect on the rate of addition. The high rate of attack of radicals on acrylonitrile probably also arises from resonance stabilization, which in this case is reinforced by polar effects since the ethyl radical is nucleophilic. Unsymmetrical orbitals in nitrile or carbonyl groups do not generally participate in delocalization to a major extent, on account of the higher electron affinities of the N and O atoms. Substituents with non-bonding electron pairs, such as F, Cl and –OR, have much smaller influences on radical additions, although they may be important in ionic reactions. Here delocalization would involve unfavourable charge separation

$$RCH_2\dot{C}HCl \longleftrightarrow RCH_2\bar{C}H\overset{+}{Cl} \qquad (5)$$

$$RCH_2\dot{C}HOR \longleftrightarrow RCH_2\bar{C}H\overset{+}{O}R \qquad (6)$$

The classic example of a monomer yielding highly stabilized radicals is styrene; note that styrene and butadiene are at the head of Mayo and Walling's sequence of monomer reactivities (Section 17.2.1), followed by carbonyls, nitrile and chloride. This sequence is not inconsistent with the data on small radical reactions discussed, but, of course, it cannot reveal all the factors which play a significant role in determining rates and regioselectivity.

17.2.1.2 Steric effects

Steric effects, specifically the degree of steric compression associated with forming the new bond, appeared in the course of the discussion in Section 17.2.1.1 as a major influence determining the regioselectivity of addition. The general rule 'addition occurs preferentially at the least-substituted carbon atom', formulated by Kharasch and his colleagues[27] in the mid 1940s, is rarely broken. Notable exceptions are the addition of $\cdot CF_3$ and $\cdot CCl_3$ to $CF_2 = CHCl$ which occurs preferentially at the CF_2 end;[28] here the steric effect of the chlorine atom predominates and the radicals add at the least sterically crowded position. Tedder and Walton[29] have studied the dependence of regioselectivity of addition on the size of the attacking radical. Such a dependence would be anticipated in view of our discussion in Section 17.2.1.1 and some results are presented in Table 2.[29] The constancy of the rates of addition is consistent with our earlier discussion, in which regioselectivity was attributed to steric deactivation of the 1-position. The increasing size of the radical is reflected in the increasing preference for addition to the unsubstituted 2-position.

Table 2 Regioselectivity and Rates of Addition of Fluoroalkyl Radicals to Vinyl Fluoride at 164 °C

Radical	$C(2)H_2 = C(1)HF$ *Orientation ratio* $[(2):(1)]$	$2k_2/k_e{}^a$
$\cdot CF_3$	1:0.1	0.5
CF_3CF_2	1:0.06	0.6
$(CF_3)_2CF$	1:0.02	0.5
$(CF_3)_3C\cdot$	1:0.005	0.5

^{a} k_2 and k_e = rate coefficients for addition to C-2 in $CH_2 = CHF$ and ethylene, respectively.[29]

Szwarc and his co-workers[30] have investigated the rates of addition of $\cdot CF_3$ radicals to symmerically substituted alkenes in the gas phase and thus assessed the influence of substitution at the reaction site. Some results are given in Table 3. These measurements present a technical difficulty in that, with alkyl-substituted alkenes, abstraction as well as addition occurs and must be allowed for. It seems clear that replacement of H by Cl or Me at the reaction site reduces the frequency factor by approximately a factor 5; F has slightly less effect. The changes are of steric origin and are attributable to the restriction of rotation of CF_3 in the transition state and consequent reduction in the entropy of activation and the frequency factor.

In the polymerization field it is well known that unsymmetrically disubstituted alkenes are relatively reluctant to polymerize. Indeed, in many cases, such as stilbene and 1,2-dichloroethylene, the compounds will not homopolymerize although they may copolymerize quite readily. Examples of the effects of disubstitution quoted by Walling[15] include comparisons of the additions to methyl acrylate $CH_2 = CHCO_2Me$ and ethyl fumarate $EtO_2CCH = CHCO_2Et$, the former being 12 times

Table 3 Frequency Factors (*A*) for Addition of $\cdot CF_3$ to Symmetrically Substituted Alkenes[30]

Alkene	*A (ethylene)/A (alkene)*
$CF_2 = CF_2$	4.4
cis-CHCl=CHCl	3.4
trans-CHCl=CHCl	5.0
$CCl_2 = CCl_2$	6.0
cis-MeCH=CHMe	5.5
trans-MeCH=CHMe	4.6
$Me_2C = CMe_2$	8.0

more reactive towards $\sim\sim CH_2\dot{C}Cl_2$ and 2.5 times more reactive towards $\sim\sim CH_2\dot{C}HPh$, and of the additions to acrylic acid $CH_2{=}CHCO_2H$ and crotonic acid $MeCH{=}CHCO_2H$ in which the former is 12 times more reactive towards $\sim\sim CH_2\dot{C}HPh$. These and many other examples arise from steric hindrance at the site of radical attack by a substituent in the 2-position. A more subtle example is the steric inhibition of resonance in the transition states of alkyl maleates on radical addition, which arises from the difficulty in attaining the necessary coplanarity. Models show that considerable steric hindrance is encountered with alkyl maleates, but not with fumarates, which are consequently more reactive.[15] In agreement with this view is the fact that the half esters show little difference in reactivity.[15] A comparison of the reactivities of *cis* and *trans* disubstituted alkenes has been given by Walling.[15]

The evidence we have been considering clearly favours tail addition to vinyl monomers. Although this mode predominates, many vinyl polymers carry steric strain, as is apparent from the data in Table 4[31] on heats of polymerization $(-\Delta H_p)$ of some common vinyl monomers; the values presented also contain minor contributions from monomer stabilization arising from conjugation. The 1,1-disubstituted monomers have values of $(-\Delta H_p)$ lower than those for monosubstituted monomers by 20–40 kJ mol^{-1}, this representing the steric strain built into the chains. The high C–C bond strength in fluorocarbons probably contributes to the high value of $-\Delta H_p$ for tetrafluoroethylene.

Table 4 Heats of Polymerization of Some Alkenic Monomers[31]

Monomer	$-\Delta H_p$ (kJ mol^{-1})
α-Methylstyrene	35.1
Methyl methacrylate	54.4
Vinylidene chloride	60.2
Styrene	68.6
Acrylonitrile	72.4
Methyl acrylate	78.2
Vinyl acetate	89.1
Vinyl chloride	108.8
Tetrafluoroethylene	138.1

17.2.1.3 *Polarity in the transition state*

Data on small radical reactions substantiate the proposal of Mayo and Walling[14] that polar contributions to the transition states of radical reactions play an important role in influencing the rates. Some very interesting results have been reported by Tedder and Walton,[29] who compared the rates of addition at 164 °C of the radicals $\cdot CH_3$, $\cdot CH_2F$, $\cdot CHF_2$ and $\cdot CF_3$ to tetrafluoroethylene and ethylene. The radicals form a series with steadily increasing electrophilicity, from the nucleophilic methyl to the strongly electrophilic trifluoromethyl. Addition of methyl is, therefore, accelerated, and that of trifluoromethyl retarded, by the presence of electron-attracting substituents in the alkene. The experimental data are set out in Table 5 and show the anticipated trend. The differences in rate arise from differences in activation energy, the frequency factors remaining approximately constant.

James and his colleagues[32] investigated the addition of ethyl radicals to a series of polar alkenes with the results shown in Table 6. These data refer to gas phase reactions at 100 °C. The rate of addition is enhanced by polar substituents in the order $-CN > Ph \sim CH_2{=}CH- > MeCO_2- > n\text{-}C_4H_9O- > $ alkyl, which is approximately that expected for a nucleophilic radical. The effect of

Table 5 Ratio of Rates of Addition of Methyl and Fluoromethyl Radicals to Tetrafluoroethylene and Ethylene at 164 °C[25]

Radical	$\cdot CH_3$	$\cdot CH_2F$	$\cdot CHF_2$	$\cdot CF_3$
$k(CF_2{=}CF_2)/k(CH_2{=}CH_2)$	9.5	3.4	1.1	0.1

steric hindrance arising from the β methyl substituent in the MeCH=CHCN isomers is apparent, and stabilization of the adduct radicals probably increases the rates of addition to styrene and the dienes. The order of monomers in Table 6 is in satisfactory agreement with Szwarc's data[33] on the rates of addition of methyl in the liquid phase at 65 °C, *viz.* acrylonitrile > methyl methacrylate > styrene > vinyl acetate > ethylene > alkyl acetate. It is interesting to compare these results, obtained with alkyl radicals, with those of Bevington[34,35] referring to additions of the benzoyloxy radical at 60 °C. The order of monomer reactivities in this case was found to be acrylonitrile < methyl methacrylate < vinyl acetate < 2,4,6-trimethylstyrene < styrene < 2,5-dimethylstyrene. The differing reactivity sequences for alkyl and benzoyloxy radicals are a consequence of the polar natures of the radicals, the alkyl radicals being nucleophilic and the benzoyloxy radicals strongly electrophilic, with $\sigma_p \approx 1$ according to the patterns treatment (Section 17.4.2).

Table 6 Addition of the Ethyl Radical to Polar Alkenes[32] Et$\cdot$ + CH$_2$=CHR → EtCH$_2\dot{C}$HR

Alkene	k(rel.)(100 °C)	Log A (dm^3 mol^{-1} s^{-1})	E_A (kJ mol^{-1})
Oct-1-ene	1.0	8.0	31.9
2-Methylpent-1-ene	1.6	8.0	30.7
n-Butyl vinyl ether	1.5	7.3	25.6
Vinyl acetate	1.8	7.8	29.0
Cyclohexa-1,3-diene	14.1	7.7	21.8
2,3-Dimethylbuta-1,3-diene	45.9	7.8	18.9
Styrene	34.1	7.5	17.2
cis-MeCH=CHCN	4.9	7.1	21.0
trans-MeCH=CHCN	6.3	7.4	21.8
Acrylonitrile	140	7.7	14.3

The above discussion illustrates the influence of polarity in the transition state on rates of addition. Polar factors can also influence regiospecificity, as is indicated by data in Table 7 due to Tedder and Walton.[24] Addition to vinyl fluoride occurs preferentially to the tail end, owing to the predominance of steric hindrance at the head site. In vinylidene fluoride additions the polar influence is apparent, the proportion of head addition decreasing with increasing electrophilicity of the radical. Finally, with trifluoroethylene the polar effects are large enough to bring about preferential head addition of $\cdot$CH$_3$ and $\cdot$CH$_2$F, in opposition to the steric hindrance. The proportion of head addition decreases along the series from $\cdot$CH$_3$ to $\cdot$CF$_3$.

Table 7 Regiospecificity [(2):(1)] of Addition of Methyl and Fluoromethyl Radicals to Fluorinated Alkenes.[24]

Radical	$C(2)H_2$=$C(1)HF$	CH$_2$=CF$_2$	CHF=CF$_2$
$\cdot$CH$_3$	1:0.2	—	1:2.1
$\cdot$CH$_2$F	1:0.3	1:0.4	1:2.0
$\cdot$CHF$_2$	1:0.2	1:0.1	1:0.9
$\cdot$CF$_3$	1:0.1	1:0.04	1:0.5

17.2.2 The Transition State in Radical Additions

A few examples have been quoted (Section 17.2.1.1) which illustrate that head and tail additions of $\cdot$CCl$_3$ to unsymmetrical alkenes have different activation energies. A more extended list (Table 8), taken from the work of Tedder and Walton and their colleagues,[36] illustrates this point more fully. These results demonstrate that two different transition states must be involved, corresponding to tail and head addition, respectively, which may be represented as in (2) and (3). These are σ transition states, and they must represent the highest points of the potential barriers for the two processes.

$$R_3C \text{---} CH_2 \text{====} CHX \qquad\qquad R_3C \text{---} CHX \text{====} CH_2$$

(2) σ transition state for tail addition   (3) σ transition state for head addition

Table 8 Activation Energies (kJ mol^{-1}) for Tail and Head Additions

Alkene Mode of addition	$CH_2=CHF$		$CHF=CF_2$		$CH_2=CF_2$	
	Tail	Head	Tail	Head	Tail	Head
$\cdot CCl_3$	26.8	35.2	38.5	42.7	32.3	47.8
$\cdot C_3F_7$	16.3	30.6	25.9	30.2	24.3	49.0

There is, however, evidence that the properties of the double bond as a whole influence the rates of addition. Thus, for additions to alkyl-substituted alkenes, linear correlations may be obtained between log k ($k =$ rate coefficient for addition) and the ionization potential.[37] The slopes of the plots for $\cdot CH_3$ and $\cdot CF_3$ are opposite in sign ($+$ and $-$, respectively). Again, the rate of addition is influenced by the total number of electron-donating or electron-releasing groups in the alkene.[20] It has, therefore, been suggested that the association between radical and π bond first gives a π complex; for reasons already given the latter must be lower in energy than the σ complexes which ultimately must be formed from it. Münger and Fischer[38] have shown that rates of addition of t-butyl radicals to alkenes depend on the electron affinities of the alkenes; this is consistent with the idea that the initial interaction between a nucleophilic alkyl radical and an alkene involves the singly occupied molecular orbital of the former and the π^* alkene LUMO.[39]

Clark[40] has recently reported the results of *ab initio* calculations, which show that the activation energy for addition of $\cdot CH_3$ to ethylene is significantly lowered when the alkene is complexed with Li$^+$. The overall heats of reaction are little affected, so that the result is a strong acceleration of the addition, without significant thermodynamic change. It is concluded that the acceleration should be general for all nucleophilic radicals and a variety of metal cations, especially those like Ag$^+$ which form stable alkene complexes.

An aspect of particular interest is the relevance of this work to the alternating copolymerization of electron donor and electron acceptor monomers in the presence of Lewis acids (see Volume 4, Chapter 23). The two types of monomer are members of the A and B groups of Hirooka;[41] generally the former are hydrocarbons or non-polar monomers, *e.g.* styrene and butadiene, while the latter have double bonds conjugated with electron-attracting substituents such as $-CO_2H$ or $-CN$ (*e.g.* methyl acrylate). Copolymerization of mixtures of A and B monomers produces copolymers showing some alternation; in the presence of a Lewis acid such as triethylaluminum sesquichloride the reaction is often greatly accelerated, and the product copolymer becomes strongly alternating with a 1:1 ratio in monomer units over a wide range of monomer feed compositions.[42] In some systems studied in detail (*e.g.* styrene, methyl acrylate and Et$_3$Al$_2$Cl$_3$) both cross-propagation reactions are considerably accelerated by the Lewis acid, especially that between the donor radical (*e.g.* ⁓S·) and the acceptor monomer, which enters into complex formation with the Lewis acid.[42] This latter propagation is normally relatively slow, and its enhancement by the Lewis acid is responsible for the high degrees of alternation. The $\cdot CH_3$ to C_2H_4 addition accelerated[40] by Li$^+$ would seem to be the prototype of this propagation reaction.

17.2.3 Wrong-way Additions

According to our discussions in Sections 17.2.1.1, 17.2.1.2 and 17.2.1.3 the structural features in common vinyl monomers are strongly conducive to head–tail propagation, but, in view of the stochastic nature of the propagation process, exceptions in the form of wrong-way additions would be anticipated and in fact are well known. They would naturally be expected to occur most frequently with monomers in which some or all of the structural features are less well developed.

The existence of 1,2-diglycol units in poly(vinyl alcohol) prepared by hydrolysis of poly(vinyl acetate) was demonstrated in 1948 by Flory and Leutner.[43] These workers cleaved poly(vinyl alcohol) chains at the 1,2-diglycol units by oxidation with periodic acid and measured the decrease in viscosity of the polymer in aqueous solution. It was concluded that the chains contained approximately 1% of 1,2-units; since this would apply to the poly(vinyl acetate) precursor, the results probably indicate the occurrence of reaction (2) to a small extent in the polymerization of vinyl acetate. NMR techniques now provide the most detailed and reliable information about polymer structures. Poly(vinyl fluoride),[44] poly(vinylidene fluoride)[45] and their copolymers[46] have received considerable attention and have been found to contain head–head units. It has been estimated that about 5–10% of the monomer units in the chains of poly(vinylidene fluoride) are reversed.

In the examples quoted, both the extent of stabilization of the propagating radicals by the substituent groups (Section 17.2.1.1) and the steric barriers of head–head addition (Section 17.2.1.2) are relatively small, although the polar influences in the fluorine compounds are significant (Section 17.2.1.3). Reference to the relevant small radical additions in Table 7 will show that significant head–head addition in the propagation reactions of $CH_2=CHF$ and $CH_2=CF_2$ is not surprising.

17.3 TRANSFER PROCESSES

Following the plan adopted for propagation reactions we shall discuss information from small radical abstraction reactions likely to assist in the interpretation of transfer processes with polymer radicals.

17.3.1 Evans–Polanyi Relations

The Evans–Polanyi relation (equation 7) formulated in 1938 still provides the most general basis for discussion of abstraction reactions.[47]

$$E_A = \alpha \Delta H^\circ + C \tag{7}$$

This equation, which relates the activation energy E_A to the total enthalpy change ΔH°, was derived from a consideration of the potential energy curves of a series of closely similar reactions. If these curves do not change shape basically from one reaction to another, equation (7) follows from geometric arguments, α and C being constants. The parameter α is less than unity and is expected to be larger for series of reactions with greater activation energies.

The assumptions underlying the relation (7) are probably valid for reactions between a radical and a series of alkanes. Trotman-Dickenson[48] showed that the relation holds for $\cdot CH_3$ and Jones and Whittle[49] reached a similar conclusion for $\cdot CF_3$

$$E_A(\cdot CH_3) = 0.49\Delta H^\circ + 62 \text{ (kJ mol}^{-1}) \tag{8}$$

$$E_A(\cdot CF_3) = 0.53\Delta H^\circ + 52 \text{ (kJ mol}^{-1}) \tag{9}$$

However, as is apparent from equations (8) and (9), the two series of reactions give different straight lines when E_A is plotted against ΔH°. In a series of reactions such as those discussed, the enthalpy change ΔH° is equal to the difference between the dissociation energies of the bond being broken and that being formed; for a given radical the latter is constant so that we obtain from equation (7)

$$E_A = \alpha[(D(R-H)] + C' \tag{10}$$

where R–H is the bond being broken and C' is a constant for the series. Equation (10) was confirmed for $\cdot CH_3$ by Trotman-Dickenson,[48] who assumed the same equation for abstraction by $\cdot CH_3$ from cycloalkanes and hence calculated bond dissociation energies in the latter. Whittle[49] calculated these dissociation energies in a similar manner, using his results for abstraction by $\cdot CF_3$ and Br$\cdot$ and obtained comparable results. The Evans–Polanyi relation, therefore, appears to be valid for cycloalkanes.

The strengths of the bonds being broken and of those being formed determine ΔH° and hence influence the activation energy, as would be supposed intuitively. For abstraction of hydrogen atoms the ease of abstraction is in the order tertiary > secondary > primary for all radicals, corresponding to the order of increasing bond strength (primary > secondary > tertiary). A rather extreme example

Table 9 Relative Selectivities in Abstraction from Ethane, Propane and Isobutane by Halogen Atoms (X$\cdot$) in the Gas Phase[25]

Atom	T (°C)	CH_3-	$-CH_2-$	$>CH-$	$D(X-H)$ (kJ mol^{-1})
F$\cdot$	25	1	1	2	569
Cl$\cdot$	25	1	4	7	431
Br$\cdot$	150	1	80	2000	364
I$\cdot$	150	1	1000	97000	297

of the influence of the strength of the new bond being formed is provided[25] by the rates of hydrogen abstraction by halogen atoms, which decrease strongly in the series $F\cdot > Cl\cdot > Br\cdot > I\cdot$, the order of decreasing halogen–hydrogen bond strength (Table 9). Table 9 also presents data for simple alkanes. The data illustrate a point of general importance, *viz.* that the selectivity increases as the rate of reaction decreases. Hydrogen abstraction by bromine and iodine atoms can evidently be highly selective.

17.3.1.1 *Polar effects*

Relations of the Evans–Polanyi type are obscured if the potential energy curves of reactions are dissimilar in shape. This may arise as a result of delocalization changes from one reaction to another, or of steric or polarity changes, *i.e.* the factors previously discussed in Sections 17.2.1.1, 17.2.1.2 and 17.2.1.3.

Polar effects are as important in transfer reactions as in propagation processes, and are well documented in small radical chemistry. A simple illustration is provided by data for hydrogen abstraction from halomethanes. The C–H bond energy probably decreases along the series CH_4, CH_3X, CH_2X_2, CHX_3 (X = halogen). Hydrogen abstraction by $\cdot CF_3$ from the compounds CH_4, CH_3Cl, CH_2Cl_2 and $CHCl_3$ shows the expected decrease in activation energy, the values being[50] 47.0, 44.5, 31.9 and 27.7 kJ mol^{-1}, respectively. Similarly, hydrogen abstraction by $\cdot CH_3$ from CH_4, CH_3F and CH_2F_2 has activation energies[51] of 60.9, 47.9 and 42.8 kJ mol^{-1}, respectively. However, abstraction by $\cdot CF_3$ from CH_4, CH_3F and CH_2F_2 has $E_A = 47.0$ kJ mol^{-1} in each case. This unexpected result probably arises from a fortuitous compensation of decreases in E_A (arising from decreasing C–H bond energy) by increases originating in polar repulsive effects, which would be expected to be strong with CH_3F and CH_2F_2.

The importance of polar effects in the transition state is illustrated by hydrogen abstraction from CH_4 and HCl by $\cdot CH_3$ and $\cdot CF_3$.[25] These reactions, which are almost thermoneutral, have the activation energies shown in Table 10. The low values of E_A for the reactions with HCl are attributable to polarization in the transition state

$$\cdot CF_3 + H{-}Cl \rightarrow \overset{\longleftarrow}{[\overset{\delta+}{CF_3} {-}{-}{-} H {-}{-}{-} \overset{\delta-}{Cl}]^{\ddagger}} \rightarrow CF_3H + Cl\cdot \qquad (11)$$

$$(4)$$

$$\cdot CH_3 + H{-}Cl \rightarrow \overset{\longrightarrow}{[\overset{\delta+}{CH_3} {-}{-}{-} H {-}{-}{-} \overset{\delta-}{Cl}]^{\ddagger}} \rightarrow CH_4 + Cl\cdot \qquad (12)$$

$$(5)$$

Note that the activation energies are reversed in order in passing from CH_4 to HCl. The formation of the polar transition state (**4**) for the HCl reaction with $\cdot CF_3$ is opposed by $\cdot CF_3$, which is nucleophilic, whereas that for the corresponding reaction with $\cdot CH_3$ (**5**) is favoured by the nucleophilicity of the latter. Consequently, abstraction from HCl by a methyl radical has the lower activation energy. The same phenomenon is operative in the reverse reactions, for which the activation energies are 33.5 and 14.7 kJ mol^{-1} for $\cdot CF_3$ and $\cdot CH_3$, respectively. Polar effects of this type are not significant in abstraction from CH_4.

Table 10 Activation Energies for Hydrogen Abstraction from CH_4 and HCl by $\cdot CH_3$ and $\cdot CF_3$[25]

Reaction	E_A (kJ mol^{-1})
$CH_4 + \cdot CH_3 \rightarrow \cdot CH_3 + CH_4$	58.6
$CH_4 + \cdot CF_3 \rightarrow \cdot CH_3 + CF_3H$	46.9
$HCl + \cdot CH_3 \rightarrow Cl\cdot + CH_4$	10.5
$HCl + \cdot CF_3 \rightarrow Cl\cdot + CF_3H$	20.9

The influence of polar factors in transfer reactions of polymer radicals is presented in Table 11.[52-54] The relatively low reactivity of polystyrene radicals towards triethylamine arises partly from the resonance stabilization of the radical, which weakens the bond being formed, but mainly from polar effects, which are here unfavourable as indicated in the transition state (**6**). On the other

Table 11 Reactivities of Polystyrene and Polyacrylonitrile Radicals[52-54]

Reaction	Rate coefficient at 60 °C ($mol^{-1} dm^3 s^{-1}$)	
∿∿AN• + NEt_3	1450	(in DMF)
∿∿S• + NEt_3	0.12	(in styrene)
∿∿AN• + $FeCl_3$	6.5×10^3	(in DMF)
∿∿S• + $FeCl_3$	9.4×10^4	(in DMF)

hand, the electrophilic polyacrylonitrile radicals and the nucleophilic triethylamine cooperate in forming the polar transition state (7). The situation is reversed with iron(III) chloride, a strong electrophile, polarity in the transition state with polystyrene radicals (8) being more highly developed than with polyacrylonitrile radicals (9).

$$[\overrightarrow{NEt_3} \cdot \overleftarrow{S∿∿}]^+ \qquad [\overleftarrow{NEt_3} \cdot \overrightarrow{AN∿∿}]^+ \qquad [\overleftarrow{FeCl_3} \cdot \overleftarrow{S∿∿}]^+ \qquad [\overleftarrow{FeCl_3} \cdot \overrightarrow{AN∿∿}]^+$$

(6)   (7)   (8)   (9)

Polar effects in systems which are closely related have been treated with some success by use of the Hammett equation, *e.g.* hydrogen abstraction from substituted toluenes by Ph• and •CCl_3 with ρ values -0.1 and -1.46, respectively.[55, 56, 20]

17.3.1.2 Steric effects

As in propagation, steric effects are important in abstraction reactions and may be divided into steric hindrance, impeding the approach of a radical to the reaction site, steric inhibition of resonance and steric compression. The last is of general importance; according to Rüchardt and colleagues[57] the release of steric compression on radical formation contributes to the progressive decrease in C–H bond strength in the series primary, secondary, tertiary to which reference has already been made (Section 17.3.1). Release of compression in CBr_4 on forming •CBr_3 probably contributes to the high transfer constant of carbon tetrachloride. A penultimate unit effect in chain transfer[58] may also be attributed to steric compression. The transfer constants of the styryl-type radicals (10, 11 and 12) in Table 12[59] towards carbon tetrabromide show a marked dependence on the nature of the penultimate unit, being very much smaller when the latter is methyl methacrylate (radical 11). No such effect is noticeable with 1-butanethiol[59] as transfer agent. Models show that the end portions of the chains formed by transfer of radical (11) with carbon tetrabromide have a close contact between the α methyl of the penultimate unit and the terminal bromine (see 13; X = Br, Y = Me); the resulting compression is not found in the polymers from radicals (10) or (12) which have styrene or methyl acrylate as penultimate units (13; X = Br, Y = H) or in the radicals with 1-butanethiol (13; X = H; Y = H, Me).

$$∿CH_2\underset{\underset{CO_2Me}{|}}{\overset{\overset{Y}{|}}{C}}-CH_2-\overset{\overset{X}{|}}{CH}$$

(13)

Table 12 Transfer Constants of Styryl-type Radicals at 60 °C[59]

Radical/Transfer Agent	CBr_4	1-BuSH
(10) ∿∿S–S•	337	22
(11) ∿∿MMA–S•	60	22
(12) ∿∿MA–S•	302	20

17.3.1.3 *Electron delocalization*

Electron delocalization in the incipient radical or the attacking radical has a major influence on the strengths of the new bond and the bond being broken and so affects rates of abstraction. Delocalization in the radical product reduces the strength of the bond being broken and enhances transfer. The activation energies for hydrogen abstraction by $\cdot CH_3$ (at 164 °C) from the molecules Et–H, $PhCH_2$–H, $MeCH{=}CHCH_2$–H and $MeCOCH_2$–H have been reported[25] as 43.5, 39.7, 31.8 and 40.6 kJ mol^{-1}, respectively. These values indicate the expected 'resonance effect', although perhaps it is surprisingly small. Electron delocalization in the attacking radical weakens the new bond and so reduces the rate of abstraction. We may refer here to the absolute rate coefficients for hydrogen abstraction from toluene presented in Table 13. These quantities, used as measures of the 'general reactivities' of radicals in the empirical patterns treatment (Section 17.4.2), were originally considered to be determined by resonance stabilization of the radicals. The order of radicals in Table 13 shows similarities to the order of monomers in Table 4, only styrene being seriously out of place. Since the values in the latter table are essentially measures of steric compression, it is difficult to resist the conclusion that steric factors play a prominent role in determining the transfer activities. From this point of view, the position of styrene is understandable; the extent of delocalization is significantly larger in the polystyrene radical than in the other radicals in Table 4 and this accounts for the low position of styrene. On the other hand, the steric compression in polystyrene is 'normal' for a polymer of a substituted ethylene with a single bulky substituent. These considerations do not invalidate use of the data in Table 13 as measures of general reactivities of radicals, but indicate that the latter term requires redefinition. Quantum-mechanical studies on hydrogen abstraction from alkanes have been published by Brown[60] and by Fukui *et al.*[61]

Table 13   Transfer Coefficients for Standard Radicals with Toluene at 60 °C[53]

Radical	$k_{3,T}$ (mol^{-1} dm^3 s^{-1})	$4 + log\ k_{3,T}$	σ_p
Styrene	2.11×10^{-3}	1.324	−0.01
Methyl methacrylate	1.25×10^{-2}	2.301	0.28
Methacrylonitrile	2×10^{-2}	2.097	0.49
Methyl acrylate	0.56	3.748	0.45
Acrylonitrile	0.785	3.895	0.66
Vinyl acetate	7.73	4.888	0.31

17.3.2 Halogen Abstraction

Transfer to halogen derivatives is commonly encountered in polymerization, and halogen abstraction has been known in small radical reactions since the work of Kharasch.[27] The ease of abstraction for the different halogens follows the inverse order of C–halogen bond strengths and is[27] I > Br ~ H > Cl > F. The position of hydrogen in the series is anomalous. Szwarc[62] studied abstraction from halogen by $\cdot CH_3$ in toluene and isooctane solutions and found that the Evans–Polanyi relation (equation 7) applies approximately; hydrogen does not fit in the series. Szwarc concluded that the approaching $\cdot CH_3$ experiences a polar repulsion by the filled orbitals of the halogen atom in X–R which would be negligible with hydrogen in H–R. Thus the rate of hydrogen abstraction is greater than expected from bond energy considerations alone. Support for this view comes from Szwarc's observations on halides containing electron-attracting substituents: for example, $\cdot CH_3$ abstracts iodine from CF_3I about 500 times as rapidly as from CH_3I.

Alcock and Whittle[63] investigated halogen abstraction by $\cdot CF_3$ in the gas phase. In the case of the chloromethanes, evidence for polar factors is encountered similar to that already discussed in connection with the work of Whittle and colleagues in hydrogen abstraction from fluoromethanes by $\cdot CF_3$ (Section 17.3.1.1).

17.4 EMPIRICAL RELATIONS

Prediction of the rates and regioselectivities of radical reaction is a very worthy theoretical objective which has attracted attention since the early work of Evans *et al.*[64] in 1948. Notable contributions have been made more recently by Yonezawa *et al.*,[65,66] who used the frontier electron density approach of Fukui and by Kawabata *et al.*[67] Although the treatments have yielded results of interest they are subject to criticism (*e.g.* see Dewar and Thompson[68]) and it must be admitted that a satisfactory comprehensive discussion has yet to be formulated. *Ab initio* calculations of reaction energy surfaces as functions of the geometric variables involved are impossible at present. It is, of course, not necessary that a comprehensive theory should be formulated in terms of the concepts used in the preceding pages; indeed, in the frontier electron technique, no formal division of substituent influences into polar and resonance effects is made.

Empirical treatments seek to quantify the concepts we have discussed. They do not necessarily have a firm theoretical basis, but they can be valuable in practice, both in summarizing experimental data and making predictions about reactions not yet examined.

17.4.1 The Q–e Scheme

We have referred in Section 17.2.1 to the alternating tendency in copolymerization which was accounted for by Mayo and Walling[14] in terms of differing electrical polarities. Further, in a radical–monomer addition each partner may be assumed to have a 'general reactivity', which, as discussed in Section 17.2.1, was originally correlated with the resonance stabilization of the species. These ideas are the basis of the Q–e scheme formulated by Price[18] and Alfrey and Price.[19] The two reactants $A\cdot$, B in a radical/monomer addition were assigned e values, e_A and e_B, respectively, to represent their charges, identical charges being assumed for a monomer and its derived radical, and general reactivities of radicals and monomers were denoted by P and Q, respectively. The rate of reaction was considered to be determined by the four quantities P_A, Q_B, e_A and e_B, as indicated by equation (13), in which k_{AB} is the rate coefficient for propagation. Similarly, equations (14)–(16) may be written for the other three propagation reactions in the copolymerization of monomers A and B.

$$k_{AB} = P_A Q_B e^{-e_A e_B} \tag{13}$$

$$k_{AA} = P_A Q_A e^{-e_A^2} \tag{14}$$

$$k_{BA} = P_B Q_A e^{-e_A e_B} \tag{15}$$

$$k_{BB} = P_B Q_B e^{-e_B^2} \tag{16}$$

The two reactivity ratios, r_A and r_B, and their product are then given by the relations

$$r_A = k_{AA}/k_{AB} = (Q_A/Q_B)e^{-e_A(e_A - e_B)} \tag{17}$$

$$r_B = k_{BB}/k_{BA} = (Q_B/Q_A)e^{-e_B(e_B - e_A)} \tag{18}$$

$$r_A r_B = e^{-(e_A - e_B)^2} \tag{19}$$

The Q–e scheme is intended to enable predictions to be made about reactivity ratios in systems which have not been studied experimentally. For this purpose it is necessary to assign characteristic Q and e values to monomers on the basis of copolymerization experiments with a limited number of reference monomers. If this can be carried through successfully, 'new' (*i.e.* unstudied) systems may be treated by application of equations (17) and (18). Thus from the binary copolymerization data of one reference monomer of known Q and e with n other monomers, the resulting Q and e assignments permit reactivity ratios for $n(n-1)/2$ other monomer combinations to be predicted. Evidently the advantages of a scheme of this type increase rapidly with the number of monomers classified. Styrene has been chosen as reference monomer and assigned $Q = 1$. Choice of e for styrene is complicated by the non-linear relation between the values of the r and e in equations (17) and (18). Originally styrene was assigned $e = -1.0$, but nowadays $e = 0.8$ is favoured. The matter has stimulated considerable debate.[67,69-73]

Three aspects of the Q–e scheme have attracted criticism: (a) the assumption of equal charges for monomer and derived radical; (b) the assumption of permanent charges rather than polarization phenomena in the transition state which would account for the alternating effect; and (c) reactivity ratios are sensibly independent of the dielectric constant of the reaction medium, which is unexpected if the reactants carry charges. Although these criticisms are mainly valid, they do not reduce the practical use of the Q–e scheme, which is undoubtedly consistent with a great deal of experimental data.[74]

In the early days of the Q–e scheme, Wall[75] drew attention to some discrepancies encountered in the copolymerization of dienes (butadiene, isoprene and chloroprene) with vinyl monomers (styrene and acrylonitrile) and suggested that radicals derived from the dienes might have polarities different from those of the monomers. He proposed introducing a third parameter e^* to accommodate this behaviour. For general use the Q–e–e^* scheme would be unmanageable unless some arbitrary features were removed. Hoyland[76] has remarked that 'the Q–e–e^* scheme should be regarded only as a mathematical model whose parameters cannot be related to any simple molecular property'.

An extensive list of Q and e values has been compiled by Young.[77]

17.4.2 Patterns of Radical Reactivity

This treatment (abbreviated to 'patterns') was devised[53, 70, 78] as an empirical scheme in which arbitrary features were reduced to a minimum. As in the Q–e approach, reaction rates were considered to depend on general reactivities of radicals and substrates together with polar factors, but these properties were defined in such a way that, as far as possible, they could be determined in independent experiments. Patterns applies to both propagation and transfer reactions (see equations 28, 29 and 31 below).

The general reactivity of a radical $R\cdot$, originally considered to be determined by its resonance stabilization, was measured[53] by the rate coefficient of its transfer reaction with toluene at 60 °C:

$$R\cdot + PhCH_3 \xrightarrow{k_{3,T}} RH + Ph\dot{C}H_2 \tag{20}$$

it will be denoted by $k_{3,T}$ for conformity with the original literature. Table 13 shows values of $k_{3,T}$ for some 'standard' polymer radicals. We have already given reasons (Section 17.3.1.3) for believing that these values include steric contributions and, since polar effects in reaction (20) are unlikely to be very significant, $k_{3,T}$ may be regarded as including all contributions to the reactivity arising from non-polar sources. In agreement with this view, linear relations were observed between log $k_{3,T}$ and log $k_{3,HC}$, where $k_{3,HC}$ is the rate coefficient for transfer to hydrocarbon HC at 60 °C. The hydrocarbons used included cyclohexane, benzene, ethyl-, isopropyl- and t-butyl-benzene, and the linear plots have slopes within 10% of unity. Hence, to this approximation, relation (21) holds,

$$k_{3,HC} = \gamma k_{3,T} \tag{21}$$

γ being constant; thus there is nothing 'special' about transfer to toluene, and transfer to other hydrocarbons could serve as measures of general reactivity.

Transfer reactions which might be expected to possess more highly polar transition states gave completely different results. No simple relation such as equation (21) appeared; instead two distinct and characteristic types of reactivity pattern emerged, which we shall refer to as types 1 and 2, respectively. The linear relation (equation 21) may be called type 0. Types 1 and 2 are illustrated in Figures 1 and 2, in which the (decadic) logarithms of the rate coefficients for reactions of the standard polymer radicals with iron(III) chloride and triethylamine are plotted against log $k_{3,T}$ for the radicals. All the transfer agents examined (other than hydrocarbons) give one or other of these patterns; for example, type 1 for carbon tetrahalides and thiols and type 2 for N,N-dimethylformamide.

Propagation reactions show similar regularities when values of log k_p for additions of a given monomer with the standard radicals are plotted against log $k_{3,T}$. Styrene and vinyl acetate give linear (type 0) plots and many familiar monomers yield type 1 patterns, *e.g.* acrylonitrile, methyl acrylate, methyl methacrylate and methacrylonitrile. Monomers giving type 2 are less common, but include p-methoxystyrene, p-diethylaminestyrene and N-vinylurethane.[70]

Monomers and transfer agents may thus be classified according to the polarities of the transition states in their reactions with the standard radicals; the reactivity patterns may be rationalized as in Table 14.

Table 14 Classification of Reactivity Patterns for Reactions of Substrates with Standard Radicals

Type of pattern	Polarization in transition state	Sign of α
0	Insignificant, substrate 'non-polar'	$\alpha = 0$
1	Substrate electron-attracting	$\alpha - ve$
2	Substrate electron-donating	$\alpha + ve$

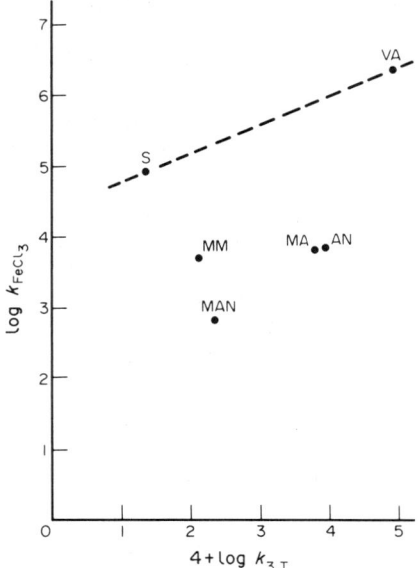

Figure 1 Reactivity pattern, type 1: substrate iron(III) chloride.[53] The styrene and vinyl acetate points are joined by a dashed line to illustrate the general disposition of the points

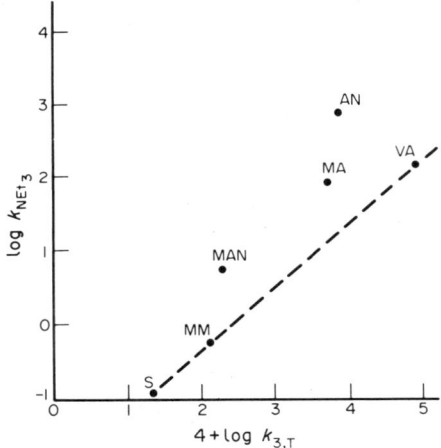

Figure 2 Reactivity pattern, type 2: substrate triethylamine.[53] The styrene and vinyl acetate points are joined by a dashed line to illustrate the general disposition of the points

The contribution of polar structures to the transition state lowers the activation energy. The reduction ΔE_R for radical $R\cdot$ relative to the standard reaction with toluene (equation 20) is a measure of the polarity of the transition state; p_R defined by equation (22) may be called the 'polarity parameter' of the reaction. Bamford *et al.*[53] developed relation (23) in which k_S is the rate coefficient for reaction (addition or transfer) of radical $R\cdot$ with substrate S, and evaluated p_R (actually *differences* in p_R) from the experimental patterns. The order of polarities for a series of radicals with electron-attracting substrates (type 1) appeared to be independent of the substrate, being the same for iron(III) chloride, carbon tetrachloride, 1-butanethiol, acrylonitrile, methyl methacrylate, methacrylonitrile and methyl acrylate without significant inversions. With triethylamine (type 2) the order is reversed, and for reactions with styrene, vinyl acetate and hydrocarbons (type 0) differences in p_R are small and there is no definite sequence. This analysis achieves a separation of non-polar and polar contributions and indicates that the polarity of the transition state depends on both partners in the reaction.

$$p_R = \frac{\Delta E_R}{RT} \tag{22}$$

$$\log k_S = \log k_{3,T} + p_R \tag{23}$$

For a given substrate, the polarity of the transition state depends on the substituent(s) on the α carbon of the radical, and indeed a linear correlation (equation 24) was found between p_R and the Hammett σ_p values for the substituent(s)

$$p_R = \alpha\sigma_p + \beta \tag{24}$$

When the radical has two substituents the algebraic sum of the σ_p values may be used. (However, although this procedure is satisfactory for substituents having opposite signs it overestimates the effective σ_p when both have the same sign.[70] One way of avoiding this difficulty is to use equation (25) for calculating σ_p from α.) Table 13 shows values of σ_p for the standard radicals.

Combination of equations (23) and (24) leads to the final expression of the patterns treatment

$$\log k_S = \log k_{3,T} + \alpha\sigma_p + \beta \tag{25}$$

This contains two parameters, $k_{3,T}$ and σ_p, characteristic of the radical and two parameters, α and β, characteristic of the substrate. The sign of α determines the direction of polarization in the transition state (*cf.* equation 24) and the type of reactivity pattern as shown in Table 14. The interrelations may be further illustrated by calculating 'idealized' patterns from equation (25) (see ref. 78). For a radical with an electron-attracting substituent (σ_p positive), polar factors influencing the rate become more favourable as α increases, and conversely for radicals with electron-donating substituents.

Two types of diagram can be constructed[70,78] which elaborate the role of polar contributions. The first indicates the reactivities of radicals towards substrates of known α. A linear plot of $(k_S - \beta)$ *vs.* α may be drawn for any radical of known $k_{3,T}$ and σ_p by use of equation (25). The intercepts on a vertical line at any α value give the relative rate coefficients for reactions of the different radicals with a substrate having this α (since β is constant for a given substrate). A complementary type of diagram, indicating relative reactivities of different substrates towards a radical of specified σ_p, is obtained by plotting $(\log k_S - \log k_{3,T})$ *vs.* σ_p. Here the lines have slope α and intercept β on $\sigma_p = 0$. A vertical at given σ_p shows the relative reactivities of the substrates towards the radical. Thus this diagram enables reactivity ratios (see Volume 3, Chapter 13) and transfer constants (see Volume 3, Chapter 2) to be estimated simply and rapidly. Clearly, the two plots summarize a great deal of data on copolymerization and chain transfer.

By applying equation (25) to the four propagations in a binary copolymerization of monomers A and B we derive the following relations for the reactivity ratios

$$\log r_A = \sigma_{pA}(\alpha_A - \alpha_B) + \beta_A - \beta_B \tag{26}$$

$$\log r_B = -\sigma_{pB}(\alpha_A - \alpha_B) - (\beta_A - \beta_B) \tag{27}$$

These may be compared to the corresponding expressions equations (17) and (18) of the $Q-e$ scheme which, written in logarithmic form, become

$$\ln r_A = -e_A(e_A - e_B) + \ln Q_A - \ln Q_B \tag{28}$$

$$\ln r_B = e_B(e_A - e_B) + (\ln Q_A - \ln Q_B) \tag{29}$$

While the two sets are formally similar, comparison reveals the effect of the assumption made in the $Q-e$ scheme that a monomer and its radical have the same e values. Equations (26) and (27) are, in fact, formally similar to those resulting from Wall's $Q-e-e^*$ scheme[75] if e^* is replaced by σ_p, but this correspondence does not, of course, expose them to Hoyland's criticism[76] of the $Q-e-e^*$ treatment that the parameters are not related to any simple molecular property (Section 17.4.1). On the contrary, the patterns treatment illustrates the possibility of attaching physical significance to the parameters of Wall's scheme. The equivalent in patterns to equations (28) and (29) would be obtained by assuming proportionality between α and σ_p.

Since the polar properties of a monomer $CH_2=CHX$ and its derived radical $\sim\sim\sim CH_2\dot{C}HX$ originate in X, some relation between α and σ_p indeed seems likely and Bamford and Jenkins[78] found that equation (30)

$$\alpha = -5.3\sigma_p \tag{30}$$

holds approximately for a number of monomers, although with considerable scatter. Introduction of equation (30) into the basic patterns equation (25) yields an expression similar to equation (13) in logarithmic form, in which P, $\ln Q$ and e correspond to $k_{3,T}$, β and α, respectively. Note, however, that relation (30) is not a necessary part of the patterns treatment, but is sometimes useful when experimental data for determining α are not available. Jenkins[79] has examined data for binary copolymerizations of a number of monomers in groups of three in connection with a proposal by

Ham,[80] and shown that the patterns equation (25) gives a good fit to the experimental findings. However, the simplified relation incorporating equation (30) and the (equivalent) $Q-e$ scheme are unsatisfactory.

Expression (31) for the transfer constant C_S of a given radical towards a substrate S, analogous to the relations for reactivity ratios (equations 26 and 27), may easily be derived from equation (25).

$$\log C_S = \sigma_{pR}(\alpha_S - \alpha_M) + \beta_S - \beta_M \tag{31}$$

Here α_M, β_M refer to the monomer M conjugate to the radical R•.

Tedder, Walton and their colleagues[81,82] have extended the patterns treatment to include addition to both ends of an alkene by using two sets of parameters α, α' and β, β' to characterize the two ends. For present purposes the problem is identical to that of a single radical species reacting with two different monomers or transfer agents, as encountered in formulating expressions for reactivity ratios or transfer constants. McMurray, Tedder, Vertommen and Walton[81] obtained the equation

$$\log OR = \log \frac{k'_2}{k_2} = \sigma_p(\alpha' - \alpha) + \beta' - \beta \tag{32}$$

in which k'_2 and k_2 are rate coefficients for head and tail additions, respectively, and OR the orientation ratio, an expression analogous to equations (26), (27) or (31). These workers studied the free-radical addition of chloroiodo- and diiodo-methane to fluoroalkenes at 150 °C, so that the chain-carrying radicals were chloromethyl and iodomethyl. The orientation ratio was determined from analysis of the products formed by addition of these radicals to the two ends of the alkenes. Values of $\alpha' - \alpha$ and $\beta' - \beta$ were obtained from plots of $\log OR$ *vs.* σ_p according to equation (32); actually the Taft σ^* was employed (which has values very similar to the Hammett σ_p except for fluorine atoms). It was found that a plot of $\log OR$ *vs.* $\sigma_p(\alpha' - \alpha) + (\beta' - \beta)$ according to equation (32) gave an excellent straight line with a gradient close to unity and a correlation coefficient of 0.98. Moreover, the experimental values of $\alpha' - \alpha$ were in satisfactory agreement with those calculated from equation (30). As the authors[81] point out, the patterns approach developed for radical polymerization in solution 'can thus be adapted to the problem of radical addition to alkenes in the gas phase with considerable success. The correlations indicate that both the polarity of the alkene and the polar character of the radical are important.'

17.4.3 The 'Electronegativity' and the 'Charge Transfer' Models

Two models have been proposed by Hoyland.[83,84] In the 'electronegativity model' electronegativities were assigned to radical and monomer and the localization energy (the resonance stabilization) of the monomer was employed. (Localization energy has been used as a 'reactivity index' in radical additions by numerous workers.)[20,85] Each monomer was characterized by three parameters: localization energy and the monomer and radical electronegativities. Localization energy and radical electronegativity were arbitrarily assumed to be zero for styrene. Good agreement with experiment was obtained, except for systems containing acrylonitrile; that is, with these exceptions, a satisfactorily self-consistent set of parameters could be computed.

The 'charge transfer' model also employs three parameters for each monomer, related to the energies of the highest occupied orbital and the lowest lying virtual orbital of the monomer and the singly occupied radical orbital. This model gives similar agreement with experimental data.

Hoyland also developed additional mathematical models and discussed the relative merits of these and others.[76] On the basis of a numerical least-squares formulation taking into account experimental errors he concluded that 'in view of the relatively large uncertainties in experimental data, it is felt that either the charge transfer or the electronegativity model represents a good compromise between quantitative accuracy, number of parameters per monomer and ease of routine usage'.

17.5 REFERENCES

1. G. Moad and D. H. Solomon, in 'Comprehensive Polymer Science', ed. G. Allen and J. C. Bevington, Pergamon Press, Oxford, 1988, vol. 3, chap. 8.
2. H. Staudinger and A. Steinhoffer, *Justus Liebigs Ann. Chem.*, 1935, **517**, 35.
3. P. J. Flory, 'Principles of Polymer Chemistry', Cornell University Press, Ithaca, NY, 1953.

4. C. S. Marvel and C. E. Denoon, *J. Am. Chem. Soc.*, 1938, **60**, 1045; C. S. Marvel and C. L. Levesque, *J. Am. Chem. Soc.*, 1938, **60**, 280; C. S. Marvel, J. H. Sample and M F. Roy, *J. Am. Chem. Soc.*, 1939, **61**, 3241; C. S. Marvel, E. H. Riddle and J. O. Corner, *J. Am. Chem. Soc.*, 1942, **64**, 92.
5. C. S. Marvel, in 'The Chemistry of Large Molecules', ed. R. E. Burke and O. Grummitt, Interscience, New York, 1943.
6. P. J. Flory, *J. Am. Chem. Soc.*, 1939, **61**, 1518; 1942, **64**, 177.
7. See, for example, D. J. E. Ingram, 'Free Radicals as Studied by Electron Spin Resonance', Butterworths, London, 1958.
8. H. Fischer, *J. Polym. Sci., Part B*, 1964, **2**, 529.
9. H. Fischer and G. Giacometti, *J. Polym. Sci., Part C*, 1965, **16**, 2763.
10. C. Corvaja, H. Fischer and G. Giacometti, *Z. Phys. Chem. (Frankfurt-am-Main)*, 1965, **45**, 1.
11. S. E. Bresler, E. N. Kozbekov, V. N. Fomichev and V. N. Shadrin, *Makromol. Chem.*, 1972, **157**, 167.
12. S. E. Bresler, E. N. Kozbekov and V. N. Shadrin, *Makromol. Chem.*, 1974, **175**, 2875.
13. M. Kamachi, M. Kohno, Y. Kuwae and S. Nozakura, *Polym. J.*, 1982, **14**, 749.
14. F. R. Mayo and C. Walling, *Chem. Rev.*, 1950, **46**, 191.
15. C. Walling, 'Free Radicals in Solution', Wiley, New York, 1957.
16. M. S. Kharasch, H. Englemann and F. R. Mayo, *J. Org. Chem.*, 1937, **2**, 288.
17. D. H. Hey and W. A. Waters, *Chem. Rev.*, 1937, **21**, 169.
18. C. C. Price, *J. Polym. Sci.*, 1946, **1**, 83; 1948, **3**, 772; *Discuss. Faraday Soc.*, 1947, **2**, 304.
19. T. Alfrey, Jr. and C. C. Price, *J. Polym. Sci.*, 1947, **2**, 101.
20. See, for example, D. C. Nonhebel and J. C. Walton, 'Free-radical Chemistry', Cambridge University Press, London, 1974, chap. 8.
21. M. Feld, A. P. Stefani and M. Szwarc, *J. Am. Chem. Soc.*, 1962, **84**, 4451.
22. A. F. Stefani, Lan-Yuh Yang Chung and H. E. Todd, *J. Am. Chem. Soc.*, 1970, **92**, 4168.
23. O. P. Strausz, I. Safarik, W. B. O'Callaghan and H. E. Gunning, *J. Am. Chem. Soc.*, 1972, **94**, 1828.
24. J. M. Tedder and J. C. Walton, *Tetrahedron*, 1980, **36**, 701.
25. J. M. Tedder, *Angew. Chem., Int. Ed. Engl.*, 1982, **21**, 401.
26. J. M. Tedder, in 'Reactivity, Mechanism and Structure in Polymer Chemistry', ed. A. D. Jenkins and A. Ledwith, Wiley, London, 1974, chap. 2.
27. M. S. Kharasch, E. V. Jensen and W. H. Urry, *Science (Washington, D.C.)*, 1945, **102**, 128; *J. Am. Chem. Soc.*, 1947, **69**, 1100; M. S. Kharasch, O. Reinmuth and W. H. Urry, *J. Am. Chem. Soc.*, 1947, **69**, 1105.
28. R. Gregory, R. N. Haszeldine and S. E. Tipping, *J. Chem. Soc. C*, 1970, 1750 and earlier papers.
29. J. M. Tedder and J. C. Walton, *Adv. Free-Radical Chem.*, 1980, **6**, 155.
30. P. S. Dixon and M. Szwarc, *Trans. Faraday Soc.*, 1963, **59**, 112; J. M. Pearson and M. Szwarc, *Trans. Faraday Soc.*, 1964, **60**, 564; G. E. Owen, J. M. Pearson and M. Szwarc, *Trans. Faraday Soc.*, 1965, **61**, 1722.
31. See R. W. Lenz, 'Organic Chemistry of Synthetic High Polymers' Wiley Interscience, New York, 1967, chap. 11.
32. J. E. Bloor, A. C. R. Brown and D. G. L. James, *J. Phys. Chem.*, 1966, **70**, 2191 and earlier papers.
33. M. Szwarc, *J. Polym. Sci.*, 1955, **16**, 367.
34. J. C. Bevington, *Proc. R. Soc. London, Ser. A*, 1957, **239**, 420.
35. J. C. Bevington, 'Radical Polymerization', Academic Press, New York, 1961, chap. 3.
36. J. Gibb, M. J. Peters, J. M. Tedder, J. C. Walton and K. D. R. Winton, *Chem. Commun.*, 1970, 978; D. P. Johari, H. W. Sidebottom, J. M. Tedder and J. C. Walton, *J. Chem. Soc. B*, 1971, 95; H. W. Sidebottom, J. M. Tedder and J. C. Walton, *Int. J. Chem. Kinet.*, 1972, **4**, 249.
37. See Figure 8-2 in D. C. Nonhebel and J. C. Walton, 'Free-radical Chemisty', Cambridge University Press, London, 1974, chap. 8.
38. K. Munger and H. Fischer, *Int. J. Chem. Kinet.*, 1985, **17**, 809.
39. See, for example, K. N. Houk, in 'Frontier of Free Radical Chemistry', ed. W. A. Pryor, Academic Press, New York, 1980.
40. T. Clark, *J. Chem. Soc., Chem. Commun.*, 1986, 1774.
41. M. Hirooka, H. Yabuuchi, S. Morita, S. Kawasumi and K. Nakaguchi, *J. Polym. Sci., Part B*, 1967, **5**, 47.
42. For review see C. H. Bamford, in 'Alternating Copolymers', ed. J. M. G. Cowie, Plenum Press, New York, 1985, chap. 3.
43. P. J. Flory and L. S. Leutner, *J. Polym. Sci.*, 1948, **3**, 880; 1950, **5**, 267.
44. C. W. Wilson III and E. R. Santee, *J. Polym. Sci., Part C*, 1964, **8**, 97.
45. R. E. Naylor and S. W. Lasoski, *J. Polym. Sci.*, 1960, **44**, 1.
46. R. C. Ferguson, *J. Am. Chem. Soc.*, 1960, **82**, 2416.
47. M. G. Evans and M. Polanyi, *Trans. Faraday Soc.*, 1938, **34**, 11; E. T. Butler and M. Polanyi, *Trans. Faraday Soc.*, 1943, **39**, 19.
48. A. F. Trotman-Dickenson, *Chem. Ind. (London)*, 1965, 379.
49. S. H. Jones and E. Whittle, *Int. J. Chem. Kinet.*, 1970, **2**, 479.
50. R. D. Giles and E. Whittle, *Trans. Faraday Soc.*, 1965, **61**, 1425; R. D. Giles, L. M. Quick and E. Whittle, *Trans. Faraday Soc.*, 1967, **63**, 662.
51. G. O. Pritchard, J. T. Bryant and P. L. Thommarson, *J. Phys. Chem.*, 1965, **69**, 664.
52. C. H. Bamford, A. D. Jenkins and R. Johnston, *Proc. R. Soc. London, Ser. A*, 1957, **239**, 217; **241**, 364.
53. C. H. Bamford, A. D. Jenkins and R. Johnston, *Trans. Faraday Soc.*, 1959, **55**, 418.
54. C. H. Bamford and E. F. T. White, *Trans. Faraday Soc.*, 1956, **52**, 716.
55. R. F. Bridger and G. A. Russell, *J. Am. Chem. Soc.*, 1963, **85**, 3754.
56. E. S. Huyser, *J. Am. Chem. Soc.*, 1960, **82**, 394.
57. C. Rüchardt, *Angew. Chem.*, 1970, **82**, 845; *Angew. Chem., Int. Ed. Engl.*, 1970, **9**, 830; C. Rüchardt and H. D. Beckhaus, *Angew. Chem.*, 1980, **92**, 417; *Angew Chem., Int. Ed. Engl.*, 1980, **19**, 429.
58. C. H. Bamford and S. N. Basahel, *J. Chem. Soc., Faraday Trans. 1*, 1978, **74**, 1020.
59. C. H. Bamford and S. N. Basahel, *J. Chem. Soc., Faraday Trans. 1*, 1980, **76**, 107, 112.
60. R. D. Brown, *J. Chem. Soc.*, 1953, 2615.
61. K. Fukui, in 'Modern Quantum Chemistry', ed. O. Sinanoglu, Academic Press, New York, 1965, pt. 1, p. 49.
62. F. W. Evans and M. Szwarc, *Trans. Faraday Soc.*, 1961, **57**, 1905; R. J. Fox, F. W. Evans and M. Szwarc, *Trans. Faraday Soc.*, 1961, **57**, 1915.
63. W. G. Alcock and E. Whittle, *Trans. Faraday Soc.*, 1965, **61**, 244; 1966, **62**, 134, 664.

64. M. G. Evans, J. Gergely and E. S. Seamen, *J. Polym. Sci.*, 1948, **3**, 866.
65. T. Yonezawa, K. Hayashi, C. Nagata, S. Okamura and K. Fukui, *J. Polym. Sci.*, 1954, **14**, 312.
66. K. Hayashi, T. Yonezawa, C. Nagata, S. Okamura and K. Fukui, *J. Polym. Sci.*, 1966, **20**, 537.
67. T. Kawabata, T. Tsuruta and J. Furukawa, *Makromol. Chem.*, 1962, **51**, 70, 80.
68. M. J. S. Dewar and C. C. Thompson, *J. Am. Chem. Soc.*, 1965, **87**, 4414.
69. F. R. Mayo, *Ber. Bunsenges. Phys. Chem.*, 1966, **70**, 233.
70. C. H. Bamford and A. D. Jenkins, *Trans. Faraday Soc.*, 1963, **59**, 530.
71. C. H. Bamford and A. D. Jenkins, *J. Polym. Sci., Part B*, 1963, **1**, 609.
72. R. D. Burkhart and N. C. Zutty, *J. Polym. Sci., Part A*, 1963, **1**, 1137.
73. C. C. Price, *J. Polym. Sci., Part B*, 1963, **1**, 433.
74. T. Alfrey, J. J. Bohrer and H. Mark, 'Copolymerization', Interscience, New York, 1952.
75. L. A. Wall, *J. Polym. Sci.*, 1947, **2**, 542.
76. J. R. Hoyland, *J. Polym. Sci., Part A-1*, 1970, **8**, 1863.
77. L. J. Young, 'Polymer Hand Book', Wiley, New York, 1965, Vol. II, p. 341.
78. C. H. Bamford and A. D. Jenkins, *J. Polym. Sci.*, 1961, **53**, 149.
79. A. D. Jenkins, *Eur. Polym. J.*, 1965, **1**, 177.
80. G. E. Ham, *J. Polym. Sci., Part A*, 1964, **2**, 4169.
81. N. McMurray, J. M. Tedder, L. L. T. Vertommen and J. C. Walton, *J. Chem. Soc., Perkin Trans. 2*, 1976, 63.
82. J. M. Tedder and J. C. Walton, *ACS Symp. Ser.*, 1978, **66**, 107.
83. J. R. Hoyland, *J. Polym. Sci., Part A-1*, 1970, **8**, 885.
84. J. R. Hoyland, *J. Polym. Sci., Part A-1*, 1970, **8**, 901.
85. See, for example, F. H. Burkitt, C. A. Coulson and H. C. Longuet-Higgins, *Trans. Faraday Soc.*, 1951, **47**, 553; M. J. S. Dewar, 'The Molecular Orbital Theory of Organic Chemistry', McGraw-Hill, New York, 1969.

18

Molecular Weight Distributions

ALFRED RUDIN

University of Waterloo, Ontario, Canada

18.1 INTRODUCTION

Molecular weight distribution is a key item of information in tailoring polymer structures for different end uses. Manufacturers offer very many different grades of commodity polymers like poly(vinyl chloride), polyethylenes and polystyrenes. These grades differ significantly between themselves. Since each polymer species has essentially the same chemical composition, the differences between commercial varieties are usually ascribed to variations in molecular weight distribution and, in some cases, to changes in branching character as well. The ability to optimize polymers for particular applications hinges, then, on understanding of the influence of polymer synthesis conditions on the molecular weight distribution of the product as well as knowledge of the influence of the molecular weight distribution on the mechanical and processing properties of the material. In addition to aiding the achievement of important practical objectives, molecular weight distribution data may provide insights into the mechanism and details of the polymerization process itself. This chapter is a review of the current (1987) state of knowledge of molecular weight distributions in polymers that are made by free radical polymerizations. Some comments on the fitting of kinetic models to molecular weight distribution data are also included.

18.2 POLYMERIZATION KINETICS AND MOLECULAR WEIGHT DISTRIBUTIONS

Measurement of fine details of the molecular weight distributions of polymers is limited at present by analytical difficulties. The preferred analytical method is gel permeation chromatography (size exclusion chromatography), in which the central region (*i.e.* $\bar{M}_w$) is measured most accurately, while measurement noise, base-line uncertainties and so on make the extremes of the distributions ($\bar{M}_n$, $\bar{M}_z$, $\bar{M}_{z+1}$) less certain.[1,2] Practical (*i.e.* high conversion) free radical polymerizations are very complex processes and must necessarily by described by a large number of kinetic parameters. The kinetic rate constants that are required in any comprehensive polymerization model outnumber the few reliable molecular weight characteristics. (This would not be so if molecular weight distributions could be measured very accurately. In principle, then, there would be available an infinite number of moments and averages of the distribution.) Thus, any realistic model of such reactions is badly overdetermined; the input parameters are far more numerous than the predicted quantities. Hence, nearly every reasonable kinetic model can be made to fit the corresponding polymer molecular weight data that can be measured by current methods and an objective preference for one high conversion polymerization model over another cannot be established.

Under these circumstances, this chapter will be confined to a summary of molecular weight distributions that can be expected under instantaneous reaction conditions, as a result of changes in process variables. Only a general survey will be attempted of engineering modelling of polymerization reactions, where concern is often with non-isothermal processes and a wide range of monomer conversions.

The most direct approach to estimation of molecular weight distributions from polymerization kinetics involves the quasi-steady state assumption.[3] That is, the time rate of change of radical concentration is effectively negligible. The assumption implies a constant rate of initiation, isothermal conditions, reaction rate constants that are invariant with conversion and the restriction that the overall reaction rate has reached a constant value before significant quantities of reactants have been consumed. Clearly, most commercial free radical polymerizations will not satisfy these criteria over significant ranges of conversion. Laboratory conditions can be adjusted fairly readily to accommodate these requirements, however, and the steady state assumption has been used successfully to account for experimental observations in many free radical polymerizations.

Instantaneous polymer size distributions under steady state conditions can be calculated by relatively straightforward methods. One assumes that initiator and monomer concentrations remain essentially constant during the polymerization. Any dependence of termination rate constants on macroradical sizes or concentrations[4] is neglected.

Any given monomer-ended radical may add monomer or undergo chain transfer or termination. The probability S that it will grow by monomer addition is

$$S = \frac{R_p}{R_p + R_{tr} + R_t} \tag{1}$$

where R_p, R_{tr} and R_t are the respective rates of propagation, transfer and termination. Now

$$R_p = k_p[M][P\cdot] \tag{2}$$

where $[M]$ and $[P\cdot]$ are the concentrations of monomer and monomer-ended radicals respectively. Also

$$R_{tr} = C_I k_p[I][P\cdot] + C_M k_p[P\cdot][M] + C k_p[TH][P\cdot] \tag{3}$$

where C_I, C_M and C are transfer constants (ratios of propagation rate constant k_p to transfer rate constant k_f) to initiator, monomer and other species (labelled TH) respectively. (Note that chain transfer effects of solvent are included in the general term for the transfer agent TH, and that transfer to polymer is excluded from these considerations.) Also

$$R_t = (k_{tc} + k_{td})[P\cdot]^2 \tag{4}$$

Substitution of equations (2)–(4) into equation (1) yields

$$\frac{1}{S} = 1 + C_I \frac{[I]}{[M]} + C_M + C \frac{[TH]}{[M]} + \frac{(k_{tc} + k_{td})}{k_p}[P\cdot] \tag{5}$$

With $[P\cdot] = R_p/k_p[M]$ from equation (2) we obtain

$$\frac{1}{S} = 1 + C_I \frac{[I]}{[M]} + C_M + C \frac{[TH]}{[M]} + \frac{(k_{tc} + k_{td})}{k_p^2[M]^2}R_p \tag{6}$$

We first consider the polymerization where each kinetic chain yields one polymer molecule. This is the case for termination of the growth of macroradicals by disproportionation or chain transfer ($k_{tc} = 0$). If an initiator residue at the end of a macromolecule is selected at random, the probability that the monomer residue which was captured by this primary radical has added another monomer is S, and the probability that this end is attached to a macromolecule which contains at least i monomers is S^{i-1}. The probability that this macromolecule contains exactly i monomers equals the product of S^{i-1} and the probability of a termination or transfer step. The latter probability must be equal to $(1-S)$ since it is certain that the last monomer under consideration will either add monomer, terminate or transfer. That is, the probability that a randomly selected molecule contains i monomer units is $S^{i-1}(1-S)$. This probability is equal to the corresponding mole fraction of this size molecule x_i. That is to say

$$x_i = (1 - S)S^{i-1} \tag{7}$$

for the number distribution function. The weight distribution function w_i is derived in a similar fashion as

$$w_i = i(1 - S)^2 S^{i-1} \tag{8}$$

The distribution described is a random one, with average degrees of polymerization

$$\overline{DP}_n = 1/(1 - S) \tag{9}$$

$$\overline{DP}_w = (1 + S)/(1 - S) \tag{10}$$

$$\overline{DP}_w/\overline{DP}_n = 1 + S \tag{11}$$

The distribution and molecular weight averages calculated here bear an evident similarity to those that can be derived for equilibrium step growth polymerizations.[5] Note, however, that the distribution functions in the step growth case apply to the whole reaction mixture, whereas in free radical polymerizations the distribution describes only the polymer that has been formed.

For most addition polymerization, $R_p \gg R_t + R_{tr}$ (or high molecular weight polymer would not be formed). In that case $S \cong 1$ and $\overline{DP}_w/\overline{DP}_n$ will be $\cong 2$, if the growth of macroradicals is terminated entirely by disproportionation or chain transfer.

When radical growth is terminated by a combination of disproportionation and combination reactions, the instantaneous distribution of degrees of polymerization is given by[5,6]

$$w_i = Fi(1 - S)^2 S^{i-1} + (1 - F)i(i - 1)(1 - S)^3 S^{i-2}/2 \tag{12}$$

where F is the fraction of product formed by chain disproportionation and/or transfer and

$$F = \frac{C_I[I] + C_M[M] + C[TH] + k_{td}R_p/k_p^2[M]}{C_I[I] + C_M[M] + C[TH] + \{(k_{tc} + k_{td})R_p/k_p^2[M]^2\}} \tag{13}$$

Equation (12) reduces to equation (8) when $k_{tc}=0$ ($F=1$) and to equation (14) when $k_{td}=0$ and transfer is negligible ($F=0$)

$$w_i = i(i - 1)(1 - S)^3 S^{i-2}/2 \tag{14}$$

In the latter case

$$\overline{DP}_n = 2/(1 - S) \tag{15}$$

$$\overline{DP}_w = (2 + S)/(1 - S) \tag{16}$$

and

$$\overline{DP}_w/\overline{DP}_n = (2 + S)/2 \tag{17}$$

The ratio in equation (17) has a limiting value of 1.5 at high polymer molecular weights when S approaches unity. This is narrower than the instantaneous distribution produced in the absence of termination by coupling.

This extremely simple scheme is typical only of isothermal polymerizations that are carried to low conversions and in which chain transfer to polymer is negligible. Synthesis of low molecular weight poly(o-vinylbenzophenone) is a recent example of such a free radical polymerization.[7]

The effects of radical chain transfer are general in the sense that this process always tends to shift the molecular weight distribution towards a more random ($\overline{M}_w/\overline{M}_n=2$) character. For example, the growth of macroradicals in the polymerization of vinyl chloride is limited primarily by chain transfer to monomer. This is a heterogeneous reaction that is carried to quite high monomer conversions. Nevertheless, the molecular weight distributions that are observed are approximately random. If free radical polymerizations are compared with and without the use of a chain transfer agent it can be expected that the molecular weight of the polymer produced under the former conditions will be smaller and the molecular weight distribution will be altered toward a random character. (Random distributions are sometimes called 'most probable' or 'Flory–Schulz' distributions in the polymer literature.)

Radical chain transfer to polymer does not change the number average degree of polymerization of the polymer. This is evident since the chain transfer event itself does not alter the number of radicals in the system nor does it change the number of monomer units that have been polymerized to that point. Now $\overline{M}_n$ is by definition the quotient of the total number of monomers that have been incorporated into polymers and macroradicals divided by the number of such polymer molecules and radicals. Its value remains the same before and after transfer of the radical site from a growing macroradical to what was a dead polymer molecule. However, $\overline{M}_w$ and higher averages can be expected to increase because preformed macromolecules are now increased in size by polymerization onto the new radical sites. Thus chain transfer to polymer will produce molecular weight distributions that are broader and more skewed to high molecular weights than would be observed under the same conditions in the absence of this process. Free radical polymerizations that yield reactive macroradicals may be characterized by significant amounts of chain transfer to polymer. The high pressure process for the manufacture of polyethylene is an example. Addition of a chain transfer agent to such a reaction will oppose such a broadening of the molecular weight distribution, for reasons given above.

If the monomer concentration decreases during the course of the polymerization while the initiator concentration and rate of decomposition remain constant, it is intuitively obvious that the molecular weight of the polymer produced will decay with increasing monomer conversion. The final distribution is the result of the superposition of a series of molecular weight distributions in which the mean molecular weight declines gradually. The distribution is therefore wider than it would be if the monomer concentration remained constant. Bamford and co-workers[8] have summarized the molecular weight distributions to be expected, in terms of the parameter S. Tompa[9] has described various mathematical techniques to solve the equations that result from free radical kinetics, both with and without the steady state assumption. The statistical characteristics of various molecular weight distributions are recorded in a book by Peebles.[10]

High conversion free radical polymerizations are often accompanied by gel effects. This results in autoacceleration, an increase in polymer molecular weight and a broadening of the molecular weight distribution. This phenomenon is very important in most commercial free radical polymerizations. It occurs when the polymer concentration and size are both high enough that macroradicals can become entangled with other polymer molecules. A decrease in the diffusion rate of radical ends is the result, along with a concurrent decrease in the frequency of encounter of radicals and therefore in the rate of termination. The rate of monomer addition to radicals is less affected, since this reaction depends on encounters between macroradicals and mobile monomers. A net increase in molecular weight and rate of polymerization result. Also, since polymerizations of vinyl monomers are exothermic, there may be a temperature increase in the reaction mass, along with a further increase in the rate of polymerization because of faster decomposition of initiator. Polystyrene provides a useful example of the effects that have been mentioned. In this polymerization termination appears to occur exclusively by combination and the expected molecular weight distribution is random, with $\bar{M}_w/\bar{M}_n = 1.5$, according to equation (11). This is indeed observed in isothermal, low conversion polymerizations. Industrial scale processes normally produce polystyrenes with $\bar{M}_w/\bar{M}_n$ slightly greater than three, however, because of their non-isothermal nature and the gel effect at later stages of the reaction, which is taken to very high conversions.

All kinetic models must become quite complicated in order to simulate real polymerizations that are carried to high conversions. This is because actual free radical polymerizations may involve many different reactions. The resulting complex kinetic schemes require a large number of kinetic parameters whose numerical values may not be known. In practice, these unknown rate constants can be estimated by fitting measurable features of the overall reaction to a particular model. This expedient is a frequent feature in engineering modelling of polymerization reactions.

Mathematical modelling of polymerization reactions is undertaken primarily to develop reactor designs to optimize control systems and to explore a range of operational variables. The model may also provide some estimates of molecular weight distributions, although this is usually an incidental reason for its development. Many mathematical expedients have been used for the solution of equations derived from free radical kinetic schemes. These have been reviewed by others[11] and are outside the scope of this chapter. Generally these methods restrict the polymerization model so that relatively simple mechanisms are required. Also, the steady state assumption is assumed to apply to all radical types in order to solve the kinetic equations. More recent computational techniques relax the requirement for steady state radical concentrations. A direct method involves integration of the differential mass balance equations that are derived from the particular kinetic scheme. It is necessary to assume a limit to the degree of polymerization of the polymer molecules, however, in order to reduce the model to a finite (but large) number of differential equations.[12]

There are relatively few instances in the current literature in which predicted molecular weight distributions of commercial polymers are compared with experimental data for products made by free radical reactions. In the same vein, it is also rare that measured molecular weight distributions are used to modify polymerization reaction models that have been proposed.

A brief scrutiny of the free radical polymerization of ethylene indicates the complexity of the reaction scheme which must be taken into account when attempting to model the polymerization and predict molecular weight distributions from reaction parameters.[1]

Ethylene is polymerized continuously in tubular or stirred autoclave reactors. Monomer conversion is about 25–30% per pass at temperatures between about 150 °C and 300 °C, and pressures of 1200 to 3000 atm. The reaction is not isothermal, proceeding as it does from a cool gas feed to a hot exit temperature. The reaction pressure can be considered to be uniform in autoclave reactors but varies with reactor length or residence time in tubular reactors. Reactions which must be considered include the usual initiation, propagation, termination and chain transfer processes. In addition, the growing macroradicals generate short (ethyl, butyl and pentyl) branches through a sequence of back-biting reactions.[13,14] Long branches are produced by radical chain transfer from macroradicals to dead polymer.[15,16]

The ethylene–polyethylene mixture may be homogeneous or it may consist of two phases: an ethylene-rich and a polyethylene-rich region. The phase behavior in the reactor is a function of the temperature and pressure. Low density polyethylene synthesized under two-phase conditions has been reported to have a lower frequency of long chain branching because the polyethylene in the polymer-rich phase is protected from attack by macroradicals.[17] The possible existence of a heterogeneous reaction mixture must be taken into account when the process is modelled, because the occurrence of long branching will affect the molecular weight distribution, as well as other properties of the polymer.

Radical scission reactions can also occur, generating volatile branched chain alkenes and vinylidene unsaturation in the polymer. The alkenes produced in this reaction can copolymerize with ethylene and they may also function as active chain transfer agents. Chain transfer tends to reduce the mean molecular weight and move the distribution to a random character (see above). This serves to narrow the breadth of the distribution compared to the situation in which chain transfer to alkenes is absent (as when the feed gas to the reactor does not contain recycled ethylene).

While high pressure ethylene polymerization is a complex process, polymer characterization data do shed some light on the mechanism of the polymer synthesis. Long chain branching frequency can be measured as a function of molecular weight by modern gel permeation chromatography techniques, subject to certain assumptions that are detailed elsewhere.[18] Long branches are formed in free radical polymerization by chain transfer to polymer.[15,16] Since larger polymer molecules provide bigger targets for macroradicals, it may be assumed that reactions that generate long chain branches will be more frequent on higher molecular weight polymers. This reasoning is sound for isothermal polymerizations in homogeneous reaction media. Recent analyses indicate, however, that long chain branch frequency is a decreasing function of molecular weight[19,20] in polyethylene. These observations can be rationalized by recalling that low density polyethylene polymerizations are carried out over a range of temperatures, with the molecular weight of the polymers produced decreasing with increasing temperature, as is general for free radical polymerizations. The activation energy for chain transfer to polymer is greater than for the propagation reaction,[21] however, and thus chain transfer to nascent polymer and long branching are most frequent under conditions where lower molecular weight polyethylenes are being formed.

It should be recalled also that higher molecular weight polyethylenes are formed in the cooler inlet regions of the continuous reactors and are carried through the hotter zones where branching reactions are increasingly more favored. That these larger molecules emerge from the reactors relatively unbranched provides information on the state of mixing in these polymerizations. It would appear that polyethylene polymerization is normally carried out in segregated reactors. That is, the reaction mixture consists of many small elements, each of which behaves as a batch reactor. The elements themselves may be more or less mixed, of course, depending on the macro conditions in the polymerization vessel. It has been suggested, in fact, that the majority of the bulk and solution polymerizations proceed in segregated systems.[22] This feature should be taken into account in any reaction model that is intended to predict polymer microstructure and molecular weight distribution in full scale systems.

18.3 REFERENCES

1. A. Rudin, *Makromol. Chem., Macromol. Symp.*, 1987, **10/11**, 273.
2. S. Balke, 'Quantitative Column Liquid Chromatography', Elsevier, Amsterdam, 1984, p. 170.
3. P. E. M. Allen and C. R. Patrick, 'Kinetics and Mechanisms of Polymerization Reactions', Wiley, New York, 1974, p. 131.
4. H. K. Mahabadi and K. F. O'Driscoll, *Macromolecules*, 1977, **10**, 55.
5. P. J. Flory, 'Principles of Polymer Chemistry', Cornell University Press, Ithaca, NY, 1953.
6. W. B. Smith, J. A. May and C. W. Kim, *J. Polym. Sci., Part A-2*, 1966, **4**, 365.
7. M. F. Tchir, A. Rudin and C. J. B. Dobbin, *Polymer*, 1983, **24**, 409.
8. C. H. Bamford, W. G. Barb, A. D. Jenkins and P. F. Onyon, 'The Kinetics of Vinyl Radical Polymerizations', Butterworths, London, 1958.

9. H. Tompa, in 'Comprehensive Chemical Kinetics', ed. C. H. Bamford and C. F. H. Tippett, Elsevier, New York, 1976, vol. 14A, chap. 7.
10. L. H. Peebles, Jr., 'Molecular Weight Distributions in Polymers', Wiley-Interscience, New York, 1971.
11. W. H. Ray, *J. Macromol. Sci., Rev. Macromol. Chem.*, 1972, **8**, 1.
12. R. D. Skeirik and E. A. Grulke, *Chem. Eng. Sci.*, 1985, **40**, 535.
13. M. J. Roedel, *J. Am. Chem. Soc.*, 1953, **75**, 6110.
14. A. H. Willbourn, *J. Polym. Sci.*, 1959, **34**, 569.
15. P. J. Flory, *J. Am. Chem. Soc.*, 1942, **69**, 2894.
16. P. J. Flory, *J. Am. Chem. Soc.*, 1937, **59**, 241.
17. C. Bergstram, A. Honkanen and M. Villanen, *Eur. Polym. J.*, 1979, **15**, 301.
18. A. Rudin, in 'Modern Methods of Polymer Analysis', ed. H. G. Barth, Wiley, New York, in press, 1988.
19. A. Rudin, V. Grinshpun and K. F. O'Driscoll, *J. Liq. Chromatogr.*, 1984, **7**, 1809.
20. M. Hert and C. Strazielle, *Makromol. Chem.*, 1983, **184**, 135.
21. G. Luft, *Chem. Eng. Tech.*, 1979, **51**, 960.
22. E. B. Nauman, *J. Macromol. Sci., Rev. Macromol. Chem.*, 1974, **10**, 75.

19

Template Polymerization

Y. YONG TAN

State University of Groningen, The Netherlands

19.1 INTRODUCTION

Any reaction implying propagation of polymer chains along template macromolecules during at least part of its lifetime may be called template (or matrix or replica) polymerization. This peculiar mode of propagation is made feasible by cooperative interaction between complementary groups of the growing (daughter) chain and the template, which is usually a linear macromolecule. The concept of template polymerization is derived from such natural and highly sophisticated processes as the self-replication of DNA and the biosynthesis of proteins.

Early examples of the template polymerization of synthetic systems include the ring-opening polymerization of the *N*-carboxyanhydride (NCA) of D,L-phenylalanine on a polysarcosine template, discovered by Ballard and Bamford,[1] and the polymerization of 4-vinylpyridine along a polyacid template by Kabanov and co-workers.[2] Many other systems have been investigated since then, either by addition, ring-opening or condensation polymerization.[3-5]

Since template polymerization generally yields a stoichiometric interpolymer complex, or polycomplex, of daughter and parent (template) polymer, one can, alternatively, devise a template system from the basis that two different polymers are able to form a polycomplex.

19.1.1 Template Effects

To study the influence of a template, the polymerization system is generally compared to a system without a template, usually called the blank, control, normal, conventional or nontemplate polymerization system. When compared to a blank polymerization, the presence of a template may affect: (a) the reaction kinetics (*i.e.* polymerization rate or reaction order); (b) the structural features of the daughter polymer, *viz.* average molecular weight, molecular weight distribution and microstructure; and (c) reactivity ratios, in the case of copolymerization. Such manifestations are called template effects. Ideally, since a template effect is a polymer (chain) effect, a template polymerization should be compared with a blank polymerization in which the template is replaced by a low molecular weight nonpolymerizable analogue. The magnitude of a template effect depends, not only on the usual reaction variables, but also on the chemical nature and physical state of the template, such as its microstructure, chain length and conformation.

19.1.2 Classification of Template Systems

If we designate the monomer molecule by M and the template by T, two idealized types of template system may be discerned, depending on whether the monomer molecules are preadsorbed on the template or not.[6, 7] This is determined by the equilibrium constant K_M (in $dm^3 \, mol^{-1}$) for the reaction $M + -T- \rightleftharpoons -T(M)-$, in which $-T-$ denotes a template site. Based on the magnitude of K_M, one can distinguish two extreme types of template polymerization.

When $K_M = \infty$ and $[M] \leq [T]$ ($[T]$ in base mol dm^{-3}), all monomer molecules are adsorbed in arrays on the template, and propagation consists of linking adjacent monomer molecules to one another (step polymerization) or of adding consecutive molecules to the active chain ends (Figure 1). This will be called a type 1 or 'zip' mechanism.

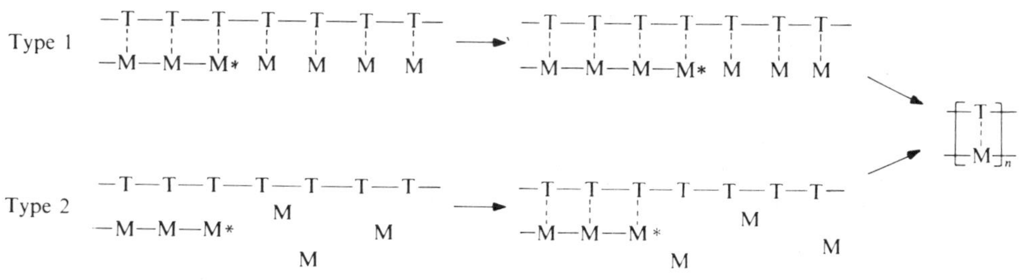

Figure 1 Diagrammatic representation of mechanistic types of template polymerization; $-T-T-T- =$ template macromolecule, $M =$ monomer molecule, $-M-M-M* =$ growing daughter chain

When $K_M = 0$, there is no preferential adsorption of monomer onto the template; in this case, template polymerization will take place only when the growing oligomer, which is created in the bulk solution, can complex with the template. Since complexation is a cooperative process, the oligomer should attain a minimum (critical) chain length prior to its attachment to the template. Template-associated propagation proceeds in this case by reaction with monomer from the surrounding solution (Figure 1). This will be called a type 2 or 'pick-up' mechanism.

The strength of the interactions between the template and, respectively, the monomer or its growing oligomer determines the mechanistic type of the process. Thus, strong ionic (Coulombic) and charge transfer interactions may favour type 1, whereas the weaker hydrogen bond tends to favour type 2. Van der Waals' forces do not lead to template systems, unless some additional driving force of cooperative nature is involved, in particular, double helix formation by stereochemical complementarity. However, factors such as solvent and temperature significantly affect the strength of any interactions and, hence, the mechanism of template polymerization. When working in aqueous solutions, one has to take pH and ionic strength into account as additional variables, and, in some cases, the possible existence of hydrophobic interactions.

True type 1 and type 2 systems are rare. One should rather speak of type 1-like and type 2-like systems, possessing large and small values of K_M respectively. Unfortunately, in many cases K_M values are unknown, so that the classification of the template system must be deduced from other experimental results, *e.g.* kinetic data.

19.2 GENERAL CONSIDERATIONS

19.2.1 Polymerization Rate

The mechanistic type of a template system can often be inferred from, or confirmed by, kinetic studies.[7] This can be done by examining the rate enhancement (with respect to blank polymerization) on varying the template concentration $[T]_0$ while keeping the monomer concentration $[M]_0$ constant. The reasons for rate enhancement are different for type 1 and type 2 systems. In type 1 it seems to be due primarily to acceleration of the propagation rate; in type 2, to retardation of the termination rate. These will be discussed in more detail in the following sections.

When a type 1 mechanism is operating (Figure 1), and given that a template usually invokes a rate enhancement, increasing $[T]_0$ should lead to an increase in the polymerization rate, up to a maximum at $[T]_0/[M]_0 = 1$, provided all monomer is bound to the template in a 1:1 base molar ratio. Beyond this ratio, the average length of the monomer arrays becomes shorter, resulting in a decline in rate (Figure 2a, curve i).

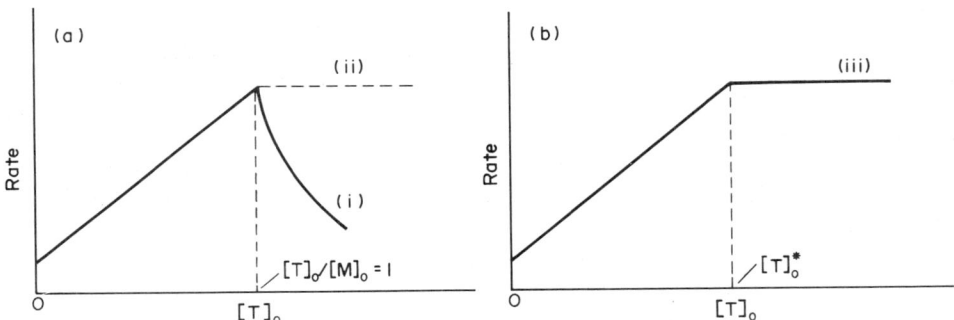

Figure 2 Schematic representation of initial overall polymerization rate *vs.* template concentration, $[T]_0$, in base mol dm^{-3}, at constant monomer concentration, $[M]_0$, in mol dm^{-3}; (a) type 1 systems, (b) type 2 systems, $[T]_0^*$ = critical coil overlap concentration of the template

When a type 2 mechanism is involved (Figure 1), the rate should increase with increasing $[T]_0$ until it reaches a critical concentration $[T]_0^*$, where the template coils begin to overlap one another. Beyond this concentration the entire reaction volume is homogeneously occupied by template segments, and, provided all growing chains are capable of complexing with template molecules at this critical concentration, the rate remains constant at this level (Figure 2b, curve iii).

In place of experiments where $[T]_0$ is varied, one can vary $[M]_0$ while keeping $[T]_0$ constant. A change in the slope of the curve at $[M]_0/[T]_0 = 1$ may be predicted for type 1 systems, since beyond this ratio the excess of monomer will polymerize at the rate of the blank polymerization.

19.2.2 Preferential Monomer Adsorption

Knowledge of preferential monomer adsorption, qualitatively or quantitatively through K_M, is essential, not only for identifying the type of template polymerization, but also for calculating propagation rate constants. In type 1-like systems, monomer adsorption automatically creates a high local monomer concentration on the template. In order to evaluate rate enhancement due to monomer alignment, one has to correct for this concentration effect. In type 2-like systems, where propagation takes place by mechanism 1 as well as by mechanism 2, K_M should be known in order to assess the individual contributions of the propagation modes.

K_M has been determined in a variety of ways, mostly by means of equilibrium dialysis,[7-10] and, occasionally, by spectroscopy,[11] calorimetry[12] or kinetic analysis.[13]

19.2.3 Critical Chain Length

The critical chain length is very important in type 2-like template polymerizations, since it affects the magnitude of the template effect: the longer it is, the less the template effect. This quantity is not always thermodynamically determined, but is often kinetically determined, especially when the interaction is rather strong, *e.g.* when preferential monomer adsorption occurs.

Critical chain lengths can be estimated from viscometric or calorimetric complexation studies between prefabricated daughter polymers of varying molecular weights (prepared under blank polymerization conditions) and the template polymer of constant length. To simulate initial template polymerization conditions, such experiments should be performed in a mixture of solvent and monomer or a suitable analogue.

Methods used to estimate the critical chain length under template polymerization conditions are based on the extrapolation to zero template effect of rate, average molecular weight or tacticity of the daughter polymer with increasing concentration of initiator or chain transfer agent to shorten the kinetic chain length,[14,15] or with increasing weight fraction of a noncomplexing comonomer to reduce the complexation ability of the growing chain in solution.[16]

Complementary to the critical chain length is the minimum template length needed for incipient complexation of the growing oligomer in the bulk solution. This can be estimated in a similar fashion, by extrapolation to zero template effect of rate as a function of the template's average degree of polymerization.[17-19]

It should be noted that these critical quantities are not constants; their values are dependent on reaction conditions.

19.2.4 Structural Features

Besides a kinetic study, it may be desirable to obtain supporting evidence of a template effect from the production of the appropriate type of polycomplex or from complementary structural features of daughter and template polymer, such as molecular weight and microstructure. However, such structural data are generally hard to acquire, since characterization of a daughter polymer is often hampered by difficulties in isolating it quantitatively from its template. So far, only partial isolations have been accomplished, probably because of grafting of the daughter polymer onto the template[20] or for hitherto unknown reasons. Sometimes indirect characterization is feasible, if a solvent or conditions can be found in which the daugher–template polycomplex is fully dissociated.

One criterion for a template effect is that the degrees of polymerization of daughter and template polymers should at least run parallel. Another criterion is that the template polymerization of prochiral monomers should yield daughter polymers possessing a microstructure distinctly different from that of their blank counterparts.

19.3 CHAIN TEMPLATE POLYMERIZATION

The chain template polymerization systems most usually studied are of the free-radical type, and only a few are of a nonradical nature.

19.3.1 Type 1-like Systems

The majority of these systems consist of an acid monomer and a polybase, and *vice versa*, in which ionic interactions play a dominant role.[2,6,21-32] Monomers used include acrylic acid, methacrylic acid and *p*-styrenesulfonic acid, on templates such as poly(ethyleneimine), ionenes and poly(L-lysine). The basic monomers 4- and 2-vinylpyridine on various polyacid templates, such as poly(acrylic acid), poly(*p*-styrenesulfonic acid) and poly(phosphoric acid), initiate spontaneously and proceed according to a nonradical mechanism.[2,32] This might also be true for the polymerization of 4-vinylpyridine by poly(4-vinyloxycarbonylphthalic anhydride) in acetone solution.[33] The template polymerization of 4-vinyl pyridine by poly(maleic anhydride) might be based on a charge transfer interaction.[34] However, systems in which hydrogen bonding plays a role, *viz.* methacrylic acid/poly(*N*-vinyl-5-methyloxazolidone) and *N*-vinyloxazolidone/poly(methacrylic acid), also fall into the category of type 1.[35,36]

Except for a couple of systems for which K_M is known, the mechanistic type is deduced from a kinetic study showing a maximum polymerization rate at $[T]_0/[M]_0$ around unity (see Figure 2a, curve i). An exception is the system acrylamide/poly(*p*-styrenesulfonic acid), with a maximum rate at $[T]_0/[M]_0 = 3$, despite a 1:1 interaction.[37]

A maximum rate is obtained only when the adsorbed monomer molecules are bound to sites on the template. When the monomer molecules possess a certain mobility along the template, the rate will drop less drastically; if mobility is substantial the rate can even remain at the maximum level beyond $[T]_0/[M]_0 = 1$ (Figure 2a, curve ii), as exemplified by the spontaneous polymerization of 4-vinylpyridine along a polyacid template in aqueous solutions at a constant pH of 5.6.[28] Monomer mobility was also found for the *p*-styrenesulfonate counterions along oligomeric ionene templates.[23] In this case, monomer mobility was deduced from the fact that, on using ionenes with greater charge separations, the polymerization rate did not drop abruptly but decreased proportionately with charge density. Monomer mobility is probably common to systems in aqueous solutions where strong ionic interactions are relevant. Under such conditions, it would be more appropriate to substitute the stoichiometric site binding model for the Manning counterion condensation model.[23, 38, 39]

A rather simple method of verifying whether mechanism 1 is operating is to add a saturated monomer analogue to the template system. This substance will compete with the monomer for the template sites, reducing the overall rate.[21]

Template polymerizations of type 1 are invariably performed in (very) dilute solutions in order to minimize intertemplate reactions. An 'external' radical-forming initiator, such as azobis(isobutyronitrile) (AIBN) or potassium persulfate, is commonly used, and, occasionally, AIBN is used in combination with UV irradiation.[21] They generally lead to random initiation of the adsorbed monomer arrays, although initiation may start at one end of short oligomeric templates.[40] The latter is ensured when an initiating group ('internal' initiator) is fixed at one end of the template macromolecules.[41]

Water and aqueous mixtures are favourite solvents. In aqueous solutions, the pH may drastically change the interaction of the monomer or its daughter polymer with the template, leading to a shift in mechanism from, say, type 1 to type 2, or even to blank polymerization.[36] A change of pH may also cause a conformation change of the template, such as the helix to coil transition of poly(L-lysine) on decreasing pH; this also affects the rate of the template polymerization of methacrylic acid.[26, 27] The effect of salt addition, and changing the dielectric constant by varying the content of a water-miscible organic solvent, such as 2-propanol or *p*-dioxane, was studied extensively for the template system *p*-styrenesulfonate/ionene.[23–25] Decreasing the dielectric constant of the medium increases K_M, and raises the rate accordingly.

19.3.1.1 Kinetics

Reasons for rate enhancement in type 1 polymerizations seem rather obvious at first sight, but there appears to be no consensus as to its true cause. The following suggestions have been or may be put forth: (a) the ordering of the monomer molecules;[2, 3, 21, 23] (b) a high local monomer concentration;[23, 36] (c) a change in the intrinsic reactivity of the reacting species on adsorption;[6] (d) a microenvironment different from that of the bulk solution; (e) a reduction of electrostatic repulsion between reacting ionic species as a consequence of adsorption on charged polyelectrolyte templates;[6] and (f) the heterogeneity of the system.[22] The last point is due to the fact that template polymerizations are often accompanied by gradual precipitation of insoluble polycomplex particles, giving rise to the possibility of entrapment of template-bound chain radicals. Such a monomolecular termination is supported by polymerization rates that are (nearly) first order in initiator concentration,[6, 20, 23] although this outcome has been attributed to degradative chain transfer to solvent.[23] Although occlusion of the polymer radicals might contribute to the overall rate enhancement by retardation of the termination step (which could be revealed at excess monomer concentration) there is, to date, no consensus about it. It should also be noted that a normal half order in initiator concentration was found, despite precipitate formation.[36]

While (c) through (e) might be ignored, and the contribution due to local monomer concentration can be allowed for through knowledge of K_M, the net rate enhancement should reflect the rate increase of the propagation step and, consequently, the degree of alignment of monomer molecules along the template. Important parameters are the distance between consecutive monomer molecules and their mutual steric orientation, assuming immobility of the adsorbed monomer molecules. This is partially determined by the template conformation.[27] Rigid templates possessing an extended conformation, such as the (2,2,2)-4-ionene template,[23–25] would be more appropriate than flexible ones, because of a smaller loss of conformational entropy on monomer adsorption and on polymerization of the adsorbed monomer.

Generally, lower overall activation energies (usually measured at $[T]_0/[M]_0 = 1$) when compared with those for blank systems,[22,23,37] were found and were ascribed primarily to a lowering of the activation energy of the propagation step.

19.3.1.2   Structural features

Very little is known about the influence of template length on polymerization rate and molecular weight of the daughter polymer. With regard to the latter, a slight parallel increase of the daughter and template chain length has been observed.[36] Approximately identical degrees of polymerization of daughter and template polymers were found in a couple of systems.[33,34]

In the template polymerization system of vinyl sulfonate (2,2,2)-4-ionene oligomer, the isolated daughter oligomer had a microstructure very different from that of the blank (tactic triads were $mm = 0.13$, $mr = 0.64$ and $rr = 0.23$ *vs.* $mm = 0$, $mr = 0.35$ and $rr = 0.65$).[42] This should reflect the best stereochemical fit between the daughter and template oligomer. Indeed, the polycomplex obtained by recomplexation of the daughter oligomer with ionene showed crystallinity, in contrast to the polycomplex between blank oligo(vinyl sulfonate) and ionene.

A peculiar result was obtained when methacrylic acid was polymerized, in the form of its polysalt, with chitosan as template in the pH range 2.60 to 6.48. The isolated daughter poly(methacrylic acid)s were claimed to be optically active, especially in the lower pH regions.[43] This was explained by the conformational structure of the polymer chain, though its microstructure ($mm = 0.29$, $mr = 0.19$ and $rr = 0.52$) was not stereoregular.

19.3.1.3   Special systems

The spontaneous polymerization of 4-vinylpyridine along a polyacid template[8,28-31] and of 2-vinylpyridine along a poly(methacrylic acid) template[32] are supposed to proceed by a stepwise mechanism *via* a zwitterionic intermediate; this hypothesis is based on extensive research on the template and the blank polymerization of the former system. Polymerization of 4-vinylpyridine does not produce poly(4-vinylpyridine), as originally thought, but its isomeric ionene, having the structural unit $-[C_5H_4N-CH_2CH_2]_n$ (ionene-1,6).[8,28,44,45] Crystalline polycomplexes between this ionene and isotactic poly(acrylic acid) were obtained by template polymerization and by mixing, whereas only amorphous polycomplexes were formed with atactic and syndiotactic poly(acrylic acid),[31] proving a stereochemical preference. In contrast, daughter poly(2-vinyl-pyridine)s isolated from polymerizations of 2-vinylpyridine along poly(methacrylic acid) possessed microstructures identical with that of the blank polymer, irrespective of the template's tacticity.[32]

An unusual template system involves the polymerization of 3-chloroprop-1-yne along poly-(4-vinylpyridine) in methanol.[45,46] The monomer reacts with the template by quaternizing the pyridine rings (the Menshutkin reaction) prior to spontaneous polymerization to a black gel. A very elaborate anionic mechanism was proposed, involving a zipping process of the attached monomer molecules, sandwiched between template chains. A network structure is eventually formed, containing highly unsaturated polyene chain segments.

Systems in which 'monomer' is covalently attached to a template (a 'perfect' type 1 system) have been developed by Kämmerer *et al.*,[40,47-49] the first example being a multiacrylate, synthesized by complete esterification of a monodisperse template oligomer of a condensation product of *p*-cresol and formaldehyde, with acrylic acid. After cyclo-oligomerization in a dilute solution of benzene, with a relatively high amount of AIBN to ensure initiation at one template end and primary termination at the other end, the ladder oligomer produced was hydrolyzed to isolate the daughter oligoacrylic acid.[48] The operation of a template effect was deduced from the equality in degree of polymerization of the daughter oligomer and its template.

The above method was extended to prepare long heterodisperse multimethacrylates, based on poly(vinyl alcohol).[41,50] These compounds are very sensitive to light and air. In such long multimers, random initiation occurs, leading to isolated monomer sequences and an increasing probability of intertemplate reactions, so that exact chain length copying of the template is lost.[50] There is, as yet, no exact description of the microstructure of daughter oligo and poly acids.

In template polymerizations, the template is commonly added to the polymerizing system as a prepolymer. It is conceivable, however, that the template could be formed *in situ* in certain instances. Such a situation has been postulated for the radiation-induced polymerization of methacrylic acid[51,52] and acrylonitrile,[51,53] in order to account for their unusual kinetics. It is deemed possible that, in the early stages of the reaction, monomer molecules associate with the polymer as it is formed, through hydrogen bonding or dipolar interaction respectively, followed by polymerization according to a zip mechanism. The poly(acrylic acid) was claimed to be more syndiotactic than a conventional one.[51]

19.3.2 Type 2-like Systems

There are several systems which may be considered to approximate to pure type 2 systems, *viz.* those based on cooperative hydrogen bond interactions between growing chain and template, *e.g.* polymerization of N-vinylpyrrolidone along tactic poly(methacrylic acid)[17,18] or poly(acrylic acid)[54] in DMF, polymerization of methacrylic acid along poly(N-vinylpyrrolidone) in DMF[55] or along poly(ethylene oxide)[14,56-58] or poly(vinyl alcohol) in water;[19] and those based on Van der Waals' interactions combined with stereochemical complementarity, *e.g.* methyl methacrylate (MMA) along isotactic or syndiotactic PMMA[7,15,59-63] and methacrylic acid along isotactic PMMA in DMF and in dioxane/methoxyethanol (3:7 v/v).[64]

In some of these systems, where absence of preferential adsorption of monomer on the template was established,[7,18] rate *vs.* $[T]_0/[M]_0$ profiles conforming to curve (iii) in Figure 2(b) were found. In the case of the system MMA/isotactic PMMA, an extra rate enhancement was observed at template concentrations near the critical coil overlap concentration, which could be rationalized by invoking hopping of the growing chains from one template to another template macromolecule.[60] In these systems, critical chain lengths of growing oligomer for complexation were also found, as well as in others.[14,64]

The following systems are considered to be type 2-like: methacrylic acid and maleic acid along poly(N-vinylpyrrolidone) (PVP) in aqueous solutions,[20,65-68] the N-β-methacroyloxyethyl (MAO) derivative of adenine along poly(MAO derivative of thymine or uracil)[69-71] and methacrylic acid along poly(4-vinylpyridine)[9,72,73] or along poly(2-vinylpyridine) in various organic solvents.[10,12,13,74-76] In the latter two systems, small but distinct K_M values were measured, pointing to some preferential monomer adsorption on the template. For the system acrylic acid/PVP, a workable type 2-like mechanistic model was suggested, from which a K_M value could be derived.[66] However, in the similar system of methacrylic acid/PVP, no preferential monomer adsorption was claimed.[67]

A simple experiment which can be used to verify a type 2-like mechanism involves adding a saturated monomer analogue to the polymerization system. This should have very little influence on the result (*cf.* type 1 systems, Section 19.3.1).

Some polymerizations were carried out in dilute solutions with constant $[M]_0$ of the order of 10^{-2} mol dm^{-3}, others in semidilute solutions of constant $[M]_0$ of 0.5 up to 0.75 mol dm^{-3}. It is not known how far concentration effects play a role in the mechanistic course of template polymerizations. Changing $[T]_0$, while keeping the ratio $[T]_0/[M]_0$ constant at unity, showed an increase in polymerization rate that approached a limit around the critical coil overlap concentration of the template,[73] giving a curve similar to curve (iii) in Figure 2(b).

External initiators are commonly used. In kinetic analyses, it is generally assumed that the initiator decomposition rate is not influenced by the template.

19.3.2.1 Kinetics

In contrast to type 1 template polymerization, rate enhancement in type 2 systems is not expected. The fact that it does occur has been generally attributed to retardation of the termination step. Evidence for this was given in a study of the template polymerization of MMA along isotactic PMMA in DMF at 5 °C (which yields syndiotactic PMMA in the initial stages) using the rotating sector technique.[60] When compared to the blank polymerization, a five-fold reduction in the propagation rate coefficient, k_p, was found which is, however, more than compensated for by an 82-fold reduction in the termination rate coefficient, k_t. The latter should correspond with an increase in the activation energy of the termination step, leading to the observed decrease in overall activation energy. Retarded termination may be caused by a strong hindrance of the segmental mobility of the template-bound chain radicals. A second rate enhancement, observed at higher conversion, might be due to hindered translational motion of growing chains in a physical network built up during polymerization.[62] This so-called second template effect was more pronounced in the nonpolar solvent, toluene, in which stereocomplexation is weaker than in DMF (see also Section 19.3.2.3).

A decrease in the rate of the propagation step indicates that the chain end radical is firmly attached to the template site, thus diminishing its activity or sterically hindering the approach of a monomer molecule. The latter seems more probable, since this should lead to a lower activation entropy of the propagation step and, hence, to the observed lower overall activation entropy.[60]

A strong decrease in the rate of the propagation step was proposed for polymerizations of methacrylic acid along poly(ethylene oxide), which showed a negative template effect,[56-58] *i.e.* the polymerization rate was reduced by the presence of the template. The effect was accompanied by a higher overall energy of activation. The lowering of the propagation rate was explained by the less favourable template environment as compared to water (a medium effect). This explanation was prompted by the fact that a decrease in rate was also observed on addition of methanol to aqueous solutions of the monomer. Pol(vinyl alcohol) also diminished the polymerization rate.[19]

The formation of gels or precipitates does not seem to affect the half order in initiator concentration of the polymerization rate greatly,[17,60,66,67,73] indicating that gel and occlusion effects do not play a role. The chains are probably terminated before a precipitate is formed, or the template-bound chain end radicals remain protruding in the solution. A kinetic explanation has been given by Smid *et al.*[74]

In order to describe type 2 systems, attempts have been made to set up kinetic schemes from which kinetic rate formulas were derived.[7,67] A generalized mechanistic model was suggested by Smid *et al.*[7,13,74,75] to include the implications of preferential adsorption of monomer. In this scheme are considered, not only the usual reaction steps in the blank and template-bound regions together with the complexation of blank chain radicals to template molecules, but also monomer adsorption with the constant K_M, and the zip propagation rate that is assumed to be independent of the length of the monomer array. Since there is already a specific interaction at the monomeric level, complexation of oligomeric chain radicals is practically irreversible and therefore largely kinetically determined. Therefore, complexation of blank radicals is visualized as a unidirectional process characterized by a 'rate coefficient' k_c. The overall propagation rate of template-bound chain radicals is characterized by a composite rate coefficient, $k_{p,T}(M)$, which is dependent on the monomer concentration and incorporates contributions of type 1 and type 2 propagations. Finally, any explicit influence of the critical coil overlap concentration of the template is ignored, but the complexation is still dependent on $[T]_0$.

This model was successfully applied to the polymerization of methacrylic acid along poly-(2-vinylpyridine) in DMF and in *p*-dioxane at 30 °C, using rate coefficients which could all be estimated.[12,13,74,75] (In DMSO there is no template effect, owing to the absence of polycomplex formation in this solvent.) The rate enhancements appear to be primarily caused by a thousand-fold decrease of the termination step, which more than compensates for the rather low fraction of chain radicals that can complex with the template. The latter derives from the unexpectedly low k_c values. The larger rate enhancements in a solvent with low dielectric constant, *p*-dioxane, seem to be mainly due to a higher concentration of template-bound radicals as a consequence of a higher k_c; $k_{p,T}(M)$ was only slightly raised despite the higher K_M of *ca.* 10 (in DMF, K_M was *ca.* 3). This system behaved rather peculiarly, in that it showed a slow build-up of radicals[10,76] (revealed by an induction period) up to concentrations high enough to be detected by ESR.[12]

This generalized model can also be applied to type 2 systems and, presumably, to type 1-like systems.[7] A somewhat similar, but simpler, kinetic model was proposed by Muramatsu *et al.*[66]

19.3.2.2 *Molecular weight*

Since both rate and molecular weight are governed by the rate coefficients k_p and k_t in the same manner, we can expect a parallel tendency of these quantities with a change in these coefficients. Thus, a rate enhancement should be accompanied by an enhancement of the molecular weight of the daughter polymer. This is indeed borne out in several systems, where the average molecular weight of the formed polymer was studied as a function of $[T]_0$ at constant $[M]_0$. The curve profiles were nearly identical with that of curve (iii) in Figure 2(b).[18,61]

There is also a parallel increase in rate and average molecular weight, or intrinsic viscosity, of the formed polymer with increasing template length.[7,17,18,60,61,65,67] This may be due to either an increase in the fraction of chain radicals that can complex with template molecules,[67] or to prolonged growth of template-bound radicals along templates of greater length,[18] *i.e.* $k_{t,T}$ decreases with template chain length, or to a combination of both.

19.3.2.3 Microstructure

Template systems of type 2 have generally led to daughter polymers whose microstructures (tacticity) differ little from those of the corresponding blank polymers. Some of them had a somewhat higher syndiotactic triad fraction,[56,67] one was claimed to contain a distinct increase in isotacticity,[65] whereas, in a few cases, no difference with the blank polymer was observed.[17,18,69]

A drawback of template polymerization proceeding according to mechanism 2 is that the daughter polymer is not entirely created on the template, its initial part being formed in the bulk solution. This means that the microstructure depends on the critical chain length of the growing chain and on the ultimate length of the daughter polymer. The shorter the former and the longer the latter are, the greater the influence on the microstructure of the daughter polymer, as was demonstrated by experiment.[15,67]

By and large, it is not yet entirely clear which factors govern the stereochemical influence of a template on its associated growing polymer. One possible factor is the firmness with which the growing chain end is attached to the template site, another one is steric hindrance which the template would exert on the adding monomer molecule in its transition state. Preferential adsorption of solvent might be detrimental to the first requirement.[18]

It might be anticipated that, when there is an influence of template tacticity on the polymerization rate, there would also be a stereochemical effect on the daughter polymer. Thus, different rate enhancements were found in the polymerization of the N-β-methacroyloxyethyl (MAO) derivative of adenine in the presence of isotactic, syndiotactic and atactic poly(MAO derivative of uracil).[69,70] Surprisingly, the isolated daughter polymers had microstructures no different from that of the blank polymer. A similar result was obtained in the template polymerization of N-vinylpyrrolidone in the presence of various tactic poly(methacrylic acid)s,[17,18] despite the fact that the syndiotactic template forms the strongest polycomplex, in accordance with kinetic findings.

A special type 2 system is the template polymerization of MMA along isotactic PMMA, which leads to the formation of syndiotactic PMMA, particularly in the early stages of the reaction.[59,77-79] The reversed system, employing syndiotactic PMMA, led to the formation of isotactic PMMA,[15,78] though the template effect was much weaker, in accordance with the weaker interaction between atactic chain radicals and the syndiotactic PMMA template. In other words, long growing chains exceeding a sizeable critical chain length are required to reveal a template effect. This requirement was probably not met in recent experiments, where no enhancement of isotactic triads in the formed PMMAs could be established on employing deuterated and undeuterated syndiotactic PMMA.[80] The driving force of the stereospecific template polymerization is the stereocomplexation of isotactic and syndiotactic PMMA to give a double helix, which is composed of an inner isotactic helix and a larger outer syndiotactic helix, such that the base molar ratio of isotactic to syndiotactic PMMA is 1:2.[81] The template effect is most pronounced in polar solvents (type A), such as DMF, DMSO and acetone, and less in nonpolar aromatic solvents (type B), such as benzene and toluene, in which stereocomplexation is weaker.[78,79] In chlorinated solvents (type C), such as chloroform and dichloromethane, where complexation does not occur, the template effect is absent. In all solvents the template effect disappears at elevated temperatures, however. Interestingly, stereoblock copolymers were also formed during these template polymerizations in type A solvents, indicating that previously formed syndiotactic PMMA (or isotactic PMMA) was also able to function as a template.[15,78,82]

In the bulk polymerization of MMA, in the presence of semisolid isotactic PMMA (or syndiotactic PMMA), unusually high isotactic (or syndiotactic) triads of newly formed PMMAs were found; this phenomenon is called the 'reverse replica effect'.[83] This was also explained in terms of previously produced syndiotactic PMMA (or isotactic PMMA) operating in its turn as a template.

Stereospecific template polymerization of MMA along tactic PMMA could also be effected anionically with n-butylmagnesium chloride in toluene at $-50\,°C$.[63] Even in the blank polymerization system, the first formed isotactic PMMA appeared to act as a template for the production of polymer chains consisting of substantial fractions of syndiotactic triads.

19.3.2.4 Selective and displacement reactions

When a monomer is polymerized in the presence of two sorts of templates, the rate is determined by the template with which the growing chain can form the stronger complex. This is exemplified by the polymerization of methacrylic acid in the presence of a mixture of poly(ethylene oxide) and poly(N-vinylpyrrolidone) of sufficient chain lengths; the initial rate was identical with that obtained in the presence of the latter template alone.[19]

Related to the above selective template polymerizations are the displacement polymerizations, in which the growing polymer of one kind can displace a polymer of another kind from its template. Thus, when methacrylic acid (MAA) was polymerized in the presence of a poly(ethylene oxide)–poly(acrylic acid) (PEO·PAA) complex in dilute aqueous solution, a template effect ensued, though reduced, according to:[19,84]

$$MAA + PEO \cdot PAA \rightarrow PMAA \cdot + PEO \cdot PAA \rightarrow PEO \cdot PMAA \cdot + PAA \tag{1}$$

19.3.2.5 Template structure

Apart from its chain length and stereochemical structure, a template can be modified through copolymerization or cross-linking. Copolymeric templates built up of a template-active and template-inactive comonomer were used by various researchers to study the effect of template gaps on the rate and the structure of the formed complex.[34,72,85] The rate is invariably slowed down. Rate vs. $[T]_0$ curves were obtained, for which the interpretation is rather obscure. The polycomplex composition depends on the polymerization type and the sort of template-inactive comonomer, which may, for example, be hydrophobic or hydrophilic.[85] Cross-linked templates reduce the polymerization rate.[37]

19.3.3 Copolymerization

Copolymerizations were performed between two sorts of multimethacrylates, in which methacrylate units are covalently linked to two different oligomeric templates, resulting in block copolymers consisting of two different ladder blocks, but yielding the poly(methacrylic acid) homopolymer after hydrolysis.[50] Copolymerization of a multimethacrylate on an oligomeric template ($K_{M,1} = \infty$) with a comonomer such as styrene or acrylonitrile ($K_{M,2} = 0$)[86] resulted in reactivity ratios corresponding to the formation of block copolymers, which contain methacrylic acid blocks after their isolation *via* hydrolysis.

Very few systems have been studied in which two comonomers are polymerized in the presence of a template.[7,34,87] An example is the copolymerization of methacrylic acid ($K_{M,1} > 0$) and styrene ($K_{M,2} = 0$) in the presence of poly(ethylene oxide) in benzene.[87] For such systems, consisting of a mixture of a template-interacting and a template-noninteracting comonomer, a mechanistic model has been proposed based on mechanism 2. It is obvious that on raising the fraction of the noninteracting comonomer, a stage will be reached where complexation of the copolymer is completely suppressed and, consequently, the template effect is suppressed as well.[16] Copolymerization can thus be used to estimate the critical chain length (see also Section 19.2.3).

Copolymerization also appears to be useful for deciding between type 1 or type 2 mechanisms in cases where the template polymerization rate is not very different from that of the blank reaction. Thus, in the nonaccelerated polymerization of the N-β-methacroylaminoethyl derivative of uracil (MAE-U) in the presence of poly(MAE derivative of adenine), preferential adsorption of monomer on the template could be indirectly deduced from the fact that addition of the comonomer, the MAE derivative of adenine (MAE-A), produced a nearly 1:1 polycomplex of poly(MAE-U) and poly-(MAE-A) (the template) and newly formed poly (MAE-A), whereas the blank polymerization gave a random copolymer.[88]

19.4 RING-OPENING TEMPLATE POLYMERIZATION

Template polymerizations *via* ring-opening are, to date, limited to N-carboxy acid anyhydrides (NCAs) of α-amino acids,[1,3,89] along suitable polypeptides, yielding polypeptides with liberation of carbon dioxide. In general, any α-amino acid NCA monomer can be used which has an unsubstituted NH group, $R^1R^2\overline{CC(O)O(O)CNH}$, where R^1 and $R^2 =$ alkyl, aryl or H, in combination with a template possessing hydrogen-accepting groups, such as a polypepetide and poly(2-vinylpyridine).

19.4.1 Polypeptides at Templates

The polypeptide must not contain unsubstituted amide groups, in order to avoid intramolecular hydrogen bonding which competes with monomer adsorption. The first system to be examined consisted of D,L-phenylalanine NCA ($R^1 = CH_2Ph$, $R^2 = H$) and poly(sarcosine)dimethylamide, $R^3R^4N\text{---}[\text{---}COC(R^5R^6)NR^7\text{---}]_n\text{---}H$ with $R^3 = R^4 = R^7 = Me$ and $R^5 = R^6 = H$, carrying the –NHMe initiating base attached to the template chain end, in nitrobenzene.[1] The general requirement of an initiating end group (an external initiator does not lead to rate enhancement) is not yet fully understood.

Since K_M values are of the order of 5 cm^3 mol^{-1},[11] these systems are typically type 2 like. The dependence of rate on the chain length of the template is evidence for a template effect. Because of the peculiarity of dealing with a template bearing an initiating group, the dependence of the rate on the template's degree of polymerization (DP) can be viewed in two ways:[1] at constant molar concentration of the template (which implies an increase in base molar concentration with increasing DP), or at constant base molar concentration of the template (which implies a decrease in the initiator concentration with increasing DP). The latter leads to a rate *vs.* DP curve which shows a conspicuous maximum; its location reflects the maximum probability of reaction (initiation) between the terminal initiating group and NCA molecules adsorbed in the various positions along the template chain. Using a kinetic model, resembling the Michaelis–Menten scheme for enzymic reactions,[1,3] it could be estimated that, in case of the polysarcosine template, the maximum probability for initiation occurs with an NCA molecule on the seventh site, counting along the template chain from the terminal base. Reaction with more remote NCA molecules will be less and less probable because the greater relative displacement for reaction will meet with increasing resistance and obstruction from adjacent parts of the chain (including the adsorbed NCA molecules).[1] This is in agreement with the larger activation energy of the overall rate with larger DP. This should also hold qualitatively for sites closer to the initiating group.

It is clear that the above rationalization is related to the flexibility of the template chain. This was demonstrated with the series of diethylamides of polysarcosine, poly(N-ethylglycine) and poly(N-propylglycine), where the decreasing template effect is not caused by a decreasing NCA adsorption but by decreasing chain flexibility.[90] If still stiffer chains were utilized, such as poly(proline)dimethylamide,[91] poly(N-methyl-D,L-alanine)diethylamide and poly(N-benzylglycine)diethylamide,[92] carrying bulky substituents, there was no rate enhancement. The overall chain flexibility was also reduced by incorporating D,L-phenylalanine units in the polysarcosine template chains.[93]

By virtue of their living character, daughter chains can grow further with another kind of NCA by normal solution polymerization, to give ultimately a triblock copolymer, including the template chain to which the daughter chain is linked through the initiating base.[41,94]

The template effect is much reduced in DMF because this solvent can strongly solvate NCA molecules through hydrogen bonding and compete with the template.[1] The rate is also lowered when the template is blockaded with a preadsorbed polymer such as poly(D,L-phenylalanine).[1]

19.4.2 Vinyl Polymers as Templates

Poly(2-vinylpyridine) and poly(N-vinyl-2-ethylimidazole) are also suitable initiating templates in methyl benzoate or acetophenone at 35 °C.[95] As the side groups in these polymers are tertiary bases, they have a dual function, *viz.* that of initiator and that of binding site. With the following monomers, NCAs of D,L-phenylalanine, p-nitro-D,L-phenylalanine and o,p-dinitro-D,L-phenylalanine, the rate enhancement increases in the above order due to an increase in the effective monomer concentration along the template chains by additional charge transfer interaction between the nitrophenyl groups and the basic moieties of the templates.

Rates increase significantly with the degree of polymerization of poly(2-vinylpyridine), especially in the oligomeric region, in spite of the relative independence of K_M. This reflects the ease with which an activated NCA molecule (presumably formed through abstraction of a proton from the NH group of the NCA) reacts with adsorbed NCA molecules on the same template molecule (intramolecularly).[96]

Small differences in rate and K_M with the tacticity of poly(2-vinylpyridine) were found, decreasing in the order isotactic, atactic and syndiotactic poly(2-vinylpyridine). It signifies that adsorption of NCA is somewhat more sterically hindered on the syndiotactic template.[96]

19.4.3 Copolymers as Templates

By introducing a noninteracting comonomeric unit in the template structure, as in 2-vinyl-pyridine/styrene copolymers, the rate is obviously lowered.[95] However, introduction of a comonomeric unit which primarily functions as an adsorption site appears to be beneficial to the rate enhancement. Thus, copolymers of N-vinyl-2-ethylimidazole and N-vinylpyrrolidone (41 mol %) are much better initiating templates than the basic homopolymer of N-vinyl-2-ethylimidazole. On the other hand, copolymers with N,N-diethylacrylamide residues as binding sites were less effective.[95] Apparently the pyrrolidone ring is more effective than the acrylamide group in adsorbing the phenylalanine NCA monomer. This was confirmed by means of IR spectroscopy.

19.5   STEP TEMPLATE POLYMERIZATION

Template polymerization systems using a step mechanism are of recent date, if earlier attempts at so-called template-directed syntheses of oligonucleotides are not considered. When two complementary monomers are employed, the usual way to achieve template polymerization is to have at least one of the monomers attached to the template. Several methods have been devised to accomplish this goal.

One method provides one monomer with groups such as hydroxyl or sulfoxy or atoms such as O or S for hydrogen bonding with a suitable template. Polymerizing such diesters with hexamethylenediamine in the presence of a template, such as poly(vinyl alcohol), poly(N-vinylpyrrolidone), or poly(4-vinylpyridine), led to an enhanced rate, a decrease in overall activation energy and an increase in the reduced viscosity of the daughter polyamide, which become more pronounced with increasing chain length of the template.[97, 98] In the case of diethyl mucate, containing four hydroxyl groups, four vinylpyridine units of the poly(4-vinylpyridine) template were required per diester molecule under optimal conditions.

Another method transforms a diacid into a derivative that interacts strongly with the template. In the polymerization involving an aromatic dicarboxylic acid, such as terephthalic acid, with the aid of triphenyl phosphite, it was postulated that the diacid monomer is bonded to the poly(4-vinylpyridine) template as an N-phosphonium salt.[99] The reaction was carried out with a diamine in N-methylpyrrolidone as solvent containing LiCl. The production of polyterephthalamides was greatly facilitated under these conditions. High molecular weights were obtained, reaching a maximum value at approximately equibase molar amounts of template and terephthalic acid. By using poly-(N-vinylpyrrolidone) as a template in the presence of triphenyl phosphite, α-amino acids could be polymerized to polypeptides of high molecular weight in high yields.[99]

A third method provides a diester with a nucleic acid base moiety for 'hooking up' on a template having complementary nucleic acid base moieties. Thus, a (di-p-nitrophenyl) methyl succinate derivative of thymine (Thy) is attached through hydrogen bonding *via* Thy to polystyrene templates having various contents of adenine (Ade) moieties, and then reacted with a diamine in a solvent mixture of pyridine and dichloromethane to form the polyamide.[100] With a template containing 54 base mol % of Ade, a maximum rate was observed at $[Ade]_0/[M]_0 = 1$ (or $[Ade]/[Thy] = 1$), when the monomer was reacted with piperazine as diamine. With the 'longer' diamine, 4-aminomethylbenzylamine, the best template contained *ca.* 25 base mol % of Ade, demonstrating the importance of the average distance between the attached diester molecules. The rate enhancement is promoted on inserting a spacer between the Ade moiety and the template backbone, indicating the importance of the flexibility of the adsorption site.

A simpler system is the template polymerization of urea and formaldehyde along poly(acrylic acid) in water.[101] It is based on the polycomplex formation of linear urea–formaldehyde (UF) oligomers with hydrogen accepting polymers. UF polymers are insoluble in water, so that polycomplexes can only be formed *in situ* by template polymerization. The polycomplexes have a stoichiometric composition of 1:1 and show interesting morphological structures.

19.6   APPLICATIONS

Template polymerization has not yet found an actual application, although various potentialities are beginning to emerge. These should be found in the properties of the new materials which are the polycomplexes formed *in situ*. Moreover, they may differ from those produced by mixing of the ready-made polymer components, in that the polycomplexes formed *in situ* are expected to have a better mutual fitting and consequently more ordered structures. As seen from the foregoing sections, structural features such as molecular weight and microstructure of the daughter polymers can, in many instances, be obtained only by means of template polymerization. Their isolation from the template forms a major obstacle, however. The following recent examples are illustrative of the promising utilization of template polymerization.

One application comprises the preparation of semi and full interpenetrating polymer networks (IPNs)[102] by polymerizing a mixture of acrylic acid and a cross-linking agent in the presence of linear or cross-linked poly(ethylene oxide) as template. The full IPNs so obtained prevent dissolution on decomplexation at pH = 7. This advantage has been put to use in membranes functioning as chemical valves, *i.e.* the permeability of the membrane can be controlled reversibly by varying the pH or ionic strength of treated aqueous solutions.

Another form of application is imaging,[103, 104] which is based on the formation of insoluble polycomplexes. A thin coating of a template polymer, dissolved in a mixture of monomer and photoinitiator, is applied onto a transparent film carrier. On UV irradiation through a mask, the exposed parts become insoluble due to formation of insoluble polycomplex by template polymerization. The unexposed areas are removed by washing and immersion in a dye bath produces the final image.

ACKNOWLEDGEMENT

The author is very grateful to Professor G. Challa for a careful perusal of the manuscript and stimulating and fruitful discussions.

19.7 REFERENCES

1. D. G. H. Ballard and C. H. Bamford, *Proc. R. Soc. London, Ser. A*, 1956, **236**, 384, 495.
2. V. A. Kargin, V. A. Kabanov and O. V. Kargina, *Dokl. Akad. Nauk SSSR*, 1965, **161**, 1131 (*Chem. Abstr.*, 1965, **63**, 3068h); V. A. Kabanov, *Pure Appl. Chem.*, 1967, **15**, 391; V. A. Kabanov, T. I. Patrikeeva, O. V. Kargina and V. A. Kargin, *J. Polym. Sci., Polym. Symp.*, 1968, **23**, 357.
3. C. H. Bamford, in 'Developments in Polymerization', ed. R. N. Haward, Applied Science, London, 1979, vol. 2, p. 215.
4. C. H. Bamford, in 'Macromolecular Chemistry', ed. A. D. Jenkins and J. F. Kennedy, Royal Society of Chemistry, London, 1980, vol. 1, p. 52.
5. G. Challa and Y. Y. Tan, in 'Polymer Complexes', ed. E. Tsuchida, K. Hori and K. Abe, Japan Scientific Society, Tokyo, 1983, p. 159.
6. E. Tsuchida and Y. Osada, *J. Polym. Sci., Polym. Chem. Ed.*, 1975, **13**, 559.
7. G. Challa and Y. Y. Tan, *Pure Appl. Chem.*, 1981, **53**, 627; Y. Y. Tan and G. Challa, *Makromol. Chem., Macromol. Symp.*, 1987, **10/11**, 215; Y. Y. Tan, in 'Recent Advances in Mechanistic and Synthetic Aspects of Polymerization', ed. M. Fontanille and A. Guyot, Reidel, Dordrecht, 1987, p. 281.
8. V. A. Kabanov, O. V. Kargina and M. V. Ulyanova, *Vysokomol. Soedin., Ser. A*, 1980, **22**, 1038 (*Polym. Sci. USSR (Engl. Transl.)*, 1980, **22**, 1143).
9. K. Fujimori, G. T. Trainor and M. J. Costigan, *J. Polym. Sci., Polym. Chem. Ed.*, 1984, **22**, 2479.
10. J. Smid, Y. Y. Tan and G. Challa, *Eur. Polym. J.*, 1983, **19**, 853.
11. C. H. Bamford and R. C. Price, *Trans. Faraday Soc.*, 1965, **61**, 2208.
12. J. Smid, J. Speelman, Y. Y. Tan and G. Challa, *Eur. Polym. J.*, 1985, **21**, 141.
13. J. Smid, Y. Y. Tan, G. Challa and W. R. Hagen, *Eur. Polym. J.*, 1985, **21**, 757.
14. Ts. I. Nedyalkova, V. Yu. Baranovskii, I. M. Papisov and V. A. Kabanov, *Vysokomol. Soedin., Ser. B.* 1975, **17**, 174 (*Chem. Abstr.*, 1975, **83**, 43 860m).
15. R. Buter, Y. Y. Tan and G. Challa, *J. Polym. Sci., Polym. Chem. Ed.*, 1973, **11**, 1003.
16. K. F. O'Driscoll and I. Capek, *J. Polym. Sci., Polym. Lett. Ed.*, 1981, **19**, 401.
17. T. Bartels, Y. Y. Tan and G. Challa, *J. Polym. Sci., Polym. Chem. Ed.*, 1977, **15**, 341.
18. D. W. Koetsier, Y. Y. Tan and G. Challa, *J. Polym. Sci., Polym. Chem. Ed.*, 1980, **18**, 1933; *Polymer*, 1981, **22**, 1709.
19. I. M. Papisov, Ts. I. Nedyalkova, N. K. Avramchuk and V. A. Kabanov, *Vysokomol. Soedin., Ser. A*, 1973, **15**, 2003 (*Polym. Sci. USSR (Engl. Transl.)*, 1973, **15**, 2259).
20. J. Ferguson and S. A. O. Shah, *Eur. Polym. J.*, 1968, **4**, 343.
21. C. H. Bamford and Z. Shikii, *Polymer*, 1968, **9**, 595.
22. J. Ferguson and S. A. O. Shah, *Eur. Polym. J.*, 1968, **4**, 611.
23. A. Blumstein and S. R. Kakivaya, in 'Polymerization in Organized Systems', ed. H. G. Elias, Gordon and Breach, London, 1977, p. 189; A Blumstein and G. Weill, *Macromolecules*, 1977, **10**, 75.
24. A. Blumstein, S. Ponrathnam and E. Bellantoni, *J. Polym. Sci., Polym. Lett. Ed.*, 1980, **18**, 299.
25. S. Ponrathnam, M. Milas and A. Blumstein, *Macromolecules*, 1982, **15**, 1251.
26. O. A. Aleksina, I. M. Papisov, K. I. Bolyachevskaya and A. B. Zezin, *Vysokomol. Soedin., Ser. A*, 1973, **15**, 1463 (*Polym. Sci. USSR (Engl. Transl.)*, 1973, **15**, 1636).
27. T. Kawai and A. Fujie, *J. Macromol. Sci., Phys.*, 1980, **B17**, 653.
28. V. A. Kabanov, O. V. Kargina and V. A. Petrovskaya, *Vysokomol. Soedin., Ser. A*, 1971, **13**, 348 (*Polym. Sci. USSR (Engl. Transl.)*, 1971, **13**, 394).
29. A. N. Gvozdetskii, V. O. Kim, V. I. Smetanyuk, V. A. Kabanov and V. A. Kargin, *Vysokomol. Soedin., Ser. A*, 1971, **13**, 2409 (*Polym. Sci. USSR (Engl. Transl.)*, 1971, **13**, 2704).
30. O. V. Kargina, L. A. Mishustina, V. I. Svergun, G. M. Ludovkin, V. P. Yerdakov and V. A. Kabanov, *Vysokomol. Soedin., Ser. A*, 1974, **16**, 1755 (*Polym. Sci. USSR (Engl. Transl.)*, 1974, **16**, 2033).
31. V. A. Kabanov, O. V. Kargina and M. V. Ulyanova, *J. Polym. Sci., Polym. Chem. Ed.*, 1976, **14**, 2351.

32. C. Klein, *Makromol. Chem.*, 1972, **161**, 85.
33. Y. Miura, Y. Kakui and M. Kinoshita, *Makromol. Chem.*, 1975, **176**, 1567.
34. K. Shima, Y. Kakui, M. Kinoshita and M. Imoto, *Makromol. Chem.*, 1972, **154**, 247; *Makromol. Chem.*, 1972, **155**, 299.
35. T. Endo, R. Numazawa and M. Okawara, *Kobunshi Kagaku*, 1971, **28**, 541 (*Chem. Abstr.*, 1971, **75**, 141 246v).
36. T. Endo, R. Numazawa and M. Okawara, *Makromol. Chem.*, 1971, **146**, 247.
37. S. Wang, F. Xi and Z. Li, *Polym. Commun. (Peking)*, 1983, 11; *Kao Fen Tzu T'ung Hsun*, 1981, 224 (*Chem. Abstr.*, 1982, **96**, 20 498d); *Kao Fen Tzu T'ung Hsun*, 1982, 241 (*Chem. Abstr.*, 1983, **98**, 107 815v).
38. G. S. Manning, *Acc. Chem. Res.*, 1979, **12**, 443.
39. A. Blumstein, M. Milas, Y. Ozcayir and E. Bellantoni, *J. Polym. Sci., Polym. Phys. Ed.*, 1983, **21**, 2159.
40. W. Kern and H. Kämmerer, *Pure Appl. Chem.*, 1967, **15**, 421.
41. C. H. Bamford, in 'Reactions on Polymers', ed. J. A. Moore, Reidel, Dordrecht, The Netherlands, 1973, p. 54.
42. A. Blumstein, S. R. Kakivaya, R. Blumstein and T. Suzuki, *Macromolecules*, 1975, **8**, 435.
43. S. Kataoka and T. Ando, *Polym. Commun.*, 1984, **25**, 24; *Kobunshi Ronbunshu*, 1980, **37**, 185 (*Chem. Abstr.*, 1980, **92**, 198 833q); *Kobunshi Ronbunshu*, 1981, **38**, 821 (*Chem. Abstr.*, 1982, **96**, 52 760z).
44. J. C. Salamone, B. Snider and W. L. Fitch, *J. Polym. Sci., Part A-1*, 1971, **9**, 1493; *Macromolecules*, 1970, **3**, 707.
45. V. A. Kabanov, *Makromol. Chem., Suppl.*, 1979, **3**, 41.
46. V. A. Kabanov, K. V. Aliev and J. Richmond, *J. Macromol. Sci., Chem.*, 1975, **A19**, 273.
47. H. Kämmerer, J. Shukla, N. Oender and G. Scheuermann, *J. Polym. Sci., Polym. Symp.*, 1967, **22**, 213.
48. H. Kämmerer and S. Ozaki, *Makromol. Chem.*, 1966, **91**, 1; H. Kämmerer and A. Jung, **101**, 284; H. Kämmerer and N. Oender, *Makromol. Chem.*, 1968, **111**, 67.
49. H. Kämmerer and G. Hegemann, *Makromol. Chem.*, 1970, **139**, 17; *Makromol. Chem.*, 1984, **185**, 635.
50. R. Jantas and S. Połowinski, *J. Polym. Sci., Polym. Chem. Ed.*, 1986, **24**, 1819.
51. A. Chapiro, *Pure Appl. Chem.*, 1972, **30**, 77; *Pure Appl. Chem.*, 1981, **53**, 643.
52. A. Chapiro, Z. Mankowski and S. Ali-Miraftab, *Eur. Polym. J.*, 1981, **17**, 259; A. Chapiro, Z. Mankowski and M. Ansarian, *Eur. Polym. J.*, 1981, **17**, 823.
53. G. Burillo, A. Chapiro and Z. Mankowski, *Eur. Polym. J.*, 1984, **20**, 653; *J. Polym. Sci., Polym. Chem. Ed.*, 1980, **18**, 327.
54. V. S. Rajan and J. Ferguson, *Eur. Polym. J.*, 1982, **18**, 633.
55. G. O. R. Alberda van Ekenstein and Y. Y. Tan, *Polym. Commun.*, 1984, **25**, 105.
56. I. M. Papisov, V. A. Kabanov, Y. Osada, M. Leskano-Brito, J. Richmond and A. N. Gvozdetskii, *Vysokomol. Soedin., Ser. A*, 1972, **14**, 2462 (*Polym. Sci. USSR (Engl. Transl.)*, 1972, **14**, 2871).
57. Y. Osada, A. D. Antipina, I. M. Papisov, V. A. Kabanov and V. A. Kargin, *Dokl. Akad. Nauk SSSR*, 1970, **191**, 399 (*Chem. Abstr.*, 1970, **73**, 4220b).
58. Y. Osada, *Kagaku no Ryoiki*, 1971, **25**, 625 (*Chem. Abstr.*, 1972, **76**, 4175c).
59. O. V. Orlova, Yu. B. Amerik, B. A. Krentsel and V. A. Kargin, *Dokl. Akad. Nauk SSSR*, 1968, **178**, 889 (*Chem. Abstr.*, 1968, **68**, 96 222b).
60. J. Gons, E. J. Vorenkamp and G. Challa, *J. Polym. Sci., Polym. Chem. Ed.*, 1975, **13**, 1699; *J. Polym. Sci., Polym. Chem. Ed.*, 1977, **15**, 3031.
61. J. Gons, L. J. P. Straatman and G. Challa, *J. Polym. Sci., Polym. Chem. Ed.*, 1978, **16**, 427.
62. J. Gons, W. O. Slagter and G. Challa, *J. Polym. Sci., Polym. Chem. Ed.*, 1977, **15**, 771.
63. T. Miyamoto and H. Inagaki, *Polym. J.*, 1970, **1**, 46; T. Miyamoto, H. Inagaki and S. Tomoshige, *Makromol. Chem.*, 1975, **176**, 3035.
64. J. H. G. M. Lohmeyer, Y. Y. Tan and G. Challa, *J. Macromol. Sci., Chem.*, 1980, **A14**, 945.
65. J. Ferguson, S. Al-Alawi and R. Granmayeh, *Eur. Polym. J.*, 1983, **19**, 475.
66. R. Muramatsu and T. Shimidzu, *Bull. Chem. Soc. Jpn.*, 1972, **45**, 2538.
67. N. Shavit and J. Cohen, in 'Polymerization in Organized Systems', ed. H. G. Elias, Gordon and Breach, London, 1977, p. 213.
68. T. Sato, K. Nemoto, S. Mori and T. Otsu, *J. Macromol. Sci., Chem.*, 1979, **A13**, 751.
69. M. Akashi, H. Takada, Y. Inaki and K. Takemoto, *J. Polym. Sci., Polym. Chem Ed.*, 1979, **17**, 747.
70. K. Takemoto, Y. Inaki and M. Akashi, *J. Macromol. Sci., Chem.*, 1979, **A13**, 519.
71. K. Takemoto and Y. Inaki, *Adv. Polym. Sci.*, 1981, **41**, 1.
72. K. Fujimori, *Polym. Bull. (Berlin)*, 1982, **8**, 207; K. Fujimori and G. Trainor, *Polym. Bull. (Berlin)*, 1983, **9**, 204.
73. K. Fujimori, *Makromol. Chem.*, 1979, **180**, 1743.
74. J. Smid, Y. Y. Tan and G. Challa, *Eur. Polym. J.*, 1984, **20**, 887.
75. J. Smid, Y. Y. Tan and G. Challa, *Eur. Polym. J.*, 1984, **20**, 1095.
76. J. Smid, G. O. R. Alberda van Ekenstein, Y. Y. Tan and G. Challa, *Eur. Polym. J.*, 1985, **21**, 573.
77. R. Buter, Y. Y. Tan and G. Challa, *J. Polym. Sci., Part A-1*, 1972, **10**, 1031.
78. R. Buter, Y. Y. Tan and G. Challa, *J. Polym. Sci., Polym. Chem. Ed.*, 1973, **11**, 1013, 2975.
79. G. Challa, A. de Boer and Y. Y. Tan, *Int. J. Polym. Mater.*, 1976, **4**, 239.
80. K. Matsuzaki, T. Kanai, C. Ichijo and M. Yuzawa, *Makromol. Chem.*, 1984, **185**, 2291.
81. F. Bosscher, G. ten Brinke and G. Challa, *Macromolecules*, 1982, **15**, 1442.
82. R. Buter, Y. Y. Tan and G. Challa, *Polymer*, 1973, **14**, 171.
83. H. Yau and S. I. Stupp, *J. Polym. Sci., Polym. Chem. Ed.*, 1985, **23**, 813.
84. I. M. Papisov, N. A. Nekrasova, V. D. Pautov and V. A. Kabanov, *Dokl. Akad. Nauk SSSR*, 1974, **214**, 861 (*Chem. Abstr.*, 1974, **81**, 13 906x).
85. J. Ferguson and C. McLeod, *Eur. Polym. J.*, 1974, **10**, 1083.
86. S. Połowinski and G. Janowska, *Eur. Polym. J.*, 1975, **11**, 183; S. Połowinski, *Eur. Polym. J.*, 1978, **14**, 563.
87. S. Połowinski, *Eur. Polym. J.*, 1983, **19**, 679; *J. Polym. Sci., Polym. Chem. Ed.*, 1984, **22**, 2887.
88. Y. Inaki, K. Ebisutani and K. Takemoto, *J. Polym. Sci., Polym. Chem. Ed.*, 1986, **24**, 3249.
89. Y. Imanishi, *Pure Appl. Chem.*, 1981, **53**, 715.
90. M. Sisido, Y. Imanishi and S. Okamura, *Polym. J.*, 1970, **1**, 198.
91. D. G. H. Ballard, *Biopolymers*, 1964, **2**, 463.
92. M. Sisido, Y. Imanishi and S. Okamura, *Biopolymers*, 1969, **7**, 937; *Biopolymers*, 1970, **9**, 791.
93. Y. Imanishi, T. Sugihara and T. Higashimura, *Biopolymers*, 1973, **12**, 1505.

94. C. H. Bamford, H. Block and Y. Imanishi, *Biopolymers*, 1966, **4**, 1067.
95. K. Suzuoki, Y. Imanishi, T. Higashimura and S. Okamura, *Biopolymers*, 1969, **7**, 917, 925.
96. Y. Imanishi, T. Higashimura, S. Nagaoka and K. Suzuoki, *Biopolymers*, 1973, **12**, 181; Y. Imanishi, T. Higashimura, Y. Amimoto and Y. Hashimoto, *Polym. J.*, 1976, **8**, 585.
97. N. Ogata, K. Sanui, H. Nakamura and H. Kishi, *J. Polym. Sci., Polym. Chem. Ed.*, 1980, **18**, 933; N. Ogata, K. Sanui, H. Nakamura and M. Kuwahara, *J. Polym. Sci., Polym. Chem. Ed.*, 1980, **18**, 939.
98. N. Ogata, K. Sanui, H. Tanaka, H. Matsuo and F. Iwaki, *J. Polym. Sci., Polym. Chem. Ed.,* 1981, **19**, 2609; N. Ogata, K. Sanui, and A. Kato, *J. Polym. Sci., Polym. Chem. Ed.*, 1982, **20**, 227; N. Ogata, K. Sanui, F. Iwaki and A. Nomiyama, *J. Polym. Sci., Polym. Chem. Ed.,* 1984, **22**, 793.
99. N. Yamazaki and F. Higashi, *Adv. Polym. Sci.*, 1981, **38**, 1.
100. M. Hattori, H. Nakagawa, and M. Kinoshita, *Makromol. Chem.*, 1980, **181**, 2325; H. Nakagawa, M. Muraki, Y. Miura and M. Kinsoshita, *Makromol. Chem.*, 1982, **183**, 2065.
101. I. M. Papisov, O. E. Kuzovleva, S. V. Markov and A. A. Litmanovich, *Eur. Polym. J.*, 1984, **20**, 195.
102. S. Nishi and T. Kotaka, *Macromolecules*, 1985, **18**, 1519; *Macromolecules*, 1986, **19**, 978.
103. M. Rätzsch, *Abstracts*, 30th IUPAC Symposium on Macromolecules, The Hague, Netherlands, 1985, p. 37.
104. S. R. Turner and R. C. Daly, in 'Comprehensive Polymer Science', ed. G. Allen and J. C. Bevington, Pergamon Press, Oxford, 1988, vol. 6, chap. 7.

20
Polymerization in Aqueous Solution

COLIN J. HAMILTON and BRIAN J. TIGHE
University of Aston, Birmingham, UK

20.1 INTRODUCTION

The unique position that water occupies, both in terms of physical properties and biological significance, has been recognized for many years and has been widely discussed.[1-4] The consequences of this position are reflected not only in the inherent behaviour of water as a polymerization solvent, but also in the fact that the interactions of the resultant polymers with the polymerization solvent have a very marked influence on their subsequent behaviour and utility. In a number of commercially and biologically important applications, therefore, water plays a special role, both in polymerization and in controlling polymer properties. It is important to recognize, however, that not all polymers that are used in solutions of, or swollen by, water are formed by aqueous polymerization and that this chapter is primarily concerned with polymerization in aqueous solution, rather than water soluble polymers.

The range of monomers whose polymerization in aqueous solution has been studied or exploited is wide. The last two decades have seen a considerable expansion of interest in hydrophilic polymers, with a corresponding increase in the availability of water soluble vinyl monomers. Much of the early work, which laid the foundations of the kinetic aspects of aqueous polymerization, was, in contrast, carried out on conventional vinyl monomers (such as methyl methacrylate) which have a relatively low water solubility. An important conclusion that can be drawn from such work is that it is inappropriate to divorce polymerization in aqueous systems from the broader aspects of polymer science, since the unusual features of the subject largely reflect the unusual solvent characteristics of water, rather than the existence of polymerization reactions that are in themselves unique. Thus many aspects of conventional free radical polymerization are directly applicable to aqueous systems. It is a consequence of this fact that some aspects of the subject area are dealt with in other chapters. Heterogeneous polymerization, for example, is obviously important in aqueous systems and is dealt with in greater detail elsewhere (Volume 4, Part 2). Similarly, redox initiation, photoinitiation and electroinitiation, which are considered individually in separate chapters (Volume 3, Chapter 9; Volume 4, Chapters 20 and 24) have particular relevance to aqueous systems. The influence of polar and ionically charged water soluble monomers on polymerization reactions and on the nature of the reaction products has, additionally, significance for topics such as template polymerization (Volume 3, Chapter 19) and polyelectrolytes (Volume 1, Chapter 11). The aim of this chapter, therefore, is to provide an overview of the subject as a whole, further detail being contained in the chapters dealing with the individual topics referred to above.

20.2 INITIATORS AND KINETIC PHENOMENA

It was mentioned at the beginning of this chapter that vinyl polymerization in the aqueous phase occurs by a free radical addition process. Consequently, we are concerned with the type of initiators that can generate *in situ* the atoms, free radicals and ion radicals that are the initiation species in such polymerizations. The requirement that initiators be water soluble obviously presents a limitation but this is more than offset by the advantages such as the ability of water to act as a solvation medium to non-organic soluble species (H_2O_2, Fenton's reagent, $K_2S_2O_8$, *etc.*). The use of water soluble species to initiate the chain growth process is not unique to this area — many are suitable as initiators for emulsion polymerization (see Volume 4, Chapter 11). For the purpose of discussion it is useful to divide the various initiation techniques into five basic types: radiolysis, photochemical, thermal, redox and electrochemical. Further subdivision of some of these five categories is possible and appropriate since some of the most interesting and unique initiator systems are contained in these subdivisions. Examples of these key categories are shown in Table 1. More detailed information may be found elsewhere in this work (Volume 3, Chapters 8–10; Volume 4, Chapters 19–21 and 24). In addition, a comprehensive tabulation of water soluble initiators has been compiled by Palit, Guha, Das and Konar.[5]

20.2.1 Radiation-induced Initiation

Any residual initiator that is present after polymerization may act as an undesirable contaminant. For example, with prosthetic implants the presence of leachable small molecular weight solutes, such

Table 1 Key Categories and Examples of Initiators for Aqueous Polymerization

Initiation	Monomer	Ref.
Radiolysis		
γ-Radiation of H_2O	Acrylonitrile	a, b
γ-Radiation of H_2O	Acrylamide	c
X-Rays or γ-radiation of H_2O_2	Acrylonitrile, methacrylonitrile	b
Photoinitiation		
Riboflavin	Acrylamide	d
$Fe^{3+}OH^-$	Methyl methacrylate	e
Thermal		
$K_2S_2O_8$	Acrylonitrile	f
LiP_2O_8	Acrylonitrile	g
H_2O_2	Methacrylic acid	h
Redox		
Peroxydiphosphate–Ag^+	Acrylonitrile	g
H_2O_2–$FeSO_4$	Methyl methacrylate	i
Ce^{4+}–glycolic acid	Methacrylamide	j
Electroinitiation		
$ZnCl_2$	Acrylamide	k
Dilute sulfuric acid	Acrylic acid, methacrylic acid and their methyl derivatives	l

[a] J. C. Arthur, Jr., R. J. Denimt and R. A. Pittman, *J. Phys. Chem.*, 1959, **63**, 1366.
[b] F. S. Dainton, *J. Phys. Colloid. Chem.*, 1948, **52**, 490.
[c] R. Schulz, G. Renner, A. Hanglein and W. Kem, *Makromol. Chem.*, 1954, **12**, 20.
[d] G. K. Oster, G. Oster and G. Prati, *J. Am. Chem. Soc.*, 1957, **79**, 595.
[e] B. Atkinson and G. R. Cotten, *Trans. Faraday Soc.*, 1958, **54**, 877.
[f] L. B. Morgan, *Trans. Faraday Soc.*, 1946, **42**, 169.
[g] S. S. Hariharan and A. Meenakshi, *J. Polym. Sci., Polym. Lett. Ed.*, 1977, **15**, 1.
[h] A. Katchalsky and G. Blauer, *Trans. Faraday Soc.*, 1951, **47**, 1360.
[i] J. H. Baxendale, M. G. Evans and J. K. Kilham, *Trans. Faraday Soc.*, 1946, **42**, 668.
[j] G. S. Misra, B. D. Arya and S. L. Abrol, *J. Macromol. Sci., Chem.*, 1986, **A23**, 923.
[k] J. F. Sobiesky and M. C. Zemer, *US Pat.* 3 464 960 (1969) (*Chem. Abstr.*, 1969, **71**, 103 294).
[l] E. Dineen, T. C. Shwan and C. L. Wilson, *Trans. Electrochem. Soc.*, 1949, **96**, 226.

as residual initiator and its side products, may create an inflammatory response by the body.[6,7] One way of avoiding this problem involves the use of γ-radiation, X-rays and ultrasonic waves to generate H$\cdot$ and OH$\cdot$ radicals as the initiating species *via* the radiolysis of the water.[8,9,10] γ-Radiation and X-rays have also been used to decompose hydrogen peroxide to hydroxyl radicals.[11]

20.2.2 Photoinitiation

It is more common to use less energetic wavelengths to bring about the initiation of aqueous polymerization, for example by the photolysis of water soluble species such as hydrazine[12] and water soluble azo compounds.[13] The use of dyes such as rose bengal, erythrosin, acriflavin and methylene blue in the presence of a reducing agent (ascorbic acid, phenylhydrazine hydrochloride, *etc.*) has also been found to produce photoinitiated polymerization.[14,15,16] The use of a reducing agent is not always necessary, as in the case of riboflavin. When used in conjunction with *N*-vinylpyrrolidone or acrylamide, for example, the monomers themselves appear to act as the reducing agent.[14]

Metal salts and complexes such as $Ce^{4+}OH^-$, $Fe^{3+}F^-$, $Fe^{3+}Cl^-$, $Pb^{2+}Cl^-$ and $Fe^{3+}(C_2O_4)^{2-}$ have been known to produce photoinitiated aqueous polymerization of monomers such as acrylonitrile and methacrylonitrile for many years.[17] The initiating radical is produced by charge transfer processes that may involve metal, ligand or solvent.

20.2.3 Thermal Initiation

Although thermal initiations are often categorized as being either dissociative or redox, the distinction can be somewhat blurred when the mechanism is examined. A dissociative initiator can be defined as a species that decomposes into the reactive intermediates that initiate the polymerization process. However, some 'dissociative' initiators may form their radicals by way of redox reactions, either with the monomer or the water. The conditions for thermal initiation to occur will, naturally, be dependent upon the system employed, but the temperature need be no higher than 20 °C.[18] One of the best known groups of thermal initiator are the peroxydisulfates, whose application to aqueous polymerization was reported by Bacon as far back as 1946.[19] The structurally similar peroxydiphosphate has been more recently employed in the aqueous polymerization of acrylonitrile.[20]

20.2.4 Redox Initiation

Solutes such as peroxydisulfates and peroxydiphosphates may be suitable initiators in their own right, but both form excellent redox initiators when used in combination with metal cations such as Ag^+, Fe^{2+}, Cu^{2+} and Ti^{3+}. Peroxydiphosphate–reducing agent redox systems were found to be better initiators for the polymerization of acrylonitrile than peroxydiphosphate alone, with the peroxydiphosphate–Ag^+ redox pair being the most effective.[20] The important features of the components constituting a redox pair for aqueous polymerization are their solubility in water, together with a sufficiently fast and constant liberation of initiating radicals in the aqueous phase. Cerium(IV) salts are commonly used as components of aqueous redox initiation systems that fulfil these criteria, and have been extensively used in the polymerization of monomers such as acrylonitrile, acrylamide and methacrylamide.[21-26] This is one area where mechanistic controversy remains (see Section 20.2.6). It has been suggested that the manganese(III) ion behaves similarly to the cerium(IV) ion. In redox pairs with glycols, aldehydes, ketones and carboxylic acids and phenols it provides good initiation systems for the polymerization of acrylonitrile and methyl methacrylate, for example.[27]

20.2.5 Electroinitiation

Electroinitiated polymerization differs from the other types discussed above in that initiation is a heterogeneous process. The initiator is formed by electron transfer at an electrode surface, thus initiation is remote from propagation, chain transfer and, usually, termination. It is conveniently defined as the generation of a polymerization initiator by electron transfer processes at an electrode in a conducting solution of a monomer. One of the advantages of this technique is that it is an electrode potential, rather than heat, which provides the activation energy for the generation of the initiator. This allows comparatively mild, low temperature conditions to be employed, thus enabling potentially undesirable side reactions to be minimized. The monomer may be polymerized in bulk, in aprotic polar solvents or in aqueous media and whereas only free radical addition mechanisms are effective for aqueous polymerization systems, electropolymerization in aprotic solvents enables propagation by ionic mechanisms to be exploited. This area is dealt with in detail in Volume 4, Chapter 24, and has been the subject of reviews by various authors.[28,29,30]

One of the earliest examples of aqueous polymerization by an electrochemical process involved aqueous solutions of acrylic acid, methacrylic acid and their methyl derivatives, which were observed to undergo polymerization at the cathode when subjected to electrolysis in the presence of dilute H_2SO_4.[31] Acrylamide behaves somewhat differently, and in aqueous $ZnCl_2$ solution electro-initiated polymerization can create an electrochemically deposited hydrogel coating on the metal cathode.[32,33] The potential of this process as a one-stage technique for the formation of polymer coatings for metal protection has been reviewed by Mengoli.[34] The author concluded that the '*in situ*' polymerization of traditional acrylic monomers gave unfavourable results in terms of high current consumption and unsatisfactory coatings. This was explained in terms of the low production of radical initiator and the contribution of desorption phenomena. In addition, polymer deposition may complicate the electrolytic process by limiting transport of electroactive species to the electrode. Anodic processes such as the Kolbe reaction with potassium acetate can also be exploited for the aqueous polymerization of monomers such as acrylic acid[35] and *N*-vinyl-2-pyrrolidone.[36] The methyl radical generated during the electron transfer process between the $MeCO_2^-$ ion and the anode acts as an initiator for the polymerization of the monomer.[34] The aqueous polymerization of vinyl ketones as a result of cathodic electrolysis has been observed with monomers such as methyl vinyl ketone (buten-3-one) using electrolytes ranging from the simple inorganic such as KCl[37,38] to species such as tetraethylammonium *p*-toluene sulfonate.[39] Although electroinitiated polymerization does have the advantages previously described, the problems should not be overlooked. These include the fact that control of the electrode potential can be restricted by the deposition of polymer on the working electrode,[40] and that the increase in solution viscosity caused by the polymerization process leads to a reduction in electrode reactions due to the resultant restrictions on the transport of electroactive species to and from the electrode.

20.2.6   Kinetic Phenomena in Aqueous Media — General Observations

Several general observations can be made from the results of early kinetic work on aqueous polymerizations, much of which was carried out by Dainton and his co-workers.[9,41-48] The monomers studied were usually of a type that exhibited solubility in both aqueous and non-aqueous solvents, allowing comparisons between solvent systems to be made. Acrylamide attracted a great deal of attention, due partially to its ability to undergo polymerization to high molecular weight in aqueous media, the degrees of polymerization obtained being substantially greater than those obtained in organic media.[10] This is in part due to the fact that water possesses a chain transfer constant of virtually zero.[47,48] In addition, other factors such as the existence of polymer–water interactions which produce a strongly bound hydration shell may help protect the propagating radical centre and hinder termination.[10] The ability of water to have a significant effect on DP is found even in systems that are essentially non-aqueous but to which traces of water have been added.[49]

Some detailed aspects of kinetic behaviour in aqueous systems still produce controversy. Recent kinetic studies[24] on the aqueous polymerization of acrylamide by a Ce^{IV}–2-chloroethanol redox pair were prompted by the differing views on the mode of initiation and termination on cerium(IV) salt initiation of vinyl polymerization.[25,26] The authors concluded that cerium(IV) ions do not participate directly in the initiation process and that termination occurs exclusively through the oxidative termination of cerium(IV) ions only — that is, there is no termination by mutual combination of growing chain radicals. This mechanism for termination was also suggested by Rout and his co-workers for the aqueous polymerization of acrylonitrile by a Ce^{IV}–thiourea redox system.[23] This is in contradiction to the observations of other workers,[21] who found in their studies of the aqueous polymerization of methacrylamide using a Ce^{IV}–glycolic acid redox pair that the cerium(IV) ions are not involved in either the initiation or termination reactions. Termination, they concluded, was due to the bimolecular reaction of the propagating polymer radicals. This proposed mechanism for the mode of termination is in agreement with earlier studies on the polymerization of methyl methacrylate, which employed Ce^{IV}–malonic acid as the redox initiator.[26]

One of the key features of aqueous polymerization systems is that they allow the study of ionizable pendant groups, and in particular the effect of degree of dissociation on the chain growth processes. Polymerization of acrylic and methacrylic acids, for example, using initiators such as hydrogen peroxide and peroxydisulfates, shows significant pH dependence.[50,51] The rate of polymerization was found to fall with increasing pH, reaching a minimum between pH 6 and 7. The rate then increased to a relatively steady value at pH 9 because of a reduction in the rate of termination due to coulombic repulsion between the fully ionized propagating chains. Another important feature of aqueous systems is that in the copolymerization of monomers, of which one or both are ionizing monomers, the relative reactivities will depend on the degree of ionization. The consequence of this phenomenon in terms of the relative effect of aqueous and non-aqueous solvents on monomer reactivities is observed in the copolymerization of acrylic acid with acrylamide, where the relative reactivity ratios, r_1 and r_2, are 1.43 and 0.60 respectively in water, and 0.36 and 1.38 in acetone.[52] These general points are discussed in terms of monomer structure in Section 20.3.

A considerable body of literature exists on the subject of inhibition and retardation of aqueous polymerization systems. This, together with information on related topics such as post-irradiation polymerization effects, is summarized in earlier reviews.[5,53] Recent studies involving polymerization in aqueous solution have drawn attention to the importance of initiation processes occurring at solid surfaces (other than electrodes).[54,55]

20.3 MONOMER STRUCTURE AND POLYMERIZABILITY

It is self-evident that a degree of water solubility is an essential requirement for the use of a monomer in aqueous polymerization studies. Several vinyl monomers that are more widely associated with reactions in organic media do, however, show a surprisingly high solubility in water and have also been used in fundamental studies of aqueous polymerization. Table 2 lists some of the monomer systems whose high water solubility has led to their use in aqueous polymerization work, together with the solubilities in water of a selected group of more conventional vinyl monomers. There are, in addition, many compounds of intermediate solubility in which the balance of hydrophilic and hydrophobic groups within the molecule has been changed. Examples include the hydroxypropyl and hydroxybutyl methacrylates and the N-alkyl-acrylamides and -methacrylamides. The ability to modulate the water-structuring ability of individual monomer units has important consequences for the properties of the polymers derived from this broad group of compounds. Some of the special features of these monomers and their interactions with water are reflected in variations of simple dilute aqueous solution polymerization with which this chapter is primarily concerned. Further reference to these types of polymerization is contained in a subsequent section. A useful overview may be achieved, however, by taking the monomers listed in Table 2 as a basis for comment on the novel features of polymerization in aqueous solution and relationships between monomer structure and polymerizability. The two volume compilation 'Functional Monomers', edited by Yocum and Nyquist, provides a substantial body of additional information relating to monomers of interest in this field.[56]

Acrylamide and substituted acrylamides have been widely studied over the last 30 years. In part, this reflects the range of industrial applications (water treatment, oil and mineral recovery, coatings and adhesives, chromatography, cosmetics, *etc.*) for this group of polymers, but is principally related to the unusual features of acrylamide polymerization. The literature reflects the facts that both homogeneous and heterogeneous polymerization of the monomers can be achieved using a wide range of free radical sources and that the fundamental mechanistic understanding has advanced little since the publication of Dainton's studies some 25 years ago.[9,41,45,46,57–59] These, and other, papers[60–65] discuss the kinetics of free radical polymerization of acrylamide and related monomers in considerable detail. The system provides something of a cornerstone in the subject of polymerization in aqueous solution. An excellent collection of information relating to acrylamide and other unsaturated amides has been compiled by MacWilliams.[66]

Although the kinetic dependence of the rate of acrylamide polymerization upon monomer (M) and initiator (I) is represented by the relationship

$$R_p = k[M][I]^{0.5} \tag{1}$$

over a wide range of conditions, the order with respect to monomer has frequently been reported to exceed one. This behaviour is most prominent at higher initiator concentrations and is related to the

Table 2 Typical Water Soluble Monomers

Monomers with pendant heterocyclic groups	*Monomers with pendant neutral hydroxyl functions*
N-Vinyl-2-pyrrolidone	Allyl alcohol
2-Vinylpyridine	2-Hydroxyethyl acrylate
3-Vinylpyridine	2-Hydroxyethyl methacrylate
4-Vinylpyridine	2-Hydroxypropyl methacrylate
4-Methylenehydantoin	
1-Vinyl-2-methyl-2-imidazoline	
4-Vinyl-3-morpholine	
Monomers with pendant basic functions	*Conventional vinyl monomers (room temperature solubility)*
Acrylamide	Styrene (0.07 g l^{-1})
Methacrylamide	Butyl methacrylate (4 g l^{-1})
N-Hydroxymethylacrylamide	Methyl methacrylate (15 g l^{-1})
N,N-Dimethylacrylamide	Vinyl acetate (25 g l^{-1})
Aminoethyl methacrylate	Acrylonitrile (118 g l^{-1})
Diethylaminoethyl methacrylate	Acrolein (180 g l^{-1})
Monomers with pendant acidic functions	
Acrylic acid	
Itaconic acid	
Methacrylic acid	
Allenesulfonic acid	
Ethylenesulfonic acid	
Styrenesulfonic acid	
2-Sulfoethyl methacrylate	

type of initiating system used. It does not occur with photolytic initiation but is marked in initiating species such as bisulfite, where the polymer chain can reduce the initiator anion to a radical. This process has been described as chain transfer to catalyst. Radiation studies have provided valuable mechanistic information, confirming that initiation with X-rays or γ-rays is a free radical process involving the reaction of a solvated electron (H· or HO·) with acrylamide. Radical lifetimes vary from microseconds in dilute solution to hours in more concentrated systems. The reversible formation of cyclic glutarimide radicals at the end of the growing chain and the very high viscosity of the polymer gel have both been cited as contributory factors. It is the high solubility of both monomer and polymer in aqueous solution, coupled with the ready formation of extremely high molecular weights ($> 10^6$) that underpin interest both in the kinetic behaviour of acrylamide polymerization and in the commercial applications of resultant polymers. The unusual aspects of the behaviour of acrylamide are best illustrated in comparison with the structural variants methacrylamide and N,N-dimethylacrylamide.

The pioneering work of Dainton and his co-workers in this field has already been mentioned. It is convenient to take the data from the 1957 Dainton and Tordoff papers, which deal with the photopolymerization of acrylamide in 0.12 M perchloric acid, as a basis for discussion.[45] These are listed below.

Propagation rate constant (k_p) $1.8 \times 10^4 \text{ M}^{-1}\text{s}^{-1}$

Termination rate constant (k_t) $14.5 \times 10^6 \text{ M}^{-1}\text{s}^{-1}$

Transfer to monomer constant (k_{trm}) $0.22 \text{ M}^{-1}\text{s}^{-1}$

Rate constant ratio ($k_p/k_t^{0.5}$) $4.2 \text{ M}^{-0.5}\text{s}^{-0.5}$

The uniquely high values of k_p and $k_p/k_t^{0.5}$ for acrylamide coupled with very low transfer to monomer and solvent result in the ready formation of polymer having a number average molecular weight in excess of one million. Under the same conditions methacrylamide has a similar termination rate constant, whereas the propagation constant is lower by almost two orders of magnitude. As a result the value of $k_p/k_t^{0.5}$ is only $0.2 \text{ M}^{-0.5} \text{ s}^{-0.5}$ (in comparison to $4.2 \text{ M}^{-0.5} \text{ s}^{-0.5}$ for acrylamide) and methacrylamide polymerization proceeds more slowly with lower molecular weight products (which, nonetheless, frequently precipitate from solution during polymerization). Some comparisons may be drawn with N-substituted derivatives of acrylamide, although polymerizations have not been reported under identical conditions. N-Alkyl groups exert a predominantly steric suppression of both k_p and k_t with the result that the polymerization of N-t-butylacrylamide proceeds at approximately one tenth the rate of acrylamide polymerization with a $k_p/k_t^{0.5}$ value of $0.6 \text{ M}^{-0.5} \text{ s}^{-0.5}$. Although N-t-butylacrylamide has a relatively low water solubility (7.0 g l^{-1},

30 °C), that of N,N-dimethylacrylamide is greater and it has been more extensively studied.[64] The propagation rate constant at 50 °C (1.1×10^4 M^{-1} s^{-1}) is around 60% of that for acrylamide at 25 °C and is thus much higher than that associated with the polymerization of conventional monomers in organic media. The major difference between acrylamide and N,N-dimethylacryl-amide polymerizations, however, lies in their relative susceptibility to chain transfer reactions. Whereas such reactions are virtually non-existent in acrylamide, they exert a marked influence on N,N-dimethylacrylamide chain growth processes, seemingly because of the lability of hydrogen atoms on the N-methyl groups. As a result of the reduction in kinetic chain length, attainable molecular weights in N,N-dimethylacrylamide polymerizations are less than 5% of those achieved with acrylamide. N-Substituted methacrylamides are relatively resistant to homopolymerization in the form of the free base, although nitrogen complexation enhances their reactivity.[65]

Solvent effects in acrylamide polymerization are pronounced, and illustrate some of the advantages of work in aqueous systems as compared to aprotic solvents. The belief that the rapid polymerization rates attainable in water are the result of the formation of a protonated radical species is supported by the effect of pH upon polymerization rate. Thus the propagation rate constant for acrylamide polymerization at 25 °C is reported to fall from 1.72×10^4 M^{-1} s^{-1} at pH 1 through 0.6×10^4 M^{-1} s^{-1} at pH 5.5, to 0.4×10^4 M^{-1} s^{-1} at pH 13.[58] The termination rate constant also declines over this pH range with the result that the value of $k_p/k_t^{0.5}$ remains substantially constant. Other workers have compared kinetic parameters for both acrylamide and methacrylamide in water, dimethyl sulfoxide and tetrahydrofuran.[63] The initiator used was AIBN and although its rate of decomposition increases thirty-fold in moving from water to tetrahydro-furan, the overall rate of polymerization was observed to decrease, at the same time, from 19.3×10^4 M^{-1} s^{-1} to 1.6×10^4 M^{-1} s^{-1}. The position of dimethyl sulfoxide is intermediate, as reflected in the progressive decrease in the value of $k_p/k_t^{0.5}$ from 3.3 M$^{-0.5}$ s$^{-0.5}$ (water) through 0.35 M$^{-0.5}$ s$^{-0.5}$ (dimethyl sulfoxide) to 0.05 M$^{-0.5}$ s$^{-0.5}$ (tetrahydrofuran). Parallel effects are observed in methacrylamide polymerization and the consequence in both cases is that attainable molecular weights in organic solvents are much lower than those obtained in water.

A progressive increase in basicity is encountered in moving from the acrylamides through the vinylpyridines to the aminoalkyl acrylates and methacrylates. Both classes undergo free radical polymerization in aqueous solution, in the form either of the free base or its quaternized derivatives. A useful overview of this topic has been given by Luskin.[67] The copolymerization of such basic monomers as diethylaminoethyl methacrylate with methacrylic acid in aqueous media is important for two reasons. In the first place, the products are polyampholytes and in the second, the compositions of the resultant copolymers show marked pH dependence. This arises, not from the effect of changing acidity on the rate of basic monomer polymerization, which is relatively small, but from the fact that the methacrylate anion is much less reactive than the free acid.[68,69] The reactivity ratios for diethylaminoethyl methacrylate (r_1) and methacrylic acid (r_2) are shown as a function of pH below.

$$\text{pH } 1.2, r_1 = 0.90; r_2 = 0.98$$

$$\text{pH } 7.2, r_1 = 0.65; r_2 = 0.08$$

A similar pH dependence is observed with itaconic acid,[70-72] whose rate of polymerization has been observed to remain constant from pH 2.5 to 3.5 and then drop to zero above pH 8. This corresponds to a structural change from the undissociated form through the monoanion (which two forms exhibit similar polymerizability) to the unreactive dianion which is the dominant form at pH 8. The point is reinforced by reactivity ratio data for acrylonitrile (r_1) and itaconic acid (r_2) expressed below as a function of pH.[73]

$$\text{pH } 1.5, r_1 = 0.25; r_2 = 1.57$$

$$\text{pH } 5.0, r_1 = 0.30; r_2 = 0.60$$

$$\text{pH } 9.5, r_1 = 0.43; r_2 = 0.1$$

In contrast, the stronger sulfonic acids, typified by ethylenesulfonic acid, allenesulfonic acid, styrenesulfonic acid and 2-sulfoethyl methacrylate, are much less affected by pH, in this respect.[74,75] This point is nicely illustrated by the comparative reactivity ratio data for acrylonitrile (r_1) and the sodium salt of p-styrenesulfonic acid (r_2) shown here under neutral and acid conditions.[74]

$$\text{pH } 7, r_1 = 0.05; r_2 = 1.40$$

$$\text{pH } 3, r_1 = 0.10; r_2 = 1.20$$

The polymerizabilities of styrenesulfonic acid and 2-sulfoethyl methacrylate are markedly influenced by the aromatic ring and ester groups respectively. Direct attachment of an acidic carboxyl or sulfonate group onto the double bond (as in acrylic acid and ethylenesulfonic acid) produces water soluble monomers of much lower reactivity than the styrenesulfonate. Whereas the reactivity of acrylic acid lies between that of itaconic and styrenesulfonic acids, the rate of homopolymerization of ethylenesulfonic acid is very low. Copolymerization activity of the sulfonated monomers may be assessed from the collected data in Table 3.[76-82] A useful overview of ionic effects in the polymerization of 2-sulfoethyl methacrylate and styrenesulfonic acids has been compiled by Kangas.[83] Both monomers polymerized approximately 10 times faster in aqueous solution than did the monomer itself, or a non-ionic equivalent, in an organic solvent. Similarly, the value of the ratio of the polymerization rate constants $k_p/k_t^{0.5}$ increases with decreasing ionic strength. The effects are attributed to the fact that increased ionic dissociation of the pendant group produces greater electrostatic repulsion between the two growing polymer radicals than between monomer and the growing polymer chain. As a consequence the termination step is retarded more markedly than is the propagation reaction.

Table 3 Reactivity Ratios of Sulfonated Monomers in Aqueous solution

M_1	r_1	M_2	r_2	Ref.
Sodium ethylenesulfonate	0	Acrylamide	14.9	a
Sodium ethylenesulfonate	0.12	Acrylonitrile	4.5	b
Sodium ethylenesulfonate	0	Sodium acrylate	5.8	a
Sodium ethylenesulfonate	1.0	N-Vinyl-2-pyrrolidone	1.0	c
Sodium allenesulfonate	0.07	Acrylonitrile	4.9	d
Sodium methylenesulfonate	0.5	Acrylonitrile	0.85	e
Sodium styrenesulfonate	2.3	Sodium acrylate	0.34	f
2-Sulphoethyl methacrylate	0.7	2-Hydroxyethyl methacrylate	1.6	g

[a] D. S. Breslow and A. Kutner, *J. Polym. Sci.*, 1958, **27**, 295.
[b] T. Alfrey, Jr. and C. R. Pfeifer, *J. Polym. Sci., Part A-1*, 1966, **4**, 2447.
[c] A. Kutner and D. S. Breslow, *J. Polym. Sci.*, 1959, **28**, 274.
[d] K. Miyamichi, A. Suzuki, S. Harada and M. Katayama, *Kobunshi Kagaku*, 1964, **21**, 79.
[e] Z. Izumi and H. Kitagawa, *Kobunshi Kagaku*, 1969, **26**, 153.
[f] C. E. Grabiel and D. L. Decker, *J. Polym. Sci.*, 1962, **59**, 425.
[g] D. A. Kangas and R. R. Pelletier, *J. Polym. Sci., Part A-1*, 1970, **8**, 3543.

In contrast to the effects described above in connection with basic and acidic monomers, the monomers that contain neutral hydroxy groups as the hydrophilic function are relatively insensitive to such changes. Indeed, it is the stability of the water-structuring centre to changes in pH, temperature and tonicity (ionic strength) that has made the water-swollen gels derived from this group of monomers so attractive for biomedical applications. As a result there has been less interest in the aqueous polymerization of hydroxyalkyl methacrylates and related compounds than in their use in gel formation. Nonetheless, the solution polymerization of 2-hydroxymethyl and 2-hydroxypropyl acrylates and of 2-hydroxyethyl methacrylate in water has been examined using a variety of initiator systems. Additionally the hydroxyl-containing polymers may be used as substrates for aqueous grafting reactions in conjunction with Ce^{IV}-based initiators.[84] The reactivity of the vinyl group in these monomers is largely influenced by the acrylate function, as illustrated by the similarity of the reactivity ratios of 2-hydroxymethyl methacrylate and 2-sulfoethyl methacrylate shown in Table 3. The behaviour of allyl alcohol on the other hand is quite different. The ready abstraction of an allylic hydrogen by the radical chain end of the growing polymer leads to the formation of a stable allyl radical, substantially incapable of reinitiation. The transfer reaction thus leads to termination of the chain reaction and is referred to as degradative chain transfer. Despite this, relatively high molecular weight copolymers of allyl alcohol and acrylonitrile have been produced in aqueous solution, using dye-sensitized photopolymerization, with acriflavin as the sensitizer and ascorbic acid as the reducing agent.[85]

A range of monomers that contain heterocyclic groups have been used in aqueous polymerization studies.[86] These include the related N-substituted compounds N-vinyl-2-pyrrolidone and 4-vinyl-3-morpholine, 1-vinyl-2-methyl-2-imidazoline and 5-methylenehydantoin. Of this group N-vinyl-2-pyrrolidone is by far the most widely studied and used, although polymerization in aqueous solution is not the normal commercial production technique. Both monomer and homopolymer are very water soluble and although homopolymerization proceeds smoothly, the monomer shows relatively low reactivity in copolymerization. This may be judged from reactivity ratio data in Table 3, and data listed below,[86,87] for copolymerization of N-pyrrolidone (M_1) with acrylamide, acrylonitrile and methyl methacrylate (M_2).

$$r_1 = 0.17; r_2 = 0.66 \quad \text{(acrylamide)}$$

$$r_1 = 0.06; r_2 = 0.18 \quad \text{(acrylonitrile)}$$

$$r_1 = 0.02; r_2 = 5.0 \quad \text{(methyl methacrylate)}$$

The behaviour of 4-vinyl-3-morpholine is in many ways analogous, although the additional ring oxygen atom does enhance the complexation and solubility behaviour of the polymer; a comprehensive body of information related to the formation and properties of copolymers exists.[86] Of the vast range of polymerizable heterocyclic monomers compiled by Tomalia,[86] relatively few have been studied in aqueous solution polymerization. Two such examples are of interest, both containing two ring nitrogens. 1-Vinyl-2-methyl-2-imidazoline is readily homopolymerized and copolymerized with aqueous persulfate initiation after first quaternizing the ring nitrogens. The imidazoline polymers thus produced are cationic in nature. Polymers and copolymers of 5-methylenehydantoin are produced with similar ease. The interesting property that poly(5-methylenehydantoin) possesses is that on opening the ring with alkali, the ampholytic product poly(α-aminoacrylic acid) is formed.

20.4 APPLICATIONS AND PROCESSES

Aqueous polymerization aroused a great deal of academic interest over a period of 20 years or so following the Second World War. The fundamental understanding of the subject developed during that time and it has changed relatively little since then. The succeeding years did, however, bring greatly increased commercial activity in the field. Increased concern over atmospheric pollution by industrial solvents, a developing interest in oil recovery and more stringent control of water pollution have all contributed to this trend. In addition pharmaceutical and medical interest in hydrophilic polymers generally has grown sharply in recent years. This trend can be readily discerned in a group of publications dealing with production and applications of the more commercially important of the water soluble polymers discussed in this chapter.[52,88-92]

Important technological consequences follow from situations in which the polymer does not remain in solution during polymerization. One such situation arises when monomers of limited solubility (Table 2) are used, and the products of polymerization precipitate from solution in the course of the reaction. Since much of the early work in the field of polymerization in aqueous solution involved such monomers, these processes (sometimes discussed as occlusion phenomena)[53] have been recognized for many years. Various aspects of their exploitation are described in Volume 4, Part 2 of this work.

Another important aspect of product insolubility in water arises when a water soluble monomer (such as acrylamide) is copolymerized with a monomer of greater functionality [such as methylene-bis(acrylamide)] leading to the onset of gelation. In the case cited the pendant double bonds have the same reactivity as that incorporated in the backbone, except for mobility restrictions. The properties of the gel may be varied by independent variation of monomer concentration in water and of monomer:monomer ratio. The polymerization is otherwise carried out in identical fashion to acrylamide homopolymerization. It was an adaptation of this process to produce soft contact lenses by the polymerization of an aqueous solution of 2-hydroxyethyl methacrylate in a rotating mould that revolutionized the contact lens industry and led to the expansion of commercial and scientific interest in 'hydrogel' polymers.[93] The significant difference between the two types of gel described above is that, in contrast to the hydroxyethyl methacrylate polymers, acrylamide gels show poor dimensional stability in conditions of changing pH and temperature. Related effects have already been described in connection with the polymerization behaviour of monomers that contain water-structuring groups which are sensitive to changes in the aqueous environment. Current knowledge in the general area of hydrogel chemistry has been recently discussed,[94] together with the way that many of the monomers listed in Table 2 influence water structure in such systems.[95,96] This dual role in both influencing polymer properties and acting as a polymerization medium, coupled with its biological and environmental significance, give water a unique position in polymer science.

20.5 REFERENCES

1. F. Franks (ed.), 'Water—a Comprehensive Treatise', Plenum Press, London, 1972–79, vols. 1–6.
2. W. Drost-Hansen and W. S. Clegg (eds.), 'Cell-associated Water', Academic Press, London, 1979.

3. S. P. Rowland, *ACS Symp. Ser.*, 1980, **127**.
4. H. H. G. Jellinek (ed.), 'Water Structure at the Water–Polymer Interface', Plenum Press, New York, 1975.
5. S. R. Palit, T. Guha, R. Das and R. S. Konar, in 'Encyclopaedia of Polymer Science and Technology', ed. H. F. Mark, N. G. Gaylord and N. Bikales, Wiley-Interscience, New York, 1965, vol. 2, p. 229.
6. A. S. Hoffman, *Radiat. Phys. Chem.*, 1977, **9**, 207.
7. Y. Ikada, T. Mita, F. Horri, I. Sakurada and M. Hatada, *Radiat. Phys. Chem.*, 1977, **9**, 633.
8. F. S. Dainton, *Nature (London)*, 1947, **160**, 268.
9. E. Collison, F. S. Dainton and G. S. McNaughton, *J. Chim. Phys.*, 1955, **52**, 556.
10. R. Schulz, G. Renner, A Henglein and W. Kem, *Makromol. Chem.*, 1954, **12**, 20.
11. F. S. Dainton, *J. Phys. Colloid. Chem.*, 1948, **52**, 490.
12. N. Uri, *Chem. Rev.*, 1952, **50**, 375.
13. S. Abuhantasch, W. Duismann and C. Rüchardt, *Makromol. Chem.*, 1976, **177**, 395.
14. G. Oster, *Nature (London)*, 1954, **173**, 300.
15. A. Watanabe and M. Koizumi, *Bull. Chem. Soc. Jpn.*, 1961, **34**, 1086.
16. A. Shepp, S. Chaberek and R. MacNeill, *J. Phys. Chem.*, 1962, **66**, 2563.
17. M. G. Evans and N. Uri, *Nature (London)*, 1949, **164**, 404.
18. O. Nuyken and R. Kerber, *Makromol. Chem.*, 1978, **179**, 2845.
19. R. G. R. Bacon, *Trans. Faraday Soc.*, 1946, **42**, 140.
20. S. S. Hariharan and A. Meenakshi, *J. Polym. Sci., Polym. Lett. Ed.*, 1977, **15**, 1.
21. G. S. Misra, B. D. Arya and S. L. Abrol, *J. Macromol. Sci., Chem.*, 1986, **A23**, 923.
22. G. S. Misra and B. D. Arya, *J. Macromol. Sci., Chem.*, 1983, **A19**, 253.
23. A. Rout, S. P. Rout, B. C. Singh and M. Santappa, *Eur. Polym. J.*, 1977, **13**, 497.
24. K. C. Gupta and K. Behari, *J. Polym. Sci., Polym. Chem. Ed.*, 1986, **24**, 767.
25. S. V. Anathanarayanan and M. Santappa, *Proc. Indian Acad. Sci.*, 1965, **62**, 150.
26. S. V. Subramanian and M. Santappa, *J. Polym. Sci., Part A-1*, 1968, **6**, 493.
27. A. Y. Drummond and W. A. Waters, *J. Chem. Soc.*, 1953, 2836.
28. B. M. Tidswell, in 'Macromolecular Chemistry', ed. A. D. Jenkins and J. F. Kennedy, Specialist Periodical Reports, No. 17, Royal Society of Chemistry, London, 1984, vol. 3, p. 72.
29. G. Mengoli, *Adv. Polym. Sci.*, 1979, **33**, 1.
30. J. W. Breitenbach, O. F. Olaj and F. Sommer, *Adv. Polym. Sci.*, 1972, **9**, 47.
31. E. Dineen, T. C. Shwan and C. L. Wilson, *J. Electrochem. Soc.*, 1943, **96**, 226.
32. Ya. D. Zytner and K. A. Makarov, *Vysokomol. Soedin., Ser. A*, 1980, **22**, 2612.
33. Ya. D. Zytner, K. A. Makarov and O. K. Lebedkina, *Izv. Vyssh. Uchebn. Zaved., Khim. Khim. Tekhnol.*, 1980, **23**, 1532.
34. G. Mengoli, *Adv. Polym. Sci.*, 1979, **33**, 1.
35. W. B. Smith and D. T. Manning, *J. Polym. Sci.*, 1962, **59**, 545.
36. B. L. Funt and F. D. Williams, *J. Polym. Sci., Part B*, 1963, **1**, 181.
37. E. I. Fulmer, J. J. Kolfenbach and L. A. Underkofler, *Ind. Eng. Chem.*, 1944, **16**, 469.
38. C. W. Johnson, C. G. Overberger and W. J. Seagers, *J. Am. Chem. Soc.*, 1953, **75**, 1495.
39. M. M. Baizer, J. D. Anderson and E. J. Prill, *J. Org. Chem.*, 1965, **30**, 3138.
40. B. L. Funt and J. Tanner, in 'Techniques of Electro-Organic Synthesis', ed. N. L. Weinberg, Wiley-Interscience, New York, 1975, part 2, p. 559.
41. E. Collison, F. S. Dainton and G. S. McNaughton, *Trans. Faraday Soc.*, 1957, **63**, 357, 476, 489.
42. F. S. Dainton and P. H. Seaman, *J. Polym. Sci.*, 1959, **39**, 279.
43. F. S. Dainton and D. G. L. James, *J. Polym. Sci.*, 1959, **39**, 299.
44. F. S. Dainton and R. S. Eaton, *J. Polym. Sci.*, 1959, **39**, 313.
45. F. S. Dainton and M. Tordoff, *Trans. Faraday Soc.*, 1957, **53**, 499, 666.
46. E. Collison, F. S. Dainton, D. R. Smith, G. L. Trundel and S. Tazuke, *Discuss. Faraday Soc.*, 1960, **29**, 188.
47. F. S. Dainton, *J. Chem. Soc.*, 1952, 1533.
48. F. S. Dainton and E. Collison, *Discuss. Faraday Soc.*, 1952, **12**, 212.
49. U. S. Nandi, P. Ghosh and S. R. Palit, *Nature (London)*, 1962, **195**, 1197.
50. A. Katchalsky and G. Blauer, *Trans. Faraday Soc.*, 1951, **47**, 1360.
51. S. H. Pinner, *J. Polym. Sci.*, 1952, **9**, 282.
52. M. L. Miller, in 'Encyclopaedia of Polymer Science and Technology', ed. H. F. Mark, N. G. Gaylord and N. Bikales, Wiley-Interscience, New York, 1965, vol. 1, p. 197.
53. A. D. Jenkins, in 'Vinyl Polymerization', ed. G. E. Ham, Edward Arnold, London, 1967, vol. 1, part 1, p. 369.
54. A. B. Moustafa, A. A. Abd El-Hakim, A. S. Badran, M. A. Abd El-Ghaffar and M. A. I. E. Sakr, *J. Appl. Polym. Sci.*, 1986, **31**, 1403.
55. M. Kuzuya, T. Kawaguchi, Y. Yanagihara, S. Nakai and T. Okuda, *J. Polym. Sci., Polym. Chem. Ed.*, 1986, **24**, 707.
56. R. H. Yocum and E. B. Nyquist (eds.), 'Functional Monomers', Dekker, New York, 1974, vols. 1 and 2.
57. F. S. Dainton and W. D. Sisley, *Trans. Faraday Soc.*, 1963, **59**, 1369, 1377.
58. D. J. Currie, F. S. Dainton and W. S. Watt, *Polymer*, 1965, **6**, 451.
59. K. W. Chambers, E. Collison and F. S. Dainton, *Trans. Faraday Soc.*, 1970, **66**, 142.
60. E. A. S. Cavell, *Makromol. Chem.*, 1962, **54**, 70.
61. J. P. Riggs and F. Rodriquez, *J. Polym. Sci., Part A-1*, 1967, **5**, 3151, 3167.
62. G. Delzenne, W. Dewinter, S. Toppet and G. Smets, *J. Polym. Sci., Part A-1*, 1964, **2**, 1069.
63. V. F. Gromov, A. V. Matveeva, A. D. Abkin, P. M. Khomikovskii and E. I. Mirokhina, *Dokl. Akad. Nauk SSSR*, 1968, **179**, 374.
64. A. M. North and A. M. Scallan, *Polymer*, 1964, **5**, 447.
65. J. Moens and G. Smets, *J. Polym. Sci.*, 1957, **23**, 931.
66. D. C. MacWilliams, in 'Functional Monomers', ed. R. H. Yocum and E. B. Nyquist, Dekker, New York, 1974, vol. 1, p. 1.
67. L. S. Luskin, in 'Functional Monomers', ed. R. H. Yocum and E. B. Nyquist, Dekker, New York, 1974, vol. 2, p. 555.
68. G. Ehrlich and P. Doty, *J. Am. Chem. Soc.*, 1954, **76**, 3764.
69. T. Alfrey, Jr., C. G. Overberger and S. H. Pinner, *J. Am. Chem. Soc.*, 1953, **75**, 4221.
70. L. S. Luskin, in 'Functional Monomers', ed. R. H. Yocum and E. B. Nyquist, Dekker, New York, 1974, vol. 2, p. 357.

71. C. S. Marvel and T. H. Shepherd, *J. Org. Chem.*, 1959, **24**, 599.
72. S. Nagai and K. Yoshida, *Kobunshi Kagaku*, 1960, **17**, 748.
73. S. Nagai, *Bull. Chem. Soc. Jpn.*, 1963, **36**, 1459.
74. Z. Izumi, H. Kiuchi and M. Watanabe, *J. Polym. Sci., Part A-1*, 1963, **1**, 705.
75. G. F. D'Alelio and T. F. Huemmer, *J. Polym. Sci., Part A-1*, 1967, **5**, 77.
76. D. S. Breslow and A. Kutner, *J. Polym. Sci.*, 1958, **27**, 295.
77. T. Alfrey, Jr. and C. R. Pfeifer, *J. Polym. Sci., Part A-1*, 1966, **4**, 2447.
78. A. Kutner and D. S. Breslow, *J. Polym. Sci.*, 1959, **38**, 274.
79. K. Miyamichi, A. Suzuki, S. Harada and M. Katayama, *Kobunshi Kagaku*, 1964, **21**, 79.
80. Z. Izumi and H. Kitagawa, *Kobunshi Kagaku*, 1969, **26**, 153.
81. C. E. Grabiel and D. L. Decker, *J. Polym. Sci.*, 1962, **59**, 425.
82. D. A. Kangas and R. R. Pelletier, *J. Polym. Sci., Part A-1*, 1970, **8**, 3543.
83. D. A. Kangas, in 'Functional Monomers', ed. R. H. Yocum and E. B. Nyquist, Dekker, New York, 1974, vol. 1, p. 489.
84. E. B. Nyquist, in 'Functional Monomers', ed. R. H. Yocum and E. B. Nyquist, Dekker, New York, 1974, vol. 1, p. 299.
85. G. Oster and Y. Mizutani, *J. Polym. Sci.*, 1956, **22**, 173.
86. D. A. Tomalia, in 'Functional Monomers', ed. R. H. Yocum and E. B. Nyquist, Dekker, New York, 1974, vol. 2, p. 1.
87. A. M. Chatterjee and C. M. Burns, *Can. J. Chem.*, 1971, **49**, 3249.
88. N. M. Bikales (ed.), 'Water-Soluble Polymers', Plenum Press, New York, 1973.
89. Y. L. Meltzer, 'Water-Soluble Polymers; Developments Since 1978', Noyes Data Corporation, Park Ridge, New Jersey, 1981.
90. R. L. Davidson and M. Sittig (eds.), 'Water-Soluble Resins', Reinhold, New York, 2nd. edn., 1968.
91. W. M. Thomas and D. W. Wang, in 'Encyclopaedia of Polymer Science and Engineering', ed. H. F. Mark, N. Bikales, C. G. Overberger and G. Meyes, Wiley-Interscience, New York, 2nd. edn., 1985, vol. 1, p. 169.
92. J. W. Nemec and W. Bauer, Jr., in 'Encyclopaedia of Polymer Science and Engineering', ed. H. F. Mark, N. Bikales, C. G. Overberger and G. Meyes, Wiley-Interscience, New York, 2nd. edn., 1985, vol. 1, p. 211.
93. B. J. Tighe, in 'Hydrogels in Medicine and Pharmacy', ed. N. A. Peppas, C. M. C. Press, New York, 1987, vol. 3, p. 53.
94. N. A. Peppas (ed.), 'Hydrogels in Medicine and Pharmacy', C. M. C. Press, New York, 1987.
95. P. C. Corkhill, A. M. Jolly, C. O. Ng and B. J. Tighe, *Polymer*, 1987, **28**, 1758.
96. D. A. Baker, P. C. Corkhill, C. O. Ng, P. J. Skelly and B. J. Tighe, *Polymer*, 1988, **29**, 691.

21

Polymerization at High Pressure

JOHN K. BEASLEY

*E. I. Du Pont de Nemours & Co., Inc., Wilmington, DE, USA**

21.1 INTRODUCTION

Most polymerization reactions proceed at a faster rate under high pressure. Higher molecular weights are obtained at higher pressures. High pressure polymerizations of a wide variety of vinyl type monomers and copolymerizations of two or more monomers have been investigated by many researchers around the world.[1-8] Such studies have been conducted with free radicals derived from oxygen and organic peroxides, photoinitiation,[9,10] and high energy radiation.[11] High pressure polymerizations have also been conducted with anionic[12] and cationic[13] initiators (catalysts) and with transition metal catalysts.[14]

High pressure polymerizations of bulk monomers have been conducted in the supercritical fluid state (ethylene), liquid state (for example styrene[15], vinyl acetate[16], methyl methacrylate[11,17]) and even in the glassy state (2-hydroxyethyl methacrylate and glycidyl methacrylate[18]) and the solid state (perfluorostyrene and several others).[19,20] Monomers have also been polymerized at high pressure in solution and in suspension or emulsion.[21,22] High pressure polymerizations have been conducted in small batch reactors in the laboratory and in very large continuous commercial reactors.

In general, any addition polymerization of vinyl type monomers can be studied as a function of pressure, just as it can be studied as a function of temperature. High pressures (above about 500 atm) do present experimental difficulties and most polymerization studies have not included effects of high pressure. Nevertheless, some general conclusions can be drawn about the effects of high pressure.

---

* Now retired.

21.2 ACTIVATION VOLUME FOR POLYMERIZATION

In general, increases in the pressure of polymerization result in higher polymerization rates and higher molecular weights. While most investigators have not done so, these increases can be described in terms of an activation volume, just as the increase in polymerization rates with temperature can be described in terms of an activation energy. The influence of temperature and pressure on a reaction rate constant can be written as shown in equation (1), where A is the frequency factor; E_A is the activation energy in kJ mol^{-1} (kcal mol^{-1}); R is the gas constant; P is the pressure; ΔV^* is the activation volume in cm^3 mol^{-1}; and T is the temperature. Table 1 shows some of the values for activation volumes reported in the literature for various free radical polymerizations.

$$k = A \exp\{-E_A/RT - (P\Delta V^*/RT)\} \tag{1}$$

Table 1 Activation Volume for High Pressure Polymerization of Various Monomers

Monomer	Activation volume (cm^3 mol^{-1})			Refs.
	Propagation	Termination	Overall	
Ethylene			−23	a, b, c
Ethylene			−20	a, b, c
Ethylene			−17.5	d
Styrene	−18.6	+ 5.8	−20.5	e
Styrene	−17.9	+13.1	−17.1	f
Vinyl acetate	−24.0	+16.3	−17.2	g
Methyl methacrylate	−19.0	+25.0		h
Butyl acrylate	−22.5	+20.8	−26.3	i
Butyl methacrylate	−23.2	+17.8	−17.4	j
Octyl methacrylate	−24.7	+20.8	−19.2	k

[a] S. Goto, K. Yamamoto, S. Furui and M. Sugimoto, *J. Appl. Polym. Sci., Appl. Polym. Symp.*, 1981, **36**, 21.
[b] P. Ehrlich and G. A. Mortimer, *Adv. Polym. Sci.*, 1970, **7**, 386.
[c] K. H. Lee and J. P. Marano, *ACS Symp. Ser.*, 1979, **104**, 221.
[d] M. Buback and H. Lendle, *Makromol. Chem.*, 1983, **184**, 193.
[e] P. W. Moore, J. G. Clouston and R. P. Chaplin, *J. Polym. Sci., Polym. Chem. Ed.*, 1983, **21**, 2503.
[f] Y. Ogo, M. Yokawa and T. Imoto, *Makromol. Chem.*, 1973, **171**, 123.
[g] M. Yokawa and Y. Ogo, *Makromol. Chem.*, 1976, **177**, 429.
[h] M. Yokawa, Y. Ogo and T. Imoto, *Makromol. Chem.*, 1974, **175**, 179.
[i] M. Yokawa, Y. Ogo and T. Imoto, *Makromol. Chem.*, 1974, **175**, 2913.
[j] M. Yokawa, Y. Ogo and T. Imoto, *Makromol. Chem.*, 1974, **175**, 2903.
[k] M. Yokawa, J. Yoshida and Y. Ogo, *Makromol. Chem.*, 1977, **178**, 443.

The negative value for the activation volume for propagation implies that the activated reaction complex for the addition of a monomer molecule to a growing free radical occupies less volume than the two components. Therefore, an increase in reaction pressure promotes the formation of this complex and increases the polymerization rate. A positive value for the activation volume for termination implies the opposite: the activated complex occupies a greater volume than the two reacting components. Therefore, an increase in pressure results in a reduced termination rate. In combination with the increased propagation rate, this will result in an increase in the molecular weight of the polymer at higher pressures.

Of course, there are exceptions. The activation volumes may not be constant over the entire pressure range. If the pressure becomes high enough to cause a phase change in the monomer to a glassy or solid state, the polymerization rate may decline drastically. No polymerization was observed in crystalline styrene,[15] while very slow polymerization was observed in crystalline methyl methacrylate.[23]

Polymerization rates can be increased by increasing the temperature of polymerization. Activation energies for polymerization are usually large, but, in most cases, an increase in polymerization temperature alone will also cause a decrease in molecular weight, sometimes to a level below that which is desired. (Activation energies for chain transfer are positive.) If *both* temperature and pressure are increased, then very large increases in polymerization rates can be achieved while maintaining molecular weight within the desired range. It is this combination which makes high pressure polymerization commercially useful.

21.3 COMMERCIAL USES OF HIGH PRESSURE POLYMERIZATION

Although many monomers have been polymerized in the laboratory under high pressure, only the free radical polymerization of ethylene and ethylene copolymers has had sufficient utility to be used commercially on a large scale. Several commercial processes operate at pressures from one to a few atmospheres to contain a monomer which has a high vapor pressure or to increase the concentration of a gaseous monomer. But high pressures, typically 1000–3000 atm, and elevated temperatures, typically 150–300 °C, are required to obtain commercially useful polymerization rates and molecular weights with ethylene and free radical initiators.

21.4 POLYMERIZATION OF ETHYLENE

Ethylene was first polymerized in 1933 by Fawcett in the research laboratories of Imperial Chemical Industries (ICI) in the UK.[24] The white waxy solid obtained from ethylene at high pressures was quickly recognized as polyethylene (PE). It gained its first commercial importance because of its remarkable properties as an electrical insulator. Since then, commercial use of PE made in high pressure processes has grown rapidly. United States consumption was 2.8×10^6 tons in 1984[25] and worldwide consumption was more than twice that amount.

ICI's first process used an autoclave, or stirred tank reactor. Ethylene was fed continuously and mixed with the contents of the reactor by mechanical stirring. Later, both Du Pont and Union Carbide in the United States developed tubular reactors in which agitation was provided by fluid flow inside the tube. Variations on these two processes, and combinations thereof, were developed later. For example, autoclaves have been used in series.[26] Internal baffles have been used to form a series of reaction zones within a single autoclave pressure vessel.[27] In another case the initial reactor was an autoclave followed in series by a tubular reactor.[28, 29] Details of the processes can be found in the vast patent literature on polyethylene processes.[30–44] Mathematical models have been constructed to describe these processes in great detail.[45–50]

21.4.1 Heat of Reaction

The polymerization of ethylene is exothermic, releasing about 94 kJ mol^{-1} (22.5 kcal mol^{-1}). For each one percent conversion of ethylene to polymer, the temperature rises 12–13 °C if none of the heat of polymerization is removed through the reactor walls by heat transfer. In the autoclave process, ethylene is injected at a temperature well below the polymerization temperature and is heated by the exothermic heat of polymerization. Relatively little heat can be removed through the thick walls in the autoclave. The difference between feed temperature and reaction temperature limits the conversion of ethylene to polymer to about 15% or less.

In the tubular process, much more surface area is available for heat transfer and external water cooling of the tube walls is used, thus permitting higher conversions.[51] Even with some heat being removed through the walls, the polymerization reaction is generally so fast that there is an increase in temperature as the reacting fluid flows down the tube. To achieve even higher conversions without excessive temperature rise, some tubular processes employ multiple injections of cold ethylene and/or inert diluents at several points along the tube.[52] Additional polymerization initiator must also be injected. A major portion of the ethylene remains unconverted and must be recovered, cooled and recycled to the process.

21.4.2 Decomposition

At high temperatures (above about 300 °C) ethylene can undergo a decomposition reaction (equation 2). This decomposition is also exothermic, releasing about 125 kJ mol^{-1} (30 kcal mol^{-1}). Once initiated, this reaction proceeds very rapidly, consumes most of the available ethylene, and results in a very rapid rise in temperature and pressure until the pressure is released, either intentionally or accidentally. Decomposition reactions are apparently initiated by adventitious hot spots in the polymerization reactor. Ethylene can undergo thermal (spontaneous) initiation of polymerization without the addition of other initiators.[53] At high temperatures (above 300 °C), this spontaneous polymerization, together with polymerization caused by added initiator, can become so fast that a runaway (accelerating) reaction occurs. When the temperature becomes high enough, the decomposition reaction starts releasing heat at a still faster rate and the reaction goes to completion very rapidly.

$$CH_2 = CH_2 \rightarrow CH_4 + C + H_2 + \text{other species} \qquad (2)$$

Industrial processes utilize carefully designed rupture discs to discharge the high pressure gases very rapidly whenever the designed release pressure is exceeded in order to quickly remove unreacted ethylene from the reactor.[54] Much work has been done on the careful control of industrial polymerization processes so that these undesirable ethylene decompositions do not often occur.

21.5 MECHANISM OF FREE RADICAL POLYMERIZATION (see also Volume 3, Chapter 6 and related chapters)

The mechanism of free radical polymerization at high pressures is generally believed to be the same as the mechanism at low pressures. Because of its commercial importance, the polymerization of ethylene has been studied more extensively than other high pressure polymerizations.[55-60] The discussion that follows describes the free radical polymerization of ethylene at high pressures. Other monomers are believed to behave in a similar manner, though some reactions such as chain transfer to polymer and the decomposition reaction may not be important for many other monomers.

An initiator (I) (generally a peroxide or hydroperoxide) decomposes to produce two free radicals $R_0^{\bullet}$ (equation 3). An ethylene molecule adds to $R_0^{\bullet}$ to produce another free radical $R_1^{\bullet}$ equation (4). Equation (3), sometimes combined with equation (4), is called the initiation reaction. Equation (5) is called the propagation reaction, in which ethylene monomer molecules are added successively to free radicals. Equations (6) and (7) are called the termination reaction, in which two free radicals combine to form dead polymer. Equation (6) shows termination by combination, in which both free radicals combine to form one polymer molecule. Equation (7) shows termination by disproportionation or disassociation in which two free radicals combine to form two polymer molecules. Equations (8) and (9) show chain transfer to telogen, in which S–H represents a molecule having an active hydrogen; such molecules are called chain transfer agents or telogens. Telogens may be added as solvents, comonomers, or as molecular weight control agents. In equation (8), a hydrogen atom is transferred from the telogen to a growing free radical, thereby terminating its growth and producing a dead polymer molecule. This reaction also results in a free radical on the telogen molecule which can add ethylene (equation 9) and undergo further growth by the polymerization reaction (equation 5).

$$I \rightarrow 2R_0^{\bullet} \tag{3}$$

$$R_0^{\bullet} + CH_2=CH_2 \rightarrow R_0CH_2CH_2^{\bullet}(=R_1^{\bullet}) \tag{4}$$

$$R_N^{\bullet} + CH_2=CH_2 \rightarrow (R_{N+1}^{\bullet}) \tag{5}$$

$$R_N^{\bullet} + R_M^{\bullet} \rightarrow P_{N+M} \tag{6}$$

$$R_N^{\bullet} + R_M^{\bullet} \rightarrow P_{N-1}CH_2CH_3 + P_{M-1}CH=CH_2 \tag{7}$$

$$R_NCH_2CH_2^{\bullet} + S-H \rightarrow P_NCH_2CH_3 + S^{\bullet} \tag{8}$$

$$S^{\bullet} + CH_2=CH_2 \rightarrow S-CH_2CH_2^{\bullet} \tag{9}$$

$$R_NCH_2CH_2^{\bullet} + CH_2=CH_2 \rightarrow R_NCH=CH_2 + CH_2CH_3^{\bullet} \tag{10}$$

In much of the early work, the initiator was oxygen, present as an impurity in the ethylene or intentionally added. Oxygen combines with hydrocarbons to form hydroperoxides which decompose at high temperatures to produce free radicals.[61] Later, ethylene was better purified to effectively remove the oxygen. When peroxides and hydroperoxides became commercially available, they became widely used because they offer advantages in handling and process control. Because they decompose at lower temperatures and can initiate polymerization at lower temperatures, polymers having different properties can be made.[62-65]

All organic molecules which contain hydrogen can be thought of as telogens, since, in principal, any hydrogen atom can participate in equation (8). However, the rate of this hydrogen transfer can vary greatly with the type of bond and the molecular structure of the telogen molecule. Hydrogens attached to carbons in an aromatic ring react very slowly, if at all. Primary hydrogens in a methyl group react more slowly than secondary hydrogens in a methylene group which, in turn, are slower than tertiary hydrogens on a carbon atom which has three other carbon atoms attached. Other atoms can also enter into a telogenic chain transfer. Generally, polyethylene processes are operated with no added telogens or with hydrocarbons as telogens. Ethylene itself can enter into telogenic chain transfer (equations 8 and 9). When ethylene enters into a telogenic chain transfer reaction, the end group on one polymer molecule is unsaturated as shown in equation (10).

21.5.1 Short Chain Branching (Intramolecular Chain Transfer) (see also Volume 3, Chapter 13)

The growing polymer molecule can itself act as a telogen. In equation (11), the growing free radical can assume conformations in which the free radical on the carbon at the end of the chain comes close to a hydrogen on a carbon that is part of the same chain. When that hydrogen transfers, a different free radical forms (equation 12). When this free radical adds ethylene through the propagation reaction (equation 5), the net result is a side chain on the PE molecule (equation 13). Steric factors favor six-membered rings and four-carbon side chains, though other structures can also be formed.[66-68] Higher polymerization temperatures result in a greater frequency of short chain branches. Since these branches interfere with the crystallization of PE, the density of the PE decreases as short chain branching increases. Conversely, higher pressures increase the rate of the propagation reaction and the ratio of short chain branching to polymer decreases as pressure increases, resulting in PE with higher densities. Both the temperature and pressure of polymerization must be controlled to obtain the desired degree of short chain branching, crystallinity, density, and those product properties which are affected by crystallinity.

$$R_NCH_2CH_2CH_2CH_2CH_2CH_2^{\cdot} \longrightarrow \quad (11)$$

$$\rightarrow R_NCH_2\overset{\cdot}{C}HCH_2CH_2CH_2CH_3 \qquad (12)$$

$$\xrightarrow{CH_2=CH_2} R_NCH_2CHCH_2CH_2^{\cdot} \qquad (13)$$
$$\underset{C_4H_9}{|}$$

21.5.2 Long Chain Branching (Intermolecular Chain Transfer) (see also Volume 3, Chapter 13)

Dead polymer molecules can also act as telogens. Previously formed polymer molecules contain many carbon–hydrogen bonds which can undergo telogenic chain transfer. In equations (14) and (15), a polymer chain is represented by a line and a free radical by a dot.

After the hydrogen transfers from a dead polymer molecule to a growing free radical (equation 14), the dead polymer molecule becomes a free radical which, by the addition of ethylene, can then grow a new PE chain. The length of this chain depends on how many ethylenes are added before growth is terminated, either by equation (6) or equation (7), or by a telogenic chain transfer (equation 8). Thus, side chains are formed having lengths comparable to the lengths of polymer molecules. This phenomenon is called long chain branching. Each of these long chains also has short chain branches, and long chain branches can form on polymer molecules which already have one or more long chain branches. The amount of long chain branching relative to the total amount of polymer increases with polymerization temperature, decreases with polymerization pressure, and increases with the dead polymer to ethylene ratio. This ratio is essentially constant in a stirred autoclave, but continually increases along a tubular reactor as ethylene is converted to polymer.

$$(14)$$

$$\xrightarrow{CH_2=CH_2} \qquad (15)$$

21.5.3 Molecular Weight and Molecular Weight Distribution

Detailed mathematical analyses[69-71] have been made of the effects of reaction kinetics and long chain branching on molecular weight and molecular weight distribution (see Volume, 3, Chapter 18). Only a few qualitative conclusions are presented here.

The number average molecular weight of the polymer is determined by the ratio of the rate of the propagation reaction (equation 5) to that of all reactions that result in new molecules (initiation and chain transfer to telogen). Chain transfer to polymer (both short and long chain branching) which does not result in additional polymer molecules will not affect number average molecular weight. However, the probability that a polymer molecule will grow a long chain branch increases as the size of the molecule increases (more carbon–hydrogen bonds become available for transfer). This has a large effect on the weight average molecular weight (and higher moment averages). Molecular weight distributions are broadened by long chain branching. Long chain branching also has important effects on melt rheology, or the melt flow characteristics of the polymer. In general, increasing long chain branching will make the polymer melt in a more non-Newtonian way; the apparent viscosity will change more rapidly with shear rate.

In a well-stirred autoclave reactor, there is a very broad distribution of residence times. Those polymer molecules which remain in the reactor for long times have the opportunity to participate in intermolecular chain transfer and to grow long chain branches many times before being removed from the reactor. A very broad distribution of molecular sizes and number of long chain branches per molecule results from these reactions.

In a tubular reactor approximating plug flow, the polymer molecules produced in the earlier portion of the reactor are exposed to more opportunities to grow long chain branches than the polymer molecules produced in the latter portions of the reactor. Ethylene conversions and polymer concentrations are higher in tubular reactors. Since long chain branching depends on polymer concentration, this higher concentration will result in more such branches being formed. Again, very broad distributions of molecular sizes and number of long chains per molecule can be formed. The changing reaction temperature in the tubular reactor also has an important effect on the molecular structures which are formed. The net result is that the molecular structures formed in each type of reactor are very complex, but are different in the two types of reactor. Two autoclaves in series, which can be operated at different temperatures and conversion levels (polymer concentrations), have been used to produce PE products having properties that match those of tubular reactor products.[72] Molecular structures of these products must differ in detail, but product properties, which depend on an averaging of molecular structures, could be matched.

21.6 DETERMINATION OF MOLECULAR STRUCTURE OF POLYETHYLENE

Because polyethylenes can be produced under a wide range of conditions, the molecular structures obtained also vary over a wide range. Published results of measurements of molecular structures do not show good agreement, probably because the samples measured were made under different conditions.

Short chain branches have been measured by IR and ^{13}C NMR.[73] Structures so identified are ethyl, *n*-butyl (which predominates), *n*-pentyl, 2-ethylhexyl and 1,3-diethyl. In 13 commercial samples, 7–17 short chain branches were found per 1000 carbon atoms.[74] The number of long chain branches (more than six carbon atoms) varied between 0.5 and 2.2 per 1000 carbon atoms.

Free radical polymerized PE made at low temperature (50–80 °C) and very high pressures (505–707 MPa) (5000–7000 atm) has a density of about 0.955 cm^{-3} and less than 0.8 branches per 1000 C atoms based on IR determination of methyl groups.[75] Commercial PEs (made at 200–300 °C and 200–300 MPa) have densities of 0.912–0.935 with 10–35 methyl groups per 1000 carbon atoms. Short chain branches interfere with crystallinity; therefore, density varies inversely with short chain branching. Since density is so much easier to measure, it is usually used as an indication of the amount of short chain branching. The number of long chain branches is small compared to the number of short chain branches; the effect of long chain branching on crystallinity and density is too small to measure in the presence of the much larger effect of the greater number of short chains.

21.6.1 High Pressure Polyethylene *vs.* Linear Polyethylene

Ethylene can be polymerized at low to moderate pressures with transition metal catalysts which operate by an entirely different mechanism (see also Volume 4, Chapters 1 and 2). These catalysts have been used at high pressures[76, 77] but normally, for economic reasons, only sufficient pressure is used to maintain the desired concentration of ethylene monomer in the gas phase or in solution in an inert solvent. Such catalysts produce very few short chain or long chain branches (densities are about 0.95 to 0.965). PE made with these catalysts is called linear polyethylene or high density polyethylene (HDPE). PE made with free radicals at high pressure is often called low density polyethylene (LDPE), but, as cited above, by operating at low temperatures and at very high pressures, high pressure polyethylene can be made which is essentially linear and which has a density comparable to HDPE.

When these transition metal catalysts are used to copolymerize ethylene with α-alkenes, side chains are introduced into the structure and crystallinity and density are reduced. Such copolymers have been made with densities down to 0.90 and even lower. Polymers in the lower part of the density range are often referred to as linear low density polyethylene (LLDPE). Because of other differences in molecular structure and because of important property differences, density and molecular weight (or melt viscosity) cannot adequately characterize a PE or fully describe its end use properties. No simple set of measurements can fully characterize all important end use properties. Commercial products are defined by product names and codes and quality control is assured by a combination of process conditions, manufacturing practices and standard quality control tests on the finished products. Different kinds of PE are preferred for different end uses and the several different processes continue to be used commercially.

21.7 COPOLYMERS (see Volume 3, Chapter 15)

Ethylene has been copolymerized with many vinyl type monomers at high pressure. In general, transition metal catalysts can be used with hydrocarbon monomers but cannot be used with polar type comonomers because they poison or kill the catalyst. In contrast, free radical initiators work well with polar type comonomers but with α-alkene comonomers, only very small amounts of comonomers can be incorporated into the polymer. Comonomers used in the greatest quantities commercially are vinyl acetate, methyl acrylate, ethyl acrylate, acrylic acid, and methacrylic acid. Another commercial copolymer, ethylene/vinyl alcohol copolymer, cannot be made directly because vinyl alcohol (CH_2=CHOH) does not exist. This copolymer is obtained by hydrolyzing ethylene/vinyl acetate copolymers so that the acetate groups are converted to hydroxyl groups. The ethylene/acrylic acid and ethylene/methacrylic acid copolymers can be partially or fully neutralized to make salts with metal ions, the so-called ionomers.

Some monomers which cannot be homopolymerized can be copolymerized with ethylene at high pressures. Examples are maleic anhydride, carbon monoxide[78] and sulfur dioxide.

In general, the incorporation of a comonomer into polyethylene leads to reduced crystallinity and stiffness. The chemical nature of the comonomer can also have a large effect on polymer properties. By choosing the type and amount of comonomer, copolymers can be optimized for a wide variety of end uses.

21.7.1 Comonomer Reactivity Ratios (see also Volume 3, Chapter 2)

When two monomers, M_1 and M_2, are polymerized together, they may both enter the chain to form a copolymer. The number of each and their arrangement are controlled by four reaction rate constants. The two types of polymer radicals with different monomer units at the end of the growing chains are designated as $M_1^\bullet$ and $M_2^\bullet$; each kind of monomer can add to each kind of radical, though the reaction rates can be different. [This simplified reaction scheme (equations 16–20) neglects the effects of the polymer chain composition beyond the terminal radical-containing monomer unit.]

$$M_1^\bullet + M_1 \xrightarrow{k_{11}} M_1\text{–}M_1^\bullet \tag{16}$$

$$M_1^\bullet + M_2 \xrightarrow{k_{12}} M_1\text{–}M_2^\bullet$$

$$M_2^\bullet + M_1 \xrightarrow{k_{21}} M_2\text{–}M_1^\bullet \tag{17}$$

$$M_2^\bullet + M_2 \xrightarrow{k_{22}} M_2\text{–}M_2^\bullet \tag{18}$$

$$r_1 = k_{11}/k_{12} \tag{19}$$

$$r_2 = k_{22}/k_{21} \tag{20}$$

If ethylene is designated as M_1, then the reactivity of ethylene relative to that of a comonomer toward the ethylene type radical is r_1. This relative reactivity toward the comonomer type radical is r_2^{-1}. The values of the reactivity ratios indicate the distribution of monomers within the chains.

If the product $r_1 r_2$ is close to unity, then the two comonomers are arranged randomly along the chain. If the product $r_1 r_2$ is less than unity, the monomers tend to alternate along the chains. In this case, if r_1 is greater than unity, the amount of M_1 in the copolymer is greater than its relative amount in the feed; if r_2 is greater than unity, the amount of M_2 in the copolymer is greater than its relative amount in the feed. If both r_1 and r_2 are zero, then the alternation is perfect. Finally, if the product $(r_1 r_2)$ is greater than one, block polymerization is favored. Such block copolymerization has not been clearly established in high pressure free radical copolymerization.

High pressures can have different effects on the reactivity of different monomers.[79-82] The reactivity ratios can change considerably with pressure, especially when ethylene is one of the monomers. Table 2 shows some reactivity ratios for ethylene and other monomers.

Table 2   Reactivity Ratios for Ethylene (M_1) and Comonomers (M_2)

Comonomer (M_2)	r_1	r_2	Pressure (MPa)	Temperature (°C)	Refs.
Propylene	3.1 ± 0.2	0.77 ± 0.05	103–172	130–220	a, b
1-Butene	3.4 ± 0.3	0.86 ± 0.02	103–172	130–220	a, b
Vinyl acetate	1.07 ± 0.06	1.09 ± 0.2	101	90	c, d
Acrylic acid	0.09 ± 0.02	6			d
	0.02	4	118–207	140–226	c, d, e
Methacrylic acid	0.1	6			d
	0.008	4	122–127	202–231	c, d, e
Methyl acrylate	0.042 ± 0.004	5.5 ± 1.5			d
	0.05	8	138	130–152	c, d, e
Ethyl acrylate	0.19 ± 0.04	2.2 ± 0.7			d
	0.04	15	207	180	c, d, e
N-Butyl acrylate	0.03 ± 0.01	11.9 ± 2.5	101	70	c, d
Vinyl chloride	0.24 ± 0.07	3.6 ± 0.3	101	90	c, d
Carbon monoxide	0.15	0.0	92	130	f

[a] L. Bogetich, G. A. Mortimer and G. W. Daues, *J. Polym. Sci.*, 1962, **61**, 3.
[b] G. A. Mortimer, *J. Polym. Sci., Part B*, 1965, **3**, 343.
[c] R. D. Burkhart and N. L. Zutty, *J. Polym. Sci., Part A*, 1963, **1**, 1137.
[d] P. Ehrlick and G. A. Mortimer, *Adv. Polym. Sci.*, 1970, **7**, 386.
[e] L. J. Young, *J. Polym. Sci.*, 1961, **54**, 411.
[f] D. D. Coffman, P. S. Pinkney, F. T. Wall and H. S. Young, *J. Am. Chem. Soc.*, 1952, **74**, 3391.

Ethylene and vinyl acetate have nearly equal reactivities at pressures within the commercial range. Copolymers of ethylene and vinyl acetate have nearly the same relative composition as feed streams. This means that such copolymers can be produced in either tubular or autoclave reactors.

However, most comonomers have different reactivity ratios. This difference in reactivity presents a special problem in tubular reactors. In commercial products, a comonomer will usually be present in a minor amount (less than about 20%). If it reacts faster than ethylene, it will be present in the feed stream in an even smaller ratio. As copolymer is produced, the comonomer will be depleted more rapidly than the ethylene. Consequently, the composition of the copolymer produced as the reaction mixture advances in a tubular reactor keeps changing toward lower comonomer content. This change in composition is undesirable for many uses.

In a well-stirred autoclave reactor, the feed stream is rapidly mixed with the contents of the reactor and steady state is rapidly reached. The comonomer ratio is nearly constant throughout the reactor and the polymer has a uniform comonomer composition when a uniform feed is maintained.[83] The autoclave process is generally preferred for making copolymers from monomers which do not have similar reactivity ratios.

21.7.2 Chain Transfer Activity of Comonomers (see also Volume 3, Chapter 17)

Most comonomers contain hydrogen atoms which can enter into telogenic, or chain transfer reactions (see above). At the same polymerization temperature and pressure, the molecular weight of the polymer would be reduced by this chain transfer activity.[84] To maintain the desired molecular weight, the polymerization temperature must be reduced as more comonomer is used; this results in a reduction of short chain branching.

When α-alkenes are used as comonomers, their telogenic activity, combined with a polymerization rate slower than that of ethylene, severely restricts the amount that can be incorporated into the copolymer in high pressure free radical polymerizations. With the more reactive polar comonomers, sufficient amounts can be readily copolymerized so that the copolymer is less crystalline than the homopolymer. However, the density of a polar copolymer is no longer a reliable measure of crystallinity since it is directly affected by the density contributions of the polar comonomer itself.

This reduction in polymerization temperature for copolymers means that the exothermic conversion must be limited. This reduction in conversion will add to the cost of production of copolymers, in addition to the cost of the comonomer itself.

21.8 REFERENCES

1. R. Raff and K. W. Doak (eds.), 'Crystalline Olefin Polymers', Wiley, New York, 1965, chap. 7, p. 267.
2. K. W. Doak and J. I. Kroschwitz (eds.), 'Encyclopedia of Polymer Science and Engineering', Wiley, New York, 1986, vol. 6, p. 386.
3. Y. Ogo, *J. Macromol. Sci., Rev. Macromol. Chem.*, 1984, **24**, 1.
4. J. P. Marano, Jr. and J. M. Jenkins, III, in 'High Pressure Technology', ed. I. L. Spain, Dekker, New York, vol. 2, p. 61.
5. G. Luft, *Chem.-Ing.-Tech.*, 1979, **51**, 960 (*Chem. Abstr.*, 1979, **91**, 211 846x).
6. A. A. Zharov, *Itogi Nauki Tekh. Khim. Tekhnol. Vysokomol. Soedin.*, 1974, **5**, 89 (*Chem. Abstr.*, 1974, **81**, 121 054b).
7. S. Terasawa and H. Itsuki, *Yuki Gosei Kagaku Kyokai Shi*, 1971, **29**, 567 (*Chem. Abstr.*, 1971, **75**, 152 097y).
8. Y. Tanaka, A. Okada and S. Ueda, *Sen-i To Kogyo*, 1969, **2**, 633 (*Chem. Abstr.*, 1970, **72**, 44 163w).
9. H. Brackemann, M. Buback and H. P. Voegele, *Makromol. Chem.*, 1986, **187**, 1977.
10. M. Buback, H. Hippler, J. Schweer and H. Voegele, *Makromol. Chem., Rapid Commun.*, 1986, **7**, 261.
11. T. Sasuga and M. Takehisa, *J. Macromol. Sci. Chem.*, 1978, **12**, 1307, 1333, 1343.
12. N. S. Isaacs and A. V. George, *Polym. Commun.*, 1984, **25**, 268.
13. M. Okamoto, M. Sasaki and J. Osugi, *Rev. Phys. Chem. Jpn.*, 1977, **47**, 33 (*Chem. Abstr.*, 1977, **87**, 118 162d).
14. J. P. Machon, *MMI Press Symp. Ser.*, 1983, **4**, 639 (*Chem. Abstr.*, 1985, **102**, 149 840f).
15. T. Sasuga, S. Kawanishi and M. Takehisa, *J. Phys. Chem.*, 1979, **83**, 3290.
16. T. Sasuga, S. Kawanishi and M. Takehisa, *Macromolecules*, 1983, **16**, 545.
17. M. Yokawa, Y. Ogo and T. Imoto, *Makromol. Chem.*, 1974, **175**, 179.
18. I. Kaetsu, F. Yoshii and Y. Watanabe, *J. Polym. Sci., Polym. Phys. Ed.*, 1978, **16**, 2645.
19. L. A. Wall, D. W. Brown, *J. Fluorine Chem.*, 1972, **2**, 73 (*Chem. Abstr.*, 1972, **77**, 102 279e).
20. M. G. Bradbury, S. D. Hamann and M. Linton, *Aust. J. Chem.*, 1970, **23**, 511 (*Chem. Abstr.*, 1970, **72**, 121 950e).
21. Y. Ogo and T. Sano, *Colloid Polym. Sci.*, 1976, **254**, 470.
22. P. W. Moore, J. G. Clouston and R. P. Chaplin, *J. Polym. Sci., Polym. Chem. Ed.*, 1983, **21**, 2491 and 2503.
23. T. Sasuga and M. Takehisa, *J. Macromol. Sci., Chem.*, 1978, **12**, 1307.
24. Brochure, Pennwalt Lucidol Chemicals, Buffalo, NY, 1982.
25. 'Plastiscope: News and Interpretation', ed. R. D. Leaversuch, *Modern Plastics*, 1987, **64** (5), 63.
26. H. Sutter, K. U. Haas and W. P. Ledet (Du Pont), *US Pat.* 4 607 086 (1986) (*Chem. Abstr.*, 1986, **105**, 209 618p).
27. I. Tincul, A. Grigorescu and D. Vlaheli (CPP) *Br. Pat.* 1 558 833 (1980) (*Chem. Abstr.*, 1980, **93**, 27 087r).
28. J. L. Adriaans and P. G. Dees (Dow) *US Pat.* 4 496 698 (1985) (*Chem. Abstr.*, 1985, **102**, 115 295s).
29. T. J. Van der Molen, in 'High Pressure, Science and Technology, Proceedings of the 7th International AIRAPT Conference', ed. B. Vodar and P. Marteau, Pergamon, Oxford, 1980, vol. 2, p. 815 (*Chem. Abstr.*, 1981, **95**, 187 861h).
30. I. Moked, R. H. Handwerk and H. J. Goettler (Union Carbide), *US Pat.* 4 452 956 (1984) (*Chem. Abstr.*, 1984, **101**, 91 730f).
31. K. Boettcher, H. G. Hoerdt, W. Zacker and O. Buecher (BASF), *US Pat.* 4 153 774 (1979) (*Chem. Abstr.*, 1979, **91**, 40 118e).
32. C. C. Clemmer and W. C. Alcorn, III (Dart Industries) *Br. Pat.* 1 509 008 (1978) (*Chem. Abstr.*, 1979, **90**, 7050h).
33. J. P. Koerner, *High Temp-High Pressures*, 1977, **9**, 587 (*Chem. Abstr.*, 1978, **89**, 130 218a).
34. A. M. Gemassmer, *High Temp.-High Pressures*, 1977, **9**, 507 (*Chem. Abstr.*, 1978, **89**, 129 967n).
35. A. M. Gemassmer, *Erdoel Kohle, Erdgas, Petrochem.*, 1978, **31**, 221 (*Chem. Abstr.*, 1978, **89**, 44 267j).
36. E. C. Ballard and J. R. Priest (Du Pont), *US Pat.* 3 988 509 (1976) (*Chem. Abstr.*, 1977, **86**, 30 491d).
37. C. Mancini and R. Gaspari (Societa Italiana Resine SPA) *US Pat.* 3 714 213 (1973) (*Chem. Abstr.*, 1973, **78**, 137 117y).
38. M. Hess and L. Saroff (Sinclair-Koppers Co.), *US Pat.* 3 577 224 (1971) (*Chem. Abstr.*, 1971, **75**, 37 085f).
39. Imhico A G, *Br. Pat.* 1 213 416 (1970) (*Chem. Abstr.*, 1971, **74**, 32 169f).
40. H. Eilbracht, H. Friedenreich, S. Goebel, M. Haeberle and B. Lehmann (BASF AG), *Fr. Pat.* 1 326 984 (1962) (*Chem. Abstr.*, 1963, **59**, 4062a).
41. BASF AG, *Br. Pat.* 1 311 984 (1973) (*Chem. Abstr.*, 1971, **74**, 126 443t).
42. J. P. Marano, Jr. (Union Carbide), *US Pat.* 4 000 357 (*Chem. Abstr.*, 1977, **86**, 90 882).
43. M. Erchak, Jr., K. W. Doak and R. M. Douglas (Rexall) *US Pat.* 3 293 233 (1966) (*Chem. Abstr.*, 1965, **63**, 3073g).
44. G. M. McDonald, M. M. Douglas and I. M. Liebson (Dart) *US Pat.* 3 842 060 (1974) (*Chem. Abstr.*, 1975, **82**, 440 541).
45. H. Mavridis and C. Kiparissides, *Polym. Process Eng.*, 1985, **3**, 263.

46. B. J. Yoon and H. K. Rhee, *Chem. Eng. Commun.*, 1985, **34**, 253.
47. S. K. Gupta, A. Kumar and M. V. G. Krishnamurthy, *Polym. Eng. Sci.*, 1985, **25**, 37.
48. W. Hollar and P. Ehrlich, *Chem. Eng. Commun.*, 1983, **24**, 57.
49. D. Stoiljkovich and S. Jovoanovich, *J. Polym. Sci., Polym. Chem. Ed.*, 1981, **19**, 741.
50. Th. J. Van der Molen and C. Van Heerden, *Adv. Chem. Ser.*, 1972, **109**, 92 (*Chem. Abstr.*, 1972, **77**, 127 119r).
51. K. W. Doak and A. Schrage, in ref. 1, chap. 8, p. 301.
52. C. D. Han and Ta-Jo Liu, *Hwahak Konghak*, 1977, **15**, 249 (*Chem. Abstr.*, 1978, **88**, 236 69f).
53. M. Buback, *Makromol. Chem.*, 1980, **181**, 373.
54. H. Van den Bossche (Exxon) *Br. Pat.* 2 093 045 (1983) (*Chem. Abstr.*, 1983, **98**, 17 245e).
55. M. Buback, H. Hippler, J. Schweer and H. P. Voegele, *Makromol. Chem., Rapid Commun.*, 1986, **7**, 261.
56. D. Stoiljkovic and S. Jovanovic, *Makromol Chem.*, 1985, **186**, 671.
57. D. Stoiljkovic and S. Jovanovic, *Br. Polym. J.*, 1984, **16**, 291.
58. P. C. Lim and G. Luft, *Inst. Chem. Eng. Symp. Ser.*, 1984, **87**, 643.
59. M. Bubach and H. Lendle, *Makromol. Chem.*, 1983, **184**, 193.
60. D. Constantin and J. P. Mackon, *Eur. Polym. J.*, 1978, **14**, 703.
61. Y. Tatsukami, T. Takahashi and H. Yoshioka, *Makromol. Chem.*, 1980, **181**, 1107.
62. G. Luft and H. Seidl, *Angew. Makromol. Chem.*, 1985, **129**, 61.
63. H. Seidl and G. Luft, *J. Macromol. Sci., Chem.*, 1980, **15**, 1.
64. G. Luft, H. Bitsch and H. Seidl, *J. Macromol. Sci., Chem.*, 1977, **11**, 1089.
65. F. J. Mercx, Th. J. Van der Molen and M. DeSteenwinkel, in 'Proceedings of the 5th European Symposium on Chemical Reaction Engineering', Elsevier, Amsterdam, 1972, Vol. B7, p. 41 (*Chem. Abstr.*, 1975, **83**, 147 814r).
66. M. J. Roedel, *J. Am. Chem. Soc.*, 1953, **75**, 6110.
67. J. Vile, P. J. Hendra, H. A. Willis, M. E. A. Cudby and A. Bunn, *Polymer*, 1984, **25**, 1173.
68. W. L. Mattice, *Macromolecules*, 1983, **16**, 487.
69. J. K. Beasley, *J. Am. Chem. Soc.*, 1953, **75**, 6123.
70. R. V. Mullikin and G. A. Mortimer, *J. Macromol. Sci., Chem.*, 1972, **6**, 1301.
71. R. V. Mullikin and G. A. Mortimer, *J. Macromol. Sci., Chem.*, 1970,´**4**, 1495.
72. H. Sutter, K. U. Haas and W. P. Ledet, (Du Pont) *US Pat.* 4 607 086 (1986) (*Chem. Abstr.*, 1986, **105**, 209 618p).
73. D. J. Cutler, P. J. Hendra, M. E. A. Cudby and H. A. Willis, *Polymer*, 1977, **18**, 1005.
74. D. E. Axelson, G. C. Levy and L. Mandelkern, *Macromolecules*, 1979, **12**, 41.
75. R. A. Hines, W. M. D. Bryant, A. W. Larcher and D. C. Pease, *Ind. Eng. Chem.*, 1957, **49**, 1071.
76. J. P. Machon, *MMI Press Symp. Ser.*, 1983, **4**, 639 (*Chem. Abstr.*, 1985, **102**, 149 840f).
77. J. C. Greaves and W. G. Oakes (ICI), *Br. Pat.* 1 205 635 (1970) (*Chem. Abstr.*, 1970, **73**, 110 322w).
78. M. Buback and H. Tups, *Physica B + C (Amsterdam)*, 1986, **139** and 626.
79. R. L. Hemmings and K. E. Weale, *Polymer*, 1986, **27**, 1819.
80. D. W. Brown and R. E. Lowry, *J. Polym. Sci., Polym. Chem. Ed.*, 1975, **13**, 1677.
81. G. Enomoto, Y. Ogo and T. Imoto, *Makromol. Chem.*, 1970, **138**, 19.
83. H. D. Anspon, W. H. Byler and G. E. Ham (Gulf Oil Corp.), *US Pat.* 3 350 372 (1964) (*Chem. Abstr.*, 1967, **67**, 117 541b).
84. G. Luft, *High Temp.-High Pressures*, 1977, **9**, 501 (*Chem. Abstr.*, 1979, **90**, 6717u).

22

Ring-opening Polymerization

WILLIAM J. BAILEY

University of Maryland, College Park, MD, USA

22.1 INTRODUCTION

Although the ionic ring-opening polymerization of heterocyclic compounds, such as ethylene oxide, tetrahydrofuran, ethyleneimine, β-propiolactone and caprolactam, as well as the Ziegler–Natta ring opening of cyclic alkenes, such as cyclopentene and norbornene, are well known, free radical ring-opening polymerizations are rather rare. In fact, unstrained five- and six-membered carbocyclic rings are usually involved in ring-closing reactions rather than ring-opening ones. For example, Butler and Angelo[1] reported in 1957 that when diallyldimethylammonium bromide was polymerized by a free radical mechanism, a soluble polymer containing five-membered rings was obtained by an inter–intramolecular polymerization as indicated in Scheme 1. Apparently the reaction is kinetically controlled to form the five-membered ring rather than the thermodynamically favored six-membered ring. The few examples of free radical ring-opening polymerization that are reported in the literature involve cyclopropane derivatives or highly strained bicyclic alkenes. For example, Takahashi[2] studied the free radical polymerization of vinylcyclopropane and reported that the cyclopropane ring opened to give a polymer containing about 80% 1,5-units and 20% of undetermined structural units. Apparently the radical adds to the vinyl group to give the intermediate cyclopropylmethyl radical which opens at a rate faster than the addition to the double bond of another monomer as in Scheme 2. Somewhat similar results were obtained with the chloro derivatives. More recently, Cho and Ahn[3] studied the related malonic ester derivative with similar results. Errede[4] found the dimer of *o*-xylylene would undergo free radical ring-opening polymerization to give the corresponding poly(*o*-xylylene). In this case the driving force for the ring-opening step is the formation of the aromatic ring (Scheme 3). Hall and co-workers[5] showed, as in equation (1), that derivatives of bicyclo[1.1.0]butane would polymerize *via* free radicals by cleavage of the highly strained central bond. Of course, the ring-opening polymerization of S_8 has been postulated to involve free radicals.[6]

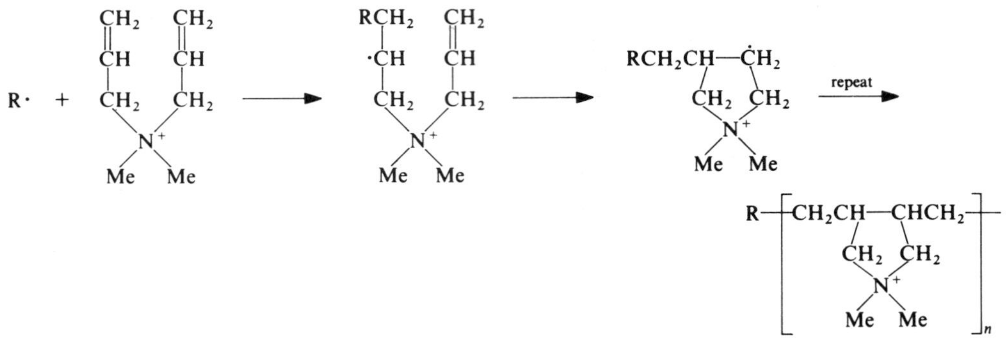

Scheme 1

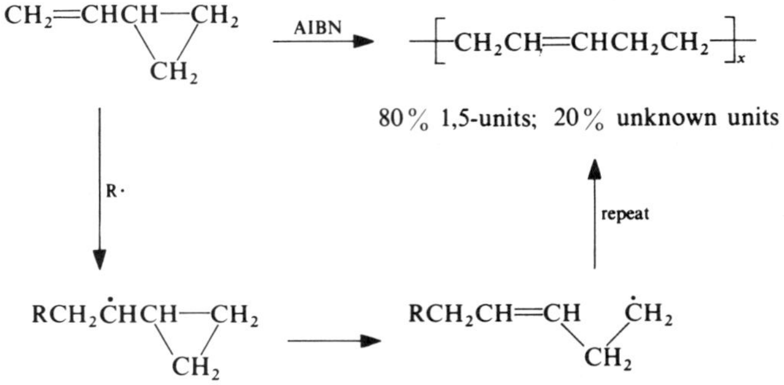

80% 1,5-units; 20% unknown units

Scheme 2

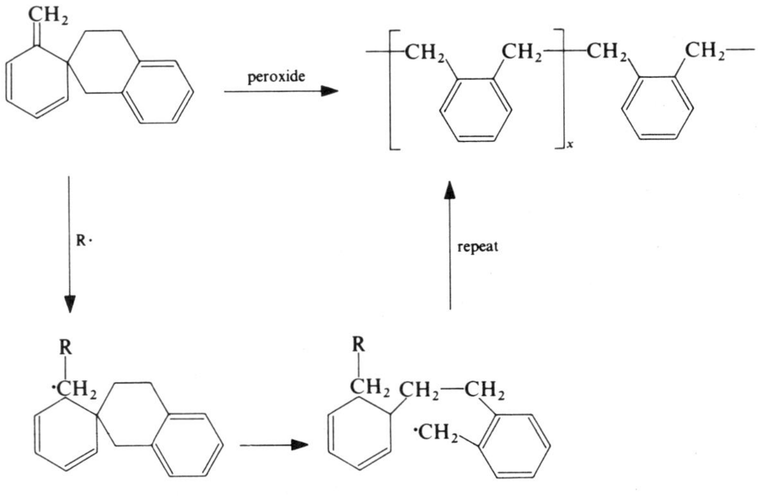

Scheme 3

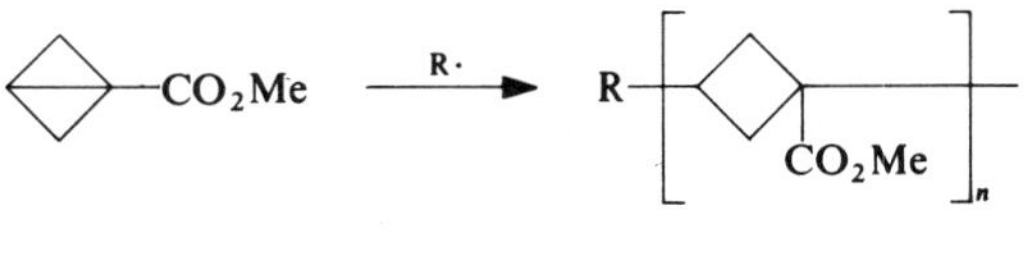

(1)

The course of some of these ring-opening and ring-closing polymerizations can be explained by the recent data of Maillard, Forrest and Ingold,[7] who studied the transformation in the cyclopropylmethyl and the cyclopentylmethyl series by electron spin resonance with the results listed in Table 1. In the case of the three-membered cyclic radical the reaction involves ring opening since the energy is favorable and the rate of the reaction is very high. In the case of the five-membered ring system the reaction proceeds in the direction of ring closure since the energetics of that reaction are favorable and the rate of the ring closure is also moderately high. With the assumption that the carbon–oxygen double bond is at least 50–60 kcal mol^{-1} (200–240 kJ mol^{-1}) more stable than the carbon–carbon double bond,[8] one can calculate that the introduction of an oxygen atom in place of a carbon atom in the cyclopentylmethyl radical would favor the reverse reaction or the ring opening. In other words the ring-opening reaction would be favored by at least 40 kcal mol^{-1} by producing the more stable carbonyl double bond.

Table 1 ESR Measurement of Radical Ring Opening and Closing

Reaction	$K_{250}(s^{-1})$	Enthalpy E (kcal mol^{-1})	log (As)
	1.3×10^8	5.94	12.48
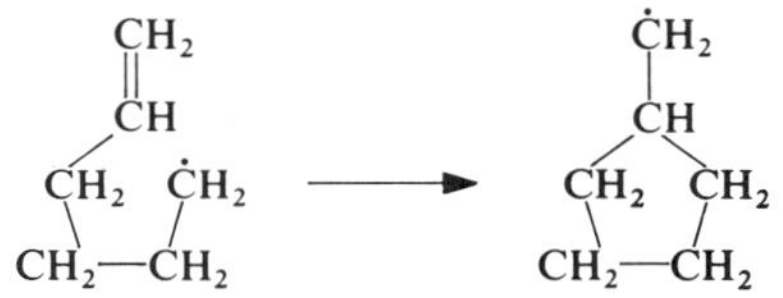	1.0×10^5	7.8	10.7

A search of the literature indeed revealed many ring systems containing an oxygen atom that would undergo a ring-opening reaction but not polymerization in the presence of free radical catalysts. One such case was the cyclic formal, ethylene formal. Maillard, Cazaux and Lalande[9] found that when ethylene formal was heated at 160 °C it rearranged to ethyl formate. The reaction could be rationalized as indicated in Scheme 4 where the driving force for the ring-opening reaction in the chain reaction was the formation of the stable carbon–oxygen double bond in the final ester. With the knowledge that such a ring system would undergo cleavage, it seemed to be a fairly straightforward process to synthesize a monomer that would undergo ring-opening polymerization by introducing a double bond at the carbon atom flanked by the two oxygens.

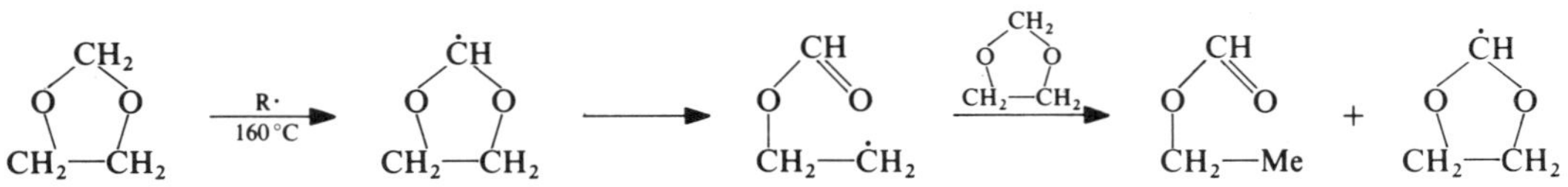

Scheme 4

22.2 POLYMERIZATION OF CYCLIC KETENE ACETALS

The monomer desired for this ring-opening polymerization had been prepared by McElvain and Curry[10] in 1948 as depicted in Scheme 5. Although Johnson, Barnes and McElvain[11] had treated diethylketene acetal with peroxide and had reported that there was no reaction, no such study was reported for the 2-methylene-1,3-dioxolane (1). A reinvestigation of the cyclic ketene acetal (1) was therefore undertaken. The cyclic ketene acetal (1) was found to be extremely sensitive to acid.[12] Exposure of (1) to the atmosphere would cause immediate polymerization to give a polymer with little or no ring opening. In order to handle this monomer conveniently it was necessary to have some base, such as potassium *t*-butoxide or a tertiary amine, always present; under these conditions no spontaneous polymerization would occur.

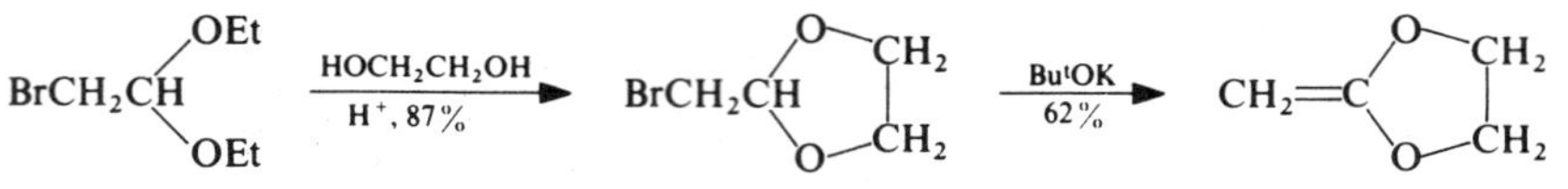

Scheme 5

Treatment of this monomer (**1**) with benzoyl peroxide at high temperatures gave a high molecular weight polyester by a free radical ring-opening polymerization which can be rationalized by the accompanying Scheme 6. This polyester is difficult to synthesize with high molecular weight from the γ-hydroxybutyric acid or the very stable γ-butyrolactone, which is isomeric with monomer (**1**). The structure of the polyester was established by analysis as well as IR and NMR spectroscopy. An alternative method of analysis of the extent of ring opening was the basic hydrolysis of the copolymer (**2**), which cleaved the ester groups but left the cyclic ketals intact.

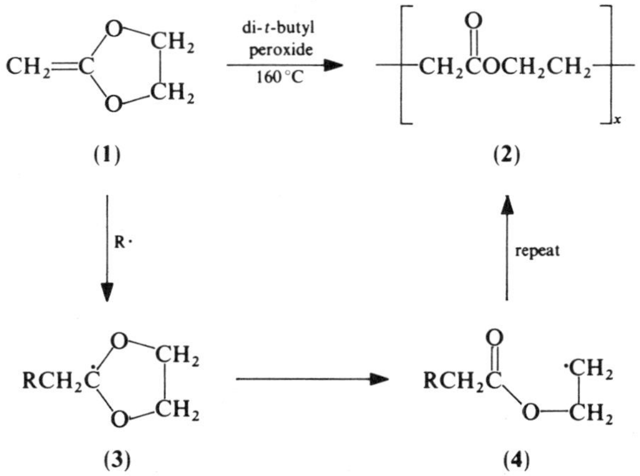

Scheme 6

At lower temperatures the ring opening was not complete. Thus at 60 °C only 50% of the rings were opened to give a random copolymer with the structure given in formula (**5**).[12] Even at 120 °C only 87% of the rings are opened. Apparently the unopened radical (**3**) can add directly to the monomer (**1**) in competition with the ring-opening process to form the open-chain polymer. High dilution was found to favor the ring-opening process since the addition of the radical (**3**) to the monomer (**1**) is a second order reaction while the conversion of the radical (**3**) to the open chain appears to be some other order. The extent of ring opening appears to be kinetically controlled with a direct competition between the rate of direct addition k_{11} and the rate of ring opening k_{iso} as indicated in Scheme 7. Any change in the system that would increase the steric hindrance of the cyclic radical would similarly increase the extent of ring opening by decreasing k_{11}. Similarly any change in the system that would produce a lower energy transition state upon ring opening would be expected to favor the ring-opening process by increasing k_{iso}.

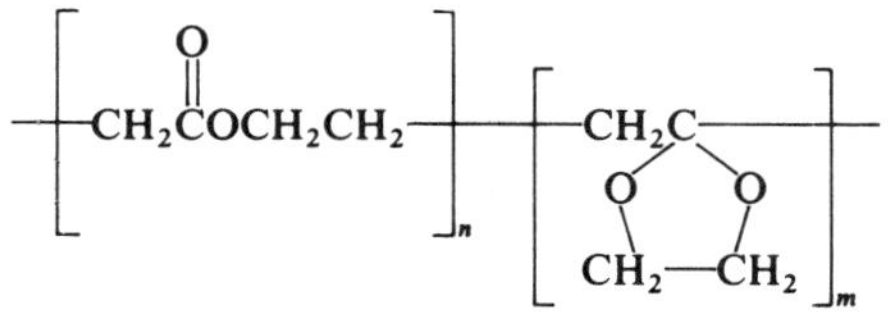

(**5**)

In a program to find other cyclic ketene acetals that would undergo quantitative ring opening even at room temperature we prepared the seven-membered cyclic ketene acetal, 2-methylene-1,3-dioxepane (**5**), which underwent essentially complete ring opening at room temperature as indicated in Scheme 8.[13] Photoinitiated polymerization of monomer (**5**) also gives complete ring opening.[14] This process makes possible the quantitative introduction of an ester group in the backbone of an addition polymer. Apparently the seven-membered ring increases the steric hindrance in the intermediate free radical (**9**) to eliminate practically all of the direct addition and also introduces a small amount of strain so that the ring opening to the radical (**9**) is accelerated.

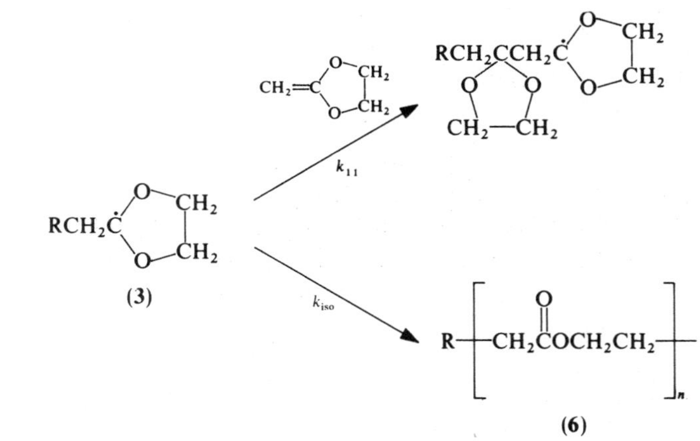

Scheme 7

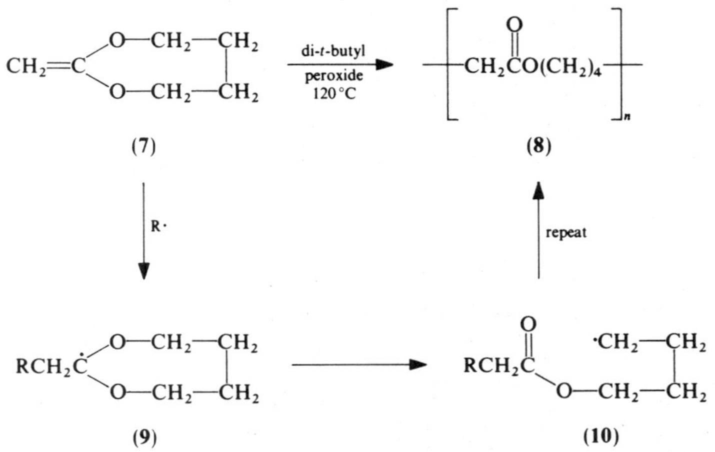

Scheme 8

Additional cyclic ketene acetals[15-18] that have been studied include the 4-phenyl-2-methylene-1,3-dioxepane (**11**) which undergoes quantitative ring opening to give the polyester (**12**). Apparently, as indicated in Scheme 9, the ring-opening step from radical (**13**) to the open-chain radical (**14**) is greatly enhanced by the formation of the relatively stable benzyl radical (**14**) even though the cyclic radical (**13**) is a five-membered ring analogous to the radical (**3**).

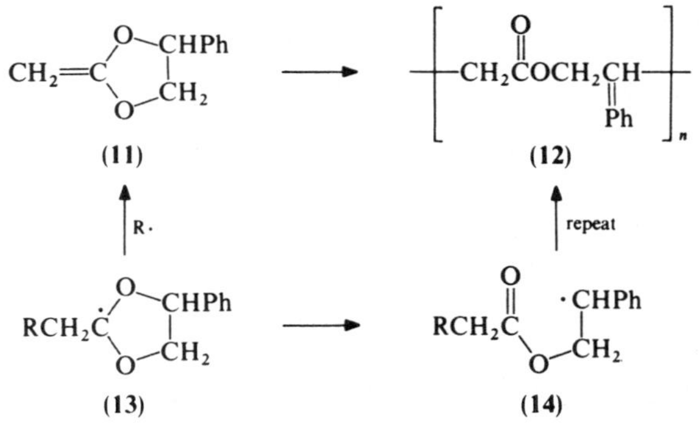

Scheme 9

A variety of other ketene acetals have been prepared and polymerized as indicated in formulae (**15**)–(**23**). In general an alkyl group in the 4-position of a cyclic ketene acetal promotes the extent of ring opening but not as much as a phenyl group. Also, a six-membered ring promotes the extent of ring opening and in that sense is intermediate between the five-membered ketene acetals, such as monomer (**1**), and the seven-membered ketene acetals, such as monomer (**7**). The effects also tend to be additive; for example, the 4,6-dimethyl-2-methylene-1,3-dioxane (21), when treated in dilute benzene at 120 °C in the presence of *t*-butyl peroxide, gave an essentially quantitative ring-opened polymer. In the case of the *cis*-4,5-diphenyl-2-methylene-1,3-dioxalane (18) the presence of the *cis*-phenyl groups ensure quantitative ring opening in both homopolymerization and copolymerization; however, the presence of these big groups makes the dehydrohalogenation step in the synthesis very difficult and makes the reactivity of monomer (18) very sluggish in homopolymerization, resulting in low conversions and in copolymerization with styrene and methyl methacrylate giving a very low incorporation into the copolymer.

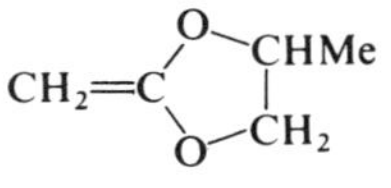

(15) 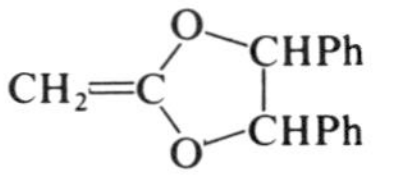 (18) 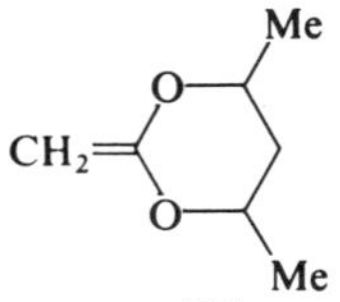 (21)

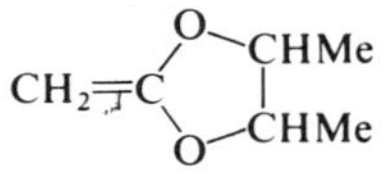

(16) 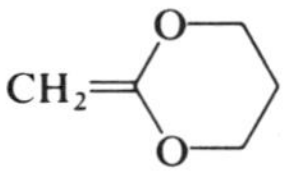 (19) 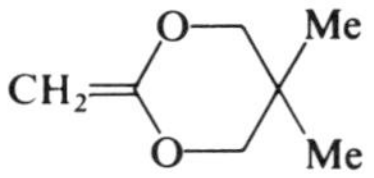 (22)

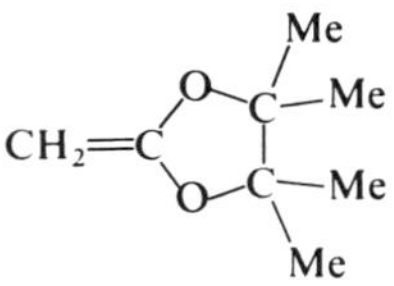

(17) 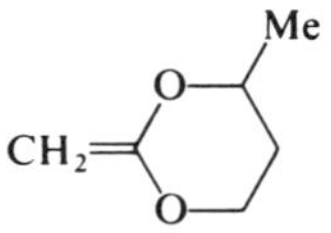 (20) 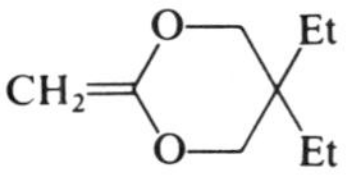 (23)

22.3 MECHANISM OF FREE RADICAL RING OPENING OF CYCLIC KETENE ACETALS

In Schemes 6, 8 and 9, the ring-opening process was written as proceeding from a cyclic radical to an open-chain radical. This process originally appeared to be consistent with the fact that substituents, such as a phenyl group in the 4-position of the cyclic ketene acetals that would stabilize the open-chain radical, tend to increase the extent of the ring opening. It also appeared to be consistent with the fact that the dilution tends to increase the extent of ring opening; that is, the direct addition of the cyclic radical (3) to the monomer (1) is bimolecular and would be expected to be reduced by dilution; but the ring-opening step going from the cyclic radical (3) to the open-chain radical (4) is unimolecular and therefore should not be strongly affected by dilution, with the result that ring-opening polymerization would be favored by dilution. However, this mechanism for the free radical ring-opening polymerization now appears to be greatly oversimplified. Evidence now seems to indicate that the ring-opened radical is never free and that the process of the cyclic radical undergoing ring opening and the addition to the next monomer to form a new cyclic radical is a concerted process.

It was found that when 2-methylene-1,3-dioxe-5-pene (24) was treated with di-*t*-butyl peroxide (DTBP) at 115 °C in benzene for 30 h, a regiospecific free radical ring-opening polymerization took place according to equation (2) to give 100% ring opening with only the vinyl-containing structure (100% 1,2-addition) being formed. The structure of the polymer (25) was clearly established by ^{1}H NMR, ^{13}C NMR and IR spectra.

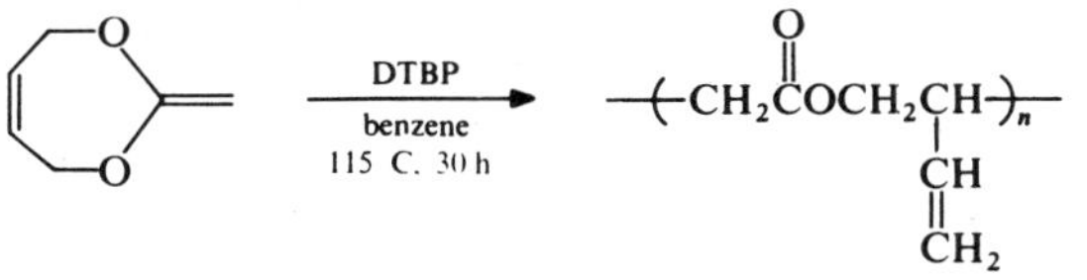

$$-(CH_2\overset{\overset{\textstyle O}{\|}}{C}OCH_2CH)_n- \quad (2)$$

(24) (25)

This result was surprising since according to the traditional mechanism of the free radical ring opening, the radical first adds to the ketene acetal double bond to form a stabilized tertiary radical, then the ring opens with the formation of ester group to generate a new allylic radical. As soon as the allylic radical forms, the end of the growing polymer chain would be expected to be similar to that present in the polymerization of butadiene as indicated in Scheme 10. At 50 °C in an emulsion polymerization of butadiene and styrene, a random copolymer consisting of about 20% 1,2-addition units is obtained.[19]

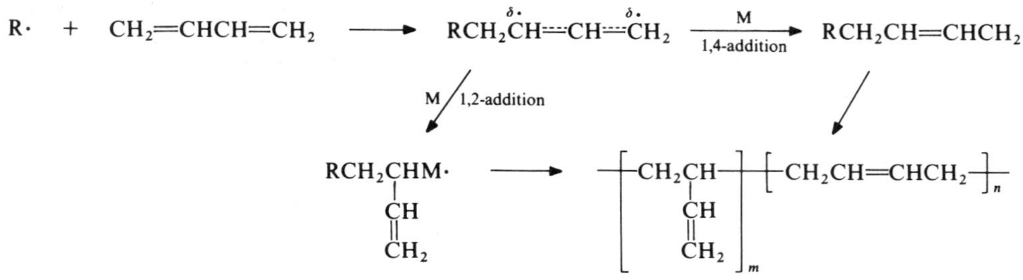

$$R\cdot \; + \; CH_2{=}CHCH{=}CH_2 \longrightarrow RCH_2\overset{\delta\cdot}{C}H{\cdots}CH{\cdots}\overset{\delta\cdot}{C}H_2 \xrightarrow[\text{1,4-addition}]{M} RCH_2CH{=}CHCH_2M\cdot$$

M / 1,2-addition

$$RCH_2CHM\cdot \longrightarrow \left[CH_2CH\right]\left[CH_2CH{=}CHCH_2\right]_n$$
with CH / CH_2 side groups

Scheme 10

In the case of the cyclic ketene acetal polymerization, one would also expect the allylic radical to undergo both 1,4-addition and 1,2-addition as depicted in Scheme 11. The result obtained is in obvious contradiction to the expected mechanism not only because of the regioselectivity but also because the thermodynamically least stable isomer is obtained. One must therefore conclude that the radical formed by ring opening is never really free and that the process is somewhat concerted. In other words, the bond-making step involving a second monomer is taking place at nearly the same time as the bond-breaking step involving the ring-opening process in some process similar to Scheme 12.

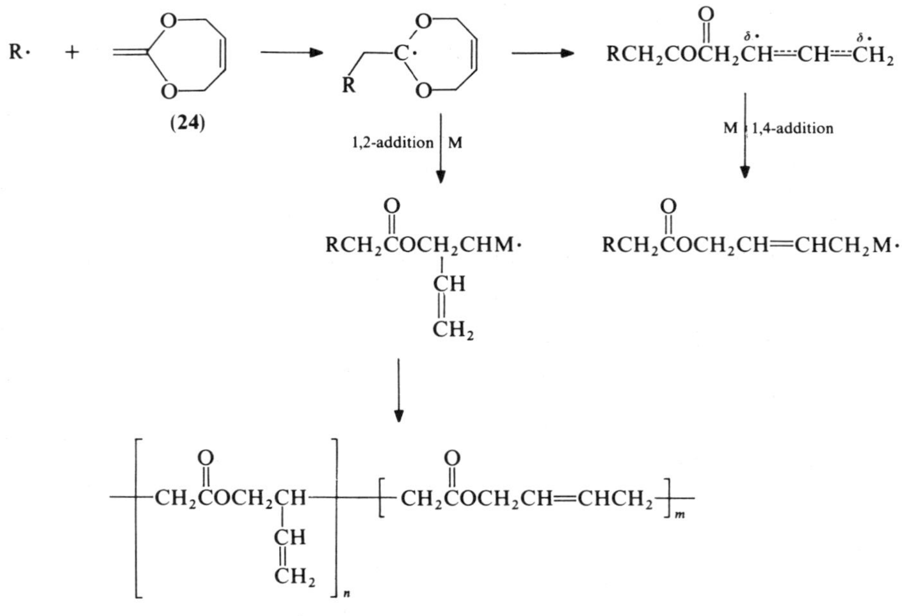

(24)

$$RCH_2\overset{O}{\overset{\|}{C}}OCH_2\overset{\delta\cdot}{C}H{\cdots}CH{\cdots}\overset{\delta\cdot}{C}H_2$$

1,2-addition | M

M ∤ 1,4-addition

$$RCH_2\overset{O}{\overset{\|}{C}}OCH_2CHM\cdot$$

$$RCH_2\overset{O}{\overset{\|}{C}}OCH_2CH{=}CHCH_2M\cdot$$

$$\left[CH_2\overset{O}{\overset{\|}{C}}OCH_2CH\right]\left[CH_2\overset{O}{\overset{\|}{C}}OCH_2CH{=}CHCH_2\right]_m$$

Scheme 11

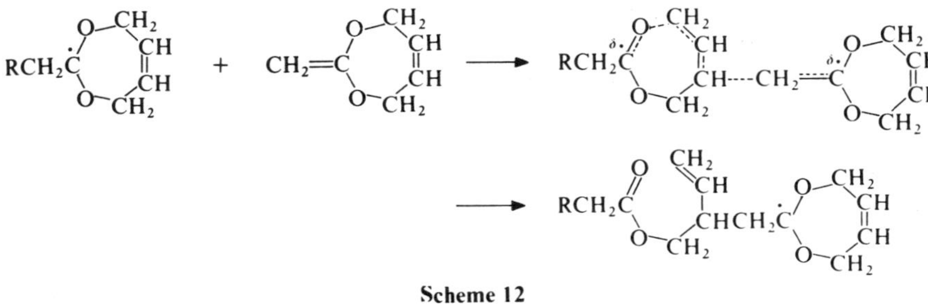

Scheme 12

In order to check this theory the isomeric 4-vinyl-2-methylene-1,3-dioxalane (26) was prepared and polymerized at 120 °C in benzene with *t*-butyl peroxide as the initiator. The spectral evidence indicated the polymerization was again regioselective with 100% ring opening to give the repeating unit with a 1,4-structure as indicated in equation (3). Thus it appears that the free radical ring-opening process for monomer (26) is again a concerted one in which the ring-opened radical is never free.

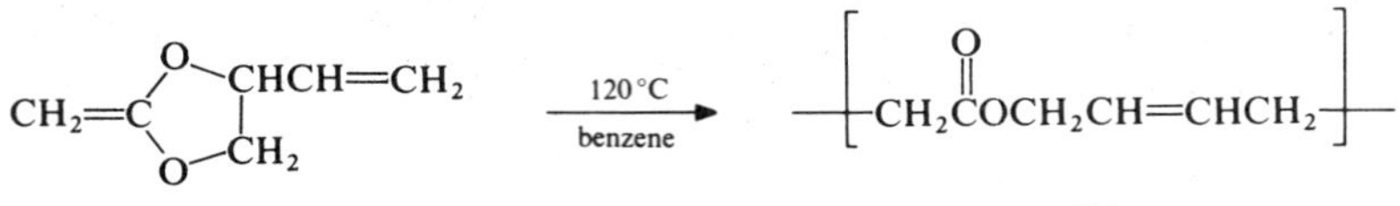

$$CH_2=C\underset{O-CH_2}{\overset{O-CHCH=CH_2}{}} \quad \xrightarrow[\text{benzene}]{120\,°C} \quad \left[CH_2\overset{O}{\overset{\|}{C}}OCH_2CH=CHCH_2 \right] \tag{3}$$

(26)   (27)

Finally an optically active cyclic ketene acetal *R*(−) 4-phenyl-2-methylene-1,3-dioxalane (28) was polymerized at 120 °C in benzene with *t*-butyl peroxide as the initiator to give an optically active polyester according to equation (4). If the ring-opened radical were really free, one would expect complete racemization of the resulting polymer (29). Work is in progress to determine if any racemization takes place in this ring-opening polymerization.

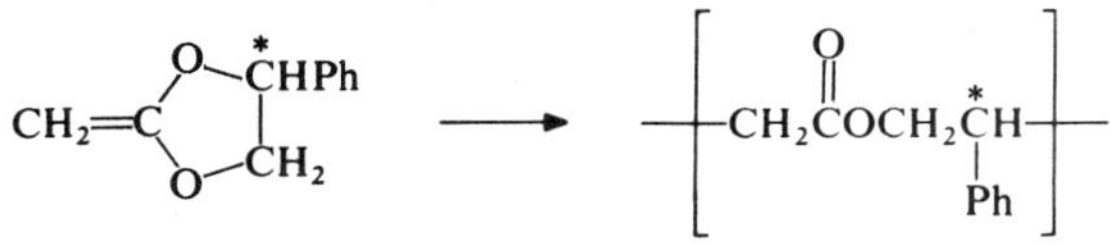

$$CH_2=C\underset{O-CH_2}{\overset{O-\overset{*}{C}HPh}{}} \quad \longrightarrow \quad \left[CH_2\overset{O}{\overset{\|}{C}}OCH_2\underset{Ph}{\overset{*}{C}H} \right] \tag{4}$$

optically active, *R* (−)   optically active (−)

(28)   (29)

A recent publication by Tsang, *et al.*[20] appears to be at odds with some of these results although the conditions of the two studies are somewhat different. They found that, in competition with a free radical ring-closing process of a five-membered ring between addition to an aldehyde and a carbon-to-carbon double bond as indicated in equation (5), the ring closure to the alkenic double bond was favored three to one at 80 °C. On the other hand in a six-membered ring competition as in equation (6) the addition product involving the aldehyde was obtained in an 86% yield.

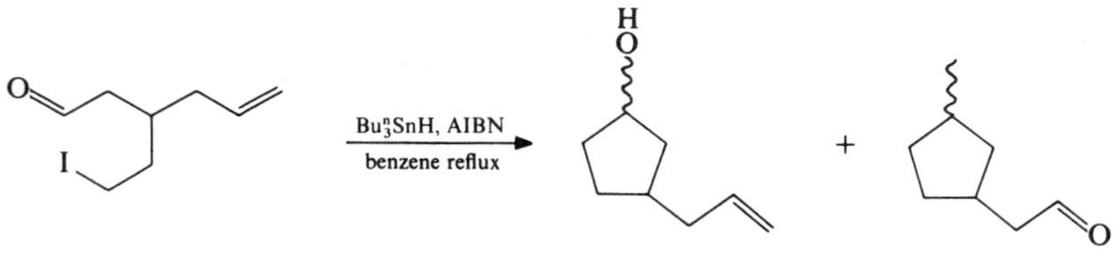

(5)

3:1

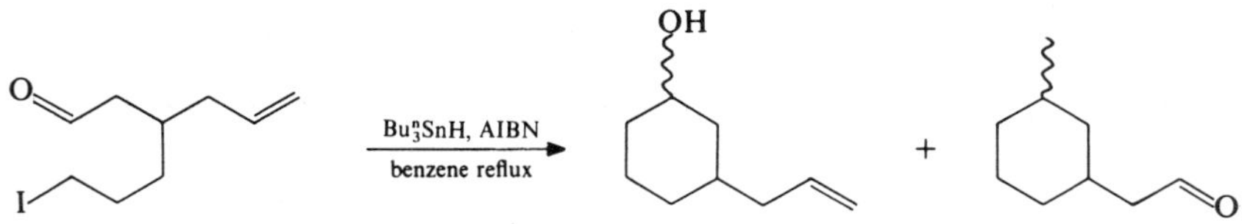

(6)

86%   0%

However, all of our free radical ring-opening reactions involve an elimination reaction with the formation of an ester, a carbonate, an amide, a thioester or a ketone group, and none involve an aldehyde group. Presumably steric effects will be much greater with an ester group than with an aldehyde. Furthermore our reactions appear to involve a concerted process and it is not clear whether any of Tsang's reactions involve such a process.

It may be possible that many radical reactions, including ring-closing (if the principle of microscopic reversibility applies) as well as ring-opening reactions, involve a concerted process. In fact many diverse free radical reactions may be concerted and involve regioselective and/or stereoselective processes.

22.4 NITROGEN AND SULFUR ANALOGS OF CYCLIC KETENE ACETALS

The nitrogen analog of 2-methylene-1,3-dioxolane (**1**) was prepared by substituting 2-*N*-methyl-aminoethanol for the ethylene glycol according to the reactions in Scheme 13.[21] Thus 2-methylene-*N*-methyl-1,3-oxazoline (**30**), which was prepared in an overall yield of 60%, was treated with peroxide at 80 °C to give the polyamide (**31**) as depicted in Scheme 14 with 100% ring opening. Even at room temperature the extent of ring opening was quantitative. Apparently the amide group is sufficiently more stable than the ester group that, in contrast with 2-methylene-1,3-dioxolane (**1**), the nitrogen analog (**30**) undergoes essentially quantitative ring opening even at low temperatures.

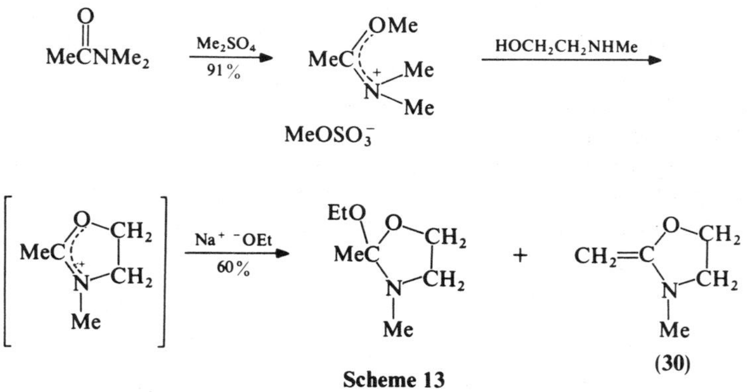

Scheme 13

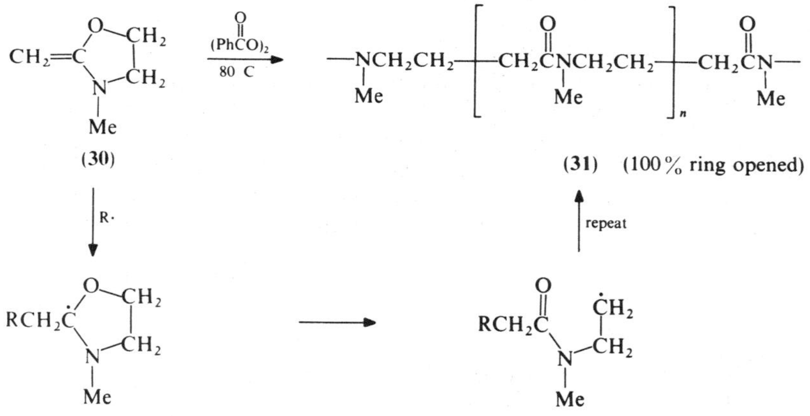

Scheme 14

The sulfur analog of monomer (**1**) was prepared by the process in Scheme 15 as the 2-methylene-1,3-oxathiolane (**32**) in an overall yield of 49%. When the sulfur derivative (**32**) was treated with peroxide at 120 °C as indicated in equation (7), a polymer (**33**) was obtained in which only 15% of the rings had been opened. Apparently the resulting thioester is of higher energy than the ordinary ester and therefore retards the extent of ring opening. Even at 160 °C only 45% of the rings were opened.[22]

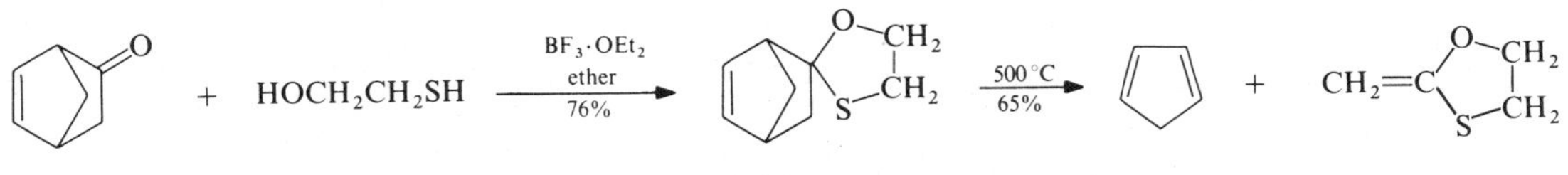

Scheme 15

(**32**)

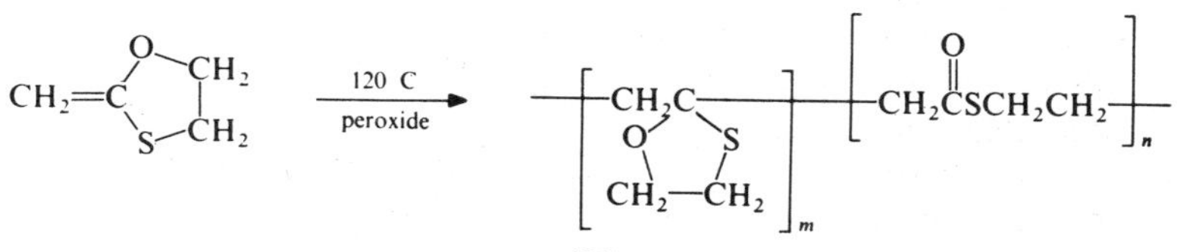

(**33**)

(7)

22.5 COPOLYMERIZATION OF CYCLIC KETENE ACETALS WITH VARIOUS COMONOMERS

22.5.1 Reactivity

The cyclic ketene acetals, in general, will copolymerize with all the common monomers to give copolymers with an ester, an amide or a thioester in the backbone of the chain. (Later, in Section 22.10, it will be shown that the use of an unsaturated spiro orthocarbonate in a copolymerization will permit the introduction of a carbonate linkage in the backbone of a copolymer.) Although copolymerization of various monomers with carbon monoxide will introduce a keto group in the backbone, copolymerization with oxygen will introduce a peroxide linkage and copolymerization with sulfur will introduce a polysulfide linkage, this is the first example of a hydrolyzable linkage being introduced into the backbone of an addition polymer. For example, the 2-methylene-1,3-dioxepane (7) will copolymerize with styrene to produce the copolymer (34) as indicated in equation (8).

(8)

One of the deficiencies of the cyclic ketene acetals is their rather low reactivity relative to monomers such as styrene. Thus in the copolymerization of 2-methylene-1,3-dioxepane (7) with styrene, $r_1 = 0.021$ and $r_2 = 22.6$ at 120 °C.[22,23] That means that in order to introduce 10% of the ester-containing units into the styrene copolymer (34) a monomer mixture consisting of 80% of the cyclic ketene acetal (7) and 20% styrene must be utilized.

22.5.2 Synthesis of Functionally Terminated Oligomers

Most functionally terminated oligomers are made commercially by ionic addition polymerization or condensation reactions rather than by the convenient and inexpensive free radical process. One of the few exceptions is the hydroxy-terminated polybutadiene produced by a free radical process by Arco Chemical Company. Also, the polymerization of butadiene and sulfur by a free radical mechanism that involves a ring opening of the S_8 ring followed by reduction of the resulting polysulfide groups gives a mercapto-terminated polymer which has found only limited use.[24] Since free radical ring-opening polymerization made it possible to introduce functional groups such as esters, carbonates, thioesters and amides into the backbone of an addition polymer, it was reasoned that simple hydrolysis of these carboxylic acid derivatives would produce the desired oligomers that could be terminated with various combinations of hydroxyl, amino, thio and carboxylic acid groups.[23,25-28]

The seven-membered ketene acetal (7) was easily copolymerized with styrene (as indicated in equation 8), and with 4-vinylanisole, vinyl acetate, ethylene and vinyl chloride, to give copolymers with ester groups in the main chain, all with quantitative ring opening. Thus by the use of equal amounts of styrene and the ketene acetal (7), followed by hydrolysis of the resultant polymer (34), an oligomer (35) of styrene was produced that was capped with a hydroxyl group at one end and a carboxylic acid group at the other as indicated in equation (9). Thus a very general method has been developed for the synthesis of a wide variety of oligomers with any desired molecular weight range. Of course, since the copolymers are random, the molecular weight distribution of the oligomers is quite broad. However, these oligomers should prove quite useful for the synthesis of polyurethanes and block polyesters.

(9)

In a similar fashion, when the copolymer (36) of ethylene and 2-methylene-1,3-dioxepane (7) (as depicted in Scheme 16) was hydrolyzed, it gave oligomers (37) of polyethylene which were capped with a hydroxyl group at one end and a carboxylic acid group at the other. For material with an ester-containing unit content of 2.1 mol %, the value of m in Scheme 16 was approximately 47 and for material with content of 10.4 mol %, the value of m was approximately nine.

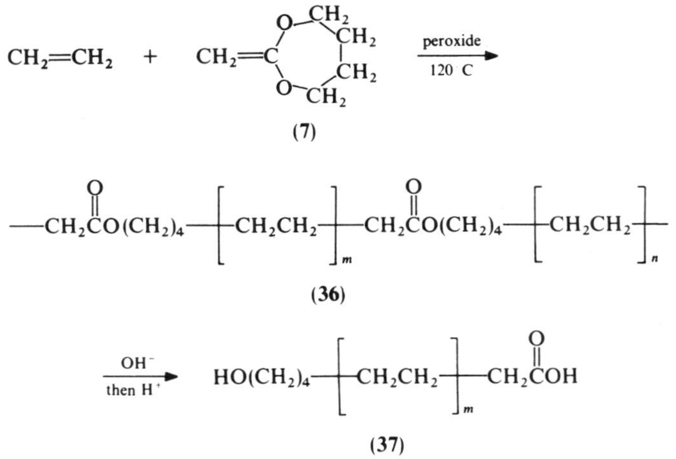

Scheme 16

Since related work had shown that the nitrogen analogs of the cyclic ketene acetals would polymerize with essentially 100% ring opening, their copolymerization with a variety of monomers was undertaken. Thus copolymerization of the aminal (30) with styrene as in Scheme 17 proceeded with essentially quantitative ring opening. The resultant copolymer (38) was readily hydrolyzed to give an oligomer (39) of polystyrene capped with an aminomethyl group and a carboxylic acid group. Even though the sulfur analog (32) of the cyclic ketene acetal (1) gives, on free radical polymerization at 120 °C, a product with only 45% of the rings opened as indicated in Scheme 18, copolymerization of the monomer (32) with styrene gave a copolymer (40) containing thioester groups, and hydrolysis of this copolymer (40) gave an oligomer (41) capped with a thiol and a carboxylic acid group. (As will be discussed later in Section 22.10, copolymerization of an unsaturated spiro orthocarbonate with styrene will give copolymer containing carbonate groups which on hydrolysis will give an oligomer capped with a hydroxyl group at each end.) Thus copolymerization followed by hydrolysis of the resulting copolymer has been shown to be a convenient method for the synthesis of functionally terminated oligomers.

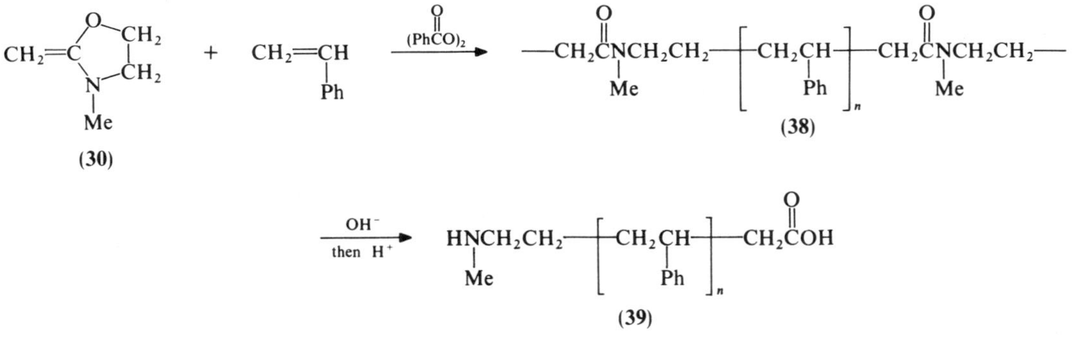

Scheme 17

22.5.3 Biodegradable Polymers

For many medical, agricultural and ecological applications it is desirable to have a biodegradable polymer that can be easily fabricated by injection molding, melt spinning and melt extrusion into films. Of course, all naturally occurring polymers such as starch, cellulose, proteins, nucleic acids and lignin are biodegradable, since they have been existing on earth for a length of time sufficient for enzymes and microorganisms to evolve that can degrade or utilize these materials as food.

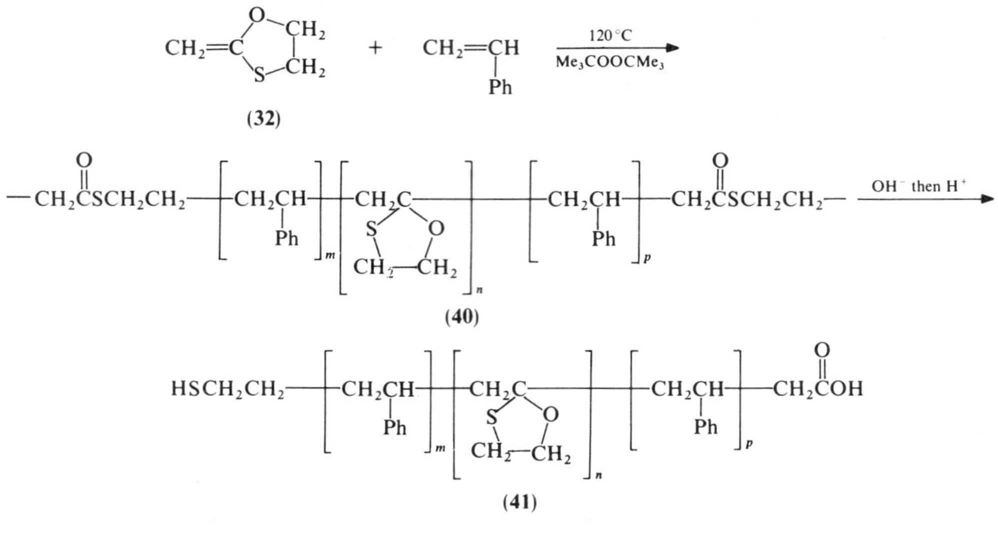

$$CH_2=C \overset{O-CH_2}{\underset{S-CH_2}{\big<}} \quad + \quad CH_2=CH\underset{Ph}{|} \quad \xrightarrow[Me_3COOCMe_3]{120\,°C}$$

(32)

$$-CH_2\overset{O}{\overset{\|}{C}}SCH_2CH_2-\left[CH_2\underset{Ph}{\overset{|}{C}}H\right]_m\left[CH_2\underset{\underset{CH_2-CH_2}{S\,\,\,\,O}}{C}\right]_n\left[CH_2\underset{Ph}{\overset{|}{C}}H\right]_p-CH_2\overset{O}{\overset{\|}{C}}SCH_2CH_2- \xrightarrow{OH^- \text{ then } H^+}$$

(40)

$$HSCH_2CH_2-\left[CH_2\underset{Ph}{\overset{|}{C}}H\right]_m\left[CH_2\underset{\underset{CH_2-CH_2}{S\,\,\,\,O}}{C}\right]_n\left[CH_2\underset{Ph}{\overset{|}{C}}H\right]_p-CH_2\overset{O}{\overset{\|}{C}}OH$$

(41)

Scheme 18

Unfortunately, all of these materials which evolved through an aqueous medium contain very polar groups and therefore decompose upon heating before they melt. By contrast, synthetic polymers have not, in general, been in existence long enough to have enzymes or microorganisms evolve that can utilize them as food. There are, however, a few biodegradable synthetic polymers currently known. For example, low melting and low molecular weight polyesters show a reasonable degree of biodegradation.[29] It is really not surprising that some synthetic polyesters are biodegradable since poly(β-hydroxybutyrate) is a naturally occurring material that many bacteria and fungi use for energy storage in the same way that animals use fat.

On the other hand, addition polymers have been generally quite resistant to biodegradation since the carbon-to-carbon bonds in the backbone are not very susceptible to biological cleavage. As a consequence, many hydrophilic addition polymers have been banned from medicinal use because the high molecular weight polymers are not degraded or eliminated from the body in a reasonable length of time. It was reasoned that if an easily hydrolyzable group, such as an ester group, could be introduced into an addition copolymer by a free radical process, a wide variety of biodegradable polymers could be prepared.[30-34]

Although there are microorganisms that will degrade linear hydrocarbons, the degradation of polyethylene is very slow.[35] Recently, Corbin[36] monitored the biodegradation of ^{14}C-labelled polyethylene in soil by means of the carbon dioxide evolved. Since he observed a minimum conversion rate of 2% per year, he concluded that polyethylene film did not degrade to any significant extent in the bioactive system studied. This slow degradation is because the mechanism of degradation of linear hydrocarbons involves the oxidation of the terminal methyl group to a carboxylic acid group and then degradation of the resulting fatty acid by stepwise β-oxidation, two carbon units at a time. In linear polyethylene of high molecular weight, there are relatively few methyl groups and even they are located in the bulk of a hydrophobic medium not readily accessible to the microorganism. However, if ester groups could be introduced into the backbone of poly-ethylene, even it should become biodegradable.

Copolymerization of ethylene and the ketene acetal (7) at 120 °C produced a copolymer (37) containing ester groups in the backbone of the copolymer also with quantitative ring opening as in Scheme 16. In order to ensure that the copolymers were reasonably homogeneous, in view of the fact that the 2-methylene-1,3-dioxepane (7) is less reactive than ethylene, all conversions were held to below 2% as indicated in Table 2. A series of copolymers containing from 2.1 to 10.4 mol % of the ester-containing units resulted. All of these copolymers had a molecular weight of about 5000 and had melting points varying from 100–105 °C for the copolymer containing 2.1 mol % of (7) to 84–88 °C for the copolymer containing 10.4 mol %.

The rate of biodegradability was determined by a method based on the determination of the carbon dioxide produced by the microbial metabolism for the polymer samples used in this study. Our screening test,[31-34] which was the method developed by Kramer and Ennis,[37,38] consisted of feeding a standard amount of hydrolyzed casein and the polymer to a culture containing a mixed

Table 2 Copolymerization of Ethylene with 2-Methylene-1,3-dioxepane (7)

Amount of (7) (mol %)		Conversion (wt. %)	Copolymer M.P. (°C)	Intrinsic[a] viscosity (dl g^{-1})	Analysis (wt. %)			
					Calculated		Found	
in feed	out feed				C	H	C	H
5.00	2.1	0.80	100–105	0.13	83.90	13.84	83.78	14.00
10.00	4.8	1.6	95–99	0.14	81.90	13.35	81.85	13.62
15.00	6.7	1.2	90–95	0.122	80.61	13.04	80.55	13.38
19.00	9.3	1.5	89–96	0.16	79.25	12.71	78.98	13.03
22.00	10.4	1.7	84–88	0.144	78.46	12.51	78.18	12.80

[a] In xylene at 75 °C.

microflora from soil in a modified Warburg apparatus. The amount of CO_2 liberated was monitored by a Fisher–Hamilton gas partitioner.

The increase in the amount of carbon dioxide in polymer-containing cultures over that of the control was used as a measure of the rate of biodegradation on the time scale used. Statistical analysis involved analysis of variance and the statistical difference was determined with a Duncan's test at a 5% level. The results are listed in Table 3. It is obvious that the copolymers containing at least 6.7 mol % of the ester-containing units are highly degradable, producing CO_2 at a rate 108 to 118% of that of the hydrolyzed casein control. The copolymers containing the lower amount of the ester groups biodegrade at quite low rates but greater than polyethylene. A variety of other addition polymers have been made biodegradable by related processes.

Table 3 Biodegradation of Copolymers of Ethylene and 2-Methylene-1,3-dioxepane (7)

Amount of (7) in copolymer (mol %)	Cumulative[a] CO_2 after	
	7 days	20 days
2.1	103	98
4.8	107	103
6.7	116	108
9.3	116	113[b]
10.4	121	118[b]

[a] Expressed as percent of control.
[b] Significantly different from the control.

The copolymer (34) of the cyclic ketene acetal (7) and styrene is also rendered biodegradable by the introduction of the ester groups into the backbone of the copolymer but at a slower rate since the branches tend to reduce the rate of degradation. Other copolymers, such as the copolymer of acrylic acid and the acetal (7), are likewise made biodegradable by the presence of the ester group in the main chain. [Later in this chapter it will be pointed out that the use of an unsaturated spiro orthocarbonate to introduce a carbonate group (Section 22.10) and a cyclic α-alkoxyacrylate to introduce a pyruvate group (Section 22.7) into a copolymer will also promote biodegradability.]

22.5.4 Increased Thermal Stability of Copolymers

Many polymers such as poly(methyl methacrylate) (PMMA) tend to decompose by a reverse polymerization process upon heating. It was reasoned that, if a small amount of a cyclic monomer were to be copolymerized with ring opening into the backbone of PMMA, the unzipping process could be greatly reduced. As indicated in equation (10), a polymer molecule can cleave at some weak spot to give the polymeric free radical (42). This radical (42) would at 225 °C unzip with the elimination of monomeric methyl methacrylate molecules until the end radical reaches an ester-containing unit. At this point it would be highly unlikely that this open-chain radical would cyclize to continue the elimination process. Thus the presence of the ring-opened unit will impart an element of thermal stability to the poly(methyl methacrylate) copolymer. This method of increasing the thermal stability is reminiscent of the way Celanese–Hoechst uses ethylene oxide to stabilize polyformaldehyde in their Celcon.

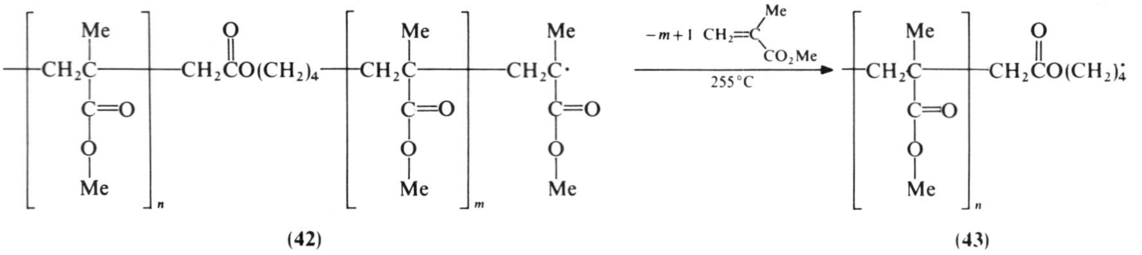

(42)   (43)

$$(10)$$

In order to test the theory discussed above, a series of copolymers of methyl methacrylate and 2-methylene-1,3-dioxepane (**7**) were prepared, the copolymers and PMMA were heated at 225 °C under nitrogen for 30 min and the weight losses of the samples are recorded in Table 4. It is obvious that as little as 1% of the cyclic monomer (**7**) gives a significant reduction in the weight loss. [The use of the unsaturated spiro orthocarbonates (Section 22.10) also produces copolymers with enhanced thermal stabilities.[39]]

Table 4   Weight Loss in Copolymers of 2-Methylene-1,3-dioxepane (**7**) and Methyl Methacrylate

Amount of (**7**) in copolymer (wt. %)	Amount of copolymer remaining[a] (wt. %)
0	74
1	90
2	93
4	96

[a] After heating for 30 min at 225 °C under nitrogen.

22.6   FREE RADICAL RING-OPENING POLYMERIZATION OF CYCLIC VINYL ETHERS

Since the nitrogen and sulfur analogs of the cyclic ketene acetal (**1**) gave interesting results, the study was extended to the carbon analogs, the cyclic vinyl ethers. Thus 2-methylenetetrahydrofuran (**44**), when treated with di-*t*-butyl peroxide at 120 °C as indicated in Scheme 19, gave a polymer (**45**) in which only about 5% of the rings had opened.[20] Apparently the ketone group is sufficiently less stable than the ester group that the extent of ring opening decreases from 87% for (**1**) to 5% for monomer (**44**). Even when the cyclic vinyl ether (**44**) was diluted with an equal volume of benzene, the ring opening increased only to 7%.

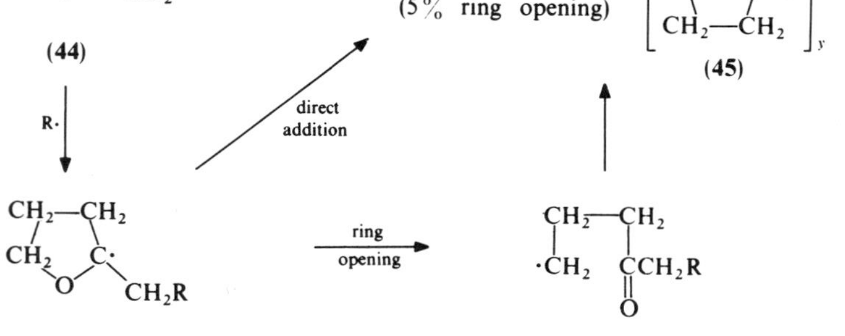

Scheme 19

Since the extent of ring opening of the cyclic vinyl ethers is less than that of the cyclic ketene acetals, this series appeared to be an ideal system to study the effect of steric hindrance and the presence of radical-stabilizing substituents on the extent of ring opening. Thus the monomers (**46**) and (**47**) were prepared and polymerized at 120 °C in the presence of di-*t*-butyl peroxide. The percentages in parentheses indicate the extent of ring opening. The highest extent of ring opening in this series was observed for the 4-phenyl-2-methylenetetrahydrofuran (**48**).[20] In this case the formation of the relatively stable benzyl radical helps promote ring cleavage so that about one-half of the rings are opened in the polymer (**49**) as indicated in Scheme 20.

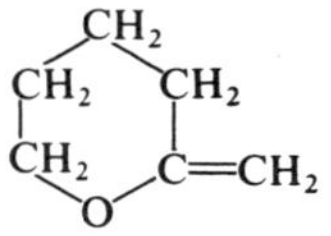

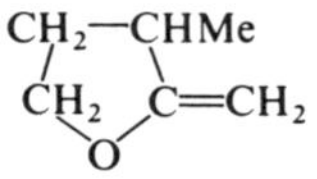

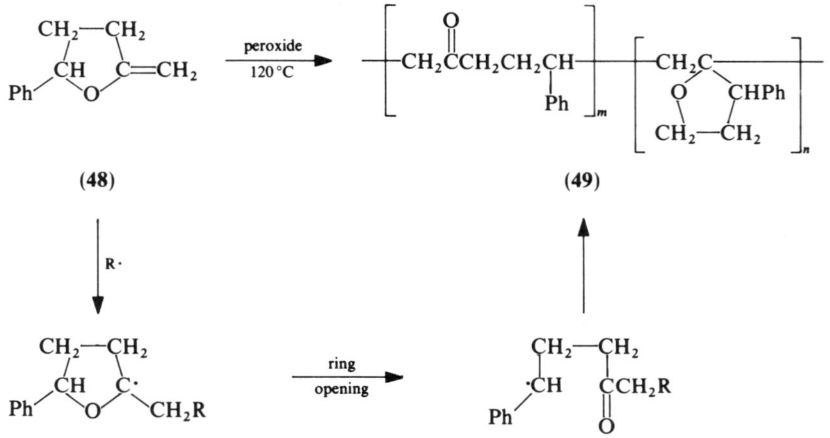

Scheme 20

Even ring strain will not produce quantitative ring opening in this series. Although the oxetane ring possesses considerable ring strain, the corresponding radical does not appear to open at a rapid rate. For example, 2-oxetanyl gives no signals in the ESR spectrum for the ring-opened product, which is in contrast to the five- and six-membered acetals which give the ring-opened radical signals at room temperature.[40] The polymerization of 2-methyleneoxetane (**50**)[41] at 120 °C gave a copolymer in which only about 40% of the rings were opened as indicated in equation (11).[21] Apparently the small size of the ring reduces the steric hindrance so that the direct addition can effectively compete with the sluggish ring-opening step.

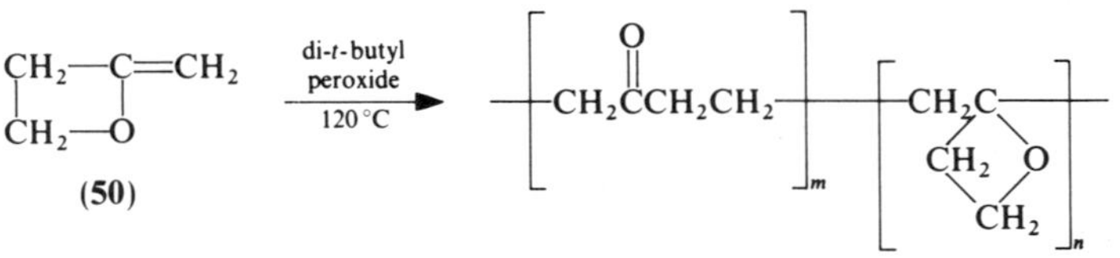

Ketene dimer (**52**), which is an analog of the 2-methyleneoxetane (**50**), will undergo at higher extent of ring opening on polymerization to give copolymers with different structures depending on whether the reaction is run in a sealed tube or open to the air as indicated in Scheme 21. The extent of ring opening is higher because the acyl radical (**53**) is more stable than the primary radical that one would expect to obtain from the 2-methylene-1,3-oxetane (**50**). In the sealed tube the acyl radical does not lose very much carbon monoxide to give a copolymer (**54**) which contains a large amount of the 1,3-diketone structure. On the other hand, the acyl radical (**53**), when the reaction is open to the air, can lose a substantial proportion of the acyl groups to give the copolymer (**55**) that is largely the 1,4-diketone.

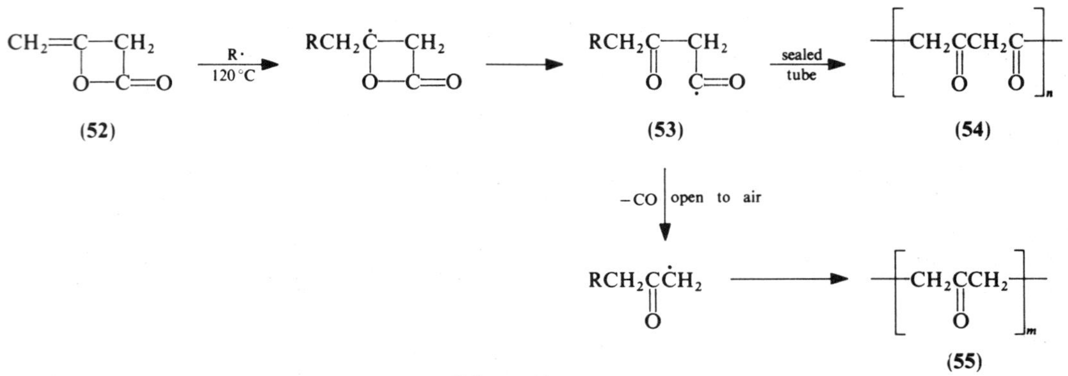

Scheme 21

A surprising result was obtained from the polymerization of the 2-methylene-3,4-benzotetra-hydrofuran (**56**), which gives a polymer (**57**) with little or no ring opening as illustrated in Scheme 22. Apparently the intermediate radical (**58**), which is a tertiary benzyl radical with additional stabiliz-ation from the oxygen, is sufficiently more stable than the open-chain primary benzyl radical (**59**) that only direct addition takes place.

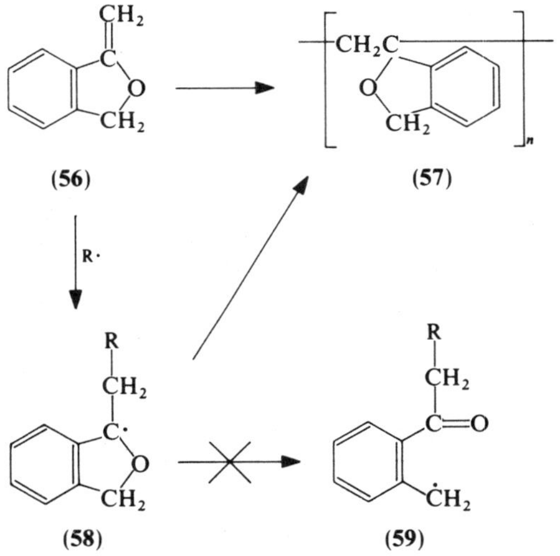

Scheme 22

Copolymerization of several of these unsaturated cyclic ethers, particularly the 2-methylene-5-phenyltetrahydrofuran (**48**) with various monomers, is a convenient way to introduce a ketone group into the backbone of the addition polymer. Such polymers containing the keto groups can be expected to be photodegradable. The competitive process involves the copolymerization of the monomer with carbon monoxide under high pressure.[42]

22.7 FREE RADICAL RING-OPENING POLYMERIZATION OF CYCLIC α-ALKOXYACRYLATES

Since cyclic ketene acetals have very low reactivity in copolymerization with common monomers (Section 22.5.1), it was of interest to find cyclic monomers with high reactivity which would still undergo a high degree of ring opening. For this reason the synthesis of a series of cyclic α-alkoxyacrylates was investigated and their polymerization characteristics were studied. By the procedure indicated in Scheme 23 various cyclic acrylates were prepared in good yields.[43] The acetal formation with bromolactic acid (**60**) to give the cyclic intermediate (**61**) proceeds easily, and since the hydrogen atom in intermediate (**61**) is activated by the carbonyl group, a strong base is not needed for the dehydrobromination to monomer (**62**). 1,5-diazabicyclo[5.4.0]undec-5-ene (DBU) is strong enough to promote the elimination in good yields but apparently not strong enough to promote polymerization.

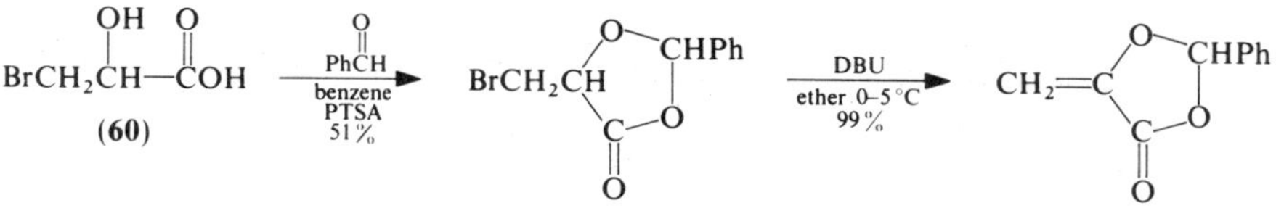

Scheme 23

When a ketone was used in this sequence of reactions to produce a disubstituted intermediate, a different approach was necessary. Thus, the direct ketal formation in this series does not proceed in as good a yield as in the acetal formation. However, the use of a ketal interchange reaction gives good yields of the cyclic intermediate (63) as indicated in Scheme 24. [A Russian article[44] also recently reported the synthesis of compounds (62) and (64) in 60–80% yields from the chloro derivatives with triethylamine without any experimental details or reported attempts at polymerization.]

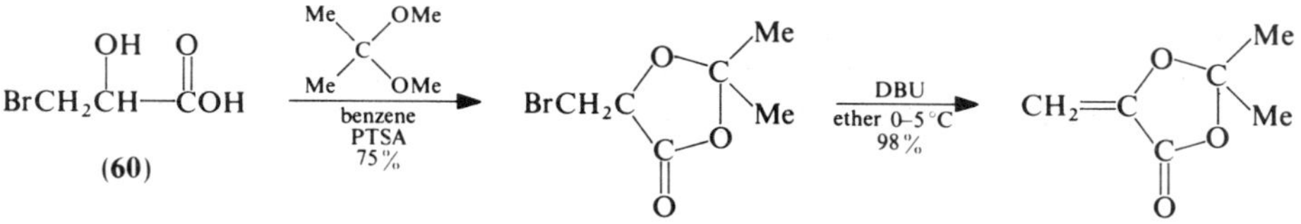

Scheme 24

When the polymerization of monomer (62) was carried out at 120 °C in bulk, a polymer consisting almost entirely of units that were nonring-opened was obtained. When the temperature was raised to 140 °C and the polymerization was carried out in a solution of *t*-butylbenzene, a limited amount of ring opening (10%) occurred. As indicated in Scheme 25 a radical can add to the cyclic acrylate (62) to give the very stable growing radical (65) in which the free radical is stabilized by resonance with both the carbonyl group and the ether oxygen. When the radical (65) undergoes ring opening, a radical (66) of about the same reactivity results, giving very little driving force for the ring-opening process. As a consequence a polymer (67) containing relatively few ring-opened units results. With monomer (64) at 120 °C in bulk the polymerization gave a mixture containing about 15% ring-opened units and 85% nonring-opened units. Again, at 140 °C and in *t*-butylbenzene solvent, polymerization of monomer (64) gave about 50% ring opening. These cyclic acrylates copolymerized readily with styrene and methyl methacrylate to give copolymers that should be both photodegradable and biodegradable. The homopolymers of monomers (62) and (64) prepared at high temperatures and in solution were shown to be biodegradable. This biodegradability probably results from the ease of hydrolysis of the cyclic acetal containing the ester group and the presence of even a small amount of the pyruvate grouping in the backbone of the polymer. [Earlier in this chapter the effect of the introduction of an ester group into the backbone of an addition polymer in rendering a polymer biodegradable was discussed (Section 22.6).] It was not surprising that the introduction of a pyruvate ester group into the backbone of an addition polymer was even more effective than a simple ester unit since pyruvates are common intermediates in many biological systems.

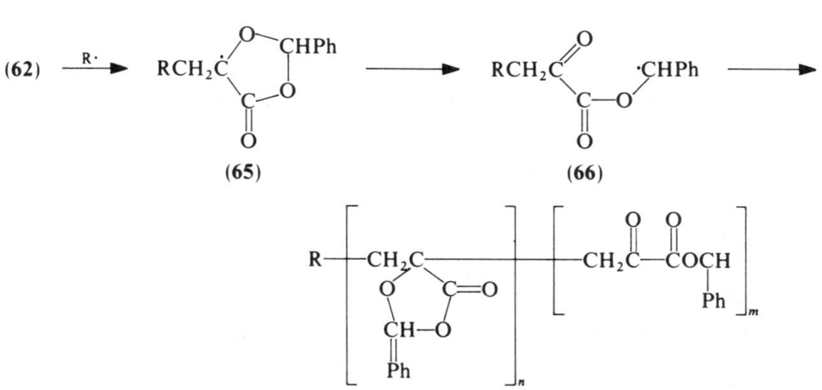

Scheme 25

An interesting extension of this work is the preparation of biodegradable polymerized vesicles. Vesicles have attracted much attention lately because they mimic body membranes and liposomes. As a result they appear very useful for the delivery of drugs and encapsulating sensitive biologically active materials. However, since many of the simple vesicles are not very stable under many biological conditions, attempts have been made to stabilize the vesicles by incorporating a polymerizable entity into the lipid-like molecules, formation of the vesicle, followed by polymerization of the monomers in the vesicle. Although this approach works, it usually renders the polymerized vesicle nonbiodegradable and thus limits its use in the body.[45-52] Only the polymerized vesicles of Samuel *et al.*[48] and Cho *et al.*,[52] who crosslinked through a disulfide group, appear possibly to be biodegradable. Since we had demonstrated that the polymers from the cyclic alkoxyacrylates were highly biodegradable, it appeared attractive to introduce a cyclic acrylate into the lipid-like molecules of the vesicles, which upon polymerization would produce polymerized vesicles with pyruvate ester groups in the polymer backbone. Thus the vesicles should be stable enough to deliver a drug or to protect a sensitive material but yet should be biodegradable in the body. The cyclic acrylates appeared especially attractive since the conditions for the synthesis are quite mild.

The required lipid-like molecule containing the cyclic acrylate (**68**) was prepared by the series of reactions indicated by Scheme 26.[53] The cyclic acrylate (**68**) will form vesicles that can be polymerized to form the polymerized vesicles containing the polymer (**69**) as schematically illustrated in Scheme 27. Upon standing in the presence of water, the polymerized vesicle containing polymer (**71**) was shown to slowly hydrolyze to generate the free lipid (**70**) and the polymer (**71**). Furthermore the polymer (**71**) was shown separately to be biodegradable. Research is in progress to determine the physical characteristics of these polymerized vesicles and to determine the effectiveness of these systems to deliver drugs in the body.

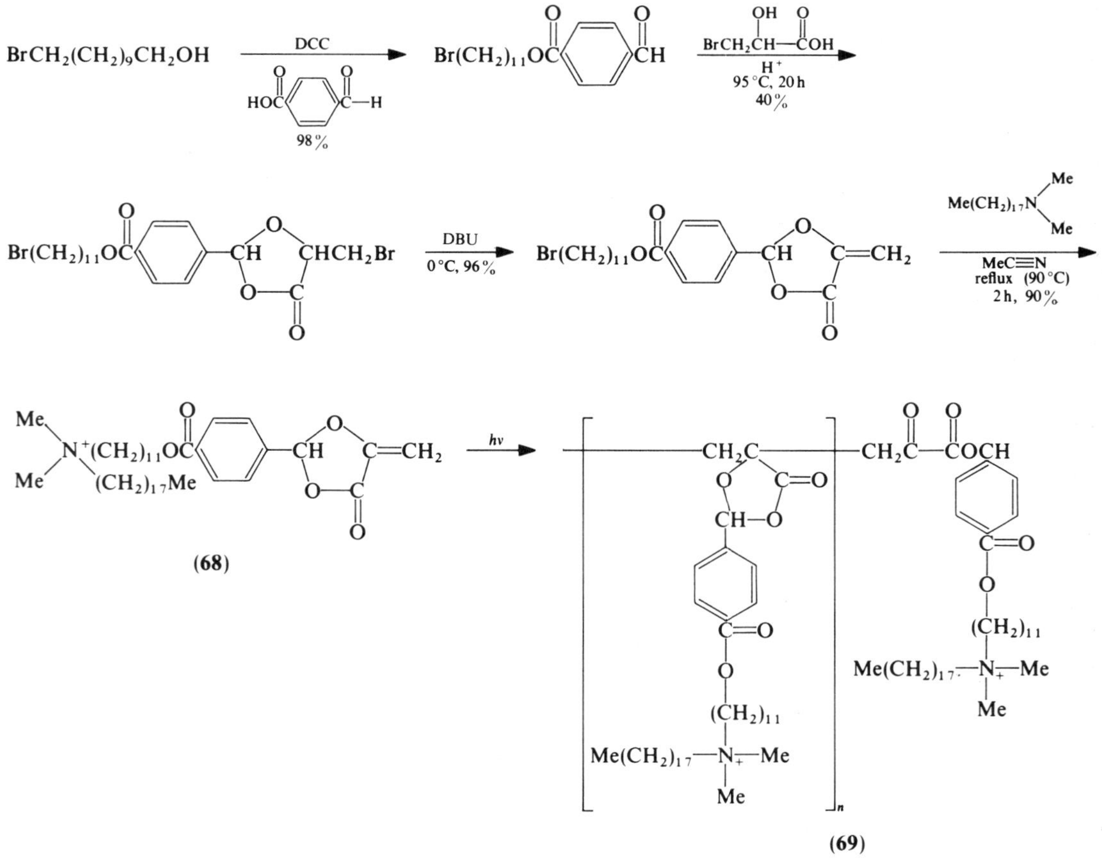

Scheme 26

An alternative procedure to polymerizing the preformed vesicle is to make a polymer and subsequently form the vesicle from this polymer. This is possible by the introduction of hydrophilic spacer groups to give very flexible linkages between the polymer chain and the amphiphilic side

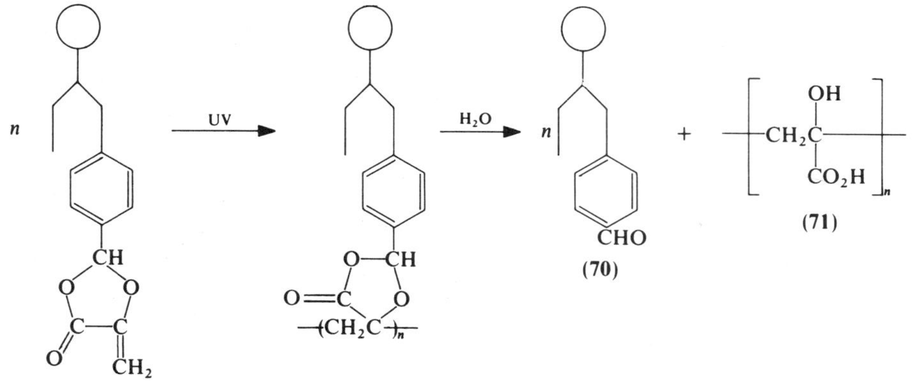

Scheme 27

groups and thus decouple the motion of the polymer main chain from the membrane-forming side groups.[54,55] The synthesis of a new polymeric lipid which contains a hydrophilic spacer group derived from ethylene glycol units between a monomeric group of a hydrolyzable cyclic acrylate and the main amphiphilic structure is described in Scheme 28.[56] The lipid (72) could be polymerized at 60 °C in the presence of AIBN to give the prepolymerized lipid, which upon ultrasonication gave the desired vesicles. On standing, the cyclic acrylate was slowly hydrolyzed to separate the polymer chain from the amphiphile and generate a water-soluble biodegradable polymer.

In an effort to find a synthetic process to prepare cyclic acrylates containing six- and seven-membered rings which would be expected to undergo a higher extent of ring opening than those containing five-membered rings, we adapted the procedure of Kirchmeyer *et al.*[57] for the insertion of a carbonyl group into a cyclic acetal. The required 2-halomethyl-1,3-dioxolanes were available from our earlier work on the synthesis of the 2-methylene-1,3-dioxolanes.[58]

For example, in Scheme 29, when the 2-bromomethyl-1,3-dioxolane (73) was treated with trimethylcyanosilane and zinc iodide at room temperature, a 62% yield of the ring-opened intermediate (74) was obtained. Hydrolysis of (74) followed by ring closure gave a 70% yield of the cyclic intermediate (75). Since (75) is a β-bromoacrylate, the elimination can be carried out with a base that is not strong enough to polymerize the resulting cyclic acrylate (76). Thus the treatment of (75) with DBU at 0–5 °C produced the cyclic acrylate (76) in a 90% yield. When the monomer (76) was treated with benzoyl peroxide in a benzene solution, a polymer (77) was obtained in which only 20% of the rings had been opened.

When the analogous dimethyl derivative (78) was polymerized under very similar conditions as indicated in equation (12), a polymer (79) was obtained in which 30% of the rings had been opened. Apparently the secondary radical that results during ring opening of monomer (78) is slightly more stable than the primary radical formed during ring opening of monomer (76) to give a somewhat higher extent of ring opening.

Since the introduction of a phenyl group greatly increased the extent of ring opening in the 2-methylene-1,3-dioxolane series, the diphenyl derivative monomer (80) was prepared by the series of reactions indicated in Scheme 30. When (80) was treated with benzoyl peroxide in benzene at 80 °C, a polymer (81) was obtained in which most of the rings had been opened. Thus the benzyl free radical formed during ring opening of monomer (80) was sufficiently more stable than the primary radical formed from monomer (76) that nearly complete ring opening occurred.

22.8 FREE RADICAL RING-OPENING POLYMERIZATION WITH AROMATIZATION AS THE DRIVING FORCE

One of the earliest examples of free radical ring-opening polymerization was reported by Errede[4] in 1961, who found that spirodi-*o*-xylylene would homopolymerize to give the corresponding poly-(*o*-xylyene) as depicted in Scheme 3. In related research, it was reasoned that the homolytic opening of a five- or six-membered ring would cost less than 8 kcal mol^{-1} (32 kJ mol^{-1}) of energy[7] while the formation of the aromatic ring would gain about 36 kcal mol^{-1} (145 kJ mol^{-1}); thus the process should be quite thermodynamically favorable. Therefore a research program was undertaken to study the polymerization of a series of methylenespirohexadienes to determine their tendency to undergo ring opening and their reactivity toward polymerization and copolymerization.

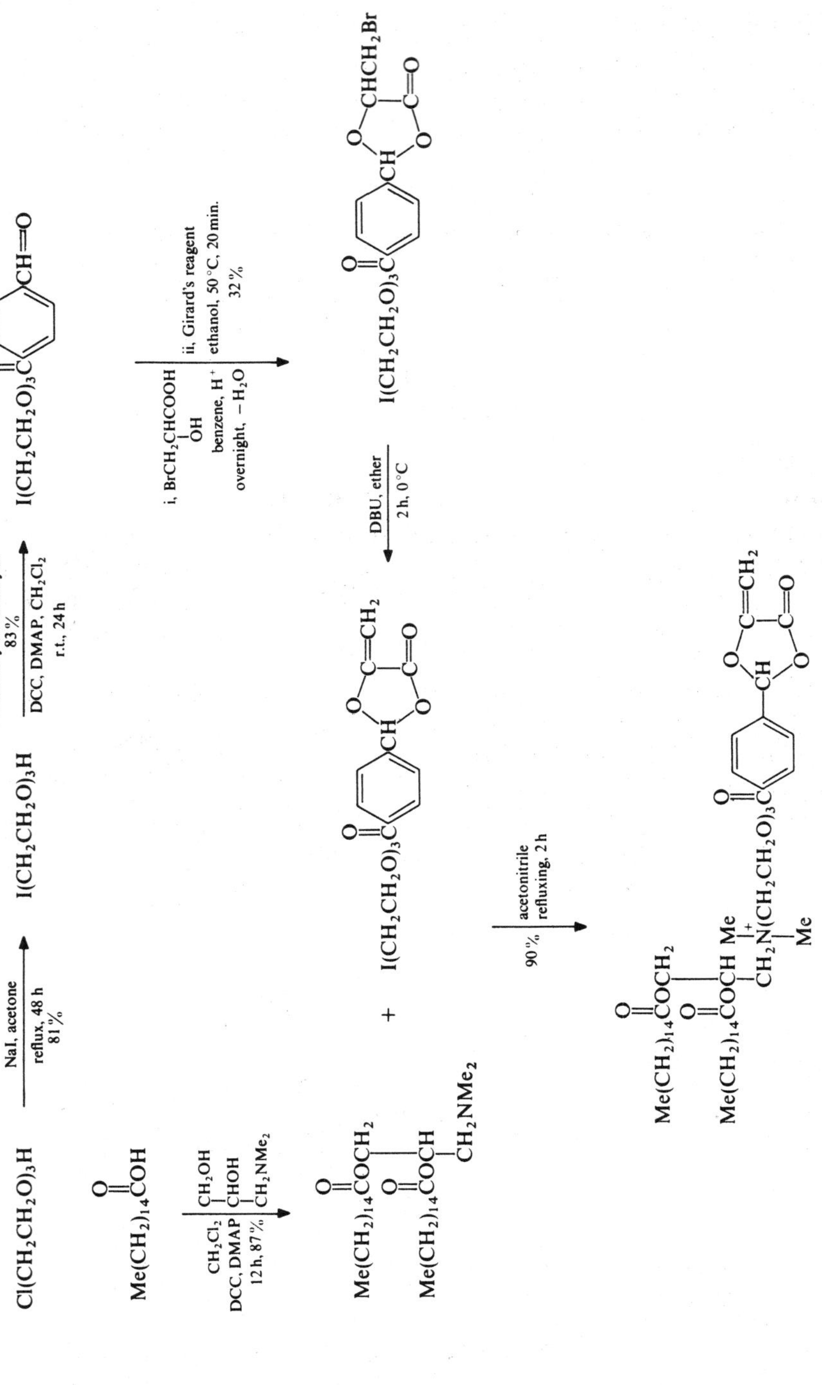
Cl(CH2CH2O)3H
NaI, acetone
reflux, 48 h
81%
I(CH2CH2O)3H
4-carboxybenzaldehyde
83%
DCC, DMAP, CH2Cl2
r.t., 24 h
I(CH2CH2O)3C
CH=O
i, BrCH2CHCOOH
OH
benzene, H+
overnight, −H2O
ii, Girard's reagent
ethanol, 50 °C, 20 min.
32%
I(CH2CH2O)3C
CHCH2Br
CH
C=O
DBU, ether
2 h, 0 °C
I(CH2CH2O)3C
C=CH2
Me(CH2)14COH
CH2Cl2
DCC, DMAP
12 h, 87%
CH2OH
CHOH
CH2NMe2
Me(CH2)14COCH2
Me(CH2)14COCH
CH2NMe2
+
90%
acetonitrile
refluxing, 2 h
Me(CH2)14COCH2
Me(CH2)14COCH Me
CH2N(CH2CH2O)3C
Me

(72)

Scheme 28

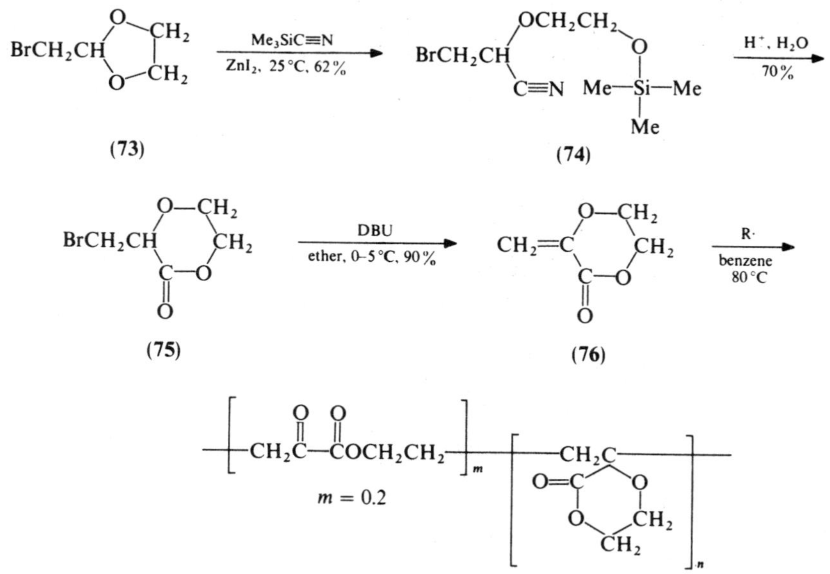

Scheme 29

(12)

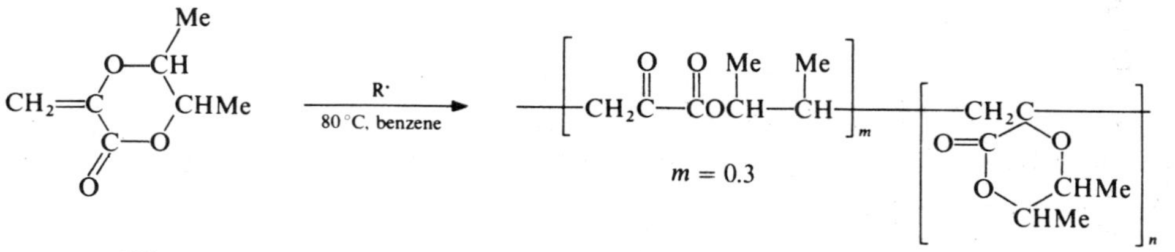

Scheme 30

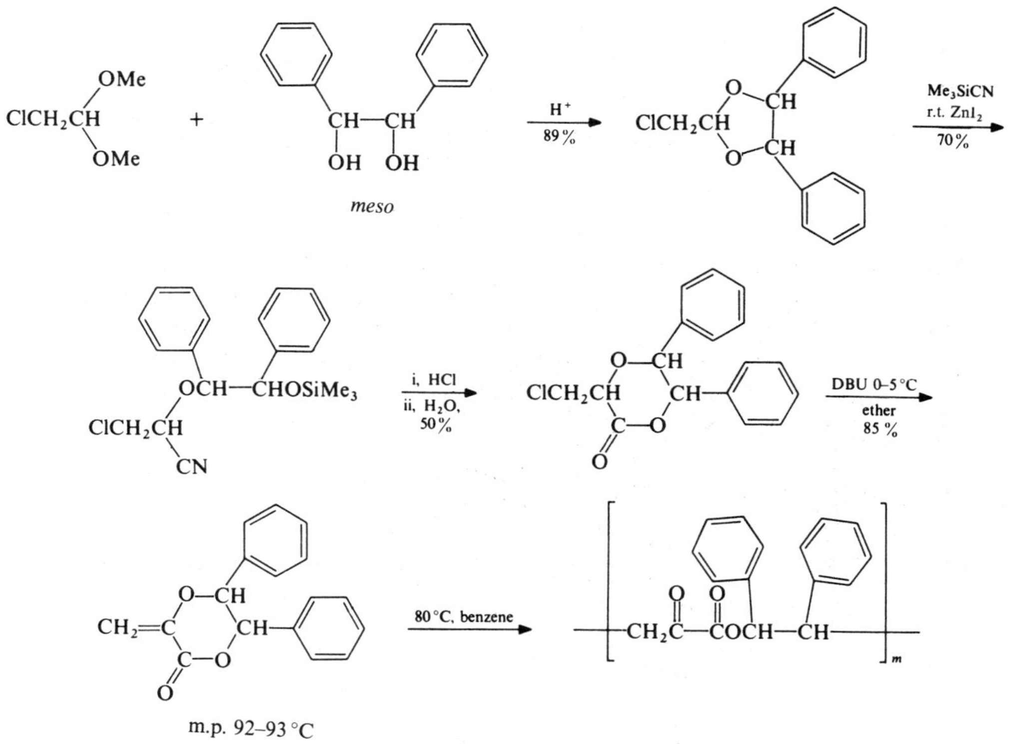

The desired monomer, 3-methylenespiro[5,5]undeca-1,4-diene (**82**) was prepared in an overall yield of 24% by the series of reactions given in Scheme 31. The intermediate cyclohexadienone, spiro[5,5]undeca-1,4-dien-3-one (**83**) was prepared on a large scale by the procedure of Kane[59] in a 36% overall yield. A Wittig reaction[60] gave the corresponding methylene derivative (**82**) in a 75% yield. In agreement with the results of Murray[61] it was found that when 1.5 equivalents of methyltriphenylphosphonium bromide and two equivalents of sodium hydride were used, the major product was the 3-ethylidenespiro[5,5]undeca-1,4-diene. The methylene monomer (**82**) was a colorless liquid that was stable under nitrogen in a freezer, but at room temperature in air the material gradually became brown and underwent spontaneous polymerization to a low molecular weight polymer.

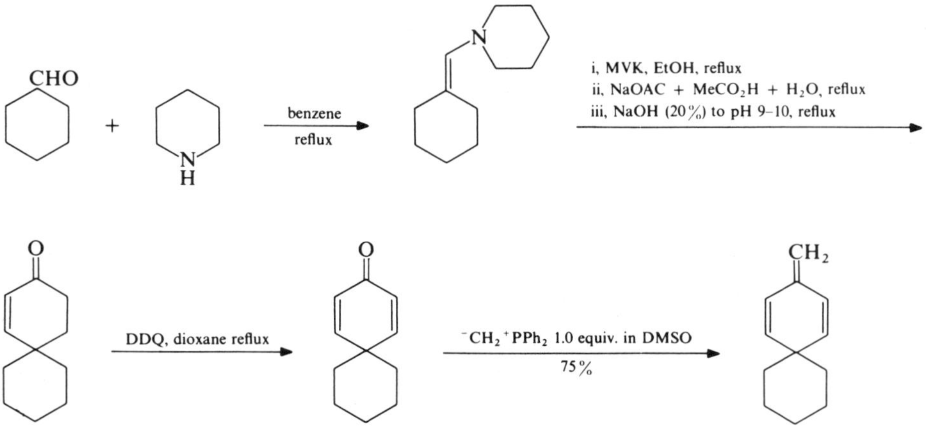

CHO

i, MVK, EtOH, reflux
ii, NaOAC + MeCO$_2$H + H$_2$O, reflux
iii, NaOH (20%) to pH 9–10, reflux

DDQ, dioxane reflux

$^-$CH$_2$ $^+$PPh$_2$ 1.0 equiv. in DMSO
75%

(**83**)   (**82**)

Scheme 31

Homopolymerizations of monomer (**82**) were carried out at different temperatures to determine the effect of temperature on the extent of ring opening. For example, when the triene (**82**) was polymerized at 85 °C over a perod of 72 h with 2 mol % of benzoyl peroxide as the initiator, a white powder was isolated after purification. The 200 MHz ^{1}H NMR spectrum showed signals at $\delta = 6.75$–7.20, corresponding to the aromatic protons, $\delta = 2.55$ corresponding to benzylic methylene protons (both indicative of ring opening); signals at $\delta = 5.25$ *vs.* those at $\delta = 5.25$ and 5.61 could be used to determine the extent of ring opening. The results of this study are listed in Table 5.

Table 5 Homopolymerizations of Triene (**82**) at Various Temperatures

Initiator (mol %)	T (°C)	Time (h)	Polymer yield (%)	Solubility	Extent of ring opening[a] (mol %)
Benzoyl peroxide	85	72	42	Soluble in CHCl$_3$, C$_6$H$_6$	43
Benzoyl peroxide	100	48	80	33% Soluble in CHCl$_3$	61
					Average 67
	100	48	80	67% Insoluble in common solvents[b]	70
t-Butyl peroxide	130	12		Insoluble in common solvents[b]	79
t-Butyl peroxide[c]	130	12		Insoluble in common solvents[b]	98

[a] From 200 MHz ^{1}H study.
[b] Below 80 °C.
[c] Solution polymerization (3:1 benzene/monomer by weight).

It can be seen from Table 5 that at 85 °C only 43% of the rings had opened, but as the temperature of polymerization was increased to 130 °C, the extent of ring opening increased to 79%. Furthermore, when the polymerization was carried out in a solvent at 130 °C, the extent of ring opening was nearly quantitative. These results are consistent with Scheme 32. Even though the ring-opening step is thermodynamically favorable, the rate of the competitive direct addition of the intermediate radical to the monomer (**82**) at 85 °C is faster. Carrying out the polymerization in solution favors the ring-opening process since it is a unimolecular reaction while the direct addition to monomer is a bimolecular reaction.

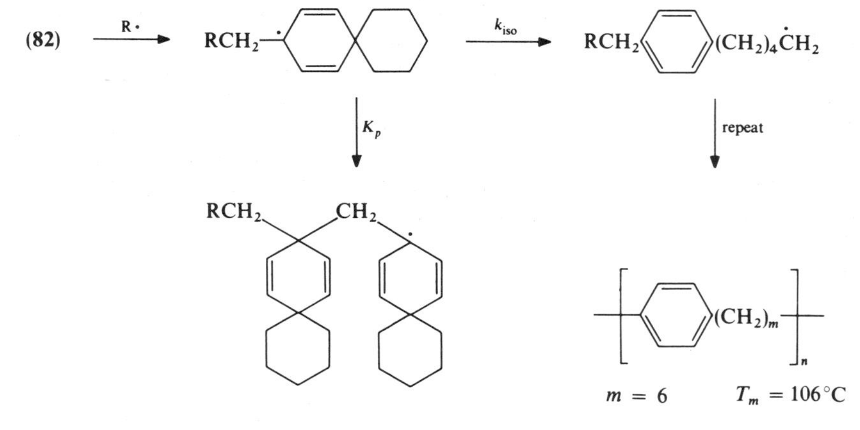

Scheme 32

The polymer (**83**) with nearly quantitative ring opening was shown to be insoluble in all common solvents below 80 °C. The NMR spectrum was obtained by dissolving the polymer in phenyl ether at 180 °C (about 10 g^{-1}) and cooling to obtain a supersaturated solution at room temperature; the resulting cold solution was then added to DCCl$_3$ to obtain the spectrum. X-Ray diffraction on an unoriented film showed a sharp ring pattern indicative of crystallinity. A further indication of the regular structure of the polymer (**83**) was shown by a DSC scan which showed a $T_m = 106$ °C and a $T_g = 6$ °C. Thus decreasing the concentration of the monomer (**82**) favors the free radical ring-opening process.

Monomer (**82**) also was shown to have relatively high reactivity in copolymerization. Thus, when equimolar amounts of monomer (**82**) and styrene were copolymerized at 130 °C in the presence of 2 mol % of *t*-butyl peroxide as the initiator for 4 h, a solid copolymer was obtained which contained 42% units from monomer (**82**) of which 92% were ring opened.

A research program to determine some of the factors that influence the extent of ring opening in this new unsaturated spiro system was then undertaken. Since the introduction of an aromatic group into the 2-methylene-1,3-dioxolane (**1**) system increased the extent of ring opening to nearly 100% even at room temperature during free radical polymerization, it was of interest to synthesize 3′,4′-dihydro-4-methylenespiro[2,5-cyclohexadiene-1,2′(1′H)-naphthalene] (**84**), which would be expected to produce a benzyl radical upon ring opening. As depicted in Scheme 33, the starting material for the synthesis of monomer (**84**) was 1,2,3,4-tetrahydro-2-naphthaldehyde (**85**), which was prepared by the method of Alder and Fremery[62] in a 44% yield from *o*-xylylene dibromide and acrolein. By the general method of Kane[59] described previously, the aldehyde (**85**) was converted to the ketone (**86**) in a 45% yield by the treatment of the enamine derived from the aldehyde (**85**) with methyl vinyl ketone (MVK). Oxidation of the tricyclic ketone (**86**) with 2,3-dichloro-5,6-dicyano-1,4-benzoquinone (DDQ) gave the dienone (**87**) in 20% yield. Finally, the conversion of the dienone (**87**) to the triene (**84**) was accomplished in a 48% yield in a Wittig reaction.

When the monomer (**84**) was treated in the bulk at 100 °C with 2 mol % of benzoyl peroxide, a white solid polymer (**88**), [η] = 0.253, was obtained in nearly quantitative yield. The extent of ring opening was determined by a 200 MHz NMR study to be in excess of 95% since only a trace of absorption in the $\delta = 5.0$–6.0 region corresponding to vinyl protons was detected. Under comparable conditions the monomer (**82**) would be expected to undergo only 67% ring opening. Thermodynamic data would indicate that the benzyl radical is about 13 kcal mol^{-1} (52 kJ mol^{-1}) more stable than the corresponding primary radical to give an additional driving force for ring opening of monomer (**84**). The extent of ring opening is, however, determined by the relative rates of the direct addition k_p and the ring opening k_{iso}.

DSC analysis showed that the polymer (**88**) had a melting point $T_m = 107$ °C and a T_g of 61 °C. For comparison, poly(*p*-xylylene) has a T_m above 400 °C[63] while poly (*o*-xylylene) was reported to melt at 110 °C.[4] Apparently the *o*-xylylene unit prevents the polymer from fitting in a compact crystal lattice. The *ortho* unit also increases solubility since the polymer (**88**) is fairly soluble in benzene and chloroform at room temperature. TGA analysis of polymer (**88**) under nitrogen at a heating rate of 10 °C min^{-1} showed that more than 90% of the polymer remained at 425 °C and 50% remained at 460 °C.

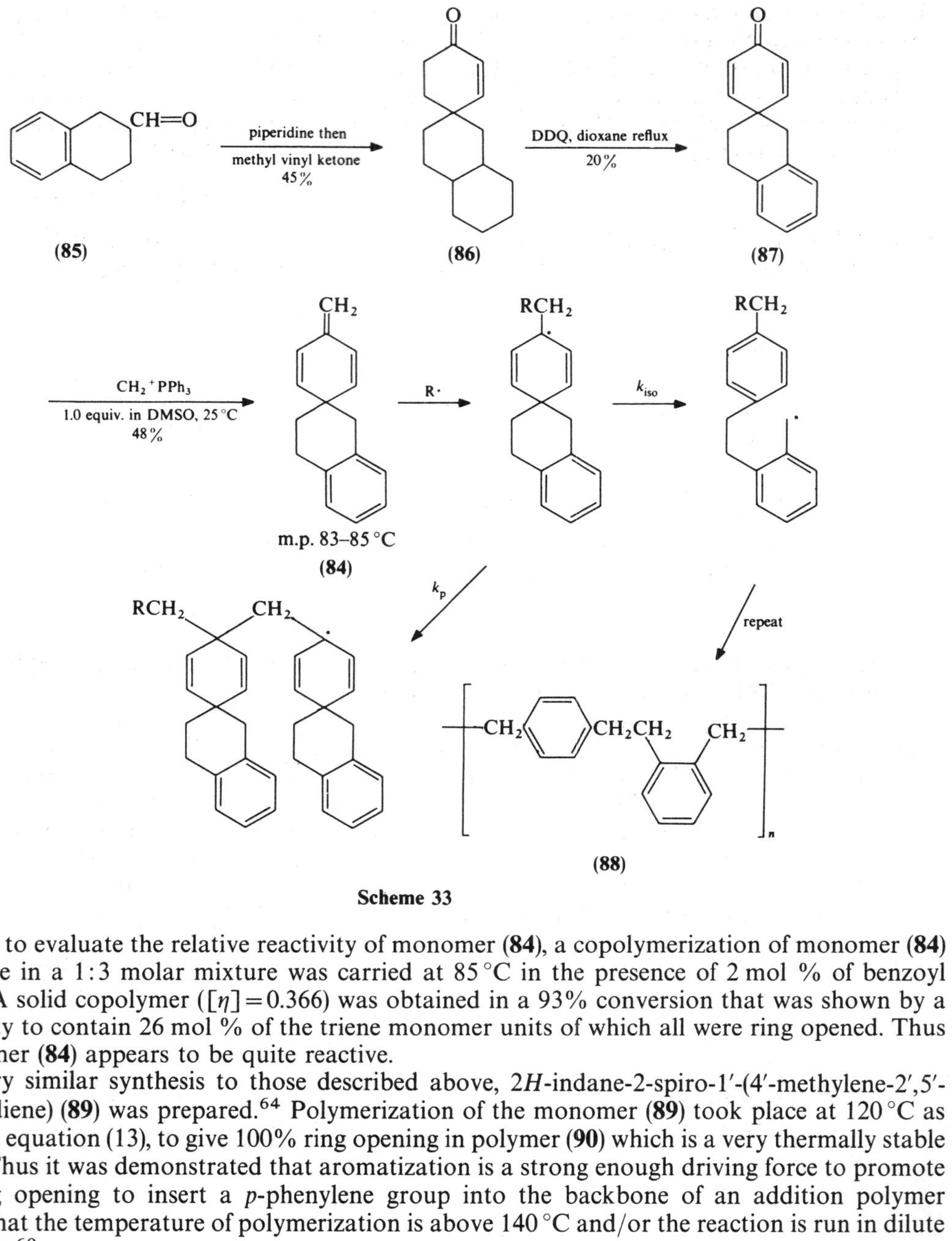

Scheme 33

In order to evaluate the relative reactivity of monomer (**84**), a copolymerization of monomer (**84**) and styrene in a 1:3 molar mixture was carried at 85 °C in the presence of 2 mol % of benzoyl peroxide. A solid copolymer ($[\eta] = 0.366$) was obtained in a 93% conversion that was shown by a NMR study to contain 26 mol % of the triene monomer units of which all were ring opened. Thus the monomer (**84**) appears to be quite reactive.

In a very similar synthesis to those described above, 2*H*-indane-2-spiro-1'-(4'-methylene-2',5'-cyclohexadiene) (**89**) was prepared.[64] Polymerization of the monomer (**89**) took place at 120 °C as depicted in equation (13), to give 100% ring opening in polymer (**90**) which is a very thermally stable material. Thus it was demonstrated that aromatization is a strong enough driving force to promote 100% ring opening to insert a *p*-phenylene group into the backbone of an addition polymer provided that the temperature of polymerization is above 140 °C and/or the reaction is run in dilute solution.[65-69]

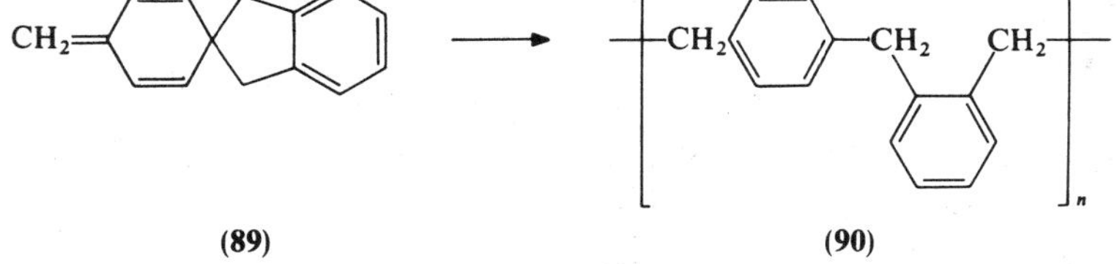

(13)

22.9 FREE RADICAL RING-OPENING POLYMERIZATION OF MISCELLANEOUS MONOCYCLIC MONOMERS

22.9.1 4-Methylene-1,3-dioxolanes

The 4-methyl-1,3-dioxolane (**91**), which is related to the cyclic vinyl ether (**44**), was shown to undergo some (30%) ring opening during polymerization at 130 °C but with elimination of formaldehyde as indicated in Scheme 34. Apparently when the radical (92) undergoes any ring opening to form the open chain radical (93), it loses a molecule of formaldehyde which is eliminated to give the more stable radical (94) resulting in a copolymer (95) containing the keto groups. (Solid paraformaldehyde was isolated from the polymerization mixture.) The cyclic monomer (91) undergoes spontaneous polymerization at room temperature with maleic anhydride to give a copolymer with very little ring opening.

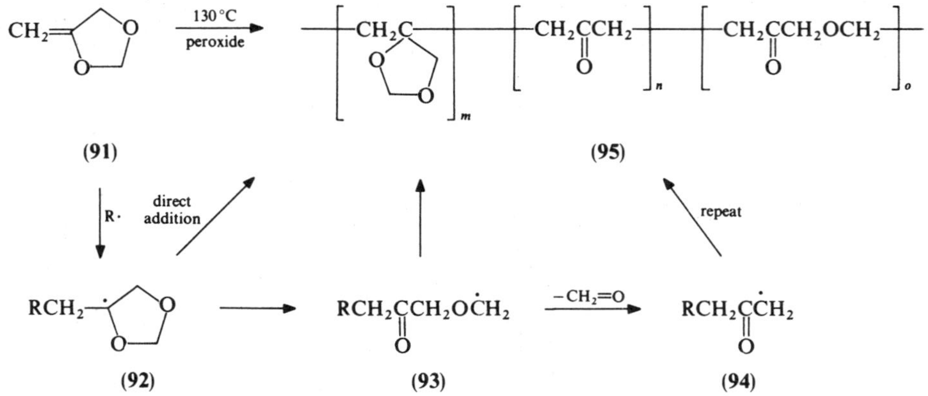

Scheme 34

When a phenyl group is introduced into the 2-position to obtain the 2-phenyl-4-methylene-1,3-dioxolane (96), this monomer (96) is much more reactive than monomer (91). As indicated in Scheme 35, a low molecular weight terpolymer (97) is obtained containing the units, I, II and III. The proportions of I, II and III vary according to the experimental conditions as indicated in Table 6. Thus at 65 °C in the bulk with an azo initiator, the polymer contains 63% of the ring-opened unit II, while at 80 °C in the bulk with a peroxide initiator the polymer contains 47% of the nonring-opened units I. In chlorobenzene solution (1:3) at 120 °C the terpolymer contained 51% of the ring-opened and benzaldehyde-eliminated units III. One can rationalize these results by assuming that a radical (98) is formed, which can add another monomer (96) or undergo ring opening to the radical (99). The radical (99) can either add a monomer (96) or undergo loss of benzaldehyde to give the keto radical (100). Finally, the radical (100) can add a monomer (96) to give the eliminated unit III in the terpolymer (97). Thus higher temperatures and higher dilution tend to favor ring opening and elimination.

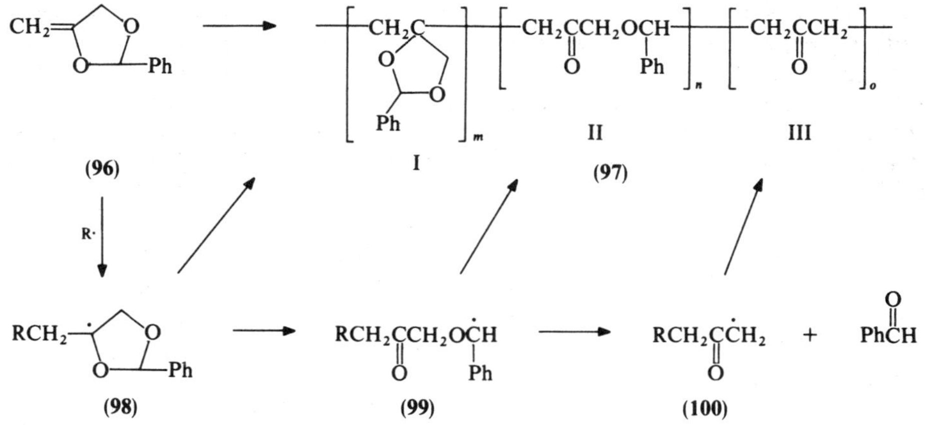

Scheme 35

Copolymerization of the 2-phenyl-1,3-dioxolane (96) with methyl methacrylate or acrylonitrile can occur either spontaneously or with a radical initiator; however, no significant amount of ring opening occurs at 55 °C or below. The monomer (96) has a very low reactivity in copolymerization with styrene.

Table 6   Polymerization of 2-Phenyl-4-methylene-1,3-dioxolane[a]

No.	Initiator (mol %)	Solvent monomer (7) solvent (7)	Temp. (°C)	Yield (%)	$\bar{M}_n^b$ (%)	Distribution in polymers		
						I	II	III
1	DTBP 3%	Bulk	120	72	1726	27	47	26
2	DTBP 3%	Chlorobenzene (1:1)	120	70	1726	14	52	36
3	DTBP 3%	Chlorobenzene (1:3)	120	51	—	14	36	51
4	BPO 3%	Bulk	80	29	1211	47	40	13
5	AIBN 3%	Bulk	65	21	1426	25	63	12

[a] Reaction time: 72 h.
[b] $\bar{M}_n$ was measured by VPO on a Colora Model 117 with toluene as solvent.

Hiraguri and Endo[72] reported that the polymerization of 2,2-diphenyl-4-methylene-1,3-dioxolane (**101**) gives only the ring-opened and eliminated product (**102**) at 120 °C as indicated in Scheme 36. These results can be rationalized by assuming that a radical adds to the monomer (**101**) to give the nonring-opened radical (**103**). At 120 °C the driving force for the ring opening to give the very stable radical (**104**) is large since there are two phenyl rings to stabilize the radical. The ring-opened radical (**104**) is sterically hindered so that addition of a monomer molecule is slow; therefore there is time for the elimination of the benzophenone molecule to form the stabilized but unhindered radical (**105**).

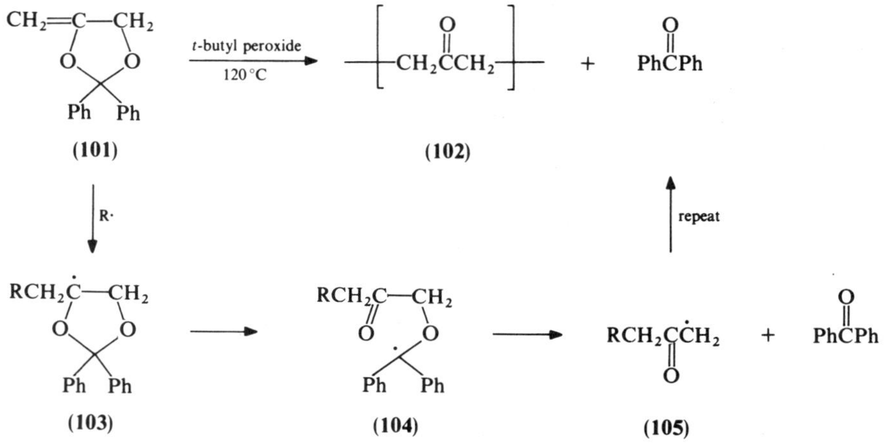

Scheme 36

In a series of patents, Endo and co-workers[73–75] describe a number of derivatives of 2-methyl-2-phenyl-4-methylene-1,3-dioxolane (**106**) which they claim undergo free radical ring-opening polymerization to give solid polymers at temperatures from 70–120 °C. However, they do not report the structure of the polymers or the extent of ring opening or elimination.

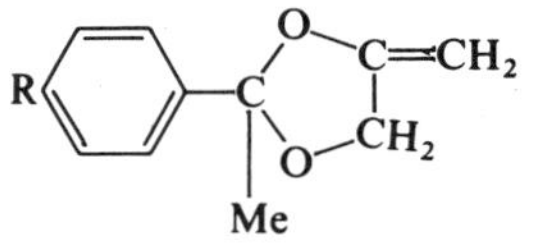

a, R = H

b, R = C₃

c, R = C≡N

(**106**)

22.9.2   2-Vinyloxiranes

In 1983, Cho and Kim[76] described the synthesis and free radical ring opening of 3-phenyl-2-vinyloxirane (**107**) as indicated in Scheme 37. The synthesis is relatively straightforward from allyl bromide, dimethylsulfide and benzaldehyde to give the 3-phenyl-2-vinyloxirane (**107**) in a 55% yield. Free radical ring-opening polymerization of the monomer (**107**) resulted in addition of a radical to the double bond followed by breakage of the carbon-to-carbon bond in the three-membered ring to produce an unsaturated ether-containing polymer (**108**).

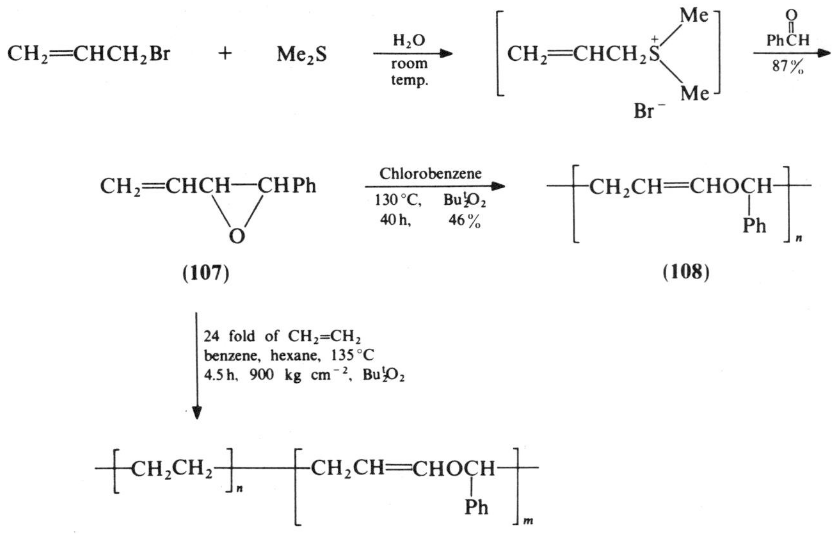

(107) **(108)**

white powder $[\eta] = 0.39$

(109)

Scheme 37

Endo and co-workers[76–78] extended this study to a series of 3-substituted-2-vinyloxiranes where the substituent could be a phenyl, substituted phenyl, α-naphthyl, β-naphthyl, nitro, acyl, acyloxy urethano or alkyl. They also copolymerized the vinyloxirane (**107**) with ethylene to produce a polyethylene containing a small amount of the ring-opened structure (**109**).

22.9.3 2-Vinyl Cyclic Sulfones

Cho and co-workers[79,80] have described the synthesis and free radical ring-opening polymerization of a series 2-vinyl cyclic sulfones. For example, as illustrated in Scheme 38, 2-vinylthiolane 1,1-dioxide (**110**) polymerized with a 100% ring opening to give polymer (**111**). The rational for this process is that a radical will add to the monomer (**110**) to give the cyclic radical (**112**), which has some ring strain. Because of this strain radical (**112**), before any addition to monomer can take place, undergoes ring opening to the linear radical (**113**). There is no indication that any sulfur dioxide is eliminated.

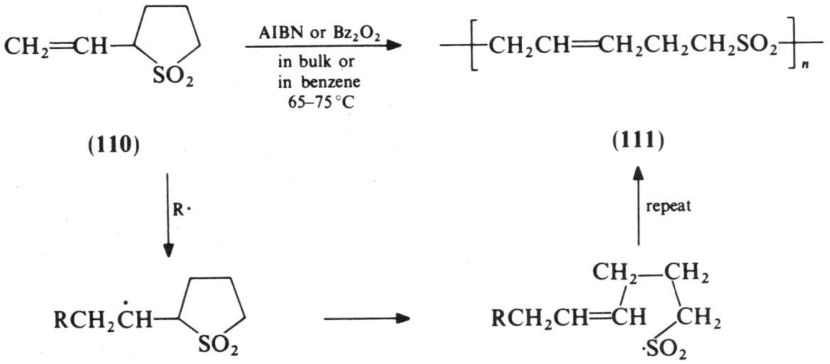

(110) **(111)** **(112)** **(113)**

Scheme 38

In a very similar reaction the 2-vinylthiane 1,1-dioxide (**114**) will undergo free radical polymerization with a very clean 1,7-ring-opening process and the 2-vinylthiepane 1,1-dioxide (**115**) will undergo a 1,9-ring-opening polymerization.

$$CH_2{=}CH \qquad\qquad\qquad CH_2{=}CH$$

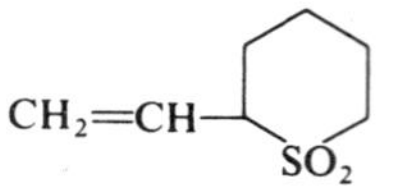

(**114**) 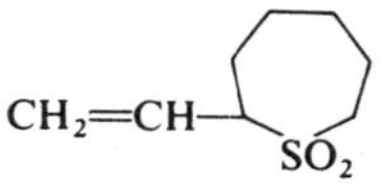   (**115**)

22.9.4   Carbocyclic Compounds

In the previous examples discussed, the three-membered rings had sufficient ring strain to undergo ring opening even in carbocyclic systems, but a five- or six-membered ring would open under radical conditions only if some other driving force was present to promote the ring opening. Since four-membered rings represent systems in between these two extremes, they appeared to be interesting cases for study. The ring opening of cyclobutylcarbinyl is exothermic [$\Delta H° = -4.0$ kcal mol^{-1} (-16.1 kJ mol^{-1}) compared to $\Delta H°$ of -5.1 kcal mol^{-1} (-20.1 kJ mol^{-1}) for the ring opening of cyclopropylcarbinyl], and the enthalpy E has been estimated at 13.0 kcal mol^{-1} (52.3 kJ mol^{-1}) corresponding to a k of about 4.5×10^3 s^{-1} at 25 °C, which is less than the values listed in Table 1 for cyclopropylcarbinyl ring opening. Finally, cyclobutylcarbinyl radicals can be observed in ESR at temperatures[82, 83] at which cyclopropylcarbinyl radicals undergo ring opening. Studies with β-pinene (**116**) show that ring opening can compete with a relatively slow abstraction reaction but cannot compete with relatively fast abstraction reactions. For example as shown in Scheme 39, β-pinene will react with carbon tetrachloride to give a ring-opened product (**117**)[84] but with thioacetic acid to give the nonring-opened product (**118**).[85]

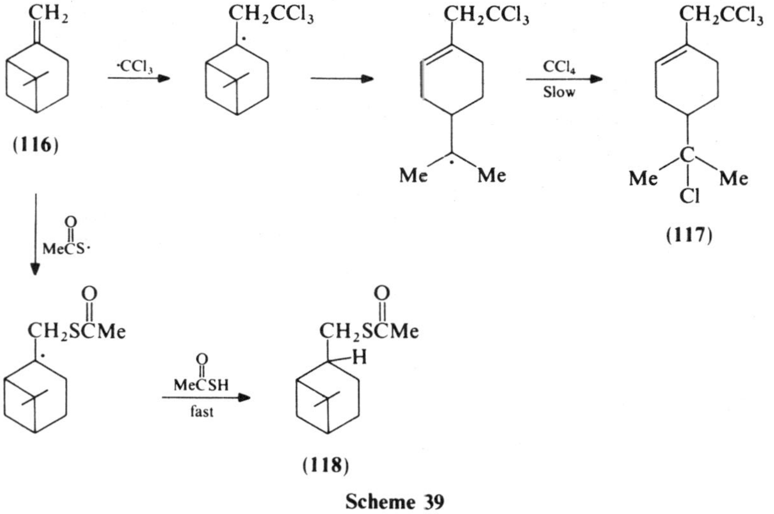

Scheme 39

It was reasoned that if β-pinene was copolymerized with a monomer in which the addition step was relatively slow and involved some steric hindrance, there was a good chance that a copolymer containing ring-opened units could be obtained. Thus maleic anhydride appeared to be an attractive choice since it does not have a high tendency to homopolymerize and is bulky enough to give considerable steric hindrance in the addition step. When equivalent quantities of β-pinene (**116**) and maleic anhydride were heated at 130 °C in the presence of di-*t*-butyl peroxide as illustrated in Scheme 40, an essentially regular alternating copolymer (**119**) containing the ring-opened β-pinene units was obtained.

22.10   FREE RADICAL DOUBLE RING-OPENING POLYMERIZATION

In a program to develop monomers that expand upon polymerization, we had prepared a series of spiro ortho esters, spiro orthocarbonates, trioxabicyclooctanes, and ketal lactones that could be

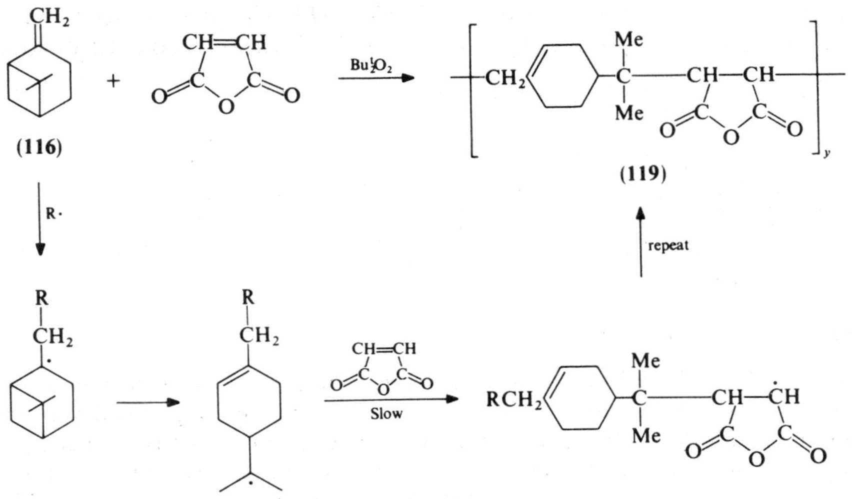

Scheme 40

polymerized with ring opening by ionic processes with expansion in volume.[86] Since a large proportion of industrial polymers are prepared by free radical polymerization, it was desirable to have available a series of monomers that would polymerize by a free radical process with no change in volume or slight expansion. Such monomers could be added to common monomers to produce copolymers with reduced shrinkage or an expansion in volume. It was reasoned that the introduction of unsaturation into a spiro orthocarbonate would permit double ring opening with expansion in volume. For that reason we undertook the synthesis of 3,9-dimethylene-1,5,7,11-tetraoxaspiro [5.5]undecane (**120**) by the set of reactions given in Scheme 41.[87]

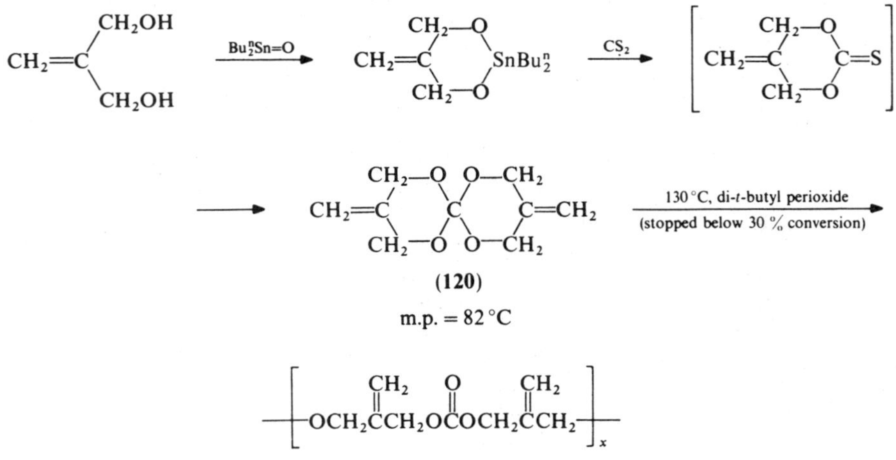

(**120**)

m.p. = 82 °C

(**121**)

Scheme 41

It was found when this monomer (**120**) was treated with di-*t*-butyl peroxide at 130 °C and the reaction stopped below 30% conversion, a soluble polymer (**121**) was obtained having the structure of a polycarbonate with pendant methylene groups. The structure of the polymer was established by elemental analysis as well as IR and NMR spectroscopy. A very similar polymer could be attained by treatment of the monomer with boron trifluoride etherate at low conversions. The mechanism of the polymerization appeared to involve a radical double ring opening according to the mechanism outlined in Scheme 42. The driving force for the double ring-opening polymerization apparently is the relief of the strain at the central spiro atom as well as the formation of the stable carbonyl group.[88] The process may be concerted all the way from the initial bicyclic radical (**122**) through to the addition of another monomer unit to form the next bicyclic radical. We know that the bicyclic radical (**122**) has a reasonable life time because it can be trapped with a thiol in making the polymer (**123**) indicated in equation (14). At 25 °C, no ring opening occurred; but at 100 °C, 50% of the rings had opened.

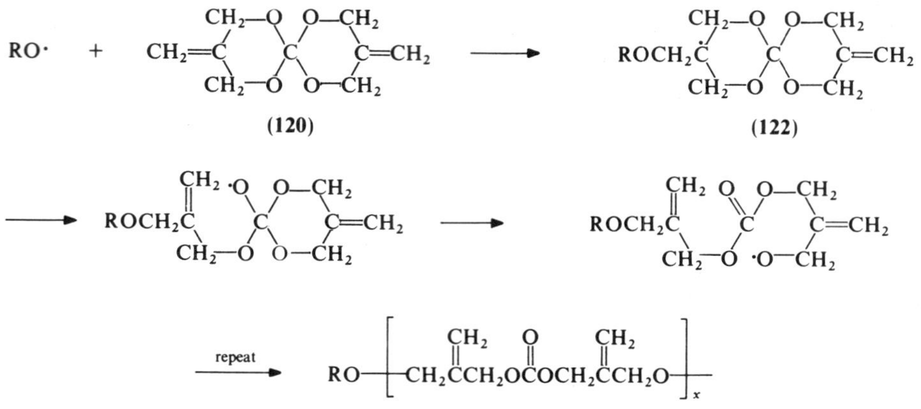

$$
RO\cdot \;+\; CH_2{=}C\!\!\underset{CH_2{-}O}{\overset{CH_2{-}O}{\diamond}}\!\!C\!\!\underset{O{-}CH_2}{\overset{O{-}CH_2}{\diamond}}\!\!C{=}CH_2 \quad\longrightarrow\quad ROCH_2\dot{C}\!\!\underset{CH_2{-}O}{\overset{CH_2{-}O}{\diamond}}\!\!C\!\!\underset{O{-}CH_2}{\overset{O{-}CH_2}{\diamond}}\!\!C{=}CH_2
$$

(120)   (122)

$$
\longrightarrow\quad ROCH_2\underset{CH_2{-}O}{\overset{CH_2\;\cdot O}{C}}\!\!C\!\!\underset{O{-}CH_2}{\overset{O{-}CH_2}{\diamond}}\!\!C{=}CH_2 \quad\longrightarrow\quad ROCH_2\underset{CH_2{-}O}{\overset{CH_2\;O}{C}}\!\!C\!\!\underset{\cdot O{-}CH_2}{\overset{O{-}CH_2}{\diamond}}\!\!C{=}CH_2
$$

$$
\xrightarrow{\text{repeat}}\quad RO{-}\!\left[CH_2\overset{CH_2}{\underset{\;}{\overset{\|}{C}}}CH_2O\overset{O}{\overset{\|}{C}}OCH_2\overset{CH_2}{\underset{\;}{\overset{\|}{C}}}CH_2O\right]_x
$$

(121)

Scheme 42

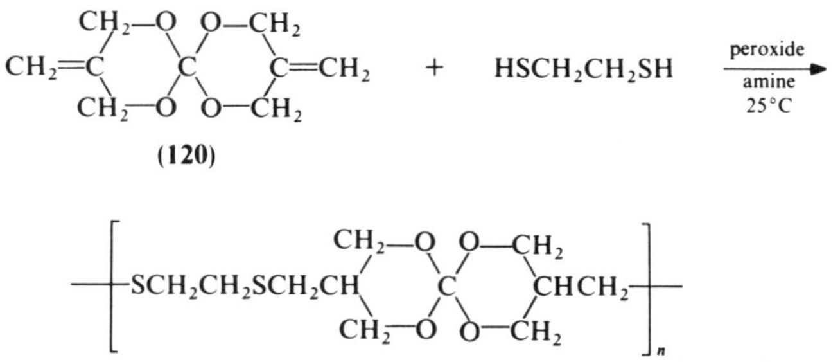

$$
CH_2{=}C\!\!\underset{CH_2{-}O}{\overset{CH_2{-}O}{\diamond}}\!\!C\!\!\underset{O{-}CH_2}{\overset{O{-}CH_2}{\diamond}}\!\!C{=}CH_2 \;+\; HSCH_2CH_2SH \quad\xrightarrow[\substack{\text{amine}\\25\,^\circ C}]{\text{peroxide}}
$$

(120)

$$
\left[-SCH_2CH_2SCH_2CH\!\!\underset{CH_2{-}O}{\overset{CH_2{-}O}{\diamond}}\!\!C\!\!\underset{O{-}CH_2}{\overset{O{-}CH_2}{\diamond}}\!\!CHCH_2-\right]_n \tag{14}
$$

(123)

At high conversion this monomer (120) produced a highly crosslinked resin, very similar in appearance to the material produced from the polymerization of diallyl carbonate. The volume change that occurred during homopolymerization was quite unusual. A room temperature, a 4.3% expansion in volume occurred, while just below its melting point at 70 °C, a 7% expansion in volume occurred; at 85 °C, a 2% expansion took place; and the expansion decreased until above 115 °C a slight shrinkage occurred. It is obvious from these data that the large expansion in volume that occurs below the melting point involves not only the increase in volume due to the double ring opening but also a change in volume of 3–6% due to the process of going from a crystalline monomer to a liquid monomer. Since the monomer is a crystalline solid, it is difficult to find examples of homopolymerization in which the full 7% expansion in volume can be utilized. However, in copolymerization it is possible to use a slurry of the crystalline monomer in liquid monomer so that as copolymerization progresses, the crystalline monomer dissolves with some expansion and also polymerizes with expansion.[89-94]

The dimethylene spiro orthocarbonate (DSOC, 120) was shown to copolymerize with a variety of monomers, including methyl methacrylate, hydroxyethyl methacrylate, styrene and diallyl carbonate. With methyl methacrylate, monomer (120) was shown to be less reactive with reactivity ratios of $r_1 = 0.87$ and $r_2 = 16.4$. Since $r_1 r_2$ is greater than unity, this fact was interpreted as involving association of DSOC (120) in the monomer mixture rather than existing in a homogeneous solution. In this copolymerization, if the monomer mixture contained 10% of the bicyclo monomer (120) and 90% methyl methacrylate and the reaction were stopped at 69% conversion, the copolymer contained only 1% of the carbonate units.

As discussed earlier, the free radical ring-opening copolymerization of the dimethylene monomer (120) with methyl methacrylate introduces a functional group into the backbone of an addition polymer as indicated in equation (15). It was obvious that the introduction of even a small amount of this unit would greatly affect the chemical and physical properties of the base polymer. In fact, the copolymers (124) of methyl methacrylate and the bicyclic monomer (120) gave data for their thermal decomposition almost indentical to those listed in Table 4.

Since the carbonate group is a relatively easy group to hydrolyze, several other uses of these copolymers were suggested. As depicted in equation (16), a copolymer (125) of the dimethylene monomer (120) and styrene was hydrolyzed to produce a hydroxyl-terminated polystyrene (126).

$$CH_2=C \underset{CH_2-O}{\overset{CH_2-O}{\underset{|}{\overset{|}{C}}}} \underset{O-CH_2}{\overset{O-CH_2}{\underset{|}{\overset{|}{C}}}} C=CH_2 \quad + \quad CH_2=C \overset{CO_2Me}{\underset{Me}{}} \xrightarrow{\text{peroxide}}$$

(120)

$$\left[-CH_2\underset{\underset{C=O}{\overset{Me}{|}}}{\overset{Me}{\underset{|}{C}}} - \right]_x -CH_2\overset{CH_2}{\underset{}{\overset{||}{C}}}CH_2O\overset{O}{\underset{}{\overset{||}{C}}}OCH_2\overset{CH_2}{\underset{}{\overset{||}{C}}}CH_2O- \left[-CH_2\underset{\underset{C=O}{\overset{Me}{|}}}{\overset{Me}{\underset{|}{C}}} - \right]_y \qquad (15)$$

(124)

This process is an extension of the work discussed in Section 22.5.2 on the synthesis of functionally terminated oligomers.

$$-CH_2\overset{CH_2}{\overset{||}{C}}CH_2O\overset{O}{\overset{||}{C}}OCH_2\overset{CH_2}{\overset{||}{C}}CH_2O-\left[-CH_2CH-\right]_x\overset{}{\underset{Ph}{}}-CH_2\overset{CH_2}{\overset{||}{C}}CH_2O\overset{O}{\overset{||}{C}}OCH_2\overset{CH_2}{\overset{||}{C}}CH_2O- \xrightarrow{\text{hydrolyze}}$$

(125)

$$\text{(16)}$$

$$HOCH_2\overset{CH_2}{\overset{||}{C}}CH_2O-\left[-CH_2CH-\right]_x\overset{}{\underset{Ph}{}}-CH_2\overset{CH_2}{\overset{||}{C}}CH_2OH$$

(126)

Still another use of the presence of a functional group within the backbone of an addition polymer is to render the copolymer biodegradable. When the dimethylene monomer (**120**) and hydroxyethyl methacrylate (HEMA) were copolymerized to produce a copolymer (**127**) containing 14 mol % of the ring-opened units, hydrolysis of the copolymer in an alcoholic solution containing 1% sodium hydroxide gave, within 3 h at room temperature, a copolymer (**128**) with one fifth of the original viscosity average molecular weight as indicated in equation (17). Preliminary results indicate that this copolymer is biodegraded by microorganisms. For use in the human body, complete biodegradability may not be necessary since polymeric materials with molecular weights below 5000 can be eliminated from the body. Although some hydrophilic character is necessary for fast biodegradation, the introduction of a sufficient number of such functional groups in the backbone may render even hydrophobic polymers, such as polystyrene, slowly biodegradable.

A potential use of this monomer is in the area of dental fillings in which a slurry containing 20% of very fine crystals of the unsaturated spiro orthocarbonate (**120**) and 60% of the adduct of methacrylic acid to bisphenol-A diglycidyl ether (Bis-GMA) plus 20% trimethylolpropane trimethacrylate produces a material with essentially no change in volume upon polymerization. An investigation of a bubble test on tooth enamel showed that this copolymer had nearly double the adhesion to the tooth structure that the base resin had without the addition of the unsaturated spiro orthocarbonate. The copolymer also had improved impact strength but yet essentially the same modulus, and filled composites appeared to have somewhat improved abrasion resistance.[95]

Since the synthesis of the spiro orthocarbonates through the tin compounds could be modified to produce unsymmetrical materials, we undertook the synthesis of the unsymmetrical 2-methylene-1,5,7,11-tetraoxaspiro[5.5]undecane (**129**) by the set of reactions listed in Scheme 43. The resulting monomer (**129**) was a crystalline solid with a melting point of 61–62 °C. When the polymerization was carried out in the presence of di-*t*-butyl peroxide and the reaction was stopped at low conversion, a linear polycarbonate containing pendant methylene groups was obtained. The structure of the polymer (**130**) which was established by elemental analysis as well as IR and NMR spectroscopy, was very similar to that of the polymer that could be obtained by the ionic polymerization of this same monomer at low conversions. Bulk polymerization of monomer (**129**) with peroxide catalyst gave a material (**130**) at 25 °C with an expansion of 4.5% and at 60 °C an

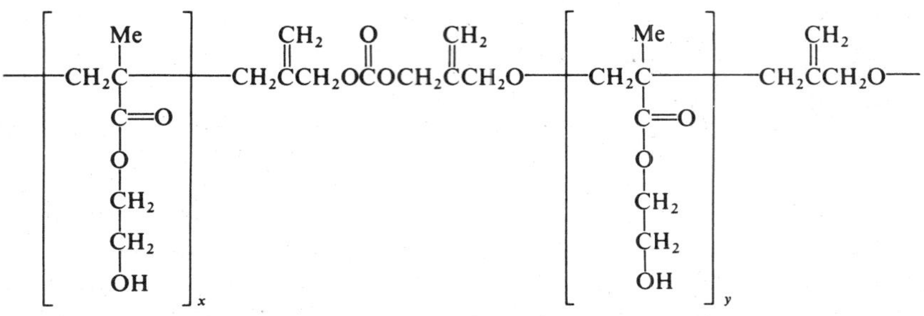

(127)

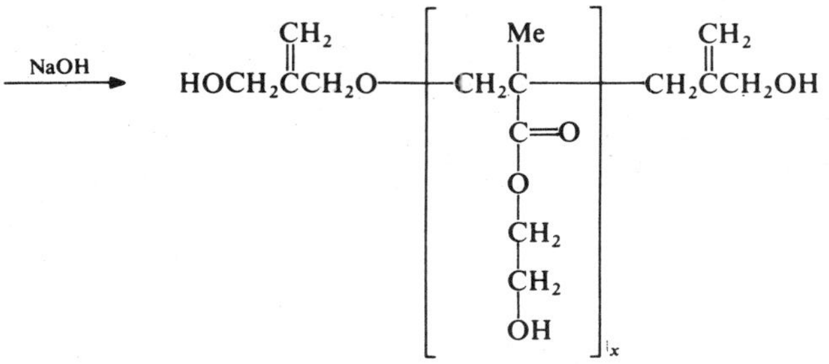

$$\xrightarrow{\text{NaOH}}$$ (17)

(128)

expansion of 5.5%; above the melting point of the polymer (130) (61–62 °C) the expansion decreased until at 111 °C the density of the monomer and the density of the polymer were the same.

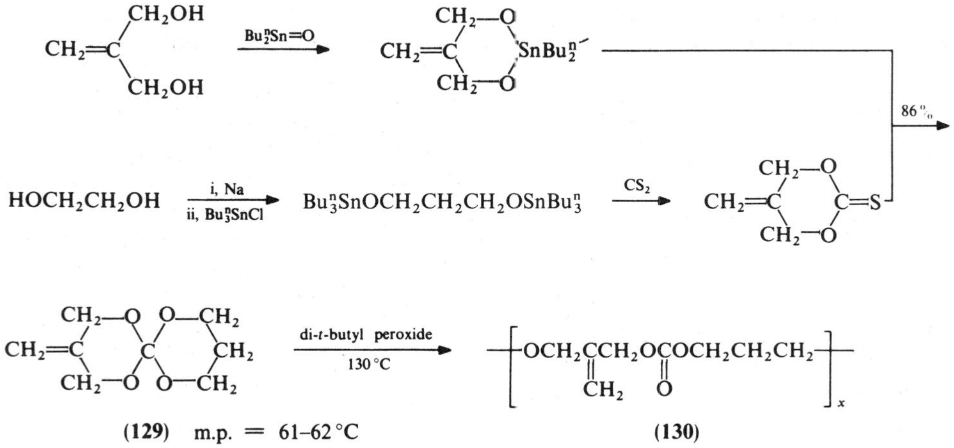

$$\xrightarrow{\text{Bu}_2^n\text{Sn=O}}$$

86%

$$\text{HOCH}_2\text{CH}_2\text{OH} \xrightarrow[\text{ii, Bu}_3^n\text{SnCl}]{\text{i, Na}} \text{Bu}_3^n\text{SnOCH}_2\text{CH}_2\text{CH}_2\text{OSnBu}_3^n \xrightarrow{\text{CS}_2}$$

$$\xrightarrow[130\,^\circ\text{C}]{\text{di-}t\text{-butyl peroxide}} \left[\text{OCH}_2\overset{\underset{\text{CH}_2}{|}}{\text{C}}\text{CH}_2\text{OCOCH}_2\text{CH}_2\text{CH}_2 \right]_x$$

(129)   m.p. = 61–62 °C   (130)

$$[\eta]^{25\,^\circ\text{C}}_{\text{CHCl}_3} = 0.11$$

Scheme 43

More recently an alternative method for the synthesis of unsymmetrical unsaturated spiro orthocarbonate has been developed as indicated in Scheme 44.[96-101] In this procedure a diol is treated with thiophosgene in the presence of dimethylaminopyridine (DMAP) to give the intermediate cyclic thiocarbonate (131). Treatment of this thiocarbonate (131) with a cyclic tin ester (132) gives the desired unsymmetrical spiro orthocarbonate (133). A variety of these spiro orthocarbonates have been prepared by this procedure in yields up to 90%. When the monomer (133) was polymerized at 120 °C, a polymer (134) was obtained in which 96% of the rings had opened. Use of this material (133) in dental composite resins containing Bis-GMA reduced the shrinkage and greatly increased the adhesion.

Since a variety of unsaturated heterocyclic monomers will undergo ring-opening polymerization quite readily, it appeared undesirable to prepare a related spiro ortho ester for use as a monomer with either no change in volume or slight expansion. Thus an unsaturated spiro ortho ester was synthesized as illustrated in Scheme 45. Polymerization of the 2-methylene-1,4,6-trioxaspiro[4.4] nonane (135) with benzoyl peroxide gave a polymer (136) with a complex structure that indicated that at 120 °C only about 10% ring opening had occurred. Apparently the radical adds to the double bond to give the intermediate (137) which can add to another monomer or undergo ring opening to

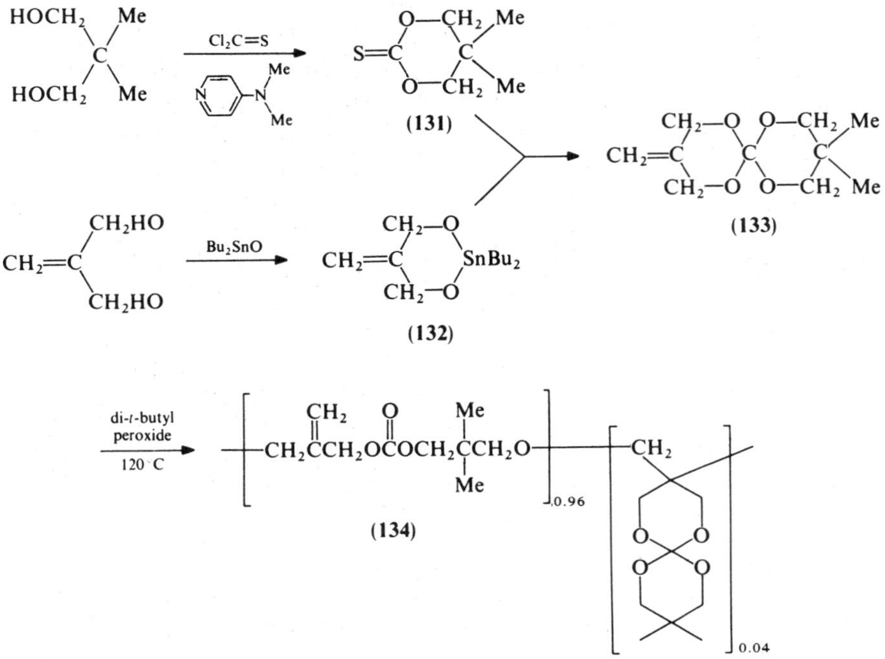

Scheme 44

give the semiopened radical (138). The radical (138), in turn, can either add to a monomer or undergo a second ring opening to give the double ring-opened radical (139). In fact the final polymer (136) has about 90% unopened units and about equal quantities of the semiopened and double-opened units.

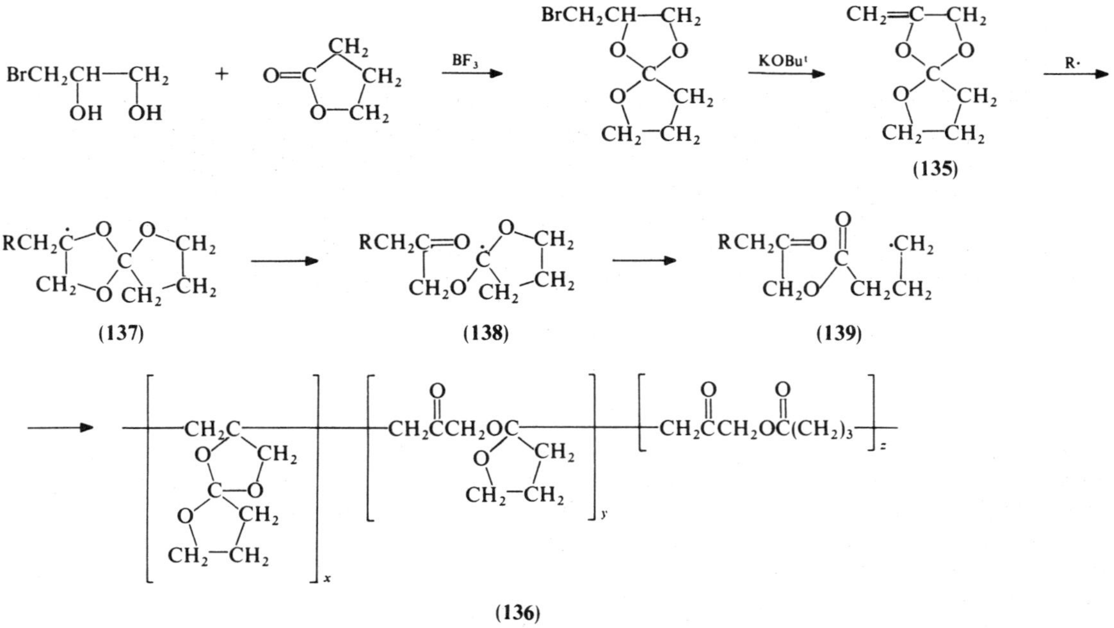

Scheme 45

When the related unsaturated spiro ortho ester (140) derived from caprolactone was polymerized, it gave a similar structure containing slightly more of the double ring-opened units but still about 90% of the unopened units. However, when the strained unsaturated spiro ortho ester (141) was polymerized, as indicated in Scheme 46, essentially complete double ring opening occurred. Apparently the intermediate radical (142) has sufficient strain to promote the ring opening. The resulting polymer (143) is intriguing since it contains both ester groups and keto groups. Such a polymer should be both biodegradable and photodegradable.

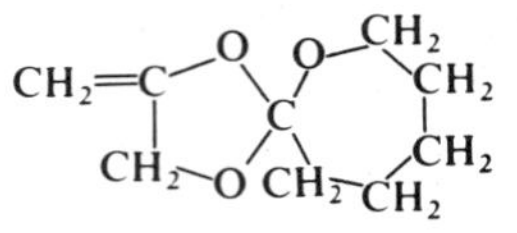

(140)

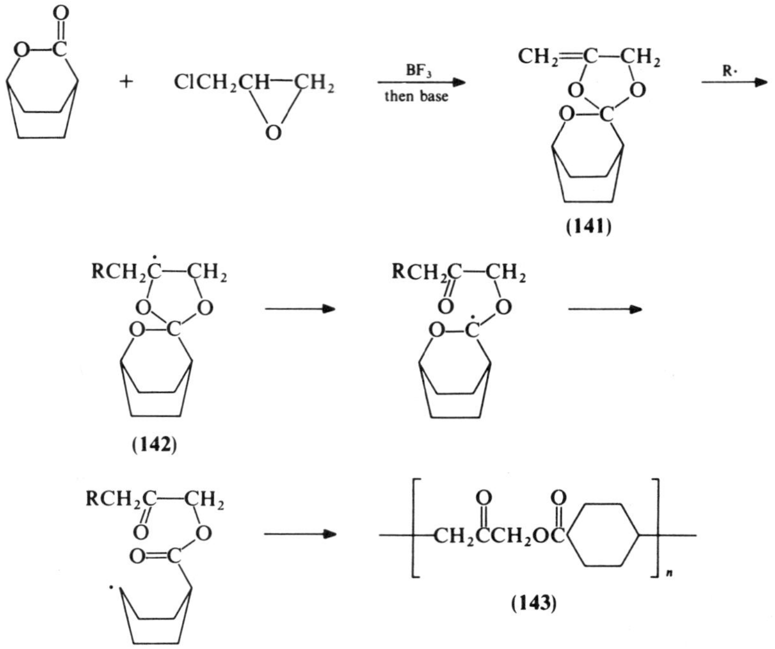

(141)

(142)

(143)

Scheme 46

A very closely related monomer (**144**), apart from the fact that it possessed a spiro orthocarbonate linkage rather than the spiro ortho ester linkage, gave essentially quantitative ring opening to yield the polymer (**145**), as indicated in Scheme 47.[101] Thus the chlorodiol was treated with thiophosgene to give the cyclic thione carbonate (**146**) which was then treated with the cyclic tin ester to give the chloro spiro compound (**147**), which on treatment with base gave the monomer (**144**). Treatment of the spiro orthocarbonate (**144**) with a peroxide at 120 °C gave the linear polymer (**145**) with 100% ring opening. Apparently the additional strain of the spiro orthocarbonate linkage was enough to promote the high extent of ring opening. Since polymer (**145**) has both a keto group and a carbonate group, it should be both photo- and bio-degradable.

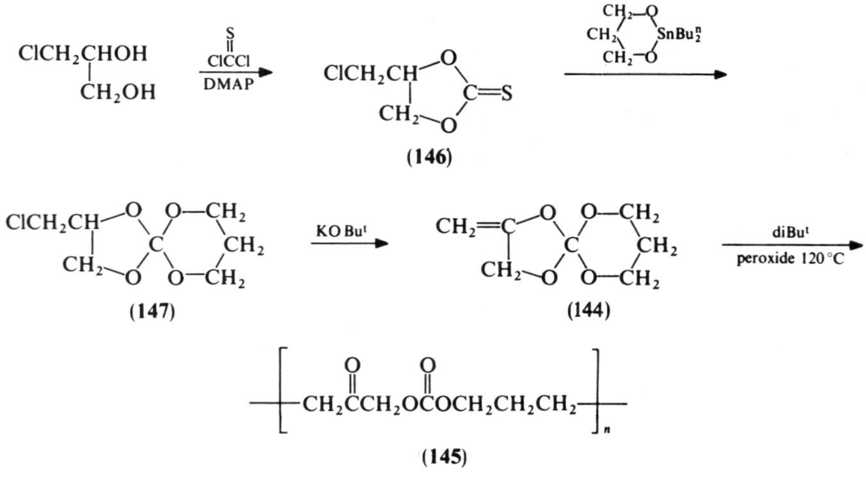

(146)

(147)   **(144)**

(145)

Scheme 47

Recently Han and Choi[102] reported that free radical polymerization of 4′-methylenespiro-[2-benzofuran-2,2′-(1,3-dioxolane)] (**148**) underwent partial ring opening to give the polymer (**149**) as indicated in Scheme 48. However, on reinvestigation we established the structure of the product

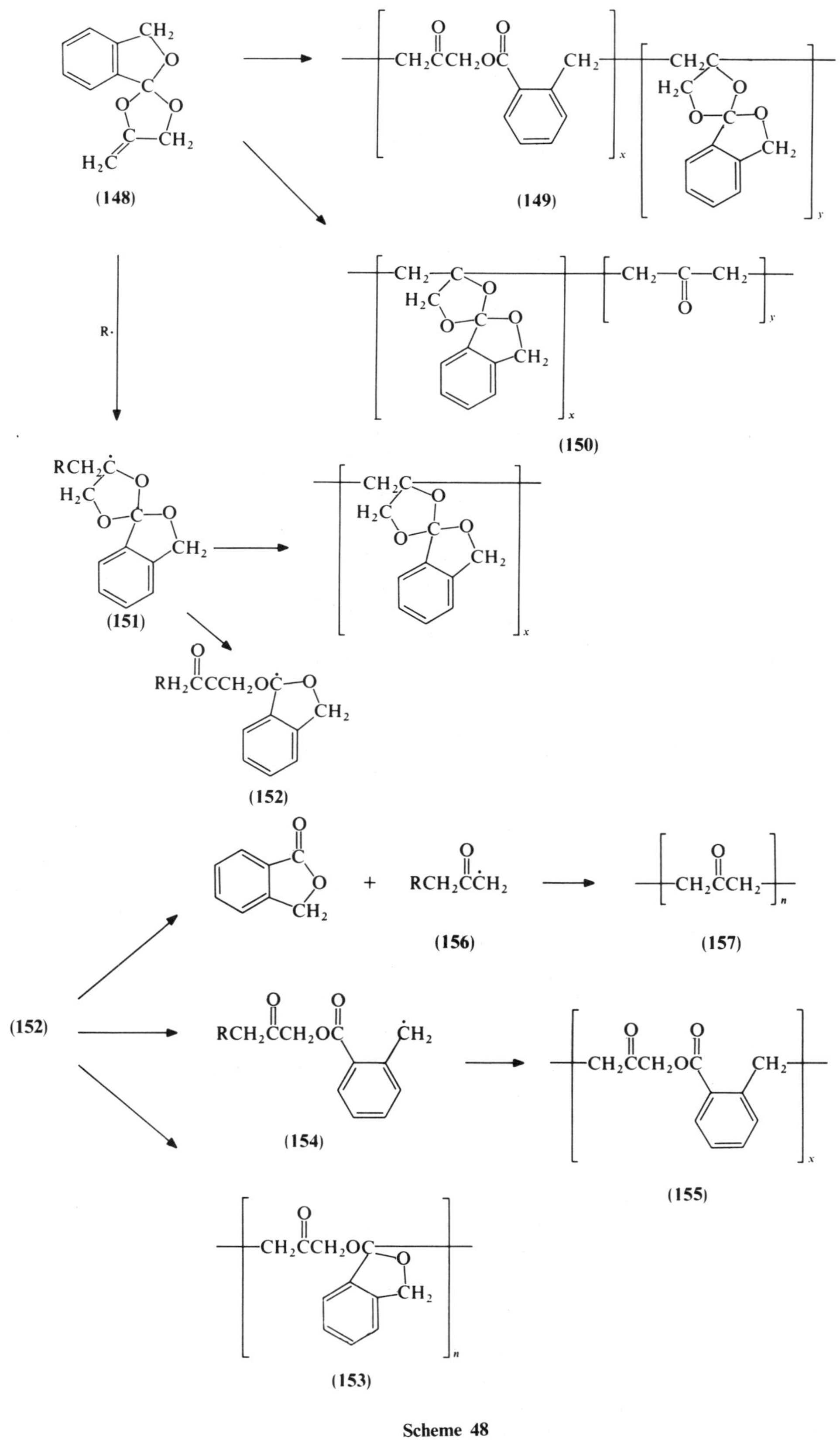

Scheme 48

as a material (**150**) which resulted from a partial ring opening followed by elimination of a molecule of phthalide.[103] The sequence of reactions can be rationalized by assuming that a radical can add to the monomer (**148**) to give the nonring-opened radical (**151**) which either can add another monomer or undergo partial ring opening to radical (**152**). Radical (**152**), which is a benzyl radical stabilized by the two adjacent oxygen atoms, has three choices: (1) it can add a monomer to introduce a partially ring-opened unit (**153**) into the polymer; (2) it can undergo a second ring opening to produce the linear radical (**154**) which will introduce the double ring unit (**155**) into the polymer chain; or (3) it can eliminate a molecule of phthalide to produce the stabilized ketone radical (**156**) which will introduce the keto unit (**157**) into the polymer chain. Spectral evidence has indicated that choice (3) is the dominant pathway.

Since it was previously demonstrated that aromatization could be used to promote free radical ring-opening polymerization, it was of interest to find a tricyclic monomer that would undergo double ring opening with the formation of an aromatic ring as the driving force for the ring opening. For this reason the synthesis of the dispiro monomer (**161**) was undertaken by the procedure indicated in Scheme 49.[104-106] Thus the dispiro acetal (**158**) was converted to the dichloro derivative (**159**) in an 89% yield. Bromination of the acetal (**159**) gave a mixture of bromides from which a 14% yield of the crystalline dibromide (**156**) could be isolated.[107] Treatment of the dibromide (**160**) with potassium *t*-butoxide and Aliquot 336 gave an 81% yield of the desired monomer (**161**). Treatment of the crude mixture of bromides gave a much higher overall yield of the monomer (**161**). Polymerization of monomer (**161**) either in bulk at 130 °C or in refluxing benzene solution with *t*-butyl peroxide gave the crystalline polymer (**162**) with an $[\eta] = 0.32$ (DMF at 82 °C). The bulk polymerization was shown to undergo an expansion in volume at 130 °C of about 2%. The spectral evidence indicated that complete double ring opening had taken place.

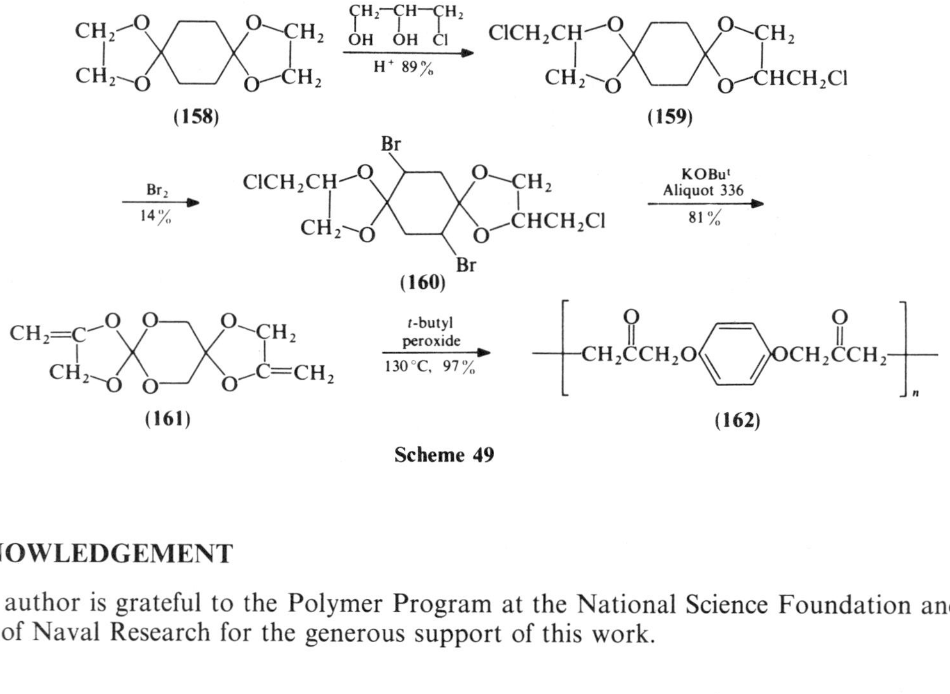

Scheme 49

ACKNOWLEDGEMENT

The author is grateful to the Polymer Program at the National Science Foundation and to the Office of Naval Research for the generous support of this work.

22.11   REFERENCES

1. G. B. Butler and R. J. Angelo, *J. Am. Chem. Soc.*, 1957, **79**, 3128.
2. T. Takahashi, *J. Polym. Sci., Part A*, 1968, **6**, 403.
3. I. Cho and K. D. Ahn, *J. Polym. Sci., Polym. Lett. Ed.*, 1977, **15**, 751.
4. L. A. Errede, *J. Polym. Sci.*, 1961, **49**, 253.
5. H. J. Hall, Jr. and P. Ykman, *Macromol. Rev.*, 1976, **11**, 1.
6. A. V. Tobolsky and A. Eisenberg, *J. Am. Chem. Soc.*, 1959, **81**, 780.
7. B. Maillard, D. Forrest and K. U. Ingold, *J. Am. Chem. Soc.*, 1976, **98**, 7024.
8. A. Streitwieser, Jr. and C. H. Heathcock, 'Introduction to Organic Chemistry', 2nd ed., Macmillan, New York, 1981, p. 1195.
9. B. Maillard, M. Cazaux and R. Lalande, *Bull. Soc. Chim. Fr.*, 1973, 1368.
10. S. M. McElvain and M. J. Curry, *J. Am Chem. Soc.*, 1948, **70**, 3781.

11. P. R. Johnson, H. M. Barnes and S. M. McElvain, *J. Am. Chem. Soc.*, 1940, **62**, 964.
12. W. J. Bailey, *Polym. J.*, 1985, **17**, 85.
13. W. J. Bailey, Z. Ni and S. -R. Wu, *J. Polym. Sci., Polym. Chem. Ed.*, 1982, **20**, 2420.
14. T. Endo, M. Okawara, W. J. Bailey, K. Azuma, K. Nate and H. Yokono, *J. Polym. Chem., Polym. Lett. Ed.*, 1983, **21**, 373.
15. W. J. Bailey, S. -R. Wu and Z. Ni, *J. Macromol. Sci., Chem.*, 1982, **18**(6), 973.
16. W. J. Bailey, Z. Ni and S. -R. Wu, *Macromolecules*, 1982, **15**, 711.
17. W. J. Bailey, S. -R. Wu and Z. Ni, *Makromol. Chem.*, 1982, **183**, 1913.
18. Z. Ni and W. J. Bailey, *Kao Fer Tzu Hsueh Pao*, 1987, 379 (*Chem. Abstr.*, 1988, **108**, 132 401q).
19. F. W. Billmeyer, Jr., 'Textbook of Polymer Science', Wiley-Interscience, New York, 1971, p. 395.
20. R. Tsang, J. K. Dickson, Jr., H. Pak, R. Walton and B. Fraser-Reid, *J. Am. Chem. Soc.*, 1987, **109**, 3484.
21. W. J. Bailey, A. Arfaei, P. Y. Chen, S. -C. Chen, T. Endo, C. -Y. Pan, Z. Ni, S. Shaffer, L. Sidney, S. -R. Wu and N. Yamazaki, *Proc. IUPAC 28th Macromol. Symp.*, Amherst, MA, 1982, July 12–16, 214.
22. L. Sidney, S. E. Shaffer and W. J. Bailey, *Polym. Prepr., Am. Chem. Soc., Div. Polym. Chem.*, 1981, **22**(2), 373.
23. W. J. Bailey, *ACS Symp. Ser.*, 1985, **286**, 47.
24. A. H. Weinstein, A. J. Constanza and G. E. Meyer (Goodyear Tire and Rubber Co.), *Fr. Pat.* 1 434 167 (1966) (*Chem. Abstr.*, 1967, **66**, 76 764k).
25. W. J. Bailey, B. Gapud, Y. -N. Lin, Z. Ni and S. -R. Wu, *Polym. Preper., Am. Chem. Soc., Div. Polym. Chem.*, 1984, **25**(1), 142.
26. W. J. Bailey, T. Endo, B. Gapud, Y. -N. Lin, Z. Ni, C. -Y. Pan, S. E. Shaffer, S. -R. Wu, N. Yamazaki and K. Yonezawa, *J. Macromol. Sci., Chem.*, 1984, **21**, 979.
27. W. J. Bailey, *Makromol. Chem., Suppl.*, 1985, **13**, 171.
28. W. J. Bailey, B. Gapud, Y. -N. Yin, Z. Ni and S. -R. Wu, *ACS Symp. Ser.*, 1985, **282**, 147.
29. J. E. Potts, R. A. Clendinning, W. B. Ackart and W. D. Niegisch, in 'Polymers and Ecological Problems', ed. J. Guillet, Plenum Press, New York, 1973, p. 61.
30. W. J. Bailey, *Proc. 3rd Int. Conf. Adv. Stab. Control. Deg. of Polym.*, Lucerne, Switzerland, 1981, June 1, 12.
31. W. J. Bailey and B. Gapud, *Polym. Prepr., Am Chem. Soc., Div. Polym. Chem.*, 1984, **27**(1), 58.
32. W. J. Bailey, *Proc. 6th Int. Conf. Adv. Stab. Control. Deg. Polym.*, Lucerne, Switzerland, 1984, May 22–4, 38.
33. W. J. Bailey and B. Gapud, *Ann. N. Y. Acad. Sci.*, 1985, **446**, 42.
34. W. J. Bailey and B. Gapud, *ACS Symp. Ser.*, 1985, **280**, 423.
35. A. -C. Albertson and B. Ranby, in 'Proceedings of the Third International Biodegradation Symposium', ed. J. M. Sharpley and A. M. Kaplan, Applied Science, Essex, 1976, p. 743.
36. D. G. Corbin, cited by T. J. Henman, *Proc. 3rd. Int. Conf. Adv. Stab. Control. Deg. Polym.*, Lucerne, Switzerland, 1981, June, 116.
37. D. Ennis and A. Kramer, *Lebensm.-Wiss. Technol.*, 1974, **7**(4), 214.
38. D. Ennis and A. Kramer, *J. Food Sci.*, 1974, **40**, 181.
39. W. J. Bailey, P. Y. Chen, S. -C. Chen, W. -B. Chiao, T. Endo, B. Gapud, Y. -N. Lin, Z. Ni, C. -Y. Pan, S. E. Shaffer, L. Sidney, S. -R. Wu, N. Yamamoto, N. Yamazaki and K. Yonezawa, *J. Macromol. Sci., Chem.*, 1984, **21**, 1511.
40. A. J. Dobbs, B. C. Gilbert and R. O. C. Norman, *J. Chem. Soc. A*, 1971, 124; *J. Chem. Soc., Perkin Trans. 2*, 1972, 786.
41. P. Hudrlik, A. Hudrlik and C. Wan, *J. Org. Chem.*, 1975, **40**, 1116.
42. P. Columbo, M. Steinberg and J. Fontana, *J. Polym. Sci., Polym. Lett. Ed.*, 1973, **11**, 447.
43. W. J. Bailey and P. -Z. Feng, *Polym. Prepr., Am. Chem. Soc., Div. Polym. Chem.*, 1987, **28**(1), 154.
44. V. R. Likhterov, V. S. Ellis and L. A. Ternovskii, *Khim. Geterotsikl. Soedin.*, 1985, (10), 1316 (*Chem. Abstr.*, 1986, **104**, 186 333e).
45. P. Tundo, *J. Am. Chem. Soc.*, 1982, **104**, 457.
46. D. Kippenberger, K. Rosenquist, L. Odberg, P. Tundo and J. H. Fendler, *J. Am. Chem. Soc.*, 1983, **105**, 1129.
47. K. Dorn, R. T. Klingbiel, D. P. Specht, P. N. Tyminski, H. Ringsdorf and D. F. O'Brien, *J. Am. Chem. Soc.*, 1984, **106**, 1627.
48. M. Singh, K. Yamaguchi and S. L. Regen, *J. Am. Chem. Soc.*, 1985, **107**, 42.
49. J. Serrano, S. Mucino, S. Millan, R. Reynoso, L. A. Fucugauchi, W. Reed, F. Nome, P. Tundo and J. H. Fendler, *Macromolecules*, 1985, **18**, 1990.
50. M. F. M. Roks, R. S. Dezentje, V. E. M. Kaats-Richters, W. Drenth, A. J. Verkleij and R. J. M. Nolte, *Macromolecules*, 1987, **20**, 920.
51. H. Ohno, Y. Ogata and E. Tsuchida, *Macromolecules*, 1987, **20**, 929.
52. S. K. Chang, D. H. Kim and I. Cho, *Chem. Lett.*, 1987, **7**, 1385.
53. W. J. Bailey and L. -L. Zhou, *Polym. Prepr., Am. Chem. Soc., Div. Polym. Chem.*, 1988, **29**, in press.
54. R. Elbert, A. Laschewsky and H. Ringsdorf, *J. Am. Chem. Soc.*, 1985, **107**, 4134.
55. A. Laschewsky, H. Ringsdorf, G. Schmidt and J. Schneider, *J. Am. Chem. Soc.*, 1987, **109**, 788.
56. W. J. Bailey and L. -L. Zhou, *Polym. Prepr., Am. Chem. Soc., Div. Polym. Chem.*, 1988, **29**, in press.
57. S. Kirchmeyer, A. Mertens, M. Arvanaghi and G. A. Olah, *Synthesis*, 1983, 498.
58. W. J. Bailey, P. Y. Chen, S. -C. Chen, W. -B. Chiao, T. Endo, B. Gapud, V. Kuruganti, Y. -N. Lin, Z. Ni, C. -Y. Pan, S. E. Shaffer, L. Sidney, S. -R. Wu, N. Yamamoto, N. Yamazaki, K. Yonezawa and L. -L. Zhou, *Makromol. Chem., Macromol. Symp.*, 1986, **6**, 81.
59. V. V. Kane, *Synth. Commun.*, 1976, **6**, 237.
60. R. Greenwald, M. Chaykovsky and E. J. Corey, *J. Org. Chem.*, 1963, **28**, 1128.
61. D. F. Murray, *J. Org. Chem.*, 1983, **48**, 4860.
62. K. Alder and M. Fremery, *Tetrahedron*, 1961, **34**, 1651.
63. L. A. Auspos, C. W. Burman, L. A. R. Hall, J. K. Hubbard, W. Kirk, Jr., J. R. Schaefgen and S. B. Speck, *J. Polym. Sci.*, 1955, **15**, 19.
64. W. J. Bailey, M. J. Amone and J. L. Chou, *Polym. Prepr., Am. Chem. Soc., Div. Polym. Chem.*, 1988, **29**(1), 178.
65. W. J. Bailey and J. L. Chou, in 'Applications of Polymers', ed. R. B. Seymour and H. F. Mark, Plenum Press, New York, 1988, p. 79.
66. W. J. Bailey, J. L. Chou, P. -Z. Feng, V. Kuruganti and L. -L. Zhou, *Acta Polym.*, 1988, **39**, in press.
67. W. J. Bailey, *IUPAC 32nd Int. Symp. Macromol.*, Kyoto, Japan, Prepr., 1988, Aug. 1–6.

68. W. J. Bailey and J. L. Chou, *Polym. Mater. Sci. Eng.*, 1987, **55**, 30.
69. W. J. Bailey, J. L. Chou, P. -Z. Feng, V. Kuruganti and L. -L. Zhou, *IUPAC Macromol. Symp., Merseberg, East Germany, Prepr.*, 1987, July 1–4.
70. C. -Y. Pan, Z. Wu and W. J. Bailey, *J. Polym. Sci., Polym. Lett. Ed.*, 1987, **25**(6), 243.
71. C. -Y. Pan, Z. Wu, and W. J. Bailey, *J. Macromol. Sci., Chem.*, 1987, **25**(1), 27.
72. Y. Hiraguri and T. Endo, *J. Am. Chem. Soc.*, 1987, **109**, 3779.
73. K. Azuma, K. Nate and T. Endo (Hitachi), *Jpn. Pat.* 61 68 485 (1986) (*Chem. Abstr.*, 1986, **105**, P227 596j).
74. K. Azuma, K. Nate and T. Endo (Hitachi), *Jpn. Pat.* 61 249 979 (1986) (*Chem. Abstr.*, 1987, **106**, P214 561g).
75. K. Azuma, K. Nate and T. Endo (Hitachi), *Jpn. Pat.* 61 118 379 (1986) (*Chem. Abstr.*, 1987, **107**, P40 598k).
76. I. Cho and J. -B. Kim, *J. Polym. Sci., Polym. Lett. Ed.*, 1983, **21**, 433.
77. N. Kanda and T. Endo (Toa Gosei Chemical Industry Co.), *Jpn. Pat.* 61 85 374 (1986) (*Chem. Abstr.*, 1986, **105**, P173 212k).
78. Y. Origasa, S. Kojima, K. Suga and T. Endo (Nippon Petrochemicals Co.), *Jpn. Pat.* 62 260 821 (1987) (*Chem. Abstr.*, 1988, **108**, P132 506c).
79. Y. Origasa, S. Kojima, K. Suga and T. Endo (Nippon Petrochemicals Co.), *Jpn. Pat.* 62 280 218 (1987) (*Chem. Abstr.*, 1988, **108**, P151 166h).
80. I. Cho, S. K. Kim and M. H. Lee, *J. Polym. Sci., Polym. Symp.*, 1986, **74**, 219.
81. I. Cho and M. H. Lee, *J. Polym. Sci., Polym. Lett. Ed.*, 1987, **25**(8), 309.
82. P. M. Blum, A. G. Davies and R. A. Henderson, *J. Chem. Soc., Chem. Commun.*, 1978, 569.
83. E. A. Hill, A. T. Chen and A. Doughty, *J. Am. Chem. Soc.*, 1976, **98**, 167.
84. G. du Pont, R. Dulov and G. Clement, *Bull. Soc. Chim. Fr.*, 1950, 1056.
85. F. G. Bordwell and W. A. Hewett, *J. Am. Chem. Soc.*, 1957, **79**, 3493.
86. W. J. Bailey, R. L. Sun, H. Katsuki, T. Endo, H. Iwama, R. Tsushima, K. Saigo and M. M. Bitritto, *ACS Symp. Ser.*, 1977, **59**, 38.
87. W. J. Bailey, H. Katsuki and T. Endo, *Polym. Prepr., Am. Chem. Soc., Div. Polym. Chem.*, 1974, **15**, 445.
88. T. Endo and W. J. Bailey, *J. Polym. Sci., Polym. Lett. Ed.*, 1975, **13**, 193.
89. T. Endo and W. J. Bailey, *J. Polym. Sci., Polym. Chem. Ed.*, 1975, **13**, 2525.
90. W. J. Bailey, *Kobunshi*, 1981, **30**(5), 331.
91. W. J. Bailey and T. Endo, *J. Polym. Sci., Polym. Symp.*, 1978, **64**, 17.
92. T. Endo and W. J. Bailey, *Makromol. Chem.*, 1975, **176**, 2897.
93. W. J. Bailey and T. Endo, *J. Polym. Sci., Polym. Chem. Ed.*, 1976, **14**, 1735.
94. T. Endo, M. Cai-Song, M. Okawara and W. J. Bailey, *Polym. J.*, 1982, **14**(6), 485.
95. V. P. Thompson, E. F. Williams and W. J. Bailey, *J. Dent. Res.*, 1979, **58**, 522.
96. W. J. Bailey, B. Issari, Y. -N. Lin, K. No, C. -Y. Pan, K. Saigo, J. Stansbury, S. -R. Tan, N. Yamazaki and J. Zhou, *Proc. Polym. Mater. Sci. Eng.*, 1985, **51**, 250.
97. W. J. Bailey, Y. -N. Lin, K. No, C. -Y. Pan, K. Saigo, J. W. Stansbury, S. -R. Tan, N. Yamazaki and J. Zhou, *Japan–U. S. Polymer Symposium, Kyoto, Japan, Prepr.*, 1985, Oct. 25–30.
98. J. W. Stansbury and W. J. Bailey, *Prepr., Am. Assoc. Dent. Res.*, Washington, DC, 1986, March 13, 452.
99. W. J. Bailey, M. J. Amone, B. Issari, Y. -N. Lin, K. No, C. Y. Pan, K. Saigo, J. W. Stansbury, S. R. Tan, C. Wu, N. Yamazaki and J. Zhou, *Proc. Polym. Mater. Sci. Eng.*, 1986, **54**, 23.
100. J. W. Stansbury and W. J. Bailey, *J. Dent. Res.*, 1986, **65**, 452.
101. J. W. Stansbury and W. J. Bailey, *Proc. Polym. Mater. Sci. Eng.*, 1988, **61**, in press.
102. Y. K. Han and S. K. Choi, *J. Polym. Sci., Polym. Chem. Ed.*, 1983, **21**, 353.
103. C. Y. Pan, S. Lu and W. J. Bailey, *Makromol. Chem.*, 1987, **188**(7), 1651.
104. W. J. Bailey, J. L. Chou, P. -Z. Feng, B. Issari, V. Kuruganti and L. -L. Zhou, *J. Macromol. Sci., Chem.*, 1988, **24**, in press.
105. B. Issari and W. J. Bailey, *Polym. Prepr., Am. Chem. Soc., Div. Polym. Chem.*, 1988, **29**(1), 217.
106. W. J. Bailey, M. J. Amone and B. Issari, *Proc. Div. Polym. Mater. Sci. Eng.*, 1988, **61**, in press.
107. J. E. Heller, A. S. Dreiding, B. R. O'Conner, H. E. Simmons, G. L. Buchanon, R. A. Raphael and R. Taylor, *Helv. Chim. Acta*, 1973, **56**, 272.

23

Polymerization of Fluoro Monomers

ROBERT E. PUTNAM

Washington Technical College, Marietta, OH, USA

23.1 INTRODUCTION

For the purposes of this chapter the term 'fluoro monomers' is meant to include only perfluorinated alkenes. While this definition excludes a number of partially fluorinated alkenes, *e.g.* vinyl fluoride and vinylidene difluoride, several important tetrafluoroethylene copolymers and a few monomers of other types (notably hexafluoropropylene oxide and thiocarbonyl difluoride), it has the advantage of focusing attention on factors peculiar to the perfluorinated structure. The foremost of these factors is polymerization mechanism. Perfluoroalkenes can be converted to high molecular weight products only under free radical conditions. No unambiguous examples of cationic or coordination polymerization appear to have been reported, nor do any commercial processes of this type exist. Anionic polymerizations provide only products of extremely low degree of polymerization,[1] the result of facile loss of a β fluoride ion from all perfluorinated carbanions.

23.2 HAZARDS IN TETRAFLUOROETHYLENE POLYMERIZATION

Because tetrafluoroethylene (TFE) is of such importance in this discussion, a note of caution about its explosiveness is appropriate. Not only does TFE burn vigorously in air to the very toxic carbonyl difluoride (equation 1), but it is thermodynamically unstable with respect to CF_4 and carbon (equation 2). Deflagration of TFE to these products is initiated readily by heat and the reaction is exothermic to the extent of $276\ kJ\ mol^{-1}$. Uncontrolled combustion of TFE can easily lead to a violent monomer deflagration.[2]

$$CF_2{=}CF_2 + O_2 \rightarrow 2CF_2{=}O \tag{1}$$

$$CF_2{=}CF_2 \rightarrow CF_4 + C \tag{2}$$

Of more importance is the fact that inadequate heat removal from the very exothermic TFE polymerization ($172\ kJ\ mol^{-1}$) also can initiate monomer deflagration. Poly(tetrafluoroethylene) (PTFE) is a very poor conductor of heat. Thus entrapment of rapidly polymerizing monomer in a large particle of PTFE can lead to a hot spot, to localized monomer decomposition and to the possibility of deflagration of the entire monomer supply in the system. Clearly great care must be taken in any TFE polymerization to avoid this sequence of events.

There is a third mechanism by which decomposition of TFE monomer can occur. It is likely that most fluoroalkenes can form polymeric peroxides of 1/1 monomer/oxygen composition. Such products have been isolated from reactions of both TFE[3] and 1,1,4,4-tetrafluorobutadiene[4] with oxygen (equations 3 and 4) and these products have been thoroughly characterized. They can be of quite high molecular weight and are explosive at any molecular weight. Furthermore, they are sensitive to both heat and shock. Hexafluoropropylene (HFP) forms a low molecular weight polyperoxide as an intermediate in its oxidation to the corresponding epoxide.[5] Finally, tetra-fluorobutatriene[6] is highly explosive in the presence of trace amounts of air, a fact that is possible evidence for reaction of the same type.

$$nCF_2{=}CF_2 + nO_2 \rightarrow n(CF_2CF_2{-}O{-}O) \tag{3}$$

$$nCF_2{=}CHCH{=}CF_2 + nO_2 \rightarrow n(CF_2CH{=}CHCF_2{-}O{-}O) \tag{4}$$

Combination of polyperoxide formation with the thermodynamic instability of TFE leads to a very dangerous set of circumstances. Mixtures of TFE and air readily form the peroxide even at temperatures as low as $-80\,°C$. The insoluble products, often found in proximity to leaking valves, can explode when subjected to heat or shock, thereby initiating monomer decomposition. This effect is thought to have caused several serious explosions in industrial handling of TFE. But even when these specific circumstances are avoided, the hazard remains. The polyperoxides are excellent free radical initiators so their presence in TFE monomer can lead to uncontrolled polymerization with sufficient heat generation to cause monomer decomposition. The net result is that, to be handled safely, TFE monomer must be kept scrupulously free of oxygen contamination.

23.3 HOMOPOLYMERIZATION OF TETRAFLUOROETHYLENE

Homopolymerization of TFE can be accomplished in aqueous media by two different free radical processes. On the other hand solution polymerization, while feasible, is of no practical value since the physical form of the precipitated polymer makes it useless in subsequent processing operations. The two aqueous processes are called 'granular' and 'dispersion', terms that derive from the nature of the polymer rather than from the reaction technique.[7] Although the processes are quite different in detail, they have several points in common and these will be discussed first.

To be useful in mechanical operations, PTFE must be produced in a molecular weight range of at least 10^6 to 10^7 g mol^{-1}. Termination by radical disproportionation does not occur at all and radical coupling is relatively inefficient.[7] In fact, a large proportion of PTFE chains remain 'alive' after polymerization is completed and these react with oxygen as the polymer is isolated.[7] Impurities or additives capable of reacting with the growing polymer$-\dot{C}F_2$ radical must be excluded very effectively. Of particular importance is exclusion of trace amounts of oxygen which would otherwise copolymerize to give a thermally unstable link in the polymer chain. Thus, a major requirement for all TFE homopolymerizations is this need for extreme purity in the monomer, in the water and in the various additives employed.

Since there is no oil phase in aqueous TFE polymerizations (temperature is above the critical temperature for TFE and pressure is below the critical pressure), initiators are usually water soluble, the most common being ammonium peroxydisulfate. Dispersing agents are perfluoro- or ω-hydroperfluoro-alkanoic acid salts. A good example is ammonium perfluorooctanoate. This type of agent has the advantages of compatibility with the polymer and, more importantly, complete resistance to attack by perfluorocarbon radicals.

Granular polymerization of TFE was the first procedure used to make PTFE[8] and is, even after 40 years, of great industrial significance. Ostensibly the process appears quite simple. Gaseous TFE at 1.4–2.8×10^6 Pa is added to an aqueous charge consisting of highly purified water, about 10 p.p.m. of ammonium peroxydisulfate and 20 p.p.m. of ammonium perfluorooctanoate. Temperature is maintained at about $60\,°C$ and the mixture is vigorously agitated. Polymerization is rapid and the reaction is terminated at a time determined by reactor volume. The initial polymerization rate is limited by the rate of solution of TFE. The vigorous agitation is intended to maximize TFE solution but also has a second objective noted below.

This simple system is, in fact, far more complex than it appears. To begin with, the polymerization proceeds in two stages. Initially the polymer is formed as an unstable suspension. This polymer is identified as α-PTFE in the final product and normally constitutes from 0.5 to 2.0% of the total product.[7] The vigorous agitation is controlled so as to break the suspension to a coagulate in which all growing radicals are trapped in polymer particles. At this point the unwetted polymer begins to react directly with gaseous monomer and the reaction rate increases dramatically. Very little initiation of new polymer chains occurs from this point on. This circumstance results in very high molecular weight since the termination rate under these conditions is low.

The coagulation process carries with it a serious safety complication. If conditions are such that adhesions of polymer to the vessel wall are permitted to grow, there exists a significant potential for monomer deflagration. The mechanism is that described in Section 23.2. Propagating radicals and unreacted monomer are trapped within the adhesions. Their environment causes impaired heat transfer to the aqueous medium. If the adhesion is large enough, heat transfer is virtually eliminated. Eventually the monomer will reach a temperature that is high enough to cause its disproportionation. Depending on circumstances this point will be recognized by a black spot in an otherwise white polymer, by a batch of polymer containing much carbon, or, in the ultimate catastrophe, by a ruptured vessel. From this discussion it should be evident that much of the technology developed for granular polymerization during the last 40 years is the result of the need to balance productivity, polymer properties and safety in industrial operation.

Dispersion polymerization of TFE is carried out under conditions quite different from those described for the granular system. The same initiator and dispersing agent can be used, though in concentrations much higher than in the granular polymerization. Nevertheless, the dispersing agent is still used at a concentration below the 'critical micelle concentration'.[7] In addition, a pure hydrocarbon wax may be employed.[9] This wax must be completely insoluble in water, completely saturated and must be molten at reaction temperature and solid when the dispersion is worked up. Its functions are to reduce coagulation tendency during polymerization as well as to trap coagulated particles in order to retard granular-type polymerization. Temperatures of 60–100 °C are employed, agitation is mild and TFE pressure is maintained in the range of $2.0–3.5 \times 10^6$ Pa. The product is a semistable dispersion of particles having a variety of shapes with dimensions of 0.1–0.3 μm. In comparison, granular particles are of the order of 500 μm in their greatest dimension.[7]

The dispersion polymerization mechanism is similar to but certainly not identical to that of a classical emulsion system.[7] There is no liquid monomer phase and there is no evidence that the highly crystalline polymer has significant solubility in TFE monomer. Furthermore, the function of the dispersing agent is not identical to that of agents used in emulsion polymerization. In consequence, the rate and molecular weight equations developed for emulsion polymerizations do not apply to TFE dispersion polymerization. Fortunately, if premature coagulation is avoided, there is little chance of inadequate heat transfer and monomer decomposition is not the potential hazard it is in granular polymerization.

23.4 HOMOPOLYMERIZATION OF PERFLUOROALKENES OTHER THAN TETRAFLUOROETHYLENE

While partially fluorinated alkenes such as vinyl fluoride and vinylidene difluoride polymerize with the same facility as TFE, the latter is unique in its class with respect to ease of polymerization. Substituted monomers such as HFP[10] and hexafluorobutadiene[11] polymerize only with great difficulty as the result of steric inhibition in the propagation step. HFP provides an unambiguous example of this phenomenon.[10] This monomer can be converted to high molecular weight polymer only at pressures above 10^8 Pa. The polymerization is most conveniently carried out in a perfluorinated solvent using perfluorinated free radical initiators. Equally laborious efforts are required for polymerization of hexafluorobutadiene[11] with reported conditions being similar to those used with HFP.

23.5 COPOLYMERIZATION OF TETRAFLUOROETHYLENE

TFE copolymerizes readily with a variety of non- and partially-fluorinated alkenes but only under free radical conditions. Success with this type of polymerization is, however, strongly dependent on monomer structure and on the susceptability of the hydrocarbon segments of the polymer to transfer with the polymer–$\dot{C}F_2$ radical. Notable commercial success has been achieved with ethylene,[12] propylene[13] and mixtures of vinylidene difluoride and HFP.[14]

Copolymerization with perfluorinated monomers, on the other hand, has been limited by the lack of availability of many such monomers as well as by the steric effects noted in the previous section. Nevertheless, important examples of this class of copolymerizations are to be found with the monomers HFP,[15] perfluoroalkyl perfluorovinyl ethers[16] and perfluorovinyl ethers containing functional groups.[17] It is with these interesting and important systems that the remainder of this chapter is concerned.

The first copolymer of TFE to achieve commercial success was that with HFP.[18] This copolymerization is carried out under dispersion conditions similar to those used with TFE alone as well as by several other techniques. Because the product is useful in a molecular weight range much lower than that of PTFE, modifications of the normal dispersion process have taken the form of changes in the polymerization conditions as well as the use of transfer agents to control molecular weight and/or end group distribution.

In the most common technique, aqueous dispersions are prepared by batch polymerization of the two monomers in water using water-soluble free radical initiators such as peroxydisulfates and dispersing agents of the perfluorinated alkanoate type. Recently sulfonic acids containing an aromatic ring have also been reported to be very effective dispersing agents for TFE/HFP copolymerization.[19] Termination of growing chains appears to occur entirely by coupling, so molecular weights are controlled by initiator concentration. Since end groups play a significant role in polymer post treatments and physical properties, it is important to note that they are not the sulfate ions anticipated from the initiator. Rather, these initially formed end groups are rapidly hydrolyzed to carboxylic acid ends with a concomitant decrease in the pH of the medium (equation 5).[18]

$$\text{Polymer–CF}_2\text{OSO}_3^- + 2\text{H}_2\text{O} \rightarrow \text{Polymer–CO}_2^- + 2\text{HF} + \text{HSO}_4^- \qquad (5)$$

A further complication lies in the low rate of addition of HFP to the propagating radicals, nearly zero in the case of the HFP-derived radical (recalling the difficulty of making HFP homopolymer). This results in a limitation of the reaction rate as well as of the amount of HFP that can be incorporated in the polymer chain at reasonable pressures. To force the desired 8 mol % HFP in the product requires that a large excess of HFP be used in the initial charge. TFE (and optionally HFP) must be fed continuously at a rate such that TFE partial pressure remains constant during polymerization. Reaction temperatures of 80–100 °C are below the critical temperature of HFP and the total pressure is in the range of 3.0–5.0×10^6 Pa. This combination of conditions results in the presence of an HFP-rich liquid phase. Polymerization is terminated before the liquid HFP is consumed, though for practical purposes reactor volume is usually the determining factor. Because of end group hydrolysis the reaction medium slowly decreases in pH and this leads to some loss of HFP by hydrolysis (equation 6). No buffer is employed, however, since both the pH decrease and monomer loss are small enough to be tolerated. The product is obtained as a semistable dispersion that is used as such or is coagulated in preparation for carboxylic acid end group removal by a variety of techniques.

$$\text{CF}_3\text{CF}{=}\text{CF}_2 + 2\text{H}_2\text{O} \rightarrow \text{CF}_3\text{CFHCO}_2\text{H} + 2\text{HF} \qquad (6)$$

These carboxylic acid ends can be avoided by use of a combination of water-insoluble chloro-fluoroacyl peroxides together with a large number of chain transfer agents.[20] In this case the end groups are those derived from the alkyl portion of the initiator and from the transfer agent. Use of these agents also affects the initiator/molecular weight relationship and polymerization rate, so somewhat different conditions are required to obtain products equivalent in molecular weight to those from peroxydisulfate initiation. The different end groups also lead to different post treatments and to different thermal stabilities of the polymer. In principle, solution processes employing a transfer-resistant solvent would be possible for preparation of TFE/HFP copolymers. However, the need to achieve an HFP mole fraction of about 0.08 could not be met because the HFP concentrations would be too low at reasonable pressures.

Unlike HFP and other substituted alkenes the perfluoroalky perfluorovinyl ethers are a class of monomers that copolymerize readily with TFE.[18] (They also homopolymerize readily but the polymers are of very limited interest.) Several members of this class of copolymers have achieved significant commerical success, notably copolymers of TFE with perfluoropropyl perfluorovinyl ether (PPVE),[18] perfluoromethyl perfluorovinyl ether (PMVE)[21] and perfluorosulfonylethyl perfluorovinyl ether (PSEPVE).[17] These polymers are prepared either by the aqueous dispersion technique or by a solution process.

Copolymers of TFE with PPVE are unusual in that the benefits of the comonomer are obtained at only 1–2 mol % comonomer content. When made by the two different processes (*i.e.* aqueous dispersion and solution) the polymers can have identical compositions except for end groups. These differences, though subtle, lead to some specificity in applications, making them worthy of detailed discussion.

Dispersion polymerization of mixtures of TFE and PPVE is similar to that of TFE/HFP but the differences are important enough to be noted. Propagating radical reactivity with PPVE is high; thus there is no need for the large excess of comonomer seen with HFP. Peroxydisulfate initiator and ammonium perfluorooctanoate surfactant are used. In addition, several modifications are needed for economic viability or to obtain products with the desired properties. Small quantities of 1,1,2-trifluoro-1,2,2-trichloroethane (Freon 113) can be added to ensure polymer homogeneity.[22] Also, an ammonium hydroxide or carbonate buffer is used to prevent a drop in pH during reaction. This precaution retards hydrolysis of the very expensive PPVE.

Two other modifications are of even more importance. All perfluoroalkyl perfluorovinyl ethers undergo an unusual free radical-catalyzed rearrangement shown in equations (7) and (8) for PPVE. This reaction, if not controlled, would lead to unacceptably low molecular weight in the copolymers. Fortunately, the reaction is quite temperature dependent so control of polymerization temperature in the range of 70–95 °C keeps the unimolecular transfer to a tolerable level. The second important modification is the use of chain transfer agents. Ethane[23] is one of several agents that act in a unique fashion. Not only does it provide molecular weight control but its use results in elimination of a serious shrinkage problem in polymer-molding applications. This is thought to be due either to a molecular weight or comonomer distribution effect but the cause has not been established conclusively. As a result of the unimolecular transfer and the ethane reaction a complex mixture of end groups is found. These include carboxylic acids from initiator and unimolecular transfer, CF_2H- and CH_3CH_2- from ethane and traces of several chlorine-containing ends from the Freon 113.

$$\dot{R}_f + CF_2=CFOC_3F_7 \rightarrow R_fCF_2\dot{C}FOC_3F_7 \tag{7}$$

$$R_fCF_2\dot{C}FOC_3F_7 \rightarrow R_fCF_2CFO + \cdot C_3F_7 \tag{8}$$

The solution process used for synthesis of TFE/PPVE copolymers employs Freon 113 as solvent and perfluoropropionyl peroxide (3P) as initiator at temperatures of 40–65 °C.[24] Polymerization is possible under either batch or continuous conditions. Were it not for the unimolecular transfer reaction shown in equations (7) and (8) as well as transfer with the solvent, this polymerization could give polymers completely free of non-fluorinated ends. However, control of molecular weight requires use of a transfer agent since use of 3P for this purpose would be prohibitively expensive. For the solution process ethane is ineffective and methanol is the transfer agent of choice. Again, a complex mixture of end groups results. These include ends derived from methanol (CF_2H- and $HOCH_2-$) together with methyl ester ends from reaction of methanol with acid fluoride ends (equation 8). The 3P initiator contributes only perfluorinated alkyl ends.

Copolymerization of TFE with PMVE is accomplished under the same general conditions as used for TFE/PPVE mixtures. However in this case, much higher PMVE concentrations are required since the polymer is intended for elastomeric applications.[21] Also, a termonomer is added in very small amounts to provide crosslinking sites. Useful termonomers are, in general, perfluorovinyl ethers with reactive side chains. Whereas TFE/PPVE copolymers contain only 1–2 mol % PPVE, the TFE/PMVE copolymers contain about 30–40 mol % PMVE.

Copolymerization of TFE with PSEPVE can also be accomplished in the manner described for TFE/PPVE. In this polymer family PSEPVE content is varied widely depending on the ion exchange capacity desired in the final products.[17]

In addition to the copolymers described above, a variety of terpolymers based on mixtures of TFE, HFP and PPVE can be prepared by superficial modification of conditions already described.[25] These polymers provide useful compromises between the properties offered by PPVE and the lower cost of HFP. However, polymerization conditions are so close to those already described that they need not be covered separately.

In closing it seems appropriate to reemphasize the safety aspects of work with TFE. Although the hazards in copolymerization are not as great as in granular polymerization, in all cases pure TFE must be handled at some point in the process. Thus, rigorous exclusion of oxygen and avoidance of other sources of heat and polymerization initiation must be ensured. These generalizations apply equally to laboratory and industrial practice.

23.6 REFERENCES

1. W. J. Brehm (Du Pont), *US Pat.* 2 918 501 (1959) (*Chem. Abstr.*, 1960, **54**, 20 875).
2. P. Thistleton, Du Pont de Nemours, Washington Laboratory, Parkersburg, W. VA, personal communication.
3. D. P. Carlson, Du Pont de Nemours, Experimental Station Laboratory, Wilmington, DE, personal communication.

4. R. E. Putnam and W. H. Sharkey, *J. Polym. Sci., Part A-1*, 1966, **4**, 2289.
5. D. P. Carlson and R. E. Putnam, unpublished work.
6. E. L. Martin and R. E. Putnam (Du Pont), *US Pat.* 2 888 447 (1959) (*Chem. Abstr.*, 1959, **53**, 18 862).
7. C. A. Sperati, 'High Performance Polymers: Their Origin and Development', Elsevier, Amsterdam, 1986.
8. T. R. Doughty, C. A. Sperati and H. H. W. Un (Du Pont), *US Pat.* 3 855 191 (1974) (*Chem. Abstr.*, 1975, **82**, 73 855).
9. S. G. Bankoff (Du Pont), *US Pat.* 2 612 484 (1952) (*Chem. Abstr.*, 1953, **47**, 3618).
10. H. S. Eleuterio (Du Pont), *US Pat.* 2 958 685 (1960) (*Chem. Abstr.*, 1960, **54**, 35 223).
11. W. T. Miller (US Atomic Energy Commission), *US Pat.* 2 567 956 (1951) (*Chem. Abstr.*, 1952, **46**, 1808).
12. D. P. Carlson (Du Pont), *US Pat.* 3 624 250 (1971) (*Chem. Abstr.*, 1970, **73**, 67 223).
13. G. Kojima and M. Hisasue (Asahi Glass Co.), *US Pat.* 4 463 144 (1984) (*Chem. Abstr.*, 1982, **97**, 24 993; 1984, **101**, 131 335).
14. J. R. Pailthrop and H. E. Schroeder (Du Pont), *US Pat.* 2 968 649 (1961) (*Chem. Abstr.*, 1961, **55**, 13 894).
15. M. I. Bro and B. W. Sandt (Du Pont), *US Pat.* 2 946 763 (1960) (*Chem. Abstr.*, 1960, **54**, 26 015).
16. R. L. Johnson, 'Encyclopedia of Polymer Science and Technology', Supplement, Wiley, New York, 1976, Vol. 1.
17. *ACS Symp. Ser.*, 1982, **180**.
18. R. E. Putnam, 'High Performance Polymers: Their Origin and Development', Elsevier, Amsterdam, 1986.
19. A. A. Khan and R. A. Morgan (Du Pont), *US Pat.* 4 380 618 (1983) (*Chem. Abstr.*, 1986, **105**, 153 709; 1986, **105**, 153 728).
20. S. Nakagawa, K. Asano, S. Sakata, T. Adachi and S. Kawachi (Daikin Kogyo Co.), *US Pat.* 4 552 925 (1985) (*Chem. Abstr.*, 1983, **99**, 213 801).
21. A. L. Barney, W. J. Keller and N. M. van Gulick, *J. Polym. Sci., Part A-1*, 1970, **8**, 1091; H. E. Schroeder, 'High Performance Polymers: Their Origin and Development', Elsevier, Amsterdam, 1986.
22. S. Nakagawa, T. Nakagawa, S. Yamaguchi, K. Ihara, T. Amano, M. Omori and K. Asano (Daikin Kogyo Co.), *US Pat.* 4 499 249 (1985) (*Chem. Abstr.*, 1984, **100**, 52 217).
23. W. F. Gresham and A. F. Vogelpohl (Du Pont), *US Pat.* 3 635 926 (1972) (*Chem. Abstr.*, 1971, **75**, 37 034).
24. D. P. Carlson (Du Pont), *US Pat.* 3 528 954 (1970); *US Pat.* 3 642 742 (1972) (*Chem. Abstr.*, 1971, **74**, 54 360).
25. D. P. Carlson (Du Pont), *US Pat.* 4 029 868 (1977) (*Chem. Abstr.*, 1977, **87**, 85 753).

24

Polymer Reaction Engineering

KARL-HEINZ REICHERT and HANS-ULRICH MORITZ
Technische Universität Berlin, FRG

24.1 INTRODUCTION

A polymerization process consists in general of three parts which can be classified as preparation, polymerization and separation. Preparation means in general purification and mixing of reactants. Intensive purification is important since very small amounts of radical scavengers or chain transfer agents in the reacting system will affect polymerization rate and molecular weight of the polymer formed, because the stationary radical concentration of the reacting system is of the order of only 10^{-8} mol l^{-1}.

Proper mixing of reactants before and during reaction may also play an important role. It may influence reactor performance as well as product quality. In the case of polymerization the following mixing processes may be distinguished: homogenization, emulsification and mixing gases and liquids. The reactants may be mixed before entering the reactor or just within the reactor.

The second step of a polymerization process is the polymerization reaction itself. Free radical polymerizations are complex chain reactions taking place in homogeneous or heterogeneous media, usually in the liquid phase. Reaction engineering problems may be caused by the increase of viscosity of homogeneous systems which may change by many orders of magnitude during the course of polymerization. The increase in viscosity can greatly affect the kinetics of reaction, the transport processes of heat, mass and momentum, the quality of mixing and finally the residence time distribution of continuous processes. Since almost all polymerization reactions are exothermic reactions which in general are conducted at constant temperature, a great amount of heat has to be removed from the polymerization reactor. This can be done by various techniques. The most general technique is by heat transfer through the reactor wall, but heat is also removed by evaporation and by the feed stream in the case of continuous processes.

The reactor design is of special importance in the case of polymerization processes. Product parameters such as molecular weight distribution, branching and crosslinking of polymers are influenced by the reactor type. Polymers once prepared with a set of data cannot be changed afterwards by any unit operation. Polymer properties of a given system are predetermined by reaction conditions and type of reactor.

The last step of a polymerization process is concerned with separation processes to achieve a polymer of certain purity and state. Usually thermal and mechanical unit operations are applied. Attention must also be paid to the flow behavior of polymer melts, solutions or dispersions. Usually they behave as non-Newtonian fluids. In Figure 1 some of the more important chemical engineering aspects of a polymerization process are schematically summarized.

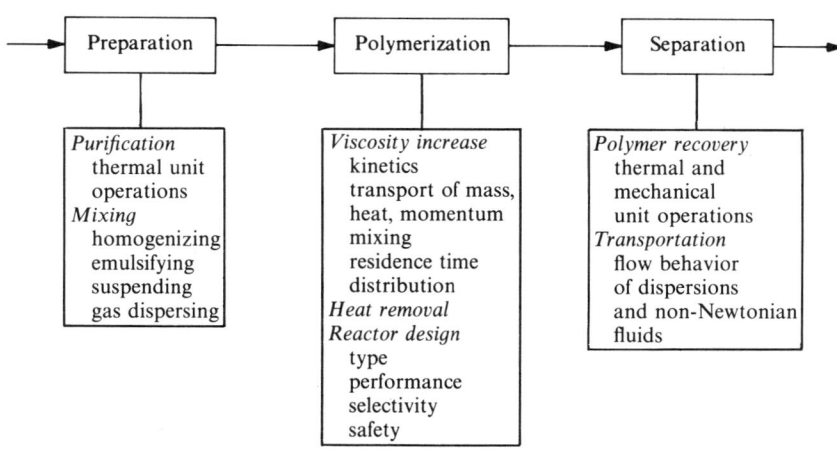

Figure 1   Major parts of a polymerization process and some reaction engineering aspects

Polymers can be produced by different polymerization techniques which are classified in four major groups and listed in Table 1 (See Section 24.5 for abbreviations of polymer names). The classification is more or less based on the historical development of these techniques. The notation is not always logical, especially in the case of suspension and emulsion polymerization. In general one can subdivide polymerization processes into homogeneous and heterogeneous systems. While suspension and emulsion polymerization are always heterogeneous systems, bulk polymerization can be homogeneous (PS, PMMA, polyester) as well as heterogeneous (PVC, HIPS). Polymerization in solution, however, is always homogeneous. If the polymer formed is not soluble in the solvent used, it will precipitate during the course of polymerization and is specified in this case as a kind of suspension polymerization (PAN).

As far as kinetics are concerned, the polymerizations in bulk, suspension and solution are usually identical whereas the kinetics of polymerization in emulsion is quite different and determined by the nature of the disperse system.

Which polymerization technique should be used depends mainly on the desired purpose of the polymer being produced. The main advantages and disadvantages of the different polymerization techniques are listed in Table 1. The most important polymerization technique is polymerization in emulsion, but polymerization in bulk and suspension are also widely used techniques for polymer production.

Most of the free radical polymerizations are run in the liquid phase. Therefore, the most often used reactor is the stirred tank reactor which can be operated batch and semibatch wise, as well as in a continuous way in the general form of a cascade of stirred tank reactors. This can also be seen from Table 2 by the number of examples of technical polymers produced in stirred tank reactors.

The reader is referred to some review articles and books which, taken together, provide a good detailed coverage of the essential features of polymer reaction engineering.[1-27]

Table 1 Polymerization Techniques

Polymerization in:	Advantages	Disadvantages	Technical polymers[a]
Bulk			
(Polymerization in absence of any inert solvent or dispersion medium)	High reactor performance; low separation cost; high purity of the product; no transfer reactions to solvents or additives	High viscosity with problems of heat removal and mixing; pumping problems; reactor wall fouling by film formation	LDPE, LLDPE, PVC, PS, HIPS, PMMA, PA, polyester
Suspension			
(Polymerization of monomer droplets dispersed in an inert phase with monomer soluble initiator, or precipitation of polymer from a polymerizing monomer solution)	Low dispersion viscosity; good heat transfer; low separation costs compared to emulsion	Smaller reactor capacity than bulk; only discontinuous operation; waste water problems; reactor wall fouling by film formation	Bead-type: PS, PMMA, SAN Non-bead-type: PVC, PTFE, PAN, HIPS Slurry-process: HDPE, PP
Emulsion			
(Formation of small polymer particles *via* micellar or homogeneous nucleation in a disperse system of monomer droplets in an inert phase with dissolved initiator)	Low dispersion viscosity compared to bulk; good heat transfer; high polymerization rate and high molecular weights; direct application of the latex	High separation costs in case of polymer isolation; waste water problems; reactor wall fouling by film formation; emulsifier as impurity of the polymer product	PVC, ABS, PVA, PMMA, SAN, PTFE, SBR, CR, NBR
Solution			
(Polymerization of monomer dissolved in an inert solvent)	Lower vicosity than bulk with better heat transfer and mixing; direct application of the solution; less reactor wall fouling than bulk	Smaller reactor capacity than bulk; high separation costs in case of polymer isolation, often inflammable and toxic solvents; transfer reactions to solvent and lower molecular weights	SBR, PVA, PAN PS, PVAL, BR, IR, EPDM

[a] For abbreviations, see Section 24.5.

Table 2 Reactor Types Classified According to their Operation Modes and Width of Residence Time Distribution[a]

Operation	Reactor type	Technical polymers
Batch and semibatch	Stirred tank reactor	PVC (B, S, E), [b] PMMA (B, S, E), PS (S), ABS (E), PVA (E), PTFE (S, E), SAN (E), polyester (B)
	Mold (RIM technique)	PUR (B)
Continuous (with broad residence time distribution)	Continuous stirred tank reactor (CSTR)	LDPE (B), HDPE (S), PP (S), PVC (E), PS (B + Sol), IR (Sol), PAN (S)
	Fluidized-bed	LLDPE (B), PP (B)
	Loop-reactor	HDPE (S)
Continuous (with narrow residence time distribution)	Tubular reactor	LDPE (B)
	Tower reactor	PA (B)
	Continuous stirred tank reactor-cascade	ABS (E), SBR (E), SAN (E), CR (E), NBR (E), HIPS (S), BR (Sol), IR (Sol), Polyester (B), EPDM (Sol)
	Tower-cascade	PS (E), PA (B)
	CSTR-tower-cascade	PS (B), PA (B)
	Endless belt reactor	PIB (B), PUR (B)
	Extruder reactor	PU (B), PA (B)

[a] Some polymers of technical importance are listed with the polymerization techniques in parentheses
[b] B: bulk, E: emulsion, S: suspension, Sol: solution polymerization

24.2 REACTOR PERFORMANCE

One of the major features of polymer reaction engineering is the proper design of a polymerization reactor which adequately meets the requirements of reactor performance, selectivity and safety. It must be evaluated what type and size of reactor and method of operation are best in order to obtain the polymer in the desired quality and quantity. A profound knowledge of the problems which may arise from increasing viscosity and decreasing heat transfer during the course of polymerization, as described briefly in the following paragraph, and the 'non-ideal' kinetics including phenomena like inhibition, gel and glass effects, particle nucleation and coagulation processes, phase separation, *etc.* are prerequisites for designing a polymerization reactor.

24.2.1 Viscosity Increase and its Influence on the Polymerization Process

One of the most important features of polymerization reactions is the large increase in viscosity with monomer conversion. The viscosity of the reaction mass increases by several orders of magnitude, particularly during polymerization in homogeneous systems such as bulk and solution polymerization. With emulsion polymerization stabilized by low molecular weight emulsifiers, the increase in viscosity is moderate in comparison, and with suspension polymerization the dispersion viscosity virtually does not change. The relative change in viscosity with conversion in bulk, solution, emulsion and suspension polymerization is qualitatively sketched in Figure 2.

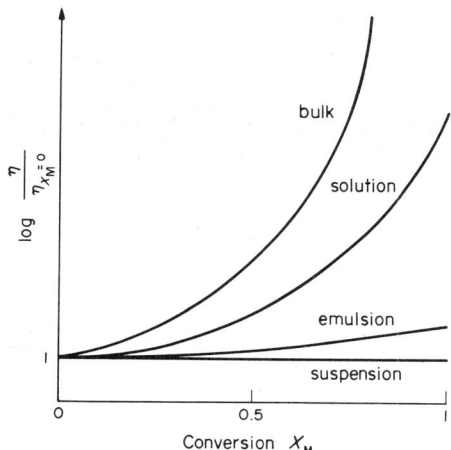

Figure 2   Relative change in viscosity $\eta/\eta_{X_M=0}$ with monomer conversion X_M for different polymerization techniques (qualitative sketch)

The viscosity is a dominant parameter of the polymerization process. The reaction kinetics as well as the heat, mass and momentum balance of the polymerization reactor are significantly affected by a change in viscosity. For a quantitative description of the complex effect of increasing viscosity on the whole polymerization process, it is necessary to analyze the process in terms of the balance equations and to quantify its influence on every term seperately. In Figure 3 the influence of the viscosity on the polymerization process is shown schematically in a flow diagram. An increasing viscosity results in a decrease of the heat transfer coefficient and an increase of the power consumption of the stirrer, and consequently the cooling capacity of the reactor decreases. Difficulties in reactor control and stability may therefore arise. Other effects of higher viscosity of the reaction mass are reduced molecular diffusion and mass transfer coefficients. The mixing time required to reach a desired degree of homogeneity increases and the degrees of macromixing and micromixing are reduced. Consequently the reaction mass becomes segregated and the residence time distribution in continuous processes is affected, for example by increasing dead water zones and bypassing in continuous stirred tank reactors. As a result of these changes monomer conversion and hence the reactor performance are influenced.

With regard to polymerization kinetics, an increasing viscosity leads to a reduction in radical formation (cage effect, reduced initiator efficiency). The polymer radical termination reaction

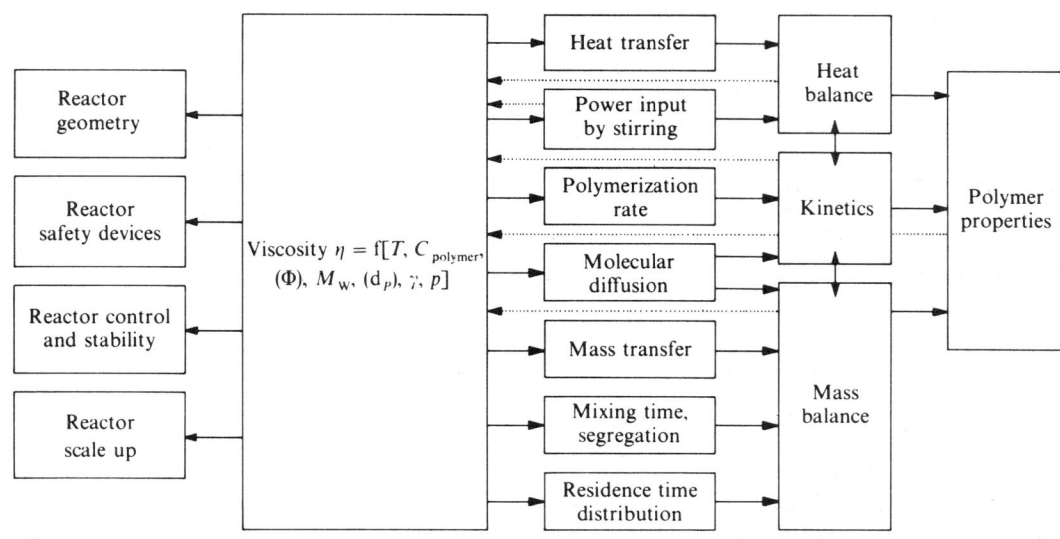

Figure 3 Schematic diagram of the influences of viscosity on the polymerization process

becomes diffusion controlled (Trommsdorff or gel effect) and at high monomer conversions so does the chain propagation reaction (glass effect). Diffusion limitation of the termination reaction involves an increase in polymer radical concentration and subsequently the reaction rate, which is an essential part of both heat and mass balance, accelerates. The increasing rate of heat generation in combination with a reduced heat transfer is a problem of major importance in running polymerization reactors. Although the overall dispersion viscosities can be low, gel and glass effects have to be taken into consideration in suspension and emulsion as well as in bulk polymerization, because these effects are related to the viscosity at the locus of reaction.

In conclusion, heat and mass balances including the polymerization kinetics are strongly affected by an increasing viscosity. Consequently selectivity, which covers the whole spectrum of polymer properties, and reactor performance are influenced. In addition, varied problems may arise by running the polymerization process under conditions of increasing viscosity and the reactor geometry, scale up aspects, reactor control and safety devices may have to be modified (Figure 3).

For running a polymerization reactor, knowledge of the flow behavior and the increase in viscosity with monomer conversion is important. One empirical approach to calculate the dependence of the specific viscosity η_{sp} on polymer concentration, intrinsic viscosity $[\eta]$ and temperature is that of Lyons and Tobolsky[28]

$$\frac{\eta_{sp}}{C_{polymer}[\eta]}=\exp\left(\frac{k_H[\eta]C_{polymer}}{1-bC_{polymer}}\right) \quad \text{or} \quad \eta=\eta_{solvent}\left[1+C_{polymer}[\eta]\exp\left(\frac{k_H[\eta]C_{polymer}}{1-bC_{polymer}}\right)\right] \quad (1)$$

where η and k_H are the solution viscosity and the Huggins constant[29] respectively; b is an adjustable parameter that can be positive or negative; $C_{polymer}$ is the concentration of polymer, which is usually substituted for by the monomer conversion X_M according to equation (2)

$$C_{polymer}=\frac{X_M C_{M,0} M_M}{10^3(1+\varepsilon_M X_M)} \ (\text{g cm}^{-3}) \quad (2)$$

where M_M and ε_M are the molecular weight of monomer and the volume contraction constant (see Section 24.3.3) respectively (definitions of symbols for all equations are given in Section 24.5). While the concentration dependence of the viscosity of concentrated polymer solutions is explicitly expressed by the Lyons–Tobolsky equation (1), the molecular weight, temperature and solvent dependences are contained in the parameters $[\eta]$, k_H, b and $\eta_{solvent}$. For a number of polymer solutions investigated over a range of molecular weights, the intrinsic viscosity $[\eta]$ strictly follows the Mark–Houwink equation[30,31] and the Huggins constant k_H decreases with increasing molecular weight as generally observed.[32] The parameter b is reported to be positive for $M < M_{cr}$, equal to zero for $M = M_{cr}$ and negative for $M > M_{cr}$.[33,34] Figure 4 shows an example of the application of the Lyons–Tobolsky equation (1), which may be quite useful for predicting solution viscosities over the

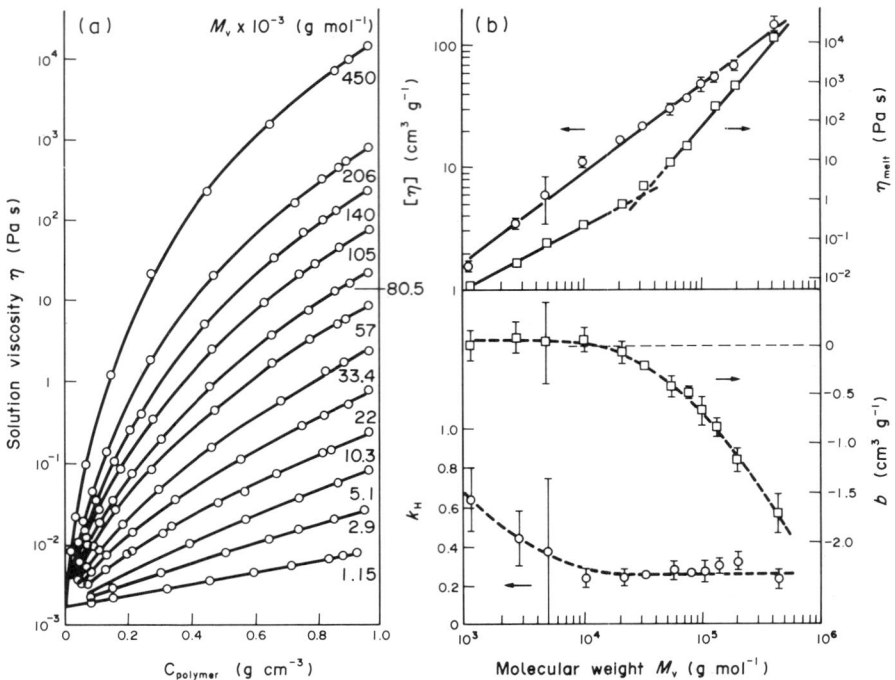

Figure 4 (a) Zero shear viscosity of polydimethylsiloxane of different molecular weights in dodecamethylpentasiloxane at 30 °C *vs.* polymer concentration. Solid lines are calculated using equation (1), while symbols represent experimental data from ref. 37. (b) Values of the parameters in the Lyons–Tobolsky equation, $[\eta]$, k_{H} and b, as functions of molecular weight M_{v}. For comparison, the experimental melt viscosity η_{melt} is also plotted *vs.* M_{v} (adapted from ref. 38)

entire concentration range from limited experimental data. Similar equations have been proposed involving free volume concepts.[35,36] In general, the reliability of interpolation will of course depend on the accuracy of the experimental data, but extrapolation for a polymerizing system with changing polymer concentration, molecular weight and possibly average shear rate should be done with caution.[39,40]

Detailed information about the flow behavior of homogeneous and heterogeneous polymer systems is delegated to numerous survey articles and books.[41–52]

24.2.2 'Ideal' First Order Free Radical Polymerization

Since reaction conditions in the reactor may vary with time as well as location, the integration of the reaction rate equation for the operation and certain boundary conditions like heat transfer through the reactor wall is necessary for the prediction of selectivity and performance. Using simple reactor models like the ideal batch reactor (BR), the ideal continuous stirred tank reactor (HCSTR), where no temperature and concentration gradients are assumed, the continuous plug flow tubular reactor (PFTR) without any back-mixing in axial direction as well as gradients in the radial direction, and the completely segregated continuous stirred tank reactor (SCSTR), where each fluid lump retains its identity and acts as an individual batch reactor, certain general recommendations and conclusions for the design of polymerization reactors can be made. Often 'non-ideal' reactors can be approximated by these 'ideal' reactor types or their combinations in multiparameter models where plug, mixed and dead water regions are interconnected by bypass, recycle and crossflow.[53–55]

The most widely used reactor for free radical polymerization in the liquid phase is the stirred tank reactor. It can be run batch-wise, semibatch-wise and continuously, either as a single reactor or in series as a cascade.[2]

Conversion–time data are the basis for the calculation of reactor performance. The ratio between the reaction time required for obtaining a certain conversion and the time constant of the reaction is a characteristic dimensionless number of the reaction, namely the Damköhler number[56] *Da*, which enables comparison of different reactor types and sizes as well as reactions and gives information about the reactor capacity. Conversion–Damköhler number and reaction volume correlations are

Table 3 Mass Balance of a First Order Polymerization at Constant Temperature, Reaction Volume and Initiator Concentration in Different Ideal Stirred Tank Reactors

	Conversion	*Reaction volume*
Stirred tank reactor BR	$X_M = 1 - \exp\{-Da\}$	$V_R = \dfrac{\dot{m}_P(t + t_{dead})}{M_M C_{M,0} X_M}$
 HCSTR	$X_{M,st} = 1 - \dfrac{1}{(1 + Da)}$	$V_R = \dfrac{\dot{m}_P}{k M_M C_{M,0}(1 - X_M)}$
 Cascade	$X_{M,st} = 1 - \dfrac{1}{\left(1 + \dfrac{Da}{N}\right)^N}$	$V_R = \dfrac{\dot{m}_P N[(1 - X_M)^{-1/N} - 1]}{k M_M C_{M,0} X_M}$

given in Table 3 for a first order polymerization at constant temperature, volume and initiator concentration in three different types of ideal stirred tank reactors.

$\dot{m}_P$, M_M and t_{dead} are the reactor performance in kg polymer s^{-1}, the molecular mass of monomer and the time required for filling, heating, cooling, discharging and cleaning a batch reactor. The dimensionless Damköhler number Da is defined for batch-wise and continuous free radical polymerization by

$$Da = kt C_{M,0}^{n-1} \quad \text{or} \quad Da = k\tau C_{M,0}^{n-1} \quad \text{with} \quad k = k_p\left(\frac{2fk_d C_{I,0}}{k_t}\right)^{0.5} \quad (3)$$

respectively. For a cascade of stirred tank reactors of equal size, τ is the total mean residence time of the whole reactor train. It has to be remembered that corresponding values of time or mean residence time and conversion which fit the conversion–Damköhler number correlations have to be used for calculating the reaction volume. Furthermore, it is assumed that in the case of free radical polymerization all polymer formed is the desired product and hence selectivity is unity.

For a first order polymerization at constant temperature, volume and initiator concentration conversion may be plotted against Damköhler number for a batch reactor, an HCSTR and a cascade of two and four HCSTRs (Figure 5).

The performances of these reactors are different since the reaction rates vary due to their different concentration–time profiles. The greatest performance can be achieved with the batch reactor, the lowest with a single HCSTR. The performance of the cascade depends on the number of vessels and lies in between the values for the continuous stirred tank reactor and the batch reactor. For an infinite number of vessels the conversion in the cascade approaches the conversion obtained in the batch reactor or the ideal plug flow reactor.[57] If the course of polymerization corresponds to a zero order reaction, conversion equals Damköhler number for all the reactor types mentioned.[58]

Since reactor performance depends on the reaction volume, the reactor capacity, defined as the number of moles of the desired product formed per unit reaction volume and time, shall be used for a comparison of different reactor types. For example the reactor capacity of an HCSTR may be obtained by rearranging the correlation in Table 3 to

$$\frac{\dot{m}_P}{M_M V_R} = r = k C_{M,0}(1 - X_M) = \frac{C_{M,0} X_M}{\tau}\left(\frac{\text{mol}}{\text{m}^3\text{s}}\right) \quad \text{with} \quad k = k_p\left(\frac{2fk_d C_{I,0}}{k_t}\right)^{0.5} \quad (4)$$

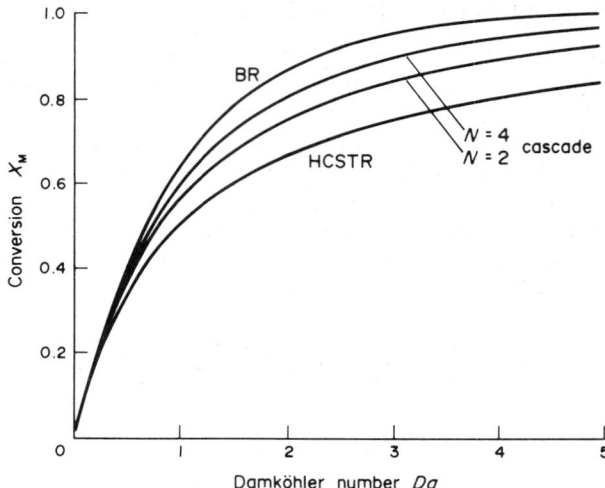

Figure 5 Conversion *vs.* Damköhler number for a first order polymerization at constant temperature, reaction volume and initiator concentration in a BR, in a cascade of four and two equal sized stirred tank reactors, and in an HCSTR

Figure 6 illustrates the decrease in reactor capacity with increasing conversion and time, t or τ respectively, for an arbitrary first order polymerization at constant temperature, volume and initiator concentration in three different types of ideal stirred tank reactors. Generally, a low reactor capacity has to be accepted when high conversion must be attained. Therefore separation and recycling of unreacted monomer or nonisothermal reaction conditions at high conversions may be suitable operations to obtain a higher reactor capacity.

A constant reactor capacity is only obtained if the course of polymerization corresponds to a zero order reaction.

First order polymerizations with constant reaction volume and constant initiator concentration are usually limited to very dilute systems initiated with initiators of low activity.

In concentrated systems, complications such as gel and glass effects, as well as a decrease of reaction volume due to increasing density of the reaction mass while converting monomer to polymer, have to be considered.

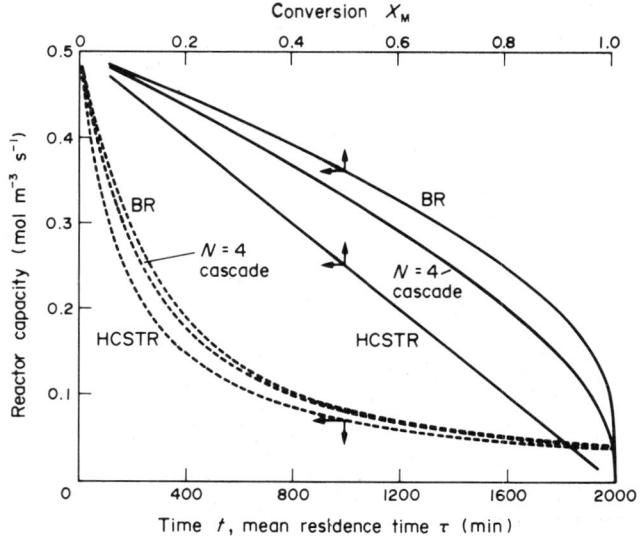

Figure 6 Reactor capacity as a function of conversion and reaction time or mean residence time for different stirred tank reactors and a first order polymerization at constant temperature, reaction volume and initiator concentration with $C_{M,0} = 5 \, \text{mol} \, l^{-1}$ and $k = 1 \times 10^{-4} \, s^{-1}$

24.2.3 Effects of Volume Contraction on Reactor Performance

A linear correlation for the volume contraction as a function of monomer conversion has been observed for many monomer/polymer systems according to

$$V_R = V_{R,0}(1 + \varepsilon_M X_M) \qquad \text{with} \qquad \varepsilon_M = \frac{V_{R,X_M=1} - V_{R,0}}{V_{R,0}} \tag{5}$$

where ε_M is the volume contraction constant of monomer which is in the range of $-0.1 > \varepsilon_M > -0.4$ for most monomers and only a function of temperature. In Table 4 the volume contraction constants of several monomers for bulk polymerization are summarized.

Table 4 Volume Contraction Constants of Several Monomers for Bulk Polymerization at 20 °C[a]

Monomer	ε_M
Styrene	−0.14
Methyl methacrylate	−0.21
Vinyl acetate	−0.22
Isoprene	−0.25
Tetrafluoroethylene	−0.34
Vinyl chloride	−0.35
Butadiene	−0.39

[a] J. Brandrup and E. H. Immergut (eds.), 'Polymer Handbook', Wiley, New York, 1975.

Consequently, monomer and initiator concentrations as well as reaction rate have to be substituted for by

$$C_M = C_{M,0}\frac{1 - X_M}{1 + \varepsilon_M X_M} \qquad C_I = C_{I,0}\frac{\exp(-k_d t)}{1 + \varepsilon_M X_M} \qquad r = \frac{C_{M,0}}{1 + \varepsilon_M X_M}\frac{dX_M}{dt} \tag{6}$$

if a first order decomposition of the initiator is assumed.[59,60] The resultant effects of volume contraction on conversion–Damköhler curves for zero and first order reactions in different reactor types is shown in Figure 7. With regard to zero order reactions, the same conversion–Damköhler

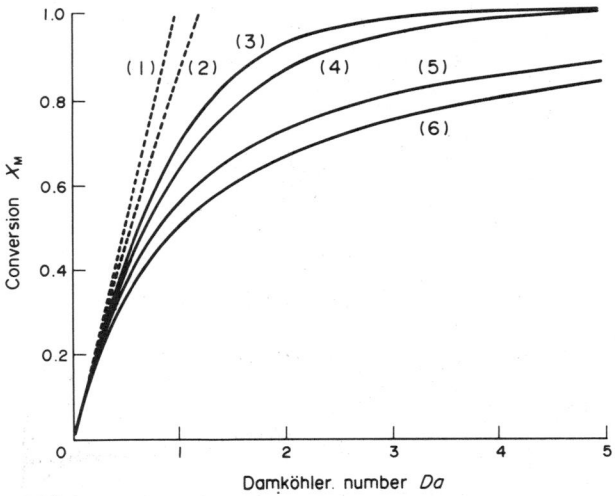

Figure 7 Conversion *vs.* Damköhler number for zero and first order reactions in different reactor types. Constant volume reactions are compared with variable volume reactions at $\varepsilon_M = -0.35$; (1) zero order, HCSTR, PFTR, cascade, constant volume and $\varepsilon_M = -0.35$, BR, constant volume; (2) zero order, BR, $\varepsilon_M = -0.35$; (3) first order, PFTR, $\varepsilon_M = -0.35$; (4) first order, BR, constant volume and $\varepsilon_M = -0.35$, PTFR, constant volume; (5) first order, HCSTR, $\varepsilon_M = -0.35$; and (6) first order, HCSTR, constant volume

number correlations hold for constant as well as variable volume reactions in the HCSTR, the PFTR and any cascade of HCSTRs. However, lower conversion is obtained in the batch reactor when the reaction volume decreases during the course of the reaction. In the case of a first order polymerization at constant temperature and initiator concentration, the batch reactor performance is independent of volume changes while HCSTR, cascade and PFTR yield higher performance when the volume contracts.

As an approximation, effects of volume contraction on reactor performance can be neglected for first order polymerizations if the desired conversion is low. Particularly in dilute systems such as solution polymerization, volume contraction is unimportant for reactor design. However, at high conversions or in case of a dead end polymerization considerable variations in the Damköhler number may occur. For example, in order to reach a conversion of $X_{M,st} = 0.95$, a Damköhler number of $Da = 19$ is required for a first order polymerization at constant volume in the HCSTR (Figure 7, curve 6) while $Da = 12.7$ for $\varepsilon_M = -0.35$ (Figure 7, curve 5). The Damköhler number decreases by a factor of $(1 + \varepsilon_M)$ at high conversion if volume contraction is taken into account. With regard to zero order batch-wise polymerizations volume contraction affects conversion significantly (Figure 7, curves 1 and 2) according to

$$X_M = \frac{\exp(\varepsilon_M Da)}{\varepsilon_M} - \frac{1}{\varepsilon_M} \approx Da + \frac{\varepsilon_M Da^2}{2} \quad \text{for} \quad Da \le \frac{\ln(1 + \varepsilon_M)}{\varepsilon_M} \approx 1 - \frac{\varepsilon_M}{2} \quad (7)$$

In emulsion and suspension polymerization only small changes in total dispersion volume are observed, which seem to be negligible. However, the effect on polymerization rate is comparable to that in bulk polymerization.

24.2.4 'Nonideal' Polymerization Kinetics

Although each set of reactant composition and reaction conditions of a polymerizing system results in an individual conversion–time curve, some common aspects may be represented by a 'typical' polymerization curve as illustrated in Figure 8.

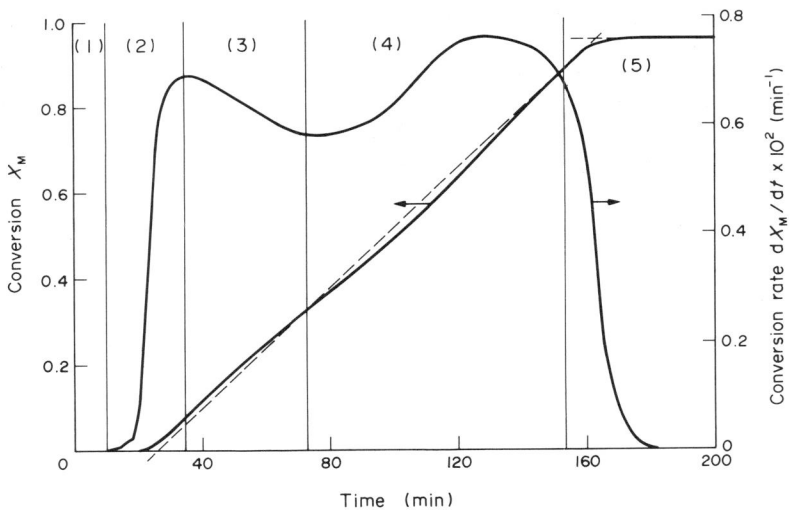

Figure 8 Conversion–time and rate–time curves of a typical polymerization at constant temperature with (1) inhibition, (2) induction, (3) period of ideal kinetics, (4) gel effect, and (5) glass effect

During the course of a typical polymerization, reaction rate and conversion may pass through the following stages:[61] (1) inhibition period where small amounts of strong radical scavengers inhibit the formation and growth of polymer radicals; (2) induction period where the reaction rate increases due to the decreasing amount of radicals which are consumed by retarding agents. In emulsion polymerization the reaction rate increases as a result of an increasing number of latex particles which contain a growing polymer radical during the latex particle nucleation period; (3) period of ideal kinetics which is observed in diluted systems and at low conversions, where the rate of reaction obeys almost first order kinetics. In emulsion polymerization a formal zero order kinetics is observed during the period of latex particle growth; (4) depending on the monomer and the monomer/polymer concentration an autoacceleration of the reaction rate occurs. This phenomenon is described as the gel or Trommsdorff effect and is observed in homogeneous as well as in heterogeneous polymerization systems; and (5) if the reaction temperature is lower than the glass transition temperature T_g of the reaction mixture, the rate of polymerization approaches zero. This phenomenon is called the glass effect.

No general kinetic model is yet available which enables *a priori* calculation of conversion–time curves of polymerization reactions.

The first step for designing a reactor for polymerization is to determine the conversion–time curve experimentally. For a rough estimate of reactor performance the conversion–time curve can often be approximated by a linear correlation which formally corresponds to zero order kinetics (dashed line in Figure 8). With this simplification, acceptable results can be obtained with regard to reactor volume calculation and comparison of reactor capacity of different reactor types. However, no information about the polymer properties can be deduced by this simplification.

If the influence of gel and glass effects on the course of polymerization can be included in the model of 'ideal' polymerization kinetics, certain additional data of polymer properties are attainable, *e.g.* molecular weight averages. The autoacceleration of the reaction rate can be described by a conversion-dependent decrease of the rate constant of the chain termination reaction. Exponential equations are often used, *e.g.*

$$k_t = k_{t,0} \exp(A_1 X_M + A_2 X_M^2 + A_3 X_M^3) \qquad \text{with} \quad A_i = f(T, C_M, C_I, \text{initiator type, } etc.) \tag{8}$$

where the constants A_i are determined by curve fitting to the experimental data.[62-67] Generally, the application of a set of values A_i is limited to a particular investigated system and range of reaction conditions. Equally, the glass effect may be described by a conversion-dependent decrease of the rate constant of the chain propagation reaction.

Without any calculation, reaction time for batch-wise or mean residence time in the case of continuous polymerization and consequently reactor volume as well as reactor performance may be determined graphically by plotting

$$\frac{n_{M,0}}{r V_R} \quad vs. \quad X_M \quad \text{or} \quad \frac{n_{M,0}}{r V_R} = \frac{C_{M,0} V_{R,0}}{\dfrac{C_{M,0}}{(1+\varepsilon_M X_M)} \dfrac{dX_M}{dt} V_{R,0}(1+\varepsilon_M X_M)} = \frac{1}{dX_M/dt} \quad vs. \quad X_M \tag{9}$$

when equations (5) and (6) are used for reactions with variable volume. The reaction time of a batch-wise polymerization is obtained as the area under the curve in Figure 9 according to

$$t_{BR} = C_{M,0} \int_0^{X_M} \frac{dX_M}{r} = \int_0^{X_M} \frac{1}{dX_M/dt} dX_M \tag{10}$$

and can be compared with the mean residence time of a continuous polymerization in an ideal homogeneous stirred tank reactor

$$\tau_{HCSTR} = C_{M,0} \frac{X_M}{r} = \frac{X_M}{dX_M/dt} \tag{11}$$

which is equal to the value of the rectangular area in Figure 9. In contrast to first order polymerizations (see Figures 5 and 6) the capacity of an HCSTR may be higher than the capacity of a BR in the case of autoaccelerating reactions. The capacity of a reactor depends on the actual polymerization kinetics and is not of fixed order.

The dashed line in Figure 9 stands for the approximation by a zero order reaction and results in the same performance for both reactor types.

It has to be kept in mind that the rate of reaction at a certain conversion in a free radical polymerization depends on the history of the system up to that point. Since the reaction environment differs in different reactor types this simple graphical method cannot be applied in general. If the initial charge or feed stream contains an inhibitor, for example, the batch-wise polymerization starts at the end of the inhibition period, when the inhibitor is completely consumed by initiator radicals, without any influence on the kind of the conversion–time curve. However, in the continuous stirred tank reactor unreacted inhibitor is constantly fed into the vessel, which reduces the

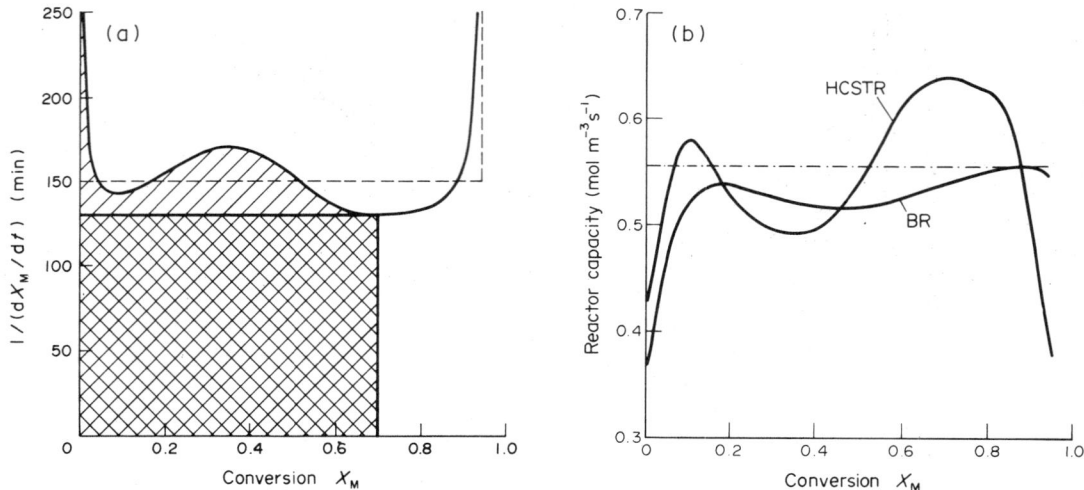

Figure 9 (a) Reciprocal conversion rate as a function of conversion for a typical polymerization with a gel effect at constant temperature. (b) Reactor capacity *vs.* monomer conversion for the batch reactor and the HCSTR with an initial monomer concentration $C_{M,0} = 5 \text{ mol l}^{-1}$. Data are taken from Figure 8 but without an inhibition period. The dashed lines represent the zero order approximation.

radical concentration, and hence the reaction rate is also reduced even at high conversion.[68–70] Therefore batch kinetics cannot be used unambiguously for the design of continuous reactors.

24.2.5 Effects of Segregation on Reactor Performance

Proper mixing of the reactants during reaction is important for fast polymerizations in homogeneous systems and for all heterogeneous systems. In particular, segregation is an important phenomenon for free radical polymerization. The degree of segregation indicates whether mixing occurs on a microscopic or macroscopic level.

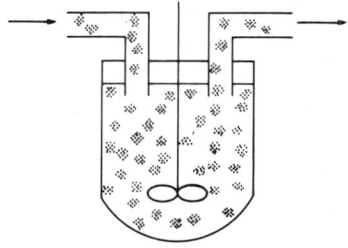

Figure 10 Completely segregated polymerization system (*e.g.* continuous suspension polymerization)

Since a fluid lump of a completely segregated fluid retains its identity, it can be regarded as a small batch reactor, which passes through the reaction volume and leaves the reactor with the exit stream according to its residence time distribution. Consequently the average conversion $\bar{X}_M$ of a completely segregated system is obtained by

$$\bar{X}_M = \int_{F=0}^{F=1} X_{M,\text{batch}}(t)\, \mathrm{d}F(t) = \int_{t=0}^{t=\infty} X_{M,\text{batch}}(t)\, E(t)\, \mathrm{d}(t) \tag{12}$$

where $F(t)$ and $E(t)$ are the cumulative and differential residence time distributions, respectively. Segregation has no effect on conversion in batch or plug flow reactors. However, it increasingly affects reactor performance as the residence time distribution shifts towards mixed flow.[71] A graphical method for the determination of average conversion in completely segregated systems was derived by Hofmann[72] and Schoenemann.[73] Its application on a typical polymerization (example from Figure 8) in a SCSTR and a cascade of four equal sized SCSTRs is illustrated in Figure 11.

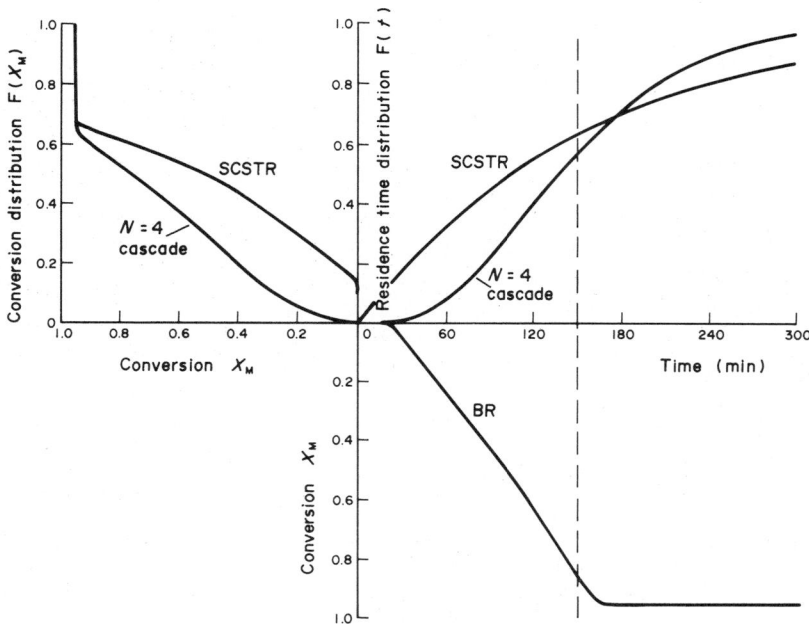

Figure 11 Average conversion and conversion distribution for a completely segregated continuous polymerization in a segregated stirred tank reactor compared with four equal sized segregated stirred tank reactors in series. The mean residence time for both reactor types is $\tau = 150$ min (dashed line) (according to refs. 72 and 73)

If the segregated aggregates keep their identity even in the final polymer particles, as is the case in many heterogeneous polymerization systems, the distribution of conversion of the individual polymer particles is important. In most cases a uniform conversion or at least a narrow conversion distribution is desired. The cumulative conversion distribution for the given example is shown in the second quadrant of the Hofmann–Schoenemann diagram in Figure 11.[74-79]

The effect of segregation on reactor performance depends strongly on kinetics. While for first order reactions conversion is unaffected by segregation, higher conversion is obtained in completely segregated systems for reaction orders greater than one and lower conversion for reaction orders smaller than one at the same Damköhler number. Especially when the course of polymerization obeys zero order kinetics, considerable reactor performance deviations are observed for microfluids and macrofluids over a wide range of conversion.[80,81] In Figure 12 conversion–Damköhler number curves for micromixed and completely segregated zero and first order reactions are compared.

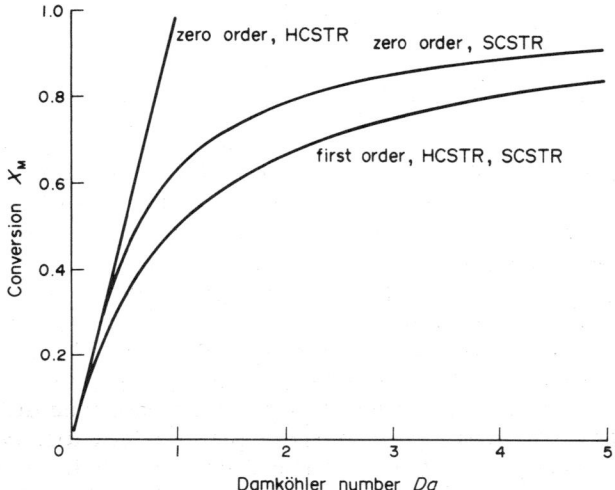

Figure 12 Conversion as a function of Damköhler number for completely segregated (SCSTR) as well as micromixed (HCSTR) and zero as well as first order polymerizations in a single continuous stirred tank reactor

In practice, polymerization systems are often partially segregated or the degree of segregation is changing during the course of polymerization, *e.g.* due to increasing viscosity in homogeneous systems[71] or reduced coalescence between polymerizing particles in heterogeneous systems.[82,83] Therefore, the actual stationary state conversion in a continuous stirred tank reactor will lie between the extremes of a micromixed and a completely segregated fluid. Numerous models for partial segregation have been suggested for calculating conversion at intermediate amounts of segregation.[82-93]

The flow pattern through a continuous stirred tank reactor may deviate considerably from the perfectly backmixed ideal HCSTR when the viscosity increases by orders of magnitude during the course of polymerization in homogeneous systems. Depending on the type and geometry of the stirrer, the reactor geometry and the position of inlet and outlet flow behavior may lead to the occurrence of crossflow, stagnancy, side capacity, channelling, bypassing, short-circuiting and recycle.[53,94-96] A large number of mixed or multiparameter models have been proposed to quantify the quality of mixing and to evaluate residence time distribution functions in terms of operating variables like the mean residence time and the stirrer speed as well as flow conditions characterized by the radial Reynolds number.[97-102] Furthermore, reactor performance of nonideal stirred tank reactors may be affected by type and size of the model compartments as well as by segregation. Figure 13 gives an example for a simple two parameter model of a nonideal stirred tank reactor with dead zone V_d and channeling V_p[97]

$$E(\Theta) = 0 \quad \text{for} \quad 0 \leq \Theta < \frac{V_p}{V_R} \quad \text{and} \quad E(\Theta) = \frac{V_R}{V_m} \exp\left\{ -\frac{V_R}{V_m}\left(\Theta - \frac{V_p}{V_R}\right) \right\} \quad \text{for} \quad \Theta \geq \frac{V_p}{V_R}$$

where Θ is the reduced residence time, $\Theta = t/\tau$; V_m, V_p and V_d are the perfectly mixed, the plug flow and the dead water volume respectively. V_R is the total reaction volume.

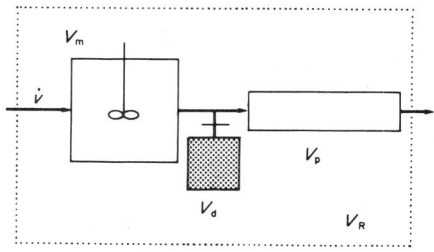

Figure 13

For a first order polymerization, average conversion of the multiparameter model $\bar{X}_{M,m}$ is calculated by using equation (12) and Table 3 as

$$\bar{X}_{M,m} = \int_{\Theta = \frac{V_p}{V_R}}^{\Theta = \infty} [1 - \exp(-Da\,\Theta)]\left[\frac{V_R}{V_m}\exp\left\{ -\frac{V_R}{V_m}\left(\Theta - \frac{V_p}{V_R}\right)\right\}\right] d\Theta = 1 - \frac{\frac{V_R}{V_m}\exp\left\{ -Da\frac{V_p}{V_m}\right\}}{Da + \frac{V_R}{V_m}} \tag{13}$$

The ratio between conversion of the nonideal stirred tank reactor $\bar{X}_{M,m}$ and the ideal one $X_{M,HCSTR}$ depends on the plug flow volume fraction, the dead water volume fraction and the Damköhler number or conversion. The latter two are corresponding values of a certain conversion–Damköhler number correlation which depends on the order of reaction (Figure 14).

Increasing the plug flow volume fraction yields higher performance of the multiparameter model and yield increases with increasing Damköhler numbers. Dead water zones lower the performance of the nonideal stirred tank reactor which again increases with increasing Damköhler numbers.

Reactor performance of polymerization reactions is a complex function of polymerization kinetics, reactor type and its residence time distribution as well as of degree of segregation and earliness of mixing. At low conversion levels, polymerization kinetics predominatly affect reactor performance, while X_M is relatively insensitive to changes in residence time distribution and degree of segregation. At medium conversion levels, the influence of residence time distribution is growing, while segregation still has little effect. At high conversion all these factors may be important. Since

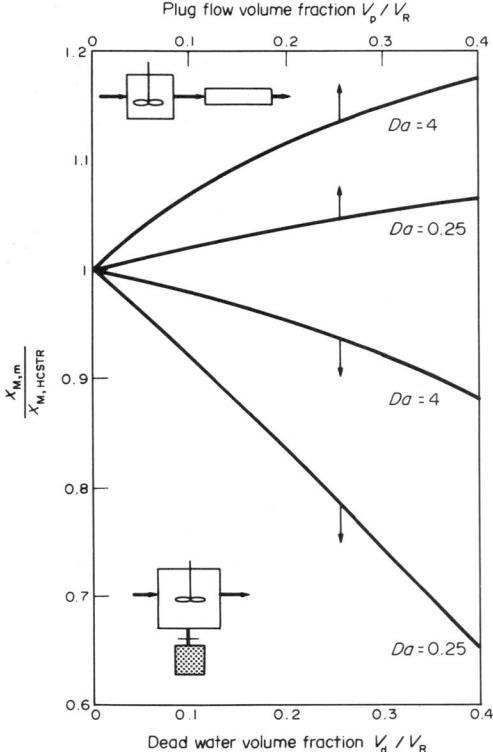

Figure 14 Ratio of average conversion of the nonideal stirred tank reactor (multiparameter model) to conversion in the HCSTR *vs.* plug flow and dead water volume fractions respectively. The curves are valid for a first order polymerization and Damköhler numbers of 0.25 and 4, which correspond to conversions of $X_{M,HCSTR} = 0.2$ and 0.8 respectively

polymerization kinetics, residence time distribution as well as degree of segregation may vary during the course of polymerization, the problem of calculating reactor performance may be rather complicated for the individual case.

24.3 REACTOR SELECTIVITY

One of the most important aims of reactor design is to optimize the selectivity of the polymerization. The molecular structure of polymers is not exclusively determined by the mechanism of polymerization (steps of reaction) but also influenced by reaction engineering parameters, *viz.* concentration and temperature gradients, feed conditions, residence time distribution and degree of segregation. Although the average life time of a growing polymer radical τ^* is of the order of seconds, usually $0.1 < \tau^* < 10$ s, and mean residence times of polymerization reactors τ are of the order of hours, the reactor type and operation mode considerably affect the properties of the polymer formed. Under isothermal operation conditions, time and location-dependent variation of reactant concentrations cause these differences in reactor selectivity.

Since in each polymerization an almost infinite number of parallel and consecutive reactions occur rather than a molecularly uniform product, polymers with a statistical molecular weight distribution are obtained. Together with other molecular parameters such as branching and crosslinking as well as chemical composition and sequence length distribution in the case of copolymerization, the molecular weight distribution determines polymer quality. In addition, particle size distribution and polymer morphology, *e.g.* porosity of polymer particles, phase separation, *etc.*, are affected by reactor type and operation conditions in heterogeneous systems. One of the greatest difficulties is to achieve quality control of the polymer product with regard to end use properties which are often nonmolecular quantities such as thermal stability, tensile strength, impact and stress crack resistance, color and clarity, plasticizer uptake, *etc.* The quantitative relationship between these product properties and molecular properties and therewith the operating conditions of the polymerization reactor may be the least understood area in polymer reaction engineering. On-line measuring devices for molecular as well as end use properties of the polymer are needed in order to achieve a better control of reactor selectivity.

The essentials of interdependences of polymerization reactions and polymerization reactors with respect to molecular weight distribution and copolymer composition have been summarized by Gerrens[1,4] and others.[6,10,12,16,24,25] The mathematical treatment of reactor selectivity has been reviewed by Ray.[14]

24.3.1 Molecular Weight Distribution

The simplest reaction scheme of free radical polymerization contains three reaction steps, *viz.* radical initiation, propagation and termination by disproportionation or combination. If volume contraction and initiator consumption are neglected and monomer concentration is assumed to be constant, the Schulz–Flory distribution is obtained for the molecular weight distribution, $W(P)$.[103,104] In the case of chain termination by disproportionation $W(P)$ is derived as

$$W(P) = P(1-\alpha)^2 \alpha^{P-1} \qquad \text{with} \qquad \alpha = \frac{k_p C_M}{k_p C_M + (2 f k_d k_{t,d} C_{I,0})^{0.5}} \tag{14}$$

where P is the degree of polymerization and α the probability of propagation.[105] $k_{t,d}$ is the rate constant of radical termination by disproportionation. Number and weight average degree of polymerization, P_n and P_w, and the width of the molecular weight distribution, which is commonly expressed by the dispersion index Q, are given by

$$P_n = \frac{1}{1-\alpha} \qquad P_w = \frac{1+\alpha}{1-\alpha} \qquad Q = \frac{P_w}{P_n} = 1 + \alpha = \frac{\lambda_2 \lambda_0}{\lambda_1^2} \approx 2 \tag{15}$$

where λ_i is the *i*th moment of the molecular weight distribution.[106]

These equations hold for the HCSTR at steady state operation and any conversion, as well as for batch-wise polymerization within small conversion increments where the monomer concentration can be regarded as constant (instantaneous molecular weight distribution).[107–110]

Monomer concentration decreases as a function of reaction time in the batch reactor and is dependent on location in the plug flow reactor. Therefore, the balance equations have to be integrated for the entire conversion interval. The resulting cumulative molecular weight distribution is broader than the Schulz–Flory distribution (Figure 15).[107–117]

The fluid lumps in a completely segregated continuous stirred tank reactor may be regarded as small batch reactors which leave the reactor according to its residence time distribution following

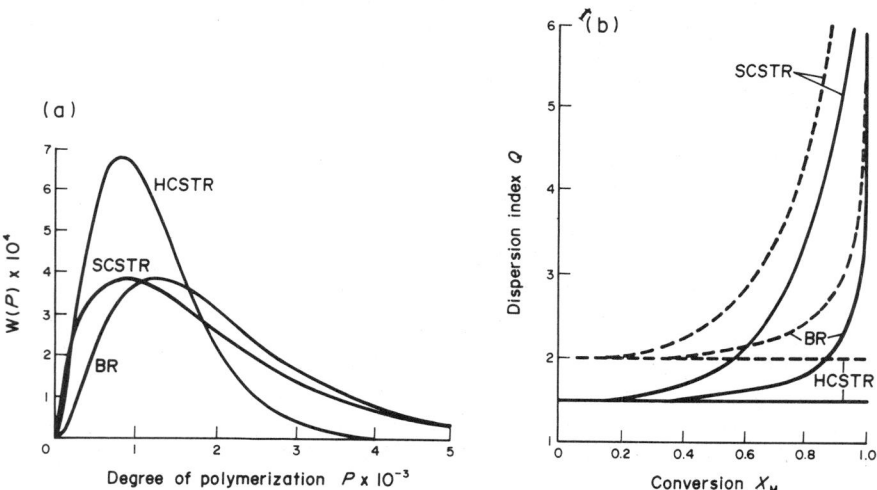

Figure 15 (a) Molecular weight distribution in three different types of ideal reactors for a simple reaction scheme (initiation, propagation, termination by combination). Conversion $X_M = 0.6$, $P_{n,0} = 1000$ (adapted from ref. 113). (b) Dispersion index as a function of monomer conversion in BR, HCSTR and SCSTR, ––––– termination by disproportionation; ——— termination by combination (adapted from refs. 1 and 4)

the relationship

$$E(t) = \frac{1}{\tau} \exp\left\{-\frac{t}{\tau}\right\} \qquad (16)$$

Consequently, a broad conversion distribution, $F(X_M)$, is obtained (see Hofmann–Schoenemann diagram, Figure 11) and the molecular weight distribution is broader than in the case of batch-wise polymerization and broadens with increasing conversion.[112,113]

Dispersion index and molecular weight distributions are compared in Figure 15. The narrowest possible molecular weight distribution, the Schulz–Flory most probable distribution, is obtained in the HCSTR, a broader one in the BR or PFTR, and the broadest in the SCSTR.

However, most free radical polymerizations are much more complex and even in the case of homogeneous homopolymerization further reaction steps must be considered, such as transfer reactions to monomer, solvent and polymer, thermal initiation, primary radical termination and terminal double bond polymerization.[23,105,106] In addition, reactor selectivity is affected by decreasing initiator concentration during the course of polymerization which in the limiting case results in dead end polymerization.[118] Diffusion control of propagation (glass effect), termination (gel effect) and terminal double bond polymerization (gel effect) severely influence molecular weight distribution and chain branching.[81,119–129]

The effects of decreasing initiator concentration and various dead end conversions on number and weight average degree of polymerization, P_n and P_w, are illustrated for isothermal batch polymerizations in Figure 16.[23,118] Compared with a constant rate of initiation (curve 1), both P_n and P_w decrease only moderately with increasing conversion if the initiator consumption is considered (curve 2). They increase during the course of polymerization at low values of dead end polymerization (curve 4). The molecular weight distribution for the BR and the SCSTR may be narrower at decreasing initiator concentration than at a constant rate of initiation. On the other hand, the dispersion index obtained in these reactor types increases considerably as dead end conversion is approached.

At constant monomer and initiator concentration (HCSTR) primary radical termination as well as transfer reactions to monomer or solvent do not change the shape of the molecular weight distribution in the case of radical termination by disproportionation, but they do change it if radical termination occurs by combination. Figure 17 shows the dispersion index Q for a polymerization in

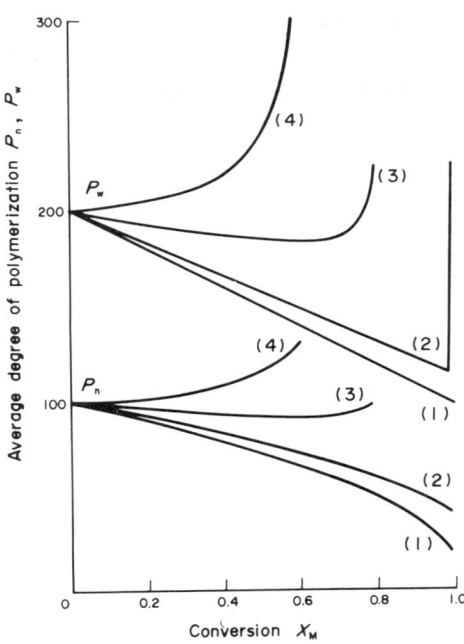

Figure 16 Number and weight average degrees of polymerization as a function of conversion. Several dead end polymerizations are compared with a batch-wise polymerization at a constant rate of initiation. No transfer reactions occur and radicals terminate by disproportionation. Adjustment of the rate constants has been made to obtain a constant initial value $P_n = 100$. (1) Constant rate of initiation (no dead end polymerization); (2) dead end conversion $X_{M,\infty} = 0.99$; (3) dead end conversion $X_{M,\infty} = 0.8$; and (4) dead end conversion $X_{M,\infty} = 0.6$ (adapted from ref. 23)

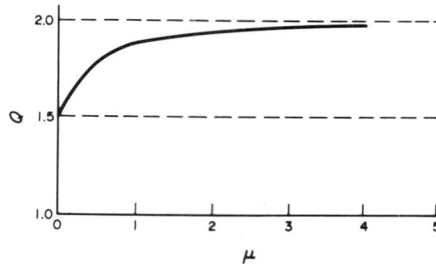

Figure 17   Dispersion index as a function of the parameter μ (see equation 17) at constant monomer concentration (HCSTR) (adapted from ref. 23)

a HCSTR as a function of the parameter μ

$$\mu = \left(1 + \frac{k_{t,d}}{k_{t,c}}\right)^{0.5} \times \frac{k_{tr,M}C_M + k_{tr,S}C_S}{(2fk_dk_{t,c}C_I)^{0.5}} + \frac{k_{t,d}}{k_{t,c}} \qquad (17)$$

which represents the ratio between radical termination by disproportionation and transfer reactions to monomer and solvent on one side and radical termination by combination on the other side. $k_{tr,M}$ and $k_{tr,S}$ are the rate constants of transfer reactions to monomer and solvent respectively, while $k_{t,c}$ is the rate constant of radical termination by combination. The dispersion index approaches $Q = 2$ at relatively low values of the parameter μ.[23]

During batch-wise polymerization at variable monomer concentration but constant rate of initiation the breadth of the molecular weight distribution is unaffected whether or not transfer reactions to solvent occur if polymer radicals are terminated by disproportionation (Figure 18, curve 1). In the presence of transfer reactions to monomer there is only little change in the distribution at low conversion while at high conversion it is narrower with transfer reactions than without (Figure 18, curve 2). In the case of radical termination by combination transfer reactions to solvent shift the entire Q–X_M curve to higher values of the dispersion index. The initial breadth of the distribution Q ($X_M = 0$) corresponds to the dispersion index at $\mu = 0.25$ in Figure 17, which has been obtained for the HCSTR. Therefore the influences of varying monomer concentration and different rate constants of chain transfer reactions to solvent may be evaluated by comparing the graphs for batch-wise polymerization when $\mu = 0$ (Figure 18, curve 5) and for variable μ at constant C_M (Figure 17).

Polymer properties for the SCSTR, such as the degree of polymerization, $P_{i,\text{SCSTR}}$ may be derived from a corresponding property in the batch reactor, $P_{i,\text{BR}}$, similar to Equation (12).[71,81]

$$P_{i,\text{SCSTR}} = \int_0^\infty P_{i,\text{BR}} E(\Theta) d\Theta = \frac{1}{\tau} \int_0^\infty P_{i,\text{BR}} \exp\left\{-\frac{t}{\tau}\right\} dt \qquad (18)$$

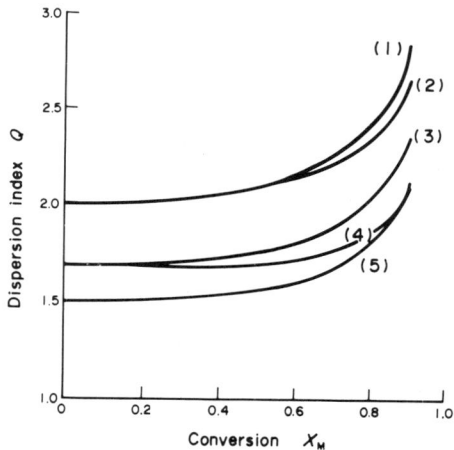

Figure 18   Dispersion index as a function of conversion for the BR. The rate of initiation is kept constant. (1) Termination by disproportionation, transfer to solvent may or may not occur; (2) termination by disproportionation, transfer to monomer, $\mu = 0.25$; (3) termination by combination, transfer to solvent only, $\mu = 0.25$; (4) termination by combination, transfer to monomer only, $\mu = 0.25$; and (5) termination by combination, no transfer reaction (adapted from ref. 23)

Peebles[23] has analyzed in detail the effects of various reactions on the molecular weight distribution obtained in free radical polymerization.

In addition to the effects discussed previously, diffusion limitation of radical termination and propagation reactions have to be considered in concentrated polymerization systems. Numerous models have been derived describing the molecular weight distribution or at least certain moments of this distribution when gel and glass effects are involved.[130-142] For example, in Figure 19 experimental and calculated number and weight average molecular weights, M_n and M_w, are given as a function of monomer conversion for batch-wise bulk polymerization of methyl methacrylate.[130] With the onset of gel effect at a conversion of about 0.3 the weight average molecular weight increases drastically, while there is only a slight increase in number average molecular weight. For comparison, M_n and M_w obtained from conventional kinetics at constant $k_{t,d}$ are shown.

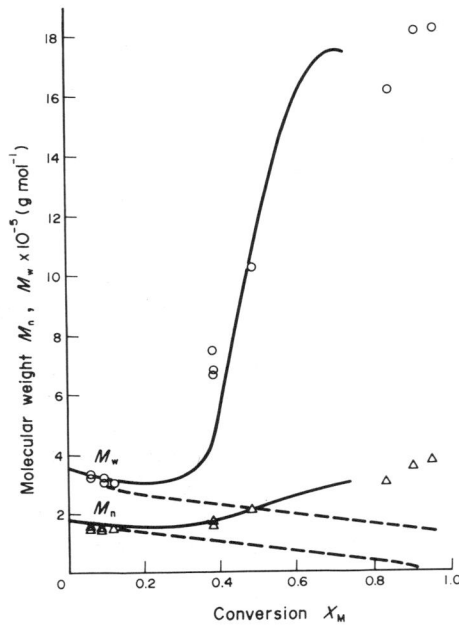

Figure 19 Number and weight average molecular weights, M_n and M_w, *vs.* monomer conversion for the bulk polymerization of methyl methacrylate at 70 °C. ——— $C_I = 0.3$ wt. % AIBN, M_w, M_n; ------ conventional kinetics (constant $k_{t,d}$) (experimental data adapted from ref. 130, calculated curves from ref. 131)

With regard to selectivity in different reactor types, equation (18) may be used to calculate the molecular properties for the SCSTR. Although theoretical estimations of the molecular weight distribution in an HCSTR are possible as gel and glass effects occur, in practice most bulk polymerizations are segregated at higher conversions.

24.3.2 Long Chain Branching and Terminal Double Bond Polymerization

Transfer reaction to polymer, as well as the polymerization of a terminal double bond of a polymer which is produced through radical termination by disproportionation and transfer to monomer, lead to long chain branching. Experimental and theoretical results for the branching of poly(vinyl acetate), which has a high rate constant of transfer reaction to polymer *via* the acetate group, show that this transfer reaction especially, as well as terminal double bond polymerization, is severely sensitive to residence time distribution and mixing effects.[81,119-129] This can be explained by the sensitivity of the second reaction in a series of consecutive reactions.

In a batch-wise polymerization the transfer reaction to polymer by itself cannot lead to gelation. On the other hand, terminal double bond polymerization causes P_w to increase faster than P_n, finally forming gel at high conversion.

In the HCSTR the concentration of polymer is larger than the average polymer concentration in the BR at a given conversion. In addition, it is possible for an individual molecule to remain in the reactor for a considerably longer period of time than the mean residence time. Consequently, if transfer to polymer occurs, some molecules grow extremely large and as a result infinite weight average molecular weights are obtained at intermediate conversions. Thus, in contradiction to Denbigh's theory[110] the HCSTR may produce a broader molecular weight distribution than the BR and even a broader one than obtained in the SCSTR, although the lifetime of the growing radicals is short compared with the mean residence time.

In Figure 20 experimental and theoretically predicted values of the polydispersity during vinyl acetate polymerization in the three reactor types BR, HCSTR and SCSTR are shown. This is one of the most complex homogeneous homopolymerization systems. Taylor and Reichert[81] have considered chain transfer reactions to monomer, solvent and polymer as well as terminal double bond polymerization in addition to initiation, propagation and termination by disproportionation. Furthermore, gel effect correlations for radical termination reactions and terminal double bond polymerization have been implemented in the model. Since terminal double bond polymerization requires two polymer molecules to occur, the contribution from this reaction should decrease as viscosity increases during bulk or suspension polymerization. Due to different initiator types and concentrations and different monomer concentrations the breadth of the molecular weight distribution in the HCSTR can be greater or less than that of the SCSTR (compare Figure 18a and b). If the SCSTR polymerizations are performed at higher monomer concentrations than those of the

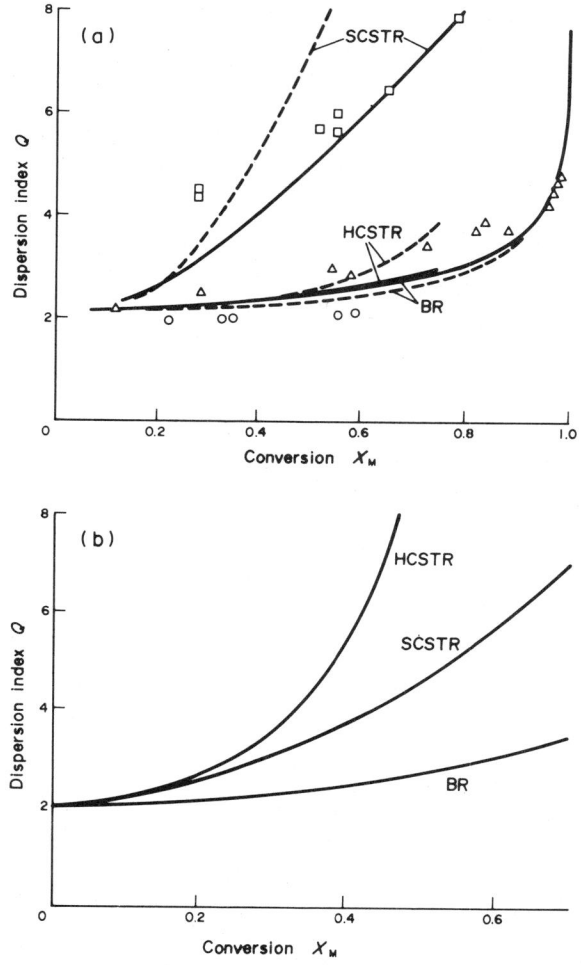

Figure 20   (a) Dispersion index as a function of monomer conversion in three different types of ideal reactors, for suspension (BR, SCSTR) and for solution polymerization (HCSTR) of vinyl acetate (VA). Predictions with (––––) and without (———) terminal double bond polymerization gel effect; all reactors with the same value of polymer chain transfer rate constants, $C_{tr,P}$ $= 2.36 \times 10^{-4}$, $C_{tr,S} = 0.34 \times 10^{-4}$, $k_d = 2.3 \times 10^{-4}\,\mathrm{s}^{-1}$, $C_{I,0} = 1.627 \times 10^{-2}\,\mathrm{mol\,l^{-1}}$, $C_{TDBP} = 0.27$. BR ($\triangle$); SCSTR ($\square$); HCSTR ($\bigcirc$) (adapted from refs. 81 and 119). (b) Experimentally and theoretically observed polydispersity as a function of monomer conversion in a BR, HCSTR, and an SCSTR for bulk and solution polymerization of VA. $k_d = 9 \times 10^{-6}\,\mathrm{s}^{-1}$, $C_{TDBP} = 0.66$, $C_{tr,P}$ $= 2.36 \times 10^{-4}$, $C_{tr,S} = 0.34 \times 10^{-4}$ (adapted from refs. 120–129)

HCSTR, or if a high concentration of a slow decomposing initiator is used, Q in the SCSTR can be higher than in the HCSTR. Under comparable reaction conditions a broader molecular weight distribution in the HCSTR is obtained than in the batch reactor. The order of the molecular weight distribution in the three ideal reactor types investigated is not fixed. The most important factors in determining the order of polydispersity in different reactor types are the degree of chain branching and the monomer and initiator concentration (Table 5).

Table 5 Effect of Concentration and Rate Constant on the Molecular Weight Distribution in Vinyl Acetate Polymerization[119]

Variable increased	Effect on Q
Monomer concentration	Increases
Initiator concentration	Decreases
Propagation rate constant	Decreases
Initiator decomposition rate constant	Decreases
Polymer chain transfer rate constant	Increases
Terminal double bond polymerization rate constant	Increases

It has to be carefully considered that one of the most important aspects of any investigation involving mathematical modelling of radical polymerization is a proper selection of the constants used.

24.3.3 Copolymer Composition

The most important polymer property for copolymers is usually the copolymer composition distribution. The instantaneous composition of the monomers M_1 and M_2 in the produced copolymer is determined for the HCSTR and the batch reactor within small conversion intervals by the following well-known copolymer equation[143–146]

$$\frac{dC_{M,1}}{dC_{M,2}} = \frac{C_{M,1}}{C_{M,2}} \frac{r_1 C_{M,1} + C_{M,2}}{r_2 C_{M,2} + C_{M,1}} \quad \text{with} \quad r_1 = \frac{k_{11}}{k_{12}} \quad r_2 = \frac{k_{22}}{k_{21}} \tag{19}$$

where r_1 and r_2 are the copolymerization parameters. If a certain copolymer composition is desired, the comonomer concentration ratio in the feed of an HCSTR has to be adjusted according to

$$\frac{C_{M,1}}{C_{M,2}} = \frac{1}{2r_1} \left\{ \left(\frac{m_1}{m_2} - 1 \right) + \left[\left(\frac{m_1}{m_2} - 1 \right)^2 + 4r_1 r_2 \frac{m_1}{m_2} \right]^{0.5} \right\} \quad \text{with} \quad F_1 = \frac{m_1}{m_1 + m_2} = 1 - F_2 \tag{20}$$

where F_1 and F_2 are the molar fractions of M_1 and M_2 in the copolymer produced.[149]

The instantaneous composition of the copolymer is identical to that of the monomer mixture only for azeotropic compositions or ideal copolymerization systems where $r_1 = r_2 = 1$. During copolymerization of all other binary systems in a batch reactor or in an SCSTR one of the monomers will be copolymerized more rapidly than the other and therefore its molar fraction in the monomer mixture will decrease gradually with increasing conversion. As a result, a nonuniform copolymer will be formed and phase separation of incompatible polymers may occur. In general, this is unwanted for technical copolymers. Although these problems are avoided by copolymerization in an HCSTR, it can be difficult to reach micromixing in concentrated systems at high conversion. Due to the high viscosity the reaction mixture will be partly segregated.

The influence of different reactor types on the copolymer composition distributions for the system methyl methacrylate/vinyl acetate is illustrated in Figure 21.[147–148] The HCSTR produces a completely uniform copolymer composition independent of conversion. The average copolymer compositions in the batch reactor and the SCSTR, given by the open circles in Figure 21, deviate little from those obtained in the HCSTR. On the other hand, the SCSTR produces a much broader copolymer composition distribution than the BR at a given conversion. Even at low average conversion there are some fluid lumps in the SCSTR reaching almost complete conversion so that almost pure monomer M_2 (vinyl acetate) is polymerized. Consequently, the copolymer composition

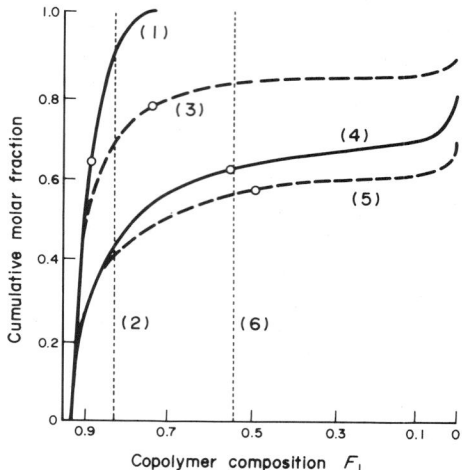

Figure 21 Copolymer composition *vs.* copolymerization distribution for the copolymerization system MMA ($r_1 = 20$)/VA ($r_2 = 0.015$) at two levels of conversion in the three reactor types BR, HCSTR and SCSTR. Conversion for curves (1), (2) and (3) $X_M = 0.35$, and for curves (4), (5) and (6) $X_M = 0.73$. The feed composition in all cases is $f_{1,0} = C_{M,1,0}/(C_{M,1,0} + C_{M,2,0}) = 0.4$. SCSTR (-----), batch reactor (———), HCSTR (····), average copolymer composition (○) (adapted from ref. 148)

distribution broadens with increasing conversion. At high conversion copolymer which contains large amounts of either monomer M_1 (methyl methacrylate) or monomer M_2 (vinyl acetate) is obtained from both reactor types BR and SCSTR. Finally, at almost complete conversion the copolymer composition distribution in the batch reactor approaches that obtained in the SCSTR.

A widely used operation mode which provides copolymers with a narrow copolymer composition distribution is to feed the faster reacting comonomer into the reactor during semibatch-wise polymerization.

In heterogeneous copolymerization systems Mayo's copolymerization equation (19) has to be modified because of the varying solubility of the two comonomers in the water and oil phases during the course of polymerization. Therefore Nernst's partition coefficients of the comonomers between the water and oil phases, α and β, and the phase ratio, ξ, of monomer phase volume V^* to water phase volume V^{**} are introduced and the copolymer equation can be written as

$$\frac{dC_{M,1}}{dC_{M,2}} = \frac{1 + r_1 \dfrac{1 + 1/(\beta\xi)}{1 + 1/(\alpha\xi)} \dfrac{C_{M,1}}{C_{M,2}}}{1 + r_2 \dfrac{1 + 1/(\alpha\xi)}{1 + 1/(\beta\xi)} \dfrac{C_{M,2}}{C_{M,1}}} \quad \text{with} \quad \alpha = \frac{C^*_{M,1}}{C^{**}_{M,1}} \quad \beta = \frac{C^*_{M,2}}{C^{**}_{M,2}} \quad \xi = \frac{V^*}{V^{**}} \tag{21}$$

where C^*_M and C^{**}_M are the monomer concentration in the latex particles and in the water phase respectively.[150] During batch-wise polymerization, the phase ratio ξ does not change significantly so that it is convenient to use the modified copolymerization parameters, r'_1 and r'_2, for practical calculation

$$r'_1 = r_1 \frac{1 + \dfrac{1}{\beta\xi}}{1 + \dfrac{1}{\alpha\xi}} \qquad r'_2 = r_2 \frac{1 + \dfrac{1}{\alpha\xi}}{1 + \dfrac{1}{\beta\xi}} \tag{22}$$

The copolymer composition depends strongly on the oil/water ratio ξ. Figure 22 illustrates the variation of the copolymer composition for several oil/water ratios for a certain set of copolymer parameters and Nernst's partition coefficients. In this example, comonomer M_1 is equally distributed between the oil and water phases ($\alpha = 1$), while comonomer M_2 is almost insoluble in the water phase ($\beta = \infty$). In homogeneous copolymerization as well as at high oil/water phase ratios in emulsion polymerization the molar fraction of M_1 in the copolymer is always greater than in the monomer mixture ($F_1 > f_1$). In the case of an oil/water ratio of $\xi = 1$, which is of practical interest for copolymerization in an emulsion, an almost ideal copolymerization system with $F_1 \approx f_1$ and $r'_1 \approx r'_2 \approx 1$ is obtained and as a result an almost uniform copolymer can be produced independent of

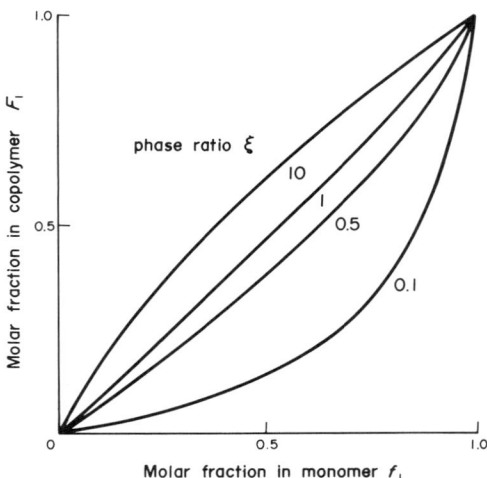

Figure 22 Copolymer composition diagram, F_1 *vs.* f_1, for a copolymerization in emulsion at different oil/water ratios from $0.1 < \xi < 10$. The copolymerization parameters are $r_1 = 1.5$, $r_2 = 0.5$; and the partition coefficients $\alpha = 1$, $\beta = \infty$ (adapted from ref. 150)

the reactor type. Consequently, narrow copolymer composition distributions for nonazeotropic monomer mixtures may not only be obtained by varying the operation mode but the oil/water phase ratio can be used as an additional parameter.

24.4 REACTOR STABILITY AND SAFETY

Reactor stability and reactor safety are by no means synonymous concepts of reaction engineering. Thus in practice appropriate reactor control can permit safe operation of polymerization reactors at operating points which are unstable in the sense of stability theory of chemical processes.[151,152] Conversely, global stability does not guarantee safe operating behavior of reactions; runaway reactions may possibly occur. The final operating state of batch reactors is always definite and stable, which is in contradiction with the behavior of continuous stirred tank reactors. Nevertheless, safe operation can be severely disturbed by parametric sensitivity, particularly in the case of strong exothermic polymerizations. For practical consideration it is quite unimportant if a runaway reaction has to be interpreted as a problem of stability or extreme parametric sensitivity.

Problems of reactor stability and safety become more important with increasing reaction enthalpy, reactor size and in particular for polymerizations with increasing viscosity and decreasing heat removal during the course of polymerization.

24.4.1 Thermal Stability

Reaction temperature severely affects the dynamic behavior of polymerization reactors due to an exponential increase in reaction rate with increasing temperature according to Arrhenius' law. Compared with temperature effects, nonlinear influences of reactant concentrations on the reaction rate are of minor importance for the analysis of thermal reactor stability. Therefore a suitable approach of the reactor dynamics is obtained by assuming simple zero or first order kinetics as an approximation of the polymerization kinetics.

Crossing certain stability limits may give rise to unstable reactor behavior. Both static and dynamic instabilities may occur. The reactor is called static unstable if reaction temperature and conversion rapidly jump from low to high values or *vice versa* in the sense of 'ignition' and 'extinction' of the polymerization reactor. On the other hand, dynamic instabilities give rise to periodic changes of temperature and conversion where certain phase shifts are observed.[153-158]

Whether or not instabilities will occur and what kind of instability will be encountered depends on reactor type and operating conditions. Thermal instabilities have been observed in a number of continuous stirred tank reactors. Examples are the perfectly mixed reactors as well as reactors with nonideal flow patterns, adiabatic and cooled reactors, and reactors in series configuration. Thermal instability may also occur in nonideal tubular reactors with axial dispersion, in tubular reactors with recycle or in autothermally operating tubular reactors (thermal recycle).[159-161]

Since temperature and conversion are independent of location (lumped parameters), the dynamic behavior of the ideal continuous stirred tank reactor is described by two ordinary differential equations, the heat and mass balance, which are coupled by the reaction term. In order to simplify the balance equations, volume contraction is neglected and the feed temperature T_0 is assumed to equal the mean value of the cooling temperature T_{jacket}. For an irreversible first order exothermic polymerization in bulk using constant initiator concentration the mass balance of HCSTR is given by

$$\tau\frac{dX_M}{dt} = Da(1-X_M) - X_M \quad \text{or} \quad \frac{dC_M}{dt} = \frac{C_{M,0} - C_M}{\tau} - C_M k_\infty \exp\left\{\frac{-E_A}{RT_R}\right\} \quad \text{with} \quad k_\infty \exp\left\{\frac{-E_A}{RT_R}\right\} = k = k_p\left(\frac{2f k_d C_{I,0}}{k_t}\right)^{0.5} \quad (23)$$

where k_∞ and E_A are the preexponential factor and the overall activation energy according to Arrhenius' law respectively. The heat balance is given by

$$V_R\rho\frac{d(c_P T_R)}{dt} = (\dot{v}\rho c_P + hA)(T_{jacket} - T_R) + V_R(-\Delta H_R)kC_M \quad \text{or} \quad \tau\frac{dT_R}{dt} = (1+St)(T_{jacket} - T_R) + \Delta T_{ad}Da(1-X_M)$$

$$\text{with} \quad St = \frac{hA}{\dot{v}\rho c_P} \quad \text{and} \quad \Delta T_{ad} = \frac{(-\Delta H_R)C_{M,0}}{\rho c_P} \quad (24)$$

where St and ΔT_{ad} are the dimensionless Stanton number (a measure of cooling intensity) and the adiabatic temperature rise respectively.[162] For constant operating conditions $dT_R/dt = 0$ and $dC_M/dt = 0$ are required and hence the reacting state is obtained by equating the mass (23) and heat balance (24) to give

$$\underbrace{(\dot{v}\rho c_P + hA)(T_{jacket} - T_{R,st})}_{\dot{Q}_{cooling}} = \underbrace{\dot{v}C_{M,0}(-\Delta H_R)\frac{\tau k_\infty \exp\left\{-\frac{E_A}{RT_{R,st}}\right\}}{1 + \tau k_\infty \exp\left\{-\frac{E_A}{RT_{R,st}}\right\}}}_{\dot{Q}_{chem.}} \quad (25)$$

where the rate constant k is expressed by Arrhenius' law. In general, the reaction temperature $T_{R,st}$ is determined graphically (Figure 23a). The rates of heat production $\dot{Q}_{chem}$ and heat removal $\dot{Q}_{cooling}$ are plotted vs. temperature. At an intersection of both curves, equation (25) is obeyed, so that steady operation is possible at the corresponding reaction temperature $T_{R,st}$. Furthermore, it is seen from Figure 23a that several solutions may exist, which is called the multiplicity of steady states. In the case of three points of intersection, those at the lowest and highest reaction temperatures represent stable conditions. The stability of these two operating points is related to the fact that the slope of the heat removal line is greater than that of the heat production curve. Thus, a negative deviation from the intersection temperature will cause stronger heat production than heat removal. Consequently, the temperature will return to the original steady state. Similarly, a positive deviation indicates stronger heat removal than heat production, thus decreasing the polymerization temperature. In the intermediate intersection point the situation is reversed, since the slope of the heat removal line is smaller than that of the heat production curve. Any positive temperature deviation will be amplified until the reactor operates in the upper stable point ('ignition' of the reaction). The reaction will 'extinguish' if the reaction temperature falls below the steady state value and finally the reactor operates on the lower branch. This $\dot{Q}$ vs. T_R diagram provides no information about the dynamic behavior of the reactor.

The rate of heat production is closely related to monomer conversion. Thus, equation (25) can be rewritten as equation (26)[163]

$$\underbrace{(\dot{v}\rho c_P + hA)(T_j - T_{R,st})}_{\text{transport line}} = \dot{v}(-\Delta H_R)(C_{M,0} - C_{M,st}) = \underbrace{\dot{v}C_{M,0}(-\Delta H_R)\frac{\tau k_\infty \exp\left\{\frac{-E_A}{RT_{R,st}}\right\}}{1 + \tau k_\infty \exp\left\{\frac{-E_A}{RT_{R,st}}\right\}}}_{\text{reaction curve}} \quad (26)$$

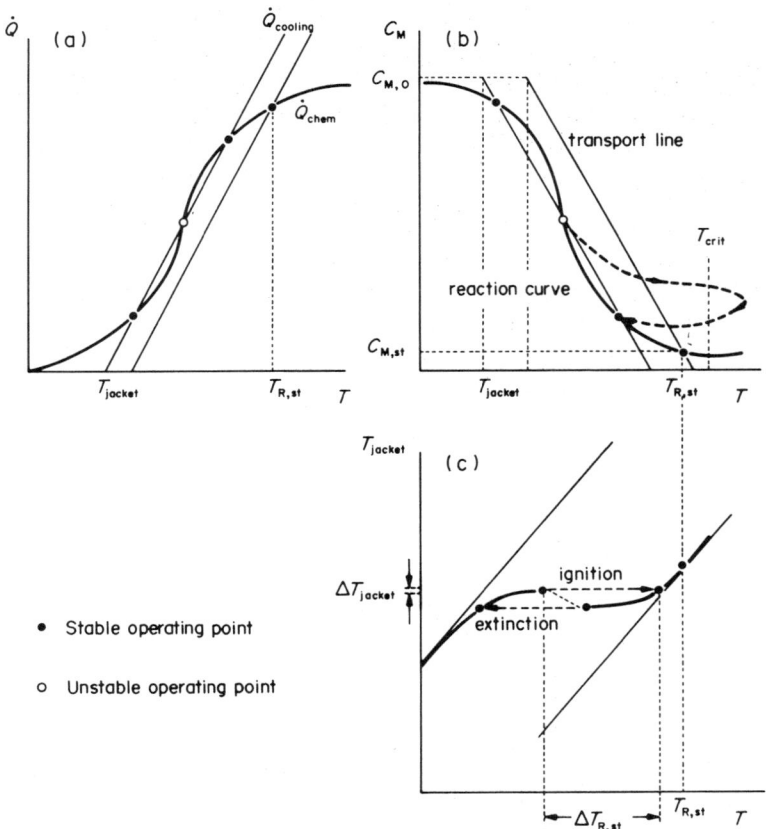

Figure 23 Steady state behavior of a HCSTR with cooling. (a) Rate of heat production $\dot{Q}_{chem}$ and heat removal $\dot{Q}_{cooling}$, as a function of reaction temperature for an exothermic first order polymerization. (b) Transport line and reaction curve in the phase plane, C_M *vs.* T_R.[163] (c) Ignition and extinction of the reactor for multiplicity of steady states. The sensitivity of the reaction temperature to a small variation in jacket temperature is indicated

which directly leads to the discussion of the dynamic reactor behavior in the phase plane, C_M *vs.* T (Figure 23b). Corresponding steady state values of reaction temperature $T_{R,st}$ as well as monomer concentration $C_{M,st}$ can be read from the intersection of the transport line and the reaction curve.

Furthermore, the trajectories, which represent the change of the reactive state as a function of time can be shown as in Figure 23b. For example, corresponding operation values of C_M and T_R during the transition from an unstable to a stable operation point are given as a dashed line. The reactor does not directly approach steady state, but may considerably overshoot steady state temperature and monomer concentration. This is illustrated in Figure 24 for a bulk polymerization of styrene, where the reaction temperature accelerates from $T_{R,st} = 130\,°C$ to $T_R = 240\,°C$ and almost complete conversion is obtained. The criterion for static stability for this first order polymerization is

$$\frac{1 + St}{\Delta T_{ad}} > \frac{E_A}{RT_{R,st}^2} \frac{Da}{(1 + Da)^2} \tag{27}$$

Temporarily, temperatures might exceed critical values during runaway reactions involving unwanted consecutive reactions and depolymerization. In order to describe the transient and dynamic behavior of the reactor, mass and heat balance have to be solved simultaneously. Only for large scale reactors is it permissible to neglect the heat capacity of the vessel itself.

Changes of viscosity which affect both the heat transfer coefficient and power input by stirring must be considered when temperature and conversion are allowed to vary.[165-171] The decrease of the heat transfer coefficient with the increase in viscosity for a stirred tank reactor with an anchor type stirrer is shown in Figure 25.[164] For this moderate increase in viscosity from 35 to 720 mPa s the heat transfer coefficient on the product side, α_i, falls to almost one third of its original value.

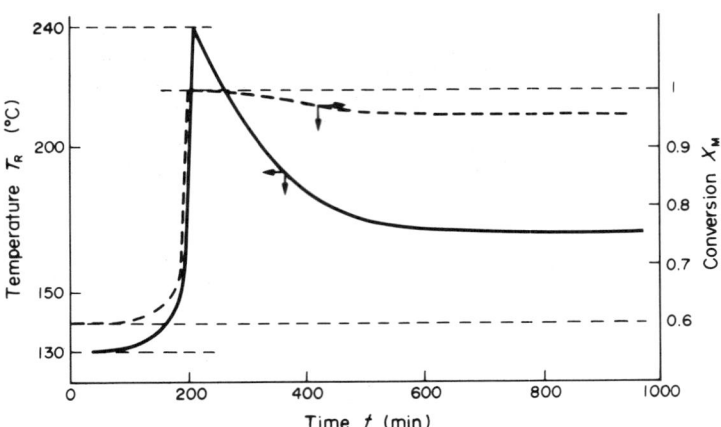

Figure 24 Dynamic reaction behavior during the transition from an unstable to a stable operating point during bulk polymerization of styrene in a HCSTR. $k_\infty = 1.39 \times 10^9$ min^{-1}, $E_A = 89$ kJ mol^{-1}, $-\Delta H_R = 74$ kJ mol^{-1} (adapted from ref. 155)

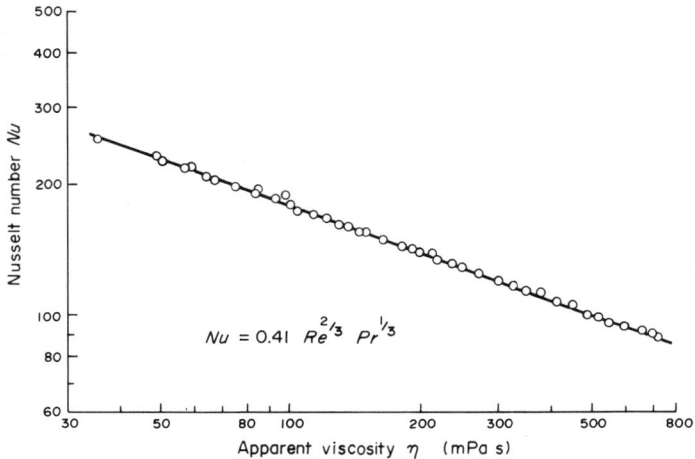

$$Nu = 0.41\ Re^{2/3}\ Pr^{1/3}$$

Figure 25 Nusselt number Nu vs. apparent viscosity η during crosslinking of PVAL in aqueous solution in a laboratory-stirred tank reactor with anchor type stirrer. Reactor geometry: $V_R = 1.31$, $D = 0.1$ m, $d/D = 0.95$, $H/D = 1.9$, $\rho = 1000$ kg m^{-3}, $\lambda = 0.65$ W m^{-1} K^{-1}, $Nu \sim \eta^{-1/3}$, $Nu = \alpha_i D/\lambda$, $Pr = C_p \eta / \lambda$, $Re = d^2 N \rho / \eta$ (adapted from ref. 164)

Therefore empirical equations which correlate the Stanton number with the temperature and conversion dependent viscosity have to be implemented into the balance equations (23) and (24).

It is important for safe operation of polymerization reactors to analyze the sensitivity of the final reactive state with respect to both start up and shut down operations and unintentional external perturbations such as variations in feed conditions, cooling temperature, concentration of impurities, *etc.* For example, a small perturbation of the jacket temperature ΔT_{jacket} may cause gradual change in reaction temperature or result in the 'ignition' of the reaction. As a consequence, reaction temperature (ΔT_R) considerably increases as is shown in Figure 23c.

The influence of the mean residence time or Damköhler number on reactor temperature and on monomer conversion is illustrated in Figure 26a for the solution polymerization of methyl methacrylate in ethyl acetate.[154,158] Both multiplicity of reactive states and hysteresis effects are predicted. For certain values of the Damköhler number limit cycle oscillations may occur with a pronounced phase shift between monomer conversion and reaction temperature. This example represents a dynamically unstable operating point of an HCSTR with cooling. In Figure 26b the oscillations of temperature and monomer conversion are plotted *vs.* reduced time. The criterion for dynamically stable reactor behavior, where the polymerization obeys first order kinetics is given by

$$St > \frac{\dfrac{E_A \Delta T_{ad}}{R T_{R,st}^2} X_{M,st} + X_{M,st} - 2}{1 - X_{M,st}} \simeq \frac{E_A \Delta T_{ad}}{R T_{R,st}^2} X_{M,st} - Da - 2 \qquad (28)$$

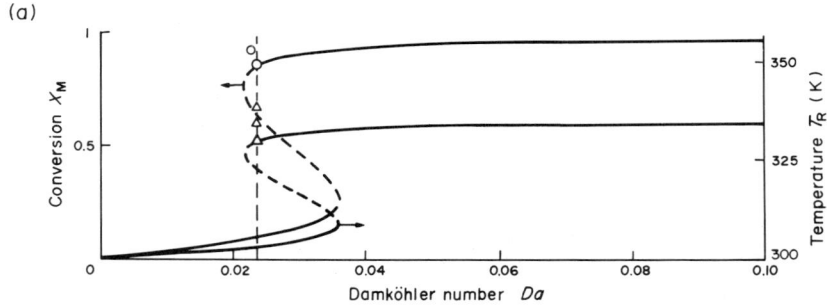

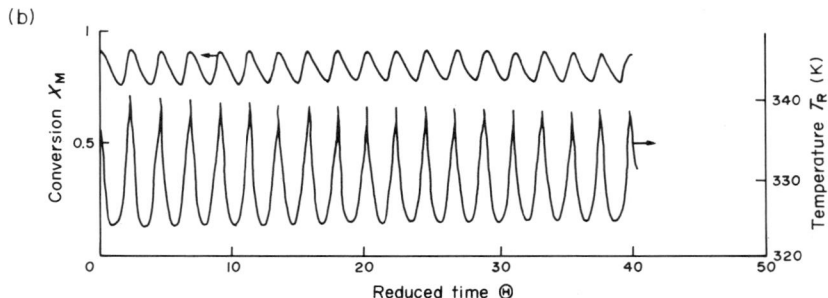

Figure 26 (a) Steady state monomer conversion and reaction temperature for the solution polymerization of MMA in ethyl acetate as a function of the Damköhler number. The circles show the region where stable limit cycles occur. (b) Periodic oscillation of temperature and conversion at $Da = 0.0239$ as a function of reduced time $\Theta = t/\tau$ (adapted from ref. 154)

For adiabatic conditions exclusively static instabilities (ignition–extinction phenomena) may occur in the HCSTR. However, both static and dynamic instabilities can be observed in the HCSTR with cooling. Obviously, the coupling of heat generation in the reaction mass and cooling by heat transfer through the reactor walls make dynamic instabilities possible. Thus limit cycle oscillations can only appear when the Stanton number surpasses a certain level.

The static and dynamic behavior of adiabatic HCSTRs and the HCSTRs with cooling has been extensively investigated both theoretically and experimentally. However, investigations considering variations of heat transfer, power input due to stirring, residence time distribution and degree of segregation as functions of the changing viscosity of the reaction mass have not yet been published.

The analysis of the thermal stability of tubular reactors is much more complicated, since temperature and conversion depend on location (distributed parameters) and partial differential balance equations have to be solved.

24.4.2 Concentration Stability

For isothermal reaction conditions multiple steady states and dynamic instabilities have also been observed in various continuous stirred tank reactors provided that autocatalytic reactions occur.[156,172–174]

The autoacceleration of the reaction rate due to the gel effect in concentrated systems as well as the particle nucleation period in emulsion polymerization can be interpreted as autocatalytic effects. The mass balance of a HCSTR for steady state operation is given by

$$\dot{v}(C_{M,0} - C_{M,st}) = V_R r_P \tag{29}$$

or

$$\frac{C_{M,0} X_{M,st}}{\tau} = r_P \tag{30}$$

if volume contraction is neglected. This balance equation can be solved graphically, similar to the determination of steady state operating points for a nonisothermal HCSTR (Figure 23). Therefore the polymerization rate curve together with the transport line for convection, $C_{M,0}/\tau$ vs. X_M, are plotted as functions of monomer conversion. At points of intersection steady state operation is possible. Figure 27 illustrates this method for the example of styrene emulsion polymerization in a

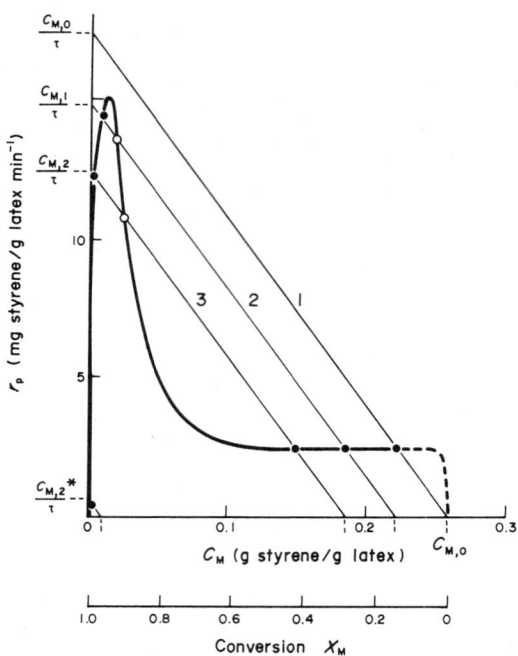

Figure 27 Steady state operating points of a cascade of three stirred tank reactors which are equal in size for an isothermal emulsion polymerization of styrene. ● stable and ○ unstable operating points (adapted from ref. 173)

cascade of three stirred tank reactors which are equal in size. Due to the strong gel effect six stable and two unstable operating points are obtained for the whole cascade. Theoretically, a cascade consisting of N CSTRs can operate at $N+1$ stable states, while $N+[N(N+1)]/2$ stable and N unstable operating points exist for all the CSTRs together.[172-174]

Dynamic instabilities have been observed for the same emulsion polymerization system. In Figure 28 temporal oscillations with a period length of 9 to 11 mean residence times are shown for monomer conversion and surface tension. The reason for these oscillations is the alternation of nucleation and wash out of latex particles. Parallel to the fast increase in surface tension, a new particle generation is formed. Subsequently, the specific surface in the polymerizing system increases and free emulsifier is needed to cover the latex particle surface. Thus the emulsifier concentration decreases. When it falls below the critical micelle concentration, particle generation stops. According to the residence time distribution of the reactor the actual particle population is washed out and the particle number N_P

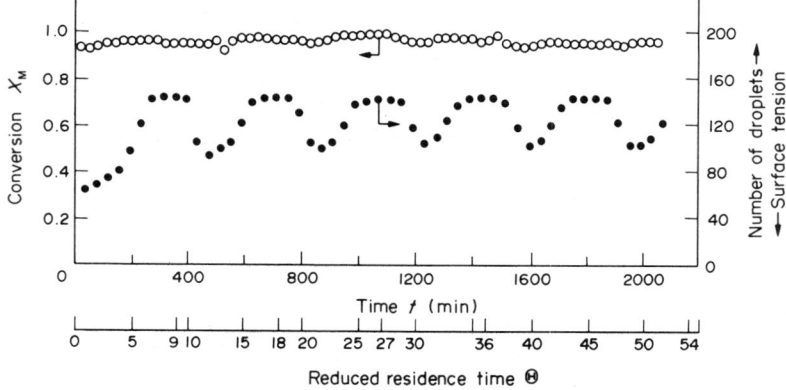

Figure 28 Periodic oscillations of conversion (○) and surface tension (●) during the isothermal emulsion polymerization of styrene in a continuous stirred tank reactor with a mean residence time of $\tau = 40$ min $vs.$ time and reduced residence time t/τ (adapted from ref. 173)

decreases following the relationship

$$N_P = N_{P,0} \exp\left\{-\frac{t}{\tau}\right\} \tag{31}$$

As a result the specific surface decreases and the emulsifier concentration increases again until it surpasses the level of the critical micelle concentration which starts the formation of a new particle generation and hence the next period of oscillation.

This kind of oscillation phenomenon has been observed in several emulsion polymerization systems as well as for continuous precipitation polymerization.

In order to avoid such instabilities small 'seed' reactors (CSTR or tubular reactors) are used as the first reactor of a cascade. A better control of particle generation is achieved in the small seed reactor.[175-177] The main portions of monomer, water and initiator are fed into the second vessel, where the concentration of emulsifier is kept below the level of critical micelle concentration. Thus uncontrolled particle formation is suppressed.

24.4.3 Nonsteady State Operation

The behavior of continuous polymerization reactors can be altered significantly by forced oscillation of the reaction parameters such as feed concentration of the reactants, flow rate and temperature. Unlike the previously discussed autonomous oscillations, the forced oscillations are controlled by the periodic perturbation applied. Therefore the later perturbation provides additional parameters for process optimization, namely perturbation amplitude and frequency.

In Table 6 some experimental results are compared, obtained for steady state and forced periodic operation of a HCSTR during solution polymerization of styrene in toluene.[178-180] The reactor capacity is considerably improved (up to 44%) by periodic operation, while the average molecular weight M_w and the polydispersity of the molecular weight distribution Q remain constant.

Table 6 Average Reactor Capacities of a HCSTR, $\dot{n}_{PS}$, for Steady State and Forced Periodic Operation during Solution Polymerization of Styrene[a]

Operation mode	τ (h)	t_P[b] (h)	M_w (g mol^{-1})	Q (—)	$\dot{n}_{PS}$ (mol m^{-3} h^{-1})	$\Delta\dot{n}_{PS}$ (%)
Periodic	1	0.5	30×10^3	2.52	203	+4
Steady state	1		30×10^3	2.52	195	
Periodic	1	3	33×10^3	2.55	193	+19
Steady state	2		32×10^3	2.53	163	
Periodic	2	4	34×10^3	2.53	170	+23
Steady state	3		34×10^3	2.53	138	
Periodic	2	6	35.5×10^3	2.54	175	+44
Steady state	4		35.5×10^3	2.53	121	

[a] Adapted from ref. 178. [b] t_P is the duration of a period.

An increase of the dispersion index of molecular weight distribution has been reported if either the initiator feed concentration or the jacket temperature is periodically perturbed during thermal styrene polymerization.[181,182]

24.4.4 Safe Operation by Reactor Control

Although the stability analysis is an essential part of safety considerations, reactor instabilities are by no means synonymous with unsafe operation. Actually, many technical polymerization reactors are run at or close to unstable operating points. Generally, stable operation at low conversion and low temperature is uneconomical, while on the other hand limited cooling capacity and high viscosity often rule out an operation at high conversion. Consequently, an intermediate level of conversion has to be adjusted, which can be unstable. Appropriately manipulated variables and controller are required in order to keep the reactor at the desired operating point. The stabilization of the operating variables by simple proportional control can be interpreted as an increase in slope of the transport line (see Figure 23b).[183]

Generally, the reaction temperature is controlled by using the temperature of the coolant in the jacket as a manipulated variable. A reliable concept is the installation of a cascade of two controllers, where the master controller, in response to changes in reaction temperature, adjusts the set point of the secondary controller. This slave controller then holds the cooling temperature at the value called for by the master controller. This type of controller provides an effective means of dealing with disturbances in the jacket loop of the reactor. Cascade control is particularly used if the reaction temperature follows changes in the jacket temperature only with strong delay due to reduced heat transfer and bad mixing. Other proven cooling concepts for polymerization reactors are summarized in refs. 184–186.

Particularly during the course of polymerizations reactor dynamics are affected severely by decreasing heat transfer as a result of increasing viscosity and the formation of polymer deposits on the reactor walls. Therefore the control parameters have to be adapted for changes of the control system, *e.g.* by changing the control parameters according to the increase in power consumption of the stirrer. Advanced process control strategies are described by Ray.[187,188]

Further complications are to be expected in batch-wise or semibatch-wise polymerizations, since no steady state operation is possible. The large amount of unreacted monomer in batch-wise operation involves a latent risk, particularly during the start up procedure. Figure 29 shows the unsteady heat release rate during batch-wise emulsion polymerization of vinyl acetate. A cooling failure which causes an adiabatic runaway reaction during a period of seven minutes leads to a doubling in heat production rate and an increase in reaction temperature of almost 9 °C. The reaction conditions during this runaway period have been evaluated in good agreement with the experimental data by assuming zero order kinetics.[189]

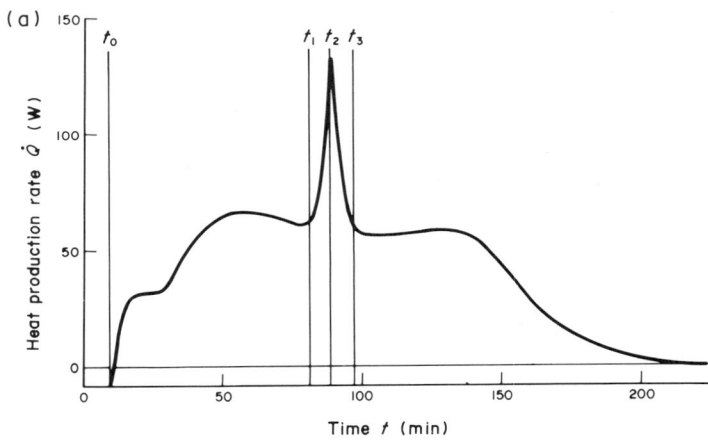

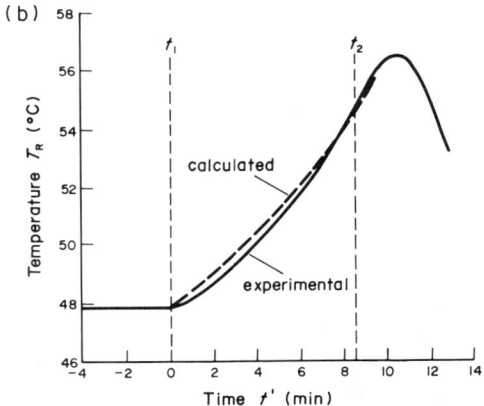

Figure 29 Specific heat production rate *vs.* time for a batch-wise emulsion polymerization of VA. The polymerization is initiated at time t_0. At time t_1 jacket cooling fails and an adiabatic runaway reaction starts. At t_2 maximum cooling is turned on and finally at t_3 the required reaction conditions are reached again. Measured and calculated values of the temperature during runaway are shown separately (adapted from ref. 189)

The occurrence of reactor wall fouling and the increasing viscosity during a runaway reaction may hinder bringing the reactor under control. Lowering the cooling temperature may not be an adequate countermeasure in the case of runaway polymerizations since the viscosity of the reaction mixture at reactor wall temperature can be very high and thus heat transfer and mixing are reduced by the formation of a stagnant region near the reactor wall. Under these conditions, however, the addition of a radical scavenger which stops the polymerization and hence the heat production may be an appropriate countermeasure. In this case it has to be considered that polymerization starts again when the scavenger is consumed.

The effects of hazardous runaways in strongly exothermic polymerizations can be limited by semibatch-wise operation, since the potential driving force, the concentration of unreacted monomer, is controlled by the feed rate and kept at a low level. This strategy is commonly used in order to control the heat release rate.

There are two dangerous periods in semibatch operation which are illustrated in Figure 30. The first dangerous situation is the start of monomer feed after successful initiation of the polymerization process. If feeding of cold monomer is started too early, the polymerization reaction may slow down and the fed monomer accumulates in the reactor. When the polymerization finally starts to accelerate, the amount of monomer is large enough to bring the reactor out of control. If the monomer feed is started too late the initially charged monomer is almost converted into polymer, the reaction rate decreases and thus the cold feed slows down the polymerization. Subsequently, monomer accumulates with the latent risk of hazardous runaway reaction when the polymerization finally ignites. The molar flow rates during the starting period of semibatch-wise polymerization are shown in Figure 30d. A further dangerous situation may occur if radical scavengers or retarders temporarily stop the polymerization during the feeding period. Under these conditions monomer also accumulates and safety problems may arise when the inhibitor is consumed and the polymerization continues with a much higher reaction rate (Figure 28a–c).[163]

Safe operation of polymerization reactors with strongly exothermic reactions requires measuring devices which are able to detect hazardous situations in the early stages. Only then can appropriate countermeasures successfully work and avoid runaway reactions. One simple criterion for detecting dangerous situations is taken from the curvature of the time-dependent course of reaction

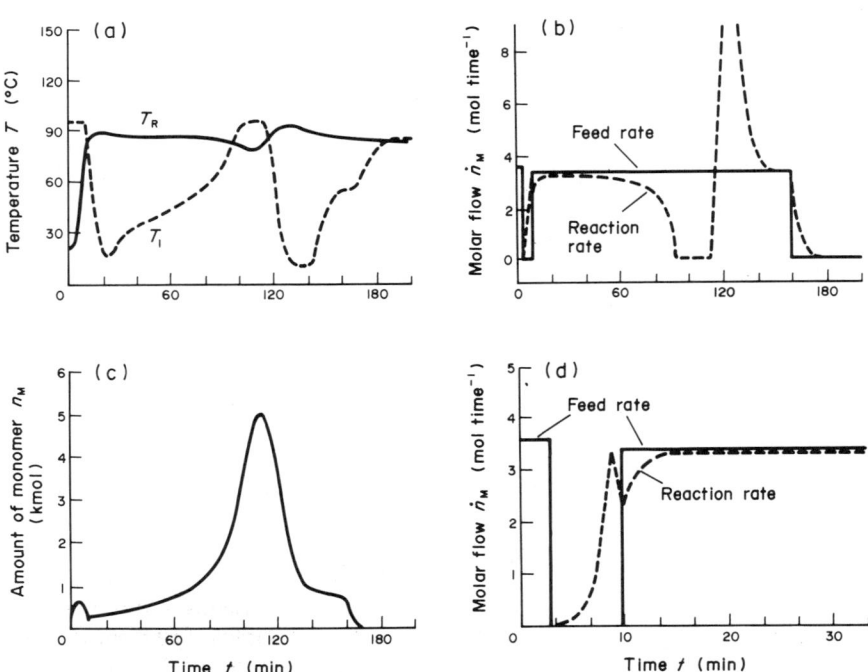

Figure 30 Reactor and coolant temperatures (a), molar flow rates (b) and accumulated amount of monomer *vs.* time (c) during a typical semibatch-wise polymerization. The polymerization is temporarily inhibited and ignites after the inhibitor is consumed, which results in a runaway reaction. The start of monomer feed during the initial period of the polymerization is shown in (d) (adapted from ref. 163)

temperature.

$$\frac{\mathrm{d}T_{\mathrm{R}}}{\mathrm{d}t} \gg 0 \quad \text{and} \quad \frac{\mathrm{d}^2 T_{\mathrm{R}}}{\mathrm{d}t^2} > 0 \tag{32}$$

Modern control theory provides model-based reconstruction methods for solving this detection problem. These methods enable dangerous situations to be avoided in advance and allow predictive calculations of future operation conditions of the reactor provided that appropriate process models are available. Methods for the detection of hazardous states of reactors in early runaway stages are summarized by Gilles and Schuler.[190]

24.5 NOTATION AND ABBREVIATIONS

24.5.1 Notation

A	m^2	heat or mass exchange area
b	$\mathrm{cm}^3\,\mathrm{g}^{-1}$	parameter in the Lyons–Tobolsky equation
C_{I}	$\mathrm{mol}\,\mathrm{l}^{-1}$	initiator concentration
C_{M}	$\mathrm{mol}\,\mathrm{l}^{-1}$	monomer concentration
C_{polymer}	$\mathrm{g}\,\mathrm{l}^{-1}$	polymer concentration
C_{S}	$\mathrm{mol}\,\mathrm{l}^{-1}$	solvent concentration
$C_{\mathrm{tr,M}}$	—	relative rate constant of transfer to monomer
$C_{\mathrm{tr,S}}$	—	relative rate constant of transfer to solvent
$C_{\mathrm{tr,P}}$	—	relative rate constant of transfer to polymer
C_{TDBP}	—	relative rate constant of terminal double bond polymerization
c_{p}	$\mathrm{kJ}\,\mathrm{kg}^{-1}\,\mathrm{K}^{-1}$	specific heat
D	m	reactor diameter
Da	—	Damköhler number
d	m	stirrer diameter
d_{P}	m	particle diameter
E_{A}	$\mathrm{kJ}\,\mathrm{mol}^{-1}$	Arrhenius activation energy
$\mathrm{E}(t)$	—	density function of residence time distribution (age distribution)
$\mathrm{F}(t)$	—	cumulative residence time distribution
F_1, F_2	—	molar fractions of monomer M_1 or M_2 in the produced copolymer
f	—	initiator efficiency
f_1, f_2	—	molar fractions of monomer M_1 or M_2 in the monomer mixture
H	m	height of reactor
ΔH_{R}	$\mathrm{kJ}\,\mathrm{mol}^{-1}$	reaction enthalpy
h	$\mathrm{W}\,\mathrm{m}^{-2}\,\mathrm{K}^{-1}$	overall heat transfer coefficient
k	$\mathrm{l}^{n-1}\,\mathrm{mol}^{1-n}\,\mathrm{s}^{-1}$	overall reaction rate constant
k_{∞}	$\mathrm{l}^{n-1}\,\mathrm{mol}^{1-n}\,\mathrm{s}^{-1}$	Arrhenius preexponential factor of a reaction rate constant
k_{d}	s^{-1}	rate constant of initiator decomposition
k_{p}	$\mathrm{l}\,\mathrm{mol}^{-1}\,\mathrm{s}^{-1}$	rate constant of polymer chain propagation reaction
k_{t}	$\mathrm{l}\,\mathrm{mol}^{-1}\,\mathrm{s}^{-1}$	rate constant of polymer chain termination reaction
$k_{\mathrm{t,c}}$	$\mathrm{l}\,\mathrm{mol}^{-1}\,\mathrm{s}^{-1}$	rate constant of polymer chain termination by combination
$k_{\mathrm{t,d}}$	$\mathrm{l}\,\mathrm{mol}^{-1}\,\mathrm{s}^{-1}$	rate constant of polymer chain termination by disproportionation
$k_{\mathrm{tr,M}}$	$\mathrm{l}\,\mathrm{mol}^{-1}\,\mathrm{s}^{-1}$	rate constant of polymer chain transfer to monomer
$k_{\mathrm{tr,S}}$	$\mathrm{l}\,\mathrm{mol}^{-1}\,\mathrm{s}^{-1}$	rate constant of polymer chain transfer to solvent
k_{H}	—	Huggins constant
M	$\mathrm{g}\,\mathrm{mol}^{-1}$	molecular weight
M_{cr}	$\mathrm{g}\,\mathrm{mol}^{-1}$	critical molecular weight
M_{M}	$\mathrm{g}\,\mathrm{mol}^{-1}$	molecular weight of monomer
M_{n}	$\mathrm{g}\,\mathrm{mol}^{-1}$	number average molecular weight
M_{w}	$\mathrm{g}\,\mathrm{mol}^{-1}$	weight average molecular weight
m	—	exponent in the power law of Ostwald–de Waele
$\dot{m}_{\mathrm{p}}$	$\mathrm{kg}\,\mathrm{s}^{-1}$	reactor performance of polymer
m_1, m_2	mol	number of moles of monomer M_1 or M_2 in the copolymer
N	s^{-1}	stirrer speed
N	—	number of stirred tank reactors in a cascade
N_{P}	m^{-3}	number of latex particles per unit volume
n	mol	number of moles of a component
$\dot{n}$	$\mathrm{mol}\,\mathrm{s}^{-1}$	molar flow rate of a component
P	—	degree of polymerization
P_{n}	—	number average degree of polymerization

P_w	—	weight average degree of polymerization
p	Pa	pressure
Q	—	dispersion index of the molecular weight distribution
$\dot{Q}$	W	heat flow rate
R	$kJ\,mol^{-1}\,K^{-1}$	ideal gas constant
r	$mol\,l^{-1}\,s^{-1}$	rate of reaction
r_1, r_2	—	copolymerization parameter
St	—	Stanton number
T	K or °C	temperature
T_J	K or °C	reactor jacket temperature
T_R	K or °C	reaction temperature
ΔT_{ad}	K or °C	adiabatic temperature rise
t	s	time, residence time
t_{dead}	s	reactor dead time required for other operations than reaction
t_p	s	duration of a period
V	m^3	volume
V_R	m^3	reaction volume
V^*	m^3	monomer phase volume in emulsion polymerization
V^{**}	m^3	water phase volume in emulsion polymerization
V_d	m^3	deadspace volume
V_m	m^3	perfectly mixed volume
V_p	m^3	piston flow volume
$\dot{v}$	$m^3\,s^{-1}$	volumetric flow rate
$W(P)$	—	polymer chain length distribution
X_M	—	monomer conversion
$\bar{X}_{M,m}$	—	monomer conversion calculated for a nonideal continuous stirred tank reactor (multiparameter model)
$X_{M,st}$	—	steady state value of monomer conversion in continuous reactor
$X_{M,\infty}$	—	dead end conversion of monomer
$\bar{X}_M$	—	average monomer conversion in segregated reaction volumes
α	—	probability of propagation
α_i	$W\,m^{-1}\,K^{-1}$	heat transfer coefficient at the product side
$\dot{\gamma}$	s^{-1}	shear rate
ε	—	volume contraction constant
ζ	—	segmental friction coefficient
η	Pa s	viscosity
η_{sp}	—	specific viscosity
$\eta_{solvent}$	Pa s	solvent viscosity
η_{wall}	Pa s	viscosity at the wall temperature
η_0	Pa s	zero shear viscosity
$[\eta]$	$cm^3\,g^{-1}$	instrinsic viscosity, Staudinger index
Θ	—	reduced residence time
λ	$W\,m^{-1}\,K^{-1}$	thermal conductivity
λ_i	—	ith moment of the molecular weight distribution
μ	—	parameter defined in equation (17)
ξ	—	volumetric phase ratio
ρ	$kg\,m^{-3}$	density
τ	s	mean residence time
τ^*	s	average life time of a growing radical
Φ	—	volume fraction

24.5.2 Dimensionless Numbers

conversion of monomer $X_M = \dfrac{n_{M,0} - n_M}{n_{M,0}}$ or $\dfrac{\dot{n}_{M,0} - \dot{n}_M}{\dot{n}_{M,0}}$ Damköhler number $Da = \dfrac{v_i k t}{C_i^{1-n}}$ or $Da = \dfrac{v_i k \tau}{C_i^{1-n}}$

dispersion index $Q = \dfrac{M_w}{M_n}$ reduced residence time $\Theta = \dfrac{t}{\tau}$ Stanton number $St = \dfrac{hA}{\dot{v}\rho c_P}$ volume con-

traction constant $\varepsilon = \dfrac{V_R - V_{R,0}}{V_{R,0}} = \dfrac{\rho_0 - \rho}{\rho}$

24.5.3 Abbreviations

ABS   acrylonitrile–butadiene–styrene
AIBN   2,2′-azobis(2-nitrilopropane)
BR   batch reactor
CR   polychloroprene
CSTR   continuous stirred tank reactor
EPDM   ethylene–propylene diene monomer
HCSTR   homogeneous continuous stirred tank reactor (perfectly micromixed)
HDPE   high density polyethylene
HIPS   high impact polystyrene
IR   polyisoprene
LDPE   low density polyethylene
LLDPE   linear low density polyethylene
NBR   nitrile–butadiene rubber
PA   polyamide
PAN   polyacrylonitrile
PFTR   plug flow tubular reactor
PIB   poly(isobutylene)
PMMA   poly(methyl methacrylate)
PP   polypropylene
PS   polystyrene
PTFE   poly(tetrafluoroethylene)
PU   polyurethane
PUR   polyurethane rubber
PVAC   poly(vinyl acetate)
PVAL   poly(vinyl alcohol)
PVC   poly(vinyl chloride)
RIM   reaction injection molding
SAN   styrene–acrylonitrile
SBR   styrene–butadiene rubber
SCSTR   completely segregated continuous stirred tank reactor
TDBP   terminal double bond polymerization

24.6 REFERENCES

1. H. Gerrens, in 'Polymerisationstechnik', in 'Ullmanns Enzyklopädie der technischen Chemie', 4th edn., Verlag Chemie, Weinheim, 1980, vol. 19, p. 107–165.
2. H. Gerrens, *Chem. Ing. Tech.*, 1980, **52**, 477, translation in *Ger. Chem. Eng.* 1981, **4**, 1.
3. H. Gerrens, *Chem. Tech.* 1982, June, Part 1, 380; 1982, July, Part 2, 434.
4. H. Gerrens, *Proc. 4th Int. 6th Eur. Symp. Chem. React. Eng. (ISCRE)*, 1976, **2**, 585.
5. C. E. Schildknecht (ed.), 'Polymer Processes', Interscience, New York, 1956.
6. C. E. Schildknecht and I. Skeist (eds.), 'Polymerization Processes', Wiley, New York, 1977.
7. G. Daumiller, *Chem. Ing. Tech.*, 1968, **40**, 673.
8. N. Platzer, *Ind. Eng. Chem.*, 1970, **62**, 6.
9. H. Gerstenberg, P. Sckuhr and R. Steiner, *Chem. Ing. Tech.*, 1982, **54**, 541.
10. N. Platzer (ed.), *Adv. Chem. Ser.*, 1972, **128**.
11. J. N. Henderson and T. C. Bouton, *ACS Symp. Ser.*, 1978, **104**.
12. T. C. Bouton and D. C. Chappelear, *AIChE Symp. Ser.*, 1976, **160**.
13. W. H. Ray and R. L. Laurence, in 'Chemical Reactor Theory', eds. L. Lapidus and N. R. Amundson, Prentice Hall, Englewood Cliffs, NJ, 1977, chap. 9.
14. W. H. Ray, *J. Macromol. Sci, Rev. Macromol. Chem.*, 1972, **8**, 1.
15. W. H. Ray, *ACS Symp. Ser.*, 1983, **226**, 101.
16. W. H. Ray, *Ber. Bunsenges. Phys. Chem.*, 1986, **90**, 947.
17. W. H. Ray, 'Dynamic Behavior of Polymerization Reactors' in 'Modelling of Chemical Reaction Systems', eds. K. H. Ebert *et al.*, Springer Verlag, Berlin, 1981.
18. D. H. Sebastian and J. A. Biesenberger, 'Polymerization Engineering', Wiley, New York, 1983.
19. H. F. Mark, N. G. Gaylord and N. M. Bikales (eds.), 'Encyclopedia of Polymer Science and Technology', Interscience, New York, 1964–1977.
20. H. F. Mark, N. M. Bikales, C. G. Overberger and G. Menges (eds.), 'Encyclopedia of Polymer Science and Engineering', Wiley–Interscience, New York, 1985.
21. I. Piirma and J. L. Gordon, 'Emulsion Polymerization', *ACS Symp. Ser.* 1976, **24**.
22. D. R. Bassett and A. E. Hamielec, 'Emulsion Polymers and Emulsion Polymerization', *ACS Symp. Ser.*, 1982, **165**.
23. L. H. Peebles, 'Molecular Weight Distributions in Polymers', Interscience, New York, 1971.
24. K. H. Reichert and W. Geiseler (eds.), 'Polymer Reaction Engineering. Influences of Reaction Engineering on Polymer Properties', Hanser, Munich, 1983.
25. K. H. Reichert and W. Geiseler (eds.) 'Polymer Reaction Engineering. Emulsion Polymerization — High Conversion Polymerization — Polycondensation', Hüthig & Wepf, Heidelberg, 1986.

26. J. Brandrup and E. H. Immergut (eds.), 'Polymer Handbook', Wiley, New York, 1975.
27. D. W. van Krevelen and P. J. Hoftyzer, 'Properties of Polymers. Correlations with Chemical Structure', Elsevier, Amsterdam, 1972.
28. P. F. Lyons and A. V. Tobolsky, *Polym. Eng. Sci.*, 1970, **10**, 1.
29. M. L. Huggins, *J. Am. Chem. Soc.*, 1942, **64**, 2716.
30. R. Houwink, *J. Prakt. Chem.*, 1941, **157**, 15.
31. H. Mark, 'Der feste Körper', Hirzel, Leipzig, 1938.
32. N. Sütterlin, in 'Polymer Handbook', eds. J. Brandrup and E. H. Immergut, Wiley, New York, 1975, p. IV-139.
33. F. Rodriguez, *Polym. Lett.*, 1972, **10**, 455.
34. O. Quadrat, *Proc. VIIIth Int. Congr. Rheol. Gothenburg, Sweden*, 1976, p. 604.
35. H. Fujita and A. Kishimoto, *J. Chem. Phys.*, 1961, **34**, 393.
36. F. N. Kelley and F. Bueche, *J. Polym. Sci.*, 1961, **50**, 549.
37. T. Kataoka and S. Ueda, *J. Polym. Sci., Part A-2*, 1967, **5**, 973.
38. M. Rink, A. Pavan and S. Roccasalvo, *J. Polym. Eng. Sci.*, 1978, **18**, 755.
39. Ch. Gabel and K. H. Reichert, *Chem. Ing. Tech.*, 1985, **57**, 612.
40. W. Loth, Ph.D. Thesis, Technical University of Berlin, 1986.
41. G. C. Berry and T. G. Fox, *Adv. Polym. Sci.*, 1968, **5**, 261.
42. W. W. Graessley, *Adv. Polym. Sci.*, 1974, **16**, 1.
43. W. W. Graessley, *Adv. Polym. Sci.*, 1982, **47**, 67.
44. V. Semjonow, *Adv. Polym. Sci.*, 1968, **5**, 387.
45. J. D. Ferry, 'Viscoelastic Properties of Polymers', Wiley, New York, 1980.
46. W. M. Kulicke, 'Fließverhalten von Stoffen und Stoffgemischen', Hüthig & Wepf, Heidelberg, 1986.
47. R. B. Bird, R. C. Amstrong and O. Hassager, 'Dynamics of Polymer Liquids', Wiley, New York, 1977.
48. G. N. Bohdanecky and J. Kovar, 'Viscosity of Polymer Solutions', Elsevier, Amsterdam, 1982.
49. G. V. Vinogradov and A. Y. Malkin, 'Rheology of Polymers', Springer Verlag, Berlin, 1977.
50. I. R. Rutgers, *Rheol. Acta.* 1962, **2**, 305.
51. V. V. Jinescu, *Int. Chem. Eng.*, 1974, **14**, 397.
52. F. R. Eirich, 'Rheology, Theory and Applications', Academic Press, New York, 1956–1960, vol. 1–3.
53. C. Y. Wen and L. T. Fan, 'Models for Flow Systems and Chemical Reactors', Dekker, New York, 1975.
54. E. B. Nauman, *J. Macromol. Sci., Rev. Macromol. Chem.*, 1974, **10**, 75.
55. E. B. Nauman and B. A. Buffham, 'Mixing in Continuous Flow Systems', Wiley, New York, 1983.
56. G. Damköhler, 'Technische Reaktionsgeschwindigkeit', in 'Der Chemie-Ingenieur', eds. A. Eucken and M. Jakob, Akademische Verlagsgesellschaft, Leipzig, 1937, vol. 3, part 1, p. 359.
57. O. Levenspiel, 'Chemical Reaction Engineering', Wiley, New York, 1962.
58. K. R. Westerterp, W. P. M. van Swaaij and A. A. C. M. Beenackers, 'Chemical Reactor Design and Operation', Wiley, New York, 1984.
59. M. Stickler, *Makromol. Chem.*, 1979, **180**, 2615.
60. N. Nishimura, *J. Macromol. Chem.*, 1966, **1**, 257.
61. G. P. Gladyshev and U. A. Popov, 'Radikalische Polymerisation bis zu hohen Umsätzen', Akademie-Verlag, Berlin, 1978.
62. N. Friis and A. E. Hamielec, *ACS Symp. Ser.*, 1976, **24**, 284.
63. A. W. Hui and A. E. Hamielec, *J. Appl. Polym. Sci.*, 1972, **16**, 749.
64. G. Weickert and R. Thiele, *Plaste Kautsch.*, 1981, **27**, 1.
65. G. Weickert and R. Thiele, *Plaste Kautsch.*, 1983, **30**, 432.
66. A. D. Schmidt and W. H. Ray, *Chem. Eng. Sci.*, 1981, **36**, 1401.
67. K. Kirchner and T. Katzenmayer, in 'Polymer Reaction Engineering. Emulsion Polymerization — High Conversion Polymerization — Polycondensation', Hüthig & Wepf, Heidelberg, 1986, p. 287–292.
68. T. Rintelen, K. Riederle and K. Kirchner, in 'Polymer Reaction Engineering. Influences of Reaction Engineering on Polymer Properties', eds. K. H. Reichert and W. Geiseler, Hanser, Munich, 1983, p. 269–286.
69. K. Stolzenberg and K. Kirchner, *Angew. Makromol. Chem.*, 1981, **95**, 185.
70. C. C. Price and D. A. Durham, *J. Am. Chem. Soc.*, 1943, **65**, 757.
71. M. Harada, K. Tanaka, W. Eguchi and S. Nagata, *J. Chem. Eng. Jpn.*, 1968, **1**, 148.
72. H. Hofmann, Ph.D. Thesis, TH Darmstadt, 1955.
73. K. Schoenemann, *DECHEMA-Monogr.*, 1952, **21**, 203.
74. H.-U. Moritz and K.-H. Reichert, *Chem. Ing. Tech.*, 1981, **53**, 386.
75. G. Deiringer and K.-H. Reichert, *Chem. Ing. Tech.*, 1985, **57**, 137.
76. K.-H. Reichert and H.-U. Moritz, *J. Appl. Polym. Sci.; Appl. Polym. Symp.*, 1981, **36**, 151.
77. K.-H. Reichert, H.-U. Moritz, Ch. Gabel and G. Deiringer, in 'Polymer Reaction Engineering. Influences of Reaction Engineering on Polymer Properties', eds. K. H. Reichert and W. Geiseler, Hanser, Munich, 1983, p. 152–174.
78. H.-U. Moritz, Ph.D. Thesis, Technical University of Berlin, 1982.
79. G. Deiringer, Ph.D. Thesis, Technical University of Berlin, 1985.
80. W. Baade, H.-U. Moritz and K.-H. Reichert, *J. Appl. Polym. Sci.*, 1982, **27**, 2249.
81. T. W. Taylor and K.-H. Reichert, *J. Appl. Polym. Sci.*, 1985, **30**, 227.
82. R. L. Curl, *AIChEJ.*, 1963, **9**, 175.
83. L. A. Spielman and O. Levenspiel, *Chem. Eng. Sci.*, 1965, **20**, 247.
84. P. V. Danckwerts, *Chem. Eng. Sci.*, 1958, **8**, 93.
85. D. Vollmerhausen, Ph.D. Thesis, University Dortmund, 1978.
86. H. Weinstein and R. J. Adler, *Chem. Eng. Sci.*, 1967, **22**, 65.
87. J. Villermaux and A. Zoulalian, *Chem.. Eng. Sci.*, 1969, **24**, 1513.
88. M. S. K. Chen and L. T. Fan, *Can. J. Chem. Eng.*, 1971, **49**, 704.
89. J. P. Klein, R. David and J. Villermaux, *Ind. Eng. Chem. Fundam.*, 1980, **19**, 373.
90. J. Villermaux, *ACS Symp. Ser.*, 1983, **226**, 135.
91. P. Sahm, Ph.D. Thesis, Institut National Polytechnique de Lorraine, Nancy, France, 1978.

92. E. B. Nauman and B. A. Buffham, 'Mixing in Continuous Flow Systems', Wiley, New York, 1983, p. 165.
93. T. T. Szabo and E. B. Nauman, *AIChE J.*, 1969, **15**, 575.
94. W. Gestrich and E. Otto, *Chem. Ing. Tech.*, 1971, **43**, 1241.
95. B. Fu, H. Weinstein, B. Bernstein and A. B. Shaffer, *Ind. Eng. Chem. Process Des. Dev.*, 1971, **10**, 501.
96. L. G. Gibilaro, *Chem. Eng. Sci.*, 1971, **26**, 299.
97. P. Zaloudik, *Br. Chem. Eng.*, 1969, **14**, 657.
98. A. Cholette and L. Cloutier, *Can. J. Chem. Eng.*, 1959, **37**, 105.
99. Y. K. Ahn, L. T. Fan and M. S. Chen, *Can. J. Chem. Eng.*, 1969, **47**, 141.
100. D. Wolf and W. Resnick, *Ind. Eng. Chem. Fundam.*, 1962, **2**, 287.
101. J. G. van de Vusse, *Chem. Eng. Sci.*, 1962, **17**, 507.
102. M. Moo-Young and K. W. Chan, *Can. J. Chem. Eng.*, 1971, **49**, 187.
103. G. V. Schulz, *Z. Phys. Chem. Abt. B*, 1935, **30**, 379.
104. P. J. Flory, *J. Am. Chem. Soc.*, 1936, **58**, 1877.
105. L. Küchler, 'Polymerizationskinetik', Springer, Berlin, 1951.
106. C. H. Bamford, W. G. Barb, A. D. Jenkins and P. F. Onyon, 'The Kinetics of Vinyl Polymerization by Radical Mechanisms', Butterworth, London, 1958.
107. K. G. Denbigh, *Trans. Faraday Soc.*, 1944, **40**, 352.
108. K. G. Denbigh, *Trans. Faraday Soc.*, 1947, **43**, 648.
109. K. G. Denbigh, *J. Appl. Chem.*, 1951, **1**, 227.
110. K. G. Denbigh, 'Chemical Reactor Theory', Cambridge University Press, Cambridge, 1965.
111. J. A. Biesenberger and Z. Tadmor, *J. Appl. Polym. Sci.*, 1965, **9**, 3409.
112. J. A. Biesenberger and Z. Tadmor, *Polym. Eng. Sci.*, 1966, **6**, 299.
113. Z. Tadmor and J. A. Biesenberger, *Ind. Eng. Chem. Fundam.*, 1966, **5**, 336.
114. Z. Tadmor and J. A. Biesenberger, *J. Polym. Sci., Polym. Lett. Ed.*, 1965, **3**, 753.
115. S. Katz and G. M. Saidel, *AIChE J.*, 1967, **13**, 319.
116. R. Thiele, *Chem. Tech. (Leipzig)*, 1967, **19**, 221.
117. R. Thiele, K. D. Herbrich, and J. Fischmann, *Chem. Tech. (Leipzig)*, 1970, **22**, 445.
118. A. V. Tobolsky, R. H. Gobran, R. Böhme, and R. Schaffhauser, *J. Phys. Chem.*, 1963, **67**, 2336.
119. T. W. Taylor and K.-H. Reichert, *1st Int. Berlin Workshop Polym. React. Eng.*, Berlin, 1983.
120. A. Chatterjee, K. Kabra, and W. W. Graessley, *J. Appl. Polym. Sci.*, 1977, **21**, 1751.
121. A. Chatterjee, W. S. Park and W. W. Graessley, *Chem. Eng. Sci.*, 1977, **32**, 167.
122. K. Nagasubramanian and W. W. Graessley, *Chem. Eng. Sci.*, 1970, **25**, 1549.
123. K. Nagasubramanian and W. W. Graessley, *Chem. Eng. Sci.*, 1970, **25**, 1559.
124. K. Nagasubramanian and W. W. Graessley, *Adv. Chem. Ser.*, 1972, **109**, 81.
125. J. Villermaux, L. Blavier and M. Pons, in 'Polymer Reaction Engineering. Influences of Reaction Engineering on Polymer Properties', eds. K. H. Reichert and W. Geiseler, Hanser, Munich, 1983, p. 1–20.
126. L. Blavier, Ph.D. Thesis, Institut National Polytechnique de Lorraine, Nancy, France, 1979.
127. P. Verlaine, Ph.D. Thesis, Institut National Polytechnique de Lorraine, Nancy, France, 1982.
128. J. Villermaux and L. Blavier, *Chem. Eng. Sci.*, 1984, **39**, 87.
129. L. Blavier and J. Villermaux, *Chem. Eng. Sci.*, 1984, **39**, 101.
130. S. T. Balke and A. E. Hamielec, *J. Appl. Polym. Sci.*, 1973, **17**, 905.
131. T. J. Tulig and M. Tirrell, *Macromolecules*, 1981, **14**, 1501.
132. J. N. Cardenas and K. F. O'Driscoll, *J. Polym. Sci., Polym. Chem. Ed.*, 1976, **14**, 883.
133. J. N. Cardenas and K. F. O'Driscoll, *J. Polym. Sci., Polym. Chem. Ed.*, 1977, **15**, 1883.
134. J. N. Cardenas and K. F. O'Driscoll, *J. Polym. Sci., Polym. Chem. Ed.*, 1977, **15**, 2097.
135. J. M. Dionisio, H. K. Mahabadi and K. F. O'Driscoll, *J. Polym. Sci., Polym. Chem. Ed.*, 1979, **17**, 1891.
136. J. M. Dionisio and K. F. O'Driscoll, *J. Polym. Sci., Polym. Chem. Ed.*, 1980, **18**, 241.
137. K. F. O'Driscoll, J. M. Dionisio and H. K. Mahabadi, *ACS Sypm. Ser.*, 1979, **104**, 361.
138. D. Panke, M. Stickler and W. Wunderlich, *Makromol. Chem.*, 1983, **184**, 175.
139. F. L. Marten and A. E. Hamielec, *ACS Sypm. Ser.*, 1979, **104**, 43.
140. F. L. Marten and A. E. Hamielec, *J. Appl. Polym. Sci.*, 1982, **27**, 489.
141. M. Stickler, *Makromol. Chem.*, 1983, **184**, 2563.
142. M. Stickler, D. Panke and A. E. Hamielec, *J. Polym. Sci., Polym. Chem. Ed.*, 1984, **22**, 2243.
143. F. R. Mayo and F. M. Lewis, *J. Am. Chem. Soc.*, 1944, **66**, 1594.
144. T. Alfrey and G. Goldfinger, *J. Chem. Phys.*, 1944, **12**, 205.
145. F. T. Wall, *J. Am. Chem. Soc.*, 1944, **66**, 2050.
146. G. E. Ham, 'Copolymerization', Interscience, New York, 1964.
147. K. F. O'Driscoll and R. Knorr, *Macromolecules*, 1968, **1**, 367.
148. K. F. O'Driscoll and R. Knorr, *Macromolecules*, 1969, **2**, 507.
149. P. Wittmer, *Makromol. Chem., Suppl.* 1979, **3**, 129.
150. H. Schuller in 'Polymer Reaction Engineering. Emulsion Polymerization — High Conversion Polymerization — Polycondensation', Hüthig & Wepf, Heidelberg, 1986, p. 137–146.
151. D. D. Perlmutter, 'Stability of Chemical Reactors', Prentice-Hall, Englewood Cliffs, NJ, 1972.
152. W. Oppelt and E. Wicke, 'Grundlagen der chemischen Prozeßregelung', Oldenbourg Verlag, München, 1964.
153. H. Hofmann and U. Hoffmann, *Ind. Chim. Belge.* 1967, **T32**, 326.
154. J. W. Hamer, T. A. Akramov and W. H. Ray, *Chem. Eng. Sci.*, 1981, **36**, 1897.
155. P. Wittmer, T. Ankel, H. Gerrens and H. Romeis, *Chem. Ing. Tech.*, 1965, **37**, 392.
156. R. S. Knorr and K. F. O'Driscoll, *J. Appl. Polym. Sci.*, 1970, **14**, 2683.
157. A. D. Schmidt and W. H. Ray, *Chem. Eng. Sci.*, 1981, **36**, 1401.
158. J. W. Hamer, Ph.D. Thesis, University of Wisconsin, 1983.
159. A. A. Butakov and E. I. Maksimov, *Dokl. Akad. Nauk SSSR*, 1973, **209**, 643.
160. J. A. Biesenberger and D. H. Sebastian, *Polym. Eng. Sci.*, 1976, **16**, 117.
161. D. H. Sebastian and J. A. Biesenberger, *Polym. Eng. Sci.*, 1979, **19**, 190.

162. K. R. Westerterp, W. P. M. van Swaaij and A. A. C. M. Beenackers, 'Chemical Reactor Design and Operation', Wiley, New York, 1984.
163. G. Eigenberger and H. Schuler, *Chem. Ing. Tech.*, 1986, **58**, 655.
164. Ch. U. Schmidt and K. H. Reichert, *Chem. Ing. Tech.*, 1987, **59**, 739.
165. M. Zlokarnik, *Chem. Ing. Tech.*, 1967, **39**, 539.
166. V. Novak and F. Rieger, *Trans. Inst. Chem. Eng.*, 1969, **47**, T335.
167. K. W. Norwood and A. B. Metzner, *AIChE J.*, 1960, **6**, 432.
168. M. Zlokarnik, *Chem. Ing. Tech.*, 1969, **41**, 1195.
169. M. Zlokarnik, in 'Ullmanns Enzyklopädie der technischen Chemie', 4th edn., Verlag Chemie, Weinheim, 1980, vol. 12, p. 260.
170. S. Nagata, '*Mixing*', Wiley, Kodansha, Tokyo, 1975.
171. K. Kipke, *Chem. Ing. Tech.*, 1979, **51**, 430.
172. H. Gerrens, K. Kuchner and G. Ley, *Chem. Ing. Tech.*, 1971, **43**, 693.
173. G. Ley and H. Gerrens, *Makromol. Chem.*, 1974, **175**, 563.
174. H. Gerrens and K. Kuchner, *Br. Polym. J.*, 1970, **2**, 18.
175. M. Nomura and M. Harada, *ACS Sypm. Ser.*, 1981, **165**, 121.
176. A. Penlidis, A. E. Hamielec and J. F. MacGregor, *J. Vinyl Technol.*, 1984, **6**, 134.
177. M. Nomura, in 'Polymer Reaction Engineering. Emulsion Polymerization — High Conversion Polymerization — Polycondensation', Hüthig & Wepf, Heidelberg, 1986, p. 41–50.
178. A. Renken, *Chem. Ing. Tech.*, 1982, **54**, 571.
179. G. Crone and A. Renken, *Chem. Ing. Tech.*, 1979, **51**, 42.
180. G. Crone and A. Renken, *Ger. Chem. Eng. (Engl. Transl.)*, 1979, **2**, 337.
181. D. Konopnicki and J. L. Kuester, *J. Macromol. Sci., Chem.*, 1974, **8**, 887.
182. J. J. Spitz, R. L. Laurence and D. C. Chappelear, *AIChE Symp. Ser.*, 1976, **72**, 86.
183. R. Aris and N. R. Amundson, *Chem. Eng. Sci.*, 1957, **7**, 121.
184. L. Schupmehl and E. Schäfer, *Chem. Ing. Tech.*, 1984, **56**, 377.
185. H. Amrehn, *Automatica*, 1977, **13**, 533.
186. Th. Ankel in 'Messen, Steuern und Regeln in der Chemischen Technik', ed. J. Hengstenberg, B. Sturm and O. Winkler, Springer, Berlin, 1981, vol. 3, p. 378.
187. W. H. Ray, 'Proceedings of the International Symposium on Process Systems Engineering', Kyoto, 1982.
188. W. H. Ray, 'Advanced Process Control', McGraw-Hill, New York, 1981.
189. Ch. U. Schmidt, Ph.D. Thesis, Technical University of Berlin, 1988.
190. E. D. Gilles and H. Schuler, *Chem. Ing. Tech.*, 1981, **53**, 673.

25

Carbanionic Polymerization: General Aspects and Initiation

MICHEL FONTANILLE

Université de Bordeaux I, Talence, France

25.1 GENERAL CHARACTERISTICS OF CARBANIONIC POLYMERIZATION

25.1.1 Definitions and Main Properties of Living Systems

Anionic polymerizations can be represented by equation (1), in which $\sim\sim M_n^-$ is the growing chain bearing an anion or negative polarization, and Mt^+ is generally an alkali or alkaline-earth metal cation. In the case of polymerization of ethylenic monomers, the active centres are species of the carbon–metal type.

$$\sim\sim M_n^-, Mt^+ \quad + \quad M \longrightarrow \quad \sim\sim M_{n+1}^-, Mt^+ \tag{1}$$

Anionic polymerization has been known for a long time; it allowed the polymerization of dienes[1] and styrene,[2] but its characteristics were not developed until the work of Szwarc published in 1956.[3] Szwarc called 'living' systems those for which all macromolecules grow proportionally to the amount of monomer consumed. One may also speak of individual living macromolecules if they grow without transfer and termination. Moreover, such living systems can be killed by appropriate reactions or die a natural death by spontaneous termination. The discovery of anionic living polymers has aroused considerable interest and, at the present time, this method of polymerization can be considered as the best known among the various methods. There are two main reasons for this; first, the stability of active centres allows their storage in relatively high concentration, thus favouring the study of their detailed structure by most methods of structural analysis, and more particularly by spectroscopic methods. Secondly, the high reactivity of active centres and their living character open the way to new macromolecular structures which are inaccessible by other methods of polymerization: monodisperse polymers, well-defined block copolymers, polymers with one or two functional end-groups, star-shaped polymers, ... These structures can be obtained with an accuracy never reached before, and can be used as standards for physicochemical studies. The manifold potential applications of anionic polymerization have drawn numerous research teams to work in this field.

25.1.2 Polymerizability of Ethylenic Monomers

The polymerizability must be considered from two viewpoints: (i) the reactivity of the double bond towards carbanionic active centres; and (ii) the necessary absence of sites reactive towards carbanionic species which would lead to inactivation of the latter.

The first aspect is related to the ability of the double bond of the monomer to undergo nucleophilic addition of the carbanionic ion-pairs produced by insertion of monomer molecules; it is related to both the intrinsic reactivity of the monomer and that of the corresponding active centres.

The intrinsic reactivity of the monomer is mainly determined by its electron structure, *i.e.* by the nature of the substituent on the double bond; electron-withdrawing substituents increase the electrophilic character of the double bond and stabilize the final state in the form of regenerated active centres. The relative reactivity of various monomers has been compared by measuring the addition rate constants of different monomers on to given living species. Thus, the kinetics of addition of butadiene, isoprene and dimethylbutadiene on to a solution of (polystyryl)sodium in tetrahydrofuran (THF), shows the strong deactivating effect on the monomer of electron-donating substituents.[4] The same phenomenon occurs in the addition of styrene derivatives: *p*-methoxystyrene adds more slowly than *p*-methylstyrene does, because the electron-donating effect of the methyl group is less than that of methoxy.[5]

As with other methods of polymerization, steric hindrance by substituents diminishes the ability of monomers to polymerize. This structural effect was studied using different substituted styrenes.[6]

The intrinsic reactivity of active centres is determined by numerous structural parameters, in particular by the structure of the corresponding monomer. Monomers bearing electron-withdrawing substituents generate active centres for which the electron density on the carbanion is lowered by the substituent, thus diminishing the reactivity. This was established by measurement of the rate constants of addition of 1,1-diphenylethylene on to active centres resulting from different *para*-substituted styrenes.[7] In order to prevent a variation in solvation state for the different active species, this study was carried out with the cesium derivative of the active centres.

Thus, two opposite effects determine the polymerizability of a given monomer, but the effect of structure on the intrinsic reactivity of the monomer surpasses that on the derived active centre.

Nevertheless, the polymerizability of monomers is variable, because, besides monomer structure, it is related to the structural state of the active ion-pair. The convenient activation of the latter can sometimes counterbalance the low reactivity of the monomer; for example, ethylene was polymerized by Langer[8] by convenient activation of active centres. The different possibilities of activation will be discussed in Volume 3, Chapter 26.

The ability of a monomer to polymerize anionically is also related to the absence in the molecule of either atoms or groups reactive towards the propagating active species, which are strong Lewis bases. Monomers bearing protic groups cannot undergo anionic polymerization, because they lead immediately to a termination reaction by formation of a new base which is too weak to reinitiate the polymerization process. Nevertheless, several monomers leading to active centres which are unstable at room temperature can be polymerized through a living mode at low temperature, because the activation energies for propagation and termination, respectively, are very different and favour propagation; this is true for methacrylic esters,[9] whose living polymerization can be obtained at temperatures lower than $-70\,°C$.[10,11]

It is also possible to protect the reactive group during polymerization and to regenerate it when the reaction is complete. For example, hydroxyethyl methacrylate, whose hydroxy group is protected by a trimethylsilyl group, can be polymerized anionically and the protic group easily regenerated by hydrolysis (equation 2).[12]

$$H_2C=\underset{\underset{O}{|}}{\overset{Me}{\underset{|}{C}}}-O\diagup\diagdown OSiMe_3 \quad \xrightarrow[\text{in THF at }-78°C]{\text{anionic initiator}} \quad +CH_2CMe +_n \overset{}{\underset{O}{\diagup\diagdown}} O\diagup\diagdown OSiMe_3 \tag{2}$$

25.1.3 Structure of Organometallic Compounds — Models of Living Species — According to the Solvent and to the Presence of Polar Additives

The reactivity of active centres towards monomers is quite variable, and relates to the structure of the carbanionic ion-pairs. It is mainly affected by the distance between the carbanion and the counter-cation, which is dependent on the electron structure of the carbanion, the nature of the counter-ion and the solvent, the possible presence of additives in the reaction medium and the temperature. These different parameters have been studied on fluorenyl and related salts of alkali and alkaline-earth metals; these salts can be considered as satisfactory models of active centres resulting from the polymerization of vinylarene monomers. Such salts were chosen because they exhibit different absorption spectra in the visible–UV region, reflecting the structure of the corresponding ion-pair.[13] Moreover, these modifications are perceptible at concentrations as low as those required for anionic polymerization conditions, and the titration of each type of species is possible. The main part of this work was performed by Smid and co-workers.[13]

In non-polar solvents, organolithium salts are aggregated, as revealed by optical spectra of 9-(n-propyl)fluorenyllithium in toluene solution (equation 3).[14] Aggregation number n is equal to 2 and disaggregation constant K_D is found to be equal to 2.9×10^{-6} mol l^{-1} at 25 °C. This is a very low value, which is even lower in cyclohexane, a solvent with solvating properties poorer than those of toluene. For fluorenyllithium in the above solvents, the steric hindrance is less than that for the 9-(n-propyl) derivative and the disaggregation process is not perceptible.

$$(PFl^-, Li^+)_n \overset{K_D}{\rightleftharpoons} nPFl^-, Li^+ \tag{3}$$

Disaggregation can also be obtained without marked modification of the anion–cation bond length, by addition of complexing additives which solvate the ion pairs externally. Thus, addition of N,N,N',N'-tetramethylethylenediamine (TMEDA) to a solution of 9-(n-propyl)fluorenyllithium will result in the filling of free orbitals of lithium responsible for the aggregation process, and form contact ion-pairs whose absorption peak is located at the same wavelength as non-aggregated ion-pairs.[15]

Optical spectra of fluorenyl salts reveal an effect of the ionic radius of the cation. An increase in the radius of the counter-ion causes the anion–cation bond in the contact ion-pair to stretch. One can observe a bathochromic shift from Li$^+$ to Cs$^+$ for 9-fluorenyl salts studied in THF solution at 25 °C.[13] One can see a linear variation of the wavenumber of the absorption peak *vs.* the inverse of the ionic radius of the cation; this allows prediction of the value of the absorption maximum for other types of ion pairs and for free ions.[13,16] A similar phenomenon was observed with ion pairs of the bicarbanionic dimer of 1,1-diphenylethylene.[17]

The nature of the cation also affects the solvation equilibrium; in THF, small cations (Li$^+$ and Na$^+$) are strongly solvated, giving loose ion-pairs (Fl$^-$//Mt$^+$) for which an overstretching results from the insertion of molecules of solvent between anion and cation. In the case of divalent alkaline-earth metal cations, loose ion-pairs are less easily formed from Fl$_2^{2-}$, Mt^{2+} than with alkali metal cations, because the cation is hindered by the two fluorenyl carbanions.[18]

The structure of the ion pair is also strongly affected by that of the anion. In the case of markedly localized carbanions, the carbon–metal bond is very strong; it may be partially covalent and only polarized as in alkyllithiums; these compounds display particular physical properties (solubility, volatility, . . .) and are capable of giving aggregates. With other alkali metal alkyls, the carbanion is very unstable and leads to easy hydride elimination. The delocalization of the negative charge favours the ionization of the bond and lowers the energy of the anion–cation interaction. Hence a stretching of the bond occurs, making the formation of solvent-separated ion-pairs easier. Thus, the comparison of the spectrum of fluorenylsodium with that of 2,3-benzofluorenylsodium in THF shows that the proportion of loose ion-pairs is higher for the more delocalized species.[13] To confirm this rule, numerous examples have been proposed with other types of carbanions or radical ions.[19,20] The presence of bulky substituents on the carbanion hinders a close approach of the cation and produces the same effect as delocalization.[16]

The solvating power of the solvent plays an important role in the structure of carbanionic species. In hydrocarbon solvents, contact ion-pairs are mainly aggregated.[14] In dioxane, disaggregation is complete and leads to externally solvated ion-pairs. With fluorenyllithium, a significant fraction of solvent-separated ion-pairs exists in THF at room temperature; in 1,2-dimethoxyethane, pyridine and dimethyl sulfoxide, all the species are loose at any temperature.[13] The solvation phenomenon results not only from an electrostatic interaction between a positively charged ion and a polarized or polarizable solvent molecule, but also — and more particularly — from the capacity of the cations to accept the donation of electron pairs from solvents considered as Lewis bases. If interactions are weak, the solvation might be mainly external. The cation–solvent interaction slightly diminishes the energy of the cation–carbanion bond, thus producing a small bathochromic shift in the optical spectrum of the fluorenyl salts. It occurs with the lithiated salt of 9-(n-propyl)fluorene when small amounts of THF are added to solutions in hydrocarbons.[14] In the case of a strong interaction between solvent and cation, the solvation energy can be high enough to stretch the anion–cation bond, and form solvent-separated ion-pairs by filling the coordination sites of the cation. Such a phenomenon is particularly important with fluorenyllithium in THF.[13]

Solvents with high dielectric constants are likely to induce the dissociation of ion pairs into free ions. The thermodynamics of this phenomenon was studied by different authors and applied to fluorenyl salts by Smid and co-workers. There is a satisfactory agreement between experimental values[21] and those calculated from both the Fuoss model[22] and the Denison and Ramsey equations[23] for contact ion-pairs such as Fl^-,Cs^+ in THF. This agreement is closely related to the fact that interionic distance is hardly affected by the solvent. Interactions are strong between solvents possessing high solvating power and cations with small ionic radii, hence lowering the anion–cation interaction energy. The corresponding values of the dissociation constants are significantly increased. One can observe that Fl^-,Li^+ in THF forms mainly solvent-separated ion-pairs; this salt is relatively strongly dissociated at low concentrations ($K_d = 3 \times 10^{-6}$ mol l^{-1} at 25 °C) as could be expected from a high ionic-radius cation. The solvent shell surrounding the cation is tightly bonded to the cation and affects the ionic mobility of the cation-plus-solvating-solvent couple. Under such conditions, the solvent influences the dissociation constant in two different ways: the dielectric characteristics are responsible for the dissociation phenomenon and the solvating power produces the lowering of the electrostatic attraction between anion and cation. It is obvious that this lowering is more marked for solvent-separated ion-pairs than for externally solvated ones.

The addition, in small amounts, of compounds possessing a strong solvating power to solutions of fluorenyl salts allows observation of important effects on the structure of ion pairs without modification of the dielectric properties of the medium. Several effects can be observed relating to both the structure of the additive and that of the solvent. In non-polar or low polarity media, the effect can only be a disaggregating one; thus, addition of TMEDA to 9-(n-propyl)fluorenyllithium (Pr^nFl^-,Li^+) in cyclohexane induces the formation of monomeric species externally complexed by one molecule of TMEDA.[15] Crown ethers and crown amines, whose cavity is adjusted to the ionic radius of the cation, exhibit a high solvating power which allows them to insert between anion and cation and form loose ion-pairs. This is the case for addition of N,N',N'',N'''-tetramethyl-1,4,8,11-tetraazacyclotetradecane (1) to Pr^nFl^-,Li^+ in cyclohexane,[15] and also for addition of cryptands to Fl^-,Li^+.[24] With an additive such as hexamethyltriethylenetetramine (HMTT; 2,5,8,11-tetramethyl-2,5,8,11-tetraazadodecane), an intermediate situation can be observed: external solvation for $r = HMTT/Li < 1$, with aggregation by means of the complexing agent, (2), and loose ion-pairs for $r > 1$.[15]

(1)   (2)

In solvents with dielectric properties allowing dissociation, the addition of strongly solvating compounds considerably increases the dissociation constant, as the dissociation energy of loose ion-pairs is much lower than that of contact ion-pairs. Such an effect has been observed with fluorenyl salts in the presence of crown ethers[25] and with carbazyllithium in THF solution, by addition of cryptand.[26]

The temperature plays a complex role in both the structure and reactivity of carbanionic species, by means of the solvation and dissociation phenomena. Solvation being an exothermic process, one observes an increasing proportion of loose ion-pairs on cooling carbanionic solutions such as Fl^-, Na^+ in THF;[13] this produces an increase in the dissociation constants of ion pairs into free ions. Moreover, a number of solvents with a medium dielectric constant exhibit an increase in dielectric constant on lowering the temperature. The relations established by Denison and Ramsey[23] allow the prediction, at a given solvation state, of higher dissociation as the temperature decreases; these relations were experimentally confirmed on Fl^-,Li^+ solutions in THF.[21] Thus, at 25 °C $K_d = 3 \times 10^{-7}$ mol l^{-1} and, at -70 °C, $K_d = 155 \times 10^{-7}$ mol l^{-1} for this salt.

The carbanionic salts of fluorene and related salts are only imperfect models of species initiating or propagating anionic polymerizations; nevertheless, from the possibility of quantifying some structural effects, their study makes the interpretation of some unexpected results easier, especially concerning the reactivity of polymerizing systems.

25.2 INITIATION OF ANIONIC POLYMERIZATION

Two methods can be used to initiate the anionic polymerization of ethylenic monomers; both involve alkali or alkaline-earth metal derivatives and differ only by the mechanism of formation of the primary carbanionic species.

25.2.1 Initiation by Electron-transfer Reaction

25.2.1.1 Direct transfer from metals

This method is based on the possibility of transfer of external electron(s) from alkali and alkaline-earth metals to organic molecules, thus leading to radical anions (equation 4).

$$\text{metal}_{(\text{solid})} + A_{(\text{solution})} \rightleftharpoons A^-, \text{metal}^+_{(\text{solution})} \tag{4}$$

The reduction of polycyclic aromatic hydrocarbons by alkali metals has been known for a long time[27] but was considered as an electron-transfer reaction after the works of Schlenk *et al.*[28] and Willstäter *et al.*[29] This reaction has been intensively studied, and knowledge of the parameters determining the enthalpy of this exothermal reaction has allowed prediction of the qualitative influence of the various structural factors on the position of equilibrium (4). This is shown in equation (5), where $\Delta H_{\text{sub(alk)}}$ is the heat of sublimation of the metal; $I^z_{(\text{alk})}$ is the ionization potential of the metal; $EA_{(A)}$ is the electron affinity of hydrocarbon (A); r is the interionic distance; and $\Delta H_{\text{solv}(A^{\mp}, \text{alk}+)}$ is the heat of solvation of the radical-ion pair.

$$-\Delta H = -\Delta H_{\text{sub(alk)}} - I^z_{(\text{alk})} + EA_{(A)} + \frac{e^2}{r} - \Delta H_{\text{solv}(A^{\mp}, \text{alk}+)} \tag{5}$$

The parameters considered are:

(i) The nature and the physical state of the metal: in a given family, both the heat of sublimation and the ionization potential of the metal decrease with increasing atomic radius. In Table 1 are given the relevant values for metals capable of transferring to hydrocarbons. Thus, it can be seen that transfer is easier for alkali metals rather than alkaline-earth ones, and for heavier metals rather than lighter ones.

Table 1 Capacity of Electron Transfer from Alkali and Alkaline-earth Metals

Metal	First ionization potential (eV)	Heat of sublimation (kJ mol^{-1})	Ref.
Li	5.392	137.7	31
Na	5.139	92.2	31
K	4.341	79.4	31
Rb	4.177	71.3	31
Cs	3.894	68.1	31
Mg	7.646	140.9	32
Ca	6.113	183.9	32
Sr	5.695	150.5	32
Ba	5.212	171.8	32

The physical state of the metal also seems to play a role in the position of equilibrium (4); for example, mirrors of alkaline-earth metals can lead to a high concentration of reduced hydrocarbons in various solvents, whereas the corresponding metals in bulk are inefficient.[33] Nevertheless, this could be due to the absence of protective impurities on the mirror surface.

(ii) Another parameter is the electron affinity of the accepting organic molecule: Electron affinity can be measured either by potentiometry[34, 35] or by polarography,[36] both techniques leading to values in good agreement. In Table 2 are collected values obtained by the polarographic method for several polycyclic aromatic hydrocarbons and for some vinylarene monomers. On the whole, they show that the more marked the electron delocalization, the higher the electron affinity.

(iii) The solvating power of the reaction medium also plays an important role in the position of equilibrium (4). While in hydrocarbon solutions the electron transfer from metals is very difficult, whatever the ionization potential of the latter, highly solvating media allow transfer from metals

Table 2 Half-wave Potentials for Various Compounds in 0.175 mol l^{-1} Tetrabutylammonium Iodide, 75% Dioxane[37,38]

Compound	$-\pi_{1/2}(V)$	Compound	$-\pi_{1/2}(V)$
Biphenyl	2.70	1,1-Diphenylethylene	2.26
Fluorene	2.65	*trans*-Stilbene	2.14
Acenaphthene	2.57	Triphenylethylene	2.12
1,2-Dihydronaphthalene	2.57	Tetraphenylethylene	2.05
α-Methylstyrene	2.54	1,2-Benzanthracene	2.03
Naphthalene	2.50	1,4-Diphenylbutadiene	1.98
Phenanthrene	2.45	Anthracene	1.94
Styrene	2.35	3,4-Benzopyrene	1.88

possessing a relatively high ionization potential. Thus, in hexamethylphosphoramide [HMPA; phosphoric tris(dimethylamide)], radical anions associated to alkaline-earth metal cations are easily formed.[39] The lower the ionic radius, the more intense the solvation of the associated cation. On the other hand, the ionization potential generally decreases as the atomic radius of the metal increases; both of these opposing effects can be responsible for the variable capacity of metals to transfer electrons, according to the nature of the solvent in which the reaction is performed.

Polycyclic aromatic radical-ions do not dimerize from the radical site, because the stabilization by resonance of the resulting species would be lower than that of the initial radical ion.[40] It is different in the case of vinyl, vinylidene, diene and related monomers, for which the dimerization process is extremely fast and leads mainly to a dimeric bicarbanion; this point will be discussed later. The direct electron-transfer from metals to monomers has not received much detailed study because the corresponding reaction is slow (solid–liquid reaction) and is concealed by the very fast dimerization of the formed species which occurs before the polymerization process. With monomers capable of polymerizing, the initiation and propagation stages overlap, thus eliminating the main advantage of anionic polymerization, namely the control of molecular weights.

Nevertheless, various methods have been proposed, allowing the production of active oligomers which can be used as initiators for subsequent polymerization:

(a) The reaction of monomers such as styrene or α-methylstyrene with finely divided metals, such as alkaline-earths or lithium, results in bifunctional active oligomers.[41,42] These fine powders are obtained by rapid condensation of vapours of metals heated slightly above their melting temperature; the very high specific area of such systems significantly increases their activity and allows the production of high concentrations of active centres.

(b) In order to prepare dicarbanionic species directly from an alkali metal mirror and polymerizing monomer, Szwarc proposed a method[37] which consists in condensing a mixture of vapours from the monomer and an adequate solvent on to an alkali metal mirror and storing the product of the reaction; the low instantaneous concentration of the monomer favours the initiation step with regard to the propagation step and limits the $\overline{DP}_n$ obtained to several units.

(c) It is also possible to obtain living oligomers of α-methylstyrene[43] by initiating the polymerization at temperatures higher than the ceiling temperature of the monomer in order to prevent any propagation step.

In studying the copolymerization of styrene and methyl methacrylate initiated by lithium and sodium metals, Overberger and co-workers[44] showed that, under given conditions, there is incorporation of a relatively important amount of styrene as a block in a copolymer. This would result from the predominant adsorption of styrene molecules on the surface of the positively charged alkali metal. A similar conclusion can be reached from the study of the competitive reaction of monomers and alkyl halides with species resulting from electron transfer.[45] Richards proposed,[46] in agreement with Overberger's results,[44] that direct initiation from metals proceeds in the following way: The first stage involves adsorption on to the metal surface; then, the monomer molecules rotate until the methylene groups of vicinal molecules are sufficiently close to allow a coupling reaction to take place, producing an adsorbed dimer dianion without transitory formation of the radical anion. Finally, the metal cation is removed from the lattice, the attractive force of the carbanion being strong enough to overcome interactions between the metal atoms. The latter stage is greatly favoured by the solvating power of the solvent.

Blue solutions of alkali metals can be obtained in a limited number of solvents (amines and ethers) and have been used to initiate polymerizations. These solutions are believed to result from the three equilibria shown in equations (6)–(8). The formation of solvated electrons is easier when the ionization potential of the metal is low and the solvating power of the solvent is high. In a solvent such as THF, the concentration of the dissolved metal is relatively low. Thus, with potassium, it has been found that the maximum concentration in THF is $\sim 5 \times 10^{-4}\,\mathrm{mol\,l^{-1}}$ and reaches $\sim 10^{-3}\,\mathrm{mol\,l^{-1}}$ in dimethoxyethane.[48]

$$2\mathrm{Mt_{(solid)}} \rightleftharpoons \mathrm{Mt^+} + \mathrm{Mt^-} \tag{6}$$

$$\mathrm{Mt^-} \rightleftharpoons \mathrm{Mt} + \mathrm{e^-_{(solv.)}} \tag{7}$$

$$\mathrm{Mt} \rightleftharpoons \mathrm{Mt^+} + \mathrm{e^-_{(solv.)}} \tag{8}$$

Dissolution of alkali and alkaline-earth metals in hydrocarbons can be obtained in the presence of strongly solvating additives, such as crown ethers[49] and cryptands.[50] All these systems initiate the polymerizations of various vinylic and dienic monomers. In spite of the homogeneity of the process, it is not possible to relate the number of active centres formed to the initial concentration in solvated electrons.[48–50]

In order to increase the activity of metals as initiators, lamellar compounds of alkali metal graphitides have been studied; such derivatives are obtained by direct reaction of the metal on to the graphite. The best results were obtained by heating and compressing the system under vacuum or argon.[51] Intercalation compounds are thus formed, e.g. $\mathrm{LiC_6}$, $\mathrm{LiC_{12}}$, $\mathrm{LiC_{18}}$, ..., which are capable, after grinding, of initiating the polymerization of vinyl and diene monomers, either in hydrocarbons or in ethers. The efficiency of these initiators is low and mainly determined by the diffusion rate of the monomers inside the graphite layers.[52,53]

In spite of the great industrial impact that it had in the past (synthesis of buna rubbers), direct initiation from metals is used very little now, except when the storage of bicarbanionic oligomers is possible.

25.2.1.2 *Transfer from polycyclic aromatic radical-anions*

Homogeneous initiation by electron transfer can be achieved by using ethereal solutions of hydrocarbon radical-ions (equation 9). Even when the electron affinity of the monomer is lower than that of hydrocarbon A, transfer may be fast because the rapid dimerization of the $\mathrm{M^{\overline{\cdot}}}$, $\mathrm{Mt^+}$ species shifts equilibrium (9) towards transfer. Thus, naphthalene–sodium in THF solution initiates easy and complete polymerization of α-methylstyrene whose electron affinity is slightly lower than that of naphthalene. The lower the electron affinity of the polycyclic hydrocarbon, the more efficient the electron transfer to the monomer. The rate constants of the electron-transfer bimolecular reaction between two non-polymerizing hydrocarbons (equation 10) have been measured and found to be in the range of $10^9\,\mathrm{l\,mol^{-1}\,s^{-1}}$.[54] It can be reasonably assumed that rate constants corresponding to equilibrium (9) are in the same range.

$$\mathrm{A^{\overline{\cdot}},Mt^+} + \mathrm{M} \rightleftharpoons \mathrm{M^{\overline{\cdot}},Mt^+} + \mathrm{A} \tag{9}$$

$$\mathrm{A_1^-} + \mathrm{A_2} \rightleftharpoons \mathrm{A_1} + \mathrm{A_2^-} \tag{10}$$

The following stage is the dimerization of radical anions, equation (11). In the case of vinyl and related monomers, it might eventually lead to different species, as shown in Scheme 1. With monomers capable of polymerizing anionicaly, the stabilizing effect of X and Y groups highly favours the head-to-head linkage corresponding to equilibrium (12), compared to (13) and (14).

$$2\mathrm{M^{\overline{\cdot}},Mt^+} \xrightarrow{k_d} \mathrm{Mt^+,^-M{-}M^-,Mt^+} \tag{11}$$

$$\mathrm{Mt^+,^-CXYCH_2CH_2XYC^-,Mt^+} \tag{12}$$

$$2(\mathrm{H_2C{=}CXY})^{\overline{\cdot}},\mathrm{Mt^+} \rightleftharpoons \mathrm{Mt^+,^-CXYCH_2CXYH_2C^-,Mt^+} \tag{13}$$

$$\mathrm{Mt^+,^-CH_2CXYCXYH_2C^-,Mt^+} \tag{14}$$

Scheme 1

A comprehensive study of dimerization has been performed by means of the radical anion of 1,1-diphenylethylene (D) whose polymerization, for steric hindrance reasons, cannot go beyond the dimerization step, equation (15). Studies have also been performed on dicarbanionic dimers of α-methylstyrene (MS), equation (16). Flash photolysis[55] of the latter species and that of Mt^+, $^-DD^-$, Mt^+, allowed the determination of accurate values for the rate constant of dimerization (Table 3). Values have also been obtained by pulsed radiolysis.[57] Dimerization is slower from $D^{\bar{\cdot}}$ free ions.[58]

$$2D^{\bar{\cdot}},Mt^+ \overset{k_d}{\rightleftharpoons} Mt^+,{}^-DD^-,Mt^+ \tag{15}$$

$$2MS^{\bar{\cdot}},Mt^+ \overset{k'_d}{\rightleftharpoons} Mt^+,{}^-MS{-}MS^-,Mt^+ \tag{16}$$

Table 3 Rate Constants of Dimerization of $D^{\bar{\cdot}},Mt^+$ and $MS^{\bar{\cdot}},Mt^+$ in THF at 25 °C, Measured from Flash Photolysis of Dimer Solutions[56]

Mt^+	$D^{\bar{\cdot}},Mt^+$ $10^{-8}k_d(1\,mol^{-1}\,s^{-1})$	$MS^{\bar{\cdot}}\cdot Mt^+$ $10^{-7}k'_d(1\,mol^{-1}\,s^{-1})$
Li^+	1.2	—
Na^+	3.5	0.2
K^+	10.0	1.0–1.2
Cs^+	30.0	—

Taking into account the relative values of the rate constants of transfer and dimerization, Szwarc[56] has shown that the rate-determining stage in the transfer-from-aromatic radical-anions-dimerization process is dimerization. However, commonly used techniques of initiation do not allow instantaneous mixing of the reactants, and the overlapping of the initiation and propagation stages is mainly determined by the rate of mixing of initiator and monomer.

Equilibrium of dimerization (11) is shifted markedly to the right because the dimeric dianion is highly stabilized.

The dissociation rate constants of dimeric dianions into monomeric radical-ions have been measured on two systems: with $^-DD^-$ it has been found to be equal to $8 \times 10^{-7}\,s^{-1}$ at 30 °C, the respective E_d being about 48 kJ mol^{-1}. The method allowing such a determination is based on the study of the exchange between the dianion and the corresponding radioactive monomer in THF solution.[59] This method cannot be applied to polymerizing monomers. The dissociation of the dimer of α-methylstyrene was studied from exchange between a hydrogenated dimer and a deuterated one.[60] It corresponds to a dissociation rate constant equal to $6 \times 10^{-8}\,s^{-1}$ at 25 °C.

In the presence of excess monomer molecules, the addition reaction of monomer molecules on to radical ions (17) competes with the dimerization of the radical ions (11). Reaction (17) was studied with D as monomer;[59] the rate constant k_1 is much lower than k_d (about 10^6 times) and the competition between (11) and (17) should be perceptible at low concentrations of initiator and high concentrations of monomer. Yet, whatever the monomer, the dimeric radical-anions are not significantly involved in the initiation stage because these species are very unstable and are destroyed by disproportionation;[57] their instantaneous concentration in the reaction medium is too low to be measurable.

$$M^{\bar{\cdot}},Mt^+ \;+\; M \overset{k_1}{\rightleftharpoons} {}^\cdot M{-}M^-,Mt^+ \tag{17}$$

Whatever the method used, the initiation of polymerization by electron transfer leads to bicarbanionic species. The propagation can occur simultaneously on both sites located at the chain ends; for a complete conversion of monomers into polymer, the $\overline{DP}_n$ of the latter can be calculated from equation (18), [C] being the concentration of active centres. It must be mentioned that, due to the difficulty of performing electron transfer in hydrocarbon solvents, this method of initiation has been used mainly in polar media.

$$\overline{DP}_n \;=\; 2([M]/[C]) \tag{18}$$

25.2.2 Initiation by Nucleophilic Addition of Organometallic Compounds

Prompt and complete initiation is closely related to the initiator chosen, the activity of which is primarily determined by its overall reactivity.

An efficiency close to unity implies the use of very strong bases which are organometallic (alkali or alkaline-earth) compounds. It must be kept in mind that the intrinsic basicity of these compounds is related to the pK_a of the associated conjugated acid, *i.e.* that of the corresponding hydrocarbon, which is very high (pK_a generally > 35).

The initiation mechanism for organometallic derivatives is similar to that of propagation: it is a nucleophilic addition on a substrate whose electrophilic character is low, thus requiring a reactant of marked nucleophilic character.

Depending on the nature of the solvent in which the reaction is performed, two main types of initiation can be considered: (i) initiation in hydrocarbon solvents (initiators available in such media are generally more or less covalent); and (ii) initiation in polar solvents (all types of initiators are available, but ionized organometallic compounds are often preferred, such as $Mt^+,^-(\alpha\text{-methyl-styrene})_4^-,Mt^+$; $Mt^+,^-DD^-,Mt^+$; $Ph\bar{C}Me_2,Mt^+$; and fluorenyl$^-,Mt^+$).

25.2.2.1 *Initiation in hydrocarbon solvents*

Organolithium compounds, particularly alkyllithium ones, are widely used to initiate polymerization in hydrocarbon media. This is due to both their solubility and good stability in such solvents, which is not the case for other alkali metal derivatives. Such properties are closely related to the partial covalency of the carbon–lithium bond as well as to the trend of these compounds to aggregate in hydrocarbons.

Calculations[61] and spectroscopic studies[62,63] assign sp^3 hybridization to the carbon of the carbon–lithium bond. From crystallographic data,[64] one can assign marked sp^2 character to the lithium atom, a result corroborated by calculation.[61] The degree of the covalent character of the carbon–lithium bond, which should result from both the relatively high electronegativity of lithium compared to other alkali metals and the small size of the corresponding cation, is a much debated question and depends on the method used to evaluate it. In the case of methyllithium, for example, the various methods of calculation result either in a strongly ionic character[66] or in a strongly covalent one.[67] Alkyllithium derivatives are generally volatile, which would seem to be in favour of covalency.

The ionicity of the bond seems to depend on the degree of aggregation of organometallic species[61] — the more marked the ionic character, the lower the degree of aggregation. As a result, the observation methods which study all of organometallic species, particularly ^{7}Li NMR,[68] cannot be used to account for the intrinsic reactivity of non-aggregated species, which are the only ones to be active in the initiation process. The aggregation of organolithium compounds in hydrocarbons is due to the fact that they are electron-deficient compounds. Indeed, lithium atoms have more low-energy orbitals than they have electrons. This leads to aggregation which allows the delocalization of bonding electron pairs. As a result, multiple bonding centres are generated. Each lithium atom is associated to two or more organic moieties and participates in two or more orbitals, which allows use of the low-energy orbitals available.

The aggregation degree of alkyllithiums depends on the physical state of the species, the nature of the solvent in which they are dissolved, their concentration, the structure of the alkyl group, the temperature and the possible presence of additives. Thus, solid ethyllithium is tetrameric,[69] whereas it is hexameric in benzene, as are most alkyllithiums in hydrocarbon solutions unless steric interactions due to branching at either the α or β carbon lead to the formation of tetramers.[70] For example, *s*-butyllithium and *t*-butyllithium are tetrameric in hydrocarbons.[71] With very bulky alkyl moieties, one can observe dimeric aggregates, as with menthyllithium.[72]

Even in non-polar and non-basic solvents, the degree of aggregation depends on the nature of the solvent; thus, trimethylsilylmethyllithium is hexameric in cyclohexane and tetrameric in benzene.[70] Concentration is a determining factor in the reaction behaviour of organolithium compounds. Derivatives revealing such an effect have steric hindrances intermediate between those leading to hexameric structures and those leading to tetrameric ones. This is the case for isopropyllithium, which is hexameric in highly concentrated solutions and tetrameric in dilute ones,[70] thus showing that the free energies of the two forms are only slightly different. The dissociation of aggregates into monomeric species is difficult to measure directly by conventional methods because the corresponding equilibria are largely shifted towards aggregated forms.

The effect of temperature was shown using 2-methylbutyllithium: its degree of aggregation changes from 6 at 30 °C to 7.6 at -12 °C, in pentane solution.[73]

The structure of aggregates was established by crystallography on crystalline alkyllithiums. It was shown, for example, that tetrameric methyllithium exhibits a structure in which lithium atoms are located at the corners of a tetrahedron, alkyl groups being oriented perpendicularly to each face of the tetrahedron.[74] An analogous structure can be proposed for the tetramer of *t*-butyllithium.

The solubility of alkyllithiums in hydrocarbons is closely related to the structure of aggregates. Indeed, the partially ionized carbon–lithium bonds are surrounded by a shell of alkyl moieties which gives a hydrophobic character to the species.[75] Only the tetrameric methyllithium is insoluble in hydrocarbons, the small size of the alkyl moiety not offsetting the hydrophobic character of the carbon–lithium group.

The structure and reactivity of arylmethyllithiums are somewhat different from those of alkyllithiums; indeed, the delocalization of the carbanionic charge on to the aromatic nucleus favours the sp^2 character of the carbanionic carbon atom. Nevertheless, Waack *et al.* showed by ^{13}C NMR study of benzyllithium in benzene solution, that the α carbon has substantial sp^3 character and the carbon–lithium bonding is probably largely σ in nature.[76] Arylmethyllithium compounds are less aggregated than alkyllithiums, and benzyllithium, for example, is only dimeric in hydrocarbons.

The dissociation of organolithium aggregates has received much attention from both theoretical and experimental viewpoints. From ^7Li and ^1H NMR spectra of mixtures of (*t*-butyllithium)$_4$ [(Bu^{t}Li)$_4$] and (trimethylsilylmethyllithium)$_4$ [(TSMLi)$_4$] in cyclopentane or toluene solutions, Brown and co-workers[77] have demonstrated that the disaggregation of tetrameric species proceeds *via* dimeric ones. Indeed, an equilibrium between the different components of the system is reached in cyclopentane after several hours—and more rapidly in toluene—which shows the formation of mixed aggregates with structure [(Bu^{t}Li)$_2$, (TSMLi)$_2$]. The rate-determining step in the process is the formation of (Bu^{t}Li)$_2$ from tetramer.

The dissociation energy of alkyllithium aggregates has not been experimentally measured. It has been estimated from various calculation methods and for different processes. The direct dissociation of aggregates into monomeric species would require so much energy that this process is very unlikely. For example, the process shown in equation (19) needs 453 kJ per mol of tetramer.[61] It is more likely that dissociation proceeds with formation of lower aggregates, a process which requires less energy and is in better agreement with the initiation temperatures. For example, equation (20) would require only 121 kJ mol^{-1},[61] and equation (21) would consume 150 kJ mol^{-1}.

$$(\text{MeLi})_4 \rightleftharpoons 4\text{MeLi} \tag{19}$$

$$(\text{MeLi})_4 \rightleftharpoons (\text{MeLi})_3 + \text{MeLi} \tag{20}$$

$$(\text{MeLi})_4 \rightleftharpoons 2(\text{MeLi})_2 \tag{21}$$

These values are markedly higher than those estimated by Bywater and co-workers[78] for the disaggregation of (polyisoprenyl)lithium in hydrocarbons. In the latter case, carbanionic delocalization could explain the difference.

In all likelihood, disaggregation would proceed according to equation (21)[65] followed by equation (22).

$$(\text{MeLi})_2 \rightleftharpoons 2\text{MeLi} \tag{22}$$

The influence of structural effects on the behaviour of organolithium derivatives is comprehensively discussed in the literature.[79]

Initiation of the polymerization of dienes and vinyl monomers reflects the phenomenon of aggregation.

Thus, the kinetics of styrene initiation by *n*-butyllithium in benzene was studied by Worsfold and Bywater[80] and showed that the rate of formation of (polystyryl)lithium species follows the kinetic law given in equation (23).

$$R_i = k_i[\text{Bu}^n\text{Li}]^{1/6}[\text{M}] \tag{23}$$

This means that only non-aggregated species are active in the process; the successive equilibria, such as (21) and (22) for methyllithium, can be summarized in equations (24) and (25).

$$(\text{Bu}^n\text{Li})_6 \rightleftharpoons 6\text{Bu}^n\text{Li} \tag{24}$$

$$\text{Bu}^n\text{Li} + \text{H}_2\text{C}{=}\text{CHPh} \longrightarrow \text{Bu}^n\text{CH}_2\bar{\text{C}}\text{HPh}, \text{Li}^+ \tag{25}$$

The reactivity of aggregated species towards the monomer is null or negligible.

Equation (23) was confirmed by a kinetic study of the addition of Bu^{n}Li to 1,1-diphenylethylene[81] which results in the formation of 1,1-diphenylhexyllithium (equation 26).

$$\text{Bu}^n\text{Li} + \text{H}_2\text{C}{=}\text{CPh}_2 \longrightarrow \text{Bu}^n\text{CH}_2\bar{\text{C}}\text{Ph}_2, \text{Li}^+ \tag{26}$$

When *t*-butyllithium is added to 1,1-diphenylethylene in benzene[82] and *s*-butyllithium is used to initiate styrene and isoprene in the same solvent,[83, 84] the corresponding kinetic law shows a 1/4 order which is in agreement with the tetrameric structure of the organometallic derivatives whose monomeric species only are reactive. To compare the intrinsic reactivity of organolithium initiators, early initiation rates only must be considered; indeed, when two types of organolithium species are in comparable concentration, mixed aggregates are formed and, as a result, the kinetics are disturbed. The overall reactivity of alkyllithium derivatives depends on the stability of the mixed aggregates *vs.* both the chemical reactions and the dissociative process.

It is widely accepted that, whatever their structure and the conditions of their use, monomeric alkyllithiums are very reactive towards diene and vinyl monomers, and overall reactivity is determined by either the value of the equilibrium constant of aggregation or the kinetics of the disaggregation process.[85] It is difficult to predict the behaviour of a given system, because initiator and initiated species form mixed aggregates whose disaggregation constant is unknown.

The polymerization of styrene initiated by *t*-butyllithium in benzene solution leads to unexpected kinetics.[86] Over a wide concentration range, the initiation rates are independent of monomer concentration and first order in initiator. The same phenomenon is observed with isoprene over a more restricted concentration range. Bywater and co-workers[86] explained such a behaviour by a slow disaggregation phenomenon which becomes the rate-determining step of the process. This assumption is supported by the fact that the observed first-order rate constant is close to the rate constant for intermolecular exchange found by Brown and co-workers[87] with the tetramer dissociation rate.

This explanation has been questioned by Szwarc in a recent review.[75] This author suggests a complexation of the monomer on to the tetramer (equation 27) followed by an internal initiation by the tetramer, this step being rate determining (equation 28).

$$(Bu^tLi)_4 \; + \; \text{monomer} \; \rightleftharpoons \; [(Bu^tLi)_4(\text{monomer})] \tag{27}$$

$$[(Bu^tLi)_4(\text{monomer})] \; \rightleftharpoons \; [(Bu^tLi)_3(Bu^t\text{monomer}^-,Li^+)] \tag{28}$$

The activity of alkyllithium initiators is difficult to measure in aliphatic hydrocarbons, because it is very low in the early stages of the process. That is the reason why the order of relative reactivity is controversial. From a general viewpoint, in aromatic solvents, the sigmoidal shape of the kinetic curves indicates that the initiation rate depends not only on the reactants, monomer and alkyllithium, but also on the lithiated product.[88] In aliphatic media, either the disaggregation constant is particularly low or the rate of disaggregation is very slow as a result of the poor complexing power of the solvent compared to aromatic media. By way of compensation, the intrinsic reactivity of these non-stabilized species might be extremely high.

Though the detailed mechanism of the initiation by alkyllithiums in aliphatic hydrocarbons is not fully known, it is interesting to compare, for the various butyllithiums, the overall efficiencies defined as more complete initiation at lower monomer consumption: Guyot *et al.*[88] reported the order *s-* > *t-* > *n*-butyllithium for the system isoprene–cyclohexane. Hsieh[84] found the order *s-* > *t-* > *n*-butyllithium for butadiene and isoprene in cyclohexane and *s-* > *n-* > *t*-butyllithium for styrene in the same solvent. Hsieh also checked the effect of the solvent on the rate of initiation by *n*-butyllithium and established the order toluene > hexane > cyclohexane for the three above-mentioned monomers.

Bywater and co-workers[86] reported that, in *n*-hexane, the efficiency is *t-* > *s-* > *n*-butyllithium with isoprene as monomer, and *s-* > *t-* > *n*-butyllithium with styrene in cyclohexane. They observed that initial rates of initiation are in the order *s-* > *t-* > *n*-butyllithium.

According to the solubility characteristics of the corresponding polymers, cylohexane and aromatics are preferred as solvents in the polymerization of styrene, whereas the polymerization of dienes can be performed in all types of hydrocarbons.

Delocalized carbanions associated to lithium are sometimes used either to initiate polymerization or as models of propagating species. Due to both the steric hindrance of the substituent and the delocalization of the charge, the corresponding aggregation degree is markedly lower than that of alkyllithiums; thus, in spite of their low absolute reactivity, they generally show a satisfactory efficiency to initiate polymerization. Benzyllithium[89] — a model of (polystyryl)lithium — and fluorenyl derivatives[90] are associated into dimers in hydrocarbon solvents. 1,1-diphenylhexyllithium can be used to initiate the polymerization of acrylics[91] because its reactivity is relatively low and therefore reduces the importance of side reactions on the carbonyl groups. A similar behaviour is achieved using fluorenyllithium as initiator.[92]

(i) Difunctional dilithium initiators

The synthesis of difunctional dilithium initiators to be used in hydrocarbon solvents is of great interest. Indeed, such compounds allow the easy preparation of triblock copolymers with a central polydiene block showing a high 1,4-*cis* content. Conventional alkyldilithium compounds are easy to synthesize by common methods but, due to the strong association of these compounds in hydrocarbons, they form completely insoluble and unreactive polymeric aggregates. The insolubility of dilithium compounds is due not only to the high apparent molecular weight of aggregates, but also to the strong proportion of organometallic polarized groups in the molecules. Research workers interested in the synthesis of dilithium initiators soluble in hydrocarbons tried to both reduce the degree of aggregation by steric effects and increase the solubilization energy of the organic moiety by increasing its size.

Dilithium initiators are prepared by two general methods: the first one involves the coupling of radical anions resulting from an electron transfer from lithium metal to a monomer whose polymerization can be controlled. To be efficient, the process must generally be performed in the presence of basic additives. If the latter are not removed, they will more or less modify the microstructure of the obtained polydiene block. For example, Fetters *et al.*[93] obtained lithiated dimers of 1,1-diphenylethylene, namely 1,4-dilithio-1,1,4-tetraphenylbutane, by reacting the monomer with a finely divided lithium powder dispersed in a hydrocarbon solvent, in the presence of anisole.[94] Even in the absence of polar additives, electron transfer to monomer can be achieved when pyrophoric lithium metal is used, but control of the molecular weights of the resulting polymer is difficult.[95]

The other way to synthesize dilithium initiators consists in reacting monolithium derivatives with precursors possessing two non-conjugated reactive double bonds and showing a low ceiling temperature in order to prevent polymerization. The difunctional initiators used are either insoluble or soluble: when they are soluble, rigorous stoichiometric conditions between precursor and alkyllithium are necessary and not always easy to achieve; when they are insoluble, they should react easily with the monomer in order to be useful. The insoluble initiators may be prepared using an excess of butyllithium that is then eliminated by washing,[102,103,104] but some of them lead to complete reaction with a stoichiometric amount of *s*-butyllithium.[96] *m*-Divinyl benzene has been proposed as a precursor,[97] but some homopolymerization can arise when it is reacted with butyllithium, and oligomers of higher functionality are formed.[98] When *m*-diisopropenyl benzene is used instead of *m*-divinylbenzene,[99,100] addition of *s*-butyllithium in stoichiometric conditions leads to an adduct which is capable of initiating styrene and isoprene polymerization, leading to a narrow distribution of molecular weights.

The reaction of *s*-butyllithium with analogues of 'double' 1,1-diphenylethylene, such as structures (**3**) and (**4**), leads to fine dispersions of dilithium derivatives. Initiation of butadiene by these compounds proceeds in two steps, the first one corresponding to solubilization by oligomerization.[101] Due to the strong stabilization of carbanionic species, the reactivity of these initiators is low.

(**3**)   (**4**)

More reactive are the lithium derivatives resulting from the reaction of *s*- or *t*-butyllithium with analogues of 'double' α-methylstyrene;[96,102,103] such as structures (**5**) and (**6**).

(**5**)   (**6**)

Solubilized seeds are obtained by addition of a small amount of butadiene, and subsequent block copolymerization leads to monodisperse triblock copolymers.

Other precursors have been proposed which are based on the same principles: Tung,[101] Höcker,[105] McGrath[106] and co-workers have prepared compound (7), which leads to polymeric derivatives with satisfactory difunctionality but whose low reactivity induces wide polydispersity.

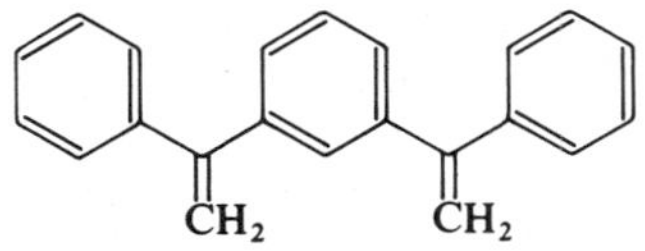

(7)

Compounds (8)[98,107] and (9),[108] when reacted with alkyllithiums, form a mixture of mono- and di-lithium derivatives. Initiators obtained from compound (10)[109] are readily soluble in toluene,[110] but produce polybutadienes with a bimodal distribution of molecular weights.

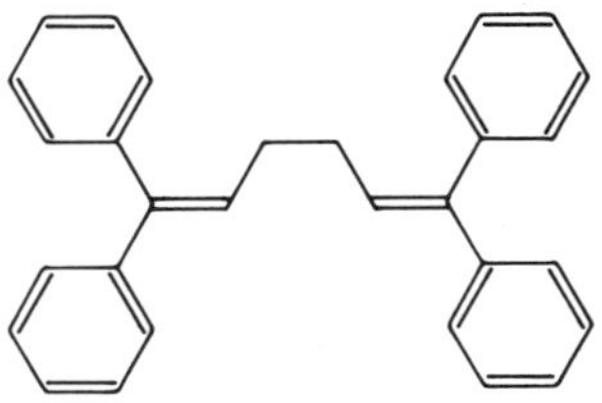

(8) 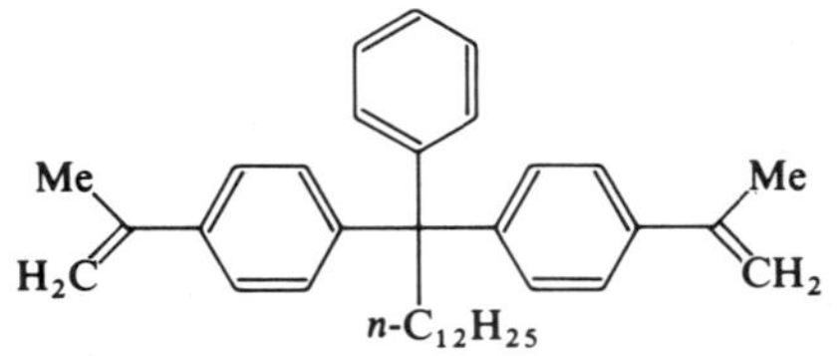   (9)

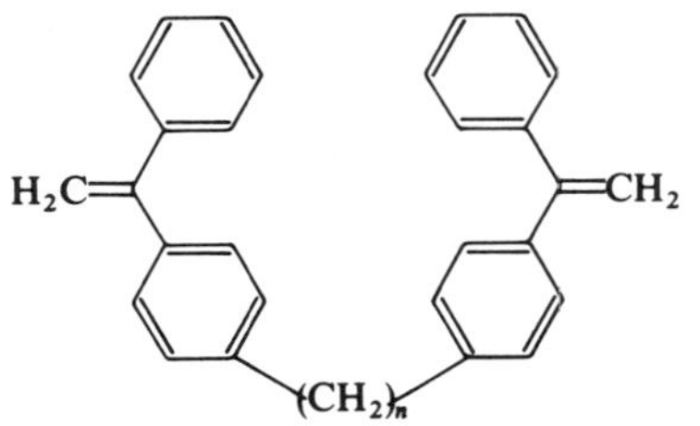

(10) $n = 2, 4, 10$

Thus, various solutions have been found to the problem of difunctional dilithium initiators but they require the synthesis of precursors with complex structures.

(ii) Other alkali metal derivatives

Alkyl derivatives of alkali metals other than lithium are insoluble in hydrocarbons. Active centres with Na^+, K^+, Rb^+ and Cs^+ as counter-ions have been obtained by oligomerization of styrene on metallic mirrors in benzene solution.[111] However, this reaction does not occur in aliphatic hydrocarbons. Studies on the propagation stage in such solvents require replacement of benzene by the selected solvent.

Another method has been used to produce organosodium initiators soluble in pure hydrocarbons; it consists of a complexation of n-butylsodium with n,s-dibutylmagnesium (sodium tributyl-magnesiate).[112] The resulting product is soluble in both cycloaliphatic and aromatic solvents, stable at room temperature and able to initiate polymerization of styrene and isoprene. It has been established that carbon–magnesium sites do not participate directly in the polymerization process. Nevertheless, the reactivity of this initiator is lower than that of the corresponding lithium derivatives. The latter characteristics would be due to the marked covalent character of the carbon–sodium bond in the magnesiate complex.

(iii) Organomagnesium compounds

Organomagnesium compounds are less reactive than the corresponding organolithium ones because, due to the higher electronegativity of magnesium compared to lithium, the carbon–

magnesium bond is less polarized than the carbon–lithium one. Nevertheless, magnesium derivatives in hydrocarbons can initiate the polymerization of monomers which are more reactive towards carbanions than styrene and dienes such as acrylic and methacrylic monomers and also vinylpyridines. The marked withdrawing effect of the substituents favours the attack of the double bond. Grignard reagents, such as phenylmagnesium bromide or *n*-butylmagnesium bromide, initiate acrylic and related monomers and produce stereoregular polymers.[113] Symmetrical organomagnesium derivatives are also efficient in the polymerization of acrylics in toluene,[114] but they are less stereospecific than Grignard reagents.[115] Natta and co-workers[116] have performed the stereospecific polymerization of 2-vinylpyridine, initiating the reaction with various ether-free Grignard reagents dispersed in toluene. Elimination of ether is achieved by heating the compounds under vacuum; nevertheless, these authors did not discover the living character of the system. Soum *et al.*[117] found that initiation by dibenzylmagnesium, prepared from the corresponding mercury derivatives, produces isotactic living poly(2-vinylpyridine), but also produces side products resulting from nucleophilic attack on the pyridine ring by the benzylic carbanion. They also showed that initiation occurs from one of the two benzylic groups, the second one being inactivated by the reaction of the first one. Polymerization proceeds with the benzylmagnesium cation as countercation. When benzylpicolylmagnesium is used as initiator,[117] side reactions are absent and well-defined living isotactic poly(2-vinylpyridine) is obtained in toluene (equation 29).

It is worthy of note that these compounds, like other magnesium derivatives,[118] are strongly aggregated in hydrocarbon solutions, but that the corresponding living chain is non-aggregated due to the complexing power of the pyridine rings.[119]

Benzylpicolylmagnesium has also been used to polymerize methyl methacrylate; the resulting polymer is atactic, whereas initiation by dibenzylmagnesium is isospecific, thus showing the influence of the pyridine ring on the stereoregulating mechanism.[120]

(iv) Effect of Lewis bases

Amines, ethers and alkoxides have been used as additives to modify the reaction behaviour of organolithium initiators. These additives coordinate by filling the vacant orbitals of the lithium atoms, thus inducing competition between the complexation and the aggregation phenomena which use the same orbitals.

The effect of additives on structure and reactivity of the corresponding systems depends on the nature of the organolithium derivative, initiated monomer and additive. The relative concentration of the latter species is also important, and additives whose concentrations are in the same order of magnitude as those of the active centres will be the only ones discussed here. Under such conditions, it can be considered that the macroscropic dielectric constant of the medium does not vary and that, in hydrocarbons, dissociation of ion pairs into free ions is excluded.

The effect of Lewis bases as additives was first mentioned by Dolgoplosk *et al.*,[121] who noticed that addition of minute amounts of THF to hydrocarbon solutions of alkyllithiums induces an acceleration of initiation. Bywater *et al.* confirmed this phenomenon[122] and showed that initiation of styrene by *s*-butyllithium is first order in lithium derivatives in the presence of THF. An analogous phenomenon was observed with dioxane and alkoxides as additives.

Guyot and Vialle[88] revealed the extreme complexity of systems including alkoxides. Thus, by initiating isoprene in cyclohexane, they observed that either an increase or a decrease in the rate of intiation can be obtained, depending on the structures of the butyllithium isomer and the alkoxide. In some cases, they observed an accelerating effect at the early stages of the process, followed by a slowing down of the reaction. Such complex behaviour was explained by the simultaneous existence, in the reaction medium, of single, binary and ternary aggregates, the latter resulting from an association between initiator and (polyisoprenyl)lithium formed. Each of these species reacts with its own reactivity and influences the overall reactivity of the system.

Heats of interaction between various Lewis bases and different alkyllithiums, allow classification of the bases according to their decreasing power of coordination.[124] This classification is independent of the structure of the alkyl moiety and of the corresponding initial aggregation degree. The order is: THF > 2-methylTHF > 2,5-dimethylTHF > diethyl ether > triethylphosphine > triethylamine > tetrahydrothiophene.

Enthalpies of coordination vary from about $1.6 \, kJ \, mol^{-1}$ for tetrahydrothiophene solvating hexameric *n*-butyllithium to $45 \, kJ \, mol^{-1}$ for THF solvating trimethylsilylmethyllithium hexamers in cyclohexane.

It must be emphasized that alkyllithiums are aggregated, generally in the tetrameric form, in monoether and monoamine solvents,[125] and it must be considered that aggregates are solvated under such conditions. Solvation would make the disaggregation process easier, accounting for the increasing reactivity.

Comprehensive studies have been performed on the effect of multiamines and multiethers on the reactivity of alkyllithium derivatives. These multidentates can produce large increases in reactivity; amines have been more studied than ethers because the latter are reactive towards activated carbanionic species and are transformed into inactive alkoxides.

Relevant results have been obtained in the polymerization of ethylene initiated by the system butyllithium–N,N,N',N'-tetramethylethylenediamine (BuLi–TMEDA).[126] Among all the alkyllithiums, only methyllithium remains tetrameric in the presence of TMEDA because the aggregation energy is higher than that of the complexation by the diamine. From ^{1}H and ^{7}Li NMR spectra of BuLi–TMEDA systems, Langer and co-workers[127] suggested the formation of a five-membered 1:1 complex (**11**). The contact ion-pair would be externally solvated by the diamine.

(**11**)

The kinetic study of ethylene initiation by various butyllithiums in hydrocarbons is very controversial. Only the most recent results are presented here. With *n*-butyllithium, a model of the growing polyethylene chain, the degree of aggregation, measured by cryometry on $\sim 10^{-1} \, mol \, l^{-1}$ solutions, depends on the structure of the complexing agent;[128] with TMEDA, organometallic species are dimeric and the corresponding kinetic order is 0.5.[129] Similar results are obtained with *s*-butyllithium;[130] with N,N,N',N'-tetraethylethylenediamine[131] and dimethyldiethylenediamine, alkyllithiums are non-aggregated and the kinetic laws are first order with respect to the initiator. Complexation with pentamethyldiethylenetriamine (2,5,8-trimethyl-2,5,8-triazanonane) produces unstable species,[131] and the system is not convenient for initiation of ethylene.

The effect of bidentates on the overall kinetics of initiation is difficult to generalize because several authors have observed, according to the monomer studied, either an increase or a decrease in reactivity when the tertiary diamine is added.[132,133] Such behaviour can be explained by a phenomenon analogous to that studied by Hélary *et al.*[134] in the polymerization of styrene by lithiated species in the presence of TMEDA. They showed that the reactivity of non-complexed non-aggregated species is higher than that of non-aggregated ones complexed by TMEDA. At a high concentration of lithiated active centres, the influence of complexed species is prominent in the kinetics and the overall effect is an activating one. At a low concentration of active centres, the complexed species formed are less reactive than would be the non-aggregated non-complexed ones whose relative proportion is increased by the high dilution.[135] If the phenomenon were the same with alkyllithiums, the equilibria in equations (30) and (31) could be written:

$$\underset{\text{inactive}}{(RLi)_n} \; \rightleftharpoons \; \underset{\text{highly active}}{n \, RLi} \tag{30}$$

$$\underset{\text{inactive}}{(RLi)_n} \; + \; \text{bidentate} \; \rightleftharpoons \; \underset{\text{moderately active}}{n \, RLi, \text{bidentate}} \tag{31}$$

Strong activation of organolithium systems can be obtained by addition of molecules showing a strong solvating power, such as crown ethers,[136,137] crown amines[138] and cryptands.[139] These additives change contact ion-pairs in loose ion-pairs, whose reactivity is much higher than that of non-aggregated contact ion-pairs. This high reactivity induces a carbanionic attack of the com-

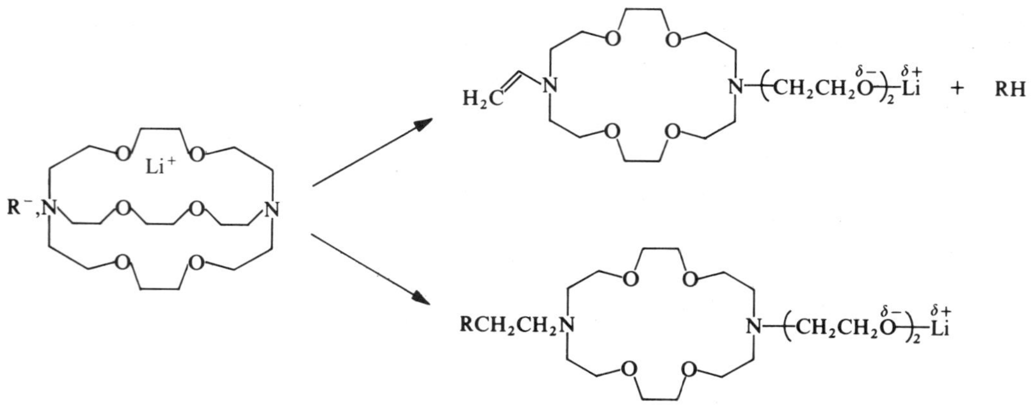

Scheme 2

plexing additive and leads to inactivation of the species. Scheme 2 has been proposed to explain the deactivation of alkyllithiums by cryptands.[140]

Asymmetric Lewis bases added to conventional initiators in hydrocarbons are capable of inducing asymmetric polymerization. Yuki *et al.* performed selective polymerization of (R,S)-methylbenzyl methacrylate by using cyclohexylmagnesium chloride (or bromide)/(−)sparteine systems as initiator.[141] Similar results were obtained with the *n*-butyllithium/(−)sparteine system,[142] and with an asymmetric initiator such as lithium (R)-N-(1-phenylethyl)aniline (**12**). With ethylmagnesium bromide–(−)sparteine complexes used in toluene at −78 °C, stereoselective polymerization of 1,2-diphenylethyl methacrylate has also been obtained.[143]

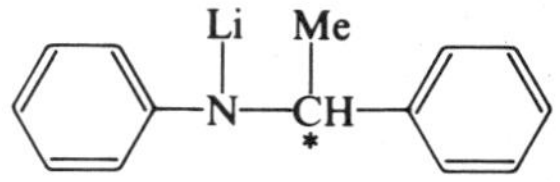

(**12**)

25.2.2.2 Initiation in polar solvents

Both the solvating power and the dielectric constant of the solvent considerably increase the reactivity of initiators, compared to that in hydrocarbons, by disaggregation of aggregates, solvation of active species and even by dissociation of ion pairs into free ions.

Difunctional initiators are easily obtained by dimerization of radical ions of monomers whose polymerization can be avoided. Such is the case for diluted solutions of α-methylstyrene reacting with alkali metals at room temperature. Due to the relatively low ceiling temperature of this monomer and depending on the conditions of the reaction, one can obtain either dicarbanionic dimers (**13**) produced from the monomeric radical-ion $M^{\bar{\cdot}}$, Mt^+, or dicarbanionic tetramers whose structure has been questioned. Those resulting from the dimerization of dimeric radical-ions (equations 32 and 33) should give a compound with structure (**14**)[144,145] whereas those formed by dimerization of $M^{\bar{\cdot}}$ followed by addition of two monomer molecules (equation 34) must have structure (**15**).

$$M^{\bar{\cdot}},Mt^+ \; + \; M \longrightarrow MM^{\bar{\cdot}},Mt^+ \tag{32}$$

$$2MM^{\bar{\cdot}},Mt^+ \longrightarrow Mt^+,^-MMMM^-,Mt^+ \tag{33}$$

$$Mt^+,^-MM^-,Mt^+ \; + \; 2M \longrightarrow Mt^+,^-MMMM^-,Mt^+ \tag{34}$$

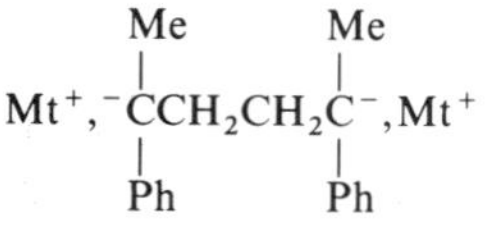

(**13**) 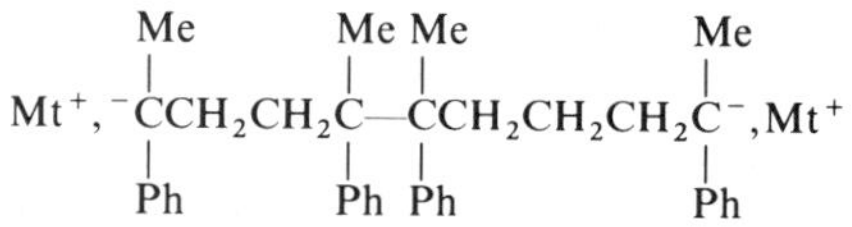   (**14**) 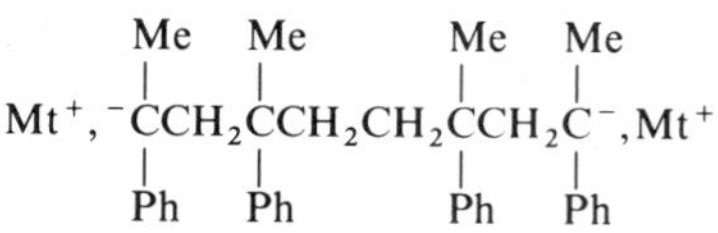   (**15**)

The latter species have been identified by their protonation derivatives.[146,147]

Whatever their structure, these species have sufficient reactivity to induce rapid initiation of diene, vinyl and related monomers.

Dimerization of 1,1-diphenylethylene (DPE) produces species with reactivity less than that of the previous examples, but which are preferred for the initiation of acrylics.

As monofunctional initiators, alkyl- and aryl-lithiums are preferably used at low temperature because, at room temperature, they react with ethers to give inactive alkoxides.[148] The only quantitative information available on these compounds concerns addition of DPE to various lithium derivatives in THF solution.[149] For all the systems studied, the kinetics are first order in alkene but of variable order in initiator. Table 4 gives kinetic orders corresponding to initial rates measured at room temperature.

Table 4 Kinetic Behaviour for Addition of Organolithium Derivatives RLi to DPE in THF at 22 °C[149]

R	Kinetic order
Methyl	0.27 ± 0.03
Phenyl	0.66 ± 0.04
Vinyl	0.34 ± 0.1
n-Butyl	0.4
Allyl	1
Benzyl	1.1 ± 0.2

Resonance-stabilized compounds, such as benzyllithium and allyllithium, have an order equal to unity which corresponds to the absence of aggregation in THF, whereas alkyllithiums and compounds with sp^2 hybridized carbanions (vinyllithium and phenyllithium) show a fractional order. In the latter case it appears that solvation by THF is not sufficient to overcome forces promoting aggregation, and the situation is analogous to that arising in hydrocarbons but with a higher overall reactivity.

In order to reduce secondary transfer and termination reactions in the polymerization of acrylates, Teyssié and co-workers[150] tried to reduce the nucleophilicity of alkyllithiums. By hindering both the carbonyl sites (t-butylacrylates) and the carbanionic centres, they obtained polyacrylates with well-defined structures, indicating the absence of side reactions. The reactivity of s-butyllithium in THF at −78 °C, was lowered by coordination with a μ-type hindering ligand such as LiCl, whose counter-anion has a strong coordinating power.

McGrath and co-workers[151] obtained similar results by using s-butyllithium as initiator and pyridine as solvent. The reversible reaction of lithiated species with pyridine would prevent side reactions on carbonyl groups.

Alkyl alkali metal compounds other than lithium ones are very unstable, and resonance-stabilized derivatives are often preferred to initiate polymerization with Na^+, K^+, Rb^+ and Cs^+ as counter-ions.

For example, benzyl⁻,Na^+ and cumyl⁻,K^+ have been used to initiate the polymerization of styrene in THF.[152] Nevertheless, benzyl⁻,Na^+ is unstable as shown by the modification of its UV spectrum. Cumyl⁻,K^+, does not change when kept under similar conditions. This indicates, with the benzyl derivative, a degradation process involving an α hydrogen.[152] Cumyl derivatives are stable and display high reactivity. They are commonly used as monofunctional initiators in polar media. They are prepared *via* the route shown in equation (35).[153] The methoxide is insoluble in THF and is removed by filtering.

$$\langle\!\!\!\bigcirc\!\!\!\rangle CMe_2OMe \ + \ 2Mt \ \longrightarrow \ MeOMt \ + \ \langle\!\!\!\bigcirc\!\!\!\rangle \bar{C}Me_2, Mt^+ \tag{35}$$

Dimsyl sodium, $MeSOCH_2^-,Na^+$, synthesized by reacting sodium hydride with dimethyl sulfoxide, initiates the polymerization of methyl methacrylate and acrylonitrile in dimethyl sulfoxide solution[154] and of styrene in HMPA.[155] Highly dissociating reaction media are necessary to activate this initiator.

Numerous initiators involving active groups other than carbon–metal bonds have been proposed for the initiation of anionic polymerization. Sodium amide was used to initiate styrene in liquid ammonia[156] but, due to the presence of a protic group at one end of the polymeric chain, the process is not a living one.

Lithium N,N-dialkylamides are capable of reacting with isoprene and with styrene.[158] In THF, one observes an induction period which has been attributed to the existence of aggregated species.[157] On the other hand, an intramolecular coordination leading to a six-membered ring (16) would complicate the polymerization kinetics.[158]

(16)

Activation of alkali metal amides has been obtained by addition of either the corresponding alkoxides or salts as $NaNO_2$. These 'complex bases' are able to initiate a large variety of monomers.[159-161]

In spite of their relatively low reactivity, alkali metal alkoxides initiate the polymerization of highly reactive monomers. Trekoval found that the initiation of methyl methacrylate (MMA) can be achieved by using lithium t-butoxides as initiators in toluene solution.[162] He determined a complex order for the initiation reaction with respect to monomer (about 2) which was attributed to the ability of the ester group in the monomer to activate lithium alkoxide. Other investigators claimed that highly polar media are necessary to activate these initiators. Zilkha and co-workers[163] performed oligomerization of MMA in dimethyl sulfoxide (DMSO), using MeO^-, Mt^+ as initiator and MeOH as transfer agent. They showed that the activity of the various methoxides varies in the order MeOLi > MeONa > MeOK.

Tomoi *et al.*[164] established that, in various polar solvents (HMPA, DMSO, dimethylacetamide, . . .) the presence of a heteroatom in the vicinity of the oxanion strongly increases the activity of alkoxides. For example, they used $MeOCH_2CH_2O^-, Na^+$ (or Li^+) and $Me_2NCH_2CH_2O^-, Na^+$ (or Li^+) and compared their activity to that of BuO^-, Mt^+.

They explained the higher activity of the former species by an internal coordination as shown in structures (17) and (18).

(17)   **(18)**

Methyl methacrylate has been initiated by alkali metal salts of methanol and t-butanol, in toluene, in the presence of cryptands and macrocyclic crown ethers.[165]

The mechanism of initiation of acrylonitrile by lithium alkoxides in dimethylformamide solution, has been studied by Berger *et al.*[166] They interpreted the complex behaviour of this system assuming the formation of donor–acceptor complexes of lithium alkoxide with DMF and with the monomer; they proposed a reaction scheme with two types of active species.

Alkali metal alkoxides are also capable of initiating monomers possessing strongly electron-withdrawing substituents, other than acrylics. Berry *et al.*[167] studied the initiation kinetics of β-nitrostyrene (eventually substituted on the phenyl ring) in THF. They observed that the kinetic law is first order with respect to monomer and initiator. It is important to point out that the carbanion generated is stabilized by the $—NO_2$ group, as shown in equation (36).

Carbanionic species associated to cations other than alkali metals, present a reactivity generally lower than that of the corresponding alkali metal derivatives. As previously mentioned for initiation performed in hydrocarbons, organomagnesium (Grignard or symmetrical) derivatives readily initiated acrylic monomers and vinyl pyridines.

Typical Lewis bases, even used alone, are capable of adding on to $>C=C<$ activated double bonds. Thus, methacrylonitrile is initiated by triethylphosphine in the presence[168] (or not[169]) of an electrolyte. The occurrence of an initiation forming zwitterionic species was demonstrated by means of spectroscopic end-group analysis for acrylonitrile polymerized in dimethylformamide.[170] With triethyl phosphite, initiation takes place in benzene and in acetonitrile; when performed in bulk, the polymerization initiated by triethyl phosphite is explosive.[171] α-Cyanoacrylates are among the most reactive monomers known and can be initiated in THF by relatively weak bases, such as tribenzylamine, pyridine and triphenylphosphine, which form zwitterionic species. With triphenylphosphine as initiator, the process is living, whereas it is more complex when tertiary amines are used.[172] LiBr, NaI, $Bu_4^n NI$, NaCN and aqueous solutions of NaOH, NH_4OH and Bu_4NOH also readily initiate these monomers.[173]

25.2.3 Electrochemically Initiated Polymerizations

This method consists of the generation of radical ions of the monomer by passing an electric current through a polar solvent containing both an electrolyte and the monomer. Thus, Funt and co-workers were successful in initiating the polymerization of acrylonitrile[174] and styrene[175] in the presence of alkali metal salts, but evidence of living anions was not obtained, probably due to the nature of the solvents used. Anderson[176] showed that quaternary ammonium salts can be used as electrolytes, thus demonstrating the direct electroinitiation process suggested by Funt[174] and illustrated in Scheme 3.

$$M + e^- \longrightarrow M^{\bar{\cdot}}$$

$$2M^{\bar{\cdot}} \longrightarrow {}^-M\!\!-\!\!M^-$$

Scheme 3

The first reported electrolytic production of living anions was that described by Yamazaki and co-workers for α-methylstyrene.[177] These authors dissolved $LiAlH_4$ or $NaAlR_4$ in dry THF in order to obtain a solution with convenient resistivity. Polymerization was carried out in a divided electrolytic cell, to prevent the deactivation of the generated carbanionic centres by the species resulting from anodic oxidation.

The living character of the system was demonstrated by the excellent agreement between experimental and theoretical molecular weights calculated from equation (37), where the amount of monomer is in grams and Q is the quantity of current in Faradays. Polymerization and copolymerization of various monomers have been performed by electrochemical initiation,[174-182] but the stability of the active centres produced is very dependent on the nature of the system components.

$$\bar{M}_n = \frac{2\,(\text{monomer})}{Q} \tag{37}$$

25.3 REFERENCES

1. C. Harries, *Justus Liebigs Ann. Chem.*, 1911, **383**, 213.
2. W. Schlenk, J. Appenrodt, A. Michael and A. Thal, *Chem. Ber.*, 1914, **47**, 473.
3. M. Szwarc, *Nature (London)*, 1956, **178**, 1168; M. Szwarc, M. Levy and R. Milkovich, *J. Am. Chem. Soc.*, 1956, **78**, 2656.
4. M. Shima, J. Smid and M. Szwarc, *J. Polym. Sci., Part B*, 1964, **2**, 735.
5. M. Shima, D. N. Bhattacharyya, J. Smid and M. Szwarc, *J. Am. Chem. Soc.*, 1963, **85**, 1306.
6. D. N. Bhattacharyya, C. L. Lee, J. Smid and M. Szwarc, *J. Am. Chem. Soc.*, 1963, **85**, 533.
7. J. C. Favier, P. Sigwalt and M. Fontanille, *J. Polym. Sci., Polym. Chem. Ed.*, 1977, **15**, 2373.
8. A. W. Langer, Jr., *Trans. N. Y. Acad. Sci.*, 1965, **27**, 741.
9. F. Wenger, *Chem. Ind. (London)*, 1959, 1094.
10. R. K. Graham, D. L. Dunkelberger and E. S. Cohn, *J. Polym. Sci.*, 1960, **42**, 501.
11. D. L. Glusker, E. Stiles and B. Yoncoskie, *J. Polym. Sci.*, 1961, **49**, 297.
12. A. Hirao, H. Kato, K. Yamaguchi and S. Nakahama, *Macromolecules*, 1986, **19**, 1294.
13. T. E. Hogen-Esch and J. Smid, *J. Am. Chem. Soc.*, 1965, **87**, 669; *J. Am. Chem. Soc.*, 1966, **88**, 307.
14. U. Takaki, G. L. Collins and J. Smid, *J. Organomet. Chem.*, 1978, **145**, 139.
15. G. Hélary, L. Lefèvre-Jenot, M. Fontanille and J. Smid, *J. Organomet. Chem.*, 1981, **205**, 139.
16. J. Smid, in 'Ions and Ion pairs in Organic Reactions', ed. M. Szwarc, Wiley-Interscience, New York, 1972, vol. 1, p. 98.
17. J. C. Favier, M. Fontanille and P. Sigwalt, *Bull. Soc. Chim. Fr.*, 1971, 526.
18. T. E. Hogen-Esch and J. Smid, *J. Am. Chem. Soc.*, 1969, **91**, 4580.
19. N. H. Velthorst and G. J. Hoijtink, *J. Am. Chem. Soc.*, 1965, **87**, 4529; *J. Am. Chem. Soc.*, 1967, **89**, 209.
20. J. W. Burley and R. N. Young, *J. Chem. Soc. B*, 1971, 1018.
21. T. Ellingsen and J. Smid, *J. Phys. Chem.*, 1969, **73**, 2712.
22. R. M. Fuoss and F. Acascina, 'Electrolytic Conductance', Interscience, New York, 1959.
23. J. T. Denison and J. B. Ramsey, *J. Am. Chem. Soc.*, 1955, **77**, 2615.

24. S. Hubert, C. Momtaz, P. Hémery, S. Boileau and J. P. Kintzinger, *Polym. Prepr., Am. Chem. Soc., Div. Polym. Chem.,* 1986, **27**, 134.
25. T. E. Hogen-Esch and J. Smid, *J. Phys. Chem.,* 1975, **79**, 233.
26. B. Vidal, Doctoral thesis, University of Paris VI, 1975.
27. M. Berthelot, *Ann. Chim.,* 1867, **12**, 155.
28. W. Schlenk and E. Bergmann, *Justus Liebigs Ann. Chem.,* 1928, **464**, 1.
29. R. Willstätter, F. Seitz and E. Bumm, *Chem. Ber.,* 1928, **61**, 871.
30. M. Szwarc and J. Jagur-Grodzinski, in 'Ions and Ion Pairs in Organic Reactions', ed. M. Szwarc, Wiley-Interscience, New York, 1974, vol. 2, p. 51.
31. R. C. Weast (ed.), 'Handbook of Chemistry and Physics', 66th edn., CRC Press, Boca Raton, FL, 1985.
32. 'Selected Values of Chemical Thermodynamic Properties', US National Bureau of Standards, Washington, DC, 1959.
33. M. Fontanille and P. Sigwalt, unpublished results.
34. G. J. Hoijtink, E. de Boer, P. H. van der Meij and W. P. Weijland, *Recl. Trav. Chim. Pays-Bas,* 1956, **75**, 487.
35. J. Jagur-Grodzinski, M. Feld, S. L. Yang and M. Szwarc, *J. Phys. Chem.,* 1965, **69**, 628.
36. H. A. Laitinen and S. Wawzonek, *J. Am. Chem. Soc.,* 1942, **64**, 1765.
37. M. Szwarc, 'Carbanions, Living Polymers and Electron Transfer Processes', Interscience, New York, 1968.
38. S. Wawzonek and H. A. Laitinen, *J. Am. Chem. Soc.,* 1942, **64**, 2365.
39. M. Fontanille and P. Sigwalt, *C. R. Hebd. Seances Acad. Sci.,* 1966, **262**, 1208; *C. R. Hebd. Seances Acad. Sci.,* 1966, **263**, 316.
40. B. J. MacClelland, *Chem. Rev.,* 1964, **64**, 301.
41. J. Minoux, B. François and Ch. Sadron, *Makromol. Chem.,* 1961, **44**, 519.
42. C. Mathis, L. Christmann-Lamandé and B. François, *J. Polym. Sci., Polym. Chem. Ed.,* 1978, **16**, 1285.
43. F. Wenger, *J. Am. Chem. Soc.,* 1960, 82, 4281.
44. C. G. Overberger and N. Yamamoto, *J. Polym. Sci., Part B,* 1965, **3**, 569; *J. Polym. Sci., Part A-1,* 1966, **4**, 3101.
45. A. Davis, D. H. Richards and N. F. Scilly, *Makromol. Chem.,* 1972, **152**, 121.
46. D. H. Richards, *Polymer,* 1978, **19**, 109.
47. J. L. Dye, M. T. Lok, F. J. Tehan, R. B. Coolen, N. Papadakis, J. M. Ceraso and M. G. Debacker, *Ber. Bunsenges. Phys. Chem.,* 1971, **75**, 659.
48. F. S. Dainton, D. M. Wiles and A. N. Wright, *J. Polym. Sci.,* 1960, **45**, 111.
49. S. Alev, F. Schué and B. Kaempf, *J. Polym. Sci., Polym. Lett. Ed.,* 1975, **13**, 397.
50. J. Lacoste, F. Schué, S. Bywater and B. Kaempf, *J. Polym. Sci., Polym. Lett. Ed.,* 1976, **14**, 201.
51. D. Guérard and A. Herold, *C. R. Hebd. Seances Acad. Sci., Ser. C,* 1972, **275**, 571.
52. E. Loria, G. Merle, J. P. Pascault and I. B. Rashkov, *Polymer,* 1981, **22**, 95.
53. I. B. Rashkov, G. Merle, Q. T. Pham, V. C. Shishkova, J. P. Pascault and I. M. Panayotov, *Eur. Polym. J.,* 1982, **18**, 37.
54. G. Rämme, M. Fisher, S. Claesson and M. Szwarc, *Chem. Phys. Lett.,* 1971, **9**, 306.
55. H. C. Wang, G. Levin and M. Szwarc, *J. Phys. Chem.,* 1979, **83**, 785.
56. M. Szwarc, *ACS Symp. Ser.,* 1983, **212**, 419.
57. C. Schneider and A. J. Swallow, *Makromol. Chem.,* 1968, **114**, 155.
58. H. C. Wang, E. D. Lillie, S. Slomkowski, G. Levin and M. Szwarc, *J. Am. Chem. Soc.,* 1977, **99**, 4612.
59. G. Spach, H. Monteiro, M. Levy and M. Szwarc, *Trans. Faraday Soc.,* 1962, **58**, 1809.
60. M. Szwarc and S. Asami, *J. Am. Chem. Soc.,* 1962, **84**, 2269.
61. G. Graham, S. Richtsmeiek and D. A. Dixon, *J. Am. Chem. Soc.,* 1980, **102**, 5759.
62. D. McKeever, R. Waack, M. A. Doran and E. B. Baker, *J. Am. Chem. Soc.,* 1969, **91**, 1057.
63. T. L. Brown, D. W. Dickerhoof and D. A. Bafus, *J. Am. Chem. Soc.,* 1962, **84**, 1371.
64. R. Zerger, W. Rhine and G. Stucky, *J. Am. Chem. Soc.,* 1974, **96**, 6048.
65. G. Graham, S. Richtsmeir and D. A. Dixon, *J. Am. Chem. Soc.,* 1980, **102**, 5759.
66. A. Streitwieser, Jr., *J. Organomet. Chem.,* 1978, **156**, 1.
67. G. D. Graham, D. S. Marynick and W. N. Lipscomb, *J. Am. Chem. Soc.,* 1980, **102**, 4572.
68. P. A. Scherr, R. J. Hogan and J. P. Oliver, *J. Am. Chem. Soc.,* 1974, **96**, 6055.
69. H. Dietrich, *Acta Crystallogr.,* 1963, **16**, 681.
70. H. L. Lewis and T. L. Brown, *J. Am. Chem. Soc.,* 1970, **92**, 4664.
71. M. Weiner, C. Vogel and R. West, *Inorg. Chem.,* 1962, **1**, 654.
72. W. H. Glaze and C. H. Freeman, *J. Am. Chem. Soc.,* 1969, **91**, 7198.
73. G. Fraenkel, W. E. Beckenbaugh and P. P. Yang, *J. Am. Chem. Soc.,* 1976, **98**, 6878.
74. E. Weiss and G. Hencken, *J. Organomet. Chem.,* 1970, **21**, 265.
75. M. Szwarc, *Adv. Polym. Sci.,* 1983, **49**, 1.
76. R. Waack, L. D. McKeever and M. A. Doran, *J. Chem. Soc., Chem. Commun.,* 1969, 117.
77. G. E. Hartwell and T. L. Brown, *J. Am. Chem. Soc.,* 1966, **88**, 4625.
78. J. E. L. Roovers and S. Bywater, *Polymer,* 1973, **14**, 594.
79. R. N. Young, R. P. Quirk and L. J. Fetters, *Adv. Polym. Sci.,* 1984, **56**, 1.
80. D. J. Worsfold and S. Bywater, *Can. J. Chem.,* 1960, **38**, 1891.
81. A. G. Evans and D. B. George, *J. Chem. Soc.,* 1961, 4653.
82. R. A. H. Casling, A. G. Evans and N. H. Rees, *J. Chem. Soc., Part B,* 1966, 519.
83. S. Bywater and D. J. Worsfold, *J. Organomet. Chem.,* 1967, **10**, 1.
84. H. L. Hsieh, *J. Polym. Sci., Part A,* 1965, **3**, 163.
85. F. Schué and S. Bywater, *Macromolecules,* 1969, **2**, 458.
86. J. E. L. Roovers and S. Bywater, *Macromolecules,* 1975, **8**, 251.
87. M. Y. Darensbourg, B. Y. Kimura, G. E. Hartwell and T. L. Brown, *J. Am. Chem. Soc.,* 1970, **92**, 1236.
88. A. Guyot and J. Vialle, *J. Macromol. Sci., Chem.,* 1970, **A4**, 79.
89. T. L. Brown, *Acc. Chem. Res.,* 1968, **1**, 23.
90. M. M. Exner, R. Waack and E. C. Steiner, *J. Am. Chem. Soc.,* 1973, **95**, 7009.
91. W. Fowells, C. Schuerch, F. A. Bovey and F. P. Hood, *J. Am. Chem. Soc.,* 1967, **89**, 1396.
92. B. Wesslén, G. Gunneby, G. Hellström and P. Svedling, *J. Polym. Sci., Polym. Symp.,* 1973, **42**, 457.

93. L. J. Fetters and M. Morton, *Macromolecules*, 1969, **2**, 453.
94. M. Morton, L. J. Fetters, J. Inomata, D. C. Rubio and R. N. Young, *Rubber Chem. Technol.*, 1976, **49**, 303.
95. B. François, M. Vernois and E. Franta, *Ger. Pat.* 2603946 (1976) (*Chem. Abstr*, 1976, **85**, 178205).
96. I. Obriot, J. C. Favier and P. Sigwalt, *Polymer*, 1987, **28**, 2093.
97. P. Lutz, E. Franta and P. Rempp, *C. R. Hebd. Seances Acad. Sci., Ser. C*, 1976, **283**, 123.
98. F. Bandermann, H. D. Speikamp and L. Weigel, *Makromol. Chem.*, 1985, **186**, 2017.
99. G. Beinert, P. Lutz, E. Franta and P. Rempp, *Makromol. Chem.*, 1978, **179**, 551.
100. P. Lutz, E. Franta and P. Rempp, *Polymer*, 1982, **23**, 1953.
101. L. H. Tung, G. Y.-S. Lo and D. E. Beyer, *Macromolecules*, 1978, **11**, 616.
102. P. Guyot, J. C. Favier, H. Uytterhoeven, M. Fontanille and P. Sigwalt, *Polymer*, 1981, **22**, 1724.
103. M. Fontanille, P. Guyot, P. Sigwalt and J. P. Vairon, *Fr. Pat.* 2313389 (1975).
104. P. Guyot, J. C. Favier, M. Fontanille and P. Sigwalt, *Polymer*, 1982, **23**, 73.
105. G. H. Schulz and H. Höcker, *Makromol. Chem.*, 1977, **178**, 2589.
106. A. D. Broske, T. L. Huang, J. M. Hoover, R. D. Allen and J. E. McGrath, *Polym. Prepr., Am. Chem. Soc., Div. Polym. Chem.*, 1984, **25** (2), 85.
107. K. H. Bauer and H. Herzog, *J. Prakt. Chem.*, 1936, **147**, 4.
108. G. Beinert, J. G. Zilliox and J. Herz, *Makromol. Chem.*, 1985, **186**, 1351.
109. G. Lattermann and H. Höcker, *Makromol. Chem.*, 1974, **175**, 2865.
110. T. Bastelberger and H. Höcker, *Angew. Makromol. Chem.*, 1984, **125**, 53.
111. J. E. L. Roovers and S. Bywater, *Trans. Faraday Soc.*, 1966, **62**, 701.
112. M. Liu, C. Kamienski, M. Morton and L. J. Fetters, *J. Macromol. Sci., Chem.*, 1986, **A23**, 1387.
113. W. E. Goode, F. H. Owens and W. L. Myers, *J. Polym. Sci.*, 1960, **47**, 75.
114. K. Hatada, K. Ute, K. Tanaka, T. Kitayama and Y. Okamoto, *Polym. Prepr., Am. Chem. Soc., Div. Polym. Chem.*, 1986, **27** (1), 151.
115. H. Yuki and K. Hatada, *Adv. Polym. Sci.*, 1979, **31**, 1.
116. G. Natta, G. Mazzanti, P. Longi, G. Dall'Asta and F. Bernardini, *J. Polym. Sci.*, 1961, **51**, 487.
117. A. Soum and M. Fontanille, *Makromol. Chem.*, 1980, **181**, 799.
118. E. C. Ashby and F. Walker, *J. Organomet. Chem.*, 1967, **7**, 17.
119. A. Soum and M. Fontanille, *Makromol. Chem.*, 1981, **182**, 1743.
120. A. Soum, N. D'Accorso and M. Fontanille, *Makromol. Chem., Rapid. Commun.*, 1983, **4**, 471.
121. E. N. Kropacheva, B. A. Dolgoplosk and E. M. Kuznetsova, *Dokl. Akad. Nauk. SSSR*, 1960, **130**, 253.
122. D. J. Worsfold and S. Bywater, *J. Chem. Soc.*, 1960, 5234.
123. J. E. L. Roovers and S. Bywater, *Macromolecules*, 1968, **1**, 328.
124. R. P. Quirk and D. E. Kester, *J. Organomet. Chem.*, 1977, **127**, 111.
125. H. L. Lewis and T. L. Brown, *J. Am. Chem. Soc.*, 1970, **92**, 4664.
126. A. W. Langer, Jr., *Trans. N. Y. Acad. Sci.*, 1965, 741; *Adv. Chem. Ser.*, 1974, **130**, 1.
127. M. T. Melchior, L. P. Klemann, T. A. Whitney and A. W. Langer, *Polym. Prepr., Am. Chem. Soc., Div. Polym. Chem.*, 1972, **13**, 649.
128. G. Crassous and F. Schué, *Polymer*, 1983, **24**, 1203.
129. H. Magnin, F. Rodriguez, M. Abadie and F. Schué, *J. Polym. Sci., Polym. Chem. Ed.*, 1977, **15**, 875.
130. M. Aldissi, F. Schué, K. Geckeler and M. Abadie, *Makromol. Chem.*, 1980, **181**, 1413.
131. G. Crassous, M. Abadie and F. Schué, *Eur. Polym. J.*, 1979, **15**, 747.
132. A. Davidjan, N. Nikolaew, V. Sgonnik, B. Belenkii, V. Nesterow and B. Erussalimsky, *Makromol. Chem.*, 1976, **177**, 2469.
133. N. Smirnowa, V. Sgonnik, K. Kalninsch and B. Erussalimsky, *Makromol. Chem.*, 1977, **178**, 773.
134. G. Hélary and M. Fontanille, *Eur. Polym. J.*, 1978, **14**, 345.
135. M. Fontanille, G. Hélary and M. Szwarc, *Macromolecules*, 1988, **21**, 1532.
136. J. P. Pascault, F. Chastrette and Q. T. Pham, *Eur. Polym. J.*, 1976, **12**, 273.
137. S. Alev, A. Collet, M. Viguier and F. Schué, *J. Polym. Sci., Polym. Chem. Ed.*, 1980, **18**, 1155.
138. G. Hélary and M. Fontanille, *Polym. Bull. (Berlin)*, 1980, **3**, 159.
139. J. Lacoste, F. Schué, S. Bywater and B. Kaempf, *J. Polym. Sci., Polym. Lett. Ed.*, 1976, **14**, 201.
140. Th. Lanoye, Doctoral thesis, University of Paris VI, 1975.
141. Y. Okamoto, K. Suzuki, K. Ohta and H. Yuki, *J. Polym. Sci., Polym. Lett. Ed.*, 1979, **17**, 293.
142. Y. Okamoto, K. Suzuki and H. Yuki, *J. Polym. Sci., Polym. Chem. Ed.*, 1980, **18**, 3043.
143. Y. Okamoto, E. Yashima, K. Hatada, H. Yuki, H. Kageyama, K. Miki and N. Kasai, *J. Polym. Sci., Polym. Chem. Ed.*, 1984, **22**, 1831.
144. C. L. Lee, J. Smid and M. Szwarc, *J. Phys. Chem.*, 1962, **66**, 904.
145. A. Vrancken, J. Smid and M. Szwarc, *Trans. Faraday Soc.*, 1962, **58**, 2036.
146. R. L. Williams and D. H. Richards, *J. Chem. Soc., Chem. Commun.*, 1967, 414.
147. D. H. Richards and N. F. Scilly, *J. Chem. Soc., Chem. Commun.*, 1968, 1641.
148. A. Rembaum, S. P. Siao and N. Indictor, *J. Polym. Sci.*, 1962, **56**, S 17.
149. R. Waack and M. A. Doran, *J. Am. Chem. Soc.*, 1969, **91**, 2456.
150. R. Fayt, R. Forte, C. Jacobs, R. Jérome, T. Ouhadi, Ph. Teyssié and S. K. Varshney, *Macromolecules*, 1987, **20**, 1442.
151. H. B. Gia and J. E. McGrath, *Polym. Prepr., Am. Chem. Soc., Div. Polym. Chem.*, 1986, **27** (1), 179.
152. R. Asami, M. Levy and M. Szwarc, *J. Chem. Soc.*, 1962, 361.
153. K. Ziegler and H. Dislich, *Chem. Ber.*, 1957, **90**, 1107.
154. J. E. Mulvaney and R. L. Markham, *J. Polym. Sci., Polym. Lett. Ed.*, 1966, **4**, 343.
155. A. Priola and L. Trossarelli, *Makromol. Chem.*, 1970, **139**, 281.
156. J. J. Sanderson and C. R. Hauser, *J. Am. Chem. Soc.*, 1949, **71**, 1595.
157. A. C. Angood, S. A. Hurley and P.-J.-T. Tait, *J. Polym. Sci., Polym. Chem. Ed.*, 1975, **13**, 2437.
158. S. A. Hurley and P.-J.-T. Tait, *J. Polym. Sci., Polym. Chem. Ed.*, 1976, **14**, 1565.
159. G. Coudert, G. Ndebeka, P. Caubère, S. Raynal, S. Lécolier and S. Boileau, *J. Polym. Sci., Polym. Lett. Ed.*, 1978, **16**, 413.
160. S. Raynal, S. Lécolier, G. Ndebeka and P. Caubère, *J. Polym. Sci., Polym. Lett. Ed.*, 1980, **18**, 13.
161. S. Raynal, *Eur. Polym. J.*, 1986, **22**, 559.

162. J. Trekoval, *J. Polym. Sci., Part A-1, 1971*, **9**, 2575.
163. S. Freireich and A. Zilkha, *J. Macromol. Sci., Chem.*, 1972, **A6**, 1383.
164. M. Tomoi, K. Sekiya and H. Hakiuchi, *Polym. J.*, 1974, **6**, 438.
165. M. Viguier, A. Collet, F. Schué and B. Mula, *Polym. Prepr., Am. Chem. Soc., Div. Polym. Chem.*, 1986, **27** (1), 157.
166. W. Berger, S. Riedel, H. J. Adler, G. Wunderlich, D. Lehmann and O. Vogl, *J. Macromol. Sci., Chem.*, 1983, **A20**, 299.
167. R. W. H. Berry, R. J. Mazza and S. F. Sullivan, *Makromol. Chem.*, 1984, **185**, 559.
168. M. A. Markevich, E. V. Kochetov, F. Ranogaets and N. S. Enikolopyan, *Vysokomol. Soedin., Ser. A*, 1973, **15**, 2063.
169. M. A. Markevich, E. V. Kochetov, F. Ranogaets and N. S. Enikolopyan, *Vysokomol. Soedin., Ser. A*, 1973, **15**, 2489.
170. C. D. Eisenbach, V. Jaacks, H. Schnecko and W. Kern, *Makromol. Chem.*, 1974, **175**, 1329.
171. T. Ogawa and P. Quintana, *J. Polym. Sci., Polym. Chem. Ed.*, 1975, **13**, 2517.
172. D. C. Pepper, *Polym. J.*, 1980, **12**, 629.
173. D. C. Pepper, *J. Polym. Sci., Polym. Symp.*, 1978, **62**, 65.
174. B. L. Funt and F. D. Williams, *J. Polym. Sci.*, 1964, **A2**, 865.
175. B. L. Funt and S. W. Laurent, *Can. J. Chem.*, 1964, **42**, 2728.
176. J. D. Anderson, *J. Polym. Sci., Part A-1*, 1968, **6**, 3185.
177. N. Yamazaki, S. Nakahama and S. Kambara, *J. Polym. Sci., Part B*, 1965, **3**, 57.
178. D. Laurin and G. Parravano, *J. Polym. Sci., Polym. Lett. Ed.*, 1966, **4**, 797.
179. S. N. Bhadani and G. Parravano, *J. Polym. Sci., Part A-1*, 1970, **8**, 225.
180. N. Yamazaki, *Adv. Polym. Sci.*, 1969, **6**, 377.
181. N. Yamazaki, I. Tanaka and S. Namahama, *J. Macromol. Sci., Chem.*, 1968, **A2**, 1121.
182. B. L. Funt and S. N. Bhadani, *J. Polym. Sci., Part A*, 1965, **3**, 4191.

26

Carbanionic Polymerization: Kinetics and Thermodynamics

AXEL H. E. MÜLLER
University of Mainz, FRG

26.1   GENERAL KINETIC EQUATIONS

For a kinetic analysis, the process of anionic polymerization has to be divided into at least three main reactions common to all types of polymerization (equations 1–3). I* denotes initiator, M monomer, P_i^* and P_i' an active or inactive polymer chain of degree of polymerization i, respectively, and X a terminating agent.

$$\text{Initiation} \quad \text{I*} \;+\; \text{M} \xrightarrow{\;k_i\;} P_i^* \tag{1}$$

$$\text{Propagation } P_i^* \; + \; M \; \xrightarrow{\; k_p \;} \; P_{i+1}^* \qquad (2)$$

$$\text{Termination } P_i^* \; (+X) \; \xrightarrow{\; k_t \;} \; P_i' \qquad (3)$$

'Living' polymerizations are characterized by the absence of termination (as well as transfer) reactions. In most cases, it is also possible to find initiators which are reactive enough to give instantaneous initiation, *i.e.* $k_i \geq k_p$. This implies that the concentration of active centres $c^* = \Sigma[P_i^*]$ remains constant during the polymerization process. For such an 'ideal' type of polymerization, only propagation has to be taken into account. Thus, for the rate of polymerization, simple pseudo-first-order kinetics can be applied

$$R_p \; = \; -\frac{d[M]}{dt} \; = \; k_p[M]c^* \; = \; k_{app}[M] \qquad (4)$$

where k_{app} is the 'apparent' pseudo-first-order rate constant. Integration leads to

$$\ln\frac{[M]_0}{[M]} \; = \; k_p c^* t \; = \; k_{app} t \qquad (5)$$

or

$$k_{app} \; = \; -d\ln[M]/dt \; = \; k_p c^* \qquad (6)$$

A constant slope of the first-order time–conversion curve indicates fast initiation (otherwise c^* would increase initially) as well as the absence of chain termination (which is accompanied by a decrease of c^* at longer reaction times).

For complete initiation ($c^* = [I]_0$), the propagation rate constant can be calculated directly from the slope of the first-order plot. However, as initiation is not always complete, it is usually necessary to determine c^* either directly (*e.g.* by UV spectroscopy) or indirectly from the number-average degree of polymerization, P_n, which is given by

$$P_n \; = \; \frac{\text{concentration of polymerized monomers}}{\text{concentration of polymer chains}} \; = \; \frac{[M]_0 x_p}{c_{tot}} \qquad (7)$$

$$c^* \; = \; k c_{tot} \qquad (8)$$

Here, c_{tot} is the total concentration of polymer chains, *including terminated ones*, and k is the number of active centres per polymer chain. A plot of P_n *vs.* monomer conversion x_p gives the constant slope $[M]_0/c_{tot}$. Any deviation from constancy indicates chain-generating reactions (*e.g.* slow initiation or transfer) or chain-coupling reactions (*e.g.* termination between two polymer molecules). Spontaneous termination, or termination by reaction with monomer or impurities during polymerization, will *not* change c_{tot} and thus will *not* harm the linearity of a plot of P_n *vs.* x_p. However, equation (8) will be invalid in that case, and equation (9) may be used

$$c_0^* \; = \; k c_{tot} \; = \; f[I]_0 \qquad (9)$$

where c_0^* denotes the initial concentration of active centres (directly after initiation) and f is the initiation efficiency.

26.1.1 Kinetics for Slow Initiation

Slow initiation ($k_i < k_p$) is frequently encountered in non-polar solvents. Because of the curved nature of the first-order time–conversion plots (corresponding to an increasing k_{app} in equations 5 and 6) it is often difficult to determine the rate constants of initiation and propagation. Kinetic evaluations for this problem were given by Gee *et al.*,[100] Beste and Hall[11] and Bawn *et al.*[12] They were reviewed and extended by Pepper.[13]

The differential equations resulting from equations (1) and (2) are

$$\frac{d[I]}{dt} \; = \; -\frac{dc^*}{dt} \; = \; -k_i[M][I] \; = \; -k_i[M]([I]_0 \; - \; c^*) \qquad (10)$$

$$\frac{d[M]}{dt} = -\{k_i[I] + k_p c^*\}[M] = \{k_i[I]_0 + (k_p - k_i)c^*\}[M] \tag{11}$$

(assuming $c^* = [I]_0 - [I]$, *i.e.* no termination). Integration of equations (10) and (11) leads to

$$\ln\frac{[I]}{[I]_0} = -k_i \int_0^t [M]dt \tag{12}$$

$$\ln\frac{[M]}{[M]_0} = -\{k_i[I]_0 + (k_p - k_i)c^*\}t = -k_{app}t \tag{13}$$

$$k_{app} = -d\ln[M]/dt = k_i[I]_0 + (k_p - k_i)c^* \tag{14}$$

Combination of equations (12) to (14) gives

$$\ln(k_p - k_{app}/[I]_0) = \ln(k_p - k_i) - k_i \int_0^t [M]dt \tag{15}$$

where the integral has to be evaluated numerically from the experimental data, and a first choice of k_p has to be estimated from the maximum slope of the first-order time–conversion curve.

In order to avoid evaluation of the integral, solving the full differential equation for monomer consumption (which can be expressed in terms of k_{app}) has also been suggested

$$dk_{app}/dt + k_i[M]k_{app} - k_i k_p[I]_0[M] = 0 \tag{16}$$

which after several transformations leads to

$$\ln\left(1 - \frac{k_{app}}{k_p[I]_0}\right) + \frac{k_{app}}{k_p[I]_0} = \frac{k_i}{k_p[I]_0}([M] - [M]_0) = -\frac{k_i[M]_0}{k_p[I]_0}x_p = -\frac{k_i}{k_p}P_n \tag{17}$$

This equation can be used for a fit of k_p and k_i. A first estimate of k_p is again obtained from the maximum value of k_{app}:

$$k_{app,max} \le k_p[I]_0 \tag{18}$$

It must be noted, however, that, even for full monomer conversion, the initiator is not always completely converted to living polymer chains. As was shown by Litt,[14] the fraction of initiator converted to polymer chains at complete monomer conversion, $f = c^*/[I]_0$, is given by the solution of equation (19):

$$[M]_0 = -[I]_0\{(k_p/k_i)[\ln(1 - f) + f] + f\} \tag{19}$$

Strictly, f can never reach unity, but for practical purposes it is sufficient to have $f > 0.99$, which is reached at a ratio of $k_i[M]_0/k_p[I]_0 \approx 4$ (or $k_i P_{n,max} \approx 4k_p$). For higher ratios, initiator will be practically consumed *before* complete monomer conversion. For most practical applications, the effect of slow termination on kinetics (and also on the molecular weight distribution, MWD) can be neglected for $k_i P_{n,max} > 4k_p$.

Litt's calculations were extended by Szwarc *et al.*[15,16] to the more general case where initiation and propagation are of different fractional orders with respect to $[I]$ and c^*, respectively. This situation is encountered in the polymerization in non-polar solvents, due to association phenomena (*cf.* Section 26.5.1). Here, initiator is completely consumed when $[M]_0$ exceeds a certain critical value, $[M]_{cr}$.

26.2 MOLECULAR WEIGHT DISTRIBUTIONS

26.2.1 'Ideal' Polymerization

The molecular weight distribution (MWD) for a 'living' polymerization with fast initiation was calculated by Flory[17] as early as 1940. It is identical with a Poisson distribution. The frequency distribution (mole fraction of *P*-mers) and the weight distribution (weight fraction of *P*-mers), respectively, are given by

$$n(P) = \frac{\exp(-v)v^{P-1}}{(P-1)!} \tag{20}$$

$$w(P) = \frac{P\exp(-v)v^{P-1}}{(v+1)(P-1)!} \tag{21}$$

where $v = P_n - 1$ is the kinetic chain length.

The weight-average degree of polymerization is given by

$$P_w = \frac{\Sigma P w(P)}{\Sigma w(P)} = v + 1 + \frac{v}{v+1} = P_n + 1 - \frac{1}{P_n} \tag{22}$$

The polydispersity ratio P_w/P_n is given by

$$P_w/P_n = 1 + \frac{1}{P_n} + \frac{1}{P_n^2} = 1 + \frac{P_n - 1}{P_n^2} \tag{23}$$

For further derivations it is practical to use the non-uniformity introduced by Schulz[18]

$$U \equiv P_w/P_n - 1 = \frac{P_n - 1}{P_n^2} \tag{24}$$

For high degrees of polymerization ($P_n \gg 1$), equations (22)–(24) can be simplified:

$$P_w = P_n + 1 \tag{25}$$

$$P_w/P_n = 1 + 1/P_n \approx 1 \tag{26}$$

$$U = 1/P_n \ll 1 \tag{27}$$

This shows that 'living' polymerization has the capability to form polymers of extremely narrow MWD.

26.2.2 Effect of Experimental Deficiencies

A variety of influences may broaden the MWD. One effect which is of practical importance in many cases is broadening due to experimental deficiencies. As anionic polymerizations are often very fast reactions, it is sometimes difficult to mix initiator and monomer solutions in a time much shorter than the half-life of the polymerization. An expression for the MWD for continuous monomer feed and finite time of mixing was derived by Figini and Schulz[19,20] and by Litt.[14] When conducting the polymerization in a flow-tube reactor, it is important to work under conditions of turbulent flow. It was shown by Löhr and Schulz[21] that laminar flow may lead to a considerable broadening of the MWD. The excess non-uniformity U_{hydr} introduced by turbulent flow alone is in the order of magnitude of 10^{-3}.

A further experimental problem encountered in anionic polymerizations is residual impurities in the monomer. This is a severe problem when monomer is added as a continuous feed or if mixing is slow compared to polymerization. The MWDs corresponding to an instantaneous reaction of the chain with impurities were calculated by Szwarc and Litt,[22] Figini[23] and Orofino and Wenger[24] for a stirred-tank reactor and by Panke[25] for a flow-tube reactor with finite time of mixing. The MWD resulting from a *slow* temination by impurities was calculated by Coleman *et al.*[26] and by Yan and Yuan.[27] The combined effects of termination by impurities and transfer to monomer,[28,29] as well as the effect of transfer to an impurity, were also calculated.[30,31]

Other factors which affect the MWD can give additional information on the polymerization mechanism. Some of these factors and the resulting distributions are discussed below. The MWDs generated from reversible polymerizations will be discussed in Section 26.6.

26.2.3 Effect of Interconversion between Different Species

Generally, anionic polymerization proceeds *via* various coexisting species of different reactivity (*cf.* Section 26.3) which are in a dynamic equilibrium. An example is given for a two-state mechanism in Scheme 1. $(P_i^*)^{(1)}$ and $(P_i^*)^{(2)}$ denote a living i-mer in states 1 and 2, respectively. If the rate constants of interconversion, k_{12} and k_{21}, are much larger than $k_{p,1}[M]$ and $k_{p,2}[M]$, respectively, the effect on the MWD will be negligible. For the opposite case, the MWD will comprise two independent Poisson distributions. For all intermediate cases, the MWD for an n-state polymerization will be a function of monomer concentration and $(3n - 2)$ rate constants. Several authors have derived expressions for the polydispersity ratio (or non-uniformity), and for the MWD resulting from a two-state mechanism.[32-35] The corresponding expressions for a multi-state polymerization were derived by Böhm.[36,37]

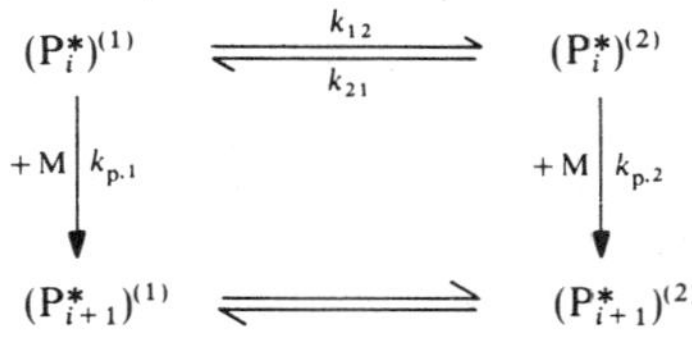

$$(P_i^*)^{(1)} \underset{k_{21}}{\overset{k_{12}}{\rightleftharpoons}} (P_i^*)^{(2)}$$

$$+M \downarrow k_{p,1} \qquad\qquad +M \downarrow k_{p,2}$$

$$(P_{i+1}^*)^{(1)} \rightleftharpoons (P_{i+1}^*)^{(2)}$$

$$K_{12} = k_{12}/k_{21}$$

Scheme 1

The experimental non-uniformity U_{app} is composed of the value given by the Poisson distribution, U_{pois}, and excess values stemming from interconversion:

$$U_{app} = U_{pois} + U_{12} + U_{23} + \dots \tag{28}$$

For a two-state mechanism, U_{12} is given by

$$U_{12} = \frac{k_1^2 K_{12}}{k_{21} k_p} c^* \frac{2 - x_p}{2 x_p} \tag{29}$$

When the rates of interconversion are comparable to those of propagation, the frequency distribution resulting from a multi-state mechanism is Gaussian:[37]

$$n(P) = (2\pi P_n^2 U_{app})^{-1/2} \exp\left[-\frac{(P - P_n)^2}{2 U_{app} P_n^2} \right] \tag{30}$$

A special case of the two-state polymerization is when one of the two species in equilibrium is inactive ('dormant'). The corresponding expressions for P_w and P_n were derived by Szwarc and Hermans[38] and by Krause et al.[39]

26.2.4 Effect of Non-Bernoullian Statistics in Stereospecific Monomer Addition

Yet another type of multi-state polymerization is encountered in some stereospecific polymerizations. The active chain-end can exist in two different states which only interconvert on monomer addition, as shown in Scheme 2. States 1 and 2 can stand for chain ends which are preceded by a *meso* and *racemic* placement of monomer units, respectively. When the rate of monomer addition is dependent on the state of the chain end (penultimate monomer placement), i.e. $k_{21} \neq k_{11}$ and $k_{12} \neq k_{22}$, non-Bernoullian statistics have to be used to describe the polymer's microstructure.[40] Scheme 2 conforms to first-order Markoff statistics. It was shown by Figini[41] that this may lead to a broadening of the MWD, especially when k_{11} or k_{22} are larger than the other rate constants.

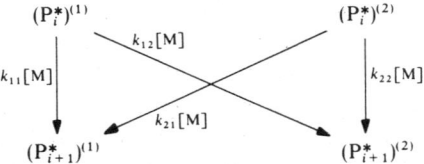

Scheme 2

The mechanism is more complex when the property which distinguishes states 1 and 2 is directly located at the chain end, because here an interconversion can occur by monomer addition *and* by an equilibrium. Examples are *cis* and *trans* chain-ends in diene polymerization[42] and *E* and *Z* chain-ends in the polymerization of polar monomers.[43] However, for the latter, it was shown that interconversion by equilibrium is much slower than by monomer addition. At the time of writing, a calculation of the MWD corresponding to such a mechanism has not yet been published.

26.2.5   Effect of Slow Initiation

Slow initiation can lead to a significant broadening of the MWD, because new polymer chains are generated during polymerization. The MWD for such a process was calculated by various authors for monofunctional[44−48] and bifunctional polymers.[49,50] The more general case, for which the rate constants of the first *n* monomer additions depend on *P*, was calculated by Figini.[51] Typically, the MWD has a tailing towards lower molecular weights. The excess non-uniformity is limited to $0 \leq U_{ex} \leq 0.375$.

In certain cases, the 'seeding technique' introduced by Wenger[52,53] can improve the MWD. The concept is to first add only a small fraction of the monomer to the initiator, in order to quantitatively convert the latter to active oligomers. Only after a while is the bulk of the monomer added. However, it was pointed out by Szwarc *et al.*[15,16] that it is generally not possible to quantitatively consume the initiator in the first step and to produce narrower MWDs using this technique. Thus, the method (at least for the 'seeding' monomer) is restricted to monomers which are susceptible to reversible polymerization, *e.g.* α-methylstyrene (*cf.* Section 26.6). Here, enough monomer is provided by depolymerization to allow for complete initiator consumption.

In practice, the seeding technique seems to have led to narrowing of the MWD in several cases. This may be explained if, in the 'classical' technique, some impurities are not consumed before a polymer of relatively high molecular weight is formed, but are consumed completely *during* the formation of seeds.

26.2.6   Effect of Termination and Transfer

Chain termination and transfer are reactions which are frequently encountered in 'real' polymerizations, especially with polar monomers. This effect can lead to distributions of the Schulz–Flory type, or to even broader ones. Numerous publications have dealt with MWDs corresponding to various types of termination and transfer and combinations thereof, combined in part with slow initiation: (a) spontaneous termination;[54,55] (b) termination by the monomer;[56,57] (c) spontaneous transfer[58−60] ($P_i^* \rightarrow P_i' + I$); and (d) transfer to monomer.[28,54,59,61−70]

The effect of termination and transfer by impurities has been dealt with in Section 26.2.2.

26.3   POLYMERIZATION OF HYDROCARBON MONOMERS IN POLAR SOLVENTS

In the following sections, the most important results on the kinetics and mechanisms of anionic vinyl polymerization will be reviewed. Recent (as well as some fundamental older) books and reviews giving more detailed information are referenced at the end of this chapter.[1−10,296]

Most of the early work on the mechanism of anionic polymerization was done on styrene and α-methylstyrene in ethereal solvents such as tetrahydrofuran (THF), 1,2-dimethoxyethane (DME), tetrahydropyran (THP) or 1,4-dioxane. These monomers (together with dienes) are perfect for kinetic investigations, because the active centres are stable for longer times than are necessary for complete polymerization. In 1958, it was shown by Worsfold and Bywater[71] that the polymerization of α-methylstyrene in THF leads to linear first-order plots (*cf.* equations 5 and 6), the apparent (*i.e.* pseudo-first-order) rate constant being a linear function of *c**. During the 1960s and early 1970s, thorough investigations, mainly by the groups of Szwarc (Syracuse) and Schulz (Mainz), gave deep insight not only into the elementary reactions of anionic polymerization, but also into the properties of ions and ion pairs in non-aqueous solvents.

26.3.1   Ions and Ion Pairs

Independently, the groups of Szwarc[72,73] and Schulz[74,75] reported that in the anionic polymerization of styrene in THF the propagation rate constant k_p decreases with increasing concentration of active centres c^*. This was attributed to the coexistence of two different kinds of active species, namely free anions and ion pairs, propagating at different rates (Scheme 3).

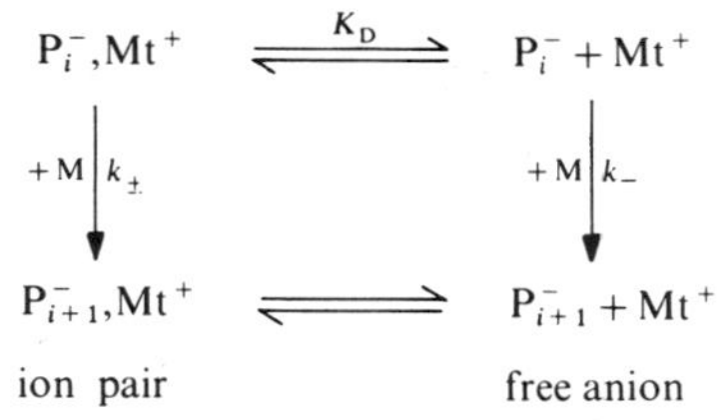

$$P_i^-, Mt^+ \xrightleftharpoons[]{K_D} P_i^- + Mt^+$$

$$+M \Big| k_\pm \qquad\qquad +M \Big| k_-$$

$$P_{i+1}^-, Mt^+ \rightleftharpoons P_{i+1}^- + Mt^+$$

ion pair $\qquad\qquad$ free anion

Scheme 3

Thus, the experimental rate constant is given by

$$k_p = \alpha k_- + (1 - \alpha)k_\pm = k_\pm + (k_- - k_\pm)\alpha \qquad (31)$$

α being the degree of dissociation. Due to the low dielectric constants of the ethereal solvents, the dissociation constants are very low ($K_D < 10^{-6}\,mol\,l^{-1}$) and consequently $\alpha \ll 1$. Thus, the mass-action law can be simplified:

$$K_D = \frac{\alpha^2 c^*}{(1 - \alpha)} \approx \alpha^2 c^* \qquad (32)$$

and

$$\alpha = (K_D/c^*)^{1/2} \qquad (33)$$

leading to

$$k_p = k_\pm + (k_- - k_\pm)K_D^{1/2}(c^*)^{-1/2} \qquad (34)$$

Figure 1 shows that a plot of k_p vs. $(c^*)^{-1/2}$ ('Szwarc–Schulz plot') is indeed linear and has a positive intercept, allowing the determination of $k_\pm$.

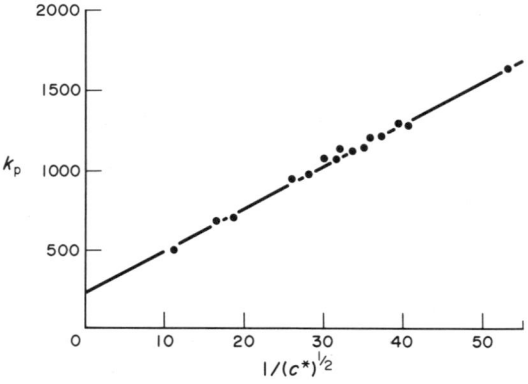

Figure 1 Dependence of propagation rate constants on the concentration of active centres in the anionic polymerization of styrene in THF, using Na^+ as the counterion, at 25 °C[9]

Two different methods were used to determine K_D and k_-. The Syracuse group used conductivity measurements in order to calculate K_D. The experimental problem is that the limiting conductivity $\Lambda_0 = \lambda_0^+ + \lambda_0^-$ can only be determined to a low degree of accuracy from the corresponding Fuoss–Krauss plot[76] because of its very low value. Whereas λ_0^+ can be measured accurately by using stable salts, λ_0^- is usually estimated from the diffusion coefficient of the polymer chain. Some problems arising from that estimation were pointed out by Schmitt and Schulz.[77]

A different approach was used by the Mainz group. Addition of a common-ion salt having a dissociation constant much larger than that of the polymeric ion-pairs decreases the degree of dissociation of the latter. Tetraphenyl borates (*e.g.* $NaBPh_4$, $CsBPh_3CN$) are suitable for this purpose. The concentration of free anions is then given by

$$[P^-] = \frac{K_D c^*}{[Mt^+]} \tag{35}$$

As the added salt is much more dissociated than the polymeric ion-pairs

$$[Mt^+] = [Mt^+]_{add} + [Mt^+]_{pol} \approx [Mt^+]_{add} \tag{36}$$

$[Mt^+]_{add}$ is calculated from the stoichiometric concentration of the salt c_s and its dissociation constant K_s is determined from conductivity measurements:

$$[Mt^+]_{add} = (K_s/2)[-1 + (1 + 4c_s/K_s)^{1/2}] \tag{37}$$

If the amount of added salt is very low, or its dissociation constant is comparable to that of the living polymer, $[Mt^+]_{pol}$ cannot be neglected. It is given by

$$[Mt^+]_{pol} = -[Mt^+]_{add}/2 + ([Mt^+]_{add}^2/4 + K_D c^*)^{1/2} \tag{38}$$

The degree of dissociaton is

$$\alpha = [P^-]/c^* = K_D/[Mt^+] \tag{39}$$

Combination with equation (31) gives

$$k_p = k_\pm + (k_- - k_\pm)K_D/[Mt^+] \tag{40}$$

For purposes of high accuracy, it should be noted that equations (37)–(40) neglect the effect of $[Mt^+]_{pol}$ on the dissociation of the salt. A precise equation is given by Schmitt and Schulz.[77]

Figure 2 shows that a plot according to equation (40) has good linearity. The values of $k_\pm$ determined from Figures 1 and 2 agree, within the limits of experimental error. Combination of the slopes of both plots renders K_D and k_-.

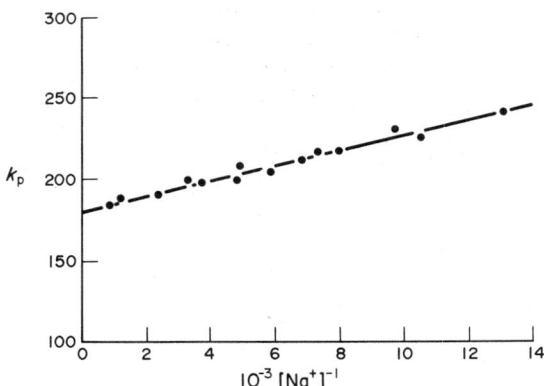

Figure 2 Dependence of propagation rate constants on the concentration of added sodium ions in the anionic polymerization of styrene in THF at 25 °C[9]

By adding a large excess of common-ion salt, it is possible to suppress the dissociation of the polymeric ion-pairs sufficiently to allow direct determination of $k_\pm$ from only one kinetic experiment. Some problems which may arise from the presence of high concentrations of electrolytes were pointed out by Szwarc.[78]

Table 1 gives the propagation rate constants of the pairs in various solvents differing in dielectric constant ε. The propagation rate constant for the free anions is nearly invariant with solvent ($k_- \approx 1.3 \times 10^5 \, l \, mol^{-1} \, s^{-1}$). In contrast, there is a strong dependence of $k_\pm$ on the polarity of the solvent, especially for the smaller cations. This is a first indication that the nature of the solvent influences the structure of the ion pair. It should be noted that although their dielectric constants are comparable, the solvating power of DME is higher than that of THF because it has a bidentate character.

Table 1 Propagation Rate Constants[a] of the Ion Pairs in the Anionic Polymerization of Styrene at *ca.* 25 °C
$$(l\,mol^{-1}\,s^{-1})$$

Solvent	ε^n	Li^+	Na^+	K^+	Cs^+
DME	7.20	1295[d]	3600[d], 5000[e]	—	340[l]
THF	7.39	176[b], 160[f]	80[f], 200[h]	60–80[f]	22[f], 130[m]
THP	5.61	11.6[b], 19.5[g]	14[i]	30[i]	53i
Oxepane	5.06	4.9[b]	10[j]	—	—
1,4-Dioxane	2.25	0.9[c]	3.4[c], 6.5[k]	20[c], 28[k]	24.5[c], 15[k]

[a] This table contains only selected data. A more complete survey is given in ref. 4.
[b] At 20 °C; S. Peeters, Ph. D. Thesis, University of Leuven, Belgium, 1982; S. Peeters and M. van Beylen, in preparation.
[c] D. N. Bhattacharyya, J. Smid and M. Szwarc, *J. Phys. Chem.*, 1965, **69**, 624.
[d] T. Shimomura, J. Smid and M. Szwarc, *J. Am. Chem. Soc.*, 1967, **89**, 5743.
[e] At 20 °C; G. Löhr and G. V. Schulz, *Makromol. Chem.*, 1968, **117**, 283.
[f] D. N. Bhattacharyya, C. L. Lee, J. Smid and M. Szwarc, *J. Phys. Chem.*, 1965, **69**, 612.
[g] At 30 °C; A. Parry, J. E. L. Roovers and S. Bywater, *Macromolecules*, 1970, **3**, 355.
[h] H. Hostalka and G. V. Schulz, *Z. Phys. Chem. (Frankfurt)*, 1965, **45**, 286.
[i] F. S. Dainton, K. J. Ivin and R. T. LaFlair, *Eur. Polym. J.*, 1969, **5**, 379.
[j] G. Löhr and S. Bywater, *Can. J. Chem.*, 1970, **48**, 2031.
[k] F. S. Dainton, G. C. East, G. A. Harpbell, N. R. Hurworth, K. J. Ivin, R. T. LaFlair, R. H. Pallen and K. M. Hui, *Makromol. Chem.*, 1965, **89**, 257.
[l] M. Bunge, G. Löhr, H. Höcker and G. V. Schulz, *Eur. Polym. J.*, 1977, **13**, 283.
[m] G. Löhr and G. V. Schulz, *Eur. Polym. J.*, 1975, **11**, 259.
[n] Dielectric constant of solvent.

The dependence of $k_{\pm}$ on the size of the counterion for a given solvent is also peculiar; for the less polar solvents such as 1,4-dioxane, $k_{\pm}$ increases with increasing ionic radius. This is expected, because a large interionic distance in the ion pair favours the charge separation in the transition state and thus leads to an increase in reaction rate. A plot of $\log k_{\pm}$ vs. the reciprocal interionic distance *a* (as calculated from the sum of the coulombic radius of the anion and the crystal radius of the cation) is shown in Figure 3. It reveals a fair linearity for 1,4-dioxane as the solvent. On the other hand, for solvents of higher polarity the dependence is reversed. The reason for this observation will be discussed below.

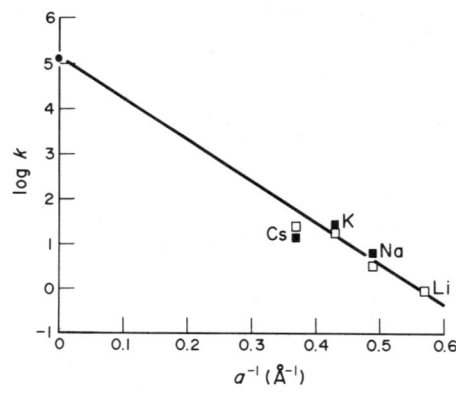

Figure 3 Dependence of the propagation rate constants on the interionic distance in the anionic polymerization of styrene in dioxane at 25 °C, data taken from Table 1: (□) ref. c, Table 1; (■) ref. k, Table 1; (●) k_-, ref. 77

26.3.2 Contact and Solvent-separated Ion Pairs

The most conclusive information on the nature of the ion pairs is obtained from the temperature dependence of the propagation rate constants. As can be seen in Figure 4, the rate constants for the free anion give one straight line in the Arrhenius plot, irrespective of counterion and solvent[77] ($\log A_- = 8.0$; $E_{a,-} = 16.7\,kJ\,mol^{-1}$).

An Arrhenius plot for the ion-pair rate constants (counterion Na^+) is shown in Figure 5. Here, the Arrhenius lines deviate significantly from linearity. For DME and THF in certain temperature intervals, this even leads to 'negative activation energies'. The reason for this observation is the

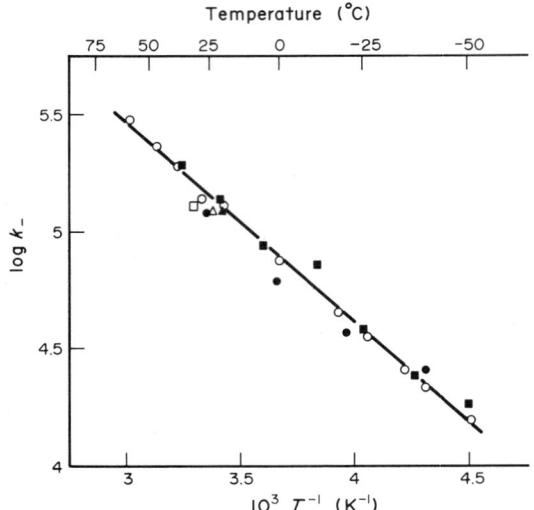

Figure 4 Arrhenius plot of the propagation rate constants of the free anions in the anionic polymerization of styrene in THF using various solvents[77]

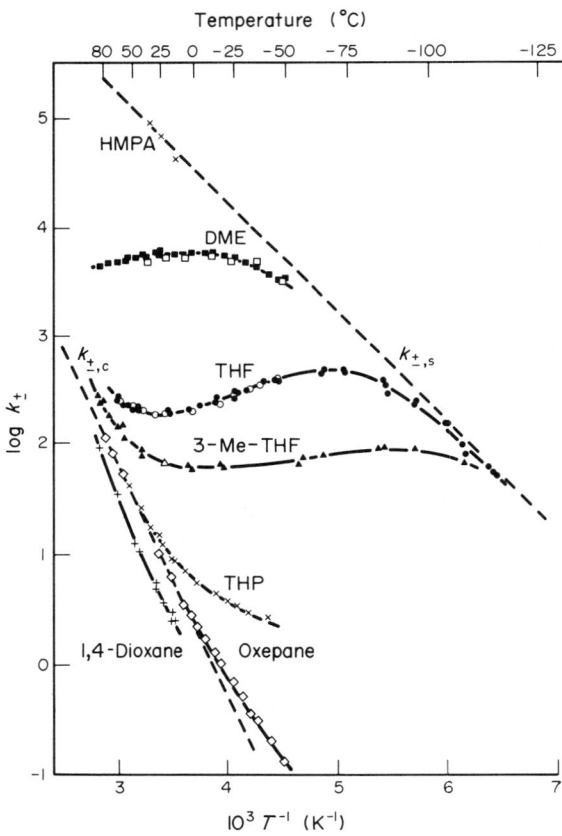

Figure 5 Arrhenius plot[77] of the propagation rate constants of the ion pairs in the anionic polymerization of styrene; counterion = Na^+

coexistence of two different types of ion pairs, *i.e.* contact (or tight) and solvent-separated (or loose) ion pairs.[79,80] This results in a three-state mechanism (Scheme 4).

In the solvent-separated ion pair, the cation is completely surrounded by solvent molecules S, and is thus separated from the carbanion. Consequently, its interionic distance and reactivity are greatly increased.

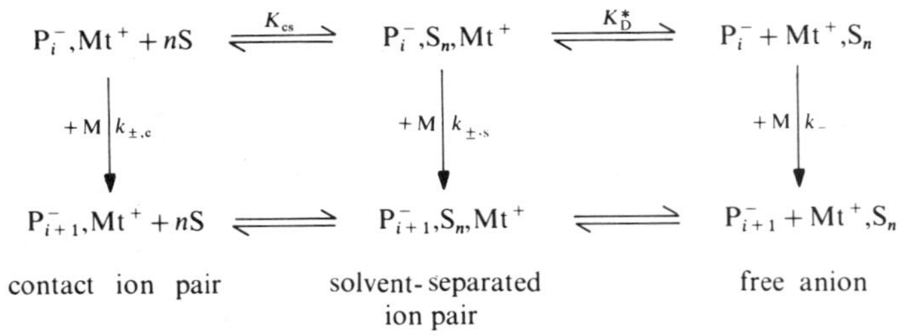

$$P_i^-,Mt^+ + nS \xrightleftharpoons{K_{cs}} P_i^-,S_n,Mt^+ \xrightleftharpoons{K_D^*} P_i^- + Mt^+,S_n$$

contact ion pair   solvent-separated   free anion
ion pair

Scheme 4

From the equilibrium constant of contact and solvent-separated ion pairs

$$K_{cs} = [P_{\pm,s}]/[P_{\pm,c}] \qquad (41)$$

the fraction of solvent-separated ion pairs can be calculated:

$$\alpha_s = [P_{\pm,s}]/[P_\pm] = \frac{K_{cs}}{1 + K_{cs}} \qquad (42)$$

The overall ion-pair propagation rate constant $k_\pm$ is given by

$$k_\pm = \alpha_s k_{\pm,s} + (1 - \alpha_s)k_{\pm,c} \qquad (43)$$

$$= \frac{k_{\pm,s}K_{cs} + k_{\pm,c}}{1 + K_{cs}} \qquad (44)$$

As $k_\pm$ is composed of three different constants, it depends on temperature in a very complex way:

$$k_\pm = \frac{A_s\exp(\Delta S_{cs}/R)\exp[-(E_s + \Delta H_{cs})/RT] + A_c\exp(-E_c/RT)}{1 + \exp[(T\Delta S_{cs} - \Delta H_{cs})/RT]} \qquad (45)$$

Here, A_s and A_c denote the frequency factors of the Arrhenius equations for $k_{\pm,s}$ and $k_{\pm,c}$, respectively, E_s and E_c denote the corresponding activation energies, and ΔS_{cs} and ΔH_{cs} denote the parameters of the van't Hoff equation for K_{cs}. By using a curve-fitting technique, it is possible to determine all the thermodynamic and activation parameters.[81]

Alternatively, K_{cs} and its van't Hoff parameters can be determined from conductance measurements. The overall dissociation constant K_D is connected to the dissociation constant of the solvent-separated ion pairs K_D^* by

$$K_D = K_D^*\alpha_s = K_D^*\frac{K_{cs}}{1 + K_{cs}} \qquad (46)$$

The temperature dependence of $\log K_D$ is given by a function composed of the van't Hoff relations for K_D^* and K_{cs} which is not linear with $1/T$ (*cf.* Figure 6):

$$K_D = \frac{\exp[(\Delta S_D^* + \Delta S_{cs})/R - (\Delta H_D^* + \Delta H_{cs})/RT]}{1 + \exp[(T\Delta S_{cs} - \Delta H_{cs})/RT]} \qquad (47)$$

By using equation (47), the thermodynamic parameters can be fitted to the experimental dissociation constants.[82]

Table 2 gives the activation and thermodynamic parameters for the polymerization of styrene using Na^+ as the counterion. It can be seen that the kinetic and conductometric determinations of ΔS_{cs} and ΔH_{cs} are in close agreement. Table 2 shows that the activation parameters for the contact and, especially, the solvent-separated ion pairs do not significantly depend on the nature of the solvent. This indicates that the solvation of the transition state does not much differ from that of the initial state. The strong influence of the solvent on the overall ion-pair rate constants is conveniently explained by changes in the position of the equilibrium between contact and solvent-separated ion pairs.

The existence of solvent-separated ion pairs has been corroborated by various direct investigations on living polymers as well as on model compounds. Some experimental evidence is reviewed in Volume 3, Chapter 25. All these investigations showed that the fraction of solvent-separated ion

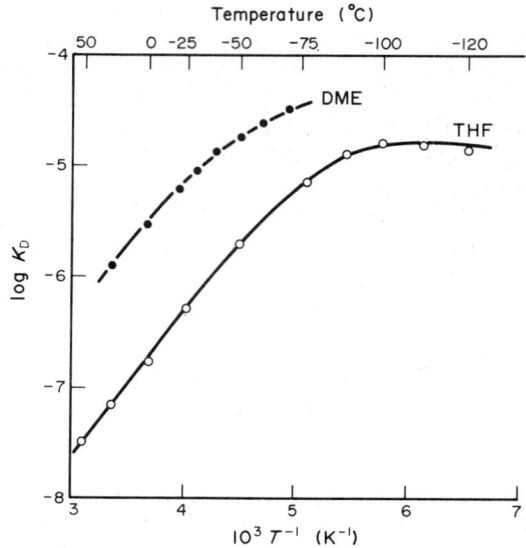

Figure 6 Van't Hoff plot of the dissociation constants of (polystyryl)sodium in DME and THF[9]

Table 2 Frequency Exponents and Activation Energies for Contact and Solvent-separated Ion Pairs in the Anionic Polymerization of Styrene using Na^+ as the Counterion, as well as Entropies and Enthalpies for the Conversion of Contact to Solvent-separated Ion Pairs

Solvent	$log A_c$	E_c (kJ mol^{-1})	$log A_s$	E_s (kJ mol^{-1})	ΔS_{cs} (J mol^{-1} K^{-1})	ΔH_{cs} (kJ mol^{-1})
DME[c]	7.8	38.5	7.8	17.6	-94,[a] -92[b]	-23.0,[a] -22.2[b]
THF[d]	7.8	36.0	8.3	19.7	-142,[a] -134[b]	-27.2,[a] -25.5[b]
THP[e]	8.1	40.6	8.0	18.8	-117[a]	-12.6[a]
Oxepane[f]	8.2	41.0	—	—	—	—
1,4-Dioxane[g]	8.4	43.9	—	—	—	—

[a] From kinetic measurements.
[b] From conductometric measurements.
[c] J. M. Alvariño, M. Chmelíř, B. J. Schmitt and G. V. Schulz, *J. Polym. Sci., Polym. Symp.*, 1973, **42**, 155.
[d] B. J. Schmitt and G. V. Schulz, *Makromol. Chem.*, 1971, **142**, 325.
[e] L. L. Böhm and G. V. Schulz, *Makromol. Chem.*, 1972, **153**, 5.
[f] G. Löhr and S. Bywater, *Can. J. Chem.*, 1970, **48**, 2031.
[g] J. Komiyama, L. L. Böhm and G. V. Schulz, *Makromol. Chem.*, 1971, **148**, 297.

pairs increases with increasing polarity of the solvent, decreasing ionic radius of the counterion, and decreasing temperature ($\Delta H_{cs} < 0$).

Whereas the formation of solvent-separated ion pairs is an exothermic process, the formation of free ions is nearly athermic (-5 kJ mol$^{-1} \leq \Delta H_D^* \leq 0$), indicating that solvation in the solvent-separated ion pair and in the free cation are similar.

26.3.3 Rates of Interconversion between Different Species

In Section 26.2.3, it was shown that polymerization *via* a multi-state mechanism may lead to a broadening in the MWD. The excess non-uniformities U_{cs} and U_d introduced by the interchange between contact and solvent-separated ion pairs and by dissociation, respectively, are only in the order of magnitude of 10^{-2} for styrene in THF and thus unimportant for practical purposes. However, by carefully determining the non-uniformity U_{app} of polymers prepared under kinetically controlled conditions, and correcting for hydrodynamic effects[21] (U_{hydr}) using

$$U_{app} = U_{pois} + U_{hydr} + U_{cs} + U_d \tag{48}$$

it was possible to determine U_{cs} and U_d with enough precision to calculate the interconversion rate constants for the three-state mechanism.[83-85] First, the contribution of free ions was suppressed by addition of common ion salt. Thus the expression for a two-state mechanism[35] could be applied, rendering k_{sc}. In a second experiment, the polymerization was conducted in the presence of free anions. Application of the corresponding equations for a three-state mechanism[36] rendered k_a.

Thus, the anionic polymerization of styrene can be described kinetically in terms of *elementary processes*. The complete set of rate constants for the three-state polymerization of styrene is given in Table 3. It can be seen that association of anion and cation is a very fast, nearly diffusion-controlled process, whereas dissociaton is fast, but somewhat slower than propagation. The rate constant for the conversion of contact to solvent-separated ion pairs, k_{cs}, is rather low and strongly dependent on the polarity of the solvent, whereas the rate constant for the reverse reaction, k_{sc}, appears to be independent of the solvent.

Table 3 Rate Constants of Propagation *via* Different Species and Rate Constants of Interconversion in the Anionic Polymerization of Styrene at 25 °C[a,b]

Solvent	$k_\pm$	$k_{\pm,c}$	$k_{\pm,s} \times 10^{-4}$	$k_- \times 10^{-4}$	k_{cs}	$k_{sc} \times 10^{-4}$	$k_d \times 10^{-4}$	$k_a \times 10^{-10}$
DME	5000	13	7.3	11.6	≥ 2000	≥ 1.5	—	—
THF	200	30	7.3	11.6	105	4.7	—	—
THP	12	9	7.3	11.6	3.5	2.4	1	1.5

[a] L. L. Böhm, G. Löhr and G. V. Schulz, *Ber. Bunsenges. Phys. Chem.*, 1974, **78**, 1064.
[b] Counterion = Na$^+$; all values in $l \, mol^{-1} \, s^{-1}$ or s^{-1}.

The corresponding activation parameters were also determined. It was found that formation of the solvent-separated ion pairs has a small activation enthalpy ($\Delta H_{cs}^\ddagger = -6 \ldots -16 \, kJ \, mol^{-1}$), but an extraordinarily high negative activation entropy ($\Delta S_{cs}^\ddagger = -170 \, J \, mol^{-1} \, K^{-1}$).

26.3.4 Triple Ions and Associates

Besides ions and ion pairs, the active centres can exist as various higher aggregates. Important species are triple ions and associates:

$$P_i^-, Mt^+ \; + \; P_i^- \overset{K_T^-}{\rightleftharpoons} P_i^-, Mt^+, P_i^- \qquad \text{negative triple-ions} \qquad (49)$$

$$P_i^-, Mt^+ \; + \; Mt^+ \overset{K_T^+}{\rightleftharpoons} Mt^+, P_i^-, Mt^+ \qquad \text{positive triple-ions} \qquad (50)$$

$$2P_i^-, Mt^+ \overset{K_A}{\rightleftharpoons} (P_i^-, Mt^+)_2 \qquad \text{dimeric associate} \qquad (51)$$

Triple ions were first detected by Kraus and Fuoss by conductance measurements of inorganic salts in water/1,4-dioxane mixtures.[86] Typically, deviations from linearity of a plot of $\log \Lambda$ *vs.* $\log c$ are found at higher concentrations which may even lead to minima. Quantitative relations were given by Wooster[87] for the formation of only one kind of triple ion, and by Fuoss and Kraus[88] for the formation of both kinds of triple ion.

Kinetic evidence was given by Szwarc and co-workers for the polymerization of bifunctional (polystyryl)cesium in THF.[73,89] Here, the formation of negative triple-ions is favoured by the fact that the two end groups are much closer than the end groups of two linear polymer chains and that the cesium ion has little peripheral solvation which might hinder the additon of the anion to the ion pair. It was found that the triple ions propagate *ca.* 100 times faster than the ion pairs, but *ca.* 30 times slower than the free anions.

Conductometric measurements show that triple ions and associates also exist for monofunctional (polystyryl)lithium in media of lower polarity, such as THP and oxepane[90] and especially in benzene/DME mixtures.[91-94] In the latter solvent mixtures, triple ions manifest kinetically by a decrease of k_p with $(c^*)^{-1/2}$. However, triple ions and associates are much more abundant in the polymerization of polar monomers (*cf.* Section 26.4).

26.3.5 Polymerization of Dienes

The kinetics of polymerization of dienes in polar solvents have been investigated in great detail by Bywater and coworkers. They are complicated by two facts:

(i) The active centres are unstable at ambient temperature, the free anions especially tending to decompose rapidly. It is thus advisable to work at lower temperatures and to add common-ion salt in order to suppress dissociation.[95]

(ii) The active centres can exist as *cis* or *trans* isomers which are in equilibrium.[96,97] Monomer can be added either in *cis*-1,4, *trans*-1,4 or (preferentially) in 1,2 (*i.e.* vinyl) fashion.[98] This leads to a modified two-state mechanism (Scheme 5). As 1,2 addition produces a vinyl group only in the *penultimate* monomer unit, it need not be included in this simplified scheme which only shows the *last* unit (corresponding to Bernoullian statistics).

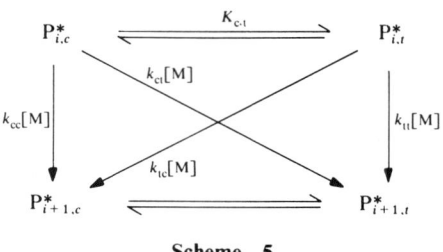

Scheme 5

The geometrical isomer formed under kinetic control may not be the thermodynamically most stable one. Competition between monomer addition and isomerization to the stable form determines which isomer reacts with the monomer.[99] In polar solvents, the *cis* form is the thermodynamically most stable one ($K_{c,t} < 1$), whereas the *trans* isomer is formed in the transition state ($k_t > k_c$). The consequences for the microstructure of the resulting polymers are outlined in Volume 3, Chapter 28. However, there are also consequences for the kinetics of polymerization.[42]

In order to adequately describe the mechanism of polymerization, it is necessary to determine the rates of isomerization of *cis* and *trans* forms. This was done spectroscopically for models of living polybutadiene.[101,102] It was found that the rates of isomerization are high for Li^+ (faster than or comparable to propagation) and very low for K^+. For Li^+ in THF it was also found that the *trans* isomers are more reactive than the *cis* ones by a factor of 2.7.[42] Consequently, polymerization based on Li^+ proceeds almost exclusively *via* the (slow) *cis* form and that based on K^+ *via* the (fast) *trans* form. Na^+ is intermediate: at high temperatures and low monomer concentrations the rates of isomerization are faster than propagation, leading to a preference for the *cis* isomer; at lower temperature and higher monomer concentrations, the *trans* structure is frozen. This is reflected in Figure 7 which shows an Arrhenius plot for the overall ion-pair propagation rate constants of budadiene in THF.[42]

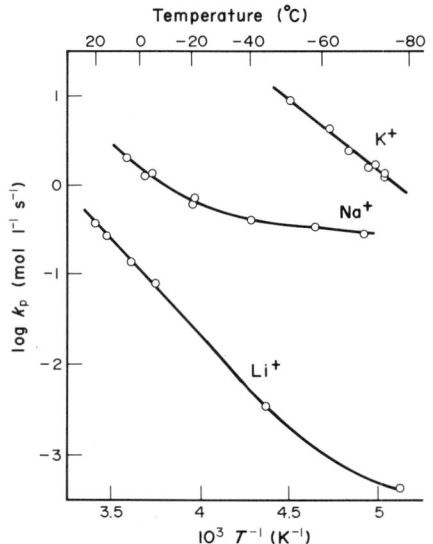

Figure 7 Arrhenius plot of the propagation rate constants of the ion pairs in the anionic polymerization of butadiene in THF[42]

26.4 POLYMERIZATION OF POLAR MONOMERS IN POLAR SOLVENTS

26.4.1 Potential Problems

In contrast to non-polar monomers, the polymerization mechanism is complicated by the presence of the polar side-group (*e.g.* ester, ketone, nitrile, pyridyl), which gives rise to chemical and physical interactions with the living chain-end. The most important potential complications are:

(a) Side reactions of the initiator or living end with the carbonyl group of the monomer or polymer. This results in low initiator efficiency or chain termination (*e.g.* equation 52).

$$R^- + CH_2{=}C\underset{\substack{\diagup COR' \\ \diagdown O}}{\overset{\diagup Me}{}} \longrightarrow CH_2{=}C\underset{\substack{\diagup C R \\ \diagdown O}}{\overset{\diagup Me}{}} + RO \qquad (52)$$

(b) Activation of protons in the α position to the carbonyl group leading to chain transfer (equation 53).

$$\text{\raise2pt\hbox{$\sim\sim$}}\bar{C}HCO_2R + RO_2C\overset{\text{\large\}}{C}H \longrightarrow \text{\raise2pt\hbox{$\sim\sim$}}CH_2CO_2R + RO_2C\overset{\text{\large\}}{C} \qquad (53)$$

(c) Due to the bidentate character of the active centres, they may attack the monomer not only by the carbanion (1,2 addition) but also by the enolate oxygen (1,4 addition) (equations 54 and 55).

$$\left[\text{\raise2pt\hbox{$\sim\sim$}}C\underset{R}{\overset{Me}{\underset{\displaystyle C{=}O}{\big|}}}^{-} \longleftrightarrow \text{\raise2pt\hbox{$\sim\sim$}}C\underset{R}{\overset{Me}{\underset{\displaystyle C{-}O^{-}}{\big|}}} \right] \equiv \text{\raise2pt\hbox{$\sim\sim$}}C\underset{R}{\overset{\overset{\delta-}{Me}}{\underset{\displaystyle C{=}O}{\big|}}}^{\delta-} \qquad (54)$$

$$\text{\raise2pt\hbox{$\sim\sim$}}C\underset{R}{\overset{\overset{\delta-}{Me}}{\underset{\displaystyle C{=}O}{\big|}}}^{\delta-} + CH_2{=}C\underset{COR}{\overset{Me}{}} \quad \substack{\xrightarrow{\;1,2\;}\\[12pt] \xrightarrow{\;1,4\;}} \quad \begin{array}{l} \text{\raise2pt\hbox{$\sim\sim$}}CCH_2C\cdots \\[6pt] \text{\raise2pt\hbox{$\sim\sim$}}C{-}OCH_2C\cdots \end{array} \qquad (55)$$

(d) The carbonyl group may coordinate with the counterion of the living chain-end ('intramolecular solvation') or may lead to association of ion pairs.

(e) The chain end may interact with electron donors.

As a result of these complications (especially in non-polar solvents), non-ideal polymerizations of acrylic monomers have been reported frequently[103] (*cf.* Section 26.4.2.3).

However, under appropriate conditions, it is possible to obtain polymers of narrow MWD and to observe 'ideal' polymerization kinetics which were characterized in the introduction (Section 26.1), *i.e.* instantaneous initiation, pseudo-first-order kinetics and a linear dependence of the number-average degree of polymerization on conversion.

26.4.2 Methacrylates

Most of the mechanistic work reported has been done on methacrylates, especially methyl methacrylate (MMA). Kinetic investigations on other methacrylates were reported elsewhere.[104, 163] The optimum conditions needed for an 'ideal' polymerization of MMA are as follows: (a) polar solvents, DME > THF ≫ THP; (b) low temperatures, $T \leq -75\,°C$ for THF; (c) large counterions, $Cs^+ > K^+ > Na^+ > Li^+$; (d) monofunctional initiators which are reactive enough to give fast initiation but not too reactive towards the carbonyl group of the monomer, such as diphenylmethyl, diphenylhexyl and benzyl [oligo(α-methylstyryl)] salts, but especially metallated esters (*e.g.* methyl α-lithioisobutyrate) which are models of the active centre;[105] and (e) addition of common-ion salts to suppress the dissociation of ion pairs into free ions.

26.4.2.1 *Structure of active species*

The first kinetic investigations stating 'ideal' kinetics for MMA polymerizaton were published independently by Mita *et al.*[106] and by Löhr and Schulz[107,108] in 1973. From the concentration dependence of the propagation rate constants, the coexistence of ions and ion pairs was shown. The kinetics were investigated in detail by Schulz, Müller and co-workers using different counterions, solvents and monomers.[109,104] When using the sodium ion complexed by the bicyclic cryptand [2.2.2] (Na$^+$,[2.2.2]) as the counterion, it was possible to independently determine the rate constants of the cryptated ion-pairs and the free anions.[110] Figure 8 shows an Arrhenius plot of the rate constants for all counterions investigated in THF.[104,111] This plot is informative in two ways: (i) the linearity indicates only one kind of active species; and (ii) except for Na$^+$, K$^+$ and Cs$^+$, a distinct dependence of the rate constants on the counterion is found. This indicates that the active species are contact ion pairs, because a large interionic distance in the contact ion pairs favours the charge separation in the transition state and thus leads to an increase in reaction rate (*cf.* Section 26.2.1).

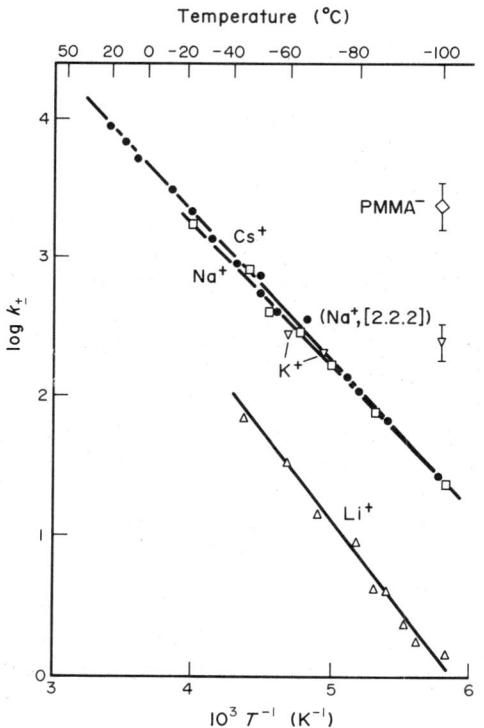

Figure 8   Arrhenius plot of the propagation rate constants of the ion pairs in the anionic polymerization of MMA in THF[111]

This is seen clearly in Figure 9, which plots $\log k_p$ *vs.* the reciprocal interionic distance a (calculated as the sum of the crystal radius of the cation and the electrostatic radius of the anion which is assumed to be 1.5 Å). The deviation of Na$^+$ and K$^+$ from the straight line indicates that the interionic distances are larger than the values expected from the crystal radii.

These results conform to a model of the active centre described as a contact ion pair peripherally solvated by THF molecules (**1**; S denotes solvent).

The increased interionic distance of Na$^+$ and K$^+$ ion pairs seems to be caused by peripheral solvation spreading the positive charge over a larger area and thus weakening the electrostatic interaction of the cation with the α carbon and the partially enolated carbonyl oxygen. On the other

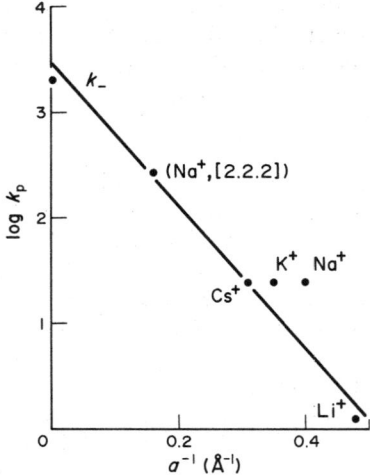

Figure 9 Dependence of the propagation rate constants on the interionic distance in the anionic polymerization of MMA in THF at $-100\,°C$[111]

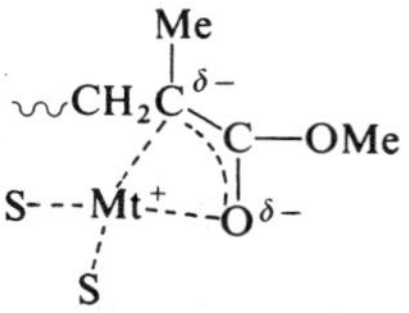

$$\begin{array}{c} \overset{\text{Me}}{\underset{|}{}} \\ \text{wCH}_2\text{C}\overset{\delta-}{}\diagdown \\ \qquad\qquad \text{C—OMe} \\ \text{S---Mt}^{+}\text{---O}^{\delta-} \\ \qquad | \\ \qquad \text{S} \end{array}$$

(1)

hand, Li^+ is on the straight line again because of a strong p–π overlap leading to a very tight bond between anion and cation.

It is interesting to note that, due to the strong interaction of the counterions with the enolate oxygen, no solvent-separated ion pairs are found.

At higher concentrations, the ion pairs tend to form dimeric or higher associates of low reactivity. This leads to decreasing rate constants. The kinetic scheme for such a process is given in Scheme 6, assuming dimeric associates only.[112-114]

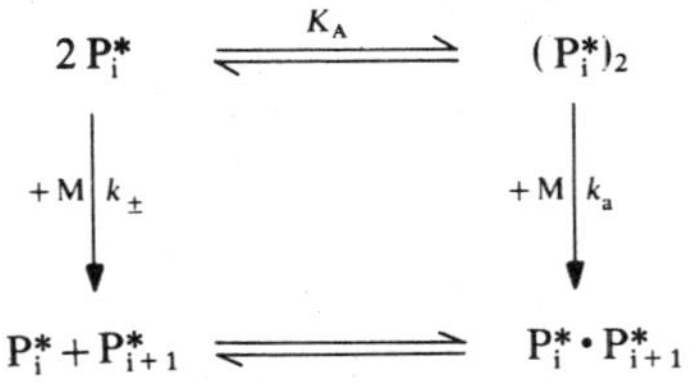

$$2\,P_i^* \quad \underset{}{\overset{K_A}{\rightleftharpoons}} \quad (P_i^*)_2$$

$$\Big\downarrow +M\;\Big|\;k_{\pm} \qquad\qquad\qquad \Big\downarrow +M\;\Big|\;k_a$$

$$P_i^* + P_{i+1}^* \quad \rightleftharpoons \quad P_i^* \cdot P_{i+1}^*$$

Scheme 6

The dependence of the rate constant k_p on the concentration of active centres is given by the rate constant of non-associated ion pairs $k_{\pm}$, the rate constant of the associates k_a and the fraction of non-associated ion pairs α:

$$k_p = \alpha k_{\pm} + [(1 - \alpha)/2]k_a = \tfrac{1}{2}k_a + (k_{\pm} - \tfrac{1}{2}k_a)\alpha \tag{56}$$

$$\alpha = [-1 + (1 + 8K_A c^*)^{1/2}]/4K_A c^* \tag{57}$$

For $K_A c^* \gg 1$

$$\alpha = (2K_A c^*)^{-1/2} \tag{58}$$

and for $K_A c^* \ll 1$

$$\alpha = 1 - 2K_A c^* \tag{59}$$

When working with bifunctional initiatiors, *e.g.* oligo(α-methylstyryl)sodium, the chain ends are very close to each other. Even at very low initiator concentrations, associates were found in these systems. The degree of association was observed to decrease with increasing degree of polymerization.[113,114]

Coordination of the counterion with the penultimate or ante-penultimate ester groups leads to an active species different from the peripherally solvated contact ion pair (equation 60). Similar structures were proposed by Fowells *et al.*[115] on the basis of the stereostructure of partially deuterated poly-acrylates and -methacrylates. There is IR-spectroscopic evidence of the existence of these structures.[116-118] However, NMR and conductivity studies on living oligomers conducted at lower temperatures did not give unequivocal results.[119,120] It appears that 'intramolecular solvation' is an endothermic process, due to the lower solvating power of the ester group compared to THF.[121] Thus, the driving force seems to be the entropy gain due to desolvation of THF, favouring this process at higher temperatures only.

$$\text{(60)}$$

26.4.2.2 *Mechanism of monomer addition*

When working in more polar solvents (*e.g.* DME), external solvation of the ion pairs presumably increases. Accordingly, the interionic distance in the ion pairs should increase, leading to higher rate constants. The opposite is expected for less polar solvents (*e.g.* THP). Figure 10 shows that this is true for Na^+ and Cs^+ as counterions but not for Li^+.[122-124] Here, the rate constants are even smaller in DME, as compared to THF (*cf.* Table 4). To understand this effect, it is necessary to have a closer look at the mechanism of monomer addition. Two possible mechanisms are discussed below. In mechanism I (Scheme 7), the monomer carbonyl oxygen first coordinates with the counterion displacing a solvent molecule (only one is shown in the scheme). Ion pair–monomer complexes of this type have been proposed for various monomer/solvent systems in order to explain the formation of highly stereoregular polymers.[115,220-222,264-266,278-280] In the transition state, the counterion is moved from the active centre to the newly formed one; then the counterion is solvated again.

Li^+ strongly coordinates with the anion as well as with the peripheral solvent molecules. Thus, it is difficult for the incoming monomer to displace the solvent, especially for DME which is a bidentate ligand. In this case, an alternative mechanism II is more likely (Scheme 8). Here, the monomer vinyl group directly attacks the carbanion without coordinating with the shielded counterion. After the transition state, the solvated counterion is transferred to the newly formed anion.

This step is energetically unfavourable. However, mechanism II is the only choice for polymerizations *via* free anions or ion pairs involving large, non-solvated counterions, such as cryptated

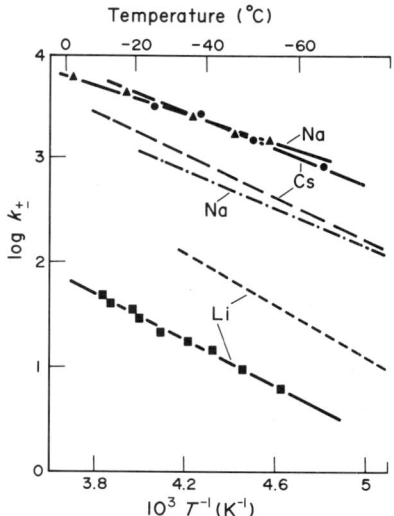

Figure 10 Arrhenius plot of the propagation rate constants of the ion pairs in the anionic polymerization of MMA in DME (●, ▲, ■, experimental points) compared to the values for THF (broken lines)[104]

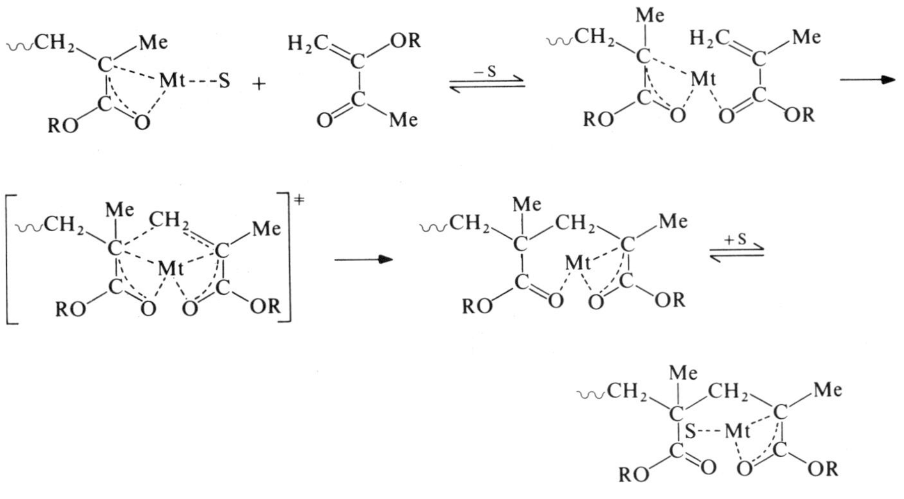

Scheme 7 Mechanism I

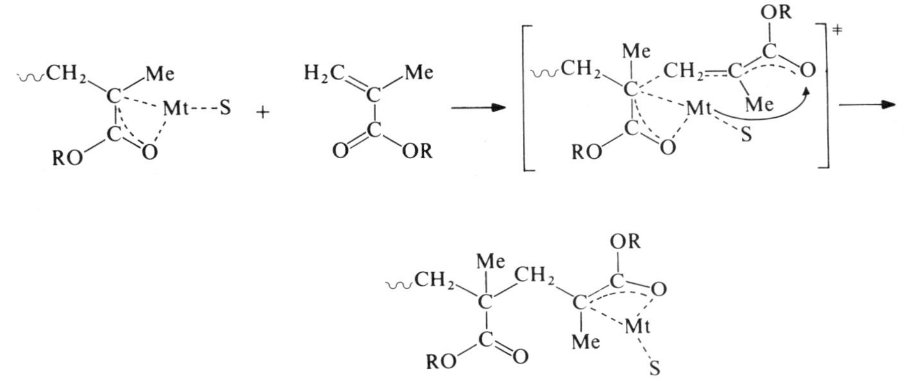

Scheme 8 Mechanism II

Table 4 Rate Constants (at $-40\,°C$) and Activation Parameters in the Anionic Polymerization of MMA *via* Different Ion Pairs[a]

Solvent		Cs^+	K^+	Na^+	Li^+	
DME	E_a (kJ mol^{-1})	15.1	—	13.8	21.4
	$\log A$	6.8	—	6.5	6.0
	$k_\pm$ (l mol^{-1} s^{-1})	2680	—	2550	16
THF	E_a (kJ mol^{-1})	19.5	19.3	18.3	24.0
	$\log A$	7.3	7.2	7.0	7.4
	$k_\pm$ (l mol^{-1} s^{-1})	860	750	800	100
THP[b]	E_a (kJ mol^{-1})	—	—	21.5	—
	$\log A$	—	—	7.5	—
	$k_\pm$ (l mol^{-1} s^{-1})	—	—	480	—

[a] A. H. E. Müller, in 'Recent Advances in Anionic Polymerization', ed. T. E. Hogen-Esch and J. Smid, Elsevier, New York, 1987, p. 205.
[b] Due to extensive deactivation, error is approximately 30%.

sodium (Na^+, [2.2.2]). Evidence from polymer microstructures (*vide infra*) indicates that this is also the preferred mechanism for Li^+.

Table 4 shows that, for all solvents investigated, Li^+ has the highest activation energies. However, the lower rate constants for Li^+ in DME seem to be due to a higher absolute activation entropy (lower frequency exponent $\log A$), which is in accordance with mechanism II.

Mechanisms I and II should differ significantly in their stereoregulating effect (*cf.* Volume 3, Chapter 28). Fowells *et al.*[115] proposed that mechanism I should lead to a *meso* placement. For mechanism II (which is valid for Li^+, cryptated Na^+ and the free anion) the stereochemistry is not pre-determined. For steric reasons, predominantly racemic placements are found,[104] as is also observed in radical polymerization.[125,126]

For the larger alkali metal cations, the fraction of racemic dyads decreases with increasing interionic distance as well as with decreasing solvating power of the solvent.[104] It is thus probable that, in these systems, both mechanisms I and II are effective to varying extents, the fraction of monomer additions *via* mechanism I increasing with interionic distance and decreasing with polarity of the solvent. These assumptions are corroborated by the fact that the statistics of the tactic placements deviate from Bernoullian ones in the same order.

26.4.2.3 Side reactions

Side reactions are frequently observed when working at temperatures above $-75\,^\circ C$. They manifest in a broadening of the MWDs (*cf.* Section 26.2.6) and deviations from 'ideal' kinetics (*cf.* Section 26.1, equations 5 and 6). A downward curvature of the first-order time–conversion curve is observed, which indicates a decrease in the concentration of active centres c^* due to chain termination. Generally, P_n is proportional to monomer conversion (*cf.* equation 7), indicting the absence of transfer reactions.

Based on these effects, it was seen qualitatively that the relative amount of termination increases with decreasing polarity of the solvent and decreasing size of the counterion.[109] In the polymerization of *t*-butyl methacrylate (TBMA), no termination was detected even at ambient temperature.[104,127]

Three termination reactions have been discussed widely in the literature. They were first proposed by Schreiber:[128]

(a) A reaction of the active centre with the carbonyl group of the monomer, resulting in a vinyl ketone ('monomer termination', equation 61).

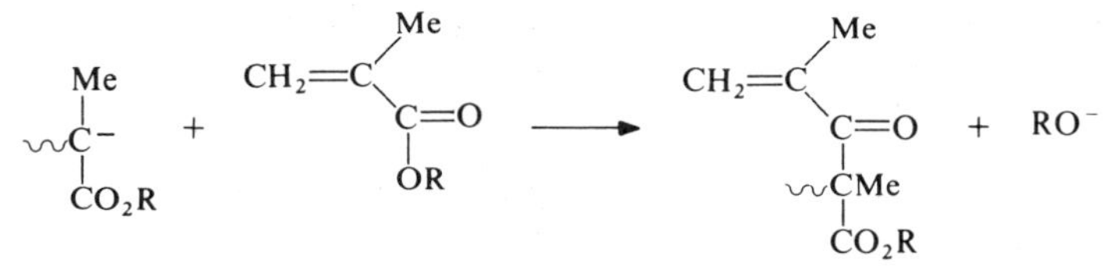

$$\text{(61)}$$

(b) Reaction of the living end with the ester group of another polymer, resulting in chain coupling ('intermolecular polymer termination', equation 62).

$$\text{(62)}$$

(c) Reaction of the active centre with the ante-penultimate ester group of its own polymer chain, resulting in a cyclic β-ketoester structure ('intramolecular polymer termination', 'backbiting', equation 63).

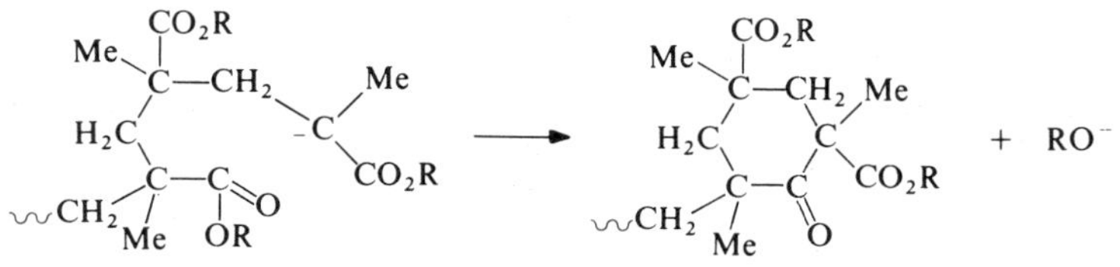

$$\text{(63)}$$

All these reactions produce alkoxide as a by-product. Wiles and Bywater[129-131] and Mita *et al.*[106] showed that most of the methoxide is produced in the *initial* stage of the polymerization of MMA. This can be explained in two ways. Either the methoxide is formed by the initiator attacking the carbonyl group of the monomer *or* by an increased tendency of living *oligomers* to undergo any type of termination. Both explanations are correct, as will be seen below.

The vinyl ketone resulting from the attack of initiator on the carbonyl group of the monomer (*cf.* equation 61), as well as the alcohol formed by subsequent addition of a second initiator molecule to the keto group, was detected by Schreiber,[128] Kawabata and Tsuruta,[132] and Hatada et al.[133-135] The vinyl ketone may also add to a living polymer *via* its vinyl group. This decreases the reactivity of the active centre leading to a 'dormant' species which will only eventually be reactivated by adding another methacrylate monomer.[134-136] There is no evidence, so far, that the living polymers are able to undergo 'monomer termination' by attacking the monomer's carbonyl group, although this was postulated by Löhr et al.[137] Consequently, the choice of the initiator is of the utmost importance. Highly reactive initiators, *e.g.* *n*-butyllithium, lead to side products due to the attack of the monomer carbonyl group. Thus, metalloesters or initiators of higher delocalization and steric hindrance, such as 1,1-diphenylhexyllithium, are preferable.

Evidence is also lacking for the existence of an *inter*molecular termination reaction (*cf.* equation 62) under normal polymerization conditions. Even at ambient temperature, Rempp et al.[138] detected only a slow grafting reaction of the much more reactive (polystyryl)sodium on to preformed PMMA. Trekoval and Kratochvil[139] reported a grafting reaction of PMMA–Li leading to one branch per 10^4 monomer units after 18 h at 20 °C in toluene. There is also no high molecular weight tailing in the GPC of polymers prepared in THF at 0 °C.[140]

In contrast, the product of *intra*molecular termination (or 'backbiting'; *cf.* equation 63), *i.e.* the cyclic β-ketoester end-group, could be detected by its characteristic IR absorption.[141-144] The termination product of the living trimer was isolated.[144] This makes 'backbiting' the most probable termination mechanism, and is a reasonable one, since the carbonyl reactivity of saturated esters, *i.e.* the polymer ester-groups, is usually higher than that of unsaturated ones, *i.e.* the monomer.

A first direct kinetic investigation of termination was published by Allen et al.,[145] who measured a first-order decrease of the maximum of the UV spectrum of PMMA–Na in toluene/THF mixtures at + 18 °C. Later, Warzelhan et al.[146] applied the method of labelling the active polymers with tritiated acetic acid ($CH_3CO_2{}^3H$) introduced by Glusker et al.[147] Again, a first-order decrease of c^* was found when taking samples during polymerization using Na^+ as the counterion in THF at temperatures below -40 °C. A linear Arrhenius plot was obtained.

By kinetically analyzing the product distribution of linear and cyclic oligomers formed in the oligomerization of MMA using Li^+ as the counterion in THF at 25 °C, Müller et al.[148] determined the rate constants of cyclization as a function of the degree of polymerization. It was found that the rate of cyclization drops from the trimer to the tetramer by a factor of *ca.* 50 and does not change significantly for the higher oligomers. This was attributed to steric effects of the pendant monomer units in the formation of the cyclic β-ketoester structure. This effect explains the above observation that most of the methoxide is formed in the initial stage of the polymerization.

Earlier investigations showed that addition of alkoxides, especially *t*-butoxide, decreases the effect of termination.[130,149] Recently, it was shown that this effect is due to a decrease of the cyclization rate constants by a factor of *ca.* 100 and of the polymerization rate constants by a factor of only 10.[150] It was assumed that alkoxide addition changes the structure of the active centre by complex formation (*cf.* Section 26.5.1.2).

A similar effect of pyridine was reported by Andrews et al.[151] and by Huynh-ba and McGrath.[152] It is reasonable to assume solvation or solvent separation by the pyridine molecules. The latter authors assume a σ complex of the PMMA α carbon with the α carbon of pyridyllithium to be the active species.

A very thorough reinvestigation and extension of Warzelhan's kinetic work by Gerner et al.[140] showed that the termination mechanism at temperatures below -40 °C is more complicated. Determination of c^* a long time after complete monomer consumption unexpectedly revealed that only a fraction of the polymer chains become terminated. This fraction, as well as the rate of termination, depends on the *initial* monomer concentration. None of the termination mechanisms discussed above (including combinations thereof) is able to explain these results. A new mechanism for termination was proposed, which is based on the assumption that a deactivating species is formed in the initial step of the polymerization. This species reacts with the living polymers during polymerization, leading to termination in a second-order reaction. The corresponding Arrhenius plots are linear and do not significantly differ for both THF and THP as solvents, indicating that the higher extent of termination in the latter solvent is due to the lower propagation rate constants.

Unfortunately, the exact nature of the deactivating species is unknown. Its initial concentration is directly proportional to the initial concentrations of both monomer and active centres. GPC and GC/MS analysis of oligomers prepared at -35 °C and end-labelled with ^3H revealed the existence of several by-products for each regular oligomer, one of which is possibly the product of termination by the deactivating species.[153,154] From the GC/MS data, it was not possible to assign a molecular structure to this series. However, it seems to be linear and to contain one initiator fragment per chain.

There is very little evidence for transfer reactions occurring in methacrylate polymerization. Müller *et al.*[155] reported that the cesium methoxide formed in the termination of PMMA–Cs in THF at ambient temperature is able to initiate the polymerization of MMA. The initiation rate, however, is very low, so that this effect only gains importance when all the polymer chains started initially have been terminated before complete monomer conversion. The newly initiated polymer chain can again suffer from termination, producing methoxide, *etc.* This is equivalent to a spontaneous transfer reaction.

26.4.3 Other Acrylic Monomers

Kinetic investigations on acrylic monomers other than methacrylates is hampered by a variety of side reactions as well as the other problems described in Section 26.4.1. The reactivity of the monomer, as well as the tendency to undergo side reactions, increases with the polarity of the side group: ester < ketone < aldehyde < nitrile. Substitution of the α-methyl group by hydrogen has a similar effect. In addition, the acidic proton α to the carbonyl group makes the monomer, and especially the polymer, susceptible to transfer (*cf.* equation 53). Thus, until recently, there were no reports on the preparation of polyacrylates having narrow MWDs. For methyl acrylate in THF, Busfield and Methven[156] reported rapid initial polymerization up to 10% conversion, followed by propagation at very low rate up to a maximum yield of 50% resulting in broad MWDs. Kitano *et al.*[157] reported first-order kinetics in the polymerization of *t*-butyl acrylate (TBA) initiated by butyllithium in THF at $-78\,°C$. The polymerization is faster than that of TBMA under the same conditions. A polydispersity of 1.4 was estimated. Investigations of Mai,[158] however, did not corroborate these findings, although various initiators were used, including metalloesters.

Recently, Teyssié and co-workers[159] reported a dramatic effect of lithium chloride on the MWD of poly(TBA). [Oligo(α-methylstyryl)]lithium was used as the initiator in THF and THF/toluene mixtures. In THF at $-78\,°C$ the polydispersity was $M_w/M_n = 3.6$ without additive, but $M_w/M_n = 1.2$ for $[\text{LiCl}]/[\text{I}]_0 = 5$. No kinetic data are available, so far. The authors suggest that the LiCl acts as a μ-type hindering ligand, preventing termination and transfer reactions. ^7Li NMR spectroscopy of living PMMA–Li confirms the formation of a complex having a tighter C—Li bond.

In the polymerization of vinyl ketones, additional problems result from the acidic α-protons of the side groups which may lead to transfer or to base-catalyzed aldol condensation (equation 64). The water produced can act as a terminating agent. Consequently, the polymerization of methyl isopropenyl ketone leads to very unsatisfactory results.[160] *t*-Butyl vinyl ketone (TBVK) and *t*-butyl isopropenyl ketone should be preferred. However, Overberger and Schiller[161] reported non-first-order kinetics for the polymerization of TBVK in THF at $0\,°C$ initiated by butyllithium and biphenyllithium. Johann[162,163] thoroughly investigated the polymerization of TBVK using Li^+, Na^+ and Cs^+ as counterions in THF. As for MMA, the rate constants, the extent of side reactions and the polydispersities decrease from Li^+ to Cs^+. For Cs^+ counterion, $M_w/M_n = 1.2$ was observed. Linear first-order time–conversion curves are obtained for Na^+ and Cs^+, but association phenomena lead to curved Arrhenius plots for non-corrected rate constants. Evidence for various interactions of the active centre with carbonyl groups present in the system also comes from IR studies of model compounds.[164]

$$(64)$$

The polymerization of (meth)acrolein can proceed *via* 1,2 and 1,4 addition (*cf.* equation 55) as well as *via* 3,4 addition leading to significant unsaturation within the main chain.[165-171] Andreyeva *et al.*[172,173] studied the kinetics of polymerization of methacrolein initiated by naphthalene sodium and *t*-butyllithium in THF. Similar studies were performed on acrolein by Gulino *et al.*[174-176] using naphthalene radical anions, polybutadienyl and poly(ethylene oxide) anions as initiators. The reaction is first order with respect to monomer and initiator. The rates are not dependent on the nature of the initiator; however, they decrease dramatically from K^+ to Li^+. It was proposed that the active centre can exist in two different forms, *i.e.* enolate and carbanion. However, as these forms are mesomeric (*cf.* equation 54) it is more likely that the nature of the counterion influences the charge distribution in the anion. Especially for Li$^+$, polymerization is strongly disturbed by transfer to both monomer and polymer. This leads to molecular weights which are lower than expected and to broad and multimodal MWDs.

In the polymerization of (meth)acrylonitrile a further complication is introduced by the insolubility of the polymers in ethereal solvents. Thus, three strategies can be followed to investigate the kinetics: (i) investigations can be restricted to the study of oligomers; (ii) a polymeric initiator [*e.g.* (polystyryl)lithium] can be used in order to solubilize the polymer; and (iii) the polymerization can be conducted in a solvent of higher dielectric constant, *e.g. N,N*-dimethylformamide (DMF).

The first two pathways were used in spectral, conductometric and quantum-mechanical studies of Erussalimsky *et al.*[177,178] and Tsvetanov *et al.*,[120,179-181] which showed that there is a strong tendency of the active centre to interact with nitrile groups of the monomer or polymer. Termination by Ziegler–Thorpe cyclization (similar to equation 63) occurs, leading to cyclic iminonitrile structures which form enamines upon protonation. A further side reaction can occur along the chain leading to ladder structures[182] (equation 65).

$$\text{(65)}$$

The polymerization of acrylonitrile in DMF was investigated by Adler *et al.*[183,184] Recently, the first kinetic results using *n*-butyllithium as initiator were reported.[185,186] The apparent rate constants of propagation strongly decrease with monomer conversion and temperature, indicating termination. This is corroborated by initiator efficiencies lower than 5%. Only below −30 °C was a certain 'livingness' of the active centres reported. Based on the dependence of polymer intrinsic viscosity on conversion, intra- or inter-molecular chain transfer (*cf.* equation 63) was also postulated.

26.4.4 Vinylpyridine

The kinetics of the polymerization of 2- and 4-vinylpyridine (2VP and 4VP, respectively) were studied by Szwarc and co-workers[187,188] and, more thoroughly, by Sigwalt, Fontanille and co-workers.[189-191] The very early investigations of Szwarc showed that VP is much more reactive than styrene. However, the conclusions were overshadowed by the lack of understanding about the coexistence of ions and ion pairs at that time. In Szwarc's second paper, this effect was taken into account. The polymerization of 2VP in THF, THP and 1,4-dioxane was studied, using Na$^+$ as the counterion. Linear first-order time–conversion plots were observed. The apparent rate constant of propagation is more or less independent of initiator concentration. Fisher and Szwarc assigned this effect to the very low dissociation constants of P2VP$^-$, Na$^+$, but Sigwalt and co-workers showed that the most probable explanation is the existence of triple ions (*vide infra*). In THP and 1,4-dioxane the ion-pair rate constants are higher than those in THF. Szwarc attributed this fact to external solvation, which is strongest in THF and may sterically hinder the addition of monomer.

Conclusive evidence on the nature of the active centres comes from conductance measurements performed in Sigwalt's[192-194] and Szwarc's[188] laboratories. The dissociation constants of P2VP$^-$, Na$^+$ ($K_D = 8 \times 10^{-10}$ mol l^{-1} at 25 °C) are 60 times lower than those of the corresponding polystyryl ion pairs. However, if (polystyryl)sodium (PS$^-$,Na$^+$) is 'capped' by only one 2VP monomer unit, the dissociation constant is considerably larger than for pure P2VP$^-$,Na$^+$. Moreover, the dissociation constant of P4VP$^-$,Na$^+$ ($K_D = 4 \times 10^{-8}$ mol l^{-1}) is of the same order of magnitude as that of PS$^-$,Na$^+$. A convenient explanation for these effects was offered by Sigwalt.

It is assumed that the counterion in P2VP$^-$,Na$^+$ is intramolecularly coordinated with the penultimate pyridyl nitrogen (2). This effect results in a stronger binding of the cation and was also

(2)

discussed in the polymerization of MMA (*cf.* equation 60). It is excluded in P4VP$^-$,Na$^+$ and in 2VP-capped PS$^-$,Na$^+$. Stereochemical investigations of Hogen-Esch *et al.*[195,196] gave additional evidence for this kind of interaction.

In their later kinetic work, Sigwalt, Fontanille and co-workers[190,197] showed that the intramolecularly coordinated ion-pair and a second cation form positive triple-ions, equation (66) (*cf.* equation 50). The higher monomer reactivity of VP as compared to styrene is accompanied by a decreased reactivity of the active centre. A kinetic investigation of the addition of 1,1-diphenylethylene to living PS and P2VP in THF[190] showed that k_- is decreased by a factor of 36, and $k_\pm$ is decreased by a factor of 18 for Cs$^+$ counterion.

$$\text{(66)}$$

26.5 POLYMERIZATION IN NON-POLAR SOLVENTS

26.5.1 Hydrocarbon Monomers

Dissociation of ion pairs is very unlikely to occur in non-polar solvents (*e.g.* cyclohexane, benzene, toluene) due to their inability to solvate the free cations. For the same reason, solvent-separated ion pairs are not present in these systems. In contrast, the polymeric ion-pairs tend to stabilize themselves by association. Thus, association phenomena play a most important role in the anionic polymerization in these media. Electron donors, such as alkoxides and Lewis bases, interact with the associated polymers leading to new active species. Often, initiators are more associated than the living polymers, leading to slow initiation. However, kinetics of initiation will not be dealt with here.

26.5.1.1 Association phenomena

The kinetics of the styrene and diene polymerization was investigated by many research groups, but most thoroughly by Worsfold and Bywater. In the propagation of styrene in benzene,[198] toluene[199] and cyclohexane,[200] a kinetic order of 0.5 with respect to living-end concentration was found for the apparent rate constants. This was attributed to the existence of dimeric associates being in equilibrium with the non-associated species, only the latter being able to propagate. This equilibrium is very much shifted to the dimeric side. A similar problem was dealt with in Section 26.4.2 (*cf.* equations 56 and 58). By setting $K_A c^* \gg 1$ and $k_a = 0$ one obtains:

$$k_p = k_\pm (2K_A c^*)^{-1/2} \tag{67}$$

Combination with equation (6) gives

$$k_{app} = k_p c^* = k_\pm (2K_A)^{-1/2}(c^*)^{1/2} \tag{68}$$

Denominating the concentration of dimeric polymers, $[P_2^*] \approx c^*/2$, and the equilibrium constant of dissociation for dimers into ion pairs, $K_d = 1/K_A$, leads to the familiar equation:

$$k_{app} = k_\pm K_d^{1/2}[P_2^*]^{1/2} \tag{69}$$

The existence of associates was confirmed by comparing the viscosities[201,202] or light-scattering intensities[203,204] of 'living' polymer solutions before and after termination. The UV spectra also depend on the state of association.[205]

A kinetic order of 1/6 to 1/4 is found for butadiene polymerization.[199,206] For isoprene, the kinetic order was first reported to be 1/4.[199,207-209] Recent kinetic measurements[210] show that the order is 1/4 at $c^* > 10^{-4}$ mol l^{-1} and 1/2 at $c^* < 10^{-4}$ mol l^{-1} (in benzene at 30 °C). Direct methods have led to contradictory results. Although at first indicating dimeric associates,[201,203] they later revealed the tetrameric nature of (polybutadienyl)lithium, which was described as the aggregation of two dimers to form a tetramer.[202,204,205,211]

$$P_4^* \underset{K_t}{\rightleftharpoons} 2P_2^* \tag{70}$$

For $K_t \ll 1$ the concentration of dimers is given by

$$[P_2^*] = K_t^{1/2}[P_4^*]^{1/2} \tag{71}$$

Introduction into equation (69) gives

$$k_{app} = k_{\pm}K_d^{1/2}K_t^{1/4}[P_4^*]^{1/4} \tag{72}$$

Spectroscopic measurements[210] also revealed the dependence of the equilibrium constant K_t (and thus of k_{app}) on the molecular weight of the 'living' polymer, complicating the kinetic analysis.

Whereas K_t can be determined spectroscopically to be $4 \times 10^{-5}\,\mathrm{mol\,l^{-1}}$ for high molecular weight (polyisoprenyl)lithium in cyclohexane at $31\,°C$ and $35 \times 10^{-5}\,\mathrm{mol\,l^{-1}}$ in benzene,[210] the dissociaton constants of the dimers, K_d, can only be estimated. Determination by means of viscosity measurements[212-214] were questioned by other authors.[4,215-217] As equation (69) is only valid for $c^*/K_d \gg 1$ and kinetic orders were established to be constant for $c^* > 10^{-6}\,\mathrm{mol\,l^{-1}}$, a value of $K_d \leq 10^{-6}\,\mathrm{mol\,l^{-1}}$ may be an upper estimate.[7]

For the larger counterions, K_d is more easily elucidated, making the propagation rate constants accessible. The propagation rates were determined for styrene in benzene and cyclohexane using all alkali metal counterions.[218,219] Whereas (polystyryl)sodium is still strongly associated, the kinetic order with respect to initiator concentration varies from unity at low concentrations to 0.5 at high concentrations for K^+. For Rb^+ and Cs^+, only non-associated ion pairs seem to exist. The results are summarized in Table 5. Except for Li^+, the rate constants are in the same order of magnitude as those measured in 1,4-dioxane (*cf.* Table 1). The low rate constant for Li^+ in 1,4-dioxane may be due to peripheral solvation hindering the attack of monomer. Various authors suggested the existence of a complex between the ion pair and a monomer molecule[220-222] (*cf.* MMA and VP polymerization). However, more precise rate constants for Li^+ and Na^+ in hydrocarbons are necessary to draw further conclusions.

Table 5 Propagation Rate Constants of the Ion Pairs and Equilibrium Constants for the Dissociation of Dimers in the Anionic Polymerization of Styrene in Benzene at $30\,°C$

	$Li^{+a,b}$	Na^{+c}	K^{+c}	Rb^{+c}	Cs^{+c}
$10^2 k_{\pm}(K_d/2)^{1/2}$	1.55	17	180	—	—
$10^4 K_d\,(\mathrm{mol\,l^{-1}})$	≤ 0.01	—	6	—	≥ 100
$k_{\pm}\,(\mathrm{l\,mol^{-1}\,s^{-1}})$	≥ 20	—	47	24	18

[a] D. J. Worsfold and S. Bywater, *Can. J. Chem.*, 1960, **38**, 1891.
[b] M. Morton, *Polym. Prepr., Am. Chem. Soc., Div. Polym. Chem.*, 1964, **5**, 1092.
[c] J. E. L. Roovers and S. Bywater, *Trans. Faraday Soc.*, 1966, **62**, 701.

In the polymerization of *o*-methoxystyrene, the methoxy group can coordinate with the counterion, thus stabilizing the ion pair. Kinetic investigations in toluene using Li^+ counterion showed that the kinetic order varied from 0.51 at high concentrations to 0.67 at the lowest measurable concentration, indicating a larger degree of dissociation.[223] The dissociation constant was estimated to have the value $K_d = 10^{-3}\,\mathrm{mol\,l^{-1}}$; the propagation rate constant is $50\,\mathrm{l\,mol^{-1}\,s^{-1}}$ at $20\,°C$.

Results of Zgonnik and co-workers[224] indicate the activity of dimers and tetramers in the polymerization of butadiene. The rate constants of the dimeric species were estimated to be *ca.* 1000 times lower than those of the ion pairs. This should result in a considerable contribution of the former at higher concentrations.

26.5.1.2 *Effect of alkoxides*

Lithium butoxides decrease the rate of polymerization for both styrene[225] and isoprene[226] (*cf.* Figure 11). The rates depend on the ratio $r = [\mathrm{BuOLi}]/c^*$, reaching their minimum value at $r \geq 1$. For $r = 1$ the reaction order is still 0.5, indicating that the species formed is not $P^* \cdot (\mathrm{BuOLi})_n$, but $(P^* \cdot \mathrm{BuOLi})_2$, the latter dissociating to the active $P^* \cdot \mathrm{BuOLi}$.

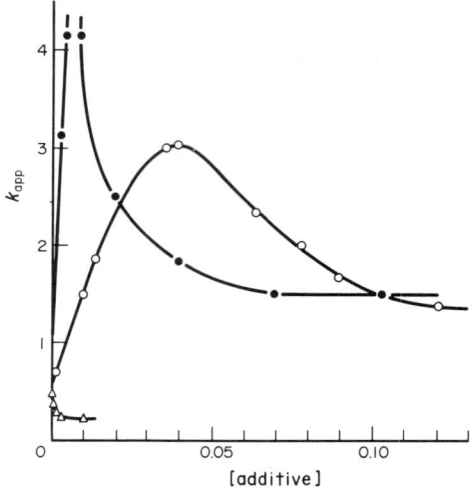

Figure 11 Effect of additives on the propagation rate in the polymerization of styrene in benzene:[7] (●) THF, (○) 1,4-dioxane, (△) lithium *t*-butoxide. Concentration of (polystyryl)lithium, $c^* \approx 10^{-3} \, \text{mol} \, \text{l}^{-1}$

26.5.1.3 *Effect of Lewis bases*

The addition of Lewis bases, such as ethers or amines, has a dramatic effect on the kinetics of propagation. This is seen in Figure 11. The addition of THF to (polystyryl)lithium in benzene strongly increases the rate of propagation.[227] When adding more than one equivalent, the rates level off to a value which is higher than the original one. This was explained by the formation of two new species; the reactive monoetherate and the less reactive dietherate:

$$P_2^* \; + \; 2E \; \overset{K_1}{\rightleftharpoons} \; 2P^* \cdot E \tag{73}$$

$$P^* \cdot E \; + \; E \; \overset{K_2}{\rightleftharpoons} \; P^* \cdot 2E \tag{74}$$

where E denotes the ether. The rate constants for the dietherate are lower than those for the 'naked' ion pair. This is further evidence for the steric hindrance exerted by peripheral solvation. The effect of ethers is less pronouned for 1,4-dioxane,[228] and especially for aromatic ethers, such as anisole[229] and diphenyl ether.[230] Viscosity measurements[203,231] confirmed the dissociation caused by THF and the weak effects of anisole and diphenyl ether.

With isoprene, the effect of THF is much smaller when compared to styrene, and no maximum is observed.[213,232] This is another indication of the stronger association in this system.

The addition of oligoethers, crown ethers or cryptands, though being favourable in the polymerization of methacrylates (*cf.* Section 26.5.2) and cyclic monomers (*cf.* Volume 3, Chapter 32), leads to chain termination for hydrocarbon monomers.[233] This was attributed to the high reactivity towards the ether bonds of the active species formed.[234] Qualitative reports on the polymerization of styrene and the dienes in the presence of crown ethers and cryptands state that the polymerization is extremely fast and that the MWDs are broad[235-240] (which may also be due to mixing problems). The polydiene microstructure is drastically altered.

Instead of oligoethers, bidentate tertiary amines, such as *N,N,N',N'*-tetramethylethylenediamine (TMEDA), have been used successfully. Figure 12 shows that the addition of TMEDA in the polymerization of styrene in cyclohexane can lead to a decrease or to an increase of the apparent rate constant, depending on concentration.[241] The effect is complete for a ratio $r = [\text{TMEDA}]/c^* = 1$. It can be seen from Figure 13 that a 1:1 addition of TMEDA reverts the reaction order from 0.5 to 1, the two straight lines intersecting at a critical concentration above which the amine addition accelerates the rate of polymerization. The equilibrium expressed in equation (73) was suggested as being responsible for this effect (with $K_1 \gg 1$). A quantitative description was given recently.[242] Calorimetric measurements of solvation enthalpy confirm the formation of a 1:1 complex.[243,244]

The rate constant of the complexed ion pairs ($k_\pm = 0.131 \, \text{mol}^{-1} \, \text{s}^{-1}$ at 25 °C) is much lower than that of the 'naked' ones and even lower than that found in 1,4-dioxane. Again, this can be understood on the basis of steric hindrance exerted by the bidentate ligand.

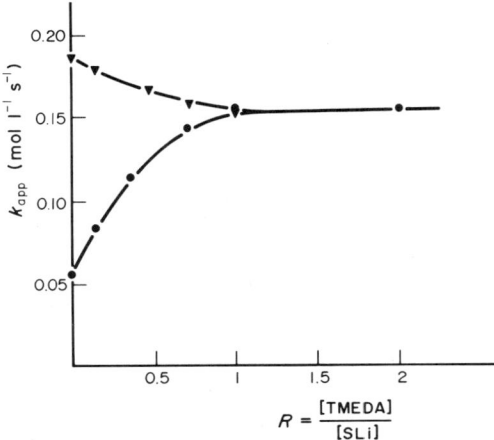

Figure 12 Effect of TMEDA on the propagation rate in the polymerization of styrene in cyclohexane,[242] (▼) $c^* = 9.2 \times 10^{-4}$; (●) $c^* = 8.3 \times 10^{-3} \, \text{mol} \, l^{-1}$

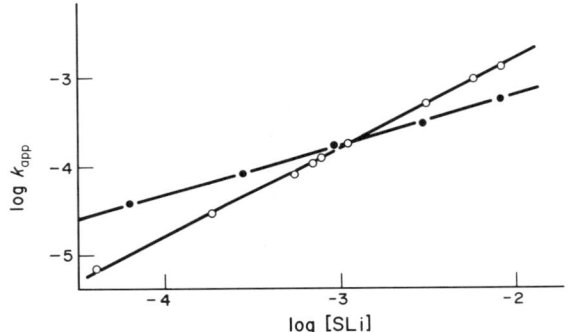

Figure 13 Reaction orders in the polymerization of styrene in cyclohexane, with (○) and without (●) added TMEDA $(r = 1)^{241}$

The addition of cyclic tertiary polyamines ('crown amines'), *e.g.* N,N',N'',N'''-tetramethyltetra-azacyclotetradecane, results in much higher rate constants ($k_{\pm} = 750 \, l \, \text{mol}^{-1} \, \text{s}^{-1}$ at 25 °C), indicating an increased interionic distance.[234,245] This is corroborated by a shift of the UV absorption maximum from 326 to 345 nm.

The effect of tertiary amines in the polymerization of dienes is similar to that in styrene polymerization. Erussalimsky and co-workers[246] found a strong increase in rate on adding TMEDA to the polymerization of isoprene in hexane. A plateau was reached for a ratio of $r \approx 4$. The content of 3,4 units in the polymer also drastically increases. When adding only 'subcatalytic' amounts of TMEDA ($r = 0.01$), bimodal MWDs were found, the peaks of which showed different microstructures.[246,247] This was attributed to a slow exchange of TMEDA molecules between tetrameric associates and ion pairs (*cf.* Section 26.2.2). It was assumed that associated as well as non-associated polymers contributed to polymerization, TMEDA increasing the reactivity of both species.

Kinetic investigations of Schué and co-workers[248-250] show that the rate effect in the polymerization of isoprene in cyclohexane depends on the concentration of active centres, similar to styrene. In contrast to Erussalimsky's results, at low concentrations a ratio of $r = 0.5$ is sufficient to give full conversion to TMEDA-complexed ion pairs. At high concentrations the plot of k_{app} *vs.* r is sigmoidal, reaching the maximum level at a ratio of unity. The same effect is seen in calorimetric measurements of the enthalpies of complex formation.[243,244] The kinetic order for the complexed species ($r = 1$) is equal to 1.1. The 1:2 stoichiometry may be best explained by assuming the following equilibrium:

$$P_4^* + 2\text{TMEDA} \underset{}{\overset{K_3}{\rightleftharpoons}} 2P_2^* \cdot \text{TMEDA} \quad (K_3 \gg 1) \tag{75}$$

but the kinetic order near to unity makes it necessary to either postulate the complexed dimer to be the active species or to assume a further equilibrium

$$P_2^* \cdot TMEDA \quad + \quad TMEDA \overset{K_4}{\rightleftharpoons} 2P^* \cdot TMEDA \tag{76}$$

producing the active complexed monomeric species.

NMR and UV experiments[250] show drastic changes in the structure of the active species for $r = 0.5$. The *trans/cis* ratio changes from two to zero, indicating the formation of a *cis*-1,4 configuration of the end group. Simultaneously, an upfield shift of the γ carbon demonstrates an increase in electron density. The UV spectrum shows a shift from 273 to 257 nm at $r = 0.5$ and, additionally, a new shoulder at 325 nm for $r > 0.5$. The former corresponds to the *trans–cis* transition and the latter to the changes in the association state. These results agree well with the equilibria given in equations (75) and (76).

Recent investigations using 1,2-dipiperidinoethane (DIPIP) as the amine[251,252] were interpreted by assuming two equilibria instead of the one in equation (75):

$$P_4^* \quad + \quad TMEDA \overset{K_5}{\rightleftharpoons} P_4^* \cdot TMEDA \qquad (K_5 \gg 1) \tag{77}$$

$$P_4^* \cdot TMEDA \quad + \quad TMEDA \overset{K_6}{\rightleftharpoons} 2P_2^* \cdot TMEDA \qquad (K_6 \gg 1) \tag{78}$$

26.5.2 Methacrylates

The polymerization of methacrylates in non-polar solvents is complicated by severe side reactions. This is manifested in complex kinetics and broad and multimodal MWDs, indicating the coexistence of various active species. On the other hand, isotactic polymers are only formed in these solvents.

Most of the earlier kinetic and mechanistic investigations in non-polar solvents focused on the nature of the side reactions,[103,135] thus rendering little evidence on the mechanism of the propagation reaction. Using fluorenyllithium for the polymerization of MMA in toluene (containing 10% diethyl ether) Glusker *et al.*[142] obtained linear first-order time–conversion plots having a finite intercept at $t = 0$. This was attributed to the rapid formation of cyclic trimer.

Using 1,1-diphenylhexyllithium (DPH–Li) in pure toluene, Wiles and Bywater[131] found first-order kinetics with respect to both monomer and initiator concentration. The rate constants are two orders of magnitude lower than those found later in polar solvents (*cf.* Section 26.4.2). However, the formation of up to 30% of methoxide during polymerization and the very broad MWDs ($M_w/M_n \approx 35$) make conclusions on the nature of the active species difficult. Experiments of Piejko[253] show that, though the distributions are broad and multimodal (M_w/M_n up to 80), a linear relationship between P_n and monomer conversion (*cf.* equation 7) is still maintained, indicating the absence of transfer reactions. Fractionation of a poly(ethyl methacrylate) prepared in toluene showed that the low molecular weight fractions are predominantly isotactic, whereas the high molecular weight fractions are predominantly syndiotactic.[254] This proves the coexistence of different active species in the polymerization in toluene.

The addition of alkoxides has a positive influence on the rates of polymerization and on the MWDs of the resulting polymers.[130,149] The microstructure is also slightly changed.[255] It was supposed that alkoxides transform associated polymers to mixed associates, similar to non-polar monomers (*cf.* Section 26.5.1). Experiments in THF show that the rate constant of PMMA$^-$,Li$^+$ ion pairs is decreased by a factor of 10 on addition of the alkoxide.[150]

Allen *et al.*[145,256] used [oligo(α-methylstyryl)]sodium as the initiator in toluene/THF mixtures. Again, a first-order dependence on monomer and initiator concentrations was established. The rate constants linearly increase with THF content (5 to 19 mol %). Even at 19% THF the rate constants (based on the assumption of full initiation efficiency) are five orders of magnitude lower than those found later in pure THF. Also, the Arrhenius parameters have unreasonably low values. The results might be understood on the basis of formation of intramolecular associates within the bifunctional polymer chain, which was discovered later in pure THF.[113,114]

Grignard reagents have been used frequently as initiators for the polymerization of MMA. Watanabe and co-workers[257-261] found the rates of propagation to be second order with Bu^n- and Bu^i-MgBr. They also investigated the effect of solvent, butyl group and halogen on rates and tacticity. Allen *et al.*[262-266] investigated the polymerization initiated by PhMgBr, Bu^n-, Bu^i- and Bu^t-MgBr, and Bu$_2^n$Mg in toluene/THF mixtures and in pure toluene. The kinetics are very complex, and the MWDs are generally very broad and multimodal, becoming narrower at higher THF contents. Multiple immutable active species were considered responsible for these effects, the active centres being associated or coordinated with THF and monomer.

Recently, Hatada et al.[267,268] reported that the polymerization of MMA in toluene initiated with Bu^tMgBr leads to narrow MWDs of 97–99% isotacticity. Initiation is fast and propagation follows first-order kinetics with respect to monomer and initiator. The rate constants are extremely low ($k_p = 4.3 \times 10^{-4}\,l\,mol^{-1}\,s^{-1}$). As narrow MWDs are related to the bulkiness of the alkyl group, it may be concluded that the active species is associated or coordinated with initiator. The former assumption is corroborated by the observation that the viscosity of the solution drops after quenching. The structure (3) was tentatively attributed to the initiator ($MgBr_2$ stemming from the Schlenk equilibrium $2RMgX \rightleftharpoons R_2Mg + MgX_2$).

$$Et_2O \underset{Br}{\overset{Bu^t}{\diagdown}} Mg \underset{Br}{\overset{Br}{\diagdown}} Mg \underset{Br}{\overset{Br}{\diagdown}} Mg \underset{Br}{\overset{Br}{\diagdown}} Mg \underset{Bu^t}{\overset{OEt_2}{\diagdown}}$$

(3)

Narrow MWDs and highly isotactic polymers were also observed in the polymerization of isopropyl acrylate initiated with mesitylmagnesium bromide in toluene.[269]

The polymerization of t-butyl methacrylate (TBMA) in toluene is different from that of MMA. When using DPH–Li or ethyl α-lithioisobutyrate as initiators, Kilz et al.[104,163,270] found first-order kinetics with respect to monomer and initiator and narrow MWDs. The rate constants are one order of magnitude *higher* than those in THF, indicating that all species exist as unsolvated contact ion pairs lacking the steric hindrance of peripheral solvation. Consequently, addition of lithium t-butoxide has no effect on kinetics, MWD and tacticity. In contrast to MMA which gives only ca. 90% isotactic polymers under these conditions, the PTBMAs prepared are 100% isotactic. The latter observation conforms with mechanism I outlined in Section 26.4.2, i.e. complex formation between monomer and cation prior to addition.

Crown ethers and cryptands have been used successfully in order to increase reactivity in the polymerization of methacrylates. Polymerization was initiated by alkali metals dissolved in toluene containing crown ethers[271] or cryptands,[235] or by common initiators complexed with crown ethers[272] or cryptands.[236,273] Alkoxides are activated by crown ethers[274] or cryptands[237,274,275] to initiate the polymerization of MMA in toluene. Tacticities resemble those found in polar solvents, indicating that the active species are complexed ion pairs. No kinetic experiments were performed in non-polar solvents. Kinetic experiments in THF using Na^+ counterion, cryptated by cryptand [2.2.2], show that the active species is a contact ion pair, the cation being the cryptated Na^+.[110] The increase in reactivity is easily understood on the basis of the charge separation in the ion pair exerted by the cryptand.

26.5.3 Vinylpyridine

The polymerization of 2-vinylpyridine (2VP) can be initiated by organomagnesium compounds in hydrocarbon solvents to give highly isotactic polymers.[276–278] Kinetic studies of Soum and Fontanille[279,280] using benzyl(2-methylpyridyl)magnesium as initiator in benzene showed that the reaction is first-order with respect to monomer and initiator. The number-average molecular weights agree well with the theoretical ones; however, no MWDs were communicated. The rate constants ($k_p = 3.0\,l\,mol^{-1}\,s^{-1}$ at 20 °C) agree with a monomeric ion pair ($P2VP^-,MgBenzyl^+$) as the active species. Viscometric experiments in THF give evidence for the monomeric nature of the active species. In contrast, cryoscopic measurements show that the initiator exists as a dimer in benzene. The lack of a tendency for the living polymers to associate was explained in terms of an intramolecular coordination of the benzylmagnesium ion by pyridyl groups (cf. equation 66).

26.6 POLYMERIZATION THERMODYNAMICS

General aspects of polymerization thermodynamics have been described in several reviews.[4,281–283] Here, the parts relating to anionic equilibrium polymerization, its kinetics and MWDs will be outlined briefly.

Generally, polymerization, *i.e.* conversion of a dissolved monomer molecule into a monomeric segment of a polymer chain, has to be regarded as an equilibrium reaction:[284]

$$P_i^* \ + \ M \ \underset{k_d}{\overset{k_p}{\rightleftharpoons}} \ P_{i+1}^*, \quad K_p \ = \ k_p/k_d \tag{79}$$

According to Flory's principle, the equilibrium constant and the rate constants should be independent of the degree of polymerization, provided the latter is high enough. Thus, equation (79) can be rewritten as

$$M_{sol} \ \underset{k_d}{\overset{k_p}{\rightleftharpoons}} \ M_{pol}, \quad K_p \ = \ k_p/k_d \tag{80}$$

Here, M_{sol} and M_{pol} denote a monomer in solution and in a polymer molecule, respectively. In contrast to k_p and k_d, the equilibrium constant K_p should be independent of the nature of the active centre. However, the ability for depropagation is related to a certain 'livingness' of the system. Thus, anionic polymerization renders some of the most important examples for polymer/monomer equilibria.

Applying the mass-action law, it is easily seen that

$$K_p \ = \ 1/[M]_{eq} \tag{81}$$

where $[M]_{eq}$ denotes the monomer concentration at equilibrium. Consequently, measurement of $[M]_{eq}$ is the most common way to determine K_p.

It must be stressed, however, that equations (80) and (81) are only valid for higher degrees of polymerization. Only here can the thermodynamic properties and the concentrations of an *i*-mer and an $(i + 1)$-mer be regarded as nearly equal.

It can be seen from Figure 14 that the equilibrium constant for the polymerization of α-methylstyrene (αMS) initiated by electron transfer strongly decreases from the dimer (the product of initiation) to the polymer.[285-288] The constant slope indicates that this effect is not due to energetic, but to entropic reasons. An effect of four orders of magnitude was reported for MMA and attributed to different stabilities of P_1^* and P_2^*.[121]

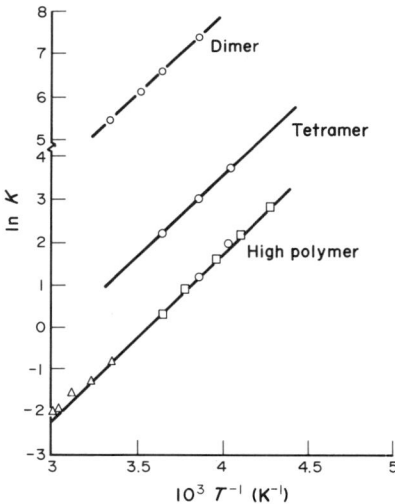

Figure 14 Equilibrium constant of polymerization of α-methylstyrene as a function of temperature and degree of polymerization in THF[4]

As most polymerizations are exothermic, the driving force for depolymerization is the negative entropy of polymerization. Thus it becomes evident that measurable depolymerization is mostly found for α,α-disubstituted vinyl monomers, *e.g.* αMS[289] ($K_p = 0.8$ mol l^{-1} at 25 °C), 2-isopropenyl-pyridine[290] ($K_p = 26$ mol l^{-1}), and, to a lesser extent, MMA[291] ($K_p = 2 \times 10^3$ mol l^{-1}). For monosubstituted monomers *e.g.* styrene[292] ($K_p = 5 \times 10^6$ mol l^{-1}), the equilibrium monomer concentration at ambient temperature becomes immeasurably low, rendering the polymerization irreversible for practical purposes.

According to equation (80), K_p only depends on the thermodynamic properties of the monomer in solution and the monomeric segment in the polymer in solution. Thus, all parameters influencing the thermodynamic properties of the polymerizing system will have an effect on K_p, e.g. temperature, pressure and composition of the solution. From this it becomes understandable that K_p depends on the nature of the solvent and on the concentration of polymer in the system. Adopting the Flory–Huggins theory for polymer soluions to ternary systems, the following equation was derived by Ivin and Leonard:[289]

$$\Delta G_{l,c}/RT = 1 + \ln \Phi_m + \Phi_s(\chi_{ms} - \chi_{sp} V_m/V_s) + \chi_{mp}(\Phi_p - \Phi_s) \tag{82}$$

Here, $\Delta G_{l,c}$ is the free energy of conversion of 1 mole of liquid monomer into 1 base-mole of amorphous polymer; Φ_m, Φ_p and Φ_s are the volume fractions of monomer, polymer and solvent; χ_{ms}, χ_{sp} and χ_{mp} are the interaction parameters for monomer–solvent, solvent–polymer and monomer–polymer; and V_m and V_s are the molar volumes of monomer and solvent, respectively.

In order to have spontaneous polymerization, the condition $\Delta G_p \leq 0$ must be fulfilled and the limiting or 'ceiling' temperature is given by

$$T_c = \Delta H_p/\Delta S_p \tag{83}$$

From equation (82) it becomes evident that T_c is not a universal constant for a given monomer. A universal constant may be obtained by extrapolating the K_p values to $\Phi_m = 1$ (bulk monomer) and $\Phi_p = \Phi_s = 0$. For solutions, the thermodynamic parameters for the standard state are frequently used ($\Delta G_p^\circ = 0$), i.e. T_c corresponds to the temperature where $[M]_{eq} = 1 \text{ mol l}^{-1}$ (extrapolated to zero polymer concentration).

Reversible polymerization has important consequences for kinetics and MWDs. The differential equation for the consumption of monomer (cf. equation 4) has to be extended for the effect of depropagation:

$$R_p = -\frac{d[M]}{dt} = (k_p[M] - k_d)c^* \tag{84}$$

leading to non-linear first-order time–conversion plots, which level off at $[M] = [M]_{eq}$. Substitution of k_d by $k_p[M]_{eq}$ gives

$$R_p = -\frac{d[M]}{dt} = k_p([M] - [M]_{eq})c^* \tag{85}$$

and integration results in the modified first-order relation (cf. equation 6)

$$\ln \frac{[M]_0 - [M]_{eq}}{[M] - [M]_{eq}} = k_p c^* t = k_{app} t \tag{86}$$

When starting from a non-equilibrated living polymer (e.g. by rapid heating of an equilibrated solution) the rate constant of depropagation can be measured independently. Substitution of k_p in equation (84) by $k_d/[M]_{eq}$ and integration leads to

$$\ln \frac{[M]_{eq}}{[M]_{eq} - [M]} = k_d c^* t \tag{87}$$

The Arrhenius plot for k_p and k_d in the polymerization of 2-isopropenylpyridine in toluene[290] is shown in Figure 15. The equilibrium constants calculated from the rate constants agree well with the values obtained from $[M]_{eq}$.

The MWD resulting from equilibrium polymerization must obviously be the most probable one, i.e. a Schulz–Flory distribution ($M_w/M_n = 2$). However, this type of distribution is only reached after full equilibration. MWDs resulting from normal polymerizations are determined by kinetic control ($M_w/M_n \approx 1$, cf. Section 26.2). As the rates of depropagation are usually very low, it may take a very long time to reach full equilibrium. The kinetics of the transformation from a Poisson to a Schulz–Flory distribution were discussed by Brown and Szwarc[293] and by Mikaye and Stockmayer.[294]

As the number of polymer chains, and thus P_n, is invariant, the kinetics of equilibrium can be described in terms of the increase in P_w, the value of which is finally doubled.

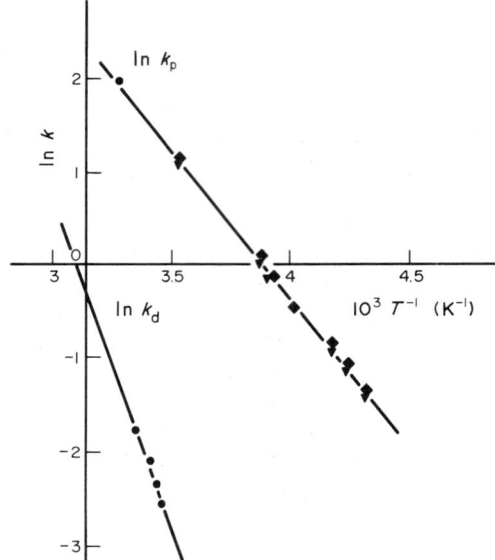

Figure 15 Arrhenius plot of the rate constants of propagation k_p and depropagation k_d in the polymerization of 2-isopropenylpyridine in toluene; counterion = Li$^+$ [290]

The approximate treatment of Brown and Szwarc gives the initial rate

$$\frac{dP_w}{dt} = \frac{2k_d}{P_n}\left(1 - \frac{[P_1^*]}{c^*}P_n\right) \approx \frac{2k_d}{P_n}. \tag{88}$$

Introducing the polydispersity P_w/P_n or better the non-uniformity $U = P_w/P_n - 1$ (equation 22) we find that the initial rate of transformation is proportional to the square of P_n:

$$\frac{dU}{dt} = \frac{2k_d}{P_n^2} \tag{89}$$

The more rigorous treatment of Mikaye and Stockmayer shows that the problem is identical to the kinetics of multilayer adsorption according to the BET (Brünaüer–Emmett–Teller) model.[295] The rate of redistribution is given as

$$\frac{dU}{dt} = \frac{2}{K_p[M]_0 P_n^2} \tag{90}$$

The time needed for complete redistribution is

$$= \frac{1}{(dU/dt)K_p[M]_0} = \frac{2P_n^2}{k_d} \tag{91}$$

As an example, the time for complete redistribution of (polystyryl)sodium of $P_n = 1000$ ($k_d = 2 \times 10^{-4}\,\mathrm{s}^{-1}$) is *ca.* 80 years in THF at room temperature, whereas the equilibrium monomer concentration is reached within seconds.

26.7 REFERENCES

1. T. E. Hogen-Esch and J. Smid (eds.), 'Recent Advances in Anionic Polymerization', Elsevier, New York, 1987.
2. B. L. Erusalimskii, 'Mechanisms of Ionic Polymerization. Current Problems', Plenum Press, New York, 1986.
3. R. N. Young, R. P. Quirk and L. J. Fetters, *Adv. Polym. Sci.*, 1984, **56**, 1.
4. M. Szwarc, *Adv. Polym. Sci.*, 1983, **49**, 1.
5. M. Morton, 'Anionic Polymerization: Principles and Practice', Academic Press, New York, 1983.
6. J. E. McGrath (ed.), *ACS Symp. Ser.*, 1981, **166**.
7. S. Bywater, in 'Comprehensive Chemical Kinetics', ed. C. H. Bamford and C. F. H. Tipper, Elsevier, Amsterdam, 1976, vol. 15, p. 1.

8. M. Szwarc (ed.), 'Ions and Ion Pairs in Organic Reactions', Wiley Interscience, New York, 1972, vol. 1; 1974, vol. 2.
9. L. L. Böhm, M. Chmeliř, G. Löhr, B. J. Schmitt and G. V. Schulz, *Adv. Polym. Sci.*, 1972, **9**, 1.
10. M. Szwarc, 'Carbanions, Living Polymers and Electron Transfer Processes', Interscience, New York, 1968.
11. L. F. Beste and H. K. Hall, *J. Phys. Chem.*, 1964, **68**, 269.
12. C. E. H. Bawn, C. Fitzsimmons, A. Ledwith, J. Penfold, D. C. Sherrington and J. A. Weightman, *Polymer*, 1971, **12**, 119.
13. D. C. Pepper, *Eur. Polym. J.*, 1980, **16**, 407.
14. M. Litt, *J. Polym. Sci.*, 1962, **58**, 429.
15. M. Szwarc, in 'Recent Advances in Anionic Polymerization', ed. T. E. Hogen-Esch and J. Smid, Elsevier, New York, 1987, p. 93.
16. M. Szwarc, M. van Beylen and D. van Hoyweghen, *Macromolecules*, 1987, **20**, 445.
17. P. J. Flory, *J. Am. Chem. Soc.*, 1940, **62**, 1561.
18. G. V. Schulz, *Z. Phys. Chem., Abt. B*, 1940, **47**, 155.
19. R. V. Figini and G. V. Schulz, *Z. Phys. Chem. (Frankfurt)*, 1960, **23**, 233.
20. R. V. Figini and G. V. Schulz, *Makromol. Chem.*, 1960, **41**, 1.
21. G. Löhr and G. V. Schulz, *Z. Phys. Chem. (Frankfurt)*, 1969, **65**, 170.
22. M. Szwarc and M. Litt, *J. Phys. Chem.*, 1958, **62**, 568.
23. R. V. Figini, *Makromol. Chem.*, 1961, **44-46**, 497.
24. T. A. Orofino and F. Wenger, *J. Chem. Phys.*, 1961, **35**, 532.
25. D. Panke, *Makromol. Chem.*, 1981, **182**, 3207.
26. B. D. Coleman, F. Gornick and G. Weiss, *J. Chem. Phys.*, 1963, **39**, 3233.
27. D. -Y. Yan and C. -M. Yuan, *J. Macromol. Sci., Chem.*, 1986, **A23**, 781.
28. V. S. Nanda and R. K. Jain, *J. Polym. Sci., Part A-1*, 1967, **5**, 2269.
29. C. -M. Yuan and D. -Y. Yan, *Makromol. Chem.*, 1986, **187**, 2629.
30. J. Largo-Cabrerizo and J. Guzmán, *Macromolecules*, 1979, **12**, 526.
31. D. -Y. Yan and C. -M. Yuan, *Makromol. Chem.*, 1987, **188**, 333.
32. B. D. Coleman and T. G. Fox, *J. Am. Chem. Soc.*, 1963, **85**, 1241.
33. R. V. Figini, *Makromol. Chem.*, 1964, **71**, 193.
34. M. Szwarc and J. J. Hermans, *J. Polym. Sci., Part B*, 1964, **2**, 815.
35. R. V. Figini, *Makromol. Chem.*, 1967, **107**, 170.
36. L. L. Böhm, *Z. Phys. Chem. (Frankfurt)*, 1970, **72**, 199.
37. L. L. Böhm, *Z. Phys. Chem. (Frankfurt)*, 1974, **88**, 297.
38. M. Szwarc and J. J. Hermans, *J. Polym. Sci., Part B*, 1964, **2**, 815.
39. S. Krause, L. Defonso and D. L. Glusker, *J. Polym. Sci., Part A*, 1965, **3**, 1617.
40. F. A. Bovey, 'High Resolution NMR of Macromolecules', Academic Press, New York, 1980.
41. R. V. Figini, *Makromol. Chem.*, 1965, **88**, 272.
42. S. Bywater, in 'Recent Advances in Anionic Polymerization', ed. T. E. Hogen-Esch and J. Smid, Elsevier, New York, 1987, p. 187; *Polym. Prepr., Am. Chem. Soc., Div. Polym. Chem.*, 1986, **27** (1), 149.
43. I. M. Khan and T. E. Hogen-Esch, in 'Recent Advances in Anionic Polymerization', ed. T. E. Hogen-Esch and J. Smid, Elsevier, New York, 1987, p. 261; *Polym. Prepr., Am. Chem. Soc., Div. Polym. Chem.*, 1986, **27** (1), 159.
44. M. Magat, *J. Chim. Phys. Phys.-Chim. Biol.*, 1950, **47**, 841.
45. L. Gold, *J. Chem. Phys.*, 1958, **28**, 91.
46. M. Litt, *J. Polym. Sci.*, 1962, **58**, 429.
47. V. S. Nanda and R. K. Jain, *J. Chem. Phys.*, 1963, **39**, 1363.
48. V. S. Nanda and R. K. Jain, *J. Polym. Sci., Part A*, 1964, **2**, 4583.
49. H. Höcker, *Makromol. Chem.*, 1972, **157**, 187.
50. D. -Y. Yan, G. -Y. Li and Y. -S. Jiang, *Sci. Sin. (Engl. Ed.)*, 1981, **24**, 46.
51. R. V. Figini, *Z. Phys. Chem. (Frankfurt)*, 1963, **38**, 341.
52. F. Wenger, *Makromol. Chem.*, 1960, **36**, 200.
53. F. Wenger, *Makromol. Chem.*, 1963, **64**, 151.
54. R. Chiang and J. J. Hermans, *J. Polym. Sci., Part A-1*, 1966, **4**, 2843; R. Chiang and H. N. Friedlander, *J. Polym. Sci., Part A-1*, 1966, **4**, 2857.
55. D. -Y. Yan and C. -M. Yuan, *J. Macromol. Sci., Chem.*, 1986, **A23**, 769.
56. G. Löhr, *Makromol. Chem.*, 1973, **172**, 151.
57. D. -Y. Yan, *J. Chem. Phys.*, 1984, **80**, 3434.
58. V. S. Nanda and S. C. Jain, *Eur. Polym. J.*, 1970, **6**, 1605.
59. V. S. Nanda and S. C. Jain, *Eur. Polym. J.*, 1977, **13**, 137.
60. C. -M. Yuan and D. -Y. Yan, *Makromol. Chem.*, 1987, **188**, 341.
61. W. T. Kyner, R. M. Radock and M. Wales, *J. Chem. Phys.*, 1959, **30**, 363.
62. M. Litt and M. Szwarc, *J. Polym. Sci.*, 1960, **42**, 159.
63. V. S. Nanda, *Trans. Faraday Soc.*, 1964, **60**, 949.
64. L. H. Peebles, *J. Polym. Sci., Part B*, 1969, **7**, 75.
65. L. H. Peebles, *J. Polym. Sci., Part A-2*, 1970, **8**, 1235.
66. S. C. Jain and V. S. Nanda, *J. Polym. Sci., Part B*, 1970, **8**, 843.
67. J. Largo-Cabrerizo and J. Guzmán, *Macromolecules*, 1979, **12**, 526.
68. Zhou Pu, D. -Y. Yan and W. -F. Tang, *Makromol. Chem.*, 1985, **186**, 159.
69. C. -M. Yuen and D. -Y. Yan, *Makromol. Chem.*, 1986, **187**, 2629.
70. C. -M. Yuen and D. -Y. Yan, *Makromol. Chem.*, 1986, **187**, 2641.
71. D. J. Worsfold and S. Bywater, *Can J. Chem.*, 1958, **36**, 1141.
72. D. N. Bhattacharyya, C. L. Lee, J. Smid and M. Szwarc, *Polymer*, 1964, **5**, 54.
73. D. N. Bhattacharyya, C. L. Lee, J. Smid and M. Szwarc, *J. Phys. Chem.*, 1965, **69**, 612.
74. H. Hostalka, R. V. Figini and G. V. Schulz, *Makromol. Chem.*, 1964, **71**, 198.
75. H. Hostalka and G. V. Schulz, *Z. Phys. Chem. (Frankfurt)*, 1965, **45**, 286.
76. F. Accascina and R. M. Fuoss, 'Electrolytic Conductance', Wiley, New York, 1965.

77. B. J. Schmitt and G. V. Schulz, *Eur. Polym. J.*, 1975, **11**, 119.
78. M. Szwarc, *Adv. Polym. Sci.*, 1983, **49**, 108.
79. W. K. R. Barnikol and G. V. Schulz, *Z. Phys. Chem. (Frankfurt)*, 1965, **47**, 89.
80. T. Shimomura, K. J. Tölle, J. Smid and M. Szwarc, *J. Am. Chem. Soc.*, 1967, **89**, 796.
81. B. J. Schmitt and G. V. Schulz, *Makromol. Chem.*, 1971, **142**, 325.
82. M. Chmelíř and G. V. Schulz, *Ber. Bunsenges. Phys. Chem.*, 1971, **75**, 830.
83. G. Löhr and G. V. Schulz, *Makromol. Chem.*, 1964, **77**, 240.
84. R. V. Figini, G. Löhr and G. V. Schulz, *J. Polym. Sci., Part B*, 1965, **3**, 985.
85. L. L. Böhm, G. Löhr and G. V. Schulz, *Ber. Bunsenges. Phys. Chem.*, 1974, **78**, 1064.
86. C. A. Kraus and R. M. Fouss, *J. Am. Chem. Soc.*, 1933, **55**, 21.
87. C. B. Wooster, *J. Am. Chem. Soc.*, 1937, **59**, 377.
88. R. M. Fuoss and C. A. Kraus, *J. Am. Chem. Soc.*, 1933, **55**, 2387.
89. D. N. Bhattacharyya, J. Smid and M. Szwarc, *J. Am. Chem. Soc.*, 1964, **86**, 5024.
90. S. Peeters, Ph. D. Thesis, University of Leuven, Belgium, 1982; S. Peeters and M. van Beylen, in preparation.
91. N. Ise, H. Hirohara, T. Makino, K. Takaya and M. Nakayama, *J. Phys. Chem.*, 1970, **74**, 606.
92. H. Hirohara and N. Ise, *J. Polym. Sci., Part D*, 1972, **6**, 295.
93. K. Takaya, S. Yamauchi and N. Ise, *J. Chem. Soc., Faraday Trans. 1*, 1974, **70**, 1330; K. Takaya and N. Ise, *J. Chem. Soc., Faraday Trans. 1*, 1974, **70**, 1338.
94. N. Ise, *J. Macromol. Sci., Chem.*, 1975, **A9**, 1047.
95. A. Gourdenne and P. Sigwalt, *Eur. Polym. J.*, 1967, **3**, 481.
96. F. Schué, D. J. Worsfold and S. Bywater, *Macromolecules*, 1970, **3**, 509.
97. S. Brownstein, S. Bywater and D. J. Worsfold, *Macromolecules*, 1973, **6**, 715.
98. A. Garton, R. P. Chaplin and S. Bywater, *Eur. Polym. J.*, 1976, **12**, 697.
99. S. Bywater, *ACS Symp. Ser.*, 1981, **166**, 71.
100. G. Gee, W. C. E. Higginson and G. T. Merrall, *J. Chem. Soc.*, 1961, 1345.
101. R. T. McDonald and S. Bywater, *Organometallics*, 1986, **5**, 1529.
102. R. T. McDonald, S. Bywater and P. Black, *Macromolecules*, 1987, **20**, 1196.
103. For reviews see: D. M. Wiles, in 'Structure and Mechanisms in Vinyl Polymerization', ed. T. Tsuruta and K. F. O'Driscoll, Dekker, New York, 1969, p. 233; M. Szwarc, in 'Ions and Ion Pairs in Organic Reactions', ed. M. Szwarc, Wiley, New York, 1974, vol. 2, p. 401; S. Bywater in 'Comprehensive Chemical Kinetics', ed. C. H. Bamford and C. F. H. Tipper, Elsevier, Amsterdam, 1976, vol. 15, p. 40; M. Szwarc, *Adv. Polym. Sci.*, 1983, **49**, 133.
104. A. H. E. Müller, in 'Recent Advances in Anionic Polymerization', ed. T. E. Hogen-Esch and J. Smid, Elsevier, New York, 1987, p. 205.
105. L. Lochmann, M. Rodova, J. Petranek and D. Lim, *J. Polym. Sci., Polym. Chem. Ed.*, 1974, **12**, 2295.
106. I. Mita, Y. Watabe, T. Akatsu and H. Kambe, *Polym. J.*, 1973, **4**, 271.
107. G. Löhr and G. V. Schulz, *Makromol. Chem.*, 1973, **172**, 137.
108. G. Löhr and G. V. Schulz, *Eur. Polym. J.*, 1974, **10**, 121.
109. A. H. E. Müller, *ACS Symp. Ser.*, 1981, **166**, 441; A. H. E. Müller, F. J. Gerner, R. Kraft, H. Höcker and G. V. Schulz, *Polym. Prepr., Am. Chem. Soc., Div. Polym. Chem.*, 1980, **21** (1), 36.
110. C. Johann and A. H. E. Müller, *Makromol. Chem., Rapid Commun.*, 1981, **2**, 687.
111. H. Jeuck and A. H. E. Müller, *Makromol. Chem., Rapid Commun.*, 1982, **3**, 121.
112. Ch. B. Tsvetanov, A. H. E. Müller and G. V. Schulz, *Macromolecules*, 1985, **18**, 863.
113. V. Warzelhan and G. V. Schulz, *Makromol. Chem.*, 1976, **177**, 2185.
114. V. Warzelhan, H. Höcker and G. V. Schulz, *Makromol. Chem.*, 1980, **181**, 149.
115. W. Fowells, C. Schuerch, F. A. Bovey and F. P. Hood, *J. Am. Chem. Soc.*, 1967, **89**, 1396.
116. L. Lochmann and D. Lim, *J. Organomet. Chem.*, 1973, **50**, 9.
117. L. Lochmann and J. Trekoval, *Makromol. Chem.*, 1982, **183**, 1361.
118. C. B. Tsvetanov, D. T. Petrova and I. M. Panayotov, *Makromol. Chem.*, 1982, **183**, 517.
119. L. Vancea and S. Bywater, *Macromolecules*, 1981, **14**, 1776.
120. C. B. Tsvetanov, D. T. Dotcheva, D. K. Dimov, E. B. Petrova and I. M. Panayotov, in 'Recent Advances in Anionic Polymerization', ed. T. E. Hogen-Esch and J. Smid, Elsevier, New York, 1987, p. 155.
121. A. H. E. Müller, L. Lochmann and J. Trekoval, *Makromol. Chem.*, 1986, **187**, 1473.
122. R. Kraft, A. H. E. Müller, H. Höcker and G. V. Schulz, *Makromol. Chem., Rapid Commun.*, 1980, **1**, 363.
123. F. J. Gerner, Ph. D. Thesis, University of Mainz, FRG, 1982.
124. P. Kilz, M. Sc. Thesis, University of Mainz, FRG, 1982.
125. M. Reinmöller and T. G. Fox, *Polym. Prepr., Am. Chem. Soc., Div. Polym. Chem.*, 1966, **7** (2), 999.
126. M. A. Müller and M. Stickler, *Makromol. Chem., Rapid Commun.*, 1986, **7**, 575.
127. A. H. E. Müller, *Makromol. Chem.*, 1981, **182**, 2863.
128. H. Schreiber, *Makromol. Chem.*, 1960, **36**, 86.
129. D. M. Wiles and S. Bywater, *Chem. Ind. (London)*, 1963, 1209.
130. D. M. Wiles and S. Bywater, *J. Phys. Chem.*, 1964, **68**, 1983.
131. D. M. Wiles and S. Bywater, *Trans. Faraday Soc.*, 1965, **61**, 150.
132. N. Kawabata and T. Tsuruta, *Makromol. Chem.*, 1965, **86**, 231.
133. K. Hatada, T. Kitayama, K. Fujikawa, K. Ohta and H. Yuki, *Polym. Bull. (Berlin)*, 1978, **1**, 103.
134. K. Hatada, T. Kitayama, M. Nagakura and H. Yuki, *Polym. Bull. (Berlin)*, 1980, **2**, 125.
135. K. Hatada, T. Kitayama, K. Fumikawa, K. Ohta and H. Yuki, *ACS Symp. Ser.*, 1981, **166**, 327; *Polym. Prepr., Am. Chem. Soc., Div. Polym. Chem.*, 1980, **21** (1), 59.
136. A. T. Bullock, G. G. Cameron and J. M. Elsom, *Eur. Polym. J.*, 1977, **13**, 751.
137. G. Löhr, A. H. E. Müller, V. Warzelhan and G. V. Schulz, *Makromol Chem.*, 1974, **175**, 497.
138. P. Rempp, V. I. Volkov, J. Parrod and Ch. Sadron, *Bull. Soc. Chim. Fr.*, 1960, 919.
139. J. Trekoval and P. Kratochvil, *J. Polym. Sci., Part A-1*, 1972, **10**, 1391.
140. F. J. Gerner, H. Höcker, A. H. E. Müller and G. V. Schulz, *Eur. Polym. J.*, 1984, **20**, 349.
141. W. E. Goode, F. H. Owens and W. L. Myers, *J. Polym. Sci.*, 1960, **47**, 75.

142. D. L. Glusker, I. Lysloff and E. Stiles, *J. Polym. Sci.*, 1961, **49**, 315.
143. F. H. Owens, W. L. Myers and F. E. Zimmerman, *J. Org. Chem.*, 1961, **26**, 2288.
144. L. Lochmann, M. Rodova, J. Petranek and D. Lim, *J. Polym. Sci., Polym. Chem. Ed.*, 1974, **12**, 2295.
145. P. E. M. Allen, R. P. Chaplin and D. O. Jordan, *Eur. Polym. J.*, 1972, **8**, 271.
146. V. Warzelhan, H. Höcker and G. V. Schulz, *Makromol. Chem.*, 1978, **179**, 2221.
147. D. L. Glusker, E. Stiles and B. Yoncoskie, *J. Polym. Sci.*, 1961, **49**, 297.
148. A. H. E. Müller, L. Lochmann and J. Trekoval, *Makromol. Chem.*, 1986, **187**, 1473.
149. L. Lochmann, J. Kolarík, D. Doskocilová, S. Vozka and J. Trekoval, *J. Polym. Sci., Polym. Chem. Ed.*, 1979, **17**, 1727.
150. L. Lochmann and A. H. E. Müller, 'IUPAC International Symposium on Macromolecules, Merseburg 1987', preprints, p. 78; *Makromol. Chem.*, in preparation.
151. G. D. Andrews, B. C. Anderson, P. Arthur, H. W. Jacobson, L. R. Melby, A. J. Playtis and W. H. Sharkey, *Macromolecules*, 1981, **14**, 1599.
152. G. Huynh-ba and J. E. McGrath, in 'Recent Advances in Anionic Polymerization', ed. T. E. Hogen and J. Smid, Elsevier, New York, 1987, p. 173; *Polym. Prepr., Am. Chem. Soc., Div. Polym. Chem.*, 1986, **27** (1), 179.
153. F. J. Gerner, A. H. E. Müller, H. Höcker and G. V. Schulz, 'IUPAC International Symposium on Macromolecules, Strasbourg, 1981', preprints, p. 213.
154. F. J. Gerner, Ph. D. Thesis, University of Mainz, FRG, 1982.
155. A. H. E. Müller, V. Warzelhan, H. Höcker, G. Löhr and G. V. Schulz, 'IUPAC International Symposium on Macromolecules, Dublin, 1977', preprints, p. 31.
156. W. K. Busfield and J. M. Methven, *Polymer*, 1973, **14**, 137.
157. T. Kitano, T. Fujimoto and M. Nagasawa, *Polym. J.*, 1977, **9**, 153.
158. P. M. Mai, M. Sc. Thesis, University of Mainz, FRG, 1985.
159. R. Fayt, R. Forte, C. Jacobs, R. Jerôme, T. Ouhadi, Ph. Teyssié and S. K. Varshney, *Macromolecules*, 1987, **20**, 1442; R. Jerôme, R. Forte, S. K. Varshney, R. Fayt and Ph. Tessié, in 'Recent Advances in Mechanistic and Synthetic Aspects of Polymerization', ed. M. Fontanille and A. Guyot, Reidel, Dordrecht, 1987, p. 101.
160. A. R. Lyons and E. Catterall, *Eur. Polym. J.*, 1971, **7**, 839.
161. C. G. Overberger and A. M. Schiller, *J. Polym. Sci., Part C*, 1963, **1**, 325.
162. C. Johann, Ph. D. Thesis, University of Mainz, FRG, 1985.
163. A. H. E. Müller, H. Jeuck, C. Johann and P. Kilz, *Polym. Prepr., Am. Chem. Soc., Div. Polym. Chem.*, 1986, **27** (1), 153.
164. C. B. Tsvetanov, D. T. Petrova and I. M. Panayotov, *Makromol. Chem.*, 1982, **183**, 517.
165. R. C. Schulz and W. Passmann, *Makromol. Chem.*, 1963, **60**, 139.
166. R. C. Schulz, G. Wegner and W. Kern, *J. Polym. Sci., Part C*, 1967, **16**, 989.
167. I. V. Andreeva, M. M. Koton and Yu. V. Medvedev, *Vysokomol. Soedin., Ser. A*, 1973, **15**, 852 [*Polym. Sci. USSR (Engl. Transl.)*, 1973, **15**, 959].
168. I. B. Rashkov and I. M. Panayotov, *Makromol. Chem.*, 1972, **151**, 275.
169. P. Thivollet and J. Golé, *J. Polym. Sci., Polym. Chem. Ed.*, 1973, **11**, 1615.
170. H. Calvayrac, P. Thivollet and J. Golé, *J. Polym. Sci., Polym. Chem. Ed.*, 1973, **11**, 1631.
171. D. Gulino, J. P. Pascault and Q. T. Pham, *Makromol. Chem.*, 1981, **182**, 2321.
172. I. V. Andreeva and Yu. V. Medvedev, *Vysokomol. Soedin., Ser. A*, 1969, **11**, 2244.
173. I. V. Andreeva, M. M. Koton and Yu. V. Medvedev, *Vysokomol. Soedin., Ser. A*, 1973, **15**, 852 [*Polym. Sci. USSR (Engl. Transl.)*, 1973, **15**, 959].
174. D. Gulino, J. Golé and J. P. Pascault, *Eur. Polym. J.*, 1979, **15**, 469.
175. D. Gulino, J. Golé and J. P. Pascault, *Makromol. Chem.*, 1981, **182**, 1215.
176. D. Gulino, Q. T. Pham, J. Golé and J. P. Pascault, *ACS Symp. Ser.*, 1981, **166**, 307; *Polym. Prepr., Am. Chem. Soc., Div. Polym. Chem.*, 1980, **21** (1), 67.
177. B. L. Erussalimsky and A. V. Novosselova, *Acta Polym.*, 1975, **26**, 293.
178. I. G. Krasnosel'skaya and B. L. Erussalimsky, *Vysokomol. Soedin., Ser. A*, 1983, **25**, 1961.
179. Ch. Tsvetanov, I. Panayotov and B. L. Erussalimsky, *Eur. Polym. J.*, 1974, **10**, 557.
180. Ch. Tsvetanov and I. Panayotov, *Eur. Polym. J.*, 1975, **11**, 209.
181. Ch. Tsvetanov, Yu. Ye. Einzer and B. L. Erussalimsky, *Eur. Polym. J.*, 1980, **16**, 219.
182. H. Venkerckhoven and M. van Beylen, *Eur. Polym. J.*, 1978, **14**, 273.
183. S. Riedel, D. Lehmann, G. Wunderlich, H. J. Adler, R. Dreyer and W. Berger, *Acta Polym.*, 1981, **32**, 227.
184. W. Berger, S. Riedel, H. J. Adler, G. Wunderlich, D. Lehmann and O. Vogl, *J. Macromol. Sci., Chem.*, 1983, **A20**, 299.
185. H. J. Adler, U. Warmuth, W. Berger and B. L. Erussalimsky, *Acta Polym.*, 1986, **37**, 169.
186. W. Berger and H. J. Adler, *Makromol. Chem., Macromol. Symp.*, 1986, **3**, 301.
187. C. L. Lee, J. Smid and M. Szwarc, *Trans. Faraday Soc.*, 1963, **59**, 1192.
188. M. Fisher and M. Szwarc, *Macromolecules*, 1970, **3**, 23.
189. D. Honnoré, J. C. Favier, P. Sigwalt and M. Fontanille, *Eur. Polym. J.*, 1974, **10**, 425.
190. J. C. Favier, P. Sigwalt and M. Fontanille, *Eur. Polym. J.*, 1974, **10**, 717.
191. A. Soum, M. Fontanille and P. Sigwalt, *J. Polym. Sci., Polym. Chem. Ed.*, 1977, **15**, 659.
192. M. Tardi, D. Rougé and P. Sigwalt, *Eur. Polym. J.*, 1967, **3**, 85.
193. M. Tardi and P. Sigwalt, *Eur. Polym. J.*, 1972, **8**,151.
194. M. Tardi and P. Sigwalt, *Eur. Polym. J.*, 1973, **9**, 1369.
195. T. E. Hogen-Esch, W. L. Jenkins, R. A. Smith and C. F. Tien, *ACS Symp. Ser.*, 1981, **166**, 231; *Polym. Prepr., Am. Chem. Soc., Div. Polym. Chem.*, 1980, **21** (1), 15.
196. C. C. Meverden and T. E. Hogen-Esch, *Makromol. Chem., Rapid Commun.*, 1983, **4**, 563.
197. D. Honnoré, J. C. Favier, P. Sigwalt and M. Fontanille, *Eur. Polym. J.*, 1974, **10**, 425.
198. D. J. Worsfold and S. Bywater, *Can. J. Chem.*, 1960, **38**, 1891.
199. Yu. L. Spivin, A. R. Gantmakher and S. S. Medvedev, *Dokl. Akad. Nauk SSSR*, 1962, **146**, 368.
200. A. F. Johnson and D. J. Worsfold, *J. Polym. Sci., Part A*, 1965, **3**, 449.
201. M. Morton, E. E. Bostick and R. Livigny, *Rubber Plast. Age*, 1961, **42**, 397.
202. H. L. Hsieh and A. G. Kitchen, *ACS Symp. Ser.*, 1983, **212**, 291.
203. M. Morton, L. J. Fetters, R. A. Pett and J. F. Meier, *Macromolecules*, 1970, **3**, 327.

204. D. J. Worsfold and S. Bywater, *Macromolecules*, 1972, **5**, 393.
205. J. E. L. Roovers and S. Bywater, *Polymer*, 1973, **14**, 594.
206. A. F. Johnson and D. J. Worsfold, *J. Polym. Sci., Part A*, 1965, **3**, 449.
207. H. Sinn and F. Patat, *Angew. Chem.*, 1963, **75**, 805.
208. D. J. Worsfold and S. Bywater, *Can. J. Chem.*, 1964, **42**, 2884.
209. A. Guyot and J. Vialle, *J. Macromol. Sci., Chem.*, 1970, **A4**, 79.
210. S. Bywater and D. J. Worsfold, in 'Recent Advances in Anionic Polymerization', ed. T. E. Hogen-Esch and J. Smid, Elsevier, New York, 1987, p. 109; *Polym. Prepr., Am. Chem. Soc., Div. Polym. Chem.*, 1986, **27** (1), 140.
211. A. Hernandez, J. Semel, H. -Ch. Broeker, H. G. Zachmann and H. Sinn, *Makromol. Chem., Rapid Commun.*, 1980, **1**, 75.
212. M. Morton, L. J. Fetters and E. E. Bostick, *J. Polym. Sci., Part C*, 1963, **1**, 311.
213. M. Morton and L. J. Fetters, *J. Polym. Sci., Part A*, 1964, **2**, 3311.
214. M. Morton, *Polym. Prepr., Am. Chem. Soc., Div. Polym. Chem.*, 1964, **5**, 1092.
215. S. Bywater, *Adv. Polym. Sci.*, 1965, **4**, 66.
216. M. Szwarc, *ACS Symp. Ser.*, 1981, **166**, 1.
217. D. J. Worsfold, *J. Polym. Sci., Polym. Phys. Ed.*, 1982, **20**, 99.
218. J. E. L. Roovers and S. Bywater, *Trans. Faraday Soc.*, 1966, **62**, 701.
219. J. E. L. Roovers and S. Bywater, *Can. J. Chem.*, 1968, **46**, 2711.
220. R. S. Stearns and L. Forman, *J. Polym. Sci.*, 1959, **41**, 381.
221. Yu. L. Spirin, A. R. Gantmakher and S. S. Medvedev, *Vysokomol. Soedin.*, 1959, **1**, 1258.
222. H. Sinn, C. Lundborg and O. T. Onsager, *Makromol. Chem.*, 1964, **70**, 222.
223. J. Geerts, M. van Beylen and G. Smets, *J. Polym. Sci., Part A-1*, 1969, **7**, 2859.
224. V. V. Shamanin, E. Yu. Melenevskaya and V. N. Zgonnik, *Acta Polym.*, 1982, **33**, 175.
235. J. E. L. Roovers and S. Bywater, *Trans. Faraday Soc.*, 1966, **62**, 1876.
226. A. Guyot and J. Vialle, *J. Macromol. Sci., Chem.*, 1970, **A4**, 107.
227. S. Bywater and D. J. Worsfold, *Can. J. Chem.*, 1962, **40**, 1564.
228. I. J. Alexander and S. Bywater, *J. Polym. Sci., Part A-1*, 1968, **6**, 3407.
229. J. Geerts, M. van Beylen and G. Smets, *J. Polym. Sci., Part A-1*, 1969, **7**, 2805.
230. A. Yamagishi, M. Szwarc, L. Tung and G. -Y. S. Lo, *Macromolecules*, 1978, **11**, 607.
231. L. J. Fetters and R. N. Young, *ACS Symp. Ser.*, 1981, **166**, 95; *Polym. Prepr., Am. Chem. Soc., Div. Polym. Chem.*, 1980, **21** (1), 34.
232. D. J. Worsfold and S. Bywater, *Can. J. Chem.*, 1964, **42**, 2884.
233. A. Okamoto, E. Watanabe and I. Mita, *J. Polym. Sci., Polym. Chem. Ed.*, 1979, **17**, 2483.
234. G. Hélary, V. Tskhovrebashvili and M. Fontanille, *Eur. Polym. J.*, 1984, **20**, 157.
235. S. Boileau, B. Kaempf, J. M. Lehn and F. Schué, *J. Polym. Sci., Polym. Lett. Ed.*, 1974, **12**, 203.
236. S. Boileau, B. Kaempf, S. Raynal, J. Lacoste and F. Schué, *J. Polym. Sci., Polym. Lett. Ed.*, 1974, **12**, 211.
237. S. Boileau, P. Hémery, B. Kaempf, F. Schué and M. Viguier, *J. Polym. Sci., Polym. Lett. Ed.*, 1974, **12**, 217.
238. T. C. Cheng and A. F. Halasa, *J. Polym. Sci., Polym. Chem. Ed.*, 1976, **14**, 583.
239. J. Lacoste, F. Schué, S. Bywater and B. Kaempf, *J. Polym. Sci., Polym. Lett. Ed.*, 1976, **14**, 201.
240. S. Alev, A. Collet, M. Viguier and F. Schué, *J. Polym. Sci., Polym. Chem. Ed.*, 1980, **18**, 1155, 1163.
241. G. Hélary and M. Fontanille, *Eur. Polym. J.*, 1978, **14**, 345.
242. M. Fontanille, G. Hélary and M. Szwarc, *Macromolecules*, 1988, **21**, 1532.
243. R. P. Quirk and D. McFay, *J. Polym. Sci., Polym. Chem. Ed.*, 1981, **19**, 1445.
244. R. P. Quirk, *ACS Symp. Ser.*, 1981, **166**, 117; *Polym. Prepr., Am. Chem. Soc., Div. Polym. Chem.*, 1980, **21** (1), 38.
245. G. Hélary and M. Fontanille, *Polym. Bull. (Berlin)*, 1980, **3**, 159.
246. A. Davidjan, N. Nikolaew, V. Sgonnik, B. Belenkii, V. Nesterow and B. Erussalimsky, *Makromol. Chem.*, 1976, **177**, 2469.
247. A. Davidjan, N. Nikolaew, V. Sgonnik, B. Belenkii, V. Nesterow, V. Krasikow and B. Erussalimsky, *Makromol. Chem.*, 1978, **179**, 2155.
248. S. Dumas, V. Marti, J. Sledz and F. Schué, *J. Polym. Sci., Polym. Lett. Ed.*, 1978, **16**, 81.
249. S. Dumas, J. Sledz and F. Schué, *ACS Symp. Ser.*, 1981, **166**, 463.
250. V. Collet-Marti. S. Dumas, J. Sledz and F. Schué, *Macromolecules*, 1982, **15**, 251.
251. S. Bywater, P. E. Black, D. J. Worsfold and F. Schué, *Macromolecules*, 1985, **18**, 335.
252. P. Brès, M. Viguier, J. Sledz, F. Schué, P. E. Black, D. J. Worsfold and S. Bywater, *Macromolecules*, 1986, **19**, 1325.
253. K. E. Piejko, Ph. D. Thesis, University of Bayreuth, FRG, 1982.
254. K. Hatada, T. Kitayama, S. Okahata and H. Yuki, *Polym. J.*, 1982, **14**, 971.
255. L. Lochmann, D. Doskočilová and J. Trekoval, *Collect. Czech. Chem. Commun.*, 1977, **42**, 1355.
256. P. E. M. Allen, D. O. Jordan and M. A. Naim, *Trans. Faraday Soc.*, 1967, **63**, 234.
257. A. Nishioka, H. Watanabe, K. Abe and Y. Sono, *J. Polym. Sci.*, 1960, **48**, 241.
258. H. Watanabe and Y. Sono, *Kogyo Kagaku Zasshi*, 1961, **64**, 720.
259. H. Watanabe, *Nippon Kagaku Zasshi*, 1961, **82**, 362.
260. H. Watanabe and A. Nishioka, *Kogyo Kagaku Zasshi*, 1962, **65**, 976.
261. H. Watanabe, *Kogyo Kagaku Zasshi*, 1962, **65**, 1104.
262. P. E. M. Allen and A. G. Moody, *Makromol. Chem.*, 1965, **81**, 234.
263. P. E. M. Allen and A. G. Moody, *Makromol. Chem.*, 1965, **83**, 220.
264. B. O. Bateup and P. E. M. Allen, *Eur. Polym. J.*, 1977, **13**, 761.
265. P. E. M. Allen and B. O. Bateup, *Eur. Polym. J.*, 1978, **14**, 1001.
266. P. E. M. Allen, M. C. Fisher, C. Mair and E. H. Williams, *ACS Symp. Ser.*, 1981, **166**, 185; *Polym. Prepr., Am. Chem. Soc., Div. Polym. Chem.*, 1980, **21** (1), 65.
267. K. Hatada, K. Ute, K. Tanaka, T. Kitayama and Y. Okamoto, *Polym. J.*, 1985, **17**, 977.
268. K. Hatada, K. Ute, K. Tanaka, T. Kitayama and Y. Okamoto, in 'Recent Advances in Anionic Polymerization', ed. T. E. Hogen-Esch and J. Smid, Elsevier, New York, 1987, p. 195; *Polym. Prepr., Am. Chem. Soc., Div. Polym. Chem.*, 1986, **27** (1), 151.
269. K. F. Mück, H. Rolly and K. Burg, *Makromol. Chem.*, 1977, **178**, 2773.
270. P. Kilz, Ph. D. Thesis, University of Mainz, FRG, 1988.

271. S. Alev, A. Collet, M. Viguier and F. Schué, *J. Polym. Sci., Polym. Chem. Ed.*, 1980, **18**, 1155.
272. J. P. Pascault, F. Chastrette and Q. T. Pham, *Eur. Polym. J.*, 1976, **12**, 273.
273. M. Viguier, M. Abadie, F. Schué and B. Kaempf, *Eur. Polym. J.*, 1977, **13**, 213.
274. M. Viguier, A. Collet, F. Schué and B. Mula, in 'Recent Advances in Anionic Polymerization', ed. T. E. Hogen-Esch and J. Smid, Elsevier, New York, 1987, p. 249; *Polym. Prepr., Am. Chem. Soc., Div. Polym. Chem.*, 1986, **27** (1), 157.
275. M. Viguier, A. Collet and F. Schué, *Polym. J.*, 1982, **14**, 137.
276. G. Natta, G. Mazzanti, G. Dall'asta and P. Longi, *Makromol. Chem.*, 1960, **37**, 160.
277. G. Natta, G. Mazzanti, P. Longi and G. Dall'asta, *J. Polym. Sci.*, 1961, **51**, 487.
278. A. Soum and M. Fontanille, *Makromol. Chem.*, 1980, **181**, 799.
279. A. Soum and M. Fontanille, *Makromol. Chem.*, 1981, **182**, 1743.
280. A. Soum and M. Fontanille, *ACS Symp. Ser.*, 1981, **166**, 239; *Polym. Prepr., Am. Chem. Soc., Div. Polym. Chem.*, 1980, **21** (1), 23.
281. H. Sawada, 'Thermodynamics of Polymerization', Dekker, New York, 1976.
282. K. J. Ivin, in 'Reactivity, Mechanism and Structure in Polymer Chemistry', ed. A. D. Jenkins and A. Ledwith, Wiley, London, 1974, p. 514.
283. R. M. Joshi and B. J. Zwolinski, in 'Vinyl Polymerization', ed. G. E. Ham, Dekker, New York, 1967, vol. 1, part I, p. 501.
284. F. S. Dainton and K. J. Ivin, *Nature (London)*, 1948, **162**, 705.
285. H. W. McCormick, *J. Polym. Sci.*, 1957, **25**, 488.
286. D. J. Worsfold and S. Bywater, *J. Polym. Sci.*, 1957, **26**, 299.
287. A. Vrancken, J. Smid and M. Szwarc, *Trans. Faraday Soc.*, 1962, **58**, 2036.
288. C. L. Lee, J. Smid and M. Szwarc, *J. Am. Chem. Soc.*, 1963, **85**, 912.
289. K. J. Ivin and J. Léonard, *Eur. Polym. J.*, 1970, **6**, 331.
290. A. Aboudalle, A. Soum, M. Fontanille and T. E. Hogen-Esch, in 'Recent Advances in Anionic Polymerization', ed. T. E. Hogen-Esch and J. Smid, Elsevier, New York, 1987, p. 137; *Polym. Prepr., Am. Chem. Soc., Div. Polym. Chem.*, 1986, **27** (1), 143.
291. S. Bywater, *Trans. Faraday Soc.*, 1955, **51**, 1256.
292. S. Bywater and D. J. Worsfold, *J. Polym. Sci.*, 1962, **58**, 571.
293. W. B. Brown and M. Szwarc, *Trans. Faraday Soc.*, 1958, **54**, 416.
294. A. Miyake and W. H. Stockmayer, *Makromol. Chem.*, 1965, **88**, 90.
295. S. Brunauer, P. H. Emmett and E. Teller, *J. Am. Chem. Soc.*, 1938, **60**, 309.
296. M. van Beylen, S. Bywater, G. Smets, M. Szwarc and D. J. Worsfold, *Adv. Polym. Sci.*, 1988, **86**, 87.

27

Carbanionic Polymerization: Termination and Functionalization

MICHEL FONTANILLE

Université de Bordeaux, France

27.1 SPONTANEOUS TERMINATION

Growing anionic active centres are often unstable. In addition to the attack by the solvent or by impurities contained in the reaction medium, so-called 'isomerization' reactions can occur, leading to new strongly stabilized species. These species are generally unable to reinitiate the polymerization process.

Szwarc and coworkers[1,2] observed that polystyrylsodium in THF is slowly transformed into new species absorbing at higher wavelength, in the visible region. The mechanism established by Szwarc proceeds in two steps: hydride elimination (equation 1) and reaction of residual polystyryl carbanions with the generated phenylallyl proton (equation 2).

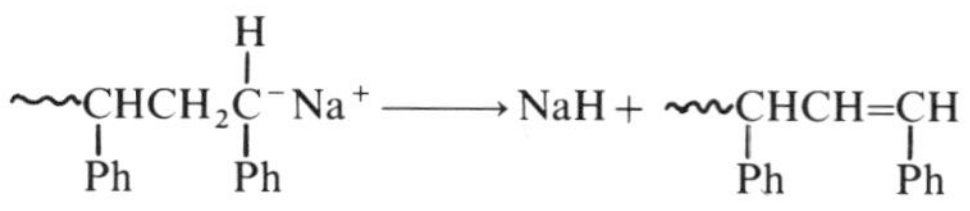

$$\text{\textasciitilde CHCH}_2\overset{\text{H}}{\underset{\text{Ph}}{\text{C}^-}}\text{Na}^+ \longrightarrow \text{NaH} + \text{\textasciitilde CHCH=CH} \qquad (1)$$

$$\text{\textasciitilde CH}_2\overset{\text{H}}{\underset{\text{Ph}}{\text{C}^-}}\text{Na}^+ + \text{\textasciitilde CHCH=CH} \longrightarrow \text{\textasciitilde CH}_2\text{CH}_2 + \text{\textasciitilde C CH=CH} \qquad (2)$$

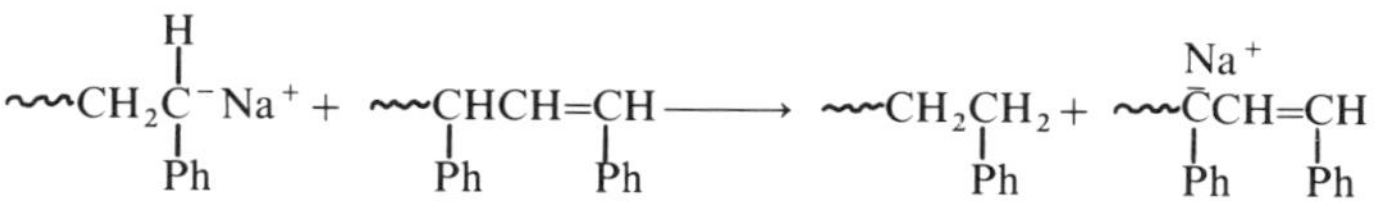

The strong stabilization of the newly formed carbanion prevents reinitiation of the styrene monomer. It has been reported[2] that this reaction is catalyzed by the presence of the alkali metal, and the mechanism shown in Scheme 1 has been suggested to explain this effect.

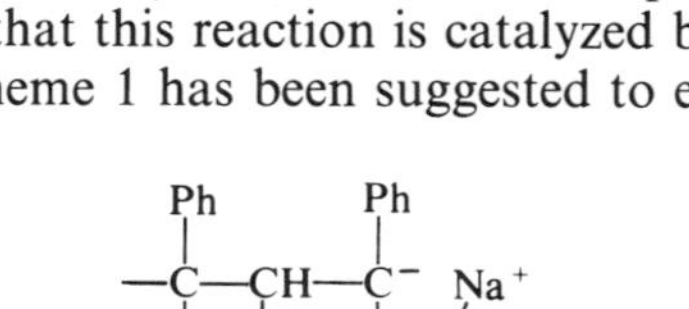

Scheme 1

An analogous transformation occurs in any polar solvent: the lower the concentration of living ends and the higher the polarity of the solvent, the faster the transformation rate. For example, isomerization is almost instantaneous in hexamethyl phosphoramide,[3] a solvent with high dissociating power. These observations are in agreement with a stability much lower for free ions than for ion pairs.

With carbanionic poly(α-methylstyrene), a mechanism identical to that established for styrene cannot be operative. In THF solution and with Na$^+$ as counterion, a wide variety of results are obtained;[4-9] these have been analyzed in a paper by Comyn and Glasse.[10] It was previously suggested[4,5] that degradation proceeds by a hydride elimination followed by a proton abstraction (Scheme 2); species (1) absorbs at 430 nm.

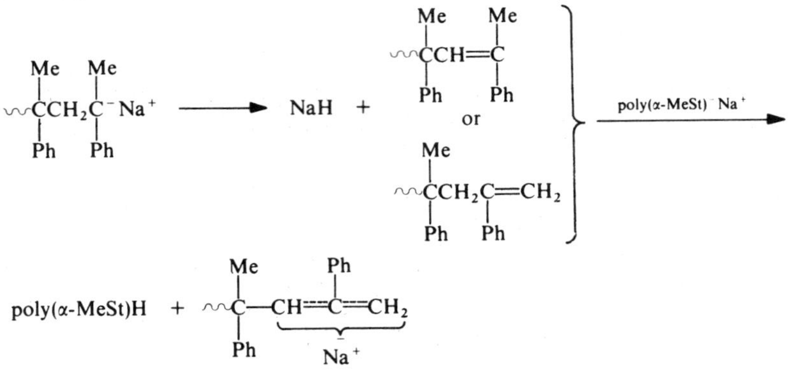

Scheme 2

Comyn and Glasse[10,11] observed that UV light and residual monomer influence the process and interpreted their results as follows (see Scheme 3). Photochemical dimerization of the monomer, catalyzed by the presence of carbanionic species, takes place. Then reaction of the normal living centres with the dimer molecule occurs, followed by transfer to monomer.

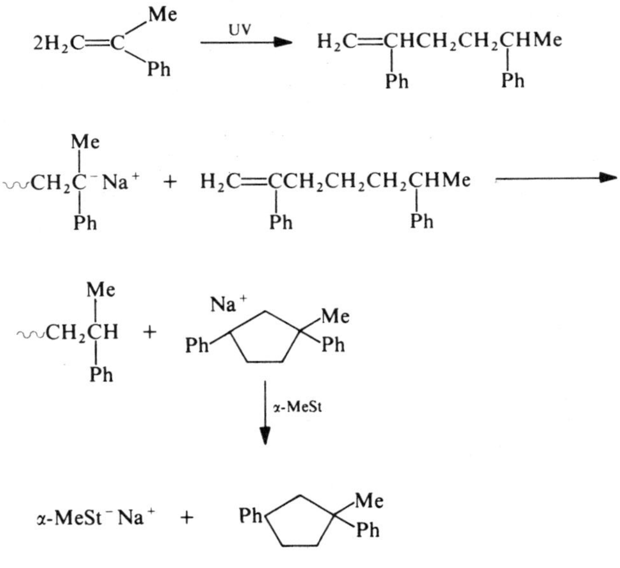

Scheme 3

Simultaneously with this process, a further reduction in the concentration of carbanionic species would result from a hydride elimination, giving indanyl-terminated polymeric chains as previously proposed by Margerison and Nyss (equation 3)[12] with Li$^+$ as counterion.

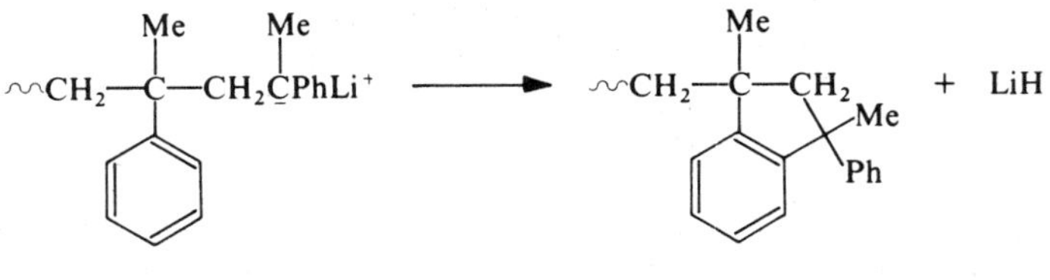

(3)

When polymerized in bulk, poly(α-MeSt)$^-$Li$^+$ (St = styrene) is unstable, even at room temperature.[13] It has been suggested that the reaction also proceeds *via* a hydride elimination. It is important to point out that the stability of the corresponding living species is much greater in the presence of Lewis bases such as tetramethylethylenediamine.[13]

While polystyryllithium and polydienyllithium are stable in hydrocarbons at toom temperature, they decompose on heating. The reaction was studied using polybutadienyllithium. Antkowiak[14] proposed a disproportionation reaction (equation 4). The dilithium derivative was identified by reaction with D$_2$O (equation 5).

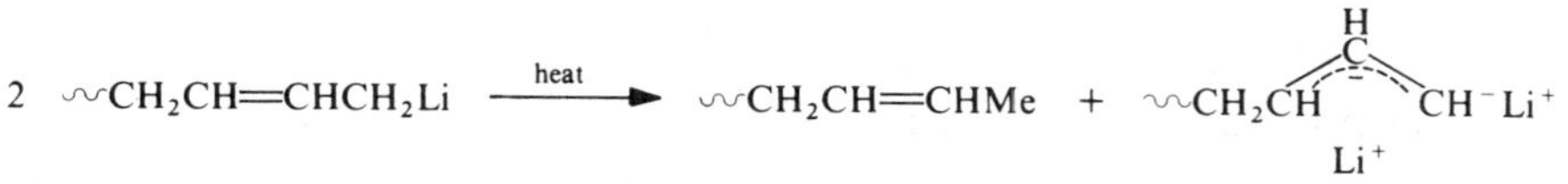

$$2 \ \text{\textasciitilde}CH_2CH=CHCH_2Li \xrightarrow{\text{heat}} \text{\textasciitilde}CH_2CH=CHMe \ + \ \text{\textasciitilde}CH_2\underset{Li^+}{\overset{\overset{\displaystyle H}{C}}{CH}} \diagup \diagdown CH^-Li^+ \tag{4}$$

$$\text{\textasciitilde}CH_2\underset{Li^+}{\overset{\overset{\displaystyle H}{C}}{CH}}\diagup\diagdown CHLi \ + \ D_2O \ \overset{\nearrow \ \text{\textasciitilde}CH_2CH=CHCHD_2}{\searrow \ \text{\textasciitilde}CH_2CHDCH=CHD} \tag{5}$$

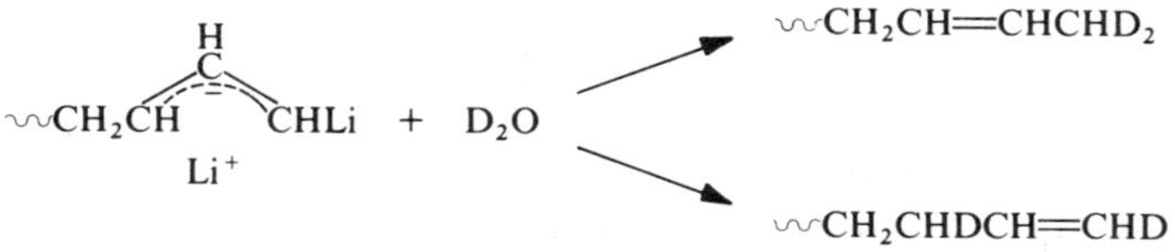

The mechanism proposed by Anderson *et al.* is different.[15] A hydride elimination occurs on heating, as well as metallation and addition reactions (Scheme 4). At higher temperatures, addition of polybutadienyllithium to isolated double bonds in the polymer backbone occurs (equation 6). Metallation of allylic sites (equation 7) is also proposed.

$$\text{\textasciitilde} CH_2CH=CHCH_2Li \rightarrow \text{\textasciitilde} CH=CHCH=CH_2 + LiH$$

$$\text{\textasciitilde} CH=CHCH=CH_2 + \text{\textasciitilde} CH_2CH=CHCH_2Li \rightarrow \text{\textasciitilde} \underset{\underset{\text{\textasciitilde}CH_2CH=CHCH_2}{|}}{CHCH=CHCH_2Li}$$

Scheme 4

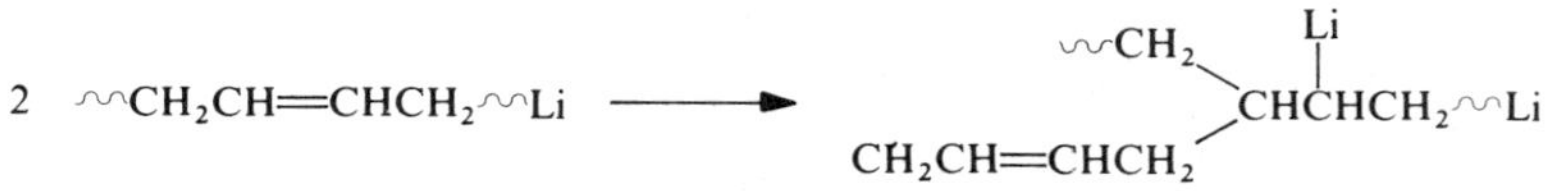

$$2 \ \text{\textasciitilde}CH_2CH=CHCH_2\text{\textasciitilde}Li \longrightarrow \underset{CH_2CH=CHCH_2}{\overset{\text{\textasciitilde}CH_2}{\diagdown}}\overset{Li}{\underset{|}{CHCHCH_2\text{\textasciitilde}Li}} \tag{6}$$

$$\text{\textasciitilde} CH_2CH=CHCH_2\text{\textasciitilde} Li \xrightarrow{\text{poly(Bu}^i)\text{Li}} \text{\textasciitilde} \underset{\underset{Li}{|}}{CHCH=CH}\text{\textasciitilde} Li \tag{7}$$

Nentwig *et al.*[16] confirmed that at 80 °C hydride elimination and reaction of the newly formed diene occur. They showed that if additional monomer is introduced into the reaction medium, it is initiated by the new active centres, leading to a three-armed polydiene, which has been identified.

With polybutadienyl-Li and -Na in THF, isomerization occurs, which is reversible when further monomer is added.[17] This change has been related to a variation in the *cis* and *trans* populations of the active centres. In addition to this isomerization, an unidentified irreversible transformation was also observed,[17] which has never been explained satisfactorily,[18] but could be a reaction with THF.[19-21]

When anionic polymerizations of acrylic and methacrylic monomers are performed at temperatures higher than -75 °C, side reactions occur, which can modify strongly both the molecular weight distribution (MWD) and the polymerization kinetics. These reactions are discussed in Volume 3, Chapter 26.

27.2 TERMINATION AND TRANSFER ON REACTIVE MOLECULES

The very strong basicity of carbanionic active centres is responsible for their termination with electrophilic sites. All proton donors react instantaneously to give the polymeric hydrocarbon (equation 8).

$$\text{\textasciitilde}M_n^-Mt^+ + XH \rightarrow \text{\textasciitilde}M_nH + X^-Mt^+ \tag{8}$$

M = monomer; Mt = metal; X $\equiv$ OH, OR, CO$_2$R

Organolithium derivatives are unstable in ethers due to the too high acidity of protons in the α position being too high with respect to oxygen. A mechanism similar to that proposed by Rembaum et al.[19] for decomposition of ethyllithium in THF is generally accepted (Scheme 5).

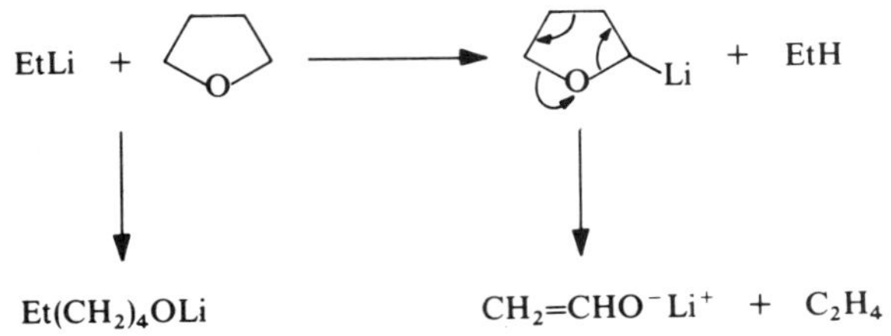

Scheme 5

Polymerizations performed in liquid ammonia and initiated by either Na blue solutions or sodium amide, lead to transfer to solvent (equations 9 and 10).[22] It has been reported[21] that for polymerizations performed in toluene alone, transfer to solvent can also take place (equation 11), the benzyllithium formed reinitiating polymerization.

$$polySt^- Na^+ + NH_3 \rightarrow polyStH + Na^+ H_2N^- \tag{9}$$

$$Na^+ H_2N^- + nSt \rightarrow NH_2(St)_n^- Na^+ \tag{10}$$

$$\tag{11}$$

Such a reaction has been used to synthesize low molecular weight polybutadiene. Polymerizations initiated by lithium or sodium derivatives, in the presence of toluene and THF, lead to phenyl-terminated polybutadiene with $\bar{M}_n$ from 700 to 10^4 (measured by end group analysis) depending on the conditions of polymerization.[23]

Living anionic centres react readily with atmospheric impurities. With oxygen, a complex reaction occurs[24] which has been comprehensively studied by Brossas and co-workers.[25] A coupling reaction, revealed by GPC,[26,27] occurs simultaneously with peroxidation, the mechanism of which was identified using low molecular weight models (Scheme 6). The proportions of various species formed are closely related to the oxidation conditions.[25]

$$\sim M_nLi + O_2 \rightarrow \sim M_n^{\bullet} + O_2^{\bullet-} Li^+$$

$$\sim M_n^{\bullet} + O_2 \rightarrow \sim M_nOO\bullet$$

$$\sim M_n^{\bullet} + O_2^{\bullet-} Li \rightarrow M_nOO^- Li^+$$

$$2\sim M_n^{\bullet} \rightarrow \sim (M_n)_2$$

$$\sim M_n-OO\bullet + \sim M_mLi \rightarrow \sim M_nOOLi + \sim M_m^{\bullet}$$

$$\sim M_nOOLi + \sim M_nLi \rightarrow 2\sim M_nOLi$$

Scheme 6

With carbon dioxide, a nucleophilic addition of carbanionic centres to carbonyl groups occurs, producing the corresponding carboxylic acids and various side products,[28] the proportions of which are dependent on the conditions of carbonation.[28,29] With polystyryllithium in hydrocarbons, the reactions shown in Scheme 7 were considered. On hydrolysis, (a) leads to carboxylated polystyrene, (b) leads to dimeric ketone, and (c) leads to trimeric tertiary alcohol.

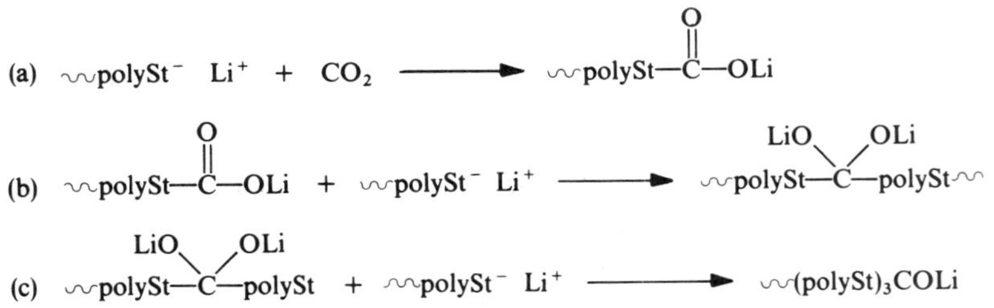

Scheme 7

With potassium as counterion and THF as solvent, Pannel[30] claimed to obtain pure 'monomeric' carboxylated polystyrene. This would also be the case for carbonation of lithiated polymers in benzene–THF mixtures.[31]

27.3 FUNCTIONALIZATION OF CHAIN ENDS

Functionalization of chain ends is of great interest because it permits the preparation of a wide variety of functional oligomers which can be used as prepolymers for subsequent polycondensations. It also enables molecular groups with specific properties to be attached to chain ends. In the preceding section, reactions carried out in the presence of atmospheric components (CO_2, O_2, H_2O) were discussed. By reacting anionic polymers with CO_2 under specific conditions, it is possible to obtain fully end-carboxylated polymers. Such species can also be obtained by reaction of carbanionic species with a cyclic anhydride (equation 12).[32] If a large excess of oxygen is added, quantitative peroxidation can occur.[33] Then coupling by a diacid chloride leads to the corresponding diperester (Scheme 8).

$$\sim M_n^-\,Mt^+ \quad + \quad \text{(cyclic anhydride)} \quad \longrightarrow \quad \sim M_n-\overset{O}{\overset{\|}{C}}-CH_2CH_2-\overset{O}{\overset{\|}{C}}O^-\,Mt^+ \tag{12}$$

$$\sim M_nLi \xrightarrow[\text{excess}]{O_2} \sim M_nO_2Li \xrightarrow{\text{ClCRCCl}} \sim M_n-O-O-\overset{O}{\overset{\|}{C}}-R-\overset{O}{\overset{\|}{C}}-O-O-M_n\sim$$

Scheme 8

Brossas *et al.* have also studied the reaction of carbanionic species with elemental sulfur.[25] They proposed the following reaction scheme (Scheme 9),

$$\sim M_nLi + S_8 \rightarrow [\sim M_nS_8Li]$$

$$[\sim M_nS_8Li] + \sim M_nLi \rightarrow \sim M_nS_xM_n + Li_2S_y$$

$$\sim M_nS_xM_n\sim + RLi \rightarrow \sim M_nS_zM_n\sim + \sim M_nS_{(x-z)}Li$$

$$x + y = 8; \quad x > z$$

Scheme 9

Hydroxy-terminated polymers can be easily obtained by reacting polymeric living-ends with ethylene oxide (Scheme 10).[34] With lithium as counterion, the propagation reaction of ethylene oxide is very slow and it is possible to limit the addition of the heterocycle to one unit. This reaction is nearly quantitative and allows the production, from difunctional initiatiors, of hydroxytelechelic polymers with a functionality close to two.[35] Another method of producing hydroxy-terminated polymers involves reaction with lactones (equation 13).[36]

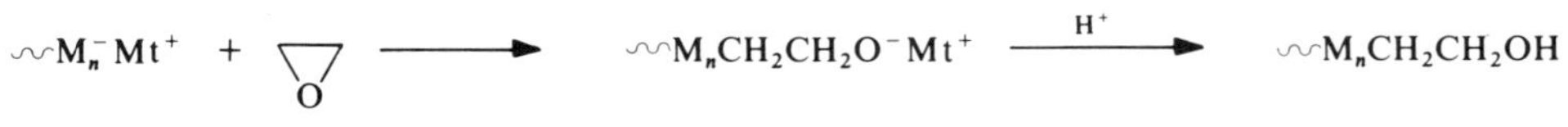

$$\sim M_n^-\,Mt^+ \quad + \quad \text{(ethylene oxide)} \quad \longrightarrow \quad \sim M_nCH_2CH_2O^-\,Mt^+ \xrightarrow{H^+} \sim M_nCH_2CH_2OH$$

Scheme 10

$$\sim M_n^-\,Mt^+ \quad + \quad \text{(lactone)} \quad \longrightarrow \quad \sim M_n \quad O^-\,Mt^+ \tag{13}$$

Primary amine terminal groups are more difficult to obtain. Two methods can be used. The first method entails initiation of polymerization using an initiator bearing a protected amine function, which is then regenerated when polymerization is achieved (Scheme 11).[37] To produce diamino

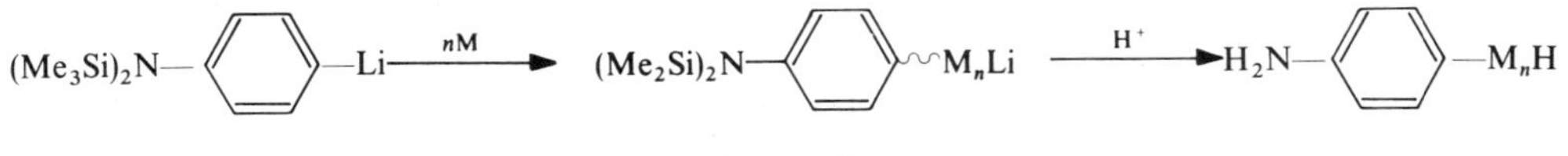

$$(Me_3Si)_2N-\!\!\left\langle\!\!\bigcirc\!\!\right\rangle\!\!-Li \xrightarrow{nM} (Me_2Si)_2N-\!\!\left\langle\!\!\bigcirc\!\!\right\rangle\!\!\sim M_nLi \xrightarrow{H^+} H_2N-\!\!\left\langle\!\!\bigcirc\!\!\right\rangle\!\!-M_nH$$

Scheme 11

telechelic polymers, coupling can be performed using Cl_2SiMe_2 to react with the living ends before deactivation.[37]

The second method involves reaction of carbanionic centres with precursors of amino groups; for example, reaction of a silylated aldimine, which adds to carbanionic centres and can be hydrolyzed to primary amino groups.[38] This method gives excellent yields with lithium derivatives (equation 14).

$$\sim M_n^- Li^+ + \begin{array}{c} Ph \\ H \end{array}\!\!\!>\!\!C=NSiMe_2 \xrightarrow{H_3O^+} \sim M_n\underset{\underset{Ph}{|}}{C}HNH_2 \tag{14}$$

It is also possible to apply to carbanionic living polymers the method described for amination of simple organolithium compounds.[39] Quirk *et al.*[40] used this method to aminate low molecular weight polystyrene and obtained a 92% yield (equation 15).

$$\sim M_n Li \xrightarrow[\text{ii, } H_2O]{\text{i, } MeONH_2-MeLi} \sim M_n NH_2 \tag{15}$$

Various attempts to halogenate polymers have been carried out because halogenated derivatives are often interesting intermediates. Direct halogenation of carbanionic active centres competes with the Wurtz reaction, which produces dimeric chains (equation 16).[41] As already mentioned for carbonation, the extent to which side reactions occur is closely related to reaction conditions. Halogenation can also be obtained by reacting an excess of dihalomethyl derivative with organometallic living ends (equation 17).[42]

$$\sim M_n Li \xrightarrow{X_2} \sim M_n X + (\sim M_n)_2 \tag{16}$$

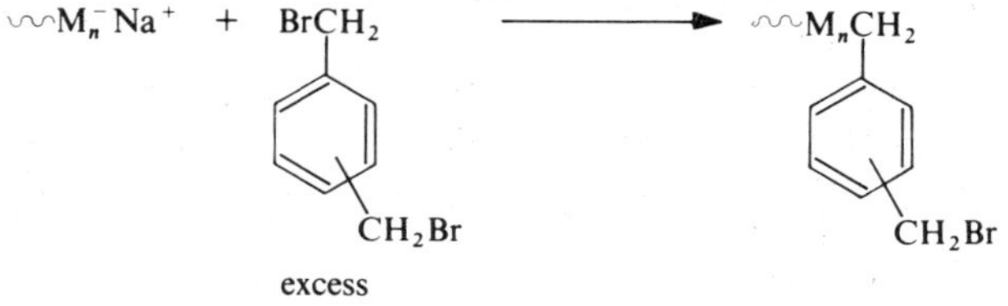

$$\sim\!\!\sim M_n^- Na^+ + BrCH_2\text{---} \longrightarrow \sim\!\!\sim M_n CH_2\text{---} \tag{17}$$

Acid chloride functions can be fixed on chain ends by reaction with a large excess of phosgene (equation 18).[43] Stoichiometry favours coupling.

$$\sim M_n^- Mt^+ + COCl_2 \rightarrow \sim M_n COCl \tag{18}$$

Covalent organometallic end groups have been generated as shown in Scheme 12.[44,45]

$$\sim M_n^- Na^+ + Et_3PbCl \rightarrow \sim M_n PbEt_3$$

$$\sim M_n Li + HgBr_2 \rightarrow \sim M_n HgBr$$

Scheme 12

These functional-ended polymers have been transformed into initiators for subsequent polymerization in order to produce various block copolymers.

ω-Ketone functions have been generated by reacting carbanionic centres with esters,[46] anhydrides, acid chlorides and nitriles (Scheme 13).[47]

Several phosphonated functions have been fixed to one or two ends of living polymers by Brossas and co-workers.[48] The reaction scheme is as shown in Scheme 14.

An N-fixed pyrrolidone group has been obtained by reaction of a carbanion, sufficiently stabilized to prevent the side reactions commonly encountered with acrylics, with 1-methacryloyl-pyrrolidone.[49]

Reaction of carbanionic sites with chlorides allows attachment of various molecules bearing another functional group. Thus, the vinylsilane group has been fixed by reaction of several vinyl polymers on $ClSiMe_2CH=CH_2$.[50] Yields of about 90% have been obtained.

Stoichiometric reaction of carbanionic species with multifunctional molecules such as $COCl_2$, $SOCl_2$, PBr_3, alkylene dihalides, *etc.* produces coupling of polymeric chains.[43]

In the field of functional polymers, special attention has to be given to ending by polymerizable groups.[51] Anionic methods are among the most suitable of those used to produce oligomers

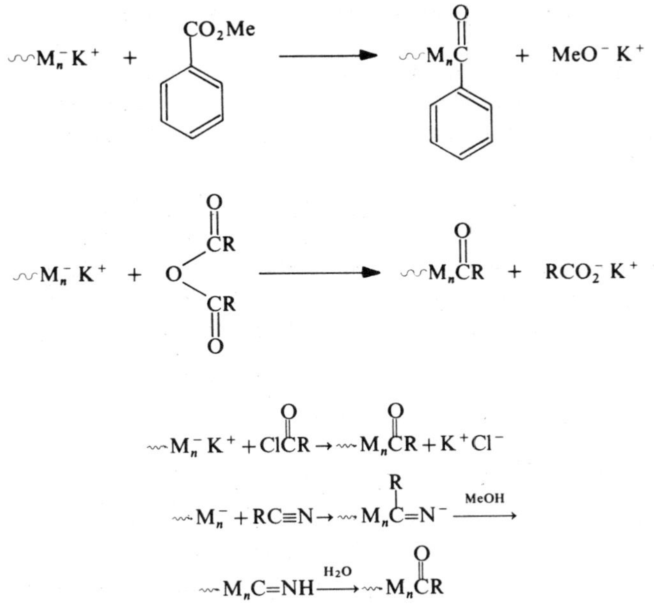

$$\text{\small{$\sim\sim$}}M_n^- K^+ \;+\; \text{[PhCO}_2\text{Me]} \longrightarrow \text{\small{$\sim\sim$}}M_n\overset{\displaystyle O}{\overset{\|}{C}}\text{[Ph]} \;+\; MeO^- K^+$$

$$\text{\small{$\sim\sim$}}M_n^- K^+ \;+\; \left[O\Big\langle\begin{matrix}\overset{\displaystyle O}{\overset{\|}{C}}R\\[4pt]\underset{\displaystyle O}{\underset{\|}{C}}R\end{matrix}\right] \longrightarrow \text{\small{$\sim\sim$}}M_n\overset{\displaystyle O}{\overset{\|}{C}}R \;+\; RCO_2^- K^+$$

$$\text{\small{$\sim\sim$}}M_n^- K^+ + Cl\overset{\displaystyle O}{\overset{\|}{C}}R \rightarrow \text{\small{$\sim\sim$}}M_n\overset{\displaystyle O}{\overset{\|}{C}}R + K^+ Cl^-$$

$$\text{\small{$\sim\sim$}}M_n^- + RC\equiv N \rightarrow \text{\small{$\sim\sim$}}M_n\overset{\displaystyle R}{\overset{|}{C}}{=}N^- \xrightarrow{\text{MeOH}}$$

$$\text{\small{$\sim\sim$}}M_nC{=}NH \xrightarrow{\text{H}_2\text{O}} \text{\small{$\sim\sim$}}M_n\overset{\displaystyle O}{\overset{\|}{C}}R$$

Scheme 13

$$\text{\small{$\sim\sim$}}M_n Li \;+\; ClP(O)(OR)_2 \longrightarrow \text{\small{$\sim\sim$}}M_n P(O)(OR)_2 \;+\; LiCl$$

$$\swarrow \text{\small{M_nLi}} \qquad\qquad \text{\small{M_nLi}} \searrow$$

$$\text{\small{$\sim\sim$}}M_n\overset{\displaystyle OR}{\underset{\displaystyle O}{\overset{|}{\underset{\|}{P}}}}{-}O^- Li^+ \;+\; \text{\small{$\sim\sim$}}M_n R \qquad\qquad \text{\small{$\sim\sim$}}M_n\overset{\displaystyle OR}{\underset{\displaystyle O}{\overset{|}{\underset{\|}{P}}}}{-}M_n\text{\small{$\sim\sim$}} \;+\; ROLi$$

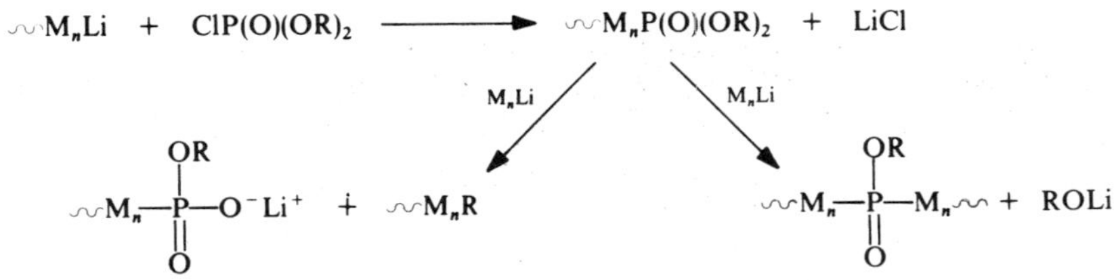

Scheme 14

called macromonomers or macromers by Milkovich *et al.*[52] and which have been defined by Rempp and Franta[53] as linear macromolecules carrying a polymerizable function at their chain end. Several examples of this reaction scheme are given below (equations 19–22). In order to prevent side reactions, lowering of the nucleophilicity of the carbanionic species is sometimes necessary. This can be achieved using an intermediate reaction with either oxirane or 1,1-diphenylethane.[56]

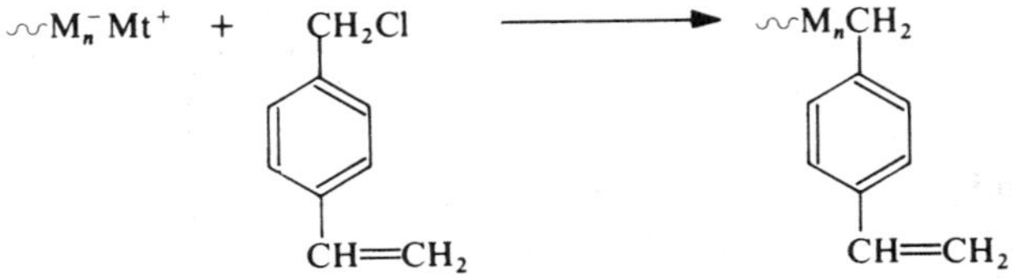

$$\text{\small{$\sim\sim$}}M_n^- Mt^+ \;+\; \left[\begin{matrix}CH_2Cl\\ \big|\\ \text{(C}_6\text{H}_4)\\ \big|\\ CH{=}CH_2\end{matrix}\right] \longrightarrow \text{\small{$\sim\sim$}}M_n\left[\begin{matrix}CH_2\\ \big|\\ \text{(C}_6\text{H}_4)\\ \big|\\ CH{=}CH_2\end{matrix}\right] \qquad (19)[54]$$

$$\text{\small{$\sim\sim$}}M_n^- Mt^+ + Cl{-}\overset{\displaystyle O}{\overset{\|}{C}}{-}CH{=}CH_2 \rightarrow \text{\small{$\sim$}}M_n{-}\overset{\displaystyle O}{\overset{\|}{C}}{-}CH{=}CH_2 \qquad (20)[55]$$

$$\text{\small{$\sim\sim$}}M_n^- Mt^+ + BrCH_2CH{=}CH_2 \rightarrow \text{\small{$\sim$}}M_nCH_2CH{=}CH_2 \qquad (21)[32]$$

$$\text{\small{$\sim\sim$}}M_n^- Mt^+ + Cl{-}\overset{\displaystyle Me}{\underset{\displaystyle Me}{\overset{|}{\underset{|}{Si}}}}{-}CH{=}CH_2 \rightarrow \text{\small{$\sim$}}M_n{-}\overset{\displaystyle Me}{\underset{\displaystyle Me}{\overset{|}{\underset{|}{Si}}}}{-}CH{=}CH_2 \qquad (22)[50]$$

27.4 REFERENCES

1. M. Levy, M. Szwarc, S. Bywater and D. J. Worsfold, *Polymer*, 1960, **1**, 515.
2. G. Spach, M. Levy and M. Szwarc, *J. Chem. Soc.*, 1962, 355.
3. M. Fontanille, unpublished results.

4. A. Okamoto, E. Watanabe and I. Mita, *J. Polym. Sci., Polym. Chem. Ed.*, 1979, **17**, 2483.
5. D. Decker, J. P. Roth and E. Franta, *Makromol. Chem.*, 1972, **162**, 279.
6. B. J. Schmitt, *Makromol. Chem.*, 1972, **156**, 243.
7. J. Comyn and M. D. Glasse, *J. Polym. Sci., Polym. Lett. Ed.*, 1980, **18**, 703.
8. J. Audureau, M. Fontanille and P. Sigwalt, *C.R. Hebd. Seances Acad. Sci., Ser. C*, 1972, **275**, 1487.
9. A. F. Podolsky and A. A. Taran, *J. Polym. Sci., Polym. Chem. Ed.*, 1974, **12**, 2187.
10. J. Comyn and M. D. Glasse, *J. Polym. Sci., Polym. Chem. Ed.*, 1983, **21**, 209.
11. J. Comyn and M. D. Glasse, *J. Polym. Sci., Polym. Chem. Ed.*, 1983, **21**, 227.
12. D. Margerison and V. A. Nyss, *J. Chem. Soc. C*, 1968, 3065.
13. D. Adès, M. Fontanille and J. Léonard, *Can. J. Chem.*, 1982, **60**, 564.
14. T. A. Antkowiak, *Polym. Prepr., Am. Chem. Soc., Div. Polym. Chem.*, 1971, **12** (2), 393.
15. J. A. Anderson, W. J. Kern, T. W. Bethea and H. E. Adams, *J. Appl. Polym. Sci.*, 1972, **16**, 3133.
16. W. Nentwig and H. Sinn, *Makromol. Chem., Rapid Commun.*, 1980, **1**, 59.
17. A. Garton and S. Bywater, *Macromolecules*, 1975, **8**, 694.
18. L. V. Vinogradova, N. I. Nikolaev, V. N. Sgonnnik, B. L. Erussalimsky, G. V. Simtsina, Ch. B. Tsvetanov and I. M. Panayotov, *Eur. Polym. J.*, 1981, **17**, 517.
19. A. Rembaum, S.-P. Siao and N. Indictor, *J. Polym. Sci.*, 1962, **56**, S 17.
20. L. J. Fetters, *J. Polym. Sci., Part B*, 1964, **2**, 425.
21. L.-S. Wang, Ph.D. Thesis, Paris, 1985.
22. J. J. Sanderson and C. R. Hauser, *J. Am. Chem. Soc.*, 1949, **71**, 1595.
23. A. Proni, C. Corno, A. Roggero, G. Santi and A. Gandini, *Polymer*, 1979, **20**, 116.
24. M. Szwarc, *Nature (London)*, 1956, **178**, 1168.
25. J. M. Catala, J. F. Boscato, E. Franta and J. Brossas, *ACS Symp. Ser.*, 1981, **166**, 483.
26. J. M. Catala, G. Reiss and J. Brossas, *Makromol. Chem.*, 1977, **178**, 1249.
27. L. J. Fetters and E. R. Firer, *Polymer*, 1977, **18**, 306.
28. D. J. Wyman, V. R. Allen and T. Altares, *J. Polym. Sci., Part A*, 1964, **2**, 4545.
29. P. Månsson, *J. Polym. Sci., Polym. Chem. Ed.*, 1980, **18**, 1945.
30. J. Pannell, *Polymer*, 1971, **12**, 547.
31. R. P. Quirk and W.-C. Chen, *Makromol. Chem.*, 1982, **183**, 2071.
32. P. Rempp and M. H. Loucheux, *Bull. Soc. Chim. Fr.*, 1958, 1497.
33. J. Brossas, J. M. Catala, G. Clouet and Z. Gallot, *C.R. Hebd. Seances Acad. Sci., Ser. C*, 1974, **278**, 1031.
34. D. H. Richards and M. Szwarc, *Trans. Faraday Soc.*, 1959, **55**, 1644.
35. M. Morton, L. J. Fetters, J. Inomata, D. C. Rubio and R. N. Young, *Rubber Chem. Technol.*, 1976, **49**, 303.
36. P. Rempp, personal communication.
37. D. N. Schulz and A. F. Halasa, *J. Polym. Sci., Polym. Chem. Ed.*, 1977, **15**, 2401.
38. A. Hirao, I. Hattori, T. Sadagawa, K. Yamaguchi and S. Nakahama, *Makromol. Chem., Rapid. Commun.*, 1982, **3**, 59.
39. P. Beak and B. J. Kokko, *J. Org. Chem.*, 1982, **47**, 2822.
40. R. P. Quirk and P.-L. Cheng, *Polym. Prepr., Am. Chem. Soc., Div. Polym. Chem.*, 1983, **24**, 461.
41. F. J. Burgess and D. H. Richards, *Polymer*, 1976, **17**, 1020.
42. F. J. Burgess, A. V. Cunliffe, D. H. Richards and D. P. Sherrington, *J. Polym. Sci., Polym. Lett. Ed.*, 1976, **14**, 471.
43. G. Finaz, Y. Gallot, P. Rempp and J. Parrod, *J. Polym. Sci.*, 1962, **58**, 1363.
44. M. Abadie, F. J. Burgess, A. V. Cunliffe and D. H. Richards, *J. Polym. Sci., Polym. Lett. Ed.*, 1976, **14**, 471.
45. A. V. Cunliffe, G. F. Hayes and D. H. Richards, *J. Polym. Sci., Polym. Lett. Ed.*, 1976, **14**, 483.
46. P. Rempp, V. I. Volkov, J. Parrod and Ch. Sadron, *Bull. Soc. Chim. Fr.*, 1960. 919.
47. V. Lazarewska and P. Rempp, *C.R. Hebd. Seances Acad. Sci., Ser. C*, 1969, **268**, 1841.
48. G. Clouet and J. Brossas, *Eur. Polym. J.*, 1981, **17**, 407.
49. M. Schmitt, E. Franta, P. Rempp and D. Froelich, *Makromol. Chem.*, 1981, **182**, 1695.
50. Ph. Chaumont, J. Herz and P. Rempp, *Eur. Polym. J.*, 1979, **15**, 537.
51. Y. Yamashita, *J. Appl. Polym. Sci.*, 1981, **36**, 193.
52. G. O. Schulz and R. Milkovich, *J. Appl. Polym. Sci.*, 1982, **27**, 4473; G. O. Schulz and R. Milkovich, *J. Polym. Sci., Polym. Chem. Ed.*, 1984, **22**, 1633; R. Milkovich and M. T. Chiang, *US Pat.* 3 989 768 (1976) (*Chem. Abstr.*, 1977, **86**, 55 925).
53. P. F. Rempp and E. Franta, *Adv. Polym. Sci.*, 1984, **58**, 1.
54. R. Asami, M. Takaki and H. Hanahata, *Macromolecules*, 1983, **16**, 628.
55. P. Masson, E. Franta and P. Rempp, *Makromol. Chem., Rapid. Commun.*, 1982, **3**, 499.
56. R. Milkovich, *Polym. Prepr., Am. Chem. Soc., Div. Polym. Chem.*, 1980, **21**, 40.

28

Carbanionic Polymerization: Polymer Configuration and the Stereoregulation Process

STANLEY BYWATER

National Research Council, Ottawa, Canada

28.1 INTRODUCTION

The earliest systematic study of anionic polymerization was carried out in the nineteen thirties by Karl Ziegler and his colleagues. By that time they had already attempted to determine the structures of oligomeric polybutadienes produced by butyllithium initiation in hydrocarbon and ether solvents.[1] The results were, however, later found to be not typical of those of high molecular weight, whose analysis had to await the development of IR[2] and NMR[3] methods of structure determination. Further development of the subject was catalyzed by the announcement in 1956 of a highly *cis*-1,4-polyisoprene produced by lithium metal or alkyl initiation in hydrocarbon solvents.[4] In addition to work on other diene polymers, examination of the structure of vinyl polymers soon led to the discovery of ordered forms of poly(methyl methacrylate)[5] and polystyrene,[6] both produced by lithium alkyl initiation. In the latter case it was later found that in strictly homogeneous solutions no ordered polymer was formed.[7] Since the unique characteristic of anionic polymerization is its ability to exert stereocontrol under homogeneous reaction conditions consideration will generally be restricted to those conditions.

28.2 SPECTROSCOPIC INVESTIGATIONS ON ACTIVE OLIGOMERS AND RELATED ORGANOALKALI METAL DERIVATIVES

The lack of termination or transfer reactions in anionic polymerization, particularly of styrene and diene monomers under suitable conditions, enables the actual active centres at the chain ends to be examined at leisure by spectroscopic techniques. The active centres of these monomers show a characteristic near UV absorption band in the region 280–350 nm.[8] In the diene series λ_{max} depends on counterion and solvent and, particularly important in the context of stereoregulation, it is sensitive to geometrical isomerism (Figure 1).[9] Diene active centres exist in *cis* (**1b**) and *trans* (**1a**) forms as can be easily ascertained by NMR analysis of model compounds (Figure 2).[10, 11]

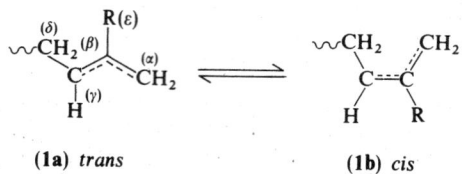

(**1a**) *trans*   (**1b**) *cis*

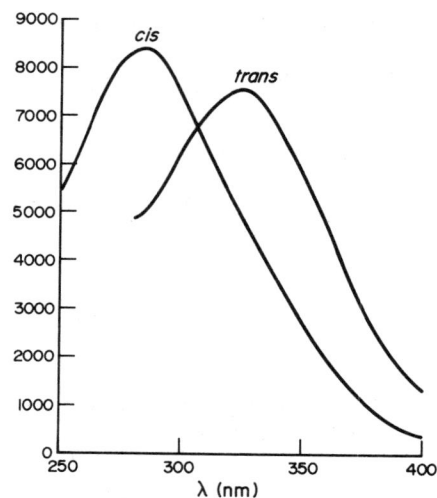

Figure 1   Estimated spectra of *cis* and *trans* forms of the active centres of poly(butadienyl)lithium at $-40\,^{\circ}$C

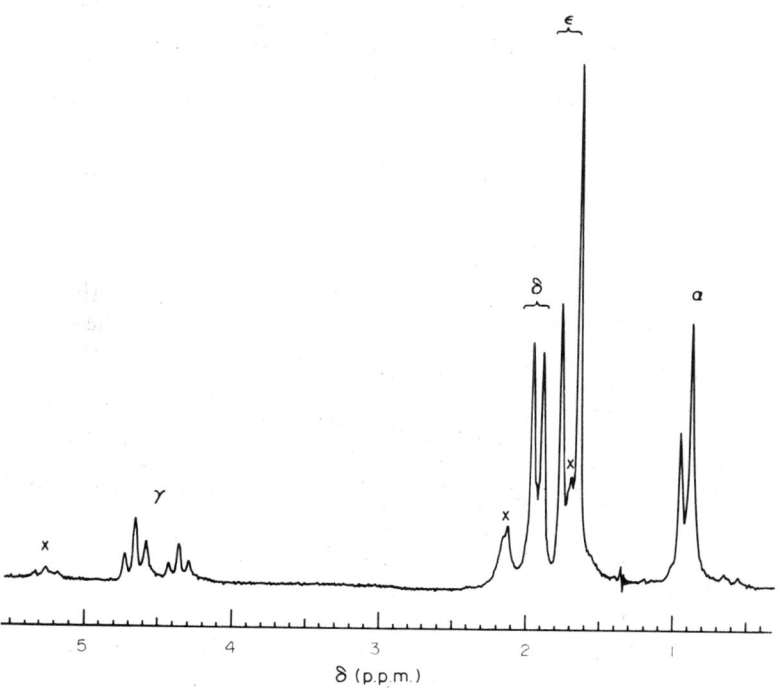

Figure 2 ^{1}H NMR spectrum of Bu^{t}CH$_{2}$CH ===C(R) ===CH$_{2}$Li in benzene (*t*-butyl group deuterated). Separated signals can be seen from *cis* and *trans* forms. Triplets are from the γ proton and singlets from the α protons (rotating). x, traces of dimer. *Cis: trans* ratio, 1:2 (from S. Brownstein, S. Bywater and D. J. Worsford, *Macromolecules*, 1973, **6**, 715)

It is therefore possible to follow the prevailing structure during polymerization and when it is complete. They are not necessarily the same. While optical spectroscopy can be used on real polymerizing systems, NMR spectroscopy can only be used on model oligomers at considerably higher concentrations generally than are used in polymerization. Much of the work has been carried out on models of the type (**2**), simply made in hydrocarbon solvents by the addition of near stoichiometric amounts of *t*-butyllithium to the diene ($Mt^+ = Li$).[10,11] The counterion can be exchanged *via* the corresponding mercury compound[12] and the solvent can be replaced by a more polar one. Individual signals can be seen in either ^1H or ^{13}C NMR corresponding to *cis* and *trans* forms, and the chemical shifts of the allylic α and γ positions shift markedly as counterion and solvent change. In hydrocarbon solvents lithium-based model compounds exist largely in the *trans* form. The preference of models of different counterion is not measurable in such solvents since they are insoluble. In polar solvents, however, the effect of a whole range of counterions can be investigated, but all prefer the *cis* form. As counterion size increases ($Li^+ \rightarrow Cs^+$) in these solvents the chemical shifts of the γ position move upfield and those of the α position move downfield. These observations can be rationalized by the suggestion[12] that charge is being delocalized from the α carbon increasingly as counterion size is increased, the process being essentially complete at K^+. Similar effects are observable for Li^+ as counterion even in non-polar solvents if a chelating diamine such as tetramethylethylenediamine is added in about an equivalent amount to lithium.[13] Corresponding UV spectroscopic changes occur, all these observations being consistent with increased charge delocalization and usually changes in the predominant geometrical isomer.

$$Mt^+$$
$$Bu^tCH_2CH{=\!=\!=}\bar{C}(R){=\!=\!=}CH_2$$

(**2**)

Information derived from model compounds (essentially one-unit active polymers) is, therefore, of great value in the assessment of structural and charge changes at active centres, for on these factors the stereoregulation process must depend to a large extent. If based on *t*-butyl-substituted diene models, the end group is of course related poorly to the penultimate monomer unit in a real growing chain. NMR studies on models with other end groups[14] show that the *cis/trans* preference is much the same for all similar models with a different alkyl group unless a *t*-butyl group is directly attached to the γ carbon, which is not the case studied. The high concentrations necessary for NMR measurements pose a different problem since the aggregation state may be different from that normal in polymerization. In THF at least, however, experiments with ^{13}C-enriched monomers down to $\sim 10^{-3}$ M concentrations show no changes in *cis/trans* preference of lithium compounds. Optical spectroscopy of this and other compounds at much lower concentrations is also consistent generally with the NMR data.

28.3 CONFIGURATION OF POLYDIENES PREPARED IN HYDROCARBON SOLVENTS

The microstructure in lithium-based polymerization of isoprene under these conditions is, as noted earlier, highly *cis* 1,4. About 5% of 3,4 structures are always present. The other alkali metals or their alkyls give a more mixed structure. Early reports were not always in agreement[15] partly since analytical methods were not as well developed, and suffer from the possibility that the reaction was not always homogeneous. More recent results, in which the initiator used was an oligomeric isoprene compound prepared separately in cyclohexane, are more likely to give results appropriate to solution polymerization. As shown in Table 1 the structure is indeed mixed, but 1,4 contents, which decrease abruptly from Li to Na, recover somewhat with use of heavier alkali metal-based initiators.

Table 1 also shows that the polyisoprene is not always as highly *cis* 1,4 as desired. Only at higher monomer to initiator ratios is the polymer produced basically *cis* 1,4 with 5% of 3,4 structures. It can also be seen that polybutadienes produced by lithium-based initiation under similar experimental conditions do not have such high fractions of *cis* units. The original simple stereospecific mechanism,[15] whereby a diene monomer coordinates in the *s-cis* form to the lithium cation while adding to the terminal carbon, is obviously inadequate. Subsequent isomerization of the *cis* active centre thus formed was therefore suggested[16] to occur. It can be shown[17] that for isoprene the new active centre immediately formed is indeed *cis* in configuration, but that it can relax to the more stable

Table 1 Configuration of Polybutadiene and Polyisoprene as a Function of Counterion and Solvent

| Polymerization Conditions | | | | Microstructure | | | | |
Monomer	Solvent	Counterion	Temperature (°C)	cis	trans	1,2	3,4	Ref.
Butadiene	hexane	Li$^+$	20[1]	0.68	0.28	0.04	—	18
Butadiene	hexane	Li$^+$	20[2]	0.30	0.62	0.08	—	18
Butadiene	THF	Li$^+$	0	0.06	0.06	0.88	—	23
Butadiene	THF	Li$^+$	−78	∼ 0	0.08	0.92	—	23
Butadiene	THF	Na$^+$	0	0.06	0.14	0.80	—	23
Butadiene	THF	Na$^+$	−78	∼ 0	0.14	0.86	—	23
Butadiene	THF	K$^+$	0 or −78	0.05	0.28	0.67	—	23
Butadiene	diethyl ether	Li$^+$	0	0.08	0.17	0.75	—	23
Butadiene	diethyl ether	K$^+$	0	0.11	0.34	0.55	—	23
Isoprene	cyclohexane	Li$^+$	30[1]	0.94	0.01	—	0.05	17
Isoprene	cyclohexane	Li$^+$	30[2]	0.76	0.19	—	0.05	17
Isoprene	cyclohexane	Na$^+$	15	—0.44—		0.06	0.50	a
Isoprene	cyclohexane	K$^+$	15	—0.59—		0.05	0.36	a
Isoprene	cyclohexane	Cs$^+$	15	—0.69—		0.04	0.27	a
Isoprene	THF	Li$^+$	30	—0.12—		0.29	0.59	b
Isoprene	THF	Na$^+$	0	—0.11—		0.19	0.70	c
Isoprene	diethyl ether	Li$^+$	20	—0.35—		0.13	0.52	d
Isoprene	diethyl ether	K$^+$	20	− 0.38—		0.19	0.43	d
Isoprene	diethyl ether	Cs$^+$	20	− 0.52—		0.16	0.32	d

Ionization suppressant added in the case of THF.
[1] At monomer/initiator ratio 5×10^4.
[2] At monomer/initiator ratio ~ 17.

[a] A. Essel, Thesis, Université de Lyon, 1974.
[b] S. Bywater and D. J. Worsfold, *Can. J. Chem.*, 1967, **45**, 1821.
[c] Authors unpublished results.
[d] C. J. Dyball, D. J. Worsfold and S. Bywater, *Macromolecules*, 1979, **12**, 819.

trans form if there is a sufficient delay before the next monomer adds. This is likely to occur at low monomer to initiator ratios. In other words the structure of each active centre is only converted to a real in-chain structure at the point where the next monomer adds on. A plausible scheme can be suggested (Scheme 1) which repeats indefinitely. At each stage a competition occurs between isomerization of the active centres (denoted*) and monomer addition. The plausible assumption is made that a *cis** active centre always produces a *cis* 1,4 in-chain polymer structure a small side reaction produces 5% 3,4 structures from either type of chain end. The scheme is susceptible to confirmation by independent estimates of the isomerization constants k_1 and k_{-1}, and the propagation constants k_p^{cis} and k_p^{trans}, the *cis* form reacting eight times faster. The former constants were measured on (a) Bu^tCH_2CH===CH===CH_2Li and (b) $Bu^t(C_5H_8)$-CH_2CH===CH===CH_2Li, model compounds which have markedly different isomerization rates ($\sim 100 \times$). This is the only type of experiment in which the *t*-butyl group was found to have a large effect. In fact (b) isomerizes so rapidly to the stable form ($k \sim 1 \times 10^{-3} s^{-1}$ at $-20\,°C$) that its rate at room temperature had to be estimated from the data on (a), which, as expected, show a typical unimolecular preexponential factor. Figure 3 shows the comparison of experimental microstructure with that calculated as described above. The agreement suggests the mechanism is plausible.

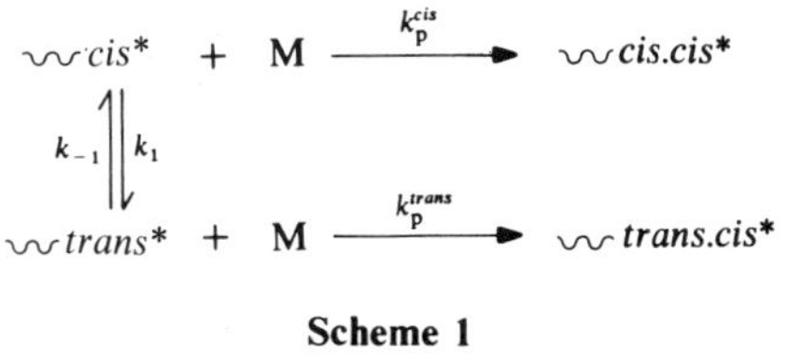

$$\text{\textbackslash\textbackslash} cis^* \quad + \quad M \quad \xrightarrow{k_p^{cis}} \quad \text{\textbackslash\textbackslash} cis.cis^*$$

$$k_{-1} \updownarrow k_1$$

$$\text{\textbackslash\textbackslash} trans^* \quad + \quad M \quad \xrightarrow{k_p^{trans}} \quad \text{\textbackslash\textbackslash} trans.cis^*$$

Scheme 1

Attempts to repeat the investigation using butadiene as monomer failed because the model compounds isomerized too rapidly to measure by the techniques used. This does suggest, however, that the same basic mechanism would hold, because a faster isomerization rate on the above mechanism would automatically produce, under comparable conditions, a higher fraction of *trans* 1,4 units in the polymer as observed. Polybutadienes having almost 90% *cis* 1,4 structural units can only be prepared using bulk monomer and [initiator] $\sim 10^{-5}$ M.[18] A plot of percentage of *trans* structures resembles that shown in Figure 3. The plateau at low monomer to initiator ratios reaches

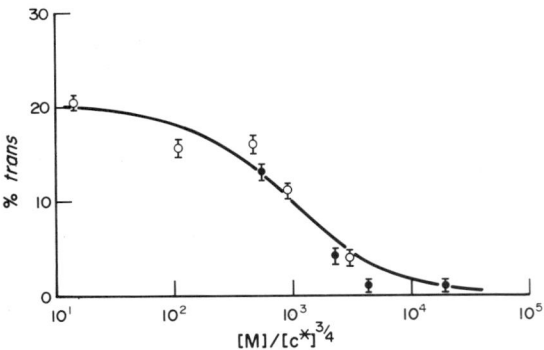

Figure 3 Percentage of *tráns* 1,4 units in polyisoprene prepared by *s*-butyllithium in cyclohexane at 30 °C (10% conversion): ●, experiments at constant initiator; [M], 0.2 to 6 molar. Monomer/initiator ratio increases from left to right (for the significance of the 3/4 power in initiator see the original reference)[17]

~ 62% *trans* units. This is consistent with a lower k_p^{cis} to k_p^{trans} ratio (~ 2) so that the generally higher *trans* fraction in polybutadienes is probably caused by more competition kinetically from *trans* active centres plus a faster isomerization rate of the *cis* centres. The effect of temperature on microstructure is small over the useful temperature range (0–50 °C), but the 1,2 content in butadiene polymerization (but not for isoprene) is sensitive to the lithium alkyl concentration used. Nearly 50% 1,2-polybutadiene can be obtained by working at an initiator concentration of 0.5 M.[19] Naturally this means the polymer molecular weight must be quite low to observe this effect. It has been suggested that this may be caused by a direct reaction of monomer with the aggregated growing centres, which are always the major form of the organolithium compounds in hydrocarbon solvents.[19] At more conventional initiator concentrations the reaction occurs predominantly with dissociated species.

28.4   CONFIGURATION OF POLYDIENES PREPARED IN POLAR SOLVENTS

As can be seen from Table 1, lithium alkyl-based polymerization produces mostly vinyl unsaturation in THF or diethyl ether, in contrast to the almost pure 1,4 polymers formed in non-polar solvents. Polymerizations based on compounds of the heavier alkali metals, however, give products whose microstructure changes far less drastically with solvent type and are more mixed. Similar trends are observed in dioxane,[20] where polymers having as high as 60% 1,4 content can be obtained using Cs^+ as counterion.

Since NMR results on model compounds reported above suggest increased charge delocalization to the γ carbon in polar solvents, higher vinyl contents in these solvents could be due to higher charge at this carbon. The trend with counterion within a solvent type is difficult to explain on this basis, since delocalization in THF is greater, but the vinyl content is lower with the large counterions. Steric blocking of the γ position is a possible complication in that case.

It should be noted that it is possible to interpret the NMR data as indicating covalent bonding of lithium in hydrocarbons and more ionic bonding in polar solvents.[21,22] Presumably again even in non-polar solvents more ionic bonding would occur with heavier alkali metals. The covalent bond localized at the terminal carbon atom would then produce 1,4 polymer and the ionic more delocalized structure, vinyl polymer. The presence of a strong electronic absorption band in either type of solvent shifting only moderately in position and intensity suggests, however, that the fundamental nature of the active centres does not change drastically with solvent. With either interpretation it is not clear why the vinyl content of the polymers should decrease in the series Li^+ to Cs^+, which should be in the direction of a more ionic or a more delocalized ion pair according to the hypothesis used.

Data from the most detailed study of microstructures of polybutadienes prepared in THF[23] are reported in Table 1. Most earlier data in this solvent were obtained without the addition of a common ion salt of high dissociation constant to suppress the free anion contribution to rates and microstructure. The results therefore do not necessarily refer to a reaction on the stated ion-pair. Together with a kinetic[9] and spectroscopic[24] study of the polymerization of butadiene using sodium as counterion, more evidence on the influence of ion pair structure could be obtained. As in the study of polyisoprene microstructure described earlier, the rates of active centre isomerization could be measured and taken into account[25,26] in THF. In the sodium-based system at $\sim 0\,°C$ and above, the active centre near UV absorption was independent of the presence or absence of polymerization. At lower temperatures, however, a long wavelength shift was noted during polymerization which relaxed to the normal absorption when polymerization was completed.[9] This shift can be interpreted as the preferential formation of *trans* active centres in the transition state since measurements on known structures always indicate *trans* ion pairs absorb at longer wavelengths.[9,25,26] The situation is, therefore, exactly opposed to that observed in non-polar solvents, with *trans* kinetically preferred but *cis* thermodynamically stable. Moreover the results imply that around about $-40\,°C$ and below with Na$^+$ as counterion, relaxation to the stabler *cis* forms between individual monomer addition steps was not possible at the particular monomer concentration used. It can be confirmed, by artificially generating oligomeric butadienyl sodium compounds in the *trans* form and measuring their relaxation rate, that this would indeed be so.[26] Experiments with similar Li$^+$ and K$^+$ compounds show that the isomerization rates decrease in the series Li$^+$ > Na$^+$ > K$^+$; potassium compounds relaxing so slowly that under no practical polymerization conditions would any relaxation occur at all. One would therefore expect microstructure changes with temperature of polybutadienes if counterions Li$^+$ or Na$^+$ were used but not for K$^+$. This is in fact what is observed with the first two counterions, *cis* 1,4 in-chain units practically disappear at $-78\,°C$, but not with the latter. In parallel with these observations, the amount of 1,2 structures increases in the first case but not in the last. These results are all consistent with the occurrence of increasing fractions of *trans* active centres as temperature is lowered; these for a given counterion are slightly more 1,2 specific than are *cis* active centres. Superposed, of course, is a counterion effect which, as noted above, increases the 1,4 content of polydienes as the counterion size is increased.

28.5 EFFECT OF SMALL AMOUNTS OF CATION COMPLEXING AGENTS ON POLYDIENE STRUCTURE

Quite small amounts of cation chelating agents, present in amounts approximately equal to lithium concentration, can have a drastic effect on polydiene microstructure even where the bulk medium is non-polar.[27] Most effective, as can be seen from Table 2, are tertiary diamines such as tetramethylethylenediamine (TMEDA)[27,28] and for butadiene polymerization, 1,2-dipiperidino-ethane (DIPIP).[29] Crown ethers and cryptands are somewhat less effective in inducing the formation of high vinyl polydienes. NMR measurements on dienyllithium oligomers at room temperature in the presence of TMEDA or DIPIP show the usual upfield movement of chemical shifts of the γ carbon or hydrogen and corresponding downfield α movements as when polar solvents are added.[30-34] These changes are, however, complete at much lower concentrations of additive, in most cases at an amine to lithium ratio (R) of about two. Only isoprenyllithium oligomers complexed with DIPIP require more diamine.[33] The ^{13}C chemical shifts and optical spectra at this point are similar to those observed with a highly delocalized ion pair (*e.g.* dienylpotassium compounds in THF). The aggregates normally present in non-polar solvents ($\sim M^- Li^+$)$_n$ have then been completely dissociated to monomers solvated by diamine, d, ($\sim M^- Li^+ : d$). This is clear from viscosity measurements on the poly(butadienyl)lithium/TMEDA system[35] and from the first-order dependence of the propagation rate on polymer chain concentration in the poly(isoprenyl)-lithium/TMEDA system.[36] The observations that more than a 1:1 diamine to lithium ratio is often required to complete spectral changes or attain a limiting propagation rate[37-39] does not imply that multiply solvated chain ends, *e.g.* $\sim M^- Li^+ : d_2$, are present at R values ~ 2,[37] but that the mono-diamine complex formation is not complete at $R = 1$. Since the ^{13}C chemical shifts of the diamine are sensitive to the complexing process, a semiquantitative measure of the amount of free diamine can be obtained.[34] Free diamine can be detected by this method at $R = 0.6$ to 0.7. At R values below two, α and γ ^{13}C shifts are averages over a number of rapidly equilibrating species in solution at room temperature. By cooling to $-20\,°C$ in one case[33] signals corresponding to intermediate complexes can be observed. From this and UV spectroscopic measurements at lower concentrations it can be inferred that the series of equilibra involved is shown by Scheme 2, where the tetramer, dimer and monomer referred to are aggregates of the growing active chains. Species T:d and M:d have been suggested previously from other evidence.[40]

While the formation of the dissociated chains M:d requires more than a diamine to lithium ratio of one, the microstructure of the polymer formed has changed to a quite high vinyl content at much lower ratios (Figure 4). This suggests that other solvates as well as M:d are responsible for the production of vinyl units in the polymer chain. In fact it could be shown[38] that the assumption of a

Table 2 Configuration of Polybutadiene and Polyisoprene Prepared in Non-polar Solvents in Presence of Cation Chelating Agents

Monomer	Solvent	Counterion	Temperature (°C)	Complex with	$[Mt^+]$	Fraction			Ref.
						1,4	1,2	3,4	
Butadiene	bulk	Li$^+$	0	[2.1.1][1]	0.2	0.40	0.60	—	a
Butadiene	benzene	Na$^+$	10	[2.2.2][1]	—	0.28	0.72	—	b
Butadiene	benzene	K$^+$	10	[2.2.2][1]	—	0.27	0.73	—	b
Butadiene	cyclopentane	Li$^+$	5	TMEDA	1×10^{-3}	0.16	0.84	—	34
Butadiene	cyclopentane	Li$^+$	5	DIPIP	1×10^{-3}	0.03	0.97	—	38
Isoprene	bulk	Li$^+$	20	[2.1.1][1]	0.2	0.30	0.20	0.50	a
Isoprene	benzene	Na$^+$	20	DCHE[2]	—	0.32	0.24	0.44	b
Isoprene	benzene	K$^+$	20	DCHE[2]	—	0.21	0.60	0.19	b
Isoprene	benzene	Cs$^+$	20	DCHE[2]	—	0.42	0.41	0.17	b
Isoprene	cyclohexane	Li$^+$	22	PMDT[3]	4.5×10^{-4} M	0.52	0.06	0.42	30
Isoprene	cyclohexane	Li$^+$	22	TMEDA	3.5×10^{-4} M	0.31	0.13	0.55	30
Isoprene	cyclohexane	Li$^+$	22	TMEDA	10^{-2} M	0.21	0.12	0.67	c
Isoprene	cyclohexane	Li$^+$	22	DIPIP	10^{-2} M	0.20	0.12	0.68	c

At 1:1 cation/complexing agent. [1] refer to cryptands with conventional numbering. [2] DCHE = dicyclohexylcrown-6. [3] PMDT = pentamethyldiethylenetriamine.

[a] S. Raynal, *J. Macromol. Sci., Chem.*, 1983, **A19**, 1049.
[b] S. Alev, A. Collet, M. Viguier and F. Schué, *J. Polym. Sci., Polym. Chem. Ed.*, 1980, **18**, 1163.
[c] P. Bries *et al.*, *Macromolecules*, 1986, **19**, 1325.

Anionic Polymerization

$$(1) \quad T \ + \ d \ \overset{K_1}{\rightleftharpoons} \ T{:}d \qquad (T{:}d = \text{monosolvated tetramer})$$

$$(2) \quad T{:}d \ + \ d \ \overset{K_2}{\rightleftharpoons} \ 2D{:}d \qquad (D{:}d = \text{monosolvated dimer})$$

$$(3) \quad 2D{:}d \ + \ 2d \ \overset{K_3}{\rightleftharpoons} \ 4M{:}d \qquad (M{:}d = \text{monosolvated monomer})$$

Scheme 2

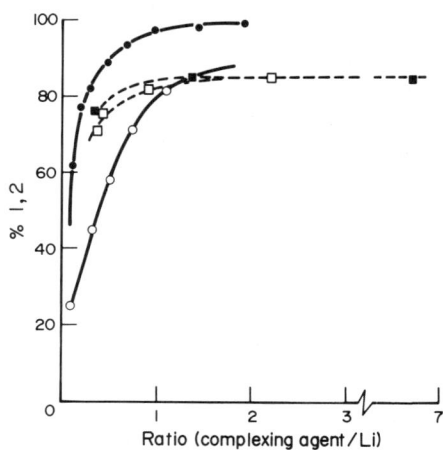

Figure 4 Microstructure of polybutadiene prepared at 5 °C in cyclopentane as a function of diamine to lithium ratio: ——, Li/cyclopentane/DIPIP; – – –, Li/cyclopentene/TMEDA; ■ ●, [Li] ~ 10^{-3} M; ○ □, [Li] ~ 10^{-4} M[34, 38]

single 1,2-producing species in the butadiene:DIPIP system was incapable of explaining the experimental results. A simpler version of Scheme 2, using equilibria (2) and (3) only, was computer fitted to allow D:d and M:d to produce 1,2 polymer units. Two rate constants and two equilibrium constants plus a fifth parameter (% of 1,2 produced by D:d) have to be estimated. Hence alternative schemes cannot be excluded. Nevertheless the choice of $K_2 = 100$ and $K_3 = 7$ allows a fit to the observed spectroscopic (UV), rate and microstructure measurements.

The question arises as to why DIPIP is so effective in producing 1,2-polybutadiene. A plausible answer to this question can be obtained by examining the *cis/trans* population of model butadienyl-lithium oligomers in the presence of TMEDA and DIPIP. In the latter case, the predominantly (75%) *trans* structure of the active centres in non-polar solvents becomes *more trans* as DIPIP is added.[31] TMEDA induces a more mixed structure in butadienyllithium oligomers and a highly *cis* structure in isoprenyllithium oligomers. These results are consistent again with the previously described deduction that *trans* active centres are the best for the formation of vinyl units in the chain.

28.6 STEREOREGULARITY OF POLYSTYRENE AND ITS DERIVATIVES

The formation of, in part, isotactic polystyrene from low temperature butyllithium-initiated polymerization in non-polar solvents was described in the introduction. This requires, however, initiator destruction by moisture.[6, 7] Presumably the isotactic fraction is formed in the presence of lithium hydroxide (or oxide). This could be colloidally dispersed to promote a heterogeneous reaction or homogeneously by simple complexing with the growing active centres. The detailed microstructure of the 'atactic' polymer produced normally by lithium alkyl initiation or by derivatives of the other alkali metals in polar or non-polar solvents is difficult to determine with exactitude. The simple spectrum of highly isotactic polystyrene is well known and even recently that of highly syndiotactic polystyrene.[41] Conventional polymerization methods, however, give polymers showing complex spectra of many overlapping signals. Most early studies suggested that polystyrene prepared by homogeneous anionic polymerization was moderately syndiotactic in structure, generally similar to that produced by classical free radical initiation. Most recent investigations based on studies on model compounds, theoretical calculations, and epimerization studies on isotactic polystyrene, have led to the conclusion that the free radically produced polymer is, in fact, rather random in microstructure,[42] with a racemic diad fraction (P_r) of 0.55. With this base a reinvestigation of a number of polystyrenes produced anionically under homogeneous conditions shows that they have similar structures[43] having P_r values lying between 0.54 and 0.48 if prepared in THF at $-78\,°C$ with initiators based on lithium to cesium. The butyllithium/toluene system gave similar polymers of $P_r = 0.53$. Only a cesium naphthalene-initiated polymer prepared in toluene (which may not be a homogeneous system) had a microstructure contaning more *meso* diads ($P_r = 0.28$).

The microstructure of poly(α-methylstyrene), on the other hand, is easily determined by NMR methods and generally found to be a mixture of hetero- and syndio-tactic triads.[44,45] Even the classical Ziegler–Natta ($AlEt_3/TiCl_4$) type catalyst failed to give isotactic polymer. All configurations are somewhat strained as can be seen by examination of molecular models and as expected from the abnormally low heat of polymerization. It appears that the formation of more than single isolated *meso* diads is particularly difficult. Some polymers are quite highly syndiotactic, P_r approaching 0.95, but these are obtained from cationic initiation. Of the anionic systems studied, the butyllithium/non-polar solvent system initiates polymer having the highest degree of stereoregularity ($P_r \sim 0.8$; Table 5). Initiation of polymerization in THF at $-78\,°C$ with any alkali metal-based compound gives polymer composed largely of hetero- and syndio-tactic triads, the former in slightly larger amount. Since no ionization suppressant was added deliberately, propagation could be largely by the much faster reacting free anions, but this needs to be checked. The lack of microstructure changes between Li^+- and Cs^+-based polymers is certainly consistent with this hypothesis.

In polymerizations in THF initiated by lithium compounds, the polymer microstructure depends on reaction conditions, particularly on initial monomer concentration and temperature.[46] A study of microstructures of polymers produced between $+50$ and $-100\,°C$ with monomer concentrations between 6 M and 0.1 M, led to the suggestion that the changes could be explained in terms of variations in the effective concentrations of racemic and *meso* diads of chain end and penultimate units. The scheme is similar to that described above for diene polymerization, equilibration to the favoured (racemic) structure occurring at low monomer concentrations, but kinetic control being dominant at higher monomer concentrations. The lifetimes calculated for the two types of structure (tens of seconds at $-70\,°C$, seconds at $-50\,°C$) seem much too long for this type of interchange, and indeed no NMR evidence can be obtained[47] from a dimeric model compound of any exchange on that time scale. Indeed counterion shifts, which are all that are required to interchange *meso* and racemic terminal diads, are expected to be orders of magnitude faster.

28.7 STEREOREGULARITY IN ACRYLATE POLYMERIZATION

Most data available refer to methacrylate esters since their microstructure is easily obtained in great detail (Tables 3, 4). Highly isotactic poly(methyl methacrylate) is formed in non-polar solvents with lithium-based initiators and some Grignard reagents. With both types of initiator the isotacticity is dependent on reaction conditions, in the first case since side reactions of monomer and initiator give rise to lithium alkoxides whose presence can modify polymer structure. The figures given in Table 3 refer to initiators whose use leads to the smallest amount of side reactions.[48,49] In the second case, besides alkoxide producing side reactions, the polymer structure depends on the treatment of the Grignard compounds to remove the diethyl ether or THF in which they were made.[50-52] The initiator stoichiometry is seldom that simply corresponding to RMgX. Magnesium alkyls themselves produce rather syndiotactic polymers so the halide component is important. The most useful Grignard-type initiators produce polymers whose structures are relatively insensitive to ether removal. These are, however, in the minority. Many magnesium-based initiators produce more than one independently propagating species, whose relative amounts depend on initiator preparation.[53] The molecular weight distributions are wide and polymer microstructure is not uniform over them. The classical initiator for isotactic acrylate preparation was for many years phenylmagnesium bromide, but more recently *t*-butylmagnesium bromide has gained interest since it can also give near monodisperse polymers.[54,55]

Alkyls of alkali metals other than lithium are insoluble in non-polar solvents so that some heterogeneous reactions may occur when they are used. The microstructure of poly(methyl methacrylate) does become less isotactic if they are used[56] — probably this, however, is a characteristic of

Table 3 Stereoregularity of Poly(methacrylate)s Prepared in Toluene at −78 °C

| | | Microstructure (triads) | | | |
Ester	Initiator	i	h	s	Ref.
Methyl	DPHL[4]	0.87	0.10	0.03	48
Ethyl	DPHL[4]	0.89	0.10	0.01	57
Diphenylmethyl	BuLi	0.99	0.01	0.00	59
Trityl	BuLi	0.96	0.02	0.02	59
Methyl	Bu^{t}MgBr[1]	0.97	0.03	0.00	54
Methyl	Bu^{t}MgBr[1,2]	0.95	0.05	0.00	54
Methyl	Bu$_2^t$Mg	0.02	0.20	0.78	a
Methyl	Bu^{n}MgBr[1]	0.11	0.15	0.74	a
Methyl	Bu^{n}MgCl[1,3]	0.31	0.21	0.48	52
Methyl	Bu^{n}MgCl[1,2]	0.03	0.19	0.78	52
Methyl	Bu^{t}MgCl[1]	0.68	0.09	0.23	52
Methyl	Bu^{t}MgCl[1,2]	0.23	0.16	0.61	52
Methyl (−70 °C)	(PhCH$_2$)$_2$Mg	0.63	0.24	0.13	60

In general oligomeric products removed. [1]'RMgX' added as a solution in diethyl ether. [2]Ether removed. [3]Separable into isotactic and syndiotactic polymers. [4]DPHL = diphenylhexyllithium.

[a] K. Hatada *et al.*, in 'Recent Advances in Anionic Polymerization', ed. T. E. Hogen-Esch and J. Smid, Elsevier, New York, 1987, p. 195.

the counterion. Small amounts of polar solvents added to the mixture destroy the stereoregulating process[57] in lithium-based polymerizations and in solvents such as THF quite syndiotactic polymers can be produced[58] as shown in Table 4. With each methacrylate, the lithium-based initiator gives the most syndiotactic polymer. Exceptionally, however, the bulky trityl ester gives rise to isotactic polymer,[59] but since even free radical initiation produces fairly highly isotactic polymers, this must be caused by steric properties intrinsic to the bulky ester group. Normally, however, the stereostructure in THF under conditions where the counterion is well solvated (or with the free anion itself) resembles that of classical free radically initiated polymers prepared at low temperatures. This is commonly supposed to be determined by steric interferences so that the bulky ester groups on chain end and monomer tend to avoid each other. The presence of more bulky alkali metal counterions, which are less easily solvatable, appears to induce more heterotactic triads, with characteristic non-

Table 4 Stereoregularity of Poly(methacrylate)s Prepared in Polar Solvents

| | | | | Microstructure (triads) | | | |
Ester	Solvent	Counterion	Temperature (°C)	i	h	s	Ref.
Methyl	THF	Li	−85	0.01	0.15	0.84	a
Methyl	THF	Li	−45	0.01	0.22	0.77	58
Methyl	THP	Li	−35	0.06	0.32	0.62	58
Methyl	DME	Li	−57	0.01	0.16	0.83	58
Diphenylmethyl	THF	Li	−78	0.02	0.11	0.87	58
Trityl	THF	Li	−78	0.94	0.04	0.02	58
t-Butyl	THF	Li	−40	0.12	0.49	0.39	58
Methyl	THF	Na	−100	0.05	0.30	0.65	b
Methyl	THF	Na	−51	0.04	0.38	0.58	58
Methyl	THP	Na	−47	0.22	0.52	0.26	58
Methyl	DME	Na	−55	0.02	0.21	0.77	58
Methyl	DME	Na[2.2.2]	−98	0.01	0.19	0.80	58
Methyl	DME	none	−98	0.01	0.20	0.79	58
t-Butyl	THF	Na	−48	0.06	0.65	0.29	58
Methyl	THF	K	−60	0.09	0.52	0.39	58
Methyl	THF	Cs	−53	0.05	0.52	0.42	58
Methyl	DME	Cs	−66	0.03	0.37	0.60	58
Methyl	THF	Cs	−100	0.00	0.40	0.60	c
t-Butyl	THF	Cs	−42	0.04	0.51	0.45	58

[a] H. Jeuck and A. H. E. Müller, *Makromol. Chem., Rapid Commun.*, 1982, **3**, 121.
[b] V. Warzelhahn, H. Höcker and G. V. Schulz, *Makromol. Chem.*, 1978, **179**, 2221.
[c] A. H. E. Müller, H. Höcker and G. V. Schulz, *Macromolecules*, 1977, **10**, 1086.

Bernoullian sequence distributions. Microstructures similar to the latter case are also obtainable in THF by Grignard initiation.[50]

For isotactic polymer formation stereoregulating mechanisms involving interactions of polar groups on monomer, chain end and penultimate groups with the counterion are usually suggested.[8,57,60] Bridging coordination of ester groups on chain end and monomer would align the two to produce a *meso* placement which would be held on bond formation, becoming interaction between chain end and penultimate monomer unit. A repeat monomer addition would transfer the coordination to chain end and monomer *etc.* A static picture of this type would produce a diisotactic polymer if a specifically deuterated monomer was used (*e.g.* 3). The polymer would have all the deuterium atoms on the 'same side' of the polymer (in the extended planar zigzag conformation). This can be termed a *threo*-diisotactic polymer and was indeed observed in low temperature polymerizations with lithium-based initiators in toluene.[57] The addition of small amounts of THF (~ 7 per Li$^+$) which competes for lithium coordination, however, largely produced an inversion of structure to give an *erythro*-diisotactic polymer with deuterium atoms opposed to the ester groups. A more natural approach of the monomer ('esters opposed') must then predominate, followed by chain end rotation to repair the configuration of a *meso* diad (Figure 5). Lithium coordination at the chain end provides the ultimate control mechanism on this scheme, a particular conformation being thermodynamically most stable. NMR experiments on model compounds, monomer or dimer equivalents of growing chains, indicate that a near planar carbanion exists at the chain end which may have a *cis* or a *trans* configuration as observed with diene active centres.[61,62] These interconvert only slowly so that the form produced in the transition state is likely to persist throughout the polymerization. Experiments with a lithio dimer of ethyl methacrylate in toluene (4) show a single ^{13}C signal from the *gem*-dimethyl group on the back monomer unit. This broadens as the temperature is reduced finally separating into two signals at $-55\,^{\circ}$C.[63] If the counterion moves rapidly on the NMR time scale between the two faces of the carbanion a single signal is expected. If the 'flip' takes longer than milliseconds then an asymmetric centre is 'seen' at the carbanion and the Me$_2$ signal will split into two, as indeed it is in the protonated product. These results, if applicable to longer chains, indicate that in toluene slowing of counterion exchange does occur at low temperatures but probably is not slow enough to avoid a number of changes of counterion position between monomer addition steps, *i.e.* the chain end cannot be defined as *meso* or racemic. Only *pro-meso* and *pro-racemic* faces can be described, attack of monomer being at a favourable conformation, an even more dynamic picture than that implied in the figure. Similar schemes to that illustrated have been

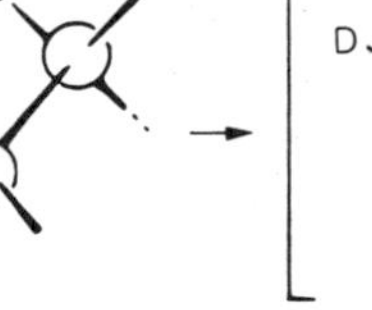

(3)

[Me$_2$C(CO$_2$R)CH$_2$C(Me)COOR] Li$^+$

(4)

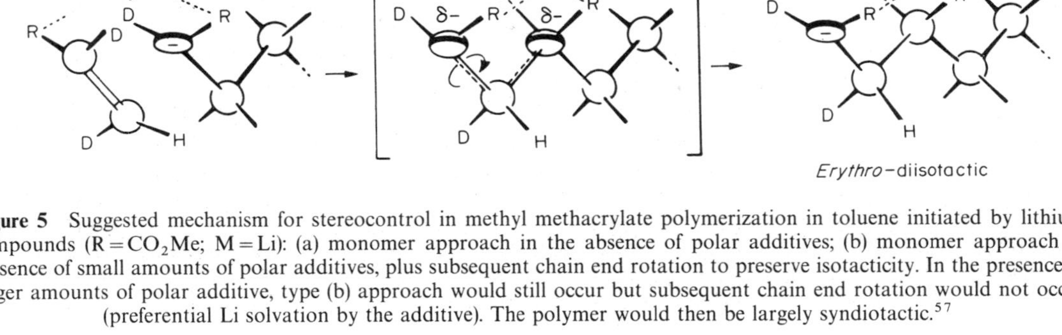

Figure 5 Suggested mechanism for stereocontrol in methyl methacrylate polymerization in toluene initiated by lithium compounds (R = CO$_2$Me; M = Li): (a) monomer approach in the absence of polar additives; (b) monomer approach in presence of small amounts of polar additives, plus subsequent chain end rotation to preserve isotacticity. In the presence of larger amounts of polar additive, type (b) approach would still occur but subsequent chain end rotation would not occur (preferential Li solvation by the additive). The polymer would then be largely syndiotactic.[57]

suggested for magnesium-based initiation.[60] In that case a more covalent type of active centre, and hence fixed stereochemistry of the chain end, is possible. The possible influences of variation of halogen, ether content, and type and stoichiometry of the Grignard reagent ensure, however, that a detailed understanding of this type of polymerization is unlikely to be easily attained.

In ether solvents and with alkali metals as counterions their coordination with carbonyl groups and monomer has also been suggested.[58] If solvation is not particularly strong, displacement of some complexed solvent by monomer has been suggested as it approaches the chain end to produce a transition state with the counterion largely solvated by monomer and chain end (type I). Solvent resolvates the chain end when addition is complete. A different mechanism is suggested for better solvated counterions in solvents such as dimethoxyethane, which are bidentate, or in presence of cryptands (type II). Here a direct attack of monomer without displacement of the solvent is envisaged. This is suggested to produce 'classical free radical' stereochemistry and would be applicable to Li⁺/THF systems as well as to polymerizations in dimethoxyethane (DME). For larger alkali metal cations, not well solvated, mixtures of type I and type II to varying extents are envisaged, the former, being similar to the mechanisms suggested in non-polar solvents, producing more *meso* diads, whereas type II would of course produce racemic diads. No permanent penultimate unit solvation of the counterion is suggested (see Volume 3, Chapter 26 for more details). NMR experiments on the lithio dimer of methyl methacrylate in THF, unlike the situation described above in toluene as solvent, show a single signal from the *gem*-dimethyl groups down to $-70\,°C$.[62] This again suggests that no long term (lifetime > milliseconds) coordination with the back ester group exists.

28.8 STEREOREGULARITY IN VINYLPYRIDINE POLYMERIZATION

Polymers of 2-vinylpyridine prepared even in THF have a quite high *meso* diad content if the counterion is Li⁺ (Table 5). Those of 4-vinylpyridine in contrast have a quite random microstructure. Coordination of the alkali metal with nitrogen atoms of the chain end and penultimate monomer unit have been suggested for the sodium salt in THF, to explain the more than two orders of magnitude lower ion pair dissociation constant compared with polystyryl salts.[64] Evidence for interaction of the lithium cation with in-chain pyridyl units has been obtained more recently by studies of the equilibrium in equation (1), where $M_1^- X^+$ is metallated 2-pyridylethane and $M_N H$ are inactive oligomers of 2-vinylpyridine.[65] Equilibrium constants of the order of 15–30 in THF in favour of $M_N^- X^+$ were obtained if $X^+ = Li^+$, whereas values about 2–4 were obtained if $X^+ = Na^+$, K^+. Even for Li⁺, addition of the strongly Li⁺ complexing cryptand[2.1.1] dropped the K value to ~ 1.5. The most plausible source of the extra stability of the lithio oligomers in THF is indeed cation interaction with a penultimate 2-pyridyl unit.

$$M_1^- X^+ \; + \; M_N H \; \rightleftharpoons \; M_1 H \; + \; M_N^- X^+ \tag{1}$$

Experiments of the type described previously for methyl methacrylate, involving stereospecific deuteration of 2-vinylpyridine, have been used to obtain the mode of monomer approach. For lithio oligomers having up to three monomer units, these were consistent with an ordered monomer approach involving Li⁺ coordination of nitrogen on monomer and active centre, although a variation between 65% and 90%[66] formal *trans* opening was observed. Experiments with 4-vinylpyridine, however suggested a random monomer approach. A plausible mechanism for dimer formation can be written where the predominant *E* (*trans*) isomer of 2-pyridylethane is attacked by monomer in the *s-cis* form, with a preferred approach of the type (5). This should produce a *Z* (*cis*)

(5)

(2)

dimer ion pair (equation 2), which with methyl iodide produces an *erythro*-H product. (The deuteration of 2-vinylpyridine is different from that described earlier in methyl methacrylate so that terms *threo* and *erythro* are interchanged.)

The situation is, however, less clear for the polymerization to trimer and tetramer since NMR evidence indicates that the lithio ion-pairs are formed almost equally in *E* and *Z* forms.[67] This should indicate that *s-cis* and *s-trans* forms of the monomer are equally involved in monomer addition. Lithium coordination of the type illustrated above seems less likely for the *s-trans* rotamer. The *Z* and *E* (*cis* and *trans*) forms of the active centres interconvert extremely slowly,[68] certainly the time scale is orders of magnitude slower than the time scale of monomer addition. This has raised the possibility of obtaining complex (non-Bernoullian) sequence distributions without the normal assumption that centres of asymmetry adjacent to the carbanion affect the monomer addition process. All that is required is interconversion of *E* and *Z* forms of the anion pairs by monomer addition *e.g.*

$$\text{Anion}(E) \quad + \quad \text{Monomer}(\textit{s-cis}) \quad \rightarrow \quad \text{Anion}(Z)$$

and
$$\text{Anion}(Z) \quad + \quad \text{Monomer}(\textit{s-trans}) \quad \rightarrow \quad \text{Anion}(E)$$

Non-Bernoullian sequence distributions are observed in the Li⁺-based polymerization of 2-vinylpyridine in THF[69] and indeed with most other counterions.[70,71] 4-Vinylpyridine and 2-isopropenylpyridine, when polymerized in THF with Li⁺ as counterion, give polymers having simple Bernoullian sequence distributions, but their microstructure is largely atactic and syndiotactic respectively. 2-Isopropenylpyridine, in fact, behaves in a very similar manner to its analogue α-methylstyrene, so that steric interferences obviously predominate in the polymerization of this monomer.

The mechanism of polymerization of 2-vinylpyridine initiated by organomagnesium compounds has also received detailed study.[72] Use of dibenzylmagnesium prepared from the corresponding mercury compound in non-polar solvents, and benzylpicolylmagnesium (**6**) prepared by its reaction with α-picoline, obviate many of the problems associated with Grignard reagents. Side reactions are largely avoided. The mechanisms of stereoregulation suggested, however, still involve the classical mechanisms involving coordination of the MgR group to pyridyl nitrogen atoms on chain end and penultimate unit plus coordination to monomer on its approach. Fixed chain end structures are more plausible with magnesium compounds since NMR experiments on benzylpicolylmagnesium, which is essentially a one-unit growing chain, show far less ionic character than do the lithium analogues.[72] Highly isotactic poly(2-vinylpyridine) is formed in aromatic solvents (Table 5). As with lithium-based systems, even in THF some tendency to retain isotactic triads is observed.

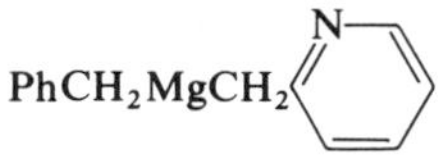

$$\text{PhCH}_2\text{MgCH}_2\text{—}$$

(6)

Table 5 Stereoregularity of Polymers Prepared from Vinylpyridine and its Derivatives and of Poly(α-methylstyrene)

Monomer	Solvent	Counterion	Temperature (°C)	*i*	*h*	*s*	Ref.
			Microstructure (triads)				
2-Vinylpyridine	toluene	Li	−78	0.69	0.21	0.10	a
2-Vinylpyridine	THF	Li	−78	0.46	0.46	0.08	b
2-Vinylpyridine	THF	Li¹	−78	0.56	0.36	0.08	a
2-Vinylpyridine	benzene	Mg	+20	0.92	0.07	0.01	b
4-Vinylpyridine	THF	Li	−78	0.26	0.51	0.23	a
2-Isopropenylpyridine	THF	Li	−78	0.03	0.08	0.89	b
2-Isopropenylpyridine	toluene	Li	−78	0.01	0.15	0.84	b
2-Isopropenylpyridine	toluene	Mg	0	0.05	0.15	0.80	b
α-Methylstyrene	cyclohexane	Li	+4	0.03	0.03	0.67	45
α-Methylstyrene	THF	Li	−78	0.13	0.48	0.39	45
α-Methylstyrene	THF	Cs	−78	0.11	0.49	0.40	45

¹ Lithium tetraphenylboride added.

ᵃ A. H. Soum, S. S. Huang and T. E. Hogen-Esch, *Polym. Prepr., Am. Chem. Soc., Div. Polym. Chem.*, 1983, **24**(2), 149.
ᵇ A. H. Soum, C.-F. Tien and T. E. Hogen-Esch, *Makromol. Chem., Rapid Commun.*, 1983, **4**, 243.

28.9 REFERENCES

1. K. Ziegler, *Angew. Chem.*, 1936, **49**, 499.
2. H. W. Thompson and P. Torkington, *Trans. Faraday Soc.*, 1945, **41**, 246.
3. F. A. Bovey and G. V. D. Tiers, *J. Polym. Sci.*, 1960, **44**, 173.
4. F. W. Stavely *et al.*, *Ind. Eng. Chem.*, 1956, **48**, 778.
5. T. G. Fox *et al.*, *J. Am. Chem. Soc.*, 1958, **80**, 1768.
6. D. Braun, W. Betz and W. Kern, *Makromol. Chem.*, 1960, **42**, 89.
7. D. J. Worsfold and S. Bywater, *Makromol. Chem.*, 1963, **65**, 245.
8. S. Bywater, in 'Comprehensive Chemical Kinetics', ed. C. H. Bamford and C. F. H. Tipper, Elsevier, Amsterdam, 1976, vol. 15, p. 1.
9. A. Garton and S. Bywater, *Macromolecules*, 1975, **8**, 694.
10. W. H. Glaze and P. C. Jones, *Chem. Commun.*, 1969, 1434.
11. F. Schué, D. J. Worsfold and S. Bywater, *J. Polym. Sci., Polym. Lett. Ed.*, 1969, **7**, 821.
12. S. Bywater and D. J. Worsfold, *J. Organomet. Chem.*, 1978, **159**, 229.
13. S. Bywater and D. J. Worsfold, in 'Advances in Elastomers and Rubber Elasticity', eds. J. Lal and J. E. Mark, Plenum Press, New York, 1987, p. 37.
14. M. Schlosser and J. Hartmann, *J. Am. Chem. Soc.*, 1976, **98**, 4674.
15. S. Bywater, *Adv. Polym. Sci.*, 1965, **4**, 66.
16. W. Gerber, J. Hinz and H. Sinn, *Makromol. Chem.*, 1971, **144**, 97.
17. D. J. Worsfold and S. Bywater, *Macromolecules*, 1978, **11**, 582.
18. M. Morton and J. R. Rupert, *ACS Symp. Ser.*, 1983, **212**, 283.
19. S. Bywater, D. J. Worsfold and G. Hollingsworth, *Macromolecules*, 1972, **5**, 389.
20. R. Salle and Q.-T. Pham, *J. Polym. Sci., Polym. Chem. Ed.*, 1977, **15**, 1799.
21. M. Morton and L. J. Fetters, *Rubber Chem. Technol.*, 1975, **48**, 359.
22. G. Fraenkel, A. F. Halasa, V. Mochel, R. Stumpe and D. P. Tate, *J. Org. Chem.*, 1985, **50**, 4563.
23. S. Bywater, Y. Firat and P. E. Black, *J. Polym. Sci., Polym. Chem. Ed.*, 1984, **22**, 669.
24. A. Garton and S. Bywater, *Macromolecules*, 1975, **8**, 697.
25. R. T. McDonald and S. Bywater, *Organometallics*, 1986, **5**, 1529.
26. R. T. McDonald, S. Bywater and P. Black, *Macromolecules*, 1987, **20**, 1196.
27. A. W. Langer, Jr., *Polym. Prepr., Am. Chem. Soc., Div. Polym. Chem.*, 1966, **7**(1), 132.
28. T. A. Antkowiak, A. E. Oberster, A. F. Halasa and D. P. Tate, *J. Polym. Sci., Part A-1*, 1972, **10**, 1319.
29. A. F. Halasa, D. F. Lohr and J. E. Hall, *J. Polym. Sci., Polym. Chem. Ed.*, 1981, **19**, 1357.
30. S. Dumas *et al.*, *Polymer*, 1983, **24**, 1291.
31. D. J. Worsfold *et al.*, *Makromol. Chem., Rapid Commun.*, 1982, **3**, 239.
32. S. Dumas, V. Marti, J. Sledz and F. Schué, *J. Polym. Sci., Polym. Lett. Ed.*, 1978, **16**, 81.
33. S. Bywater, P. Black, D. J. Worsfold and F. Schué, *Macromolecules*, 1985, **18**, 335.
34. R. Werbowyj, S. Bywater and D. J. Worsfold, *Eur. Polym. J.*, 1986, **22**, 707.
35. R. Milner, R. N. Young and A. R. Luxton, *Polymer*, 1983, **24**, 543.
36. S. Dumas, J. Sledz and F. Schué, *ACS Symp. Ser.*, 1981, **166**, 463.
37. J. N. Hay, J. F. McCabe and J. C. Robb, *J. Chem. Soc., Faraday Trans. 1*, 1972, **68**, 1.
38. S. Bywater, D. H. MacKerron, D. J. Worsfold and F. Shué, *J. Polym. Sci., Polym. Chem. Ed.*, 1985, **23**, 1997.
39. L. V. Vinogradova, N. I. Nikolayev and V. N. Zgonnik, *Polym. Sci. USSR (Engl. Transl.)*, 1976, **18**, 2008.
40. A. Davidjan *et al.*, *Makromol. Chem.*, 1978, **179**, 2155.
41. N. Ishihara, T. Seimiya, M. Kuramoto and M. Uoi, *Macromolecules*, 1986, **19**, 2464.
42. H. J. Harwood, T.-K. Chen and F.-T. Lin, *ACS Symp. Ser.*, 1984, **247**, 197.
43. T. Kawamura, T. Uryu and K. Matsuzaki, *Makromol. Chem., Rapid Commun.*, 1982, **3**, 661.
44. S. Brownstein, S. Bywater and D. J. Worsfold, *Makromol. Chem.*, 1961, **48**, 127.
45. T. Kawamura, T. Uryu, T. Seki and K. Matsuzaki, *Makromol. Chem.*, 1982, **183**, 1647.
46. R. Wicke and K.-F. Elgert, *Makromol. Chem.*, 1977, **178**, 3075.
47. Y. Firat and S. Bywater, *Eur. Polym. J.*, 1982, **18**, 265.
48. D. M. Wiles and S. Bywater, *Trans. Faraday Soc.*, 1965, **61**, 150.
49. S. Bywater, *Prog. Polym. Sci.*, 1975, **4**, 27.
50. P. E. M. Allen and C. Mair, *Eur. Polym. J.*, 1984, **20**, 697.
51. P. E. M. Allen, C. Mair, M. C. Fisher and E. H. Williams, *J. Macromol. Sci., Chem.*, 1982, **A17**, 61.
52. K. Hatada *et al.*, *Polym. J.*, 1983, **15**, 771.
53. P. E. M. Allen, *J. Macromol. Sci., Chem.*, 1980, **A14**, 11.
54. K. Hatada *et al.*, *Polym. J.*, 1985, **17**, 977.
55. K. Hatada *et al.*, *Polym. J.*, 1986, **18**, 1037.
56. D. Braun, M. Herner, U. Johnsen and W. Kern, *Makromol. Chem.*, 1962, **51**, 15.
57. W. Fowells, C. Schuerch, F. A. Bovey and F. P. Hood, *J. Am. Chem. Soc.*, 1967, **89**, 1396.
58. A. H. E. Müller, in 'Recent Advances in Anionic Polymerization', ed. T. E. Hogen-Esch and J. Smid, Elsevier, New York, 1987, p. 205.
59. H. Yuki, K. Hatada, T. Niinomi and Y. Kikuchi, *Polym. J.*, 1970, **1**, 36.
60. A. Soum, N. D'Accorso and M. Fontanille, *Makromol. Chem., Rapid Commun.*, 1983, **4**, 471.
61. L. Vancea and S. Bywater, *Macromolecules*, 1981, **14**, 1321.
62. L. Vancea and S. Bywater, *Macromolecules*, 1981, **14**, 1776.
63. S. Bywater, in 'Recent Advances in Anionic Polymerization', ed. T. E. Hogen-Esch and J. Smid, Elsevier, New York, 1987, p. 187.
64. M. Tardi, D. Rougé and P. Sigwalt, *Eur. Polym. J.*, 1967, **3**, 85.
65. C. C. Meverden and T. E. Hogen-Esch, *Makromol. Chem., Rapid Commun.*, 1983, **4**, 563.
66. T. E. Hogen-Esch and W. L. Jenkins, *Macromolecules*, 1981, **14**, 510.

67. I. M. Khan and T. E. Hogen-Esch, *Makromol. Chem., Rapid Commun.*, 1983, **4**, 569.
68. I. M. Khan and T. E. Hogen-Esch in 'Recent Advances in Anionic Polymerization', ed. T. E. Hogen-Esch and J. Smid, Elsevier, New York, 1987, p. 261.
69. S. Strong Huang and T. E. Hogen-Esch, *J. Polym. Sci., Polym. Chem. Ed.*, 1985, **23**, 1203.
70. S. Strong Huang, A. H. Soum and T. E. Hogen-Esch, *J. Polym. Sci., Polym. Lett. Ed.*, 1983, **21**, 559.
71. A. H. Soum and T. E. Hogen-Esch, *Macromolecules*, 1985, **18**, 690.
72. A. H. Soum and M. Fontanille, *Makromol. Chem.*, 1980, **181**, 799; 1981, **182**, 1743; 1982, **183**, 1145.

29

Carbanionic Polymerization: Copolymerization

MICHAEL K. MARTIN

3M Company, St Paul, MN, USA

Because of factors beyond the editors' control, the submission of the manuscript for this chapter was delayed. In order to minimize any delay in publishing 'Comprehensive Polymer Science' as a whole, the coverage of Carbanionic Polymerization: Copolymerization appears at the end of this Volume, commencing on page 861.

30

Carbanionic Polymerization: Hydrogen Migration Polymerization

YUKIO IMANISHI

Kyoto University, Japan

30.1 HYDROGEN MIGRATION POLYMERIZATION OF UNSATURATED CARBOXYLIC ACID AMIDES WITH BASIC INITIATORS

30.1.1 Discovery of Hydrogen Migration Polymerization and Mechanistic Investigations

In 1957 Breslow *et al.*[1] reported the production of poly(β-alanine) in the polymerization of acrylamide with basic initiators. They proposed an acid–base reaction (1) and a Michael-type addition reaction (2) for the initiation reaction, and an intramolecular hydrogen migration reaction (3) and an intermolecular hydrogen migration reaction (4) for the propagation reaction.

$$CH_2{=}CHCONH_2 \; + \; B^- \; \rightleftharpoons \; CH_2{=}CHCON\bar{H} \; + \; BH \tag{1}$$

$$CH_2{=}CHCONH_2 \; + \; B^- \; \rightleftharpoons \; BCH_2\bar{C}HCONH_2 \; \rightleftharpoons \; BCH_2CH_2CON\bar{H} \tag{2}$$

$$\text{\small w}CH_2CON\bar{H} + CH_2{=}CHCONH_2 \longrightarrow \text{\small w}CH_2CONH{-}CH_2\bar{C}HCONH_2 \rightleftharpoons \text{\small w}CH_2CONHCH_2CH_2CON\bar{H} \tag{3}$$

$$CH_2{=}CHCON\bar{H} \; + \; CH_2{=}CHCONH_2 \rightarrow CH_2{=}CHCONH{-}CH_2\bar{C}HCONH_2 \tag{4a}$$

$$CH_2{=}CHCONH{-}CH_2\bar{C}HCONH_2 \; + \; CH_2{=}CHCONH_2 \rightleftharpoons CH_2{=}CHCONH{-}CH_2CH_2CONH_2$$
$$+ \; CH_2{=}CHCON\bar{H} \rightarrow CH_2{=}CHCONH{-}CH_2\bar{C}HCONH{-}CH_2CH_2CONH_2 \tag{4b}$$

Based on experimental evidence it was proposed that in polymerization initiated by Bu^{t}ONa or Bu^{t}OK the main initiation reaction is reaction (1) and that of propagation is reaction (4).[2]

Ogata investigated the polymerization of acrylamide initiated with metal alkoxides[3] or Grignard reagents,[4] and proposed that the main initiation reaction is reaction (2).[5]

Tani *et al.*[6] showed that the differences in the mechanisms proposed by Breslow *et al.* and Ogata depend upon the different types of initiators used. When Bu^{t}ONa is used as initiator, reaction (1) dominates as Breslow *et al.* reported. However, with MeONa as initiator reaction (2) was important, as Ogata proposed.

Trossarelli *et al.*[7-9] showed that the polymerization of acrylamide with Bu^{t}ONa at 100 °C is initiated by reaction (1) and propagated through the stepwise reaction (3) in aprotic solvents, whereas it is initiated by reaction (2) when the conjugate acid of the initiator is used as solvent.

Ogata observed the occurrence of anionic polymerization of acrylamide without the involvement of hydrogen migration (vinyl polymerization, reaction 5) with Bu^{t}ONa in chlorobenzene containing water (1%)[3] or with Grignard reagents in tetrahydrofuran.[4] His observation shows the possibility of concurrent hydrogen migration and vinyl polymerization in base-catalyzed polymerization of acrylamide.

$$\underset{\underset{CONH_2}{|}}{\text{\small$\sim\sim$}\bar{C}H} \quad + \quad \underset{\underset{CONH_2}{|}}{CH_2{=}CH} \quad\longrightarrow\quad \underset{\underset{CONH_2}{|}\qquad\underset{CONH_2}{|}}{\text{\small$\sim\sim$}CH{-}CH_2{-}\bar{C}H} \tag{5}$$

Okamura *et al.*[10] reported concurrent hydrogen migration and vinyl polymerizations of acrylamide initiated by Mg(OR)$_2$, which were dependent on polymerization conditions. Nakayama *et al.*[11-14] showed that the hydrogen migration polymerization competes with the vinyl polymerization, and determined the relative rate quantitatively.

Furthermore, the effects of monomer concentration, initiator concentration, type and polarity of solvent and polymerization temperature were investigated.

Glickson and Applequist[15] showed that branching in poly(β-alanine) was produced in the polymerization of acrylamide with metal *t*-butoxide in aprotic solvent.

30.1.2 Hydrogen Migration Polymerization of Acrylamide Derivatives with Basic Initiators

Okamura *et al.*[16] reported that the hydrogen migration polymerization of methacrylamide with basic initiators proceeded to a high yield of poly(α-methyl-β-alanine) under suitable conditions. Nakayama *et al.*[17] showed that both hydrogen migration (equation 4) and vinyl (equation 5) polymerizations occur in the polymerization of methacrylamide with basic initiators, and that the contribution of the former type of polymerization was higher in the methacrylamide polymerization than in acrylamide polymerization. Yamaguchi and Minoura[18] used optically active alcoholates for the initiation of hydrogen migration polymerization of methacrylamide and obtained optically active poly(α-methyl-β-alanine).

Yokota *et al.*[19] investigated the Bu^{t}ONa-initiated polymerization of *N*-substituted acrylamides at 105 °C, and observed that *N*-cyclohexylacrylamide underwent vinyl polymerization and other *N*-substituted acrylamides yielded poly(α-methyl-β-alanine).

Fujii and Kudo[20] showed that the polymerization of crotonamide with basic initiators was initiated by the mechanism of equation (1) to yield poly(β-methyl-β-alanine) with a terminal unsaturation. Iwakura *et al.*[21] showed that the initiation of base-catalyzed polymerization of cinnamide is of the type shown in equation (1), and that unsaturated oligomers of β-phenyl-β-alanine were formed by hydrogen migration polymerization.

Two types of nucleophilic addition reactions, 1-addition (equation 6) and 2-addition (equation 7), of acrylamide derivatives carrying a β-substituent that is as electrophilic as an amide group are possible.

$$\text{\small$\sim\sim$}CO\bar{N}H \quad + \quad \underset{\underset{CONH_2}{|}}{\overset{(1)\ \ (2)}{CH{=}CHX}} \quad\xrightarrow{\text{1-addition}}\quad \underset{\underset{CONH_2}{|}}{\text{\small$\sim\sim$}CONH{-}\bar{C}HCHX} \tag{6}$$

$$\text{\small$\sim\sim$}CO\bar{N}H \quad + \quad \underset{\underset{CONH_2}{|}}{\overset{(2)\ \ (1)}{XCH{=}CH}} \quad\xrightarrow{\text{2-addition}}\quad \underset{X\ \ \ CONH_2}{\text{\small$\sim\sim$}CONH{-}CH\bar{C}H} \tag{7}$$

There are two types of intramolecular hydrogen migration reactions. One of them is α-migration (equation 8) in which a proton migrates from the amide substituent of an alkenic carbon to which the nucleophilic addition occurred. The other is β-migration (equation 9) in which a proton migrates from the amide substituent of an alkenic carbon adjacent to the nucleophilic addition site.

$$\text{\textbackslash\textbackslash}CONH-\underset{\underset{CONH_2}{|}}{C}H\bar{C}HX \quad \underset{\Longleftarrow}{\overset{\alpha\text{-migration}}{\longrightarrow}} \quad \text{\textbackslash\textbackslash}CONH-\underset{\underset{CO\bar{N}H}{|}}{C}HCH_2X \qquad (8)$$

$$\text{\textbackslash\textbackslash}CONH-\underset{\underset{X\ \ CONH_2}{|\ \ \ |}}{C}H\bar{C}H \quad \underset{\Longleftarrow}{\overset{\beta\text{-migration}}{\longrightarrow}} \quad \text{\textbackslash\textbackslash}CONH-\underset{\underset{X\ \ CO\bar{N}H}{|\ \ \ |}}{C}HCH_2 \qquad (9)$$

Bamford *et al.*[22] polymerized maleamide (**1**; MA), mesaconic acid α-methyl ester β-amide (**2**; MEA) and mesaconamide (**3**; MSA) with Bu^{n}Li or Bu^{t}ONa. The amino acids produced by the hydrolysis of the polymerization products depend on the mechanism of the polymerization. They are summarized in Table 1 and compared with experimental findings. The experimental results show the occurrence of both 1-addition and 2-addition reactions and the formation of α-peptide bonds by α-migration and β-peptide bonds by β-migration.

$$\underset{\underset{CONH_2\ \ CONH_2}{|\qquad\quad|}}{CH=CH} \qquad\qquad \underset{\underset{CONH_2\ \ H}{|\qquad\quad|}}{\overset{\overset{Me\qquad CO_2Me}{|\qquad\quad|}}{C=C}} \qquad\qquad \underset{\underset{CONH_2\ \ H}{|\qquad\quad|}}{\overset{\overset{Me\qquad CONH_2}{|\qquad\quad|}}{C=C}}$$

$$\text{(1)} \qquad\qquad\qquad \text{(2)} \qquad\qquad\qquad \text{(3)}$$

Table 1 Polymerization Mechanism and Experimental Results

Monomer	Polymerization mechanism	Peptide bond	Hydrolysate expected	Found
MA	α-Migration	α	Aspartic acid	Yes
	β-Migration	β	Aspartic acid	Yes
MEA	1-Addition	α	α-Methylaspartic acid	Yes
	2-Addition	β	β-Methylaspartic acid	Yes
MSA	1-Addition/α-Migration	α	α-Methylaspartic acid	Yes
	1-Addition/β-Migration	β	α-Methylaspartic acid	Yes
	2-Addition/α-Migration	α	β-Methylaspartic acid	Yes
	2-Addition/β-Migration	β	β-Methylaspartic acid	Yes

Imanishi *et al.* attempted to synthesize α-polypeptide by the hydrogen migration polymerization of acrylamide derivatives carrying a strongly electron-withdrawing β-substituent, and observed that about one third of the repeating units in poly(*trans-p*-nitrocinnamide) were peptide bonds containing α-peptide bonds,[23,24] that *trans*-cinnamide produced polypeptide without containing α-peptide bonds,[23] and that about one fourth of the repeating units in poly(*trans-β*-chloroacrylamide) were peptide bonds without definite evidence for the presence of α-peptide bonds.[23]

Maleimides are regarded as β-substituted acrylamides. Tawney *et al.*[25] polymerized maleimide in ethanol with metal ethoxides as initiators, and showed the occurrence of vinyl polymerization. Kojima *et al.*[26] reported that the polymerization of maleimide with basic initiators in dimethylformamide consists of hydrogen migration (75–85%) and vinyl (15–25%) polymerizations. Nakayama and Smets[27] reported that the polymerization of maleimide with Bu^{t}ONa in dimethylformamide at 20 °C yielded copolymers containing repeating units resulting from hydrogen migration polymerization (70–75%).

30.2 HYDROGEN MIGRATION POLYMERIZATION OF UNSATURATED CARBOXYLIC ACID AMIDES OTHER THAN ACRYLAMIDE DERIVATIVES WITH BASIC INITIATORS

Asahara and Yoda[28] reported that in the polymerization of *p*-vinylbenzamide with basic initiators copolymers consisting of hydrogen migration polymerization and vinyl polymerization units were formed, and that the former polymerization was completely suppressed by the addition of LiCl to the polymerization solution. Nakayama *et al.*[29] also observed that the contribution of the hydrogen migration polymerization to the base-catalyzed polymerization of *p*-vinylbenzamide was low, and that it did not occur under suitable conditions.

Nakayama *et al.*[30] found that the base-catalyzed polymerization of vinylacetamide involved the isomerization of the allyl group, yielding poly(2-methyl-β-alanine).

30.3 HYDROGEN MIGRATION POLYMERIZATION OF UNSATURATED COMPOUNDS WITHOUT AMIDE GROUPS WITH BASIC INITIATORS

Iwatsuki *et al.*[31] reported the formation of polyketones in the hydrogen migration polymerization of methyl vinyl ketone with metal alkoxides as initiators.

Saegusa *et al.*[32] discovered the formation of poly(β-propiolactone) (polyester) in the polymerization of acrylic acid with triphenylphosphine *via* the formation of a macrozwitterion and probably intermolecular hydrogen migrations. Saegusa *et al.*[33] further found the formation of poly(ester–ether) in the base-catalyzed polymerization of 2-hydroxyethyl methacrylate involving intra- or intermolecular hydrogen migrations.

30.4 HYDROGEN MIGRATION POLYMERIZATION IN ALTERNATING COPOLYMERIZATION OF NUCLEOPHILIC AND ELECTROPHILIC MONOMERS

The discovery by Saegusa and his coworkers that on mixing a nucleophilic compound M_N, *e.g.* 2-oxazolines, and an electrophilic compound M_E, *e.g.* β-propiolactone a betaine is formed that induces uncatalyzed alternating copolymerization by repeating the addition to macrozwitterion is important and far reaching. If M_E bears a mobile hydrogen, the uncatalyzed alternating copolymerization should be accompanied by hydrogen migrations. This type of copolymerization has been observed for the following pairs of M_N and M_E: 2-oxazoline (M_N) and acrylic acid (M_E), yielding poly(amine–ester);[34] 2-benzyliminotetrahydrofuran (M_N) and acrylic acid (M_E), yielding poly-(amide–ester);[35] 1,3,3-trimethylazetidine (M_N) and acrylic acid (M_E), yielding poly(amine–ester);[36] several cyclic imino ethers (M_N) and acrylamide (M_E), yielding poly(amide–imidate);[37] ethylene phenylphosphonite (M_N) and acrylic acid or acrylamide (M_E), yielding poly(phosphinate–ester) or poly(phosphinate-amide);[38] and several cyclic imino ethers (M_N) and acrylic acid (M_E), yielding poly(amide–ester).[39]

30.5 REFERENCES

1. D. S. Breslow, G. E. Hulse and A. S. Matlack, *J. Am. Chem. Soc.*, 1957, **79**, 3760.
2. L. W. Bush and D. S. Breslow, *Macromolecules*, 1968, **1**, 189.
3. N. Ogata, *Bull. Chem. Soc. Jpn.*, 1960, **33**, 906.
4. N. Ogata, *J. Polym. Sci.*, 1960, **46**, 271.
5. N. Ogata, *Makromol. Chem.*, 1960, **40**, 55.
6. H. Tani, N. Oguni and T. Araki, *Makromol. Chem.*, 1964, **76**, 82.
7. L. Trossarelli, M. Guaita and G. Camino, *Makromol. Chem.*, 1967, **105**, 285.
8. L. Trossarelli, M. Guaita and G. Camino, *J. Polym. Sci., Part C*, 1969, **22**, 721.
9. G. Camino, M. Guaita and L. Trossarelli, *Makromol. Chem.*, 1970, **136**, 155.
10. S. Okamura, T. Higashimura and T. Seno-o, *Kobunshi Kagaku*, 1963, **20**, 364.
11. H. Nakayama, T. Higashimura and S. Okamura, *Kobunshi Kagaku*, 1966, **23**, 433.
12. H. Nakayama, T. Higashimura and S. Okamura, *Kobunshi Kagaku*, 1966, **23**, 439.
13. H. Nakayama, T. Higashimura and S. Okamura, *Kobunshi Kagaku*, 1966, **23**, 537.
14. H. Nakayama, T. Higashimura and S. Okamura, *Kobunshi Kagaku*, 1967, **24**, 42.
15. J. P. Glickson and J. Applequist, *Macromolecules*, 1969, **2**, 628.
16. S. Okamura, Y. Oishi, T. Higashimura and T. Seno-o, *Kobunshi Kagaku*, 1962, **19**, 323.
17. H. Nakayama, Y. Yamasawa, T. Higashimura and S. Okamura, *Kobunshi Kagaku*, 1967, **24**, 296.
18. K. Yamaguchi and Y. Minoura, *J. Polym. Sci., Part A-1*, 1972, **10**, 1217.
19. K. Yokota, M. Shimizu, Y. Yamashita and Y. Ishii, *Makromol. Chem.*, 1964, **77**, 1.
20. K. Fujii and S. Kudo, *Kobunshi Kagaku*, 1964, **21**, 613.
21. Y. Iwakura, N. Nakabayashi, K. Sagara and Y. Ichikura, *J. Polym. Sci., Part A-1*, 1967, **5**, 675.
22. C. H. Bamford, G. C. Eastmond and Y. Imanishi, *Polymer*, 1967, **8**, 651.
23. Y. Imanishi, T. Andoh and S. Okamura, *Kobunshi Kagaku*, 1968, **25**, 708.
24. Y. Imanishi, T. Andoh and S. Okamura, *J. Polym. Sci., Part A-1*, 1969, **7**, 773.
25. P. O. Tawney, R. H. Snyder, R. P. Conger, K. A. Leibbrand, C. H. Stiteler and A. R. Williams, *J. Org. Chem.*, 1961, **26**, 15.

26. K. Kojima, N. Yoda and C. S. Marvel, *J. Polym. Sci., Part A-1*, 1966, **4**, 1121.
27. Y. Nakayama and G. Smets, *J. Polym. Sci., Part A-1*, 1967, **5**, 1619.
28. T. Asahara and N. Yoda, *Polym. Lett.*, 1966, **4**, 921.
29. H. Nakayama, T. Sogo, T. Higashimura and S. Okamura, *Bull. Chem. Soc. Jpn.*, 1968, **41**, 520.
30. H. Nakayama, T. Higashimura and S. Okamura, *J. Macromol. Sci., Chem.*, 1968, **A2**, 53.
31. S. Iwatsuki, Y. Yamashita and Y. Ishii, *Polym. Lett.*, 1963, **1**, 545.
32. T. Saegusa, S. Kobayashi and Y. Kimura, *Macromolecules*, 1974, **7**, 256.
33. T. Saegusa, S. Kobayashi and Y. Kimura, *Macromolecules*, 1975, **8**, 950.
34. T. Saegusa, S. Kobayashi and Y. Kimura, *Macromolecules*, 1974, **7**, 139.
35. T. Saegusa, Y. Kimura, K. Sano and S. Kobayashi, *Macromolecules*, 1974, **7**, 546.
36. T. Saegusa, Y. Kimura, S. Sawada and S. Kobayashi, *Macromolecules*, 1974, **7**, 956.
37. T. Saegusa, S. Kobayashi and Y. Kimura, *Macromolecules*, 1975, **8**, 374.
38. T. Saegusa, Y. Kimura, N. Ishikawa and S. Kobayashi, *Macromolecules*, 1976, **9**, 724.
39. T. Saegusa, Y. Kimura and S. Kobayashi, *Macromolecules*, 1977, **10**, 236.

31

Anionic Ring-opening Polymerization: General Aspects and Initiation

TEIJI TSURUTA
Science University of Tokyo, Japan
and
YUHSUKE KAWAKAMI
Nagoya University, Japan

31.1 INTRODUCTION AND GENERAL ASPECTS

Many cyclic compounds can be polymerized *via* a cationic mechanism, but relatively few by an anionic or a nucleophilic mechanism. Most research on ring-opening polymerization has been focused on the modes of polymerization, and some comprehensive reviews have already appeared.[1-3]

The ring-opening polymerization of a heterocyclic monomer can be generally described as shown in equation (1).

$$n \quad \bigcirc_X \longrightarrow \left(\diagdown \diagup X \right)_n \tag{1}$$

Polymerizability of cyclic monomers is influenced by the ring size, the atoms constituting the ring, the initiator and the reaction conditions. Thermodynamic considerations should be examined first in order to discuss the polymerizability of monomers. Kinetic considerations are also crucial, especially in discussing the polymerizability of cyclic monomers by an anionic or nucleophilic mechanism. Typical examples of the thermodynamic polymerizability of monomers are shown in Table 1.[1-21]

The polymerizability of a monomer is basically determined by the free energy change in the polymerization reaction, namely

$$\Delta G_p = \Delta H_p - T\Delta S_p \tag{2}$$

where ΔG_p, ΔH_p, ΔS_p and T are free energy, enthalpy and entropy changes in the polymerization, and absolute temperature (in K), respectively.

The free energy change must be negative to cause polymerization. When the ring size is small, the polymerizability is mainly controlled by the enthalpy term, which is determined by (a) strain in the bond angle; (b) repulsion between neighboring atoms; and (c) steric repulsion between quasi-axial atoms. In the case of three- or four-membered ring systems, the strain in bond angle is the principal factor which determines the polymerizability. In five-membered ring systems, the repulsion between neighboring eclipsed hydrogens is the principal factor that opens the ring. In six-membered ring

Table 1 Thermodynamic Polymerizability of Cyclic Monomers $\lfloor(CHR)_nX\rfloor$ Containing Functional Group X

Functional group X	Ring size[a] 3	4	5	6	7	8	9	Ref.
—O—	P	P	P	N	P			3, 7, 11
—S—	P	P	N	N				3, 7
—NH—	P	P	N	N				3, 7
—OCH$_2$O—	U	U	P	N	P	P	P	3, 4, 12
—SCH$_2$S—	U	U	P	N				2
—CO— (C=O)	U	P	P	P	P			3, 4, 10
—CNH— (C=O)	U	P	P	P	P	P	P	3, 9, 13
—OCO— (C=O)	U	U	N	P	P			4, 14
—HNCO— (C=O)	U	U	N	P	P			10
—COC— (O O)	U	U	N	N	P	P	P	8
—CNHC— (O O)	U	U	N	N				3, 10
—HNCNH— (C=O)	U	U	P	N	P			10
—COCNH— (O O)	U	U	P					3
—SS—	U	P	P	P	P	P	P	5, 11
—CH=CH—	U	P	P	N	P	P		3
—SiMe$_2$OSiMe$_2$—		U	P					21
—OP— (R)		U	P					15
—OPO— (R)		U	P					16
—OPO— (O—)		U	P	P				17
—OPO— (OR / O)	U	U	P	P	P	P		18—20

[a] P: polymerized; N: not polymerized; U: compounds are unknown; blank: not determined; R: H or alkyl.

systems, these strains are negligibly small, and monomers usually lack polymerizability since the free energy change does not become negative. In the case of cyclic siloxanes, the six-membered ring has some strain and can be polymerized. Steric repulsion between quasi-axial atoms becomes important in ring systems having more than six atoms. In systems with rings bigger than fourteen- or fifteen-membered, this enthalpy term becomes unimportant; instead, entropy change by ring opening becomes the controlling factor.[6–8, 11]

When the free energy change is small, the equilibrium between monomer and polymer cannot be completely on the polymer side and the concentration of monomer at equilibrium becomes high.

The equilibrium concentration of monomer M_e can be expressed by equation (3).[8]

$$\ln M_e \;=\; \Delta H_p / RT \;-\; \Delta S_p / R \qquad\qquad (3)$$

In the case of cyclic siloxane polymerization, attack of propagating silanolate on an already formed siloxane bond in the polymer molecule, called a back-biting reaction, would lead to formation of the cyclic oligomer, and finally an equilibrium between monomer, polymer and cyclic oligomer would be reached. This type of equilibrium between a monomer and its polymer or cyclic oligomer is a consequence of the small free energy difference between monomer and polymer or cyclic oligomer.[8] In the initiation, the fragmentation of the monomer to the silanolate unit may also take place concurrently, depending on the ratio of the initiator to the monomer concentration (Scheme 1).[22]

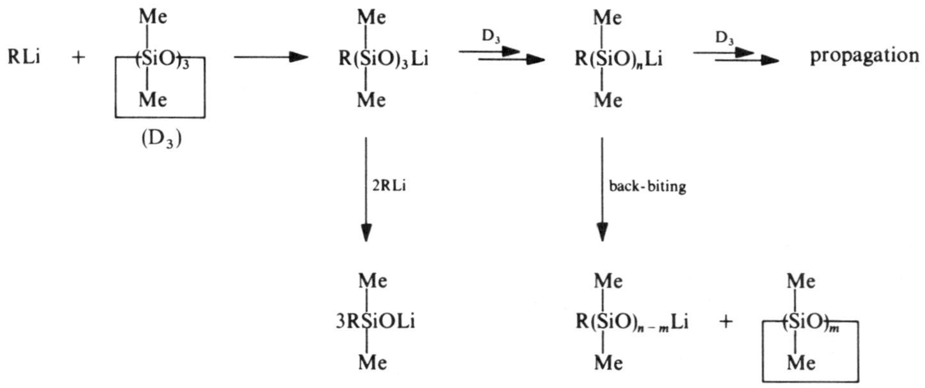

Scheme 1

The chain–ring equilibrium can be treated by the Jacobson–Stockmayer theory.[23] There are extensive studies by Flory and Semlyen[24] on the polymerizability of cyclic siloxanes and the chain–ring equilibrium. Such a situation is often seen for five- or six-membered cyclic monomers whose ΔH_p values are considerably smaller than those of smaller ring monomers.[25]

Bicyclic monomers generally show higher reactivity in the polymerization because of the higher strain energy of the bicyclic system than that of the ordinary monocyclic system. Bicyclo[1.1.0]-butane derivatives were reported to react with various reagents to give cyclobutane derivatives through the release of high strain energy by ring opening, and they can even be polymerized radically or anionically.[26,27] The initiation reaction of bicyclo[1.1.0]butane carbonitrile by butyl-lithium is the nucleophilic attack of the butyl group to open the ring.[26]

In bicyclo[2.2.1], [2.2.2] and [3.2.2] systems, the cyclohexane ring is forced to take a boat form in the monomer. The release of strain by the change in conformation from boat form to chair form is the driving force for the polymerization. Less strained bicyclo[3.2.1] and [3.3.1] systems usually show low or no polymerizability.[2,3] Bicyclic lactones or lactams of [2.2.2] and [3.2.1] systems containing six-membered rings show considerably higher polymerizability.[28]

Spiro monomers containing three-, four-, or five-membered rings are known to be polymerizable.[2,3,29]

In the anionic or nucleophilic ring-opening polymerization of a heterocyclic monomer, the initiation reaction is generally a bimolecular nucleophilic substitution reaction by an initiator leading to the formation of an anion or nucleophile of the heteroatom. The formed anion or nucleophile must have a high enough reactivity toward the monomer in order to enable it to propagate. In this sense, anions or nucleophiles of heteroatoms such as alkoxides, amides, and sulfides can be used as initiators. Alkali metals are mostly used as countercations and polar solvents are generally used. Zinc and aluminum are also used as countercations.

In order to realize a clean nucleophilic initiation reaction through substitution, the strong basic character of the initiatior must be suppressed. Metal alkyls are generally stronger initiators than the corresponding metal alkoxides, amides and sulfides. Caution must be taken to avoid side reactions caused by metal alkyls acting as basic initiators.

A coordination type initiator is suitable for achieving a stereospecific initiation and polymerization. Typical examples of these initiators are alkoxides of aluminum or zinc, and bimetallic alkoxide systems.

Polymerization of aziridines such as ethylenimine can be explained by thermodynamic considerations, and actually ethylenimine can be easily polymerized by cationic initiators. However, it is well
known that it cannot be polymerized by an anionic mechanism. Meanwhile, ethylamine is known to
open up the ethylenimine ring under vigorous conditions, although the reaction is very slow.[30]
Moreover, some of the *N*-acyl-substituted ethylenimines are claimed to be polymerized by an
anionic mechanism.[2] The nonpolymerizability of ethylenimine through an anionic mechanism
should be explained by kinetic considerations. The lower electronegativity of the nitrogen atom
compared to the oxygen atom results in the lower reactivity of the aziridine ring compared with that
of the oxirane ring toward nucleophilic attack.

Potassium hydroxide, which polymerizes ethylene oxide according to a living anionic mode, is not
suitable for the polymerization of propylene oxide. Hydrogen abstraction from the substituent
methyl group leading to the formation of allyl alcoholate occurs in this combination (equation 4).[2, 3]
This reaction would be a chain transfer reaction if it occurred in the propagation step. The presence
of double bonds in the polymer is also reported in the polymerization of isobutylene oxide by
sodium metal.[31]

$$\text{Me} \;+\; \text{KOH} \;\longrightarrow\; \text{CH}_2^-\text{K}^+ \;\longrightarrow\; \text{KOCH}_2\text{CH}\!=\!\text{CH}_2 \tag{4}$$

Polymerizations of δ-valerolactone and ε-caprolactone by lithium alkoxide were reported to
consist of (i) fast propagation, and (ii) slow back-biting reactions.[32, 33] Care must also be taken with
the reaction conditions in order to achieve a clean polymerization reaction without undesirable side
reactions.

Recently, interest has been shown in the development of new polymerization systems which give
controlled structures in the polymers and copolymers formed. Living polymerization systems, in
which no termination and transfer reactions occur, are the most suitable for achieving a narrow
molecular weight distribution of the formed polymer, and eventually for well-defined block
copolymer synthesis. In order to realize a narrow molecular weight distribution of the polymer, the
initiation reaction must be effectively faster than the propagation reaction. Control of the structure
and reactivity of the initiating and propagating species is very important. This is discussed in detail
in Chapters 32, 34, and 37 of this volume. Control of the stereochemical structure of the polymer is
another important feature of the polymerization. This is discussed in Chapter 33.

31.2 INITIATION REACTION OF MONOMERS WITH ONLY ONE HETEROATOM

Initiators used for the anionic polymerization of oxiranes are summarized in Table 2.[2, 3, 34–42]
These initiators can also initiate thiiranes and thietanes.

Table 2 Anionic Initiators for Oxirane Polymerization

Derivatives of	Examples[a]
Alkali metal	Na, K, NaNH$_2$, NaOR, KOH, NaOH, NaNaph.[34–36]
Alkaline earth metal	CaO, SrO
Other metals	ZnO, R$_2$Zn/H$_2$O,[2, 3, 37] R$_2$Zn/R'OH, [Zn(OMe)$_2$·(EtZnOMe)$_6$],[38] [Zn(OCH$_2$CH$_2$OMe)$_2$·(EtZnOCH$_2$CH$_2$OMe)$_6$],[38] [{MeOCH$_2$CH-(Me)OZnOCH(Me)CH$_2$OMe}$_2$·{EtZnOCH(Me)CH$_2$OMe}$_2$],[38] R$_3$Al, R$_3$Al/H$_2$O, R$_3$Al/H$_2$O/acetylacetone, Al(OR)$_3$–ZnCl$_2$,[39] (RO)$_2$-AlOZnOAl(OR)$_2$,[40] porphinatoaluminum,[41] FeCl$_3$, Fe(OR)$_3$, Ph$_3$SnBr$_2$/Ph$_3$P[42]
Amines	Me$_3$N

[a] Taken from refs. 2 and 3 unless otherwise noted.

Typical initiators are alkali metal hydroxides and alkoxides, and metal alkyls and their reaction
products with alcohol or water. The former initiators are mainly used to obtain polymers of well-
controlled structure, and the latter to obtain stereochemically controlled polymer structures.

Potassium hydroxide in water and sodium alkoxide or phenoxide initiate the polymerization of
ethylene oxide by a nucleophilic attack of OH^- or RO^- on the CH_2—O carbon. The oligomeric

metal alkoxides formed have the ability to react further with monomer. The reaction was shown to proceed at different rates for the first and second steps. Generally, the initiation reaction is faster than the following second step reaction or propagation reaction, and the requirement for living anionic polymerization can be fulfilled. It has also been pointed out that free alcohol or phenol in the reaction system has significant effects on the kinetics.[2,3]

Poly(ethylene oxide) having controlled molecular weight with a narrow molecular weight distribution, which is usually synthesized with sodium methoxide as initiator, is used as a standard sample in GPC analysis of aqueous systems.

Polymerization of ethylene oxide by KOH in dimethyl sulfoxide is much faster than that in ordinary solvents, and the degree of polymerization can be controlled by the concentration ratio of monomer to initiator.[43] Dimsyl anion $MeS(O)CH_2^-$, which is formed by the reaction of KOH with dimethyl sulfoxide, has been shown to play an important role in the initiation reaction.[44,45]

Anhydrous potassium hydroxide, sodium ethoxide and sodium amide give the metal thiolate as the propagating species in the initiation reaction of thiiranes.[46] Effects of water have been reported.

Alkali metal and alkaline earth metal alkyls are also strong initiators for the ethylene oxide polymerization. Fluorenyl salt is a typical initiator, acting as a divalent initiator with sodium as counterion, and as a monovalent initiator with barium.[47] The detailed mechanism has been discussed.[48]

Not only metal alkyls but also sodium naphthalide can be used as an initiator. The initiation reaction by sodium naphthalide was reported to be a direct addition of the naphthalene radical anion to ethylene oxide, followed by further electron transfer from sodium naphthalide, leading to the formation of bifunctional alkoxide propagating species with low monomer to initiator ratio (Scheme 2).[34-36]

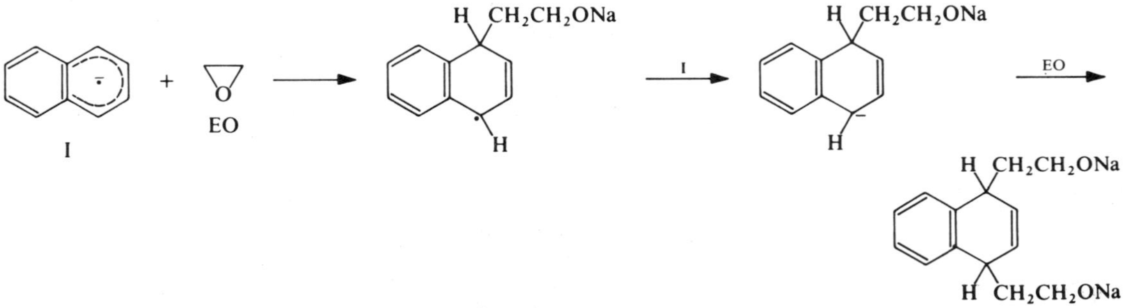

Scheme 2

Caution must be taken with strong initiators in the polymerization of propylene oxide to avoid side reactions such as hydrogen abstraction from the α-methyl group.[2,3] The porphinatoaluminum system reported by Inoue has been found to be an effective initiator for the polymerization of propylene oxide in living anionic mode.[41] The details are described in Chapter 37 of this volume.

Thiiranes are more easily polymerized than oxiranes. For instance, $ZnEt_2$ itself, which is inactive in initiating the polymerization of oxirane, can polymerize thiirane. Even ammonia, amines[49] and trialkylphosphines,[50] phosphonium and ammonium salts, alkali metal alkoxides[2,3] (needless to say), hydroxides and amides,[46] alkyllithium,[46] sodium triphenylmethide,[51] sodium naphthalide[52] and tetrabutylammonium fluorenyl[53] can initiate the polymerization. Among these, tertiary amines, phosphines and $ZnEt_2/H_2O$[54] are specially good initiators. A zwitterionic mechanism has been proposed as a possible mechanism in the initiation with tertiary amine.[49]

The initiation reaction by alkyllithium or sodium naphthalide is not a simple nucleophilic ring opening of the thiirane ring. Formation of lithium ethanethiolate was proposed for the initiation reaction of propylene sulfide by ethyllithium by the loss of propylene from the monomer as shown in equation (5).[55]

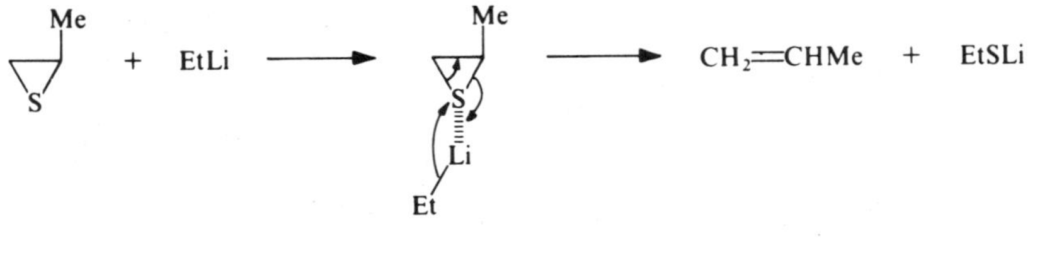

$$(5)$$

Propylene sulfide was reported to be initiated by sodium naphthalide in tetrahydrofuran through electron transfer from the initiator.[56] This is in strong contrast with the case of oxirane polymerization, where direct addition of the 6,9-dihydronaphthalene-6,9-diyl group was observed.[34–36] Bifunctional thiolate living polymers were formed, but there was no incorporation of groups derived from naphthalene in the polymer. Although the formation of the dianion through the coupling of the electron-transferred radical anion of S⁻ was proposed,[52] a more complex mechanism is likely.[57]

According to the hard–soft acid base (HSAB) concept,[58, 59] thiolate anion is a softer base than the alkoxide anion. Since the soft thiolate anion does not abstract a hard proton from the substituent methyl group of propylene sulfide, the polymerization of propylene sulfide tends to be easily carried out according to a living anionic mode compared with that of propylene oxide.

Metal alkyl/water or metal alkyl/alcohol systems such as the $ZnEt_2/H_2O$ system are good initiating systems not only for the polymerization of oxiranes but also for thiiranes.[54] Cadmium and zinc thiolates are reported to be active in the polymerization of thiiranes.[60, 61] Cadmium carbonate is active even in water[60, 62] or in the presence of amine.[63] These initiator systems are of interest in connection with the stereospecific polymerization of oxiranes and thiiranes. This is discussed in Section 31.3, and in Chapter 33 of this volume.

Less reactive thietanes can be polymerized by anionic initiators like ethyllithium. The propagating species was reported to be a carbanion formed by the ring opening, but not a thiolate anion.[55] The formation of carbanion species can also be rationalized by the HSAB principle (equation 6).

$$
\underset{\text{Me}}{\overset{}{\square}}\text{S} \;+\; \text{EtLi} \;\longrightarrow\; \text{EtS}\underset{\text{H}}{\overset{\text{Me}}{\text{C}}}\text{CH}_2\text{CH}_2\text{Li} \tag{6}
$$

There are two possible atoms for ethyl anion (harder base than thiolate anion) to attack on the thietane ring: the methylene carbon or the sulfur atom. The attack on methylene carbon is less favorable, since lithium cation is a hard acid and the thiolate anion to be formed by the attack is a softer base than the carbanion to be formed by the attack on sulfur. Thus the attacks by carbanions in initiation and propagation steps may proceed regioselectively on the sulfur atom, regenerating the carbanion at a sufficiently low temperature.

Although slow ring-opening reactions of the oxetane ring by base hydrolysis,[64] by thiol (base catalyzed or not), amine, or sodium phenoxide,[65] and ready opening by a Grignard reagent[66] have been reported, it is well known that oxetane cannot be polymerized with anionic initiators under ordinary reaction conditions. The low reactivity of the alcoholate formed in the initial reaction towards the oxetane ring, whose strain is much smaller than that of the oxirane ring, seems to be the reason for the non-polymerizability of oxetanes by an anionic mechanism. It is worthwhile to note that oxetane is reported to be polymerized by the coordination type initiator $Zn(OMe)_2$.[67] Larger ring size cyclic ethers cannot be polymerized according to anionic or nucleophilic modes.

In the ring opening of oxirane by an anionic mechanism, the configuration of the carbon attacked by the initiator was shown to be inverted[68, 69] (*cf.* Chapter 33 of this volume), which is stereochemical evidence of an S_N2 mechanism for the ring opening. Regiospecificity arises in the ring opening of substituted oxiranes. α-Scission (ring opening at the carbon–oxygen bond of the more substituted carbon) and β-scission (ring-opening at the less substituted carbon–oxygen bond) are possible. The regiospecificity of the attack of the propagating alkoxide on the monomer and the stereochemistry of the attacked carbon are very important in determining the constitutional and configurational regularities of the formed polymers. The model ring-opening reaction of propylene oxide by an alcohol in the presence of KOH was shown to proceed almost exclusively (95% or more) *via* β-scission to give a secondary alcoholate.[70] The concentration of primary alcoholate propagating species formed by α-scission would be quite small in the polymerization system, since the primary alcoholate has a higher reactivity than the secondary alcoholate, and the primary alcoholate formed would be consumed to ring open the monomer in 95% β-scission mode, and converted to the secondary alcoholate. It has been shown that the concentration of the primary alcoholate is as low as 0.5% of the propagating species by measuring ^{19}F NMR of trifluoroacetylated propagating species.[2, 3] A similar situation is also true in the ring-opening polymerization of styrene oxide,[71] *i.e.* it is opened rather randomly (65% α- and 35% β-scission) by KOH at the initiation reaction, but the propagating species, after several steps of the reaction, is converted into the secondary alcoholate.

(Tetraphenylporphinato)chloroaluminum ring opens propylene oxide exclusively according to the β-mode.[41] Although $Al(OPr^i)_3$ itself opens propylene oxide rather randomly in the α- and β-modes, the combination of $Al(OPr^i)_3$ and $ZnCl_2$ selectively opens the ring in the β-mode.[39]

Styrene oxide is opened exclusively in the α-mode by $Al(OPr^i)_3$,[72] but selectively in the β-mode by $ZnEt_2/H_2O$.[73] α,α-Disubstituted ethylene sulfide is opened in the β-mode.[74] The mode of ring opening is further discussed in Chapter 33 of this volume.

31.3 INITIATION REACTION OF MONOMERS CONTAINING CARBONYL GROUPS

Lactones and lactams are typical monomers containing carbonyl groups. Carothers, Gresham and Hall systematically studied the polymerization of lactones including the cyclic dimeric ester of lactic acid, β-propiolactone and other lactones.[2,3,9] Typical anionic initiators for lactones polymerization are shown in Table 3.[2,3,9,10,40,75-88]

Table 3 Anionic Initiators for Lactone Polymerization

Derivatives of	Examples[a]
Alkali metals	Na, NaNaph, Li_2 (or K_2) benzophenone,[75] LiR,[75] LiH, NaH, K_2CO_3,[10] KOH, NaOR,[2,3,76] LiOR, Li (or Na or K) acetate (active only for β-propiolactone)
Other metals	MgR_2,[75] RMgX,[75] $Mg(OR)_2$, ZnR_2,[2,3,75] $ZnEt_2/H_2O$,[75] AlR_3,[2,3,75] R_2AlX, R_2AlOR', $Al(OR)_3$, $AlEt_3/H_2O$,[75] (porphinato)aluminum,[77] $R_2AlOZnOAlR_2$,[40,80] $CdEt_2/H_2O$,[75] titanium phosphate
Amines	tertiary amines (active only for β-propiolactone)[78,79]

[a] Taken from refs. 2 and 3 unless otherwise noted.

Sodium ethoxide and acetate are good initiators for β-propiolactone polymerization. The ring opening was reported to proceed with different regiospecificity in response to the initiators used. Strong nucleophiles like sodium ethoxide (hard base) attack the hard carbonyl carbon; on the other hand, weak nucleophiles like sodium carboxylate (soft base) attack the soft methylene carbon as shown in equations (7) and (8).[76,77,81]

$$EtONa \; + \; \underset{O}{\square}O \longrightarrow EtOCCH_2CH_2ONa \qquad (7)$$

$$RCO_2Na \; + \; \underset{O}{\square}O \longrightarrow RCO_2CH_2CH_2CONa \qquad (8)$$

In the case of α,α-bis(chloromethyl)-β-propiolactone, it was noticed that extent of alkyl–carbon bond scission was much higher compared with β-propiolactone, and this was ascribed to the steric effect of the two α-chloromethyl groups.[81]

Recently, Penczek[76] reinvestigated the reaction of alkoxide with β-propiolactone quantitatively, and concluded that the principal propagating species, even by alkoxide initiator, was the carboxylate anion, which was converted from the alkoxide with its consumption in the early stage of the polymerization. Formation of a betaine followed by the propagation *via* carboxylate species was proposed for the polymerization of β-propiolactone by a tertiary amine like pyridine.[78,79]

Porphinatoaluminum opens β-propiolactone through alkyl–oxygen cleavage leading to the formation of an aluminum carboxylate propagating species,[77] whereas the ring-opening mode by μ-oxoalkoxide is acyl–oxygen cleavage.[80]

The polymerizations initiated by $AlEt_3$ or $ZnEt_2$, as well as their modified initiators with water and $Al(OR)_3$, are considered to proceed according to the coordinate anionic mechanism. $ZnEt_2$ opened β-functionalized β-propiolactones, such as β-vinyl-β-propiolactone, through selective opening of the propiolactone ring with the functional group intact.[2,3] α- and/or β-substituted β-propiolactones can also be polymerized.[82,83] The ring opening usually proceeds through alkyl–oxygen bond cleavage;[84-86] however, it is reported that the C—C bond between α and β carbons was cleaved in the case of β-methyl-β-propiolactone.[83] The propagation reaction was considered to proceed according to the same regiospecificity as the initiation reaction. A detailed discussion of the propagation reaction, including that by the coordination mechanism, appears in Chapter 34 of this volume.

Five-membered γ-butyrolactone cannot be polymerized under ordinary conditions. However, it was claimed that the monomer could be polymerized under high pressure at elevated temperature to give a polymer that is thermally stable at room temperature.[87] Six- or seven-membered lactones are opened through acyl–oxygen cleavage, even with lithium trimethylsilanolate,[88] but sodium acetate or pyridine, which cleave β-propiolactone at the methylene–oxygen bond, cannot polymerize these monomers.

Glycolide or lactide and other alkyl-substituted analogues can be polymerized by cationic or anionic initiators,[2,3] and the copolymers with other cyclic monomers are of interest as biodegradable polymers. Polymerizations of four-membered thiol lactones[89] and seven-membered ε-thiol caprolactones[90] were also reported. Ballard *et al.* reported that cyclic anhydrosulfites of α-hydroxy acids could be polymerized to polyester by the loss of sulfur dioxide.[91]

Cyclic amides (lactams) can be polymerized by alkali metals, alkali metal hydroxides, alkoxides and alkali metal lactam salts.[2,3] The four-membered β-lactam structure can be polymerized by alkali metal lactam salts, tetraalkylammonium benzoate and amino compounds.[2,3,92,93] Non-substituted five-membered lactam can be polymerized. Six-membered lactam does not give high polymers.

Seven-membered ε-caprolactam is an important monomer of nylon 6. This monomer can be polymerized by water, amine, or strong bases. Alkali metal, alkali metal halide, carbonate, or hydroxide can initiate fast polymerization of the monomer in the absence of water.[2,3] The initiation reaction is the attack of *N*-metalated lactam anion on the lactam monomer (see Scheme 3).

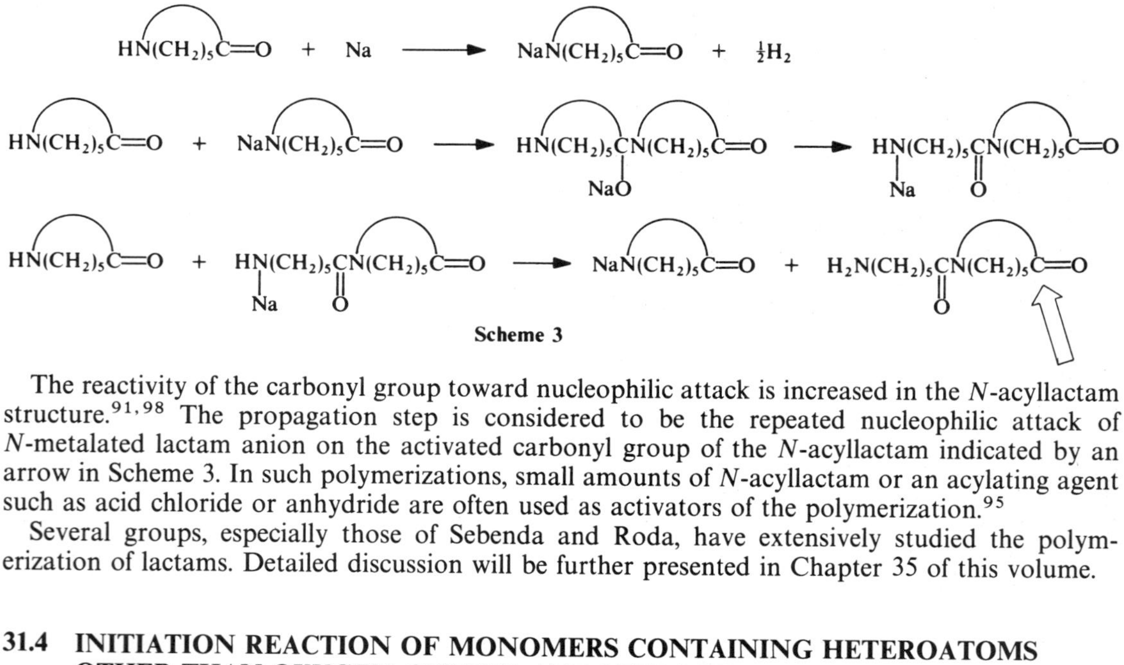

$$HN(CH_2)_5C{=}O + Na \longrightarrow NaN(CH_2)_5C{=}O + \tfrac{1}{2}H_2$$

$$HN(CH_2)_5C{=}O + NaN(CH_2)_5C{=}O \longrightarrow HN(CH_2)_5\underset{NaO}{C}N(CH_2)_5C{=}O \longrightarrow HN(CH_2)_5\underset{Na}{\underset{\|}{C}}\underset{O}{\,}N(CH_2)_5C{=}O$$

$$HN(CH_2)_5C{=}O + HN(CH_2)_5\underset{Na}{C}\underset{O}{\|}N(CH_2)_5C{=}O \longrightarrow NaN(CH_2)_5C{=}O + H_2N(CH_2)_5\underset{O}{\underset{\|}{C}}N(CH_2)_5C{=}O$$

Scheme 3

The reactivity of the carbonyl group toward nucleophilic attack is increased in the *N*-acyllactam structure.[91,98] The propagation step is considered to be the repeated nucleophilic attack of *N*-metalated lactam anion on the activated carbonyl group of the *N*-acyllactam indicated by an arrow in Scheme 3. In such polymerizations, small amounts of *N*-acyllactam or an acylating agent such as acid chloride or anhydride are often used as activators of the polymerization.[95]

Several groups, especially those of Sebenda and Roda, have extensively studied the polymerization of lactams. Detailed discussion will be further presented in Chapter 35 of this volume.

31.4 INITIATION REACTION OF MONOMERS CONTAINING HETEROATOMS OTHER THAN OXYGEN, SULFUR AND NITROGEN

Six-membered cyclic trisiloxane hexamethylcyclotrisiloxane (D₃) can be polymerized by alkyllithium,[96-98] alkali metal hydroxides[99] and silanolate,[100,101] as initiators. The initiation reaction is a nucleophilic attack of the initiator on the silicon atom of the monomer to form the silanolate, shown in the typical case of alkyllithium in equation (9). The propagation reaction is fast in THF, but slow in hydrocarbon solvents.

$$D_3 + RLi \longrightarrow R\underset{Me}{\overset{Me}{Si}}{-}O\underset{Me}{\overset{Me}{Si}}{-}O\overset{Me}{Si}OLi \xrightarrow{D_3} R(SiO)_{3n}\underset{Me}{\overset{Me}{Li}} \quad (9)$$

Larger ring analogues of the six-membered ring system can be polymerized. Hexaphenylcyclo-trisiloxane cannot be polymerized to a high polymer under ordinary conditions. Five-membered 1-oxa-2-silacyclopentane can be polymerized by KOH. The initiation reaction seems to be nucleophilic attack by the initiator to form silanolate growing species.[102]

Hexachlorocyclotriphosphonitrile (cyclophosphazene) can be polymerized by heating or in the presence of suitable initiators including sulfur and organometallic compounds at elevated temperature.[103] Even though a radical or anionic mechanism has been proposed for the polymerization, a cationic mechanism seems most plausible.

Five- and six-membered cyclic phosphates can be polymerized by anionic initiators.[17-20] The initiators used are $(C_5H_5)_2Mg$, Bu^i_3Al, Li metal, Pr^i_3Al and others. The initiation reaction is the attack of a nucleophile on the phosphorus atom leading to the formation of alkoxide anion, which will propagate. 2-Hydro-2-oxo-1,3,2-dioxaphosphorinane is also reported to be polymerized by anionic and coordination initiators. The polymer was further transformed into poly(phosphoric ester).[104] Detailed discussions on these polymerizations may be found in Chapter 50 of this volume and in Chapters 26 and 27 of Volume 4.

31.5 REFERENCES

1. K. C. Frisch and S. L. Reegen, (eds.), 'Ring-Opening Polymerization', Dekker, New York, 1969.
2. T. Saegusa, 'Ring-Opening Polymerization' (Kaikan Jyugo in Japanese), Kagaku Dojin, Kyoto, 1971, vols. 1 and 2.
3. K. J. Ivin and T. Saegusa (eds.), 'Ring-Opening Polymerization', Elsevier, London, 1984, vols. 1-3.
4. J. W. Hill and W. H. Carothers, *J. Am. Chem. Soc.*, 1935, **57**, 925.
5. J. G. Affleck and G. Dougherty, *J. Org. Chem.*, 1950, **15**, 865.
6. F. S. Dainton, T. R. E. Devlin and P. A. Small, *Trans. Faraday Soc.*, 1955, **51**, 1710.
7. P. A. Small, *Trans. Faraday Soc.*, 1955, **51**, 1717.
8. F. S. Dainton and K. J. Ivin, *Q. Rev., Chem. Soc.*, 1958, **12**, 61.
9. H. K. Hall, Jr., *J. Am. Chem. Soc.*, 1958, **80**, 6404.
10. H. K. Hall, Jr. and A. K. Schneider, *J. Am. Chem. Soc.*, 1958, **80**, 6409.
11. F. S. Dainton, K. J. Ivin, D. A. G. Walmsley, *Trans. Faraday Soc.*, 1960, **56**, 1784.
12. F. S. Dainton, J. A. Davies, P. P. Manning and S. A. Zahir, *Trans. Faraday Soc.*, 1957, **53**, 813.
13. J. Sebenda, *Makromol. Chem., Macromol. Symp.*, 1986, **6**, 1.
14. H. Keul, R. Bacher and H. Höcker, *Makromol. Chem.*, 1986, **187**, 2579.
15. S. Kobayashi, M. Suzuki and T. Saegusa, *Polym. Bull. (Berlin)*, 1981, **4**, 315.
16. T. Mukaiyama, T. Fujisawa, Y. Tamura and Y. Yokota, *J. Org. Chem.*, 1964, **29**, 2572.
17. W. Vogt and R. Pfluger, *Makromol. Chem., Suppl.*, 1975, **1**, 97.
18. S. Penczek, P. Kubisa and K. Matyjaszewski, *Adv. Polym. Sci.*, 1980, **37**, 1.
19. H. Yasuda, M. Sumitani and A. Nakamura, *Macromolecules*, 1981, **14**, 458.
20. S. Penczek, T. Biela, P. Klosinski and G. Lapienis, *Makromol. Chem., Macromol. Symp.*, 1986, **6**, 123.
21. W. A. Piccoli, G. G. Haberland and R. L. Merker, *J. Am. Chem. Soc.*, 1960, **82**, 1883.
22. C. L. Frye, R. M. Salinger, F. W. G. Fearon, J. M. Klosowski and T. DeYoung, *J. Org. Chem.*, 1970, **35**, 1308.
23. H. Jacobson and W. H. Stockmayer, *J. Chem. Phys.*, 1950, **18**, 1600.
24. P. J. Flory, 'Statistical Mechanics of Chain Molecules', Wiley-Interscience, New York, 1969, pp. 2, 18, 39, 42, 174-180, 394-396.
25. M. P. Dreyfuss and P. Dreyfuss, *J. Polym. Sci., Part A-1*, 1966, **4**, 2179.
26. H. K. Hall, Jr., E. P. Blanchard, Jr., S. C. Cherkovsky, J. B. Sieja and W. A. Sheppard, *J. Am. Chem. Soc.*, 1971, **93**, 110.
27. E. P. Blanchard and A. Cairncross, *J. Am. Chem. Soc.*, 1966, **88**, 487.
28. K. Hashimoto and H. Sumitomo, *Macromolecules*, 1980, **13**, 786.
29. T. Saegusa, S. Kobayashi and A. Nakamura, *Macromolecules*, 1975, **8**, 593.
30. J. R. Malpass, in 'Comprehensive Organic Chemistry', ed. D. Barton and W. D. Ollis, Pergamon Press, Oxford, 1979, vol. 2, p. 154.
31. S. L. Malhotra and L. P. Blanchard, *J. Macromol. Sci., Chem.*, 1977, **11**, 1809.
32. K. Ito and Y. Yamashita, *Macromolecules*, 1978, **11**, 68.
33. K. Ito, M. Tomida and Y. Yamashita, *Polym. Bull. (Berlin)*, 1979, **1**, 569.
34. D. H. Richards and M. Szwarc, *Trans. Faraday Soc.*, 1959, **55**, 1644.
35. K. S. Kazanskii, A. A. Solovyanov and S. G. Entelis, *Eur. Polym. J.*, 1971, **7**, 1421.
36. I. Cabasso and A. Zilkha, *J. Macromol. Sci., Chem.*, 1974, **8**, 1313.
37. F. M. Rabagliati and F. L. Cariasquero, *Eur. Polym. J.*, 1985, **21**, 1061.
38. T. Tsuruta, *Makromol. Chem., Macromol. Symp.*, 1986, **6**, 23.
39. Z. Jedlinski, A. Dworak and M. Bero, *Makromol. Chem.*, 1979, **180**, 949.
40. Ph. Teyssié, J. P. Bioul, A. Hamitou, J. Heuschen, L. Hochs, R. Jerome and T. Ouhadi, *ACS Symp. Ser.*, 1977, **59**, 165.
41. T. Aida and S. Inoue, *Makromol. Chem., Macromol. Symp.*, 1986, **6**, 217.
42. R. Nomura, H. Hisada and A. Ninagawa, *Makromol. Chem.*, 1982, **183**, 1073.
43. C. C. Price and D. D. Carmelite, *J. Am. Chem. Soc.*, 1966, **88**, 4039.
44. K. S. Kazanskij, A. A. Solovyanov and S. A. Dubrovsky, *Makromol. Chem.*, 1978, **179**, 969.
45. A. Stolarzewicz, *Makromol. Chem.*, 1986, **187**, 745.
46. S. Boileau and P. Sigwalt, *C. R. Hebd. Seances Acad. Sci.*, 1961, **252**, 882.
47. I. U. Berlinova and I. M. Panayotov, *Eur. Polym. J.*, 1980, **16**, 769.

48. D. Lassalle, S. Boileau and P. Sigwalt, *Eur. Polym. J.*, 1977, **13**, 591.
49. D. R. Morgan, G. T. Williams and R. T. Wragg, *Eur. Polym. J.*, 1970, **6**, 309.
50. B. E. Jennings (ICI Ltd.), *Br. Pat.* 1 077 958 (1967) (*Chem. Abstr.*, 1967, **67**, 82 542).
51. S. Boileau, P. Sigwalt and N. D'Haeyer, *Bull. Soc. Chim. Fr.*, 1968, 1054.
52. S. Boileau, G. Champetier and P. Sigwalt, *Makromol. Chem.*, 1963, **69**, 180.
53. G. Tersac, S. Boileau and P. Sigwalt, *Makromol. Chem.*, 1971, **149**, 153.
54. J. P. Machon and P. Sigwalt, *C. R. Hebd. Seances Acad. Sci.*, 1965, **260**, 549.
55. M. Morton and R. F. Kammereck, *J. Am. Chem. Soc.*, 1970, **92**, 3217.
56. S. Boileau, G. Champetier and P. Sigwalt, *J. Polym. Sci., Part C*, 1967, **16**, 3021.
57. J. C. Favier, S. Boileau and P. Sigwalt, *Eur. Polym. J.*, 1968, **4**, 3.
58. R. G. Pearson, 'Hard and Soft Acids and Bases', Dowden, Hutchinson & Ross, Inc., Stroudsburg, 1973.
59. T.-L. Ho, 'Hard and Soft Acids and Bases Principles in Organic Chemistry', Academic Press, New York, 1977.
60. W. Cooper, D. R. Morgan and R. J. Wragg, *Eur. Polym. J.*, 1969, **5**, 71.
61. P. Guerin, S. Boileau, F. Subira and P. Sigwalt, *Eur. Polym. J.*, 1980, **16**, 121, 129.
62. Y. B. Amerik, L. A. Shirokova, I. M. Toltchinsky and B. A. Krestsel, *Makromol. Chem.*, 1984, **185**, 899.
63. D. R. Morgan and R. T. Wragg, *Makromol. Chem.*, 1969, **125**, 220.
64. J. G. Pritchard and F. A. Long, *J. Am. Chem. Soc.*, 1958, **80**, 4162.
65. S. Searles, *J. Am. Chem. Soc.*, 1951, **73**, 4515.
66. S. Searles, *J. Am. Chem. Soc.*, 1951, **73**, 124.
67. T. Hirano, S. Nakayama and T. Tsuruta, *Makromol. Chem.*, 1975, **176**, 1897.
68. C. C. Price and R. Spector, *J. Am. Chem. Soc.*, 1965, **87**, 2069.
69. P. H. Khanh, T. Hirano and T. Tsuruta, *Makromol. Chem.*, 1972, **153**, 331.
70. L. C. Case and N. H. Rent, *J. Polym. Sci., Part B*, 1964, **2**, 417.
71. Z. Jedlinski, J. Kasperczyk, A. Dworak and B. Matuszewska, *Makromol. Chem.*, 1982, **183**, 587.
72. Z. Jedlinski, J. Kasperczyk and A. Dworak, *Eur. Polym. J.*, 1983, **19**, 899.
73. J. Kasperczyk and Z. Jedlinski, *Makromol. Chem.*, 1986, **187**, 2215.
74. S. A. Bhatti and E. J. Goethals, *Makromol. Chem.*, 1985, **186**, 317.
75. S. Inoue, Y. Tomoi, T. Tsuruta and J. Furukawa, *Makromol. Chem.*, 1961, **48**, 229.
76. A. Hofman, S. Slomkowski and S. Penczek, *Makromol. Chem.*, 1984, **185**, 91.
77. T. Yasuda, T. Aida and S. Inoue, *J. Macromol. Sci., Chem.*, 1984, **21**, 1035.
78. K. Boehlke, M. J. Han, V. Jaacks, N. Mathes and K. Zimmerschied, *Angew. Chem.*, 1969, **81**, 336.
79. Y. Yamashita, K. Ito and T. Nakakita, *Makromol. Chem.*, 1969, **127**, 292.
80. A. Hamitou, T. Ouhadi, R. Jerome and Ph. Teyssié, *J. Polym. Sci., Polym. Chem. Ed.*, 1977, **15**, 865.
81. Y. Yamashita, T. Tsuda, H. Ishida, A. Uchikama and Y. Kuriyama, *Makromol. Chem.*, 1968, **113**, 139.
82. G. Perego, A. Melis and M. Cesari, *Makromol. Chem.*, 1972, **157**, 269.
83. Z. Jedlinski, P. Kurcok, M. Kowalczuk and J. Kasperczyk, *Makromol. Chem.*, 1986, **187**, 1651.
84. R. E. Prud'homme, *Makromol. Chem., Macromol. Symp.*, 1986, **6**, 189.
85. S. C. Arnold and R. W. Lenz, *Makromol. Chem., Macromol. Symp.*, 1986, **6**, 285.
86. Ph. Guérin, J. Francillette, C. Braud and M. Vert, *Makromol. Chem., Macromol. Symp.*, 1986, **6**, 305.
87. F. Korte and W. Glet, *J. Polym. Sci.*, 1966, **4**, 685.
88. S. Sosnowski, S. Slomkowski and S. Penczek, *J. Macromol. Sci., Chem.*, 1983, **20**, 979.
89. D. Krilov, Z. Veksli and D. Fles, *J. Polym. Sci., Polym. Chem. Ed.*, 1976, **14**, 777.
90. C. G. Overberger and J. Weise, *J. Polym. Sci., Part B*, 1964, **2**, 329.
91. M. D. Thomas, D. G. H. Ballard and B. J. Tighe, *J. Chem. Soc. B*, 1970, 1039.
92. R. W. Lenz, E. Bigdeli and H. Sekiguchi, *Polym. Bull. (Berlin)*, 1980, **3**, 91.
93. T. Sebenda and J. Haver, *Polym. Bull. (Berlin)*, 1981, **5**, 529.
94. N. Yoda and A. Miyake, *J. Polym. Sci.*, 1960, **43**, 117.
95. J. Stehlicek, B. Valter and J. Sebenda, *Makromol. Chem.*, 1986, **187**, 513.
96. J. G. Zilliox, J. E. L. Roovers and S. Bywater, *Macromolecules*, 1975, **8**, 573.
97. P. M. Lefebvre, R. Jerome and Ph. Teyssié, *Macromolecules*, 1977, **10**, 871.
98. J. C. Saam, D. J. Gordon and S. Lindsey, *Macromolecules*, 1970, **3**, 1.
99. D. T. Hurd, R. C. Osthoff and M. L. Corrin, *J. Am. Chem. Soc.*, 1954, **76**, 249.
100. H. J. Hölle and B. R. Lehnen, *Eur. Polym. J.*, 1975, **11**, 663.
101. Y. Kawakami, Y. Miki, T. Tsuda, R. A. N. Murthy and Y. Yamashita, *Polym. J.*, 1982, **14**, 913.
102. J. Chojnowski and M. Mazurek, *Makromol. Chem.*, 1975, **176**, 2999.
103. H. R. Allcock, *Makromol. Chem., Macromol. Symp.*, 1986, **6**, 101.
104. K. Kaluzynski, J. Libiszowski and S. Penczek, *Makromol. Chem.*, 1977, **178**, 2943.

32

Anionic Ring-opening Polymerization: Epoxides and Episulfides

SYLVIE BOILEAU

Collège de France, Paris, France

32.1 INTRODUCTION

Extensive studies of the mechanism and kinetics of the anionic polymerization of epoxides and episulfides have been reported during the last few decades, and they were reviewed quite recently by several authors.[1-5] Most of the experiments were carried out with ethylene oxide (oxirane) and propylene sulfide (2-methylthiirane) because the living character of the polymerization of these two monomers, under specific conditions, has been well established;[6,7] propylene oxide (2-methyloxirane) polymerizes anionically with transfer to monomer, and poly(ethylene sulfide) is insoluble. In this chapter, the kinetics and mechanisms of the ring-opening reactions of epoxides and episulfides with nucleophilic initiators, as well as with the living ends during propagation, will be presented.

32.2 RING-OPENING REACTIONS OF EPOXIDES AND EPISULFIDES

Alkali metals, classical bases and radical anions, as well as anions and their ion pairs, are efficient initiators for the polymerization of epoxides and episulfides in aprotic media such as dioxane, tetrahydrofuran (THF) or hexamethylphosphoramide [HMPA; phosphoric tris(dimethyl-amide)].[3-5] It must be noted that a much wider range of nucleophilic initiators can be used for episulfides than for epoxides.

32.2.1 Initiation by Radical Anions

Radical anions can react with a given monomer in two ways, either by electron transfer to monomer followed by recombination of monomer radical-anions or by direct addition of the monomer to the radical anion. The latter mechanism was established for the reaction of sodium napthalene with ethylene oxide (Scheme 1).[8] The first step was found to be rate determining.[9] Its kinetics were investigated by spectrophotometry and electron spin resonance (ESR). The process was found to be first order with respect to monomer and initiator, and the bimolecular rate constant was about 11 mol^{-1} s^{-1} in THF at 25 °C. The reduction of the adduct (1) by sodium naphthalene gives the dianion (2) which has been isolated and characterized in THF[10] and in dimethyl sulfoxide (DMSO).[11] Addition of another molecule of ethylene oxide yields the *para* and/or *ortho* diadduct bearing two alkoxide groups (3) that are responsible for the subsequent propagation.

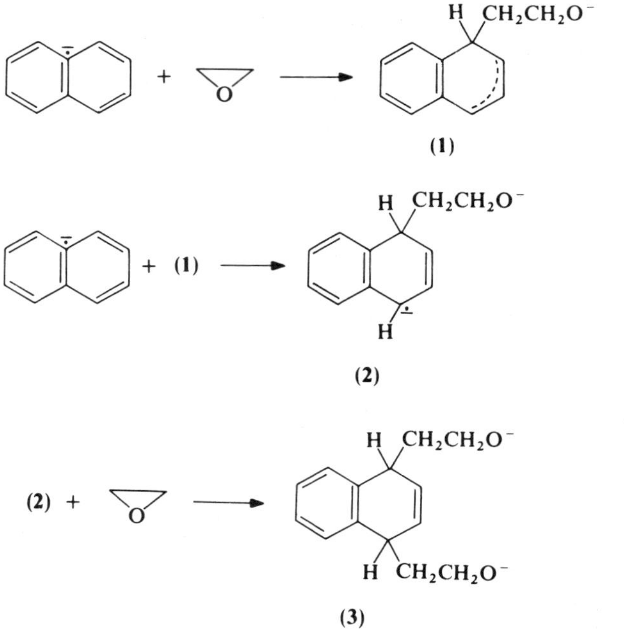

Scheme 1

Such a mechanism is supported by two observations: (a) the presence of a dihydronaphthalenic moiety in the resulting polymer and (b) the quantitative analysis of the solution recovered after precipitation of the polymer which showed that only half of the sodium naphthalene utilized was converted into naphthalene. It has been claimed that, at very low concentrations of the radical anion, polymers with only one growing group are formed.[10] Hydrogen abstraction from solvent was proposed as an explanation. However, terminating impurities become significant at very low concentrations of initiators and their action may account for the observations.

Another heterocyclic compound whose polymerization by naphthalene complexes has been studied in some detail is propylene sulfide. It was found that sodium naphthalene leads to the formation of polymers with two living ends per macromolecule.[7] The initiation mechanism was later shown to be quite different from that for epoxides.[12] With propylene sulfide, the sodium naphthalene utilized was recovered quantitatively as free naphthalene at the end of the reaction. Since propylene was formed in the course of this reaction, the mechanism shown in Scheme 2 was proposed. NaS· can dimerize and the subsequent sodium disulfide initiates the propagation. The presence of S—S bonds in the resulting polymer was confirmed by reducing them selectively with 9-methylfluorenylsodium, and the molecular weight of the polymer decreased by half.[13]

Naph = naphthalene

Scheme 2

Potassium complexes of aromatic nitriles and ketones have also been used for polymerizing episulfides.[14] The reaction in THF of radical anions of 4-cyanobiphenyl, 1-cyanopyrene and benzophenone was interpreted by a mechanism similar to that shown for sodium naphthalene

(Scheme 2), involving Na_2S_2 formation and regeneration of the initial nitrile or ketone. However, the percentage of regenerated compound recovered varied with the initiator and with the episulfide and was always lower than 100%. This was attributed to more or less extensive coupling of monomer radical-anions with the initiator (for propylene and styrene sulfides) or to a direct attack on the monomer by the radical anions (ethylene sulfide). In the same study,[14] it was assumed that initiation with dianions of nitriles proceeds mainly by electron transfer to monomer.

Electroinitiated anionic polymerization of propylene sulfide in dimethylformamide (DMF) was also explained by an electron-transfer reaction to monomer.[15]

32.2.2 Initiation by Anions and their Ion Pairs

Initiation of anionic polymerization by anions is conceptually similar to propagation in which every polymer living-end is an initiator for its own monomer. The behaviour of ethylene oxide in anionic homopolymerization is rather complicated because associations of ion pairs subsist down to low living-end concentrations and occur even in high polarity solvents such as HMPA.[6] Moreover, dissociation constants of alkoxide ion-pairs into free ions are very low in solvents of medium polarity (such as THF). All this makes the determination of reactivities of free ions and ion pairs difficult. A possible complication may come from solvation involving monomer and polymer which may influence ion associations. This was the reason why kinetic studies of the nucleophilic cleavage of ethylene oxide by various carbanion and nitranion salts were investigated, for which the proportion of the different ionic species is known with precision.[16-23] Such reactions are conveniently studied by UV and visible spectrophotometry.

An elegant way to overcome the problem of aggregates and to simplify the kinetics is to use crown ethers (*e.g.* compound **4**) or cryptands (*e.g.* compound **5**) that could encapsulate cations and greatly affect their behaviour. Cryptands are macrobicyclic ligands discovered by Lehn *et al.*[24,25] that form extremely stable cation-inclusion complexes, called cryptates, in which the cation is completely surrounded by the ligand and well hidden inside the molecular cavity, and this leads to a considerable increase in the interionic distance in the ion pairs. Thus cryptands, and to a lesser extent crown ethers, can convert tight ion-pairs, or some still larger aggregates, into loose ion-pairs. Activation of some initiators by addition of these ligands was shown in the case of potassium carboxylates and carbonates with [18]-crown-6 (**4**) for the ring opening of epoxides.[26]

(4) [18]-Crown-6 (18-C-6)

$$CH_2CH_2 \text{–}\!\!\left[OCH_2CH_2\right]_m$$
$$N\text{–}CH_2CH_2\text{–}\!\!\left[OCH_2CH_2\right]_n\text{–}N$$
$$CH_2CH_2\text{–}\!\!\left[XCH_2CH_2\right]_p$$

(5) a: [2.2.2]Cryptand; X = O, $m = n = p = 2$

    b: [2.2.1]Cryptand; X = O, $m = n = 2, p = 1$

    c: [3.2.2]Cryptand; X = O, $m = 3, n = p = 2$

    d: [2.1.1]Cryptand; X = O, $m = 2, n = p = 1$

    e: $[2_O.2_O.2_S]$Cryptand; X = S, $m = n = p = 2$

The kinetics of initiation of ethylene oxide polymerization by salts of anions derived from polystyrene,[18] fluoradene (7*bH*-indeno[1,2,3-*jk*]fluorene),[17] 9-methylfluorene,[19-22] carbazole[16,19,21,27,120] and their derivatives were studied in detail. In some systems, such as carbazyl (N^-) and polystyryl (PS^-) salts, free anions are more reactive than the corresponding ion pairs. In more delocalized systems, such as 9-methylfluorenyl (9-MeF⁻) and fluoradenyl (FD⁻) salts, the results are opposite. Reactivities of tight ion-pairs decrease on increasing the bulkiness of the cation. Moreover, free ions and loose ion-pairs are less reactive than tight ion-pairs.

In the case of fluoradenylsodium with pure ethylene oxide, the addition of sodium tetraphenylboride (that depresses the dissociation of ion pairs into free ions) speeds up the cleavage, whereas the addition of a complexing agent (crown ether, cryptand) slows it down.[17] Thus, it was suggested that electrophilic activation of the epoxide by cation plays an important role in the ring-opening reaction. Such a 'push–pull' mechanism may be written as equation (6).

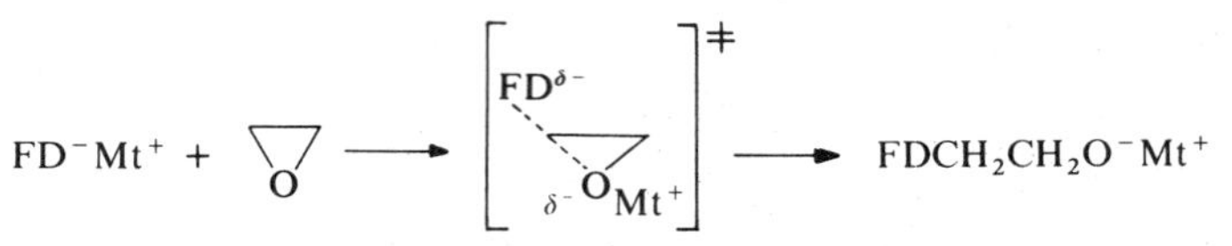

$$FD^-Mt^+ \; + \; \triangle_O \; \longrightarrow \; \left[\begin{array}{c} FD^{\delta -} \\ \diagdown \\ {}_{\delta^-}O_{Mt^+} \end{array} \right]^{\neq} \; \longrightarrow \; FDCH_2CH_2O^-Mt^+ \qquad (6)$$

FD⁻ = fluoradenyl anion

The cleavage of ethylene oxide by alkali metal salts of 9-methylfluorenyl anion was investigated in several ethereal solvents.[19–22] It was shown that it is a simple process of addition, whereas side reactions occur for fluorenyl alkali metal salts,[28] according to the mechanism shown in Scheme 3. In the first step, there is an addition of one molecule of monomer to one molecule of fluorenylsodium to give (6) which forms a new carbanion (7) by reaction with fluorenylsodium. During this process, half of the initial fluorenylsodium is consumed, giving inactive fluorene. Solvation by the monomer of carbanion (7) is instantaneous. It reacts slowly afterwards with ethylene oxide to give a dialkoxide. In the case of 9-methylfluorenylsodium, reactions (7) and (8) are impossible since there is no hydrogen in the 9-position. The reaction with ethylene oxide is first order in monomer and in salt. Here again, the tight ion-pairs were found to be more reactive than the cryptated pairs or free anions. For example, the bimolecular rate constant of the cleavage, determined in tetrahydropyran (THP) at $-30\,^{\circ}C$, is $8.5 \times 10^{-1}\,mol\,l^{-1}\,min^{-1}$ for a tight sodium salt, whereas it is only $3.0 \times 10^{-5}\,mol\,l^{-1}\,min^{-1}$ for the cryptated salt. The coordination of the cation with the cryptand makes it inaccessible for interaction with ethylene oxide, therefore the cleavage becomes much slower in the presence of the ligand. Interestingly, the rate constant for cleavage is about three times lower in THF than in THP, presumably because ethylene oxide, which interacts strongly with the cation, has to compete with the solvent, and the better the solvating power of the solvent the more difficult the competition.

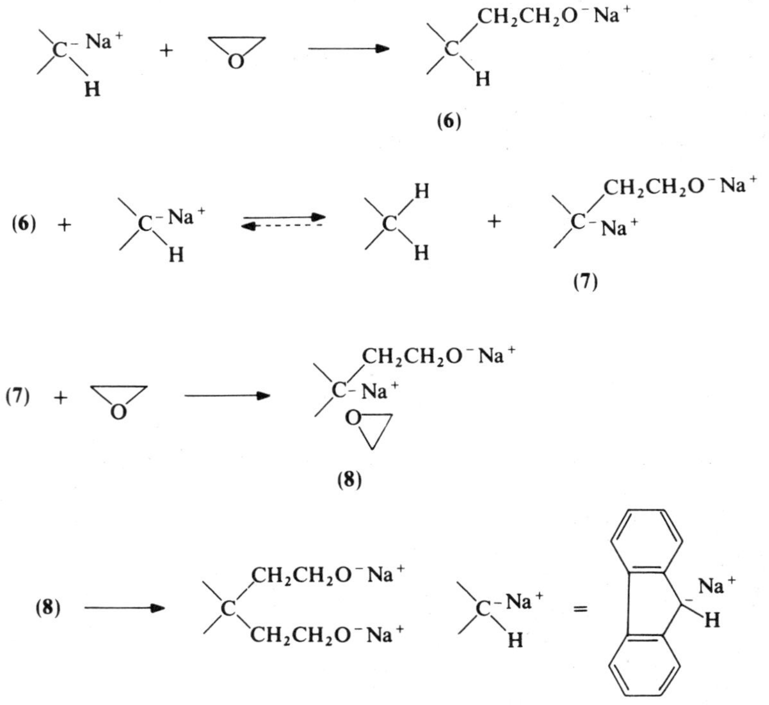

$$\underset{H}{\overset{}{>}}C^-Na^+ \; + \; \triangle_O \; \longrightarrow \; \underset{H}{\overset{CH_2CH_2O^-Na^+}{>}}C \qquad (7)$$
$$(6)$$

$$(6) \; + \; \underset{H}{\overset{}{>}}C^-Na^+ \; \rightleftharpoons \; \underset{H}{\overset{H}{>}}C \; + \; \underset{H}{\overset{CH_2CH_2O^-Na^+}{>}}C^-Na^+ \qquad (8)$$
$$(7)$$

$$(7) \; + \; \triangle_O \; \longrightarrow \; \underset{O\triangle}{\overset{CH_2CH_2O^-Na^+}{>}}C^-Na^+ \qquad (9)$$
$$(8)$$

$$(8) \; \longrightarrow \; \underset{CH_2CH_2O^-Na^+}{\overset{CH_2CH_2O^-Na^+}{>}}C \; \quad \underset{H}{\overset{}{>}}C^-Na^+ \; = \qquad (10)$$

Scheme 3

According to Scheme 4, Mt^+ is, of course, attracted by A^- but it can also interact with the solvent (S), ethylene oxide (EO) or with a ligand (L) such as crown ethers (CE) or cryptands, the strength of interaction following the sequence: THP < THF < EO < CE < (**5a**). This coordination will be stronger if the negative charge on the anion is more delocalized. For a given anion, it will also be stronger with smaller cations of high field strength, such as Li^+ or Na^+. Thus, if ethylene oxide can act as a coordinating agent (delocalization of the anion, absence of cryptand), the predominant effect is the electrophilic activation of ethylene oxide by the cation. The specific cation assistance is completely inhibited on addition of a cryptand, thus leading to a decrease in reactivity.

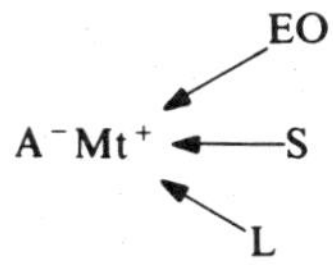

$$A^- Mt^+ \longleftarrow S \quad\substack{\nearrow EO \\ \\ \searrow L}$$

Scheme 4

For more localized anions, such as carbazyl salts, the interaction of the anion with Mt^+ is stronger and that of Mt^+ with ethylene oxide is smaller, weakening the electrophilic activation of the monomer.[16,19,21,27,120] The classical order — free ions more reactive than ion-pairs — is observed in that system. Introduction of two additional benzene rings in the carbazyl anion (**9**) leads to a delocalization of the negative charge on the nitrogen in the dibenzocarbazyl anion (**10**). Kinetic studies of the epoxide cleavage in THF at 20 °C led to the following bimolecular rate constants. $N^- K^+ : k_{i_\pm} = 21 \times 10^{-3}\,l\,mol^{-1}\,min^{-1}$; $N^- : k_{i_-} = 78 \times 10^{-3}\,l\,mol^{-1}\,min^{-1}$; $DBN^- K^+ : k_{i_\pm} = 5.7 \times 10^{-3}\,l\,mol^{-1}\,min^{-1}$; $DBN^- : k_{i_-} = 2.3 \times 10^{-3}\,l\,mol^{-1}\,min^{-1}$.

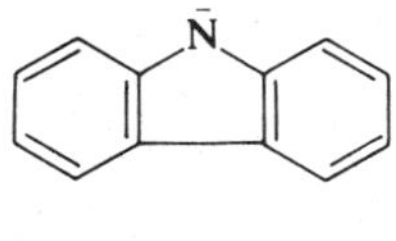

(9) N^- = carbazyl$^-$ 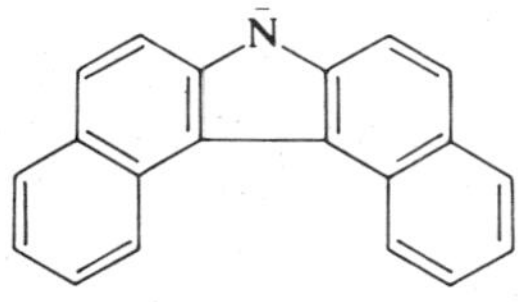   **(10)** DBN^- = dibenzocarbazyl$^-$

The rate constants k_{i_-} were derived from experiments made with cryptated salts. As a consequence of this delocalization, free ions as well as ion pairs in the dibenzo derivative are less reactive than the similar species of the carbazyl salt and, significantly, the order of reactivities of the free ions and ion pairs is reversed.

Interesting initiators for the polymerization of ethylene, propylene or 1,2-butene oxides were reported by Inoue *et al.*[4,29–31] Following their earlier studies on tetraphenylporphyrin–aluminum complexes, which were efficient initiators for propylene oxide polymerization and its copolymerization with carbon dioxide, they modified the original complex by substituting $AlEt_2Cl$ for $AlEt_3$. The active species is (**11**).

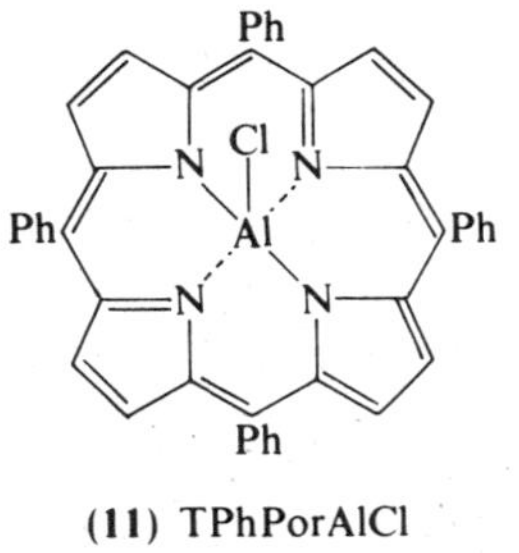

(11) TPhPorAlCl

The initiation reaction proceeds rapidly, involving insertion into the Al—Cl bond according to equation (11), followed by the analogous propagation, since $ClCH_2CHMeCH_2OH$ was isolated when an equimolar mixture of the initiator (**11**) and propylene oxide was hydrolyzed.

$$\text{TPhPorAlCl} + \overset{O}{\triangle} \longrightarrow \text{TPhPorAlOCH}_2\text{CH}_2\text{Cl} \tag{11}$$

Numerous anions and their ion pairs have been used for the ring opening of episulfides. Unexpectedly, it was found that triphenylmethylsodium in THF leads to the formation of poly-(propylene sulfide) having two living ends per macromolecule and a degree of polymerization which was double the theoretical value $\overline{DP}_n = [M]/[I]$.[32] The mechanism shown in Scheme 5 was proposed. A rapid electron-transfer from the carbanion to the monomer gives the triphenylmethyl radical which slowly dimerizes; NaS· dimerizes also and the subsequent sodium disulfide initiates the propagation.

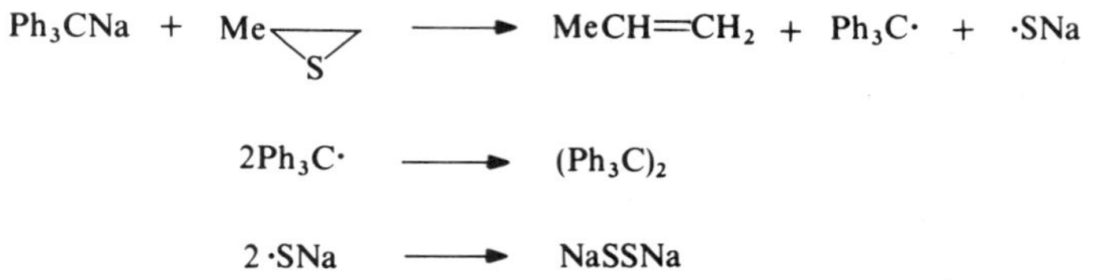

$$Ph_3CNa \ + \ Me\text{—}\underset{S}{\triangle} \quad \longrightarrow \quad MeCH{=}CH_2 \ + \ Ph_3C\cdot \ + \ \cdot SNa \qquad (12)$$

$$2Ph_3C\cdot \quad \longrightarrow \quad (Ph_3C)_2 \qquad (13)$$

$$2\cdot SNa \quad \longrightarrow \quad NaSSNa \qquad (14)$$

Scheme 5

The fluorenyl tetrabutylammonium salt also leads to the formation of poly(propylene sulfide), the molecular weights of which obey the relationship $\overline{DP}_n = 2[M]/[I]$; only half of the fluorene residues are incorporated in the polymer.[33] The reactions shown in Scheme 6 can take place. The reaction of (14) with propylene sulfide is slow and goes to completion only when a large excess of monomer is present. A similar reaction of fluorenyl salts was studied in more detail with ethylene oxide, but in that case there is a straightforward nucleophilic attack on the CH_2 of the monomer.

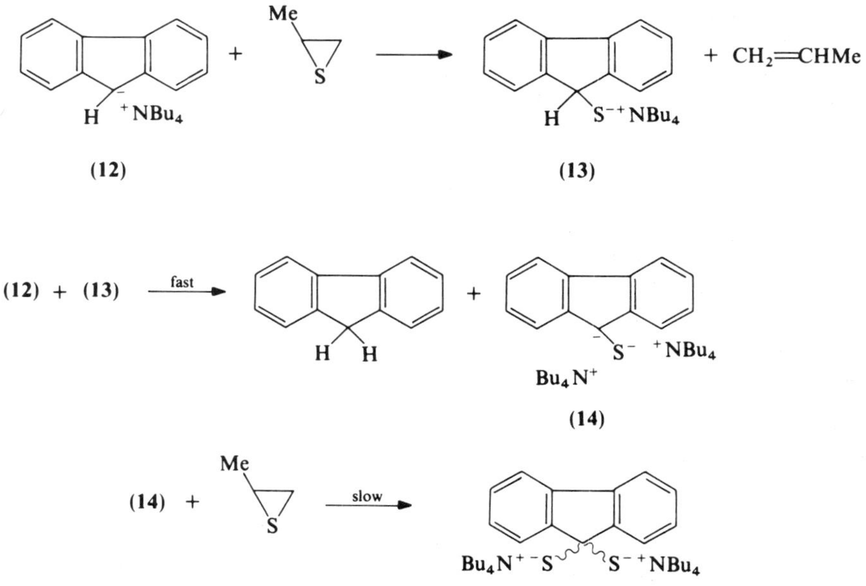

$$(15)$$

(12)   (13)

$$(16)$$

$$(17)$$

Scheme 6

Carbazyl salts were used for the preparation of monofunctional living polymers of propylene sulfide.[22] The effect of the nature of the counterion on the initiation rate in THF[34] will be examined in Section 32.3.3 after the discussion of similar results for the propagation reaction. In the same manner, the kinetic results obtained for the ring opening of 2-methyl-2-ethylthiirane with carbazyl salts in THP[35] are close to those relative to the propagation of propylene sulfide under similar conditions.[36]

Further discussion about the ion-pairing effect can be found in the section dealing with the propagation steps. As pointed out by Szwarc,[3] 'the usefulness of an initiator is determined by two factors. Firstly, how readily it reacts with a monomer in producing its respective anion. Secondly, the anion derived from a monomer may or may not interact with the cation introduced by the initiator. Hence, such cations that produce the most reactive propagating species are the most desired. Here the effect of cryptates is of a great importance since in many propagation reactions the cryptated ion-pairs are more reactive than the tight ion-pairs'.

32.2.3 Initiation by Alkyllithiums

Alkyllithiums are extremely reactive and versatile reagents. However, attempts to polymerize ethylene oxide *via* organolithium initiators were reported to be unsatisfactory.[11] The use of *N,N,N',N'*-tetramethylethylenediamine (TMEDA) with Bu^{n}Li as well as with 1,1'-dilithioferrocene leads to the formation of poly(ethylene oxide). Complexation of the lithium counterion partly destroys the aggregates and greatly enhances the basic character of the alkoxide anions, allowing them to propagate *via* ring opening.[38] However, no propagation of ethylene oxide was observed when (polystyryl)lithium with TMEDA or the lithium salt of poly(methyl methacrylate) with [12]-crown-4 were used as initiators.[39]

Organolithium initiators are effective in initiating polymerization of cyclic sulfides, but the mechanism differs from that of ethylene oxide since a desulfuration process may occur (equations 18 and 19).

$$RLi \ + \ \triangle_{Me}^{S} \ \longrightarrow \ CH_2{=}\underset{\underset{Me}{|}}{CH} \ + \ RSLi \tag{18}$$

$$RSLi \ + \ \triangle_{Me}^{S} \ \xrightarrow{\text{several steps}} \ RS{-}[CH_2\underset{\underset{Me}{|}}{CH}{-}S]_xLi \tag{19}$$

With propylene sulfide and ethyllithium in equimolar amounts at $-78\,°C$ in THF, the products after hydrolysis are propylene and ethanethiol only, and it is concluded that the S atom is the site of attack by the carbanion.[40] The thiolates formed in this first step may then initiate the polymerization of the thiirane. The initiation of propylene sulfide polymerizaton by α-sulfonyl type carbanions, $-SO_2CH_2Li$, proceeds in a conventional way.[41]

32.3 KINETICS OF POLYMERIZATION

32.3.1 Ethylene Oxide

The living character of ethylene oxide polymerization initiated by KOH was recognized in 1940.[42] Extensive studies of this reaction have been reported over the last 40 years. However, early investigations of polymerizations initiated by alkoxides dealt with systems involving alcohols which solubilized the scarcely soluble alkoxides. Since alcohols act as chain-transfer agents, living polymers are not formed in those systems. These difficulties were overcome by performing the polymerization in powerfully solvating aprotic media and, more recently, in aprotic solvents using complexing agents of the alkali metal cations.

32.3.1.1 Polymerization in the presence of alcohols

A study of ethylene oxide polymerization initiated by alkoxides in dioxane[43] was coupled with a study of the reaction of ethylene oxide with alkoxides dissolved in their parent alcohols.[44] The extremely low solubility of sodium alkoxides in dioxane made it necessary to add some alcohol to the medium, which acted as a transfer agent. The propagation is described in equation (20), and this process is coupled with chain transfer (equation 21).

$$\sim{}CH_2CH_2O^-,Na^+ \ + \ \triangle^{O} \ \longrightarrow \ \sim{}CH_2CH_2OCH_2CH_2O^-,Na^+ \tag{20}$$

$$P_nCH_2CH_2O^-, Na^+ \ + \ P_mCH_2CH_2OH \ \leftrightharpoons \ P_nCH_2CH_2OH \ + \ P_mCH_2CH_2O^-, Na^+ \tag{21}$$

The system is composed of two types of chains: the growing ones having the $-CH_2O^-,Na^+$ end-group and the 'dormant' ones possessing the $-CH_2OH$ end-group as noted by Szwarc.[45] The reversible chain-transfer equilibrium converts the growing polymers into dormant ones and *vice versa*. The propagation reaction was found to be first order in alcohol. The range of dielectric constant was 5–13 and the active species were considered to be the ion pairs of the alkali metal alkoxides. Moreover, the rate constant of propagation increased with the size of the counterion, from Na$^+$ to K$^+$.

In the associated study in pure methanol, the alkoxide was thought to be completely dissociated into ions, and the active species to be the free alkoxide ions.[44] Their reactivity was found to be higher than that of sodium ion-pairs determined in dioxane–methanol mixtures. In alcohols of lower dielectric constant, the reaction order in alkoxide fell below one. This was attributed to a lower degree of ionization.

Polymerization of ethylene oxide has been studied in chlorobenzene, anisole and diethyl ether at 100–110 °C with dry KOMe as the initiator.[46] The reaction was found to be first order in monomer, but second order in the alkoxide. A mechanism involving the binding of KOMe to the O of the oxide as a step facilitating the ring-opening stage was proposed. The role of alcohols when added to the medium[47] was suggested to be either solvating alkoxide, thus reducing its nucleophilic activity or taking part in the reaction as proton donors instead of the second alkoxide molecule.

32.3.1.2 *Polymerization in aprotic solvents*

Polymerizaton of ethylene oxide initiated by the sodium or potassium salts of the monomethyl ether of diglyme (2,2'-oxybisethanol), $MeOCH_2CH_2OCH_2CH_2O^-$, Mt^+, has been investigated in HMPA.[6,48] However, the polymerization was complex and the results were not readily explained. It was found that termination and chain-transfer reactions were avoided in this aprotic solvent which interacts strongly with cations.

Investigation of the kinetics of propagation in the case of the sodium salt revealed first-order character with respect to the monomer. However, the rate of propagation was found to be independent of the concentration of living ends [C] over a hundredfold range (7×10^{-5} mol l^{-1} $<$ [C] $< 10^{-2}$ mol l^{-1}). The first-order propagation rate constant, $-d\ln[M]/dt$, was found to be $\sim 3 \times 10^{-4}$ s^{-1} at 40 °C with $[M]_0 = 0.09$ mol l^{-1}.

Conductance measurements made on solutions of the initiator showed that the equivalent conductance was low and nearly constant in the range 1.4×10^{-2} down to 10^{-3} mol l^{-1}, although its value increased fivefold on further dilution to 6×10^{-5} mol l^{-1}. Thus, the sodium alkoxide behaved like a weakly dissociated electrolyte in HMPA in spite of the relatively high dielectric constant of this solvent, 26 at 40 °C. In contrast, the dissociation of sodium alkoxides such as sodium methoxide is virtually quantitative at 10^{-3} mol l^{-1} in methanol.[49] This can be explained by the fact that methanol is able to solvate both the anion and the cation through either the hydrogen or oxygen atom. On the other hand, sodium tetraphenylboride was found to behave as a strong electrolyte in HMPA. Its addition to polymerizing solutions caused a decrease in the rate of propagation. This means that the free alkoxide anions propagate faster than their sodium ion-pairs.

The viscosity of living poly(ethylene oxide) solutions decreased on termination with a slight excess of hydrochloric acid, indicating the presence of some unspecified aggregates. The authors suggested that the aggregation is responsible for the complex kinetics of propagation. An interesting explanation of these results was provided by Szwarc.[45] The low degree of direct dissociation of alkoxide ion-pairs, hindered by the lack of stabilization of alkoxide anions, does not preclude the formation of triple ions through the reaction shown in equation (22). Such a process could account for the independence of polymerization rate on the concentration of living ends, provided that the RO^- anions, coming from the buffered dissociation $RO^-,Na^+ \leftrightarrows RO^- + Na^+$, are the main contributors to the propagation. The dissociation of the $(RO^-,Na^+)_n$ aggregates, taking place at higher dilution, would then account for the increase in the equivalent conductance without affecting the rate of propagation.

$$2RO^-,Na^+ \rightleftharpoons RO^-,Na^+,{}^-OR \; + \; Na^+ \tag{22}$$

In the case of the potassium salt, propagation is again first order in monomer, but the order with respect to the potassium alkoxide is between 0 and 1. The rates are near to first order at lower concentrations of living ends, but nearly independent of concentration at the highest. Addition of potassium tetraphenylboride leads to a decrease in the rate of propagation. It seems that the aggregation of ion pairs is less pronounced in the potassium system than in the sodium, while their direct dissociation into free ions as well as their capacity to form triple ions is enhanced.

Simple behaviour is observed in the case of the cesium salt.[50] The reaction was found to be first order in monomer. The observed bimolecular propagation constant, $k_p = (-d\ln[M]/dt)/[C]$ with [C] = concentration of living ends, was found to be a linear function of $1/[C]^{1/2}$ in the absence of cesium tetraphenylboride, whereas it is a linear function of $1/[Cs^+]$ in the presence of cesium tetraphenylboride. These relationships are shown in Figure 1. They represent a classical behaviour, in which propagation occurs through ion pairs in equilibrium with a small amount of free ions which are much more reactive.[45] From the slopes of the lines shown in the figure, the propagation constant of the free anions was determined as $k_{p-} = 22$ l mol^{-1} s^{-1} and the dissociation constant of the ion pairs as 5×10^{-5} mol l^{-1}, at 40 °C. The latter value was confirmed by conductance measurements. The propagation constant of ion pairs was deduced from the intercept of the lines (Figure 1), $k_{p\pm} = 0.2$ l mol^{-1} s^{-1}.

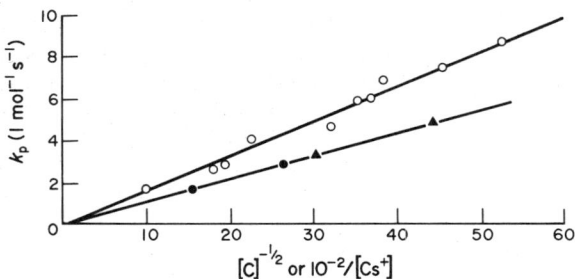

Figure 1 Plots of bimolecular propagation rate constant k_p of ethylene oxide polymerization in HMPA at 40 °C, with Cs$^+$ counterion, *vs.* reciprocal of square-root concentration of living ends $1/[C]^{1/2}$ (○) or reciprocal of Cs$^+$ concentration, $1/[Cs^+]$ (●), [C] = 1.90×10^{-3} mol l^{-1}; ▲, [C] = 8.07×10^{-4} mol l^{-1} (reproduced from ref. 50)

Several papers have been published on the anionic polymerization of ethylene oxide in DMSO, initiated by alkali metal alkoxides[51-55] and by dimsylpotassium or dimsylcesium (dimsyl =MeSOCH$_2^-$).[56] It was shown that ethylene oxide was readily converted to living polymer by potassium *t*-butoxide at room temperature without any noticeable chain-transfer reactions.[51] In all cases, the propagation was found to be first order in the monomer.

A reaction which was less than first order with respect to the reactive species was noticed in the case of Na$^+$ counterion.[56] This was attributed to the presence of ion pairs in equilibrium with a relatively small amount of free ions. The respective propagation rate constants were determined as $k_{p\pm} = 0.63 \times 10^{-2}$ l mol^{-1} s^{-1} and $k_{p-} = 1.2 \times 10^{-2}$ l mol^{-1} s^{-1} with $K_D = 6.15 \times 10^{-4}$ mol l^{-1} at 50 °C. In the case of K$^+$ and Cs$^+$ counterions, conductance measurements performed on meth-oxides[57] as well as on living poly(ethylene oxide)s[53, 56] showed that the active species are fully dissociated into free ions in the concentration range examined. Thus, K_D is equal to 4.71 $\times 10^{-2}$ mol l$^-$ for K$^+$ and to 9.43×10^{-2} mol l^{-1} for Cs$^+$, for living poly(ethylene oxide)s in DMSO at 25 °C.[56]

A high order in initiator, namely 1.9, was reported for the polymerization of ethylene oxide initiated by EtOC$_2$H$_4$OC$_2$H$_4$O$^-$,Cs$^+$ in DMSO at 50 °C,[53] as shown in Figure 2. On the other hand, first-order dependence on the concentration of alkoxides was claimed by other auth-ors.[52, 55, 56] An interesting attempt to reconcile these contradictory findings has been reported.[54] As revealed by conductance studies made on alkali metal alkoxides in DMSO,[57] the complex equilibria shown in Scheme 7 were established in the alkoxide–DMSO systems; where S$^-$ denotes free dimsyl anion; Mt$^+$, the alkali counterion; S$^-$,Mt$^+$, the ion pair of the dimsyl anion; RO$^-$, the free alkoxide

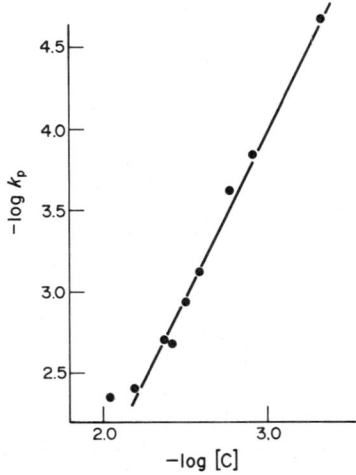

Figure 2 Plot of log of bimolecular propagation rate constant k_p of ethylene oxide polymerization in DMSO at 50 °C *vs.* log of concentration of the initiator [C], EtOC$_2$H$_4$OC$_2$H$_4$O$^-$,Cs$^+$; slope = 1.9 (reproduced by permission of Hüthig & Wepf from ref. 53)

anion; RO^-, Mt^+, the alkali metal alkoxide ion-pair; SH, dimethyl sulfoxide; ROH, the alcohol; and, finally, $RO(HOR)^-$, RO^-, $Mt^+(HOR)$, $RO(HOR)_2^-$ and $RO^-, Mt^+(HOR)_2$, aggregates of different order between dissociated and undissociated alkoxide and its respective alcohol. Addition of ethylene oxide to such a system leads to its polymerization through the formation of new alkoxides by reaction with RO^- and S^- if they are assumed to be active initiating species. These new alkoxides will attack DMSO to form the corresponding alcohols according to equilibrium (23), which will associate according to equilibrium (24), and so on . . . At any time, the rate of propagation is given by equation (30).

$$S^- \ + \ ROH \overset{K_0}{\rightleftharpoons} SH \ + \ RO^- \tag{23}$$

$$RO^- \ + \ ROH \overset{K_1}{\rightleftharpoons} RO(HOR)^- \tag{24}$$

$$S^- \ + \ Mt^+ \overset{K_2}{\rightleftharpoons} S^-, Mt^+ \tag{25}$$

$$RO^- \ + \ Mt^+ \overset{K_3}{\rightleftharpoons} RO^-, Mt^+ \tag{26}$$

$$RO(HOR)^- \ + \ ROH \overset{K_4}{\rightleftharpoons} RO(HOR)_2^- \tag{27}$$

$$RO(HOR)^- \ + \ Mt^+ \overset{K_5}{\rightleftharpoons} RO^-, Mt^+(HOR) \tag{28}$$

$$RO(HOR)_2^- \ + \ Mt^+ \overset{K_6}{\rightleftharpoons} RO^-, Mt^+(HOR)_2 \tag{29}$$

Scheme 7

$$R_p \ = \ [M] \Sigma k_i [\text{active species } i] \tag{30}$$

The problem arises when looking for the order with respect to the initiator. It seems that the free alkoxide anions are the most reactive species contributing to the propagation. The activity of aggregates is negligible compared to other species, since the addition of small amounts of alcohol strongly decreases the rate of propagation.[52] Moreover, the dimsyl anions do not play an important role otherwise the oligomers would have a significant sulfur content which is contrary to the experimental findings.[52] Thus, equation (30) would simplify to equation (31). Using the equilibrium constants reported in the conductance studies on alkali metal alkoxides,[57] it is possible to calculate the concentration of RO^- anions as a function of the initiator concentration $[RO^-, M^+]_0$ as shown in Figure 3. The results reveal two regions of alkoxide concentration: at higher dilution, the concentration of $[RO^-]$ depends on the initiator concentration to the power of 1.95, whereas, at higher concentrations, it is proportional to the initiator concentration to the power of 1. Since the rate of propagation is assumed to be proportional to $[RO^-]$, the contradictory claims seem to be reconciled.

$$R_p \approx k_{p-}[RO^-][M] \tag{31}$$

The kinetics of polymerization of ethylene oxide have also been reported in ethers, mainly in THF.[9,10,21,22,27,58-62] THF solutions of sodium, potassium and cesium naphthalene were used as the initiators and they gave living polymers.[9,10,58,59] An increase in the propagation rate was observed on increasing the size of the counterion. The study was complicated by strong association of alkoxide end-groups. This phenomenon was clearly shown by the gelation observed on addition of ethylene oxide to THF solutions of (polystyryl)sodium with two active end-groups.[63] The association was manifested by the low fractional kinetic order of propagation,[10] as illustrated in Figure 4. The rate of propagation was shown to be order 0.25 in concentration of sodium alkoxide over the complete range of concentration examined, and about 0.33 for the potassium and cesium salts.[10] For concentrations lower than 10^{-4} mol l^{-1}, the process tended towards first order in the case of K^+ and Cs^+ counterions. On this basis, the propagation constants of the unassociated alkoxide ion-pairs were estimated to be 0.94 l mol^{-1} s^{-1} for the potassium salt and 3.5 l mol^{-1} s^{-1} for the cesium salt at 70 °C.[10] The contribution of free alkoxide ions was discarded in view of the extremely low dissociation constant of alkoxide ion-pairs in THF: $K_D = 10^{-11}$ mol l^{-1} for K^+ and 7×10^{-10} mol l^{-1} for Cs^+ at 20 °C.[59] The association of alkoxide end-groups was also confirmed by

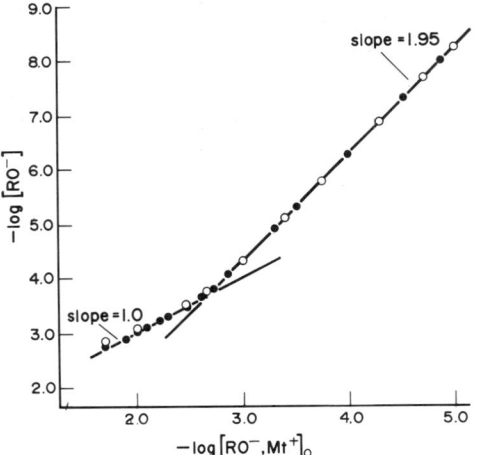

Figure 3 Change in free alkoxide anion concentration $[RO^-]$ with alkali metal alkoxide concentration $[RO^-,Mt^+]_0$ for potassium *t*-butoxide (○) and cesium *t*-butoxide (●) in DMSO at 25 °C (reproduced by permission of Hüthig & Wepf from ref. 54)

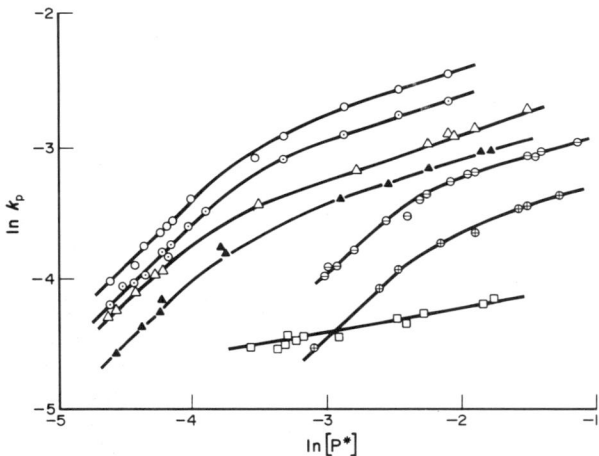

Figure 4 Logarithmic plots of propagation rate constant (s^{-1}) of ethylene oxide polymerization *vs.* concentration of active centres $[P^*]$: (□) Na^+, 99.4 °C, THF; (⊕) K^+, 30 °C; (⊖) K^+, 40 °C, in bulk; (▲) K^+, 70 °C; (△) K^+, 80 °C; (⊙) Cs^+. 60 °C; (○) Cs^+, 70 °C, all in THF, $[M]_0 \simeq 2 \, mol\,l^{-1}$ (reproduced from ref. 10)

viscometric measurements on living and terminated poly(ethylene oxide)s.[10] The stability of these aggregates decreases from Li^+ to Cs^+.

The propagation was found to be first order in ethylene oxide except at the very beginning of the reaction, where the rate constant k_p increased with the chain length, reaching a limiting value at $n = 4\text{–}6$ (see equation 32).[58,62] This autoacceleration process was observed even at low concentrations of active centres ($< 10^{-4} \, mol\,l^{-1}$). A possible explanation is that this is the result of progressive solvation of the ion pair by the growing macromolecule. Several indirect proofs of this phenomenon have been afforded.

$$R(OC_2H_4)_nO^-,Mt^+ \quad + \quad \triangle\!\!\!\!O \quad\longrightarrow\quad R(OC_2H_4)_{n+1}O^-,Mt^+ \tag{32}$$

The poly(ethylene glycol) dimethyl ethers with the general formula $MeO\text{---}(C_2H_4O)_x\text{---}Me$, referred to as glymes, are very effective alkali-metal-solvating agents.[64] Quantitative studies of solvation of ion pairs with poly(ethylene oxide) have been performed using different physicochemical techniques.[65-67] It is also well known that remarkable rate accelerations are observed in several nucleophilic reactions on adding poly(ethylene oxide) to the medium.[68] This was explained by a cooperative coordination of the oxygen atoms of poly(ethylene oxide) with metal cations, promoting ion dissociaton of the reagent.

Several authors have tried to obtain more direct experimental proofs of this self-solvation effect during the anionic polymerization of ethylene oxide. They examined the dissociation constant, K_D, of several mono-[60] and bi-functional[69] oligomers of ethylene oxide. Though there are some important discrepancies between the two sets of results, K_D was found to increase with the number of ether units in the chain. This was attributed to the intramolecular solvation of the counterion K^+ by the oxygen of the macromolecule nearest to the chain end. The mechanism of participation in conductivity of the cation entrapped by the polymer chain is still unclear. Thus, the conductance results on living poly(ethylene oxide)s in THF should be considered as tentative. Moreover, no attention was paid to the perturbations afforded by the glass walls of the conductance apparatus. Poly(ethylene oxide) in THF can dissolve a non-negligible amount of Na^+ from the glass since it is a powerful solvating agent; the conductivity of these species can be higher than that of the studied electrolyte.

The influence of different ethereal solvents upon the rate of propagation of ethylene oxide has been examined.[59] It was found to be nearly the same in THF ($k_p = 0.94 \, \mathrm{l \, mol^{-1} \, s^{-1}}$), in 1,2-dimethoxyethane (DME) ($k_p = 0.71 \, \mathrm{l \, mol^{-1} \, s^{-1}}$) and in the dimethyl ether of diethylene glycol ($k_p = 0.8 \, \mathrm{l \, mol^{-1} \, s^{-1}}$), at 70 °C, with K^+ as the counterion. This insensitivity of the anionic polymerization of ethylene oxide to an 'external' solvent was explained by the strong shielding effect of the polymer.

32.3.1.3 Polymerization in the presence of complexing agents

Addition of only two equivalents of DMSO to potassium hydroxide formed *in situ* by addition of water to a solution of potassium naphthalene in THF was found to increase the rate of polymerization of ethylene oxide threefold.[70] This increase cannot be due to an increase in the bulk dielectric constant of the medium since the amounts of added DMSO were small, but solvation effects of the potassium cation should be important.

The results concerning the effects of crown ethers, such as [18]-crown-6 (4) and its analogues, upon the rate of propagation of ethylene oxide are somewhat divergent. The sodium hydroxide catalyzed oligomerization of ethylene oxide in 1-butanol at 100 °C was found to be accelerated more than tenfold with (4).[71] A similar acceleration was noticed for the polymerization of ethylene oxide initiated by potassium benzoate in the presence of (4).[26] However, an important decrease of the rate of propagation of this monomer, in THF at 80 °C with K^+ as the counterion, was observed on adding dibenzo-[18]-crown-6.[60] The authors assumed that the reactivity of crowned ion-pairs was lower than that of the self-solvated species.

The association of alkali metal alkoxide end-groups of living poly(ethylene oxide) can be avoided by complexing the cations with suitable cryptands.[21,22,27,61,72,73] Polymerization of ethylene oxide initiated by carbazylpotassium complexed by the [2.2.2] ligand (5a) gave living polymers,[72] termination and transfer being prevented if an excess of carbazole is avoided.[74] Kinetic measurements were performed at concentrations of living ends lower than $7 \times 10^{-4} \, \mathrm{mol \, l^{-1}}$, since associations of ion pairs became too important and could not be neglected at higher concentrations. Use of cryptates led to the formation of a simple equilibrium between cryptated alkoxide ion-pairs and free ions. The dissociation constant of the cryptated potassium salt was measured by conductivity, $K_D = 3 \times 10^{-7} \, \mathrm{mol \, l^{-1}}$ in THF at 20 °C. This value was in good agreement with that determined from two sets of kinetic experiments made in the absence and in the presence of potassium tetraphenylboride complexed by (5a). The dissociation constant of this salt was determined separately in THF: $K_D = 8.1 \times 10^{-5} \, \mathrm{mol \, l^{-1}}$ at 20 °C.[75] A plot of the observed bimolecular rate constant, k_p *vs.* α, the fraction of free ions, is shown in Figure 5. The values of $k_{p\pm}$ and k_{p-} were found to be $1.5 \, \mathrm{l \, mol^{-1} \, min^{-1}}$ and $100 \, \mathrm{l \, mol^{-1} \, min^{-1}}$, respectively, at 20 °C in THF (Table 1).

These investigations were extended later to the cesium salt, using the spheroidal cryptand SC 24 (15) as the complexing agent.[73] Its synthesis and properties were reported in 1975.[76] The conductance study of cesium tetraphenylboride complexed by this ligand led to its dissociation constant, which was equal to $1.1 \times 10^{-4} \, \mathrm{mol \, l^{-1}}$ at 20 °C. A linear relation was obtained for k_p plotted *vs* α, the fraction of free anions, in the case of experiments made in the absence as well as in the presence of the cryptated cesium tetraphenylboride, on assuming that k_{p-} has the same value whatever the nature of the counterion (Figure 5). Thus, $k_{p\pm}$ was found to be $5.6 \, \mathrm{l \, mol^{-1} \, min^{-1}}$ with k_D equal to $1 \times 10^{-6} \, \mathrm{mol \, l^{-1}}$ at 20 °C.

Kinetic measurements were also made with K^+ and Cs^+ as counterions in the absence of cryptand, for concentrations in living ends [C] lower than $1.5 \times 10^{-4} \, \mathrm{mol \, l^{-1}}$ in order to avoid the associations of ion pairs. This point was verified by viscometric measurements made on living and

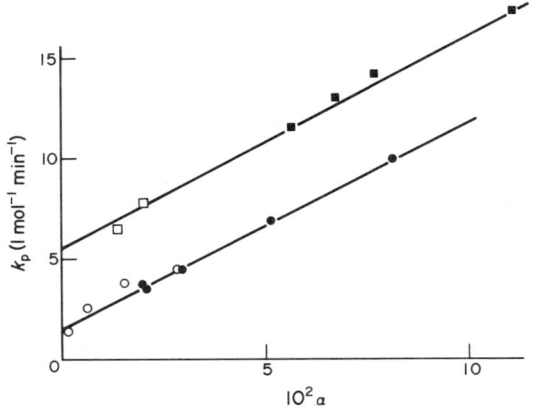

Figure 5 Plot of propagation rate constant k_p of living poly(ethylene oxide) propagation *vs.* molar fraction of free anions α with cryptates as counterions, in THF at 20 °C: ($\bullet$) K$^+$ + (**5a**); ($\bigcirc$) with Ph$_4$BK + (**5a**); ($\blacksquare$) Cs$^+$ + (**15**); ($\square$) with Ph$_4$BCs + (**15**) (reproduced by permission of the American Chemical Society from ref. 61)

Table 1 Propagation Rate Constants of Ion Pairs $k_{p\pm}$ and Free Ions k_{p-} for the Anionic Polymerization of Ethylene Oxide in THF at 20 °C[a]

Counterion	K_D (mol l^{-1})	$k_{p\pm}$ ($l\,mol^{-1}\,min^{-1}$)	k_{p-}
Na$^+$	—	0	—
K$^+$	1.8×10^{-10} [b]	2.9	—
Cs$^+$	2.7×10^{-10} [b]	7.3	—
Na$^+$ + (**5a**)	—	0	—
K$^+$ + (**5a**)	3×10^{-7}	1.5	100
Cs$^+$ + (**15**)	1×10^{-6} [c]	5.6	100[d]

[a] Reproduced by permission of the American Chemical Society from ref. 61.
[b] A. A. Solovyanov and K. S. Kazanskii, *Vysokomol. Soedin., Ser A*, 1972, **14**, 1063.
[c] Determined from kinetic data by assuming that $k_{p-} = 100\,l\,mol^{-1}\,min^{-1}$.
[d] k_{p-} is assumed to be the same as for K$^+$ + (**5a**).

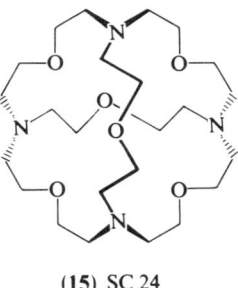

(**15**) SC 24

terminated poly(ethylene oxide)s. The results obtained with different counterions are listed in Table 1. Free alkoxide ions are about 70 times more reactive than K$^+$ cryptated ion-pairs. It is difficult to compare the values of k_{p-} obtained in different solvents since they were measured at different temperatures. However, free ions are much more reactive in THF ($k_{p-} = 100\,l\,mol^{-1}\,min^{-1}$ at 20 °C) than in methanol ($k_{p-} \simeq 1.2 \times 10^{-2}\,l\,mol^{-1}\,min^{-1}$ at 30 °C)[44] and DMSO ($k_{p-} = 0.7\,l\,mol^{-1}\,min^{-1}$ at 50 °C).[56] This can be explained by solvation of oxanions by these latter solvents.[77]

Cryptated ion-pairs are, surprisingly, less or as reactive as the corresponding non-complexed ion-pairs though the charges are more separated: K_D is 1700 times higher in the case of K$^+$(**5a**) than for K$^+$ alone. Solvation of K$^+$ by the oxygen atoms of the terminal units of the living poly(ethylene oxide) chains, described as the self-solvation effect,[62] would make the negative charge more available for the ring opening of the incoming monomer and, thus, would increase the reactivity of the potassium ion-pair compared to that of a simple alkoxide. Another additional effect can explain the surprisingly low reactivity of cryptated ion-pairs. The dissociation constant found for cryptated potassium alkoxide ion-pairs in THF ($K_D = 3 \times 10^{-7}$ mol l^{-1}) led to an interionic distance a equal to 4.6 Å according to the classical Fuoss equation. This value was too small compared to the results obtained for cryptated living poly(propylene sulfide)[78] as well as for cryptated tetraphenylborides in THF.[75] This might mean that either K$^+$ was not located inside the cavity of the ligand or the oxanion could penetrate into the cavity of the cryptand. This last explanation was consistent with comparative conductance data obtained for model compounds,[27,61] as shown in Table 2. The K_D of cryptated potassium 2,6-dimethylphenoxide is higher (about four times) than that of the 3,5-dimethylphenoxide cryptated salt. This is due to the methyl groups, which hinder the penetration of the oxanion into the cavity of the cryptate when they are located in the 2- and 6-positions. Moreover, a good agreement between the theoretical value of the interionic distance and the value found for the cryptated 2,6-dimethylphenoxide was observed.

Table 2 Conductance Data for Cryptated Alkoxides and Phenoxides in THF at 20 °C [a]

Salt	K_D (mol l^{-1})	a_K^F (Å)[b]
∿CH$_2$CH$_2$O$^-$,K$^+$ + (5a)	3×10^{-7}	4.6
2,6-Me O$^-$,K$^+$ + (5a)	1.9×10^{-6}	5.3
3,5-Me O$^-$,K$^+$ + (5a)	6.8×10^{-6}	6.0
Theoretical value	—	5.9

[a] Reproduced by permission of the American Chemical Society from ref. 61.
[b] Interionic distance according to Fuoss' equation (ref. 75).

32.3.2 Propylene Oxide and Higher Epoxides

Depending on the type of monomer and initiator system, different chain-transfer reactions may take place in the anionic polymerization of propylene oxide and higher epoxides. Chain transfer to the monomer leading to the formation of new active centres containing aliphatic double bonds was found to occur in the case of propylene oxide (2-methyloxirane), 2,3-epoxypropyl phenyl ether and 2-neopentyloxirane according to equations (33)–(35).

$$\sim O^-,Mt^+ + Me\text{---}\triangle_O \xrightarrow{k_{tr}} \sim OH + CH_2\!\!=\!\!CHCH_2O^-,Mt^+ \qquad (33)$$

$$\sim O^-,Mt^+ + PhOCH_2\text{---}\triangle_O \xrightarrow{k_{tr}} \sim OH + PhOCH\!\!=\!\!CHCH_2O^-,Mt^+ \qquad (34)$$

$$\sim O^-,Mt^+ + Bu^tCH_2\text{---}\triangle_O \xrightarrow{k_{tr}} \sim OH + Bu^tCH\!\!=\!\!CHCH_2O^-,Mt^+ \qquad (35)$$

The problem of the structure, mechanism and kinetics of formation of C=C double bonds in the synthesis of polyethers has been the subject of large number of papers over the last 30 years, for propylene oxide,[51,79–89] for 2,3-epoxypropyl phenyl ether and its derivatives[87,90–92] as well as for 2-neopentyloxirane.[93] The influence of crown ethers such as (4) on the course of polymerization of propylene oxide in bulk, initiated by potassium alkoxides in the presence of alcohols, has been examined.[89] An increase in the rate of propagation was observed together with an increase in the molecular weight of the polymers. Very interestingly, the percentage of double bonds was found to decrease on increasing the concentration of crown ether. The results were explained by assuming that the propagation occurred mainly through the free alkoxide anions, the proportion of which was increased on adding (4). The chain transfer to the monomer was thought to occur through the ion pairs with the intervention of K^+ cations, which was lowered when they were complexed by crown ether (Scheme 8).

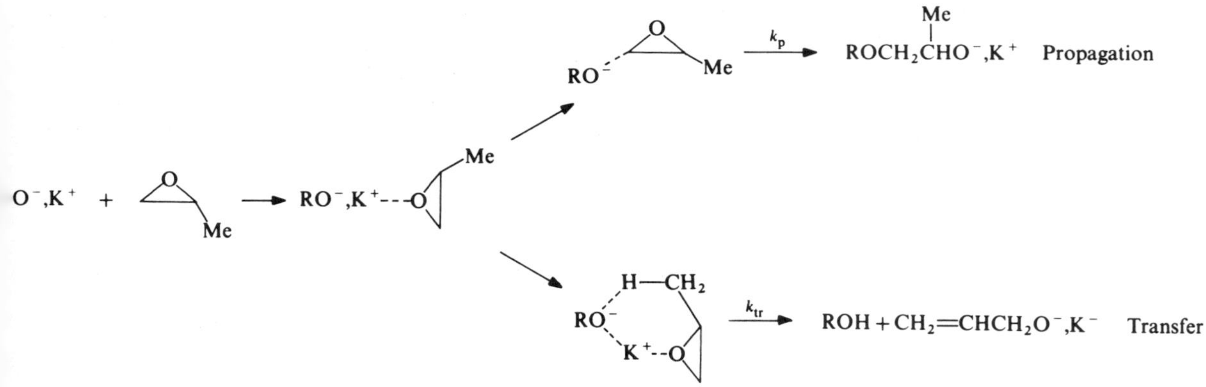

$$O^-,K^+ \; + \; \underset{Me}{\triangle\!\!\!-\!\!\!O} \; \longrightarrow \; RO^-,K^+ \text{---} O$$

Propagation:
$$RO^- + \overset{O}{\triangle}\!\!-Me \;\xrightarrow{k_p}\; ROCH_2\overset{\underset{|}{Me}}{C}HO^-,K^+$$

Transfer:
$$RO^-,K^+\text{---}O \;\xrightarrow{k_{tr}}\; ROH + CH_2{=}CHCH_2O^-,K^-$$

Scheme 8

As in the case of ethylene oxide polymerization, it was assumed that a dimsyl anion, formed in the reaction of alkoxide with the solvent (equation 36) and by chain transfer to the solvent (equation 37) was responsible for initiating the polymerization of propylene oxide and higher epoxides when using potassium *t*-butoxide as the initiator in DMSO.[51,84,86,92,94,95] The above assumption was not confirmed, however, by experiments in the case of 2,3-epoxypropyl phenyl ether polymerization since no significant amounts of sulfur were found in the polymer.[91]

$$Bu^tO^-,K^+ \; + \; Me\underset{\underset{O}{\|}}{S}Me \;\underset{}{\overset{K_i}{\rightleftharpoons}}\; Me\underset{\underset{O}{\|}}{S}CH_2^-,K^+ \; + \; Bu^tOH \tag{36}$$

$$\leadsto O^-,K^+ \; + \; Me\underset{\underset{O}{\|}}{S}Me \;\rightleftharpoons\; \leadsto OH \; + \; Me\underset{\underset{O}{\|}}{S}CH_2^-,K^+ \tag{37}$$

A detailed analysis of polymers of 2,3-epoxypropyl phenyl ether was made recently.[96] They were prepared with potassium *t*-butoxide in THF and in DMSO at 25 °C, in the absence and in the presence of (4) with a molar ratio [4]/[K$^+$] equal to 1. It was shown that, irrespective of the chain-transfer reaction to the monomer depicted in equation (34), another reaction leading to the formation of carbonyl groups occurred according to Scheme 9. The carbanion formed in equation (38) acted as a new polymerization initiator, which was transformed into an alkoxide active centre after the addition of a monomer molecule.

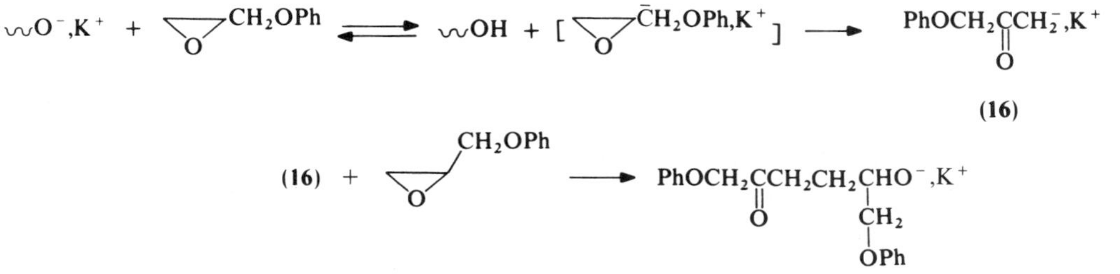

$$\leadsto O^-,K^+ \; + \; \underset{O}{\triangle}\!\!CH_2OPh \;\rightleftharpoons\; \leadsto OH \; + \; [\underset{O}{\triangle}\!\!\overset{\cdot}{C}H_2OPh,K^+] \;\longrightarrow\; PhOCH_2\underset{\underset{O}{\|}}{C}CH_2^-,K^+ \tag{38}$$

$$\textbf{(16)}$$

$$\textbf{(16)} \; + \; \underset{O}{\triangle}\!\!\overset{CH_2OPh}{} \;\longrightarrow\; PhOCH_2\underset{\underset{O}{\|}}{C}CH_2CH_2\underset{\underset{\underset{OPh}{|}}{CH_2}}{C}HO^-,K^+ \tag{39}$$

Scheme 9

Kinetic studies of the anionic polymerization of propylene oxide and higher epoxides were performed in bulk and in polar solvents. In all cases, the propagation was first order in the monomer. In DMSO, some discrepancies appeared in the order with respect to initiator. First order was observed for the polymerization of propylene oxide,[51,92] 2-*t*-butyloxirane,[94] and 2,3-epoxypropyl phenyl ether and its derivatives,[92] initiated by potassium *t*-butoxide. However, nearly second-order was found in the case of propylene oxide (1.7)[84] as well as for 2-ethyloxirane (1.8)[95] with the same initiator in DMSO. Some authors tried to reconcile these controversial findings[53] (see Section 32.3.1.2). It must be noted that the studies on 2-ethyloxirane polymerization were complicated by phase separation, the polymer being insoluble in DMSO. On performing the experiments in mixtures of DMSO with THF, in which the polymer was soluble, the order of the reaction was equal to one.

The effect of different additives on the bulk anionic polymerization of propylene oxide has been examined.[97] The rate of propagation increased in the following order: MeOH < bulk polymerization < DMF < HMPA, which is in agreement with the donor–acceptor properties of these compounds. The results found for the polymerization of propylene oxide in HMPA and in DMSO show that the rate constant of propagation was higher in the former solvent than in the latter by a factor of about three.[92]

Kinetic studies on the polymerization of propylene oxide were made in bulk at 120 °C using potassium salts of poly(ethylene glycol) monobutyl ethers, $Bu\{-OCH_2CH_2\}_xO^-,K^+$ with $x = 1, 4, 5$ and 13.[98] An increase in the rate constant of propagation was observed on increasing x. This was attributed to a specific solvation of potassium cations. In the same manner, noticeable increases in the rate of polymerization of epoxides were found on adding crown ethers.[26,89,92,99] For instance, the rate constant of propagation of the polymerization of propylene oxide initiated by potassium *t*-butoxide, in HMPA at 25 °C, was multiplied by two on adding dicyclohexyl-[18]-crown-6 with [crown ether]/[K^+] equal to one.[92]

The following order of reactivity of epoxides was found: ethylene oxide > propylene oxide > 2-ethyloxirane in HMPA[92] and in DMSO[94] using potassium *t*-butoxide as the initiator. Moreover, it was found that 2-neopentyloxirane polymerized between 70 to 30 times more slowly than propylene oxide at 30 °C with potassium as the counterion.[93] These results are consistent with steric hindrance considerations.

The relative rates of propagation to transfer, k_p/k_{tr}, have been estimated for several epoxides under various conditions.[51,82,83,91–93] The values of this ratio were very similar for propylene oxide and 2-neopentyloxirane except for those measured in DMSO. This solvent is known to be excellent for proton abstraction reactions and thus gave a low value for k_p/k_{tr}. Transfer reactions were found in polymerizations of 2,3-epoxypropyl phenyl ether initiated by potassium *t*-butoxide in bulk at 50 °C,[92] and by sodium methoxide in DMSO at 50 °C[91] ($k_p/k_{tr} \simeq 75$ in both cases). However, these reactions did not occur significantly for experiments made at 25 °C in THF/DMSO mixtures (1v/2v) using potassium naphthalene as the initiator.[100]

32.3.3 Propylene Sulfide

The absence of transfer or termination reactions has allowed a thorough investigation of the propylene sulfide propagation reaction and of the nature and reactivities of the active centres.[5,7,22,61] This is probably the most complete kinetic study for the polymerization of any ring compound.

The configuration of poly(propylene sulfide) samples prepared with anionic initiators was studied by ^{13}C NMR.[101,102] No significant amount of head-to-head or tail-to-tail irregularities could be detected. Moreover, the polymers are atactic, except those obtained with soluble cadmium thiolates as initiators, which are isotactic.[103]

It was shown that in polar (or even weakly polar) media and also in the presence of electron-donor additives, the propylene sulfide polymerization initiated by various initiating systems yields only poly(propylene sulfide) without formation of propylene and SS bonds in the polymer chain (except during the initiation step with radical anions and triphenylmethylsodium: see Section 32.2.1). However, in bulk or in non-polar media, lithium alkoxides and their complexes with alkyl lithiums and thiolates give polymers containing up to 100% SS bonds.[104–106] The results were explained according to the HSAB principle.[107]

Kinetic studies were performed in THF at −30 °C, using sodium naphthalene as the initiator.[108] Living polymers form aggregates at concentrations higher than 10^{-3} mol l^{-1}, but at lower concentrations the system is simple, being composed of free ions and contact ion-pairs only. The absence of association for these concentrations was shown by comparative viscometric measurements on living and 'killed' polymer solutions. The first dissociation constant values K_D were determined on bifunctional living polymers prepared from sodium naphthalene and were rather approximate.[109]

This was due to the rather arbitrary values assumed for Λ_0 and to irreproducibilities when high initiator concentrations were used. A good agreement was observed later between the K_D of monofunctional polymers (obtained with carbazylsodium as the initiator) and that of bifunctional polymers, between $-40\,°C$ ($K_D = 5 \times 10^{-8}$ mol l^{-1}) and $0\,°C$ ($K_D = 0.7 \times 10^{-8}$ mol l^{-1}).[110,111] The presence of the monomer slightly affects the dielectric constant of the solution, causing some increase in K_D.

The apparent bimolecular rate constants k_p for the propagation reaction were measured by dilatometry in high vacuum systems.[108,110,111] The order of the reaction with respect to monomer is about one for propylene sulfide concentrations lower than 1 mol l^{-1}, but is a little higher at higher concentrations due to the influence of the dielectric constant of the monomer (12.9 at $-30\,°C$ compared to 9.4 for THF).[111] The values of k_p vary with the living-end concentration, in agreement with a simple equilibrium between ion pairs and free ions. Using these data as well as the kinetic data obtained from runs involving sodium tetraphenylboride, the individual rate constants were determined, viz. $k_{p+} = 1 \times 10^{-3}$ l mol^{-1} s^{-1} and $k_{p-} = 1.7$ l mol^{-1} s^{-1} at $-40\,°C$, and $k_{p\pm} = 3.2 \times 10^{-2}$ l mol^{-1} s^{-1} and $k_{p-} \simeq 40$ l mol^{-1} s^{-1} at $0\,°C$.[111] Propagation takes place mainly through free ions which are much more reactive than the tight ion-pairs.

Van't Hoff plots for K_D between -50 and $+20\,°C$ give $\Delta H_D = -30.1$ kJ mol^{-1} at $-30\,°C$ (-34.3 kJ mol^{-1} at $20\,°C$), with no evidence for the presence of non-intimate ion-pairs.[112] The same conclusion was drawn from the Arrhenius plots for k_{p-} and $k_{p\pm}$. The activation energy of the free ion propagation was estimated as 40 kJ mol^{-1}, corresponding to the frequency factor of $\simeq 10^9$ l mol^{-1} s^{-1}, while the respective values for the Na$^+$ ion-pairs were reported as 46 kJ mol^{-1} and 2×10^7 l mol^{-1} s^{-1}.[111] The main difference in reactivities comes from the difference in the frequency factors. The ratio $k_{p-}/k_{p\pm}$ is of the same order of magnitude as for styrene (with the same counterion) but there is no contribution of solvent-separated ion-pairs, unlike the case of styrene in THF where these are the dominant non-dissociated kinetic species, the negative charge on the thiolate end-groups being more localized than on the carbanions of living polystyrene.

A kinetic study was also performed in THP with Na$^+$ as the counterion.[113] The overall rate of propagation is much smaller than in THF. K_D does not change significantly between -30 and $22\,°C$ and has a value of 3×10^{-12} mol l^{-1}. At $20\,°C$, using low living-end concentrations ($< 4 \times 10^{-4}$ mol l^{-1}), the same rate law as in THF is obeyed, with $k_{p\pm} = 5.8 \times 10^{-2}$ l mol^{-1} s^{-1} and $k_{p-} = 250$ l mol^{-1} s^{-1}. The lower rate of reaction in THP compared with THF is due mainly to the lower concentration of free ions in THP at a given living-end concentration.

At higher living-end concentrations ($> 10^{-3}$ mol l^{-1} in THF, 4×10^{-4} mol l^{-1} in THP), the rate becomes independent of thiolate concentration.[113,114] Moreover, viscometric measurements show that living polymers have a higher molecular weight than the corresponding 'killed' polymers. However, it is not possible to measure the aggregation coefficient n of equation (40). The independence of the rate of propagation on the living-end concentration [C] seems to indicate that there is a concentration above which the excess living-ends form inactive aggregates.

$$n \sim\!\!\sim S^-,Na^+ \quad \rightleftharpoons \quad (\sim\!\!\sim S^-,Na^+)_n \qquad (40)$$

With lithium counterion in THF, associations between polymer chains become much stronger. Gel formation has been observed at high concentrations and was attributed not only to associations of living ends but also to their interactions with bivalent sulfur atoms located in the polymer chains.[115] Upon addition of 5 to 10 mol% of tetrahydrothiophene, which provides 'monomeric' bivalent sulfur necessary to prevent the formation of a three-dimensional network, the gel was broken up, but the viscosity of the polymer solution still remained at abnormally high levels.[116]

Kinetic studies have been extended to other counterions such as Bu$_4$N^{+} [33] and Cs$^+$.[78] The most interesting results were obtained with cryptated cations.[61,78,117,118] The values of the apparent propagation rate constant $k_p = R_p/[M][C]$ (with R_p = rate of polymerization, [M] and [C] = monomer and living-end concentrations) were determined for a given counterion, at a fixed temperature. Since the main ionic species are ion pairs in equilibrium with free ions, k_p depends on the fraction of free ions according to equation (41).

$$k_p = (1 - \alpha)k_{p\pm} + \alpha k_{p-} \qquad (41)$$

Examples of plots of k_p vs. α are shown in Figure 6, for Na$^+$ + (5a) in THF and in THP, at $-30\,°C$, giving $k_{p\pm}$ and k_{p-}. It should be mentioned that the cryptands can be used in nearly stoichiometric amounts (excess of about 10%) since the complexation constant giving the 'cryptates' is extremely high.

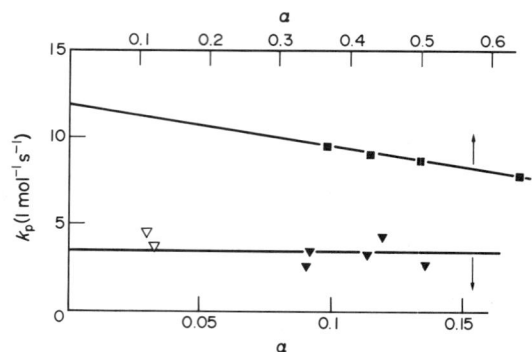

Figure 6 Linear dependence of the apparent bimolecular rate constant k_p of living poly(propylene sulfide) propagation on the fraction of free ions α with Na$^+$ + (**5a**) as counterion at $-30\,°C$: (■) in THF; (▼) in THP; (▽) with Ph$_4$BNa + (**5a**) in THP (reproduced by permission of the American Chemical Society from ref. 61)

The values of $k_{p\pm}$ and k_{p-} found in THF at $-30\,°C$ are listed in Table 3, together with the values of the dissociation constant K_D and of the interionic distance a_K^F for the ion pairs. The values of k_{p-} may be considered to be the same, within experimental error. The $k_{p\pm}$ values increase rapidly with the size of the counterion. With Cs$^+$, $k_{p\pm}$ is already about 100 times higher than for Na$^+$. In the case of Bu$_4$N$^+$, $k_{p\pm}$ is about half k_{p-}; however, there is also some termination reaction, probably of the type shown in equation (42) since most of the amine can be recovered after a long time.[119]

Table 3 Propagation Rate Constants of Free Ions and Ion Pairs for the Anionic Polymerization of Propylene Sulfide in THF at $-30\,°C^a$

Counterion	$10^5 K_D\,(\mathrm{mol\,l^{-1}})$	$a_K^F\,(\text{Å})^e$	$k_{p\pm}\,(\mathrm{l\,mol^{-1}\,s^{-1}})$	$k_{p-}\,(\mathrm{l\,mol^{-1}\,s^{-1}})$
Na$^+$	0.0054^b	3.7	0.0025	3.8
Cs$^+$	—	4.3	0.23^c	—
Bu$_4$N^{+d}	1.90	6.5	1.7	3.5
Li$^+$ + (**5d**)	2.53	6.7	4.9	2.8
Na$^+$ + (**5b**)	3.35	7.0	8.9	3.0
Na$^+$ + (**5e**)	4.55	7.3	8.3	3.3
Cs$^+$ + (**5c**)	5.50	7.5	8.7	4.7

^{a} Reproduced by permission of the Americal Chemical Society from ref. 61.
^{b} Obtained from kinetic experiments made with and without Ph$_4$BNa.
^{c} Extrapolated from the value at $+15\,°C$ using $E_\pm = 46\,\mathrm{kJ\,mol^{-1}}$.
^{d} Obtained from kinetic data in ref. 33 using a calculated K_D.
^{e} See Table 2, footnote b.

$$\sim\!\!\sim\!\!\mathrm{CH_2CHMeS^-,\,^+NBu_4} \longrightarrow \sim\!\!\sim\!\!\mathrm{CH_2CHMeSBu} + \mathrm{NBu_3} \tag{42}$$

It is interesting to note that the polymerization of propylene sulfide can be initiated electrochemically between -10 and $50\,°C$, in the presence of tetrabutylammonium perchlorate.[15] The observed rate can be very satisfactorily explained by a propagation process occuring mainly *via* free ions with a k_{p-} value in agreement with that derived from 'chemically initiated' polymerizations.

The most unexpected results are those for cryptated cations,[61, 78, 117, 118] since the propagation of cryptated ion-pairs turned out to be *higher* than that of the free anions. A plot of log $k_{p\pm}$ *vs.* $1/a$ gives approximately a straight line (Figure 7), and it can be seen that the experimental value for k_{p-} is quite far from the extrapolated value for $a = \infty$. This seems to show that the mechanism for ring opening differs for free ions and ion pairs. It is possible that a polarization of the CH$_2$—S bond in the monomer occurs before the ring opening and is induced by the ion-pair dipole of the living end. Such a modification should increase with the magnitude of the ion-pair dipole and then with the interionic distance a. This leads to an increase in the interaction with the monomer, together with an increase in the reaction rate. It is understandable that modification of the polarization of the CH$_2$—S bond may be quite different when the interaction occurs with free ions. A similar behaviour was observed in the case of the ring opening of propylene sulfide by carbazyl salts in THF as shown in Figure 7.

The activation energy for the cryptated Na$^+$ + (**5a**) ion-pairs was estimated as 33 kJ mol^{-1},[118] whereas the values obtained for free ions and for sodium ion-pairs were equal to 40 and 46 kJ mol^{-1},

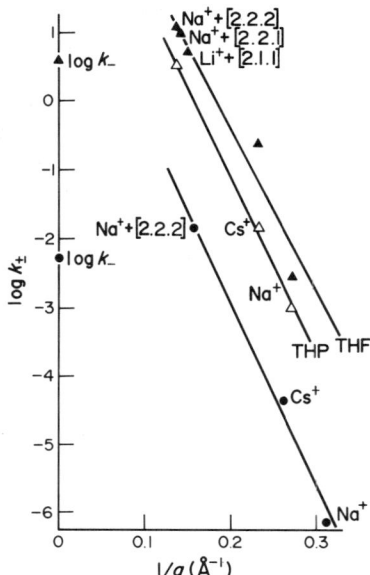

Figure 7 Variations of the thiolate and carbazyl ion-pairs' reactivities with the reciprocal of the interionic distance parameter of ion pairs for the ring opening of propylene sulfide at $-30\,°C$ (▲) propagation with thiolate ion-pairs in THF; (△) in THP; (●) initiation with carbazyl ion-pairs in THF

respectively.[111] The frequency factor in the case of cryptated ion-pairs is close to that for sodium ion-pairs: $5 \times 10^7\,l\,mol^{-1}\,s^{-1}$[118] compared to $2 \times 10^7\,l\,mol^{-1}\,s^{-1}$,[111] whereas it is quite different from that of free ions ($10^9\,l\,mol^{-1}\,s^{-1}$). These observations are in agreement with the assumption of two different mechanisms for the ring opening of propylene sulfide according to the type of ionic species (ion pairs or free ions).

The influence of the solvent was examined in the case of Cs^+ counterion.[78] The propagation constant of ion pairs increases from 0.27 in dioxane to 0.5 in THP and to $7–8\,l\,mol^{-1}\,s^{-1}$ in THF or DME at 15 °C. Similarly, the reactivity of cryptated $Na^+ + (5a)$ ion-pairs is lower in THP than in THF.[36] The plot of $\log k_{p\pm}$ vs. $1/a$ for the results found in THP at $-30\,°C$, in the case of Na^+, Cs^+ and $Na^+ + (5a)$, gives a straight line nearly parallel to the line observed in THF. In general, the ion pairs' reactivity decreases with the dielectric constant of the medium for a given counterion because separation of the charges in the transition state is more difficult in a medium of lower dielectric constant. The reactivity of free ions is the same in THP and in THF: $k_{p-} = 3.4\,l\,mol^{-1}\,s^{-1}$ in THP at $-30\,°C$,[36] whereas the average value of k_{p-} in THF at 30 °C is $3.8\,l\,mol^{-1}\,s^{-1}$ (see Table 3). Moreover, these values are very close to that found in DMF at $-30\,°C$ from data of propylene sulfide polymerization, initiated electrochemically: $k_{p-} = 4\,l\,mol^{-1}\,s^{-1}$.[15]

32.4 REFERENCES

1. S. Bywater, *Prog. Polym. Sci.*, 1975, **4**, 27.
2. N. C. Billingham, in 'Developments in Polymerization', ed. R. N. Haward, Applied Science, London, 1979, vol. 1, p. 147.
3. M. Szwarc, *Adv. Polym. Sci.*, 1983, **49**, 1.
4. S. Inoue and T. Aida, in 'Ring Opening Polymerization', ed. K. J. Ivin and T. Saegusa, Elsevier Applied Science, London and New York, 1984, vol. 1, p. 185.
5. P. Sigwalt and N. Spassky, in 'Ring Opening Polymerization', ed. K. J. Ivin and T. Saegusa, Elsevier Applied Science, London and New York, 1984, vol. 2, p. 603.
6. J. E. Figueruelo and D. J. Worsfold, *Eur. Polym. J.*, 1968, **4**, 439.
7. S. Boileau, G. Champetier and P. Sigwalt, *Makromol. Chem.*, 1963, **69**, 180.
8. D. H. Richards and M. Szwarc, *Trans. Faraday Soc.*, 1959, **55**, 1644.
9. T. J. Dudek, Ph. D. Thesis, University of Akron, 1961.
10. K. S. Kazanskii, A. A. Solovyanov and S. G. Entelis, *Eur. Polym. J.*, 1971, **7**, 1421.
11. I. Cabasso and A. Zilkha, *J. Macromol. Sci., Chem.*, 1974, **A8**, 1313.
12. S. Boileau, G. Champetier and P. Sigwalt, *J. Polym. Sci., Part C*, 1967, **16**, 3021.
13. L. Sousselier, Thèse Dt. 3° Cycle, Université P. et M. Curie, Paris, 1974.
14. I. M. Panayotov and I. V. Berlinova, *Makromol. Chem.*, 1972, **154**, 139.
15. G. Mengoli and S. Daolio, *J. Polym. Sci., Polym. Lett. Ed.*, 1975, **13**, 743.
16. D. Lassalle, Thèse Dt. d'Etat, Université P. et M. Curie, Paris, 1972.

17. C. J. Chang, R. F. Kiesel and T. E. Hogen-Esch, *J. Am. Chem. Soc.*, 1973, **95**, 8446.
18. A. A. Solovyanov and K. S. Kazanskii, *Vysokomol. Soedin., Ser. A*, 1974, **16**, 595.
19. P. Sigwalt, *J. Polym Sci., Polym. Symp.*, 1975, **50**, 95.
20. D. Lassalle, S. Boileau and P. Sigwalt, *Eur. Polym. J.*, 1977, **13**, 599.
21. S. Boileau, A. Deffieux, D. Lassalle, F. Menezes and B. Vidal, *Tetrahedron Lett.*, 1978, 1767.
22. P. Sigwalt and S. Boileau, *J. Polym. Sci., Polym. Symp.*, 1978, **62**, 51.
23. A. Deffieux, P. Sigwalt and S. Boileau, *Eur. Polym. J.*, 1984, **20**, 77.
24. B. Dietrich, J. M. Lehn and J. P. Sauvage, *Tetrahedron Lett.*, 1969, 2885, 2889.
25. For a review, see J. M. Lehn, in 'Structure and Bonding', Springer, Berlin, 1973, vol. 16, p.1.
26. H. Koinuma, K. Naito and H. Hirai, *Makromol. Chem.*, 1982, **183**, 1383.
27. A. Deffieux, Thèse Dt. d'Etat, Université P. et M. Curie, Paris, 1978.
28. D. Lassalle, S. Boileau and P. Sigwalt, *Eur. Polym. J.*, 1977, **13**, 591.
29. N. Takada and S. Inoue, *Makromol. Chem.*, 1978, **179**, 1377.
30. T. Aida and S. Inoue, *Macromolecules*, 1981, **14**, 1162, 1166.
31. S. Inoue, *Makromol. Chem., Macromol. Symp.*, 1986, **3**, 295.
32. S. Boileau and P. Sigwalt, *Bull. Soc. Chim. Fr.*, 1968, 1054.
33. G. Tersac, S. Boileau and P. Sigwalt, *Makromol. Chem.*, 1971, **149**, 153.
34. P. Hémery, V. Warzelhan and S. Boileau, *Polymer*, 1980, **21**, 77.
35. A. Rakotomanga, Thèse Dt. 3° Cycle, Université P. et. M. Curie, Paris, 1977.
36. H. Dang Ngoc, Thèse Dt. d'Etat, Université P. et M. Curie, Paris, 1979.
37. M. Morton, 'Anionic Polymerization: Principles and Practice', Academic Press, New York, 1983, p. 102.
38. K. Gonsalves and M. D. Rausch, *J. Polym. Sci., Polym. Chem. Ed.*, 1986, **24**, 1419.
39. R. P. Quirk and N. S. Seung, *ACS Symp. Ser.*, 1985, **286**, 37.
40. M. Morton and K. F. Kammereck, *J. Am. Chem. Soc.*, 1970, **92**, 3217.
41. A. Roggero, T. Salvatori, A. Proni and A. Mazzei, *J. Polym. Sci., Polym. Lett. Ed.*, 1979, **17**, 125.
42. S. Perry and H. Hibbert, *J. Am. Chem. Soc.*, 1940, **62**, 2599.
43. G. Gee, W. C. E. Higginson and G. T. Merall, *J. Chem. Soc.*, 1959, 1345.
44. G. Gee, W. C. E. Higginson, P. Levesley and K. T. Taylor, *J. Chem. Soc.*, 1959, 1338.
45. M. Szwarc, 'Carbanions, Living Polymers and Electron Transfer Processes', Interscience, New York, 1968.
46. N. N. Lebedev and Yu. I. Baranov, *Vysokomol. Soedin.*, 1966, **8**, 198.
47. N. N. Lebedev and Yu. I. Baranov, *Kinet. Katal.*, 1966, **7**, 619.
48. J. E. Figueruelo and A. Bello, *J. Macromol. Sci., Chem.*, 1969, **A3**, 311.
49. G. E. M. Jones and O. L. Highes, *J. Chem. Soc.*, 1934, 1197.
50. S. Nenna and J. E. Figueruelo, *Eur. Polym. J.*, 1975, **11**, 511.
51. C. C. Price and D. D. Carmelite, *J. Am. Chem. Soc.*, 1966, **88**, 4039.
52. C. E. H. Bawn, A. Ledwith and N. McFarlane, *Polymer*, 1969, **10**, 653.
53. S. Nenna and J. E. Figueruelo, *Makromol. Chem.*, 1975, **176**, 3377.
54. M. Rodriguez and J. E. Figueruelo, *Makromol. Chem.*, 1975, **176**, 3107.
55. K. S. Kazanskii, A. A. Solovyanov and S. A. Dubrovsky, *Makromol. Chem.*, 1978, **179**, 969.
56. A. A. Solovyanov and K. S. Kazanskii, *Vysokomol. Soedin., Ser. A*, 1972, **14**, 1072.
57. J. H. Exner and E. C. Steiner, *J. Am. Chem. Soc.*, 1974, **96**, 1782.
58. A. A. Solovyanov and K. S. Kazanskii, *Vysokomol. Soedin., Ser. A*, 1970, **12**, 2114.
59. A. A. Solovyanov and K. S. Kazanskii, *Vysokomol. Soedin., Ser A*, 1972, **14**, 1063.
60. N. V. Ptitsyna, V. K. Kazakevich and K. S. Kazanskii, *Vysokomol. Soedin., Ser. A*, 1977, **19**, 2787.
61. S. Boileau, *ACS Symp. Ser.*, 1981, **166**, 283.
62. K. S. Kazanskii, *Pure Appl. Chem.*, 1981, **53**, 1645.
63. H. Brody, D. H. Richards and M. Szwarc, *Chem. Ind. (London)*, 1958, 1473.
64. J. Smid, in 'Ions and Ion Pairs in Organic Reactions', ed. M. Szwarc, Wiley-Interscience, New York, 1972, vol. 1, p. 86.
65. D. K. Dimov, I. M. Panayotov, V. N. Lazarov and C. B. Tsvetanov, *J. Polym. Sci., Polym. Chem. Ed.*, 1982, **20**, 1389.
66. G. N. Arkhipovitch, S. A. Dubrovskii, K. S. Kazanskii, N. V. Ptitsina and A. N. Shupik, *Eur. Polym. J.*, 1982, **18**, 576.
67. G. N. Arkhipovitch, Ye. A. Ugolkova and K. S. Kazanskii, *Polym. Bull. (Berlin)*, 1984, **12**, 181.
68. See, for instance, A. Hirao, S. Nakahama, M. Takahashi and N. Yamazaki, *Makromol. Chem.*, 1978, **179**, 915.
69. I. V. Berlinova, I. M. Panayotov and C. B. Tsvetanov, *Eur. Polym. J.*, 1977, **13**, 757.
70. A. Bar-Ilan and A. Zilkha, *J. Macromol. Sci., Chem.*, 1970, **A4**, 1727.
71. J. A. Orvik, *J. Am. Chem. Soc.*, 1976, **98**, 3322.
72. A. Deffieux and S. Boileau, *Polymer*, 1977, **18**, 1047.
73. A. Deffieux, E. Graf and S. Boileau, *Polymer*, 1981, **22**, 549.
74. C. D. Eisenbach and M. Peuscher, *Makromol. Chem., Rapid Commun.*, 1980, **1**, 105.
75. S. Boileau, P. Hémery and J. C. Justice, *J. Solution Chem.*, 1975, **4**, 873.
76. E. Graf and J. M. Lehn, *J. Am. Chem. Soc.*, 1975, **97**, 5022.
77. A. J. Parker, *Chem. Rev.*, 1969, **69**, 1.
78. P. Hémery, S. Boileau and P. Sigwalt, *J. Polym. Sci., Polym. Symp.*, 1975, **52**, 189.
79. L. E. St. Pierre and C. C. Price, *J. Am. Chem. Soc.*, 1956, **78**, 3432.
80. G. J. Dege, R. L. Harris and J. S. MacKenzie, *J. Am. Chem. Soc.*, 1959, **81**, 3374.
81. D. M. Simons and J. J. Verbanc, *J. Polym. Sci.*, 1960, **44**, 303.
82. G. Gee, W. C. E. Higginson, K. J. Taylor and M. W. Trenholme, *J. Chem. Soc.*, 1961, 4298.
83. E. C. Steiner, R. R. Pelletier and R. O. Trucks, *J. Am. Chem. Soc.*, 1964, **86**, 4678.
84. L. P. Blanchard, V. Hornof, J. Moinard and F. Tahiani, *J. Polym. Sci., Polym. Chem. Ed.*, 1972, **10**, 3089.
85. G. A. Gladkovskii, L. P. Golovina, G. F. Vedeneyeva and V. S. Lebedev, *Vysokomol. Soedin., Ser. A*, 1973, **15**, 1221.
86. C. C. Price, *Acc. Chem. Res.*, 1974, **7**, 294.
87. A. Stolarzewicz, H. Becker and G. Wagner, *Acta Polym.*, 1981, **32**, 483.
88. H. Becker, G. Wagner and A. Stolarzewicz, *Acta Polym.*, 1982, **33**, 34.
89. H. Becker and G. Wagner, *Acta Polym.*, 1984, **35**, 28.

90. C. C. Price, Y. Atarashi and R. Yamamoto, *J. Polym. Sci., Part A-1*, 1969, **7**, 569.
91. P. Banks and R. H. Peters, *J. Polym. Sci., Part A-1*, 1970, **8**, 2595.
92. C. C. Price and M. K. Akkapeddi, *J. Am. Chem. Soc.*, 1972, **94**, 3972.
93. W. H. Snyder and A. E. Meisinger, Jr., *Polym. Prepr., Am. Chem. Soc., Div. Polym. Chem.*, 1968, **9** (1), 382.
94. C. C. Price and H. Fukutani, *J. Polym. Sci., Part A-1*, 1969, **6**, 2653.
95. L. P. Blanchard, K. T. Dinh, J. Moinard and F. Tahiani, *J. Polym. Sci., Part A-1*, 1972, **10**, 1353.
96. A. Stolarzewicz, *Makromol. Chem.*, 1986, **187**, 745.
97. N. P. Doroshenko and Yu. L. Spirin, *Vysokomol. Soedin., Ser. A*, 1970, **12**, 2481.
98. M. J. Béchet, *Bull. Soc. Chim. Fr.*, 1971, 3593, 3596.
99. Z. Jedlinsky, A. Stolarzewicz, P. Szewczyk and R. Tymczynski, *Polym. Bull. (Berlin)*, 1980, **2**, 555.
100. G. Ezra and A. Zilkha, *J. Polym. Sci., Part A-1*, 1970, **8**, 1343.
101. S. Boileau, H. Chéradame, P. Guérin and P. Sigwalt, *J. Chim. Phys. Phys.-Chim. Biol.*, 1972, **69**, 1420.
102. S. Boileau, H. Chéradame, W. Lapeyre, L. Sousselier and P. Sigwalt, *J. Chim. Phys. Phys.-Chim. Biol.*, 1973, **70**, 879.
103. P. Guérin, S. Boileau and P. Sigwalt, *Eur. Polym. J.*, 1974, **10**, 13.
104. A. D. Aliev, B. A. Krentsel, G. M. Mamediarov, I. P. Solomatina and E. P. Tiurina, *Eur. Polym. J.*, 1971, **7**, 1721.
105. A. D. Aliev, I. P. Solomatina and B. A. Krentsel, *Macromolecules*, 1973, **6**, 797.
106. A. D. Aliev, I. P. Solomatina, A. Yu. Koshevnik, Zh. Zhumabayev and B. A. Krentsel, *Vysokomol. Soedin., Ser. A*, 1977, **19**, 173.
107. A. D. Aliev, B. A. Krentsel and S. L. Alieva, *Eur. Polym. J.*, 1983, **19**, 71.
108. J. P. Favier, S. Boileau and P. Sigwalt, *Eur. Polym. J.*, 1968, **4**, 3.
109. S. Boileau and P. Sigwalt, *Eur. Polym. J.*, 1967, **3**, 57.
110. P. Guérin, P. Hémery, S. Boileau and P. Sigwalt, *Eur. Polym. J.*, 1971, **7**, 953.
111. P. Hémery, S. Boileau and P. Sigwalt, *Eur. Polym. J.*, 1971, **7**, 1581.
112. P. Sigwalt, in 'Kinetics and Mechanism of Polyreactions', Akadémiai Kiado, Budapest, 1971, p. 251.
113. G. Tersac, S. Boileau and P. Sigwalt, *J. Chim. Phys. Phys.-Chim. Biol.*, 1968, **65**, 1141.
114. P. Guérin, S. Boileau and P. Sigwalt, *Eur. Polym. J.*, 1971, **7**, 1119.
115. E. Campos-Lopez, A. Leon-Gross and M. A. Ponce Velez, *J. Polym. Sci., Polym. Chem. Ed.*, 1973, **11**, 3021.
116. M. Morton, R. F. Kammereck and L. J. Fetters, *Br. Polym. J.*, 1971, **3**, 120.
117. P. Hémery, S. Boileau, P. Sigwalt and B. Kaempf, *J. Polym. Sci., Polym. Lett. Ed.*, 1975, **13**, 49.
118. P. Hémery, Thèse Dt. d'Etat, Université P. et M. Curie, Paris, 1976.
119. G. Tersac, S. Boileau and P. Sigwalt, *Bull. Soc. Chim. Fr.*, 1970, 2537.
120. B. Vidal, Thèse Dt. 3° Cycle, Université P. et M. Curie, Paris, 1975.

33

Anionic Ring-opening Polymerization: Stereospecificity for Epoxides, Episulfides and Lactones

TEIJI TSURUTA
Science University of Tokyo, Japan
and
YUHSUKE KAWAKAMI
Nagoya University, Japan

33.1 INTRODUCTION

The first report of the stereospecific polymerization of propylene oxide (PO) was published by Pruitt and Baggett[1] in 1955. Their catalyst was a reaction product of iron(III) chloride with propylene oxide. A few years later, various types of stereospecific catalyst for PO polymerization were found and developed almost simultaneously: aluminum isopropoxide/zinc chloride by Price and Osgan;[2] zinc alkyl/water (or alcohol) by Furukawa and Tsuruta;[3] aluminum alkyl/water by Colclough, Gee *et al.*;[4] and aluminum alkyl/water/acetylacetone by Vandenberg.[5] A zinc alkyl/water system was used by Sigwalt and co-workers in 1965[6] for the stereospecific polymerization of propylene sulfide. In 1967, Osgan and Teyssie[7] reported the synthesis of a bimetallic catalyst, μ-oxoalkoxide $(RO)_2AlOZnOAl(OR)_2$, and noted its high activity.

The first example of asymmetric selective polymerization of (*RS*)-propylene oxide was given by Tsuruta, Inoue *et al.*[8,9] using the catalyst system $ZnEt_2$–(+)-borneol. A few years later, Sigwalt and Spassky[10,11] and Furukawa *et al.*[12] reported that (*RS*)-propylene sulfide (PS) also underwent a stereoelective polymerization with the use of similar organozinc catalyst systems to those for PO. The term 'stereoelective' is a synonym of 'asymmetric selective' as will be discussed later (see Section 33.2).

Details of studies carried out in and before the early years of the 1960s have been reviewed elsewhere.[13]

The enormous progress in methodology for structural analyses of polymers has made it much easier than before to have an insight into mechanistic features of the stereospecific polymerization. In this chapter, studies on stereospecificity in the anionic ring-opening polymerizations of epoxides and episulfides and also of some lactones will be reviewed, with the emphasis on recent progress in this field.

33.2   SOME STEREOCHEMICAL DEFINITIONS RELATING TO POLYMERS

The prerequisite condition for formation of a stereoregular polymer from an epoxide or episulfide is the regioselective ring-opening of monomer molecules in the polymerization process, which results in the formation of a regular polymer. According to the IUPAC documents,[14, 15] a regular polymer is a polymer whose molecules can be described by only one species of constitutional unit in a single sequential arrangement. Regular poly(propylene oxide) (1) is an example. Every constitutional unit shown by the adjacent dashed lines is linked together with its neighboring units in a single sequential (head to tail) arrangement.

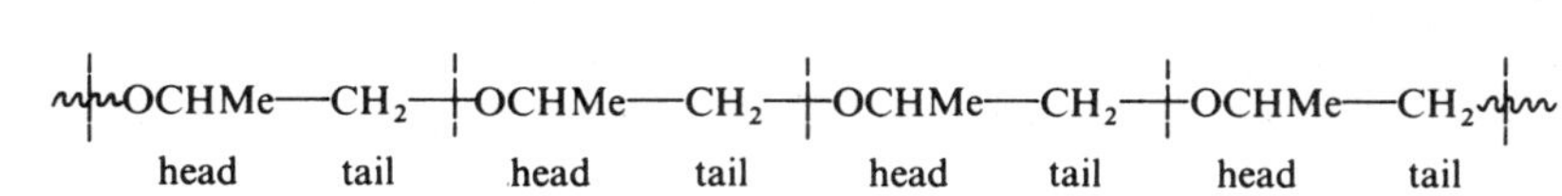

$$\text{\textasciitilde}\text{OCHMe—CH}_2\text{—}\big|\text{OCHMe—CH}_2\text{—}\big|\text{OCHMe—CH}_2\text{—}\big|\text{OCHMe—CH}_2\text{\textasciitilde}$$

   head   tail   head   tail   head   tail   head   tail

(1)

When racemic propylene oxide [(RS)-PO] is polymerized by KOH (or KOR), the ring opening takes place predominantly ($\sim 95\%$) at the O—CH$_2$ bond to form a 'regular' polymer similar to (1). The alkali metal catalyst system, however, cannot recognize the difference between R and S monomer, resulting in the polymer being regular but atactic.

The polymerization of (RS)-PO with ZnEt$_2$/MeOH (or Zn(OMe)$_2$[16-21]) as initiator was proved[22, 23] to be regioselective (O—CH$_2$ bond scission) and stereoselective to form the isotactic polymer[23] as shown in Scheme 1. Stereoselective polymerization is defined[14] as 'a polymerization in which a polymer molecule is formed from a mixture of stereoisomeric monomer molecules (*e.g.* (R)- and (S)-PO) by incorporation of only one stereoisomeric species (*e.g.* (R)-PO) into a growing polymer chain'. In Scheme 1, (RS)-PO is polymerized by the optically inactive initiator, so that an equal number of moles of poly[(R)-PO] and poly[(S)-PO] are formed. When an optically active initiator such as the ZnEt$_2$–(+)-borneol system[8, 9] is used, preferential polymerization of (R)-PO over the S enantiomer takes place, unreacted PO monomer being enriched with the S enantiomer. This is a typical example of asymmetric selective (or stereoelective) polymerization.

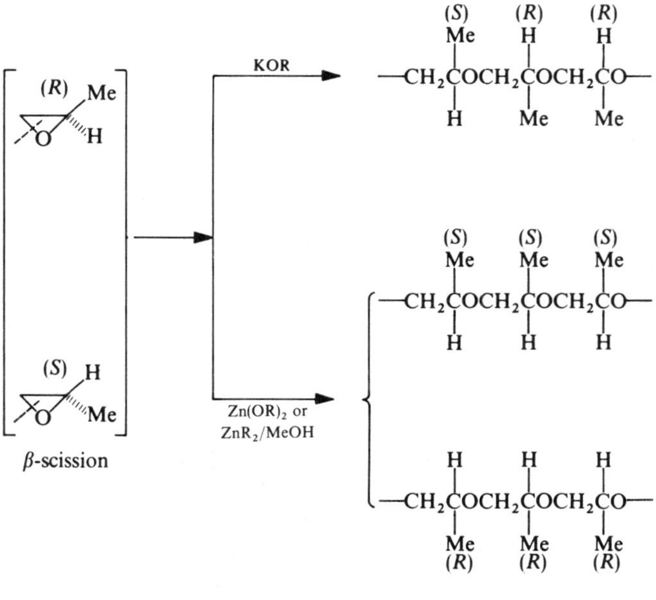

Scheme 1

The IUPAC Macromolecular Nomenclature Commission has recently recommended[25] the terms *enantiosymmetric* and *enantioasymmetric* polymerizations for the more specific cases of *stereoselective* polymerization. The above-mentioned polymerization (see Scheme 1) in which an equal number of moles of poly[(*R*)-PO] and poly[(*S*)-PO] are formed is obviously an example of enantiosymmetric polymerization, while asymmetric selective (or stereoelective) polymerization is one of enantioasymmetric polymerization.

Stereospecific polymerization[14,24] is 'a polymerization in which a tactic polymer is formed'. A *tactic* polymer is defined as 'a regular polymer, the molecules of which can be described in terms of only one species of configurational repeating unit in a single sequential arrangement'. The polymerizations of (*RS*)-PO with optically inactive (Scheme 1) and also with optically active organozinc catalyst systems[8,9] are *stereospecific* because *isotactic* (or tactic in broader definition) polymers are formed in all of these polymerization systems. It is to be noted that the stereospecific polymerization of propylene cannot be a stereoselective polymerization because the starting monomer, propylene, has no chiral center.

33.3 REGIOSELECTIVITY AND STEREOCHEMISTRY AT THE SITE OF BOND CLEAVAGE

Some previous papers[2] reported that the regioselectivity in the oxirane polymerizations depended strongly upon the nature of the initiator systems. The general conclusion was that anionic and anionic-coordinate catalysts mostly cleaved the CH_2—O bond (β-bond in 2) to form regular head-to-tail linkages, while cationic catalysts also cleaved the CHMe—O bond (α-bond in 2) concurrently to a significant extent to form irregular polymer chains with head-to-head and tail-to-tail linkages. Most of the catalyst systems proposed for stereospecific polymerization of PO are binary or ternary systems, so that active centers are formed having a varying catalytic nature in terms of regioselectivities and stereoselectivities, which are changeable according to the conditions under which the catalyst systems are prepared. It has long been known that the whole polymer obtained by polymerizing racemic PO with these catalyst systems could be fractionated into crystalline and amorphous polymers.[13] The crystalline fraction was proved to be an isotactic polymer many years ago.[26-28] The structure of the amorphous fraction, however, was not fully elucidated until recently. Owing to progress in the NMR technique, it is now clear that some amorphous poly(PO) prepared by $ZnEt_2$/MeOH[23] or $Al(OPr^i)_3$/$ZnCl_2$[30] consists of regular head-to-tail linked units, but the percentage of isotactic diads is lower than 60%[23] (example of an atactic polymer in the broad sense). Some other amorphous poly(PO) samples prepared with $ZnEt_2$/H_2O (1:0.5),[30,31] $AlEt_3$/H_2O (1:1)[33] and $Al(OPr^i)_3$[30] have been shown to have head-to-head and tail-to-tail structures along their polymer chains (example of an irregular polymer).

(2)

It is to be noted in this connection that styrene oxide is polymerized by $Al(OPr^i)_3$ under selective cleavage of the O—CHPh bond (α-bond).[34] With the $ZnPh_2$/H_2O (1:1) system,[35] on the other hand, polymerization of styrene oxide proceeds under β-bond cleavage.

Tsuruta *et al.*[22,23] prepared poly(*trans*-propylene oxide-β-*d*) (3) with $ZnEt_2$/MeOH (1:1.7), and fractionated it into crystalline and amorphous parts. The crystalline fraction was shown to have an *erythro*-diisotactic structure as in (4) and (5). It was concluded, therefore, that the ring-opening took place at the β-bond with *inversion* of the configuration at the methylene carbon. The inversion mechanism at the ring cleavage coincides with those reported by Vandenberg[36-38] for butene 2-oxide, by Price[39-42] for ethylene oxide and propylene oxide, by Tani[29,44,45] for propylene oxide and by Tsuruta[46] for cyclohexene oxide.

racemic *trans*-(*RR,SS*)-propylene oxide-β-*d*

(3)   (4)   (5)

33.4   ENANTIOMORPHIC CATALYST SITES MODEL FOR THE STEREOSPECIFIC POLYMERIZATION OF PROPYLENE OXIDE

The reaction mechanism for the stereospecific polymerization of propylene oxide with zinc dialkoxide (the active species in the $ZnEt_2/MeOH$ system)[16-21] has been satisfactorily explained by the enantiomorphic catalyst sites model, in which the presence of $R*$ and $S*$ sites is assumed to be the origin of the steric control mechanism. The $R*$ sites accept (R)-propylene oxide in preference to the S isomer, resulting in the formation of —RRR · · · RRR— isotactic sequences, and *vice versa* (see Section 33.2). Results of copolymerization between R and S monomers were found to be a useful experimental tool for elucidation of the mechanism of stereoregulation. The simple and important copolymerization formula (equation 1) can be derived from the concept of the enantiomorphic catalyst sites model having a symmetrical distribution of $R*$ and $S*$ sites. Equation (1) implies that the relative rate of incorporation of R monomer against S monomer is exactly the same as the ratio of concentrations of the two isomers. More details about phenomenological approaches to the enantiomorphic mechanism were summarized elsewhere[47] in 1972.

$$d[R]/d[S] \;=\; [R]/[S] \tag{1}$$

The ^{13}C NMR technique has given information on the triad tacticity of poly(PO).[48-50] Matsuzaki and Uryu[51] reported that the triad tacticity of poly(PO) prepared with $ZnEt_2/MeOH$ catalyst accorded with values calculated on the basis of the enantiomorphic model[52-54] shown in equations (2)–(4), where parameter σ_2 is the probability of entering of an R monomer at an $R*$ catalyst site. The triad tacticity data of another poly(PO) sample prepared with $Zn(OMe)_2$ were also in good agreement with the enantiomorphic model.[55]

$$I \;=\; 1 \;-\; 3\sigma_2(1 \;-\; \sigma_2) \tag{2}$$

$$H \;=\; 2\sigma_2(1 \;-\; \sigma_2) \tag{3}$$

$$S \;=\; \sigma_2(1 \;-\; \sigma_2) \tag{4}$$

33.5   MOLECULAR LEVEL ELUCIDATION OF THE ENANTIOMORPHIC MECHANISM USING WELL DEFINED ORGANOZINC COMPLEXES

As stated in the foregoing sections, the mechanism of the stereospecific polymerization of PO could be explained by assuming the presence of $R*$ and $S*$ sites in the catalyst system. No information, however, was available concerning the chiral structure of $R*$ and $S*$ catalyst sites, because none of the active catalysts possessed a well defined structure.

To elucidate the stereocontrol mechanism in terms of molecular level considerations, a series of studies[32, 56-58] were carried out in homogeneous systems using well defined organozinc complexes such as $[EtZnOMe]_6[MeOZnOMe]$ (6),[59-61] $[EtZnOCH_2CH_2OMe]_6[Zn(OCH_2CH_2OMe)_2]$ (7),[62] and $[EtZnOCHMeCH_2OMe]_2[Zn(OCHMeCH_2OMe)_2]_2$ (8)[63] as initiator. All of these complexes were isolated in the form of single crystals. Complex (6) consists of two enantiomorphic distorted cubes (*i.e.* R cube and S cube), which share a six-coordinated central zinc atom. Each cube has three inner methoxy groups and one outer methoxy group. The single crystal is soluble in benzene, and the benzene solution induces the polymerization of PO at 80 °C. Cryoscopic and NMR measurements of the benzene solution have shown that the organozinc complex exists as a monomeric form and retains its structure even in benzene in the temperature range from 5 to 80 °C. Hagiwara, Ishimori and Tsuruta[61] carried out ^{13}C-PFT NMR analysis of a reaction system in which (6) and (RS)-PO in excess were allowed to react in benzene at 80 °C. They found that all of the observed signals in the reaction system, including those that had newly appeared or disappeared could reasonably be explained in terms of the initiation mechanism by one of the inner methoxy groups. This result indicates that the principal framework of the complex is retained even after the start of the propagation reaction. It was also demonstrated that every complex molecule of (6) consumes only one inner methoxy group for the initiation reaction. The triad tacticities (%) of poly(PO) (unfractionated) formed were found to be in good agreement with those anticipated from the enantiomorphic catalyst sites model ($\sigma_2 = 0.72$). The chiral structure around the central zinc atom in (6) was considered to be responsible for the formation of the $R*$ (or $S*$) site.

Complex (7) has a similar spatial structure to (6) but exhibits inferior steric control ability to that of (6).

Complex (8) is a newly found organozinc complex,[63] the structure of which was shown to be a 'chair' type in contrast with the 'cube' type for (6) or (7). It was confirmed that (8) exists as a monomeric form in benzene and that the stereochemical structure of the chair framework is retained in benzene at 80 °C. Complex (**8**) exhibited much higher activity and stereospecificity in PO polymerization than either of complexes (**6**) or (**7**). The population of the triad tacticities of poly(PO) prepared with (**8**) fitted well with those predicted from the enantiomorphic catalyst sites model [$\sigma_2 = 0.91$ for unfractionated poly(PO)]. As seen in Figure 1, Zn(1) is surrounded by the three types of (*S*)-methoxypropoxy group (*endo*-coordinated, *exo*-coordinated and non-coordinated), and Zn(2) by the three types of (*R*)-methoxypropoxy group. By using a partly deuterated complex (**8**) and PO-α-*d*, Hasebe and Tsuruta[64] obtained a result which showed the initiation reaction to be caused by an attack with the non-coordinated methoxypropoxy group onto a monomer molecule. It was also confirmed that only one methoxypropoxy group for one mole of (**8**) participated in the initiation reaction.

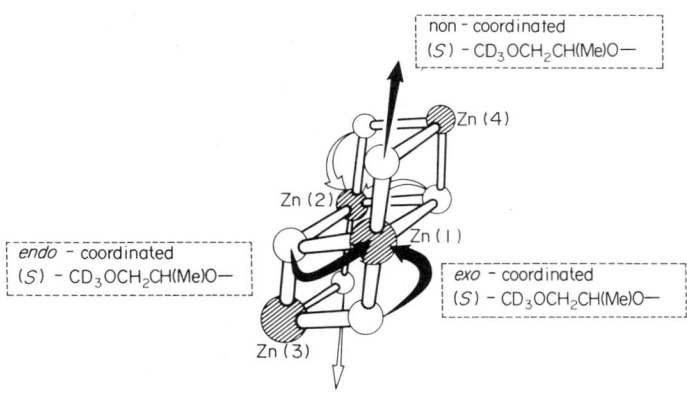

non - coordinated
(*S*) – CD$_3$OCH$_2$CH(Me)O—

Zn (4)

Zn (2)

Zn (1)

endo – coordinated
(*S*) – CD$_3$OCH$_2$CH(Me)O—

exo – coordinated
(*S*) – CD$_3$OCH$_2$CH(Me)O—

Zn (3)

Figure 1 A simplified illustration of complex (**8**)

Another experiment using (*S*)-PO as monomer showed that the coordination of the *S* monomer took place at Zn(2), and was attacked by the non-coordinated (*R*)-methoxypropoxy group. A molecular model study showed that there is formation of a *chiral hole* which can accommodate only enantiomeric molecules with one sense. The chiral structure of the hole is determined by the chair type Zn—O framework together with the coordinated methoxypropoxy groups. In the PO polymerization, the chirality of the non-coordinated methoxypropoxy group has nothing to do with the enantioselection. In the propagation stage, an R^* chiral hole receives predominantly (*R*)-PO, while an S^* chiral hole accepts (*S*)-PO, to form isotactic linkages. Since spatial allowance at the chiral hole decreases with increasing bulkiness of substituent of oxirane monomer, the parameter σ_2 becomes larger the bulkier the substituent of the monomer molecule is. Actually, σ_2 was found to be unity for poly(*t*-butylethylene oxide) which was prepared with complex (**8**).

It is to be noted that the chiral hole of (**8**) did not recognize the mode of orientation of a *cis*-disubstituted epoxide such as cyclohexene oxide, which is an achiral monomer. Complex (**8**) seems to serve as a simple bulky group, which facilitates syndiotactic addition of cyclohexene oxide monomer molecule to the active site, in cooperation with strong steric effects exerted from the terminal and penultimate units of the growing chain.[46]

33.6 GROWING CHAIN CONTROL MECHANISM IN THE POLYMERIZATION OF *t*-BUTYLETHYLENE OXIDE

When racemic PO is polymerized with potassium alkoxide as initiator, *R* and *S* monomers are randomly incorporated into a polymer chain to form an atactic polymer (see Section 33.2). Results obtained in copolymerizations between (*R*)- and (*S*)-PO were in good agreement with equation (1), no change at all being observed in the optical purity of the monomer phase during the course of the copolymerization. The physical meaning here is obviously different from that of the enantiomorphic catalyst sites model.

When the *R* and *S* copolymerization of *t*-butylethylene oxide was carried out with a monomer mixture consisting of $R/S = 76/24$ with Bu^{t}OK as initiator, *R* monomer was incorporated into polymer chain preferentially over *S* monomer.[65] As a consequence, the optical purity in the recovered monomer became smaller than that of the starting mixture as the copolymerization reaction proceeded. This was explained by a growing chain control mechanism, the *t*-butyl group being bulky enough to make the chiral structure of the growing polymer chain able to undergo stereoselection of the chiral monomer. Sato, Hirano and Tsuruta[65] analyzed the curve of *R* content in unreacted monomer *vs.* conversion for *t*-butylethylene oxide by dividing the curve into *j*-stages. From experimental data at every stage, they estimated the value of a parameter, α_j, defined as equation (5)

$$(d[R]/d[S])_j = \alpha_j([R]/[S])_j \tag{5}$$

The parameter α_j was found to become greater as the conversion increased. Since the polymerization of *t*-butylethylene oxide initiated by Bu^{t}OK was proved to form a living system, the increase of the α_j value should be ascribed to a surplus contribution from a chiral secondary structure of the growing chain, along with the primary chiral structure of the chain end.

Price[42] reported previously the IsoSyn mechanism for the stereochemistry of polymerization of (*RS*)-*t*-butylethylene oxide (Bu^{t}EO), initiated with Bu^{t}OK. For the formation of IsoSyn poly-(Bu^{t}EO), —*RRSSRRSSRR*—, $P_{SR/R}$ (or $P_{RS/S}$) should be larger than $P_{RR/R}$ (or $P_{SS/S}$), where $P_{SR/R}$ means the conditional probability for *R* monomer to add on a growing chain end which has *S* and *R* units respectively as the penultimate and terminal units. The other probabilities are defined correspondingly.

From results of NMR studies[66, 67] on racemic poly(Bu^{t}EO) prepared with Bu^{t}OK as initiator, the conditional probabilities, $(P_{SR/R} + P_{RS/S})$ and $(P_{RR/R} + P_{SS/R})$ were evaluated to be 0.55 and 0.43 respectively. This result indicates that the terminal unit of a growing chain reacts with an incoming monomer without any significant influence from the penultimate unit, because $P_{SR/R}$ (or $P_{RS/S}$) is almost equal to $P_{RR/R}$ (or $P_{SS/S}$).

In the above-mentioned *R* and *S* copolymerizations the ratio of the initial concentrations of *R* and *S* monomer was 76/24, so that there was an enhanced chance to form isotactic linkages, which would result in the formation of a chiral secondary structure. It is to be noted in this connection that Spassky *et al.*[68] reported a much larger rate of polymerization of (*R*)-(−)-Bu^{t}EO in comparison with that of the *RS* monomer under identical conditions.

The stereochemical behavior of *t*-butylethylene sulfide (Bu^{t}ES) is in sharp contrast with that of Bu^{t}EO, no stereoelection being observed in the reaction systems in which (*S*)-(−)-Bu^{t}ES in excess was copolymerized with the (*R*)-(+) isomer in bulk or in dimethyl sulfoxide using Bu^{t}OK as initiator.[69]

A similar result was reported[70] for the polymerization of styrene oxide of different enantiomeric compositions using a potassium alkoxide initiator. The stereoelective effect was almost completely suppressed by addition of complexing agents such as crown ethers or cryptands. These observed phenomena were explained in terms of polymer chain effects.

Another example[46] of the growing chain control mechanism has already been given in the last part of Section 33.5.

33.7 STEREOSELECTIVE AND STEREOELECTIVE POLYMERIZATIONS OF EPISULFIDES: COMPARISON WITH EPOXIDES

Extensive studies have been carried out by Sigwalt, Spassky and co-workers on the stereoselective and stereoelective polymerizations of episulfides and epoxides.[71-78] The enantiomorphic catalyst sites control mechanism was found to be valid also for the stereospecific polymerization of episulfides. Sigwalt *et al.* found that (*R*)-(−)-3,3-dimethyl-1,2-butanediol (DMBD)/ZnEt$_2$ (1:1) had a high stereoelective nature both in epoxide and episulfide polymerizations.[71, 74] Using a 'stepwise procedure' they succeeded in isolating almost optically pure monomer as the unreacted monomer.[73, 77] They found that equation (6) was valid for this stereoelective polymerization

$$d[R]/d[S] = r([R]/[S]) \tag{6}$$

Equation (6), however, was not valid for the polymerization of *t*-butylethylene sulfide with the ZnEt$_2$/(*R*)-(−)-DMBD system.[71, 78-80] Sigwalt *et al.* proposed equation (7) for this system

$$d[R]/d[S] = \rho([R]^2/[S]^2) \tag{7}$$

To explain the second-order law, they assumed that incorporation of an *R* monomer molecule into a polymer chain took place only on the site already complexed by another *R* monomer molecule, and *vice versa*.

It appears also that *t*-butylthiirane gives pure isotactic chains with most of the stereospecific initiators and that when polymerizing an enantiomerically enriched *t*-butylthiirane with $ZnEt_2/H_2O$ initiator one can separate the polymer into a pure polyenantiomer partly soluble in $CHCl_3$ and a racemic stereocomplex which is insoluble in $CHCl_3$.[81]

Spassky, Sigwalt et al.[83,88] revealed also that some cadmium compounds, including simple salts, were excellent initiators for stereoselective polymerization of propylene sulfide. For instance, the percentage of isotactic diad in the poly(propylene sulfide) sample prepared using Cd (*R*)-tartrate catalyst was more than 95%, in contrast with 69% that was found in a polymer sample prepared using Zn (*R*)-tartrate as catalyst. Cd (*RS*)-tartrate also exhibited the same level of stereoselectivity as Cd (*R*)-tartrate. The superior stereoselectivity of the Cd tartrate was shown also by an experiment in which the more effective separation into fractions having opposite optical rotation was performed in poly(propylene sulfide) prepared by Cd tartrate, compared with that by Zn tartrate.[84]

In the case of stereoelectivity, precisely the opposite situation was found in the two tartrates; only very slight optical activity was associated with the poly(propylene sulfide) sample prepared by Cd (*R*)-tartrate, whereas $[\alpha]_D^{25} = -5.8$ (in benzene) for the polymer sample prepared by Zn (*R*)-tartrate.

The $CdMe_2/(-)$-DMBD system, however, was found to exhibit a stereoelectivity lower than that of $ZnEt_2/(-)$-DMBD system in the polymerizations of propylene sulfide, *cis*-butene sulfide and cyclohexene sulfide, but the elected chirality was opposite to that found with the Zn system.[85,86]

Stereoselective and stereoelective polymerizations of propylene sulfide have also been studied in a homogeneous phase using two series of chiral cadmium derivatives of cysteine and methionine.[87-89] The first series (series I) are the cadmium thiolates of cysteine esters, and the second (series II) are the cadmium carboxylates of cysteine and methionine.

(9) Series I

(10) Series II

Poly(propylene sulfide) samples prepared by the series I Cd compounds (thiolates) are highly isotactic and optically active though the value of the optical rotation is not large. Polymer samples from the series II compounds (carboxylates) are also isotactic but not optically active. The lack of stereoelectivity of the carboxylates should be noted in connection with the behavior of Cd (*R*)-tartrate. Further studies[89] with the Cd thiolate catalyst (R = Pr^i) revealed that stereospecificity in a wide sense (both stereoselection and stereoelection) appeared only for molecular weights higher than 6000 and also depended on the temperature of propagation: with decreasing temperature, the isotacticity increased and stereoelection could be inverted.

Spassky and his co-workers[90,91] reported recently that an optically active atropoisomeric system exhibited a remarkable stereoelection almost one order of magnitude higher than that obtained with the best previously known initiator. For instance, zinc (*S*)-binaphtholate polymerized methyl- or ethyl-substituted episulfide with very high stereoelectivity, k_S/k_R being 15–20, where k_S and k_R denote the rate constants for polymerizations of S and R monomer respectively. Sepulchre et al.[92] reported the formation of disulfide linkages when using the atropoisomeric catalyst and the particular role of additions of tetrahydrothiophene which allows the most crystalline, isotactic products to be obtained in the case of most substituted thiiranes including *cis*-butene sulfide. It was found also that the resolution efficiency decreased significantly when the episulfide ring carried a bulky group such as *t*-butyl.[90,91]

On the basis of the results obtained from stereoelective polymerizations of epoxides and episulfides with a variety of chiral catalysts, Spassky et al.[72,77,85] proposed configurational rules governing the stereochemical control of chiral initiators in the ring-opening polymerization. They

(11) (+)-(*R*)-
3,3-dimethyl-2-butanol

(12) (−)-(*R*)-
3,3-dimethyl-1,2-butanediol

(13) (+)-(*R*)-
propylene sulfide

classified substituents into three categories, (i) CH_2OH, CH_2SH, CH_2S; (ii) OH, OR, SH, SR; (iii) Me, Bu^t, and any alkyl group. For instance, molecules (11)–(13) were considered to have the 'same' spatial configuration.

When initiator systems prepared from these chiral alcohols with $ZnEt_2$ were used, the enantiomer having the same configuration as the alcohol was preferentially chosen for the polymerization. More generally, the chiral initiator preferentially polymerizes the enantiomer with the same spatial configuration. Spassky *et al.* confirmed that this rule could be extended to a number of experimental results. They named this type of polymerization as 'homosteric' stereoelective polymerization.[93] It is to be noted that all of the homosteric initiators were chiral zinc systems, the composition of which was expressed as $[RZnOR^*]_x [R^*OZnOR^*]_y$, $(x/y < 1)$.

In contrast with the homosteric initiator, an initiator derived from the reaction between $CdMe_2$ and $(-)$-DMBD polymerized preferentially (S)-$(-)$-propylene sulfide which is the enantiomer with the opposite spatial configuration to that of the chiral diol in the initiator system. This was named by Spassky *et al.* as 'antisteric' stereoelection. Some examples of antisteric stereoelection were also found in polymerization systems with $ZnEt_2$–$(-)$-DMBD as initiator. A common compositional feature of the 'antisteric' initiator $[RMtOR^*]_x [R^*OMtOR^*]_y$ is $x/y > 2$, where Mt is Zn or Cd.

The nature of homosteric and antisteric stereoelection has not yet been fully elucidated at the molecular level because the structure of the operating species and the reaction mechanism with these initiator systems are not clearly established.

33.8 STEREOSPECIFICITY IN THE POLYMERIZATION OF LACTONES

There are a great number of papers available which deal with lactone polymerizations (see Chapter 34). Stereospecificity in the polymerizations, however, has been discussed only in relatively few papers compared with those for epoxides and episulfides. This section will concentrate on stereochemical behavior of mono- and di-substituted β-propiolactones in the polymerization reactions induced by anionic or coordinate anionic initiators.

The mode of ring cleavage of lactones with anionic as well as coordinate anionic initiators was discussed in Chapter 34. From an overview of results so far reported in the literature, it may be possible to deduce the following criteria regarding the mode of ring cleavage: (i) the more free anionic character the attacking reagent possesses, the higher the regioselectivity observed at the β-carbon for the reaction site which causes the C—O bond cleavage to form a carboxylate anion; while (ii) the more important the role of the coordination process of the ester group of the lactone onto the metallic moiety of initiator, the higher is the regioselectivity observed at the carbonyl carbon which causes the O=C—O bond cleavage to form an alkoxy group.

(i) Anionic type attack   (ii) Coordination type attack

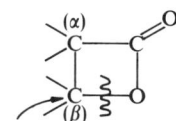   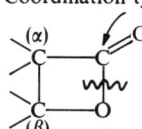

α,α-disubstituted β-propiolactone β-substituted and β,β-disubstituted β-propiolactone

⟶   preferential site of attack

〜〜〜   preferential site of ring cleavage

Scheme 2

It is understandable from points (i) and (ii) above that β- and β,β-substituted propiolactones are generally not polymerized with typical anionic initiators owing to the steric environment around the β-carbon. On the other hand, coordination type initiators should be successfully applied to β-, β,β- and α,α-substituted propiolactones (substituted PLs).

It is to be noted that α-monosubstituted PL is basically not an appropriate monomer for stereospecific polymerization, because steric configuration at the α-carbon is liable to be racemized during the polymerization process owing to its active hydrogen.

Most studies on the stereochemistry of polymers from α- and β-substituted PLs have been carried out, first by Tani *et al.*, and by Spassky, Leborgne and co-workers.

33.8.1 α,α-Disubstituted Propiolactones

Copolymerizations of (R)-$(+)$- and (S)-$(-)$-α-ethyl-α-methyl-β-propiolactone (MEPL) using a $1:1$ $MeCO_2K$/dicyclohexyl-18-crown-6 complex as initiator were carried out with different enantiomeric compositions.[94] The optical rotation of the poly(MEPL) obtained was found to be linear with the optical purity of the starting monomer.

If $p = [S]/[R]$ is defined as the enantiomeric ratio in the initial monomer with $[S] + [R] = 1$, the Bernoullian statistics predict triad populations as shown in equations (8) and (9)

$$I = (p^3 + 1)/(p + 1)^3 \tag{8}$$

$$H_i = H_s = S = p(p + 1)/(p + 1)^3 \tag{9}$$

where I, H_i, H_s and S are triad contents of $([RRR] + [SSS])$, $([RRS] + [SSR])$, $([RSS] + [SRR])$ and $([RSR] + [SRS])$ respectively.[96]

A study with high field NMR showed that triad as well as tetrad populations observed in poly-(MEPL) were in good agreement with the Bernoullian statistics (*cf.* ref. 47). Since the initiator attacks exclusively at the β-carbon of the monomer (see above) to break the CH_2—O bond, the monomer molecules are incorporated into polymer chains with the steric configuration unchanged. In consequence, it is concluded that the rate constant k_{RS} for reaction of S monomer with a growing chain end having an R monomeric unit should be equal to the rate constant k_{RR}, defined correspondingly. This situation is just the same as that obtained from R,S copolymerization of propylene oxide with potassium alkoxide as initiator (see Section 33.6).

The Bernoullian statistics are also applicable to the triad as well as the tetrad populations in poly(α-methyl-α-n-propyl-β-propiolactone) [poly(MPPL)], which was prepared with the $MeCO_2K$ initiator complexed with 18-crown-6.

Racemic MEPL gave only an atactic polymer ($I = 0.25$; $H_s = 0.26$; $H_i = 0.26$; $S = 0.24$) even when the $ZnEt_2$/MeOH system was used as initiator.[96] In other words, the enantiomorphic catalyst sites (Section 33.4) in the zinc-coordinated initiator were not able to recognize the chirality of MEPL, a fact which forms a sharp contrast with the case of racemic propylene oxide (see Section 33.2). It was reported that some irregularities, not yet identified, appeared in the spectra of poly(MEPL) prepared with the zinc-coordinated initiators.

Polymerization[95] of racemic MEPL using the $ZnEt_2$/(R)-$(-)$-DMBD system was found to obey the first-order law (see Section 33.7, equation 6) to form an optically active polymer. The stereo-election was a homosteric type, though the k_R/k_S ($= r_R$) value was very low (1.02–1.07).

A stereoelective polymerization[98] of racemic MPPL using $ZnEt_2$/(R)-$(-)$-DMBD gave a similar result (homosteric, $r_R = 1.25$) to that of MEPL. The high field NMR analysis showed a slight enrichment of isotactic tetrads (*ca.* 7%) in the stereochemical sequence of the poly(MPPL) formed. This result is compatible with the previous finding[97] that the optically active poly(MPPL) had a higher equilibrium melting point than the racemic polymer prepared with an anionic initiator.

Antisteric type stereoelective polymerization of racemic MPPL was also reported[98] when $CdMe_2$/(R)-$(-)$-DMBD was used as initiator (see Section 33.7).

33.8.2 β-Monosubstituted and β,β-Disubstituted β-Propiolactones

Agostini, Lando and Shelton[99] polymerized (RS)-β-butyrolactone (β-BL) by using initiator systems derived from zinc and aluminum alkyls. Poly(β-BL) prepared with $AlEt_3$/H_2O ($1:1$ in molar ratio) as initiator was fractionated by chloroform. The chloroform insoluble fraction showed, in the X-ray diffraction pattern, the practically identical crystalline structure to that of the naturally occurring poly[(R)-β-hydroxybutyrate]. In comparison with this, polymers prepared with the $ZnEt_2$/H_2O system were less crystalline.

Leborgne and Spassky reviewed a number of initiator systems so far proposed in terms of their stereospecificity.[100] They concluded that in most cases Al coordinate type initiators gave higher values of the index of stereospecificity (IS), defined by Tani *et al.* as the percent crystalline fraction with respect to the unfractionated polymer. The IS was generally evaluated by the solvent fractionation method. Among the initiators reviewed, the highest IS value (72%) was found for the $AlEt_3$/H_2O/epichlorohydrin system which was prepared *via* a high-vacuum drying process.[101,102]

It is to be noted that the proposed mechanism of polymerization of β-BL by the $AlEt_3$/H_2O initiator was assumed by Shelton *et al.*[103] to involve an oxonium ion as the propagating species, as previously proposed by Yamashita *et al.*[104] This is considered to relate to the strong acidic character

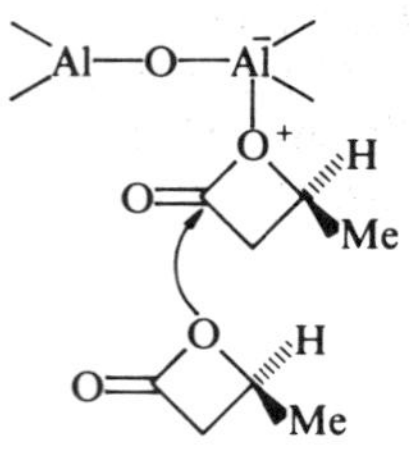

Scheme 3

of aluminum, and initiation may be written as shown in Scheme 3. In contrast with this, it has been confirmed that $ZnEt_2/H_2O$ (or $ZnEt_2/MeOH$) systems possess much less cationic character than aluminum systems (see Section 33.3).

Araki, Tani et al.[102] reported that diad and triad tacticities of β-alkyl- or β-chloroalkyl-PL are difficult to observe in 100 MHz ^{1}H NMR, but diad tacticity may be observed in ^{13}C NMR. Leborgne et al.[105] succeeded in the stereoelective polymerization of (RS)-β-BL by using $ZnEt_2$–(R)-(−)-DMBD as initiator. Unreacted monomer was enriched in S enantiomer (enantiomeric excess, e.e. = 46%) at 84% conversion. This corresponds to k_R/k_S (= r_R) = 1.6, which is not too far from the value observed in the stereoelective polymerization of propylene oxide (r_R = 1.8). The stereoelection is of the homosteric type. The whole polymer was fractionated into two parts by methanol. The methanol insoluble fraction (25% by weight) was shown by ^{13}C NMR to contain 72% isotactic diads. The stereoselective behavior of $ZnEt_2$–(R)-(−)-DMBD contrasted with that of $ZnEt_2/H_2O$ (or $ZnEt_2/MeOH$) system, the latter of which had no stereoselectivity as stated above. According to these results, it may be suggested that the mechanisms of stereoselection or stereoelection of the lactones must fit the concept of the enantiomorphic sites model (Section 33.4).

The effect of substituents on stereospecific polymerization of monosubstituted β-alkyl-PL and β-chloroalkyl-PL was studied extensively by Araki, Tani et al.[102] The values of IS (%) for polymers obtained with $(EtAlO)_n$ as initiator were 66% for methyl, 73% for ethyl, 44% for isopropyl, 0% for t-butyl, 22% for chloromethyl, 48% for dichloromethyl and 0% for trichloromethyl substituents. The non-formation of crystalline polymers from t-butyl and trichloromethyl-substituted PL was attributed to the presence of highly crowded substituents. It is to be noted, however, that the IS values may not always parallel the tacticities of the relevant macromolecules. Organozinc initiator systems again gave only amorphous polymers from these substituted PLs.

Recently, Prud'homme, Spassky et al.[106] compared polymerizations of the optically active and racemic β-chloroalkyl monosubstituted and disubstituted monomers (CCl_3-PL; CF_3,Me-PL; and CF_3,Et-PL). They found in the polymerizations of optically active CCl_3-PL and CF_3,Me-PL, with tetraphenylporphyrin/$AlEt_2Cl$ and $ZnEt_2/H_2O$ initiators respectively, that (i) the rotatory power of the polymer varies linearly as a function of the enantiomeric excess of the monomer, and (ii) the specific rotation of the residual monomer is equal to that of the starting monomer. These results indicate that the R,S copolymerization of these lactones takes place similarly to that of α,α-disubstituted PL (see above), which is in the same category as the propylene oxide polymerization induced by alkali catalyst, where $k_{RS} = k_{RR}$ (see Scheme 1, and also Section 33.6).

Prud'homme, Spassky et al.[106] studied 100.6 MHz ^{13}C NMR spectra of poly(CCl_3-PL) samples which were prepared starting from monomers with enantiomeric excesses of 0, 50 and 69%. They observed that the population of isotactic sequences increased with the enantiomeric excess, but found it very difficult to integrate isotactic, heterotactic and syndiotactic sequences separately or even to group them into four patterns for the corresponding triads, owing to the overlapping of some of the peaks.

The mode of ring cleavage of β-substituted lactones was established when the substituent was a polar ester group. Guerin, Vert et al.[107] showed that polymerization of (R)-benzyl malolactonate with NEt_3 as initiator led to poly[(S)-malic acid benzyl ester], which means that the ring cleavage takes place at the alkyl–oxygen bond with inversion of configuration at the asymmetric carbon (Scheme 4).

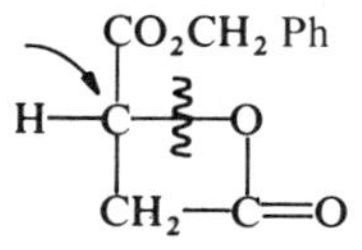

Scheme 4

γ-Ray polymerization of racemic β-dichloromethyl-PL in the solid phase was found to result in the formation of a syndiotactic polymer. On the other hand, β-trichloromethyl-PL formed an isotactic polymer under the same conditions. These results were explained by the particular molecular packing in the monomer crystals.[108]

33.9 REFERENCES

1. M. E. Pruitt and J. M. Baggett (Dow Chemical Co.), *US Pat.* 2 706 181 (1955) (*Chem. Abstr.*, 1955, **49**, 9325*f*).
2. C. C. Price and M. Osgan, *J. Polym. Sci.*, 1959, **34**, 153.
3. J. Furukawa, T. Tsuruta, R. Sakata, T. Saegusa and A. Kawasaki, *Makromol. Chem.*, 1959, **32**, 90.
4. R. O. Colclough, G. Gee and A. H. Jagger, *J. Polym. Sci.*, 1960, **48**, 273.
5. E. J. Vandenberg, *J. Polym. Sci.*, 1960, **47**, 485.
6. J. P. Machon and P. Sigwalt, *C.R. Hebd. Seances Acad. Sci.*, 1965, **260**, 549.
7. M. Osgan and Ph. Teyssie, *J. Polym. Sci., Part B*, 1967, **5**, 789; 1970, **8**, 319.
8. S. Inoue, T. Tsuruta and J. Furukawa, *Makromol. Chem.*, 1962, **53**, 215.
9. T. Tsuruta, S. Inoue, N. Yoshida and J. Furukawa, *Makromol. Chem.*, 1962, **55**, 230.
10. N. Spassky and P. Sigwalt, *C. R. Acad. Sci., Paris, Ser. C*, 1967, **265**, 624.
11. N. Spassky and P. Sigwalt, *Eur. Polym. J.*, 1971, **7**, 7.
12. J. Furukawa, N. Kawabata and A. Kato, *Polym. Lett.*, 1967, **5**, 1073.
13. (a) J. Furukawa and T. Saegusa, 'Polymerization of Aldehydes and Oxides', Wiley, New York, 1963; (b) J. Furukawa and T. Saegusa, in 'Encyclopedia of Polymer Science and Technology', Interscience, New York, 1967, vol. 6, p. 103; (c) T. Tsuruta, in 'Stereochemistry in Macromolecules', ed. A. D. Ketley, Dekker, New York, 1967, vol. 2, p. 117.
14. IUPAC Macromolecular Division, Nomenclature Commission, 'Basic Definitions of Terms Relating to Polymers', *Pure Appl. Chem.*, 1974, **40**, 477.
15. IUPAC Macromolecular Division, 'Nomenclature of Regular Single-Strand Organic Polymers', *Pure Appl. Chem.*, 1976, **48**, 373.
16. M. Ishimori and T. Tsuruta, *Makromol. Chem.*, 1963, **64**, 190.
17. T. Tsuruta, S. Inoue, M. Ishimori and N. Yoshida, *J. Polym. Sci., Part C*, 1964, **4**, 267.
18. S. Inoue, T. Tsuruta and N. Yoshida, *Makromol. Chem.*, 1964, **79**, 34.
19. M. Ishimori, T. Tomoshige and T. Tsuruta, *Makromol. Chem.*, 1968, **119**, 161.
20. M. Ishimori, G. Hsiue and T. Tsuruta, *Makromol. Chem.*, 1969, **124**, 143.
21. M. Ishimori, G. Hsiue and T. Tsuruta, *Makromol. Chem.*, 1969, **128**, 52.
22. P. H. Khanh, T. Hirano and T. Tsuruta, *J. Macromol. Sci., Chem.*, 1971, **5**, 1287.
23. T. Hirano, P. H. Khanh and T. Tsuruta, *Makromol. Chem.*, 1972, **153**, 331.
24. IUPAC Macromolecular Division, Nomenclature Commission, 'Stereochemical Definitions and Notations Relating to Polymers'; *Pure Appl. Chem.*, 1981, **53**, 733.
25. IUPAC Macromolecular Division, Nomenclature Commission, 'Polymerization Reactions Involving Chiral Monomer Molecules or Giving Optically Active Polymers', provisional document.
26. C. C. Price and R. Spector, *J. Am. Chem. Soc.*, 1965, **87**, 2069.
27. C. C. Price and M. Osgan, *J. Am. Chem. Soc.*, 1956, **78**, 4787.
28. A. Kawasaki, J. Furukawa, T. Tsuruta, T. Saegusa, G. Kakogawa and R. Sakata, *Polymer*, 1960, **1**, 315.
29. H. Tani, N. Oguni and S. Watanabe, *J. Polym. Sci., Part B*, 1968, **6**, 577.
30. Z. Jedlinski, A. Dowark and M. Bero, *Makromol. Chem.*, 1979, **180**, 949.
31. S. Tsuchiya and T. Tsuruta, *Makromol. Chem.*, 1967, **110**, 123.
32. T. Tsuruta, *Pure Appl. Chem.*, 1981, **51**, 1745.
33. N. Oguni, K. Lee and H. Tani, *Macromolecules*, 1972, **5**, 819.
34. Z. Jedlinski, J. Kasperczyk and A. Dowark, *Eur. Polym. J.*, 1983, **19**, 899.
35. F. M. Rubogliati and J. M. Contreras, *Eur. Polym. J.*, 1987, **23**, 63.
36. E. J. Vandenberg, *J. Am. Chem. Soc.*, 1961, **83**, 3538.
37. E. J. Vandenberg, *J. Polym. Sci., Part B*, 1964, **2**, 1085.
38. E. J. Vandenberg, *J. Polym. Sci., Part A-1*, 1969, **7**, 525.
39. C. C. Price and R. Spector, *J. Am. Chem. Soc.*, 1966, **88**, 4171.
40. C. C. Price and A. L. Tumolo, *J. Polym. Sci., Part A-1*, 1967, **5**, 175.
41. C. C. Price, R. Spector and A. L. Tumolo, *J. Polym. Sci., Part A-1*, 1967, **5**, 407.
42. C. C. Price, M. K. Akkapeddi, B. T. DeBona and B. C. Furie, *J. Am. Chem. Soc.*, 1972, **94**, 3964.
43. M. Yokoyama, H. Ochi, H. Tadokoro and C. C. Price, *Macromolecules*, 1972, **5**, 690.
44. N. Oguni, S. Watanabe, M. Maki and H. Tani, *Macromolecules*, 1973, **6**, 195.
45. N. Oguni, S. Maeda and H. Tani, *Macromolecules*, 1973, **6**, 459.
46. (a) Y. Hasebe and T. Tsuruta, *Makromol. Chem.*, 1987, **188**, 1403; (b) T. Tsuruta, *Makromol. Chem., Makromol. Symp.*, 1986, **6**, 23.
47. T. Tsuruta, *J. Polym. Sci., Part D*, 1972, **6**, 179.
48. J. Schaefer, *Macromolecules*, 1969, **2**, 533.
49. N. Oguni, K. Lee and H. Tani, *Macromolecules*, 1972, **5**, 819.
50. W. Lapeyre, H. Cheradame, N. Spassky and P. Sigwalt, *J. Chim. Phys. Phys.-Chim. Biol.*, 1973, **5**, 838.
51. T. Uryu, H. Schimazu and K. Matsuzaki, *J. Polym. Sci., Part B*, 1973, **11**, 275.
52. R. A. Sheldon, T. Fueno, T. Tsunetsugu and J. Furukawa, *J. Polym. Sci., Part B*, 1965, **3**, 23.
53. T. Fueno and J. Furukawa, *J. Polym. Sci., Part A*, 1964, **2**, 3681.
54. T. Fueno, R. A. Sheldon and J. Furukawa, *J. Polym. Sci., Part A*, 1965, **3**, 1279.
55. M. Ishimori, K. Tsukikawa and T. Tsuruta, *J. Macromol. Sci., Chem.*, 1977, **11**, 379.
56. T. Tsuruta, *J. Polym. Sci., Polym. Symp.*, 1980, **67**, 73.

57. T. Tsuruta, *Makromol. Chem., Suppl.*, 1981, **5**, 230.
58. T. Tsuruta, T. Hagiwara and M. Ishimori, in 'Coordination Polymerization', ed. C. C. Price and E. J. Vandenberg, Plenum Press, New York, 1983, p. 45.
59. M. Ishimori, T. Hagiwara, T. Tsuruta, Y. Kai, N. Yasukawa and N. Kasai, *Bull. Chem. Soc. Jpn.*, 1976, **49**, 1165.
60. M. Ishimori, T. Hagiwara and T. Tsuruta, *Makromol. Chem.*, 1978, **179**, 2337.
61. T. Hagiwara, M. Ishimori and T. Tsuruta, *Makromol. Chem.*, 1981, **182**, 501.
62. H. Hasegawa, K. Miki, N. Tanaka, N. Kasai, M. Ishimori, T. Heki and T. Tsuruta, *Makromol. Chem., Rapid Commun.*, 1982, **3**, 947.
63. H. Kageyama, Y. Kai, N. Kasai, C. Suzuki, N. Yoshino and T. Tsuruta, *Makromol. Chem., Rapid Commun.*, 1984, **5**, 89.
64. Y. Hasebe and T. Tsuruta, *Makromol. Chem.*, to be submitted.
65. A. Sato, T. Hirano and T. Tsuruta, *Makromol. Chem.*, 1975, **176**, 1187.
66. T. Tsuruta, *ACS Symp. Ser.*, 1977, **59**, 178.
67. A. Sato, T. Hirano and T. Tsuruta, *Makromol. Chem.*, 1977, **178**, 609.
68. M. Sepulchre, A. Khalil, N. Spassky and M. Vert, *Makromol. Chem.*, 1979, **180**, 131.
69. Ph. Dumas, N. Spassky and P. Sigwalt, *Eur. Polym. J.*, 1977, **13**, 713.
70. C. Kazanskij, M. Reix and N. Spassky, *Polym. Bull.*, 1979, **1**, 793.
71. P. Sigwalt, *Pure Appl. Chem.*, 1976, **48**, 257, and references therein.
72. N. Spassky, in 'Ring-Opening Polymerization', *ACS Symp. Ser.*, 1977, **59**, 191.
73. M. Sepulchre, N. Spassky and P. Sigwalt, *Israel J. Chem.*, 1976/77, **15**, 33.
74. C. Coulon, N. Spassky and P. Sigwalt, *Polymer*, 1976, **17**, 821.
75. P. Sigwalt and N. Spassky, in 'Ring-Opening Polymerization', ed. K. J. Ivin and T. Saegusa, Elsevier, New York, 1984, vol. 2, p. 603.
76. A. Momtaz, N. Spassky and P. Sigwalt, *Polym. Bull.*, 1979, **1**, 267.
77. N. Spassky, A. Leborgne and M. Sepulchre, *Pure. Appl. Chem.*, 1977, **53**, 1735.
78. Ph. Dumas, N. Spassky and P. Sigwalt, *J. Polym. Sci., Polym. Chem. Ed.*, 1979, **17**, 1583.
79. Ph. Dumas, N. Spassky and P. Sigwalt, *J. Polym. Sci., Polym. Chem. Ed.*, 1979, **17**, 1595.
80. Ph. Dumas, N. Spassky and P. Sigwalt, *J. Polym. Sci., Polym. Chem. Ed.*, 1979, **17**, 1605.
81. Ph. Dumas, N. Spassky and P. Sigwalt, *Makromol. Chem.*, 1972, **156**, 55; N. Spassky, Ph. Dumas, M. Sepulchre and P. Sigwalt, *J. Polym. Sci., Polym. Symp.*, 1975, **52**, 327.
82. K. J. Ivin, E. D. Little, P. Sigwalt and N. Spassky, *Macromolecules*, 1971, **4**, 345.
83. N. Spassky and P. Sigwalt, *Bull. Soc. Chem. Fr.*, 1968, 4617.
84. M. Marchetti, E. Chiellini, M. Sepulchre and N. Spassky, *Makromol. Chem.*, 1979, **180**, 1305.
85. N. Spassky, A. Leborgne, A. Momtaz and M. Sepulchre, *J. Polym. Sci., Polym. Chem. Ed.*, 1980, **18**, 3089.
86. N. Spassky, A. Momtaz and P. Sigwalt, in 'Coordination Polymerization', ed. C. C. Price and E. J. Vandenberg, Plenum Press, New York, 1983, p. 111.
87. Ph. Dumas, Ph. Guerin and P. Sigwalt, *Nouv. J. Chim.*, 1980, **4**, 95.
88. Ph. Dumas, P. Sigwalt and P. Guerin, *Makromol. Chem.*, 1981, **182**, 2225.
89. Ph. Dumas, P. Sigwalt and Ph. Guerin, *Makromol. Chem.*, 1984, **185**, 1317.
90. M. Sepulchre and N. Spassky, *Makromol. Chem., Rapid Commun.*, 1981, **2**, 261.
91. M. Sepulchre, *Makromol. Chem.*, 1987, **188**, 1583.
92. M. Sepulchre, A. Momtaz and N. Spassky, in 'Recent Advances in Anionic Polymerization', ed. T. Hogen-Esch and J. Smid, Elsevier, 1987, 297; reported in ACS Meeting in New York, *Polym. Prepr.*, 1986, 173.
93. A. Deffieux, M. Sepulchre, N. Spassky and P. Sigwalt, *Makromol. Chem.*, 1974, **175/2**, 339.
94. D. Grenier, R. E. Prud'homme, A. Leborgne and N. Spassky, *J. Polym. Sci., Polym. Chem. Ed.*, 1981, **19**, 1781.
95. A. Leborgne, D. Grenier, R. E. Prud'homme and N. Spassky, *Eur. Polym. J.*, 1981, **17**, 1103.
96. N. Spassky, A. Leborgne and W. E. Hull, *Macromolecules*, 1983, **16**, 608.
97. N. Spassky, A. Leborgne, M. Reix, R. E. Prud'homme, E. Bigdeli and R. W. Lenz, *Macromolecules*, 1978, **11**, 716.
98. A. Leborgne, N. Spassky and P. Sigwalt, *Polym. Bull.*, 1979, **1**, 825.
99. D. E. Agostini, J. B. Lando and J. R. Shelton, *J. Polym. Sci., Polym. Chem. Ed.*, 1971, **9**, 2775.
100. A. Leborgne and N. Spassky, private communication.
101. K. Teranishi, M. Iida, T. Araki, S. Yamashita and H. Tani, *Macromolecules*, 1974, **7**, 421.
102. M. Iida, T. Araki, K. Teranishi and H. Tani, *Macromolecules*, 1977, **10**, 275.
103. J. R. Shelton, D. E. Agostini and J. B. Lando, *J. Polym. Sci., Polym. Chem. Ed.*, 1971, **9**, 2789.
104. Y. Yamashita, Y. Ishikawa, T. Tsuda and S. Miura, *Kogyo Kagaku Zasshi*, 1963, **63**, 110.
105. A. Leborgne, N. Spassky and P. Sigwalt, '27th International Symposium on Macromolecules (July, 1981, Strasbourg)', Preprint, vol. 1, p. 152.
106. C. Lavallee, A. Leborgne, N. Spassky and R. E. Prud'homme, *J. Polym. Sci., Polym. Chem. Ed.*, 1987, **25**, 1315.
107. P. Guerin. J. Francillette, C. Braud and M. Vert, *Makromol. Chem., Makromol. Symp.*, 1986, **6**, 305.
108. Y. Chatani, M. Yokouchi and H. Tadokoro, *Macromolecules*, 1979, **12**, 822.

34

Anionic Ring-opening Polymerization: Lactones

ROBERT JEROME and PHILIPPE TEYSSIÉ
University of Liège, Belgium

34.1 RING-OPENING REACTIONS OF LACTONES

The ring-opening reactions of lactones, in particular β-lactones, are amongst the most striking examples of the diversity of mechanistic pathways that can be induced by minute and subtle variations in the orbital, inductive and steric characteristics of a molecule.

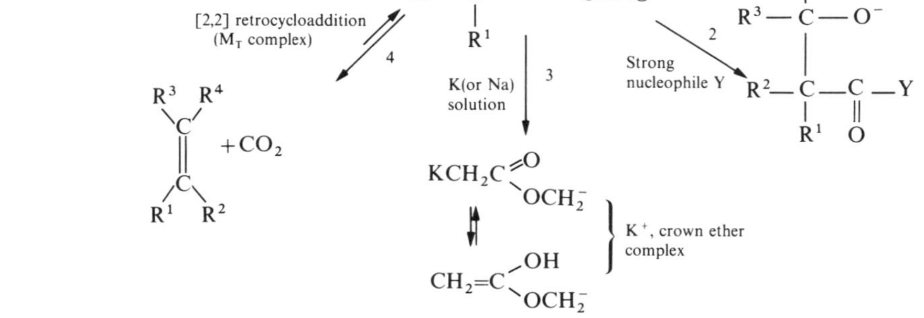

Scheme 1

It has been shown that β-propiolactone, and its substituted derivatives, can undergo five types of ring-opening reaction as illustrated in Scheme 1. All of these are important in polymerization reactions: paths 1 and 2 represent classical initiation reactions for the formation of linear polyesters, through either oxygen–alkyl (path 1) or oxygen–carbonyl (path 2) bond scission;[1] path 3 postulates a C–C bond scission in the initiation step, although the propagation proceeds through active carboxylic centers;[2] the reverse of path 4 (direct cycloaddition) leads to polyesters through intermediate lactones that are not always isolated;[3,4] and the metal-assisted [1,3] prototropic shift, (path 5) yields a carboxylic butadiene monomer that polymerizes very easily under mild conditions.[5,6]

34.2 KINETICS OF POLYMERIZATION INITIATED BY CHARGED NUCLEOPHILES

Polymerizations promoted by the usual anionic initiators, *i.e.* alkoxide and carboxylate anions, are the main source of information. The availble kinetic data will be discussed separately for the most extensively studied lactones, *i.e.* β-propiolactone, α,α-dialkyl-β-propiolactone and ε-caprolactone.

34.2.1 β-Propiolactone

Sodium acetate has been used as a weak nucleophile in the anionic polymerization of β-propiolactone (β-PL).[7] Its reactivity at room temperature was especially low, and transfer reactions were usually reported. Nowadays, the addition of macrocyclic ligands, which act as complexing agents for the alkali metal cations, allows β-PL to be polymerized by carboxylate anions under conditions that would not usually lead to polymers.[8] As well as providing a noticeable rate enhancement, crowned cations also eliminate transfer reactions, resulting in a living process.[9] Another advantage of complexing the alkali metal cations is that it provides systems in which well-defined active species are present, *i.e.* ion pairs and free ions. This particular situation has enabled Penczek and co-workers to quantify the contribution of specific active species, rather than to report global propagation rate constants.[9-15] From the dissociation constant of the crowned potassium carboxylate ion pairs (calculated approximately from conductometric data), the proportion of macroion pairs and macroions has been established, and the rate constant of propagation on these species determined.

In dichloromethane, the dissociation constant is found to be largely independent of temperature, therefore the enthalpy of dissociation of the macroion pairs is close to zero. Moreover, the entropy of dissociation is equal to -92 and -75 J k^{-1} mol^{-1} at initial β-PL concentrations ($[β\text{-PL}]_0$) of 1 and 3 mol l^{-1}, respectively. These values correlate with the existence of loose ion pairs which are subjected to only a small change in their solvation when becoming free ions. This behavior is most probably due to the bulkiness of the complexed potassium counterion.[9]

As would be expected, macroions are more reactive propagating species than are macroion pairs.[9] The ratio of the related rate constants, $k_p^-/k_p^\pm$, which amounts to 150 at 35 °C, falls, however, down to 5.6 at -20 °C. From the dependence of $\ln(k_p^-)$ and $\ln(k_p^\pm)$ on T^{-1} (Figure 1), activation parameters have been calculated (Table 1) and interpreted as follows. The solvation efficiency of the free carboxylate anion is much higher than that of the activated complex, in which the negative charge is greatly delocalized (Figure 2a). Therefore, the formation of the activated complex from the carboxylate anion requires a desolvation step, which increases the activation energy and makes the entropy of activation less negative. When the propagation takes place on macroion pairs, the charge distribution and the related solvation ability are quite similar in both the ground state and the activated complex (Figure 2b). The enthalpy of activation is, thus, essentially independent of solvation effects, and lower than that for free anions. A much more negative entropy of activation should result from the necessity of proper orientation of both ion and monomer molecules in the activated complex. In order to be consistent with the experimental results, and especially with the dependence of the activation parameters on $[β\text{-PL}]_0$ (Table 1), the aforementioned reaction scheme must include the participation of β-PL in the solvation of the active species. Therefore, $\Delta H_p^{\ddagger\,-}$ strongly depends on $[β\text{-PL}]_0$, while $\Delta H_p^{\ddagger\,\pm}$ is influenced much less. Since an increase in temperature (at constant $[β\text{-PL}]_0$) is expected to have the same effect on the solvation state of the active species as a decrease in $[β\text{-PL}]_0$ (at constant temperature), it is easy to account for the experimental dependence of $\ln(k_p^-)$ and $\ln(k_p^\pm)$ on T^{-1} (Figure 1). Similarly, the reactivity of macroion pairs might become higher than that of macroions ($k_p^-/k_p^\pm < 1$) under extremely low temperature conditions.

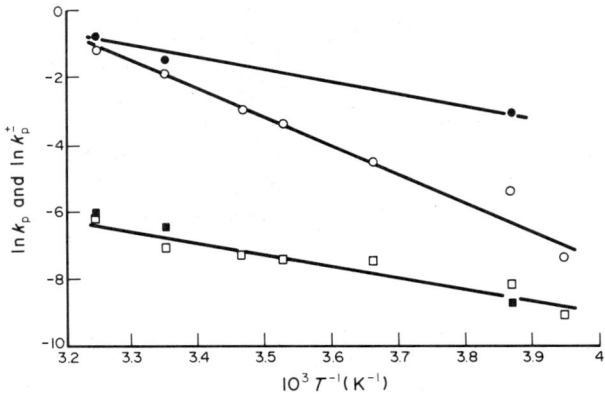

Figure 1 Dependence of $\ln k_p^-$ ($\bullet$, $[\beta\text{-PL}]_0 = 1 \text{ mol l}^{-1}$; $\bigcirc$, $[\beta\text{-PL}]_0 = 3 \text{ mol l}^{-1}$) and $\ln k_p^{\pm}$ ($\blacksquare$, $[\beta - \text{PL}]_0 = 1 \text{ mol l}^{-1}$; $\square$, $[\beta\text{-PL}]_0 = 3 \text{ mol l}^{-1}$) on T^{-1}, (k_p^- and $k_p^{\pm}$ in $\text{mol}^{-1} 1 \text{ s}^{-1}$) (reproduced by permission of the American Chemical Society from *Macromolecules*, 1980, **13**, 229)

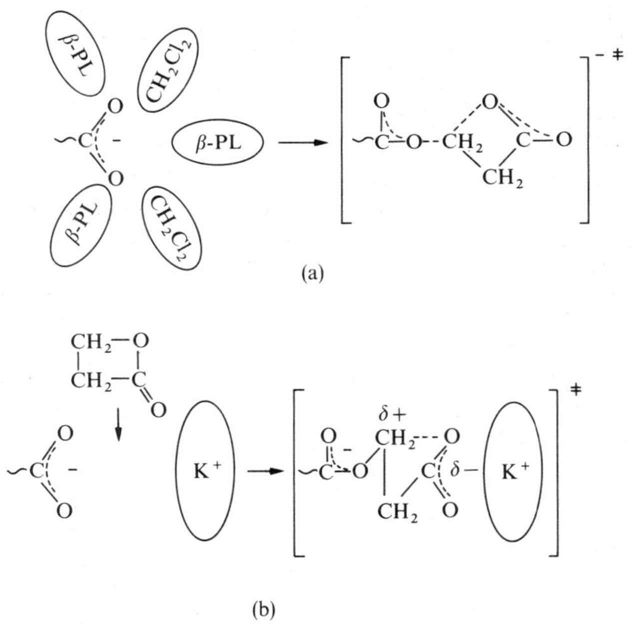

(a)

(b)

Figure 2 (a) Formation of the activated complex for propagation on macroions; monomer and solvent molecules participate in the solvation shell of the macroions, but in unknown numbers. (b) Formation of the activated complex for propagation on macroion pairs

Table 1 Activation Parameters[a] for the Polymerization of β-PL with K^+ Dibenzo-[18]-crown-6-ether Counterion in Dichloromethane[9]

$[\beta\text{-}PL]_0$ (mol l^{-1})	$\Delta H_p^{\ddagger -}$ (kJ mol^{-1})	$\Delta S_p^{\ddagger -}$ (J K^{-1} mol^{-1})	$\Delta H_p^{\ddagger \pm}$ (kJ mol^{-1})	$\Delta S_p^{\ddagger \pm}$ (J K^{-1} mol^{-1})
3	67±4	−42±17	—	—
1	25±4	−167±21	—	—
1 and 3	—	—	25±4	−218±17

When β-PL is polymerized with crowned potassium acetate in dimethylformamide (DMF) instead of dichloromethane, the predicted inversion of reactivity of macroions and macroion pairs is actually reported.[15] At $[\beta\text{-PL}]_0 = 1 \text{ mol l}^{-1}$, macroion pairs are less reactive than macroions at $35\,°C$ ($k_p^-/k_p^\pm = 3$), but the situation is quite reversed at $20\,°C$, since $k_p^-/k_p^\pm = 0.75$. It is worth noting that β-PL has a dipole moment ($\mu = 4.18$) higher than that of dichloromethane ($\mu = 1.14$) and DMF ($\mu = 3.86$). These values conform with the suggestion that the solvation shell of the propagating macroions contains a higher fraction of β-PL when compared with the bulk solution.

In conclusion, when the monomer has a higher solvating power than the solvent the solvation effects govern the kinetics of polymerization, and the relative reactivity of macroions and macroion pairs can be changed, and even reversed, when temperature and/or initial monomer concentration are varied over a sufficiently large range.

34.2.2 α,α-Disubstituted β-Propiolactones

The anionic polymerization of α,α-dialkyl-β-propiolactones, initiated by carboxylate anions, has been extensively studied in relation to the ion pair structure, the nature of the substituents and the polarity of the solvent.

34.2.2.1 *Ion pair structure*

Any increase in the nucleophilicity of the carboxylate anion results in a decrease of the ratio of the propagation rate constant to that of initiation.[16,17]

Complexing the alkali metal counterion,[18] or replacing it by a tetraalkylammonium cation,[19,20] has a beneficial effect on the rate of polymerization and endows the propagation step with the main features of a living process.

In the anionic polymerization of α-methyl-α-propyl-β-propiolactone (THF, $-20\,°C$), the reactivity of the cryptated carboxylate ion pairs is found to be higher than that of the free ions, as supported by the ratio $k_p^-/k_p^\pm$ equalling 0.1.[18] Although discussed in the previous section, this was the first time that this particular situation had been observed. Since α-methyl-α-propyl-β-propiolactone may be assumed to have the same dipole moment as β-PL, it is possible that a reaction scheme based on Figure 2 is followed, involving the participation of the monomer in the solvation of the carboxylate anion. At the time of their observation Haggiage *et al.*[18] proposed a push–pull mechanism, only operating for cryptated ion pairs, in order to account for the comparatively lower reactivity of the free carboxylate ions (Figure 3). According to this reaction pathway, the very high dipole moment of the lactone favours a strong dipole–dipole interaction with the carboxylate ion pairs. The attack on the alkyl–oxygen bond of the lactone is then assisted by a transfer of charge through the cation. The face-to-face explanations are both founded on the high polarity of β-lactones, but Penczek's proposal emphasizes the solvation of the carboxylate anions by the monomer, whereas Haggiage *et al.* stress the dipole–dipole interaction between the monomer and the ion pairs.

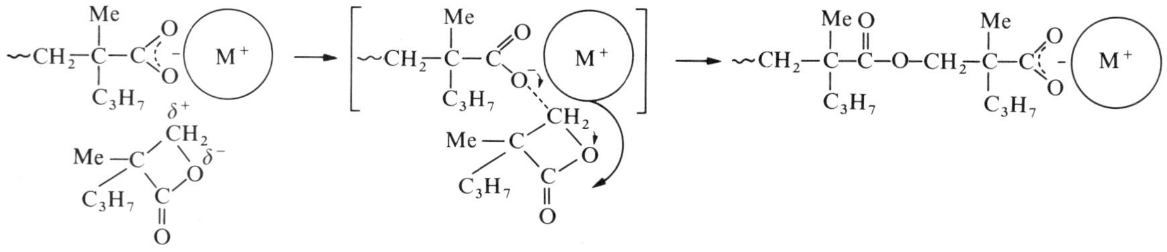

Figure 3 Push–pull mechanism for the propagation of α-methyl-α-propyl-β-propiolactone on cryptated ion pairs

34.2.2.2 *Substitution effect*

Within a series of α-methyl-α-(n-alkyl)-β-propiolactones, the apparent propagation rate constant unexpectedly increases with increasing size of the n-alkyl group (Table 2). This effect is observed

Table 2 Propagation Rate Constants and Activation Parameters[a] for the Polymerization of α,α-Disubstituted β-PL in DMSO, Initiated by Tetraethylammonium Benzoate[22]

Monomer Substituents α,α	k_p (mol l min^{-1}) at T (°C)				ΔE_p (kJ mol^{-1})	ΔS_p (J K^{-1} mol^{-1})
	22	30	37	40		
Me, Me	10.2					
Me, Et	11.2		19.0		26.8	−177
Me, Pr^n	11.8		20.4		27.6	−174
Me, Bu^n	13.0		22.9		28.9	−169
Et, Bu^n		9.5		13.8	31.3	−141

independently of solvent, counterion and temperature.[21] The increase in the activation energy with increasing substituent size is an expression of a deactivation effect, due to either steric hindering of attack or increased ring stability. The deactivation effect is, however, offset by a favorable change in the entropic contribution.[22]

Table 2 shows that k_p for α-ethyl-α-(n-butyl)-β-PL is lower than for α-methyl-α-(n-butyl)-β-PL, all conditions of solvent and initiator being unchanged. The so-called 'steric acceleration' reported in the α-methyl-α-(n-alkyl)-β-PL series is, thus, no longer observed when two larger alkyl groups are simultaneously present. In this instance, the decrease in activation entropy is not sufficient to offset the increase in activation energy. So far, the observed changes in the activation parameters have not been found to be consistent with any mechanistic scheme.

34.2.2.3 Solvent effect

In general, the apparent propagation rate constant for the anionic polymerization of α,α-dialkyl-β-PL is higher when a solvent of lower polarity is used. This is true when dimethyl sulfoxide (DMSO) or acetonitrile (AN) are compared with the less polar tetrahydrofuran (THF).[20,21,23] For the polymerization of α-methyl-α-butyl-β-PL at 36 °C, k_p is more than twice as large in THF when compared with DMSO.[20] This observation could mean that THF is a much better dissociating agent of the tetraethylammonium benzoate used as the initiator than are DMSO or AN.

In order to cope with this unexpected solvent effect, Bigdeli and Lenz[21] suggest that the polymerization reaction might be controlled by some type of specific solvation of the carboxylate anions. The acceptor number, as defined by the coordinate bond energy of a given solvent with a specific electron donor, is higher for DMSO (19.3) than for THF (8.0). Moreover, DMSO is a much better solvent for cations than THF, as supported by its higher donor number (29.8 for DMSO and 20.0 for THF). Thus, either DMSO strongly stabilizes solvated ion pairs by a type of dual coordination (Figure 4), or the anion solvation by DMSO is more important than the cation solvation, resulting in both cases in a less reactive nucleophile.

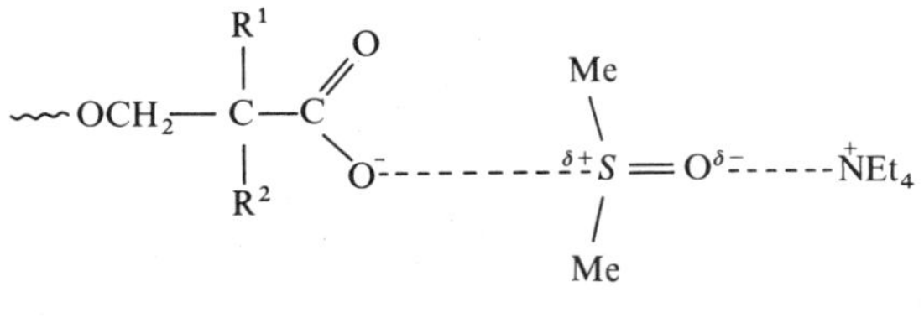

Figure 4 Structured ion pair solvation with specific coordination of carboxylate anion to DMSO

Finally, Bigdeli and Lenz suggest that an alternative explanation could involve the desolvation of the free carboxylate anions which is required to reach the transition state characteristic of the ion–dipole S_N2 polymerization reaction.[21] Due to the higher solvation efficiency of DMSO, the reactivity of carboxylate anions should be lower in DMSO than in THF. A very similar effect has been discussed by Slomkowski[15] who attributes the very low dissociation of crowned potassium carboxylate ion pairs to stabilizing solvation by dimethylformamide (DMF). Owing to the zwitter-ionic nature of its resonance structure, DMF might be able to increase the tightness of the solvated ion pairs rather than decrease it, thereby decreasing the dissociation constant (Figure 5).

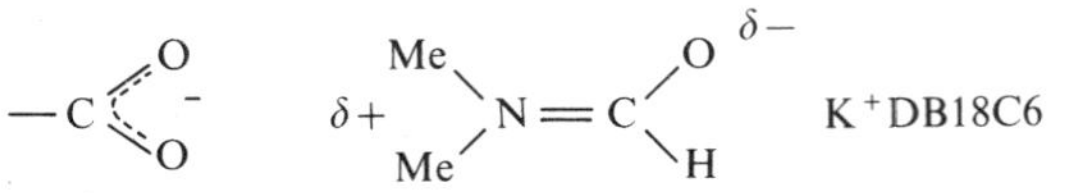

Figure 5   Stabilizing solvation of the K$^+$ DB18C6 carboxylate ion pair by DMF (DB18C6 = dibenzo-[18]-crown-6-ether)

34.2.3   ε-Caprolactone

When polymerized by alkali metal alkoxides (or stronger organoalkali initiators), ε-caprolactone (ε-CL) is subject to a ring–chain equilibrium which is more rapidly attained as the alkali metal cation is more electropositive.[17] Once equilibrium is reached, the reaction mixture is still reactive, since additional monomer is completely converted to give a different product distribution.[24] According to Yamashita and co-workers,[24] linear chains are formed in the first stages of polymerization (equation 1) but then break down to cyclic oligomers by a back-biting reaction of the active chain ends. The cyclization process can actually proceed, *via* back biting, to either an optional chain unit (equation 2) or to the last unit of the chain (end-to-end ring closure, equation 3).[25]

$$A–M_x–B + M_1 \underset{k_d}{\overset{k_{p_1}}{\rightleftharpoons}} A–M_{x+1}–B \tag{1}$$

$$A–M_x–B \underset{k_{p(y)}}{\overset{k_{b(y)}}{\rightleftharpoons}} A–M_{x-y}–B + M_y \tag{2}$$

$$A–M_x–B \overset{k_{e(x)}}{\longrightarrow} A–B + M_x \tag{3}$$

M_y and M_x designate cyclic oligomers with a polymerization degree of y and x, respectively. A is the ester end group, B the growing center, A–B the released initiator and M_1 the monomer.

34.2.3.1   Kinetic control of macroring and linear chain formation

In order to avoid the formation of cyclic oligomers as much as possible, several authors have considered kinetic control of the living reaction mixture.[25-27] With reference to the kinetic scheme described above (equations 1–3), the formation of linear polymer can be kinetically enhanced at the early stages of polymerization, provided that $k_{e(x)} < k_{b(y)} \ll k_p[M_1]$. Morton and Wu[27] have defined experimental conditions in which the polymerization of ε-CL is complete in three minutes (in benzene, with Li$^+$ as the counterion), while oligomers are formed only when the active polymer solution is allowed to stand for longer periods of time. Advantage has been taken of this kinetic control of the ring–chain competition in order to synthesize polyesters containing block polymers.[26, 27]

34.2.3.2   Thermodynamics of the ring–chain equilibrium

The thermodynamic aspect was first analyzed by Jacobson and Stockmayer,[28] who neglected the enthalpy contribution and assumed the formation of unstrained macrocycles obeying Gaussian statistics. The macrocyclic distribution was therefore expressed by:

$$[M_y]_e = Py^{-5/2} \tag{4}$$

where P is a constant and $[M_y]_e$ is the equilibrium concentration of cyclic oligomers with a polymerization degree y. The theory actually relies upon two competing reactions: the formation of macrocycles by back biting with a rate constant $k_{b(y)}$ and the disappearance of macrocycles with a rate constant $k_{p(y)}$ (equation 2). According to theoretical expectations, $k_{p(y)}$ should be proportional to y, and $k_{b(y)}$ to $y^{-3/2}$.

Penczek and co-workers[29] have studied the anionic polymerization of ε-CL, initiated by sodium trimethylsilanolate in THF, in the temperature range 10–40 °C. They have determined $k_{p(y)}$ and $k_{b(y)}$ for monomer and cyclic oligomers with $y \le 7$. From a kinetic treatment based on equations (1) and

(2), assuming a partial aggregation of the sodium alkoxide ion pairs, they have concluded that

$$k_{b(y)}^{app} = k_{p(y)}^{app}[M_y]_e \tag{5}$$

$$k_{b \text{ or } p(y)}^{app} = k_{b \text{ or } p(y)} K^{1/m} \tag{6}$$

where K is the equilibrium constant of aggregation and m the degree of aggregation.

For sodium alkoxide ion pairs, m is found to be close to 3. Within the limits of experimental error, the apparent energy of activation for propagation and back biting are very close to each other in value and do not vary with ring size. In agreement with Yamashita's previous results,[26] the equilibrium concentration of cyclic oligomers is independent of temperature, illustrating their unstrained conformation.

Furthermore, the theoretical predictions of Jacobson and Stockmayer, for chains unperturbed by long range interactions, are confirmed, since, starting from the tetramer, $[M_y]_e$ is found to depend on $y^{-2.2\pm0.2}$ and $k_{b(y)}^{app}$ on $y^{-1.4\pm0.1}$ in the region 10–40 °C. Figure 6 shows the experimental dependences observed at 40 °C. Only the dependence of $k_{p(y)}^{app}$ on $y^{+0.70\pm0.15}$ is significantly different from the theoretical prediction. This discrepancy is attributed to 'conformational hindrance', which makes some of the ester groups in the macrocycles less accessible to attack by the growing alkoxide ion pair. Since the hindrance can be expected to increase with ring size ($y \leq 7$), the exponent in the dependence law of $k_{p(y)}^{app}$ on y may decrease from 1 to 0.7.

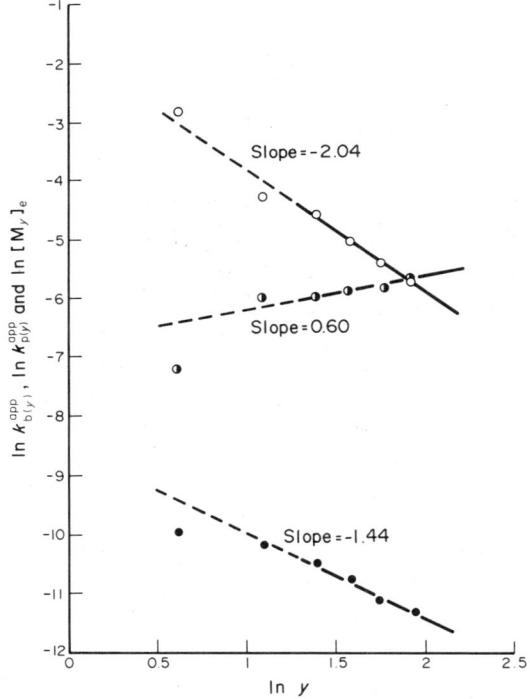

Figure 6 Dependence of $\ln[M_y]_e$ ($\bigcirc$), $\ln k_{p(y)}^{app}$ ($\circleddash$) and $\ln k_{b(y)}^{app}$ ($\bullet$) on $\ln y$. $[\varepsilon\text{-CL}]_0 = 0.5$ mol l^{-1}, $T = 40$ °C, solvent = THF. $k_{p(y)}^{app}$ in mol$^{-1/3}$ l$^{1/3}$ s^{-1} and $k_{b(y)}^{app}$ in mol$^{2/3}$ l$^{-2/3}$ s^{-1} (reproduced by permission of Hüthig and Wepf Verlag from *Makromol. Chem.*, 1983, **184**, 2159)

ε-CL is not usually detected in the equilibrated polymerization mixture, due to a much higher reactivity than those of the other cyclic oligomers. As a kinetic consequence of the dependence of $k_{p(y)}^{app}$ and $k_{b(y)}^{app}$ on y (Figure 6), the higher the ring size of the cyclic oligomer, the higher is its reactivity towards polymerization and the lower its equilibrium concentration.

34.2.3.3 *Aggregation of alkoxide ion pairs*

From the experimental data relating to sodium alkoxides as growing species in THF, Penczek and co-workers conclude that these ion pairs are aggregated ($m \sim 3$).[29] In contrast to Na$^+$, no aggregation is reported in THF with K^+ as the counterion,[12] whereas, for Li^+, Yamashita[17] observes a kinetic order close to zero with respect to the initiator, suggesting a very high degree of association of the active lithium alkoxide species.

The big difference in association exhibited by potassium and lithium alkoxide anions could account for the much slower propagation (*ca.* 300 times slower) when ε-CL is polymerized in THF at 0 °C with Li^+ instead of K^+ as the counterion.[17] Similarly, in the presence of Li^+, the rate of back biting becomes slower by several orders of magnitude, and allows the formation of living polymer chains to be kinetically controlled at the initial stage of polymerization. The cation size governs the cation charge density and, accordingly, the strength of the ion pairs formed with the alkoxide anions. With lithium alkoxides, the very small size of the lithium atom allows the cation of one ion pair to interact with the anion alkoxide of another one, leading to a multifold coordination and aggregation. In the particular case of K^+ (THF, 20 °C), the propagation of ε-CL proceeds exclusively on the non-associated ion pairs which have, moreover, no significant tendency to dissociate (dissociation constant $= 4 \times 10^{-10}$ mol l^{-1}). Under these conditions, the experimental value of k_p^{app} can be identified with $k_p^{\pm}$.[12]

34.3 KINETICS OF POLYMERIZATION INITIATED BY NON-CHARGED NUCLEOPHILES

The polymerization of β-PL and α,α-disubstituted-β-PL can be initiated by tertiary amines and phosphines with formation of zwitterions[30-33] (Scheme 2). Growing ion pairs may be formed by either cyclization of individual chains or their pairing with each other.[32] k_i is expected to be much lower than k_p, since initiation involves the interaction of two neutral molecules, leading to a charge separation within one molecule, whereas propagation proceeds on both ion pairs and free ions.

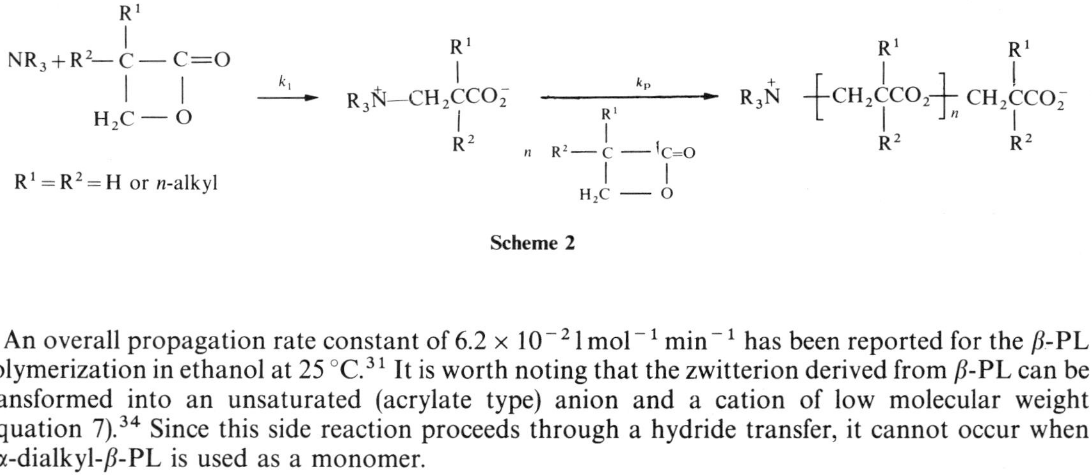

Scheme 2

An overall propagation rate constant of 6.2×10^{-2} l mol^{-1} min^{-1} has been reported for the β-PL polymerization in ethanol at 25 °C.[31] It is worth noting that the zwitterion derived from β-PL can be transformed into an unsaturated (acrylate type) anion and a cation of low molecular weight (equation 7).[34] Since this side reaction proceeds through a hydride transfer, it cannot occur when α,α-dialkyl-β-PL is used as a monomer.

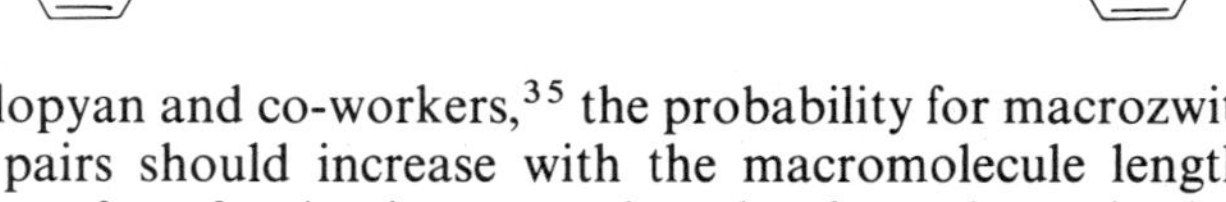

$$ \text{equation 7} \tag{7} $$

According to Enikolopyan and co-workers,[35] the probability for macrozwitterions to exist as free ions rather than ion pairs should increase with the macromolecule length. Assuming that the propagation rate constant for a free ion is greater than that for an ion pair, the propagation rate of a given polymer chain is expected to increase with the molecular weight. Although not documented in the polymerization of β-PL, this self-accelerated chain propagation effect has been reported in the polymerization of methacrylonitrile initiated by triethylphosphine in DMF at 41 °C, and all the elementary rate constants have been determined.[35]

It is also known that conventional active hydrogen compounds, such as alcohols, amines and carboxylic acids, can initiate lactone polymerization. High temperatures are generally required for quantitative conversion into low molecular weight species, at a very low rate. It is interesting to note that Rozenberg has reported that ε-CL can be polymerized by aniline (150–200 °C) with formation of carboxylic acid telechelic poly(ε-caprolactone).[36] A reaction scheme and the kinetic parameters for the initiation and chain propagation have been reported.

34.4 KINETICS OF POLYMERIZATION INITIATED BY ANIONIC COORDINATION CATALYSTS

In anionic polymerization, various types of propagating species can be found in equilibrium in the reaction medium. Classified in order of decreasing reactivity, these species may be: free ions $\rightleftharpoons$ solvated ion pairs $\rightleftharpoons$ contact ion pairs $\rightleftharpoons$ polarized covalent bonds. When the equilibrium is displaced in favor of the less ionized species, a pseudoionic reaction takes place.

Derivatives of aluminum alkyls and alkoxides are believed to behave in this way, since aluminum is much less electropositive than the commonly used alkali metals. Aluminum porphyrins,[37,38] bimetallic-μ-oxo-alkoxides[39–41] and aluminum alkoxides[42,43] are the most representative coordinate anionic catalysts used in lactone polymerization. The performances of aluminum porphyrins and their main mechanistic features are described in Volume 3, Chapter 37.

Bimetallic-μ-oxo-alkoxides are well defined compounds based on aluminum alkoxides, having the general formula $(RO)_2AlOM_tOAl(OR)_2$. M_t is a metal in a bivalent form such as Zn, Co, Fe or Mo. The striking solubility of both bimetallic-μ-oxo-alkoxides and aluminum alkoxides in most organic solvents is due to their association, which increases as the polarity of the solvent decreases.

Since the propagation step is known to proceed through the insertion of the monomer into an Al–OR bond,[41] the association of the catalyst by intra- and inter-molecular oxygen–metal coordinative bonds governs the mean number of active alkoxide groups per catalyst molecule. The mean degree of the catalyst association has a further decisive effect on the kinetics of polymerization, since the ε-CL polymerization is second order in catalyst for higher aggregates and first order with the less associated catalysts.[40]

It is worth noting that, all other conditions being unchanged, the polymerization rate decreases with increasing polarity of the solvent.[40] For di-μ-oxo-bis[bis(n-butoxy)Al]Zn at 20 °C, the half-polymerization time is 180 s in toluene but rises to 1470 s in chloroform.

Similarly, in the presence of alcohol, which completely dissociates the catalyst, the global propagation rate constant is about 17 times lower than the value at the same temperature in the absence of that ligand. These kinetic effects stress the strong competition between the monomer and the solvent for coordination to the catalyst sites, and support the so-called 'coordinate anionic mechanism'. Indeed, the monomer seems to be first coordinated at the catalyst site prior to propagation, a feature demonstrated by ^{13}C NMR in the polymerization of oxiranes by the same catalytic systems. The metal alkoxide, which is essentially nucleophilic in nature, then attacks the coordinated monomer, and enables the chain growth to proceed. At first glance, these data cannot support the idea that the addition of solvating agents leads to more ionized (and thus more reactive) species, as generally observed for alkali metal alkoxides. Bimetallic-μ-oxo-alkoxides[39–41] and aluminum alkoxides[42,43] have the great advantage of polymerizing β-PL, ε-CL and some methylated ε-CL's in a living manner. Furthermore, the equilibrium of the ε-CL polymerization is shifted towards the open chains to the extent of more than 99%. Due to these very favorable conditions, novel multiphase polyesters containing block polymers have been synthesized with well-controlled molecular parameters.[44]

When catalyzed by aluminum isopropoxide, the polymerization of ε-CL, ε-methyl-ε-CL (MCL) and β,δ-methyl-ε-CL (mixed isomers; XCL) obeys the same living anionic-type coordinated insertion mechanism previously reported for bimetallic-μ-oxo-alkoxides.[43] The methyl substitution of MCL and XCL is responsible for a substantial decrease in the propagation rate constant as compared to that of ε-CL. ($k_p = 243$, 2.12 and 13.8 mol^{-1} l min^{-1} for ε-CL, MCL and XCL, respectively, in toluene at 25 °C.) Nevertheless, the copolymerization of ε-CL with MCL and XCL closely approaches a random sequence distribution of monomer units, since $r_1 r_2 = 0.90$ and 0.88 for the ε-CL/MCL and ε-CL/XCL copolymerization (toluene, 25 °C), respectively.[43] From the experimental values of k_{11}, k_{22}, r_1 and r_2, the k_{12} and k_{21} apparent rate constants of cross-propagation have been calculated for the ε-CL (monomer 1)/MCL (monomer 2) pair. A comparison of the elementary propagation rate constants ($k_{11} = 243$, $k_{12} = 162$, $k_{21} = 2.65$ and $k_{22} = 2.12$ mol^{-1} l min^{-1}) sheds light on the kinetic consequence of the methyl substitution. The substitution of the carbon atom in the α position of the insertion site is responsible for a dramatic decrease in the overall insertion rate as measured by a comparison of $k_{11}[-CH_2OAl\langle +\varepsilon$-CL] and $k_{21}[-CHMeOAl\langle +\varepsilon$-CL]. On the other hand, k_{11} and k_{12} (or k_{21} and k_{22}) values seem to indicate a relatively small effect of the methyl substitution of ε-CL on the global propagation rates. It is interesting to note that the crystallinity of poly(ε-CL-co-MCL) and poly(ε-CL-co-XCL) samples can be largely controlled in relation to the ε-CL content. When the latter is lower than *ca.* 50% amorphous materials are obtained, which, however, retain a remarkable property of poly(ε-caprolactone), in that they are miscible with poly(vinyl chloride).

34.5 REFERENCES

1. A. Hofman, S. Slomkowski and S. Penczek, *Makromol. Chem.*, 1984, **185**, 91.
2. Z. Jedlinski, P. Kurcok and M. Kowalczuk, *Macromolecules*, 1985, **18**, 2679.
3. J. H. McCain and E. Marcus, *J. Org. Chem.*, 1970, **35**, 2414.
4. A. F. Noels and P. Lefebvre, *Tetrahedron Lett.*, 1973, **32**, 3035.
5. A. F. Noels, J. J. Herman and Ph. Teyssie, *J. Org. Chem.*, 1976, **41**, 2527.
6. A. F. Noels, J. J. Herman, Ph. Teyssie, J. M. Andre, J. Delhalle and J. G. Frippiat, *J. Mol. Struct.*, 1984, **109**, 293.
7. T. Shiota, Y. Goto and K. Hayashi, *J. Appl. Polym. Sci.*, 1967, **11**, 753.
8. A. Deffieux and S. Boileau, *Macromolecules*, 1976, **9**, 371.
9. S. Slomkowski and S. Penczek, *Macromolecules*, 1980, **13**, 229.
10. S. Penczek, S. Slomkowski and D. Kotynska, *Polym. Prepr., Am. Chem. Soc., Div. Polym. Chem.*, 1980, **21**, 53.
11. S. Slomkowski and S. Penczek, *ACS Symp. Ser.*, 1981, **166**, 271.
12. S. Sosnowski, S. Slomkowski and S. Penczek, *J. Macromol. Sci., Chem.*, 1983, **A20**, 979.
13. S. Penczek, P. Kubisa, S. Slomkowski and K. Matyjaszewski, *ACS Symp. Ser.*, 1985, **286**, 117.
14. S. Penczek and S. Slomkowski, *Polym. Prepr., Am. Chem. Soc., Div. Polym. Chem.*, 1986, **27**, 171.
15. S. Slomkowski, *Polymer*, 1986, **27**, 71.
16. Y. Yamashita and T. Hane, *J. Polym. Sci., Polym. Chem. Ed.*, 1973, **11**, 425.
17. Y. Yamashita, *Polym. Prepr., Am. Chem. Soc., Div. Polym. Chem.*, 1980, **21** (1), 51.
18. J. Haggiage, P. Hemery, S. Boileau and R. W. Lenz, *Polymer*, 1983, **24**, 578.
19. J. Cornibert, R. H. Marchessault, A. E. Allegrezza, Jr. and R. W. Lenz, *Macromolecules*, 1973, **6**, 676.
20. C. D. Eisenbach and R. W. Lenz, *Makromol. Chem.*, 1976, **177**, 2539.
21. E. Bigdeli and R. W. Lenz, *Macromolecules*, 1978, **11**, 493.
22. R. W. Lenz, E. M. Minter, D. B. Johns and S. Hvilsted, *ACS Symp. Ser.*, 1985, **286**, 105.
23. R. W. Lenz, C. G. D'Hondt and E. Bigdeli, *ACS Symp. Ser.*, 1977, **59**, 210.
24. K. Ito, Y. Hashizuka and Y. Yamashita, *Macromolecules*, 1977, **10**, 821.
25. K. Matyjaszewski, M. Zielinski, P. Kubisa, S. Slomkowski, J. Chojnowski and S. Penczek, *Mackromol. Chem.*, 1980, **181**, 1469.
26. Y. Yamashita, *ACS Symp. Ser.*, 1981, **166**, 199.
27. M. Morton and M. Wu, *ACS Symp. Ser.*, 1985, **286**, 175.
28. H. Jacobson and W. H. Stockmayer, *J. Chem. Phys.*, 1950, **18**, 1600.
29. S. Sosnowski, S. Slomkowski and S. Penczek, *Makromol. Chem.*, 1983, **184**, 2159.
30. T. L. Gresham, J. E. Jansen, F. W. Shaver, R. A. Bankert and F. T. Fiedorek, *J. Am. Chem. Soc.*, 1951, **73**, 3168.
31. V. Jaacks and N. Mathes, *Makromol. Chem.*, 1970, **131**, 295; *Makromol. Chem.*, 1971, **142**, 209.
32. N. R. Mayne, *Chem. Technol.*, 1972, 728.
33. G. Bier and N. Vollkommer, *Br. Polym. J.*, 1979, **11**, 28.
34. Y. Yamashita, K. Ito and F. Nakakita, *Makromol. Chem.*, 1969, **127**, 292.
35. M. A. Markevich, E. V. Kochetov, F. Ranogajec and N. S. Enikolopyan, *J. Macromol. Sci., Chem.*, 1974, **A8**, 265.
36. B. A. Rozenberg, *Pure Appl. Chem.*, 1981, **53**, 1715.
37. T. Yasuda, T. Aida and S. Inoue, *Macromolecules*, 1983, **16**, 1792.
38. S. Asano, T. Aida and S. Inoue, *Macromolecules*, 1985, **18**, 2057.
39. T. Ouhadi and J. M. Heuschen, *J. Macromol. Sci., Chem.*, 1975, **A9**, 1183.
40. T. Ouhadi, A. Hamitou, R. Jerome and Ph. Teyssie, *Macromolecules*, 1976, **9**, 927.
41. A. Hamitou, T. Ouhadi, R. Jerome and Ph. Teyssie, *J. Polym. Sci., Polym. Chem. Ed.*, 1977, **15**, 865.
42. T. Ouhadi, C. Stevens and Ph. Teyssie, *Makromol. Chem., Suppl.*, 1975, **1**, 191.
43. J. M. Vion, R. Jerome, Ph. Teyssie, M. Aubin and R. E. Prud'homme, *Macromolecules*, 1986, **19**, 1828.
44. J. Heuschen, R. Jerome and Ph. Teyssie, *Macromolecules*, 1981, **14**, 242.

35

Anionic Ring-opening Polymerization: Lactams

JAN ŠEBENDA

Institute of Macromolecular Chemistry, Prague, Czechoslovakia

35.1 INTRODUCTION

The anionic polymerization of lactams differs from the anionic polymerizations of most unsaturated and heterocyclic monomers, because the growth center at the chain end is not represented by an anionically activated group but by a neutral N-acylated lactam; the anionically activated species is the incoming monomer (Scheme 1). For the sake of simplicity, the anionically activated lactam is written as an N anion without counterion, although it may be present in various ionized states, such as a contact or solvent-separated ion pair, as well as complexed, solvated or free anion. Depending on the reaction medium and monomer structure, the negative charge is distributed between the three atoms of the amide group.

The high activity of the anionic system is due to the fact that the two types of active species involved in the propagation reaction (1) are lactam units activated in opposite directions: the anionically activated lactam of increased nucleophilicity and the N-acylated lactam of increased

$$\text{\sim\sim CON—CO} + \text{}^-\text{N—CO} \longrightarrow \text{\sim\sim CON}^- \quad \text{CON—CO} \tag{1}$$

$$\text{\sim\sim CON}^- \quad \text{CON—CO} + \text{HN—CO} \longrightarrow \text{\sim\sim CONH} \quad \text{CON—CO} + \text{}^-\text{N—CO} \tag{2}$$

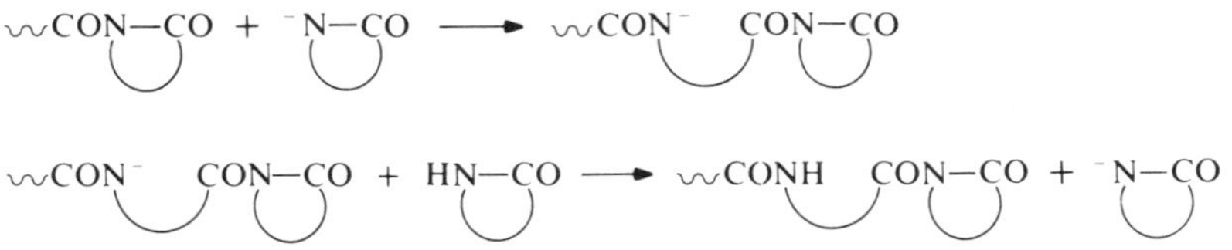

Scheme 1

electrophilicity. The effective cooperation of these two oppositely activated species provides a favorable pathway for cleavage of the otherwise highly stabilized amide bond.

35.2 CATALYTIC COMPONENTS

Anionic lactam polymerizations are usually started by a two-component catalytic system,[1-6] composed of lactamates (1), or their precursors, and N-acyllactams (2) or similarly acting compounds (3) and (4). The specific nature of the propagation reaction also poses a nomenclature problem. Various terms have been used in the literature for the two catalytic components, *e.g.* catalyst, initiator, activator, coinitiator, accelerator, promoter, cocatalyst and chain initiator.

$$ Mt\left[N{-}CO \right] \qquad RCON{-}CO \qquad RNHCON{-}CO \qquad R_2NCON{-}CO $$

(1) Mt = metal   (2)   (3)   (4)

Since the actual growth center at the chain end is represented by an N-acyllactam, the term chain initiator, suggested by Sekiguchi,[1] properly expresses the role of N-acyllactams and their precursors. When using this term, however, it should be noted that these substances are incapable of starting the propagation reaction alone, *i.e.* in the absence of lactamate.

The term initiator would be appropriate for lactamates and their precursors, because these substances are capable of initiating alone an anionic polymerization, by forming the necessary N-acyllactam in the initiation reaction (3; Scheme 2). However, in order to avoid confusion with the term chain initiator, these strongly basic compounds and their precursors will be designated as catalysts.

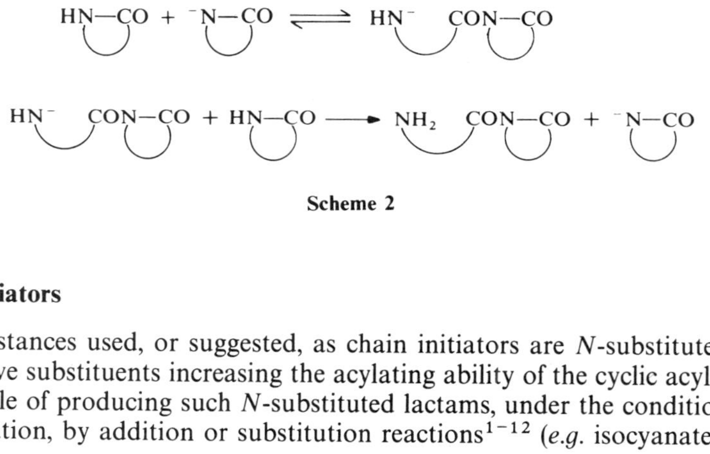

$$ HN{-}CO + {}^{-}N{-}CO \rightleftharpoons HN^{-} \quad CON{-}CO \tag{3} $$

$$ HN^{-} \quad CON{-}CO + HN{-}CO \longrightarrow NH_2 \quad CON{-}CO + {}^{-}N{-}CO \tag{4} $$

Scheme 2

35.2.1 Chain Initiators

Among the substances used, or suggested, as chain initiators are N-substituted lactams (2)–(4) with electronegative substituents increasing the acylating ability of the cyclic acyl group, as well as compounds capable of producing such N-substituted lactams, under the conditions of the anionic lactam polymerization, by addition or substitution reactions[1-12] (*e.g.* isocyanates, acid halides or esters).

35.2.2 Catalysts

Lactamates (1), or strong bases yielding lactamates in sufficiently clean reactions, such as alkali metal alkoxides, metal hydrides, organometallic compounds and hydroxides, may serve as catalysts.[1] They may also be generated from quaternary ammonium salts[13] and other strong bases, *e.g.* pentamethyl guanidine.[14] Lactamates may also be generated from low basicity compounds by the thermal decomposition of easily decomposed salts[15] or by the electrolysis of lactam solutions of neutral salts.[16] The properties of both catalytic components may even be inherent in a single compound[17] or a pair of neutral compounds,[18] from which lactamate and growth center are generated by thermal decomposition, *e.g.* alkali metal salts of N,2-disubstituted 3-oxoamides,[17] or a mixture of potassium fluoride and N-trimethylsilylcaprolactam.[18] For kinetic studies, recrystallized lactamates[19] or those prepared from tertiary butoxides[20] are recommended. Lactamates prepared from alkali metals are inadvisable, because they are contaminated by amines and water,[21] which destroy the N-acyllactam growth centers. Some contradictory results encountered in the literature[22,23] may be explained by the use of such contaminated lactamates. When using sodium caprolactamate prepared from metallic sodium, the initial rate of polymerization is lower[24] than that found with the catalyst prepared from alkoxides.[25] The peculiar decrease in the rate of polymerization with increasing concentration of sodium caprolactamate[24] (at a constant concentration of N-acetylcaprolactam) can be explained by the fact that, with increasing lactamate concentration, increasing concentrations of amine and water are introduced into the reaction mixture. In this way, an increasing fraction of the active species is destroyed, thus inevitably decreasing the rate of polymerization in proportion to the concentration of lactamate. The poor reproducibility of the reaction between an alkali metal and caprolactam is illustrated by the fact that, in one communication, the activity of lithium lactamate was reported to be higher than that of sodium lactamate,[22] whereas the reverse was stated in a subsequent communication[23] by the same group of authors.

For the five-membered lactam, the effect of impurities and of the method used in the preparation of the alkali metal salt is not as pronounced as in the case of caprolactam, because of the lower acidity of the latter.[26, 27]

35.2.3 Dissociation of Lactamates

Under polymerization conditions, lactamates (1), and the corresponding salts of polymer amide groups, enter into complicated equilibria involving various aggregates, ion pairs, free ions and triple ions, which may interact with other donors and acceptors present. Hence, the conductometrically estimated dissociation constant, based on the simple ionization equilibrium (5) and equation (6), includes contributions from various types of ions (where $[ML]_0$ is the initial lactamate concentration). At the catalytic concentrations used in bulk polymerizations, aggregated lactamates and triple ions predominate.[28] Among the various types of ionic species, the contribution of free lactam anions starts to predominate at 150 °C only for concentrations of the alkali metal salts of caprolactam below 0.0005 mol 1^{-1}.[28] Except at high temperatures, most lactamates are only weakly dissociated,[29] so that the value of $[L^-]$ may be approximated by $[L^-] = (K_d[ML]_0)^{1/2}$. Accordingly, the initial rate of initiation and assisted polymerization of caprolactam with sodium caprolactamate increases with the square root of the lactamate concentration[19, 25] (see Section 35.3.2).

$$\left[\overset{N=\!=\!=C=\!=\!=O}{\underset{\smile}{}} \right]^{-} Mt \; \rightleftharpoons \; \left[\overset{N=\!=\!=C=\!=\!=O}{\underset{\smile}{}} \right]^{-} + \; Mt^+ \tag{5}$$

$$[L^-] = (K_d[ML]_0 + K_d^2/4)^{1/2} - K_d/2 \tag{6}$$

In the absence of other solvating or complexing agents, the dissociation of alkali metal lactamates (1) in the molten lactam depends on the acidity and permittivity of the lactam, as well as on its donor–acceptor properties (Table 1). Substances of the same acidity as caprolactam, but of a substantially higher permittivity, strongly increase the dissociation of lactamates[29, 30] (Figure 1). The same effect is brought about by cryptands,[31] which increase both the dissociation of lactamates[31] as well as the rate of polymerization.[32] For various cations, the dissociation (in the corresponding lactam) increases with decreasing electronegativity of the cation[33] (Table 1). Also, the size and structure of substituents affect the ionization of lactamates and, consequently, the rate of polymerization.[34, 35]

Table 1 Effect of Lactam Acidity (pK_a) and Permittivity (ε) on the Dissociation Constant (K_d) of Lactamates in the Corresponding Lactam (of Ring Size n)

Lactam n	pK_a^a at 25 °C	ε_T^b	T	Li^+	Na^+	K^+	Me_3N^+	Ref.
				\multicolumn{4}{c}{$10^4 K_d$ (mol 1^{-1})}	
5	24.5	27.1	31	—	—	1600	—	31
6	26.7	17.5	45	—	1.3	2.6	4.5	c
7	27.2	12.5	150	0.7	1.0	1.9	—	23
13	27.2	35.0	160	—	47	—	—	d

[a] Ref. 48. [b] Ref. 68. [c] B. Coutin and H. Sekiguchi, *J. Polym. Sci., Polym. Chem. Ed.*, 1977, **15**, 2539. [d] V. V. Korshak, T. M. Frunze, V. A. Kotel'nikov, M. P. Ivanov, V. V. Kurashev, S. P. Davtyan and L. B. Danilevskaya, *Dokl. Akad. Nauk SSSR*, 1981, **257**, 641.

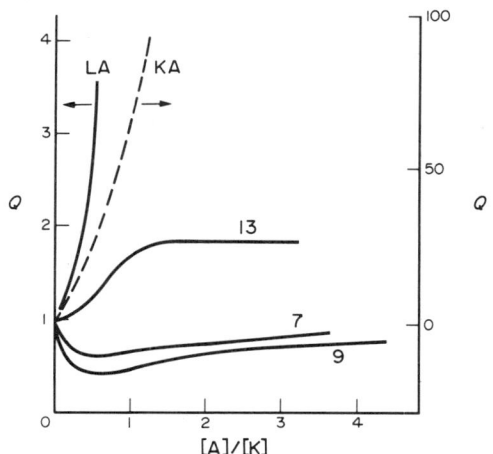

Figure 1 Effect of added compounds (A) on the relative conductivity (Q) of potassium lactamates (K) in THF at 25 °C (ref. 30). Concentration of K $= 0.06$–0.08 mol l^{-1}. Addition of the seven-, nine- and thirteen-membered lactam to the corresponding potassium lactamate (7, 9 and 13), and addition of a β-ketoamide (KA) or linear amide (LA) to potassium caprolactamate

35.3 REACTION SCHEME

So far, the following types of reactions have been found under the conditions of anionic lactam polymerization: acylation, condensation, dissociation, hydrolysis and neutralization.[3] The chemistry of the complicated set of reactions taking place during the polymerization may be derived from the interaction of the first two components present, *i.e.* lactam and lactamate.

35.3.1 Initiation

Although the acylating ability of the resonance-stabilized amide group is fairly weak, it is sufficiently strong to acylate the strongly nucleophilic lactamate in reaction (3), which is followed by the neutralization reaction (4; Scheme 2). The net result of this disproportionation between two amide groups is the formation of an amine and an acyllactam moiety. From this moment there are four types of reactive species present in the reaction mixture. The strongest acylating agent is now the *N*-acyllactam, which reacts with the two nucleophiles (lactamate and amine) in propagation, condensation and addition reactions (see Sections 35.3.2 and 35.3.6). As soon as linear amide groups are present, these may be involved in similar disproportionation reactions.

The neutralization reaction (4) is extremely fast, *e.g.* for pyrrolidone at 30 °C the rate constant of proton exchange[36] $k_e = 10^5$ l mol^{-1} s^{-1}. Therefore, the initial rate of disproportionation follows equation (7).[19]

$$v_d = d[NH_2]/dt = d[I]/dt = k_d[L][L^-] = k_d(K_d[ML]_0)^{1/2}[L] \tag{7}$$

Because *N*-acyllactam as well as other diacylamine structures undergo fast secondary reactions (see Section 35.3.6), the rate of disproportionation can be estimated only from the rate of amine group formation. So far, the rate of formation of amine groups by reaction (3) has been determined only for caprolactam[19] in the presence of its sodium salt at 160–190 °C, yielding $k_d = 1.6 \times 10^7$ exp $(-112\,000/RT)$ (kg mol^{-1} s^{-1}). For the disproportionation of a linear model alkylamide[19] (*N*-butylcaproamide) in the presence of its sodium salt, the value of k_d at 190 °C was found to be the same as that for caprolactam. Generally, however, the rate of disproportionation depends not only on the counterion[33] and reaction medium, but also on the ring size and substitution of the lactam, as well as on the structure of the open chain monomer unit. Whereas for caprolactam (in the presence of its sodium salt) the initiation reaction (3) is perceptible above 100 °C,[37] in the case of four-membered lactams (in the presence of their lithium salt) the initiation reaction (3) is already effective at room temperature.[38]

Depending on the electrophilicity and steric effects of the substitutent X on the *N*-substituted lactam, the rate of addition of the first lactamate in reaction (8) is generally different from that of the subsequent additions (equation 9; Scheme 3 and Figure 2). From published data[41] we can see that reaction (8) proceeds with *N*-benzoylcaprolactam 40 times faster than with *N*-propionylcapro-

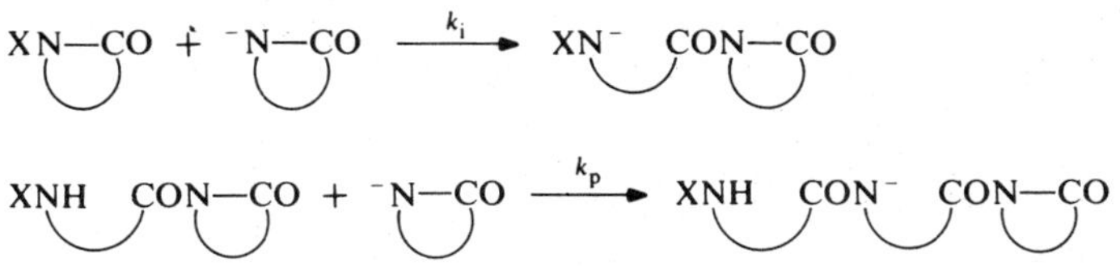

XN—CO + ⁻N—CO
k_i
XN⁻ CON—CO
XNH CON—CO + ⁻N—CO
k_p
XNH CON⁻ CON—CO

(8)

(9)

Scheme 3

C4H9NHCO-
PhCO-
MeCO-
Me3CO-
PhNHCO-
Polymer (mol kg⁻¹)
4
3
2
1
0
10
20
Time (min)

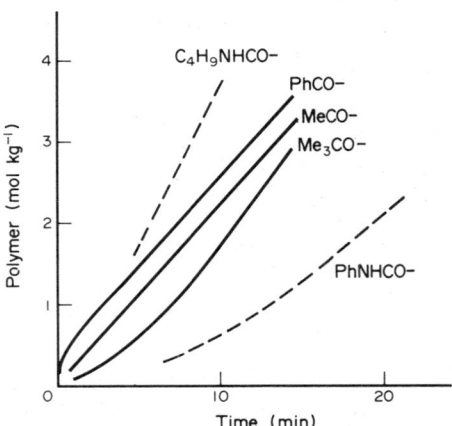

Figure 2 Effect of the substitutent in *N*-substituted caprolactam (I) on the polymerization of caprolactam with equimolar concentrations of sodium caprolactam and I (0.0177 mol kg^{-1})[39,40]

lactam. Therefore, reaction (8) must also be included among initiation reactions, although its rate is much higher than that of the disproportionation reaction (9). In this sense, all reactions yielding *N*-acyllactams with acyl residues derived from the polymerizing lactam should be considered as initiation reactions.

Polymerizations with added chain initiators are designated as assisted[1] or activated[3] polymerizations, whereas polymerizations triggered by lactamates alone are designated as non-assisted or non-activated.

35.3.2 Propagation and Depolymerization

Chain growth involves acylation of the lactamate by the cyclic acyl group of the *N*-acyllactam and, subsequently, very fast proton exchange (Scheme 1). Thus, the incorporation of one monomer unit results in the regeneration of both active species, *i.e.* lactamate and *N*-acyllactam growth center. Several hypotheses have been suggested for the detailed course of the propagation step.

The free ion mechanism[42] (Scheme 4) suggests attack of the free lactam anion at the endocyclic carbonyl group of the growth center, and ring-opening of the anionic intermediate (**5**) in reaction (11), the product (**6**) entering, subsequently, into the neutralization equilibrium (12).

According to the lactamolytic mechanism[43,44] (Scheme 5), ring-opening only occurs after protonation of the ionic intermediate (**5**) in reaction (13), thus complexation of the cation by the growth center plays an important role.[1] This mechanism is based on the observation that the addition of an *N*-acyllactam to a solution of a lactamate increases its conductivity. This effect has been attributed to an enhanced dissociation of the lactamate, resulting from the coordination of the cation to the two carbonyl oxygen atoms of the growth center, as shown in equation (14). This hypothesis is supported by the fact that the carbonyl vibrations of *N*-acyllactams are shifted to lower frequencies by addition of potassium lactamate.[30] Nevertheless, in the case of the seven- and nine-membered lactams the conductivity decreased, whereas with the thirteen-membered lactam, or with non-cyclic components (*i.e. trans*-amides), the conductivity increased (Figure 1). It may be inferred from these results that there is no simple relationship between conductivity and interaction of lactamates with growth centers.

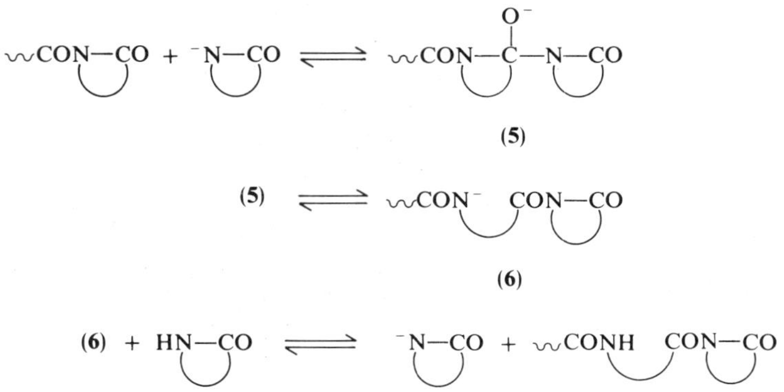

$$\text{\footnotesize vvCON—CO} + \text{\footnotesize ¯N—CO} \rightleftharpoons \text{\footnotesize vvCON—}\overset{\overset{\displaystyle O^-}{|}}{C}\text{—N—CO} \qquad (10)$$

(5)

$$\text{\footnotesize (5)} \rightleftharpoons \text{\footnotesize vvCON}^- \quad \text{\footnotesize CON—CO} \qquad (11)$$

(6)

$$\text{\footnotesize (6)} + \text{\footnotesize HN—CO} \rightleftharpoons \text{\footnotesize ¯N—CO} + \text{\footnotesize vvCONH} \quad \text{\footnotesize CON—CO} \qquad (12)$$

Scheme 4

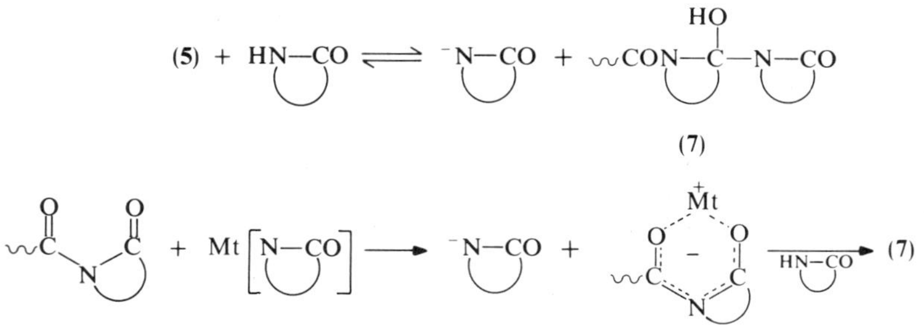

$$\text{\footnotesize (5)} + \text{\footnotesize HN—CO} \rightleftharpoons \text{\footnotesize ¯N—CO} + \text{\footnotesize vvCON—}\overset{\overset{\displaystyle HO}{|}}{C}\text{—N—CO} \qquad (13)$$

(7)

$$\text{\footnotesize vvC} + \text{Mt}\left[\text{N—CO}\right] \rightarrow \text{¯N—CO} + \cdots \xrightarrow{\text{HN—CO}} (7) \qquad (14)$$

Scheme 5

The ion-coordinative mechanism[22,45,46] suggests formation of a complex **(8)** between the lactamate and both carbonyl groups of the growth center in reaction (15), with propagation involving both free anions and ion pairs within complex **(8)**.

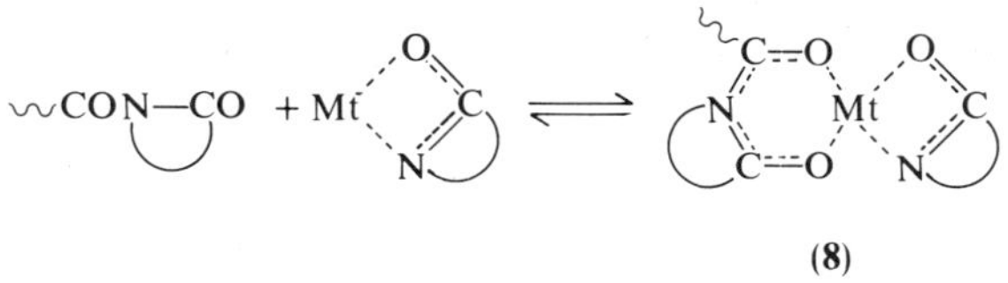

$$\text{vvCON—CO} + \text{Mt} \rightleftharpoons \qquad \qquad (15)$$

(8)

At present, there is no unambiguous evidence for any of these mechanisms. Free anions will play a decisive role at high temperatures and in media of high permittivity and donicity. At low temperatures and in less polar media, the participation of ion pairs might be of increased importance.[1] The predominant role of ions, even at low temperatures and in less polar media, is illustrated by the dramatic increase in the rate of polymerization on addition of a cryptand[32] (Figure 3).

Nevertheless, there are some results which indicate that ion pairs could participate in the propagation reaction to some extent.[22,23,28,45,46] From the depression of the initial rate of caprolactam polymerization by increasing counterion concentration, it was estimated that the contribution of ion pairs to the overall rate could be 3–5% for potassium caprolactamate and 10–15% for sodium caprolactamate at 150 °C. These data were obtained from a plot of the apparent rate constant (k_p^{app}) against the concentration of lactam anions ($[L^-] = K_d/[Mt^+]$), the values of propagation rate constants for free anions (k_p^-) and ion pairs ($k_p^{\pm}$) calculated from the relationship $k_p^{app} = k_p^{\pm} + (k_p^- - k_p^{\pm})[L^-]$. Results summarized in Table 2 indicate that the contribution of ion pairs increases in the series $K^+ < Na^+ < Li^+$. However, it can be estimated from published data[47] that, at 150 °C, a large number of the growth centers are converted into derivatives of β-ketoamides within only a few seconds. These products of secondary reactions (see Section 35.3.6) are much more acidic than caprolactam[48,49] and appreciably decrease the concentration of lactamate, so that the real values of the rate constants must be very much higher than those given in Table 2. The suppression of secondary reactions by the addition of lithium chloride[50] could result in an increase in the actual

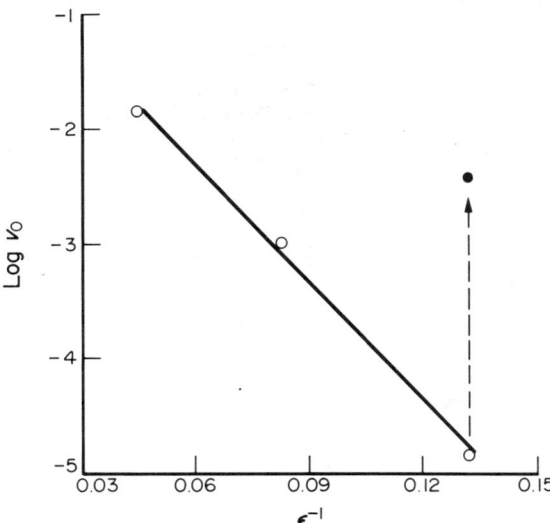

Figure 3 Effect of the permittivity of the reaction medium (ε) on the initial rate of reaction of equimolar concentrations of potassium caprolactam and N-propionylcaprolactam at 25 °C (○) and after addition of an equimolar amount of cryptand (●). Concentration of reactants = 0.08 mol l^{-1} (ref. 32)

Table 2 Propagation Rate Constants (k_p^- and $k_p^\pm$) and Concentration of Lactam Anions ([L$^-$]) in the Polymerization of Caprolactam with Equimolar Concentrations of Alkali Metal Caprolactamate and N-Acetylcaprolactam (0.0191 mol l^{-1}) at 150 °C[28]

Counterion	[L$^-$] (mol l^{-1})	k_p^- (l mol^{-1} s^{-1})	$k_p^\pm$ (l mol^{-1} s^{-1})
K$^+$	0.00188	66.5	0.15
Na$^+$	0.00152	68.0	0.95
Li$^+$	0.00109	67.3	1.45

concentration of both active species with increasing counterion concentration, and thus change the value of k_p^{app} and, accordingly, of k_p^- and $k_p^\pm$.

The rate of depolymerization in the reverse reaction (1) strongly depends on the ring size and substitution of the lactam, and on temperature. Its impact has to be taken into account, especially for the five-, six- and seven-membered lactams at conversions approaching the monomer–polymer equilibrium. However, at the present state of knowledge, it is useless to derive a complete rate equation incorporating propagation, depolymerization and disproportionation, as well as other reactions in which monomer is consumed or formed. The main obstacle is the lack of appropriate equations describing the decay and regeneration of active species in secondary reactions (see Section 35.3.6). Moreover, a term implying changes of the permittivity during polymerization should also be included (see Section 35.3.4).

Irrespective of these difficulties, semiempirical rate equations have been suggested, which are valid in a given range of temperature and concentration of catalytic components[3, 19, 25, 51−58] and, in some cases, for a given lactam only. Most of these equations are based on the free ion mechanism, so that the rate of polymerization is proportional to the square root of the catalyst concentration. For caprolactam, the assisted and non-assisted polymerizations follow equations (16) and (17), where k_p, k_d, K_d, [I] and [C] are rate constants of propagation and disproportionation, the dissociation constant and the concentrations of chain initiator and catalyst, respectively.

$$-\mathrm{dL}/\mathrm{d}t = k_p K_d^{1/2} [\mathrm{I}][\mathrm{C}]^{1/2} = k[\mathrm{I}][\mathrm{C}]^{1/2} \tag{16}$$

$$-\mathrm{dL}/\mathrm{d}t = k_d k_p K_d [\mathrm{C}][\mathrm{L}]t, \quad k_d k_p K_d = 10^{19} \exp(-215\,000/RT) \quad (\mathrm{kg\,mol^{-1}\,s^{-2}}) \tag{17}$$

35.3.3　Effect of the Structure of the Chain Initiator

Certain types of chain initiators affect not only the rate of the first lactamate addition in reaction (8), but also influence the whole course of polymerization. This is particularly the case with aryl isocyanates and the corresponding *N*-arylcarbamoyllactams (3). The disubstituted urea unit (ArNHCONH⌇) incorporated into the polymer chain in the initiation reaction (8) has a much higher acidity[59] than the lactam or polymer amide groups (Table 3), and decreases the concentration of lactamate during the polymerization. The NH-acidity of such urea units increases with the electronegativity of the substituents, and the rate of polymerization decreases accordingly.[51,60,61] On the other hand, the acidity of *N,N′*-dialkylurea units (RNHCONH⌇), introduced by alkyl isocyanates or alkylcarbamoyllactams, is comparable with that of caprolactam (Table 3). The polymerization with such chain initiators proceeds much faster than that with arylcarbamoyllactams or aryl isocyanates,[51] and even proceeds faster than that with *N*-acyllactams (Figure 2). In the series of *N,N*-diarylcarbamoyllactams (4) having no acidic NH group, the initial rate of polymerization increases with the electronegativity of the substituents.[4]

Table 3　Acidity (pK_a in DMSO at 25 °C) of Substituted Ureas and Alkylamide[59]

Compound	PhNHCONHEt	EtNH CONHEt	EtCONHC$_3$H$_7$
pK_a	22.4	26.7	27.2

A variation of the rate of polymerization with the size and structure of the acyl group in *N*-acyllactams (2) has been explained by polar and steric effects.[62-64] The effect of the structure of the chain initiator on the reactivity ratio observed in the copolymerization of the seven- and thirteen-membered lactams[65] has not yet been explained (see Section 35.4).

35.3.4　Reaction Medium and Kinetics

Anions (or salts) derived from polymer amide groups (6) formed in the propagation reaction (1) enter into a neutralization equilibrium (12) with lactam amide groups. The position of this equilibrium is determined by the concentrations and acidities of lactam and polymer amide groups. From seven-membered lactams upwards, the acidity of the lactam is equal or comparable to that of the corresponding polymer amide group[48] (Table 4). In these cases, the concentration of lactamate is proportional to the fraction of lactam amide groups, and the rate of polymerization should be proportional to the lactam concentration. As a matter of fact, the polymerization of the thirteen-membered lactam has been found to be of first order with respect to monomer at equimolar concentrations of catalyst and chain initiator ([I]).[66]

Table 4　Acidity (pK_a in DMSO at 25 °C) of Lactams and Alkylamide[48]

Ring size	4	5	6	7	9	13	C$_3$H$_7$CONHC$_8$H$_{17}$
pK_a	22.7	24.5	26.7	27.2	27.3	27.2	27.1

On the other hand, the rates of polymerization of caprolactam and pyrrolidone have been found to be independent of the monomer concentration.[24,25,67] So far, the zero order in monomer, sometimes attributed to autocatalysis,[51,54-58] has not been satisfactorily explained. It has been suggested[58] that the autoacceleration is due to the formation of additional growth centers in the initiation reaction (3). This explanation cannot be correct, because of the very low rate of this reaction[29] at temperatures around 150°C at which autoacceleration has been observed.

An explanation of this phenomenon should be sought in some particular property in which these two monomers differ substantially. It has been established[68] that the permittivity of caprolactam is much lower than that of its polymer (Figure 4) so that the polymerization is accompanied by a large increase in permittivity. On the other hand, the permittivities of the thirteen-membered lactam

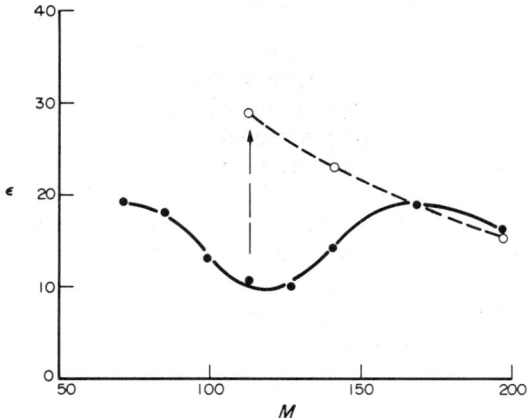

Figure 4 Permittivity (ε) of lactams ($\bullet$) of molecular weight M and of the corresponding polymers ($\bigcirc$) at 240 °C (ref. 68)

and of its polymer are very close (Figure 4). It was found that the dissociation of sodium capro-lactamate[29] and the rate of polymerization of caprolactam[32] strongly increase with increasing permittivity of the medium (Figure 3). Thus, the increasing permittivity increases the rate of caprolactam polymerization in such a manner that the propagation, which should be first order in monomer, seems to be formally independent of the lactam concentration.

Such complications should not occur in solution polymerizations of α,α-disubstituted four-membered lactams (**9**). Polymers of these lactams are soluble in aprotic solvents so that in dilute solution the permittivity remains essentially constant during polymerization. Substitution of both hydrogen atoms in the vicinity of the carbonyl group prevents secondary reactions consuming active species, and the polymerization proceeds at a constant concentration of growth centers as a living polymerization.[69] Moreover, the acidity of four-membered lactams is several orders of magnitude higher than that of a linear amide group (Table 4), so that equilibrium (12) is shifted almost entirely

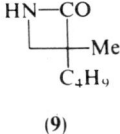

(**9**)

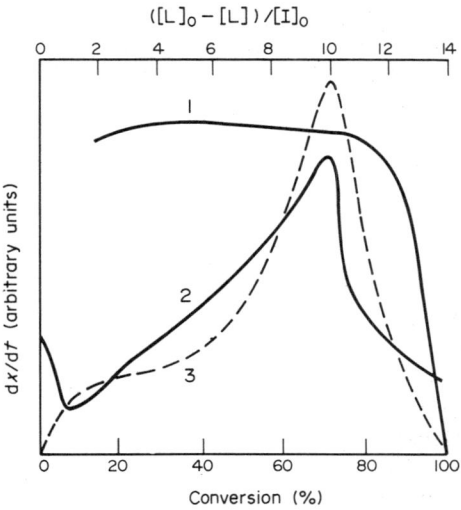

Figure 5 Polymerization of lactam (**9**) with its *N*-pivaloyl derivative (0.007 mol kg^{-1}) and lithium salt (0.002 mol kg^{-1}) at 30 °C (ref. 70). Concentration of (**9**) $= 0.095$ mol kg^{-1}. Rate of polymerization (1) and changes of the permittivity ($2 = \mathrm{d}\varepsilon/\mathrm{d}t$) and conductivity ($3 = \mathrm{d}G/\mathrm{d}t$) plotted against conversion

in favor of the lactamate. Therefore, the lactamate concentration, as well as the rate of polymerization, should remain constant until the monomer concentration approaches that of the lactamate. Curiously, the polymerization of lactam (**9**) has been found to be zero order with respect to monomer only up to certain conversions.[38, 70] It turned out that as soon as approximately 10 monomer units have been added per growth center the polymerization becomes gradually first order with respect to monomer (Figure 5). The changing character of the polymerization kinetics is accompanied by similar anomalies in the course of permittivity and conductivity (Figure 5). These phenomena indicate that interaction of growth centers and lactamate, as well as solvation of the latter by lactam and polymer amide groups, play an important role in this particular polymerization. Under these circumstances, the growing chain end could be viewed as an ionized rather than as a neutral structure. Similar anomalies observed in DMSO have been assigned to impurities[34] or separation of the polymer from solution during polymerization.[71]

35.3.5 Monomer Structure and Reactivity

The fundamental properties affecting the reactivity of lactams in anionic polymerization include ring strain, substitution, acidity and permittivity of the lactam and its polymer, as well as nucleophilicity of the lactam anion or lactamate and electrophilicity of the *N*-acyllactam.[72, 73] Although these properties can be estimated separately, their correlation with the specific requirements of the initiation and propagation steps is difficult. For example, the rate of the base-catalyzed hydrolysis of *N*-acyllactams (**2**)[74] or *N*-substituted carbamoyllactams (**3**)[40] does not reflect the rate of propagation reaction (1), although both reactions involve a nucleophilic attack on the same substrate. The acidity of lactam and polymer amide groups may be used to calculate the relative proportion of the corresponding salts or anions, but does not tell us very much about their nucleophilicities. The higher the acidity of the amide group, the lower the nucleophilicity of the anion or ion pair.

A comparison of lactam reactivities from the rates of bulk polymerization is circumvented by their different permittivities (Figure 4). Owing to the insolubility of most lactam polymers in aprotic solvents, the investigation of polymerization in dilute solution is confined to low conversions. But even under these conditions, the initial rate of acyllactam conversion cannot be used as a measure of reactivity, because the dependence of reactivities on the ring size varies with the concentration of lactamate (Figure 6). Therefore, it is not too surprising that contradictory conclusions concerning the effect of substitution on lactam reactivities have been published.[34, 35, 73, 75] Such results are mostly correct, and the contradictory deductions result from different reaction conditions, *e.g.* solvent and relative concentrations of reactants.

In order to measure the rates of propagation of all lactams under identical conditions, living anionic lactam polymers that are soluble in aprotic solvents have been used.[73] This technique takes advantage of the fact that living polymers of the four-membered lactam (**9**) remain in solution even after addition of a few monomer units of another lactam, the polymer of which is insoluble in the given solvent. In this way, the rate of addition of the first and second monomer unit of any

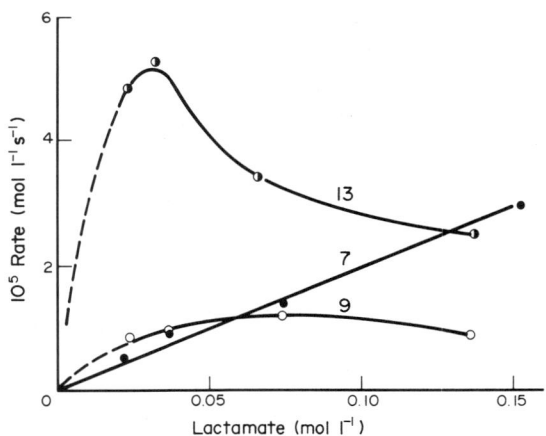

Figure 6 Effect of the ring size and concentration of lactamate on the initial rate of reaction in the system *N*-propionyllactam/potassium lactamate in THF at 25 °C (ref. 32); figures indicate the ring size

polymerizable lactam can be estimated under equal conditions, provided that the exchange of terminal lactam units in reaction (18) is not faster than propagation. Although this transacylation (18) is usually faster than the propagation reaction (1), favorable conditions may be achieved by adjusting the reaction temperature (Table 5). In the series of various α,α-disubstituted four-membered lactams, the rate of polymerization in THF has been found to decrease with the size of the substitutent,[73] whereas the reverse has been found in DMSO.[35] The latter result has been attributed to increased shielding of the lactam anion with increased length of the substituent, resulting in increased dissociation.[35]

$$\text{\small vCON—CO} \underset{A}{} + \text{\small ⁻N—CO} \underset{B}{} \rightleftharpoons \text{\small vCON—CO} \underset{B}{} + \text{\small ⁻N—CO} \underset{A}{} \qquad (18)$$

Table 5 Second Order Rate Constant of Propagation (k_p) and Transacylation (k_t) Between *N*-Acyllactams and Lactamates in THF

N-Acyllactam	Lactamate	T (°C)	k_p (l mol^{-1} s^{-1})	k_t (l mol^{-1} s^{-1})	*Ref.*
MeCON—CO (5)	[N—CO (7)]MgBr	25	0.02	0.20	89
MeCON—CO (7)	[N—CO (5)]MgBr	25	0.05	0.60	89
Me₃CCO—N—CO | Me | C₄H₉	[N—CO | Me | C₆H₁₃]Li	−20	0.08	0.21	a
		−60	0.0004	0.004	a

[a] J. Šebenda, J. Hauer and J. Světlik, in preparation.

35.3.6 Secondary Reactions

The enhanced reactivity of the anionically activated monomer and of the *N*-acylated lactam provides suitable pathways for polymerization reactions but also for undesirable secondary reactions. The complicated sequence of secondary reactions[3] starts with *C*-, *N*- and, in some cases, *O*-acylations, which are followed by condensation, cleavage and hydrolytic reactions. These secondary reactions and their products have a great impact on the polymerization process and polymer structure.[72]

35.3.6.1 Acylation

The carbonyl group of the growth center (**10**), as well as that of diacylamine branching points (**11**), behaves in many respects like an ester or ketone carbonyl group. The acidity of C–H bonds in the vicinity of imide groups is sufficient[49] to promote ionization under conditions of anionic lactam polymerization (Scheme 6).

Although the concentration of the strongly basic anions (**12**)–(**14**) will be low under the polymerization conditions (excess of lactam), *C*- and *O*-acylations at this anion will compete with the propagation reaction (1). There are two reactive carbonyl groups and up to two labile C–H bonds in each imide group (**10**) and (**11**). *C*-Acylation reactions (19)–(21) of the three types of anions (**12**)–(**14**) by the cyclic or exocyclic carbonyl group yield three types of structures, denoted as β-ketoimides (**15**)–(**17**) (Scheme 7).

The structures of these β-ketoimides have been derived from those of products formed by the hydrolysis of polymers[76–80] and products of model reactions.[81] As expected, the reactivities of C–H bonds and of the two carbonyl groups in *N*-acyllactams (**10**) are different, and depend on the ring size[78,80] and substituents of the lactam[79] and chain initiator.[78] Model reactions with *N*-propionyl-

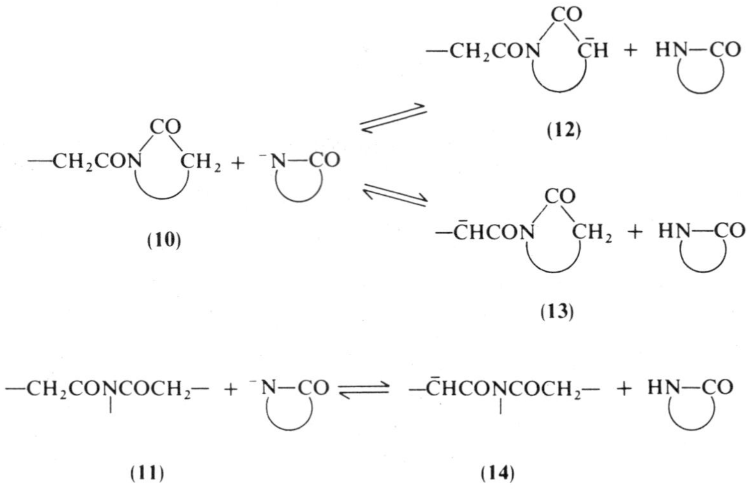

$$-CH_2CON{\overset{CO}{\underset{\smile}{}}}\overset{..}{C}H + HN-CO$$

(12)

$$-CH_2CON{\overset{CO}{\underset{\smile}{}}}CH_2 + {}^-N-CO$$

(10)

$$-\overset{..}{C}HCON{\overset{CO}{\underset{\smile}{}}}CH_2 + HN-CO$$

(13)

$$-CH_2CON\!COCH_2- + {}^-N-CO \rightleftharpoons -\overset{..}{C}HCON\!COCH_2- + HN-CO$$

(11)   (14)

Scheme 6

$$-CH_2CON{\overset{CO}{\underset{\smile}{}}}CHCO- + {}^-N-CO \qquad (19)$$

(15)

$$-CON-CO \xrightarrow[\ (13)\ ]{} -\overset{|}{C}HCON{\overset{CO}{\underset{\smile}{}}}CH_2 + {}^-N-CO \qquad (20)$$
$$\underset{-\overset{|}{C}O}{}$$

(16)

$$-\overset{|}{C}HCON\!COCH_2- + {}^-N-CO \qquad (21)$$
$$\underset{-\overset{|}{C}O}{}$$

(17)

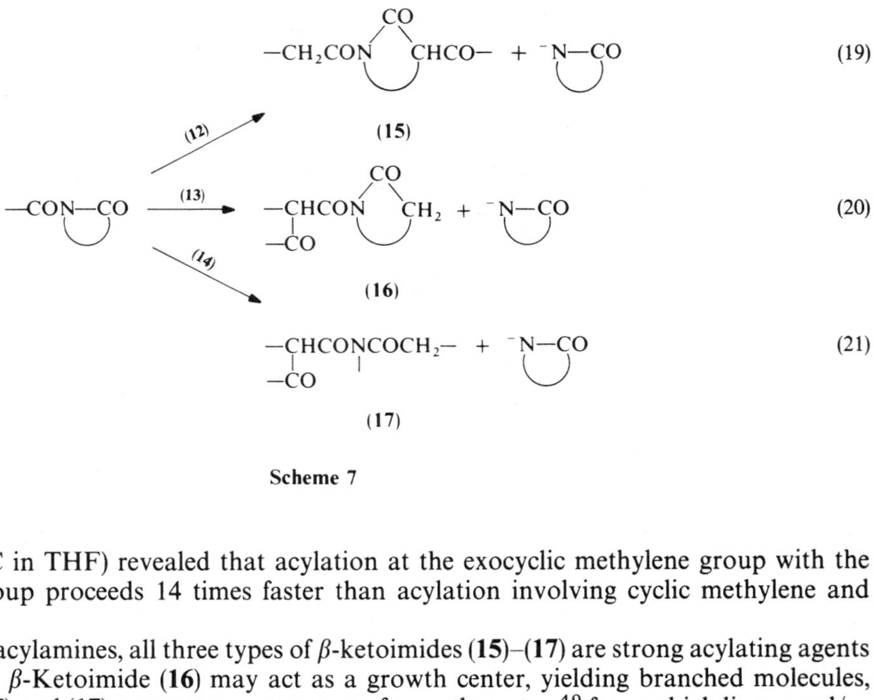

Scheme 7

caprolactam (at 25 °C in THF) revealed that acylation at the exocyclic methylene group with the exocyclic carbonyl group proceeds 14 times faster than acylation involving cyclic methylene and carbonyl groups.[41]

As derivatives of diacylamines, all three types of β-ketoimides (15)–(17) are strong acylating agents in their neutral form. β-Ketoimide (16) may act as a growth center, yielding branched molecules, whereas structures (15) and (17) may act as precursors of growth centers[49] from which linear and/or branched molecules are formed (Scheme 8).

In reactions (22)–(24), followed by proton exchange (2), β-ketoimides (15)–(17) are converted into β-ketoamides of the general structure (18; Scheme 8). Both β-ketoimides and β-ketoamides are many orders of magnitude more strongly acidic than lactam or polymer amide groups[49] (Tables 4 and 6). Their formation is accompanied by a decrease in the lactamate concentration in the neutralization reactions (25) and (26). The net result of C-acylation and neutralization reactions is a decrease in the concentration of both active species and, consequently, a decrease in the rate of polymerization, as well as a decrease in the rate of the C-acylation reactions. Therefore, the C-acylation proceeds very fast until a temporary equilibrium (27) is attained.

Subsequent secondary reactions consuming β-ketoimides distort this equilibrium and thus induce further C-acylations of imides, as long as these, and the strongly basic lactamates (or polymer amide anions), are present. The subsequent secondary reactions are much slower than the C-acylation reactions, therefore the decay of growth centers and strong base proceeds in two distinct stages. At the usual polymerization temperatures, the first stage of C-acylation proceeds too fast to enable it to be followed. Model experiments with diacylamine and the sodium salt of alkylamide revealed that

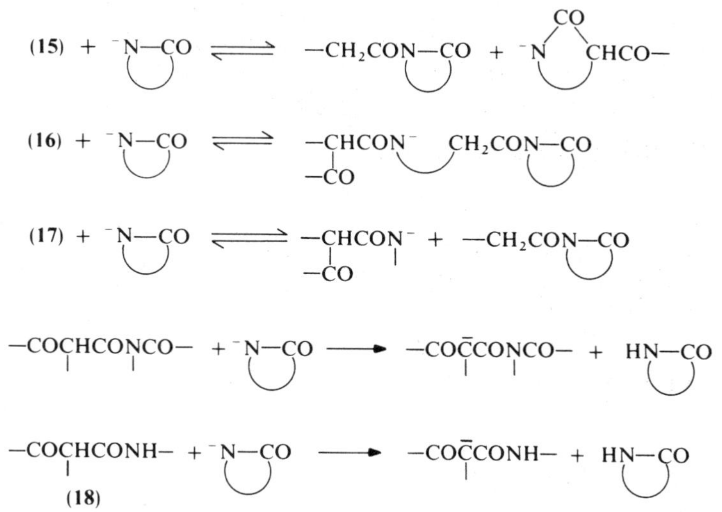

$$(15) + \ ^-N{-}CO \rightleftharpoons \ -CH_2CON{-}CO \ + \ ^-N \overset{CO}{\diagdown} CHCO- \tag{22}$$

$$(16) + \ ^-N{-}CO \rightleftharpoons \ -\underset{-\overset{|}{C}O}{CHCON^-} \quad CH_2CON{-}CO \tag{23}$$

$$(17) + \ ^-N{-}CO \rightleftharpoons \ -\underset{-\overset{|}{C}O}{CHCON^-} \ + \ -CH_2CON{-}CO \tag{24}$$

$$-CO\overset{|}{C}H\overset{|}{C}ONCO- \ + \ ^-N{-}CO \longrightarrow \ -CO\overset{|}{\bar{C}}\overset{|}{C}ONCO- \ + \ HN{-}CO \tag{25}$$

$$-CO\overset{|}{C}HCONH- \ + \ ^-N{-}CO \longrightarrow \ -CO\overset{|}{\bar{C}}CONH- \ + \ HN{-}CO \tag{26}$$
(18)

Scheme 8

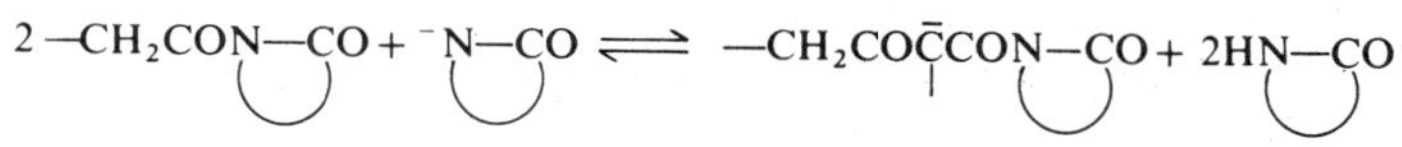

$$2\ -CH_2CON{-}CO + \ ^-N{-}CO \rightleftharpoons \ -CH_2CO\overset{|}{\bar{C}}CON{-}CO + 2HN{-}CO \tag{27}$$

Table 6 Acidity (pK_a in DMSO at 25 °C)[49] of β-Ketoimides and β-Ketoamides

Compound[a]	$RCON \overset{CO}{\diagdown} CHCOR$	$HN \overset{CO}{\diagdown} CHCOR$	$\underset{COC_3H_7}{CHCONHEt}$
pK_a	13.2	19.7	18.4

[a] R = Ph.

the rapid stage of decay of the imide groups is finished within a few minutes at 80 °C, and only the slow stage could be accurately followed.[47]

It has been estimated that at equimolar concentrations of *N*-propionylcaprolactam and potassium caprolactamate (at 25 °C in THF) the rate of *C*-acylation is comparable with the rate of polymerization.[32,41] A similar situation is also met at elevated temperatures, because the overall activation energies of the polymerization and *C*-acylation reactions are very close,[32] *e.g.* for caprolactam, 40 and 30 kJ mol^{-1}, respectively. In view of these facts, it is not too surprising that with increasing concentration of chain initiator the polymerization slows down substantially, long before the monomer–polymer equilibrium is attained[82] (Figure 7).

Along with *C*-acylation, *O*-acylation of the ambident anions derived from *N*-acyllactams has also been observed[49] under certain conditions, yielding enol esters (19) of *N*-acyllactams in reaction (28).

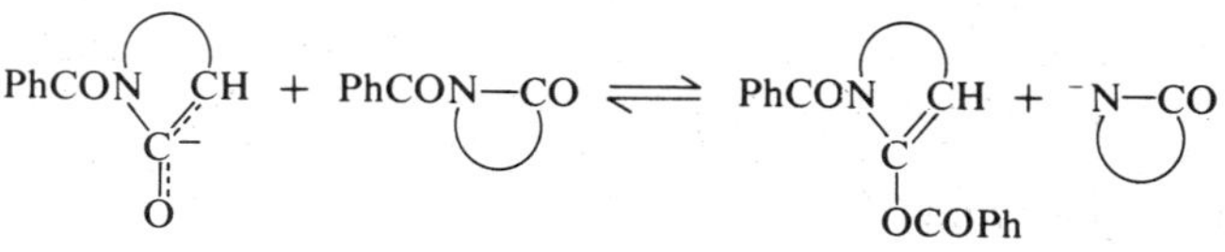

$$PhCON \diagup CH \ + \ PhCON{-}CO \rightleftharpoons PhCON \diagup CH \ + \ ^-N{-}CO \tag{28}$$
(19)

Anions derived from polymer amide groups are acylated by the cyclic or exocyclic acyl group of the *N*-acyllactam (10) in reactions (29) and (30) (Scheme 9). The resulting imide groups (11) represent easily hydrolyzable branching points, as well as sources of growth centers (10). Reaction (29), proceeding with ring-opening, can be considered as a propagation reaction.

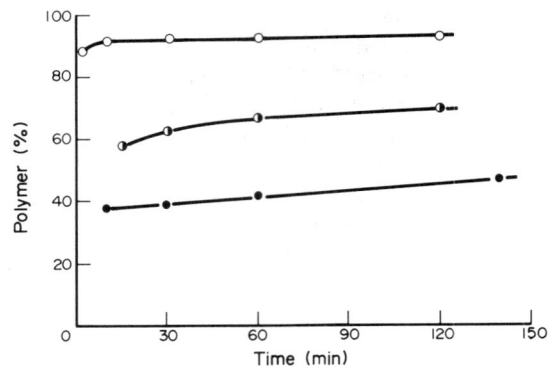

Figure 7   Polymerization of caprolactam with sodium caprolactam $(0.044 \text{ mol kg}^{-1})$ and N-butyrylcaprolactam (I) at $210\,^{\circ}\text{C}$ (ref. 82). Concentration of I: 0.088 ($\bigcirc$), 0.262 ($\pmb{\circ}$) and $0.437 \text{ mol kg}^{-1}$ ($\bullet$)

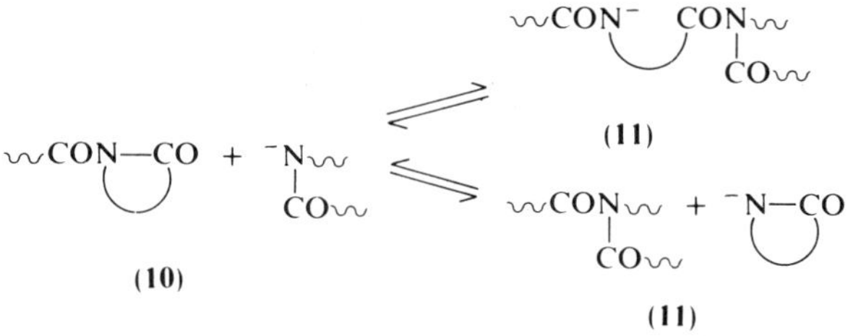

$$\text{wCON—CO} + {}^{-}\text{N}_{\text{w}} \quad \rightleftharpoons \quad \text{wCON}^{-}\ \text{CON}_{\text{w}} \atop \text{CO}_{\text{w}} \tag{29}$$

$$\textbf{(11)}$$

$$\begin{array}{c} \text{wCON—CO} \\ \textbf{(10)} \end{array} + {}^{-}\text{N}_{\text{w}} \atop \text{CO}_{\text{w}} \quad \rightleftharpoons \quad \text{wCON}_{\text{w}} + {}^{-}\text{N—CO} \atop \text{CO}_{\text{w}} \tag{30}$$

$$\textbf{(11)}$$

Scheme 9

Table 7   Initial Rates of Polymerization (v_p), C-Acylation (v_c) and N-Acylation (v_n) in Reactions of Equimolar Concentrations of Potassium Caprolactamate and N-Propionyl-caprolactam $(0.081 \text{ mol l}^{-1})$ or N-Benzoylcaprolactam $(0.055 \text{ mol l}^{-1})$ in THF at $25\,^{\circ}\text{C}$

Acyl	v_p (mol l^{-1} s^{-1})	v_c (mol l^{-1} s^{-1})	v_n (mol l^{-1} s^{-1})	Ref.
EtCO–	8.9×10^{-6}	4.7×10^{-6}	3.0×10^{-7}	41
PhCO–	2.3×10^{-4}	4.0×10^{-7}	2.9×10^{-5}	41

The rates of the first step of propagation, C-acylation and N-acylation of polymer amide groups have been found to depend on the nature of the exocyclic acyl group (Table 7).

The effects of the counterion, ring size and structure of the chain initiator on the extent of C-acylation are represented in Tables 8–10. At high temperatures, the concentration of diamino ketone units (**20**) formed in C-acylation and subsequent reactions strongly depends on the concentration of the catalyst, rather than on that of the chain initiator, and to a lesser extent, on the counterion (Table 8) and on the ring size (Table 9). It must be pointed out that the content of irregular structural units (**20**) can be even higher than that of the added chain initiator, because

$$-\text{NH(CH}_2)_n\text{CO(CH}_2)_n\text{NH}-$$
$$\textbf{(20)}$$

Table 8   Content of Diamino Ketone Units (**20**) in Caprolactam Polymers Prepared with N-Benzoylcaprolactam (I) and Lactamates (C) at $225\,^{\circ}\text{C}$ ($t = 150$ min)[78]

[C] (mol kg^{-1})	[I] (mol kg^{-1})	K^+	[(**20**)] (mol kg^{-1}) Na^+	Li^+	$BrMg^+$
0.052	0.013	0.062	0.062	0.066	0.044
0.051	0.153	0.044	0.043	0.040	0.019

Table 9 Effect of Ring Size of the Lactam on the Content of Diamino Ketone Units (**20**) in Polymerizations with Sodium Lactamate $(0.035 \, \text{mol kg}^{-1})$ and *N*-Benzoyllactam $(0.018 \, \text{mol kg}^{-1})$ at 225 °C $(t = 150 \, \text{min})$[78]

Ring size, n	7	9	13
[(**20**)] (mol kg^{-1})	0.056	0.017	0.020

Table 10 Effect of the Chain Initiator (I) on the Content of Diamino ketone Units (**20**) and Number of Polymer Molecules (N) in Caprolactam Polymers Prepared with Equimolar Concentrations of Sodium Caprolactamate and *N*-Substituted Caprolactam (I) $(0.044 \, \text{mol kg}^{-1})$ at 175 °C $(t = 30 \, \text{min})$[77]

I	$PhCOL^7$	$PhNHCOL^7$	$C_3H_7COL^7$
[(**20**)] (mol kg^{-1})	0.030	0.008	0.030
[N] (mol kg^{-1})	0.036	0.040	0.035

additional growth centers (**10**) are generated in the initiation reaction (3) as long as lactam or polymer amide anions are present. At a high chain-initiator to catalyst ratio, these anions are consumed very rapidly and, consequently, the concentration of ketone units (**20**) is always lower than that of the catalyst (Table 8). The same effect is shown by acidic chain initiators such as *N*-phenylcarbamoyllactams (Table 10).

35.3.6.2 *Condensation and cleavage of β-keto compounds*

Like other β-keto compounds, β-ketoimides (**15**)–(**17**) and β-ketoamides (**18**) are also very reactive. At elevated temperature and/or in the presence of a strong base, they undergo condensation reactions (31) and (32) yielding water, carbon dioxide, amine and ketone, as well as oxypyridone[83] (**21**) and uracil structures[84] (**22**; Scheme 10). The latter two, rather stable, heterocyclic structures represent centers of branching and cross-linking, as well as some of the final products of the decay of growth centers.[84]

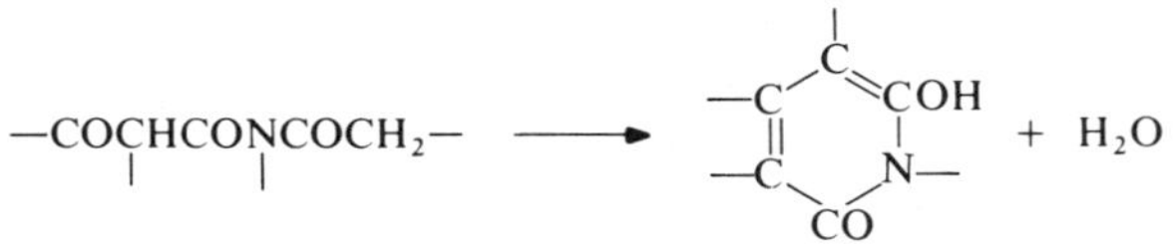

$$ -COCHCONCOCH_2- \longrightarrow \quad \text{(21)} \quad + \; H_2O \tag{31} $$

$$ 2 \; -COCHCONH- \longrightarrow \quad \text{(22)} \quad + \; (H_2O + CO_2 + -CO-) \tag{32} $$

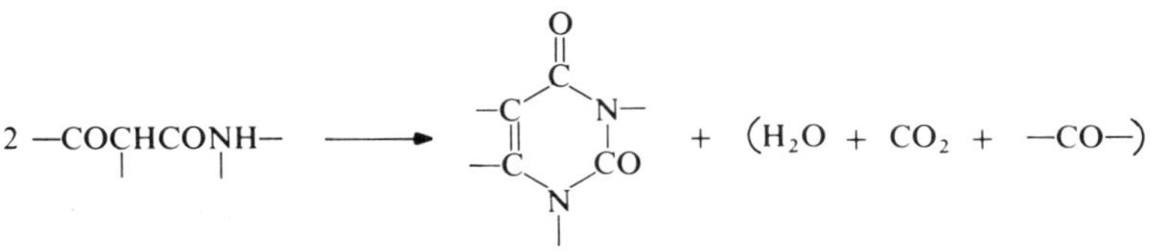

Scheme 10

$$ \text{\textasciitilde}\overline{N}COCHCO- \longrightarrow \text{\textasciitilde}N{=}C{=}O + \overline{C}HCO- \xrightarrow[HN-CO]{} CH_2CO- \tag{33} $$

$$ \text{\textasciitilde}NHCOCHCO- \longrightarrow \text{\textasciitilde}N{=}C{=}O + CH_2CO- \tag{34} $$

Scheme 11

Base-catalyzed (equation 33) or thermal cleavage (equation 34) of β-ketoamides (**18**) yields ketones and isocyanates[85-87] (Scheme 11), which are very efficient chain initiators.

35.3.6.3 *Hydrolytic reactions*

Water formed in reactions (31) and (32) hydrolyzes growth centers (**10**) and imide branching points (**11**), as well as β-ketoimides (**15**)–(**17**) and β-ketoamides (**18**) with formation of ketone and carbon dioxide,[47] as well as amine and carboxyl groups.[88] The latter then convert lactam and other strongly basic anions into carboxylate[88] and carbonate,[47] which represent final products of the decay of catalytic components (Scheme 12).

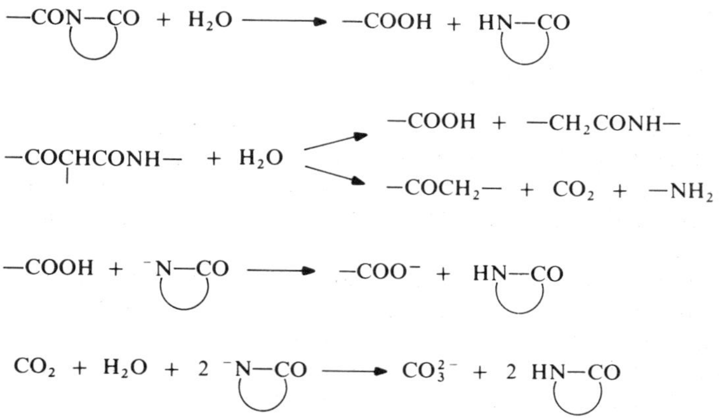

$$-\text{CON}-\text{CO} + \text{H}_2\text{O} \longrightarrow -\text{COOH} + \text{HN}-\text{CO}$$

$$-\text{COCHCONH}- + \text{H}_2\text{O} \begin{cases} -\text{COOH} + -\text{CH}_2\text{CONH}- \\ -\text{COCH}_2- + \text{CO}_2 + -\text{NH}_2 \end{cases}$$

$$-\text{COOH} + {}^-\text{N}-\text{CO} \longrightarrow -\text{COO}^- + \text{HN}-\text{CO}$$

$$\text{CO}_2 + \text{H}_2\text{O} + 2\ {}^-\text{N}-\text{CO} \longrightarrow \text{CO}_3^{2-} + 2\ \text{HN}-\text{CO}$$

Scheme 12

With the exception of α,α-disubstituted lactams, the decay of catalytic components is very rapid in all anionic lactam polymerizations. For example, in the polymerization of caprolactam with an initial chain-initiator/catalyst ratio of four, the effective concentration of lactamate drops to 1% of its initial value within 30 minutes at 210 °C.[82] With respect to the numerous fast consecutive and reversible secondary reactions, reasonable kinetic measurements are restricted to low temperatures only and very low conversions.[30, 32, 41]

35.4 COPOLYMERIZATION

When deriving the reaction scheme for the anionic copolymerization of lactams, two more effects must be taken into account,[71] in addition to the four rate constants of homo- and cross-propagation reactions (33)–(36) (Scheme 13). Due to the different acidity of the copolymerizing lactams, the

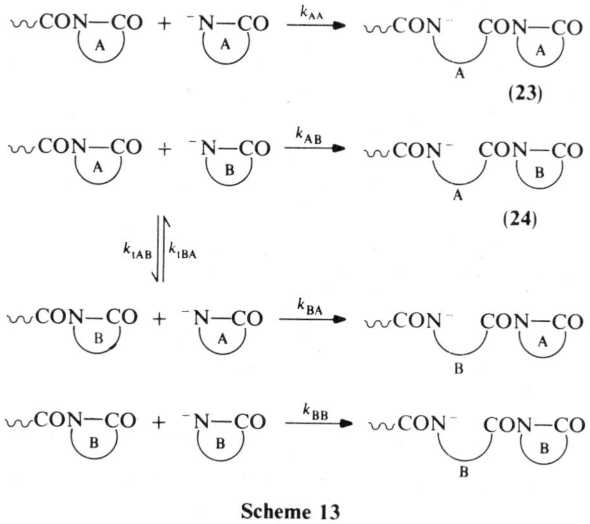

$$\text{wCON}-\text{CO} + {}^-\text{N}-\text{CO} \xrightarrow{k_{AA}} \text{wCON}^- \quad \text{CON}-\text{CO} \tag{33}$$

(**23**)

$$\text{wCON}-\text{CO} + {}^-\text{N}-\text{CO} \xrightarrow{k_{AB}} \text{wCON}^- \quad \text{CON}-\text{CO} \tag{34}$$

(**24**)

$k_{tAB} \Big\| k_{tBA}$

$$\text{wCON}-\text{CO} + {}^-\text{N}-\text{CO} \xrightarrow{k_{BA}} \text{wCON}^- \quad \text{CON}-\text{CO} \tag{35}$$

$$\text{wCON}-\text{CO} + {}^-\text{N}-\text{CO} \xrightarrow{k_{BB}} \text{wCON}^- \quad \text{CON}-\text{CO} \tag{36}$$

Scheme 13

representation of the corresponding anions (L_A^- and L_B^-) depends on both their concentration and acidity (Table 4). Also, the representation of the two kinds of growth centers (23) and (24) depends not only on the concentration of both lactams, as well as on the propagation rate constants (k_{AA}, k_{AB}, k_{BB} and k_{BA}), but also on the rates of transacylation (equation 18). It has been established that the rate constants of transacylation (k_{tAB} and k_{tBA}) are usually much higher than those of propagation (Table 5). Consequently, the composition of the copolymer formed at the beginning of the polymerization is controlled by the rates of the two transacylation reactions, which depend on the transacylation rate constants, the acidities of the lactams and the dissociation constants of the corresponding lactamates. Hence, the relative rates of incorporation of the two lactams are given by equation (37), in which K_a is the equilibrium constant determining the distribution of anions L_A^- and L_B^-, and K_t is the transacylation equilibrium constant.[89] In the copolymerization of the five- and seven-membered lactams (L^5 and L^7), the former is eight times more reactive,[89] irrespective of the higher ring strain of L^7.

$$d[L_A]/d[L_B] = K_a K_t [L_A]/[L_B] \tag{37}$$

The temperature affects the acidity of lactams, dissociation of lactamates and the rates of propagation and transacylation. Due to the different activation parameters of transacylation and propagation, the values of apparent reactivities depend on temperature. For a pair of five- and seven-membered lactams, (L^5 and L^7), the latter becomes the more reactive monomer with increasing temperature.[90]

With respect to the different permittivities of lactams (Figure 4), the initial ratio of monomers affects the dissociation equilibria of lactamates, as well as the apparent reactivities. In the copolymerization of L^7 with L^{13}, the increasing fraction of the latter raised not only the rate of copolymerization, but also the relative reactivities.[91] Unfortunately, an inadequate evaluation of the experimental data resulted in the wrong conclusion being drawn. It was found that, with an increasing fraction of L^{13}, the initial rate of monomer consumption increased faster for L^{13} than for L^7. Korshak *et al.* inferred from this result that, with an increasing fraction of L^{13} in the reaction mixture, the reactivity of the latter increases faster than that of L^7.[91] However, the opposite is true, because the rates have to be related to the initial lactam concentrations (Table 11).

Table 11 Initial Rates of Monomer Consumption (v_7 and v_{13}) and Relative Reactivities ($v_n/[L^n]$) in the Copolymerization of L^7 and L^{13} at Different Initial Molar Ratios (L^7/L^{13}) at 150 °C

L^7/L^{13}	$10^4\,v_7^a$ $10^4\,v_{13}^a$ (mol l^{-1} s^{-1})		$10^4 v_7/[L^7]$	$10^4 v_{13}/[L^{13}]$	$\varepsilon L^7 + L^{13b}$
0.85/0.15	10	1.7	1.5	1.4	16.6
0.60/0.40	16	5.7	3.9	2.1	19.4
0.25/0.75	22	26.7	15.5	6.3	28.0

[a] Ref. 91. [b] M. Provazník, Ph.D. Thesis, Institute of Macromolecular Chemistry, Prague, 1987.

At high temperatures and/or with less strained lactams, the reversible nature of propagation must be taken into account. In the later stages of polymerization, the composition of the copolymer is not determined kinetically, but thermodynamically. The same also holds for the distribution of monomer units in the copolymer.[90]

35.5 NON-ISOTHERMAL POLYMERIZATION

Due to the rapid heat evolution in the bulk polymerizations of lactams, isothermal conditions can be maintained only at low conversions and, therefore, most polymerizations proceed under non-isothermal conditions. Assuming that the heat transfer and the specific heat capacity of the polymerizing system are independent of conversion (β), the latter may be expressed in terms of temperature differences. The simplest situation is met under adiabatic conditions, when $\beta = ([L]_0 - [L])/([L]_0 - [L]_e) = (T - T_0)/(T_e - T_0)$, where T_0, T_e and T are the initial, equilibrium and instantaneous temperature, respectively. Under non-adiabatic conditions, a term accounting for heat transfer has to be included,[92] and the polymerization can be described by equation (38), where c_p

represents the specific heat capacity of the polymerizing system, ΔH_p is the heat of polymerization, α is the heat transfer constant, T is the instantaneous temperature of the reacting system and T_w is the wall temperature.

$$-d[L]/dt = \frac{c_p}{\Delta H_p}\left[\frac{dT}{dt} + \alpha(T - T_w)\right] \tag{38}$$

With respect to rapid changes in the catalytic activity during polymerization (see Section 35.3.6.3), the kinetic equation cannot be derived on the basis of the reaction mechanism. Instead, an inverse method has been used to find the kinetic function and the corresponding constants of the thermokinetic model.[93-95] It was found that the assisted non-isothermal polymerization of caprolactam can only be described satisfactorily if a term accounting for the autoacceleration is included in the rate equation (see Section 35.3.4). The autoacceleration is considered to be proportional to conversion, so that, together with the second term accounting for the increasing temperature (also proportional to conversion), equation (39) was obtained.[94] In this equation, $E = 65\ \text{kJ mol}^{-1}$, and the values of k and m depend on the kind of chain initiator; for N-acetylcaprolactam, $k = 1.3 \times 10^8$ ($1\ \text{mol}^{-1}\ \text{s}^{-1}$) and $m = 0.93$ (mol l^{-1}).

$$d\beta/dt = k\frac{[C][I]}{[L]_0}(1-\beta)\left[1 + \frac{m\beta}{[C][I]^{1/2}}\right]\exp(-E/RT) \tag{39}$$

Adiabatic polymerizations of caprolactam in the presence of increasing concentrations of lithium chloride revealed that the value of ΔH_p decreases from 134 to about 60 kJ mol^{-1}. This has been attributed to the decrease in the enthalpy of the monomeric system as a result of ion–dipole interactions between Li$^+$ ions and carbonyl groups.[96,97] Surprisingly, addition of lithium chloride up to 3% increased the initial rate of polymerization, although a decrease would be expected, due to the suppression of the dissociation of lithium caprolactamate.[96] The increasingly negative entropy term with increasing concentration of LiCl is responsible for the increasing equilibrium monomer content.[97]

35.6 REFERENCES

1. H. Sekiguchi, in 'Ring-opening Polymerization', ed. K. J. Ivin and T. Saegusa, Elsevier, London, 1984, vol. 2, p. 832.
2. T. M. Frunze, V. V. Kurashev, V. A. Kotel'nikov and T. V. Volkova, *Usp. Khim.*, 1979, **48**, 1856; *Russ. Chem. Rev. (Engl. Transl.)*, 1979, **48**, 991.
3. J. Šebenda, in 'Comprehensive Chemical Kinetics', ed. C. H. Bamford and C. F. H. Tipper, Elsevier, Amsterdam, 1976, vol. 15, p. 279.
4. G. Falkenstein and H. Dörfel, *Makromol. Chem.*, 1969, **34**, 127.
5. I. Tanaka, A. Suzuki and T. Yoshida, in 'Preprints of Scientific Papers of the IUPAC International Symposium on Macromolecular Chemistry, Tokyo–Kyoto, 1966', Preprint I–255.
6. J. Stehlíček, J. Šebenda and O. Wichterle, *Collect. Czech. Chem. Commun.*, 1964, **29**, 1236.
7. A. Mattiussi and G. B. Gechele, *Eur. Polym. J.*, 1968, **4**, 695.
8. T. Yasumoto, *J. Polym. Sci., Part A*, 1965, **3**, 3877.
9. H. W. Tesch and R. C. Schulz, *Makromol. Chem., Rapid Commun.*, 1981, **2**, 667.
10. S. W. Shalaby and H. K. Reimschuessel, *J. Polym. Sci., Polym. Chem. Ed.*, 1977, **15**, 1349.
11. J. Roda, P. Sysel and J. Králíček *Polym. Bull. (Berlin)*, 1981, **5**, 9.
12. J. H. Park, B. Jung and S. K. Choi, *Taehan Hwahakhoe Chi*, 1980, **24**, 167 (*Chem. Abstr.*, 1980, **93**, 132897).
13. H. Sekiguchi, P. Rapacoulia Tsourkas and B. Coutin, *J. Polym. Sci., Polym. Symp.*, 1973, **42**, 51.
14. J. Králíček and F. Fiala, *Angew. Makromol. Chem.*, 1978, **71**, 29.
15. O. Wichterle, J. Králíček and J. Šebenda, *Collect. Czech. Chem. Commun.*, 1959, **24**, 755.
16. H. Gilch and D. Michael, *Makromol. Chem.*, 1966, **99**, 103.
17. Z. Bukač, J. Tomka and J. Šebenda, *Collect. Czech. Chem. Commun.*, 1969, **34**, 2057.
18. H. S.-I. Chao and P. P. Policastro, *J. Polym. Sci., Polym. Lett. Ed.*, 1986, **24**, 253.
19. E. Šittler and J. Šebenda, *Collect. Czech. Chem. Commun.*, 1968, **33**, 3182.
20. J. Šebenda, A. Stiborová, L. Lochman and Z. Bukač, *Org. Prep. Proced. Int.*, 1980, **12**, 289 (*Chem. Abstr.*, 1981, **94**, 16 165).
21. A. Ciaperoni, L. Mariani and G. B. Gechele, *Chim. Ind. (Milan)*, 1968, **50**, 771.
22. T. M. Frunze, V. A. Kotel'nikov, T. V. Volkova and V. V. Kurashev, *Eur. Polym. J.*, 1981, **17**, 1079.
23. V. V. Korshak, V. A. Kotel'nikov, Yu. A. Avakyan, V. V. Kurashev, T. M. Frunze and S. P. Davtyan, *Dokl. Akad. Nauk SSSR*, 1982, **266**, 896.
24. T. M. Frunze, S. P. Davtyan, V. A. Kotel'nikov, T. V. Volkova and V. V. Kurashev, *Vysokomol. Soedin., Ser. B*, 1981, **23**, 388.
25. E. Šittler and J. Šebenda, *Collect. Czech. Chem. Commun.*, 1968, **33**, 270.
26. J. Roda, M. Kušková and J. Králíček, *Makromol. Chem.*, 1977, **178**, 247.
27. J. Roda, I. Kmínek and J. Králíček, *Makromol. Chem.*, 1978, **179**, 353.

28. T. M. Frunze, V. A. Kotel'nikov, V. V. Kurashev, Yu. A. Arakyan, L. B. Danilevskaya and S. P. Davtyan, in 'Abstracts, IUPAC Macro 83 Bucharest, 1983', Section I, Polymer Chemistry, Commun. I 72, p. 233.
29. E. Šittler and J. Šebenda, *J. Polym. Sci., Part C*, 1967, **16**, 67.
30. J. Stehlíček and J. Šebenda, *Eur. Polym. J.*, 1987, **23**, 237.
31. B. Coutin and H. Sekiguchi, in '1st European Discussion Meeting on Polymer Science: New developments in ionic polymerizations, Strasbourg, France, 1978', Abstracts, p. 88.
32. J. Stehlíček and J. Šebenda, *Eur. Polym. J.*, 1986, **22**, 769.
33. R. Puffr and J. Šebenda, *Eur. Polym. J.*, 1972, **8**, 1037.
34. C. D. Eisenbach and R. W. Lenz, *Macromolecules*, 1976, **9**, 227.
35. C. D. Eisenbach, R. W. Lenz and H. Sekiguchi, *J. Polym. Sci., Polym. Lett. Ed.*, 1977, **15**, 83.
36. S. Barzakay, M. Levy and D. Vofsi, *J. Polym. Sci., Part B*, 1965, **3**, 601.
37. J. Šebenda, P. Čefelín and O. Wichterle, *Chem. Prum.*, 1962, **12**, 41.
38. J. Šebenda, J. Hauer and J. Světlík, *J. Polym. Sci., Polym. Symp.*, 1986, **74**, 303.
39. J. Stehlíček, J. Labský and J. Šebenda, *Collect. Czech. Chem. Commun.*, 1987, **32**, 545.
40. J. Stehlíček, K. Gehrke and J. Šebenda, *Collect. Czech. Chem. Commun.*, 1967, **32**, 370.
41. J. Stehlíček and J. Šebenda, *Eur. Polym. J.*, 1986, **22**, 5.
42. J. Šebenda, *J. Macromol. Sci., Chem.*, 1972, **A6**, 1145.
43. H. Sekiguchi, *Bull. Soc. Chim. Fr.*, 1960, 1835.
44. G. Champetier and H. Sekiguchi, *J. Polym. Sci.*, 1960, **48**, 309.
45. T. M. Frunze, V. A. Kotel'nikov, T. V. Volkova, V. V. Kurashev, S. P. Davtyan and V. V. Korshak, *Dokl. Akad. Nauk SSSR*, 1980, **255**, 612.
46. T. M. Frunze, V. A. Kotel'nikov, T. V. Volkova, V. V. Kurashev, S. P. Davtyan and I. V. Stankevich, *Acta Polym.*, 1981, **32**, 31.
47. J. Šebenda, B. Masař and Z. Bukač, *J. Polym. Sci., Part C*, 1967, **16**, 339.
48. B. Valter, M. I. Terekhova, E. S. Petrov, J. Stehlíček and J. Šebenda, *Collect. Czech. Chem. Commun.*, 1985, **50**, 834.
49. J. Stehlíček, B. Valter and J. Šebenda, *Makromol. Chem.*, 1986, **187**, 513.
50. G. Bontá, A. Ciferri and A. Russo, *ACS Symp. Ser.*, 1972, **59**, 216.
51. A. Ya. Malkin, V. G. Frolov, A. N. Ivanova, Z. S. Andrianova and L. A. Alekseichenko, *Vysokomol. Soedin., Ser. A*, 1980, **22**, 995.
52. R. Z. Greenley, J. C. Stauffer and J. E. Kurz, *Macromolecules*, 1969, **2**, 561.
53. A. Rigo, G. Fabbri and G. Talamini, *J. Polym. Sci., Polym. Lett. Ed.*, 1975, **13**, 469.
54. S. A. Bolgov, V. P. Begishev, A. Ya. Malkin and V. G. Frolov, *Vysokomol. Soedin., Ser. A*, 1981, 23, 1341.
55. P. W. Sibal, R. E. Camargo and C. W. Macosko, *Polym. Process Eng.*, 1984, **1**, 147.
56. D. J. Lin, J. M. Ottino and E. L. Thomas, *Polym. Eng. Sci.*, 1985, **25**, 1153.
57. S. A. Iobst, *Polym. Eng. Sci.*, 1985, **25**, 425.
58. R. A. Cimini and D. C. Suddberg, *Polym. Eng. Sci.*, 1986, **26**, 560.
59. B. Valter, M. I. Terekhova, E. S. Petrov, J. Stehlíček and J. Šebenda, *Collect. Czech. Chem. Commun.*, 1985, **50**, 840.
60. V. G. Frolov, V. P. Pshenitsyna and I. A. Krasnova, *Vysokomol. Soedin., Ser. B*, 1980, **22**, 758.
61. V. G. Frolov, *Vysokomol. Soedin., Ser. B*, 1983, **25**, 134.
62. J. Králíček, M. Kušková and J. Roda, *Makromol. Chem.*, 1977, **178**, 3203.
63. N. I. Vasil'iev, B. V. Kholodenko and G. P. Andrianova, *Vysokomol. Soedin., Ser. B*, 1984, **26**, 72.
64. R. P. Scelia, S. E. Schonfeld and L. G. Donaruma, *J. Appl. Polym. Sci.*, 1967, **11**, 1299.
65. T. M. Frunze, V. A. Kotel'nikov, M. P. Ivanov, T. V. Volkova, V. V. Kurashev and S. P. Davtyan, *Vysokomol. Soedin., Ser. A*, 1981, **23**, 2675.
66. A. Ya. Malkin, S. L. Ivanova and M. A. Korchagina, *Vysokomol. Soedin., Ser. A*, 1977, **19**, 2224.
67. H. Sekiguchi, *Bull. Soc. Chim. Fr.*, 1960, 1831.
68. M. Provazník, R. Puffr and J. Šebenda, *Eur. Polym. J.*, submitted for publication.
69. J. Šebenda and J. Hauer, *Polym. Bull. (Berlin)*, 1981, **5**, 529.
70. M. Provazník, R. Puffr, J. Hauer, J. Světlík and J. Šebenda, in preparation.
71. J. Šebenda, *Prog. Polym. Sci.*, 1978, **6**, 123.
72. J. Šebenda, *Pure Appl. Chem.*, 1976, **48**, 329.
73. J. Šebenda, *Makromol. Chem., Macromol. Symp.*, 1986, **6**, 1.
74. H. Sekiguchi and B. Coutin, *J. Polym. Sci., Polym. Chem. Ed.*, 1973, **11**, 1601.
75. R. Graf, G. Lohaus, K. Börner, E. Schmidt and H. Bestian, *Angew. Chem.*, 1962, **74**, 523.
76. J. Stehlíček, P. Čefelín and J. Šebenda, *Collect. Czech. Chem. Commun.*, 1972, **37**, 1926.
77. P. Čefelín, J. Stehlíček and J. Šebenda, *Collect. Czech. Chem. Commun.*, 1972, **37**, 3861.
78. P. Čefelín, J. Stehlíček and J. Šebenda, *Collect. Czech. Chem. Commun.*, 1974, **39**, 2212.
79. J. Stehlíček, P. Čefelín and J. Šebenda, *J. Polym. Sci., Polym. Symp.*, 1973, **42**, 89.
80. J. Stehlíček, J. Roda, Z. Votrubcová and S. Pokorný, *Angew. Makromol. Chem.*, 1980, **91**, 117.
81. J. Šebenda, *Collect. Czech. Chem. Commun.*, 1966, **31**, 1501.
82. J. Šebenda and V. Kouřil, *Eur. Polym. J.*, 1972, **8**, 437.
83. O. Wichterle, *Makromol. Chem.*, 1960, **35**, 174.
84. Z. Bukač and J. Šebenda, *Collect. Czech. Chem. Commun.*, 1967, **32**, 3537.
85. Z. Bukač and J. Šebenda, *J. Polym. Sci., Polym. Symp.*, 1973, **42**, 345.
86. F. Wiloth and E. Schindler, *Chem. Ber.*, 1970, **103**, 757.
87. F. Wiloth and E. Schindler, *Chem. Ber.*, 1967, **100**, 2373.
88. P. Čefelín, J. Stehlíček and J. Šebenda, *Eur. Polym. J.*, 1974, **10**, 227.
89. S. Barzakay, M. Levy and D. Vofsi, *J. Polym. Sci., Part A-1*, 1967, **5**, 965.
90. H. R. Kricheldorf, B. Coutin and H. Sekiguchi, *J. Polym. Sci., Polym. Chem. Ed.*, 1981, **20**, 2353.
91. V. V. Korshak, T. M. Frunze, V. A. Kotel'nikov, M. P. Ivanov, V. V. Kurashev, S. P. Davtyan and L. B. Danilevskaya, *Dokl. Akad. Nauk SSSR*, 1981, **257**, 641.
92. H. K. Reimschuessel, in 'Ring-opening Polymerization', ed. K. C. Frisch and S. L. Reegen, Dekker, New York, 1969, p. 303.

93. A. Ya. Malkin, V. G. Frolov, A. N. Ivanova and Z. S. Andrianova, *Vysokomol. Soedin., Ser. A*, 1979, **21**, 632.
94. A. Ya. Malkin, V. P. Begishev, and S. A. Bolgov, *Polymer*, 1982, **23**, 385.
95. A. Ya. Malkin, S. L. Ivanova, V. G. Frolov, A. N. Ivanova and Z. S. Andrianova, *Polymer*, 1982, **23**, 1791.
96. G. C. Alfonso, G. Bontá, S. Russo and A. Traverso, *Makromol. Chem.*, 1981, **182**, 929.
97. E. Biagini, B. Pedemonte, E. Pedemonte, S. Russo and A. Turturro, *Makromol. Chem.*, 1982, **183**, 2131.

36

Anionic Ring-opening Polymerization: *N*-Carboxyanhydrides

HANS R. KRICHELDORF
Universität, Hamburg, FRG

36.1 APPLICATIONS OF *N*-CARBOXYANHYDRIDES AND THEIR PEPTIDE DERIVATIVES

Since their discovery by Leuchs in the year 1906,[1] α-amino acid *N*-carboxyanhydrides (α-NCAs) have found increasing interest as monomers for the synthesis of oligopeptides and polypeptides. All *N*-carboxyanhydrides (NCAs) have the basic disadvantage that they are not stable on storage. The mechanisms of the spontaneous polymerization have not yet been elucidated. A detailed discussion of this aspect is presented in a recent monograph.[2] Despite their instability, α-NCAs are highly useful for a variety of preparative purposes. Syntheses of various different low molecular weight amino acid derivatives such as *N*-ω-protected α,ω-diamino acids, amino acid alkyl esters, amino acid amides, amino acid hydroxamic acids, amino acid hydrazides, α-aminoacetophenones, hydantoic acids and α-isocyanatocarboxylic acid chlorides, have been described.[2] In addition to pure oligopeptides, and polypeptides, a broad variety of block and graft copolymers have also been synthesized containing peptide chains attached to blocks or backbones derived from numerous different monomers. For instance two- or three-block copolymers have been prepared from poly(ethylene oxide) bearing one or two functionalized end groups.[3-6] Three-block copolymers consisting of a central block of polystyrene and two flanks of peptide chains, or of central peptide blocks with flanks of polystyrene, have been reported.[7-11] Similar block copolymers have been described based on blocks of anionically initiated poly(butadiene).[9,12-15] These block copolymers have been studied as models of proteins with hydrophobic or hydrophilic domains, and their compatibility with blood has been investigated.

In addition to block copolymers, numerous graft copolymers have been synthesized by means of NCAs. Peptide chains have been grafted onto poly(vinyl alcohol) activated by means of phosgene.[3] Various copolymers with pendant amino groups that serve as polymeric initiators for grafting of NCAs have been prepared by radical homopolymerization and copolymerization of *N*-(2-benzyl-oxycarbonylaminoethyl)acrylamide.[16] Other workers have used partially saponified cellulose acetate with alcoholate anions randomly scattered along the chain as the initiator for the grafting process.[17,18] Several authors[19-26] have devoted intensive studies to synthesis and characterization of so-called peptidyl proteins. This term means that NCAs are used to graft peptide chains on naturally occuring polypeptides and proteins such as insulin, trypsin, chymotrypsin, ribonucleases, bovine serum albumin, hen egg-white albumin, gelatin and even tobacco mosaic virus. These peptidyl proteins are characterized with regard to their enzymatic activity or stability against enzymatic attack, and also with regard to their solubility and stability on storage and, in particular, with regard to their immunological properties.

For the synthesis of oligopeptides by means of NCAs two quite different strategies may be used. Firstly, beginning with an amino acid, amino acid amides or ester oligopeptides are built up in a stepwise manner, so that exactly defined sequences may be obtained at will. For such stepwise syntheses, either NCAs containing a protecting group attached to the nitrogen are used, or N-unsubstituted NCAs under sophisticated reaction conditions.[2] Despite the instability of NCAs, these stepwise peptide syntheses have been successful enough to enable the preparation of enzymes.[27, 28] Secondly, NCAs have been polymerized by means of highly nucleophilic initiators at low monomer/initiator ratios (M/I) so that the degree of polymerization (DP) of the resulting peptides is limited by the product of M/I ratio and conversion. This procedure is quite simple but yields oligopeptides with a broader molecular weight distribution (MWD). Nonetheless, even these simple homooligopeptides may be useful for studying the dependence of solubility and conformation on DP.[5]

α-NCAs are, of course, most widely used for the preparation of polypeptides. Four classes of polypeptides can be obtained from NCAs: (i) homopolypeptides with varying molecular weights and varying secondary structure; (ii) two- or three-block copolypeptides consisting of two or three different amino acids; (iii) copolypeptides with more or less random sequences of two or more different amino acids; and (iv) graft copolypeptides with a backbone prepared from the NCA of a trifunctional amino acid (*e.g.* L-lysin).

The only class of polypeptides which cannot be synthesized by means of α-NCAs are copolypeptides with an exactly alternating sequence of two or more different amino acids. In addition to their versatility α-NCAs have the significant advantage that neither synthesis nor polymerization involve racemization if conducted under appropriate conditions. For this reason polypeptides prepared from NCAs have been the object of numerous investigations, in particular of conformational studies in solution or in the solid state.[2, 5, 6, 29] They have served as substrates of enzymes, as components of chiral catalysts and as synthetic immunogens. Their compatibility with blood and various living tissues has been investigated. Furthermore, they have been manufactured as textile fibers as synthetic analogues of silk or wool.[30]

α-NCAs can be polymerized by heating above their melting point and by initiation with nucleophilic or basic catalysts. However, they are rather insensitive to radicals and do not polymerize in contact with protic acids or other cationic initiators. Because the nature of the initiator has a considerable influence on the polymerization mechanism and on the properties of the resulting polypeptides (*e.g.* molecular weights, end groups, sequences) the following mechanistic discussion is subdivided according to classes of initiators. A simplified overview on initiators and proposed reaction mechanisms is given in Table 1. Finally, it is worth noting that all methods suitable for the synthesis of NCAs have been summarized and discussed in a recent monograph.[2] The same work also contains a list of almost all NCAs reported so far, including their melting points, optical rotations and the corresponding references.

36.2 INITIATION WITH PROTIC NUCLEOPHILES

α-NCAs possess four reactive sites, namely the two carbonyl groups and the NH and α-CH groups. The C(2)=O carbonyl is significantly less electrophilic than C(5)=O owing to the delocalization of the nitrogen electron pair. Nucleophilic attacks at C(2)=O have never been proven so far, although they were many times postulated by several authors. Neither NH nor α-CH groups are nucleophilic in the ground state, yet after deprotonation highly nucleophilic anions are formed. Because the NH group is more acidic than α-CH, deprotonation of the carbon is only expected when the nitrogen is substituted (*e.g.* in Sar-NCA or Pro-NCA).

Due to these four reactive sites of different reactivities nearly all initiation steps involving N-unsubstituted NCAs (the vast majority of NCAs) may be classified either as nucleophilic attack onto C(5)=O (equation 1) or as deprotonation of the NH group (equation 2). Which initiation step is dominant depends, of course, on the reaction conditions, but it depends mainly on the chemical structure of the initiator. Thus, the most important property of initiators in this regard is their nucleophilicity/basicity ratio (N/B), which depends on three parameters: electron density, polarizability of electron pairs, and steric demands. With regard to the propagation steps it is reasonable to subdivide initiators into protic and nonprotic ones. Also important for the classification of initiators is their ability to form stable end groups (*i.e.* dead chain ends). This property is characteristic for anionic and protic initiators, unless high steric demands prevent a nucleophilic attack at C(5)=O. In contrast tertiary amines never form stable end groups.

Table 1 Proposed Mechanisms of Initiation and Propagation Reactions with Various Initiators

Initiator	N-Unsubstituted N-carboxyanhydrides		N-Substituted N-carboxyanhydrides	
	Initiation	*Propagation*	*Initiation*	*Propagation*
Primary amine	Nucleophilic attack on C-5 (amine mechanism)	Nucleophilic attack on C-5 (amine mechanism)	Nucleophilic attack on C-5 (amine mechanism)	Nucleophilic attack on C-5 (amine mechanism)
Secondary amine	Competition of nucleophilic attack and deprotonation	Competition of amine carbamate and activated monomer mechanism	Nucleophilic attack on C-5 (amine mechanism)	Nucleophilic attack on C-5 (amine mechanism)
Trialkylamine	Deprotonation	Competition of activated monomer and carbamate mechanism	Deprotonation at C-4	Carbamate mechanism
Pyridine	Deprotonation and activated monomer mechanism or nucleophilic attack and formation of zwitterions	Activated monomer mechanism and carbamate mechanism	Nucleophilic attack on C-5	Zwitterion and carbamate mechanism
Alcoholate and alcohol	Deprotonation and activated monomer mechanism followed by nucleophilic attack	Mainly carbamate and amine mechanism activated monomer mechanism	Nucleophilic attack on C-5	Carbamate mechanism
Hydride ion	Deprotonation and activated monomer mechanism	Carbamate mechanism or activated monomer mechanism	Not investigated so far	
Carbamate and carboxylate	Nucleophilic attack on C-5	Mainly carbamate mechanism, partially activated monomer mechanism	Not investigated so far	

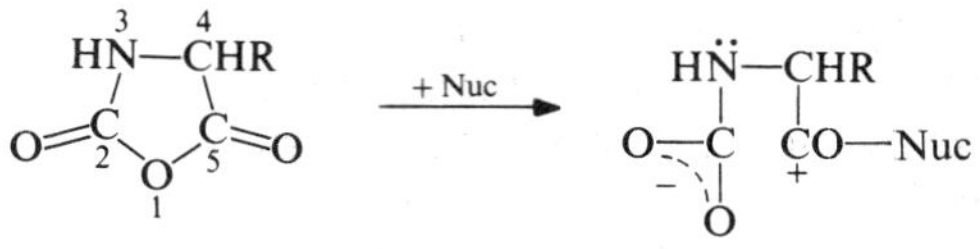

$$\text{(NCA)} \xrightarrow{+\,\text{Nuc}} \text{(ring-opened adduct)} \tag{1}$$

$$\text{(NCA)} \xrightarrow[-\,\overset{+}{B}H]{+\,B} \text{(deprotonated NCA)} \tag{2}$$

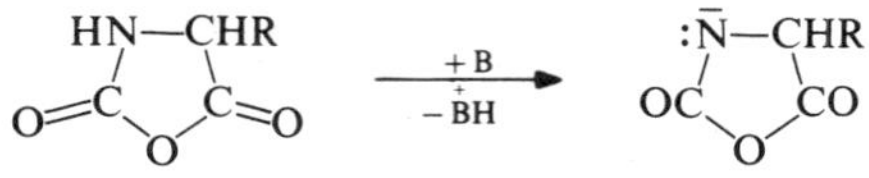

The most widely used protic initiators are water and primary amines. Water (like alcohols) is not basic enough to deprotonate NCAs, so that the nucleophilic attack at $C(5){=}O$ is the only possible initiation step. The hydrolysis of NCAs has been carefully investigated by several research groups,[1,31–35] and the following results were obtained. The reaction of water with NCAs can take two extreme courses: low NCA/H_2O ratios ($< 1{:}1000$) result in complete hydrolysis, whereas high NCA/H_2O ratios ($> 10{:}1$) yield polypeptides in nearly quantitative yields. Intermediate NCA/H_2O ratios favor the formation of oligopeptides. Decreasing pH increases the yield of free amino acids at the expense of polypeptides, whereas increasing pH favors polypeptides at the expense of amino acid. These findings and careful kinetic investigations conducted at various pH values[35,36] indicate the following reaction mechanism. In neutral and acidic water it is the H_2O molecule itself which attacks $C(5){=}O$ (equation 3). The resulting carbamic acid is detectable by the increasing acidity of the solution. Addition of calcium or barium hydroxide to the ice-cold solution of NCAs yields the so-called 'Siegfried salts' (equation 4). The carbamate ions are stable in aqueous solution in contrast to the carbamic acid, which decarboxylates above $0\,°C$ yielding a free amino acid (equation 5).

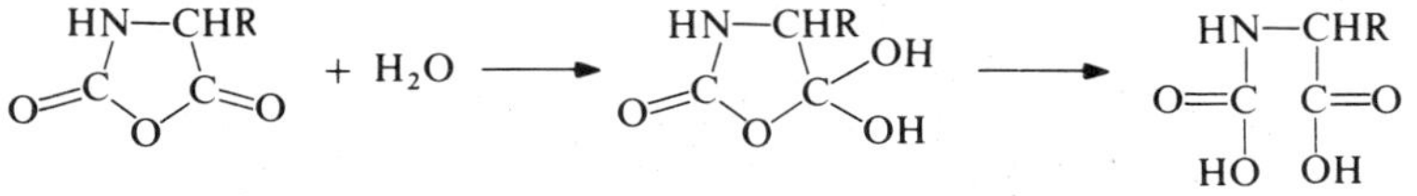

$$\text{(NCA)} + H_2O \longrightarrow \text{(tetrahedral intermediate)} \longrightarrow \text{(carbamic acid)} \tag{3}$$

$$\text{(carbamic acid)} + Ba(OH)_2 \xrightarrow{-2H_2O} \text{(carbamate)} + Ba^{2+} \tag{4}$$

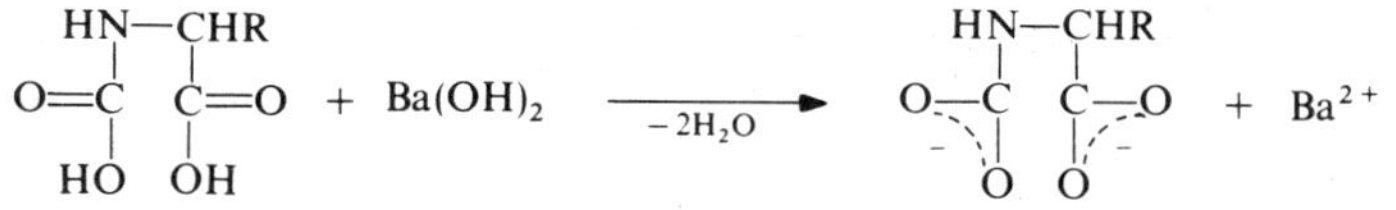

$$\text{(carbamate)} \xrightarrow{+H^+} \text{(carbamic acid)} \xrightarrow{-CO_2} NH_2CHRCO_2H \tag{5}$$

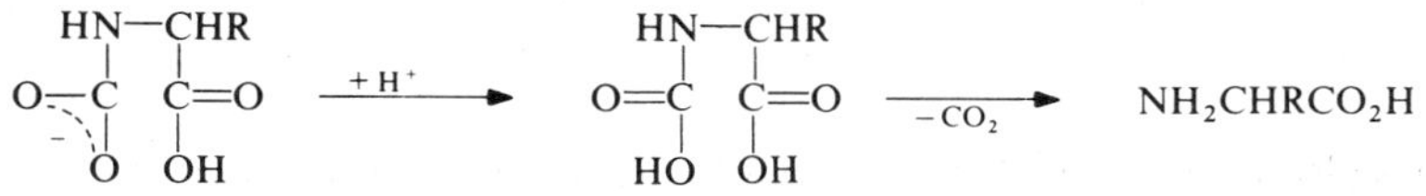

The amino group generated by hydrolysis can attack another NCA, inasmuch as it is more nucleophilic than H_2O, and this initiate a chain growth *via* amino end groups (equation 6). Such a chain growth is called the amine mechanism. Alkaline solutions prevent the protonation of amino groups and thus favor the polymerization. However, an alkaline solution also stabilizes the carbamate ions, which are more nucleophilic (due to higher polarizability) than the carboxylate ions, therefore another chain growth mechanism involving carbamate end groups, the so-called carbamate mechanism (equation 7), may also be operating. It is not yet known to what extent polymerizations of NCAs in neutral or alkaline aqueous solutions proceed *via* the amine or the carbamate mechanism. Water-initiated polymerizations of NCAs have been used by several research groups for preparative purposes. However high molecular weights ($M_n > 20\,000$) are not attainable in this way, as evidenced by titration of amino or carboxyl end groups.[37–41] Several authors[42,43]

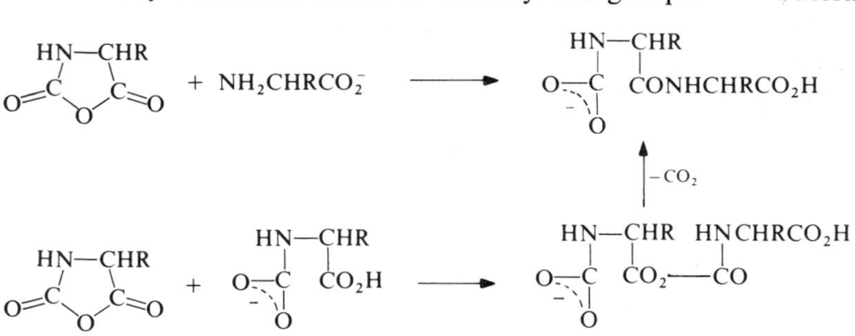

$$\text{(NCA)} + NH_2CHRCO_2^- \longrightarrow \text{(CONHCHRCO}_2\text{H product)} \tag{6}$$

$$\Big\downarrow -CO_2$$

$$\text{(NCA)} + \text{(carbamate)} \longrightarrow \text{(adduct)} \tag{7}$$

initially reported molecular weights of the order of 500×10^3 g mol^{-1} for water-initiated poly(D,L-phenylalanine) and copolymers of L-leucine and D,L-phenylalanine on the basis of membrane osmometry in benzene. However, these authors did not take into account that polypeptides heavily associate in nonpolar solvents due to H bonds. A reinvestigation[44] by means of light scattering and end group titrations has revealed M_n and M_w values below 20 000.

The treatment of NCAs with dilute acids prevents the formation of polypeptides and may be used to prepare free amino acids in quantitative yields. This approach is useful for the synthesis of N-protected basic amino acids according to Scheme 1.[45-48] The reaction of NCAs with alcohols largely resembles that with water. In the presence of at least equimolar amounts of a strong acid, amino acid esters are formed in nearly quantitative yields (Scheme 1).[46,48] When NCAs are heated in neutral alcohols, polypeptides with ester end groups are formed. Because this procedure neither yields high molecular weights nor enables a systematic variation of the molecular weights, it has never been investigated in more detail. The reaction of NCAs with small amounts of alcoholate ion is discussed in Section 36.4.

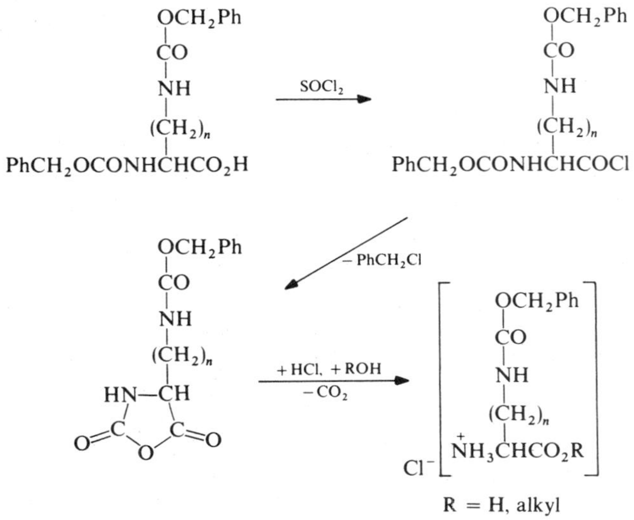

Scheme 1

Primary amines also possess a sufficiently high N/B ratio for nucleophilic attack at C(5)=O to be the dominant initiation step. Amino acid amines with carbamate groups (equation 6), or after decarboxylation with free amino groups, are the first reaction products. These nucleophilic end groups continue the chain growth when excess NCA is present. The course of the polymerization largely depends on the nuelcophilicity of the amine and on the M/I ratio. When the initiator is less nucleophilic than the active chain end, as is true for aromatic amines, propagation is faster than initiation ($k_p > k_i$ in equations 8–10) and the resulting DP is higher than expected from the M/I ratio (equation 11), in close analogy to initiation with water or alcohols. When the primary amine is more nucleophilic, the initiation step is faster ($k_p < k_i$ in equations 8–10) and the DP is predictable according to equation (11), at least for M/I ratios below 100. Therefore n-alkylamines and benzylamine have been widely used for syntheses of polypeptides with DPs below 200. Although the reactivity of primary alkylamines is high enough to initiate a living polymerization of α-NCAs, the results do not fit in with the living pattern for two reasons. First, the low solubility of most oligopeptides and polypeptides in the reaction medium along with their secondary structure may cause a 'physical death' of the growing chains, and may also yield bimodal MWDs (see Section 36.5). Second, termination steps may occur.

$$V_i = k_i[\text{NCA}][\text{RNH}_2] \tag{8}$$

$$V_{p1} = k_{p1}[\text{NCA}][\text{pol}\text{\textasciitilde}\text{NH}_2] \tag{9}$$

$$V_{p2} = k_{p2}[\text{NCA}][\text{pol}\text{\textasciitilde}\text{NH}-\text{CO}_2^-] \tag{10}$$

$$DP = \frac{M}{I} \times \frac{\% \text{ conversion}}{100} \tag{11}$$

Two kinds of termination steps have been reported so far. When the amino end group of the growing chain can react with the side chain of the ultimate (or penultimate) monomer unit (*e.g.* by transacylation), an intramolecular termination occurs. Such termination steps are most likely for polyglutamates (equation 12).[49-51] poly(*O*-acylserine)s (equation 13), poly(*O*-acylthreonine)s and poly(*S*-acylcysteine)s. The formation of hydantoic acid end groups is independent of the nature of the side chain. Most authors who have investigated[39] or discussed the formation of hydantoic acids from NCAs originally postulated that they result from the nucleophilic attack of an amino (end) group on $C(2)=O$ (equation 14). However, any experimental evidence in favor of this reaction is lacking. In contrast, hydantoic acids have never been obtained from N-substituted NCAs, even when electron-withdrawing substituents are attached to the nitrogen.[52,53] Hence a reaction sequence involving deprotonation of an NCA followed by intermediate formation of an α-isocyanate carboxylate ion is the most likely explanation, in connection with chain growth *via* amino end groups (Scheme 2).[53,54] Whenever propagation *via* carbamate chain ends takes place, an alternative mechanism needs be taken into account (equation 15): again any experimental evidence is lacking.

$$
\begin{array}{c}
RO \overset{\displaystyle CO}{\diagdown} CH_2 \\
| \\
CH_2 \\
| \\
\ddot{N}H_2CHCO\text{\textasciitilde pol}
\end{array}
\quad\xrightarrow{\ -\,ROH\ }\quad
\begin{array}{c}
O=C \overset{\displaystyle CH_2}{\diagdown} CH_2 \\
| \\
HN-CHCO\text{\textasciitilde pol}
\end{array}
\tag{12}
$$

$$
\begin{array}{c}
R-CO \\
\diagdown O \\
| \\
CH_2 \\
| \\
NH_2CHCO\text{\textasciitilde pol}
\end{array}
\quad\longrightarrow\quad
\begin{array}{c}
OH \\
| \\
CH_2 \\
| \\
RCONH\dot{C}HCO\text{\textasciitilde pol}
\end{array}
\tag{13}
$$

$$
pol\text{\textasciitilde}NH_2 \;+\;
\begin{array}{c}
HN-CHR \\
O=C \qquad C=O \\
\diagdown O \diagup
\end{array}
\quad\longrightarrow\quad pol\text{\textasciitilde}NHCONHCHRCO_2H
\tag{14}
$$

$$
\begin{array}{c}
HN-CHR \\
O=C \qquad C=O \\
\diagdown O \diagup
\end{array}
\;+\; RNH_2 \;\longrightarrow\; R\overset{+}{N}H_3 \;+\;
\begin{array}{c}
{}^{-}\!:\!\ddot{N}-CHR \\
O=C \qquad C=O \\
\diagdown O \diagup
\end{array}
\;\rightleftharpoons\;
\begin{array}{c}
N-CHR \\
\| \quad\quad | \\
C \quad C=O \\
\| \quad | \\
O \quad O
\end{array}
$$

$$
RNH_2 \;+\; O=C=NCHRCO_2^{-} \xrightarrow{\ +H^{+}\ } RNHCONHCHRCO_2H
$$

Scheme 2

$$
\begin{array}{cc}
CHRCO_2 & \\
| \quad\quad | & \\
NH_2 \quad CONHCHRCO\text{\textasciitilde pol} &
\end{array}
\quad\longrightarrow\quad
\begin{array}{c}
HO_2CCHRNHCONHCHR \\
| \\
CO\text{\textasciitilde pol}
\end{array}
\tag{15}
$$

The question of to what extent primary or secondary amine-initiated polymerizations involve a chain growth *via* carbamate ions (equation 7) cannot be answered quantitatively. The decarboxylation equilibrium (equation 5) certainly depends on the CO_2 pressure and thus it is obvious that the rate of polymerization should be influenced by the CO_2 pressure. Regardless of whether an acceleration is found upon increasing the CO_2 pressure or not,[55-58] such an observation can never prove that the majority of growing steps occur *via* carbamate ions. However, several observations militate against there being a significant role for the carbamate mechanism. First, the carbamate chain ends need a protonated amine as counterion for their stabilization. But end group analyses (*e.g.* by means of NMR spectroscopy) demonstrate that aliphatic amines used as initiators are almost completely incorporated as dead end groups even at low conversion. Second, polymerization is prevented when excess of primary or secondary amine is used as low temperatures ($\leq 0\,^{\circ}C$) so that the carbamate ion is stabilized. Polymerization is also suppressed when the carbamate group is stabilized in alkaline water. Obviously the carbamate ion is less nucleophilic than a primary aliphatic amino group and thus its stabilization has been used for the stepwise synthesis of peptides.[27,59,60] Third, end group analyses by means of 2,4-dinitrofluorobenzene always reveals amino and never carbamate[61-63] end groups.

Incorporation of primary or secondary amines as dead chain ends is easily detectable by high resolution ^{1}H or ^{13}C NMR spectroscopy, yet ^{14}C-enriched amines have also been used by several authors.[64-66] For the purpose of ^{1}H NMR spectroscopic end group analyses it is advisable to use aliphatic amines with aromatic protons (*e.g.* benzylamine) when aliphatic NCAs are to be polymerized, or aliphatic amines with numerous identical protons, such as isopropylamine or *t*-butylamine, in the case of aromatic amino acids.[67,68] Quantification of signal intensities allows calculation of the DP, when termination steps are almost absent, *i.e.* for M/I ratios $\leq$ 100. ^{1}H NMR measurements have the additional advantage that they enable the determination of the molar composition of copolymers.[69,70]

Secondary aliphatic amines react like primary amines when the N/B ratio is high, *i.e.* when the steric demands of both substituents are low, as is true for dimethylamine, piperidine or morpholine. Secondary amines with slightly bulkier substituents such as *N*-methylbenzylamine, diethylamine or di-*n*-propylamine may react as both nucleophile and base. In the latter case, deprotonation of the NCA is the first step. At high M/I ratios this first step may be followed by an anionic polymerization according to the activated monomer mechanism (AM mechanism) as discussed for trialkylamines in Section 36.3. High concentrations of secondary amines and higher temperatures not only favor deprotonation but also favor the formation of hydantoic acids (according to Scheme 2) at the expense of polymerization.[52,71] Secondary amines with two bulky substituents, such as diisopropylamine or dicyclohexylamine, exclusively react as bases. Whether a secondary amine mainly reacts as a nucleophile or as a base is easy to determine. The incorporation of the secondary amine is detectable by NMR spectroscopy and addition of strong electrophiles such as *N*-acetylglycine NCA (AcGly NCA) or isocyanates stops the chain growth proceeding *via* nucleophilic chain ends.[72] When the amine exclusively reacts as a base, it does not form amide end groups and addition of the aforementioned electrophiles has a cocatalytic effect, *i.e.* it enhances the rate of polymerization.[72]

36.3 INITIATION WITH TERTIARY AMINES

The tertiary amines used as initiators may be subdivided into two groups with largely differing N/B ratios, namely trialkylamines and pyridines. Trialkylamines, such as triethylamine or tributylamines, possess pK_a values around 11 and a low nucleophilicity due to the steric demands of the three alkyl groups. Any experimental evidence for nucleophilic attack of these tertiary amines on the carbonyl group of (amino) acid derivatives is lacking so far. In contrast pyridines possess pK_a values in the range of 4–7 and a high nucleophilicity when C-2 and C-6 are unsubstituted. Their high nucleophilicity is not only a result of lacking steric hindrance, but also results from the delocalization of positive charges by the π electrons. Careful kinetic studies of the hydrolysis of acid anhydrides has revealed[73] that pyridine can cleave anhydride groups under formation of *N*-acylpyridinium ions (equation 16).

$$\text{MeCO}{-}\text{O}{-}\text{OCMe} + :N\langle \rangle \rightleftharpoons \text{MeCON}^+\langle \rangle + \text{MeCO}_2^- \qquad (16)$$

Two different initiation steps (equations 1–2) and different propagation mechanisms have been proposed for tertiary amine-initiated polymerizations. The earliest mechanistic concept, published in 1951,[74] postulates a nucleophilic attack of the tertiary amine followed by condensation polymerization of the zwitterions formed as intermediates (equation 17). However, all experimental results published so far demonstrate that this so-called zwitterion mechanism is not operating when N-unsubstituted NCAs are polymerized, because the energy of activation of the deprotonation is lower. Nonetheless, pyridine-initiated polymerizations of N-substituted NCAs may involve zwitterions as discussed below.

$$2 \begin{array}{c} \text{HN}{-}\text{CHR} \\ | \quad\quad | \\ \text{O}{=}\text{C} \quad \text{CONR}_3 \\ \diagdown | \\ \text{O} \end{array} \xrightarrow[-\text{NR}_3]{} \begin{array}{c} \text{HN}{-}\text{CHR} \;\; \text{HN}{-}\text{CHR} \\ | \quad\quad | \quad\quad\;\; | \quad\quad | \\ \text{O}{=}\text{C} \quad \text{CO}_2{-}\text{CO} \quad \text{CONR}_3 \\ \diagdown | \\ \text{O} \end{array} \xrightarrow{-\text{CO}_2} \begin{array}{c} \text{HNCHRCONHCHR} \\ | \quad\quad\quad\quad\quad\quad | \\ \text{O}{=}\text{C} \quad\quad\quad\quad\quad \text{CONR}_3 \\ \diagdown | \quad\quad\quad\quad\quad\quad + \\ \text{O} \end{array} \qquad (17)$$

When an N-unsubstituted NCA is deprotonated, the resulting NCA anion (equation 2) may attack another NCA, so that a dimer with an *N*-acyl-NCA group on one side and a carbamate ion on the other side is formed (equation 18). Because the *N*-acyl-NCA is more electrophilic than an unsubstituted NCA, the polymerization may proceed by further reaction of NCA anions with *N*-acyl-

NCA chain ends (equation 19). This reaction pathway, which closely resembles the anionic polymerization of lactams, is an example of the activated monomer (AM) mechanism.[49,56,76,77] A third propagation mechanism which may follow the formation of reactive dimers (equations 18, 20) is a condensation polymerization (equation 21). However, the kinetic and molecular weight measurements of several research groups[49,51,56,75–82] militate against this mechanism. The crystallization of oligopeptides in the early stages of base-initiated polymerizations is another argument against predominance of condensation steps. A fourth potential propagation mechanism is again the carbamate mechanism (equation 7), which may follow an initiation involving nucleophilic attack of the initiator (equation 1) or deprotonation (equations 2 and 18).[83–85] So, the long-standing debate on the polymerization mechanisms of NCAs is mainly concentrated on the following two questions: (i) whether the deprotonation of NCAs (*e.g.* by trialkylamines) initiates a propagation *via* the AM mechanism or *via* the carbamate mechanism; and (ii) whether the nucleophilic attack of an aprotic initiator (*e.g.* pyridine) onto an N-substituted NCA is mainly followed by condensation of zwitterions or again by the carbamate mechanism.

$$\tag{18}$$

$$\tag{19}$$

$$\tag{20}$$

$$\tag{21}$$

Doubts about the existence of the AM mechanism for a long time included doubts about the deprotonation of NCAs.[86] Meanwhile four reactions were reported which demonstrate that substitution of NCAs is feasible when trialkylamines are added both as catalysts and HCl acceptors (equations 22–25).[53,87–89] In this connection the sodium methoxide-catalyzed cyclization of α-amino-δ-O-benzyladipate NCA (equation 25)[89] is noteworthy. Model reactions using the non-polymerizable 3-methylhydantoins[49,90] confirmed that NCAs and similar heterocycles are easily deprotonated by trialkylamines and stronger bases, although their pK_a values are higher[91] than those of protonated tertiary amines.

$$\tag{22}$$

$$\tag{23}$$

$$\tag{24}$$

$$\tag{25}$$

The existence of the AM mechanism requires that the *N*-acyl-NCA chain end is significantly more electrophilic than a monomer, so that its reaction with NCA anions (equation 19) is considerably more rapid than the initiation step of equation (18). Consequently, addition of an *N*-acyl-NCA, such as *N*-acetyl-Gly-NCA, should enhance the rate of polymerization, because the relatively slow initiation step of equation (18) is replaced by the faster step of equation (26). This acceleration effect is indeed found.[53,92] Furthermore, it has been demonstrated by means of ^{1}H NMR spectroscopy that the coinitiator is incorporated into the growing peptide in the form of a dead acetylglycyl end group (Figure 1).[53] Moreover, both IR and ^{1}H NMR spectroscopy have revealed that peptides initiated with a combination of *N*-acetyl-Gly-NCA and triethylamine (or diisopropylamine) possess active *N*-acyl-NCA chain ends (Figure 1).[53,72] In addition to *N*-acetyl-Gly-NCA various other compounds have been tested as potential coinitiators.[72,93] All these results clearly prove that the AM mechanism does indeed operate when appropriate reaction conditions are chosen. Yet it should be kept in mind that all these results have been obtained under conditions yielding low molecular weight reaction products.

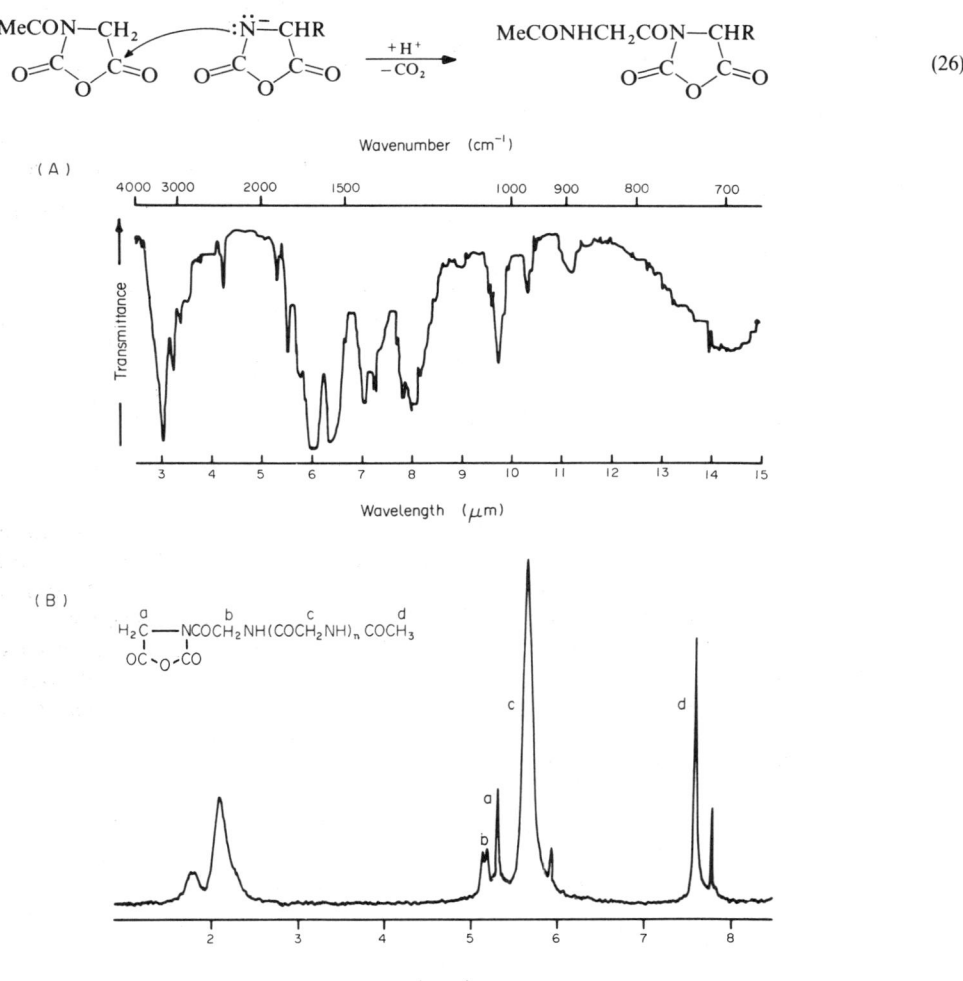

Figure 1 Oligoglycine (DP = 9, initiated by means of diisopropylamine and *N*-acetyl-Gly-NCA) in dioxane at 20 °C (A) IR spectrum (in KBr) exhibiting the CO bands of the *N*-acyl-Gly-NCA endgroup at 5.8, 5.5 and 5.8 μm. (B) 100 MHz ^{1}H NMR spectrum of the same sample dissolved in trifluoroacetic acid[71]

The trialkylamine-initiated polymerizations of NCAs are of particular interest because they enable the synthesis of high molecular weight polypeptides (up to 500×10^3)[50] in contrast to polymerizations initiated by protic nucleophiles. Reaction conditions favoring high molecular weights are characterized by high M/I ratios and by the absence of electrophilic coinitiators. Even if the initial formation of few *N*-acyl-NCA chain ends occurs, these chain ends may be lost in the early stages of the polymerization due to side reactions. It is well known that base-initiated polymerizations of NCAs yield cyclic byproducts such as hydantoin-3-carboxylic acids, 2,5-dioxopiperazines or larger cyclopeptides, which may be formed according to Scheme 3.[72,94-97] Furthermore, N-acyl-NCAs are seemingly able to attack amide groups when trialkylamines are present. Regardless of the nature of the side reaction, if the N-acyl-NCA chain ends disappear, further chain growth must involve nucleophilic end groups (*i.e.* carbamate or amino groups). Kinetic measurements do not allow a clearcut determination of the propagation mechanism and any reliable evidence for the existence of N-acyl-NCA chain ends in high molecular weight polypeptides is lacking so far. In other words, the nature of the mechanism yielding high molecular weight polypeptides is not yet established, even when the initial formation of NCA anion is accepted.

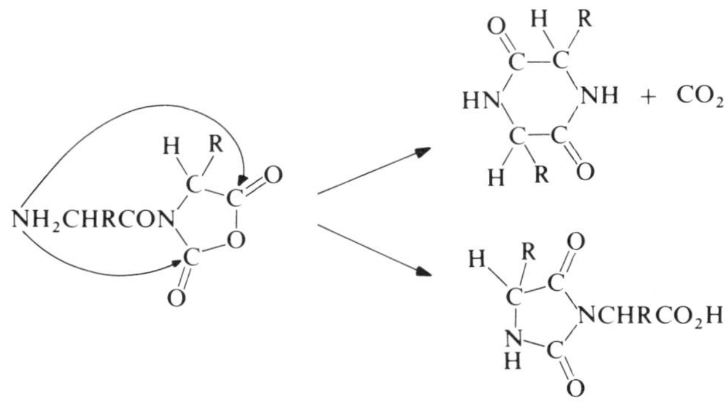

Scheme 3

In the case of N-unsubstituted NCAs not only trialkylamines but also pyridines cause initiation by deprotonation. Accordingly the rate of initiation increases with increasing basicity of substituted pyridines, whereas the steric demands of the substituents do not affect the reaction rate.[76] Furthermore, addition of N-acetyl-Gly-NCA causes acceleration and formation of an acetyl end group.[53] In the case of N-substituted NCAs, *e.g.* Sar-NCA or Pro-NCA, neither deprotonation nor the AM mechanism can be operating, and the results reported so far indicate that trialkylamines and pyridines initiate according to different polymerization mechanisms. Careful kinetic studies of trialkylamine-initiated polymerizations of Sar-NCA revealed that this monomer is rather insensitive to tri-*n*-butylamine at room temperature under extremely pure reaction conditions.[56] Addition of a protic coinitiator causes rapid polymerization, because its deprotonation generates a reactive anion (Scheme 4). Sar-NCA seems to be more sensitive against sterically less hindered trialkylamines, such as triethylamine and N-methylpiperidine, possibly because deprotonation of the α-CH$_2$ group takes place (equation 27). Convincing experimental evidence for such an initiation step is lacking, yet in the case of N-(*o*-nitrophenylsulfenyl) NCAs even 1% solutions of triethylamine or N-methyl-morpholine suffice to effect rapid racemization (equation 27).[87]

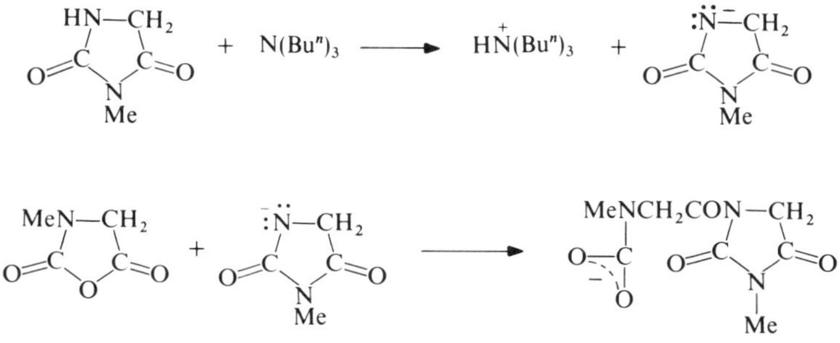

Scheme 4

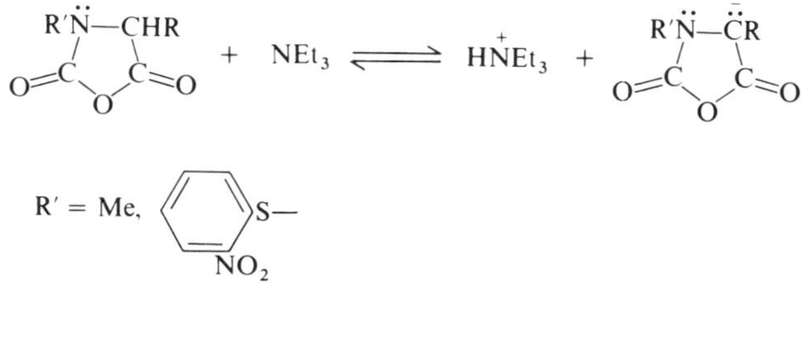

$$(27)$$

Several authors agree[36,67,98] that pyridines cause rapid polymerization of Sar-NCA or other N-alkyl-NCAs regardless of their purity. In this case a nucleophilic attack (equations 16, 17), resulting in the formation of zwitterions, is the most likely initiation step for two reasons. First, pyridines are not basic enough to deprotonate the α carbon. Second, 2,5-dioxopiperazines and higher cyclopeptides are frequent byproducts,[34,98,99] although the mechanism of Scheme 3 cannot be operating. The occurrence of cyclization and decarboxylation of dimeric zwitterions is thus a conceivable explanation (equation 28). The initial formation of zwitterions may entail two different propagation steps, namely the condensation of zwitterions (equation 17) or the carbamate mechanism. Unfortunately, detailed studies of this problem are scarce. When the DP of a pyridine-initiated polymerization of Sar-NCA is plotted *vs.* conversion (Figure 2) high DPs are found even at low conversion, along with an upward curvature at higher conversion.[67] This result fits in with a combination of both mechanisms. The carbamate mechanism predominates in the beginning when the monomer concentration is high (see equation 29), whereas the zwitterion mechanism gains in importance with increasing concentration of zwitterions and decreasing concentration of monomers (equation 30). Figure 2 also indicates that in general pyridines provide higher yields and molecular weights than trialkylamines when N-substituted NCAs are to be polymerized, whereas the opposite is true for N-unsubstituted NCAs.

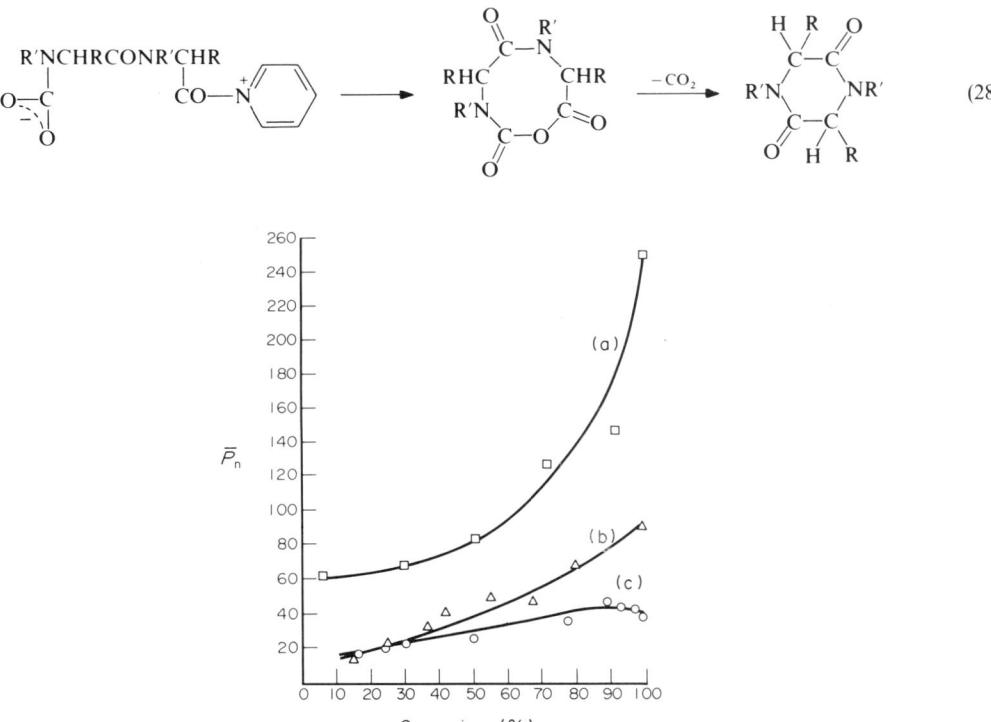

$$(28)$$

Figure 2 Plots of DP *vs.* conversion for tertiary amine-initiated polymerizations of sarcosine NCA: (a) in pyridine; (b) with γ-picoline in DMF; (c) with triethylamine in DMF[67]

$$V_z = k_z[\text{zwitterion}]^2 \qquad (29)$$

$$V_c = k_c[\text{zwitterion}] - [\text{NCA}] \qquad (30)$$

The (co)polymerization of Pro-NCA is of particular interest because poly(L-proline) can adopt two unusual helical conformations (3_1 helix or 10_3 helix) and because proline-containing polypeptides are useful models of collagen and elastin.[100-101]

36.4 INITIATION WITH METAL SALTS AND ORGANOMETALLIC COMPOUNDS

Various metal salts have been successfully used as initiators in dry aprotic solvents, for example lithium chloride, sodium hydride, sodium methoxide, sodium N-benzylcarbamate, sodium acetate and 9-fluorenylpotassium. The N/B ratios of these salts differ greatly, as, for instance, the basicities of hydride or methoxide ions are more than 15 orders of magnitude larger than that of lithium chloride when compared in water. Therefore it is not reasonable to classify all these salts as strong bases. In water, neutral salts such as lithium or sodium chloride, sodium sulfate or sodium phosphate do not have a significant influence on rates of hydrolysis or polymerization of NCAs.[102] Basic salts such as sodium acetate or benzoate have a catalytic effect on the polymerization, and evidence for a mechanism based on the nucleophilic attack of a benzoate ion onto C(5)=O of an NCA have been presented.[102] Despite this catalytic effect, polymerization in water has never yielded high molecular weight polypeptides, and detailed investigation of propagation and termination steps has never been conducted.

When lithium chloride is dissolved in a dry aprotic solvent such as DMF, it reacts as a highly effective initiator, which can produce polypeptides of fairly high molecular weight. Careful kinetic studies with various NCAs and 3-methylhydantoin revealed that the chloride ion is basic enough to deprotonate NCAs.[103] This finding substantially contributed to the first formulation of the AM mechanism. Yet again any convincing evidence that the AM mechanism is operating over the whole course of the polymerization is lacking.

The most widely used salt initiator is sodium methoxide, usually applied as a concentrated solution in methanol. The methoxide/methanol initiator possesses a couple of interesting properties. First, the rate of polymerization is higher than that caused by most other aprotic initiators (*e.g.* tertiary amines), and second, it can produce high molecular weight polypeptides.[50] The mechanism of alcoholate-initiated polymerizations was and still is the object of controversial discussion. One group[64, 65] attempted to differentiate between the nucleophilic attack of methoxide anions (equation 1) and deprotonation (equation 2) by measuring the formation of ^{14}C-labeled methyl ester end groups. From the low degree of radioactivity found in the isolated poly(γ-*O*-benzylglutamate) samples the deprotonation mechanism was derived. However solid sodium methoxide was used so that the initiation only occured at the surface of the solid particles and the poly(γ-*O*-benzylglutamate) was incompletely precipitated with methanol. Therefore these experiments are not conclusive. Another group has found that NCAs can be grafted in partially saponified cellulose acetate containing alcoholate ions on its surface.[104] Two further groups.[84, 105] detected methyl ester end groups by means of ^{1}H NMR spectroscopy in methoxide/methanol-initiated poly(L-alanine) and poly(L-leucine). Nevertheless, the existence of ester end groups does not prove that initiation exclusively occurs by nucleophilic attack (equation 1). Methoxide ions are basic enough to deprotonate NCAs and prototropic reactions are more rapid than any kind of nucleophilic attack. The initially formed oligomers with *N*-acyl NCA end groups (equation 31) can in turn easily react with alcohols or alcoholate ions yielding oligomers with inactive ester groups at one end and a carbamate ion at the other end (equation 31). The ratio of deprotonation and direct nucleophilic attack (equations 1, 2) may depend on the experimental conditions.[106] In any case, the detection of ester end groups is compatible with both initiation mechanisms, and initiation of the AM mechanism is in this case also compatible with a further chain growth *via* carbamate chain ends.

$$\text{(31)}$$

There remains the question of why methoxide/methanol-initiated polymerizations are more rapid than primary or tertiary amine-initiated ones when compared under similar conditions. This difference is possibly not only a result of different polymerization mechanisms but also a result of microphase separation. The chain ends, monomer and methanol can interact *via* H bonds and thus form a phase with a relatively high concentration of NCAs close to the active chain end. Clearcut experimental evidence for this hypothesis is lacking, yet several authors[103, 107, 108] have demonstrated that the NCAs associate with dissolved and precipitated peptide chains.

When 9-fluorenylpotassium is used as initiator,[109] fluorenyl end groups are not detectable by UV spectroscopy in the isolated polypeptide, and initiation obviously results from deprotonation of the NCAs. In contrast, when the less basic nitrophenolate ion is used as initiator, nitrophenolate end groups are clearly detectable.[102] No end group analyses have been conducted with lactam anions used as initiators.[110] A controversial debate exists on the reaction mechanism of sodium hydride and sodium *N*-benzylcarbamate. Sodium hydride is a highly effective initiator[111] and when added to a slow tertiary amine-initiated polymerization causes acceleration.[84] Unfortunately the high efficiency of sodium hydride may be interpreted in terms of a higher concentration of NCA anions or in terms of a stabilization of carbamate ions. Sodium hydride is thus a rather useless tool for the elucidation of the propagation mechanism.

Sodium *N*-benzylcarbamate-initiated polymerizations of γ-*O*-benzylglutamate NCA have been conducted by means of a ^{14}C labeled initiator. In analogy to the sodium methoxide experiment, a low radioactivity is found in the isolated poly(γ-*O*-benzylglutamate),[64] but again the evaluation of this result is affected by incomplete precipitation of the poly(γ-*O*-benzyl-L-glutamate). Recent ^1H NMR spectroscopic analyses[85] of sodium *N*-benzylcarbamate-initiated poly(L-leucine) and poly(L-valine) clearly demonstrate the incorporation of the initiator in the form of benzylamide end groups. If a nearly quantitative incorporation of a carbamate could be confirmed by another group, these results would be a good argument in favor of the carbamate mechanism.

In addition to metal salts, various organometallic compounds have been used as initiators, in particular diethyl- and dibutyl-zinc,[112] triethyl-[113,114] and tributyl-aluminum[112,115] and butyltin alkoxides.[116] Table 2[117] summarizes results obtained with D,L-alanine NCA and numerous organometallic initiators. Mechanistic sudies based on alkylzinc and alkylaluminum have revealed that the first reaction step in all these cases is a deprotonation of the NCA, which is easily detectable by the evolution of ethane or butane (equation 32). IR spectroscopic examination of the reaction products and model reactions of sarcosine NCA with zinc oxazolidinate suggest that the second reaction step is a nucleophilic attack of the metalated NCA onto C(5)=O of another NCA (equation 33). Unfortunately the further course of the propagation has not been satisfactorily elucidated. It has been proposed that transmetalation between carbamate group and NCA regenerates N-metalated NCAs (equation 34) which react with another NCA, but this hypothesis does not agree with the relatively high stability of zinc carbamates. An alternative mechanism, which resembles the AM mechanism, is the reaction of N-metalated NCAs with the *N*-acyl-NCA chain end generated according to equation (33). Such a propagation mechanism has been observed for N-silylated NCAs (equation 35).[118] However, N-silylated NCAs also form an equilibrium with α-isocyanatocarboxylic acid silyl esters and the isomeric isocyanates can in turn polymerize *via* the C=N double bond (equation 36). Because both chain growth processes occur simultaneously, copolymers with biuret and peptide segments are the result. Regardless of the nature of the initiator, organometallic compounds are seemingly not suitable for the synthesis of high molecular weight polypeptides.

$$2 \begin{array}{c} HN-CHR \\ \diagup \quad \diagdown \\ O=C \quad C=O \\ \diagdown O \diagup \end{array} \quad \xrightarrow[-2BuH]{+ZnBu_2} \quad \begin{array}{c} RHC-N \diagdown \quad \diagup N-CHR \\ \diagup \quad \diagdown Zn \diagup \quad \diagdown \\ O=C \quad CO \quad OC \quad C=O \\ \diagdown O \diagup \qquad \diagdown O \diagup \end{array} \qquad (32)$$

Table 2 Polymerization of *N*-carboxy-DL-alanine Anhydride by Various Catalyst Systems[a]

Catalyst	Polymerization time (days)	Yield of polymer[b]		(%) total	sp/C[c] (dl g^{-1})	
		I	*II*		*I*	*II*
LiC$_4$H$_9$	5	40.5	0	40.5	0.051	—
C$_6$H$_5$MgBr	5	61.3	0	61.3	0.086	—
ZnEt$_2$	5	69.7	0	69.7	0.065	—
ZnEt$_2$/MeOH	4	8.3	0	8.3	0.034	—
ZnEt$_2$/2MeOH	5	25.3	0	25.3	0.049	—
ZnEt$_2$/H$_2$O	2	18.5	0	18.5	0.043	—
CdEt$_2$	2	100	0	100	0.073	—
CdEt$_2$/MeOH	2	54.9	0	54.9	0.058	—
CdEt$_2$/2MeOH	5	72.9	0	72.9	0.055	—
CdEt$_2$/H$_2$O	4	60.5	0	60.5	0.021	—
B(Bu^n)$_3$	14	0	0	0	—	—
AlEt$_3$	2	trace	84.2	84.2	—	0.14
AlEt$_3$	2	trace	51.0	51.0	—	0.10
AlEt$_3$/MeOH	2	20.5	14.2	34.7	0.063	0.087
AlEt$_3$/2MeOH	2	18.3	11.6	29.9	0.071	0.085
AlEt$_3$/3MeOH	5	23.6	10.3	33.9	0.067	0.092
AlEt$_3$/H$_2$O	2	46.5	5.8	52.3	0.037	0.052
MeOH	13	70.6	0	70.6	0.13	—
H$_2$O	7	51.2	0	51.2	0.21	—
(+)-Borneol	5	38.4	10.6	49.0	0.15	0.20

[a] *N*-Carboxy-DL-alanine anhydride 0.035 mol (4 g); solvent, tetrahydrofuran 15 ml; catalyst, 5 mol % for monomer; reaction under nitrogen atmosphere at 30 °C.
[b] I: water soluble part; II: water insoluble part.
[c] Reduced viscosity at 30 °C of 0.5% of solution in dichloroacetic acid.

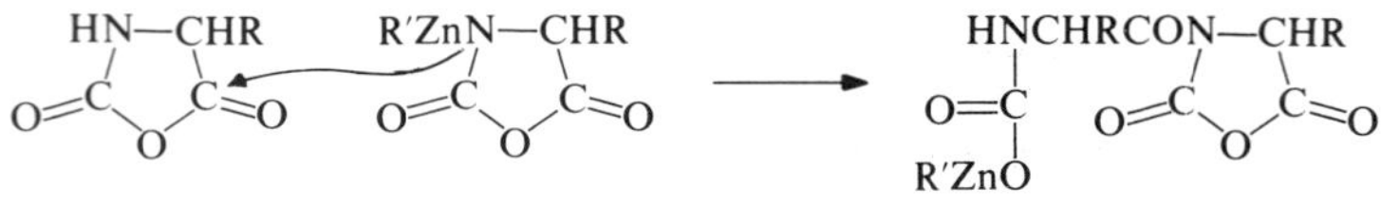

$$\text{(33)}$$

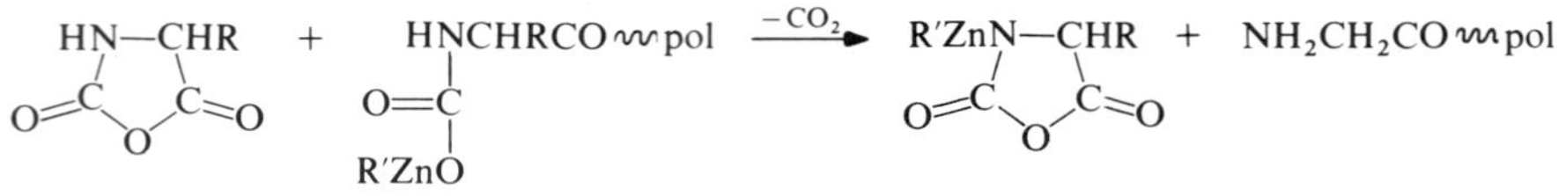

$$\text{(34)}$$

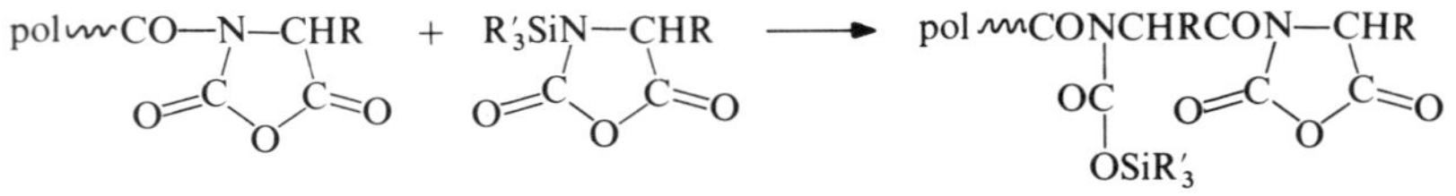

$$\text{(35)}$$

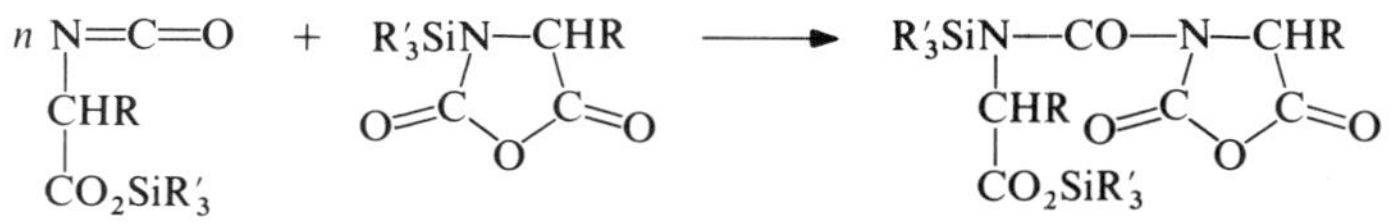

$$\text{(36)}$$

36.5 CHAIN EFFECTS AND CRYSTAL GROWTH

The term chain effect was coined[119] when an unexpected acceleration was found for poly-(sarcosine)-initiated polymerizations of D,L-Phe-NCA in less polar solvents. Whereas the reactivity of sarcosine N,N-dimethylamide is comparable to that of other secondary amines with a similar N/B ratio, increasing chain length of oligo(sarcosine)dimethylamides enhances the rate of polymerization until a maximum is reached at DP = 7–8. Acetylation of the oligo(sarcosine)dimethylamides reduces their activity to zero. From these observations the following polymerization mechanism has been derived. The polymeric initiator attacks C(5)=O of the monomer (equation 37) and the chain growth continues according to the amine mechanism so that finally a two-block copolypeptide is formed.[119,120] The acceleration results from association of monomers with the backbone of the initiator, which increases the concentration of NCAs in the neighborhood of the active chain end. NCAs associated with the seventh and eighth monomer unit of the initiator seems to be in a favorable position for a cyclic transition state as illustrated in Figure 3.[121,122]

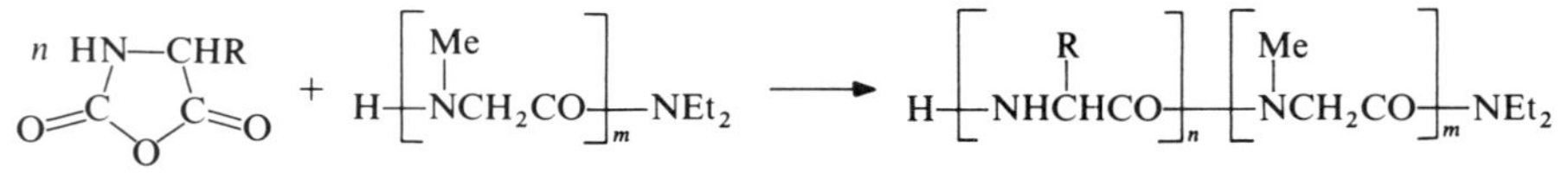

$$\text{(37)}$$

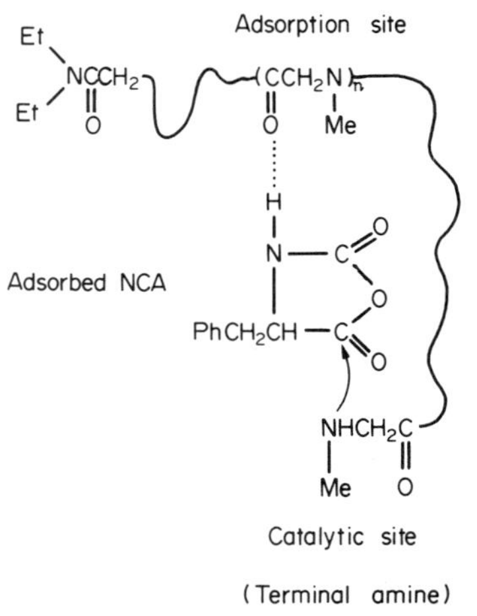

Figure 3 Schematic reaction mechanism of the polymerization of D,L-phenylalanine NCA initiated by poly(sarcosine) dimethylamide[121]

This mechanistic scheme is supported by the following observations. IR spectroscopic studies confirm that poly(sarcosine) acts as an H-bond acceptor for NCAs or phenols in less polar solvents such as dioxane or chlorinated hydrocarbons. No acceleration is detectable when polymerizations are conducted in DMF, which can itself act as an H-bond acceptor. The efficiency of other poly-(N-alkylamino acid)amides as initiators decreases with increasing bulkiness of their substituents.[121]

Another kind of chain effect, yet again an acceleration of the polymerization, is found when NCAs are polymerized by means of primary aliphatic amines under conditions that lead to precipitation or at least strong aggregation of the growing peptide chains. During the first stage of the polymerization resulting in the formation of oligopeptides, a relatively low rate of polymerization is observed, whereas the second stage is characterized by a higher rate of polymerization.[107,122-127] A typical example of the so-called two-stage kinetics is shown in Figure 4. Several authors who have investigated the two-stage kinetics have also investigated the secondary structure of their reaction products. When NCAs of helicogenic amino acids are used, oligopeptides forming crystalline β-sheets are found in the first stage, whereas α-helical polypeptides seem to be characteristic for the second stage.[5,6,51,79,123] Therefore the most common interpretation relates the faster propagation with a so-called helical chain growth, and the acceleration is attributed to the β-sheet α-helix transition.[124,127]

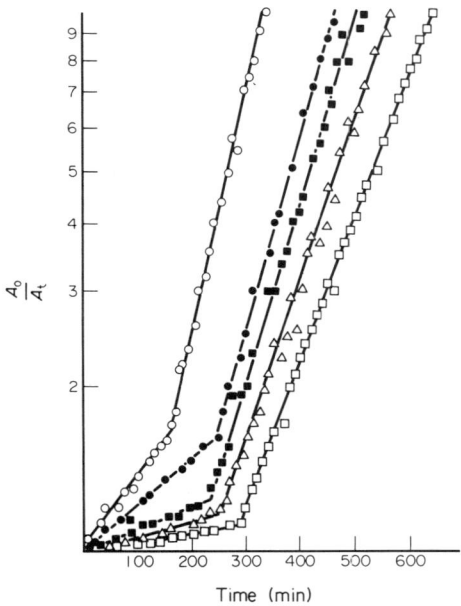

Figure 4 First order plot for *n*-hexylamine-initiated copolymerizations of γ-O-benzyl-D- and L-glutamate NCA in dioxane at 25 °C[107] (M/I = 22): ○ = 100% L; ● = 95% L; ■ = 85% L; △ = 75% L; □ = 50% L

Unfortunately, this interpretation, which might seem obvious at first glance, implies two misunderstandings. First, a β-sheet → α-helix transition does not occur in such a way that peptide chains change their conformation after they adopt the β-sheet structure. The β-sheets, once they are formed in aprotic solvents, are stable over the whole course of the polymerization. The α-helices which seem to predominate at higher conversions either grow on the surface of β-sheet lamellae or (more probably, see below) result from the initially dissolved fraction of oligopeptides. At this point it is worth noting that a minimum DP of 10–12 (depending on the nature of the amino acids) is required before the α-helical conformation is sufficiently stable. Second, two independent groups[123,128,129] have reported that even non-α-helix-forming NCAs such as Gly-NCA, Pro-NCA and Sar-NCA display two-stage kinetics in reaction media where aggregation and/or precipitation occur. Furthermore, the acceleration is barely detectable when γ-O-benzyl-L-Glu-NCA is polymerized in DMF which prevents association of NCAs and peptide chains.[78,125] Therefore it is obvious that the origin of the two-stage kinetics is not the helical chain growth, but phase separation owing to aggregation of the growing oligopeptides. The concentration of NCAs increases in the peptide phase due to association *via* H bonds. The appearance of growing α-helices may enhance the acceleration, because active chain ends of α-helices are sterically less hindered than those of β-sheet lamellae.

As already demonstrated by these chain effects, it is characteristic for polymerizations of NCAs, in contrast to those of most other cyclic monomers, that the chemical aspects of the polymerization mechanisms are closely interrelated with several physical aspects. The physical aspects are association equilibria between growing peptide chains, between monomers or between monomers and polymers, and also crystallization equilibria and conformational changes. Because most oligopeptides or polypeptides are almost insoluble in the common reaction media, their solubility and secondary structure are of particular importance not only for the kinetic course of the polymerization but also for the molecular weight distribution and for the crystal growth. Molecular weight distributions are determined in two ways: either dissolved polypeptides are subjected to chromatographic analyses,[80, 81, 130, 131] or relatively soluble fractions are extracted from less soluble polypeptides.[5, 6, 51, 78, 79] All these analyses agree in that at least primary amine-initiated polymerizations lead to bimodal molecular weight distributions with a first minimum at DP values of 6–8. Regardless of the nature of the amino acid, oligopeptides of DP ≤ 10 adopt the β-sheet structure, and thus the reactivity of chain ends on the surface of β-sheet lamellae is an important factor for the understanding of the polymerization process.

The initial formation of crystalline β-sheets by association and precipitation of the growing oligopeptides has been reported by several groups.[5, 51, 78, 79] The main problem was to find out what happens to these β-sheet lamellae and their active end groups in the further course of the polymerizations. A systematic comparison of helicogenic and nonhelicogenic NCAs under identical reaction conditions[132, 133] has revealed that polymerizations of nonhelicogenic NCAs rapidly slow down and stop far below 100% conversion, although unreacted NCAs are present in the reaction mixture. From this and other spectroscopic observation, it has been concluded that a 'physical death' is responsible for the incomplete conversions. The chains growing on the surface of antiparallel β-sheets need to fold-back to form H bonds and thus hinder the further chain growth for steric reasons (Figure 5a).[132, 133]

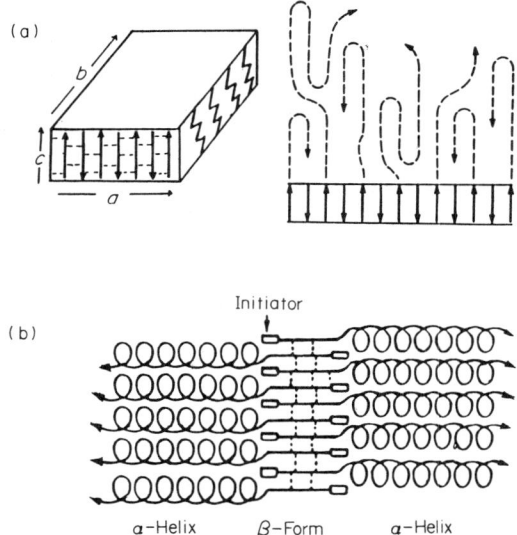

Figure 5   Models of the chain growth on the surface of antiparallel β-sheets; (a) chain growth of nonhelicogenic NCAs leading to 'physical death'; (b) chain growth of helicogenic NCAs[132–135]

In the case of α-helix-forming peptides two crystal growth models have been proposed. The older model postulates that the α-helices grow from the surface of the initially formed β-sheet lamellae (Figure 5b).[133–135] This model is not compatible with the fact that most oligopeptides can be separated from the helical high molecular weight fraction either by extraction[5, 6, 51, 78, 79] or by chromatographic methods.[80, 81, 130, 131] The second model assumes that the chain growth of the β-sheets slows down due to the physical death and the α-helices grow from the soluble fraction of oligopeptides when their DP has reached the order of 10–12. The faster chain growth of the α-helices enables complete conversions and high molecular weights. This model of crystal growth (Figure 6) is also compatible with recent ^{13}C NMR cross polarization/magic angle spinning (CP/MAS) studies of various polypeptides.[5, 6, 136] The two crystal growth models of Figures 5b and 6 are certainly simplifications; for a more detailed discussion see ref. 2.

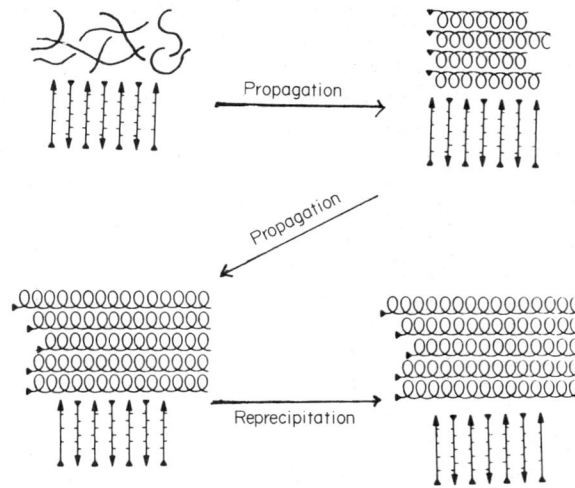

Figure 6 Model of crystal growth and effect of reprecipitation in the case of α-helix-forming NCAs[5]

36.6 STEREOSPECIFICITY AND TACTICITY

The polymerization of a racemic D,L-amino acid NCA may involve four stereochemically different growing steps (equations 38–41). When all four propagation steps occur at identical rates an atactic sequence is the result. When the polymerization is stereospecific so that the crossover steps (equations 40 and 41) do not occur, isotactic blocks are formed, whereas a syndiotactic sequence is obtained when the homopropagation (equations 38 and 39) does not take place. In the absence of stereospecificity the sum of all four propagation rates ($V_{LL} + V_{DD} + V_{LD} + V_{DL}$) is the same as the rate of polymerization of a pure D- or L-NCA when conducted under identical conditions. Therefore a high degree of stereospecificity (either $V_{LL} = V_{DD} = 0$ or $V_{LD} = V_{DL} = 0$) must reduce the rate of polymerization.

$$V_{LL} = k_{LL}[P^*_L][\text{L-NCA}] \tag{38}$$

$$V_{DD} = k_{DD}[P^*_D][\text{D-NCA}] \tag{39}$$

$$V_{LD} = k_{LD}[P^*_L][\text{D-NCA}] \tag{40}$$

$$V_{DL} = k_{DL}[P^*_D][\text{L-NCA}] \tag{41}$$

P^*_L: polypeptide with L unit as active chain end

P^*_D: polypeptide with D unit as active chain end

Several authors[78,107,126,127,137] have conducted polymerizations of D,L-NCAs under a variety of conditions and have indeed found significantly reduced rates of polymerization (Figure 4). For such kinetic studies, based on CO_2 measurements, a high degree of stereospecificity is derived, in particular for primary amine-initiated polymerizations. Furthermore, D,L-NCAs with varying D/L ratios have been polymerized by means of optically active amines or amino acid esters and the optical rotations of the resulting polypeptides and/or of the unreacted NCAs have been measured.[78,125,138−140] These measurements indicate a certain degree of enantioselectivity for primary amine-initiated polymerizations of helicogenic NCAs so that L-chain ends favor the incorporation of L-NCAs ($k_{LL} > k_{LD}$ and $k_{DD} > k_{DL}$). These results, along with a sophisticated enzymatic degradation of poly(D,L-leucine)[141,142] seemingly agree in that primary amine-initiated polymerizations of D,L-NCA normally involve a high degree of isotactic stereospecificity due to the helical chain growth.[123] However, the rate of CO_2 evolution depends on various parameters including association equilibria between two NCAs and between peptides and NCAs. These equilibria are different for D/L associations and D/D or L/L associations.[106−108,137] The ambiguity of the kinetic approach is best illustrated by the fact that two groups using kinetic and optical rotation measurements have come to almost opposite conclusions with regard to the stereospecificity.[78,123,137,140,143−147] Furthermore, it is worth noting that optical rotation measurements do not enable a quantification of the D/L ratio of isolated polypeptides, because the molar rotation of an L unit flanked by other L units is different from that of an L unit flanked by D units.[147]

Moreover, more or less stereospecific polymerizations are also found for nonhelicogenic NCAs.[148,151]

Several research groups have attempted to characterize stereosequences by means of IR spectroscopy and X-ray diffraction.[126,144,152-154] When such spectroscopic studies are combined with other methods, for instance measurements of solubilities or optical rotation, reasonable results are obtained. These results indicate that polymerizations of D,L-Ala-NCA initiated by protic nucleophiles involve a low degree of stereospecificity and yield nearly atactic sequences. Nonetheless, X-ray and IR spectroscopic studies alone do not allow a reliable characterization of stereosequences, because they mainly reflect the secondary structure of the polypeptides and not their tacticity. In contrast to polymers of alkenes and vinyl monomers, no direct relationship exists between tacticity and secondary structure or crystallinity of polypeptides. Due to H bonds and dipolar forces even atactic polypeptides prefer the β-sheet structure or α-helical conformation to an amorphous state. Therefore the detection of a high α-helical content may not serve as a proof of the presence of long isotactic blocks.[155,156]

Nevertheless, it is obvious that atactic sequences reduce the stability of both β-sheets and α-helices, and thus atactic poly(D,L-amino acid)s show a considerably better solubility in organic solvents and poorer mechanical properties than the corresponding poly(L-amino acid)s. Systematic comparisons of the solubilities of poly(L-amino acid)s and poly(D,L-amino acid)s clearly demonstrate that primary amine-initiated poly(D,L-amino acid)s possess a good solubility in various organic solvents, in agreement with a nearly atactic stereosequence.[151,157] Tacticity and properties of poly(D,L-amino acid)s are also of technical interest, because D,L-amino acids are less expensive than L-amino acids, and poly(D,L-amino acid)s with long isotactic blocks and good mechanical properties would be useful as synthetic textile fibres. However, comprehensive studies of this aspect have revealed[30] that poly(D,L-amino acid)s, in contrast to poly(L-amino acid)s, are not useful for fibre making, because of insufficient molecular weights, poor mechanical properties and good solubilities in organic solvents. These unfavorable properties are obviously a consequence of more or less atactic stereosequences.

The only spectroscopic method which enables direct characterization of stereosequences is NMR spectroscopy of dissolved poly(D,L-amino acid)s. Unfortunately, ^{1}H NMR spectroscopy is not useful, because the broad signals do not display tacticity splittings. The application of ^{13}C NMR spectroscopy is limited to amino acids with branched side chains, such as D,L-leucine, D,L-isoleucine and D,L-valine.[151-160] Only ^{15}N NMR spectroscopy can be applied to a broader variety of poly (D,L-aminoacid)s. Clearcut assignments of syndiotactic and isotactic triad peaks have been obtained with ^{15}N enriched guest–host copolypeptides.[159] Under suitable conditions quantification of signal intensities allows the calculation of average lengths of isotactic blocks.

The results obtained from ^{13}C and ^{15}N NMR spectroscopy agree with each other and also agree with the solubilities of the poly(D,L-amino acid)s under investigation and fit in with the kinetic and IR spectroscopic measurements of four groups.[143-146,148-150,152-155] These results may be summarized as follows: (i) primary amine initiated polymerizations of D,L-NCAs yield nearly atactic

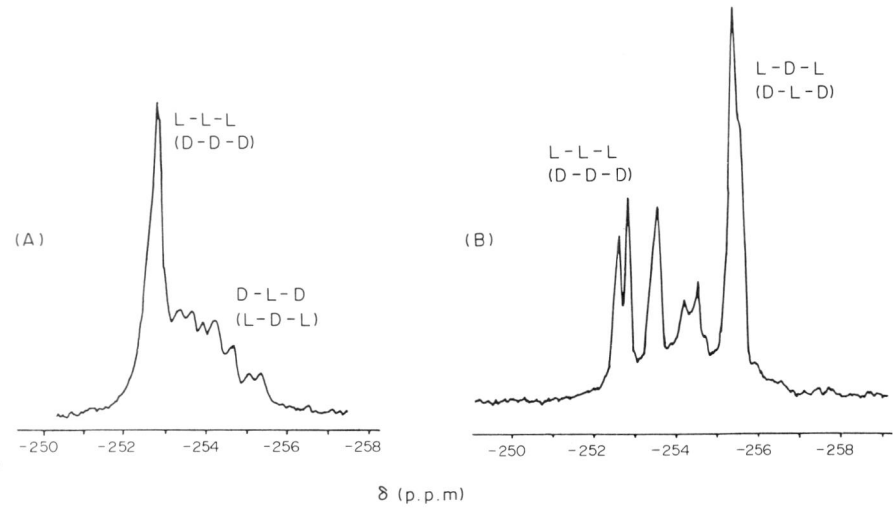

Figure 7   40.5 MHz ^{15}N NMR spectra of poly(γ-*O*-Me-L-glutamate) measured in trifluoroacetic acid; (A) initiation with triethylamine and *N*-acetyl-Gly-NCA in DMF; (B) polymerization in pure pyridine[159]

sequences. Decreasing temperatures favor the formation of isotactic dyads; (ii) reaction conditions favoring the AM mechanism also favor isotactic sequences when helicogenic NCAs are polymerized (Figure 7a). Yet they favor syndiotactic sequences when applied to nonhelocogenic NCAs; and (iii) when helicogenic NCAs are polymerized in pyridine or mixtures of pyridine and benzene, syndiotactic dyads are preferentially formed (Figure 7b).

Finally, it is worth noting that a catalyst system which enables the production of highly isotactic polypeptides from racemic D,L-NCAs has not yet been found.

36.7 REFERENCES

1. H. Leuchs, *Ber. Dtsch. Chem. Ges.*, 1906, **39**, 857.
2. H. R. Kricheldorf 'α-Amino Acid N-Carboxyanhydrides and Related Heterocycles', Springer Verlag, Berlin, 1987.
3. L. Reibel and G. Spach, *Bull. Soc. Chim. Fr.*, 1972, 1025.
4. L. Reibel, G. Spach and C. Dufour, *Biopolymers*, 1973, **12**, 2391.
5. H. R. Kricheldorf, M. Mutter, F. Maser, D. Müller and H. Förster, *Biopolymers*, 1983, **22**, 1357.
6. H. R. Kricheldorf and D. Müller, *Int. J. Biol. Macromol.*, 1983, **5**, 171.
7. Y. Yamashita, Y. Iwaya and K. Ito, *Makromol. Chem.*, 1975, **176**, 1207.
8. J. P. Billot, A. Douy and B. Gallot, *Makromol. Chem.*, 1976, **177**, 1889.
9. J. P. Billot, A. Douy and B. Gallot, *Makromol. Chem.*, 1977, **178**, 1641.
10. Y. Imanishi, M. Tanaka and C. H. Bamford, *Int. J. Biol. Macromol.*, 1985, **7**, 89.
11. M. Tanaka, A. Mori, Y. Imanishi and C. H. Bamford, *Int. J. Biol. Macromol.*, 1985, **7**, 173.
12. B. Perly, A. Douy and B. Gallot, *Makromol. Chem.*, 1976, **177**, 2569.
13. A. Nakajima, K. Kugo and T. Hayashi, *Polym. J.*, 1979, **11**, 995.
14. G.-W. Chen, T. Hayashi and A. Nakajima, *Polym. J.*, 1981, **13**, 433.
15. K. Kugo, M. Murashima, T. Hayashi and A. Nakajima, *Polym. J.*, 1983, **15**, 267.
16. Y. Imanishi, T. Kimura and T. Higashimura, *Polymer*, 1981, **22**, 1407.
17. Y. Avny and A. Zilkha, *Eur. Polym. J.*, 1966, **2**, 355, 367.
18. Y. Avny and A. Zilkha, *J. Appl. Polym. Sci.*, 1965, **9**, 3503.
19. M. A. Stahmann and R. Becker, *J. Am. Chem. Soc.*, 1952, **74**, 2695.
20. M. A. Stahmann in 'Polyamino Acids, Polypeptides and Proteins', ed. M. A. Stahmann, University of Wisconsin Press, Madison, 1962, p. 329 (and references therein).
21. H. Fraenkel-Contrat, *Biochem. Biophys. Acta* 1953, **10**, 180.
22. R. Becker, in 'Polyamino Acids, Polypeptides and Proteins', ed. M. A. Stahmann, University of Wisconsin Press, Madison, 1962, p. 301 (and references therein).
23. E. Katchalski, in 'Polyamino Acids, Polypeptides and Proteins', ed. M. A. Stahmann, University of Wisconsin Press, Madison, 1962, p. 283 (and references therein).
24. M. Sela and R. Arnon, *Biochem. J.*, 1960, **75**, 91.
25. R. Arnon and G. E. Perlmann, *J. Biol. Chem.*, 1963, **233**, 963.
26. J. P. Cooke, C. B. Anfirnsen and M. Sela, *J. Biol. Chem.*, 1963, **238**, 2034.
27. R. G. Denkewalter, H. Schwam, R. G. Strachan, T. E. Beesley, D. F. Veber, E. F. Schoenewaldt, A. Barkmeyer, W. J. Paleveda, Jr., T. A. Jacobs and R. Hirschmann, *J. Am. Chem. Soc.*, 1966, **88**, 3163.
28. R. G. Denkewalter, D. F. Veber, F. W. Holly and R. Hirschmann, *J. Am. Chem. Soc.*, 1969, **91**, 502.
29. F. A. Bovey, *J. Polym. Sci., Macromol. Rev.*, 1974, **9**, 1.
30. J. Noguchi, S. Tokura and N. Nishi, *Angew. Makromol. Chem.*, 1972, **22**, 107.
31. F. Wessely, *Z. Physiol. Chem.*, 1925, **146**, 72.
32. F. Wessely and F. Sigmund, *Z. Physiol. Chem.*, 1926, **159**, 102.
33. F. Wessely and M. John, *Z. Physiol. Chem.*, 1927, **170**, 38.
34. F. Wessely, K. Riedl and K. Tuppy, *Monatsh. Chem.*, 1950, **81**, 86d.
35. P. D. Bartlett and R. H. Jones, *J. Am. Chem. Soc.*, 1957, **79**, 2153.
36. P. D. Bartlett and D. C. Dittmer, *J. Am. Chem. Soc.*, 1957, **79**, 2159.
37. C. J. Brown, D. Coleman and A. C. Farthing, *Nature (London)*, 1949, **163**, 834.
38. A. C. Farthing, *J. Chem. Soc.*, 1950, 3222.
39. M. Sela and A. Berger, *J. Am. Chem. Soc.*, 1955, **77**, 1893.
40. L. A. Sluytermann and B. Laruyere, *Recl. Trav. Chim. (Pays Bas)*, 1954, **73**, 347.
41. A. B. Meggy, *J. Chem. Soc.*, 1956, 1444.
42. R. B. Woodward and C. H. Schramm, *J. Am. Chem. Soc.*, 1947, **69**, 1552.
43. J. W. Breitenbach and F. Richter, *Makromol. Chem.*, 1950, **4**, 262.
44. J. W. Breitenbach and K. Allinger and A. Koref, *Monatsh. Chem.*, 1955, **86**, 269, 278.
45. E. Katchalski and P. Spitnik, *J. Am. Chem. Soc.*, 1951, **73**, 2946.
46. M. Bergmann, L. Zervas and M. F. Ross, *J. Biol. Chem.*, 1935, **111**, 245.
47. D. T. Gish and F. H. Carpenter, *J. Am. Chem. Soc.*, 1953, **75**, 5872.
48. A. Patchornik, M. Sela and E. Katchalski, *J. Am. Chem. Soc.*, 1954, **76**, 299.
49. C. H. Bamford and H. Block, *J. Chem. Soc.*, 1961, 4989, 4992.
50. E. R. Blout and R. H. Karlson, *J. Am. Chem. Soc.*, 1956, **78**, 941.
51. M. Idelson and E. R. Blout, *J. Am. Chem. Soc.*, 1957, **79**, 3948.
52. K. D. Kopple, *J. Am. Chem. Soc.*, 1957, **79**, 662.
53. H. R. Kricheldorf, *Makromol. Chem.*, 1977, **178**, 905.
54. K. D. Kopple, *J. Am. Chem. Soc.*, 1957, **79**, 6442.
55. D. G. H. Ballard and C. H. Bamford, *Proc. R. Soc. London., Ser. A*, 1954, **223**, 495.
56. C. H. Bamford, H. Block and A. C. P. Pugh, *J. Chem. Soc.*, 1961, 2057.

57. D. Thunig, J. Semen and H.-G. Elias, *Makromol. Chem.*, 1977, **178**, 603.
58. M. El Sabbah and H.-G. Elias, *Makromol. Chem.*, 1981, **182**, 1617.
59. J. L. Bailey, *Nature (London)*, 1949, **164**, 889.
60. J. L. Bailey, *J. Chem. Soc.*, 1950, 3461.
61. E. Katchalski, I. Grossfeld and M. Frankel, *J. Am. Chem. Soc.*, 1948, **70**, 2094.
62. G. Schramm and H. Restle, *Makromol. Chem.*, 1954, **13**, 103.
63. T. Makino, S. Inoue and T. Tsuruta, *Makromol. Chem.*, 1971, **150**, 137.
64. M. Goodman and J. Hutchison, *J. Am. Chem. Soc.*, 1966, **88**, 3627.
65. M. Goodman and E. Peggion, *Vysokomol. Soedin, Ser. A*, 1967, **9**, 247.
66. E. Peggion, M. Terbojevich, A. Cosani and C. Colombini, *J. Am. Chem. Soc.*, 1966, **88**, 3630.
67. H. R. Kricheldorf and K. Bösinger, *Makromol. Chem.*, 1976, **177**, 1243.
68. H. R. Kricheldorf, D. Müller and W. E. Hull, *Biopolymers*, 1985, **24**, 2113.
69. H. R. Kricheldorf, W. E. Hull and D. Müller, *Macromolecules*, 1985, **18**, 2135.
70. H. R. Kricheldorf, D. Müller and W. E. Hull, *Int. J. Biol. Macromol.*, 1986, **8**, 20.
71. H. R. Kricheldorf, *Makromol. Chem.*, 1977, **178**, 1959.
72. H. R. Kricheldorf, *J. Polym. Sci., Polym. Chem. Ed.*, 1979, **17**, 97.
73. V. Gold and E. G. Jefferson, *J. Chem. Soc.*, 1953, 1409, 1416.
74. T. H. Wieland, *Angew. Chem.*, 1951, **63**, 7.
75. E. Katchalski and M. Sela, *Adv. Protein Chem.*, 1957, **23**, 243.
76. M. Szwarc, *Adv. Polym. Sci.*, 1965, **4**, 1.
77. Y. Shalitin, in 'Ring-opening Polymerization', ed. K. C. Frisch and S. Reegen, Dekker, New York, 1969, p. 471.
78. R. D. Lundberg and P. Doty, *J. Am. Chem. Soc.*, 1957, **79**, 3961.
79. P. Doty, J. H. Bradbury and A. M. Holtzer, *J. Am. Chem. Soc.*, 1956, **78**, 947.
80. M. Rinaudo and A. Domard, *Biopolymers*, 1976, **15**, 2185.
81. E. Peggion, E. Scofone, A. Cosani and A. Portolan, *Biopolymers*, 1966, **4**, 695.
82. J. B. Milstein and A. Ferreti, *Biopolymers*, 1973, **12**, 2335.
83. M. Idelson and E. R. Blout, *J. Am. Chem. Soc.*, 1958, **80**, 2387.
84. C. E. Seeney and J. Harwood, *J. Macromol. Sci., Chem.*, 1975, **179**, 779.
85. D. Giannakidis and H. J. Harwood, *J. Polym. Sci., Polym. Lett. Ed.*, 1978, **16**, 491.
86. H. R. Kricheldorf, *Angew. Chem.*, 1973, **85**; *Angew. Chem., Int. Ed. Engl.*, 1973, **12**, 73.
87. H. R. Kricheldorf and M. Fehrle, *Chem. Ber.*, 1974, **107**, 3533.
88. H. R. Kricheldorf and G. Greber, *Chem. Ber.*, 1971, **104**, 3131.
89. N. S. Choi and M. Goodman, *Biopolymers*, 1972, **11**, 67.
90. Y. Hashimoto and Y. Imanishi, *Biopolymers*, 1980, **19**, 655.
91. A. Caillet, D. Bauer, G. Froyer and H. Sekiguchi, *C.R. Hebd. Seances Acad. Sci., Ser. C*, 1973, **277**, 1211.
92. M. Amouyal, B. Coutin and H. Sekiguchi, *J. Macromol. Sci., Chem.*, 1986, **23**, 451.
93. M. Sekiguchi and G. Froyer, *C.R. Hebd. Seances Acad. Sci., Ser. C*, 1974, **279**, 623.
94. D. G. H. Ballard, C. H. Bamford and F. J. Weymouth, *Proc. R. Soc. London., Ser. A*, 1955, **227**, 155.
95. C. H. Bamford and F. J. Weymouth, *J. Am. Chem. Soc.*, 1955, **77**, 6368.
96. L. Bilek, J. Derkosch, A. Michl and F. Wessely, *Monatsh. Chem.*, 1953, **84**, 717.
97. M. Rothe and D. Mühlhausen, *Angew. Chem.*, 1976, **88**, 338; *Angew. Chem., Int. Ed. Engl.*, 1976, **15**, 307.
98. F. Sigmund and F. Wessely, *Z. Physiol. Chem.*, 1926, **157**, 91.
99. H. R. Kricheldorf, *Makromol. Chem.*, 1974, **175**, 3325.
100. L. Mandelkern, in 'Poly-α-Amino Acids', ed. G. D. Fasman, Dekker, New York, 1967, p. 675.
101. H. R. Kricheldorf and D. Müller, *Int. J. Biol. Macromol.*, 1984, **6**, 145.
102. T. K. Miwa and M. A. Stahmann, in 'Polyamino Acids, Polypeptides and Proteins', ed. M. A. Stahmann, University of Wisconsin Press, Madison, 1962, p. 81.
103. D. G. H. Ballard, C. H. Bamford and A. Elliot, *Makromol. Chem.*, 1960, **35**, (Suppl. 2), 222.
104. Y. Anvy, S. Migdal and A. Zilkha, *Eur. Polym. J.*, 1966, **2**, 355.
105. H. R. Kricheldorf, *Makromol. Chem., Macromol. Symp.*, 1987, **6**, 165.
106. M. Amouyal, B. Coutin and H. Sekiguchi, *J. Macromol. Sci., Chem.*, 1983, **20**, 675.
107. F. D. Williams, M. Eshaque and R. D. Brown, *Biopolymers*, 1971, **10**, 753.
108. N. Oguni, H. Kuboyama and A. Nakamura, *Makromol. Chem.*, 1983, **31**, 1559.
109. M. Goodman and U. Arnon, *J. Am. Chem. Soc.*, 1964, **86**, 3384.
110. H. Sekiguchi and J. F. Doussin, *C.R. Hebd. Seances Acad. Sci., Ser. C*, 1975, **281**, 433.
111. M. Goodman, E. Peggion, M. Szwarc and C. H. Bamford, *Macromolecules*, 1977, **10**, 1299.
112. K. Matsuura, S. Inoue and T. Tsuruta, *Makromol. Chem.*, 1967, **103**, 140.
113. T. Tsuruta, K. Matsuura and S. Inoue, *Makromol. Chem.*, 1965, **83**, 289.
114. T. Makino, S. Inoue and T. Tsuruta, *Makromol. Chem.*, 1965, **83**, 316.
115. T. Makino, S. Inoue and T. Tsuruta, *Makromol. Chem.*, 1971, **150**, 137.
116. S. Freireich, D. Gertner and A. Zilkha, *Eur. Polym. J.*, 1974, **10**, 439.
117. K. Matsuura, S. Inoue and T. Tsuruta, *Makromol. Chem.*, 1964, **80**, 149.
118. H. R. Kricheldorf and G. Greber, *Chem. Ber.*, 1971, **104**, 3131.
119. D. G. H. Ballard and C. H. Bamford, *Proc. R. Soc. London, Ser. A*, 1956, **236**, 384.
120. C. H. Bamford, H. Block and Y. Imanishi, *Biopolymers*, 1960, **4**, 1067.
121. Y. Imanishi, T. Sugihara and T. Higashimura, *Biopolymers*, 1973, **12**, 1505.
122. Y. Imanishi, *Pure Appl. Chem.*, 1981, **53**, 715.
123. H. Weingarten, *J. Am. Chem. Soc.*, 1958, **80**, 352.
124. R. E. Nylund and W. G. Miller, *Biopolymers*, 1969, **2**, 131.
125. S. Inoue, K. Matsuura and T. Tsuruta, *J. Polym. Sci., Part C*, 1968, **23**, 721.
126. Y. Iwakura, K. Uno and M. Oya, *J. Polym. Sci., Part A*, 1968, **15**, 2165.
127. H.-G. Elias and M. M. El-Sabbah, *Makromol. Chem.*, 1981, **182**, 1629.
128. D. G. H. Ballard and C. H. Bamford, *J. Chem. Soc.*, 1959, 1039.

129. D. G. H. Ballard, C. H. Bamford and A. Elliot, *Makromol. Chem.*, 1960, **35** (Suppl. 2), 222.
130. J. N. Stewart and M. A. Stahmann, in 'Polyamino Acids, Polypeptides and Proteins', ed. M. A. Stahmann, University of Wisconsin Press, Madison, 1962, p. 95.
131. E. Scoffone, E. Peggion, A. Cosani and M. Terbojevich, *Biopolymers*, 1965, **3**, 535.
132. T. Komoto, K. Y. Kim, Y. Minoshima, M. Oya and T. Kawai, *Makromol. Chem.*, 1973, **168**, 261.
133. T. Komoto, K. Y. Kim, M. Oya and T. Kawai, *Makromol. Chem.*, 1975, **174** (Suppl. 1), 283, 301.
134. T. Komoto and T. Kawai, *Makromol. Chem.*, 1973, **172**, 221.
135. A. Fujie, T. Komoto, M. Oya and T. Kawai, *Makromol. Chem.*, 1973, **169**, 301.
136. H. R. Kricheldorf, D. Müller and J. Stulz, *Makromol. Chem.*, 1983, **184**, 1407.
137. F. D. Williams and R. D. Brown, *Makromol. Chem.*, 1973, **169**, 191.
138. K. Matsuura, S. Inoue and T. Tsuruta, *Makromol. Chem.*, 1964, **80**, 149.
139. T. Tsuruta, S. Inoue and K. Matsuura, *Biopolymers*, 1967, **5**, 343.
140. H. G. Bührer and H.-G. Elias, *Makromol. Chem.*, 1973, **169**, 145.
141. J. Semen and H.-G. Elias, *Makromol. Chem.*, 1978, **179**, 463.
142. F. G. Fick, J. Semen and H.-G. Elias, *Makromol. Chem.*, 1978, **179**, 579.
143. Y. Hashimoto, A. Aoyama, Y. Imanishi and T. Higashimura, *Biopolymers*, 1976, **15**, 2407.
144. Y. Imanishi, A. Aoyama, Y. Hashimoto and T. Higashimura, *Biopolymers*, 1977, **16**, 187.
145. Y. Hashimoto, Y. Imanishi and T. Higashimura, *Biopolymers*, 1978, **17**, 2561.
146. H. Onishi, Y. Hashimoto and Y. Imanishi, *Int. J. Biol. Macromol.*, 1981, **3**, 327.
147. J. Schlechter and A. Berger, *Biochemistry*, 1966, **5**, 3362.
148. T. Akaike, Y. Aogaki and S. Inoue, *Biopolymers*, 1975, **14**, 2577.
149. F. D. Williams and R. D. Brown, *Biopolymers*, 1973, **12**, 647.
150. N. F. Blair and W. A. Bonner, *Origins of Life*, 1980, **10**, 255.
151. H. R. Kricheldorf and T. Mang, *Makromol. Chem.*, 1981, **182**, 3077.
152. K. Matsuura, S. Inoue and T. Tsuruta, *Makromol. Chem.*, 1964, **80**, 149.
153. K. Itoh, T. Nakahara, T. Shimanouchi, M. Oya, K. Uno and Y. Iwakura, *Biopolymers*, 1968, **6**, 1759.
154. K. Itoh and T. Shimonouchi, *Biopolymers*, 1970, **9**, 383.
155. M. Tsuboi, I. Mitsui, A. Wada, T. Mazawa and N. Nagashima, *Biopolymers*, 1963, **1**, 297.
156. M. Amouyal-Ruderman and H. Sekiguchi, *Polym. Bull.*, 1981, **6**, 69.
157. H. R. Kricheldorf and W. E. Hull, *Biopolymers*, 1983, **22**, 1635.
158. W. E. Hull and H. R. Kricheldorf, *J. Polym. Sci., Polym. Lett. Ed.*, 1978, **16**, 215.
159. H. R. Kricheldorf and W. E. Hull, *Biopolymers*, 1982, **21**, 359.
160. H. R. Kricheldorf and W. E. Hull, *Makromol. Chem.*, 1979, **180**, 1715.

37

Anionic Ring-opening Polymerization: Copolymerization

SHOHEI INOUE and TAKUZO AIDA

University of Tokyo, Japan

37.1 INTRODUCTION

Anionic ring-opening copolymerizations of heterocyclic compounds have not been systematically studied much, primarily because of a rather narrow range of monomers that polymerize with nucleophilic character. Moreover, numerous side reactions together with multiplicities of active species make it difficult to accurately evaluate the reactivities of monomers and active species. Among the limited studies reported to date, copolymerizations of epoxides, episulfides and lactones have been investigated in detail,[1] and indicate the dominant factors to be the electronic and steric effects of the substituents of monomers associated with the balance between nucleophilicity and Lewis acidity of initiators or catalysts. Thus, monomers bearing electron-withdrawing substituents are preferred by initiators of high nucleophilicities, while monomers with electron-donating substituents are preferred by initiators of Lewis acidic character.[2] Stereochemical preference for geometrically isomeric monomers is also observed.[3] These features are due to the coordinative interactions between heterocyclic monomers and active species, which are quite essential and operate to varying extents, resulting in scattered values being reported for copolymerization parameters.

Thus, statistical anionic ring-opening copolymerizations have only a limited importance, and the present chapter focuses attention on recent developments in the copolymerizations between different types of comonomers, mainly from a synthetic point of view.

37.2 CYCLIC ETHER

37.2.1 Cyclic Acid Anhydride

The alternating copolymerization between epoxide (1) and cyclic acid anhydride (2) proceeds in the presence of a catalyst, such as inorganic and organic salts, organometallic compounds and Lewis bases, producing polyester (3) under rather rigorous conditions (equation 1).[4] A recent development of particular interest is the discovery of the catalyst system composed of aluminum porphyrin [4; (TPP)AlX, where (TPP) = 5,10,15,20-tetraphenylporphinato)] and a quaternary organic salt, which gives the alternating copolymer (3) of controlled molecular weight with narrow molecular weight distribution under mild conditions.[5,6]

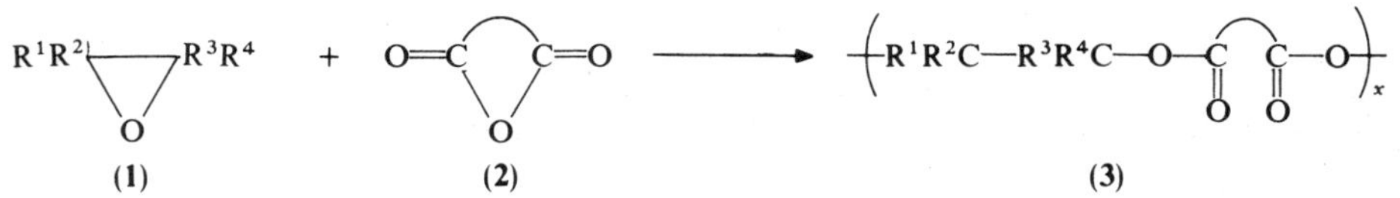

$$R^1R^2\overset{\triangle}{\underset{O}{}}R^3R^4 \quad + \quad O=C\underset{O}{\overset{\frown}{}}C=O \quad \longrightarrow \quad \{R^1R^2C-R^3R^4C-O-\overset{O}{\underset{O}{C}}\overset{O}{\underset{O}{C}}-O\}_x \quad (1)$$

$$(1) \qquad\qquad (2) \qquad\qquad\qquad (3)$$

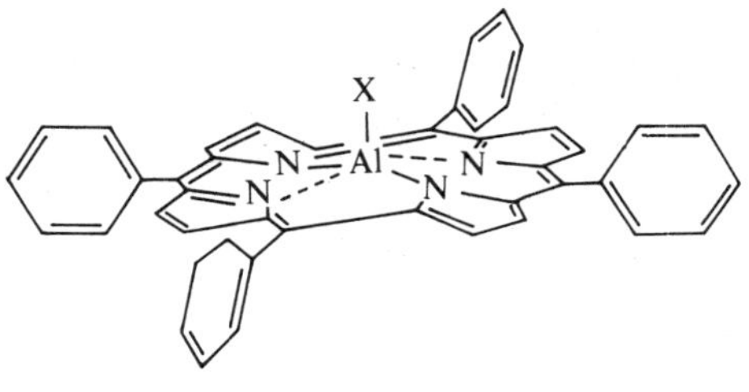

(4) (TPP)AlX

For example, the copolymerization of 1,2-epoxypropane (1; R^1 = Me and R^2 = R^3 = R^4 = H) and phthalic anhydride (25 equiv./25 equiv.) by the system of (TPP)AlCl (4; X = Cl) coupled with ethyltriphenylphosphonium bromide (EtPh$_3$PBr) (1 equiv./1 equiv.) proceeds quantitatively at room temperature to give the corresponding polyester, which shows a unimodal, sharp elution curve in the gel permeation chromatogram (GPC). The number average molecular weight ($\bar{M}_n$) of the copolymer increases linearly with the conversion of comonomers, while the ratio of the weight and number average molecular weights ($\bar{M}_w/\bar{M}_n$) remains constant at close to unity. This catalyst system is widely applicable to the copolymerization of phthalic anhydride and unsubstituted, monosubstituted and disubstituted epoxides as listed below:

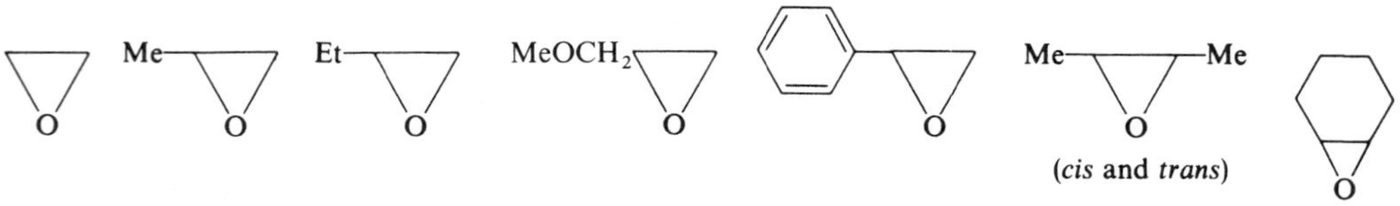

(cis and trans)

Aluminum porphyrin (4) was originally found to be a very effective catalyst for the 'living' polymerization of epoxide (equation 2) and β-lactone (equation 3), which proceeds *via* (porphinato)aluminum alkoxide (5) and carboxylate (6), respectively, as the growing species.[7,8] On the other hand, the copolymerization of epoxide and cyclic acid anhydride by aluminum porphyrin alone proceeds very slowly, and the product contains a considerable amount of ether linkages. The development of 'living' and 'alternating' natures when using aluminum porphyrin as a catalyst, coupled with a quaternary organic salt (Q$^+$Y$^-$), is due to the formation of a six-coordinate aluminum porphyrin (7), in which the reaction proceeds on both sides of the aluminum porphyrin

$$\text{(2)}$$

$$\text{(3)}$$

$$\text{(4)}$$

plane (equation 4.)[5] Accordingly, the number of the molecules of copolymer (N_p) produced is twice the number of the molecules of aluminum porphyrin (N_{Al}).

37.2.2 Carbon Dioxide

Carbon dioxide is able to copolymerize with epoxide (three-membered) and oxacyclobutane (oxetane, four-membered) to give, in some cases, the alternating copolymer (aliphatic polycarbonate) under rather mild conditions.[9, 10]

37.2.2.1 Epoxide

A representative example of an effective catalyst for the alternating copolymerization of carbon dioxide and epoxide is the system of dialkylzinc combined with a protic compound such as water, dihydric phenol or aromatic hydroxycarboxylic acid (equation 5).[11] The system, based on an inorganic zinc compound such as zinc oxide or zinc dihydroxide coupled with carboxylic acid, was later also found to be effective as a catalytic system without any cumbersome association with air sensitive metal alkyls.[12] In particular, the system composed of zinc oxide, isophthalic acid and propionic acid (2/1/2) is of an activity comparable to the organozinc-based system. This catalytic system is considered to consist of structure (9) containing zinc phthalate and propionate bonds, as indicated by elemental and IR analyses.

$$\text{(5)}$$

The copolymers of carbon dioxide and 1,2-epoxypropane obtained respectively by the systems of diethylzinc coupled with isophthalic acid, terephthalic acid and resorcinol (10–12) carry an aromatic residue covalently incorporated at the terminal, as evidenced by the gel permeation chromatography monitored by UV absorption detector (GPC–UV) and the quantitative end group analysis by UV

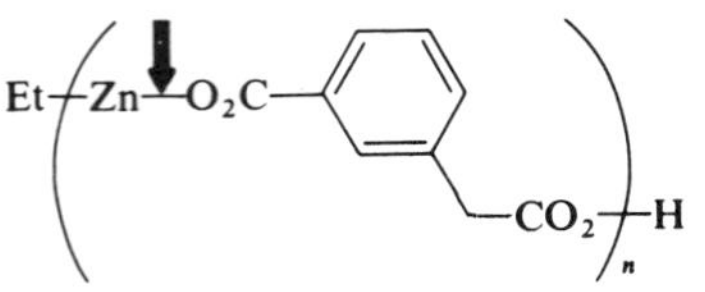

(10) 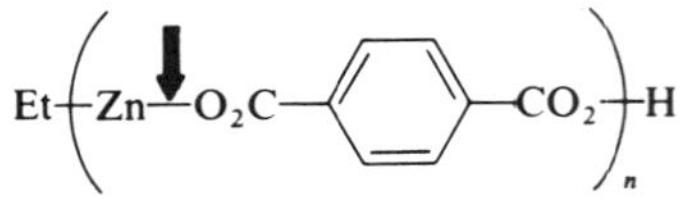   (11) 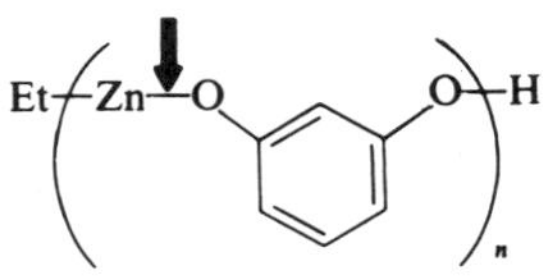   (12)

spectroscopy.[13] Thus, the copolymerization takes place between the zinc–oxygen bond of the catalyst. The same result is obtained for the copolymer by using (9) as catalyst and indicates that the copolymerization proceeds exclusively at the zinc–phthalate bond, although there exist in (9) two different zinc–carboxylate bonds of a potential reactivity to epoxide, *i.e.* zinc–phthalate and zinc–propionate.

In the ternary copolymerization of carbon dioxide, 1,2-epoxybutane (1; $R^1 = Et$, $R^2 = R^3 = R^4 = H$) and 2,3-epoxybutane (1; $R^1 = R^3 = Me$, $R^2 = R^4 = H$) by the diethylzinc–water system, *cis*-2,3-epoxybutane is incorporated in the copolymer, while the *trans* isomer is hardly incorporated.[14] Thus, the coordinative activation of an epoxide molecule by a zinc species is considered essential prior to being incorporated into the growing terminal, taking into account the less steric hindrance for the *cis* form than the *trans*.

The copolymer (13) from carbon dioxide and 1,2-epoxycyclohexane (*cis*) (1; $R^1 = R^3 = +CH_2)_4$, $R^2 = R^4 = H$) or *cis*-2,3-epoxybutane by the diethylzinc–water system gives a *trans* (*threo*) diol upon alkaline hydrolysis (equation 6), indicating the inversion of configuration at the carbon atom of the epoxide ring where the ring is cleaved.[14, 15]

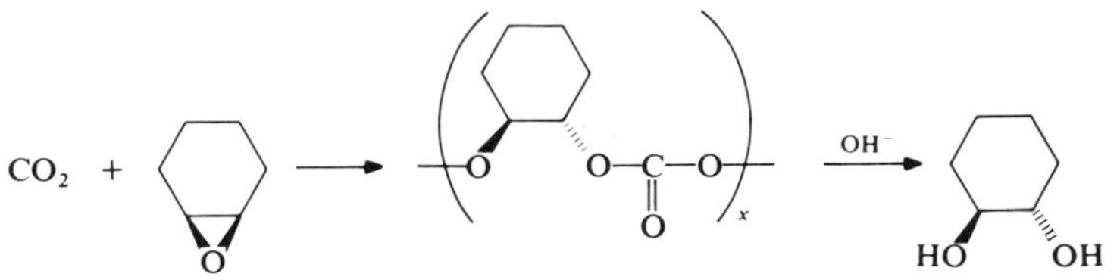

(6)

In the copolymerization of carbon dioxide and epoxyethane by the system of diethylzinc and water as catalyst, epoxyethane is hardly detected throughout the copolymerization, even at the very early stage. A major detectable product formed initially is the oligoether, which is suggested to gradually depolymerize to epoxyethane, which is subsequently incorporated into the copolymer by the copolymerization with carbon dioxide.[16] Thus, the molecules of oligoether formed initially (14) are suggested to have an oxonium terminal which is depolymerizable to epoxide by a 'back-biting' reaction, and a zinc alkoxide terminal on the other side at which the copolymerization takes place.

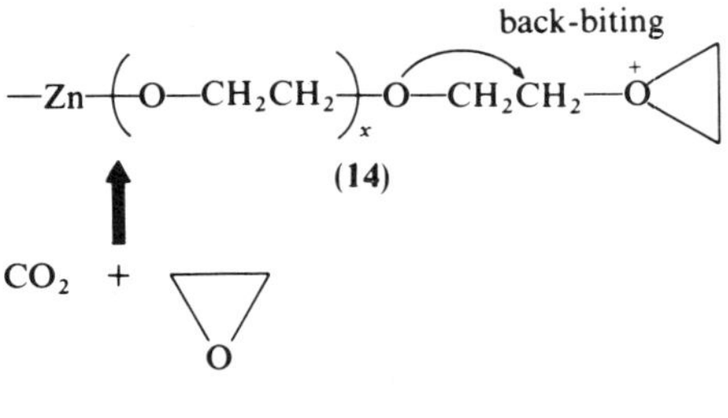

(14)

Of particular interest to note from a synthetic viewpoint is the first successful example for the control of molecular weight of the copolymer. The catalyst system is the equimolar mixture of aluminum porphyrin (4) and a quaternary organic salt, which is also effective for the 'living' and 'alternating' copolymerization of cyclic acid anhydride and epoxide,[5, 6] as described in Section 37.2.1. For example, the copolymerization of 1,2-epoxycyclohexane (100 equiv.) and carbon dioxide by the (TPP) AlCl (4; $X = Cl$)–$EtPh_3P^+Br^-$ (1 equiv./1 equiv.) system proceeds at room temperature under the CO_2 pressure of 50 kg cm^{-2} to give quantitatively the alternating copolymer (13) of very narrow molecular weight distribution ($\bar{M}_w/\bar{M}_n = 1.06$).[17] A sharp elution curve of the copolymer thus obtained (a) in the GPC chromatogram is in clear contrast with a much broader one (b) typical of the copolymer obtained by usual catalyst system, such as the diethylzinc–isophthalic acid system

(Figure 1). The number average molecular weight ($\bar{M}_n$) of the copolymer, 6800, which is almost a half of that calculated from the mole ratio of comonomers and aluminum porphyrin, 14 200, indicates that the copolymerization proceeds similarly to equation (4). Triphenylphosphine can be used in place of the quaternary organic salt for the present copolymerization. On the other hand, in the absence of any such additives, epoxide is consumed preferentially over carbon dioxide to give the copolymer which contains ether linkages to a considerable extent.[18]

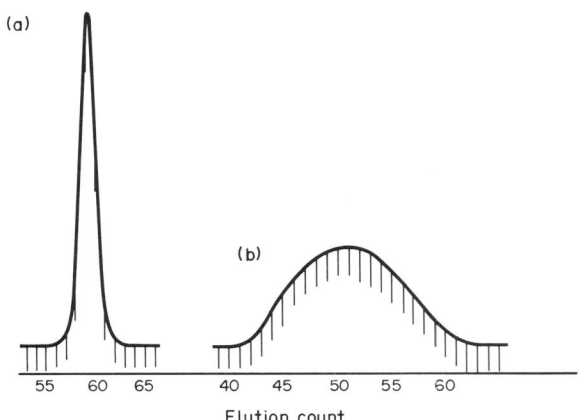

Figure 1 GPC curves of the copolymer from carbon dioxide (CO_2) and epoxide (**1**): (a) CO_2–1,2-epoxycyclohexane (**1**; R^1, $R^3 = -(CH_2)_4$, R^2, $R^4 = H$) by the (TPP)AlCl(**4**; $X = Cl$)–$EtPh_3P^-Br^+$ (1/1) system,[17] $[\mathbf{1}]_0/[\text{cat}]_0 = 100$, $(P_{CO_2})_0 = 50\ \text{kg cm}^{-2}$, 100% conversion, $\bar{M}_n = 6800$, $\bar{M}_w/\bar{M}_n = 1.06$; (b) CO_2–1,2-epoxypropane (**1**; $R^1 = Me$, $R^2 = R^3 = R^4 = H$) by the diethylzinc–isophthalic acid (1/0.9) system,[13] $[\mathbf{1}]_0/[\text{cat}]_0 = 43$, $(P_{CO_2})_0 = 40\ \text{kg cm}^{-2}$, 22% yield, $\bar{M}_n = 7400$

37.2.2.2 *Oxacyclobutane*

Among the cyclic ethers other than epoxide, only oxacyclobutane (oxetane, **15**), a four-membered cyclic ether, has been shown to copolymerize with carbon dioxide (equation 7).

$$CO_2 + \ \ \square\!\!-\!O \quad \longrightarrow \quad \left(\!O\!-\!C\!-\!C\!-\!C\!-\!O\!-\!\underset{\underset{O}{\|}}{C}\!\right)_x \tag{7}$$

$$(\mathbf{15}) \qquad\qquad\qquad\qquad (\mathbf{16})$$

This copolymerization does not take place in the presence of organozinc catalyst, which is generally effective for the copolymerization of carbon dioxide and epoxide. On the other hand, the system composed of triethylaluminum, water and acetylacetone as catalyst gives the copolymer with an intrinsic viscosity $[\eta]$ of 0.4, though with a low CO_2 content (17%).[19]

The first example of the formation of alternating copolymer (**16**) was developed by the system of organotin halide coupled with a Lewis base as catalyst.[20] For example, the copolymerization of carbon dioxide and oxetane (50 equiv.) by the catalyst tributyltin iodide–tributylphosphine ($Bu_3^nSnI–PBu_3$) (1 equiv./1 equiv.) under the CO_2 pressure of 50 kg cm^{-2} proceeds up to 100% conversion in 24 h at 100 °C to give the alternating copolymer with $\bar{M}_n$ of 1200. $Bu_2^nSnI_2$ and $Bu_2^nSnBr_2$ give a similar result when coupled with trialkylphosphine or triarylphosphine, while $Bu_2^nSnCl_2$ and $SnCl_4$ with an enhanced Lewis acidity or organotin compound without any halide such as Bu_4^nSn are quite ineffective under the same conditions. In the absence of Lewis base, organotin halide does not bring about the reaction. The reaction of carbon dioxide and $Bu_3^nSn–O-(CH_2)_3I$, a possible intermediate for the reaction between Bu_3^nSnI and oxetane, gives, in the presence of triphenylphosphine (PPh_3) at 50 °C for 4 h under the CO_2 pressure of 6 kg cm^{-2}, trimethylene carbonate (**17**) in 24% yield. Moreover, with the ring-opening polymerization of trimethylene carbonate it is possible to give (**16**) at 80 °C in the presence of the system composed of $Bu_2^nSnI_2$ and PPh_3 as catalyst. Thus, in the copolymerization of carbon dioxide and oxetane catalyzed by the organotin halide–Lewis base system, the polycarbonate would not be formed directly from the alternating reaction of comonomers but by the ring-opening polymerization of trimethylene carbonate (**17**) formed *in situ*, as in equation (8).

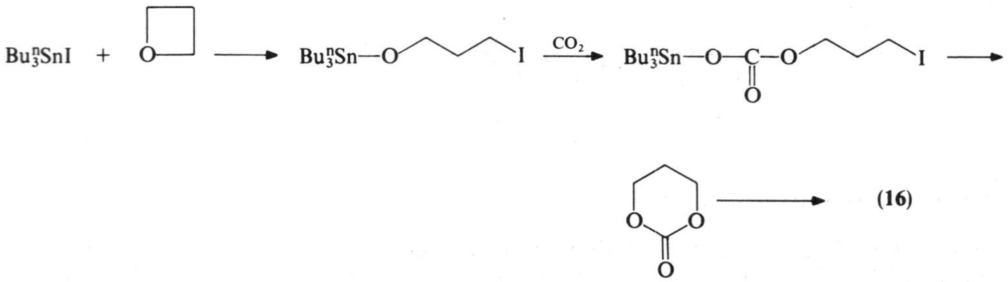

$$Bu_3^nSnI \; + \; [\text{oxetane}] \longrightarrow Bu_3^nSn{-}O{\diagdown\diagup}I \xrightarrow{CO_2} Bu_3^nSn{-}O{-}\underset{O}{\underset{\|}{C}}{-}O{\diagdown\diagup}I \longrightarrow$$

$$[\text{cyclic carbonate (17)}] \longrightarrow (16) \tag{8}$$

(17)

In connection with this, the ring-opening polymerization of trimethylene carbonate, either substituted or unsubstituted, has been known to give polycarbonate by some anionic catalysts such as potassium carbonate or *n*-butyllithium without any loss of CO_2.[21,22] On the other hand, the polymerization of five-membered cyclic carbonate (18), which proceeds only at a rather high temperature (100–170 °C), is accompanied by a considerable extent of decarboxylation ($>70\%$) to give a poly(ether carbonate) (equation 9).[23] Thus, cyclic carbonate is not the precursor for polycarbonate in the case of the copolymerization of carbon dioxide and epoxide.

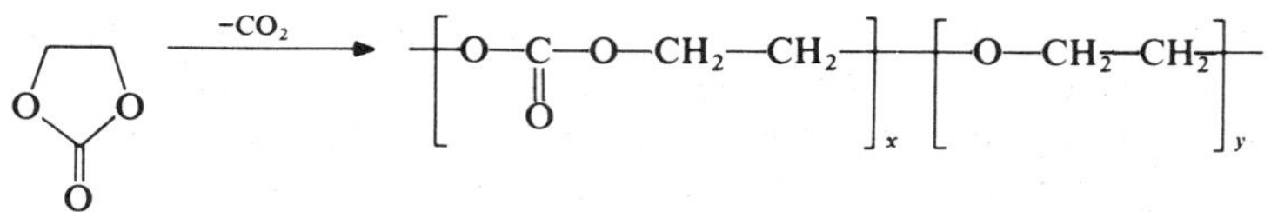

$$[\text{(18)}] \xrightarrow{-CO_2} {+}{\Big[}O{-}\underset{O}{\underset{\|}{C}}{-}O{-}CH_2{-}CH_2{\Big]}_x{\Big[}O{-}CH_2{-}CH_2{\Big]}_y{+} \tag{9}$$

(18)

37.2.3 Sulfur Dioxide

The copolymerization of epoxide and sulfur dioxide (SO_2) takes place in the presence of a catalyst such as Lewis acid[24] or Lewis base.[25,26] When a Lewis base such as aromatic tertiary amine is used, the copolymerization takes place alternately to give poly(alkylene sulfite) (19 in equation 10).

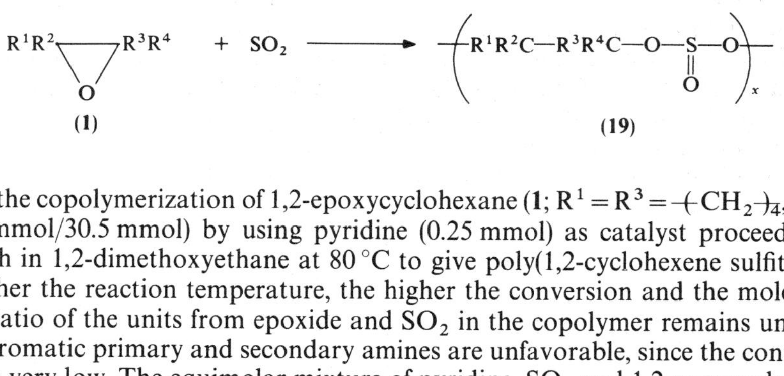

$$\underset{\text{(1)}}{R^1R^2{\diagup\diagdown}R^3R^4} \; + \; SO_2 \longrightarrow {\Big(}R^1R^2C{-}R^3R^4C{-}O{-}\underset{O}{\underset{\|}{S}}{-}O{\Big)}_x \tag{10}$$

(1)   (19)

For example, the copolymerization of 1,2-epoxycyclohexane (1; $R^1 = R^3 = {+}CH_2{+}_4$, $R^2 = R^4 = H$) and SO_2 (29.8 mmol/30.5 mmol) by using pyridine (0.25 mmol) as catalyst proceeds up to 86% conversion in 3 h in 1,2-dimethoxyethane at 80 °C to give poly(1,2-cyclohexene sulfite) with $\bar{M}_n$ of 6300.[26] The higher the reaction temperature, the higher the conversion and the molecular weight, while the mole ratio of the units from epoxide and SO_2 in the copolymer remains unchanged, and close to unity. Aromatic primary and secondary amines are unfavorable, since the content of SO_2 in the copolymer is very low. The equimolar mixture of pyridine, SO_2 and 1,2-epoxycylohexane, upon stirring at 20 °C for 1 h, gives a crystalline precipitate which is identified to be a zwitterion (20). Thus, copolymerization proceeds as in equation (11).

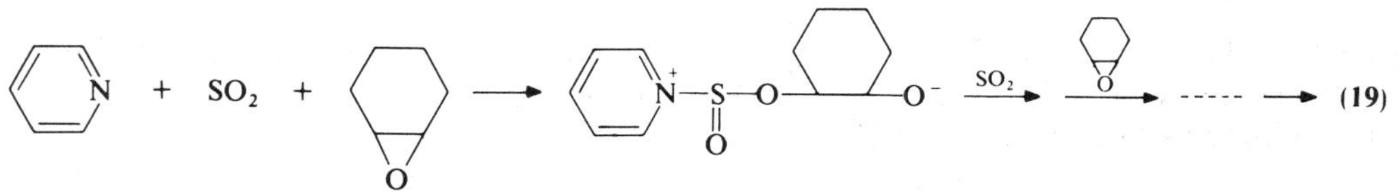

$$[\text{pyridine}] + SO_2 + [\text{epoxycyclohexane}] \longrightarrow [\text{zwitterion (20)}] \xrightarrow{SO_2} [] \text{-----} \longrightarrow (19) \tag{11}$$

(20)

With respect to the thermal properties of the copolymer, poly(alkylene sulfite) obtained from epoxyethane and SO_2, for example, decomposes at 218 °C, as in (equation 12).[25]

$$\left(C-C-O-\underset{\underset{O}{\overset{\parallel}{S}}}{\overset{}{S}}-O \right)_x \quad \xrightarrow[218\,°C]{\Delta} \quad \underset{\underset{O}{\overset{\parallel}{S}}}{\overset{}{O}}\,O\,O \quad + \quad \underset{O}{\triangledown} \quad + \quad SO_2 \quad + \quad residue \qquad (12)$$

37.2.4 Other Unsaturated Compounds

The copolymerizations of epoxide with carbon disulfide,[27] carbon monoxide,[28] α-amino acid N-carboxyanhydride,[29] benzonitrile[30] and phenyl isocyanate[31] have been attempted to date, although the products are not alternating copolymer but are more complicated than expected.

37.3 CYCLIC THIOETHER

37.3.1 Carbon Dioxide or Carbon Disulfide

Episulfides, such as propylene sulfide (**21**, R = Me), copolymerize with carbon dioxide to give the copolymer (**22**) containing a thiocarbonate linkage (equation 13). For example, the copolymerization of propylene sulfide (50 equiv.) and carbon dioxide under a pressure of $30\,kg\,cm^{-2}$ proceeds by the system of triethylaluminum-1,2,3-trihydroxybenzene (1 equiv./2 equiv.) in dioxane at 45 °C for 96 h to give the copolymer in 11.6% yield with respect to episulfide.[32] The content of carbon dioxide, as determined by the elemental analysis, is 42% ($x = 0.16$, $y = 0.84$ in structure **22**). In contrast to the case of the copolymerization of epoxide and carbon dioxide, the organozinc-based catalyst gives virtually the homopolymer of episulfide in the present copolymerization.

$$R-\underset{S}{\triangledown} \quad + \quad CO_2 \quad \longrightarrow \quad \left[CH_2-\underset{\overset{|}{R}}{CH}-S \right]_x \left[CH_2-\underset{\overset{|}{R}}{CH}-S-\underset{\overset{\parallel}{O}}{C}-O \right]_y \qquad (13)$$

(**21**)   (**22**)

Carbon disulfide has been reported to copolymerize with an episulfide, such as ethylene sulfide (**21**, R = H), in the presence of triethylamine at room temperature for four days, to give the alternating copolymer (**23**) in 63% yield, with a concomitant formation of ethylene trithiocarbonate (**24**).[33] With a metal thiolate, such as mercury bis(n-butanethiolate), as catalyst, carbon disulfide and episulfide give a copolymer with a thiocarbonate content of 50–70%.[34] The thiocarbonate content is 0–18% when an organometallic compound such as triethylaluminum, diethylzinc or diethylcadmium is used as catalyst. The cyclic trithiocarbonate (**24**) is not the precursor for (**23**), since the ring-opening polymerization of (**24**) does not take place by the catalyst, which is effective for the copolymerization of episulfide and carbon disulfide under the same conditions.

$$R-\underset{S}{\triangledown} \quad + \quad CS_2 \quad \longrightarrow \quad \left(CH_2-\underset{\overset{|}{R}}{CH}-S-\underset{\overset{\parallel}{S}}{C}-S \right)_x \quad + \quad R-\underset{\underset{\overset{\parallel}{S}}{C}}{S\,\,\,\,S} \qquad (14)$$

(**21**)   (**23**)   (**24**)

37.3.2 Elemental Sulfur

Elemental sulfur (S_8, **25**) with an eight-membered ring has been known to undergo free radical homopolymerization upon heating above 160 °C to eventually give a viscous liquid which contains repeating sulfur linkages, with $\bar{M}_n$ of 10^6 (equation 15).[35]

(**25**)

$$\tfrac{n}{8} \cdot S_8 \;\rightleftharpoons\; \text{----} -S_n- \text{-----} \tag{15}$$
(25)

Below the floor temperature for radical homopolymerization (equation 15), the copolymerization of episulfide (21') and elemental sulfur (25) takes place in the presence of $CdCO_3$ or with the alkali metal thiophenoxide–crown ether system as catalyst in aromatic hydrocarbon solvent (equation 16).[36–39] The copolymer produced has molecular weight ($\bar{M}_n$) and sulfur content, in some cases, up to 50 000 and 85%, respectively. For example, the copolymerization of isobutylene sulfide (21'; $R^1 = R^2 = Me$) and S_8 catalyzed by crowned sodium thiophenoxide proceeds in benzene at 80 °C, as shown in Figure 2, to give a rubber-like material (26), which has a statistical distribution of polysulfide units with $\bar{y}$ of 2.6, as determined by NMR and Raman spectroscopies.[38] Cyclic polysulfide (27) is initially formed, but gradually disappears with the progress of reaction. Thus, the chain growth (equations 17 and 18) is accompanied by side reactions, such as the intra- and inter-molecular attacks of the growing thiolate anion onto the already formed polysulfide bond, resulting in chain reshuffling (equation 19) and cyclization (equation 20). Based on kinetic analysis without considering the scrambling reaction (equation 19), elemental sulfur (S_8) is 7.5 times more reactive than propylene sulfide (21'; $R^1 = Me$, $R^2 = H$) towards the growing thiolate anion.[39]

The copolymers thus formed are very reluctant to depolymerize, and give transparent films by casting the solution.

$$\underset{\underset{\displaystyle(21')}{S}}{\triangle}\!\!R^1R^2 \;+\; \underset{(25)}{S_8} \;\longrightarrow\; \underset{(26)}{\left(\!CH_2\!-\!CR^1R^2\!-\!S_y\!\right)_x} \tag{16}$$

$$\text{\small$\sim$}C\!-\!C\!-\!S_m^- \;+\; S_8 \;\rightleftharpoons\; \text{\small$\sim$}C\!-\!C\!-\!S_{m+8}^- \tag{17}$$

$$\text{\small$\sim$}C\!-\!C\!-\!S_m^- \;+\; \underset{S}{\triangledown} \;\longrightarrow\; \text{\small$\sim$}C\!-\!C\!-\!S_m\!-\!C\!-\!C\!-\!S^- \tag{18}$$

$$\text{\small$\sim$}C\!-\!C\!-\!S^- \;+\; -S_x- \;\longrightarrow\; \text{\small$\sim$}C\!-\!C\!-\!S_m^- \;+\; \text{\small$\sim$}S_{x-m+1}\text{\small$\sim$} \tag{19}$$

$$\text{\small$\sim$}C\!-\!C\!-\!S_m\!-\!C\!-\!C\!-\!S^- \;\longrightarrow\; \text{\small$\sim$}C\!-\!C\!-\!S_{m-n+1}^- \;+\; \underset{\underset{\displaystyle(27)}{S_n}}{\triangledown} \tag{20}$$

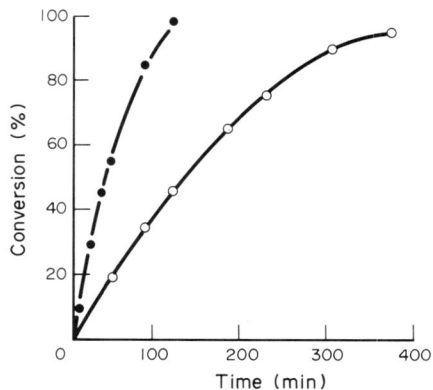

Figure 2 Time–conversion curve for the copolymerization of isobutylene sulfide (21'; $R^1 = R^2 = Me$; ○) and elemental sulfur (S_8; ●) catalyzed by crowned sodium thiophenoxide. $[21']_0/[S_8]_0/[cat]_0/1000/250/1$, in benzene at 80 °C[38]

37.4 LACTIDE (GLYCOLIDE)

37.4.1 Lactone

A cyclic dimer of an α-hydroxy acid, such as glycolide (**28**; R=H, 1,4-dioxacyclohexane-2,5-dione) or lactide (**28**; R=Me, 3,6-dimethyl-1,4-dioxacyclohexane-2,5-dione), gives the polymer (**29**) consisting of ester linkages (equation 21),[40] which is potentially very useful for biomedical and pharmaceutical applications owing to the favorable biodegradability and low toxicity. In view of developing controlled mechanical properties and biodegradability, the copolymerization of (**28**) and lactone (**30**) in equation (22) is of practical importance as well as of fundamental interest. For the copolymerization with β-propiolactone ($n=2$), δ-valerolactone ($n=4$) and ε-caprolactone ($n=5$), an anionic catalyst such as quaternary ammonium or phosphonium salt, and a coordinate anionic catalyst such as aluminum isopropoxide or dibutyltin dimethoxide are rather effective.[41-43] In any case, (**28**) is incorporated in the copolymer more preferentially over (**30**) to form a block sequence. For example, the copolymerization of an equimolar mixture of glycolide (**28**, R=H) and ε-caprolactone (**30**, $n=5$) catalyzed by aluminum isopropoxide proceeds up to 25% for 2 h at 100 °C without solvent to give the copolymer with the glycolide content of 85%.[41] The average degrees of polymerization of the homoblock sequences for glycolide and ε-caprolactone are 14.0 and 2.5 respectively.

$$(21)$$

$$(22)$$

In the polymerization of glycolide by aluminum isopropoxide in the presence of the homopolymer of ε-caprolactone, the transesterification to give the glycolide-ε-caprolactone sequence does not take place (equation 23).[41] This is in sharp contrast with the case using cationic catalysts such as $BF_3 \cdot OEt_2$, where the product contains 84% of the glycolide-ε-caprolactone sequence and 16% of the ε-caprolactone-ε-caprolactone sequence.

$$(23)$$

In the copolymerization of L-lactide (**28**; R = Me) and ε-caprolactone by aluminum isopropoxide, the number average molecular weight of the copolymer linearly increases with the mole ratio of the comonomers reacted with the catalyst, indicating the absence of appreciable side reactions.[43] The reactivity ratios for L-lactide and ε-caprolactone are 17.9 and 0.58, respectively, indicating the tendency to form block sequences in the copolymerization by aluminum isopropoxide.

37.4.2 Morpholinedione

Morpholine-2,5-dione (**31**) is considered both as a six-membered lactone as well as a six-membered lactam, and reported to polymerize in the presence of coordinate anionic catalysts such as diethylzinc, zinc oxide[44] and tin bis(2-ethylhexanoate).[45] The product is a polydepsipeptide consisting of alternating sequences of hydroxy acid and amino acid (**32**), and thus very interesting as a biocompatible material.

$$(24)$$

The copolymerization of (31) and lactide (28; R = Me) provides the polymer (33) with an increased content of hydroxy acid units (equation 25).[44,46] For example, the copolymerization of an equimolar mixture of 6-methylmorpholine-2,5-dione (31'; R = H, R' = Me) and D,L-lactide by using tin bis(2-ethylhexanoate) as catalyst, with a mole ratio of comonomers to the catalyst of 2500, proceeds at 135 °C without solvent to give the copolymer in 70% yield.[46] The number and weight average molecular weights are 16 000 and 28 000, respectively, and the content of amino acid units is 24% in agreement with that in the feed (25%). With varying mole ratios of morpholinedione and lactide in the feed, the composition of the copolymer can be controlled. The glass transition temperature of the copolymer changes almost linearly from 109 °C for the homopolymer of (31') (R = H, R' = Me) to 53 °C for the D,L-lactide homopolymer on increasing the content of hydroxy acid units.

$$(25)$$

The N-substituted morpholinedione is less copolymerizable than the unsubstituted one with lactide.[44] In fact, in the copolymerization of (31') (R' = CHMe₂) with an equimolar amount of D,L-lactide catalyzed by zinc oxide at 175 °C, the N-unsubstituted one (31'; R = H) is incorporated in the copolymer with almost the same mole ratio to lactide, while for the N-methyl derivative (31'; R = Me) the mole ratio is only 2 mol % under the same conditions.

37.5 BLOCK COPOLYMERIZATION

The synthesis of block copolymers composed of two or more different backbone structures with well-defined block lengths has been very limited to date, since an initiator or catalyst is required which will be widely effective at bringing about the 'living' polymerizations.[47] Following are examples achieved by two representative systems, aluminum porphyrin[48] and bimetallic μ-oxoalkoxide,[49] which have recently been found to be of exceptionally wide potential utility.

37.5.1 Aluminum Porphyrin System

As described above, an aluminum porphyrin such as (4) is capable of initiating the 'living' polymerization and copolymerization of a wide variety of monomers as listed below:

(30) $n = 2^{8.51}$

(30) $n = 4^{52}$

(30) $n = 5^{53}$

(34)[54]

Thus, the synthesis of block copolymers with well-defined structures is possible by a simple combination of these polymerizations.

37.5.1.1 Polyether–polyester

The combinations of epoxide (1') with lactone (30) such as β ($n=2$), δ ($n=4$) and ε ($n=5$) lactone or lactide (28, R = Me) give polyether–polyester block copolymers. For example, when the living polymerization of β-butyrolactone by (TPP)AlCl (4, X = Cl) with the mole ratio of 50 is followed at 100% conversion by the addition of 400 equiv. 1,2-epoxypropane (1'; R = Me), the GPC chromatogram of the poly(β-butyrolactone) formed at the first stage (Figure 3(a); $\bar{M}_n = 3200$, $\bar{M}_w/\bar{M}_n = 1.09$) clearly shifts toward the higher molecular weight region without any broadening to give the block copolymer (35) (Figure 3b); $\bar{M}_n = 12\,100$, $\bar{M}_w/\bar{M}_n = 1.17$, at 50% conversion).[51b] The present block copolymerization is accompanied by the conversion of the structure of the growing species from a (porphinato)aluminum carboxylate (6) to the alkoxide (5'), as shown in equation (26).

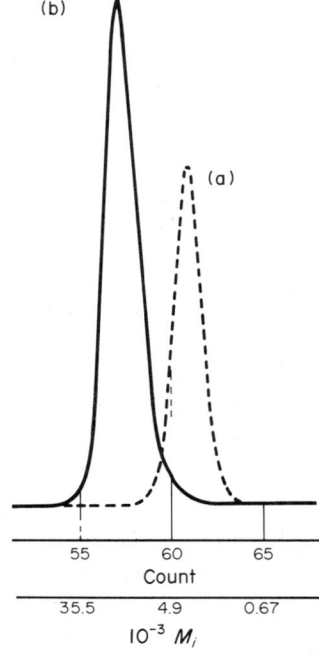

(b)

(a)

55   60   65
Count

35.5   4.9   0.67
$10^{-3}\,M_i$

Figure 3 Block copolymerization of 1,2-epoxypropane (1'; R = Me) from the 'living' polymer of β-butyrolactone (6; R = Me, X = Cl) prepared by the polymerization with (TPP)AlCl (4; X = Cl)[51b]: (a) prepolymer of (30; $n=2$, R = Me), $\bar{M}_n = 3200$, $\bar{M}_w/\bar{M}_n = 1.17$; (b) block copolymer; [1'; R = Me]$_0$/[6]$_0$ = 400, 50% conversion, $\bar{M}_n = 12\,100$, $\bar{M}_w/\bar{M}_n = 1.17$

$$\begin{array}{c} \text{Me} \qquad\qquad \text{Me} \\ \text{—}(\text{CH—CH}_2\text{—CO}_2)_m(\text{CH}_2\text{—CH—O})_n\text{—} \end{array}$$

(35)

$$\text{(Al}(\text{O—C(O)—CH}_2\text{—CH(Me)})_m\text{—X} + \text{Me}\triangle\text{O} \longrightarrow \text{(Al—O—CH(Me)—CH}_2(\text{O—C(O)—CH}_2\text{—CH(Me)})_m\text{—X} \quad (26)$$

(6) R = Me   (1') R = Me   (5')

A block copolymer of polyether and polyester of the type (36) can be synthesized by the polymerization of epoxide from the 'living' copolymer of epoxide and phthalic anhydride (7' in equation 4) prepared by the system of aluminum porphyrin coupled with a quaternary organic salt.[6] Further addition of an equimolar mixture of epoxide and phthalic anhydride to the above reaction system gives the ABA type of block copolymer. Throughout these successive block copolymerizations, the ratio of the weight and the number average molecular weights ($\bar{M}_w/\bar{M}_n$) of the block copolymer remains constant at about 1.1–1.2.

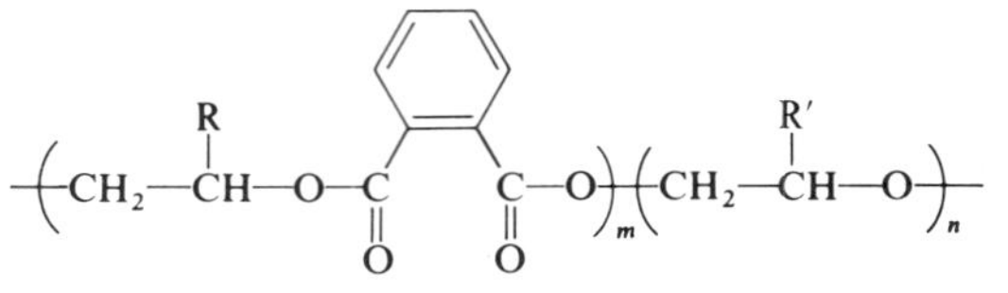

$$\left(CH_2-\underset{\underset{R}{|}}{CH}-O-\underset{\underset{O}{\|}}{C}-\underset{}{}-\underset{\underset{O}{\|}}{C}-O\right)_m\left(CH_2-\underset{\underset{R'}{|}}{CH}-O\right)_n$$

(36)

37.5.1.2 Polyester–polyester'

The block copolymerization of lactone (30) from the living copolymer of phthalic anhydride and epoxide (7'), described above, gives a novel class of polyester–polyester' block copolymer.[6] For example, when the polymerization of β-butyrolactone (100 equiv.) is carried out from the 'living' copolymer of phthalic anhydride and 1,2-epoxypropane ($\bar{M}_n = 3000$, $\bar{M}_w/\bar{M}_n = 1.07$), the block copolymer (37; R = R' = Me) with $\bar{M}_n$ and $\bar{M}_w/\bar{M}_n$, respectively, of 6800 and 1.07, is formed at 100% conversion.

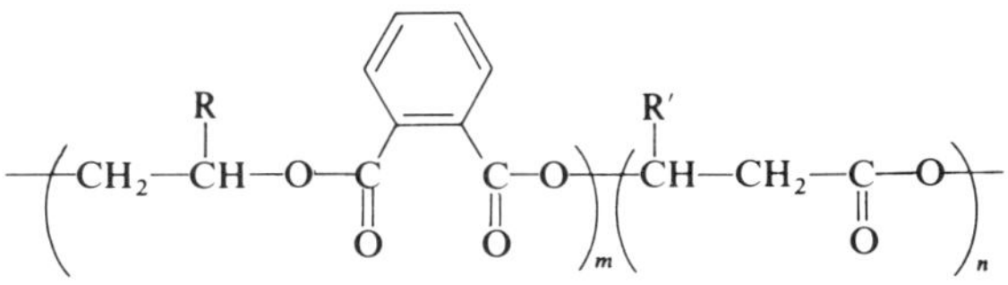

$$\left(CH_2-\underset{\underset{R}{|}}{CH}-O-\underset{\underset{O}{\|}}{C}-\underset{\underset{O}{\|}}{C}-O\right)_m\left(\underset{\underset{R'}{|}}{CH}-CH_2-\underset{\underset{O}{\|}}{C}-O\right)_n$$

(37)

37.5.1.3 Polyester–polycarbonate

The 'living' and 'alternating' characters of the copolymerization of epoxide and cyclic acid anhydride by the system of aluminum porphyrin (4) coupled with a quaternary organic salt is successfully followed by the copolymerization of epoxide and carbon dioxide to give a novel polyester–polycarbonate block copolymer (38) of controlled molecular weight.[17] For example, the copolymerization of 1,2-epoxypropane with phthalic anhydride (25 equiv./25 equiv.) to afford the polyester (3) ($\bar{M}_n = 2500$, $\bar{M}_w/\bar{M}_n = 1.09$), followed by the addition of 1,2-epoxycyclohexane (113 equiv.) and carbon dioxide under the pressure of 50 kg cm^{-2} gives the block copolymer (38; R = Me, R' = $-(CH_2)_4-$) with $\bar{M}_n$ and $\bar{M}_w/\bar{M}_n$, respectively, of 6800 and 1.11. The block lengths can be controlled by the mole ratio of comonomers reacted with the catalyst. The ABA type of block copolymer consisting of polyester (A) and polycarbonate (B) can be obtained similarly by the third stage copolymerization of epoxide and phthalic anhydride with the above reaction system.

$$\left(CH_2-\underset{\underset{R}{|}}{CH}-O-\underset{\underset{O}{\|}}{C}-\underset{\underset{O}{\|}}{C}-O\right)_m\left(\underset{\underset{R'}{|}}{CH}-\underset{\underset{R'}{|}}{CH}-O-\underset{\underset{O}{\|}}{C}-O\right)_n$$

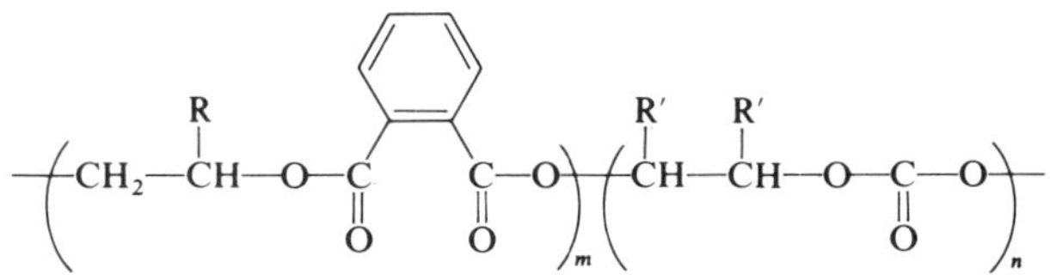

(38)

37.5.1.4 Polyvinyl–polyether

Alkylaluminum porphyrins such as methylaluminum porphyrin (4; X = Me) initiate the polymerization of alkyl methacrylate (34) with 'living' nature under the irradiation with visible light to give the polymer (39) with narrow molecular weight distribution (equation 27).[54] The degree of polymerization of (39) is in excellent agreement with the mole ratio $[34]_0/[4]_0$.

$$xCH_2=C\overset{\displaystyle R}{\underset{\displaystyle \underset{O}{\overset{\|}{C}}-R'}{|}} \longrightarrow \left(CH_2-\overset{\displaystyle R}{\underset{\displaystyle \underset{O}{\overset{\|}{C}}-OR'}{|}}\right)_x \qquad (27)$$

$$(34) \qquad\qquad\qquad (39)$$

The block copolymerization of epoxide from the 'living' polymer of (34) affords a novel polyvinyl–polyether block copolymer of controlled molecular weight (40).[55] An example is shown by the 'living' polymerization of methyl methacrylate (34; R = R' = Me) by aluminum porphyrin (4; X = Me) with the mole ratio $[34]_0/[4]_0$ of 50 at 100% conversion, followed by the addition of 100 equiv. of 1,2-epoxypropane, where the molecular weight of the polymer at the first stage ($\bar{M}_n = 5500$, $\bar{M}_w/\bar{M}_n = 1.16$) clearly increases without broadening the molecular weight distribution to give finally the block copolymer of poly(methyl methacrylate) and poly(oxy(methyl)ethylene) (40; R = R' = R'' = Me) ($\bar{M}_n = 12\,800$, $\bar{M}_w/\bar{M}_n = 1.14$) at the second stage.

$$\left(CH_2-\overset{\displaystyle R}{\underset{\displaystyle \underset{O}{\overset{\|}{C}}-OR'}{|}}\right)_m \left(CH_2-\overset{\displaystyle R''}{\underset{\displaystyle }{\underset{}{C}H}-O}\right)_n$$

$$(40)$$

37.5.1.5 'Immortal' block copolymerization

In sharp contrast to the 'living' polymerization, the polymerization of epoxide by aluminum porphyrin (equation 2) proceeds even in the presence of a protic compound such as HCl, carboxylic acid or alcohol, to give a polymer with very narrow molecular weight distribution. This finding built up a new concept of 'immortal' polymerization,[56,57] which involves a reversible chain transfer reaction (exchange reaction 28) taking place much more rapidly than propagation (equation 2), resulting in uniform chain growth from all the molecules of aluminum porphyrin and the protic compound (equation 29).

$$\left(Al-OR\right) + HOR' \rightleftharpoons \left(Al-OR'\right) + HOR \qquad (28)$$

$$(4)\ X=OR \qquad\qquad (4)\ X=OR'$$

$$\begin{array}{c}\left(Al-OR\right)\\ \uparrow\downarrow H-OR'\\ \text{Exchange}\quad\Big|\quad H-OR'\\ \downarrow H-OR'\end{array} \quad \overset{Growth}{\underset{O}{\triangledown}} \quad \begin{array}{c}\left(Al\text{+}O-C-C\text{+}_xOR\right)\\ \uparrow\downarrow H\text{+}O-C-C\text{+}_xOR'\\ H\text{+}O-C-C\text{+}_xOR'\\ \downarrow H\text{+}O-C-C\text{+}_xOR'\end{array} \qquad (29)$$

By taking advantage of the 'immortal' nature of polymerization, not only the homopolymer but also the block copolymer can be synthesized with the number of the molecules (N_p) more than that of the molecules of aluminum porphyrin (4) (N_{Al}) without broadening the molecular weight distribution.[57] For example, when the 'immortal' polymerization of epoxyethane (200 equiv.) is carried out at the first stage by the system of (TPP) AlCl (4; X = Cl) and methanol (1 equiv./9 equiv.), the reaction mixture containing poly(oxyethylene) with $\bar{M}_n$, $\bar{M}_w/\bar{M}_n$ and N_p/N_{Al}, respectively of 960, 1.05 and 9.2 is formed. The successive polymerization of 200 equiv. of ε-caprolactone (30; $n = 5$) initiated with the reaction mixture thus obtained gives the block copolymer of poly(oxyethylene) and poly(ε-caprolactone) (41) with $\bar{M}_n$, $\bar{M}_w/\bar{M}_n$ and N_p/N_{Al}, respectively, of 3800, 1.20 and 8.3. When a bifunctional protic compound such as ethylene glycol or bisphenol A (2,2-bis(4-hydroxyphenyl) propane) is used, the formation of an ABA type of block copolymer is the result.[52]

$$\left(CH_2-CH_2-O\right)_m\left(\overset{}{\underset{O}{\overset{\|}{C}}}-CH_2-CH_2-CH_2-CH_2-CH_2-O\right)_n$$

$$(41)$$

The block copolymer composed of lactide (**28**; R = Me) and δ-valerolactone (**30**; $n = 4$) with narrow molecular weight distribution (**42**) can be synthesized similarly by the 'immortal' block copolymerization with the aluminum porphyrin–protic compound system.[58]

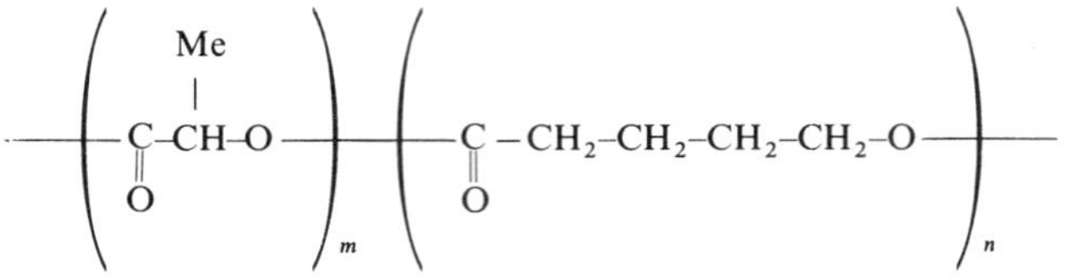

$$\left(\begin{array}{c} Me \\ | \\ -C-CH-O- \\ \| \\ O \end{array}\right)_m \left(\begin{array}{c} -C-CH_2-CH_2-CH_2-CH_2-O- \\ \| \\ O \end{array}\right)_n$$

(**42**)

37.5.2 Al$_2$Zn Bimetallic μ-Oxoalkoxide

The bimetallic μ-oxoalkoxide containing aluminum and zinc (**43**) is an excellent catalyst, in particular, for the 'living' polymerization of ε-caprolactone (**30**; $n = 5$) to give the polyester with narrow molecular weight distribution. The chain growth takes place uniformly at every fourth Al–O–Bu^n bond.[49, 59] This catalyst system is applicable to the synthesis of poly(ε-caprolactone)-based block copolymer of a well-defined structure.

$$(Bu^nO)_2Al-O-Zn-O-Al(OBu^n)_2$$

(**43**)

37.5.2.1 Polyester–polyester'

The block copolymerization of D,L-lactide (**28**; R = Me) from the above-mentioned 'living' poly-(ε-caprolactone) gives the corresponding AB type block copolymer. For example, when the polymerization of ε-caprolactone (**30**; $n = 5$) by (**43**) ($[30]_0/[43]_0 = 200$) at 90 °C in toluene to give poly-(ε-caprolactone) ($\bar{M}_n = 10\,300$, $\bar{M}_w/\bar{M}_n = 1.12$) is followed by the addition of 176 equiv. of D,L-lactide, the block copolymer (**44**) with $\bar{M}_n$ and $\bar{M}_w/\bar{M}_n$, respectively, of 22 400 and 1.12 is formed.[60]

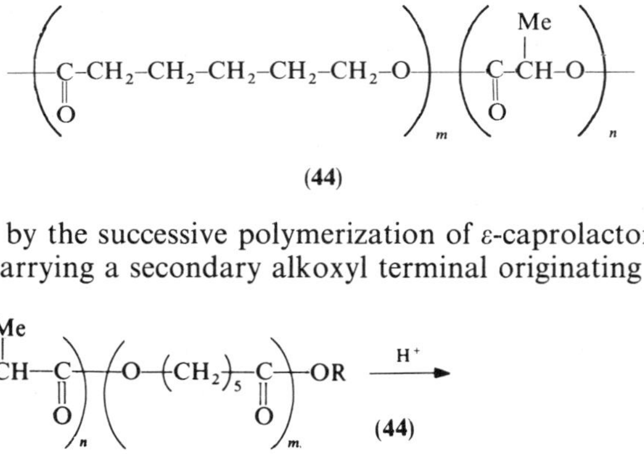

$$\left(\begin{array}{c} -C-CH_2-CH_2-CH_2-CH_2-CH_2-O- \\ \| \\ O \end{array}\right)_m \left(\begin{array}{c} Me \\ | \\ -C-CH-O- \\ \| \\ O \end{array}\right)_n$$

(**44**)

The ABA type analog by the successive polymerization of ε-caprolactone is not formed directly from (**44'**) (equation 30) carrying a secondary alkoxyl terminal originating from lactide, but can be

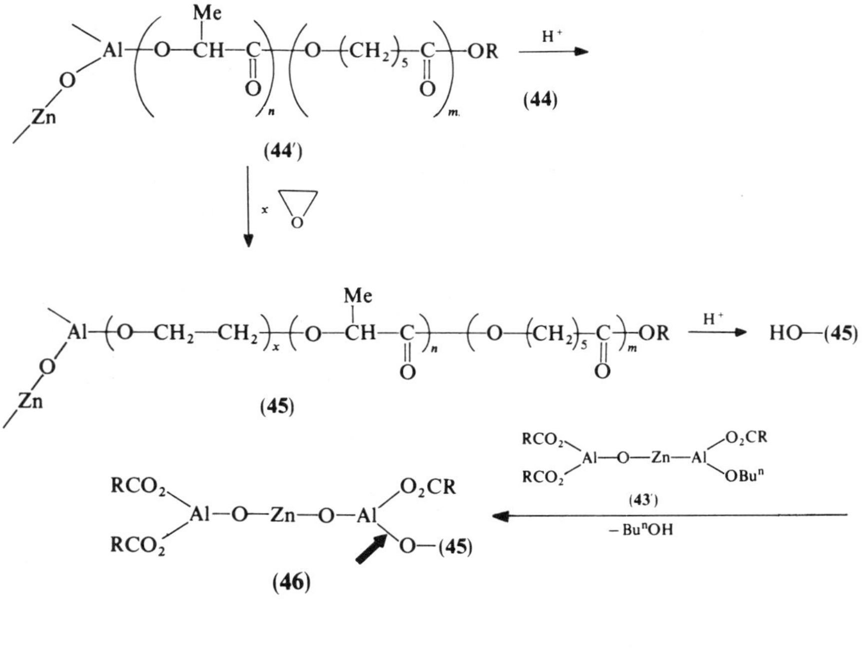

(30)

formed when this terminal is converted into the primary alkoxyl structure by reaction with a small excess of epoxyethane.[61] In order to avoid the formation of unfavorable side products, the isolation of the precopolymer as the alcohol form (HO–**45**), followed by remounting on the aluminum atom (**46**) by the exchange reaction with (**43'**) is necessary, according to the scheme illustrated by equation (30).

37.5.2.2 Polyvinyl–polyester

ω-Hydroxypolystyrene can be used for a similar exchange reaction with (**43'**) to give the macromolecular initiator (**47**), which affords the AB type polystyrene–poly(ε-caprolactone) block copolymer (**48**) by the polymerization of ε-caprolactone.[62] The molecular weight of poly(ε-caprolactone) block can be controlled by the mole ratio of ε-caprolactone reacted to (**47**). However, about 10% of the initial polystyrene always remains unreacted, indicating the essential difficulty in completing the exchange reaction (31).

$$\text{polystyrene—OH} \quad + \quad (\mathbf{43'}) \quad \xrightarrow{\text{Bu}^n\text{OH}} \quad \underset{(\mathbf{47})}{\overset{\text{RCO}_2}{\underset{\text{RCO}_2}{\diagup}}\text{Al—O—Zn—O—Al}\overset{\text{O}_2\text{CR}}{\underset{\text{O—polystyrene}}{\diagdown}}} \tag{31}$$

$$\left(\text{CH}_2\text{—CH}\right)_m\text{—O}\left(\overset{\text{O}}{\underset{\|}{\text{C}}}\text{—CH}_2\text{—CH}_2\text{—CH}_2\text{—CH}_2\text{—CH}_2\text{—O}\right)_n$$

(**48**)

37.5.2.3 Polydiene–polyester

When α,ω-dihydroxypolybutadiene (HO–PBD–OH) is subjected to a similar exchange reaction with (**43'**), the macromolecular initiator (**49**), in which two molecules of (**43'**) are bridged together by a polybutadiene chain, is formed on average.[62] The polymerization of 300 equiv. of ε-caprolactone initiated with (**49**) (HO–PBD–OH; $\bar{M}_n = 4400$, $\bar{M}_w/\bar{M}_n = 1.6$) gives the block copolymer (**50**) with $\bar{M}_n$ and $\bar{M}_w/\bar{M}_n$ of 35 000 and 1.5, respectively. Upon careful treatment of (**50**) with hydroperoxide

$$\underset{\text{RCO}_2}{\overset{\text{RCO}_2}{\diagup}}\text{Al—O—Zn—O—Al}\overset{\text{O}_2\text{CR}}{\underset{\text{O}\sim\text{PBD}\sim\text{O}}{\diagdown}}\underset{\text{RCO}_2}{\overset{\text{RCO}_2}{\diagup}}\text{Al—O—Zn—O—Al}\overset{\text{O}_2\text{CR}}{\underset{\text{O}_2\text{CR}}{\diagdown}}$$

(**49**)

$$\left(\text{O—CH}_2\text{CH}_2\text{CH}_2\text{CH}_2\text{CH}_2\overset{\text{O}}{\underset{\|}{\text{C}}}\right)\text{—O—PBD—O}\left(\overset{\text{O}}{\underset{\|}{\text{C}}}\text{—CH}_2\text{CH}_2\text{CH}_2\text{CH}_2\text{CH}_2\text{—O}\right)$$

(**50**)

coupled with osmium tetroxide, the polybutadiene block degrades to leave exclusively the poly-(ε-caprolactone) components without any hydrolytic cleavage of the ester linkage. The molecular weight, 17 000, observed for the poly(ε-caprolactone) thus obtained meets the value expected from the ABA structure of the block copolymer (**50**), 15 300.

37.6 REFERENCES

1. Reviews for the ring-opening polymerization and copolymerization: for example: (a) K. C. Frisch and S. Reegen, (eds), 'Ring-Opening Polymerization', Dekker, New York, 1967; (b) A. Noshay and J. E. McGrath, 'Block Copolymers. Overview and Critical Survey', Academic Press, New York, 1977; (c) Y. Yamashita, *Adv. Polym. Sci.*, 1978, **28**, 1; (d) K. J. Ivin and T. Saegusa (eds), 'Ring-Opening Polymerization', Elsevier, London, 1984; (e) J. E. McGrath, (ed), 'Ring-Opening Polymerization', *ACS Symp. Ser.*, 1985, vol. 286.

2. (a) C. C. Price, Y. Atarashi and R. Yamamoto, *J. Polym. Sci., Polym. Chem. Ed.*, 1969, **7**, 569; (b) C. C. Price and L. R. Brecker, *J. Polym. Sci., Polym. Chem. Ed.*, 1969, **7**, 575.
3. (a) E. J. Vandenberg, *J. Polym. Sci., Polym. Chem. Ed.*, 1969, **7**, 525; (b) T. Aida, K. Wada and S. Inoue, *Macromolecules*, 1987, **20**, 237.
4. J. Lustoň and F. Vašš, *Adv. Polym. Sci.*, 1984, **56**, 91.
5. T. Aida and S. Inoue, *J. Am. Chem. Soc.*, 1985, **107**, 1358.
6. T. Aida, K. Sanuki and S. Inoue, *Macromolecules*, 1985, **18**, 1049.
7. T. Aida and S. Inoue, *Macromolecules*, 1981, **14**, 1166.
8. T. Yasuda, T. Aida and S. Inoue, *Macromolecules*, 1983, **16**, 1792.
9. S. Inoue and T. Aida, in 'Ring-Opening Polymerization', ed. K. J. Ivin and T. Saegusa, Elsevier, London, 1984, vol. 1, p. 283.
10. S. Inoue, in 'Organic and Bio-organic Chemistry of Carbon Dioxide', ed. S. Inoue and N. Yamazaki, Kodansha, Tokyo, 1981, p. 167.
11. (a) M. Kobayashi, S. Inoue and T. Tsuruta, *Macromolecules*, 1971, **4**, 658; (b) M. Kobayashi, Y. L. Tang, T. Tsuruta and S. Inoue, *Makromol. Chem.*, 1973, **169**, 69; (c) M. Kobayashi, S. Inoue and T. Tsuruta, *J. Polym. Sci., Polym. Chem. Ed.*, 1973, **11**, 2383.
12. (a) S. Inoue, T. Takada and H. Tatsu, *Makromol. Chem., Rapid Commun.*, 1980, **1**, 775; (b) K. Soga, E. Imai and I. Hattori, *Polym. J.*, 1981, **13**, 407.
13. Y. Hino, Y. Yoshida and S. Inoue, *Polym. J.*, 1984, **16**, 159.
14. S. Inoue, K. Matsumoto and Y. Yoshida, *Makromol. Chem.*, 1980, **181**, 2287.
15. S. Inoue, H. Koinuma, Y. Yokoo and T. Tsuruta, *Makromol. Chem.*, 1971, **178**, 241.
16. D. D. Dixon and M. E. Ford, *J. Polym. Sci., Polym. Lett. Ed.*, 1980, **18**, 599.
17. T. Aida, M. Ishikawa and S. Inoue, *Macromolecules*, 1986, **19**, 8.
18. T Aida and S. Inoue, *Macromolecules*, 1982, **15**, 682.
19. H. Koinuma and H. Hirai, *Makromol. Chem.*, 1977, **178**, 241.
20. A. Baba, H. Meishou and H. Matsuda, *Makromol. Chem., Rapid Commun.*, 1984, **5**, 665.
21. W. H. Carothers and F. J. Van Natta, *J. Am. Chem. Soc.*, 1930, **52**, 314.
22. H. Keul, R. Bächer and H. Höcker, *Makromol. Chem.*, 1986, **187**, 2579.
23. (a) K. Soga, Y. Tazuke, S. Hosoda and S. Ikeda, *J. Polym. Sci., Polym. Chem. Ed.*, 1977, **15**, 219; (b) W. Kuran and P. Gorecki, *Makromol. Chem.*, 1983, **184**, 907.
24. (a) R. J. Kern and J. Schaefer, *J. Am. Chem. Soc.*, 1967, **89**, 6; (b) J. Schaefer, *Macromolecules*, 1968, **1**, 111; (c) M. Matsuda, F. Akiyama and Y. Hara, *Kobunshi Ronbunshu*, 1975, **32**, 660.
25. K. Soga, I. Hattori, J. Kinoshita and S. Ikeda, *J. Polym. Sci., Polym. Chem. Ed.*, 1977, **15**, 745.
26. K. Soga, K. Kiyohara, I. Hattori and S. Ikeda, *Makromol. Chem.*, 1980, **181**, 2151.
27. S. Inoue, H. Koinuma and T. Tsuruta, *Makromol. Chem.*, 1969, **2**, 110; (b) N. Adachi, Y. Kida and K. Shikata, *J. Polym. Sci., Polym. Chem. Ed.*, 1977, **15**, 937.
28. J. Furukawa, Y. Iseda, T. Saegusa and H. Fujii, *Makromol. Chem.*, 1965, **89**, 263.
29. T. Tsuruta, K. Matsuura and S. Inoue, *Makromol. Chem.*, 1965, **81**, 258.
30. K. Ree, M. Yanagida, J. Kanesaka and Y. Minoura, *Polymer*, 1977, **18**, 308.
31. J. Furukawa, S. Yamashita, M. Maruhashi and K. Harada, *Makromol. Chem.*, 1965, **85**, 80.
32. W. Kuran, A. Rokicki and W. Wielgopolan, *Makromol. Chem.*, 1978, **179**, 2545.
33. G. A. Razuvaev, V. S. Etlis and L. N. Gribov, *Zh. Obshch. Khim*, 1963, **33**, 1366.
34. K. Soga, H. Imamura, M. Sato and S. Ikeda, *J. Polym. Sci., Polym. Chem. Ed.*, 1976, **14**, 677.
35. B. Meyer, in 'Inorganic Sulfur Chemistry', ed. G. Nickless, Elsevier, New York, 1968.
36. S. Penczek, R. Ślazak and A. Duda, *Nature (London)*, 1978, **273**, 738.
37. S. Penczek and A. Duda, *Pure Appl. Chem.*, 1981, **53**, 1679.
38. A. Duda and S. Penczek, *Makromol. Chem.*, 1980, **181**, 995.
39. A. Duda and S. Penczek, *Macromolecules*, 1982, **15**, 36.
40. (a) J. Kleine and H. H. Kleine, *Makromol. Chem.*, 1959, **30**, 23; (b) T. Tsuruta, K. Matsuura and S. Inoue, *Makromol. Chem.*, 1964, **75**, 211; (c) W. Dittrich and R. C. Schulz, *Angew. Makromol. Chem.*, 1971, **15**, 109; (d) F. E. Kohn, J. G. Van Ommen and J. Feijen, *Eur. Polym. J.*, 1983, **19**, 1081; (e) L. Trofimoff, T. Aida and S. Inoue, *Chem. Lett.*, 1987, 991.
41. H. R. Kricheldorf, T. Mang and J. M. Jonté, *Macromolecules*, 1984, **17**, 2173.
42. H. R. Kricheldorf, T. Mang and J. M. Jonté, *Makromol. Chem.*, 1985, **186**, 955.
43. J.-M. Vion, R. Jérôme, Ph. Teyssié, M. Aubin and R. E. Prud'homme, *Macromolecules*, 1986, **19**, 1828.
44. N. Yonezawa, F. Toda and M. Hasegawa, *Makromol. Chem., Rapid Commun.*, 1985, **6**, 607.
45. J. Helder, F. E. Kohn, S. Sato, J. W. Vanden Berg and J. Feijen, *Makromol. Chem., Rapid Commun.*, 1985, **6**, 9.
46. J. Helder, J. Feijen, S. J. Lee and S. W. Kim, *Makromol. Chem., Rapid Commun.*, 1986, **7**, 193.
47. M. Szwarc, *Adv. Polym. Sci.*, 1983, **49**, 1.
48. S. Inoue and T. Aida, *ACS Symp. Ser.*, 1985, **286**, 137.
49. Ph. Teyssié, J. P. Bioul, A. Hamitou, J. Heuschen, L. Hocks, R. Jérôme and T. Ouhadi, *ACS Symp. Ser.*, 1977, **59**, 165.
50. (a) T. Aida, R. Mizuta, Y. Yoshida and S. Inoue, *Makromol. Chem.*, 1981, **182**, 1073; (b) T. Aida and S. Inoue, *Makromol. Chem., Rapid Commun.*, 1980, **1**, 677; (c) T. Aida and S. Inoue, *Macromolecules*, 1981, **14**, 1162; (d) K. Sanuki, T. Aida and S. Inoue, *Macromolecules*, 1985, **18**, 1049, and Ref. 7.
51. (a) T. Yasuda, T. Aida and S. Inoue, *Makromol. Chem., Rapid Commun.*, 1982, **3**, 585; (b) T. Yasuda, T. Aida and S. Inoue, *Macromolecules*, 1984, **17**, 2217; (c) T. Yasuda, T. Aida and S. Inoue, *J. Macromol. Sci., Chem.*, 1983, **A21**, 1035; (d) T. Yasuda, T. Aida and S. Inoue, *Bull. Chem. Soc. Jpn.*, 1986, **59**, 3931.
52. M. Endo, T. Aida and S. Inoue, *Macromolecules*, 1987, **20**, 2982.
53. K. Shimasaki, T. Aida and S. Inoue, *Macromolecules*, 1987, **20**, 3076.
54. K. Kuroki, T. Aida and S. Inoue, *J. Am. Chem. Soc.*, 1987, **109**, 4737.
55. K. Kuroki, S. Nashimoto, T. Aida and S. Inoue, *Polym. Prepr. Jpn.*, 1987, **36**, 242.
56. S. Asano, T. Aida and S. Inoue, *J. Chem. Soc., Chem. Commun.*, 1985, 1148.
57. T. Aida, Y. Maekawa, S. Asano and S. Inoue, *Macromolecules*, 1988, in press.
58. K. Shimasaki, T. Aida and S. Inoue, unpublished results.

59. (a) A. Hamitou, R. Jérôme, A. Hubert and Ph. Teyssié, *Macromolecules*, 1973, **6**, 651; (b) A. Hamitou, T. Ouhadi, R. Jérôme and Ph. Teyssié, *J. Polym. Sci., Polym. Chem. Ed.*, 1977, **15**, 865; (c) Ph. Teyssié, J. P. Bioul, P. Condé, J. Druet, J. Heuschen, R. Jérôme, T. Ouhadi and R. Warin, *ACS Symp. Ser.*, 1985, **286**, 97.
60. X. D. Feng, C. X. Song and W. Y. Chen, *J. Polym. Sci., Polym. Lett. Ed.*, 1983, **3**, 177.
61. C. X. Song and X. D. Feng, *Macromolecules*, 1984, **17**, 2764.
62. J. Heuschen, R. Jérôme and Ph. Teyssié, *Macromolecules*, 1981, **14**, 242.

38

Experimental Aspects

MICHEL FONTANILLE
University of Bordeaux, France
and
AXEL H. E. MÜLLER
University of Mainz, FRG

38.1 INTRODUCTION

Experimental methods used in the preparation of polymers, as well as the classical methods used to follow the kinetics of polymerization reactions (*e.g.* gravimetry, dilatometry) are reviewed in Chapters 5 and 7 of this volume. These methods can be applied to anionic polymerization, provided that they allow for (or can be modified for) high purity conditions necessitated by the high reactivity of the active centres. In particular, experiments performed at low concentrations of active centres (so as to determine the corresponding kinetic order) require techniques using sealed vessels.

Besides the high nucleophilicity of the active centres involved, anionic polymerization (especially in polar solvents) differs from radical polymerization by the fact that for anionic polymerization the concentration of active centres is much higher and as a consequence the rates can be much higher (or half-lives much shorter). Thus among the various possible techniques, the choice is mainly determined by the half-life of the polymerization to be investigated. The methods reviewed in this chapter are especially suitable for fast reactions having half-lives of 0.05 to 2 s (flow-tube reactor), 1 to 20 s (calorimetry), 1 s to several minutes (UV spectrophotometry), and 2 s to hours (automated tank reactor).

38.2 UV SPECTROPHOTOMETRY

This method consists of following the variation of the monomer concentration in the reaction medium by measuring the optical density at the absorption maximum of the monomer *vs.* time. It is therefore of prime importance that the solvent should not absorb in the same region of the spectrum as the monomer. As a result, aromatic solvents are excluded as reaction media.

Different systems have been proposed by Szwarc[1] allowing reactions with half-life times higher than 10 s to be followed; for lower values, the mixing of reactants is not conveniently performed during the early stages of the process. A reactor is schematically represented in Figure 1. The vessel is placed under vacuum and sealed off at P; the solution of initiator (or living oligomers) and the monomer solution are each transferred to the corresponding finger by crushing the break seals; the mixture of reactants is produced in the optical cell A by inverting the vessel. The cell is placed in the

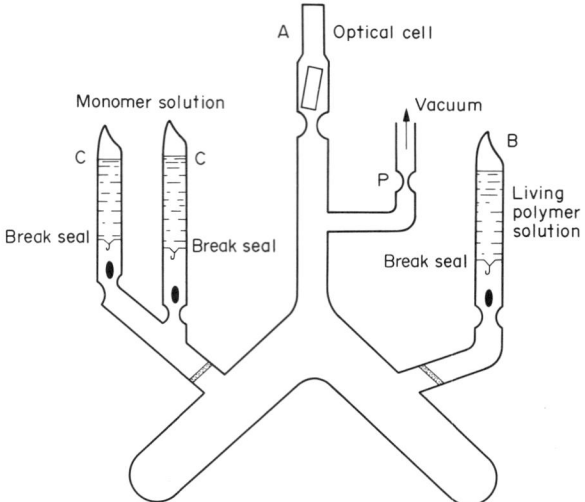

Figure 1 Reactor to follow the kinetics of slow polymerizations near room temperature by spectrometry (reproduced from ref. 1 by permission of Wiley Interscience)

cell compartment of the spectrophotometer and the optical density measured. For slow kinetics, the concentration in active centres is measured spectrometrically at regular intervals. For fast ones, it is determined only at the end of the reaction, requiring an extreme purity of both the reactor and the reactants.

An alternative to this device has also been proposed by Szwarc[1] to study polymerization kinetics at low temperature. The principle of operation is similar but the shape of the vessel (Figure 2) allows it to be put into a Dewar flask, equipped with quartz optical windows, which may be placed in the cell compartment of a spectrophotometer.

The mixing of reactants can be improved in order to allow the kinetics of polymerizations with half-lives as short as one second to be followed. Favier *et al.*[2] have proposed a device, which is

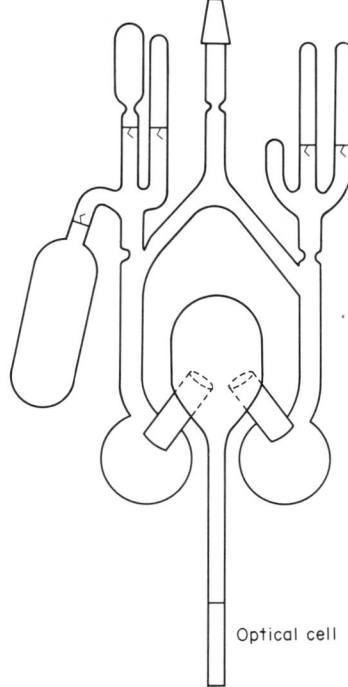

Figure 2 Reactor used to study the kinetics of polymerization at low temperatures (reproduced from ref. 1 by permission of Wiley Interscience)

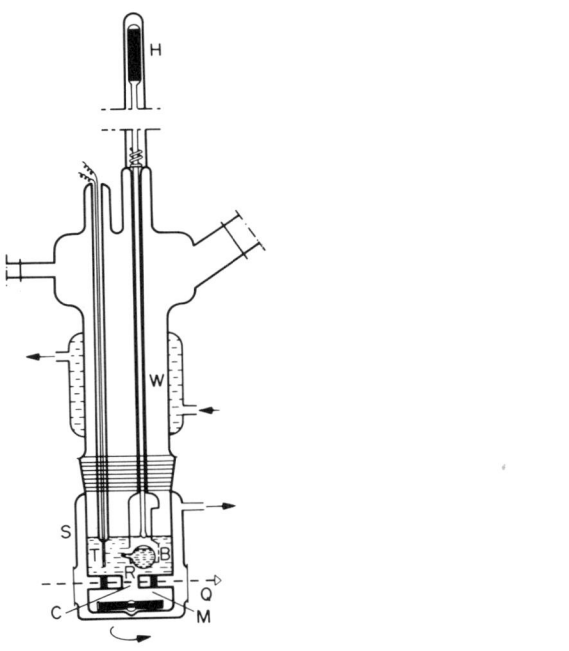

Figure 3 Reactor to follow the kinetics of relatively fast polymerizations at temperatures lower than room temperature by spectrometry

represented in Figure 3. All the following operations have to be performed, the whole vessel being placed in the cell compartment of the spectrophotometer. The solution of initiator is contained in reactor R and its temperature is regulated from room temperature down to $-80\,°C$ by condensing the vapours of the solvent on wall W, which is cooled by circulation of a freezing liquid. To avoid condensation of water on the windows of optical cell C, the bottom of the reactor is placed in a surrounding wall provided with quartz windows Q and put under either vacuum or nitrogen. The temperature of the reaction medium is measured by means of thermistor T and concentration in active species is determined by spectrometric titration before the polymerization starts. Bulb B contains the solution of monomer; the living solution is vigorously stirred by means of magnetic barrel M. The mixing of the reactants results in the crushing of bulb B by dropping magnetic hammer H. The time of complete mixing of the reactants is about 0.2 s.

This device has also been used[2] to measure the rate constants of copolymerization by the appearance of the signal corresponding to new species formed by addition of the comonomer. To neglect the homopropagation of the added comonomer, its concentration must be much lower than that of active centres, or better still, it should not be able to homopolymerize (*e.g.* 1,1-diphenyl-ethylene).

38.3 CALORIMETRY

The method consists of measuring the temperature rise resulting from the variation of enthalpy of the system during the polymerization. This variation is obviously proportional to the conversion of monomer into polymer.

To be of any actual use, the calorimetric technique implies that the reactor may be considered as adiabatic during the greater part of the polymerization process. Moreover, the response time of the thermic probe is about 0.5 s. Consequently, the present calorimetric technique can be used only for values of half time of reaction from 1 to 20 s.

The principle of this technique has been given by Biddulph and Plesch[3] and modified for completely sealed apparatus by Cheradame *et al.*[4] Applications to anionic polymerizations have been developed by Honnoré *et al.*[5]

The apparatus used for calorimetric measurements can be described by means of Figure 3. Indeed, this apparatus allows simultaneous calorimetric and spectrophotometric determinations. A simpler version without the optical cell C and circulation device around W, is described in ref. 4. To obtain excellent control of temperature in the reactor, the surrounding wall S is replaced by another one allowing external circulation of a thermoregulated fluid. A satisfactory adiabaticity of the system is obtained by evacuation of the enclosed space surrounding the reactor. The fast mixing of the reactants is obtained by crushing bulb B, which contains the solution of monomer, in the stirred solution, as previously described for the calorimetric technique. The evolution of heat due to the polymerization is measured by means of a platinum thermistor T immersed in the reaction medium; the thermal inertia of the thermistor and of its support must be sufficiently low not to increase the response time of the system. The amount of monomer contained in the bulb must be low enough to induce a temperature rise no higher than several tenths of a degree; under such conditions the kinetics of the reaction are not modified by the variation of temperature.

For polymerizations to be performed at low temperature, the body of the reactor is cooled in a cooling bath at a temperature slightly lower than that of the reaction; then surrounding wall E is put under vacuum and the temperature of the living solution rises slowly up to the reaction temperature. A deficient adiabaticity of the system is easily perceptible from the shape of the recording of temperature *vs.* time.

38.4   FLOW-TUBE AND BATCH REACTORS

Flow-tube reactors are suitable for following the kinetics of fast reactions having half-lives between 0.05 and 2 s. Originally invented by Hartridge and Roughton,[6] they were first applied to the kinetics of anionic polymerization by Szwarc and co-workers.[7] The method has been improved and extensively used by Schulz and co-workers.[8-10] The principle of Schulz's apparatus is illustrated in Figure 4.

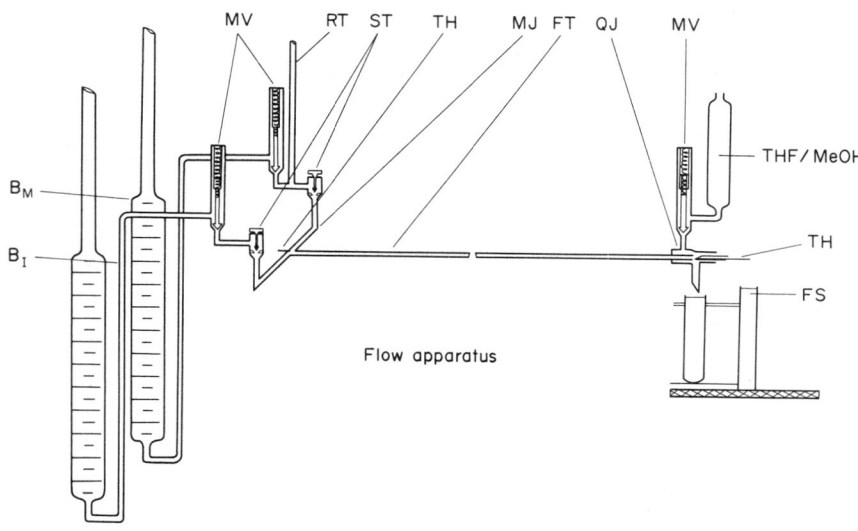

Figure 4 Flow-tube apparatus, B_M, B_I are burettes for monomer and initiator solutions, respectively; MV, magnetic valve; ST, Teflon valve; RT, side arm for rinsing the flow tube; FT, flow tube; MJ, mixing jet; QJ, quenching jet; FS, fraction sampler; TH, thermocouple. The complete apparatus is immersed in a thermostatted bath

A kinetic experiment is performed as follows. With the magnetic valves closed, nitrogen pressure is applied to the monomer and initiator burettes. After opening the valves, monomer and initiator solutions are mixed within a few milliseconds and the reacting polymer solutions flow through the flow tube. After passing the quenching jet the solution is sampled in a fraction sampler, discarding the first two tube fillings to allow for nonstationary flow and to rinse the tube.

An alternative set up has been given by Adler *et al.*[11] Here, the flow is established and controlled by motor-driven glass pistons. The flow tube as well as the mixing and quenching jets are made of stainless steel.

The reaction time τ, *i.e.* the residence time of the 'living' solution in the tube, is given by

$$\tau = V_R/v \qquad (1)$$

where V_R is the tube volume and $v = \Delta V/\Delta T$ is the velocity of the solution in the tube. The velocity v

is measured as follows. By applying fibreoptics, light beams are sent through the monomer burette at various volume intervals ΔV. The light is passed through phototransistors. Each time the meniscus passes a light beam, an electronic clock is stopped, to give the corresponding value of Δt. Monomer conversion can be measured either gravimetrically (by precipitating the polymer) or by determining the residual monomer concentration (*e.g.* by GC). In order to establish a time–conversion curve, several runs have to be performed by varying the flow velocity (*i.e.* by varying the nitrogen pressure) and/or by varying the tube volume. The accessible range of residence times is determined by two factors. The lower limit (*ca.* 0.05 s) is set by the volumes of the mixing and quenching jet, which have to be small compared to the tube volume. Decreasing the jet volumes leads to a higher flow resistance and thus lower flow velocities. The upper limit is set by a minimum flow rate which has to be maintained in order to establish turbulent flow. Laminar flow seriously affects the molecular weight distribution of the resulting polymer.[12,13] For turbulent flow, a Reynolds number of

$$Re \quad = \quad 2\rho v/\pi r\eta > 3000 \tag{2}$$

is necessary,[14] where ρ is the density and η the viscosity of the solution, and r is the tube radius.

Due to the high reaction rates, the heat of polymerization is only partially transferred to the cooling bath. Thus it is important to measure the temperature shortly after the mixing jet and before the quenching jet. A function to correct for the temperature rise has been given by Löhr *et al.*[9]

For kinetic investigations of reactions with half-lives ≥ 2 s an automated stirred tank reactor has been introduced by Warzelhan *et al.*[15] It is illustrated in Figure 5. A kinetic experiment, which is controlled by an electronic timer, is performed as follows. By opening magnetic valve MV1, nitrogen pressure is applied to the monomer and initiator solutions. In this way the reactor is filled within less than 0.4 s. At predetermined intervals of ≥ 1 s samples are taken by opening the reactor outlet MV2 and the quenching jet valve MV3. After closing MV2 and MV3, the reactor outlet is rinsed with pure

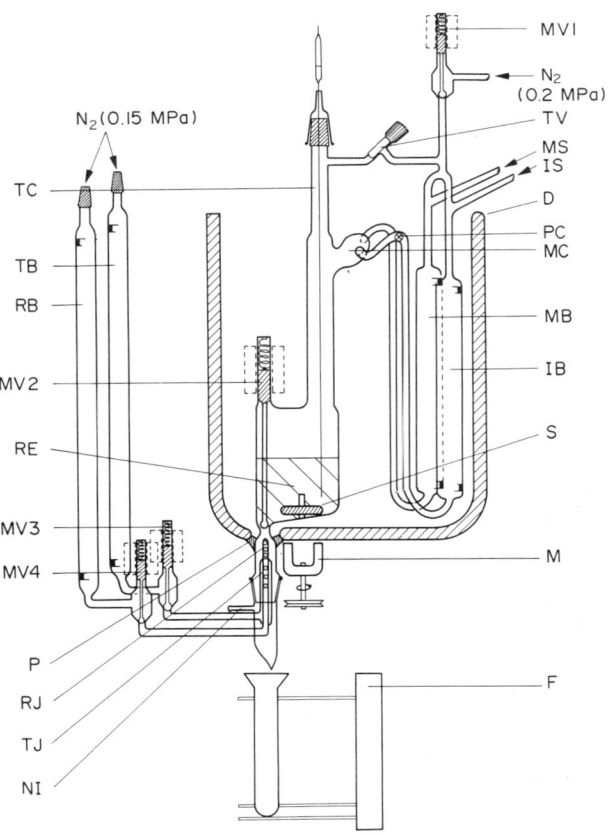

Figure 5 Discontinuous stirred tank reactor. TC, thermocouple; TB, RB, MB and IB, burettes with quenching, rinsing, monomer and initiator solutions, respectively; MS and IS, monomer and initiator solution inlet, respectively; NI, nitrogen inlet; MV, magnetic valves; RE, reactor; P, silicone rubber packing; RJ and TJ, rinsing and quenching jet, respectively; TV, teflon valve; D, dewar vessel; PC, photo cell; MC, mixing chamber, S, magnetic stirrer; M, magnet; F, fraction collector.

solvent by opening MV4, and the fraction collector is then stepped into its next position. Determination of monomer conversion is identical to that using the flow-tube reactor. The advantage of the tank reactor as compared to the flow-tube reactor is that a complete time–conversion curve is obtained within only one experiment. Moreover, it spans the largest range of half-lives (2 s to hours) of all the methods reviewed.

The advantage of both methods over methods which directly monitor monomer conversion (*cf.* Sections 38.2, 38.3) is that for each point on the time–conversion curve a polymer sample is obtained which can be characterized with respect to molecular weight distribution, tacticity, *etc.* Both methods are especially advantageous for monomers which do not absorb in the accessible range of wavelengths, *e.g.* acrylic monomers.

38.5 DETERMINATION OF THE CONCENTRATION OF ACTIVE SPECIES

As pointed out in Chapter 26 (equation 6) of this volume, kinetic experiments only render *apparent* rate constants of propagation. These have to be divided by the concentration of active centres c^* in order to obtain the 'true' bimolecular rate constants

$$k_{app} = -d\ln[M]/dt = k_p c^* \tag{3}$$

For complete initiation c^* is equal to the initial initiator concentration which can be determined spectrometrically or by titration. For incomplete initiation or when termination takes place simultaneously, c^* has to be determined independently. For nonpolar monomers this can be done spectrometrically (*cf.* Section 38.2). For acrylic monomers or when working in aromatic solvents, the absorption maximum does not lie within the accessible range of wavelengths.

Alternatively, the number average degree of polymerization P_n can be used to determine c^*

$$P_n = \frac{\text{concentration of polymerized monomers}}{\text{concentration of polymer chains}} = \frac{[M]_0 x_p}{c_{tot}} \tag{4}$$

$$c^* = k c_{tot} \tag{5}$$

Here c_{tot} is the total concentration of polymer chains, including terminated ones, and k is the number of active centres per polymer chain. A plot of P_n *vs.* monomer conversion x_p gives the constant slope $[M]_0/c_{tot}$, where $[M]_0$ is the initial monomer concentration.

Equation (5) can only be applied when the total number of chains remains constant during polymerization (fast initiation, no transfer). For simultaneous termination, equation (5) will render the initial concentration of active centres c_0^*.

In order to determine c^* as a function of time (for systems inaccessible to spectrometry) it is necessary to label the living polymers. This can be done by quenching the living polymer solution at various times using a labelled quenching agent (*e.g.* a strongly absorbing dye or a radioactive compound). Triteated acetic acid ($MeCO_2{}^3H$) has been used as a labelling agent in the anionic polymerization of methyl methacrylate.[16,17] By using this method, only those polymers bearing active centres are labelled. The activity of a constant volume taken at various reaction times is proportional to c^*. However, due to the kinetic isotope effect it is difficult to obtain the *absolute* concentrations of active centres. The latter can be obtained from equations (4) and (5).

38.6 REFERENCES

1. M. Szwarc, 'Carbanions, Living Polymers and Electron Transfer Processes', Wiley, New York, 1968.
2. J. C. Favier, P. Sigwalt and M. Fontanille, *Eur. Polym. J.*, 1974, **10**, 717.
3. R. H. Biddulph and P. H. Plesch, *Chem. Ind. (London)*, 1959, 1482.
4. H. Cheradame, J. P. Vairon and P. Sigwalt, *Eur. Polym. J.*, 1968, **4**, 13.
5. D. Honnoré, J. C. Favier, P. Sigwalt and M. Fontanille, *Eur. Polym. J.*, 1974, **10**, 425.
6. H. Hartridge and F. J. W. Roughton, *Proc. R. Soc. London, Ser. A*, 1923, **104**, 376.
7. C. Geactinov, J. Smid and M. Szwarc, *J. Am. Chem. Soc.*, 1961, **83**, 1283.
8. R. V. Figini, H. Hostalka, K. Hurm, G. Löhr and G. V. Schulz, *Z. Phys. Chem. (Frankfurt)*, 1965, **45**, 269.
9. G. Löhr, B. J. Schmitt and G. V. Schulz, *Z. Phys. Chem. (Frankfurt)*, 1972, **78**, 177, 334.
10. G. Löhr and G. V. Schulz, *Eur. Polym. J.*, 1974, **10**, 121.
11. H.-J. Adler, L. Lochmann, S. Pokorny, W. Berger and J. Trekoval, *Makromol. Chem.*, 1982, **183**, 2901.
12. G. Löhr and G. V. Schulz, *Makromol. Chem.*, 1964, **77**, 240.
13. G. Löhr and G. V. Schulz, *Z. Phys. Chem. (Frankfurt)*, 1969, **65**, 170.

14. L. Prandtl, 'Führer durch die Strömungslehre', Vieweg, Braunschweig, 1957.
15. V. Warzelhan, G. Löhr, H. Höcker and G. V. Schulz, *Makromol. Chem.*, 1978, **179**, 2211.
16. D. L. Glusker, E. Stiles and B. Yonkoskie, *J. Polym. Sci.*, 1961, **49**, 297.
17. F. J. Gerner, H. Höcker, A. H. E. Müller and G. V. Schulz, *Eur. Polym. J.*, 1984, **20**, 349.

39

Carbocationic Polymerization: General Aspects and Initiation

GEORGES SAUVET
Université Paris-Nord, France
and
PIERRE SIGWALT
Université Pierre et Marie Curie, Paris, France

39.1 INTRODUCTION

39.1.1 General Features of Cationic Polymerizations[1, 2, 3]

Polymerizations are generally considered as being 'cationic' when they involve a positively charged active species acting as an electrophile* towards the monomer. These species may be identified ionic species or may appear only in a transition state. Generally, they are located at the end of a growing polymer molecule and react with the monomer acting as nucleophile, but in some cases the situation may be reversed (if the monomer is activated by an acid).

Two main categories of monomers have been found to be polymerized by cationic polymerization. The first one is that of ethylenic monomers, for which the reactive intermediate has been assumed to

* The expression 'electrophilic polymerization' has been recently proposed as more general.[8]

be a carbocation (also called 'carbenium ion'), the propagation being written schematically as in equation (1).

$$\text{mmCH}_2\overset{+}{\underset{\underset{R}{|}}{\text{CH}}}\text{X}^- \quad + \quad \underset{\underset{R}{|}}{\text{CH}_2}{=}\text{CH} \quad \longrightarrow \quad \text{mmCH}_2\underset{\underset{R}{|}}{\text{CH}}\text{CH}_2\overset{+}{\underset{\underset{R}{|}}{\text{CH}}},\text{X}^-$$
(1)

The secondary category is that of heterocyclic monomers containing at least one heteroatom (O, S or P) in the ring, for which propagation has been shown in various cases to involve onium ions, for example oxonium ions in the case of cyclic ethers (equation 2).

$$\text{mmCH}_2\overset{+}{\underset{\text{X}^-\ \ \text{CH}_2}{\text{O}}}\Big) \quad + \quad \text{O}\Big) \quad \longrightarrow \quad \text{mmCH}_2\text{O}\quad \text{CH}_2\overset{+}{\underset{\text{X}^-}{\text{O}}}\Big)$$
(2)

For a third category of monomers including aldehydes and acetals, propagation resembles that of carbocationic polymerization, but strongly delocalized carbocations may be involved (equations 3 and 4).

$$\text{mmO}{\overset{+}{{=}{=}}}\text{CH}_2 \quad + \quad \text{H}_2\text{C}{=}\text{O} \quad \longrightarrow \quad \text{mmOCH}_2\text{O}{\overset{+}{{=}{=}}}\text{CH}_2$$
(3)

$$\text{mmCH}_2\text{CH}_2\text{O}{\overset{+}{{=}{=}}}\text{CH}_2 \quad + \quad \text{O}\underset{}{\Diamond}\text{O} \quad \longrightarrow \quad \text{mmCH}_2\text{CH}_2\text{OCH}_2\text{CH}_2\text{CH}_2\text{O}{\overset{+}{{=}{=}}}\text{CH}_2$$
(4)

The types of positively charged intermediates have not been well identified in most cases, particularly for carbocationic polymerizations, and this is mainly the consequence of the high reactivity of carbenium ions and of the large number of reactions competitive with the propagation step. Whereas in anionic polymerizations many systems are known in which active species are stable for times much longer than the duration of the polymerization (from minutes to hours), such systems were until recently very rare in carbocationic polymerization. Being more stable, onium ions have been identified more easily in several ring-opening polymerizations (see Volume 3, Chapters 45, 48).

In carbocationic polymerization, the most important reactions competitive with propagation are transfer reactions, particularly those involving the monomer, and termination reactions. Both have been found to decrease in non-reactive solvents of sufficiently high dielectric constants, particularly at low temperatures, the most used being chlorinated hydrocarbons (MeCl, CH_2Cl_2, EtCl, $(CH_2Cl)_2$). Various systems in which carbocationic polymerization occurs have recently been described as living (see particularly Chapters 40 and 42 of this volume), but two remarks may be made. First, the presence of a living polymer (with terminationless and transferless propagation) does not necessarily imply the existence of stable active species, and macromolecules may eventually grow on stable end groups which are activated for short periods; second, the complete absence of transfer could not yet be established in these systems, since the molecular weights of the polymers formed were generally rather low.

Another particular feature of cationic polymerizations is the influence of impurities, particularly on the initiation step. Small quantities of polar compounds, and particularly of (even weak) acidic compounds, have been found to modify strongly the reactivity of Friedel–Crafts acids, which are generally the most efficient initiators for carbocationic polymerization, also giving polymers of high molecular weight. This explains why particular care about the purification of the reagents and about polymerization conditions (high vacuum, sealed apparatus) has been mandatory in order to obtain reproducible results.

39.1.2 Monomers Polymerized by Carbocationic Polymerization

A large number of monomers are polymerizable by such a mechanism, and an extensive list may be found for example in ref. 2, p. 37 and in ref. 4. However, the number of usual monomers polymerized by carbocationic polymerization is much more limited, and particularly the number of those having given polymers of high molecular weight. The main ones are shown in Table 1, which also includes a few important monomers that give only oligomers. The absence of polymerization may result from the difficulties of initiation or from the occurrence of a very rapid termination step. Monomers most easily polymerized are those bearing on the double bond electrorepulsive substituents increasing the monomer reactivity, and also groups that stabilize the carbocation by resonance such as an aromatic ring, a double bond, at least one ether group, or two methyl groups.

Table 1 Main Ethylenic Monomers Polymerized by a Cationic Mechanism

Monomer	Type of polymer	Refs.
Ethylene	Low molecular weight oligomers[a,b]	4, p. 53
Propylene	Medium molecular weight[b] (5×10^3)	9
Isobutylene	High molecular weight (10^6)	4, p. 10 and 129[d]
Styrene	Medium or high molecular weight (2×10^3 to 10^5)	4, p. 243[e]
α-Methylstyrene	High molecular weight (10^6)	1, p. 313[e]
p-Methoxystyrene	High molecular weight (10^6)	10
trans-β-Methyl-p-methoxystyrene (anethole)	High molecular weight (10^5)	11
cis-1,2 Dimethoxyethylene	High molecular weight (10^6)	12
Vinyl ethers	High molecular weight (10^5)	1, p. 393[f]
Butadiene, isoprene	High molecular weight (10^5)[c]	4, p. 154[d]
Cyclopentadiene	High molecular weight (10^5)	13
Indene	High molecular weight (10^6)	14
Acenaphthylene	High molecular weight (10^5)	15
N-Vinylcarbazole	High molecular weight (10^6)	1, p. 543[g]

[a] With high Friedel–Crafts acid concentration. [b] Numerous isomerizations giving irregular structures. [c] Often cyclized (low unsaturation) and crosslinked. [d] See also Volume 3, Chapter 40. [e] See also Volume 3, Chapter 41. [f] See also Volume 3, Chapter 42. [g] See also Volume 3, Chapter 43.

39.1.3 Formation of Carbocationic Active Species

In a carbon–carbon double bond, the electrons are more distant from the axis of the bond than in a single bond. Thus double bonds are generally sensitive to electrophilic attack and are subject to electrophilic addition.

The electrophilic reagent (*viz.* as a positively charged species E^+) adds to the double bond, forming intermediately a *carbocation* which may react with all nucleophilic compounds (*viz.* as a negative species Nu^-) present in the medium (equations 5 and 6).

$$E^+ \; + \; {\gt}C{=}C{\lt} \longrightarrow E{-}\overset{|}{\underset{|}{C}}{-}\overset{|}{\underset{|}{C}}{}^+ \tag{5}$$

$$E{-}\overset{|}{\underset{|}{C}}{-}\overset{|}{\underset{|}{C}}{}^+ \; + \; Nu^- \longrightarrow E{-}\overset{|}{\underset{|}{C}}{-}\overset{|}{\underset{|}{C}}{-}Nu \tag{6}$$

When the double bond is enriched in electrons by electron-donating substituents, the alkene may compete with other nucleophiles (equation 7).

$$E{-}\overset{|}{\underset{|}{C}}{-}\overset{|}{\underset{|}{C}}{}^+ \; + \; {\gt}C{=}C{\lt} \longrightarrow E{-}\overset{|}{\underset{|}{C}}{-}\overset{|}{\underset{|}{C}}{-}\overset{|}{\underset{|}{C}}{-}\overset{|}{\underset{|}{C}}{}^+ \tag{7}$$

If conditions are found in which step (7) is faster than step (6), a cationic polymerization may be observed, that is, a chain reaction in which equation (5) represents the initiation step, equation (7) the propagation step and equation (6), or a similar one, the termination step.

In cationic polymerization of ethylenic monomers, the initiator plays a prominent role because each step of the reaction depends on it: (1) initiation depends on its electrophilicity (*i.e.* its ability to produce a carbocation by opening the double bond), and (2) propagation, chain transfer and termination depend on the nucleophilicity of the counterion (the negative moiety derived from the initiator).

39.1.3.1 *Terminology*

Initiation of a cationic polymerization is rarely a simple process. Nevertheless, in some cases the formation of the carbocationic active centre results from the addition of a positive fragment from the compound used as initiator (*e.g.* a proton from a Brönsted acid, or the cation from an organic salt such as $Ph_3C^+SbCl_6^-$, *etc.*; equations 8 and 9).

$$H^{\delta+}A^{\delta-} \quad + \quad {>}C{=}C{<} \quad \longrightarrow \quad H{-}\overset{|}{\underset{|}{C}}{-}\overset{|}{\underset{|}{C}}{}^+A^- \tag{8}$$

$$Ph_3C^+A^- \quad + \quad {>}C{=}C{<} \quad \longrightarrow \quad Ph_3C{-}\overset{|}{\underset{|}{C}}{-}\overset{|}{\underset{|}{C}}{}^+A^- \tag{9}$$

This electrophilic addition may be called *cationation* in general and *protonation* in the particular case of equation (8). The term 'catalyst' often used in the literature for initiator should be avoided because — except in rare cases — the initiator is consumed during the initiation and is not regenerated in the termination step.

In many cases, the nature of the true electrophilic moiety responsible for the cationation of the double bond is not known. For instance, in systems involving Friedel–Crafts halides which are Lewis acids, it is very difficult to know whether they are capable of adding directly to the double bond (equation 10) or if the intervention of a proton donor is necessary (equation 11).

$$2AlCl_3 \quad + \quad {>}C{=}C{<} \quad \longrightarrow \quad Cl_2AlC{-}\overset{|}{\underset{|}{C}}{}^+AlCl_4^- \tag{10}$$

$$AlCl_3 \quad + \quad H_2O \quad + \quad {>}C{=}C{<} \quad \longrightarrow \quad HC{-}\overset{|}{\underset{|}{C}}{}^+AlCl_3OH^- \tag{11}$$

In the last case, water acting as a proton source was traditionally referred to as the 'cocatalyst' or the coinitiator. It has been pointed out[2,4] that the word initiator would better apply to the proton donor (or protogen) and coinitiator to the Lewis acid because the role of the latter is only to enhance the acidity of water. However this terminology may lead to confusion when processes (10) and (11) are simultaneously operating (AlCl$_3$ should be called initiator in equation 10 and coinitiator in equation 11).

Also, the term 'coinitiator' has been used to describe either the protogen or cationogen[5,160] or the Lewis acid.[2] In the present chapter cationogen or protogen will be preferred to 'cocatalyst' when there is no doubt about their role. But we have felt it sometimes necessary to use the expressions 'cocatalytic effect' or 'cocatalysis' even if they are not really adequate. They are widely used in the literature and may be useful in order to refer to the experimental evidence demonstrating the intervention of a potential protogen or cationogen, whatever is the real mechanism occurring.

As stated above, initiation of the cationic polymerization of an ethylenic compound may be defined as the first step of the chain reaction, consisting of the formation of a carbocation (or strongly polarized ester), the structure of which is identical to that of the propagating active centre.

Identification of the initiation step in chain polymerizations has been based either on measurement of incorporation of initiator fragments into the polymer or on identification of the active species formed and of their concentration. These methods could not be used successfully in most carbocationic polymerizations. The first method is difficult for initiation involving protonic acids — and also for all reactions cocatalyzed by various proton donors — since protons may be incorporated in many ways other than initiation into the polymer. Attempts at using tritiated compounds were not really conclusive, and active species have not generally been identified in most carbocationic polymerizations. It follows from this that in many cases initiation reactions cannot be discussed separately from the problem of the type of active species involved in propagation and from that of the whole polymerization scheme. Particularly, the question of the active species concentration has been extremely controversial, evaluations differing sometimes by a factor as high as 10^4.

Cationic initiation may result from different mechanisms which may be classified in two main categories: (1) addition of an electron-deficient moiety on the double bond (this fragment may already exist as such in the compound used as initiator or it may be produced *in situ* by a chemical reaction between the components of an initiating system), and (2) removal of one electron from the double bond yielding a radical cation. This may be achieved by chemical means (strong oxidants, formation of charge-transfer complexes) or by physical means (electromagnetic radiation, electric field, electrochemical oxidation).

Since the first extensive review of the literature up to 1963,[1] general reviews of carbocationic polymerization including the initiation reactions have been published recently.[2-6] A more comprehensive analysis and discussion of the data on initiation reactions has been made by Gandini and Cheradame.[7]

39.2 ELECTROPHILIC ADDITION TO THE CARBON–CARBON DOUBLE BOND

39.2.1 Initiation Involving Protonic Acids

39.2.1.1 *Initiation by protonic acids used alone*

(i) Classification of protonic acids (Brönsted acids)

Brönsted acids HA used alone may polymerize ethylenic monomers, following the electrophilic addition of a proton to the monomer as pictured symbolically in equation (8) but in fact involving several acid molecules (see below). The lifetime of the carbocation formed (and the possibility for a polymer to be formed) depends both on its intrinsic stability and on the nucleophilicity of the counterion A^-. Propagation also competes more easily with termination and transfer reactions when solvation by polar solvents is possible, and when the temperature is lowered.

The most reactive monomers generally give the most stable carbocationic species since the latter are stabilized by resonance (*e.g.* by aromatic rings) and by the possible electrodonating effect of the substituent. It has been found that even weak protonic acids such as acetic acid or HCl (weak in non-aqueous solvents) may polymerize a very reactive monomer such as vinylcarbazole.

The most reactive acids for initiation are those with the highest acidities (*i.e.* the lowest pK_a) which also generally correspond to the lower nucleophilicity of their counterions A^-. Acids with high pK_a either do not initiate the polymerization or give extensive termination by recombination of the nucleophilic A^- with the carbocation. This happens for the reaction of propylene and isobutylene with HCl for which the corresponding chloride (monoadduct) is formed. With a stronger acid such as sulfuric acid (used at high concentration) and isobutylene, some propagation occurs and dimers and trimers are formed. But the abstraction of the proton by the counterion is preferred (also linked to its nucleophilicity) corresponding to a transfer reaction (equations 12–14).

$$H_2SO_4 + CH_2{=}CMe_2 \longrightarrow Me_3C^+HSO_4^- \tag{12}$$

$$Me_3C^+HSO_4^- + CH_2{=}CMe_2 \longrightarrow Me_3CCH_2\overset{+}{C}Me_2HSO_4^- \tag{13}$$

$$Me_3C\overset{+}{C}H_2CMe_2HSO_4^- + H_2SO_4 \underset{\diagdown}{\overset{\diagup}{}}
\begin{matrix} Me_3CCH{=}CMe_2 \\[6pt] Me_2CCH_2CMe \\ \underset{CH_2}{\overset{\|}{}} \end{matrix}
\; + \; 2H_2SO_4 \tag{14}$$

With ethylenic monomers substituted by an aromatic group (such as styrene) higher polymers may be formed (*vide infra*).

A classification of protonic acids according to their pK_a values corresponds roughly to their initiating ability. Some pK_a values are given in Table 2 and vary similarly in different solvents, showing that trifluoromethanesulfonic acid (triflic acid) and perchloric acid should be the most effective protonic initiators, as has been verified experimentally. With them, both recombination of the anion with the carbocation (termination) or abstraction of a proton on the β carbon (transfer) should be less important than with weaker acids. Significant values of the pK_a in the most used solvents for cationic polymerization, alkyl chlorides, are unfortunately not available.[7]

It should be noted that the pK_a values found in the literature did not always take into account the homoconjugation phenomenon, that is, the association of the anion (and eventually of the proton) with the undissociated acid. The matter is discussed in ref. 7, p. 8–9, where it is suggested that a dissociation equilibrium should in fact involve three acid molecules (equations 15–17).

$$HB + HB \rightleftharpoons (HB)_2 \tag{15}$$

$$(HB)_2 + HB \rightleftharpoons H_2B^+ + HB_2^- \tag{16}$$

Table 2 pK_a Values of Some Brönsted Acids[a]

Acid	Solvents	
	$MeCO_2H$	$MeCN$
CF_3SO_3H	4.7	2.6
$HClO_4$	4.9	1.6
FSO_3H	6.1	3.4
H_2SO_4	7.0	7.3
HCl	8.4	8.9
$MeSO_3H$	8.6	8.4
CF_3CO_2H	11.4	10.6
CCl_3CO_2H	12.2	12.7
$MeCO_2H$	12.8	22.5

[a] A. Gandini and H. Chéradame, *Adv. Polym. Sci.*, 1980, **34/35**.

or

$$3HB \rightleftharpoons H_2B^+ \;+\; HB_2^- \tag{17}$$

For CF_3SO_3H, dimerization in acetonitrile was reported by Kolthoff *et al.*[16] but was not observed later in dilute solution.[17] It might occur in less polar CH_2Cl_2 or $(CH_2Cl)_2$, but Chmelir[18] observed a linear dependence of the molar fraction of free ions α as a function of concentration which may be explained by the self-ionization of two acid molecules (equations 18 and 19).

$$HA \;+\; HA \rightleftharpoons HA_2^- \;+\; H^+ \tag{18}$$

$$HA \;+\; HA \rightleftharpoons H_2A^+ \;+\; A^- \tag{19}$$

For the reasons stated above, trifluoromethanesulfonic acid and perchloric acid (and its derivatives) have been the most studied in recent years and are those for which some conclusions could be reached. They will be discussed first.

(ii) Perchloric acid

Extensive studies have been made of the polymerization of styrene in ethylene dichloride (EDC) or methylene chloride (MC), which gave similar results.[19, 20, 21] Polymerizations realized near room temperature have given only oligomers but apparently these occurred without termination reactions since they resumed with the same rate after a new monomer addition. The reaction mixture was colourless during polymerization (with a very low conductivity) and active centres could not be detected. A yellow colour ($\lambda_{max} = 415$ nm) that has been ascribed to indanyl cations appeared after total monomer consumption. Propagation has been assumed to involve the perchloric ester and to occur by a 'pseudocationic' mechanism[19] with either a four-membered transition state, or more likely a six-membered one (equation 20):

$$\tag{20}$$

The main arguments in favour of a propagation involving non-ionic species (see Chapter 41 of this volume) have been the apparent absence of any visible cations and of a significant effect of water on the rate. This theory has also been supported by the bimodality of the polymer made in CH_2Cl_2[22] and by some results of block copolymerization showing the presence of reactive end groups in the polystyrene.[23]

But the absence of UV absorption does not exclude the possibility of a mechanism[6] involving ions in very low concentration if they are very reactive (*e.g.* $k_p^+ \approx 10^4$ to 10^5) and reinitiation by the acid resulting from hydrolysis of the ester might be possible. It has also been shown recently that the model ester 1-phenylethyl perchlorate reacted rapidly with water even at $-70\,^\circ C$ to give the acid[24] and it is unlikely it would be unreactive near room temperature, even if stabilized by the monomer.[20]

It was also found that at $-40\,°C$ and below, polymerization of styrene occurs in two successive stages.[22,25] During stage 1, which is very rapid and lasts only a few seconds, a transient UV absorption was observed with a maximum at 340 nm, which was attributed to the polystyryl cation and allowed the calculation of ionic propagation rate constants. A simultaneous transient conductivity was observed. Below $-80\,°C$ (*e.g.* $-97\,°C$) the reaction practically stopped before complete conversion. At $-80\,°C$ and higher temperatures a much slower polymerization occurred during stage 2, with a stationary concentration of active species as deduced from the kinetics. Stage 2 seems analogous to what happens near room temperature, and no UV absorption is observed.

Carbocationic species P* are formed during stage 1 proportionally to $[HA]_0^2$ but with an approximately inverse dependence on $[M]_0$ (from 0.05 M to 0.2 M).[25,26] To explain the second order in acid, the virtually instantaneous initiation was suggested as involving dimeric forms of the initiator that then ionize. The yields of P* were very low and of the order of 1% of the acid, and the explanation given for its decrease with increasing monomer concentration was the competitive formation of the ester. The polymers formed at the monomer concentrations used were bimodal.[25,26] The oligomeric fraction was assumed to result from a non-ionic propagation on esters (mainly during phase 2 but also during phase 1) and the polymer with higher molecular weight (10^4) to result from an ionic reaction. In a later article,[27] the order two in $(HClO_4)_0$ for the reaction rate at $0\,°C$ in CH_2Cl_2 was confirmed for $[M_0]/[C_0]$ ratios of five to ten, but for higher ratios, the order was equal to one, which was ascribed to the predominance of the 'non-ionic' contribution.

(iii) Trifluoromethanesulfonic (triflic) acid

The reaction with styrene carried out in CH_2Cl_2 solution was found to show several analogies to the initiation with perchloric acid, with an initial rapid rate at $-14\,°C$ or $-52\,°C$ followed by a slower rate.[28,29] No UV absorption was observed except that at 415 nm resulting from the slow appearance of indanyl ions formed after the termination reaction. A reaction order between 2.6 and 3 was found for the slow rate with respect to $[HA]$ and the initial polymerization rate depended on monomer concentration in a complex manner, passing through a maximum around a concentration of 0.1 M. The decrease in rate at higher concentration was attributed to the formation of an inactive complex between the monomer and the acid.[28] However, this led to a rather unlikely value of the assumed association constant ($K_m = 23.1\ \mathrm{mol}^{-1}$ at $-15\,°C$).

There was a correlation between the rates and the conductivities of polymerization solutions when $[M]_0$ was varied.[29] Polymerization was assumed to occur through ionic species formed from two or three molecules of acid. From a controversy about the existence of the complex between monomer and acid and about the interpretation of the rate decrease when $[M]_0$ increases,[30-32] it may be concluded that an apparent internal first order in monomer is observed from -15 to $-60\,°C$ for low monomer concentrations but is verified less satisfactorily for high monomer concentration, and that the complex formation hypothesis is not satisfactory to explain the quantitative data. Another analogy with perchloric acid is the polymodality of the polymer samples obtained in CH_2Cl_2 for high monomer concentrations either at $0\,°C$[33] or at -15 or $-60\,°C$[30] (see Volume 3, Chapter 41).

The kinetics of initiation and propagation of the fast polymerization of styrene giving ionic species (UV maximum at 340 nm) have been studied by Kunitake *et al.*[34] using a stopped-flow technique with rapid scan spectroscopy. Experiments were made between $-1\,°C$ and $30\,°C$, with acid concentrations (~ 2.5 to 10×10^{-3} M) similar to those used by Léon *et al.* with perchloric acid.[26] The results are similar to those observed at lower temperatures with perchloric acid. There is a very rapid formation of carbocations, their maximum concentration being reached in 20 to 100 ms, followed by their disappearance.

The maximum yields of carbocations (calculated on the assumption of an ε_{340} of 10^4) were low, between 1% and 10%, and increased strongly with acid concentration. The initiation rate was proportional to initiator concentration and roughly first order in monomer, and was assumed to be a simple protonation of styrene (equation 8). Initiation rate constants k_i were calculated assuming that $[HA]_0$ remained constant until the maximum was reached and gave rather low values: $k_i \approx 10\ \mathrm{M}^{-1}\mathrm{s}^{-1}$ at $-1\,°C$ and $\approx 20\ \mathrm{M}^{-1}\mathrm{s}^{-1}$ at $30\,°C$. However, results obtained later for polymerizations made at $-78\,°C$[35] have shown that reformation of acid from the growing end was extremely rapid. The low yields in active species are probably the consequence of the balance between a very fast initiation and a very fast deactivation by proton abstraction, and a scheme based on this assumption could explain satisfactorily the decrease of P* concentration when $[M]$ or temperature increased.[35]

α-Methylstyrene initiation has also been followed by stopped-flow techniques.[36] The initiation rate constant k_i for the formation of a carbocation ($\lambda_{max} = 326$ nm) was evaluated to be of the order of 1 to 3×10^3, much higher than for styrene.

A similar stopped-flow technique was used by Sawamoto *et al.*[37] for a kinetic study of the polymerization of *p*-methoxystyrene by triflic acid in CH_2Cl_2 at 30 °C. A very rapid formation of ionic species with a UV peak at 380 nm was observed (reaching a maximum in 20–40 ms) followed by a rapid decrease of their concentration. The 380 nm peak was assumed to be that of the expected *p*-methoxystyryl carbocation, and using an ε of 2.8×10^4 for this species, the initiation yields were calculated to be much higher than in the styrene case (from 36 to 42% based on the acid). Initiation rate constants k_i were calculated assuming a linear increase with time during the time t_d needed to reach the maximum, and a first order in monomer and acid (equation 21).

$$R_i = \frac{\Delta OD_{380}}{\varepsilon_{380}} \frac{1}{t_d} = k_i[M]_0 [HA]_0 \tag{21}$$

Values of k_i of the order of 50 000 $M^{-1} s^{-1}$ were calculated, much larger than those obtained by Kunitake with styrene. From monomer conversion and variation of optical density with time, k_p values could also be obtained (see Chapter 41 of this volume).

However, it seems unlikely that the peak at 380 nm corresponds to the expected 'classical' growing *p*-methoxystyryl cation since the corresponding monomeric cation formed with the alcohol has a maximum around 348 nm. But it is true that there is a strong correlation between the peak intensity and the polymerization rate, and the growing species, in fact much more reactive, might be in equilibrium with the species absorbing at 380 nm.

The main difficulties in the identification of carbocationic species are their instability and short lifetimes. This was the reason for the use of model compounds such as 1,1-diphenylethylene (DPE) that gives well identified carbocations stable at −30 °C and below. The most conclusive data have been obtained with DPE and CF_3SO_3H in combined ^1H and ^{19}F experiments.[39] It was found that there was formation of only one carbocation per three acid molecules (between −30 °C and −80 °C in CH_2Cl_2),[38,39] the two unreacted molecules being trapped as conjugate acid in the counterion (equation 22).

$$CH_2{=}CPh_2 \;+\; 3CF_3SO_3H \longrightarrow Me\overset{+}{C}Ph_2(CF_3SO_3^- 2CF_3SO_3H) \tag{22}$$

This 33% yield in carbocations did not change with monomer concentration.

A similar homoconjugation was found for the counterion of triphenylmethyl cation, stable at +20 °C, and formed by reaction of Ph_3COH with triflic anhydride (giving $Ph_3C^+H(OSO_2CF_3)_2^-$). On the other hand, reaction of the anhydride with $MeCH_2OH$ or $PhCH_2OH$ gave the corresponding stable triflic esters.[39] With the secondary alcohol $Ph(Me)CHOH$, the NMR data were first interpreted by the presence of the corresponding ester $PhMeCHOSO_2CF_3$, a model for the ester hypothetically formed during styrene polymerization. But this was not confirmed in later studies in which various attempted syntheses of this ester led to ethers or to secondary Friedel–Crafts reactions even at low temperatures.[35]

A well identified carbocation was also formed in the reaction of 3-isopropylindene with CF_3SO_3H at −70 °C[40] and by that of 3-phenylindene.[38] For this last monomer the yield was 50% at −60 °C, in agreement with a homoconjugation giving the $(CF_3SO_3^-, CF_3SO_3H)$ counterion. The higher initiation yield (compared to DPE) probably results from the higher reactivity of the monomer. More recently, well identified carbocations (by ^1H NMR) were also observed for some *p*-substituted α-methylstyrenes (with OMe[41] or CMe$_3$).[42] The monomeric cations (with λ_{max} respectively at 363 and 350 nm) were formed with an excess of the acid and the dimeric carbocations had similar spectra. However the higher oligomeric (or polymeric) species exhibited changes in the UV and NMR spectra (see Chapter 40 of this volume) showing the presence of 'isomerized' species.

The initiation rates were too rapid to be followed using these monomers with the technique used. But for *p*-isopropyl-α-methylstyrene in CH_2Cl_2 at −40 °C, for lower M and HA concentration, stopped-flow experiments showed that the yield in 'isomerized' species was complete in less than 10 ms showing that k_i was higher than 5×10^3.[43]

In conclusion, initiation of ethylenic monomers with triflic acid did not give a detectable concentration of covalent ester species, the difference with perchloric acid probably resulting from the higher nucleophilicity of the ClO_4^- ion. With triflic acid, only ionic species have been detected and sometimes identified when the carbocations were stable enough.

(iv) Sulfonic acids other than CF_3SO_3H

$ClSO_3H$ and FSO_3H both rapidly initiate styrene polymerization, but give limited yields, showing also an important termination reaction. Initiation occurs in liquid SO_2,[44] and in various solvents[45] (CH_2Cl_2, $PhNO_2$, benzene).

A more detailed study of the initiation step has been made with $MeSO_3H$ for *p*-methoxystyrene polymerization at $30\,°C$ in $(CH_2Cl)_2$, using a stopped-flow technique.[46] The peak observed in the UV at 380 nm reached a maximum ($[P^*]_{max}$) in 0.5 to 1 s, but with a yield of about 10^{-3} based on the acid (polymer with a molecular weight of 6.5×10^3, showing a dominant transfer). However, in $(CH_2Cl)_2$ mixed with CCl_4,[47] $[P^*]_{max}$ decreased considerably, nearly disappearing for a volume fraction of 40% CCl_4 while the global rate decreased only moderately. This implied the presence of propagating species not absorbing at this wavelength or in the neighbourhood ('invisible' species). Propagation involving activated esters was suggested.

The rates of formation of carbocationic species in $(CH_2Cl)_2$ were found to be intermediate between those obtained with I_2 and $BF_3 \cdot OEt_2$.[48] Apparent initial rate constants of initiation have been calculated assuming a reaction which is first order in monomer and in acid, and also lifetimes τ of the 'propagating' species, but this approach has been criticized.[49] Since steady state conditions are not operative, a calculation of k_i and τ is not possible if k_t is unknown.

(v) Sulfuric acid

Polymerization of styrene by sulfuric acid was one of the first examples of polymer formation through cationic polymerization[50] but was not much studied later on. A study by Pepper and co-workers[51-54] was also one of the first kinetic treatments of a chain polymerization without a stationary state. Polymerizations were rapid at $25\,°C$ in $(CH_2Cl)_2$ and led (in 10 to 20 min) to limited yields (for $H_2SO_4 \leq 4 \times 10^{-3}$ M) that could reach 100% for higher H_2SO_4 concentration. Initiation was assumed to be fast and to correspond to the formation of a concentration of ionic active species equal to that of the acid. The progressive decrease of the rate was interpreted by a termination reaction giving polystyryl bisulfate. However, it was observed later by Gandini that no absorption in the UV above 320 nm was seen during the reaction,[55] and also that the kinetics could not be fully explained by Pepper's scheme.[56] It was also observed that a polymerization with limited yield could start again by addition of an excess of a weaker acid such as CF_3CO_2H or CCl_3CO_2H, which did not polymerize styrene at the concentrations used. The general explanation proposed[56] was that an ester of sulfonic acid was formed that was activated by H_2SO_4 (or eventually by the other acid) (Scheme 1).

Initiation: $\begin{cases} M + HA \longrightarrow HMA \\ HMA + HA \longrightarrow HMA \cdot HA \end{cases}$

Propagation: $HM_nA \cdot HA + M \longrightarrow HM_{n+1}A \cdot HA$

Termination: $HM_{n+1}A \cdot HA + M \longrightarrow HM_{n+1}A + HMA$

Scheme 1

The esters would be inactive, but might be reactivated by addition of a weaker acid in high concentration.

Polymerization of α-methylstyrene by H_2SO_4 in CH_2Cl_2 solution[57] has given high molecular weights but limited yields even at $-70\,°C$ after more than 6 h. A stopped-flow study was made at $30\,°C$.[58] With an excess of acid ($[HA]/[M] \approx 20$), a carbocationic species was formed rapidly ($\lambda_{max} = 336$ nm) with a maximum intensity after 20 ms that disappeared after 100 ms. The pseudo-first order $k_{i,app}/[HA]_0$ was of the order of 2×10^3 $M^{-1}\,s^{-1}$.

(vi) Other inorganic acids

Hydrogen halides initiate the polymerization of reactive monomers such as vinyl ethers, but they are not very efficient.[59] *N*-vinylcarbazole polymerization in CCl_4 at $20\,°C$ is initiated in HCl, HBr and HI.[60] The initiation of styrene has been shown to occur in CH_2Cl_2 with fairly high concentrations (0.1–0.2 M) of dry HCl at $-78\,°C$,[61] giving high molecular weight polymers, but did not take place in liquid SO_2.[62]

HI is a very efficient initiator for the polymerization of isobutyl vinyl ether in toluene or in CH_2Cl_2 (at $-15\,°C$, 100% yield in 1–2 min) but initiation (and polymerization) in *n*-hexane was very slow.[63] HI also initiated the polymerization of *trans*-anethole in $(CH_2Cl)_2$.[64]

(vii) Haloacetic acids

Unsubstituted carboxylic acids are very weak cationic initiators, and only the polymerization of *N*-vinylcarbazole initiated by $MeCO_2H$ and various other acids has been reported,[65] giving oligomers including the corresponding acetates.

Monochloroacetic acid initiates the polymerization in chlorinated solvents of vinylcarbazole[66] and of α-methylstyrene,[67] but not that of styrene.

Di- and tri-chloroacetic acids initiate styrene polymerization (at room temperature).[68] In these studies, very large initiator concentrations were used (2 to 5 M!). The reaction was slow, but occurred more rapidly in more polar solvents. In bulk, addition of water was necessary to have a significant and reproducible rate. The initial rates were second order in styrene and of varying order in acid. A discussion of the complex kinetics is made in ref. 7, p. 55–57, and explained on the assumption of a pseudocationic mechanism.

Trifluoroacetic acid is the most effective initiator among the haloacetic acids, but only in solvents with high dielectric constants. For example, it was found that addition of CF_3CO_2H in benzene to styrene gives only oligomeric esters, while styrene introduced in CF_3CO_2H gives a high polymer (molecular weight $\sim 15\,000$).[69] This last result was attributed to the anion stabilization by acid in excess.

Polymerizations of styrene by CF_3CO_2H at 50 °C in solutions of various dielectric constants[70] were realized with rather high initiator concentrations ($[CF_3CO_2H] = 0.2$ M, $[M] = 1$ M). Even in nitrobenzene the rates were small and decreased together with the solvent polarity, the molecular weight distribution being bimodal in $(CH_2Cl)_2$ and in mixed solvent with benzene. ^{19}F NMR spectra showed the progressive transformation of the acid into the ester. This one did not initiate the polymerization when it was used independently. The kinetics could be explained on the assumption of an initiation rate that was first order in acid and by the complete inactivity of the esters,[70] with a propagation on ionic chain carriers, but this has been challenged.[71]

Polymerization of styrene at 40 °C in CH_2Cl_2, with similar concentrations of reagents ($[M]_0 = 1$ M, $[HA]_0 = 0.3$ to 1 M) has given very different results according to the water content (even low) of the reagents.[72] The rate decreased when $[H_2O]$ increased, even if the global order in acid was equal to three in all cases. The oligomers of small molecular weight ($\sim 3 \times 10^2$) in the polymer disappeared in the absence of water, only the high polymer (molecular weight $\approx 10^4$) remaining. The order in monomer was equal to one in the presence of H_2O and to two in dry systems. However, the general conclusions are similar to those of ref. 70: the high polymer is formed through ionic species and the oligomers through non-ionic species, resulting from the activation of the ester (probably by the acid) which is inactive when alone.

Rather different results and conclusions were reached in the study of the much more reactive isobutyl vinyl ether (IBVE) polymerization by CF_3CO_2H in $(CH_2Cl)_2$[73] (between -2.5 and 35 °C). Much lower acid concentrations (2×10^{-4} to 3×10^{-3}) than for styrene were needed, even if polymerization did not take place for $[HA]_0$ equal to 10^{-4} M. The initial rates were found to be first order in acid and second order in monomer. The effect of water concentration was slight (except for the occurrence of an induction period), even when it was in small excess over the acid. The salt effect of $NEt_4^+ CF_3CO_2^-$ was also very small. The initiation step was considered to be the addition of HA to IBVE to give the ester, which would then become solvated by the monomer to give the active species, and the termination step to involve the reformation of the acid.

(viii) Initiation with protonic acids: general conclusions

The problem of the type of active species responsible for polymerization and that of their formation cannot be considered to be completely solved, and the answers are different according to the monomer and acid used. For monomers giving stable or relatively stable carbocationic species. (*e.g.* by reacting DPE and various *p*-substituted styrenes with triflic acid) carbocations have been shown to be formed exclusively. Their formation may also explain the data observed with styrene with this acid, for which also no evidence for the presence of esters could be found.

With perchloric acid, the ester derived from styrene has been identified at -40 °C and below,[24] and might eventually contribute to propagation. But it is unstable in the presence of water, and the occurrence of a 'pseudocationic' polymerization near room temperature seems unlikely since it still occurs readily in the presence of H_2O.

For weaker acids such as trifluoroacetic acid, esters with styrene have been isolated and shown to be inactive for initiation when they are alone, but there are good arguments (and also for sulfuric acid) in favour of an initiation involving the interaction of the ester with the acid. But the type of active species (ionic or not) is still unknown, the situation being similar to that described for other esters activated by Friedel–Crafts acids or iodine (see Section 39.2.2.2).

39.2.1.2 Activation of weak Brönsted acids by Friedel–Crafts acids

(i) The 'cocatalysis' phenomenon and its general explanation

It will be seen in Section 39.2.3 that some Friedel–Crafts halides (denoted F–C acids[2]) such as $AlCl_3$ or $TiCl_4$ may initiate the polymerization of various monomers without intervention of a third component, originally called a 'cocatalyst'. However, these 'direct initiations' are strongly dependent on general experimental conditions (such as temperature) and their contribution to initiation may be small, particularly with the less acid metal halides, for which the presence of a third component, such as a weak Brönsted acid, may be necessary for initiation to occur.

The distinction between these two modes of initiation is not easy since very low concentrations of third components have been shown to be still active (this is particularly the case for water) and it is not possible to prove their total absence. In order to have significant and reproducible results, very stringent polymerization conditions are necessary, such as purification of the reagents on several films of alkali metals and polymerization carried out in apparatus completely sealed under high vacuum. Some clear-cut data about the occurrence of cocatalysis phenomena could however be obtained by the so-called 'stopping experiments' described first by Evans, Plesch, Polanyi and co-workers,[74] for which polymerization either did not occur in the presence of the F–C acid, or occurred very slowly, or stopped before complete conversion. This first reaction might result either from direct initiation or from the presence of residual cocatalyst. But if the further addition of a third substance led to rapid polymerization, this was proof of a cocatalytic effect.

There are several examples of reactions of this type for various monomers and F–C acids, but most of them are purely qualitative since the active centres concentration (and the efficiency of the initiating system) could not be measured. It is generally admitted that 'cocatalysis' results from an enhancement of the acidity of weak Brönsted acids by the metal halide, permitting the attack of the monomer by a proton. Only a few experiments in which stable carbocationic species could be observed have given quantitative data on initiation.

Another type of 'cocatalysis' will be discussed in Section 39.2.2.2.i involving alkyl or arylalkyl halides.

The mechanism of initiation involving weak protonic acids is still conjectural and we shall see that the experimental data vary considerably according to the type of initiation system. Originally, it was assumed that initiation always resulted from the reaction of the alkene with a complex of the protogen with the Friedel–Crafts acid (Scheme 2).

$$HX \ + \ MtX_n \ \rightleftharpoons \ HX \cdot MtX_n \ \ (\text{or} \ \ H^+MtX_{n+1}^-) \tag{23}$$

$$H^+MtX_{n+1}^- \ + \ M \ \longrightarrow \ HM^+MtX_{n+1}^- \tag{24}$$

Scheme 2

However, the existence of stable complexes $HX \cdot MtX_n$ (where Mt = metal) has been shown only in some rare cases (*e.g.* for $H_2O \cdot BF_3$) and they are generally not formed with hydrogen halides (*e.g.*, from $TiCl_4$ and HCl) so that a second possibility was considered (Scheme 3).

$$M \ + \ MtX_n \ \rightleftharpoons \ M \cdot MtX_n \tag{25}$$

$$M \cdot MtX_n + HX \ \longrightarrow \ HM^+MtX_{n+1}^- \tag{26}$$

Scheme 3

Both Schemes 2 and 3 lead to a rate of initiation of the type $R_i = k_i[M] [MtX_n] [HX]$ which has sometimes been observed, particularly for $[HX] \ll [MtX_n]$. But the situation is generally more complex for Scheme 2 since several types of complexes may be formed between MtX_n and HX (particularly with water).

A third scheme has also been proposed,[75] for which the initiation step involves both types of complexes formed in equilibria (23) and (25) (see Scheme 4).

It may be occurring in reactions for which $R_i = k_i[M] [MtX_n, HX] [MtX_n]$.

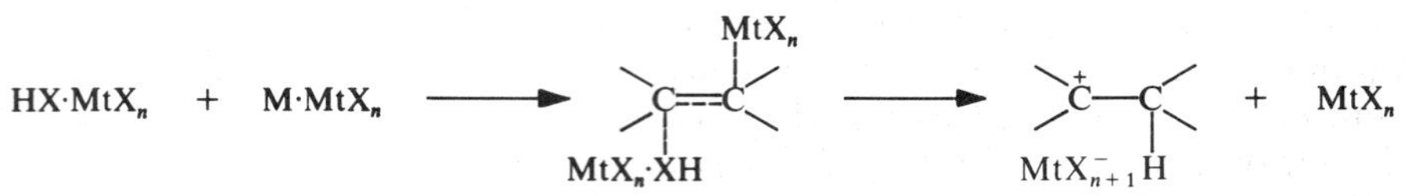

$$\text{HX·MtX}_n \ + \ \text{M·MtX}_n \ \longrightarrow \ \underset{\text{MtX}_n\text{·XH}}{\overset{\text{MtX}_n}{\text{C=C}}} \ \longrightarrow \ \underset{\text{MtX}_{n+1}^-\text{H}}{\overset{+}{\text{C—C}}} \ + \ \text{MtX}_n$$

Scheme 4

(ii) *Efficiencies of the Friedel–Crafts acids and of the weak Brönsted acids*

The efficiencies of F–C acids for the initiation of carbocationic polymerization are not easy to compare since only data relative to yields or global polymerization rates and molecular weights are available, and it is not generally known what are the respective contributions of 'cocatalyzed' and 'direct' initiation. Moreover, a higher yield (or polymerization rate) may result from either a faster initiation (and propagation) or from a slower termination reaction.

The following order (based on yields) has been found for isobutylene (IB) at $-80\,°\text{C}$;[76] $\text{BF}_3 > \text{AlBr}_3 > \text{TiCl}_4 > \text{TiBr}_4 > \text{SnCl}_4 > \text{BCl}_3 > \text{BBr}_3$. The molecular weights followed the same trend except for SnCl_4 (giving the lowest value.) For styrene in $(\text{CH}_2\text{Cl})_2$/benzene mixtures at $30\,°\text{C}$, the rates varied in the order:[77] $\text{TiCl}_4 > \text{SnCl}_4 > \text{FeCl}_3 > \text{BF}_3,\text{OEt}_2$. However, the effect of proton donors was not considered for the two series mentioned.

Similar considerations have been applied to relative cocatalyst efficiencies. For example, for the polymerization of IB by SnCl_4 at $-78\,°\text{C}$ in EtCl, the rates fell in the order:[78] $\text{Cl}_3\text{CCO}_2\text{H} > \text{ClCH}_2\text{CO}_2\text{H} > \text{MeCO}_2\text{H} > \text{MeNO}_2 > \text{PhOH} > \text{H}_2\text{O}$ which was interpreted by a decreasing sequence of the dissociation constants, but might result also from the termination reaction. These results may also be misleading since it was found later[79] with styrene that the rates with the system $\text{SnCl}_4/\text{Cl}_3\text{CCO}_2\text{H}$ were extremely low when the reagents were purified and used under vacuum, and that H_2O or MeOH added in low concentration with respect to $\text{Cl}_3\text{CCO}_2\text{H}$ (*e.g.* 10 mol %) increased the rate considerably and suppressed the induction period.

For IB polymerization initiated with AlEt_2Cl, the following order of decreasing activity of proton donors was also reported.[80] $\text{HCl} \sim \text{HBr} > \text{HF} \sim \text{H}_2\text{O} > \text{MeCO}_2\text{H} \gg \text{MeOH} > \text{MeCOMe}$. But we shall see that for other Friedel–Crafts acids (BCl_3, AlCl_3, AlBr_3, TiCl_4...) water is generally much more efficient than HCl, which is often inactive. Many examples have been given of the effect of water on cationic polymerization, in which it may play the role of a terminating (or transfer) agent, or that of a 'cocatalyst'. H_2O is always present in some amount in the reagents or on the walls of the vessels in which they are contained, which explains the difficulty of obtaining definite proof, and the evidence for cocatalysis could be best established for F–C acids which did not lead to direct initiation, such as boron trifluoride. We shall examine successively the main examples of 'cocatalysis' with F–C acids, starting from the most active ones.

(iii) *Boron trifluoride*

The first example in which a cocatalytic effect of H_2O was demonstrated was that of the polymerization of IB by BF_3. When these two reagents were sufficiently purified under high vacuum[81] they could be mixed either in the gas phase (at room temperature) or condensed in the liquid phase (at $-80\,°\text{C}$) with either a very slow polymerization or no polymerization at all (according to the purification of the reagents). The addition of small quantities of H_2O (added either to the monomer or to its mixture with BF_3) led to rapid polymerization, giving oligomers ($\overline{\text{DP}}_n \sim 4$ to 30) in the gas phase, and high polymer in the liquid phase. Similar results were observed with the addition of *t*-butyl alcohol or acetic acid. These experiments[81] were realized with relatively high concentrations of BF_3 ($[\text{BF}_3]/[\text{IB}] \approx 10$ to 50 mol %). A very small amount of water (*e.g.* that adsorbed on the walls) was enough to bring about the polymerization, but more quantitative experiments were also made (with $[\text{ROH}]/[\text{BF}_3] \approx 1$ to 30%).

Attempts at proving the addition of protons in the initiation step have been made by using $^3\text{H}_2\text{O}$[82] ($\sim 1\%$ of the BF_3). Polymerizations in heptane solution at $-78\,°\text{C}$ proceeded very rapidly even without added water, and in the presence of $^3\text{H}_2\text{O}$, the quantity of tritium incorporated was about 1% of the number of macromolecules. If water is active, it is so in 'subanalytical' concentration. But no evidence for incorporation during initiation could be obtained since tritium incorporation in the polymer was the same when $^3\text{H}_2\text{O}$ was added before the polymerization or *after* it had taken place,[82] showing that exchange reactions could occur rapidly between some H atoms of the polymer and tritiated water. Earlier data about the small incorporation of D atoms with the use of D_2O are not conclusive either.[83]

The *cis–trans* isomerization of 2-butene by $BF_3 + H_2O$ has been reasonably assumed to involve the formation of carbenium ions and might be considered as a model of an initiation reaction. In $(CH_2Cl)_2$ solution at 25 °C, with carefully purified reagents, Eastham *et al.*[84] observed a strong effect of water with a maximum rate for $[H_2O]/[BF_3]$ between 1/3 and 1/2. They showed later[85] that two hydrates, $H_2O \cdot BF_3$ and $2H_2O \cdot BF_3$ were formed in equilibrium with free BF_3, and the maximum in the rate may be explained if the formation of the dihydrate is faster than that of the monohydrate. The rate of isomerization R_i followed the law: $R_i = k [IB] [BF_3 \cdot H_2O] [BF_3]$. This is compatible with a mechanism involving the reaction of the monohydrate with the complexed alkene such as in Scheme 4. Similar results were observed with other proton donors such as methanol[86] or acetic acid,[87] which were much less effective than water.

In the cationic oligomerization of 3-methyl-1-butene in hexane at 0 °C, it was shown later that the complex $H_2O \cdot BF_3$ is very active and $2H_2O \cdot BF_3$ inactive.[88]

The effect of water on the formation of carbocations from DPE and BF_3 has been studied in CH_2Cl_2 at 0 °C.[89] These cations are not stable at this temperature and extrapolations to zero time were made to measure their concentration from the UV absorption. This initial concentration showed a maximum for $[H_2O]/[BF_3] \approx \frac{1}{2}$ which could be expected from the above equation.

(iv) Boron trichloride

BCl_3 has been long considered to be a weak cationic initiator. However, it was found later that it could be very effective at low temperatures in the presence of specific cationogens such as H_2O[90] and arylalkyl halides.[159] Using purified reagents and sealed apparatus, no polymerization occurred at -78 °C with pure IB and $[BCl_3] = 0.12$ M even with further water addition. But in CH_2Cl_2 or MeCl solution polymerization occurred with $[BCl_3] = 2 \times 10^{-2}$ M, with small yields that increased considerably if water was premixed with the monomer, being nearly total when $[H_2O]/[BCl_3] \approx 1$.[90]

A more detailed investigation of the effect of H_2O and of temperature[91] showed that the yield passed through a maximum for $[H_2O]/[BCl_3] = 1$ at both -50 °C and -78 °C and that polymerization was negligible at -40 °C and higher, which was attributed to a strong termination reaction.

The effect of HCl has also been investigated,[91] and it was found to be less efficient than H_2O with $[HCl]/[BCl_3]$ equal to unity. The increase of rate in its presence was significant at -78 °C, but not at -50 °C. This was attributed again to a rapid termination at higher temperatures, following the initiation step (equation 27).

$$HCl + BCl_3 + CH_2{=}CMe_2 \longrightarrow Me_3C^+ BCl_4^- \longrightarrow Me_3CCl + BCl_3 \qquad (27)$$

The absence of a cocatalytic effect of HCl has been confirmed in high purity conditions at -30 °C[92] ($[IB] = 1$ M; $[BCl_3] = 10^{-2}$ M; $[HCl] = 1.6 \times 10^{-3}$ M).

(v) Aluminum trichloride and chloroalkylaluminums

$AlCl_3$ is the most important F–C acid and it is used in the industrial synthesis of butyl rubber and of 'polybutene's.[93] Unfortunately there has been little work done in high purity conditions and consequently very few clear-cut data about the effect of proton donors. One difficulty of the studies comes from the low solubility of $AlCl_3$, even in some alkyl halides.

(a) Effect of water. Styrene in bulk did not react with (insoluble) sublimed $AlCl_3$, but polymerization occurred after addition of water vapour.[94] It has also been mentioned that styrene in MeCl solution was not initiated by $AlCl_3$ at -45 °C.[95]

There are some arguments in favour of a possible direct initiation of IB in chlorinated solvents (see Section 39.2.3.1.i). It is usual however, in industrial polymerizations at -100 °C in MeCl, to premix some water with the $AlCl_3$ solution in order to have higher and more reproducible yields. On the other hand, with IB in CH_2Cl_2 at -35 °C, an increase in water concentration had no effect on the rate, but led to the formation of an appreciable quantity of oligomers.[96] Experiments at -60 °C showed also that water had no effect on the initial rate, but, strangely enough, it inhibited the polymerization at -94 °C. It was concluded that water is not required as cocatalyst, and is an effective chain breaker.

The cocatalytic effect of water could be measured quantitatively for the model system $DPE, AlCl_3, H_2O$ from the yield of stable carbocations measured by spectrophotometry.[97] At -30 °C in CH_2Cl_2, direct initiation is relatively slow. When the monomer solution contains some water, initiation is much more rapid and complete in about 1 min. The quantity of rapidly formed carbocations R$^+$ is proportional to [AlCl$_3$] up to [AlCl$_3$]/[H$_2$O] = 2 and becomes independent of [H$_2$O] for higher ratios. The stoichiometry of the reaction is: H$_2$O + 2AlCl$_3$(+ DPE)→$\frac{4}{3}$R$^+$. This means that more than one proton of water but not all the protons resulting from the hydrolysis of AlCl$_3$ are active (which is confirmed by the absence of a cocatalytic effect with dry HCl) and the hypothesis was made that starting from 3H$_2$O and the stoichiometric amount of AlCl$_3$ (equation 28) four HCl molecules only (out of six) would become active, through coordination with autoassociated tetrachloroaluminoxanes such as (Cl$_2$Al)$_2$O --- AlCl$_2$OAlCl$_2$ --- O(AlCl$_2$)$_2$ which has only four free orbitals.

$$6AlCl_3 \ + \ 3H_2O \longrightarrow 3Cl_2AlOAlCl_2 \ + \ 6HCl \tag{28}$$

It has also been shown that the presence of the proton trap 2,6-di-*t*-butyl-4-methylpyridine did not modify the yield of carbocations formed by water cocatalysis.[98] This seems to show that the protons derived from water never appear and are involved in a concerted mechanism.

(b) Effect of HCl. HCl was found not to activate the polymerization of styrene[99] in CCl$_4$ solution (which is similar to the above case with DPE), but led to an increase of the yields for the polymerization of IB in pentane.[100]

(c) Aluminium alkyl chlorides. The cocatalytic effect of various proton donors on polymerizations initiated using AlEt$_2$Cl could be shown, since this compound alone did not initiate IB polymerization or copolymerization with dienes in MeCl.[80] The rates were not measured with precision but they decreased together with the yields in the following order: HCl ≈ HBr > HF ≈ H$_2$O > MeCO$_2$H.

Isobutyl vinyl ether was not polymerized by AlEt$_2$Cl alone at −78 °C in toluene but did so when some water was added, giving a non-stereoregular polymer. On the other hand, additions of O$_2$ led to the formation of a highly stereoregular polymer.[101] The possible initiating compound was suggested to be Al(O$_2$Et)EtCl.

With AlEtCl$_2$, polymerizations of several monomers (*e.g.* IB, styrene, α-methylstyrene *etc.*) occur without addition of a proton donor, but reaction in pure conditions have not been realized and direct initiation is possible but not proven.[102]

(vi) Aluminum bromide

In one of the earliest kinetic studies of carbocationic polymerization, that of propylene in butane solution at −78 °C using the initiating system AlBr$_3$/HBr,[103] which is very active and gives medium to high molecular weight polymer (containing fractions up to 8 × 10^5), it is mentioned that no polymerization occurred in the absence of cocatalyst and that rates were not reproducible in non-anhydrous solutions but became so when HBr was added (usually [HBr]/[AlCl$_3$] = $\frac{1}{4}$).

Polymerization of IB in heptane[104] initiated by AlBr$_3$ (−60 °C to 20 °C) proceeds with a rate proportional to [AlBr$_3$]^{2,99} and this has been assigned to a direct initiation (see Section 39.2.3.1). The addition of H$_2$O (*e.g.* 10^{-4} M for 3 × 10^{-4} M of AlBr$_3$) led to a strong decrease of the initial rate. However, the rates also obtained at −10 °C by other authors without water addition were lower by a factor of eight[105] and addition of HCl 3 × 10^{-4} M for AlCl$_3$ ≈ 7 × 10^{-4} led to a very strong increase of the rate. It seems then that a possible cocatalytic effect of traces of HBr present in AlBr$_3$ cannot be excluded.

A confirmation of the probable inactivity of water for initiation with AlBr$_3$ may be deduced from the addition of ^{3}H$_2$O where results[82] were similar to those obtained with BF$_3$: tritium incorporation was the same if ^{3}H$_2$O was added after polymerization. The possibility of an effect of 'subanalytical' concentration of water on the autodissociation of AlBr$_3$ was considered however.[82]

(vii) Titanium tetrachloride and butoxytrichlorotitanium

In polymerizations of IB initiated by TiCl$_4$ in CH$_2$Cl$_2$ solutions ([TiCl$_4$] ≈ 2 × 10^{-3}; [IB] ≈ 0.1) Plesch *et al.*[106] showed that when the reagents had been carefully purified, the reaction stopped for limited conversions, that the yield increased when the temperature was lowered, and was complete below −60 °C. For the 'stopped reactions' the introduction of water could lead to complete conversions for concentrations higher than a 'critical water concentration' [H$_2$O]$_c$, which decreased with temperature. This demonstrated the 'cocatalysis' by water, the total conversions below −60 °C being attributed to residual water.

These results were confirmed by experiments made in all-sealed apparatus[107,108] at $-72\,°C$ in CH_2Cl_2, for which partial conversions (8% to 38%) were observed, which became total by 'polymerization by condensation' (see Section 39.2.3.1). With IB and $TiCl_4$ in CH_2Cl_2, H_2O has been confirmed to be, and HCl has been shown not to be,[109] a cocatalyst at $-30\,°C$, but HCl is a cocatalyst at $-70\,°C$.

The role of HCl (and of water) in these experiments with IB has not been clearly established. However, similar experiments made with 1-butene[110] showed that HCl was not a cocatalyst at any temperature (and did not interact with $TiCl_4$), while water was. It was concluded that for HCl, Scheme 3 was probably operative, while it might be Scheme 2 for H_2O.

The reason for the 'stopping' of IB polymerization may be either a capture of a Cl of the counterion giving a CMe_2Cl end group, (Me_3CCl is not a cocatalyst[109]) or, if the solvent is not carefully purified (*e.g.* by treatment with oleum) a termination involving traces of impurities that could be identified.[111] Their inhibiting effect was shown for polymerizations of indene and α-methylstyrene in CH_2Cl_2 at $-70\,°C$.[112,113] With these stopped polymerizations, addition of water ($[H_2O]/[TiCl_4] = 0.8$ or 2) or of HCl ($[HCl]/[TiCl_4] = 0.15$) led to complete conversions, showing the cocatalytic effect.[111,113] However, with carefully purified solvent, total conversions were always observed, and also reproducible rates that were first order in monomer. The rates became slightly lower when water (for $[H_2O]/[TiCl_4] = 0.3$ and 1) was added showing that it may also play the part of an inhibitor.[111] For styrene polymerization, the results were similar. The initial rates increased with water concentration at $-25\,°C$ but were independent of it between -33 and $-90\,°C$.[114] It seems that for those three aryl-substituted monomers, cocatalysis by water is not important at low temperatures.

With DPE, the initiation yields for its reaction with $TiCl_4$ and H_2O could be measured from the carbocation concentration.[115,117] At $-30\,°C$, whatever the order of addition was, they were formed in concentrations corresponding to 160% of the H_2O used, showing that more than one proton was active.[117] An explanation similar to that given for $AlCl_3$ may be proposed. But with $TiCl_4$, the direct initiation yields based on $TiCl_4$ were much lower (0.5 to 10%). Another difference with $AlCl_3$ was that HCl was a very active cocatalyst.[115]

HF in excess was found to be a very efficient activator of the polymerization of IB in heptane at $-20\,°C$[118] by $TiCl_4$ and also by $TiBr_4$, VCl_4 and BCl_3 (*e.g.* for $[HF]/[TiCl_4] = 5$). HCl had no effect, and also HF with TiF_4. The formation of precipitates of fluorides was observed for the active systems, and initiation has been attributed to the formation of transitory mixed fluorohalides.

Polymerizations of isoprene with $TiCl_4$, associated with CF_3CO_2H, CCl_3CO_2H, and some of their esters (in concentrations similar to that of $[TiCl_4] = 1.5 \times 10^{-2}\,M$), has been studied at $20\,°C$ in various solvents such as heptane and 1,2-dichlorobenzene.[119] The order of activity was $CF_3CO_2H > CCl_3CO_2H > CF_3CO_2\text{-}s\text{-Bu} > CCl_3CO_2\text{-}s\text{-}\tilde{B}u$.

The *n*-alkyl esters were inactive. It was shown that with the acids, the following reaction occurred

$$TiCl_4 \; + \; Cl_3CCO_2H \; \rightleftharpoons \; Cl_3CCO_2TiCl_3 \; + \; HCl \qquad (29)$$

the equilibrium being reached in a few minutes. No cocatalytic influence of added HCl was noted.

n-Butoxytrichlorotitanium has been used for the polymerization of cyclopentadiene because usual F–C acids such as $TiCl_4$, $SnCl_4$ and BF_3,OEt_2 give insoluble cross-linked polymers whereas soluble high molecular weight polycyclopentadienes are obtained rapidly with Cl_3TiOBu^n in CH_2Cl_2 between $0\,°C$ and $-90\,°C$.[120] Even if a possible direct initiation cannot be excluded, an important effect of water has been observed on yields and rates,[121] the latter being first order in H_2O up to the stoichiometry with Cl_3TiOBu^n. The results obtained with α-methylstyrene at $-70\,°C$ in CH_2Cl_2[122] are similar, and addition of HCl led to slightly higher rates than that of H_2O.

(viii) Tin(IV) chloride

Polymerizations in the absence of added proton donor and in non-polar solvents either do not occur or do so after an inhibition period.[123] They generally take place in the presence of water and also sometimes on HCl addition. In solvents of higher dielectric constants, rates are higher, but induction periods may also be observed.

For IB in EtCl at $-78\,°C$, the rate is proportional to $[H_2O]$ up to $[H_2O]/[SnCl_4] = 0.5$.[124] When phenol was used as proton donor[125] the rate passed through a maximum for $[ROH]/[SnCl_4] = 0.1$. With styrene and related compounds, there are several examples of a maximum rate for a specific $[H_2O]/[SnCl_4]$ ratio, depending on the solvent. At room temperature with styrene,[126,127] this ratio varied from 2×10^{-3} to 2.5 (see Table 3). Two types of hypothesis have been

Table 3   Maximum Rate of Polymerization According to the $[H_2O]/[SnCl_4]$ Ratio

Monomer	Solvent	Temperature (°C)	$SnCl_4$ concentration (mol l^{-1})	$[H_2O]/[SnCl_4]$ at the maximum rate	Ref.
Styrene	CCl_4	25	8×10^{-2}	2×10^{-3}	126
Styrene	CH_2ClCH_2Cl	25	2×10^{-3}	2.5	126
Styrene	$CCl_4/PhNO_2$				
	70/30	25	2×10^{-2}	1	127
	85/15	25	2×10^{-2}	0.3	127
1,1-Diphenylethylene	Benzene	25	9×10^{-3}	2	128
Indene	CH_2Cl_2	-30	2.6×10^{-2}	0.25	129
Indene	CH_2Cl_2	-30	9×10^{-3}	0.25	129

offered to explain these maxima: either a variation in the proportion of lower (active) and higher (inactive) hydrates[124,126,129] or a variation of the solubility of the active hydrate.[130]

The effect of HCl differs according to the monomer and experimental conditions. With styrene, it may play the part of an inhibitor, probably on account of phenylethyl chloride formation.[130] But it is a very efficient proton donor for the polymerization of indene in CH_2Cl_2 at $-30\,°C$.[129,131] With DPE, the rate of dimerization increased continuously with HCl addition.[132] Using the related (non-polymerizable) model, 1,1-diphenylpropene (DPP), Bywater *et al.*[133] were able to measure the stable carbocation concentrations for the reaction in CH_2Cl_2 at $-30\,°C$, which are in agreement with the following equilibria (equation 30).

$$SnCl_4 \; + \; HCl \; + \; MeCH{=}CPh_2 \; \underset{}{\overset{K_1}{\rightleftharpoons}} \; MeCH_2\overset{+}{C}Ph_2 \cdot SnCl_5^- \; \underset{}{\overset{K_2}{\rightleftharpoons}} \; MeCH_2\overset{+}{C}Ph_2 \; + \; SnCl_5^- \qquad (30)$$

K_1 is small (cations yields $\sim 2\%$) and if $K_2 \sim 5 \times 10^{-4}$, most of the species are in the form of free ions.

(ix) Initiation involving weak protonic acids: general conclusions

Unambiguous results have been obtained with BF_3 and IB, and with some of the less active F–C acids such as $TiCl_3OBu$, $SnCl_4$ and $AlEt_2Cl$ with various monomers, for which clear cocatalytic effects have been shown to occur with both HCl and H_2O and with various other proton donors. But the yields in carbocations are not known except for DPP[133] and $SnCl_4$ for which they are very small.

For more reactive F–C acids, published data have sometimes been contradictory, which may result from both the effect of traces of proton donors and from the possibility of direct initiation (see Section 39.2.3.1). The most reliable data have been obtained in studies with HCl and H_2O and are summarized in Table 4 which is limited to the available experiments (generally made in CH_2Cl_2) realized under high vacuum for BF_3, $AlCl_3$ and $TiCl_4$. With IB, water always showed a cocatalytic effect with BF_3 and $TiCl_4$ whereas the effect of HCl with $TiCl_4$ depended on the temperature. H_2O seems to be an inhibitor for $AlCl_3$ (and also for $AlBr_3$ in heptane[104]). The effect depends on the monomer. With DPE, H_2O is a 'cocatalyst' with both $TiCl_4$ and $AlCl_3$, but HCl at $-30\,°C$ is not a 'cocatalyst' with $AlCl_3$, whereas it is one for $TiCl_4$.

The role of water as a proton donor is not necessarily the same with the various F–C acids, as shown in the examples given above, and also by the addition of hindered bases. Their use as 'proton traps' has been suggested in order to discriminate between initiation involving proton donors and 'direct' initiation. For example, it was found that polymerizations of IB initiated by BF_3 in EtCl at $-40\,°C$ that occurred in the presence of residual moisture, were slowed down by adding 2,6-di-*t*-butyl-4-methylpyridine (DBMP) for concentrations lower than $3 \times 10^{-3}\,M$ and stopped for higher concentrations.[134] It was concluded that the transient acid $H^+BF_3OH^-$ was neutralized by the base and that direct initiation did not occur with BF_3. On the other hand, with α-methylstyrene (α-MeSt) in CH_2Cl_2 and in the presence of a base, polymerization still occurred but the yields were lower, the effect being much stronger at $-30\,°C$ than at $-70\,°C$. This was given as evidence for two types of initiation, one of which is not inhibited (direct initiation) and the other which is 'cocatalyzed' initiation.[134] Similar results have been observed with α-MeSt and BCl_3.[135] However, it was found with this system that in the presence of a hindered base the yields decreased (without change of the

Table 4 Effect of H_2O and HCl on Initiation by Various Friedel–Crafts Acids

F–C acid	Monomer	HCl[b]	H_2O	Number of carbocations per H_2O molecule	Ref.
BF_3	IB[a]	—	Yes	—	81
$AlCl_3$	DPE	—	Yes	$1 C^+ (-70\,°C)$	89
$AlCl_3$	IB	?	No		106
$AlCl_3$	Styrene	—	No		96
$AlCl_3$	DPE	No	Yes	$1.33 C^+ (-30\,°C)$	98
$TiCl_4$	IB	No at $-30\,°C$ Yes at $-60\,°C$	Yes		106, 109
$TiCl_4$	1-Butene	No	Yes		110
$TiCl_4$	DPE	Yes	Yes	$1.6 C^+ (-30\,°C)$	117
$TiCl_4$	α-Methylstyrene	Yes	Yes		113
$TiCl_4$	Indene	Yes	Yes		112
$TiCl_4$	Styrene	—	Yes at $-25\,°C$ No at $-90\,°C$		114

Reactions done in CH_2Cl_2. [a] Polymerization realized in bulk. [b] 'Yes' and 'no' indicate that initiation or polymerization occur or do not occur on protogen addition.

molecular weight) when the purification of the reagents increased. The authors concluded that initiation involving water still occurred in the presence of the base, and suggested a concerted process not involving free protons.[135] A similar interpretation[98] may explain the results observed with DPE and $AlCl_3$, for which water strongly increases the initiation rate, which is, however, unchanged when DBMP is present. Other possibilities linked to different reactivities with the base of active centres differing by the type of counterion have been discussed.[98]

39.2.2 Initiation by Organic Cations

39.2.2.1 Stable carbocations

Stable organic cations are interesting initiators for fundamental kinetic studies of cationic polymerizations, because they lead generally to simpler reactions than other types of initiators. Indeed, a compromise should be found between cations which are too stable and therefore non-reactive with ethylenic compounds (quaternary ammonium, xanthylium, *etc.*) and cations which are too unstable and lead to side reactions and are difficult to handle (alkyldiaryl, *t*-butyl, *etc.*) . In between, the best compounds are triphenylmethyl (trityl, Ph_3C^+) and cycloheptatrienyl (tropylium, $C_7H_7^+$). Provided certain precautions are taken to exclude moisture and light, solutions are indefinitely stable and lead to reproducible results. Instability reported in some instances[136] is probably due to impurities.

(i) Trityl cation

One of the most useful carbocationic salts is triphenylmethyl cation, Ph_3C^+, which possesses a strong electronic absorption double band near 410–435 nm. In association with counterions derived from Friedel–Crafts halides ($SbCl_6^-$, SbF_6^-, AsF_6^-, PF_6^-, *etc.*) or from Brönsted acids (ClO_4^-), trityl salts show apparent molar extinction coefficients in halogenated solvents, $\varepsilon_{410} \approx \varepsilon_{435} \approx 4 \times 10^4\,l\,mol^{-1}\,cm^{-1}$, which are similar to that of Ph_3COH in sulfuric acid. This indicates that these salts are completely ionized. This property allows a direct spectrometric measurement of the initiation rate to be made.

In several cases, kinetics of the reaction of a trityl salt with an alkenic compound have been found to obey a second order law (first order with respect to each reactant) and the rate constants and the activation energies have been calculated (Table 5). Different counterions have been used, but their effect on the rate constants is apparently very weak (compare styrene with $SnCl_5^-$ and $SbCl_6^-$ in Table 5). The monomer reactivities with respect to the electrophilic attack by trityl cation varies similarly to the rates of homopolymerization. *N*-Vinylcarbazole (NVC) is the most reactive monomer, followed by isobutyl vinyl ether (IBVE), *p*-methoxystyrene (*p*-MeOSt) and *p*-methyl-styrene (*p*-MeSt). A little less reactive are cyclopentadiene (CPD) and α-methylstyrene (α-MeSt), and then finally styrene and indene. It can be seen that 1,2-disubstitution on the double bond strongly decreases its reactivity (compare anethole and *p*-MeOSt in Table 5). For alkyl vinyl ethers, a comparison of the monomer reactivities as a function of the alkyl group is available.[146] The order is the same for initiation as for the propagation rate: methyl < ethyl < isobutyl < isopropyl, which shows that the reactivity is mainly governed by the monomer structure.

Table 5   Initiation Rate Constants of Ethylenic Monomers by Trityl Salts at 20 °C

Monomer[a]	Counterion	Solvent	T (°C)	$10^2 \times k_i$ $(\mathrm{l\,mol^{-1}\,s^{-1}})$	E_i $(\mathrm{kcal\,mol^{-1}})$[b]	Ref.
Indene	$SnCl_5^-$	CH_2Cl_2	20	0.07	18.4	137
Styrene	$SnCl_5^-$	$(CH_2Cl)_2$	30	0.09	6.7	138
Styrene	$SbCl_6^-$	CH_2Cl_2	30	0.09		139
Anethole	SbF_6^-	CH_2Cl_2	25	0.1–0.2		140
α-MeSt	$SnCl_5^-$	$(CH_2Cl)_2$	30	2.2	5.0	138, 141
CPD	$SbCl_6^-$	CH_2Cl_2	22	1.1	9.2	142
p-MeSt	$SnCl_5^-$	$(CH_2Cl)_2$	30	6.5		141
p-MeOSt	$SbCl_6^-$	CH_2Cl_2	20	63	13.3	143
IBVE	$SbCl_6^-$	CH_2Cl_2	20	1600	9.4	144
NVC	SbF_6^-	CH_2Cl_2	20	13000	≃10.6	145
MVE	$SbCl_6^-$	CH_2Cl_2	0	60	8.7	146
EVE	$SbCl_6^-$	CH_2Cl_2	0	230	6.7	146
IBVE	$SbCl_6^-$	CH_2Cl_2	0	540	9.3	146
IPVE	$SbCl_6^-$	CH_2Cl_2	0	1500	5.6	146

[a] MVE = methyl vinyl ether; EVE = ethyl vinyl ether; IBVE = isobutyl vinyl ether; IPVE = isopropyl vinyl ether. For other abbreviations, see text. [b] kcal = 4.2 kJ.

Table 5 does not list monomers which are polymerized by trityl salts, such as isoprene in nitrobenzene,[147] for which no kinetic results are available. On the other hand, alkenes such as isobutylene are not polymerized by $Ph_3C^+SbF_6^-$.[148] Nevertheless it should be mentioned that this is not due to a lack of reactivity in the initiation process, but rather to side reactions. It has been shown that addition of the trityl cation onto isobutylene did occur but that indanic termination by an intramolecular Friedel–Crafts reaction followed immediately (equation 31).[149]

$$Ph_3C^+ \;+\; CH_2{=}CMe_2 \;\longrightarrow\; \cdots \;\longrightarrow\; \cdots \;+\; H^+ \tag{31}$$

High yields of 1,1-dimethyl-3,3-diphenylindane were obtained (28%) when isobutylene was added slowly to a solution of trityl perchlorate in acetonitrile. However, when isobutylene was added rapidly, a viscous oil was formed, presumably oligoisobutylene. On the other hand, $(p\text{-}ClC_6H_4)_3C^+SbF_6^-$ and diphenylmethyl cation, $Ph_2CH^+SbF_6^-$, have been found to initiate the polymerization of isobutylene.[148]

In many other cases, the initiation step has been shown to occur by a direct electrophilic addition of the carbocation to the double bond, according to the kinetics observed. For example, incorporation of trityl fragments in the polymer has been shown by IR in the case of cyclopentadiene[142] and vinyl ethers,[150] by NMR in the case of *p*-methoxystyrene[143] and by GLC in the case of styrene.[139] Evidence for the addition on the double bond has been also obtained by using a bifunctional trityl salt (**1**) to initiate the polymerization of cyclopentadiene.[151] This monomer was chosen because it had been shown previously to give no transfer.[142]

$$SbCl_6^-\,Ph_2C^+ \!\!\!\!\!\!\!\!\!\!\!\! \text{⟨⟩} \!\!\!\!\!\!\! CH_2CH_2 \text{⟨⟩} \!\!\!\!\!\!\! \overset{+}{C}Ph_2\,SbCl_6^-$$

(**1**)

In the case of α-methylstyrene at low monomer/initiator ratio (M/I = 2–13) at 30 °C in dichloroethane, the presence of Ph_3CH was detected by ^1H NMR and GLC and attributed to a different mechanism of initiation by hydride abstraction from the monomer (equations 32 and 33).[152]

$$Ph_3C^+ \; + \; CH_2{=}CMePh \longrightarrow Ph_3CH \; + \; PhC\overset{CH_2}{\underset{CH_2}{\big\langle}}{+} \tag{32}$$

(2)

$$PhC\overset{CH_2}{\underset{CH_2}{\big\langle}}{+} \; + \; CH_2{=}CMePh \longrightarrow CH_2{=}CCH_2CH_2\overset{+}{C}\overset{Me}{\underset{Ph}{\big\langle}} \tag{33}$$

$$\underset{Ph}{|}$$

(3)

Unfortunately, no direct evidence for the allylic cation (**2**) and compound (**3**) has been given. The same authors also used $(p\text{-}ClC_6H_4)_3C^+SbCl_6^-$ as initiator for α-methylstyrene and found no chlorinated trityl groups in the oligomers (elemental analysis and fluorescence spectroscopy). It is worth noting that in the same conditions the fluorescence spectrum of oligo(isobutyl vinyl ether) presented a strong emission band at 430 nm assigned to p-chlorotrityl groups, which points to the possibility of different mechanisms for different monomers. Recently hydride abstraction from the monomer has also been proposed for the system $NVC/Ph_3C^+SbCl_6^-/$nitrobenzene, but no evidence was given for this assumption.[153]

The reactivities of the ion pairs and free ions seem almost identical, as shown by the fact that the addition of a common anion electrolyte has no effect on the initiation rate in the system p-methoxystyrene/$Ph_3C^+SbCl_6^-/CH_2Cl_2$ at 10 °C, whereas the proportion of free trityl cation was varied from 0.36 to 0.96.[154] This peculiar behaviour was attributed to the bulkiness of the ion pair by comparison to the situation prevailing in anionic polymerization where the ratio $k_p^-/k_p^\pm$ decreases strongly when the size of the counterion increases.[155]

(ii) Tropylium ion

Another organic cation of interest as an initiator for cationic polymerization is cycloheptatrienyl (tropylium, $C_7H_7^+$). This cation is more stable than trityl since it can be formed by exchange between tropylidene and trityl cation (equation 34).

$$C_7H_8 \; + \; Ph_3C^+ \; \rightleftharpoons \; C_7H_7^+ \; + \; Ph_3CH \tag{34}$$

It is thus expected to be less reactive than the corresponding trityl salt. Nevertheless it is still capable of initiating the polymerization of reactive monomers such as isobutyl vinyl ether,[150] N-vinyl-carbazole[156] and p-methoxystyrene.[157] In the last two cases, incorporation of tropylium groups in the polymer was shown, probably as a result of electrophilic addition of tropylium to the double bond. However, the assumption that initiation is complete soon after the beginning of the polymerization might be wrong, since this is not the case with trityl salts which are stronger electrophiles. This might cause an overestimation of the concentration of active centres, which in turn leads to underestimation of the propagation rate constant. The difficulty comes from the impossibility of getting a direct measurement of the initiation rate with this initiator, because it absorbs at 217 and 273 nm, wavelengths at which most of the monomers also absorb. It is thus less suitable for spectroscopic and kinetic studies.

39.2.2.2 Carbocations generated in situ

Carbocations more reactive than Ph_3C^+ or $C_7H_7^+$ have generally not been isolated, on account of their high reactivity. They have often been supposed to be reaction intermediates for initiation involving various cationogens reacting either with a F–C acid (equation 35)

$$RX \; + \; MtX_n \; \rightleftharpoons \; R^+MtX_{n+1}^- \tag{35}$$

or with a salt of a complex acid, preferably an Ag salt that precipitates (equation 36).

$$RCl \; + \; AgSbF_6 \longrightarrow AgCl(\downarrow) \; + \; R^+SbF_6^- \tag{36}$$

The RX cationogens have been either halides (esters of HX) or esters of oxygenated acids.

(i) Alkyl or arylalkyl halides and F–C acids

The stopping experiments described in Section 39.2.1.2 show that primary halides used as solvents, such as CH_2Cl_2, are no cationogens under the usual conditions. However, there was some indication of a possible effect of isopropyl chloride on styrene polymerization with $SnCl_4$ in nitrobenzene solution at $25\,°C$,[126] even if the slow rate might result from a very low H_2O concentration in the solvent. A much larger effect of *t*-butyl chloride was shown in the same conditions, the rate being proportional to its concentration.[126] There are other examples of the effect of Bu^tCl on the polymerization of IB with various F–C acids. With $AlCl_3$, a complex described as a donor–acceptor could be isolated[158] (stable at $-78\,°C$ but not at $-30\,°C$) that led to higher polymerization yields than for $AlCl_3$ or for the same components added separately, which suggested that $AlCl_3$ complexation with IB was faster than with Me_3CCl. On the other hand, with $TiCl_4$ in CH_2Cl_2, Bu^tCl had no cocatalytic effect, both at $-30\,°C$ and $-70\,°C$.[109]

Bu^tCl and also arylalkyl chlorides such as benzyl chloride have also been found to give strong cocatalytic effects[159] with various alkylaluminums and with the corresponding monochlorides (such as $AlEt_2Cl$). The advantage of these Lewis acids is that they do not initiate the polymerization of IB or styrene, even when the purity of the reagents is not extremely high, but may do so on addition of cationating agents. From model studies of addition of Bu^tCl to 2,4,4-trimethyl-1-pentene (TMP) in $MeCl$,[161] the following order of decreasing reactivity of the Lewis acids was deduced: $Me_2AlCl > Et_2AlCl > Me_3Al > Et_3Al$.

From the yields of IB polymerization using Et_2AlCl in $MeCl$ at $50\,°C$, the efficiencies of various alkyl, allyl and arylalkyl chlorides have been compared, and went through a maximum for Bu^tCl when plotted against the sequence of carbocation stabilities. Similar results have been observed with styrene.[159, 160] It was proposed that efficiency for initiation results from two opposing effects:[159,160] the carbenium ion availability linked to the ionization of the halide, and its stability, which increases when ionization is easier. The maximum efficiency would be for halides that give a sufficient concentration of carbocations by ionization but that are also sufficiently reactive for cationation (equations 37 and 38).

Ion generation:

$$RCl \ + \ MtCl_n \rightleftharpoons R^+MtCl_{n+1}^- \tag{37}$$

Cationation:

$$R^+MtCl_{n+1}^- \ + \ \underset{|}{\overset{|}{C}}{=}\underset{|}{\overset{|}{C}} \longrightarrow R{-}\underset{|}{\overset{|}{C}}{-}\underset{|}{\overset{|}{C}}{}^+MtCl_{n+1}^- \tag{38}$$

However, the above scheme has been challenged and it was suggested[162] that initiation involving $AlEt_2Cl$ might result from an exchange reaction with the halide giving higher halides (equations 39 and 40).

$$AlEt_2Cl \ + \ RCl \longrightarrow AlEtCl_2 \ + \ REt \tag{39}$$

$$AlEtCl_2 \ + \ RCl \longrightarrow AlCl_3 \ + \ REt \tag{40}$$

Such types of reactions have been shown to occur readily in CH_2Cl_2[163] and it was pointed out[162] that Ph_3CCl addition was quite effective for IB polymerization[160] with $AlEt_2Cl$ even if Ph_3C^+ does not lead to polymerization (see Section 39.2.2.1.i). The higher halides, $AlEtCl_2$ or $AlCl_3$, might then initiate the polymerization by direct initiation. This hypothesis has been discussed in detail in ref. 7.[162] The problem is that the formation of the higher halides may occur through the intermediate of the ionized species, and nothing is known of their reactivity compared to that of the higher aluminum halides, supposed to be larger in the nascent form.[162]

A partial answer to this dilemma has been given by the results of carbocationic grafting on chlorinated backbones. Earlier results with chlorinated butyl rubber and $AlEt_2Cl$ had shown high grafting efficiencies for styrene[164] and for indene,[165] but the occurrence of regrafting of trans-ferred chains on $\underset{}{>}C{=}CH_2$ groups complicated the interpretation. Using a backbone bearing $—C_6H_4CH_2Cl$ groups, the grafting efficiencies remain high.[166] They are limited only by transfer reactions, and it could be shown that grafting did *not* occur by reaction of growing chains with aromatic nuclei.[167] The grafting observed is as yet the best argument for an initiation reaction directly involving the halide, but it is still not known whether reaction (38) is involved, or the reaction of the halide with a monomer complexed with $AlEt_2Cl$. Another argument in favour of the possibility of an initiation according to reactions (37) and (38) is the observation of the racemization of optically active 1-phenylethyl chloride by $SnCl_4$ in benzene (at $25\,°C$),[168] but other explanations of this result have been offered.[162]

Some of the most interesting recent developments in cationic polymerizations have been the use of arylalkyl halides with BCl_3 as initiators for IB polymerizations, that has led to the synthesis of polymers with well defined end groups and predicted molecular weights. It was mentioned that IB polymerization in CH_2Cl_2 is not initiated by BCl_3 at $-78\,^{\circ}C$ in conjunction with Bu^tCl, or with allyl and benzyl chlorides, but that substituted allyl and benzyl chlorides are quite effective[169] (it was found later that initiation could occur with Bu^tCl at $-20\,^{\circ}C$, but in the presence of very large concentrations of BCl_3: 0.2 to 0.8 mol l^{-1}).[170] The most used initiators have been cumyl chloride (CC) and related halides such as the bifunctional [1,4-bis(1-chloro-1-methylethyl)benzene, (p-DCC)][171-173] or trifunctional [1,3,5-(2-chloro-2-methylethyl)benzene, (TCC)],[174] which have led to the synthesis of polyfunctional polyisobutylenes bearing terminal —CMe_2Cl groups. p-DCC and TCC are much more stable than CC and lead to fewer secondary reactions (such as indan formation). They may give macromolecules each bearing one aromatic group, in agreement with an initiation step that would involve a cationic species —$C_6H_4\overset{+}{C}Me_2$. Indan formation could also be reduced by using sterically hindered dichlorides such as 1,3-di(2-chloro-2-propyl)5-t-butyl-benzene.[175]

In order to have a unimodal distribution with TCC and to exclude other types of initiation, the use of a proton trap was found to be necessary.[176] Unfortunately no data are available about the initiation step with these initiators, but arguments based on the addition of a common-ion salt are in favour of a propagation involving mainly free ions.[176]

The use of BF_3 instead of BCl_3 led to rapid polymerizations even with Bu^tCl, allyl and benzyl chlorides,[177] but the molecular weights were much higher and there was no correlation with the halide concentration.

The efficiency of a series of chlorides as cationating agents has been recently compared, and the data confirm the hypothesis of Kennedy[159] about the necessity of a large enough dissociation (equation 37) and also of a sufficient reactivity of the carbocations (equation 38). For the model reaction with 2-methyl-1-pentene (with $[SnCl_4] = [halide]$ in CH_2Cl_2) of a series of diarylmethyl chlorides $(R^1Ph)(R^2Ph)CHCl$ bearing different R^1 and R^2 groups,[178] an increase of electron donation (favouring reaction 37) increased the relative reactivity by several orders of magnitude, but there was a maximum in the variation, with a decrease for the most stable carbocations (such as $(MeOPh)_2\overset{+}{C}H$) (equation 41).

$$(41)$$

In another study[179] in which the cationating agents were completely ionized (*e.g.* with the corresponding tetrachloroborates with eventually an excess of BCl_3) their rate constants of addition to the same alkene were measured and decreased by a factor up to 10^5 when the electron-releasing ability of R^1 and R^2 increased. The activation entropy was found to be similar (120 ± 5 J mol^{-1} K^{-1}), whereas the activation enthalpy increased from 11.6 ($R^1 = R^2 = Me$) to 30 kJ mol^{-1} (for $R^1 = R^2 = OMe$).

$Ph_2CH^+SbF_6^-$ prepared *in situ* has been used to initiate the polymerization of acenaphthylene.[180] The monomeric carbocation $Me_3CC_6H_4\overset{+}{C}Me_2\ SbF_5Cl^-$ has been prepared by reaction of SbF_5 with the chloride at $-70\,^{\circ}C$ in CH_2Cl_2. It was stable up to $0\,^{\circ}C$ and was used to initiate the oligomerization of p-t-butyl-α-methylstyrene at $-70\,^{\circ}C$.[181]

Polymerization of vinyl ethers initiated by $HI + I_2$[182-184] proceeds with the formation of one macromolecule per HI molecule, in hexane, toluene or CH_2Cl_2 ($T = -40\,^{\circ}C$ to $-15\,^{\circ}C$). Narrow molecular weight distributions have been observed ($\bar{M}_w/\bar{M}_n = 1.04$ to 1.08) even when I_2 concentration was lower than $[HI]$ and has been attributed to the presence of living polymers resulting from propagation on non-ionic species. In comparison with the usual F–C acids-based initiator systems, there is certainly either a strong reduction or absence of termination and transfer. But it is difficult to conclude yet, on account of the low values ($\sim 10^4$) of the molecular weights up to which their linear growth with conversion has been observed.

It has been shown that the first reaction is the addition of HI to give the 1-iodo ether $MeCH(OR)I$ and the growth has been assumed to occur by activation by iodine of the covalent C—I bond which is reformed in each propagation step (equation 42)[183] (for a more detailed review with a large variety of ethers, see Volume 3, Chapter 42).

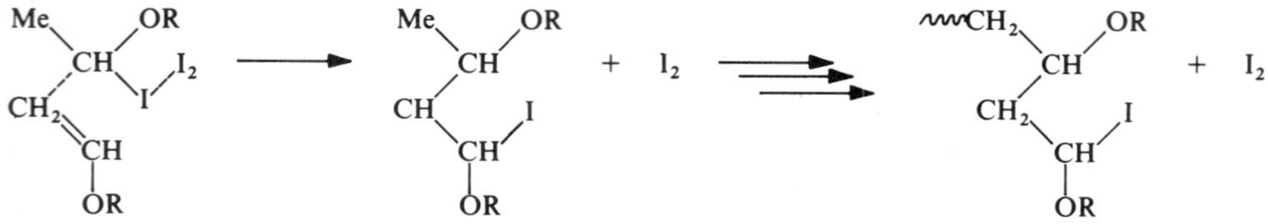

$$\text{(structures)} \qquad (42)$$

However, the above scheme does not explain the kinetics observed in *n*-hexane[185] or toluene[186] which are more in agreement with a zero (or at least very low) order in monomer for both IBVE[185] and for chloroethyl vinyl ether.[186] These kinetics have been explained by initiation and growth involving a monomer/iodine complex (equation 43).

$$\underset{\mid\;\;\text{OR}}{\text{MeCHI}} \;+\; \underset{\mid\;\;\text{OR}}{\text{CH}_2\!=\!\text{CH}\cdot\text{I}_2} \longrightarrow \underset{\mid\quad\;\;\mid\;\;\text{OR}\;\;\;\text{OR}}{\text{MeCHCH}_2\text{CHI}} \;+\; \text{I}_2 \qquad (43)$$

More recently, similar results have been obtained with a (weak) F–C acid-activated system, HI/ZnI_2, for the polymerization of vinyl ethers[187] and of *p*-methoxystyrene[188] in toluene, and this even for reactions carried out at 25 °C. The reactions were much more rapid than with I_2 systems, (*e.g.* complete in 10 min for IBVE instead of 4 h at 25 °C in toluene), and the molecular weight continued to grow linearly after a new monomer addition (up to $\bar{M}_n = 1.2 \times 10^4$).

(ii) Carbocations from esters of oxygenated acids and Friedel–Crafts acids

The possible formation of perchlorates by reaction between the monomer and the acid and their reactivity have been discussed in Section 39.2.1.1.ii. The formation *in situ* of the perchlorate or of the corresponding carbocation has also been realized by reaction of the halide with silver perchlorate. This was the case with 1-phenylethyl bromide in order to initiate the polymerization of styrene.[189] The results were similar to those observed with an equivalent amount of HClO_4. The relative reactivities as initiators of a series of organic halides reacting with AgClO_4 have been compared.[190] For styrene in bulk, the global rates decreased in the order: benzyl bromide > 1-phenylethyl chloride > *t*-butyl chloride. For the same organic group, the activities decreased in the order iodide > bromide > chloride.

Polymerizations supposed to involve esters of carboxylic acids as cationating agents have been mentioned in Section 39.2.1.1 for activation of the ester by protonic acids. Recently, initiating systems for IB based on the association of esters[191] (or ethers, see below), and BCl_3 have been described, for which the number of macromolecules is very near to that of the added ester, with a linear growth of the molecular weight with conversion from the origin, up to molecular weights of about 10^4. These systems have also been described as 'living', and even if this is not clearly established, there is no doubt that they are very interesting for the synthesis of functional polymers, with low, predictable molecular weights, and that the transfer reactions are strongly reduced (or suppressed?) compared to more classical systems, even at temperatures as high as -30 °C in various solvents such as MeCl, CH_2Cl_2 and EtCl. The esters giving the best results are those which might give potentially more stable cationic species, such as cumyl acetate or *t*-butyl acetate.[191–192] Initiation and propagation have both been supposed to occur by activation of the ester — or of ester end groups — by BCl_3, giving active ionic species (equation 44).

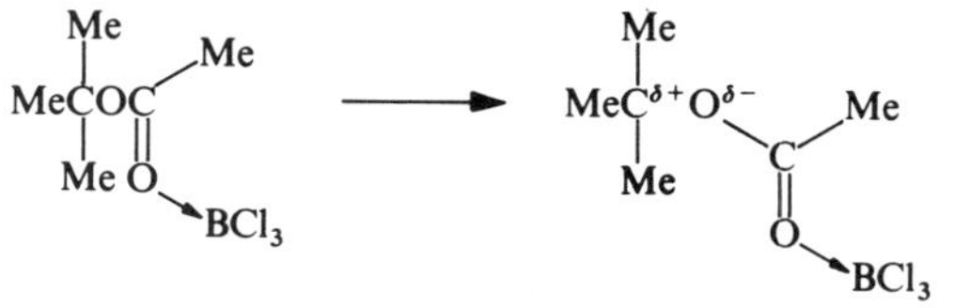

$$\text{(structures)} \qquad (44)$$

The addition of a common ion (formed from $\text{Ph}_3\text{COAc} + \text{BCl}_3$) strongly reduced the rate but led to a narrow molecular weight distribution (1.1 instead of 2) which was interpreted as the suppression of more active species and by propagation occurring on 'less-dissociated species'.[192] Diesters such as $p\text{-AcOCMe}_2\text{C}_6\text{H}_4\text{CMe}_2\text{OAc}$ have been found to give similar results.[193] In all cases, the end groups of the recovered polymer were not acetate groups but —CMe_2Cl groups. This has been attributed to the quenching reaction (with methanol). By using a cyclic lactone (γ-phenyl-γ-butyrolactone) as the ester, arguments for the formation of a cyclic polymer have been given when pyridine was used as a quenching agent.[194]

Very recently, similar results have been obtained for the polymerization of *p*-methylstyrene[195] (PMS) and of 2,4,6-trimethylstyrene (TMS)[196] with cumyl acetate (and other acetates in MeCl at —30 °C), but with the last monomer, the linear growth of molecular weight was also observed in the absence of the acetate (initiation was then assumed to be by the 'H_2O/BCl_3 system'). With styrene,[197] the above esters did not give a number of macromolecules near to their concentration. This was achieved only with 1-(*p*-methylphenyl)ethyl acetate, but the molecular weight distributions were much broader ($\bar{M}_w/\bar{M}_n \approx 5$–6).

Other interesting data have been obtained with IBVE initiated by $AlEtCl_2$ in the presence of a large excess of an ester (*e.g.* 10% by volume of ethyl acetate in toluene).[198] There was no apparent correlation between polymer molecular weight and $AlEtCl_2$ or ester concentrations, which suggests that initiation might result from an unknown cocatalytic effect. But contrary to reactions in the absence of ester, the molecular weights grew linearly with conversion from the origin, for temperatures between —40 and +25 °C, the polymers having narrow distributions (*e.g.* $\bar{M}_w/\bar{M}_n = 1.15$). The role of the ester as a solvating and stabilizing agent (formation of oxonium ions?) has been suggested. Similar results have been observed with the addition of smaller concentrations of ethers such as THF or Et_2O.[199]

(iii) Ethers as cationogens

The stable complex $BF_3 \cdot OEt_2$, soluble in most solvents, has been much used as an initiator, but few systematic studies have been made, particularly in high purity conditions. The mechanism of initiation is unknown, even if it has been generally written

$$BF_3 \cdot OEt_2 \rightleftharpoons (BF_3 \cdot OEt)^- Et^+ \tag{45}$$

$$BF_3 \cdot OEt^- Et^+ + M \longrightarrow (BF_3 \cdot OEt^-)EtM^+ \tag{46}$$

It was shown that a small excess of Et_2O (and also of THF or Et_2S) had both an inhibitory effect and a retarding effect for the polymerization of styrene in benzene at 30 °C.[200] The effect of water on its polymerization in CH_2Cl_2 at 25 °C has been studied by Giusti and co-workers,[201] who showed a cocatalytic effect with a maximum rate for $[H_2O]/[BF_3 \cdot OEt_2]$ between 0.1 and 0.6, this value increasing with styrene concentration. However, the rate was still high in the absence of added water and proportional to $[BF_3 \cdot OEt_2]^2$. Initiation involving only the etherate was also likely.

Polymerization of IB initiated by ethers such as $Me_3CCH_2CMe_2OMe$ or $PhCMe_2OMe$ with an excess of BCl_3 in MeCl or CH_2Cl_2 at —30 °C has given results similar to those described above with the cumyl esters, with a number of polymer chains equal to that of the ether, but with wider molecular weight distributions ($\bar{M}_w/\bar{M}_n \approx 1.7$ to 2).[202] Similar interpretations to those given for the ester may be considered. Bifunctional[203] or trifunctional[204] polymers with —CMe_2Cl end groups have also been prepared using diethers of triethers.

39.2.2.3 Oxocarbenium ions and acetyl perchlorate

The reaction of an acyl or aryl halide with a F–C acid may give an oxocarbenium salt. For example, such salts were prepared by reaction of an acylfluoride with SbF_5 in Freon 113 at low temperature (equation 47).[205] They precipitate as solid salts.

$$\underset{\underset{O}{\|}}{R}CF + SbF_5 \longrightarrow \underset{\underset{O}{\|}}{R}C^+ SbF_6^- \tag{47}$$

When they are put into solution, there is a partial dissociation into the components and the conductivity in CH_2Cl_2 obeys the binary ionogenic equilibrium (equation 48).[206]

$$MeCOF + SbF_5 \rightleftharpoons Me\overset{+}{C}O + SbF_6^- \quad K_D = 5 \times 10^{-2} \quad \text{at } 0 °C \tag{48}$$

They may also be prepared by reaction of an aryl chloride with an Ag salt (equation 49) and this method has been used for the *in situ* initiation of tetrahydrofuran polymerization.[207]

$$MeCOCl + AgSbF_6 \longrightarrow RCO^+ SbF_6^- + AgCl(\downarrow) \tag{49}$$

If the acyl halide is reacted with an Ag salt of a protonic acid, a mixed anhydride may be formed, such as acetyl perchlorate (equation 50).

$$MeCOCl + AgOClO_3 \longrightarrow MeCO_2ClO_3 + AgCl(\downarrow) \tag{50}$$

This compound has been shown to be mainly covalent in relatively concentrated solutions, but has never been isolated. It is assumed to initiate polymerizations following the dissociation equilibrium: $MeCO_2ClO_3 \rightleftharpoons MeCO^+ClO_4^-$.

(i) Initiation by oxocarbenium ions

Polymerization of acenaphthylene in nitrobenzene at 25 °C has been initiated using $EtCOSbF_6$ and $PhCOPF_6$,[178] which were soluble in CH_2Cl_2. The initiation was said to be 'fast' on account of the observed internal first order in monomer. Another set of experiments in the same solvent used $PhCO^+SbF_6^-$ for the polymerization of acenaphthylene and styrene.[208]

(ii) Initiation with acetyl perchlorate

The mixed anhydride acetyl perchlorate has been initially used as a substitute for perchloric acid, which is difficult to obtain and use perfectly pure since it is dangerous when dry (risk of explosion). For polymerization reactions, acetyl perchlorate has usually been prepared by reaction of silver perchlorate with acetyl chloride. However, the compound was not isolated, an excess of acetyl chloride was generally present[209] and the stability of the mixtures in various solvents has not been studied.

As pointed out in a recent and complete review (see ref. 7, p. 212), its general behaviour and reactivity are very similar or even identical to those of perchloric acid in cases where comparison has been made, e.g. for the polymerization of styrene in CH_2Cl_2 at 0 °C[210] or cyclopentadiene in CH_2Cl_2/toluene at -78 °C.[211] When relatively low concentrations ($< 10^{-4}$ M) were used, perchloric acid (resulting from hydrolysis by water in the system) was present in significant and sometimes even predominant concentration, and it is difficult to ascertain the part really played by the mixed anhydride. This was the case for the stopped-flow experiments realized with p-methoxy-styrene[212] in CH_2Cl_2 for which high yields of carbocationic species P* ($\lambda_{max} = 380$ nm) were observed, about twice as high for the anhydride as for triflic acid. The initiation rate constants were estimated from the average rate of formation of carbocations and were very high for both $MeCOClO_4$ ($k_i = 1.2 \times 10^5$ M^{-1}s^{-1}) and CF_3SO_3H ($k_i \approx 5 \times 10^4$ M^{-1}s^{-1}), these values being underestimated on account of the method used. They are much higher than those observed with the same monomer with $SnCl_4$, BF_3OEt or $Ph_3C^+SbCl_6^-$. Addition of the common ion salt $Bu_4^nN^+ClO_4^-$ to the monomer solution caused a decrease of [P*]. However, the rate did not change significantly and this was attributed to the presence of 'invisible propagating species' leading to an increase of the apparent k_p calculated from [P*].

For the polymerization of a less reactive monomer such as styrene,[210] higher concentrations of initiator [C]$_0$ were used (e.g. 4×10^{-4} to 6×10^{-3} M) and the proportion of residual water was much lower. The effect of added water concentration on the rate (in CH_2Cl_2 at 0 °C) was found to be insignificant when it was mixed with the monomer and small when it was added to the initiator solution (even for [H$_2$O]$_0 = 6 \times 10^{-3}$ and [C]$_0 = 4 \times 10^{-4}$). The very small effect of added water concentration on the rates and also on the molecular weight distribution may be explained in two different ways, between which it does not seem possible to choose yet: either an initiation step by $HClO_4$ or $MeCO_2ClO_3$, which would give the same type of species (ester or ions) still reactive in the presence of water, or an initiation by only $HClO_4$ formed by hydrolysis of the ester by water in concentration higher than 4×10^{-4} M. The apparent rate constant was proportional to [C]$_0$ but this might result from the presence of $HClO_4$ in the 'perchlorate'.

Addition of $Bu_4N^+ClO_4^-$ in large excess suppressed the high polymer fraction, the formation of which was ascribed to 'ionic species', but did not considerably decrease the whole rate. The author's conclusion was that the rate of propagation of the non-dissociated species (at 0 °C) was not very different from that of the dissociated ones. As in Pepper's experiments with $HClO_4$, polymerizations made at -78 °C occurred with a very rapid initial stage 1 followed by a slow stage 2, the high polymer fraction being made mainly during stage 1.

Initiation rate constants have also been deduced from conductance stopped-flow experiments[213] with p-methoxystyrene and styrene, permitting a comparison between the two monomers. As with perchloric acid, k_i (and k_p) decreased strongly with an increase of monomer concentration, which is not consistent with the explanation given by Pepper for the similar observation made with $HClO_4$ (ester formation). Values of k_i for styrene were about 100 times lower than for *p*-methoxystyrene.

Attempts at determining the initiation efficiency of this initiator were made with styrene at $0\,°C$ in CH_2Cl_2 or in mixtures of CH_2Cl_2 with $PhNO_2$.[214] Quenching of perchlorate end groups was realized by reaction with a suspension of sodium *β*-naphthyloxide as end-capping agent. Naphthyloxy end groups were incorporated in a large part of the oligomer and of the higher polymer ($\bar{M}_n \approx 2000$) formed, as shown by GPC with UV detection. For high ratios $[M]_0/[C]_0$ (*e.g.* ~ 2000) this should correspond to 100% formation of first the perchlorates and then of naphthyloxy end-groups. However, the apparent k_p calculated from the corresponding 'active centres' concentration were very low and changed strongly with the above ratio. It is not clear whether the perchlorate end groups really corresponded to *active* species and whether they had been formed by initiation (for example involving perchloric acid) or through a transfer reaction to the acetyl perchlorate. The same question arises for experiments using aryl perchlorates as initiators (formed *in situ*), with aryl groups absorbing in the UV.[215] For these last reactions carried out at $-25\,°C$, there was a satisfactory agreement between the $\bar{M}_n$ measured directly by VPO and those calculated from the end group concentration.

A study of the polymerization of *p*-methylstyrene at $-78\,°C$ has given important results[216] permitting an evaluation of the initiator efficiency. In CH_2Cl_2, high polymer (molecular weight $\sim 10^5$) was formed rapidly, with a single GPC peak. In mixtures with toluene, bimodal distributions were observed for a volume ratio CH_2Cl_2/toluene of two and only a low molecular weight fraction was formed for a ratio of $\frac{1}{4}$ with a large decrease of the rate and a narrowing of the molecular weight distribution. This is quite similar to what was observed with $HClO_4$. For polymerization realized with the solvent ratio $\frac{1}{4}$, an increase in the molecular weight of the polymer with conversion was observed at $-78\,°C$, but not at $-40\,°C$. In the presence of an excess of $Bu_4^n NClO_4$ (equal to five times $[C]_0$) in CH_2Cl_2 alone at $-78\,°C$, only the low polymer fraction was formed, the molecular weight of which increased regularly with conversion up to $\bar{M}_n \approx 30\,000$. A similar linear increase was observed after new monomer additions (after 5 and 10 h) and the observed $\bar{M}_n$ were in good agreement with the calculated ones based on $[M]_0/[C]_0$ and with the formation of one polymer chain per initiator molecule. These results have been interpreted by the existence of long-lived 'non-dissociated species'. They might be the consequence of the stability of the ester at low temperature[23,24] and of the suppression of free ion propagation in the presence of added salt. They seem to show that initiation with the perchlorate is relatively rapid and complete even at low temperature, but it is still not possible to know whether initiation and propagation involve the perchlorate ester alone or if it is activated by the acid (which is also probably homoconjugated with the salt).

39.2.2.4 *Dialkoxycarbenium ions and dioxolenium ions*

Dialkoxycarbenium ions and dioxolenium ions are known in organic chemistry as alkylating agents. With counterions such as BF_4^- or $SbCl_6^-$, they are able to initiate ring-opening polymerizations of cyclic ethers and acetals by alkylation.[217] Recently, it has been reported that these compounds are also capable of initiating the cationic polymerization of ethylenic monomers, such as styrene, *α*-methylstyrene and isobutyl vinyl ether.[218] Nevertheless in the case of 1,1-diphenylethylene, it was found that initiation occurs by proton transfer and not by alkylation as in the case of heterocycles (equation 51).

$$HC^{+}\begin{array}{c}OEt\\\\OEt\end{array}\quad BF_4^- \;+\; CH_2{=}CPh_2 \;\longrightarrow\; HCO_2Et \;+\; CH_2{=}CH_2 \;+\; Me\overset{+}{C}Ph_2 \qquad (51)$$

39.2.3 Initiation by Inorganic Lewis Acids

39.2.3.1 *'Direct initiation' by Friedel–Crafts acids*

We have said that Friedel–Crafts acids (F–C acids) are efficient activators of weak protogens such as H_2O or HX (Section 39.2.1.2) and of weak cationogens such as organic halides (Section 39.2.2.2). However a large number of systems have been reported in which F–C acids can initiate the polymerization without the participation of any added cationogen (so-called 'direct initiation'). In most cases, the claim is based on the observation that a very careful purification of the reactants and solvent, by the best high-vacuum techniques of purification and drying, is not sufficient to stop the polymerization. In a few cases, more direct proof could be obtained through the measurement of the yield of stable carbocationic species.

(i) Experimental evidence

A first series of arguments in favour of 'direct initiation' is given by the observation of high polymerization rates (and complete conversions), in the 'purest' experimental conditions. Some systems presenting this behaviour are: styrene/TiCl$_4$/CH$_2$Cl$_2$/$-90\,°$C;[219] α-methylstyrene/ TiCl$_4$/CH$_2$Cl$_2$/$-30\,°$C, $-70\,°$C;[113] indene/TiCl$_4$/CH$_2$Cl$_2$/$-30\,°$C, $-70\,°$C;[220] isobutylene/ TiCl$_4$/CH$_2$=CHCl/$-138\,°$C;[221] isobutylene/AlCl$_3$/CH$_2$Cl$_2$/0\,°C, $-90\,°$C[96] isobutylene/ AlBr$_3$/MeBr/$-78\,°$C;[222] isobutylene/AlBr$_3$/heptane/$-10\,°$C;[104] and isobutylene/BCl$_3$/CH$_2$Cl$_2$/ $-30\,°$C.[92] In some cases, reproducible rates have been obtained with different batches of reactants.

A second set of arguments is based on the study of hindered alkenes which do not give high polymers, but rather stable carbocations. For example concentrations of 1,1-diphenylalkyl cation much higher than the concentration of residual water were obtained in the system 1,1-diphenyl-ethylene (DPE)/TiCl$_4$/CH$_2$Cl$_2$/$-30\,°$C.[117] By using AlCl$_3$ instead of TiCl$_4$, it is possible to form about one carbocation per two aluminums.[223] Similarly the system 1,1-diphenylpropene/ TiCl$_4$/CH$_2$Cl$_2$ at $-30\,°$C also gave one carbocation for two TiCl$_4$ molecules, whereas SnCl$_4$ was almost inactive under the same conditions[133] and the system 3-isopropylindene/TiCl$_4$/CH$_2$Cl$_2$/$-70\,°$C yielded more than one carbocation per two TiCl$_4$ molecules.[224]

A third set of arguments is brought about by the experiments of 'polymerization by condensation'. If a quiescent mixture of isobutylene, TiCl$_4$ and CH$_2$Cl$_2$ in which polymerization has stopped at a low conversion is distilled under vacuum in a side arm of the reactor, maintained at a lower temperature, a rapid and complete polymerization occurs.[108]

Polymerization was observed even when the side vessel had been covered by a sodium mirror, but no polymerization took place when the mixture was poured into the side vessel. Similar phenomena were reported for the indene/TiCl$_4$/CH$_2$Cl$_2$ system[111] and the cyclopentadiene/TiCl$_3$OBu^n/ CH$_2$Cl$_2$ system.[121] With the system 1,1-diphenylethylene (DPE)/TiCl$_4$/CH$_2$Cl$_2$,[117] carbocations in concentrations ten times higher than in the main vessel could be obtained by bulb-to-bulb distillation. It was shown that the simultaneous vaporization of DPE and TiCl$_4$ was necessary for the initiation to occur.

Recently, a fourth set of arguments has been derived from the use of sterically hindered pyridines such as 2,6,-di-t-butyl-4-methylpyridine (DBMP). According to earlier work,[225,226] these bases are assumed to quantitatively trap the protons without interfering with F–C acids and carbocations (see Section 39.2.1.2). A series of experimental results could be interpreted in this way.[227] Systems were found which were not affected by DBMP (*e.g.* isobutylene/AlCl$_2$Et/EtCl/$-60\,°$C) and others for which polymerization was inhibited (isobutylene/BF$_3$/EtCl/$-40\,°$C). In between, systems were also reported in which the presence of DBMP caused a decrease in the yield, but could not totally stop the polymerization (α-methylstyrene/TiCl$_4$/CH$_2$Cl$_2$/$-30\,°$C, $-70\,°$C). The authors concluded that 'direct initiation' was predominant in the first case, cocatalysis by residual water in the second one, and that both mechanisms were simultaneously operating in the third case. The same investigators found that the polymerization of various monomers (isobutylene, styrene, vinyl ethers) initiated by Al(OSO$_2$CF$_3$)$_3$ in nitromethane was not altered by the addition of DBMP in equimolar amounts.[228] However, the role of DBMP in probably much more complicated than anticipated. Using the system DPE/AlCl$_3$/CH$_2$Cl$_2$/$-30\,°$C as a model for initiation by F–C halides, it has been shown that (1) initiation coinitiated by water is not suppressed by DBMP; (2) free DBMP reacts readily with carbocations; and (3) DBMP forms a strong complex with AlCl$_3$.[98] Though generalization is not possible owing to the diversity of the systems under discussion, the fact that, in some cases, 'cocatalysis' is not perturbed does not allow the non-inhibition by DBMP to be considered as a proof for direct initiation.

(ii) Kinetics of 'direct initiation'

As stated above, it is not possible to observe 'direct initiation' by a Friedel–Crafts acid without interference with the reaction cocatalyzed by residual cationogens. Moreover, in most cases, data concerning the real concentration of active centres are not available, and kinetic studies are limited to the dependence of the overall polymerization rate on the initial concentrations.

This explains the attention paid to the non-polymerizable alkenes giving stable carbocations, which provide a simplified model for initiation by F–C acids. The system DPE/AlCl$_3$/CH$_2$Cl$_2$ has been of particular interest.[98,223] After a very fast, but limited initiation due to a residual cationogenic impurity (probably water, see Section 39.2.1.2.v) or eventually to free ions formed by self-ionization of $AlCl_3$ (see Section 39.2.3.1.iii), a much slower reaction yielding one carbocation for two aluminums can be followed by spectrophotometry. The initiation rate during this stage is first order with respect to the monomer and $AlCl_3$ (equation 52).

$$R_i = d[R^+]/dt = k_i[DPE][AlCl_3] \qquad (52)$$

This allowed the first direct estimation of an initiation rate constant by a F–C acid and that of the corresponding activation energy: $k_i = (2.5 \pm 0.3) \times 10^{-2} \, l \, mol^{-1} \, s^{-1}$ at $-30\,^\circ C$; and $E_i = 10 \pm 1.5 \, kcal \, mol^{-1}$ $(-30\,^\circ C, -65\,^\circ C)$ $(1 \, kcal = 4.2 \, kJ)$. It is worth noting that preliminary results obtained under the same conditions showed that BCl_3 was not an initiator.[229]

On account of the diversity of the systems examined and of the diversity of the experimental conditions, it seems unlikely, at the present time, that the experimental facts summarized above might be rationalized by a unique mechanism. Nevertheless, among the various hypotheses reviewed below, some appear more probable because they are better argued and find a wider field of application.

(iii) Mechanisms of 'direct initiation'

(a) Solvent cocatalysis. Most of the polymerizations in which a 'direct initiation' mechanism is likely have been carried out in halogenated solvents. This observation suggested the hypothesis that the Lewis acid could be able to induce the heterolytic cleavage of a carbon–halogen bond in the solvent. The mechanism could be written as in equation (53).

$$RX + MtX_n + {>}C{=}C{<} \longrightarrow R\overset{|}{\underset{|}{C}}{-}\overset{|}{\underset{|}{C}}{}^+ MtX_{n+1}^- \qquad (53)$$

In fact, in only a few instances have solvent fragments been identified in the polymer and this cannot be considered as evidence for solvent cocatalysis since these fragments could have been incorporated by transfer as well. It is unlikely that the solvent is directly involved in direct initiation, since in the reaction of DPE with $AlCl_3$, the same yields of carbocations (50%) were obtained in CH_2Cl_2 and in MeCl.[230]

(b) Allylic hydride abstraction. It is known that metal halides which are strong oxidants (*e.g.* $SbCl_5$) are able to produce carbocations by abstraction of hydride from a C—H bond, when the carbocation is strongly stabilized, *e.g.* Ph_3C^+ from Ph_3CH.[231] The assumption has been made in the case of monomers containing allylic hydrogens that allylic cations could be formed by this mechanism (equation 54).[232]

$$CH_2{=}CH\overset{|}{\underset{|}{C}}H + MtX_n \longrightarrow CH_2{=}CH\overset{|}{\underset{|}{C}}{}^+ MtX_nH^- \qquad (54)$$

Indeed the hypothesis cannot explain the cases of 'direct initiation' which concern monomers such as styrene and 1,1-diphenylethylene.

(c) Self-ionization of the F–C acid. Conductometric studies have shown that metal halides in solvents of medium dielectric constant behave as weak electrolytes. When the metal halide is monomeric in solution, ions are formed through a 2:2 ionogenic equilibrium (equation 55).

$$2MtX_n \rightleftharpoons MtX_{n-1}^+ + MtX_{n+1}^- \qquad (55)$$

This situation is generally found in dichloromethane, *e.g.* for $TiCl_4$,[223] $SbCl_5$,[234] $AlCl_3$ and $AlBr_3$.[235] It is worth noting in this context that mixtures of F–C acids of different strength show enhanced conductivities.[236]

It has been postulated that the cationic moiety resulting from self-ionization could initiate the polymerization of some ethylenic monomers by addition to the double bond (equation 56).[222,237]

$$MtX_{n-1}^+ + {>}C{=}C{<} \longrightarrow MtX_{n-1}\overset{|}{\underset{|}{C}}{-}\overset{|}{\underset{|}{C}}{}^+ \qquad (56)$$

A similar scheme had been previously formulated for the polymerization of isobutylene by mixtures of F–C halides such as $TiCl_4$ and $AlBr_3$ in heptane solution.[238] It was found that $TiCl_4$ alone was inactive and that $AlBr_3$ was a very weak initiator. Nevertheless when $AlBr_3$ was added to quiescent mixture of isobutylene and $TiCl_4$ (or $TiCl_4$ to isobutylene/$AlBr_3$), a rapid and complete polymerization occurred. The strong activation was attributed to the higher ionization caused by mixing two Lewis acids of different strengths. However, this might be due to a lower termination rate or a higher propagation rate rather than to a higher initiation rate. This is supported by the fact that a mixture of $TiCl_4$ and $AlBr_3$ failed to produce high concentrations of carbocations in the case of DPE.[105]

An important set of experiments have been carried out concerning the polymerization of isobutylene by aluminum compounds.[239, 240] These authors found that the strongest Lewis acids ($AlCl_3$ and $AlCl_2Et$) led to direct initiation in MeCl: polymerizations were usually rapid, without an acceleration period (initiation faster than propagation), but incomplete (important termination). On the other hand, weaker Lewis acids ($AlClEt_2$ and $AlEt_3$) required a cationogen (Cl_2, HCl or Me_3CCl) and presented an acceleration period, but they led invariably to complete conversion. When $AlClEt_2$ or $AlEt_3$ was added to a mixture of isobutylene with $AlCl_3$ or $AlCl_2Et$ in which polymerization had stopped at a low yield, polymerization started again and went to completion after an acceleration period. This suggests that the second polymerization was due to a coinitiator formed *in situ* by a termination reaction occurring during the first step and that the role of the second aluminum compound, $AlClEt_2$ or $AlEt_3$, is to prevent the termination. This constitutes an alternative explanation for the enhanced polymerization rates observed with mixed F–C halides.

It has been shown recently that a very small amount of $FeCl_3$ could initiate the polymerization of isobutylene in the presence of BCl_3, $TiCl_4$ or $VOCl_3$, these compounds alone not being initiators or being very weak initiators.[236] In this particular case, the observation of Fe^{2+} in the medium suggested that initiation proceeded by electron transfer with formation of a radical cation from the monomer (see Section 39.3.1.1).

Self-ionization of the initiator was also suggested in the case of aluminum triflate, $Al(OSO_2CF_3)_3$ (equation 57).[241]

$$2Al(OSO_2CF_3)_3 \rightleftharpoons Al(OSO_2CF_3)_2^+ + Al(OSO_2CF_3)_4^- \tag{57}$$

This compound was used to initiate polymerizations of various ethylenic monomers in nitromethane: isobutylene, butadiene, styrene, α-methylstyrene, indene, 1,1-diphenylethylene, vinyl ethers, NVC and 1,2-dimethoxyethylene. An alternative explanation of the presence of an acidic impurity acting as a protogen was ruled out by the authors who showed that the addition of a hindered pyridine did not affect the polymerization rate in the case of isobutylene. An argument in favour of the direct addition of the $Al(OSO_2CF_3)_2^+$ moiety on the double bond was brought about by the use of the ionic salt, $Al(OSO_2CF_3)_2^+ SbF_6^-$, which is also a very efficient initiator.[228, 241] It has been mentioned that this aluminum salt reacted with a non-polymerizable alkene, 3-phenylindene, and give rise to the formation of one carbocation for two moles of $Al(OSO_2CF_3)_3$.[228]

In the same papers, a large number of metal salts (perchlorates and triflates) were shown to be efficient initiators in nitromethane solution. Their reactivity decreased in the order: $Co(ClO_4)_2 > Ni(ClO_4)_2 \approx Al(ClO_4)_3 > Al(OSO_2CF_3)_3 > Ga(OSO_2CF_3)_3 > Mg(ClO_4)_2$. It is worth mentioning that the same metal salts are also active in heterogeneous conditions. For example, isobutylene is slowly polymerized by $Mg(ClO_4)_2$ in bulk, in CH_2Cl_2 or in hexane solution, whereas $Co(ClO_4)_2$ under the same conditions gives flash polymerizations.[242]

(d) Direct addition onto the double bond. The phenomenon of 'polymerization by condensation' which implies that initiation takes place in the vapour phase (see above) and the observation that the concentration of carbocations is inversely proportional to the monomer concentration in the system $DPE/TiCl_4/CH_2Cl_2$[115, 117] are not easily explained by the self-ionization theory. To account for these findings, it has been postulated that initiation could proceed from a direct acid–base interaction between the Lewis acid and the double bond.[243] The zwitterionic intermediate (equation 58) would probably isomerize to covalent species (equation 59) which could alternately act as a cationogen (equation 60) or eliminate HX, a potential protogen (equation 61).

$$MtX_n + {>}C{=}C{<} \longrightarrow MtX_n^- {-}\overset{|}{\underset{|}{C}}{-}\overset{|}{\underset{|}{C}}{}^+ \tag{58}$$

$$MtX_n^- {-}\overset{|}{\underset{|}{C}}{-}\overset{|}{\underset{|}{C}}{}^+ \longrightarrow MtX_{n-1}^- {-}\overset{|}{\underset{|}{C}}{-}\overset{|}{\underset{|}{C}}X \tag{59}$$

$$MtX_{n-1} \overset{|}{\underset{|}{C}}\overset{|}{\underset{|}{CX}} \; + \; MtX_n \longrightarrow MtX_{n-1}\overset{|}{\underset{|}{C}}\overset{|}{\underset{|}{C}}{}^+ MtX_{n+1}^- \tag{60}$$

$$MtX_{n-1}\overset{|}{\underset{|}{C}}\overset{|}{\underset{|}{CX}} \longrightarrow MtX_{n-1}\overset{|}{C}{=}C{<} \; + \; HX \tag{61}$$

A similar mechanism was also proposed for the polymerization of isobutylene by $AlBr_3$ and SbF_5.[244]

It should be stressed that mechanism (*c*) (self-ionization) and mechanism (*d*) (direct addition onto the double bond) lead to the formation of a metal–carbon bond. Identification of this type of bond has been attempted by terminating polymerizations with tritiated water.[245] In the case of the systems isobutylene/$AlBr_3$/MeBr and isobutylene/$AlCl_3$/EtCl, the incorporation of 0.66 to 1.5 moles of tritium per mole of polymer was found. As a large number of macromolecules are initiated by proton transfer, it was necessary to assume that most of these Al—C bonds resulted from a quantitative post-alumination of the terminal double bonds produced by proton transfer. In the case of styrene/$AlBr_3$/MeBr, only 4–5% of the chains contained a tritium atom, in agreement with the fact that saturated, indanic end groups are formed in this case.

The identification by elemental analysis, IR and NMR of a compound of formula $MeOCPh_2CH_2SbCl_4(NH_3)$ during the reaction of DPE with an excess of $SbCl_5$ in CH_2Cl_2 at $-80\,°C$ (after quenching by a mixture of methanol and liquid ammonia) was attributed to the direct addition of $SbCl_4^+$ to the alkene.[246]

An attempt to identify organoaluminum compounds by ^{27}Al NMR was made with the system DPE/$AlCl_3$/CH_2Cl_2.[230] After concentrating the reaction mixture by distillation of the solvent under vacuum at low temperature, the ^{27}Al NMR spectrum of the solution presented a weak and broad signal at -98.2 p.p.m. (from Al^{3+}) which might be assigned to a compound Cl_2AlR. Due to the extreme reactivity of this species, a precise characterization was not possible.

To justify the low yields of carbocations often observed, the hypothesis has been made of the formation of strong inactive complexes between Lewis acids and monomers. In fact, complexes have been identified in some cases, but they are very weak.[247] For instance, the equilibrium constant for the formation of the isobutylene $TiCl_4$ complex is about 0.1 at $25\,°C$ in heptane. Another explanation of the low yields might be the transformation of intermediate compounds (*e.g.* formed in equations 58 and 59) into active compounds.

39.2.3.2 *Electrophilic halogens*

In this section, we shall examine briefly initiation which involves (or might involve) halonium ions. A first case, relatively well documented, concerns the coinitiation by halogens of some polymerizations in the presence of F–C acids. Initiation by iodine alone, though the mechanism is probably of a different nature, will also be considered in this section.

(i) *Halogens and F–C acids*

Halogens have been reported to be active 'cationogens' in the presence of Lewis acids. For example, $AlMe_3$ alone does not initiate the polymerizations of isobutylene or styrene; on addition of chlorine, a rapid, but incomplete polymerization occurs.[248] In the same conditions (solvent: methyl chloride, temperature range $= -40$ to $-100\,°C$), bromine is very weakly active and iodine is completely inactive. To establish the mechanism of initiation, a study of the reaction of Cl_2/$AlMe_3$ with a non-polymerizable alkene showed that the first step is probably the addition of an electrophilic chlorine onto the double bond, followed by the expulsion of a proton (5) and (6) or the capture of a methyl group from the counterion (7) (equation 62).

$$CH_2{=}C\overset{Me}{\underset{R}{<}} \xrightarrow{Cl_2/AlMe_3} ClCH_2C\overset{CH_2}{\underset{R}{<}} \; + \; ClCH{=}C\overset{Me}{\underset{R}{<}} \; + \; ClCH_2\overset{Me}{\underset{R}{\underset{|}{C}}}Me \tag{62}$$

     (4)   (5)   (6)   (7)

R = 2,2-dimethylpropyl

The same authors found that reaction of chlorine with trimethylaluminum led to the formation of $AlMe_2Cl$ even at $-50\,°C$ and that the system X_2/$AlEt_2Cl$ was more active than the corresponding

$X_2/AlMe_3$. Thus participation by $AlMe_2Cl$ cannot be ruled out in the systems involving $Cl_2/AlMe_3$. Chlorine has been shown to be a better cationogen than HCl or Me_3CCl for the polymerization of isobutylene initiated by $AlEt_2Cl$ or $AlEt_3$.[239] All collected data, including kinetics and conductivity measurements, have been interpreted by the hypothesis that initiation is due to a minute amount of chloronium ions in equilibrium (equation 63).[240]

$$Cl_2 \; + \; AlEt_2Cl \rightleftharpoons Cl^+ \; + \; AlEt_2Cl_2^- \tag{63}$$

In the presence of isobutylene, the rate-determining step of initiation would be the rate of formation of Cl^+ from Cl_2 and $AlEt_2Cl$ (zero order in monomer for initiation, in agreement with the external overall first order found for the polymerization rate). The authors did not consider the possibility that $AlEtCl_2$ and $AlCl_3$, formed by an exchange reaction, might be the true initiators. This hypothesis has been discussed (see ref. 7, p. 176).

The system $AlEt_2I/I_2/$isobutylene has been shown to behave similarly.[249] In pentane, efficiency is very low (high concentrations of iodine are necessary to get high conversions, which lowers the molecular weight). In CH_2Cl_2, the rate of polymerization is high, but somewhat irreproducible.

The addition of a chloronium ion from a Friedel–Crafts chloride to a double bond has been shown to occur in the particular case of a non-polymerizable alkene (di-*p*-methoxyphenylethylene) in the presence of a large excess of $SbCl_5$ at $-70\,°C$.[250] After quenching by ammonia, $ClCH=C(C_6H_4OMe)_2$ was identified. Nevertheless, two routes may lead to this compound, either the direct addition of a chloronium ion (equation 64)

$$SbCl_5 \; + \; CH_2=C(C_6H_4OMe)_2 \longrightarrow ClCH_2\overset{+}{C}(C_6H_4OMe)_2 \cdot SbCl_4^- \tag{64}$$

or the addition of chlorine (formed by decomposition of $SbCl_5$ to $SbCl_3$ and Cl_2), followed by the ionization of the chloride by $SbCl_5$ (equations 65 and 66).

$$SbCl_5 \; + \; CH_2=C(C_6H_4OMe)_2 \longrightarrow ClCH_2C(C_6H_4OMe)_2Cl \; + \; SbCl_3 \tag{65}$$

$$ClCH_2C(C_6H_4OMe)_2Cl \; + \; SbCl_5 \longrightarrow ClCH_2\overset{+}{C}(C_6H_4OMe)_2SbCl_6^- \tag{66}$$

(ii) Iodine

Iodine initiates the polymerization of aromatic nonomers, vinyl ethers and *N*-vinylcarbazole. The initiation mechanism proposed for styrene,[251] acenaphthylene[252] and anethole[253] has been accepted also for vinyl ethers[254–256] and *N*-vinylcarbazole,[257,258] which leads to narrow molecular weight distributions at $-78\,°C$. The common features of this complex reaction are the following (equations 67 and 68): (1) iodine adds to the double bond and forms the corresponding diiodide; (2) the diiodide eliminates hydrogen iodide; (3) hydrogen iodide adds to the monomer; and (4) in several cases, propagation is thought to occur on these iodide derivatives activated by solvation with free iodine.

$$\begin{array}{c} \diagup \\ C=C \\ \diagdown \end{array} + \; I_2 \longrightarrow \begin{array}{c} | \;\; | \\ -C-C- \\ | \;\; | \\ I \;\; I \end{array} \longrightarrow \begin{array}{c} \diagup \\ C=C \\ \diagdown \end{array} + \; HI \tag{67}$$

$$\begin{array}{c} \diagup \\ C=C \\ \diagdown \end{array} + \; HI \longrightarrow \begin{array}{c} | \;\; | \\ HC-CI \\ | \;\; | \end{array} \xrightarrow{\; I_2 \;} \begin{array}{c} | \;\; | \\ HC-CI---I_2 \\ | \;\; | \end{array} \longrightarrow polymer \tag{68}$$

Direct evidence concerning points (1) to (3) has been gathered, whereas the fourth point remains more speculative (see Section 39.2.2.2.i). No effect of an electric field has been found in the case of acenapthylene, which is evidence in favour of a non-ionic propagation.[252] However the bimodal molecular weight distributions observed for *p*-methoxystyrene (*p*-MOS) and *p*-methylstyrene[259] point to the intervention of at least two kinds of active centres, one of which probably involves free ions since the addition of a small amount of Bu_4^nNI depressed the formation of the high molecular weight portion in the system *p*-MOS/I_2/CH_2Cl_2/CCl_4. Stopped-flow spectroscopy performed on the system *p*-MOS/I_2/1,2-dichloroethane showed an absorption at 320 nm assigned to a π complex and an absorption at 380 nm assigned to the propagating species.[46] On this basis, a very low initiation efficiency was deduced ($\sim 0.1\%$, see, however, Section 39.2.1.1.iii).

39.3 OXIDATION OF THE CARBON–CARBON DOUBLE BOND

Besides the addition of an electrophilic species to a carbon–carbon double bond that was examined in the preceding section, an alternative way to create a carbocation is to remove one electron from the double bond. This can be achieved either by chemical means (strong oxidants, formation of charge-transfer complexes) or by physical means (electromagnetic radiation, electric field, electrochemical oxidation). We shall briefly survey each of these possibilities.

39.3.1 Chemical Oxidation

39.3.1.1 Single electron transfer

Strong oxidants may be able to directly transfer one electron from a carbon–carbon double bond of sufficiently low ionization potential, yielding an unstable monomer cation radical. The subsequent coupling of this cation radical leads to a tail-to-tail bicationic dimer (equation 69).

$$\text{ox} \; + \; CH_2{=}C{\prec} \longrightarrow \text{red} \; + \; [CH_2{=}C{\prec}]^{\ddot{+}} \longrightarrow {}^+\overset{\mid}{C}CH_2CH_2\overset{\mid}{C}{}^+ \qquad (69)$$

This oxidation process is symmetric to the reduction of some alkenic monomers which occurs in anionic polymerization in the presence of an alkali metal (reduction of the double bond with formation of an anion radical and coupling in a bicarbanion which leads to a bifunctional polymer).

Initiation of cationic polymerization has been proved to occur through this pathway in some particular cases, but the same mechanism has been proposed speculatively for many other systems.

The clearest evidence for that mechanism has been reported in the case of 1,1-diphenylethylene (D)[246] that gives only dimers by a chain reaction. By using this model compound under very specific conditions (a large excess of SbCl$_5$ in dichloromethane at $-80\,°C$), the tail-to-tail bicationic dimer was obtained, which upon deactivation yielded 1,1,4,4-tetraphenylbutadiene. The reaction was written as involving a charge transfer complex (CTC) (equation 70).

$$D \; + \; SbCl_5 \rightleftharpoons CTC \rightleftharpoons D^{\ddot{+}}SbCl_5^- \longrightarrow SbCl_5^- \, {}^+DD^+ SbCl_5^- \qquad (70)$$

This type of reaction is probably limited to Lewis acids of very high electron affinity, such as SbCl$_5$, since the same experiment repeated with TiCl$_4$ failed to produce a tail-to-tail dimer.[117]

The same explanation has been claimed to hold for the system DPE/AlCl$_2$Et/CH$_2$Cl$_2$.[260] Recently, the same authors have found that the polymerization of isobutylene by AlEtCl$_2$ was totally inhibited by the simultaneous presence of a hindered pyridine (to prevent cocatalysis) and of oxygen. This effect was claimed to prove that the mechanism of direct initiation involved the formation of a cation radical.[261]

Recently, the polymerization of isobutylene by mixtures of F–C halides has been reinvestigated.[236] BCl$_3$ or FeCl$_3$ alone are not initiators in methylene chloride at $-20\,°C$, whereas their mixture leads to complete polymerizations. From spectrophotometric and conductometric studies, it was concluded that the ionogenic equilibrium (71) takes place.

$$FeCl_3 \; + \; BCl_3 \rightleftharpoons FeCl_2^+ \; + \; BCl_4^- \qquad (71)$$

It was shown that about 80% of the Fe^{3+} was reduced to Fe^{2+} during initiation, from which it was concluded that initiation proceeded by redox reaction between FeCl$_2^+$ and the alkene

$$FeCl_2^+ BCl_4^- \; + \; CH_2{=}C{\prec} \longrightarrow FeCl_2 \; + \; {\cdot}CH_2\overset{\mid}{C}{}^+ BCl_4^- \qquad (72)$$

In this context, it is noteworthy that cation radicals have been identified by ESR in the reaction of alkenes with F–C halides[262] by a method consisting of trapping intermediate cation radicals by 2,4,6-tri-t-butylnitrosobenzene (TBN), giving a stable nitroxide radical. Thus cation radicals were identified in the system styrene (or α-methylstyrene)/BF$_3$–PhOH (or BF$_3 \cdot$Et$_2$O or AlCl$_3$)/TBN/ benzene at 15–20 °C. In most cases, α-radicals (**8**) were formed. Only with the system α-methylstyrene/AlCl$_3$ was the expected β-radical (**9**) obtained. Unfortunately, these experiments have not been carried out under high purity conditions.

$$\text{(8)} \qquad\qquad\qquad \text{(9)}$$

In the case of NVC, which is a very nucleophilic monomer easily giving charge-transfer complexes, initiation by trityl salts has been also claimed to involve an electron transfer (equations 73 and 74).[263]

$$Ph_3C^+X^- + NVC \rightleftharpoons CTC \longrightarrow Ph_3C\cdot + NVC^{\overset{+}{\cdot}}X^- \tag{73}$$

$$2NVC^{\overset{+}{\cdot}}X^- \longrightarrow {}^-X\overset{+}{N}VCNVC^+X^- \tag{74}$$

The fate of the radical $Ph_3C\cdot$ is obscure. Since the amount of Ph_3C^+ consumed during the polymerization was very small, it was assumed that Ph_3C^+ was spontaneously regenerated from $Ph_3C\cdot$. A simpler explanation would have been to consider that initiation is much slower than propagation, as shown for other systems.

Direct oxidation of an alkenic double bond by a stable oxidizing cation radical, tri(*p*-bromophenyl)aminium perchlorate, has also been observed in the case of 1,1-bis[di(*p*-dimethylaminophenyl)]ethylene (equations 75 and 76).[264]

$$Ar_3'N^{\overset{+}{\cdot}}ClO_4^- + Ar_2C{=}CH_2 \longrightarrow Ar_3'N + [Ar_2\overset{+}{\underset{\cdot}{C}}CH_2]ClO_4^- \tag{75}$$

$$2[Ar_2\overset{+}{\underset{\cdot}{C}}CH_2]ClO_4^- \longrightarrow ClO_4^-[Ar_2\overset{+}{C}CH_2CH_2CAr_2^+]ClO_4^- \tag{76}$$

39.3.1.2 Charge transfer complexes

It has been known for many years that very nucleophilic vinyl monomers of low ionization potential, such as NVC, *p*-MOS or vinyl ethers, give charge transfer complexes (CTCs) with electron acceptors such as chloranil, benzoquinone, tetracyanoethylene and tetranitromethane (TNM). CTC in the ground state may be excited and the excited state (sometimes called an exciplex) may dissociate into ion radicals able to initiate a cationic polymerization; in principle, the energy necessary for the transition can be provided by heat or by an electromagnetic radiation of convenient wavelength, generally UV. An excellent review of thermally and photochemically induced charge transfer polymerizations may be found in ref. 265.

(i) Thermally induced polymerization

'Spontaneous' (*i.e.* thermally induced) polymerizations have been observed under the action of CTCs, but the mechanisms are not fully established, because of a controversy about the role of impurities. For instance, it was demonstrated that very pure chloranil is almost inactive in the polymerization of NVC at room temperature and that an impurity, namely 2-hydroxy-3,5,6-trichloro-*p*-benzoquinone, was responsible for the rapid polymerizations previously observed.[266]

In other cases, an acidic compound may be formed *in situ* by decomposition of the CTC. For example, the NVC/TNM complex decomposes at $+10\,°C$, leading to nitroform, which initiates the cationic polymerization of NVC.[267] The same conclusion was reached for the system DPE/TNM.[268] In the same way, chloranil-induced polymerization of NVC in benzene at $80\,°C$ is probably due mainly to HCl eliminated by the zwitterion intermediate formed (equations 77 and 78).[266]

$$\tag{77}$$

$$\tag{78}$$

Cz = *N*-carbazolyl

The similarity of this mechanism with the mechanism of 'direct initiation' by F–C halides described in Section 39.2.3.1.iii. *d* is worth emphasizing.

Recently, it has been reported that alkenes bearing strong electron acceptor substituents A and a good leaving group X such as triflate[269] or iodide[270] were efficient initiators of the cationic polymerization of nucleophilic vinyl monomers of low ionization potential, such as NVC or *p*-MOS. High molecular weights were obtained. The suggested mechanism involves the formation of a zwitterion and the rearrangement of the carbanionic end by *β*-elimination of X$^-$ (equation 79).

$$\text{(79)}$$

When no leaving group is available (*e.g.* with tetracyanoethylene), the *gauche* tetramethylene zwitterion (**10**) is in equilibrium both with cyclobutane (**11**) and with a *trans* conformer (**12**) which can lead to a cationic propagation (equation 80).[271]

$$\text{(80)}$$

(**11**)   (**10**)   (**12**)

(ii) Photo-induced polymerization

When the difference between donor and acceptor strength in a CTC is not very large, thermal dissociation is not observed. In that case, the energy necessary to dissociate the CTC can be brought by UV radiation (photo-induced cationic polymerization). For example, by irradiating a tetracyanoethylene–*α*-methylstyrene complex at 363 nm in 1,2-dichloroethane and cyclohexane solvent at room temperature and below, polymerization occurs.[272] Dissociation occurs essentially from the triplet state (equation 81).

$$^3(DA)^* \longrightarrow D^+_{solv} + A^-_{solv} \tag{81}$$

Using laser flash photolysis in the nanosecond range, the same authors were able to determine the rate constant k_1 for the addition of monomer to the cation radical and its lifetime ($k_{p1} = 1 \times 10^{-5}$ l mol^{-1} s^{-1} and $t_{1/2} = 2 \times 10^{-5}$ s).[273]

A photo-induced charge-transfer initiation has also been proposed in the case of isobutylene in the presence of VCl$_4$ to explain the fact that no polymerization occurred in the dark in heptane solution, but that a rapid polymerization was induced by UV, visible and even IR illumination.[274]

39.3.2 Physical Methods

39.3.2.1 Ionizing radiations

When an ethylenic compound is submitted to *γ*-ray irradiation, a variety of unstable species are formed (excited states, ion radicals, anions, cations, *etc.*) and they may contribute to the formation of a high polymer. For monomers which can be polymerized by radical as well as by anionic or cationic polymerization, such as styrene, radical polymerization was observed in 'wet' monomer (*i.e.* not thoroughly purified) and the contribution of a cationic mechanism increased with improved drying. In the best experimental conditions, only cationic polymerization was obtained. Similarly the polymerization of isobutylene (which is not radical polymerized) can be achieved only in 'dry' conditions. Monomers very sensitive to cationic polymerization such as cyclopentadiene and vinyl ethers have also been polymerized by ionizing radiation. Very high molecular weights are usually obtained.

Comparison of *γ*-ray-induced cationic polymerization with chemically induced polymerization is very useful, particularly concerning the propagation rate constants (although their determination relies on a number of questionable hypotheses) which are reference values since they are related to free cations (see Volume 4, Chapter 19).

The initiation rate depends on the dose rate I according to

$$R_i = \Phi_i I / 100 \tag{82}$$

where Φ_i is the yield of ionization (in cations for 100 eV). Φ_i is usually taken to be equal to 0.1 by hypothesis, since it is very difficult to evaluate in conditions close to those used for polymerizations.

In continuous irradiation, a steady-state concentration of active species is established at a very low level (typically $[M^+] \approx 10^{-10}\,\text{mol}\,l^{-1}$ at $\sim 1\,\text{Mrad}\,h^{-1}$) which is demonstrated by the dependence of the polymerization rate on the square root of the dose rate, but the characterization of the cationic intermediates is very difficult. Solvent participates in the first step of the initiation process, which then proceeds with the formation of larger cation radicals (equation 83).[275,276]

$$\text{Solv}^{\ddagger} \; + \; M \; \longrightarrow \; \text{Solv} \; + \; M^{\ddagger} \xrightarrow{\; M \;} \cdot MM^+ \; \longrightarrow \; \text{polymer} \tag{83}$$

By using pulse radiolysis coupled with rapid scanning spectrophotometry, direct identification of some intermediates has been attempted. For example, in a study of the irradiation of styrene in a mixture of isopentane and *n*-butyl chloride at $-165\,°C$ UV spectra were given that were attributed to monomer and dimer cation radicals.[277] For polymerizations done in CH_2Cl_2 at $-23\,°C$, the dimers formed from tail-to-tail cation radicals have been identified.[278]

39.3.2.2 *UV radiation*

UV radiation of short wavelength has been shown to initiate the polymerization of isobutylene. Photoinitiation has been observed in solution at very low temperature ($-153\,°C$) using vacuum UV of wavelength 1236 Å (10.04 eV) or 1470 Å (8.4 eV).[279] In the second case, the energy is lower than the ionization energy of isobutylene in the vapour phase (9.25 eV), which implies that initiation took place in the liquid phase. Very low conversions were obtained, but molecular weights were high ($\sim 5 \times 10^5$). Low quantum yields of the order of 0.074 to 1.5×10^{-2} ions per quantum were calculated.

Very high molecular weights ranging from 1 to 4×10^6 have also been reported for experiments carried out in neat isobutylene at temperatures between $-115\,°C$ and $-145\,°C$ with irradiation at 1165 and 1235 Å (Kr source).[280] Similar molecular weights were obtained in isopentane or toluene solution in which the polymer is soluble, whereas a much lower molecular weight ($\sim 60\,000$) was obtained in ethyl chloride in which the polymer precipitates.[281] The independence of the molecular weight in the presence of 1,1-diphenyl-2-picrylhydrazyl (DPPH) proved that radical species did not participate in the polymer growth. A right-angle cell was designed to produce ions in the vapour phase by means of a horizontal UV beam and to transfer them in the liquid monomer using a vertical electric field, the negative electrode being immersed in the liquid.[281] The mechanism probably involves the formation of a *t*-butyl cation in the gas phase through a cation radical intermediate (equations 84 and 85).

$$CH_2=CMe_2 \; + \; h\nu \; \longrightarrow \; [CH_2=CMe_2]^{\ddagger} \; + \; e^- \tag{84}$$

$$[CH_2=CMe_2]^{\ddagger} \; + \; CH_2=CMe_2 \; \longrightarrow \; Me_3C^+ \; + \; CH_2=C(Me)CH_2\cdot \tag{85}$$

Photoinitiation of cationic polymerization also includes photolysis of compounds which lead to acidic by-products. This is probably the way by which diaryliodonium or diarylsulfonium salts initiated the polymerization of ethylenic monomers (styrene, α-methylstyrene or 2-chloroethyl vinyl ether) as well as heterocyclic monomers (oxiranes, THF, cyclic acetals, *etc.*)[296-298] For example, in the case of $Ar_2I^+MtX_{n+1}^-$, the true initiator is probably $H^+MtX_{n+1}^-$ ($MtX_{n+1}^- = BF_4^-$, ArF_6^-, PF_6^-, *etc.*) formed through equations (86) and (87):

$$Ar_2I^{\cdot+}MtX_{n+1}^- \xrightarrow{\; h\nu \;} ArI^{\cdot+}MtX_{n+1}^- \; + \; Ar\cdot \tag{86}$$

$$ArI^{\cdot+}MtX_{n+1}^- \; + \; RH \; \longrightarrow \; ArI \; + \; R\cdot \; + \; H^+MtX_{n+1}^- \tag{87}$$

where RH is a protic molecule (solvent, monomer, protogenic impurity, *etc.*). Conversions are usually high, but molecular weights obtained at room temperature are very low in the case of ethylenic monomers.

39.3.2.3 *Electric field*

The effect of high intensity electric fields has been shown to induce the cationic polymerization of some ethylenic monomers. The electrodes used are different in shape, one of them being generally a sharp point or a thin blade. They are immersed in the pure monomer, separated by a short distance (about 5 mm) and a potential difference of at least 5 kV is applied. According to the polarities, two phenomena are possible, both leading to the formation of cation radical intermediates.

If the sharp electrode is positively charged, it exerts an electric field which is strong enough to abstract an electron from the monomer (field ionization) (equation 88).

$$\diagup C{=}C\diagdown \longrightarrow [\diagup C{=}C\diagdown]^{\ddagger} + e^- \tag{88}$$

If the sharp electrode is negatively charged, there is emission of electrons of sufficient energy to abstract electrons from the molecules in the beam (field emission) (equation 89).

$$e^- + \diagup C{=}C\diagdown \longrightarrow [\diagup C{=}C\diagdown]^{\ddagger} + 2e^- \tag{89}$$

The first experiments reported used both techniques to polymerize styrene.[282] This was possible because styrene polymerizes by a cationic as well as by a radical mechanism. On the other hand, in similar conditions THF was polymerized only when the point was positive (cationic mechanism) and methyl methacrylate only when the point was negative (anionic or radical mechanism). As in any cationic polymerization, purification and drying is crucial to obtain reproducible results. The molecular weight of the polystyrene obtained is also indicative of the type of mechanism: when the point is positive, molecular weights are rather low ($\sim 5 \times 10^4$), probably as a consequence of important transfer reactions when the active centre is a carbocation, whereas molecular weights as high as 1×10^6 may be obtained with a negative point.[283]

From the bimodal molecular weight distributions and from the effect of specific scavengers (water and pyridine for the cationic process, DPPH for the radical process), it was shown that both radical and cationic chain ends participate independently in the propagation in the case of styrene.[284] α-Methylstyrene and isobutyl vinyl ether have also been polymerized using a field ionization technique.[285]

39.3.2.4 *Electrochemical initiation*

Passage of a current through a solution containing a supporting electrolyte and an ethylenic monomer, generally in a polar solvent, may lead to a polymerization of the monomer. According to the many complex reactions that may occur at electrodes, involving the electrolyte and/or the monomer (and eventually impurities), the nature of the mechanism may be anionic, radical or cationic (see Volume 4, Chapter 24).[286]

When the oxidation potential of the monomer is lower than that of the anion of the supporting electrolyte, initiation is thought to occur by direct monomer oxidation (equation 90).

$$M \longrightarrow M^{\cdot+} + e^- \tag{90}$$

In the opposite case, the primary process may consist of the oxidation of the anion, with production of an acidic substance capable of initiating a cationic polymerization. In this case, electroinduced polymerization is nothing more than the electrochemical generation of a chemical initiator. It is often difficult to distinguish between both mechanisms, which may also coexist.

Examples in which initiation has been claimed to result from a direct oxidation of the monomers are relatively scarce. They include the case of *n*-butyl vinyl ether with tetrabutylammonium perchlorate as electrolyte in acetonitrile or nitrobenzene[287] and the case of isobutyl vinyl ether with tetraethylammonium perchlorate in mixtures of acetonitrile and isopropyl ether.[288] Also in the case of styrene (anodic peak at $+1.8$ V with respect to an Ag^0/Ag^+ reference electrode) with tetrabutyl-ammonium fluoroborate (oxidation of BF_4^- at a potential higher than 2.5 V), constant potential electrolyses (at $+1.8$ V) were interpreted by direct oxidation of the monomer absorbed on the anode surface.[289] Recently, the same mechanism was proposed for the cationic polymerization of isoprene in constant potential electrolyses at $+2.3$ V *vs.* Ag^0/Ag^+.[290]

Examples of electrolytic generation of an initiator are more abundant. This is the case with vinyl ethers, *n*-butyl vinyl ether,[291] ethyl vinyl ether and isobutyl vinyl ether,[292] when sodium tetra-phenylboride is used instead of $Bu_4^nN^+ \cdot ClO_4^-$. BPh_4^- is supposed to be oxidized in BPh_2^+ which might either add to the double bond (though it could not be detected by UV and IR in the polymer) or initiate through electron transfer or after subsequent reactions leading to the formation of protons or other acidic intermediates. Other examples concern the electropolymerization of styrene: with lithium perchlorate as electrolyte in propylene carbonate, the formation of perchloric acid at the anodic compartment has been documented.[293] With sodium tetrafluoroborate in sulfolane, the true initiator is thought to be boron fluoride (eventually with HF or water acting as cationogens).[294] Also in the case of styrene with tetraethylammonium hexachloroantimoniate in nitrobenzene, the polymerization would result from the formation of $SbCl_5 \cdot HCl$.[295]

39.4 REFERENCES

1. P. H. Plesch (ed.), 'The Chemistry of Cationic Polymerization', Pergmon Press, Oxford, 1963.
2. J. P. Kennedy and E. Maréchal, 'Carbocationic Polymerization', Wiley, New York, 1982.
3. A. Gandini and H. Chéradame, in 'Encyclopedia of Polymer Science and Technology', Wiley, New York, 1985, vol. 2, p. 729.
4. J. P. Kennedy, 'Cationic Polymerization, a Critical Inventory', Wiley–Interscience, New York, 1975.
5. D. J. Dunn, in 'Developments in Polymerization', R. N. Haward, Applied Science, London, 1979, chap. 2, p. 45.
6. P. Sigwalt, *Polym. J.*, 1985, **17**, 57.
7. A. Gandini and H. Chéradame, *Adv. Polym. Sci.*, 1980, **34/35**.
8. T. Saegusa and S. Kobayashi, *Makromol. Chem., Macromol. Symp.*, 1986, **1**, 23.
9. C. M. Fontana, *J. Phys. Chem.*, 1959, **63**, 1167.
10. S. Matsushita, T. Higashimura and S. Okamura, *Kobunshi Kagaku*, 1960, **17**, 456.
11. P. Sigwalt, *C. R. Hebd. Seances Acad. Sci.*, 1961, **252**, 3998.
12. V. V. Stepanov, S. I. Klenin, A. V. Troitskaya and S. S. Skorokhodov, *Vysokomol. Soedin., Ser. A*, 1976, **18**, 821; *ibid., J. Polym. Sci. USSR*, 1976, **A18**, 933.
13. J. P. Vairon and P. Sigwalt, *Bull. Soc. Chim. Fr.*, 1964, 482.
14. P. Sigwalt, *J. Polym. Sci.*, 1961, **52**, 15.
15. M. Imoto and T. Takemoto, *J. Polym. Sci.*, 1955, **15**, 271.
16. I. M. Kolthoff and M. K. Chantooni, Jr., *J. Am. Chem. Soc.*, 1973, **95**, 8539.
17. T. Fujinaga and I. Sakamoto, *J. Electroanal. Chem.*, 1977, **85**, 185.
18. M. Chmelir, Communication to the International Symposium on Cationic Polymerization, Preprints C7.1, Rouen, 1973.
19. A. Gandini and P. H. Plesch, *J. Chem. Soc.*, 1965, 4826.
20. A. Gandini and P. H. Plesch, *Eur. Polym. J.*, 1968, **4**, 55.
21. D. C. Pepper and P. J. Reilly, *Proc. R. Soc. London, Ser A*, 1966, **291**, 41.
22. D. C. Pepper, *Makromol. Chem.*, 1974, **175**, 1077.
23. P. K. Bossaer, E. J. Goethals, P. J. Hackett and D. C. Pepper, *Eur. Polym. J.*, 1977, **13**, 489.
24. K. Matyjaszewski, *Polym. Prepr., Am. Chem. Soc., Div. Polym. Chem.*, 1986, **27**[2], 112.
25. D. C. Pepper, *J. Polym. Sci., Polym. Symp.*, 1975, **50**, 51.
26. J. P. Lorimer and D. C. Pepper, *Proc. R. Soc. London, Ser. A*, 1976, **351**, 551.
27. L. M. Léon, P. Altuna and D. C. Pepper, *Eur. Polym. J.*, 1980, **16**, 929.
28. M. Chmelir, *Makromol. Chem.*, 1975, **176**, 2099.
29. M. Chmelir, N. Cardona and G. V. Schulz, *Makromol. Chem.*, 1977, **178**, 169.
30. N. Cardona-Sütterlin, *Polym. Bull. (Berlin)*, 1979, **1**, 307.
31. M. Chmelir and G. V. Schulz, *Polym. Bull. (Berlin)*, 1979, **1**, 355.
32. N. Cardona-Sütterlin, *Polym. Bull. (Berlin)*, 1979, **1**, 361.
33. M. Sawamoto, T. Masuda and T. Higashimura, *Makromol. Chem.*, 1976, **77**, 2995.
34. T. Kunitake and K. Takarabe, *Macromolecules*, 1979, **12**, 1061.
35. K. Matyjaszewski and P. Sigwalt, *Makromol. Chem.*, 1986, **187**, 2299.
36. K. Takarabe and T. Kunitake, *Makromol. Chem.*, 1981, **182**, 1587.
37. M. Sawamoto and T. Higashimura, *Macromolecules*, 1979, **12**, 581.
38. A. Le Borgne, G. Sauvet and P. Sigwalt, *Eur. Polym. J.*, 1980, **16**, 855.
39. D. Souverain, A. Le Borgne, G. Sauvet and P. Sigwalt, *Eur. Polym. J.*, 1980, **16**, 861.
40. A. H. Nguyen, F. Subira, H. Chéradame and P. Sigwalt, *Tetrahedron*, 1978, **34**, 335.
41. M. Moreau, K. Matyjaszewski and P. Sigwalt, *Macromolecules*, 1987, **20**, 1456.
42. K. Matyjaszewski and P. Sigwalt, *Macromolecules*, 1987, **20**, 2679.
43. D. Teyssié, M. Villesange and J. P. Vairon, *Polym. Bull. (Berlin)*, 1984, **11**, 459.
44. R. Asami and N. Tokura, *J. Polym. Sci.*, 1960, **42**, 553.
45. T. Masuda, M. Sawamoto and T. Higashimura, *Makromol. Chem.*, 1976, **177**, 2981.
46. M. Sawamoto and T. Higashimura, *Macromolecules*, 1978, **11**, 328.
47. M. Sawamoto and T. Higashimura, *Macromolecules*, 1978, **11**, 501.
48. T. Higashimura and M. Sawamoto, *Polym. Bull. (Berlin)*, 1978, **1**, 11.
49. Ref. 7, p. 66.
50. M. Berthelot, *Bull. Soc. Chim. Fr.*, 1866, **6**, 294.
51. R. E. Burton and D. C. Pepper, *Proc. R. Soc. London, Ser. A*, 1961, **263**, 58.
52. M. J. Hayes and D. C. Pepper, *Proc. R. Soc. London, Ser. A*, 1961, **263**, 63.
53. A. Albert and D. C. Pepper, *Proc. R. Soc. London, Ser. A*, 1961, **263**, 75.
54. D. H. Jenkinson and D. C. Pepper, *Proc. R. Soc. London, Ser. A*, 1961, **263**, 82.
55. Ref. 7, p. 64.
56. C. Peniche and A. Gandini, *Rev. CENIC (Cuba)* 1973, **4**, 59 (*Chem. Abstr.*, 1976, **84**, 5472).
57. R. G. Heiligmann, *J. Polym. Sci.*, 1951, **6**, 155.
58. K. Takarabe and T. Kunitake, *Makromol. Chem.*, 1981, **182**, 1587.
59. Ref. 7, p. 51.
60. J. Pielichowski, *J. Polym. Sci., Polym. Symp.*, 1973, **42**, 451.
61. Y. Tsuda, *Makromol. Chem.*, 1960, **36**, 102.
62. N. Tokura and T. Kawakara, *Bull. Chem. Soc. Jpn.*, 1962, **35**, 1902.

63. M. Miyamoto, H. Sawamoto and T. Higashimura, *Macromolecules*, 1984, **17**, 265.
64. F. Andruzzi, P. Censi and P. Giusti, *Chim. Ind. (Milan)*, 1970, **52**, 466.
65. A. Gandini and S. Pineto, *J. Polym. Sci., Polym. Lett. Ed.*, 1977, **15**, 337.
66. Ref. 7, p. 54.
67. C. D. Brown and A. R. Mathieson, *J. Chem. Soc.*, 1958, 3445.
68. C. D. Brown and A. R. Mathieson, *J. Chem. Soc.*, 1957, 3612, 3631.
69. J. J. Throssell, S. P. Sood, M. Szwarc and V. Stannett, *J. Am. Chem. Soc.*, 1956, **78**, 1122.
70. M. Sawamoto, T. Masuda, T. Higashimura, S. Kobayashi and T. Saegusa, *Makromol. Chem.*, 1977, **178**, 389.
71. Ref. 7, p. 60.
72. N. Obrecht and P. H. Plesch, *Makromol. Chem.*, 1981, **182**, 1459.
73. F. Bolza and F. E. Treloar, *Makromol. Chem.*, 1980, **181**, 839.
74. A. G. Evans, D. Holden, P. Plesch, M. Polanyi, H. A. Skinner and M. A. Weinberger, *Nature (London)*, 1946, **157**, 102.
75. Ref. 7, p. 153.
76. F. Fairbrother and E. L. Seymour, Ph. D. Thesis (Seymour), University of Manchester, 1943; D. C. Pepper, in 'Friedel–Crafts and Related Reactions', ed. G. A. Olah, Interscience, New York, 1964, vol. 2, part 2, chap. 30.
77. Y. Sakurada, T. Higashimura and S. Okamura, *J. Polym. Sci.*, 1958, **33**, 496.
78. K. E. Russell, in 'Cationic Polymerization and Related Complexes', Heffer, Cambridge, UK, 1953, p. 114.
79. T. Higashimura and I. Moribe, *J. Polym. Sci., Polym. Lett. Ed.*, 1974, **12**, 391.
80. J. P. Kennedy, *J. Polym. Sci., Part A-1*, 1968, **6**, 3139.
81. A. G. Evans and G. W. Meadows, *J. Polym. Sci.*, 1949, **4**, 359; *Trans. Faraday Soc.*, 1950, **46**, 327.
82. N. A. Ghanem and M. Marek, *Eur. Polym. J.*, 1972, **8**, 999.
83. F. S. Dainton and G. B. B. M. Sutherland, *J. Polym. Sci.*, 1949, **4**, 37.
84. A. M. Eastham, *J. Am. Chem. Soc.*, 1956, **78**, 6040.
85. J. M. Clayton and A. M. Eastham, *J. Am. Chem. Soc.*, 1957, **79**, 5368.
86. J. M. Clayton and A. M. Eastham, *Can. J. Chem.*, 1961, **39**, 138.
87. J. M. Clayton and A. M. Eastham, *J. Chem. Soc.*, 1963, 1636.
88. A. Priola, G. Gozzelino and F. Ferrero, *Polym. Bull. (Berlin)*, 1985, **13**, 245.
89. S. Bywater and D. J. Worsfold, *Can. J. Chem.*, 1977, **55**, 85.
90. J. P. Kennedy, S. Y. Huang and S. C. Feinberg, *J. Polym. Sci., Polym. Chem. Ed.*, 1977, **15**, 2869.
91. J. P. Kennedy and F. J. Y. Chen, *Polym. Bull. (Berlin)*, 1986, **15**, 201.
92. L. Bui, H. A. Nguyen and E. Maréchal, *Polym. Bull. (Berlin)*, 1987, **17**, 157.
93. Ref. 2, p. 469, 481.
94. D. O. Jordan and F. E. Treloar, *J. Chem. Soc.*, 1961, 737.
95. M. Di Maina, P. Narducci and G. Pizzirani, *Chim. Ind. (Milan)*, 1981, **63**, 263.
96. J. H. Beard, P. H. Plesch and P. P. Rutherford, *J. Chem. Soc.*, 1964, 2566.
97. M. Masure, G. Sauvet and P. Sigwalt, *Polym. Bull. (Berlin)*, 1980, **2**, 699.
98. M. Masure, P. Sigwalt and G. Sauvet, *Makromol. Chem., Rapid. Commun.*, 1983, **4**, 269.
99. D. O. Jordan and A. R. Mathieson, *J. Chem. Soc.*, 1952, 611.
100. J. P. Kennedy and R. G. Squires, *J. Macromol. Sci., Chem.*, 1967, **1**, 995.
101. H. Hirata and H. Tani, *Polymer*, 1968, **9**, 60.
102. J. Maslinka-Solich, M. Chmelir and M. Marek, *Collect. Czech. Chem. Commun.*, 1969, **34**, 2611.
103. C. M. Fontana and G. A. Kidder, *J. Am. Chem. Soc.*, 1948, **70**, 3745; see also ref. 4, p. 19.
104. M. Chmelir, M. Marek and O. Wichterle, *J. Polym. Sci., Part C*, 1967, **16**, 833.
105. M. Masure, N. A. Hung, G. Sauvet and P. Sigwalt, *Makromol. Chem.*, 1981, **182**, 2695.
106. R. H. Biddulph, P. H. Plesch and P. P. Rutherford, *J. Chem. Soc.*, 1965, 275.
107. H. Chéradame and P. Sigwalt, *C. R. Hebd. Seances Acad. Sci.*, 1964, **259**, 4273.
108. H. Chéradame and P. Sigwalt, *Bull. Soc. Chim. Fr.*, 1970, 843.
109. P. M. Chau, Thèse 3è cycle, Paris, 1973.
110. P. Sigwalt, W. Lapeyre and H. Chéradame, *Int. Symp. Cationic Polym., Prepr.*, Rouen (France), 1973, Comm. C-33.
111. H. Chéradame, M. Mazza, A. H. Nguyen and P. Sigwalt, *Eur. Polym. J.*, 1973, **9**, 375.
112. H. Chéradame, A. H. Nguyen and P. Sigwalt, *C. R. Hebd. Seances Acad. Sci.*, 1969, **268**, 476.
113. R. Bourne Branchu, H. Chéradame and P. Sigwalt, *C. R. Hebd. Seances Acad. Sci.*, 1969, **268**, 1292.
114. W. R. Longworth, C. J. Panton and P. H. Plesch, *J. Chem. Soc.*, 1965, 5579.
115. G. Sauvet, J. P. Vairon and P. Sigwalt, *Bull. Soc. Chim. Fr.*, 1970, 4031.
116. G. Sauvet, J. P. Vairon and P. Sigwalt, *J. Polym. Sci., Polym. Symp.*, 1975, **52**, 173.
117. G. Sauvet, J. P. Vairon and P. Sigwalt, *J. Polym. Sci., Polym. Chem. Ed.*, 1978, **16**, 3047.
118. P. Lopour, J. Pecka and M. Marek, *Makromol. Chem.*, 1973, **174**, 1.
119. B. Matyska, L. Petrusova, K. Mach and M. Svestka, *Collect. Czech. Chem. Commun.*, 1979, **44**, 1262.
120. P. Sigwalt and J. P. Vairon, *Bull. Soc. Chim. Fr.*, 1964, 482.
121. J. P. Vairon and P. Sigwalt, *Bull. Soc. Chim. Fr.*, 1971, 559.
122. M. Villesange, G. Sauvet, J. P. Vairon and P. Sigwalt, *J. Macromol. Sci., Chem.*, 1977, **11**, 391.
123. S. Okamura and T. Higashimura, *J. Polym. Sci.*, 1956, **21**, 289.
124. R. G. W. Norrish and K. E. Russell, *Trans. Faraday Soc.*, 1952, **48**, 91.
125. R. F. Bauer, R. T. La Flair and K. E. Russell, *Can. J. Chem.*, 1970, **48**, 1251.
126. R. O. Colclough and F. S. Dainton, *Trans. Faraday Soc.*, 1958, **54**, 886; **54**, 894.
127. C. G. Overberger, R. J. Ehrig and R. A. Marcus, *J. Am. Chem. Soc.*, 1958, **80**, 2456.
128. A. G. Evans and J. Lewis, *J. Chem. Soc.*, 1957, 2975.
129. A. Polton and P. Sigwalt, *Bull. Soc. Chim. Fr.*, 1970, 131.
130. Ref. 7, p. 132, 147.
131. Ref. 7, p. 134, 5.
132. A. G. Evans and E. D. Owen, *J. Chem. Soc.*, 1959, 4123.
133. S. Bywater and D. J. Worsfold, *Can. J. Chem.*, 1978, **56**, 2093.

134. J. M. Moulis, J. Collomb, A. Gandini and H. Chéradame, *Polym. Bull. (Berlin)*, 1980, **3**, 197.
135. J. P. Kennedy and R. T. Chou, *J. Macromol. Sci., Chem.*, 1982, **18**, 47.
136. G. Heublein, S. Spange and P. Hallpap, *Makromol. Chem.*, 1979, **180**, 1935.
137. F. Subira, J. P. Vairon, A. Polton and P. Sigwalt, *Bull. Soc. Chim. Fr.*, 1974, **12**, 2903.
138. T. Higashimura, T. Fukushima and S. Okamura, *J. Macromol. Sci., Chem.*, 1967, **1**, 683.
139. A. F. Johnson and D. A. Pearce, *J. Polym. Sci., Polym. Symp.*, 1976, **56**, 57.
140. J. M. Rooney, *Makromol. Chem.*, 1978, **179**, 2419.
141. K. Yamamoto and T. Higashimura, *Polymer*, 1975, **16**, 815.
142. G. Sauvet, J. P. Vairon and P. Sigwalt, *J. Polym. Sci., Part A-1*, 1969, **7**, 985.
143. R. Cotrel, G. Sauvet, J. P. Vairon and P. Sigwalt, *Macromolecules*, 1976, **9**, 931.
144. F. Subira, G. Sauvet, J. P. Vairon and P. Sigwalt, *J. Polym. Sci., Polym. Symp.*, 1976, **56**, 221.
145. J. M. Rooney, *J. Polym. Sci., Polym. Symp.*, 1976, **56**, 47.
146. F. Subira, J. P. Vairon and P. Sigwalt, *Proc. 28th IUPAC Macromol. Symp.*, 1982, 151.
147. N. G. Gaylord and M. Svestka, *J. Macromol. Sci., Chem.*, 1969, **3**, 897.
148. S. D. Pask, Ph. D. Thesis, University of Keele, 1980.
149. H. G. Richey, Jr., R. K. Lustgarten and J. M. Richey, *J. Org. Chem.*, 1968, **33**, 4543.
150. C. E. H. Bawn, C. Fitzsimmons, A. Ledwith, J. Penfold, D. C. Sherrington and J. A. Weightman, *Polymer*, 1971, **12**, 119.
151. M. Villesange, G. Sauvet, J. P. Vairon and P. Sigwalt, *Polym. Bull. (Berlin)*, 1980, **2**, 131.
152. R. Velichkova, I. M. Panayotov, J. Doicheva, G. Heublein, H. Schutz, P. Adler, S. Spange and R. Wondraczek, *J. Polym. Sci., Polym. Chem. Ed.*, 1982, **20**, 2895.
153. E. Bilbao, M. Rodriguez and L. M. Leon, *Polym. Bull. (Berlin)*, 1983, **10**, 483.
154. G. Sauvet, M. Moreau and P. Sigwalt, *Makromol. Chem., Macromol. Symp.*, 1986, **3**, 33.
155. P. Hémery, S. Boileau and P. Sigwalt, *J. Polym. Sci., Polym. Symp.*, 1975, **52**, 189.
156. P. M. Bowyer, A. Ledwith and D. C. Sherrington, *Polymer*, 1971, **12**, 509.
157. A. M. Goka and D. C. Sherrington, *Polymer*, 1975, **16**, 819.
158. S. Cesca, A. Priola and G. Ferraris, *Makromol. Chem.*, 1972, **156**, 325.
159. J. P. Kennedy, in 'Polymer Chemistry of Synthetic Elastomers', ed. J. P. Kennedy and E. Törnquist, Wiley-Interscience, New York, 1968, vol. I, chap. 5, p. 301.
160. J. P. Kennedy and J. K. Gilham, *Adv. Polym. Sci.*, 1972, **10**, 1.
161. J. P. Kennedy and S. Rengachary, *Adv. Polym. Sci.*, 1974, **14**, 1.
162. Ref. 7, p. 172.
163. A. Priola, S. Cesca and G. Ferraris, *Makromol. Chem.*, 1972, **160**, 41; see also *J. Polym. Sci., Polym. Symp.*, 1976, **56**, 162.
164. J. P. Kennedy and J. J. Charles, *J. Polym. Sci., Polym. Symp.*, 1977, **30**, 1.
165. P. Sigwalt, A. Polton and M. Miskovic, *J. Polym. Sci., Polym. Symp.*, 1976, **56**, 13; C. Baudin, M. Tardi, A. Polton and P. Sigwalt, *Eur. Polym. J.*, 1980, **16**, 695.
166. B. Pary, M. Tardi, A. Polton and P. Sigwalt, *Eur. Polym. J.*, 1985, **21**, 393.
167. M. Tazi, M. Tardi, A. Polton and P. Sigwalt, *Eur. Polym. J.*, 1986, **22**, 451.
168. R. M. Evans and R. S. Satchell, *J. Chem. Soc. B*, 1970, 300.
169. J. P. Kennedy, S. Y. Huang and S. C. Feinberg, *J. Polym. Sci., Polym. Chem. Ed.*, 1977, **15**, 2869.
170. L. Toman, S. Pokorny, J. Spevacek and J. Danhelka, *Polymer*, 1986, **27**, 1121.
171. J. P. Kennedy and R. A. Smith, *J. Polym. Sci., Polym. Chem. Ed.*, 1980, **18**, 1523.
172. R. Santos, A. Fehervari and J. P. Kennedy, *J. Polym. Sci., Polym. Chem. Ed.*, 1984, **22**, 2685.
173. O. Nuyken, S. D. Pask, A. Vischer and M. Walter, *Makromol. Chem.*, 1985, **186**, 173.
174. J. P. Kennedy, L. R. Ross, J. E. Lackey and O. Nuyken, *Polym. Bull. (Berlin)*, 1981, **4**, 67.
175. R. Santos, J. P. Kennedy and M. Walters, *Polym. Bull. (Berlin)*, 1984, **11**, 261.
176. S. D. Pask and O. Nuyken, *Polym. Bull. (Berlin)*, 1982, **8**, 457.
177. R. M. Wondraczek, J. P. Kennedy and R. F. Storey, *J. Polym. Sci., Polym. Chem. Ed.*, 1982, **20**, 43.
178. H. Mayr and R. Schneider, *Makromol. Chem., Rapid Commun.*, 1984, **5**, 43; H. Mayr, R. Schneider and R. Pock, *Makromol. Chem., Macromol. Symp.*, 1986, **3**, 19.
179. R. Schneider, U. Grabis and H. Mayr, *Angew. Chem., Int. Ed. Engl.*, 1986, **25**, 89.
180. S. D. Pask, P. H. Plesch and S. B. Kingston, *Makromol. Chem.*, 1981, **182**, 3031.
181. K. Matyjaszewski and P. Sigwalt, *Macromolecules*, 1987, **20**, 2679.
182. M. Miyamoto, M. Sawamoto and T. Higashimura, *Macromolecules*, 1984, **17**, 265.
183. T. Higashimura and M. Miyamoto, *Macromolecules*, 1985, **18**, 611.
184. T. Enoki, M. Sawamoto and T. Higashimura, *J. Polym. Sci, Polym. Chem. Ed.*, 1986, **24**, 2261.
185. C. G. Cho and J. E. McGrath, *Polym. Prepr. Am. Chem. Soc., Div. Polym. Chem.*, 1987, **28**[1], 455.
186. V. Heroguez, A. Deffieux and M. Fontanille, *Polym. Bull. (Berlin)*, 1987, **18**, 287.
187. M. Sawamoto, C. Okamoto and T. Higashimura, *Macromolecules*, 1987, **20**, 2693.
188. T. Higashimura, K. Kojima and M. Sawamoto, *Polym. Bull. (Berlin)*, 1988, **19**, 7.
189. A. Gandini and P. H. Plesch, *J. Chem. Soc.*, 1965, 4826.
190. T. Kagiya, M. Izu, H. Maruyama and K. Fukui, *J. Polym. Sci., Part A-1*, 1969, **7**, 917.
191. R. Faust and J. P. Kennedy, *Polym. Bull. (Berlin)*, 1986, **15**, 317.
192. R. Faust and J. P. Kennedy, *J. Polym. Sci., Polym. Chem. Ed.*, 1987, **25**, 1847.
193. R. Faust, A. Nagy and J. P. Kennedy, *J. Macromol. Sci., Chem.*, 1987, **24**, 595.
194. A. F. Fehervari, R. Faust and J. P. Kennedy, *Polym. Prepr., Am. Chem. Soc., Div. Polym. Chem.*, 1987, **28**[1], 382.
195. R. Faust and J. P. Kennedy, *Polym. Bull. (Berlin)*, 1988, **19**, 29.
196. R. Faust and J. P. Kennedy, *Polym. Bull. (Berlin)*, 1988, **19**, 35.
197. R. Faust and J. P. Kennedy, *Polym. Bull. (Berlin)*, 1988, **19**, 21.
198. S. Aoshima and T. Higashimura, *Polym. Bull. (Berlin)*, 1986, **15**, 417.
199. S. Aoshima and T. Higashimura, *Polym. Prepr. Jpn.*, 1987, **36**, 5–10, E284.
200. M. Imoto and S. Aoki, *Makromol. Chem.*, 1963, **63**, 141.
201. P. Giusti, F. Andruzzi, P. Cerrai and G. L. Possanzini, *Makromol. Chem.*, 1970, **136**, 97.
202. M. K. Mishra and J. P. Kennedy, *J. Macromol. Sci., Chem.*, 1987, **24**, 933.

203. B. Wang, M. K. Mishra and J. P. Kennedy, *Polym. Bull. (Berlin)*, 1987, **17**, 213.
204. M. K. Mishra, B. Wang and J. P. Kennedy, *Polym. Bull. (Berlin)*, 1987, **17**, 307.
205. G. A. Olah, S. J. Kuhn, W. S. Tolgyesi and E. B. Baker, *J. Am. Chem. Soc.*, 1962, **84**, 2733.
206. O. Nuyken and P. H. Plesch, *Chem. Ind. (London)*, 1973, 379.
207. E. Franta, L. Reibel, J. Lehmann and S. Penczek, *J. Polym. Sci., Polym. Symp.*, 1976, **56**, 139.
208. G. E. Holdcroft and P. H. Plesch, *Makromol. Chem.*, 1984, **185**, 27.
209. T. Masuda and T. Higashimura, *J. Macromol. Sci., Chem.*, 1971, **5**, 550.
210. T. Higashimura and O. Kishiro, *J. Polym. Sci., Polym. Chem. Ed.*, 1974, **12**, 967.
211. S. Kohjiya and S. Yamashita, *Chem. Lett.*, 1972, 671; 1973, 1007.
212. M. Sawamoto and T. Higashimura, *Macromolecules*, 1979, **12**, 581.
213. M. Sawamoto, T. Higashimura, A. Enokida and T. Okubo, *Polym. Bull. (Berlin)*, 1980, **2**, 309.
214. M. Sawamoto, A. Furukawa and T. Higashimura, *Macromolecules*, 1983, **16**, 518.
215. H. Kämmerer and A. Seyed-Mozaffari, *Makromol. Chem.*, 1984, **185**, 509.
216. A. Tanizaki, M. Sawamoto and T. Higashimura, *J. Polym. Sci., Polym. Chem. Ed.*, 1986, **24**, 87.
217. U. Seitz, R. Hoene and K. H. W. Reichert, *Makromol. Chem.*, 1975, **176**, 1689.
218. M. L. Hallensieben and K. Möller, *Polym. Bull. (Berlin)*, 1984, **11**, 7.
219. W. R. Longworth, C. J. Panton and P. H. Plesch, *J. Chem. Soc.*, 1965, 5579.
220. A. H. Nguyen, H. Chéradame and P. Sigwalt, *Eur. Polym. J.*, 1973, **9**, 385.
221. M. Marek, J. Pecka and M. Maleska, in 'Cationic Polymerization and Related Processes', ed. E. J. Goethals, Academic Press, London, 1984, p. 17.
222. D. W. Grattan, Ph. D. Thesis, University of Keele, 1973.
223. M. Masure, G. Sauvet and P. Sigwalt, *J. Polym. Sci., Polym. Chem. Ed.*, 1978, **16**, 3065.
224. H. Chéradame, A. H. Nguyen and P. Sigwalt, *J. Polym. Sci., Polym. Symp.*, 1976, **56**, 335.
225. H. C. Brown and B. Kanner, *J. Am. Chem. Soc.*, 1953, **75**, 3865.
226. H. C. Brown and B. Kanner, *J. Am. Chem. Soc.*, 1966, **88**, 986.
227. J. M. Moulis, J. Collomb, A. Gandini and H. Chéradame, *Polym. Bull. (Berlin)*, 1980, **3**, 197.
228. J. Collomb, A. Gandini and H. Chéradame, *Makromol. Chem., Rapid. Commun.*, 1980, **1**, 489.
229. M. Masure and P. Sigwalt, unpublished data.
230. M. Masure, Doctorate Thesis, University of Paris-VI, 1979.
231. J. Holmes and R. Pettit, *J. Org. Chem.*, 1963, **28**, 1695.
232. J. P. Kennedy, *J. Macromol. Sci., Chem.*, 1972, **6**, 329.
233. W. R. Longworth and P. H. Plesch, *J. Chem. Soc.*, 1959, 1887.
234. D. D. Eley, D. F. Monk and D. F. Rochester, *J. Chem. Soc., Faraday Trans. 1*, 1976, 1584.
235. D. W. Grattan and P. H. Plesch, *J. Chem. Soc., Dalton, Trans.*, 1977, 1734.
236. M. Marek, J. Pecka and V. Halaska, *Makromol. Chem., Makromol. Symp.*, 1988, **13–14**, 443.
237. P. H. Plesch, *Makromol. Chem.*, 1974, **175**, 1065.
238. M. Marek and M. Chmelir, *J. Polym. Sci., Part C*, 1968, **23**, 223.
239. M. Di. Maina, S. Cesca, P. Giusti, G. Ferraris and P. L. Magagnini, *Makromol. Chem.*, 1977, **178**, 2223.
240. P. L. Magagnini, S. Cesca, P. Giusti, A. Priola and M. Di. Maina, *Makromol. Chem.*, 1977, **178**, 2235.
241. J. Collomb, P. Arlaud, A. Gandini and H. Chéradame, in 'Cationic Polymerization and Related Processes', ed. E. J. Goethals, Academic Press, London, 1984, p. 49.
242. J. Collomb, B. Morin, A. Gandini and H. Chéradame, *Eur. Polym. J.*, 1980, **16**, 1135.
243. P. Sigwalt, *Makromol. Chem.*, 1974, **175**, 1077.
244. G. A. Olah, *Makromol. Chem.*, 1974, **175**, 1039.
245. D. W. Grattan and P. H. Plesch, *Makromol. Chem.*, 1980, **181**, 751.
246. B. E. Fleischfresser, W. J. Cheng, J. M. Pearson and M. Szwarc, *J. Am. Chem. Soc.*, 1968, **90**, 2172.
247. V. Halaska, J. Pecka and M. Marek, *Makromol. Chem., Makromol. Symp.*, 1986, **3**, 3.
248. J. P. Kennedy and S. Sivaram, *J. Macromol. Sci., Chem.*, 1973, **7**, 969.
249. P. Giusti, A. Priola, P. L. Magagnini and P. Narducci, *Makromol. Chem.*, 1975, **176**, 2303.
250. W. Bracke, W. J. Cheng, J. M. Pearson and M. Szwarc, *J. Am. Chem. Soc.*, 1969, **91**, 203.
251. P. Giusti and F. Andruzzi, *J. Polym. Sci., Part C*, 1968, **16**, 3797.
252. P. Cerrai, F. Andruzzi and P. Giusti, *Makromol. Chem.*, 1968, **117**, 128.
253. P. Cerrai, P. Giusti, G. Guerra and M. Tricoli, *Eur. Polym. J.*, 1974, **10**, 1141.
254. A. Ledwith and D. C. Sherrington, *Polymer*, 1971, **12**, 344.
255. K. M. Janjua and A. F. Johnson, *Int. Symp. Cationic Polym., Prepr.*, Rouen (France), 1973, Comm. C 15.
256. A. F. Johnson and R. N. Young, *J. Polym. Sci., Polym. Symp.*, 1976, **56**, 211.
257. T. Higashimura, H. Teranishi and M. Sawamoto, *Polym. J.*, 1980, **12**, 393.
258. T. Higashimura, Y. -X. Deng and M. Sawamoto, *Polym. J.*, 1983, **15**, 385.
259. T. Higashimura, O. Kishiro and T. Takeda, *J. Polym. Sci., Polym. Chem. Ed.*, 1976, **14**, 1089.
260. F. M. Nasirov and F. R. Khalafov, *Int. Symp. Cationic Polym., Prepr.*, Rouen (France), 1973, Comm. C 29.
261. F. R. Khalafov, F. M. Nasirov, N. E. Melnikova, B. A. Krentsel and T. N. Shakhtakhtinsky, *Makromol. Chem., Rapid Commun.*, 1985, **6**, 29.
262. K. Yamada, H. Tanaka and H. Kawazura, *J. Polym. Sci., Polym. Lett. Ed.*, 1976, **14**, 517.
263. M. Rodriguez and L. M. León, *Eur. Polym. J.*, 1983, **19**, 589.
264. C. E. H. Bawn, F. A. Bell and A. Ledwith, *Chem. Commun.*, 1968, 599.
265. Y. Shirota and H. Mikawa, *J. Macromol. Sci., Rev. Macromol. Chem.*, 1977–1978, **16**, 129.
266. T. Natsuume, M. Nishimura, M. Fujimatsu, M. Shimazu, Y. Shirota, H. Hirata, S. Kusabayashi and H. Mikawa, *Polym. J.*, 1970, **1**, 181.
267. R. Gumbs, S. Penczek, J. Jagur-Grodzinski and M. Szwarc, *Macromolecules*, 1969, **2**, 77.
268. S. Penczek, J. Jagur-Grodzinski and M. Szwarc, *J. Am. Chem. Soc.*, 1968, **90**, 2174.
269. H. K. Hall, Jr. and H. A. A. Rasoul, *Contemp. Top. Polym. Sci.*, 1984, **4**, 295.
270. A. B. Padias and H. K. Hall, Jr., *Polym. Bull. (Berlin)*, 1985, **13**, 329.
271. T. Gotoh, A. B. Padias and H. K. Hall, Jr., *J. Am. Chem. Soc.*, 1986, **108**, 4920.

272. M. Irie, S. Tomimoto and K. Hayashi, *J. Phys. Chem.*, 1972, **76**, 1419.
273. M. Irie, H. Masuhara, K. Hayashi and N. Mataga, *J. Phys. Chem.*, 1974, **78**, 341.
274. J. Pilar, L. Toman and M. Marek, *J. Polym. Sci., Polym. Chem. Ed.*, 1976, **14**, 2399.
275. Y. Yoshida and K. Hayashi, *Adv. Polym. Sci.*, 1969, **6**, 401.
276. K. Hayashi, *Actions Chim. Biol. Radiat.*, 1971, **15**, 7.
277. Y. Tabata, *J. Polym. Sci., Polym. Symp.*, 1976, **56**, 409.
278. T. Gotoh, M. Yamamoto, Y. Nishijima, *J. Polym. Sci., Polym. Chem. Ed.*, 1981, **19**, 1047.
279. C. Vermeil, M. Matheson, S. Leach and F. Muller, *J. Chim. Phys.*, 1964, **61**, 596.
280. E. W. Schlag and J. J. Sparapany, *J. Am. Chem. Soc.*, 1964, **86**, 1875.
281. J. J. Sparapany, *J. Am. Chem. Soc.*, 1966, **88**, 1357.
282. M. Lambla, G. Schreibling and A. Banderet, *C. R. Hebd. Seances Acad. Sci., Ser. C*, 1970, **271**, 924.
283. M. Brendlé, *C. R. Hebd. Seances Acad. Sci., Ser. C*, 1971, **272**, 734.
284. M. Lambla, R. Koenig and A. Banderet, *Eur. Polym. J.*, 1972, **8**, 1.
285. M. Brendlé and A. M. Ilvoas, *Br. Polym. J.*, 1976, 11.
286. O. F. Olaj, *Makromol. Chem., Macromol. Symp.*, 1987, **8**, 235.
287. G. Mengoli and G. Vidotto, *Makromol. Chem.*, 1970, **139**, 293.
288. J. W. Breitenbach, F. Sommer and J. Unger, *Monatsh. Chem.*, 1976, **107**, 359.
289. U. Akbulut, J. E. Fernandez and R. L. Birke, *J. Polym. Sci., Polym. Chem. Ed.*, 1975, **13**, 133.
290. U. Akbulut, L. Toppare and B. Yurttas, *J. Polym. Sci., Polym. Lett. Ed.*, 1986, **24**, 185.
291. G. Mengoli and G. Vidotto, *Eur. Polym. J.*, 1972, **8**, 671.
292. G. Mengoli and G. Vidotto, *Makromol. Chem.*, 1972, **153**, 57.
293. G. Pistoia, *Eur. Polym. J.*, 1974, **10**, 279.
294. B. M. Tidswell and A. G. Doughty, *Polymer*, 1971, **12**, 431.
295. A. Ghose and S. N. Bhadani, *Indian J. Technol.*, 1975, **13**, 172.
296. J. V. Crivello and J. H. W. Lam, *Macromolecules*, 1977, **10**, 307.
297. J. V. Crivello, in 'Cationic Polymerization and Related Processes', ed. E. J. Goethals, Academic Press, London, 1984, p. 289.
298. J. V. Crivello, *Adv. Polym. Sci.*, 1984, **62**, 1.

40

Carbocationic Polymerization: Alkenes and Dienes

O. NUYKEN
University of Bayreuth, FRG
and
St. D. PASK
Bayer A G, Dormagen, FRG

40.1 INTRODUCTION

Although probably the first polymerizations ever to be observed were what are now known to be processes involving chain growth *via* positively charged active-species (cationic polymerizations), the cationic polymerization of alkenes and dienes remains a topic beset by the difficulties of its experimental study.

One of the first reports of a cationic polymerization in the scientific literature is that from Bishop Watson in his Chemical Essays,[1] describing the resinification of oil of turpentine by sulfuric acid. During the following century, it was realized[2,3] that metal halides, such as BF_3, could also be used to yield similar products. The cationic polymerization of terpenes, and especially α-pinene, using metal halides ($AlCl_3$, $SnCl_4$) remains an important industrial application of cationic polymerization.[4]

619

In the 19th century, many reactions which are now acknowledged as cationic polymerizations were discovered. For example, it was during this period that methylpropene (isobutylene, IB)[5] and styrene[6] were first polymerized using sulfuric acid.

During the 1930s, the development of organic chemistry, and the growing necessity to utilize the unsaturated by-products from the refining and cracking of crude oil, led to the recognition that the reactions of Brønsted and Lewis acids with $C{=}C$ double bonds involves positively charged active-species. This period also heralded the development of what remains the only industrial application of cationic polymerization of an alkene to a polymer with a high molecular weight: the polymerization of IB with metal halides (see Section 40.4.1).

In order to appreciate the difficulties which have hampered the elucidation of carbocationic polymerizations, it is important to recognize that the initiation and propagation steps involve exceedingly reactive carbocations; the more reactive a species is, the less specific its reactions. In particular, carbocations undergo transfer reactions and react with a wide variety of impurities, especially with the ubiquitous water. The low concentration of such compounds and the reactivity of the growing species makes the direct experimental observation of carbocations very difficult.

This chapter will endeavour to emphasize the most modern developments in this field, while outlining some of the essential aspects of the cationic polymerization of alkenes and dienes. The emphasis on more modern literature should not be taken to indicate that all the questions which earlier workers hoped to answer have been adequately explained; this is not the case.

Since the first international conference concerned with cationic polymerization[7] there have been regular meetings with published proceedings.[8-11] Milestones in the appreciation of the theory of cationic polymerization are represented by the reviews of Plesch[12] and Gandini and Cheradame.[13]

Other useful introductions to the profuse literature can be found in the detailed critical review of the cationic polymerization of vinyl monomers by Dunn[14] and in two books.[15,16] Lastly, the cationic polymerization of alkenes and dienes with metal-halide initiation-systems is a special case of a Friedel–Crafts reaction; an exhaustive review of which has also been published.[17]

40.2 TERMINOLOGY

The term carbenium ion is restricted in this chapter to the trivalent trigonal sp^2-hybridized carbocation derived from CH_3^+.

The cationic polymerization of an alkene (or diene) can be envisaged to involve the reactions shown in Scheme 1. In addition to the transfer to monomer, which although terminating polymer-chain growth does not lead to a kinetic termination, transfer and termination reactions with impurities can occur; terminations in carbocationic polymerization usually involve the formation of relatively stable carbocations which are no longer capable of adding monomer.

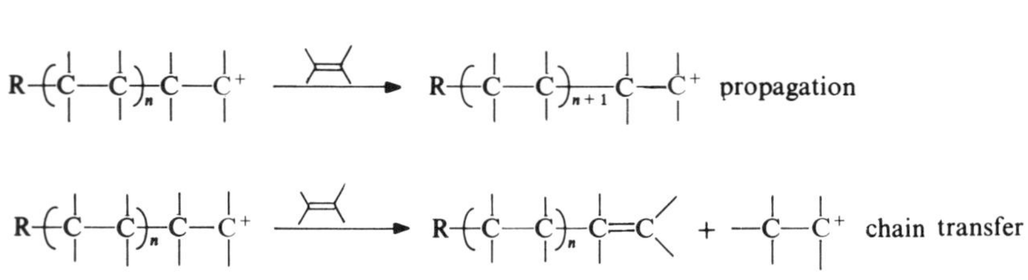

Scheme 1

Considerable confusion arises from the use of the word initiator; the true initiating species is R^+, which may, for example, be a proton, a carbenium ion or a metal-halide cation. The species from which the initiating cation is derived is a cationogen. For many cationic polymerizations, the precursor of the growing chain has not been unambiguously identified, and in these cases it is preferable to name the initiating system or cationogen. Particularly misleading is the use of the word catalyst; metal halides have traditionally been called Friedel–Crafts catalysts. Metal halides generally require an additional reagent, such as H_2O or a Brønsted acid, in order for polymerization to take place. Thus, here too, it is preferable to describe an initiating system.

Lastly, the literature contains many examples of reaction schemes involving a free proton. Such a species is far too reactive in solution to be considered as a viable reality. A better understanding of such reactions is almost always promoted by an appreciation of the form in which the protons actually exist, namely as H_3O^+ or H_2X^-.

40.3 GENERAL CONSIDERATIONS

All molecules which can be cationated, and thereby give rise to an ion which is capable, in turn, of an electrophilic attack on the neutral molecule, can, in principle, be polymerized cationically. Thus, suitable monomers can be considered as nucleophiles or bases.

Within the scope of this chapter, a cationically polymerizable C=C double bond requires an adjacent electron-releasing substituent. However, not all molecules which fulfill these requirements will polymerize cationically.

The reasons for this can be summarized in terms of the molecular structure or additional functional-groups. The presence of, for example, a nitrile, ester, hydroxy or an amino group in a molecule will generally preclude its carbocationic polymerization.

1,1-Diphenylethylene and 2,4,4-trimethyl-1-pentene can both be readily cationated, but the resulting cations are so sterically hindered that they are unable to propagate; at best, dimers or low oligomers, respectively, can be synthesized. For this reason, 1,1-diphenylethylene has often been used as a model for the initiation of cationic polymerization.[18-20]

A reduction in the rate of polymerization is also observed with 1,2-substituted monomers, such as β-substituted styrenes.[21] This effect can, however, be reduced or reversed if the two substituents form part of a ring system. Thus, indene[22,23] and acenaphthylene[24] both polymerize readily with cationic initiating-systems.

If the initially formed cation is not in the most stable possible conformation, or if several isomers have similar free energies of formation, rearrangement accompanies the propagation step. An example of such is the cationic polymerization of propene, which has seldom led to high molecular weight products;[25] generally, only low molecular weight oils can be prepared from a cationic polymerization of propene, and these materials contain highly branched structures.[26] For some monomers, a rearrangement is an essential step in their polymerization. An example is the polymerization of β-pinene (1; 2(10)-pinene), during which the rearrangement of the sterically hindered endocyclic carbenium-ion to an exocyclic ion is the sole reaction-pathway leading to high molecular weight products[27] (Scheme 2).

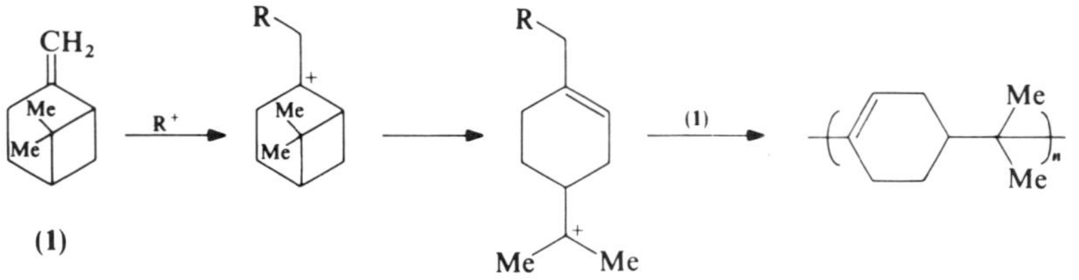

Scheme 2

The classical example of an isomerization accompanying a polymerization and leading to a unique polymer structure is that of the polymerization of 3-methyl-1-butene (2). With Lewis-acid initiating-systems in hexane ($T < -100\,^\circ$C) the crystalline polymer products are almost exclusively the product of a 1,3-addition reaction (Scheme 3).[28]

$$CH_2{=}CHCHMe_2 \xrightarrow{R^{\cdot}} RCH_2\overset{+}{C}HCHMe_2 \longrightarrow RCH_2CH_2\overset{+}{C}Me_2 \xrightarrow{(2)} (CH_2CH_2CMe_2)_n$$

(2)

Scheme 3

40.4 INITIATION

Within the scope of this chapter, only some of the basic principles can be discussed. The interested reader should consult the excellent critical review of Gandini and Cheradame[13] for more detail.

'In order to define the kinetics of any system it is essential to have a precise knowledge of what the reaction mixture contains'.[29] The identification of the binary ionogenic equilibria associated with metal halides[30] and metal halides with organic halides[31] in organic solvents (equations 1 and 2, respectively), the preparation of carbocation salts[32] and the development of high-energy radiation initiation-techniques[33] have all helped towards a more detailed understanding of the cationic polymerization of alkenes. These four topics, as well as the more classical cationogens, the Brønsted acids, will be dealt with in some detail, but it should not be forgotten that cationic polymerization can also be started electrochemically,[34] photochemically[35] and *via* charge-transfer complexes.[36-40]

$$2MtX_n \rightleftharpoons MtX_{n+1}^- MtX_{n-1}^+ \rightleftharpoons MtX_{n+1}^- + MtX_{n-1}^+ \tag{1}$$

$$RX + MtX_n \rightleftharpoons R^+ MtX_{n+1}^- \rightleftharpoons R^+ + MtX_{n+1}^- \tag{2}$$

40.4.1 Metal Halides

Metal halides, such as $SnCl_4$, $TiCl_4$, $AlCl_3$, $AlBr_3$ and BF_3, form the basis of the most widely studied group of initiating systems for carbocationic polymerization Two modes of initiation are possible starting with a metal halide and a C=C double bond, as shown in equations (3) and (4). The first mode (equation 3), elucidated during the now-classic work of Polanyi and his co-workers,[41] and since confirmed by many other studies,[42] implies the necessity of a third component, a protogen (H_2O, HX) and is analogous to the alkyl halide/metal halide systems which have recently been exploited to prepare telechelic polymers *via* cationic polymerization (see Section 40.7). However, in contrast to the metal halide/alkyl halide systems, the water/metal halide systems are not simple. Water can react with the metal halide to yield a variety of species depending on the concentration ratios and the nature of the metal halide. In particular, the corresponding Lewis acid may form (equation 5). The extent to which HX can act as a protogen, especially in conjunction with MtX_n, varies from system to system (monomer/solvent/metal halide).

$$MtX_n + H_2O + \text{C=C} \longrightarrow -\overset{|}{\underset{|}{C}}-\overset{|}{\underset{|}{C}}^+ + MtX_nOH^- \tag{3}$$

$$MtX_n + \text{C=C} \longrightarrow X_{n-1}Mt-\overset{|}{\underset{|}{C}}-\overset{|}{\underset{|}{C}}^+ + MtX_{n+1}^- \tag{4}$$

$$MtX_n + H_2O \longrightarrow MtX_{n-1}OH + HX \tag{5}$$

Many workers have also found that the order of addition of the reagents is critical in terms of reproducibility, and the actual mechanism for most systems remains unclear.[43-45] Indeed, using the the so-called 'condensation technique', Cheradame and Sigwalt[46] have claimed that IB can be initiated using $TiCl_4$ without a protogen. This controversial 'direct initiation' has also been described for IB using $AlCl_3$ and $AlBr_3$,[47] or $AlEtCl_2$[48] initiating-systems. Most recently, a direct initiation of an IB polymerization by BCl_3 has been proposed.[49] This paper exemplifies the complex problems involved. The authors carried out several types of experiments in 'all glass' vacuum apparatus, with and without added impurities (*i.e.* H_2O or HCl) and under varying degrees of purity. However, in this work it is stated incorrectly that a direct initiation following a self-ionization (Scheme 4) would lead to yields being proportional to $[BCl_3]_0^2$ or $[BCl_3]_0^3$. On the contrary, the concentration of growing chains will reflect the initial concentration of BCl_3. The yield will be determined by the nature of the termination reactions, into which this work gives no insight. Furthermore, a direct initiation (cationation) of an IB polymerization would be in conflict with the results of Plesch[41] and Kennedy.[50]

$$2BCl_3 \rightleftharpoons BCl_2^+ + BCl_4^- \xrightarrow{CH_2=Me_2} Cl_2BCH_2\overset{+}{C}Me_2 BCl_4^-$$

Scheme 4

The careful work of Plesch,[51] taken in conjunction with the more recent work by Priola,[52] which excludes the possibility that CH_2Cl_2 acts as a cationogen, seems to confirm that $AlCl_3$ can lead to a 'direct initiation' of the polymerization of IB without the intervention of a third component. The yields in this case are close to 100%. That 'direct initiation' does not always lead to complete monomer conversion is shown by the result of Di Maina.[53] This work also appears to support a 'direct initiation' mechanism by $AlCl_3$ or $AlEtCl_2$ in MeCl, but the yields in these cases are limited.

One last example which supports a 'direct initiation' mechanism is the excellent work of Sigwalt's group.[54] This paper exemplifies the exceedingly complex experiments which are required in order to approach an elucidation of initiation mechanism, and that some 2 molecules of $AlCl_3$ are required to yield one cation from 1,1-diphenylethylene. Each system must, however, be judged alone; earlier work by Sigwalt's group[55] showed that the cationation of 3-isopropylindene (3) involves a 1:1 stoichiometry under certain specific reaction conditions (CH_2Cl_2, $-70\,^\circ C$). In order to explain this result these authors proposed a zwitterionic structure for the product (equation 6). Such initiating systems are complicated by the possibility of charge-transfer complexes forming between the monomer and the metal halide. That such complexes exist is corroborated by a large body of literature; the strength of such complexes and their influence on the course of cationic polymerizations remains to be ascertained. Furthermore, the simple binary ionogenic equilibrium of metal halides described by Grattan and Plesch for $AlCl_3$ and $AlBr_3$ in alkyl halide solvents[30] cannot easily be distinguished from the complex equilibria associated with higher aggregates (Scheme 5). That such species exist is also well documented in the literature.[56,57]

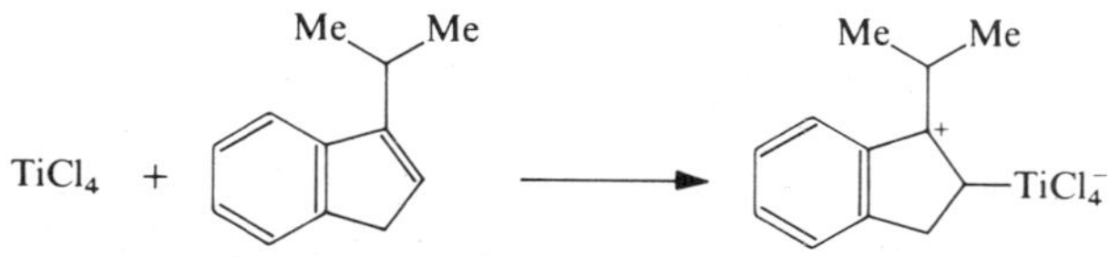

$$\text{(6)}$$

$$2MtX_n \rightleftharpoons MtX_{n-1}^+ + MtX_{n+1}^-$$

$$\Big\downarrow\Big\uparrow MX_n \qquad\qquad \Big\downarrow\Big\uparrow MX_n$$

$$Mt_2X_{2n-1}^+ + Mt_2X_{2n+1}^-$$

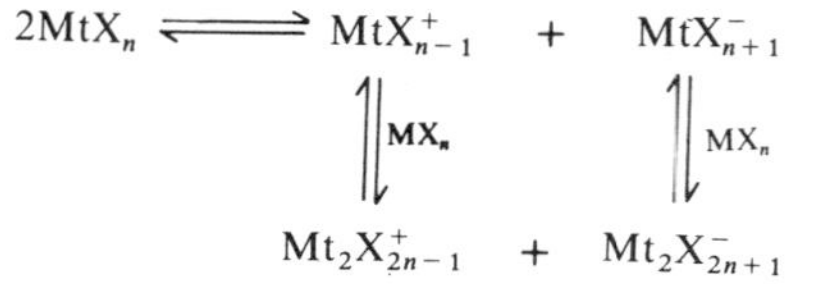

Scheme 5

One approach to the investigation of the nature of initiating cations involves the use of hindered pyridines, such as 2,6-dibutylpyridine or its 4-methyl derivative. The latter cannot be attacked by an electrophile at its 4-position.[58,59] The principle is based on the rapid reaction of these 'proton traps' with protonic acids and inactivity towards Lewis acids. It has, however, been demonstrated[60] that these pyridines form π complexes with $AlCl_3$ (but apparently not with $TiCl_4$) and also that these pyridines do not inhibit the protonation of 1,1-diphenylethylene by H_2O and $AlCl_3$.[60] Obviously, the protonation of this monomer by H_2O and $AlCl_3$ does not involve a 'free' proton.

Lastly, the use of hindered pyridines to restrict the cationation process involving Brønsted acids can also lead to changes in the course of transfer and termination reactions.[61]

40.4.2 Brønsted Acids

As with the apparently simple metal-halide initiating-systems, those involving Brønsted acids are also not simple, and the true mechanism of initiation, and even more so the nature of the propagating species, remains a source of controversy. The simple mechanism shown in equation (7) could no longer be accepted as a general case following the failure of Gandini and Plesch[62] to observe any carbenium ions during the polymerization of styrene with perchloric acid ($HClO_4$) in CH_2Cl_2. These authors proposed the so-called 'pseudocationic' polymerization, whereby the propagating species is an ester.

$$HA + \diagdown\!\!=\!\!\diagup \rightarrow H\overset{|}{\underset{|}{C}}-\overset{|}{\underset{|}{C}}{}^+ A^- \qquad\qquad \text{(7)}$$

In the light of the work by Higashimura[63] and that of Chmelir,[64,65] it now seems well established that esters are present during the cationic polymerization of styrene by strong Brønsted acids. Perhaps the strongest argument for a propagation *via* species other than a carbenium ion is the insensitivity of the $HClO_4$/styrene polymerization to added water, even in large excess.[66] Although there is considerable evidence[63-65] that esters are present during the cationic polymerization of styrene by strong Brønsted acids, more recent evidence[67] suggests that the insensitivity of the $HClO_4$/styrene system to water[66] cannot be taken as an indication that esters are responsible for propagation; 1-phenylethyl perchlorate reacts with water to form bis(1-phenylethyl) ether. Furthermore, it has recently[68] been suggested that a classical ionic mechanism can be used to explain the phenomena associated with the polymerization of styrene with trifluoromethanesulfonic acid (triflic acid, CF_3SO_3H). A summary of the arguments for and against a pseudocationic mechanism can be found in the literature.[69]

An unsolved problem associated with polymerizations initiated by Brønsted-acid systems is the nature of these acids in organic solvents. In CH_2Cl_2, $HClO_4$ forms stable solutions with very low conductivities,[70] suggesting that any proposed ionization scheme, *e.g.* equation (8), must lie essentially on the molecule side. It has, however, been demonstrated that such an ionization for CF_3SO_3H lies further towards the ion side.[65]

$$2HA \rightleftharpoons H_2A^+ + A^- \tag{8}$$

In view of the similarity of such systems with the recently described 'living' polymerization of IB with ROCOMe/BCl_3,[71] it is interesting to note the proposed mechanism for these systems (Scheme 6). If one accepts that the BCl_3-complexed ester (such species have been studied in their own right[72]) can propagate, then the formal similarity to a protonated or acid-complexed ester is obvious. That both types of initiating system are rather insensitive to water makes this approach reasonable.

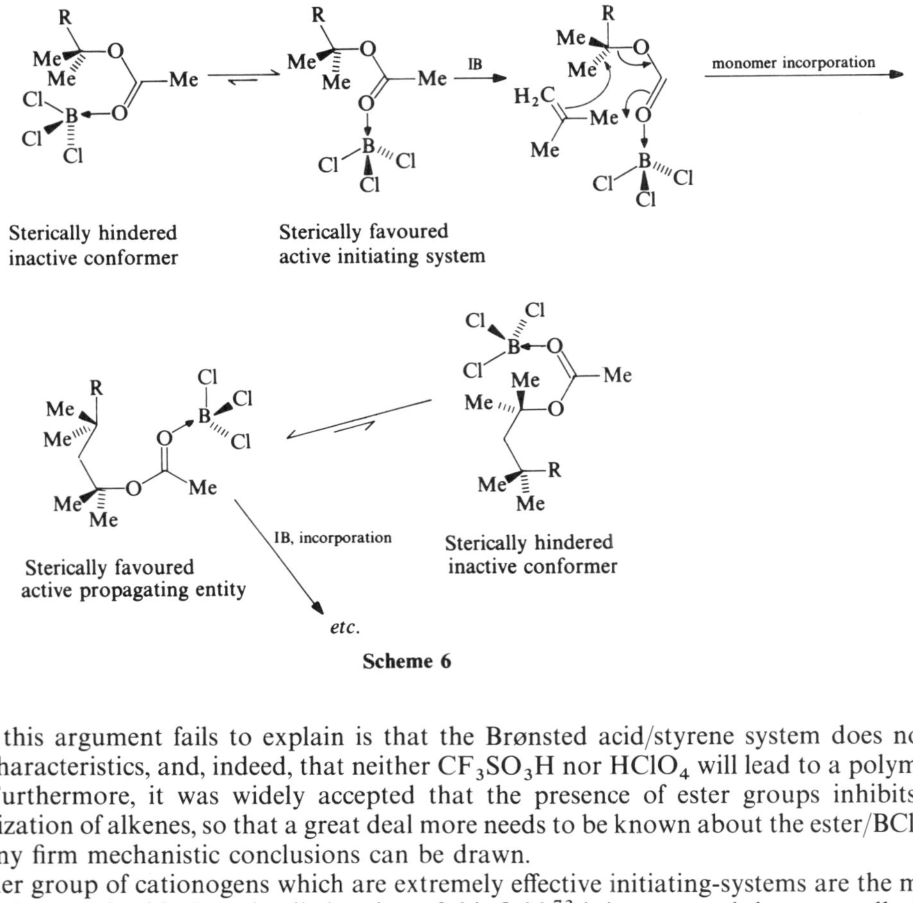

Scheme 6

What this argument fails to explain is that the Brønsted acid/styrene system does not exhibit 'living' characteristics, and, indeed, that neither CF_3SO_3H nor $HClO_4$ will lead to a polymerization of IB. Furthermore, it was widely accepted that the presence of ester groups inhibits cationic polymerization of alkenes, so that a great deal more needs to be known about the ester/BCl_3 systems before any firm mechanistic conclusions can be drawn.

Another group of cationogens which are extremely effective initiating-systems are the metal salts of strong Brønsted acids. In a detailed review of this field,[73] it is suggested that a metallation of the double bond is the initiation reaction. In particular, aluminum triflate, which is very easily prepared from aluminum metal and triflic acid, is a very convenient and clean initiating-system for the preparation of polymers.[74]

40.4.3   Carbocation Salts

The initiation of cationic alkene polymerization with preformed carbocation-salts has been thoroughly reviewed by Ledwith and Sherrington[32] up to 1974. A facile approach suggests that a carbocation salt (either preformed or prepared *in situ*) should provide the most simple initiating system (Scheme 7).

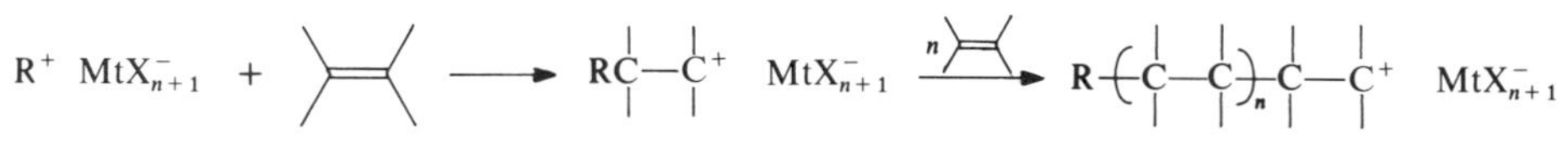

$$R^+ \ MtX_{n+1}^- \ + \quad \diagdown\!\!=\!\!\diagup \quad \longrightarrow \quad R\overset{|}{\underset{|}{C}}\!-\!\overset{|}{\underset{|}{C}}{}^+ \ MtX_{n+1}^- \quad \overset{n}{\diagdown\!\!=\!\!\diagup} \quad R\!\left(\!\overset{|}{\underset{|}{C}}\!-\!\overset{|}{\underset{|}{C}}\!\right)_{\!\!n}\!\!\overset{|}{\underset{|}{C}}\!-\!\overset{|}{\underset{|}{C}}{}^+ \ MtX_{n+1}^-$$

Scheme 7

However, the importance of determining the fate of the carbocation during the polymerization has been neglected by many of the earlier workers in this area. Carbocations can react with alkenes in at least two ways: by direct addition or by proton abstraction. In addition, these reactions may be more or less rapid than the propagation reaction. Even if the cationation of the double bond is rapid and complete, it is important to realize that the carbocation salt comprising the growing-chain end and the anion from the original salt may be considerably less stable than the original salt. Furthermore, the ionization/dissociation equilibrium-constants of the original salts will almost certainly be different from those associated with the growing chain. The whole complex of problems associated with this type of initiation has been discussed.[13]

An excellent example of what is required when using carbocation salts as initiating systems is given by the work of Sigwalt's group.[75,76] These authors show, using a combination of dilatometry, spectroscopy and product analysis, that the initiation of the polymerization of cyclopentadiene with trityl hexachloroantimonate involves a 1:1 addition (direct initiation), and that initiation is relatively rapid and complete. The results of this work were also used to calculate k_p and k_t for this system, and to show that both these reaction steps have negative activation-energies. An explanation for this was proposed in terms of a shift of equilibrium at lower temperatures towards 'tight ion-pairs' (covalent molecules?), which are suggested to be inactive toward propagation.

A thermodynamic treatment of carbocation salts and some guidelines on how to choose the right salt has been published by Pask and Plesch.[29] This work, as well as rationalizing, in thermodynamic terms, why some salts are effective initiators while others are not, reflects the lack of fundamental thermodynamic data which has plagued the study of the cationic polymerization of alkenes in the past. The situation is, however, rapidly improving, not least through the work of Arnett and co-workers,[77,78] who have recently obtained heterolysis bond-energies *via* the study of the reaction of cations with anions in solution.

One of the most pleasing developments of the past decade, because it has provided a large body of fundamental data, is the growing interest of non-polymerization chemists in the individual reaction-steps associated with a cationic polymerization. Thus, for example, the excellent pulse-radiolysis studies of Dorfmann's school[79] have provided absolute reaction-rates for the addition of benzyl and diphenylmethyl cations to a number of alkenes in CH_2Cl_2. Another approach, which is proving equally fruitful, is demonstrated by the detailed systematic work of Mayr *et al.*[80,81] These authors have studied the reaction of diphenylmethyl chlorides with alkenes in the presence of BCl_3 using conductimetric, spectroscopic and standard product analyses. From their results, as well as providing a sound basis for a number of novel initiation-systems, these authors suggest that the free energy of activation for cationation of the alkene is the same for both free and paired cations; an important advance for a consideration of reaction rates in different solvents. This finding, if it can be substantiated, may also lead to an end in the long-running discussion about the importance of knowing the concentrations of free and paired ions before comparing reaction rates.

That Mayr has turned his attention to BCl_3 (his earlier work is of no less importance for an appreciation of his methods) reflects the interest in this metal halide due to the inifer technique developed by Kennedy's school[82,83] (see Section 40.7.2). The principle of the inifer technique is the formation of a carbocation salt *in situ*, and as such it is not new. The importance lies in the use of BCl_3, since the anion BCl_4^- leads to growing chains of, for example, IB which undergo little or no transfer to monomer (below 0 °C) and no irreversible termination reactions.[84]

40.4.4 High-energy Radiation

Since the discovery by Davison[85] in 1957 that IB could be polymerized both with high-energy electrons and ^{60}Co radiation, there has been a considerable effort by a number of schools to develop these techniques.

A detailed discussion of the techniques and theory involved has been given by Williams.[86] The initiation process (for IB, for example) can be envisaged as[87] Scheme 8, whereby an n-electron is ejected from the double bond by the incoming radiation. The allyl radical formed after reaction of the cation radical with a second molecule is a relatively stable entity, and the highly reactive *t*-butyl cation starts a polymerization. Although this mechanism is corroborated by mass spectrometric studies,[88] it is interesting to note that, in systems containing an anion, the initiating capability of the *t*-butyl cation with respect to IB is a matter of controversy.[47, 84, 89]

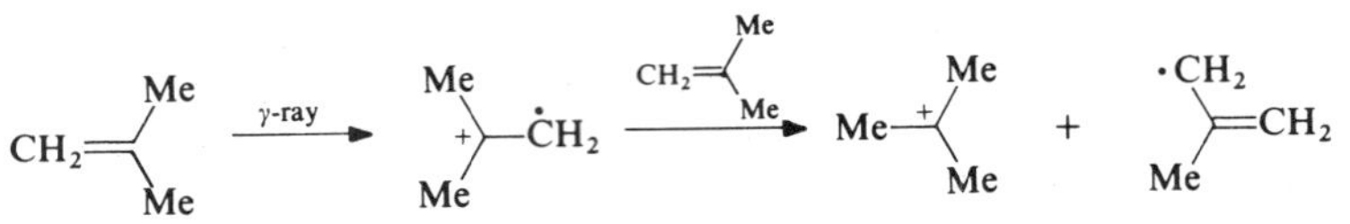

Scheme 8

Cyclopentadiene (CPD) and isoprene have been polymerized in devitrifying *t*-butyl chloride[90] after irradiation of the glass with ^{60}Co γ-rays. Since, after bleaching of the CPD/Bu^{t}Cl glass, no polymerization occurs on warming, the authors assume that the polymerization is ionic. Interestingly, the products of the 'free ion' polymerization were soluble materials with $\bar{M}_w \sim 8 \times 10^4$ (*cf.* Section 40.6.5). Another interesting point is that, at these low temperatures (77 K), the effects of small amounts of water are apparently negligible. At this temperature, the water present is probably 'frozen' out of the solution and thus inactive.

The advantages of a radiation-induced polymerization lie in the possibility of measuring the rates of polymerization without the influence of a counterion. Unfortunately, the highly reactive 'bare' cations are extremely sensitive to water (*vide supra*) and other impurities, so that the most exacting experimental techniques are required. The efficiency of the techniques used can be estimated by the approach of the dependence of the rate of polymerzation to a half power order, since, assuming that the negative and positive charged species annihilate each other on contact, equation (9) is valid (R_p = rate of propagation, R_i = rate of initiation). Using conductimetric methods, it is possible to obtain a good measure of the average lifetimes of the active species. With these and estimates of ion mobilities,[86, 90, 91] values of k_p^+ can be obtained. Comparison of the values obtained with those from carbocation-salt-initiated polymerizations[92] or those for bulk and solution processes[87] have led to some interesting, if as yet only weakly substantiated, propositions concerning ion solvation. Until more exact information is available concerning the actual mechanism involved, such comparisons, in particular in absolute terms, should be treated with due scepticism.

$$R_p = (R_i^{\frac{1}{2}}/k_t^{\frac{1}{2}})k_p \text{ [Monomer]} \tag{9}$$

An alternative approach to radiation-initiated cationic polymerization, which, despite its apparent usefulness, seems not to have been followed up, is that reported by Akulov *et al.*[93] These authors used low concentrations of tritium to initiate the polymerization of bulk IB at $-78\,^\circ$C. The decay of tritium (equation 10) provides a convenient source of low-energy 'bare' cations. In this work, extremely high molecular weight products were obtained, a typical feature of radiation-induced polymerizations in bulk media.

$$^3H_2 \rightarrow HeT^+ + \beta^- + \bar{\nu} \tag{10}$$

40.5 PROPAGATION AND TERMINATION

Studies of the propagation and termination reactions in cationically polymerizing systems have been comprehensively reviewed up to 1979 by Dunn[14] and an update can be found in the work of Kennedy and Maréchal.[16]

It should be made clear that the purpose of elucidating the mechanism of a reaction, and thereafter determining the parameters for individual reaction steps, is to be able to predict the products of further experiments and, indeed, to control such experiments. With respect to cationic alkene polymerization, considerable effort has been expended trying to distinguish between covalent species, ion pairs and free ions. Rather than review the more recent trends in this discussion, this section will focus on some principal considerations. Indeed, if the result of Mayr[81] can be generalized (see Section 40.4.3), then it may be superfluous to distinguish between ion pairs and free ions from a kinetic point of view. Whether the different ion-states lead to stereochemically different polymers remains to be demonstrated for a cationic polymerization. Furthermore, the recent demonstration by Kennedy[71] that BCl$_3$-activated esters can propagate a 'living' polymerization of IB exemplifies the need to examine the older literature with extreme care.

The propagation of a cationic alkene polymerization can be formally represented by a triedic system (*i.e.* there are three types of growing species) in which the total rate of polymerization (monomer consumption) can be represented by[94] equation (11), where k_p^+, $k_p^{\pm}$ and k_p^E are the rate constants of propagation for the 'free ions', the ion pairs and the covalent esters, respectively. Even for a group of similar monomers with an identical anion and in the same solvent, the proportions of the three species will vary. Indeed, the proportions may vary during any polymerization. Thus, a detailed knowledge of the equilibrium constants defined in equation (12) is required if individual constants are to be obtained (K_i = ionization constant, K_D = dissociation constant). All three rate constants are, however, affected by the experimental environment in known or, at least, qualitatively predictable ways: (i) the more polar the medium or the more polarizable the solvent, the greater the degree of dissociation (more polar solvents also lead to slower polymerization);[95, 96] and (ii) the more electronegative or the smaller the anion, the less viable it is as a 'free' species; the equilibria will be shifted towards the molecules.

$$-\frac{d[\text{monomer}]}{dt} = (k_p^+[P_n^+] + k_p^{\pm}[P_n^+A^-] + k_p^E[E])[\text{Monomer}] \tag{11}$$

$$RA \underset{}{\overset{K_i}{\rightleftharpoons}} R^+A^- \underset{}{\overset{K_D}{\rightleftharpoons}} R^+ + A^- \tag{12}$$

There remains much to be done with respect to propagation rates in alkene polymerizations, and, apart from those from radiation-induced polymerizations, there are few reliable measurements in the literature. It can, however, be expected that, with an increasing industrial interest and the development of 'living' alkene polymerizations, this situation will soon improve.

40.6 DIENES

It is not the purpose of this section to bring up to date the compendium published by Kennedy.[15] The authors have selected, more or less subjectively, several monomers which they hope will give an insight into the trends and problems in this field. A review up to 1963 can be found in the literature.[97]

40.6.1 2-Methyl-1,3-butadiene (Isoprene)

The polymerization of isoprene with Ziegler–Natta catalysts can apparently lead to polymers having a variety of microstructures, depending on the nature of the initiating system. From studies of such systems, Gaylord *et al.*[98–100] have suggested that the more acidic the catalyst, the more prevalent are structures produced from a carbocationic propagation mechanism in the product. Based on a kinetic study, Ghandi *et al.*[101] inferred that the system VCl$_4$/AlEt$_2$Br leads to a cationic propagation mechanism. Unfortunately, the latter authors fail to give details of the molecular weights of the product, so that their conclusions are somewhat dubious.

An extremely interesting and apparently clean polymerization of isoprene to a fully cyclized product by constant potential electrolysis in CH$_2$Cl$_2$ has been published by Akbulut *et al.*[102] These authors suggest that the cyclization takes place after termination of polymer growth through an electron transfer followed by cyclization (Scheme 9). Although this paper lacks full experimental details, the rate of polymerization ($3–20 \times 10^{-3}$ mol l^{-1} min^{-1}) seems to be exceptionally slow.

A very interesting study of the homopolymerization of isoprene using a TiCl$_4$/halogenated acetic acid/t-butyl chloride initiating-system has been reported by Matyska *et al.*[103] These exceedingly complicated systems lead to polymers having very broad molecular weight distributions and predominantly cyclic products. It was shown that primary and secondary esters are inactive as components of an initiating system, a finding that corresponds well with that of Kennedy[71] for his BCl$_3$/ester initiating-systems (see Section 40.4.2). Matyska's detailed study suggests that carbocations (as measured by conductivity) are not the active propagating-species; rather the active centres are complexes of TiCl$_4$ and the esters. The contrast in terms of the products between Kennedy's results with IB and Matyska's results with isoprene are intriguing and warrant further study.

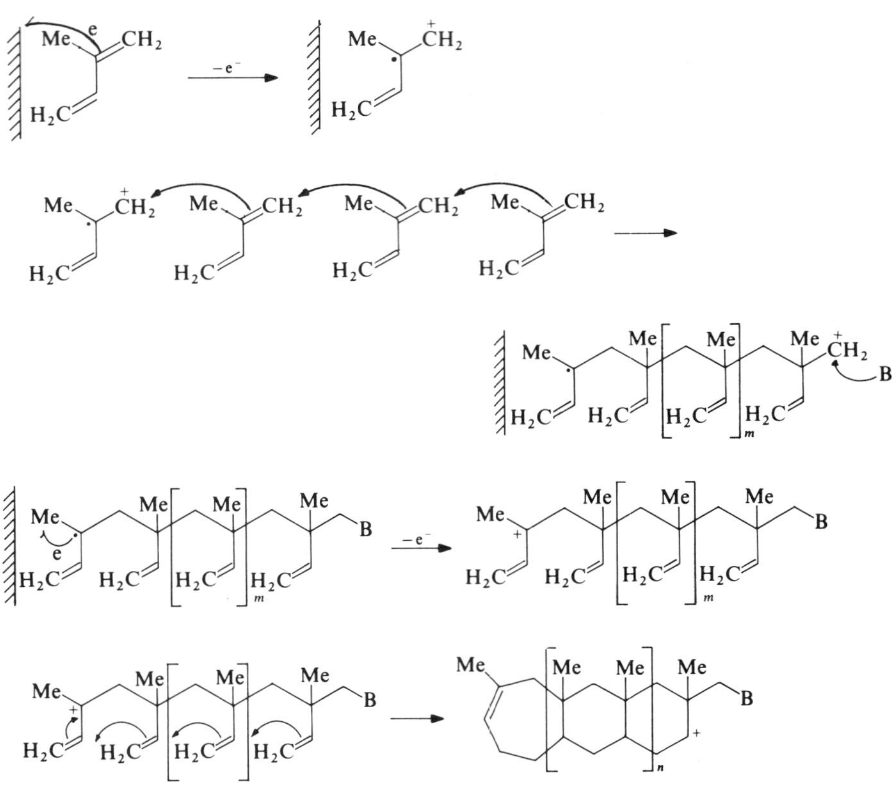

Scheme 9

The report by Katsuda *et al.*[104] on the polymerization of 4-acylisoprene seems to be the only study of a non-alkyl-substituted isoprene. In comparison to radical and anionic initiators, $BF_3 \cdot Et_2O$ and $AlEt_2Cl/Bu^tCl$ led to higher yields of lower molecular weight products. The products were essentially completely cyclic, according to IR spectroscopy.

40.6.2 Butadiene

The low level of interest in the cationic polymerization of butadiene, one of the most important monomers as far as commercial polymers are concerned, is due to the fact that the products are inferior to these produced by free radical or coordination techniques in terms of the mechanical properties of compounds made from these polymers. In particular, the products generally contain less than 40% of the theoretical unsaturation due to cyclization reactions.[97] The only commercial product to date is Budium® (Du Pont), which is an oligomer ($M_n \sim 1500$) used for lining tin-cans, prepared using $BF_3 \cdot Et_2O/H_2O$ in hexane at between 0 and 5 °C.[105]

The available literature up to 1974 has been thoroughly reviewed by Kennedy.[15] In view of the increasing interest in telechelics (see Section 40.7), it may well be interesting to look again at this monomer using an alkyl halide/BCl_3 or a BCl_3/ester initiating-system.

40.6.3 Aryl-substituted Butadienes

Asami *et al.* have reported studies of phenyl-1,3-butadienes,[106-109] diphenyl-1,3-butadiene[110] and methylphenyl-1,3-butadiene.[111] The results of these studies, employing a variety of Friedel–Crafts-type initiating-systems in CH_2Cl_2, C_6H_{14}, benzene, cyclohexane and toluene, suggest that: (a) up to one half of the double bonds are invariably consumed by cyclization reactions during the polymerization of 1- and 2-substituted butadienes; (b) the cationic polymerization of diphenylbutadienes leads only to oligomers, due to steric hindrance by the phenyl ring; and (c) steric considerations are generally crucial for an understanding of the microstructure of the polymers.

Another Japanese group[112,113] has studied the cationic polymerization of 1- and 2-phenylbutadiene. From the latter work, it appears that the loss of double bonds during polymerization (presumably due to cyclizations) can be controlled to some extent by a judicious choice of initiating system. Thus, $AcClO_4$ leads to only a 16% loss of double bonds, whereas the $SnCl_4/CCl_3CO_2H$ system leads to a 35% loss of double bonds. This may, however, reflect the lower yields obtained with the former system; *i.e.* that the degree of cyclization increases during the polymerization (*cf.* Section 40.6.4). A comparison of the 1- and 2-substituted phenylbutadienes shows that, whereas the 1-substituted monomer polymerizes predominantly (75%) *via* a 3,4-addition mechanism, with virtually no evidence for a 1,2-addition, the 2-phenyl derivative polymerizes almost exclusively *via* a 1,4-addition. These authors also report an interesting analysis of the copolymerization of these two monomers and some ring-substituted derivatives with styrene. They compare the results with the ^{13}C NMR spectra of the monomers. Thus, for 2-phenylbutadiene, only C-1 showed an upfield shift with an electron-donating substituent (Cl). This compares very well with the predominant 1,2-microstructure of the polymers of the 2-phenyl derivative, which indicates an almost exclusive attack of the growing chain at this carbon atom. Whereas the 1-phenyl derivative is more reactive than styrene, the 2-phenyl derivative is only some 0.66 times as reactive with respect to cationic polymerization.

40.6.4 Alkyl-substituted Butadienes

A comparison of a number of alkyl-substituted butadienes — of which isoprene (see Section 40.6.1) is a special case — with respect to homopolymerization and their copolymerization with styrene has been published by Hasegawa *et al.*[114] In particular, these authors emphasize the excessive chain-transfer which occurs when $AcClO_4$ is used as an initiating system (equation 13).

$$\sim\!\sim\!CH_2CH\!\!=\!\!\overset{+}{CH}\!\!=\!\!CHR \xrightarrow{\text{monomer}} \sim\!\sim\!CH\!\!=\!\!C\!\!=\!\!CHR + H\ Monomer^+ \qquad (13)$$

For the 1-substituted systems, the transfer reaction is enhanced with more bulky R-groups. For the 2-substituted systems, the prevalence of transfer is controlled, not only by the steric hindrance presented to the attacking growing-chain, but also by the charge distribution in the growing-chain end. The copolymerization parameters of these monomers and styrene can also be rationalized in the same way. These considerations cannot, however, explain the very different yield obtained with different initiating systems. For example, with $SnCl_4/CCl_3CO_2H$ the yield of poly(1,3-pentadiene) was 78.5%, whereas with $AcClO_4$ it was only 15.7%. For 2-*t*-butyl-1,3- butadiene, the yields were 29.2% and 96.6%, respectively.

Kamachi *et al.*[115] have reported on the polymerization of 2,4-hexadiene in toluene and nitroethane using both $BF_3 \cdot Et_2O$ and WCl_6 as initiating systems. Their rather bold conclusion, on the basis of kinetic measurements, that the more rapid polymerization using WCl_6 is due to greater concentration of free ions is not adequately substantiated, since there is no direct evidence that the concentration of growing chains in the two systems was identical.

More recently,[116] the same authors have reported that, at relatively high temperatures ($\sim 30\,^\circ$C) with $TiCl_4/CH_2Cl_2$ or Et_2AlCl/benzene, high yields of an amorphous polymer can be obtained. Although the molecular weights obtained with these cationic systems were greater than those obtained using coordination catalysts, the authors were interested particularly in the crystalline products obtained with the latter systems. They do, however, state that no crosslinking was observed during the polymerization and that, predominantly, the polymers were of a 1,4-*trans* structure, despite a 1:1 mixture of the *trans,trans* and *cis,trans* isomers being used as monomer. Obviously, an isomerization takes place during the polymerization.

40.6.5 Cyclopentadiene

Despite a large number of studies of this monomer, which is one of the most reactive monomers with respect to cationic polymerizaton, no commercial development of such products has been undertaken. The most obvious reason for this is the lack of stability of the products with respect to oxidation;[117] it is also partly because, using traditional Friedel–Crafts initiating-systems, some crosslinking invariably accompanies polymerization, although soluble high-polymer ($\bar{M}_n$ up to 2×10^5) has been achieved with initiation by $TiCl_3(OBu)$.[118,119] A resonance stabilization of the growing-chain cation can be used to rationalize the predominance of 1,4-structures (**4**) and 1,2-structures (**5**) in the polymers[120,121] (Scheme 10).

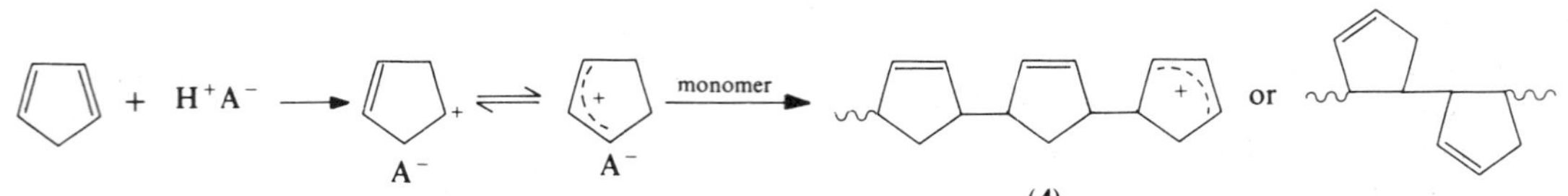

Scheme 10

More recently,[76,120,122] it has been shown that, with trityl carbocation salts, a direct initiation of CPD occurs to give a polymerization in which transfer can be neglected and which yields totally soluble products. These authors demonstrate in their more recent work that the molecular weight of the product obtained with a bifunctional initiator (**6**) is almost twice that obtained with the trityl salt. The assumption that the broadening of the molecular weight distribution when $SbCl_5OH^-$ was employed as anion is due to this system containing more ion pairs would be more easily acceptable had the authors made a fuller spectroscopic analysis of the product. (There is some evidence[123,124] that ion pairs and free ions yield different proportions of the 1,2- and 1,4-microstructures.) The $SbCl_5OH^-$ anion is almost certainly less stable than $SbCl_6^-$, and may well lead to some termination reaction. This would also lead to a broadening of the molecular weight distribution. Interestingly, the initiation of a CPD polymerization by either $HClO_4$ or $AcClO_4$ leads to a very rapid but incomplete polymerization;[125,126] these authors give no explanation of this phenomenon, but one explanation might be that a cyclization reaction is more favourable for an ester-terminated chain than for an ionically growing chain.

$$SbCl_6^- \ Ph_2\overset{+}{C}\text{———}CH_2CH_2\text{———}\overset{+}{C}Ph_2 \ SbCl_6^-$$

(**6**)

40.6.6 Copolymers

40.6.6.1 *Isoprene*

The copolymerization of isoprene and IB is one of the most important industrial cationic-polymerization processes. The industrial synthesis of these copolymers (Butyl Rubber®) started in 1943, and some 500 000 tonnes were produced in 1980.[127] Butyl Rubber is synthesized from IB with between 1 and 5% isoprene in MeCl using an $AlCl_3$/Brønsted acid initiating-system. As the polymerization proceeds, the polymer precipitates and the molecular weight of the product drops.[128] The polymerization is usually carried out at *ca.* 173 K in order to reduce transfer reactions and thus achieve acceptable molecular weights. The copolymerization parameters at low temperature are (IB = monomer 1, isoprene = monomer 2) $r_1 = 2.5 \pm 0.5$ and $r_2 = 0.4 \pm 0.1$. Thus $r_1 r_2 \approx 1$ and more or less statistical copolymers can be prepared. The limiting factor for the inclusion of isoprene is the regulation effect of conjugated dienes on cationic polymerizations. The structure of Butyl Rubber (**7**) indicates an almost exclusive head-to-tail addition of IB to isoprene, the latter being present essentially only in the *trans*-1,4-configuration.[129]

$$\text{\textbackslash\textbackslash}CH_2 CMe_2\text{——}(CH_2CMe=CHCH_2)\text{——}CH_2CMe_2\text{\textbackslash\textbackslash}$$

(**7**)

Using cumyl chloride and $TiCl_4$ at $-50\,°C$ in a 60/40 hexane/CH_2Cl_2 solvent mixture, up to 10% isoprene could be incorporated into an IB/isoprene copolymer using a so-called 'quasi-living' technique.[130] The results in this publication show, however, that even with the very low stationary monomer concentrations used in this technique, the increase in molar mass with further monomer addition seems to tail off at *ca.* 1.9×10^4 Daltons. This effect is indicative of the regulating effect of dienes and its extent is inversely proportional to the concentration of isoprene in the monomer feed. At higher isoprene concentrations the molecular weight distribution also becomes broader.

40.6.6.2 *Cyclopentadiene*

In analogy to Butyl Rubber, the copolymerization of dienes with other monoalkenes should enable the polymers of the latter to be more efficiently crosslinked. Thus, Kohjiya *et al.*[131] have studied the copolymerization of CPD with 2-chloroethyl vinyl ether using $BF_3 \cdot Et_2O$ as the initiating system. They showed that, as the solvent was changed from toluene to CH_2Cl_2, the value of $r_1 r_2$ increases. Interestingly, the value of $r_1 r_2$ in toluene was close to 1, so that statistical copolymers and, thus, useful networks should be obtainable from these systems.

Random copolymers have also been obtained from CPD and butyl vinyl ether with a $BF_3 \cdot Et_2O$/toluene system.[132] It appears that the major termination-reaction in these systems involves a cyclization reaction after two CPD units have been added to a single polymer chain.

40.6.6.3 *Pentadiene and hexadiene*

1,3-Pentadiene (piperylene) is usually obtained from hydrocarbon fractionation as a mixture of the *cis* and *trans* isomers in a 40:60 ratio (structures **8** and **9**).

(**8**) *Trans*-1,3-pentadiene   (**9**) *Cis*-1,3-pentadiene

In their study of the copolymerization of *cis*- and *trans*-1,3-pentadiene, Priola *et al.*[133] report that homopolymers of the *trans* isomer, obtained using Et_2AlCl/Cl_2, or Me_3CCl in MeCl, contain some 61% 1,4-*trans*, 24% 1,2 *trans* and 15% cyclic structures. The *cis* isomer gives a polymer under similar conditions which contains some 35% cyclic moieties, some 36% *cis* and 12% vinyl double-bonds. No other details (*e.g.* $\bar{M}_w$ or $\bar{M}_n$) are given about the homopolymer products.

As a comonomer with IB, the *trans* isomer is almost twice as reactive as the *cis* isomer in terms of propagation and transfer to monomer at $-55\,°C$. The individual transfer coefficients, obtained from experiments at high conversion, $k_{tr, cis} \sim 50$, $k_{tr, trans} \sim 113$, are notably different from those reported by Kennedy.[134] The differences may be due to differences in solvent and initiator system (Kennedy and Canter employed $AlCl_3$ in pentane), but it should be noted that in both publications the data exhibit considerable scatter. Indeed, Priola *et al.*[133] showed that a change of catalyst or temperature has little effect on the amount of diene incorporated into the copolymer. A decrease in temperature does, however, lead to a decrease in conversion; this effect being more extreme for the *trans* than for the *cis* isomer.

Both isomers are incorporated in the polymers with a *trans* configuration, predominantly 1,4-*trans*, so that an isomerization of the *cis* isomer, similar to that observed during a coordination polymerization,[135] presumably occurs. A very detailed study of the NMR spectra of poly-(IB-*co*-*cis*-1,3-pentadiene) has also been published by Priola *et al.*[136]

The application of detailed NMR analyses of copolymers for determining relative rate-constants has been reported by Corno *et al.*[137] From this study the parameters given in Table 1 were calculated.

Table 1 Copolymerization Parameters for IB with Various Comonomers

Comonomer	r_1	r_2
Isoprene (high conversion)	1.49 ± 0.02	0.98 ± 0.2
Isoprene (low conversion)	1.56 ± 0.17	0.94 ± 0.15
Trans-1,3-pentadiene	0.59 ± 0.09	1.03 ± 0.19
2,3-Dimethylbutadiene	3.20 ± 1.10	0.98 ± 0.53

As a continuation of their systematic studies of the copolymerization of IB with conjugated dienes, Priola *et al.*[138] have also studied the three geometric isomers of 2,4-hexadiene (*trans,trans*, *cis,trans* and *cis,cis*). The work was carried out in a 1:1 vol/vol mixture of pentane and CH_2Cl_2 with $EtAlCl_2$ as ionogen. The authors made a very detailed analysis of the ^{13}C NMR spectra of the products in order to determine the copolymerization parameters (Table 2). The *cis,trans* isomer is more reactive than 2,3-dimethylbutadiene (*vide supra*), and it would seem that the geometry of the isomer is more important than the position of substitution for the copolymerization behaviour of such substituted butadienes.

Table 2   Copolymerization Parameters for IB with Isomers
of 2,4-Hexadiene

Comonomer	r_1	r_2
Cis,trans	0.74 ± 0.15	1.8 ± 0.59
Cis,cis	2.3 ± 0.46	1.5 ± 0.89
Trans,trans	6.0 ± 1.39	0.88 ± 0.78

40.6.6.4   Dimethylfulvene

6,6-Dimethylfulvene can be easily prepared from the condensation of acetone and CPD.[139] It can be polymerized to high molecular weight soluble homopolymers *via* a cationic polymerization.[140] The products are, however, very sensitive to oxidation and rapidly crosslink. From model studies,[113] the probable structure of such a crosslink is as shown in equation (14).

$$(14)$$

Recently this monomer has been copolymerized with styrene, vinyl ethers and CPD, using $SnCl_4$ and CCl_3CO_2H or a mixture of these in CH_2Cl_2 and bulk.[141] In all cases, the fulvene was more readily polymerized; the comonomer and the molar masses (M_w) obtained were of the order of 20–60 $\times 10^3$ Daltons. Since the authors used large concentrations of initiator ($100 \le [M]/[I] \le 200$) and carried out the reactions at 0 °C it may be assumed that higher molecular weight products should be obtainable. Such air-sensitive polymers, which crosslink under atmospheric conditions, are interesting for a number of applications, such as joint fillers.

40.7   TELECHELICS *VIA* CATIONIC POLYMERIZATION

With the growing industrial interest in well-defined polymeric materials, the interest in prepolymers has grown. One type of prepolymer are the telechelics; relatively low molecular weight materials ($M_n \le 20000$) with well-defined end-groups which can be utilized for further chemical reactions, such as the synthesis of networks or block copolymers.

In general reviews on telechelics,[142, 143] carbocationic polymerization of vinyl monomers has received little attention. This is, in part, due to high reactivity and low selectivity of unstabilized carbocations. Thus, until recently, it was widely accepted that the molecular weight and end-group control, necessary for the synthesis of useful telechelics, would not be achievable *via* a carbocationic polymerization. This situation has, as has been mentioned in the previous sections of this chapter, changed appreciably in the last 10 years.

40.7.1   Poly(divinylbenzene)

Linear polymers with C=C double bonds in the backbone and/or at the chain ends are attractive as monomers or as telechelics because of the potential reactivity of the double bonds. Such polymers can be prepared from divinylbenzene (**10**; DVB) using, for example, CF_3SO_3H or $AcClO_4$[144–147] initiating-systems in CH_2Cl_2 or benzene at relatively high temperatures (*e.g.* 70 °C). In particular, it is worth noting that the mechanism shown in Scheme 11[136] involves a proton, despite the use of $AcClO_4$ as initiating system, which suggests that the system indeed contained enough water to produce essentially a Brønsted-acid direct protonation. However, it should be noted that the $BF_3 \cdot Et_2O$ system led to some crosslinking and cyclization. The molecular weight increase in such polymerizations indicates (as shown in the scheme) a step-growth rather than a linear-addition type of mechanism.

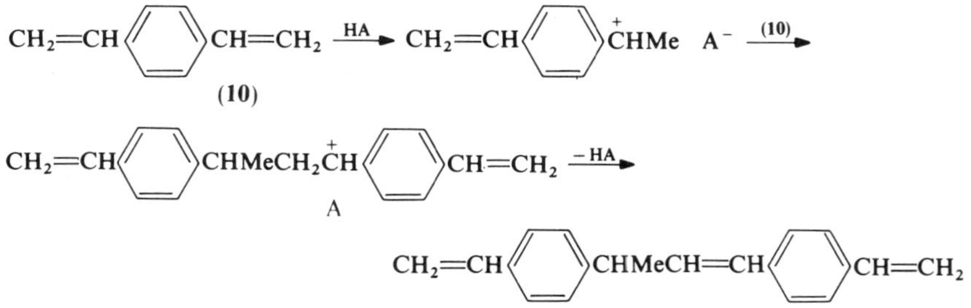

$$CH_2=CH\!\!\left\langle\!\!\bigcirc\!\!\right\rangle\!\!CH=CH_2 \xrightarrow{\;HA\;} CH_2=CH\!\!\left\langle\!\!\bigcirc\!\!\right\rangle\!\!\overset{+}{C}HMe \;\; A^- \xrightarrow{\;(10)\;}$$
(10)

$$CH_2=CH\!\!\left\langle\!\!\bigcirc\!\!\right\rangle\!\!CHMeCH_2\overset{+}{C}H\!\!\left\langle\!\!\bigcirc\!\!\right\rangle\!\!CH=CH_2 \xrightarrow{\;-HA\;}$$
A

$$CH_2=CH\!\!\left\langle\!\!\bigcirc\!\!\right\rangle\!\!CHMeCH=CH\!\!\left\langle\!\!\bigcirc\!\!\right\rangle\!\!CH=CH_2$$

Scheme 11

40.7.2 Polyisobutylene (PIB) and Polypinene

On the basis of Scheme 12, involving an initiation by a metal halide (MX_n) and an alkyl halide (RX) initiating-system, one can predict under which conditions the formation of telechelics should be possible. Control of ionization and cationation gives rise to specific head-groups, and control of propagation, termination and transfer allows the molecular weight of the products and their terminal groups to be regulated.

$$RX + MtX_n \rightleftharpoons R^+ + MtX_{n+1}^- \qquad \text{ionization}$$

$$R^+ + n\text{Monomer} \longrightarrow R\!\!\sim\!\!\sim\!\!+ \qquad \text{cationation/propagation}$$

$$R\!\!\sim\!\!\sim^+ \; + \; MtX_{n+1}^- \longrightarrow R\!\!\sim\!\!\sim X \; + \; MtX_n \qquad \text{termination}$$

$$R\!\!\sim\!\!\sim^+ \; + \; RX \longrightarrow R\!\!\sim\!\!\sim X \; + \; R^+ \qquad \text{transfer}$$

Scheme 12

The earliest report of a deliberate exploitation of this principle for the synthesis of telechelics *via* a cationic polymerization is that from VerStrate and Baldwin.[148,149] Since then, the technique has been considerably refined and has come to be called the 'inifer' technique (*initiator transfer*).[82,83] The reaction scheme for a typical inifer polymerization is as shown in Scheme 13. This reaction scheme was, in part, elucidated by Kennedy and Smith,[82,83] and has since been examined in considerable detail by Nuyken *et al.*[61,84,150-152] In particular, it has now been demonstrated that the reactions are not limited to isopropylbenzene systems (**11, 18–20**), and that there is only negligible termination if the systems are clean and dry enough.

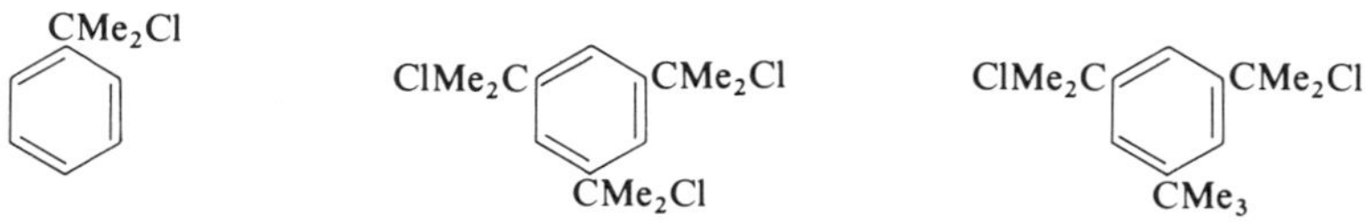

(18) (19) (20)

The tertiary chloro end-groups of such polymers can be easily converted into other more interesting end-groups, such as vinyl and primary alcohol groups. Another very interesting possibility is to employ the chloro-terminated telechelic in conjunction with a Friedel–Crafts catalyst for the alkylation of, for example, phenol. Friedel–Crafts alkylation of benzene or toluene followed by nitration and then reduction yields amine telechelics.[153,154] Both phenol or amino end-groups can then be used either for polycondensation reactions to yield, for example, polycarbonates (equation 15) or polyamides or for further derivatization, for example with epichlorohydrin to yield oxirane-terminated telechelics.[155,156]

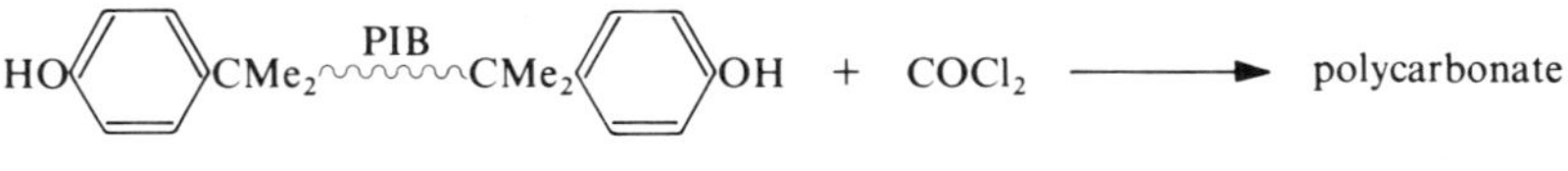

$$HO\!\!\left\langle\!\!\bigcirc\!\!\right\rangle\!\!CMe_2\overset{PIB}{\sim\!\!\sim\!\!\sim}CMe_2\!\!\left\langle\!\!\bigcirc\!\!\right\rangle\!\!OH \; + \; COCl_2 \longrightarrow \text{polycarbonate} \qquad (15)$$

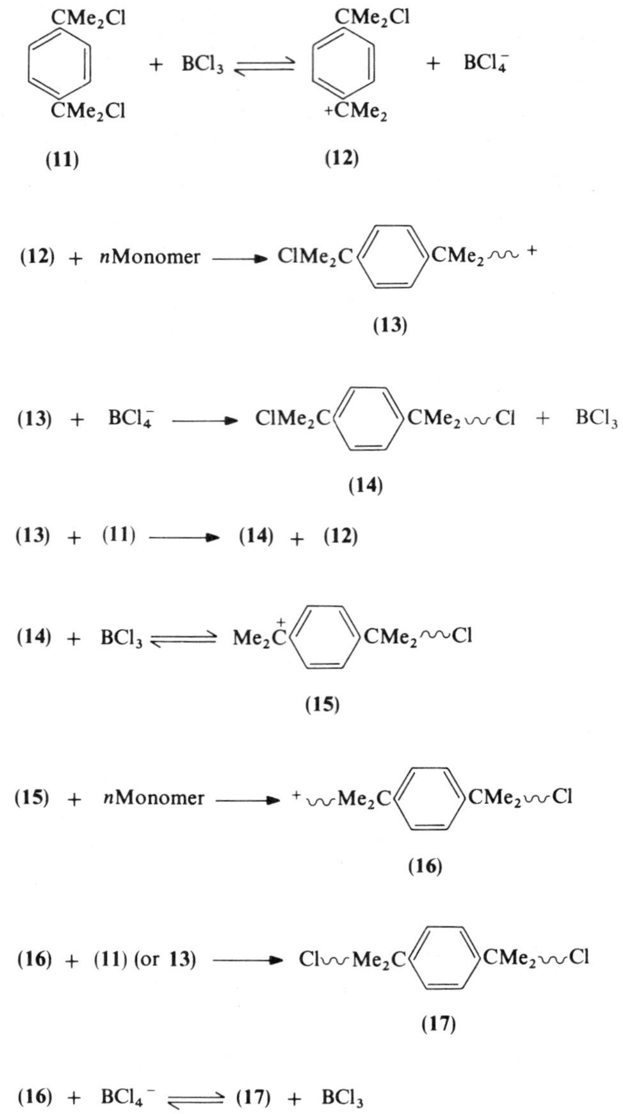

(11)   **(12)**

(12) + *n*Monomer ⟶ ClMe$_2$C⟨☐⟩CMe$_2$〰 $^+$

(13)

(13) + BCl$_4^-$ ⟶ ClMe$_2$C⟨☐⟩CMe$_2$〰Cl + BCl$_3$

(14)

(13) + (11) ⟶ (14) + (12)

(14) + BCl$_3$ ⇌ Me$_2$Ċ⟨☐⟩CMe$_2$〰Cl

(15)

(15) + *n*Monomer ⟶ $^+$〰Me$_2$C⟨☐⟩CMe$_2$〰Cl

(16)

(16) + (11) (or 13) ⟶ Cl〰Me$_2$C⟨☐⟩CMe$_2$〰Cl

(17)

(16) + BCl$_4^-$ ⇌ (17) + BCl$_3$

Scheme 13

Alternatively, the sulfonation of an α,ω-alkenic telechelic with, for example, acetyl sulfate, quantitatively yields the corresponding sulfonate which is an ionomer,[157,158] a new and industrially interesting group of polymeric materials. Such materials (equation 16) can be moulded in machines designed for thermoplastics, but have the properties of crosslinked rubber at ambient temperatures; they are a class of thermoplastic elastomers. Further examples of derivatization of such IB telechelics include the synthesis of SiCl,[159] —CONC, and —CH$_2$CN[160,161] terminal groups.

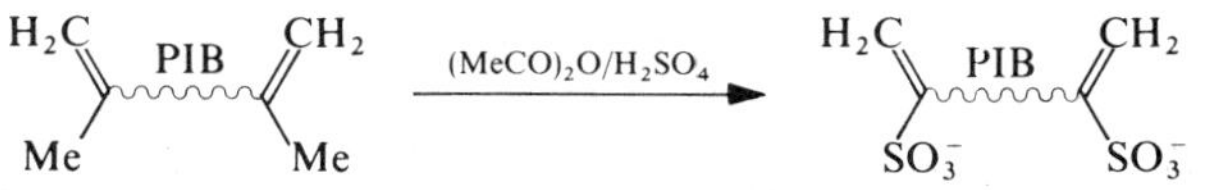

$$(16)$$

Using a monofunctional inifer, macromonomers can be synthesized, *e.g.* equation (17).[162] Such prepolymers can be employed for the synthesis of block and comb copolymers.

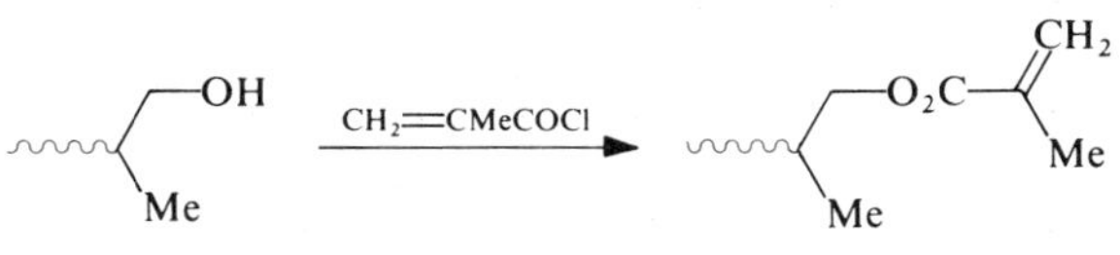

$$(17)$$

There have, as yet, been few attempts to use the inifer technique for the synthesis of telechelics from monomers other than IB. However, one such report is the systematic study of β-pinene with a 1,4-dichloroisopropylbenzene (**11**)/BCl$_3$ initiator-system.[163]

With the development of the activated-ester polymerizations (see Section 1.3.2, ref. 71) dramatic advances in this field can be expected. The inifer technique gave a new impulse to the study of cationic alkene polymerization. This technique has not proved as useful as it first appeared, and for the production of commercial quantities of telechelics with acceptable functionality there remain several problems to be solved, such as the influence of water and solvent on these polymerizations. From the initial publications, the activated-ester polymerizations appear to be closer to an ideal living system, so that molecular weight and end-group control should be possible. However, it is worth remembering that, where telechelics are the goal of a synthesis, the amount of initiating system used is considerable and the cost is important for the cost of the end product. With the inifer technique, a recycling of the BCl$_3$ is possible; for the activated-ester polymerization this will not be so facile.

40.8 REFERENCES

1. Bishop Watson, 'Chemical Essays', 5th edn., printed for T. Evans and sold by J. Evans, London, 1789, vol. III.
2. M. Deville, *Ann. Chim. (Paris)*, 1839, **75**, 66.
3. M. Berthelot, *C. R. Hebd. Seances Acad. Sci.*, 1858, **47**, 266.
4. 'Ullmans Enzyklopädie der Technischen Chemie', 4th edn., Verlag Chemie, Weinheim, 1976, vol. 12, p. 542; vol. 22, p. 553.
5. A. M. Butlerov and V. Gorianov, *Justus Liebigs Ann. Chem.*, 1873, **169**, 146.
6. M. Berthelot, *Bull. Soc. Chim. Fr.*, 1866, **6**, 294.
7. P. H. Plesch (ed.), 'Cationic Polymerization and Related Complexes', Heffer, Cambridge, 1953.
8. *Makromol. Chem.*, 1973, **175**.
9. *J. Polym. Sci., Polym. Symp.*, 1977, **56**
10. E. J. Goethals (ed.), 'Cationic Polymerization and Related Processes', Academic Press, London, 1984.
11. G. Heublein (ed.), *Makromol. Chem., Macromol. Symp.*, 1986, **3**.
12. P. H. Plesch, in 'Progress in High Polymers 2', ed. J. C. Robb and F. W. Peaker, Iliffe Books, London, 1968, p. 137.
13. A. Gandini and H. Cheradame, *Adv. Polym. Sci.*, 1980, **34/35**.
14. D. J. Dunn, in 'The Cationic Polymerization of Vinyl Monomers, Development in Polymerization 1', ed. R. N. Haward, Applied Science, London, 1979.
15. J. P. Kennedy, 'The Cationic Polymerization of Olefins: A Critical Inventory', Wiley, New York, 1975.
16. J. K. Kennedy and E. Marechal, 'Carbocationic Polymerization', Wiley, New York, 1982.
17. G. A. Olah (ed.), 'Friedel–Crafts and Related Reactions', Interscience, New York, 1963.
18. B. Elliott, A. G. Evans and E. D. Owen, *J. Chem. Soc.*, 1962, 689.
19. G. Heublein, D. Bauerfeind, S. Spange, G. Knoeppel and R. Wondraczek, in 'Cationic Polymerization and Related Processes', ed. E. J. Goethals, Academic Press, London, 1984, p. 69.
20. A. Leborgne, D. Souverain, G. Sauvet and P. Sigwalt, *Eur. Polym. J.*, 1980, **16**, 855.
21. T. Higashimura and M. Hoshino, *J. Polym. Sci., Part A-1*, 1972, **10**, 673.
22. E. Marechal, *J. Polym. Sci., Part A-1*, 1970, **3**, 2867.
23. P. Sigwalt, *J. Polym. Sci.*, 1961, **52**, 15.
24. H. J. Prosser and R. N. Young, *Eur. Polym. J.*, 1975, **11**, 403.
25. C. M. Fontana, *J. Phys. Chem.*, 1959, **63**, 1167.
26. A. D. Ketley and M. C. Harvey, *J. Org. Chem.*, 1961, **26**, 4649.
27. M. Modena, R. B. Bates and C. S. Marvel, *J. Polym. Sci., Part A*, 1965, **3**, 949.
28. J. P. Kennedy, W. W. Schulz, R. G. Squires and R. M. Thomas, *Polymer*, 1965, **6**, 287.
29. S. D. Pask and P. H. Plesch, *Eur. Polym. J.*, 1982, **18**, 839.
30. D. W. Grattan and P. H. Plesch, *J. Electroanal. Chem. Interfacial Electrochem.*, 1979, **103**, 81.
31. P. H. Plesch and O. Nuyken, *Chem. Ind. (London)*, 1973, 379.
32. A. Ledwith and D. C. Sherrington, *Adv. Polym. Sci.*, 1975, **19**, 1.
33. A. Chapiro, *Makromol. Chem.*, 1974, **175**, 1181.
34. P. Giusti, *J. Polym. Sci., Polym. Symp.*, 1975, **50**, 133.
35. S. I. Schlesinger, *Photogr. Sci. Eng.*, 1974, **18**, 387.
36. R. F. Tarvin, S. Aoki and J. K. Stille, *Macromolecules*, 1972, **5**, 663.
37. T. Natsuume, Y. Shirota, M. Hirata, S. Kusabayashi and M. Mikawa, *J. Chem. Soc., Chem. Commun.*, 1969, 289.
38. H. Gilbert, F. F. Miller, S. J. Averill, E. J. Carlson, V. L. Folt, H. J. Heller, F. D. Stewart, R. F. Stewart, R. F. Schmidt and H. L. Trumbull, *J. Am. Chem. Soc.*, 1956, **78**, 1669.
39. J. K. Stille and D. C. Chung, *Macromolecules*, 1975, **8**, 114.
40. T. Gotoh, A. B. Padias and H. K. Hall, *J. Am. Chem. Soc.*, 1986, **108**, 4920.
41. P. H. Plesch, M. Polanyi and H. A. Skinner, *J. Chem. Soc.*, 1947, 257.
42. R. H. Biddulph, P. H. Plesch and P. P. Rutherford, *J. Chem. Soc.*, 1965, 275.
43. M. Chmelir, M. Marek and O. Wichterle, *J. Polym. Sci., Part C*, 1967, **16**, 833.
44. A. G. Evans and G. W. Meadows, *Trans. Faraday Soc.*, 1950, **46**, 327.
45. N. A. Ghanem and M. Marek, *Eur. Polym. J.*, 1972, **8**, 999.
46. H. Cheradame and P. Sigwalt, *Bull. Soc. Chim. Fr.*, 1970, 843.
47. D. W. Grattan and P. H. Plesch, *Makromol. Chem.*, 1980, **181**, 751.

48. F. R. Khalafov, F. M. Nasirov, N. E. Melnikova, B. A. Krentsel and T. N. Shakhtakhtinsky, *Makromol. Chem., Rapid Commun.*, 1985, **6**, 29.
49. L. Bui, H. A. Nguyen and E. Marechal, *Polym. Bull. (Berlin)*, 1987, **17**, 157.
50. J. P. Kennedy, S. Y. Huang and S. C. Feinberg, *J. Polym. Sci., Polym. Chem. Ed.*, 1977, **15**, 2801.
51. J. H. Beard, P. H. Plesch and P. P. Rutherford, *J. Chem. Soc.*, 1964, 2566.
52. A. Priola, S. Cesca and G. Ferraris, *Makromol. Chem.*, 1972, **160**, 41.
53. P. L. Magagnini, S. Cesca, P. Guisti, A. Priola and M. DiMaina, *Makromol. Chem.*, 1977, **178**, 2235.
54. M. Masure, P. Sigwalt and G. Sauvet, *Makromol. Chem., Rapid Commun.*, 1983, **4**, 269.
55. H. Cheradame, A. H. Nguyen and P. Sigwalt, *J. Polym. Sci., Polym. Symp.*, 1976, **56**, 335.
56. J. Bacon, P. A. W. Dean and R. J. Gillespie, *Can. J. Chem.*, 1971, **49**, 1276.
57. M. I. Ignatov, I. G. Ilian and Y. A. Buslaev, *Dokl. Akad. Nauk SSSR*, 1979, **245**, 604.
58. J. M. Moulis, J. Collomb, A. Gandini and H. Cheradame, *Polym. Bull. (Berlin)*, 1980, **3**, 197.
59. J. P. Kennedy and R. T. Chou, *J. Macromol. Sci., Chem.*, 1982, **A18**, 1.
60. M. Masure, G. Sauvet and P. Sigwalt, *Polym. Bull. (Berlin)*, 1980, **2**, 699.
61. O. Nuyken, S. D. Pask and M. Walter, *Polym. Bull. (Berlin)*, 1982, **8**, 451.
62. A. Gandini and P. H. Plesch, *Proc. Chem. Soc., London*, 1964, 240.
63. M. Sawamoto, T. Masuda, T. Higashimura, S. Kobayashi and T. Saegusa, *Makromol. Chem.*, 1977, **178**, 389.
64. M. Chmelir, *Makromol. Chem.*, 1975, **176**, 2099.
65. M. Chmelir, *J. Polym. Sci., Polym. Symp.*, 1976, **56**, 311.
66. D. C. Pepper and P. J. Reilly, *Proc. R. Soc. London, Ser. A*, 1966, **291**, 41.
67. K. Matyjaszewski, *Polym. Prepr., Am. Chem. Soc., Div. Polym. Chem.*, 1986, **27** (2), 112.
68. K. Matyjaszewski and P. Sigwalt, *Markomol. Chem.*, 1986, **187**, 2299.
69. P. H. Plesch and K. Matyjaszewski, *Makromol. Chem., Macromol. Symp.*, 1988, **13/14**, 345, 389, 393.
70. A. Gandini and P. H. Plesch, *Eur. Polym. J.*, 1968, **4**, 55.
71. R. Faust and J. P. Kennedy, *Polym. Bull. (Berlin)*, 1986, **15**, 317.
72. M. F. Lappert, *J. Chem. Soc.*, 1961, 817.
73. J. Collomb, B. Morin, A. Gandini and H. Cheradame, *Eur. Polym. J.*, 1980, **16**, 1135.
74. J. Collomb, A. Gandini and H. Cheradame, *Makromol. Chem., Rapid Commun.*, 1980, **1**, 489.
75. G. Sauvet, J. P. Vairon and P. Sigwalt, *J. Polym. Sci., Part A-1*, 1969, **7**, 983.
76. G. Sauvet, J. P. Vairon and P. Sigwalt, *Eur. Polym. J.*, 1974, **10**, 501.
77. E. M. Arnett and K. E. Molter, *Acc. Chem. Res.*, 1985, **18**, 339.
78. E. M. Arnett and K. E. Molter, *J. Phys. Chem.*, 1986, **90**, 383.
79. Y. Wang and L. M. Dorfman, *Macromolecules*, 1980, **13**, 63.
80. R. Schneider and H. Mayr, *Angew. Chem., Int. Ed. Engl.*, 1986, **25**, 1016.
81. H. Mayr, R. Schneider and U. Grabis, *Angew. Chem., Int. Ed. Engl.*, 1986, **25**, 1017.
82. J. P. Kennedy and R. A. Smith, *Polym. Prepr., Am. Chem. Soc., Div. Polym. Chem.*, 1979, **20** (2), 316.
83. J. P. Kennedy and R. A. Smith, *J. Polym. Sci., Polym. Chem. Ed.*, 1980, **18**, 1523.
84. O. Nuyken, S. D. Pask, A. Vischer and M. Walter, *Makromol. Chem.*, 1985, **186**, 173.
85. W. H. T. Davison, S. H. Pinner and R. Worrall, *Chem. Ind. (London)*, 1957, 1274.
86. F. Williams, 'Fundamental Processes in Radiation Chemistry', ed. P. Ausloos, Interscience, New York, 1968.
87. V. Stannett and J. Silverman, *ACS Symp. Ser.*, 1983, **212**, 435.
88. F. W. Lampe, *J. Phys. Chem.*, 1959, **63**, 1986.
89. L. Toman, S. Pokorny, J. Spevacek and J. Danhelka, *Polymer*, 1986, **27**, 1121.
90. K. Hayashi, K. Hayashi and S. Okamura, *J. Polym. Sci., Part A-1*, 1971, **9**, 2305.
91. K. Hayashi, Y. Yamazawa, T. Takagaki, F. Williams, K. Hayashi and S. Okamura, *Trans. Faraday Soc.*, 1967, **63**, 1489.
92. V. Stannett, H. Garreau, C. C. Ma, J. M. Rooney and D. R. Squire, *J. Polym. Sci., Polym. Symp.*, 1976, **56**, 233.
93. A. Akulov, N. Geller, V. Kropatchev, V. Nefedov, E. Sinotova, S. Skorochodov, V. Stepanov and M. Toropova, *Makromol. Chem.*, 1978, **179**, 2775.
94. P. H. Plesch, in 'Cationic Polymerization and Related Processes', ed. E. J. Goethals, Academic Press, London, 1984, p. 1.
95. S. D. Pask, P. H. Plesch and S. Kingston, *Makromol. Chem.*, 1981, **182**, 3031.
96. G. E. Holdcroft and P. H. Plesch, *Makromol. Chem.*, 1984, **185**, 27.
97. W. Cooper, in 'The Chemistry of Cationic Polymerization', ed. P. H. Plesch, Macmillan, New York, 1963, p. 349.
98. N. G. Gaylord, I. Kössler and M. Stolka, *J. Macromol. Sci., Chem.*, 1968, **A2**, 1105.
99. N. G. Gaylord, I. Kössler, B. Matyska and K. Mach, *J. Polym. Sci., Part A-1*, 1968, **6**, 125.
100. N. G. Gaylord, I. Kössler and M. Stolka, *J. Macromol. Sci., Chem.*, 1968, **A2**, 421.
101. V. G. Ghandi, A. B. Deshpande and S. L. Kapur, *J. Polym. Sci., Polym. Chem. Ed.*, 1974, **12**, 1257.
102. U. Akbulut, L. Toppare and B. Yurttas, *J. Polym. Sci., Polym. Lett. Ed.*, 1986, **24**, 185.
103. B. Matyska, L. Petrusova, K. Mach and M. Svesta, *Collect. Czech. Chem. Commun.*, 1979, **44**, 1262.
104. M. Katsuta, Y. Ito and M. Taniguchi, *J. Polym. Sci., Polym. Lett. Ed.*, 1986, **24**, 463.
105. L. A. Eldib, *Hydrocarbon Process. Pet. Refiner*, 1963, **42**, 164.
106. R. Asami, K. Hasegawa and T. Onoe, *Polym. J.*, 1976, **8**, 43.
107. R. Asami and K. Hasegawa, *Polym. J.*, 1976, **8**, 67.
108. R. Asami, K. Hasegawa, N. Asai, I. Moribe and A. Doi, *Polym. J.*, 1976, **8**, 74.
109. K. Hasegawa and R. Asami, *Polym. J.*, 1976, **8**, 276.
110. R. Asami and K. Hasegawa, *Polym. J.*, 1976, **8**, 53.
111. K. Hasegawa, R. Asami and T. Higashimura, *Macromolecules*, 1977, **10**, 585.
112. T. Masuda, M. Otsuki and T. Higashimura, *J. Polym. Sci., Polym. Chem. Ed.*, 1974, **12**, 1385.
113. T. Masuda, T. Mori and T. Higashimura, *J. Polym. Sci., Polym. Chem. Ed.*, 1974, **12**, 2065.
114. K. Hasegawa, R. Asami and T. Higashimura, *Macromolecules*, 1977, **10**, 592.
115. M. Kamachi, K. Matsumura and S. Murahashi, *Polym. J.*, 1970, **1**, 499.
116. M. Kamachi, N. Wakabayashi and S. Murahashi, *Macromolecules*, 1974, **7**, 744.
117. U. Schädeli and M. Neuenschwander, *Chimia*, 1986, **40**, 239.
118. J. P. Vairon and P. Sigwalt, *Bull. Soc. Chim. Fr.*, 1964, 482.

119. J. P. Vairon and P. Sigwalt, *Bull. Soc. Chim. Fr.*, 1971, **559**, 569.
120. C. Aso, T. Kunitake and Y. Ishimoto, *J. Polym. Sci., Part A-1*, 1968, **6**, 1163, 1175.
121. A. Schmidt and G. Kolb, *Makromol. Chem.*, 1969, **130**, 90.
122. M. Villesange, G. Sauvet, J. P. Vairon and P. Sigwalt, *Polym. Bull. (Berlin)*, 1980, **2**, 131.
123. C. Aso and O. Ohara, *Makromol. Chem.*, 1969, **127**, 78.
124. G. Heublein and B. Adelt, *Plaste Kautsch.*, 1972, **19**, 728.
125. S. Kohjiya, A. Terada and S. Yamashita, *Chem. Lett.*, 1972, 671.
126. S. Kohjiya and S. Yamashita, *Chem. Lett.*, 1973, 1007.
127. K. Dinges, in 'Polymere Werkstoffe', ed. H. Batzer, Thieme, Stuttgart, 1984, III Technologie 2, p. 330.
128. G. Marwede, G. Sylvester and J. Witte, in 'Winnacker + Küchler-Chemische Technologie', 4th edn., Hanser, München, 1982, vol. 6, p. 550.
129. J. Rehner, *Ind. Eng. Chem., Ind. Ed.*, 1944, **36**, 46.
130. M. Györ, J. P. Kennedy, T. Kelen and F. Tüdös, *J. Macromol. Sci., Chem.*, 1984, **A21**, 1323.
131. S. Kohjiya, K. Nakamura and S. Yamashita, *Angew. Makromol. Chem.*, 1972, **27**, 189.
132. S. Kohjiya, M. Matsuzuki, M. Iwatani and S. Yamashita, *Kobunshi Ronbunshu*, 1980, **37**, 115.
133. A. Priola, S. Cesca, G. Ferraris and M. Bruzzone, *Makromol. Chem.*, 1975, **176**, 1969.
134. J. P. Kennedy and N. H. Canter. *J. Polym. Sci., Part A-1*, 1967, **5**, 2712.
135. F. Ciampelli, M. P. Lachi, M. Tacchi-Venturi and L. Porri, *Eur. Polym. J.*, 1967, **3**, 353.
136. A. Priola, C. Corno and S. Cesca, *Macromolecules*, 1981, **14**, 475.
137. C. Corno, A. Priola, G. Spallanzani and S. Cesca, *Macromolecules*, 1984, **17**, 37.
138. A. Priola, C. Corno and S. Cesca, *Polym. Bull. (Berlin)*, 1982, **7**, 599.
139. P. Yates, in 'Advances in Alicyclic Chemistry', ed. H. Hart and G. J. Karabatsos, Academic Press, New York, 1968, vol. 2.
140. C. Rentsch, M. Slongo, S. Schönholzer and M. Neuenschwander, *Makromol. Chem.*, 1980, **181**, 19.
141. U. Schädeli and M. Neuenschwander, *Chimia*, 1986, **40**, 241.
142. R. D. Athey, Jr., *J. Coat. Technol.*, 1982, **54**, 47.
143. R. D. Athey, Jr., *Prog. Org. Coat.*, 1979, **7**, 289.
144. H. Hasegawa and T. Higashimura, *Macromolecules*, 1980, **13**, 1350.
145. S. Aoshima and T. Higashimura, *J. Polym. Sci., Polym. Chem. Ed.*, 1984, **22**, 2443.
146. T. Higashimura and M. Sawamoto, *Adv. Polym. Sci.*, 1984, **62**, 49.
147. T. Higashimura, S. Aoshima and H. Hasegawa, *Macromolecules*, 1982, **15**, 1221.
148. G. VerStrate and F. P. Baldwin, *Polym. Prepr., Am. Chem. Soc., Div. Polym. Chem.*, 1976, **17** (2), 808.
149. G. VerStrate and F. P. Baldwin, *US Pat.* 4 278 822 (1981) (*Chem. Abstr.*, 1981, **95**, 187 880r).
150. O. Nuyken, S. D. Pask and A. Vischer, *Makromol. Chem.*, 1983, **184**, 553.
151. O. Nuyken, S. D. Pask, A. Vischer and M. Walter, *Makromol. Chem., Macromol. Symp.*, 1986, **3**, 129.
152. C. V. Freyer, H. P. Mühlbauer and O. Nuyken, *Angew. Makromol. Chem.*, 1986, **145/146**, 69.
153. J. P. Kennedy and M. Hiza, *J. Polym. Sci., Polym. Chem. Ed.*, 1983, **21**, 3573.
154. B. Keszler, V. S. C. Chang and J. P. Kennedy, *J. Macromol. Sci., Chem.*, 1984, **A21**, 307.
155. J. P. Kennedy, S. Guhaniyogi and V. Percec, *Polym. Bull. (Berlin)*, 1982, **8**, 571.
156. J. P. Kennedy, V. S. C. Chang and W. P. Francik, *J. Polym. Sci., Polym. Chem. Ed.*, 1982, **20**, 2809.
157. J. P. Kennedy and R. F. Storey, *Org. Coat. Appl. Polym. Sci. Proc.*, 1981, **46**, 182.
158. J. P. Kennedy, R. F. Storey, Y. Mahajer and G. L. Wilkes, in 'IUPAC, Macro 82, Amherst', p. 905.
159. T. R. Fang and J. P. Kennedy, *Polym. Bull. (Berlin)*, 1983, **10**, 82.
160. R. H. Wondraczek and J. P. Kennedy, *Polym. Bull. (Berlin)*, 1981, **4**, 445.
161. V. Percec and J. P. Kennedy, *Polym. Bull. (Berlin)*, 1983, **10**, 31.
162. J. P. Kennedy and M. Hiza, *J. Polym. Sci., Polym. Chem. Ed.*, 1983, **21**, 1033.
163. J. P. Kennedy, T. P. Liao, S. Guhaniyogi and V. S. C. Chang, *J. Polym. Sci., Polym. Chem. Ed.*, 1982, **20**, 3219.

41

Carbocationic Polymerization: Styrene and Substituted Styrenes

KRZYSZTOF MATYJASZEWSKI

Carnegie-Mellon University, Pittsburgh, PA, USA

41.1 INTRODUCTION

Although the cationic polymerization of styrene and substituted styrenes (referred to here collectively as 'styrenes') belongs to one of the oldest and most studied polymerization systems, it is not yet completely understood. In this chapter we shall try to review selectively and critically the existing literature on the cationic polymerization of styrene and its derivatives as well as other chemical reactions involving similar intermediates, and we shall attempt to give an overall picture of these systems by discussing the chemistry of the participating elementary reactions. First we shall describe the polymerizability of different derivatives of styrene in kinetic and thermodynamic terms. Then we shall review the peculiarities of different initiation systems and attempt to characterize the resulting chain carriers. We shall focus on the discussion of the chemistry of initiation, propagation, transfer and termination reactions and different factors influencing their relative importance. Finally we shall briefly describe the copolymerization of styrenes, emphasizing the reactivities of monomers and growing species. Because of the limited size of this chapter we cannot cover all phenomenological observations. The interested reader should consult the comprehensive reviews which describe this field extensively.[1-5]

41.1.1   Historical Background

Styrene (S) was among the first monomers to be polymerized cationically. Deville[6] polymerized styrene with tin(IV) chloride in 1839. Next, in 1866 Berthelot polymerized styrene using either iodine or sulfuric acid.[7] α-Methylstyrene (αMS) was polymerized with different Lewis acids by Staudinger.[8] First quantitative studies of polymerization of S were carried out by Williams,[9] who used tin(IV) chloride as the initiator.

In the 1950s and 1960s, the polymerization of styrene derivatives by strong protonic acids was investigated intensively by Evans,[10] Pepper[11] and Plesch.[12] Later this system was also studied by Higashimura and Sawamoto,[13] Sigwalt and Sauvet,[14] Kunitake and Takarabe[15] and in other laboratories. Weak UV absorption was the main advantage of these initiators because it allowed the observation of the growing carbenium cations and the estimation of reactivities of active species by a stopped flow method, which was applied for the first time in cationic systems by Pepper and Szwarc.[16] The polymerization of styrenes by different Lewis acids (MtX_n) was first critically reviewed by Mathieson[17] who concluded that most MtX_n are active when cocatalyzed. Later, some evidence for the direct initiation (without cocatalyst) was discussed (see also Chapter 39).[18]

There is a large number of initiation systems for styrenes and it is impossible to discuss all of them in this chapter. We suggest some earlier reviews in cationic polymerization which extensively described all initiation systems[3,4] (the latter one in a very critical way).

41.1.2   Commercial Importance

Polystyrenes are prepared industrially by radical polymerization. Polymers with a narrow molecular weight distribution used for some special applications are synthesized by anionic living polymerization. Cationic polymerization is usually limited to the synthesis of ill-defined resins. The aromatic resins obtained from refinery fractions C_8 to C_{10} by the action of trichloroaluminum or other Lewis acids contain large amounts of polyindenes. In the fraction boiling between 140 and 200 °C there is also styrene, α-methylstyrene, some dimethylstyrenes and alkyl aromatics. Therefore, besides polymerization *via* double bonds, alkylation of the aromatic rings plays a very important role. Oligomers of polystyrene and poly(α-methylstyrene) ($M_n < 1000$) are commercially available and used as overprint varnishes and additives. For example, Hercules Inc. produces polystyrene with a trade name Piccoelastic and Amoco produces poly(α-methylstyrene), called Resin 18, which softens at 144, 119 and 99 °C depending on the molecular weight.[2]

Preparation of well-defined polystyrenes *via* a cationic mechanism can open a new route towards block and graft copolymers. For example, triblock copolymers of polystyrene with poly(isobutylene) as the central soft segment should have better properties as thermoplastic elastomers than triblock copolymers with unsaturated blocks [polybutadiene or poly(isoprene)].

41.1.3   Model Reactions

Reactions similar to the elementary reactions of the polymerization process, *i.e.* initiation, transfer, termination, propagation and isomerization have been intensively studied in organic chemistry. Reactive intermediates in polymerization — carbenium ions — appear as intermediates in the solvolysis of alkyl esters and halides. Thus, considerable information relevant to cationic polymerization can be obtained from studies of nucleophilic and electrophilic aromatic substitution, elimination and electrophilic addition to double bonds. We will recommend some excellent reviews on these reactions and will try to discuss the chemistry of the elementary reactions along with analogous reactions from the organic chemistry of low molecular weight compounds.[19-25]

41.2 MONOMERS

First we must define which monomers we consider as derivatives of styrene. This will include styrenes substituted both in the ring and at the C=C bond. In this chapter we will use short abbreviations for the following monomers [see formulae (1) and (2)]: styrene (S, $R_1 = R_2 = H$), α-methylstyrene (αMS, $R_1 = Me$, $R_2 = H$), 1,1-diphenylethene (DPE, $R_1 = Ph$, $R_2 = H$), p-methoxy-styrene (pMOS, $R_1 = H$, $R_2 = p$-OMe), p-methoxy-α-methylstyrene (pMOαMS, $R_1 = Me$, $R_2 = p$-OMe), p-methyl-α-methylstyrene (pMαMS, $R_1 = Me$, $R_2 = p$-Me), p-t-butyl-α-methylstyrene (pBαMS, $R_1 = Me$, $R_2 = p$-Bu^t), p-isopropyl-α-methylstyrene (pPαMS, $R_1 = Me$, $R_2 = p$-Pr^i). The polymerization of indene (3; IN), which is a combination of the above structures, will also be covered.

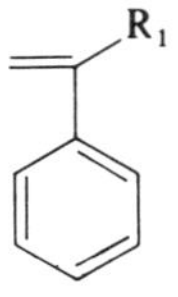

(1) 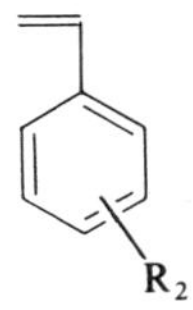   (2) 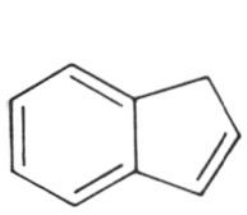   (3)

We will discuss the polymerizability of different styrenes in terms of the thermodynamic feasibility of the formation of high polymers and then we will consider some reasons for difficulties in syntheses of this type of polymers.

41.2.1 Thermodynamic Polymerizability

In this section polymerizability means the thermodynamic possibility of the formation of a high polymer and will be purely related to the change of the free enthalpy of polymerization (ΔG_p°):

$$\Delta G_p^\circ = \Delta H_p^\circ - T\Delta S_p^\circ \qquad (1)$$

Thus, polymerizability is independent of the reaction mechanism and will be the same for the cationic, anionic and radical systems, provided that the same type of polymer is formed with respect to conformation, tacticity, and possible isomerization of repeating units.

The thermodynamics of the polymerization of styrenes has been studied quantitatively for only a few systems.[26-28] The standard enthalpy of polymerization of S is $\Delta H_{p(lc)}^\circ = -73.1$ kJ mol^{-1} and the standard entropy $\Delta S_{p(lc)}^\circ = 104$ J K^{-1} mol^{-1}.[29] These values, calculated for polymerization in benzene, indicate that at room temperature the equilibrium monomer concentration is $[M]_e = 10^{-7}$ mol l^{-1}. Therefore, at any practically usable concentration, S should polymerize quantitatively.

Ring substituted styrenes behave similarly. The polymerization remains very exothermic[30] even for the trisubstituted monomer for which $\Delta H_p^\circ = -70.4$ kJ mol^{-1}.[31] Pentamethylstyrene,[32] as well as vinyl naphthalenes[33] can also be polymerized quantitatively to a high polymer.

Substitution at the double bond changes the polymerizability of styrenes dramatically. The thermodynamics of the polymerization of αMS was intensively studied in anionic systems and precise values of the enthalpy and entropy of polymerization as well as of oligomerization are known.[34, 35] This allows us to calculate not only the ceiling temperature for any concentration but also the equilibrium monomer concentration for a given temperature:

$$[M]_e = K^{-1} = \exp(\Delta G_p^\circ/RT) = \exp(\Delta H_p^\circ/RT - \Delta S_p^\circ/R) \qquad (2)$$

The standard enthalpy and entropy for the polymerization of αMS, pPαMS and pBαMS are $\Delta H_p^\circ(ss) = -10 \pm 2$ kJ mol^{-1} and $\Delta S_p^\circ(ss) = -35 \pm 10$ J K^{-1} mol^{-1} in THF solvent.[27, 28, 36] These parameters lead to the following values of $[M]_e$ (in mol l^{-1}) ≈ 2 (20 °C), 0.8 (0 °C), 0.3 (−20 °C) and 0.01 (−80 °C).

It is obvious that no high polymer can be formed at 0 °C from αMS, using for example $[M]_0 = 0.1$ mol l^{-1}, but dimers, and probably some higher oligomers, are thermodynamically allowed. Of course, a suitable mechanism must exist for this type of reaction. Keeping in mind these values of $[M]_e$, some of the results reported in the literature are quite surprising. For example, the quantitative polymerization of αMS to a polymer with a molecular weight $M_n > 10^3$ using

$[M]_0 = 0.6 \, \text{mol} \, l^{-1}$ at $-20\,°C$ contradicts the thermodynamic predictions (maximum yield could be only 62%).[37] However, differences in tacticity of polymers formed by cationic and anionic mechanism can change values of $[M]_e$.

β-Methylstyrene gives a lower yield of polymer than the α-isomer.[38] Anethole (*p*-methoxy-β-methylstyrene) forms a high polymer at low temperatures.[39] No information is available on the thermodynamics of polymerization of *cis* and *trans* isomers (or *Z* and *E*).

Neither α- nor β-ethyl substituted styrenes have been polymerized. Similar behavior was observed for α-methoxystyrene,[40] which was copolymerized successfully *via* cationic and radical mechanisms but could not be homopolymerized. Thus, a small increase in the size of the α- or β-substituent from a methyl to an ethyl group prevents monomers from polymerizing. In agreement with this explanation, aryl substituted styrenes, *e.g.* stilbenes or 1,1-diphenylethylenes, provide only linear and cyclic dimers. Cyclization is the Friedel–Crafts alkylation of aromatic rings and can sometimes lead even to higher oligomers of irregular structure. It is important to remember that cyclization, proceeding by the formation of an indan-type structure, is practically irreversible.[42] Therefore, the large energetic gain from the formation of stable indan can overbalance the small loss in the free energy for oligomerization. In this case the structure of the end group influences the thermodynamics of short chain oligomer formation. Hence, reliable data on the thermodynamics of polymerization of αMS can be obtained only by anionic polymerization in which indans are not formed.

Indene is a special case. Although it resembles β-methylstyrene, indene polymerizes very well, preserving its bicyclic structure.[43] The same is true for acenaphthylene[44] and 1-methylindene. Other indenes such as 2- and 3-methyl yield only oligomers,[45,46] but dimethylindenes substituted in the five-membered ring yield high polymers.[46]

Substituents in the *m* and *p* positions have no influence on the polymerization thermodynamics of αMS. Thus, *p*-phenyl-αMS (**4**)[47] and β-isopropenylnaphthalene (**5**)[47] were successfully polymerized. *Ortho* substitution prevents polymerization. Thus, neither α-isopropenylnaphthalene nor 2,4-dimethyl-αMS could be polymerized.[47]

In summary, we may conclude that α or β substitution in styrenes very strongly increases repulsive interactions in polymer chains and any substituent larger than methyl prevents polymerization. The only exception can be 1-methylindene, due to its cyclic structure. Ring substitution has smaller effect but *o*-substituted αMS cannot be homopolymerized.

41.2.2 Kinetic Polymerizability

Conversion of a monomer, which is thermodynamically polymerizable into a high polymer, is possible if a mechanism exists which allows this conversion at a suitable rate. Some strained monomers were polymerized only by an anionic or a radical mechanism (*e.g.* acrylates, tetrafluoroethylene), and some others only by a cationic mechanism (isobutylene). The main criterion is the formation of sufficiently stable and sufficiently reactive growing species as well as the absence of groups leading to termination and extensive transfer. Therefore, substituents can strongly affect the

(3)

kinetics of polymerization and may make a monomer inert towards a certain initiation system. For example, amino groups inhibit the cationic polymerization of styrene by forming onium ions (equation 3). A *p*-dimethylamino substituent strongly decreases the overall rate of polymerization but allows the formation of a polymer with $M_n = 4000$.[48]

41.2.3 Polymerizability of Some Oligomers

Extensive transfer reactions lead to linear oligomers which contain potentially reactive alkene end groups. Chances for the homopolymerization of styrenes with an α or β substituent larger than methyl are very small, but copolymerization with *p*-chlorostyrene was observed for α-ethyl-, β-ethyl- and β-propyl-styrenes.[50] A similar reaction may lead to the incorporation of unsaturated oligomers into polymer chains (equation 4).

$$-CH_2CHPhCH=CHPh + -CH_2CHPh^+ \rightarrow -CH_2CHPhCH-CHPh^+ \qquad (4)$$
$$\underset{-CH_2-CHPh}{\overset{\mid}{}}$$

Unsaturated dimers of styrene and *p*-methylstyrene have been oligomerized in 80% yield to trimers and tetramers, confirming their polymerizability.[51] The copolymerization of dimers with monomers has not yet been studied. However, the rapid increase of the molecular weight observed in the cationic polymerization of pentamethylstyrene and accompanied by the considerable decrease of polydispersity at higher conversion may suggest chain coupling reactions.[32] Because poly(pentamethylstyrene) can hardly be attacked at the ring, this coupling should involve the incorporation of the unsaturated oligomers into the polymer chain in a way similar to equation (4). This reaction is less possible for αMS, since attempts at copolymerization of α,β-dimethylstyrene were unsuccessful.[52]

41.3 INITIATORS AND REACTION CONDITIONS

The cationic polymerization of styrene derivatives must be carried out in dry and non-nucleophilic solvents. Best are halogenated solvents, such as dichloromethane or methyl chloride. Oxygen- and nitrogen-containing solvents are too nucleophilic. Aromatic solvents are sometimes used but Friedel–Crafts alkylation is possible.[53] Nitro compounds might be alkylated at the nitro group leading to nitronium cations.[54] However, since high polymers are formed and conversion is complete,[55] this reaction must be reversible.

Ionic reactions are sensitive to changes in solvent polarity. The rate constant of recombination of ions (termination) decreases strongly in more polar solvents. Rate constants for the reaction between an ion and a polar molecule (propagation, transfer) also decrease in solvents with higher dielectric constants, but to a smaller extent.[56, 57] Transition states in these reactions have more dispersed charge and are destabilized in more polar solvents. Reactions in which ions are formed (ionization of covalent species, dissociation of ion pairs) are accelerated in more polar solvents because the transition states have stronger ionic character than the starting compounds. Synthesis of high molecular weight polymers often requires low temperatures because the energy of activation of transfer is higher than that of propagation.

The sensitivity of the polymerization of styrenes to nucleophiles calls for pure solvents uncontaminated by nucleophilic impurities (moisture included). Therefore, reliable and reproducible studies require special purification of solvents. For example, CH_2Cl_2 should first be treated with oleum to remove vinyl compounds, then neutralized, carefully distilled, and dried with several P_2O_5 films and Na mirrors.[58]

Styrenes have been polymerized by a wide variety of initiators. Practically any Lewis or protonic acid may, under some conditions, produce polystyrene. Below we discuss briefly the main types of initiation systems. An excellent discussion of different initiation systems was published by Gandini and Cheradame a few years ago.[4] A more detailed discussion of initiation can be found in Section 41.5.1 and also in Chapter 39 in this volume.

In this section we will list different initiation systems and discuss their efficiencies towards different monomers. The mechanism and kinetics of initiation will be discussed later together with other elementary reactions.

41.3.1 Protonic Acids

The efficiency of protonic acids depends on their strength, the nucleophilicity of monomers, polarity of solvent and temperature. An idealized (in real systems side reactions could not be neglected, *cf.* Scheme 3) scheme of polymerization (Scheme 1) can explain this behavior.

$$M + HA \rightleftharpoons M_1^+, A^- \tag{a}$$

$$M_1^+, A^- + M \rightleftharpoons M_2^+, A^- \tag{b}$$

$$M_n^+, A^- + M \rightleftharpoons M_{n+1}^+, A^- \tag{c}$$

$$M_n^+, A^- \rightleftharpoons M_n A \tag{d}$$

Scheme 1

If a monomer, M, has a very low basicity and an acid HA is weak, the addition reaction (a) is very slow and may be possible only in more polar media. On the other hand, if the anion A^- is very nucleophilic, recombination of the counterions (d) would occur before addition of the first monomer molecule to the ion (b). For strong acids and reactive monomers reaction (c) leads to high polymer. In cationic polymerization there are examples of all these cases. Addition of hydrogen halides to styrenes in low polar solvents leads to the quantitative formation of the corresponding 1-phenylethyl halides.[59,60] The same was observed for haloacetic acids.[61,62] Reaction of trifluoroacetic acid with styrene in the non-polar CCl_4 yielded a mixture of 1-phenylethyl trifluoroacetate and some higher oligomeric acetates.[63,64] Also linear oligomers with C=C bond were found as a result of transfer or reversible protonation (a reaction similar to (−a) but for higher oligomers). In the more polar CH_2Cl_2, a higher polymer was formed because the relative rate of addition of monomer to active centers was higher than the rate of collapse of ion pairs.[65,66] The addition of styrene to bulk trifluoroacetic acid (which has a quite high dielectric constant) yields a high polymer.[67] High polymers are easily formed using a more nucleophilic monomer (with a *p*-methoxy group).[68]

The strongest protonic acids, such as triflic and perchloric, polymerize styrenes rapidly and quantitatively even in less polar solvents and at low concentrations.[69−75] For example, the ratio of acid to monomer concentrations can be decreased from 0.1 to 0.0001 for trifluoroacetic and triflic acid respectively to achieve similar overall rates of polymerization.

Triflic, perchloric and sulfuric acids lead to the quantitative polymerization of styrenes and to relatively high molecular weight products. Among other efficient initiators are chlorosulfonic and fluorosulfonic acids, but they participate in irreversible termination reactions.[76]

41.3.2 Lewis Acids

Many Lewis acids have been used as initiators for the polymerization of styrenes. The main problem which should be cleared up is the way these acids add to styrenes: directly or with coinitiators? In the latter case the Lewis acid is not a true cationogen. A detailed analysis of initiation with Lewis acids can be found in refs. 4 and 18 and a more updated discussion in Chapter 39 of this volume.

The direct initiation by the Lewis acid may involve either zwitterionic species or disproportionation (self ionization) of two molecules of the Lewis acid (equation 5).

$$MtX_n + CH_2=\underset{\underset{Ph}{|}}{CH} \rightarrow {}^-MtX_n-CH_2-\underset{\underset{Ph}{|}}{CH}{}^+ \tag{5a}$$

$$2MtX_n \rightleftharpoons MtX_{n-1}^+, MtX_{n+1}^- \xrightarrow{\overset{Ph}{\overset{\|}{}}} MtX_{n-1}-CH_2\underset{\underset{Ph}{|}}{CH}{}^+, MtX_{n+1}^- \tag{5b}$$

The former mechanism was proposed originally by Hunter and Yohe.[77] Electrochemical studies showed that the latter mechanism (equation 5b) operates for $AlBr_3$.[78] Direct initiation has been proved only for a few Lewis acids such as $TiCl_4$, $TiCl_3OBu$,[79,80,81] and $AlCl_3$ (see Table 2, ref. n).

Organic and inorganic compounds containing halogen or oxygen atoms may react with Lewis acids forming intermediate carbenium ions or protonic acids. Alcohols, water, alkyl and hydrogen halides were found as efficient components of the initiating system, as shown in equation (6).

$$HOH + MtX_n \rightleftharpoons [HOMtX_n^-, H^+] \xrightarrow{\underset{Ph}{\overset{\overset{Ph}{\|}}{}}} \underset{Ph}{\overset{}{MeCH^+}}, HOMtX_n^- \qquad (6)$$

$$RX + MtX_n \rightleftharpoons [R^+, MtX_{n+1}^-] \xrightarrow{\underset{Ph}{\overset{\overset{Ph}{\|}}{}}} R\text{–}CH_2\underset{Ph}{\overset{}{CH^+}}, MtX_{n+1}^-$$

Lewis acids such as aluminoorganic compounds and boron halides cannot initiate polymerization of styrenes in rigorously purified and dried solvents. Antimony halides can either form complexes with styrenes which, *via* electron transfer, can rearrange to cationic species or serve as Cl^+ donors.[82] Some Lewis acids are less efficient initiators of styrene polymerization than strong protonic acids because of what may be ascribed to irreversible termination by the formation of inactive alkyl halides (equation 7).

$$-CH_2\text{–}\underset{Ph}{\overset{}{CH^+}} BF_4^- \rightarrow -CH_2\text{–}\underset{Ph}{\overset{}{CH}}\text{–}F + BF_3 \qquad (7)$$

41.3.3 Carbenium Salts and Esters

There are two carbenium ions which are relatively stable and available commercially: triphenylmethylium and tropylium. Both have been used in the polymerization of most reactive styrenes.[83-84] The most efficient are salts containing the most stable anion SbF_6^-.[85] This suggests that other complex anions cannot survive during the entire polymerization process and terminate it by the formation of alkyl halides and Lewis acids unable to reinitiate polymerization alone (*cf.* equation 7).

Carbenium ions participate easily in hydride transfer reactions, especially at higher temperatures (hydride transfer has a higher energy of activation than direct initiation). This may change for more bulkily substituted monomers due to crowding from trityl moieties. Hydride transfer has been observed for αMS[86] but under conditions different than in polymerization. For *p*MOS initiation proceeds by the addition of the trityl cation to the C=C bond.[87]

Another group of initiators used successfully for the polymerization of styrenes is oxocarbenium salts.[88] Here again the best results were obtained for the SbF_6^- anion. Styrene and acenaphthylene produced high molecular weight polymers.[55]

Reactive perchlorate esters were formed *in situ*. Different alkyl halides have been treated with silver perchlorate in the presence of styrene, leading to its polymerization.[89] It is difficult to establish whether they reacted in covalent form or after ionization. In general, esters and halides do not initiate the polymerization of styrenes. Even quite unstable and reactive 1-phenylethyl trifluoroacetate cannot induce the polymerization of styrene without any additives.[66] More reactive esters and mixed anhydrides—perchlorates—prepared *in situ*, have been successfully used in the polymerization of styrenes. They were prepared from 1-phenylethyl bromide or acetyl chloride and silver perchlorate directly in the solution containing styrene as the monomer.[90,91] Only very recently was 1-phenylethyl perchlorate observed directly by NMR at a low temperature ($-78\,°C$).[92] This confirmed the covalent nature of a majority of the end groups in polymerization with perchloric acid. Perchlorates containing strong chromophoric groups were also used as initiators and were found in the polymer chains.[93,94] Triflate esters were unstable at $-78\,°C$ in CH_2Cl_2.[95]

Some esters, inactive alone, initiate polymerization in the presence of a Lewis or a protonic acid. For example, the addition of sulfuric acid was claimed to reactivate inactive trifluoroacetate esters.[96] The addition of deuterated styrene to the mixture of 1-phenylethyl trifluoroacetate and trifluoroacetic acid led to the much more rapid consumption of the ester than of the acid.[97] Lewis acids are more efficient coinitiators, and even weaker chloro and unsubstituted acetates have been activated by BCl_3 and BBr_3.[98] Systems of this type may also be considered as activated Lewis acids (*cf.* previous section).

41.3.4 Other Systems

We have already discussed chemical types of initiation. Styrenes can also be polymerized by radiolysis, field ionization, field emission and electrodialysis.[1,4] Photochemical initiation and nuclear chemical initiation have also been described. In all these systems the structure of the real initiating species is not known because the concentration of active species is too low to allow the structural analysis of the end groups. Nevertheless, some results are of primary importance. For example, the use of radiation in the polymerization of vinyl compounds for the first time suggested high rate constants of propagation (10^6 mol^{-1}1s^{-1}), contrary to the low apparent rate constants observed at that time in the majority of systems.[99,100]

Nuclear chemical initiation employing tritium as the initiator led to the highest molecular weight observed in the polymerization of styrene ($\bar{M}_n = 1.7 \times 10^6$).[101] This would suggest that transfer reactions depend on the nature of anions because transfer was apparently suppressed in the system in which there was no anion. However, the cationic nature of this polymerization has not yet been sufficiently proven.

41.4 ACTIVE SPECIES

In addition to propagating carbenium ions as the active species, in this section we will consider esters and isomerized species which may not be active *per se*, but which under some conditions may produce reactive growing species. With such a broad definition the scheme of propagation should read as shown in Scheme 2.

$$-CH_2CHR\text{-}A \rightleftharpoons -CH_2CHR^+,A^- \rightleftharpoons -CH_2CHR^+ \| A^- \rightleftharpoons -CH_2CHR^+ + A^-$$
$$(M_n^c) \downarrow k_p^c \qquad\qquad \downarrow k_p^{\pm} \qquad\qquad \downarrow k_s^{+\|-} \qquad\qquad \downarrow k_p^+ \qquad + M$$
$$M_{n+1}^c \quad \rightleftharpoons \quad M_{n+1}^{\pm} \quad \rightleftharpoons \quad M_{n+1}^{+\|-} \quad \rightleftharpoons \quad M_{n+1}^+$$

Scheme 2

Scheme 2 resembles any scheme of solvolysis of esters or halides. The active species in the polymerization of styrenes are analogs of the species derived from 1-phenylethyl and cumyl esters or halides. We will therefore discuss polymeric cations and esters simultaneously with low molecular weight models.

41.4.1 Growing Ions

Growing carbenium ions in the polymerization of styrenes have a strongly stabilizing α-phenyl group. There are only a few theoretical calculations for these cations. Two limiting structures are possible: classical carbenium ions and non-classical bridged cations. Calculations refer to the gas phase and they give only a general idea about relative stabilities and charge distribution in cations because cations have a strong affinity to nucleophiles, even so weak as solvents and monomer. The *ab init o* calculations with a small basis set favor the classical ions, whereas the semiempirical methods overemphasize the stability of the non-classical, bridged cations. Near the Hartree–Fock limit both structures have similar energies for aliphatic cations (*e.g.* ethylium).[102]

Calculations show that the bridged structure is more stable than the β-phenylethylium ion.[102,103] However, the γ-phenylpropylium ion prefers to take an antiperiplanar, perpendicular conformation as a classical cation (**6**) (stabilization through hyperconjugation).[104] The corner protonated cyclopropane has higher energy by more than 80 kJ mol^{-1}. The α-phenylpropylium ion is the most stable among different phenylpropyl cations due to better charge delocalization. The α-phenylethylium cation was found experimentally to be more stable than the bridged cation.[105]

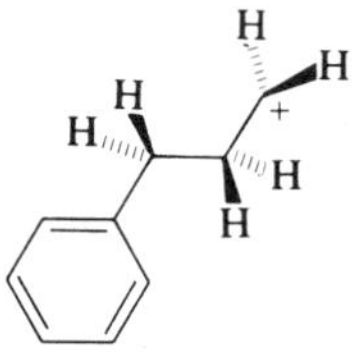

(6)

We have done some approximate calculations (STO-3G and MNDO) for dimethylphenylmethylium and methylphenylmethylium ions and found that a considerable amount of charge was on the β-H atom (8 to 12%). The tertiary carbenium C atom has only 20 to 30% of the charge, depending on the substituents. The carbon atoms in the ring have negative charge (*ipso* −4.5%, *meta* −5.4%) or small positive charge (*para* 1.7%, *ortho* 0.39%). These results are in qualitative agreement with the ^{13}C NMR chemical shifts.[106,107,108]

The *p*-methoxy substituent decreases the charge on the tertiary atom to 18%, shifts it toward the *p*-C atom, and leads to higher stability of cations, which was confirmed experimentally. The calculations for the models of the growing dimers indicate that the second aromatic ring interacts weakly with the carbenium center as shown by structure (**7**). Although the calculations cannot account for the solvent effects and could have limited credibility, these interactions were also detected in the NMR spectra of dimeric cations formed from *p*MOαMS.[115]

(**7**)

The ions involved in the polymerization of S in the usual solvents (*e.g.* CH$_2$Cl$_2$) have a very short life time.[15,75,109] At room temperature, this life time is less than one second and is not much longer at lower temperatures. Therefore only stopped flow methods could be used for the detection of short living cations. Initially, in the early sixties, strong absorption observed at 420 nm (indanylium cations) was ascribed to growing cations. However, it was proposed on the basis of molecular orbital theory that the maxima of absorption of benzyl cations and anions should coincide.[73,110] The latter absorb at 340 nm.[111] The subsequent stopped flow studies of Pepper and Szwarc proved that a transient absorption at 340 nm could be correlated with the rate of polymerization,[16,75] which indicated that this absorption was related to the active species. UV studies of the corresponding models derived from tertiary alcohol in superacid media confirmed the proposed similarity of absorptions of cations and anions.[112] In those studies the extinction coefficients (ε) were calculated assuming the quantitative formation of the cations from tertiary alcohols. It would be safer to state that the lower limit of ε was calculated in this way.

The absorption maxima and extinction coefficients of the most important growing and model cations are shown in Table 1. The absorption of dimeric cations formed from S in strong acids at shorter wavelengths (315 nm) is very unexpected, as is also the high stability of these species.[113] Attempts to study these cations using NMR were unsuccessful. It is possible that the UV absorption was due to some kind of a complex (σ or π) or indanic species.[114]

The dimeric cation derived from *p*MOαMS in CH$_2$Cl$_2$ solvent has the same UV absorption as the monomeric cation.[115] At higher monomer concentrations a red shift is observed for cations derived from different αMS. This shift can be ascribed to isomerized cations or complexes with alkenes.[119] 3-Alkyl indanylium cations absorb at shorter wavelengths than styryl cations, whereas 3-phenyl substituted indanylium cations absorb at a maxima higher by 100 nm.[116]

Carbenium ions were also studied using ^{1}H NMR and ^{13}C NMR.[106] Because of the low stability of cations, superacid media were used again, in which only monomeric cations could be formed (large excess of acid over monomer). Chemical shifts of cations derived from ring-substituted S and αMS were correlated with the σ parameters of the substituents. Recently a new σ_{c+} scale was derived which is based on ^{13}C NMR shifts of the charged C atoms in dimethylarylmethyl cations.[117] However, the stability of cations is still better correlated with the classical σ^+ parameter.[118]

Very recently monomeric (**8a**), dimeric (**8b**) and trimeric (**8c**) cations were observed in CD$_2$Cl$_2$ in the polymerization of *p*MOαMS.[115] This was the first direct NMR observation of styryl-type

(**8a**)   (**8b**)   (**8c**)

Table 1 Maxima of Absorption and Extinction Coefficients of Model and Growing Carbenium Ions

Monomer	Cation	Maximum	ε	Comments	Ref.
S	$-CH_2CHPh^+$	340	10^4	1	a, b
αMS	$-CH_2CMePh^+$	336	10^4	$\overline{DP}_n < 2$, 1	c
	$-CH_2CMePh^+$	350	10^4	polymer, 1	a
	Me_2CPh^+	333	2.6×10^4		d
	Me_2CPh^+	326	1.1×10^4	2	e
pMS	$-CH_2CHAr^+$	334	—		f
pMOS	$-CH_2CHAr^+$	340	—		a
	$-CH_2CHAr^+$	380	2.8×10^4	2	g
	$MeCHAr^+$	348	2.8×10^4	2	h
pClS	$-CH_2CHAr^+$	325	—		a
2,4,6MS	$-CH_2CHAr^+$	325	—		a
pMαMS	$MeCHAr^+$	345	$> 2.5 \times 10^4$	4	d
	$-CH_2CHAr^+$	358	$> 2.2 \times 10^4$	3	d
pPαMS	$-CH_2CHAr^+$	362	$> 1.6 \times 10^4$	3	i
pBαMS	$MeCHAr^+$	350	3.5×10^4	4	d
	$-CH_2CHAr^+$	363	$> 2.2 \times 10^4$	3	d
pMOαMS	$MeCHAr^+$	368	2.9×10^4	4	d
	$-CH_2CHAr^+$	382	$> 2.5 \times 10^4$	3	d
IN	$-CHRCHAr^+$	404	—		f
3iPIN	$-CHRCHAr^+$	320	2.9×10^4		j
3MIN	$-CHRCHAr^+$	312	3×10^4		j
3PhIN	$-CHRCHAr^+$	412	3.2×10^4		j
DPE	$MeCPh_2^+$	435	2.6×10^4	5	k, f, l

^{1} mol^{-1} $l\,cm^{-1}$; ε assumed from the model tertiary alcohol in superacid media. ^{2} ε assumed quantitative formation of ions. ^{3} Assumed A^- structure, ε should be higher for HA_2^- counterions. ^{4} Confirmed by NMR. ^{5} Additional bands were observed at 520 nm (and 395) and at 456 nm (and 330) and were ascribed to the cation solvated internally by the neighboring aromatic ring or by monomer.

^{a} M. DeSorgo, D. C. Pepper and M. Szwarc, *J. Chem. Soc., Chem. Commun.*, 1973, 419.
^{b} T. Kunitake and K. Takarabe, *Macromolecules*, 1979, **12**, 1061.
^{c} K. Takarabe and T. Kunitake, *Makromol. Chem.*, 1981, **182**, 1587.
^{d} K. Matyjaszewski and P. Sigwalt, *Macromolecules*, 1987, **20**, 2679.
^{e} G. A. Olah, C. U. Pittman, Jr., R. Waack and M. Doran, *J. Am. Chem. Soc.*, 1966, **88**, 1488.
^{f} T. Kunitake and K. Takarabe, *J. Polym. Sci., Polym. Symp.*, 1976, **56**, 33.
^{g} M. Sawamoto, T. Masuda and T. Higashimura, *J. Polym. Sci., Polym. Chem. Ed.*, 1978, **16**, 2675.
^{h} T. Higashimura, N. Kanoh and S. Okamura, *J. Macromol. Sci., Chem.*, 1966, **1**, 109.
^{i} D. Teyssie, M. Villesange and J. P. Vairon, *Polym. Bull.*, 1984, **11**, 459.
^{j} H. Cheradame, N. Hung and P. Sigwalt, *J. Polym. Sci., Polym. Symp.*, 1976, **56**, 335.
^{k} G. Sauvet, J. P. Vairon and P. Sigwalt, *J. Polym. Sci., Polym. Symp.*, 1975, **52**, 173.
^{l} B. E. Fleischfresser, W. J. Cheng, J. M. Pearson and M. Szwarc, *J. Am. Chem. Soc.*, 1968, **90**, 2172.

cations in a solvent usually applied in cationic polymerization. Monomeric cations were also studied for other α-methylstyrenes.[119] The ^1H NMR spectrum of the mixture of monomeric, dimeric and trimeric cations in the polymerization of pMOαMS is shown in Figure 1. These species were also studied by UV (*cf.* Table 1).

'Free' carbenium ions were detected conductometrically. The increase of the conductivity was proportional to the increase of the absorption at 420 nm (absorption of the final non-reactive indanylium cations).[120] Growing carbenium cations absorbing at 340 nm have very short life times and are present at very low concentrations. Hence, stopped flow conductometry was used and successfully correlated with the results found by the stopped flow UV studies.[121]

Carbenium ions, which can be models for growing species, have life times that are too short to be studied by the other methods which have been successfully used for more stable cations (tritylium and tropylium ions) such as magnetic circular dichroism,[122] IR and Raman spectroscopy,[123] calorimetry,[124] X-ray crystallography,[125] ESCA,[126] cryoscopy,[127] *etc.*

41.4.2 Isomerized and Inactive Ions

Very recently in the polymerization of pMOαMS, isomerized ions were observed by NMR and UV.[115] These ions are present at much higher concentrations than the usual linear growing species.

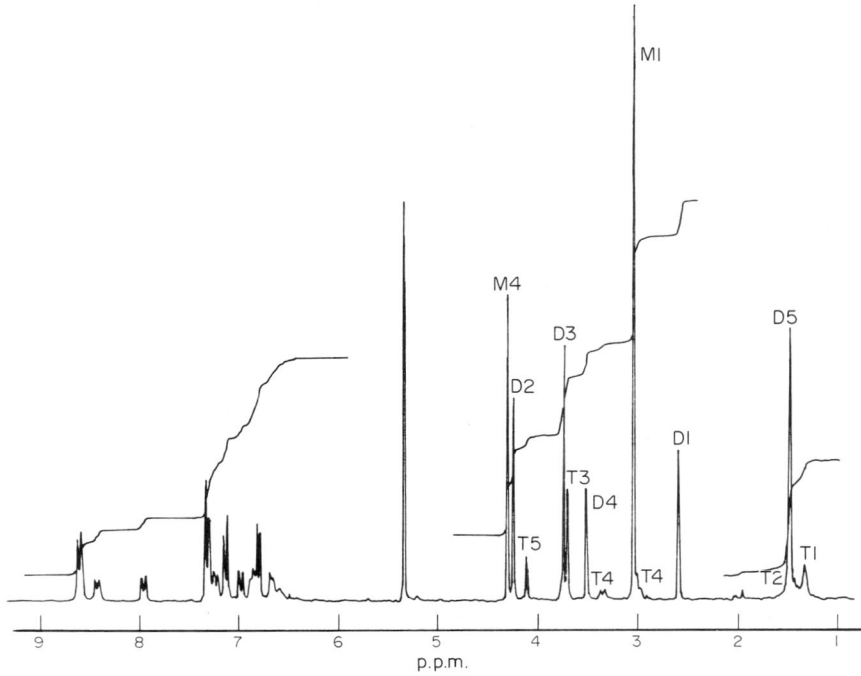

Figure 1 250 MHz ^{1}H NMR spectrum after reaction of $[p\text{MO}\alpha\text{MS}]_0 = 0.031 \text{ mol l}^{-1}$ and $[\text{HOSO}_2\text{CF}_3]_0 = 0.037 \text{ mol l}^{-1}$ in CD_2Cl_2 at equilibrium ($T = -63\,^\circ\text{C}$)

It has been found that the isomerized ions are in equilibrium with the linear species (equation 8).

$$-CH_2-CR^1R^2-CH_2-\overset{+}{C}R^1R^2 \rightleftharpoons -CH_2-\overset{+}{C}R^2-CH_2-CR^1R^1R^2 \tag{8}$$

The mechanism of this interconversion is discussed in Section 41.5.6. The UV absorptions of isomerized cations are different from those of the usual linear dimeric and monomeric cations. The main absorption is shifted from 368 nm to 382 nm and (additionally) the second absorption appears at 460 nm to 480 nm. These changes can be ascribed to the interactions with two neighboring aromatic rings. Indeed, carbenium ions are so reactive that they stabilize themselves by interactions with any donor, such as an aromatic ring from a polymer chain, a conjugated alkene (monomer or unsaturated oligomer) or a solvent molecule. In the dimerization of DPE, different absorptions were explained by the first two interactions.[128] On the other hand, the polymodality of MWD observed in the polymerization of *p*MOS was ascribed to species complexed by monomer, solvent and polymer chains.[129]

The polymerization of styrenes (except 2,6-disubstituted) is accompanied by Friedel–Crafts cycloalkylation, *i.e.* the formation of an indan. Under acidic conditions an indan is transformed to an indanyl cation.[113] This reaction may proceed by hydride transfer for styrenes and by transfer of methide anion or protonation of the aromatic rings for α-methylstyrenes (*cf.* Section 41.5.6). These cations were observed by NMR and by UV.[113,119] The spectral characteristics were given in Table 1.

At later stages an absorption at 430 nm was observed. This maximum was ascribed to triaryl-methylium cations.[113] In some systems the ions are finally converted to allyl cations with a broad charge delocalization. These ions (inactive) absorb at 450 nm.

41.4.3 Esters

Pseudocationic polymerization assumes propagation involving esters as active chain carriers. It was found, however, that alone even the strongest acetates (trifluoroacetates) cannot polymerize styrene.[66] In a number of systems esters were detected spectroscopically. ^{1}H NMR was used for trifluoroacetates,[63] trichloroacetates[98] and perchlorates.[92] ^{19}F NMR was used for trifluoroacetates as well.[66] The ^{1}H NMR absorption of the methine proton in the macromolecular ester (triplet at 5.30 p.p.m.) shifts upfield with respect to the starting 1-phenylethyl perchlorate (quartet at 5.83

p.p.m.).[92] This is not due to the change in the electron density at the α-C atom but was ascribed to diamagnetic shielding by the neighboring aromatic nucleus. A similar effect was found for the corresponding chlorides and trifluoroacetates. Triflates, esters of the strongest protonic acid, were too unstable to be observed by NMR even at $-70\,^\circ$C in CH_2Cl_2.[95] Esters and halides have very weak UV absorptions and were not studied by this method.

41.4.4 Activated Esters

Pseudocationic polymerization might also involve covalent species activated by protonic or Lewis acids. If these acids polymerize styrene it will be practically impossible to distinguish between covalent and ionic polymerization. However, styrene cannot be polymerized by BCl_3 alone. This Lewis acid forms complexes with acetates. Complexes with less nucleophilic haloacetates are much less stable and decompose even at low temperatures. Complexes exchange with the excess ester leading to the line broadening of the corresponding signals at $-30\,^\circ$C. Simultaneously the complex decomposes slowly to form 1-phenylethyl chloride as illustrated in equation (9).[98]

$$\underset{\underset{Ph}{|}\underset{Me}{|}}{Me-CH-O-C=O} + BCl_3 \rightleftarrows \underset{\underset{Ph}{|}\underset{Me}{|}}{Me-CH-O-C=O} \cdot BCl_3 \rightarrow \underset{\underset{Ph}{|}}{Me-CH-Cl} + \underset{\underset{O}{\|}}{BCl_2-O-C-Me} \qquad (9)$$

Interactions of esters with protonic acids led to the preferential incorporation of the 1-phenylethyl trifluoroacetate activated by the trifluoroacetic acid. However, no complex between the ester and the acid was observed by NMR.[97] This indicates weak complexation and fast exchange between the free and complexed ester.

41.5 ELEMENTARY REACTIONS

To describe the mechanism of a polymerization process and to control the properties of the final polymer it is necessary to know the chemistry and kinetics of the elementary reactions involved in the polymerization. This requires knowledge of the structure and concentration of different active species and monomer, the structure of the final product, the microstructure of the polymer chain, the structures of the end groups and any isomerized units, and the rate constants involved. At present we have some information on the structure of the active species in the cationic polymerization of styrenes but we do not know precisely their concentrations and reactivities. We do not know whether the monomer is complexed (*e.g.* by cations) before addition to the growing center and whether such a complexed monomer molecule can react with the growing species, or if only an external molecule of monomer has sufficient kinetic energy to propagate. We are not even sure what the structure of the initiator is — does some Lewis acid react alone or with the coinitiator, what is the state of association of some protonic acids, *etc.*

In this chapter, a sketch of the polymerization process of styrenes based on different qualitative and quantitative observations will be attempted.

41.5.1 General Scheme

The overall reaction scheme, for apparently the most studied system initiated with protonic acids, can be described by Scheme 3.

We will try to discuss these elementary reactions in more detail, showing the peculiarities of the polymerization of styrenes.

41.5.2 Initiation

There is some information on the chemistry and kinetics of the addition of protonic acids to alkenes and, in particular, to styrenes. A few studies were performed for non-polymerizable monomers such as DPE[131] and in media which trap cations (*e.g.* trifluoroacetic acid/ethanol),[132,133] while some others were carried out directly in polymerization systems.[16,75,109,134]

$$n\text{HA} \rightleftharpoons (\text{HA})_n \text{ (in the simplest case } n=2) \tag{a}$$

$$(\text{HA})_2 + \text{M} \rightleftharpoons \text{M}_1^+, \text{HA}_2^- \tag{b}$$

$$\text{M}_1^+, \text{HA}_2^- + \text{M} \rightleftharpoons \text{M}_2^+, \text{HA}_2^- \tag{c}$$

$$\text{M}_n^+, \text{HA}_2^- + \text{M} \rightleftharpoons \text{M}_{n+1}^+, \text{HA}_2^- \tag{d}$$

$$\text{M}_n^+, \text{HA}_2^- \rightleftharpoons \text{M}_n^+ + \text{HA}_2^- \tag{e}$$

$$\text{M}_n^+, \text{HA}_2^- \rightleftharpoons \text{M}_n^= + (\text{HA})_2 \tag{f}$$

$$\text{M}_n^+, \text{HA}_2^- + \text{M} \rightleftharpoons \text{M}_n^= + \text{M}_1^+, \text{HA}_2^- \tag{g}$$

$$\text{M}_n^+, \text{HA}_2^- \rightleftharpoons \text{M}_n^I + (\text{HA})_2 \tag{h}$$

$$\text{M}_n^+, \text{HA}_2^- \rightleftharpoons \text{M}_{ni}^+ + \text{HA}_2^- \tag{i}$$

$$\text{M}_n^+, \text{HA}_2^- \rightleftharpoons \text{M}_n^E + \text{HA} \tag{j}$$

$$\text{M}_n^E + \text{M} \rightleftharpoons \text{M}_{n+1}^E \tag{k}$$

$$\text{M}_n^+, \text{HA}_2^- + \text{M}_m^= \rightleftharpoons \text{M}_{(n+m)i}^+, \text{HA}_2^- \tag{l}$$

Scheme 3

The kinetics of initiation of polymerization was interpreted on the assumption of a simple bimolecular reaction between non-associated acid and monomer molecules.

The overall orders of the kinetics of initiation should depend on the relative rates of reactions (a) and (b) from Scheme 3 as well as the degree of association of the acid and conjugation of the anions. This may change with solvent and temperature. There is no hard data on the association of strong protonic acids in solvents typical for cationic polymerization. The acids may form cyclic dimers or open dimers in a way similar to carboxylic acids as shown by structures (9) and (10).[135,136] The dimers can have different reactivities, the latter being much more reactive; it may be considered as the only active form.

(9)   (10)

A little more is known about the conjugation of the anions with acid molecules. This conjugation was initially proposed for carboxylic acids and carboxylate anions, and was later observed for phenols as well as the perchlorate anion and perchloric acid.[137,138,139] The conjugation of the triflate anion with one or two molecules of triflic acid was detected by ^{19}F and ^{1}H NMR[140] and confirmed by the 33 to 50% yields of carbocations found in reactions between the acid and alkenes as well as in the formation of carbocation from carbinols.[141] Of course, the degree of conjugation will depend on the ratio of the concentrations of acid to monomer and for a very large excess of the monomer lower conjugation is expected. Anions are quite basic and they complex with acid molecules much more strongly than alkenes do. In the presence of more basic monomers such as pMOαMS or 3-phenylindene[115,140] the structure of the anions is $\text{CF}_3\text{SO}_3\text{H} \cdot \text{CF}_3\text{SO}_3^-$. For DPE and less basic monomers,[140] one more acid molecule is complexed and the structure changes to $(\text{CF}_3\text{SO}_3\text{H})_2 \cdot \text{CF}_3\text{SO}_3^-$. The larger aggregates are probably not stable.

The rate constants of initiation for some systems are summarized in Table 2. The highest rate constants were reported for strong protonic acids, the lowest for stable carbenium salts. Most reactive were basic monomers such as pMOS, and stabilized by resonance such as DPE and αMS. The stopped flow kinetic data for the protonation of styrenes by strong protonic acids (triflic and perchloric) apparently indicate first order with respect to the acid and the monomer. However, the change in concentration was usually limited to less than one order of magnitude and the results are not very reliable. Earlier studies indicated higher kinetic orders in acid (from two up to four).[142,143,144] In order to explain the observed second orders in both perchloric acid and 1,3-diphenylbutene-1 (U, unsaturated styrene dimer; $\text{M}_2^=$, species from Scheme 3), Scheme 4 was proposed[145] (this scheme has been slightly modified to account for the dimerization of the acid prior to the reaction with U). If both equilibria are rapidly established, the indan (I) formation will be the rate determining step (equation 10). (The chemistry of this reaction is discussed in detail in Secton 41.5.4).

Table 2 Rate Constants of Initiation in the Cationic Polymerization of Styrenes

Monomer	Initiatior	Solvent	Temperature (°C)	Rate Constant[1]	Comments	Ref.
S	$Ph_3C^+SnCl_5^-$	$C_2H_4Cl_2$	30	5×10^{-2}	2	a
S	$Ph_3C^+SbCl_6^-$	CH_2Cl_2	30	9×10^{-4}	3	b
S	CF_3SO_3H	$C_2H_4Cl_2$	30	2×10^1	4	c
S	CF_3CO_2H	CH_2Cl_2	40	1×10^{-2}	5	d
S	CF_3CO_2H	CF_3CO_2H	25	1.3×10^{-3}	6	e
S	CF_3CO_2H	CCl_4	25	1.0×10^{-5}	6	f
αMS	CF_3SO_3H	$C_2H_4Cl_2$	30	2×10^3	4	g
αMS	H_2SO_4	$C_2H_4Cl_2$	30	2×10^3	4	g
αMS	$Ph_3C^+SnCl_5^-$	$C_2H_4Cl_2$	30	2×10^{-2}	2	a
αMS	$BuOTiCl_3$	CH_2Cl_2	−30	1.3×10^3	2	h
βMS	CF_3CO_2H	CF_3CO_2H	25	6×10^{-5}	6	e
pMOS	$MeCOClO_3$	$C_2H_4Cl_2$	30	10^5	3, 4	i, j
pMOS	CF_3SO_3H	$C_2H_4Cl_2$	30	5×10^4	3, 4	i, j
pMOS	$MeSO_3H$	$C_2H_4Cl_2$	30	0.64	4	k
pMOS	I_2	$C_2H_4Cl_2$	30	10^{-1}	4, 6	k
pMOS	$SnCl_4$	$C_2H_4Cl_2$	30	10^1	4, 6	k
pMOS	$BF_3 \cdot OEt_2$	$C_2H_4Cl_2$	30	10^1	4, 6	k
pMOS	$Ph_3C^+SbCl_6^-$	CH_2Cl_2	25	0.89	2	l
DPE	CF_3SO_3H	$C_2H_4Cl_2$	30	3.4×10^3	4	m
DPE	$AlCl_3$	CH_2Cl_2	−30	2.5×10^{-2}	2	n
TPB[7]	CF_3SO_3H	CH_2Cl_2	−71	0.67	2, 8	o
1-PhCH[9]	Ph_2CH^+	$C_2H_4Cl_2$	25	7×10^7	10	p
pPαMS	CF_3SO_3H	CH_2Cl_2	−20	$>5 \times 10^3$	6, 8	q
3,4DMOS	Ph_3C^+	CH_2Cl_2	25	0.4	6	r
IN	$Ph_3C^+SnCl_5^-$	CH_2Cl_2	20	6×10^{-4}	2	s

[1] In $mol^{-1} l s^{-1}$. [2] UV Studies. [3] Conductivity. [4] Stopped flow. [5] In $mol^{-4} l^4 s^{-1}$. [6] Assumed bimolecular reaction. [7] Dimer of DPE. [8] Calculated from the experimental data. [9] 1-Phenylcyclohexene. [10] Pulsed radiolysis.

[a] T. Higashimura, T. Fukushima and S. Okamura, *J. Macromol Sci., Chem.*, 1967, **1**, 683.
[b] A. F. Johnson and D. A. Pearce, *J. Polym. Sci., Polym. Symp.*, 1976, **56**, 57.
[c] T. Kunitake and K. Takarabe, *Macromolecules*, 1979, **12**, 1061.
[d] W. Obrecht and P. H. Plesch, *Makromol. Chem.*, 1981, **182**, 1459.
[e] A. D. Allen, M. Rosenbaum, N. O. L. Seto and T. T. Tidwell, *J. Org. Chem.*, 1982, **47**, 4234.
[f] A. D. Allen and T. T. Tidwell, *J. Am. Chem. Soc.*, 1982, **104**, 3145.
[g] K. Takarabe and T. Kunitake, *Makromol. Chem.*, 1981, **182**, 1587.
[h] M. Villesange, G. Sauvet, J. P. Vairon and P. Sigwalt, *J. Macromol. Sci., Chem.*, 1977, **A11**, 391.
[i] M. Sawamoto and T. Higashimura, *Macromolecules*, 1979, **12**, 581.
[j] M. Sawamoto, T. Higashimura, A. Enokida and T. Okubo, *Polym. Bull.*, 1980, **2**, 309.
[k] M. Sawamoto and T. Higashimura, *Macromolecules*, 1978, **11**, 328.
[l] B. Cotrel, G. Sauvet, J. P. Vairon and P. Sigwalt, *Macromolecules*, 1976, **9**, 931.
[m] K. Takarabe and T. Kunitake, *Polym. J.*, 1980, **12**, 239.
[n] M. Masure, G. Sauvet and P. Sigwalt, *J. Polym. Sci., Polym. Chem. Ed.*, 1978, **16**, 3065.
[o] D. Souverain, A. Leborgne, G. Sauvet and P. Sigwalt, *Eur. Polym. J.*, 1980, **16**, 861.
[p] Y. Wang and L. M. Dorfman, *Macromolecules*, 1980, **13**, 63.
[q] D. Teyssie, M. Villesange and J. P. Vairon, *Polym. Bull.*, 1984, **11**, 459.
[r] J. M. Rooney, *Polym. Bull.*, 1983, **10**, 414.
[s] F. Subira, A. Polton and P. Sigwalt, *Int. Symp. Cationic Polym., Rouen, 1973*, **C36**.

$$2HA \overset{K_1}{\rightleftharpoons} (HA)_2$$

$$(HA)_2 + U \overset{K_2}{\rightleftharpoons} HU^+, HA_2^-$$

$$HU^+, HA_2^- + U \overset{k_3}{\rightarrow} I + HU^+, HA_2^-$$

Scheme 4

$$\text{Rate} = \frac{d[I]}{dt} = k_3[U][HU^+, HA_2^-] = k_3 K_2 K_1 [U]^2 [HA]^2 \qquad (10)$$

The conjugation of the acid with the anion was observed in model systems and also deduced from stopped flow experiments on the polymerization of styrenes with triflic acid. Self association of the acid molecules is relatively weak in these experiments. At the usually employed concentration range

(10^{-4} mol l^{-1}), even with an equilibrium constant of dimerization assumed to be as large as $K_1 = 10^3$ mol^{-1} l (measured in acetonitrile at room temperature;[146] however, the equilibrium constant for the dimerization of CF_3CO_2H[147] in CH_2Cl_2 at 35 °C is only $K_1 = 2.1$ mol^{-1} l), only 10% of the acid will be in the form of dimers. At higher concentrations of acid the proportion of dimer increases.

Thus, kinetics can be either first or second order (or higher if larger aggregates are involved) with respect to acid, depending on the relative values of the rate constants of the first and second reactions in Scheme 4. For example, in the polymerization of S with trifluoroacetic acid, third external order was found, which might be in agreement with the formation of aggregates of three acid molecules.[65] The order depends on the solvent, and in more polar solvents the order of the addition of e.g. hydrogen halide to alkenes changes from third to second.[144]

Qualitatively, the rate of initiation increases with the strength of the acid: $CF_3CO_2H < MeSO_3H < H_2SO_4 < HClO_4 < CF_3SO_3H$, with the nucleophilicity (or rather basicity) of the monomer: S < αMS < DPE, pMOS, and with the polarity of the reaction mixture: $CCl_4 < C_2H_4Cl_2$, $CH_2Cl_2 < MeNO_2$.[134, 148, 149] Initiation proceeds *via* a transition state more polar than the ground state. Thus, the reaction accelerates in solvents of higher dielectric constants.

The addition of protonic acids as well as halides to α,β-disubstituted alkenes proceeds predominantly with Z stereochemistry. This suggests the complexation of the anion by one or two acid molecules, which at a later stage add from the same side of the half space to the intermediate cation (equation 11). The proportion of Z addition is usually above 70%. Originally, π complexes were proposed as reactive intermediates[150] but careful studies of the rates and stereochemistry of hydration showed that the reaction proceeds by an Ad_E2 or Ad_E3 mechanism.

$$ \tag{11} $$

When relatively weak acids are used, protonation is irreversible. Hence hydration of β,β-dideuterostyrene proceeded without loss of deuterons at partial conversion when acetic acid was used alone.[151, 152] The same is true for the hydration of 2-methyl-1-butene which does not isomerize to more stable 2-methyl-2-butene.[153] However, when stronger acids are used, such as triflic acid, the protonaton becomes reversible. For example, the addition of acetic acid to *cis* and *trans* 2-butene in the presence of triflic acid proceeded with isomerization and hydrogen–deuterium exchange.[154] In the oligomerization of the *endo* isobutene dimer, initial isomerization to the more stable *exo* isomer was observed.[155] These results, as well as a recent reinvestigation of the polymerization of styrene with triflic acid, indicate that the protonation of alkenes with strong acids is reversible.[95] The triflate anion may not be *nucleophilic enough* to form a covalent bond with a tertiary or secondary (with an adjacent phenyl group) carbon atom, but it may be *basic enough* to abstract a β-hydrogen atom having a considerable positive charge, shown by equation (12).

$$ \tag{12} $$

Initiation with stable carbenium ions is usually slow (more than 10^5 times slower than k_p) because of steric hindrances. With αMS it proceeds not by addition but by a hydride transfer process and subsequent protonation.

Initiation with Lewis acids alone is also slow and the mechanism still uncertain. It may involve the direct addition of a Lewis acid (in neutral or ionized form) or it may require the presence of a protonogenic or cationogenic additive (see Chapter 39).

41.5.3 Propagation

41.5.3.1 *Ions and ion pairs*

Propagation is the reaction between a monomer and a growing species with the subsequent reformation of the same active species (equation 13). The active center 'z' may be ionic or covalent and may be complexed with a counterion, solvent or a monomer molecule. Formally, the complexation state before and after addition should be the same. Rate constants of propagation for a first and a macromolecular growing species may be different (equations b, c in Scheme 3).

$$M_n^z + M \rightleftarrows M_{n+1}^z \tag{13}$$

The most general scheme of propagation involving four types of species (covalent, contact and separated ion pairs, and ions) was shown in Scheme 2.

In such systems the apparent propagation rate constant is the sum of the products of the reactivities (k_p^i) and proportions (α^i) of all of the species involved:

$$\bar{k}_p = \sum_{i=1}^{\infty} (\alpha_i k_p^i) \tag{14}$$

We discussed previously a possible approach for determining the proportion of each of the species (*cf.* equations d, e, k in Scheme 3). For example, in the polymerization of styrenes with strong acids studied by stopped flow UV measurements, the total concentration of propagating ionic species was estimated by means of the extinction coefficients derived from the model compounds. Then the measurements of the overall rate of ionic propagation (with the assumption of the complete inactivity of the covalent species) as a function of the degree of ionization might lead to the absolute values of the rate constants of propagation on ions and ion pairs.[156] This approach was successfully used for the polymerization of heterocycles[157,158] but its transfer to the polymerization of styrene gave less reliable data. Namely, dissociation constants K_D of paired ions are known neither for model nor growing species. K_D was estimated for the polymerization of styrene initiated by triflic acid in the presence of different amounts of a salt with a common anion (Bu_4N^+, $CF_3SO_3^-$).[156] The rate constants k_p^+ and $k_p^\pm$ were calculated assuming a simple mass law action of triflate anion. However, we know that anions homoconjugate with acid molecules and the structure of complexed and 'free' anions is different, hence the simple second salt effect cannot be applied. No homoconjugation was found for systems with heterocyclic monomers.

Moreover, the addition of the salt with triflate anion to the acid solution will lead to the trapping of the acid to form the complexed anion, which decreases the proportion of the acid capable of initiation. Therefore, the observed decrease of the rate in the presence of the salt has two origins: first, the reduction of the concentration of the acid, and second, the shift of the equilibrium between ions and ion pairs to the side of the ion pairs. The overall results published in ref. 156 indicate that the cations' reactivity is 6 to 24 times higher than the reactivity of the ion pairs. The analysis of the rate constants did not take into account the complexation of the acid by anion. Thus, because the overall decrease of the rate of polymerization was partially due to the complexation, the ratio of the reactivities of unpaired to paired ions should be lower than calculated in ref. 156.

The activation parameters calculated in ref. 156 were quite unexpected. The energy of activation of propagation on ions was as high as $56\,kJ\,mol^{-1}$. For ion pairs a more reasonable value ($E = 20\,kJ\,mol^{-1}$) was given. Thus, at temperatures below $-30\,°C$ ion pairs should have been more reactive than ions. We believe that the origin of the above discrepancies lies in the complexation of the acids by the anions. Namely, the energy of activation of propagation by unpaired ions is a composite value (the real energy of activation and *e.g.* the enthalpy of the complexation). The same is true for the activation entropy of propagation by ions which was calculated to be $\Delta S^{\neq} = 40\,J\,mol^{-1}\,K^{-1}$, a value undoubtedly too high for a bimolecular reaction in which three degrees of translational freedom are lost (propagation by ion pairs gives the expected value $\Delta S^{\neq} = -88\,J\,mol^{-1}\,K^{-1}$).

Although we may doubt the interpretation of the kinetic data and the precision of the calculations of the rate constants, we still have a very solid conclusion that the reactivities of unpaired and paired ions are similar. Recent kinetic results on the addition of alkenes to benzhydryl cations showed that k^+ was, within 10% error, the same as $k^{\pm}$.[130]

In contrast to the anionic polymerization of vinyl monomers (where reactivities of ions are sometimes 10^5 times higher than reactivities of ion pairs), in cationic systems the reactivities of ions and ion pairs are comparable. In heterocyclic systems similar behavior was observed. Moreover, in the polymerization of cyclic ethers, tetrahydrofuran or oxepane, as well as in the polymerization of cyclic amines, equal reactivities of ions and ion pairs were found.[157,158]

What are the similarities and differences between the cationic polymerization of styrenes, heterocycles and the anionic polymerization of vinyl monomers? We discuss them below.

(i) *Counterions.* In anionic polymerization very small counterions such as Li^+ or Na^+ are used. They can interact very strongly with growing species, forming tight contact ion pairs. The Coulombic interactions are strong due to the small ionic radii (less than 1 Å). On the other hand, the counterions in cationic polymerization are quite large. Even the smallest F^- anion (1.36 Å) is still larger than the K^+ cation (1.33 Å). In cationic polymerization counterions usually have radii larger than 3 Å. As a typical anion we may consider $CF_3SO_3^-$ (or even larger species in the homo-conjugated form $CF_3SO_3 \cdot CF_3SO_3H^-$), $SbCl_6^-$, SbF_6^- or $Sb_2F_{11}^-$, etc.[159, 160] In all of these anions the charge is delocalized and interactions with a cation are weak.

(ii) *Growing species.* The steric structures of growing carbocations and carbanions may be similar. The effective ionic radii can be quite small for cations derived from styrene and probably higher for those from α-substituted styrenes. The carbenium ions are nearly planar. The distance from the edge of the anion to the positively charged carbenium center can be as low as 1 Å. In comparison, onium ions are larger.[158] The ion pairs in the cationic polymerization of vinyl monomers may be intermediate in size, between very loose pairs typical for the cationic polymerization of heterocycles, and tight pairs in anionic polymerization. The electrostatic interactions influencing charge distribution and reactivity of ionic growing species decrease with the second power of the interionic distance. This is in agreement with the observed ratio $\rho = k_p^+/k_p^\pm$. For the onium ions ρ is close to 1,[158, 160] for carbanions (with Na^+, or Li^+) $\rho > 10^{3, 161, 162}$ and for cationic polymerization of styrenes $1 < \rho < 10$.[156]

We discussed the relative reactivities of ions and ion pairs without commenting on the reported values of the average rate constants of ionic propagation (Table 3). The rate constants of propagation on ions estimated from the γ radiation experiments are one order of magnitude higher than the rate constants calculated from experiments with chemical initiation. The calculations of the propagation rate constants in systems initiated chemically were based on the extinction coefficients determined for model compounds ($\varepsilon = 10^4\ l\,mol^{-1}\,cm^{-1}$) and on the assumption that at the range ≈ 350 nm only growing ions absorb. However, if the extinction coefficients were three times larger (*cf.* Table 1) and in addition to the growing ions the isomerized ions were present (*cf.* Section 41.5.6), then the k_p would be a few times larger and closer to k_p^+ calculated in the γ radiation experiments.

Values calculated for the polymerization in $C_6H_5NO_2$ are 10^3 times lower than those in less polar CH_2Cl_2. The rate constant of the reaction between ion and molecule should decrease with the increase of the dielectric constant because the transition state is less polar than the ground state and

Table 3 Rate Constants of Propagation in Cationic Polymerization of Styrenes

Monomer	Initiator	Solvent	Temperature (°C)	Rate Constant[1]	Comments	Ref.
S	$PhCO^+SbF_6^-$	$C_6H_5NO_2$	25	1.9×10^2	2	a
S	$HClO_4$	CH_2Cl_2	-80	5×10^3	3, 2	b
S	$HClO_4$	CH_2Cl_2	25	10^5	3, 2	b
S	CF_3SO_3H	$C_2H_4Cl_2$	30	2×10^5	3	c
S	γ radiation	—	15	3×10^6		d
αMS	γ radiation	—	0	4×10^6		e
αMS	CF_3SO_3H	$C_2H_4Cl_2$	30	3×10^4	3, 4, 5	f
αMS	H_2SO_4	$C_2H_4Cl_2$	30	$>10^6$	3, 4	f
αMS	$BuOTiCl_3$	CH_2Cl_2	-70	2×10^4		g
pMOS	$MeCOClO_4$	$C_2H_4Cl_2$	30	10^5		h
pMOS	CF_3SO_3H	$C_2H_4Cl_2$	30	10^5		h
pMOS	$MeSO_3H$	$C_2H_4Cl_2$	30	2×10^4		h
pMOS	Ph_3C^+	CH_2Cl_2	10	4×10^4		i
pPαMS	CF_3SO_3H	CH_2Cl_2	-58	8×10^3	5	j

[1] In $mol^{-1}\,l\,s^{-1}$. [2] Assumed to be k_p^+, probably underestimated. [3] Based on the assumed $\varepsilon = 10^4\ mol^{-1}\,l\,cm^{-1}$. [4] No propagation was possible. [5] Probably underestimated due to slow transfer.

[a] G. E. Holdcroft and P. H. Plesch, *Makromol. Chem.*, 1984, **185**, 27.
[b] J. P. Lorimer and D. C. Pepper, *Proc. R. Soc. London, Ser. A*, 1976, **351**, 551.
[c] T. Kunitake and K. Takarabe, *Macromolecules*, 1979, **12**, 1061.
[d] F. Williams, K. Hayashi, K. Ueno and S. Okamura, *Trans. Faraday Soc.*, 1967, **63**, 1501.
[e] E. Hubman, R. B. Taylor and F. Williams, *Trans. Faraday Soc.*, 1966, **62**, 88.
[f] K. Takarabe and T. Kunitake, *Makromol. Chem.*, 1981, **182**, 1587.
[g] M. Villesange, G. Sauvet, J. P. Vairon and P. Sigwalt, *J. Macromol. Sci., Chem.*, 1977, **A11**, 391.
[h] M. Sawamoto and T. Higashimura, *Macromolecules*, 1979, **12**, 581.
[i] B. Cotrel, G. Sauvet, J. P. Vairon and P. Sigwalt, *Macromolecules*, 1976, **9**, 931.
[j] D. Teyssie, M. Villesange and J. P. Vairon, *Polym. Bull.*, 1984, **11**, 459.

desolvation should favor less polar solvents. The thousandfold decrease of k_p may also be ascribed to the reversible deactivation of the carbenium ions by the nitro group from the solvent.

The reactivity of loose ion pairs is usually not influenced by the structure of the counterion. The observed differences may rather be ascribed to the different proportions of growing species. The higher value of the rate constant of propagation in the polymerization of αMS with H_2SO_4 than with CF_3SO_3H is even more surprising because of the larger size and better charge delocalization in the anion derived from the latter acid. The reaction conditions described in ref. 163 indicate that no high polymer could be formed ($[M]_0 < [M]_e$). Therefore, after initiation and formation of a dimer, the consumption of the monomer can proceed only by the transfer of the proton to a new monomer molecule as shown in Scheme 5. If transfer is much slower than propagation, then the overall rate constant of consumption of monomer will be equal to k_{tr}. This may well happen for the triflate anion. With sulfuric acid, k'_{tr}, the rate constant of transfer to counterion (more basic), may be much larger and (at the larger excess of the acid) propagation may become the rate limiting step. This can explain the unexpectedly much larger apparent rate constant of propagation found with H_2SO_4. This rate constant is close to the values found by γ radiation.

$$HA + M \overset{k_1}{\rightleftarrows} M_1^+, A^-$$

$$M_1^+, A^- + M \overset{k_p}{\rightleftarrows} M_2^+, A^-$$

$$M_2^+, A^- + M \overset{k_{tr}}{\rightleftarrows} M_2^= + M_1^+, A^-$$

$$M_2^+, A^- \overset{k'_{tr}}{\rightleftarrows} M_2^= + HA$$

Scheme 5

The same can be true for $pP\alpha MS$ which was also polymerized below $[M]_e$.[164] According to the original article 'the propagation rate observed at higher $[M]$ is larger and deviates significantly from linearity.' Under these conditions, in addition to the usually observed mixture of dimer and trimer, higher oligomers were also formed and transfer was no longer the rate determining step.

We may conclude that the reactivities of ions and ion pairs in the propagation of styrenes are similar. The values of the rate constants range from 10^5 to 10^6 mol^{-1} l s^{-1} at room temperature (energies of activation are low, $E < 20$ kJ mol^{-1}). Reactivities of monomers and growing species are also discussed in Section 41.6.

41.5.3.2 *Stereochemistry of propagation*

The stereochemistry of the propagation step could be studied by the overall chain microstructure or, in more detail, at the model level for β-substituted styrenes. Comparison of propagation with the electrophilic addition reactions requires application of the correct models of carbenium ions, because the hydration and addition of halides proceeds by a different mechanism (*e.g.* bromonium ions). Very recent studies of the stereochemistry of the addition of diphenylmethylium cations in the presence of $ZnCl_2/Et_2O$ to *cis* and *trans* β-methylstyrenes show the very high E/Z ratio (ρ) for the *trans* isomer ($\rho_t = 62$) and low ratio for the *cis* isomer ($\rho_c = 0.24$).[165] This was explained by weakly bridged intermediates. Although in the original paper bridging in the three-membered cycle was suggested, we think that complexation of the cation with the adjacent aromatic group is also probable.[119] Similar complexes have already been described in the discussion of the UV spectra of dimeric cations (*cf*. Section 41.4.1).

Stereochemical studies indicate that the life time of the carbenium ion derived from the *cis* alkene is long enough for a conversion to a less sterically hindered isomer and for eventual recombination with the chloride, which produces a Z adduct. The *trans* isomer hardly interconverts to the more hindered isomer and gives a high proportion of the E product. Because $\rho_t \neq 1/\rho_c$, the thermodynamically most favorable adduct ratio is not formed, which indicates that either the complexation (or bridging) prevents free rotation in the carbenium ion, or that recombination occurs before equilibrium is reached.

The microstructure of polystyrenes also provides information on the stereochemistry of propagation. The stereoregularity of monomer addition is very low in the polymerization of S ($P_r = 0.63$).[166]

However, in the polymerization of different *p*-substituted αMS the tendency to form syndiotactic triads is quite high (up to 92%).[167, 168] This tendency depends on the structure of the *p*-substituent, the temperature, and very strongly on the polarity of the solvent; it is usually not influenced by the structure of the counterion.

The classical reaction scheme accounting for the observed tacticity is based on the assumption of front-side attack in more polar solvents and back-side attack in less polar solvents (equation 15).[169] The high proportion of front-side attack in CH_2Cl_2 solvent is surprising. What prevents the back-side attack? The substituents at *β*-C atoms seem to be too far from the approaching monomer to hinder the back-side attack. The explanation could be based again on interactions with an adjacent aromatic ring. These interactions would not allow the free approach of a monomer molecule from the side opposite the anion. Thus, instead of the extended zigzag conformation, the real chain is perhaps twisted to maximize the interactions between the *γ* aromatic ring and the carbenium center. The presence of the aromatic ring in the immediate neighborhood and the relatively weak interactions with the anion may facilitate the front attack which apparently requires an unfavorable separation of counterions. In more polar solvents (CH_2Cl_2) the front attack is the preferred pathway. In mixtures with hexanes (lower polarity) the separation of ions becomes more difficult and the decomplexation can be faster, which favors the back-side attack.

$$\text{(15)}$$

41.5.3.3 *Complexation of ions*

Carbenium ions are extremely reactive species that need to be stabilized in one way or another. Thus, they can interact with anion, solvent, polymer and monomer. The first interaction may lead to the formation of an ion pair or a covalent species (active or not). The second will depend on the nucleophilicity of the solvent; even weakly nucleophilic solvents such as CH_2Cl_2 or SO_2 can interact with carbenium ions and Lewis acids.[170, 171] The interactions with polymer may be either intramolecular or intermolecular. The polymerization of monomers with nucleophilic groups (*e.g.* OMe) may lead to the formation of oxonium ions. These ions were detected in model systems[172] but were usually neglected when the real polymerization was discussed. In addition to the typical nucleophilic groups, polystyrenes possess aromatic rings which are known to form complexes with electrophiles.[173, 174, 175, 176] Interactions with the remote group are especially important due to high neighboring group participation (local concentration or anchimeric assistance).[177]

Eventually, cations can interact with monomers. These interactions will lead to propagation, transfer or complexation. Although these considerations may be treated as purely theoretical, there is recent evidence which supports the participation of the distinct species solvated by CH_2Cl_2, polymer and monomer in the polymerization of *p*MOS.[129] Namely, a tri-modal MWD was correlated with different proportions of the discussed compounds. The most striking conclusion is that the life times of these species had to be long enough to build the entire polymer chain (≈ 1 s) before they could exchange. Otherwise broadening of the MWD, not polymodality, would be observed. High yield of linear polymer in cyclopolymerization of α,ω-bis(4-vinylphenyl)alkenes also confirms intramolecular interactions.[197]

41.5.3.4 *Covalent propagation*

Carbenium ions in the polymerization of styrenes seem to be so reactive that they will interact with any nucleophilic reagent present in the system. Interactions with a counterion may lead either to the formation of a contact ion pair, to transfer of the *β*-H atom, or to termination, by formation of the inactive covalent species. In some systems these species, later observed directly by NMR[92] (*cf.* Section 41.4.3), were claimed to be active.[4, 70]

Covalent species may react with monomer *via* multicenter rearrangement or through intermediate ionic species.[178] For the first case the theoretical considerations of covalent propagation in terms of orbital symmetry theory[179] and parity rules[180] reject the four-membered transition state but allow the six-membered transition state (equation 16). Reactions with an even number of electron pairs involved (two for the four-membered transition state) should have antarafacial stereochemistry

$$-CH_2-CH-O\cdots Cl=O \quad \times\!\!\longrightarrow \quad -CH_2-CH \quad O-Cl=O$$

with Ph, CH$_2$=CH, O, Ph substituents   (16)

(first scheme structures)

(second scheme)

(inversion of configuration and *trans* addition). Reactions involving an odd number of electron pairs (three for the six-membered transition state) should proceed with suprafacial stereochemistry (retention and *cis* addition). Covalent multicenter addition requires *cis* addition and retention of configuration. This is only possible in the six-membered transition state.[178]

Thus, covalent propagation cannot proceed with halide anions unless they are additionally activated. This is in agreement with recent results reported in the polymerization of vinyl ethers.[181,182] There are also possible two other mechanisms of propagation which involve covalent species. The first resembles electrophilic addition to alkenes (*e.g.* bromination or hydration), proceeding *via* intermediate carbenium ions (equation 17). The second involves preliminary intra-molecular ionization of the covalent species, true ionic propagation, and the recombination of counterions back to the covalent species (equation 18). The last mechanism is purely ionic, although a majority of the growing chain ends have covalent structures.

$$-CH_2-CH-OClO_3 + CH_2=CH \rightleftharpoons -CH_2-CH \cdots \overset{CH_2}{\underset{CH}{\|}} {}^-ClO_4 \rightleftharpoons -CH_2-CH-CH_2-CHOClO_3$$

(with Ph substituents)   (17)

$$-CH_2-CHOClO_3 \rightleftharpoons -CH_2-CH^+, ClO_4^- \xrightarrow{nM,\, k_p^i} \{CH_2-CH\}_n CH_2-CH^+, ClO_4^-$$

(with Ph substituents)

$$\rightleftharpoons \{CH_2-CH\}_n CH_2-CHOClO_3$$

(with Ph substituents)   (18)

As we discussed in Section 41.4.3, the covalent species were observed in the polymerization of styrene with perchloric acid and with different haloacetic acids. The activity of these esters has not been established, however. Quite often these esters require the presence of a strong electrophile to react with a monomer. The activated covalent species can propagate *via* a multicenter mechanism (equation 16) or through an ionic intermediate in a reaction similar to equation (18). An example of this system is the polymerization of S in the presence of 1-phenylethyl trifluoroacetate and trifluoroacetic acid. When deuterated S was used, two types of end groups were found. One was formed by the direct addition of the acid to S d_8 and the other by the addition of the activated ester (ester alone is inactive),[66] as in equation (19). In spite of the excess acid, the end groups of the second type prevail (>90%). The reaction with the optically active ester was accompanied by a racemization of the ester that was faster than the incorporation of this ester into the polymer chain. In covalent propagation proceeding *via* multicenter rearrangement, the optically active 1-phenylethyl ester should yield an optically active 1-phenylethyl end groups. The complete loss of activity found in these systems suggests the presence of ionic intermediates which racemize before the reaction with monomer. Because carbenium ions are very reactive, monomer, at least in this system, may be consumed exclusively in ionic polymerization although nearly all species have covalent structures.[97]

$$HOC(O)CF_3 + CD_2=CD \rightarrow HCD_2-CD-$$

(with Ph substituents)   (19)

$$Me-CH-O-C-CF_3 + CD_2=CD \rightarrow Me-CH-CD_2-CD-$$

(with Ph, O·HA, Ph, Ph, Ph substituents)

Perchlorates are much more reactive than trifluoroacetates and the chance of direct reaction between monomer and covalent species is higher. The bimodal molecular weight distributions were explained by the contribution of ionic species (high MW fraction) and covalent species (low MW fraction).[4] The high MW fraction was completely suppressed by the addition of the salt with the common anion, indicating that this fraction was formed by the free ionic species.[11,183] The low molecular weight polymer could be formed by either ion pairs or covalent esters. There is no proof that an ester reacts with a monomer directly. It may also ionize prior to a monomer addition (equation 18). The correlation of the rates with proportions of ions may indicate some contribution from the species which do not absorb in the UV, *i.e.* covalent esters.[184] Reliable data on the rate constants of covalent propagation are not available. Unfortunately, preparation of optically active perchlorate was unsuccessful and studies similar to those for trifluoroacetate could not be done.

Bimodality of the MWD was also observed for polymerization initiated by triflic acid.[185] This was also initially explained by the simultaneous propagation on ionic and covalent species.[4] However, taking into account very low proportions of ions (10^{-7} mol l^{-1}), their high reactivities ($k_p \approx 10^6$ mol^{-1} l s^{-1}), and diffusion controlled association ($k_a \approx 10^{10}$ mol^{-1} l s^{-1}), the high MW fraction ($DP_n = 10^3$ at $[M]_0 = 1$ mol l^{-1}) can be formed exclusively on ions and the low MW fraction on ion pairs.[95]

Thus, the activity of covalent species in the polymerization of styrenes is still an open question and the acceptance of the pseudocationic propagation requires some additional experimental proof.

41.5.4 Transfer

Semiempirical calculations indicate that in carbenium ions a considerable amount of charge is present on the β-hydrogen atoms.[186] Hence, in addition to the carbenium center, these atoms can be attacked by a basic monomer, by the anion, or the solvent. The result of these reactions is the formation of an unsaturated end group and a protonated monomer molecule (*i.e.* a new growing species), protonic acid, or the protonated solvent. In all these cases the growth of the material chain is terminated. Transfer may or may not decrease the overall propagation rate depending on the reactivity of the new electrophilic species. These reactions correspond to equations (f) and (g) in Scheme 3.

It is not clear what the mechanism of transfer is in the polymerization of styrenes. Does it involve predominantly monomer or counterion? The preparation of very high molecular weight polystyrene by nuclear chemical initiation (in the absence of any gegenion)[101] could suggest that the direct transfer to monomer is not important, if polymer had been formed only by the cationic mechanism. On the other hand, the oligomerization of S is kinetically first and sometimes even second order in monomer (*cf.* Section 41.5.2 and equation 10). To fit these kinetics the rate-determining step must be either initiation or transfer to monomer.[145] A very interesting approach to the determination of the relative proportion of transfer to monomer and to counterion is based on cationic grafting.[187,188,189] When the backbone is a cationating agent [for example, poly(isobutylene-*co*-chloromethylstyrene)], then in the presence of a Lewis acid (*e.g.* AlEt$_2$Cl, inactive alone), transfer would lead to a homopolymer and without transfer a graft copolymer would be formed. Separation and characterization of the products gives proportions of different transfer reactions. Transfer to monomer as a bimolecular reaction will depend on conversion whereas unimolecular transfer to anion will not. Usually a Schultz–Harboth or modified plot is used.[190,191] The unimolecular transfer may be based on the transfer not only to counterion but to solvent as well.[189]

In the polymerization of αMS at $-70\,^\circ$C, the ratio of transfer to propagation rate constants is $k_{trm}/k_p = 10^{-5}$ and $k_{tru}/k_p = 5 \times 10^{-5}$ mol l^{-1}. Therefore, unimolecular transfer dominates at concentrations of monomer $[M] < 5$ mol l^{-1}. In the polymerization of IN only transfer to monomer was observed. This may be ascribed to the less hindered approach of a molecule of IN to the more exposed active center. In the polymerization of αMS this reaction is more sterically hindered. Earlier work on the polymerization of S by perchloric acid led to similar ratios of k_{trm}/k_p and k_{tru}/k_p.[190] These ratios were strongly dependent on the composition of the reaction mixture.

As we discussed before, aromatic rings interact with carbenium ions by complex formation but they may also react *via* Friedel–Crafts alkylation. The π complex can be considered as the intermediate in this alkylation. As discussed by Olah,[192] the proportion of an outer π complex, the tetracoordinated π complex, and of a σ complex may vary with the structure of the reagents and reaction conditions (see equation (20), where E is the γ-phenylpropyl cation). In some systems π complexes may be so stable that they can be isolated, and in other systems the σ complex can be

$$(20)$$

stable enough to be detected by spectroscopic methods. Nevertheless, the final thermodynamic product is always the indan.[42]

Intermolecular reactions of this kind lead to branched polymers and an intramolecular reaction within the last polymer segment gives indan. The formation of indans is one of the most typical features of the cationic polymerization of styrenes. Indans are preferentially formed at higher temperatures and extremely easily from αMS.[193] Alkylation proceeds with a higher activation energy than propagation. Hence at higher temperatures, especially below the equilibrium concentration of monomer when no high polymer can be formed, dimeric or trimeric cations cyclize to form indans. This reaction proceeds through the formation of the intermediate benzenonium cation and expulsion of the proton which gives rise to a new monomeric cation, and the addition of monomer yields a dimeric cation which cyclizes, *etc.*

The formation of indans slows down in the presence of a *p*-methoxy group which stablizes the cation and also decreases the nucleophilicity of the corresponding C atom in the second ring.[115] Thus, the dimeric cations involved in the polymerization of *p*MOαMS are stable up to $-30\,°C$. Above this temperature indans are slowly formed but, in contrast to the other indans, they are not stable in the presence of protonic acid and are converted quantitatively at room temperature to spirobiindans and anisole (equations 21). The proportion of the unsaturated oligomers and indans depends on the monomer structure, temperature and time. In the polymerization of different α-methylstyrenes at low temperatures, unsaturated oligomers are initially formed when the concentration of the initiator is low. At later stages they can be protonated and can cyclize, forming more stable indans.[194, 195] The same process operates for DPE (equation 22).

$$(21)$$

$$(22)$$

The oligomerization of αMS may lead to either an *exo* or *endo* alkene (equation 23). Apparently the *exo* alkene is formed more rapidly, at least at low acid concentration. Steric considerations may explain the more facile approach of the transfer agent (monomer or counterion) to the H atoms of the methyl than to the methylene group. The *exo* alkenes are later protonated and transformed to more stable *endo* isomers. Styrene can form only *endo* alkene.

$$(23)$$

Recently, the selective preparation of unsaturated and indan oligomers from styrenes and α-methylstyrenes was reviewed.[195] For styrene monomers high temperature, low polar solvent and protonic acid favor the formation of unsaturated dimers with yields as high as 95% in some cases.[196] α-Methylstyrenes form indans under similar conditions. However, as we discussed previously, a nearly quantitative yield of the unsaturated oligomers may be obtained at $[M]_0 < [M]_e$ and at short reaction times, with indan being formed later.

The last aspect of the formation of indans is the isomerization step growth polymerization of diisopropenylbenzenes.[197,198] In this case indan formation is a reaction that couples aromatic rings (equation 24).

(24)

Transfer is the main side reaction in the cationic polymerization of other monomers as well as styrenes. It is the main reaction limiting the formation of high molecular weight polymer *via* a cationic mechanism. Transfer is usually suppressed at low temperatures because it has a higher enthalpy of activation than propagation. For example, the synthesis of poly(isobutylene) with a molecular weight above 200 000 requires temperatures below $-100\,^\circ$C. Is changing the temperature the only possible way to suppress transfer? It seems that strong interactions with the anion (or some additives) may change the charge distribution in the growing species. This may lead to the stronger relative decrease of the reactivity of β-H atoms in comparison with the carbenium center. For example, MNDO calculations show that the charge on the C^+ atom is reduced from 28% for phenylethylium to 21% for 1-phenylethyl acetate, and on the β-H atoms from 9% to 2%.[202] Formation of the covalent species will lead to an overall decrease in the reactivity of the active species. If this decrease is more pronounced for β-H atoms than for the C atom, the contribution of transfer will diminish. Thus, the overall rate of polymerization will decrease, but the drop in the rate of transfer will be more significant than in the rate of propagation.

It has been recently found that this approach (covalent species) in the polymerization of vinyl ethers $(HI + I_2)$[199,200] and of isobutylene (activated acetates)[201] leads to well-defined polymers with narrow molecular weight distributions. With regard to the polymerization of styrenes two approaches are worth noting. One is based on the application of different electron acceptors to stabilize the active species and to decrease their activity to an extent which is larger for transfer than for propagation.[203] Polystyrenes with MWD as low as $\rho = 1.8$ and $M_n = 24\,000$ were prepared. In the second approach, covalent acetates were activated by Lewis and protonic acids. Polystyrenes with molecular weights in the range from $M_n = 2 \times 10^3$ to $M_n = 2 \times 10^4$ and polydispersities as low as $\rho = 1.15$ were synthesized.[98] Better results were obtained for *p*MS and 2,4,6-trimethylstyrene than for S.[202]

41.5.5 Termination

The termination of ionic polymerization usually proceeds either by the recombination of counterions or by the formation of sterically hindered ions with a very broad charge distribution. For example, the indanylium cations, allyl type cations and branched cations are inactive in polymerization of S.

Indanylium cations are usually formed at room temperature (or very slowly at lower temperatures) from indan end groups and acid (see equation 25).[119] Three types of indanylium cations are formed. Cation I_1^+, produced by the expulsion of the aromatic ring, was formed the most rapidly. Indanylium cations are so stable that they decompose slowly at room temperature only in open systems, *i.e.* in the presence of moisture.

UV absorption in the range above 400 nm suggested the formation of cations (**11**) with a large degree of charge delocalization (an allyl type).[120] The absorptions at 500 nm and 650 nm observed at later stages in the polymerization of acenaphthylene were ascribed to a similar type of ions (**12**).[204]

The polymerization of both styrenes and α-methylstyrenes is accompanied by the formation of unsaturated end groups. These macromolecules (macromonomers) may react with growing species but for the thermodynamic (steric) reasons they cannot homopolymerize (*cf.* equation (1) in Scheme 3). Unsaturated dimers and trimers of S and *p*MS were oligomerized in low yield.[51] This system may be considered as the model for consumption of polymeric chains with alkene groups.

Recombination of the counterions is not possible for anionic polymerization where alkali metals are used but in cationic polymerization it is the most important termination process. When anions capable of forming covalent bonds are used, the corresponding ester or halide is formed. The stability of complex anions was studied in detail for trityl cations. Most stable was the anion SbF_6^-, which also gave the best results in the polymerization of styrene;[85] the order of anion stabilities is $SbF_6^- > BF_4^- > SbCl_6^- > PF_6^-$.

Chlorosulfonate and fluorosulfonate anions decompose yielding macromolecular halides and SO_3[76] (see equation 26); in both cases termination is irreversible.

$$-CH_2-CH-OSO_2Cl \rightarrow -CH_2-CH-Cl + SO_3 \qquad (26)$$
$$\quad\quad\;\; Ph \qquad\qquad\qquad Ph$$

Termination also occurs with any nucleophilic impurities in the system such as moisture, alcohols and amines.

41.5.6 Isomerization

One of the possible side reactions in the cationic polymerization of vinyl monomers is isomerization of growing carbenium ions to less reactive and more stable cations. If this reaction is reversible it may lead to a rate decrease and broadening of the molecular weight distribution. If it is irreversible it must be considered as termination.

Carbocations rearrange especially easily when they have exclusively alkyl substituents. A polymer with a completely isomerized structure was formed in the polymerization of 3-methyl-1-butene at low temperatures.[205]

The 1,2-hydride anion transfer should be extremely easy and rapid.[206] However, in the polymerization of S, the formation of the cations which are more stable than 'normal' growing species requires two consecutive 1,2-H⁻ shifts or one 1,3 shift (equation 27).[207]

This reaction should leave an imprint on the chain microstructure. Unfortunately, neither the model reactions nor the microstructure of cationic polystyrenes was investigated in sufficient detail to support or reject this isomerization.

However, in the polymerization of α-methylstyrenes there is a strong indication of the formation of isomerized species. They were observed directly by ^{1}H NMR and by UV in the polymerization of *p*MOαMS.[115] It cannot be at present established whether these cations are formed by the intramolecular 1,3-methide shift or by a bimolecular reaction with the *exo* alkenes (equation 28).[115] Very recently a similar 1,3-alkenyl shift was found to proceed rapidly at low temperatures.[208] UV studies on the polymerization of *p*PαMS revealed a similar absorption pattern (two maxima, the longer wavelength maximum at approximately 490 nm and with 20% intensity of the main maximum at 360 nm). This absorption appeared immediately after mixing the monomer with triflic acid when no *exo* alkenes were present.[164] Therefore, either transfer proceeds intramolecularly or the second absorption comes from the complex which was not formed intramolecularly but intermolecularly with a monomer molecule.

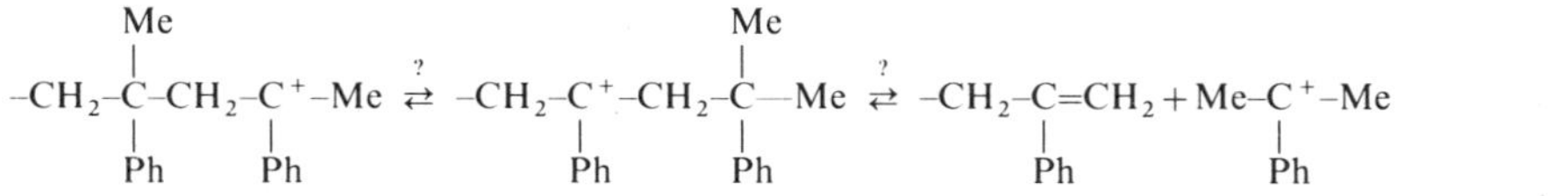

$$-CH_2-\underset{\underset{Ph}{|}}{\overset{\overset{Me}{|}}{C}}-CH_2-\overset{+}{C}-Me \underset{Ph}{\overset{?}{\rightleftarrows}} -CH_2-\overset{+}{C}-CH_2-\underset{\underset{Ph}{|}}{\overset{\overset{Me}{|}}{C}}-Me \overset{?}{\rightleftarrows} -CH_2-\underset{Ph}{C}=CH_2 + Me-\overset{+}{C}-Me \qquad (28)$$

Some of the cations discussed in the previous section may also be formed by rearrangement.

41.6 COPOLYMERIZATION AND REACTIVITIES

In this section we will describe briefly the most important examples of styrene copolymers. Three different types of cationic copolymers are known: random, graft and block copolymers. The copolymerization of different styrenes was extensively studied in the 1950s and 1960s and seems to have given (mostly) random structures. Unfortunately, the reactivity ratios determined for a large majority of these systems are meaningless. They were calculated using the simplest classical scheme with four irreversible reactions as in Scheme 6.

$$M_1^+ + M_1 \overset{k_{11}}{\rightarrow} M_{11}^+$$

$$M_1^+ + M_2 \overset{k_{12}}{\rightarrow} M_{12}^+$$

$$M_2^+ + M_1 \overset{k_{21}}{\rightarrow} M_{21}^+$$

$$M_2^+ + M_2 \overset{k_{22}}{\rightarrow} M_{22}^+$$

Scheme 6

The reactivity ratios $r_1 = k_{11}/k_{12}$ and $r_2 = k_{22}/k_{21}$ were derived by different methods without taking into account two major considerations:

(i) The reversibility of the polymerization of highly substituted alkenes (α- and β-methylstyrenes, DPE, *etc.*). For these monomers the apparent ratios depend on the composition of the reaction mixture and temperature, and may reflect either the kinetic or thermodynamic polymerizability of these monomers.

(ii) Multiplicity of the active species. Depending on the initiator, solvent and temperature, more or less reactive species can be formed. Thus, the ionic or unreactive covalent species may be formed by protonic acids; with Lewis acids more or less extensive termination depends on the stability of the cations. Often oligomeric products were formed. Therefore comparison of the monomer reactivities can be done only on the qualitative and not on the quantitative level, referring to the same type of active species, *e.g.* ionic or covalent. We will discuss reactivities of monomers and active species later for different model systems.

The preparation of pure block copolymers from styrenes by cationic process is an extremely difficult and generally unsuccessful task. Transfer to monomer, anion, solvent and polymer prevents the preparation of block copolymers in high yields and with high molecular weights. Better results were obtained for low molecular weight products when transfer was less important. Two approaches are worth citing. One involves so-called quasi-living polymerization with a continuous supply of monomer, the concentration of which is maintained at a very low level.[209,210] This suggests that the external order with respect to monomer for the transfer to monomer is higher than for propagation. Apparently, transfer was reduced under these conditions, but an alternative explanation was also proposed.[18]

The other approach is based on the application of the activated covalent species (HI + I₂, etc.). This approach was successfully used for more basic monomers such as *p*MOS and vinyl ethers.[211,212]

Block copolymers were also reported for the stepwise polymerization of S and heterocyclic monomers such as aziridines[213] and oxazolines. First, polymerization of S was carried out in the presence of a salt with a common anion to suppress transfer to the anion, and later the heterocyclic monomer was added. The block structure of the copolymers was confirmed by the solubility behavior and by the higher MW of the product compared with the starting polystyrene.

Triblock copolymers having a soft central block [poly(isobutylene)] and hard terminal blocks (polyαMS) were synthesized from difunctional poly(isobutylene) with two terminal chlorine atoms as the starting material, which was later activated by $AlEt_2Cl$ in the presence of αMS.[214,215] The blocking efficiency was low but the product had promising properties as a thermoplastic material.

The same approach was used in grafting polystyrenes from copolymers of isobutylene with chlorinated monomers (*e.g.* chloromethylstyrene). This method can be used for grafting from PVC, chlorinated rubber, *etc.*[217] $AlEt_2Cl$ is a very convenient coinitiator because it does not initiate polymerization of S and it activates different chlorinated compounds.[217,218]

In addition to typical grafting from, grafting onto and grafting through were also applied in the polymerization of styrenes.[187]

41.6.1 Reactivities of Monomers

Styrenes and α-methylstyrenes were used as classical monomers in studies of the influence of ring substituents on reactivities. The correlation was usually expressed in terms of Hammett's σ parameter and Brown's σ^+ parameter. As shown in Table 4, the thermodynamic as well as kinetic parameters were correlated. The gas phase basicities gave higher ρ values ($\rho = -23$) than the kinetic experiments ($\rho \approx -5$).[219] The substituents at the double bond are important. The gas phase basicity of S is approximately 20 kJ mol⁻¹ lower than that of αMS. Usually a proton was used as the electrophilic reagent, however, values of ρ obtained with cations carrying more bulky substituents, such as the benzhydryl cation formed *in situ* from the chloride, are similar.[220] This may suggest that in solution the nucleophilic reactivities of styrenes follow the same sequence as the basicities. Perhaps the acid reacts as the soft dimeric species in a way similar to the other electrophiles. The reactivities of monomers calculated from copolymerizations agree well with model studies.[2]

Table 4   Reactivities and Gas Phase Basicities of *p*-Substituted (R) Styrenes

R	$S^{1,a}$	$\alpha MS^{1,a}$	$H_2SO_4^{2,c}$ S	$H_2SO_4^{2,c}$ αMS	$CF_3CO_2H^{3,b}$ S	σ^+	σ
*p*MeO	867	876	6.4×10^{-5}	8×10^{-3}	0.112	−0.78	−2.02
*p*Me	835	851	2.5×10^{-6}	1×10^{-3}	3.8×10^{-3}	−0.31	−0.67
*p*F		834				−0.07	−0.40
H	816	834	3.2×10^{-7}		1.4×10^{-4}	0	0
*p*Cl		827	1.0×10^{-7}	6×10^{-5}		0.11	−0.24
*p*Br	812		2.0×10^{-7}		2.0×10^{-5}	0.15	−0.19
*p*CF₃		803				0.61	0.79

[1] Basicities in kJ mol⁻¹. [2] Hydration rate constants in mol⁻¹ s⁻¹, at 25 °C in aqueous acids. [3] Addition rate constants in mol⁻¹ s⁻¹ at 25 °C in 20% CF_3CO_2H in CCl_4.

[a] A. G. Harrison, R. Houriet and T. T. Tidwell, *J. Org. Chem.*, 1984, **49**, 1302.
[b] P. Knittel and T. T. Tidwell, *J. Am. Chem. Soc.*, 1977, **99**, 3408.
[c] A. D. Allen, M. Rosenbaum, N. O. L. Seto and T. T. Tidwell, *J. Org. Chem.*, 1982, **47**, 4234.

The reactivities of β-methylstyrenes differ for the Z and E isomers (see Table 5). In hydration and bromination the relative reactivities are similar.[221,222] The much higher rate constant in the addition of benzhydryl cation to the E isomer than to the Z isomer may suggest the preliminary coordination of the carbenium ion with a phenyl substituent prior to the addition.[223] In the Z isomer, the position of the methyl group on the same side as the phenyl group decreases the chance of this interaction.

Table 5 Rate Constants of Reactions of Some Electrophiles with Different Monomers

Monomer	$Ph_2CH^{+1,a}$	$H_3O^{+1,b}$	$Br_2^{1,c}$	$CF_3CO_2H^{2,d}$	$Tol_2CH^{+2,3,e}$	$An_2CH^{+2,4,e}$
$CH_2=CHMe$	1.8×10^{-4}	1.3×10^{-4}	0.01			
$CH_2=CMe_2$	1	1	1			
$CH_2=CMeC_3H_7$	1.16		0.976		1×10^3	3×10^{-2}
$CH_2=CHC_6H_5$	0.43	1×10^{-3}	0.041	1.75×10^{-2}	1×10^3	9×10^{-3}
$MeCH=CHC_6H_5(E)$	2.0	3×10^{-4}	0.086	3×10^{-4}		
$MeCH=CHC_6H_5(Z)$	0.06	3×10^{-4}	0.021	3×10^{-4}		

[1] Relative rate constants assuming bimolecular reaction. [2] Rate constants in $mol^{-1}1s^{-1}$ assuming bimolecular reactions. [3] $(pMePh)_2CH^+$.
[4] $(pMeOPh)_2CH^+$.
[a] H. Mayr and R. Pock, *Tetrahedron Lett.*, 1983, **24**, 2155.
[b] P. Knittel and T. T. Tidwell, *J. Am. Chem. Soc.*, 1977, **99**, 3408.
[c] E. Bienvenue and J. E. Dubois, *Tetrahedron*, 1978, **34**, 2021.
[d] A. D. Allen, M. Rosenbaum, N. O. L. Seto and T. T. Tidwell, *J. Org. Chem.*, 1982, **47**, 4234.
[e] R. Schneider and H. Mayr, *Angew. Chem., Int. Ed. Engl.*, 1986, **25**, 1016.

The determination of the relative reactivities of monomers by means of a standard electrophilic species may give access to the order of reactivities of different classes of cationically polymerizable monomers. This approach, used initially by Szwarc[225] and later developed by Giese[226] and Tirrell[227] in radical polymerization, allows the explanation of the influence of the penultimate unit on the reactivity of the active center. It is hoped that the same level of understanding of the propagation step will be reached in cationic systems.

41.6.2 Reactivities of Active Species

We discussed in previous sections the absolute values of the rate constants of some elementary reactions and found that the rate constants of propagation range from 10^5 to 10^6 $mol^{-1}1s^{-1}$. These values were calculated by assuming that the extinction coefficients of the growing cations were close to those for the model compounds and that in the observed range of UV only active species absorb.

The rate constants obtained with γ radiation as the initiation source led to values similar to those cited above when they were correlated with conductivity measurements, but to values of about 10^8 $mol^{-1}1s^{-1}$ when the scavenger method was applied. Are such high values of the rate constants of propagation reasonable? Some light was shed on this problem by pulsed radiolysis studies of reactions of different cations with nucleophiles, bases and alkenes.[229]

In Table 6 the rate constants of the recombination of anions, nucleophilic substitution and electrophilic addition to double bonds are listed for the benzyl, benzhydryl and trityl cations. The reactivity of the styryl cation should be between those of the first two and the reactivity of the cation derived from αMS should be similar to that of benzhydryl (*cf.* also Table 7). The monomer with the reactivity closest to styrene is isobutylene (*cf.* previous section). Thus, at room temperature the rate constant of this model reaction equals $k \approx 10^7$ $mol^{-1}1s^{-1}$. 1-Phenylcyclohexene can serve as a model of αMS or IN (more precisely of α,β-dimethylstyrene). It reacts with the benzhydryl cation with a rate constant $k = 7 \times 10^7$ $mol^{-1}1s^{-1}$. Are these rate constants reliable as estimates of the rate constants of propagation? The growing cation can be additionally stabilized by either the monomer or its own polymer chain. Thus, the values in Table 6 give rather the upper limit of the rate constants of propagation on carbenium ions stabilized at most by a solvent molecule.

It is interesting to compare the relative reactivities of alkenes and different nucleophilic species. For example, the reactivity of S is similar to H_2O! This means that even in the presence of a small amount of water $(10^{-4}$ $mol\,l^{-1})$, a polymer with a degree of polymerization $DP_n = 10^3$ can be formed before termination occurs in bulk. Although methanol and ethanol may be more efficient terminating agents, termination will not be immediate unless a large excess of an alcohol is added.[230] The

Table 6 Rate Constants of Addition of Carbenium Ions to Different Nucleophiles

Nucleophiles	$PhCH_2^{+1}$	Ph_2CH^{+1}	Ph_3C^{+1}	$ArCH^+(Me)^2$ pH	pMe	pMeO
Br^-	5×10^{10}					
I^-	5×10^{10}					
Cl^-			$\approx 10^{7\,3}$			
NH_3	4.2×10^9	4.3×10^9	2.4×10^7			
NEt_3	2×10^9	1.2×10^9	7×10^7			
H_2O	1.8×10^7	1.3×10^6		3×10^9	1.3×10^8	1.5×10^6
$MeOH$	7×10^7				2×10^8	1×10^7
$EtOH$	1×10^8					
Et_2O	1.5×10^7					
THF	1.4×10^8		$<5 \times 10^5$			
$CH_2=CHMe$	1.9×10^6	$<10^5$				
$CH_2=CMe_2$	1.9×10^7	9×10^6				
1-phenylcyclohexene	—	7×10^7				
N_3^-				5×10^9	5×10^9	5×10^9
CF_3CH_2OH				3×10^9	1×10^8	8×10^5

[1] Rate Constants in $mol^{-1}1s^{-1}$ measured by pulse radiolysis, at 30 °C in $C_2H_4Cl_2$.[a,b,c] [2] Rate constants in $mol^{-1}1s^{-1}$ measured in 50% aqueous trifluoroethanol at 25 °C. Rate constant of the reaction of cations with N_3^- assumed to be diffusion controlled $(5 \times 10^9 \, mol^{-1}1s^{-1})$.[e,f,g] A similar rate constant has been recently measured directly. [3] Extrapolated from low temperatures to 30 °C, measured by dynamic NMR.[d]

[a] Y. Wang and L. M. Dorfman, *Macromolecules*, 1980, **13**, 63.
[b] L. M. Dorfman and V. M. DePalma, *Pure Appl. Chem.*, 1979, **51**, 123.
[c] R. Sujdak, R. L. Jones and L. M. Dorfman, *J. Am. Chem. Soc.*, 1976, **98**, 4875.
[d] H. Kessler and M. Feigel, *Acc. Chem. Res.*, 1982, **15**, 2.
[e] J. P. Richard, M. E. Rothenberg and W. P. Jencks, *J. Am. Chem. Soc.*, 1984, **106**, 1361.
[f] J. P. Richard and W. P. Jencks, *J. Am. Chem. Soc.*, 1984, **106**, 1373.
[g] J. P. Richard and W. P. Jencks, *J. Am. Chem. Soc.*, 1984, **106**, 1383.
[h] R. A. McClelland, N. Banait and S. Steenken, *J. Am. Chem. Soc.*, 1986, **108**, 7023.

Table 7 Reactivities of Some Cations and Related Species

X =	$Cl^{1,a}$	$Cl^{2,b}$	Gas Phase Stabilities of Cations[3,c]	Stabilities of Cations in Solution[4,d]	Addition of Cations to IB[5,e]
Me_2CH-X	2×10^{-9}	2×10^{-12}	1044		
Me_3C-X	9×10^{-6}	8.6×10^{-8}	981	148	
$PhCH_2-X$			999		1.9×10^7
$PhCHMe-X$	1×10^{-5}	2.2×10^{-7}	958		
$PhCMe_2-X$	2×10^{-3}	3.94×10^{-4}		180	
Ph_2CH-X	2×10^{-3}	5.75×10^{-4}		164	9.5×10^6
Ph_2CMe-X		2×10^{-2}		157	
Ph_3C-X	2×10^2	5.8×10^{-1}		205	

[1] Rate constants of solvolysis measured in 80% ethanol at 25 °C (in s^{-1}). [2] The same but in pure ethanol. [3] In $kJ\,mol^{-1}$, measured as hydride anion affinities at 25 °C. [4] Enthalpies (in $kJ\,mol^{-1}$) of the reaction of formation of cations from carbinols with 1:1 SbF_5/FSO_3H in SO_2ClF. [5] Measured by pulsed radiolysis in $C_2H_4Cl_2$ at 30 °C, in $mol^{-1}1s^{-1}$.

[a] H. Mayr and W. Striepe, *J. Org. Chem.*, 1983, **48**, 1159.
[b] H. C. Brown and M. Rei, *J. Am. Chem. Soc.*, 1964, **86**, 5008.
[c] C. J. Wolf, R. H. Staley, I Koppel, M. Taagepera, R. T. McIvor, Jr., J. L. Beauchamp and R. W. Taft, *J. Am. Chem. Soc.*, 1977, **99**, 5417.
[d] E. M. Arnett and T. C. Hofelich, *J. Am. Chem. Soc.*, 1983, **105**, 2889.
[e] Y. Wang and L. M. Dorfman, *Macromolecules*, 1980, **13**, 63.

halide anions are most reactive. The rate constants for these anions exceed the limit of the diffusion-controlled rate constant under similar conditions $(k \approx 5 \times 10^9 \, mol^{-1}1s^{-1})$.[231,232] The latter value was assumed to be the upper limit of the rate constant of the different 1-phenylethyl cations with the most nucleophilic azide anion (N_3^-).[233] Recently, a very similar value $(4 \times 10^9 \, mol^{-1}1s^{-1})$ was measured for the recombination of the azide anion with the trityl cation.[234]

The results shown in Table 7 indicate that halides derived from styrene and isobutylene should have similar reactivities; the same is true for diphenylmethyl (benzhydryl) and dimethylphenylmethyl (cumyl) derivatives. One may generalize that the influence of one phenyl is similar to two methyl groups (but this is not always the case).[118]

Thus, by looking at Tables 6 and 7 we can conclude that the reactivities of cations should decrease in the order S > αMS > DPE, and for *p*-substituted rings H > Me > MeO. Recent studies of the disubstituted benzhydryl cations confirmed the above order.[235,236]

The next step in the correlation of the structure of monomers and cations with their reactivities should be the estimation of their relative contribution to the rate constant of propagation, *i.e.* the determination of whether the structure of monomer or cation is the rate determining factor. This was done for the cationic polymerization of heterocyclic monomers,[237] radical polymerization of vinyl monomers,[238] and anionic polymerization of styrenes.[239] In the first two systems the electrophilic onium ions and radicals, and in the last one the electrophilic monomers, determined the propagation rate constants. Very recently 'late' transition state for the addition of alkenes to benzhydryl cations was proposed.[130] It may be anticipated that in the cationic polymerization of styrenes the same will be true, *i.e.* the structure of the electrophilic carbenium ions will be the rate determining factor, but this still remains to be experimentally proven.

41.7 OVERVIEW

The most condensed picture of the cationic polymerization of styrenes is given below. Initiation is usually slower than the subsequent propagation but this may change for the most basic monomers such as *p*MOS and *p*MOαMS (they will also form the least reactive growing ions). Protonation may be reversible. Propagation involving the complexed gegenion, HA_2^-, occurs with similar rate constants for ions and ion pairs ($k_p \approx 10^5$ to $10^6 \, mol^{-1} \, s^{-1}$). Counterions capable of forming covalent bonds recombine with cations. Covalent species may be in dynamic equilibrium with ions, they may be completely inactive alone, and under some conditions they may participate in propagation (probably in low polarity solvents). The mechanism of the possible covalent propagation is not yet clear but it may involve multicenter rearrangement (pseudocationic process).

Carbenium ions, being very reactive, self stabilize by the formation of complexes with gegenions, monomer, solvent or polymer. In growing ions the positive charge is also distributed on the β-H atoms which are easily and rapidly attacked by different basic components leading to transfer reactions. Transfer to monomer seems to be the most important reaction, but for α-methylstyrenes unimolecular transfer is also important. Another transfer reaction is the formation of indans in unimolecular Friedel–Crafts cycloalkylation.

Growing carbenium ions rearrange to more stable cations. These ions may be inactive and if the isomerization is irreversible it becomes termination. Termination may also occur by the formation of inactive covalent species and by reaction with the nucleophilic impurities in the system (*e.g.* moisture).

Lewis acids as initiators usually require the addition of cationogenic or protonogenic compounds (water, alcohols, alkyl halides). Initiation is slow and irreversible termination occurs by the decomposition of complex gegenions. Stable carbenium salts are sluggish initiators.

Complete elimination of transfer, termination and isomerization is a difficult, if not impossible, task under usual reaction conditions when active species have ionic structures. The best chance for the synthesis of well-defined polystyrenes might be polymerization with covalent species (activated) in which the proportion of the positive charge on the β-H atoms is significantly reduced.[202] This reaction, however, could occur *via* short-living intermediate ion-pairs which recombine in the solvent cage before dissociation.

ACKNOWLEDGEMENT

Valuable discussions with Professors P. Sigwalt and P. H. Plesch are gratefully acknowledged.

41.8 REFERENCES

1. A. Gandini and H. Cheradame, in 'Encyclopedia of Polymer Science and Engineering', 2nd edn., Wiley, New York, 1984, vol. 2, p. 729.
2. J. P. Kennedy and E. Maréchal, in 'Carbocationic Polymerization', Wiley, New York, 1982.
3. J. P. Kennedy, in 'Cationic Polymerization of Olefins: A Critical Inventory', Wiley, New York, 1975.
4. A. Gandini and H. Cheradame, *Adv. Polym. Sci.*, 1980, **34/35**, 1.
5. P. H. Plesch, in 'The Chemistry of Cationic Polymerization', Pergamon Press, Oxford, 1963.
6. M. Deville, *Ann. Chim. Phys.*, 1839, **75**, 66.

668 *Cationic Polymerization*

7. M. Berthelot, *Bull. Soc. Chim. Fr.*, 1866, **6**, 294.
8. H. Staudinger and F. Breusch, *Ber. Dtseh. Chem. Ges.*, 1929, **62B**, 442.
9. G. Williams, *J. Chem. Soc.*, 1938, 246.
10. A. G. Evans and J. Lewis, *J. Chem. Soc.*, 1957, 2975.
11. D. C. Pepper, *Makromol. Chem.*, 1974, **175**, 1077.
12. P. H. Plesch, *Adv. Polym. Sci.*, 1971, **8**, 137.
13. T. Higashimura and M. Sawamoto, *Adv. Polym. Sci.*, 1984, **62**, 49.
14. P. Sigwalt and G. Sauvet, *Polym. J.*, 1980, **9**, 651.
15. T. Kunitake and K. Takarabe, *Macromolecules*, 1979, **12**, 1061.
16. M. DeSorgo, D. C. Pepper and M. Szwarc, *J. Chem. Soc., Chem. Commun.*, 1973, **3**, 419.
17. A. R. Mathieson, in 'The Chemistry of Cationic Polymerization', Pergamon Press, Oxford, 1963, p. 235.
18. P. Sigwalt, *Polym. J.*, 1985, **17**, 57.
19. G. A. Olah and P. v. R. Schleyer, 'Carbonium Ions', Wiley, New York, 1970.
20. G. A. Olah, 'Friedel–Crafts and Related Reactions', Wiley, New York, 1963.
21. P. Vogel, 'Carbocation Chemistry', Elsevier, Amsterdam, 1985.
22. P. B. D. De la Mare and R. Bolton, 'Electrophilic Addition to Unsaturated Systems', Elsevier, Amsterdam, 1982.
23. C. K. Ingold, 'Structure and Mechanism in Organic Chemistry', 2nd edn., Cornell University Press, Ithaca, New York, 1969.
24. M. Jones, Jr., and R. A. Moss (eds.), 'Reactive Intermediates', Wiley, New York, 1978, vol. I; 1981, vol. II; 1985, vol. III.
25. M. Szwarc, 'Ions and Ion Pairs in Organic Reactions', Wiley, New York, 1974.
26. S. Bywater, *Makromol. Chem.*, 1962, **52**, 120.
27. K. J. Ivin and J. Leonard, *Eur. Polym. J.*, 1970, **6**, 331.
28. S. L. Malhotra, J. Leonard and P. E. Harvey, *J. Macromol. Sci., Chem.*, 1977, **A11**, 2199.
29. S. Bywater and D. Worsfold, *J. Polym. Sci.*, 1962, **58**, 571.
30. F. S. Dainton, K. J. Ivin and D. A. G. Walmsley, *Trans. Faraday. Soc.*, 1960, **56**, 1784.
31. L. K. J. Tong and W. O. Kenyon, *J. Am. Chem. Soc.*, 1947, **69**, 1402.
32. D. J. Worsfold, S. Bywater and P. Black, 'Cationic Polymerization and Related Processes', ed. E. J. Goethals, Academic Press, London, 1984, p. 43.
33. P. Blin, C. Bunel and E. Maréchal, *J. Chem. Res.*, 1978, **6**, 2619.
34. A. Vrancken, J. Smid and M. Szwarc, *Trans. Faraday Soc.*, 1962, **58**, 2036.
35. J. Leonard and S. L. Malhotra, *J. Polym. Sci.*, 1971, **9**, 1983.
36. R. Asami, 'Kobunshi No Tenbo 1970', Maruzen, Tokyo, 1970, p. 53.
37. J. P. Kennedy and R. T. Chou, *J. Macromol. Sci., Chem.*, 1982, **A18**, 47.
38. A. Shimizu, T. Otsu and M. Imoto, *Bull. Chem. Soc. Jpn.*, 1968, **41**, 953.
39. P. Sigwalt, *C. R. Hebd. Seances Acad. Sci.*, 1961, **252**, 3998.
40. H. Lussi, *Makromol. Chem.*, 1967, **103**, 68.
41. M. Sonntag and W. Funke, *Makromol. Chem.*, 1970, **137**, 23.
42. R. M. Roberts and A. A. Khalaf, 'Friedel–Crafts Alkylation Chemistry', Marcel Dekker, New York, 1984.
43. P. Sigwalt, *J. Polym. Sci.*, 1961, **52**, 15.
44. M. Imoto and K. Takemoto, *J. Polym. Sci.*, 1955, **15**, 271.
45. E. Maréchal, P. Evrard and P. Sigwalt, *Bull. Soc. Chim. Fr.*, 1969, 1981.
46. A. G. Heilbrunn and E. Maréchal, *C.R. Hebd. Seances Acad. Sci.*, 1972, **274**, 1149.
47. H. Hopff and H. Lussi, *Makromol. Chem.*, 1963, **62**, 31.
48. G. Heublein and H. Dawczynski, *J. Prakt. Chem.*, 1972, **314**, 557.
49. C. G. Overberger, L. H. Arond, D. H. Tanner, J. J. Taylor and T. Alfrey, Jr., *J. Am. Chem. Soc.*, 1952, **74**, 4848.
50. C. G. Overberger, D. H. Tanner and E. M. Pearce, *J. Am. Chem. Soc.*, 1958, **80**, 4566.
51. M. Sawamoto and T. Higashimura, *Macromolecules*, 1981, **14**, 467.
52. C. G. Overberger, E. M. Pearce and D. H. Tanner, *J. Am. Chem. Soc.*, 1958, **80**, 1761.
53. S. D. Hamann, A. J. Murphy, D. H. Solomon and R. I. Willing, *J. Macromol. Sci., Chem.*, 1972, **6**, 771.
54. G. A. Olah and J. R. DeMember, *J. Am. Chem. Soc.*, 1970, **92**, 2562.
55. G. E. Holdcroft and P. H. Plesch, *Makromol. Chem.*, 1984, **185**, 27.
56. C. Reichardt, 'Solvent Effects in Organic Chemistry', Verlag Chemie, New York, 1979.
57. E. S. Amis, 'Solvent Effects on Reaction Rates and Mechanisms', Academic Press, New York, 1966.
58. H. Cheradame, M. Mazza, N. A. Hung and P. Sigwalt, *Eur. Polym. J.*, 1973, **9**, 375.
59. Y. Pocker and K. D. Stevens, *J. Am. Chem. Soc.*, 1969, **91**, 4205.
60. P. Giusti and F. Andruzzi, *J. Polym. Sci., Part C*, 1968, **16**, 3797.
61. C. P. Brown and A. R. Mathieson, *J. Chem. Soc.*, 1958, 3445.
62. C. P. Brown and A. R. Mathieson, *J. Chem. Soc.*, 1957, 3625.
63. T. Hamaya, *Makromol. Chem., Rapid Commun.*, 1982, **3**, 953.
64. T. Hamaya and S. Yamada, *Makromol. Chem.*, 1979, **180**, 2979.
65. W. Obrecht and P. H. Plesch, *Makromol. Chem.*, 1981, **182**, 1459.
66. M. Sawamoto, T. Masuda, T. Higashimura, S. Kobayashi and T. Seagusa, *Makromol. Chem.*, 1977, **178**, 389.
67. J. J. Throssell, S. P. Sood, M. Szwarc and V. Stannett, *J. Am. Chem. Soc.*, 1956, **78**, 1122.
68. P. H. Plesch and A. Gandini, *SCI Monogr.* 1966, **20**, 107.
69. C. Pepper and P. J. Reilly, *J. Polym. Sci.*, 1962, **58**, 639.
70. A. Gandini and P. H. Plesch, *J. Chem. Soc.*, 1965, 4765.
71. M. Sawamoto, A. Furukawa and T. Higashimura, *Macromolecules*, 1983, **16**, 518.
72. T. Masuda, M. Sawamoto and T. Higashimura, *Makromol. Chem.*, 1976, **177**, 2981.
73. S. Bywater and D. Worsfold, *Can. J. Chem.*, 1966, **44**, 1671.
74. M. Chmelir, N. Cardona and G. V. Schulz, *Makromol. Chem.*, 1977, **178**, 169.
75. J. P. Lorimer and D. C. Pepper, *Proc. R. Soc. London, Ser. A.*, 1976, **351**, 551.
76. T. Masuda, M. Sawamoto and T. Higashimura, *Makromol. Chem.*, 1976, **177**, 2981.
77. W. H. Hunter and R. V. Yohe, *J. Am. Chem. Soc.*, 1933, **55**, 1248.

78. P. H. Plesch, *Makromol. Chem.*, 1974, **175**, 1065.
79. N. A. Hung, H. Cheradame and P. Sigwalt, *Eur. Polym. J.*, 1973, **9**, 375.
80. W. R. Longworth, C. J. Panton and P. H. Plesch, *J. Chem. Soc.*, 1965, 5579.
81. M. Villesange, G. Sauvet, J. P. Vairon and P. Sigwalt, *J. Macromol. Sci., Chem.*, 1977, **A11**, 391.
82. W. Bracke, W. J. Cheng, J. M. Pearson and M. Szwarc, *J. Am. Chem. Soc.*, 1969, **91**, 203.
83. G. Sauvet, J. P. Vairon and P. Sigwalt, *C.R. Hebd. Seances Acad. Sci.*, 1967, **C265**, 1090.
84. A. D. Eckard, A. Ledwith and D. C. Sherrington, *Polymer*, 1971, **12**, 444.
85. A. F. Johnson and D. A. Pearce, *J. Polym. Sci., Polym. Symp.*, 1976, **56**, 57.
86. R. Velichkova, I. M. Panayotov, J. Doicheva, G. Heublein, H. Schultz, P. Alter, S. Spang and R. Wondraczek, *J. Polym. Sci.*, 1982, **20**, 2895.
87. R. Cotrel, G. Sauvet, J. P. Vairon and P. Sigwalt, *Macromolecules*, 1976, **9**, 931.
88. W. Obrecht and P. H. Plesch, *Makromol. Chem.*, 1981, **182**, 1459.
89. T. Kagiya, M. Izu, H. Maruyama and K. Fukui, *J. Polym. Sci., Part A-1*, 1969, **7**, 917.
90. T. Masuda and T. Higashimura, *J. Polym. Sci., Polym. Lett. Ed.*, 1971, **9**, 783.
91. T. Takeda, M. Sawamoto and T. Higashimura, *Polym. J.*, 1977, **9**, 377.
92. K. Matyjaszewski, *Polym. Prepr. Am. Chem. Soc., Div. Polym. Chem.*, 1987, **27**(2), 112; K. Matyjaszewski, *Macromolecules*, 1988, **21**, 933.
93. H. Kammerer and A. S. Mozafarri, *Makromol. Chem.*, 1983, **184**, 1143.
94. H. Kammerer and A. S. Mozafarri, *Makromol. Chem.*, 1984, **185**, 509.
95. K. Matyjaszewski and P. Sigwalt, *Makromol. Chem.*, 1986, **187**, 2299.
96. C. Peniche and A. Gandini, *Rev. CENIC Cienc. Fis.*, 1973, **4**, 59.
97. K. Matyjaszewski and C. H. Lin, *Polym. Prepr., Am. Chem. Soc., Div. Polym. Chem.*, 1987, **28** (2), 224.
98. K. Matyjaszewski and C. H. Lin, *Polym. Prepr., Am. Chem. Soc., Div. Polym. Chem.*, 1987, **28**(1), 176.
99. F. Williams, K. Hayashi, K. Ueno and S. Okamura, *Trans. Faraday Soc.*, 1967, **63**, 1501.
100. E. Hubman, R. B. Taylor and F. Williams, *Trans. Faraday Soc.*, 1966, **62**, 88.
101. G. O. Akulov, N. Gellez, V. Kropatchev, V. Nefedor, E. Sinotova, S. Skozokhodov, V. Stepanov and U. Toropova, *Makromol. Chem.*, 1978, **179**, 2775.
102. W. J. Hehre, *Acc. Chem. Res.*, 1975, **8**, 369.
103. W. J. Hehre, *J. Am. Chem. Soc.*, 1972, **94**, 5919.
104. H. Griengl and P. Schuster, *Tetrahedron*, 1974, **30**, 117.
105. R. J. Spear, D. A. Forsyth and G. A. Olah, *J. Am. Chem. Soc.*, 1976, **98**, 2493.
106. D. P. Kelly and R. J. Spear, *Aust. J. Chem.*, 1978, **31**, 1209.
107. G. A. Olah, A. L. Berrier and G. K. S. Prahash, *Proc. Natl. Acad. Sci. USA*, 1981, **78**, 1998.
108. H. C. Brown, M. Periasamy and P. T. Perumal, *J. Org. Chem.*, 1984, **49**, 2754.
109. M. Sawamoto and T. Higashimura, *Macromolecules*, 1978, **11**, 328.
110. N. C. Deno, *Prog. Phys. Org. Chem.*, 1964, **2**, 129.
111. S. Bywater, A. F. Johnson and D. J. Worsfold, *Can. J. Chem.*, 1964, **42**, 1255.
112. G. A. Olah, C. U. Pittman, R. Waack and M. Doran, *J. Am. Chem. Soc.*, 1966, **88**, 1488.
113. V. Bertoli and P. H. Plesch, *J. Chem. Soc. (B)*, 1968, 1500.
114. D. G. Farnum, *J. Am. Chem. Soc.*, 1967, **89**, 2970.
115. M. Moreau, K. Matyjaszewski and P. Sigwalt, *Macromolecules*, 1987, **20**, 1456.
116. H. Cheradame, N. A. Hung and P. Sigwalt, *J. Polym. Sci., Polym. Symp.*, 1976, **56**, 335.
117. H. C. Brown, D. P. Kelly and M. Periasamy, *Proc. Natl. Acad. Sci. USA*, 1980, **77**, 6956.
118. E. M. Arnett and T. C. Hofelich, *J. Am. Chem. Soc.*, 1983, **105**, 2889.
119. K. Matyjaszewski and P. Sigwalt, *Macromolecules*, 1987, **20**, 2679.
120. A. Gandini and P. H. Plesch, *Eur. Polym. J.*, 1968, **4**, 55.
121. M. Sawamoto, T. Higashimura, A. Enokida and T. Okubo, *Polym. Bull.*, 1980, **2**, 309.
122. H. P. J. M. Dekkers and E. C. M. Kielman van Luyt, *Mol. Physics*, 1976, **32**, 899.
123. J. C. Evans, in 'Carbonium Ions', ed. G. A. Olah and P. v. R. Schleyer, Wiley, New York, 1968, vol. 1, p. 223.
124. E. M. Arnett and B. Chawla, *J. Am. Chem. Soc.*, 1979, **101**, 7141.
125. M. I. Watkins, W. M. Ip, G. A. Olah and R. Bau, *J. Am. Chem. Soc.*, 1982, **104**, 2365.
126. G. A. Olah, *Angew. Chem., Int. Ed. Engl.*, 1973, **12**, 173.
127. R. J. Gielespi and E. A. Robinson, in 'Carbonium Ions', ed. G. A. Olah and P. v. R. Schleyer, Wiley, New York, 1968, vol. 1, p. 111.
128. G. Sauvet, J. P. Vairon and P. Sigwalt, *J. Polym. Sci., Polym. Symp.*, 1975, **52**, 173.
129. G. Sauvet, M. Moreau and P. Sigwalt, *Makromol. Chem., Macromol. Symp.*, 1986, **3**, 33.
130. H. Mayr, *Makromol. Chem., Macromol. Symp.*, 1988, **13/14**, 43.
131. K. Takarabe and T. Kunitake, *Polym. J.*, 1980, **12**, 239.
132. A. D. Allen, M. Rosenbaum, N. O. L. Seto and T. T. Tidwell, *J. Org. Chem.*, 1982, **47**, 4234.
133. V. J. Nowlan and T. T. Tidwell, *Acc. Chem. Res.*, 1977, **10**, 252.
134. T. Kunitake and K. Takarabe, *J. Polym. Sci., Polym. Symp.*, 1976, **56**, 33.
135. M. Kriszenbaum, J. Corset and M. Josien, *J. Phys. Chem.*, 1971, **75**, 1327.
136. T. Sano, N. Tatsumoto, Y. Mende and T. Yasunaga, *Bull. Chem. Soc. Jpn.*, 1972, **45**, 2673.
137. I. M. Kolthoff, *Anal. Chem.*, 1974, **46**, 1992.
138. M. M. Krevoy, T. Liang and K. Chang, *J. Am. Chem. Soc.*, 1977, **99**, 5207.
139. D. M. Coutagne, *J. Am. Chem. Soc.*, 1971, **93**, 1518.
140. D. Souverain, A. Leborgne, G. Sauvet and P. Sigwalt, *Eur. Polym. J.*, 1980, **16**, 861.
141. A. Leborgne, D. Souverain, G. Sauvet and P. Sigwalt, *Eur. Polym. J.*, 1980, **16**, 855.
142. Y. Pocker and K. D. Stevens, *J. Am. Chem. Soc.*, 1969, **91**, 4205.
143. J. Haugh and D. R. Dalton, *J. Am. Chem. Soc.*, 1975, **97**, 5674.
144. Y. Pocker, K. D. Stevens and J. J. Champoux, *J. Am. Chem. Soc.*, 1969, **91**, 4199.
145. J. M. Barton and D. C. Pepper, *J. Chem. Soc.*, 1964, 1573.
146. I. M. Kolthoff and M. K. Chantooni, Jr., *J. Am. Chem. Soc.*, 1973, **95**, 8539.

147. L. Wilczek and J. Chojnowski, *Macromolecules*, 1981, **14**, 9.
148. M. Sawamoto and T. Higashimura, *Macromolecules*, 1978, **11**, 501.
149. M. Sawamoto and T. Higashimura, *Macromolecules*, 1979, **12**, 581.
150. R. W. Taft, Jr., *J. Am. Chem. Soc.*, 1952, **74**, 5372.
151. W. M. Schubert and J. R. Keeffe, *J. Am. Chem. Soc.*, 1972, **94**, 559.
152. W. M. Schubert and B. Lamm, *J. Am. Chem. Soc.*, 1966, **88**, 120.
153. J. B. Levy, R. W. Taft, Jr. and L. P. Hammett, *J. Am. Chem. Soc.*, 1953, **75**, 1253.
154. D. J. Pasto, G. R. Meyer and B. Lepeska, *J. Am. Chem. Soc.*, 1974, **96**, 1858.
155. H. Hasegawa and T. Higashimura, *J. Appl. Polym. Sci.*, 1982, **27**, 171.
156. T. Kunitake and T. Takarabe, *Macromolecules*, 1979, **12**, 1067.
157. K. Matyjaszewski, S. Slomkowski and S. Penczek, *J. Polym. Sci., Polym. Chem. Ed.*, 1979, **17**, 2413.
158. K. Matyjaszewski, *Makromol. Chem.*, 1984, **185**, 51.
159. E. R. Nightingale, Jr., *J. Phys. Chem.*, 1959, **63**, 1381.
160. S. Penczek, P. Kubisa and K. Matyjaszewski, *Adv. Polym. Sci.*, 1980, **37**, 1.
161. S. Bywater, in 'Comprehensive Chemical Kinetics', ed. C. H. Bamford and F. H. Tipper, vol. 15, Elsevier, Amsterdam, 1976.
162. M. Szwarc, *Adv. Polym. Sci.*, 1983, **49**, 1.
163. K. Takarabe and T. Kunitake, *Makromol. Chem.*, 1981, **182**, 1587.
164. D. Teyssie, M. Villesange and J. P. Vairon, *Polym. Bull.*, 1984, **11**, 459.
165. R. Pock, H. Mayr, M. Robow and E. Wilhelm, *J. Am. Chem. Soc.*, 1986, **108**, 7767.
166. T. Uryu, T. Seki, T. Kawamura, A. Funamoto and K. Matsuzaki, *J. Polym. Sci., Polym. Chem. Ed.*, 1976, **14**, 3035.
167. N. Koide and R. W. Lenz, *J. Polym. Sci., Polym. Symp.*, 1976, **56**, 283.
168. R. W. Lenz, D. J. Fisher, and J. M. Jonte, *Macromolecules*, 1985, **18**, 1659.
169. T. Kunitake and C. Aso, *J. Polym. Sci., Part A-1*, 1970, **8**, 665.
170. E. M. Arnett and C. Petro, *J. Am. Chem. Soc.*, 1978, **100**, 5402.
171. E. M. Arnett and C. Petro, *J. Am. Chem. Soc.*, 1978, **100**, 5408.
172. G. A. Olah and T. Ohyama, *J. Am. Chem. Soc.*, 1984, **106**, 5284.
173. H. A. Benesi and J. H. Hildebrand, *J. Am. Chem. Soc.*, 1949, **71**, 2703.
174. R. S. Mulliken, *J. Am. Chem. Soc.*, 1952, **74**, 811.
175. M. Feldman and S. Winstein, *J. Am. Chem. Soc.*, 1961, **83**, 3338.
176. G. Heublein, *J. Macromol. Sci., Chem.*, 1981, **16**, 563.
177. M. I. Page, *Chem. Soc. Rev.*, 1973, **2**, 295.
178. K. Matyjaszewski, *J. Polym. Sci., Polym. Chem. Ed.*, 1987, **25**, 765.
179. R. B. Woodward and R. Hoffman, *J. Am. Chem. Soc.*, 1965, **87**, 395.
180. J. Mathieu and A. Rassat, *Tetrahedron*, 1974, **30**, 1753.
181. T. Higashimura, M. Miyamoto and M. Sawamoto, *Macromolecules*, 1985, **18**, 611.
182. M. Sawamoto, T. Enoki and T. Higashimura, *Macromolecules*, 1987, **20**, 1.
183. D. C. Pepper, *J. Polym. Sci., Polym. Symp.*, 1975, **50**, 51.
184. L. M. Leon, P. Altuna and D. C. Pepper, *Eur. Polym. J.*, 1980, **16**, 929.
185. M. Sawamoto, T. Masuda and T. Higashimura, *J. Polym. Sci., Polym. Chem. Ed.*, 1978, **16**, 2675.
186. M. S. Dewar and D. Landman, *J. Am. Chem. Soc.*, 1977, **99**, 7439.
187. P. Sigwalt, A. Polton and M. Miskovic, *J. Polym. Sci., Polym. Symp.*, 1976, **56**, 13.
188. M. Tazi, M. Tardi, A. Polton and P. Sigwalt, *Eur. Polym. J.*, 1986, **22**, 451.
189. M. Tazi, M. Tardi, A. Polton and P. Sigwalt, *Br. Polym. J.*, 1987, **19**, 369.
190. D. C. Pepper and P. J. Reilly, *Proc. R. Soc. London*, 1966, **291**, 41.
191. G. V. Schultz and G. Harboth, *Makromol. Chem.*, 1947, **1**, 104.
192. G. A. Olah, *Acc. Chem. Res.*, 1971, **4**, 240.
193. Y. Kawakami, N. Toyoshima and Y. Yamashita, *Chem. Lett.*, 1980, 13.
194. A. R. Taylor, G. W. Keen and E. J. Eisenbraun, *J. Org. Chem.*, 1977, **42**, 3477.
195. T. Higashimura and M. Sawamoto, *Adv. Polym. Sci.*, 1984, **62**, 49.
196. T. Higashimura, M. Hiza and H. Hasegawa, *Macromolecules*, 1984, **17**, 217.
197. J. Furukawa and J. Nishimura, *J. Polym. Sci., Polym. Symp.*, 1976, **56**, 437.
198. J. Nishimura, N. Tanaka, N. Hayashi and S. Yamashita, *J. Polym. Sci., Polym. Chem. Ed.*, 1980, **18**, 515.
199. M. Miyamoto, M. Sawamoto and T. Higashimura, *Macromolecules*, 1984, **17**, 265.
200. T. Enoki, M. Sawamoto and T. Higashimura, *J. Polym. Sci., Polym. Chem. Ed.*, 1986, **24**, 2261.
201. R. Faust and J. P. Kennedy, *Polym. Bull.*, 1986, **15**, 317.
202. K. Matyjaszewski, *Makromol. Chem., Macromol. Symp.*, 1988, **13/14**, 433.
203. G. Heublein and S. Spange, *Markromol. Chem., Macromol. Symp.*, 1986, **3**, 65.
204. S. D. Pask and P. H. Plesch, *Eur. Polym. J.*, 1982, **18**, 939.
205. J. P. Kennedy, L. S. Minckler, G. Wanless and R. M. Thomas, *J. Polym. Sci., Part A.*, 1964, **2**, 2093.
206. M. Sanders, J. Chandrasekhar and P. v. R. Schleyer, in 'Rearrangements in Ground and Excited States', Academic Press, New York, 1980, vol. 1, p. 1.
207. M. Saunders and J. J. Stofko, Jr., *J. Am. Chem. Soc.*, 1973, **95**, 252.
208. E. Bauml, G. Kolberg and H. Mayr, *Tetrahedron Lett.*, 1987, 387.
209. J. P. Kennedy and T. Kellen, *J. Macromol. Sci., Chem.*, 1982, **A18**, 1189.
210. J. Puskas, G. Kaszas, J. P. Kennedy, T. Kelen, and F. Tudos, *J. Macromol. Sci., Chem.*, 1982, **A18**, 1315.
211. T. Higashimura, M. Mitsuhashi and M. Sawamoto, *Macromolecules*, 1979, **12**, 178.
212. M. Sawamoto, T. Ohtoyo, T. Higashimura, K.-H. Gühzs and G. Heublein, *Polym. J.*, 1985, **17**, 929.
213. P. K. Bossaer, E. J. Goethals, P. J. Hackett and D. C. Pepper, *Eur. Polym. J.*, 1977, **13**, 489.
214. J. P. Kennedy and R. A. Smith, *J. Polym. Sci., Polym. Chem. Ed.*, 1980, **18**, 1539.
215. J. P. Kennedy and R. A. Smith, *J. Polym. Sci., Polym. Chem. Ed.*, 1980, **18**, 1523.
216. M. Samsani, M. Tardi, A. Polton and P. Sigwalt, *Eur. Polym. J.*, 1983, **19**, 287.
217. J. P. Kennedy, *J. Appl. Polym. Sci., Appl. Polym. Symp.*, 1977, **30**, 1
218. Y. Jolivet and J. Peyrot, *Int. Symp. Cationic Polym., Rouen, 1973*, C18.

219. A. G. Harrison, R. Houriet and T. T. Tidwell, *J. Org. Chem.*, 1984, **49**, 1302.
220. R. Pock and H. Mayr, *Chem. Ber.*, 1986, **119**, 2497.
221. P. Knittel and T. T. Tidwell, *J. Am. Chem. Soc.*, 1977, **99**, 3408.
222. M. F. Rausse, J. E. Dubois and A. Argile, *J. Org. Chem.*, 1979, **44**, 1173.
223. H. Mayr and R. Pock, *Tetrahedron Lett.*, 1983, **24**, 2155.
224. H. Mayr, R. Schneider and R. Pock, *Makromol. Chem., Macromol. Symp.*, 1986, **3**, 19.
225. M. Szwarc, *J. Polym. Sci.*, 1955, **16**, 367.
226. B. Giese, *Angew. Chem.*, 1983, **22**, 753.
227. S. Jones, G. Prementine and D. Tirrell, *J. Am. Chem. Soc.*, 1985, **107**, 5275.
228. Y. Wang and L. M. Dorfman, *Macromolecules*, 1980, **13**, 63.
229. L. M. Dorfman and V. M. DePalma, *Pure Appl. Chem.*, 1979, **51**, 123.
230. R. Sujdak, R. L. Jones and L. M. Dorfman, *J. Am. Chem. Soc.*, 1976, **98**, 4875.
231. J. P. Richard and W. P. Jencks, *J. Am. Chem. Soc.*, 1982, **104**, 4689.
232. J. P. Richard, M. E. Rothenburg and W. P. Jencks, *J. Am. Chem. Soc.*, 1984, **106**, 1361.
233. J. P. Richard and W. P. Jencks, *J. Am. Chem. Soc.*, 1984, **106**, 1373.
234. R. A. McClelland, N. Banait and S. Steenken, *J. Am. Chem. Soc.*, 1986, **108**, 7023.
235. R. Schneider, U. Grabis and H. Mayr, *Angew. Chem., Int. Ed. Engl.*, 1986, **25**, 89.
236. R. Schneider and H. Mayr, *Angew. Chem., Int. Ed. Engl.*, 1986, **25**, 1016.
237. K. Matyjaszewski, *J. Macromol. Sci., Rev. Macromol. Chem.*, 1986, **C26**, 1.
238. N. L. Billingham and A. D. Jenkins, in 'Comprehensive Chemical Kinetics', ed. C. H. Bamford and C. F. H. Tipper, Elsevier, Amsterdam, 1976, vol. 14a.
239. M. Szwarc, 'Carbanions, Living Polymers and Electron Transfer Processes', Wiley, New York, 1968.

42

Carbocationic Polymerization: Vinyl Ethers

TOSHINOBU HIGASHIMURA and MITSUO SAWAMOTO

Kyoto University, Japan

42.1 INTRODUCTION

Vinyl ethers (**1**), which have a strongly electron-donating alkoxy substituent, readily form polymers on treatment with an acidic compound (initiator). The polymerization is thus cationic in nature; no anionic or radical polymerizations give high polymers from vinyl ethers (except for the alternating radical copolymerization with an electron-deficient monomer). The existence of an ether oxygen in the pendant group renders the cationic polymerization of vinyl ethers considerably different from those of nonpolar hydrocarbon monomers, such as styrene and isobutylene.

$$CH_2\!=\!\underset{\underset{OR}{|}}{CH}$$

(**1**)

It was in 1878 that Wislicenus reported the transformation of ethyl vinyl ether into a viscous material in the presence of iodine.[1] The first systematic investigations of vinyl ether polymerization began in 1928 at I.G. Farbenindustrie in Germany. An excellent and comprehensive review of these early developments is available.[2] Two decades later, in 1947, Schildknecht *et al.*[3] in the US recognized that poly(isobutyl vinyl ether) obtained with boron trifluoride or its etherate (BF_3OEt_2) is either crystalline (nontacky) or amorphous (tacky) depending on polymerization conditions. This finding is important as it is the first experimental evidence for the stereoisomerism of vinyl polymers (see Section 42.5).

Despite such a long history of research, which provided many interesting findings and even heralded the stereochemical study of polymers, vinyl ether polymerization attracted limited interest among polymer scientists and engineers, primarily because vinyl ether polymers had thus far failed to find extensive industrial use.[4-6] However, the recent developments in this field have been sufficient to revitalize intensive investigations that are directed particularly toward the synthesis of novel functional polymers based on the polymerization chemistry of vinyl ethers. For example, a variety of functionally substituted vinyl ethers (*e.g.* 2-(vinyloxy)ethyl cinnamate[7]) have been found cationically polymerizable (see Sections 42.6.2 and 42.7.2). Another important development is the discovery of 'living' cationic polymerization of vinyl ethers, which has allowed hitherto unattainable control of the molecular weight and structure of poly(vinyl ethers) (see Section 42.6).[8]

42.2 MONOMERS AND INITIATORS

42.2.1 Vinyl Ether Monomers

A number of vinyl ethers have been prepared since the Reppe method was developed. The major methods include: (i) reactions of acetylene with alcohols (Reppe's method); (ii) dealcoholation of acetals; and (iii) transetheration of existing vinyl ethers. Excellent reviews are available on the synthesis,[2] reactions[2] and physical properties[2,9] of vinyl ether monomers; the properties of poly(vinyl ethers) are also reviewed.[2,4,5]

The vinyl ethers studied so far are classified as follows:[9] (i) alkyl vinyl ethers (CH_2=CHOR), R = alkyl (linear, branched or cyclic); (ii) aryl vinyl ethers (CH_2=CHOAr), Ar = phenyl, substituted phenyl, naphthyl; (iii) functionally substituted vinyl ethers (CH_2=CHOX), X = alkyl group with a heteroatom (halogen, silicon, *etc.*), ether, ester, amine or their derivatives (see Section 42.6.2); (iv) divinyl ethers (CH_2=CHOCH=CH_2 and CH_2=CHOXOCH=CH_2), X = $-(CH_2)_n$, $-(CH_2CH_2O)_n CH_2CH_2-$, $-CH_2CH_2OCMe_2C_6H_4CMe_2OCH_2CH_2-$, *etc.* (see Section 42.7.4); (v) α-substituted vinyl ethers (CH_2=CR'OR), R = alkyl; R' = methyl, alkoxy, chlorine; and (vi) β-substituted vinyl ethers (R'CH=CHOR), R = alkyl; R' = alkyl, alkoxy, chlorine.

42.2.2 Initiators for Vinyl Ether Polymerization

Because of their high reactivity, vinyl ethers can be polymerized by a variety of acidic compounds, including those which are too weakly acidic to initiate cationic polymerization of styrene, isobutene and other less reactive vinyl monomers. Typical examples are: (i) protonic acids (*e.g.* CF_3CO_2H, CF_3SO_3H); (ii) metal halides (*e.g.* $SnCl_4$, $FeCl_3$, BF_3, BF_3OEt_2); (iii) halogenated metal alkyls (*e.g.* $RAlCl_2$, R_2AlCl, RMgX; R = alkyl, X = halogen); (iv) cation-forming salts (*e.g.* $Ph_3C^+SnCl_5^-$, $C_7H_7^+SbCl_6^-$, $MeCO^+ClO_4^-$); (v) halogens (*e.g.* I_2, IBr); (vi) modified Ziegler–Natta catalysts (*e.g.* Et_3Al–VCl_4–Bu_3^iAl–THF); (vii) solid acids (*e.g.* Cr_2O_3, $Al_2(SO_4)_3$–H_2SO_4); and (viii) high-energy radiation (*e.g.* γ-ray).

42.3 ELEMENTARY REACTIONS IN VINYL ETHER POLYMERIZATION

The systematic study of the kinetics of vinyl ether polymerization was started in 1947 by Eley and co-workers[10] who polymerized *n*-butyl vinyl ether with $SnCl_4$,[11] iodine,[12] $AgClO_4$[13] and Ph_3CCl.[13] The British research group eventually established the cationic chain polymerization mechanism for vinyl ethers, consisting of initiation, propagation, chain transfer and termination.

42.3.1 Initiation

The initiation step in the cationic polymerization of vinyl ethers is an electrophilic addition of a cation (A^+), derived from an initiator (A^+B^-), across a vinyl ether double bond, to form a monomeric carbocation (equation 1). The initiation processes of typical cationic initiators are briefly discussed below.

$$A^+B^- + CH_2=\underset{\underset{OR}{|}}{CH} \rightarrow ACH_2\underset{\underset{OR}{|}}{CH^+}B^- \qquad (1)$$

42.3.1.1 Protonic acids

Polymerization of vinyl ethers by a protonic acid hardly ever produces high polymers because of accompanying side reactions (see Section 42.3.3), and little study has been devoted to this field. Only recently, the polymerization of isobutyl vinyl ether with CF_3CO_2H was kinetically investigated in detail.[14] High polymers may form at lower temperatures; for instance, at $-78\,°C$ the CF_3SO_3H-initiated polymerization of 2-chloroethyl vinyl ether in CH_2Cl_2 gives polymers of molecular weight $>10^5$.[15]

42.3.1.2 Metal halides and halogenated metal alkyls

In general, metal halides (MtX_n; Mt = metal, X = halogen) need a cationogen or protogen (AB; formerly called 'cocatalyst') in order to initiate cationic polymerization (Scheme 1). The high reactivity of vinyl ethers, however, severely hampered the unambiguous demonstration of the necessity of a cationogen in the MtX_n-mediated polymerization of these monomers. For example, in spite of early studies,[16–18] it still remains unestablished as to whether BF_3OEt_2 is capable of directly

$$MX_n + AB \rightarrow A^+(MX_nB)^-$$

$$A^+(MX_nB)^- + \underset{\underset{OR}{|}}{CH_2{=}CH} \rightarrow ACH_2\underset{\underset{OR}{|}}{CH^+}(MX_nB)^-$$

Scheme 1

initiating vinyl ether polymerization *via* Et^+ or whether it needs water as an impurity for initiation in the absence of a deliberately added cationogen.

Among metal halides, BF_3OEt_2 is most frequently employed for alkyl vinyl ether polymerization, because it readily produces high polymers at low temperature. For more nucleophilic monomers, such as ketene acetals $[CH_2{=}C(OR)_2]$, weakly acidic metal halides (*e.g.* $ZnCl_2$) are more effective for the preparation of high polymers.[19] Metal alkyls also polymerize vinyl ethers either by themselves or in conjunction with an appropriate cationogen (protogen).[20] Typical examples include Et_3Al/H_2O (1:1), Et_2Zn/H_2O (1:1) and $EtMgBr$, among which the first is the most active.

42.3.1.3 Cation-forming salts

Organic salts of suitably reactive carbocations or acyl cations initiate vinyl ether polymerization.[21] Some onium salts decompose upon irradiation by UV and visible light to form a cation (or cation radical) that in turn polymerizes vinyl ethers.[22] Among these, iodonium (**2**) and sulfonium (**3**) salts are useful as thermally stable photoinitiators.

$$Ar_2I^+X^-$$

(**2**) $Ar = Ph$, C_6H_4OMe, *etc.*, $X^- = SbF_6^-$, AsF_6^-, PF_6^-, BF_4^-, *etc.*

$$Ar_3S^+X^-$$

(**3**) $Ar = Ph$, C_6H_4OMe, *etc.*, $X^- = SbF_6^-$, AsF_6^-, PF_6^-, BF_4^-, *etc.*

42.3.1.4 Halogens

Historically, iodine was the first initiator employed for vinyl ether polymerization,[1] for which kinetic studies were carried out by Eley and Richards.[12] Other halogens and interhalogen compounds (*e.g.* IBr) also induce polymerization of vinyl ethers. The rate of polymerization of *n*-butyl vinyl ether with these initiators decreases in the order: $I_2 > ICl > IBr > Br_2$; the rate for bromine is only one thousandth of that for iodine.[23]

42.3.1.5 Heterogeneous initiators[5]

Metal oxides and related solid acids have been extensively studied as initiators for the synthesis of stereoregular or high molecular weight poly(vinyl ethers) at room temperature; see Section 42.5.

42.3.1.6 High-energy radiation

See Section 42.3.2.2.

42.3.2 Propagation

The propagation reaction in vinyl ether polymerization, in general, proceeds *via* a carbocation (**4**) as the growing species (equation 2, where B^- is the counteranion derived from an initiator). The kinetics and mechanism of the propagation process for vinyl ethers have been investigated in detail;[10,24] see also Section 42.4.5.

$$\sim CH_2\underset{\underset{OR}{|}}{CH^+}B^- + \underset{\underset{OR}{|}}{CH_2{=}CH} \overset{k_p}{\rightarrow} \sim CH_2\underset{\underset{OR}{|}}{CH}CH_2\underset{\underset{OR}{|}}{CH^+}B^- \qquad (2)$$

$$(\textbf{4})(\textbf{4})$$

42.3.2.1 Propagation rate constant

In an organic solvent, the carbocationic growing species (4) is either a (solvated) free ion or an ion pair or an equilibrium mixture of both. The free-ion propagation rate constant k_p^+ may be obtained in the polymerization by high-energy radiation[25, 26] or cation-forming salts[21, 27] under suitable conditions. The ion-pair propagation rate constant $k_p^{\pm}$ has not yet been determined separately.

Table 1 lists the representative k_p^+ values for a series of alkyl vinyl ethers; a more comprehensive compilation is found in Kennedy and Maréchal's book.[24] It must be kept in mind that all these k_p^+ values involve an assumption in the determination of the concentration of the propagating species. The dependence of k_p^+ on the alkyl groups in vinyl ethers, as seen in Table 1, is discussed in Section 42.4.1.

Table 1 Representative Free Ion Propagation Rate Constants $(k_p^+)^a$ for Alkyl Vinyl Ethers (CH$_2$=CHOR) at 0 °C

	Radiation (γ-ray, bulk)			Carbocation salt (CH$_2$Cl$_2$)	
Monomer (R)	k_p^+	Ref.	Initiator	k_p^+	Ref.
Methyl	—[b]	d	C$_7$H$_7^+$SbCl$_6^-$	$1.4 \times 10^{2\,c}$	g
Ethyl	7.2×10^3	e	C$_7$H$_7^+$SbCl$_6^-$	$1.5 \times 10^{3\,c}$	g
			Ph$_3$C$^+$SbCl$_6^-$	5.1×10^3	h
Isopropyl	9.0×10^5	e	Ph$_3$C$^+$SbCl$_6^-$	1.6×10^3	i
Isobutyl	3.8×10^4	e	Ph$_3$C$^+$SbCl$_6^-$	4.0×10^3	j
	3.8×10^4	f	Ph$_3$C$^+$SbCl$_6^-$	7.0×10^3	k
			Ph$_3$C$^+$SbCl$_6^-$	9.2×10^3	h
			Ph$_3$C$^+$BF$_4^-$	2.8×10^3	j
			C$_7$H$_7^+$SbCl$_6^-$	6.8×10^3	j
t-Butyl	5.0×10^4	e	C$_7$H$_7^+$SbCl$_6^-$	3.5×10^3	g
Cyclohexyl	—		C$_7$H$_7^+$SbCl$_6^-$	3.3×10^3	g
2-Chloroethyl	—		C$_7$H$_7^+$SbCl$_6^-$	$2.0 \times 10^{3\,c}$	g

[a] 1 mol^{-1} s^{-1}. See ref. 24 for more comprehensive data.
[b] No polymerization due to the radiolysis of methyl vinyl ether.
[c] Not k_p^+ but a composite of k_p^+ and $k_p^{\pm}$.
[d] V. R. Desai, Y. Suzuki and V. Stannett, *J. Macromol. Sci., Chem.*, 1977, **A11**, 133.
[e] A. M. Goineau, J. Kohler and V. Stannett, *J. Macromol. Sci., Chem.*, 1977, **A11**, 99.
[f] Ka. Hayashi, Ko. Hayashi and S. Okamura, *J. Polym. Sci., Part A-1*, 1971, **9**, 2305.
[g] A. Ledwith, E. Lockett and D. C. Sherrington, *Polymer*, 1975, **16**, 31.
[h] Y. J. Chung, J. M. Rooney, D. R. Squire and V. Stannett, *Polymer*, 1975, **16**, 527.
[i] C. C. Ma, H. Kubota, J. M. Rooney, D. R. Squire and V. Stannett, *Polymer*, 1979, **20**, 317.
[j] C. E. H. Bawn, C. Fitzsimmons, A. Ledwith, J. Penfold, D. C. Sherrington and J. A. Weightman, *Polymer*, 1971, **12**, 119.
[k] F. Subira, G. Sauvet, J. P. Vairon and P. Sigwalt, *J. Polym. Sci., Polym. Symp.*, 1976, **56**, 221.

42.3.2.2 Radiation-induced polymerization in solution

Because of the difficulty in purifying and drying solvents, radiation-induced polymerization is usually performed in bulk monomer.[25, 26] For comparison with these bulk processes, Stannett and co-workers studied γ-ray-induced polymerizations of ethyl and isopropyl vinyl ethers in various solvents under the so-called super-dry conditions.[29–34] With both monomers, the polymerization rate and k_p^+ sharply decrease on going from bulk to solution: bulk > C$_6$H$_6$ > Et$_2$O- ≫ CH$_2$Cl$_2$ ≫ MeNO$_2$. Stannett concluded that the decrease in k_p^+ is due to the solvation (or stabilization) of the free ionic propagating carbocation, either by solvent or by a pendant ether oxygen.[34]

42.3.2.3 Polymerization initiated by iodine

In an attempt to determine the growing center concentration, Higashimura et al.[35] measured, by iodometry, the amount of iodine consumed during the early stages of the polymerization of vinyl ethers by this halogen. Later studies, however, showed that the reaction of iodine with a vinyl ether produces not only an active propagating species but an inactive adduct. For example, Ledwith and Sherrington reported the rapid formation of the iodine–vinyl ether adduct (5) in methylene chloride at $-40\,^\circ$C, and proposed an initiation mechanism in which the $^\alpha$CH–I bond of (5) undergoes heterolysis with the assistance of free iodine (Scheme 2).

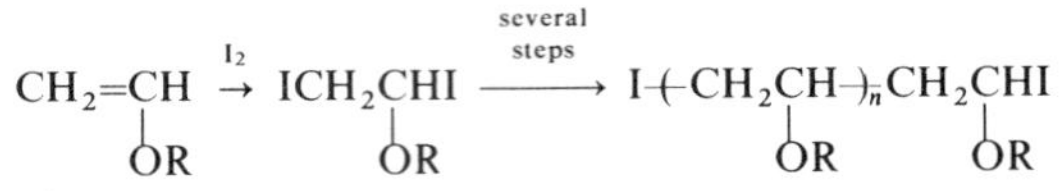

$$CH_2=CH \xrightarrow{I_2} ICH_2CHI \xrightarrow[\text{steps}]{\text{several}} I(CH_2CH)_nCH_2CHI$$

with OR groups:

(5)   (6)

Scheme 2

The formation of the diiodide (5) or its higher homologs (6) was also observed in the polymerization of *n*-butyl vinyl ether with iodine in *n*-hexane.[37] It was proposed that propagation might proceed *via* polymer (6) with a terminal C–I bond (Scheme 2).[37–39] The involvement of such a terminal C–I bond is now known to also play an important role in the living polymerization of vinyl ethers by the hydrogen iodide/iodine initiating system (see Section 42.6.1).[40]

42.3.3 Chain Transfer and Termination

Cationic polymerization of vinyl ethers hardly ever produces high polymers unless carried out at low temperature. The low molecular weight of the product is due to the chain transfer that frequently occurs, particularly at or above room temperature.

42.3.3.1 Termination

Compared with chain transfer, termination is much less important as a chain-breaking reaction in vinyl ether polymerization, except when a basic compound (*e.g.* pyridine[41]) is present in the reaction mixture.

42.3.3.2 Chain transfer

Chain transfer to bases (or transfer agents) in vinyl ether polymerization was first recognized by Eley *et al.*[11,12] Even in the absence of a transfer agent, it is usually difficult to obtain high polymers from vinyl ethers at room temperature. This difficulty is attributed to the instability of the propagating carbocation, which tends to transfer its β proton either to the counteranion or to the incoming monomer. In most cases,[12,13,27,42,43] the more frequently chain transfer occurs, the higher are the amounts of terminal double bond ($\sim$CH=CHOR) and acetal end group [$\sim$CH$_2$CH(OR)$_2$] in the product polymers.

The polymerization mixtures of vinyl ethers in polar media or at high temperatures often exhibit an intense color that changes from yellow through blue-green to dark violet with the progress of the reaction. This coloration long eluded a consistent explanation, but has recently been established to result from delocalized polymeric carbocations that form *via* protonation of conjugated polyene terminals such as (7) and (8).[44] Scheme 3 illustrates how (7) and (8) evolve. The formation of polyene terminals was also reported for ketene acetals.[45]

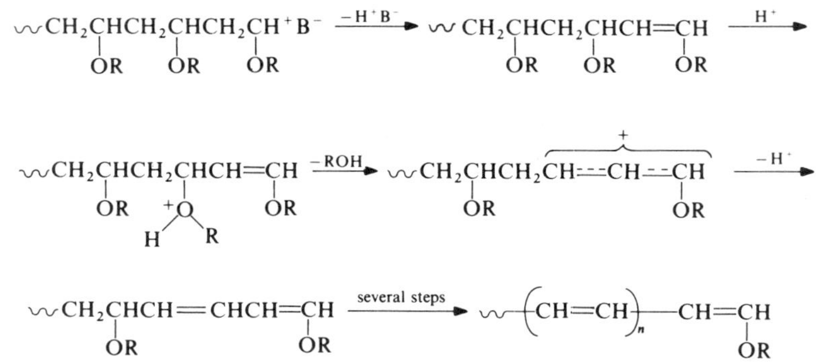

(7)   (8)

Scheme 3

42.4 STRUCTURE AND REACTIVITY IN VINYL ETHER POLYMERIZATION

42.4.1 Alkyl Vinyl Ethers

42.4.1.1 *Effects of alkyl substituents*

Even in the late 1940s, when the first systematic study on vinyl ether polymerization began, it was well known that the polymerization reactivity of alkyl vinyl ethers depended on their pendant alkyl groups. The reactivity orders for homopolymerization[27,28,41,46,47] roughly agree with those for copolymerization,[48-50] although there are some discrepancies from one to another depending on research groups. Inspection of these orders indicates that the reactivity of alkyl vinyl ethers depends primarily on the electron-donating power of the alkyl substituents through the inductive effect (*e.g.* $Me < Et < Pr^i < Bu^t$); in fact, the $\log(1/r_1)$ values in copolymerization correlate linearly to Taft's σ^* constants for the pendant groups (r_1 = monomer reactivity ratio).[48,49] Similar side-chain effects were found in the related electrophilic reactions of alkyl vinyl ethers, such as acid-catalyzed hydrolysis,[51,52] cycloaddition to diphenylketene ($Ph_2C=C=O$),[28] and π-complex formation with molecular iodine.[53]

The alkyl substituents in vinyl ethers may affect their reactivity not only through the inductive effect but through the resonance effect, mediated by the intervening ether oxygen ($CH_2CH-O-R \leftrightarrow {}^-CH_2CH=O^+-R$). The importance of the latter was shown in the copolymerization among a series of ring-substituted phenyl vinyl ethers ($CH_2=CHOC_6H_4X$) by BF_3OEt_2[54] and also in their acid-catalyzed hydrolysis,[55] where the relative reactivity of these vinyl ethers correlates linearly to Hammett's σ constants for X.

42.4.1.2 *π-Electron density of the vinyl ether double bond*

Yuki and co-workers[56] found that the π-electron density (estimated by ^{13}C NMR) on the vinyl β carbon is *lower* in a more reactive vinyl ether that carries a more electron-donating alkyl pendant group. An explanation of this unexpected fact involves the characteristic conformation of vinyl ether monomers. Because of a partial double-bond character of the C–O linkage (see the resonance structures shown above), vinyl ethers may assume the *s-cis* and *gauche* (or *s-trans*) conformations. According to IR[57] and ^1H NMR[58] structural analyses, methyl vinyl ether assumes almost entirely the *s-cis* structure;[57] the contribution of the *gauche* form increases with increasing bulkiness of the alkyl group; and the latter conformer predominates in *t*-butyl vinyl ether.[58]

The *s-cis* structure permits an effective overlap between the π orbital of the vinyl group and the nonbonded electron-pair orbital of the ether oxygen, whereas such a resonance interaction is sterically prohibited in the *gauche* form. Thus, the π-electron density on the β carbon may increase in methyl vinyl ether (*s-cis*).

The characteristic structure–reactivity relationship in vinyl ether polymerization, described above, provides insights into the pathway of the propagation reaction (see Section 42.4.5).

42.4.2 Functionalized Vinyl Ethers

In the course of their investigations on living cationic polymerization (see Section 42.6), Higashimura *et al.* found that the reactivity of functionally substituted vinyl ethers ($CH_2=CHOCH_2CH_2X$) in homo- and co-polymerizations is affected dramatically by their pendant groups (X).[59-61] For example, the polymerization rates change with X in the following order: $O(CH_2CH_2O)_nEt > H \sim alkyl > O_2CMe$; the rate difference between the poly(oxyethylene)- and ester-containing monomers reaches two orders of magnitude.[60] Such a great change in reactivity is rather surprising in view of the fact that the polar substituents of these monomers are isolated from the vinyl ether moiety by an ethylene spacer. It is proposed that the substituent (X) may affect the reactivity of the growing end through an intramolecular solvation.[34,60]

42.4.3 α-Substituted Vinyl Ethers

The reactivity of α-substituted vinyl ethers (**9**)[2] is influenced, electronically and sterically, by the α-substituents, R. In homopolymerization, for instance, the monomers with an α-methyl (**9a**)[62] or α-ethoxy (**9d**; ketene acetals)[19] group are more reactive than their nonsubstituted counterparts (ethyl vinyl ether), whereas those with a bulky α-substituent (*e.g.* **9b** and **9c**) have no homopolymerizability

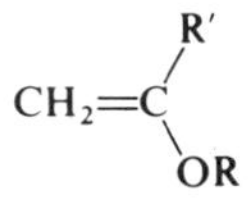

$$CH_2{=}\underset{\underset{\displaystyle OR}{|}}{\overset{\overset{\displaystyle R'}{|}}{C}}$$

(9) a: R = Et, R′ = Me
   b: R = Et, R′ = Et
   c: R = Et, R′ = Ph
   d: R = Et, R′ = OEt

at all.[62] However, all of (9a)–(9d) undergo copolymerization with ethyl vinyl ether, in which they are invariably more reactive than the parent vinyl ether.[62] Thus, the introduction of an electron-donating group on to the α carbon increases the reactivity of alkyl vinyl ethers, unless the α-substituent exerts a severe steric hindrance.

42.4.4 β-Substituted Vinyl Ethers

42.4.4.1 *Alkenyl ethers and their derivatives*[63]

Higashimura and co-workers[64, 65] copolymerized a series of propenyl ethers (10; R = Me, Et, Bu^n, Bu^t; R′ = Me; *cis* and *trans*) with the corresponding alkyl vinyl ethers. Table 2 shows the typical results for the *cis* isomers. Except when R = Bu^t, these *cis*-propenyl ethers are consistently 2–4 times more reactive (with respect to both $1/r_1$ and r_2) than the vinyl ether counterparts; *trans* isomers exhibit a comparable reactivity. When R is very bulky, such as *t*-butyl, the reactivities of both *cis*- and *trans*-propenyl ethers are lower than those of the corresponding vinyl ethers. Similar results were obtained for alkenyl ethers with a longer straight-chain β substituent.[66, 67]

$$R'CH{=}CHOR$$

(10) *cis, trans*

Table 2 Relative Reactivity of *cis*-Propenyl Ethers in Cationic Copolymerization with the Corresponding Vinyl Ethers Initiated by BF$_3$OEt$_2$ at $-78\,°C^a$

M_1	$M_2\,(cis)$	$1/r_1\,(k_{12}/k_{11})$	$r_2\,(k_{22}/k_{21})$	*Ref.*
CH$_2$=CHOMe	MeCH=CHOMe	3.8	3.8	c
CH$_2$=CHOEt	MeCH=CHOEt	2.9	4.0	c
CH$_2$=CHOBu^n	MeCH=CHOBu^n	2.0	4.0	b, d
CH$_2$=CHOBu^t	MeCH=CHOBu^t	0.45	0.28	e

^{a} Solvent = toluene.
^{b} Solvent = CH$_2$Cl$_2$.
^{c} M. Mizote, S. Kusudo, T. Higashimura and S. Okamura, *J. Polym. Sci., Part A-1*, 1967, **5**, 1727.
^{d} T. Higashimura, S. Kusudo, Y. Ohsumi and S. Okamura, *J. Polym. Sci., Part A-1*, 1968, **6**, 2523.
^{e} T. Higashimura and K. Yamamoto, unpublished results (1974).

The introduction of a β-alkoxy group (10; R′ = OMe, OEt) to an alkyl vinyl ether remarkably increases its reactivity in cationic polymerization.[68–70] On the other hand, the reactivity of a β-chloro derivative (ClCH=CHOBu^i) is 1/26 that of isobutyl vinyl ether.[71] The electronic nature of the β substituents, therefore, is the primary factor that determines the reactivity of β-substituted vinyl ethers.

42.4.4.2 *Effect of monomer geometry*

An interesting feature of β-substituted vinyl ethers is their geometrical isomerism. In the cationic copolymerization of a pair of *cis* and *trans* isomers,[65–69, 71, 72] in general, the *cis* form is more reactive than its *trans* counterpart, independent of the structure of the α- and β-substituents. The higher reactivity of the *cis* isomer is attributed, at least partly, to its lower thermodynamic stability.[73]

42.4.4.3 *π-Electron density of the double bond*

Higashimura *et al.*[74] compared the π-electron densities of the carbon–carbon double bonds in *cis*-alkenyl ethers and the corresponding vinyl ethers on the basis of ^{13}C NMR chemical shifts. On going from ethyl vinyl ether to ethyl propenyl ether, for example, the α-carbon shifts slightly upfield while the β carbon shifts considerably downfield. Namely, the introduction of an electron-donating β-alkyl group *decreases* the π-electron density on the vinyl β carbon. The same conclusion was reached from quantum chemical calculation of the atomic orbital populations for *cis*- and *trans*-alkenyl ethers.[75,76]

42.4.5 Pathway of the Propagation Reaction

In the propagation step in cationic polymerization, the electrophilic growing end combines with the β carbon of a vinyl monomer (see equation 2). Despite this fact, the higher the polymerization reactivity of a vinyl ether, the lower the π-electron density on its β carbon and the higher is that of the α carbon (see above). This apparently puzzling trend suggests that the π-electron density on the β carbon is not the sole factor that determines the reactivity of vinyl ethers and their analogs.

Several models[21,28,69,75,76] have been proposed for the transition state in vinyl ether polymerization to account for these experimental results. A general conclusion drawn from all these studies is that the propagating species interacts not only with the β carbon but also with the α carbon of the incoming monomer, probably *via* a π-complex-type transition state, illustrated in model (11).

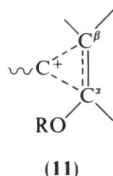

(11)

42.5 STEREOCHEMISTRY OF VINYL ETHER POLYMERIZATION

42.5.1 Alkyl Vinyl Ethers

42.5.1.1 *Discovery of stereospecific polymerization*[2]

Poly(vinyl ether) was the first vinyl polymer for which the stereoisomerism of the main chain was discovered. In 1947, Schildknecht noted that the nontacky[3] and crystalline[77] polymer, obtained in the 'polyphase' ('proliferous')[46] polymerization of isobutyl vinyl ether (IBVE) with BF$_3$OEt$_2$ at about $-80\,°C$, differed in appearance and physical properties from the tacky and amorphous[77] poly(IBVE) then commercially available as Oppanol C. It is important that the crystalline and amorphous poly(IBVE) can be obtained separately by simply regulating polymerization procedure.[2,46]

Unfortunately, the structure of this apparently stereoregular polymer was not pursued further at that time. It was not until 1956 that, after establishing the stereochemistry of polyalkenes, Natta *et al.* recognized the crystalline poly(IBVE) as 'isotactic'.[78] Schildknecht suggested that the formation of crystalline polymers might require a heterogeneous polymerization proceeding slowly at low temperature in a viscous gel phase.[79]

42.5.1.2 *Homogeneous stereoregular polymerization*

In 1958, Higashimura and co-workers[80] found that crystalline poly(IBVE) could be obtained even in a homogeneous polymerization where IBVE was added to a cooled solution of vigorously stirred fine suspension of BF$_3$OEt$_2$ in a nonpolar solvent (*e.g.* toluene). The polymerization occurred in a transparent solution free from a polymer gel phase. The produced polymer was more stereoregular (isotactic) than the poly(IBVE) prepared by corresponding polyphase polymerization.[81,82] The Japanese group also obtained crystalline polymers from methyl vinyl ether (MVE) in a homogeneous polymerization with BF$_3$OEt$_2$ in a toluene–*n*-hexane mixed solvent.[80] These findings demonstrate that the heterogeneity of a polymerization system is not always mandatory for the formation of crystalline isotactic poly(vinyl ethers).

42.5.1.3 Influence of reaction conditions

(i) Polymerization solvent

In the polymerzation of vinyl ethers at low temperature, a polymer rich in the isotactic structure forms in a nonpolar solvent in general, and the syndiotactic and heterotactic structures increase with increasing solvent polarity. The dependence of polymer stereoregularity on solvent polarity is small with MVE[83] but is considerable with vinyl ethers carrying a bulky substituent (*e.g.* trimethylsilyl).[84, 85]

(ii) Initiator (counteranion)

The nature of initiators may affect the stereoregularity of poly(vinyl ethers), because it determines the nature of the counteranion associating with the propagating species. For example, in the polymerization of IBVE in a nonpolar solvent at low temperature, BF_3OEt_2,[3] $EtAlCl_2$[86] and Et_2AlCl[86] lead to crystalline polymers, but $SnCl_4$ does not.[81]

(iii) Polymerization temperature

Highly isotactic poly(vinyl ethers) form only at low temperature in a homogeneous polymerization (*cf.* Section 42.5.2); elevating polymerization temperature results in an increase of random sequences in the backbone.

(iv) Monomer side-chains

The isotacticity of a series of poly(alkyl vinyl ethers) obtained under identical conditions increases, though not dramatically, with increasing bulkiness of the pendant groups.[87]

42.5.1.4 Mechanism of stereoregular polymerization

Cationic polymerization in nonpolar media often gives rise to isotactic-rich polymers, whereas radical polymerization generally yields syndiotactic-rich polymers. There is no doubt, therefore, that the counteranion of a propagating species plays a crucial role in the formation of isotactic polymers.

Several mechanisms have been developed which interpret the isotactic propagation of vinyl ethers.[88-91] All these mechanisms account for, at least qualitatively, the observed variation of the isotacticity of polymers with reaction conditions such as solvent polarity. However, they do not suffice to discuss the direction of the double-bond opening (see Section 42.5.5) and the penultimate-group effects.[92]

42.5.2 Stereoregular Polymerization by Solid Initiators

The homogeneous polymerization of vinyl ethers by soluble initiators, such as BF_3OEt_2 and $EtAlCl_2$, affords isotactic polymers only at low temperature (see above). This limitation is also the case for binary Ziegler-type initiators (*e.g.* $TiCl_4$–R_3Al[93] and $(CPD)_2TiCl_2$–R_3Al (CPD = cyclopentadienyl)[85]). In contrast, another class of initiators, mostly insoluble solid acids, produce isotactic poly(vinyl ethers) at higher temperatures above 0 °C.

42.5.2.1 Modified Ziegler–Natta catalysts

It was Vandenberg[94] who first obtained in 1959 a series of highly crystalline poly(vinyl ethers) at room temperature. He employed so-called modified Ziegler–Natta catalysts, such as a VCl_4–Et_3Al mixture treated with Bu^i_3Al–THF complex in *n*-heptane.[95] In particular, this vanadium-based initiator led to highly crystalline poly(MVE) (m.p. 144 °C) insoluble in water as well as in methanol.

42.5.2.2 Solid-supported protonic acids

Simple protonic acids such as hydrochloric and sulfuric acids, which *per se* cannot produce high polymers from vinyl ethers, may lead to effective initiators when bound to appropriate solid supports. Examples of such heterogeneous initiators include $Al_2(SO_4)_3$–H_2SO_4[96] and Ni_2O_3–HCl[97] complexes. Higashimura and co-workers[98] found that a series of metal sulfate–sulfuric acid complexes efficiently polymerize vinyl ethers at room temperature to give crystalline (isotactic) polymers. The best results are obtained with sulfates of Al^{3+}, Cr^{3+} and Fe^{3+};[98,99] in particular, the aluminum sulfate–sulfuric acid complex yields stereoregular poly(IBVE) with higher crystallinity than those obtained in the homogeneous polymerization with BF_3OEt_2 at low temperature. The yield and molecular weight of the crystalline polymers vary with the metal cations in the sulfates.[100]

42.5.2.3 Metal oxides

Cationic polymerization of vinyl ethers can be initiated by metal oxides such as Cr_2O_3. As solid acids, they consist of Lewis-acid sites (electron-deficient metal atoms) and Brönsted-acid sites (active protons derived from adsorbed water *etc.*). The latter have been shown to be responsible for the initiation of cationic polymerization.

Iwasaki[101] obtained highly stereoregular poly(IBVE) in high yield using Cr_2O_3 as initiator in toluene at a temperature as high as 80 °C; in contrast, the use of MgO_2 and Al_2O_3 resulted in amorphous products. It was proposed that catalysts for stereoregular polymerization should have a tetrahedral structure with a short 'active edge', as visualized by the thick line in (12) for Cr_2O_3.[102]

(12)

42.5.3 Asymmetric Polymerization of Vinyl Ethers[103]

42.5.3.1 Asymmetric-selective polymerization

Vinyl ethers may carry a substituent with a chiral carbon(s). When a chiral initiator is employed with a racemic mixture of such a monomer, one enantiomer (either *R*- or *S*-monomer) may be polymerized selectively, with another enantiomer left unchanged, to give an optically active polymer. For vinyl ethers, however, a highly asymmetric-selective polymerization has not yet been attained,[104,106] although a small enantiomer selection was observed in the polymerization of a chiral propenyl ether (13; *cis*) with chiral initiators.[106]

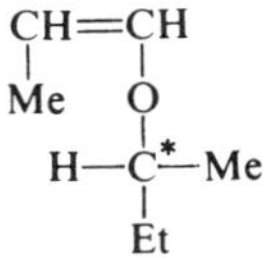

(13) *cis*

42.5.3.2 Asymmetric-induction polymerization

This polymerization gives a chiral (optically active) polymer from an achiral monomer with a prochiral carbon(s). A classic example is the polymerization of benzofuran with $EtAlCl_2$ in conjunction with chiral cationogens, the chirality of which determines the optical activity of the product polymers; *e.g.* $[\alpha]_D$ (in benzene) $= -24.1$ to -33.1 with $(-)$-β-phenylalanine and $+13.1$ with $(+)$-β-phenylalanine.[107] Despite some success, neither asymmetric-selective nor asymmetric-induction cationic polymerization of vinyl ethers so far exceeds the corresponding anionic and coordination processes in the extent of enantiomer selectivity and in the induced optical activity of polymers.

42.5.4 α-Methylvinyl Ethers

α-Methylvinyl ethers are considerably more reactive than the parent vinyl ethers and can be polymerized with weakly acidic initiators, such as $FeCl_3$ and iodine.[108] The ^1H and ^{13}C NMR study by Matsuzaki *et al.*[109] showed that, in contrast to poly(vinyl ethers), the syndiotactic (or racemic) structure is favored for poly(α-methylvinyl alkyl ethers), irrespective of solvent polarity and of the nature of initiators (Table 3).

Table 3 Steric Structure (Syndiotacticity)[a] of Poly(α-methylvinyl alkyl ethers) Obtained at $-78\,°C$[b]

Polymerization conditions		Poly(α-methylvinyl methyl ether)		Poly(α-methylvinyl isobutyl ether)	
Initiator	Solvent	r	rr	r	rr
I_2	CH_2Cl_2	0.90	0.80	0.76	0.38
I_2	Toluene	0.80	0.64	0.70	0.49
$FeCl_3$	Toluene	0.75	0.57	0.66	0.49

[a] By ^{13}C NMR spectroscopy; r = the fraction of racemic dyads.
[b] K. Matsuzaki, S. Okuzono and T. Kanai, *J. Polym. Sci., Polym. Chem. Ed.*, 1979, **17**, 3447.

42.5.5 β-Methylvinyl Ethers (Propenyl Ethers)

Ample studies are available on the stereochemistry of the cationic polymerization of propenyl ethers.[110-116] It is uncertain, however, that the conclusions drawn for these β-methyl derivatives can immediately be applied without reserve to vinyl ethers, because the β-methyl substituent affects cationic polymerization not only electronically but sterically as well.

The stereochemistry of the polymerization of propenyl ethers, as α,β-disubstituted ethylenes, depends on both the direction of monomer addition to the growing end and the direction of the subsequent double-bond opening (*cis* and *trans* openings). The steric structures of poly(β-substituted vinyl ethers) obtained in homogeneous polymerization indicate that *cis* opening occurs in all cases.[110,111] The steric course of propenyl ether polymerization is also discussed on the basis of model reactions[112,113] and the dependence of polymer steric structure on reaction conditions.[114-116]

42.6 LIVING CATIONIC POLYMERIZATION OF VINYL ETHERS[8,117]

42.6.1 Living Polymerization of Alkyl Vinyl Ethers

42.6.1.1 *Hydrogen iodide/iodine initiating system*

In 1984, Higashimura and co-workers demonstrated, for the first time in this field, that an initiating system consisting of hydrogen iodide and molecular iodine (HI/I_2) induces truly living polymerization of IBVE in nonpolar media.[118] When hydrogen iodide is added to excess IBVE in a nonpolar solvent at low temperature, below $-15\,°C$, no polymerization occurs, but instead the adduct (**14**) forms between the acid and IBVE (see Scheme 4 below). However, addition of a small amount of iodine to this quiescent mixture leads to a rapid polymerization, which is faster than that by iodine alone under the same conditions. Detailed studies demonstrated the perfectly living nature of the HI/I_2-initiated polymerization of IBVE.[118,119] For example, the number-average molecular weight ($\bar{M}_n$) of the polymers increases in direct proportion to monomer conversion, is inversely proportional to the initial concentration of hydrogen iodide and is in good agreement with the calculated value assuming one polymer chain per unit hydrogen iodide (not iodine) (Figure 1). The molecular weight distribution (MWD) of the polymers stays very narrow ($\bar{M}_w/\bar{M}_n < 1.1$) at any conversion ($\bar{M}_w$ = weight-average molecular weight).

The living polymerization by the HI/I_2 system was soon extended to include a variety of alkyl vinyl ethers (methyl to *n*-hexadecyl,[120] 2-chloroethyl[121] and benzyl[59]) and propenyl ethers.[122,123] Figure 2 shows the very narrow MWDs of the living poly(alkyl vinyl ethers).[120] Although these living polymerizations are usually carried out in nonpolar solvents (*n*-hexane, methylcyclohexane and toluene),[118-120] a recent study showed that a relatively polar solvent (CH_2Cl_2) can also be used,

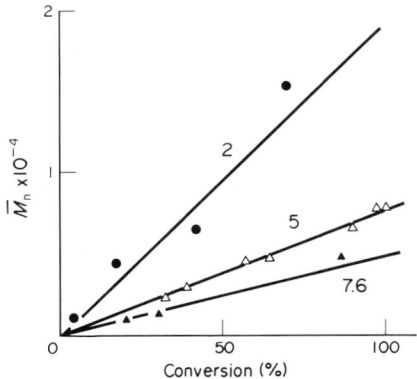

Figure 1 $\bar{M}_n$ *vs.* conversion plots for poly(isobutyl vinyl ether) obtained by the HI/I_2 initiating system in *n*-hexane at $-15\,^{\circ}C$: $[\text{monomer}]_0 = 0.38\ \text{mol}\,l^{-1}$; $[HI]_0 = [I_2]_0$ (as indicated in $\text{mmol}\,l^{-1}$).[118] The straight lines show the calculated $\bar{M}_n$ values for living polymers: $\bar{M}_n = (\text{weight of consumed monomer per l})/[HI]_0$

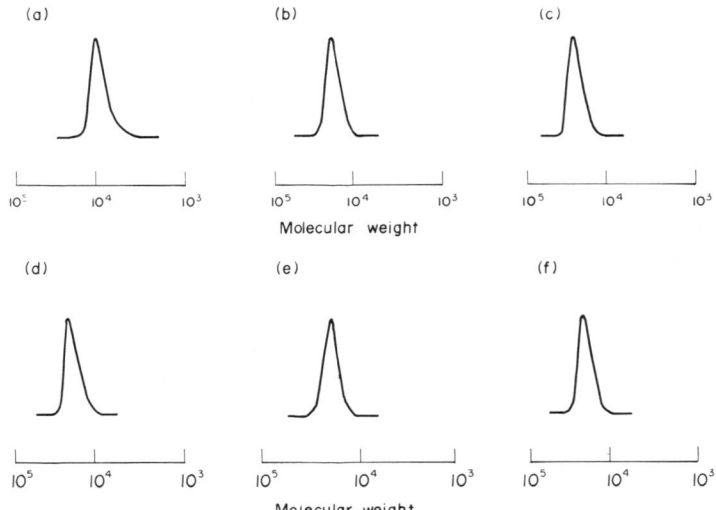

Figure 2 Molecular weight distribution of polymers obtained from alkyl vinyl ethers ($CH_2{=}CHOR$) by the HI/I_2 initiating system in nonpolar media at -5 to $-15\,^{\circ}C$:[120] (a) $R = Me$, $[M]_0/[HI]_0 = 100$, $\bar{M}_w/\bar{M}_n = 1.10$; (b) $R = Et$, $[M]_0/[HI]_0 = 190$, $\bar{M}_w/\bar{M}_n = 1.06$; (c) $R = Pr^i$, $[M]_0/[HI]_0 = 168$, $\bar{M}_w/\bar{M}_n = 1.04$; (d) $R = Bu^n$, $[M]_0/[HI]_0 = 267$, $\bar{M}_w/\bar{M}_n = 1.07$; (e) $R = Bu^i$, $[M]_0/[HI]_0 = 134$, $\bar{M}_w/\bar{M}_n = 1.04$; (f) $R = n\text{-}C_{16}H_{33}$, $[M]_0/[HI]_0 = 111$, $\bar{M}_w/\bar{M}_n = 1.05$

where living polymerization is much faster than in *n*-hexane at the same concentrations of hydrogen iodide and iodine.[124]

42.6.1.2 Mechanism of living polymerization

Higashimura *et al.*[40] analyzed the living polymerization system initiated by HI/I_2 *in situ* by low-temperature ^1H NMR and UV spectroscopy. On the basis of these and other studies, the Kyoto school proposed the initiation/propagation mechanism shown in Scheme 4.

The C–I bond of (**14**) and its higher homologs (**16**) are *activated* (dissociated or polarized) by added iodine as in (**15**), and the incoming monomer reacts with this activated bond. The vinyl ether polymerization by iodine alone, which yields long-lived polymers,[37, 118, 125] most probably follows a similar mechanism.[40]

According to this mechanism, a stable C–I bond is formed at the growing end in *each* propagation step. The formation of this bond is attributed to the strong interaction of the iodide anion with the propagating carbocation.[8] The stability of the propagating species, in turn, suppresses chain transfer, termination and other side reactions. It is concluded that a suitable combination of the

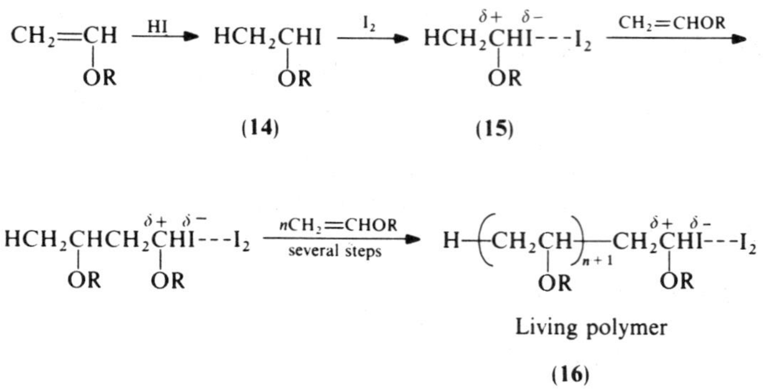

$$CH_2{=}CH \xrightarrow{\text{HI}} HCH_2CHI \xrightarrow{I_2} HCH_2\overset{\delta+}{C}HI\overset{\delta-}{---}I_2 \xrightarrow{CH_2{=}CHOR}$$
$$\quad\underset{OR}{\quad}\qquad\qquad \underset{OR}{\quad}\qquad\qquad\quad \underset{OR}{\quad}$$

$$\qquad\qquad\qquad\qquad\qquad (14)\qquad\qquad\qquad (15)$$

$$HCH_2CHCH_2\overset{\delta+}{C}HI\overset{\delta-}{---}I_2 \xrightarrow[\text{several steps}]{nCH_2{=}CHOR} H{-}{\Big(}CH_2CH{\Big)}_{n+1}{-}CH_2\overset{\delta+}{C}HI\overset{\delta-}{---}I_2$$
$$\underset{OR}{\quad}\ \underset{OR}{\quad}\qquad\qquad\qquad\qquad\qquad\qquad\qquad\ \underset{OR}{\quad}\qquad \underset{OR}{\quad}$$

$$\text{Living polymer}$$

$$(16)$$

Scheme 4

stability of the growing carbocation and the nucleophilicity of its counteranion is the key to living cationic polymerization.[8,126]

42.6.2   Living Polymerization of Functionalized Vinyl Ethers

An advantage of vinyl ethers as monomers is that a variety of functional groups (*e.g.* ester) can readily be incorporated into their pendant groups.[2,9,59] The recent studies by Higashimura *et al.*[59] revealed that such functionalized vinyl ethers, similarly to the alkyl derivatives, undergo living polymerization in the presence of HI/I_2, although esters and other polar functions often induce chain transfer and termination in cationic polymerization initiated by metal halides and protonic acids.

Figure 3 lists functionally substituted vinyl ethers (**17–27**) that are known thus far to give living polymers when polymerized by the HI/I_2 system. The functional pendant groups of these monomers fall into the following four groups: (i) saturated esters (**17–20**)[59,127–129] (ii) unsaturated esters (**21–24**);[130–131] (iii) oxyethylene or poly(oxyethylene) (**25** and **26**);[60,61] and (iv) imide (**27**).[132] Although vinyl ethers with an oxyethylene unit(s) (**25**[60] and **26**[61]) or an imide function (**27**)[132] form living polymers, the conditions for the HI/I_2-initiated living polymerizations are limited compared

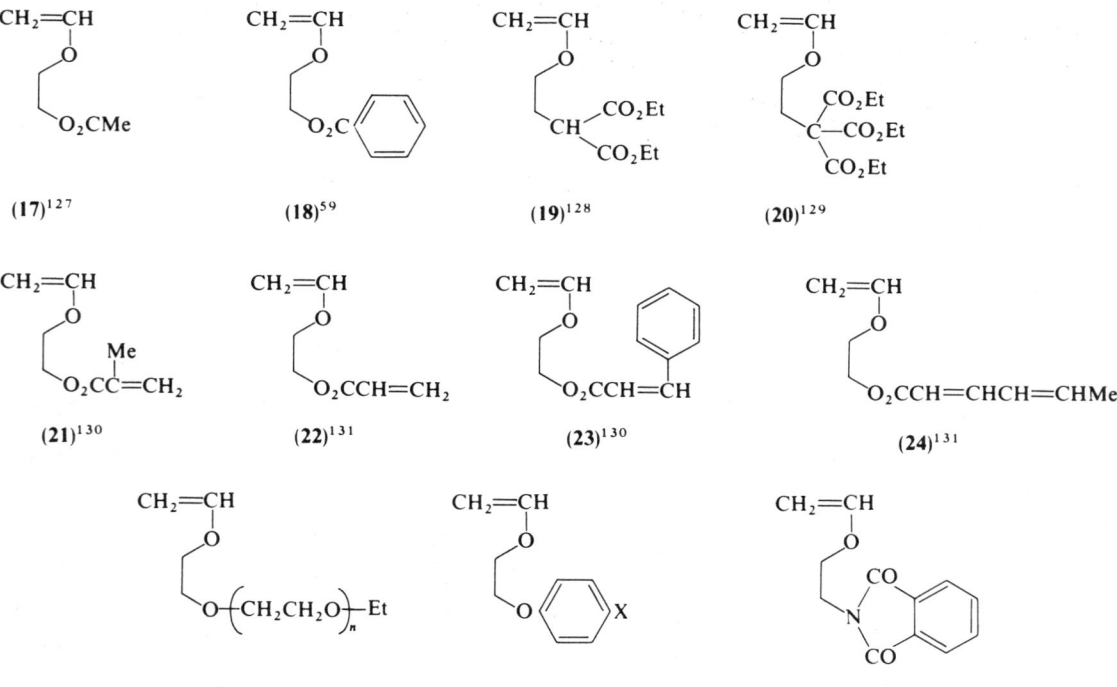

Figure 3   Functionalized vinyl ethers for living cationic polymerization

with those for alkyl vinyl ethers and the ester-containing derivatives. The propagating species derived from (25)–(27) are apparently less stable.[60]

Vinyl ethers (28) and (29) were employed for the synthesis of thermotropic liquid-crystalline polymers with pendant mesogens.[133] The polymerization was induced by BF_3OEt_2 and $(CF_3SO_2)_2O$ but was not living.

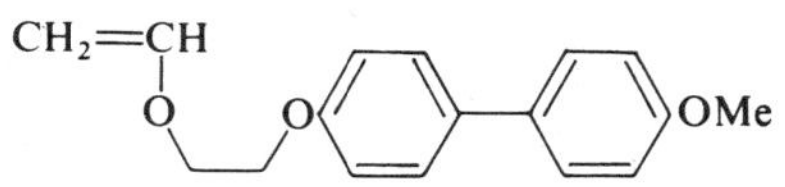

(28)

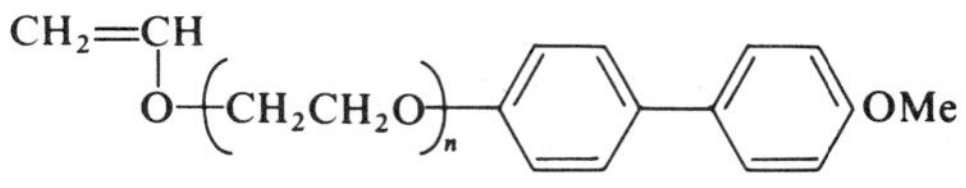

(29) $n = 1–3$, *cis* and *trans*

42.6.3 Polymer Synthesis by Living Polymerization of Vinyl Ethers

42.6.3.1 End-functionalized polymers

The absence of chain transfer and termination in living polymerization provides straightforward and versatile methods to attach a variety of functional groups to polymer chain ends. The synthesis of end-functionalized polymers by living polymerization is usually achieved *via* either an *initiation* or a *termination* reaction.[134]

(i) Synthesis by initiation reaction (Scheme 5)

This approach is based on the initiation in Scheme 4. Thus, the hydrogen iodide–vinyl ether adduct (31), obtained from (30), is employed as an initiator that induces living polymerization of vinyl ethers (other than 30) in the presence of iodine *via* the activation of its C–I bond. The product (32) carries a functional 'head' (α-end) group X derived from the initiator.

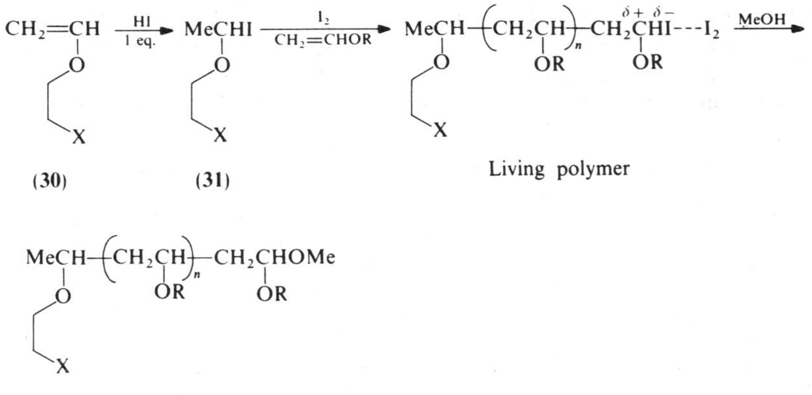

(30)

(31)

Living polymer

(32)

Scheme 5

(ii) Synthesis by termination reaction (Scheme 6)

In this end-capping method, the HI/I_2-initiated living poly(vinyl ether) (16) is quenched with a nucleophile (:NuX) to yield polymer (33) in which the 'tail' (ω-end) group carries the function X derived from the quencher. Under suitable conditions, the living end of (16) is so stable and selective that it combines to :NuX without forming an alkenic by-product ($\sim$ CH=CHOR) *via* β-proton elimination.

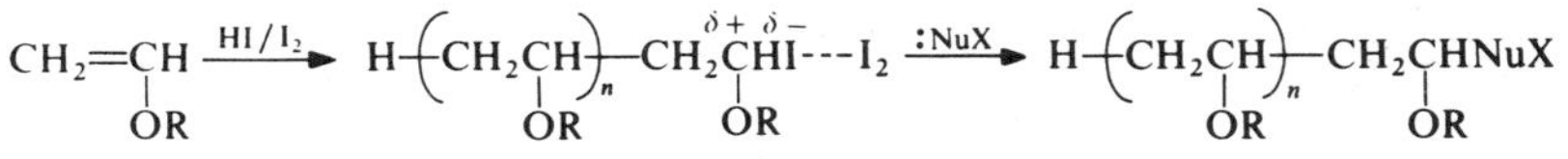

(16) Living polymer

(33)

Scheme 6

(iii) Poly(vinyl ethers) with terminal carboxyl or amino groups

As described in Section 42.6.2, vinyl ether (**19**) with a malonate pendant group undergoes living polymerization in the presence of HI/I$_2$.[128] On treatment with an equimolar amount of hydrogen iodide at room temperature, this vinyl ether quantitatively forms the adduct (**34**) that, in conjuction with iodine, initiates living polymerization of vinyl ethers to give a malonate-capped living polymer (**35**; Scheme 7).[134] Quenching its active end with methanol and with sodiomalonic ester (**36**), respectively, leads to monofunctional (**37**) and bifunctional (telechelic; **38**) polymers with malonate end group(s). The malonate terminals of (**37**) and (**38**) can be converted into the corresponding carboxylic acids (**39 and 40**, respectively) *via* the conventional saponification followed by thermal decarboxylation.[134] Alternatively, termination of the HI/I$_2$-initiated living polymer (**16**; Scheme 6) with (**36**) gives another monofunctional poly(vinyl ether) with a 'tail' malonate group that can also be transformed into a carboxylic acid form [$\sim$CH$_2$CH(OR)CH$_2$CO$_2$H].[134]

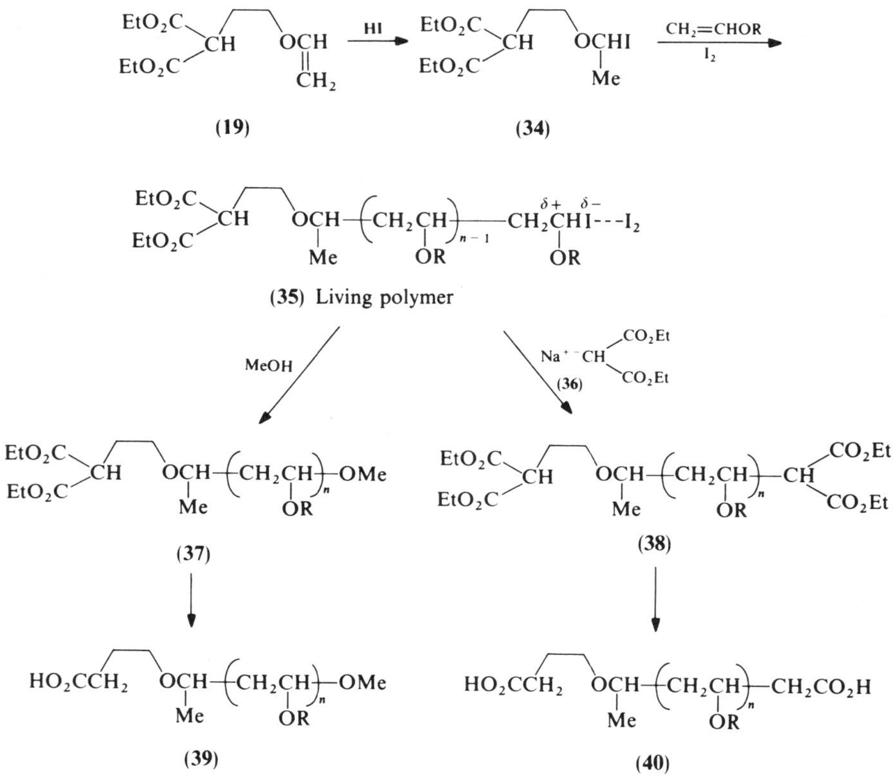

Scheme 7

Similarly to Scheme 7, a poly(vinyl ether) (**41**) with an aliphatic primary amino end-group is obtained from imide vinyl ether (**27**).[132,135] Terminal amino functions can also be attached by termination reactions with the use of *n*-butylamine,[119] aniline[136] or phenylenediamine[136] as an end-capping agent.

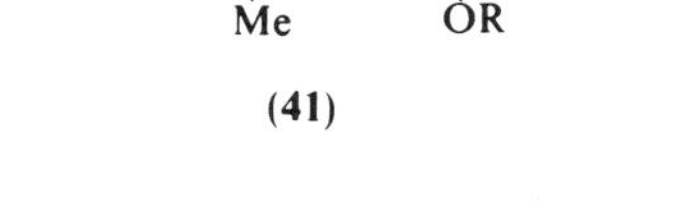

(41)

(iv) Macromonomers

A polymerizable functional group can be attached to the end of the poly(vinyl ether) chain either by initiation or termination. For example, macromonomer (**43**) with a methacrylate end is prepared by living polymerization of vinyl ethers using the initiator (**42**) derived from methacrylate vinyl monomer (**21**; equation 3).[137] Quenching the HI/I$_2$-initiated living polymer (**16**) with functional anion (**44**) leads to vinyl ether-capped poly(vinyl ethers) (**45**; equation 4);[138] (**44**) is prepared by

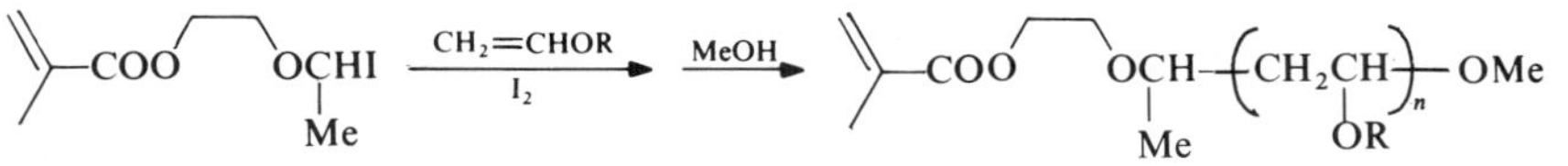

$$\text{(42)} \qquad \text{(43)} \tag{3}$$

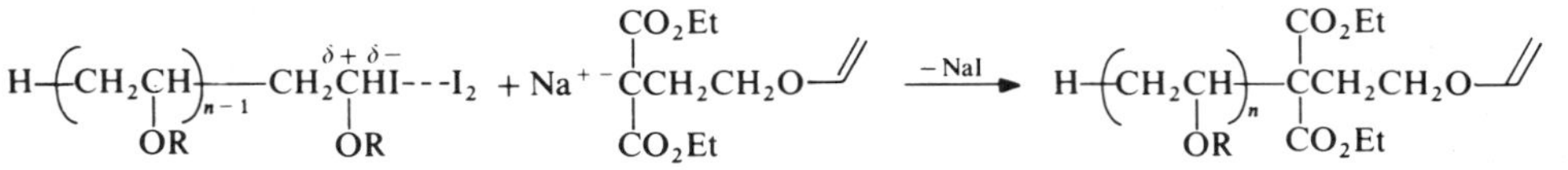

$$\text{(16) Living polymer} \qquad \text{(44)} \qquad \text{(45)} \tag{4}$$

treatment of malonate vinyl ether (19) with sodium hydride. The macromonomers (43) and (45) are characterized by their perfect number-average end-functionality ($\bar{F}_n \simeq 1.0$), narrow MWD and controlled molecular weight.

42.6.3.2 Poly(vinyl ethers) with functional pendant groups

(i) Water-soluble polymers with a narrow MWD

Although numerous water-soluble synthetic polymers are known, few possess controlled molecular weight and narrow MWD. Polymers of vinyl ethers (25) with a poly(oxyethylene) pendant group are *per se* soluble in water at room temperature, provided that they have enough ether oxygens per-pendant group ($n \geq 1$; see Figure 3).[60] They can be regarded as comb-like polymers consisting of a monodisperse poly(vinyl ether) backbone and poly(oxyethylene) grafts with a predetermined length (n). Recently, one of them (46) was found to act as a carrier for the transport of alkali metal ions (K^+ in particular) through a liquid membrane.[139]

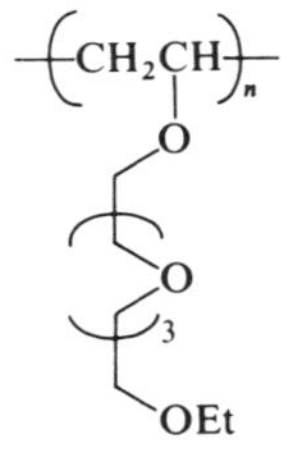

(46)

Alkaline hydrolysis of the ester pendant groups in poly(17) and poly(18) gives polyalcohols (47) which are soluble in water and inherit the narrow MWD of the precursors obtained by living polymerization (see Section 42.6.2).[127]

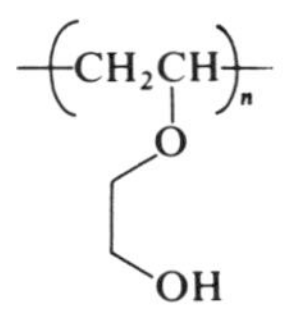

(47)

(ii) Polycarboxylic acids and polyamines

By the conventional saponification and subsequent decarboxylation of the malonate pendant groups, polymers of (19) are converted into polycarboxylic acid (48), a vinyl ether-based polyelectrolyte with a narrow MWD.[128] A poly(vinyl ether) with aliphatic primary amino pendant groups (49) can be obtained from poly(27).[132] Both (48) and (49) are readily soluble in water at room temperature.

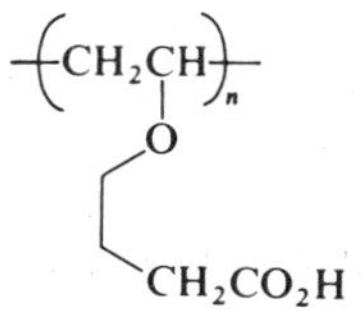

$$-\left(CH_2CH\right)_n-$$
$$O$$
$$CH_2CO_2H$$

(48)

$$-\left(CH_2CH\right)_n-$$
$$O$$
$$NH_2$$

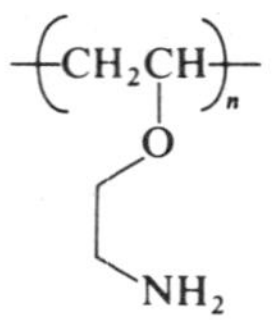

(49)

42.6.3.3 Block copolymers

Living polymerization is most frequently employed for the synthesis of block copolymers where two or more monomers are polymerized sequentially, and this is not exceptional for the HI/I$_2$ initiated living polymerization of vinyl ethers.[8,117d] The reported examples include: alkyl vinyl ether–alkyl vinyl ether[120] (*e.g.* methyl vinyl ether–hexadecyl vinyl ether; **50**); alkyl vinyl ether–alkyl propenyl ether;[122] alkyl vinyl ether–*p*-methoxystyrene (**51**);[120,140] alkyl vinyl ether–2-hydroxyethyl vinyl ether (**52**);[141] and methyl vinyl ether–hexadecyl vinyl ether–methyl vinyl ether (**53**).[119]

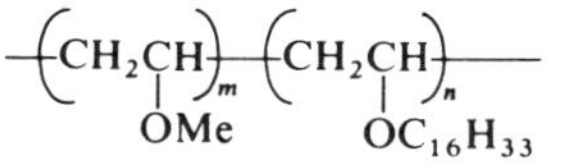

$$-\left(CH_2CH\right)_m\left(CH_2CH\right)_n-$$
$$OMe \qquad OC_{16}H_{33}$$

(50)

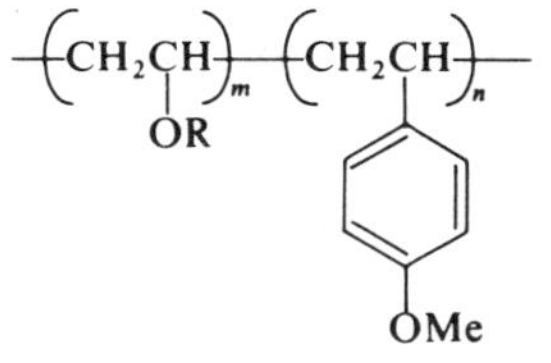

$$-\left(CH_2CH\right)_m\left(CH_2CH\right)_n-$$
$$OR$$
$$OMe$$

(51) R = alkyl

$$-\left(CH_2CH\right)_m\left(CH_2CH\right)_n-$$
$$O \qquad\qquad OR$$
$$OH$$

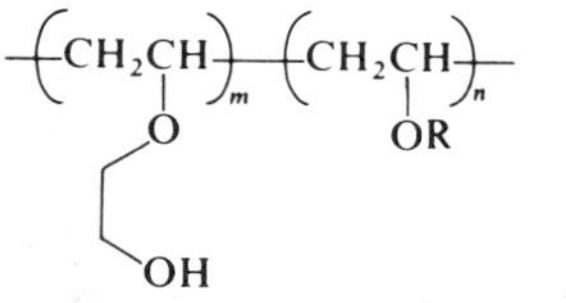

(52) R = *n*-C$_4$H$_9$, iso-C$_8$H$_{17}$, *n*-C$_{16}$H$_{33}$

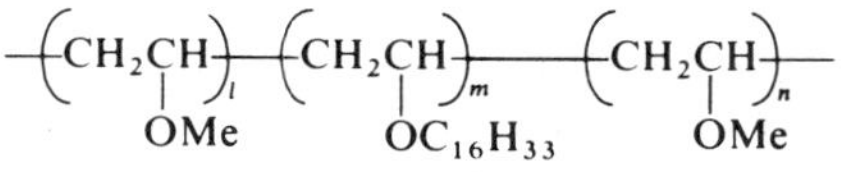

$$-\left(CH_2CH\right)_l\left(CH_2CH\right)_m\left(CH_2CH\right)_n-$$
$$OMe \qquad OC_{16}H_{33} \qquad OMe$$

(53)

All these block copolymers, except for (**51**), were obtained in a very high blocking efficiency because of the living nature of the polymerization of each comonomer. The block polymers (**50**), (**52**) and (**53**) are amphiphilic, among which (**52**) exhibits an excellent surface activity, and reduces the surface tension of water to 30 dyn cm^{-1} (3×10^{-2} N m^{-1}) or below at room temperature.[141]

42.6.4 New Initiators and New Living Processes

42.6.4.1 Hydrogen iodide/zinc iodide initiating system

In the living polymerization by the HI/I$_2$ system, iodine acts as a weak Lewis acid that activates electrophilically the terminal C–I bond of the propagating species (**16**; Scheme 4).[38] This mechanism implies that weak Lewis acids other than iodine may serve to activate the terminal C–I linkage for living propagation. In fact, Sawamoto *et al.*[142] recently found that zinc iodide (ZnI$_2$) in conjunction with hydrogen iodide induces living polymerizations of IBVE and 2-acetoxyethyl vinyl ether (**17**) in methylene chloride or toluene in a wide temperature range from -40 to $+25\,°C$. The advantages of the HI/ZnI$_2$ system over HI/I$_2$ include the higher maximum temperature ($+25\,°C$ *vs.* $-15\,°C$) and higher rate of the living process. Like iodine, ZnI$_2$ apparently interacts with the C–I bond for its activation, *i.e.* $\sim$ CH$_2$CH(OR)I --- ZnI$_2$.

42.6.4.2 EtAlCl$_2$/ester and EtAlCl$_2$/ether initiating systems

EtAlCl$_2$ by itself efficiently induces cationic polymerization of vinyl ethers, but the product polymers are not living and have broad MWDs. However, according to Higashimura and his co-workers, combinations of EtAlCl$_2$ with a large excess of inert esters (*e.g.* MeCO$_2$Et)[143] or ethers (*e.g.* dioxane)[144] as initiating systems lead to living polymerization of IBVE. It is notable that living polymers with a high molecular weight ($\bar{M}_n \sim 10^5$) and a narrow MWD ($\bar{M}_w/\bar{M}_n = 1.1$–1.2) are prepared in a wide temperature range from -40 to even $+70\,°C$. For ester-containing vinyl ethers (*e.g.* **17** and **18**; Figure 3), the presence of an ester additive is unnecessary; the polymerizations of such monomers with EtAlCl$_2$ alone are already living.

As another Lewis acid/ester combination for living cationic polymerization, the BCl$_3$/tertiary carboxylate systems are reported,[145] which are thus far applicable to isobutene only.

The discovery of the HI/ZnI$_2$, EtAlCl$_2$/ester, and EtAlCl$_2$/ether initiating systems is particularly significant in that, for the first time, living cationic polymerization of vinyl monomers has been realized at or above room temperature.

42.6.4.3 Quasi-living polymerization

When a vinyl monomer (α-methylstyrene, vinyl ethers, *etc.*) is added slowly but continuously to an initiator solution, a polymerization occurs which is phenomenologically similar to living polymerization. Kennedy *et al.*, who found this phenomenon, named it 'quasi-living' carbocationic polymerization.[117e] For vinyl ethers, Sawamoto and Kennedy studied the quasi-living polymerizations of isobutyl[146] and methyl[147] derivatives, using a bifunctional initiating system consisting of *p*-dicumyl chloride (ClCMe$_2$C$_6$H$_4$CMe$_2$Cl) and AgSbF$_6$. The $\bar{M}_n$ of the resulting polymer ($\bar{M}_w/\bar{M}_n = 1.2$–1.5) increased in direct proportion to the cumulative amount of the added monomer. ABA triblock copolymers of α-methylstyrene (A) with isobutyl[148] or methyl[147] vinyl ether (B) were obtained.

42.6.4.4 Group transfer polymerization

Sogah and Webster quite recently reported living polymerization of trialkylsilyl vinyl ethers by the aldehyde (or its precursor)/Lewis acid initiating systems.[149] A typical example utilizes benzaldehyde, in conjunction with zinc bromide (ZnBr$_2$), to polymerize *t*-butyldimethylsilyl vinyl ether at -60 to $+30\,°C$. The growing end is proposed to be an aldehyde, probably associated by ZnBr$_2$, to which the trialkylsilyl group of the incoming monomer transfers to effect propagation. Details of this new class of 'group transfer polymerization' through aldol condensation are described in Volume 4, Chapter 10.

42.6.4.5 Polymerization of cis-1,2-dimethoxyethylene

According to Skorokhodov and co-workers, this β-substituted vinyl ether (MeOCH=CHOMe, *cis*) forms a polymer ($\bar{M}_n = 10^5$–10^6, $\bar{M}_w/\bar{M}_n = 1.2$–1.5) of long lifetime when polymerized by BF$_3$OEt$_2$ in toluene at $-78\,°C$.[150] Block copolymers with benzyl vinyl ether[151] or styrene[152] were prepared.

42.7 RECENT TOPICS IN VINYL ETHER POLYMERIZATION

42.7.1 Photoinitiated Interfacial Polymerization

Willson and co-workers allowed gaseous vinyl ethers (methyl vinyl ether and α-methylvinyl methyl ether) to come into contact with a polystyrene or poly(*p*-methoxystyrene) film that was doped with an arylsulfonium salt (*e.g.* Ph$_3$S$^+$SbF$_6^-$) and then irradiated with UV light.[153] A protonic acid, photochemically generated from the salt, polymerized the monomers on or inside the polymer film by a cationic mechanism. This photoinitiated interfacial polymerization appears interesting as a method for polymer surface modification.

42.7.2 Crosslinkable Poly(vinyl ethers) for Lithography

Polymer (**54**) of 2-(vinyloxy)ethyl cinnamate is perhaps the first poly(vinyl ether) successfully commercialized as a specialty material.[7] Upon exposure to UV light, this polymer crosslinks *via* cyclodimerization of the cinnamate pendant groups[7] and serves as a negative photoresist material with a high sensitivity and resolution (1.0 μm).[154] It was applied to the fabrication of integrated circuits for various artificial satellites and for the Viking mission for the exploration of Mars (1976).

A random copolymer (55) between 2-chloroethyl vinyl ether and 2-(vinyloxy)ethyl acrylate was studied as an X-ray developable resist material (sensitivity, 18 mJ cm^{-1}; resolution, 1.0 μm).[155] In addition, a number of crosslinkable homopolymers and random copolymers of vinyl ethers are known.[156]

(54)   (55)

42.7.3   Photoconductive and Semiconductive Poly(vinyl ethers)

Vinyl ethers (56–59) with a pendant π-conjugating system (a fused aromatic ring or carbazolyl group) were polymerized cationically (mostly with BF$_3$OEt$_2$) to study the photoconductivity[157,158] and complex formation[159,160] of the product polymers. Simionescu et al. recently reported the cationic grafting on to poly(vinyl chloride)[161] and electroinitiated polymerization[162] of the carbazole derivative (59).

(56)[157]   (57)[157]   (58)[158]   (59)[159-162]

According to Nakamura and co-workers, vinyl ether (60) carrying an organometallic complex is readily polymerized by metal halides (BF$_3$OEt$_2$, SnCl$_4$, etc.).[163] The as-formed sample of the polymer ($\bar{M}_n \sim 2000$) is an insulator ($\sigma \leq 10^{-10}$ S cm^{-1}), but upon doping with iodine *in vacuo* it exhibits semiconductivity ($\sigma = 3.2 \times 10^{-3}$ S cm^{-1}).[164]

(60)   Mt = Fe, Ru

42.7.4   Bifunctional Vinyl Ethers

More than 30 bifunctional vinyl ethers (61)[165-174] and propenyl ethers (62)[175] have been prepared and cationically polymerized. For example, Crivello et al. polymerized (61)[165,166] and (62)[175] using photo and thermal cationic initiators. The resultant crosslinked materials were comparable with commerical epoxy resins in performance.

Under suitable conditions, bifunctional vinyl ethers (61) do not crosslink but undergo cationic cyclopolymerization.[169-174] When the middle link R is a poly(oxyethylene) chain or its derivative,

$$CH_2{=}CH \qquad CH{=}CH_2$$
$$| \qquad\qquad |$$
$$O{-}R{-}O$$

(61) R = $-(CH_2)_n-$,[165] $-(CH_2CH_2O)_n{-}CH_2CH_2{-}$,[168,169]

,[166,167] *etc.*

$$MeCH{=}CH \qquad CH{=}CHMe$$
$$| \qquad\qquad\qquad |$$
$$O{-}R{-}O$$

(62) R =

,[175] *etc.*

the cyclopolymerization leads to a backbone consisting of cyclic repeat units similar to crown ethers.[171-174] The complexation of metal ions by these cyclopolymers,[172-174] as well as by poly(vinyl ethers) carrying pendant crown ether units[176-178] or poly(oxyethylene) chains,[139] was studied.

42.8 CONCLUDING REMARKS

The basic research into the cationic polymerization of vinyl ethers has now entered the third era in its history that spans over a century since 1878 (Section 42.1): the first era (1940s and 1950s) involved the pioneering kinetic studies (Section 42.3) and the first discovery of stereoregular polymers in macromolecular science in 1947 (Section 42.5); the second era (1960s and 1970s) featured the prolific research on structure–reactivity relationships (Section 42.4); and the third era was no doubt triggered by the discovery of living polymerization of alkyl and functionalized vinyl ethers in the 1980s (Section 42.6). This last breakthrough has rejuvenated the interest in vinyl ether polymerization by providing the methodology for fine control of polymer molecular weight, MWD and end-group structure. The current investigations in this field, as in basic polymer research as a whole,[179] are thus directed towards the synthesis of well-designed polymers with specific functions and physical properties.

42.9 REFERENCES

1. J. Wislicenus, *Justus Liebigs Ann. Chem.*, 1878, **92**, 106.
2. C. E. Schildknecht, 'Vinyl and Related Polymers', Wiley, New York, 1952, p. 593.
3. C. E. Schildknecht, A. O. Zoss and C. McKinley, *Ind. Eng. Chem.*, 1947, **39**, 180.
4. J. Lal, in 'Polymer Chemistry of Synthetic Elastomers', ed. J. P. Kennedy and E. Törnqvist, Wiley–Interscience, New York, 1968, part 1, p. 331.
5. N. M. Bikales, in 'Encyclopedia of Polymer Science and Technology', Wiley–Interscience, New York, 1971, vol. 14, p. 511.
6. J. P. Kennedy and E. Maréchal, 'Carbocationic Polymerization', Wiley–Interscience, New York, 1982, p. 499.
7. M. Kato, T. Ichijo, K. Ishii and M. Hasegawa, *J. Polym. Sci., Part A-1*, 1971, 9, 2109; S. Watanabe, M. Kato and S. Kosakai, *J. Polym. Sci., Polym. Chem. Ed.*, 1984, **22**, 2801.
8. T. Higashimura and M. Sawamoto, *Adv. Polym. Sci.*, 1984, **62**, 49.
9. D. H. Lorenz, in 'Encyclopedia of Polymer Science and Technology', Wiley–Interscience, New York, 1971, vol. 14, p. 504.
10. D. D. Eley, in 'The Chemistry of Cationic Polymerisation', ed. P. H. Plesch, Pergamon Press, London, 1963, chap. 9.
11. D. D. Eley and D. C. Pepper, *Trans. Faraday Soc.*, 1947, **43**, 112.
12. D. D. Eley and A. W. Richards, *Trans. Faraday Soc.*, 1949, **45**, 425.
13. D. D. Eley and A. W. Richards, *Trans. Faraday Soc.*, 1949, **45**, 436.
14. F. Bolza and F. E. Treloar, *Makromol. Chem.*, 1980, **181**, 839.
15. S. Aoshima and T. Higashimura, unpublished results (1983).
16. J. D. Coombes and D. D. Eley, *J. Chem. Soc.*, 1957, 3700.
17. J. P. Kennedy, *J. Polym. Sci.*, 1959, **38**, 262.
18. V. A. Kruglova, V. G. Liporich and N. A. Tyukavkina, *Vysokomol. Soedin., Ser. A*, 1967, **9**, 932.
19. S. M. McElvain, *Chem. Rev.*, 1949, **45**, 453.
20. H. Imai, T. Saegusa and J. Furukawa, *Makromol. Chem.*, 1965, **61**, 92.

21. A. Ledwith and D. C. Sherrington, *Adv. Polym. Sci.*, 1975, **19**, 1.
22. J. V. Crivello, *Adv. Polym. Sci.*, 1984, **62**, 1.
23. D. D. Eley and J. Saunders, *J. Chem. Soc.*, 1954, 1668.
24. J. P. Kennedy and E. Maréchal, 'Carbocationic Polymerization', Wiley–Interscience, New York, 1982, chap. 5.
25. F. Williams, Ka. Hayashi, K. Ueno, Ko. Hayashi and S. Okamura, *Trans. Faraday Soc.*, 1967, **63**, 1501.
26. Ka. Hayashi, Ko. Hayashi and S. Okamura, *J. Polym. Sci., Part A-1*, 1971, **9**, 2305.
27. C. E. H. Bawn, C. Fitzsimmons, A. Ledwith, J. Penfold, D. C. Sherrington and J. A. Weightman, *Polymer*, 1971, **12**, 119.
28. A. Ledwith, E. Lockett and D. C. Sherrington, *Polymer*, 1975, **16**, 31.
29. H. Kubota, V. Ya. Kobanov, D. R. Squire and V. Stannett, *J. Macromol. Sci., Chem.*, 1978, **A12**, 1299.
30. W. C. Hsieh, H. Kubota, D. R. Squire and V. Stannett, *J. Polym. Sci., Polym. Chem. Ed.*, 1980, **18**, 2773.
31. A. Deffieux, W. C. Hsieh, D. R. Squire and V. Stannett, *Polymer*, 1981, **22**, 1575.
32. A. Deffieux, W. C. Hsieh, D. R. Squire and V. Stannett, *Polymer*, 1982, **23**, 65.
33. W. C. Hsieh, A. Deffieux, D. R. Squire and V. Stannett, *Polymer*, 1982, **23**, 427.
34. A. Deffieux, J. A. Young, W. C. Hsieh, D. R. Squire and V. Stannett, *Polymer*, 1983, **24**, 573.
35. S. Okamura, N. Kanoh and T. Higashimura, *Makromol. Chem.*, 1961, **47**, 35.
36. A. Ledwith and D. C. Sherrington, *Polymer*, 1971, **12**, 344.
37. A. F. Johnson and R. N. Young, *J. Polym. Sci., Polym. Symp.*, 1976, **56**, 211.
38. A. Gandini and P. H. Plesch, *J. Polym. Sci., Part B*, 1965, **3**, 127; *J. Chem. Soc.*, 1965, 4826; *Eur. Polym. J.*, 1968, **4**, 55.
39. P. Giusti, G. Puce and F. Andruzzi, *Makromol. Chem.*, 1966, **98**, 170; P. Giusti and F. Andruzzi, *J. Polym. Sci., Part C*, 1968, **16**, 3797; P. Cerrai, F. Andruzzi and P. Giusti, *Makromol. Chem.*, 1968, **117**, 128.
40. T. Higashimura, M. Miyamoto and M. Sawamoto, *Macromolecules*, 1985, **18**, 611.
41. D. D. Eley and J. Saunders, *J. Chem. Soc.*, 1954, 1672.
42. Y. Imanishi, T. Higashimura and S. Okamura, *Kobunshi Kagaku*, 1962, **19**, 154.
43. Y. Imanishi, H. Nakayama, T. Higashimura and S. Okamura, *Kobunshi Kagaku*, 1962, **19**, 565.
44. S. Aoshima and T. Higashimura, *Polym. J.*, 1984, **16**, 249.
45. D. J. Dunn and P. H. Plesch, *Makromol. Chem.*, 1974, **175**, 2821.
46. C. E. Schildknecht, A. O. Zoss and F. Grosser, *Ind. Eng. Chem.*, 1949, **41**, 2891.
47. A. M. Goineau, J. Kohler and V. Stannett, *J. Macromol. Sci., Chem.*, 1977, **A11**, 99.
48. T. Higashimura, J. Masamoto and S. Okamura, *Kobunshi Kagaku*, 1968, **26**, 702.
49. H. Yuki, K. Hatada and M. Takeshita, *J. Polym. Sci., Part A-1*, 1969, **7**, 667.
50. S. Nozakura, M. Kitamura and S. Murahashi, *Polym. J.*, 1970, **1**, 736.
51. D. M. Jones and N. F. Wood, *J. Chem. Soc.*, 1964, 5400.
52. A. Ledwith and H. J. Woods, *J. Chem. Soc. (B)*, 1966, 753.
53. N. Kanoh, K. Ikeda, A. Gotoh, T. Higashimura and S. Okamura, *Makromol. Chem.*, 1965, **86**, 200.
54. T. Fueno, T. Okuyama, I. Matsushita and J. Furukawa, *J. Polym. Sci., Part A-1*, 1969, **7**, 1447.
55. T. Fueno, I. Matsushita, T. Okuyama and J. Furukawa, *Bull. Chem. Soc. Jpn.*, 1968, **41**, 818.
56. H. Yuki, K. Hatada, K. Nagata and T. Emura, *Polym. J.*, 1970, **1**, 269.
57. N. L. Owen and N. Shepherd, *Trans. Faraday Soc.*, 1964, **60**, 634.
58. K. Hatada, M. Takeshita and H. Yuki, *Tetrahedron Lett.*, 1968, 4621.
59. T. Higashimura, S. Aoshima and M. Sawamoto, *Makromol. Chem., Macromol Symp.*, 1986, **3**, 99.
60. T. Nakamura, S. Aoshima and T. Higashimura, *Polym. Bull (Berlin)*, 1985, **14**, 515.
61. W. O. Choi, M. Sawamoto and T. Higashimura, *Polym. J.*, 1987, **19**, 889.
62. T. Masuda and T. Higashimura, *Makromol. Chem.*, 1973, **167**, 191.
63. T. Higashimura and K. Yamamoto, *Makromol. Chem.*, 1974, **175**, 1139.
64. A. Mizote, S. Kusudo, T. Higashimura and S. Okamura, *J. Polym. Sci., Part A-1*, 1967, **5**, 1727.
65. T. Higashimura, S. Kusudo, Y. Ohsumi and S. Okamura, *J. Polym. Sci., Part A-1*, 1968, **6**, 2523.
66. T. Higashimura, S. Kusudo and S. Okamura, *Kobunshi Kagaku*, 1968, **25**, 694.
67. T. Okuyama, T. Fueno, J. Furukawa and K. Ueno, *J. Polym. Sci., Part A-1*, 1968, **6**, 1001.
68. T. Higashimura, J. Masamoto and S. Okamura, *Polym. J.*, 1971, **2**, 153.
69. T. Okuyama and T. Fueno, *J. Polym. Sci., Part A-1*, 1971, **9**, 629.
70. Yu. E. Eizner, S. S. Skorokhodov and T. P. Zubova, *Eur. Polym. J.*, 1971, **7**, 869.
71. T. Okuyama, T. Fueno and J. Furukawa, *J. Polym. Sci., Part A-1*, 1969, **7**, 2433.
72. T. Okuyama, T. Fueno and J. Furukawa, *J. Polym. Sci., Part A-1*, 1968, **6**, 993.
73. T. Okuyama and T. Fueno, *Tetrahedron*, 1969, **25**, 5409.
74. T. Higashimura, S. Okamura, I. Morishima and T. Yonezawa, *J. Polym. Sci., Part B*, 1969, **7**, 23.
75. T. Higashimura, T. Masuda, S. Okamura and T. Yonezawa, *J. Polym. Sci., Part A-1*, 1969, **7**, 3129.
76. T. Fueno, T. Okuyama and J. Furukawa, *J. Polym. Sci., Part A-1*, 1969, **7**, 3219.
77. C. E. Schildknecht, S. T. Gross, H. R. Davidson, J. M. Lambert and A. O. Zoss, *Ind. Eng. Chem.*, 1948, **40**, 2104.
78. G. Natta, I. Bassi and P. Corradini, *Makromol. Chem.*, 1956, **18/19**, 455.
79. C. E. Schildknecht and P. H. Dunn, *J. Polym. Sci.*, 1956, **20**, 597.
80. S. Okamura, T. Higashimura and H. Yamamoto, *J. Polym. Sci.*, 1958, **33**, 510.
81. S. Okamura, T. Higashimura and I. Sakurada, *J. Polym. Sci.*, 1959, **39**, 507.
82. T. Higashimura, T. Kodama and S. Okamura, *Kobunshi Kagaku*, 1960, **17**, 163.
83. Y. Ohsumi, T. Higashimura and S. Okamura, *J. Polym. Sci., Part A-1*, 1967, **5**, 849.
84. M. Sumi, S. Nozakura and S. Murahashi, *Kobunshi Kagaku*, 1967, **24**, 424.
85. M. Delfini, F. Conti, M. Paci, A. L. Segre, R. Solaro and E. Chiellini, *Macromolecules*, 1983, **16**, 1212.
86. G. Natta, G. Dall'Asta, G. Mazzanti, U. Giannini and S. Cesca, *Angew. Chem.*, 1959, **71**, 205.
87. K. Hatada, T. Kitayama, N. Matsuo and H. Yuki, *Polym. J.*, 1983, **15**, 719.
88. D. J. Cram and K. R. Kopecky, *J. Am. Chem. Soc.*, 1959, **81**, 2748.
89. C. E. H. Bawn and A. Ledwith, *Q. Rev., Chem. Soc.*, 1962, **16**, 361.
90. T. Higashimura, T. Yonezawa, S. Okamura and K. Fukui, *J. Polym. Sci.*, 1959, **39**, 487.
91. T. Kunitake and C. Aso, *J. Polym. Sci., Part A-1*, 1970, **8**, 655.
92. Y. Ohsumi, T. Higashimura and S. Okamura, *J. Polym. Sci., Part A-1*, 1967, **5**, 849.

93. J. Lal, *J. Polym. Sci.*, 1958, **31**, 179.
94. E. J. Vandenberg, R. F. Heck and D. S. Breslow, *J. Polym. Sci.*, 1959, **41**, 519.
95. E. J. Vandenberg, *J. Polym. Sci., Part C*, 1963, **1**, 207.
96. S. A. Mosley (Union Carbide Co.) *US Pat.*, 2 549 921 (1951) (*Chem. Abstr.*, 1951, **45**, 5972d).
97. S. Aoki, K. Nakamura and T. Otsu, *Makromol. Chem.*, 1968, **115**, 282.
98. S. Okamura, T. Higashimura and T. Watanabe, *Makromol. Chem.*, 1961, **50**, 137.
99. J. Lal and J. E. McGrath, *J. Polym. Sci., Part A*, 1964, **2**, 3369.
100. T. Higashimura, T. Watanabe and S. Okamura, *Kobunshi Kagaku*, 1963, **20**, 680.
101. K. Iwasaki, *J. Polym. Sci.*, 1962, **56**, 27.
102. S. Nakano, K. Iwasaki and H. Fukutani, *J. Polym. Sci., Part A*, 1963, **1**, 3277.
103. P. Pino, *Adv. Polym. Sci.*, 1965, **4**, 393.
104. E. Chiellini, G. Montagnolli and P. Pino, *J. Polym. Sci., Part B*, 1969, **7**, 121.
105. E. Chiellini, *Macromolecules*, 1970, **3**, 527.
106. T. Higashimura and Y. Hirokawa, *J. Polym. Sci., Polym. Chem. Ed.*, 1977, **15**, 1137.
107. G. Natta, M. Farina, M. Peraldo and G. Bressan, *Makromol. Chem.*, 1961, **43**, 68; *Makromol. Chem.*, 1963, **61**, 79.
108. M. Goodman and Y.-L. Fan, *J. Am. Chem. Soc.*, 1964, **86**, 4922, 5712.
109. K. Matsuzaki, S. Okuzono and T. Kanai, *J. Polym. Sci., Polym. Chem. Ed.*, 1979, **17**, 3447.
110. G. Natta, *J. Polym. Sci.*, 1960, **48**, 219.
111. G. Natta, M. Peraldo, M. Farina and G. Bressan, *Makromol. Chem.*, 1962, **55**, 139.
112. T. Higashimura, Y. Hirokawa, K. Matsuzaki and T. Uryu, *J. Polym. Sci., Polym. Chem. Ed.*, 1980, **18**, 1489.
113. K. Matsuzaki, H. Morii, N. Inoue, T. Kanai and T. Higashimura, *Makromol. Chem.*, 1981, **182**, 2421.
114. T. Higashimura, S. Kusudo, Y. Ohsumi, A. Mizote and S. Okamura, *J. Polym. Sci., Part A-1*, 1968, **6**, 2511.
115. T. Higashimura, M. Hoshino, Y. Hirokawa, K. Matsuzaki and T. Uryu, *J. Polym. Sci., Polym. Chem. Ed.*, 1977, **15**, 2691; *J. Polym. Sci., Polym. Chem. Ed.*, 1979, **17**, 1473.
116. Y. Hirokawa, T. Higashimura, K. Matsuzaki, T. Kawamura and T. Uryu, *J. Polym. Sci., Polym. Chem. Ed.*, 1979, **17**, 3923; *Polym. Bull. (Berlin)*, 1979, **1**, 365.
117. For reviews of living cationic polymerization by HI/I_2, see: (a) T. Higashimura and M. Sawamoto, in 'Cationic Polymerization and Related Processes', ed. E. J. Goethals, Academic Press, London, 1984, p. 77; (b) M. Sawamoto, M. Miyamoto and T. Higashimura, in 'Cationic Polymerization and Related Processes', ed. E. J. Goethals, Academic Press, London, 1984, p. 89; (c) T. Higashimura and M. Sawamoto, *Makromol. Chem., Suppl.*, 1985, **12**, 153; (d) M. Sawamoto and T. Higashimura, *Makromol. Chem., Macromol. Symp.*, 1986, **3**, 83. For the so-called quasi-living polymerization, see: (e) J. P. Kennedy *et al.*, *J. Macromol. Sci., Chem.*, 1982–83, **A18**, 1185–1418.
118. M. Miyamoto, M. Sawamoto and T. Higashimura, *Macromolecules*, 1984, **17**, 265.
119. M. Miyamoto, M. Sawamoto and T. Higashimura, *Macromolecules*, 1985, **18**, 123.
120. M. Miyamoto, M. Sawamoto and T. Higashimura, *Macromolecules*, 1984, **17**, 2228.
121. T. Higashimura, Y.-M. Law and M. Sawamoto, *Polym. J.*, 1984, **16**, 401.
122. M. Sawamoto, K. Ebara, A. Tanizaki and T. Higashimura, *J. Polym. Sci., Polym. Chem. Ed.*, 1986, **24**, 2919.
123. T. Higashimura, A. Tanizaki and M. Sawamoto, *J. Polym. Sci., Polym. Chem. Ed.*, 1984, **22**, 3173.
124. T. Enoki, M. Sawamoto and T. Higashimura, *J. Polym. Sci., Polym. Chem. Ed.*, 1986, **24**, 2261.
125. T. Ohtori, Y. Hirokawa and T. Higashimura, *Polym. J.*, 1979, **11**, 471.
126. M. Sawamoto, J. Fujimori and T. Higashimura, *Macromolecules*, 1987, **20**, 916.
127. S. Aoshima, T. Nakamura, N. Uesugi, M. Sawamoto and T. Higashimura, *Macromolecules*, 1985, **18**, 2097.
128. T. Higashimura, T. Enoki and M. Sawamoto, *Polym. J.*, 1987, **19**, 515.
129. M. Minoda, M. Sawamoto and T. Higashimura, *Polym. Bull. (Berlin)*, 1987, **17**, 107.
130. S. Aoshima, O. Hasegawa and T. Higashimura, *Polym. Bull. (Berlin)*, 1985, **13**, 229.
131. S. Aoshima, O. Hasegawa and T. Higashimura, *Polym. Bull. (Berlin)*, 1985, **14**, 417.
132. M. Sawamoto, H. Ibuki, T. Hashimoto, T. Enoki and T. Higashimura, *Polym. Prepr., Jpn.*, 1986, **35**, 1316; T. Hashimoto, H. Ibuki, M. Sawamoto and T. Higashimura, *J. Polym. Sci., Polym. Chem. Ed.*, 1988, in press.
133. J. M. Rodriguez–Parada and V. Percec, *J. Polym. Sci., Polym. Chem. Ed.*, 1986, **24**, 1363.
134. M. Sawamoto, T. Enoki and T. Higashimura, *Macromolecules*, 1987, **20**, 1.
135. T. Hashimoto, M. Sawamoto and T. Higashimura, *J. Polym. Sci., Polym. Chem. Ed.*, submitted.
136. M. Sawamoto, T. Enoki and T. Higashimura, *Polym. Bull. (Berlin)*, 1987, **18**, 117.
137. S. Aoshima, K. Ebara and T. Higashimura, *Polym. Bull. (Berlin)*, 1985, **14**, 425.
138. M. Sawamoto, T. Enoki and T. Higashimura, *Polym. Bull. (Berlin)*, 1986, **16**, 117.
139. T. Higashimura, T. Nakamura and S. Aoshima, *Polym. Bull. (Berlin)*, 1987, **17**, 389.
140. T. Higashimura, M. Mitsuhashi and M. Sawamoto, *Macromolecules*, 1979, **12**, 178.
141. M. Minoda, M. Sawamoto and T. Higashimura, *Macromolecules*, 1987, **20**, 2045.
142. M. Sawamoto, C. Okamoto and T. Higashimura, *Macromolecules*, 1987, **20**, 2693.
143. S. Aoshima and T. Higashimura, *Polym. Bull. (Berlin)*, 1986, **15**, 417; *Macromolecules*, submitted.
144. T. Higashimura, Y. Kishimoto and S. Aoshima, *Polym. Bull. (Berlin)*, 1987, **18**, 111.
145. R. Faust and J. P. Kennedy, *Polym. Bull. (Berlin)*, 1986, **15**, 317; *J. Polym. Sci., Polym. Chem. Ed.*, 1987, **25**, 1847.
146. M. Sawamoto and J. P. Kennedy, *J. Macromol. Sci., Chem.*, 1982–83, **A18**, 1275.
147. M. Sawamoto and J. P. Kennedy, *J. Macromol. Sci., Chem.*, 1982–83, **A18**, 1301.
148. M. Sawamoto and J. P. Kennedy, *J. Macromol. Sci., Chem.*, 1982–83, **A18**, 1293.
149. D. Y. Sogah and O. W. Webster, *Macromolecules*, 1986, **19**, 1775.
150. V. V. Stepanov, S. I. Klenin, A. V. Troitskaya and S. S. Skorokhodov, *Vysokomol. Soedin., Ser. A*, 1976, **18**, 821; S. S. Skorokhodov, V. V. Stepanov, V. V. Nesterov and V. D. Krasikov, *Acta Polym.*, 1986, **37**, 583.
151. V. V. Stepanov, A. R. Barantseva, V. D. Krasikov, V. V. Nesterov and S. S. Skorokhodov, *Acta Polym.*, 1985, **36**, 605.
152. V. V. Stepanov, A. R. Barantseva, V. D. Krasikov, V. V. Nesterov and S. S. Skorokhodov, *Acta Polym.*, 1985, **36**, 609.
153. A. Hult, S. A. MacDonald and C. G. Willson, *Macromolecules*, 1985, **18**, 1804.
154. M. Kato and H. Nakane, *Photogr. Sci. Eng.*, 1979, **23**, 207.
155. S. Imamura, S. Sugawara and K. Murase, *J. Electrochem. Soc.*, 1977, **124**, 1139.
156. T. Nishikubo, M. Kishida, T. Ichijo and T. Takaoka, *Nippon Kagaku Kaishi*, 1974, 1581.

157. S. Yoshimoto, K. Okamoto, H. Hirata, S. Kusabayashi and H. Mikawa, *Bull. Chem. Soc. Jpn.*, 1973, **46**, 358.
158. K. Okamoto, A. Itaya and S. Kusabayashi, *Polym. J.*, 1975, **7**, 622.
159. K. Okamoto, A. Itaya and S. Kusabayashi, *Chem. Lett.*, 1974, 1167.
160. K. Okamoto, M. Ozeki, A. Itaya, S. Kusabayashi and H. Mikawa, *Bull. Chem. Soc. Jpn.*, 1975, **48**, 1362.
161. C. I. Simionescu, I. Rabia and M. Grigoras, in 'Cationic Polymerization and Related Processes', ed. E. J. Goethals, Academic Press, London, 1984, p. 373.
162. C. I. Simionescu, M. Grovu-Ivanoiu and M. Grigoras, *Eur. Polym. J.*, 1986, **22**, 71.
163. Y. Morita, M. Yamauchi, H. Yasuda and A. Nakamura, *Kobunshi Ronbunshu*, 1980, **37**, 677; H. Yasuda, Y. Morita, I. Noda, K. Sugi and A. Nakamura, *J. Organomet. Chem.*, 1981, **205**, C5.
164. H. Yasuda, I. Noda, S. Miyanaga and A. Nakamura, *Macromolecules*, 1984, **17**, 2453.
165. J. V. Crivello, J. L. Lee and D. A. Conlon, *J. Radiat. Curing*, 1983, **10**, 6.
166. J. V. Crivello and D. A. Conlon, *J. Polym. Sci., Polym. Chem. Ed.*, 1983, **21**, 1785.
167. R. R. Gallucci and R. C. Going, *J. Org. Chem.*, 1983, **48**, 342.
168. H. M. Teeter, E. J. Defek, C. B. Coleman, C. A. Glass, E. H. Melvin and J. C. Cowan, *J. Am. Oil Chem. Soc.*, 1956, **33**, 99.
169. L. J. Mathias and J. B. Canterberry, *Polym. Prepr., Am. Chem. Soc., Div. Polym. Chem.*, 1981, **22**(1), 38; L. J. Mathias, J. B. Canterberry and M. South, *J. Polym. Sci., Polym. Lett. Ed.*, 1982, **20**, 473.
170. S. L. N. Seung and R. N. Young, *J. Polym. Sci., Polym. Lett. Ed.*, 1978, **16**, 367.
171. G. B. Butler and Q. S. Lien, *Polym. Prepr., Am. Chem. Soc., Div. Polym. Chem.*, 1981, **22**(1), 54.
172. K. Yokota, M. Matsumura, K. Yamaguchi and Y. Tanaka, *Makromol. Chem., Rapid Commun.*, 1983, **4**, 721.
173. T. Kakuchi and K. Yokota, *Makromol. Chem., Rapid Commun.*, 1985, **6**, 551.
174. S. S. Skorokhodov, *Makromol. Chem., Macromol. Symp.*, 1986, **3**, 153.
175. J. V. Crivello and D. A. Conlon, *J. Polym. Sci., Polym. Chem. Ed.*, 1984, **22**, 2105.
176. M. Shirai, T. Orikata and M. Tanaka, *Makromol Chem., Rapid Commun.*, 1983, **4**, 65.
177. M. Shirai, T. Orikata and M. Tanaka, *J. Polym. Sci., Polym. Chem. Ed.*, 1985, **23**, 463.
178. M. Shirai, A. Ueda and M. Tanaka, *Makromol. Chem.*, 1985, **186**, 2519; *J. Polym. Sci., Polym. Chem. Ed.*, 1987, **25**, 1811.
179. N. M. Bikales, *Polym. J.*, 1987, **19**, 11.

43

Carbocationic Polymerization: *N*-Vinylcarbazole

JOHN M. ROONEY
Loctite Corporation, Newington, CT, USA

43.1 INTRODUCTION

Two features which distinguish *N*-vinylcarbazole (9-vinylcarbazole, NVC) from most cationically polymerizable compounds are its physical form and its reactivity. Under ambient conditions, NVC is a white crystalline solid, purified readily, in principle, by repeated crystallization. Both the crystallinity and basicity of NVC are explained by reference to the molecular structure, in which the carbon–carbon double bond is attached directly to a large planar aromatic amine. The electronegative nitrogen atom withdraws electrons from the double bond through induction, but this depletion is more than offset by the tendency of nitrogen to donate its unshared electron-pair, thus creating an electron-rich double bond.[1]

Numerous synthetic routes to NVC have been devised, including direct vinylation of carbazole and dehydrohalogenation of *N*-2-chloroethylcarbazole.[2] Among the first groups to prepare NVC was that of Reppe, who also reported the first cationic polymerizations of the monomer.[3]

43.2 SOLID-STATE POLYMERIZATION

Bulk polymerizations of NVC conducted at temperatures below 337 K proceed ostensibly in the solid state. Although early kinetic measurements[4] and most subsequent studies support a free-radical mechanism for the solid-state polymerization of NVC, the chemistry of initiation remains obscure. Electron spin resonance (ESR) spectra of NVC irradiated at ambient temperatures, or irradiated at 77 K and warmed to temperatures above 90 K, showed a multiplet ascribed to the $\geq$NĊHCH$_2$– group.[5] Irradiation followed by scanning at 77 K produced a triplet spectrum speculatively assigned to the radical cation formed by removal of an electron from the NVC nitrogen. An initiation mechanism was proposed in which this radical cation served as a transient intermediate, rearranging to a primary carbocation (the structure shown on p. 1312 in ref. 5 is incorrect) which underwent a highly improbable addition to the α carbon of a monomer molecule.

The radical end of the radical cation was presumed to remain dormant, accounting for the presence of terminal radicals in this putatively cationic polymerization. The tendency to infer propagation by carbocations in solid-state radiation-induced polymerizations of NVC and related monomers results in part from the relative insensitivity of these systems to the presence of oxygen.[6] Diagnostic tests of this type have led to the proposal of a zwitterion as the initiator of cationic chain-growth in the solid-state photoinduced polymerization of 9-ethyl-3-vinylcarbazole.[7] As yet, no conclusive evidence has been adduced to support the contention that these polymerizations are cationic.

The physical form and reactivity of NVC have been exploited in an interesting series of nominally solid-state polymerizations induced by the addition of electrophilic gases to the reaction vessel. Scavenging experiments demonstrated that cations were the propagating species, and the participation of radical cations was advanced on the basis of ambiguous ESR spectra.[8] However, consideration of the rates and exothermicity of these reactions led to the conclusion that polymerization proceeded in the melt.[9]

43.3 SOLUTION POLYMERIZATION

43.3.1 Protogenic Acids

Possibly the most obvious method of cationically polymerizing NVC involves treating the monomer with acids. Control over the rate of addition and molar ratio of strong acids, such as sulfuric acid, to monomer must be exercised, since the carbazole rings are susceptible to alkylations which eventually crosslink the polymer chains.[10]

Due to the nucleophilicity of NVC, even relatively weak carboxylic acids will cause polymerization.[11] In a solvent of intermediate polarity (1,2-dichloroethane), oligomers bearing a terminal acetate ester were isolated from reactions between NVC and large excesses of acetic acid.[12] This observation suggests that proton addition to the β carbon of the monomer double bond is followed by addition of the counteranion to the positively charged α carbon, forming a polarized ester. The lability of the ester linkage was amply demonstrated by rapid hydrolysis of the monomeric adduct in the presence of atmospheric moisture.

The proximity of a chiral counteranion (derived from a protogenic acid) to the propagating species can apparently be used to induce optical activity in NVC polymers.[13] Unfortunately, the design of the experiments left the question of the identity of these species unresolved. Although solvents of different dielectric constant were employed in the reactions, these solvents differ widely in electron-donor activity as well, conceivably altering any ionic or ion–ester equilibria. The dependence of the relatively small angles of optical rotation on polymer molecular weights suggests that the optical activity may be due entirely to terminal groups.

43.3.2 Aprotogenic Organic Compounds

Since the mid-1960s, a large number of publications dealing with the polymerization of NVC by organic electron-acceptors have appeared in the literature.[14] In most of these reports, the mechanism of propagation is shown unequivocally to be cationic, and the remaining questions concern the exact identity of the initiator or its mode of interaction with the monomer. In other cases, both the mechanism of initiation and propagation are unclear

Certain organic halides, for example benzoyl chloride, were found to initiate the cationic polymerization of NVC.[15] A charge-transfer complex was presumed to be an intermediate in the initiation process, but the relatively high reaction temperature (333 K), long reaction times (12 h) and high halide concentrations, coupled with the reactivity of NVC, leave room for doubt. Attempts have also been made to invoke a cationic mechanism in describing NVC polymerizations in the presence of benzoyl peroxide.[16] Again, a charge-transfer complex between monomer and initiator was postulated, with kinetic and spectroscopic measurements adduced in support of the theory. However, a case for conventional free-radical polymerization under similar conditions was established.[17] One effort to resolve the conflict[18] creates further uncertainty, since a kinetic calculation designed to probe mechanistic changes appears to involve a misplaced decimal point (see Table 4 of ref. 20).

Studies dealing with the polymerization of NVC by chloranil (tetrachloro-1,4-benzoquinone) and related compounds must be interpreted with caution, since compounds of this type are notoriously difficult to free from traces of protogenic acid derivatives.[19] A different problem arose when NVC solutions were treated with tetranitromethane (TNM).[20] Although the TNM was apparently pure, this compound and nucleophilic alkenes were subsequently shown to undergo a sequence of reactions leading to formation of nitroalkenes and trinitromethane (nitroform).[21] Nitroform, a protogenic acid, proved to be an effective initiator of NVC polymerization. Consequently, any analysis of the kinetics of the TNM–NVC system would require an understanding of the rates of formation and consumption of nitroform as well as insight into the dissociation equilibria involving the trinitromethyl counteranion.

The complicated reaction pattern which ensues upon mixing NVC and electrophilic alkenes has only recently been elucidated.[22] Zwitterionic tetramethylene intermediates were trapped by treatment with methanol as linear 1-methoxybutanes, indicating that the addition of electrophilic alkenes to NVC occurs in a stepwise manner. Once these intermediates are formed, the course of the reaction is determined largely by experimental conditions. In the presence of a large excess of NVC, cationic polymerization occurs. If the concentrations of NVC and alkene are comparable, collapse of the intermediate to the corresponding cyclobutane is favored, as shown in Scheme 1. An interesting aspect of this chemistry is that the cyclobutane formation is reversible, and the isolated product is capable of initiating the cationic polymerization of NVC. Experiments in which the cyclobutane was employed as an initiator demonstrated that neither an electron donor–acceptor complex or a radical-ion pair was essential to initiation, as had previously been hypothesized.[23,24]

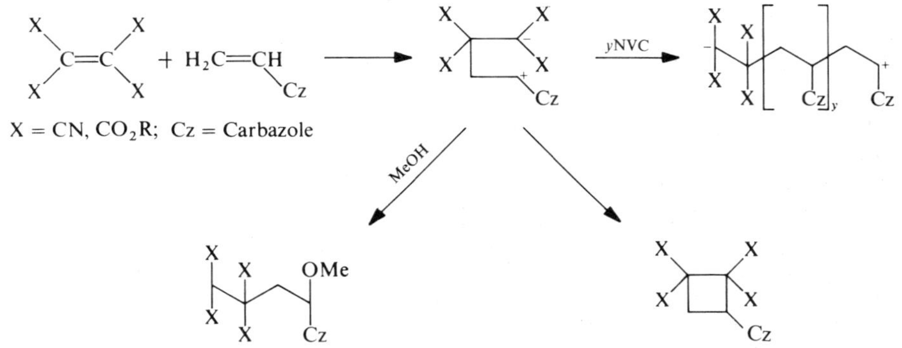

Scheme 1

43.3.3 Metal Halides and Metal Salts

Polymerizations of NVC induced by arsenic trichloride (AsCl$_3$) were shown to propagate by a cationic mechanism.[25] An initiation mechanism was proposed in which the AsCl$_3$ adds directly to the NVC double bond, generating a carbocation with a chloride counteranion. In support of this contention, analytical results were presented which showed the presence of one arsenic atom and two chlorine atoms per polymer chain. However, these results create more problems than they solve. Polymer chain lengths were short in these experiments and appeared to be governed by chain-breaking reactions. Moreover, the concentration of AsCl$_3$ did not seem to influence chain lengths either through a simple monomer/initiator ratio or by participation in the chain-breaking reactions. Consequently, it is difficult to see how approximately one arsenic atom could be bonded to each chain.

When metal nitrates were used in place of metal halides, the resulting NVC polymerizations were again propagated by cationic active centers.[26,27] In certain solvents, the nitrates were insoluble or only partially soluble, complicating interpretation of kinetic measurements. ESR spectra demonstrated the absence of radical species in solutions of the nitrates alone and their presence in solutions containing both nitrate and monomer, but also their presence in solutions of the nitrates and saturated analogs of the monomer. By a process of elimination, an electron-transfer mechanism was proposed for initiation. The metal nitrates were thought unlikely to initiate by direct addition or through the participation of hydrolysis products. Therefore, transfer of an electron from the NVC to the metal nitrates was postulated.

A number of interesting effects of additives were observed when sodium tetrachloroaurate (NaAuCl$_4$) was used to polymerize NVC.[28] This salt was selected for examination since Lewis-acid behavior was felt to be less likely than in the case of the metal nitrates. In general, reducing agents seemed to accelerate the polymerization process, raising the possibility that the initiation involved a redox reaction. The devised mechanism envisaged electron transfer from NVC to Au[II], formed by electron transfer from NVC to Au[III]. Obviously, both reactions generate NVC radical-cations, and the selection of Au[II] as the true initiating species was based on consideration of the probable relative rates of the two reactions.

Metal oxyhalides as well as metal halides initiate the cationic polymerization of NVC effectively.[29] However, the activity of both types of compound is altered dramatically by the addition of metal alkyls to the reaction medium. The ability of the resulting complexes to induce stereoregularity in NVC polymers has led to the hypothesis that propagation proceeds through a 'cationic coordination mechanism'.[30]

43.3.4   Organic Cation Salts

Despite the fact that diagnostic tests indicate that most of the NVC polymerizations discussed in the preceding sections are cationic in nature, a disappointingly small amount of kinetic and mechanistic information can be gleaned from them. The principal difficulty is that the concentration, if not the identity, of the true initiating species is usually unknown. Furthermore, the nature and extent of dissociation of the counteranion is also often a mystery. Lack of knowledge of these parameters prevents reliable estimates of the rate constants of the fundamental chemical processes (initiation, propagation transfer and termination) from being drawn. One approach to rectifying this lack of knowledge involves the use of preformed stable organic cation salts as initiators.

Not all organic cations are capable of initiating NVC polymerizations. For example, careful analysis of polymer formed in the presence of isotopically labeled triethyloxonium hexafluorophosphate ($Et_3O^+PF_6^-$) showed that the cation was not participating in initiation.[31] Similarly, cationically polymerized 'living' poly(tetrahydrofuran) failed to add to NVC monomer. These results corroborated the earlier observation that NVC and cyclic ethers do not cross-propagate in cationic systems.[21]

While stable preformed ammonium salts (quaternary salts, such as benzyltrimethylammonium hexafluoroantimonate) do not initiate the cationic polymerization of NVC, stable amminium salts (electron-deficient tertiary salts) will. Tris-*p*-bromophenylamminium hexachloroantimonate $[(BrC_6H_4)_3N^+SbCl_6^-]$ is isolated in the form of blue crystals. When a solution of these crystals was added to NVC in dichloromethane, the blue color was discharged and rapid polymerization ensued.[32] In this case, an electron-transfer initiation mechanism was not only proposed but verified. Electron transfer between the amminium salt and NVC generated a monomer radical-cation which could add a second monomer unit to give a dimeric radical-cation. Further oxidation by a second amminium radical-cation would yield a dication. Alternatively, the monomer radical-cations could combine to form the dication. The intermediacy of the dication was demonstrated by quenching the reaction with methanol and isolating the dimethoxybutane.

An extensive kinetic study of the polymerization of NVC in dichloromethane initiated by stable preformed tropylium (cycloheptatrienylium) hexachloroantimonate and perchlorate salts provided the first genuine measurements of the rate constant of cationic propagation for this monomer.[33] A methanol trapping experiment showed that initiation involved direct addition of the tropylium cation to the monomer double bond. Thermograms obtained by adiabatic calorimetry corresponded to a kinetic scheme in which initiation was rapid and complete, giving rise to a fixed concentration of active species. Significantly, kinetic termination appeared to be negligible during the observed reaction time. Since this concentration of active centers could be equated with the level of initiator added, and the rate of polymerization and monomer concentration could be calculated from the thermograms, estimation of the rate constant of propagation was possible.

One preliminary observation made in the course of these experiments was that initiation of NVC polymerization by tropylium salts did not seem to occur at temperatures below 223 K. Later work, however, employing tropylium hexachloroantimonate, hexafluoroantimonate (SbF_6^-) and hexafluoroarsenate (AsF_6^-) salts produced polymerization at temperatures as low as 203 K.[34] Reactions at these low temperatures were characterized by slow initiation, and the rectilinear portion of the kinetic first-order plots was used to evaluate rate constants of propagation. Neither variation of the initiator concentration nor addition of large excesses of nonreactive common counteranion salts served to alter appreciably these rate constants, suggesting that propagation rates for paired and unpaired poly(NVC) cations were similar.

A detailed investigation of the influence of ion-pairing equilibria on the rate of cationic propagation for NVC in dichloromethane was conducted with trityl (triphenylmethyl) hexafluoroantimonate as initiator.[35] Measurements of initiator consumption demonstrated that initiation was slow but virtually complete during the lifetime of the polymerizations. The validity of the techniques and assumptions was subsequently verified by an alternative kinetic treatment.[39] Rate constants for initiation, propagation and chain transfer to monomer were estimated. The propagation rate constant showed a dependence on initiator concentration and on the level of added common counteranion, which was also reflected in polymer molecular weights. Both the rate constants and the molecular weight decreased as initiator levels increased, implying an influence of ion pairing. Two interesting conclusions emerged from this analysis: the rate constants of propagation estimated for paired and unpaired ions differed by only about one order of magnitude, and the rate constant of chain transfer to monomer was higher for unpaired ions than for paired ions.

Preconceived ideas, derived from familiarity with similar studies of anionic polymerizations, led to criticism of the proposal that the reactivity of unpaired poly(NVC) cations exceeded that of paired ions by a factor of 10 or less, depending on temperature.[40] However, the anionic polymerizations, involving different monomers, solvents and counterions, bore little relevance to the discussion, and this aspect of reactivity in cationic polymerizations has since been noted with other vinyl monomers.[38] Indeed, later work confirmed the essential features of the equilibrium shift and its effect on the propagation rate constant,[42] although results obtained at very high salt concentrations could be influenced by multiple ion aggregation. A further observation drawn from these later experiments was that the trityl cation appeared to be regenerated at the end of the polymerizations.[40] This behavior was explained in terms of electron-transfer reactions. Initiation was presumed to occur through transfer from NVC to the trityl cation, yielding a monomer radical-cation and a trityl radical. As the remaining monomer was consumed, slow reconversion of the trityl radical to the cation would occur. However, no demonstration of the presence or absence of trityl residues in NVC polymers has yet been made.

Since all of the previous rate determinations had been conducted in dichloromethane, an important advance in the understanding of cationic reactivity was achieved when NVC was polymerized by trityl salts in nitrobenzene.[41] The dielectric constant for this solvent is high (about 36 at 293 K) compared to that of dichloromethane (approximately 9 at the same temperature), but the electron-donor activities of both solvents are low. Dissociation of paired ions to unpaired ions would be enhanced by an increase in the dielectric constant of the reaction medium. However, according to transition-state rate theory, the rate constant for the reaction between an ion (propagating cation) and a neutral molecule (monomer) resulting in an ionic product should be depressed. This effect is precisely what was observed in the experiments. A summary of the kinetic parameters measured in polymerizations of NVC by organic cation salts is presented in Table 1.

One related monomer, 9-ethyl-3-vinylcarbazole, has been subjected to the same type of kinetic analysis using an organic cation salt with a well-defined counteranion. Polymerizations of this monomer in dichloromethane at 273 K were characterized by a lower rate constant of propagation ($20\,000\ \mathrm{dm^3\,mol^{-1}\,s^{-1}}$) than corresponding NVC polymerizations, presumably due to the higher stability of the propagating cations.[44] Although polymer molecular weights appeared to be

Table 1 Kinetic Parameters for the Cationic Polymerization of *N*-Vinylcarbazole

			$10^{-4} \times$ Rate constants ($\mathrm{dm^3\,mol^{-1}\,s^{-1}}$)				
Initiator	Solvent	Temperature (K)	Unpaired ion propagation	Paired ion propagation	Initiation	Enthalpy of polymerization ($\mathrm{kJ\,mol^{-1}}$)	Ref.
$Ph_3C^+AsF_6^-$	CH_2Cl_2	293	95	—	—	−92.6	37, 42
$Ph_3C^+SbF_6^-$	CH_2Cl_2	293	60	5	0.013	−87.1	35
$Ph_3C^+AsF_6^-$	$PhNO_2$	293	8.5	—	—	—	41
$Ph_3C^+SbCl_6^-$	$PhNO_2$	293	9.5	—	—	—	41
$Ph_3C^+SnCl_5^-$	$PhNO_2$	293	8.9	—	—	—	41
$C_7H_7^+ClO_4^-$	$CH_2Cl_2^{\,a}$	273	16	—	—	−95.1	33
$C_7H_7^+SbCl_6^-$	CH_2Cl_2	273	46	—	—		33
$Ph_3C^+AsF_6^-$	$CH_2Cl_2^{\,a}$	273	48	—	—	−95.1	39, 42
$Ph_3C^+AsF_6^-$	CH_2Cl_2	273		—	—		39, 42
$Ph_3C^+SbF_6^-$	CH_2Cl_2	273	45	5	0.0034	−99.6	35
$C_7H_7^+ClO_4^-$	$CH_2Cl_2^{\,a}$	248	6.6	—	—	−104.8	33
$C_7H_7^+SbCl_6^-$	CH_2Cl_2	248	16	—	—		33
$Ph_3C^+AsF_6^-$	CH_2Cl_2	248	12	—	—	—	43
$C_7H_7^+SbF_6^-$	$CH_2Cl_2^{\,b}$	233	2	2	—	−77.9	34
$Ph_3C^+AsF_6^-$	CH_2Cl_2	233	5	—	—	—	43

^{a} Containing 1% v/v MeCN. ^{b} Containing 1% v/v $MeNO_2$.

governed by the monomer/initiator ratio, implying the absence of chain termination, spectroscopic evidence indicated that at least one kinetically important side-reaction of the propagating ions did occur.

43.3.5 Iodine

In recent years, criteria for generating 'living' (no chain transfer, no termination) cationic polymerizations have been developed. Reaction conditions required include low temperatures, nucleophilic counteranions and solvents of low dielectric constant and electron-donor activity. These requirements were satisfied when iodine was used to polymerize NVC in toluene and mixtures of dichloromethane with carbon tetrachloride at low temperatures.[45] Polymer molecular weights increased in a linear fashion with monomer consumption during reactions and upon the addition of successive aliquots of monomer. The influence of counterion dissociation on the course of these polymerizations was demonstrated in a series of experiments conducted at 195 K in dichloromethane involving the addition of tetrabutylammonium iodide.[46]

43.3.6 Radiation-induced Solution Polymerizations

The nature of the solvent in radiation-induced polymerizations of NVC is crucial. Upon exposure to high-energy electron beams, NVC formed a cyclic dimer, *trans*-1,2-dicarbazylcyclobutane, in non-deaerated solvents of low electron-donor activity, regardless of dielectric constant. However, in non-deaerated solvents of high electron-donor activity, again over a wide range of dielectric constants, polymerization occurred.[47] In the absence of air, polymer was formed in both types of solvent. These observations were explained in terms of ease of electron capture and radical-anion formation: solvents of low donor activity formed a radical anion readily whether oxygen was present or not, while in the case of solvents with high donor activity, oxygen competed effectively for electrons and prevented formation of the organic radical-anion. The nature of the radical counteranion dictated the course of the polymerization, with oxygen radical-anions tending to promote cyclization.

When the electron accelerator was replaced by a lower-energy radiation source (UV light at 366 nm), no polymerization occurred in irradiated benzene solutions.[48] Addition of 2,2'-azobis-(2-methylpropionitrile) [azobis(isobutyronitrile), AIBN] caused polymer formation during irradiation, apparently by quenching excited monomer singlets and triplets. A mechanism was postulated in which quenching of singlets led to radical polymerization, and triplet quenching led to simultaneous cationic polymerization. In dichloromethane solutions, NVC polymerization occurred during irradiation by UV light whether AIBN was present or not.[49] No identification of the initiating species was possible, but analysis of the polymer molecular weight distributions again led to the conclusion that radical and cationic active centers propagated simultaneously. Depending on the nature of the solvent, photosensitized reactions of NVC with organic electron-acceptors were found to give rise to: (i) cationic NVC polymerization; (ii) cyclodimerization of NVC; (iii) radical NVC polymerization; and (iv) radical copolymerization.[50]

Irradiation of NVC solutions containing inorganic electron-acceptors resulted in cationic polymerization alone.[51,52] Initiation in nitrobenzene solutions of sodium tetrachloroaurate and NVC was attributed to stable species generated by redox reactions similar to those postulated for the corresponding thermal polymerizations. When the sodium cation was replaced by a tetra-*n*-butylammonium cation, flash photolysis indicated that halogen atoms or their derivatives were formed and reacted with NVC. An initiation mechanism was proposed in which complex formation between the halogen and NVC was followed by successive electron transfer and proton transfer to the halogen, releasing the acid HX. However, the proton transfer step presents difficulties as depicted (ref. 51, p. 823) since it requires removal of either a vinyl or aromatic proton from the NVC monomer radical-cation.

43.4 HETEROGENEOUS POLYMERIZATION

Several reports have outlined the salient features of electroinitiated polymerizations of NVC.[53-55] Polymerization occurred in the anodic compartment of divided cells, consistent with a cationic mechanism of chain propagation. A soluble supporting electrolyte was required, and polymer molecular weights were higher in dichloromethane than in either nitrobenzene or acetone. Polymerizations of NVC in the presence of very high ratios of electrolyte to monomer resulted in the deposition of crosslinked conductive polymer on the anode.[61] In order to increase the current-carrying ability of the reaction medium without reducing polymer solubility, a two-phase system was developed with formamide as the electrolysis phase and toluene as the polymer solvent.[62] Since contact between the electrolyte and the monomer was required for polymerization, initiation apparently took place at the interface between the solvents. Electrolysis of iron(III) chloride solutions, for example, was presumed to yield $\cdot FeCl_4$ radicals, which migrated to the interface and reacted with NVC to give the monomer radical-cation together with an $FeCl_4^-$ counteranion.

Polymerizations of NVC have also been induced by exposing solutions of the monomer to carbon black.[58] While scavenging tests showed that these polymerizations are cationic in nature, disagreement as to the initiation mechanism exists. Carboxylic acid residues on the carbon surface are claimed to be protogenic initiators,[59] and evidence of termination by grafting to the surface counteranion could also be interpreted as the result of propagation through esters. Model studies have been adduced to support the contrary hypothesis that initiation proceeds by electron transfer from the monomer to unpaired electrons on the solid carbon.[60,61]

A detailed kinetic scheme for NVC polymerization in the presence of 13X molecular sieves was derived.[62] The activity of the sieves was enhanced dramatically by metal exchange with both copper(II)[63] and cobalt(II),[64] and initiation was presumed to involve both protons and the metal ions. The most intriguing feature of these polymerizations was the apparently linear dependence of polymer molecular weight on the level of transition metal in the sieves.

43.5 REFERENCES

1. V. P. Naidenov, N. E. Kruglyak and V. G. Syromyatnikov, *Ukr. Khim. Zh. (Russ. Ed.)*, 1979, **45**, 1083 (*Chem. Abstr.*, 1980, **92**, 110 372).
2. J. Pielichowski, R. Popielarz and R. Chrzaszcz, *J. Polym. Sci., Polym. Lett. Ed.*, 1985, **23**, 387.
3. W. Reppe, E. Keyssner and E. Dorrer (I. G. Farbenindustrie AG), US Pat. 2 072 465 (1937) (*Chem. Abstr.*, 1937, **31**, 2717).
4. A. Chapiro and G. Hardy, *J. Chim. Phys. Phys.-Chim. Biol.*, 1962, **59**, 993.
5. P. B. Ayscough, A. K. Roy, R. G. Croce and S. Munari, *J. Polym. Sci., Part A-1*, 1968, **6**, 1307.
6. W. Pekala and A. Lesinski, *Nukleonika*, 1974, **19**, 769 (*Chem. Abstr.*, 1975, **82**, 98 480).
7. S. Tazuke, O. Supakorn and T. Inoue, *J. Polym. Sci., Polym. Chem. Ed.*, 1982, **20**, 2239.
8. K. Tsuji, K. Takakura, M. Nishii, K. Hayashi and S. Okamura, *J. Polym. Sci., Part A-1*, 1966, **4**, 2028.
9. R. A. Meyers and E. M. Christman, *J. Polym. Sci., Part A-1*, 1968, **6**, 945.
10. J. Polaczek, *Nuova Chim.*, 1973, **49**, 93 (*Chem. Abstr.*, 1973, **79**, 5689).
11. A. Gandini and P. H. Plesch, *J. Chem. Soc. (B)*, 1966, 7.
12. A. Gandini and S. Prieto, *J. Polym. Sci., Polym. Lett. Ed.*, 1977, **15**, 337.
13. J. Asakura, M. Yoshihara and T. Maeshima, *Makromol. Chem., Rapid Commun.*, 1983, **4**, 103.
14. L. P. Ellinger, *Adv. Macromol. Chem.*, 1968, **1**, 169.
15. M. Ko, T. Nakanishi, T. Sato and T. Otsu, *Mem. Fac. Eng., Osaka City Univ.*, 1973, **14**, 153 (*Chem. Abstr.*, 1975, **82**, 73 529).
16. J. C. Bevington, C. J. Dyball and J. Leech, *Makromol. Chem.*, 1977, **178**, 2741.
17. T. Sato, M. Abe and T. Otsu, *Makromol. Chem.*, 1977, **178**, 1259.
18. P. K. Sengupta and G. Mukhopadhyay, *Makromol. Chem.*, 1982, **183**, 1093.
19. T. Natsuume, Y. Shirota, H. Hirata, S. Kusabayashi and H. Mikawa, *J. Chem. Soc., Chem. Commun.*, 1969, 289.
20. J. Pac and P. H. Plesch, *Polymer*, 1967, **8**, 237.
21. R. Gumbs, S. Penczek, J. Jagur-Grodzinski and M. Szwarc, *Macromolecules*, 1969, **2**, 77.
22. T. Gotoh, A. B. Padias and H. K. Hall, Jr., *J. Am. Chem. Soc.*, 1986, **108**, 4920.
23. C. E. H. Bawn, A. Ledwith and M. Sambhi, *Polymer*, 1971, **12**, 209.
24. T. Nakamura, M. Soma, T. Onishi and K. Tamaru, *Makromol. Chem.*, 1970, **135**, 241.
25. M. Biswas and D. Chakravarty, *J. Polym. Sci., Polym. Chem. Ed.*, 1973, **11**, 7.
26. S. Tazuke, T. B. Tjoa and S. Okamura, *J. Polym. Sci., Part A-1*, 1967, **5**, 1911.
27. T. Kawamura and K. Matsuzaki, *Makromol. Chem.*, 1981, **182**, 3003.
28. S. Tazuke, M. Asai and S. Okamura, *J. Polym. Sci., Part A-1*, 1968, **6**, 1809.
29. M. Biswas and G. C. Mishra, *Makromol. Chem.*, 1980, **181**, 1629.
30. M. Biswas and G. C. Mishra, *J. Polym. Sci., Polym. Chem. Ed.*, 1981, **19**, 3081.
31. G. Turchi, F. Matera and P. L. Magagnini, *Makromol. Chem.*, 1973, **170**, 75.
32. A. Ledwith and D. C. Sherrington, *Macromol. Synth.*, 1972, **4**, 183.
33. P. M. Bowyer, A. Ledwith and D. C. Sherrington, *Polymer*, 1971, **12**, 509.
34. J. M. Rooney, *Makromol. Chem.*, 1978, **179**, 165.
35. J. M. Rooney, *J. Polym. Sci., Polym. Symp.*, 1976, **56**, 47.
36. D. C. Pepper, *Eur. Polym. J.*, 1980, **16**, 407.
37. A. Gandini and H. Cheradame, *Adv. Polym. Sci.*, 1980, **34/35**, 1.
38. T. Kunitake and K. Takarabe, *Macromolecules*, 1979, **12**, 1067.
39. M. Rodriguez and L. M. Leon, *Eur. Polym. J.*, 1983, **19**, 585.
40. M. Rodriguez and L. M. Leon, *Eur. Polym. J.*, 1983, **19**, 589.
41. E. Bilbao, M. Rodriguez and L. M. Leon, *Polym. Bull. (Berlin)*, 1984, **12**, 359.

42. M. Rodriguez and L. M. Leon, *J. Polym. Sci., Polym. Lett. Ed.*, 1983, **21**, 881.
43. M. Rodriguez and L. M. Leon, *Makromol. Chem., Rapid Commun.*, 1983, **4**, 601.
44. A. H. DeMola, A. Ledwith, J. F. Yanus, W. W. Limburg and J. M. Pearson, *J. Polym. Sci., Polym. Chem. Ed.*, 1978, **16**, 761.
45. T. Higashimura, H. Teranishi and M. Sawamoto, *Polym. J.*, 1980, **12**, 393.
46. T. Higashimura, Y. X. Deng and M. Sawamoto, *Polym. J.*, 1983, **15**, 385.
47. Y. Tabata, *J. Polym. Sci., Polym. Symp.*, 1976, **56**, 409.
48. R. G. Jones and R. Karimian, *Polymer*, 1980, **21**, 832.
49. D. R. Terrell, *Polymer*, 1982, **23**, 1045.
50. K. Tada, T. Shirota and H. Mikawa, *Macromolecules*, 1973, **6**, 9.
51. M. Asai and S. Tazuke, *Macromolecules*, 1973, **6**, 818.
52. M. Asai, S. Tazuke and S. Okamura, *J. Polym. Sci., Polym. Chem. Ed.*, 1974, **12**, 45.
53. J. W. Breitenbach and C. Srna, *Pure Appl. Chem.*, 1962, **4**, 245.
54. D. C. Phillips, D. H. Davies and J. D. B. Smith, *Macromolecules*, 1972, **5**, 674.
55. E. B. Mano and B. A. L. Calafate, *J. Polym. Sci., Polym. Chem. Ed.*, 1983, **21**, 829.
56. Y. Shirota, N. Noma, H. Kanega and H. Mikawa, *J. Chem. Soc., Chem. Commun.*, 1984, 470.
57. S. Sanyal, R. C. Bhakta and B. Nayak, *Macromolecules*, 1985, **18**, 1314.
58. K. Ohkita, M. Uchiyama and N. Nishioka, *Carbon*, 1978, **16**, 195.
59. N. Tsubokawa, H. Maruyama and Y. Sone, *Polym. Bull. (Berlin)*, 1986, **15**, 209.
60. M. Biswas and S. A. Haque, *J. Polym. Sci., Polym. Chem. Ed.*, 1983, **21**, 1861.
61. S. A. Haque and M. Biswas, *J. Polym. Sci., Polym. Chem. Ed.*, 1985, **23**, 2567.
62. M. Biswas, M. Banerjee and M. M. Maiti, *J. Polym. Sci., Polym. Chem. Ed.*, 1985, **23**, 2631.
63. M. Biswas, M. Banerjee and M. M. Maiti, *J. Polym. Sci., Polym. Chem. Ed.*, 1984, **22**, 1997.
64. M. Biswas, M. Banerjee and M. M. Maiti, *Polymer*, 1985, **26**, 625.

44

Carbocationic Polymerization: Copolymerization

JOSEPH P. KENNEDY
University of Akron, OH, USA

44.1 DEFINITIONS AND FUNDAMENTALS

Carbocationic copolymerizations are random polyaddition processes in which both propagating sites exhibit carbenium ion character.[1,2,3] After having established the cationic polymerizability of a pair of monomers M_1 and M_2, the critical steps for copolymer synthesis are the crosspropagation reactions

$$\sim M_1^+ + M_2 \xrightarrow{k_{12}} \sim M_1 M_2^+ \tag{1}$$

and

$$\sim M_2^+ + M_1 \xrightarrow{k_{21}} \sim M_2 M_1^+ \tag{2}$$

where $\sim M^+$ are propagating cations. The rate constants k_{12} and k_{21} must not be too different, otherwise meaningful random copolymerization will not arise. In practical terms copolymerization can occur only when the ratio of crosspropagation constants k_{12}/k_{21} is not more than about an order of magnitude different from unity, *i.e.* when the stabilities of the propagating species are not very different.[4] In contrast, the magnitudes of the rate constants of the individual homopolymerization steps

$$\sim M_1^+ + M_1 \xrightarrow{k_{11}} \sim M_1 M_1^+ \tag{3}$$

and

$$\sim M_2^+ + M_2 \xrightarrow{k_{22}} \sim M_2 M_2^+ \tag{4}$$

are much less important in determining the outcome of the copolymerization. The ratios of rate constants k_{11}/k_{12} and k_{22}/k_{21} are termed the reactivity ratios r_1 and r_2 for M_1 and M_2 respectively and, preferentially in their reciprocal forms $1/r_1$ and $1/r_2$, are measures of monomer reactivities.

Although the number and variety of carbocationically polymerizable monomers is very large,[1] owing to the relatively large differences in monomer reactivities (carbenium ion stabilities) the number of monomer pairs that can be transformed into random copolymers by carbocationic techniques is rather limited. Unfortunately many incorrect and meaningless reactivity ratios can be found in the literature. According to a recent comprehensive analysis,[2] out of the 643 comonomer pairs that have been examined and described in the literature, only 155 can be regarded as true random copolymers whose composition can be described by the conventional two-parameter copolymerization equation.

In general, substituents α to the propagating ion (or radical) determine the reactivity of the growing species. Efforts have been made to correlate the electron-donating effect of substituents on monomer reactivity of styrenes by means of the Hammett σ–ρ relationship[3] and more recently by ^{13}C NMR chemical shifts of the β carbons of substituted styrenes.[4] The substituents tend to increase the reactivity of styrene in the order *p*-OMe > *p*-Me > *p*-H > *p*-Cl > *m*-Cl > *m*-NO$_2$. For monomers other than styrene comparative data are sparse and the general order of reactivities is vinyl ethers > α-methylstyrene > isobutylene > styrene > isoprene > butadiene > propylene.

Since inductive or conjugative interactions between ion and substituent are much stronger than between radical and substituent, the effect of the nature of the substituent will be much larger in ionic copolymerizations than in radical systems. Thus ionic copolymerizations are more selective and restrictive than radical systems. Indeed true ionic random copolymerization is rather the exception than the rule.[5]

Notwithstanding this generalization copolymerization has been demonstrated in several seemingly surprising monomer pairs, for example 2-chloroethyl vinyl ether/4-methylstyrene,[6] benzofuran/benzothiophene,[7] 1-phenyl-1,3-butadiene/2-chlorostyrene,[8] 4-vinylbiphenyl/2-vinylfluorene[9] and α-methylstyrene/isobutyl vinyl ether,[10] *i.e.* in systems in which monomer reactivity and cation stability are fortuitously balanced. Even more surprising are copolymerizations between heterocyclic and vinyl compounds, *e.g.* styrene/trioxane[11] and styrene/dioxolane.[12] In these systems the heterocyclic compounds probably propagate by carboxonium ions ($\sim$O–$\overset{+}{C}$H$_2$ $\leftrightarrow$ $\sim$$\overset{+}{O}$=CH$_2$); indeed copolymerizability with styrene may be viewed as evidence for the presence of carbocationic intermediates.

In contrast to free radical systems, the reaction conditions (*i.e.* the nature of solvent, counter anion and temperature) greatly affect the outcome of carbocationic copolymerizations and the effects of these reaction parameters on the reactivity ratios are complex and interdependent.[2] For example, by changing the solvent,[13–18] the counter anion[19] or in the presence of various additives, *e.g.* π electron acceptors,[20–24] r_1 and r_2 will be significantly changed.

Seminal information regarding the various ionicities that prevail during copolymerization has been generated mainly by Yamamoto and Higashimura, who investigated the effect of common salts on copolymerization parameters.[25–27]

The molecular weights of copolymers will be lower than those of the corresponding homopolymers synthesized under the same conditions.[2] The extent of the molecular weight depression of copolymers relative to homopolymers has been analyzed for a large number of systems.[28–31] The phenomenon of molecular weight depression may become quite large at low temperatures.[28]

Sequence distribution analyses[32] of various cationic copolymers have been carried out.[28–31] In the case of the isobutylene/β-pinene system this method helped to establish the validity of the two-parameter copolymerization equation, *i.e.* the random nature of the copolymer.[31] In contrast, analysis of isobutylene/styrene systems indicated 'blocky' (non-random) copolymers.[31]

44.2 IMPORTANT COPOLYMERS AND TERPOLYMERS

44.2.1 Isobutylene–Conjugated Diene Copolymers

From a technological point of view, copolymers of isobutylene and conjugated dienes, particularly isoprene (*i.e.* a copolymer of isobutylene plus a small but critical amount of isoprene, so called butyl rubber, or isobutene–isoprene rubber IIR), are far more important than carbocationic homopolymers.[2] Commercially available carbocationically prepared copolymers are high molecular weight ($\bar{M}_n = 100\,000$–$150\,000$) isobutylene–isoprene rubbers comprising from $\sim$0.5 to $\sim$2.5% isoprene in the chain. This family of low unsaturation general purpose elastomers are mainly used as inner tubes or (in their halogenated form) inner liners, and also in various other applications such as cable insulation, vibration dampers, curing bladders, automotive body mounts, ditch and tank liners, gaskets and sealants, blending agents and adhesive bases.[2]

The desirable combination of physical, mechanical and chemical properties of IIR is largely due to the unique head-to-tail linked isobutylene and *trans*-1,4 linked isoprene structure (**1**) of these copolymers.[33, 34] The seven allylic hydrogens in the *trans*-1,4 isoprene unit provide sulfur vulcaniz-

$$-\left(CH_2\underset{\underset{Me}{|}}{\overset{\overset{Me}{|}}{C}}\right)_{\sim 98.5\%}\left(CH_2\overset{\overset{Me}{|}}{C}{=}CHCH_2\right)_{\sim 1.5\%}-$$

(1)

ability, the numerous configurational possibilities of the polyisobutylene sequences explain the low glass transition temperature ($T_g \sim -73\,°C$), the absence of tertiary hydrogens results in high oxidative environmental stability, and the low free volume due to the interlocking methyl groups goes a long way to account for the excellent barrier properties.

The properties and structure/property relationship of IIR are the subjects of numerous treatises.[35-43] The most detailed recent characterization data of isobutylene–conjugated diene copolymers has been generated by Cesca and co-workers (see refs. 34, 44–48 and the review, ref. 49, and refs. therein). Analyses have involved mainly high resolution ^{1}H and ^{13}C NMR spectroscopy, and older structural information together with reactivity ratio data have been critically examined.

Isobutylene–isoprene copolymerizations are carried out on a commercial scale by various international companies (Exxon, Polysar *etc.*).[2] Total world capacity (1986) was $\sim 600\,000$ tons. The rubbers are produced by a slurry process invariably employing the 'H$_2$O' AlCl$_3$-initiating system and MeCl diluent (butyl rubber is insoluble in MeCl) at approximately $-100\,°C$. Cryogenic temperatures are needed to 'freeze out' the molecular-weight-limiting event, *i.e.* chain transfer to isoprene, and thus to obtain high molecular weight products.[2] In spite of sustained research since the 1940s a satisfactory process for the synthesis of high molecular weight IIR at higher than $-100\,°C$ has not yet been introduced in commercial practice (see, for example, refs. 50–52).

Low molecular weight (liquid) butyl rubber is also a commercial commodity and is used for sealants, caulking agents, adhesives and blending agents.[53-55] It is produced by shear degradation (extrusion) of the high molecular weight product.[53] The structure of these materials, *e.g.* the nature of the end groups, has not been investigated in detail.

In view of the high chain transfer activity of isoprene it is difficult to produce high molecular weight IIR containing more than ~ 1.5 mol % isoprene in the copolymer. Also, higher amounts of isoprene in the feed, *i.e.* beyond ~ 3 vol. %, may cause gelation. Promising leads for the preparation of high molecular weight, high unsaturation (~ 5–40 mol %), soluble isobutylene–conjugated diene (isoprene, cyclopentadiene) copolymers have been developed by the use of cosolvents.[56]

The most recent research in this field concerns the 'living' copolymerization of isobutylene–isoprene charges.[57] Thus it has been demonstrated that gel-free copolymers containing from 1.2 to 7.8 mol % 1,4-isoprene units can be prepared by novel ester/BCl$_3$-initiating systems at $-30\,°C$; however, the molecular weights of these products are still low ($\bar{M}_n = 2400$–$12\,100$).[57]

A considerable amount of research has also been carried out with isobutylene–butadiene copolymers (see, for example, ref. 58), but a commercial product has not yet been developed. Butadiene incorporation is mainly by *trans*-1,4 units but minor ($\sim 20\%$) amounts of 1,2 units are also present.[29,30] The reactivity of butadiene is higher in nonpolar than in polar solvents and it is affected by temperature. Since these copolymers contain at least three kinds of repeat units (isobutylene, *trans*-1,4- and 1,2-butadiene), these products are in fact terpolymers and should not be described by the two-component copolymerization model.[59] The microstructure of these copolymers has been investigated in detail by ^{1}H and ^{13}C NMR spectroscopy[44] and evidence for blocky *trans*-1,4-butadiene units was found.

Efforts have also been made to develop high molecular weight isobutylene–butadiene rubbers containing up to 12 mol % butadiene in the chain by the use of BF$_3$-based initiating systems at cryogenic temperatures.[60]

Soluble high molecular weight isobutylene–cyclopentadiene copolymers containing up to 40 mol % cyclopentadiene have been prepared under essentially homogeneous conditions at $-120\,°C$ by the use of solvent mixtures (*e.g.* *n*-heptane and carbon disulfide) and AlCl$_3$ dissolved in methyl chloride.[56] The copolymers presumably contain 1,4-cyclopentadiene enchainments. They exhibit excellent ozone resistance and good compatibility with high-unsaturation conventional rubbers (nitrile rubber, styrene–butadiene rubber, polybutadiene) under sulfur-curing conditions.[56]

In addition to these simple conjugated dienes, copolymerization of isobutylene has also been investigated with *trans*-1,3,5-hexatriene,[61] 2,4,6-octatriene,[62] 1,3,5-heptatriene,[62] 2,5-dimethyl-1,3,5-hexatriene,[62] alloocimene,[62] with the three geometric isomers of 2,4-hexadiene,[63] 1,1,4,4-tetramethyl-1,3-butadiene,[64] *trans*-1,3-pentadiene,[45,46] *cis*-1,3-pentadiene[47] and 2,3-dimethyl-1,3-butadiene.[48] The products have been characterized in detail by ^{1}H and ^{13}C NMR spectroscopy.

44.2.2 Isobutylene–β-Pinene Copolymers

Ozone resistance combined with sulfur vulcanizability was the incentive for the development of high molecular weight rubbery isobutylene–β-pinene copolymers.[65] Isobutylene/β-pinene monomer charges readily copolymerize (for example with 'H$_2$O'/EtAlCl$_2$/EtCl/−50 °C to −130 °C) to give random structures containing the β-pinene moiety (2), *i.e.* a structural element that may be regarded as consisting of a cyclohexene ring inserted into an isobutylene unit.[28] The six allylic hydrogens provide sulfur vulcanizability and due to the 'protected' nature of the unsaturation, ozone resistance occurs.

(2)

Interestingly, the isobutylene/β-pinene pair leads to azeotropic copolymerizations at cryogenic temperatures. β-Pinene is more reactive than isobutylene (as judged by 1/r data) in the −50 to −100 °C range; however, the difference in reactivities diminishes with decreasing temperatures and at about −110 °C and below, $r_{IB} = r_{\beta P} = 1.0$, *i.e.* azeotropic conditions prevail.[28] Azeotropic copolymerizations have also been mentioned in connection with isobutylene–isoprene copolymerizations at −120 °C in homogeneous systems.[56]

44.2.3 Isobutylene–Styrene Copolymers

Both isobutylene and styrene (and styrene derivatives) are readily polymerizable by cationic means.[2] High molecular weight isobutylene–styrene copolymers have been prepared and have been evaluated for various applications. On account of their excellent barrier properties, films made of these copolymers were found to exhibit outstanding packaging characteristics for fresh and dried fruits.[66]

44.2.4 Terpolymerizations

Carbocationic terpolymerizations have also been investigated, *e.g.* the terpolymerization of isobutylene–styrene–α-methylstyrene.[67] Some terpolymers, notably partially insoluble random terpolymers of isobutylene–isoprene–divinylbenzene, are used commercially in blends with conventional IIR in some inner tubes where they impart green strength during manufacture.[68]

44.3 IMPORTANT DERIVATIVES OF CARBOCATIONIC COPOLYMERS

Halogenated butyl rubbers, *i.e.* chlorinated and brominated IIR, are also important commercial commodities and used mainly as inner liners for passenger tires, inner tubes for trucks, hoses, *etc.* These products are produced by elemental halogenation of butyl rubbers.[69-75] The (overall) structure of chlorinated IIR is a composite of the units (3), (4) and (5). These structural elements reflect the advantageous combination of physical–chemical properties of the rubber, *i.e.* improved blending and covulcanizability with other rubbers, a faster cure rate than butyl rubber and good heat resistance.

(3)   (4)   (5)

Treatment of chlorobutyl rubber, for example, with zinc carboxylates[76,77] yields isobutylene chains containing conjugated diene units[78] (6) and possibly (7). The conjugated units are active for

$$\underset{\underset{\text{Me}}{|}}{\overset{\overset{\text{Me}}{|}}{-\text{CH}_2\text{C}}}\text{CH}_2\text{CH}_2\underset{}{\overset{\overset{\text{CH}_2}{\|}}{\text{C}}}\text{CH}=\text{CH}-$$

$$\underset{\underset{\text{Me}}{|}}{\overset{\overset{\text{Me}}{|}}{-\text{CH}_2\text{C}}}\text{CH}=\overset{\overset{\text{Me}}{|}}{\text{C}}\text{CH}=\text{CH}-$$

exo–cis; exo–trans   *cis–cis; cis–trans; trans–trans; trans–cis*

(6)   (7)

Diels–Alder additions with maleic anhydride, acrylate esters, *etc.* which lead to a great variety of modification possibilities.[2]

44.4 REFERENCES

1. J. P. Kennedy, 'Cationic Polymerization of Olefins', Wiley, New York, 1975.
2. J. P. Kennedy and E. Maréchal, 'Carbocationic Polymerization', Wiley, New York, 1982.
3. R. B. Cundall, in 'The Chemistry of Cationic Polymerization', ed. P. H. Plesch, Macmillan, New York, 1963, chap. 15.
4. K. Hatada, K. Nagata, T. Hasegawa and H. Yuki, *Makromol. Chem.*, 1977, **178**, 2413.
5. J. P. Kennedy, T. Kelen and F. Tüdös, *J. Polym. Sci., Polym. Chem. Ed.*, 1975, **13**, 2277; T. Kelen., F. Tüdös, B. Turcsanyi and J. P. Kennedy, *J. Polym. Sci., Polym. Chem. Ed.*, 1977, **15**, 3047.
6. T. Masuda and T. Higashimura, *Polym. J.*, 1971, **2**, 29.
7. C. Zaffran and E. Maréchal, *Bull. Soc. Chim. Fr.*, 1970, 3523.
8. P. Borg and E. Maréchal, *J. Macromol. Sci., Chem.*, 1977, **11**, 897.
9. S. Cohen and E. Maréchal, *J. Polym. Sci., Polym. Symp.*, 1975, **52**, 83.
10. P. Trivedi, *J. Macromol. Sci., Chem.*, 1980, **14**, 589.
11. W. Kern, H. Cherdron and V. Jaacks, *Angew. Chem.*, 1961, **73**, 177.
12. M. Okada, Y. Yamashita and Y. Ishii, *Kogyo Kagaku Zasshi*, 1965, **68**, 364.
13. C. G. Overberger and V. G. Kamath, *J. Am. Chem. Soc.*, 1959, **81**, 2910.
14. T. Masuda and T. Higashimura, *Polym. J.*, 1971, **2**, 29.
15. J. P. Kennedy, 'Copolymerization' in 'High Polymers', ed. G. E. Ham, Wiley Interscience, New York, 1964, vol. 18, p. 293.
16. Y. Imanishi, T. Higashimura and S. Okamura, *J. Polym. Sci., Part A-1*, 1965, **3**, 2455.
17. F. Laval and E. Maréchal, *J. Polym. Sci., Polym. Chem. Ed.*, 1977, **15**, 149.
18. T. Masuda and T. Higashimura, *J. Macromol. Sci., Chem.*, 1971, **5**, 549.
19. A. V. Tobolsky and R. J. Boudreau, *J. Polym. Sci.*, 1961, **51**, S53.
20. I. M. Panayotov, I. K. Dimitrov and I. E. Bakerdjiev, *J. Polym. Sci., Part A-1*, 1969, **7**, 2421.
21. I. M. Panayotov, R. S. Velichkova and N. Matev, *Dokl. Bolg. Akad. Nauk*, 1974, **27**, 1679.
22. V. Toncheva, R. S. Velichkova and I. M. Panayotov, *Bull. Soc. Chim. Fr.*, 1974, 103.
23. G. Heublein and O. Barth, *J. Prakt. Chem.*, 1974, **316**, 649.
24. G. Heublein and B. Heublein, *Faserforsch. Textiltech.*, 1975, **26**, 107.
25. T. Higashimura and K. Yamamoto, *J. Polym. Sci., Polym. Chem. Ed.*, 1977, **15**, 301.
26. K. Yamamoto and T. Higashimura, *J. Polym. Sci., Chem. Ed.*, 1974, **12**, 613.
27. K. Yamamoto and T. Higashimura, *J. Polym. Sci., Polym. Chem. Ed.*, 1976, **14**, 2621.
28. J. P. Kennedy and T. Chou, *Adv. Polym. Sci.*, 1976, **21**, 1.
29. Y. Imanishi, Z. Momiyama, T. Higashimura and S. Okamura, *Kobunshi Kagaku*, 1963, **20**, 369.
30. S. Okamura, T. Higashimura and K. Takeda, *Kobunshi Kagaku*, 1961, **18**, 389.
31. S. Okamura, T. Higashimura, Y. Imanishi, R. Yamamoto and K. Kimura, *J. Polym. Sci., Part C*, 1967, **16**, 2365.
32. H. J. Harwood and W. M. Ritchey, *J. Polym. Sci., Polym. Lett. Ed.*, 1964, **2**, 601.
33. C. Y. Chu and R. Vukov, *Macromolecules*, 1985, **18**, 1423.
34. C. Corno, A. Priola and S. Cesca, *Macromolecules*, 1980, **13**, 1092.
35. D. J. Buckley, in 'Encyclopedia of Polymer Science and Technology', Wiley, New York, 1964, vol. 2, p. 754.
36. R. M. Thomas and W. J. Sparks, in 'Synthetic Rubber', ed. G. S. Whitby, Wiley, New York, 1954, chap. 24, p. 838.
37. D. J. Buckley, *Rubber Chem. Technol.*, 1959, **32**, 1475.
38. F. P. Baldwin and R. H. Schatz, in 'Kirk-Othmer: Encyclopedia of Chemical Technology', 3rd edn., Wiley, New York, 1979, vol. 8, p. 470.
39. R. M. Thomas, I. E. Lightbown, W. J. Sparks, P. K. Frolich and E. V. Murphree, *Ind. Eng. Chem.*, 1940, **32**, 1283.
40. R. M. Thomas, *Rubber World*, May 1954.
41. R. J. Adams and E. J. Buckler, *Kautsch. Gummi*, 1953, **6**, 225.
42. R. M. Thomas, *Rubber World*, 1954, **130**, 203.
43. P. J. Flory, *Ind. Eng. Chem.*, 1946, 417.
44. C. Corno, A. Priola and S. Cesca, *Macromolecules*, 1979, **12**, 411.
45. C. Corno, A. Priola and S. Cesca, *Macromolecules*, 1980, **13**, 1099.
46. C. Corno, A. Priola and S. Cesca, *Macromolecules*, 1980, **13**, 1314.
47. A. Priola, C. Corno and S. Cesca, *Macromolecules*, 1981, **14**, 475.
48. C. Corno, A. Priola and S. Cesca, *Macromolecules*, 1982, **15**, 840.
49. S. Cesca, in 'Cationic Polymerization and Related Processes', IUPAC Proceedings, Ghent, 1983; Proceedings: Academic Press, 1984, p. 105.
50. S. Cesca, A. Priola, M. Bruzzone, G. Ferraris and P. Giusti, *Makromol. Chem.*, 1975, **176**, 2339.
51. M. Baccaredda, M. Bruzzone, S. Cesca, M. DiMaina, G. Ferraris, P. Giusti, P. L. Magagnini and A. Priola, *Chim. Ind. (Milan)*, 1973, **55**, 109.
52. S. Cesca, M. Bruzzone, A. Priola, G. Ferratis and P. Giusti, *Rubber Chem. Technol.*, 1976, **49**, 937.
53. Kalene 800, Data Sheet of Hardman Inc., Belleville, NJ.

54. A. J. Berejka and N. E. Stucker, *Rubber Age*, 1983, **105**, 33.
55. J. S. Glazman, lecture, Society of Plastic Engineers, New York, 1970, April.
56. W. A. Thaler and D. J. Buckley, *Rubber Chem. Technol.*, 1976, **49**, 960.
57. R. Faust, A. Fehérvári and J. P. Kennedy, *Br. Polym. J.*, 1987, **19**, 379.
58. C. E. Schildknecht, 'Vinyl and Related Polymers', Wiley, New York, 1952, p. 571.
59. T. Kelen, F. Tüdös, B. Turcsányi and J. P. Kennedy, *J. Polym. Sci., Polym. Chem. Ed.*, 1977, **15**, 3047.
60. K. Lee and J. Oziomek (Firestone Tire & Rubber Co.), *US Pat.* 4 390 673 (1984).
61. A. Priola, C. Corno, M. Bruzzone and S. Cesca, *Polym. Bull.*, 1981, **4**, 735.
62. A. Priola, C. Corno, M. Bruzzone and S. Cesca, *Polym. Bull.*, 1981, **4**, 743.
63. A. Priola, C. Corno and S. Cesca, *Polym. Bull.*, 1982, **7**, 599.
64. C. Corno, A. Priola and S. Cesca, *Polym. Bull.*, 1983, **9**, 132.
65. J. P. Kennedy and T. Chou (University of Akron), *US Pat.* 3 923 759 (1975).
66. R. G. Newberg, J. R. Briggs and W. A. Fairclough, *American Chemical Society, Division of Rubber Chemistry*, Los Angeles, 1948, July.
67. M. Györ, J. P. Kennedy, T. Kelen and F. Tüdös, *J. Macromol. Sci., Chem.*, 1984, **21**, 1339.
68. J. Walker, G. J. Wilson and K. J. Kumbhani, lecture, Society of Chemical Industry, Plastics & Polymer Group, London, UK, 1972, October 3.
69. I. Kuntz, R. L. Zapp and R. J. Pancirov, *Rubber Chem. Technol.*, 1984, **57**, 813.
70. A. van Tongerloo and R. Vukov, *Proc. Int. Rubber Conf.*, 1979, 70.
71. D. A. Patterson and R. G. Schwammberger, *Proc. Int. Rubber Conf.*, 1982, 723.
72. F. P. Baldwin, D. J. Buckley, I. Kuntz and S. B. Robinson, *Rubber Plast. Age*, 1961, **42**, 500.
73. N. E. Odam, *J. Inst. Rubber Ind.*, 1971, **5**, 49.
74. J. Walker, R. H. Jones and G. Feniak, Philadelphia Rubber Group, Technical Meeting, 1972, September 22.
75. G. C. Blackshaw and K. J. Robinson, American Chemical Society, Division of Rubber Chemistry, Meeting, Detroit, 1973, May 1.
76. F. P. Baldwin (Exxon Research & Engineering Co.), *US Pat.* 3 775 387 (1973).
77. F. P. Baldwin and A. Malatesta (Exxon Research & Engineering Co.), *US Pat.* 3 965 213 (1976).
78. F. P. Baldwin, I. J. Gardner, A. Malatesta and J. A. Rae, American Chemical Society, Division of Rubber Chemistry, 108th Meeting, 1975, October 7.

45

Cationic Ring-opening Polymerization: Introduction and General Aspects[1-11]

ERIC J. GOETHALS
State University of Ghent, Belgium
and
STANISLAW PENCZEK
Polish Academy of Sciences, Lodz, Poland

45.1 INTRODUCTION

Ring-opening polymerization is an important method for the synthesis of macromolecules and is currently used for the production of a variety of polymers on a commercial scale.

Cationic ring-opening polymerization proceeds either by nucleophilic attack of the monomer molecule on the onium ion at the end of the growing macromolecule (equation 1, where Z = heteroatom), or by nucleophilic attack of the chain end on the monomer molecule bearing a positive charge (*e.g.* protonated), as in equation (2).

activated chain-end mechanism
$$\tag{1}$$

activated (here, protonated) monomer mechanism
$$\tag{2}$$

Besides these two general mechanisms, pseudocationic polymerization may also proceed, where in equation (1), instead of the growing macroion, a highly polarized bond is present at the end of the macromolecule. These mechanisms are briefly discussed in this chapter and then treated in detail in the subsequent chapters in this volume.

45.2 CATIONIC RING-OPENING POLYMERIZATION

Cationic ring-opening polymerization has provided two important industrial polymers: polyformaldehyde, based on cationically homo- or co-polymerized 1,3,5-trioxane (with a few percent of comonomer) and poly(tetrahydrofuran). The former is an engineering thermoplastic material, produced in the USA, West Germany and Japan (260×10^3 ton year^{-1}). The latter is available as an α,ω-hydroxy-ended telechelic ($M_n = 1000$ or 2000) used as a soft segment for polyurethanes (including Biomer® for the artificial heart Jarvic 7) or elastoplastic polyether–polyesters (Hytrel®, Arnitel®, *etc.*).

711

There are a number of other polymers, produced on a smaller scale, based on cationic ring-opening polymerization. A recently produced polymer which is particularly interesting is poly(dichlorophosphazene), obtained by polymerization of hexachlorotriphosphazene, which is subsequently modified by substituting the chlorine atoms by perfluoroalkoxy groups. In this way, a heat-stable rubber is obtained. Cationic polymerization of cyclic amines, lactams, cyclic siloxanes, α-epichlorohydrin, 3,3-bis(chloromethyl)oxetane and some other monomers that already have found industrial applications, complete an impressive list of technological achievements for cationic ring-opening polymerization.

Another aspect, equally important and perhaps even more spectacular, is the contribution of this field to the development of polymer science. The real milestone in this area was set up using ring-opening polymerization: Staudinger established the macromolecular features of the products of polymerization by using polyformaldehyde and poly(ethylene oxide) as the major models. Simultaneous measurements of the end-group concentrations and molecular weights provided evidence for the proposal that these were not merely associates of lower molecular weight products, but long chains, each with distinctive end-groups.

Cationic ring-opening polymerization has also made an important contribution to the modern theory of polymer chemistry. With regard to reaction mechanisms, several systems, particularly the polymerization of tetrahydrofuran (THF), cyclic sulfides (particularly those which are substituted and four-membered) and some aziridines and oxazolines, have provided living systems, in which the elementary reactions have been studied quantitatively and the relevant rate constants determined.

In cationic ring-opening polymerization, the propagation reaction can be described as a nucleophilic substitution reaction in which the positively charged active species is the electrophile and the monomer is the nucleophile. As in classical organic chemistry, the reaction can be of the S_N2 type, in which the new bond is formed and the old bond is broken simultaneously, or of the S_N1 type, in which the old bond is broken first, with the formation of a carbenium ion, which then immediately reacts with the monomer. The latter mechanism will be favoured if the structure of the monomer is such that the resulting carbenium ion is stabilized and if the monomer is a weak nucleophile.

The driving force for polymerization is the relief of strain as a consequence of the ring-opening reaction. For three- and four-membered rings this strain is high (60–90 kJ mol^{-1}) and therefore, from a thermodynamic point of view, these rings are always polymerizable. Monomers with five or more atoms in the ring have low ring-strains and therefore not all these monomers are polymerizable and, if they are, they show a low ceiling temperature (see Volume 3, Chapter 46) above which the polymerization is thermodynamically impossible.

In cationic ring-opening polymerization, the important step in the propagation reaction involves the breaking of a carbon–onium bond. This is in contrast with anionic ring-opening polymerization where a carbon–heteroatom bond is broken. It is well known that carbon–onium bonds provide

Table 1 Some Parent Heterocycles which can be Polymerized Cationically

[a] These compounds can also be polymerized anionically.

better leaving groups compared with the corresponding carbon–heteroatom bonds (*e.g.* ammonium–amine, oxonium–ether . . .). Therefore, it is not surprising that there are more heterocycles which can be polymerized by a cationic mechanism than by an anionic mechanism. Table 1 shows some of the most common parent heterocycles which can be polymerized cationically.

In the cationic ring-opening polymerization with active species at the chain end Y, the major Y driving force of the reaction is the ring opening of the active species. The monomer only provides the nucleophilic heteroatom and does not contribute to the negative reaction enthalpy since it remains cyclic. In fact, any other nucleophile present in the reaction mixture is a potential reagent for the active species. Even if the experiments are carried out under the purest conditions, there will always be two potential competitors for the monomer to react with the active species: (1) the counteranion and (2) the heteroatoms of the resulting polymer chain.

The first reaction transforms the cationic active species into an uncharged chain-end (equation 3). Depending on the nature of X and of Z, the covalent carbon–X bond can be reactivated to an active species; for example, equation (4).

$$\text{~}\overset{+}{Z}\langle\hspace{0.5em} X^- \longrightarrow \text{~}Z\text{~}X \tag{3}$$

$$\text{~}Z\text{~}X + Z\langle \longrightarrow \text{~}Z\text{~}\overset{+}{Z}\langle\hspace{0.3em} X^- \tag{4}$$

Correspondingly, the formation of a covalent bond by reaction with the counterion can be a real termination or a temporary termination. In the former case, the uncharged chain-end is called 'dead'; in the latter case, it is called 'dormant'. If initiators are selected which produce anions which have little or no nucleophilicity, the (temporary) termination with counterion is negligible or non-existent. Typical counterions with low nucleophilicity are FSO_3^-, $CF_3SO_3^-$, ClO_4^- and trinitrobenzene-sulfonate. Counterions with no nucleophilic reactivity are BF_4^-, SbF_6^-, PF_6^- and AsF_6^-. However, some of these counterions have been found to react with strong electrophilic groups, with the formation of fluoride and the corresponding uncharged fluoro derivative; for example, equation (5) with BF_4^-. This reaction is even more pronounced with some chlorides such as $SbCl_6^-$ and $AlCl_4^-$.

$$R_2\overset{+}{O}\text{-}R + F\text{-}BF_3^- \rightarrow ROR + RF + BF_3 \tag{5}$$

The ring opening of the active species by a heteroatom of the polymer chain results in a non-strained onium ion, equation (6). This reaction can also be irreversible or reversible, leading to 'dead' or 'dormant' species. In the former case, the polymerization will stop before all monomer is consumed and the final yield will be determined by the initiator concentration and the values of the rate constants of polymerization and termination. In the latter case, an equilibrium between active and dormant species will be established, resulting in a decrease in the observed rate of polymerization.

$$\text{~}\overset{+}{Z}\langle + \hspace{0.3em} :Z \rightleftharpoons \text{~}Z\text{~}\overset{+}{Z} \tag{6}$$

If the active species reacts with a heteroatom of another polymer molecule, a branched structure is formed. If it reacts with a heteroatom of its own polymer chain, a (macro)cyclic polymer end-group will form (equation 7). If the latter reaction predominates and the termination is reversible, cyclic oligomers will be formed (equation 8).

$$\text{~}\overset{.}{Z}\hspace{0.5em}\overset{+}{Z}\langle \longrightarrow \text{~}\overset{+}{Z}\text{~}Z \tag{7}$$

$$\text{~}\overset{+}{Z}\bigcirc + Z\langle \longrightarrow \text{~}\overset{+}{Z}\langle + Z\bigcirc \tag{8}$$

In many cationic ring-opening polymerizations, cyclic oligomers are an important product. The thermodynamics of the ring–chain equilibria are described by the Jacobson–Stockmayer equation, relating the concentration of the given macrocyclic compound $[M_n]$ to the number of repeating units n: $[M_n] = An^{-5/2}$. However, before this thermodynamically inevitable concentration is attained, various kinetically controlled states are possible. Thus, systems are known in which the linear polymer is formed first, practically free of cyclic compounds, even at complete monomer conversion, and then, slowly, the thermodynamically controlled concentration of macrocyclics is built up.

In the general case of cationic ring-opening polymerization, the reaction scheme for propagation is as shown in Scheme 1.

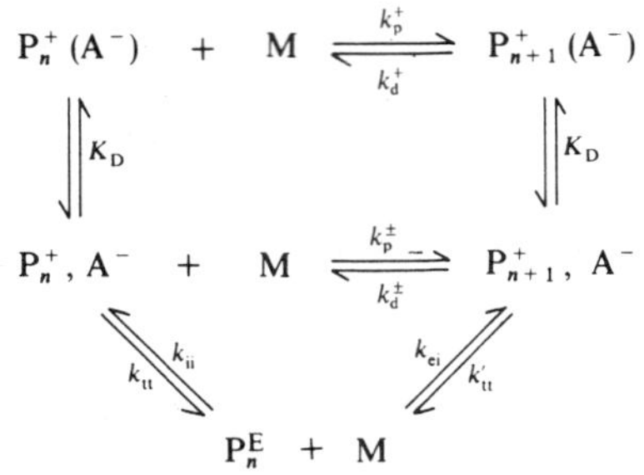

Scheme 1

This scheme is discussed in detail in Volume 3, Chapter 48. The rate constants and dissociation constants (K_D) shown in this scheme have been determined for the polymerization of THF and oxepane. These are two systems which are understood on the level approaching the present knowledge of the anionic polymerization of styrene, which is far superior to the present understanding of any vinyl cationic polymerization.

This interpretation of kinetics and mechanism has been made possible because of the early recognition and then direct observation of the growing species. On top of this, some systems exhibit a living nature after careful choice of the polymerization conditions, particularly after finding stable anions. Some examples of active species, observed directly by ^{1}H NMR, are given in structures **(1)**–**(5)**, counterions omitted.

$$-OCH_2CH_2CH_2CH_2OSO_2CF_3$$

(1)

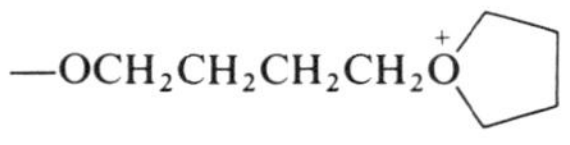

(2)

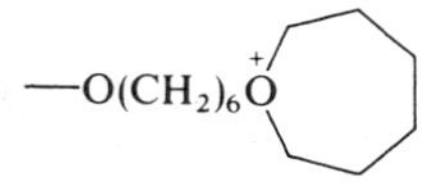

(3)

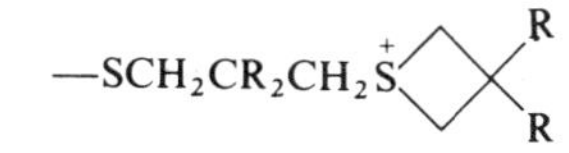

(4)

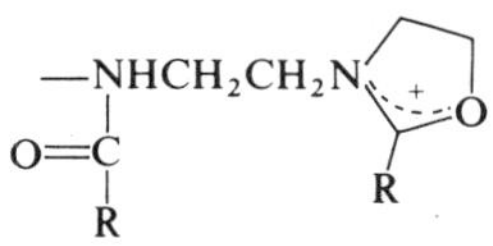

(5)

The covalent active species, such as triflate in the cationic polymerization of THF, are formed when the anion can provide a covalent bond.

The following generalities can be formulated for cationic chain-growth in ring-opening polymerization.

(i) Active species are highly solvated by the constituents of the systems; in contrast to vinyl polymerization, monomers and formed polymers are the strongest solvating agents. In cationic polymerization, the growing cations are mostly solvated (in contrast to anionic vinyl polymerization, for example, where counterions are solvated).

(ii) Macrocations and macrocation ion-pairs react with monomers with similar rate constants (*i.e.* $k_p^+ \approx k_p^\pm$). This is explained by the high solvation of the growing species, mostly by monomer and polymer units. The particular stereochemical course of the propagation step, in which the anion is not significantly altering the reacting macrocation, may also be responsible for this equality.

(iii) Covalent species are usually less reactive than their ionic counterparts; however, sterically hindered onium ions may add the puckered cyclic monomers more slowly than the respective covalent species, if these are less sterically demanding (*e.g.* oxepane or some P-containing monomers).

(iv) Equilibria between covalent species and macroion-pairs (ionization–collapse) as well as equilibria between the macroion-pairs and macrocations (dissociation–ion recombination) are both mostly governed by solvation. Thus, both ionization and dissociation, leading to the more ionic states, are exothermic, due to the enhanced solvation of the more ionic species formed.

Reactivity in cationic ring-opening polymerization can be described by the rate constant of propagation. However, when reactivities are expressed in this way, it is not clear whether the difference between the observed rate constants is due to the difference in reactivities of the monomers or of the corresponding active species. Therefore, in order to compare reactivities in the correct way, one has to take a certain (standard) active species and measure the rate of addition to it of the monomers under consideration. By the same token, in order to correlate structures and reactivities of active species, the rate constants of addition of the same (standard) monomer to these species has to be measured. This is illustrated schematically in Scheme 2.

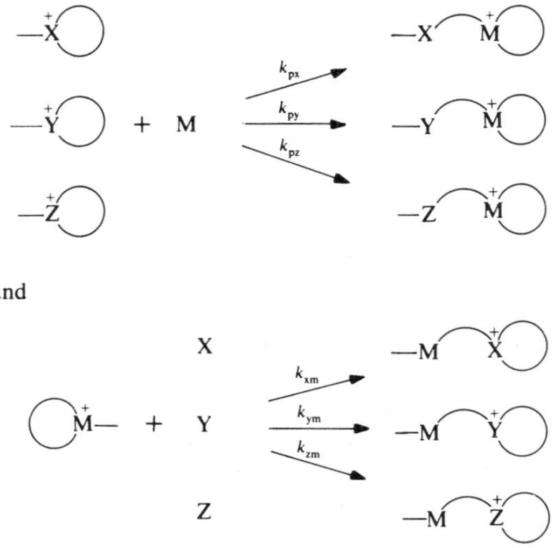

(9)

and

(10)

X, Y, Z and M are heterocyclic monomers

Scheme 2

Similar information can be obtained from copolymerization studies, but the four-reactions scheme (two homo- and two cross-propagations) can seldom be applied to the cationic polymerization of heterocyclics, because usually some additional reactions, such as reversible propagation or reaction of the exocyclic group in the growing species, cannot be neglected. Nevertheless, at least in some systems (*i.e.* cyclic sulfides, measured by Goethals *et al.*), the copolymerization approach gave quantitative results.

More recently, kinetics of addition of various heterocyclic monomers to the same active species have been studied. Similar measurements have been performed for various active species, in the form of covalent and ionic species, adding to the same monomer. The 1-methyltetrahydrofuranium cation was used as a model of the ionic species, while ethyl triflate was employed as a model of the covalent ones (Scheme 3). The values of the corresponding rate constants determined for these systems are given in Table 2, which also lists the pK_a values of the monomers.

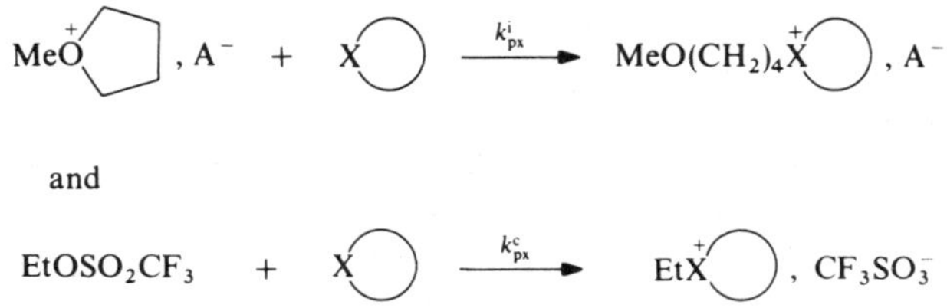

(11)

and

(12)

Scheme 3

The data of Table 2 indicate that the rate constants of reactions between various monomers and model active species, both ionic and covalent, are a function of monomer basicity and not of the ring strain.

Table 2 Correlation of Structure and Reactivity of the Heterocyclic Monomers in Cationic Ring-Opening Polymerization[a]

Monomer	k_{px}^{i} [b]	k_{px}^{c} [b]	pK_a
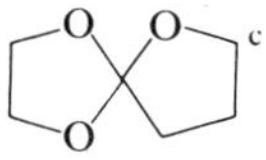 [structure: dioxaspiro] [c]	$\sim 4 \times 10^{-3}$	10^{-5}	-7.3
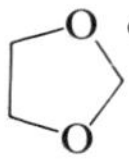 [structure: 1,3-dioxolane] [c]	10^{-4}	6×10^{-5}	-6.5
 [structure: tetrahydrofuran]	4×10^{-2}	2×10^{-4}	-2.0
 [structure: tetrahydrothiophene]	20	4.4×10^{-2}	1.2
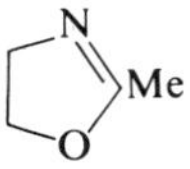 [structure: 2-methyl oxazoline]	120	4×10^{-1}	3.4
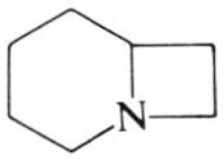 [structure: conidine]	500	170	11

[a] S. Penczek, P. Kubisa, S. Slomkowski and K. Matyjaszewski, *ACS Symp. Ser.*, 1985, **286**, 117.
[b] Rate constants of addition of heterocyclic monomers to the models of active species, according to equations (11) and (12), in $mol\,l^{-1}\,s^{-1}$, measured in $PhNO_2$ solvent at 35 °C.
[c] Rate constants not certain, but not lower than the given values.

Table 3 Correlation of Structure and Reactivity of the Onium Ions in Cationic Ring-opening Polymerization[a]

Active center	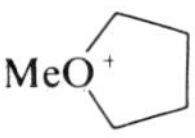 MeO^+ [structure]	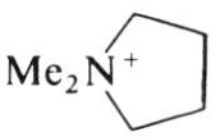 Me_2N^+ [structure]	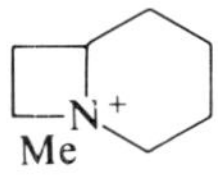 [structure: N-Me bicyclic]$^+$	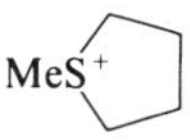 MeS^+ [structure]
	SbF_6^-	$CF_3SO_3^-$	$CF_3SO_3^-$	$CF_3SO_3^-$
k_p ($l\,mol^{-1}\,s^{-1}$)[b]	5×10^2	9×10^{-2}	7×10^{-3}	1×10^{-3}

[a] Ref. a from Table 2.
[b] Rate constants of conidine addition.

In order to correlate the reactivities of various active species (onium ions) with their structures, the reactions between the corresponding ion-pairs, modelling these active species, and a standard monomer, conidine, were studied. The measured rate constants are given in Table 3.

According to Table 3, the higher the basicity of the parent monomer, the lower the reactivity of active species from this monomer towards the standard monomer. Thus, the observed order of rate constants of homopropagation, Scheme 4, is proportional to the reactivities of active species and *inversely* proportional to the reactivities determined for different monomers with standard active species.

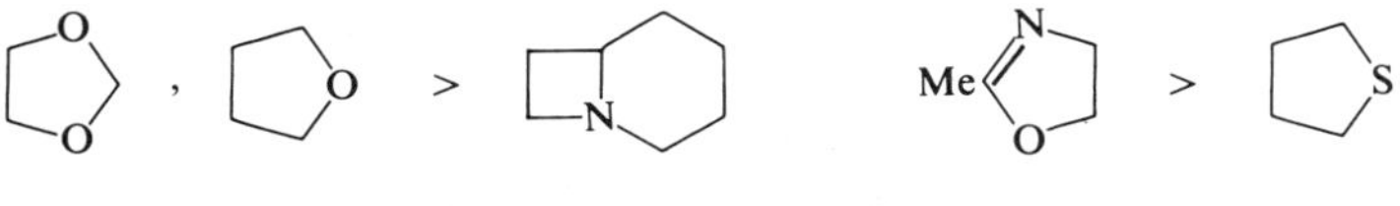

Scheme 4

This is a clear demonstration that, in passing from the ground state to the transition state, the bond-breaking is more advanced than the bond-making (borderline S_N2).

With a further shift towards the direction of still more advanced breaking of the bond within active species, this borderline S_N2 mechanism could eventually convert into the S_N1 mechanism, in which bond-breaking in the onium ion, giving the carbenium ion (unassisted by monomer), becomes a rate-determining step.

It has been shown that, besides the mechanism of growth involving the onium ions at the ends of the macromolecules, another mechanism may operate in which the growing macromolecules are not fitted with ions. In this mechanism, it is a monomer in the form of an ion, or charge-transfer complex, that adds to the electrically neutral macromolecule. The expression 'activated-monomer propagation' has been coined for these systems, because it is the chemically transformed monomer, activated in the ground state, that adds to the growing chain.

Thus, the activated (*e.g.* protonated) monomer propagation proceeds as shown in Scheme 5 (for ethylene oxide with a protonic acid as catalyst and an alcohol as initiator).

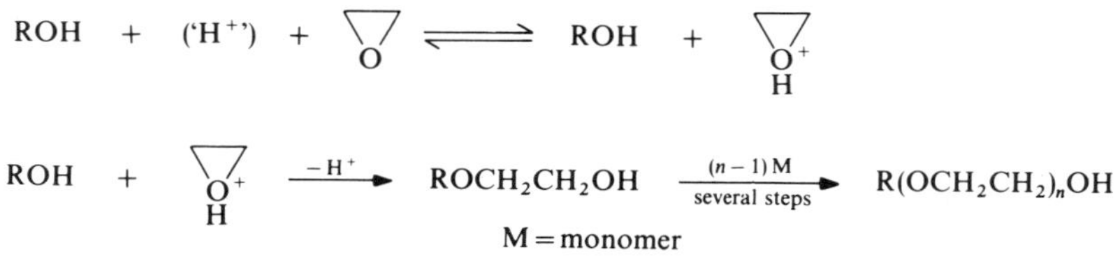

Scheme 5

Since the growing macromolecule in this process does not contain ions, as in active chain-end propagation, formation of cyclic compounds is highly hampered. However, as is discussed in Volume 3, Chapter 48, chains may also be protonated. This provides a route to the formation of cyclic compounds in activated-monomer polymerization. However, the rate of this process is far inferior to that in the active-chain process, when compared with the rate of propagation. A mechanism, similar to the one shown above for ethylene oxide, has also been proposed in the polymerization of lactams and cyclic amines.

This introduction merely describes the major phenomena observed in cationic ring-opening polymerization. These and more specific problems are discussed in subsequent chapters.

45.3 GENERAL REFERENCES

1. P. H. Plesch (ed.), 'The Chemistry of Cationic Polymerization', Pergamon Press, Oxford, 1963.
2. M. Szwarc, 'Anionic Polymerization', Wiley, New York, 1969.
3. S. Penczek (ed.), 'Polymerization of Heterocycles (Ring-Opening)', Pergamon Press, Oxford, 1975.
4. T. Saegusa and E. Goethals (eds.), 'Ring-Opening Polymerization', *ACS Symp. Ser.*, 1977, **59**.
5. S. Penczek, P. Kubisa and K. Matyjaszewski, *Adv. Polym. Sci.*, 1980, **37**.
6. E. J. Goethals (ed.), 'Cationic Polymerization and Related Processes', Academic Press, London, 1984.
7. K. J. Ivin and T. Saegusa (eds.), 'Ring-Opening Polymerization', Elsevier, Essex, 1984, vols. I–III.
8. J. E. McGrath (ed.), *ACS Symp. Ser.*, 1985, **286** (Proceedings of St. Louis ACS Meeting, 1984).
9. S. Penczek, in 'Polymer Year Book', ed. H.-G. Elias and R. A. Pethric, Harwood, London, 1984.
10. S. Penczek, P. Kubisa and K. Matyjaszewski, *Adv. Polym. Sci.*, 1985, **68, 69**.
11. S. Penczek, P. Kubisa and R. Szymanski, *Makromol. Chem., Macromol. Symp.*, 1986, **3**, 203.

46

Cationic Ring-opening Polymerization: Thermodynamics

Polish Academy of Sciences, Lodz, Poland

and

ERIC J. GOETHALS

State University of Ghent, Belgium

46.1 INTRODUCTION

There are a number of general reviews on the thermodynamics of polymerization[1,2,3] and several reviews specifically devoted to ring-opening (mostly cationic) polymerization, published[4,5,6] since Flory's 'Principles . . .' appeared.[7]

The present review is mostly based on refs. 2 and 4–6, although the thermodynamics of polymerization in real systems is treated in a modified way.

The thermodynamics of conversion of a monomer molecule into a polymer chain unit does not depend, at least under ideal conditions, on the polymerization mechanism. However, ideal conditions are seldom, if ever, met. Thus, in this chapter, some general thermodynamic dependences are discussed first, and then peculiarities of cationic ring-opening polymerization in real systems are reviewed in more detail.

46.2 POLYMERIZABILITY AND REACTIVITY

A clear distinction is made throughout this chapter between polymerizability and reactivity. Polymerizability refers to the thermodynamic features of a given monomer, and can be expressed by the extent to which the free energy of the polymerizing system changes when monomer converts into polymer.

Reactivity is related to kinetics and can be expressed by the rate of polymerization (more properly, by the rate of propagation). It may happen that a monomer of low polymerizability is highly reactive. In this case, a small fraction of the monomer comes rapidly to equilibrium with its polymer. At equilibrium, however, a large proportion of unreacted monomer is left. The opposite phenomenon can also be envisaged. However, it is more typical that high polymerizability corresponds to high reactivity (and *vice versa*); *e.g.* ethylene oxide (both high) and tetrahydrofuran (both low).

46.3 BASIC RELATIONSHIPS

46.3.1 Ideal Systems

Conversion of one mole of monomer into one mole of a linear unit is accompanied by a certain change in free energy. This conversion can be described as

$$n\text{M} \rightleftharpoons \text{-(M-)}_n \tag{1}$$

Provided that no interactions involving monomer M are broken, and/or no new interactions involving polymer units -(M-)- emerge, the change of free energy ΔG_p, due to polymerization of a given *system*, in which monomer converts into polymer, is described by a difference between the free energies of the initial and final states:

$$\Delta G_p = \Delta G_{p,a} - \Delta G_{m,1} = \Delta H_p - T\Delta S_p \tag{2}$$

For comparing various polymerizations, it is better to use the normalized (standard) thermodynamic functions. In the standard state (indicated with a superscript, *e.g.* ΔG_p°), concentration of monomer $[\text{M}]_0 = 1 \text{ mol l}^{-1}$ and pressure $P = 1 \text{ atm} = 101\,325 \text{ Pa}$. The change in enthalpy does not depend on $[\text{M}]_0$ under ideal conditions, but the change in entropy does. The entropy of the monomer increases with dilution by $-RT\ln[\text{M}]_0$. The corresponding term related to polymer, namely $\ln([\text{M}]_0 - [\text{M}]_e)/n$, can be omitted, because it is relatively small for large n.[8] This term stems from treating a macromolecule as one unit. Thus, we have for non-standard conditions:

$$\Delta G_p = \Delta G_p^\circ - RT\ln[\text{M}]_0 \tag{3}$$

At equilibrium $\Delta G_p = 0$, and hence:

$$\Delta G_p^\circ = \Delta H_p^\circ - T_c\Delta S_p^\circ = RT_c\ln[\text{M}]_e \tag{4}$$

thus

$$R\ln[\text{M}]_e = -\frac{\Delta H_p^\circ}{T_c} + \Delta S_p^\circ \tag{5}$$

or

$$T_c = \Delta H_p^\circ/(\Delta S_p^\circ + R\ln[\text{M}]_e) \tag{6}$$

where T_c is a temperature at which the equilibrium monomer concentration equals $[\text{M}]_e$. Depending on the signs of ΔH_p° and ΔS_p°, this temperature is called either the 'ceiling' (T_c) or 'floor' (T_f) temperature. Remembering that $\Delta G_p^\circ = -RT\ln K_e$, we have $K_e = 1/[\text{M}]_e$.

The same relationship follows from the kinetics of polymerization. Indeed, at equilibrium the rates of propagation and depropagation are equal to each other:

$$k_p[\text{P*}]_n[\text{M}]_e = k_d[\text{P*}]_{n+1} \tag{7}$$

Since, for sufficiently long chains, $[\text{P*}]_n = [\text{P*}]_{n+1}$ (P* are active species), then $k_p/k_d = K_e = 1/[\text{M}]_e$.

Equilibria are usually studied when liquid monomer (in bulk) is converted into a solution of the polymer in monomer or when a solution of monomer is converted into a solution of polymer in a mixture of solvent and monomer (left at equilibrium). Some data are also available for polymerization of a gaseous monomer into condensed polymer. To distinguish between these data the related subscripts are used. Thus, we have, correspondingly, $\Delta G_{l,s}$, $\Delta G_{s,s}$, and $\Delta G_{g,c}$.

The change of enthalpy and entropy of polymerization can be determined from equation (5) by plotting the experimentally measured $\ln[\text{M}]_e$ as a function of $1/T$.

46.3.2 Non-ideal (Real) Systems

When the conversion of monomer into a polymer unit is measured at high $[\text{M}]_0$, the monomer–polymer interaction cannot be neglected. The same is true when the measurements are made in a solvent interacting differently with monomer and polymer. Under these conditions, the corresponding interactions have to be taken into account. If the molecular weight of polymer is high enough, the following equation can be used to determine ΔG_p:[9]

$$\Delta G_p = RT[\ln\phi_1 + 1 + \chi(\phi_2 - \phi_1)] \qquad (8)$$

where ϕ_1 and ϕ_2 are volume fractions of monomer and polymer and χ is the polymer–monomer interaction parameter.

46.3.2.1 *Influence of initial monomer concentration*

It was observed in the polymerization of THF in benzene solvent that the equilibrium monomer concentration $[THF]_e$ depends linearly on $[THF]_0$.[9] A method was elaborated to determine ΔH_{lc}° and ΔG_{lc}° from the experimental data for such non-ideal thermodynamics.

It has also been observed that the extent of dependence of $[THF]_e$ on $[THF]_0$ is a function of the acidity of the solvent used.[10] This is shown in Figure 1, where $[THF]_e$ is plotted as a function of $[THF]_0$ for CCl_4, C_6H_6, CH_2Cl_2 and $MeNO_2$ solvents. Some of these solvents interact with THF, making its polymerization more difficult. This is because THF is a stronger base than its open-chain counterpart and, thus, interaction with polymer units is relatively weaker. These dependences can be envisaged as if the actual momentary concentration of THF available for polymerization were lowered by a fraction of complexed monomer, provided that the complexed monomer propagates less rapidly, requiring desolvation. Indeed, the heat of mixing of THF with CCl_4 (lowest line in Figure 1) is only -2.9 kJ mol^{-1}, whereas the heat of mixing of THF with CH_2Cl_2 (highest line) is -5.0 kJ mol^{-1}.[11]

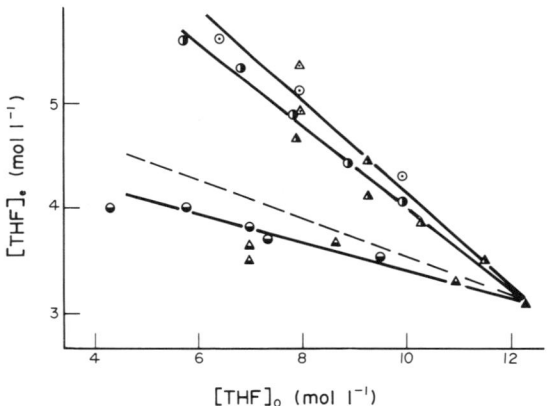

Figure 1 Dependence of the monomer equilibration concentration $[THF]_e$ on initial monomer concentration $[THF]_0$ at 25 °C; circles = measured by NMR, triangles = measured by gravimetry. Solvent: (⬤, ▲) CCl_4, (◗, ▲) CH_2Cl_2, (⊙, △) $MeNO_2$, (---) benzene;[10] (▲) neat THF

1,3-Dioxolane (DXL) is much less basic than THF and, besides, in contrast to THF, DXL is less basic than its polymer. It follows that the dependence of its equilibrium concentration $[DXL]_e$ on $[DXL]_0$ and solvent acidity is much less pronounced than in the polymerization of THF.[12]

A knowledge of the dependence of $[M]_e$ on $[M]_0$ is of primary importance in the studies of kinetics of polymerization. The use of the wrong value of $[M]_e$ may lead to erroneous values of k_p, especially when polymerization is studied starting from $[M]_0$ close to $[M]_e$.[13, 14, 15] Therefore, the studies of equilibrium polymerization, when $[M]_e$ is relatively high, require determination of $[M]_e$ for every run.

For low degrees of polymerization (a few units) $[M]_e$ *increases* with $[M]_0$, as required by the Tobolsky equation.[16] Therefore, the complete dependence of $[M]_e$ on $[M]_0$ goes from the 0,0 coordinates to a certain maximum, and then *decreases* according to the discussion given above.

46.3.2.2 Influence of degree of polymerization

The equilibrium monomer concentration $[M]_e$ depends on DP_n for shorter chains because, in the neighborhood of the end groups, the chemical potentials of the polymer units are different from those in the middle of the chain.

According to Tobolsky,[16] the relationship between $[M]_e$ and DP_n is given by

$$\frac{[M]_0 - [M]_e}{[I]_0} = \frac{K[M]_e}{1 - K[M]_e} \tag{9}$$

Recently, these relationships have been confirmed experimentally for THF polymerization.[17] When practically no initiator is left at equilibrium (*i.e.* if $[I]_0 \gg [I]_e$), then $([M]_0 - [M]_e)/[I]_0 = DP_n$. If $K \approx 1/[M]_e$ (*cf.* equation 7), then $DP_n = (1 - K[M]_e)^{-1}$. This is true, however, only when the equilibrium constant K is the same for all the propagation/depropagation steps, including the first one, namely, the reaction of initiator with monomer.

When the equilibrium constants are not the same, they can be determined by Szwarc's treatment, based on the reduction of the number of steps.[18]

Recently, Heitz *et al.*[19] have studied the cationic oligomerization of THF and established that only the first equilibrium constant K_0 differs from K_n, *i.e.* the equilibrium constant for every next step. Thus, $K_0 = 1.7 \text{ mol}^{-1} \text{ l}$ and $K_n = 0.25 \text{ mol}^{-1} \text{ l}$. This study has been made possible by using high resolution HPLC, allowing measurement of the concentration of all of the chains involved. It is also based on the assumption that the distribution of the growing chains is the same as that of the dead macromolecules.

46.3.2.3 Influence of phase separation

When liquid monomer in solution is converted into an insoluble polymer, the following possibilities arise: (a) solid amorphous polymer precipitates out of the saturated solution; (b) solid crystalline polymer precipitates from the saturated solution; and (c) polymerization proceeds in the solid crystalline state.

Only polymerization proceeding in the crystalline state differs thermodynamically from the polymerization of the same monomer in homogeneous solution. In the first two instances, growing species may still remain in solution and $[M]_e$ would be the same for both homogeneous and apparently heterogeneous systems. However, this does not mean that the polymer yield, *i.e.* $([M]_0 - [M]_e)/[M]_0$, would be the same. Although $[M]_e$ remains the same, in the precipitating system the total volume of liquid polymer–monomer mixture becomes smaller, so the *amount* of monomer at equilibrium is smaller anyway when polymer precipitates.

The change in free energy of the polymerization proceeding in the crystalline state differs from that taking place in the homogeneous solution by the heat of crystallization. Therefore, we have:

$$[M]_{e,s}/[M]_{e,ss} = \exp\Delta G_{cryst}/RT \tag{10}$$

It follows from this equation that $[M]_e$, observed in the homogeneous polymerization, may be greatly reduced by performing polymerization under conditions when the propagation step proceeds at the crystalline sites of a solid polymer.

Cationic polymerization of 1,3,5-trioxane (TXN) takes place in the crystalline state. The hypothetical equilibrium monomer concentration calculated for the homogeneous polymerization is $[TXN]_e = 2.5 \text{ mol l}^{-1}$ (CH_2Cl_2, 20 °C),[20] whereas $[TXN]_e$ for polymerization from dissolved monomer directly to crystalline polymer is 0.11 mol l^{-1} (at 25 °C). Thus, if one starts from a 30% solution of TXN ($\sim 3.3 \text{ mol l}^{-1}$), then, in the homogeneous solution, 72% of monomer would be left at equilibrium, whereas the real polymerization, proceeding in the crystalline state, leaves only $\sim 3\%$ of unreacted monomer.

46.4 FACTORS AFFECTING POLYMERIZABILITY: ENTHALPY OF POLYMERIZATION

In polymerization at normal pressure, the $P\Delta V$ term in $\Delta H = \Delta E - P\Delta V$ is negligible and the enthalpy change is almost equivalent to the change in the internal energy of the monomer. Thus, the heat of polymerization may be used as a measure of the strain energy in the cyclic compound. The major sources of ring strain are: (i) bond-angle distortion (angular strain); (ii) bond stretching or compression; (iii) repulsion between eclipsed hydrogen atoms (conformational strain, bond torsion, bond opposition); and (iv) non-bonded interaction between atoms or substituents attached to different parts of the ring (transannular strain, compression of the van der Waals radii).

In some groups of monomers, there are additional sources of strain, such as inhibition or reduction of amide-group resonance in lactams.[21]

The contribution of each type of strain depends on the chemical structure of the cyclic monomer, ring size and its substitution.

Distortion in bond angles is the major source of ring strain in three- and four-membered cyclic monomers. In five-membered cyclic compounds, the strain is mostly due to bond opposition forces, arising from eclipsed conformations. Tetrahydrofuran is an example. In medium-sized rings, strain arises primarily from non-bonded interactions and bond oppositions. Any kind of strain, including transannular interactions, can be removed completely in very large rings by arranging the ring atoms into two almost parallel chains (1).[22]

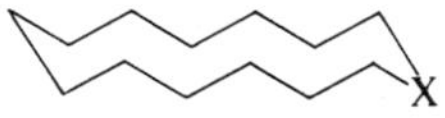

(1)

The dependence of ΔH_p on ring size for cationically polymerizable cyclic ethers is illustrated in Table 1. The same dependence for lactam polymerization is given in Table 2.[23]

Table 1 Dependence of ΔH_p on the Ring Size for Cyclic Ethers[a]

Monomer	Ring size	$-\Delta H_p$ (kJ mol^{-1})
Ethylene oxide (oxirane)	3	94.5
Trimethylene oxide (oxetane)	4	81
Tetrahydrofuran (oxolane)	5	15
Tetrahydropyran (oxane)	6	~ 0
1,4-Dioxane	6	~ 0
Hexamethylene oxide (oxepane)	7	33.5[b]

[a] H. Sawada, *J. Macromol. Sci., Rev. Macromol. Chem.*, 1970, **C5** (1), 151.
[b] W. K. Busfield, R. M. Lee and D. Merigold, *Makromol. Chem.*, 1972, **156**, 183.

Table 2 Dependence of ΔH_p on the Ring Size for Lactams

Monomer	Ring size	$-\Delta H_p$ (kJ mol^{-1})
4-Butanelactam	5	4.6
5-Pentanelactam	6	7.1
6-Hexanelactam (ε-caprolactam)	7	13.8
7-Heptanelactam	8	22.6
8-Octanelactam	9	35.1
9-Nonanelactam	10	23.4
10-Decanelactam	11	11.7
11-Undecanelactam	12	-2.1
12-Dodecanelactam	13	2.9

46.5 EQUILIBRIA INVOLVING MACRORINGS

In the preceding sections of this chapter, only the monomer–polymer equilibrium has been discussed. This equilibrium stems mostly from a back-biting process, involving the penultimate unit in the polymer chain. Thus, for instance, for THF polymerization we have the process shown in equation (11).[24]

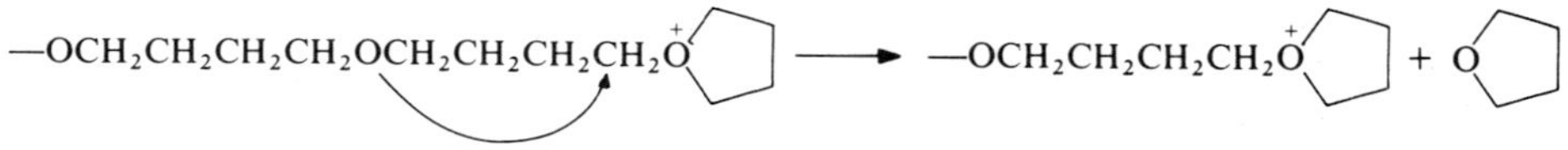

(11)

However, in equation (11), any oxygen atom from the chain can attack either the endo- or exo-cyclic carbon atom of the growing species, producing larger size cyclic compounds at the end, displaced by subsequent attack on the exocyclic carbon atom.

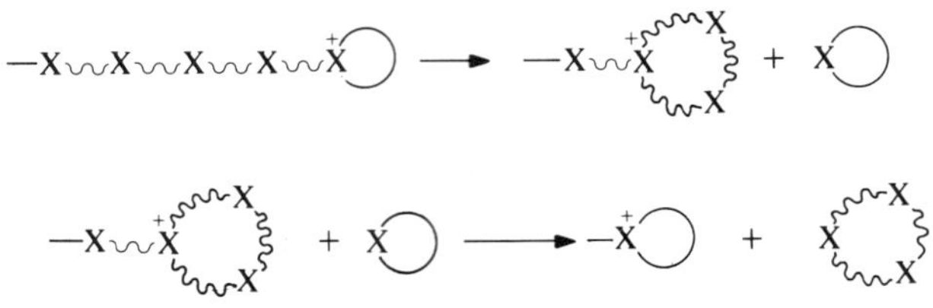

Scheme 1

Thus, in Scheme 1, formation of a cyclic trimer is illustrated. In the same manner, as well as by end-to-end cyclization, a complete range of macrocyclic compounds can be formed. The kinetics and thermodynamics involving these larger rings are discussed separately in Volume 3, Chapter 47.

46.6 REFERENCES

1. F. S. Dainton and K. J. Ivin, *Q. Rev., Chem. Soc.*, 1958, **12**, 61.
2. M. Szwarc, 'Carbanions, Living Polymers and Electron Transfer Processes', Interscience, New York, 1968, chap. 3.
3. H. Sawada, 'Thermodynamics of Polymerization', Dekker, New York, 1976.
4. K. J. Ivin and T. Saegusa (eds.), 'Ring Opening Polymerization', Elsevier, Essex, 1984, chap. 1.
5. S. Penczek, P. Kubisa and K. Matyjaszewski, 'Cationic Ring Opening Polymerization', *Adv. Polym. Sci.*, 1985, **68/69**, Chapter 2, Springer, Heidelberg, 1985.
6. J. Sebenda, in 'Polymerization of Heterocycles', ed. S. Penczek, Pergamon Press, Oxford, 1976, p. 329.
7. P. J. Flory, 'Principles of Polymer Chemistry', Cornell University Press, Ithaca, NY, 1953.
8. S. Penczek, P. Kubisa and K. Matyjaszewski, *Adv. Polym. Sci.*, 1985, **68/69**, 3.
9. K. J. Ivin and J. Leonard, *Eur. Polym. J.*, 1970, **6**, 331.
10. S. Penczek and K. Matyjaszewski, *J. Polym. Sci., Polym. Symp.*, 1976, **56**, 255.
11. S. Dincer and H. C. van Ness, *J. Chem. Eng. Data*, 1971, **16**, 378.
12. L. I. Kozub and N. S. Enikolopian, *Vysokomol. Soedin., Ser. A*, 1968, **10**, 2007.
13. P. Bourdauducq and D. J. Worsfold, *Macromolecules*, 1975, **8**, 562.
14. K. Brzezinska, K. Matyjaszewski and S. Penczek, *Makromol. Chem.*, 1978, **179**, 2387.
15. K. Matyjaszewski, S. Slomkowski and S. Penczek, *J. Polym. Sci., Polym. Chem. Ed.*, 1979, **17**, 69, 2413.
16. A. V. Tobolsky and A. Eisenberg, *J. Am. Chem. Soc.*, 1960, **82**, 289.
17. M. Hirota and H. Fukuda, *Makromol. Chem.*, 1987, **188**, 2259.
18. A. Vrancken, J. Smid and M. Szwarc, *Trans. Faraday Soc.*, 1962, **58**, 2036.
19. H. J. Kress, W. Stix and W. Heitz, *Makromol. Chem., Macromol. Symp.*, 1988, **13/14**, 507.
20. Al. Al. Berlin and N. S. Enikolopian, *Vysokomol. Soedin., Ser. A*, 1973, **15**, 555.
21. H. C. Brown, *J. Chem. Soc.*, 1956, 1248.
22. H. Sawada, *J. Macromol. Sci., Rev. Macromol. Chem.*, 1970, **C5** (1), 151.
23. K. J. Ivin, in 'Polymer Handbook', ed. J. Brandrup and E. H. Immergut, Wiley, New York, 1975, vol. 2, p. 421.
24. V. V. Korshak, V. A. Kotelnikov, V. V. Kurashev and T. M. Frunze, *Usp. Khim.*, 1976, **45**, 1671.

47
Cationic Ring-opening Polymerization: Formation of Cyclic Oligomers

STANISLAW PENCZEK and STANISLAW SLOMKOWSKI

Polish Academy of Sciences, Lodz, Poland

47.1 GENERAL SCHEME OF POLYMERIZATION WITH CYCLIZATION: CHEMISTRY, THERMODYNAMICS AND KINETICS

47.1.1 Mechanism of Formation of Cycles

In ring-opening polymerizations, and particularly in the cationic processes, electrophilic active centres can react with heteroatoms in their own chains. This leads to formation of cyclic oligomers. For example, in this way 1,4-dioxane (**1**) is formed during the cationic polymerization of oxirane (Scheme 1).

Similarly, cycles are also produced during the cationic polymerization of other cyclic ethers as well as in the polymerization of cyclic acetals, esters, sulfides, aziridines and siloxanes. Scheme 2 describes the polymerization with cyclization: *n* denotes the number of monomer units in the given macrocycle, and X is used for a nucleophilic heteroatom. Formation of cyclic oligomers is a reversible process; monomers and macrocycles react with active centres in a similar way.

Another potential method (not yet proved experimentally) of cycle formation is a direct ring expansion, from monomer to higher cycles.

The equilibrium concentration of any given cyclic oligomer is determined by the rates of its formation (*e.g.* back-biting and/or end-biting) and by the rate of its addition to the active species. For polymers with long chains and when the contribution of end-biting becomes negligible, the equilibrium concentration of cycles equals $k_b(n)/k_p(n)$ (*cf.* Scheme 2). As an example, the distribution

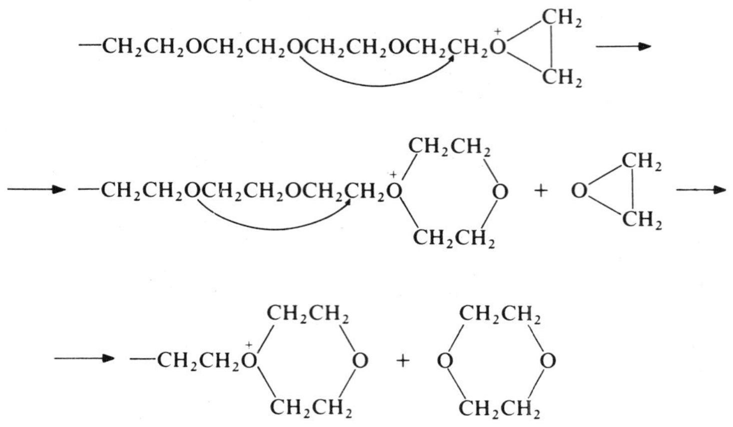

$$-CH_2CH_2OCH_2CH_2OCH_2CH_2OCH_2CH_2\overset{+}{O} \begin{array}{c} CH_2 \\ | \\ CH_2 \end{array} \longrightarrow$$

$$\longrightarrow -CH_2CH_2OCH_2CH_2OCH_2CH_2\overset{+}{O} \begin{array}{c} CH_2CH_2 \\ \diagup \quad \diagdown \\ O \\ \diagdown \quad \diagup \\ CH_2CH_2 \end{array} + O \begin{array}{c} CH_2 \\ | \\ CH_2 \end{array} \longrightarrow$$

$$\longrightarrow -CH_2CH_2\overset{+}{O} \begin{array}{c} CH_2CH_2 \\ \diagup \quad \diagdown \\ O \\ \diagdown \quad \diagup \\ CH_2CH_2 \end{array} + O \begin{array}{c} CH_2CH_2 \\ \diagup \quad \diagdown \\ O \\ \diagdown \quad \diagup \\ CH_2CH_2 \end{array}$$

(1)

Scheme 1

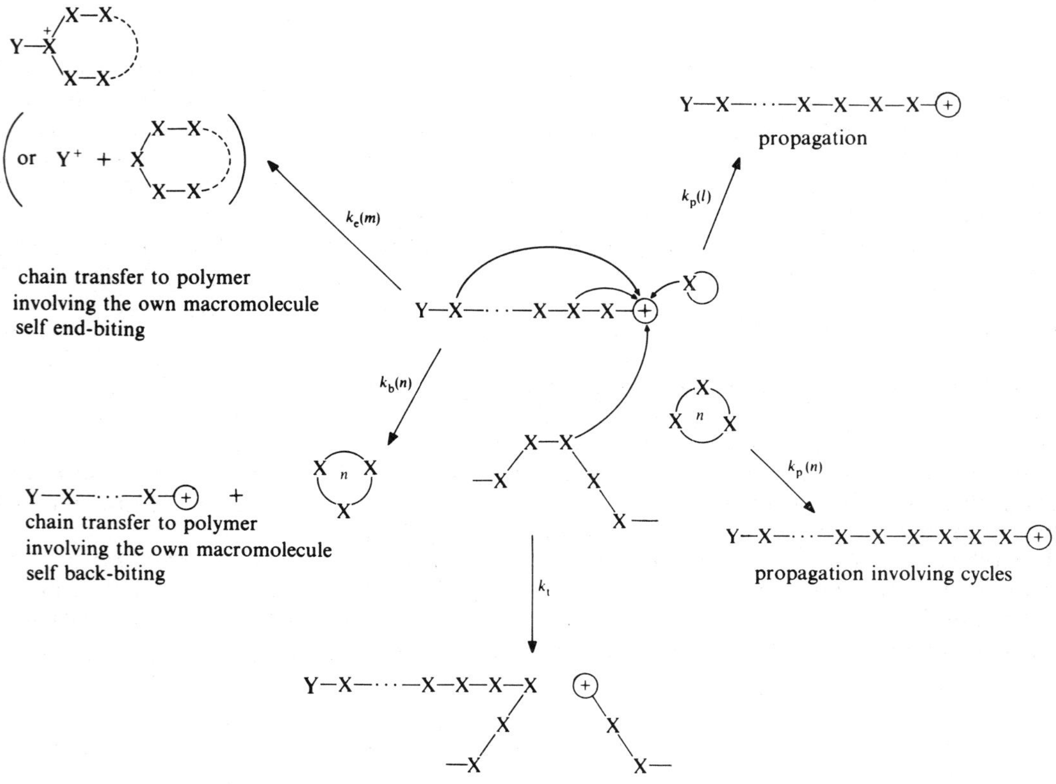

Scheme 2

of cyclic oligomers obtained in the cationic polymerization of hexamethylcyclotrisiloxane to poly(dimethylsiloxane) is given in Figure 1.

Distribution of unstrained cyclic oligomers at equilibrium is described by the dependence $[M_n] = An^{-5/2}$, derived theoretically by Jacobson and Stockmayer for polymers with unperturbed chain conformations.[1] However, the character of changes of the concentration of cycles with time and the rate with which the ring–chain equilibrium is attained are influenced by rate constants of all reactions contributing to the general scheme (Scheme 2), and particularly by the rate of propagation to linear macromolecules.

Competition between back-biting, leading to macrocycles, and propagation depends on the following factors: (i) nucleophilicities of the monomer and polymer units; (ii) steric hindrance for these two reactions; and (iii) the dependence of the rates of these two reactions on monomer

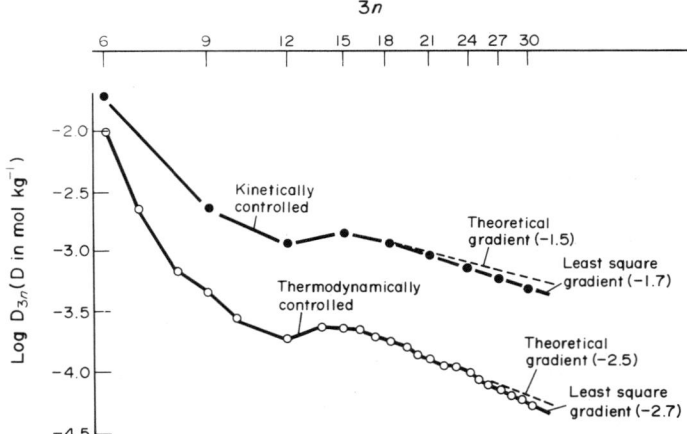

Figure 1 Comparison of the kinetically and thermodynamically controlled distributions of hexamethylcyclotrisiloxane (D_3) cyclic oligomers. $[D_3] \approx 2 \, mol \, kg^{-1}$, $[CF_3SO_3H]_0 \approx 7 \times 10^{-4} \, mol \, kg^{-1}$, heptane, 30 °C; ●, at 30% conversion of D_3; ○, at equilibrium (reproduced from ref. 2 by permission of Hüthig and Wepf Verlag, Basel)

concentration. The following examples can be given. In THF polymerization, oxygen atoms in the $-CH_2OCH_2-$ groups in the polymer unit are *much* less nucleophilic than those in the monomer and thus the competition of back-biting with propagation is efficiently decreased.[3-6] Conversely, for 1,3-dioxolane, nucleophilicity of the oxygen atoms in the $-OCH_2O-$ groups in both monomer and polymer units are comparable and in effect cyclization is already more important at the early stages of monomer conversion.[3,7] Another difference stems from steric hindrance. Thus, in the polymerization of oxirane, mostly cyclic dimer (1,4-dioxane) is formed, whereas in the polymerization of methyloxirane, cyclization to a dimer is reduced due to the steric factors, and formation of cyclic tetramers prevails.[8-12]

Cyclization by end- or back-biting is a unimolecular reaction, virtually independent of monomer concentration. Propagation is a monomer dependent bimolecular process (Scheme 2). Therefore, competition between these two reactions depends on the instantaneous monomer concentration. Decreasing the initial monomer concentration should favour formation of cyclic oligomers (Figure 2), provided that a usual mechanism of polymerization takes place, *i.e.* a process in which the macromolecule is fitted with cationic active species at the end (active chain end polymerization). In the activated monomer mechanism the opposite is true (peculiarities of macrocycle formation in the activated monomer propagation are discussed in Section 47.1.4.4).

The following problems, related to the ring–chain equilibria, have been studied so far: (i) factors governing the kinetically and thermodynamically controlled distributions of cycles and the numerical values of the coefficients in the Jacobson–Stockmayer equation, relating the concentration of a given macrocycle to the number of repeating units in the cycle ($[M_n] = An^{-\delta}$); (ii) methods enabling

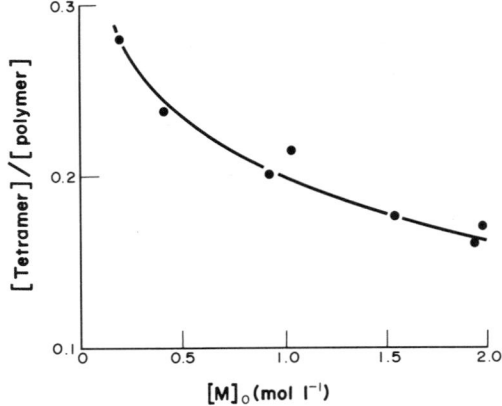

Figure 2 Influence of an initial monomer concentration $[M]_0$ on the ratio of tetramer to polymer in the polymerization of 3,3-dimethyloxetane with $Et_3O^+BF_4^-$, $[Et_3O^+BF_4^-]_0 = 10^{-2} \, mol \, l^{-1}$ (reproduced from ref. 61 by permission of Hüthig and Wepf Verlag, Basel)

the proportions of macrocycles in the product (kinetically controlled distributions) to be influenced; and (iii) special systems giving high concentrations of cyclic oligomers (*e.g.* synthesis of crown ethers by ethylene oxide polymerization).

47.1.2   Thermodynamics of Cyclization

The thermodynamics of cyclization to small cycles (constituted from less than approximately 20 skeletal bonds) is strongly influenced by the ring strain. Strain depends not only on the size but also on the chemical structure of the formed cycles. Thus distribution of small cycles often differs significantly for various systems. For example, as already mentioned, oxirane is converted with all of the studied initiators predominantly into a cyclic dimer,[8,9,13] from 3,3-bis(chloromethyl)oxetane mainly cyclic trimer is obtained,[14] whereas ethyloxirane yields predominantly a cyclic tetramer.[11] Extensive formation of a cyclic tetramer has also been found for the polymerization of chloro-methyloxirane (epichlorohydrin).[12,13] For the polymerization of THF, cycles from trimer through to octamer were observed.[6]

Large rings are strainless. Thus equilibrium of cyclization involving these rings becomes solely governed by the entropic factor, related to the probability that active centres and appropriate nucleophilic groups along the chains would be located at a distance necessary to close a cycle. This has been analysed by Kuhn,[15] Jacobson and Stockmayer,[1] and by Flory.[16] According to their theories the equilibrium constant of cyclization K_n can be described in the following way

$$K_n = \frac{P_n(0)}{N_A \sigma_{Rn}} \tag{1}$$

where $P_n(0)$ represents the probability that segments of polymeric chain, forming a cycle, are in a close contact, N_A denotes the Avogadro number, and $\sigma_{Rn} = 2n$ is a symmetry number for cyclic oligomers (n equals the number of bonds that can be broken by opening a cycle in propagation). For polymeric chains with conformations described by Gaussian statistics

$$P_n(0) = \lim_{r=0} \left(\frac{3}{2\pi \langle r_n^2 \rangle}\right)^{\frac{3}{2}} \exp\left(-\frac{3r}{2\langle r_n^2 \rangle}\right) = \left(\frac{3}{2\pi \langle r_n^2 \rangle}\right)^{\frac{3}{2}} \tag{2}$$

$$K_n = \left(\frac{3}{2\pi \langle r_n^2 \rangle}\right)^{\frac{3}{2}} \bigg/ (N_A \sigma_{Rn}) \tag{3}$$

where $\langle r_n^2 \rangle$ denotes the mean-square end-to-end distance for macromolecules containing n segments. For the unperturbed polymer conformations (without taking into account the excluded volume effects) $\langle r_n^2 \rangle = 2Cl^2 n$, where l is an average bond length in the cycle and C is the characteristic constant for a polymer with a given chemical structure. Thus.

$$K_n = \frac{\left(\frac{3}{\pi C}\right)^{\frac{3}{2}}}{2^4 l^3 N_A n^{\frac{5}{2}}} \qquad \left[P_n(0) = \left(\frac{3}{4\pi Cl^2}\right)^{\frac{3}{2}} \left(n^{-\frac{3}{2}}\right)\right] \tag{4}$$

For polymers in good solvents the dependence of K_n on n is slightly changed and $K_n \approx n^{-2.8}$. Generally, $P_n(0) \approx n^{-(\delta-1)}$ and $K_n \approx n^{-\delta}$, where $2.5 \leq \delta \leq 2.8$. The above theoretically derived dependences of K_n on n have been experimentally verified for many systems, including poly-(dimethylsiloxane)[17-19] and poly(1,3-dioxolane).[20,21] For poly(dimethylsiloxane), a good agreement between theoretical and experimental dependences of K_n on n was observed for cycles containing more than 15 —SiO— units (*cf.* Figure 1).

47.1.3   Kinetics of Cyclization

47.1.3.1   *General decription of the cyclization process*

Two consecutive steps are required for a cyclization process to occur. First, conformational changes of a linear macromolecule, consisting of diffusion of the appropriate groups and leading to their close approach (characteristic time $\tau_c^D(n) = 1/k_c^D(n)$ take place. In the second step, chemical reaction proceeds between these groups, assuming that they are properly oriented at a sufficiently short distance (characteristic time $\tau_R(n) = 1/(P_n(0)k)$ where k denotes the rate constant of the chemical reaction).

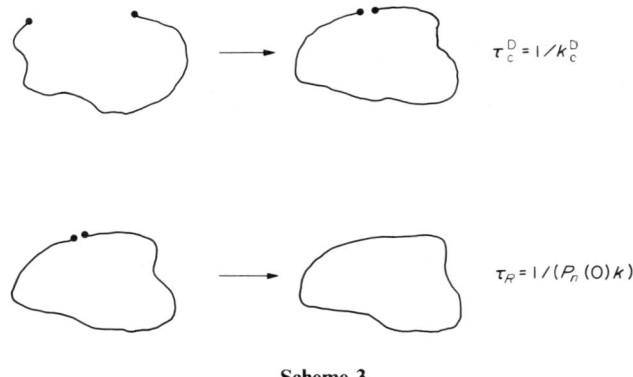

$$\tau_c^D = 1/k_c^D$$

$$\tau_R = 1/(P_n(0)k)$$

Scheme 3

Thus the measured overall characteristic time of cyclization $\tau_c(n) = 1/k_c(n)$ equals

$$\tau_c = \tau_c^D + \tau_R \qquad \left(\frac{1}{k_c} = \frac{1}{k_c^D} + \frac{1}{P_n(0)k} \right) \tag{5}$$

For systems with slow diffusion and fast chemical reaction $[\tau_c^D(n) \gg \tau_R(n)]$, $\tau_c(n) = \tau_c^D(n)$, $k_c(n) = k_c^D(n)$ and cyclization is diffusion controlled. For fast diffusion and slow chemical reaction $[\tau_c^D(n) \ll \tau_R(n)]$, $\tau_c(n) = \tau_R(n)$, $k_c(n) = P_n(0)k$ and cyclization is controlled by the equilibrium distribution of polymer conformations. Chemical processes, leading to cyclization in the ionic ring-opening polymerization, are slow (in comparison with diffusion) and hence belong to the second group of cyclization reactions, namely those controlled by the chemical change. For these reactions the dependence of $k_c(n)$ on n is determined only by $P_n(0)$ and is the same for all cycles, regardless of the nature of the chemical reaction responsible for closing a cycle (*e.g.* cationic or anionic). For small cycles $P_n(0)$ depends on many factors related to the ring size, and cycles with a given size can be produced preferentially. Only for large rings, starting from more than 20 skeletal bonds, does the general statistical description become possible.

47.1.3.2 *Formation of small cycles*

Cyclization leading to rings composed from less than 15 skeletal bonds is strongly affected by the energy terms. These include unfavourable interactions when the linear molecule or its fragment assumes the form of the cyclic one. These interactions involve strains resulting from bond length and bond angle deformations, transannular interactions, and 'Pitzer strain' due to the *gauche* interactions in the cyclic molecule.[22, 23] Cyclization to small rings was studied not only for polymerization reactions but also for many model systems. Some of these processes, related to formation of epoxides and lactones, will be discussed later in this section.

Experimental and theoretical studies of the ring strain in hydrocarbons revealed that the ring strain changes in the following order: cyclobutane $\geqslant$ cyclopentane > cyclohexane < cycloheptane < cyclooctane < cyclononane < cyclodecane > cyclododecane[23] (*cf.* Figure 3). Similar dependences were observed for monomers polymerizing cationically, such as cyclic ethers, esters or amides, although the six-membered lactone gives a polymer whereas the five-membered one does not.

Cyclization rate constants depend strongly on the chemical structure, influencing the flexibility of the cyclizing chains. It has been found that rates of cyclization leading to small rings (three- and four-membered) are strongly enhanced by substitution at the vicinal carbon atoms (Thorpe–Ingold effect).[22] This can be exemplified by measurements of rate constants for the formation of epoxides from chlorohydrins (Table 1).

Rate constants of cyclization of ω-bromoalkanecarboxylate ions to lactones are lowest for eight-membered rings (Figure 4).[26] This reflects some additional strain in small lactone rings, in comparison with cycloalkanes, for which maximum strain in medium rings corresponds to cyclo-

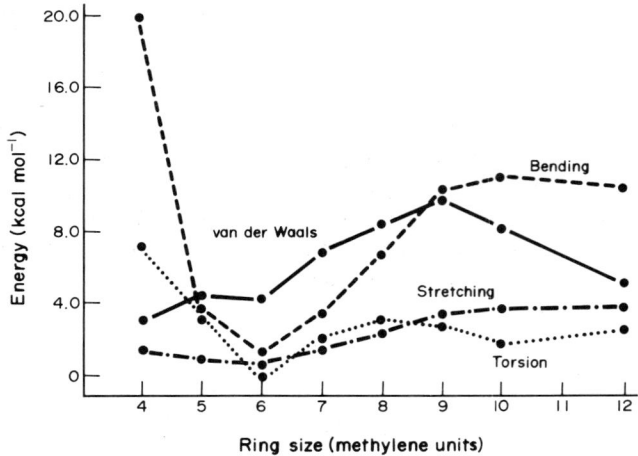

Figure 3 Contribution to the total steric energies in the C-4 to C-12 rings (1 cal = 4.18 J) (reproduced from ref. 23 by permission of The American Chemical Society)

Table 1 Formation of Epoxides from Chlorohydrins

Compound	k (cyclization) (s⁻¹)	Temperature (°C)	Ref.
(structure)	1.05×10^{-2}	18	24
(structure)	3.42	18	25
(structure)	4.1×10^{2}	18	25

decane. It has been proposed that for small ring lactones, strain related to the *cis* conformation of the fragment involving the ester group becomes important.[27] For linear esters of carboxylic acids, energetically the *trans* conformation (**2**) is most convenient, and thus preferred. However in rings smaller than eight-membered ones only *cis* conformations are possible (**3**). Twelve-membered lactones and lactones with larger rings are puckered and assume exclusively the energetically favoured *trans* conformation.[27]

trans

(**2**)

cis

(**3**)

Differences in the ring strain manifest themselves in the polymerization process through the preferential formation of cyclic oligomers with a certain size. In the polymerization of ethylene oxide (oxirane) the nonstrained six-membered cyclic dimer (1,4-dioxane) is obtained almost exclusively.[8-10] Substitution in the oxirane ring leads to additional interactions and for methyl-oxirane (1,2-propylene oxide) and *t*-butyloxirane tetramers are the predominant cyclic

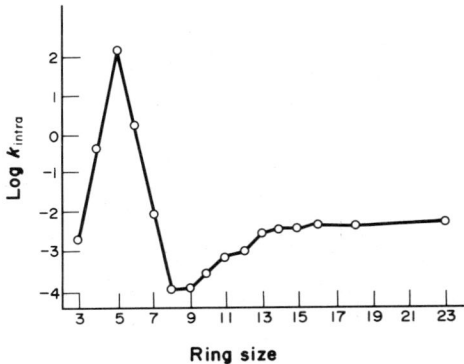

Figure 4 Reactivity profile for lactone formation: $Br(CH_2)_m CO_2^- \rightarrow \overline{(CH_2)_m CO_2} + Br^-$ (reproduced from ref. 26 by permission of The American Chemical Society)

products.[11,28] In the polymerization of α-epichlorohydrin, cycles larger than a trimer constitute the main fraction.[29]

The detailed ^1H and ^{13}C NMR spectroscopic studies indicate that the cyclic tetramer from (R)-t-butylethylene oxide assumes the energetically favourable *gauche$^+$–gauche$^+$–trans* conformation (Figure 5).[28]

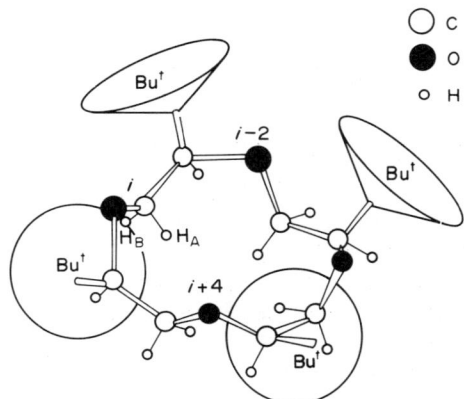

Figure 5 Structure of the cyclic tetramer of (R)-t-butylethylene oxide (reproduced from ref. 28 by permission of The Society of Polymer Science, Japan)

In the polymerization of oxetane, the cyclic trimer and tetramer dominate in the fraction of macrocycles.[30] Formation of the eight-membered cyclic dimer is hindered because of the ring strain resulting from transannular interactions.

The higher ring strain of the ten- and fifteen-membered rings reduces the formation of the cyclic dimer and trimer in the cationic polymerization of THF (Figure 6). Relatively higher concentrations of cyclics containing four and five monomeric units are observed.[6]

47.1.3.3 Formation of large cycles

Rate constants of cyclization $[k_c(n) = P_n(0)k$, *cf.* Section 47.1.3.1] for large strainless cyclic oligomers (containing more than 20 skeletal bonds) depend exclusively on the entropic factors. In Section 47.1.2 we indicated that for polymers with conformations described by Gaussian statistics the following dependence is valid: $P_n(0) \approx n^{-\kappa}$ ($1.5 \leq \kappa \leq 1.8$; $\kappa = 1.5$ for θ solvents and $\kappa = 1.8$ for good solvents respectively). A similar dependence should be observed for $k_c(n)$.

Kinetically controlled distribution of cyclic oligomers for the cationic polymerization of hexamethylcyclotrisiloxane, initiated with CF_3SO_3H, is shown in Figure 1. For cycles containing more than 15 skeletal bonds the dependence of $\ln[M_n]$ ($[M_n]$ denotes concentrations of cyclics with n monomer units) on $\ln(n)$ gives a straight line graph with slope -1.7, *i.e.* within the region predicted

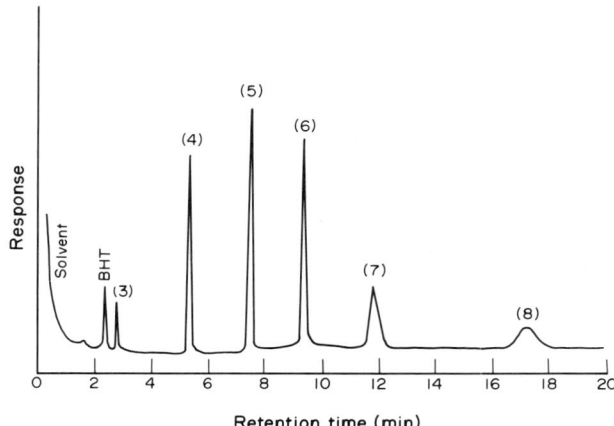

Figure 6 Gas chromatograms of the polymerizing mixture of THF in $MeNO_2$ initiated with CF_3SO_3H: (3)–(8) indicate the cyclic oligomers (reproduced from ref. 6 by permission of The American Chemical Society)

theoretically. For comparison, the equilibrium distribution of cyclic oligomers is also presented in Figure 1. In agreement with theoretical predictions (*cf.* Section 47.1.2) a plot with a slope equal to -2.7 was obtained.

47.1.3.4 Formation of cycles by cyclization and by ring expansion

There are some systems for which propagation by ring expansion has been suggested. According to this mechanism, developed originally by Plesch for the polymerization of 1,3-dioxolane initiated with protonic acids,[31-33] propagation consists of addition (by an insertion mechanism) of the monomer molecules (M_1) into macrocyclic oligomers with active centres on the ring (C_n). Neutral macrocycles (M_n) are produced by exchange reactions, predominantly with monomer, at the beginning of polymerization (Scheme 4). Later, however, Penczek *et al.*[40] proved, by a direct NMR analysis of the end groups of the growing and dead macromolecules, that in the cationic poymerization of cyclic acetals, active centres on linear macromolecules are responsible for cyclization. Kubisa[41] has also shown, by analyzing the chemical shift of protons introduced with protonic acids into cyclic acetals, that with conversion of monomer into a polymer a substantial upfield shift of these protons indicates formation of the hydroxyl end groups. In the ring expansion polymerization, protons should merely sit as such on cycles, growing in size with monomer conversion. The chemical shift should not therefore depend substantially on conversion. ^{31}P NMR of ions trapped with phosphines gave similar results, showing that the larger the average degree of polymerization ($\overline{DP_n}$), the higher the proportion of linear growing macromolecules (end-to-end closure decreases).[70]

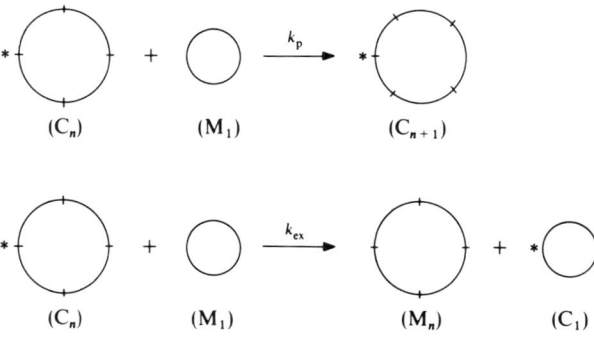

Scheme 4

Chojnowski *et al.*[2] have shown that polymerization with formation of cyclics by cyclization (back-biting and/or end-biting) and polymerization proceeding by ring expansion should produce two different kinetically controlled distributions of cyclic oligomers. Formation of cyclic oligomers by cyclization should result in the distribution of cycles described by equation (6).

$$k_c(n) = kP_n(0) = k\left(\frac{3}{4\pi Cl^2}\right)^{\frac{3}{2}} n^{-\frac{3}{2}} \qquad (k_c(n) \sim n^{-\frac{3}{2}}) \qquad (6)$$

On the other hand, for ring expansion the system of differential equations in Scheme 5 holds

$$\frac{d[C_1]}{dt} = -k_p[C_1][M_1] + k_{ex}[M_1]\Sigma[C_n]$$

$$\frac{d[C_n]}{dt} = k_p[C_{n-1}][M_1] - k_p[C_n][M_1] - k_{ex}[C_n][M_1]$$

$$\frac{d[M_n]}{dt} = k_{ex}[C_n][M_1]$$

Scheme 5

For the stationary state, $d[C_n]/dt = 0$ and the following equations are valid

$$\frac{[C_n]}{[C_{n-1}]} = \frac{k_p}{k_p + k_{ex}} = q$$

$$[C_n] = q[C_{n-1}] = q^2[C_{n-2}] = \ldots = q^{n-1}[C_1]$$

$$[M_n] = k_{ex}[C_n]\int[M_1]dt = k_{ex}q^{n-1}[C_1]\int[M_1]dt$$

$$[M_n] \sim q^{n-1}$$

Scheme 6

Thus, when macrocyles are formed by back- or end-biting, $\ln[M_n]$ is a linear function of $\ln(n)$ (*cf.* equation 6), whereas for polymerization by ring expansion $\ln[M_n]$ is a linear function of n (*cf.* Scheme 6). As mentioned in Section 47.1.3.3, in the cationic polymerization of hexamethylcyclotrisiloxane a linear dependence of $\ln[M_n]$ on $\ln(n)$ was observed (*cf.* Figure 1). This observation strongly supports formation of cycles by end-biting.

Sigwalt recently analyzed the cationic polymerization of cyclic siloxanes and indicated that perhaps conditions can also be found for ring-expansion.[99] As indicated above, direct proof (*e.g.* NMR observation of the active species) for this mechanism is still lacking.

47.1.4 Kinetics of Polymerization with Macrocyclization

47.1.4.1 *Overall kinetic scheme of polymerization with macrocyclization*

Scheme 7 is a general scheme describing polymerization including macrocyclization. In Scheme 7, L_n and L_{n+m} denote linear macromolecules, M_m and M_n are cyclic oligomers, and C_n and C_m are cyclic oligomers with an active centre on the ring. For example, in the polymerization of 1,3-dioxolane initiated with protonic acid, C_n corresponds to the protonated cyclic oligomers (structure **4**).

$$L_n + M_m \underset{k_b(m)}{\overset{k_p(m)}{\rightleftharpoons}} L_{n+m}$$

Propagation and depropagation (back-biting)

$$L_n \underset{k_o(n)}{\overset{k_e(n)}{\rightleftharpoons}} C_n$$

End-biting leading to macrocycles with an active center on the ring and ring opening

$$C_n + M_m \underset{k_{ex}(n)}{\overset{k_{ex}(m)}{\rightleftharpoons}} M_n + C_m$$

Exchange

Scheme 7

$$
\begin{bmatrix}
\overset{\text{H}}{\underset{\mid}{}} \\
-(CH_2)_2OCH_2O^+ - \\
\\
-[OCH_2O(CH_2)_2]_{n-1}-
\end{bmatrix}
$$

(4)

Indexes n, m, and $n + m$ in Scheme 7 indicate the number of monomeric units in the species mentioned. How the cyclization rate constants $[k_b(n)$ and $k_e(n)]$ depended on the ring size was chosen according to the analysis given in Section 47.1.3. It was also assumed that for propagation involving large, nonstrained cycles, rate constants are simply proportional to the number of these sites in the macrocycle that can be attacked. Due to the ring strain in small monomeric cycles, the rate constants involving the monomer were chosen arbitrarily, unlike those resulting from the general dependence.

There are two particular solutions of differential kinetic equations corresponding to Scheme 7.[34, 35] For systems with very fast propagation and slow back-biting to monomer and without any end-biting, a kinetic enhancement in linear polymer is observed. This means that first monomer is converted into the linear macromolecules and then, when the monomer is virtually consumed, these macromolecules are converted (partially or completely, depending on the initial monomer concentration) into cycles of various sizes. Thus the total concentration of the monomeric units incorporated into the linear polymer, may temporarily exceed its final equilibrium concentration (*e.g.* can go through a maximum when the concentration is plotted as a function of monomer conversion). This behaviour is illustrated in Figure 7.

On the other hand, for systems with fast propagation and slow end-biting giving monomer (for example, due to the high ring strain) and with fast end-biting giving other cycles, monomer is

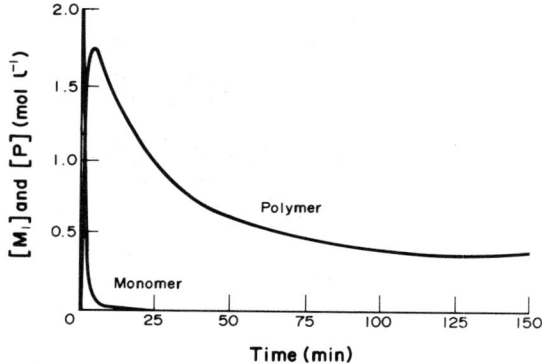

Figure 7 Polymerization with kinetic enhancement in linear macromolecules. Dependence of monomer and polymer concentrations (in moles of monomeric units per litre) on time. Initial monomer concentration $[M_1]_0 = 2\ mol\,l^{-1}$ (reproduced from ref. 35 by permission of Hüthig and Wepf Verlag, Basel)

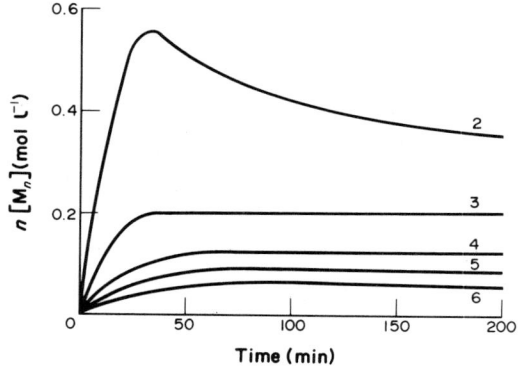

Figure 8 Polymerization with kinetic enhancement in cyclic oligomers. Dependence of the concentrations of cyclic oligomers $n[M_n]$ (in moles of monomeric units per litre) on time. Numbers assigned to each curve indicate the degree of polymerization of an oligomer (reproduced from ref. 35 by permission of Hüthig and Wepf Verlag, Basel)

converted first (*via* short linear oligomers) into cycles, and only later are these cyclic molecules partially or completely converted into linear macromolecules. In such systems the kinetic enhancement in cyclic oligomers is observed. As is shown in Figure 8, the temporary concentrations of cyclic oligomers can exceed their corresponding equilibrium concentrations.

Polymerizations involving formation of cyclic oligomers should formally be treated as reversible copolymerizations and cyclic oligomers should be treated as comonomers. Each particular system is characterized by an equilibrium concentration of monomeric units, incorporated into cyclic oligomers (involving cyclic monomers), called a critical concentration $[M_1]_{crit} = \Sigma n[M_n]_e$. When $[M_1]_0 < [M_1]_{crit}$ then the linear macromolecules are practically absent at equilibrium.[1,35,36] One of the first mathematical treatments of this situation was given by Enikolopyan and Rosenberg.[100] Thus at equilibrium the proportion of cyclic polymers in the product depends on the initial monomer concentration (as shown in Figure 2).

47.1.4.2 Kinetic enhancement in linear polymers

In the cationic polymerization of heterocycles, clear evidence of kinetic enhancement in linear polymers was obtained for substituted thiiranes by Goethals.[37-39] Polymerizations of methylthiirane, ethylthiirane, phenylthiirane, and 2,3-dimethylthiirane, initiated with $(MeCH_2)O^+BF_4^-$, lead to a rapid formation of a linear polymer. Only at the later stages are polymers converted into cycles.[37-39] Examples of such a polymerization are illustrated in Figures 9 and 10, in which GPC chromatograms of polymerizing *cis*- and *trans*-2,3-dimethylthiirane are given.[39] However, in this system, after initial polymer formation, complex cyclization follows, leading eventually not only to the cyclic tetramer (**5**) (in the polymerization of *cis*-2,3-dimethylthiirane) but also to isomers of 3,4,6,7-tetramethyl-1,2,5-trithiepane (**6**) and *cis*- and *trans*-butene (**7**), (**8**) for *cis* and *trans* monomers respectively.

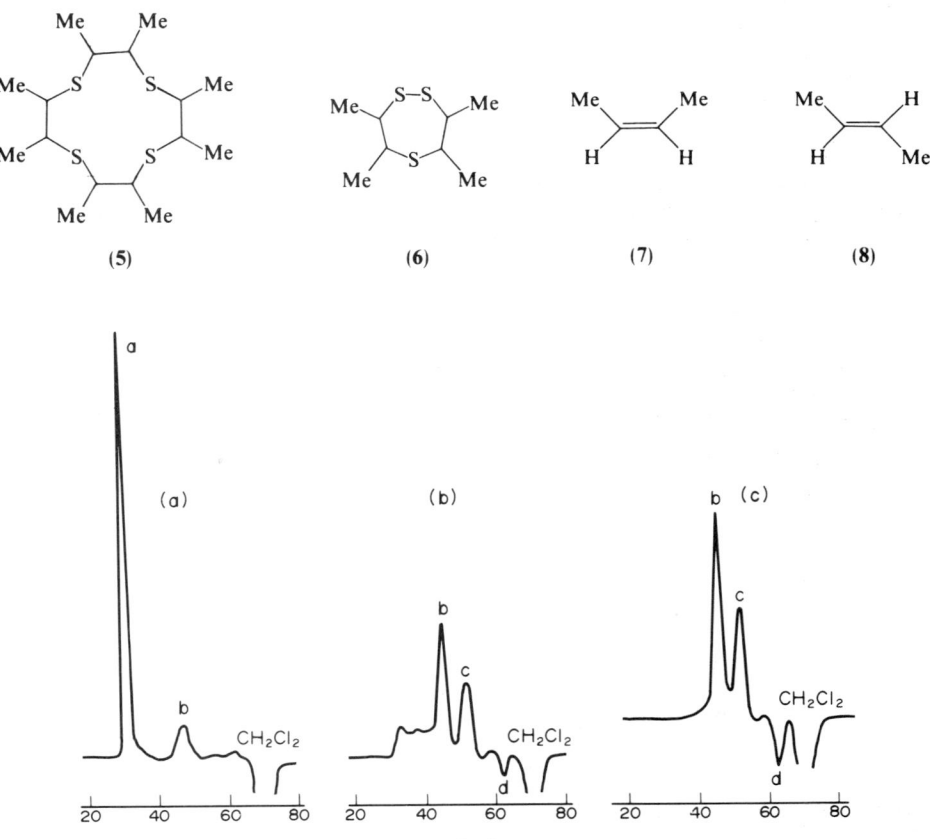

Figure 9 Gel permeation chromatograms of the reaction mixture obtained from *cis*-2,3-dimethylthiirane; $[M]_0 = 0.991 \text{ mol l}^{-1}$, $[Et_3O^+ BF_4^-]_0 = 3 \times 10^{-3} \text{ mol l}^{-1}$. (a), 10 min; (b), 75 min; (c), 475 min after initiation (at 20 °C). Assignment of peaks: a, polymer; b, cyclic tetramer (**5**); c, 3,4,6,7-tetramethyl-1,2,5-trithiepane (**6**); d, *cis*-butene (**7**) (reproduced from ref. 39 by permission of Pergamon Press)

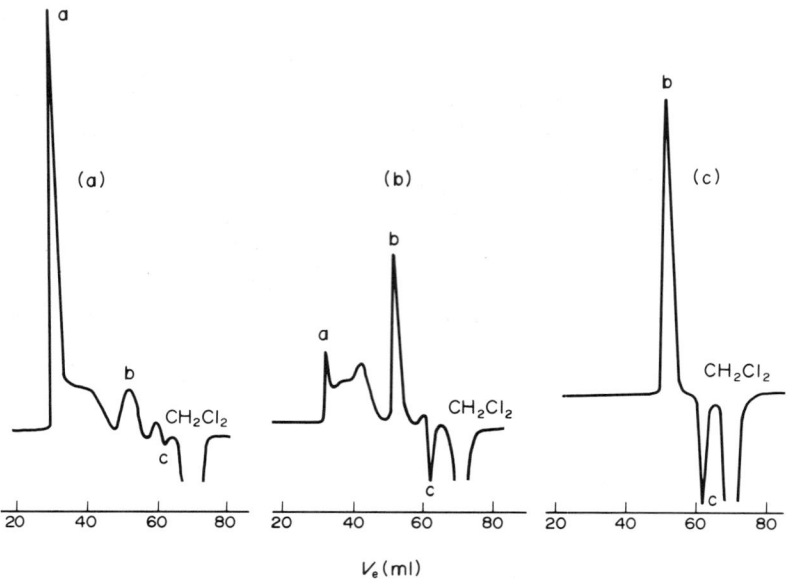

Figure 10 Gel permeation chromatograms of the reaction mixture obtained from *trans*-2,3-dimethylthiirane; $[M]_0 = 1.018 \, mol \, l^{-1}$, $[Et_3O^+BF_4^-]_0 = 10^{-2} \, mol \, l^{-1}$. (a), 20 min; (b), 80 min; (c), 1400 min after initiation (at 20 °C). Assignment of peaks: a, polymer; b, 3,4,6,7-tetramethyl-1,2,5-trithiepane (**6**); c, *trans*-butene (**8**) (reproduced from ref. 39)

47.1.4.3 *Kinetic enhancement in cyclic oligomers*

Kinetic enhancement in cyclic oligomers was noticed for the polymerization of cyclic acetals [1,3-dioxolane (**9**), 1,3,6,9-tetraoxacycloundecane (**10**) and 1,3,6,11-tetraoxacyclotridecane (**11**)], initiated with protonic acids and studied extensively by Schulz and Yamashita.[40–45] In these systems, linear oligomers with OH terminal groups are finally formed. Fast end-biting, reaction of cationic active centres with —OH groups, and subsequent proton transfer (exchange reactions in Scheme 7) lead to rapid formation of cyclic oligomers. According to the GPC traces, in the cationic polymerization of 1,3,6,9-tetraoxacycloundecane initiated with CF_3SO_3H (*cf.* Figure 11) only cyclic oligomers are present at low monomer conversion (24%). The conversion–time curves (Figures 12 and 13) for the polymerization of 1,3,6,9-tetraoxacycloundecane and 1,3,6,11-tetraoxacyclotridecane clearly indicate the kinetic enhancement in cyclic dimers. During the initial stage of polymerization there is a period when the concentration of dimer exceeds its corresponding equilibrium concentration.

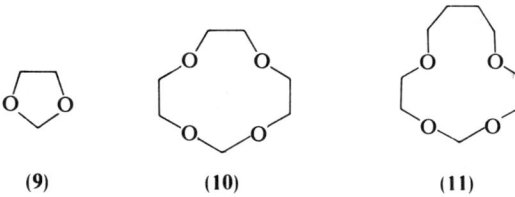

47.1.4.4 *Reduction of cyclic oligomers in cationic polymerization by the activated monomer mechanism*

Formation of cyclic oligomers can kinetically be decreased by increasing the ratio of the rate constants of propagation to the rate constants of reactions leading to cyclization.[35] In the cationic polymerization of heterocycles this has been achieved by switching from 'active chain end propagation' to 'activated monomer propagation'.[46–48] According to this mechanism (Scheme 8), monomer is first activated in its ground state, *e.g.* by protonation. Protonated monomer adds to the macromolecule hydroxyl end groups. Since the charge is mostly located on the monomer molecules the end-to-end as well as the back-biting reactions are highly suppressed. Thus the usual source of the formation of cycles is eliminated.

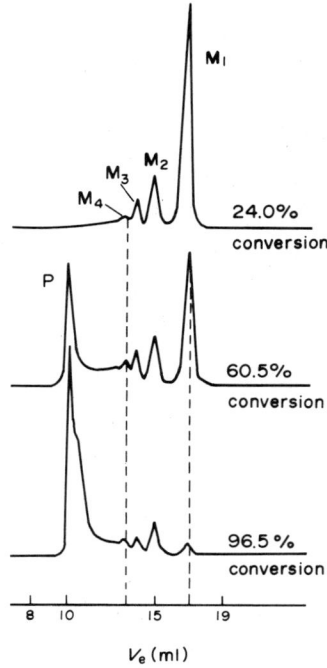

V_e (ml)

Figure 11 GPC chromatograms of polymerizing 1,3,6,9-tetraoxacycloundecane initiated with CF_3SO_3H. Conditions of polymerization: $[M]_0 = 0.5$ mol1^{-1}, $[CF_3SO_3H]_0 = 5 \times 10^{-4}$ mol1^{-1}, solvent CH_2Cl_2, temperature $0\,°C$. P = polymer, M_1 = monomer, M_2 = cyclic dimer, *etc.* (reproduced from ref. 42 by permission of Hüthig and Wepf Verlag, Basel)

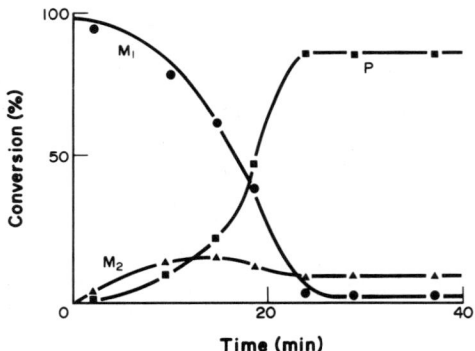

Figure 12 Conversion–time curves for the polymerization of 1,3,6,9-tetraoxacycloundecane. Polymerization conditions given in caption for Figure 11 (reproduced from ref. 42 by permission of Hüthig and Wepf Verlag, Basel)

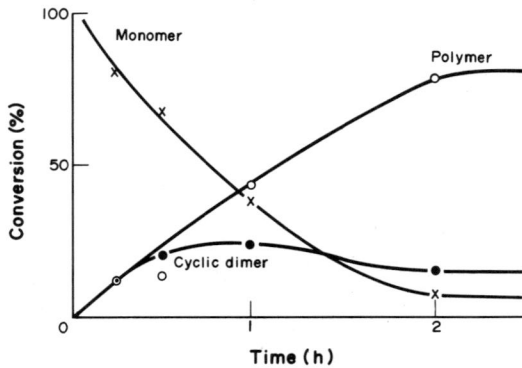

Figure 13 Conversion–time curves for the polymerization of 1,3,6,11-tetraoxacyclotridecane initiated with CF_3SO_3H. Polymerization conditions: $[M]_0 = 0.25$ mol1^{-1}, $[CF_3SO_3H]_0 = 0.5$ mol %, solvent CH_2Cl_2, temperature $0\,°C$ (reproduced from ref. 45 by permission of The Society of Polymer Science, Japan)

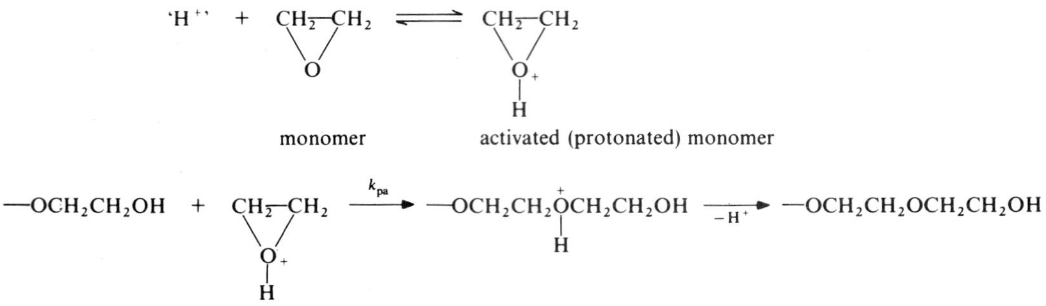

$$\cdot H^{+} + \underset{\underset{O}{\diagdown\diagup}}{CH_2\!-\!CH_2} \;\rightleftharpoons\; \underset{\underset{O_+}{\underset{|}{H}}}{CH_2\!-\!CH_2} \xrightarrow{\hspace{1cm}}$$

monomer   activated (protonated) monomer

$$-OCH_2CH_2OH \;+\; \underset{\underset{O_+}{\underset{|}{H}}}{CH_2\!-\!CH_2} \xrightarrow{k_{pa}} -OCH_2CH_2\overset{+}{O}CH_2CH_2OH \xrightarrow[-H^+]{} -OCH_2CH_2OCH_2CH_2OH$$

with H on the oxonium.

Scheme 8

For example, in the polymerization of ethylene oxide, initiated with protonic acids and carried out in the presence of alcohols, polymerization involving the activated monomer mechanism prevails and formation of 1,4-dioxane may be almost completely excluded.[47] At the very beginning of polymerization, protonated monomer can participate in one of the two reactions, namely with alcohol and/or with monomer.

Addition of the protonated monomer to alcohol (Scheme 9) regenerates —OH groups, reacting further with protonated monomer and eventually forming linear macromolecules by the activated monomer mechanism (*cf.* Scheme 8). The second reaction given in Scheme 9 leads to tertiary oxonium active centres, capable of participation in 'normal' cationic propagation (equation 7). These active centres can also participate in macrocyclization, *e.g.* by back-biting (*cf.* Scheme 1).

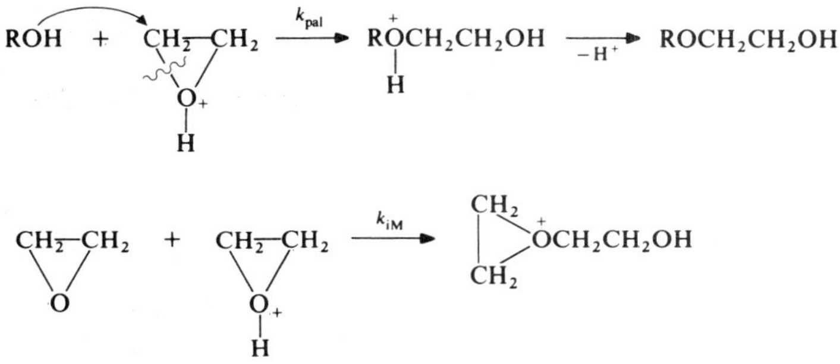

$$ROH \;+\; \underset{\underset{O_+}{\underset{|}{H}}}{CH_2\!-\!CH_2} \xrightarrow{k_{pal}} R\overset{+}{O}CH_2CH_2OH \xrightarrow[-H^+]{} ROCH_2CH_2OH$$

$$\underset{\underset{O}{\diagdown\diagup}}{CH_2\!-\!CH_2} \;+\; \underset{\underset{O_+}{\underset{|}{H}}}{CH_2\!-\!CH_2} \xrightarrow{k_{iM}} \underset{\underset{CH_2}{}}{\overset{CH_2}{}}\!\!>\!\overset{+}{O}CH_2CH_2OH$$

Scheme 9

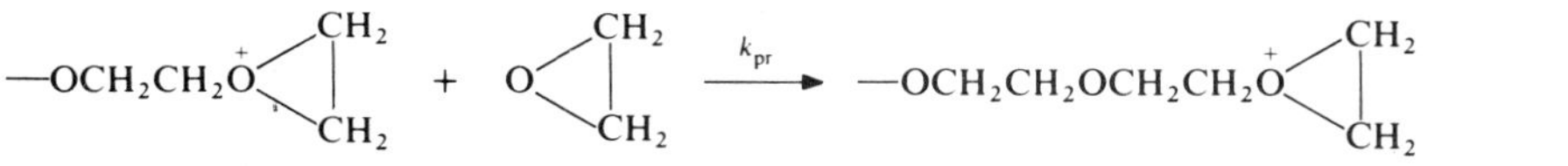

$$-OCH_2CH_2\overset{+}{O}\!\!<\!\!\underset{CH_2}{\overset{CH_2}{}} \;+\; O\!\!<\!\!\underset{CH_2}{\overset{CH_2}{}} \xrightarrow{k_{pr}} -OCH_2CH_2OCH_2CH_2\overset{+}{O}\!\!<\!\!\underset{CH_2}{\overset{CH_2}{}} \qquad (7)$$

The decrease of the instantaneous monomer concentration decreases the proportion of chains growing *via* active chain ends and thus increases the contribution of the activated monomer mechanism, reducing the overall rate of cycle formation. This is contrary to the polymerization proceeding according to the active chain end mechanism, where decreasing monomer concentration facilitates cyclization (*cf.* discussion in Section 47.1.1 and Figure 2).

Details of the propagation proceeding according to the activated monomer mechanism are discussed in Section 47.2.1.

47.2 MAJOR CASES OF CATIONIC RING-OPENING POLYMERIZATION WITH FORMATION OF CYCLIC OLIGOMERS

47.2.1 Cyclic Ethers

Formation of cyclic oligomers has been observed in the cationic polymerization of unsubstituted and substituted oxiranes,[8-13, 28, 29, 49-59] oxetanes,[14, 30, 60-62] and in the polymerization of THF,[5, 6] as has already been discussed.

In the polymerization of oxirane, initiated with BF_3 and/or $Et_3O^+BF_4^-$, predominantly the six-membered cyclic dimer (1,4-dioxane) is formed (*cf.* Scheme 1). It has been established that in the polymerization of the nondeuterated monomer, carried out in the presence of perdeuterated 1,4-dioxane, cyclic oligomers containing two —CD_2CD_2O— groups are produced.[49] Therefore 1,4-dioxane is involved in the formation of cycles according to Scheme 10.

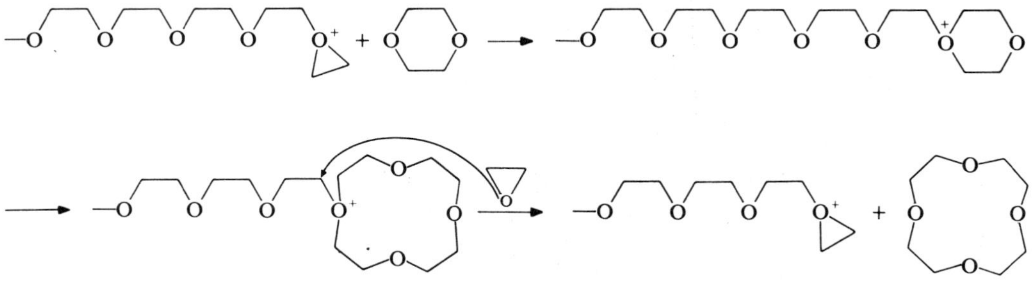

Scheme 10

Polymerization of substituted oxiranes yields products containing not only cyclic dimers but also higher cyclic oligomers. This was observed for methyloxirane (propylene oxide),[11,12,63] ethyloxirane (butylene oxide),[11,12,63-65] phenyloxirane (styrene oxide),[57-59] 1,1-dimethyloxirane (isobutylene oxide)[52,66] and chloromethyloxirane (epichlorohydrin).[11,29,53,54,67]

Cyclooligomerization of (R)-t-butyloxirane, initiated with Et_2OBF_3, proceeds with retention of configuration and gives mainly the cyclic tetramer.[28] It was proposed that the selective formation of cyclic tetramer is favoured by the energetically convenient conformation of this oligomer (cf. Figure 5).

Cyclic oligomers can be selectively produced in good yield in the polymerization carried out in the presence of various salts with cations forming complexes with cycles of a certain size (template polymerization).[68] Some examples of such processes, useful for the synthesis of crown ethers, are illustrated in Table 2.

Table 2 Distribution of Cyclic Oligoethers from Ethylene Oxide Obtained with BF_3 in the Presence of Metal Salts[a]

Salt	Tetramer (%)	Pentamer (%)	Hexamer (%)
$LiBF_4$	30	70	0
$NaBF_4$	25	50	25
KBF_4	0	50	50
KPF_6	20	40	40
$KSbF_6$	40	20	40
$RbBF_4$	0	0	100
$CsBF_4$	0	0	100
$Ca(BF_4)_2$	50	50	0
$Sr(BF_4)_2$	10	45	45
$Ba(BF_4)_2$	10	30	60
$AgBF_4$	35	30	35
$Hg(BF_4)_2$	20	70	10
$Ni(BF_4)_2$	20	80	0
$Cu(BF_4)_2$	5	90	5
$Zn(BF_4)_2$	5	90	5

[a] Reproduced from ref. 68 by permission of the Chemical Society.

Polymerization of oxetanes (unsubstituted and substituted), initiated with BF_3, BF_3OEt_2, $Et_3O^+BF_4^-$, Et_3Al and $EtOSO_2CF_3$, is accompanied with extensive cyclization, especially at elevated temperatures.[14,30,60-62]

For reasons explained above, activated monomer polymerization cannot be induced for THF. This monomer, initiated with $HOSO_2CF_3$, slowly gives cyclic oligomers.[5,6] The distribution of these oligomers is shown in Figure 6. In the polymerization initiated with $MeOSO_2CF_3$, the low molecular weight fraction contains predominantly linear oligomers (Figure 14).

This is because formation of cycles by the end-to-end cyclization is faster when the head end group is a hydroxyl group. Hydroxyl end groups react faster with the tertiary oxonium ions (*i.e.* during the

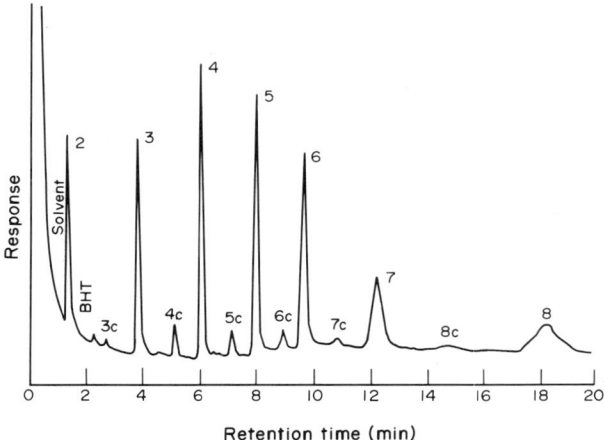

Figure 14 Gas chromatograms of the polymerizing mixture of THF in MeNO$_2$ initiated with CF$_3$SO$_3$Me; 3c, 4c, *etc.* refer to cyclic oligomers; 2, 3, 4, *etc.* to linear oligomers (reproduced from ref. 6 by permission of The American Chemical Society)

reaction responsible for cyclization) than do the ether groups (Scheme 11). Indeed, transfer of the proton (breaking of the —OH bond) is much easier than transfer of the carbenium ion (breaking of the —OR bond).

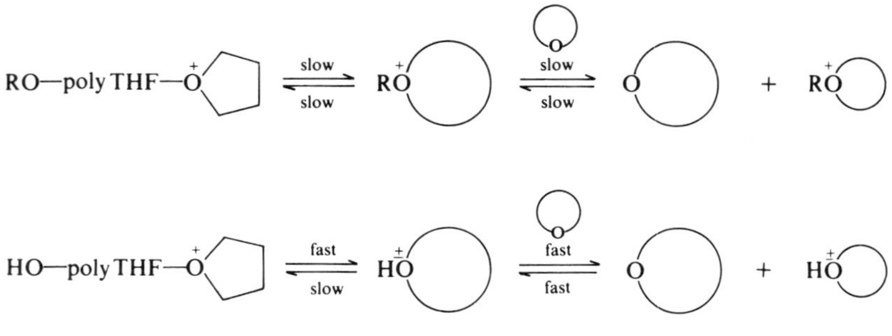

Scheme 11

Polymerization of diepoxides may proceed by cyclopolymerization: polymers with rings in the chain are formed. This topic is discussed below.

47.2.2 Cyclic Acetals

Cyclic oligomers were detected in the polymerization of 1,3-dioxolane,[20,21,31-33,40,69,70] 1,3-dioxane,[71] 1,3-dioxacycloundecane,[45] 1,3,5-trioxane[72,73] and cyclic formals of di-, tri-, tetra- and penta-ethylene glycols.[42,45,75,76] For the polymerization of 1,3-dioxolane, initiated with BF$_3$OEt$_2$, the equilibrium concentrations of cyclic oligomers up to the cyclic octamer were determined.[20,21] In the reaction of 1,3-dioxane ('nonpolymerizable' compound) with cationic initiators, cyclic dimer and tetramer were produced.[71]

Polymerization of 1,3,5-trioxane is accompanied by formation of a number of cyclic oligomers, not merely repeating 1,3,5-trioxane (*i.e.* [(CH$_2$O)$_3$]$_n$), but rather cyclic oligomers derived from formaldehyde (containing 4, 5, 6, *etc.* —CH$_2$O— units in the ring.[72,73] These cycles are produced either by back-biting or by end-biting.

In the polymerization of 1,3-dioxacycloundecane (**12**), 1,3,6,9-tetraoxacycloundecane (**11**), 1,3,6,9,12-pentaoxacyclotetradecane (**13**), and 1,3,6,9,12,15-hexaoxacycloheptadecane (**14**), cyclic oligomers were obtained.[42-45,75-78] These cycles contain 2,3,4 . . . monomeric units and are formed by end-biting reactions. Back-biting seems not to be important in these systems, because otherwise cycles with irregular structure (containing fragments of monomeric units) would also have been produced. Polymerization of 1,3,6,9-tetraoxacycloundecane and 1,3,6,11-tetraoxacyclotridecane initiated with CF$_3$SO$_3$H proceed with kinetic enhancement in cyclic oligomers. Analysis of this problem has been presented in Section 47.1.4.3.

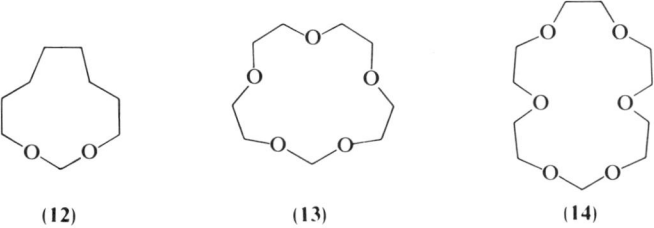

(12)   (13)   (14)

The efficient end-biting results in a specific behaviour of the cationic copolymerization of 1,3,6,9-tetraoxacycloundecane with styrene.[44] During the initial period only cycles are formed. Later, these cycles and comonomers are involved in copolymerization and finally copolymers with a statistical distribution of styrene and 1,3,6,9-tetraoxacycloundecane units are produced. The equilibrium distribution of cycles, obtained from cyclic formals of oligoethylene glycols, is shown in Figure 15.[75]

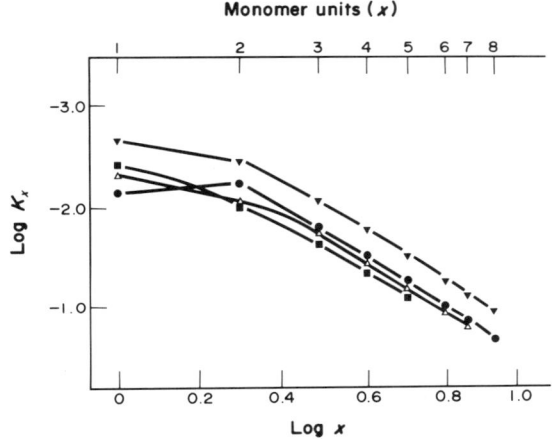

Figure 15 Log–log plots of equilibrium concentrations of cycles against degree of polymerization x for different polymerization systems: ▼, 1,3,6-trioxocane; ●, 1,3,6,9-tetraoxacycloundecane; △, 1,3,6,9,12-pentaoxacyclotetradecane; and ■, 1,3,6,9,12,15-hexaoxacycloheptadecane (reproduced from ref. 75 by permission of The Americal Chemical Society)

Cycles with more than two monomeric units are strainless and the dependences of $\ln[M_n]_e$ on $\ln(n)$ are linear with slopes close to -2.5, as predicted by the Jacobson–Stockmayer theory (*cf.* Table 3).

47.2.3 Cyclic Esters

Formation of cyclic oligomers was observed in the cationic polymerization of ε-caprolactone initiated with CF_3SO_3H.[79] The equilibrium distribution of cycles was the same as previously found for the anionic process.[80] The dependence $[M_n]_e n^{-5/2}$ indicates that cyclization conforms to the Jacobson–Stockmayer theory. Critical monomer concentration ($[M_1]_{crit}$) equals 0.25 mol l^{-1}; thus below $[M_1]_0 = 0.25$ mol l^{-1} only cycles are formed.[80]

47.2.4 Cyclic Sulfides

In the cationic polymerization of substituted thiiranes, initiated with $Et_3O^+BF_4^-$ and Et_2OBF_3, extensive cyclization is usually preceded by complete monomer conversion into a linear polymer.[37-39] Preferentially, cyclic dimers (*e.g.* for phenylthiirane and 1,1-dimethylthiirane) and cyclic tetramers (*e.g.* for ethylthiirane and isopropylthiirane) are produced from the degraded polymers.[39]

In the polymerization of *cis*- and *trans*-1,2-dimethylthiirane, cyclization is accompanied by side reactions, giving for the *cis* monomer a mixture of a cyclic tetramer, substituted 1,2,5-trithiepane (**6**) and *cis*-butene (**7**).[39] In the oligocyclization of the *trans* isomer only substituted 1,2,5-trithiepane (**6**)

Table 3 Measured Slopes of the Dependence of $\ln[M_n]_e$ on $\ln(m)$[a]

Monomer	Range of m	Slope	Ref.
$[OSiMe_2]_3$	15–40	−2.50	18
	16–40	−2.50	19
	40–200	$−2.86 \pm 0.15$	17
	40–200	−2.69	19
	16–202	−2.48	19
	18–30	−2.7	2
$(CH_2)_5CO$	14, 28, 35, 42,	−2.50	80
	28, 35, 42, 49	$−2.2 \pm 0.2$	90
$OCH_2(OCH_2CH_2)_3$	22, 44, 55, 66, 77, 88	−2.50	75
	22–165	$−2.45 \pm 0.05$	76
$OCH_2(OCH_2CH_2)_4$	28, 42, 56, 70, 84 98	−2.34	75
$OCH_2(OCH_2CH_2)_5$	34, 51, 68, 85	−2.17	75

[a] m = Number of bonds in the cyclic oligomer.

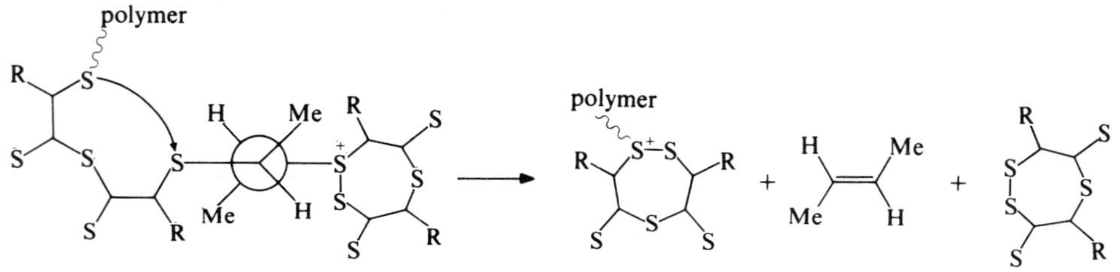

Scheme 12 Cylization and formation of *trans*-butene during polymerization of *trans*-2,3-dimethyl thiirane

and *trans*-butene (**8**) are obtained. Scheme 12 was proposed for cyclization, resulting in formation of butene and 1,2,5-trithiepane derivatives.[39]

47.2.5 Aziridines

1-Alkylaziridines, initiated with HCl, HI, HBr, NaI or $Et_3O^+BF_4^-$ in solvents of low polarity (*e.g.* heptane, acetone, CH_2Cl_2), are converted efficiently into 1,4-dialkylpiperazine cyclic dimers.[81-83]

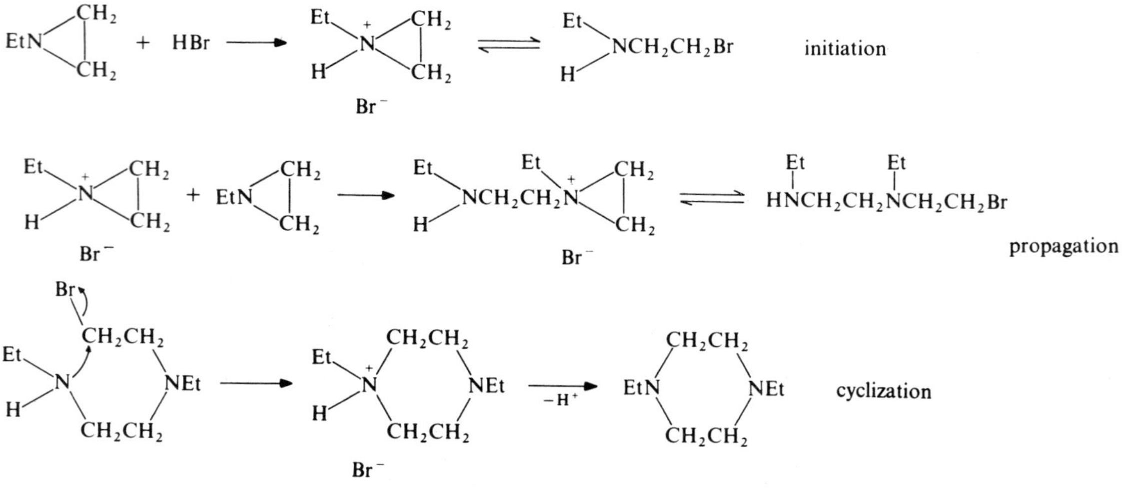

Scheme 13

Table 4 Oligomer Formation During Cationic Ring-opening Polymerization

Monomer	Initiator	Solvent	Temperature (°C)	Structure of main cyclic oligomer	Refs.
$CH_2{-}CH_2$ (epoxide)	BF_3, $SnCl_4$ Et_3OBF_4	$MeCH_2Cl$	20	Dimer	8, 9, 13
$CH_2{-}CHMe$ (epoxide)	BF_3, Et_3OBF_4 $AlMe_3$	bulk CH_2Cl_2	35 0–80	Dimer, tetramer, pentamer Dimer, trimer, tetramer	11, 12 63
$CH_2{-}CHEt$ (epoxide)	BF_3, Et_3OBF_4	CH_2Cl_2	41	Dimer, tetramer, pentamer	11
$CH_2{-}CHCH_2Cl$ (epoxide)	BF_4, Et_3OBF_4 $AlEt_3$	CCl_4 Et_2O	77 −50	Dimer, tetramer, pentamer Dimer	11 53, 54
$CH_2{-}CHPh$ (epoxide)	Ph_3CSbF_6 $AlCl_3$, $FeCl_3$, $SbCl_5$ $BeCl_2$, $SnCl_4$	bulk bulk	20, 50 70–100	Dimer Dimer	59 58, 91
$CH_2{-}CMe_2$ (epoxide)	BF_3	—	—	Dimer	66
$CH_2{-}CHCH_2ONO_2$ (epoxide)	BF_3	—	—	Tetramer	28
$CH_2{-}CHCMe_3$ (epoxide)	BF_3OEt_2	C_6H_6	Reagents mixed at 0 °C	Tetramer	28
$(CH_2)_3O$	BF_3, Et_3OBF_4 CF_3SO_3Et	Bulk or EtCl C_6H_6	−80±50 70	Tetramer Trimer, tetramer	60 30
$CH_2CMe_2CH_2O$	BF_3, Et_3OBF_4	EtCl, CH_2Cl_2	0–40	Tetramer	60, 61
$CH_2C(CH_2Cl)_2CH_2O$	Et_3Al	Bulk	180	Trimer	14

Table 4 *(continued)*

Monomer	Initiator	Solvent	Temperature (°C)	Structure of main cyclic oligomer	Refs.
⌐$(CH_2)_4O$⌐	CF_3SO_3H	$MeNO_2$	35, 50	From trimer to octamer	5, 6
⌐$OCH_2O(CH_2)_2$⌐	BF_3OEt_2	Bulk, CH_2Cl_2	60	From dimer to octamer	20, 21
⌐$OCH_2O(CH_2)_3$⌐	$HClO_4$	Bulk, CH_2Cl_2	−78 to 0	Dimer, trimer	71
⌐$(CH_2O)_3$⌐	BF_3OEt_2 $FeCl_3$	Bulk, $C_2H_4Cl_2$, C_6H_{12}, C_6H_6, $PhNO_2$	30	⌐$(CH_2O)_4$⌐	72, 73
⌐$OCH_2O(CH_2)_2O(CH_2)_4O(CH_2)_2$⌐	CF_3SO_3H	CH_2Cl_2	20	Dimer	45
⌐$OCH_2O(CH_2)_8$⌐	CF_3SO_3H	CH_2Cl_2	20	Dimer	45
⌐$OCH_2(OCH_2CH_2)_3$⌐	CF_3SO_3H BF_3OEt_2 Et_3OBF_4	CH_2Cl_2	0	From dimer to cyclooligomers containing 14 monomeric units	42, 43 44, 45 75, 76
⌐$OCH_2(OCH_2CH_2)_4$⌐	BF_3OEt_2	CH_2Cl_2	0	From dimer to heptamer	45, 75 77
⌐$OCH_2(OCH_2CH_2)_5$⌐	BF_3OEt_2	CH_2Cl_2	0	From dimer to pentamer	75
⌐$(CH_2)_5CO_2$⌐	CF_3SO_3H	Xylene	100	From dimer to hexamer	79
$CH_2{-}CHMe$ (S)	BF_3OEt_2 Et_3OBF_4	CH_2Cl_2	0–20	Tetramer, pentamer dimethyl-1,2,5,-tri-thiacycloheptane	37
$CH_2{-}CHCH_2Me$ (S)	Et_3OBF_4	—	20	Tetramer	98

Monomer	Catalyst	Solvent	Temp.	Product	Ref.
CH_2—CHPh (thiirane)	Et_3OBF_4	—	0–20	Dimer	98
CH_2—$CHMe_2$ (thiirane)	Me_2SO_4	Bulk	Room temp.	Dimer	92
CH_2—$CHMe_2$ (thiirane)	Et_3OBF_4	—	0–20	Dimer	98
MeCH—CHMe *cis* (thiirane)	Et_3OBF_4	CH_2Cl_2	0–20	Tetramer, substituted 1,2,5-trithiepane	39
MeCH—CHMe *trans* (thiirane)	Et_3OBF_4	CH_2Cl_2	0–20	Substituted 1,2,5-trithiepane	39
CH_2—$CHCH_2CMe_3$ (thiirane)	Et_3OBF_4	—	0–20	Tetramer	98
CH_2—CH_2 N-Et (aziridine)	HI, HBr, HCl	Acetone	25–30	Dimer	81, 82
CH_2—CH_2 N-Bu^n	HI, NaI	Acetone	25–30	Dimer	81
CH_2—CH_2 N-Ph	NaI	H_2O/EtOH	80	Dimer	83
CH_2—CH_2 N-C_6H_4Me (*para*)	NaI	H_2O/EtOH	80	Dimer	83
CH_2—CH_2 N-$CH_2CH=CH_2$	HI, NaI	Acetone	25–30	Dimer	81

Table 4 (continued)

Monomer	Initiator	Solvent	Temperature (°C)	Structure of main cyclic oligomer	Refs.
$CH_2\!-\!CH_2$ $\diagdown N \diagup$ CH_2CH_2Ph	HI, NaI Et$_3$OBF$_4$	Acetone CH$_2$Cl$_2$	25–30 20	Dimer Dimer	81 93
$CH_2\!-\!CH_2$ $\diagdown N \diagup$ CH_2CH_2CN	Et$_3$OBF$_4$	CH$_2$Cl$_2$	20	Tetramer	93
$CH_2\!-\!CH_2$ $\diagdown N \diagup$ CH_2Ph	Et$_3$OBF$_4$ MeC$_6$H$_4$SO$_3$H	CH$_2$Cl$_2$ EtOH	20 75	Dimer Tetramer	93 94
$CH_2\!-\!CH_2$ $\diagdown N \diagup$ $CH_2CH_2SiEt_3$	—	Bulk	250–300	Dimer	95
$CH_2\!-\!CHEt$ $\diagdown N \diagup$ CH_2Ph	BF$_3$OEt$_2$	Bulk C$_6$H$_6$, EtOH	Room Reflux	Tetramer	96
$CH_2\!-\!CHMe$ $\diagdown N \diagup$ CH_2Ph	HI	Acetone	—	Dimer	97
$\overline{(OSiMe_2)_3}$	CF$_3$SO$_3$H	Heptane CH$_2$Cl$_2$	30	Cyclic oligomers containing from 6 to 30 dimethylsiloxane units	2 84–99, 99

The contribution of cyclic oligomers is substantially suppressed in solvents of high polarity and high solvating power.[81] It has been proposed that ionic active centres are responsible for propagation leading to linear polymers, whereas covalent species, being in equilibrium with ion pairs, facilitate cyclization.[82]

Depression of cyclization in polar solvents is due to the shifting of the position of equilibrium between ionic and covalent species to the ionic side.

47.2.6 Cyclic Siloxanes

Cyclization has been observed in the cationic polymerization of hexamethylcyclotrisiloxane (D_3, 15) and octamethylcyclotrisiloxane (D_4, 16) initiated with CF_3SO_3H.[2,84-89,100] Cycles containing up to 30 dimethylsiloxane units (D) have been identified. At the initial stage of the D_3 polymerization (monomer conversion lower than 30%) predominantly D_{3n} cycles, containing an integral number of hexamethylene units, are formed.[2,99]

(15)   (16)

There are different opinions as to whether these cycles are formed due to the end-to-end cyclization[2] or according to the ring expansion mechanism.[99] Whereas the first hypothesis is substantiated by the character of the kinetically controlled distribution of cycles (*cf.* Section 47.1.3.4) the second one is supported by kinetic enhancement in D_3 in the polymerization of D_6. Nevertheless, this does not mean, in our opinion, that the splitting off of the smaller ring from a larger one (monomer) necessarily indicates that the chain growth proceeds by ring expansion of this monomer.

At equilibrium there is no preference for D_{3n} macrocycles, because cyclic oligomers with D_{3n+1} and D_{3n+2} are produced by subsequent back-biting reactions.[2] Kinetically and thermodynamically controlled distributions of cyclic D_n are illustrated in Figure 1 and were analyzed earlier in Sections 47.1.2 and 47.1.3.

A list of the main cyclic oligomers formed from the cyclic monomers discussed in Section 47.2 is given in Table 4, together with references.

47.3 REFERENCES

1. H. Jacobson and W. H. Stockmayer, *J. Chem. Phys.*, 1950, **18**, 1600.
2. J. Chojnowski, M. Scibiorek and J. Kowalski, *Makromol. Chem.*, 1977, **178**, 1351.
3. E. M. Arnett, in 'Progress in Physical Organic Chemistry', ed. S. G. Cohen, R. Streitwieser, Jr. and R. W. Taft, Interscience, New York, 1963, vol. 1, p. 223.
4. J. M. Andrews and J. A. Semlyen, *Polymer*, 1971, **12**, 642.
5. J. M. McKenna, T. K. Wu and G. Pruckmayr, *Macromolecules*, 1977, **10**, 877.
6. G. Pruckmayr and T. K. Wu, *Macromolecules*, 1978, **11**, 265.
7. S. Penczek, P. Kubisa and K. Matyjaszewski, *Adv. Polym. Sci.*, **68/69**, 1.
8. D. U. Worsfold and A. M. Eastham, *J. Am. Chem. Soc.*, 1957, **79**, 900.
9. G. A. Latremouille, G. T. Merrall and A. M. Eastham, *J. Am. Chem. Soc.*, 1960, **82**, 120.
10. S. Kobayashi, K. Morikawa and T. Saegusa, *Polym. J.*, 1979, **11**, 405.
11. R. J. Kern, *J. Org. Chem.*, 1968, **33**, 388.
12. R. J. Katnik and J. Schaefer, *J. Org. Chem.*, 1968, **33**, 384.
13. D. J. Worsfold and A. M. Eastham, *J. Am. Chem. Soc.*, 1957, **79**, 897.
14. Y. Arimatsu, *J. Polym. Sci., Part A-1*, 1966, **4**, 728.
15. N. Kuhn, *Kolloid Z.*, 1934, **68**, 2.
16. P. J. Flory, 'Statistical Mechanics of Chain Molecules', Interscience, New York, 1969, chap. 11, sect. 2.
17. J. F. Brown and G. M. J. Slusarczuk, *J. Am. Chem. Soc.*, 1965, **87**, 931.
18. J. A. Semlyen and P. V. Wright, *Polymer*, 1969, **10**, 643.
19. P. V. Wright, *J. Polym. Sci., Polym. Phys. Ed.*, 1973, **11**, 51.
20. J. M. Andrews and J. A. Semlyen, *Polymer*, 1972, **13**, 142.
21. J. A. Semlyen, *Adv. Polym. Sci.*, 1976, **21**, 41.

22. E. L. Eliel, 'Stereochemistry of Carbon Compounds', McGraw-Hill, New York, 1962, chaps. 7–9.
23. N. L. Allinger, M. T. Tribble, M. A. Miller and D. W. Wertz, *J. Am. Chem. Soc.*, 1971, **93**, 1637.
24. A. C. Knipe, *J. Chem. Soc., Perkin Trans. II*, 1973, 589.
25. H. Nilsson and L. Smith, *Z. Phys. Chem., Abt. A*, 1933, **166**, 136.
26. G. Illuminati and L. Mandolini, *Acc. Chem. Res.*, 1981, **14**, 95.
27. C. Galli, G. Illuminati and L. Mandolini, *J. Am. Chem. Soc.*, 1973, **95**, 8374.
28. A. Sato, T. Hirano, M. Suga and T. Tsuruta, *Polym. J.*, 1979, **9**, 209.
29. K. Ito, N. Usami and Y. Yamashita, *Polym. J.*, 1979, **11**, 171.
30. P. Dreyfuss and M. P. Dreyfuss, *Polym. J.*, 1976, **8**, 81.
31. P. H. Plesch and P. H. Westerman, *J. Polym. Sci., Part C*, 1968, **16**, 3837.
32. P. H. Plesch, *Br. Polym. J.*, 1973, **5**, 1.
33. P. H. Plesch, *Pure Appl. Chem.*, 1976, **46**, 287.
34. K. Matyjaszewski, M. Zielinski, P. Kubisa, S. Slomkowski, J. Chojnowski and S. Penczek, *Makromol. Chem.*, 1980, **181**, 1469.
35. S. Slomkowski, *Makromol. Chem.*, 1985, **186**, 2581.
36. S. Slomkowski, *J. Macromol. Sci., Chem.* 1984, **21**, 1383.
37. J. L. Lambert, D. Van Ooteghem and E. J. Goethals, *J. Polym. Sci., Part A-1*, 1971, **9**, 3055.
38. D. Van Ooteghem and E. J. Goethals, *Macromol. Chem.*, 1974, **175**, 1513.
39. W. Van Craeynest and E. J. Goethals, *Eur. Polym. J.*, 1976, **12**, 859.
40. R. Szymanski, P. Kubisa and S. Penczek, *Macromolecules*, 1983, **16**, 1000.
41. P. Kubisa, *J. Polym. Sci., Polym. Chem. Ed.*, 1987, **25**, 873.
42. C. Rentsch and R. C. Schulz, *Macromol. Chem.*, 1977, **178**, 2535.
43. Y. Yamashita, Y. Kawakami and K. Kitano, *J. Polym. Sci., Polym. Lett. Ed.*, 1977, **15**, 213.
44. Y. Kawakami and Y. Yamashita, *Macromolecules*, 1979, **12**, 399.
45. R. C. Schulz, K. Albrecht, Q. V. Tran Thi, J. Nienburg and D. Engel, *Polym. J.*, 1980, **12**, 639.
46. K. Brzezinska, R. Szymanski, P. Kubisa and S. Penczek, *Makromol. Chem., Rapid Commun.*, 1986, **7**, 1.
47. S. Penczek, P. Kubisa and R. Szymanski, *Makromol. Chem., Makromol. Symp.*, 1986, **3**, 203.
48. M. Wojtania, P. Kubisa and S. Penczek, *Makromol. Chem., Makromol. Symp.*, 1986, **6**, 201.
49. J. Dale, K. Daasvatn and T. Grønneberg, *Makromol. Chem.*, 1977, **178**, 873.
50. S. Kobayashi, K. Morikawa and T. Saegusa, *Macromolecules*, 1975, **8**, 952.
51. Y. Yamashita, K. Iwao and K. Ito, *Polym. Lett.*, 1979, **17**, 1.
52. Y. Yamashita and I. Ito, *Polym. Bull. (Berlin)*, 1978, **1**, 73.
53. S. G. Entelis, G. V. Korovina and A. I. Kuzayev, *Vysokomol. Soedin, Ser. A*, 1971, **13**, 1438.
54. Y. I. Estrin and S. G. Entelis, *Vysokomol. Soedin, Ser. A*, 1971, **13**, 1654.
55. E. J. Vandenberg, *J. Polym. Sci., Polym. Chem. Ed.*, 1972, **10**, 329.
56. V. Hornof, G. Gabra and L. P. Blanchard, *J. Polym. Sci., Polym. Chem. Ed.*, 1973, **11**, 1825.
57. R. K. Summerbell and M. J. Kland-English, *J. Am. Chem. Soc.*, 1955, **77**, 5095.
58. R. O. Colclough, G. Gee, W. C. E. Higginson, J. B. Jackson and M. Litt, *J. Polym. Sci.*, 1959, **34**, 171.
59. W. M. Pasika, *J. Polym. Sci., Part A*, 1965, **3**, 4287.
60. J. B. Rose, *J. Chem. Soc.*, 1956, 542, 546.
61. M. R. Bucquoye and E. J. Goethals, *Makromol. Chem.*, 1978, **179**, 1681.
62. M. R. Bucquoye and E. J. Goethals, *Makromol. Chem.*, 1981, **182**, 3379.
63. R. O. Colclough and K. Wilkinson, *J. Polym. Sci., Part C*, 1966, **4**, 322.
64. J. L. Down, J. Lewis, B. Moore and G. Wilkinson, *J. Chem. Soc.*, 1959, 3767.
65. I. M. Robinson and G. Pruckmayr, *Macromolecules*, 1979, **12**, 1043.
66. A. M. Eastham, in 'The Chemistry of Cationic Polymerization', ed. P. H. Plesch, Pergamon Press, New York, 1963, chap. 10.
67. K. Weissermel and E. Nölken, *Makromol. Chem.*, 1963, **68**, 140.
68. J. Dale and K. Daasvatn, *J. Chem. Soc., Chem. Commun.*, 1976, 295.
69. K. Boehlke and V. Jaaks, *Mákromol. Chem.*, 1971, **145**, 219.
70. P. Kubisa and S. Penczek, *Makromol. Chem.*, 1979, **180**, 1821.
71. P. H. Plesch and P. H. Westermann, *Polymer*, 1969, **10**, 105.
72. T. Miki, T. Higashimura and S. Okamura, *J. Polym. Sci., Part A-1*, 1967, **5**, 95, 2977, 2389.
73. V. Jaacks, *Makromol. Chem.*, 1966, **99**, 300.
74. R. C. Schulz, K. Albrecht, C. Rentsch and Q. V. Tran Thi, *ACS Symp. Ser.*, 1977, **59**, 77.
75. Y. Yamashita, J. Mayumi, Y. Kawakami and K. Ito, *Macromolecules*, 1980, **13**, 1075.
76. C. Rentsch and R. C. Schulz, *Makromol. Chem.*, 1978, **179**, 1403.
77. Y. Kawakami, J. Suzuki and Y. Yamashita, *Polym. J.*, 1977, **9**, 519.
78. Y. Yamashita, J. Mayumi, Y. Kawakami and K. Ito, *Kobunshi Ronbunshu*, 1978, **35**, 615.
79. Y. Yamashita, *Polym. Prepr., Am. Chem. Soc., Div. Polym. Chem.*, 1979, **20**(1), 126.
80. K. Ito, Y. Mashizuka and Y. Yamashita, *Macromolecules*, 1977, **10**, 821.
81. C. R. Dick, *J. Org. Chem.*, 1967, **32**, 72.
82. C. R. Dick, *J. Org. Chem.*, 1970, **35**, 3950.
83. H. W. Heine, W. G. Kenyon and E. M. Johnson, *J. Am. Chem. Soc.*, 1961, **83**, 2570.
84. J. Chojnowski and L. Wilczek, *Makromol. Chem.*, 1979, **180**, 117.
85. L. Wilczek and J. Chojnowski, *Macromolecules*, 1981, **14**, 9.
86. L. Wilczek and J. Chojnowski, *Makromol. Chem.*, 1983, **184**, 77.
87. J. Chojnowski, L. Wilczek and S. Rubinsztajn, in 'Cationic Polymerization and Related Processes', ed. E. J. Goethals, Academic Press, London, 1984, p. 253.
88. J. J. Lebrun, G. Sauvet and P. Sigwalt, *Makromol. Chem., Rapid Commun.*, 1982, **3**, 757.
89. G. Sauvet, J. J. Lebrun and P. Sigwalt, in 'Cationic Polymerization and Related Processes', ed. E. J. Goethals, Academic Press, London, 1984, p. 237.
90. S. Sosnowski, S. Slomkowski, S. Penczek and L. Reibel, *Makromol. Chem.*, 1983, **184**, 2159.

91. S. Kondo and L. P. Blanchard, *J. Polym. Sci., Polym. Lett. Ed.*, 1969, **7**, 621.
92. A. Noshay and C. C. Price, *J. Polym. Sci.*, 1961, **54**, 533.
93. E. J. Goethals, E. H. Schacht and P. Bruggeman, in 'International Symposium on Cationic Polymerization', Rouen, 1973, prepr. C-30.
94. G. R. Hansen and T. E. Burg, *J. Heterocycl. Chem.*, 1968, **5**, 305.
95. N. N. Nametkin, I. A. Grushevenko and V. N. Perchenko, *Dokl. Akad. Nauk SSSR, Chem. Sect.*, 1965, **162**, 347.
96. S. Tsuboyama, K. Tsuboyama, I. Higashi and M. Yanagita, *Tetrahedron Lett.*, 1970, **16**, 1367.
97. M. Yanagita, S. Tsuboyama and K. Tsuboyama, *Jpn. Pat.* 74 109 386 (1975) (*Chem. Abstr.*, 1975, **83**, 10 142a).
98. E. J. Goethals, *Adv. Polym. Sci.*, 1977, **23**, 103.
99. P. Sigwalt, *Polym. J.*, 1987, **19**, 567.
100. B. A. Rosenberg, W. I. Irzak and N. S. Enikolopyan, 'Interchain Exchange Reactions in Polymers' (Russian), Khimia, Moscow, 1975, and references therein.

48

Cationic Ring-opening Polymerization: Ethers

STANISLAW PENCZEK and PRZEMYSLAW KUBISA
Polish Academy of Sciences, Lodz, Poland

48.1 INTRODUCTION

There are several comprehensive reviews, covering the cationic polymerization of cyclic ethers, which have been published recently.[1-4] In the present survey we summarize the major findings of the field, giving priority to the new advances (*e.g.* polymerization involving the activated monomer mechanism) as well as the problems overlooked by previous reviews (*e.g.* cationic cyclopolymerization of diepoxides or application of cationic initiators in curing of epoxy resins).

In the present chapter we will discuss separately, and in more detail, polymerization of the five-membered cyclic ether oxolane (tetrahydrofuran — THF), because all of the features of this process are understood on a level highly superior to other monomers.

48.2 MONOMERS

48.2.1 Ring Strains

Three-, four- and five-membered cyclic ethers are of major interest for polymer chemists. These monomers are sufficiently strained to polymerize easily. Six-membered rings do not undergo polymerization, because of the low ring strain.[2,4] Polymerization of seven-membered and higher rings, as well as polymerization of bicyclic ethers, has been studied less extensively.

The ring strains of typical, unsubstituted cyclic ethers are listed in Table 1.[5]

Table 1

Unsubstituted cyclic ether	Ring strain (kJ mol^{-1})	Unsubstituted cyclic ether	Ring strain (kJ mol^{-1})
Oxirane (ethylene oxide)	114	Oxane (tetrahydropyran)	5
Oxetane (trimethylene oxide)	107	Oxocane	42
Oxolane (tetrahydrofuran)	23	7-oxabicyclo[2.2.1]heptane	44

The geometry of cyclic ethers is governed mainly by the number of atoms in the ring. Ring strain is caused by the difference between the angles resulting from a normal orbital overlap and the angles resulting from the requirements imposed by the ring size.[7] The other source of strain arises from the interactions of the non-bonded atoms located in close proximity one to another.

Molecules of cyclic ethers, with the exception of ethylene oxide, are not planar, although the deviation from planarity for the four- and five-membered rings is very limited. The free electron pairs on the oxygen atoms behave sterically like a medium size substituent. This is the result of the sp^3 hybridization of the oxygen atom supported by nearly tetrahedral angles.

Five-membered cyclic ethers, *e.g.* THF, exists in two interconverting conformations: envelope and half chair (see equation 1).[8] The energy of this interconversion has not been determined, but taking into account the energy barriers for related compounds it can be estimated as rather low.

$$\text{(1)}$$

The distortion from planarity can be characterized by the torsional angles. This angle for THF is equal to 21°,[8] as in the THF molecule a considerable repulsion exists between the non-bonded electron pairs on the oxygen atom and the electrons engaged in bonding of the hydrogen atoms to the nearby carbon atoms.

48.2.2 Nucleophilicity and Basicity of Cyclic Ethers

Nucleophilicity and basicity reflect the same property, namely the ability to share the lone electron pair (or pairs) with an electron acceptor; thus, to combine with an electrophilic substrate. Nucleophilicity is determined at kinetically controlled conditions whereas basicity is determined at conditions controlled thermodynamically (from the studies of equilibria).

Linear and cyclic ethers belong to the group of weak organic bases.[9, 10] Basicities of cyclic ethers decrease in the order illustrated in Table 2.

Table 2

pK_a	2.02[a]	2.1[a] 2.08[b]	—	3.59[b]	3.7[a]
Ref.	10	10, 12	—	10	11

[a] From MeOD shift in IR. [b] Calorimetrically from the heat of mixing with $CHCl_3$.

The low basicity of ethylene oxide is, at least partly, attributed to the ring strain. The angle of the unperturbed C—O—C bond is close to 110°, whereas in ethylene oxide it equals 60°. For tricoordinated oxygen having sp^2 hybridization, the C—O—C angle is equal to 120°; thus protonation (as well as alkylation) of ethylene oxide leads to an increase of the ring strain.

Oxetane is more basic than THF in spite of higher inductive effect in the latter compound. This stems from the lower steric hindrance and stronger interaction between the lone electron pair on the oxygen with δ electrons of the C—H bonds in THF.

Knowledge of the order of basicities of cyclic and linear ethers is important for understanding certain phenomena in cyclic ether polymerization. Thus, chain transfer to the polymer is a general feature of the cationic polymerization of cyclic ethers, because the nucleophilic site of the monomer molecule (ethereal oxygen) is being transferred to the polymer chain.

To what extent this reaction competes with propagation depends on the relative nucleophilicities of monomer and polymer unit. Thus, for THF, the polymer unit is a weaker base than the monomer. This, combined with a lack of strain in the linear polyether, makes the polymer less reactive than the monomer in nucleophilic substitution type reactions. Consequently, for this monomer, chain transfer to polymer is slow and does not interfere with propagation. In contrast to THF, in the polymerization of ethylene oxide the polymer unit is more basic than the monomer. Therefore, reactions involving the polymer chain are important in this system, as will be discussed in following sections.

48.2.3 Polymerizability of Cyclic Ethers

Polymerizability, in a thermodynamic sense, is related to the free energy change (ΔG) associated with a conversion of a monomer molecule into a polymer unit. ΔG of polymerization is a function of enthalpy (ΔH) and entropy (ΔS) of polymerization (equation 2).

$$\Delta G_p = \Delta H_p - T\Delta S_p \qquad (2)$$

The relationship between ΔG of polymerization and the ceiling temperature, T_c, and/or monomer equilibrium concentration, $[M]_e$, is discussed in Chapter 46 of this volume, which is devoted to thermodynamics (*cf.* also refs. 2 and 4).

The thermodynamic polymerizability is governed primarily by the enthalpy of polymerization (ΔH), closely related to the ring strain of a cyclic monomer. Thus, polymerization of highly strained three- and four-membered cyclic ethers is practically irreversible, whereas polymerization of the five-membered THF is a typical example of reversible polymerization, in which a significant concentration of monomer remains in equilibrium with the growing polymer chains.

48.3   GENERAL FEATURES OF THE POLYMERIZATION OF CYCLIC ETHERS

The most comprehensively studied cationic polymerization of cyclic ethers proceeds with tertiary oxonium ions as the growing species. It has generally been accepted that propagation then proceeds predominantly as a S_N2 type substitution (equation 3).

$$\text{(structure)} \qquad (3)$$

In some systems, the existence and possible participation in propagation of branched or macrocyclic oxonium ions, resulting from chain transfer to the polymer (inter- or intra-molecular), has also to be considered (equation 4).

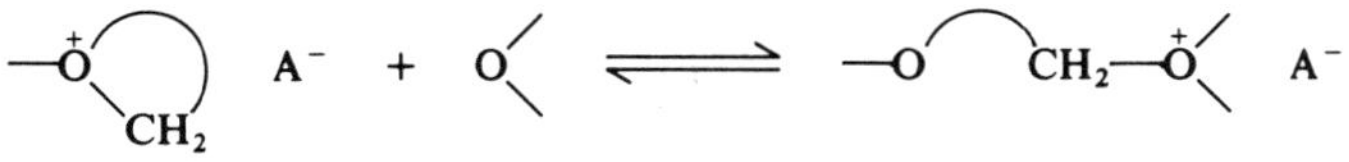

$$\text{(structure)} \qquad (4)$$

There is no direct evidence of the participation of carbenium ions in propagation of the four- and five-membered cyclic ethers. Analysis of the isomerized side products, as well as studies of stereochemistry of polymerization, indicate, however, that carbenium ions may play a certain rôle in the polymerization of highly strained three-membered cyclic ethers (equation 5), where the ring strain may facilitate the unimolecular opening of the oxonium ion ring (*cf.* Section 48.4.2.1.ii).

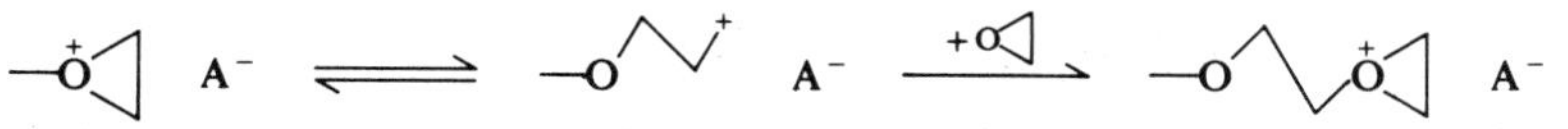

$$\text{(structure)} \qquad (5)$$

When ionic active species are associated with counterions that can form covalent bonds, such as $A^- = OSO_2CF_3^-$ or $OClO_3^-$, isomeric covalent species may be reversibly formed (equation 6). These have been shown to participate in propagation (*cf.* Section 48.6.3.2.iii).

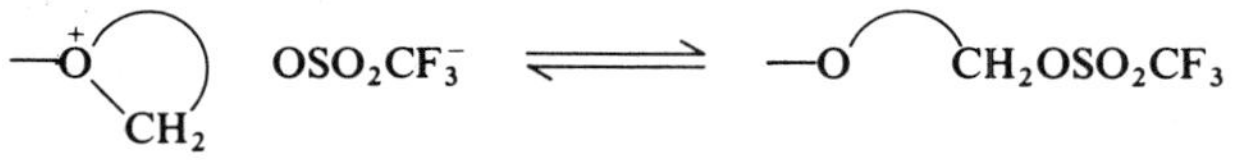

$$\text{(structure)} \qquad OSO_2CF_3^- \rightleftharpoons \qquad CH_2OSO_2CF_3 \qquad (6)$$

In this generally accepted mechanism of cyclic ether polymerization, the active species, *i.e.* ions and/or macroesters (covalent species), are located at the end of the growing macromolecule. Thus, we coined the term: 'Active Chain End' (ACE) to describe this mechanism.

It has only recently been shown that, at certain conditions, charge can be located not at the polymer end but on the monomer itself.[13, 14] Growth then proceeds with the non-ionic chain end, fitted with hydroxyl groups, which attack the cationated (*e.g.* protonated) monomer (equation 7).

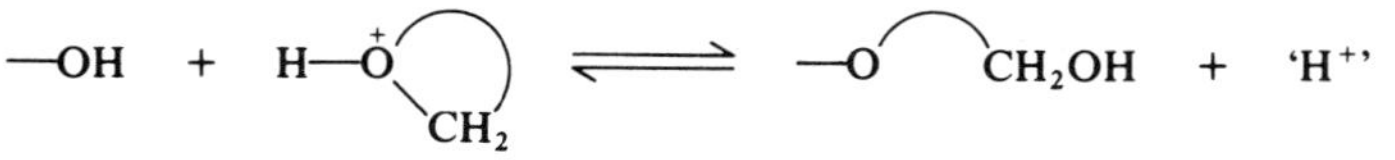

$$-OH + H-\text{(structure)} \rightleftharpoons \text{(structure)} CH_2OH + \text{'}H^+\text{'} \qquad (7)$$

This mechanism, which we call 'Activated Monomer' (AM) mechanism, is discussed in more detail in Section 48.4.

48.3.1   Formation of Active Species — Initiation

There are three possible routes of initiation of cyclic ether polymerization *i.e.* formation of the first growing species (equation 8) by: (a) direct addition of initiator to monomer molecule; (b) abstraction of hydride ion (H^-) from monomer; and (c) formation of zwitterion between initiator and monomer.

$$I + M \longrightarrow M_I^+ \qquad (8)$$

Typical examples of each of these systems will be shown below, with a more detailed discussion presented in the paragraphs devoted to the specific groups of monomers.

Protonic acids, their derivatives (anhydrides, esters), oxycarbenium and oxonium salts generally initiate polymerization of cyclic ethers by direct addition (protonation, acylation, alkylation), as shown schematically in Scheme 1.

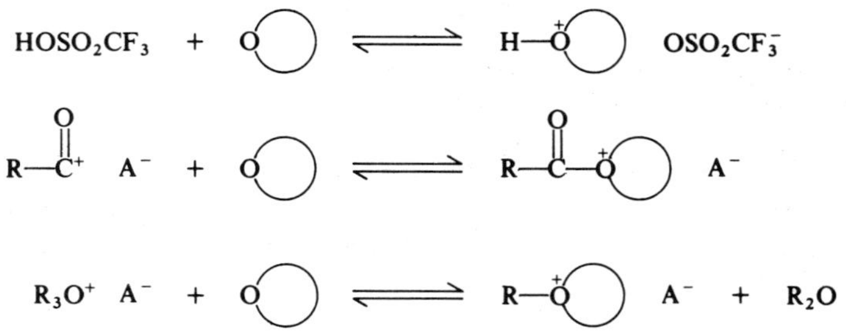

Scheme 1

Carbenium salts produce sterically hindered addition products, *e.g.* triphenylmethylium (trityl) salts, which may lead to hydride ion abstraction (Scheme 2).[15,16]

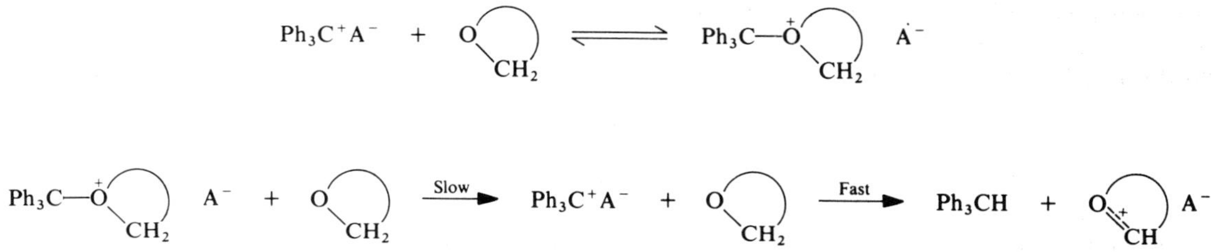

Scheme 2

Oxycarbenium ions thus formed serve as the actual initiators (*cf.* Section 48.6.3.1.i). In these systems initiation may involve additional side reactions like proton expulsion and formation of a double bond within a monomer molecule (*e.g..* as in Scheme 3 for THF).[15,17]

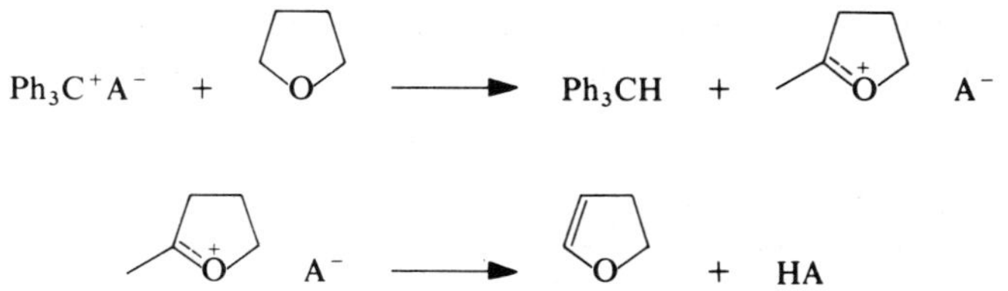

Scheme 3

Friedel–Crafts type initiators (*e.g.* BF$_3$), in the presence of a suitable coinitiator (H$_2$O, RX), may form protonic acids or carbenium ions which then act as actual initiators. This mode of initiation has been shown to operate in the polymerization of four-membered cyclic ethers (Scheme 4).[18-20]

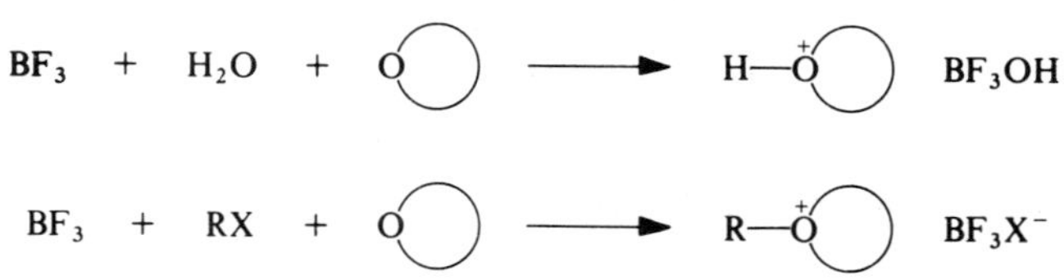

Scheme 4

It has been claimed that some of the cyclic ethers, particularly the strained ones, may give directly a zwitterion with a tertiary oxonium ion at one of its ends, when initiated with BF$_3$ (equation 9).[21]

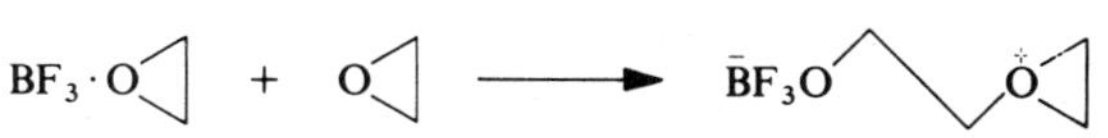

$$BF_3 \cdot O \triangleleft \ + \ O \triangleleft \ \longrightarrow \ \bar{B}F_3 O \wedge \overset{+}{O} \triangleleft \tag{9}$$

THF is known to give a stable 1:1 crystalline complex with BF_3,[22] which, when mixed with excess THF, does not initiate polymerization.[23]

PF_5 behaves differently; it does directly initiate polymerization of THF, proceeding through a transient state involving a zwitterion. This is a well-documented system[24] (by ^{1}H and ^{31}P NMR).

Special features of the different initiation mechanisms are discussed in more detail in the sections dealing with polymerization of the particular groups of monomers.

48.3.2 Reactions of Active Species — Propagation *vs.* Chain Transfer to Polymer

As stated in the introduction to this chapter, propagation in cyclic ether polymerization proceeds either by the active chain (ACE) mechanism, in which active species are located at the end of the growing macromolecule (Scheme 5) and/or by the activated monomer (AM) mechanism. In the latter the positive charge is located on the monomer and the growing macromolecule is terminated with a non-ionic *nucleophilic* group (*e.g.* HO— group), as shown in equation (10).

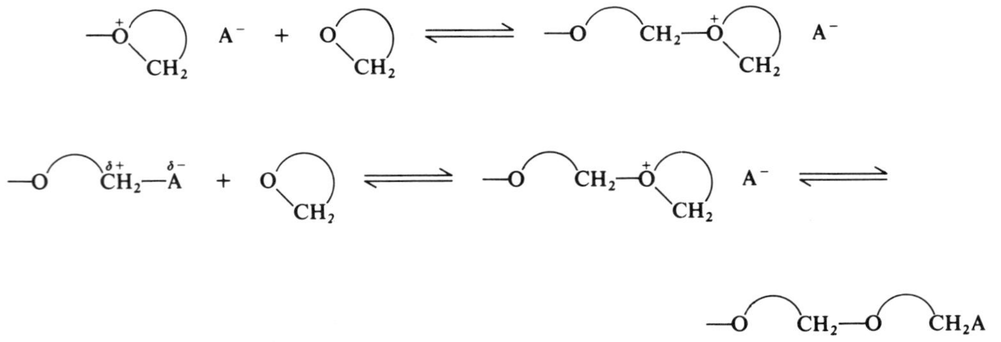

Scheme 5

$$\text{—OH} + \text{H—O}^+\text{(CH}_2\text{)} \longrightarrow \text{—O}\text{—CH}_2\text{—OH} + \text{'H}^+\text{'} \tag{10}$$

The ACE and AM mechanisms may coexist in some systems; methods allowing the determination of the contribution of both mechanisms to the chain growth are discussed in Section 48.4, which is devoted to the AM polymerization of oxiranes.

Knowledge of the kinetics of propagation by the ACE mechanism mainly results from studies of THF polymerization and is discussed in detail in Section 48.6.3.2.

Polymerization of THF proceeds as a living process; active species, once formed, are stable and do not participate appreciably in any side reactions within the time period needed to achieve equilibrium conversion of monomer. This system differs from the living systems defined by Szwarc[25] because, due to the reversibility of the propagation, the Poisson distribution cannot be achieved.

In polymerization of the majority of other cyclic ethers, side reactions are very common (*i.e.* transfer or termination). Although termination may be avoided in some systems, chain transfer to polymer is more pronounced; this results in cyclization and/or scrambling. Thus, with some monomers and conditions, cyclic oligomers are found to be the only product of cationic polymerization. For instance, nearly quantitative conversion of ethylene oxide into its cyclic dimer, 1,4-dioxane, has been reported.[26]

Chain transfer to polymer is a general phenomenon in the cationic polymerization of heterocyclic monomers. The nucleophilic site of the monomer molecule (ethereal oxygen) is transferred to the polymer chain by reaction with a growing oxonium ion.

Chain transfer to polymer may involve either foreign (intermolecular reaction) or the own (intramolecular reaction) chain, as shown in Scheme 6.

The extent of competition of these reactions with propagation depends on the ratio k_{tr}/k_p. This, in turn, is related to the relative nucleophilicities of a monomer molecule and of the polymer chain unit as well as to the ability of the growing chain end (for the intramolecular reaction) to attain the conformation required for closing the ring.

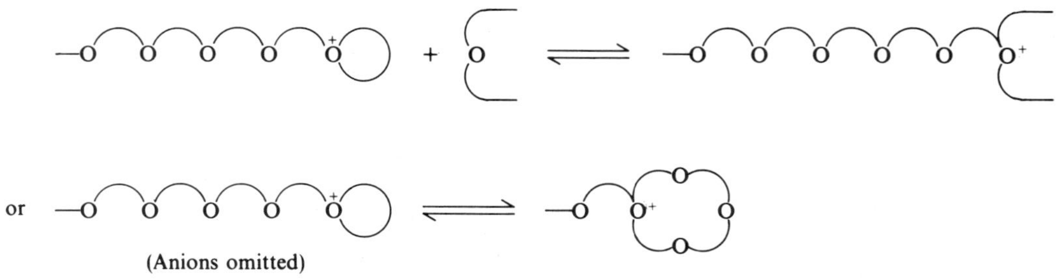

(Anions omitted)

Scheme 6

From the point of view of polymer synthesis, the contribution of the intramolecular reaction is more critical, because this process leads to formation of a cyclic fraction.

Problems related to cyclization phenomena in the cationic polymerization of heterocycles are discussed in more detail in Chapter 47 of this volume which deals with cyclization. Following from this discussion, the differences in rates and yields of cyclic oligomers formation, observed for different classes of cyclic ethers (negligible for THF, but significant for oxiranes) may be explained on the basis of a combination of kinetic (relative rates) and thermodynamic (conformational) factors.

48.3.3 Transfer and Termination Reactions

Here two groups of problems related to transfer and/or termination reactions will be discussed separately: (a) transfer and/or termination reactions as inherent features of the system; and (b) transfer and/or termination intentionally induced in order to introduce the desired end groups.

In some systems undesirable transfer or termination reactions may be avoided, provided the impurities content does not exceed a certain level. Due to the lower reactivity of oxonium ions and basic character of monomers this factor is not as critical as in vinyl (both anionic and cationic) polymerization.

The best documented system, without involving any termination is, as indicated above, polymerization of THF with stable counterions (SbF_6^-, AsF_6^-, PF_6^-) or noncomplex anions like $CF_3SO_3^-$ for which recombination occurs but is fully reversible. For these systems the dependence in equation (11) holds up to the high $\overline{DP}_n$ values.[27, 28]

$$\overline{DP}_n = \{[THF]_0 - [THF]_e\}/[I]_0 \tag{11}$$

With the less stable counterions, like $AlCl_4^-$ (equation 12), termination due to counterion fragmentation was observed even in THF polymerization.[29] Chlorine end groups were detected in the resulting polymers.

$$-O(CH_2)_4\overset{+}{O}\bigcirc \quad AlCl_4^- \longrightarrow -O(CH_2)_4O(CH_2)_4Cl \;+\; AlCl_3 \tag{12}$$

The H^+ transfer, which is a typical route of transfer in cationic vinyl polymerization, does not seem to be important in the polymerization of THF or in any other polymerization which proceeds by oxonium ions. It seems that it is the presence of carbenium ions that is required for proton transfer (equation 13).

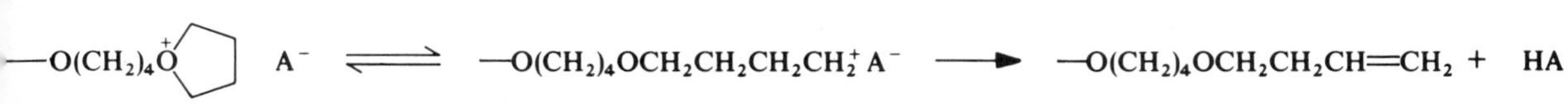

$$-O(CH_2)_4\overset{+}{O}\bigcirc \quad A^- \rightleftharpoons -O(CH_2)_4OCH_2CH_2CH_2CH_2^+\,A^- \longrightarrow -O(CH_2)_4OCH_2CH_2CH{=}CH_2 \;+\; HA$$

$$\tag{13}$$

The situation is less straightforward in the polymerization of three- and four-membered cyclic ethers. For these monomers transfer and/or termination reactions are obscured by effective chain transfer to the polymer. This leads to cyclization, lowering in this way $\overline{M}_n$ of the linear fraction, when compared with a value calculated for a living system.

Very high molecular weight polymers ($\overline{M}_n$ up to 10^6 and even higher) were obtained in the polymerization of the three-membered (*e.g.* epichlorohydrin)[30] and four-membered (*e.g.* 3,3-bis-(chloromethyl)oxetane)[31] cyclic ethers with organoaluminum compounds.

In these systems propagation seems to proceed not by oxonium ions but rather by covalent species (coordinate–'pseudocationic'? polymerization). There is no detailed work on the mechanism of polymerization with these initiators.

As indicated in Section 48.4.2.1.ii, in the cationic polymerization of three-membered cyclic ethers carbenium ions may exist in equilibrium with oxonium ions (equation 14) although their concentration is probably low.

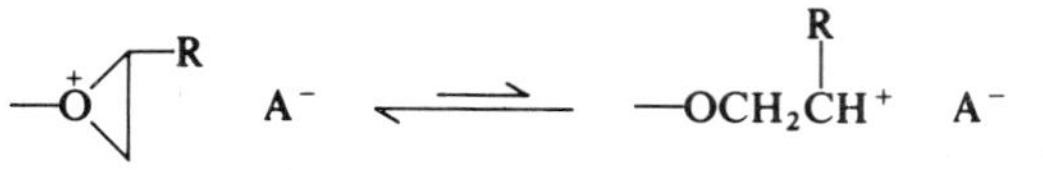

$$-\overset{+}{O}\overset{R}{\diagdown}\quad A^- \quad \rightleftharpoons \quad -OCH_2\overset{R}{\underset{|}{C}}H^+ \quad A^- \tag{14}$$

Thus, in oxirane polymerization, transfer (*e.g.* proton transfer) and/or termination involving carbenium ions, cannot be excluded, Indeed, Entelis and co-workers found both unsaturated (detected by bromination) and —CHRF (by ^{19}F NMR) end groups in the copolymerization of THF with nitroglycidyl ether and epichlorohydrin respectively.[32]

The second group of transfer and termination processes includes systems where specific chain transfer or terminating agents are introduced into the system with the aim of preparing polymers with desired end groups. It is required for this purpose that no other transfer and/or termination reactions proceed than these leading to the proper structure of the end groups. In other words, active species should inherently be 'long-living'. This approach has most advantageously been used in the functionalization of PTHF.

Saegusa's 'phenoxy end-capping' method[33,34] and the phosphine 'ion-trapping' method[35,36] which was developed in the present authors' laboratory, both employ the principle of quantitative transformation of the active species into end groups which are detectable spectroscopically. This allows the quantitative determination of the active species concentration (Scheme 7). For the phenoxy end-capping method, —OPh was determined by UV ($\lambda_{max} = 272\,m\mu$; $\varepsilon_{max} = 1.93 \times 10^3\,mol^{-1}\,l\,cm^{-1}$). For the ion-trapping method, —$\overset{+}{P}Ph_3$ was determined by ^{31}P NMR ($\delta = 23.4$ p.p.m. in CH_2Cl_2); the chemical shift depends on the structure of ligand and on the solvent.

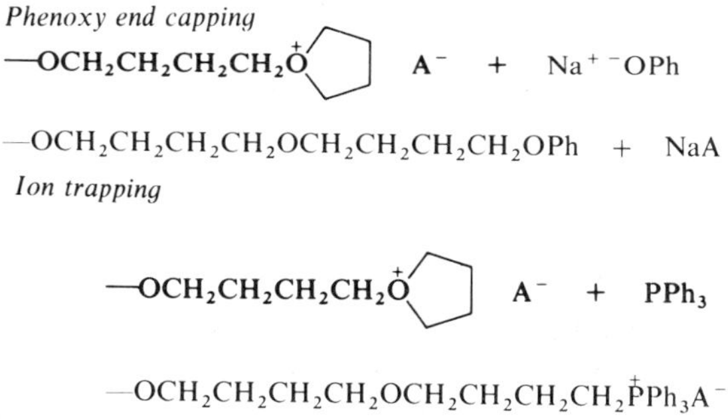

Phenoxy end capping

$$-OCH_2CH_2CH_2CH_2\overset{+}{O}\diagup\quad A^- \quad + \quad Na^+\,{}^-OPh$$

$$-OCH_2CH_2CH_2CH_2OCH_2CH_2CH_2CH_2OPh \quad + \quad NaA$$

Ion trapping

$$-OCH_2CH_2CH_2CH_2\overset{+}{O}\diagup\quad A^- \quad + \quad PPh_3$$

$$-OCH_2CH_2CH_2CH_2OCH_2CH_2CH_2CH_2\overset{+}{P}Ph_3A^-$$

Scheme 7

Both reactions are fast and quantitative. Ion trapping has been used in the polymerization of THF as well as in other systems, permitting one not only to measure the concentration of active species down to $10^{-4}\,mol\,l^{-1}$, but also to distinguish between different possible structures of active species (*e.g.* in copolymerization). This is made possible by taking advantage of the sensitivity of the chemical shift of phosphorus in phosphonium ions to the nature of the ligand from the polymer chain.[36]

Various aspects of application of transfer or termination reactions for preparing polymers having one or two required end groups will be discussed in more detail in Section 48.6.4.

48.4   THREE-MEMBERED CYCLIC ETHERS—OXIRANES

Oxirane (ethylene oxide) and methyloxirane (propylene oxide) are prepared by direct oxidation of the corresponding alkenes. Chloromethyloxirane (epichlorohydrin) is produced industrially from the corresponding chlorohydrin (*e.g.* propylene + $Cl_2 \rightarrow$ allyl chloride + HOCl → glycerol dichlorohydrin → epichlorohydrin).

In spite of the availability of monomers, cationic polymerization of oxiranes is of little industrial interest, due to the side reactions (mostly cyclization) which prohibit the preparation of well-defined, high molecular weight polymers. The two important exceptions are: preparation of poly(epichloro-hydrin) elastomers and curing of the epoxy resins, discussed in Sections 48.4.3.1 and 48.4.3.2.

Activated monomer polymerization of oxiranes opens new perspectives for preparation of reactive oligomers (macromonomers, telechelics) by cationic polymerization.

Finally, cyclopolymerization of oxiranes is studied as a method for preparing crown ethers.

48.4.1 Polymerizability

Due to the high strain of the three-membered rings, propagation of oxiranes can be practically treated as an irreversible reaction. Although substitution generally decreases the polymerizability (*cf.* Chapter 46 of this volume on thermodynamics), release of the ring strain in three-membered cyclic ethers more than compensates the conformational strain. Thus, even polymers from fully substituted oxiranes, like 1,1,2,2-tetramethyloxirane, have been prepared.[37]

48.4.2 Mechanism of Polymerization

As discussed earlier in this chapter, polymerization of oxiranes may proceed by the Active Chain End (ACE) mechanism, or by the Activated Monomer (AM) mechanism (Scheme 8).

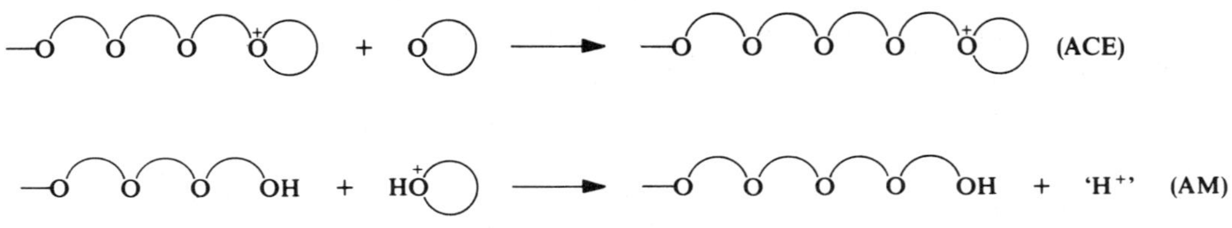

Scheme 8

In the following sections, we will present the characteristic features of both mechanisms.

48.4.2.1 Active chain end mechanism

(i) Generation of active species — initiation

From three possible general mechanisms of initiation, described in previous sections, the route *via* hydride ion abstraction from the monomer molecule is not important for oxiranes, which are poor H^- donors. Thus, active species are generated either through addition of the initiating cation or by formation of a zwitterion. Protonic acids, carbenium, oxonium or oxocarbenium salts add to the monomer molecules, whereas Lewis acid type initiators (BF_3, PF_5) are believed to form zwitterions. However, in the presence of a suitable coinitiator (*e.g.* water) initiation may be caused also by the formed '*in situ*' protonic acid.

There are not too many results in the literature actually proving that initiation of oxirane polymerization with initiators of the first of the two mentioned groups indeed proceeds by addition. It was shown conclusively that triphenylmethylium cation initiates oxirane polymerization by addition (equation 15).[38]

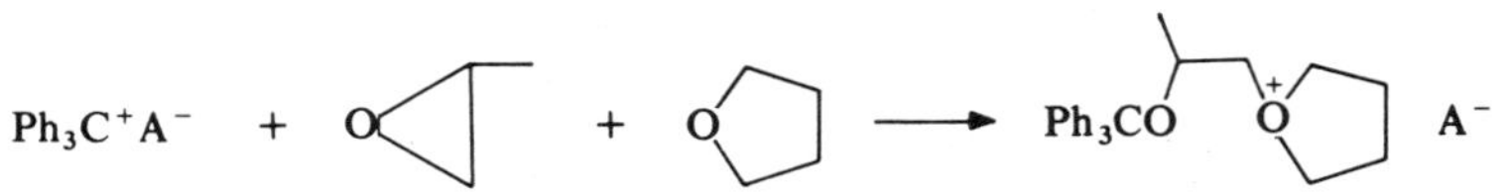

Formation of the transient zwitterionic species in the reaction of some oxiranes with Lewis acids has already been postulated by Meerwein,[39] in his scheme describing the formation of trialkyl-oxonium salts. This type of initiation reaction has further been advocated by Entelis for the BF_3–epichlorohydrin system (equation 16).[40]

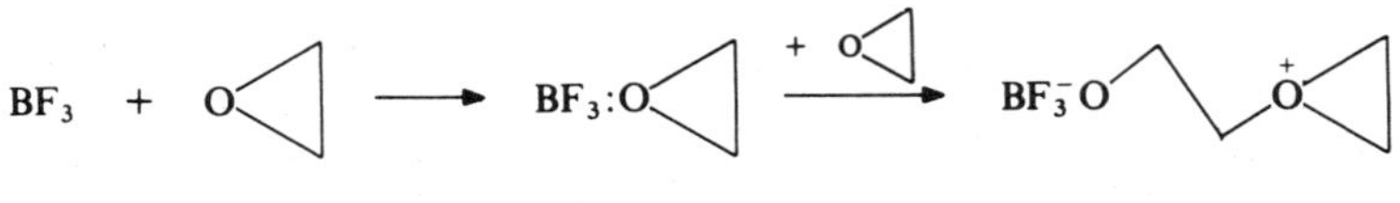

The initially formed zwitterionic species may, during the course of reaction, be converted into a usual ion pair by reaction with another molecule of Lewis acid (equation 17).

$$BF_3^-O \diagdown \diagdown O^+ \triangleleft \; + \; BF_3 \longrightarrow BF_2O \diagdown \diagdown O^+ \triangleleft \; BF_4^- \tag{17}$$

These reactions are still obscure and not as well documented as in the polymerization of THF with PF_5.[24]

(ii) Structure of active species

By analogy with cationic polymerization of THF, the structure of the oxonium ion active species in the polymerization of oxiranes is frequently shown as structure (**1**).

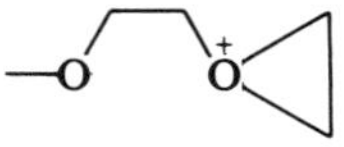

(**1**)

In contrast to the tetrahydrofuranium cation, the oxiranium cation has never been directly observed in any polymerization process. On the contrary, it has been shown recently that strained oxonium ion (**1**) is partly converted into the less strained oxonium ion involving a six-membered ring (equation 18)[41] and to the strainless branched oxonium ion (equation 19).

(18)

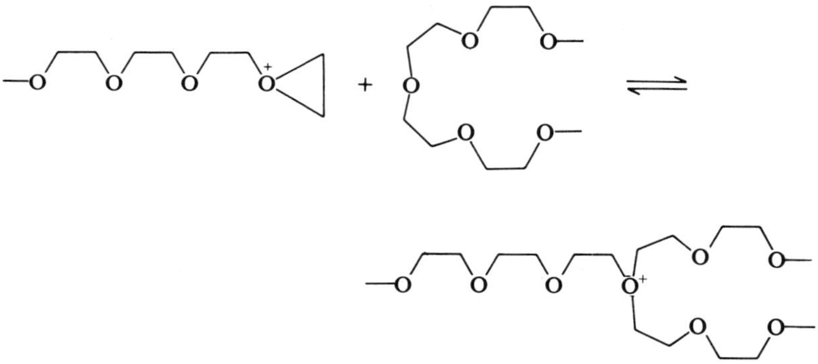

(19)

Both species have been identified directly in ^1H and ^{13}C NMR spectra,[41] the corresponding spectrum is shown in Figure 1. Active species in the ACE polymerization of other oxiranes, although frequently presented as oxiranium cations (**2**) have never been shown experimentally to have such a structure.

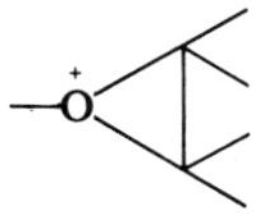

(**2**)

(iii) Reactions of active species — propagation vs. cyclization

In the ACE polymerization of three-membered cyclic ethers propagation is inevitably accompanied by cyclization. Cyclic oligomers that are formed can be divided into two series: (i) 'main' series, composed of cyclic oligomers containing exclusively repeating units of the same structure as that of the monomer (cyclic *n*-mers); (ii) 'side' series composed of cyclic oligomers containing one or more isomerized units.

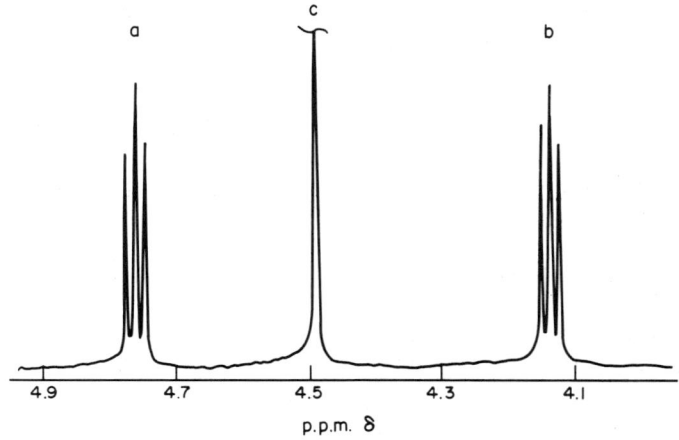

Figure 1 ^{1}H NMR spectrum of 1-methyl-1,4-dioxanium cation (solvent: SO_2, 0 °C; counterion: SbF_5I^-)

Formation of the 'main' series cyclic oligomers may be analyzed in terms of the classical cyclization theory, based on the Jacobson–Stockmayer treatment,[42] involving additionally, whenever required, kinetic and steric factors (*cf.* Chapter 47 on cyclization).

Formation of the 'side' series cyclic products indicates the occurence of the side reactions, leading to isomerization of active species. Thus, analysis of this fraction provides a valuable diagnostic tool for studies of the mechanism of polymerization. Small amounts of isomerized structures, which could be undetectable if built into the polymer chain, may easily be analyzed when present as the low molecular weight cyclic oligomers.

It has been shown that 'side' series cyclic oligomers are formed in the cationic polymerization of various oxiranes.[43–46] Typical examples are given in Scheme 9.

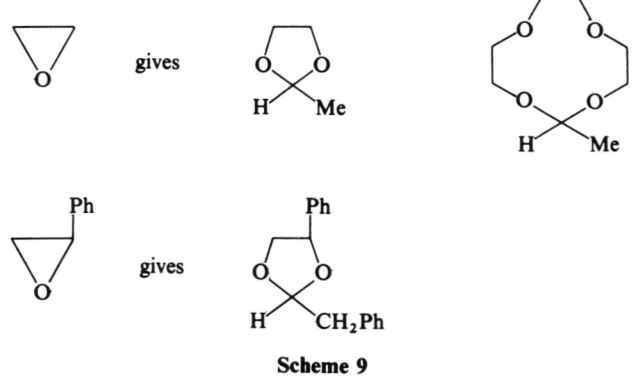

Scheme 9

Formation of the isomerized products shows that tertiary oxonium ion type active species may propagate not only by the usual S_N2 type reaction (equation 20) but in addition to this by a mechanism resembling an S_N1 process (Scheme 10).

$$-\overset{+}{O}\triangleleft \ + \ O\triangleleft \ \longrightarrow \ -O\diagdown\diagup\overset{+}{O}\triangleleft \tag{20}$$

The carbenium ion, having then a definite lifetime, can isomerize by a 1,2-hydride shift to the corresponding oxycarbenium ion, as shown in equation (21).

$$-OCH_2CHRCH_2CHR^+ \ \longrightarrow \ -OCH_2CHR\overset{+}{O}{=\!=\!=}CHCH_2R \tag{21}$$

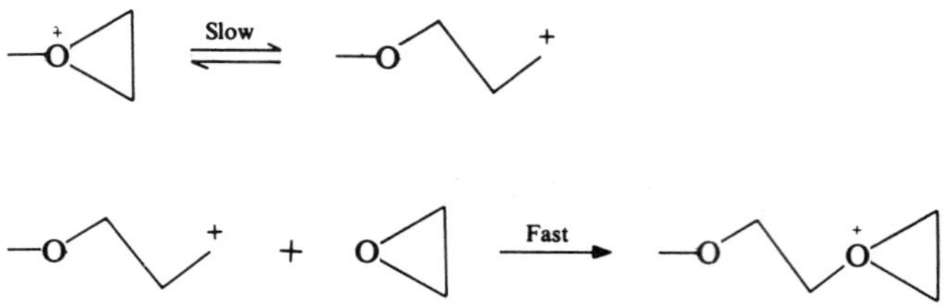

Scheme 10

Further back-biting reactions, involving isomerized active species, produce cyclic oligomers of the 'side' series.

Participation of carbenium ion type active species in the polymerization of oxiranes has also been proposed by Schuerch who studied the polymerization of 1,2-anhydrosugars.[47] It has particularly been shown (Scheme 11) that the corresponding polymers contain both α- and β-linkages.

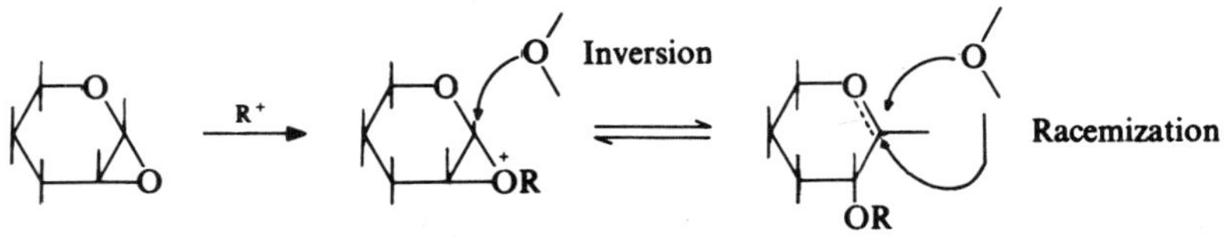

Scheme 11

Racemization implies participation of the flat carbenium ion with sp^2 hybridization and the possibility of an attack of the incoming monomer molecule from both sides of a plane. Propagation on oxonium ions should lead exclusively to the inversion of configuration but not to racemization.

Not only the difference in structure but also stereochemistry of cyclic oligomers may provide valuable information on the reaction mechanism. Thus, it was shown that in polymers obtained from *cis*- and *trans*-2,3-butene oxide, configuration on one carbon atom in each unit is different to that in the starting monomer[48] (**3**; where r = retention and i = inversion)

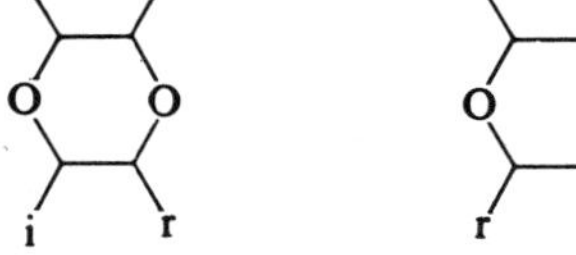

(3)

On the other hand, the cyclic dimer formed in this system has configuration (**4a**) and not (**4b**).

(4a) 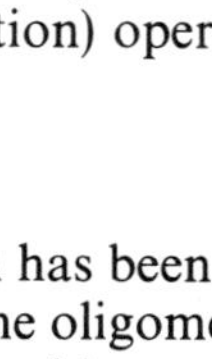   **(4b)**

This shows, that from possible routes (in Scheme 12) for cyclic dimer formation, route (b) involving double inversion (resulting formally in retention) operates exclusively.

(iv) Synthesis of cyclic oligomers

The high tendency of oxiranes to undergo cyclization has been used for synthesis of crown ethers and their analogues. Dale increased the proportion of the oligomers with desired ring size by taking advantage of the template effect.[49] Thus, ethylene oxide was polymerized cationically in the presence of inorganic cations (Scheme 13).

Folding of the chain end around the cation leads to the preferential formation of the cyclic oligomer, which is crown ether specific for the cation used. Representative examples are given in Table 3.

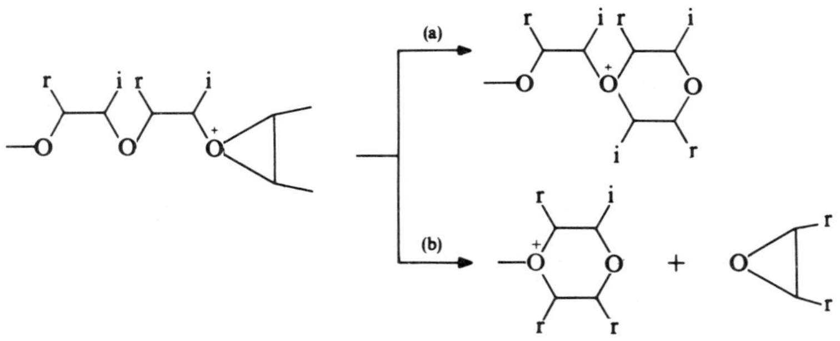

Scheme 12

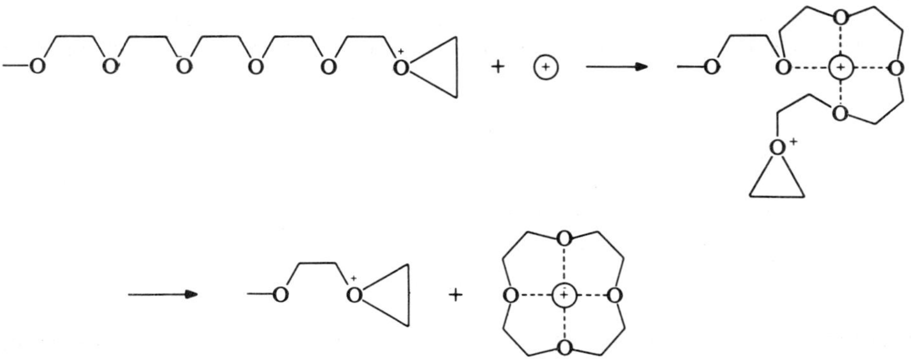

Scheme 13

Table 3 Composition of Cyclic Oligomer Mixtures in the Polymerization of Ethylene Oxide in the Presence of Various Complexing Salts

Salt	Tetramer (wt %)	Pentamer (wt %)	Hexamer (wt %)
LiBF$_4$	30	70	
NaBF$_4$	25	25	50
KBF$_4$	—	50	50
RbBF$_4$	—	—	100
Ca(BF$_4$)$_2$	50	50	—
Cu(BF$_4$)$_2$	5	90	5

Crown ether type cyclic oligomers may be prepared also by cationic cyclopolymerization of diepoxides. 1,2-bis[2-(2,3-epoxypropoxy)ethoxy]benzene polymerized with a variety of cationic initiators gave soluble, linear polymers containing no unreacted epoxy groups.[50] The mechanism in equation (22) was proposed.

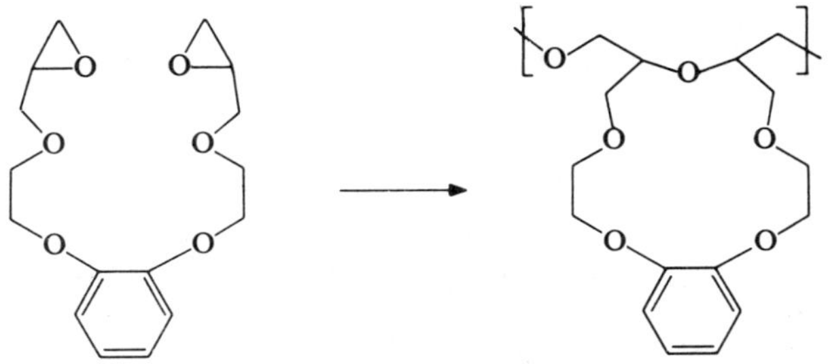

(22)

These polymers show cation-binding properties with the highest selectivity for K$^+$ and Rb$^+$ cations.

48.4.2.2 Activated monomer mechanism

The activated monomer (AM) mechanism of propagation was first formulated in our laboratory and has been continually developed here since then.[13,14,51-53] This mechanism in its simplest form can be described by equation (23).

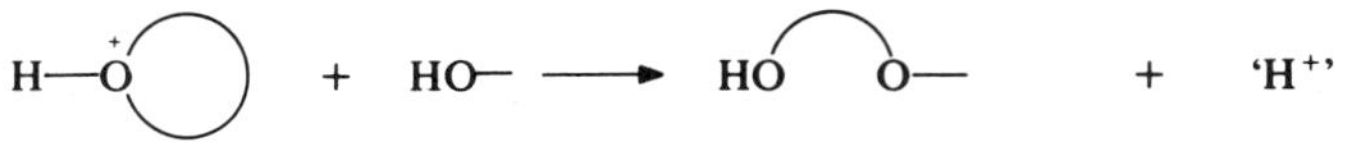

$$\text{H}\!-\!\overset{+}{\text{O}}\bigcirc \;\; + \;\; \text{HO}\!- \;\; \longrightarrow \;\; \text{HO}\frown\text{O}\!- \;\; + \;\; \text{`H}^{+}\text{'} \tag{23}$$

It follows that the AM mechanism may operate effectively providing the concentration of the nucleophilic end group (reproduced in each propagation step) is high enough to make this route competitive with respect to the ACE mechanism.[13,14] This may be illustrated by Scheme 14.

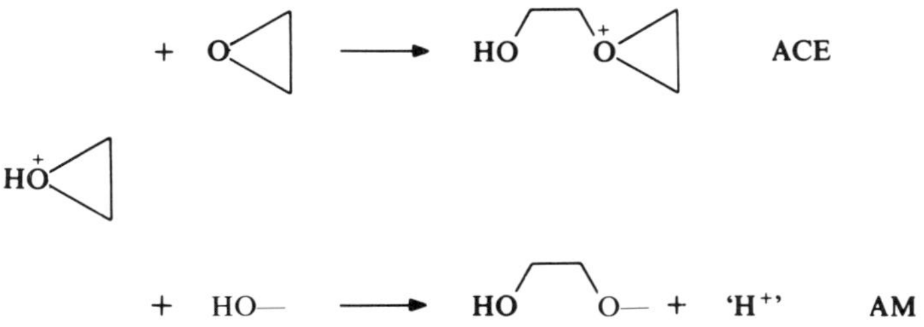

Scheme 14

In all the systems studied so far that operate by the AM mechanism, it was observed that macrocycle formation was suppressed.

The direct proof of the participation of the AM mechanism should be given by finding the concentrations of tertiary oxonium ions and showing kinetically that their contribution to the chain growth can be neglected. This approach is under study but the final data are not yet available. Another proof comes from the studies of distribution of linear and cyclic oligomers in the polymerization of ethylene oxide (EO), carried out in the presence of the monomethyl ether of ethylene glycol.[14] As shown in Scheme 15, the same structures, namely a series of homologues of the monomethyl ethers of oligo(ethylene oxide) glycols may result either from the AM type propagation or from reaction proceeding through the tertiary oxonium ion species, followed by chain transfer to A_1, as in Scheme 15.

Formation of A_2, as shown in Scheme 15, is possible in both mechanisms, but in ACE, for every molecule of A_2 one molecule of 1,4-dioxane (DXN) should be formed. This is not required for the AM mechanism. Formation of A_2 coupled with absence of DXN, as shown experimentally, strongly suggests the occurence of the AM mechanism of growth.[14] Results are shown in Table 4. Thus, the ratio of $[A_2]/[EO]_{consumed}$ gives directly the contributions of both competing reactions. These results show that the contribution of the AM type reaction increases with decreasing $[EO]/[HO\!-\!]$ ratio and approaches 100% at a sufficiently low ratio of $[EO]$ to $[HO\!-\!]$.

The chain end, from which a macromolecule grows in the AM type propagation, does not bear a charge. Therefore, one can expect chain transfer to polymer to be considerably suppressed, as compared to the ACE mechanism. In particular, this type of propagation should lead to much less cyclization, as already mentioned.

It was indeed shown that the amount of cyclic dimer, 1,4-dioxane, formed in the cationic polymerization of ethylene oxide, decreases to nearly zero with increasing contribution of the AM mechanism of propagation, caused by increasing the $[HO\!-\!]/[EO]$ ratio.[51] Results presented in Table 5 show that it is possible to convert oxiranes into essentially linear chains with the proportion of the cyclic fraction being below a few molar percent. Thus formulation of the AM mechanism, and understanding of its consequences, opens important synthetic possibilities.

As shown in Table 5, to suppress the formation of the cyclic fraction, relatively high $[HO\!-\!]/[EO]$ ratios are required. On the other hand, if the products of relatively high polymerization degree are to be prepared, lower $[HO\!-\!]/[EO]$ ratios are needed. This is because $\overline{DP}_n$ (assuming that only AM type propagation takes place) is given by equation (24).

$$\overline{DP}_n = \frac{[EO]_{consumed}}{[HO\!-\!]} \tag{24}$$

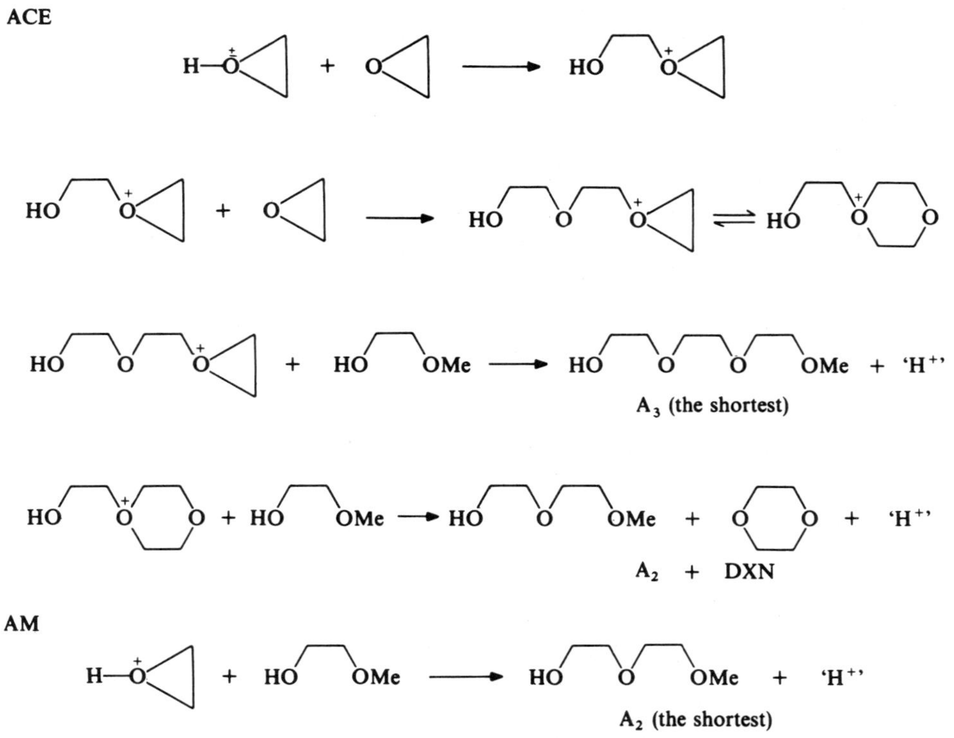

ACE

A₃ (the shortest)

A₂ + DXN

AM

A₂ (the shortest)

Scheme 15

Table 4 Contribution of Activated Monomer Mechanism
($C_{AM} = [A_2]/[EO]_{consumed}$) to the Chain Growth in the Polymerization
of Ethylene Oxide[a]

$[EO]$ (mol l^{-1})	$[EO]/[HO-]$	$[A_2]/[EO]_{consumed} \times 10^2$ (%)
2.0	1.5	99
3.6	2.7	96
2.0	4.0	96
3.0	4.0	93
4.0	4.0	91
6.1	4.6	87

[a] Conditions: $[HSbF_6]_0 = 7 \times 10^{-4}$ mol l^{-1}; solvent CH_2Cl_2; 0 °C

Table 5 Proportion of 1,4-Dioxane Formed in the Polymerization of Ethylene Oxide in the
Presence of MeOH[a]

Catalyst	$[EO]$ (mol l^{-1})	$[EO]/[MeOH]$	Yield of 1,4-dioxane (mol %)
HOSO₂CF₃	0.96	2.2	2.9
(0.1 mol l^{-1})	0.80	6.15	5.2
	0.89	10.7	7.0
	0.97	35	13.5
	0.80	103	16.2
BF₃·MeOH	1.1	2.0	1.1
(0.03 mol l^{-1})	1.1	4.2	13.4
	1.2	6.0	39.5
	1.3	7.8	43.0
	1.3	17.0	66.5
	1.2	120	100

[a] Conditions; solvent CH_2Cl_2; 25 °C; measured at the complete conversion of EO.

The contradiction between these two requirements is only apparent because $\overline{DP}_n$ is governed by the overall concentration of the monomer consumed, whereas the contribution of the AM mechanism depends on the instantaneous concentration of monomer in the system. Thus, in order to produce a polymer of the required molecular weight, one should introduce the needed amount of monomer into the reaction mixture with the rate close to the rate of its consumption. In this way the instantaneous monomer concentration can be kept low enough. The influence of $[M]_{inst.}$ on the proportion of cyclic and linear fraction is illustrated by the data of Table 6 for polymerization of propylene oxide (PO).[52]

Table 6   Influence of the Instantaneous Monomer Concentration (Propylene Oxide, PO) on the Amount of Cyclic Tetramer Formed

Conditions	$[PO]_{inst}$ $(mol\,l^{-1})$	Conversion of PO (wt %)	Tetramer formed (wt %)
PO introduced at the beginning	$2.0 \to 0$	100	4.0
PO continuously charged			
(a) faster	0.3–0.7	82	2.35
(b) slower	0.05–0.14	100	0.95

Conditions: $[PO]$ total $= 2.0\,mol\,l^{-1}$, $[HO—] = 4 \times 10^{-1}\,mol\,l^{-1}$, $[HBF_4 \cdot Et_2O] = 2.5$–$5 \times 10^{-3}\,mol\,l^{-1}$; CH_2Cl_2 solvent; 25 °C.

(i) Kinetics of activated monomer polymerization

Applying the normally used formalism for the description of the AM type polymerization, we should treat the reaction between the original alcohol and the monomer as initiation, and subsequent reactions of the same kind with the next linear oligomers as propagation.

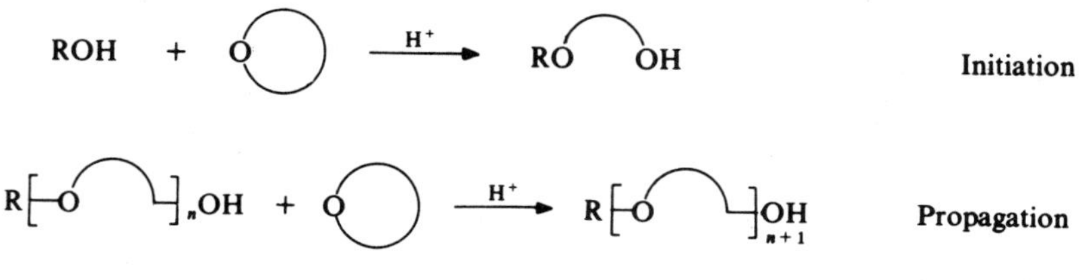

Scheme   16

According to this scheme, ROH should be treated as an initiator and 'H$^+$' as a catalyst. Both initiation and propagation reactions actually proceed as reactions of the protonated monomer with a terminal hydroxyl group (equation 25). The rate of this reaction may be expressed by equation (26). Thus, to analyze the kinetics of AM polymerization one should know the concentration of the protonated monomer.[14]

$$R = [H—\overset{+}{O}\bigcirc][HO—]k \qquad (26)$$

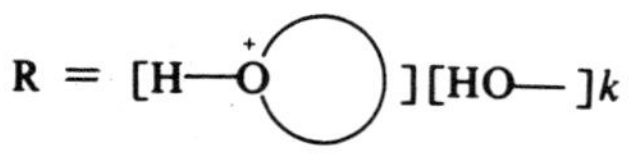

When catalyst (e.g. protonic acid) is added to a system consisting of a cyclic monomer and ROH, the equilibrium in equation (27) is established.

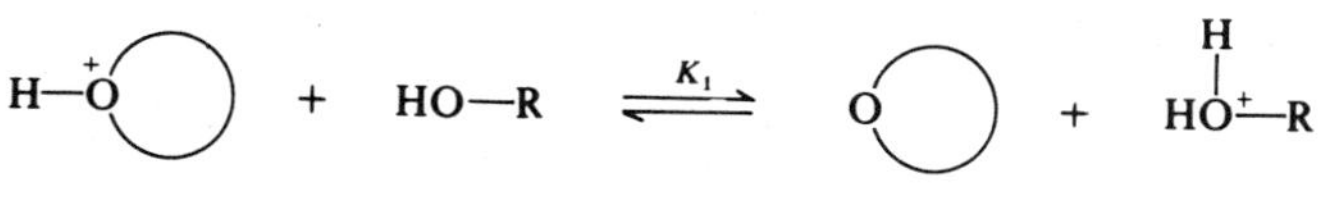

With the progress of the polymerization, when the polymer starts to accumulate, the protonated linear polyether also participates in the equilibrium (equation 28). Thus, the concentration of the protonated monomer changes throughout the reaction.

$$H—\overset{+}{O} \bigcirc \; + \; O\!\!< \quad \underset{\rightleftharpoons}{\overset{K_2}{\longrightarrow}} \quad H—\overset{+}{O}\!\!< \; + \; O\bigcirc \tag{28}$$

Studies of the kinetics of AM polymerization are at the preliminary stage and no quantitative data are available yet. The following features have however been noted in the polymerization of oxirane.

The apparent rate constant of reaction increases with increasing $\overline{DP}_n$ of the growing oligomer.[14] This is surprising in the light of the equilibria discussed above (*cf.* equations 27 and 28) because concentration of the protonated monomer should *decrease* with increasing concentration of the polymer chain units. Thus, there should be another process able to overcome this effect. It has eventually been shown that this is the polymer chain reacting (besides the —OH end group) with the protonated monomer (as in equation 29). In this way, the higher the $\overline{DP}_n$ of a given macromolecule, the higher its overall rate of reaction (and thus apparent k_p) with monomer.

$$-O\diagup\diagdown_O\diagup\diagdown_O- \; + \; H—\overset{+}{O}\!\!\triangleleft \quad \longrightarrow \quad -O\diagup\diagdown\underset{\underset{HO}{|}}{\overset{+}{O}}\diagdown_O- \tag{29}$$

Tertiary oxonium ions formed in reaction (29) may further participate in side reactions, leading to scrambling and cyclization. This results finally in the broadening of the molecular weight distribution, observed experimentally in the later stages of AM polymerization of ethylene oxide.[14]

Steric hindrance may prevent the protonated monomer from reacting with a polymer chain, thus reactions similar to reaction (29) are much less important in the AM polymerization of substituted oxiranes like propylene oxide and epichlorohydrin.[53]

In these systems there is, however, an additional complication due to the possible presence of two nonequivalent hydroxyl end groups, namely primary and secondary ones. This may affect both rate constants (in equation 26) and equilibrium constant (equation 27). Studies of this and related dependences are now under progress.

(ii) Synthetic aspects of the activated monomer polymerization

The AM polymerization is a convenient synthetic method for preparation of telechelics and macromonomers. Thus, if a difunctional alcohol (a diol) is used as initiatior, an oligomeric α,ω-oligodiol results. $\overline{M}_n$ of the product can easily be regulated. The required $\overline{DP}_n$ is equal to the starting [monomer]/[HO—] ratio. This is shown in Figure 2 for polymerization of epichlorohydrin in the presence of ethylene glycol.[53] $\overline{M}_n$ increases linearly with conversion and is equal to the calculated value over the whole concentration range studied.

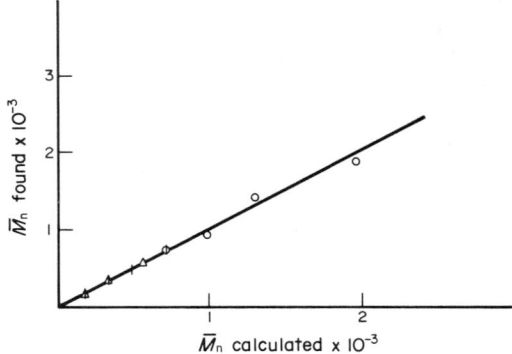

Figure 2 Dependence of $\overline{M}_n$ found on $\overline{M}_n$ calculated in the polymerization of epichlorohydrin (ECH) in the presence of ethylene glycol (EG) using GPC (|), VPO (○) and ¹H NMR (△) ($\overline{M}_n$ calculated = {[ECH]₀ − [ECH]ₜ}/[EG]₀; polymerization in bulk, catalyst $BF_3 \cdot EG$ (0.1 mol%; 25 °C)

The behaviour of the AM polymerization allows preparation of macromonomers of the required $\overline{DP}_n$. Initiators containing, simultaneously in one molecule, an hydroxyl group and a polymerizable double bond, such as hydroxyethyl methacrylate, have to be used. This method is applicable however only for systems with suppressed scrambling (*i.e.* for substituted oxiranes), otherwise macromonomers may contain a certain proportion of bifunctional products. Table 7 lists typical examples of products prepared by an AM polymerization according to the principle discussed above.[14, 53]

48.4.3 Industrial Applications of Cationic Polymerization of Oxiranes

There are two main areas in which cationic polymerization of three-membered cyclic ethers have found industrial application.

Curing of epoxy resins with cationic catalysts belongs to the better developed area. Traditionally, latent cationic initiators, like complexes of BF_3 with amines, were used.[54] More recently, processes based on cationic photoinitiators have been developed.[55, 56]

The other industrial application of cationic polymerization of three-membered cyclic ethers is the production of elastomers, based on epichlorohydrin homo- and co-polymers (with ethylene oxide).[57] These elastomers are known under the trade names Herclor (Hercules Co.) or Hydrin (B.F. Goodrich Chemical Co.). The actual mechanism of the polymerization process applied industrially (*i.e.* the structure of active species) is not known; it is possible that this polymerization, in which organoaluminum initiators or tin derivatives are used,[30, 57] proceeds by pseudocationic mechanism on polarized covalent type active species.

48.4.3.1 *Epichlorohydrin elastomers*

At least in one of the industrial processes, an initiating system based on the alkylaluminum compounds is used. The original process is related to Vandenberg's discovery of the catalytic activity of the reaction products of alkylaluminum compounds with water in the polymerization of various oxiranes.[30, 57] The aluminum trialkyls alone were shown to be inactive.

There is a number of patents suggesting a possibility of using quite a large number of different organometallic initiators. Recently, diphenylcalcium modified by monomer itself was found to be effective in producing an amorphous polymer.[58]

In the family of Vandenberg's initiators, the proper adjustment of the components of the system (AlR_3/H_2O or AlR_3/H_2O/acetylacetone) gives a polymer with a low content of the crystalline fraction and predominantly ($> 97\%$) head-to-tail sequences.

Commercial products are either epichlorohydrin homopolymers (ECH, $\bar{M}_w \sim 450\,000$) or epichlorohydrin–ethylene oxide copolymers ($\bar{M}_w \sim 1\,400\,000$, 68% weight of ECH), soluble in toluene, acetone or methylene chloride. They may be vulcanized by different difunctional compounds (for example diamines) reacting with the chloromethyl group of the rubber.

In addition to the excellent oil and heat-ageing resistance, the ECH homopolymer especially has particularly good air retention properties, by a factor of two better than butyl rubber.

48.4.3.2 *Curing of epoxy resins*

Epoxy resins are catalytically cured primarily by means of basic (anionic) curing agents like amines at room temperature or close to it. Cationic curing finds only limited applications. BF_3 induces fast polymerization of epoxy groups, even at room temperature, and curing proceeds too quickly to be controlled, particularly in larger masses when heat transfer is difficult.[54]

Thus, for practical applications, complexes of BF_3 with bases such as ethylamine, are used. This complex, known commercially as BF_3:400, gives a pot life of about four months at room temperature. The complex dissociates on heating, freeing BF_3, and curing (as in Scheme 17) occurs. The main use of BF_3 complexes is in the catalysis of curing with acid anhydrides or in the preparation of laminates.

Photochemically initiated cationic hardening of epoxy coatings is another example of a technical application of cationic polymerization. This may be accomplished with the aid of aryldiazonium ($ArN_2^+A^-$), diaryliodonium ($Ar_2I^+A^-$) and triarylsulfonium ($Ar_3S^+A^-$) salts, mostly studied by Ledwith[55] and Crivello.[56]

Table 7   Reactive Oligomers Prepared by Activated Monomer Polymerization of Oxiranes

Monomer	Initiator	Structure of the chain	$\bar{M}_n$ calculated	$\bar{M}_n$ found	Cyclic fraction (wt %)
PO	HO(CH₂)₆OH	$\overset{\text{Me}}{\underset{}{+\!\!\left[\text{CH}_2-\text{CH}-\text{O}\right]_n}}$	705 1300	710 1390	1.0 1.2
ECH	HO(CH₂)₂OH	$\underset{\text{CH}_2\text{Cl}}{+\!\!\left[\text{CH}_2-\text{CH}-\text{O}\right]_n}$	1280 2280	1270 1970	0.1 0.2
ECH + PO	HO(CH₂)₂OH	$\left\{\overset{\text{Me}}{+\!\!\left[\text{CH}_2-\text{CH}-\text{O}\right]_x}\underset{\text{CH}_2\text{Cl}}{\left[\text{CH}_2-\text{CH}-\text{O}\right]_y}\right\}_n$			
ECH + THF	HO(CH₂)₂OH	$\left\{\underset{\text{CH}_2\text{Cl}}{+\!\!\left[\text{CH}_2-\text{CH}-\text{O}\right]_x}\left[(\text{CH}_2)_4\text{O}\right]_y\right\}_n$			
ECH	$\text{HO}\!\left[\text{THF}\right]_n\text{OH}$ $\bar{M}_n = 1120$	$\underset{\text{CH}_2\text{Cl}}{+\!\!\left[\text{CH}-\text{CH}_2\text{O}\right]_x}\left[(\text{CH}_2)_4\text{O}\right]_y\underset{\text{CH}_2\text{Cl}}{\left[\text{CH}_2-\text{CH}-\text{O}\right]_x}$	2190	2150	
ECH	$\text{HO}\!\left[\text{Bu}\right]_n\text{OH}$ $\bar{M}_n = 2100$	$\underset{\text{CH}_2\text{Cl}}{+\!\!\left[\text{CH}-\text{CH}_2\text{O}\right]_x}\left[\text{CH}_2\text{CH}=\text{CHCH}_2\right]_y\underset{\text{CH}_2\text{Cl}}{\left[\text{OCH}_2-\text{CH}\right]_x}$	3350	3940	
ECH	HOCH₂CH₂OCOCH=CH₂	$\underset{\text{CH}_2\text{Cl}}{+\!\!\left[\text{CH}_2-\text{CH}-\text{O}\right]_n}$	990	1040	

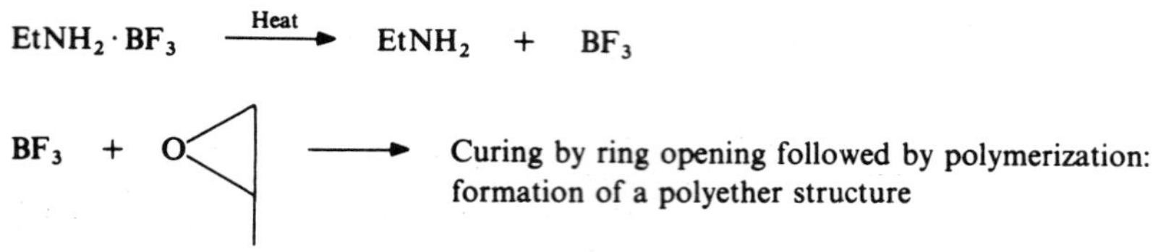

$$EtNH_2 \cdot BF_3 \xrightarrow{\text{Heat}} EtNH_2 + BF_3$$

BF$_3$ + O⟨△ ⟶ Curing by ring opening followed by polymerization: formation of a polyether structure

Scheme 17

It is generally assumed that the primary process for initiation by iodonium and sulfonium salts involves photolytic dissociation to an aryl radical and the appropriate cation radical (Scheme 18). Further reaction of the cation radical with solvent, monomer, *etc.* (RH) produces protonic acids which act as the actual initiators (equation 30).

$$Ar_2I^+ \longrightarrow ArI^{\cdot +} + Ar\cdot$$
$$Ar_3S^+ \longrightarrow Ar_2S^{\cdot +} + Ar\cdot$$

Scheme 18

$$ArI^{\cdot +} + RH \longrightarrow ArI + R\cdot + H^+ \tag{30}$$

In order to increase sensitivity to the longer wavelengths (300–600 nm) these systems may be modified by introducing substituents or combining with dyes and electron donor molecules.

48.5 FOUR-MEMBERED CYCLIC ETHERS — OXETANES

The general method of synthesis of oxetanes is based on the cyclization of the corresponding 1,3-disubstituted compounds, mostly halogenohydrins or their esters (see equation 31 where: R^1, R^2 = —H, alkyl, haloalkyl or aryl group; X = —H or —COMe; Y = —Cl, —Br, —OH or —OCOMe).

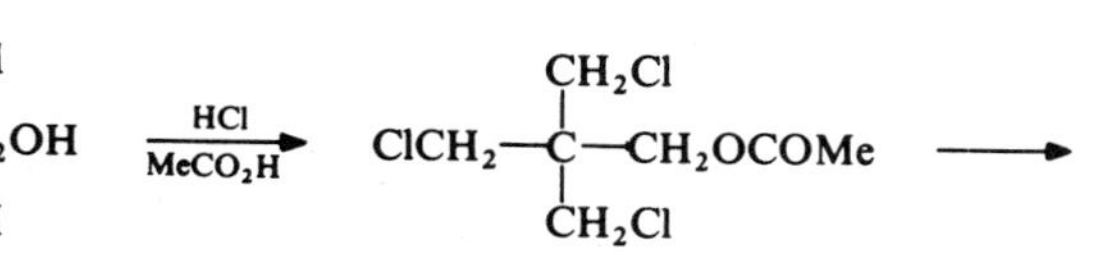

$$\text{R}^1\text{, R}^2\text{C(CH}_2\text{OX)(CH}_2\text{Y)} \longrightarrow \text{R}^1\text{, R}^2\text{ oxetane} + XY \tag{31}$$

The only polymer of industrial importance obtained by cationic polymerization of four-membered cyclic ethers is poly(3,3-bis(chloromethyl)oxetane), known under the trade name Penton (USA) or Pentaplast (USSR).

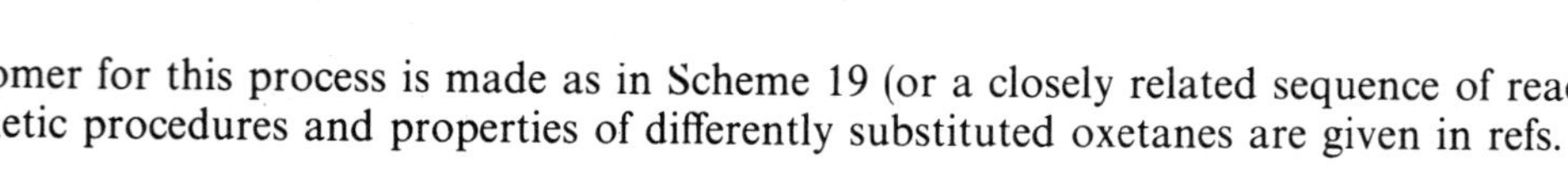

$$HOCH_2-\underset{\overset{|}{CH_2OH}}{\overset{\overset{|}{CH_2OH}}{C}}-CH_2OH \xrightarrow[\text{MeCO}_2\text{H}]{\text{HCl}} ClCH_2-\underset{\overset{|}{CH_2Cl}}{\overset{\overset{|}{CH_2Cl}}{C}}-CH_2OCOMe \longrightarrow \text{(ClCH}_2\text{)}_2\text{ oxetane}$$

Scheme 19

Monomer for this process is made as in Scheme 19 (or a closely related sequence of reactions). Synthetic procedures and properties of differently substituted oxetanes are given in refs. 59–61.

48.5.1 Polymerizability

Ring strain of the four-membered oxetane ring (107 kJ mol^{-1}) is not much lower than that of the three-membered oxirane ring (114 kJ mol^{-1}). Thus, polymerization of oxetanes proceeds practically irreversibly, even for monomers containing two bulky substitutents (*e.g.* 3,3-bis(chloromethyl) oxetane).

48.5.2 Mechanism of Polymerization

48.5.2.1 Initiation

Initiation of oxetane polymerization apparently proceeds by direct protonation or alkylation of the oxetane ring. Thus, for $Ph_3C^+PF_6^-$ and $Et_3O^+PF_6^-$, addition of the initiation species has been assumed,[62] but definite proof has not been presented. With Friedel–Crafts type initiators (*e.g.* BF_3) polymerization can not be initiated in the absence of water.[18-20] Farthing concluded, on the basis of kinetic measurements, that the real initiator in the BF_3–oxetane system is a protonic acid, formed from BF_3 and water present as an impurity.[18]

The AlR_3/H_2O system is even more complex, and was the most thoroughly studied.[63-65] There is no doubt that like the epichlorohydrin polymerization described in the preceding section, this system is better classified as cationic coordinate, because oxetanes do not polymerize by an anionic mechanism (there is however some conflicting information on the anionic copolymerization of oxetane itself).[66] The Bu^i_3Al/H_2O system has been found, at least for BCMO, to be the most effective at the 1:1 molar ratio of components and at these conditions, approximately ten molecules of water are needed to start one polymer chain.[64] This means that a longer structure, comprising of a few Al atoms, is required to accommodate one growing macromolecule.

The rate constant of initiation for the Bu^i_3Al/H_2O–BCMO system has been reported as equal to $k_i = 1.6 \times 10^{-3} mol^{-1}1s^{-1}$ (PhCl, 70 °C).[64] Kropachev *et al.*, in their studies of 3-methyl-3-chloromethyloxetane polymerization, confirmed the general features of BCMO polymerization, although the activity of the Bu^i_3Al/H_2O system was the highest for a 2:1 molar ratio.[65] Cationic polymerization of BCMO, initiated by radiation, also leads to high molecular weight polymers ($\bar{M}_n > 10^5$).[67,68]

48.5.2.2 Propagation

Propagation in the polymerization of oxetanes can be represented as an S_N2 type reaction, like in other cationic polymerizations of cyclic ethers, proceeding by an active chain end mechanism (equation 32).

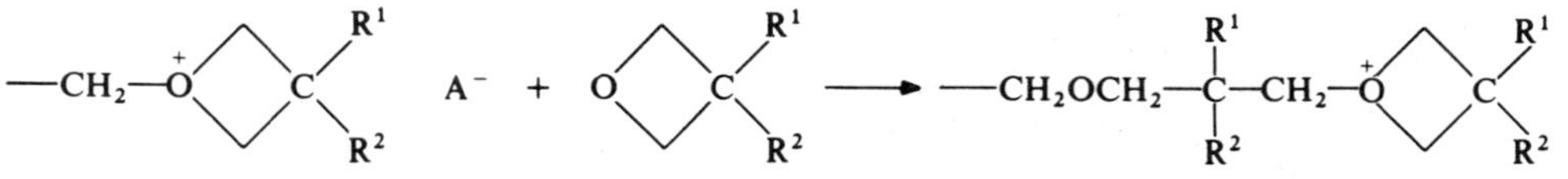

$$\tag{32}$$

The kinetics of cationic propagation of oxetanes has been studied by Saegusa (using the endcapping method),[69] Worsfold[62] and the present authors.[63,64] Propagation rate constants have been reported as equal to: $k_p = 5.7 \times 10^{-2} mol^{-1}1s^{-1}$ (oxetane, BF_3, CH_2Cl_2; -10 °C);[69] $k_p = 6.8$ $mol^{-1}1s^{-1}$ (3,3-dimethyloxetane, BF_3, CH_2Cl_2, -9 °C);[69] and $k_p = 8.5 mol^{-1}1s^{-1}$ (3,3-bis(chloromethyl)oxetane, Bu^i_3Al/H_2O (1:1) PhCl, 70 °C).

The values reported by Saegusa may be too low because, as pointed out by Worsfold,[62] the endcapping method gives the sum of the concentrations of active (*i.e.* monomer holding) and dormant (polymer holding) ionic species (Scheme 20).

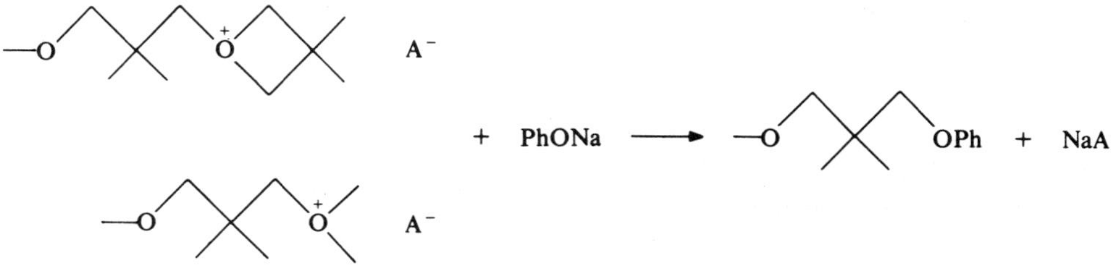

Scheme 20

48.5.2.3 Transfer and termination

Cationic polymerization of oxetanes proceeds without appreciable transfer, except for the chain transfer to polymers discussed above. Thus, high molecular weight polymers were prepared in a number of systems. For the polymerization of substituted oxiranes (BCMO, 3-methyl, 3-chloro-methyloxetane) with the R_3Al/H_2O initiating system, it has been shown that $\bar{M}_n$ increases almost linearly, with conversion reaching values close to 10^6.[64,65]

48.5.2.4 Industrial applications of the cationic oxetane polymerization

Poly(3,3-bis(chloromethyl)oxetane) [poly(BCMO)] has been one of the large scale production plastics based on cationic ring-opening polymerization.[70] Production was started in the USA in the 1950s (Penton by Herculus Co.) but was discontinued in the 1970s. This polymer has also been produced in the USSR (Pentaplast). Poly(BCMO) is a thermoplastic (m.p. = 181 °C), having mechanical properties close to Nylon 6, but showing outstanding thermal and chemical resistance (*e.g.* stable up to 120 °C in concentrated H_2SO_4). Thus, Penton has competed with some fluorinated polymers, especially in the field of construction of chemical equipment.[70]

48.6 FIVE-MEMBERED CYCLIC ETHERS — OXOLANES

48.6.1 Monomer

The unsubstituted five-membered cyclic ether, *i.e.* tetrahydrofuran (THF), occupies a special position among cationically polymerized cyclic ethers. This is the most thoroughly studied monomer of this group, and the kinetics and mechanism of THF polymerization have been understood at a level comparable to that of the living anionic polymerization of styrene.

Thus, this polymerization is to be discussed in the following sections in detail as an exemplary system. The kinetics and thermodynamics of THF polymerization have already been reviewed by other authors[3,4,17] and by ourselves.[1,2] Thus, we discuss here the newer data and some features overlooked in other reviews (*e.g.* kinetics of the ester–ion exchange or reactivity of the *endo* and *exo* groups in oxolanium cation).

On the laboratory scale THF is easily purified by distillation (b.p. = 66 °C) from lithium aluminum hydride, blue or violet sodium–benzophenone solution or sodium–potassium alloy. Special care should be taken to remove peroxides (formed upon storage) prior to distillation as distillation of THF containing peroxides may cause violent explosions.[71]

48.6.2 Polymerizability

General discussion on the thermodynamics of reversible polymerization is presented in Chapter 46 of this volume. In this section we will discuss only some specific features of the thermodynamics of THF polymerization; their understanding is indispensable for the proper understanding of the kinetics.

Ring strain of the five-membered cyclic ethers is considerably lower ($23\,kJ\,mol^{-1}$) than that of three- and four-membered rings. Consequently, the entropy term $-T\Delta S$ becomes equal to the enthalpy term (*i.e.* $\Delta G = 0$) at 80 °C for bulk polymerization ([THF]$_0$ = 12.8 mol l^{-1}).[4] Thus, polymerization of THF is highly reversible and its monomer equilibrium concentration is relatively high.

In ideal systems, the monomer equilibrium concentration [THF]$_e$ is given by equation (33), thus [THF]$_e$ should not depend on conditions, such as solvent properties or concentrations. In real systems however, the value of [THF]$_e$ strongly depends on the experimental conditions.[72,73] In Chapter 46 of this volume the reasons for these deviations are discussed and the methods allowing one to quantitatively treat real systems are presented.

$$RT \ln[M]_e = \Delta H_p - T\Delta S_p \tag{33}$$

Knowledge of the dependence of [THF]$_e$ on [THF]$_0$ and on solvent properties is important for both synthesis of poly(THF) and for kinetic studies. The use of a wrong value of [THF]$_e$ may lead to erroneous values of k_p, especially when polymerization is studied at [THF]$_0$ close to [THF]$_e$.

48.6.3 Mechanism of Polymerization

48.6.3.1 *Initiation*

Dozens of initiators were used for THF polymerization and more or less adequately tested. They were classified according to their chemical structure and reviewed.[2,3,17] The large majority of the reported systems have no or merely historical importance. There are however a few ones that are well characterized and give fast and quantitative initiation.

(i) Initiators giving quantitative fast initiation

It is desirable to have an initiator which reacts faster with THF than THF propagates, thus conforming to the inequality $k_i > k_p$, where k_i and k_p are the corresponding rate constants of initiation and propagation. It is also important that they initiate quantitatively, *i.e.* neither is the cation involved in side reactions, nor is the anion unstable. Apparently, only 1,3-dioxolenium salts with stable anions fit all of these requirements.[74-76] These salts can be prepared conveniently by H^- abstraction from 1,3-dioxolane (equation 34).[77]

$$Ph_3C^+A^- + \text{(1,3-dioxolane)} \longrightarrow Ph_3CH + \text{(1,3-dioxolenium)} A^- \tag{34}$$

Triphenylmethyl (trityl) salts with various anions are commercially available, reaction proceeds smoothly at room temperature and 1,3-dioxolenium salts can be obtained in high yield.[77] They are stable in solution and in the solid state if stored under vacuum, preferably in the dark.

Reaction of unsubstituted 1,3-dioxolenium salts with THF proceeds by direct addition,[78] in contrast to reaction of THF with trityl salts, involving H^- transfer[16] (equation 35). Due to the ambient reactivity of the dioxolenium salts, a transient kinetic product may precede the formation of the thermodynamic product shown in equation (35)[79] although this is irrelevant for further polymerization.

$$\text{(dioxolenium)} A^- + \text{(THF)} \longrightarrow \text{H—C(=O)—OCH}_2\text{CH}_2\text{—O}^+\text{(ring)} \; A^- \tag{35}$$

Initiation by dioxolenium salts, coupled with AsF_6^-, SbF_6^-, PF_6^-, BF_4^- and other anions, has been described by the present authors; thus, $k_i(AsF_6^-) = 1.0 \times 10^{-3}\,mol^{-1}\,s^{-1}$ at 25 °C with $MeNO_2$ as solvent.[78] This was measured by ^1H NMR, monitoring the disappearance of the acidic proton in the 1,3-dioxolenium salt (5.7 p.p.m. δ) and the appearance of the formate proton (8.1 p.p.m. δ), after the dioxolenium salt ring opening and alkylating of THF.

1,3-Dioxolenium salts with noncomplex anions like ClO_4^- and $CF_3OSO_2^-$ have also been prepared in the same way.[77,80] Special care is required when working with dry perchlorate, as we have observed that this is a shock-sensitive explosive compound. Yamashita prepared bis(dioxolenium)salt, used subsequently in preparation of the dicationically growing poly(THF) and ABA block copolymer thereof.[81]

Dioxolenium salts provide both fast and quantitative initiation. They are suitable to prepare high polymers but are not easily available. There are a number of other initiators, commercially available, providing high molecular weight poly(THF) but they initiate slowly and are not as eminently suited for the kinetic studies as dioxolenium salts. These include, first of all, strong protonic acids (superacids) and their derivatives (anhydrides or esters),[83-89] trialkyloxonium salts[90-93] and some oxycarbenium salts.[28] All of these should be coupled with stable anions (SbF_6^-, AsF_6^-, PF_6^-). There are several reports showing that initiation with the acids and organic cations listed above proceeds quantitatively by addition of the intiating species. With some carbenium salts, however (*e.g.* triphenylmethylium), hydride transfer reaction occurs[16] and today these initiators have only historical interest in the polymerization of THF.

Anhydrides or esters of superacids initiate THF polymerization by direct acylation or alkylation. The former ones (*e.g.* triflic anhydride) are acting as bifunctional initiators, as discussed in the following section.

The situation is more complex with superacids themselves. Due to the reaction between the head HO— group and active species (ionic or covalent), chain coupling and/or end-to-end cyclization occurs, leading to polymers with $\bar{M}_n$ higher than those expected from the $\{[THF]_0-[THF]_e\}/[HA]$ ratio (Scheme 21).[94]

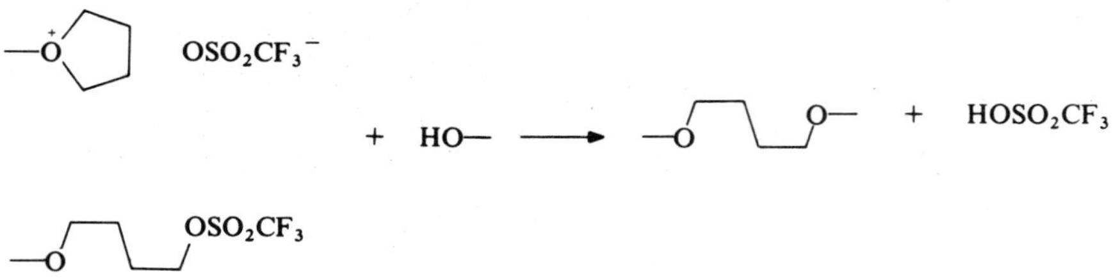

Scheme 21

When fluorosulfonic acid was used as initiator, formation of HF in concentrations approaching that of fluorosulfonic acid itself was observed by ^{19}F NMR.[95] This was attributed to the chain coupling, proceeding as in Scheme 22.

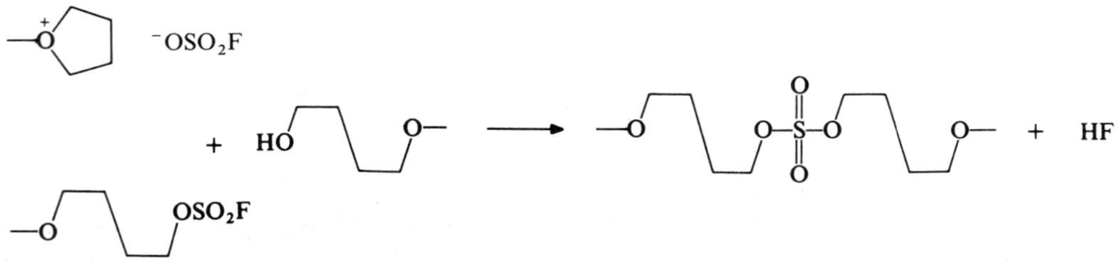

Scheme 22

Thus, although superacids do give high molecular weight polymers, the mechanism of polymerization is not as straightforward as in the case of other initiators described earlier. We discovered recently that this difficulty can be avoided by using heteropolyacids. Their acid strength is even higher than that of superacids and the anions do not form covalent bonds.[96]

(ii) Multifunctional initiators

Some initiators may yield macromolecules growing in two or more directions. The most commonly used bifunctional initiators is triflic anhydride (*cf.* Scheme 23).[82] The interesting feature of this system is that finally no initiator fragment is incorporated into the polymer chain.

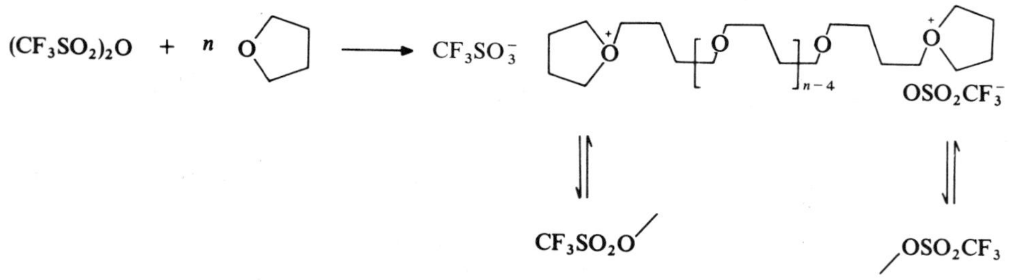

Scheme 23

This easily available initiator leads to macromolecules terminated at both ends with ionic and/or covalent active species. They may further be converted into a variety of functional end groups by means of suitable terminating agents.

Other bifunctional initiators used for THF polymerization include the already mentioned bis(dioxolenium) salts (**5**),[81] silicenium salts (**6**; which is a schematic representation only as the existence of silicenium cation has not been proven),[97] or oxycarbenium salts.[28] This latter group is

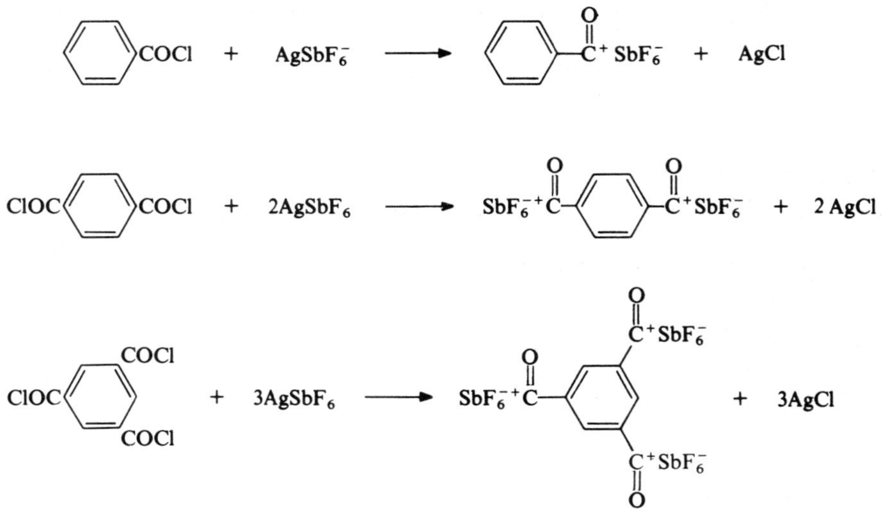

$$\text{C}_6\text{H}_5\text{COCl} + \text{AgSbF}_6^- \longrightarrow \text{C}_6\text{H}_5\overset{O}{\underset{}{\text{C}}}{}^+ \text{SbF}_6^- + \text{AgCl}$$

$$\text{ClOC-C}_6\text{H}_4\text{-COCl} + 2\text{AgSbF}_6 \longrightarrow \text{SbF}_6^-\,{}^+\text{C}(\text{O})\text{-C}_6\text{H}_4\text{-C}(\text{O}){}^+\text{SbF}_6^- + 2\,\text{AgCl}$$

$$\text{ClOC-, COCl, COCl (C}_6\text{H}_3) + 3\text{AgSbF}_6 \longrightarrow \text{trifunctional acylium} + 3\,\text{AgCl}$$

Scheme 24

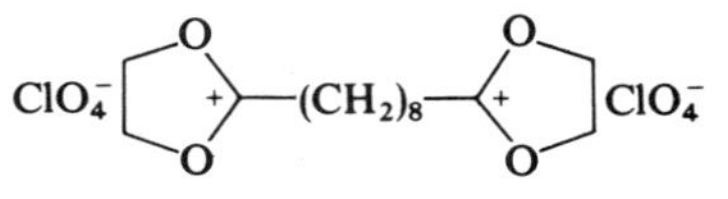

(5) ClO_4^- ... (CH$_2$)$_8$ ... ClO_4^- (bis-dioxolanylium)

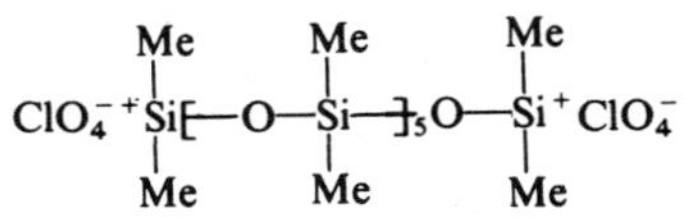

(6) $ClO_4^-\,{}^+\text{Si(Me)}_2\text{-[O-Si(Me)}_2\text{]}_5\text{-O-Si(Me)}_2{}^+\,ClO_4^-$

especially useful, because, depending on the nature of the precursor, mono-, di- or tri-functional initiating systems may be obtained (Scheme 24).

As shown in our work with Franta,[28] each of the $-\overset{+}{\text{C}}{=}\text{O}$ species initiated the polymerization, thus three-armed stars were readily prepared from a trifunctional initiator. Some typical data are shown in Table 8.

Table 8 The $\overline{DP}_n$ Values of Poly(THF) Prepared with Mono-, Di- and Tri-functional Oxycarbenium Ions

Initiator	$\overline{DP}_n$ calculated	$\overline{DP}_n$ found UV	IR	Br analysis[a]	Osmometry
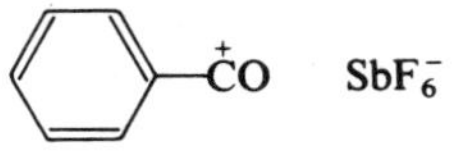 $C_6H_5\overset{+}{C}O \; SbF_6^-$	175	170	170	150	185
	370	360	355	330	365
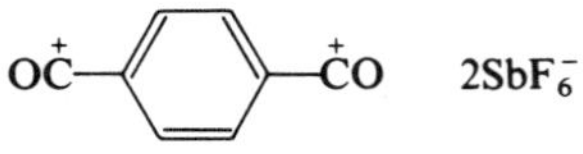 $O\overset{+}{C}\text{-}C_6H_4\text{-}\overset{+}{C}O \; 2SbF_6^-$	255	230		266	245
	355	319		331	340
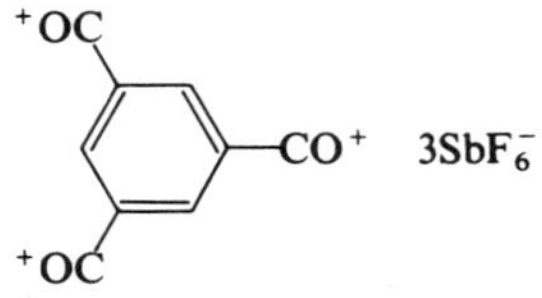 ${}^+OC\text{-}C_6H_3(\text{-}CO^+)(\text{-}OC^+) \; 3SbF_6^-$	366			348	340
	660			625	630

[a] Polymerizations were terminated with LiBr.

(iii) Other initiation mechanisms

Until now we have only discussed the mechanisms of initiation of THF polymerization that involve direct addition of the initiator moiety to the monomer molecule. With some initiators the mechanism of initiation is more complex.

Triphenylmethylium (trityl) salts cause H⁻ abstraction from a THF molecule (equation 36).[15,16]

$$Ph_3C^+A^- \quad + \quad \text{(THF ring with O)} \quad \longrightarrow \quad Ph_3CH \quad + \quad \text{(cyclic oxonium ion)} \quad A^- \tag{36}$$

Further reaction may involve proton expulsion, with the protonic acid thus formed acting as the actual initiator.

Thus, initiation with $Ph_3C^+A^-$ salts leads to side reactions and poorly defined products. These initiators lost their original importance when dioxolenium salts and superacids were discovered to be efficient initiators.

A route through zwitterion formation is also possible in the polymerization of THF. It was shown, by using ^{31}P NMR, that the sequence of reactions in Scheme 25 operates for the THF–PF_5 system.[24]

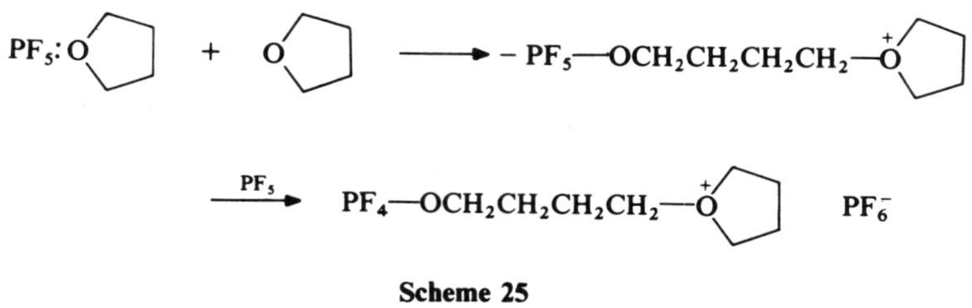

$$PF_5\!:\!O\text{(ring)} \quad + \quad O\text{(ring)} \quad \longrightarrow \quad -PF_5-OCH_2CH_2CH_2CH_2-\overset{+}{O}\text{(ring)}$$

$$\xrightarrow{PF_5} \quad PF_4-OCH_2CH_2CH_2CH_2-\overset{+}{O}\text{(ring)} \quad PF_6^-$$

Scheme 25

These examples are shown to make the picture complete. There are however such a large number of initiators available, which give clean and quantitative (although sometimes relatively slow) initiation, that initiators leading to complex or nonstoichiometric initiation should simply be avoided. Nevertheless a cheap, efficient initiator for industrial production of THF α,ω-ended telechelic is still needed. The presently used solutions are not without deficiences: FSO_3H or H_2SO_4 have to be used in high concentrations in order to secure a sufficient rate for the process, which is lowered by formation of macroesters of less reactivity than macroions.

(iv) Rate constants of initiation

The rate constants of initiation that have been measured up until now are summarized in Table 9 and taken from ref. 1.

Table 9 Rate Constants of Initiation in the Polymerization of THF

Initiator	Solvent	Temperature (°C)	$k_i \times 10^3$ $(mol^{-1}1s^{-1})$	Method	Ref.
$Et_3O^+BF_4^-$	CH_2Cl_2	25	0.7	^1H NMR	98
$MeOSO_2CF_3$	CH_2Cl_2	25	0.4	^1H NMR	99
$EtOSO_2CF_3$	CH_2Cl_2	25	0.17	^1H NMR	99
$EtOSO_2CF_3$	CCl_4	25	0.06	^1H NMR	75
$EtOSO_2CF_3$	CCl_4	25	0.04	^{19}F NMR	83
$EtOSO_2CF_3$	$MeNO_2$	25	0.6	^1H NMR	99
$EtOSO_2CF_3$	CCl_4	25	0.02	^{19}F NMR	100
PF_5	CH_2Cl_2	25	0.005	^{19}F NMR	101
$DXL^+AsF_6^-$ [a]	$MeNO_2$	25	1.0	^1H NMR	78
$Et_3O^+Br_4^-$	$(CH_2Cl)_2$	0	0.25	^{14}C NMR	92

[a] 1,3-Dioxolenium ion

There are a number of relatively simple but unanswered questions concerning initiation of THF polymerization. For instance, the rate constants are not known for either the reaction of THF with the secondary oxonium ion (initiation with protonic acids) or for the initiation *via* acyltetrahydrofuranium cation (Scheme 26).

From kinetic plots of polymerization it follows that both k_{i2} and k'_{i2} are smaller than k_p but actual values have not yet been determined.

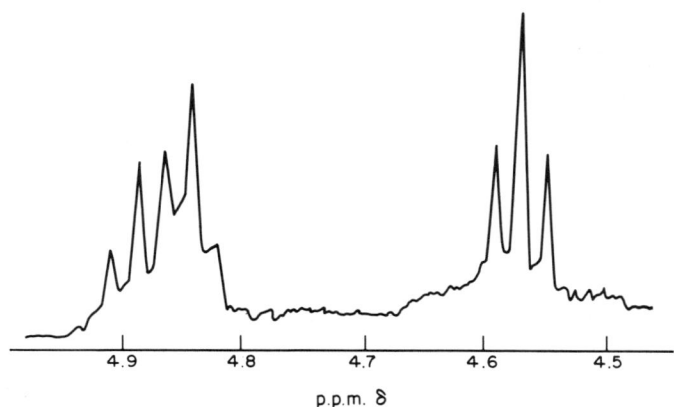

Scheme 26

48.6.3.2 Propagation

(i) Structure of ionic active species

Tertiary oxonium ions involving a THF molecule result from initiation as discussed in the previous section. These ions have been observed directly by ^1H and ^{13}C NMR.[74,85] The fine structure of the spectrum, particularly at high resolution, is compatible only with structure (7), which is taken from our work.[74]

$$-OCH_2CH_2CH_2CH_2-\overset{+}{O}\langle\ \rangle$$

(7)

In Figure 3 ^1H NMR (300 MHz) of the region of active species in the polymerization of THF at $-18\,°C$ with $CF_3SO_3^-$ anion in CH_2Cl_2 solvent ($[THF]_0 = 8.0\,mol\,l^{-1}$, $[I]_0 = 10^{-1}\,mol\,l^{-1}$) is shown.[74]

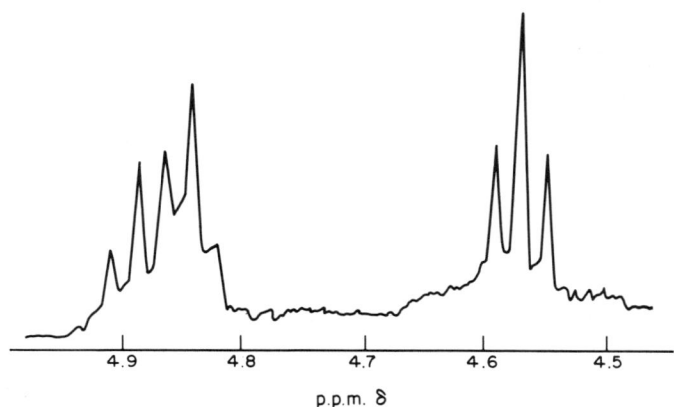

| 4.9 | 4.8 | 4.7 | 4.6 | 4.5 |

p.p.m. δ

Figure 3 ^1H NMR spectrum of the region of active species in the polymerization of tetrahydrofuran with $CF_3SO_3^-$ anion $\{[THF]_0 = 8.0\,mol\,l^{-1};\ [I]_0 = 10^{-1}\,mol\,l^{-1};$ solvent CH_2Cl_2, at $25\,°C\}$

Although the concentration of the growing species is almost 10^2 times lower than that of the monomer used, the resolution is good enough to observe both *endo-* and *exo*-cyclic methylene groups separately. Both give slightly distorted triplets, due to the splitting by two different pairs of protons (*endo-* and *exo*-cyclic ones). Their peak area ratio is close to 2:1, as required.

The chemical shifts of the *exo*-cyclic protons has moved downfield in comparison with the *endo*-cyclic ones. The same phenomenon has recently been observed in our laboratory for the dioxanium cation.[41]

The actual chemical shifts of the *endo-* and *exo*-cyclic protons (4.85 and 4.89 p.p.m. δ respectively in Figure 3) depend slightly on the solvent used, the anion and concentration. This latter dependence can be rationalized by a difference in chemical shifts of the paired and unpaired ions, although no data are available for the chemical shifts of these two forms of tetrahydrofuranium cations. It was shown in our laboratory, however, that for 1,3-dioxolenium ions, the chemical shifts of the unpaired ions has moved downfield in comparison with ion pairs.[102]

Endo- and *exo*-cyclic carbon atoms in methylene groups absorb in ^{13}C NMR at 87.5 and 89.6 p.p.m. δ respectively.

The concentration of the tetrahydrofuranium ions, determined from ^{1}H NMR spectra, was close to the concentration of initiator used. More precise determination of the concentration of the growing species became possible after end capping[33,34] and ion trapping[35,36] were elaborated.

This information on structure and concentration is not sufficient to maintain that these are the tertiary oxonium ions that are involved in the rate-determining step of propagation. Indeed, two kinetically distinguishable steps (shown as (a) and (b) in Scheme 27) are possible.

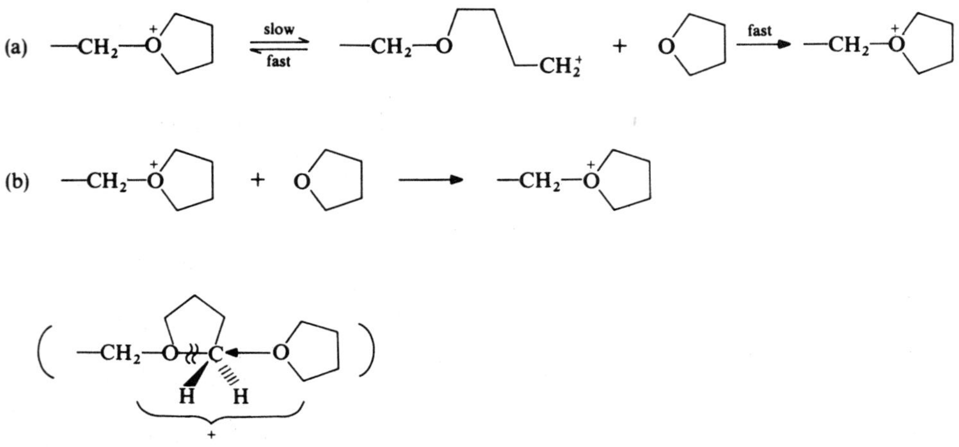

Scheme 27

The first path, resembling the S_N1 process, requires the transient presence of the primary carbenium ions. These are known to quickly isomerize by internal hydride shift to the secondary and tertiary ones (equation 37).

$$-CH_2CH_2CH_2^+ \longrightarrow -CH_2\overset{Me}{\underset{}{C}}H^+ \longrightarrow -CH_2-\overset{Me}{\underset{Me}{C}}^+ \tag{37}$$

In the cationic polymerization of unsaturated compounds this isomerization was observed in many systems and reviewed by Kennedy.[103] In THF polymerization only the tetramethylenoxy units in the main chain were detected, even if the living system was kept for a long time at the monomer–polymer equilibrium. Mostly on this basis the S_N1 path was dismissed and the second path, similar to S_N2 solvolysis, has been accepted by almost all of the researchers in the field.

(ii) Covalent active species

Smith and Hubin initiated polymerization of THF with triflic acid and assumed that covalent active species coexist with tertiary onium ions.[82] The first kinetic evidence of this phenomenon was given in the paper from our group[104] and later it was confirmed by ^{1}H NMR that macroesters do exist in these systems.[74]

Covalent active species were also directly observed by Saegusa *et al.* using ^{19}F NMR[83] and Pruckmayr *et al.* using ^{13}C NMR.[85]

^{1}H NMR clearly showed the existence of two distinct species with chemical shifts identical with model compounds, the corresponding ions and esters. These spectra, taken in our group,[74] have been reproduced several times in the review papers by ourselves[1] and by others.[3,17] The upfield signal in the spectrum shown in Figure 3 corresponds to the covalent species with the chemical shift being equal to: 4.56 p.p.m. δ (^{1}H NMR)[14], 38.0 p.p.m. δ(^{19}F NMR)[83] and 79.1 p.p.m. δ(^{13}C NMR).[85]

(iii) Equilibria between covalent and ionic active species

In the preceding sections it has been shown that, in the presence of counterions able to form covalent bonds, ionic and covalent active species coexist in equilibrium (equation 38).

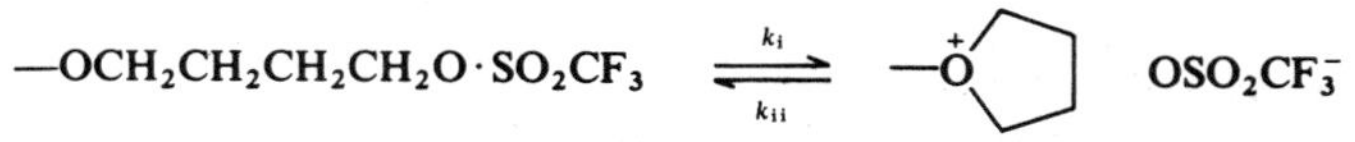

$$—OCH_2CH_2CH_2CH_2O\cdot SO_2CF_3 \quad \underset{k_{ii}}{\overset{k_i}{\rightleftharpoons}} \quad —\overset{+}{O}\hspace{-0.3em}\bigcirc \quad OSO_2CF_3^- \qquad (38)$$

Ionization of covalent species may proceed either within a macromolecule, as an intramolecular reaction involving the ultimate unit of the chain, or as an intermolecular process, involving a monomer molecule; both are shown schematically in Scheme 28. These two reactions responsible for ionization can simultaneously operate in the same system and their contributions to the chain growth depend on the rates of the internal and external ionization as well as on the rates of the back reactions.

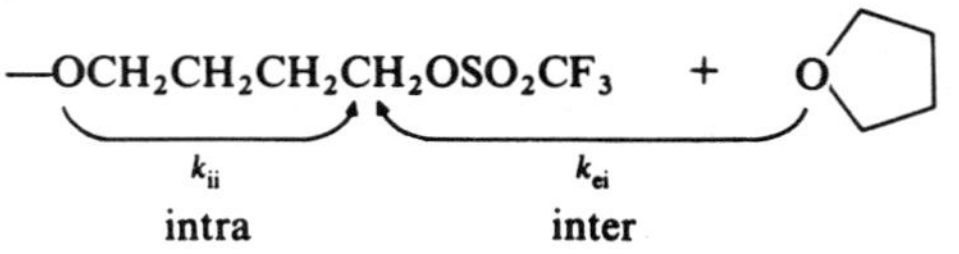

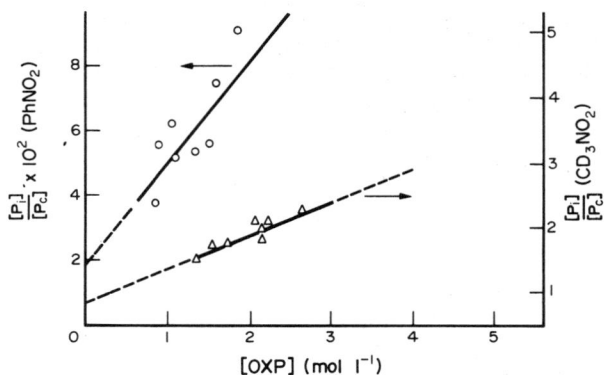

Scheme 28

The internal ionization is a unimolecular process whereas the external ionization is a bimolecular reaction with a rate depending on the monomer concentration. Therefore, if the proportion of ions does not change during the polymerization this is strong evidence that ionization proceeds mostly by the unimolecular process. On the other hand, the decrease of the proportion of ions with monomer conversion indicates the preponderance of the external ionization.[105]

It has been shown, that in the polymerization of THF, the concentration of macroions does not depend on conversion (*i.e.* on the monomer concentration).[74] Thus, in the polymerization of THF, ionization proceeds mostly as a unimolecular reaction (internal ionization).

This may be related to the low ring strain, facilitating the closure of the five-membered tetrahydrofuranium ion as required for the internal ionization. In the polymerization of a more strained seven-membered cyclic ether, oxepane, the contribution of the external ionization route is significant and as soon as the monomer concentration is equal to $1.7\,\mathrm{mol\,l^{-1}}$ (MeNO$_2$, 25 °C), this becomes a dominating route of ionization. In Figure 4 it is shown that with conversion the proportion of ions among the growing species decreases.[105,106]

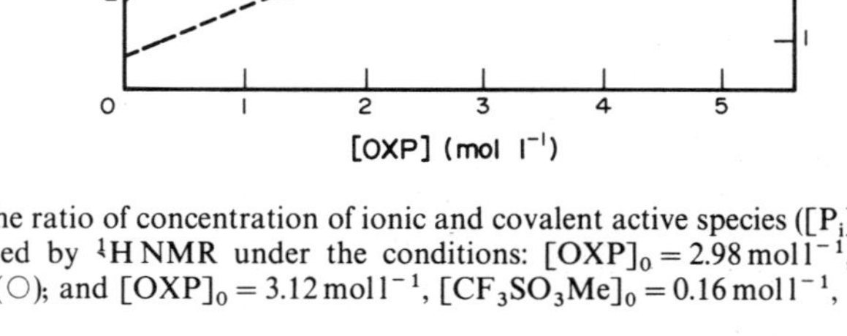

Figure 4 Dependence of the ratio of concentration of ionic and covalent active species ($[P_i]/[P_c]$) in the polymerization of oxepane (OXP), as observed by ^{1}H NMR under the conditions: $[OXP]_0 = 2.98\,\mathrm{mol\,l^{-1}}$, $[CF_3SO_3Me]_0 = 0.54\,\mathrm{mol\,l^{-1}}$, PhNO$_2$, 22 °C (○); and $[OXP]_0 = 3.12\,\mathrm{mol\,l^{-1}}$, $[CF_3SO_3Me]_0 = 0.16\,\mathrm{mol\,l^{-1}}$, CD$_3NO_2$, 22 °C (△)

The ratio k_{ii}/k_{ei} gives directly the so called 'effective monomer concentration', *i.e.* concentration of monomer at which rates of both internal and external ionizations would become equal to each other (Scheme 29).

Typical values of the ionization constant $K_e = k_{ii}/k_{tt}$ for the polymerization of THF are given in Table 10.

The rates of interconversion of covalent macromolecules and macroion pairs have been measured for the cationic polymerization of THF by using the 'temperature jump' method.[88] The system was first brought to equilibrium at the given temperature and then the equilibrium was disrupted by

THF

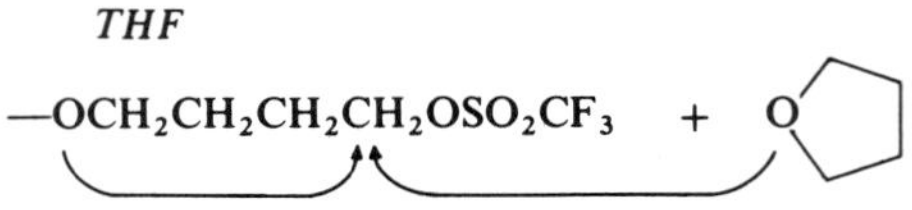

$$-OCH_2CH_2CH_2CH_2OSO_2CF_3 \quad + \quad O$$

$$k_{ii} = 1.9 \times 10^{-2}\,s^{-1}, \quad k_{ei} = 2 \times 10^{-4}\,mol^{-1}\,1s^{-1}, \quad k_{ii}/k_{ei} = 100\,mol\ 1^{-1} \qquad (CH_2Cl_2,\ 25\,°C)$$

Oxepane

$$-OCH_2CH_2CH_2CH_2CH_2CH_2OSO_2CF_3 \quad + \quad O \begin{matrix} CH_2CH_2CH_2 \\ | \\ CH_2CH_2CH_2 \end{matrix}$$

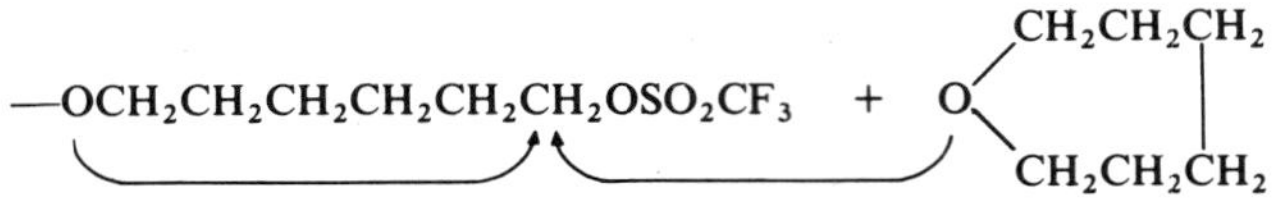

$$k_{ii} = 2.3 \times 10^{-4}\,s^{-1}, \qquad k_{ei} = 1.35 \times 10^{-4}\,mol^{-1}\,1s^{-1}, \qquad k_{ii}/k_{ei} = 1.7\,mol\ 1^{-1} \qquad (MeNO_2,\ 25\,°C)$$

Scheme 29

lowering the temperature and the rate of reestablishing the new equilibrium was measured. The change of ^1H NMR spectrum, accompanying this interconversion, is shown in Figure 5.[88]

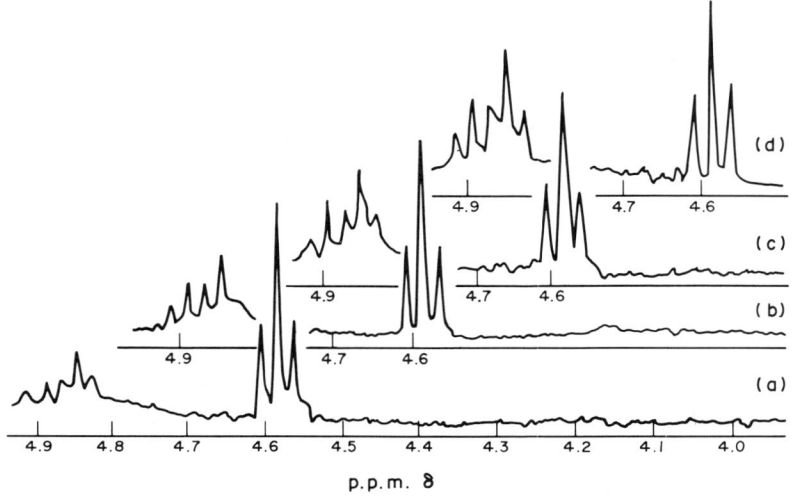

Figure 5 Dependence of ^1H NMR spectra of living ends on time in the polymerization of tetrahydrofuran (THF) with $CF_3SO_3^-$ anion at $-18\,°C$ after disrupting an equilibrium established at $+18\,°C$, with measurements at the following times ($\times\ 10^2$ s): (a) 1.5, (b) 5.7, (c) 13.2, (d) 22.8 (solvent CCl_4, $[THF]_0 = 8.0\,mol\,1^{-1}$, $[CF_3SO_3Me]_0 = 8 \times 10^{-2}\,mol\,1^{-1}$)

Treating the studied interconversion as the first order opposed reaction we have equations (39a) and (39b)

$$\ln\left(\frac{[E]_0 - [E]_e}{[E]_t - [E]_e}\right)\Bigg/\frac{1 + K_e}{K_e} = k_{ii}t \tag{39a}$$

and

$$k_{tt} = \frac{K_e}{k_{ii}} \tag{39b}$$

The values of rate constants k_{ii} and k_{tt} are also given in Table 10.

(iv) Rate constants of propagation of covalent and ionic active species

In the polymerization of THF it is possible to have conditions when unpaired macroions are excluded and the only active species participating in propagation are macroion pairs and covalent macroesters. For such a system, the apparent rate constant of propagation is given by dependence in equation (40), where: k_p^{app} = experimental apparent rate constant; α = degree of ionization; and k_{pi} and k_{pc} = rate constants of propagation of ionic and covalent active species respectively.

$$k_p^{app} = \alpha k_{pi} + (1 - \alpha)k_{pc} \tag{40}$$

Table 10 Equilibrium and Rate Constants of the Macroion–Macroester Inter-conversion in the Polymerization of THF at 25 °C

Anion	Solvent	Dielectric constant of the monomer–solvent mixture	K_e	k_{ii} (s^{-1})	k_{tt} (s^{-1})
CF$_3$SO$_3^-$	CCl$_4$	5.9	0.06	0.8	12.1
	CH$_2$Cl$_2$	8.2	0.58	1.9	3.3
	MeNO$_2$	20.6	42.0		
FSO$_3^-$	CH$_2$Cl$_2$	8.2	0.27	1.6	5.9
	MeNO$_2$	20.6	20.0		
ClO$_4^-$	CH$_2$Cl$_2$	8.2	0.09		
	MeNO$_2$	20.6	0.5		

Rearranging equation (40) one gets equation (41).

$$k_p^{app}/\alpha \;=\; k_{pi} \;+\; \frac{1 - \alpha}{\alpha} k_{pc} \tag{41}$$

The plots k_p^{app}/α *vs.* $(1-\alpha)/\alpha$ give k_{pi} from the intercept and k_{pc} from the slope. In Table 11 the values of propagation rate constants on ionic and covalent active species for THF propagation are compared with the corresponding values for the polymerization of oxepane (OX), the seven-membered cyclic ether.[105]

Table 11 Rate Constants of Propagation on Ionic and Covalent Active Species in the Polymerization of THF and Oxepane (OX)

Monomer	Anion	Solvent	Temperature (°C)	k_p^i (mol^{-1} l s^{-1})	k_p^c (mol^{-1} l s^{-1})
THF	SO$_3$CF$_3^-$	MeNO$_2$	25	2.4×10^{-2}	5×10^{-4}
OX	SO$_3$CF$_3^-$	MeNO$_2$	25	1.3×10^{-4}	3×10^{-4}

As it follows from the data shown in Table 11, for polymerization of THF $k_{pi} \gg k_{pc}$, whereas for polymerization of oxepane $k_{pi} \approx k_{pc}$. Let us analyze the reasons of this difference in more detail.

The reactivities of covalent species in the homopolymerization of THF and OX are nearly the same. This shows that the involved electronic and steric factors are either similar or irrelevant (equation 42).

$$\tag{42}$$

Ionic propagations, in contrast to the covalent process, differ remarkably in their respective rate constants; the homopropagation of THF is approximately 10^2 times faster than that of OX. The difference in k_{pi} should be connected with the different steric hindrances exerted by the two different macroions. This is best visualized by comparing the Newman projections of both cations (**8a** and **8b**), remembering that attack of the approaching monomer molecule should proceed along the C—O$^+$ bond.

(8a)   (8b)

The above discussion shows that if the steric hindrance does not decrease the reactivity of ions, the covalent active species are much less reactive than the ionic ones.

(v) Macroions and macroion pairs

Relatively high stability of oxonium ions allows the dissociation constant K_D of ion pairs to be directly measured. This in turn leads to the knowledge of the degree of dissociation (see equation 43).

$$\alpha = \frac{[\text{ions}]}{[\text{ion pairs}] + [\text{ions}]} \qquad K_D = \frac{[\text{ions}]^2}{[\text{ion pairs}]} \qquad (43)$$

The values of K_D for dissociation of tetrahydrofuranium ion pairs were measured in our laboratory. Typical values of K_D are as follows:[76]

$$K_D = 2 \times 10^{-3}\,\text{mol}\,l^{-1}\ (25\,^\circ\text{C}; [\text{THF}]_0 = 7.0\,\text{mol}\,l^{-1}\ \text{in MeNO}_2;\ \text{counterion is SbF}_6^-, \text{AsF}_6^-)$$

$$K_D = 1.5 \times 10^{-5}\,\text{mol}\,l^{-1}\ (25\,^\circ\text{C}; [\text{THF}]_0 \doteq 7.0\,\text{mol}\,l^{-1}\ \text{in CH}_2\text{Cl}_2;\ \text{counterion is SbF}_6^-, \text{AsF}_6^-)$$

These values indicate, that in the range of concentrations of active species typically used in kinetic studies (10^{-2}–$10^{-4}\,\text{mol}\,l^{-1}$), the degree of dissociation may vary within a relatively wide range, depending on the solvent used, namely from a few percent of unpaired ions (in CH_2Cl_2) to over 80% of unpaired ions at higher dilution in $MeNO_2$ solvent (the rest being ion pairs).

To determine the rate constants of propagation for macroions and macroion pairs, the apparent rate constant of propagation at different degrees of dissociation, α, were measured dilatometrically and the results were plotted according to equation (44).[76]

$$k_p^{\text{app}}/\alpha = (k_p^+ - k_p^\pm) + k_p^\pm/\alpha \qquad (44)$$

The plot of k_p^{app}/α vs. $1/\alpha$ is shown in Figure 6. From this plot, $k_p^\pm$ was determined as the slope and k_p^+ from the intercept.

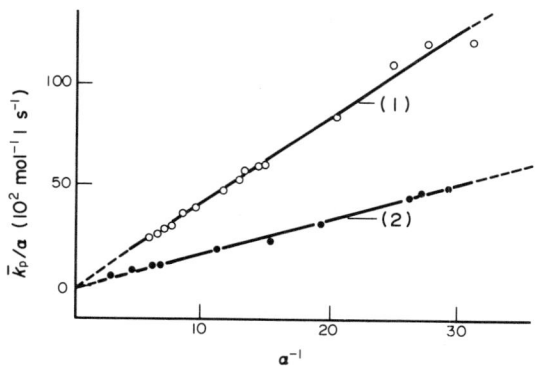

Figure 6   Dependence of the ratio k_p^{app}/α on $1/\alpha$ in the polymerization of tetrahydrofuran at (1) 25 °C; and (2) 10 °C (solvent is CH_2Cl_2)

The measured values k_p^+ and $k_p^\pm$ were found to be equal to each other within experimental error, thus it was concluded that in the polymerization of THF the reactivities of ions and ion pairs are the same. This finding seems to be a fundamental one for all of the cationic polymerizations of heterocycles.[105]

This is in marked contrast to the anionic polymerization of vinyl monomers, where, except with highly solvated ion pairs, a large difference in reactivities which favoured ions was found.[25] It should be remembered, however, that in the cationic polymerization of cyclic ethers, monomers themselves strongly solvate the growing cationic active species. Thus, in the polymerization of THF, ions are 'free' in the electrochemical sense but not free in the molecular understanding.

It has also recently been observed in our laboratory that in anionic polymerization conducted in highly solvating media, particularly when solvation involves the actual growing end (and not merely counterions, as in a majority of vinyl polymerization), the reactivity of ions and ion pairs may be equal to each other.[107]

48.6.3.3 *Transfer and termination*

As discussed already in Section 48.3.3, the distinction should be made between transfer and/or termination processes being either inherent features of the system (thus, undesirable) or arranged purposely by adding chain transfer or termination agents. This provides the required end groups in the synthesis of telechelics.

Reactions belonging to the former group can easily be avoided in the polymerization of THF, providing that the concentration of impurities (*e.g.* water) is low and stable counterions are used. For a number of systems the relationship in equation (45) holds up to high $\overline{DP}_n$ values and polymers with $\overline{M}_n$ up to 10^6 can be prepared.

$$\overline{DP}_n = \frac{[THF]_0 - [THF]_e}{[I]_0} \tag{45}$$

Transfer and/or termination reactions are purposely induced in order to fit the polymer with desired end groups. This may serve both analytical (determination of active species concentration by phenoxy end capping[33, 34] or phosphine ion trapping[35, 36]) and synthetic (preparation of telechelic oligomers) purposes. Some examples of the end groups introduced intentionally into the poly(THF) chain are listed in Scheme 30.[17]

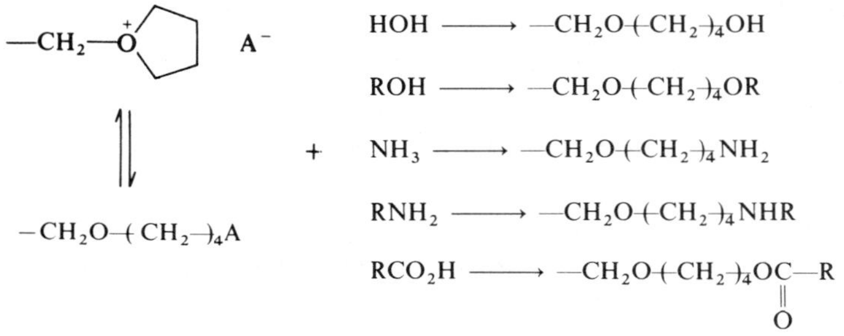

$$-CH_2-\overset{+}{O} \begin{pmatrix} \ \end{pmatrix} A^-$$

$$\Updownarrow$$

$$-CH_2O-(CH_2)_4A$$

$$+ \quad \begin{array}{ll} HOH \longrightarrow & -CH_2O-(CH_2)_4OH \\[6pt] ROH \longrightarrow & -CH_2O-(CH_2)_4OR \\[6pt] NH_3 \longrightarrow & -CH_2O-(CH_2)_4NH_2 \\[6pt] RNH_2 \longrightarrow & -CH_2O-(CH_2)_4NHR \\[6pt] RCO_2H \longrightarrow & -CH_2O-(CH_2)_4OC-R \\ & \qquad\qquad\qquad\quad \underset{O}{\overset{\parallel}{}} \end{array}$$

Scheme 30

Anionic living polymers can also be used to terminate living poly(THF) chains. This approach, leading to block copolymers, has been pioneered by Richards.[108] Reactions shown schematically in Scheme 31 proceed quantitatively.

$$Na^{+-}[St]_n^- Na^+ + 2[THF]_m^+ PF_6^- \longrightarrow [THF]_m [St]_n [THF]_m + 2NaPF_6$$

and

$$PF_6^- {}^+[THF]_n^+ PF_6^- + 2[St]_n^- Li^+ \longrightarrow [St]_m [THF]_n [St]_m + 2LiPF_6$$

Scheme 31

The quantitative transformations of living linear poly(THF) chains are possible, because in contrast to the polymerization of *e.g.* oxiranes, chain transfer to polymer does not interfere with propagation. The reasons have already been discussed in this section. Propagation is fast compared to chain transfer to the polymer, and within the time period necessary to obtain complete (equilibrium) conversion of the monomer no significant chain transfer to the polymer is observed. However, when such living systems are kept for a prolonged period of time, cyclic oligomers start to appear, indicating the occurrence of intramolecular transfer.[94] The rate of formation of cyclic oligomers is low, except for systems initiated with superacids due to the faster end-to-end cyclization to the —OH head group[95] (*cf.* Chapter 47 on Cyclization).

48.6.4 Reactive Oligomers of Tetrahydrofuran

The living character of THF polymerization allowed the reactive oligomers (macromonomers and telechelics) to be developed. Macromonomers were prepared by introducing polymerizable groups both as a head group[109] or as a tail group,[110] according to Scheme 32.

Telechelics are prepared either by termination of difunctional living poly(THF) with a suitable terminating agent, as discussed in the previous paragraph, or by conducting polymerization in the presence of a chain transfer agent, for example acetic anhydride (equation 46). Mechanism of this process has been elaborated by Rozenberg[111] and Heitz.[112]

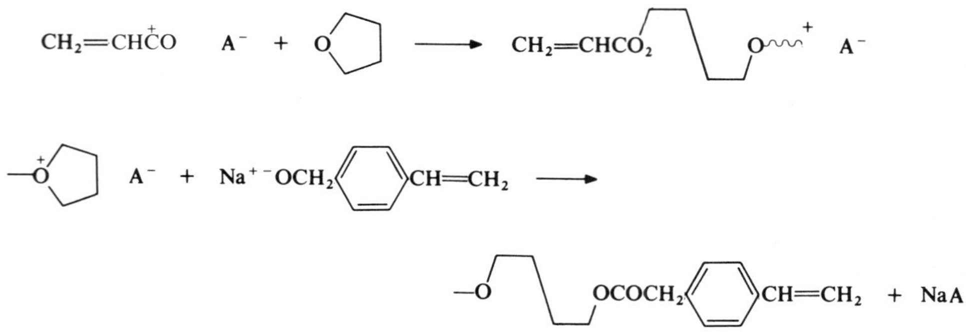

CH$_2$=CHĊO　A$^-$　+　[oxolane]　⟶　CH$_2$=CHCO$_2$[chain]O⁺ⁿⁿⁿ A$^-$

—O⁺[oxolane]　A$^-$　+　Na⁺⁻OCH$_2$[phenyl]CH=CH$_2$　⟶

—O[chain]OCOCH$_2$[phenyl]CH=CH$_2$　+　NaA

Scheme 32

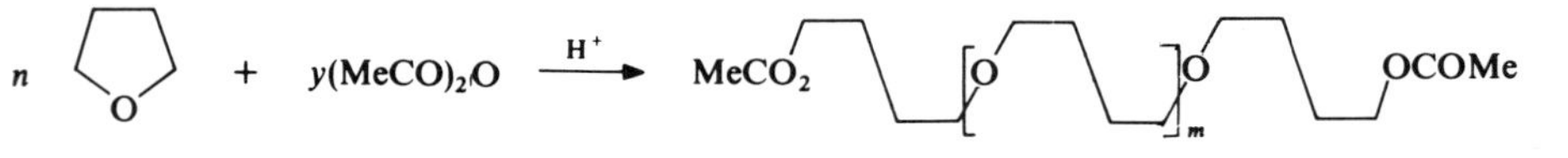

n [oxolane] + y(MeCO)$_2$O　$\xrightarrow{H^+}$　MeCO$_2$[chain][O chain]$_m$[O chain]OCOMe　　　(46)

Dihydroxy-terminated THF oligomers are used on an industrial scale as soft blocks in block copolymers of polyurethanes and poly(ethero)esters.[17] Some information on the industrial processes for making these oligodiols is given in the following section.

48.6.5　Industrial Applications of THF Polymerizations

Poly(THF) diols are available in commercial quantities as Teracol (Du Pont) or Polymeg (Quaker Oats) in a molecular weight range from 650 to 3000.

These products are prepared by protonic acid initiated polymerization of THF, followed by hydrolysis. Superacids,[113] Nafion resins[114] or heteropolyacids[115] may be used as initiators.

This process has several disadvantages; the required concentration of acid is high and its neutralization is costly. The yields are limited due to the equilibrium character of THF polymerization, enhanced at higher temperatures.

48.6.6　Substituted Oxolanes

Substitution generally decreases the thermodynamic polymerizability. Thus substituted five-membered cyclic ethers polymerize very reluctantly and give only low molecular weight products with poor yields.[4] If, however, substituents are linked, forming additional rings, the excess strain imposed on the five-membered oxolane ring may considerably facilitate the polymerization. This effect may be illustrated (in Scheme 33) by the behaviour of *cis*- and *trans*-8-oxabicyclo-[4.3.0]nonane.

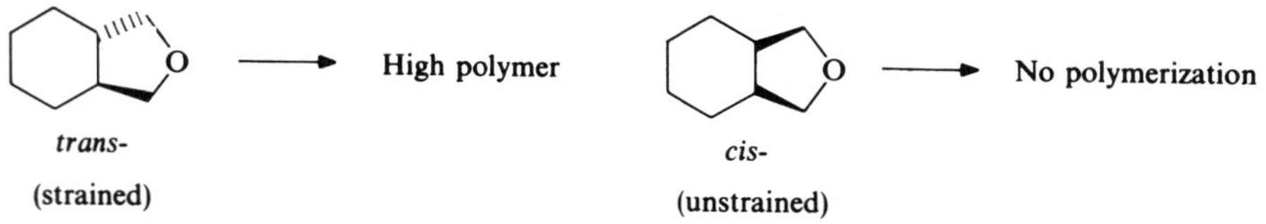

[structure] ⟶ High polymer　　　　[structure] ⟶ No polymerization

trans-　　　　　　　　　　　　　　*cis-*

(strained)　　　　　　　　　　　　(unstrained)

Scheme 33

The *trans* monomer, in which a five-membered ring is pressed into strained, twisted conformation, was polymerized to high molecular weight polymer, while almost unstrained *cis* monomer failed to polymerize. Polymerizability of various bicyclic ethers was studied by Saegusa[116] and by Kops,[117] whilst the exhaustive list of monomers of this structure, polymerized till now, is given in ref. 17.

48.7 HIGHER CYCLIC ETHERS

Six-membered cyclic ethers do not homopolymerize; their rings are not strained and consequently ΔG of polymerization is positive at any condition. The seven-membered cyclic ether, oxepane, is more strained than THF (*cf.* Section 48.6.3.2.iii) and polymerizes to a high molecular weight polymer.[106] Kinetic and mechanistic aspects of its polymerization were discussed in Section 48.6.3.2.v in relation to THF polymerization.

There have been no systematic studies on the polymerization of cyclic ethers with larger rings.

48.8 REFERENCES

1. S. Penczek, P. Kubisa and K. Matyjaszewski, *Adv. Polym. Sci.*, 1980, **37**, 1.
2. S. Penczek, P. Kubisa and K. Matyjaszewski, *Adv. Polym. Sci.*, 1985, **68/69**, 1.
3. S. Inoue and T. Aida, in 'Ring-opening Polymerization', ed. K. J. Ivin and T. Saegusa, Elsevier Applied Science, New York, 1984, vol. 1, chap. 5, p. 185.
4. K. J. Ivin and T. Saegusa, in 'Ring-opening Polymerization', ed. K. J. Ivin and T. Saegusa, Elsevier Applied Science, New York, 1984, vol. 1, chap. 1, p. 1.
5. A. S. Pell and G. Pilcher, *Trans. Faraday Soc.*, 1965, **61**, 71.
6. I. Andruzzi, G. Pilcher, Y. Virmani and P. H. Plesch, *Makromol. Chem.*, 1977, **178**, 2367.
7. C. Romers, C. Altona, H. R. Buys and E. Haringa, in 'Topics in Stereo-Chemistry', ed. E. L. Eliel and N. L. Allinger, Wiley, New York, 1969, vol. 4, p. 39.
8. J. B. Lambert, J. J. Papay, S. A. Khan, K. A. Koppauf and E. S. Magyar, *J. Am. Chem. Soc.*, 1974, **96**, 6112.
9. S. Searles, Jr. and M. Tamres, in 'The Chemistry of the Ether Linkage', ed. S. Patai, Wiley, New York, 1967, p. 243.
10. E. M. Arnett, in 'Progress in Physical Organic Chemistry', Interscience, New York, 1963, vol. 7, p. 243.
11. S. Searles, Jr., M. Tamres and E. R. Lippincott, *J. Am. Chem. Soc.*, 1953, **75**, 2775.
12. E. M. Arnett and C. Y. Wu, *J. Am. Chem. Soc.*, 1962, **84**, 1684.
13. S. Penczek, P. Kubisa, K. Matyjaszewski and R. Szymanski, *Pure Appl. Chem.*, 1984, 140.
14. S. Penczek, P. Kubisa and R. Szymanski, *Makromol. Chem., Macromol. Symp.*, 1986, **3**, 203.
15. I. Kuntz, *J. Polym. Sci., Part A-1*, 1967, **5**, 193.
16. M. P. Dreyfuss, J. C. Westphal and P. Dreyfuss, *Macromolecules*, 1968, **1**, 437.
17. P. Dreyfuss, in 'Poly(tetrahydrofuran)', Gordon and Breach, New York, 1982, p. 35.
18. A. C. Farthing and R. J. W. Reynolds, *J. Polym. Sci.*, 1954, **12**, 503.
19. J. B. Rose, *J. Chem. Soc.*, 1956, 542.
20. I. Penczek and S. Penczek, *J. Polym. Sci.*, 1970, **8**, 2465.
21. S. G. Entelis and G. V. Korovina, *Makromol. Chem.*, 1974, **175**, 1523.
22. H. C. Brown and R. M. Adams, *J. Am. Chem. Soc.*, 1942, **64**, 2557.
23. S. Browstein, A. M. Eastham and G. A. Latremouille, *J. Phys. Chem.*, 1963, **67**, 1028.
24. R. Hoene and K. -H. W. Reichert, *Makromol. Chem.*, 1976, **177**, 3545.
25. M. Szwarc, in 'Carbanions, Living Polymers and Electron Transfer Processes', Interscience, New York, 1968.
26. S. Kobayashi, K. Morikawa and T. Saegusa, *Polym. J.*, 1979, **11**, 405.
27. K. Matyjaszewski, S. Slomkowski and S. Penczek, *J. Polym. Sci.*, 1979, **17**, 2413.
28. E. Franta, L. Reibel, J. Lehman and S. Penczek, *J. Polym. Sci., Polym. Symp.*, 1976, **56**, 139.
29. T. Saegusa and S. Matsumoto, *J. Macromol. Sci., Chem.*, 1970, **A4**, 873.
30. E. J. Vandenberg, *Pure Appl. Chem.*, 1976, **48**, 295.
31. P. Kubisa and S. Penczek, *Makromol. Chem.*, 1969, **130**, 186.
32. T. V. Grinevich, A. N. Shupik, G. V. Korovina and S. G. Entelis, *Eur. Polym. J.*, 1981, **17**, 1107.
33. T. Saegusa and S. Matsumoto, *J. Polym. Sci., Part A-1*, 1968, 1559.
34. T. Saegusa and S. Kobayashi, *Prog. Polym. Sci. Jpn.*, 1973, **6**, 107.
35. K. Brzezinska, W. Chwialkowsa, P. Kubisa, K. Matyjaszewski and S. Penczek, *Makromol. Chem.*, 1977, **178**, 2491.
36. K. Matyjaszewski and S. Penczek, *Makromol. Chem.*, 1981, **182**, 1735.
37. T. L. Cairns and R. M. Joyce, *US Pat.* 2 455 912 (1948).
38. I. Kuntz and M. T. Melchior, *J. Polym. Sci.*, 1969, **7**, 1959.
39. H. Meerwein, H. Battenberg, H. Gold, E. Pfeil and G. Willfang, *J. Prakt. Chem.*, 1939, **154**, 83.
40. T. V. Grinevich, A. N. Shupik, G. V. Korovina and S. G. Entelis, *Vysokomol. Soedin.*, 1980, **22**, 1576.
41. J. Libiszowski, R. Szymanski and S. Penczek, *Makromol. Chem.*, in press.
42. W. Jacobson and W. H. Stockmayer, *J. Chem. Phys.*, 1950, **18**, 1600.
43. D. J. Worsfold and A. M. Eastham, *J. Am. Chem. Soc.*, 1957, **79**, 900.
44. J. Dale, K. Daasvatn and T. Gronneberg, *Makromol. Chem.*, 1977, **178**, 873.
45. S. Kobayashi, K. Morikawa and T. Saegusa, *Macromolecules*, 1975, **8**, 952.
46. Y. Yamashita, K. Iwao and K. Ito, *J. Polym. Sci., Polym. Lett. Ed.*, 1979, **17**, 1.
47. J. Zachoval and C. Schuerch, *J. Polym. Sci., Part C*, 1969, **28**, 187.
48. Y. Kawakami, A. Ogawa and Y. Yamashita, *J. Polym. Sci., Polym. Chem. Ed.*, 1979, **17**, 3785.
49. J. Dale and K. Daasvatn, *J. Chem. Soc., Chem. Commun.*, 1976, 295.
50. K. Yokota, H. Hashimoto, T. Kakuchi and Y. Takada, *Makromol. Chem., Rapid Commun.*, 1984, **5**, 115.
51. K. Brzezinska, R. Szymanski, P. Kubisa and S. Penczek, *Makromol. Chem., Rapid Commun.*, 1986, **7**, 1.
52. M. Wojtania, P. Kubisa and S. Penczek, *Makromol. Chem., Macromol. Symp.*, 1986, **6**, 201.
53. P. Kubisa, *Makromol. Chem., Macromol. Symp.*, 1988, **13/14**, 203.
54. W. G. Potter, 'Epoxide Resins', Iliffe Books, London, 1970, p. 83.
55. A. Ledwith, S. Al-Kass and A. Hulme-Lowe, in 'Cationic Polymerization and Related Processes', ed. E. J. Goethals, Academic Press, New York, 1984, p. 275.
56. J. V. Crivello, in 'Cationic Polymerization and Related Processes', ed. E. J. Goethals, Academic Press, 1984, p. 289.

57. E. J. Vandenberg, in 'Kirk-Othmer Encyclopedia of Chemical Technology', 3rd edn., 1979, vol. 8, p. 568.
58. R. Lamot, *Makromol. Chem.*, 1988, **189**, 45.
59. K. C. Frisch (ed.), 'Cyclic Monomers', Wiley, New York, 1972, p. 54.
60. W. R. Sorensen and T. W. Campbell, 'Preparative Methods of Polymer Chemistry', 2nd edn., Wiley, New York, 1968, p. 367.
61. E. J. Goethals, *Ind. Chem. Belge*, 1965, **30**, 559.
62. P. E. Black and D. J. Worsfold, *Can. J. Chem.*, 1976, **54**, 3325.
63. P. Kubisa, J. Brzezinski and S. Penczek, *Makromol. Chem.*, 1967, **100**, 286.
64. P. Kubisa and S. Penczek, *Makromol. Chem.*, 1969, **130**, 186.
65. G. P. Aleksiuk, V. V. Shamanin, A. F. Podolsky, L. V. Alferova and V. A. Kropachev, *Polym. J.*, 1981, **13**, 23, 33.
66. T. Hirano, S. Nakayama and T. Tsuruta, *Makromol. Chem.*, 1975, **176**, 1897.
67. A. Chapiro and S. Penczek, *J. Chim. Phys. Phys.-Chim. Biol.*, 1962, **59**, 696.
68. S. Okamura, K. Hayashi and Y. Kitanishi, *J. Polym. Sci.*, 1962, **58**, 925.
69. T. Saeusa, *J. Macromol. Sci., Chem.*, 1972, **A6**, 997.
70. H. Boardman, in 'Encyclopedia of Polymer Science and Technology', ed. H. F. Mark and N. G. Gaylord, Interscience, New York, 1968, vol. 9, p. 668.
71. J. S. Coates, *Chem. Eng. News*, 1978, **56**, 3.
72. J. Leonard and D. Maheux, *J. Macromol. Sci.*, 1973, **A7**, 1421.
73. S. Penczek and K. Matyjaszewski, *J. Polym. Sci., Polym. Symp.*, 1976, **56**, 255.
74. K. Matyjaszewski and S. Penczek, *J. Polym. Sci., Polym. Chem. Ed.*, 1974, **12**, 1905.
75. K. Matyjaszewski, P. Kubisa and S. Penczek, *J. Polym. Sci., Polym. Chem. Ed.*, 1975, **13**, 764.
76. K. Matyjaszewski, S. Slomkowski and S. Penczek, *J. Polym. Sci., Polym. Chem. Ed.*, 1979, **17**, 69.
77. A. Stolarczyk, P. Kubisa and S. Penczek, *J. Macromol. Sci., Chem.*, 1977, **25**, 627.
78. P. Kubisa, *Bull. Acad. Pol. Sci., Ser. Sci. Chim.*, 1977, **8**, 627.
79. Z. Jedlinski, J. Lukaszczyk, J. Dudek and M. Gibas, *Macromolecules*, 1976, **9**, 622.
80. K. Matyjaszewski, S. Penczek and E. Franta, *Polymer*, 1979, **20**, 1185.
81. Y. Yamashita, M. Hirota, H. Matsui, A. Hirao and K. Nobutoki, *Polym. J.*, 1971, **2**, 43.
82. S. Smith and A. Hubin, *J. Macromol. Sci., Chem.*, 1973, **A7**, 1399.
83. S. Kobayashi, H. Danda and T. Saegusa, *Macromolecules*, 1974, **7**, 415.
84. G. Pruckmayr and T. K. Wu, *Macromolecules*, 1973, **6**, 33.
85. G. Pruckmayr and T. K. Wu, *Macromolecules*, 1975, **8**, 954.
86. K. Matyjaszewski, P. Kubisa and S. Penczek, *J. Polym. Sci., Polym. Chem. Ed.*, 1974, **12**, 1333.
87. K. Matyjaszewski, A. M. Buyle and S. Penczek, *J. Polym. Sci., Polym. Lett. Ed.*, 1976, **14**, 125.
88. A. M. Buyle, K. Matyjaszewski and S. Penczek, *Macromolecules*, 1977, **10**, 296.
89. S. Penczek, *Makromol. Chem., Suppl.*, 1979, **3**, 17.
90. E. B. Lyudwig, B. A. Rozenberg, T. M. Zuereva, A. R. Gantmakher and S. S. Medvedev, *Polym. Sci. USSR (Engl. Transl.)*, 1965, **7**, 296.
91. D. Vofsi and A. V. Tobolsky, *J. Polym. Sci., Part A*, 1965, **3**, 3261.
92. T. Saegusa and S. Matsumuto, *J. Macromol. Sci, Chem.*, 1970, **A4**.
93. H. Meerwein, D. Delfs and H. Morshel, *Angew. Chem.*, 1960, **72**, 927.
94. G. Pruckmayr and T. K. Wu, *Macromolecules*, 1978, **11**, 265.
95. G. Pruckmayr and T. K. Wu, *Macromolecules*, 1978, **11**, 662.
96. K. Brzezinska, M. Bednarek, J. Stasinski, P. Kubisa and S. Penczek, *Makromol. Chem.*, submitted.
97. M. Kucera, F. Bozek and K. Majerowa, *Polymer*, 1979, **20**, 1013.
98. T. Saegusa, Y. Kimura, H. Fujii and S. Kobayashi, *Macromolecules*, 1973, **6**, 657.
99. K. Matyjaszewski, Ph. D. Thesis, Lodz, 1976.
100. S. Kobayashi, K. Morikawa and T. Saegusa, *Macromolecules*, 1975, **8**, 386.
101. F. Andruzzi, A. Prescia and G. Ceccarelli, *Makromol. Chem.*, 1974, **175**, 1253.
102. A. Stolarczyk, P. Kubisa and S. Penczek, *Bull. Acad. Pol. Sci., Ser. Sci. Chim.*, 1974, **22**, 431.
103. J. P. Kennedy, 'Cationic Polymerization of Olefins: A Critical Inventory', Wiley, New York, 1975.
104. K. Matyjaszewski, P. Kubisa and S. Penczek, paper presented at the International Symposium on Cationic Polymerization, Rouen, 1973.
105. K. Matyjaszewski, R. Szymanski, P. Kubisa and S. Penczek, *Acta Polym.*, 1984, **35**, 14.
106. T. Baran, K. Brzezinska, K. Matyjaszewski and S. Penczek, *Makromol. Chem.*, 1983, **184**, 2497.
107. S. Slomkowski, *Polymer*, 1986, **27**, 71.
108. D. H. Richards, S. B. Kingston and T. Sonel, *Polymer*, 1976, **19**, 68.
109. J. Sierra-Vargas, J. G. Ziliox, P. Rempp and E. Franta, *Polym. Bull. (Berlin)*, 1980, **3**, 83.
110. M. Takaki, R. Asami and T. Kuwabara, *Polym. Bull. (Berlin)*, 1982, **7**, 521.
111. B. A. Rosenberg, T. I. Ponomarieva, L. D. Narkevitch and N. S. Enikolopyan, *Dokl. Akad. Nauk SSSR*, 1967, **175**, 365.
112. H. J. Kress and W. Heitz, *Makromol. Chem., Rapid Commun.*, 1981, **2**, 427.
113. K. Matsuda, Y. Tanaka and T. Sakai, *J. Appl. Polym. Sci.*, 1976, **20**, 2821.
114. Patentee, *US Pat.* 4 153 786 (1979).
115. Patentee, *Jpn. Pat. Appl.*, 83/89 081 (1983).
116. T. Saegusa, M. Motoi, S. Matsumoto and H. Fuji, *Macromolecules*, 1972, **5**, 233.
117. J. Kops and H. Spangaard, *Makromol. Chem.*, 1975, **176**, 299.

49

Cationic Ring-opening Polymerization: Acetals

STANISLAW PENCZEK and PRZEMYSTAW KUBISA
Polish Academy of Sciences, Boczna, Lodz, Poland

49.1 INTRODUCTION

Cyclic acetals are five- and higher-membered monomers with at least one unit in which two oxygen atoms flank a substituted or unsubstituted methylene group (**1**).

The bridge connecting two acetal oxygen atoms may be either an all carbon chain (*e.g.* $-(CR^1R^2)_{\overline{n}}$) or may contain additional heteroatoms (*e.g.* $-(CR^1R^2)_n X(CR^1R^2)_{\overline{m}}$, where X is oxygen or sulfur). Some typical monomers belonging to this class are (**2**)–(**8**) below.

(1)   (2) 1,3-dioxolane (3) 1,3-dioxepane (4) 1,3,6,7-tetraoxacycloundecane (5) 1,3,5-trioxane

(6) 2,6-dioxabicyclo[2.2.1]heptane (7) 6,8-dioxabicyclo[3.2.1]octane (8) 1,6-anhydro-β-D-glucopyranose

Cationic polymerization of these monomers yields polyacetals, *i.e.* polymers containing acetal bonds $-OCR^1R^2O-$ in the main chain. Polyacetals were among the first synthetic polymers studied by Staudinger in the 1920s.[1] Polyformaldehydes prepared by both anionic polymerization of formaldehyde and cationic polymerization of 1,3,5-trioxane were, however, thermally unstable. Thus, they were not commercialized until the late 1950s when the reasons for the thermal instability were understood and the stabilization methods developed. The Du Pont process,[2] based on anionic formaldehyde polymerization, involves esterification of unstable hemiacetal end groups from which degradation starts, while in the Celanese–Hoechst[3] process 1,3,5-trioxane is copolymerized with a few per cent of comonomer, *e.g.* ethylene oxide or 1,3-dioxolane. Randomly distributed $-OCH_2CH_2-$ units interrupt the sequences of $-OCH_2-$ units, which prevents depolymerization (unzipping). Polyformaldehyde (Delrin®, Celcon®, Hostaform®) is the only polyacetal made on an industrial scale (over 150×10^3 tons year^{-1} worldwide).

In recent years, there has been a growing interest in the polymerization of anhydrosugars bearing two fused rings. Synthetic polysaccharides prepared by polymerization of the acetal ring in anhydrosugars cannot compete, however, with inexpensive products isolated from natural sources for use on a large scale such as food additives, paper additives, adhesives and coatings. Thus, research on synthetic polysaccharides is stimulated mainly by their potential biomedical applications. It is expected that it will be easier to meet the high requirements for standardization with synthetic products than with the natural ones.

49.2 MONOMERS

Cyclic acetals are usually prepared by condensation of formaldehyde (conveniently in the form of paraformaldehyde or 1,3,5-trioxane) or higher aldehydes with diols in the presence of 1–2 wt % of an acid catalyst (equation 1); *p*-toluenesulfonic acid or ion exchange resins are most frequently used for this purpose.[4,5]

Resulting cyclic acetals and water are distilled off, the organic product being separated from water by extraction or addition of salt (*e.g.* CaCl$_2$) and separation of organic and aqueous layers. Cyclic acetals are stable towards bases, thus basic drying agents like CaH$_2$, sodium or sodium–potassium alloy can be used for the final drying. This method can be used for the synthesis of both substituted and unsubstituted monocyclic acetals. In some cases however, when intramolecular cyclization is not possible due to steric restrictions, condensation may proceed as an intermolecular process, leading to linear oligomeric products (equation 2).[6,7]

$$RCH{=}O + HO\frown OH \xrightarrow{H^+} 1/n{-}[O\frown OCHR]_n + H_2O \qquad (2)$$

This is the case with some cyclic diols (*e.g.* 1,3-dihydroxymethylcyclobutane or 1,4-dihydroxy-cyclohexane) or long chain diols (*e.g.* triethylene glycol).

Linear oligomers, formed by condensation, may be converted into cyclic monomers by depolymerization, *i.e.* heating and distilling off cyclic monomers (equation 3).[8]

$$\text{+(OCH}_2\text{CH}_2)_3\text{OCH}_2\text{+}_n \rightarrow n \; \overline{\text{---(OCH}_2\text{CH}_2)_3\text{OCH}_2\text{---}}} \tag{3}$$

1,3,5-Trioxane, the cyclic trimer of formaldehyde, is prepared by heating aqueous solutions of formaldehyde in the presence of 1–2 wt % of H_2SO_4.[9] Water–1,3,5-trioxane azeotrope (~70% 1,3,5-trioxane) is distilled off and pure 1,3,5-trioxane is isolated by extraction. The simplest purification method involves fractionation (b.p. 115 °C), often in the presence of a solvent which forms a ternary azeotrope with water, *e.g.* n-heptane.[9]

Synthesis of bicyclic acetals is more complex; thus, for example, 6,8-dioxabicyclo[3.2.1]octane is prepared in three steps from acrolein (Scheme 1).[10]

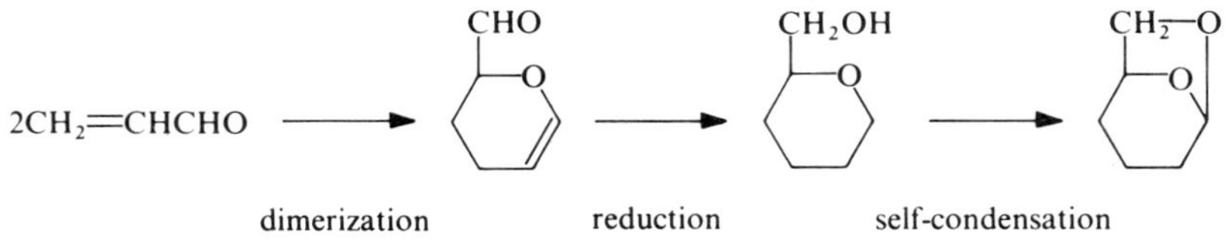

Scheme 1

Anhydrosugars are formed by elimination of water from simple sugars isolated from natural sources (Scheme 2).[11]

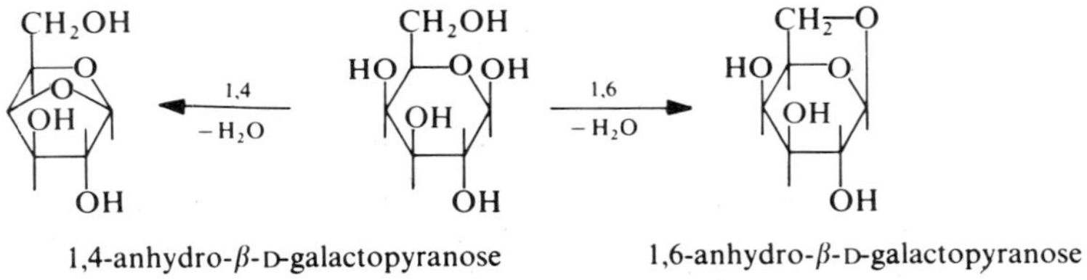

Scheme 2

Prior to cationic polymerization, the free hydroxyl functions of anhydrosugars have to be blocked by etherification or esterification.[11] Benzyl ether protecting groups are frequently used; after the polymerization they are removed using sodium in liquid ammonia.

The properties of some typical cyclic acetals are listed in Table 1.

Table 1 Properties of Cyclic Acetals

Cyclic acetal	b.p. (°C)	m.p. (°C)	d_4^{20} (g cm^{-3})	n_D^{20}	Ref.
1,3-Dioxolane	74–75	−95	1.0600	1.3974	5
2-Methyl-1,3-dioxolane	82–83		0.9811	1.4035	5
2,2-Dimethyl-1,3-dioxolane	92–93				5
4,5-Dimethyl-1,3-dioxolane	102–104				5
1,3,5-Trioxane	114.5[a]	64	1.17[b]		5
1,3-Dioxepane	119			1 4275	5
1,3,6-Trioxocane	150–155				5
1,3-Dioxacycloundecane	196		0.985	1.4564	8
1,3,6,9-Tetraoxacycloundecane	56/0.4 Torr	27	1.1314[c]	1.4541[c]	8

[a] Azeotrope 70% 1,3,5-trioxane 30% water, b.p. = 91.3 °C. [b] At 65 °C. [c] At 30 °C.

49.3 POLYMERIZABILITY OF CYCLIC ACETALS

Polymers are formed from the corresponding monomer when: (a) polymerization is thermodynamically possible; and (b) a suitable mechanism exists.

For polymerization of cyclic acetals, the thermodynamic criterion is not always fulfilled. The smallest ring size for a cyclic acetal is five atoms. Ring-opening polymerization of cyclic monomers with rings composed of more than four atoms, due to a relatively low negative ΔH_p, and consequently ΔG_p, value (see Volume 3, Chapter 46) is a reversible process.[12] Thus, monomer–polymer equilibrium is established, and if the monomer equilibrium concentration $[M]_e$ is higher than its bulk concentration, polymerization is not possible. This is the case with the six-membered cyclic acetal 1,3-dioxane.

The acetal bond is highly reactive and is easily opened in the presence of acid catalysts;[13] thus, cationic polymerization provides a suitable mechanism for converting a cyclic acetal into a linear polymer.

49.3.1 Thermodynamics of Polymerization

Thermodynamic parameters for polymerization of some unsubstituted cyclic acetals are listed in Table 2.

Table 2 Thermodynamic Parameters of Polymerization of Unsubstituted Cyclic Acetals

Monomer	Standard state[a]	ΔH_p° (kJ mol^{-1})	ΔS_p° (J mol^{-1} K^{-1})	T_c for 1 mol l^{-1} (°C)	Ref.
1,3-Dioxolane	gg	−25.9			12, 14
	ss (C$_6$H$_6$)	−15.0	−58.5		15
	ss (CH$_2$Cl$_2$)	−21.7	−77.7	1	16
	lc	−23.0	−62.7		17
1,3-Dioxepane	gg	−19.6			12, 14
	ss (C$_6$H$_6$)	−13.4	−38.9		15
	ss (CH$_2$Cl$_2$)	−15.0	−48.1	27	16
1,3,5-Trioxepane[b]	ss (CH$_2$Cl$_2$)	−6.6	−18.9		18
	ss (CH$_2$Cl$_2$)	−6.9	−31.5		19
1,3,6-Trioxocane	ss (C$_6$H$_6$)	−22.1	−38.9		15
	ss (CH$_2$Cl$_2$)	−16.7	−34.3		20
1,3,6,9-Tetraoxacycloundecane	ss (CH$_2$Cl$_2$)	−13.4	−13.0		20

[a] gg = gas–gas, ss = solution–solution, lc = liquid–condensed. [b] For detailed discussion of the thermodynamics of 1,3,5-trioxepane polymerization see ref. 19

Substitution generally decreases the thermodynamic polymerizability of cyclic monomers[12] (lower negative ΔH°). Some thermodynamic parameters for polymerization of substituted cyclic acetals are given in Table 3.

Table 3 Thermodynamic Parameters of Polymerization of Substituted Cyclic Acetals

Monomer	Standard state[a]	ΔH_p° (kJ mol^{-1})	ΔS_p° (J mol^{-1} K^{-1})	T_c for 1 mol l^{-1} (°C)	Ref.
4-Methyl-1,3-dioxolane	ss (CH$_2$Cl$_2$, 3 mol l^{-1})	−18.4	−54		21
	lc	−13.4	−53		22
		−11.7[b]			22
4-Ethyl-1,3-dioxolane	lc	−13.0	−59.4		22
4,4-Dimethyl-1,3-dioxolane		−5.85[b]			22
cis-4,5-Dimethyl-1,3-dioxolane		−10.45[b]			22
trans-4,5-Dimethyl-1,3-dioxolane		−5.85[b]			22
2-Methyl-1,3-dioxepane	ss (CH$_2$Cl$_2$)	−8.8	−37.2	−37	27
4-Methyl-1,3-dioxepane	ss (CH$_2$Cl$_2$)	−9.3	−39		24

[a] ss = solution–solution, lc = liquid–condensed. [b] Values calculated on the basis of differences between *gauche* interactions between substituents. The value 3.35 kJ mol^{-1} was assumed per one interaction and −17.5 kJ mol^{-1} for ΔH for the unsubstituted ring.

For reversible polymerization, two interrelated parameters characterize the system: monomer equilibrium concentration $[M]_e$ (at the given temperature) and ceiling temperature T_c (for a given $[M]_0$, usually equal to $1 \, mol \, l^{-1}$ or in bulk). The smaller the negative ΔG° value, the higher the $[M]_e$ value and the lower the T_c value (equation 4).

$$\ln(1/[M]_e) = -\Delta G^{\circ}/RT \qquad (4)$$

It was not always recognized in the past that in order to make polymerization of some cyclic acetals feasible, the highest possible $[M]_0$ and the lowest possible temperature should be used. This led to some unsuccessful attempts to prepare high polymers from substituted cyclic acetals. When, however, suitable conditions were used, such substituted cyclic acetals as 4-phenyl-1,3-dioxolane[25] or 2-butyl-1,3,6-trioxocane[26] gave reasonably high yields of high molecular weight polymers.

For bicyclic (*i.e.* disubstituted) acetals, the thermodynamic polymerizability is usually enhanced due to the additional strain introduced by the presence of the second ring; thus, for example, 6,8-dioxabicyclo[3.2.1]octane at concentrations of *ca.* $2.5 \, mol \, l^{-1}$ can be polymerized up to 95% conversion at $-78 \, ^{\circ}C$ in CH_2Cl_2, which means that $[M]_e < 1.25 \times 10^{-1} \, mol \, l^{-1}$ under these conditions.[27]

The influence of the additional strain on the thermodynamic polymerizability may be illustrated by the polymerization of 7,9-dioxabicyclo[4.3.0]nonane.[28] *Trans* monomer gives high molecular weight polymer in high yield, while under the same conditions *cis*-monomer gives only a cyclic dimer (Scheme 3).

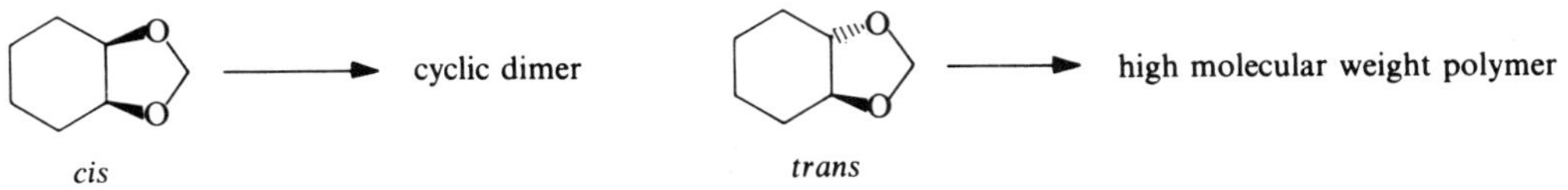

cis → cyclic dimer   *trans* → high molecular weight polymer

Scheme 3

In some instances, the excess strain can even lead to the opening of the six-membered ring instead of the five-membered one. Thus, 2,6-dioxabicyclo[2.2.1]heptane, in which the six-membered ring is forced into an unfavourable boat conformation, gives polymers containing both five- and six-membered rings (Scheme 4).[29]

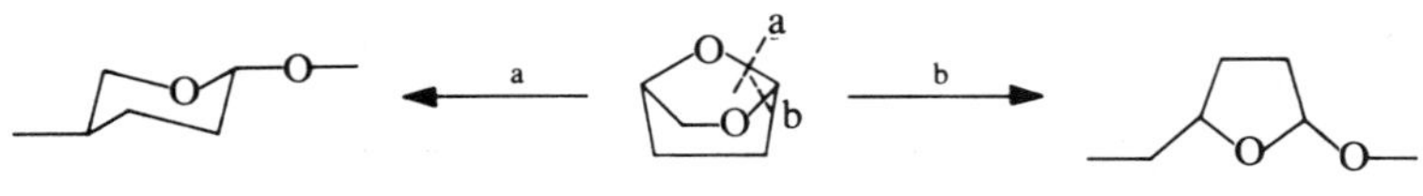

Scheme 4

The thermodynamics of 1,3,5-trioxane polymerization, studied extensively by Enikolopyan *et al.*,[30] is more complex. This is due to the fact that propagation proceeds simultaneously with phase transition; monomer molecule, when being incorporated into a polymer chain, is at the same time transferred from a liquid into the crystalline phase. Thus, the overall free energy change (ΔG°) for the propagation step is the sum of those for the chemical reaction and phase transition. Consequently, equilibrium is shifted more towards polymer and 1,3,5-trioxane (six-membered ring of low strain) can be polymerized with high conversions. 1,3,5-Trioxane is a cyclic trimer of formaldehyde. Thus, equilibria existing in polymerization can be described by Scheme 5.

Thermodynamic parameters for polymerization of 1,3,5-trioxane are given in Tables 4 and 5.

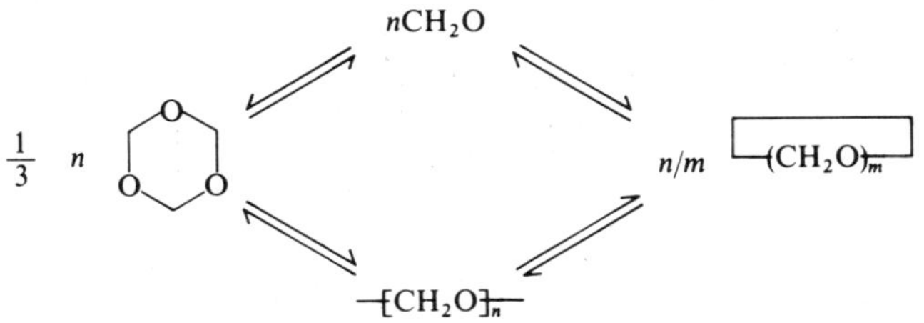

Scheme 5

Table 4 Thermodynamic Parameters of Polymerization of
1,3,5-Trioxane[50]

Standard state[a]	ΔH_p^0 (kJ mol^{-1})	ΔS_p^0 (J mol^{-1} K^{-1})
gl	-50.1	-115
gc	-62.7	-155
lc	-21.7	-48.9
sc (nitrobenzene)	-21.3	-43.1
sc (CH$_2$Cl$_2$)	-19.6	-50.1

[a] gl = gas liquid. gc = gas-condensed, lc = liquid-condensed, sc = solution-condensed.

Table 5 Calculated Concentrations of Monomers in Equilibrium with Polyoxymethylene[a]

Standard state	Monomer	Equilibrium concentration (mol l^{-1})
ss (hypothetical polymerization from dissolved monomer to dissolved polymer)	Formaldehyde	0.013
	1,3,5-Trioxane	1.41
	1,3,5,7-Tetraoxocane	0.10
sc′ (real polymerization from dissolved monomer to crystalline polymer)	Formaldehyde	0.003
	1,3,5-Trioxane	0.11
	1,3,5,7-Tetraoxocane	0.004

[a] In nitrobenzene solvent at 25 °C; calculated on the basis of data given in ref. 30.

49.4 MECHANISM OF POLYMERIZATION

49.4.1 Initiation

Cyclic acetals polymerize exclusively by a cationic mechanism. Most typical cationic initiators will initiate polymerization of cyclic acetals. Examples are (i) strong protonic acids, *e.g.* HClO$_4$, HOSO$_2$CF$_3$ and their derivatives, esters (ROSO$_2$CF$_3$), anhydrides [(CF$_3$SO$_2$)$_2$], or, as shown recently, tungsten or molybdenum heteropolyacids, *e.g.* H$_3$PMo$_{12}$O$_{40}$;[31] (ii) organic salts: carbenium, *e.g.* Ph$_3$C$^+$A$^-$, oxonium, *e.g.* Et$_3$O$^+$A$^-$, or carboxonium, *e.g.* PhCO$^+$A$^-$; and (iii) Lewis acids (Friedel–Crafts catalysts), *e.g.* BF$_3$, SbCl$_5$, PF$_5$ and their complexes with, for example, ethers.

49.4.1.1 *Influence of the counterion structure*

Organic cations (carbenium, carboxonium, oxycarbenium and oxonium) are coupled with complex counterions of the MtX$_{n+1}^-$ type (Mt = metal), for example SbF$_6^-$, SbCl$_6^-$ or BF$_4^-$, or with noncomplex ones, *e.g.* OClO$_3^-$ or OSO$_2$CF$_3^-$. It has been shown for 1,3-dioxolane polymerization that only those initiators which contain the most stable, least nucleophilic counterions, such as SbF$_6^-$ and AsF$_6^-$, lead to the quantitative formation of stable active species.[32-34] The unstable SbCl$_6^-$ and BF$_4^-$ counterions, due to their fragmentation, give rise to side reactions, resulting in incomplete initiation and decrease of active species concentration.[32,35] The fragmentation of counterions and resulting decay of active species has been studied in detail for 1,3-dioxolane polymerization in the presence of SbCl$_6^-$.[35] Noncomplex counterions, derived from protonic acids (*e.g.* OClO$_3^-$ or OSO$_2$CF$_3^-$) react with active species, giving covalent esters. This reaction, however, being reversible, does not lead to the irreversible termination of the material chain (equation 5)

$$-\text{OCH}_2^+ + \text{OSO}_2\text{CF}_3^- \rightleftharpoons -\text{OCH}_2\text{OSO}_2\text{CF}_3 \tag{5}$$

In equation (5) and subsequent reaction schemes, ionic active species are shown for simplicity in the form of alkoxycarbenium ions, although in reality the large majority of active species exist in the form of oxonium ions. There are several types of oxonium ions coexisting in the system (see Section 49.4.2.1.ii) and due to their multiplicity, schematic representation is difficult.

Equilibria like (5) have not been systematically studied for cyclic acetal polymerization. The quantitative data are available only for polymerization of cyclic ethers, mainly THF, which has been studied extensively in our laboratory.[36] In this system, the equilibrium constants of temporary termination have been measured and it has been shown that covalent species are able to react with monomer by themselves, although the rate of propagation is about 10^2 times lower than that for ionic species.[37] Other systems are known, however, in which propagation on the covalent species is faster than on the puckered sterically hindered ionic species (*cf.* polymerization of oxepane, Volume 3, Chapter 48).

Cyclic acetals do not induce significant hindrance and their polymerization can be described by the THF scheme.

For 1,3-dioxolane and 1,3-dioxepane polymerizations initiated with $(CF_3SO_2)_2O$, the degree of polymerization is given by equation (6). Equation (6) indicates that the number of growing macromolecules is equal to the number of initiator molecules used.[38] Some representative data are given in Table 6.

$$DP_n = \frac{[\text{monomer}]_0 - [\text{monomer}]_e}{[(CF_3SO_2)_2O]_0} \tag{6}$$

Table 6 Calculated and Measured Molecular Weights in the Polymerization of 1,3-Dioxolane and 1,3-Dioxepane[a]

Monomer	$\overline{M}_n$ calculated	$\overline{M}_n$ measured[b]
1,3-Dioxolane	19 350	18 250
	21 300	21 100
	35 000	44 500
1,3-Dioxepane	13 250	10 700
	55 600	57 100
	76 500	78 000

[a] Initiated with $(CF_3SO_2)_2O$ $[M]_0 = 1–3$ mol l^{-1}, $[I]_0 = 10^{-3}–10^{-2}$ mol l^{-1}, CH_2Cl_2, temperature -50 to $-78\,°C$.[38]
[b] Membrane osmometry.

The overall rates of polymerization are, however, *ca.* 10^2 times lower than for $PhCO^+SbF_6^-$ initiated polymerization. This indicates that active species exist predominantly in the form of less reactive covalent species.

49.4.1.2 *Addition* vs. *hydride transfer*

The first, fast step in initiation is the protonation or alkylation (acylation) of the monomer molecules. This leads to formation of the secondary or tertiary oxonium ions, as shown for 1,3-dioxolane in Scheme 6.

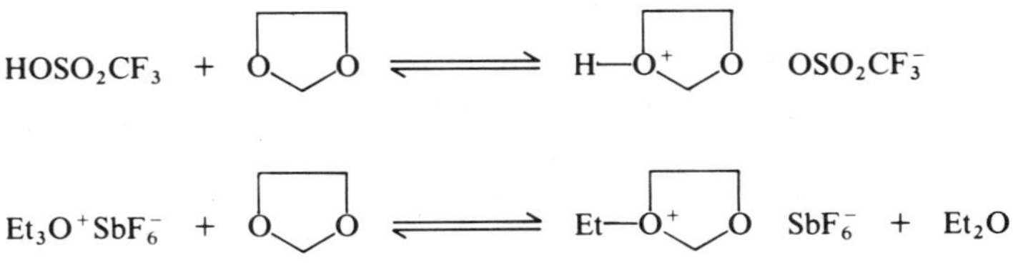

Scheme 6

The subsequent, slower step involves nucleophilic attack of the next monomer molecule, leading to the opening of the ring and formation of active species (equation 7).

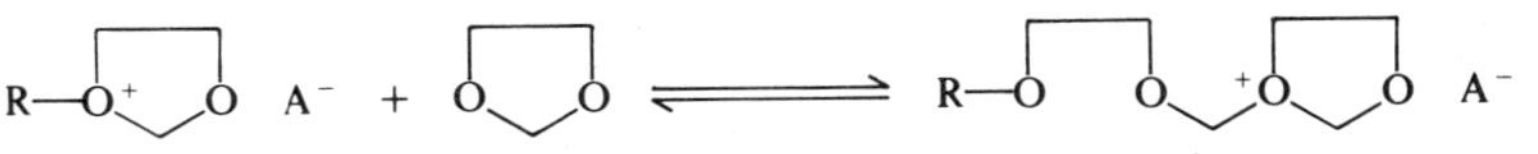

Initiation proceeds according to equation (7) for the majority of initiators used, although the rate of initiation depends on the nature of R. In the polymerization of 1,3-dioxolane initiated with $Et_3O^+SbF_6^-$ and $PhCO^+SbF_6^-$ the corresponding initiator fragments are incorporated quantitatively as the end groups.[33] These end groups were identified by ^1H NMR; perdeuterated monomer was used to minimize signals of the polymer chain. Results are shown in Table 7.

Table 7   Structure of Poly-1,3-dioxolane End-Groups[a]

Initiator	Terminating agent	DP_n calc.	DP_n found from ^1H NMR analysis of the indicated end groups	
$PhCO^+SbF_6^-$ 4.7×10^{-3} mol l^{-1}	EtONa	700	650 PhCOO–	575 EtO–
$PhCO^+SbF_6^-$ 1.95×10^{-3} mol l^{-1}	PBu$_3^n$	1690	1380 PhCOO–	—
$Et_3O^+SbF_6^-$ 5.2×10^{-3} mol l^{-1}	PBu$_3^n$	750	1000 EtO–	920 Bu$_3^n$P–

[a] Perdeuterated monomer was used; polymerization conditions: $[1,3\text{-dioxolane}]_0 = 4.05$ mol l^{-1}, CD_3NO_2 solvent, -15 °C.[33]

For some initiators, however, an alternative route of initiation is possible. For triphenylmethylium (tritylium) salts, the preinitiation step is highly reversible (equation 8).[39] ·

$$Ph_3C^+A^- \quad + \quad \text{[dioxolane]} \quad \underset{}{\overset{K}{\rightleftharpoons}} \quad Ph_3C\text{—[dioxolanium]}^+ \quad A^- \tag{8}$$

The equilibrium constant of this reaction is equal to $K = 3.2 \times 10^{-2}$ mol^{-1} l (CH_2Cl_2, 25 °C).[39] Thus, under typical polymerization conditions, a significant amount of carbenium ions exist in equilibrium and participate in hydride transfer reactions.[40] Oxonium ions are not H$^-$ ion acceptors (equation 9).

$$Ph_3C^+A^- \quad + \quad \text{[dioxolane]} \quad \longrightarrow \quad Ph_3CH \quad + \quad \text{[dioxolenium]}^+\text{H} \quad A^- \tag{9}$$

This reaction is relatively fast ($k = 8.3 \times 10^{-3}$ mol^{-1} l s^{-1}, CH_2Cl_2, 25 °C), while an attack of the next monomer molecule on oxonium ion, shown in equation (7), is apparently slow. Thus, instead of simple initiation by addition, as shown in equation (7), hydride transfer takes place and 1,3-dioxolan-2-ylium (dioxolenium) salt and triphenylmethane are formed quantitatively. 1,3-Dioxolan-2-ylium salts, which are relatively stable and can be isolated as pure crystalline compounds, are thus the true initiators, formed *in situ* in the 1,3-dioxolane–$Ph_3C^+A^-$ system.

1,3-Dioxolan-2-ylium salts initiate the polymerization according to equation (10).[41] As was indicated earlier,[42] in agreement with the known ambident reactivity of dioxolenium salts,[43] the intermediate kinetic product can sometimes be formed (equation 11).

$$\text{[dioxolenium]}^+\text{H} \quad A^- \; + \; \text{[dioxolane]} \quad \longrightarrow \quad \overset{H}{\underset{O}{\diagup}}C\text{—}OCH_2CH_2\text{—}[dioxolanium]^+ \quad A^- \tag{10}$$

$$\text{[dioxolenium]}^+\text{R} \quad A^- \quad \underset{}{\overset{+Nu}{\rightleftharpoons}} \quad \overset{}{\underset{R \; Nu^+}{\diagdown\diagup}} \quad A^- \tag{11}$$

Nu = nucleophile

This reversible reaction does not alter the actual path of initiation. Indeed, Jedlinski, *et al.* reported that 1,3-dioxolan-2-ylium salts, substituted at the 2-position, behave in exactly this way.[44]

Hydride transfer reactions to the Ph_3C^+ cation proceeds relatively fast for other cyclic and linear acetals and probably the initiation mechanism described by equations (8)–(10) also operates for other cyclic acetals. Some values of hydride transfer rate constants are given in Table 8.

Table 8   Rate Constants of Hydride Transfer Reaction from Cyclic and Linear Acetals to Ph_3C^+ Cation at 25 °C

Acetal	k_{H^-} $(mol^{-1}1s^{-1})$	*Ref.*
1,3-Dioxolane	1.2×10^{-2}	45
	9.4×10^{-3}	46
	7.7×10^{-3} (23 °C)	47
2-Methyl-1,3-dioxolane	7.9×10^{-3}	45
2-Phenyl-1,3-dioxolane	1.5×10^{-2}	45
4,5-Dimethyl-1,3-dioxolane	4.15×10^{-2} (23 °C, *trans*)	47
	7.35×10^{-2} (23 °C, *cis*)	
1,3-Dioxepane	$2.4–3.0 \times 10^{-3}$ (22 °C)	45
Dimethoxymethane	6.0×10^{-4}	45

49.4.1.3   *Friedel–Crafts type initiators: direct initiation* vs. *coinitiation*

The mechanism of initiation with Friedel–Crafts type initiators is generally obscure. For BF_3, the most commonly used initiator of this type, two mechanisms, namely direct initiation and initiation by coinitiator (*e.g.* water present as impurity), have been proposed (Scheme 7).[48,49]

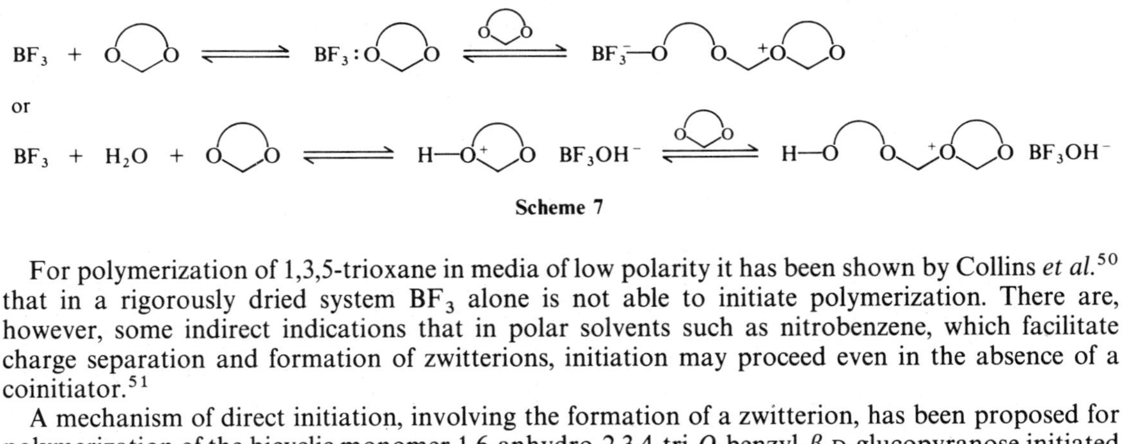

Scheme 7

For polymerization of 1,3,5-trioxane in media of low polarity it has been shown by Collins *et al.*[50] that in a rigorously dried system BF_3 alone is not able to initiate polymerization. There are, however, some indirect indications that in polar solvents such as nitrobenzene, which facilitate charge separation and formation of zwitterions, initiation may proceed even in the absence of a coinitiator.[51]

A mechanism of direct initiation, involving the formation of a zwitterion, has been proposed for polymerization of the bicyclic monomer 1,6-anhydro-2,3,4-tri-*O*-benzyl-β-D-glucopyranose initiated with PF_5 (Scheme 8).[52]

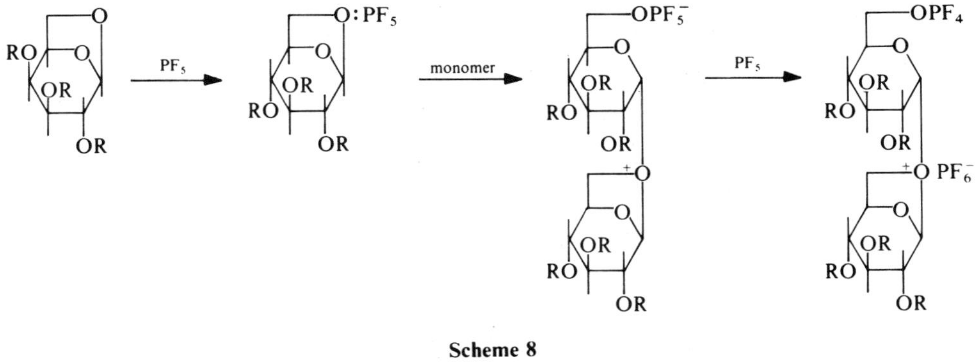

Scheme 8

All the proposed species, *e.g.* $PF_5:O=$, PF_4O- and PF_6^- were identified on the basis of ^{31}P and ^{19}F NMR spectra. In addition, formation of POF_3 was observed, indicating that terminal $-OPF_4$ groups undergo decomposition to $-CH_2F$ and POF_3.[52]

It may be concluded that, unlike protonic acids and organic salts, which under suitable conditions give clean quantitative initiation, Friedel–Crafts type initiators lead to poorly defined systems. Efficiency of initiation, structure of counterions and the structure of end groups are generally not known with any certainty for these initiators. The overall polymerization rates are usually lower than with other initiators, indicating the low efficiency of initiation. Some relevant data are given in Table 9.

Table 9 Copolymerization of 1,3,5-Trioxane (50 wt %) with 1,3-Dioxolane (50 wt %) (total 100 g) in Bulk, 60 min[53]

Initiator (mol l^{-1} × 10^{-5})	Temperature (°C)	Yield (%)
HClO$_4$ (0.07)	40	34
	50	86
	60	94
	70	93
BF$_3$·OBu$_2^n$ (21.3)	40	15
	50	43
	60	76
	70	83

49.4.1.4 Radiation-initiated solid state polymerization of 1,3,5-trioxane

Solid state polymerization of 1,3,5-trioxane can be initiated by different kinds of radiation, including γ-rays, X-rays, electron beams or α-particles.[54-56] The mechanism of initiation is not well understood. It is, however, generally accepted that radiation induces cationic polymerization. Ions or radical ions are generated by electron transfer, the loss of hydrogen ions or the heterolytic cleavage of 1,3,5-trioxane rings.[57]

Radiation-induced solid state polymerization of 1,3,5-trioxane was studied intensively in the past because it was believed that it could lead to large scale production. The process is, however, highly irreproducible and extremely sensitive to impurities.[58] Thus, although high molecular weight polymers can be obtained in high yields ($\sim 80\%$) by direct irradiation of the crystalline monomer,[55] interest in radiation-induced polymerization is vanishing.

49.4.2 Propagation

49.4.2.1 Structure of active species

(i) Oxonium–alkoxycarbenium ion equilibria

The simplest structure of active species in the polymerization of cyclic acetals, by analogy with the polymerization of other heterocyclic monomers, is an oxonium ion holding the monomer molecule (9).

$$-OCH_2-\overset{+}{O}\diagdown\diagup O \quad A^-$$
(9)

Studies of the model systems have revealed that oxonium ions coexist in equilibrium with their alkoxycarbenium counterparts, in which the positive charge is distributed between the carbon and α-oxygen atoms (equation 12).[59] Such equilibria were earlier only tacitly assumed.[60,61]

$$-OCH_2-\overset{+}{O}\diagdown\diagup O \quad A^- \rightleftharpoons -OCH_2^+ \, A^- \, (+ \, O\diagdown\diagup O) \qquad (12)$$

Equilibrium constants for the model reaction between methoxymethylium cation and dimethoxymethane (simple linear model acetal) (equation 13) have been determined in our laboratory and found to be $K = k_a/k_d = 3 \times 10^3$ mol$\,$l^{-1} (SO$_2$, $-70\,°$C).[59] This value indicates that active species in cyclic acetal polymerizations exist predominantly in the form of oxonium ions, although a small proportion exist in the form of alkoxycarbenium ions.

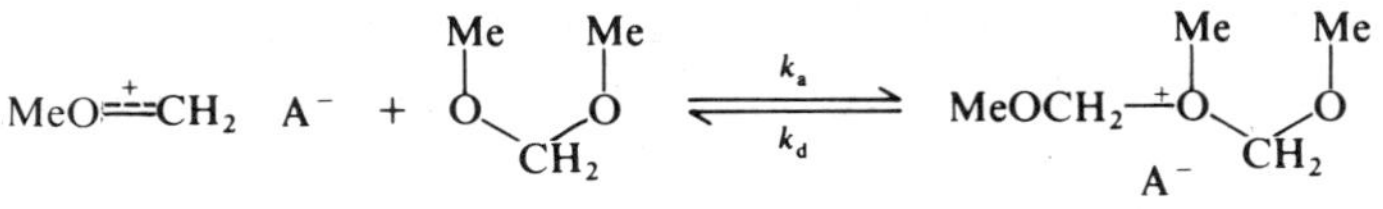

$$\text{(13)}$$

(ii) Structure of oxonium active species

Rate constants for formation and dissociation of oxonium ions (reaction 13) are high: $k_a = 2 \times 10^6$ mol$^{-1}\,$l$\,$s^{-1} and $k_d = 1.9 \times 10^4$ mol$^{-1}\,$l$\,$s^{-1} (SO$_2$, $-70\,°$C).[59] Thus, the system is in a fast dynamic equilibrium. This leads to very fast isomerization of oxonium active species, proceeding *via* alkoxycarbenium ions.

When 1,3-dioxolane was used instead of dimethoxymethane in studies of equilibria similar to (13), it has been observed that the predominant structure is not an oxonium ion holding the five-membered ring but the corresponding cyclic ion involving the seven-membered ring (Scheme 9).[62]

Scheme 9

The ratio of equilibrium constants K_7 and K_5, corresponding to the ratio of concentrations of the seven-membered to the five-membered oxonium ions, is equal to 3×10^2.

When the six-membered (1,3-dioxane) or the seven-membered (1,3-dioxepane) cyclic acetals were used in the studies of related equilibria, isomerization to the enlarged ring structures (eight- and nine-membered, respectively) was negligible because the starting rings were less strained than the enlarged ones.[62]

All these reactions, which may be described by equation (14), proceed with high rates, therefore at any stage of polymerization, equilibrium between alkoxycarbenium ions and various cyclic (formed by intramolecular reaction) and branched (formed by intermolecular reaction) oxonium ions is quickly established (Scheme 10).

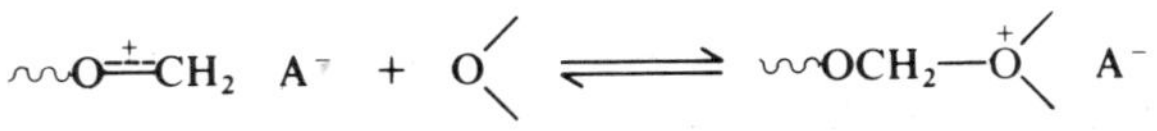

$$\text{(14)}$$

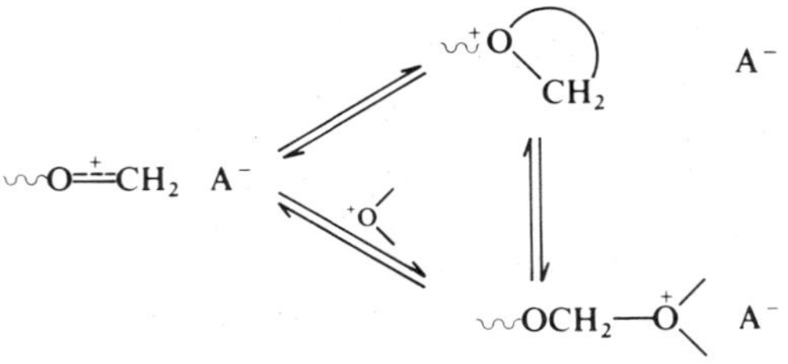

Scheme 10

(iii) Secondary vs. tertiary oxonium ions

Polymerization of cyclic acetals initiated with protonic acids may lead to kinetic enhancement in macrocycles (see Section 49.4.4.2). This effect is due to the efficient end-to-end cyclization of the short growing macromolecules (equation 15).[63]

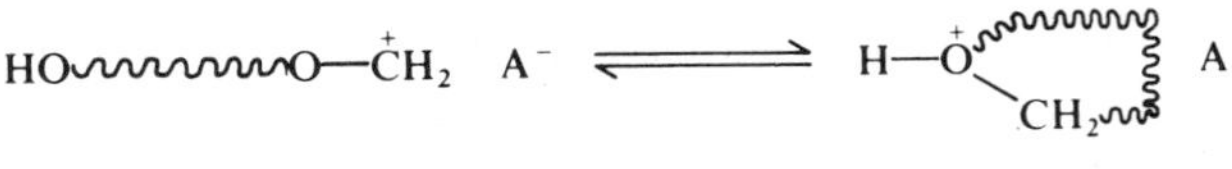

$$\text{(15)}$$

Thus, active tertiary oxonium ion species are in this process converted into secondary oxonium ions.

These equilibria have been studied in our laboratory by ion-trapping methods for the polymerization of 1,3-dioxolane initiated with $HOSO_2CF_3$. The principle of the method is shown in Scheme 11.[64]

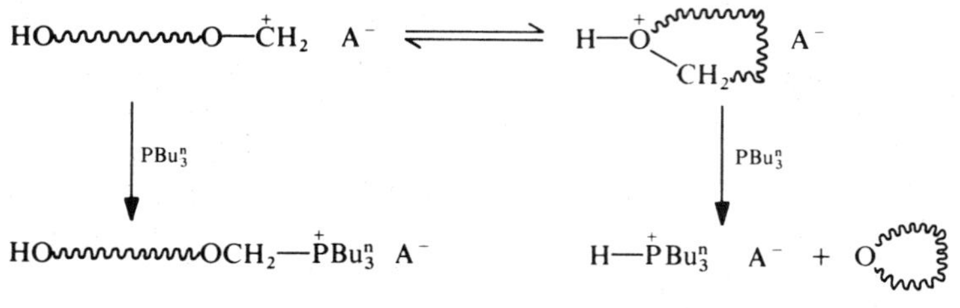

$$HO\text{———}O\overset{+}{—}CH_2 \quad A^- \rightleftharpoons H—\overset{+}{O}\underset{CH_2}{\overset{\frown}{}} \quad A^-$$

$$\downarrow PBu_3^n \qquad\qquad\qquad\qquad \downarrow PBu_3^n$$

$$HO\text{———}OCH_2\overset{+}{—}PBu_3^n \quad A^- \qquad H—\overset{+}{P}Bu_3^n \quad A^- + O\!\!\curvearrowright$$

Scheme 11

Analysis of ^{31}P NMR spectra, based on the different chemical shifts of the tertiary ($\delta = 11.9$ p.p.m.) and quaternary ($\delta = 31.4$ p.p.m.) phosphonium ions, allows the determination of the relative concentrations of the secondary and tertiary oxonium ions. It was shown that, in agreement with the end-to-end cyclization scheme, the concentration of the secondary oxonium ions is high for short growing chains, but decreases gradually with increasing chain length. This conclusion was later confirmed by studies of the ^1H NMR spectra of the same system.[65] Protons in HO– end groups and $H–\overset{+}{O}=$ species exchange fast, giving one narrow signal. The averaged chemical shift of this signal allows determination of the relative contribution of both species.

These observations prove that secondary oxonium ions are indeed formed by end-to-end cyclization, as shown in equation (15).

49.4.2.2 *Reactivity of active species*

It follows from the preceding discussion that in the polymerization of cyclic acetals small but definite concentrations of alkoxycarbenium active species exist in equilibrium with oxonium active species. The equilibrium constant, measured for a model system, indicates that alkoxycarbenium ions constitute *ca.* $10^{-2}\%$ of all active species in the polymerization of 1,3-dioxolane at $-78\,°C$ in CH_2Cl_2.[59] This proportion may vary substantially with the conditions applied and, of course, depends enormously on the structure of the monomer. To estimate to what extent the alkoxycarbenium ions participate in propagation, the rate constants of model reactions have been measured (Scheme 12).

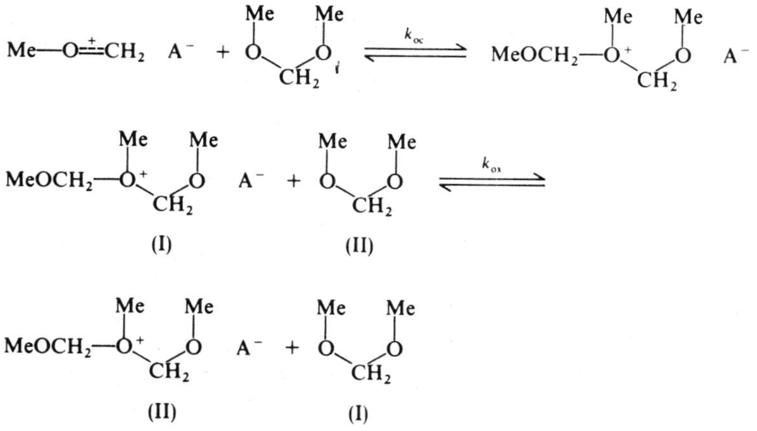

$$Me—\overset{+}{O}{=}CH_2 \quad A^- + \underset{CH_2}{O\bigvee O} \overset{k_{oc}}{\rightleftharpoons} MeOCH_2—\overset{+}{O}\underset{CH_2}{\bigvee O} \quad A^-$$

$$MeOCH_2—\overset{+}{O}\underset{CH_2}{\bigvee O} \quad A^- + \underset{CH_2}{O\bigvee O} \overset{k_{ox}}{\rightleftharpoons}$$

$$(I) \qquad\qquad (II)$$

$$MeOCH_2—\overset{+}{O}\underset{CH_2}{\bigvee O} \quad A^- + \underset{CH_2}{O\bigvee O}$$

$$(II) \qquad\qquad (I)$$

(I) and (II) denote different molecules of the same compound

Scheme 12

The corresponding values, determined by Penczek and Szymanski, using dynamic NMR line broadening, are $k_{oc} = 2 \times 10^6\ mol^{-1}\,l\,s^{-1}$ and $k_{ox} = 1.9 \times 10^4\ mol^{-1}\,l\,s^{-1}$ (both in SO_2 at $-70\,°C$).[59]

Thus, oxonium ions are only *ca.* 10^2 times less reactive than alkoxycarbenium species. Since the ratio of concentrations is *ca.* 10^4, it may be concluded that propagation proceeds predominantly on the oxonium ions.

49.4.2.3 *Propagation in the polymerization of 1,3,5-trioxane*

Polyformaldehyde, the product of 1,3,5-trioxane polymerization, is insoluble in common organic solvents, including molten monomer. Thus, polymerization of this monomer proceeds as a heterogeneous process and conclusions based on the model studies of homogeneous systems cannot be directly adopted for 1,3,5-trioxane polymerization. Studies of the topochemical aspects of polymerization, carried out by Wegner *et al.*[66-68] strongly indicate that the active species in 1,3,5-trioxane polymerization are located on the surface of the growing polymer crystal. Thus, they cannot be directly observed or identified by spectroscopic methods.

It is known that in this polymerization, even in the rigorously purified systems, the molecular weights, corresponding to the lengths of the kinetic chains, cannot be attained. This was attributed by Weissermel *et al.*[69] to the hydride transfer reaction shown in equation (16).

$$—OCH_2O—\overset{+}{C}H_2 \ A^- \ + \ \langle \text{trioxane} \rangle \ \longrightarrow \ —OCH_2OMe \ + \ \langle \text{trioxanylium} \rangle \ A^- \tag{16}$$

As a result, the polymer would aquire –OMe end groups, which have indeed been detected by Kern and Jaacks,[70] and the 1,3,5-trioxan-2-ylium cation would start a new chain. The stable –OMe end groups may be responsible, at least partially, for the presence of *ca.* 25% of the thermally stable fraction in otherwise nonstabilized polymers.[71] It has to be remembered, however, that it may also be due to the presence of a cyclic fraction (see Section 49.4.9).

The fact that hydride transfer reactions occur in 1,3,5-trioxane polymerization indicates that the fraction of the alkoxycarbenium active species may be significant, because only these species can participate in hydride transfer; no hydride transfer reaction was observed with oxonium ions.

Active species in 1,3,5-trioxane polymerization are located on the surface of the growing polymer crystals. Thus, monomer molecules, when being incorporated into the polymer chain, are at the same time transferred from the liquid into the crystalline phase. This affects the thermodynamics of polymerization, as discussed already in Section 49.3.1, and introduces an additional factor to the mechanism of chain growth.

As soon as polymer starts to form, it precipitates in the form of small hexagonal crystals.[66-68] As polymerization progresses, these crystals grow both in diameter and in height. A schematic representation of this process is given in Figure 1.

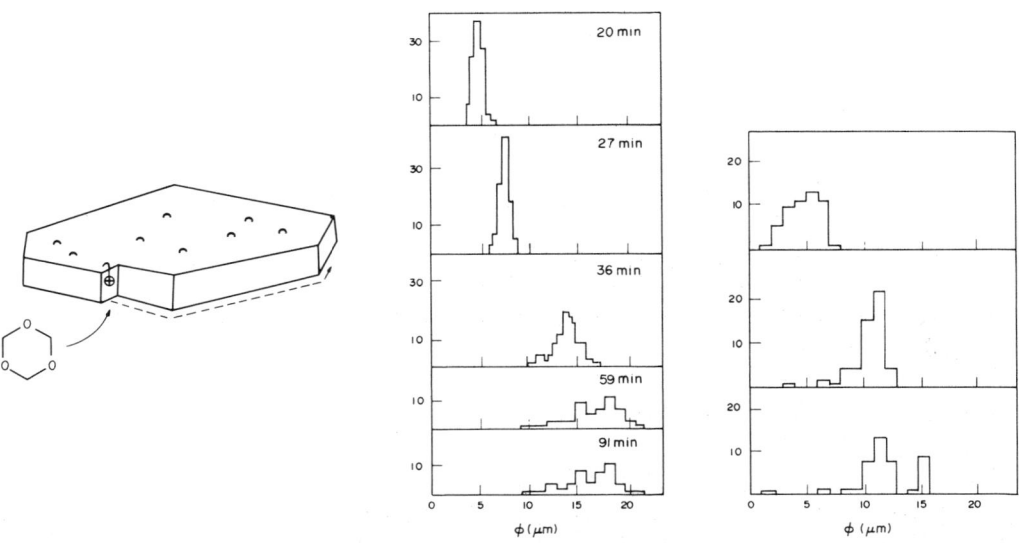

Figure 1 Model for lateral growth of polyoxymethylene crystals (left). Histograms of lateral crystal sizes during 1,3,5-trioxane polymerization initiated by (centre) $HClO_4$ and (right) $BF_3 \cdot OEt_2$ (ref. 67)

The growth of crystal height is explained in terms of insertion of monomer molecules into the chain folds, facilitated by the strain induced by these folds. The actual mechanism probably involves a multistep process (Scheme 13).

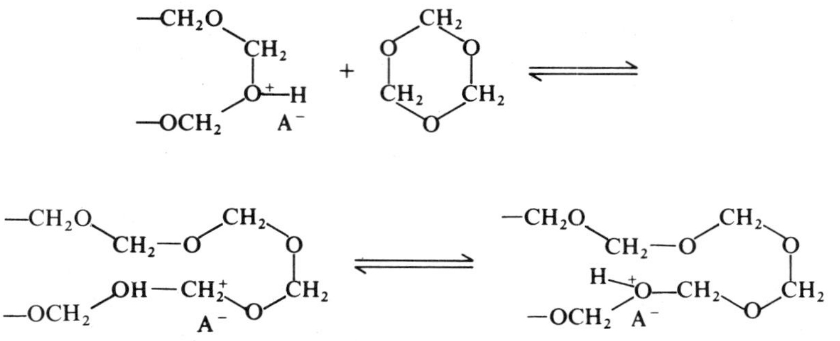

Scheme 13

Thus, the size of the crystals increases when reaction progresses; at the same time the degree of crystallinity increases. This process does not stop even after equilibrium conversion is reached. This is explained in terms of dissolution (by depolymerization) of the smaller, less perfect crystals in favour of the thermodynamically preferred larger, more perfect crystals. The changes of crystal size and degree of crystallinity in the polymerization of 1,3,5-trioxane are illustrated by the data in Table 10.

Table 10   Growth of Polyoxymethylene Crystals During Cationic Polymerization of 1,3,5-trioxane in Nitrobenzene at 35 °C[68]

Time (h)	Conversion (%)	$\bar{M}_n \times 10^{-3}$	Crystal diameter (μm)	Crystal thickness (μm)	Degree of crystallinity (%)
0.5	20	17.5	14	0.01	79.1
1.0	55	40	27	0.03	84.3
3.0	75	50	37	—	88.2
6.0	75	53	57	0.18	88.2
12.0	75	53	70	0.20	88.8

49.4.3   Transfer and Termination

49.4.3.1   *Chain transfer to polymer — transacetalization*

Active species in the polymerization of cyclic acetals undergo fast isomerization, as already discussed in Section 49.4.2.1.ii. This results in chain transfer to polymer, *i.e.* formation (by intramolecular reaction) of cyclic structures (equation 17) or formation (by intermolecular reaction) of branched oxonium ions, followed by exchange of the linear fragments of the chain (transacetalization) (equation 18).

$$\text{mmOmmO—}\overset{+}{\text{C}}\text{H}_2 \rightleftharpoons \text{mmO}^+ \underset{\text{CH}_2\text{O}}{\overset{\text{mmmm}}{\diagdown}} \qquad (17)$$

$$\text{mmO—}\overset{+}{\text{C}}\text{H}_2 + \text{O}{\diagdown} \rightleftharpoons \text{mmOCH}_2\text{—}\overset{+}{\text{O}}{\diagdown} \qquad (18)$$

Reactions (17 and (18) are facilitated by the higher basicity of oxygen atoms in the polymer chain than in the monomer molecule; this is a general feature of polyacetals, in contrast to polyethers.

Reaction (17) leads to the formation of macrocycles (see Section 49.4.4). Formation of branched structures (equation 18) results in transacetalization (also called scrambling). In branched ions (equation 18) two (or all three for 1,3,5-trioxane) of the bonds between carbon and oxygen atoms bearing the positive charge are identical. The further reaction of branched oxonium ions may thus lead to the redistribution of polymer segments, as shown in Scheme 14 for 1,3,5-trioxane polymerization.

$$A\text{\scriptsize$\sim$}OCH_2O\overset{+}{=\!\!=}CH_2 \quad A^- \quad + \quad O\!\!\begin{array}{c} {}^{CH_2O\text{\scriptsize$\sim$}B} \\ {}_{CH_2O\text{\scriptsize$\sim$}C} \end{array} \quad \rightleftharpoons$$

$$A\text{\scriptsize$\sim$}OCH_2OCH_2\!\!-\!\!\overset{+}{O}\!\!\begin{array}{c} {}^{CH_2O\text{\scriptsize$\sim$}B} \\ {}_{CH_2O\text{\scriptsize$\sim$}C} \end{array} \quad A^- \quad \rightleftharpoons$$

$$A\text{\scriptsize$\sim$}OCH_2OCH_2OCH_2\text{\scriptsize$\sim$}B \quad + \quad C\text{\scriptsize$\sim$}O\overset{+}{=\!\!=}CH_2 \quad A^-$$

Scheme 14

This reaction, which is fast, results in continuous transfer of the monomer units or longer sequences between the chains. Thus, when sequential polymerization of cyclic acetals was performed by introducing 1,3-dioxepane into a solution of living poly-1,3-dioxolane or *vice versa*, the co-polymer isolated after longer reaction times showed DXL–DXL/DXL–DXP/DXP–DXP (DXL = 1,3-dioxolane; DXP = 1,3-dioxepane) diad distribution (determined by ^{13}C NMR) in ratios of 1:2:1, *i.e.* characteristic of ideally random copolymers.[38] When copolymerization was terminated after shorter reaction times, the proportion of heterodiads was lower, but at no stage of copolymerization could pure block copolymer be isolated.

Transacetalization plays an important role in copolymerization of 1,3,5-trioxane and 1,3-di-oxolane.[72] In this system, as shown by Jaacks,[73] due to the different reactivities of both com-onomers, 1,3-dioxolane polymerizes first. Therefore, at the early stages of polymerization, soluble polymer, consisting essentially of 1,3-dioxolane units, is formed. This is illustrated in Figures 2 and 3.[73]

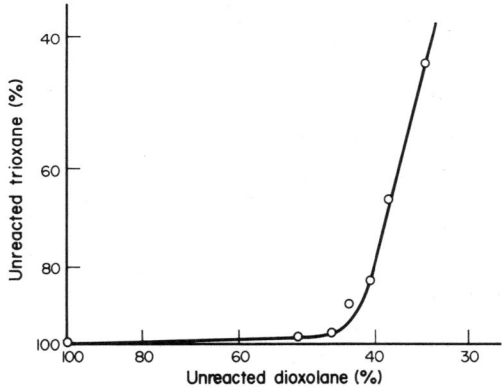

Figure 2 Rate of comonomer consumption in 1,3,5-trioxane–1,3-dioxolane copolymerization. $[1,3,5\text{-trioxane}]_0 = 2.5 \text{ mol l}^{-1}$, $[1,3\text{-dioxolane}]_0 = 4.5 \times 10^{-2} \text{ mol l}^{-1}$, $[\text{SnCl}_4]_0 = 1.7 \times 10^{-2} \text{ mol l}^{-1}$, CH_2Cl_2, 30 °C (ref. 48)

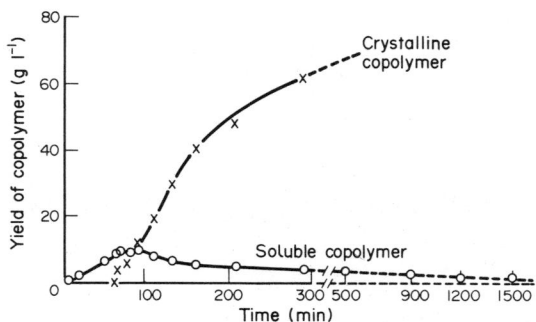

Figure 3 Rates of formation of soluble and insoluble copolymers in 1,3,5-trioxane–1,3-dioxolane copolymerization. $[1,3,5\text{-trioxane}]_0 = 2.8 \text{ mol l}^{-1}$, $[1,3\text{-dioxolane}]_0 = 2.8 \times 10^{-1} \text{ mol l}^{-1}$, $[\text{SnCl}_4]_0 = 2.5 \times 10^{-2} \text{ mol l}^{-1}$, CH_2Cl_2, 30 °C (ref. 48)

Nevertheless, the distribution of 1,3-dioxolane units in the final product is close to random. This indicates that the originally formed poly-1,3-dioxolane blocks undergo further fast and efficient scrambling.

Thus, due to the transacetalization, thermally stable polymers containing randomly distributed oxyethylene units derived from 1,3-dioxolane are formed. A small amount of the unstable fraction ($<5\%$) corresponds to oxymethylene units at the end of the chain (equation 19). This fraction is removed by heating before processing the polymer.

$$-OCH_2OCH_2OCH_2CH_2OCH_2OCH_2OH \rightarrow -OCH_2OCH_2OCH_2CH_2OH + 2CH_2O \tag{19}$$

oxyethylene   unstable
unit   fraction

49.4.3.2 *Termination*

In the absence of impurities, and when stable counterions (SbF_6^-, AsF_6^-, $OSO_2CF_3^-$) are used, polymerization of simple cyclic acetals, namely 1,3-dioxolane and 1,3-dioxepane, proceeds without termination. It has already been shown (Tables 6 and 7) that molecular weights in these systems are described by equation (20) and the only end groups, identified quantitatively, are those coming from initiator and intentionally added terminating agent.[33]

$$DP_n = \frac{[\text{monomer}]_0 - [\text{monomer}]_e}{[\text{initiator}]_0} \tag{20}$$

Polymerization of 1,3,5-trioxane was also described, by Cherdron *et al.*, as a terminationless process.[71] Hydride transfer to apparently living active species has been observed even several hours after equilibrium conversion of monomer had been reached. Some of this long-lived species may, however, become unavailable for propagation as a result of occlusion within the polymer particles.[74]

49.4.4 Formation of Cyclic Oligomers

49.4.4.1 *Back-biting*

Chain transfer to polymer leads to the formation of macrocyclic oxonium ions (equation 21), which, reacting further, may release the cyclic fragment (equation 22).

$$\tag{21}$$

$$\tag{22}$$

A thermodynamic approach to the ring–chain equilibria has been developed by Jacobson and Stockmayer (J–S theory).[75] According to this approach, the concentration of each cyclic oligomer in equilibrium with a linear chain unit is related to the probability of the required conformation of linear chain being achieved and can be calculated using the rotational isomeric state model.

For 1,3-dioxolane polymerization, these 'theoretical' equilibrium concentrations of cyclic oligomers have been calculated and compared with the values measured for real systems by GLC.[76] The data are given in Figure 4.

The experimental curve shows typical deviation from theory for oligomers with $n = 2-4$, which are still strained. (J–S theory is valid only for the strainless rings.)

1,3-Dioxolane and macrocyclic formals (10)[20, 77] are the only cyclic acetals for which cyclization has been studied quantitatively. Formation of cyclic fraction may, however, be expected in the

$$\overline{}(OCH_2CH_2)_nOCH_2\overline{}$$

(10)

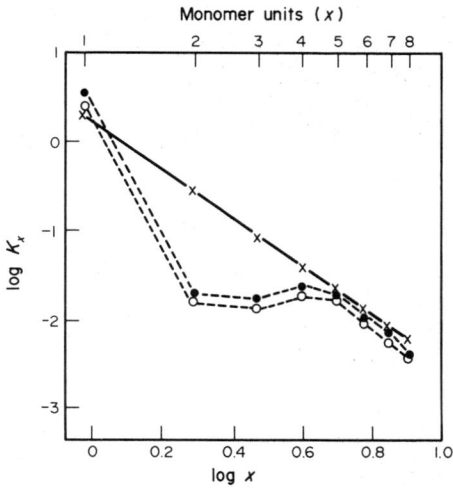

Figure 4 Comparison of experimental molar cyclization equilibrium constants $K_x (= \lceil M_x \rceil_e$ in mol l^{-1}) in polymerization of 1,3-dioxolane with values calculated according to Jacobson–Stockmayer theory: $\times$, calculated; $\bullet$, measured for bulk polymerization $[BF_3 \cdot OEt_2] \sim 10^{-2}$ mol l^{-1}; $\bigcirc$, measured for solution polymerization, $[M]_0 = 4.8$ mol l^{-1}, $C_2H_4Cl_2$, 60 °C (from ref. 76)

polymerization of any cyclic acetal. There are, for example, indications that cyclic oligomers are formed in the polymerization of 1,3,5-trioxane. Unstabilized homopolymer always contains some amount (up to 25%) of a stable fraction,[71] which, at least partly, is assumed to be cyclic (instability is related to the hemiacetal groups, which are absent in cyclic macromolecules). Cyclic oligomers are also believed to be responsible for the bimodal molecular weight distribution observed by GPC.[78, 79]

49.4.4.2 End-to-end cyclization

It has been shown recently that in certain systems,[63] for example for the protonic acid initiated polymerization of 1,3-dioxolane[64] or 1,3,6,9-tetraoxacycloundecane,[80] end-to-end cyclization may dominate over random back-biting in the early stages of polymerization, *i.e.* when the growing chains are still relatively short (Scheme 15).

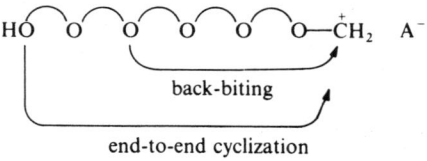

Scheme 15

This leads to the 'kinetic enhancement' of the cyclic macromolecules. It means that at the early stages of polymerization concentration of a given cyclic oligomer is higher than when the final equilibrium is attained. This, in turn, results from the contribution of the end groups of short chains, which vanish when the high polymer is finally formed. Thus, concentration of small cyclics grows, goes through a maximum when the end-to-end cyclization helps its formation the most, and then decreases with increasing chain length, falling to the equilibrium concentration; this follows the J–S theory for sufficiently long chains.

This behaviour, explained in a series of papers from our group, is illustrated in Figure 5, giving the computer-simulated concentrations of cyclic oligomers formed by the end-to-end mechanism.[63]

This type of dependence was observed experimentally by Schulz *et al.* for the polymerization of 1,3,6,9-tetraoxacycloundecane in the presence of $HOSO_2CF_3$ as initiator (Figure 6).[80]

49.4.4.3 Proportions of cyclic and linear fractions

Reactions involving polyacetal chains, including intramolecular reactions, are fast compared to propagation, resulting in the simultaneous formation of macrocycles and linear polymers.

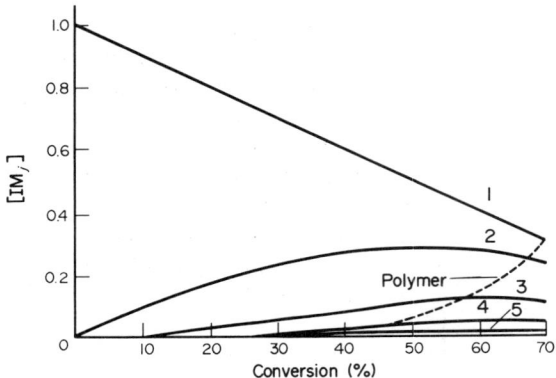

Figure 5 Computer simulation of the dependence of concentration of macrocycles $[\stackrel{\frown}{M_j}]$ (in mol l^{-1} of monomeric units) on monomer conversion for end-to-end cyclization (ref. 63)

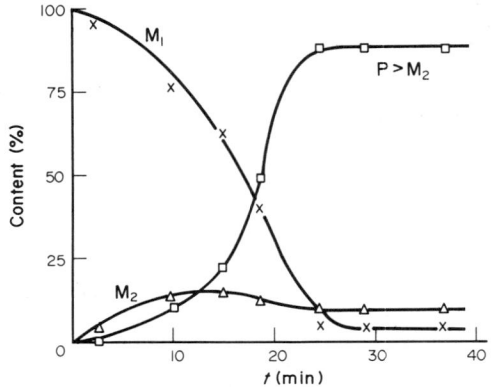

Figure 6 Time conversion curves for the consumption of monomer (M$_1$), formation of polymer (P) and formation of cyclic dimer (M$_2$) in the polymerization of 1,3,6,9-tetraoxacycloundecane (0.5 mol l^{-1}) in CH$_2$Cl$_2$ [HOSO$_2$CF$_3$]$_0$ = 5 × 10^{-4} mol l^{-1}, 0 °C (ref. 80)

Therefore, preparation of purely linear polymers is virtually impossible. The equilibrium concentration of the individual cyclic oligomers is, however, not very high (*ca.* 10^{-2} mol l^{-1} for the cyclic dimer and similar values for oligomers with $n=3$–5 in 1,3-dioxolane polymerization).[76] The sum of thus-calculated equilibrium concentrations of cyclics from $n=2$ to infinity gives a relatively high value, 0.85 mol l^{-1}, for 1,3-dioxolane polymerization. The actual measured values are somewhat lower than the calculated ones, as shown in Figure 4, thus the value of 0.85 mol l^{-1} should be taken as an upper limit. The concentration of 1,3-dioxolane in bulk equals 12.75 mol l^{-1} (at 25 °C). Thus, the lowest proportion of cyclic oligomers, starting from a dimer, that can be attained for this monomer is close to 7.5 wt %.

The percent fraction of cyclic oligomers at equilibrium is expressed as shown in equation (23).

$$f = \frac{100 \sum_{n=2}^{\infty} [1,3\text{-dioxolane units in cyclics}]_e}{[1,3,\text{-dioxolane}]_0 - [1,3\text{-dioxolane}]_e} \tag{23}$$

Thus at low [M]$_0$, polyacetals at equilibrium may contain a considerably high proportion of the cyclic fraction. At these conditions, when equation (24) holds, almost all of the polymer can be cyclic.

$$\sum_{n=2}^{\infty} [1,3\text{-dioxolane units in cyclics}]_e \geq [1,3\text{-dioxolane}]_0 - [1,3\text{-dioxolane}]_e \tag{24}$$

Some authors observed almost exclusively cyclic polymers and concluded, without considering the above-discussed consequences of the Jacobson–Stockmayer theory, that the chain growth must proceed by ring expansion.[81] With our present knowledge of the mechanism of cyclization however, all experimental observations can be interpreted in terms of the conventional propagation mechanism (*i.e.* involving linear active species) coupled with back-biting and/or end-to-end cyclization (see also Section 49.4.2.1.iii).

49.5 STEREOCHEMISTRY OF POLYMERIZATION

Unsubstituted monocyclic acetals (symmetrical molecules) are not suitable for studies of the stereochemistry of polymerization. Because there are no reliable data on the stereochemistry of polymerization of the substituted monocyclic acetals, the only available information relating to the steric course of polymerization comes from studies of bicyclic acetals. This subject has recently been reviewed by Sumitomo and Okada.[82]

For polymerization of 6,8-dioxabicyclo[3.2.1]octane, it has been shown that at $-78\,^{\circ}C$ inversion of configuration occurs in nearly 100% (equation 25).[83] This result is in full agreement with an oxonium ion mechanism involving S_N2 substitution and an attack of the incoming monomer molecule from the 'opposite' side,[84] along the carbon–oxygen bond as shown in Scheme 16.

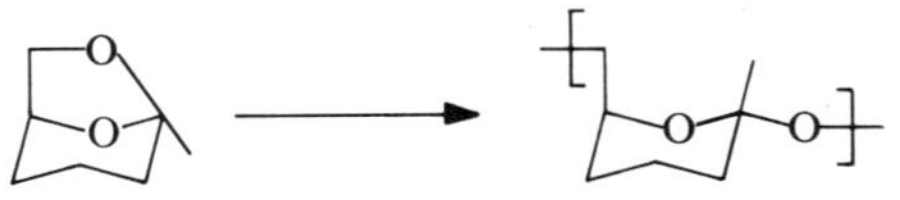

$$(25)$$

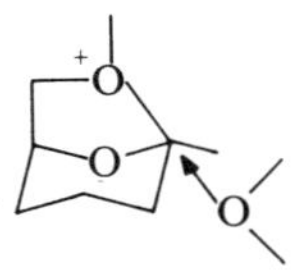

Scheme 16

At higher temperatures the stereospecificity of reaction is lower. It has been argued, however, that this effect is only apparent, *i.e.* propagation proceeds with inversion of configuration, while racemization is due to the side reactions involving polymer chains (transacetalization).

The steric course of polymerization of anhydrosugars depends on the size of the fused rings.[11,29] Thus, for polymerization of 1,2- and 1,4-anhydrosugars, racemization has been observed and explained by participation of alkoxycarbenium ions in the propagation step (*e.g.* for 1,2-anhydrosugars, see Scheme 17).

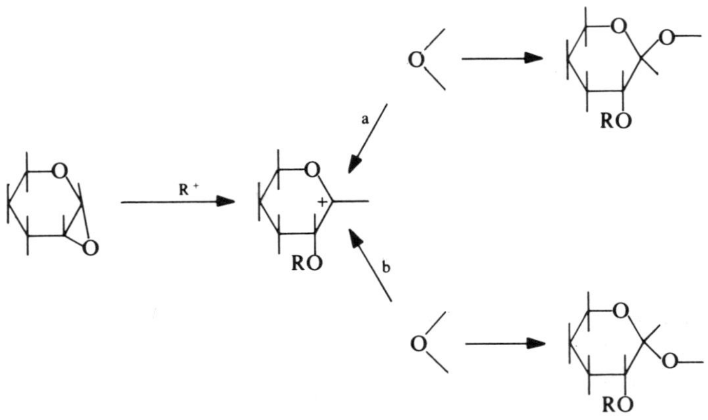

Scheme 17

1,6-Anhydrosugars, which have a 6,8-dioxabicyclo[3.2.1]octane skeleton, polymerize mainly with inversion of configuration; thus α-monomers give β-polymers and *vice versa*. For example, 1,6-anhydro-2,3,4-tri-*O*-benzyl-β-D-glucose (**11**) polymerizes at $-60\,^{\circ}C$ in CH_2Cl_2 with PF_5 initiator (1 mol %) to give a high molecular weight polymer ($\bar{M}_n$ up to 7×10^5) in 50% yield. The polymer, after debenzylation, has a specific rotation up to 197°[85] and is identical with a naturally occurring polysaccharide (dextran) for which a highly specific α structure has been established. Natural dextran has $[\alpha]_D^{25}$ values between $+196^{\circ}$ and $+199^{\circ}$.

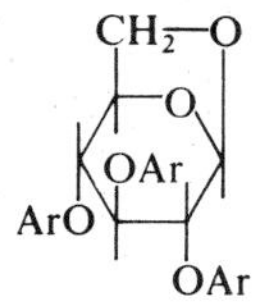

$$
\begin{array}{c}
CH_2\!\!-\!\!O \\
\diagdown\!\!-\!\!O \\
OAr \\
ArO \\
OAr
\end{array}
$$

(11)

For the polymerization of 6,8-dioxabicyclo[3.2.1]octane, it has also been shown that stereo-selection operates to some extent. Thus, polymer formed from racemic monomer is enriched in isotactic diads (up to 85%).[86]

Polymerization of a racemic mixture enriched with one of the enantiomers has also been studied, and it was shown that selection of the enantiomer present in excess leads to polymers with an enhanced proportion of this enantiomer.[87] On the basis of these observations, the mechanism of enantiomer selection by growing chain ends has been proposed. According to this theory steric repulsion is minimized when the monomer incorporated into the growing species and the incoming monomer molecule have the same chirality.

49.6 MICROSTRUCTURE OF POLYACETALS

49.6.1 End Groups

The quantitative analysis of end groups has been performed for only a few systems. Thus, when polymerization of 1,3-dioxolane-d_6 was initiated with $PhCO^+SbF_6^-$ and terminated with EtONa, it was shown in our laboratory by ^{1}H NMR (see Table 7) that the polymer had the structure (12).[33]

No end groups other than those shown in (12) were detected. Similarly, initiation with $Et_3O^+SbF_6^-$ and termination with NMe_3 gave quantitatively a polymer of structure (13).[34]

Termination of polymerization with tertiary phosphines gives a structure (14), which has phosphonium terminal end groups.[33]

$$PhCOO\!\!\sim\!\!\sim\!\! OEt$$

(12)

$$EtO\!\!\sim\!\!\sim\!\! \overset{+}{N}Me_3$$

(13)

$$EtO\!\!\sim\!\!\sim\!\! \overset{+}{P}R_3$$

(14)

The quantitative incorporation of initiator moiety and formation of the EtO– head groups has also been observed by Lyudvig and Ponomarenko in the polymerization of 1,3-dioxolane initiated with Et_3O^+ cation, labelled with ^{14}C in the ethyl group.[88]

In the polymerization initiated with protonic acid ($HClO_4$) and terminated with EtONa, Jaacks has proven the presence of the terminal –OEt groups by hydrolyzing the polymer and determining the concentration of EtOH formed.[89]

49.6.2 Microstructure of the Polymer Chain

Polymerization of cyclic acetals proceeds by opening of one of the $-O-CH_2-O-$ bonds; thus polyacetals are built up as shown in equation (26). This was confirmed by ^{1}H and ^{13}C NMR spectroscopy; thus, for example, the ^{1}H NMR spectrum of poly-1,3-dioxolane consists of two sharp singlets at $\delta = 4.76$ p.p.m. (OCH_2O group) and 3.73 p.p.m. (OCH_2CH_2O group).[90] The regular structure has also been confirmed by dipole moment measurements.[91]

$$
n\ \underset{O\diagdown_{CH_2}\diagup O}{\overset{(CH_2)_n}{\diagup\ \ \diagdown}} \longrightarrow \ \ \{O(CH_2)_nOCH_2\}_n \tag{26}
$$

For substituted cyclic acetals, opening of either the O(1)–C(2) or the O(3)–C(2) bond may lead to a polymer of irregular structure. Indeed, by analyzing the methylene group signals in ^{1}H NMR spectra of poly(4-ethyl-1,3-dioxolane), it has been shown that the polymer is composed of both types of units (equation 27).[92]

$$\text{(cyclic acetal structure)} \longrightarrow -\text{OCH}_2\text{CHOCH}_2- \quad \text{and} \quad -\text{OCHCH}_2\text{OCH}_2- \qquad (27)$$

The thermal stability of 1,3,5-trioxane copolymers, as already discussed in Section 49.4.3.1, depends on the distribution of comonomer units, thus the microstructure of copolyacetals has been studied extensively. Methods based on NMR spectroscopy have mainly been used,[93, 94] although chemical analyses of hydrolysis products have also been used.[94] ^{1}H NMR allows identification of triads composed of oxymethylene (M) and oxyethylene (E) units. More detailed analysis is possible by using ^{13}C NMR, especially when shift reagents [*e.g.* 1,1,1,2,2,3,3,-heptafluoro-7,7-dimethyl-4,6-octadionatoeuropium, Eu(fod)$_3$] are used. This method gives access to pentad and even heptad sequences. The related chemical shifts are given in Table 11.[93]

Table 11 Chemical Shifts of Oxymethylene and Oxyethylene Units in Polyacetals[93]

^{1}H NMR, CDCl$_3$, 25°C, 60 MHz	
Triad	*Chemical shift, δ(p.p.m.)*
–M–M–M–	4.89
–M–M–E– (=–E–M–M–)	4.83
–E–M–E–	4.77
–M–E–M–	3.73
–M–E–E– (=–E–E–M–)	3.68

^{13}C NMR, CDCl$_3$, 25°C, Eu(fod)$_3$/CH$_2$O = 0.0053		
Pentad	*Heptad*	*Chemical shift, δ (p.p.m.)*
–M–E–M–E–M–	–E–M–E–M–E–M–M–	96.57
	–M–M–E–M–E–M–M–	96.05
	–E–M–M–M–E–M–E–	92.53
–M–M–M–E–M–	–M–M–M–M–E–M–M–	
	–M–M–M–M–E–M–E–	93.30
	–E–M–M–M–E–M–M–	
–M–E–M–M–E	–M–M–E–M–M–E–M–	92.01
	–M–M–M–M–M–M–M–	
–M–M–M–M–M–	–M–M–M–M–M–M–E–	88.86
	–E–M–M–M–M–M–E–	
–E–M–M–M–M–	–M–E–M–M–M–M–M–	88.55
	–M–E–M–M–M–M–E–	
–E–M–M–M–E–	–M–E–M–M–M–E–M–	88.15

49.7 SOME PROPERTIES OF LINEAR, HIGH MOLECULAR WEIGHT POLYACETALS

The only polyacetal which is made on an industrial scale (150×10^3 tons year^{-1}) is polyformaldehyde. This is a highly crystalline polymer (50–60% crystallinity for commercial products Delrin® or Hostaform®) with a crystalline melting point of *ca.* 180°C and a density of 1.42 g cm^{-3}[95] (both density and melting point, however, decrease with increasing content of oxyethylene units).[96] The polymer is insoluble in common organic solvents; it dissolves well only in strongly hydrogen-bonding solvents (*e.g.* phenols) at above 100°C, or in hexafluoroacetone at room temperature.

Polyformaldehyde is perfectly stable in even strongly basic media and moderately stable in acidic media;[95] it is a high performance engineering plastic.

Other polyacetals have not found practical applications yet. Poly-1,3-dioxolane is a crystalline (50–80%) polymer of m.p. = 55 °C,[97] soluble in common organic solvents; polymers of lower molecular weight are also soluble in water.

Poly-1,3-dioxepane exhibits a lower degree of crystallinity and has a m.p. = 24 °C;[98] thus, at room temperature the high molecular weight polymer behaves like a typical elastomer. It is soluble in organic solvents (CH_2Cl_2, PhH, THF) but insoluble in water. Polymers of higher acetals of general formulae $[-(OCH_2CH_2)_nOCH_2-]_m$ are moderately crystalline materials with crystalline melting points around room temperature.[99]

49.8 LABORATORY PREPARATION OF POLYACETALS

49.8.1 Polymerization of Simple Cyclic Acetals

High molecular weight polymers can easily be obtained in the polymerization of 1,3-dioxolane or 1,3-dioxepane, providing that suitable reaction conditions are used.

Cyclic acetal monomers are conveniently purified by fractionation; CaH_2, sodium mirror and sodium/potassium alloy are recommended as drying agents.

The solvent of choice is methylene chloride; its purification involves treatment with fuming sulfuric acid to remove unsaturated impurities, washing with water, drying, fractionation over CaH_2 and final purification over a sodium mirror.

Convenient initiators are triflic acid (trifluoromethanesulfonic acid) and its anhydride, which can be purified by distillation (b.p. 162 °C and 80 °C, respectively).[100]

Polymerization of cyclic acetals proceeds rapidly, thus reaction may be carried out at low temperatures. At −78 °C the equilibrium monomer concentrations are low and high conversions may be attained, the side reactions being minimized.

Polymerization reaches equilibrium in several hours with initiator concentration in the range 10^{-2}–10^{-3} mol l^{-1}. The molecular weight of the isolated polymers corresponds to $\overline{DP}_n = ([M]_0 - [M]_e)/[I]_0$, thus polymers having $\overline{M}_n \approx 10^5$ can easily be prepared (see Table 6).

49.8.2 Polyacetal-based Telechelics

The theory of chain transfer leading to telechelics has mostly been developed by Irzak, Rozenberg and Enikolopyan and summarized in a monograph.[61] Of special interest is the polymerization of cyclic acetals in the presence of diols. Thus, it has been shown that polymerization of 1,3-dioxolane or 1,3-dioxepane with $HOSO_2CF_3$ as catalyst ($\sim 10^{-2}$ mol l^{-1}) in the presence of 1,4-butanediol (3–10 mol % with respect to monomer) at room temperature leads to low molecular weight oligomers terminated at both ends with –OH groups.[101] One molecule of macrodiol is formed from one molecule of the starting diol, thus molecular weights in the range 1000–2000 can be easily controlled (see equation 28).

$$\overline{DP}_n = \frac{[M]_0 - [M]_e}{[diol]_0} \tag{28}$$

When oligomeric diols, *e.g.* poly(ethylene oxide) diol, are used instead of 1,4-butanediol, telechelic block copolymers are formed, as shown by Franta and Reibel.[102] This process is, however, accompanied by side reactions, leading to formation of some multiblock copolymers (equation 29).[103]

$$HO\sim\sim OH + \underset{O\diagdown\diagup O}{\bigtriangleup} + HO\sim\sim OH \underset{}{\overset{H^+}{\rightleftharpoons}} HO\sim\sim O\diagdown_{CH_2}O\sim\sim OH + HOCH_2CH_2OH \tag{29}$$

These telechelics are bifunctional and contain only one type of end-group, namely $-OCH_2O(CH_2)_nOH$ groups, as shown by end-group analysis based on the complexation of –OH groups with hexafluoroacetone and determination of the ^{19}F NMR chemical shift.[104]

49.8.3 Polymerization of 1,3,5-Trioxane

Several methods of 1,3,5-trioxane polymerization have been described in the patent literature. In this section, we give a detailed description of the solution–precipitation polymerization in which the aggregation of precipitating polymer particles, which is a major shortcoming of the solution process, is avoided.[105] A typical procedure is as follows: 144 ml (168 g, 1.87 mol) of freshly distilled 1,3,5-trioxane ($[H_2O] < 50$ p.p.m.), 6.6 ml (7 g, 0.094 mol) of 1,3-dioxolane, 120 ml of cyclohexane and 3.6 g of poly(ethylene oxide) ($\bar{M}_n \approx 15\,000$) are charged into the reaction flask under nitrogen. The flask is thermostatted at 60 °C. To initiate polymerization, 0.1 mol % (with respect to 1,3,5-trioxane, *i.e.* 0.14 g) of $BF_3 \cdot OBu_2^n$ complex is added under constant stirring. Polymerization ensues after a short induction period, the temperature rises to 70–75 °C and polymer precipitates. The reaction is terminated after 60 min by adding a solution of NH_3 in MeOH, the polymer is filtered off, washed with warm water and methanol and then dried at 70 °C. A typical yield is 70–80%.

Poly(ethylene oxide) prevents aggregation of the precipitated polymer particles. In the absence of polyether, the reaction mixture containing 45 wt % of polyacetal solidifies and cannot be mixed (at least on a laboratory scale). In the presence of polyether, even a mixture containing 60 wt % of the solid polymer is fluid, like an emulsion, and can be efficiently mixed. The reason for this behaviour is a difference in the shape and size of particles precipitating in the presence and in the absence of polyether, as shown in Figures 7 and 8 respectively.

Particles obtained in the presence of poly(ethylene oxide) are smaller and more regular in shape. This result indicates that polyether forms a protective layer on the growing polymer particles and, at a certain size, this layer effectively prohibits further growth of individual particles. Poly(ethylene oxide) cannot be removed by extraction, thus it has to be chemically bonded[105] or somehow embedded in the polyacetal core.

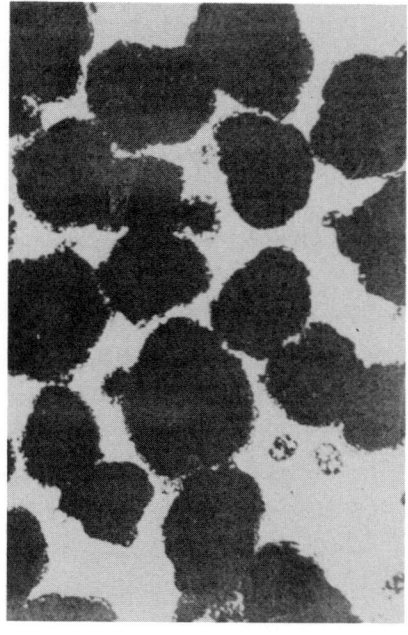

Figure 7 Particles of copolymer obtained without poly(ethylene oxide), magnification × 500, $[1,3,5\text{-trioxane}]_0 = 7.0$ mol kg^{-1}, $[1,3\text{-dioxolane}]_0 = 0.35$ mol kg^{-1}, $[BF_3 \cdot OBu_2^n]_0 = 2 \times 10^{-3}$ mol kg^{-1}, conversion 71% (ref. 105)

49.9 NOTE ON THE CATIONIC POLYMERIZATION OF ALDEHYDES

Polyacetals are also formed by C=O double bond-opening polymerization of aldehydes (equation 30).

$$n \quad \begin{array}{c} R \\ | \\ C=O \\ | \\ H \end{array} \rightarrow \begin{array}{c} R \\ | \\ \mathbf{\{}C-O\mathbf{\}}_n \\ | \\ H \end{array} \tag{30}$$

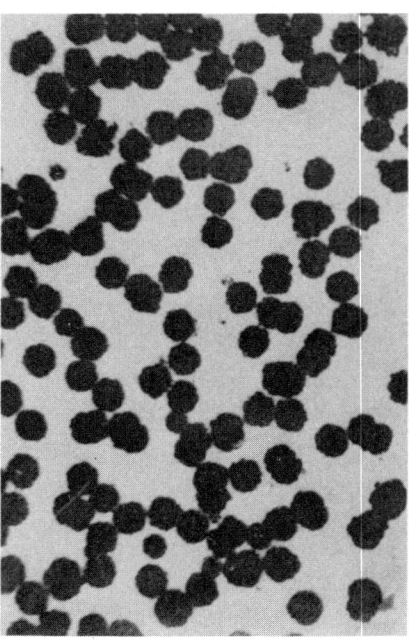

Figure 8   Particles of copolymers obtained in the presence of 1.5 wt % of poly(ethylene oxide) $\bar{M}_n = 15\,000$, magnification × 500, $[\text{1,3,5-trioxane}]_0 = 7.0 \text{ mol kg}^{-1}$, $[\text{1,3-dioxolane}]_0 = 0.35 \text{ mol kg}^{-1}$, $[\text{BF}_3 \cdot \text{OBu}_2^n]_0 = 4.0 \times 10^{-3} \text{ mol kg}^{-1}$, conversion 86% (ref. 105)

Aldehydes polymerize predominantly by ionic mechanisms, and although anionic methods are more important, several aldehydes, including formaldehyde, can be polymerized by cationic methods. Polymerization of this class of monomers has been studied extensively by Vogl's group and the subject has been reviewed.[106]

Acetaldehyde in the presence of Lewis of Brönsted acids gives, at temperatures below $-40\,°C$, amorphous high molecular weight polymers which behave like typical elastomers.[107] As with polyformaldehyde, polymers of acetaldehyde are thermally unstable and have to be end-capped.

Several halogenated aldehydes also polymerize by cationic mechanisms; polymerization of trichloroacetaldehyde–chloral has been studied most extensively.[108] Polymers of chloral are, however, intractable (insoluble, infusible), which hampers the application of these polymers, which are inflammable, chemically highly resistant and have good mechanical properties. Other halo-aldehydes which have been polymerized by cationic mechanisms include tribromoacetaldehyde–bromal[109] and trifluoroacetaldehyde–fluoral.[110] Of particular interest is formation of helical polymers by direct polymerization of chloral and related monomers in the solid state.[111]

49.10   APPENDIX A: REVIEWS ON THE POLYMERIZATION OF CYCLIC ACETALS

S. Penczek, P. Kubisa and K. Matyjaszewski, 'Cationic Ring-opening Polymerization. Volume I. Mechanisms.', *Adv. Polym. Sci.*, 1980, **37**, 1.

S. Penczek, P. Kubisa and K. Matyjaszewski, 'Cationic Ring-opening Polymerization. Volume II. Synthetic Applications.', *Adv. Polym Sci.*, 1985, **68/69**, 1.

K. J. Ivin and T. Saegusa (eds.), 'Ring-opening Polymerization', Elsevier, London, 1984, vol. I, chaps. 1, 5 and 6.

M. Okada and H. Sumitomo, 'Polymerization of Bicyclic Acetals', *Makromol. Chem. Suppl.*, 1985, **14**, 29.

R. Szymanski, P. Kubisa and S. Penczek, 'Mechanisms of Cyclic Acetal Polymerization. End of a Controversy?', *Macromolecules*, 1983, **16**, 1000.

49.11 REFERENCES

1. H. Staudinger, 'Die Hochmolekularen Organischen Verbindungen', Springer Verlag, Berlin, 1922.
2. R. N. McDonald (E.I. du Pont de Nemours and Co.) *US Pat.* 2 768 994 (1956).
3. C. T. Walling, F. Brown and K. W. Bartz (Celanese Corporation of America), *US Pat.* 3 027 352 (1962).
4. J. M. Astle, J. A. Zaslowsky and P. G. Lafyatis, *Ind. Eng. Chem.*, 1954, **46**, 787.
5. K. C. Frisch (ed.), 'Cyclic Monomers', Wiley Interscience, New York, 1972, p. 114.
6. J. W. Hill and W. H. Carothers, *J. Am. Chem. Soc.*, 1935, **57**, 925.
7. N. G. Gaylord, in 'Encyclopedia of Polymer Science and Technology', ed. H. F. Mark and N. G. Gaylord, Wiley, New York, 1969, vol. 10, p. 319.
8. K. Albrecht, D. Fleischer, A. Kane, Ch. Rentsch, Q. V. T. Thi, H. Yamaguchi and R. C. Schulz, *Markromol. Chem.*, 1977, **178**, 881.
9. J. F. Walker, 'Formaldehyde', Reinhold, New York, 1964.
10. T. P. Murray, C. S. Williams and R. K. Brown, *J. Org. Chem.*, 1971, **36**, 1311.
11. C. Schuerch, in 'Encyclopedia of Polymer Science and Technology', ed. H. F. Mark, N. G. Gaylord and N. M. Bikales, Wiley, New York, 1976, suppl. vol. 1, p. 810.
12. F. S. Dainton and K. J. Ivin, *Trans. Faraday Soc.*, 1950, **46**, 331.
13. E. H. Cordes, in 'Progress in Physical Organic Chemistry', ed. A. Streitwieser and R. W. Tafts, Interscience, New York, 1967, vol. 4, p. 1.
14. S. M. Skuratov, A. A. Strepikheev, S. M. Shtekher and A. V. Volokhina, *Dokl. Akad. Nauk SSSR*, 1957, **117**, 263.
15. Y. Yamashita, M. Okada, K. Suyama and H. Kasahara, *Makromol. Chem.*, 1968, **114**, 146.
16. P. H. Plesch and P. H. Westermann, *Polymer*, 1969, **10**, 105.
17. L. I. Kuzub, M. A. Markevich, Al. Al. Berlin and N. S. Enikolopyan, *Vysokomol. Soedin., Ser. A*, 1968, **10**, 2007.
18. R. C. Schulz, K. Albrecht, C. Rentsch and Q. V. T. Thi, *ACS Symp. Ser.*, 1977, **59**, 77.
19. M. Szwarc, *Markromol. Chem. Suppl.*, 1979, **3**, 327.
20. Y. Yamashita, J. Mayumi, Y. Kawakami and K. Ito, *Macromolecules*, 1980, **13**, 1075.
21. Y. Firat and P. H. Plesch, *Makromol. Chem.*, 1975 **176**, 1979.
22. M. Okada, K. Mita and H. Sumitomo, *Makromol. Chem.*, 1975, **176**, 859.
23. M. Okada, K. Yagi and H. Sumitomo, *Makromol. Chem.*, 1973, **163**, 225.
24. M. Okada, T. Hisada and H. Sumitomo, *Makromol. Chem.*, 1978, **179**, 959.
25. B. Krummenacher and H. G. Elias, *Makromol. Chem.*, 1971, **150**, 271.
26. B. Xu. P. Lillya and J. C. W. Chien, *Macromolecules*, 1988, in press.
27. M. Okada, H. Sumitomo and H. Komada, *Macromolecules*, 1979, **12**, 395.
28. J. Kops and H. Spanggaard, *Makromol. Chem.*, 1975, **176**, 299.
29. C. Schuerch, *Adv. Polym. Sci.*, 1972, **10**, 173.
30. Al. Al. Berlin *et al.*, *Vysokomol. Soedin., Ser. A*, 1975, **17**, 643.
31. M. Bednarek, K. Brzezińska, J. Stasiński, P. Kubisa and S. Penczek, *Makromol. Chem.*, 1988, submitted.
32. F. R. Jones and P. H. Plesch, *Chem. Commun.*, 1969, 1230, 1231.
33. P. Kubisa and S. Penczek, *Makromol. Chem.*, 1978, **179**, 445.
34. P. Kubisa and S. Penczek, *Macromolecules*, 1977, **10**, 1216.
35. S. Penczek and P. Kubisa, *Makromol. Chem.*, 1973, **165**, 121.
36. K. Matyjaszewski and S. Penczek, *J. Polym. Sci., Polym. Chem. Ed.*, 1974, **12**, 1905.
37. K. Matyjaszewski, T. Diem and S. Penczek, *Makromol. Chem.*, 1979, **180**, 1817.
38. W. Chwiałkowska, P. Kubisa and S. Penczek, *Makromol. Chem.*, 1982, **183**, 753.
39. S. Słomkowski and S. Penczek, *J. Chem. Soc., Perkin Trans. 2*, 1974, 1718.
40. S. Słomkowski and S. Penczek, *Chem. Commun.*, 1970, 1347.
41. A. Stolarczyk, P. Kubisa and S. Penczek, *J. Macromol. Sci., Chem.*, 1977, **A11**, 2047.
42. S. Penczek, P. Kubisa and K. Matyjaszewski, *Adv. Polym. Sci.*, 1980, **37**, 21.
43. C. U. Pittman, Jr, S. P. McManus and J. W. Larsen, *Chem. Rev.*, 1972, **72**, 357.
44. Z. Jedlinski and M. Gibas, *Macromolecules*, 1980, **13**, 1700.
45. S. Słomkowski, Ph.D Thesis, University of Lodz, 1974.
46. Kabir-ud-din, P. H. Plesch, *J. Chem. Soc., Perkin Trans. 2*, 1978, 937.
47. Z. Jedlinski, J. Lukaszczyk, J. Dudek and M. Gibas, *Macromolecules*, 1976, **9**, 622.
48. V. Jaacks and W. Kern, *Makromol. Chem.*, 1963, **62**, 1.
49. W. Kern, *Chem. Ztg., Chem. Appar.*, 1964, **88**, 623.
50. G. L. Collins, R. K. Greene, F. M. Berardinelli and W. V. J. Garruto, *J. Polym. Sci., Polym. Lett. Ed.*, 1979, **17**, 667.
51. S. Okamura, T. Higashimura and T. Miki, *Prog. Polym. Sci. Jpn.*, 1972, **3**, 97.
52. T. Uryu, K. Ito, K. Kobayashi and K. Matsuzaki, *Makromol. Chem.*, 1979, **180**, 1509.
53. K. Burg, H. Schlaf and H. Cherdron, *Makromol. Chem.*, 1971, **145**, 247.
54. A. Chapiro, in 'Encyclopedia of Polymer Science and Technology', ed. H. F. Mark and N. G. Gaylord, Wiley, New York, 1969, vol. 11, p. 702.
55. G. C. Eastmond, in 'Progress in Polymer Science', ed. A. D. Jenkins, Pergamon Press, Oxford, 1970, vol. II, p. 3.
56. Y. Chatani, *Prog. Polym. Sci. Jpn.*, 1974, **7**, 149.
57. V. I. Tupikov and S. Y. Pshezhetskii, *Zh. Fiz. Khim.*, 1964, **38**, 2430.
58. G. Kiss, K. Kiss, A. J. Kovacs and J. C. Wittmann, *J. Appl. Polym. Sci.*, 1981, **26**, 2485.
59. S. Penczek and R. Szymański, *Polym. J.*, 1980, **12**, 617.
60. M. Kuchera and Yu. Pichler, *Vysokomol. Soedin.*, 1965, **7**, 3.
61. B. A. Rosenberg, W. J. Irzhak and N. S. Enikolopyan, 'Interchain Exchange in Polymers', Chimia, Moscow, 1975.
62. R. Szymański and S. Penczek, *Makromol. Chem.*, 1982, **183**, 1587.
63. M. Zieliński, S. Słomkowski, K. Matyjaszewski, P. Kubisa, J. Chojnowski and S. Penczek, *Makromol. Chem.*, 1980, **181**, 1469.
64. P. Kubisa and S. Penczek, *Makromol. Chem.*, 1979, **180**, 1821.
65. P. Kubisa, *J. Polym. Sci.*, 1987, **25**, 873.

66. R. Mateva, G. Wegner and G. Lieser, *J. Polym. Sci., Polym. Lett. Ed.*, 1973, **11**, 369.
67. A. Lücke and G. Wegner, *Polym. Prepr., Am. Chem. Soc., Div. Polym. Chem.*, 1980, **21**, 197.
68. G. Wegner, M. Rodriguez-Baeza, A. Lücke and G. Lieser, *Makromol. Chem.*, 1980, **181**, 1763.
69. H. D. Hermann, E. Fischer and K. Weissermel, *Makromol. Chem.*, 1966, **90**, 1.
70. W. Kern, H. Deibig, A. Giefer and V. Jaacks, *Pure Appl. Chem.*, 1966, **12**, 371.
71. K. Weissermel, E. Fischer, K. Gutweiler, H. D. Hermann and H. Cherdron, *Angew. Chem., Int. Ed. Engl.*, 1967, **6**, 526.
72. K. Weissermel, E. Fischer, K. Gutweiler and H. D. Herman, *Kunstoffe*, 1964, **54**, 410.
73. V. Jaacks, *Adv. Chem. Ser.*, 1969, **91**, 371.
74. L. Leese and M. W. Baumber, *Polymer*, 1964, **5**, 380.
75. W. Jacobson and W. H. Stockmayer, *J. Chem. Phys.*, 1950, **18**, 1600.
76. J. M. Andrews and J. A. Semlyen, *Polymer*, 1972, **13**, 141.
77. C. Rentsch and R. C. Schulz, *Makromol. Chem.*, 1978, **179**, 1403.
78. I. Ishigaki, Y. Morita, K. Nishimura and A. Ito, *J. Appl. Polym. Sci.*, 1974, **18**, 1927.
79. A. Lücke, PhD Thesis, University of Freiburg, 1979, p. 137.
80. C. Rentsch and R. C. Schulz, *Makromol. Chem.*, 1977, **178**, 2535.
81. P. H. Plesch and P. H. Westermann, *J. Polym. Sci., Part C*, 1968, **16**, 3837.
82. M. Okada and H. Sumitomo, *Makromol. Chem. Suppl.*, 1985, **14**, 29.
83. M. Okada, H. Sumitomo and Y. Hibino, *Polym. J.*, 1974, **6**, 256.
84. Ye. L. Berman, A. A. Gorkovenko and V. A. Ponomarenko, *Vysokomol. Soedin., Ser. A*, 1984, **26**, 2165.
85. T. Uryu, H. Tachikawa, K. I. Ohaku, K. Terui and K. Matsuzaki, *Makromol. Chem.*, 1977, **178**, 1929.
86. H. Komada, M. Okada and H. Sumitomo, *Macromolecules*, 1979, **12**, 5.
87. M. Okada, H. Sumitomo and H. Komada, *Macromolecules*, 1979, **12**, 395.
88. Z. N. Nysenko, E. L. Berman, E. B. Lyudwig, A. P. Klimov, V. A. Ponomarenko and G. V. Isagulyants, *Vysokomol. Soedin., Ser. A*, 1976, **18**, 1696.
89. V. Jaacks, K. Boehlke and E. Eberius, *Markromol. Chem.*, 1968, **118**, 354.
90. D. Fleischer and R. C. Schulz, *Makromol. Chem.*, 1975, **176**, 677.
91. J. E. Mark and E. Riande, *Macromolecules*, 1978, **11**, 956.
92. M. Okada, K. Mita and H. Sumitomo, *Makromol. Chem.*, 1976, **177**, 895.
93. G. Opitz, *Plaste Kautsch.*, 1975, **22**, 951.
94. L. Höhr, H. Cherdron and W. Kern, *Makromol. Chem.*, 1962, **52**, 59.
95. S. J. Barker and M. B. Price, 'Polyacetals', Iliffe Books, London, 1970.
96. M. Dröscher, G. Lieser, H. Reimann and G. Wegner, *Polymer*, 1975, **16**, 497.
97. S. Sasaki, Y. Takahashi and H. Tadakoro, *J. Polym Sci., Polym. Phys. Ed.*, 1972, **10**, 2363.
98. S. Sasaki, Y. Takahashi and H. Tadokoro, *Polym. J.*, 1973, **4**, 172.
99. R. C. Schulz, K. Albrecht, Q. V. T. Thi, J. Nienburg and D. Engel, *Polym. J.*, 1980, **12**, 639.
100. 'Products For Synthesis', Merck–Schuchard, 1985/1986, p. 1084.
101. S. Penczek *et al.*, to be published.
102. L. Reibel, H. Zouine and E. Franta, *Makromol. Chem., Macromol. Symp.*, 1986, **3**, 221.
103. E. Franta, P. Kubisa, J. Refai and L. Reibel, *Makromol. Chem., Macromol. Symp.*, 1988, **13/14**, 127.
104. G. R. Leader, *Anal. Chem.*, 1970, **42**, 16.
105. S. Penczek, J. Fejgin, W. Sadowska and M. Tomaszewicz, *Makromol. Chem.*, 1968, **116**, 203.
106. P. Kubisa, K. Neeld, J. Starr and O. Vogl, *Polymer*, 1980, **21**, 1433.
107. O. Vogl, *J. Polym. Sci.*, 1960, **46**, 261.
108. P. Kubisa and O. Vogl, *Vysokomol. Soedin., Ser. A*, 1975, **17**, 929.
109. D. W. Lipp and O. Vogl, *Polymer*, 1977, **18**, 1051.
110. W. K. Busfield, *Polymer*, 1968, **9**, 479.
111. O. Vogl, presented at the 32nd IUPAC International Symposium on Macromolecules, Kyoto, 1988.

50
Cationic Ring-opening Polymerization: Cyclic Esters

STANISLAW PENCZEK and STANISLAW SLOMKOWSKI
Polish Academy of Sciences, Lodz, Poland

50.1 INTRODUCTION

The groups of monomers given in structures (**1**)–(**10**) will be discussed in this chapter. Among these monomers, only lactones and cyclic phosphates have been comprehensively studied.

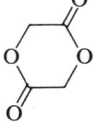

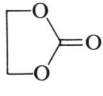

(**1**) β-propiolactone (internal ester of β-hydroxypropionic acid)

(**2**) 1,4-dioxane-2,5-dione (glycollide; cyclic dimer of α-hydroxyacetic acid)

(**3**) 1,3-dioxalan-2-one (ethylene carbonate)

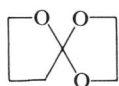

(**4**) 2,6,7-trioxabicyclo[2.2.1]heptane (bicyclic fused ortho ester)

(**5**) 1,4,6-trioxaspiro[4.4]nonane (spiro ortho ester)

(6) 2-methoxy-1,3,2-dioxaphospho-
lane (methyl ethylene phosphite)

(7) 2-methoxy-1,3,2-dioxaphospho-
lane 2-oxide (methyl ethylene phosphate)

(8) 1-phospha-2,6,7-trioxabicyclo-
[2.2.1]heptane 1-oxide (bicyclo-
glycerol phosphate)

(9) 1,3,2-dioxathiolane   2-oxide
(ethylene sulfite)

(10) 1,3,2-dioxathiolane   2,2-diox-
ide (ethylene sulfate)

50.2 ABILITY OF MONOMERS TO POLYMERIZE

Cyclic esters belonging to each of the groups mentioned above can be cationically polymerized, and some can be polymerized exclusively by this mechanism. However, when anionic or pseudo-anionic mechanisms are also operative, these would be the methods of choice when high molecular weight polymers are required. Cationic polymerization does not usually lead to polymers with high molecular weights, due to the inter- or intra-molecular (back-biting) chain transfer to polymer or other transfer processes, such as hydride ion or proton transfer. Apparently, only in a few systems can lactones be polymerized by a cationic method in nearly 'living' conditions. Thus, polymerization of β-propiolactone initiated with $MeCOCl \cdot SbCl_5$ gave polymers with a number-average molecular weight ($\bar{M}_n$) from 90 000 to 300 000, close to the calculated values, assuming a $\overline{DP}_n(calc.) = [M]_0/[I]_0$ (where $[M]_0 =$ initial monomer concentration, $[I]_0 =$ initial initiator concentration and $\overline{DP}_n =$ average degree of polymerization).[1] High molecular weight poly(β-propiolactone) ($\bar{M}_n \simeq 10^5$) was also obtained in polymerizations initiated with $AlCl_3$/trifluoroacetic anhydride, acetyl perchlorate and trifluoroacetic acid.[2]

It appears that anionic polymerization, and particularly pseudoanionic (covalent–coordinate) polymerization, provides the best results. Nevertheless, in the anionic polymerization of lactones with larger rings (*e.g.* δ-valerolactone, ε-caprolactone), as well as in the anionic polymerization of cyclic carbonates (*e.g.* neopentyl carbonate), undesired cyclic oligomers are formed.[3–8] It has only recently been shown that application of pseudoanionic initiators leads to polymerizations with kinetic suppression of rings.[9,10]

50.3 SOME PROPERTIES OF POLYESTERS

β-Propiolactone has been found to be highly carcinogenic, and is banned from any practical use, although its polymer might have been an interesting product for the plastic and fiber industry. Poly(pivalolactone) (α,α-dimethyl-β-propiolactone) was considered for some time as a potential fiber-forming crystalline material, but was abandoned because of the unfavorable crystalline interconversions. Poly(ε-caprolactone), prepared anionically, is used as a telechelic oligomer, providing soft blocks for polyurethane elastomers. Its biodegradability has a number of potential applications in the biomedical and agricultural fields. Polyglycolide and polylactide, as well as their copolymers, are used as biomaterials (including absorbable sutures). Poly(alkylene phosphates) are used as models in studies of biopolymers.

50.4 POLYMERIZATION OF LACTONES

50.4.1 Mechanism of Initiation

The ambident reactivity of lactones makes possible the attack of the cationic species on exo- or endo-cyclic oxygen atoms, followed by alkyl–oxygen or acyl–oxygen bond scission, respectively (equation 1).

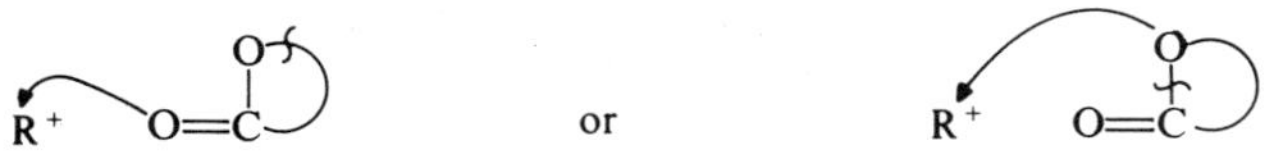

initiation with alkyl–oxygen bond scission   initiation with acyl–oxygen bond scission   (1)

Recently, the mechanism of initiation has been elaborated and explained in detail for the polymerization of β-propiolactone, δ-valerolactone and ε-caprolactone initiated with alkylating (*e.g.* $Me_2I^+SbF_6^-$, $Me_2Br^+SbF_6^-$, $MeOSO_2CF_3$, $MeOSO_2Me$, $EtOSO_2F$, $(MeO)_2SO_2$, $MeOSO_2C_6H_4Me$ and $Et_3O^+BF_4^-$) and acylating (*e.g.* $MeCO^+SbF_6^-$) initiators.[11-14] Methyl, ethyl or acyl cations are transferred from these initiators to monomer molecules. The subsequent formation of polymers with ester end-groups (for 'Me$^+$' or 'Et$^+$' cations; see, for example, equation 2) was established by ^1H and/or ^{13}C NMR spectroscropy. This observation indicated that the more basic exocyclic oxygen atoms in monomer molecules were attacked, resulting eventually in the alkyl–oxygen bond scission and formation of an oxonium ion as a growing species (**12**; for β-propiolactone).

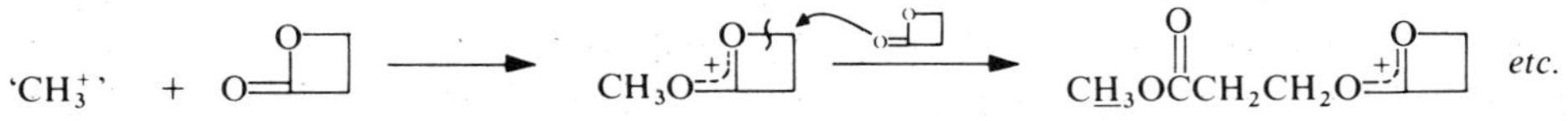

$$\delta = 3.67 \text{ p.p.m. } (^1H \text{ NMR})$$

(2)

The less probable alternative attack of cation on the less basic endocyclic oxygen atom, and ring-opening with acyl–oxygen bond scission, was thus excluded.[11] This type of attack would have yielded polymers with ether end-groups, which were not detected.

In cation (**13**) the ring-opening with alkyl–oxygen bond scission was excluded because we found in models studies that acyl–oxygen bond scission (equation 3) is $\sim 10^6$ times faster.[15]

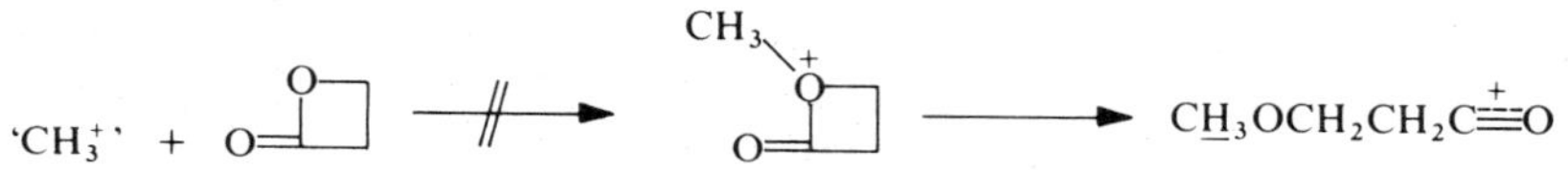

$$\delta = 3.20 \text{ p.p.m. } (^1H \text{ NMR})$$

(3)

The mechanism of initiation, described above for β-propiolactone, was confirmed in the same way for the much less strained δ-valerolactone and ε-caprolactone.[11-14] Very recently, oxonium ions (**14**), formed by transfer of a methyl cation from $MeOSO_2CF_3$ to the exocyclic oxygen atom of δ-valerolactone, were observed directly by ^1H NMR (Table 1).[14]

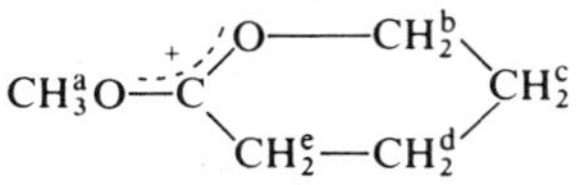

(**14**)

Table 1 Chemical Shifts in ^1H NMR Spectrum[14] of Cation (**14**)

Groups of protons in (**14**)	Chemical shift (p.p.m.)
a	4.33 (s)
b	5.16 (t)
c	2.19 (m)
d	2.13 (m)
e	3.07 (t)

In contrast to the simple initiation with alkylating initiators (equation 2), the initiation with acylium salts is more complicated. ^1H NMR spectra of poly(β-propiolactone) initiated with $MeCO^+SbF_6^-$ revealed that two kinds of macromolecules had been formed; with anhydride ($CH_3C(O)OC(O)CH_2CH_2-$, $\delta = 2.20$ p.p.m.) and ester ($CH_3C(O)OCH_2CH_2-$, $\delta = 2.08$ p.p.m.) end-groups. The ratio of macromolecules with these end-groups was close to $1:1$.[11] Similar observations were made for the polymerization of ε-caprolactone.[11] Anhydride end-groups were observed by IR spectroscopy ($\nu_{CO} = 1820$ cm^{-1}) for the polymerization of β-propiolactone, δ-valerolactone and ε-caprolactone initiated with MeCO$^+$SbF$_6^-$.[11,13] Thus, it has been concluded that, in the polymerization of β-propiolactone and ε-caprolactone, initiation with MeCO$^+$SbF$_6^-$ leads to alkyl–oxygen and acyl–oxygen bond scission with equal probability. After a limited number of monomer additions (approximately five) the growing species become oxonium ions and less than 4% of acylium cations are left as growing species. At the earlier stage of polymerization, when acylium cations are still present (before being converted into oxonium ions; (**15**)), these cations are involved in a fast equilibrium (equation 4). This equilibrium is strongly shifted to the right-hand side.

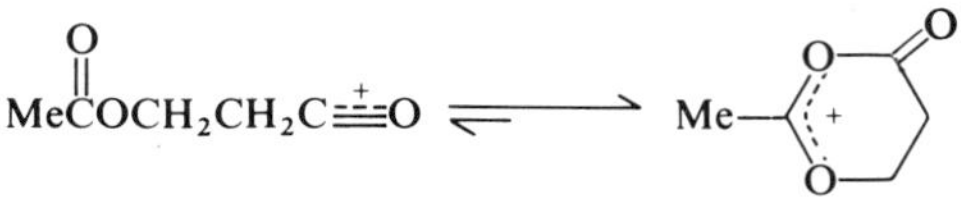

$$\text{MeCOCH}_2\text{CH}_2\text{C}\overset{+}{\equiv}\text{O} \rightleftharpoons \text{Me}\cdots \qquad (4)$$

50.4.2 Mechanism and Kinetics of Propagation of Unsubstituted Lactones

As explained in the previous section, the ambient reactivity of lactones requires, first of all, establishing the site of attack and the mode of ring opening for both initiation and propagation processes.

For the polymerization of unsubstituted lactones (β-propiolactone, δ-valerolactone and ε-caprolactone) the structure of active centers has been determined by the method of cation-trapping with triphenylphosphine.[11,13] For example, in the polymerization of β-propiolactone initiated with Me$_2$I$^+$SbF$_6^-$ and/or Me$_2$Br$^+$SbF$_6^-$, the reaction of active centers with PPh$_3$ (after complete consumption of monomer) led to the formation of alkyltriphenylphosphonium cations, *e.g.* cations (**16**), as shown in equation (5). This substantiates the hypothesis that the active centers are oxonium cations, formed from the very beginning in the course of initiation (equation 2), and that propagation proceeds with alkyl–oxygen bond scission (equation 6). Alternatively, cation (**16**) could have been formed through attack by a polymeric carbenium ion, but its existence at any kinetically meaningful and measurable concentration is doubtful in the presence of lactones, taken at concentrations used for polymerization.

$$(5)$$

(**15**)   (**16**)

$\delta = 23.8$ p.p.m. (^{31}P{^{1}H} NMR)

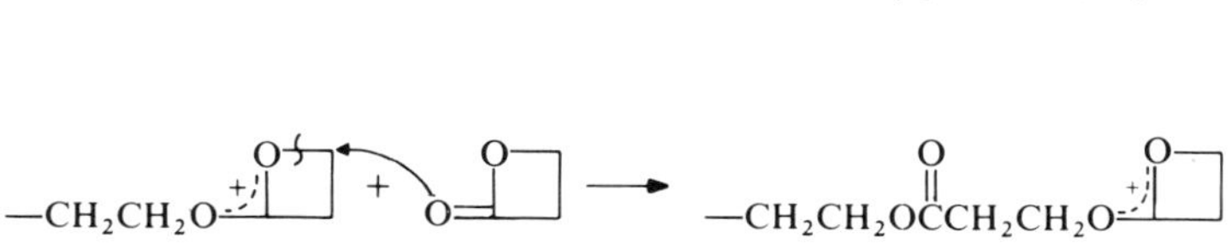

$$(6)$$

Propagation on acylium active centers (**17**) with acyl–oxygen bond scission (equation 7) has to be excluded on the following grounds: acylium active centers should react with PPh$_3$ giving cations (**18**) absorbing at ~10 p.p.m. (equation 8) and those signals were absent in the ^{31}P{^{1}H} NMR spectra of poly(β-propiolactone). The same mechanism was established for the polymerization of δ-valerolactone and ε-caprolactone initiated with alkylating initiators.[11,13]

$$(7)$$

(**17**)

$$(8)$$

(**18**)

$\delta \simeq 10$ p.p.m. (^{31}P{^{1}H} NMR)

Recently, oxonium ions, corresponding to those given in equation (6) for β-propiolactone, were observed directly by ^{1}H NMR for oligomers of poly(δ-valerolactone) [triflate (trifluoromethanesulfonate) counterion, $CDCl_3$ solvent].[14] In the polymerization with triflate counterion one also has to take into consideration an equilibrium involving the formation of macroesters (**20**). Such an equilibrium has been observed recently by ^{1}H NMR in the polymerization of δ-valerolactone.[14]

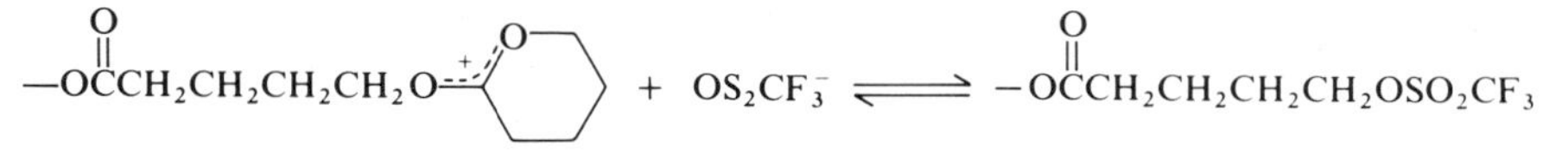

$$-O\overset{\overset{O}{\|}}{C}CH_2CH_2CH_2CH_2O \overset{+}{-} \overset{O}{\diagdown} + OS_2CF_3^- \rightleftharpoons -O\overset{\overset{O}{\|}}{C}CH_2CH_2CH_2CH_2OSO_2CF_3 \qquad (9)$$

$$(19) \qquad\qquad\qquad\qquad\qquad (20)$$

The ion-trapping method indicated that, in the polymerization of β-propiolactone initiated with $MeCO^+SbF_6^-$, the active species are also oxonium cations (**15**). Therefore alkyl–oxygen bond scission is responsible for propagation, as in the polymerization initiated by alkylating initiators (equation 6).[11] As mentioned in the previous section on initiation, acylium cations, constituting 50% of the active centers formed during initiation, react with monomer in a similar manner to an acylium initiator. This means that in every propagation step roughly half of the acylium active centers are replaced by oxonium centers. Thus, after a few propagation steps, oxonium ions (**15**) predominate, *i.e.* constitute more than 95% of all the active centers after five monomer additions. Propagation on oxonium active centers was also established for the polymerization of δ-valerolactone and ε-caprolactone, initiated with $MeCO^+SbF_6^-$.[11,13]

The kinetics of the cationic polymerization of β-propiolactone and ε-caprolactone in CH_2Cl_2 and $MeNO_2$ have been studied by Lyudvig and co-workers;[16-19] complicated kinetic behavior was observed. Polymerization in CH_2Cl_2 proceeded with acceleration. The order of propagation with respect to initiator was fractional (3/2). Initial rates of propagation were proportional to initial monomer concentrations for $[M]_0 < 2\ mol\ l^{-1}$ but became independent from monomer concentration for $[M]_0 > 2\ mol\ l^{-1}$. To explain these results ions, ion pairs and triple ions, as well as complexes of ionic species with monomer and polymer, were proposed as coexisting in the system. These analyses were based on an assumption that the active species were acylium cations (**17**), which contradicts the findings described above. Apparently none of the rate constants of propagation reported in the literature refer to the elementary reactions.

50.4.3 Mechanism of Polymerization of Substituted Lactones

In the case of monomers with electron-donating groups, one could envisage formation of carbocations as active species. This was postulated for β,β'-dimethyl-β-propiolactone, but has never been directly proved by spectroscopic or other direct methods.[20]

On the other hand, the substitution of lactones at the β position results in increased stability of cation (**13**; equation 3), due to the inductive effect of the alkyl substituent, and decreases, due to steric hindrance, the probability of nucleophilic attack on the β carbon atom pertinent to propagation with alkyl–oxygen bond scission (equation 6). Thus, for lactones substituted at the β position, the probability of propagation with acyl–oxygen as against alkyl–oxygen scission is increased.

It has been found that cationic polymerization of D(+)-β-methyl-β-propiolactone yields polymers with optical activity lower than that for the polymer built only from D polymeric units, indicating that propagation involves reactions with retention (acyl–oxygen bond scission) and inversion (alkyl–oxygen bond scission) of configuration at the β carbon atom (Scheme 1).[21,22] The cationic polymerization of L-lactide did not change the configuration at C* atoms (hydrolysis of monomer and polymer gave lactic acid with the same optical purity).[23,24] Thus, propagation

retention   inversion

Scheme 1

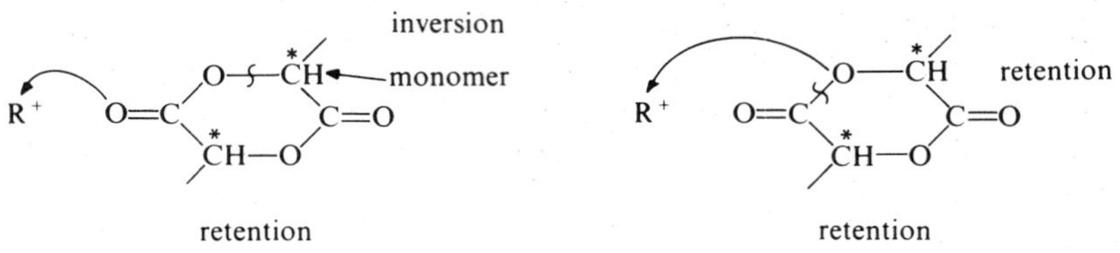

inversion

retention   retention

Scheme 2

with lactide proceeds with acyl–oxygen bond scission without any significant contribution of alkyl–oxygen bond breaking (Scheme 2).

50.4.4   Chain Transfer to Polymer

Polymerization of some lactones (*e.g.* δ-valerolactone and ε-caprolactone) is accompanied by transesterification and macrocyclization.[6,25] These reactions consist of attack by active centers on the ester group of a foreign (transesterification) or their own (macrocyclization) molecules. Due to the basicity and nucleophilicity of ester groups being higher in monomer molecules compared to linear polyesters, polymerization is faster than these side reactions. For example, Yamashita *et al.* have shown that, in the polymerization of ε-caprolactone initiated with triflic acid, monomer is converted first into a linear polymer, and that only after monomer conversion has approached 90% do cyclic oligomers appear in appreciable proportions.[25] The *rate* of formation of cyclics (unimolecular) is, of course, constant from the very beginning of polymerization, but this rate is very much lower than the rate of monomer consumption (bimolecular), even when the concentration of the monomer left in the system is quite low. However, at certain monomer concentrations the rate of macrocyclization becomes equal to and then higher than the rate of monomer consumption, until an equilibrium (linear living polymer ⇌ macrocyclics) is established. Thus, in a polymerization with initial concentrations of ε-caprolactone lower than 2.5×10^{-1} mol l^{-1} all polymer is finally degraded to cyclic oligomers. This concentration is equal to the critical monomer concentration $[M]_{crit}$, equation (10); see Chapter 47.[3]

$$[M]_{crit} = \sum_{x=1}^{\infty} \times [\overline{M_x}] \tag{10}$$

It follows that the synthesis of linear poly(ε-caprolactone) by any ionic method, including cationic polymerization, requires: (i) high initial monomer concentrations; and (ii) killing of active centers prior to formation of macrocycles. On the other hand, when macrocycles are required as the final product, the synthesis should be carried out for $[M]_0 \leq [M]_{crit}$ (2.5×10^{-1} mol l^{-1} for ε-caprolactone) during the period of time required for complete equilibration.

50.5   POLYMERIZATION OF ORTHO ESTERS

Bicyclic fused ortho esters and spiro ortho esters (*e.g.* **4** and **5**) polymerize by the opening of both rings. However, there are observed preferences in the opening of the first and second ring, depending on the conditions of polymerization and on the nature of the monomer. For example, polymerization of 2,6,7-trioxabicyclo[2.2.1]heptane (**4**), initiated with BF$_3 \cdot$OEt$_2$ and carried out in CH$_2$Cl$_2$ at -78 °C, proceeds with opening of only one ring.[26] ^{1}H NMR spectra indicated that the polymer has predominantly the structure described by formula (**21**). Contribution of units (**22**), containing six-membered rings, does not exceed 10%. With respect to 1,3-dioxolane rings, poly(ortho ester)s (**21**) have *cis* and *trans* configurations.[27] Heating the polymer solution in the presence of triflic acid leads to opening of the substituted 1,3-dioxolane rings; eventually, linear polymer (**23**) is obtained, containing two types of units (structures determined by ^{1}H NMR).[26] The first step (polymerization with opening of one ring) is accompanied by a contraction in volume: 6% from comparison of monomer and polymer densities at 25 °C. The second step results in an expansion in volume (9.7%). Thus, conversion of a bicyclic ortho ester (**4**) into a linear polymer leads to an overall expansion in volume (~4%).[26]

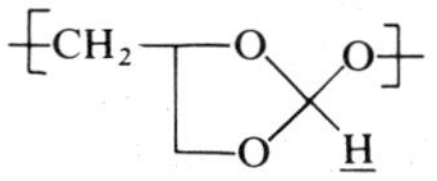

(21)   (22)

$\delta = 5.81$ p.p.m. (^{1}H NMR) $\delta = 5.24$ p.p.m. (^{1}H NMR)

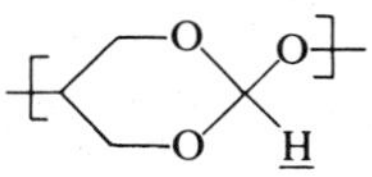

$$\text{-[CH}_2\text{CHO]-[CH}_2\text{CHCH}_2\text{O]-}$$

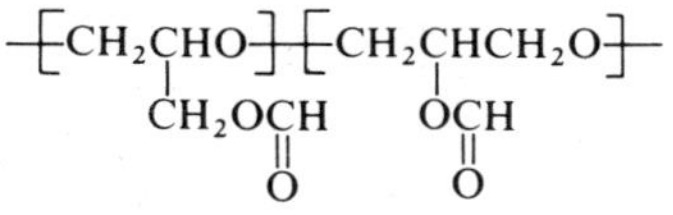

(65%)   (35%)

(23)

Linear polymer, with the structure (23), can be obtained in one step, when polymerization initiated with triflic acid is carried out in bulk at 80 °C.[26]

Hall suggested that poly(ortho ester)s (21) and polyethers with pendant ester groups (23) are the kinetically and thermodynamically controlled products, respectively.[27-29] For example, in the polymerization of 2,6,7-trioxabicyclo[2.2.2]octanes the routes shown in Scheme 3 were proposed.

Poly(ortho ester) (kinetically controlled product)

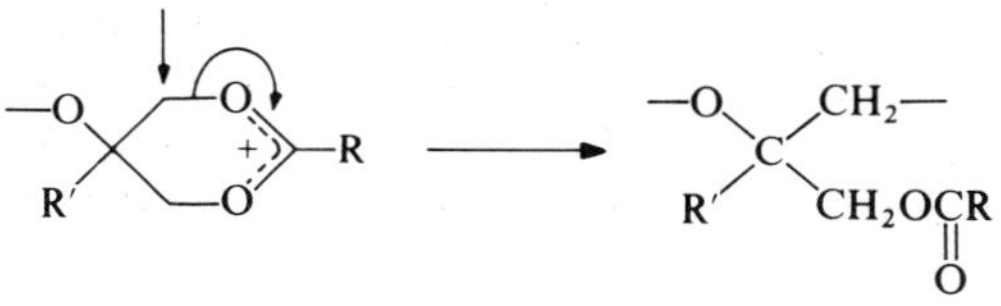

Polyether (thermodynamically controlled product)

Scheme 3

Recently, Hall and co-workers described cationic polymerization of 4-methoxycarbonyl-2,6,7-trioxabicyclo[2.2.2]octane (24).[30] Polymerization proceeds with opening of only one ring and leads to polymers with *cis* and *trans* units (with respect to 1,3-dioxane rings). The subsequent hydrolysis gave the water-soluble potassium carboxylate polymer (25).

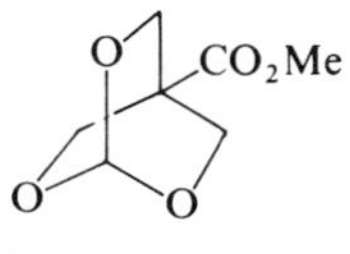

(24) 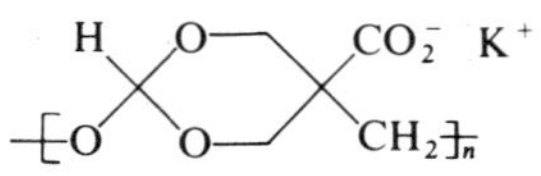   (25)

Cationic polymerization of 1,4,6-trioxaspiro[4,4]nonane (5) gives a linear polymer, with predominantly (80–90%) the structure (26) (equation 11).[31,32] Analysis of the ^{13}C NMR spectra of the polymerizing mixtures led to the conclusion that both rings are not necessarily sequentially opened, and that, at the intermediate stage, segments depicted in formula (27) contribute to the polymer chain structure. At the later stages of polymerization these rings are opened. Eventually, polymers containing not only regular units (26) but also head-to-head (28) and tail-to-tail segments (29),

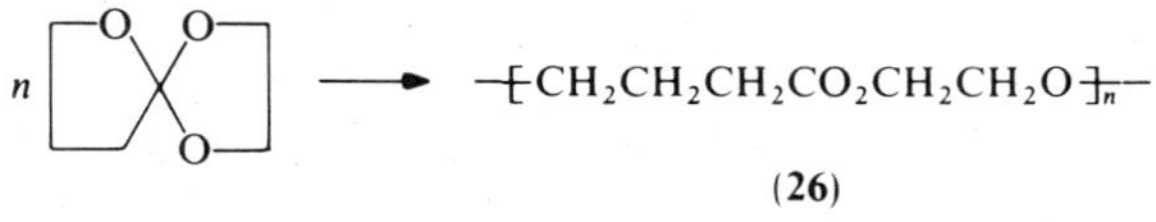

$$n \ \boxed{\text{O O}} \longrightarrow \text{-[CH}_2\text{CH}_2\text{CH}_2\text{CO}_2\text{CH}_2\text{CH}_2\text{O]}_n\text{-} \qquad (11)$$

(26)

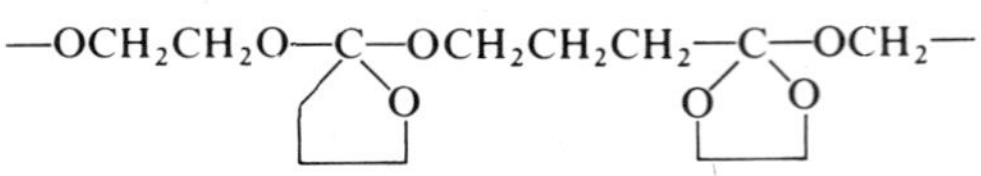

$$-OCH_2CH_2O-\underset{O}{\overset{|}{C}}-OCH_2CH_2CH_2-\underset{O}{\overset{|}{C}}-OCH_2-$$

(27)

$$-OCH_2CH_2O\overset{O}{\overset{\|}{C}}CH_2CH_2CH_2OCH_2CH_2CH_2\overset{O}{\overset{\|}{C}}OCH_2CH_2-$$

(28) head-to-head

$$-CH_2CH_2CH_2\overset{O}{\overset{\|}{C}}OCH_2CH_2OCH_2CH_2O\overset{O}{\overset{\|}{C}}CH_2CH_2O-$$

(29) tail-to-tail

$$-CH_2O\overset{O}{\overset{\|}{C}}CH_2CH_2CH_2OCH_2CH_2OCH_2CH_2CH_2\overset{O}{\overset{\|}{C}}OCH_2-$$

(30)

$$-CH_2OCH_2CH_2CH_2\overset{O}{\overset{\|}{C}}OCH_2CH_2O\overset{O}{\overset{\|}{C}}CH_2CH_2CH_2OCH_2-$$

(31)

segments with two consecutive ether groups (**30**) and segments with two consecutive ester groups (**31**) are formed.

50.6 POLYMERIZATION OF PHOSPHORUS-CONTAINING CYCLIC ESTERS

There is a much larger variety of cyclic esters of phosphorus-containing acids than of the derivatives of carboxylic and related acids. Below, some of the most representative groups of monomers are illustrated (Scheme 4; these monomers also have thio analogues); in this section only the most general results are discussed. Polymerization of cyclic phosphorus compounds has recently been reviewed.[34]

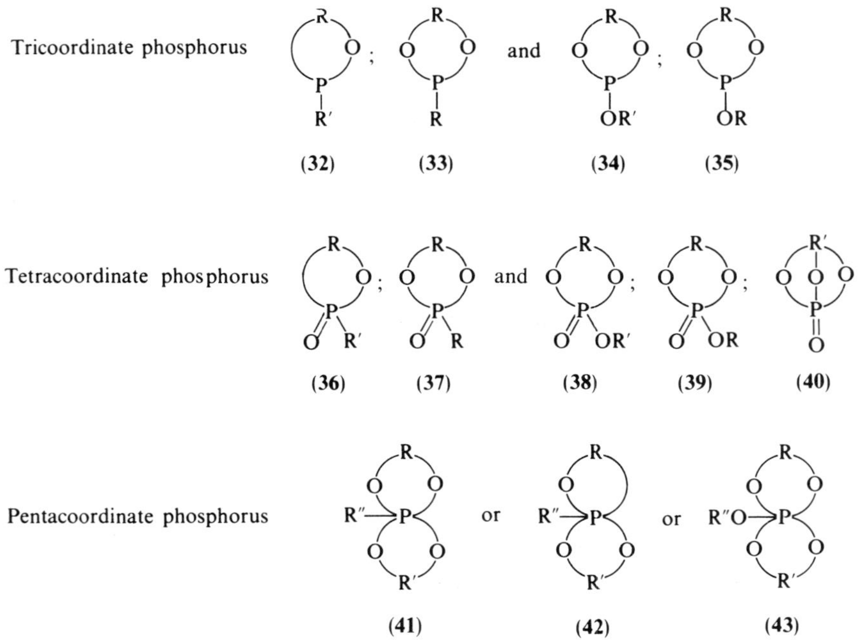

Scheme 4

Patent literature describes the polymerization of almost all of these monomers, as well as their statistical copolymers. Several of these compounds are prone to zwitterionic polymerization, playing the role of a donor monomer.[35]

Polymerization of some P-containing monomers proceeds with isomerization, *i.e.* the polymer units do not have a chemical structure reflecting that of the parent monomer. It happens mostly when Arbuzov rearrangement is possible, as was amply demonstrated by Shimidzu *et al.*[36, 37] Thus, we shall first describe a few simple cases (without rearrangements) and then the 'isomerization polymerization'.

50.6.1 Polymerization of Cyclic Esters of Phosphoric Acid

Mostly five- and six-membered monomers (**7** and **44**) were polymerized. In the earlier work of Pietrov *et al.*[38, 39] anionic initiators were used, Munoz *et al.*[40, 41] in the 1960s applied cationic initiators such as HCl and AlCl$_3$. Relatively high temperatures (60–100 °C) and long reaction times were needed for nearly complete monomer conversion (for R = Me or Et). Molecular weights did not exceed 5×10^3. Apparently, cationic polymerization is not the method of choice; indeed, polymerization of several five-membered esters was shown to proceed rapidly (within minutes), even at -70 °C, with organometallic initiators, producing polymers with $\bar{M}_n$ greater than 10^4.[42] Thermodynamic studies gave $\Delta H_{ss} = 15 \pm 2$ kJ mol^{-1} for phosphate (**7**).[43] High polymers from five-membered esters ('rubbery substances') were also described by Vandenberg.[44]

Studies of the cationic polymerization of six-membered esters (**44**; Scheme 5) provided more detailed information. It was shown by NMR (both ^{1}H and ^{31}P) that alkylation with alkylating agents gives, as the first species, the tetraalkoxyphosphonium salt (**45**). Further breaking of the O–C bond, under the nucleophilic attack of the next monomer molecule, provides actual growing species, being also tetraalkoxyphosphonium cations (cation **46** observed by NMR).[45, 46]

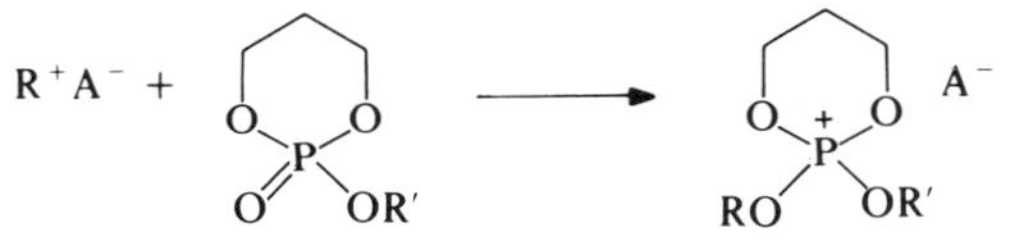

Scheme 5

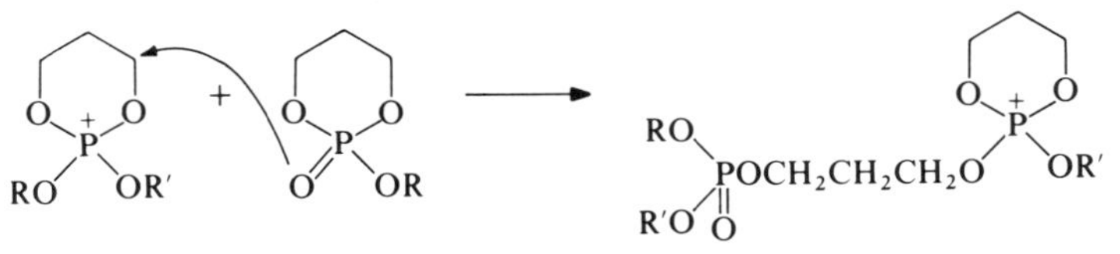

Thus, there is an anology with the polymerization of other heterocyclics, proceeding on oxonium ions. Polymerization is relatively slow ($k_p^{\pm} = 3.2 \times 10^{-3}$ mol^{-1} s l^{-1} at 85 °C bulk), six-membered monomers are either slightly strained ($\Delta H_{1c} \approx -4.0$ kJ mol^{-1} for R = Me) or not strained at all ($\Delta H_{1c} \geqslant 0$ for larger exocyclic substituents). Polymer molecular weights are less than a few thousand. On the other hand, studies of polymer microstructure by ^{1}H and ^{31}P NMR, and determination of $\bar{M}_n$ by vapor pressure osmometry (VPO) and end-group analysis (^{1}H NMR), indicated that every macromolecule has a cyclic monomer fragment (*e.g.* **47** for R = Me) as one of the end-groups. Formation of this end-group was explained by assuming a chain transfer from active species (**46**).

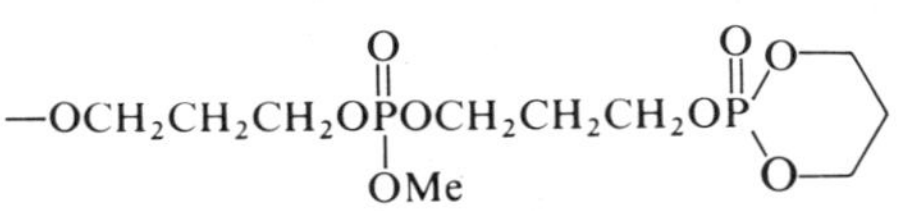

(**47**)

Indeed, breaking the O–C bond in the exocyclic group (O–R′) within a growing species gives 'R′$^+$' (for reinitiation) and the cyclic end-group (**47**). A very similar mechanism of chain transfer operates in anionic polymerization of six-membered esters of phosphoric acid.

It may be of interest to mention here that high molecular weight polymers of six-membered monomers ($\bar{M}_n \approx 10^5$) were obtained only in anionic or pseudoanionic polymerization for monomers with exocyclic groups unable to participate in the chain transfer.[42,47]

50.6.2 Isomerization (Rearrangement) in the Polymerization of Cyclic Esters of Phosphoric Acid

The best example is provided by the polymerization of methyl (or ethyl) ethylene phosphite (*i.e.* 2-alkoxy-1,3,2-dioxaphospholane; **48**). This monomer, depending on the initiator used, yields the polymeric units (**49**), (**50**) and (**51**), as shown in Scheme 6.

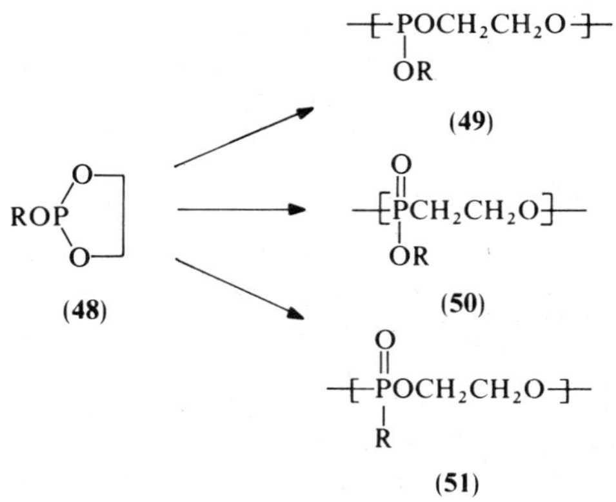

Scheme 6

Cationic polymerization (initiated, for example, with $AlCl_3$, $TiCl_4$ or BF_3, and performed above $150\,°C$) leads to structure (**51**) almost exclusively,[36] whereas structures (**49**) and (**50**) can be obtained under milder conditions with anionic initiators.[36,39,48]

Stark *et al.*[49] assumed that the actual propagation step is preceded by rearrangement (Scheme 7).

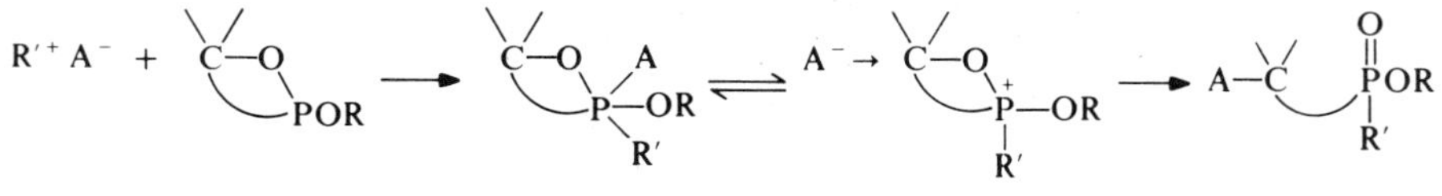

Scheme 7

A cationic mechanism operates in the polymerization of cyclic phosphazenes. This important field has been reviewed comprehensively by Allcock[50] (see also Volume 4, Chapter 27).

50.7   REFERENCES

1. A. K. Khomyakov and E. B. Lyudvig, *Dokl. Akad. Nauk SSSR*, 1971, **201**, 877.
2. H. Cherdron, H. Ohse and F. Korte, *Makromol. Chem.*, 1962, **56**, 179.
3. K. Ito, Y. Hashizuka and Y. Yamashita, *Macromolecules*, 1977, **10**, 821.
4. K. Ito and Y. Yamashita, *Macromolecules*, 1978, **11**, 68.
5. S. Sosnowski, S. Slomkowski, S. Penczek and L. Reibel, *Makromol. Chem.*, 1983, **184**, 2159.
6. K. Ito, M. Tomida and Y. Yamashita, *Polym. Bull. (Berlin)*, 1979, **1**, 569.
7. Y. Yamashita, *Polym. Prepr., Am. Chem. Soc., Div. Polym. Chem.*, 1980, **21**, 51.
8. H. Keul, R. Bächer and H. Höcker, *Makromol. Chem.*, 1986, **187**, 2579.
9. A. Hamitou, T. Ouhadi, R. Jerome and Ph. Teyssié, *J. Polym. Sci., Polym. Chem. Ed.*, 1977, **15**, 865.
10. H. L. Hsieh and I. W. Wang, *ACS Symp. Ser.*, 1985, **286**, 181.
11. A. Hofman, R. Szymanski, S. Slomkowski and S. Penczek, *Makromol. Chem.*, 1984, **185**, 655.

12. H. R. Kricheldorf, J. M. Jonté and R. Dunsing, *Makromol. Chem.*, 1986, **187**, 771.
13. A. Hofman, S. Slomkowski and S. Penczek, *Makromol. Chem.*, 1987, **188**, 2027.
14. H. R. Kricheldorf, R. Dunsig and A. Serra, *Macromolecules*, 1987, **20**, 2050.
15. S. Slomkowski, R. Szymanski and A. Hofman, *Makromol. Chem.*, 1985, **186**, 2283.
16. E. B. Lyudvig, G. S. Sanina and A. K. Khomyakov, *Vysokomol. Soedin., Ser. A*, 1974, **16**, 801.
17. A. K. Khomyakov, A. T. Gorelikov, N. N. Shapet'ko and E. B. Lyudvig, *Vysokomol. Soedin., Ser. A*, 1976, **18**, 1699.
18. B. G. Belenkaya and E. B. Lyudvig, *Vysokomol. Soedin., Ser. A*, 1978, **20**, 565.
19. B. G. Belenkaya, A. I. Levenko and E. B. Lyudvig, *Vysokomol. Soedin., Ser. A*, 1978, **20**, 559.
20. Y. Yamashita, T. Tsuda, M. Okada and S. Iwatsuki, *J. Polym. Sci., Part A-1*, 1966, **4**, 2121.
21. D. E. Agostini, J. B. Lando and J. R. Shelton, *J. Polym. Sci., Part A-1*, 1971, **9**, 2775.
22. J. R. Shelton, D. E. Agostini and J. B. Lando, *J. Polym. Sci., Part A-1*, 1971, **9**, 2789.
23. E. Lillie and R. C. Schulz, *Makromol. Chem.*, 1975, **176**, 1901.
24. A. Schindler and D. Harper, *J. Polym. Sci., Polym. Lett. Ed.*, 1976, **14**, 729.
25. Y. Yamashita, *Polym. Prepr., Am. Chem. Soc., Div. Polym. Chem.*, 1979, **20**, 126.
26. H. K. Hall, Jr. and Y. Yokayama, *Polym. Bull. (Berlin)*, 1980, **2**, 281.
27. Y. Yokoyama, A. Buyle Padias, F. De Blauwe and H. K. Hall, Jr., *Macromolecules*, 1980, **13**, 252.
28. Y. Yokoyama and H. K. Hall, Jr., *Adv. Polym. Sci.*, 1982, **42**, 107.
29. A. Buyle Padias, R. Szymanski and H. K. Hall, Jr., *ACS Symp. Ser.*, 1985, **286**, 161.
30. R. Szymanski and H. K. Hall, Jr., *J. Polym. Sci., Polym. Lett. Ed.*, 1983, **21**, 177.
31. W. J. Bailey and R. L. Sun, *Polym. Prepr., Am. Chem. Soc., Div. Polym. Chem.*, 1972, **13**, 281.
32. W. J. Bailey, H. Iwama and R. Tsushima, *J. Polym. Sci., Polym. Symp.*, 1976, **56**, 117.
33. K. Matyjaszewski, *J. Polym. Sci., Polym. Chem. Ed.*, 1984, **22**, 29.
34. G. Lapienis and S. Penczek, in 'Ring-Opening Polymerization', ed. K. J. Ivin and T. Saegusa, Elsevier, New York, 1984, vol. 2, p. 919.
35. K. J. Ivin and T. Saegusa, in 'Ring-Opening Polymerization', ed. K. J. Ivin and T. Saegusa, Elsevier, New York, 1984, vol. 1, p. 1.
36. T. Shimidzu, T. Hakozaki, T. Kagiya and K. Fukui, *Bull. Chem. Soc. Jpn.*, 1966, **39**, 562.
37. T. Shimidzu, T. Hakozaki, T. Kagiya and K. Fukui, *J. Polym. Sci., Part B*, 1965, **3**, 871.
38. K. A. Petrov, E. E. Nifant'ev and L. W. Fedorchuk, *Vysokomol. Soedin.*, 1960, **2**, 917.
39. K. A. Petrov, E. E. Nifant'ev, L. V. Khorkhyanu, M. I. Merkulova and V. F. Voblikov, *Vysokomol. Soedin.*, 1962, **4**, 246.
40. A. Munoz and J.-P. Vives, *C.R. Hebd. Seances Acad. Sci.*, 1961, **253**, 1693.
41. A. Munoz, J. Navech and J.-P. Vives, *Bull. Soc. Chim. Fr*, 1966, 2350.
42. J. Libiszowski, K. Kaluzynski and S. Penczek, *J. Polym. Sci., Polym. Chem. Ed.*, 1978, **16**, 1275.
43. S. Sosnowski, J. Libiszowski, S. Slomkowski and S. Penczek, *Makromol. Chem., Rapid Commun.*, 1984, **5**, 239.
44. E. J. Vandenberg (Hercules Powder Co.), *US Pat.* 3 520 849 (1970) (*Chem. Abstr.*, 1970, **73**, 67 314).
45. G. Lapienis and S. Penczek, *Macromolecules*, 1974, **7**, 166.
46. G. Lapienis and S. Penczek, *Macromolecules*, 1977, **10**, 1301.
47. K. Kaluzynski, J. Libiszowski and S. Penczek, *Makromol. Chem.*, 1977, **178**, 2943.
48. W. Vogt and N. U. Ahmad, *Makromol. Chem.*, 1977, **178**, 1711.
49. B. P. Stark, A. J. Duke and R. J. Martin, *J. Appl. Chem.*, 1967, **17**, 127.
50. H. R. Allcock, 'Phosphorus–Nitrogen Compounds', Academic Press, New York, 1972.

51

Cationic Ring-opening Polymerization: Sulfides

ERIC J. GOETHALS
State University of Ghent, Belgium

51.1 INTRODUCTION

In this series of compounds, the three- and four-membered rings (thiiranes and thietanes) possess enough ring strain to be polymerized. The five-membered thiolane cannot be polymerized but 1,3-oxathiolane can. Among the six- and higher-membered rings, only trithiane has been reported to polymerize (Scheme 1).

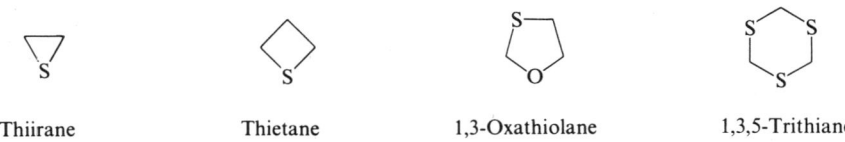

| Thiirane | Thietane | 1,3-Oxathiolane | 1,3,5-Trithiane |

Scheme 1

51.2 THIIRANES

A large number of thiiranes have been polymerized by a variety of cationic initiators. A general feature of these polymerizations is that, if no terminating agent has been added, the polymers degrade to cyclic low molecular weight compounds, the structures of which depend on the type of monomer and on the reaction conditions.

51.2.1 Mechanism of Polymerization

The initiation reaction (equation 1) is the formation of a thiiranium ion (**1**). The most straightforward way to obtain this is by addition of a (strong) alkylating agent such as a trialkyloxonium salt or

$$\triangleright S \;+\; Et_3O^+BF_4^- \;\longrightarrow\; \overset{+}{\underset{BF_4^-}{\triangleright S}}\!\!-Et \;+\; Et_2O \tag{1}$$

(1)

an ester of a super acid. Typical examples are triethyloxonium tetrafluoroborate or hexafluoroantimonate, and methyl trifluoromethanesulfonate.

Lewis acids such as boron trifluoride (etherate) are also able to initiate the polymerization. Direct addition of the boron fluoride to the sulfur atom of the monomer is not likely to be the initiation step since the kinetics of the polymerization of methylthiirane with boron trifluoride depend more on the concentration of water than on the 'initiator' concentration, the highest rates being obtained for a ratio $[BF_3]/[H_2O]$ of 2. This indicates that initiation is a protonation and suggests that *both* protons of water can be used for the initiation.[1]

Proton acids such as perchloric acid are also good initiators, giving kinetics of polymerization similar to those of the oxonium-ion-initiated ones.[1]

Photoinitiated cationic polymerization of cyclohexene sulfide with diaryliodonium salts has been reported.[2]

A combination of trialkylaluminum and water was found to produce a cationic polymerization of methylthiirane[3] if the ratio H_2O/AlR_3 was higher than 0.7.

The propagation reaction is a nucleophilic substitution reaction in which the monomer is the nucleophile and the thiiranium ion the electrophile. With a number of monomers it was possible to prove that the reaction is of the S_N2 type. This was the case for the polymerization of the 2,3-dimethylthiiranes (*cis* or *trans*) and proof came from the fact that the degradation products formed after polymerization had diastereoisomeric structures, which could only be obtained by inversion of one of the carbon configurations.[4] However, in the polymerization of 2,2-dimethylthiirane (isobutylene sulfide, IBS) there may be a partial S_N1 mechanism operating, as was shown by the observation that, in the cyclic oligomers formed by degradation, a small but non-negligible amount of a reaction product resulting from a hydride shift was formed.[5] Such a hydride shift leading to an isomerized unit can only be explained by the occurrence of a carbenium ion as an intermediate (**2**; equation 2). This hydride shift only takes place during the cationic degradation of poly-IBS and not during the polymerization, since the polymer itself does not contain isomerized (thioacetal) structural units. Apparently, the rate of polymerization is much higher than the rate of hydride transfer, whereas the rate of degradation is of the same order of magnitude.

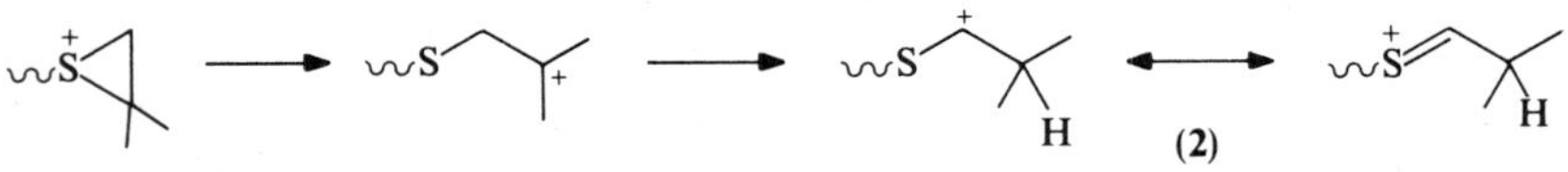

$$\mathcal{w}\overset{+}{S}\!\!\!\diagup\!\!\diagdown \;\longrightarrow\; \mathcal{w}S\!\!\diagup\!\!\!\diagup_+ \;\longrightarrow\; \underset{H}{\mathcal{w}S\!\!\diagup\!\!\diagup} \;\longleftrightarrow\; \underset{H}{\mathcal{w}\overset{+}{S}\!\!\diagup\!\!\diagdown} \tag{2}$$

(2)

Other evidence for the carbenium ion character of the active species in the case of IBS is the fact that the polymerization initiated by a triphenylcarbenium salt leads to compound (**3**), which is formed by a Friedel–Crafts type of electrophilic substitution (Scheme 2).[6]

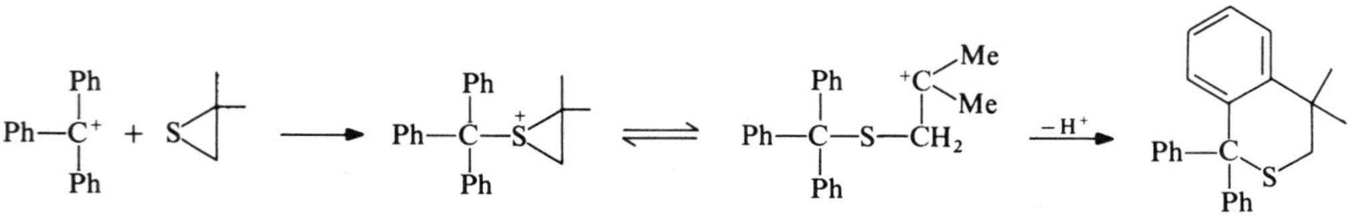

$$\underset{Ph}{\overset{Ph}{Ph-\underset{|}{\overset{|}{C}}{}^+}} + S\!\!\diagdown\!\!\!\diagup \;\longrightarrow\; \underset{Ph}{\overset{Ph}{Ph-\underset{|}{\overset{|}{C}}-\overset{+}{S}\!\!\diagdown\!\!\!\diagup}} \;\rightleftharpoons\; \underset{Ph}{\overset{Ph}{Ph-\underset{|}{\overset{|}{C}}-S-\overset{+C<^{Me}_{Me}}{CH_2}}} \;\xrightarrow{-H^+}\; \underset{Ph}{Ph-\overset{|}{\underset{|}{C}}}\diagdown_{S}$$

(3)

Scheme 2

In the case of unsymmetrically substituted thiiranes, ring-opening can occur at two positions (see structure **4**). In anionic and coordinative polymerizations the ring-opening occurs almost exclusively at β, leaving unchanged the configuration of the asymmetric carbon atom and producing pure head-to-tail polymers. When using one enantiomer, an isotactic polymer with a high optical activity is obtained. In cationic polymerization a substantial fraction of the rings are opened at α, thus leading to inversion of the configuration and producing polymers containing a similar fraction of head-to-head and tail-to-tail structures. This was demonstrated by Spassky and Sigwalt for methylthiirane.[7] These structural irregularities have also been identified by ^{13}C NMR.[8,9]

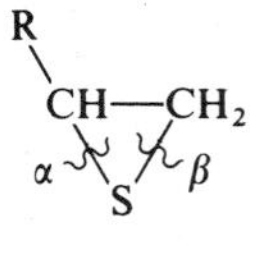

$$\overset{\displaystyle R}{\underset{\alpha \;\;\; \beta}{\underset{S}{\overset{|}{CH\!-\!CH_2}}}}$$

(4)

Kinetic studies performed with methylthiirane have shown that the active species for the polymerization are rapidly deactivated by attack of a sulfur atom of the polymer chain. This reaction forms a non-strained sulfonium ion that is unreactive for propagation.[10] From kinetic studies and also from the observation that the polymer slowly degrades mainly to cyclic tetramer, it was concluded that, in the case of methylthiirane, the temporary termination reaction is mainly the formation of a 12-membered cyclic sulfonium ion (5; equation 3) at the end of the polymer chain. The ratio of the rate constant of propagation to the rate constant of termination, k_{tt}, at 0 °C in methylene chloride is *ca.* 9 l mol^{-1}. This ratio does not vary much for different initiator systems.[1]

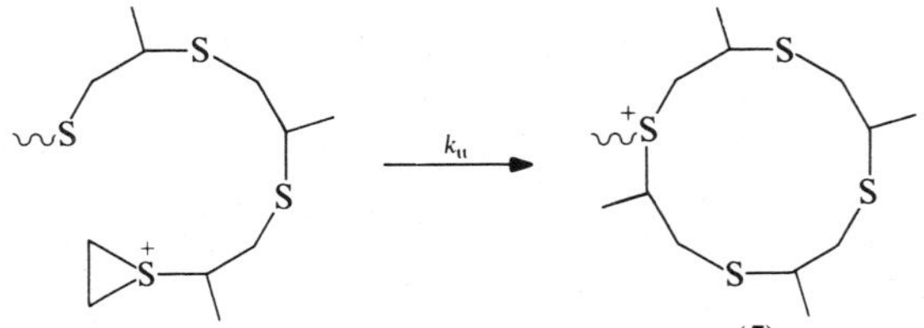

$$\xrightarrow{\;k_{tt}\;}$$

(3)

(5)

This termination, however, is only a temporary one. This is shown by the observation that with 'fast initiators' the polymerization consists of a very fast, non-stationary state A and a much slower subsequent state B which leads to quantitative conversion.

The reinitiation is proposed to be an intramolecular nucleophilic attack of the neighbouring sulfur atom on the carbon of the sulfonium ion (equation 4).

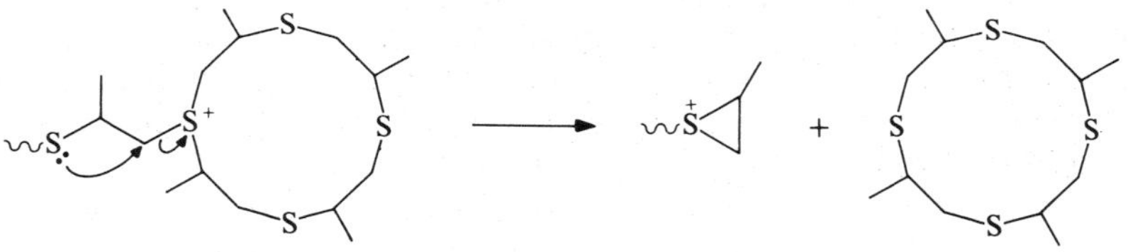

(4)

Evidence for the intramolecular mechanism of reinitiation was obtained from the observation that simple trialkylsulfonium salts are very poor initiators or completely incapable of initiating the polymerization, whereas under identical conditions, 2-thiaalkylsulfonium salts (6) give polymerizations without stage A which are similar to stage B of the oxonium-ion-initiated reactions (equation 5).

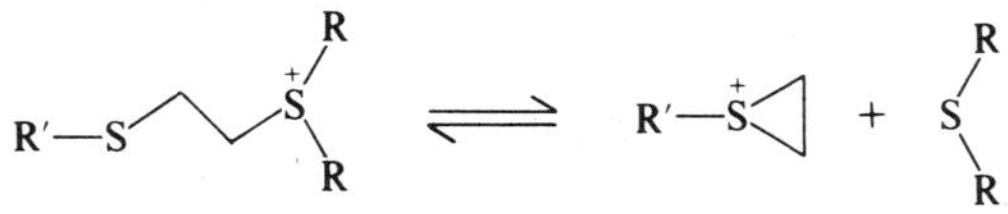

$$R'\!-\!S \cdots \overset{R}{\underset{R}{S}} \;\;\rightleftharpoons\;\; R'\!-\!\overset{+}{S} + \overset{R}{\underset{R}{S}}$$

(5)

(6)　　　　　　Initiating thiiranium ion

Irreversible termination due to reaction with the counter anion can be avoided by choosing initiating systems producing non-nucleophilic anions such as BF_4^-, PF_6^- and $CF_3SO_3^-$. However, with some of these non-nucleophilic anions termination may occur by halide displacement. This has been observed with the counter ion $SbCl_6^-$, which reacts with the active species as shown in equation (6).

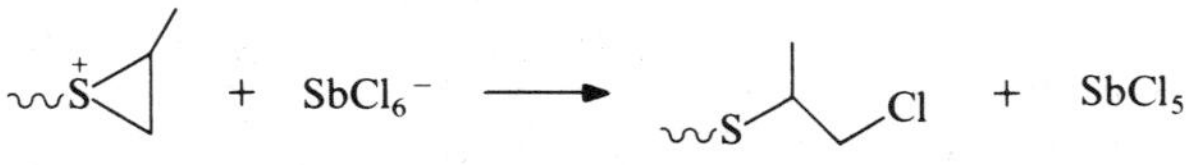

$$\text{S}^+ + SbCl_6^- \longrightarrow \text{S} \!\!\sim\!\! Cl + SbCl_5$$

(6)

51.2.2 Transalkylation Reactions

If the active species of the polymerization, the thiiranium ion, reacts with a sulfide function of another polymer chain, a branched, dormant species is obtained. From the kinetics it follows that this dormant species is reactivated by intramolecular re-formation of the thiiranium ion. There are three different possibilities for such a reactivation: one leads back to the original situation, the two others lead to two polymer chains which have other chain lengths than the original ones and the process can be considered as a transfer of an alkyl group from one chain to another, *i.e.* a transalkylation (Scheme 3).

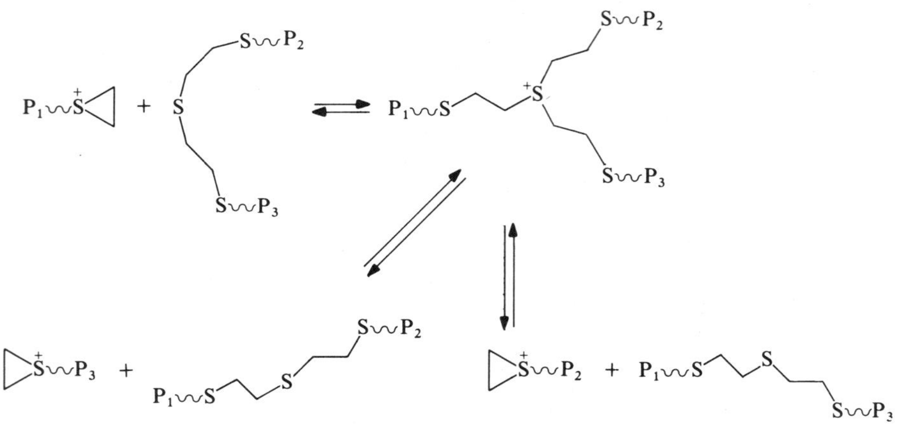

Scheme 3

If this process is repeated many times during the course of the polymerization, a broadening of the molecular weight distribution is obtained. A number of experiments have demonstrated that this transalkylation reaction is very rapid and probably occurs at a rate comparable to that of the propagation reaction. This is proved by the observation that if 'dead' poly(methylthiirane) is added to a cationically degrading solution of poly(methylthiirane) the newly added polymer degrades at the same rate as if the initiator for the degradation had been added to that newly added polymer. Furthermore, if the newly added polymer is deuterated, the cyclic tetramers formed by the degradation consist of a mixture of undeuterated, mono-, di-, tri- and tetra-deuterated products.[11] This is in agreement with a very fast exchange reaction between sulfonium ions and sulfide functions by which a mixture of undeuterated and deuterated polymer is transformed into a 'copolymer' with random structure. The low reactivity of the macrocyclic dormant sulfonium ion (5), as shown by the kinetics, compared with the high reactivity of the 'linear' dormant sulfonium ions, is in the first place due to the fact that the former can reinitiate only through the exocyclic sulfide function, whereas the latter can reinitiate through all three β sulfide functions. Also, it is reasonable to accept that the steric requirements for the thiiranium ion formation are more easily met for the linear sulfonium ion than for the cyclic one.

The occurrence of the rapid transalkylation explains that the stereoregularity of a poly(methylthiirane) (obtained with Cd tartrate) is immediatly destroyed on addition of a catalytic amount of Et_3OBF_4 at room temperature, as was shown by NMR spectroscopy.[12] The rapid transalkylation reactions also prevent the formation of block copolymers by sequential monomer addition, a phenomenon which has also been reported for cyclic acetals (see Volume 3, Chapter 49). However, block copolymers of methylthiirane and thiirane have been prepared by polymerization of the latter in the presence of poly(methylthiirane),[13] a result that is attributed to the transalkylation reaction. Complete scrambling of the two structural units is probably suppressed by the insolubility of the polythiirane.

51.2.3 Cationic Degradation of Polythiiranes

If the reaction mixture obtained after quantitative cationic polymerization of a thiirane is kept for some time at room temperature or higher, the polymer degrades completely to low molecular weight compounds. Polymers obtained with anionic or coordinative initiators are stable, but when these polymers are treated with a cationic initiator, they start to degrade to the same low molecular weight compounds as those obtained by the cationic polymerization–degradation process. For some polythiiranes this degradation is very fast even at room temperature.[14] Four types of degradation products are obtained. They are shown in Table 1.

Table 1 Degradation Products of polythiiranes[a]

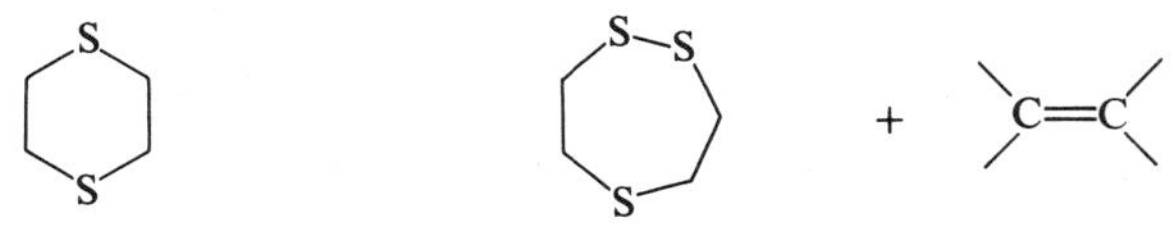

1,4-Dithiane   1,2,5-Trithiepane   +   alkene

Cyclic tetramer and higher rings   1,3-Dithiolane (isomerized dimer)

[a] Carbon substituents omitted.

The nature of the degradation products is in the first place determined by the nature and number of substituents on the carbon atoms of the starting thiirane monomer.

The active species for the degradation reactions are believed to be the same as for the polymerization, *i.e.* tertiary thiiranium ions. Several observations provide evidence for this, such as the fact that if monomer is added to a degraded (or degrading) reaction mixture, that monomer starts to polymerize immediately. Also, the stereochemistry of the degradation reaction is in agreement with the thiiranium ion intermediate.

Starting from a 'dead' polythiirane, the degradation is initiated by protonation or alkylation of a sulfide function as shown in Scheme 4. The different degradation products are obtained as a result of a reaction between the thiiranium ion and a sulfur atom of its own polymer chain. It is known that thiiranium ions can react with nucleophiles in different ways:[15] the nucleophile can attack at the α carbon in an S_N2 like reaction mechanism with ring opening. The nucleophile can also attack at the most positively charged site of the thiiranium ion, *i.e.* the positive sulfur, forming a sulfur–nucleophile bond. Simultaneously expulsion of an alkene takes place. It is also known that sulfonium ions can react with nucleophiles *via* an S_N1 type reaction if the transient carbenium ions are stabilized[16] and this is also possible with thiiranium ions. Such carbenium ions may rearrange by hydride transfer to produce isomeric (more stable) species. Several of these reactions can occur simultaneously during the degradation of polythiiranes and this explains the variety of products that may be formed.

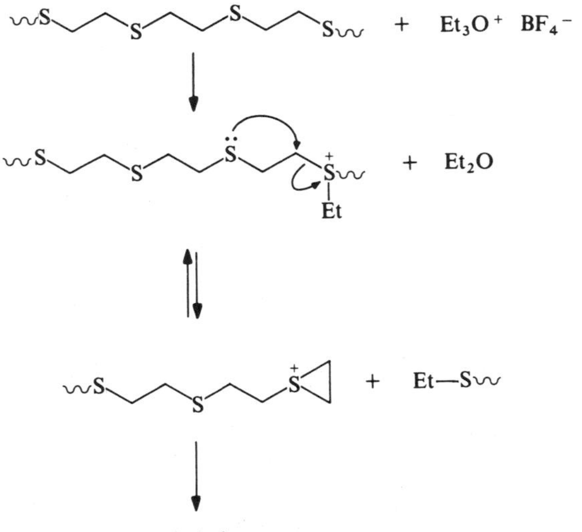

Scheme 4

The higher cyclics are formed by a back-biting reaction between the thiiranium ion and a sulfur atom further along the chain, followed by re-formation of a thiiranium ion and expulsion of the oligomer (Scheme 5). The smallest cyclic that is formed in this way is the 12-membered ring, which is always the most abundant of the higher cyclics. The absence of the nine-membered trimer is explained by the assumption that the reaction is an S_N2 reaction, which implies that the nucleophile must attack the thiiranium ion from a direction opposite to the bond that must be broken.

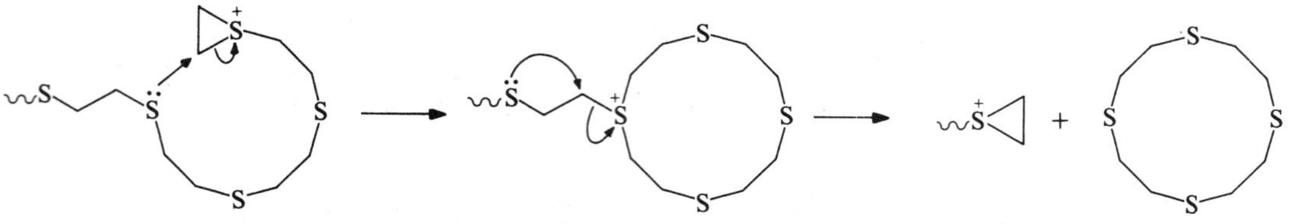

Scheme 5

The trithiepane and the corresponding alkene are formed by nucleophilic attack of the second sulfur atom next to the thiiranium ion at the positive sulfur, with expulsion of the alkene. Typical trithiepane-forming polythiiranes are the poly(2,3-dimethylthiirane)s.[4,18] *trans*-2,3-Dimethylthiirane (TDMT) is a mixture of two enantiomers which both have a *threo* configuration. During the polymerization, one of the carbon atoms undergoes an inversion so that in the polymer the repeating units have *erythro* configurations. *cis*-2,3-Dimethylthiirane (CDMT) has a (*meso*) *erythro* configuration and after the polymerization the repeating units have the *threo* configuration (Scheme 6).

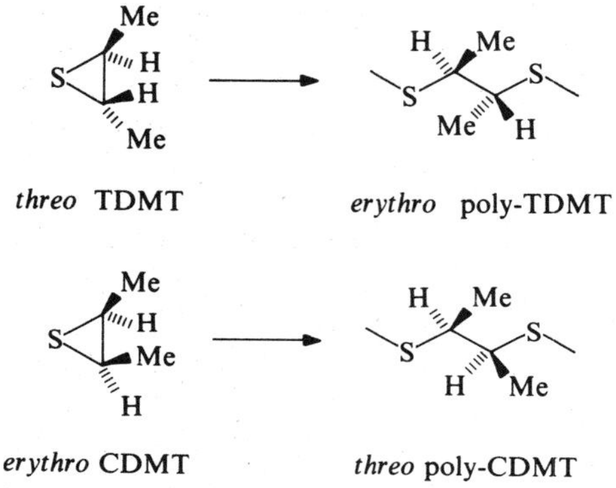

threo TDMT   *erythro* poly-TDMT

erythro CDMT   *threo* poly-CDMT

Scheme 6

The tetramethyltrithiepane obtained from the 2,3-dimethylthiiranes can occur as any of the six geometrical isomers shown in Scheme 7. By means of gas chromatography, these six isomers can be separated and quantitatively analyzed. The polymerization–degradation of poly-CDMT leads to only two out of the six isomers and the polymerization–degradation of poly-TDMT to two other isomers. By ${}^1H$ NMR analysis it was shown that the former isomers are those with two *threo* configurations (I and II) and the latter are those with two *erythro* configurations (V and VI).[17] The other two isomers (III and IV) are formed if a mixture of CDMT and TDMT is polymerized and degraded. Another important observation is that poly-TDMT leads to *trans*-butene whereas poly-CDMT leads to *cis*-butene as the alkene.

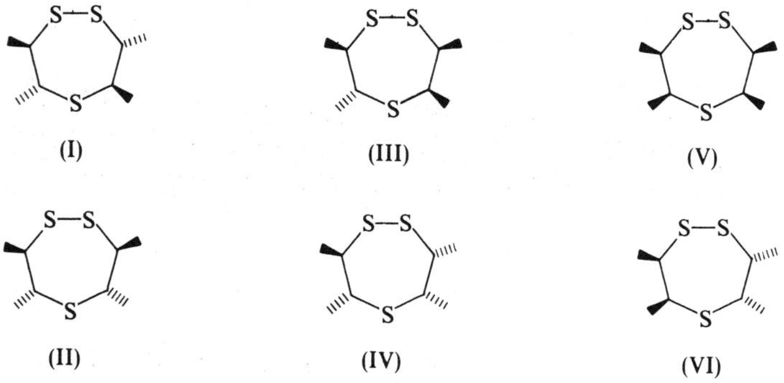

(I)   (III)   (V)

(II)   (IV)   (VI)

Scheme 7

Since the mechanism of degradation involves the formation of a sulfur–sulfur bond, there is no change of configuration of the repeating units during the degradation.

All these observations are in agreement with a reaction mechanism in which the active species is a thiiranium ion. This is illustrated for poly-TMDT in Scheme 8.

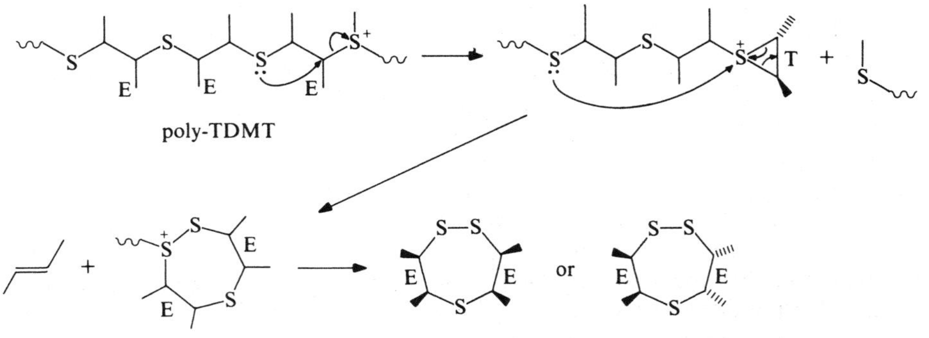

E = *erythro* unit;   T = *threo* unit

Scheme 8

The formation of cyclic dimer is observed for polymers that can give rise to a stabilized carbenium ion such as poly(styrene sulfide)[18] and poly(isobutylene sulfide).[5] Because the reaction is now of the S_N1 type, the formation of six-membered rings is possible because there is now no necessity to attack the carbon–sulfonium bond from the carbon side, as is the case in S_N2 reactions (see Scheme 9). That the reaction occurs *via* a carbenium ion is further indicated by the fact that along with the six-membered cyclic dimer an isomerized dimer is also found [at least in the depolymerization of poly(isobutylene sulfide)].[5] This isomer is the 1,3-dithiane, the formation of which can be explained by a hydride shift, which transforms the stable tertiary carbenium ion into an even more stable α-thiacarbenium ion. Cyclization by attack of the first sulfur atom of the polymer chain on this

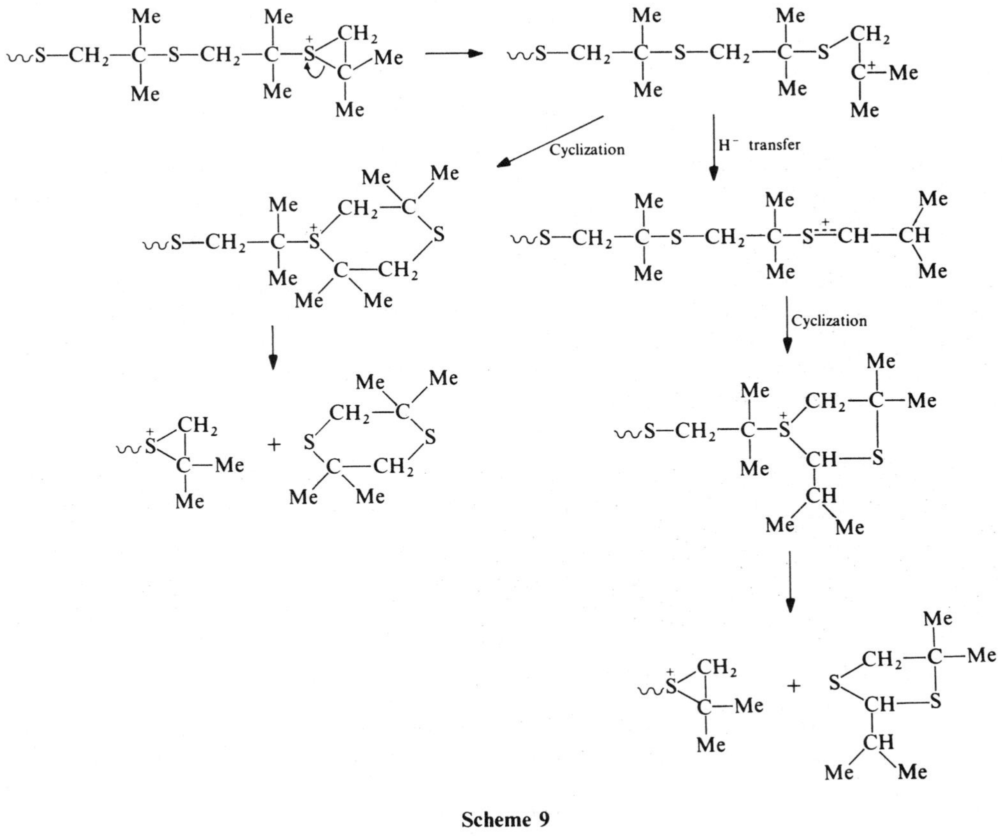

Scheme 9

carbenium ion, followed by re-formation of the thiiranium ion, leads to the 1,3-dithiane. Recently the degradation of polythiiranes by the combined action of UV light and a photoinitiator, such as a diphenyliodonium salt, has been described.[19]

51.2.4 Polysulfide Formation During the Cationic Thiirane Polymerization

In the anionic polymerization of thiiranes, it frequently happens that the polymer contains more sulfur than the monomer because during the polymerization an alkene is expelled with the formation of a disulfide bond. This phenomenon is much less frequently observed in the cationic polymerization. As was shown in the preceding section, disulfide functions do form during the degradation of polythiiranes. Disulfide and trisulfide functions in the polymer chain have been observed in the case of highly substituted thiiranes such as trimethylthiirane[20] and tetramethylthiirane.[21] It has also been reported that if the polymerization of methylthiirane is initiated with Et_3Al/H_2O, a system which is assumed to produce a cationic mechanism, polymers containing substantial fractions of disulfide functions are obtained.[22] The formation of disulfide bonds is the result of a direct attack of the monomer on the sulfur atom of the thiiranium ion (Scheme 10).

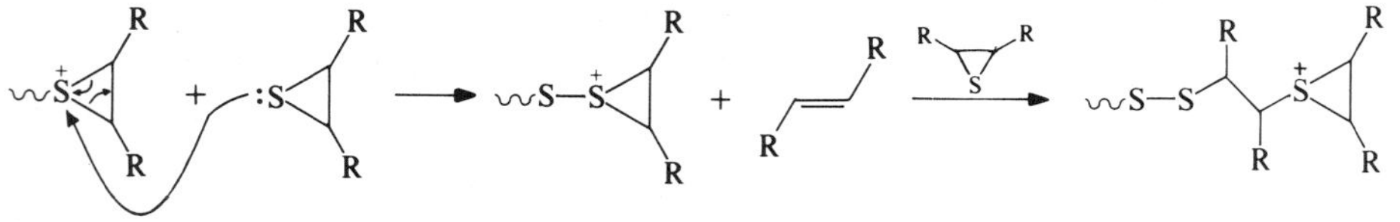

Scheme 10

Another possibility is that the disulfide bond formation occurs during the transalkylation reaction, *i.e.* when a thiiranium ion is attacked by a sulfur atom of a 'foreign' polymer chain followed by re-formation of the thiiranium ion (Scheme 11).

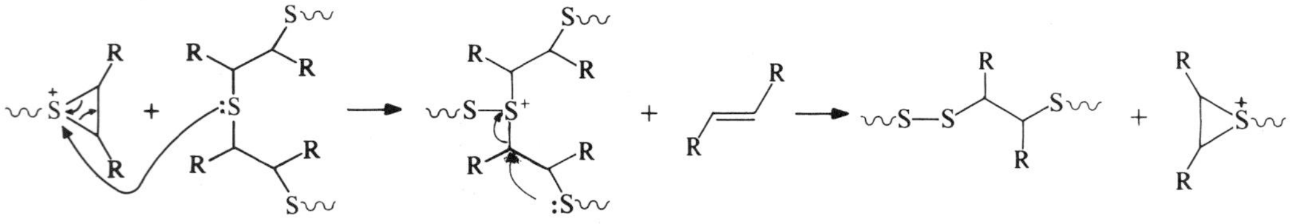

Scheme 11

51.3 THIETANES

The literature on the polymerization of these four-membered cyclic sulfides is much less abundant compared with that on the three-membered analogues. This is undoubtedly due to the more difficult synthesis of the former. The ring strains of thietanes are almost as high as those of thiiranes and, therefore, their polymerization is thermodynamically highly favourable.

The cationic polymerization of thietane was first reported in 1967 by Price and Blair[23] and by Stille and Empen[24] with initiators such as dimethyl sulfate, boron trifluoride etherate or triethyloxonium tetrafluoroborate. In the 1970s the mechanism of cationic polymerization of thietanes was studied in detail by Goethals and coworkers.[25] 3,3-Dimethylthietane (DMT) was the preferred monomer because this monomer has reasonable rates of polymerization at room temperature and because the ^1H NMR spectra of monomer and polymer are simple and different, which made it possible to use ^1H NMR for studying the kinetics.

51.3.1 Mechanism of Polymerization

Most 'classical' cationic initiators can be used for the polymerization of DMT but it was found that trialkyloxonium salts gave rather simple polymerization kinetics because the initiation with this kind of initiator is fast and quantitative.[26] This was shown by conductivity, by NMR, and especially by the finding that time–conversion curves obtained with Et_3OBF_4 are identical to those obtained with 1-ethyl-3,3-dimethylthietanium tetrafluoroborate (**7**), which can be prepared from DMT and Et$_3$OBF$_4$ (equation 7).[27] With triphenylcarbenium ion, the initiation occurred by direct addition of the trityl group to the sulfur atom (equation 8).[28]

$$ \diagup\!\!\!\diagdown S \ + \ Et_3OBF_4 \ \longrightarrow \ \overset{+}{S}\!-\!Et \ + \ BF_4^- \ + \ Et_2O \tag{7} $$

(**7**)

$$ \diagup\!\!\!\diagdown S \ + \ Ph_3C^+SbF_6^- \ \longrightarrow \ \overset{+}{S}\!-\!CPh_3 \ SbF_6^- \ \longrightarrow \ polymer \tag{8} $$

(**8**)

Although the reaction between a trityl salt and a sulfide is reversible, leading to an equilibrium, the initiation is quantitative because the tritylthietanium ion (**8**) is rapidly consumed by the propagation reaction.

Propagation occurs by nucleophilic attack of the monomer on the α carbon of the thietanium ion (equation 9). The nature of the active species was confirmed by 300 MHz ^{1}H NMR spectroscopy. The presence of an AB system at δ 3.7–4.0 p.p.m. due to the non-equivalency of the ring methylene protons and of two signals for the *gem*-dimethyl groups are in accordance with the thietanium structure as was proved by comparison with the spectrum of the model compound (**7**).[29]

$$ \diagup\!\!\!\diagdown S \ + \ \overset{+}{S}\!\!\sim \ \longrightarrow \ \overset{+}{S}\!\diagup\!\!\!\diagdown S\!\!\sim \tag{9} $$

Termination occurs when a sulfur atom of a polymer chain reacts with the active species. This reaction forms a non-strained and therefore inactive sulfonium ion (equation 10). By ^{1}H NMR spectroscopy it was possible to measure the concentration of active species during the course of the polymerization. The kinetics of the termination indicate an intermolecular reaction, *i.e.* with the formation of branched structures. This leads to molecular weights which are higher than calculated from the initial monomer and initiator concentrations.

$$ \tilde{S} \ + \ \overset{+}{S}\!\!\sim \ \longrightarrow \ \overset{+}{S}\!\!\sim \tag{10} $$

In contrast with the termination reaction in the thiirane polymerization, the termination in the thietane polymerization is not reversible because the formation of a thietanium ion is, for entropic reasons, much more difficult than the formation of a thiiranium ion.

The nature of the termination reaction has been confirmed by several techniques and the values of k_t have been determined. Because of the occurrence of an irreversible termination, the polymerizations of thietanes stop before complete conversion is attained. The monomer concentration remaining at the end of the polymerization, m_f, is determined by the ratio k_p/k_t and by the initial monomer and initiator concentrations (m_0 and c_0) according to

$$ \ln\left(\frac{m_0}{m_f}\right) - \left(\frac{m_0 - m_f}{m_0}\right) = \frac{k_p c_0}{k_t m_0} \tag{11} $$

By plotting $\ln(m_0/m_f) - [(m_0 - m_f)/m_0]$ against c_0/m_0 straight lines going through the origin with slopes equal to k_p/k_t are obtained.[30] Values of k_p/k_t for a series of thietanes are given in Table 2. The very important differences in k_p/k_t values can be explained by taking into account differences in basicity as well as differences in steric hindrance between monomer and the corresponding polymer units.

The importance of both the basicity and steric hindrance was confirmed by carrying out the polymerization of DMT in the presence of small amounts of low molecular weight sulfides.[25] When a sulfide of higher basicity (measured by IR spectroscopy) was added, the final yield of the polymerization of DMT was lowered, *e.g.* from 60% without added sulfide to 18% for thiolane. On the other hand, the basic but very bulky di-*t*-butyl sulfide did not affect the polymerization, the steric hindrance preventing the reaction with the thietanium ion.

From Table 2 it may be concluded that the polymerization of thietanes may vary from typical 'suicidal' polymerization (active species killed by its own polymer) to almost living polymerization

Table 2 Values of k_p/k_t for the
Polymerization of Different Thie-
tanes with Et$_3$OBF$_4$ Obtained
According to Equation (11)[a]

Monomer	k_p/k_t
	1.1
Me	2.4
Me Me	28
Et Et	450
Ph Ph	∞ (?)

[a] E. J. Goethals, W. Drijvers, D. Van
Ooteghem and A. M. Buyle, *J.
Macromol. Sci., Chem.*, 1973, **A7**, 1375.
[b] 1:1 mixture of *cis* and *trans* isomers.

by changing the number and the nature of the substituents on the monomer. It is also clear from this
table and from Table 3 that the introduction of substituents in the 3-position drastically decreases
the rate of propagation but, even more drastically, reduces the rate of the termination reaction. Thus,
the less reactive monomers give the more living systems.

Copolymerization studies of thietanes have shown that the decrease in 'reactivity' for the
substituted monomers is mainly due to a decrease in reactivity of the active species rather than to a

Table 3 Rate Constants (in $1\,mol^{-1}\,s^{-1} \times 10^4$) in Copolymerization of
Thietanes[a]

Propagating Species	Nucleophlic reactant		
	9000[b]	5000	8000
	1800	650	600
	9	2	2

[a] E. J. Goethals, *J. Polym. Sci., Polym. Symp.*, 1976, **56**, 271. At 20 °C in CH$_2$Cl$_2$ with
Et$_3$OBF$_4$ as initiator.
[b] Extrapolated value.

decrease of reactivity of the monomer.[31] Thus, as shown in Table 3, in homopropagation, thietane reacts about 4500 times more rapidly than 3,3-diethylthietane, but in the copolymerization of these two monomers, the reaction of thietane monomer with a 3,3-diethylthietanium ion is only 4.5 times more rapid than the reaction of diethylthietane with the same ion.

51.4 HIGHER-MEMBERED CYCLIC SULFIDES

51.4.1 Thiolane

Thiolane (tetrahydrothiophene) cannot be polymerized, apparently because of its low ring strain. The polymer can be prepared by polycondensation of 1,5-dibromopentane with pentane-1,5-dithiol. When treated with a small amount of alkylating agent such as methyl fluorosulfonate or Et_3OBF_4, the polymer degrades to thiolane (equation 12).[32]

$$\text{equation 12} \tag{12}$$

51.4.2 1,3-Oxathiolane

1,3-Oxathiolane has been reported to polymerize to low molecular weight polymers with sulfuric acid, Et_3OBF_4 with BF_3Et_2O, and $MeCO^+SbF_6^-$ as initiator. According to the NMR analysis, the polymer has a structure of a regularly alternating copolymer of 1,3-dioxolane and 1,3-dithiolane,[33,34] which can only be explained by accepting extensive and selective *trans*-alkylation reactions (equation 13).

$$\longrightarrow \quad -CH_2CH_2SCH_2SCH_2CH_2OCH_2O- \tag{13}$$

51.4.3 Trithiane

Trithiane and higher cyclic thioformaldehyde oligomers are polymerized by cationic initiators such as BF_3, SbF_5, Me_2SO_4 or MeI at high temperatures.[35] The polymerization mechanism is assumed to be similar to that for trioxane, *i.e.* through a sulfonium ion in equilibrium with an α thiacarbenium ion (equation 14).

$$\longrightarrow \quad \text{wwS} \diagup S \diagdown S\overset{+}{\cdots}CH_2 \quad X^- \tag{14}$$

Trithiane may also be polymerized in the solid state by irradiation with γ-rays at 193 °C.[36,37] The preparation and polymerization to poly(methylene sulfide) of 1,3,5,7-tetrathiacyclooctane and 1,3,5,7,9-pentathiacyclodecane under the influence of BF_3 etherate has also been described.[38,39]

Poly(methylene sulfide) has a m.p. of 250 °C, and is insoluble in common organic solvents. The polymer is quite unstable in the molten state and can therefore not be succesfully processed.

51.5 REFERENCES

1. D. Van Ooteghem and E. J. Goethals, *Makromol. Chem.*, 1976, **177**, 3389.
2. J. V. Crivello and J. H. W. Lam, *Macromolecules*, 1977, **10**, 1307.
3. P. Dumas, N. Spassky and P. Sigwalt, *Makromol. Chem.*, 1972, **156**, 65.
4. E. J. Goethals, R. Simonds, N. Spassky and A. Momtaz, *Makromol. Chem.*, 1980, **181**, 2481.
5. S. Batthi and E. J. Goethals, *Makromol. Chem.*, 1985, **186**, 317.
6. F. D'Haese, S. Florquin and E. J. Goethals, paper presented at the Ring-opening Polymerization Symposium, Blois, 1986.
7. N. Spassky and P. Sigwalt, *Bull. Soc. Chim. Fr.*, 1963, 4617.
8. S. Boileau, H. Cheradame, N. Spassky, K. J. Ivin and E. Lillie, *C. R. Hebd. Seances. Acad. Sci., Ser. C*, 1970, **271**, 1232.
9. K. J. Ivin, E. D. Lillie and I. H. Petersen, *Makromol. Chem.*, 1973, **168**, 217.
10. D. Van Ooteghem and E. J. Goethals, *Makromol. Chem.*, 1974, **175**, 1513.
11. R. Simonds, E. J. Goethals and N. Spassky, *Makromol. Chem.*, 1979, **179**, 1851.

12. M. Sepulchre, N. Spassky, D. Van Ooteghem and E. J. Goethals, *J. Polym. Sci., Polym. Chem. Ed.*, 1974, **12**, 1683.
13. L. A. Korotneva and G. V. Belonowskaya,, *Usp. Khim.*, 1972, **41**, 150.
14. R. Simonds and E. J. Goethals, *Makromol. Chem.*, 1978, **179**, 1689.
15. G. H. Schmid, *Top. Sulfur Chem.*, 1978, **3**, 101.
16. Y. Pocker and A. J. Parker, *J. Org. Chem.*, 1966, **31**, 1526.
17. W. M. Van Craeynest and E. J. Goethals, *Eur. Polym. J.*, 1976, **12**, 859.
18. W. M. Van Craeynest and E. J. Goethals, *Makromol. Chem.*, 1978, **179**, 2613.
19. E. J. Goethals, D. Van Meirvenne and R. De Clercq, *Makromol. Chem., Macromol. Symp.*, 1988, **13/14**, 175.
20. A. D. Aliev, B. A. Krentsel and S. L. Alieva, *Eur. Polym. J.*, 1983, **19**, 71.
21. E. J. Goethals, unpublished results.
22. P. Dumas, N. Spassky and P. Sigwalt, *J. Polym. Sci., Polym. Chem. Ed.*, 1976, **14**, 1015.
23. D. C. Price and E. A. Blair, *J. Polym. Sci., Part A-1*, 1967, **5**, 171.
24. J. Stille and J. A. Empen, *J. Polym. Sci., Part A-1*, 1967, **5**, 273.
25. E. J. Goethals, W. Drijvers, D. Van Ooteghem and A. M. Buyle, *J. Macromol. Sci., Chem.*, 1973, **A7**, 1375.
26. E. J. Goethals and W. Drijvers, *Makromol. Chem.*, 1970, **136**, 73.
27. W. Drijvers and E. J. Goethals, *Makromol. Chem.*, 1971, **148**, 311.
28. S. M. Florquin and E. J. Goethals, *Makromol. Chem.*, 1981, **182**, 3371.
29. E. J. Goethals and W. Drijvers, *Makromol. Chem.*, 1973, **165**, 329.
30. E. J. Goethals, *Makromol. Chem.*, 1974, **175**, 1309.
31. E. J. Goethals, *J. Polym. Sci., Polym. Symp.*, 1976, **56**, 271.
32. D. Van Ooteghem, R. Deveux and E. J. Goethals, *J. Polym. Sci., Polym. Symp.*, 1976, **56**, 459.
33. J. Guzman and E. Riande, *Macromolecules*, 1980, **13**, 1715.
34. L. Carrido, J. Guzman and E. Riande, *Makromol. Chem., Rapid Commun.*, 1981, **2**, 379.
35. E. Gipstein, E. Wellisch and O. J. Sweeting, *J. Polym. Sci., Part B*, 1963, **1**, 237.
36. J. E. Herz and V. Stannett, *J. Polym. Sci., Part B*, 1966, **4**, 995.
37. G. Carrazolo and M. Mammi, *J. Polym. Sci., Part B*, 1964, **2**, 1057.
38. M. Russo, L. Mortillaro, L. Credali and C. De Checchi, *J. Polym. Sci., Part B*, 1965, **3**, 455.
39. M. Russo, L. Mortillaro, D. De Checchi, G. Valle and M. Mammi, *J. Polym. Sci., Part B*, 1965, **3**, 501.

52

Cationic Ring-opening Polymerization: Amines and N-containing Heterocycles

ERIC J. GOETHALS
State University of Ghent, Belgium

52.1 INTRODUCTION

The polymerization of a three-membered cyclic amine was reported 100 years ago by Ladenberg and Abel.[1] These authors tried to isolate aziridine (ethylenimine) but instead of the expected volatile compound they obtained a hygroscopic solid. In the 1940s, the polymerization of aziridine was investigated in detail by Kern and Brenneisen[2] and by Jones and co-workers.[3] It was found that cyclic amines, in contrast with their sulfur and oxygen analogues, could be polymerized only by 'cationic' initiators. As in the case of the sulfides, only the three- and four-membered rings — aziridines and azetidines — contain enough ring strain to be polymerizable. Higher rings polymerize only if additional strain is introduced into the molecule, *e.g.* in bicyclic systems, or if a more stable functional group is formed by an isomerization, *e.g.* the oxazolines. Typical examples of polymerizable cyclic amines are shown in structures (1)–(5).

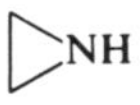

(1) Aziridine

(2) *N*-Alkylaziridine

(3) Azetidine

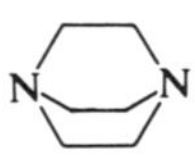

(4) 1,4-Diazabicyclo[2.2.1]octane

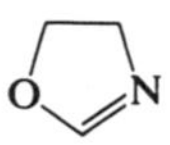

(5) Δ^2-1,3-Oxazoline

The mechanism of the polymerization is determined by the nature of the amino group, *i.e.* secondary or tertiary. With secondary amines, the active species are tertiary ammonium ions which easily transfer a proton to other amino functions thus leading to a polymerization proceeding by a combination of active chain-end and activated monomer mechanism (see Chapter 45). With tertiary amines, the active species are quaternary ammonium ions and the propagation occurs by the activated chain-end mechanism only. Polymerizations proceeding *via* non-charged electrophilic active centers have been reported for cyclic imino ethers.

52.2   AZIRIDINES

52.2.1   Aziridine (Ethylenimine)

The parent compound of the aziridine family is commonly called ethylenimine and the corresponding polymer poly(ethylenimine) (PEI).

The mechanism of polymerization involves addition of a proton acid to aziridine to produce the corresponding aziridinium ion (**6**; equation 1). If an alkylating agent is used as initiator, the alkylated monomer will transfer a proton to a monomer thus producing the same initiating species (equation 2). The first propagation step is the ring opening of (**6**) by nucleophilic attack of monomer (equation 3). The thus formed dimer (**7**) can react in two ways: a new ring-opening of the aziridinium ion by monomer leads to trimer (**8**); the second possibility is that (**7**) transfers its proton to another amino function present in the reaction mixture thus forming the uncharged dimer *N*-(2-aminoethyl)-aziridine (**9**).

$$ \text{(6)} \tag{1} $$

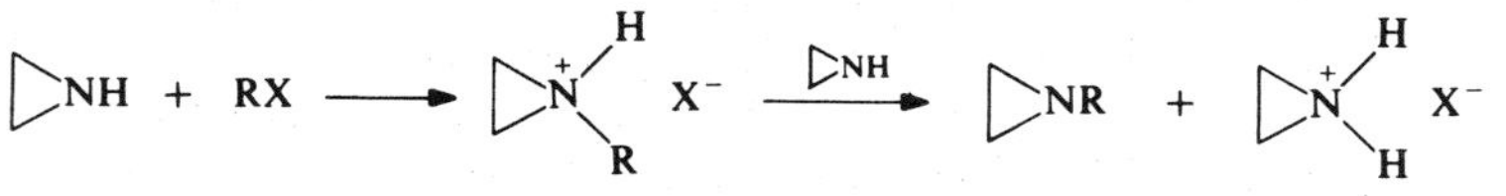

$$ \text{(6)} \tag{2} $$

$$ \text{(7)} \tag{3} $$

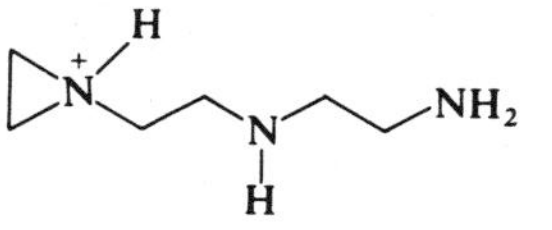

(8) 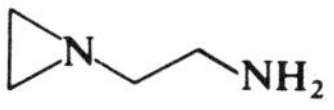   (9)

Compound (**9**) contains two new nucleophilic groups, a primary amine and an *N*-alkylaziridine, which were not present in the monomer. Chain growth can occur by reaction of these nucleophilic groups with protonated monomer, formed by the initiation or by the proton transfer from (**6**) (see Scheme 1).

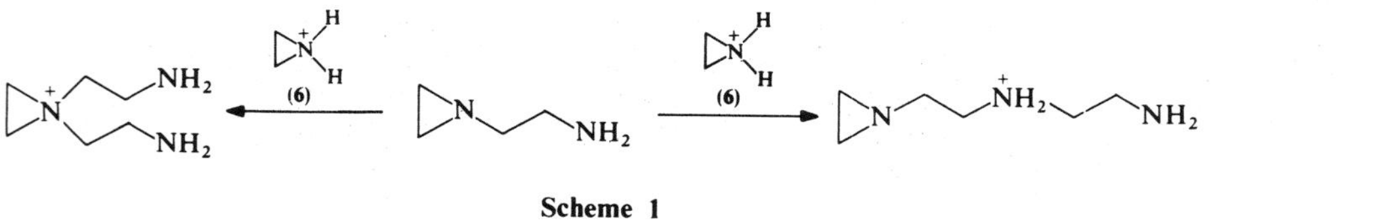

Scheme 1

These reactions are similar to the activated monomer mechanism discussed in the Chapter 45. It is clear that this proton transfer from the active center can also occur with trimer (**8**) or at any state during the polymerization. The nucleophilic aziridine group in dimer (**9**) can be reactivated by protonation and becomes then an electrophilic active chain again, as was demonstrated by Barb.[4] The concentrations of protonated aziridine species such as (6), (7) or (8) become lower as polymerization proceeds to produce the more basic acyclic amino groups. This explains the observed decrease in the rate of polymerization. The formation and disappearance of the various oligomeric species have been followed by gas chromatography.[5,6] One important consequence of the proton transfer from the original active species is that very early in the polymerization branched structures, due to tertiary amine functions, are formed. The most straightforward route for the formation of a tertiary amine is the attack of a secondary amine function on an aziridinium ion followed by proton transfer (equation 4). The importance of this reaction is shown by the observation that, in the tetramer stage, a substantial fraction already consists of the branched isomer.[5] Consequently, the final polymer contains a large fraction of tertiary and an equal fraction of primary amino groups.

$$\overset{\xi}{\underset{\xi}{NH}} + \underset{H}{\overset{H}{\underset{\displaystyle N}{\triangleright}}} \longrightarrow \overset{H}{\underset{\displaystyle N}{\underset{\displaystyle \sim}{\overset{+}{N}}}}\!\!\!\smallsmile\!\!\!NH_2 \xrightarrow{-H^+} \sim\!\!N\!\!\smallsmile\!\!NH_2 \qquad (4)$$

Much work has been devoted to the accurate measurement of the degree of branching in PEI. In 1941, Kern and Brenneisen had already shown that benzoylation of PEI is incomplete and deduced that 20 to 30% of the amino groups were tertiary. Many other methods were used but the clearest demonstration of the branched structure of PEI was given by ^{13}C NMR spectroscopy.[7,8] The spectrum contains eight lines which were attributed by Lukovkin *et al.*[7] to methylene groups having either primary, secondary or tertiary amino groups in α and β positions. As an example, Figure 1 shows the spectrum of a commercial PEI and, for comparison, the spectra of linear PEI and poly[1-(2-aminoethyl)]aziridine. Based on this ^{13}C NMR spectrum it was concluded that, for PEI obtained with 1,3-dichloropropanol as initiator (molecular weight 20 000), the fraction of primary amines is 26 ± 4, of secondary amines 49 ± 4 and of tertiary amines $25 \pm 4\%$.

Linear poly(ethyleneimine) (LPEI), containing only secondary amino groups, has been prepared in three different ways. One method consists of the isomerization–polymerization of a Δ^2-1,3-oxazoline to the corresponding poly(N-acylaziridine) followed by hydrolysis of the amide (equation 5). This method, first described in a patent by Seeliger,[9] was investigated in detail by

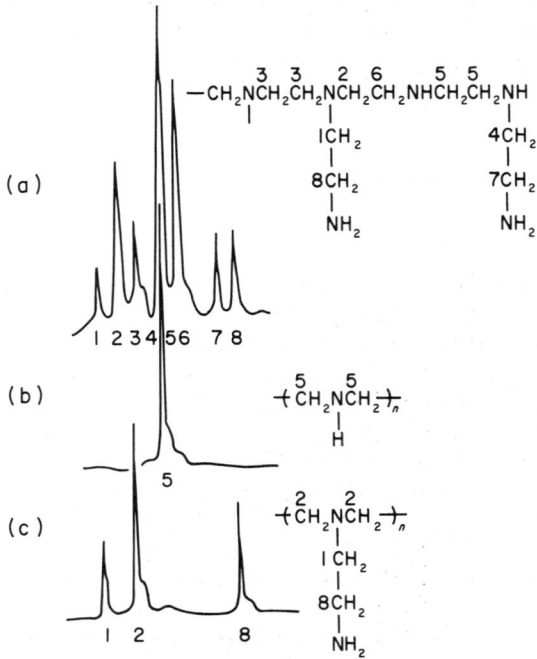

Figure 1 ^{13}C NMR spectra of (a) branched PEI, (b) linear PEI and (c) poly[1-(2-aminoethyl)aziridine]; peak 1 at 71.9 p.p.m. and peak 8 at 89.0 p.p.m., relative to benzene[7]

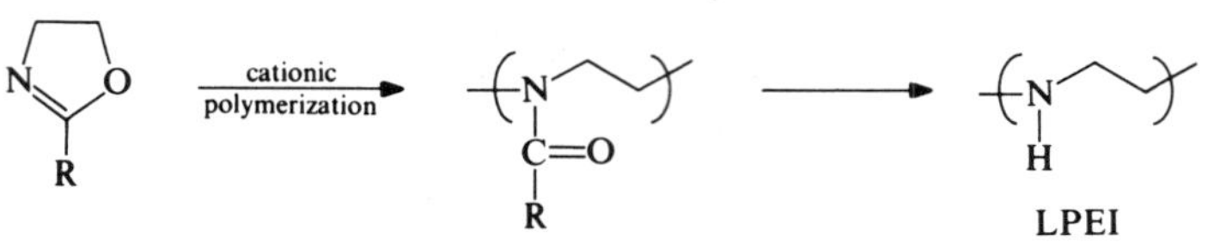

$$(5)$$

Saegusa and co-workers.[10] It was found that alkaline hydrolysis (in contrast with acid hydrolysis) gives an almost complete transformation of the amide to the amine.

In the early 1970s, Zhuk and Gembitskii reported that low temperature acid-initiated aqueous polymerization of aziridine leads to a mixture of branched polymer, which remains in solution, and linear polymer, which precipitates from solution.[11] Yields of up to 33% LPEI of molecular weight up to 13 000 were obtained using perchloric acid and a reaction time of several days. A better technique for producing LPEI is a two-stage process consisting of a limited-conversion pre-polymerization of aziridine to dimer and trimer predominantly, followed by proton initiation of the polycondensation of these oligomers in the presence of a specific amount of water.[12] The process is described as a head-to-tail coupling of the oligomers (Scheme 2). The role of the water is to form a crystalline hydrate with the polymer which is therefore physically blocked. As a consequence, the secondary amines are prevented from reacting with the aziridinium ions (which is the mechanism leading to branching), whereas the primary amino groups located at the periphery of the crystalline structure retain a degree of freedom and therefore are still able to react.[13]

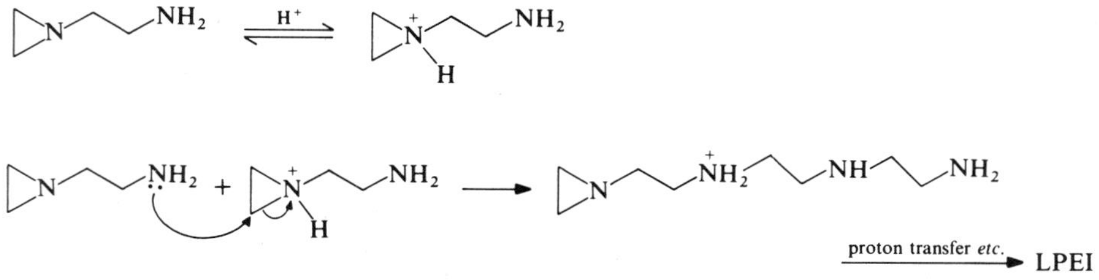

Scheme 2

Recently, a third method for the production of LPEI has been reported by Weyts and Goethals.[15] It consists of the cationic polymerization of N-(α-tetrahydropyranyl)aziridine followed by the acid hydrolysis of the resulting poly(imino ether) and neutralization with sodium hydroxide to form the free base. The starting monomer is prepared from aziridine and α-chlorotetrahydropyran (Scheme 3).

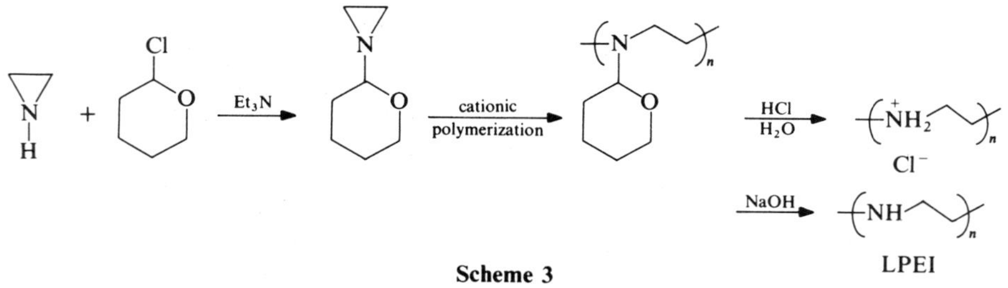

Scheme 3

In contrast with branched PEI, LPEI is a crystalline material. The melting point of dry LPEI is 58.5 °C but the polymer can form various crystalline hydrates containing up to 2 mol of water for one repeat unit.[14] The melting point of these hydrates varies from 70 to 80 °C. LPEI is insoluble in water below 50 °C. Various block and graft copolymers containing LPEI segments have been described.[15]

52.2.2 *C*-Substituted Aziridines

The polymerization tendency of aziridines is markedly influenced by the number of substituents on carbon. It was observed by Jones[16] that only those monomers having at least one CH$_2$ in the ring could be polymerized with boron trifluoride. 2,2-Dimethylaziridine gave polymer along with oligomeric products. Identification of these oligomers lead to the conclusion that the propagation reaction takes place by an S_N2 reaction at the methylene group of the aziridinium ion. The dimer, for example, had structure (**10**) and not structure (**11**) which would have been formed if an S_N1 mechanism were operative (Scheme 4).

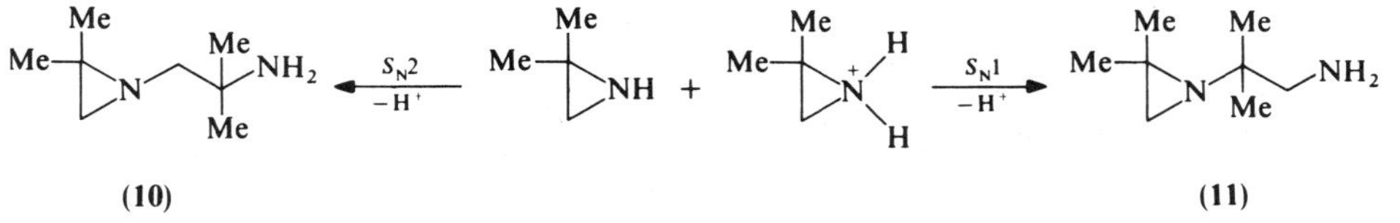

(**10**)   (**11**)

Scheme 4

Similar conclusions were drawn for the polymerization of 2-methylaziridine (propylenimine) by Price *et al.*:[17] polymerization of the racemic mixture led to a liquid, but polymerization of one of the enantiomers gave solid polymers which had strong optical rotation of opposite sign to that of the corresponding monomer. This indicated a high degree of retention of asymmetry during the polymerization, which was explained by an S_N2 ring-opening at the methylene group of the aziridinium ion.

Conclusive evidence for the S_N2 attack without inversion came from the work of Tsuboyama *et al.*[18] who showed that polymerization of (*R*)-2-ethylaziridine led to the same polymer as the polycondensation of (*R*)-2-amino-1-bromobutane.

Nothing seems to be known about the degree of branching of polymers of 2-substituted aziridines. Linear pure head-to-tail poly(2-methylaziridine) has been prepared by alkaline hydrolysis of poly(*N*-formyl-2-methylaziridine) which, in turn, was obtained by the isomerization–polymerization of 5-methyl-Δ^2-1,3-oxazoline[19] (equation 6). Starting from optically active monomer, isotactic (optically active) poly(2-methylaziridine) was obtained.[20]

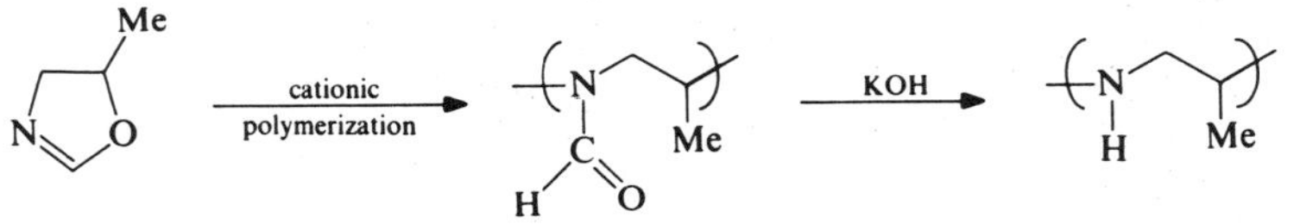

(6)

52.2.3 *N*-Substituted Aziridines

The mechanism of polymerization of tertiary aziridines is simpler than that for secondary amines because the aziridinium ions are now quaternary so that transfer of a proton is not possible. The propagation step is the classical nucleophilic attack of the nitrogen of the monomer on the active species. However, this attack can also occur with a nitrogen of the polymer molecule (Scheme 5). The thus formed branched or cyclic quaternary ammonium ion is unreactive toward propagation and this reaction is a real termination reaction. As a consequence, the polymerizations of *N*-alkyl-aziridines generally do not proceed to completion and do not lead to a high molecular weight polymers. This was first observed for the polymerization of *N*-methylaziridine by Jones[21] and more recently by the group of Ponomarenko.[22] It was shown later[23] that the nature of the nitrogen substituent plays a predominant role in the rate of the termination reaction. A number of monomers were characterized by the ratio of k_p/k_t which was taken as a measure for the 'living character' of their respective polymerizations. Table 1 gives a survey of the k_p/k_t values obtained.

A kinetic study revealed that the termination reaction of polymerizations carried out in solution follows first-order kinetics, which indicates that under the conditions used the reaction takes place between the aziridinium group and an amino function of its own polymer chain forming a

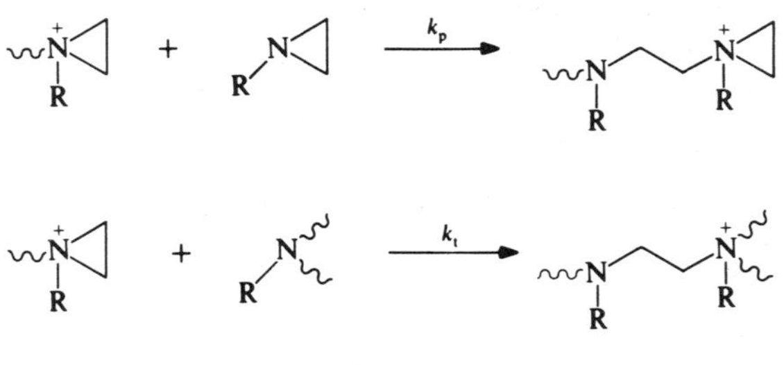

Scheme 5

Table 1 Values of k_p/k_t for the Polymerization of Various
N-Substituted Aziridines[a]

Monomer	k_p/k_t ($l\,mol^{-1}$)
▷NEt	6
▷NPr^i	21
▷NBu^t	12 000
▷NCH$_2$Ph	85
▷NCH$_2$CH$_2$Ph	14
▷NCH$_2$CH$_2$CN	82

[a] In methylene chloride solution at 10 °C; initiator triethyloxonium tetrafluoroborate.

macrocyclic ammonium end-group. This assumption was supported by the observation that the
molecular weights of the polymers were close to the expected values assuming quantitative initiation
which excludes intermolecular termination. It is, however, not excluded that such intermolecular
termination would occur at high monomer concentrations. Further evidence for the formation of
large rings as end groups was found in the fact that cyclic oligomers (mainly tetramers) are formed
when the polymerization is initiated with proton initiators (see below).

An important factor determining the living character is the bulkiness of the substituent. Compared with the nitrogen in the corresponding monomer, the nitrogens in the polymer chain are much
more sterically hindered by the free rotating polymer chain and the substituent, thus reducing their
ability to perform an S_N2 reaction with the active species.

From Table 1, it can be seen that N-t-butylaziridine has a very high k_p/k_t value so that the
polymers obtained under certain conditions are 'temporarily living'. This means that, although there
is a termination reaction, it is so slow compared with propagation that, at almost quantitative
conversion, almost all polymer chains still possess an active center. This allows the production of
polymers having predictable molecular weights and narrow molecular weight distributions if a fast
initiating system is used. The active species itself is still highly reactive and reacts readily with
nucleophiles which are not sterically hindered. This has enabled the synthesis of a series of poly(N-t-
butylaziridines) with different end groups.[24] When the nucleophile is the acrylate or methacrylate
anion a polyamine macromer with an acrylate ester end-group is obtained (equation 7).[25]

$$\underset{Bu^t}{\sim\!\sim N^+} \triangleleft \;+\; CH_2\!\!=\!\!\overset{Me}{\underset{}{C}}CO_2^- \;\longrightarrow\; \underset{Bu^t}{\sim\!\sim N}\!\!\frown\!\!O_2C\overset{Me}{\underset{}{C}}\!\!=\!\!CH_2 \qquad (7)$$

The living character of the polymerization of N-t-butylaziridine was also used to produce block
copolymers; for example, by initiation of the polymerization with living poly(tetrahydrofuran)
(PTHF; Scheme 6).[26] Block copolymers consisting of a poly(N-t-butylaziridine) (PTBA) segment
and a poly(dimethylsiloxane) segment were obtained by reaction of the cationically living polyamine
with the anionically living polysiloxane[27] and graft copolymers containing poly(N-t-butylaziridine)
segments were made by copolymerization of the methacrylate macromer[28] and by grafting the living
polymer on to cellulose derivatives.[29]

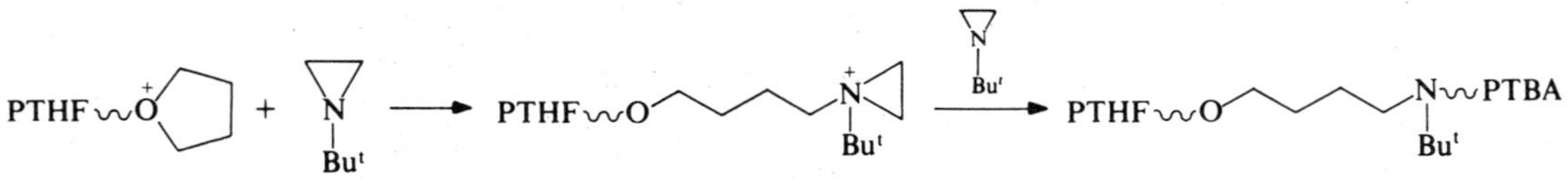

Scheme 6

Polyamine networks have been prepared from bifunctionally living poly(*N*-*t*-butylaziridine) and polyfunctional acids such as 1,2,4,5-benzenetetracarboxylic acid (pyromellitic acid; equation 8).[30] The bifunctionally living polymer is obtained if the polymerization is initiated by low molecular weight bifunctionally living PTHF which in turn is obtained by initiation with trifluoromethane-sulfonic anhydride (triflic anhydride).

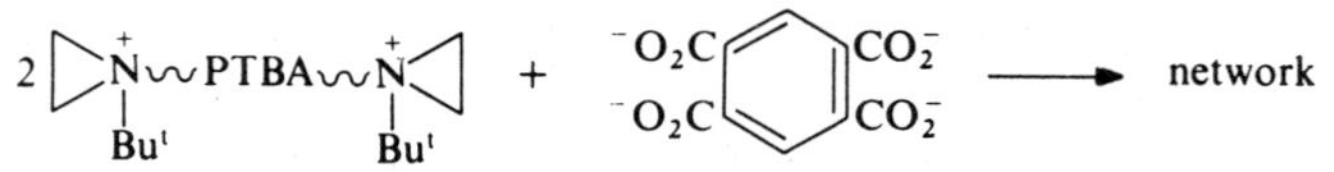

$$2 \;\; \underset{Bu^t}{\triangleright}\overset{+}{N}\!\!\sim\!\!PTBA\!\!\sim\!\!\underset{Bu^t}{\overset{+}{N}}\triangleleft \;\; + \;\; \begin{array}{c} ^-O_2C \\ ^-O_2C \end{array}\!\!\bigcirc\!\!\begin{array}{c} CO_2^- \\ CO_2^- \end{array} \longrightarrow \text{network} \tag{8}$$

N-Alkylaziridines can be polymerized with an alkyl halide as initiator if the reaction is carried out in a polar solvent.[31] In a solvent of low polarity, no polymer is formed but the corresponding 1,1,4-trialkylpiperazinium ion is formed instead. This is due to the equilibrium between aziridinium halide and covalently bonded halide that exists in the non-polar media. This favours an intramolecular S_N2 reaction at the dimer state (Scheme 7).

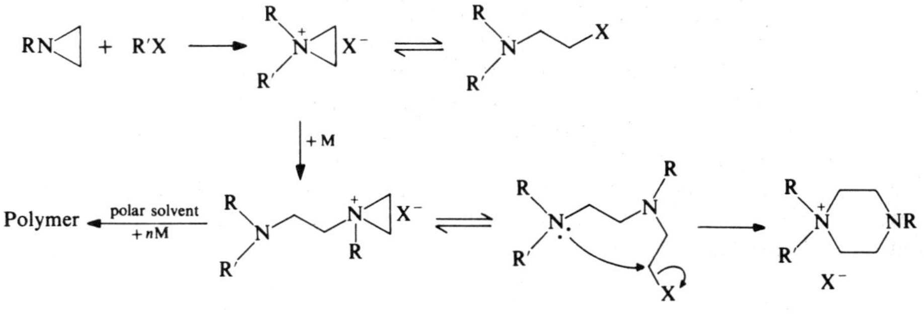

M = monomer

Scheme 7

Initiation of the polymerization with a proton acid leads to polymers having a secondary amine as head group (equation 9). Although the concentration of this secondary amino group is small compared with the tertiary amino groups (monomer and polymer nitrogens), its reactivity towards the aziridinium ion is considerably higher so that its presence has a marked effect on the polymerization. This differs from the alkyl-initiated polymerization; primarily, in the occurrence of an end-group coupling reaction of different polymer chains (equation 10). Indeed, it was observed[32] that the molecular weight of poly(*N*-*t*-butylaziridine), obtained by initiation with triflic acid, continues to increase after all the monomer has been consumed, a phenomenon that was absent in the case of alkyl initiation and that is consequently to be attributed to the presence of the secondary amino head-group.

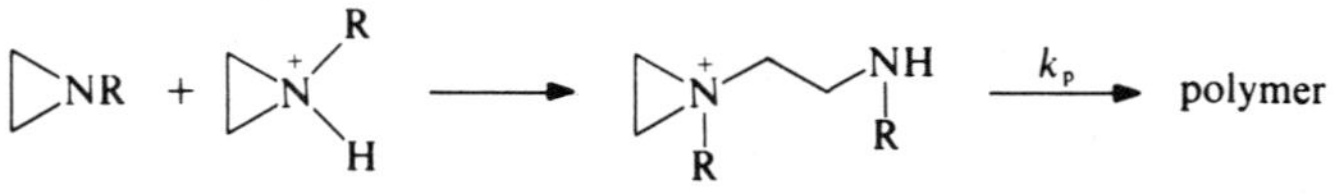

$$\triangleright NR \;\; + \;\; \underset{H}{\triangleright}\overset{R}{\overset{+}{N}} \longrightarrow \underset{R}{\triangleright}\overset{+}{N}\!\!\sim\!\!\overset{NH}{\underset{R}{}} \;\; \xrightarrow{k_p} \;\; \text{polymer} \tag{9}$$

$$\underset{R}{\triangleright}\overset{..}{N}\!\!\sim\!\!\underset{R}{\overset{..}{N}H} \;\; + \;\; \underset{R}{\triangleright}\overset{+}{N}\!\!\sim\!\!\underset{R}{NH} \longrightarrow \underset{R}{\triangleright}\overset{+}{N}\!\!\sim\!\!\underset{R}{\overset{\overset{H}{|}}{N}}\!\!\sim\!\!\underset{R}{NH} \;\; \xrightarrow{etc.} \cdots \tag{10}$$

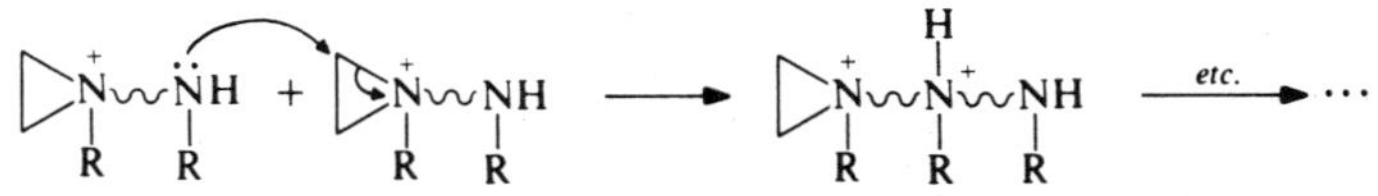

A second remarkable observation is that the proton-initiated polymerization of *N*-alkylaziridines invariably leads to the formation of substantial amounts of cyclic oligomers. This phenomenon was first reported for *N*-benzylaziridine by Hansen and Burg[33] and was then confirmed for a series of aziridines by Ham.[34] Generally, a mixture of cyclic oligomers consisting mainly of tetramer and small amounts of dimer, trimer and pentamer is obtained, but the distribution varies with the reaction conditions. The formation of these cyclic oligomers occurs by an intramolecular head-to-end coupling between the secondary amino group and the aziridinium group. Apparently, the approach between these two chain ends is most favourable when they are separated by three structural units (equation 11).

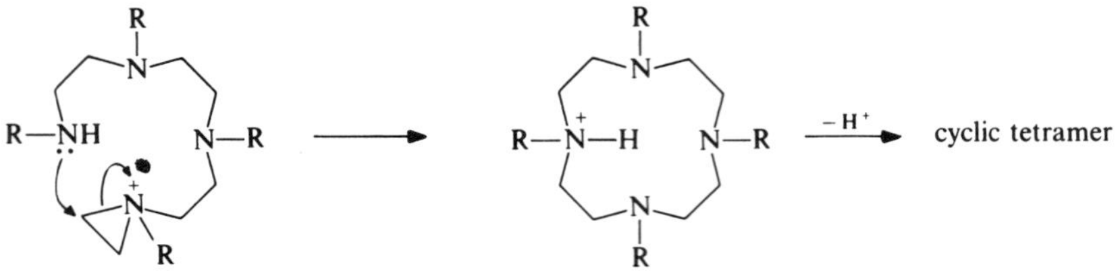

$$\text{(11)}$$

If the initiating proton acid has a nucleophilic counteranion such as halide and if the reaction is carried out in a non-polar solvent, the main reaction product is the corresponding N,N'-dialkyl-piperazine.[35] This is formed by a similar route to that described in Scheme 7 with the difference that R' is now a proton that can be transferred to another monomer thus leading to high yields of the piperazine.

A number of substituted aziridines carrying a functional group in the substituent have been polymerized. For example, polyampholytes have been prepared by polymerization of monomer (12) followed by hydrolysis (Scheme 8).[36, 37] Such polymers have been examined for their ability to form complexes with metal ions.[37]

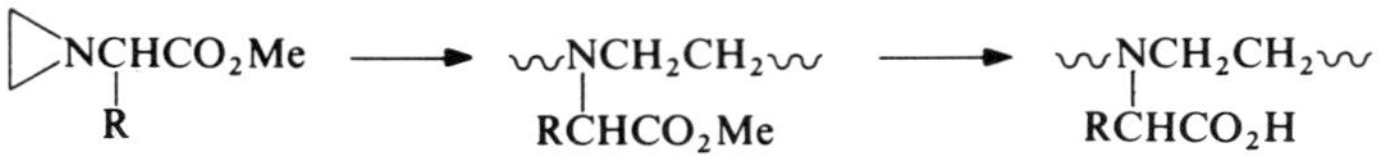

$$\triangleright\text{NCHCO}_2\text{Me} \quad \longrightarrow \quad \text{wNCH}_2\text{CH}_2\text{w} \quad \longrightarrow \quad \text{wNCH}_2\text{CH}_2\text{w}$$

```
       |                              |                              |
       R                            RCHCO₂Me                       RCHCO₂H
      (12)
                                   R = H or Me
```

Scheme 8

52.2.4 *N*- and *C*-Substituted Aziridines

The introduction of an alkyl group on to one of the carbon atoms in an N-alkylaziridine has a dramatic effect on the rate of propagation. With the C-unsubstituted aziridines the final yields are generally obtained after a few minutes at 25 °C but with the corresponding 2-methyl derivatives it takes several hours. This difference has been demonstrated for the N-methyl-,[38] N-benzyl-,[23] N-phenethyl-[23] and N-(2-cyanoethyl)-aziridine[23] and their corresponding 2-methyl homologues. The termination reaction is reduced more than the propagation reaction by the introduction of the methyl substituent, and therefore the polymerizations have a much greater living character. With fast initiators the polymerizations follow first-order kinetics up to high conversions and the molecular weights are determined by the ratio of reacted monomer over initiator concentration. This is also the case for the bicyclic aziridine (13) which was reported to give a living polymerization (equation 12).[39]

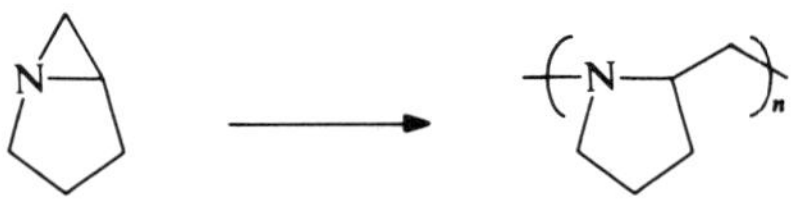

$$\text{(12)}$$

```
       (13)
```

The importance of C-substitution is further demonstrated when the polymerizations of the monomers (14)–(16) are compared.[40] At 20 °C, the polymerization of (14) is very fast but stops at limited conversions. Under the same reaction conditions (15) leads to virtually quantitative conversion in a few hours. It was not possible to polymerize (16) at temperatures between 20 and 120 °C either in solution or in bulk. (16) does copolymerize with (14) but the amount of (16) incorporated in the copolymer is small and always less than 50%.

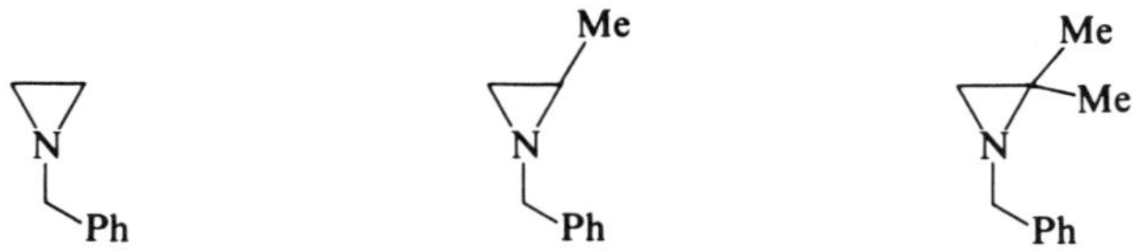

```
       (14)                (15)                (16)
```

N-Alkylaziridines with a substituent on one of the ring carbons can open in two ways (structure **17**). A ^{13}C NMR spectroscopy study of polymers obtained from (*S*)-*N*-(methoxycarbonylmethyl)-2-methylaziridine (**17**; R = —CH$_2$CO$_2$Me) showed that ring opening occurs predominantly but not exclusively at α.[41] The same was found for the polymerization of *N*-phenethyl-2-methylaziridine.[42] The polymers contain approximately 10% of head-to-head and tail-to-tail structures.

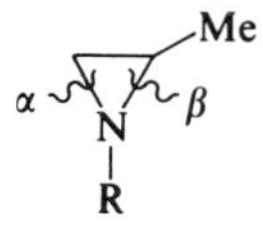

(17)

Similarly to the aziridines, 2-methylaziridines lead to large fractions of cyclic oligomers if the polymerization is initiated by proton acids.[43] The most abundant oligomers are the tetramers but, if an acid with a nucleophilic counterion is used, the piperazine is formed.[44]

52.3 AZETIDINES

52.3.1 Azetidine

The polymerization of the parent monomer of this family was first described only in 1974.[45] The mechanism is very similar to that of aziridine in that oligomeric compounds are formed first due to the possibility of transferring a proton from the original active species to monomer. In the case of azetidine this transfer is even more pronounced due to the higher basicity of the secondary amino function of the four-membered monomer, compared with the tertiary amino function of an *N*-substituted azetidine. When all monomer has reacted, 70% of the reaction mixture consists of dimer (**18**; Scheme 9).

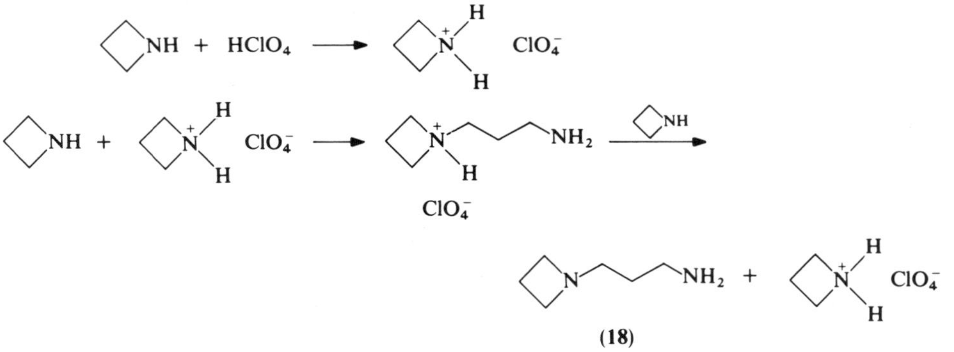

(18)

Scheme 9

Further polymerization leads to a branched polymer containing 20% primary, 20% tertiary and 60% secondary amines, as deduced from its ^{1}H NMR spectrum.[45]

Linear poly(iminotrimethylene) cannot be obtained by direct polymerization of azetidine. It has been synthesized by isomerization–polymerization of 5,6-dihydro-4*H*-1,3-oxazine (**19**) followed by hydrolysis of the polyamide.[46] In contrast with the branched polymer, the linear polymer is a crystalline solid with a melting point between 74 and 84 °C depending on the content of water of crystallization.

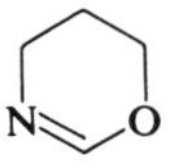

(19)

52.3.2 *N*-Alkyl Substituted Azetidines

As in the case of aziridines, the absence of a proton on the nitrogen simplifies the mechanism of polymerization of azetidines. The most studied monomers in this group are conidine (**20**;

1-azabicyclo[4.2.0]octane) and 1,3,3-trimethylazetidine (**21**), the polymerizations of which were reported to be living (equation 13).

(13)

(**20**)

(**21**)

Toy and Price[49] prepared and polymerized the two pure enantiomers of conidine. The polymers had a strong optical rotation of a sign opposite to that of the monomer from which they were formed, gave a sharp X-ray pattern and had a crystalline melting point of 94 °C. From these observations it was concluded that the ring opening of the azetidinium ion, which is the active species for the propagation, is exclusively at the methylene carbon atom.

The polymerization of 1-methylazetidine (**22**) and 1,3,3-trimethylazetidine (**21**) has been reported by Schacht *et al.*[48, 50] *N*-Methylazetidine polymerizes at room temperature. A termination reaction between the active species and an amino function of the polymer, leading to branched or macrocyclic ammonium salts, was observed.[50] The polymerization of 1,3,3-trimethylazetidine is possible only above 60 °C. At room temperature, initiation to the azetidinium salt occurs immediately and quantitatively but no polymerization takes place. Above 60 °C, the polymerization goes to completion according to first-order kinetics. Second-monomer addition experiments showed that the concentration of active species remains constant even a long time after the polymerization.[48] The quaternary azetidinium ions could be observed directly by ^{1}H NMR spectroscopy; it was also shown by this technique that their concentration did not change appreciably after several days at 78 °C.[51] Consequently, the cationic polymerization of 1,3,3-trimethylazetidine is a typical living polymerization.

(**22**)

Comparison of the polymerization behavior of 1-methylazetidine with that of 1,3,3-trimethyl-azetidine shows the importance of the presence of the *gem*-dimethyl substituents in the 3-position. The presence of these substituents retards the termination much more than the propagation which results in a slower polymerization but one having a more living character.

The copolymerization of the two azetidines, (**21**) and (**22**) has been studied.[52] The following reactivity ratios (at 80 °C in nitrobenzene) were obtained: $r_{MA} = k_{11}/k_{12} = 3.3 \pm 0.2$; $r_{TMA} = k_{22}/k_{21} = 0.3 \pm 0.1$. These results mean that 1-methylazetidine reacts three times more rapidly with both active species than 1,3,3-trimethylazetidine. The difference between the homo-propagation constants k_{11} and k_{22}, however, is a factor of at least 10^3. This shows that the differences in reactivities in the homopolymerizations are in the first place governed by differences in reactivity of the active species and to a much lesser extent by the reactivities of the monomers.

Other azetidines which have been polymerized are 1-(2-cyanoethyl)-,[53] 1-(2-ethoxycarbonyl-ethyl)-[54] and 1-benzyl-azetidine.[55] 1-Alkyl-3-hydroxyazetidines (**23**) which are synthesized from epichlorohydrin and a primary alkylamine, can be polymerized to the corresponding hydroxy-aminopolymers (equation 14).[56]

(14)

(**23**)

52.4 HIGHER-MEMBERED RINGS

Only a few reports on the polymerization of cyclic amines having more than four ring atoms have been published. Pyrrolidine and *N*-alkylpyrrolidines seem to be non-polymerizable, most probably because of the low ring strain in these molecules and hence a too low ceiling temperature. The same is true for piperidine or piperazine and their derivatives. Nevertheless, two monomers containing six-membered amine structures have been reported to undergo ring-opening polymerization,[57] namely 1-azabicyclo[2.2.2]octane (**24**; equation 15) and 1,4-diazabicyclo[2.2.2]octane (**25**; equation 16).

$$\text{(15)}$$

(24)

$$\text{(16)}$$

(25)

52.5 CYCLIC 1,3-OXAZA COMPOUNDS

52.5.1 Δ^2-1,3-Oxazolines

Unsubstituted as well as a large number of substituted Δ^2-1,3-oxazolines can be polymerized with typical cationic initiators to produce the corresponding poly(*N*-acylaziridine)s. The polymerization is a ring-opening reaction with isomerization (equation 17).[58-63] The mechanism of the polymerization is as shown in Scheme 10.

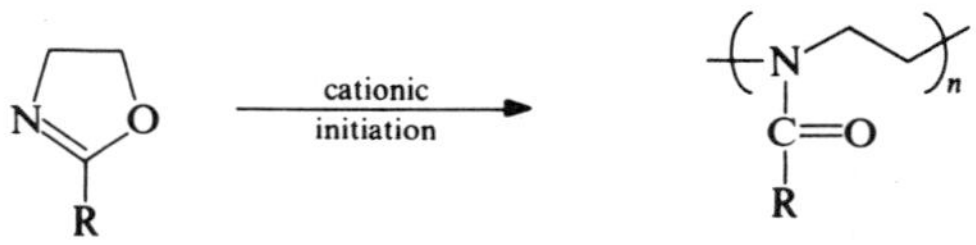

$$\text{(17)}$$

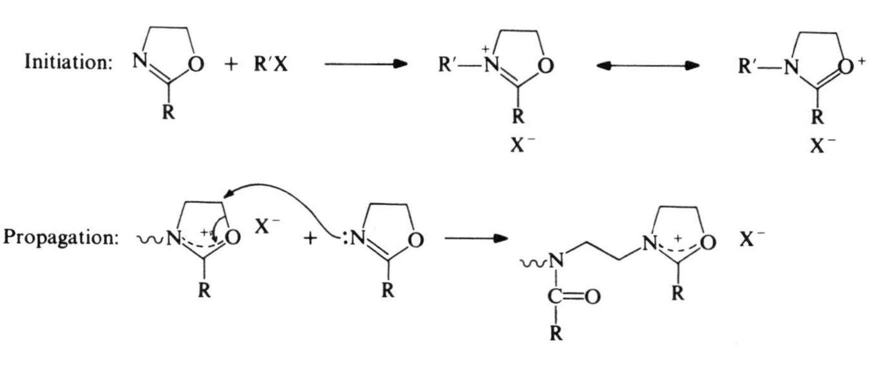

Initiation:

Propagation:

Scheme 10

A termination reaction by attack of a polymeric heteroatom on the active species, which usually occurs in the cationic ring-opening polymerization of heterocyclic monomers, is not possible in the polymerization of oxazolines because the nucleophilicity of the amide functions of the polymer chain is much lower than that of the monomer. As a consequence, the polymerization is of the living type and the molecular weight can be controlled by varying the ratio of monomer to initiator concentrations.

For the mechanism of polymerization of Δ^2-1,3-oxazoline two types of propagating species have been identified.[64] One is the oxazolinium salt, which is obtained with initiators producing counter-ions with low nucleophilicity *e.g.* *p*-toluenesulfonate (tosylate). The other is the covalent species which is formed if an initiator such as methyl iodide is used. In the latter case, the first formed oxazolinium iodide is unstable under the reaction conditions used because of the high nucleophilicity of the iodide ion. This causes a rapid ring-opening, forming the covalently bound alkyl iodide (Scheme 11).

In the polymerization of 5-methyl-Δ^2-1,3-oxazoline initiated with methyl iodide, both propagating species were found to be present in equilibrium.[65]

The living character of the oxazoline polymerization has been employed to produce a variety of graft and block copolymers. For example, poly(Δ^2-1,3-oxazoline) has been grafted on to chloro-

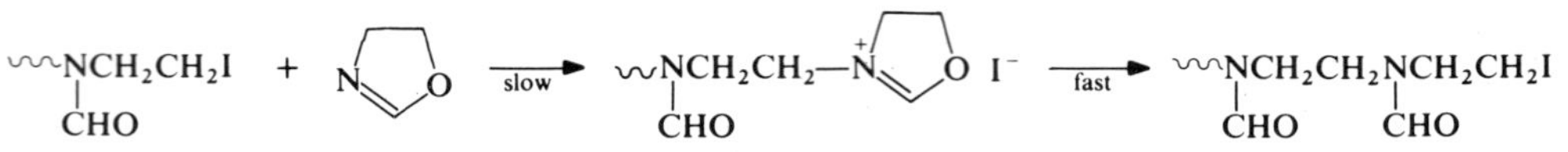

Scheme 11

methylated polystyrene[66] and on to a (butadiene)–(1-chlorobutadiene) copolymer.[67] Block copolymers containing poly(N-acetylaziridine) segments have been prepared with polystyrene,[68] poly(ethylene oxide)[69] and polybutadiene.[70] The synthesis consisted of initiating the Δ^2-1,3-oxazoline polymerization with a tosylate-terminated polymer (Scheme 12). Hydrolysis of the poly(N-acylaziridine)s leads to LPEI.[15]

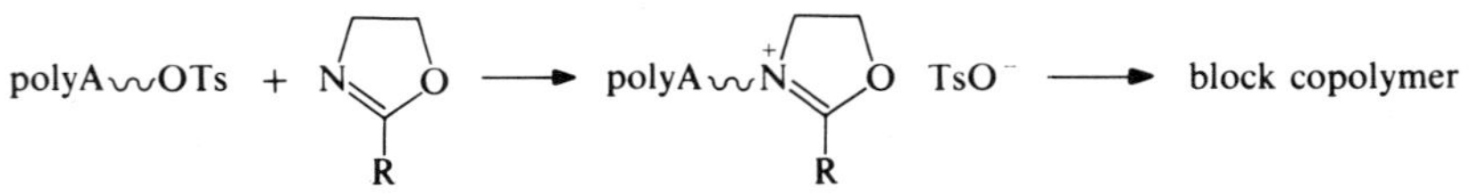

Scheme 12

Generally, the molecular weights of the LPEIs obtained by hydrolysis of the poly(N-acylaziridine)s are rather low. However, Tanaka *et al.*[71] and, more recently, Schulz[72] reported that polymerization of 2-phenyl-Δ^2-1,3-oxazoline leads to polymers which after hydrolysis yield LPEIs with molecular weights above 100 000.

52.5.2 5,6-Dihydro-4H-1,3-oxazines

Several 2-substituted 5,6-dihydro-4H-1,3-oxazines[73] as well as the unsubstituted monomer[46] have been polymerized by cationic initiators at elevated temperatures to poly[(acylimino)trimethylene] (equation 18).

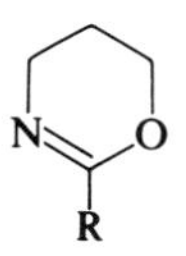

(18)

R = alkyl or H

Complete conversions were achieved in every case. The driving force of the polymerization of these six-membered rings is attributed to the conversion of the imino ether group into an amide group which has more resonance stabilization. Hydrolysis of the polymers leads to linear poly(iminotrimethylene).[46]

52.6 REFERENCES

1. A. Ladenberg and J. Abel, *Ber. Dtsch. Chem. Ges.*, 1888, **21**, 758.
2. W. Kern and E. Brenneisen, *J. Prakt. Chem.*, 1941, **159**, 193.
3. G. D. Jones, A. Langsjoen, M. M. Neumann and J. Zomlefer, *J. Org. Chem.*, 1944, **9**, 125.
4. W. G. Barb, *J. Chem. Soc.*, 1955, 2464, 2577.
5. C. R. Dick and G. E. Ham, *J. Macromol. Sci., Chem.*, 1970, **A4**, 1301.
6. G. L. Gromova, V. G. Berezkin, P. A. Gembitskii and D. S. Zhuk, *Vysokomol. Soedin., Ser. A*, 1976, **18**, 240.
7. G. M. Lukovkin, V. S. Pshezhetsky and G. A. Murtazaeva, *Eur. Polym. J.*, 1973, **9**, 559.
8. T. St. Pierre and M. Geckle, *Polym. Prepr., Am. Chem. Soc., Div. Polym. Chem.*, 1981, **22**, 128.
9. W. Seeliger, W. Thier and W. Kriesten, *Ger. Pat.* 1 720 436 (1971).
10. T. Saegusa, H. Ikeda and H. Fujii, *Macromolecules*, 1972, **5**, 108.
11. P. A. Gembitskii, A. I. Chmarin, N. A. Kheshcheva and D. S. Zhuk, *Vysokomol. Soedin., Ser. A*, 1978, **20**, 1505.
12. D. S. Zhuk, P. A. Gembitskii and A. I. Chmarin, *Br. Pat.* 1 459 809 (1976); *Ger. Pat.* 2 530 042 (1977).
13. K. Weyts and E. J. Goethals, *Polym. Bull. (Berlin)*, 1988, **19**, 13.
14. P. A. Gembitskii, N. A. Kleshcheva, A. I. Chmarin and D. S. Zhuk, *Vysokomol. Soedin., Ser. A*, 1978, **20**, 1613.
15. T. Saegusa and S. Kobayashi, in 'Polymeric Amines and Ammonium Salts', ed. E. J. Goethals, Pergamon Press, 1980, p. 55.
16. G. D. Jones, *J. Org. Chem.*, 1944, **9**, 484.

17. Y. Minoura, M. Takebayashi and C. C. Price, *J. Am. Chem. Soc.*, 1959, **81**, 4689.
18. K. Tsuboyama, S. Tsuboyama and M. Yanagita, *Bull. Chem. Soc. Jpn.*, 1967, **40**, 2954.
19. T. Saegusa, S. Kobayashi and M. Ishiguro, *Macromolecules*, 1974, **7**, 958.
20. J. G. Hamilton, K. J. Ivin, L. C. Kuan-Essig and P. Watt, *Polymer*, 1975, **16**, 763.
21. G. D. Jones, D. C. MacWilliams and N. A. Braxtor, *J. Org. Chem.*, 1965, **30**, 1994.
22. V. A. Ponomarenko *et al.*, *Vysokomol. Soedin., Ser. B.*, 1974, **16**, 815; 1981, **23**, 230.
23. E. J. Goethals, E. H. Schacht, P. Bruggeman and P. Bossaer, *ACS Symp. Ser.*, 1977, **59**, 1.
24. A. Munir and E. J. Goethals, *J. Polym. Sci., Polym. Lett. Ed.*, 1981, **19**, 1985.
25. E. J. Goethals and M. Vlegels, *Polym. Bull. (Berlin)*, 1981, **4**, 521.
26. E. J. Goethals, M. Van de Velde and A. Munir, in 'Cationic Polymerization and Related Processes', ed. E. J. Goethals, Academic Press, 1984, p. 387.
27. Y. Tezuka *et al.*, *Makromol. Chem.*, 1988, in press.
28. M. Vlegels, Ph.D. Thesis, University of Gent, 1986.
29. A. Munir and W. Daly, *Makromol. Chem., Rapid Commun.*, 1983, **4**, 589.
30. M. Van de Velde and E. J. Goethals, *Makromol. Chem., Macromol. Symp.*, 1986, **6**, 271.
31. C. R. Dick, *J. Org. Chem.*, 1967, **32**, 72.
32. E. J. Goethals, R. Deveux and L. Vandenberghe, *Makromol. Chem., Rapid Commun.*, 1982, **3**, 515.
33. G. R. Hansen and T. E. Burg, *J. Heterocycl. Chem.*, 1968, **5**, 304.
34. G. E. Ham, in 'Polymeric Amines and Ammonium Salts', ed. E. J. Goethals, Pergamon Press, Oxford, 1980, p. 1.
35. C. R. Dick, *J. Org. Chem.*, 1970, **35**, 3940.
36. BASF A. G., *Ger. Pat.* 806 992 (1951) (*Chem. Abstr.*, 1952, **45**, 9230).
37. C. Samijn and G. Smets, in 'Polymeric Amines and Ammonium Salts', ed. E. J. Goethals, Pergamon Press, Oxford, 1980, p. 25.
38. V. A. Ponomarenko, I. A. Chekulaeva, N. B. Bogacheva, E. Y. Gorodetskaya and A. V. Ignatenko, *Vysokomol. Soedin., Ser. B*, 1981, **23**, 235.
39. E. F. Razvodovskii, A. A. Berlin, A. V. Nekrasov, A. T. Ponomarenko, L. Pushchayeva, N. G. Puchkova and N. S. Enikolopyan, *Vysokomol. Soedin., Ser. A*, 1973, **15**, 2233.
40. E. J. Goethals, P. Bossaer and R. Deveux, *Polym. Bull. (Berlin)*, 1980, **6**, 121.
41. C. Samijn, S. Toppet and G. Smets, *Makromol. Chem.*, 1976, **177**, 2849.
42. G. Eekhaut, Ph.D. Thesis, University of Gent, 1987.
43. S. Tsuboyama, K. Tsuboyama, I. Higashi and M. Yanagita, *Tetrahedron Lett.*, 1970, **16**, 1357.
44. M. Yanagita, S. Tsuboyama and K. Tsuboyama, *Jpn. Pat.* 74 109 386 (1974) (*Chem. Abstr.*, 1974, **83**, 10 143a).
45. E. H. Schacht and E. J. Goethals, *Makromol. Chem.*, 1974, **175**, 3447.
46. T. Saegusa, Y. Nagura and S. Kobayashi, *Macromolecules*, 1973, **6**, 495.
47. E. F. Razvodovskii *et al.*, *Dokl. Akad. Nauk SSSR*, 1971, **198**, 894.
48. E. H. Schacht and E. J. Goethals, *Makromol. Chem.*, 1973, **165**, 329.
49. M. S. Toy and C. C. Price, *J. Am. Chem. Soc.*, 1960, **82**, 2613.
50. E. H. Schacht, P. Bossaer and E. J. Goethals; *Polym. J.*, 1977, **9**, 329.
51. E. J. Goethals and E. H. Schacht, *J. Polym. Sci., Polym. Lett. Ed.*, 1973, **11**, 497.
52. E. J. Goethals *et al.*, *Polym. J.*, 1980, **12**, 571.
53. S. Hashimoto, T. Yamashita and J. Hino, *Polym. J.*, 1977, **9**, 19.
54. J. Lucasczyk, E. H. Schacht and E. J. Goethals, *Makromol. Chem., Rapid Commun.*, 1980, **1**, 79.
55. S. Hashimoto and T. Yamashita, *J. Macromol. Sci., Chem.*, 1986, **A23**, 295.
56. K. Banthia, E. H. Schacht and E. J. Goethals, *Makromol. Chem.*, 1978, **179**, 841.
57. E. F. Razvodovskii *et al.*, *Vysokomol. Soedin., Ser. A*, 1973, **15**, 2219, 2233.
58. T. Saegusa, H. Ikeda and H. Fujii, *Polym. J.*, 1972, **2**, 35.
59. T. G. Bassiri, A. Levy and M. Litt, *J. Polym. Sci., Part B*, 1967, **5**, 871.
60. M. Litt, T. Bassiri and A. Levy, *Belg. Pat.* 666 828 (1965); *Belg. Pat.* 666 831 (1965).
61. D. A. Tomalia and D. P. Shutz, *J. Polym. Sci., Part A-1*, 1966, **4**, 2253.
62. W. Seeliger, *Ger. Pat.* 1 206 585 (1965); *Ger. Pat.* 1 215 930 (1966); *Angew. Chem.*, 1966, **78**, 613, 913.
63. T. Kagiya, S. Narisawa, T. Maeda and K. Fukui, *J. Polym. Sci., Part B*, 1966, **4**, 257, 441.
64. T. Saegusa and H. Ikeda, *Macromolecules*, 1973, **6**, 315.
65. S. Kobayashi, K. Morikawa, N. Shimizu and T. Saegusa, *Polym. Bull. (Berlin)*, 1984, **11**, 253.
66. T. Saegusa, S. Kobayashi and A. Yamada, *Macromolecules*, 1975, **8**, 390.
67. T. Saegusa, A. Yamada and S. Kobayashi, *Polym. J.*, 1979, **11**, 53.
68. K. Ishizu, T. Fukutomi and T. Kakurai, *J. Polym. Sci., Polym. Lett. Ed.*, 1983, **21**, 405.
69. C. J. Simionescu and I. Rabia, *Polym. Bull. (Berlin)*, 1983, **10**, 311.
70. T. Saegusa and H. Ikeda, *Macromolecules*, 1973, **6**, 805.
71. R. Tanaka, I. Ueko, Y. Takaki, K. Kataoka and S. Saito, *Macromolecules*, 1983, **16**, 849.
72. R. C. Schulz, *Makromol. Chem., Macromol. Symp.*, 1986, **4**, 67.
73. A. Levy and M. Litt, *J. Polym. Sci., Part B.*, 1967, **5**, 881.

53

Cationic Ring-opening Polymerization: Copolymerization

PATRICIA DREYFUSS and M. PETER DREYFUSS
Michigan Molecular Institute, Midland, MI, USA

53.1 INTRODUCTION AND GENERAL CONSIDERATIONS

Many cyclic compounds can be copolymerized by cationic ring-opening polymerization (see refs. 1–9 and references therein). Copolymers from various combinations of cyclic ethers, acetals, sulfides, amines, lactones, lactams and hydrocarbons have been prepared. Some of the copolymers have a random distribution of monomer units and some are well-defined block or graft copolymers. Some of the copolymers include units from compounds that will not homopolymerize. Sometimes attempted copolymer syntheses have failed, and mixtures of homopolymers or cyclic oligomers have formed instead. Much more work needs to be done before the cationic copolymerization of cyclic compounds of all kinds can be controlled at will.

The same thermodynamic, kinetic and steric factors that govern homopolymerizations also apply to copolymerizations, except that now pairs or groups of monomers (and the cations formed from them) rather than a single monomer (and its cation) need to be considered. Familiar considerations from homopolymerizations include temperature, pressure, initiator effects, solvent effects, relative reactivities of ions formed from monomer and from polymer, relative reactivity of covalent *vs.* ionic species, relative reactivity if more than one kind of growing species form from a single monomer, cyclic oligomer formation, and termination and transfer reactions of all kinds. These were described in earlier chapters and will not be discussed further here.

Relative basicity and ring strain are particularly important considerations in copolymerizations, although they are not equally important for all kinds of monomers. It seems fairly well established that relative basicity plays an important role in controlling the relative reactivity of cyclic ethers.[7, 10] The basicity is affected by chemical structure, ring size and substituents. The ring size affects basicity in the order $4 > 5 \cong 7 > 6 > 3$.[4, 6, 11, 12] In five-membered rings the basicity order is ether > lactone > formal. Methyl substitution increases the basicity and chloromethyl substitution decreases the basicity. The relative basicities of cyclic ethers are broadly in agreement with their reactivity in copolymerization experiments.[1] An attempt has been made to make a quantitative estimate of the contribution of ring strain and basicity to reactivity of cyclic ethers in cationic copolymerization.[13] Free energy of polymerization was used as a measure of ring strain. The relationship derived relates the logarithm of relative reactivity, $1/r_n$, of m-membered ring ethers with i substituents to n-membered ring compounds with j substituents, to a linear combination of the differences in basicity, $\Delta(pK_b)_{m,i-n,j}$ and free energy, $\Delta(\Delta G)_{m,i-n,j}$, namely

$$\log(1/r_n) = \alpha\Delta(\Delta G)_{m,i-n,j} + \beta\Delta(pK_b)_{m,i-n,j} + \gamma \tag{1}$$

where α, β and γ are constants. Calculated values of $1/r_n$ agreed well with observed values taken from the literature. In the cyclic ether series, there seems to be no correlation between ring strain and copolymerization reactivity.

Ring size effects the reactivity, and the basicity, of the cyclic acetal monomers also. Additionally, there are equilibria involving the terminal oxonium ion that further influence reactivity, or more exactly the concentration of active end groups *vs.* dormant cationic end groups (equations 2 and 3).

$$\tag{2}$$

$$\tag{3}$$

In each case, the equilibrium favors the seven-membered ring. The larger rings containing three oxygen atoms are considered to be dormant species. Thus, while the five-membered oxonium ion derived from 1,3-dioxolane is the most reactive, its presence in relatively low concentration results in a greater apparent reactivity of the larger 1,3-dioxepane monomer.[14] Whereas the six-membered cyclic formal 1,3-dioxane does not homopolymerize, the related six-membered cycle, trioxane, can be polymerized in solution or in bulk, the latter either as a molten liquid or as a crystalline solid. Information on the reactivity of trioxane relative to the five- and seven-membered cyclic formals does not seem to be available.

In contrast, basicity seems to be less important than ring strain in determining the relative reactivity of cyclic sulfides.[15] The reactivity measured in the sulfide case was in homopolymerizations. The order of basicity found for the cyclic sulfides on the basis of ring size is $5 > 6 > 4 > 3$. This order is different from that in cyclic ethers. The difference has been ascribed to the differences in heteroatom size, differences in ring size (ring strain), and also to differences in polarizability between oxygen and sulfur atoms. Basicity did not correlate well with reactivity in the sulfide series.

Products from interactions between cyclic ethers and cyclic sulfides or cyclic amines are limited by the fact that an oxonium ion will initiate the polymerization of a cyclic sulfide or cyclic amine, but the reverse does not occur. Therefore, irreversible and complete passage from oxonium ion to sulfonium ion or ammonium ion takes place and there is no possibility of statistical monomer placement in copolymers from these combinations of monomers. The number and nature of the substituents on a cyclic amine ring is of great importance. The polymerizability of the cyclic amines is determined by the nature of their substituents. Cyclic amine monomers that do not homopolymerize, do not seem to copolymerize either.

Numerous studies of the kinetics of copolymerization of cyclic compounds have been reported.[1,2,4,6] Tables of relative reactivity ratios have been collected (see refs. 1, 10 and references therein). These numbers should be used with some caution, for other studies have shown that these kinetics can be very complicated and the standard Mayo Lewis equations[16] often do not adequately describe the copolymerizations.[2] The kinetics of copolymerization must be modified to take these complications into account. Some examples of the kind of complications that can occur include the following. (i) In the presence of monomers like formaldehyde and tetrahydrofuran, there is a polymerization–depolymerization equilibrium with active centers that occurs at a significant rate. (ii) In trioxane (a six-membered ring) polymerizations, cleavage of cyclic formals from the cationic chain end occurs and oxacyclic compounds such as tetroxane (a reactive eight-membered ring) are formed. (iii) In copolymerizations of tetrahydrofuran with substituted oxetanes above the ceiling temperature for tetrahydrofuran, copolymerization occurs and the penultimate unit has a significant effect on the rate of copolymerization. (Under these conditions homopolymerization of tetrahydrofuran would not occur. The derivation of the original Mayo Lewis equations assumes there is no effect of the penultimate unit.) (iv) Some compounds, which will not homopolymerize, do copolymerize. Examples are tetrahydropyran or 2-methyltetrahydrofuran with epichlorohydrin, and 1,3- or 1,4-dioxane with 3,3-bis(chloromethyl)oxetane or tetrahydrofuran. (v) Some combinations of monomers give products that have block character or that are mixtures of copolymer and homopolymers, or just of homopolymers. Examples are given below.

53.2 CYCLIC ETHER COPOLYMERIZATION

Cyclic ethers are capable of copolymerizing with other cyclic ethers as well as with a variety of other compounds. Among cyclic compounds, this section will be restricted to discussion of copolymerization with cyclic ethers. Copolymerization with other cyclic compounds will be discussed below under the particular type of cyclic compound in question. This section will also include discussion of other copolymers, which are not related to the parent compounds discussed below.

53.2.1 Cyclic Ether Copolymerization of Monomer Mixtures

Copolymerization of one cyclic ether with another is probably the most widely studied type of cationic ring-opening copolymerization. Three-, four- and five-membered ring compounds have been copolymerized in a large variety of combinations and with just about every type of initiator mentioned in the section on homopolymerization. The products of attempted copolymerization of cyclic ethers are not necessarily statistical copolymers. The distribution of monomer units in the copolymer can vary with the relative reactivities of the monomers and sometimes with the initiator used. Mixtures of different 3,3-substituted oxetanes usually give random copolymers.[10] 2-Methyltetrahydrofuran and tetrahydropyran, which do not homopolymerize, copolymerize with 3,3-bis(chloromethyl)oxetane to give alternating copolymers.[17] Tetrahydrofuran and epichlorohydrin polymerized with $AlEt_3$ or $AlEt_3$–H_2O give block copolymers.[18] In copolymerization of ethylene oxide and tetrahydrofuran by nonhydrolyzable proton acid initiators (*e.g.* CF_3SO_3H), the product includes cyclic tetramers.[19] 2,3-Epoxy-2,4,4-trimethylpentane, 'copolymerized' with tetrahydrofuran, gives only end-capped homopolytetrahydrofuran.[20] As can be seen from Table 1, which is representative but not all-inclusive, each size ring will copolymerize with a variety of compounds of the same ring size and with a variety of rings of different sizes.

The studies of copolymerization have had a variety of objectives including property modification of specific homopolymers, structure–property relationships determination, property determination, kinetics determination and mechanism delineation. For example, a three-membered ring compound might be added to a four- or five-membered ring compound to reduce the tendency of the polymer from the larger ring size to crystallize or to lower its melting temperature. Or tetrahydrofuran might be copolymerized with an energetic monomer like 3,3-bis(azidomethyl)oxetane to improve processing and modify its glass transition temperature. More detailed discussion is beyond the intended scope of this brief review.

53.2.2 Block and Graft Copolymers by Special Techniques

Most of the methods available for the synthesis of block and graft copolymers have been applied to cyclic ethers (see refs. 5, 6, 21, 22 and references therein). The presence of living ends on cyclic ether polymers have been used to good advantage for the intentional, controlled preparation of block copolymers. Both sequential addition and polymerization of different monomers, and termination of living polyether cations with living macromolecular anions have resulted in well-defined products. Graft copolymers have been made by the macromer method, and by polymerization from, through and onto polymeric backbones using polymerization methods tailored to the functional groups already present or purposely added to the selected backbone. Star copolymers have also been prepared from living cations. By far the largest number of these copolymers have been prepared where one of the substituents is polytetrahydrofuran, although copolymers from other cyclic ethers and, for example, neoprene backbones are known. A few of the copolymers have been extensively characterized but so far none of the polymers prepared by these methods enjoys a significant commercial market.

In contrast, copolymers from homo- and co-polyols prepared from cyclic ethers by cationic polymerization methods are of considerable commercial importance.[5, 9] These polyols are fabricated into polyurethanes and polyesters, which find a wide variety of uses ranging from coatings, films and fibers, to binders for propellants and thermoplastic moldings.

Table 1 Representative Combinations of Cyclic Ethers Copolymerized[1,5,6,8-10]

M_2: Three-membered	Ref.	Four-membered	Ref.	Five-membered	Ref.
		(a) Three-membered Rings			
		M_1 = *Epichlorohydrin*			
Ethylene oxide	a	3,3-Bis(chloromethyl)oxetane	b	2,3-Dihydrofuran	c
		2-Methyloxetane	d	3,4-Dimethyltetrahydrofuran	e
				Tetrahydrofuran	f–k
		M_1 = *Ethylene oxide*			
Epibromohydrin	a	3,3-Bis(chloromethyl)oxetane	l	7-Oxabicyclo[2.2.1]heptane	m
Propylene oxide	n, o	3-Chloromethyl-3-methyloxetane	p	Tetrahydrofuran	q–s
		M_1 = *Propylene oxide*			
Allyl glycidyl ether	t	3,3-Bis(chloromethyl)oxetane	l, u	2,3-Dihydrofuran	c
Ethyl glycidate	v	3-Chloromethyl-3-methyloxetane	p	3,4-Dimethyltetrahydrofuran	e
Glycidyl nitrate	w			Tetrahydrofuran	x–mm
Styrene oxide	nn				
Substituted phenyl glycidyl ethers	oo				
Trifluoromethylethylene oxide	pp				
		(b) Four-membered Rings			
		M_1 = *3,3-Bis(chloromethyl)oxetane*			
Styrene oxide	u	3,3-Bis(fluoromethyl)oxetane	qq	2,3-Dihydrofuran	c
		3-Chloromethyl-3-methyloxetane	qq	2-Methyltetrahydrofuran	i, rr
		3,3-Dimethyloxetane	qq	Tetrahydrofuran	k, ss–
		Oxetane	qq		yy
		M_1 = *3,3-Bis(Fluoromethyl)oxetane*			
		3,3-Diethyloxetane	qq		
		3,3-Dimethyloxetane	qq		
		3,3-Bis(bromomethyl)oxetane	qq		
		3,3-Bis(iodomethyl)oxetane	qq		
		M_1 = *Dimethyloxetane*			
		Oxetane	zz, aaa		

Table 1 *(continued)*

M_2: Three-membered	Ref.	Four-membered	Ref.	Five-membered	Ref.
		(c) Five-membered Rings M_1 = *Tetrahydrofuran*		3-Methyltetrahydrofuran	eee, fff
1-Butene oxide	bbb	3,3-Bis(azidomethyl)oxetane	ccc, ddd	7-Oxabicyclo[2.2.1]heptane	jjj
2,3-Epoxyhexane	ggg, hhh	3-Chloromethyl-3-methyloxetane	iii		
4,5-Epoxy-1-hexene	kkk–nnn	3-(2,2-Dinitropropoxymethyl)-3-methyloxetane	ccc		
5,6-Epoxy-1-hexene	kkk, lll	3-Methyl-2-allyloxymethyloxetane	ooo		
2,3-Epoxy-2,4,4-trimethylpentane	ggg, hhh	2-Methyloxetane	d		
Glycidyl ethers	ppp				
Glycidyl nitrate	w, qqq–sss				
Trifluoromethylethylene oxide	pp	*(d) Six-membered Rings* M_1 = *Tetrahydropyran*			
4-Vinyl cyclohexene oxide	lll	3,3-Bis(chloromethyl)oxetane	u		
		(e) Seven-membered Rings M_1 = *Oxepane*		Tetrahydrofuran	ttt

[a] T. N. Kuren'gina, L. V. Alferova and V. A. Kropachev, *Vysokomol. Soedin., Ser. A*, 1969, **11**, 1985.
[b] Y. Yamashita, T. Tsuda, M. Okada and S. Iwatsuki, *J. Polym. Sci., Part A-1*, 1966, **4**, 2121.
[c] Y. Minoura and M. Mitoh. *Makromol. Chem.*, 1968, **119**, 104.
[d] V. A. Ponomarenko, S. N. Sakharova, G. I. Alikberova and A. S. Zharova, *Izv. Akad. Nauk SSSR, Ser. Khim.*, 1969, 687.
[e] M. Malanga and O. Vogl, *J. Polym. Sci., Polym. Chem. Ed.*, 1982, **20**, 2033.
[f] A. I. Kusaev, G. N. Komratov, G. V. Korovina and S. G. Entelis, *Vysokomol. Soedin., Ser. A*, 1970, **12**, 995.
[g] T. V. Grinevich, G. V. Korovina and S. G. Entelis, *Vysokomol. Soedin., Ser. B*, 1977, **19**, 690.
[h] V. A. Ponomarenko and M. G. Deborin, *Izv. Akad. Nauk SSSR, Ser. Khim.*, 1969, 682.
[i] A. Ishigaki, T. Shono and Y. Hachihama, *Makromol. Chem.*, 1964, **79**, 170.
[j] B. A. Rozenberg, E. B. Lyudvig, N. V. Desyatova, A. R. Gantmakher and S. S. Medvedev, *Vysokomol. Soedin., Ser. A*, 1965, **7**, 1010.
[k] T. Saegusa, T. Ueshima, H. Imai and J. Furukawa, *Makromol. Chem.*, 1964, **79**, 221.
[l] S. Aoki, K. Fujisawa, T. Otsu and M. Imoto, *Kogyo Kagaku Zasshi*, 1966, **69**, 131.
[m] M. Paci, F. Andruzzi and G. Ceccarelli, *Macromolecules*, 1982, **15**, 835.
[n] A. Rastogi and L. E. St. Pierre, *J. Appl. Polym. Sci.*, 1970, **14**, 1179.
[o] T. N. Kuren'gina, L. V. Alferova and V. A. Kropachev, *Vysokomol. Soedin., Ser. B*, 1971, **13**, 418.
[p] N. M. Geller, V. A. Kropachev and B. A. Dolgoplosk, *Vysokomol. Soedin., Ser. A*, 1968, **10**, 1878.
[q] W. J. Murbach and A. Adicoff, *Ind. Eng. Chem.*, 1960, **52**, 772.
[r] I. M. Robinson, E. Pechhold and G. Pruckmayr, *ACS Symp. Ser.*, 1981, **172**, 197.
[s] I. M. Robinson and G. Pruckmayr, *Macromolecules*, 1979, **12**, 1043.
[t] G. Corradini, G. Ghetti, M. Bruzzone and W. Marconi, *Chim. Ind. (Milan)*, 1970, **52**, 135.
[u] S. Aoki, Y. Harita, Y. Tanaka, H. Mandai and T. Otsu, *J. Polym. Sci., Part A-1*, 1968, **6**, 2585.
[v] D. Tirrell, O. Vogl, T. Saegusa, S. Kobayashi and T. Kobayashi, *Macromolecules*, 1980, **13**, 1041.
[w] S. Abe, M. Ito and K. Namba, *Makromol. Chem.*, 1970 **134**, 121.

Table 1 *(continued)*

[x] G. N. Komratov, R. A. Barzykina and G. V. Korovina, *Vysokomol. Soedin., Ser. B*, 1979, **21**, 326.

[y] M. D. Baijal and L. P. Blanchard, *J. Polym. Sci., Part C*, 1968, **23**, 157.

[z] G. N. Komratov, G. V. Korovina, G. A. Mirontseva and S. G. Entelis, *Vysokomol. Soedin., Ser. A*, 1969, **11**, 443.

[aa] L. P. Blanchard, S. Kondo, J. Moinard, J. F. Pierson and F. Tahiani, *J. Polym. Sci., Part A-1*, 1972, **10**, 399.

[bb] G. N. Komratov, R. A. Barzykina and G. V. Korovina, *Vysokomol. Soedin., Ser. A*, 1980, **22**, 2342.

[cc] L. P. Blanchard, G. G. Gabra, V. Hornof and S. L. Malhotra, *J. Polym. Sci., Polym. Chem. Ed.*, 1975, **13**, 271.

[dd] L. P. Blanchard, C. Raufast, H. H. Kiet and S. L. Malhotra, *J. Macromol. Sci., Chem.*, 1975, **9**, 1219.

[ee] S. L. Malhotra and L. P. Blanchard, *J. Macromol. Sci., Chem.*, 1975, **9**, 1485.

[ff] H. H. Kiet, S. L. Malhotra and L. P. Blanchard, *J. Macromol. Sci., Chem.*, 1976, **10**, 1317.

[gg] E. J. Alvarez, V. Hornof and L. P. Blanchard, *J. Polym. Sci., Part A-1*, 1972, **10**, 1895.

[hh] L. N. Turovskaya, N. G. Matveeva and A. A. Berlin, *Vysokomol. Soedin., Ser. A*, 1973, **15**, 1842.

[ii] A. A. Berlin, L. N. Turovskaya and N. G. Matveeva, *Vysokomol. Soedin., Ser. A*, 1976, **18**, 1322.

[jj] A. A. Berlin, L. N. Turovskaya and N. G. Matveeva, *Plaste Kautsch.*, 1977, **24**, 794.

[kk] L. A. Dickinson, *J. Polym. Sci.*, 1962, **58**, 857.

[ll] T. N. Kuren'gina, L. V. Alferova and V. A. Kropachev, *Vysokomol. Soedin., Ser. A*, 1966, **8**, 293.

[mm] L. P. Blanchard, J. Singh and M. D. Baijal, *Can. J. Chem.*, 1966, **44**, 2679.

[nn] L. P. Blanchard, A. Aghadjan and S. L. Malhotra, *J. Macromol. Sci., Chem.*, 1975, **9**, 299.

[oo] C. C. Price, Y. Atarashi and R. Yamamoto, *J. Polym. Sci., Part A-1*, 1969, **7**, 569.

[pp] V. A. Ponomarenko, A. M. Khomutov and N. A. Zadorozhnyi, *Izv. Akad. Nauk SSSR, Ser. Khim.*, 1968, 1847.

[qq] S. Okamura, H. Miyaki and N. Shimazaki, *Encycl. Polym. Sci. Technol.*, 1968, **9**, 668.

[rr] T. Tsuda, T. Nomura and Y. Yamashita, *Makromol. Chem.*, 1965, **86**, 301.

[ss] T. Saegusa, H. Imai and J. Furukawa, *Makromol. Chem.*, 1962, **54**, 218.

[tt] T. Saegusa, H. Imai and J. Furukawa, *Makromol. Chem.*, 1962, **56**, 55.

[uu] J. Furukawa, *Polymer*, 1962, **3**, 487.

[vv] L. V. Alferova and V. A. Kropachev, *Vysokomol. Soedin., Ser. A*, 1965, **7**, 1065.

[ww] T. Saegusa, H. Imai, S. Hirai and J. Furukawa, *Kogyo Kagaku Zasshi*, 1962, **65**, 699.

[xx] M. P. Dreyfuss and P. Dreyfuss, *J. Polym. Sci., Part A-1*, 1966, **4**, 2179.

[yy] I. Penczek and S. Penczek, *J. Polym. Sci., Part B*, 1967, **5**, 367.

[zz] L. Garrido, E. Riande and J. Guzman, *Makromol. Chem., Rapid Commun.*, 1983, **4**, 725.

[aaa] L. Garrido, J. Guzman, E. Riande and J. De Abajo, *J. Polym. Sci., Polym. Chem. Ed.*, 1982, **20**, 3377.

[bbb] J. M. Hammond, J. F. Hooper and W. G. P. Robertson, *J. Polym. Sci., Part A-1*, 1971, **9**, 265.

[ccc] *Industrial Chemical News*, November, 1981, cover.

[ddd] K. D. Pae, C. L. Tang and E. S. Shin, *J. Appl. Phys.*, 1984, **56**, 2426.

[eee] H. Stratmann, B. Stutzel and R. Feinauer, *Angew. Makromol. Chem.*, 1978, **74**, 105.

[fff] J. Guzman and E. Riande, *J. Polym. Sci., Polym. Chem. Ed.*, 1987, **25**, 365.

[ggg] P. Dreyfuss and J. P. Kennedy, *J. Polym. Sci., Polym. Symp.*, 1977, **60**, 47.

[hhh] P. Dreyfuss and J. P. Kennedy, *J. Appl. Polym. Sci., Appl. Polym. Symp.*, 1977, **30**, 153.

[iii] Yu. A. Gorin, K. N. Charskaya and E. I. Rodina, *Vysokomol. Soedin., Ser. A*, 1968, **10**, 405.

[jjj] E. L. Wittbecker, H. K. Hall, Jr. and T. W. Campbell, *J. Am. Chem. Soc.*, 1960, **82**, 1218.

[kkk] T. G. Shchibriya, A. V. Ignatenko, S. P. Krukovskii and V. A. Ponomarenko, *Vysokomol. Soedin., Ser. A*, 1976, **18**, 1805.

[lll] J. G. Burt and H. C. Walter (E.I. DuPont de Nemours & Co), *US Pat. 3 287 330* (1966) (*Chem. Abstr.*, 1967, **66**, 29 854e).

[mmm] M. P. Dreyfuss and J. H. Macey, *Polym. Prepr., Am. Chem. Soc., Div. Polym. Chem.*, 1979, **20**(2), 324.

[nnn] L. P. Blanchard, J. C. Asselin and S. L. Malhotra, *Polym. J.*, 1975, **7**, 326.

[ooo] Yu. A. Gorin, E. I. Rodina, N. V. Kozlova and K. V. Nelison, *Vysokomol. Soedin., Ser. A*, 1969, **11**, 1477.

[ppp] V. A. Ponomarenko, A. M. Khomutov, S. I. Il'chenko, G. N. Gorshkova and V. S. Bogdanov, *Vysokomol. Soedin., Ser. A*, 1969, **11**, 182.

[qqq] S. G. Entelis and G. V. Korovina, *Makromol. Chem.*, 1974, **175**, 1253.

[rrr] G. V. Korovina, D. Ya. Rossina, D. D. Novikov and S. G. Entelis, *Vysokomol. Soedin., Ser. A*, 1974, **16**, 1274.

[sss] T. V. Trinevich, D. Ya. Rossina, G. V. Korovina and S. G. Entelis, *Plast. Massy*, 1975, 14.

[ttt] G. Pruckmayr and T. K. Wu, *ACS Symp. Ser.*, 1979, **103**, 237.

53.3 TRIOXANE AND CYCLIC ACETAL COPOLYMERIZATION

Copolymers from cyclic acetals are numerous. Even cyclic acetal homopolymers can be considered to be copolymers in the sense that they are 1:1 alternating copolymers of the parent glycol and formaldehyde. Cyclic acetals copolymerize readily with each other[23] and have been reported to copolymerize with cyclic ethers such as 3,3-bis(chloromethyl)oxetane and tetrahydrofuran.[24] Reactivity ratios have been reported. Some studies have shown that they vary with the initiator used. For example, copolymerization proceeded normally if methyl fluorosulfonate was used, while only block copolymers or mixtures of homopolymers resulted with oxonium ion initiators.[25]

Copolymerization of 1,3-dioxolane with unsaturated cyclic lactones can involve the carbon–carbon double bond.[26] Thus, copolymerization with diketene results in two types of diketene residues in the product: $-OCH_2CH_2OCH_2-CH_2C(=O)CH_2C(=O)-$ and $-OCH_2CH_2OCH_2-OC(=CH_2)CH_2C(=O)-$. Depending on the initiator employed, copolymerization with an unsaturated γ-lactone results in polymerization either through the double bond or both through the double bond and by lactone ring-opening.[23] Copolymerization with the vinyl monomer, styrene, has also been reported.[27]

Polyformaldehyde, whether originating from formaldehyde or from trioxane, decomposes readily at elevated temperatures, 'unzipping' to regenerate monomeric formaldehyde. End capping is used to prevent this unzipping. One of the commercially important methods for stabilizing formaldehyde polymers is the introduction of copolymeric units which act, ultimately, as a means to introduce stable end groups. The product formed is deliberately degraded until a copolymeric unit is reached, which then becomes the stable end group. Ethylene oxide, the simplest and perhaps most commonly used comonomer, is also used in the form of the formaldehyde cyclic acetal, 1,3-dioxolane. The cyclic acetal derived from formaldehyde and 1,4-butanediol, dioxepane, is also used as a comonomer. The carbon–carbon bonds thus introduced provide a stable linkage beyond which unzipping cannot occur.

The cationic polymerization of trioxane with ethylene oxide is not a straightforward copolymerization. An extended induction period is observed in the presence of ethylene oxide. During this period, preferential polymerization of ethylene oxide and formation of 1,3-dioxolane and 1,3,5-trioxepane occurs. The latter two acetals result from ethylene oxide reacting with formaldehyde released from the trioxane–catalyst complex. Only when the ethylene oxide is consumed does the polymerization of trioxane and copolymerization of trioxane with the dioxolane and trioxepane commence. Nevertheless, a random copolymer does form by a process referred to as transacetalization. This is the process in which an active growing polymer chain end attacks a polymeric oxygen atom to form an intermediate branch point at an oxonium ion. The result of this process is a randomization of the polymer structures and an approach of the polydispersity towards a value of two (see equation 4).

$$-OCH_2OCH_2OCH_2CH_2OCH_2CH_2OCH_2CH_2OCH_2^+ \quad + \quad \text{\raisebox{0pt}{\wedge\wedge}} OCH_2OCH_2OCH_2^+ \quad \rightarrow$$

$$-OCH_2OCH_2OCH_2CH_2\overset{+}{O}CH_2CH_2OCH_2CH_2OCH_2^+ \quad \rightarrow$$
$$\underset{\displaystyle CH_2OCH_2OCH_2O\text{\wedge\wedge}}{|} \tag{4}$$

$$-OCH_2OCH_2OCH_2CH_2^+ \quad + \quad \text{\wedge\wedge} OCH_2OCH_2OCH_2OCH_2CH_2OCH_2CH_2OCH_2^+$$

A large number of oxirane monomers other than ethylene oxide have also been studied and patented as comonomers for trioxane. Some are used to achieve specific property variations. In one case, adding a controlled amount of diepoxide introduces a little chain branching and significantly alters the melt rheology.[28]

A number of other miscellaneous cyclic acetal copolymerization combinations have also been studied and undoubtedly many additional combinations are possible should the specific need arise. Copolymerization of nonhomopolymerizable cyclic acetals such as 1,3-dioxane and 4-methyl-1,3-dioxane with trioxane has been reported.[29] Dioxepane copolymerization with 1,3-dioxolane[30] and with isobutyl vinyl ethers has been studied.[31] Copolymerizations involving larger ring size acetals containing three or more oxygen atoms have also been reported.[31,32] A common feature of the copolymerization among these cyclic acetals is the transacetalization reaction. Random copolymers are generally obtained, especially at high conversions. Thus, an attempt to prepare a block copolymer by reacting biliving dioxepane with dioxolane only gave evidence of the desired block copolymer very early in the copolymerization.[30] A statistical copolymer was observed before all of the dioxolane was incorporated. An asymmetric-selective copolymerization was observed when a racemic bromo bicyclo acetal was copolymerized with an optically active glucopyranose, which is also a bicyclic acetal derivative.[33] The D-enantiomer of the bromo bicyclo acetal was preferentially incorporated into the copolymer and the remaining unreacted bicyclo acetal became optically active.

53.4 CYCLIC SULFIDE OR CYCLIC AMINE COPOLYMERIZATION

The cationic copolymerization of these monomers has not been studied extensively. As already mentioned, the formation of random copolymers with oxygen containing cyclic monomers is not possible. Thus, for example, when copolymerization of episulfides with epoxides was attempted, only block copolymer and episulfide homopolymer was observed.[34] Deliberate efforts to prepare block copolymers have been carried out by initiating the polymerization of a cyclic sulfide or a cyclic amine with living polytetrahydrofuran. The results indicated that the products were mixtures of block copolymers and homopolytetrahydrofuran.[35]

Copolymerization between two cyclic sulfides or between two cyclic amines is possible. The copolymerization of propylene sulfide with both cyclohexene sulfide and isobutylene sulfide has been reported.[34] Reactivity ratios indicated that the order of reactivity was cyclohexene sulfide > propylene sulfide > isobutylene sulfide. Steric factors have been shown to play an important role. Copolymerization of the *cis*- and *trans*-2,3-dimethylthiiranes showed that the *cis* isomer is markedly more reactive.[34]

Cyclic amines are highly reactive monomers that undergo a variety of interesting and unusual copolymerization reactions. However, only a few of these can be classified as cationic ring-opening polymerization. Random copolymerization among the *N*-substituted aziridines is readily observed.[36] Block copolymers of aziridine monomers have been prepared by sequential addition of the monomers. Poly(1-*t*-butylaziridine-*b*-1-phenylethyl-2-methylaziridine) was prepared starting with either monomer.[37] If living carbocationic or oxonium ion polymers are used as initiators for cyclic amine polymerization, block copolymers are also obtained. In this way, poly(styrene-*b*-1-*t*-butylaziridine)[38] and poly(tetrahydrofuran-*b*-1,3,3-trimethylaziridine)[39] have been realized. An amine containing ABA block copolymer was prepared by terminating a living poly(1-*t*-butylaziridine) with a dicarboxy terminated polybutadiene.[7]

The four-membered ring analogs, the azetidines, also copolymerize with each other.[40] In addition, the preparation of an ABA block polymer, where the B block is polytetrahydrofuran and the A blocks are derived from 1,3,3-trimethylazetidine is described.[40] Copolymerization between aziridines and azetidines does not seem to have been considered as yet.

53.5 LACTONE AND LACTAM COPOLYMERIZATIONS

The lactones seem particularly well suited for cationic copolymerization with a number of monomer types including lactones, cyclic ethers, trioxane and vinyl monomers. Rather few reports of lactone–lactone copolymerizations seem to exist. One such copolymerization provides a route for incorporating the nonhomopolymerizable γ-butyrolactone.[41] Copolymerization with β-propiolactone was possible using a triethyl aluminum–water initiator (which may not be truly cationic). A true cationic copolymerization of γ-butyrolactone was observed with 3,3-bis(chloromethyl)oxetane using boron trifluoride–etherate as initiator.[42]

In general, copolymerization studies have involved β-lactones, though a few examples with others, notably ε-caprolactone, can also be found. 3,3-Bis(chloromethyl)oxetane as comonomer was employed in most of these copolymerization studies, although more recently there have been some reports of copolymerization of β-propiolactone and ε-caprolactone with tetrahydrofuran.[43,44] Mention should also be made of an unusual graft copolymer involving carbon black. Cationic sites were introduced onto the carbon black surface and these were used to initiate the polymerization of a number of lactones, producing polyester grafts from the carbon black.[45] A number of examples of lactone copolymerization with vinyl compounds have also been observed (see below).

Lactams can be polymerized by several mechanisms. Of these, the cationic mechanism is one of lesser importance. Lactam copolymerization in general has been unimportant and cationic copolymerization has offered no particular advantages, so little effort has gone into this area. The mechanism of a lactam–phosphoric acid complex initiated copolymerization of caprolactam with dodecanolactam is reported to be cationic.[46,47]

53.6 COPOLYMERIZATION WITH VINYL MONOMERS

Generally speaking cross-propagation between a carbocation and a heterocation is not favored. In spite of this there are a number of examples of copolymerization of these cyclic heteroatom-containing monomers with vinyl monomers. Some have already been alluded to.

Very few cyclic ether copolymerizations with carbon monomers are known. One is the copolymerization of tetrahydrofuran or epichlorohydrin with cyclopentadiene. The copolymers seem to be rather block-like in character. So far the active cyclopentadienyl sites have not helped in incorporating other vinyl comonomers.[48,49]

Cationic copolymerization of cyclic formals, such as 1,3-dioxolane, 4-chloromethyl-1,3-dioxolane, trioxane and 1,3-dioxepane, with vinyl monomers has been reported on a number of occasions.[23,27] The vinyl monomers have generally been either styrene or a vinyl ether. On the other hand, β-propiolactones have been reported to copolymerize with a rather wide assortment of vinyl monomers.[42] But of these only the copolymerization with styrene was initiated cationically. A number of low molecular weight polymers produced from the heterocyclic monomers discussed here have been used to prepare macromers by attaching a vinyl monomer to one end of the chain. Usually an acrylate ester macromer is employed. Graft copolymers containing a hydrocarbon backbone and heterocyclic polymer grafts are then prepared by vinyl copolymerization of the macromer with a chosen vinyl monomer.

53.7 REFERENCES

1. K. C. Frisch and S. L. Reegen (eds.), 'Kinetics and Mechanisms of Polymerization Reactions', 'Ring-opening Polymerization', Marcel Dekker, New York, 1969, vol. 2.
2. C. H. Bamford and C. F. H. Tipper (eds.), 'Comprehensive Chemical Kinetics', Elsevier, Amsterdam, Netherlands, 1976, vol. 15.
3. T. Saegusa and E. Goethals (eds.), 'Ring-Opening Polymerization', *ACS Symp. Ser.*, 1977, **59**.
4. S. Penczek, P. Kubisa and K. Matyjaszewski, *Adv. Polym. Sci.*, 1980, **37**, 1.
5. P. Dreyfuss, 'Poly(tetrahydrofuran)', Gordon and Breach, New York, 1982.
6. K. J. Ivin and T. Saegusa (eds.), 'Ring-Opening Polymerization', Elsevier, London and New York, 1984, vols. 1 and 2.
7. J. E. McGrath (ed.), 'Ring-Opening Polymerization', *ACS Symp. Ser.*, 1985, **286**.
8. E. J. Goethals (ed.), 'Cationic Polymerization and Related Processes', Academic Press, New York, 1984.
9. M. P. Dreyfuss and P. Dreyfuss, *Encycl. Polym. Sci. Eng.*, 2nd edn. (in press).
10. S. Okamura, H. Miyake and N. Shimazaki, *Encycl. Polym. Sci. Technol.*, 1968, **9**, 668.
11. Y. Yamashita, T. Tsuda, M. Okada and S. Iwatsuki, *J. Polym. Sci., Part A-1*, 1966, **4**, 2121.
12. K. Brzezińska, K. Matyjaszewski and S. Penczek, *Makromol. Chem.*, 1978, **179**, 2387.
13. Y. Tanaka, *J. Macromol. Sci., Chem.*, 1967, **1**, 1059.
14. S. Penczek, P. Kubisa, K. Matyjaszewski and R. Szymański, in ref. 8, p. 139.
15. J. K. Stille and J. A. Empen, *J. Polym. Sci., Part A-1*, 1967, **5**, 273.
16. F. R. Mayo and F. M. Lewis, *J. Am. Chem. Soc.*, 1944, **66**, 1594.
17. T. Tsuda and Y. Yamashita, *Makromol. Chem.*, 1966, **99**, 297.
18. T. Saegusa, T. Ueshima, H. Imai and J. Furukawa, *Makromol. Chem.*, 1964, **79**, 221.
19. I. M. Robinson and G. Pruckmayr, *Macromolecules*, 1979, **12**, 1043.
20. P. Dreyfuss and J. P. Kennedy, *J. Appl. Polym. Sci., Appl. Polym. Symp.*, 1977, **30**, 153.
21. P. Dreyfuss, L. J. Fetters and D. R. Hansen, *Rubber Chem. Technol.*, 1980, **53**, 728.
22. P. Dreyfuss and R. P. Quirk, *Encycl. Polym. Sci. Eng.*, 1987, **7**, 551.
23. R. C. Schulz, W. Hellermann and J. Nienburg, in ref. 6, vol. 1, p. 369.
24. M. Okada, N. Takikawa, S. Iwatsuki, Y. Yamashita and Y. Ishii, *Makromol. Chem.*, 1965, **82**, 16.
25. Y. Yokoyama, M. Okada and H. Sumitomo, *Polym. J.*, 1979, **11**, 629.
26. M. Okada, Y. Yokoyama and H. Sumitomo, *Makromol. Chem.*, 1972, **162**, 31.
27. M. Okada, Y. Yamashita and Y. Ishii, *Macromol. Chem.*, 1966, **94**, 181.
28. K. Weissermel, E. Fischer, K. Gutweiler and H. D. Hermann, *Kunststoffe*, 1964, **54**, 410.
29. C. S. H. Chen, *J. Polym. Sci., Polym. Chem. Ed.*, 1976, **14**, 143.
30. W. Chwialkowska, P. Kubisa and S. Penczek, *Makromol. Chem.*, 1982, **183**, 753.
31. M. Okada and Y. Yamashita, *Macromol. Chem.*, 1969, **126**, 266.
32. Y. Yamashita, T. Inoue, G. Hattori and K. Ito, *Macromol. Chem.*, 1972, **151**, 91.
33. M. Okada, H. Sumitomo and T. Hirasawa, *Macromolecules*, 1985, **18**, 2345.
34. P. Sigwalt and N. Spassky, in ref. 6, vol. 2, p. 603.
35. J. L. Lambert and E. J. Goethals, *Macromol. Chem.*, 1970, **33**, 289.
36. E. J. Goethals, in ref. 6, vol. 2, p. 715.
37. E. J. Goethals, M. Van de Velde, G. Eckhaut and G. Bouquet, in ref. 7, p. 219.
38. P. K. Bossaer, E. J. Goethals, P. J. Hackett and D. C. Pepper, *Eur. Polym. J.*, 1977, **13**, 489.
39. Y. Tezuka and E. J. Goethals, *Polym. Prepr., Am. Chem. Soc., Div. Polym. Chem.*, 1981, **22**(2), 313.
40. E. J. Goethals, E. H. Schacht, Y. E. Bogaert, S. I. Ali and Y. Tezuka, *Polym. J.*, 1980, **12**, 571.
41. K. Tada, Y. Numata, T. Saegusa and J. Furukawa, *Macromol. Chem.*, 1964, **77**, 220.
42. R. D. Lundberg and E. F. Cox, in ref. 1, p. 247.

43. A. K. Khomyakov, E. B. Lyudvig and N. N. Shapet'ko, *Eur. Polym. J.*, 1981, **17**, 1089.
44. A. K. Khomyakov, E. B. Lyudvig and N. N. Shapet'ko, *Vysokomol. Soedin.. Ser. A*, 1980, **22**, 2649; *Polym. Sci. USSR (Engl. Transl.)*, 1980, **22**, 2902.
45. N. Tsubokawa, *J. Appl. Polym. Sci.*, 1985, **30**, 2041.
46. R. Gomola, R. Alijev, J. Kondelikova and J. Kralicek, *Makromol. Chem., Rapid Commun.*, 1983, **4**, 745.
47. R. Alijev, M. Budesinsky, J. Kondelikova and J. Kralicek, *Angew. Makromol. Chem.*, 1982, **105**, 107.
48. I. Kuntz, *Trans. N.Y. Acad. Sci.*, 1971, **33**, 529.
49. M. P. Dreyfuss, unpublished observations.

29

Carbanionic Polymerization: Copolymerization

MICHAEL K. MARTIN
3M Industrial Tape Division, St. Paul, MN, USA

29.1 INTRODUCTION

There have been numerous studies on anionic copolymerization and many have recently been summarized in a book by Morton.[1] As in free radical copolymerization, the copolymerization of two monomers by anionic methods involves the four basic rate equations shown in Scheme 1.

$$\sim\sim M_1^* \ + \ M_1 \ \xrightarrow{\ k_{11}\ } \ \sim\sim M_1 - M_1^* \tag{1}$$

$$\sim\sim M_1^* \ + \ M_2 \ \xrightarrow{\ k_{12}\ } \ \sim\sim M_1 - M_2^* \tag{2}$$

$$\sim\sim M_2^* \ + \ M_2 \ \xrightarrow{\ k_{22}\ } \ \sim\sim M_2 - M_2^* \tag{3}$$

$$\sim\sim M_2^* \ + \ M_1 \ \xrightarrow{\ k_{21}\ } \ \sim\sim M_2 - M_1^* \tag{4}$$

Scheme 1

The asterisks represent the active carbanionic chain ends of either monomer M_1 or M_2. In this kinetic treatment, one basic assumption is that only the terminal unit of a growing chain controls its reactivity. This assumption has been invoked in the treatment of free radical copolymerization and assumed to apply to the anionic chain growth polymerization mechanism. From these equations reactivity ratios can be defined such that

$$r_1 \ = \ \frac{k_{11}}{k_{12}} \quad \text{and} \quad r_2 \ = \ \frac{k_{22}}{k_{21}} \tag{5}$$

The ratios of these propagation rate constants represents the relative reactivity of a growing chain end with its own monomer as opposed to the other monomer. When the value of r_1 is much larger than r_2 a tendency for monomer M_1 to homopolymerize will occur. The homopolymerization will occur even in the presence of monomer M_2. Assuming a steady state concentration of active species (*i.e.* carbanions) one can derive the instantaneous copolymer equation

$$\frac{d\{M_1\}}{d\{M_2\}} = \frac{\{M_1\}}{\{M_2\}} * \frac{r_1\{M_1\} + \{M_2\}}{r_2\{M_2\} + \{M_1\}} \qquad (6)$$

Here the composition of the copolymer $d\{M_1\}/d\{M_2\}$ can be related to the composition of the monomer mixture at a given instant.

Unlike free radical copolymerization, anionic copolymerization rates and copolymer compositions show a marked solvent effect.[1,23] Likewise, metal counterion and the polymerization temperature can influence the relative reaction of monomer with the anionic chain end. Electronegative substituents on the monomer will give the double bond more electrophilic character and make the monomer more reactive. The relative chain end basicities and nucleophilic character is also an important consideration.[2] Due to the ionic character of the chain end, the chain end structure can exist as either a dissociated carbanion or a more covalent-like carbon–metal bond. The nature of the counterion and solvent control the chain end structure so that the copolymerization kinetics are dependent on the initiator and polymerization solvent. In contrast, free radical copolymerizations are generally independent of the initiator and solvent employed for the copolymerization.

A detailed explanation of methods used to determine the reactivity ratios is beyond the scope of this review; however, both conventional[3,4] and statistical[5,6] methods can be employed.

29.2 EARLY COPOLYMERIZATIONS

Many early studies[7–9] on anionic copolymerization were complicated by the fact that termination free systems were not easy to obtain. Likewise, many alkali metal initiators were somewhat heterogeneous which complicated the kinetics.[7,8] However, useful relative rates between a variety of monomers have been obtained and are summarized in Table 1.

The copolymerization data in Table 1 indicate that the more polar monomer seems to be preferred for the sodium initiator in liquid ammonia. This could be due to the heterogeneous nature of the initiator.

Table 1 Earlier Carbanionic Copolymerization Studies

M_1	M_2	Initiator	Solvent	Reactivity ratios r_1	r_2	Ref.
Styrene	Methyl methacrylate	Na	Liq. NH$_3$	0.12	6.4	7
Styrene	Vinyl acetate	Na	Liq. NH$_3$	0.01	0.01	7
Styrene	Butadiene	EtNa	Benzene	0.96	1.60	8
Styrene	Acrylonitrile	Li alkyls	—	0.20	12.5	9
Styrene	p-Bu^t styrene	s-RLi	Benzene	1.30	0.90	10
Styrene	p-Methylstyrene	s-RLi	Benzene	0.75	1.10	10
α-Methylstyrene	Isoprene	EtLi	THF	10.0	0.07	11
Methyl methacrylate	Acrylonitrile	NaNH$_2$	Liq. NH$_3$	0.25	7.90	12
Acrolein	Methyl vinyl ketone	Imidazole	THF	2.02	0.06	13

29.3 SOLUBLE ALKYLLITHIUM INITIATORS

The use of soluble alkyllithium compounds such as *n*-butyllithium and *s*-butyllithium allowed for well-defined copolymerization studies in homogeneous solutions. Both nonpolar and polar solvent systems have been studied, and these polymerizations are known to result in few chain termination reactions. A summary of copolymerization studies on styrene, butadiene and isoprene using soluble alkyllithium catalysts has been given in Table 2.

The reactivity ratios (r_1) of styrene (M_1) with isoprene or butadiene (M_2) in hydrocarbon solvents is quite low. More importantly, these copolymerizations have shown that both 'living' polystyrene chain ends or polydiene chain ends prefer to react with diene monomers as opposed to styrene monomer. The crossover reaction between a polystyryl carbanion and diene monomer is much faster than the reaction with styrene monomer. This occurs even though the homopolymerization rate of styrene is faster than dienes utilizing similar solvents and initiators.[1]

Table 2 Soluble Organolithium Vinyl Aromatic and Diene Copolymerizations

M_1	M_2	Initiator	Temperature (°C)	Solvent	r_1	r_2	Ref.
Styrene	Butadiene	Bu^{n}Li	30–50	Benzene	0.04	10.0	8
Styrene	Butadiene	Bu^{n}Li	29	Benzene	0.30	4.50	14
Styrene	Butadiene	Bu^{n}Li	40	C_6H_{12}	0.04	26.0	15
Styrene	Butadiene	Bu^{n}Li	30	Et_2O	0.11	1.78	16
Styrene	Isoprene	Bu^{n}Li	30	Benzene	0.26	10.6	14
Styrene	Isoprene	Bu^{n}Li	40	C_6H_{12}	0.046	16.6	17
Styrene	Isoprene	Bu^{n}Li	25	THF	9.00	0.10	18
Butadiene	Isoprene	Bu^{n}Li	50	Hexane	3.38	0.47	19
Butadiene	Isoprene	Bu^{n}Li	40	Benzene	3.60	0.50	20
Butadiene	Isoprene	s-RLi	40	Hexane	1.85	0.32	21
Butadiene	Isoprene	s-RLi	30	Hexane	1.60	0.35	21
Butadiene	Isoprene	s-RLi	20	Hexane	2.62	0.39	21
1,1-DPE	m-DVB	Bu^{n}Li	−78	THF	0.00	1.20	22
1,1-DPE	p-DVB	Bu^{n}Li	−78	THF	0.00	2.80	22

In the case of polar solvents, such as diethyl ether or THF, reaction of the carbanionic chain ends with styrene monomer is preferred over that with the diene monomers.

In the copolymerization of 1,1-diphenylethylene (DPE) with m-divinylbenzene (m-DVB) and p-divinylbenzene (p-DVB) the r_1 values were zero. Copolymerization, even in excess DPE, gives a highly alternating copolymer structure. The pendant double bonds of the copolymer are thus flanked by DPE units reducing the tendency of the copolymer towards gelation.[22]

29.4 TEMPERATURE AND SOLVENT EFFECTS

The s-butyllithium initiated copolymerization of styrene with butadiene has recently been studied in detail.[1,23] In these studies the effect of reaction temperature and polar solvent content on the reactivity ratios were investigated. In the case of hydrocarbon solvent, reaction temperature showed a minimal influence on reactivity ratios. However, in the case of polar solvents such as THF, the reactivity ratios varied considerably when comparing results obtained at 25 °C and −78 °C as observed in Table 3.

The results obtained in THF are rationalized by considering the solvation of the carbon–lithium bond. As the temperature is increased from −78 °C to 25 °C the extent of solvation by THF is expected to decrease. Decreased solvation results in a more covalent carbon–lithium bond suggesting that chain end association effects influence the rate of monomer addition. Apparently, more associated organolithium chain ends favor butadiene addition relative to styrene. This is clearly the case in hydrocarbon solvents.

Figure 1 illustrates the concerted four center reaction of butadiene with the dimeric associated chain ends of polybutadienyllithium. The dimeric association in hydrocarbons has been verified by several laboratories.[25-30]

The addition of butadiene occurs at either of the two carbon–lithium bonds. Subsequent sequential butadiene additions continue and likewise, the association continues to exist after each monomer addition. The exclusion of styrene monomer as opposed to butadiene based purely on steric arguments is not obvious. It is possible that diene monomer complexation to the associated chain ends, prior to addition, could favor diene propagation vs. styrene addition. Further detailed

Table 3 Temperature and Solvent Effects on Styrene/Butadiene Reactivity Ratios

M_1	M_2	Temperature (°C)	Solvent	r_1	r_2	Ref.
Styrene	Butadiene	0	n-Hexane	0.03	13.3	a, 23
Styrene	Butadiene	50	n-Hexane	0.04	11.8	a, 23
Styrene	Butadiene	25	THF	4.00	0.30	a, 23
Styrene	Butadiene	−78	THF	11.0	0.04	a, 23

a Ref. 1, p. 142.

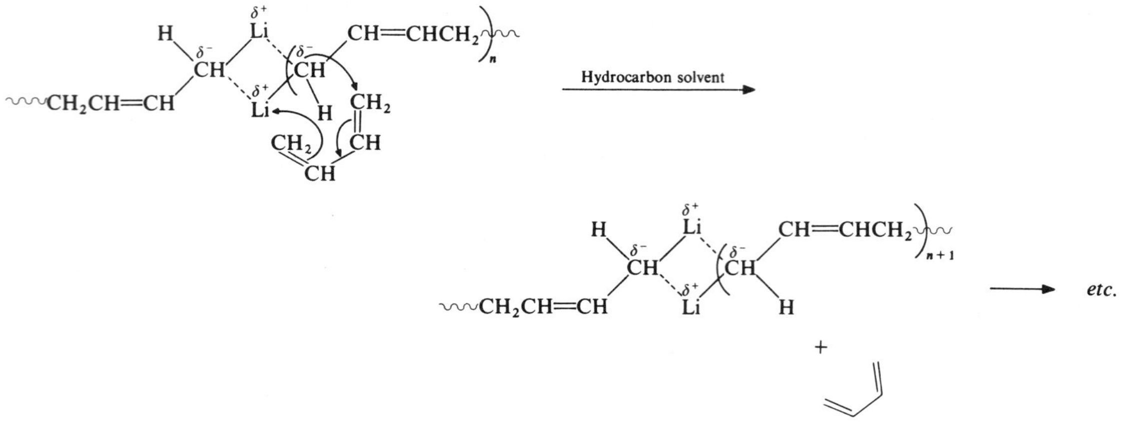

Figure 1 The *cis* approach of a butadiene monomer to a polybutadienyllithium dimeric chain end

structural characterization is necessary to determine the exact mechanism of diene-styrene copolymerization.

29.5 COPOLYMERIZATION WITH POLAR MONOMERS

The carbanionic copolymerization of polar monomers with diene or styrene type monomers is difficult and not well understood. In fact, many carbanion chain ends which have resulted from organometallic initiation of polar monomers will not initiate monomers of the stryene or diene type. Apparently, the nucleophilic strength of carbanions derived from these monomers is not high enough to participate in a crossover reaction with hydrocarbon monomers. Unlike free radical copolymerization, this fact limits the range of copolymer compositions available by carbanionic copolymerization. For example, a diblock copolymer of styrene with 2-vinylpyridine must be synthesized by first polymerizing styrene followed by the addition of 2-vinylpyridine monomer and subsequent polymerization to give the polystyrene–poly(vinylpyridine) diblock polymer. The carbanion derived from the initiation of 2-vinylpyridine will not initiate styrene monomer. This is also the case for polar monomers like the alkyl acrylates and methacrylates. Other factors complicating the carbanionic polymerization of polar monomers are the side reactions which occur between the carbanionic chain end and the pendant ester functionality of acrylic monomers. The copolymerization of polar monomers with themselves can occur however, and an extensive study on the copolymerization of methyl methacrylate with various alkyl methacrylates has been presented.[24] This work evaluated the reactivity ratios of MMA with several alkyl methacrylates at $-78\,°C$ in both toluene and THF as the reaction solvent. The organometallic initiator in all cases was *n*-butyllithium. Table 4 summarizes the monomers and their corresponding reactivity ratios with MMA. Unlike the case of styrene and butadiene copolymerizations, a pronounced solvent effect on the reactivity ratios is not evident in many MMA/alkyl methacrylate monomer pairs.

Solvent and/or monomer effects are observed for the more sterically hindered methacrylate esters. These include: diphenylmethyl, dimethylbenzyl, trityl, and to an extent, methylbenzyl methacrylate. For these monomer pairs a preference for MMA to homopolymerize is observed when toluene is used as the polymerization solvent.

Another complication of methacrylate copolymerization studies is the stereospecific nature of these polymerizations. Isotactic chains are preferred when the polymerizations are achieved in hydrocarbon solvents. This can also lead to helical chain conformations in solution which influence the steric approach of the incoming monomer reaction with the carbanionic chain end. Further detailed *in situ* structural information is needed before the mechanism of these copolymerizations is fully understood.

29.6 SUMMARY

In closing, considerable effort has been put forth to understand the copolymerization of a variety of monomers by carbanionic polymerization. The unique 'living' nature of the carbanionic polymerization allows for well-defined copolymers where copolymer composition can be controlled by

Table 4 Carbanionic Reactivity Ratios of Various Methacrylate Monomers in Polar and Nonpolar Solvents at $-78\,^{\circ}$C

$M_1{}^a$	$M_2{}^a$	Initiator	Solvent	Temperature (°C)	r_1	r_2	Ref.
MMA	DPMMA	Bu^{n}Li	Toluene	-78	0.57	0.55	24
			THF		1.11	1.57	
MMA	BzMA	Bu^{n}Li	Toluene	-78	0.59	1.60	24
			THF		0.70	1.46	
MMA	DMBMA	Bu^{n}Li	Toluene	-78	19.1	0.56	24
			THF		2.59	2.00	
MMA	TrMA	Bu^{n}Li	Toluene	-78	6.28	0.13	24
			THF		0.62	0.62	
MMA	EtMA	Bu^{n}Li	Toluene	-78	1.10	0.38	24
			THF		1.13	0.52	
MMA	Pr^{i}MA	Bu^{n}Li	Toluene	-78	2.75	0.20	24
			THF		2.29	0.42	
MMA	MBMA	Bu^{n}Li	Toluene	-78	1.68	0.78	24
			THF		2.04	1.52	
MMA	Bu^{t}MA	Bu^{n}Li	Toluene	-78	4.41	0.02	24
			THF		5.07	0.02	

^{a} Abbreviations are as follows: methyl methacrylate (MMA, M_1) copolymerized with M_2's of diphenylmethyl (DPMMA), benzyl (BzMA), dimethylbenzyl (DMBMA), trityl (TrMA), ethyl (EtMA), *iso*-propyl (Pr^{i}MA), α-methylbenzyl (MBMA), and *t*-butyl (Bu^{t}MA) methacrylates.

the nature of the solvent, initiator and reaction temperature. The polymer chain microstructure can be adjusted by the choice of reaction solvent as well. This allows for control of copolymer properties by adjusting the composition and microstructure *via* changes in polymerization solvent or temperature. The diene microstructure tends toward a more branched polymer chain when polar solvents or modifiers are part of the polymerization medium.

With the knowledge gained by the many studies[1-30] on carbanionic copolymerization a wide variety of block polymer and copolymer structures are attainable. These options broaden the scope of useful materials which can result from the limited economic raw materials and resources available to the polymer chemist.

29.7 REFERENCES

1. M. Morton, 'Anionic Polymerization: Principles and Practice', Academic Press, London, 1983, p. 134.
2. L. J. Fetters, *J. Polym. Sci., Part C*, 1969, **26**, 1.
3. F. R. Mayo and F. M. Lewis, *J. Am. Chem. Soc.*, 1944, **66**, 1594.
4. M. Fineman and S. D. Ross, *J. Polym. Sci.*, 1950, **5**, 269.
5. R. M. Joshi and S. G. Joshi, *J. Macromol. Sci., Chem.*, 1971, **A5** (8), 1329.
6. J. P. Kennedy, T. Kelen and F. Tudos, *J. Polym. Sci., Polym. Chem. Ed.*, 1975, **13**, 2277.
7. I. Landler, *C. R. Hebd. Seances Acad. Sci.*, 1950, **230**, 539.
8. A. A. Korotkov, *Angew. Chem.*, 1958, **70**, 85.
9. N. L. Zutty and F. J. Welch, *J. Polym. Sci.*, 1960, **43**, 445.
10. J. Chen and L. J. Fetters, *Polym. Bull.*, 1981, **4**, 275.
11. D. K. Polyakov, A. R. Gantmakher and S. S. Medvedev, *Vysokomol. Soedin., Ser. A*, 1967, **9**, 2329.
12. I. Landler, *J. Polym Sci.*, 1952, **8**, 63.
13. S. Monita, I. Kazuyuki, H. Inoue, N. Yamashita and T. Maeshima, *J. Macromol. Sci., Chem.*, 1982, **A17** (9), 1495.
14. M. Morton and F. R. Ells, *J. Polym. Sci.*, 1962, **61**, 25.
15. A. F. Johnson and D. J. Worsfold, *Makromol. Chem.*, 1965, **85**, 273.
16. A. A. Korotkov, S. P. Mitzengendler and K. M. Aleyev, *Polym. Sci. USSR (Engl. Transl.)*, 1962, **3**, 487.
17. D. J. Worsfold, *J. Polym. Sci., Polym. Chem. Ed.*, 1967, **5**, 2783.
18. D. J. Worsfold, *Macromolecules*, 1970, **3**, 514.
19. G. V. Rakova and A. A. Korotkov, *Rubber Chem. Technol.*, 1960, **33**, 623.
20. J. Furukaqa, T. Saegusa and K. Irako, *Kogyo Kagaku Zasshi*, 1962, **65**, 2029.
21. I. C. Wang, Y. Mohajer, T. C. Ward, G. L. Wilkes and J. E. McGrath, *ACS Symp. Ser.*, 1981, **166**, 529.
22. K. Hatada, T. Kitayama, Y. Okamoto and K. Ute, *Polym. Prepr., Am. Chem. Soc., Div. Polym. Chem.*, 1985, **26** (1), 249.
23. M. Morton, Li-K. Huang, unpublished data; Li-K. Huang, Ph. D. Dissertation, University of Akron, Ohio, 1979.
24. H. Yuki, Y. Okamoto, K. Ohta and K. Hatada, *J. Polym. Sci., Polym. Chem. Ed.*, 1975, **13**, 1161.

25. M. Morton and L. J. Fetters, *J. Polym. Sci., Part A*, 1964, **2**, 3311.
26. L. J. Fetters and M. Morton, *Macromolecules*, 1974, **7**, 552.
27. H. S. Makowski and M. Lynn, *J. Macromol. Chem.*, 1966, **1**, 443.
28. M. M. F. Al-Jarrah and R. N. Young, *Polymer*, 1980, **21**, 119.
29. W. H. Glaze, J. E. Hanicak, M. L. Moore and J. Chadhuni, *J. Organomet. Chem.*, 1972, **44**, 39.
30. A. Hernandez, J. Semel, H. C. Broeker, H. G. Zachmann and H. Sinn, *Makromol. Chem., Rapid Commun.*, 1980, **1**, 75.

Subject Index